Encyclopedia of
NANOSCIENCE and NANOTECHNOLOGY®

Volume 3

El–H

International Advisory Board

* Nobel Prize Laureate.

Encyclopedia of NANOSCIENCE and NANOTECHNOLOGY®

Volume 3

El–H

Editor

Hari Singh Nalwa, M.Sc., Ph.D.
Stanford Scientific Corporation
Los Angeles, California, USA

Formerly at
Hitachi Research Laboratory
Hitachi Ltd., Ibaraki, Japan

ASP

AMERICAN SCIENTIFIC PUBLISHERS
25650 North Lewis Way
Stevenson Ranch, California 91381-1439, USA

AMERICAN SCIENTIFIC PUBLISHERS
25650 North Lewis Way, Stevenson Ranch, California 91381-1439, USA
Tel.: (661) 254-0807
Fax: (661) 254-1207
E-mail: order@aspbs.com
WEB: www.aspbs.com

Encyclopedia of Nanoscience and Nanotechnology edited by Hari Singh Nalwa

This book is printed on acid-free paper. ♾

Library of Congress Control Number: 2002110511

International Standard Book Number: 1-58883-001-2 (Set)
International Standard Book Number: 1-58883-057-8 (Volume 1)
International Standard Book Number: 1-58883-058-6 (Volume 2)
International Standard Book Number: 1-58883-059-4 (Volume 3)
International Standard Book Number: 1-58883-060-8 (Volume 4)
International Standard Book Number: 1-58883-061-6 (Volume 5)
International Standard Book Number: 1-58883-062-4 (Volume 6)
International Standard Book Number: 1-58883-063-2 (Volume 7)
International Standard Book Number: 1-58883-064-0 (Volume 8)
International Standard Book Number: 1-58883-065-9 (Volume 9)
International Standard Book Number: 1-58883-066-7 (Volume 10)

PRINTED IN THE UNITED STATES OF AMERICA
10 9 8 7 6 5 4 3 2 1

Foreword

Nanotechnology is one of the fastest growing and most dynamic areas of research in this decade. The novel materials and devices made on a nanoscale offer unique and entirely different properties and applications compared to conventional technology, and this development has significantly surged research activities. Nanotechnology is rapidly spreading to academic institutions, industries, and government laboratories as new nanotechnology centers are being established in the United States, Western Europe, Japan, and other parts of the world. This is being viewed as an area of tremendous future growth because it has the potential to capture a multibillion dollar market in electronic, photonic, and biotechnology industries. Several research groups from academia and industry have already reported significant developments especially in the area of nanoelectronics. In this view, the encyclopedia is very timely for the scientific community.

The *Encyclopedia of Nanoscience and Nanotechnology* is the world's first and the single most comprehensive reference source ever published in the field of nanotechnology. This encyclopedia is an indispensable source for any research professional as well as for technology investors and developers seeking up-to-date information on nanotechnology among a wide range of disciplines from science to engineering to medicine. It should inspire future generations of academic and industrial researchers who endeavor to develop new nanoscale materials and devices. I hope that this 10-volume encyclopedia will support future growth is this newly emerging field of great scientific and commercial interest.

Richard E. Smalley
Nobel Prize Laureate, 1996
Rice University
Houston, Texas, USA

Preface

Over the last decade, new inventions have been reported at a prodigious rate in the growing field of nanotechnology. The *Encyclopedia of Nanoscience and Nanotechnology* records the field of nanotechnology as we understand it today where the leading experts document the relevant information on many different aspects of nanoscale science and technology. Nanotechnology is derived from the Greek words *nanos*, which means "dwarf," and *technologia*, which means a systematic treatment of an art or craft. Scientifically, "nano" means 10^{-9} where 1 nanometer is equivalent to one millionth of a millimeter and one billionth of a meter (10^{-9} m). Nanotechnology is the science and engineering of manipulation, assembling, synthesizing, characterizing, manufacturing, and viewing materials, functional structures, and devices at the atomic, molecular, or nanometer scale. The feature size is approximately in the 0.1–100 nm range. Atoms have a size of 1–4 Å. A hydrogen atom is about one-tenth of a nanometer. Nanostructured materials could hold tens of thousands of atoms together. Biomacromolecules such as the DNA double helix have a diameter of about 2 nm while ribosomes have a diameter of 25 nm. Proteins are between 1 and 20 nm in size. Therefore nanotechnology deals with atomic and molecular scale functional structures. The diameter of a human hair is 50 μm and would be on a nanometer scale if the diameter is reduced several thousand times. From our current understanding, it has become apparent that some nanoscale materials such as fine-grained minerals in rocks, DNA, proteins, nanoparticles in bacteria, smoke, etc., might have existed in nature since the inception of life on Earth, but we did not visualize in terms of nanotechnology before. Why are we now so fascinated and interested in downsizing materials, functional structures, and devices to a nanometer scale? Because the fundamental physical, chemical, and biological properties of materials are changed dramatically as the size of their constituent grains is decreased to nanometer scale owing to their distinguished size, shape, surface chemistry, and topology. It has been proved experimentally that nanocrystalline materials offer unique and significantly improved mechanical properties because they become harder and wear resistant as well as acquire incredible strength compared to their counterpart bulk materials. For example, 6 nm copper grains show five times more hardness than conventional copper. Carbon nanotubes and nanotube-composite fibers are several times tougher than steel, Kevlar, or spider silk and are the strongest materials known today. Likely, cadmium selenide (CdSe) can yield any color in the spectrum simply by controlling the size of its constituent grains. The unique and superior electrical, optical, magnetic, mechanical, and other physical properties acquired by materials at a nanoscale are related to their quantum size effect. Nanotechnology platforms offer significant potential benefits to create functional materials and devices that could not be made before. A new challenging vision of molecular nanotechnology is creating materials atom by atom where new inventions will have widespread societal impact. From a historic point of view, nanotechnology was first documented in the classic talk titled "There's Plenty of Room at the Bottom" presented by Nobel Laureate Richard Feynman to the American Physical Society in 1959. This field changed dramatically when Richard E. Smalley, Robert F. Curl, Jr., and Sir Harold W. Kroto won the 1996 Nobel Prize for their discovery of fullerenes. Many books, journals, and conference proceedings have been published since research activities in the field of nanotechnology got started in the 1990s. The five-volume *Handbook of Nanostructured Materials and Nanotechnology* published in October 1999 by the Editor (H. S. Nalwa) became a milestone reference source for researchers in the field of nanotechnology, and immediately thereafter work commenced on this encyclopedic reference.

A nanocluster can have several tens to thousands of atoms or molecules and the formation of well-ordered and uniform size metal and semiconductor nanoclusters has been visualized using the technique of scanning tunneling microscopy. Techniques such as atomic force microscopy and scanning tunneling microscopy have been used for the structural modification of materials and fabrication of devices at the atomic, molecular, and nanometer scales. Nanoscale technology offers a wide range of applications in almost every industry covering all cross-disciplinary areas of science, engineering, and medicine. The potential applications include high-density data storage devices, field effect transistors, flat screen displays, video projectors, computer monitors, energy storage, fuel cells, solar cells, sensors, detectors, actuators, transducers, photonic circuits, tunable lasers, quantum cascade lasers, nanorobots, supercomputers, molecular electronic devices, molecular motors, atom-sized chips and switches, atomic scale gears, integrated circuits, supercapacitors, lithography, nanoimprints, computing, cosmetics, adhesives, fabrics and textiles, catalysts, coatings, flame retardants, telecommunications and information technologies, anti-counterfeit and security technology, and automotive and aerospace applications. The potential applications for nanotechnology also include modification of a material's surface properties associated with surface phenomena such as catalysis, adhesion, corrosion, sintering, electrochemistry, composite formation, electronic packaging, etc. and utilizing those phenomena in applications in, for example, semiconductor processing, catalysis, paints, pigments, coatings, adhesives, vacuum technology, microelectronics, flat-panel displays, compact disks, televisions, computers, environmental monitoring of pollutants, optical coatings, energetic materials, etc. The potential applications of nanotechnology in fields of life sciences include drug and protein delivery, biomanipulations of human genes and individual living cells, tissue engineering, clinical and medical diagnostics, biomedical applications, nucleic acids, immunological assays, single-molecule analysis, health care products, pharmaceuticals, artificial muscles and organs, biomembranes, bioseparation processes, biopolymers for orthopedic and cardiovascular applications, dentistry, bone, molecular sensors, clinical diagnostic techniques, dendrimers for medicine, biomedical implantation, biomechanics, bioinstrumentation, nanoscale physiology and pathology, bioinformatics, nanoscale genomics, proteomics, biocomputing, instrumentation techniques for nanobioscience, nanoscale cellular and tissue engineering, detection and identification of cancerous tissues, pathogens and biological warfare agents as well clinical and biological diagnostics by chips and nanosensors, biomaterials, artificial joints, soft tissues, food safety and agriculture. The breakthroughs in nanoelectronics are a few years away from any other emerging nanotechnologies. Currently 64-bit chips are available for servers, desktops, notebooks, and laptops. Today, the fastest computer chips use 130 nm processors while the next generation 90 and 45 nm chips would offer 100 times faster speed and improved performance over today's 130 nm processors. We look forward to a lot of new breakthroughs in nanodevices and commercialization of these products over the coming years. Nanoscale technologies offer tremendous opportunities and promise for future generation industries to have an impact on the quality of daily life in the areas of health, energy, and environment. Nanotechnology has attracted extraordinary attention among the scientific community and institutions from around the world. There are many major research and development government programs in the United States, Europe, and Asia. In the United States, the National Nanotechnology Research and Development Act provides $3.67 billion over the next four years for research and development of new nanoscale technologies; this would eventually spur investments and accelerate more research and development activities in the field of nanotechnology. Major industrial corporations and governments around the world have started investing in nanoscale science and technology focused on building nanotechnology resources for future economic development. Nanotechnology pundits predict that the worldwide market for new emerging nanotechnologies could reach over $1 trillion by 2015. Beyond any doubt, nanotechnology is the most dominating technology of the 21st century.

The *Encyclopedia of Nanoscience and Nanotechnology*® is the world's first encyclopedia published in the field of nanotechnology. The 10-volume encyclopedia is an unprecedented single reference source that provides an ideal introduction and overview of the most recent advances and emerging new aspects of nanotechnology from science to engineering to medicine. This encyclopedia covers all theoretical and experimental aspects of nanoscale science and technology dealing with design, modeling, synthesis, fabrication, processing, characterization, and structure–property relationships of nanostructured ceramics, metals, semiconductors, alloys, polymers, and organic and biological materials in the form of nanocomposites, nanopowders, nanoparticles, nanotubes, nanowires, quantum dots, nanoclusters, nanocrystals, self-assemblies, supramolecules, monolayers, nanofilms, and their-based device applications. All kinds of nanostructured materials are now widely available on the market. The encyclopedia fills the gap to provide basic information on all fundamental and applied aspects of nanotechnology by drawing on two decades of pioneering research. It is the only scientific work of its kind chronicling the origins of nanotechnology and providing insight on core knowledge and the very latest advances. This groundbreaking encyclopedia is an indispensable source for research professionals, technology investors, and developers seeking the most up-to-date information on the current nanotechnologies among a wide range of disciplines from science to engineering to medicine. This is the world's first and only encyclopedia ever published in the field of nanoscience and nanotechnology with the most up-to-date information available as of today. The encyclopedia contains 419 review chapters contributed by over 900 of the world's leading scientists from academia, industries, and government laboratories. It has approximately 80,000 bibliographic citations offering extensive cross-referencing in each chapter. There are approximately 7500 figures, 800 tables, and hundreds of chemical structures and equations. Internationally known authoritative experts from 35 countries who are familiar with current technologies write the state-of-the-art review chapters. All state-of-the-art entries in the encyclopedia have been arranged alphabetically. The encyclopedia provides the most comprehensive coverage spanning the field of nanotechnology by compiling up-to-date physical data and recent information published by September 2003. Each state-of-the-art review chapter is self-contained with sufficient cross-references. Though some overlap may exist in a few chapters, significant efforts were made to minimize such overlap. The review chapters in the encyclopedia are written in a very clear and concise fashion.

This encyclopedia is intended for a very broad and diverse audience working in the fields of nanotechnology and nanoscience having strong connections with materials science, electrical and electronic engineering, microelectronics, solid-state physics, surface science, aerosol technology, chemistry, colloid science, ceramic and chemical engineering, polymer science and engineering, sol–gel science, supramolecular science, mechanical engineering, metallurgy and powder technology, optical science and engineering, optoelectronics, semiconductor technology, device engineering, aerospace engineering, microscopy, computer technology, information technology, environmental engineering, biology, biological physics, pharmacy, biotechnology, biomedical sciences, and food science. I hope this encyclopedia will be an invaluable reference source for academic and research libraries in universities, industrial institutions, and government and independent institutes, as well as for R&D and defense research laboratories, individual research groups, scientists, students, college and university professors, research professionals, technology investors and developers, research enterprises, and everyone working in the field of nanoscale science and technology.

I am highly grateful to all contributing authors for their tremendous efforts in writing timely and comprehensive state-of-the-art review chapters on many cross-disciplinary areas of nanoscience and nanotechnology. I cordially extend my very special thanks and congratulations to all distinguished colleagues who contributed to these volumes. This encyclopedia has been the most enormous effort of my publishing career. One of the most rewarding and remarkable editorial works undertaken has been putting together the world's first encyclopedia in the field of nanotechnology. It has been a great pleasure to work with so many wonderful and supportive colleagues from around the world over a period of three years. It is my hope that the scientific community working in this field will greatly benefit from the advanced technical information this 10-volume encyclopedia provides. I am personally very thankful to Robert McKinney at Cadmus Communications for his tireless efforts and great dedication in bringing this encyclopedia to fruition. I am highly thankful to David Tebera and Murali Krishnan for their outstanding support in completing this project. I also wish to thank Professor Satya Vir Arya, Professor Padma Vasudevan, Professor I. K. Varma, Professor Prem Vrat, Dr. Sivaram Arepalli, Rakesh Misra, Krishi Pal Raghuvanshi, Deepak Singhal, Yogesh, Jagmer Singh, Ranvir Singh Chaudhary, Ashish Kumar, and Anjali Kumari for their encouragement. I also thank all publishers and authors who granted us their copyright permissions to use their illustrations in this encyclopedia. Finally, I would like to give special acknowledgment and thanks for the continuous support and cooperation of my wife, Dr. Beena Singh Nalwa, the great enthusiasm given by my parents, Sri Kadam Singh and Srimati Sukh Devi, and the great love offered by my children, Surya, Ravina, and Eric, in this exciting endeavor.

I especially express my sincere gratitude to Professor Richard E. Smalley for his insightful Foreword and endorsements by Professor Jean-Marie Lehn and Professor C. N. R. Rao.

Hari Singh Nalwa
Los Angeles, California, USA

Endorsements

"In view of the great interest that Nanoscience and Nanotechnology are attracting, the rapid development they are experiencing and of the profound impact they will have, an Encyclopedia covering this field is most welcome. It should become a source of inspiration as well as of information for a very wide range of scientists."

Professor Jean-Marie Lehn
Nobel Prize Laureate

"The Encyclopedia of Nanoscience and Nanotechnology edited by Dr. Hari Singh Nalwa will be of great value to students, teachers and research workers interested in the area. I recommend this volume to individuals, research groups and libraries."

Professor C. N. R. Rao
Bangalore, India

Contents

Guide to the Encyclopedia

The *Encyclopedia of Nanoscience and Nanotechnology* is the world's first and only comprehensive single reference source compiling multidisciplinary research topics related to nanoscale science and technology in all disciplines of science, engineering, and medicine. It contains 419 chapters written by the world's leading experts, drawing on the past two decades of pioneering research work. It consists of 10 volumes and all chapters are organized in alphabetical order. Each chapter provides comprehensive coverage of the research activities on a selected topic starting from a very basic background for nonscientists to complete scientific and technological details including the most recent research developments in the field for experts. Therefore each chapter is designed for a very wide range of readers from students to highly trained research professionals. This guide provides a general outline on how this encyclopedia is organized and how to get the right information on a particular research topic so that readers could most benefit.

FINDING A CHAPTER

The chapters in the *Encyclopedia of Nanoscience and Nanotechnology* are organized in alphabetical order by title. Therefore readers are advised to consult the alphabetical Table of Contents to locate a particular topic. An A–Z sequence is followed. Therefore chapters having titles beginning with letters A through Ch are in Volume 1; titles from letter from Cl through E are in Volume 2. This alphabetical sequence continues through Z in Volume 10, which also has the detailed Index. The complete A–Z list of 419 chapters with appropriate volume and page numbers is located in Volume 10. All related topics are grouped together in a sequence. For example, all chapters beginning with the key terms "bio" or "carbon" appear together in Volume 1. The second method of locating a specific topic is to search the contents by Subject Area Section which follows the Table of Contents in Volume 10.

FORMAT OF CHAPTERS

All chapters in the *Encyclopedia of Nanoscience and Nanotechnology* have exactly the same format, which is as follows:

- Title
- Author and affiliation
- Contents
- Main body of the chapter
- Glossary
- References

This format is simple and easy for readers having any background in the different fields of science, engineering, and medicine. This format has been widely used in most scientific and technical books; the only difference is the inclusion of a Glossary for each chapter. We believe that our readers and librarians would like this format. The Contents section is intended to provide an overview. Therefore it lists only the first-level major headings of a chapter while second- and third-level headings are only available within the chapter. The main body of the chapter provides different levels of information about a topic. For example, the main body of a chapter dealing with "boron nitride nanotubes" will provide a full up-to-date coverage of the topic. The chapter would contain a brief historical background, all types of synthetic methods and spectroscopic techniques used for characterization, physical properties, and applications related to boron nitride nanotubes. Detailed tables extensively summarizing physical data available about a topic or a particular nanostructured material are provided, giving appropriate cross-references.

GLOSSARY

The Glossary defines the scientific and technical terms related to a particular topic that are important for readers. The terms related to the chapter are listed at the end, just before the References. The same term may appear in different chapters though the definition may vary slightly due to different writing styles. The encyclopedia includes approximately 3000 glossary entries.

REFERENCES

The sources listed in the References section are the author's recommendation of the most important and appropriate materials related to a particular topic for further detailed research. For example, a chapter on "Computational Atomic Nanodesings" includes 1026 references. The encyclopedia contains about 80,000 bibliographic citations from hundreds of different journals. Each chapter is self-contained and complete. The extensive cross-referencing in each chapter provides readers with a broader range of knowledge so that it could be consulted for further information on closely related topics.

INDEX

The Index appears as the last segment of Volume 10. The Subject Index is listed alphabetically and contains the volume and page number indicating where to locate a particular term or topic. Volume 10 includes a complete subject index for the entire work and an alphabetical list of all authors who contributed to this 10-volume encyclopedia.

ONLINE EDITION

The online edition offers the exact same chapters, format, and features as the print edition. All entries are organized in a single alphabetical sequence providing browsing capabilities within different categories. The online edition allows multiple users and fully searchable text.

COMPANION WORKS

The *Encyclopedia of Nanoscience and Nanotechnology* is part of a continuing program of nanotechnology works published by American Scientific Publishers that encompasses different nanotechnology journals and books.

JOURNALS

1. *Journal of Nanoscience and Nanotechnology*
2. *Journal of Computational and Theoretical Nanoscience*

BOOKS

1. *Handbook of Nanostructured Biomaterials and Their Applications in Nanobiotechnology*
2. *Handbook of Theoretical and Computational Nanotechnology*
3. *Handbook of Semiconductor Nanostructures and Devices*
4. *Handbook of Organic–Inorganic Hybrid Materials and Nanocomposites*
5. *Molecular Nanoelectronics*
6. *Magnetic Nanostructures*
7. *Nanoclusters and Nanocrystals*
8. *Quantum Dots and Nanowires*
9. *Synthesis, Functionalization and Surface Treatment of Nanoparticles*.

For information on American Scientific Publishers titles, please visit the website at:

www.aspbs.com

About the Editor

Dr. Hari Singh Nalwa is the President and Chief-Executive Officer of American Scientific Publishers based in Los Angeles, California, USA. Previously, he was Head of Department and R&D Manager at the Ciba Specialty Chemicals Corporation in Los Angeles (1999–2000) and a staff scientist at the Hitachi Research Laboratory, Hitachi Ltd., Japan (1990–1999). While working at Hitachi Ltd., he was also concurrently an Honorary Visiting Professor at the Indian Institute of Technology in New Delhi (1996–1998). He has authored over 150 scientific articles in journals and books. He has 18 patents to his credit on electronic and photonic materials and devices based on them.

Dr. Nalwa has published 73 books including *Ferroelectric Polymers* (Marcel Dekker, 1995), *Nonlinear Optics of Organic Molecules and Polymers* (CRC Press, 1997), *Organic Electroluminescent Materials and Devices* (Gordon & Breach, 1997), *Handbook of Organic Conductive Molecules and Polymers*, Vols. 1–4 (John Wiley & Sons, 1997), *Handbook of Low and High Dielectric Constant Materials and Their Applications*, Vols. 1–2 (Academic Press, 1999), *Handbook of Nanostructured Materials and Nanotechnology*, Vols. 1–5 (Academic Press, 2000), *Handbook of Advanced Electronic and Photonic Materials and Devices*, Vols. 1–10 (Academic Press, 2001), *Advanced Functional Molecules and Polymers*, Vols. 1–4 (Gordon & Breach, 2001), *Photodetectors and Fiber Optics* (Academic Press, 2001), *Silicon-Based Materials and Devices*, Vols. 1–2 (Academic Press, 2001), *Supramolecular Photosensitive and Electroactive Materials* (Academic Press, 2001), *Nanostructured Materials and Nanotechnology*—Condensed Edition (Academic Press, 2001), *Handbook of Thin Film Materials*, Vols. 1–5 (Academic Press, 2001). *Handbook of Surfaces and Interfaces of Materials*, Vols. 1–5 (Academic Press, 2001), *Semiconductors and Semimetals*, Vol. 73 (Academic Press, 2001), *Advances in Surface Science*, Vol. 38 (Academic Press, 2001), *Handbook of Polyelectrolytes and Their Applications*, Vols. 1–3 (American Scientific Publishers, 2002), *Magnetic Nanostructures* (American Scientific Publishers, 2002), *Nanoclusters and Nanocrystals* (American Scientific Publishers, 2003), *Handbook of Organic-Inorganic Hybrid Materials and Nanocomposites*, Vols. 1–2 (American Scientific Publishers, 2003), *Handbook of Luminescence, Display Materials, and Devices*, Vols. 1–3 (American Scientific Publishers, 2003), *Quantum Dots and Nanowires* (American Scientific Publishers, 2003). *Handbook of Photochemistry and Photobiology*, Vols. 1–4 (American Scientific Publishers, 2003), *Polymer Optical Fibers* (American Scientific Publishers, 2004), *Encyclopedia of Nanoscience and Nanotechnology*, Vols. 1–10 (American Scientific Publishers, 2004), and *Handbook of Nanostructured Biomaterials and Their Applications in Nanobiotechnology*, Vols. 1–2 (American Scientific Publishers, 2004). The *Handbook of Nanostructured Materials and Nanotechnology* edited by him received the 1999 Award of Excellence in Engineering Handbooks from the Association of American Publishers.

Dr. Nalwa is the founder and Editor-in-Chief of the *Journal of Nanoscience and Nanotechnology* (2001–). He also was the founder and Editor-in-Chief of the *Journal of Porphyrins and Phthalocyanines* published by John Wiley & Sons (1997–2000) and has served on the editorial boards of *Journal of Macromolecular Science-Physics* (1994–2000), *Applied Organometallic Chemistry* (1993–1999), *International Journal of Photoenergy* (1998–2001) and *Photonics Science News* (1995–). He has been a referee for many international journals including *Journal of American Chemical Society, Journal of Physical Chemistry, Applied Physics Letters, Journal of Applied Physics, Chemistry of Materials, Journal of Materials Science, Coordination Chemistry Reviews, Applied Organometallic Chemistry, Journal of Porphyrins and Phthalocyanines, Journal of Macromolecular Science-Physics, Applied Physics, Materials Research Bulletin*, and *Optical Communications*.

Dr. Nalwa helped organize the First International Symposium on the Crystal Growth of Organic Materials (Tokyo, 1989) and the Second International Symposium on Phthalocyanines (Edinburgh, 1998) under the auspices of the Royal Society of Chemistry. He also proposed a conference on porphyrins and phthalocyanines to the scientific community that, in part, was intended to promote public awareness of the *Journal of Porphyrins and Phthalocyanines*, which he founded in 1996. As a member of the organizing committee, he helped effectuate the First International Conference on Porphyrins and Phthalocyanines, which was held in Dijon, France in 2000. He was on the organizing committee of the BioMEMS and Smart Nanostructures, (December 17–19, 2001, Adelaide, Australia).

Dr. Nalwa has been cited in the *2000 Outstanding Intellectuals of the 21st Century, Order of Excellence, Dictionary of International Biography, Who's Who in Science and Engineering, Who's Who in America,* and *Who's Who in the World.* He is a member of the American Chemical Society (ACS), the American Physical Society (APS), the Materials Research Society (MRS), and the American Association for the Advancement of Science (AAAS). He has been awarded a number of prestigious fellowships including National Merit Scholarship, Indian Space Research Organization (ISRO) Fellowship, Council of Scientific and Industrial Research (CSIR) Senior fellowship, NEC fellowship, and Japanese Government Science & Technology Agency (STA) Fellowship.

Dr. Nalwa received a B.Sc. degree in Biosciences from Meerut University in 1974, a M.Sc. degree in Organic Chemistry from the Indian Institute of Technology at Roorkee (formerly University of Roorkee) in 1977, and a Ph.D. degree in Polymer Science from the Indian Institute of Technology in New Delhi in 1983. His thesis research focused on the electrical properties of macromolecules. Since then, his research activities and professional career have been devoted to studies of electronic and photonic organic and polymeric materials. His endeavors include molecular design, chemical synthesis, spectroscopic characterization, structure-property relationships, and evaluation of novel high performance materials for electronic and photonic applications. He was a Guest Scientist at Hann-Meitner Institute in West Berlin, Germany (1983) and Research Associate at the University of Southern California in Los Angeles (1984–1987) and State University of New York at Buffalo (1987–1988). In 1988, he moved to the Tokyo University of Agriculture and Technology in Japan as a lecturer (1988–1990), where he taught and conducted research on electronic and photonic materials. His research experience include ferroelectric polymers, electrically conducting polymers, polymer electrets, nonlinear optical materials for integrated optics, electroluminescent materials, low and high dielectric constant materials for microelectronics packaging, organic semiconductors, nanocrystalline and nanostructured materials. Langmuir–Blodgett films, high temperature resistant polymer composites, epoxy-based photopolymers, rapid modeling, and stereolithography.

Encyclopedia of Nanoscience and Nanotechnology
www.aspbs.com/enn

Electromagnetism in Nanogranular Metal Oxides

Christian Brosseau
Université de Bretagne Occidentale, Brest, France

CONTENTS

1. INTRODUCTION

The science of manipulating matter on the nanometer scale has provided scientists and engineers at the forefront of this research with a remarkable diverse set of tools for probing the behavior of nanophases. One of the strengths of this sector is the concomitance of the scientific interests in the universities with the economic interests of the markets for new nanotechnologies. In that respect, the electromagnetic characterization of dry granular materials can be considered a major part of the emerging field of advanced materials relevant to the fields of electronics, photonics, or information technology. Powder metallurgy in particular offers opportunities for engineering and ultimately trying to control novel features, functions, and properties of granular materials on different scales by using appropriate procedures, with promising commercial applications. Over the past decade or so, a substantial amount of research and development across the scientific community has been conducted in the areas of nanotechnology for materials and device structures whose characteristic feature sizes are at or near the nanometer scale and significant fundamental advances have been made; see, for example, [1–12] and references therein. In addition, the increasing demand for new materials that satisfy multiple roles, for example, electrical and mechanical functions, has provided the impetus for an important new area of research of nanomaterials science. An elaborate overview of the most recent advances and emerging new aspects of nanoscience and nanotechnology has been given by Nalwa et al. [13]. The potentially profound implications both for transport properties and optics are only beginning to be explored. Interest in nanophases has expanded as investigators have recognized that many of the properties of finely divided matter strongly depend on the interfacial properties of the constituents by virtue of the high fraction of the overall material that is in the vicinity of an interface as well as of the confinement of electrons, excitons, and photons in small volumes. This allows the exploration of a vast range of experimental regimens, from noninteracting to strongly correlated. Their highly specific surface area combined with their smallness led to the observation of many unique behaviors, such as superparamagnetism in Fe clusters embedded in a nonmagnetic medium [14]. Observe that usual microwave materials, such as the ferrites, are often heavy, and may not have very desirable mechanical properties. In contrast, micro- and nanostructured composite materials can be designed with considerable magnetic activity and can be made extremely light. One of the interesting and important issues in predicting and understanding nanostructures and their functional behaviors is whether the properties of matter evolve gradually from bulk, as system size is reduced, and what determines this evolutionary behavior. Because electronic properties of interfaces affect the carrier mobility and scattering in a material, plasmon excitations at planar interfaces and at nanoparticle interfaces have also been studied [15–17]. Recently, it has been argued that the electronic band structure of nanocrystals can be different from the band structure of the bulk semiconductor, that is, the opening of the semiconductor band gap due to electron confinement [2, 13]. From the technological point of view, the observation of giant magnetoresistance [18, 19] or large coercivities [20, 21] in various nanophase materials has attracted much attention. Despite a number of notable contributions, theory has often been outpaced by these rapid experimental advances. From an applications standpoint, powder metallurgy is a particularly attractive technique for preparation of nanophase materials in that it offers a rapid, cost-effective, and controllable route to yield materials with specific properties. To improve performance and reliability of powder metallurgy, experimenters have

ISBN: 1-58883-059-4/$35.00

developed a range of techniques for performing characterization of properties of nanophases such as structural, dielectric, magnetic, and/or coupling between of these properties. The practical importance and industrial interest in these materials demand optimization of several types of properties in these materials. These properties include polarization, magnetization, and stability of the materials to mechanical, electrical, and magnetic fields applied during processing and operation. One of the fundamental goals of this field should be the understanding of the relationships of these properties on the composition, particle size and boundaries variations, defect structure, and separation of the residual pores, but in most cases they are not well understood.

1.1. Powder Metallurgy Composites and Magnetic Properties

Although extensive literature exists on the properties of nanophase materials, only a limited number of systematic studies have been reported on the electromagnetic and magnetic properties of granular systems. It is useful to succintly summarize existing studies. Shaikh and co-workers [22] were among the first to report on the dielectric properties of ultrafine grained $BaTiO_3$: the permittivity was observed to increase with decreasing grain size, reaching a peak value at 1 μm and decreasing subsequently with any further decrease of the grain size. Panina and co-workers [23] found that the effective permeability of composite materials containing micrometer-sized, Fe particles tends to zero near the percolation threshold. Newnham and Trolier-McKinstry [24] examined ferroelectric nanocomposites consisting of particles in the nanometer scale dispersed in a matrix to get higher frequencies as well as other advantages such as miniaturization. Also noteworthy is the observation of Xiao and Chien [25] that the coercivity changes in a dramatic variation in granular Fe/silica mixtures, which was interpretated by invoking percolation effects. In recent years a number of studies [7, 19, 26–30] have focused on the behavior of fine magnetic particles in a nonmagnetic metallic matrix. Until now, the vast majority of these publications suggest that the magnetic state of the particles is influenced by the effect of the surface whose contribution becomes predominant. On another front, questions concerning the coercivity and reversal magnetization in nanophases have also come to the fore. Particles of ferromagnetic metals with dimension typically below 100 nm are too small to show well-developed domain structures [31] because the exchange correlation length is much greater than the particle size. For ultrafine magnetically ordered particles, there exists a critical size below which the grains can acquire only single magnetic domains even in zero applied magnetic field. The critical size of a single domain particle can be evaluated $D^* \cong \frac{36}{2\pi}(AK_1)^{1/2}M_S^{-2}$, where A is the exchange constant and K_J is the magnetocrystalline anisotropy constant. For Fe, Ni, and Co, D^* is on the order of a few tens of nanometers, depending on the shape of the particle. As areal recording density in magnetic storage devices approaches a few 10^2 Gbits/in^2, the medium should possess magnetically isolated fine grains (less than 10 nm in gram size) exhibiting a narrow-size distribution. The areal density remains limited by the random size of the grains that may couple to form larger magnetic switching units. Moreover, there is a fair amount of experimental evidence that shows an enhanced coercivity for nanoparticles [32]. In the sintering process, compaction occurs due to the enhanced diffusion between neighboring particles. The advantage of the hot isostatic pressing process is that the porosity of the final product is negligible, and therefore the basic properties of the product are better. The technique of mechanical alloying typically involves the high-energy ball milling of elemental powder mixtures. During the milling process, large powder particles may undergo plastic deformation and/or fragmentation, and small powder particles may coalesce by cold welding. The repeated fragmentation and coalescence of powder particles causes the intimate mixing of particles on the nanoscale. Meanwhile, the exchange coupling between neighboring magnetic particles can overcome the anisotropy and demagnetizing effect of individual particles [33]. From the theoretical point of view, Stoner and Wohlfarth (SW) were pioneers in studying the hysteretic behavior of a random distribution of noninteracting magnetic particles [34, 35]. This early, insightful, work (in 1948) has pointed toward interesting effects that are specific to single-domain particles, for example, the reduced remanence ratio $\frac{M_r}{M_s}$ is equal to 0.5 for uniaxial particles, and the coercivity is $H_c = 0.48(\frac{2K_1}{M_s})$, where K_1 is the magnetocrystalline anisotropy constant. Since its introduction, the SW model was expanded by others [36] for multiaxial cubic or competing types of magnetic anisotropy. However, the SW model is static theory and does not take into account thermal fluctuations in the magnetization process: therefore, its range of validity is restricted to low temperatures or very strong anisotropies [37]. Still in use in the current literature to explore the case of higher temperatures and/or if anisotropies are small is the superparamagnetic Néel–Brown approach eventually modified to include (magnetostatic) dipolar interactions among magnetic nanograins, which provides an important framework for the understanding and interpretation of experimental data [38–40]. Within this framework, magnetization reversal takes place by thermal activation over finite energy barriers, leading to temperature and time dependences of the magnetic behavior. However, it is worth noting that the effect of interparticle interactions cannot be easily included in this model, except as a mean-field approximation. Some authors argued that the enhanced remanence ratio (>0.5) observed experimentally in nanoscale isotropic magnetic materials may be due to the exchange coupling between neighboring nanograins [41, 42]. It should, perhaps, be emphasized in this context that one important, and as yet not completely answered, question concerns the understanding and the control of the dynamic magnetization processes in these granular materials on nanosecond time scales. A general framework for understanding the damping mechanism in nanostructures is proving elusive. Finally, we observe that these basic and applied considerations have stimulated much research concerning the effect of both dipolar and exchange interactions on the magnetic and transport properties of interacting nanoscale particles [27, 29, 35, 43]. However, it is worth noting that some of the results are controversial, and several studies reaching contradictory conclusions exist.

1.2. Electromagnetic Transport in Heterostructures

Theoreticians pointed out that the theoretical framework used in many instances to explain the electromagnetic transport in granular materials is a modelization in terms of a continuum media. As previously observed elsewhere [44], the basic assumption used is that there exist length scales above which fluctuations at the microscopic level may be ignored. In this way, useful predictions of a granular material's bulk properties can be made with no references to its discrete nature. At issue is the manner in which these local fluctuations influence the dynamics of ensemble-averaged quantities such as the effective permittivity ε and the effective permeability μ. A related basic question is to examine the links with coarse-graining models based on the effective medium approximation (EMA). This major simplifying approximation is based on a self-consistent procedure in which a grain of one component is assumed to have a convenient shape and to be embedded in an effective medium whose properties are determined self-consistently. In particular, it is natural to ask if this long-wavelength physics approach is still relevant to describe (and, if so, how) the response of the material at microwave frequencies by the materials parameters ε and μ and what happens in granular media, such as those made up of very fine grains. Particularly related to the current work are earlier model calculations which showed that the macroscopic character of the materials parameters ε and μ of two-phase systems depends sensitively on the microstructural properties such as grain size, particle shape, and grain boundaries type [45]. Early conceptualizations of EMA approaches arc given in [44–47]. Another classic review we found useful for general information is that of Bergman and Stroud [48]. Historically, one fruitful venue for studying the electromagnetic properties of random heterostructures has been the concept of percolation, which may shine light on the importance of the three-dimensional (3D) metal-insulator transition [49]. The fundamental precept of the percolation model is the prediction of a percolation threshold above which particles form a connecting network. Additionally, advances in *ab initio* computational techniques along with constantly growing computational resources now provide opportunities for the development of a much more profound understanding of the bulk magnetic and transport properties of granular magnetic solids [45, 46]. The power of these computational techniques, as a viable alternative to expensive or complicated laboratory experiments, provides theoreticians with a tool for exploring and predicting the electromagnetic properties for nanostructured materials. In the past it has been demonstrated that there are numerous questions or unresolved issues on the effective materials properties of heterostructures. One such issue is the effect of size on the materials parameters ε and μ. There are also interesting microwave μ measurements on submicrometer-size dispersed Co_x,Ni_{1-x} particles, where it was shown that losses are associated with domain structure or surface effects [50]. Additionally, there are few quantitative data on the effect of the distribution of the magnetostatic interaction fields between the particles. There is another good motivation to study this issue. In recent years, a new race for the understanding of magnetoelectric properties in condensed matter systems was initiated by the recent discovery of new types of unconventional materials, that is, bianisotropic media in which the electric displacement and the magnetic field vectors depend on both the electric field and the magnetic induction vectors [51]. However, the problem remains unsolved due to its intrinsic many-body complexity from intergranular magnetostatic and exchange interactions. The interplay between collective magnetic properties and transport properties in these systems presents a rich phenomenology and offers the prospect of new functionality for specific applications [35, 51]. This chapter is intended to introduce the reader to the field by providing some insights into some of the theoretical and experimental approaches that have been employed thus far to understand the relationships between particle size and the dynamical and hysteretic properties of granular solids consisting of heterogeneous dispersions of magnetic particles with dimensions ranging from nanometers to micrometers in a nonmagnetic host. The intriguing prospective of extending magnetic and electromagnetic studies of heterostructures into the domain of nanophases may be hindered by the collective transport properties of strongly correlated spins systems. Therefore, magnetic and electromagnetic behaviors in nanosystems are challenging in terms of both experimental observation and development of theoretical analyses.

2. NANOGRANULAR METAL OXIDES

To our knowledge, to date few experimental studies have addressed the physical mechanisms involved in the electromagnetic wave transport in nanogranular metal oxides, with only a few systems having received significant attention; see Table 1 for a summary. It should be emphasized parenthetically that the oxidation mechanisms of metal nanoparticles have considerable fundamental and technical importance and have been the focus of many experimental and theoretical studies during the past decade. Due to the high fraction of the overall material in the vicinity of the surface, nanoparticles can be extremely susceptible to oxidation. One way to ensure the chemical stability of nanoparticles is to encapsulate the reactive particles in a protective matrix. In clinical applications, the coating protects the nanoparticles from leaching in an acidic environment. In the current work, we will examine the links between the properties of the individual particles with those of the bulk materials. The study of these materials offered the opportunity to explore the use

Table 1. Composite materials studied.

Composite materials	Ref.
Fe/SiO_2	[42]
Fe/Al_2O_3	[42]
γ-Fe_2O_3/ZnO	[29, 66, this work]
$MgO/Ba_xSr_{1-x}TiO_3$	[99]
$LaAlO^3/Ba_xSr_{1-x}TiO_3$	[100]
$SrTiO^3/Ba_xSr_{1-x}TiO_3$	[101]

Note: In recent years, understanding of the electromagnetic properties of nanogranular metal oxides has advanced steadily. The materials studied along with the corresponding reference are listed in the table.

of powder metallurgy as a route to tunable magnetoelectric materials, with properties tailored to the demands of various applications. In writing this chapter, we have chosen naturally to draw on our expertise, which is electromagnetism of nanostructures. We consider the ZnO–Fe_2O_3 system as a prototype system to illustrate the magnetic and electromagnetic transport properties of particulate composite materials for at least four main reasons. First, these metal oxides are commonly found in technological applications; for example, ZnO is a remarkable II–VI semiconductor with wide applications for various devices such as optical waveguides [52], piezoelectric transducers [53], and varistors [54–56], and maghemite (γ-Fe_2O_3) is the most popular general-purpose magnetic tape material [57–59]. Second, characterization of the crystal oxides has been carried out by a variety of experimental techniques. For example, the magnetic behavior of γ-Fe_2O_3 has been studied extensively: it is a ferrimagnetic spinel with a bulk saturation magnetization of 390 kA m^{-1} [60–62]. Much work has been devoted to the magnetic properties of γ-Fe_2O_3 nanoparticles; for example, Dai et al. [62] found that the long-range magnetostatic interactions between surfactant coated γ-Fe_2O_3 nanoparticles and the blocking temperature (below which the moments are blocked) can be increased by decreasing the average interparticle distance through an increase of the applied pressure on the bulk samples. γ-Fe_2O_3 is of particular interest because of its applications in devices such as magnetic recording media. γ-Fe_2O_3 is a metastable phase at ambiant conditions and tends to transform to a stable-phase, antiferromagnetic hematite (γ-Fe_2O_3), when heated up to 400 °C. Ferrimagnetic oxides of iron combine the properties of a magnetic material with that of an insulator. Commercially available pure ZnO and γ-Fe_2O_3, powders were used as basic components of the composite materials and had different origins. Third, the low cost and ease of fabrication of these materials make study of the electromagnetic and magnetic properties of their small clusters, where the size and composition can be varied, extremely interesting. Fourth, nanostructures of these materials exhibit materials properties that can differ strongly from those of the bulk phases [62–66]. We are convinced that the comprehension of complex problems posed by nanogranular metal oxides has the potential to produce a substantial advance in electromagnetism and magnetism of condensed matter nanophases and nanostructures and will eventually shed light into other unsolved problems.

Because there is no review that covers electromagnetism of nanophases, our emphasis has been on producing a self-contained account of the basic physical pictures, while referring the reader to the original literature for details. Computerized literature searches citing "electromagnetism of nanogranular metal oxides" as key words find very few articles. As a result, it is impossible to give a comprehensive review of the electromagnetic properties of many of these materials. We shall therefore limit ourselves to discussing an overview of some of the electromagnetic and magnetic properties of heterostructures composed of metal oxide nanoparticulate powders, Samples consist of grains of the magnetic phase (γ-Fe_2O_3) homogeneously dispersed in a nonmagnetic insulating background (ZnO). In this review the starting powders had different origins. The micro oxide powders were purchased from Aldrich Chemical Co. and marked as $\geq$99% pure, whereas the nano oxide powders were supplied from Nanophase Technologies Corp. (Burr Ridge, II). The manufacturer's specifications are listed in Table 2. Control over the average particle size and shape was achieved by transmission electron microscopy (TEM). Epoxy resin (Scotchcast 265) purchased from 3M and was used as binder phase. Composite materials for this investigation were prepared by a four-step procedure. (1) The appropriate amounts of metal oxide powders were first preheated at 200 °C for dehydratation, and then, after cooling to room temperature, the powders were subsequently mixed with acetone and a certain amount of epoxy resin in a bath sonicator for 1 h; the mixture was then cast into a culture dish and the acetone was allowed to evaporate slowly at room temperature. (2) The mold was filled by the resulting mixturer. (3) The powder was compacted to a specified size and shape. (4) The dense powder compact was ejected from the mold. Room temperature-pressed ZnO/Fe_2O_3 compacts were made under a pressure of 10^7 N m^{-2} for 2 min. As a result of this procedure, test samples with a toroidal shape of 7 mm outside diameter were used for electromagnetic spectroscopy experiments. In the following we refer to the samples containing micrometer-sized particles and those containing nanometer-sized particles as M-type and N-type samples, respectively. We studied three sets of samples that nominally had volume fractions of 0.10, 0.15, and 0.25 of

Table 2. Selected physical properties of the powders investigated in this study.

Powder	ZnO	ZnO	γ-Fe_2O_3	γ-Fe_2O_3
Average particle size[a, b]	1 μm	49 nm	5 μm	23 nm
Powder color	White	White	Purple	Brown
Specific surface area BET (m^2 g^{-1})		22		51
Morphology	Elongated	Elongated	Nearly spherical, faceted	Nearly spherical, faceted
Crystal phase	Wurtzite	Wurtzite	Maghemite (cubic spinel)	Maghemite (cubic spinel)
Density[a] (g cm^{-3})	5.61	5.6	5.24	5.2

[a] From manufacturer product literature.
[b] Determined from specific surface area.

epoxy. The fractional volume of voids (porosity) is deduced knowing the volume fractions of ZnO, Fe_2O_3, and epoxy and was controlled by density measurements. Throughout the text, f denotes the volume fraction of Fe_2O_3 and ϕ indicates the volume fraction of the epoxy resin.

3. PHYSICAL PROPERTIES

In this chapter we focus on the present understanding of what we think are the most remarkable electromagnetic properties of nanogranular metal oxides, observed by state of the art experiments and interpreted with modem theories. We have organized the discussion into three main modules: (1) zero-field (ZF) permittivity and permeability of a set of mixtures having different compositions and grain sizes (see Appendix A for the measurement details), (2) gyromagnetic resonance, and (3) magnetic hysteresis loop data for these samples. We end the chapter with some concluding remarks and a brief look at some open questions that remain in the electromagnetic properties of particulate nanophases.

TEM bright field using a JEOL microscope at an accelerating voltage of 100 kV was used to reveal the grain size and shape of the as-received ZnO and Fe_2O_3 particles. The particles in solution were brought onto holey carbon films suspended over copper grids for TEM observations. From TEM micrographs of the powders, we have observed that the individual ZnO particle is elongated with an average aspect ratio of about 3:1 whereas γ-Fe_2O_3 particles are nearly spherically shaped. To obtain information on the morphology of the mixtures, scanning electron microscopy (SEM) data were collected using JEOL 6400 scanning electron microscopy operating at an accelerating voltage of 25 kV and magnification of 60,000. Samples were mounted on an aluminium stub using an adhesive carbon tab and were gold-coated. A representative SEM image of a cross-sectional view of a composite sample is displayed in Figure 1. A visual examination of Figure 1 reveals a finely dispersed structure and a large void distribution. The identification of the crystalline phases was carried out by powder X-ray diffraction (XRD) with Cu K_α, radiation. ZnO crystallizes in the hexagonal wurtzite structure whereas the crystal structure of Fe_2O_3 is maghemite

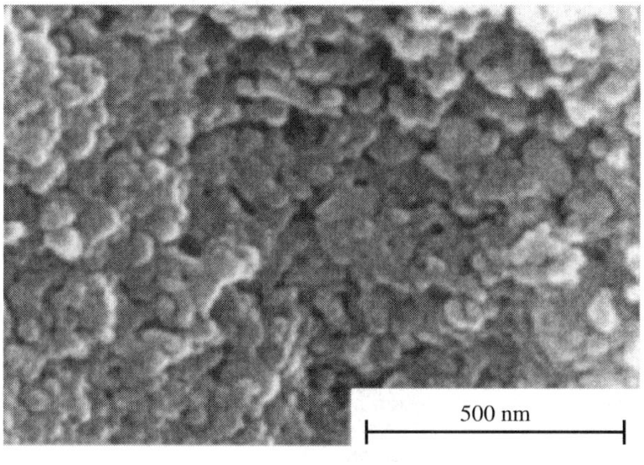

Figure 1. SEM micrographs of fracture cross-section view of a composite sample ($f = 0.53$ and $\phi = 0.25$) showing the distribution of particle sizes. The black bar in the lower right corner is 500 nm.

(cubic spinel). The average grain sizes were obtained in the XRD analysis from the line broadening of the diffraction peak and by using the Scherrer formula and are consistent with the estimates obtained from TEM.

3.1. Dielectric Properties

Representative frequency dependence curves of the real and imaginary parts of the relative effective permittivity ε for N and M composite. samples ($f = 0.53$ and $\phi = 0.25$) are plotted in Figure 2a and b. Upon inspection of these figures we observe that ε' is a slowly decreasing function for N samples and constant for M samples in the range of frequencies explored. The larger conductivity for N composite samples, shown in Figure 2b, can probably be attributed to the increase of the density of interfaces as surface-to-volume ratio increases in magnetic nanostructures. Figure 2c and d show the corresponding plots for the real and imaginary parts of the relative permeability μ. From Figure 2c and d, one can see the differences between the permeability of M samples and that of N samples. Perhaps the most intriguing fact is that whatever the volume fraction of Fe_2O_3 investigated (from 0 to 58 vol%), M samples do not show absorbing properties in the microwave range. To gain a better understanding of how the composition affects the permittivity and permeability, a set of experiments was conducted in which the volume fraction of Fe_2O_3 is varied, as illustrated in Figure 3a and d. It is noticeable that the data in Figure 3a and b, corresponding to M and N samples at $F = \frac{\omega}{2\pi} =$ 1 GHz, are almost coincident. Another important observation is that there is no evidence for a percolation threshold over the whole volume fraction of Fe_2O_3 range investigated. Figure 3c and d indicate that the $\mu'(f)$ and $\mu''(f)$ curves show substantial changes from M and N samples; for example, μ' is enhanced by a factor of 3 and μ'' is enhanced by a factor of 20 for $f \approx 0.5$

The experimental plots of $\varepsilon'(f)$ and $\varepsilon''(f)$ were compared with the results of simple calculations in Figure 3a and b using the Bruggeman symmetric medium equation; that is, $\sum_i f_i((\varepsilon - \varepsilon_i)/(2\varepsilon + \varepsilon_i)) = 0$, where f_i and ε_i denote the volume fraction and permittivity of the constituent i ($i =$ Fe_2O_3, ZnO, epoxy resin or air). See Appendix B for details. The inputs Bruggeman's equation were obtained from our measurements and the literature [66–70]. The permittivities predicted by this model are inconsistent with experiment in magnitude and in composition trend.

We believe that the discrepancies originate from two assumptions. First, the theory of Bruggeman assumes that the electric dipole–dipole interactions between the inclusions is taken into account in an average way. Unlike the dilute limit case, for which dipolar interactions provide a good description of the electrostatic interactions between the inclusions provided that the composite has the assumed geometry (in which a material of one type is embedded in an identifiable host material), such an approach fails to account for high concentrations of inclusions. Moreover, it does not depend explicitly on the size of the grains. This is because only the dipole term, which is proportional to grain volume, is retained in the series expansion of the amplitude of the magnetic field scattered by a single grain. Thus, only volume fractions appear in this approach. A second source of this

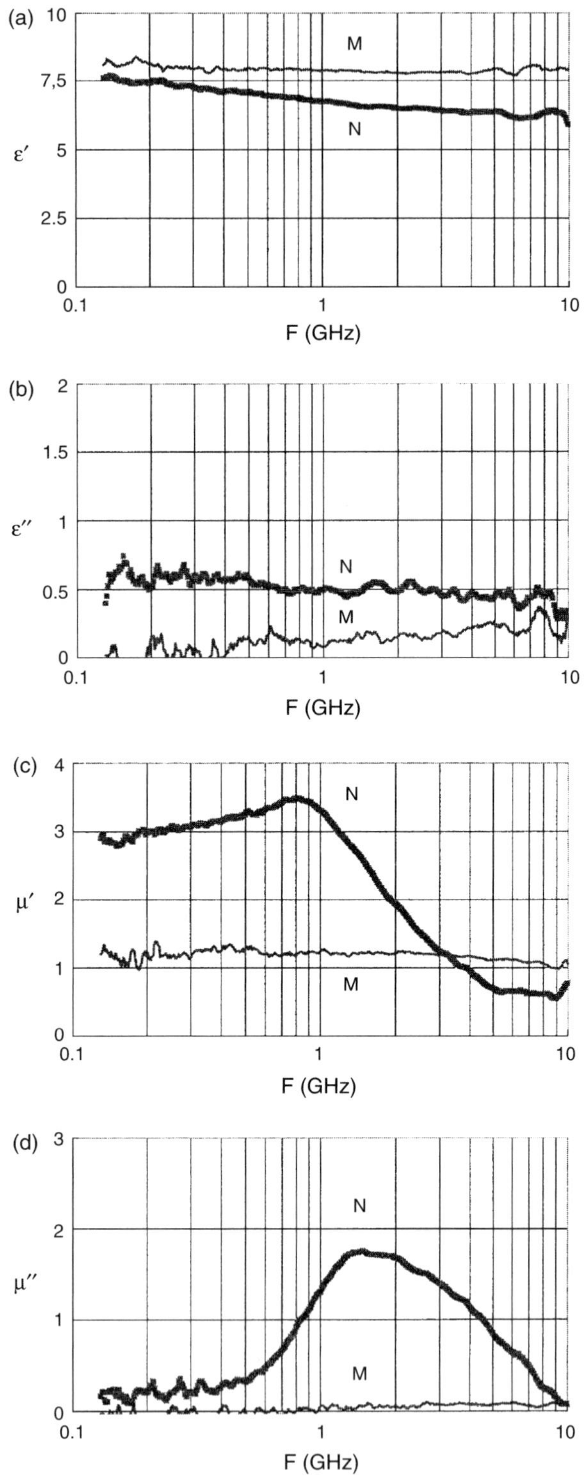

Figure 2. The frequency dependence of the relative permittivity ε and the relative permeability μ for micrometer-sized (M, red curve) and nanometer-sized (N, thick red curve) ZnO/Fe_2O_3 composite samples and a given value of the volume fraction of Fe_2O_3, $f = 0.53$. Room temperature. $\phi = 0.25$ (a) Real part ε', (b) imaginary part ε'', (c) real part μ', and (d) imaginary part μ''.

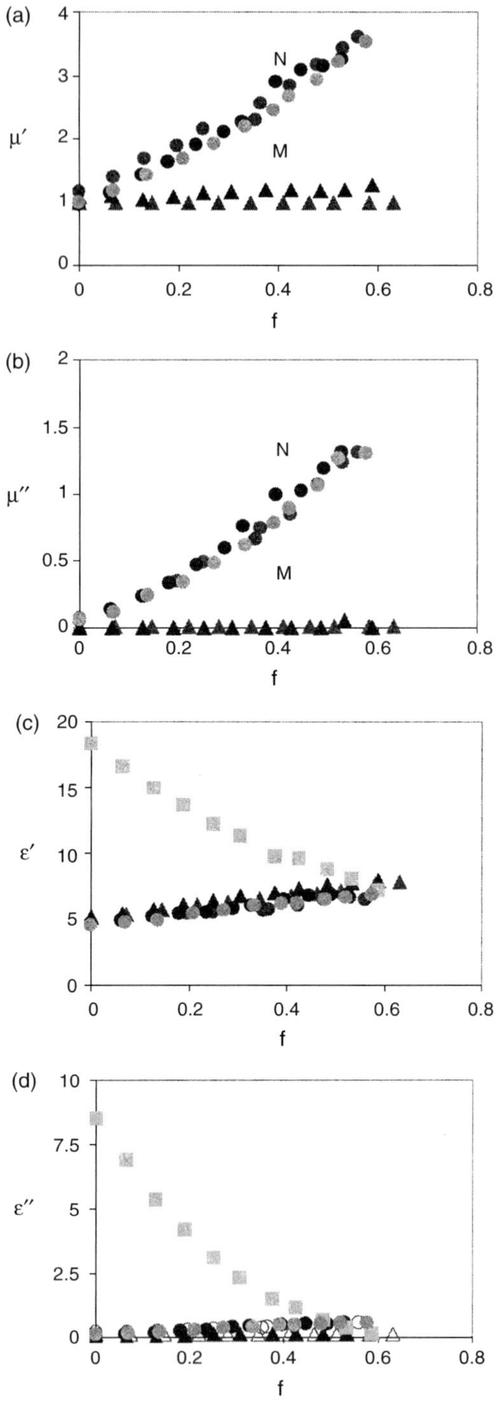

Figure 3. Variation of the relative permittivity as a function of the volume fraction Fe_2O_3, for M-type (triangles) and N-type (circles) ZnO/Fe_2O_3 composite samples at a given value of frequency, $F = 1$ GHz. Room temperature. Squares in (a) and (b) indicate the values of ε' and ε'' calculated using the mean-field Bruggeman model and taking as inputs $\varepsilon_{ZnO}(1\ \text{GHz}) = 38.2 - j0.3$, $\varepsilon_{Fe_2O_3}$ (1 GHz) $= 14.2 - j0.3$, ε_{epoxy} (1 GHz) $= 5 - j0.01$, and ε_{air} (1 GHz) $= 1 - j0$. Observe that the caluculated values take into account the porosity of the samples. Red symbols correspond to $\phi = 0.25$, green symbols correspond to $\phi = 0.15$, and light blue symbols correspond to $\phi = 0.10$. (a) real part ε' of the relative permittivity, (b) imaginary part ε'' of the relative permittivity, (c) real part μ' of the relative permeability, and (d) imaginary part μ'' of the relative permeability.

discrepancy is that the parameters for the material are not uniform, but rather they are influenced by the presence of defects such as impurities and grain boundaries. It is usually argued that EMA approaches have the disadvantages typical of all mean-field theories because they Ignore the fluctuations in the system. It assumes that the local electric field is the same in the volume occupied by each constituent in the heterostructure. The microscopic local fields near the interfaces may differ considerably from those in the bulk regions because of the existence of the scattering fields from boundaries [71]. We call attention to the fact that the standard effective (continuous) medium approximations underlying the present calculations were done under the simplifying assumption of spherical particles, even if there is a significant departure from perfect sphericity for the ZnO particles. However, the use of depolarizing factors to account for shape anisotropy of the ZnO particles still leads to a significant deviation between the mean-field theory and experiment.

Observe also that no conduction (percolation) threshold is evidenced, indicating a disconnected Fe_2O_3 particles network. In these powder compact composites, the presence of insulating grains and of the amorphous epoxy resin increases the conduction barrier height and therefore is expected to decrease the overall conductivity of the samples. However, even if the volume fraction of Fe_2O_3—and hence the intergrain spacing—and grain size were independently adjustable, we have not yet studied the local charge transport mechanisms. A difficulty is that transport through such systems often involves processes that are strongly influenced by local fluctuations, including variations in morphology and particle sizes.

3.2. Gyromagnetic Resonance

An interesting observation can be made in the results of Figure 3. In addition, a 3D rendering of the $\mu'(f, F)$ and $\mu''(f, F)$ surfaces over the f–F plane is exhibited in Figure 4a and b. It is apparent from Figure 4c that the resonance frequency F_0 is slightly increased the percentage of Fe_2O_3 is increased. One can see from Figure 4d that the resonance width at half-height ΔF decreases significantly as the percentage of Fe_2O_3 is increased. For completeness, we also show the three sets of data corresponding to $f = 25$ vol% for ε', ε'', μ', and μ'' in Figure 5a through d, where the epoxy concentration is increased from $\phi = 10$ vol% to $\phi = 15$ and 25 vol%. Careful inspection of the μ'' curves with different epoxy volume fractions reveals a fine structure of the broad resonance line. This effect is most pronounced for the curve corresponding to $\phi = 25$ vol% in Figure 5d.

We now turn our attention to a phenomenological analysis of the gyromagnetic resonance observed in N-type samples. To go further one must fit the gyromagnetic resonance to an appropriate functional form for the dynamics of magnetization. We now give technical information about the calculations and analyze the results in detail. The dynamics of an individual magnetic moment can be described by a magnetization vector $\vec{M}$. In theoretical aspect, the response of this magnetization vector to the effective field is commonly described by the Landau–Lifshitz–Gilbert (LLG) equation of motion, which has the

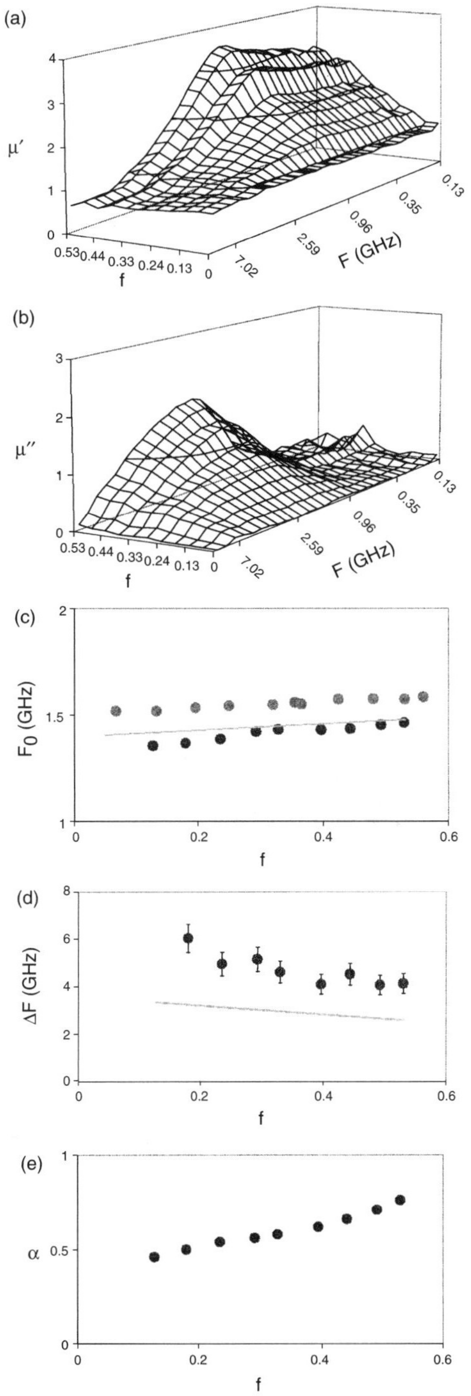

Figure 4. (a) Real part μ' of the permeability on a (f, F) surface. $f = 0.53$ and $\phi = 0.25$. (b) Same as in Figure 4a for the imaginary part μ''. $f = 0.53$ and $\phi = 0.25$. (c) The resonance frequency F_0 as a function of the volume fraction of Fe_2O_3. The blue line shows the calculated values using our modeling approach (see text). Red symbols correspond to $\phi = 0.25$ and green symbols correspond to $\phi = 0.15$. (d) The resonance width ΔF as a function of the volume fraction of Fe_2O_3. The blue line shows the calculated values using our modeling approach (see text). Same notation for the symbols as in Figure 4c. (e) The damping constant α, obtained from the fit of the gyromagnetic resonance, as a function of the volume fraction of Fe_2O_3.

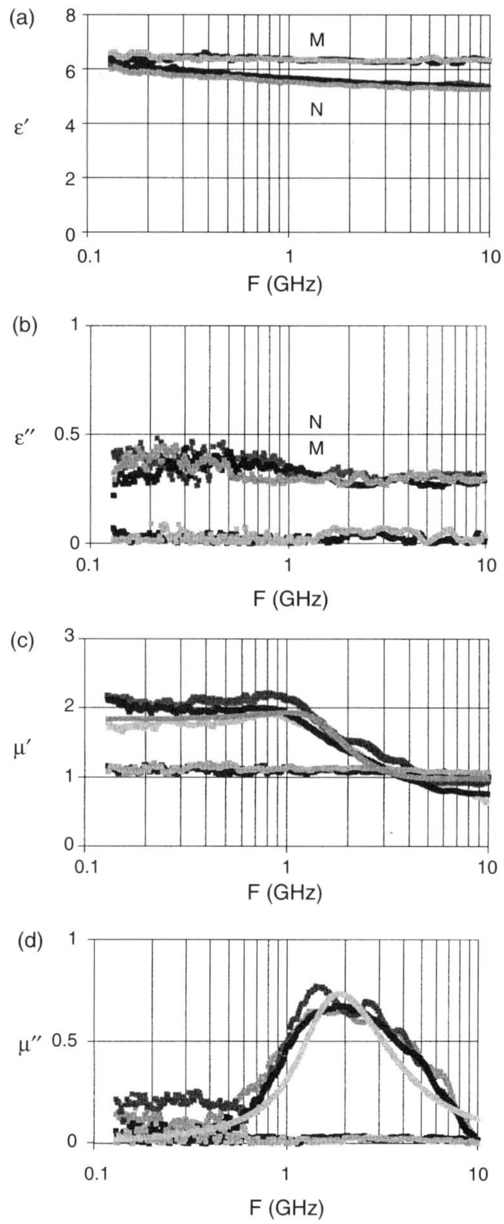

Figure 5. The frequency dependence of the relative permittivity ε and the relative permeability μ for micrometer-sized (N) ZnO/Fe_2O_3 composite samples for different volume fractions of epoxy and a given value of the volume fraction of Fe_2O_3, $f = 0.25$. Room temperature. The different colors are: (red, $\phi = 0.25$), (green, $\phi = 0.15$) and (dark blue, $\phi = 0.10$) for N-type samples, and (black, $\phi = 0.25$) and (orange, $\phi = 0.15$) for M-type samples. The light blue curves in (c) and (d) are fits of μ' and μ'' data by the Landau–Lifshitz–Gilbert equation (see text) using one free parameter: $\alpha = 0.5$. The magnetic dilution was described using the Bruggeman equation (see text). (a) Real part ε', (b) imaginary part ε'', (c) real part μ', and (d) imaginary part μ''.

form $\frac{\partial \vec{M}}{\partial t} = -\gamma \vec{M} \times \vec{H}_{\text{eff}} + \frac{\alpha}{M_s} \vec{M} \times \frac{\partial \vec{M}}{\partial t}$. The first term of the right-hand side of this equation accounts for the gyromagnetic precession of the magnetization $\vec{M}$. The second term arises from viscous damping and was introduced phenomenologically just from simple geometric considerations. At high damping, the magnetization rotates more or less directly toward the direction of the field as the second term is dominant. The term γ is the gyromagnetic ratio of the free electron spin ($\gamma = 0.0176$ Oe^{-1} ns^{-1}), and α is the Gilbert dimensionless damping constant [72, 73]. The effective field term H_{eff} follows from the variation of the magnetic Gibbs free enthalpy G with respect to the magnetization (i.e., $\vec{H}_{\text{eff}} = \frac{-\partial G}{\partial \vec{M}}$) and includes the external applied field H_0, the exchange field, the magnetocrystalline field, and the demagnetizing field. Our approach can be quantified in the following manner. Because all of our experiments were conducted at wavelengths much larger than the characteristic internal lengths, that is, interparticle spacing and particle size, and because the Fe_2O_3 particles are polycrystalline with no preferred orientation in the samples giving all possible angular variations with respect to the magnetization, it is convenient to use the Bruggeman symmetric equation (i.e., $\sum_i f_i((\mu - \mu_i)/(2\mu + \mu_i)) = 0$, where μ_i represents the permeability of the ith component with volume fraction f_i), for it can explicitly provide the composition dependence of the bulk effective permeability. As before, the summation has four terms corresponding, respectively, to γ-Fe_2O_3 ($\mu_1 = \mu_1' - j\mu_1''$), ZnO ($\mu_2 = 1 - j0$), epoxy resin ($\mu_3 = 1 - j0$), and porosity ($\mu_4 = 1 - j0$). The other ingredient of the model was to describe the dynamics of magnetization by the phenomenological LLG equation. Following Suhl [72], are can easily show that

$$\mu_1' = 1 + \frac{\omega_0\omega_M(\omega_0^2 - \omega^2) + \omega_0\omega_M\omega^2\alpha}{[\omega_0^2 - \omega^2(1+\alpha^2)]^2 + 4\omega_0^2\omega^2\alpha^2}$$

and

$$\mu_1'' = \frac{\omega_0\omega_M[\omega_0^2 + \omega^2(1+\alpha^2)]}{[\omega_0^2 - \omega^2(1+\alpha^2)]^2 + 4\omega_0^2\omega^2\alpha^2}$$

where ω_0 and ω_M are respectively, the gyromagnetic angular frequency given by γH_0 and the saturation magnetization frequency given by γM_s. The procedure described above was applied to fit the shape of the resonance line with only one adjustable parameter α. The result is given in Figure 5c and d. A reasonable agreement between theory and experiment is observed, considering the simplicity of the model; that is, the calculated $\mu'(F)$ and $\mu''(F)$ spectra are nicely reproduced but show some deviations that cannot be removed by tuning α. We found the following value for α: 0.50. Our appraisal of the literature indicates that the room temperature values of α, obtained from ferromagnetic resonance (FMR) linewidths, of about 0.33–0.45 were found for nanocomposites of iron particles dispersed in a ZnO matrix [29]. It must, however, be emphasized that smaller values of α have been reported for pure polycrystalline Fe films [74], but as explained by the authors of [74], the actual linewidths observed by FMR are somewhat larger than those attributable only to intrinsic damping. While our modeling of the gyromagnetic resonance line seems reasonable, it apparently is not the whole story. A remaining and crucial question comes from the origin of the large value of the Gilbert dimensionless damping constant α. It should be remembered that an enhanced Gilbert damping constant was found in ultrathin iron films [75] although this provides a different configuration of surface interface regions because the surface area is parallel to the sample's magnetization (assuming that the magnetization is held in the plane of the film). Lack of microscopic theories is one reason for the remaining fundamental questions and debate concerning the interpretation of a large value of α. We tend to be cautious about the actual prediction of the damping constant as discussed here because this value may be exaggerated. Knowledge of the damping constant in nanophases is minimal because it is difficult to probe it directly. Two (possibly interrelated) aspects are important here: the origin of the damping mechanism and its intensity. In any event, it would appear that, even if the notion of exchange interaction beyond nearest neighboring grains notably affecting the high-frequency spin dynamics seems reasonable, the details have not been completely disentangled. One has no rigorous calculation of the characteristic length scale on which this process ought to proceed. It is desirable to develop an analysis based on micromagnetics calculations to check the result displayed in Figure 5.

As can be shown in Figure 4c, this model reproduces satisfactorily the experimentally observed slight increase of the resonance mode F_0 as the volume fraction of Fe_2O_3 is increased within experimental uncertainty. We also note from this picture that the results obtained for the two lowest concentrations of epoxy are close to each other. It is also instructive to examine the relative importance of the resonance width at half-height ΔF shown in Figure 4d as a function of composition. The decrease of ΔF with increasing f that we obtained by the above model is also shown in Figure 4d for reference. The amount of decrease is lower than the experimental results, but is consistent with the same trend. The variations $F_0(f)$ and $\Delta F(f)$ are inconsistent with the results of previous measurements on the microwave behavior of soft ferrites for which a broadening of the resonance band was evidenced [76]. We are led to believe that the enhanced value of the resonance model F_0 we observe is due, at least partly, to the polydispersity of the nanoparticles. It is immediately apparent from Figure 4e that the damping constant, obtained from those simulations, has the interesting feature that the composition dependence of α follows an increase from 0.46 for $f = 15$ vol% to 0.73 for $f = 53$ vol%.

It is intriguing to contrast the above results for nanometric composites with experimental results for micrometric composites using a similar sample configuration. The absence of gyromagnetic resonance for M-type samples in the frequency range explored seems puzzling. We have examined the possibility that this effect could be instrumental or due to the sample preparation. This has been shown not to be the case because the results were reproductible when both the preparation procedure and the epoxy content varied. At least three factors may contribute to the differing behaviors between M and N samples that are displayed in Figure 3c and d: reduced exchange coupling between the grains, porosity, and a possible effect of the binder. In the isotropic nanocrystalline mixtures of ZnO and Fe_2O_3, exchange interactions override the anisotropy of the individual grains, and we expect that particles with this small size will avoid large demagnetizing fields. One possible explanation for the absence of resonance in M samples is that the nonmagnetic phases separating adjacent grains induce a screening effect for the electromagnetic field.

A closer look at the gyromagnetic resonance and its shape dependence, at constant volume fraction of Fe_2O_3, on the binder concentration is shown in Figure 5d. As the epoxy concentration increases, the tendency to form multiresonances becomes apparent. Actually, several ideas have been discussed in the literature relating the multiresonance observed with submicrometer-sized particles. For example, Toneguzzo et al. [77] have established that the $\mu''(F)$ spectrum of Co_xNi_{1-x} and $Fe_z[Co_xNi_{1-x}]_{1-z}$ particles systems with a mean diameter lower than 400 nm presents several resonance bands in the 0.1–18 GHz range. For this reason, the authors of [77] argued that the nonuniform resonance modes resulting from the exchange energy contribution to the magnetization precession and the particle size distribution may be possible mechanisms responsible for this multiresonant behavior in these materials. Key to these arguments is the idea that surface anisotropy is expected to play a dominant role for describing the magnetization mechanisms in nanophases. The reader may find also enlightening the article by Toneguzzo et al. [77], in which this issue is discussed.

3.3. Magnetization and Hysteretic Properties

After an investigation of the permeability has been conducted, we now turn to the magnetization data of M and N samples. The magnetization was studied as a function of particle size, composition, and epoxy content. Magnetic hysteresis measurements in magnetic field perpendicular ($H_\perp$), and parallel ($H_\parallel$) to the torus axis were performed at room temperature using vibrating sample mangetometry (VSM). No distinction was made between M–$H_\parallel$ and M–$H_\perp$, loops. Thus, we note that the shapes of these loops are isotropic in field direction. In what follows, consider only M–H loops, where H is defined as the magnetic field parallel to the torus axis. Figure 6 shows a representative example of such M–H loops, after normalizing them to saturation, for both M and N samples corresponding to a Fe_2O_3 volume fraction $f = 0.53$ and an epoxy concentration $\phi = 0.25$. From the inset of Figure 6 showing the full hysteresis loop, we observe that the magnetization saturates at a large magnetic field ($H_s > 10$ kOe) whereas the coercivity is relatively small ($H_c < 200$ Oe). The set of hysteresis curves reveals that the saturation magnetization M_s (normalized to the amount of Fe_2O_3) and the coercive field H_c do not vary substantially, within the limits of the experimental errors, when the volume fraction of Fe_2O_3 is varied in the samples at a specified particle size, as illustrated in Figure 7a through c, respectively, while the hysteresis loop squareness, that is, the remanence magnetization M_r normalized to M_s, decreases by increasing the volume fraction of $Fe2O_3$, as shown in Figure 7b. In Figure 7a through c, we also collect the values for H_c, $\frac{M_r}{M_s}$, and M_s for the three concentrations of epoxy resin investigated. Finally, over the whole volume fraction range of Fe_2O_3 investigated, it is noteworthy that the saturation magnetization of N samples is much lower (by an order of magnitude) than the saturation magnetization of M samples, whereas the opposite is true for the coercivity, that is, H_c (N samples) $\approx 2H_c$ (M samples) and the squareness $\frac{M_r}{M_s}$.

As a final point, we would like to comment on the room temperature magnetization measurements $M(H)$. There are a number of features worthy of discussion in these plots. All the loops show a smooth demagnetization curve and a

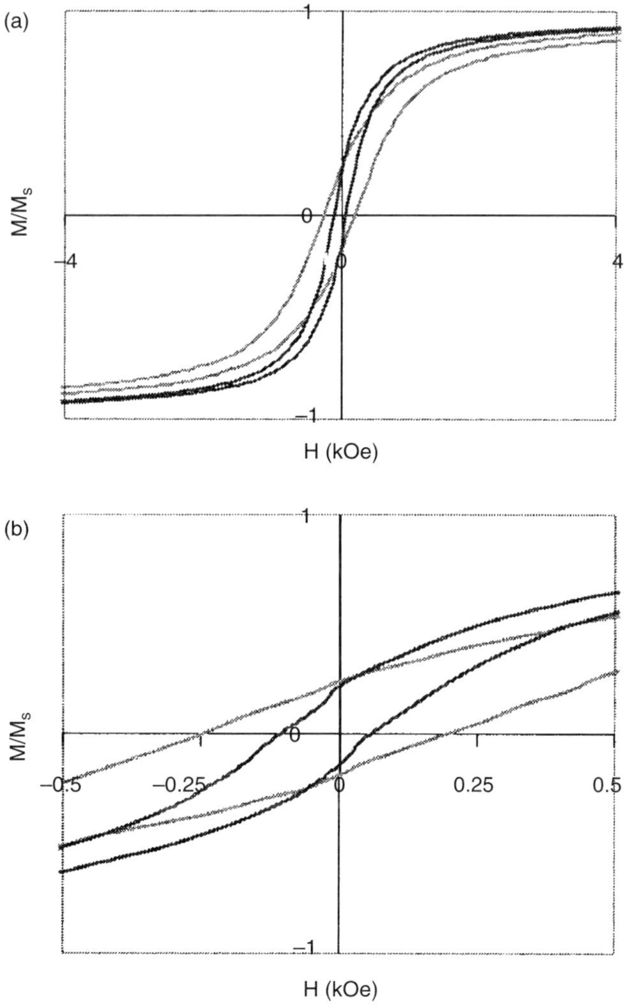

Figure 6. The magnetic hysteresis loops has a function of the applied magnetic field for N-type (red line) and M-type (green line) $ZnOFe_2O_3$ composite samples. $f = 0.53$ and $\phi = 0.25$. Room temperature. For the sake of comparison, the magnetization is normalized to the full saturation value. (a) Minor hysteresis loop and (b) major hysteresis loops.

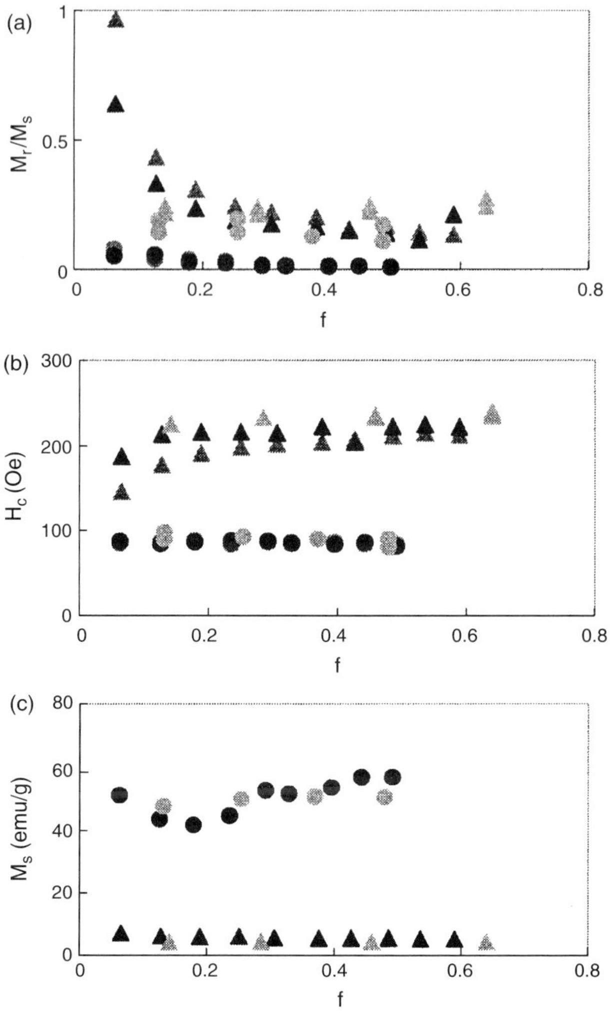

Figure 7. Dependence of the reduced remanence ratio M_r/M_s of M-type (triangles) and N-type (circles) ZnO/Fe_2O_3 composite samples as a function of the volume fraction of Fe_2O_3 for two different volume fractions of the epoxy resin $\phi = 0.10$ (blue), $\phi = 0.15$ (green), and $\phi = 0.25$ (red). Room temperature. (b) Same as in Figure 7a for the coercivity. (c) Same as in Figure 7a for the saturation magnetization M_s normalized to the mass content of Fe_2O_3.

moderate saturation magnetization ($\cong$50 em μg^{-1} for N-type samples), suggesting a very fine and uniform particle size in the samples. The hysteretic behavior shows that the saturation magnetization (resp. coercivity) takes higher (resp. lower) values for larger particles. This observation is consistent with recent reports by several groups [42, 78, 79], in which the authors reported similar particle size variations of coercivity and remanence for different types of magnetic fine particles in the size range of a few to a few tens of nanometers. For instance, Chen et al. [42] have shown that the coercivity curves of Fe/SiO_2 and Fe/Al_2O_3 granular films show a maximum at particle diameter $d \cong 20$ nm and then the coercivity decreases according to $1/d$, independently of the matrix material. The results of the above-mentioned studies and other reports [65, 80, 81] suggest that surface anisotropy and spin disorder at the particle surface play a dominant role in the magnetization mechanisms in such fine particle systems. Over the years several models have been proposed to explain this size dependence. Chen et al. [42] provided remarkable new insight into this issue by the following argument. Consider an assembly of spherical, random, noninteracting single-domain particles with uniaxial surface anisotropy constant K_s and diameter d. The magnetization reversal of fine particles should overcome an energy barrier whose height is assumed to be independent on sample volume and depends only on surface area. This barrier may be overcome by thermal agitation. After some calculations, these authors concluded that the coercivity describing the physics of the 3D random uniaxial assembly can be written as

$$H_c = \frac{5.76\,K_s}{M_s d}\left[1 - \left(\frac{25\,kT}{K_s \pi d^2}\right)^{1.155}\right]$$

where T is the temperature. Applying Chen et al.'s calculations to our N-type samples, we found that $K_s = 0.48$ erg cm^{-2}. This is similar to the values of 0.4 erg cm^{-2}

for ultrafine particle systems of α-Fe found in [42]. On general grounds, it is natural to expect that the lower values of H_c corresponding to the M-type samples may come from the multidomain structure and the bulk anisotropy of the particles. Very recently, Verdes et al. [43] suggested a model that turned out to be successful for explaining the experimental findings for the magnetic and transport properties of Co nanoparticles with uniaxial anisotropy. The physics behind the model discussed by Verdes et al. considers a system of single-domain particles coupled via magnetostatic and short-range exchange interactions. It is found that the magnetic properties are strongly dependent on the ratio $\beta = \frac{KV}{kT}$, with K being the anisotropy constant and V being the mean particle volume. For small β the interactions lead to an increase in H_c with packing density. For large β the anisotropy and magnetostatic interaction fields are comparable and the competition between these effects leads to a decrease in the coercivity H_c as the packing density is increased. Unfortunately, the results are numerical, making it difficult to use them quantitatively for a general case.

The saturation magnetization changes only slightly with f. We observed also that the coercivity exhibits a monotonic behavior insensitive to the concentration of nonmagnetic particles. A similar observation is made for the remanence ratio except for the lower values of f investigated ($f < 0.1$). Moreover, the samples have very low remanence ratio compared with the value predicted by the SW model under the assumption of uniaxial anisotropy in the reversal process; that is, $\frac{M_r}{M_s} < 0.5$. It is widely believed that the remanence ratio sheds light on the interaction between particles [82]. Although it has long been recognized that the hysteretic behavior of these multiphase heterostructures depends not only on the intrinsic properties and the grain size distribution of the different phases but also on the exchange and magnetostatic interphase coupling [83, 84], the lack of methods to accurately compute or measure the relative strength of these interactions has limited the understanding of the magnetic properties of nanophases.

The lack of a significant Fe_2O_3 volume fraction dependence of M_s, H_c, and $\frac{M_r}{M_s}$ requires additional attention. One would expect that the coercivity of samples will increase with their concentration of nonmagnetic particles because the variation of the demagnetizing field around the nonmagnetic particles, especially for large particles, causes a domain wall energy increase during the magnetization. Within the SW model, only coherent rotation of the magnetic moments can cause the magnetization and demagnetization processes in single-domain particles. However, many experimental studies revealed that the SW theory does not explain magnetic properties of fine particle systems even when the particles can be considered as single-domain regions [25, 42]. It is by now well established that the magnetization reversal in an assembly of small-sized particles with high magnetocrystalline anisotropy is controlled by nucleation of reversed domain walls at the particle surface where the anisotropy is lower than the interior of the particle due to the presence of surface defects or large localized demagnetization factors [41, 85]. The nucleation field is much higher for nanometer-sized particles because each grain has a small surface area and thus less surface defects. To close this section, we also note that in the composite materials considered here, the Fe_2O_3 particles are strongly bonded to the insulating background and may be subjected to very large stresses. Consequently, the magnetic phase–nonmagnetic phase interfaces may dominate the behavior of the coercivity as the packing is varied.

APPENDIX A: MICROWAVE MEASUREMENTS OF ε AND μ

For the purposes of this review it is not necessary to go into the details of the various ingenious schemes that have been developed over the last decades for investigations of the permittivity and permeability of materials. We will concentrate our discussion on measuring ε and μ having special relevance to isotropic materials in the microwave region of the electromagnetic spectrum. Research into the characterization of the permittivity and permeability tensors in dielectric and magnetic heterostructures having anisotropy is still very new and is in its developing stage [96]. Different experimental techniques and geometries measure different components of these tensors; for these topics, the reader is referred, for example, to [47].

The ZF effective complex permittivity $\varepsilon = \varepsilon' - j\varepsilon''$ and permeability $\mu = \mu' - j\mu''$ of the composite samples were measured using the transmission/reflection method based on the Nicolson and Ross and Weir [97] and Baker-Jarvis et al. [98] algorithms for the broadband characterization of isotropic solid materials. This experimental technique is based on the measurement of the scattering parameters (S parameters) of a toroidal sample of the test material that has been machined so that the geometrical dimensions fit well with those of the coaxial line used (to avoid the presence of air gaps between the sample and the line walls). The geometries of the cell and the material to be tested are illustrated in Figure 8a and b. The method enables us to calculate simultaneously ε and μ of the material from the measurement of the transmission S_{21} and reflexion S_{11} parameters (i.e., Fig. 8c) of a coaxial line that contains it. The electromagnetic analysis of the propagation line permits us to write

$$S_{21} = \frac{(1-\Gamma^2)T^2}{1-\Gamma^2 T^2} \tag{A1}$$

$$S_{11} = \frac{(1-T^2)\Gamma^2}{1-\Gamma^2 T^2} \tag{A2}$$

From Eqs. (A1) and (A2), one gets ε and μ [97]:

$$\mu = \mu' - j\mu'' = \left(\frac{1+\Gamma}{1-\Gamma}\right)\frac{j\ln(T)}{2\pi d}\frac{c}{F} \tag{A3}$$

$$\varepsilon = \varepsilon' - j\varepsilon'' = \frac{\left(\frac{j\ln(T)}{2\pi d}\frac{c}{F}\right)^2}{\mu} \tag{A4}$$

where c denotes the velocity of the wave in free space, d is the thickness of the material sample (Fig. 8c), and F is the frequency of the wave, respectively.

In this coaxial line measurement the fundamental transverse electromagnetic mode is the only mode that

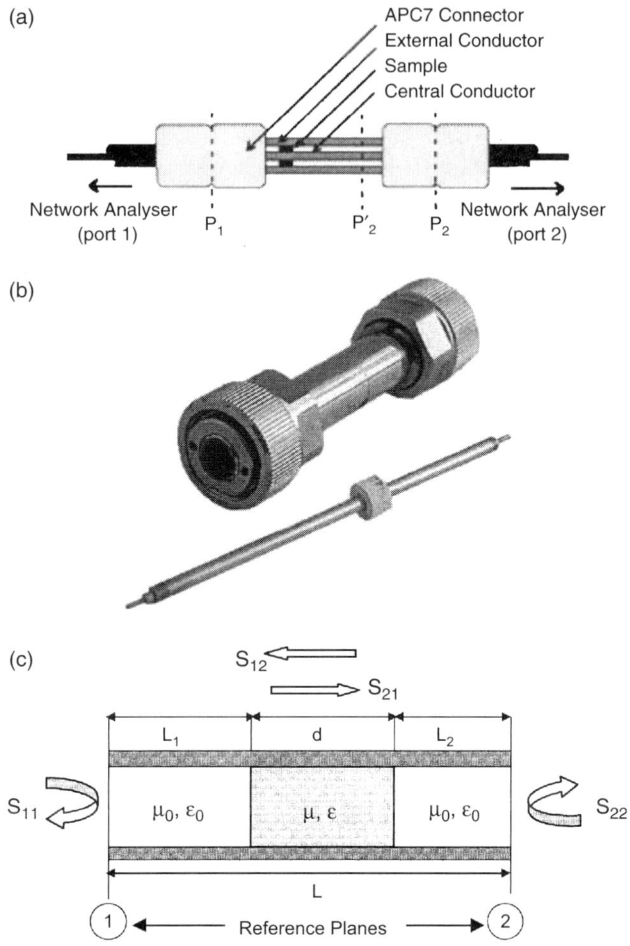

Figure 8. (a) A schematic diagram of the experimental setup showing its different parts. (b) Picture of the test cell and of the sample material indicating the scale of the experiment. (c) Principle of the measurement of the S parameters. The thickness of the material having effective complex (relative) permittivity ε and permeability μ is d.

propagates in the coaxial line. An error analysis indicates modest uncertainties in ε' (<5%), ε'' (<1%), μ' (<3%), and μ''(<1%) for the data. A Hewlett Packard HP 8510B network analyzer setup is used to measure the S parameters of the cell containing the material under test over a wide range of frequencies (10 MHz–20 GHz) with SOLT calibration. Typical dimensions of the sample are inner diameter = 3.04 mm, outer diameter = 7 mm, and thickness = 2–4 mm. All experiments were carried out at room temperature. In addition we have used VSM to measure the hysteresis loops at room temperature. In this arrangement, the sample could be oriented either parallel or perpendicular to the direction of the applied horizontal magnetic field.

APPENDIX B: BRUGGEMAN EFFECTIVE MEDIUM APPROXIMATION (EMA) THEORY

Several basic assumptions underlying the Bruggeman symmetric effective-medium theory deserve note. They are the following. (1) The applicability of this approach rests firmly on the quasistatic assumption, that is, when the typical size of inclusions (or inclusions clusters), a, is small compared with the wavelength of the electromagnetic wave probing the system. This puts an upper limit on angular frequency $\omega \ll \frac{c}{a}(\varepsilon\mu)^{-1/2}$, where c is the vacuum light velocity, ε is the effective relative permittivity, and μ is the effective relative permeability ($\mu = 1$). Within this assumption, the system can be described by an average (effective) permittivity. The no-scattering condition leads to absorption losses only. (2) The inclusions do not interact with one another. (3) The various inclusions of any single homogeneous material constituent are assumed to be embedded in an effective medium and are statistically equivalent; in particular, they exhibit the same form. (4) The constituents are distributed randomly without interstices. (5) The rationale behind this self-consistent (mean-field) approach lies in the notion that the volume average polarization in the composite material is equal to zero. It is important to recall that while this dipolar approximation is adequate to describe the dielectric properties for sufficiently low volume fraction of the inclusions, it must ultimately fail for predicting the dielectric behavior at high concentrations where quadrupoles and higher-order interactions are required, that is, near the percolation threshold. This point has been debated in the literature without, until recently, satisfactory resolution. The difficulty is addressed by the so-called generalized effective medium theory, first developed by McLachlan, that combines elements of the percolation theory with the conventional Bruggeman effective medium theory [44].

Although limited to a dipolar description, the popularity of the Bruggeman theory among the researchers is obviously due to one great virtue—its simplicity. Moreover, its prediction offers a means to get parameters with clear physical meaning, which could be related to direct experimental observations. In the most commonly used form, the Bruggeman effective medium theory for spherical inclusions (of permittivity ε_1) embedded in a host matrix (of permittivity ε_2 and treated as a continuum medium enveloping the inclusions) leads to the self-consistency condition

$$f\frac{\varepsilon - \varepsilon_1}{2\varepsilon + \varepsilon_1} + (1-f)\frac{\varepsilon - \varepsilon_2}{2\varepsilon + \varepsilon_2} = 0 \qquad \text{(B1)}$$

where f is the volume fraction of the more conducting constituent. It should be noticed that the Bruggeman formula is still valid if the properties of the constituent materials have complex values; that is, $\varepsilon = \varepsilon' - j\varepsilon''$, where j is the imaginary unit, thus enabling a prediction of complex valued properties of filled polymers. Although the definition, in a two-component system of what is the filler and what is the matrix can be confusing when in the case of certain composites the volume fraction occupied by the filler is higher than that of the matrix, the Bruggeman model assumes no explicit choice of the host material [Eq. (B1) is symmetrical with respect to the constituents]. The procedure amounts to finding the roots of a second-order polynomial and the physical root of Eq. (B1) is determined from requirements of positivity of ε' and ε''. Although this approach was originally designed for a two-component system, it can be straightforwardly extended to mixtures of any number of components

[46]. The concentration at which the system undergoes an insulator–conductor transition is designated as f_c. In the dilute limit (low concentration of the particles and $f < f_c$), one has $\varepsilon = \varepsilon_2(1 + 3f)$. This model can be useful within its range of validity. However, a basic difficulty arises in using it to describe real composite materials. The derivation of Eq. (B1) assumes that each inclusion is surrounded by a mixture of the two constituents that have the effective value for the medium. This is not the case for percolating systems in which the inclusions come in contact with each other. Thus, in practice, the range of validity is limited to $f < f_c$ and often does not encompass the actual experimental conditions. The Bruggeman symmetric medium equation is particularly appropriate for composites and polycrystals in which the grains of the various components are randomly and symmetrically distributed, so that none of the components is identifiable as a host in which the others are preferentially embedded.

4. CONCLUSIONS

This work has yielded stable and reproductible data on the composition, size, and frequency dependences of the complex permittivity and permeability of nanometer- and micrometer-sized Fe_2O_3 particles dispersed in a nonmagnetic metal oxide matrix. We described a number of previously unreported features for magnetic nanophases. We conclude two things based on the discussion and experimental evidence above. The first is that manipulations of the composition in metal oxide composite materials and the grain size can be good approximations of reduced dimensionality systems and offer a means of precisely tailoring the electromagnetic properties of these materials. The second feature of our results is that there are correlations in these multiphase nano- and microstructured systems that go far beyond a standard coarse graining description of the dielectric response, that is, the Bruggeman theory, which makes no reference to the arrangement and the size of the constituents [46, 86]. Some possible reasons for this discrepancy have been discussed above. For example, we expect that clustering effects, surface defects, size, shape, and separation of the residual pores and the fact that ZnO and Fe_2O_3 are wholly surrounded by an epoxy layer are key points to describe the polarization mechanisms. It would be interesting to test whether the combination of fractal boundaries with spheres arranged in a simple cubic array, such as that proposed in [68], is an efficient model to describe this mesostructure under confinement. On the other hand, the complex effective permeability of these particulate dispersions can be rationalized in terms of the general framework of mean-field theory, that is, the Bruggeman theory, coupled to the LLG equation describing the dynamic behavior of the magnetization. It is also important to realize that treating the effective magnetic permeability by a mean-field approach is not inconsistent with the result mentioned above for the effective permittivity. This state of affairs can be understood if we view the polarization and magnetization mechanisms in these materials as decoupled. There are several mechanisms for the damping term, which is responsible of the losses in the material [87]. What is of paramount physical relevance in ascertaining a realistic damping mechanism is the magnitude of the Gilbert constant α. In this regard, the comparison of the value of α deduced from the tit of the gyromagnetic resonance with that obtained from the FMR linewidth will be important. It is fair to say that the shape of the gyromagnetic curve in Figure 5 reflects, in part, the uncontrolled polydispersity of the nanophases. It should perhaps be emphasized once again that this information is necessary if we aim to understand the basic physics behind the magnetic properties of these materials [88]. In addition, the experimental behavior observed does not follow the general trend that had been predicted from the SW model for magnetization reversal. After several years of intense theoretical studies, it is clear that we do not have the proper skills and tools to deal with the collective transport properties of strongly correlated spin systems whose internal interactions modify the influence of external applied fields. Missing from this work is any information regarding the scale of fluctuations of the local fields experienced by the moments. For this, it is necessary to go beyond the mean-field formalism considered here, which neglects composition fluctuations. As mentioned in the Introduction, these difficulties have led numerous groups to study computational methods that address fundamental issues in nanomagnetism.

There are several directions that we could see taking our work in the future. In light of our earlier discussion concerning the likely significance of the multiresonance structure observed at high epoxy concentrations, it appears extremely interesting to further investigate composites with even smaller particles with controlled size. The biggest bottleneck for the preparation of well-defined systems is the homogeneous dispersion process of the particles. Obtaining samples with well-separated particles is far from trivial and becomes more difficult the higher the packing density. Meanwhile, the puzzling feature of the absence of gyromagnetic resonance for M-type samples has not been satisfactorily explained, and this a question to be pursued in future studies. To tackle this problem experimentally it would clearly be of general interest to investigate whether the behavior observed here persists for another choice of the dielectric matrix and the magnetic component. In future work, we plan also to give a detailed discussion of the study of the magnetic properties, for example, exchange coupling between the grains, using ^{57}Fe Mössbauer spectrometry elsewhere. We should note that this work has also highlighted the fact that refined first-principles (and realistic) micromagnetics simulations combined with experimental measurements are needed to understand the magnetic behavior of nanophases. It is an area of much current interest; for example, see [89]. Computational results can contribute to the acceptance or rejection of mean-field based theories and can also indicate directions, in which new approaches should be developed. Furthermore, we would like to point out that this work highlights the importance (and complexity) of particle characteristics (particle size and shape) in comparison of bulk (effective) electromagnetic properties. To understand the significance of these results in the study of wave transport in nanocomposites, it will be interesting in the future to see more and more experiments realized with increasingly smaller particles. The exploitation of interfacial properties, especially at the nanometer length scale will be central to developments of nano-technical devices of the future, for example, for applications in high-density magnetic storage [90]. The resolution of these issues will be a target of future investigation.

It is a conspicuous implication of the present experiments that manipulation of the composition, the preparation procedure, and the grain sizes offered a means to "tailor," to a particular need, the electromagnetic properties of a (isotropic) distribution of fine magnetic particles embedded in a nonmagnetic matrix, where complementary parameters, such as permittivity and permeability, are interestingly combined. For example, very small coercivity, high remanence, high permeability, low dielectric, and low magnetic loss soft magnetic material having a square hysteresis loop is required for microwave operation. Other requirements for a nearly ideal material eventually include fine grain, defect-free, and dense [91]. Materials that simultaneously exhibit a large permittivity and a large permeability with corresponding low losses are of significant technological interest as components of magnetoelectric sensors and actuators [24, 92, 93]. Observe that the utility foreseen for magnetoelectric materials rests on the possibility of tailoring their electric and magnetic properties on the same footing. Another related question is how all of this can be related to investigations of the electromagnetic properties of materials with simultaneously negative permittivity and permeability or left-handed materials (LHMs). In 1968 the Russian physicist Victor Veselago [94] theoretically investigated the electromagnetic consequences of a LHM and predicted that such a medium would have dramatically different propagation characteristics stemming from the sign change of the group velocity, including reversal of the Doppler shift and Cherenkov radiation, anomalous refraction, and even reversal of radiation pressure to radiation tension. However, at that time, these effects could not be experimentally verified since, as Veselago pointed out, materials with negative permeability were not available. Recently, a number of authors have made progress in preparing artificial LHMs, for example, arrays of conducting nonmagnetic split ring resonators or metallic magnetic granular composites [95]. Nanostrctured LHMs can potentially be applied to alternative technologies to be used in displays and devices such as antenna radoms, band-pass filters, microwave couplers, and high-resolution optical lithography in the semiconductor industry.

GLOSSARY

Effective medium approximation (EMA) The basic idea is to describe the system by a simple average permittivity or (permeability) and is based on a self-consistent procedure in which a grain of one of the constituents is assumed to have a convenient shape (usually taken as spherical or ellipsoidal) and to be embedded in an effective medium (homogeneous) whose properties are determined self-consistently. This requires that the wavelength of the electromagnetic radiation probing the system has to be larger than a typical scale of length that characterizes the inhomogeneities in the material medium.

Gyromagnetic resonance This is the spin of the electron that is the main cause of magnetization. In ferromagnetic materials, there exists a predominance of either left- or right-handed spins. Moreover, there is a restraining force that tends to keep electrons near each other spinning in the same direction. Ferromagnetic materials, also called cooperative materials, can retain their magnetic properties. This is due to the restraining force. Normally, in an unmagnetized state, the electron spins are randomly oriented so that the net sum of all the Bohr magnetons is zero. When a magnetic field is applied, the material changes to the magnetized state, where all the tiny magnets line up and add together. In an experiment the sample is subjected to a microwave field of fixed frequency and to an applied static field, the magnitude of which is varied. At each field value, the reflected or absorbed power from the sample is measured. These may yield one or more minima and maxima, respectively, corresponding to resonance modes of the ensemble. The field at which this gyromagnetic resonance occurs, the so-called resonance field, depends on several magnetic parameters.

Hysteresis This represents the history dependence of physical systems. When a ferromagnetic material is magnetized in one direction, it will not relax back to zero magnetization when the imposed magnetizing field is removed. It must be driven back to zero by a field in the opposite direction. If an alternating magnetic field is applied to the material, its magnetization will trace out a loop called a hysteresis loop. The lack of retraceability of the magnetization curve is the property called hysteresis, and it is related to the existence of magnetic domains in the material. Once the magnetic domains are reoriented, it takes some energy to turn them back again. This property of ferromagnetic materials is useful as a magnetic "memory."

Magnetoelectric materials These hybrid nanostructures form a fascinating melting pot for studying the interplay between dielectric and magnetic properties. In fact, the addition of the spin degree of freedom to conventional electronics know-how is paving the way to the appealing field dubbed "spintronics." This emerging area includes the active control of carrier spin dynamics and transport in electronics materials, with the final goal of nanolithic integration of the nanoscale with larger structures.

Particulate composite materials In the most general sense, a particulate composite material can be defined as the combination of many materials (or the same material in different states) present as separated components (particles) with different geometric and physical properties and mixed together to form structures that may improve a specific property.

ACKNOWLEDGMENTS

This review could not have been written without important input from many people. I would like to express my appreciation to the following for their technical contributions and discussions: J. Ben Youssef, A. M. Konn, D. Rozuel, and P. Talbot. The Laboratoire d'Electronique et Systèmes de Télécommunications is Unité Mixte de Recherche CNRS 6165.

REFERENCES

1. D. Ferry, J. R. Barker, and C. Jacoboni, Eds., "Granular Nanoelectronics," Nato ASI Series B: Physics, Vol. 251. Plenum, New York, 1990.
2. G. C. Hadjipanayis and R. W. Siegel, Eds., "Nanophase Materials: Synthesis, Properties, Applications." Kluwer Academic, New York, 1994; G. L. Trimp, "Nanotechnology." AIP Press, New York, 1998.
3. H. Gleiter, *Nanostruct. Mater.* 6, 1 (1995). See also H. Gleiter, *Prog. Mater. Sci.* 33, 223 (1989); D. Ferry, J. R. Barker, and C. Jacoboni, Eds., "Granular Nanoelectronics," Nato ASI Series B: Physics, Vol. 251. Plenum, New York, 1990; N. Ichinose, Y. Ozaki, and S. Kaghu "Superfine Particle Technology." Springer, New York, 1992.
4. R. Landauer, *J. Phys.: Condens. Matter* 1, 8099 (1989).
5. G. F. Hohl, T. Hihara, M. Sakurai, T. J. Konno, K. Sumiyama, F. Hensel, and K. Suzuki, *Appl. Phys. Lett.* 66, 383 (1995).
6. E. Nieto-Vesperinas and N. Garcia, Eds., "Optics at the Nanometer Scale." Kluwer Academic, Dordrecht, 1996. See also U. Kreibig and M. Vollmer, "Optical Properties of Metal Clusters." Springer, New York, 1995.
7. R. W. Chantrell, N. Walmsley, J. Gore, and M. Maylin, *Phys. Rev. B* 63, 024410 (2000). See also R. W. Chantrell, in "Science & Technology of Nanostructured Magnetic Materials" (G. C. Hadjipanayis and G. A. Prinz, Eds.). Plenum, New York, 1991.
8. J. Karch, R. Birringer, and H. Gleiter, *Nature (London)* 330, 556 (1987).
9. H. Hahn and R. S. Averback, *J. Am. Ceram Soc.* 74, 2918 (1991).
10. R. D. McMichael, R. D. Schull, L. J. Swartzendruber, and L. H. Bennet, *J. Magn. Magn. Mater.* 11, 29 (1992).
11. R. W. Siegel, *Springer Ser. Mater. Sci.* 27, 65 (1994).
12. F. J. Himpsel, J. E. Ortega, G. J. Mankey, and R. F. Willis, *Adv. Phys.* 47, 511 (1998).
13. P. V. Kamat and D. Meisel, Eds., "Semiconductor Nanocrystals—Physical, Chemical, and Catalytic Aspects." Elsevier, Amsterdam, 1997. See also H. S. Nalwa, Ed., "Encyclopedia of Nanoscience and Nanotechnology" Vols. 1–10. American Scientific Publishers, Stevenson Ranch, CA, 2004; A. Aviram and M. A. Ratner, Eds., "Molecular Electronics, Science and Technology." New York Academy of Sciences, New York, 1998.
14. G. F. Hohl, T. Hihara, M. Sakurai, T. J. Konno, K. Sumiyama, F. Hensel, and K. Suzuki, *Appl. Phys. Lett.* 66, 383 (1995).
15. H. Fukuda, M. Yasuda, and T. Iwabuchi, *Appl. Phys. Lett.* 61, 693 (1992).
16. M. A. Turowski, T. F. Kelly, and P. E. Batson, *J. Appl. Phys.* 76, 3776 (1994).
17. S. Munnix and M. Schmeits, *Phys. Rev. B* 32, 4192 (1985).
18. M. J. Carey et al., *Appl. Phys. Lett.* 61, 2935 (1992).
19. J. S. Jiang, J. Q. Xiao, and C. L. Chien, *Appl. Phys. Lett.* 61, 2362 (1992).
20. J. Löffler et al., *Nanostruct. Mater.* 9, 523 (1997).
21. Y. D. Yao et al., *Nanostruct. Mater.* 9, 933 (1997).
22. A. A. Shaikh, R. W. Vest, and G. M. Vest, "Proceedings of IEEE 1986 6th International Symposium on Applications of Ferroelectrics." Pennsylvania, 1986.
23. L. V. Panina et al., *J. Appl. Phys.* 76, 6365 (1994).
24. R. E. Newnham and S. E. Trolier-McKinstry, *Ceram. Dielectrics: Composition, Processing and Properties* 8, 235 (1990).
25. G. Xiao and C. L. Chien, *Appl. Phys. Lett.* 51, 1280 (1987).
26. S. Bednarek, *Appl. Phys. A: Mater. Sci. Process* 68, 63 (1999).
27. M. El-Hilo, R. W. Chantrell, and K. O'Grady, *J. Appl. Phys.* 84, 5114 (1998).
28. A. E. Berkowitz, W. J. Shuele, and P. J. Flanders, *J. Appl. Phys.* 39, 1261 (1968). See also A. E. Berkowitz, J. R. Mitchell, M. J. Carey, and G. Thomas, *Phys. Rev. Lett.* 68, 3745 (1992), 68, 3749 (1992).
29. M. Pardavi-Horvath and L. J. Swartzendruber, *IEEE Trans. Magn.* 35, 3502 (1999).
30. M. Pardavi-Horvath, G. Zheng, and G. Vertery, *Physica B* 233, 287 (1997).
31. There is much work in this area: See, for example, A. Aharoni, "Introduction of the Theory of Ferromagnetism." Oxford University Press, New York, 1996. The interested reader may also consult R. L. White, R. M. H. News, and R. F. W. Pease, *IEEE Trans. Magn.* 33, 990 (1997).
32. J. L. Dormann and D. Fiorani, Eds., "Magnetic Properties of Fine Particles." North-Holland, Amsterdam, 1992. See also J. L. Dormann, E. Tronc, and D. Fiorani, *Adv. Chem. Phys.* 98, 283 (1997); R. M. German, "Particle Packing Characteristics." Metal Powder Industries, Princeton, NJ, 1989.
33. Y. Hayakawa, A. Makino, H. Fujimori, and A. Inoue, *J. Appl. Phys.* 81, 3747 (1997).
34. E. C. Stoner and E. P. Wohlfarth, *Philos. Trans. R. Soc. London, Ser. A* 240, 599 (1948); E. H. Frei, S. Shtrikman, and T. Treves, *Phys. Rev.* 106, 446 (1957); W. F. Brown, Jr., *Phys. Rev.* 105, 1479 (1957); E. W. Lee and J. E. L. Bishop, *Proc. Phys. Soc. London* 89, 661 (1966).
35. A. S. Arrott, in "Nanomagnetism" (A. Hernando, Ed.). Kluwer, Dordrecht, 1993.
36. C. P. Bean and D. J. Livingstone, *J. Appl. Phys.* 30, 1205 (1959).
37. D. A. Dimitrov and G. M. Vysin, *Phys. Rev. B* 54, 9237 (1996).
38. J. O. Anderson, C. Djuberg, T. Jonsson, P. Svedlindh, and P. Noblad, *Phys. Rev. B* 56, 13983 (1997).
39. P. Allia, M. Corsson, M. Knobel, P. Tiberto, F. Vinai, M. A. Novek, and W. C. Nunes, *Phys. Rev. B* 64, 144420 (2001).
40. J. C. Denardin, A. L. Brandl, M. Knobel, P. Panissod, A. B. Pakhomov, H. Liu, and X. X. Zhang, *Phys. Rev. B* 65, 064422 (2002).
41. R. Fischer and H. Kronmüller, *Phys. Rev. B* 54, 7284 (1996). See also R. Fischer and H. Kronmüller, *J. Magn. Magn. Mater.* 191, 225 (1999).
42. Z. M. Chen, X. Meng-Burany, H. Okumura, and G. C. Hadjipanayis, *J. Appl. Phys.* 87, 3409 (2000).
43. C. Verdes, B. Ruiz-Diaz, S. M. Thompson, R. W. Chantrell, and Al. Stancu, *Phys. Rev. B* 65, 174417 (2002).
44. C. Brosseau, P. Queffelec, and P. Talbot, *J. Appl. Phys.* 89, 4532 (2001); C. Brosseau, *J. Appl. Phys.* 90, 3197 (2002).
45. S. Torquato, "Random Heterogeneous Materials." Springer, New York, 2001.
46. C. Brosseau and A. Beroual, *Prog. Mater. Sci.*, in press (2002).
47. A. Priou, Ed., "Dielectric Properties of Heterogeneous Materials." Elsevier, New York, 1992; R. Landauer in "Electrical Transport and Optical Properties of Inhomogeneous Media" (J. C. Garland and D. B. Tanner, Eds.), AIP Conference Proceedings No. 40, p. 2. AIP, New York, 1978.
48. D. J. Bergman and D. Stroud, "Solid State Physics" (H. Ehrenreich and D. Turnbull, Eds.), Advances in research and Applications, Vol. 46, p. 147. Academic Press, San Diego, CA, 1992.
49. A. N. Lagarkov, S. M. Matytsin, K. N. Rozanov, and A. K. Sarychev, *J. Appl. Phys.* 84, 3806 (1998). See also A. N. Lagarkov and A. K. Sarychev, *Phys. Rev. B* 53, 6319 (1998); A. N. Lagarkov, S. M. Matytsin, K. N. Rozanov, and A. K. Sarychev, *Physica A* 241, 58 (1997). For various generalizations of the ordinary percolation, see the articles in the special issue, *Physica A* 266 (1999).
50. G. Viau, F. Ravel, and O. Acher, *J. Appl. Phys.* 76, 6570 (1994).
51. O. N. Singh and A. Lakhtakia, Eds., "Electromagnetic Fields in Unconventional Materials and Structures." Willey, New York, 2000.
52. H. Karzel, W. Potzel, M. Kofferlein, W. Schiessl, M. Steiner, U. Hiller, G. M. Kalvius, D. W. Mitchell, T. P. Das, P. Blaka, K. Schwarz, and M. P. Pasternak, *Phys. Rev. B* 53, 11425 (1996). See also C. T. Troy, *Photonics Spectra* 31, 34 (1997).
53. F. C. M. Van de Pol, *Ceram. Bull.* 69, 1959 (1990).

54. T. Shiosaki, S. Ohnishi, and A. J. Kawabata, *Appl. Phys.* 50, 3113 (1979). See also T. K. Gupta, *J. Am. Ceram. Soc.* 73, 1817 (1990).
55. M. Matsuoka, *Jpn. J. Appl. Phys.* 10, 736 (1971).
56. L. M. Levinson and H. R. Philipp, *Am. Ceram. Soc. Bull.* 65, 639 (1986).
57. F. Jorgensen, "The Complete Handbook of Magnetic Recording." McGraw-Hill, New York, 1996.
58. M. P. Sharrock and R. E. Bodnar, *J. Appl. Phys.* 57, 3919 (1985).
59. R. H. Kodama and A. E. Berkowitz, *Phys. Rev. B* 59, 6321 (1999).
60. R. M. Cornell and U. Schwertmann "The Iron Oxides Structures Properties, Occurrences and Uses." VCH Verlagsgesellschaft, Weinheim, 1996.
61. B. D. Cullity, "Introduction to Magnetic Materials." Addison-Wesley, London, 1972.
62. J. Dai, J. Q. Wang, C. Sangregorio, J. Fang, E. Carpenter, and J. Tang, *J. Appl. Phys.* 87, 7397 (2000). See also J. M. D. Coey, *Phys. Rev. Lett.* 25, 1140 (1971).
63. S. Mørup and E. Tronc, *Phys. Rev. Lett.* 72, 3278 (1994).
64. F. Bødker, S. Mørup, and S. Linderoth, *Phys. Rev. Lett.* 72, 282 (1994). See also F. Bødker, M. F. Hansen, C. Bender Koch, K. Lefmann, and St. Mørup, *Phys. Rev. B* 61, 6826 (2000).
65. B. Martinez, X. Obradors, Ll. Barcells, A. Rouanet, and C. Monty, *Phys. Rev. Lett.* 80, 181 (1998).
66. P. Talbot, A. M. Komn, and C. Brosseau, *J. Magn. Magn. Mater.* 249, 483 (2002).
67. J. V. Mantese, A. L. Micheli, D. F. Dungan, R. G. Geyer, J. Baker-Jarvis, and J. Grosvenor, *J. Appl. Phys.* 79, 1655 (1996).
68. J. P. Calame, A. Birman, Y. Carmel, D. Gershon, B. Levush, A. A. Sorokin, V. E. Semenov, D. Dadon, L. P. Martin, and M. Rosen, *J. Appl. Phys.* 80, 3992 (1996).
69. P. Neelakanta, "Handbook of Electromagnetic Materials." CRC Press, New York, 1995.
70. For an overview, see H. Fukuyama, N. Nagaosa, Eds., "Physics and Chemistry of Transition Metal Oxides." Springer Series in Solid State Science 125, Proceedings of a Symposium, Kashikojima, Japan, May 1998. Springer-Verlag, New York, 1999.
71. J. A. C. Bland and B. Heinrich, "Ultrathin Magnetic Structures I and II." Springer-Verlag, Berlin, 1994.
72. H. Suhl, *IEEE Trans. Magn.* 34, 1834 (1998).
73. T. L. Gilbert, *Phys. Rev.* 100, 1243 (1955); L. D. Landau, E. M. Lifshitz, and L. P. Pitaevski, "Statistical Physics," Part 2, 3rd ed. Pergamon Press, Oxford, 1980. The interested reader may also consult H. B. Callen, *J. Chem. Phys. Solids* 4, 256 (1958), and references therein.
74. F. Schreiber, J. Pflaum, Z. Frait, Th. Mühge, and J. Pelzl, *Solid State Commun.* 93, 965 (1995). See also S. M. Bhagat and P. Lubitz, *Phys. Rev. B* 10, 179 (1974); P. Lubitz, S. F. Cheng, F. J. Rachhford, M. M. Miller, and V. G. Harris, *J. Appl. Phys.* 91, 7783 (2002).
75. R. Urban, G. Woltersdorf, and B. Heinrich, *Phys. Rev. Lett.* 87, 217204 (2001).
76. D. Polder and J. Smit, *Rev. Mod. Phys.* 25, 89 (1953).
77. P. Toneguzzo, G. Viau, O. Acher, F. Guillet, E. Bruneton, F. Fievet-Vincent, and F. Fievet, *J. Mater. Sci.* 35, 3767 (2000). See P. Toneguzzo, O. Acher, G. Viau, F. Guillet, E. Bruneton, F. Fievet-Vincent, and F. Fievet, *J. Appl. Phys.* 81, 5546 (1997).
78. Y. Chen, J. E. Snyder, C. R. Schwichtenberg, K. W. Denins, D. K. Falzgref, R. W. McCallum, and D. C. Jiles, *Appl. Phys. Lett.* 74, 1159 (1999).
79. X. N. Xu, Y. Wolfus, A. Shaulov, Y. Yeshurun, I. Felner, I. Nowik, Yu. Koltypin, and A. Gedanken, *J. Appl. Phys.* 91, 4611 (2002).
80. H. Sohl and H. Neal Bertram *J. Appl. Phys.* 82, 6128 (1997).
81. M. F. Hansen, F. Bødker, S. Mørup, K. Lefmann, K. N. Clausen, and P. Lindgärd, *Phys. Rev. Lett.* 79, 4910 (1997).
82. H. N. Bertram and A. K. Bhatia, *IEEE Trans Magn.* mag-9, 127 (1973).
83. A. Hemando, *J. Phys.: Condens. Matter* 11, 9455 (1999).
84. G. Bertotti, "Hysteresis in Magnetisn." Acadamic Press, San Diego, CA, 1998.
85. J. D. Livingston, *J. Appl. Phys.* 52, 2544 (1981).
86. See, for example, R. Pelster and U. Simon, *Colloid Polym. Sci.* 277, 2 (1999), and references therein.
87. A. S. Chakravarty, "Introduction to the Magnetic Properties of Solids." John Wiley, New York, 1980. See also P. E. Wigen, Ed., "Nonlinear Phenomena and Chaos in Magnetic Materials." World Scientific, Singapore, 1994, and references therein.
88. N. Vukadinovic, J. Ben Youssef, H. Legall, and J. Ostorero, *Phys. Rev. B* 62, 9021 (2000). See also A. Thiaville, J. Miltat, and J. Ben Youssef, *Eur. Phys. J. B* 23, 37 (2001).
89. W. F. Brown, Jr., *Phys. Rev.* 130, 1677 (1963). See also W. F. Brown, Jr., "Micromagnetics." Interscience, New York, 1963.
90. R. D. Shull, J. J. Ritter, A. J. Shapiro, L. J. Swartzendruber, and L. H. Bennett, *J. Appl. Phys.* 67, 4490 (1990).
91. See, for example, J. Tressler, S. Alkoy, and R. Newnham, *J. Electrocerm.* 2, 257 (1998); M. Shahinpoor, Y. Bar-Cohen, J. Simpson, and J. Smith, *Smart Mater. Struct.* 7, R15–30 (1998).
92. R. Edelstein, C. Tamanaha, P. Sheehan, M. Miller, D. Baselt, L. Whitman, and R. Colton, *Biosens. Bioelectron.* 14, 805 (2000).
93. M. Miller, P. Sheehan, R. Edelstein, C. Tamanaha, L. Zhong, S. Bounnak, L. Whitman, and R. Colton, *J. Magn. Magn. Mater.* 225, 138 (2001).
94. V. G. Veselago, *Sov. Phys. Usp.* 10, 509 (1968).
95. J. B. Pendry, A. J. Holden, W. J. Stewart, and I. Youngs, *Phys. Rev. Lett.* 76, 4773 (1996), and the following comment: R. M. Walser, A. P. Valanju, and P. M. Valanju, *Phys. Rev. Lett.* 87, 119701 (2001). See also J. B. Pendry, A. J. Holden, D. J. Robbins, and W. J. Stewart, *J. Phys. Conden. Matter.* 10, 4785 (1998); J. B. Pendry, D. J. Robbins, and W. J. Stewart, *IEEE Trans. Microwave Theory Tech.* 47, 2075 (1999); D. R. Smith, D. C. Vier, W. Padilla, S. C. Nemat-Nasser, and S. Schultz, *Appl. Phys. Lett.* 75, 1425 (1999); D. R. Smith, W. Padilla, D. C. Vier, S. C. Nemat-Nasser, and S. Schultz, *Phys. Rev. Lett.* 84, 4184 (2000); S. T. Chui and L. Hu, *Phys. Rev. B* 65, 144407 (2002). See also R. Marqués, F. Medina, and R. Rafii-El-Idrissi, *Phys. Rev. B* 65, 14440 (2002).
96. S. Mallegol, C. Brosseau, P. Queffelec, and A. M. Konn, *Phys. Rev. B*, in press (2003).
97. A. M. Nicolson and G. F. Ross, *IEEE Trans. Instrum. Meas.* 9, 377 (1970). See also W. B. Weir, *proc. IEEE* 62, 33 (1974).
98. J. Baker-Jarvis, E. J. Vanzura, and W. A. Kissick, *IEEE Trans. Microwave Theory Tech.* 38, 1096 (1990).
99. W. Chang and L. Sengupta, *J. Appl. Phys.* 92, 3941 (2002). See also W. Chang, J. S. Horvitz, A. C. Carter, J. M. Pond, S. W. Kirchoefer, C. M. Gilmore, and D. B. Christy, *Appl. Phys. Lett.* 74, 1033 (1999).
100. S. B. Quadri, J. S. Horvitz, D. B. Chrisey, R. C. Y. Auyeung, and K. S. Grabowski, *Appl. Phys. Lett.* 66, 1605 (1995).
101. E. J. Cukauskas, S. W. Kirchoefer, and J. M. Pond, *J. Appl. Phys.* 88, 2830 (2000).

Encyclopedia of Nanoscience and Nanotechnology
www.aspbs.com/enn

Electron Beam Writing in Nanoparticle Films

Y. Chen, R. E. Palmer

The University of Birmingham, Birmingham, United Kingdom

CONTENTS

1. INTRODUCTION

In recent years, significant progress has been made in creating a variety of functional, nanometer-scale architectures, such as molecular-scale wires and switches [1–4], molecular-scale transistors based on nanotubes [5], and semiconductor nanowires [6]. Even logic circuits, constructed from individual carbon nanotube molecules [7] and semiconductor nanowires [8], have been reported, presenting dramatic advances toward the realization of nanoelectronics and especially nanocomputers.

The main challenge to the advance of molecular-scale electronics from the single-device level to the circuit level lies in the assembling together of the component devices, each of which must be addressable. Fabrication techniques coping with the size region from the nanometer to the micrometer are essential to assemble the nanocomponents and link them to the macroscopic world. Here, we introduce one of these techniques, namely, electron beam (e-beam) writing in passivated nanoparticle films.

This chapter is divided into five sections. Passivated nanoparticles, the building materials for the films modified by the direct e-beam writing process, are surveyed in Section 2. Section 3 gives a review of the effect of electron irradiation of self-assembled monolayers and films of nanoparticles, which provides the background to the e-beam writing process. The direct e-beam writing technique is demonstrated in Section 4, which provides details of the basic experimental process and its quantitative assessment and also discusses the underlying mechanism. Section 5 gives some examples of the application of this technique to create structures of technological interest, aiming to give the reader a flavor of potential applications. Finally, the future development of the technique is discussed in Section 6.

2. PASSIVATED NANOPARTICLES

Passivated nanoparticles are nanometer sized colloidal particles chemically terminated by organic monolayers, typically alkanethiol molecules. They can be produced from many materials and show a diversity of functional behavior beyond that of the corresponding bulk materials, with many of their electronic [9], optical [10, 11], and catalytic [12–14] properties originating from their quantum-scale dimensions. The organic layer, typically bound to the nanoparticle core by a metal–sulphur (thiolate) bond, prevents the particles from aggregating together, in solution or on a surface, which is more likely to occur in the case of bare particles (clusters), so that they retain their unique, size-dependent functionality.

2.1. Self-Assembly of Passivated Nanoparticles

Passivated metal (e.g., gold) nanoclusters can be used as "building blocks" to form ordered monolayers and thin films. The diameter of these nanoparticles (1–10 m) makes them ideal for constructing nanoscale architectures and devices. X-ray diffraction of the superlattice formed by a three-dimensional assembly of passivated gold nanocrystals shows a sequence of intense, sharp Bragg peaks in the small angle region, in addition to the peaks arising from the planes of Au atoms within individual nanocrystal cores.

ISBN: 1-58883-059-4/$35.00

Encyclopedia of Nanoscience and Nanotechnology
Edited by H. S. Nalwa
Volume 3: Pages (17–28)

The low angle peaks arise from the ordered layers of nanocrystals stacked to form a giant three-dimensional superlattice [15]. The macroscopic properties of the nanoparticle superlattice are determined by both the properties of each individual particle and the coupling/interaction between the nanocrystals (via the ligands). Thus the organic ligands serve not only as a protective layer for the particles, preventing direct contact between the cores, but also regulate interparticle bonding. For example, the tuneable interparticle spacing, derived by changing the length of the molecular chain, allows control over structural, optical, and transport properties [16].

Since a variety of organic molecules can be employed to passivate the nanoparticle cores, it is possible to choose ligands with a range of functionalities, allowing attachment to particular substrates or molecules. For example, one can envisage the selective attachment of individual nanoparticles to selected substrate regions, as part of the fabrication of a nanodevice. The electronic properties of single nanoparticles are already the subject of active research [17, 18], which exploits the size-dependent properties of nanoparticles to fabricate individual single-electron devices [19, 20] and similar structures [21, 22].

2.2. Patterning the Nanoparticle Assembly

In order to manufacture nanostructures or, ultimately, devices using nanoparticles, one needs to be able to assemble these "artificial atoms" from solution onto technically interesting substrates and possibly also to process these assemblies further to create robust structures with practical functions. Exciting advances toward the self-assembly of nanoparticles into ordered structures on various substrates have been reported in recent years [23]. As an example, linear structures of width ~1–50 nm are of technological importance for potential applications and also hold great fundamental interest for quantum confinement behavior. A number of such "wires," based on gold and silver nanoparticles, have been prepared by various methods. The spontaneous alignment of silver nanoparticles to form wires several micrometers long and 20–300 nm wide was observed by Chung and co-workers at the water/air interface [24]. Simon and co-workers reported the spontaneous formation of a wire consisting of a single line of clusters [25]. A metal particle chain, which formed a bridge between a pair of nanoelectrodes, was created by Burghard and co-workers by self-assembly under directional flow conditions [26]. Another approach to the fabrication of nanowires is self-assembly of nanoparticles along linear templates. One example, reported by Hutchinson and colleagues, is the self-assembly of gold nanowires from ligand stabilized gold nanoparticles drop cast onto a corrugated amorphous carbon thin film with topographical features of nanometer dimensions, followed by heat treatment [27]. The continuous wire segments resulting from the sintering of the nanoparticles were typically 3 to 5 nm wide and up to 100 nm long. Another example of the use of linear templates was reported by Fullam and co-workers, who decorated bundles of single-walled carbon nanotubes with gold particles and obtained continuous, polycrystalline gold nanowires after sintering the particles under heat treatment [28]. Passivated gold nanoparticles have also been deposited from solution onto silicon dioxide surfaces prepatterned by photolithography and preferential cluster accumulation is observed along the edges of the resist structure after a lift-off of the photoresist [29]. A guided flow along the boundary between a resist and a silicon surface can extend for over 600 μm across the surface, producing cluster chains as narrow as ~120 nm [30].

Recently, nanoparticles of CdTe were found to spontaneously reorganize into crystalline nanowires upon removal of the protective shell of organic ligands. The stabilizer was removed by precipitating the CdTe colloids by methanol addition and redissolving in pure water. Then the obtained dispersion was allowed to age in darkness at room temperature for several days. Nanowires were formed during this period through pearl-necklace aggregates. It is believed that the strong dipole–dipole interaction is the driving force of nanoparticle self-organization. The produced nanowires have high aspect ratio (100–500) and their diameters were determined by the diameter of the nanoparticles [31].

Patterning of nanoparticle assembly can also be achieved by modifying surfaces with electrons or photons in a way such that the exposed area is functionalized with molecules to which the nanoparticles bind selectively [32–34]. For example, focused e-beam writing to a self-assembled monolayer incorporation of an aromatic nitro derivative allows the unreactive nitro functionality to be converted to a nucleaophilic amino group which could subsequently react with simple carboxylic acids in solution to form amide bonds. Nanoparticles with designed ligands can thus be selectively assembled to an e-beam modified area.

2.3. Electron Transport in Nanoparticle Film

The conductivity of nanoparticle assemblies is important for both fundamental reasons and technical applications and has been the subject of several studies in recent years. It is found that the resistance of a nanoparticle network depends on factors such as the particle size and particle separation, as well as the number of particle layers (in a thin film). Gold nanoparticle networks connected by alkanedithiol linkers demonstrated nonmetallic conductivity behavior when the temperature was varied, probably involving activated electron hopping from particle to particle [35]. Significantly different surface resistances were measured for multilayers of gold nanoparticles separated by alkanedithiol molecules of various chain lengths [36]. It has also been reported that colloidal gold multilayers undergo a metal–insulator transition at a critical particle density [37, 38]. Films of nanoparticles (diameter 11 nm) with density up to 25×10^{11} cm^{-2} were found to be insulating, with a resistance in excess of 10^7 Ω, but an increase in the number of gold particle layers resulted in an enormous decrease in the resistance. When the particle density was doubled (50×10^{11} cm^{-2}), the resistance was below 100 Ω, a change of five orders of magnitude. Reversible metal–insulator transitions in ordered silver nanoparticle monolayers have been observed when the Langmuir–Blodgett (LB) films are compressed and decompressed [39]. As the interparticle separation is reduced, the mechanism of charge transport changes from hopping to tunnelling and then to metallic transport, resulting in significant conductivity changes. In contrast to the many studies of nanoparticle films, relatively little has been reported in the area of electron transport by linear nanoparticle structures.

3. ELECTRON BEAM EFFECTS ON SELF-ASSEMBLED MONOLAYERS AND NANOPARTICLE FILMS

3.1. Electron Beam Irradiation of Self-Assembled Monolayers

Self-assembled monolayers (SAMs) are close-packed arrays of chainlike molecules which are chemically bound to a planar substrate via their head groups. SAMs are formed from the same kinds of molecules (e.g., alkanethiols) used to terminate nanoparticles. Thus the response of SAMs to electron irradiation provides clues as to the response of nanoparticle films to the same type of irradiation. SAMs also provide structurally and chemically well-defined organic substrates, relevant to biological interfaces and membranes, corrosion protection, electrochemistry, and molecular devices [40, 41]. The formation of well-defined organic surface phases by the immersion of a clean gold substrate in a dilute solution of a long-chain dialkyldisulfide was first reported in 1983 [42]. A typical example of a self-assembled monolayer is the thiol/Au system (see [43, 44] and references therein), where S–Au bonding leads to the formation of a gold thiolate. A $(\sqrt{3} \times \sqrt{3})R30°$ overlayer structure results, with domains hundreds of square nanometers in size. It is found that the chains are fully extended and tilted with respect to the surface normal by around 30°.

The response of SAMs to ion, X-ray, and electron beam irradiation is a substantial issue for scientific and technological applications. Many experimental techniques applied to characterize SAMs involve the exposure of the sample to such irradiation. On the other hand, modification of SAMs by irradiation can be used to pattern the self-assembly monolayers and thus represents a novel method of lithography. Moreover, the initial pattern generated in a SAM (e.g., by electron beam writing) can be transferred into the underlying substrate (e.g., by chemical etching or ion beam milling) [45].

An understanding of the nature of the beam damage in SAMs, especially the changes of the physical and chemical properties of the organic films in the exposed region, is important for the optimization of the lithography process and also for the selection of appropriate molecules. So far, investigations into the modification of alkanethiol SAMs by X-ray and electron exposure have explored the chemical and structural changes within the monolayers [46–52], the relation between packing density and the extent of irradiation-induced damages [53], changes in tribological properties due to cross-linking driven by electron irradiation [54], and the effect of irradiation by secondary electrons originating from the substrate [55]. Dehydrogenation of alkanethiol/Au/mica films has been reported, principally via dissociative electron attachment [56]. A well-resolved maximum in the probabilities for C–H bond cleavage was observed at $\sim$10 eV in these experiments. Electron irradiation also causes a whole series of other effects: cleavage of C–C bonds resulting in the partial desorption of fragments, a reduction in the film thickness, the appearance of C–C/C=C cross-links, the loss of conformational and orientational order within the alkanethiolate films, the breaking of Au–S bonds, the formation of disulphide bonds, and incorporation of sulfur into an alkyl matrix [57–59].

3.2. Low Energy Electron Impact on Films of Passivated Nanoparticles

In contrast to the wealth of knowledge on SAMs of alkanethiol molecules on planar metal surfaces, the nature of the adsorption of organic ligands on metal clusters is a relatively unexplored field [60, 61]. Moreover, the specific characteristics of the clusters, such as the shape, electronic structure, and surface reactivity derived from their limited size, may introduce new aspects to the interaction of the ligands with low energy electrons. The effect of electron impact on the organic shells of passivated metal nanoclusters is thus not only of relevance to practical e-beam writing but is also of fundamental scientific interest.

Chen et al. have investigated films of gold particles, passivated with alkanethiol (octanethiol, $C_8H_{17}S$) ligands and deposited from a toluene solution onto the graphite surface, using high resolution electron energy loss spectroscopy (HREELS) [62]. A typical HREELS spectrum from this film (Fig. 1) shows vibrational modes of the CH_x groups of the ligands, together with their overtones. Impact energy dependent HREELS measurements reveal an enhancement in the intensity of these vibrational modes at an energy of around 10 eV, as displayed in Figure 2, indicating the formation of a negative ion resonance state resulting from low energy electron attachment [63]. Olsen and Rowntree [56], who investigated the electron-induced dehydrogenation of alkanethiols on planar Au films, found C–H dissociation probabilities with a well-resolved maximum at 10 eV, which is close to the resonance value observed in the HREELS study of the nanoparticle film. The width of the resonance, about 7 eV, is significantly greater than found in SAM (4 eV). One possible explanation is that the resonance lifetime is reduced in the passivated cluster film (e.g., by the opening of new decay channels for the transient negative ion). In the film of passivated Au clusters, one can envisage interactions between the ligands of adjacent clusters [61, 64], for example, by interdigitation, which might lead to broadening of the resonance.

Low energy electron impact on gold nanoparticles passivated with dialkyl sulphide molecules ($H_{21}C_{10}SC_{10}H_{21}$) has also been investigated [65]. Dialkyl sulphide is an interesting passivating agent because the Au–S bond is weaker (60 kJ mol^{-1}) than in the alkanethiol case (126 kJ mol^{-1}) [66], which might result in a higher probability for Au–S bond cleavage during electron beam exposure [67, 68].

A typical HREELS spectrum obtained from a film of dialkyl sulphide passivated gold nanoparticles on graphite is shown in Figure 3a. Energy loss peaks observed in the 700–3000 cm^{-1} region are related to CH_x and C–C vibrational modes of the organic ligands (see [65] and references therein). The Au–S stretching mode at 230 cm^{-1} is also clearly visible in this case [69]. After prolonged exposure

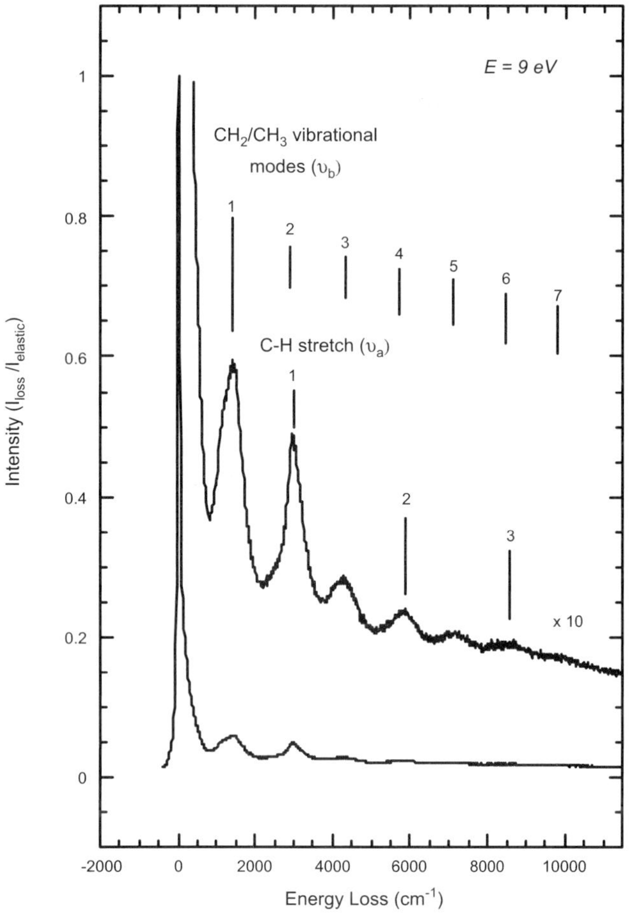

Figure 1. HREELS spectrum from a film of gold nanoparticles passivated with alkanethiol ($C_8H_{17}S$) ligands deposited on the graphite surface.

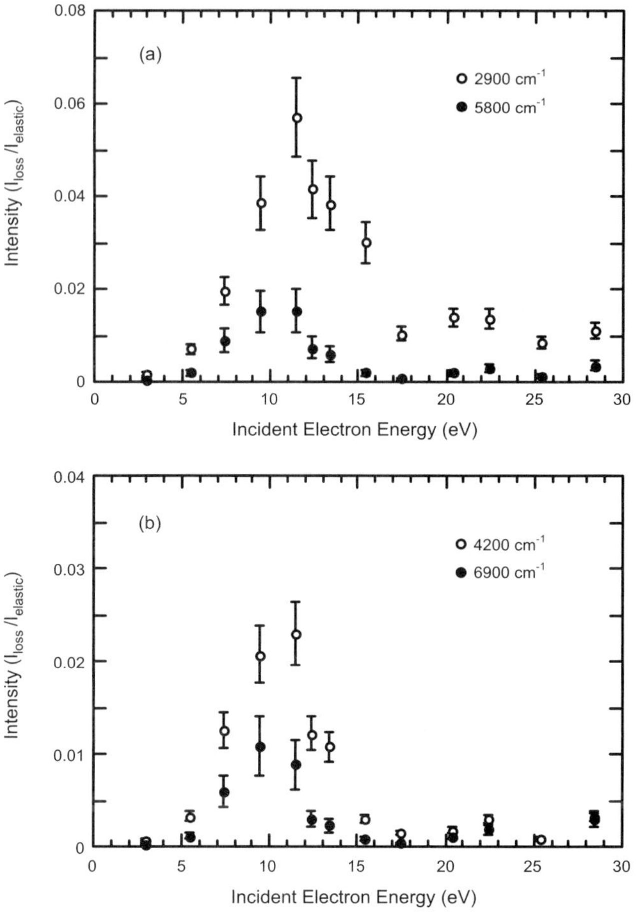

Figure 2. Incident energy dependence of the intensities of the CH stretch (υ_a) vibrational mode and first overtone (upper panel) and of two overtones of the CH_2/CH_3 vibrational modes around 1400 cm^{-1} (υ_b) (lower panel) obtained from HREELS spectra taken in the specular scattering geometry from a film of passivated gold nanoparticles on graphite.

of the film to electron beam irradiation at 50 eV, with a dosage of approximately 360 μC cm^{-2}, significant changes were found in the HREEL spectra (Fig. 3b). The Au–S stretch mode disappears and a new feature at 460 cm^{-1} appears, indicating the cleavage of the Au–S bond and, possibly, the formation of sulphoxide. Thus the HREELS measurements provide evidence of direct Au–S bond cleavage upon low energy electron irradiation, possibly related to the relatively weak Au–S bond in the dialkyl sulphide–gold system.

The picture emerging of the interaction between low energy electrons and passivated nanoparticle films resulting in cleavage of the C–H, C–C, and Au–S bonds is thus rather similar to the picture of electron beam–SAM interaction.

3.3. High Energy Electron Irradiation on Nanoparticles

The morphology changes in a nanoparticle film as a result of high energy electron beam irradiation can be imaged directly with a transmission electron microscope (TEM). Figure 4 was obtained after alkanethiol passivated gold nanoparticles, deposited onto an SiO_2 substrate, were exposed to a focused electron beam (200 kV, 36 nA), giving an approximate dose (assuming a probe of 10 nm) of 55 nC nm^2. Figure 4 shows TEM images taken before and after electron beam irradiation; only the gold cores of the nanoparticles are visible in the TEM (not the ligands). The coalescence (sintering) of local clusters within the exposed area is clearly illustrated in the TEM images. High resolution TEM imaging also show that grain boundary diffusion occurs during sintering [70].

4. DIRECT ELECTRON BEAM WRITING IN NANOPARTICLE FILMS

4.1. Electron Beam Lithography

Electron beam lithography utilizes the fact that certain chemicals (e-beam resist) change their properties after electron irradiation and creates extremely fine patterns by scanning a highly focused energetic electron beam across a surface covered with resist and thus deposits energy in the desired area. The electron beam can be generated by either thermionic sources or field emission sources and is focused on the substrate by a multilens system to a spot size on the order of nanometers. In order to expose arbitrarily shaped patterns, the electron beam is vector scanned

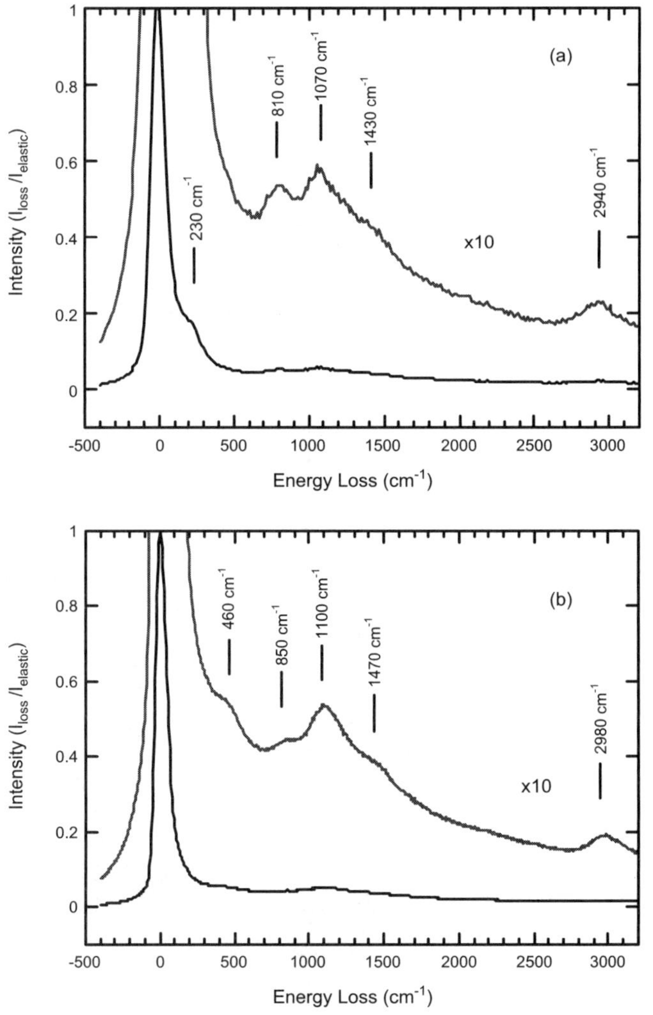

Figure 3. (a) Typical HREELS spectrum from a film of dialkyl sulphide ($H_{21}C_{10}SC_{10}H_{21}$) passivated gold nanoclusters on graphite. Incident electron energy 4.5 eV; incident angle 60°; specular scattering geometry. (b) HREELS spectrum taken after the film was exposed to a low energy electron dose of approximately 90 μC at 50 eV. Scattering conditions as for (a).

under computer control by electrostatic deflectors. A commercial e-beam lithography system can maintain a focus e-beam in a diameter of 8 nm or less throughout a typical 100 μm × 100 μm writing field with beam currents in the

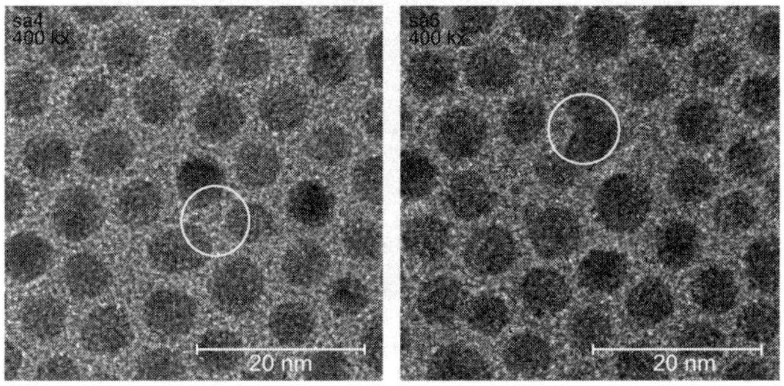

Figure 4. TEM images taken before and after the sample was exposed to a focused electron beam (200 kV, 36 nA) for 2 min (an approximate dose of 55 nC nm^2, assuming a probe diameter of 10 nm).

10 pA range at an acceleration voltage of 50 kV [71]. In addition to sophisticated and expensive commercial electron beam lithography systems, standard scanning electron microscopes or scanning transmission electron microscopes (STEM) may be used for pattern exposure. In this case, the SEM or STEM is interfaced with a pattern generation system, which controls the deflection system in the microscopes during the exposure.

Depending on the chemical structure of the resist, the electron either intersects the polymer chains (positive tone) or cross-links the small chain (negative tone), thus changing the solubility in organic developer compared to the unexposed area. For a positive tone resist the exposed material is removed, whereas for a negative tone resist only the exposed material remains after development. PMMA (polymethylmethacrylate) is a most commonly used positive tone high resolution e-beam resist for a typical exposure dose of several hundred μC cm^{-2} at 50 keV [72].

Figure 5 schematically shows the basic steps for the fabrication of a patterned material structure on a substrate using electron beam writing with a positive tone resist and lift-off process. The e-beam resist is first spin coated onto a substrate and then the e-beam writing process doses electrons to selective areas. The following development step removes the materials in the exposed area. Then a thin layer of materials (metal or insulator) is evaporated on the sample. Finally in a lift-off process, the unexposed resist together with its (metal or insulator) coating is removed with a suitable solvent. Only the materials in the area where they are directly deposited on the substrate surface remain.

The e-beam lithography technique is capable of high resolution pattern generation, flexible to create an almost infinite number of patterns, but slow in comparison to optical lithography. It is currently used in making masks for photolithography, advanced prototyping device fabrication, and research on novel physics phenomena at very small dimensions.

4.2. Direct Electron Beam Writing in Passivated Nanoparticles

Section 3 shows that the organic ligands, which prevent the aggregation of the metal cores when passivated nanoparticles are deposited on a surface, can be removed by

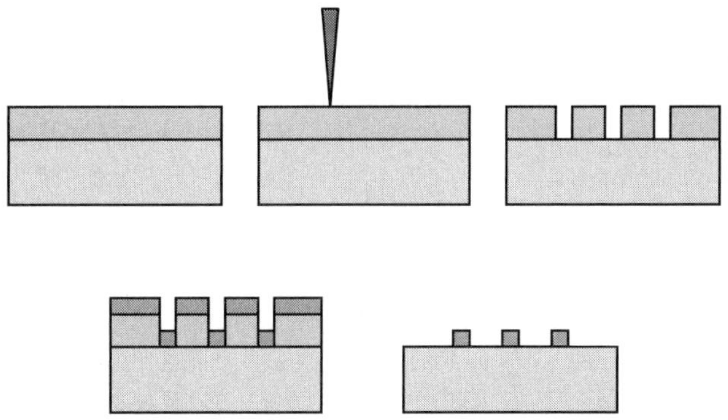

Figure 5. Basic steps for the fabrication of patterned materials on a substrate using e-beam lithography with positive tone e-beam resist.

electron beam irradiation. A metal-based nanostructure can thus be obtained, provided aggregation of the metal cluster cores takes place. Previous work on electron beam writing in films of passivated metal nanoparticles has demonstrated the creation of nanoscale architectures of different conductivity. Forty kV electron beam writing in multilayers of passivated gold clusters was employed to fabricate structures that demonstrated room temperature Coulomb blockade conduction characteristics [73–75]. The carbon residuals arising from the exposed ligands are exploited to create a series of tunnelling gaps between metal islands in 100 nm wide structures. Structures with genuinely metallic conductivity have also been observed after an additional annealing stage [76–78]. In this case, passivated Pd clusters were used to fabricate 70 nm structures with a STEM operated at 100 kV.

In general terms, the technique of direct electron beam writing in nanoparticles is a three-step process consisting of coating, writing, and rinsing, as schematically shown in Figure 6. First of all, a substrate is coated with nanoparticles, which can be achieved by spin coating, the Langmuir–Blodgett technique, liquid drop deposition, or simply dipping a substrate into a solution of nanoparticles [23, 79–87]. Then the nanoparticle film is patterned though direct electron beam writing, such that decomposition of the ligands of the nanoparticles takes place in the exposed area and may result in sintering to continuous structures. Following the patterning, a rinsing process is employed to remove unexposed nanoparticles by dissolution in a suitable solvent.

Thus direct e-beam writing is a relatively simple process for production of metal nanostructures compared with the more conventional e-beam lithography and lift-off process, which usually employs five steps as shown in Figure 5, namely, resist coating, e-beam exposure, resist development, metal deposition, and lift-off. Note that the development of the resist after exposure may result in some loss of resolution [88, 89]. Moreover, since the ultimate resolution possible in electron beam writing is limited by the molecular size of resist, a further attraction of passivated nanoparticles is the selectable size of the particles, which can be chosen to be as small as 2–3 nm in diameter.

The use of organometallic compounds for direct write e-beam production of metal-based structures is well established. As long ago as 1986, the formation of metal lines of 250 nm width was demonstrated by e-beam lithography in gold mercaptides [90]. The major difficulty with using conventional organometallic molecules as precursors is the carbon contamination, which results from the relatively low metal content [e.g., Au:C ratio of 1:7 in Me2-Au (tfa), $(CH_3)_2$–Au–$(C_5H_4F_3O_2)$] [91]. The level of carbon impurity affects the conductivity of the structures and may thus determine the effectiveness of this technique for electronic device fabrication. Passivated nanoparticles provide a greater metal loading than typical organometallics, (Au:C ratio of 1:1 for a 4 nm Au core capped with decanethiol molecules). Therefore e-beam writing in nanoparticle films may allow fabrication of nanowires with superior conductivity [92].

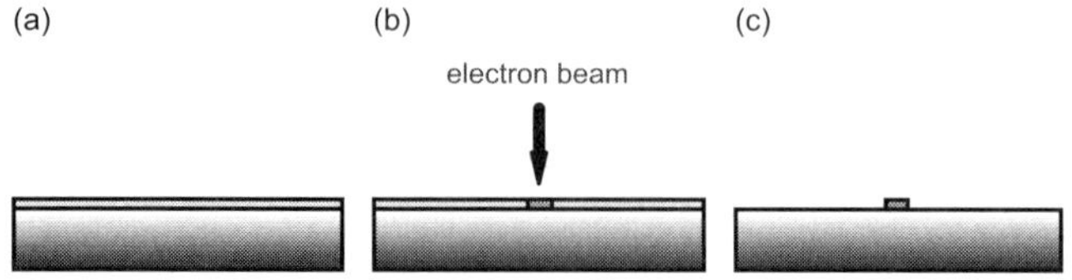

Figure 6. Schematic showing the three steps involved in the direct electron beam writing process: (a) coating, (b) writing, and (c) rinsing.

Figure 7 gives a typical example of the result of e-beam writing in thin films of passivated gold nanoparticles [93]; in this case the gold cores were 4–4.5 nm in diameter and surrounded by hexadecanethiol ($C_{16}H_{33}S$) ligands. They were deposited from solution in octane onto clean Si and a thermally grown SiO_2 film (thickness 65 nm). Thin films of monolayer coverage (~6 nm thick), calibrated by ellipsometry measurements, were exposed to an electron beam at 6 keV in an ISI DS-130 field emission scanning electron microscope equipped with an electron beam lithography system (Raith Elphy Quantum). The unexposed nanoparticles were removed by rinsing in octane. Figure 7 shows an atomic force microscopy (AFM) image (contact mode) of lines fabricated by e-beam writing with different doses on a thermally grown SiO_2 film (thickness 65 nm) on a Si wafer. It was found that the linewidth increases as the dose increases. The average linewidth of the upper most line, corresponding to the lowest e-beam dose (11.7 mC cm^{-2}), is only 26 nm [93]. This resolution lies well below the smallest feature sizes (currently) employed in integrated circuit, ~100 nm.

Direct e-beam writing of nanoparticles on Si_3N_4 has been carried out by Lin et al. [94]. It was reported that the electron beam exposure strips the passivating molecules (dodecanethiol) from the gold cores, enabling the cores to stick to the underlying substrate. A subsequent washing step removes unexposed nanoparticles, leaving behind the desired pattern.

Recently, Werts et al. reported nanometer scale patterning of Langmuir–Blodgett films of gold nanoparticles by electron beam lithography [95]. Nanowires of sub-50 nm wide with thickness controlled at the single particle level are created with e-beam doses in the mC cm^2 region. It was

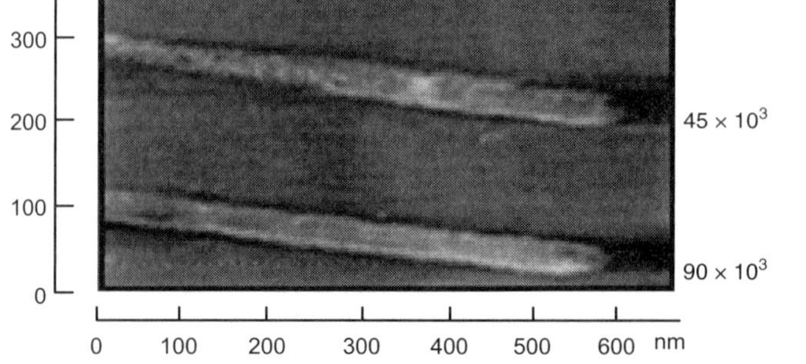

Figure 7. AFM image of lines produced by direct e-beam writing (6 keV) in a passivated gold nanoparticle film deposited on a thermal SiO_2 film (thickness 65 nm) on Si wafer.

shown that the patterns are formed by radiation-induced cross-linking of the alkyl chains. Interestingly, these wires can be "etched" away by prolonged exposure to a solution of capping agent, which is probably due to a replacement of cross-linked ligand shell by "fresh" passivating agents and therefore making the particles soluble again.

4.3. Quantitative Evaluation of Direct E-Beam Writing in Passivated Nanoparticles

The primary goals of e-beam lithography are high resolution and high speed, the latter requiring high resist sensitivity. In the direct electron beam writing technique described in the preceding section, the layer of passivated gold nanoparticles can be viewed as a negative tone e-beam resist. To investigate the sensitivity of nanoparticles as a resist, it is necessary to evaluate the e-beam writing process quantitatively.

The lines fabricated by e-beam writing can be characterized quantitatively using AFM. Indeed, the AFM image of Figure 7, showing lines of 1 μm long and 200 nm apart, was obtained as part of this procedure. The height of the lines varies little as a function of dose, always lying in the range of ~8–12 nm. This confirms the nominal thickness of the nanoparticle layer as 1 (–2) monolayers. The average linewidth of the four lines, measured by AFM, ranges from 26 to 92 nm.

For a monolayer resist, standard measures of the resist sensitivity, based on the depth of resist exposed as a function of dose, are not applicable. Thus the linewidth as a function of dose was defined as a practical, quantitative measure of the sensitivity of the cluster film resist, resulting in dose curves of the kind shown in Figure 8. In fact, this plot is effectively a measurement of the degradation of the resolution with increasing dose, so that the finest resolution is obtained just after the onset of lithography, as labelled in Figure 8. If the sensitivity of the nanoparticle film resist is

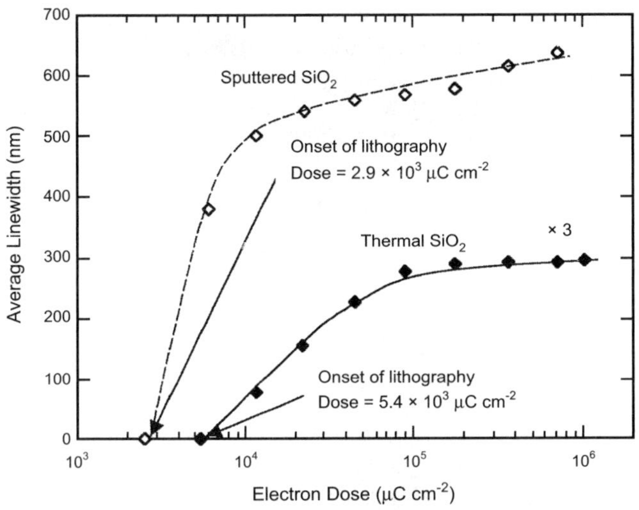

Figure 8. Response curve for passivated gold nanoparticle films (~6 nm) on thermal and sputtered SiO_2. Shown is the width of the lines produced by direct e-beam writing at 6 kV as a function of electron dose. The threshold doses obtained from the curves are marked (see text).

described by the threshold dose for production of exposed lines, then a sensitivity for the film on thermal SiO_2 of 5.4×10^3 $\mu C\,cm^{-2}$ can be quoted. It is observed from Figure 8 that the feature size (linewidth) increases with dose and after that the rate of increase slows down considerably after a linewidth of ~100 nm is reached.

Bedson's result (i.e., 5.4×10^3 $\mu C\,cm^{-2}$) on thermally grown SiO_2 on Si wafer is comparable to those recently published by other groups. Werts et al. reported 1×10^3 $\mu C\,cm^{-2}$ for a film of AuC_{12} particles on SiO_2 coated Si wafer [95] and Lin et al. found 7.7×10^3 $\mu C\,cm^{-2}$ for gold nanoparticles on Si_3N_4 substrate [94]. As will be discussed in the following section, the threshold dose and the rate of growth of the linewidth with dose rely not only on the nature of the nanoparticle film but also the substrate, since secondary electrons from the substrate play an important role in the electron exposure.

In comparison to the sensitivity of a well established e-beam resist such as PMMA (quoted as 0.13×10^3 $\mu C\,cm^{-2}$) [96], the nanoparticle film is about 40 times less sensitive. However, the sensitivity is comparable to that of other high resolution negative tone e-beam resists, for example, calixarene MC6AOAC (7×10^3 $\mu C\,cm^{-2}$) [97], C_{60} (10^4 $\mu C\,cm^{-2}$) [98], and hexapentyloxytriphenylene (2.5×10^3 $\mu C\,cm^{-2}$) [99].

One minor advantage arising from the modest sensitivity of the nanoparticle film is that SEM imaging for a short time period does not expose the film. Therefore, it is possible to locate a selected area on the specimen with the help of SEM imaging and align the writing field properly to this area before e-beam exposure.

4.4. Mechanism of Electron Beam Writing in Passivated Nanoparticles

When the incident electrons impinge on the nanoparticle film, they experience many small angle scattering events (forward scattering) and propagate through the cluster film into the substrate. Here the electrons occasionally undergo large angle scattering events and may thus return to the cluster film (possibly at some distance from the incident beam), causing additional resist exposure. The electron backscattering probability depends on the substrate material; low atomic number materials give less backscattering. Like the primary electrons, the backscattered electrons can also generate secondary electrons, with energies less than 50 eV. The cross-section for bond breaking is generally large for these low energy electrons.

In the case of a thin cluster film, these effects mean that the nature of the substrate plays an important role in the e-beam writing process, whereas the nature of the cluster film itself regulates the exposure of a thick film.

Figure 8 also illustrates results from a study intended to shed light on the mechanism of direct e-beam writing in passivated nanoparticle films. Passivated Au nanoparticles were deposited from solution onto a variety of substrates, namely, highly oriented pyrolitic graphite, silicon (with the native oxide layer intact), thermally grown silicon dioxide, and sputtered silicon dioxide. Thin films of monolayer coverage (i.e., thickness ~6 nm) were calibrated by ellipsometry measurements. Thicker films of multilayer coverage (thickness ~135 nm) were also prepared as a reference. A response curve describing the exposure of monolayer films of passivated gold nanoparticles on sputtered and thermally grown SiO_2 is shown in Figure 8, which plots the width of the exposed lines as a function of electron beam dose. The threshold for electron beam writing obtained in the case of sputtered SiO_2 is 2900 $\mu C\,cm^{-2}$. This is about half the value obtained with thermally grown SiO_2, 5400 $\mu C\,cm^{-2}$. In addition, it is evident from Figure 8 that the linewidth increases at a much faster rate above the onset of lithography in the case of the sputtered SiO_2. Moreover, saturation of the linewidth on sputtered SiO_2 starts at ~500 nm compared with ~100 nm in the case of thermal SiO_2. However, complete saturation was not observed in either case even at the highest doses used (28.5 $C\,cm^{-2}$), which is possibly due to the ability of the weakly bound nanoclusters to diffuse across the surface to the region where the line is being fabricated and thus provide a continuous supply of material for further growth of the structure.

The quantitative difference between the response curves for passivated nanoparticles on the two different (sputtered and thermally grown) SiO_2 substrates demonstrates that the substrate plays a key role in the mechanism of exposure of thin nanoparticle films. This is consistent with the generation of backscattering and secondary electrons in the vicinity of the initially irradiated region. Indeed, nanowires were first found to be formed in the layer of nanoparticles directly on the substrate. AFM imaging of nanowires created on a three layer thick AuC_6 LB film reveals that the height of the nanowire increased with exposure in the case of underexposure and structures of monolayer or multilayer thickness were observed [95].

Monte Carlo simulations of the scattering process confirm this prehension. Figure 9 shows results for (a) thin and (b) thick nanoparticle film on a thermal SiO_2 surface. The thin cluster film (~6 nm) increases the total scattering width of the primary electron beam—defined as the distance between the outermost emerging electrons—when compared with the thick nanoparticle film (~135 nm).

The predominant role of substrate electron scattering in the exposure of thin passivated nanoparticle films is confirmed by the experimental results for graphite and silicon (with native oxide) substrates. In neither case was it possible to write lines in monolayer cluster films [100]. This is consistent with the much reduced secondary electron coefficients, δ, for these materials (1 for graphite and 1.1 for silicon) compared with SiO_2 (where δ lies between 2.1 and 4) [100]. It is likely that low energy secondary electrons, generated both by the primary beam and the backscattering electrons, are the electrons which actually expose the passivated clusters, via dissociation or removal of the ligands.

5. APPLICATIONS

Direct e-beam writing in passivated nanoparticle films, like conventional e-beam lithography, is capable of creating various types of nanoscale architectures. As an example, we shall consider the fabrication of linear structures with potential applications as nanoscale electrodes and connections.

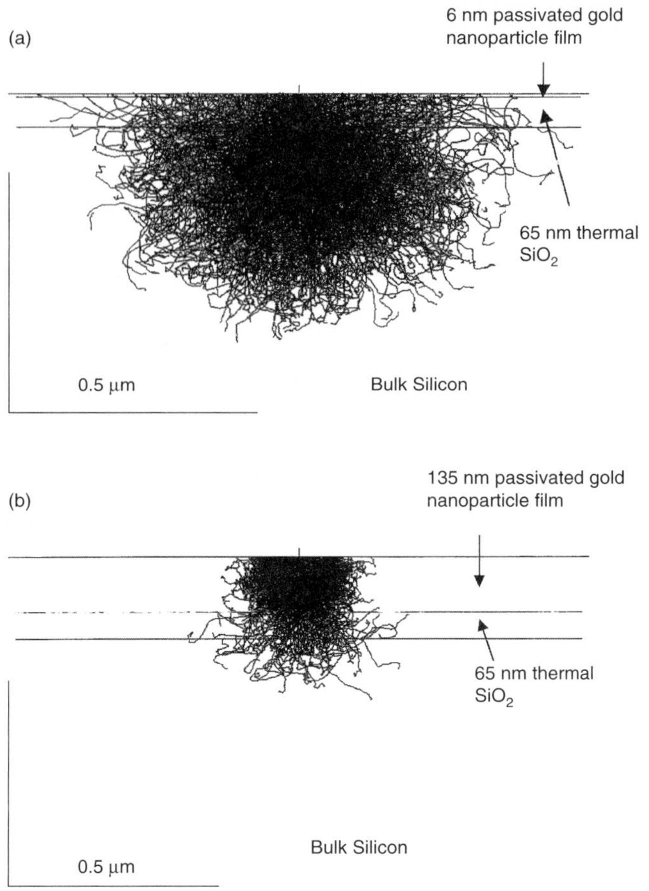

Figure 9. Monte Carlo simulations of a 6 keV electron beam scattering through (a) thin (~6 nm) and (b) thick (~135 nm) films of passivated gold nanoclusters on a thermal SiO_2 layer (65 nm thick) on bulk silicon. One thousand trajectories are shown in each case.

5.1. Fabrication of Nanowires

Figure 10 illustrates the fabrication of individual wires obtained via direct electron beam writing in thin films of alkanethiol passivated Au nanoparticles on a Si substrate with a thermally grown SiO_2 layer. The line is 20 μm long but only 40–50 nm wide and written with a 7 keV electron beam. The pattern shown in Figure 10 above the line is an alignment mark. Success in obtaining long lines lies mainly in the quality of the nanoparticle film; large voids in the film result in a discontinuous line.

One potential application of such wires is to make electrical contacts to nanometer-scale objects (e.g., nanoparticles). The creation of a gap structure provides a template for the incorporation of the molecule or nanoparticles whose transport properties are to be explored. In recent years, a number of techniques have been developed to fabricate small gaps between two electrodes, including shadow deposition [101], electrochemical deposition [102], electromigration [103], scanning probe lithography [104], and controlled e-beam lithography [105]. Direct e-beam writing can also create such gaps, as Figure 11 illustrates. A clear gap of 20 nm is visible. It is expected that the gap can be further narrowed via proximate effect correction.

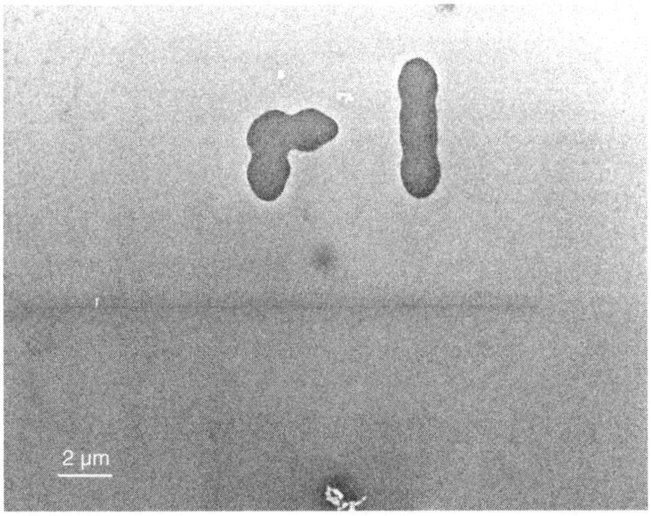

Figure 10. SEM image of a line 20 μm long and 40–50 nm wide, produced by direct electron beam writing in a film of passivated gold nanoparticles on a thermal SiO_2 substrate (7 keV, electron dose, 50×10^3 $\mu C\,cm^{-2}$). An alignment mark is shown above the wire.

5.2. Nanowires on Patterned Surfaces

The realization of a true nanoscale device involves the connection of a nanometer-scale component to the microscopic (and thence macroscopic) world. It is necessary to combine both microfabrication and nanofabrication techniques. The direct e-beam writing technique is compatible with conventional lithography as illustrated in Figure 12. Micrometer scale gold patterns were prepared via photolithography and ion beam milling on an SiO_2/Si substrate. After coating of this sample with alkanethiol passivated gold nanoparticles, direct e-beam writing was applied, as illustrated in Figure 12a, to create a long wire running across four gold microelectrodes with 5 μm gaps. The whole length of the wire is 40 μm with an average width of about 70 nm. Another example, Figure 12b, is a cross formed by two nanowires bridging two pairs of electrodes.

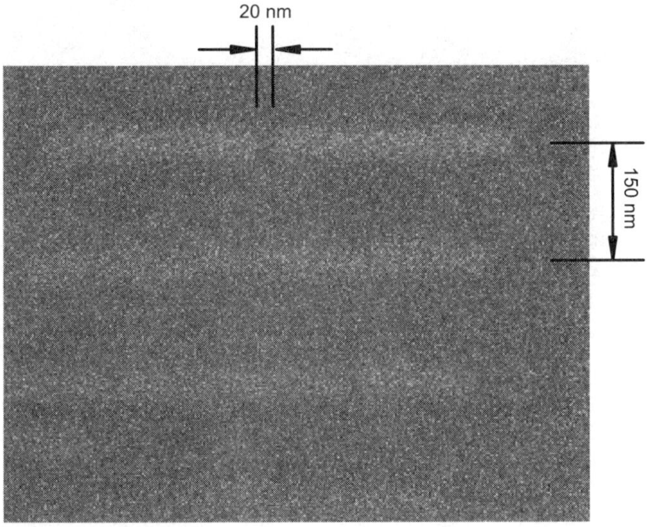

Figure 11. SEM image of a wire produced by direct electron beam writing in a film of passivated gold nanoparticles on a thermal SiO_2 substrate. The wire contains a narrow gap (generated without proximity effect correction) to create a pair of electrodes.

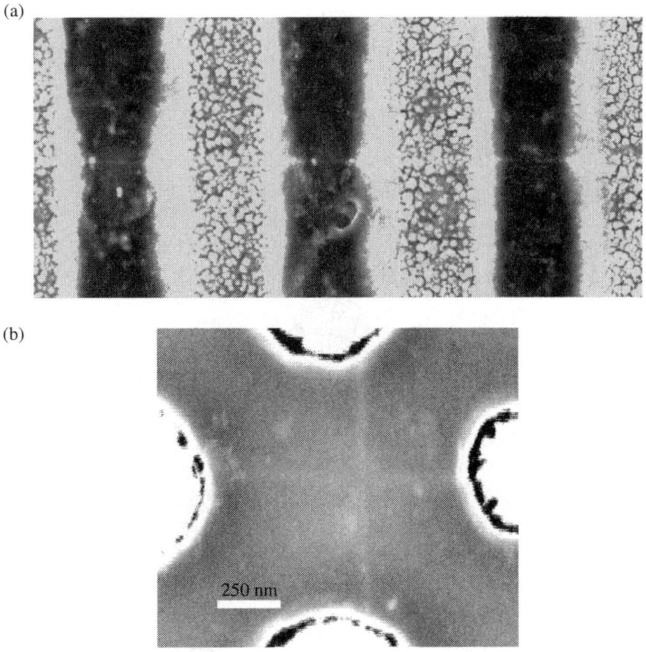

Figure 12. SEM images of nanowires fabricated on patterned substrates. (a) A long (40 μm) wire running across four Au electrodes with 5 μm gaps fabricated on SiO_2/Si substrate via photolithography and ion beam milling. The average width of the wire is ~70 nm; (b) perpendicular nanowires connecting two pairs of Au electrodes.

5.3. Conductivity Evaluation of Nanowires

One critical issue regarding the application of the nanowire as electrodes or electrical connection is the conductivity of the wires themselves. Figure 13 demonstrates that "wires" produced by e-beam writing in nanoparticle film are indeed

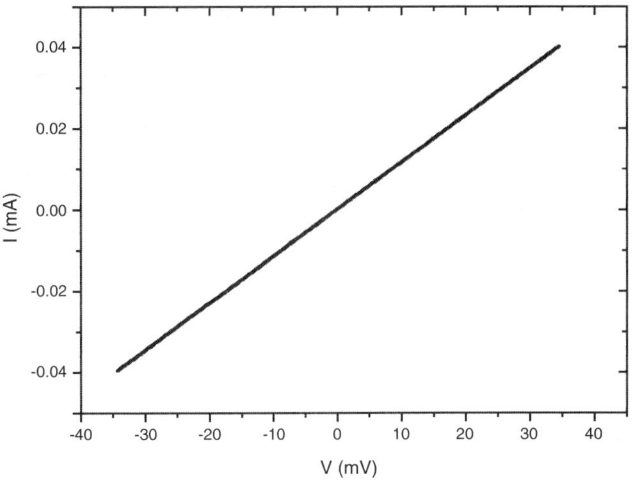

Figure 13. Linear I–V characteristic obtained from a nanowire fabricated via direct electron beam writing in a film of alkanethiol passivated gold nanoparticles on a thermal SiO_2/Si substrate.

metallic. A wire of dimension of 60 nm was produced between patterned Au/Cr electrodes. A linear I–V characteristic is observed, with a resistivity of 40 $\mu\Omega$ cm^{-1} (and scope for optimization in future). This compares with a value of 2.44 $\mu\Omega$ cm^{-1} for bulk Au [106]. The increase in resistance compared with bulk most likely originates from impurities due to fragmentation of the passivating ligands.

6. CONCLUSIONS

Direct electron beam writing in films of passivated nanoparticles is a new technique with clear potential for the creation of metallic wires and contacts of 10–100 nm in size. Future improvements include the use of nanoparticles with specially designed organic ligands. Employing shorter chain ligands would enhance the metal loading in the initial nanoparticle film, while molecules other than alkane thiols (e.g., dialkyl sulphides with a weaker S–Au bond) may lead to ligand removed at low electron beam dose. Again, the choice of ligand molecules, which produce "volatile" fragments under electron irradiation, may reduce the impurity level in the resulting wires and hence increase their conductivity. E-beam writing in the presence of oxygen or hydrogen gas may have a similar effect.

While we have so far considered the fabrication of linear structures (i.e., nanowires), the advanced capabilities of a modern e-beam writer also suggest much more complex patterns from production of arrays of nanosize dots right up to the creation or repair of integrated circuits. Further research into the interaction of electrons with nanoparticles is essential to understand and thus optimize the fabrication processes. This should involve detailed structural and compositional analysis. For example, *in-situ* high resolution TEM images of the direct e-beam writing process would give insights into the formation and growth of the nanowires as a function of the electron dose. By combining this structural information with the compositional analysis and electrical measurements, it may be possible to improve both the sensitivity and the resolution of the technique as well as to optimize the conductivity of the resulting nanometer-scale metallic structures for advanced applications.

GLOSSARY

Atomic force microscope (AFM) A microscope that produces an image by scanning a tip attached to the end of a cantilever across the sample surface while monitoring the change in cantilever deflection due to the tip-surface interaction with a split photodiode detector.

High-resolution electron energy loss spectroscopy (HREELS) A sample surface is irradiated with a monoenergetic and well-collimated electron beam and a spectrum obtained as a function of energy and momentum transfer by energy analysis of the backscattered electrons.

Highly oriented pyrolitic graphite (HOPG) A special type of graphite with highly oriented monocrystal grains prepared by pyrolysis of organic compounds.

Langmuir–Blodgett (LB) film A mechanically assembled array of amphiphilic molecules on a water surface often transferred to a solid support once the molecules are compressed into a desired phase. An LB film can consist of single or many layers.

Polymethyl methacrylate (PMMA) A polymer that changes its properties under irradiation by electron, UV photons or X-rays and can be used as a resist, providing high contrast, high resolution lithography.

Scanning electron microscope (SEM) A microscope that produces an image by scanning an electron beam across the surface of a specimen and collecting the backscattered or secondary electrons. Examination of microstructural features with a resolution below 10 nm is possible.

Scanning tunnelling microscope (STM) A microscope that produces an image by scanning a sharp tip across a conducting sample surface and (usually) recording the tip-surface distance required to keep tunnelling current constant.

Self-assembled monolayers (SAM) An ordered and highly oriented single layer of molecules spontaneously assembled on a solid surface, as governed by the molecule-surface and intermolecular interactions.

Transmission electron microscope (TEM) A microscope that produces an image by using electron beams that transmitted (passed through) a thin specimen and provides internal structural information of the specimen at high magnification.

REFERENCES

1. T. Rueckes, K. Kim, E. Joselevich, G. Y. Tseng, C. L. Cheung, and C. M. Lieber, *Science* 289, 94 (2000).
2. M. A. Reed, J. Chen, A. M. Rawlett, D. W. Price, and J. M. Tour, *Appl. Phys. Lett.* 78, 3735 (2001).
3. C. P. Collier, E. W. Wang, M. Belohradsky, F. M. Raymo, J. F. Stoddart, P. J. Kuekes, R. S. Williams, and J. R. Heath, *Science* 295, 391 (1999).
4. C. P. Collier, G. Mattersteig, E. W. Wong, Y. Luo, K. Beverly, J. Sampaio, F. M. Raymo, J. F. Stoddart, and J. R. Heath, *Science* 289, 1172 (2000).
5. S. J. Tans, A. R. M. Vershueren, and C. Dekker, *Nature* 393, 49 (1998).
6. Y. Cui and C. M. Lieber, *Science* 291, 851 (2001).
7. A. Bachtold, P. Hadley, T. Nakanishi, and C. Dekker, *Science* 294, 1317 (2001).
8. Y. Huang, S. Duan, Y. Cui, L. J. Lauhon, K. H. Kim, and C. M. Lieber, *Science* 294, 1313 (2001).
9. R. F. Khairutdinov, *Collid J.* 59, 535 (1997).
10. A. P. Alivisatos, *J. Phys. Chem.* 100, 13226 (1996).
11. M. M. Alvarez, J. T. Khoury, T. G. Schaaff, M. N. Shafigullin, I. Vezmar, and R. L. Whetten, *J. Phys. Chem. B* 101, 3706 (1997).
12. L. N. Lewis, *Chem. Rev.* 93, 2693 (1993).
13. Y. Li and M. A. El-Sayed, *J. Phys. Chem. B* 105, 8938 (2001).
14. H.-G. Boyen, G. Kastle, F. Weigl, B. Koslowski, C. Dietrich, P. Ziemann, J. P. Spatz, S. Riethmuller, C. Hartmann, M. Moller, G. Shmid, M. G. Garnier, and P. Oelhafen, *Science* 297, 1533 (2002).
15. R. L. Whetten, M. N. Shafigullin, J. T. Khoury, T. G. Schaff, J. Vezmar, M. M. Alvarez, and A. Wilkinson, *Acc. Chem. Res.* 32, 397 (1999).
16. Z. L. Wang, *Adv. Mater.* 10, 13 (1998).
17. A. P. Alivisatos, *Science* 271, 933 (1996).
18. D. L. Feldheim and C. D. Keating, *Chem. Soc. Rev.* 27, 1 (1998).

19. D. L. Klein, R. Roth, A. K. L. Kim, A. P. Alivisatos, and P. L. McEuen, *Nature* 389, 699 (1997).
20. T. Sato, H. Ahmed, D. Brown, and B. F. G. Johnson, *J. Appl. Phys.* 82, 696 (1997).
21. R. S. Ingram, M. J. Hostetler, R. W. Murray, T. G. Schaaff, J. T. Khoury, R. L. Whetten, T. R. Bigioni, D. K. Guthrie, and P. N. First, *J. Am. Chem. Soc.* 119, 9279 (1997).
22. T. Sato and H. Ahmed, *Appl. Phys. Lett.* 70, 2759 (1997).
23. A. N. Shipway, E. Katz, and I. Willner, *Chem. Phys. Chem.* 1, 18 (2000) and references therein.
24. S. M. Chung, G. Markowich, and J. R. Heath, *J. Phys. Chem. B* 102, 6685 (1998).
25. U. Simon, G. Schon, and G. Schmid, *Angew Chem. Int. Ed. Engl.* 32, 250 (1993).
26. M. Burghard, S. Philips, S. Roth, and K. von Klitzing, *Appl. Phys. A* 67, 591 (1998).
27. T. O. Hutchinson, Y. P. Liu, C. Kiely, C. J. Kiely, and M. Brust, *Adv. Mater.* 13, 1800 (2001).
28. S. Fullam, D. Cottell, H. Rensmo, and D. Fitzmaurice, *Adv. Mater.* 12, 1430 (2000).
29. A. J. Parker, P. A. Childs, R. E. Palmer, and M. Burst, *Appl. Phys. Lett.* 74, 2833 (1999).
30. A. J. Parker, P. A. Childs, R. E. Palmer, and M. Burst, *Nanotech.* 12, 6 (2000).
31. Z. Tang, N. A. Kotov, and M. Giersig, *Science* 297, 237 (2002).
32. Q. Guo, X. Sun, Y. Chen, and R. E. Palmer, *Surf. Sci.* 497, 269 (2002).
33. W. Eck, V. Stadler, W. Geyer, M. Zharnikov, A. Golzhauser, and M. Grunze, *Adv. Matter.* 12, 805 (2000).
34. A. Golzhauser W. Eck, W. Geyer V. Stadler, T. Weimann, P. Hinze, and M. Grunze, *Adv. Mater.* 13, 806 (2001).
35. M. Brust, D. Bethell, D. J. Shiffrin, and C. J. Kiely, *Adv. Mater.* 7, 795 (1995).
36. D. Bethell, M. Brust, D. J. Schiffrin, and C. Kielly, *J. Electron. Chem.* 409, 137 (1996).
37. M. D. Musick, D. J. Pena, S. L. Botsko, T. M. McEvoy, J. N. Richardson, and M. J. Natan, *Langmuir* 15, 844 (1999).
38. M. D. Musick, C. D. Keating, M. H. Keefe, and M. J. Natan, *Chem. Mater.* 9, 1499 (1997).
39. G. Markovich, D. P. Collier, and J. R. Heath, *Phys. Rev. Lett.* 80, 3807 (1998).
40. J. D. Swalen, D. L. Allara, J. D. Andrade, E. A. Chandross, and S. Garoff, *Langmuir* 3, 932 (1987).
41. A. Ulman, "An Introduction in Ultrathin Organic Films." Academic, New York, 1991.
42. R. G. Nuzzo and D. L. Allara, *J. Am. Chem. Soc.* 105, 4481 (1983).
43. R. G. Nuzzo, L. H. Dubois, and D. L. Allara, *J. Am. Chem. Soc.* 112, 558 (1990).
44. L. H. Dubois and R. G. Nuzzo, *Annu. Res. Phys. Chem.* 43, 437 (1992).
45. E. Delamarche, A. C. F. Hoole, B. Michel, S. Wilkes, M. Despont, M. E. Welland, and H. Biebuyck, *J. Phys. Chem. B* 101, 9263 (1997).
46. A. J. Wagner, S. R. Carlo, C. Vecities, and D. H. Fairbrother, *Langmuir* 18, 1542 (2002).
47. K. Heister, M. Zharnikov, and M. Grunze, *Langmuir* 17, 8 (2001).
48. M. Zhamikov, S. Frey, K. Heisterm, and M. Grunze, *Langmuir* 16, 2697 (2000).
49. K. Heister, S. Frey, A. Golzhauser, A. Ulman, and M. Zharmikov, *J. Phys. Chem. B* 103, 11098 (1999).
50. M. Zhaenikov, W. Geyer, A. Golzhauser, S. Frey, and M. Grunze, *Phys. Chem. Chem. Phys.* 1, 3163 (1999).
51. H. U. Muller, M. Zharnikov, B. Volkel, A. Schertel, P. Harder, and M. Grunze, *J. Phys. Chem. B* 102, 7949 (1998).
52. H. Kondoh and H. Nozoye, *J. Phys. Chem. B* 102, 2367 (1998).
53. S. Frey, H. T. Rong, K. Heister, Y. J. Yang, M. Buck, and M. Zhamikov, *Langmuir* 18, 3142 (2002).
54. H. W. Liu, B. Bhushan, W. Eck, and V. Stadler, *J. Vac. Sci. Technol. A* 19, 1234 (2001).
55. P. E. Laibinis, R. L. Graham, H. A. Biebuyck, and G. M. Whitesides, *Science* 254, 981 (1991).
56. C. Olsen and P. A. Rowntree, *J. Chem. Phys.* 108, 3750 (1998).
57. M. Zharnikov, W. Geyer, A. Golzhauser, S. Frey, and M. Grunze, *Phys. Chem. Chem. Phys.* 1, 3163 (1999).
58. M. Zharnikov, S. Frey, K. Heister, and M. Grunze, *Langmuir* 16, 2697 (2000).
59. K. Heister, M. Zharnikov, and M. Grunze, *Langmuir* 17, 8 (2001).
60. S. Chen, R. S. Ingram, M. J. Hostetler, J. J. Pietron, R. W. Murray, T. G. Schaaff, J. T. Khoury, M. M. Alvarez, and R. L. Whetten, *Science* 280, 2098 (1998).
61. Z. L. Wang, S. A. Harfenist, R. L. Whetten, J. Bentley, and N. D. Evans, *J. Phys. Chem. B* 102, 3068 (1998).
62. Y. Chen, R. E. Palmer, and J. P. Wilcoxon, *Surface Sci.* 454–456, 963 (2000).
63. R. E. Palmer and P. J. Rous, *Rev. Mod. Phys.* 64, 383 (1992).
64. S. Murthy, Z. L. Wang, and R. L. Whetten, *Philos. Mag. Lett.* 75, 321 (1997).
65. Y. Chen, E. J. Shelley, J. A. Preece, and R. E. Palmer, *Surface Sci.* 502, 208 (2002).
66. D. J. Lavrich, S. M. Wetterer, S. L. Bernasek, and G. Scoles, *J. Phys. Chem. B* 102, 3456 (1998).
67. M. C. Wang, J. D. Liao, C. C. Weng, R. Klauser, S. Frey, M. Zhamikkov, and M. Grunze, *J. Phys. Chem. B* 106, 6220 (2002).
68. J. D. Liao, M. C. Wang, C. C. Weng, R. Klauser, S. Frey, M. Zhamikkov, and M. Grunze, *J. Phys. Chem. B* 106, 77 (2002).
69. R. G. Nuzzo, B. R. Zegarski, and L. H. Dubois, *J. Am. Chem. Soc.* 109, 733 (1987).
70. M. Couillard, Electron Microscopy of Nanostructures Resulting from the Deposition of Size-Selected Metal Clusters, Ph.D. Thesis, University of Birmingham, UK.
71. A. Forchel, P. Ils, R. Steffen, and M. Bayer, in "Nanofabrication and Biosystems" (H. V. Hoch, L. W. Jelinski, and H. G. Craighead, Eds.), p. 3. Cambridge Univ. Press, Cambridge, UK, 1996.
72. M. A. Mccord and M. J. Rooks, in "Handbook of Microlithography, Micromaching, and Microfabrication" (P. Rai-Choudhury, Ed.), IEE Materials and Devices Series 12, Vol. 1: Microlithography, p. 205. SPIE Press/Institute of Electronical Engineers, 1997.
73. M. N. Wybourne, L. Clarke, M. Yan, S. X. Cai, L. O. Brown, J. Hutchinson, and J. F. W. Keana, *Jpn. J. Appl. Phys. 1* 36, 7796 (1997).
74. L. Clarke, M. N. Wybourne, M. Yan, S. X. Cai, and J. F. W. Keana, *Appl. Phys. Lett.* 71, 617 (1997).
75. L. Clarke, M. N. Wybourne, M. Yan, S. X. Cai, L. O. Brown, J. Hutchinson, and J. F. W. Keana, *J. Vac. Sci. Technol. B* 15, 2925 (1997).
76. M. T. Reetz and M. Winner, *J. Am. Chem. Soc.* 119, 4539 (1997).
77. J. Lohau, S. Friedrichowski, G. Dumpich, E. F. Wassermann, M. Winner, and M. T. Reetz, *J. Vac. Sci. Technol. B* 16, 77 (1998).
78. G. Dumpich, J. Lohau, E. F. Wassermann, M. Winner, and M. T. Reetz, *Mater. Sci. Forum* 287–288, 413 (1998).
79. J. R. Heath, C. M. Knobler, and D. V. Leff, *J. Phys. Chem. B* 101, 189 (1997).
80. M. Brust, N. Stuhr-Hansen, K. Norgaard, J. B. Christensen, L. K. Nielsen, and T. Bjomholm, *Nano. Lett.* 1, 189 (2001).
81. V. I. Klimov, A. A. Mikhailovsky, S. Xu, A. Malko, J. A. Hollingsworth, C. A. Leatherdale, H. J. Eisler, and M. G. Bawendi, *Science* 290, 314 (2000).
82. J. P. Bourgoin, C. Kergueris, E. Lefeven, and S. Palacin, *Thin Solid Films* 327–329, 515 (1998).

83. L. F. Chi, M. Harting, T. Drechsler, T. Schawaack, C. Seidel, H. Fuchs, and G. Schimid, *Appl. Phys. Lett.* 66, S187 (1998).
84. G. Schmid, Y. P. Liu, M. Schumann, T. Raschke, and C. Radehaus, *Nano Lett.* 1, 405 (2001).
85. M. Bruchez, M. Moronne, P. Gin, W. Weiss, and A. P. Alivisatos, *Science* 281, 2013 (1998).
86. W. C. W. Chan and S. M. Nie, *Science* 281, 2016 (1998).
87. C. B. Mayer, C. R. Kagan, and M. G. Bawendi, *Annu. Res. Mater. Sci.* 30, 545 (2000).
88. C. Vieu, F. Carcenac, A. Pepin, Y. Chen, M. Mejias, A. Lebib, L. Manin-Ferlazzo, L. Couraud, and H. Launois, *Appl. Sur. Sci.* 164, 111 (2000).
89. W. Chen and H. Ahmed, *Appl. Phys. Lett.* 62, 1499 (1993).
90. H. G. Craighead and L. M. Schiavone, *Appl. Phys. Lett.* 48, 1748 (1986).
91. I. Utke, P. Hoffmann, B. Dwir, K. Leifer, E. Kapon, and P. Doppelt, *J. Vac. Sci. Technol. B* 18, 3168 (2000).
92. R. F. Service, *Science* 290, 1524 (2000).
93. T. R. Bedson, T. E. Jenkins, D. J. Hayton, J. P. Wilcoxon, and R. E. Palmer, *Appl. Phys. Lett.* 78, 1921 (2001).
94. X. M. Lin, R. Parthasarathy, and H. M. Jaeger, *Appl. Phys. Lett.* 78, 1915 (2001).
95. M. H. V. Werts, M. Lambert, J. P. Boourgoin, and M. Brust, *Nano Lett.* 2, 43 (2002).
96. Y. M. Ham, *J. Vac. Sci. Technol. B* 15, 2313 (1997).
97. J. Fujita, Y. Ohnishi, S. Manako, and S. Matsui, *Microelectron. Eng.* 41/42, 323 (1998).
98. T. Tada and T. Kanayama, *Jpn. J. Appl. Phys. 2* 35, L63 (1996).
99. A. P. G. Robinson, R. E. Palmer, T. Tada, T. Kanayama, M. T. Allen, J. A. Preece, and K. D. M. Harris, *J. Phys. D* 32, L75 (1999).
100. T. R. Bedson, R. E. Palmer, and J. P. Wilcoxon, *Appl. Phys. Lett.* 78, 2061 (2001).
101. E. T. T. Jones, O. M. Chyan, and M. S. Wrighton, *J. Am. Chem. Soc.* 109, 5526 (1987).
102. H. He and N. Tao, *Adv. Mater.* 14, 161 (2002).
103. H. Park, A. K. L. Lim, J. Park, A. P. Alivisatos, and P. L. McEuen, *Appl. Phys. Lett.* 75, 301 (1999).
104. T. Miyazaki, K. Kobayashi, T. Horiuchi, H. Yamada, and K. Matsushige, *Jpn. J. Appl. Phys.* 40, 4365 (2001).
105. K. Liu, Ph. Avouris, J. Bucchignano, R. Martel, S. Sun, and J. Michl, *Appl. Phys. Lett.* 80, 865 (2002).
106. H. D. Young and R. A. Freedman, "University Physics," 9th ed. Addison–Wesley, Reading, MA, 1996.

ENN Encyclopedia of Nanoscience and Nanotechnology
www.aspbs.com/enn

Electron Dynamics in Metallic Nanoparticles

M. Aeschlimann
University of Kaiserslautern, Kaiserslautern, Germany

CONTENTS

1. INTRODUCTION

The excitation and relaxation of electrons in metallic nanoparticles play a key role in a number of important physical phemomena, including linear and nonlinear optical effects. The main motivation for the interest in these systems is related to the possibility of tailoring, to a considerable extent, their electrodynamical behavior on the basis of size and shape. Furthermore, by studying the temporal evolution of collective electronic excitations (e.g., after a laser excitation), on the time scale from a few femtoseconds to several picoseconds, one can improve the knowledge about basic properties of the light–matter interaction in metallic nanoparticles.

A typical signature of the optical response of metallic nanoparticles is given by the presence of a local plasmon resonance (Mie-plasmon; see Section 2.1) [1–4]. While the positions of the resonances of Mie-plasmon excitation as a function of particle size, shape, and dielectric properties are well understood (as an overview, see [1]), the ultrafast dynamics of these collective electronic excitations have remained a highly interesting topic to be revealed in much greater detail. An essential issue is how rapidly the collective excitation lose their phase coherence. Further, one would like to investigate the mechanisms responsible for dephasing. After these dephasing processes, typical internal and external thermalization processes set in to distribute the energy of the excited quasiparticles (single excited *hot* electron) into the electron gas and nanoparticle lattice, and finally to the nanoparticle–substrate complex.

With the advance of ultrashort pulsed laser systems, a transient (nonthermal) electron distribution can now be created and probed. Compared to the time scale of these laser pulses, the long-lasting nonequilibrium situation offers the unique possibility of analyzing the interactions (energy exchange) between the different subsystems (electron gas, phonon system, substrate/matrix complex).

In this chapter, the different mechanisms, which can lead to phase destruction as well as energy dissipation of the collective electron motion, are revealed, and pioneering experiments in this new field are reviewed. All different energy dissipation steps after an ultrashort laser pulse are described in a sequential approach. However, one should keep in mind that this is a straightforward view. All processes start at once, but they are finalized on different time scales, that is, electron gas energy losses occur before the electron gas is fully thermalized. It is just the time scale of the different energy exchange mechanisms which differs over several magnitudes (from 10 fs to 120 ps) that allows, with some exception, this simplified treatment.

It is also important to mention that all considerations discussed in this chapter are relevant for metallic nanoparticles larger than 10 Å. The metal is still considered as a three-dimensional system. When the size becomes smaller than 10 Å, the discrete nature of the electronic states (quantum effects) shows up [5–9]. Above that size, the width of the discrete levels becomes comparable to or larger than the separation between them. That is, the conduction band is quasicontinuous, and the behavior of the nanoparticles tends to approach that of the bulk metal, presenting, however, some significant differences concerning the electrodynamic properties.

2. ADSORPTION OF LIGHT IN METALLIC NANOPARTICLES

2.1. General Considerations

The optical properties of nanostructured systems have been extensively investigated in order to reveal their fundamental processes and to examine possible technological applications

ISBN: 1-58883-059-4/$35.00

[1, 2]. Concerning the interaction of visible light with metals, the conduction electrons are of central importance. As a useful approximation, all electrons can be described collectively as a plasma with density N_e [10]. A collective excitation of the dense electron gas in a metal is called a *plasmon*. In the most general case, the term *plasmon* describes a longitudinal collective oscillation of the electron plasma relative to the crystal lattice [11]. This excitation consists of a coherent motion of a high number of electrons in contrast to those excitations in which the external perturbation acts on a single electron. In general, one has to distinguish among bulk, surface, and particle (Mie) plasmons.

The *bulk plasmon* denotes a collective excitation of the electron gas in the bulk of the metal, which propagates as a longitudinal charge density fluctuation at a resonance frequency [12]:

$$\omega_{Pl} = \sqrt{\frac{N_e \cdot e^2}{\varepsilon_0 m}} \tag{1}$$

The quanta of these bulk plasmons possess an energy $\hbar\omega_{Pl}$ which is, for most metals, on the order of 10–15 eV. Excitation of bulk plasmons with light is not possible due to the longitudinal nature of bulk plasmons.

At *surfaces*, the mobility of the electrons in a plane parallel to the surface is high (quasifree electrons), whereas perpendicular to the surface, the mobility is limited due to the border of the metal. This results in a reduced energy of the collective mode ω_{SP} regarding the bulk value ω_{Pl} [1, 13]:

$$\omega_{SP} = \frac{\omega_{Pl}}{\sqrt{2}} \tag{2}$$

Because the phase velocity of surface plasmons is slower than that of a photon with the same energy, direct coupling of freely propagating light to an excitation of surface plasmons is forbidden in a smooth metal film. However, the coupling is possible by means of two well-known methods [1, 14, 15]: (1) the grating coupling method invokes modification of the surface with a lattice structure which adds additional reciprocal lattice vectors to the initial wavevector of the light, and (2) the attenuated total reflection method exploits the total internal reflection inside a prism which is attached to the metal film. Surface plasmons have been extensively studied with different optical methods, mainly because of their extremely high sensitivity to interface structures and adsorbates [1, 16].

In *metal nanoparticles* with sizes smaller than the wavelength of light as well as the optical penetration depth, all atoms in the particle can be collectively excited. Hence, collective electronic oscillations, the so-called Mie-plasmons, can be excited by light, and are therefore detectable as a pronounced optical resonance in the visible or UV parts of the spectrum [1, 17, 18]. The resonance frequency of the oscillation is determined by the dielectric properties of the metal and the surrounding medium, and by the particle size and shape. The collective oscillation can be interpreted as a displacement of the center of mass of all electrons in the particle against the positively charged background of the atomic cores [19]. There is no propagation of a (longitudinal) charge density fluctuation; hence, a plasmon in a metallic nanoparticle has to be distinguished from a propagating surface plasmon, not often done in the literature. This collective charge oscillation causes a large resonant enhancement f of the local field inside and near the particle [20]. This resonance dominates the linear and nonlinear response of the materials. The particle-size dependence and host-matrix dependence of the absorption spectrum have been discussed in many systems with Mie scattering theory [21]. In general, these studies took into account two contributions to the dielectric function $\varepsilon(\omega)$: a Drude term $\varepsilon_{\text{intra}}(\omega)$ originating from free electrons, and an interband term $\varepsilon_{\text{inter}}(\omega)$ reflecting the band-to-band transitions [22].

Considerable interest has been focused on silver nanoparticles as they exhibit a particularly strong size-dependent optical resonance in the visible spectral range (1.8–3 eV), that is, at energies below the interband transition threshold (~4 eV). Consequently, the absorption cross section in the visible region is dominated by the Drude term [23]. The contribution due to interband transitions is negligible. Quite the contrary, investigation on isolated copper and gold nanoparticles allows us to study the influence due to the strong overlap between the Mie-plasmon resonance (~2.3 eV) and the interband ($d \rightarrow sp$) transition at ~2.4 eV. Hence, both the intraband and interband optical processes are involved in the material response to a light excitation.

2.2. Plasmon Decay Time

The most important factor for the coherent interaction of light with nanoparticles is the dephasing time

$$T_2 = 2\hbar/\Gamma_{\text{hom}} \tag{3}$$

where Γ_{hom} is the homogeneous line width of the plasmon resonance. This dephasing time T_2 corresponds to the time scale on which the decay of the coherent electron plasma oscillation (and hence the local-field enhancement) takes place, for example, the time scale on which the electron oscillation preserves the memory of the optical phase of the excitation pulse. This dephasing time is essential because, for example, the enhancement factor f of the electric field near the particle surface due to the collective excitation is expected to be proportional to T_2 [2, 24]. Note that many authors use the expression plasmon lifetime $\tau_{sp} = \hbar/\Gamma_{\text{hom}}$. According to Eq. (3), the plasmon lifetime τ_{sp} is related to the dephasing time T_2 by $2\tau_{sp} = T_2$.

In recent years, several line width measurements and time-resolved second- and third-harmonic generation autocorrelation measurements (for the technique, see, e.g., [25–28]) on metallic nanoparticles have been published, reporting a dephasing time of the Mie-plasmon excitation on the order of 4–20 fs. All experimental studies have been performed in the low-excitation regime, where the energy optically transferred to the system is too small to strongly perturb the nanoparticle [29]. In the following, different experimental techniques will be discussed which allow us to determine the dephasing time T_2. The different mechanisms, which can lead to this phase destruction of the collective electron motion, are discussed in the next section.

First experimental knowledge of the line width of metallic nanoparticles in a two-dimensional arrangement was gained for thermally evaporated ultrathin films, which show a broad distribution in particle size and shape [2]. By this randomness, a strong inhomogeneous broadening of the absorption band results which is much broader than the resonance of a single individual particle. This inhomogeneous broadening is caused by variations in the size, shape, surface structure, and environment of the individual particles within the cluster ensemble. Since the magnitude of the inhomogeneous broadening is not known quantitatively, the homogeneous width Γ_{hom} and, hence, the dephasing time T_2 cannot be extracted [28]. The absorption spectra only provide a lower limit of T_2.

Craighead and Niklasson reported the first extinction spectra of a lithographically produced regular two-dimensional array of identical metal nanoparticles [30]. In contrast to thermally evaporated ultrathin films, by nanofabrication, the size and shape of particles can be designed and the distribution width of these parameters can be extremely narrowed.

The potential of nanodesigned optical films in order to study plasmon decay dynamics has also been nicely demonstrated by Gotschy et al. They deposited nearly identical, parallel-oriented silver particles on a transparent ITO substrate by use of electron-beam lithography [31]. Figure 1 shows SEM pictures of two-dimensional arrays of elliptic-shaped, nearly identical, parallel-oriented silver particles. As the particle size and interparticle distances can be varied independently, this method allows us to tailor the optical properties of single particles. In addition, elliptic-shaped metal nanoparticles show two different plasmon resonances which lie at different wavelengths for light polarized parallel to the short and long axes (see, e.g., Fig. 6). From the experimentally obtained line widths, Gotschy et al. expected a plasmon decay time of a few femtoseconds.

Another way to overcome the problem of inhomogeneous broadening due to variations in the size and shape is the spectroscopy of a single particle. Using SNOM spectroscopy, Klar et al. measured the homogeneous line shape of the plasmon resonance in single gold nanoparticles [24]. The measured line width varies around a value of 160 meV, which corresponds to a plasmon dephasing time T_2 of 8 fs. Besides the single-mode resonances, they also observed more complex line shapes caused by electromagnetic coupling between close-lying particles.

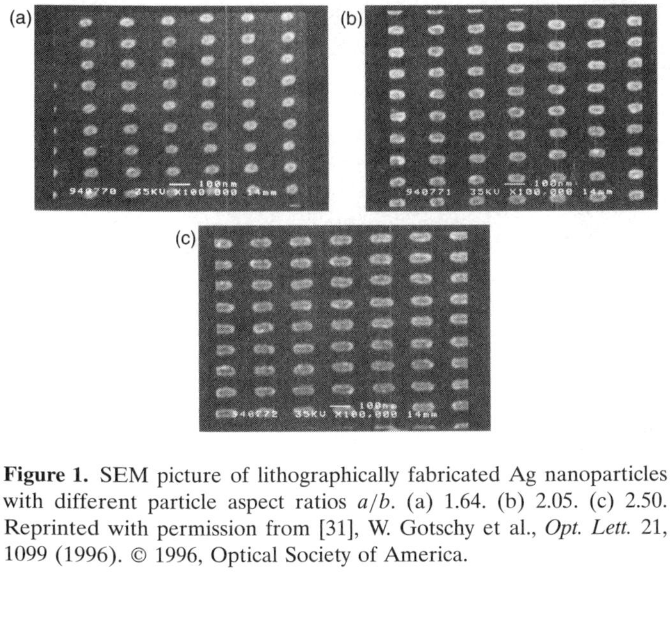

Figure 1. SEM picture of lithographically fabricated Ag nanoparticles with different particle aspect ratios a/b. (a) 1.64. (b) 2.05. (c) 2.50. Reprinted with permission from [31], W. Gotschy et al., *Opt. Lett.* 21, 1099 (1996). © 1996, Optical Society of America.

Träger and co-workers applied an interesting novel technique where the size and shape selectivity of the clusters was obtained by optical hole burning after deposition [32, 33]. They used this technique to extract dephasing times of silver nanoparticles with radii smaller than 10 nm, dimensions which are not accessible by lithographic techniques.

Spectral hole burning is a well-known technique in atomic, molecular, and solid-state physics. The achievement of the authors was to adopt the method for the specific needs in order to investigate T_2 in metallic nanoparticles. The idea of the method is as follows. First, nanoparticles with a broad size distribution are prepared on a transparent substrate. After measuring their optical absorption spectrum, the nanoparticles are irradiated with nanosecond laser pulses, the photon energy being located within the inhomogeneously broadened absorption profile. The fluence of the light is chosen such that the temperature increase of the particles induced by rapid conversion of the absorbed energy into heat is sufficiently high to stimulate evaporation of atoms. As a result, the distribution changes, and a hole is permanently burned into the absorption profile. Finally, the optical spectrum is measured a second time, and subtracted from the spectrum of the particles as grown to determine the width of the hole, that is, Γ_{hom}. From the experimental results and a theoretical model of hole burning, the line width of 260 meV corresponding to a decay time of 4.8 fs was extracted for silver nanoparticles with a radius of 7.5 nm. This T_2 value is at least a factor of 2 smaller than one would expect from the bulk dielectric function of silver. The authors concluded that additional damping mechanisms, in particular surface scattering, come into play in this size regime.

A further way to obtain accurate decay time constants is by means of time- instead of frequency-resolved measurements. In principle, all femtosecond real-time measurements use a two-pulse (pump–probe) technique: the first laser pulse drives the systems, and the second, suitably delayed, probes the actual state. Lamprecht et al. fabricated noncentrosymmetric particles in order to study the decay time of particle plasmons by means of a real-time second-order nonlinear optical (SHG) autocorrelation experiment [34]. SHG can be considered as an optimal noninvasive probe of plasmon dynamics, but a shape without centrosymmetry is an essential condition for high SHG efficiency. Figure 2 shows an interferometric autocorrelation measurement (thin line) for gold nanoparticles. Also shown is the autocorrelation function of the laser pulse (bold line) measured with a nonlinear crystal (BBO). For gold, the experiment revealed a dephasing time T_2 of 12 fs. In the case of silver, the experimentally obtained values T_2 vary between 14 and 20 fs.

Using the same time-resolved SHG technique, Rubahn and co-workers studied the plasmon lifetime of Na nanoparticles with mean radii of $5 < r_0 < 55$ nm [35]. They found a pronounced maximum in T_2 of 20 fs ($\tau_{sp} = 10$ fs) for nanoparticles of an average radius of 22 nm, with lifetimes

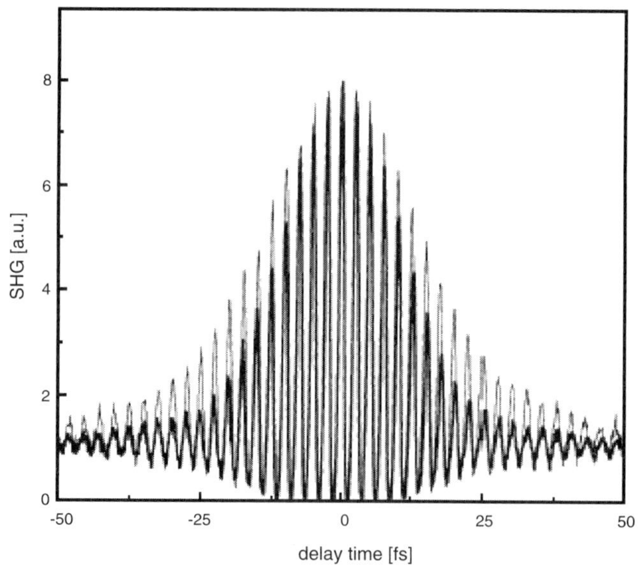

Figure 2. Interferometric autocorrelation functions for the laser pulse (bold) measured with BBO crystal and the SHG signal (thin line) for gold nanoparticles. Reprinted with permission from [34], B. Lamprecht et al., *Appl. Phys. B* 68, 419 (1999). © 1999, Springer-Verlag.

decreasing both for smaller and for larger particle radii (see Fig. 3).

The major drawback of the real-time SHG method, the restriction to noncentrosymmetric particle shapes, can be eliminated by using third-harmonic generation (THG). Lamprecht et al. used this technique to study the decay of resonant and slightly off-resonant driven plasmons in gold nanoparticles [36]. By comparing the measured third-order interferometric autocorrelation function of the plasmon field with a simulation based on a simple harmonic oscillator model, they obtained the same dephasing time T_2 of 12 fs in both (resonant and off-resonant) cases. For off-resonant excitation, the results show that the phase difference between the driving laser field and the driven plasmon oscillation is no longer constant at $\pi/2$ as in the resonant case, but varies with time (beating, see Fig. 4).

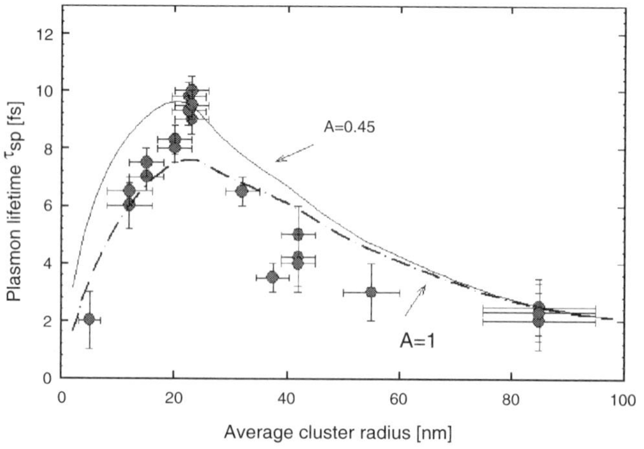

Figure 3. Measured plasmon lifetimes as a function of mean cluster radius for Na clusters adsorbed on mica.

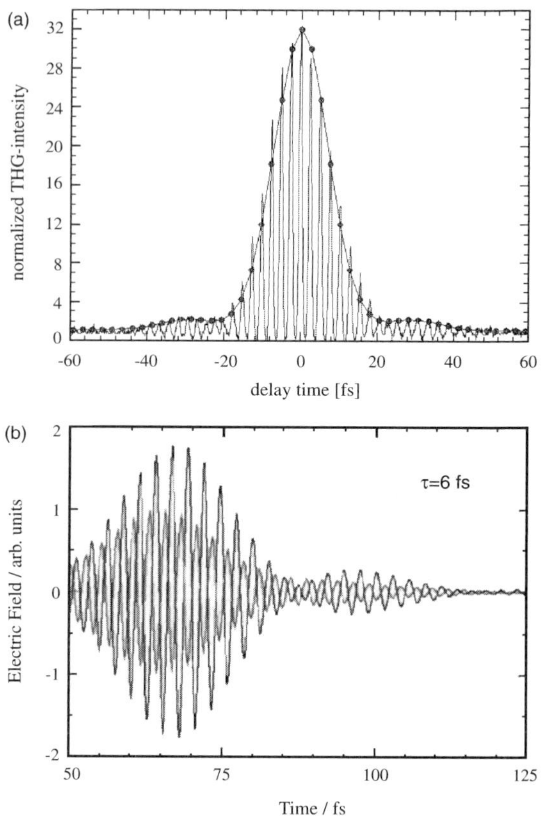

Figure 4. (a) Solid line: measured third-order autocorrelation functions for an off-resonantly excited particle plasmon, maximum normalized to 32; open circles: envelope of the calculated autocorrelation functions (the solid line serves as a guide to the eye). (b) Driving laser pulse field (red) together with the calculated off-resonant plasmon field oscillation.

2.3. Mechanisms for Damping

While the dephasing time T_2 of metallic particles as a function of particle size, shape, and dielectric properties is well investigated, the physical mechanism responsible for dephasing has remained a highly interesting and debated topic. Knowledge of the dephasing mechanisms is essential for basic science and for a large variety of applications. Depending on the size, size distribution, shape, dielectric constant of the surrounding medium, and so on, the potential mechanisms are as follows. First, the plasmon can dephase by internal decay, for example, a decay of the fixed phase correlation between the individual electronic excitations of the whole oscillator ensemble, described by the *pure dephasing time* T_2^*. A typical mechanism is the decay of the collective mode due to inhomogeneous phase velocities caused by the spread of the excitation energy or the local inhomogeneity of the nanoparticles. Second, the plasmon can also decay due to a transfer of energy into quasiparticles (electron–hole pairs, i.e., due to surface scattering, phonon scattering, or *e–e* scattering) or reemission of photons (radiation damping and luminescence [37–40]), described by T_1. The total dephasing of the plasmon is given by

$$\frac{1}{T_2} = \frac{1}{2T_1} + \frac{1}{T_2^*} \qquad (4)$$

In general, it is predicted that, for sizes below about 10 nm, surface scattering is quite essential [1, 32]. For larger particles, however, radiation damping becomes more pronounced. According to Liebsch and co-workers [23], the overall broadening $\Gamma(\omega)$ due to Drude damping γ and radiation losses is given by the function

$$\Gamma(\omega) = \gamma + \omega B(\omega)/A_i \tag{5}$$

The term $B(\omega)$ characterizes radiation damping, and is given by

$$B(\omega) = \frac{4\pi^2}{3}\frac{V}{\lambda^3} \tag{6}$$

where λ is the wavelength of the laser pulse and V is the volume of an isolated metallic particle. A_i represents the depolarization factors. For an ideal spheroid, the A_i satisfies the same rule $A_1 + A_2 + A_3 = 1$; in the case of a sphere, $A_i = 1/3$. The Drude damping γ includes all microscopic damping processes due to photons, phonons, impurities, surface scattering, and electron–electron interactions. Compared with Drude damping, the radiation damping is explicitly frequency dependent. In particular, this implies that the damping near the maximum of the plasmon resonance is not representative of the damping far from resonance. Below the resonance, radiation damping is weaker than at resonance, while above the resonance, it becomes rapidly stronger (see Fig. 5).

This theoretical prediction has been experimentally verified by Scharte et al. [23]. They used elliptically shaped metal nanoparticles with semiaxes of 40 nm × 80 nm and a height of 45 nm. Elliptic-shaped metal nanoparticles show two different plasmon resonances which lie at different wavelengths for light polarized parallel to the short (b) and long (a) axes, respectively (see Fig. 6). Tuning the laser wavelength to one of these two resonances allows us to distinguish between resonance excitation and off-resonance

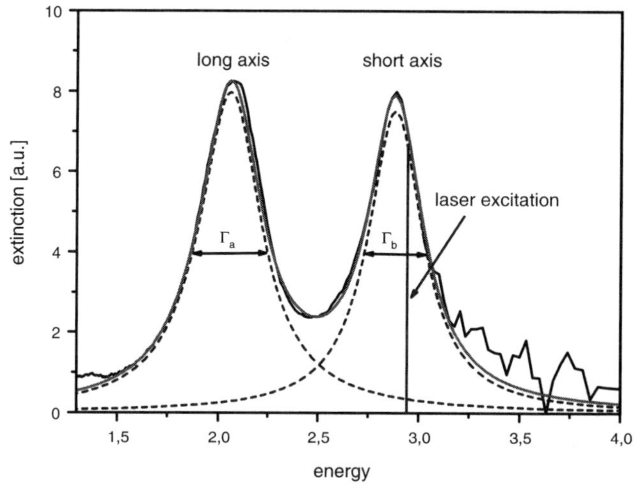

Figure 6. Measured extinction spectrum of the nanoparticle array (solid black line). The low- and high-frequency peaks correspond to the Mie plasmon, excited along the long and short axes of the elliptical Ag particles, respectively. Reprinted with permission from [23], M. Scharte et al., *Appl. Phys. B* 73, 305 (2001). © 2001, Springer-Verlag.

excitation by simply changing the polarization of the laser pulse. In both cases, the excitation is the result of an intraband process since the excitation energies of the Mie plasmons in those silver nanoparticles lie below the interband transition threshold (~4 eV). Consequently, the absorption cross section in the visible region is dominated by Drude damping. The contribution due to interband transitions is negligible.

Liebsch's theory predicts that collective plasma oscillations within the nanoparticles are subject to the same microscopic Drude damping processes due to phonons, impurities, and electron–electron interactions as off-resonance excitations [23]. None of these inelastic scattering mechanisms is reduced or absent if the electrons oscillate near the resonance excitation. The plasma mode signifies a larger amplitude of oscillation, but not the presence or absence of a particular microscopic scattering mechanism.

However, as pointed out above, radiation damping is strongly laser wavelength dependent [see Eq. (5)], which leads to a different damping along the axes of the ellipsoid, as shown in Figure 5. Hence, at first sight, it is surprising that two different resonances in the extinction spectrum at $\omega_a = 2.1$ eV and $\omega_b = 2.95$ eV exhibit about the same FWHM, that is, about the same Γ (see also Fig. 6). However, as evident from Eq. (5), the radiation damping scales inversely with the depolarization factor A_i, that is, the damping is also determined by these coefficients. Thus, for the low (high)-frequency Mie plasmon, both $\omega B(\omega)$ and A are small (or large), so that $\Gamma_a(\omega) \approx \Gamma_b(\omega)$.

An interesting question arises concerning the lifetime of the plasma oscillation in the nanoparticle near resonance and off resonance. The experimental method used to investigate the dynamical parameters T_1 and T_2 of optically excited electrons in a metal is the time-resolved two-photon photoemission (TR-2PPE). This method is known to be a highly accurate method to determine the dynamics of hot electrons in unoccupied and occupied states. The pump–probe technique enables a direct measurement of the dynamical

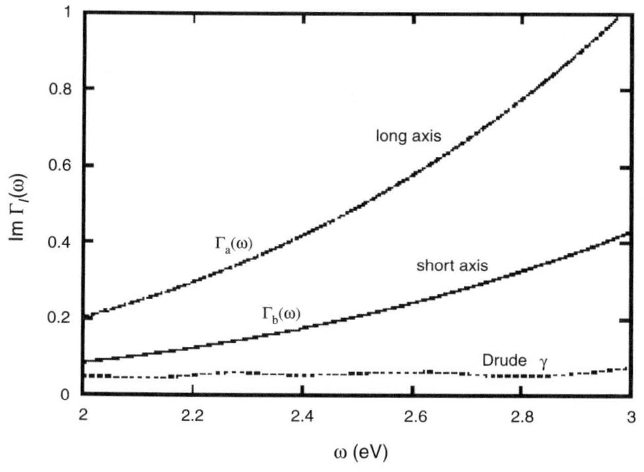

Figure 5. Plasmon broadening $\Gamma_i(\omega)$ as a function of frequency. Solid curve: damping of high-frequency mode $\Gamma_b(\omega)$ ($\omega_b = 2.9$ eV, short axis); dashed curve: damping of low-frequency mode $\Gamma_a(\omega)$ ($\omega_a = 2.1$ eV, long axis); dot–dashed curve: Drude damping γ derived from bulk optical data. Reprinted with permission from [23], M. Scharte et al., *Appl. Phys. B* 73, 305 (2001). © 2001, Springer-Verlag.

properties in the time domain with a resolution in the range of few femtoseconds [41–43]. The principle is schematically shown in Figure 7. A pump pulse excites electrons out of the valence band into usually unoccupied states (on or off a plasmon resonance) with energies between the Fermi and vacuum level. A second probe pulse subsequently photoemits the excited electrons. The photoemission yield depends on the transient population of these intermediate states as a function of the temporal delay of the probe pulse with respect to the pump pulse. The measured pump–probe signal contains information on the energy relaxation time T_1 of the population in the intermediate state, as well as its dephasing time T_2.

In Figure 8, the FWHM derived from the autocorrelation 2PPE traces is plotted as a function of the state of polarization relative to the short axis (b). For comparison, the behavior of polycrystalline tantalum is also shown. The FWHM for tantalum is not affected by rotating the polarization angle of the incoming light. Therefore, any effect caused by an increase of the dispersion due to rotation of the half-wave plate can be excluded. For the elliptical Ag nanoparticles, however, a rotation of 90° reduces the FWHM from almost 75 fs in the b direction (short-axis mode, resonant plasmon excitation) to 69 fs in the a direction (long-axis mode, off-resonant plasmon excitation). Further rotation of the polarization to the short axis b restores the long lifetime of the plasmon excitation. One should be aware that these FWHM values still include the autocorrelation of the laser pulse width. The autocorrelation traces of tantalum were measured at an intermediate state energy $E - E_F = 2.8$ eV. In this energy range, the lifetime of excited electrons in a transition metal is on the order of 1 fs, and hence, the traces obtained for tantalum represent the pure laser autocorrelation curve. Any increase in the FWHM in the autocorrelation measurement for Ag nanoparticles compared to the Ta values must be caused by a lifetime effect of the optically excited electron system.

According to Eq. (5), the different damping rates obtained under on- and off-resonance conditions can be explained by different magnitudes of radiation damping along the short and long axis of the elliptical particles, as shown in Figure 5. The increase in the FWHM of the TR-2PPE autocorrelation curve in Figure 8 is about a factor of 2

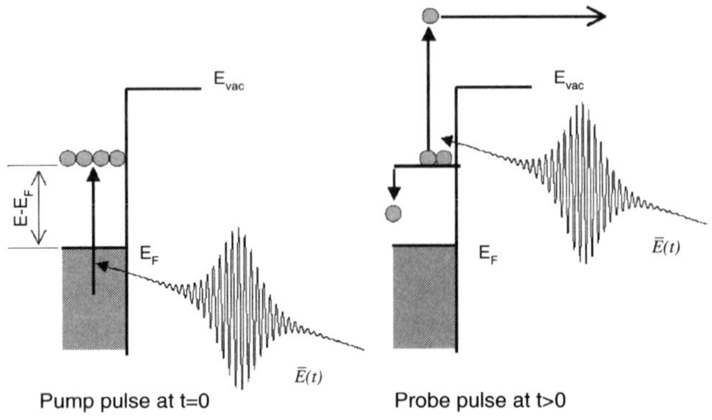

Figure 7. Principal mechanism of the time-resolved two-photon photoemission (TR-2PPE).

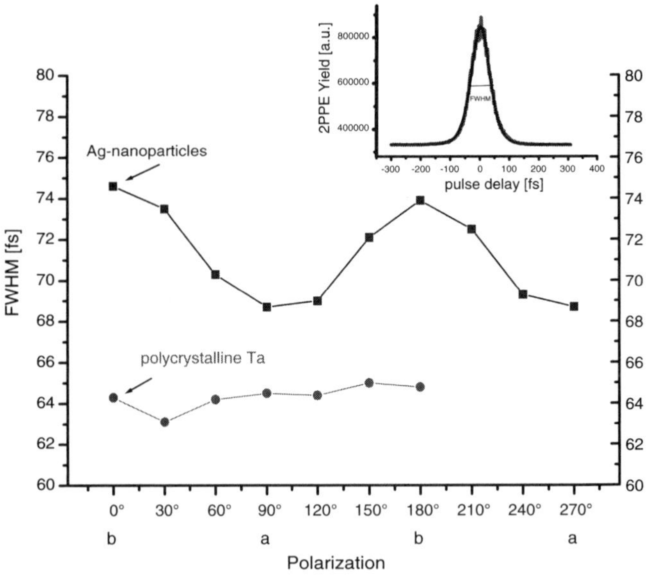

Figure 8. Nanoparticles show a variation of the FWHM of the autocorrelation over rotating states of polarization, while tantalum shows no effect. In the inset, a typical autocorrelation measurement can be seen from which the data points have been derived. Reprinted with permission from [23], M. Scharte et al., *Appl. Phys. B* 73, 305 (2001). © 2001, Springer-Verlag.

smaller for the long-axis mode as compared to the short-axis mode, taking into account that the FWHM of the Ta measurements represents the FWHM of the laser autocorrelation curve. This trend agrees well with the theoretical calculation, as shown in Figure 5. For a light polarization along the short b axis, we would expect to observe a damping given approximately by $\Gamma_b = \Gamma_b(\omega_b)$. If we now rotate the polarization direction toward the long a axis without changing the laser frequency, we excite the high-frequency tail of the ω_a plasma mode (see Fig. 5), and expect to find the damping $\Gamma_a = (\omega_b)$, which is about twice as large as Γ_b.

It might be tempting to conclude from the FWHM results of Figure 8 that on-resonance collective excitations in general have a longer lifetime than excitations far from resonance. This is not correct, however, since the exact opposite should be obtained if we perform the analogous experiment near the frequency ω_a a of the long-axis mode: for a polarization along the a axis, we observe this mode on resonance with a damping given by $\Gamma_a = \Gamma_a(\omega_a)$. If we rotate the polarization toward the short b axis, while keeping the laser frequency fixed at ω_a, we should observe a strong decrease in damping since excitations of the nanoparticle along the short axis are less subject to radiation losses. Thus, in this alternative configuration, the collective mode ω_a detected on resonance has a shorter lifetime than the excitations in the low-frequency tail of the ω_b mode. To test this theoretical prediction, it would be desirable to repeat the two-photon photoemission measurement with about 2.1 eV pump laser frequency.

On the basis of the above theoretical analysis, we can give the following tentative interpretation of the observed two-photon photoemission spectra: the initial pump pulse excites an electron from an occupied state 1 to an intermediate state 2. The effective field governing this excitation consists of a superposition of an external field and an induced field generated by the dynamical response of the electrons of the Ag particle. Accordingly, the amplitude of this excitation process is greatly enhanced if the laser frequency coincides with the Mie resonance. Because of the strong intrinsic broadening of the Mie plasmon, the induced contribution to the effective electromagnetic field exciting state 1 is coherent over rather short times on the order of 0.6–6 fs, corresponding to $\Gamma \approx 0.1$–1.0 eV (see Fig. 5). Since the energy of the intermediate state is about 2 eV above the Fermi energy, the excited electrons scatter quasielastically via phonons and impurities and inelastically via electron–electron processes, giving a typical overall lifetime of about 6–10 fs, corresponding to an imaginary part of the Ag quasiparticle self-energy of about 0.06–0.1 eV [42]. The probe beam excites the electron from the intermediate state to the final state 3 above the vacuum level such that it can be detected as an emitted electron. The probability of the second excitation process is also greatly enhanced if the probe laser is in resonance with the Mie plasma frequency. The probe pulse is subject to the same frequency-dependent damping mechanisms as the pump pulse. The measurements discussed in the present work suggest that time-resolved two-photon photoemission is sensitive not only to the decay of the intermediate electronic state, but also to the intrinsic damping of the field enhancement causing the single-electron transitions.

Similar time-resolved 2PPE experiments have been performed by Lehmann and co-workers [44–46]. They investigated the nonlinear response due to surface plasmons in silver nanoparticles on graphite. The results provide direct evidence for multiplasmon excitation and allow identification of two of their decay channels, namely, decay into one or several single-particle excitations. In another work, they reported similar studies on Au nanoparticles on graphite, although the resonance was much less pronounced compared to Ag nanoparticles [47]. The resonance of the nanoparticles (2.1 eV) was detected by the peak in the yield in the high-energy tail of the photoemission spectrum, and was compared with the resonant excitation seen in EELS (2.5 eV). The rather complicated shape of the photoemission spectra is explained by multiphoton-assisted thermionic emission from the nanoparticles and resonant emission via the image-potential state on graphite.

3. DISSIPATION OF ENERGY

3.1. Internal Thermalization

As pointed out above, Drude and radiation damping are the dominant mechanisms for the decay of the collective excitation in a metallic nanoparticle with sizes >10 nm. In the case of Drude damping, the energy is primarily transferred into one or several quasiparticle pairs (electron–hole pairs). The next step of the energy relaxation corresponds to an *internal* thermalization of the electron gas. The excited electronic states tend to a Fermi–Dirac distribution with a well-defined temperature which depends on the laser pulse intensity. It has to be noted, however, that the expression "thermalization time" has never been deeply discussed so far, for example, the question of what fraction of electrons have to lie within the Fermi distribution until the electron gas can be called thermalized.

The electron–electron scattering is the most efficient process to redistribute the energy to thermalize the electron gas. Quasielastic electron–phonon and electron–surface scattering do not play a dominant role in the dissipation of the excess energy of laser excited hot electrons; hence, no strong size effect in the thermalization process is expected. Also, surface-plasmon-assisted resonant scattering of *d* holes into the conduction band as described by Bigot and co-workers is only relevant for sizes <10 nm [48].

Fann and co-workers have shown, in a time-resolved photoemission experiment performed in Au films, that the temporal scale of this thermalization process is a few hundreds of femtoseconds [49, 50]. No comparable direct measurement of the thermalization time has been reported so far for metallic nanoparticles. However, there are few indirect indications for a transient nonthermal component of the electron gas. Bigot et al. studied the nonthermal component in Ag nanoparticles (size 6.5 nm) embedded in a transparent matrix, using the pump–probe femtosecond spectroscopy [51]. Figure 9 shows the differential transmission $\Delta T/T(t)$ excited with pulses of 30 fs duration at 800 nm, far from the plasmon resonance. The maximum of the signal is reached after a delay of $\approx$200 fs with respect to the laser pulse autocorrelation. This delay is a measure for the evolution from a nonthermal to a thermal Fermi–Dirac distribution, and is just slightly longer than the corresponding dynamics in silver film (delay $\approx$ 120 fs) as measured by del Fatti et al. [52]. This indicates that the thermalization process due to electron–electron scattering is not much slower in metallic nanoparticles compared to a metal film, in agreement with time-resolved two-photon photoionization measurements on free noble metal nanoparticles by Fierz et al. [53].

Similar electron dynamic studies in gold nanoparticles (size 22 nm) have been completed by Link et al. [54]. Taking into account that they used longer laser pulses (100 fs), the delay rise of the signal is in good agreement with the results of Bigot et al. The authors modeled their results (decay of

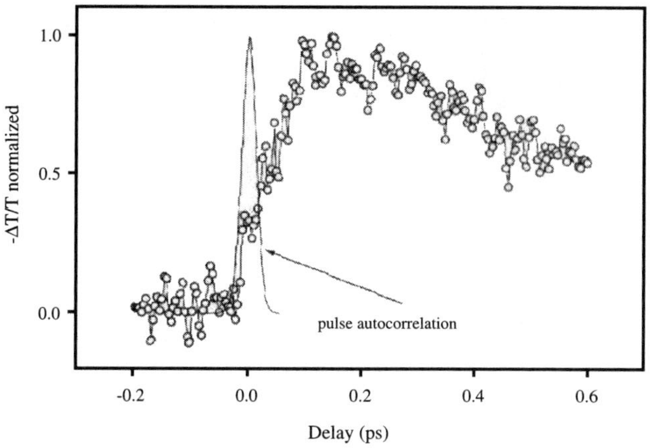

Figure 9. Differential transmission $\Delta T/T(t)$ of 6.5 nm Ag nanoparticles excited with pulses of 30 fs duration at 800 nm. The normalized pulse autocorrelation is shown as a reference. Reprinted with permission from [51], J.-Y. Bigot et al., *Chem. Phys.* 251, 181 (2000). © 2000, Elsevier.

the bleach data) by a rate equation derived by Sun et al. [55]. The model consists of a modification of the two-temperature model (see below), including a nonthermal electron distribution part. From the obtained fitting parameter, a thermalization time between 400 and 500 fs was obtained, in very good agreement with the values obtained by Sun et al. for 20 nm thin gold films.

Stella and co-workers also found a rise time in the transient reflectivity change on the order of 100 fs for tin nanoparticles embedded in an Al_2O_3 matrix [56]. They found that the build up of the hot Fermi distribution was constant in the size range from 2 to 6 nm. It is expected that transition metals have a shorter thermalization time compared to noble metals, caused by the increased phase space for electron–electron scattering due to the unfilled *d* band [42].

3.2. External Thermalization

Let us now discuss the energy transfer from the hot-electron system to the still cold lattice of the nanoparticles, often called external thermalization. In general, it takes about a few picoseconds until a quasiequilibrium state is formed between the electron system and the phonon system. This heat-exchange process is usually described by the electron–phonon coupling model developed for thin metallic films [57–59]. The electron system is characterized by an electron temperature Θ_e, and the phonon system is characterized by a lattice temperature Θ_l, where each subsystem is assumed to be in local equilibrium. The time evolution of the temperatures is obtained with the heat equations [60, 61]

$$C_e(\Theta_e)\frac{\partial \Theta_e}{\partial t} = -G(\Theta_e - \Theta_l) + p(t) \tag{7}$$

$$C_l\frac{\partial \Theta_l}{\partial t} = G(\Theta_e - \Theta_l) \tag{8}$$

where G reflects the *e–p* coupling constant, $C_e(\Theta_e)$ and C_l are electronic and lattice heat capacities, and $p(t)$ is the absorbed laser power density. The heat capacity of the electron gas is about two orders of magnitude smaller than the heat capacity of the lattice [11]. Therefore, a high maximum electron gas temperature of several thousand kelvins can be reached, although the rise of the equilibrium temperature is only a few hundred kelvins [62].

In Eqs. (7) and (8), the electron energy losses are supposed to be proportional to the difference between the electron and lattice temperatures; the *e–p* coupling constant G governs the thermal equilibrium process. Due to the reduced dimensionality of the nanoparticles, however, the surface modes of the metallic particles will influence the energy transfer between the electrons and the lattice. Hence, a major goal in many studies has been to determine how the *e–p* coupling constant G changes as the particle size changes.

Most of the published results are obtained by means either of transient transmission and reflectivity measurements or transient absorption spectra as a function of the temporal delay between the pump and the probe beam. The increase in the electronic temperature induced by the pump laser changes the real and imaginary part of the complex dielectric function $\varepsilon(\omega, t)$ of the particles, which causes transient changes in the spectrum (decrease in intensity, broadening and shift of the plasmon absorption band). The subsequent electron cooling via phonon emission results in a characteristic decay time τ of the modification of the optical constants [63]. It has to be noted that the electronic heat capacity $C_e(\Theta_e)$ depends on the electron temperature Θ_e. Hence, the effective rate constant $\tau^{-1} = G/C_e$ for the decay of the electronic temperature decreases as the initial electronic temperature increases, that is, longer decay times are obtained for higher pump laser powers [64].

The response of the transient absorption spectra with respect to a nonthermal and thermal electron distribution at higher temperature can be modeled by combining the Mie equation with a theory developed by Rosei et al. [65, 66].

Faulhaber et al. [67] measured the energy relaxation of gold particles (size 12–18 nm) by means of femtosecond absorption spectra, and obtained a decay of about 7 ps, much larger than the decay on thin ($d <$ optical penetration depth) gold films ($\tau < 2$ ps [57]). The authors explained that the slower decay observed in nanoparticles is because there is less nonequilibrium electron transport due to the spatial size confinement of the electrons in three dimensions.

In a gold nanoparticle system, Inouye and co-workers made white-light pump–probe experiments in the picosecond region too [68]. They discussed the origin of the change in the transient differential absorption spectra for various delay times by decomposing the temporal behavior into two decay components: a fast component (<3 ps), and a slow decay component (>100 ps). They suggested that the fast decay reflects the thermal equilibrium process between the electron system and the lattice system within the nanoparticles, and the slow decay comes from the heat transfer between the nanoparticles and the host matrix, as will be discussed below. Quite contrary to the work of Faulhaber et al., Inouye and co-workers reported an *e–p* coupling constant G which was about two times larger than the result of a gold film [49, 57], indicating a faster decay in the particles.

The electron–phonon coupling time constant has also been investigated by Rubahn and co-workers for Na nanoparticles [35]. They measured the relative change in the reflected second-harmonic generation (SHG) signal of a pump–probe setup. They obtained a decay of 1.1 ps, which is comparable to the values for thin metal films [57], opposite to the observation for gold nanoparticles by Inouye et al. [68] and Faulhaber et al. [67].

Also, Hodak et al. found, for Au (11 nm) and Ag (10 nm, 50 nm) nanoparticles, similar relaxation times to those of bulk metals, which implies that there are no size-dependent effects in the heat exchange between the electron and phonon system [69]. A similar result was obtained by del Fatti et al. on Ag nanoparticles for the size range 3 nm $\leq R \leq$ 13 nm [70]. Hence, one can assume that, after a few tens of femtoseconds, the electrons have no memory of the way (intraband, interband, on or off Mie resonance) they have been excited. Furthermore, Hodak et al. reported relaxation times which are strongly dependent on the pump laser power. This means that it is not possible to fit high-power (pump laser pulse) data using the simple two-temperature model calculations using a constant electronic heat capacity C_e in order to obtain G. They performed a series of measurements at successively lower pump laser powers, and extrapolated the measured time constants τ to zero power. From the obtained data, the authors drew the conclusion that surface interaction has a remarkably small effect on the electron–phonon coupling constant G for noble metal particles with diameters greater than 10 nm.

Nisoli et al. investigated the thermalization dynamics in gallium nanoparticles of different sizes in the solid and liquid phase [71]. They found clear size-dependent characteristic time constants ($\tau \sim$ 0.6–1.6 ps by increasing the size), but a quite similar temporal behavior of the electron energy relaxation in both phases (solid and liquid). In bulk, the electron energy loss mechanisms are quite different in the two phases. These results shed some light on the role of the energy exchange with surface phonons in small metallic structures.

On the basis of all published results, one can summarize that the thermal equilibrium process between an electron system and a lattice system within the nanoparticles is larger than the internal thermalization time to a Fermi distribution, but still of the same magnitude [64]. Hence, for a deeper theoretical treatment, both processes have to be simultaneously included in the calculation. At present, the reason for the rather different results for the electron–phonon coupling constant versus size is not clear. One possibility for these discrepancies has been addressed by Bigot and co-workers [51, 72]. They investigated the dynamics of Cu and Ag films in order to compare their behavior with the corresponding Cu and Ag nanoparticles, using the same experimental apparatus. The simultaneous measurement of the differential transmission and reflection as a function of the pump–probe delay allows retrieval of the time-dependent complex dielectric function $\varepsilon(t)$ of the metal. Figure 10 shows the normalized $\Delta\varepsilon_2$ and $-\Delta\varepsilon_1$ for the Cu film investigated for two probe wavelengths probed near and far from the interband optical transition. The curves show quite different relaxation times; in particular, the relaxation of the real part is much faster off resonance (0.4 ps) compared to on resonance (1.3 ps). The authors explained the differences due to the nonthermal character of the electron distribution after the pump laser pulse. This nonthermal population can persist for a long time near the Fermi level, and has a strong effect on $-\Delta\varepsilon_1$. A similar effect was reported on Cu nanoparticles by the same authors. In Figure 11, the time dependence of $\Delta T/T$ decays with a longer relaxation time when the laser energy is at the plasmon resonance. One should keep in mind that, in the pump–probe experiment, the collective character of the electrons excited by the pump is lost within a few tens of femtoseconds. Therefore, it is the plasmon that is created by the *probe*, which is sensitive to this mechanism. Hence, these effects have to be taken into account by comparing results of different groups obtained on the same metallic nanoparticles, but at different wavelengths. This counts, in particular, for Cu and Au, for which the $d \rightarrow sp$ interband transition occurs at approximately the same frequency as the plasmon resonance. The situation is somehow simpler when the dynamics is studied far from the interband transition, for example, for Ag, where the interband transition is displaced to the plasmon resonance by >0.7 eV.

Feldmann and co-workers investigated the vibrational dynamics of ellipsoidal silver nanoparticles after

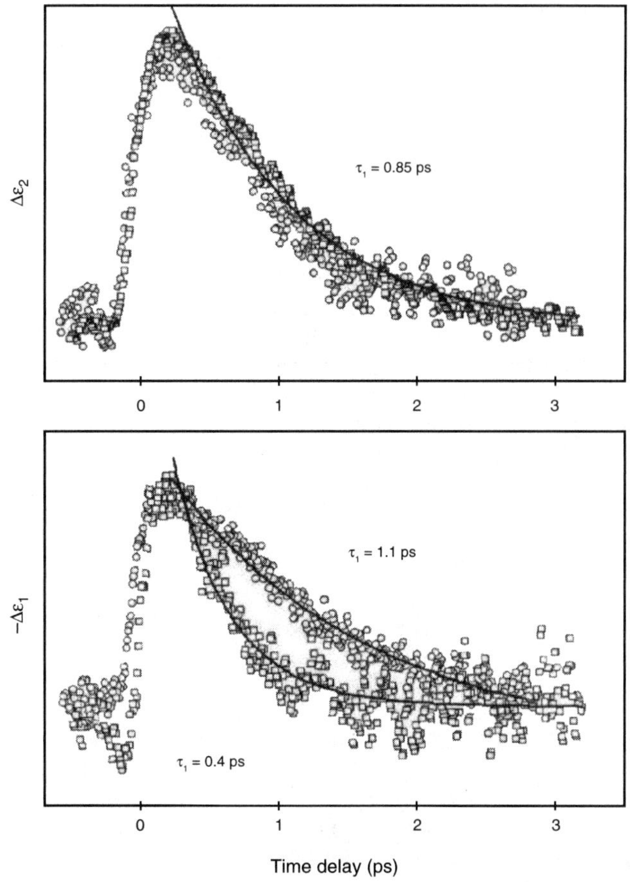

Figure 10. Dynamics of the normalized $\Delta\varepsilon_2$ and $-\Delta\varepsilon_1$ for the Cu film for two probe wavelengths near (circles: $\lambda = 570$ nm, 2.175 eV) and far (squares: $\lambda = 611$ nm, 2.03 eV) from the interband optical transition. Reprinted with permission from [51], J. Y. Bigot et al., *Chem. Phys.* 251, 181 (2000). © 2000, Elsevier.

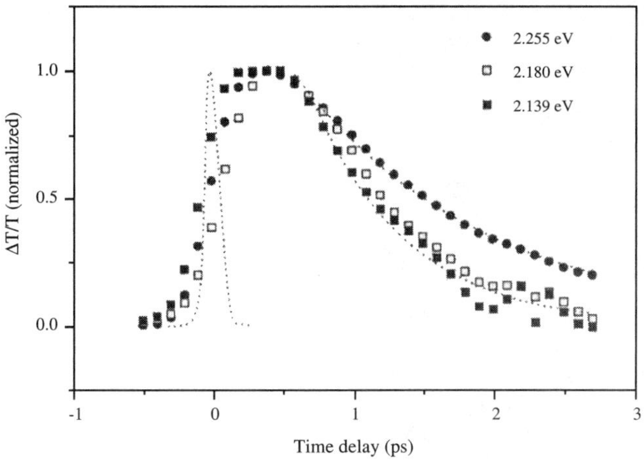

Figure 11. Time dependence of $\Delta T/T$ in Cu nanoparticles of 10 nm diameter in the vicinity of the surface plasmon resonance. The dotted line is the cross correlation between the 60 fs pump and 10 fs probe pulses. Reprinted with permission from [51], J.-Y. Bigot et al., *Chem. Phys.* 251, 181 (2000). © 2000, Elsevier.

femtosecond laser pulse excitation [73]. They measured the differential transmission $\Delta T/T$ induced by a 150 fs laser pump pulse at $h\nu = 3.1$ eV, and found the typical two-step decay (fast and slow decay) as discussed by Hodak et al. In addition, the signals exhibit periodic modulations during the second decay step. The modulation periods are in the range of 20–50 ps, and are found to be independent of pump intensity, but show a strong dependence on the particle axis probed. They explained their results in the following way. The heating by the pump pulse results in a thermal expansion of the particle. The electron pressure to this expansion is extremely large due to the high temperature of the electron gas. During this expansion, the inertia of the ions will make them overshoot their new (high-temperature) equilibrium positions. This overshoot will be followed by a reversal of the expansion, due to the elasticity of the metal, resulting in an oscillating behavior of the particle length. These particle-length oscillations are expected to induce a plasmon shift, which can even quantitatively explain the observed modulation in the differential transmission change $\Delta T/T$.

3.3. Heat Transfer to the Support

The last step in the energy relaxation is the heat transfer between the nanoparticles and the support (host matrix, substrate). A single electronic cluster–substrate interaction process happens in the femtosecond time scale, as reported by Reihl and co-workers [74]. Photoemission spectra of silver particles on a graphite surface reveal distinct deviations in the Fermi-level onset from the step-like shape typical for metals. These deviation could be explained by using a simple model that takes into account the finite lifetime of the photohole remaining on the metal cluster during the photoemission process and the cluster–substrate.

The time to establish a thermal equilibrium between the cluster and substrate takes much longer; in general, it happens in a time scale above 100 ps. This process is sensitive to the thermal conductivity of the surrounding medium. In this time region, we can consider that the lattice temperature of the nanoparticles is the same as the electron temperature.

First, we consider the thermal diffusion from the metallic nanoparticle to the substrate by a simple diffusion equation:

$$\frac{\partial \Theta_{\text{subst}}}{\partial t} = D_{\text{subst}} \nabla^2 \Theta_{\text{subst}} \tag{9}$$

where D_{subst} is the diffusivity of the heat in the substrate. From Eq. (9), we obtain the thermal distribution function outside the metal nanoparticle as a function of the distance r and time t. In the case of a dielectric substrate, thermal diffusion inside the nanoparticle can be neglected because, in general, the diffusivity of metal D_m is 10^3 time larger than D_{diel}.

As already mentioned above, Inouye and co-workers found, in a gold nanoparticle, a slow decay component (~120 ps), and elucidated it as the heat transfer between the nanoparticles and the host matrix [68]. Also, Hodak and co-workers found a slow component on a ~100 ps time scale in their transient bleach data recorded with a high-power (>1 mJ/cm^2) pump laser for Au particles (10 nm) in a solvent [69]. For sufficiently high pump power, the heat deposited into the lattice by the electrons causes the particles to expand. This shifts the plasmon band to the red, leading to a long-lived transient absorption and bleach signals on the red and blue sides of the plasmon band, respectively. For gold particles with a mean diameter of 30 nm in a sol–gel matrix, Perner et al. found a slow decay constant of 200 ps [75].

4. CONCLUSION

The impact of ultrafast excitation and relaxation of the conduction band electrons of metallic nanoparticles has been reviewed. The attention is confined to metallic nanoparticles with sizes larger than 100 Å, for example, the metal is still considered as a three-dimensional system. Quantum size effects due to the discrete nature of the electronic states [6] do not show up, and do not play a role in the energy dissipation processes. However, the size of the nanoparticles is smaller than the wavelength of light, as well as the optical penetration depth. Hence, all atoms in the particle can be collectively excited. This in turn influences the excitation (optical properties) and relaxation (electrodynamic properties) process.

Summarizing the general trend, only the interaction of light with the nanoparticles shows a strong size dependency, especially the damping (dephasing time), and hence, the resonant enhancement f of the local field depends critically on the size of the particles. But as soon as the energy is stored in the electron system (e.g., in electron–hole pairs), the following thermalization processes (electron cooling dynamics) show—if at all—only a slight size dependence. In addition, independent of whether the light is absorbed due to plasmon absorption or interband transition, no difference is observed in the femtosecond electron dynamics, as clearly shown by Link et al. [54].

So far, most of the published work has been performed on noble metal nanoparticles; only a few results were reported on transition metal particles [56, 71]. The detailed study of the electron dynamics in transition metal nanoparticles would help us understand the influence of the more localized d band to the energy dissipation dynamics. Also, from the point of catalytic reaction, the move to the more reactive transition metals would be desirable [76].

Furthermore, spin- and time-resolved measurements open the possibility to study the spin-dependent electron dynamics in magnetic nanoparticles, a new research field which has just emerged in the last years [77, 78].

GLOSSARY

Drude damping Includes all microscopic damping processes of excited electrons due to photons, phonons, impurities, surface scattering, and electron-electron interactions.

External thermalization Cooling of the optically excited electron gas via phonon emission. This heat-exchange process is usually described by an electron-phonon coupling model.

Inelastic decay time T_1 Decay of the plasmon population via transformation into photons (radiation damping) and via decay into electron-hole excitations.

Interface damping Contribution to the plasmon decay due to electron scattering off the particle surface and interaction with adsorbates.

Interferometric autocorrelation Autocorrelation of two laser pulses performed with interferometric accuracy. If the pulses are sent through an optically nonlinear material, higher order interferometric autocorrelation function can be obtained.

Internal thermalization Thermalization of optically excited electronic states towards a Fermi-Dirac distribution with a well defined temperature. Electron-electron scattering is the most efficient process to redistribute the energy to thermalize the electron gas.

ITO substrate Indium tin oxide, an often used transparent and electrically conducting substrate to investigate deposited metallic nanoparticles.

Local field enhancement The electromagnetic field in the particle near field is enhanced due to the resonantly driven charges in the nanoparticles. Resonant and nonresonant excitation result in different amounts of field enhancement.

Particle (Mie) plasmon A collective excitation of the electron plasma in a nanoparticle gives rise to polarization charges at the particle surface which act as a restoring force like an oscillator system. On the contrary to the bulk and the surface plasmon, the particle plasmon can be directly excited by freely propagating light waves.

Plasmon dephasing time T_2 Corresponds to the time scale on which the decay of the coherent electron plasma oscillation (and hence the local-field enhancement) takes place, e.g. the time scale on which the electron oscillation preserves the memory of the optical phase of the excitation pulse. T_2 is inverse proportional to the homogeneous linewidth Γ in the spectral domain and is related to the inelastic decay time T_1 and the pure dephasing time T_2^*.

Plasmon resonance A collective excitation of a dense electron gas. This excitation consists of a coherent motion of a high number of electrons in contrast to those excitations in which the external perturbation acts on a single electron. The effective field governing this excitation consists of a superposition of external field and induced field generated by the dynamical response of the electrons system. In general, one has to distinguish between bulk, surface, and particle (Mie) plasmons.

Pump-probe technique A stroboscopic method, a first laser pulse drives the systems and a second, suitable delayed, probes the actual state. This technique enables a direct measurement of the dynamical properties in the time domain.

Pure dephasing time T_2^* Elastic phase-loss process, e.g. a decay of the fixed phase correlation between the individual electronic excitations of the whole oscillator ensemble.

Radiation damping Coherent reemission of photons from the oscillating electron plasma. It is the origin of the reflected beam.

ACKNOWLEDGMENTS

The author would like to thank R. Porath, T. Ohms, M. Scharte, J. Beesley, M. Wessendorf, C. Wiemann, O. Andreyev, B. Lamprecht, H. Ditlbacher, J. R. Krenn, F. R. Aussenegg, A. Liebsch, J. Feldmann, H.-G. Rubahn, and J.-Y. Bigot for their contribution to many experimental results that are reported here. Special thanks go to M. Bauer and C. Ziegler for critical comments on various aspects of this chapter. This work was supported by the Deutsche Forschungsgemeinschaft through SPP 1093.

REFERENCES

1. H. Raether, "Surface Plasmons," Springer Tracts in Modern Physics, Vol. 111. Springer, Berlin, 1988.
2. U. Kreibig and M. Vollmer, "Optical Properties of Metal Clusters," Springer Series in materials Science, Vol. 25. Springer, Berlin, 1995.
3. C. Bohren and D. Huffmann, "Absorption and Scattering of Light by Small Particles." Wiley, New York, 1983.
4. A. A. Lushnikov and A. J. Simonov, *Z. Phys.* 224, 307 (1969).
5. D. A. Evans and K. Horn, *Surf. Sci.* 307–309, 321 (1994).
6. A. Bettac, L. Köller, V. Rank, and K. H. Meiwes-Broer, *Surf. Sci.* 402–404, 475 (1998).
7. H.-V. Roy, P. Fayet, F. Patthey, W.-D. Schneider, B. Delley, and C. Massobrio, *Phys. Rev. B* 49, 5611 (1994).
8. G. K. Wertheim, S. B. DiCenzo, and D. N. E. Buchanan, *Phys. Rev. B* 33, 5384 (1986).
9. H. Handschuh, G. Ganteför, P. S. Bechthold, and W. Eberhardt, *J. Chem. Phys.* 100, 7093 (1994).
10. B. Lamprecht, Dissertation, Karl-Franzen-University of Graz, 2000, unpublished.
11. C. Kittel, "Introduction to Solid State Physics." Wiley, New York, 1976.
12. H. Raether, "Excitations of Plasmons and Interband Transitions by Electrons." Springer, Berlin, 1980.
13. D. Pines, *Rev. Mod. Phys.* 28, 184 (1956).
14. A. Otto, *Z. Phys.* 216, 39 (1968).
15. R. H. Ritchie, *Surf. Sci.* 34, 1 (1973).
16. B. Lamprecht, J. R. Krenn, G. Schider, H. Ditlbacher, M. Salerno, N. Felidj, A. Leitner, F. R. Aussenegg, and J. C. Weeber, *Appl. Phys. Lett.* 79, 51 (2001).
17. G. Ritchie, E. Burstein, and R. B. Stephens, *J. Opt. Soc. Am. B* 2, 544 (1985).
18. M. Meier, A. Wokaun, and P. F. Liao, *J. Opt. Soc. Am. B* 2, 931 (1985).
19. F. Moresco, M. Rocca, T. Hildebrandt, and M. Henzler, *Phys. Rev. Lett.* 83, 2238 (1999).
20. H. Wenzel and H. Gerischer, *Chem. Phys. Lett.* 76, 460 (1980).
21. G. Mie, *Ann. Phys.* 25, 377 (1905).
22. B. R. Cooper, H. Ehrenreich, and H. R. Philipp, *Phys. Rev.* 138, 494 (1965).
23. M. Scharte, R. Porath, T. Ohms, M. Aeschlimann, J. R. Krenn, H. Ditlbacher, F. R. Aussenegg, and A. Liebsch, *Appl. Phys. B* 73, 305 (2001).
24. T. Klar, M. Perner, S. Grosse, G. von Plessen, W. Spirkl, and J. Feldmann, *Phys. Rev. Lett.* 80, 4249 (1998).
25. J.-C. M. Diels, J. J. Fontaine, I. C. McMichael, and F. Simoni, *Appl. Opt.* 36, 1270 (1985).
26. C. Spielmann, L. Xu, and F. Krausz, *Appl. Opt.* 36, 2523 (1997).
27. D. Meshulach, Y. Barad, and Y. Silberberg, *J. Opt. Soc. Am. B* 14, 2122 (1997).
28. T. Vartanyan, M. Simon, and F. Träger, *Appl. Phys. B* 68, 425 (1999).
29. G. Seifert, M. Kaempfe, K.-J. Berg, and H. Graener, *Appl. Phys. B* 71, 795 (2000).
30. H. G. Craighead and G. A. Niklasson, *Appl. Phys. Lett.* 44, 1134 (1984).
31. W. Gotschy, K. Vonmetz, A. Leitner, and F. R. Aussenegg, *Opt. Lett.* 21, 1099 (1996).
32. F. Stietz, J. Bosbach, T. Wenzel, T. Vartanyan, A. Goldmann, and F. Träger, *Phys. Rev. Lett.* 84, 5644 (2000).

33. T. Vartanyan, J. Bosbach, F. Stietz, and F. Träger, *Appl. Phys. B* 73, 391 (2001).
34. B. Lamprecht, A. Leitner, and F. R. Aussenegg, *Appl. Phys. B* 68, 419 (1999).
35. J.-H. Klein-Wiele, P. Simon, and H.-G. Rubahn, *Phys. Rev. Lett.* 80, 45 (1998).
36. B. Lamprecht, J. R. Krenn, A. Leitner, and F. R. Aussenegg, *Phys. Rev. Lett.* 83, 4421 (1999).
37. J. D. Jackson, "Klassische Elektrodynamik." Walter de Gruyter, Berlin, New York, 1983.
38. A. Wokaun, in "Solid State Physics" (H. Ehrenreich, T. Thurnbull, and F. Seitz, Eds.), Vol. 38, p. 223. Academic, New York, 1984.
39. F. J. Heilweil and R. M. Hochstrasser, *J. Chem. Phys.* 82, 4762 (1985).
40. M. Adelt, S. Nepijko, W. Drachsel, and H.-J. Freund, *Chem. Phys. Lett.* 291, 425 (1998).
41. C. A. Schmuttenmaer, M. Aeschlimann, H. E. Elsayed-Ali, R. J. D. Miller, D. Mantell, J. Cao, and Y. Gao, *Phys. Rev. B* 50, 8957 (1994).
42. M. Aeschlimann, M. Bauer, and S. Pawlik, *Chem. Phys.* 205, 127 (1996).
43. M. Wolf, *Surf. Sci.* 343, 377 (1997).
44. K. Ertel, U. Kohl, J. Lehmann, M. Merschdorf, W. Pfeiffer, A. Thon, S. Voll, and G. Gerber, *Appl. Phys. B* 68, 439 (1999).
45. J. Lehmann, M. Merschdorf, W. Pfeiffer, A. Thon, S. Voll, and G. Gerber, *J. Chem. Phys.* 112, 5428 (2000).
46. J. Lehmann, M. Merschdorf, W. Pfeiffer, A. Thon, S. Voll, and G. Gerber, *Phys. Rev. Lett.* 85, 2921 (2000).
47. C. Kennerknecht, H. Hövel, M. Merschdorf, S. Voll, and W. Pfeiffer, *Appl. Phys. B* 73, 425 (2001).
48. T. V. Shahbazyan, I. E. Perakis, and J.-Y. Bigot, *Phys. Rev. Lett.* 81, 3120 (1998).
49. W. S. Fann, R. Storz, H. W. Tom, and J. Bokor, *Phys. Rev. B* 46, 13592 (1992).
50. W. S. Fann, R. Storz, H. W. Tom, and J. Bokor, *Phys. Rev. Lett.* 48, 2834 (1992).
51. J.-Y. Bigot, V. Halté, J.-C. Merle, and A. Daunois, *Chem. Phys.* 251, 181 (2000).
52. N. del Fatti, R. Bouffanais, F. Vallée, and C. Flytzanis, *Phys. Rev. Lett.* 81, 922 (1998).
53. M. Fierz, K. Siegmann, M. Scharte, and M. Aeschlimann, *Appl. Phys. B* 68, 415 (1999).
54. S. Link, C. Burda, Z. L. Wang, and M. A. El-Sayed, *J. Chem. Phys.* 111, 1255 (1999).
55. C.-K. Sun, F. Vallée, L. H. Acioli, E. P. Ippen, and J. G. Fujimoto, *Phys. Rev. B* 50, 15337 (1994).
56. A. Stella, M. Nisoli, S. De Silvestri, O. Svelto, G. Lanzani, P. Cheyssac, and R. Kofmann, *Phys. Rev. B* 53, 15497 (1996).
57. H. E. Elsayed-Ali, T. B. Norris, M. A. Pessot, and G. A. Mourou, *Phys. Rev. Lett.* 58, 1212 (1987).
58. R. W. Schoenlein, W. Z. Lin, J. G. Fujimoto, and G. L. Eesley, *Phys. Rev. Lett.* 58, 1680 (1987).
59. C.-K. Sun, F. Vallée, L. Acioli, E. P. Ippen, and J. G. Fujimoto, *Phys. Rev. B* 48, 12365 (1993).
60. S. I. Anisimov, B. L. Kapeliovich, and T. L. Perel'man, *Sov. Phys. JETP* 39, 375 (1974).
61. G. L. Eesley, *Phys. Rev. B* 33, 2144 (1986).
62. S. A. Nepijko, S. A. Gorban, L. V. Viduta, and R. D. Fedorovich, *Int. J. Electron.* 73, 1011 (1992).
63. N. del Fatti and F. Vallée, *Appl. Phys. B* 73, 383 (2001).
64. T. Tokizaki, A. Nakamura, S. Kaneko, K. Uchida, S. Omi, H. Tanji, and Y. Asahara, *Appl. Phys. Lett.* 65, 941 (1994).
65. R. Rosei, F. Antongeli, and U. M. Grassano, *Surf. Sci.* 37, 689 (1973).
66. R. Rosei and D. W. Lynch, *Phys. Rev. B* 5, 3883 (1972).
67. A. E. Faulhaber, B. A. Smith, J. K. Andersen, and J. Z. Zhang, *Mol. Cryst. Liq. Cryst.* 283, 25 (1996).
68. H. Inouye, K. Tanaka, I. Tanahashi, and K. Hirao, *Phys. Rev. B* 57, 11334 (1998).
69. J. H. Hodak, I. Martini, and G. V. Hartland, *J. Phys. Chem. B* 102, 6958 (1998).
70. N. Del Fatti, F. Vallée, C. Flytzanis, Y. Hamanaka, and A. Nakamura, *Chem. Phys.* 251, 215 (2000).
71. M. Nisoli, S. Stagira, S. DeSilvestri, A. Stella, P. Tognini, P. Cheyssac, and R. Kofman, *Phys. Rev. Lett.* 78, 3575 (1997).
72. J.-Y. Bigot, J.-C. Merle, O. Cregut, and A. Daunois, *Phys. Rev. Lett.* 75, 4702 (1995).
73. M. Perner, S. Gresillon, J. März, G. von Plessen, J. Feldmann, J. Porstendorfer, K.-J. Berg, and G. Berg, *Phys. Rev. Lett.* 85, 792 (2000).
74. H. Hövel, B. Grimm, M. Pollmann, and B. Reihl, *Phys. Rev. Lett.* 81, 4608 (1998).
75. M. Perner, P. Bost, U. Lemmer, G. von Plessen, J. Feldmann, U. Becker, M. Mennig, M. Schmitt, and H. Schmidt, *Phys. Rev. Lett.* 78, 2192 (1997).
76. P. W. Jacobs, S. J. Wind, F. H. Ribeiro, and G. A. Somorjai, *Surf. Sci.* 372, 249 (1997).
77. U. Wiedwald, M. Spasova, M. Farle, M. Hilgendorff, and M. Giersig, *J. Vac. Sci. Technol.* A19, 1773 (2001).
78. F. Kronast, B. Heitkamp, H. A. Dürr, W. Eberhardt, G. Bihlmayer, S. Blügel, A. Liebsch, S. Landis, and B. Rodmacq (to be published).

Encyclopedia of Nanoscience and Nanotechnology
www.aspbs.com/enn

Electron Holography of Nanostructured Materials

Rafal E. Dunin-Borkowski
University of Cambridge, Cambridge, United Kingdom

Martha R. McCartney, David J. Smith
Arizona State University, Tempe, Arizona, USA

CONTENTS

1. INTRODUCTION

The transmission electron microscope (TEM) is a powerful and widely used tool for characterizing the microstructure and chemistry of materials. There are many different TEM imaging modes, but most suffer from the major drawback that the final recorded image is a spatial distribution of intensity. All information about the phase shift of the high-energy electron wave that has passed through the sample is lost. The technique of electron holography, which was originally described by Denis Gabor [1, 2] and is based on the formation of an interference pattern or "hologram," overcomes this limitation by allowing the phase shift of the electron wave to be recovered. As the phase shift is sensitive to both the in-plane component of the magnetic induction and the electrostatic potential in the sample, an electron hologram can provide quantitative information about magnetic and electric fields in materials to a spatial resolution that can approach the nanometer scale. The earliest attempts at electron holography were restricted by the limited brightness of the tungsten hairpin filaments used as electron sources [3], but the development of the high-brightness field-emission electron gun (FEG) for the TEM [4] now allows the practical implementation of electron holography in commercial instruments. Since the first of these instruments (the 200 kV Hitachi HF-2000 FEG-TEM) became available in 1990, FEG-TEMs have been sold widely, and electron holography has been applied to a wide variety of materials such as quantum well structures, magnetoresistive films, nanowires, and semiconductor devices.

In this chapter, the application of electron holography to the characterization of magnetic and electrostatic fields in nanostructured materials is reviewed. The properties of such materials differ significantly from those observed in larger structures. For example, magnetic interactions between closely spaced, deep-submicrometer magnetic elements depend sensitively on their size, shape, composition, and spacing. The characterization of such interactions is of great importance if these materials are to be used in future magnetic recording and sensing applications. Such applications may require specific values of the coercive fields and remanent magnetizations of individual magnetic nanostructures, as well as stable and reproducible magnetic domain states. Most of the experimental results that are presented in this chapter were obtained using the off-axis, or "sideband," mode of electron holography in the TEM, which is ideally suited to the characterization of nanostructured materials for reasons that will be discussed. As a result, this mode will be described in detail, while other modes will only be mentioned briefly or in passing.

ISBN: 1-58883-059-4/$35.00

Encyclopedia of Nanoscience and Nanotechnology
Edited by H. S. Nalwa
Volume 3: Pages (41–99)

A particular advantage of electron holography for the analysis of magnetic or electric fields in nanostructured materials is that unwanted contributions to the contrast from local variations in composition and sample thickness can be removed from a holographic phase image more easily than from images recorded using alternative phase contrast techniques such as Lorentz microscopy. As the dimensions of magnetic storage and electronic devices continue to shrink, electron holography provides an increasingly unique approach for solving industrial problems, as well as providing advances to fundamental scientific knowledge.

This chapter begins with an outline of the experimental and theoretical procedures that are used to obtain amplitude and phase information from electron holograms, as well as some of the limitations and pitfalls that may be encountered when using the technique. Applications of electron holography to the characterization of nanostructured materials are then reviewed, in three distinct sections. First, the use of measurements of mean inner potential and sample thickness to characterize the morphologies of fine particles and steps on the surfaces of materials is described. The influence of dynamical diffraction and chemical bonding on such measurements is discussed.

Second, the characterization of magnetic domain structures in a wide variety of nanostructured materials using electron holography is reviewed. Examples of "optical" holographic reconstruction of magnetic fields in nanostructures that are larger than a few hundred nanometers in size include the characterization of bubble memories, ferromagnetic wires that are wide enough to contain domain walls, fine particles, recorded magnetic tapes, and the experimental confirmation of the Aharonov–Bohm effect. In smaller magnetic structures, particularly those below 100 nm in size, the measured holographic signal is dominated by unwanted contributions to the measured phase shift from local variations in sample thickness and composition. Digital acquisition and analysis of electron holograms are then essential in order to recover the weak magnetic signal of primary interest. The procedures that can be used to characterize deep-submicrometer magnetic nanostructures are illustrated for hard magnets, magnetic nanowires, lithographically patterned elements, chains of strongly interacting single-domain nanocrystals, magnetic tunnel junctions examined in cross-section, magnetic force microscopy tips, and fluxons in superconductors. Both the need to perform micromagnetic simulations to understand the observed magnetic microstructure and the quantitative nature of the results that can be obtained using electron holography are emphasized.

Third, the application of electron holography to the characterization of electrostatic fields in nanostructured materials is reviewed. Experiments that are described include the measurement of electrostatic fringing fields around charged latex particles and field-emitting carbon nanotubes, electrostatic potentials associated with depletion regions in doped semiconductor devices, and space charge layers at interfaces in ceramics. Special attention is given to factors that currently limit the accuracy of results obtained from doped semiconductors, in particular the effect of different TEM sample preparation techniques on the contrast recorded from such samples. Recent results from the application of electron holography to working semiconductor devices, which are examined *in-situ* with a current flowing through them, are reported. Finally, several alternative modes of electron holography are described, and their respective advantages for different experiments are discussed.

The reader is referred to several books (e.g., [5–7]), review papers (e.g., [8–13]), and special journal issues (e.g., [14]) on the subject of electron holography, which offer comprehensive descriptions of other aspects of the technique that cannot be covered adequately here.

1.1. Basis of Off-Axis Electron Holography

The electron microscope geometry required for the TEM mode of off-axis electron holography is shown schematically in Figure 1a. The sample is examined using defocused, highly coherent illumination, which is usually provided by a FEG electron source. In reality, the source is never perfectly coherent, either spatially or temporally, but the degree of coherence must be such that an interference fringe pattern of sufficient quality can be recorded within a reasonable acquisition time, during which specimen and/or beam drift must be negligible. Cold FEGs have high temporal coherence ($\Delta E \sim 0.3$ eV) and high spatial coherence, but they require an ultrahigh vacuum environment for reliable operation. In addition, the beam current usually falls during operation as a result of gas molecules adhering to the tip. Schottky FEGs, which are used in JEOL and FEI FEG-TEMs, do not require such stringent vacuum conditions, and they provide a beam with a higher total current that is also more stable over time. Although electron holograms have historically been recorded on photographic film, digital acquisition using charge coupled device (CCD) cameras is now commonly used.

In order to acquire an off-axis electron hologram, the region of interest on the sample is positioned so that it covers approximately half the field of view. A positive voltage (or a negative voltage depending on the electron-optical configuration [15]) of typically between 50 and 200 V is then applied to an electrostatic biprism wire [16, 17], which is located close to a conjugate image plane in the electron microscope (most often in place of one of the conventional selected-area apertures). The biprism is usually a fine (<1 μm diameter) wire made from either Pt- or Au-coated quartz, and it may be rotatable. The optimal position for the biprism has been calculated by Lichte [18]. The voltage applied to the biprism acts to tilt the reference wave that has passed only through vacuum (or alternatively through a thin region of amorphous support film) relative to the wave that has passed through the sample, so that the two waves overlap and interfere to form a hologram in a slightly defocused image plane of the object.

The application of a voltage to the biprism results in an electron-optical configuration that is equivalent to the use of two electron sources S_1 and S_2, as shown in Figure 1a. The width of the interference region, W, is increased by applying a larger voltage to the wire, which in turn moves the two virtual sources further apart. This places a greater constraint on the spatial coherence of the source required to maintain

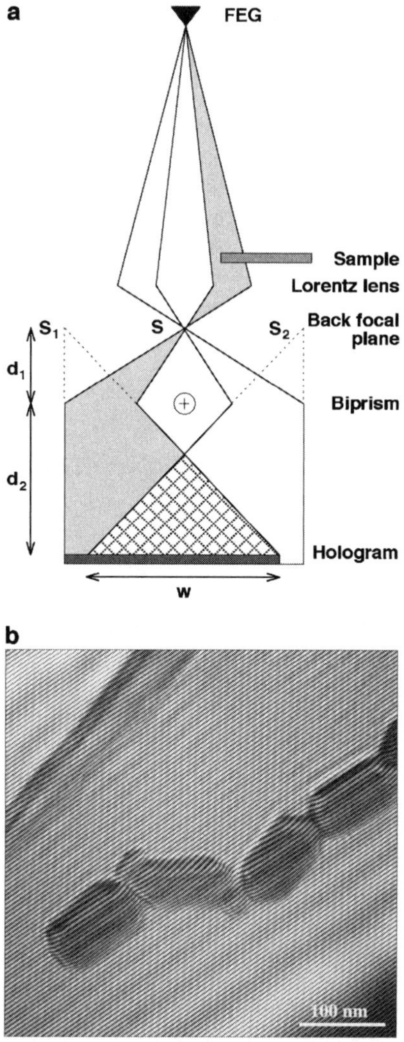

Figure 1. (a) Schematic illustration of the setup used for generating off-axis electron holograms. The sample occupies approximately half the field of view. Essential components are the field-emission electron gun electron source, which provides coherent illumination, and the positively charged electrostatic biprism, which causes overlap of the object and (vacuum) reference waves. The Lorentz lens allows imaging of magnetic materials in close-to-field-free conditions. The symbols are defined in the text. (Taken from Dunin-Borkowski, unpublished work.) (b) Off-axis electron hologram of a chain of magnetite (Fe_3O_4) crystals from a bacterial cell, recorded at 200 kV using a Philips CM200 FEG-TEM. The crystals are supported on a holey carbon film. Holographic phase changes can be seen most clearly, in the form of bending of the holographic interference fringes as they pass through the crystals, by looking along the fringes at a shallow angle to the plane of the paper. Coarsely spaced Fresnel fringes from the edges of the biprism wire are visible within the overlap region in addition to the holographic fringes. A hole in the support film is visible at the top left corner of the image. (Taken from McCartney, unpublished work.)

sufficient fringe contrast. The interference width is given by the expression

$$W = 2\left(\frac{d_1 + d_2}{d_1}\right)\left(\alpha\frac{d_1 d_2}{d_1 + d_2} - R\right) \tag{1}$$

where R is the radius of the biprism. α, the deflection angle introduced by the biprism, is directly proportional to the voltage applied to the wire [19, 20]. Equation (1) highlights the importance of making the biprism wire as narrow as possible. The voltage that must be applied to the wire to achieve overlap between the two parts of the electron wave on either side of it is directly proportional to the wire diameter. The interference fringe spacing in the overlap region, s, is inversely proportional to the biprism voltage and is given by the expression

$$s = \lambda\left(\frac{d_1 + d_2}{2\alpha d_1}\right) \tag{2}$$

where λ is the wavelength of the incident illumination. The total number of fringes in the overlap region is approximately proportional to the square of the applied biprism voltage.

A representative off-axis electron hologram from a chain of ~100 nm magnetite crystals is shown in Figure 1b. In this image, the biprism has deliberately been rotated to achieve overlap between the crystals and a hole in the carbon support film (top left). Although the hologram is superficially similar to a bright-field image, the field of view is also covered by two distinct sets of interference fringes. The coarser (and lower contrast) fringes are Fresnel fringes from the edges of the biprism wire, while the finer fringes are the holographic interference fringes. The amplitude and phase shift of the electron wave that leave the sample are recorded in the intensity and position of the holographic interference fringes, respectively. Hence, the hologram contains information about the complete wavefunction transmitted by the specimen.

When examining magnetic materials in the electron microscope, holograms are usually acquired with the conventional microscope objective lens turned off, as its strong magnetic field would saturate the sample magnetization along the electron beam direction. A high-strength minilens located below the sample may then be used to provide high magnification with the sample in a field-free environment, as well as to provide a large overlap width of 1–2 μm with a small interference fringe spacing of 1–5 nm.

1.2. Imaging Theory

For most cases of electron microscope imaging, the imaging process may be assumed, as a first approximation, to be coherent. Imaging is described in terms of the modification of the electron wavefunction by the object and the imaging objective lens. The intensity of an image is then found from the modulus squared of the corresponding wavefunction [21]. The Abbe description of the imaging process is valid for electron microscopy, and a small-angle approximation can be made since typical scattering angles rarely exceed 10^{-1} radians. For coherent imaging in the TEM, the electron wavefunction at the exit surface of the sample can be written in the form

$$\psi_s(\mathbf{r}) = A_s(\mathbf{r}) \exp[i\phi_s(\mathbf{r})] \tag{3}$$

where **r** is a two-dimensional vector in the plane of the sample with components x, y. The corresponding wavefunction in the image plane can be written as

$$\psi_i(\mathbf{r}) = A_i(\mathbf{r}) \exp[i\phi_i(\mathbf{r})] \quad (4)$$

In Eqs. (3) and (4), A and ϕ refer to amplitude and phase, and the subscripts s and i refer to the sample and image planes, respectively. In a standard bright-field TEM image, the recorded intensity distribution is given by the expression

$$I(\mathbf{r}) = |A_i(\mathbf{r})|^2 \quad (5)$$

The back focal plane of the electron microscope objective lens contains the Fraunhofer diffraction pattern (i.e., the Fourier transform) of the object wave $\psi_s(\mathbf{r})$, denoted $\psi(\mathbf{q}) = \mathrm{FT}[\psi_s(\mathbf{r})]$, where **q** is a two-dimensional vector in reciprocal space. Subsequent transfer from the back focal plane to the image plane is expressed as an inverse Fourier transform, denoted $\mathrm{FT}^{-1}[\ ]$.

For a perfect thin lens, neglecting any magnification and rotation of the image, the complex image wave would be exactly equivalent to the object wave $\psi_s(\mathbf{r})$. In practice, severe and unavoidable objective lens aberrations result in modifications to the amplitude and phase of the electron wave. These modifications are usually represented by a multiplication of the electron wavefunction in the back focal plane by a transfer function of the form

$$T(\mathbf{q}) = B(\mathbf{q}) \exp[i\chi(\mathbf{q})] \quad (6)$$

In Eq. (6), $B(\mathbf{q})$ is an aperture function that has a value of unity for values of q less than the objective aperture size and zero beyond the edge of the aperture. The effects of defocus and spherical aberration of the objective lens are included in the phase factor

$$\chi(\mathbf{q}) = \pi \Delta z \lambda q^2 + \frac{\pi}{2} C_S \lambda^3 q^4 \quad (7)$$

where Δz is the defocus of the lens and C_S is its spherical aberration coefficient. (A more complete form of contrast transfer theory for electron holography that takes partial coherence into account has also been derived [22].) The complex wave in the image plane can then be written in the form

$$\psi_i(\mathbf{r}) = \mathrm{FT}^{-1}[\mathrm{FT}[\psi_s(\mathbf{r})] \times T(\mathbf{q})] \quad (8)$$

$$= \psi_s(\mathbf{r}) \otimes t(\mathbf{r}) \quad (9)$$

where $t(\mathbf{r})$ is the inverse Fourier transform of $T(\mathbf{q})$, and the convolution $\otimes$ of the object wave $\psi_s(\mathbf{r})$ with $t(\mathbf{r})$ describes the loss of resolution that results from lens imperfections. Since both $\psi_s(\mathbf{r})$ and $t(\mathbf{r})$ are in general complex, the intensity in the image plane, which can be expressed in the form

$$I(\mathbf{r}) = A_i^2 = |\psi_s(\mathbf{r}) \otimes t(\mathbf{r})|^2 \quad (10)$$

is not normally related simply to the object structure.

The intensity in an off-axis electron hologram can be obtained from the previous expressions by adding a plane reference wave with a tilt corresponding to $\mathbf{q} = \mathbf{q}_c$ to the complex object wave in the image plane, to give

$$I_{\mathrm{hol}}(\mathbf{r}) = \left|\psi_s(\mathbf{r}) \otimes t(\mathbf{r}) + \exp\left[2\pi i \mathbf{q}_c \cdot \mathbf{r}\right]\right|^2 \quad (11)$$

$$= 1 + A_i^2(\mathbf{r}) + 2A_i(\mathbf{r}) \cos\left[2\pi \mathbf{q}_c \cdot \mathbf{r} + \phi_i(\mathbf{r})\right] \quad (12)$$

Equation (12) can be interpreted as follows. The intensity in the hologram is the sum of three terms: the intensities of the reference and image waves, and an additional set of cosinusoidal fringes with local phase shifts ϕ_i and amplitudes A_i that are equivalent to the corresponding phases and amplitudes of the image wave, respectively. These are the fine fringes visible in Figure 1b.

1.3. Reconstruction of Electron Holograms

The procedure typically used to "reconstruct" an electron hologram (i.e., to obtain amplitude and phase information about the image wave) is summarized in Figure 2a. The hologram is first Fourier transformed. The complex Fourier transform of the hologram can be expressed in the form

$$\begin{aligned}\mathrm{FT}\left[I_{\mathrm{hol}}(\mathbf{r})\right] = {} & \delta(\mathbf{q}) + FT\left[A_i^2(\mathbf{r})\right] \\ & + \delta(\mathbf{q} + \mathbf{q}_c) \otimes \left[A_i(\mathbf{r}) \exp\left[i\phi_i(\mathbf{r})\right]\right] \\ & + \delta(\mathbf{q} - \mathbf{q}_c) \otimes \left[A_i(\mathbf{r}) \exp\left[-i\phi_i(\mathbf{r})\right]\right] \quad (13)\end{aligned}$$

Equation (13) contains four terms: a peak at the origin ($\mathbf{q} = 0$) corresponding to the Fourier transform of the uniform intensity of the reference image, a second peak centered on the origin ($\mathbf{q} = 0$) comprising the Fourier transform of the intensity distribution of the normal TEM image, a peak centered on $\mathbf{q} = -\mathbf{q}_c$ comprising the Fourier transform of the desired image wavefunction, and a peak centered on $\mathbf{q} = +\mathbf{q}_c$ comprising the Fourier transform of the complex conjugate of the image wavefunction. In the Fourier transform of the hologram visible at the bottom left of Figure 2a, these contributions are visible as a central "autocorrelation" peak and two "sidebands." The two sidebands contain identical information apart from a change in the sign of the phase; experimentally, neither choosing the other sideband nor moving the biprism to the other side of the region of interest provides any additional information.

The reconstruction procedure involves using an "aperture" to select a single sideband from the Fourier transform. The chosen sideband is then inverse Fourier transformed, as shown in the lower half of Figure 2a. Complete separation of the sideband region from the central peak of the Fourier transform can only be achieved if a large enough value of $\mathbf{q}_c$ (i.e., a small enough fringe spacing) is chosen, while the optimal aperture size depends on the object being studied [23]. The aperture used to select the sideband is normally circular, and it can be given diffuse edges during processing to minimize effects due to the abrupt loss of information at the mask edge. The larger the radius of the aperture, the higher the spatial frequencies that are retained and thus the higher the spatial resolution, but paradoxically the higher the noise, in the reconstructed image. It is often the case that the highest spatial frequencies are not required, and a reduction in mask radius may remove high frequency

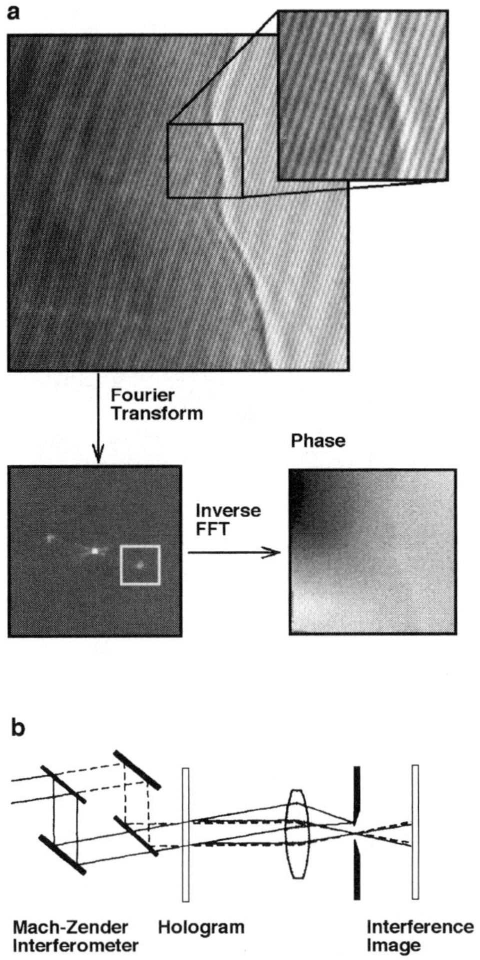

Figure 2. (a) Representative off-axis electron hologram of a thin crystal, recorded at 200 kV using a Philips CM200 FEG-TEM, with the enlargement showing interference fringes within the sample. The Fourier transform of the electron hologram and a phase image obtained after inverse Fourier transformation of one sideband are also shown. Reprinted with permission from [217], R. E. Dunin-Borkowski et al., *J. Microsc.* 200, 187 (2000). © 2000, Blackwell Science Ltd. (b) Schematic diagram showing a typical optical reconstruction setup for electron holography. Reprinted with permission from [188], N. Osakabe et al., *Phys. Rev. A* 34, 815 (1986). © 1986, American Physical Society.

noise from the reconstructed amplitude and phase. This procedure is particularly applicable to electromagnetic fields, since the phase shift may vary slowly across the image and only the lowest frequencies are required to portray the field accurately. However, care is required since the use of a small aperture can lead to extremely serious artifacts (e.g., [24]).

The reconstruction of electron holograms can be carried out either digitally with a computer [25, 26] or "optically" on a light bench. In common with many experimental techniques in the physical sciences, electron holography benefits greatly from the use of quantitative recording. Indeed, CCD cameras are now becoming the most widely used medium for recording holograms. They have a linear response over many orders of magnitude, they have a high detection quantum efficiency, and the recorded information is immediately accessible for digital processing. Moreover, fast and inexpensive computers and sophisticated image processing software (e.g., [27]) are now readily available. The convenient combination of digital recording with computer processing also facilitates novel applications of electron holography, some of which are described below. In particular, computational approaches make it straightforward to isolate the desired magnetic contribution to the holographic phase from other sources. Although many CCD cameras are still only 1024×1024 pixels in size, which limits the number of holographic fringes that can be sampled, 2048 and 4096 pixel cameras are now also available to relieve this limitation. As with any recording medium, the CCD camera has a point spread function that describes how much intensity has spread from one pixel to its neighbor, as well as a detection quantum efficiency, which is a measure of the electron dose required to separate an image detail from noise [28–31]. Although the accurate measurement of fringe contrast requires deconvolution of the point spread function from a hologram, this procedure can increase noise substantially. As a result, deconvolution approaches are rarely applied to medium resolution electron holograms.

When a hologram is reconstructed digitally, it is usually multiplied by a two-dimensional apodization function such as a Hanning window before its Fourier transform is calculated. This procedure eliminates streaks in the Fourier transform associated with the mismatch in intensity between opposite edges of the hologram. For an image of size N pixels on each side, a Hanning window of order τ is defined [32] by the expression

$$H^{\tau}(m,n) = H_x^{\tau}(m)H_y^{\tau}(n) \tag{14}$$

where

$$H_x^{\tau}(m) = \left(\frac{N}{2\pi}\right)\sum_{j=1}^{\tau}\frac{(-1)^j}{j}\sin\left[\pi j\left(\frac{N-1}{N}\right)\right]\cos\left[\left(\frac{2\pi jm}{N}\right)-1\right] \tag{15}$$

and

$$H_y^{\tau}(n) = \left(\frac{N}{2\pi}\right)\sum_{j=1}^{\tau}\frac{(-1)^j}{j}\sin\left[\pi j\left(\frac{N-1}{N}\right)\right]\cos\left[\left(\frac{2\pi jn}{N}\right)-1\right] \tag{16}$$

The Hanning window, which primarily affects the reconstructed amplitude image, can be removed from the complex image wave after a sideband has been selected and inverse Fourier transformed.

Although it is rarely used, it should be mentioned that a *real-space* digital reconstruction approach for electron holography has also been proposed [33]. A neural network approach has also been applied to this problem [34].

Optical reconstruction involves the use of a laser bench that has been configured as a Mach–Zender interferometer to illuminate a hologram that has been recorded photographically (Fig. 2b). The approach involves splitting a plane parallel light wave from a collimated laser into two separate beams, which, by proper adjustment of optical mirrors, pass through the hologram at slightly different angles and are then recombined using a lens. Similar to Figure 2a, the fringe modulation of the hologram acts as a diffraction grating and produces three primary beams in the back focal plane of the lens: the central beam and two diffracted beams, with a separation that is inversely proportional to the holographic fringe spacing. One of the diffraction maxima contains amplitude and phase information about the object, while the other contains the complex conjugate. When the region around one of the diffraction spots is selected physically using the aperture, the amplitude and phase of the resulting optical image are the same as those of the original electron image (i.e., the electron wavefronts are reproduced as optical wavefronts).

Whereas electron holograms that have been recorded digitally can be used to provide either raw phase images that have 2π phase jumps remaining or "unwrapped" phase images, optical reconstruction typically provides either "interferograms" (formed by adding a phase ramp to the image) or "interference images" (contour maps), which are equivalent to displaying the cosine of the measured phase. Two times phase amplification can be achieved optically by interfering the first order (+) spot produced by one laser beam (the reconstructed wavefront) with the first order (−) spot from the other beam (the conjugate wavefront) instead of with a plane wave. Greater phase amplification may be achieved by using nonlinear photographic reproduction of the hologram to produce a series of several diffraction orders on each side of the central spot in the back focal plane, instead of just one order on each side as given by the cosinusoidal fringes. The higher-order reflections are correspondingly more sensitive to phase changes and may be chosen to give contours at phase intervals as small as $(\pi/100)$. Alternatively, the phase amplification factor can be increased by forming a new hologram that has twice the phase sensitivity of the original. After processing, the original hologram may be replaced in the laser bench by the newly formed hologram and the procedure repeated to obtain greater phase amplification [35]. An alternative phase amplification scheme that is based on the difference between the phases of two holograms obtained using different biprism voltages has also been proposed [36].

1.4. Experimental Considerations

1.4.1. Microscope Alignment

The basic requirement for high electron beam coherence would imply demagnification of the electron source as far as practicable within the constraint of the available current density per unit solid angle imposed by the finite brightness of the source. Consideration of the imaging geometry shows, however, that the use of rotationally symmetric illumination is not required. A significant improvement in the coherence available with a given electron source and condenser lens system can be achieved by employing highly elliptical illumination. The condenser lens stigmators and focus settings are deliberately misaligned to produce illumination that is very wide in the direction perpendicular to the biprism wire when the condenser lens is overfocused but relatively narrow in the parallel direction. For example, in a typical medium resolution electron hologram obtained with the image located below the biprism, the minor axis of the ellipse may be adjusted to be 2–5 times greater than the overlap width, while the major axis may be 50–100 times greater again. The typical experimental illumination condition for acquiring an off-axis electron hologram is shown in Figure 3a [37].

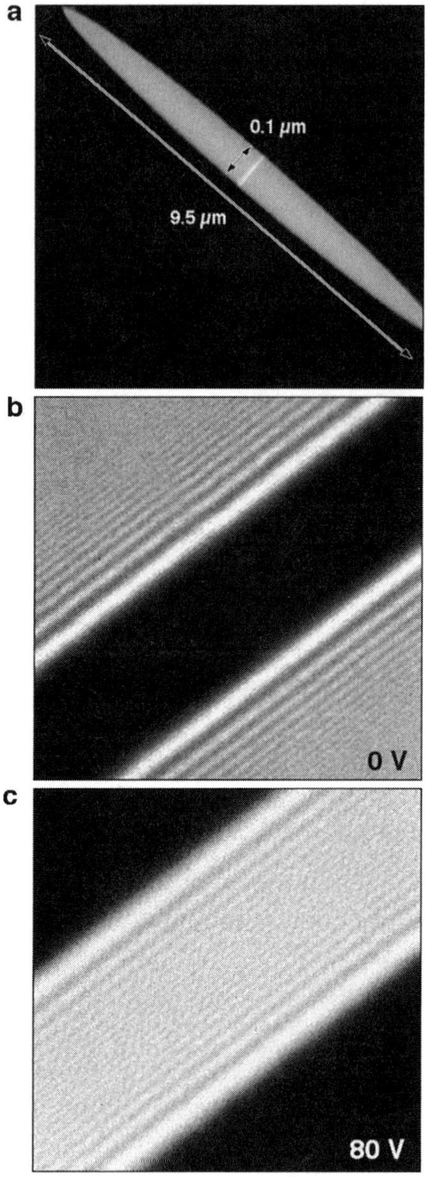

Figure 3. (a) Highly elliptical illumination formed by deliberately misadjusting the condenser stigmators, showing the typical condition used for recording electron holograms. The major and minor axis dimensions are 9.5 and 0.1 μm, respectively. Holographic interference fringes are formed within the central bright band (indicating the position and direction of the biprism wire) visible within the ellipse. The major axis of the illumination is aligned accurately perpendicular to the direction of the biprism wire. Images (b) and (c) show the effect of changes in the biprism voltage on hologram formation, recorded at 200 kV using a Philips CM200 FEG-TEM, for biprism voltages of 0 and 80 V, respectively. Fresnel fringes at the edges of the wire are visible outside the shadow of the wire in (b) and inside the overlap region in (c). Reprinted with permission from [37], D. J. Smith and M. R. McCartney, in "Introduction to Electron Holography" (E. Völkl, L. F. Allard, and D. C. Joy, Eds.), pp. 91–92. Kluwer Academic/Plenum, New York, 1998. © 1998, Kluwer Academic/Plenum.

Accurate alignment of the condenser stigmators can be achieved when the illumination is underfocused so that the major axis of the ellipse is parallel to the biprism wire direction. It is convenient to perform this adjustment with a relatively low biprism voltage, since the holographic fringes then have higher contrast and larger spacing. Even when using a FEG electron source, a small spot size, small condenser aperture size, and low gun extraction voltage are required to maximize coherence. As a result of the inevitable trade-off between intensity and coherence, the loss of signal must be balanced with a loss in coherence to achieve a reasonable hologram acquisition time [38].

Figure 3b illustrates the fact that when the biprism wire is initially inserted into the field of view, it creates a black shadow that has Fresnel fringes on either side. As the voltage is applied to the wire, the black shadow becomes progressively narrower until overlap is achieved. As the overlap region continues to expand with increasing biprism voltage, it appears bright and contains increasingly finer holographic interference fringes that run parallel to the length of the wire (in addition to the Fresnel fringes that remain from the edges of the biprism). This condition is shown in Figure 3c.

1.4.2. Vacuum Reference Wave

The requirement for a (vacuum) reference wave that can be overlapped onto the region of interest by the application of a voltage to the biprism wire usually implies that a hologram can only be recorded from near a specimen edge. The region of interest and the reference wave must be within a few μm of each other in order to achieve sufficient fringe contrast in the overlap region, and the orientation of the biprism wire may need to be optimized to satisfy this criterion. The latter restriction can be relaxed if a thin and clean region of electron-transparent support film, rather than vacuum, can be overlapped onto the region of interest. An alternative proposal involves acquiring holograms of the same area of the sample using two different biprism voltages [39].

Great care should always be taken to assess the effect on the reference wave of long-range electromagnetic fields that may extend outside the sample [40–42]. Both the object wave and the reference wave may then be affected by the field from the sample. The observed phase image may be altered from one that would be recorded using a perfect reference wave, as illustrated in Figure 4 for simulations of a small magnetic bar. Figure 4a shows the phase image that would be recorded using a perfect reference wave, while Figure 4b shows the effect of using a reference wave that is modulated by the long-range magnetic field from the object [43]. Although a possible numerical method for correcting experimental phase images for this effect has been suggested [44], care should in general be taken both to avoid using a perturbed reference wave in the first place and to recognize its presence if it cannot be avoided.

1.4.3. Reference Hologram

In most standard microscope applications, image distortions are small enough to be ignored. Off-axis electron holography, however, stores information in the lateral displacement of interference fringes, and distortions can cause similar displacements. The reconstructed phase image may therefore carry long-range phase modulations that stem from a number of factors, including inhomogeneities in the charge and the thickness of the biprism wire, distortions caused by the projector lenses and the recording system, and charging effects in the path of the electrons (e.g., at fixed apertures). In some cases, these distortions can be distinguished from true phase information if the phase information is bandwidth-limited to higher spatial frequencies. As distortions characteristically show up in the lower frequency range, a simple background filtering may be sufficient to remove their contribution. In general, however, since these distortions are geometrical in nature and invariant in time, it is a simple matter to remove them. A reference hologram is obtained by carefully removing the sample from the field of view without changing any of the optical parameters of the microscope. Correction is then possible by a complex division of the reconstructed (sample and reference) image waves in real space, to reveal the distortion-free phase of the image wave [45–47]. This procedure is illustrated dramatically in Figure 5 for a wedge-shaped crystal of InP. Figure 5a shows a reconstructed phase image obtained before distortion correction, with the vacuum region outside the edge of the sample on the left side of the image [48]. Figure 5b shows the corresponding vacuum reference phase image, which has been acquired with the sample removed from the field of view but with all other imaging parameters unchanged. The equiphase contour lines now delineate only the distortions that must be removed from the phase image of the sample. Figure 5c shows the distortion-corrected phase image of the sample, obtained by subtracting Figure 5b from Figure 5a. This subtraction is in practice carried out by dividing the complex image wave. The vacuum region in Figure 5c is flattened substantially by

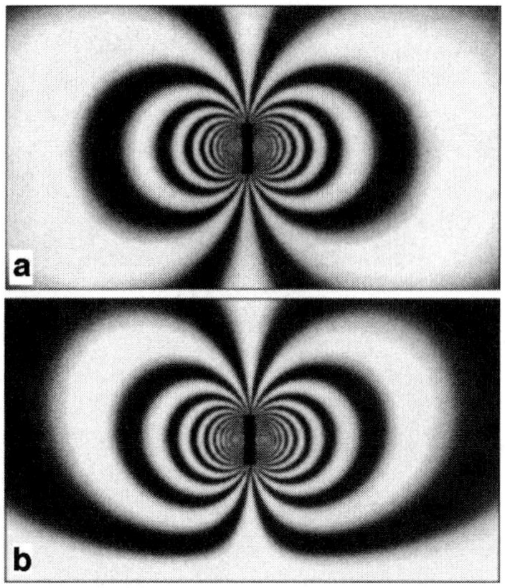

Figure 4. Simulations of phase lines around a small magnetic bar, calculated (a) assuming a perfect reference wave and (b) using a reference wave that has been perturbed by the long-range magnetic field extending into vacuum around the object. Reprinted with permission from [43], B. G. Frost and G. Matteucci, in "Introduction to Electron Holography" (E. Völkl, L. F. Allard, and D. C. Joy, Eds.), p. 193. Kluwer Academic/Plenum, New York, 1998. © 1998, Kluwer Academic/Plenum.

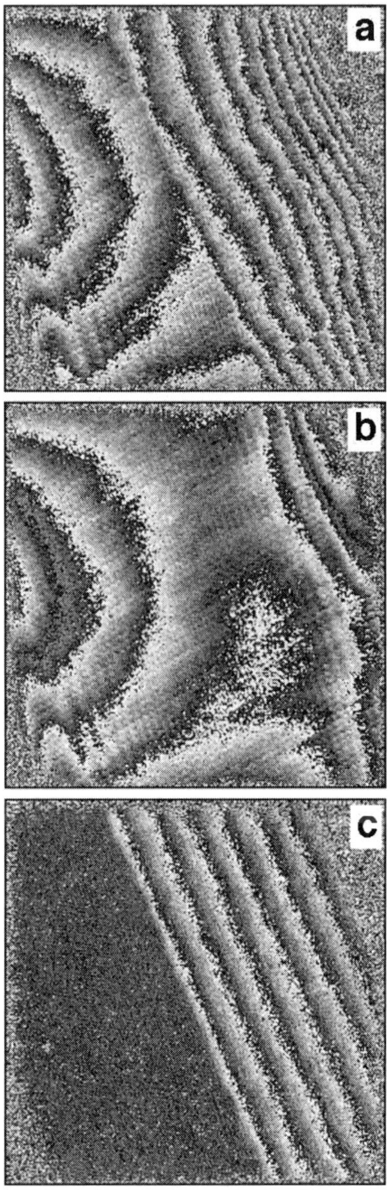

Figure 5. Illustration of distortion correction procedure for phase images acquired at 100 kV using a Philips 400ST FEG-TEM equipped with a Gatan 679 slow-scan CCD camera. (a) Initial (six-times-amplified) phase image of a wedge-shaped InP crystal. (b) Phase image obtained from vacuum, with the sample removed from the field of view, showing distortions due to the microscope imaging system. (c) Distortion-corrected phase image, obtained by subtracting phase image (b) from phase image (a). This procedure is best carried out by dividing the complex image waves obtained by inverse Fourier transforming the sideband obtained from each hologram, and by calculating the phase of the resulting wavefunction. Reprinted with permission from [48], D. J. Smith et al., in "Introduction to Electron Holography" (E. Völkl, L. F. Allard, and D. C. Joy, Eds.), p. 116. Kluwer Academic/Plenum, New York, 1998. © 1998, Kluwer Academic/Plenum.

this procedure, which allows relative phase changes within the sample to be interpreted faithfully.

The acquisition of a reference hologram has the additional advantage that it allows the true center of the sideband in Fourier space to be determined more accurately. Although the interference fringes in a hologram usually form the strongest contribution to a single spatial frequency in Fourier space, the true center of the sideband may not be the pixel that has the highest amplitude. This is a problem for automated reconstruction procedures, particularly for strong electric and magnetic fields as these can have a significant influence on the reference wave. Using a reference hologram, however, allows the position of the sideband to be determined to within ± 1 pixel accuracy by selecting the pixel with the highest amplitude from the sideband of the reference hologram. The use of the same location for the sideband in the Fourier transforms of both the sample and reference waves removes any tilt of the recorded wave introduced by the inability to locate the exact (subpixel) position of the sideband frequency in Fourier space.

1.4.4. Phase Unwrapping

When a phase image is calculated digitally it is usually evaluated modulo 2π, which means that 2π phase discontinuities that are unrelated to specimen features will appear at positions where the phase shift exceeds this amount. The phase image must therefore be "unwrapped" carefully using suitable algorithms before reliable interpretation of image features becomes possible. Although several phase unwrapping algorithms exist to locate and unwrap such discontinuities, none appears to solve all of the problems. The most straightforward approach involves searching the image by row or column for adjacent pixels with phase differences that are more than a prespecified threshold value, and then adding or subtracting 2π to the subsequent pixels [49]. This method may be unsuitable when noise is significant (e.g., for a thick sample or when the fringe contrast is poor), when the phase gradient exceeds 2π over three pixels, or when the interference field is smaller than the image. In an alternative approach [50], a second phase image may be calculated after multiplying the complex image wave by $\exp[i\alpha]$ to move the phase jumps to different positions in the image. In this manner, areas of the first phase image that were difficult to unwrap may be replaced by identical areas of the second phase image where there are no wraps. This method is more applicable for noisy images, although in practice a combination of the two approaches is often best. Advanced phase unwrapping techniques (e.g., [51]), many of which have been developed in fields outside electron microscopy, are now being introduced into the software that is used to process and analyze electron holograms.

1.4.5. Aberration Correction

If required, the aberrations of the objective lens can be removed by multiplying the complex image wave by a suitable phase plate corresponding to $T^*(\mathbf{q})$ to provide the amplitude and the phase of the *object* wave $\psi_s(\mathbf{r})$ rather than the *image* wave $\psi_i(\mathbf{r})$, and hence to improve the interpretable resolution of the image beyond the point resolution of the electron microscope. This procedure is usually only applied to high-resolution electron holograms, for which an optimal defocus can be defined in order to maximize the resolution of the reconstructed object wave after correction of aberrations [52–62]. The value of the optimal focus for recording a high-resolution electron hologram is given by the expression

$$\Delta z_{opt} = -\frac{3}{4} C_S (\lambda q_{max})^2 \tag{17}$$

where q_{max} is the maximum desired spatial frequency. The spatial frequency of the holographic fringes should be at least $3q_{max}$ for strong objects and $2q_{max}$ for weak objects [63]. At this defocus, the minimum number of fringes required in the hologram is given by the expression

$$N_{fringes} \geq 6C_S \lambda^3 q_{max}^4 \tag{18}$$

although a prefactor of 4 rather than 6 in this equation has been quoted also [64].

Figure 6 illustrates the application of aberration correction to an electron hologram of crystalline Si imaged at the $\langle 110 \rangle$ zone axis, at which characteristic "dumbbell" contrast, of spacing 0.136 nm, is expected [65]. The original hologram was acquired using an interference fringe spacing of 0.05 nm on a CM30 FEG-TEM, which has a point resolution of 0.198 nm and an information limit of 0.1 nm at 300 kV. Figure 6a and b shows, respectively, the reconstructed amplitude and phase of the hologram after aberration correction using a phase plate. The phase image reveals the expected white "dumbell" contrast, at a spatial resolution that is considerably better than the point resolution of the microscope, after the lens aberrations, which include the residual astigmatism and off-axis coma, have been measured and removed.

In contrast to high-resolution electron holography, the final focus is not so critical when characterizing electromagnetic fields at moderate resolution. In such experiments, the complex image wave can be manipulated in different ways by applying an aperture or phase plate to the sideband to reproduce images of the same sample that would normally be obtained using different techniques such as the Fresnel, Foucault, coherent Foucault, or differential phase contrast imaging modes of electron microscopy.

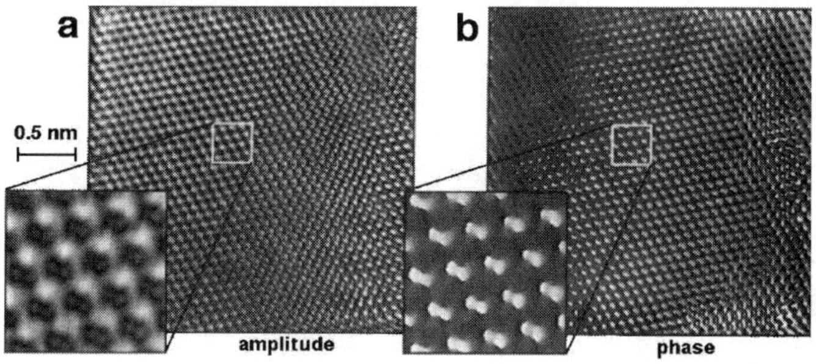

Figure 6. High-resolution (a) amplitude and (b) phase images of the aberration-corrected object wave reconstructed from an electron hologram of [110] Si, obtained at 300 kV on a CM30 FEG-TEM. The spacing of the original hologram fringes was 0.05 nm. The sideband contained {111}, {220}, {113}, and {004} reflections, corresponding to lateral information of 0.136 nm. The characteristic Si-dumbbell structure is visible only after aberration correction. Reprinted with permission from [65], A. Orchowski et al., *Phys. Rev. Lett.* 74, 399 (1995). © 1995, American Physical Society.

1.5. Magnetic and Mean Inner Potential Contributions to the Phase

The phase shift of an electron wave that has passed through a TEM sample is sensitive both to the electrostatic potential and to the in-plane component of the magnetic induction in the sample integrated in the incident beam direction. The phase changes are given by integrating along the electron paths, although it should be noted that electric and magnetic fields may extend into the surrounding vacuum and may be of considerable extent, as shown in Figure 4. For the magnetic case, the beam is deflected perpendicular to both the incident electron beam direction and the magnetic field direction.

Neglecting dynamical diffraction effects (i.e., assuming that the sample is thin and weakly diffracting), the measured phase shift can be expressed in the form

$$\phi(x) = C_E \int V(x, z)\, dz - \left(\frac{e}{\hbar}\right) \int A_z(x, z)\, dz \tag{19}$$

$$= C_E \int V(x, z)\, dz - \left(\frac{e}{\hbar}\right) \iint B_\perp(x, z)\, dx\, dz \tag{20}$$

where

$$C_E = \left(\frac{2\pi}{\lambda}\right)\left(\frac{E + E_0}{E(E + 2E_0)}\right) \tag{21}$$

z is the incident electron beam direction, x is a direction in the plane of the sample, A_z is the component of the magnetic vector potential **A** in the beam direction, $B_\perp$ is the component of the magnetic induction in the sample perpendicular to both x and z, V is the electrostatic potential, λ is the (relativistic) electron wavelength, and E and E_0 are, respectively, the kinetic and rest mass energies of the incident electron [66]. The energy-dependent constant C_E has the value 7.29×10^6, 6.53×10^6, and 5.39×10^6 rad V^{-1} m^{-1} at microscope accelerating voltages of 200 kV, 300 kV, and 1 MV, respectively. (The refractive index of a TEM sample for high-energy electrons is only slightly larger than unity; the phase shift is appreciable because of the small wavelength of the electron.)

In addition to electrostatic fields associated with the long-range redistribution of charge (e.g., at depletion layers in semiconductors [67]), the dominant contribution to the electrostatic potential is from the mean inner potential, which depends on the local composition and density in the sample, as will be discussed.

If neither V nor $B_\perp$ varies in the incident beam direction, then Eq. (20) can be simplified to

$$\phi(x) = C_E V(x) t(x) - \left(\frac{e}{\hbar}\right) \int B_\perp(x) t(x)\, dx \tag{22}$$

where t is the sample thickness. Differentiation with respect to x then leads to

$$\frac{d\phi(x)}{dx} = C_E \frac{d}{dx}\{V(x)t(x)\} - \left(\frac{e}{\hbar}\right) B_\perp(x) t(x) \tag{23}$$

In a sample of uniform thickness and composition, the first term in Eq. (23) disappears and the phase gradient is directly proportional to the in-plane component of the magnetic induction

$$\frac{d\phi(x)}{dx} = \left(\frac{et}{\hbar}\right) B_{\perp}(x) \tag{24}$$

A direct representation of the magnetic induction in the sample is then readily obtained by adding contours to the phase image, as shown in Figure 4. These contours correspond to magnetic line of force, along which $\int \mathbf{B} \cdot d\mathbf{S} = 0$. The fact that phase differences of 2π between adjacent contours correspond to $\int \mathbf{B} \cdot d\mathbf{S} = (h/e) = 4 \times 10^{-15}$ Wb provides the basis for the measurement of magnetic fields on an absolute basis. Subject to the limitations of signal-to-noise, smaller phase differences can be measured by decreasing the spacing of the contours obtained from the phase image.

An experimental phase image does not need to be unwrapped in order to evaluate its first differential for use in Eq. (24). In fact, the phase differential can be calculated directly from the reconstructed image wave (the inverse Fourier transform of the sideband selected from the Fourier transform of the hologram) without any introduction of artificial phase "wraps". If the real and imaginary parts of the reconstructed image wave are designated $\mathrm{Re}(\psi)$ and $\mathrm{Im}(\psi)$, then two of the possible ways of calculating the phase differential in the x-direction directly from the image wave are given by the expressions

$$\frac{d\phi}{dx}\left(\frac{\mathrm{Re}(\psi)\frac{d}{dx}\mathrm{Im}(\psi) - \mathrm{Im}(\psi)\frac{d}{dx}\mathrm{Re}(\psi)}{\mathrm{Re}(\psi)^2 + \mathrm{Im}(\psi)^2}\right) \tag{25}$$

and, more simply,

$$\frac{d\phi}{dx} = \mathrm{Im}\left(\frac{\frac{d}{dx}\psi}{\psi}\right) \tag{26}$$

1.6. Limitations

A major restriction of electron holography is that holograms are usually processed off-line so that dynamic events cannot be followed in real time [68–70]. Moreover, the field of view is typically limited to about 5 μm (unless a wide field, low magnification mode with the objective lens weakly excited is used [71]), either by the dimensions of the recording medium (e.g., the finite size of the CCD array) and the sampling of the holographic fringes or by the width of the fringe overlap region if the electron-optical magnification and the biprism voltage are reduced.

1.6.1. Contrast and Noise

Loss of interference fringe contrast occurs because of finite source size, beam energy spread, stray magnetic fields, the mechanical stability of the biprism wire, and the point spread function of the recording medium. The contrast of holographic interference fringes is determined largely by the lateral coherence of the electron wave at the specimen level. As a result, the contrast decreases when the voltage on the biprism wire is increased, and as the interference fringe spacing decreases. Experimentally, the biprism voltage is increased, first, to achieve a large field of view, second, to minimize the influence of the Fresnel fringes from the edges of the wire (which are independent of biprism voltage), and third, to minimize the fringe spacing to obtain the best spatial resolution in the reconstructed phase image. The decrease in fringe contrast is usually incorporated into Eq. (12) in the form

$$I_{\mathrm{hol}}(\mathbf{r}) = 1 + A_i^2(\mathbf{r}) + 2\mu A_i(\mathbf{r}) \cos[2\pi \mathbf{q}_c \cdot \mathbf{r} + \phi_i(\mathbf{r}) + \theta] \tag{27}$$

where μ is referred to as the fringe contrast or visibility and θ represents an arbitrary phase offset that is independent of position.

The standard experimental method of determining fringe visibility involves taking a line trace across a holographic interference fringe pattern that has been recorded in the absence of a sample. μ is then defined according to the relation

$$\mu = \left(\frac{I_{\mathrm{max}} - I_{\mathrm{min}}}{I_{\mathrm{max}} + I_{\mathrm{min}}}\right) \tag{28}$$

where I_{max} and I_{min} are the maximum and minimum intensities of the interference fringes, respectively, in a part of the hologram where the Fresnel fringes from the biprism can be neglected [72]. An alternative method of determining μ involves applying a Hanning window to the hologram, taking its Fourier transform, and then measuring the heights of the central peak and the two sidebands. The fringe visibility is then defined as twice the amplitude of the sidebands divided by the amplitude of the central peak [32, 73]. Although the fringe visibility should be as close to its theoretical maximum value of unity as possible, some decrease in μ is inevitable. Should it decrease too much, no reconstruction of the image wavefunction will be possible.

The spatial resolution of a phase image is determined primarily by the choice of magnification and the biprism voltage used. The interference fringe spacing varies inversely with biprism voltage and depends on the accelerating voltage, the focal length of the objective lens, and the distance between the back focal plane of the objective lens and the biprism wire [74, 75]. The phase detection limit [76] for electron holography is often determined by the signal-to-noise ratio of the recorded phase image, which may be low because of the weak intensity of the signal being measured. Off-axis electron holograms are typically recorded at electron doses of 100–500 electrons per pixel, so the recording process is dominated by Poisson-distributed shot noise [48, 77]. The recorded hologram is also affected by the detection quantum efficiency of the CCD camera (the spatial-frequency-dependent ratio of the squared signal-to-noise ratios at the output and the input of the camera, where the squared signal-to-noise ratio at the input is equal to the number of electrons incident on the pixel [28]) and the fringe visibility, which is in turn affected by parameters such as the illumination diameter, exposure time, and biprism voltage. The recording medium may have a detrimental effect on the recording process; film has a nonlinear response and a finite grain size, whereas digital recording involves a finite pixel size and strong attenuation of higher spatial frequencies due to the point spread function of the detector [78].

The minimum phase difference between two pixels that can be detected is given by the expression

$$\Delta\phi_{\min} = \frac{\text{SNR}}{\mu}\sqrt{\frac{2}{N_{\text{el}}}} \tag{29}$$

where SNR is the signal-to-noise ratio, μ is defined in Eq. (28), and N_{el} is the number of electrons collected per pixel [79]. In practice, some averaging of the measured phase image is often implemented, particularly if the features of interest vary slowly across the image or only vary in one direction (e.g., in a layered cross-sectional sample).

1.6.2. Artifacts

The effects of Fresnel diffraction at the biprism wire can cause phase and amplitude modulations of both the image and the reference wave [80]. These disturbances can be removed to some extent by using a reference hologram, although the positions of the Fresnel fringes in the sample and reference holograms are not always identical. They can also be minimized by using the undisturbed part of a broader hologram for recording and reconstruction. Residual effects can be removed by Fourier filtering the sideband before reconstructing the image wave or, alternatively, by making use of criteria such as the fact that a recorded magnetic signal must satisfy the Laplace equation [81].

A further geometrical effect known as vignetting results from the fact that a thin strip of the hologram close to the shadow of the biprism wire contains only single-sideband information about certain spatial frequencies because of screening by the biprism. The object wave in this region cannot be determined uniquely, and the reconstructed wave should be interpreted with care. The width of the hologram that is free from vignetting is

$$W_{\text{free}} = W\left(1 - \frac{q_{\max}}{q_c}\right) - 2R\frac{f}{d_1}\left(\frac{q_{\max}}{q_c}\right) \tag{30}$$

where W is the full width of the hologram [82]. In this expression, $q_{\max}$ is the maximum spatial frequency to be recorded, q_c is the carrier fringe frequency, R is the biprism radius, f is the objective lens focal length, and d_1 is defined in Figure 1a. The second term is usually much smaller than the first, and so for $q_c = 3q_{\max}$ only 2/3 of the hologram is free from vignetting. However, if the maximum frequency of interest is oversampled, as is usually the case for medium resolution electron holography of electric and magnetic fields, then this problem is easily minimized. The use of a window of finite size to select the sideband may also lead to artifacts around the outer edges of the reconstructed phase image. Analysis indicates that about three-quarters of the field of view should be reconstructed correctly [82].

2. MEASUREMENT OF MEAN INNER POTENTIAL AND SAMPLE THICKNESS

Electron holography has been widely used to obtain medium-resolution information about sample morphology and composition. In this section, we consider phase changes that are associated solely with local variations in mean inner potential and sample thickness. The characterization of magnetic and electrostatic fields is considered in the following Sections 3 and 4.

The mean inner potential, V_0, is defined as the volume average, or the zero-order Fourier coefficient, of the electrostatic potential in a material [83]. It can be written in the form

$$V_0 = \left(\frac{1}{\Omega}\right)\int_{\Omega} V(x, y, z)\,dx\,dy\,dz \tag{31}$$

where Ω is the volume of the region over which the integral is evaluated. V_0 is proportional to the second moment of the atomic charge density and can be calculated from its mean square radius [84, 85]. Exchange and correlation contributions to the potential are negligible for high-energy electrons [86], and the measured mean inner potential can be regarded as a basic property of the material and not of the electrons that are used to probe it. For low-energy electrons, exchange and correlation effects cannot be neglected, and the mean inner potential varies with the energy of the incident electrons [87–89].

In the absence of magnetic and long-range electric fields (such as those occurring at depletion regions in semiconductors), and assuming that dynamical diffraction effects can be neglected, Eq. (19) can be rewritten as

$$\phi(x) = C_E \int V_0(x, z)\,dz \tag{32}$$

where x is in the plane of the sample, z is the incident electron beam direction, and C_E is defined in Eq. (21). If the sample has uniform structure and composition in the electron beam direction, then Eq. (32) reduces further to

$$\phi(x) = C_E V_0(x) t(x) \tag{33}$$

where t is the sample thickness. Equations (32) and (33) are kinematical expressions because they ignore multiple scattering in the sample and are therefore only applicable to crystalline samples when they are tilted to a weakly diffracting orientation.

Electron holography provides the most accurate technique currently available for measuring the mean inner potential of a sample through the use of Eq. (33), but an independent measure of the sample thickness is required. The most straightforward sample that can be used is either a cleaved wedge or a spherical particle of known size and geometry, in which the thickness changes in a well-defined manner as a function of distance from the edge of the sample. This behavior is shown in Figure 7a for a line phase profile obtained from a 90° crystal wedge of GaAs that has been tilted to a weakly diffracting orientation. The procedure that is used to determine V_0 from such a profile is described in Section 2.2.

2.1. Calculation of Mean Inner Potential

The simplest approach that can be used to calculate V_0 is to treat the sample as though it contains an array of neutral, free atoms. In this "nonbinding" approximation, Eq. (31)

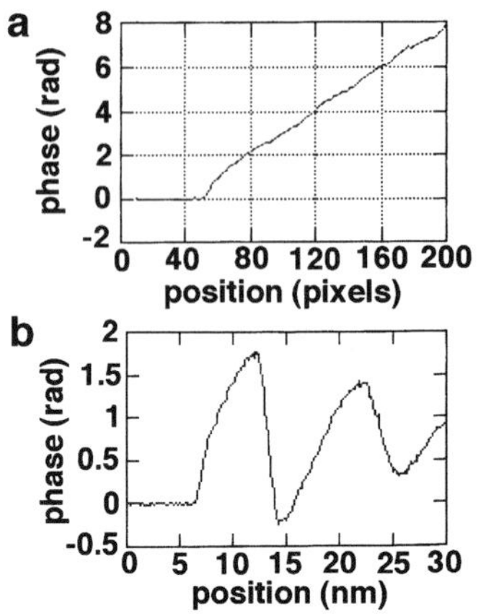

Figure 7. (a) Phase profile plotted as a function of distance into the sample, obtained from a 90° GaAs cleaved wedge tilted away from zone axes and systematic row orientations. The phase change increases approximately linearly with sample thickness, as predicted kinematically. (b) Equivalent phase profile obtained from a GaAs wedge tilted close to a [100] zone axis. Strong dynamical effects make the phase change nonlinear. Reprinted with permission from [90], M. Gajdardziska-Josifovska and A. Carim, in "Introduction to Electron Holography" (E. Völkl, L. F. Allard, and D. C. Joy, Eds.), p. 270. Kluwer Academic/ Plenum, New York, 1998. © 1998, Kluwer Academic/Plenum.

can be rewritten in the form

$$V_0 = \left(\frac{h^2}{2\pi m e \Omega}\right) \sum_{\Omega} f_{\mathrm{el}}(0) \tag{34}$$

where $f_{\mathrm{el}}(0)$ is the electron scattering factor at zero scattering angle for each atom in volume Ω [$f_{\mathrm{el}}(0)$ is approximately proportional to atomic number Z] and Ω is the summation volume (typically a single unit cell for a crystalline material) [90, 91, 101]. Values of $f_{el}(0)$ for neutral atoms to be included in Eq. (34) have been calculated from relativistic atomic wavefunctions, most recently by Doyle and Turner [92] and Rez et al. [93], in both tabulated and parameterized forms. Earlier calculations (e.g., [94, 95]) are occasionally also referred to. The electron scattering factors of Rez et al. [93], which were converted from X-ray scattering factors calculated using the multiconfiguration Dirac–Fock code of Grant et al. [96] and the Mott formula, are within 0.1% of the values calculated by Doyle and Turner [92].

2.1.1. *Ionicity and Bonding*

A limitation of the "nonbinding" approximation (i.e., the use of neutral atom scattering factors) is that calculated values of V_0 are overestimated because the redistribution of valence electrons due to bonding (which typically results in a contraction of the average electron density around each atom) is neglected. Radi [97] considered this nonbinding approximation as an upper limit to the true value of the mean inner potential. A lower limit for V_0 can be calculated by assuming that the crystal contains ions at the lattice points, with the remaining valence electrons distributed uniformly. The contribution of p free electrons to V_0 is given by the expression

$$V_0 = \left(\frac{3}{10}\right)\left(\frac{ep}{4\pi\varepsilon_0 r_0}\right) \tag{35}$$

where r_0 is chosen so that a sphere of radius r_0 has the same volume as the atom within the crystal [98]. A lower limit for V_0 is then provided by the sum of V_0 calculated for a p-times-ionized free atom (with the p outermost electrons of the neutral atom missing) and V_0 calculated for p free electrons using Eq. (35). Lower bound values calculated using this approach have been compiled by Gajdardziska-Josifovska and Carim [90] for both elemental and compound crystals and are reproduced in Tables 1 and 2, respectively. Table 1 also provides nonbinding values of V_0 calculated using the scattering factors of both Radi [93] and Doyle and Turner [92]. Differences between the two sets of values presumably arise because they are calculated using different scattering factors. The upper and lower limits for V_0 are plotted for a wide range of elements in their solid state in Figure 8a. The graph is plotted as a function of mean atomic number per unit volume because, according to Eq. (34), V_0 is expected to be approximately proportional to this parameter.

An empirical rule for evaluating the influence of bonding on V_0 has been proposed by Ross and Stobbs [91]. By analyzing the ratio between values of V_0 calculated using the scattering factors of Doyle and Turner [92] and Radi [93] (the average of the upper and lower limits given in Table 1), which is plotted as a function of mean atomic number per unit volume in Figure 8b, the following phenomenological dependence was found,

$$\frac{V_0(\mathrm{Radi})}{V_0(\mathrm{Doyle\text{–}Turner})} \approx 0.0325\frac{Z}{\Omega} + 0.6775 \tag{36}$$

where Z is atomic number and Z/Ω is given in units of 10^3 nm^3. Equation (36) provides a procedure for calculating an approximate value of V_0 that includes the effects of bonding by starting from Eq. (35), and subsequently using Eq. (36) to include the effects of charge redistribution. Figure 8b shows that the effect of bonding on V_0 decreases with increasing atomic number density.

2.2. Measurement of Mean Inner Potential

Experimental measurements of V_0 reported in the literature are sparse in number, and the scatter in the quoted values is wide. Values obtained using both refraction and phase-shifting techniques have been compiled by Gajdardziska-Josifovska and Carim [90] and are reproduced in Table 3. A similar table has been published by Spence [83]. Electron holography permits much greater accuracy to be obtained.

One approach to measure V_0 has used electron holograms of 20–40 nm diameter Si nanospheres coated in layers of amorphous SiO_2 [99]. The contribution to the phase images from the uniform SiO_2 overlayer can be separated from the contribution of the Si core if the particle is assumed to be exactly spherical, yielding measurements of the mean inner potential for both the surface layer and the core. Using

Table 1. Calculated values of mean inner potential, V_0, for elemental crystals using binding [Eqs. (34) and (35)] and nonbinding [Eq. (34) only] approximations.

Z	Element	Structure	Lattice parameter a/c (Å)	Ionicity p	V_0 (V) model: Radi [93] binding	V_0 (V) model: Radi [93] nonbinding	V_0 (V) model: D-T [92] nonbinding
6	C	diamond	3.56	4	15.93	19.75	21.30
14	Si	diamond	5.42	4	11.47	12.20	14.02
32	Ge	diamond	5.66	4	13.69	13.82	15.59
50	Sn	diamond	6.49	4	14.03	13.69	15.22
3	Li	bcc	3.46	1	2.88	6.16	7.60
11	Na	bcc	4.30	1	3.30	4.66	5.75
19	K	bcc	5.33	1	3.62	4.46	5.68
23	V	bcc	3.03	2	15.43	23.93	28.59
24	Cr	bcc	2.89	1	17.55	23.84	27.65
26	Fe	bcc	2.86	2	17.23	25.03	29.33
37	Rb	bcc	5.62	1	4.31	5.08	6.35
41	Nb	bcc	3.30	1	19.83	24.38	—
42	Mo	bcc	3.15	1	22.22	27.27	31.43
55	Cs	bcc	6.13	1	4.91	5.58	6.86
56	Ba	bcc	5.02	2	9.11	11.40	13.83
73	Ta	bcc	3.30	2	24.07	31.08	—
74	W	bcc	3.16	2	27.12	34.80	—
81	Tl	bcc	3.88	3	17.26	19.25	—
10	Ne	fcc	4.52	0	3.22	3.22	3.43
13	Al	fcc	4.04	3	10.00	14.91	17.10
18	Ar	fcc	5.43	0	4.96	4.96	5.48
20	Ca	fcc	5.58	2	6.75	8.87	10.93
25	Mn	fcc	3.70	2	16.38	24.08	28.38
26	Fe	fcc	3.63	2	16.95	24.51	28.69
27	Co	fcc	3.54	2	17.70	25.28	29.59
28	Ni	fcc	3.52	2	17.78	24.90	28.84
29	Cu	fcc	3.61	1	15.82	19.98	22.80
36	Kr	fcc	5.59	0	6.87	6.87	7.56
38	Sr	fcc	6.08	2	7.35	9.09	11.17
45	Rh	fcc	3.80	1	23.68	28.55	—
46	Pd	fcc	3.89	0	22.37	22.37	—
47	Ag	fcc	4.08	1	18.74	21.87	24.45
54	Xe	fcc	6.24	0	7.77	7.77	8.51
77	Ir	fcc	3.84	2	30.07	37.24	—
78	Pt	fcc	3.92	1	28.14	32.14	—
79	Au	fcc	4.08	1	25.02	28.20	29.81
82	Pb	fcc	4.95	4	18.25	19.41	19.89
90	Th	fcc	5.08	2	20.91	26.58	—
4	Be	hcp	2.28/3.59	2	7.41	16.32	18.08
12	Mg	hcp	3.20/5.20	2	6.53	9.24	10.81
21	Sc	hcp	3.31/5.27	2	9.99	14.69	17.82
22	Ti	hcp	2.95/4.68	2	12.92	19.77	23.83
27	Co	hcp	2.51/4.07	2	17.77	25.44	29.56
30	Zn	hcp	2.66/4.95	2	13.55	16.64	19.15
39	Y	hcp	3.65/5.73	2	11.49	15.22	—
40	Zr	hcp	3.23/5.14	2	15.49	21.01	—
43	Tc	hcp	2.73/4.39	2	23.38	31.54	—
44	Ru	hcp	2.71/4.28	1	24.30	29.55	—
48	Cd	hcp	2.98/5.62	2	15.92	18.29	20.45
57	La	hcp	3.75/6.06	2	14.60	19.51	—
58	Ce	hcp	3.65/5.96	2	14.32	20.22	—
59	Pr	hcp	3.99/5.91	2	14.20	19.93	—
60	Nd	hcp	3.65/5.89	2	14.27	19.90	—
64	Gd	hcp	3.63/5.78	2	14.81	19.21	—
65	Tb	hcp	3.60/5.70	2	14.09	19.43	—
66	Dy	hcp	3.59/5.65	2	14.08	19.37	—
67	Ho	hcp	3.58/5.62	2	14.10	19.40	—

continued

Table 1. Continued

Z	Element	Structure	Lattice parameter a/c (Å)	Ionicity p	V_0 (V) model: Radi [93] binding	V_0 (V) model: Radi [93] nonbinding	V_0 (V) model: D-T [92] nonbinding
68	Er	hcp	3.56/5.59	2	14.12	19.38	—
69	Tm	hcp	3.54/5.56	2	14.17	19.45	—
72	Hf	hcp	3.20/5.05	2	19.75	25.45	—
75	Re	hcp	2.76/4.46	2	29.08	36.95	—
76	Os	hcp	2.73/4.32	2	30.42	38.26	—
81	Tl	hcp	3.46/5.53	3	17.59	19.73	—

Source: Reprinted with permission from [90], M. Gajdardziska-Josifovska and A. Carim, in "Introduction to Electron Holography" (E. Völkl, L. F. Allard, and D. C. Joy, Eds.), p. 274. Kluwer Academic/Plenum, New York, 1998. © 1998, Kluwer Academic/Plenum.
Note the difference between the nonbinding values for V_0 resulting from the use of different neutral atom scattering factors.

this approach, the mean inner potential of crystalline Si was found to be 12.1 ± 1.3 V, that of amorphous Si 11.9 ± 0.9 V, and that of amorphous SiO_2 10.1 ± 0.6 V. Similar measurements obtained from spherical latex particles embedded in vitrified ice have provided values of V_0 of 8.5 ± 0.7 and 3.5 ± 1.2 V for the two materials, respectively [100].

In principle, V_0 can be determined from phase profiles such as that shown in Figure 7a by measuring the gradient of the phase $d\phi/dx$ and then making use of the relation

$$V_0 = \left(\frac{1}{C_E}\right)\left(\frac{d\phi/dx}{dt/dx}\right) \tag{37}$$

This approach, which is independent of any contributions to the phase shift from amorphous overlayers on the sample surfaces (for which $dt/dx = 0$ if these layers are of uniform thickness), has been used successfully to measured the mean inner potentials of cleaved wedges and cubes of Si, MgO, GaAs, PbS [101, 102], and Ge [103]. The values of V_0 determined for MgO, GaAs, PbS, and Ge using this approach are 13.0 ± 0.1, 14.5 ± 0.2, 17.2 ± 0.1, and 14.3 ± 0.2 V, respectively [101, 103]. In these studies, the value of an independent measure of sample thickness using a technique such as weak-beam dark-field imaging cannot be overemphasized. This need arises because semiconductor materials such as Si or Ge that appear to have cleaved on $\langle 110 \rangle$ planes may have different thickness profiles at their very edges [103]. Such variations in wedge angle most likely account for the anomalous value of the mean inner potential of Si reported in the literature [101].

In a similar study, wedge-shaped Si samples with stacked Si oxide layers on their surfaces have been used to measure the mean inner potentials of the oxide layers [104]. Values of 10.51 ± 0.35, 10.65 ± 0.32, 11.19 ± 0.38, and 10.58 ± 0.69 V were measured for undensified thermal oxide, densified thermal oxide, undensified deposited oxide, and densified deposited oxide, respectively. Chemical etch rate differences between the oxide layers were also used to provide thickness differences between different oxide layers, which could be recorded using electron holography. The widths of silicon oxide–nitride–oxide layers have also been measured using electron holography [105].

Table 2. Calculated values of mean inner potential, V_0, for compound crystals that have the rocksalt crystal structure.

Compound	Lattice parameter a (Å)	Ionicity p	V_0 (V) model: Radi [93] binding	V_0 (V) model: D-T [92] nonbinding
LiF	4.02	+1, −1	13.61	15.00
LiCl	5.14	+1, −1	10.82	11.48
LiBr	5.49	+1, −1	11.40	11.97
LiI	6.00	+1, −1	12.12	12.58
NaF	4.61	+1, −1	11.34	12.86
NaCl	5.63	+1, −1	9.54	10.34
NaBr	5.95	+1, −1	10.06	10.76
NaI	6.46	+1, −1	10.56	11.14
KF	5.33	+1, −1	11.38	13.64
KCl	6.28	+1, −1	9.33	10.70
KBr	6.58	+1, −1	9.57	10.79
KI	7.05	+1, −1	9.86	10.87
RbF	5.63	+1, −1	12.19	14.57
RbCl	6.56	+1, −1	9.78	11.29
RbBr	6.86	+1, −1	9.85	11.18
RbI	7.32	+1, −1	9.95	11.08
AgCl	5.55	+1, −1	14.32	15.18
AgBr	5.76	+1, −1	15.01	15.76
CsF	6.01	+1, −1	13.55	16.15
MgO[a]	4.21	+2, −2	17.94	18.44
PbS	5.94	+2, −2	16.88	16.26
PbSe	6.14	+2, −2	17.11	16.38
PbTe	6.44	+2, −2	17.74	—

Source: Reprinted with permission from [90], M. Gajdardziska-Josifovska and A. Carim, in "Introduction to Electron Holography" (E. Völkl, L. F. Allard, and D. C. Joy, Eds.), p. 275. Kluwer Academic/Plenum, New York, 1998. © 1998, Kluwer Academic/Plenum.

Note: Results calculated using the binding approximation of Radi [97] are shown in column four, while those calculated using the nonbinding approximation of Doyle and Turner [92] are shown in the last column.

[a] Radi [97] gives a nonbinding value of 16.23 V only for MgO.

2.2.1. Dynamical Effects

In practice, dynamical contributions to the phase should be evaluated and taken into account when determining V_0 from a crystalline sample such as a cleaved semiconductor wedge, even when the phase profile is as linear as that shown in Figure 7a [106]. These corrections can be calculated using either multislice or Bloch wave algorithms and are still required even when an electron hologram has been acquired at a supposedly weakly diffracting orientation for greatest

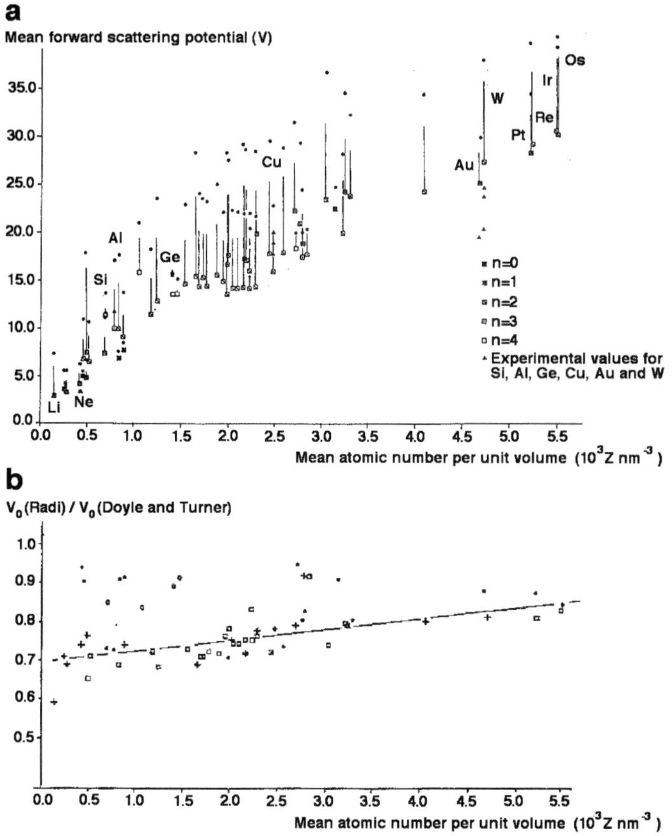

Figure 8. The effect of bonding on calculated values of mean inner potential V_0. (a) V_0 plotted as a function of mean atomic number per unit volume. All elements from $Z = 3$ to 90 are shown in their solid state. The small filled circles are "free-atom" values calculated using the scattering factors of Doyle and Turner [92], while the squares are values calculated according to Radi [97] using an approximation for bonding in which n electrons per atom (see key) are assumed to be free. Each square is joined by a line to the value calculated by Radi [97] assuming free-atom wavefunctions, which should be comparable to the Doyle and Turner [92] values. Experimentally measured values of V_0 for Si, Al, Ge, Cu, Au, and W are also shown. (b) Ratio of the values of V_0 calculated according to Radi [97] [the average of the two values given in (a)] to those calculated according to Doyle and Turner [92]. There is a trend for this ratio to increase with atomic number density. The structure of each crystal is given by the following symbols. Filled circle: face-centered cubic (fcc); open square: hexagonal close-packed (hcp); plus sign: body-centered cubic; and open circle: diamond structure. Reprinted with permission from [91], F. M. Ross and W. M. Stobbs, *Philos. Mag. A* 63, 37 (1991). © 1991, Taylor and Francis Ltd.

possible accuracy. It is essential to record the precise orientation of a crystalline sample when the hologram is acquired, using a technique such as convergent beam electron diffraction [101, 103]. For measurements obtained from crystalline Ge at several weakly diffracting orientations, dynamical corrections to V_0 were found to be typically between 0.1 and 0.2 V [103].

The fact that Eq. (34) is no longer valid when the sample is tilted to a strongly diffracting orientation is shown in Figure 7b for a GaAs 90° cleaved wedge sample tilted to a ⟨110⟩ zone axis. The measured phase shift now varies nonlinearly with sample thickness and is extremely sensitive to small changes in sample tilt from the exact zone axis

Table 3. Experimental values of mean inner potential, V_0.

Material	V_0 (V)	Method	Ref.
Be	7.8 ± 0.4	interferometry of films	[429]
C	7.8 ± 0.6	interferometry of films	[430]
Al	13.0 ± 0.4	interferometry of films	[430]
	12.4 ± 1	interferometry of films	[431]
	11.9 ± 0.7	interferometry of films	[432]
Si	9.26 ± 0.05	holography of wedges	[101]
	12.1 ± 1.3	holography of spheres	[99]
	(11.5)	(RHEED)	[433]
	11.9 ± 0.9	holography of spheres	[99]
Cu	23.5 ± 0.6	interferometry of films	[430]
	20.1 ± 1.0	interferometry of films	[432]
Ge	15.6 ± 0.8	interferometry of films	[432]
Ag	20.7 ± 2	interferometry of films	[431]
	17.0–21.8	interferometry of films	[430]
(W)	(23.4)	(RHEED)	[433]
Au	21.1 ± 2	interferometry of films	[431]
	22.1–27.0	interferometry of films	[430]
MgO	13.01 ± 0.08	holography of cleaved wedge	[101]
	13.2	interferometry of cleaved wedge	[434]
	13.5	interferometry of smoke wedge	[124]
	13.9	holography of smoke wedge	[124]
	(13.7)	(electron diffraction)	[435]
	(16)	(electron diffraction)	[436]
	(15.4)	(electron diffraction)	[437]
	(13.4)	(electron diffraction)	[438]
	(12.3)	(electron diffraction)	[439]
	(14.4 ± 0.8)	(STEM with sector detector)	[440]
SiO_2 amorph.	10.1 ± 0.6	holography of SiO_2 films on Si spheres	[99]
ZnS	10.2 ± 1	interferometry of thin films	[431]
GaAs	14.53 ± 0.17	holography of cleaved wedge	[101]
	(13.2)	(RHEED)	[441]
PbS	17.19 ± 0.12	holography of cleaved wedge	[101]
Chrysotile	11.5	interferometry of fiber	[442]
Kohlenstoff	24 ± 5 V	interferometry of thin films	[443]
(GaP)	(12.2)	(RHEED)	[441]

Source: Reprinted with permission from [90], M. Gajdardziska-Josifovska and A. Carim, in "Introduction to Electron Holography" (E. Völkl, L. F. Allard, and D. C. Joy, Eds.), p. 276. Kluwer Academic/Plenum, New York, 1998. © 1998, Kluwer Academic/Plenum.

Note: Values without brackets were obtained using holographic methods, either by direct interferometric measurements from a hologram, or by true holographic measurements from a reconstructed phase image. Values in brackets were obtained using other techniques, predominantly based on measurements of refraction effects in electron diffraction patterns.

orientation. If a strong two-beam condition is set up, then it is possible to measure both V_0 and V_g using electron holography [38].

Additional experimental factors that may affect measurements of V_0 include the presence of electrical dipoles at sample surfaces, which are sensitive to the chemical and physical state of the surface and its crystallographic orientation [107, 108] and unwanted charging of the sample surface, which has been observed directly using electron holography for a wide range of materials that include semiconductors and sapphire [109].

2.3. Measurement of Sample Thickness and Morphology

If the mean inner potential of a sample is constant, and if dynamical contributions to the contrast are small, then Eq. (33) indicates that a phase image can be used to provide a map of the projected thickness of the sample. Electron holography can be used to study surface topography and the shapes of nanoparticles.

The local sample thickness $t(x)$ can be inferred from a holographic amplitude image in units of λ_{in}, the mean free path for inelastic scattering, using the equation

$$\frac{t(x)}{\lambda_{\text{in}}} = -2\ln\left(\frac{A_i(x)}{A_r(x)}\right) \tag{38}$$

where $A_i(x)$ and $A_r(x)$ are the measured amplitudes of the sample and reference holograms, respectively [110]. (Although the information in the sideband of an electron hologram is, in principle, perfectly energy filtered, the degree to which inelastically scattered electrons interfere with each other and with a vacuum reference wave has recently been debated [111–114].) One disadvantage of using Eq. (38) to measure the sample thickness is that the recorded amplitude image is always more noisy than the corresponding phase image. In addition, the value of λ_{in} may be unknown, and this parameter is different for each material in the sample and depends on the objective aperture size used. Values of λ_{in} can, on occasion, be measured from samples of known geometry. For example silicon and polystyrene nanospheres were analyzed to give values of λ_{in} for Si of 88.6 ± 6.9 and 53.8 ± 5.5 nm at 200 and 120 kV, respectively, and corresponding values for polystyrene of 113.0 ± 5.9 and 78.1 ± 3.4 nm [115].

Whether the sample thickness profile is measured from a phase image or an amplitude image, interpretation may be complicated if the sample has nonuniform composition. Most TEM sample preparation methods yield nonuniform sample thickness profiles. For example, defects and interfaces between materials often thin at a different rate to the material that surrounds them. If necessary, the thickness dependence of the phase and amplitude images can be removed by combining Eqs. (33) and (38). This approach results in the expression

$$\frac{\phi(x)}{-2C_E\ln\left(\frac{A_i(x)}{A_r(x)}\right)} = V_0(x)\lambda_{\text{in}}(x) \tag{39}$$

Equation (39) can be used to generate an image that has contrast equal to the product of the local values of the mean inner potential and the inelastic mean free path [110]. Both parameters depend only on the local composition of the sample. This approach has been used to interpret images obtained from samples with varying compositions but unknown thickness variations, including multilayers of amorphous Si and polycrystalline Mo [116] and interfaces between amorphous Si and SiN_x [117].

Alternatively, if the composition of a sample is required, an approach based on electron holography and an (002) chemically sensitive reflection in a sphalerite-type material has been proposed [118, 119]. The phase of the (000) beam from the sideband of a high-resolution electron hologram acquired slightly away from a axis orientation, and the amplitude of the chemically sensitive (002) beam, are used together to evaluate both the local composition and the local thickness of the sample.

2.3.1. Fine Particles

The morphology of a wide range of fine particles has been studied by electron holography by making use of Eq. (33). The projected thickness of a polyhedral particle provides information about the locations and sizes of the particle's edges and faces without the need to tilt the sample to multiple zone axes, although complications may arise because of the unknown relative inclinations of the entrance and exit surfaces. A significant advantage of examining small particles (in particular those below 50 nm in diameter) is that dynamical contributions to the phase are more likely to be small. When examining magnetic particles in low-field conditions, the mean inner potential and magnetic contributions to the phase must still be separated.

Figure 9a shows a high-resolution lattice image of a 15 nm cuboctahedral ZrO_2 crystal that has {111} and {001} faces [90, 120]. Schematic diagrams of the expected shape and [110] projection of the crystal are shown in Figure 9b. A reconstructed electron holographic phase image of the particle, and line profiles obtained across different sections of the phase image, are shown in Figure 9c–g and are consistent with the expected geometry. A void within the particle is apparent in one of the profiles (Fig. 9g). Similar surface structures and internal voids have been reported for 5–15 nm Pd particles supported on amorphous silica spheres [121–124].

Electron holography has been applied to the study of cubic single crystals of MgO, which have {100} faces and are produced by burning magnesium wire or ribbon and collecting the resulting smoke (e.g., [120, 125–127]). Further applications of holographic phase imaging have included the formation of 1.3 nm thickness contours on a 200 nm Be particle by making use of 32-times phase amplification [127] and the examination of carbon nanotubes [128, 129], voids in Si [130, 131], SiGe quantum dots [132], the tobacco mosaic virus [133], bacterial flagellae [134], strands of DNA [135], and both organic polymers and mesoporous silica [136].

2.3.2. Surface Steps

A subtle use of phase imaging with electron holography involves the examination of atomic-height steps on clean surfaces. Figure 10 shows a 24-times phase-amplified interference micrograph of a molybdenite (MoS_2) thin film [137]. In this image, thickness changes due to steps of height five, three, and one atomic layers are observed at positions A, B, and C, respectively. The fringe shift at C corresponds to a thickness change of 0.62 nm, which is the height of a single growth step. Similar observations have been performed by using reflection off-axis electron holography to examine Pt (111) surfaces [138, 139] and strain fields around screw

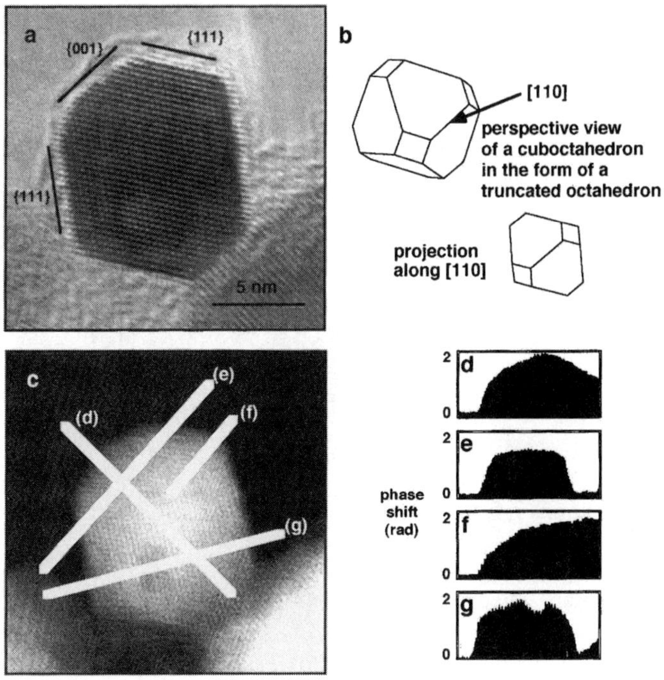

Figure 9. (a) High-resolution lattice image of an apparently cuboctahedral ZrO_2 particle. (b) Sketch of the corresponding cuboctahedral shape and its expected [110] projection. (c) Phase image reconstructed from a hologram of the particle shown in (a). Line profiles across different sections of the phase image are shown in (d), (e), (f), and (g). Each profile starts at the label end of the corresponding line in (c). All are consistent with the postulated geometry, with the exception of one area where there is a void in the particle, as seen in (g). Reprinted with permission from [90], M. Gajdardziska-Josifovska and A. Carim, in "Introduction to Electron Holography" (E. Völkl, L. F. Allard, and D. C. Joy, Eds.), p. 292. Kluwer Academic/Plenum, New York, 1998. © 1998, Kluwer Academic/Plenum.

dislocations on GaAs {110} surfaces [140]. This approach involves formation of the reference wave by the reflection of electrons from an atomically flat part of the surface. The object wave is then formed from parts of the surface that contain the features of interest, an objective aperture is used to select the specularly reflected beam, and a biprism is used to superimpose the reference and object waves. A related experiment has involved the examination of the surface potential of MgO in cross-section at atomic resolution using electron holography [141].

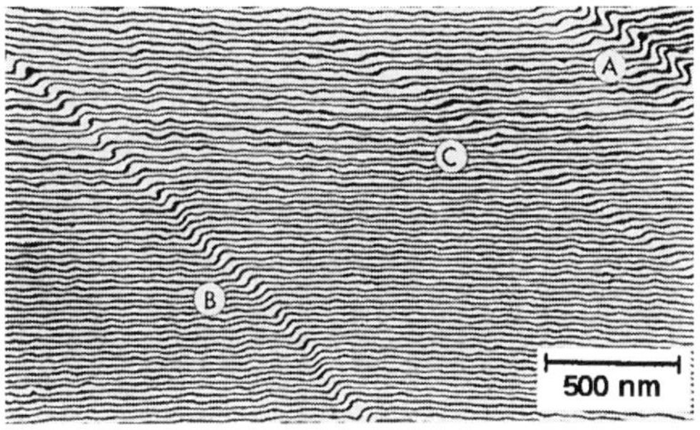

Figure 10. A 24-times-phase-amplified interference micrograph of a molybdenite (MoS_2) film. Thickness changes due to steps of height five, three, and one atomic layers are observed at positions A, B, and C, respectively. Reprinted with permission from [137], A. Tonomura et al., *Phys. Rev. Lett.* 54, 60 (1985). © 1985, American Physical Society.

2.3.3. Three-Dimensional Information

Electron tomography has been used widely in the biological sciences to recover the three-dimensional structures of bacteria, viruses, and other macromolecules from tilt series of two-dimensional images [142, 143]. The technique is starting to be applied in the physical sciences, primarily using annular dark field imaging [144, 145]. By acquiring successive holograms over a range of tilts, electron tomography can be combined with electron holography to allow sample morphology to be determined directly in three dimensions based on tilt series of phase images. Figure 11a shows an individual electron hologram of three latex particles taken from a tilt series of holograms obtained in 5° steps between ±60°. Two views of the reconstructed three-dimensional shapes of the particles obtained after applying tomographic techniques to the tilt series are shown in Figure 11b and c [146]. Despite the successful demonstration of the combination of electron tomography with electron holography shown in Figure 11, the techniques are still rarely applied together because the acquisition, processing, and alignment of a large number of holograms is necessarily time-consuming. Considerable patience and care are required since maximum sample tilt angles of at least ±60°, small tilt steps (ideally every 2° or less), and accurate alignment of the phase images are all essential to avoid reconstruction artifacts.

3. MAGNETIC FIELDS IN NANOSTRUCTURED MATERIALS

The most successful and productive applications of electron holography have involved the characterization of magnetic fields both within and surrounding materials. This work has followed primarily from the pioneering work carried out in the 1980s by the research group of Tonomura, which has been the subject of several monographs and review articles (e.g., [7, 8, 35, 147–151]). Many alternative techniques exist for the characterization of magnetic microstructure in the electron microscope [152]. These techniques include the Fresnel and Foucault modes of Lorentz microscopy

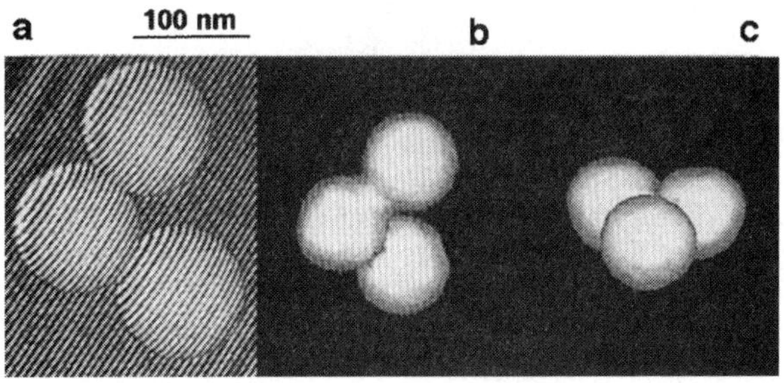

Figure 11. (a) Electron hologram of three latex particles recorded at 200 kV using a Hitachi HF-2000 TEM and a biprism voltage of 10 V. (b) and (c) Two views of the reconstructed three-dimensional shapes of the particles obtained after applying tomographic techniques to a tilt series of phase images of the three latex particles obtained in 5° steps between ±60°. Reprinted with permission from [146], G. Lai et al., *Appl. Opt.* 33, 829 (1994). © 1994, Optical Society of America.

[153–161], coherent Foucault imaging [162], and differential phase contrast imaging in the scanning transmission electron microscope (STEM) [163–167]. Each of these techniques is best applied to a particular type of magnetic material. More general literature on the subject of electron microscopy of magnetic materials should be consulted for further detailed information (e.g., [168, 169]).

3.1. Results Obtained Using Optical Reconstruction

Results for magnetic materials obtained using optical reconstruction of off-axis electron holograms acquired on photographic film are reviewed in this section. Applications using digital reconstruction are discussed separately below. This distinction between optical and digital reconstruction is made primarily to highlight the different ways in which holograms can be analyzed and interpreted using the two approaches and to emphasize the different types of results obtained.

3.1.1. Fringing Fields in Vacuum

Figure 12 illustrates two examples where electron holography has been used to characterize magnetic fringing fields that extend into vacuum. In such experiments, no complications arise from contributions to the measured phase shift from variations in sample composition or thickness. However, care is required when interpreting these types of results because magnetic leakage fields outside TEM samples are three-dimensional. In contrast, electron holography only records a projection in the electron beam direction of the components of the magnetic induction B that lie in the plane of the sample. Thus, comparisons with micromagnetic simulations may be necessary before quantitative statements can be made about the results.

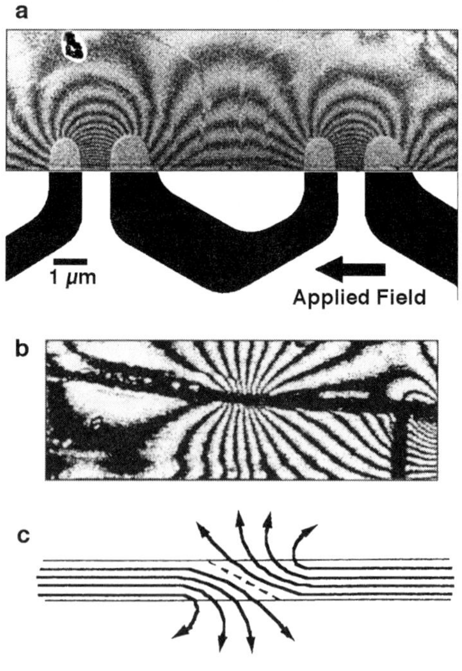

Figure 12. (a) Interference micrograph obtained from a hologram of microscopic horseshoe magnets (chevron patterns) from the propagation circuit of a bubble memory made from permalloy (Ni 80% Fe 20%), supported on a thin C film. The period of the chevron is 8 μm and its thickness is 350 nm. A 20 Oe magnetic field was applied to the sample during hologram acquisition in the direction shown in the figure. Reprinted with permission from [170], T. Matsuda et al., *J. Appl. Phys.* 53, 5444 (1982). © 1982, American Institute of Physics. (b) Interference micrograph and (c) schematic illustration of the flux distribution associated with a 0.3 μm Fe wire that contains a single domain wall. Reprinted with permission from [171], Y. Otani et al., *Mater. Res. Soc. Symp. Proc.* 475, 215 (1997). © 1997, Materials Research Society.

The two-times phase-amplified interference micrograph shown in Figure 12a was obtained from a hologram of a line of microscopic horseshoe magnets, which were part of the propagation circuit of a bubble memory [170]. Each magnet is 350 nm in thickness, made from permalloy ($Ni_{0.8}Fe_{0.2}$), and supported on a thin C film. The period of the chevron pattern is 8 μm. A 20 Oe magnetic field was applied to the sample during hologram acquisition in the direction shown. The contours, which are separated by a constant magnetic flux of $h/2e$, show the magnetic lines of force that leave one end of each magnet, being directed both to its neighbor and to a smaller extent in the opposite direction. The magnetic field strength in the gap between adjacent magnets is estimated from the contours to be in excess of 1000 Oe.

A similar interference micrograph, acquired from a 0.3 μm Fe wire that contains a single domain wall at its center [171], is shown in Figure 12b. A strong magnetic leakage field, which originates at the position of the domain wall, is visible outside the wire. A schematic diagram that illustrates the origin of the flux leakage associated with the domain wall is shown in Figure 12c.

Results equivalent to those shown in Figure 12 have been obtained from single Fe_2O_3 [172] and barium ferrite [173] particles. In the latter case, each particle could only support a single magnetic domain. As a result, the magnetic lines of force were no longer entirely contained within the particles. The leakage field was then clearly identified using electron holography.

3.1.2. Recorded Magnetic Tape

A practical application of electron holography, which required the characterization of magnetic microstructure within the sample as well as the leakage fields outside, involved the examination of magnetic fields in high-density magnetic recording media. Figure 13a shows a schematic diagram of the method used to record bits on a magnetic film. Figure 13b shows results obtained from an in-plane-recorded high-coercivity evaporated Co film that had a thickness of 45 nm, a coercive field of 340 Oe, and a bit length of 5 μm [174]. The sample was prepared by evaporating the Co film obliquely onto a glass substrate that had been coated with photoresist. The Co film was partly oxidized to control its coercive field and remanent magnetization, and the recording was performed using a ring-type magnetic head with a track width of 300 μm and gap lengths of 200 nm and 1 μm. The interference micrograph in Figure 13b shows the projected magnetic lines of force both inside and outside the film. The recorded track runs parallel to the film edge, the film is observed in the upper half of Figure 13b, and the lower half shows empty space. The magnetization

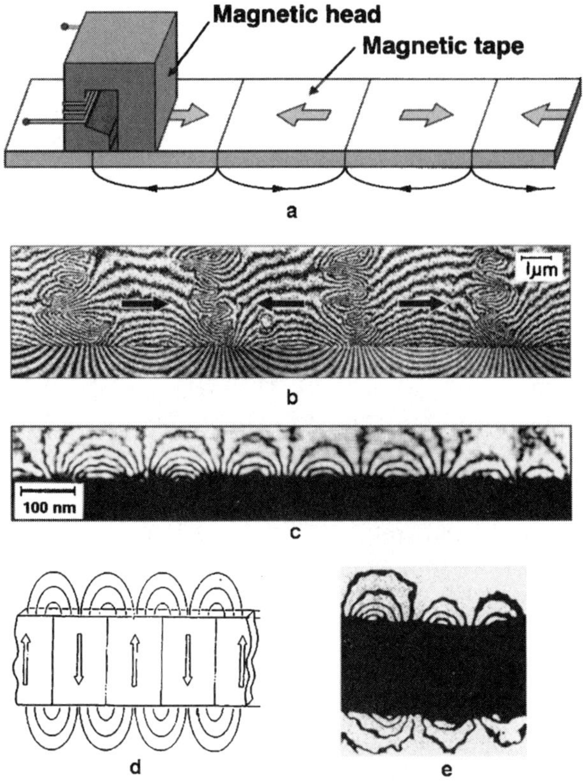

Figure 13. (a) Schematic diagram showing the method used to record bits on a magnetic film. (b) Interference micrograph formed from a holographic phase image of a 45-nm-thick recorded Co film with a bit length of 5 μm. The magnetic stray field in vacuum at the side of the thin film is visible at the bottom of the image. (c) Interference micrograph of a 30-nm-thick recorded Co film with a bit length of 0.15 μm, showing periodic stray fields in vacuum outside the edge of the sample. Reprinted with permission from [174], N. Osakabe et al., *Appl. Phys. Lett.* 42, 746 (1983). © 1983, American Institute of Physics. (d) Schematic diagram and (e) interference micrograph of a perpendicularly recorded 80-nm-thick Co–Cr film with a bit length of 0.25 μm. Reprinted with permission from [176], K. Yoshida et al., *IEEE Trans. Magn.* 23, 2073 (1987). © 1987, IEEE.

directions within the film are indicated by arrows. The maximum packing density for longitudinal recording in such films is limited by the widths of the transition regions between two oppositely magnetized regions. Oppositely magnetized regions in adjacent domains can be seen to collide with each other head-on to produce vortexlike patterns. The magnetic field then meanders to the edge of the film edge and leaks out into the vacuum. (Although stray fields above the film are read by the magnetic head, the contour lines visible at the bottom of Fig. 13b are at the side of the film.)

Figure 13c shows an interference micrograph obtained from an electron hologram of a 30-nm-thick Co film that had been recorded with a bit length of 0.15 μm [174]. Periodic stray fields are again visible in the vacuum outside the sample edge. The narrowest transition region that has been characterized using this approach is 100 nm. Corresponding results have been obtained from recorded CoCr films [175]. Figure 13d and e shows a schematic diagram and an interference micrograph, respectively, of the magnetic field emerging from both the top and the bottom surfaces of a perpendicularly recorded 80-nm-thick Co–Cr film that has a bit length of 0.25 μm [176].

3.1.3. Fine Particles

Figure 14 shows representative results obtained by applying off-axis electron holography and optical reconstruction to particles of Co and permalloy. The Co fine particles shown in Figure 14a–d were prepared by gas evaporation of Co in an inert gas atmosphere [177]. Figure 14a shows a bright-field image of a 55-nm-thick plate-shaped Co particle, which has {111} surfaces and ⟨110⟩ edges and is uniform in thickness. In this image, only the outline of the particle can be

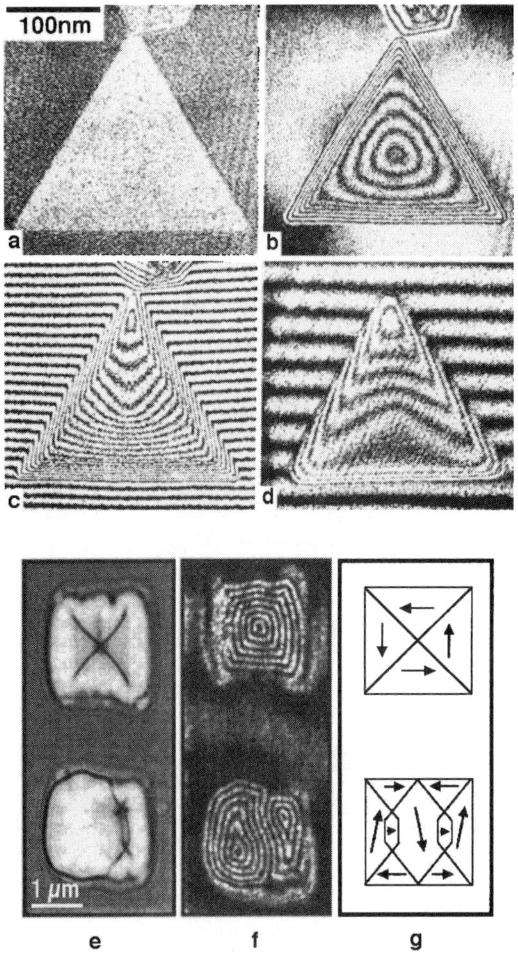

Figure 14. (a) Bright-field image of a 55-nm-thick plate-shaped Co particle with {111} surfaces and ⟨110⟩ edges. (b) Two-times-phase-amplified interference micrograph obtained from a holographic phase image of the particle obtained at 80 kV. (c) Interferogram formed from the recorded holographic phase image. Reprinted with permission from [177], A. Tonomura et al., *Phys. Rev. Lett.* 44, 1430 (1980). © 1980, American Physical Society. (d) Interferogram obtained from another particle that has the opposite rotation direction of magnetization. Reprinted with permission from [35], A. Tonomura, "Electron Holography," p. 92. Springer-Verlag, Berlin, 1999. © 1999, Springer-Verlag. (e) Lorentz micrograph, (f) interference micrograph, and (g) inferred magnetization distribution of four and seven domain remanent magnetic states of 2 μm × 2 μm × 40 nm permalloy particles fabricated onto a C film. Reprinted with permission from [178], K. Runge et al., *J. Appl. Phys.* 79, 5075 (1996). © 1996, American Institute of Physics.

seen. In contrast, in the holographic interference micrograph that has been phase-amplified two times (Fig. 14b), contour lines are visible running parallel to the three edges of the particle. The outer, more closely spaced contours arise from sample thickness variations and can be used to show that the thickness of the particle increases rapidly at its edge to 55 nm. The inner, more widely spaced contours are magnetic lines of force, which are clearly visible in the region of the particle that is uniform in thickness. A flux of $(h/2e) = 2 \times 10^{-15}$ Wb flows between adjacent contours in Figure 14b. Although the particle consists of three distinct magnetic domains, the magnetic lines of force are almost circular since the particle radius is so small. Stray magnetic fields outside the particle are estimated to be small from the contour map. The widths of the three domain walls in the particle are measured to be 40 nm.

The contours in Figure 14b cannot be used to establish whether the magnetization in the particle rotates clockwise or counterclockwise. Instead, the rotation direction can be determined from an interferogram, which is obtained by tilting the reference beam during optical reconstruction (Fig. 14c). In Figure 14c, the interference fringes are displaced downward at the particle edges and downward further inside the particle. As an electron beam travels faster in the crystal than in vacuum and consequently has a shorter wavelength, the wavefront of the transmitted beam through the particle must be retarded. Depending on whether the magnetization is clockwise or counterclockwise, the beam is either further retarded or advanced. Analysis of Figure 14c shows that the magnetization direction in the particle is counterclockwise. If the particle was inverted inside the electron microscope, then the rotation direction would be reversed and the wavefront would be retarded rather than advanced. This behavior can be observed directly in Figure 14d, which shows an interferogram recorded from a triangular Co particle that has the opposite rotation direction of magnetization to the particle shown in Figure 14a–c [35]. Equivalent results from thinner triangular Co particles of similar size, which show simpler magnetization configurations, have also been obtained [170].

Figure 14e–g shows results obtained by applying optical holographic reconstruction to an array of 2 μm $\times$ 2 μm $\times$ 40 nm permalloy particles that have been fabricated directly onto an electron-transparent C film using electron beam lithography [178]. The particles are arranged in a square array of side 4 μm and are largely magnetically isolated from each other. A Lorentz micrograph (i.e., a Fresnel defocus image) of two of the particles, which is shown in Figure 14e, reveals the positions of domain walls as bright or dark lines for a four-closure domain and a seven-closure domain. A corresponding interference micrograph is shown in Figure 14f, and a schematic diagram showing the magnetic domain structure is shown in Figure 14g. The spacing of the contours in Figure 14f has been used to confirm that the in-plane magnetization of the particles is consistent with that of bulk permalloy. Similar results have been obtained as the domain structure evolves when a range of both in-plane and out-of-plane external fields is applied to the particles [178].

The interference micrographs presented in Figure 14 from both Co and permalloy particles show strong contrast at the particle edges, where the sample thickness, and therefore the measured phase, changes rapidly. Such compositional contributions to the contrast can be removed to leave just the magnetic signal of primary interest. Some of these approaches, which are straightforward to apply if holograms are recorded and analyzed digitally, are described in section 3.2.

3.1.4. Verification of the Aharonov–Bohm Effect

One of the most elegant scientific contributions of electron holography is the experimental confirmation of the Aharonov–Bohm effect [179]. The Aharonov–Bohm effect states that when an electron wave from a point source passes on either side of an infinitely long solenoid then a relative phase shift should occur between the two parts of the wave. This phase shift should be proportional to the magnetic flux flowing inside the coil and should occur as a direct result of the presence of vector potentials. The significance of the Aharonov–Bohm effect is that it provides the only observable phenomenon that confirms the physical reality of gauge theory. It implies that vector potentials are more fundamental physical quantities in quantum mechanics than electric and magnetic fields, and that electrons can be influenced by electromagnetic fields "without touching them."

A series of increasingly complicated and difficult electron holography experiments was performed in order to confirm the Aharonov–Bohm effect unambiguously. Initial measurements were obtained from iron whiskers [180–183]. Although the measured phase shifts across the whiskers agreed with predictions, the results received a skeptical reception since it was believed that magnetic stray fields around the whiskers, rather than vector potentials, were influencing the electron beam [184]. Subsequent experiments were performed on tiny toroidal permalloy (80% Ni, 20% Fe) magnets of various sizes that exhibited negligible magnetic flux leakage [185, 186]. These magnets were fabricated lithographically by vacuum-evaporation of the permalloy onto glass substrates covered with NaCl. They were then floated off in water onto 10-nm-thick C films. Electron holography observations involved the interference of electrons that passed through toroids of different sizes with electrons that passed outside them. Figure 15a shows an interferogram obtained from an electron hologram of a toroidal magnet prepared from a 40-nm-thick film of permalloy. The arms of the toroid are approximately 640 nm in width. [185]. The phase differences measured from results such as Figure 15a were proportional to toroid width and agreed with predictions. Critics still claimed, however, that the observed phase shifts were not due to the Aharonov–Bohm effect because the electron wave could penetrate the magnetic material and therefore interact with it directly [187].

In response, 20-nm-thick permalloy toroidal magnets were entirely covered with 300-nm-thick layers of superconducting Nb. The Nb was deposited both to prevent electrons from penetrating the magnetic material and to confine the magnetic flux by exploiting the Meissner effect. A scanning electron micrograph of a toroidal permalloy magnet covered with superconducting Nb is shown in Figure 15b, and a

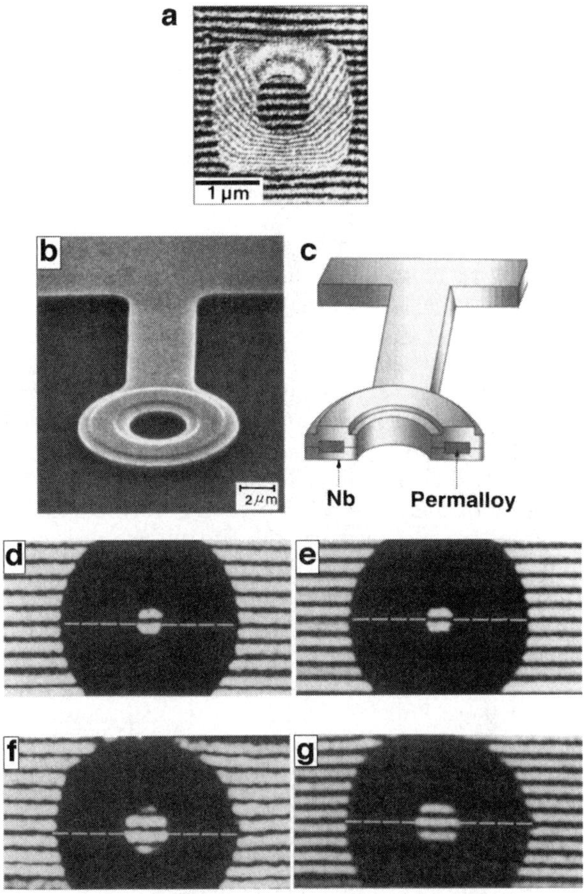

Figure 15. (a) Interferogram obtained from an electron hologram of a toroidal magnet prepared from a 40-nm-thick film of permalloy, obtained at 100 kV. The arms of the toroid are approximately 640 nm in width. Reprinted with permission from [185], A. Tonomura et al., *Phys. Rev. Lett.* 48, 1443 (1982). © 1982, American Physical Society. (b) Scanning electron micrograph of a toroidal permalloy magnet covered with superconducting Nb. (c) Schematic diagram of the structure of the sample. (d)–(g) Interferograms showing the existence of the Aharonov–Bohm effect. (d) and (e) are for a toroid in which the magnetic flux is quantized below T_c in units of $n(h/2e)$, where n is even. For (f) and (g), n is odd. For (d) and (f), $T = 15$ K, whereas for (e) and (g) $T = 5$ K (phase amplification × 1). Reprinted with permission from [188], N. Osakabe et al., *Phys. Rev. A* 34, 815 (1986). © 1986, American Physical Society.

schematic diagram of the structure is shown in Figure 15c. It was estimated that a negligible proportion ($\sim 10^{-6}$) of the electron wave should see the magnetic field coherently in such a toroid. Holograms were obtained using a liquid–helium temperature specimen stage [188, 189]. Each toroid was attached to a Nb plate to cool it effectively and covered with a thin layer of Au or Cu to discount the effects of contact potential differences in the sample. The observations, which are summarized in Figure 15d–g, showed that the measured phase difference became quantized to a value of 0 or π when the temperature was reduced below the superconducting critical temperature for Nb (5 K), that is, when a supercurrent was induced to circulate in the magnet. An even number of flux quanta ($h/2e$) trapped in the toroid corresponded to a phase shift of 0 when the Nb was superconducting (Fig. 15e), while an odd number corresponded to a phase shift of π (Fig. 15g). Figure 15d and f shows corresponding images obtained at a temperature of 15 K, when the Nb was no longer superconducting and flux quantization no longer occurred. The observed quantization of magnetic flux and the measured phase differences with the magnetic field entirely screened by the superconductor provided final confirmation of the Aharonov–Bohm effect.

3.2. Results Obtained Using Digital Reconstruction

Electron holography studies of magnetic materials that have involved the digital acquisition, reconstruction, and analysis of electron holograms are now presented. Digital approaches are often more suitable for many applications, either because of the small feature size or because the sample exhibits rapid variations in thickness or composition. Digital analysis can be readily used to separate the magnetic and mean inner potential contributions to the measured holographic phase, particularly at the edges of nanostructured elements where rapid changes in sample thickness are likely to dominate the measured phase and phase gradient. Digital analysis also facilitates the construction of line profiles from phase images, which can then provide quantitative information such as the widths of domain walls.

The magnetic contribution to the measured holographic phase shift for a nanostructured material is much smaller than the mean inner potential contribution, as inferred from Eq. (22). Neglecting any return flux associated with magnetic leakage fields, the magnetic contribution to the phase shift is proportional to the width of a small magnetic element, whereas the mean inner potential contribution to the phase shift of an element of the same thickness is independent of its lateral dimensions. For a small magnetic element of width a, the ratio of the magnetic to the mean inner potential contributions to the phase shift is given by the expression

$$\frac{\Delta\phi_{\text{MAG}}}{\Delta\phi_{\text{MIP}}} = \frac{eB_{\perp}a}{\hbar C_E V_0} \tag{40}$$

which is directly proportional to the width of the element.

3.2.1. NdFeB Hard Magnets

Hard magnetic materials such as NdFeB alloys have large magnetic remanence and high coercivity, which lead to many practical applications. Figure 16a shows a Lorentz (Fresnel defocus) image of a die-upset $Nd_2Fe_{14}B$ sample that was prepared for electron microscopy by standard dimpling and ion-milling [190–192]. The serpentine and "Y"-shaped domains that extend to the sample edge are similar to those previously observed at grain boundaries in sintered NdFeB [193]. An electron holographic phase image obtained from the same area of the sample is shown in Figure 16b. Figure 16a and b was recorded at room temperature using a Philips CM200 FEG-TEM operated at 200 kV in Lorentz mode, and the electron hologram had an interference fringe spacing of 2.5 nm.

In order to map the magnetic induction in the sample, gradients of the phase image werc calculated digitally along the $+x$ and $-y$ directions, as shown in Figure 16c and d,

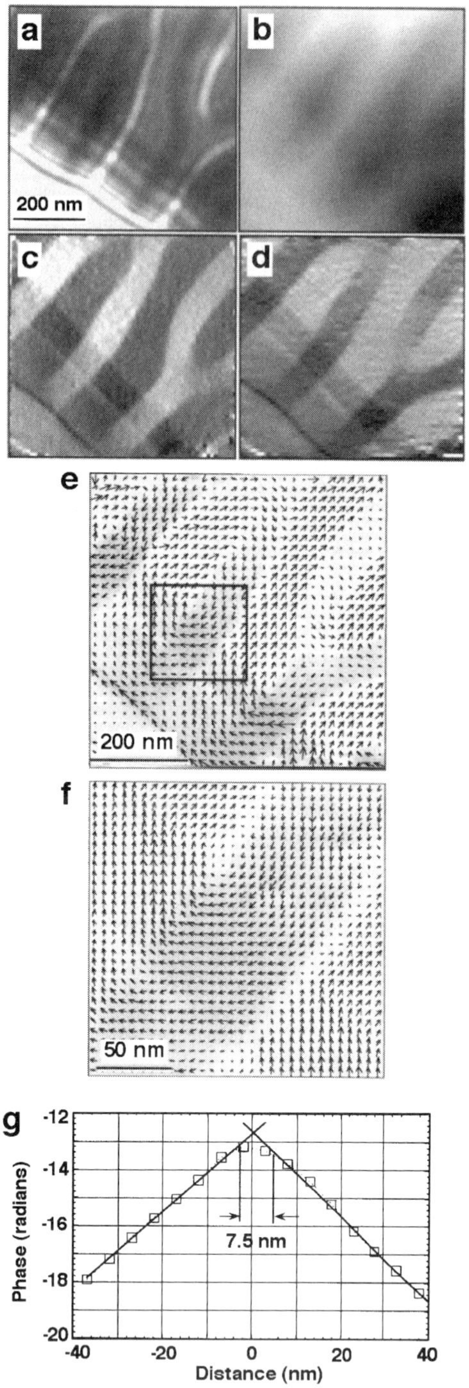

Figure 16. (a) Room temperature Lorentz (Fresnel underfocus) image of a die-upset $Nd_2Fe_{14}B$ hard magnet, recorded at 200 kV using a Philips CM200 FEG-TEM operated in Lorentz mode. (b) Phase image of the same area of the sample reconstructed from an electron hologram obtained using an interference fringe spacing of 2.5 nm. (c) and (d) Gradients of the phase image shown in (b), calculated along the $+x$ and $-y$ directions, respectively. (e) Induction map derived from the phase gradients shown in (c) and (d). (f) Enlargement of the area indicated in (e), showing singularities and domain wall character. (g) Line scan obtained from the phase image shown in (b) across a 90° domain wall that appears as a bright ridge near the center of (b). The line profile provides an upper limit for the domain wall width of 10 nm. Reprinted with permission from [190], M. R. McCartney and Y. Zhu, *Appl. Phys. Lett.* 72, 1380 (1998). © 1998, American Institute of Physics.

respectively. The two gradient images were combined to form a vector map of the magnetic induction in the domains, which is shown in Figure 16e. The vector map in Figure 16e is divided into 20 nm × 20 nm squares and has a low contrast image of one of the original phase gradient images superimposed onto it for reference purposes. The minimum vector length is zero (which is indicative of out-of-plane induction), while the maximum vector length is consistent with a measured induction $B = 1.0$ T at an estimated sample thickness of 90 nm. The vector map shows that the domains in this region of the sample are oriented at approximately 90° to each other rather than at 180°, as might be expected for a material that has such strong uniaxial anisotropy. Pairs of singularities visible in the image are reminiscent of cross-tie walls that contain periodic arrays of Bloch lines of alternating polarity [194]. A vector map of the area outlined in Figure 16e is shown at higher magnification in Figure 16f. This map is divided into 6.7 nm × 6.7 nm squares. There is an apparent tendency for the induction in the center domain to rotate toward a 180° orientation near the vortices. The vortices show Bloch-like character, with vanishingly small vector length. However, it is emphasized that care is needed when interpreting fine details in such induction maps due to the undetermined effects of the fringing fields immediately above and below the sample surfaces, as well as due to possible contributions to the vector maps from variations in sample thickness rather than from the strength of the local magnetic induction. Finally, a single pixel line scan obtained across a 90° domain wall, which appears as a bright ridge near the central part of Figure 16b, is shown in Figure 16g. This line profile places an upper limit of 10 nm on the domain wall width, which agrees well with theoretical estimates [194]. Similar observations were carried out after heating the sample to 300 °C and then cooling it to room temperature. The well-defined domains disappeared, and reduced phase gradients showed that the remaining structure had a large out-of-plane component. Domain walls were released from the thin edge of the sample, although some remaining domains were found to have interacted with structural features that were identified as planar defects and grain boundaries. Further heating to 400 °C, which is above the Curie temperature of 312 °C [195], resulted in complete disappearance of the domain structure. In a separate study, electron holography was used to examine as-quenched NdFeB permanent magnets, as well as samples that had been annealed at a range of temperatures [196]. Approaches based on noninterferometric phase reconstruction (including the transport of intensity equation) have also been used to study the magnetic microstructure of polycrystalline NdFeB samples [197].

Similar results that show both the magnetic fields within a TEM sample and the fringing fields outside it were obtained by applying digital reconstruction to holograms of continuous thin films of magnetic Fe_xPt_{1-x} [198]. Spontaneous ordering of Fe–Pt can occur during deposition, causing the development of an anisotropic ordered phase with the $L1_0$ crystal structure [199]. The tetragonal c-axis may then be predominantly out-of-plane or in-plane, depending on the substrate used for growth. As a result, thin films of Fe–Pt can have large magnetic moments with either longitudinal (in-plane) or perpendicular (out-of-plane) magnetization.

A reconstructed phase image was used to show the presence of magnetic fringing fields outside an epitaxial $Fe_{0.5}Pt_{0.5}$ ordered alloy that had been deposited using molecular beam epitaxy onto an MgO (110) single-crystal substrate. In this growth direction, the anisotropic $L1_0$ ordered alloy phase had the (001) easy axis parallel to the film. Adjacent domains within the FePt film were shown to have opposite polarity.

Magnetic fringing fields have been characterized outside ball-milled, nanocrystalline iron particles [200] using electron holography. Measurements were made at temperatures of between 300 and 1220 K and were used to provide insight into temperature-induced reversible variations in the exchange correlation length and saturation magnetization of the material. Nanocrystalline FeZrN films with a variety of grain sizes and thicknesses have also been charazterized using electron holography [201], as have CoNiAl shape memory alloys [202]. Most recently, electron holography was used to verify that charge order exists both in regions that have no net magnetization and in ferromagnetic regions in the mixed-valent manganite $La_{0.5}Ca_{0.5}MnO_3$ [203].

3.2.2. Layered Magnetic Nanowires

Low-magnification electron holography in field-free conditions has been used to study the remanent magnetic states of magnetic nanowires, prepared by electrodeposition of Co, Cu, or Ni in the cylindrical pores of polycarbonate track-etched membranes that are between 6 and 10 μm in thickness [204–206]. The composition of the wire may be changed during deposition to form multilayered wires that contain alternating metals. The wires themselves have diameters of between 20 and 150 nm. After dissolution of the track-etched membrane, the wires can be placed on standard C films for electron holography. Figure 17a shows an electron hologram of an electrodeposited, layered Co/ Cu nanowire that has a radius of 40 nm and a length of 5.3 μm. Successive Co and Cu segments in the wire have lengths of 1.2 and 0.3 μm, respectively. The holographic interference fringe spacing in this image is 80 nm. Figure 17b shows a four-times-amplified reconstructed phase image of the same wire, where the arrows indicate the local magnetization direction and the approximate lengths of the Co grains. Figure 17c shows the corresponding best-fitting simulated phase image, also four-times-phase-amplified. Because of the large holographic interference fringe spacing in these experiments, only the magnetic leakage fields outside these wires could be interpreted.

In general, considerable care is required when applying low magnification modes of electron holography to nanostructured materials, since gross artifacts may result if the magnetic fields in the sample are undersampled by the holographic fringes. As a result, more recent experiments on the remanent magnetic states of single crystalline 4-nm-diameter ferromagnetic Co wires have been carried out at high spatial resolution using a microscope equipped with a Lorentz lens [207]. In contrast to earlier holographic experiments on larger diameter magnetic nanowires [208], the magnetic and mean inner potential contributions to the measured phase shift were separated successfully in these experiments.

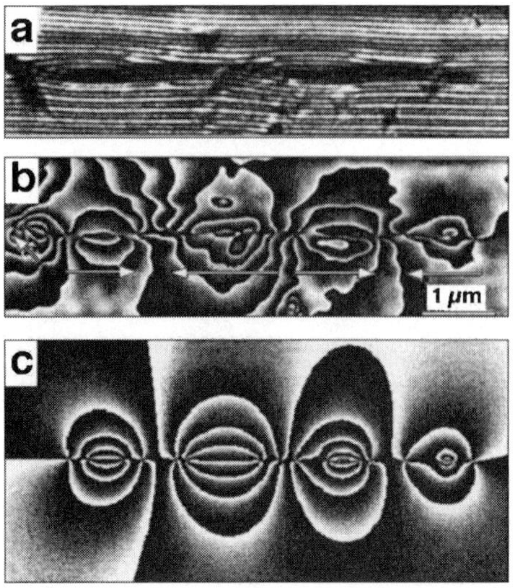

Figure 17. (a) Electron hologram of an electrodeposited, layered Co/Cu nanowire that has a radius of 40 nm and a length of 5.3 μm, obtained using a Hitachi HF-2000 FEG-TEM. Successive Co and Cu segments have lengths of 1.2 and 0.3 μm, respectively. A fragment of another nanowire is visible near the left end of the layered Co/Cu wire. (b) Four-times-amplified reconstructed phase image. Arrows indicate the local magnetization direction and the approximate lengths of the Co grains. (c) Corresponding simulated phase image, also 4-times-phase-amplified. Reprinted with permission from [204], C. Beeli et al., *Ultramicroscopy* 67, 143 (1997). © 1997, Elsevier Science.

3.2.3. Separation of Magnetic and Mean Inner Potential Contributions to the Phase

The mean inner potential contribution must be removed from the measured phase image in order to interpret the magnetic contribution to the phase shift of a nanostructured magnetic element. This separation of the two contributions to the signal becomes more important for smaller nanostructures. There are several ways to achieve this separation, all of which can be understood with reference to Eqs. (22) and (23). For example, the sample may be inverted to change the sign of the magnetic contribution to the signal and a second hologram recorded. The sum, and the difference, of the two measured phase images can then be used to provide twice the magnetic contribution and twice the mean inner potential contribution to the phase, respectively [209, 210]. An alternative method involves acquiring two holograms from the same area of the sample at two different microscope accelerating voltages. In this case, the magnetic signal is independent of accelerating voltage, and subtraction of the two phase images can be used to provide the mean inner potential contribution.

In many practical situations, it is neither feasible to turn the sample over during an experiment nor convenient to change the microscope accelerating voltage. A simpler method of removing the mean inner potential contribution involves performing a magnetization reversal experiment *in-situ* in the electron microscope and subsequently selecting pairs of holograms that differ only in the (opposite) directions of the magnetization within each element. The magnetic and mean inner potential contributions to the phase can be calculated by taking half the difference and half the sum of the phases of such pairs of holograms, respectively. The mean inner potential contribution can then be subtracted from all other phase images acquired from the same sample [211].

3.2.4. Lithographically Patterned Magnetic Nanostructures

An understanding of the interactions between magnetic nanostructures is important because of their potential applications as magnetic random access memory cells, which should have a limited number of magnetically accessible states. However, such elements are susceptible to interactions with their neighbors [212–215], particularly as switching in submicrometer cells often involves the propagation of magnetic domains that have larger stray fields than the magnetic vortices that accompany switching in larger cells [216].

For the experiments on lithographically patterned experiments described here, a Lorentz minilens ($C_S = 8$ m and 1.2 nm line resolution at 200 kV), located in the bore of the objective lens pole-piece of a Philips CM200ST FEG-TEM, allowed electron holograms to be obtained at magnifications of up to 70 k× with the main objective lens switched off and the sample located in an almost field-free environment. The objective lens could then be excited slightly and the sample tilted by angles of up to ±30° (Fig. 18a) in order to apply known magnetic fields to the sample, allowing magnetization processes to be followed *in-situ* over entire hysteresis cycles. Phase images that differed only in the opposite directions of the magnetization within the field of view could then be identified in order to subtract the mean inner potential contribution to the measured phase. Figure 18b shows a Hall probe calibration of the vertical magnetic field at the position of the sample in this microscope, plotted as a function of objective lens current. The residual vertical field of ~130 Oe was unaffected by the excitation of the Lorentz lens.

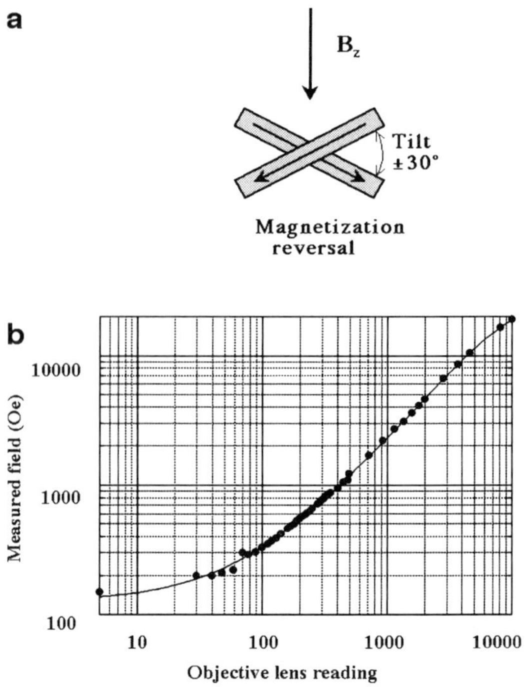

Figure 18. (a) Schematic diagram illustrating the use of specimen tilt to provide the in-plane component of the external field for *in-situ* magnetization reversal experiments. Reprinted with permission from [217], R. E. Dunin-Borkowski et al., *J. Microsc.* 200, 187 (2000). © 2000, Blackwell Science Ltd. (b) Hall probe calibration of the magnetic field in the specimen plane of a Philips CM200 FEG-TEM as a function of objective lens current in mA. The field is parallel to the incident electron beam direction and is insensitive to small changes in specimen height and to the excitation of the Lorentz lens. Reprinted with permission from [211], R. E. Dunin-Borkowski et al., *Ultramicroscopy* 74, 61 (1998). © 1998, Elsevier Science.

Bright-field images of linear arrays of magnetic elements patterned onto self-supporting electron-transparent membranes using electron beam lithography and lift-off processes are shown in Figure 19. These magnetic elements were patterned from a variety of different magnetic materials and included layered spin-valve structures. Adjacent linear arrays were separated from each other by at least 5 μm. The samples were made by coating 55-nm-thick silicon nitride windows with polymethyl methacrylate (PMMA) and then exposing them to the rastered electron beam of a scanning electron microscope. The exposed parts were removed using a solution of 11:10:1 methyl isobutyl ketonc:cellusol:methyl alcohol. An electron-beam evaporator was then used to deposit the magnetic materials at a rate of ~0.05 nm/s. Immersion in acetone was used to remove the unwanted PMMA and overlying material. Electrostatic charging of the membranes during examination in the electron microscope was reduced by depositing a further thin (~1 nm) layer of Al, as well as by using a small objective aperture. Each membrane contained at least 10 different array designs, and each design was repeated for several different exposure times during patterning to ensure that an optimal electron dose was achieved. The narrowest feature sizes produced were approximately 7 nm in width. The elements included rectangles of varying aspect ratio (Fig. 19a) and diamonds, rectangular bars, and ellipses of varying size (Fig. 19b). At least two adjacent elements could typically be included within the field of view of a single electron hologram.

Figure 20a and b shows the amplitude and unwrapped phase of a hologram of two representative rectangular magnetic elements that were patterned from 30-nm-thick Co [217, 218]. The elements are polycrystalline, with a grain size of below 5 nm, and they form one end of a linear array similar to that shown in Figure 19a. The elements in the array are 275 nm in width, between 220 and 800 nm in length, and are typically separated from their neighbors by gaps of between 170 and 350 nm. The rectangles in Figure 20 are 275 × 220 and 275 × 300 nm in size and are separated by 170 nm. The hologram was recorded at 200 kV using a rotatable biprism located in the selected-area aperture plane and a 1024 × 1024 pixel Gatan 794 multiscan CCD camera. A biprism voltage of 100 V was used, corresponding to an overlap width of about 0.6 μm at the sample level. The hologram was obtained by overlapping a clean region of the silicon nitride support film onto the magnetic elements, since no regions of vacuum were available on this sample. A reference hologram was also acquired from

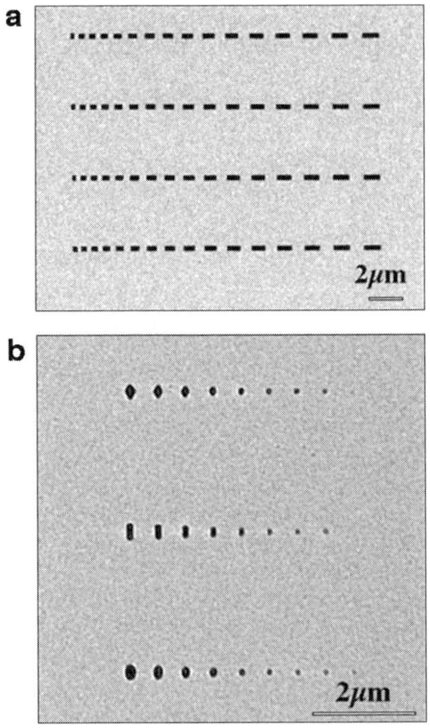

Figure 19. Low-magnification bright-field images of linear arrays of lithographically patterned magnetic elements on a 55-nm-thick electron-transparent silicon nitride membrane. (a) Rectangles with different aspect ratios. (b) Diamonds, rectangular bars, and ellipses with different sizes. Reprinted with permission from [217], R. E. Dunin-Borkowski et al., *J. Microsc.* 200, 187 (2000). © 2000, Blackwell Science Ltd.

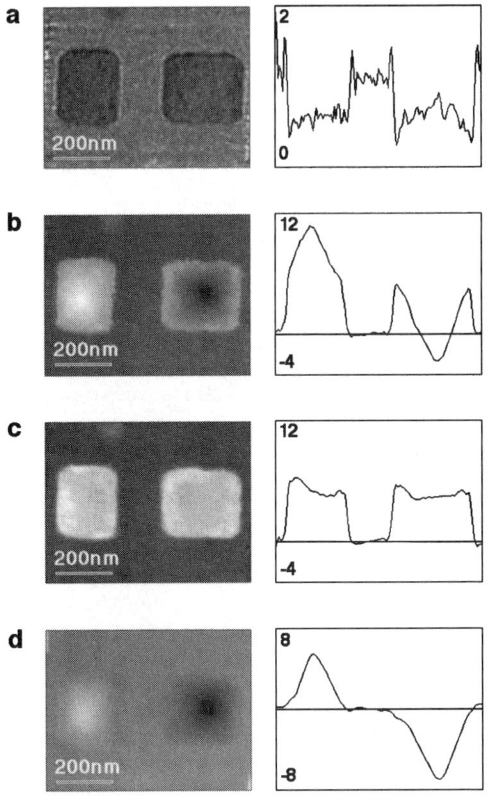

Figure 20. (a) and (b) Reconstructed amplitude and unwrapped phase images obtained from an electron hologram of two 30-nm-thick patterned rectangular Co elements shown at the extreme left end of the top array in Figure 19. (c) and (d) Mean inner potential and magnetic contributions to the holographic phase shown in (b) (see text for details). Reprinted with permission from [217], R. E. Dunin-Borkowski et al., *J. Microsc.* 200, 187 (2000). © 2000, Blackwell Science Ltd.

the adjacent silicon nitride to ensure that artifacts caused by local irregularities of the imaging/recording system were removed. Horizontal line traces that pass through the center of the amplitude and phase image of each element are also shown in Figure 20. When performing magnetization reversal experiments, the in-plane component of the applied magnetic field was always parallel to the line joining the rectangles shown in Figure 20. Figure 20c and d shows the mean inner potential and magnetic contributions to the phase, calculated using the approach described previously, from a pair of phase images similar to that shown in Figure 20b but which showed a reversed direction of the magnetic induction everywhere across the field of view. The bright areas at the edges of the elements in Figure 20c may originate from material left on the elements after the lift-off procedure. The fact that the magnetic signal is much smoother and more slowly varying than the compositional signal is also apparent from the figures.

Before obtaining electron holographic phase images such as that shown in Figure 20b, a preliminary examination of the magnetic domain structures in such elements is always performed. These images are usually obtained using the Fresnel mode of Lorentz microscopy, which relies on highly defocused images to reveal lines of dark and white contrast at domain walls [219]. Although the resolution of the technique is inherently limited, it permits real-time viewing of dynamic processes and it was used to provide an initial assessment of the micromagnetic structure in the samples. Suitable ranges for the applied magnetic fields can then be selected for more detailed examination by electron holography.

Magnetization Reversal Holograms of the Co elements were recorded over an entire magnetization reversal cycle, and the resulting magnetic contributions to the measured phase are shown in Figure 21a. The in-plane component of the applied field is indicated at the bottom left corner of each image, and the hysteresis cycle should be followed in a counterclockwise sense. The external field points toward the right in the image shown at the top of the cycle and toward the left in the image at the bottom. The images are displayed at a sampling density of ~6 nm/pixel. In each image in the montage, contours have been generated from the measured magnetic contribution to the phase and superimposed onto the original hologram. The local separation of the phase contours, 0.21π radians, is inversely proportional to the in-plane component of the magnetic induction in the sample integrated in the incident beam direction. Arrows have also been added to show the local direction of the magnetic induction in the sample. The advantage of separating the magnetic and mean inner potential contributions is clear from these figures; contours can easily be followed from within the nanostructures to outside, allowing the magnetic

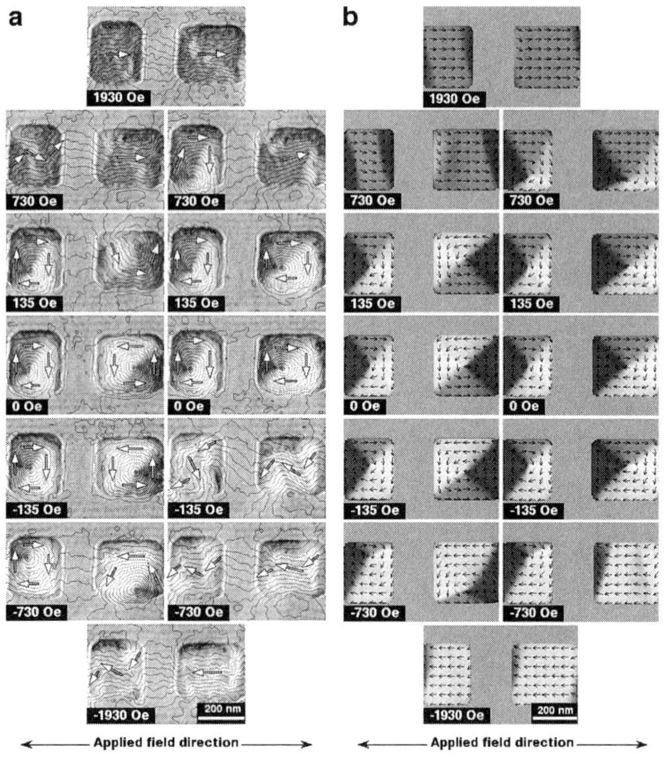

Figure 21. (a) Magnetic contributions to measured phases of holograms of 30-nm-thick Co rectangular elements for an entire magnetization reversal cycle, for the applied in-plane fields indicated. Phase contours lie parallel to lines of constant magnetic induction integrated in the incident beam direction. The contour spacing is 0.21π radians. The images, which should be followed in a counterclockwise sense, were recorded with an average out-of-plane field of 3600 Oe. Reprinted with permission from [218], R. E. Dunin-Borkowski et al., *J. Appl. Phys.* 84, 374 (1998). © 1998, American Institute of Physics. (b) Best-fitting micromagnetic simulations to hysteresis cycle shown in (a). Initial states in the top image are an S-state for the smaller rectangle and a C-state for the larger rectangle. Reprinted with permission from [224], R. E. Dunin-Borkowski et al., *Appl. Phys. Lett.* 75, 2641 (1999). © 1999, American Institute of Physics.

interactions between adjacent elements to be quantified at all points across each image. In particular, the effects of strong variations in sample thickness that were at the edges of the Co and NiFe particles in Figure 14 are now absent.

The images in Figure 21 show a variety of domain structures that include vortices and both S-shaped and C-shaped configurations. Few contours are visible between the elements when they both exhibit flux closure, whereas the strongest interactions occur when they both have S-shaped domain structures. Although the magnetic flux enclosed by adjacent contours is known, quantitative measurements of the local magnetic field strength both within and between the elements would require accurate models for the magnetic fringing fields around them. The strength of the magnetic fringing field above and below each element can be estimated by comparing the spacings of the contours within the elements between the nonsolenoidal and solenoidal configurations.

The phase contours show that complete saturation of the magnetization is never achieved at the extreme ends of the cycle, even for large applied fields at which the domain structures are nonsolenoidal and the surrounding fringing fields are strong. The initial almost saturated state transforms to a vortex, which then moves up or down the element before finally switching to an almost saturated state in the opposite direction. The stability of a single vortex appears to influence the field at which an S-shaped domain structure forms irreversibly in each element at large values of the applied in-plane field. The solenoidal domain configurations contrast with the single-domain structures reported previously for 100–300 nm wide Co elements [220].

Differences are observed between the magnetic domain structures seen in successive cycles similar to that shown in Figure 21a [217]. These differences arise in part because the magnetic microstructure is sensitive to the strength of the out-of-plane field used in the experiment, but also from the sensitivity of the domain structure to the starting magnetic state and the statistical nature of the reversal process. This lack of reproducibility is of obvious importance for potential device applications. The extent to which the magnetization is saturated at the end of a hysteresis cycle also has a strong influence on the evolution of the domain structure. For example, if a vortex is still present in the "saturated" state then it will be retained in the second half of the cycle. The persistence of such vortices during switching will produce a variability in the switching field that will seriously diminish the ability of a nanostructured magnetic device to discriminate between states [221].

Results similar to those shown in Figure 21a have been obtained for a variety of remanent magnetic states, with the two Co rectangles only in the residual field of the microscope objective lens after being subjected to a variety of different applied magnetic fields [217, 218]. The remanent states showed domain configurations that were not observed in the hysteresis cycle, including a double-vortex structure in the larger element and a "vertical" S-shape in the smaller element. These observations highlight the importance of recording remanent states, which are particularly significant for device applications in which an element is first magnetized and the external field is then removed with the objective of retaining a nonsolenoidal domain structure.

One of the primary strengths of the digital analysis of electron holograms is that quantitative measurements can be made easily from recorded phase images. For example, Figure 22 shows hysteresis loops for the two Co elements, generated from the phase images in Figure 21a by plotting the fractional magnetization M/M_s (where M_s is the saturation magnetization in the direction of the applied field) for each element. The vertical steps in the loops correspond to the irreversible steps between S-shaped and solenoidal domain structures. The graphs highlight the fact that, as the applied field is reduced, the smaller element forms a vortex before the larger element and then retains it to larger fields before switching to an S-shaped domain structure in the opposite direction.

Similar experiments have also been performed for elements of different thickness, shape, and composition [217]. Domain structures in thinner (10-nm-thick) Co rectangles were found to be much simpler than those observed in Figure 21a, with no distinct vortex states at any stage of the

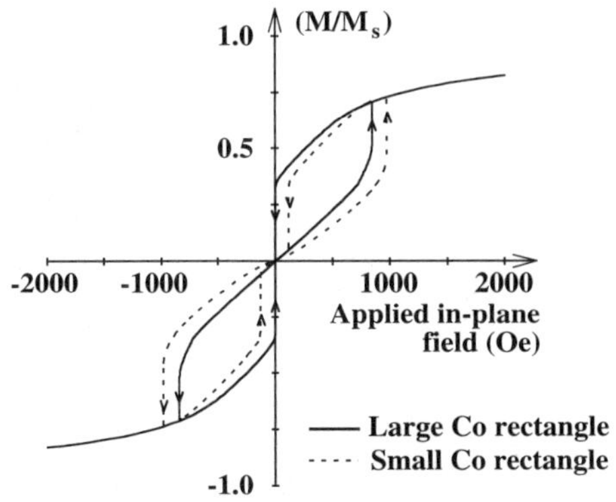

Figure 22. Best-fitting curves to measured hysteresis loops for the rectangles shown in Figure 21a. The curves were generated by measuring the fraction of each element magnetized in the direction of the applied field and are plotted as a function of the magnitude of the applied field. Reprinted with permission from [218], R. E. Dunin-Borkowski et al., *J. Appl. Phys.* 84, 374 (1998). © 1998, American Institute of Physics.

cycle. For 10-nm-thick Ni elements, it was found that a large component of the magnetization was pulled out of plane by the external field. The Ni elements were also found to oxidize over time, resulting in the presence of a magnetically "dead" layer. The general topic of magnetic domains in elements of different size has been treated comprehensively elsewhere [222].

Need for Micromagnetic Simulations The benefit of comparing electron holographic measurements, such as those shown in Figure 21a, with micromagnetic simulations is illustrated through comparisons of experimental results with solutions to the Landau–Lifshitz–Gilbert (LLG) equations [223]. The room temperature parameters used to simulate the magnetic microstructure of the Co rectangles included the exchange stiffness, $A = 1.55$ μerg/cm, and the saturation magnetization, $M_s = 1414$ emu/cm^3. The magnetocrystalline anisotropy constant, K, in the polycrystalline films was set to zero as the grain size was below 10 nm. Letting $K = 0$ is consistent with the observation that the coercivity in the submicrometer cells (ascribed to shape anisotropy) is much larger than that typical of bulk films of the same thickness, implying that magnetocrystalline anisotropy plays only a minor role in the energetics of switching. The effects of temperature fluctuations were not included. The gyromagnetic frequency, $\gamma = 17.6$ MHz/Oe, and a damping constant, $\alpha = 1$, were used in the LLG equations. Square, in-plane, discrete moments ranged in size between 5.4 and 6.9 nm on a side. Changing the out-of-plane discretization between 7.5 and 30 nm produced identical magnetization distributions; hence, a single layer of moments was used. Magnetization reversal cycles were followed by assigning an initial condition and then integrating the LLG equations in a fixed external field until equilibrium was reached. The exit criteria (equilibrium) corresponded to the largest change in the residual direction cosine of all discretized moments in the grid changing by less than 2 to 5×10^{-5}.

Effect of Initial Magnetic State Figure 21b shows a computed magnetization reversal cycle, where the initial conditions were selected to provide a best fit to the measurements shown in Figure 21a [224]. In the initial state (at the top of the figure), the smaller cell is oriented in an S-state and the larger cell is in a modified C-state. The 3600 Oe vertical field is oriented into the plane. The larger element exhibits a flowerlike divergence at its right side, coupling the flux back to the smaller bit along both the top and bottom edges of the cell. The simulated vortex helicities match those measured, and almost all of the S-shaped domain structures are reproduced. The main differences between the measurements and the simulations were the inability to match the S-shaped domain structure at one end of the reversal cycle and the fact that vortices formed at higher fields in the simulation. The most important point established from the simulations was the sensitivity of the magnetic domain structure to the detailed magnetic history of each element. Differences in the initial state that were often on a scale too small to be distinguished visually were found to be of paramount importance to the formation of the subsequent domain structure, and in particular to the sense (the handedness) with which the vortex in each element unrolled. In general, these initial states will be determined by the magnetic history of each cell and may differ even though the magnetization may appear to be saturated.

Interactions between Neighboring Elements Simulations similar to those shown in Figure 21b were used to illustrate the way in which the presence of neighboring cells affected the magnetic domain structure in the linear array [224]. In a simulated magnetization reversal cycle of the two adjacent cells, the magnetic microstructure evolved differently as a result of intercell coupling depending on whether the cells were simulated as a pair, or whether they were simulated separately and then displayed as a pair. The energy difference between the simulations of the individual cells and the simulation of the pair was found to be on the order of the thermal energy.

Effect of Out-of-Plane Component of Applied Field The dependence of the domain structure on the out-of-plane component of the applied field, which is often assumed to be unimportant in such experiments, is illustrated in Figure 23. Calculations of hysteresis cycles were performed for identical initial conditions for perpendicular fields (of 3600 Oe) both into (Fig. 23a) and out of (Fig. 23b) the plane. The simulations show that the sense of the perpendicular field affects the domain structure appreciably, with the primary effect being to define the handedness of the vortices that evolve during magnetization reversal [224].

In combination with the results shown in Figures 21 and 23, the sensitivity of the simulated magnetic domain structure to the initial magnetization state, as well as the out-of-plane component of the applied magnetic field and intercell coupling, emphasizes the need to correlate high quality experimental holographic measurements with micromagnetic simulations.

Interlayer Coupling within Individual Elements Electron holography can be used to provide information about magnetic interactions between closely separated, ferromagnetic layers within an individual lithographically defined structure. For example, spin valves and magnetic tunnel junctions [225] consist of two ferromagnetic layers separated by a thin nonmagnetic spacer layer. Changes in either

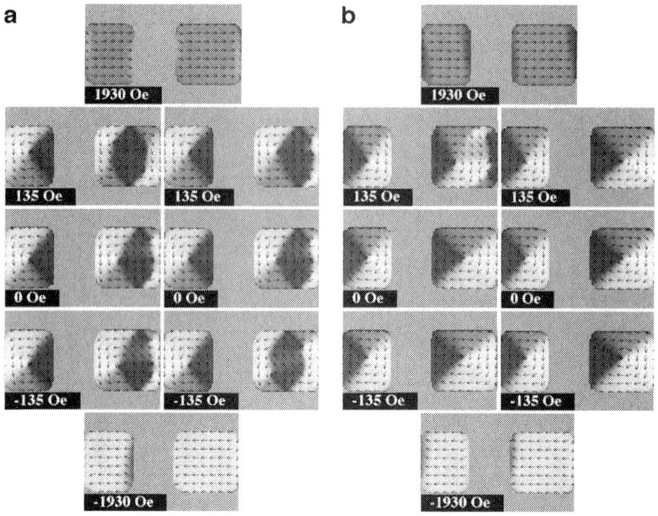

Figure 23. Micromagnetic simulations of hysteresis cycles for 30-nm-thick rectangular Co elements, for the applied in-plane fields indicated. An S-state was used as the initial state for both elements. A 3600 Oe out-of-plane field was included in the simulations, in a direction (a) into and (b) out of the plane of the page. Reprinted with permission from [224], R. E. Dunin-Borkowski et al., *Appl. Phys. Lett.* 75, 2641 (1999). © 1999, American Institute of Physics.

the separation of the ferromagnetic layers or their parallel to antiparallel magnetic alignment can cause substantial changes in electrical resistance when the magnitude and direction of an externally applied magnetic field is varied. This effect is termed giant magnetoresistance [226].

Trilayer Co (10 nm)/Au (5 nm)/Ni (10 nm) spin-valve elements were patterned into ellipses [227]. Figure 24a shows a representative off-axis electron hologram of a pair of elliptical spin-valve elements, for which the corresponding hysteresis cycle is shown in Figure 24b. The direction of the in-plane component of the applied field is now parallel to the long axis of each element (i.e., perpendicular to the line joining the elements). As a result of their narrow width, these elements do not show complicated domain structures. A solenoidal vortex state is occasionally visible during both forward and reverse cycles (at −168 and +336 Oe). In contrast, vortex states are never observed in similar rectangular bars, which have narrower dimensions (width of 70 nm) and a more dominant influence of shape anisotropy on their magnetic response. Significantly, the bars are also too thin and narrow to form end domains, which govern the reversal of larger rectangular elements. Instead, the phase contours curve at their ends by a maximum angle of ~45° just before magnetization reversal [227].

A measured hysteresis loop (Fig. 24c) was obtained from the ellipses by plotting the magnetic contribution to the measured phase difference across the midpoint of each element. The most significant observation from Figure 24b is the presence of two different contour spacings at different applied fields within each element (narrower at higher applied fields and wider close to remanence), and corresponding steps in the hysteresis loop in Figure 24c, in agreement with simulated loops [227]. The magnetization direction of the Ni layer in each element reverses well before the external field is reduced to zero as a result of flux closure associated

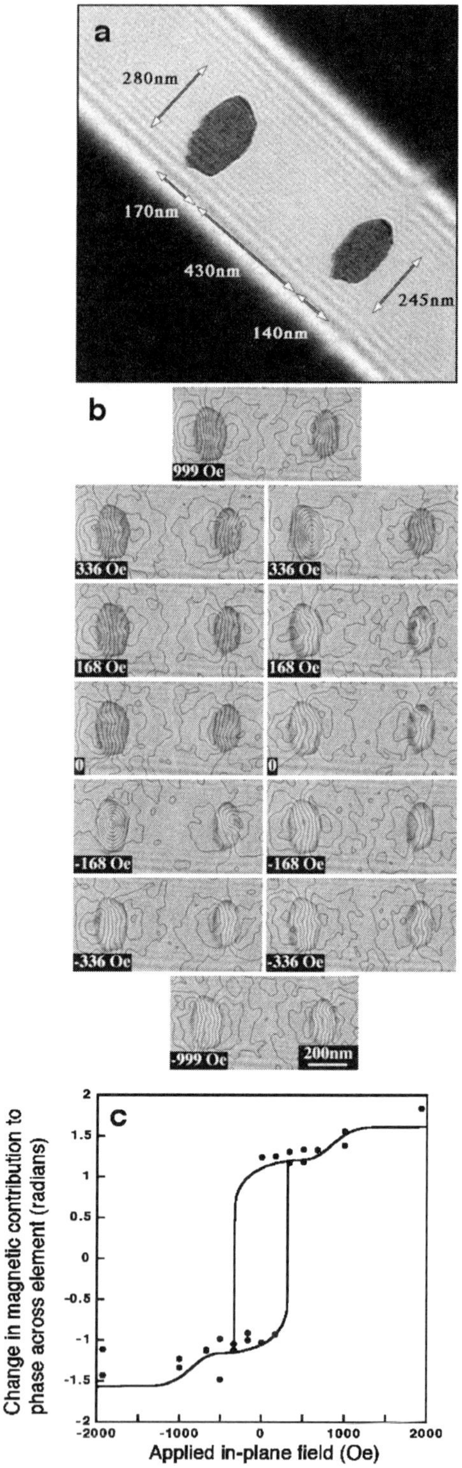

Figure 24. (a) Off-axis electron hologram of elements with a nominal layer sequence of Co (10 nm)/Au (5 nm)/Ni (10 nm) that have been patterned into ellipses. (b) Corresponding hysteresis cycle, with a contour spacing of 0.064π radians and an average out-of-plane field of 3600 Oe. The external in-plane field is applied vertically (parallel to the major axes of the ellipses). (c) Hysteresis loop deduced from magnetic contributions to the holographic phase shown in (b) for the smaller ellipse. Reprinted with permission from [217], R. E. Dunin-Borkowski et al., *J. Microsc.* 200, 187 (2000). © 2000, Blackwell Science Ltd.

with the strong fringing field of the magnetically more massive and closely adjacent Co layer. An antiferromagnetically coupled state is the normal remanent state that would be obtained after saturation of the element followed by removal of the external field. The drop in Figure 24c that occurs before 0 Oe is reached corresponds to reversal of the Ni. In the simulation, the drop occurs later and is more abrupt than observed experimentally [227], perhaps reflecting a variability in the grain size and orientation in the experimental Ni film that facilitates earlier reversal. The simulations confirm that the two different contour spacings arise because of the strength of the coupling between the Co and Ni layers within each element at low values of the in-plane applied field.

Exchange-Biased Elements Exchange-biased elements, consisting of Ti (5 nm)/Pd (15 nm)/$Fe_{54}Mn_{46}$ (10 nm)/$Co_{84}Fe_{16}$ (2.8 nm)/Al_2O_3 (2 nm)/Pd (7.5 nm), were characterized in a manner analogous that used for the spin valves [228]. In such elements, the magnetization direction of the ferromagnetic CoFe layer is pinned, or "exchange biased," by the adjacent antiferromagnetic FeMn layer [229]. In this sample, the ferromagnetic layer was very thin ($\sim$3 nm) relative to the total element thickness ($\sim$42 nm). As a result, the magnetic signal was very weak, and in addition the fringe contrast was low because of the presence of heavy Pd layers in the structure. Electron holographic phase measurements of two submicrometer-sized patterned elements were obtained over a magnetization reversal cycle. No solenoidal or vortex states formed in either particle, and the measured hysteresis loop was shifted sideways from zero by about 90 Oe due to the influence of the pinning layer. Best-fitting simulations corresponded to an exchange field of 79 Oe.

3.2.5. Chains of Magnetic Nanocrystals

The application of electron holography to closely spaced magnetic nanocrystals, such as those in the hologram shown in Figure 1b, is now described.

Magnetite in Magnetotactic Bacteria Magnetotactic bacteria synthesize intracellular, membrane-bounded ferrimagnetic crystals that are known as magnetosomes. These are composed of magnetite (Fe_3O_4) and/or greigite (Fe_3S_4), and they are usually arranged in one or more linear chains within each cell [230, 231]. The magnetosomes have a narrow size distribution, are specific to each cell type, and usually have their magnetic easy axes aligned (roughly) parallel to the chain axis. The fact that they are typically only 40–200 nm in size suggests that they should each contain a single magnetic domain [232]. The magnetic moment that they impart to the cell results in its alignment and subsequent migration along the Earth's magnetic field lines. This attribute is thought to increase the efficiency with which the cells find their optimal location in a vertical chemical or redox gradient within aquatic environments [233]. The ability to obtain high spatial resolution, quantitative information about the magnetic microstructure of bacterial magnetosomes is likely to be crucial to understanding magnetic-field-sensing mechanisms in a wide range of organisms [234, 235], as well as providing magnetic biomarkers that may be used to establish the occurrence of ancient life [236]. Such measurements may also have implications for understanding the interaction of external fields with iron biominerals that form in the human brain [237–239].

Magnetic Microstructure Holograms obtained from two chains of magnetite crystals that formed within a single bacterial cell are shown in Figure 25a and b. This bacterial cell is a magnetic coccus, which was collected from Sweet Springs Nature Reserve in Morro Bay, California. The sample was prepared by placing drops of water that had been enriched in bacteria onto a 3-mm-diameter C-coated Ni grid for TEM examination. The two chains of crystals, with morphologies that have been described as "truncated hexahedral prisms" [240, 241], lie on opposite sides of the cell body. The holograms were acquired at 300 kV in a Philips CM300ST FEG-TEM using a biprism voltage of 160 V, an acquisition time for each hologram of 32 s, and a 2048 × 2048 pixel

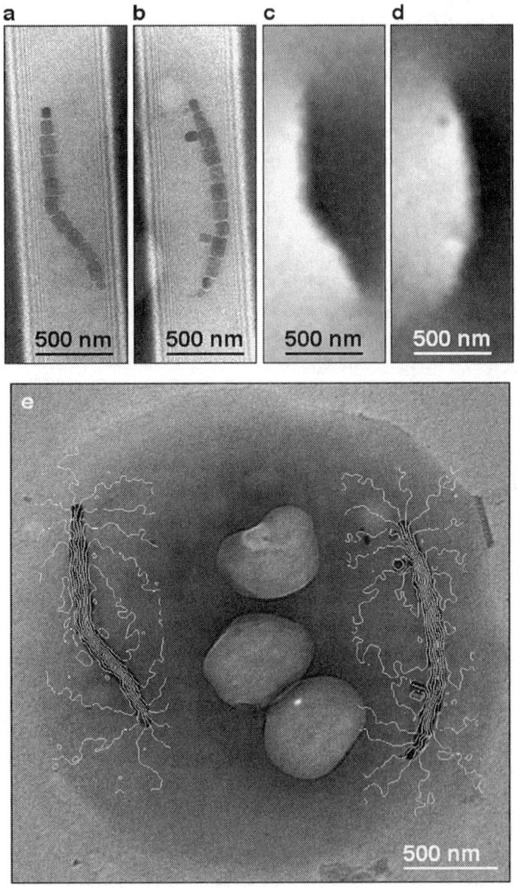

Figure 25. (a) and (b) Electron holograms of two chains of magnetite crystals from within a single bacterial cell (a coccus) collected by Professor R. B. Frankel at Sweet Springs Nature Reserve, Moro Bay, California. The holograms were acquired in Lorentz mode (in field-free conditions) at 300 kV using a Philips CM300 FEG-TEM equipped with a Gatan Imaging Filter. The crystals are elongated parallel to their ⟨111⟩ crystallographic axes. (c) and (d) Corresponding magnetic contributions to the electron holographic phase shift, obtained by acquiring holograms after reversing the direction of the magnetization in each chain (see text for details). (e) 0.5 radian contours generated from (c) and (d) and overlaid onto a bright-field image of the cell, showing the direction and the strength of the local magnetic flux both within and around each chain of crystals. (Taken from Dunin-Borkowski, unpublished work.)

CCD camera located at the end of a Gatan Imaging Filter 2000. The long acquisition time was required because the cell has a thickness of 3 μm in the electron beam direction. The biprism is oriented in such a way that the entire length of each chain is contained within a single hologram.

Figure 25c and d shows magnetic contributions to the holographic phase for the two chains. As for the patterned elements described above, the direction of magnetization in each chain was reversed *in-situ* by tilting the sample by $\pm 30°$ and turning on the conventional microscope objective lens to provide a large (>10,000 Oe) component of the applied in-plane field parallel to the chain axis. The objective lens was then turned off and the sample tilted back to 0° for hologram acquisition in field-free conditions. In this way, holograms were recorded with the chain of crystals magnetized in opposite directions. On the assumption that the magnetic field in the chain reversed exactly, the mean inner potential and magnetic contributions to the measured phase were then obtained by taking half the difference and half the sum of the phases of each pair of holograms, respectively. As for Figure 20, the magnetic signals in Figure 25c and d vary across each phase image smoothly, and the positions of the individual crystals are no longer visible. The step in the magnetic contribution to the phase across each chain is approximately 4.5 radians. The height of the corresponding mean inner potential contribution to the phase at the position of each crystal is approximately 9 radians. Contours of spacing 0.5 radians, which have been generated from the magnetic phase images shown in Figure 25c and d, are shown superimposed onto a bright-field image of the entire cell in Figure 25e. The contours provide a semiquantitative map of the strength of the local magnetic field along the chain. As predicted, all of the crystals in the chains are single magnetic domains that are magnetized approximately parallel to the chain axes. The contours are most closely spaced within the crystals, where the field is strongest, and an increase in separation occurs at the ends of the chains where the field becomes weaker. The noise in the contours is most visible where their spacing is large.

The identification of single magnetic domains in bacterial cells such as that shown in Figure 25 is consistent with the qualitative prediction of a single- to multidomain transition when the crystal size is comparable to the domain wall width of ~100 nm in bulk magnetite. The particle size at which this transition occurs, as well as the external factors that may affect it, have been uncertain. Numerical micromagnetic modeling predicts that the transition should occur at a size of ~70 nm for perfect cubes, although this value is predicted to increase with axial ratio [242]. The situation is complicated by the fact that metastable single domains may exist in the multidomain regime, particularly in the presence of interactions between closely spaced particles. Results similar to those shown in Figure 25 have been obtained [243] from chains of smaller magnetite nanocrystals originating from cells of the bacterial marine vibrioid strain MV-1 [244] and the aquatic strain *Magnetospirillum magnetotacticum* MS-1.

Magnetic Moment Although the contours in Figure 25e provide much useful information, one of the strengths of electron holography is its ability to provide quantitative information about magnetic properties. One of the most straightforward parameters that can be obtained from the magnetic contribution to the phase is the magnetic moment of a nanocrystal or, in this case, a chain of nanocrystals. Two approaches for measuring this parameter are now outlined.

By definition, the magnetic moment of a chain of magnetite crystals (or any other magnetic nanostructure) along the x and y directions is given by the relations

$$m_x = \int_{-\infty}^{+\infty} \int_{-\infty}^{+\infty} \int_{-\infty}^{+\infty} B_x(x, y, z)dx\,dy\,dz \tag{41}$$

and

$$m_y = \int_{-\infty}^{+\infty} \int_{-\infty}^{+\infty} \int_{-\infty}^{+\infty} B_y(x, y, z)dx\,dy\,dz \tag{42}$$

respectively. However, according to Eq. (20) the magnetic contribution to the phase

$$\phi_{\text{mag}}(x, y) = -\left(\frac{e}{\hbar}\right) \int_{z=-\infty}^{z=\pm\infty} \int_{x=\pm\infty}^{x} B_y(x, y, z)dx\,dz \tag{43}$$

$$\equiv -\left(\frac{e}{\hbar}\right) \int_{z=-\infty}^{z=+\infty} \int_{y=\pm\infty}^{y} B_x(x, y, z)dy\,dz \tag{44}$$

It follows from Eqs. (43) and (44) that

$$\frac{\partial}{\partial_x}\phi_{\text{mag}}(x, y) = -\left(\frac{e}{\hbar}\right) \int_{z=-\infty}^{z=+\infty} B_y(x, y, z)dz \tag{45}$$

and

$$\frac{\partial}{\partial_y}\phi_{\text{mag}}(x, y) = +\left(\frac{e}{\hbar}\right) \int_{z=-\infty}^{z=+\infty} B_x(x, y, z)dz \tag{46}$$

By combining Eqs. (41), (42), (45), and (46), the magnetic moment can be expressed in the form

$$m_x = +\left(\frac{\hbar}{e}\right) \int_{y=-\infty}^{y=+\infty} \int_{x=-\infty}^{x=+\infty} \frac{\partial}{\partial y}\phi_{\text{mag}}(x, y)dx\,dy \tag{47}$$

and

$$m_y = -\left(\frac{\hbar}{e}\right) \int_{y=-\infty}^{y=+\infty} \int_{x=-\infty}^{x=+\infty} \frac{\partial}{\partial x}\phi_{\text{mag}}(x, y)dx\,dy \tag{48}$$

The magnetic moment of a chain of crystals in a given direction can thus be obtained by measuring the area under the first differential of the measured magnetic contribution to the phase evaluated in the *perpendicular* direction. The application of Eq. (48) to the MV-1 cell mentioned above [243] gives a magnetic moment (along the average directions of the chain axis) of $(7.5 \pm 0.3) \times 10^{-16}$ A m^2 (7.5×10^{-13} emu).

An alternative, simpler approach for measuring the magnetic moment results from the fact that, to a good approximation, Eq. (48) can be rewritten in the form

$$m_y \approx +\left(\frac{\hbar}{e}\right) \int_{\text{chain}} \Delta\phi_{\text{mag}}(x, y)dl \tag{49}$$

where $\Delta\phi_{\text{mag}}$ is the local change in the magnetic contribution to the phase across the chain and l is a direction along the chain axis. The moment can therefore be calculated by multiplying the average value of $\Delta\phi_{\text{mag}}$ along the chain by its length and by $(h/2\pi e)$. This approach gives a moment for the MV-1 cell of 7×10^{-16} A m^2 (7×10^{-13} emu), in good agreement with the more accurate approach described previously.

Remanent Magnetization and Coercive Field Measurements of either the magnetic moment of a chain of crystals parallel to its axis or the parameter $\Delta\phi_{mag}$ [defined in relation to Eq. (49) above] can be used to infer the magnetization state of the constituent crystals. (For an isolated crystal, this would require a model for the magnetic fringing field that surrounds it). Electron holographic data obtained from chains of magnetite are consistent with a value for the magnetization of the crystals of 0.60 T (480 emu/cm^3) [245]. To within experimental error, this value is equal to the known magnetization of magnetite. This approach may be used to measure values of magnetization that are still not well known, such as that of greigite (Fe_3S_4) magnetosomes. However, care should be taken when examining crystals with easy axes that are not parallel to the chain axis or when examining chains that do not lie perpendicular to the incident electron beam direction.

The magnetization reversal process in a magnetosome chain may also be followed *in-situ*. For example, electron holograms may be acquired at successive sample tilts with the objective lens excited to provide a known component of the applied field in the plane of the sample. This approach was used to measure the coercive field of an MV-1 cell (the in-plane component of the applied field at which the magnetization direction of the chain reversed) [245]. By monitoring the difference in the measured phase shift across the chain as a function of applied field, the coercive field was measured to be between 300 and 450 Oe. The sensitivity of the coercive field to the size, separation, morphology, and the number of crystals in a chain [246], as well as to magnetocrystalline anisotropy, highlights the importance of studying individual magnetosome chains rather than obtaining statistical measurements from large numbers of bacteria. A comparison of the measured value with a chain-of-spheres calculation for magnetite [247] suggests that magnetization reversal proceeds by a symmetric fanning mechanism rather than by the parallel rotation of the magnetic moments of the crystals.

Morphology The experimental approach used to generate the phase images of the chain of bacterial crystals shown in Figure 25c and d provides both the mean inner potential and the magnetic contributions to the holographic phase shift. Whereas only the magnetic contribution is required to generate the contours shown in Figure 25e, the mean inner potential contribution can be used to provide additional information about the morphology and orientation of the magnetite crystals, particularly if it can be assumed that the external shape of each crystal can be formed only from a combination of {111}, {110}, and {100} faces. Figure 26a and b shows phase contours generated from, respectively, the mean inner potential and magnetic contributions to the phase shift of the end of a chain of magnetite crystals. The crystals are from the "Itaipu 1" and "Itaipu 3" strains of bacteria and were collected from a brackish lagoon at Itaipu in Brazil. In each case, the contours have been overlaid onto the measured mean inner potential contribution to the phase. The contours in Figure 26a are confined primarily to the crystals, while those in Figure 26b extend smoothly from within the crystals to the region surrounding them. Figure 26c shows line profiles that have been measured across the large and small magnetite crystals visible

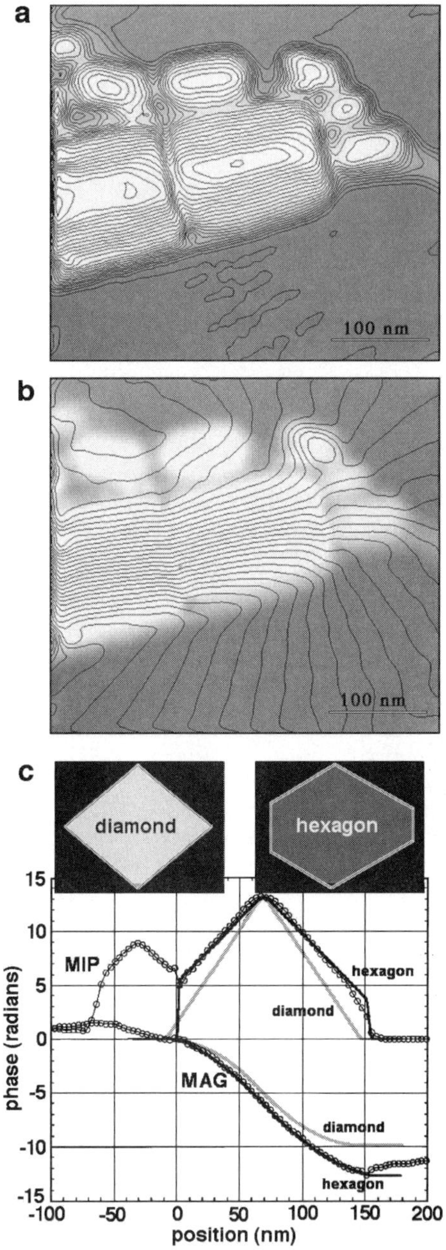

Figure 26. Phase contours showing (a) the mean inner potential and (b) the magnetic contribution to the reconstructed phase shift of the end of a chain of magnetite crystals from "Itaipu 1" and "Itaipu 3" strains of magnetotactic bacteria collected from a brackish lagoon at Itaipu in Brazil. In each case, the contours have been overlaid onto the measured mean inner potential contribution to the phase. (c) Line profiles obtained from images (a) and (b) across the large and small magnetite crystals close to the center of each image. The experimental data are shown as open circles. The darker solid line shows the best-fitting simulation to the data for the larger crystal, corresponding a distorted hexagonal cross-sectional shape (shown above the figure). The lighter line shows the worse fit provided by assuming a diamond shape in cross-section (also shown above the figure). (Taken from McCartney, unpublished work.)

close to the center of Figure 26a and b. Individual experimental data points are shown as open circles on the line profiles. Corresponding simulations are shown on the same axes. The darker solid line shows the best-fitting simulation to the data for the larger crystal, which corresponds to a distorted hexagonal shape in cross-section (shown as an inset above the figure). The lighter line shows a worse fit, provided by assuming a diamond shape in cross-section (also shown as an inset above the figure).

Small Crystals and Superparamagnetism The smallest crystals in some of the bacterial magnetite chains examined by holography should be superparamagnetic if they were isolated. At room temperature, this transition would occur when the crystals are below approximately 20 to 35 nm in size. As a result, their magnetization direction should fluctuate rapidly and their measured magnetization should be zero. However, the crystals are still found to be magnetized approximately parallel to the chain axes [245]. The expected thermal fluctuations may be overcome by interactions between the smallest crystals and the field of the larger crystals in the chain, rather than by local magnetostatic interactions between the small crystals themselves.

Modeling the Phase Shift of a Spherical Magnetic Nanocrystal The magnetization and the mean inner potential of a uniformly magnetized, isolated spherical particle [248] can be measured by fitting a line trace obtained across a phase image of the particle to an analytical expression for the expected phase shift, including the effect of magnetic fringing fields [249], which can in turn be derived from an expression for the vector potential of a uniformly magnetized sphere [250]. For a spherical particle of radius a, magnetization $B_\perp$ (along y), and mean inner potential V_0, the resulting expressions for the phase shift across the particle are

$$\phi(x, y)|_{(x^2+y^2)\le a^2} = 2C_E V_0\sqrt{a^2-(x^2+y^2)} + \left(\frac{e}{\hbar}\right)B_\perp\left(\frac{x}{x^2+y^2}\right) \times\left[1-\left(1-\left(\frac{x^2+y^2}{a^2}\right)\right)^{\frac{3}{2}}\right] \quad (50)$$

$$\phi(x)|_{(x^2+y^2)>a^2} = \left(\frac{e}{\hbar}\right)B_\perp a^3\left(\frac{x}{x^2+y^2}\right) \quad (51)$$

For a line profile obtained through the center of the particle in a direction perpendicular to that of $B_\perp$, these expressions reduce to

$$\phi(x)|_{x\le a} = 2C_E V_0\sqrt{a^2-x^2} + \left(\frac{e}{\hbar}\right)B_\perp\left[\frac{a^3-(a^2-x^2)^{\frac{3}{2}}}{x}\right] \quad (52)$$

$$\phi(x)|_{x>a} = \left(\frac{e}{\hbar}\right)B_\perp\left(\frac{a^3}{x}\right) \quad (53)$$

Figure 27a and b shows a hologram and a reconstructed phase image for a chain of spherical Co particles [251] that are suspended over a hole in a C support film. Figure 27c and d shows corresponding line traces obtained from the phase image across the centers of two particles. Each line trace is obtained in a direction perpendicular to the chain

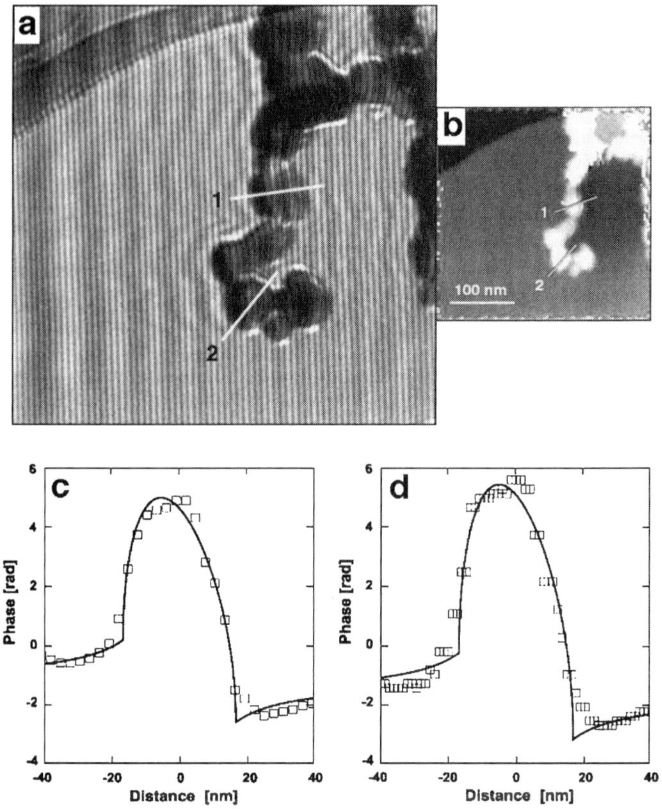

Figure 27. (a) Off-axis electron hologram of a chain of Co particles suspended over a hole in a carbon support film, acquired at 200 kV using a Philips CM200 FEG-TEM and a biprism voltage of 90 V. (b) Corresponding unwrapped phase image. (c) and (d) Experimental line profiles formed from lines 1 and 2 in (b), and fitted phase profiles generated for spherical Co nanoparticles. Reprinted with permission from [249], M. de Graef et al., *J. Microsc.* 194, 84 (1999). © 1999, Blackwell Science Ltd.

axis. Care is required when applying Eqs. (52) and (53) to chains of crystals, since these equations are strictly only applicable to isolated particles. Nevertheless, least squares fits of Eqs. (52) and (53), which are also shown in Figure 27c and d, to the data points provide respectable best-fitting values for a, $B_\perp$, and V_0 of 17 nm, 1.7 T, and 26 V, respectively. A decrease in the apparent remanent magnetization of a magnetized sphere with particle size may be attributed to the formation of nonuniform domain structures as the particle size increases [252].

As a point of interest, the height of the step in the magnetic contribution to the phase shift across a uniformly magnetized sphere of radius a can be determined from Eqs. (52) and (53) to be given by the expression

$$\Delta\phi_{\mathrm{MAG}} = \left(\frac{2\sqrt{2}\left(1-\left(1-\frac{\sqrt{3}}{2}\right)^{\frac{3}{2}}\right)}{3^{\frac{1}{4}}}\right)\left(\frac{e}{\hbar}\right)B_\perp a^2 \equiv 2.044\left(\frac{e}{\hbar}\right)B_\perp a^2 \quad (54)$$

which can be contrasted with the equivalent expression for a uniformly magnetized cylinder of radius a of

$$\Delta\phi_{\mathrm{MAG}} = \pi\left(\frac{e}{\hbar}\right)B_{\perp}a^2 \qquad (55)$$

3.2.6. Nanomagnet Arrays Patterned Using Interferometric Lithography

TEM sample preparation presents a challenge for certain samples that contain nanostructured magnetic materials. An example involves the characterization of nanomagnet arrays that have been fabricated using interferometric lithography [253] directly onto Si substrates rather than onto self-supporting electron-transparent membranes of silicon nitride, alumina, or carbon. Figure 28a shows a scanning electron microscope image of 100-nm-diameter, 20-nm-thick Co dots fabricated on Si in a square array of side 200 nm. The dots were prepared for TEM examination using focused ion beam milling in plan-view geometry, by micromachining a trench from the substrate side of the sample to leave a freestanding 10×12 μm membrane of crystalline Si, which was approximately 100 nm in thickness and contained approximately 3250 Co dots. Figure 28b shows an off-axis electron hologram from part of the electron-transparent membrane containing the dots, prepared using focused ion beam (FIB) milling. The hologram was acquired at 200 kV using a Philips CM200 FEG-TEM, a biprism voltage of 160 V, a holographic interference fringe spacing of 3.05 nm, and an overlap width of 1.04 μm. The holograms were recorded in zero applied field, with the samples tilted slightly away from zone axis orientations of the underlying Si substrate to minimize diffraction contrast. The sample edge is toward the bottom left of the figure. Although great care was taken to minimize damage from sample preparation, it is possible that the elliptical shape of the dots results from damage sustained during FIB milling [254].

Figure 28c and d shows contours that have been added to the (slightly smoothed) magnetic contribution to the holographic phase for two different remanent magnetic states of the Co dots. Subtraction of the mean inner potential contribution to the measured phase before forming the contours was essential for successful characterization of the magnetic microstructure. The contour spacing is $0.033 \approx \pi/94$ radians. In Figure 28c, which was recorded after saturating the dots upward and then removing the external field, the dots are all oriented magnetically in the direction of the applied field. In contrast, in Figure 28d, which was formed by saturating the dots upward, applying a 382 Oe downward field, and then removing the external field, the dots are magnetized in a range of directions. The experiments show that the dots are sometimes magnetized out of the plane (e.g., at the bottom left of Fig. 28d) and also that shape anisotropy does not seem to dominate their behavior. The measured saturation magnetizations are smaller than expected for pure Co, possibly because of oxidation or damage sustained during sample preparation. Similar electrodeposited 57-nm-diameter, 200-nm-high Ni pillars arranged in square arrays of side 100 nm, which were prepared for TEM examination using focused ion beam milling in cross-sectional geometry, have also been examined. Despite their shape, not all of

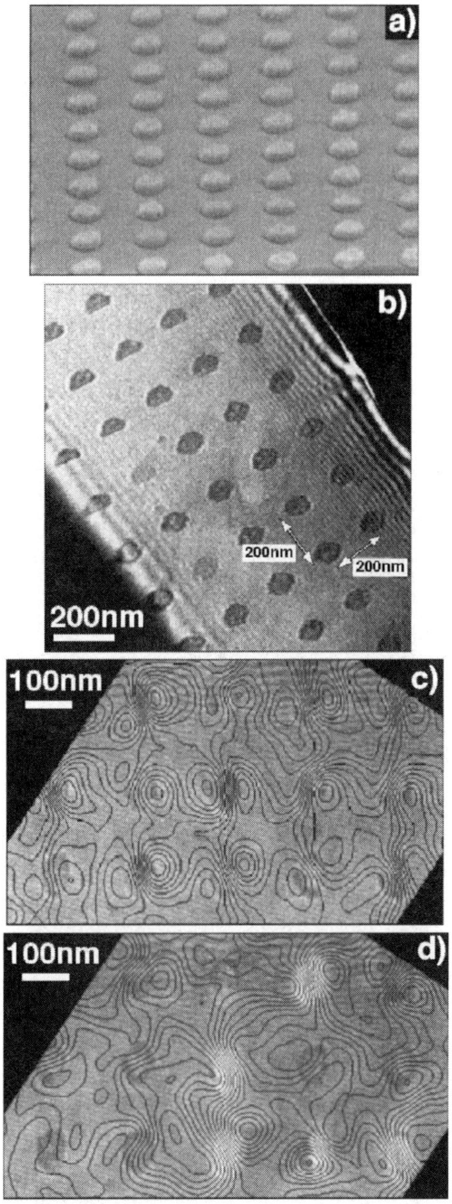

Figure 28. (a) Scanning electron microscope image of 100-nm-diameter, 20-nm-thick Co dots fabricated on Si in a square array of side 20 nm using interferometric lithography. (b) Off-axis electron hologram of part of an electron-transparent membrane containing the dots, prepared using focused ion beam milling. The hologram was acquired at 200 kV using a Philips CM200 FEG-TEM, a biprism voltage of 160 V, a holographic interference fringe spacing of 3.05 nm, and an overlap width of 1.04 μm. (c) and (d) Magnetic contributions to the measured electron holographic phase for two remanent states. The contour spacing is 0.033 radians. (c) was formed by saturating the dots upwards and then removing the external field. (d) was formed by saturating the dots upward, applying a 382 Oe downward field and then removing the external field. Reprinted with permission from [254], R. E. Dunin-Borkowski et al., *Inst. Phys. Conf. Ser.* 168, 485 (2001). © 2001, Institute of Physics Publishing.

the Ni pillars were magnetized parallel to their long axes. Instead, the pillars interacted with each other strongly, with two, three, or more adjacent pillars frequently combining to form vortices [254].

3.2.7. Magnetic Force Microscopy Tips

Magnetic force microscopy (MFM) offers an inexpensive way of measuring magnetic microstructure by scanning a magnetized tip across the surface of a sample. However, the observed contrast is a complicated function of the magnetic fields emanating from the tip and the sample, both of which may be unknown. Electron holography has been used by several groups, in combination with micromagnetic simulations, to interpret the magnetic fields surrounding MFM tips.

As an example, Figure 29a shows a schematic diagram of the coating procedure applied to a pyramidal thin-film ferromagnetic MFM tip. Figure 29b shows the magnetic leakage field measured using electron holography outside a commercially available silicon nitride tip similar to that shown in Figure 29a, which was mounted on the end of a cantilever, coated on one of its faces with a 100-nm-thick layer of $Co_{80}Ni_{20}$ and subsequently magnetized in a direction parallel to the pyramid axis [255, 256]. In Figure 29b, the magnetic material covers only the left hand side of the tip. Figure 29c shows a best-fitting simulation to the contoured phase image shown in Figure 29b. The simulation was obtained by describing the tip using a uniform distribution of magnetic dipoles, which were initially oriented randomly and were then each allowed to rotate freely until a best fit to the experimental data was obtained. In contrast, similar results obtained from a thin Ni wire [40] were modeled more simply by comparing them with the field of a 20-μm-long dipole.

Figure 30a shows a reconstructed holographic phase image from a sputter-coated MFM tip obtained using the far-out-of-focus mode of STEM electron holography in a VG HB5 STEM [257]. The nominal radius of the tip is 30 nm, and the spacing between adjacent phase contours is $\pi/2$ radians. Even though the MFM tip magnetic field is three-dimensional, in this case the experimental electron holographic data were interpreted quantitatively on the assumption that the field was cylindrically symmetrical. Figure 30b shows a best-fitting simulation, superimposed onto the part of the experimental phase image close to the tip, which was obtained by fitting a multipole expansion of the magnetic scalar potential to the phase image. The fitting coefficients include monopole, dipole, and quadrupole (tensor) terms, with no assumption made about either their size or their orientation. As in Figure 29, the clear agreement between the experimental data and the simulation is evident, and the strength of the magnetic field along the tip axis can be derived directly from the multipole fit to the data. The

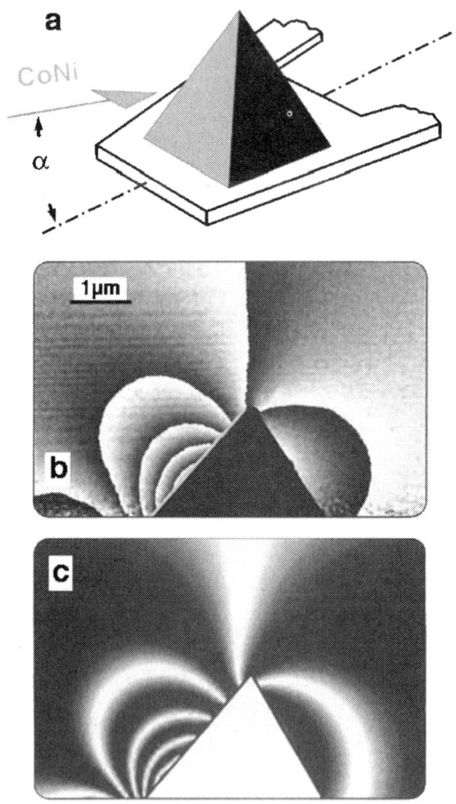

Figure 29. (a) Schematic diagram of the coating procedure applied to a pyramidal thin-film ferromagnetic tip used for magnetic force microscopy. (b) Magnetic lines of force measured using electron holography for a pyramidal tip that was coated on one of its faces with a 100-nm-thick layer of $Co_{80}Ni_{20}$ and subsequently magnetized in a direction parallel to the pyramid axis. The hologram was acquired at 200 kV using a Hitachi HF-2000 FEG-TEM. (c) Best-fitting simulation to the contoured phase image shown in (b), obtained by describing the tip as a uniform distribution of magnetic dipoles, which are each allowed to rotate freely until a best fit to the experimental data is obtained. Reprinted with permission from [255], B. G. Frost et al., *Appl. Phys. Lett.* 68, 1865 (1996). © 1996, American Institute of Physics.

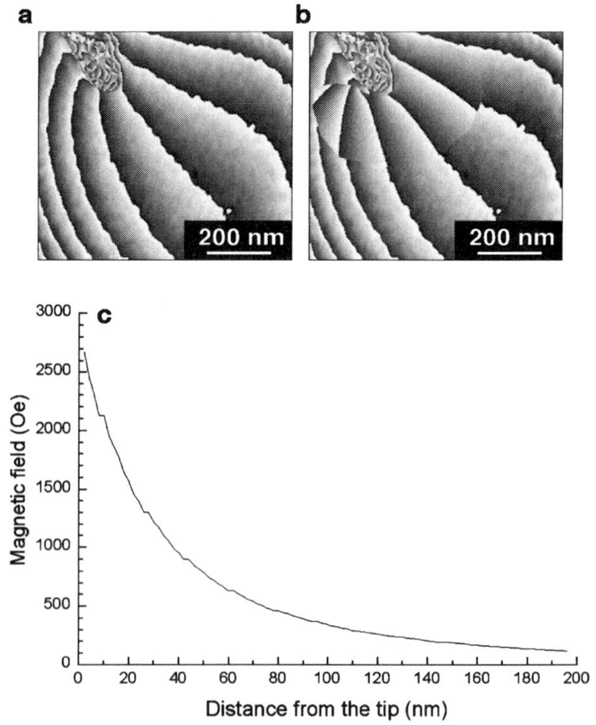

Figure 30. (a) Recorded holographic phase image of a sputter-coated magnetic force microscopy tip of nominal radius 30 nm, which was initially magnetized along its axis ($\pi/2$ phase contours). (b) Experimental phase image with multipole expansion fit superimposed. (c) Absolute value of the magnetic field directly in front of the tip. Reprinted with permission from [257], D. G. Streblechenko et al., *IEEE Trans. Magn.* 32, 4124 (1996). © 1996, IEEE.

resulting field, shown in Figure 30c, falls from about 2500 Oe near the tip surface to 500 Oe at a distance of 100 nm in front of the tip. The fitted dipole is oriented nearly along the tip (15° off-axis), which is consistent with the fact that the tip was oriented ~12° from the magnetic field direction during the magnetization process.

3.2.8. Cross-Sectional Samples—Removal of Sample Thickness Effects

One of the most challenging problems for electron holography of magnetic materials is the quantitative measurement of the magnetic properties of nm-scale magnetic layers that are examined in cross-section. The primary difficulty is the presence of rapid and unknown variations in both the composition and the thickness of the sample, from which the weak magnetic signal must be separated. In a cross-sectional sample, magnetic contributions to the phase and the phase gradient are then much smaller than the mean inner potential contribution. This point has been illustrated both experimentally and using simulations [211] for an AlO_x tunnel barrier sandwiched between CoPtCr and CoPt ferromagnetic layers. In this work, the similarity noted between the shapes of the magnetization profile and the magnetic contribution to the phase gradient evaluated across the ferromagnetic layers suggested the possibility that differences in the phase gradient between images in which the magnetization has been reversed could provide the basis for more accurate magnetization characterization. In contrast, the magnetic contribution to the phase was observed to change cumulatively across the layers. In such samples, the magnetization profile differs from the magnetic contribution to the phase gradient only because of variations in specimen thickness within the magnetic layers. These contributions due to specimen thickness cannot be eliminated by using the normalized amplitude of the hologram defined in Eq. (38), both because the mean free path λ_{in} in each material in such a cross-sectional sample is generally unknown and also because the amplitude image is in general noisy and may contain strong contributions from diffraction contrast and Fresnel contrast.

Rearrangement of Eqs. (22) and (23) indicates that sample thickness effects may be removed by plotting the difference in the phase gradient between images in which the magnetization has reversed divided by the average of their phases, multiplied by a constant and by the value of the mean inner potential of each magnetic layer separately. Formally, this process can be written in the form

$$\left(\frac{C_E \hbar V(x)}{e}\right)\left\{\frac{\Delta[d\phi(x)/dx]}{\langle\phi(x)\rangle}\right\} = \frac{\Delta[B_\perp(x)]}{(1-(e/C_E\hbar V(x))\{\langle\int B_\perp(x)t(x)dx\rangle/t(x)\})} \quad (56)$$

According to Eq. (56), by combining the phase profiles and phase gradients (evaluated in a direction perpendicular to the layers) from successive holograms between which the magnetization direction has been reversed, the specimen thickness profile is eliminated and the magnetization can be determined quantitatively. This process is equivalent to using the average phase to calculate the specimen thickness and then using this knowledge to eliminate the thickness from the magnetic contribution to the phase gradient. Most importantly, both the magnitude and the sign of $\Delta[B_\perp(x)] = 2B_\perp(x)$ are obtained exactly using Eq. (56) if the magnetization reverses exactly everywhere in the sample between the two holograms. (The denominator on the right-hand side of the equation is then unity.) Furthermore, nonzero values are returned only in regions where the magnetization has changed.

Figure 31 illustrates the application of Eq. (56) to a cross-sectional sample of a magnetic tunnel junction that contains a layer sequence of 22 nm Co/4 nm HfO_2/36 nm CoFe on a Si substrate [258]. Two holograms were obtained, similar to that shown in Figure 31a, between which the magnetization directions of the Co and CoFe layers in the sample were reversed *in-situ* in the electron microscope. Figure 31b shows an unwrapped phase profile obtained from the hologram in Figure 31a by taking a line profile in the direction perpendicular to the layers. The phase profiles from the two holograms appeared almost identical irrespective of the direction of magnetization. The application of Eq. (56) to the two phase images (i.e., dividing the difference in the gradients of the two phase images obtained perpendicular to the layers by the average of the original phases) results in the image shown in Figure 31c, which represents $2B_\perp(x)$. The line profile in Figure 31d was obtained by averaging Figure 31c along the direction of the layers. As predicted, Figure 31d, which should by now be independent of variations in composition and sample thickness, is nonzero only in the magnetic layers and yields a value for the magnetization of Co of 1.5 T (assuming a mean inner potential of 25 V). (In a similar experiment, holograms of $La_{0.5}Ca_{0.5}MnO_3$ have

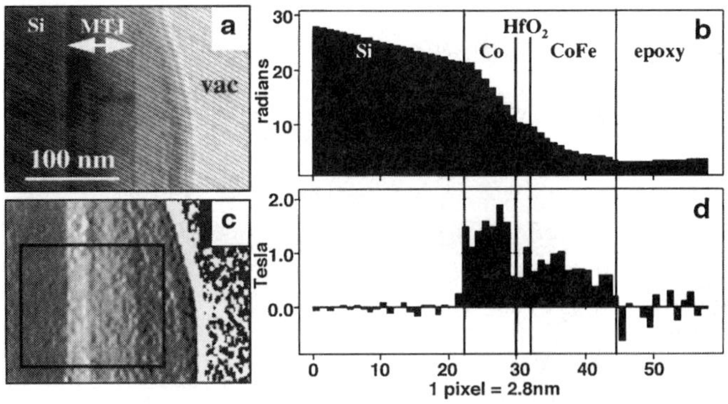

Figure 31. (a) Off-axis electron hologram obtained from a magnetic tunnel junction with a 4-nm HfO_2 tunnel barrier. (b) Measured phase profile across layers in tunnel junction structure. (c) Image formed by recording two holograms with opposite directions of magnetization in the sample, taking the difference between the recorded phase gradients (calculated in a direction perpendicular to the layers) and dividing by the average of the two phases. (d) Measured magnetic induction in the tunnel junction sample, generated by multiplying a line profile obtained from image (c) by a constant (see text for details), with the vertical scale now plotted in units of Tesla. Reprinted with permission from [258], M. R. McCartney and R. E. Dunin-Borkowski, in "Electron Microscopy 98" (H. A. Calderón Benavides and M. José Yacamán, Eds.), Vol. 2, p. 497. Paper presented at the 14th International Conference on Electron Microscopy, Cancun, Mexico, 31 August–4 September, 1998. Institute of Physics Publishing, Bristol, UK. © 1998, Institute of Physics Publishing.

recently been acquired both above and below the Curie temperature of the material to remove sample thickness and mean inner potential contributions from the measured phase [203].)

Artifacts may be present in a profile such as that shown in Figure 31d. The primary disadvantage of using Eq. (56) to determine the magnetization in a cross-sectional sample is that it utilizes differences between subtle changes in the phase gradient. Thus, holograms of high quality are required and it is important to register them to subpixel accuracy. The magnetic layers must be unbroken and uniform in cross-section along their length to minimize any magnetic fringing fields around them. If the sample is too thick or the image too far from focus, then Fresnel effects at the layers may cause additional oscillations in the phase images [211]. As a result, it is particularly important to record the two phase images both close to focus and at the same defocus. The accuracy with which the absolute magnetization can be inferred is limited by the degree to which the mean inner potential of the area of interest can be estimated. The approach is not affected by diffraction contrast if the local orientation of the sample is identical for the two holograms. However, surface contamination layers should be thin in order to validate multiplication by the local value of the mean inner potential. The sample should be sufficiently thick so that division by the average phase does not substantially increase the noise in the inferred magnetization profile.

Recently, electron holography has been used to examine spin valve structures containing oxide layers, and to infer a 3–4 V step in potential across a metal/oxide interface. This potential barrier was related to the magnetoresistive properties of the material [259].

3.3. Superconducting Thin Films

An external magnetic field can pass through a type-II superconductor in the form of thin filaments with a magnetic flux that is a multiple of $h/(2e) = 2 \times 10^{-15}$ Wb. These filaments are known as fluxons, or vortices. The critical current of the superconductor is determined by how tightly the fluxons are attached to pinning centers when a Lorentz force is exerted on them by a current passing through the sample. Electron holography allows fluxons to be observed directly, both in profile and in transmission. Dynamic observations of their motion are also possible.

Figure 32a illustrates the procedure used to study static fluxons in profile. In this case, lead was evaporated onto one side of a 30-μm-diameter tungsten wire that had been cleaned by flash-heating. (Lead is a type-I superconductor in bulk form and a type-II superconductor when it is in the form of a thin film.) The sample was examined with a magnetic field of a few Gauss applied in a direction perpendicular to the film. Whereas the Meissner effect excludes a weak magnetic field from the superconducting film, magnetic lines of force penetrate the superconductor as the strength of the applied field is increased. These magnetic lines of force can then be characterized directly using electron holography. Figure 32b shows a two-times-phase-amplified interference micrograph obtained from a 0.2-μm-thick superconducting lead film at a temperature of 4.5 K [260, 261]. The critical temperature of the film is 7.2 K. In this figure, each fringe corresponds to a single fluxon, which is observed to have a diameter of approximately 150 nm at the sample surface. The image shows a single fluxon on the right and a fluxon pair on the left. The magnetic field at the position of a fluxon near the surface of the sample is as high as 1000 Gauss, even though the applied magnetic field is only a few Gauss. As expected, the magnetic lines of force outside the sample disappear above the critical temperature. Figure 32c shows corresponding magnetic lines of force recorded from a 1-μm-thick Pb film, again with two times phase amplification. In this image, the magnetic flux always emerges from the sample surface in integral multiples of $h/(2e)$. Observations similar to those shown in Figure 32 have been made "dynamically" as the sample temperature is raised and the fluxon diameters increase [262]. Using this approach, fluxons

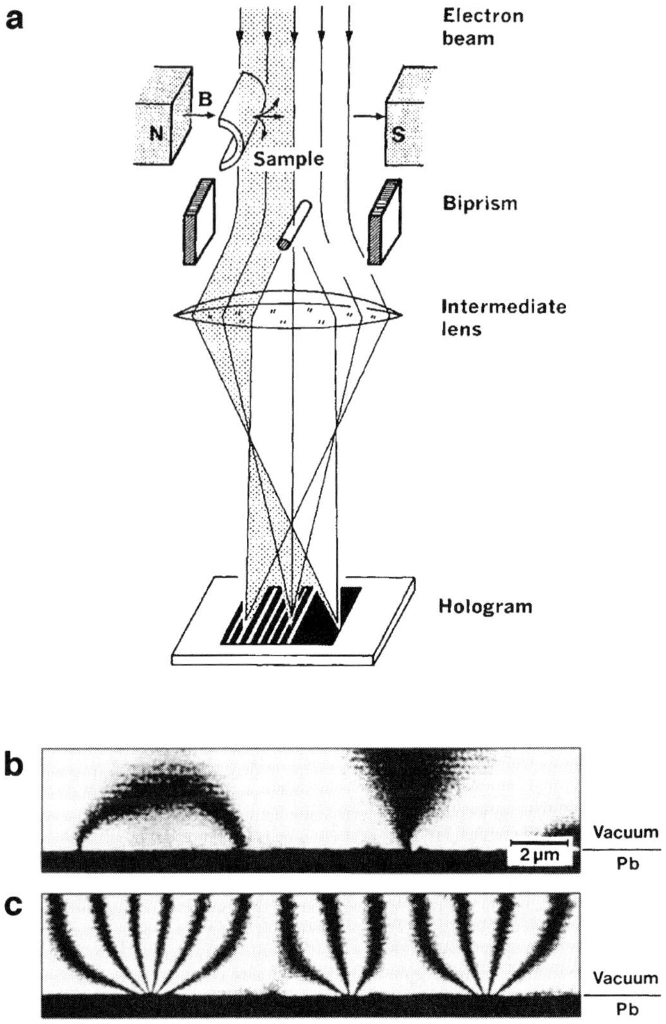

Figure 32. (a) Electron-optical configuration used for recording a hologram of a superconducting Pb film evaporated onto one side of a tungsten wire. (b) and (c) Two-times-phase-amplified interference micrographs of magnetic flux penetrating superconducting Pb films of thickness 0.2 and 1.0 μm, respectively. The holograms were recorded at a temperature of 4.5 K. The critical temperature of the films is 7.2 K. Reprinted with permission from [260], T. Matsuda et al., *Phys. Rev. Lett.* 62, 2519 (1989). © 1989, American Physical Society.

were observed to move from one pinning center to the next. The dynamics of fluxons that are driven by the Lorentz force as a current is passed through the sample have also been studied [263], allowing the pinning force in the film to be measured. Computer simulations of the phase contours of fluxons observed in profile have highlighted the importance of including the effect of the perturbed reference wave on such calculations [264].

The experiment described in Figure 32 only reveals the magnetic lines of force *outside* the superconductor. A transmission geometry is required to visualize fluxon distributions inside a superconductor. In this mode, the superconducting film has to be tilted, as shown in Figure 33a, since the phase shift associated with a fluxon is too small to be detected when the electron beam is parallel to its axis [265, 266]. The TEM sample should have uniform thickness, so that changes in film thickness do not obscure small phase changes that arise from the presence of the fluxon. Figure 33b shows a 16-times-phase-amplified interference micrograph of a vortex lattice in a 100-nm-thick superconducting film of single-crystalline Nb, recorded at a temperature of 4.5 K with a 150 Gauss magnetic field applied horizontally and the sample tilted by 45° [267]. Regions where the contours are most closely spaced, which are circled in the image, correspond to the positions of fluxons. Figure 33c, d, and e illustrates measurements of changes in the sizes of vortices as the sample temperature is changed from 4.5 to 7.0 and 8 K, respectively [268]. The average vortex diameter at these temperatures is measured to be 150, 185, and 230 nm, respectively. As the vortices broaden with increasing temperature, the flux enclosed remains constant. Simulations of the holographic contrast observed at fluxons in transmission mode [269–275] have been performed. In contrast to electron holography, Lorentz microscopy, which reveals fluxons as pairs of bright and dark contrast features, can be used to observe fluxon dynamics in real time, including flux pinning as the sample temperature and the applied magnetic field are varied [276–284].

4. ELECTROSTATIC FIELDS IN NANOSTRUCTURED MATERIALS

The characterization of magnetic fields in nanostructured materials by electron holography, as described in Section 3, has been explored in considerable depth. Key issues and problems have been addressed, and many remaining challenges are either technical or else involve the further development of existing aspects of the technique. In contrast, the application of electron holography to the characterization of electrostatic fields in nanostructured materials (e.g., [285, 286]) is less developed and requires several basic technical and conceptual problems to be resolved. In this section, work that has already been carried out in this field is reviewed. Initial examples are taken from the characterization of electrostatic fringing fields outside charged particles and electrically biased carbon nanotubes. The considerable challenges that are associated with imaging dopant contrast at depletion layers in semiconductors are described, before proceeding to other more complicated systems such as the characterization of interfaces at which both charge redistribution and changes in chemistry are possible.

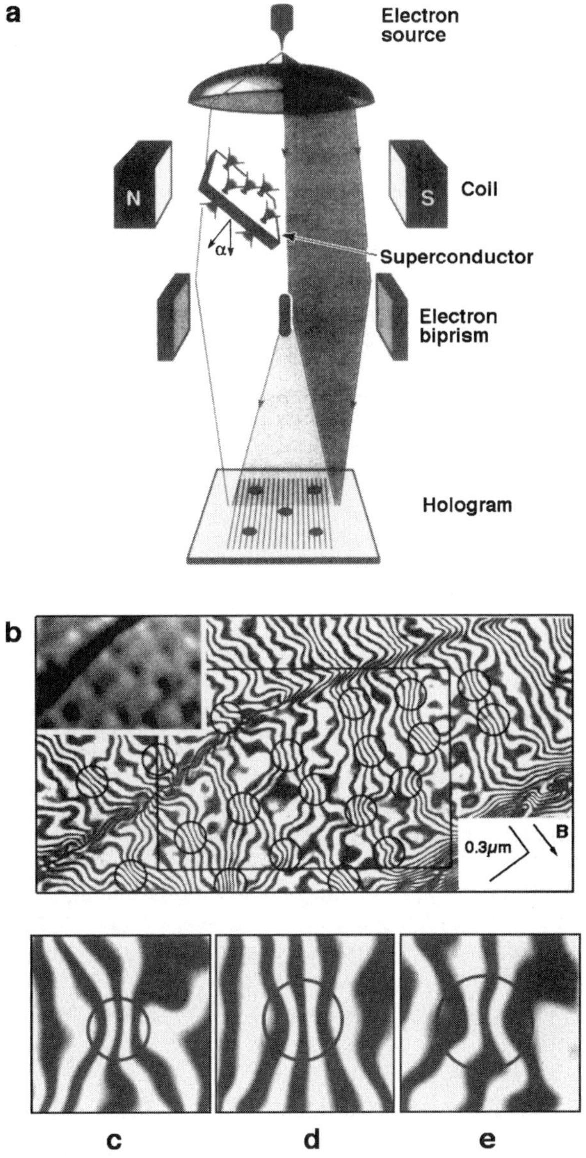

Figure 33. (a) Electron-optical configuration used for recording holograms of vortices in a tilted 100-nm-thick superconducting sample of Nb, with a magnetic field applied horizontally. (b) A 16-times-phase-amplified interference micrograph of a vortex lattice, recorded at 300 kV and at a temperature of 4.5 K. The circled regions correspond to the positions of quantized vortices. A bend contour runs diagonally across the image. Reprinted with permission from [267], J. E. Bonevich et al., *Phys. Rev. Lett.* 70, 2952 (1993). © 1993, American Physical Society. (c), (d), and (e) Vortex broadening observed with increasing temperature: 4.5, 7.0, and 8 K, respectively (12-times-amplified phase images). Reprinted with permission from [268], J. E. Bonevich et al., *Phys. Rev. B* 50, 567 (1994). © 1994, American Physical Society.

4.1. Charged Latex Particles

Latex spheres that become charged when exposed to a high-energy electron beam are often used as "standard" samples for studying electrostatic fields using electron holography (e.g., [41, 42, 285, 287]). Electron holography has also been used to measure charging effects due to electron irradiation in 250-nm-diameter SiO_2 particles [288]. For example, Figure 34 shows a simulation of equiphase contours

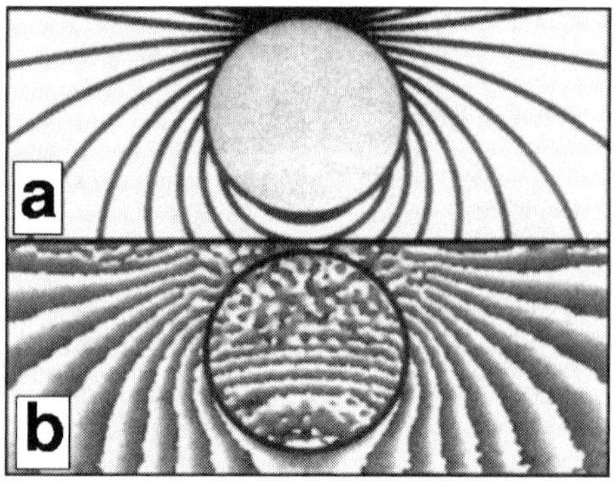

Figure 34. (a) Simulated (b) experimental equiphase lines for a charged 2.7-μm-diameter latex sphere, reconstructed using a reference hologram. Reprinted with permission from [289], B. Frost and T. L. Jenkins, *J. Microsc.* 187, 85 (1997). © 1997, Blackwell Science Ltd.

associated with a 2.7-μm-diameter latex sphere placed close to a Cu grid bar, together with the corresponding experimental phase image [289]. The simulation takes into account the image charge of the sphere in the grid bar and assumes that the charge is distributed uniformly throughout the particle. The match between the simulation and the measured phase profile in Figure 34 appears to be satisfactory, but changes in the measured phase distributions across latex spheres and other materials have been reported to depend on the condenser aperture size and the objective lens strength used when recording the holograms [290, 291]. These observations indicate the need for care when recording and interpreting such low resolution phase images. However, the origin of the differences remains to be explained. Possible reasons could include changes in the way in which either the sample or parts of the microscope have accumulated electrostatic charge as the electron optical conditions are varied.

4.2. Field-Emitting Carbon Nanotubes

Electron holography can be used *in-situ* in the TEM to measure the electrostatic potential in samples that have a voltage applied to them. Early experiments that demonstrated this capability were carried out on tungsten microtips, which exhibited a strong electrostatic fringing field [292], on pairs of parallel conducting (1-μm-diameter Pt) wires held at different potentials [293], and on single conducting wires [294]. Simulations have also been presented for electrostatic phase plates [295]. A more recent example involved the use of electron holography to map the electrostatic potential surrounding the end of a field-emitting, electrically biased, multiwalled carbon nanotube [296]. (The ability to bias carbon nanotubes in the TEM has also been demonstrated in other experiments, e.g., [297, 298].) The carbon nanotubes were mounted onto a piezo-driven three-axis manipulation electrode using conducting epoxy and positioned approximately 6 μm from a gold electrode, as shown in Figure 35a. Depending on the biasing conditions, electrons could be emitted from the nanotube to the gold collector plate, in a direction perpendicular to the incident high-energy electron beam. Electron holograms were acquired during this biasing experiment using a Philips CM200ST FEG-TEM

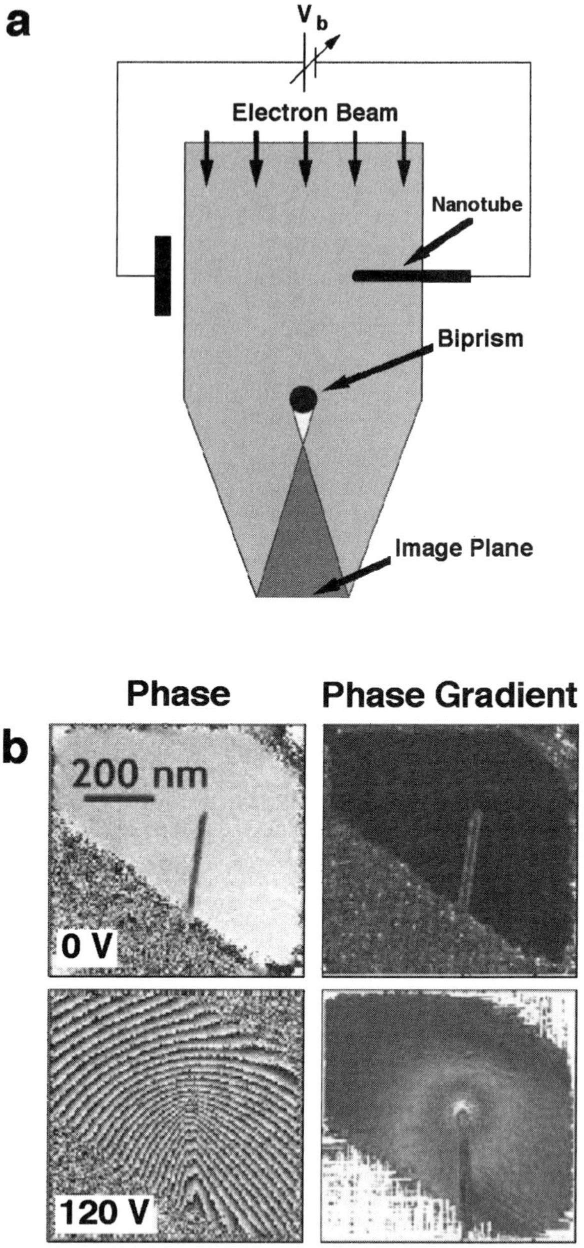

Figure 35. (a) Schematic diagram of the experimental setup used to record electron holograms of field-emitting carbon nanotubes. (b) Phase shift and phase gradient maps extracted from holograms of the same nanotube at bias voltages of 0 and 120 V. The phase gradient indicates where the electric field is strongest; note the concentration of the electric field at the nanotube tip when the bias voltage is 120 V. Reprinted with permission from [296], J. Cumings et al., *Phys. Rev. Lett.* 88, 056804 (2002). © 2002, American Physical Society.

operated at 200 kV and an interference fringe spacing of 4 nm. Figure 35b shows contoured phase images in the left hand column recorded both before any bias was applied to the sample and at a bias V_b above the threshold for electron field emission (approximately 70 V for this sample). The upper phase shift map (at $V_b = 0$) shows a featureless area around the nanotube, whereas the lower map (at $V_b = 120$ V) shows a striking distribution of closely spaced 2π

phase contours. The right hand column in Figure 35b shows the corresponding two-dimensional phase gradient for each phase image. (The phase gradients are much less affected by a perturbed reference wave than the original phase images.) When $V_b = 0$, the phase gradient is featureless around the nanotube, whereas the gradient is concentrated around the nanotube tip when $V_b = 120$ V. The phase and phase gradient images shown in Figure 35b were interpreted quantitatively by comparing them with simulated images, calculated on the assumption that the nanotube could be approximated by a line charge, on which the charge distribution was varied until a best fit to the data was found. The best fit to the 120 V phase data shown in Figure 35b gave a value of 1.22 V/nm for the electric field at the nanotube tip. This electric field was concentrated at the tip and was not observed at other defects such as sidewall imperfections. It was also stable over time, even when the field emission current varied. The fact that clear holograms could be recorded from a single field-emitting nanotube placed an upper limit on fluctuations in the electric field strength at the tip of 0.03 V/nm.

4.3. Dopant Potentials in Semiconductors

One of the most elusive but tantalizing problems for electron holography has been the quest for a reliable and quantitative approach for the characterization of electrostatic potentials associated with charge redistribution at depletion layers in doped semiconductors. Attempts to tackle this problem have been made since the 1960s using many forms of electron interferometry, both experimentally under reverse-bias conditions (e.g., [299–306]) and also theoretically (e.g., [307–312]). In most of these studies, in which chemically thinned samples were examined, virtually zero contrast was observed at the position of a *p–n* junction until it was reverse-biased *in-situ* in the TEM. Although it was predicted that external electrostatic fringing fields outside the TEM sample could, in principle, dominate the observed phase shift at the junction position, these fringing fields were rarely seen at any *p–n* junction until it was reverse-biased.

It has recently been recognized that TEM sample preparation can have a profound effect on the contrast seen in holographic phase images of doped semiconductors, either because of physical damage to the sample surface or because of the implantation of dopant ions such as Ar or Ga. Either an electrically inactive surface layer or a doped layer, with a thickness that varies depending on the sample preparation method, may then form at the sample surface. In addition, the sample may become charged during observation, to such an extent that all dopant contrast is lost. It is possible that the effects of sample preparation, and in particular the electrical state of the sample surface, could account for many of the anomalous results seen in early experiments, including the absence of electrostatic fringing fields outside unbiased *p–n* junctions. Recent studies, which are now reviewed, suggest that these problems are being successfully addressed.

4.3.1. Transistors

Following earlier experiments on one-dimensional dopant profiles (e.g., [313]), the first unequivocal demonstration of the two-dimensional mapping of the electrostatic potential in an *unbiased* doped semiconductor using electron holography was achieved for metal-oxide-semiconductor (MOS) Si transistor structures [67, 314, 315]. Figure 36 shows electron holography results obtained from several 0.35-μm-channel-length MOS transistors, examined in cross-section at 200 kV using a Philips CM200 FEG-TEM. The samples were tilted by approximately 4° from Si $\langle 110 \rangle$ to minimize dynamical diffraction contributions to the contrast. Amplitude images from these samples are shown in Figure 36a and b, and corresponding phase images are shown in Figure 36c and d. The images shown in Figure 36a and c were obtained from an *n*-MOS transistor, while Figure 36b and d was obtained from a *p*-MOS transistor. The source and drain regions in each transistor structure (marked n^+ and p^+) are clearly visible in the phase images with a spatial resolution of 10 nm and an energy resolution of 0.10 V. The effect of differential thinning across the junctions was discounted as a possible cause of the observed phase shifts, and an optimal sample thickness of between 200 and 400 nm was identified for such experiments.

Two aspects of this study deserve further comment. First, the transistor structures were prepared for TEM examination using conventional mechanical polishing and Ar-ion milling. Although this approach proved successful here, it has since been found elsewhere to result in gross charging effects in other similar samples. Second, a 25-nm-thick electrically dead layer was identified on each surface of the sample, which resulted in a measured built-in voltage across each *p–n* junction of 0.9 ± 0.1 V. This value was lower than the value of 1.0 V predicted for the specified dopant concentrations. It should, however, be noted that the voltage profile is remarkably insensitive to changes in the charge density profile across the junction. As a

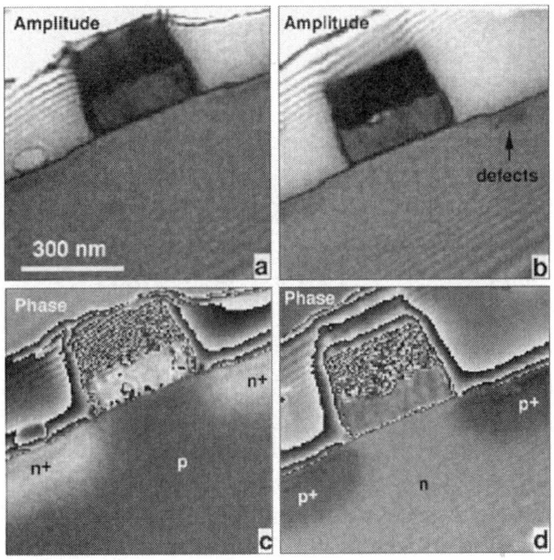

Figure 36. Electron holography results revealing the electrostatic potential distributions in metal-oxide-semiconductor transistors. Amplitude and phase images of 0.35 μm channel length (a) and (c) *n*-MOS and (b) and (d) *p*-MOS transistors, viewed in cross-section. The source and drain areas (marked n^+ and p^+) are clearly visible in the phase images. The holograms were recorded at 200 kV in a Philips CM200 FEG-TEM. Reprinted with permission from [67], W. D. Rau et al., *Phys. Rev. Lett.* 82, 2614 (1999). © 1999, American Physical Society.

result, partial deactivation of the dopant in the depletion region due to the TEM sample preparation may account for the lower than expected value of the built-in voltage, but without changing the position of each junction in the device appreciably. Further results have since been published on the examination of process-induced boron redistribution in Si/SiGe/Si heterostructures [316]. These results were compared with process simulations in order to measure parameters such as surface recombination lengths for self-interstitials.

More recently, electron holography studies of transistor structures have been compared with process simulation and secondary ion mass spectrometry results [317]. Figure 37a shows a contoured phase image of the electrostatic potential distribution in a 0.35 μm Si device structure, which was recorded at 200 kV in a Philips CM200 ST FEG-TEM. The contour step in this image corresponds to a potential step of 0.1 V. As in Figure 36, the B-doped source and drain regions in the device are clearly delineated in the phase image. In this study, the sample was prepared using tripod wedge-polishing followed by limited low-angle Ar ion milling at 3.5 kV. Significantly, no electrically dead layer needed to be taken into account to quantify the results. However, there was some evidence for charging in the near-surface regions under the sidewall spacers. Figure 37b and c shows a comparison between simulations and line profiles obtained from Figure 37a, both laterally across the junction and as a function of depth from the Si surface. Simulations for "scaled loss" and "empirical loss" models, which account for B-implant segregation into the adjacent oxide and nitride layers, are shown. The scaled-loss model, which leads to stronger B diffusion, assumes uniform B loss across the device structure, while the empirical loss model assumes segregation of the implanted B at the surface of the source and drain regions. In both Figure 37b and c, the empirical loss model provides a better match to the experimental results. Figure 37d shows a simulated electrostatic potential map for the same device structure based on the "empirical loss" model, with a contour step of 0.1 V, which matches closely with the experimental image in Figure 37a. Overall, this study demonstrated the successful two-dimensional mapping of the electrostatic potential in 0.13 and 0.35 μm device structures with a spatial resolution of 6 nm and a sensitivity of 0.17 V.

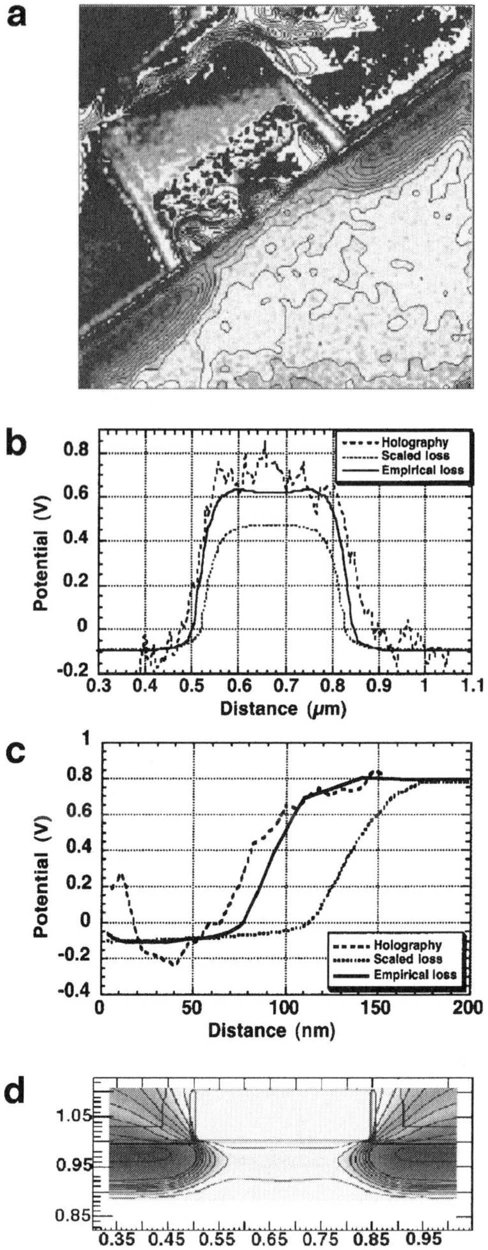

Figure 37. (a) Reconstructed maps of the electrostatic potential distribution in a 0.35 μm semiconductor device structure, with a contour step of 0.1 V, recorded at 200 kV using a Philips CM200 FEG-TEM. (b) Lateral and (c) depth profiles obtained from the image shown in (a). Predictions from process simulations for "scaled loss" and "empirical loss" models are also shown. (d) Two-dimensional simulated map of the potential based on the "empirical loss" model, with a contour step of 0.1 V. The dimensions are in μm. Reprinted with permission from [317], M. A. Gribelyuk et al., *Phys. Rev. Lett.* 89, 025502 (2002). © 2002, American Physical Society.

4.3.2. Sample Preparation and Charging

The results presented in Figure 36 and 37 highlight the fact that the nature and extent of the electrically dead layer on the sample surface of a doped semiconductor depend on the TEM sample preparation procedure. The ways in which the sample preparation technique of "wedge-polishing" affects both the dead layer thickness and sample charging are explored further in Figure 38, which shows results obtained from a one-dimensional *p–n* junction in Si [318]. This sample was prepared from a *p*-type wafer that had been subjected to a shallow B implant and a deeper P implant, resulting in the formation of an *n*-type well and a *p*-doped surface region. A 30-nm-thick $CoSi_2$ film and a further oxide (tetraethylorthosilicate—TEOS) layer were deposited onto the wafer surface before TEM sample preparation, which used wedge-polishing and brief Ar-ion milling to provide a continuous range of sample thicknesses for examination. Figure 38a and b shows typical reconstructed phase images from this sample, obtained both before and after coating one side of the sample with approximately 40 nm of carbon, respectively. The images were obtained with the sample

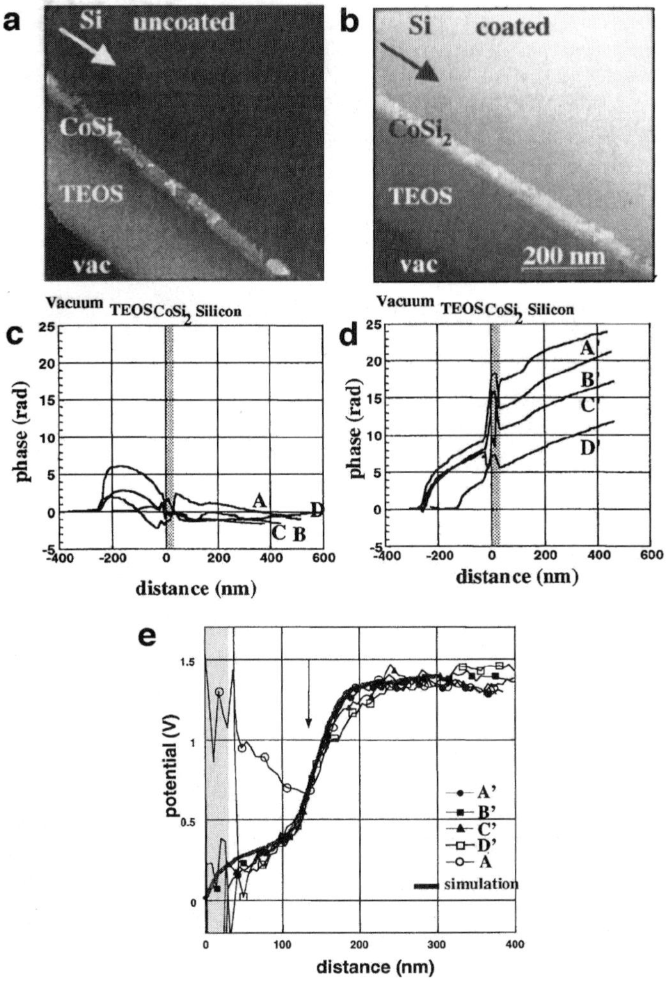

Figure 38. (a) Reconstructed phase image of a one-dimensional *p–n* junction sample prior to carbon coating. (b) Reconstructed phase image from a similar region after carbon coating. (c) Phase profiles for uncoated sample regions of decreasing thickness. At 180 nm from the top of the $CoSi_2$ (gray region), the thicknesses are A = 250 nm, B = 170 nm, C = 120 nm, and D = 50 nm, as measured from holographic amplitude images. (d) Phase profiles after carbon coating. At 180 nm, the thicknesses are A′ = 280 nm, B′ = 220 nm, C′ = 160 nm, and D′ = 85 nm. (e) Comparison of the electrostatic potential derived from simulations based on secondary ion mass spectrometry with measurements obtained using electron holography. Note the close agreement of profiles A′, B′, and C′. The position of the *p–n* junction is marked. The gray region is the $CoSi_2$ layer. Reprinted with permission from [318], M. R. McCartney et al., *Appl. Phys. Lett.* 80, 3213 (2002). © 2002, American Institute of Physics.

tilted by about 10° to a weakly diffracting orientation close to ⟨110⟩, and with the layers kept 1–2° from edge-on.

Figure 38c and d shows line profiles obtained perpendicular to the layers from the uncoated and carbon-coated sample, respectively, for a range of sample thicknesses (measured from the corresponding amplitude images). The phase profiles obtained from the uncoated sample (Fig. 38c) show an initial increase in the measured phase going from vacuum into the TEOS layer. However, the phase profiles then drop steeply and even become negative at the largest sample thicknesses. In contrast, after the sample surface was coated with carbon (Fig. 38d), the phase profiles increase with sample thickness, suggesting that the anomalous behavior observed in Figure 38c is associated with sample charging. Such charging effects may result from the electron-beam-induced emission of secondary electrons, which is negated by carbon-coating. The sample would indeed need to be positively charged along its thin edge to reproduce the features visible in Figure 38c.

Figure 38e shows electrostatic potential profiles derived for four different sample thicknesses from the phase profiles shown in Figure 38d, displayed alongside a simulation based on a secondary ion mass spectrometry measurement of the dopant profile. The position of the *p–n* junction is marked with an arrow. Close agreement between the data and the simulation is obtained for all of the profiles apart from D′, which was obtained from the thinnest area after carbon-coating. Profile A was obtained by subtracting a linear ramp from the phase profile from the thickest uncoated sample and shows a distinct change in slope in the part of the sample that lies between the *p–n* junction and the silicon–silicide interface. Significantly, after carbon-coating, depleted surface layers did not need to be taken into account to obtain quantitative matches to simulations.

In addition to charging problems, an important issue associated with the preparation of a TEM sample for electron holography studies of dopant potentials is the need for uniformly thin areas [319]. This requirement is particularly demanding for focused ion beam milling of device structures since curtaining effects (thickness corrugations) result if substantial metallization is present on the wafer surface. One solution to this problem involves focused ion beam milling of the sample from several directions to form a freestanding electron-transparent membrane. The final milling is then carried out from the substrate side of the wafer to provide an electron-transparent site-specific TEM sample that exhibits minimal thickness variations [320].

4.3.3. In-situ Electrical Biasing Experiments

The experiments illustrated by Figure 36 to 38 involved the examination of unbiased semiconductor device structures, where some part of the TEM sample may have been grounded to the electron microscope sample holder, while some other part may have been subject to current flow or to charging as a result of the generation of positive charge caused by secondary electron emission. The effect on the contrast of examining an unbiased doped semiconductor may be assessed by comparing the results with samples that have been biased electrically *in-situ* in the TEM. Electrical biasing experiments may then provide the potential distribution in a semiconductor device under "working conditions," as well as allowing unwanted effects from variations in sample thickness and composition to be removed by making use of the difference between holographic phase images acquired at different sample voltages.

The earliest applications of electron holography to dopant delineation were carried out under conditions of reverse bias [305, 306]. In these experiments, which involved the examination of chemically thinned Si samples, difference images were recorded between phase images obtained at different reverse-bias voltages. This procedure allowed an electrostatic fringing field outside the sample to be visualized at the position of the p–n junction.

Electrostatic potential profiles have recently been measured from reverse-biased Si p–n junctions that were prepared for TEM examination using focused ion beam milling [321]. A significant aspect of these experiments is the fact that focused ion beam milling is the technique of choice for preparing TEM samples from site-specific regions of integrated circuits. It is therefore essential to establish whether results obtained from unbiased samples prepared by focused ion beam milling are reliable, as well as to develop a suitable sample geometry that allows currents to be passed through TEM samples prepared using this technique. Samples for electrical biasing were prepared for TEM examination by using a 30 kV FEI 200 focused ion beam workstation to machine parallel-sided electron-transparent membranes at the corners of 1 mm × 1 mm 90° cleaved squares of wafer, as shown schematically in Figure 39a. This sample geometry allowed electrical spring contacts to be made to the front and back surfaces of the cleaved wafer in a modified single-tilt sample holder, as shown in Figure 39b. Holograms were recorded at 200 kV in a Philips CM200 ST FEG-TEM operated in Lorentz mode, using a biprism voltage of 100 V. The samples were always tilted by 1–2° from the ⟨100⟩ zone axis, while keeping the p–n junction edge-on to better than 0.2°, in order to ensure that dynamical diffraction effects were minimized. Care was taken during sample preparation to minimize physical damage and implantation of Ga ions into the samples [322, 323]. The area of interest was exposed to the focused beam of Ga ions only at a glancing angle to its surface, and the final stage of milling was performed exactly parallel to the surface of the thin membrane using a small spot size of 10 nm and a low ion beam current of 150 pA.

Figure 39c shows a representative electron holographic phase image obtained from an unbiased Si p–n junction sample prepared by focused ion beam milling. The crystalline thickness of this sample was measured to be 550 nm using convergent beam electron diffraction. The p-type and n-type regions are delineated clearly as areas of dark and bright contrast, respectively. An additional "gray" band at the sample edge is presumably associated with the presence of an electrically dead layer, which is visible in cross-section in this image but is thought to run around the entire sample surface. No electrostatic fringing field is visible in vacuum outside the sample, indicating that its surface is an equipotential. Line profiles across the p–n junction were obtained from phase images acquired with different reverse-bias voltages applied to a sample with a crystalline thickness of 390 nm (Fig. 39d), as well as from several unbiased samples of different thickness. Each profile in Figure 39d is qualitatively consistent with the expected potential profile of a p–n junction contained in a sample of uniform thickness. The height of the potential step across the junction, $\Delta\phi$, increases linearly with reverse-bias voltage V_{appl}, as shown in Figure 39e. This behavior is described by the equation

$$\Delta\phi = C_E(V_{\mathrm{bi}} + V_{\mathrm{appl}})t_{\mathrm{active}} \tag{57}$$

where C_E is defined in Eq. (21) and the p–n junction is contained within an electrically active layer of thickness t_{active}

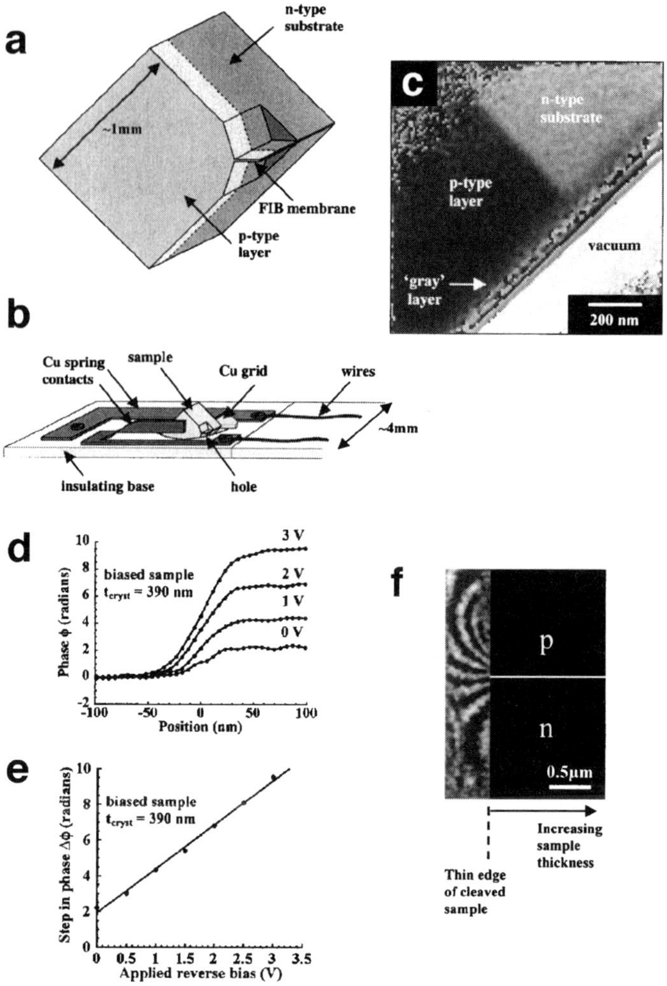

Figure 39. (a) Schematic diagram showing the sample geometry used for biasing experiments. Focused ion beam milling has been used to machine a membrane of uniform thickness that contains a p–n junction, at one corner of a 90° cleaved wedge. (b) Schematic diagram showing the sample position in a single tilt electrical biasing holder. The sample is glued to the edge of a Cu grid using conducting epoxy and then clamped between two spring contacts on an insulating base. (c) Reconstructed phase image of an unbiased Si sample containing a p–n junction. Note the "gray" layer running along the edge of the membrane, which is discussed in the text. No attempt has been made to remove the 2π phase "wraps" at the edge of the sample. (d) Phase (ϕ) measured across a p–n junction as a function of reverse bias for a single sample of crystalline thickness 390 nm (measured using convergent beam electron diffraction). (e) shows the height of the measured step in phase ($\Delta\phi$) across the junction as a function of reverse bias. (f) Four-times-amplified reconstructed phase image, showing the vacuum region outside a p–n junction in a 2 V reverse-biased cleaved wedge sample that had not been focused ion beam milled. Reprinted with permission from [321], A. C. Twitchett et al., *Phys. Rev. Lett.* 88, 238302 (2002). © 2002, American Physical Society.

within the total sample thickness t. Measurement of the gradient of Figure 3e, which is equal to $C_E t_{\mathrm{active}}$ according to Eq. (67), provides a value to t_{active} of 340 ± 10 nm, indicating that 25 ± 5 nm of the crystalline sample thickness on each surface of the TEM sample is electrically inactive. Similarly, the intercept of the graph with the vertical axis is $C_E V_{\mathrm{bi}} t_{\mathrm{active}}$, which provides the expected value for the built-in voltage across the junction of 0.9 ± 0.1 V. Depletion widths across the junction measured from the line profiles are higher than expected, suggesting that the electrically active dopant concentration in the sample is lower than the nominal value. However, these experiments also show that electrical biasing reactivates some of the dopant in the sample that has been passivated by sample preparation [324].

The electrical nature of the sample surface was investigated by examining the vacuum region just outside the edge of the sample in the recorded phase images. Figure 39f shows a four-times-amplified phase image obtained from a 90° cleaved wedge sample that had *not* been focused ion beam milled, for an applied reverse bias of 2 V, in which an external electrostatic fringing field similar to that predicted [325] is visible. Such fringing fields were never observed outside either *unbiased* cleaved wedges or any focused ion beam milled samples, indicating that the surface of a TEM sample prepared by focused ion beam milling is always an equipotential under applied bias.

The importance of minimizing and assessing damage, implantation, and curtaining problems when examining a focused ion beam milled TEM sample that contains a *p–n* junction has been highlighted by three recent papers that present results from unbiased samples [326–328]. The most elegant of these experiments involved the use of focused ion beam milling to form a 45° wedge, from which both the phase change across the junction and the absolute phase shift relative to vacuum on each side of the junction could be plotted as a function of sample thickness. The slopes of these phase change and phase profiles were then used to provide the built-in voltage across the junction and the mean inner potentials of the *p* and *n* sides of the junction, respectively, while the intercept of the phase change graph was used to determine the electrically dead layer thickness. Using this approach, the built-in voltage across a junction with a dopant concentration of approximately 10^{15} cm^{-3} was measured to be 0.71 ± 0.05 V, while the mean inner potentials of the *p* and *n* sides of the junction were measured to be 11.50 ± 0.27 and 12.1 ± 0.40 V, respectively. The electrically dead layer thickness was measured to be approximately 25 nm on each sample surface.

4.3.4. Simulations

The electrical state of the surface of a TEM sample that contains a doped semiconductor can be assessed by comparing experimental holographic results with computer simulations. Simulations have recently been performed using commercial semiconductor process simulation software [329]. These simulations considered the effect of surface states, charged surface oxide layers, and the generation of electron–hole pairs on the electrostatic potential in the sample. Figure 40a shows simulated equipotential contours for a Si sample of thickness 100 nm that has hole traps on its surfaces. The *n* and *p* sides have doping concentrations of 4.86×10^{15} and 9×10^{18} cm^{-3}, respectively. A sample with these doping concentrations has been studied experimentally [312] and has revealed anomalous contrast features in defocused images of the junction. The presence of surface states in the simulation leads to the formation of a surface depletion layer on the *p*-side of the junction, but the model cannot

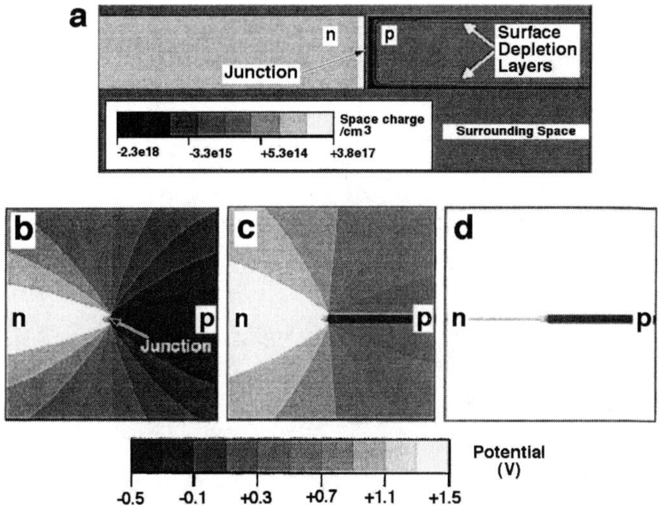

Figure 40. (a) Simulation showing equipotential contours that depict surface depletion layers on the *p*-side of a *p–n* junction in a Si sample of thickness 100 nm, created by the presence of hole traps on the sample surfaces. The *n*- and *p*-sides have doping concentrations of 4.86×10^{15} and 9×10^{18} cm^{-3}, respectively. (b), (c), and (d) Equipotential contours showing the effect of increasing positive surface charge concentrations of 1×10^{13}, 2.5×10^{13}, and 1×10^{14} cm^{-2}, respectively. The external electrostatic leakage field outside the sample is no longer present in (d). Reprinted with permission from [329], M. Beleggia et al., *Inst. Phys. Conf. Ser.* 169, 427 (2001). © 2001, Institute of Physics Publishing.

account for the experimental observations. A better model involves including electron-beam-induced charging of the silicon oxide layer that forms on the sample surfaces, at a level of approximately 10^{13} to 10^{14} cm^{-2}. The presence of this concentration of positive charge creates an inversion layer on the *p*-side of the junction, which eventually cancels out the surface potential and results in the absence of electrostatic fringing fields outside the sample surface, in agreement with observations for unbiased *p–n* junctions. (Such fringing fields, if present, would dominate the observed phase contrast [330].) The depletion width in the simulation varies with distance from the sample surfaces, in agreement with experimental measurements obtained as a function of sample thickness from a focused ion beam milled sample [321]. Figure 40b, c, and d shows the effect of increasing the surface charge concentration on the equipotential contours that form both within and surrounding the sample described in Figure 40a. The electrostatic fringing field disappears progressively with increasing surface charge density, until it is no longer visible at a concentration of 10^{14} cm^{-2}.

4.3.5. Layered Device Structures

Although questions still remain about the phase contrast observed at a simple Si *p–n* junction, successful attempts have been made to interpret electron holographic data obtained from more complicated semiconductor device structures, where changes in composition as well as doping concentration were present. Figure 41 shows results obtained from a strained n-$Al_{0.1}Ga_{0.9}N/In_{0.1}Ga_{0.9}N/p$-$Al_{0.1}Ga_{0.9}N$ heterojunction diode, in which large piezoelectric and polarization fields are used to induce high two-dimensional electron gas (2DEG) concentrations [331].

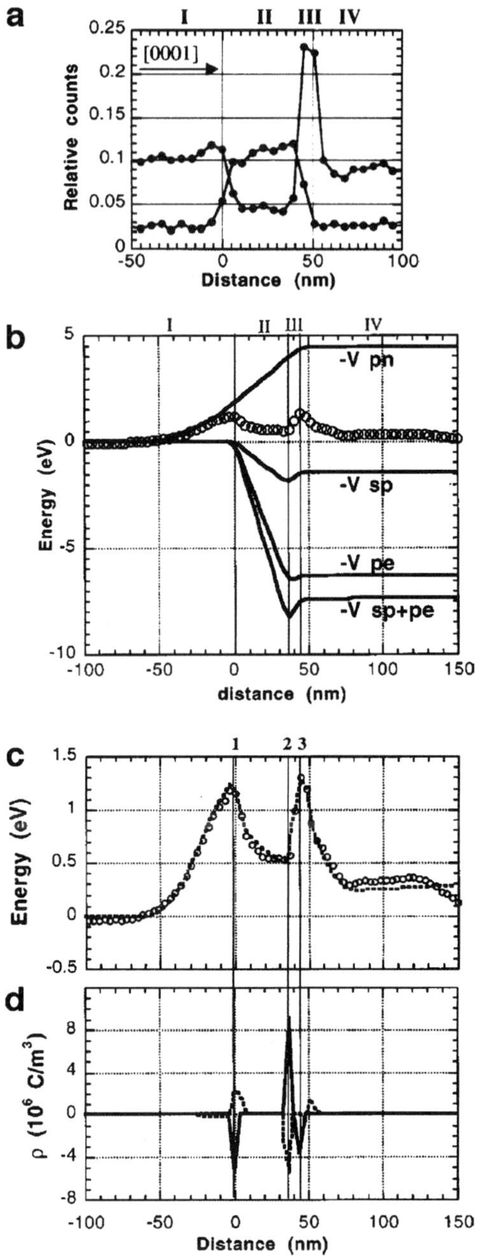

Figure 41. (a) Energy dispersive X-ray profiles showing the Al and In signals obtained from an n-$Al_{0.1}Ga_{0.9}N/In_{0.1}Ga_{0.9}N/p$-$Al_{0.1}Ga_{0.9}N$ structure with an additional thin barrier layer with an enhanced Al concentration of 30% grown on the InGaN. The n-AlGaN layer was nominally doped with 1×10^{18} cm^{-3} of Si, while the p-AlGaN layer was nominally doped with 1×10^{20} cm^{-3} of Mg. Integrated peak counts were normalized to the Ga signal. Regions I, II, III, and IV correspond to n-AlGaN, InGaN, the high concentration AlGaN barrier, and p-AlGaN, respectively. (b) Energy-profile across structure measured from an electron holographic phase image acquired at 200 kV using a Philips CM200 FEG-TEM (open circles). Calculated energy contributions for p–n junction V_{pn}, spontaneous polarization V_{sp}, and piezoelectric V_{pe} contributions to the energy are indicated. (c) Experimental (open circles) and simulated (dotted line) energy across structure, incorporating both spontaneous and piezoelectric polarization. (d) Bound polarization-induced interface charge (solid line) and additional charge (dotted line) used for energy simulation shown in (c). Reprinted with permission from [331], M. R. McCartney et al., *Appl. Phys. Lett.* 76, 3055 (2000). © 2000, American Institute of Physics.

The n-AlGaN layer in this sample was nominally doped with 1×10^{18} cm^{-3} of Si, while the p-AlGaN layer was nominally doped with 1×10^{20} cm^{-3} of Mg. An additional thin barrier layer with an enhanced Al concentration of 30% was grown on the InGaN. Figure 41a shows Al and In energy-dispersive X-ray profiles obtained from this sample, in which regions I, II, III, and IV correspond to n-AlGaN, InGaN, the high concentration AlGaN barrier, and p-AlGaN, respectively.

Electron holograms of a sample that had been prepared using conventional mechanical polishing and Ar-ion milling, and also lightly carbon-coated, were acquired at 200 kV using a Philips CM200 ST FEG-TEM operated in Lorentz mode. The open circles in Figure 41b show experimental measurements of the potential profile across the heterojunction, after correcting for changes in sample thickness (assuming a linear thickness profile and neglecting contributions to the measured phase from variations in mean inner potential). In order to interpret the experimental energy profile, simulations incorporated the effects of the p–n junction potential, the presence of both spontaneous and piezoelectric polarization, and additional charge accumulation. These contributions to the potential are shown in Figure 41b alongside the experimental data points. Additional charge had to be added to the simulations in order to account for discrepancies between the experimental data and the simulations in Figure 41b, within the constraint that the total charge in the model had to be zero. Figure 41c shows the best-fitting simulation, which incorporates both spontaneous and piezoelectric polarization, plotted alongside the experimental data. Figure 41d shows the magnitude of the bound polarization-induced interface charge and the additional charge required to obtain the match shown in Figure 41c. In particular, a sheet of negative charge was included at the bottom of the InGaN well, at a position consistent with the expected location of a 2DEG. The sheet charge density at this position was 2.1×10^{13} cm^{-2}.

Off-axis electron holography has also been used to characterize and to interpret electrostatic potentials within GaN-based quantum well structures [332–336] and to study charging effects at different types of threading dislocations in GaN [337–339].

4.4. Space Charge Layers at Grain Boundaries

Figure 41 demonstrated the use of electron holography to characterize interfaces in semiconductor devices where there were rapid changes in both composition and doping concentration. The technique has also been used to characterize space charge layers at doped and undoped grain boundaries in electroceramics such as $SrTiO_3$ and ZnO, where several possible contributions to the measured electron holographic phase complicate interpretation. The space charge distribution that is predicted to form at such a grain boundary [340, 341] is often described as a double (back-to-back) Schottky barrier, and the barrier height and width greatly influence the performance of the material in applications that include varistor devices for overvoltage protection.

Figure 42a shows an electron hologram of a Mn-doped grain boundary in $SrTiO_3$, acquired at 200 kV in a Hitachi HF-2000 FEG-TEM using an interference fringe spacing of

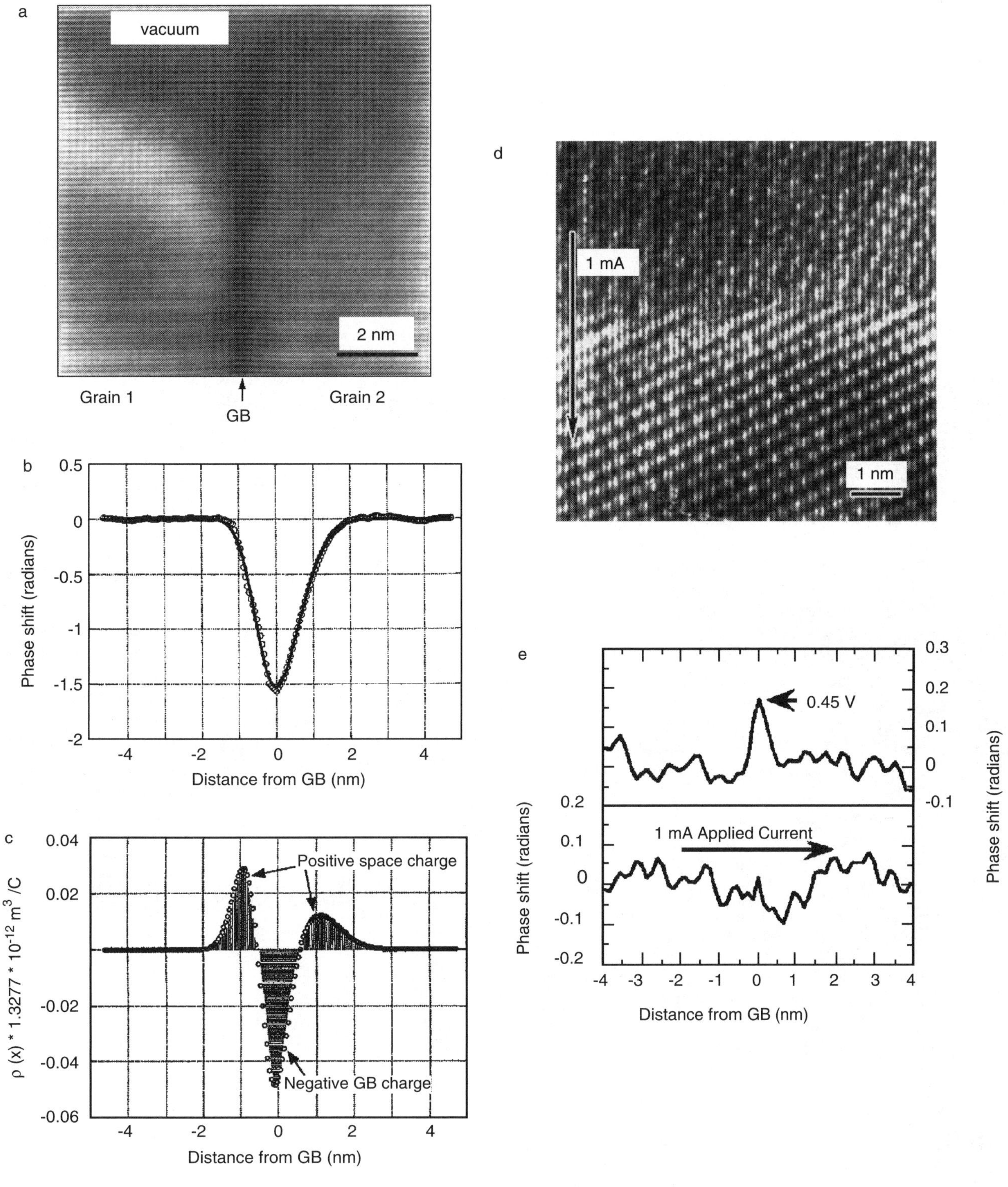

Figure 42. (a) Electron hologram of a Mn-doped grain boundary in $SrTiO_3$ acquired using a Hitachi HF-2000 FEG-TEM. The interference fringe spacing is 0.14 nm. (b) Phase profile measured across grain boundary shown in (a). (c) Charge density profile inferred by applying Poisson's equation direction to the phase profile shown in (b). The charge density profile suggests the presence of negative grain boundary charge and compensating positive space charge. The spatial extent of the boundary charge is approximately 1.4 nm. Reprinted with permission from [342], V. Ravikumar et al., *J. Phys. D: Appl. Phys.* 29, 1799 (1996). © 1996, Institute of Physics Publishing. (d) Electron hologram acquired from a Nb-doped grain boundary in $SrTiO_3$ that has a current flowing across it *in-situ* in the TEM. (e) Measured phase shift across the boundary shown in (d), both before applying a current and as a current is applied. The absence of contrast at the boundary when a current is flowing suggests barrier breakdown. Reprinted with permission from [357], K. D. Johnson and V. P. Dravid, *Appl. Phys. Lett.* 74, 621 (1999). © 1999, American Institute of Physics.

0.14 nm [342–348]. Similar holograms were obtained from several undoped and doped grain boundaries in $SrTiO_3$. All results showed a decrease in the measured phase shift at the boundary relative to that in the sample. However, the changes in phase measured at the doped boundaries were consistently larger in both magnitude and spatial extent than at similar undoped grain boundaries. Possible contributions to the contrast from changes in density, composition, sample thickness, dynamical diffraction, and electrostatic fringing fields [349, 350] were considered, and the remaining contributions to the measured phase shifts at the doped boundaries were attributed to space charge. The sign of this space charge contribution to the specimen potential was consistent with the presence of Mn^{2+} and Mn^{3+} ions on Ti sites at the boundaries. The phase profile and the corresponding charge density profile at a Mn-doped boundary, as inferred from the hologram in Figure 42a, are shown in Figure 42b and c, respectively. Both the force acting on the boundary as a result of the presence of the space charge layer and the resulting strain profile were calculated for these boundaries [345], presumably on the basis of equations published in an earlier study [351]. The results were finally interpreted in terms of a narrow (1–2 nm) region of negative grain boundary charge and a wider (3–5 nm) distribution of positive space charge.

A similar approach has recently been applied to the characterization of grain boundaries in ZnO, where a space charge layer width of approximately 150 nm has been measured [352]. In an earlier study, the defocus contrast recorded from delta-doped layers in Si and GaAs was also attributed to the presence of space charge [353]. Defocus contrast has been used to assess possible space charge contributions to electrostatic potential profiles across grain boundaries in mullite [354] and in both doped and undoped $SrTiO_3$ [355, 356]. In contrast to the results shown in Figure 42, the contrast observed in the latter experiments was not found to be consistent with a dominant contribution to the signal from space charge.

The electrical transport properties of Nb-doped and Mn-doped grain boundaries in $SrTiO_3$ bicrystals have recently been examined *in-situ* in the TEM by acquiring electron holograms while electric currents were passed across the boundaries [357–359]. In order to perform these experiments, Ni contact pads were evaporated onto the samples after they had first been prepared for TEM examination using dimpling and ion milling. Figure 42d shows an electron hologram acquired from a Nb-doped grain boundary in $SrTiO_3$ as a current of 1 mA flows across it. The measured phase shift across the boundary is shown in Figure 42e both before applying the current and as the current is applied. The absence of a detectable phase shift at the boundary when the current is flowing suggests that barrier breakdown then occurs. This suppression of the boundary potential is reversible as the current is applied and then removed.

Related experiments have been performed to measure the polarization distributions across domain boundaries in ferroelectric materials such as $BaTiO_3$ and $PbTiO_3$ [360–365] and the external electrostatic fringing fields [366]. The theory of electron holographic contrast from ferroelectric domain walls has also been presented [367]. Considerable further work is possible on this subject.

5. ALTERNATIVE FORMS OF ELECTRON HOLOGRAPHY

There are many different forms of electron holography that can be implemented both in the TEM and in the STEM [368]. Equally, there are several alternative ways in which the off-axis mode of TEM electron holography can be implemented. A full discussion of the various possible schemes that have been investigated, which include interferometry in the diffraction plane of the electron microscope [369, 370], reflection electron holography [371], and coherent convergent beam electron diffraction [372, 373], is beyond the scope of this chapter. Some developments that are historically or potentially most useful for examining nanostructured materials are reviewed.

5.1. Differential Form of TEM Electron Holography

The need for a vacuum reference wave is a major drawback of the normal off-axis mode of TEM holography since this requirement restricts the region that can be examined to the specimen edge. In many applications, the feature of interest is not so conveniently located. In addition, the presence of an edge may dominate the magnetic contribution to the measured phase shift for magnetic samples. The implementation of a differential phase contrast (DPC) mode of electron holography in the TEM enables this restriction to be overcome. DPC imaging is well established as a technique in the scanning transmission electron microscope, involving various combinations of detectors to obtain magnetic domain contrast [374–376]. It has also been shown [377] that DPC contrast can be obtained using far-out-of-focus STEM electron holography. An equivalent configuration can be achieved in the TEM by using an electrostatic biprism located in the condenser aperture plane of the electron microscope [378].

Figure 43a shows a schematic ray diagram that illustrates the electron-optical configuration of the microscope for the differential mode of off-axis TEM holography. The application of a positive voltage to the biprism wire results in the formation of two closely spaced, overlapping plane waves, which appear to originate from sources S_1 and S_2 to create an interference fringe pattern at the specimen level. When the observation plane is defocused by a distance Δz with respect to the specimen plane, the two coherent beams produced by the beamsplitter, which are labelled k_1 and k_2 in Figure 43a, impinge upon different parts of the specimen. For a magnetic material, the difference in the component of the magnetic induction parallel to the biprism wire between these two points in the specimen plane determines the relative phase shift of the holographic fringes, thus giving differential phase contrast. Since the hologram is acquired under out-of-focus conditions, it is in effect the superposition of a pair of Fresnel images. The biprism voltage must be adjusted so that the feature of interest or the desired spatial resolution is sampled by at least three interference fringes. An appropriate postspecimen magnification should be chosen to ensure that the interference fringes are properly sampled by the recording medium.

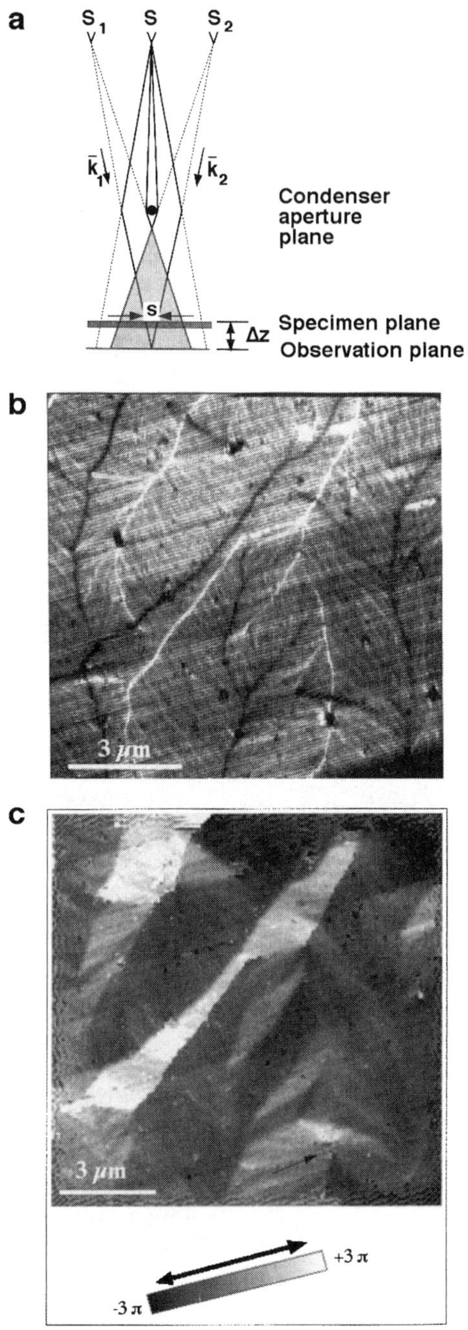

Figure 43. (a) Schematic ray diagram for the differential mode of off-axis TEM holography. The symbols are defined in the text. (b) Composite differential mode hologram formed from a series of eight holograms of a 30-nm-thick Co film. (c) Differential phase contrast image obtained from the hologram shown in (b). The arrow below the image indicates the direction of the magnetic field analyzed. The arrow at the bottom of the image indicates the position of a magnetic vortex. Reprinted with permission from [378], M. R. McCartney et al., *Ultramicroscopy* 65, 179 (1996). © 1996, Elsevier Science.

Figure 43b shows a composite phase image obtained from a series of eight DPC holograms of a 30-nm-thick Co film, where the fringe system was shifted progressively across the specimen plane between exposures. In addition to the holographic interference fringes, the image shows characteristic black and white lines that delineate walls between magnetic domains, with magnetization ripple visible within the domains. All of the image features appear doubled due to the split incident beam. Figure 43c shows the final reconstructed DPC image obtained from Figure 43b, in which the contrast is proportional to the component of the magnetic induction parallel to the holographic fringes. The arrow below the image indicates the direction of the component of the magnetic induction analyzed in this experiment. Several magnetic vortices are visible, where the measured field direction circles an imperfection in the film. One such vortex is indicated by an arrow in the lower right corner of the image. For complete characterization of both components of the in-plane induction without removing the sample from the electron microscope, either a rotating biprism or a rotating sample holder is required.

5.2. Amplitude Division Electron Holography

A scheme that is conceptually similar to the differerential mode of electron holography in the TEM, but which does not require the use of either an electron biprism or a field emission electron gun, is termed amplitude division electron holography. Whereas conventional modes of electron holography involve splitting the wavefront of the incident illumination and thus require high spatial coherence in order to form interference fringes, this coherence requirement can be removed by dividing the amplitude of the incident electron wave instead of the wavefront. This division of the amplitude of the incident electron wave can be achieved by using a crystal film located before the sample. The lattice fringes of the crystal film are then used as carrier fringes. The original configuration for this scheme involved placing the specimen in the selected-area-aperture plane of the electron microscope [379, 380]. The specimen can also be inserted into the normal object plane by using the electron-optical configuration shown in Figure 44a [381, 382]. A single-crystal thin film and the sample of interest are placed on top of each other, in close proximity, in the electron microscope. The single-crystal film is then tilted to a strong Bragg condition and used as an electron beam splitter. As a result of the separation of the crystal and sample, the hologram plane contains two defocused images of the sample that are shifted laterally with respect to each other. One of these images is carried by the direct beam and the other by the Bragg-reflected beam, as indicated by the solid and dashed lines in Figure 44a, respectively. When the distance between the two images is greater than the object size, the images separate perfectly and interfere with adjacent plane waves to form an off-axis electron hologram. The distance between the two images is given by the expression

$$l = \frac{\lambda z}{d} \tag{58}$$

where λ is the electron wavelength, z is the distance between the crystal and the specimen, and d is the lattice spacing of the crystal. Because the single crystal is in focus and the object is out of focus, a Fresnel electron hologram of the object is obtained. The defocus of the object can

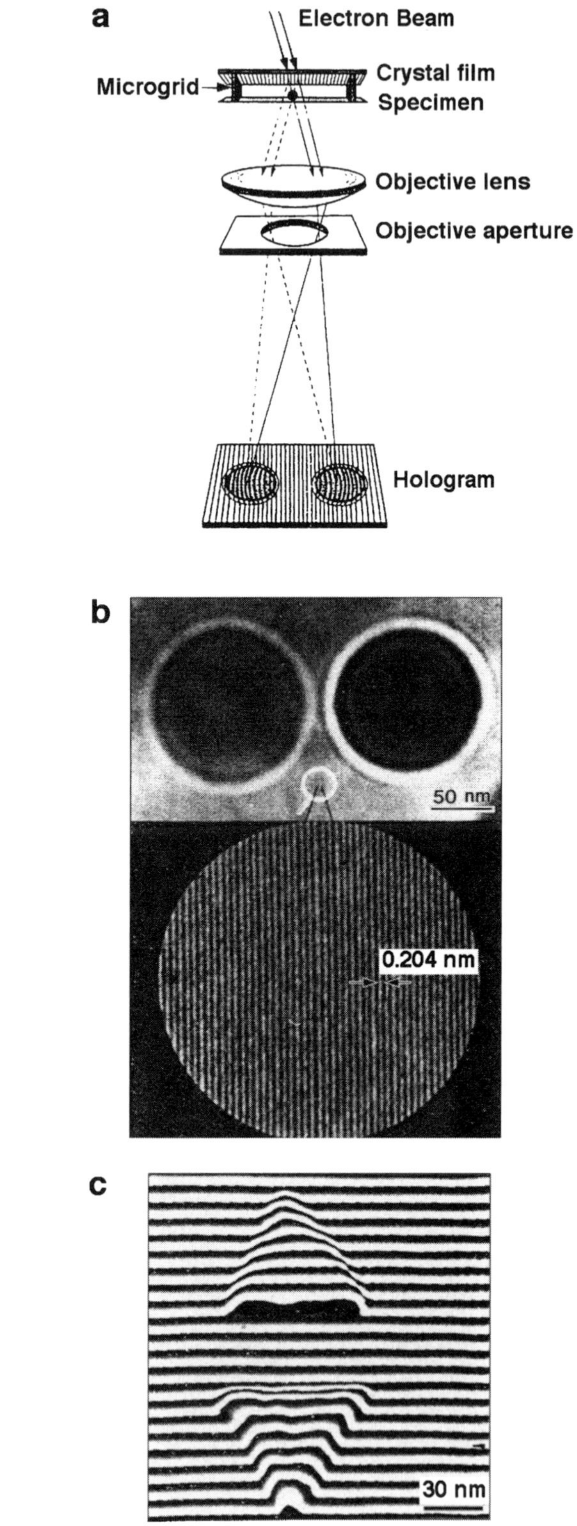

Figure 44. (a) Electron-optical setup for hologram formation using a single-crystal film as an electron beamsplitter. The solid lines indicate the direct electron rays, while the dashed lines indicate the Bragg-reflected rays. (b) Fresnel hologram of a polystytrene latex particle. Approximately 1500 Au lattice fringes are recorded. The left and right images are formed by the Bragg-reflected and direct beams, respectively. (c) Reconstructed phase map of a pair of (111) Au platelets grown epitaxially on MoS_2. Reprinted with permission from [381], Q. Ru et al., *Ultramicroscopy* 53, 1 (1994). © 1994, Elsevier Science.

be corrected at the reconstruction stage by using a phase plate, although high coherence of the incident illumination is then required. The coherence used when forming the image therefore determines the spatial resolution of the final reconstructed image. In addition, poor contrast in the lattice fringes will lead to noise in the reconstructed phase image. Although amplitude-division electron holography has several obvious disadvantages over wavefront-division holography, the final phase image is not affected by such effects as Fresnel diffraction from the edges of the biprism wire. The diffracting condition and the thickness of the crystal film also have little effect on the results [383].

Figure 42b shows a hologram of a 120-nm-diameter polystyrene latex particle, illuminated by an electron beam that has been split by a single crystal Au film with a lattice spacing of 0.204 nm. This image was recorded using convergent illumination, under conditions where no interference fringes would be produced in a conventional off-axis electron hologram. The image on the right of Figure 42b is formed by the direct beam, while the image on the left is formed by the Bragg-reflected beam. The crystal and the sample are separated by 10 nm, and approximately 1500 Au lattice fringes are recorded in the hologram. An enlarged area of the hologram, where the lattice fringes are visible, is shown at the bottom of Figure 44b. Figure 42c shows a reconstructed phase image obtained by amplitude division electron holography from a pair of (111) Au platelets grown epitaxially on a molybdenite (MoS_2) substrate.

5.3. Phase Shifting Electron Holography

One approach that can be used to increase the phase sensitivity of electron holography is termed phase-shifting electron holography. This approach is based on the acquisition of several off-axis holograms while the phase offset (the initial phase) of the image is changed, either by tilting the incident electron beam or by shifting the biprism [384, 385]. The electron optics that are required for the scheme that involves tilting the incident beam are shown in Figure 45a. Electron holograms are recorded at successive values of the incident beam tilt, such that the phase is shifted by at least 2π over the image series. The fringe shift can be monitored in the complex Fourier spectra of the holograms. In Figure 45b, the real and imaginary components given by the center band are denoted R_0 and I_0, while those given by the sideband are denoted R and I. The real and imaginary parts of the complex Fourier spectrum are shown as solid and broken lines, respectively. Since shifting the initial phase does not change the complex spectrum of the center band nor the absolute amplitude of the sideband, the parameters R_0, I_0, and $(R^2 + I^2)^{0.5}$ do not change when the phase is shifted. A series of real–imaginary pair values obtained from holograms acquired with different initial phases therefore trace out a circular trajectory, with radius $(R^2 + I^2)^{0.5}$ and center located at (R_0, I_0), as shown schematically in Figure 45c. The values of R_0 and I_0 can be determined by fitting the trajectory with a circle, while R and I are given by subtracting R_0 and I_0 from the measured spectrum data. The initial phase of each hologram is determined by calculating the inverse tangent of the ratio I/R. A series of calculated

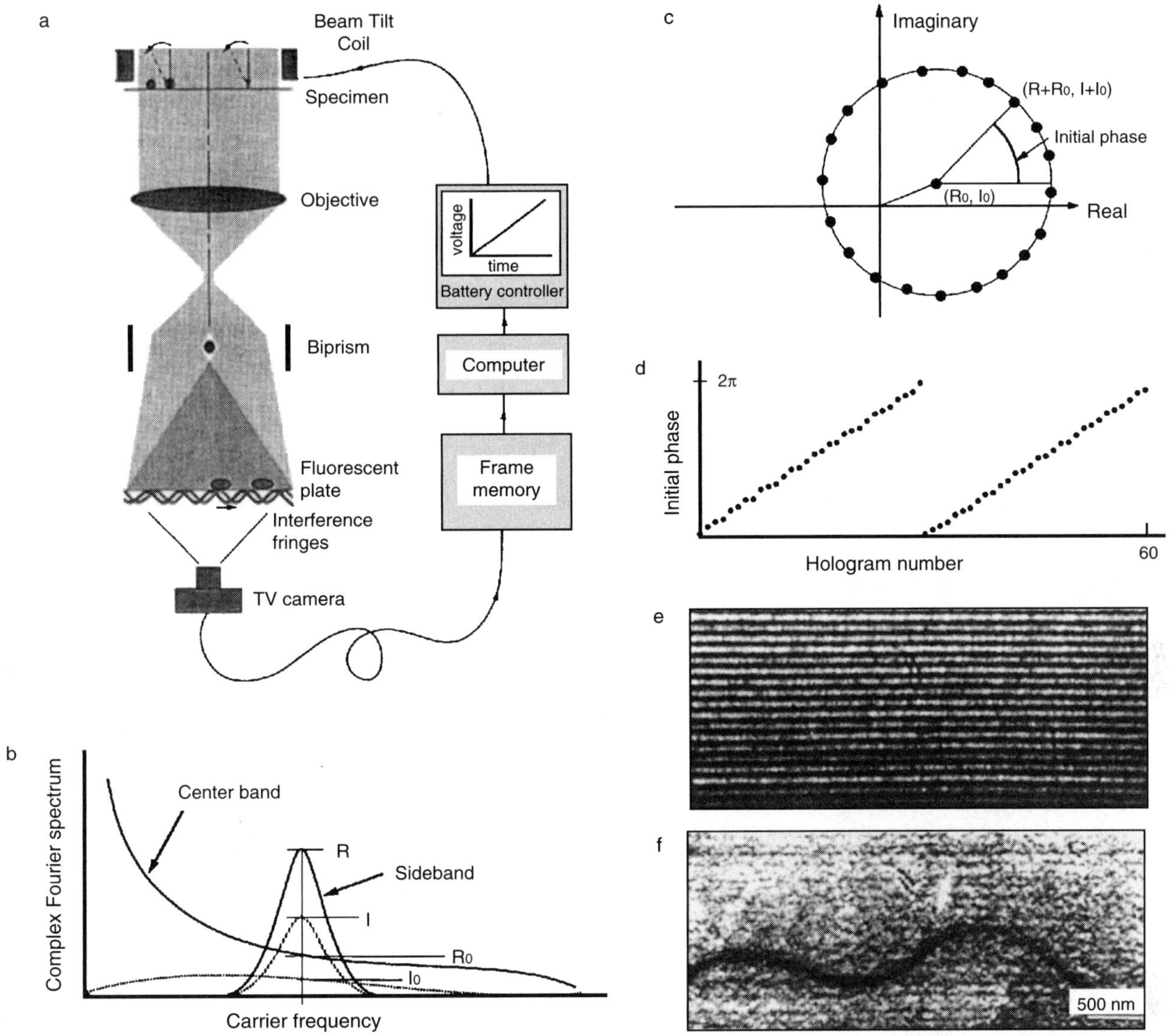

Figure 45. (a) Electron optics and equipment required for shifting the initial phase of an electron hologram. (b) Schematic diagram of the complex Fourier spectrum, with the real and imaginary parts shown as solid and broken lines, respectively. (c) Complex plane, illustrating the circular trajectory of the spectrum data at the carrier frequency. (d) Calculated initial phase data for a series of holograms. (e) Hologram of a thin flagellum filament. (f) The corresponding phase image, obtained from a series of 96 holograms with different initial phases from 0 to 4π. The maximum phase change due to the filament is $2\pi/40$. Reprinted with permission from [69], Q. Ru et al., *Ultramicroscopy* 55, 209 (1994). © 1994, Elsevier Science.

initial phases is plotted in Figure 45d for a 4π shift in the initial phase.

For N holograms recorded in equal steps between phase shifts of 0 and 2π, the procedure involves multiplying the intensity in the nth hologram in the series by $\exp(2\pi in/N)$, and adding these images together in the form

$$C_3 = \left(\frac{1}{N}\right) \sum_{n=1}^{N} I(n) \exp\left(\frac{2\pi\, in}{N}\right) \tag{59}$$

The desired phase distribution $\phi(x, y)$ is then given by the expression

$$\phi(x, y) = \tan^{-1}\left(\frac{\mathrm{Im}[C_3]}{\mathrm{Re}[C_3]}\right) - \frac{2\pi x}{T_x} - \frac{2\pi y}{T_y} \tag{60}$$

where T_x and T_y are the fringe spacings in the x and y directions, respectively. Although three holograms can in principle be used to reconstruct the object wave, in practice as many holograms as possible are used to reduce noise in the reconstructed phase.

The advantages of the phase-shifting approach are greatly improved phase sensitivity and spatial resolution. Since the object wave is reconstructed from only the intensity data at each point, defects do not propagate into adjacent areas, and the spatial resolution is equal to that of the original image, independent of the fringe spacing. Objects that are smaller than one fringe width can therefore be reconstructed. Care is required if the object is out of focus, as tilting the beam will then induce an image shift between successive images. Shifting the biprism may then be required if atomic resolution phase measurements are to be achieved. Figure 45e shows an individual hologram of a thin bacterial flagellum filament, and Figure 45f shows the corresponding phase image, obtained from a series of 96 holograms acquired with different initial phases ranging from 0 to 4π. The maximum phase change due to the filament is $2\pi/40$ [69]. Very small phase shifts have also been observed from individual unstained ferritine molecules [386] and both latex and iron particles [387].

5.4. Real-Time Electron Holography

Electron holograms can be acquired at video rate and subsequently digitized and processed individually to record dynamic events, but this procedure is very time-consuming. An alternative, real-time approach for acquiring and processing holograms has been demonstrated by using a liquid–crystal panel to reconstruct holograms in real time [388–390], in the configuration shown in Figure 46. The electron holograms are recorded at TV rate and then transferred to a liquid–crystal spatial light modulator, which is located at the output of a Mach–Zender interferometer. The liquid–crystal panel is illuminated using a He–Ne laser, and interference micrographs are observed and recorded at video rate on the monitor beside the microscope as the sample is examined. (In an alternative configuration, a liquid–crystal panel can also be used as a computer-controlled phase plate to correct for aberrations [391].) Recently, an all-digital system that allowed reconstructed (wrapped) phase images to be displayed at approximately one frame per second has been facilitated by developments in computer speed, with the number of phase wraps across the field of view minimized automatically in each image [392]. This performance represents a substantial improvement over a time of 30 s per hologram reported in 1991 [47].

5.5. In-Line Electron Holography

Whereas an off-axis electron hologram is formed by the interference of an object and a reference wave that propagate in different directions in the electron microscope, as shown schematically in Figure 47a and b [35], the simplest way of recording an electron hologram without using an electron biprism involves using the transmitted wave as the reference wave to form an in-line electron hologram. Gabor's original paper [1] describes the reconstruction of an image by illuminating an in-line hologram with a parallel beam of light using a spherical aberration correcting plate and an astigmatism corrector. The reconstructed image is, however, disturbed by a "ghost" or "conjugate" twin image. If the hologram is recorded and subsequently illuminated by a plane wave, then both the reconstructed image and a

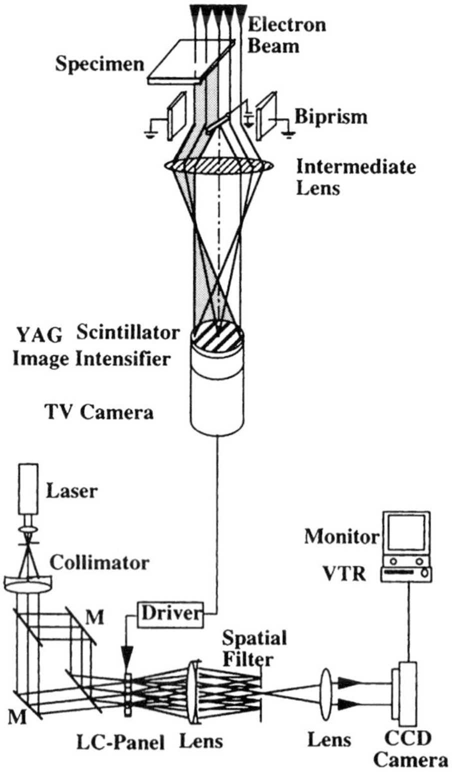

Figure 46. Schematic diagram of an on-line real-time electron holography system. The hologram formed in the microscope is transferred, as a video signal, to a liquid–crystal spatial light modulator located at the output part of a Mach–Zender interferometer. Interference micrographs are observed on the monitor beside the microscope. Reprinted with permission from [390], T. Hirayama et al., *Ultramicroscopy* 54, 9 (1994). © 1994, Elsevier Science.

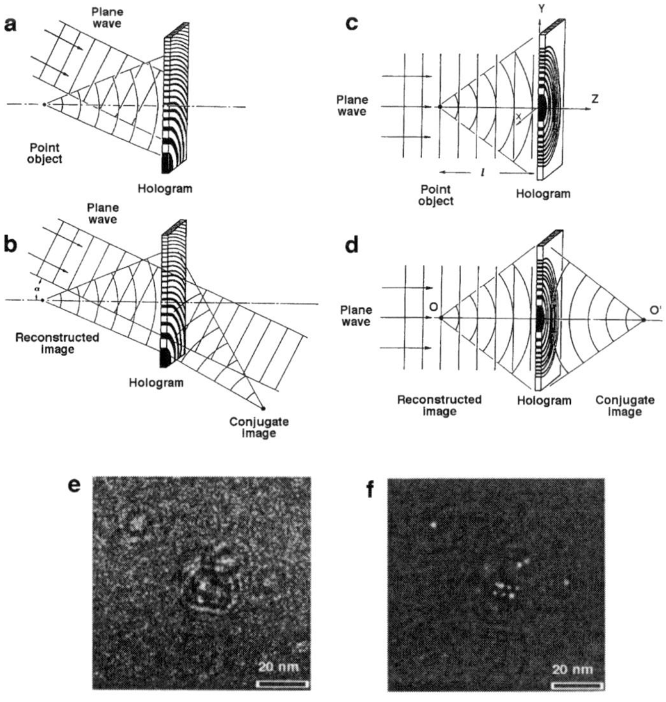

Figure 47. Schematic diagrams showing (a) hologram formation and (b) image reconstruction for off-axis electron holography of a point object. (c) and (d) As for (a) and (b) but for in-line electron holography of a point object. Reprinted with permission from [35], A. Tonomura, "Electron Holography," pp. 4 and 7. Springer-Verlag, Berlin, 1999. © 1999, Springer-Verlag. (e) Fraunhofer in-line electron hologram of an undecagold cluster, at a defocus of 13.6 μm. The corresponding Fresnel number is 0.0155. (f) Phase image reconstructed from the hologram shown in (e). Reprinted with permission from [396], T. Matsumoto et al., *Ultramicroscopy* 54, 317 (1994). © 1994, Elsevier Science.

defocused conjugate image of the object are superimposed on top of each other since both images lie on the same axis, as shown schematically in Figure 47c and d. The most effective method for separating the twin images is to use Fraunhofer in-line holography [15]. Here, in-line holograms are recorded in the Fraunhofer diffraction plane of the object [393, 394], for which the defocusing distance

$$z \gg \frac{d^2}{\lambda} \tag{61}$$

where d is the object width and λ is the wavelength of the incident illumination. Under this condition, the conjugate image is so blurred that its effect on the reconstructed image is negligible [395].

An example of a Fraunhofer in-line electron hologram of an undecagold cluster, which is supported on a thin (<5 nm) carbon film, and recorded at a defocus of 13.6 μm, is shown in Figure 47e [396]. The Fraunhofer condition described by Eq. (61) is satisfied for this object. Figure 47f shows the phase image reconstructed from the hologram shown in Figure 47e. The Fraunhofer condition would no longer be satisfied if a considerably larger object were present in the field of view at this defocus. A maximum entropy approach has also been applied to the reconstruction of in-line electron holograms [397].

It should be noted that in-line electron holography also encompasses "Fresnel imaging," which has been used to measure the structure and composition of a variety of boundaries and interfaces [398–409], including magnetic domain boundaries [410, 411].

5.6. Differential Form of STEM Electron Holography

The STEM holographic mode used for DPC imaging, which has similarities with the TEM differential mode of electron holography described above, is a point projection technique in which a *stationary* illuminating beam in a STEM is split by a biprism preceding the sample so that two mutually coherent electron point sources are formed just above the sample. The electron-optical configuration for this mode of holography is shown schematically in Figure 48. In this operating mode, the objective lens is weakly excited so that the hologram is formed in the diffraction plane rather than the image plane [412]. By greatly defocusing the objective lens, a shadow image of the object is formed, which has the appearance of a TEM hologram, although it is distorted by spherical aberration and defocus. The image magnification and the separation of the sources relative to the sample are flexible in this configuration and can easily be adjusted by changing the biprism voltage and/or the objective or postspecimen lens settings. The far-out-of-focus mode of STEM holography has been applied to the characterization of a wide range of magnetic materials [377, 413–420], where the sample had to be displaced by ~15 mm from the center of the objective lens so that it could be located in a field-free region of the electron microscope.

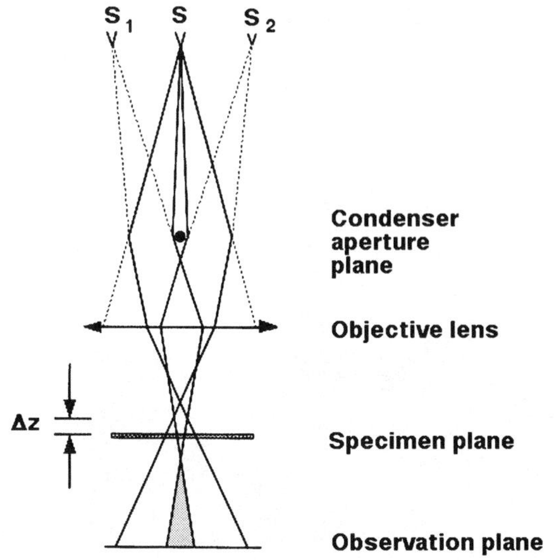

Figure 48. Schematic ray diagram for greatly defocused off-axis electron holography performed in a STEM. Reprinted with permission from [378], M. R. McCartney et al., *Ultramicroscopy* 65, 179 (1996). © 1996, Elsevier Science.

5.7. Double Exposure Electron Holography

A simple approach that can be used to visualize equiphase contours involves superimposing a hologram of the sample onto a reference hologram acquired under identical conditions, with the sample removed from the field of view [421]. Interference effects between the holographic fringes in the two images then provide widely spaced, low contrast bright and dark bands that reveal phase contours directly. Defocusing the combined image slightly can then be used to remove the unwanted, finely spaced holographic interference fringes. The technique has been applied to image both electrostatic and magnetic fields [41]. For example, Figure 49a shows an experimental double-exposure electron hologram of the magnetic leakage field outside the apex of a Ni magnetic force microscopy tip. The phase difference between neighboring black or white fringes is 2π. A computer simulation of the contrast, generated by assuming that the tip can be approximated by a single macroscopic magnetic dipole, is shown in Figure 49b. The simulation includes the effect of the perturbed reference wave on the recorded phase image [40]. Figure 49c shows a double-exposure electron hologram of crossed domain walls in a permalloy thin film [41]. A similar double-exposure hologram can be obtained after changing the biprism voltage by a small amount before recording the reference hologram [36].

A related approach involves the use of two parallel or perpendicular electron biprisms to generate an interference pattern between either three or four electron waves, respectively. Equiphase contours are displayed in the recorded hologram in the same manner as for double-exposure electron holography. This method has been used to form images of electric field lines outside charged latex and alumina particles, as well as magnetic fields outside barium ferrite particles [422–424], and even to expose a resist to fabricate a 100-nm-period two-dimensional grating lithographically [425]. Two parallel biprisms have also been used to

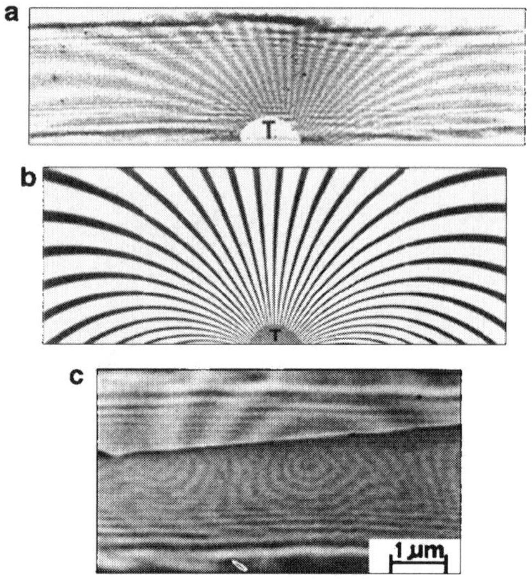

Figure 49. (a) Experimental double-exposure electron hologram of the magnetic leakage field outside the apex of a Ni magnetic force microscopy tip, recorded in a Philips EM 400T FEG-TEM. The overlap width is 4 μm. The phase difference between neighboring black or white fringes is 2π. (b) Computer simulation of the contrast in image (a), generated by assuming that the tip can be approximated by a single macroscopic magnetic dipole. The simulation includes the effect of the perturbed reference wave on the recorded phase image. Reprinted with permission from [40], G. Matteucci et al., *Phys. Rev. B* 50, 6823 (1994). © 1994, American Physical Society. (c) Double-exposure electron hologram of crossed domain walls in a permalloy film. Reprinted with permission from [41], G. Matteucci et al., *Appl. Phys. Lett.* 52, 176 (1988). © 1988, American Institute of Physics.

form a "trapezoidal" biprism, in order to record a double-exposure hologram with the biprism voltage changed [426], but the reference wave unaffected by the biprism voltage [427]. This scheme is similar conceptually to that presented by van Dyck [39].

6. CONCLUSIONS

The reconstruction of electron holograms using both optical and digital methods has been outlined, and the different types of information that can be obtained using the two approaches have been illustrated using results from a wide variety of nanostructured materials. A key point is that as nanostructured materials decrease in size so the magnetic and electrostatic signals from them also become progressively weaker. Of the different microscopy techniques that are available, it is easiest to measure these signals with high sensitivity from a phase image recorded using electron holography. For example, most other TEM techniques for measuring magnetic microstructure (such as the Fresnel and Foucault modes of Lorentz microscopy and DPC imaging) record signals that are proportional to either the first or the second differential of the phase shift, which inherently enhance contributions to the signal from rapid variations in sample thickness and composition, as compared to the weak and slowly varying magnetic signal.

Results have been presented from many different types of materials, including fine magnetic particles, the experimental confirmation of the Aharonov–Bohm effect, magnetic force microscopy tips, field-emitting carbon nanotubes, and a range of semiconductor device structures. *In-situ* experiments, which allow magnetization reversal processes to be followed and electrostatic fields in working semiconductor devices to be characterized, have been described. Although the results that are presented here are specific to the dimensions and the morphologies of the samples that are described, they illustrate the ways in which electron holography can be adapted to tackle different materials problems and the issues that remain to be answered, both about the technique and about the materials of interest.

Future developments in the field of electron holography of nanostructured materials are likely to follow several distinct paths: the development of new forms of electron holography and instrumentation, the development of new approaches that can enhance weak magnetic and electrostatic signals, the further examination of novel materials, the formulation of a complete understanding of the effect of different TEM sample preparation approaches on dopant contrast in semiconductors, and the combination of electron holography with electron tomography to record both electrostatic scalar fields and magnetic vector fields *inside* nanostructured materials (rather than simply the stray fields outside [428]) in *three* rather than just two dimensions [146]. The application of electron holographic tomography to the characterization of magnetic vector fields requires two high-tilt series of holograms to be recorded about orthogonal axes in order to reconstruct the three-dimensional distributions of two of the three components of the magnetic field in the sample B_x and B_y. The third component B_z could then be evaluated by making use of the criterion that div.**B** must equal zero everywhere, in the form

$$B_z(x, y, z) = -\int_{-\infty}^{z} \left[\frac{\partial}{\partial x} B_x(x, y, \zeta) + \frac{\partial}{\partial y} B_y(x, y, \zeta) \right] d\zeta \quad (62)$$

If this approach is ultimately successful, the primary difficulty may lie in the measurement and subtraction of the unwanted mean inner potential contribution to the measured phase shift at every one of the tilt angles required.

Electron holography has progressed through many generations of development. It has now evolved to the stage where the emphasis is shifting toward its optimization for specific challenging applications to diverse materials problems. This statement is particularly true for electron holography applications at medium resolution, for which the phase image provides information about magnetic and electric fields both in the sample and in the surrounding space. Such information is rarely available using other TEM imaging techniques.

The unique capability of the technique to provide quantitative information at resolutions approaching the nanometer scale, coupled with the increasing availability of field-emission-gun transmission electron microscopes and quantitative digital recording, suggests that the technique has a very promising future.

GLOSSARY

Electron holography An electron microscopy technique that can be used to record an interference pattern (an electron hologram) that contains information about changes in the amplitude and phase shift of an electron wave scattered by a material. The hologram can be used to provide information about local variations in electrostatic and magnetic potential within and surrounding the material at a spatial resolution that can approach the atomic scale.

Transmission electron microscope (TEM) An instrument that makes use of a beam of high energy (>50 kV) electrons, which are focused using electromagnetic lenses, to obtain information about the microstructure of a thin (<1 micrometer) sample, through which the electrons pass. Further electromagnetic lenses are used to magnify either an image or a diffraction pattern of the sample. Additional components such as spectrometers can be used to record a variety of signals from the sample in order to characterize its chemical composition and electrical and magnetic properties at a spatial resolution that can approach the atomic scale.

ACKNOWLEDGMENTS

Thanks are due to the Royal Society (R.D.B.) and the National Aeronautics and Space Administration (R.D.B. and M.R.M.) for financial support and to D. A. Bazylinski, C. B. Boothroyd, P. R. Buseck, R. B. Frankel, K. Harada, R. J. Harrison, B. Kardynal, P. A. Midgley, S. B. Newcomb, S. S. P. Parkin, M. Pósfai, G. Pozzi, A. Putnis, C. A. Ross, M. R. Scheinfein, A. Tonomura, A. C. Twitchett, and Y. Zhu for discussions and for ongoing collaborations. One of the authors (R.D.B.) is also grateful to the CNRS and the European Community for support through a European research network (GDR-E) entitled "Quantification and measurement in transmission electron microscopy".

REFERENCES

1. D. Gabor, *Proc. Roy. Soc. London Ser. A* 197, 454 (1949).
2. D. Gabor, *Proc. Phys. Soc. B* 54, 449 (1951).
3. M. E. Haine and T. Mulvey, *J. Opt. Soc. Am.* 42, 41 (1955).
4. A. Tonomura, T. Matsuda, and J. Endo, *Jpn. J. Appl. Phys.* 18, 1373 (1979).
5. A. Tonomura, L. F. Allard, G. Pozzi, D. C. Joy, and Y. A. Ono, Eds., "Electron Holography." Elsevier, Amsterdam, 1995.
6. E. Völkl, L. F. Allard, and D. C. Joy, Eds., "Introduction to Electron Holography." Plenum, New York, 1998.
7. A. Tonomura, "The Quantum World Unveiled by Electron Waves." World Scientific, Singapore, 1998.
8. A. Tonomura, *Adv. Phys.* 41, 59 (1992).
9. K.-J. Hanszen, *Adv. Electron. Electron Phys.* 59, 1 (1982).
10. G. Matteucci, G. F. Missiroli, G. Pozzi, P. Merli, and I. Vecchi, *Thin Solid Films* 62, 5 (1979).
11. A. Tonomura, *J. Electron Microsc.* 44, 425 (1995).
12. P. A. Midgley, *Micron* 32, 167 (2001).
13. H. Lichte, *Phil. Trans. Roy. Soc. London Ser. A* 360, 897 (2002).
14. *Mater. Charact.* 42, 173 (1999).
15. J. Munch, *Optik* 43, 79 (1975).
16. G. Möllenstedt and H. Düker, *Naturwissenschaften* 42, 41 (1954).
17. G. Möllenstedt and H. Düker, *Z. Phys.* 145, 377 (1956).
18. H. Lichte, *Ultramicroscopy* 64, 79 (1996).
19. G. F. Missiroli, G. Pozzi, and U. Valdré, *J. Phys. E: Sci. Instrum.* 14, 649 (1981).
20. G. Matteucci, G. F. Missiroli, and G. Pozzi, *Adv. Imaging Electron Phys.* 99, 171 (1998).
21. J. M. Cowley and J. C. H. Spence, in "Introduction to Electron Holography" (E. Völkl, L. F. Allard, and D. C. Joy, Eds.), pp. 17–56. Kluwer Academic/Plenum, New York, 1998.
22. L. D. Marks and R. Plass, *Ultramicroscopy* 55, 165 (1994).
23. K. Ishizuka, *Ultramicroscopy* 52, 1 (1993).
24. T. M. Hiller, P. A. Midgley, M. Saunders, Y. Ge, and W. Schwarzacher, *Phys. Rev. Lett.* 76, 977 (1996).
25. M. Takeda and Q. Ru, *Appl. Opt.* 24, 3068 (1985).
26. E. Völkl, L. F. Allard, and B. Frost, *J. Microsc.* 180, 39 (1995).
27. W. O. Saxton, T. J. Pitt, and M. Horner, *Ultramicroscopy* 4, 343 (1979).
28. W. J. de Ruijter, *Micron* 26, 247 (1995).
29. R. R. Meyer, A. I. Kirkland, R. E. Dunin-Borkowski, and J. L. Hutchison, *Ultramicroscopy* 85, 9 (2000).
30. R. R. Meyer and A. I. Kirkland, *Micr. Res. Techn.* 49, 269 (2000).
31. J. M. Zuo, *Ultramicroscopy* 66, 21 (1996).
32. E. Völkl and L. F. Allard, *MSA Bull.* 24, 466 (1994).
33. M. Lehmann, E. Völkl, and F. Lenz, *Ultramicroscopy* 54, 335 (1994).
34. R. R. Meyer and E. Heindl, *J. Microsc.* 191, 52 (1997).
35. A. Tonomura, "Electron Holography." Springer-Verlag, Berlin, 1999.
36. X. F. Duan, M. Gao, and L.-M. Peng, *Appl. Phys. Lett.* 72, 771 (1998).
37. D. J. Smith and M. R. McCartney, in "Introduction to Electron Holography" (E. Völkl, L. F. Allard, and D. C. Joy, Eds.), pp. 87–106. Kluwer Academic/Plenum, New York, 1998.
38. G. Ade, *Adv. Electron. Electron Phys.* 89, 1 (1994).
39. D. van Dyck, *J. Electron Microsc.* 48, 33 (1999).
40. G. Matteucci, M. Muccini, and U. Hartmann, *Phys. Rev. B* 50, 6823 (1994).
41. G. Matteucci, G. F. Missiroli, J. W. Chen, and G. Pozzi, *Appl. Phys. Lett.* 52, 176 (1988).
42. G. Matteucci, G. F. Missiroli, E. Nichelatti, A. Migliori, M. Vanzi, and G. Pozzi, *J. Appl. Phys.* 69, 1835 (1991).
43. B. G. Frost and G. Matteucci, in "Introduction to Electron Holography" (E. Völkl, L. F. Allard, and D. C. Joy, Eds.), pp. 183–200. Kluwer Academic/Plenum, New York, 1998.
44. L. Kou and J. Chen, *J. Mod. Opt.* 42, 1171 (1995).
45. W. J. de Ruijter and J. K. Weiss, *Ultramicroscopy* 50, 269 (1993).
46. E. Völkl, L. F. Allard, and B. Frost, *Mater. Characterization* 42, 307 (1999).
47. W. D. Rau, H. Lichte, E. Völkl, and U. Weierstall, *J. Comput.-Assisted Microsc.* 3, 51 (1991).
48. D. J. Smith, W. J. de Ruijter, J. K. Weiss, and M. R.. McCartney, in "Introduction to Electron Holography" (E. Völkl, L. F. Allard, and D. C. Joy, Eds.), pp. 107–124. Kluwer Academic/Plenum, New York, 1998.
49. M. Takeda, H. Ina, and S. Kobayashi, *J. Opt. Soc. Am.* 72, 156 (1982).
50. E. Völkl, L. F. Allard, A. Datye, and B. Frost, *Ultramicroscopy* 58, 97 (1995).
51. D. C. Ghiglia and M. D. Pritt, "Two-dimensional Phase Unwrapping. Theory, Algorithms and Software." Wiley, New York, 1998.
52. H. Lichte, *Ultramicroscopy* 38, 13 (1991).
53. Q. Fu, H. Lichte, and E. Völkl, *Phys. Rev. Lett.* 67, 2319 (1991).
54. H. Lichte, *Ultramicroscopy* 47, 223 (1992).
55. E. Völkl and H. Lichte, *Ultramicroscopy* 32, 177 (1990).
56. H. Lichte, E. Völkl, and K. Scheershmidt, *Ultramicroscopy* 47, 231 (1992).
57. T. Kawasaki and J. M. Rodenburg, *Ultramicroscopy* 52, 248 (1993).
58. H. Lichte, P. Kessler, F. Lenz, and W. D. Rau, *Ultramicroscopy* 52, 575 (1993).
59. H. Lichte and W. D. Rau, *Ultramicroscopy* 54, 310 (1994).
60. K. Ishizuka, T. Tanji, A. Tonomura, T. Ohno, and Y. Murayama, *Ultramicroscopy* 55, 197 (1994).

61. A. Harscher, G. Lang, and H. Lichte, *Ultramicroscopy* 58, 79 (1995).
62. Q. Fu and H. Lichte, *J. Microsc.* 179, 112 (1995).
63. K. Ishizuka, *Ultramicroscopy* 52, 1 (1993).
64. K. Ishizuka, *Ultramicroscopy* 53, 297 (1994).
65. A. Orchowski, W. D. Rau, and H. Lichte, *Phys. Rev. Lett.* 74, 399 (1995).
66. L. Reimer, "Transmission Electron Microscopy." Springer-Verlag, Berlin, 1991.
67. W. D. Rau, P. Schwander, F. H. Baumann, W. Höppner, and A. Ourmazd, *Phys. Rev. Lett.* 82, 2614 (1999).
68. Q. Ru, T. Hirayama, J. Endo, and A. Tonomura, *Jpn. J. Appl. Phys.* 31, 1919 (1992).
69. Q. Ru, G. Lai, K. Aoyama, J. Endo, and A. Tonomura, *Ultramicroscopy* 55, 209 (1994).
70. J. W. Chen, T. Hirayama, G. Lai, T. Tanji, and A. Tonomura, *Opt. Lett.* 18, 1887 (1993).
71. B. G. Frost, E. Voelkl, and L. F. Allard, *Ultramicroscopy* 63, 15 (1996).
72. E. Völkl, L. F. Allard, A. Datye, and B. Frost, *Ultramicroscopy* 58, 97 (1995).
73. W. J. de Ruijter and J. K. Weiss, *Rev. Sci. Instrum.* 63, 4314 (1992).
74. H. Lichte, *Ultramicroscopy* 20, 293 (1986).
75. H. Lichte, *Ultramicroscopy* 51, 15 (1993).
76. A. Harscher and H. Lichte, *Ultramicroscopy* 64, 57 (1996).
77. H. Lichte, K. H. Hermann, and F. Lenz, *Optik* 77, 135 (1987).
78. R. R. Meyer, A. I. Kirkland, R. E. Dunin-Borkowski, and J. L. Hutchison, *Ultramicroscopy* 85, 9 (2000).
79. H. Lichte, in "Electron Holography" (A. Tonomura, L. F. Allard, G. Pozzi, D. C. Joy, and Y. A. Ono, Eds.), pp. 11–31. Elsevier, Amsterdam, 1995.
80. K. Yamamoto, T. Tanji, and M. Hibino, *Ultramicroscopy* 85, 35 (2000).
81. R. Huhle, R. Goldberg, and H. Lichte, in "Proceedings of the 12th EUREM," Brno, Czech Republic (J. Gemperlová and I. Vávra, Eds.), Vol. 2, p. 163. Czechoslovak Society for Electron Microscopy, Brno, Czech Republic, 2000.
82. H. Lichte, D. Geiger, A. Harscher, E. Heindl, M. Lehmann, D. Malamidis, A. Orchowski, and W. D. Rau, *Ultramicroscopy* 64, 67 (1996).
83. J. C. H. Spence, *Acta Crystallogr. A* 49, 231 (1993).
84. S. Miyake, *Proc. Phys.-Math. Soc. Japan* 22, 666 (1940).
85. J. A. Ibers, *Acta Crystallogr.* 11, 178 (1958).
86. J. C. H. Spence and J. M. Zuo, "Electron Microdiffraction." Plenum Press, New York, 1992.
87. J. B. Pendry, "Low Energy Electron Diffraction." Academic Press, Boston, 1974.
88. M. A. van Hove, W. H. Weinberg, and C. M. Chan, "Low Energy Electron Diffraction." Springer-Verlag, Berlin, 1986.
89. D. K. Saldin and J. C. H. Spence, *Ultramicroscopy* 55, 397 (1994).
90. M. Gajdardziska-Josifovska and A. Carim, in "Introduction to Electron Holography" (E. Völkl, L. F. Allard, and D. C. Joy, Eds.), pp. 267–293. Kluwer Academic/Plenum, New York, 1998.
91. F. M. Ross and W. M. Stobbs, *Philos. Mag. A* 63, 37 (1991).
92. P. A. Doyle and P. S. Turner, *Acta Crystallogr. A* 24, 390 (1968).
93. D. Rez, P. Rez, and I. Grant, *Acta Crystallogr. A* 50, 481 (1994).
94. J. A. Ibers and B. K. Vainshtein, "International Crystallographic Tables," Vol. III. Kynoch Press, 1962.
95. G. H. Smith and R. E. Burge, *Acta Crystallogr.* 15, 182 (1962).
96. I. P Grant, B. J. McKenzie, P. H. Norrington, D. F. Mayers, and N. C. Pyper, *Comput. Phys. Commun.* 21, 207 (1980).
97. G. Radi, *Acta Crystallogr. A* 26, 41 (1970).
98. H. Bethe, *Ann. Phys.* 87, 55 (1928).
99. Y. C. Wang, T. M. Chou, M. Libera, and T. F. Kelly, *Appl. Phys. Lett.* 70, 1296 (1997).
100. A. Harscher and H. Lichte, in "Electron Microscopy 98" (H. A. Calderón Benavides and M. José Yacamán, Eds.), Vol. 1, p. 553. Paper presented at the 14th International Conference on Electron Microscopy, Cancun, Mexico, 31 August–4 September 1998. Institute of Physics Publishing, Bristol, UK, 1998.
101. M. Gajdardziska-Josifovska, M. R. McCartney, W. J. de Ruijter, D. J. Smith, J. K. Weiss, and J. M. Zuo, *Ultramicroscopy* 50, 285 (1993).
102. W. J. de Ruijter, M. Gajdardziska-Josifovska, M. R. McCartney, R. Sharma, D. J. Smith, and J. K. Weiss, *Scanning Microsc. Suppl.* 6, 347 (1992).
103. J. Li, M. R. McCartney, R. E. Dunin-Borkowski, and D. J. Smith, *Acta Crystallogr. A* 55, 652 (1999).
104. W. D. Rau, F. H. Baumann, J. A. Rentschler, P. K. Roy, and A. Ourmazd, *Appl. Phys. Lett.* 68, 3410 (1996).
105. G. L. Waytena, J. Hren, and P. Rez, *J. Appl. Phys.* 73, 1750 (1993).
106. G. Möllenstedt, H. Stumpp, and H. Lichte, *Ultramicroscopy* 15, 1 (1984).
107. M. O'Keeffe and J. C. H. Spence, *Acta Crystallogr. A* 50, 33 (1994).
108. M. Y. Kim, J. M. Zuo, and J. C. H. Spence, *Phys. Status Solidi A* 166, 445 (1998).
109. S. J. Lloyd, R. E. Dunin-Borkowski, and C. B. Boothroyd, *Inst. Phys. Conf. Ser.* 153, 113 (1997).
110. M. R. McCartney and M. Gajdardziska-Josifovska, *Ultramicroscopy* 53, 283 (1994).
111. J. C. H. Spence and J. M. Zuo, *Ultramicroscopy* 69, 185 (1997).
112. A. Harscher, H. Lichte, and J. Mayer, *Ultramicroscopy* 69, 201 (1997).
113. H. Lichte and B. Freitag, *Ultramicroscopy* 81, 177 (2000).
114. D. Van Dyck, H. Lichte, and J. C. H. Spence, *Ultramicroscopy* 81, 187 (2000).
115. T. M. Chou and M. Libera, *Ultramicroscopy* 94, 31 (2003).
116. J. K. Weiss, W. J. de Ruijter, M. Gajdardziska-Josifovska, D. J. Smith, E. Völkl, and H. Lichte, in "Proceedings of the 49th Annual EMSA Meeeting" (G. W. Bailey, Ed.), p. 674. San Francisco Press, San Francisco, 1991.
117. J. K. Weiss, W. J. de Ruijter, M. Gajdardziska-Josifovska, M. R. McCartney, and D. J. Smith, *Ultramicroscopy* 50, 301 (1993).
118. A. Rosenauer, D. Van Dyck, M. Arzberger, and G. Abstreiter, *Ultramicroscopy* 88, 51 (2001).
119. A. Rosenauer, D. Gerthsen, D. Van Dyck, M. Arzberger, G. Bohm, and G. Abstreiter, *Inst. Phys. Conf. Ser.* 169, 33 (2001).
120. L. F. Allard, E. Völkl, A. Carim, A. K. Datye, and R. Ruoff, *Nano. Mater.* 7, 137 (1996).
121. L. F. Allard, E. Völkl, D. S. Kalakkad, and A. K. Datye, *J. Mater. Sci.* 29, 5612 (1994).
122. A. K. Datye, D. S. Kalakkad, E. Völkl, and L. F. Allard, in "Electron Holography" (A. Tonomura, L. F. Allard, G. Pozzi, D. C. Joy, and Y. A. Ono, Eds.), pp. 199–208. Elsevier, Amsterdam, 1995.
123. E. Völkl, L. F. Allard, A. Datye, and B. Frost, *Ultramicroscopy* 58, 97 (1995).
124. R. A. Herring, T. Tanji, and A. Tonomura, in "Proceedings of the 50th EMSA Meeting" (G. W. Bailey, J. Bentley, and J. A. Small, Eds.), p. 990. San Francisco Press, San Francisco, 1992.
125. D. C. Joy, Y.-S. Zhang, X. Zhang, T. Hashimoto, R. D. Bunn, L. F. Allard, and T. A. Nolan, *Ultramicroscopy* 51, 1 (1993).
126. H. Lichte, in "Advances in Optical and Electron Microscopy" (T. Mulvey and T. R. Sheppard, Eds.), Vol. 12, p. 25. Academic Press, New York, 1991.
127. A. Tonomura, *Rev. Mod. Phys.* 59, 639 (1987).
128. L. F. Allard, E. Völkl, S. Subramoney, and R. S. Ruoff, in "Electron Holography" (A. Tonomura, L. F. Allard, G. Pozzi, D. C. Joy, and Y. A. Ono, Eds.), pp. 219–230. Elsevier, Amsterdam, 1995.
129. X. Lin and V. P. Dravid, *Appl. Phys. Lett.* 69, 1014 (1996).
130. C. Beeli, G. Matteucci, G. Lulli, P. G. Merli, and A. Migliori, *Philos. Mag. Lett.* 78, 445 (1998).
131. C. Beeli, G. Matteucci, A. Migliori, G. Lulli, and P. G. Merli, *Helv. Phys. Acta* 70, S3 (1997).

132. P. A. Midgley, J. S. Barnard, and D. Cherns, *Inst. Phys. Conf. Ser.* 157, 75 (1997).
133. K. Aoyama and Q. Ru, in "Electron Holography" (A. Tonomura, L. F. Allard, G. Pozzi, D. C. Joy, and Y. A. Ono, Eds.), pp. 239–248. Elsevier, Amsterdam, 1995.
134. K. Aoyama and Q. Ru, *J. Microsc.* 182, 177 (1996).
135. H. W. Fink, H. Schmid, E. Ermentraut, and T. Schulz, *J. Opt. Soc. Am. A* 14, 2168 (1997).
136. P. Simon, R. Huhle, M. Lehmann, H. Lichte, D. Mönter, T. Bieber, W. Reschetilowski, R. Adhikari, and G. H. Michler, *Chem. Mater.* 14, 1505 (2002).
137. A. Tonomura, T. Matsuda, T. Kawasaki, J. Endo, and N. Osakabe, *Phys. Rev. Lett.* 54, 60 (1985).
138. H. Banzhoff, K. H. Herrmann, and H. Lichte, *Microsc. Res. Techn.* 20, 450 (1992).
139. H. Banzhoff, K. H. Herrmann, and H. Lichte, *Inst. Phys. Conf. Ser.* 93, 263 (1988).
140. N. Osakabe, J. Endo, T. Matsuda, A. Tonomura, and A. Fukuhara, *Phys. Rev. Lett.* 62, 2969 (1989).
141. T. Tanji, K. Urata, K. Ishizuka, Q. Ru, and A. Tonomura, *Ultramicroscopy* 49, 259 (1993).
142. D. J. De Rosier and A. Klug, *Nature* 217, 130 (1968).
143. R. A. Crowther, L. A. Amos, J. T. Finch, and A. Klug, *Nature* 226, 421 (1970).
144. P. A. Midgley, M. Weyland, J. M. Thomas, and B. F. G. Johnson, *Chem. Commun.* 10, 907 (2001).
145. P. R. Buseck, R. E. Dunin-Borkowski, B. Devouard, R. B. Frankel, M. R. McCartney, P. A. Midgley, M. Posfai, and M. Weyland, *Proc. Nat. Acad. Sci. USA* 98, 13490 (2001).
146. G. Lai, K. Hirayama, T. Ishizuka, T. Tanji, and A. Tonomura, *Appl. Opt.* 33, 829 (1994).
147. G. Matteucci, G. F. Missiroli, and G. Pozzi, *IEEE Trans. Magn.* 20, 1870 (1984).
148. A. Tonomura, *IEEE Trans. Magn.* 29, 2488 (1993).
149. A. Tonomura, *Ultramicroscopy* 47, 419 (1992).
150. T. Hirayama, J. Chen, T. Tanji, and A. Tonomura, *Ultramicroscopy* 54, 9 (1994).
151. A. Tonomura, *Int. J. Mod. Phys.* A15, 3427 (2000).
152. R. E. Dunin-Borkowski, M. R. McCartney, and D. J. Smith, *Recent Res. Dev. Appl. Phys.* 1, 119 (1998).
153. M. S. Cohen, *J. Appl. Phys.* 38, 4966 (1967).
154. P. J. Grundy and R. S. Tebble, *Adv. Phys.* 17, 153 (1968).
155. J. N. Chapman, *J. Phys. D: Appl. Phys.* 17, 623 (1984).
156. J. P. Jakubovics, in "Electron Microscopy in Materials Science" (U. Valdré and E. Ruedl, Eds.), Part IV, p. 1303. Commission of the European Communities, Brussels, 1976.
157. A. C. Daykin and A. K. Petford-Long, *Ultramicroscopy* 58, 365 (1995).
158. J. Dooley and M. De Graef, *Ultramicroscopy* 67, 113 (1997).
159. M. De Graef and Y. Zhu, *J. Appl. Phys.* 89, 7177 (2001).
160. K. Tang, M. R. Visokay, C. A. Ross, R. Ranjan, T. Yamashita, and R. Sinclair, *IEEE Trans. Magn.* 32, 4130 (1996).
161. J. Dooley and M. De Graef, *Ultramicroscopy* 67, 113 (1997).
162. A. B. Johnston and J. N. Chapman, *J. Microsc.* 179, 119 (1995).
163. J. N. Chapman, P. E. Batson, E. M. Waddell, and R. P. Ferrier, *Ultramicroscopy* 3, 203 (1978).
164. G. R. Morrison, H. Gong, J. N. Chapman, and V. Hrnciar, *J. Appl. Phys.* 64, 1338 (1988).
165. J. N. Chapman, I. R. McFadyen, and S. McVitie, *IEEE Trans. Magn.* 26, 1506 (1990).
166. Y. Liu and R. P. Ferrier, *IEEE Trans. Magn.* 31, 3373 (1995).
167. K. J. Kirk, S. McVitie, J. N. Chapman, and C. D. W. Wilkinson, *J. Appl. Phys.* 89, 7174 (2001).
168. K. M. Krishnan, Ed., *Mater. Res. Soc. Bull.* 20 (1995).
169. *Ultramicroscopy* 47, 313 (1992).
170. T. Matsuda, A. Tonomura, R. Suzuki, J. Endo, N. Osakabe, H. Umezaki, H. Tanabe, Y. Sugita, and H. Fujiwara, *J. Appl. Phys.* 53, 5444 (1982).
171. Y. Otani, K. Fukamichi, O. Kitakami, Y. Shimada, B. Pannetier, J.-P. Nozieres, T. Matsuda, and A. Tonomura, in "Magnetic Ultrathin Films, Multilayers and Surfaces" (D. D. Chambliss, J. G. Tobin, D. Kubinski, K. Barmak, W. J. M. de Jonge, T. Katayama, A. Schuhl, and P. Dederichs, Eds.), Materials Research Society Symposium Proceedings, Vol. 475, p. 215. Materials Research Society, Warrendale, PA, 1997.
172. M. Tsutsumi, F. Kugiya, S. Hasegawa, and A. Tonomura, *IEEE Trans. Magn.* 25, 3665 (1989).
173. T. Hirayama, Q. Ru, T. Tanji, and A. Tonomura, *Appl. Phys. Lett.* 63, 418 (1993).
174. N. Osakabe, K. Yoshida, Y. Horiuchi, T. Matsuda, H. Tanabe, T. Okuwaki, J. Endo, H. Fujiwara, and A. Tonomura, *Appl. Phys. Lett.* 42, 746 (1983).
175. S. Hasegawa, T. Kawasaki, J. Endo, A. Tonomura, Y. Honda, M. Futamoto, K. Yoshida, F. Kugiya, and M. Koizumi, *J. Appl. Phys.* 65, 2000 (1989).
176. K. Yoshida, Y. Honda, T. Kawasaki, M. Koizumi, F. Kugiya, M. Futamoto, and A. Tonomura, *IEEE Trans. Magn.* 23, 2073 (1987).
177. A. Tonomura, T. Matsuda, J. Endo, T. Arii, and K. Mihama, *Phys. Rev. Lett.* 44, 1430 (1980).
178. K. Runge, Y. Nozaki, Y. Otani, H. Miyajima, B. Pannetier, T. Matsuda, and A. Tonomura, *J. Appl. Phys.* 79, 5075 (1996).
179. Y. Aharonov and D. Bohm, *Phys. Rev.* 115, 485 (1959).
180. R. G. Chambers, *Phys. Rev. Lett.* 5, 3 (1960).
181. H. A. Fowler, L. Marton, J. A. Simpson, and J. A. Suddeth, *J. Appl. Phys.* 32, 1153 (1961).
182. H. Boersch, H. Hamisch, K. Grohmann, and D. Wohlleben, *Z. Phys.* 165, 79 (1961).
183. A. Fukuhara, K. Shinagawa, A. Tonomura, and H. Fujiwara, *Phys. Rev.* B 27, 1839 (1983).
184. P. Bocchieri and A. Loinger, *Nuovo Cimento A* 66, 164 (1981).
185. A. Tonomura, T. Matsuda, R. Suzuki, A. Fukuhara, N. Osakabe, H. Umezaki, J. Endo, K. Shinagawa, Y. Sugita, and H. Fujiwara, *Phys. Rev. Lett.* 48, 1443 (1982).
186. A. Tonomura, H. Umezaki, T. Matsuda, N. Osakabe, J. Endo, and Y. Sugita, *Phys. Rev. Lett.* 51, 331 (1983).
187. P. Bocchieri, A. Loinger, and G. Siragusa, *Nuovo Cimento Soc. Ital. Fis.* 35, 370 (1982).
188. N. Osakabe, T. Matsuda, T. Kawasaki, J. Endo, A. Tonomura, S. Yano, and H. Yamada, *Phys. Rev. A* 34, 815 (1986).
189. A. Tonomura, N. Osakabe, T. Matsuda, T. Kawasaki, J. Endo, S. Yano, and H. Yamada, *Phys. Rev. Lett.* 56, 792 (1986).
190. M. R. McCartney and Y. Zhu, *Appl. Phys. Lett.* 72, 1380 (1998).
191. M. R. McCartney and Y. Zhu, *J. Appl. Phys.* 83, 6414 (1998).
192. Y. Zhu and M. R. McCartney, *J. Appl. Phys.* 84, 3267 (1998).
193. H. Kronmuller, in "Science and Technology of Nanostructured Materials" (G. C. Hadjipanayis and G. Prinz, Eds.), p. 657. Plenum Press, New York, 1990.
194. B. O. Cullity. "Introduction to Magnetic Materials." Addison–Wesley, New York, 1992.
195. J. F. Herbst and J. J. Croat, *J. Magn. Magn. Mater.* 100, 57 (1991).
196. Y. G. Park and D. Shindo, *J. Magn. Magn. Mater.* 238, 68 (2002).
197. Y. M. Zhu, V. V. Volkov, and M. De Graef, *J. Electron Microsc.* 50, 447 (2001).
198. M. R. McCartney, D. J. Smith, R. F. C. Farrow, and R. F. Marks, *J. Appl. Phys.* 82, 2461 (1997).
199. R. F. C. Farrow, D. Weller, R. F. Marks, M. F. Toney, A. Cebollada, and G. R. Harp, *J. Appl. Phys.* 79, 5330 (1996).
200. E. Bonetti, L. Del Bianco, L. Pasquini, G. Matteucci, C. Beeli, and S. Signoretti, *J. Appl. Phys.* 90, 4152 (2001).

201. J. T. M. De Hosson, N. G. Chechenin, D. H. Alsem, T. Vystavel, B. J. Kooi, A. R. Chezan, and D. O. Boerma, *Microsc. Microanal.* 8, 274 (2002).
202. Y. Murakami, D. Shindo, K. Oikawa, R. Kainuma, and K. Ishida, *Acta Mater.* 50, 2173 (2002).
203. J. C. Loudon, N. D. Mathur, and P. A. Midgley, *Nature* 420, 797 (2002).
204. C. Beeli, B. Doudin, J.-Ph. Ansermet, and P. A. Stadelmann, *Ultramicroscopy* 67, 143 (1997).
205. C. Beeli, *Nanostruct. Mater.* 11, 697 (1999).
206. C. Beeli, P. Stadelmann, and B. Doudin, *Helv. Phys. Acta* 68, 494 (1995).
207. E. Snoeck, R. E. Dunin-Borkowski, F. Dumestre, P. Renaud, C. Amiens, B. Chaudret, and P. Zurcher, *Appl. Phys. Lett.* 82, 88 (2003).
208. A. Sugawara, D. Streblechenko, M. R. McCartney, and M. R. Scheinfein, *IEEE Trans. Magn.* 34, 1081 (1998).
209. D. J. Wohlleben, "Electron Microscopy in Materials Science," Vol. 2, p. 712. Academic Press, New York, 1971.
210. A. Tonomura, T. Matsuda, J. Endo, T. Arii, and K. Mihama, *Phys. Rev. B* 34, 3397 (1986).
211. R. E. Dunin-Borkowski, M. R. McCartney, D. J. Smith, and S. S. P. Parkin, *Ultramicroscopy* 74, 61 (1998).
212. K. J. Kirk, J. N. Chapman, and C. D. W. Wilkinson, *Appl. Phys. Lett.* 71, 539 (1997).
213. A. D. Kent, T. M. Shaw, S. von Molnar, and D. D. Awschalom, *Science* 262, 1249 (1993).
214. S. Y. Chou, P. R. Krauss, and L. Kong, *J. Appl. Phys.* 79, 6101 (1996).
215. Y. Zheng and J.-G. Zhu, *J. Appl. Phys.* 81, 5470 (1997).
216. J. Shi, T. Zhu, M. Durlam, E. Chen, S. Tehrani, Y. F. Cheng, and J.-G. Zhu, *IEEE Trans. Magn.* 34, 997 (1998).
217. R. E. Dunin-Borkowski, M. R. McCartney, B. Kardynal, S. S. P. Parkin, M. R. Scheinfein, and D. J. Smith, *J. Microsc.* 200, 187 (2000).
218. R. E. Dunin-Borkowski, M. R. McCartney, B. Kardynal, and D. J. Smith, *J. Appl. Phys.* 84, 374 (1998).
219. S. Henzelmann, T. Uhlig, M. Heumann, H. Brückl, and J. Zweck, in "Proceedings of the 12th EUREM," Brno, Czech Republic (J. Gemperlová and I. Vávra, Eds.), Vol. 2, p. 167. Czechoslovak Society for Electron Microscopy, Brno, Czech Republic, 2000.
220. M. Rührig, B. Khamsehpour, K. J. Kirk, J. N. Chapman, P. R. Aitshison, S. McVitie, and C. D. W. Wilkinson, *IEEE Trans. Magn.* 32, 4452 (1996).
221. J. Shi, S. Tehrani, T. Zhu, Y. F. Zheng, and J.-G. Zhu, *Appl. Phys. Lett.* 74, 2525 (1999).
222. A. Hubert and R. Schäfer, "Magnetic Domains." Springer-Verlag, Berlin, 1998.
223. M. R. Scheinfein, J. Unguris, J. L. Blue, K. J Coakley, D. T. Pierce, R. J. Celotta, and P. J. Ryan, *Phys. Rev. B* 43, 3395 (1991).
224. R. E. Dunin-Borkowski, M. R. McCartney, B. Kardynal, D. J. Smith, and M. R. Scheinfein, *Appl. Phys. Lett.* 75, 2641 (1999).
225. Y. Lu, R. A. Altman, A. Marley, S. A. Rishton, P. L. Trouillard, G. Xiao, W. J. Gallagher, and S. S. P. Parkin, *Appl. Phys. Lett.* 70, 2610 (1997).
226. S. S. P. Parkin, *Ann. Rev. Mater. Sci.* 25, 357 (1995).
227. D. J. Smith, R. E. Dunin-Borkowski, M. R. McCartney, B. Kardynal, and M. R. Scheinfein, *J. Appl. Phys.* 87, 7400 (2000).
228. R. E. Dunin-Borkowski, M. R. McCartney, B. Kardynal, M. R. Scheinfein, D. J. Smith, and S. S. P. Parkin, *J. Appl. Phys.* 90, 2899 (2001).
229. J. C. S. Kools, *IEEE Trans. Magn.* 32, 3165 (1996).
230. R. P. Blakemore, *Science* 190, 377 (1975).
231. D. A. Bazylinski and B. M. Moskowitz, *Mineralog. Soc. Am. Rev. Mineral.* 35, 181 (1997).
232. D. J. Dunlop and Ö. Özdemir, "Rock Magnetism." Cambridge Univ. Press, Cambridge, UK, 1997.
233. R. B. Frankel, D. A. Bazylinski, M. S. Johnson, and B. L. Taylor, *Biophys. J.* 73, 994 (1997).
234. C. E. Diebel, R. Proksch, C. R. Green, P. Neilson, and M. M. Walker, *Nature* 406, 299 (2000).
235. S. Mann, N. H. Sparks, M. M. Walker, and J. L. Kirschvink, *J. Exp. Biol.* 140, 35 (1988).
236. D. S. McKay, E. K. Gibson, Jr., K. L. Thomas-Keprta, H. Vali, C. S. Romanek, S. J. Clemett, X. D. F. Chillier, C. R. Maechling, and R. N. Zare, *Science* 273, 924 (1996).
237. J. L. Kirschvink, A. Kobayashi-Kirschivink, and B. J. Woodford, *Proc. Nat. Acad. Sci. USA* 89, 7683 (1992).
238. J. R. Dunn, M. Fuller, J. Zoeger, J. P. Dobson, F. Heller, E. Caine, and B. M. Moskowitz, *Brain Res. Bull.* 36, 149 (1995).
239. P. P. Schultheiss-Grassi, R. Wesiken, and J. Dobson, *Biochim. Biophys. Acta* 1426, 212 (1999).
240. M. Hanzlik, M. Winklhofer, and N. Petersen, *Earth Planet. Sci. Lett.* 145, 125 (1996).
241. V. P. Shcherbakov, M. Winklhofer, M. Hanzlik, and N. Petersen, *Eur. Biophys. J.* 26, 319 (1997).
242. K. A. Fabian, A. Kirchner, W. Williams, F. Heider, T. Leibl, and A. Hubert, *Geophys. J. Int.* 124, 89 (1996).
243. R. E. Dunin-Borkowski, M. R. McCartney, R. B. Frankel, D. A. Bazylinski, M. Pósfai, and P. R. Buseck, *Science* 282, 1868 (1998).
244. F. C. Meldrum, S. Mann, B. R. Heywood, R. B. Frankel, and D. A. Bazylinski, *Proc. Roy. Soc. London Ser. B* 251, 231 (1993).
245. R. E. Dunin-Borkowski, M. R. McCartney, M. Pósfai, R. B. Frankel, D. A. Bazylinski, and P. R. Buseck, *Eur. J. Mineral.* 13, 671 (2001).
246. I. S. Jacobs and C. P. Bean, *Phys. Rev.* 100, 1060 (1955).
247. P. C. Kuo, *J. Appl. Phys.* 64, 5071 (1988).
248. S. A. Nepijko and R. Wiesendanger, *Appl. Phys.* A 65, 361 (1997).
249. M. De Graef, T. Nuhfer, and M. R. McCartney, *J. Microsc.* 194, 84 (1999).
250. J. D. Jackson, "Classical Electrodynamics," 2nd ed., p. 197. Wiley, New York, 1975.
251. J. Jiao, S. Seraphin, X. Wang, and J. C. Withers, *J. Appl. Phys.* 80, 103 (1996).
252. S. Seraphin, C. Beeli, J.-M. Bonard, J. Jiao, P. A. Stadelmann, and A. Chatelain, *J. Mater. Res.* 14, 2861 (1999).
253. C. A. Ross, *Ann. Rev. Mater. Res.* 31, 203 (2001).
254. R. E. Dunin-Borkowski, S. B. Newcomb, M. R. McCartney, C. A. Ross, and M. Farhoud, *Inst. Phys. Conf. Ser.* 168, 485 (2001).
255. B. G. Frost, N. F. van Hulst, E. Lunedei, G. Matteucci, and E. Rikkers, *Appl. Phys. Lett.* 68, 1865 (1996).
256. E. Lunedei, G. Matteucci, B. G. Frost, and J. Greve, *J. Magn. Magn. Mater.* 157/158, 434 (1996).
257. D. G. Streblechenko, M. R. Scheinfein, M. Mankos, and K. Babcock, *IEEE Trans. Magn.* 32, 4124 (1996).
258. M. R. McCartney and R. E. Dunin-Borkowski, in "Electron Microscopy 98" (H. A. Calderón Benavides and M. José Yacamán, Eds.), Vol. 2, p. 497. Paper presented at the 14th International Conference on Electron Microscopy, Cancun, Mexico, 31 August–4 September 1998. Institute of Physics Publishing, Bristol, UK, 1998.
259. Y. G. Wang, F. Shen, Z. Zhang, J. W. Cai, and W. Y. Lai, *Chinese Phys. Lett.* 19, 1480 (2002).
260. T. Matsuda, S. Hasegawa, M. Igarashi, T. Kobayashi, M. Naito, H. Kajiyama, J. Endo, N. Osakabe, and A. Tonomura, *Phys. Rev. Lett.* 62, 2519 (1989).
261. S. Hasegawa, T. Matsuda, J. Endo, N. Osakabe, M. Igarashi, T. Kobayashi, M. Naito, and A. Tonomura, *Phys. Rev. B* 43, 7631 (1991).
262. T. Matsuda, A. Fukuhara, T. Yoshida, S. Hasegawa, A. Tonomura, and Q. Ru, *Phys. Rev. Lett.* 66, 457 (1991).

263. T. Yoshida, T. Matsuda, and A. Tonomura, in "Proceedings of the 50th EMSA Meeting" (G. W. Bailey, J. Bentley, and J. A. Small, Eds.), p. 68. San Francisco Press, San Francisco, 1992.
264. A. Migliori and G. Pozzi, *Ultramicroscopy* 41, 169 (1992).
265. C. Colliex, B. Jouffrey, and M. Kleman, *Acta Crystallogr. A* 24, 692 (1968).
266. C. Capiluppi, G. Pozzi, and U. Valdré, *Philos. Mag.* 26, 865 (1972).
267. J. E. Bonevich, K. Harada, T. Matsuda, H. Kasai, T. Yoshida, G. Pozzi, and A. Tonomura, *Phys. Rev. Lett.* 70, 2952 (1993).
268. J. E. Bonevich, K. Harada, H. Kasai, T. Matsuda, T. Yoshida, and A. Tonomura, *Phys. Rev. B* 50, 567 (1994).
269. A. Migliori and G. Pozzi, *Ultramicroscopy* 41, 16 (1992).
270. A. Migliori, G. Pozzi, and A. Tonomura, *Ultramicroscopy* 49, 87 (1993).
271. J. E. Bonevich, D. Capacci, K. Harada, H. Kasai, T. Matsuda, R. Patti, G. Pozzi, and A. Tonomura, *Phys. Rev. B* 57, 1200 (1998).
272. M. Beleggia, S. Fanesi, R. Patti, and G. Pozzi, *Mater. Charact.* 42, 209 (1999).
273. M. Beleggia and G. Pozzi, *Ultramicroscopy* 84, 171 (2000).
274. M. Beleggia and G. Pozzi, *Phys. Rev. B* 63, 054507 (2001).
275. M. Beleggia and G. Pozzi, *J. Electron Microsc.* 51, S73 (2002).
276. K. Harada, T. Matsuda, H. Kasai, J. E. Bonevich, T. Yoshida, U. Kawabe, and A. Tonomura, *Phys. Rev. Lett.* 71, 3371 (1993).
277. T. Matsuda, K. Harada, H. Kasai, O. Kamimura, and A. Tonomura, *Science* 271, 1393 (1996).
278. K. Harada, O. Kamimura, H. Kasai, T. Matsuda, and A. Tonomura, *Science* 274, 1167 (1996).
279. K. Harada, T. Matsuda, H. Kasai, O. Kamimura, and A. Tonomura, *Ultramicroscopy* 67, 133 (1997).
280. K. Harada, H. Kasai, T. Matsuda, M. Yamasaki, and A. Tonomura, *J. Electron Microscopy* 46, 227 (1997).
281. A. Tonomura, *J. Microsc.* 190, 366 (1998).
282. A. Tonomura, H. Kasai, O. Kamimura, T. Matsuda, K. Harada, J. Shimoyama, K. Kishio, and K. Kitazawa, *Nature* 397, 308 (1999).
283. A. Tonomura, H. Kasai, O. Kamimura, T. Matsuda, K. Harada, Y. Nakayama, J. Shimoyama, K. Kishio, T. Hanaguri, K. Kitazawa, M. Sasase, and S. Okayasu, *Nature* 412, 620 (2001).
284. O. Kamimura, H. Kasai, T. Akashi, T. Matsuda, K. Harada, J. Masuko, T. Yoshida, N. Osakabe, A. Tonomura, M. Beleggia, G. Pozzi, J. Shimoyama, K. Kishio, T. Hanaguri, K. Kitazawa, M. Sasase, and S. Okayasu, *J. Phys. Soc. Japan* 71, 1840 (2002).
285. G. Matteucci, G. F. Missiroli, and G. Pozzi, *Adv. Imaging Electron Phys.* 122, 173 (2002).
286. G. Matteucci, G. F. Missiroli, and G. Pozzi, *J. Electron Microsc.* 45, 27 (1996).
287. J. W. Chen, G. Matteucci, A. Migliori, G. F. Missiroli, E. Nichelatti, G. Pozzi, and M. Vanzi, *Phys. Rev. A* 40, 3136 (1989).
288. Y. Aoyama, Y. G. Park, C. W. Lee, and D. Shindo, *Mater. Trans.* 43, 474 (2002).
289. B. Frost and T. L. Jenkins, *J. Microsc.* 187, 85 (1997).
290. B. Frost and E. Voelkl, *Ultramicroscopy* 72, 101 (1998).
291. B. Frost, *Ultramicroscopy* 75, 105 (1998).
292. G. Matteucci, G. F. Missiroli, M. Muccini, and G. Pozzi, *Ultramicroscopy* 45, 77 (1992).
293. G. Matteucci, G. F. Missiroli, and G. Pozzi, *Physica B* 151, 223 (1988).
294. T. Kawasaki, G. F. Missiroli, G. Pozzi, and A. Tonomura, *Optik* 92, 168 (1993).
295. T. Matsumoto and A. Tonomura, *Ultramicroscopy* 63, 5 (1996).
296. J. Cumings, A. Zettl, M. R. McCartney, and J. C. H. Spence, *Phys. Rev. Lett.* 88, 056804 (2002).
297. Z. L. Wang, P. Poncharal, and W. A. de Heer, *J. Phys. Chem. Solids* 61, 1025 (2000).
298. Z. L. Wang, P. Poncharal, and W. A. de Heer, *Microsc. Microanal.* 6, 224 (2000).
299. J. M. Titchmarsh, A. J. Lapworth, and G. R. Booker, *Phys. Status Solidi* 34, K83 (1969).
300. P. G. Merli, G. F. Missiroli, and G. Pozzi, *Phys. Status Solidi A* 20, K87 (1973).
301. P. G. Merli, G. F. Missiroli, and G. Pozzi, *J. Microsc.* 21, 11 (1974).
302. E. H. Darlington and U. Valdré, *J. Phys. E* 8, 321 (1975).
303. P. G. Merli, G. F. Missiroli, and G. Pozzi, *Phys. Status Solidi A* 30, 699 (1975).
304. C. Capiluppi, P. G. Merli, G. Pozzi, and I. Vecchi, *Phys. Status Solidi A* 35, 165 (1976).
305. S. Frabboni, G. Matteucci, G. Pozzi, and M. Vanzi, *Phys. Rev. Lett.* 55, 2196 (1985).
306. S. Frabboni, G. Matteucci, and G. Pozzi, *Ultramicroscopy* 23, 29 (1987).
307. C. Capiluppi, P. G. Merli, and G. Pozzi, *Optik* 47, 205 (1977).
308. G. Pozzi, *Optik* 53, 381 (1979).
309. G. Pozzi and M. Vanzi, *Optik* 60, 175 (1982).
310. M. Vanzi, *Optik* 68, 319 (1984).
311. G. Pozzi, *Phys. Status Solidi A* 156, K1 (1996).
312. M. Beleggia, D. Cristofori, P. G. Merli, and G. Pozzi, *Micron* 31, 231 (2000).
313. M. R. McCartney, D. J. Smith, R. Hull, J. C. Bean, E. Voelkl, and B. Frost, *Appl. Phys. Lett.* 65, 2603 (1994).
314. W. D. Rau, P. Schwander, and A. Ourmazd, *Solid State Phenom.* 63–4, 525 (1998).
315. W. D. Rau, P. Schwander, and A. Ourmazd, *Phys. Status Solidi B* 222, 213 (2000).
316. A. Orchowski, W. D. Rau, H. Rücker, B. Heinemann, P. Schwander, B. Tillack, and A. Ourmazd, *Appl. Phys. Lett.* 80, 2556 (2002).
317. M. A. Gribelyuk, M. R. McCartney, J. Li, C. S. Murthy, P. Ronsheim, B. Doris, J. S. McMurray, S. Hegde, and D. J. Smith, *Phys. Rev. Lett.* 89, 025502 (2002).
318. M. R. McCartney, M. A. Gribelyuk, J. Li, P. Ronsheim, J. S. McMurray, and D. J. Smith, *Appl. Phys. Lett.* 80, 3213 (2002).
319. A. E. Thesen, B. G. Frost, and D. C. Joy, *J. Vac. Sci. Technol. B* 20, 3063 (2002).
320. R. E. Dunin-Borkowski, S. B. Newcomb, D. Doyle, A. Deignan, and M. R. McCartney, in "Proceedings of the 12th EUREM," Brno, Czech Republic (J. Gemperlová and I. Vávra, Eds.), Vol. 3, p. 163. Czechoslovak Society for Electron Microscopy, Brno, Czech Republic, 2000.
321. A. C. Twitchett, R. E. Dunin-Borkowski, and P. A. Midgley, *Phys. Rev. Lett.* 88, 238302 (2002).
322. R. M. Langford and A. K. Petford-Long, *J. Vac. Sci. Technol. A* 19, 2186 (2001).
323. D. Sutton, S. M. Parle, and S. B. Newcomb, *Inst. Phys. Conf. Ser.* 168, 377 (2001).
324. R. E. Dunin-Borkowski, A. C. Twitchett, and P. A. Midgley, *Microsc. Microanal.* 8 (Suppl. 2), 42 (2002).
325. M. Beleggia, R. Capelli, and G. Pozzi, *Philos. Mag. B* 80, 1071 (2000).
326. Z. Wang, T. Hirayama, T. Kato, K. Sasaki, H. Saka, and N. Kato, *Appl. Phys. Lett.* 80, 246 (2002).
327. Z. Wang, K. Sasaki, N. Kato, K. Urata, T. Hirayama, and H. Saka, *J. Electron Microsc.* 50, 479 (2002).
328. Z. Wang, T. Kato, N. Shibata, T. Hirayama, N. Kato, K. Sasaki, and H. Saka, *Appl. Phys. Lett.* 81, 478 (2002).
329. M. Beleggia, G. C. Cardinali, P. F. Fazzini, P. G. Merli, and G. Pozzi, *Inst. Phys. Conf. Ser.* 169, 427 (2001).
330. R. E. Dunin-Borkowski and W. O. Saxton, in "Atomic Resolution Microscopy of Surfaces and Interfaces" (D. J. Smith and R. J. Hamers, Eds.), Materials Research Society Symposium Proceedings, Vol. 466, p. 73. Materials Research Society, Warrendale, PA, 1996.
331. M. R. McCartney, F. A. Ponce, J. Cai, and D. P. Bour, *Appl. Phys. Lett.* 76, 3055 (2000).
332. D. Cherns, J. Barnard, and H. Mokhtari, *Mater. Sci. Eng. B* 66, 33 (1999).

333. D. Cherns, J. Barnard, and F. Ponce, *Solid State Comm.* 111, 281 (1999).
334. D. Cherns, H. Mokhtari, C. G. Jiao, R. Averbeck, and H. Riechert, *J. Cryst. Growth* 230, 410 (2001).
335. J. Cai, F. A. Ponce, S. Tanaka, H. Omiya, and Y. Nakagawa, *Phys. Status Solidi A* 188, 833 (2001).
336. J. Cai and F. A. Ponce, *J. Appl. Phys.* 91, 9856 (2002).
337. D. Cherns and C. G. Jiao, *Phys. Rev. Lett.* 87, 205504 (2001).
338. C. G. Jiao and D. Cherns, *J. Electron Microsc.* 51, 105 (2002).
339. D. Cherns, *Mater. Sci. Eng. B* 91, 274 (2002).
340. J. Frenkel, "Kinetic Theory of Liquids." Oxford Univ. Press, Oxford, 1946.
341. J. Eshelby, C. Newey, P. Pratt, and A. Lidiard, *Philos. Mag.* 3, 75 (1958).
342. V. Ravikumar, R. P. Rodrigues, and V. P. Dravid, *Phys. Rev. Lett.* 75, 4063 (1995).
343. V. Ravikumar, R. P. Rodrigues, and V. P. Dravid, *J. Phys. D* 29, 1799 (1996).
344. V. Ravikumar, R. P. Rodrigues, and V. P. Dravid, *J. Am. Ceram. Soc.* 80, 1117 (1997).
345. V. Ravikumar, R. P. Rodrigues, and V. P. Dravid, *J. Am. Ceram. Soc.* 80, 1131 (1997).
346. H. J. Chang, J. D. Lee, R. P. Rodrigues, D. E. Ellis, and V. P. Dravid, *J. Mater. Synth. Proc.* 6, 323 (1998).
347. V. P. Dravid and V. Ravikumar, *Interface Sci.* 8, 177 (2000).
348. Y. G. Wang and V. P. Dravid, *Philos. Mag. Lett.* 82, 425 (2002).
349. R. E. Dunin-Borkowski and W. O. Saxton, *Acta Crystallogr. A* 53, 242 (1997).
350. G. Pozzi, *J. Phys. D* 29, 1807 (1996).
351. R. E. Dunin-Borkowski, W. O. Saxton, and W. M. Stobbs, *Acta Crystallogr. A* 52, 705 (1996).
352. M. Elfwing and E. Olsson, *J. Appl. Phys.* 92, 5272 (2002).
353. R. E. Dunin-Borkowski, W. M. Stobbs, D. D. Perovic, and Z. R. Wasilewski, in "Proceedings of ICEM13" (B. Jouffrey and C Colliex, Eds.), Vol. 1, p. 411 (1994).
354. F. Kara, R. E. Dunin-Borkowski, C. B. Boothroyd, W. M. Stobbs, and J. A. Little, *Ultramicroscopy* 66, 59 (1996).
355. Z. Mao, R. E. Dunin-Borkowski, C. B. Boothroyd, and K. M. Knowles, *J. Am. Ceram. Soc.* 81, 2917 (1998).
356. Z. Mao, R. E. Dunin-Borkowski, C. B. Boothroyd, and K. M. Knowles, in "Interfacial Engineering for Optimized Properties" (C. L. Briant, C. B. Carter, and E. L. Hall, Eds.), Materials Research Society Symposium Proceedings, Vol. 458, p. 109. Materials Research Society, Warrendale, PA, 1996.
357. K. D. Johnson and V. P. Dravid, *Appl. Phys. Lett.* 74, 621 (1999).
358. K. D. Johnson and V. P. Dravid, *Microsc. Microanal.* 5, 428 (1999).
359. K. D. Johnson and V. P. Dravid, *Interface Sci.* 8, 189 (2000).
360. X. Zhang, T. Hashimoto, and D. C. Joy, *Appl. Phys. Lett.* 60, 784 (1992).
361. J. C. H. Spence, J. M. Cowley, and J. M. Zuo, *Appl. Phys. Lett.* 62, 2446 (1993).
362. X. Zhang, D. C. Joy, Y. Zhang, T. Hashimoto, L. Allard, and T. A. Nolan, *Ultramicroscopy* 51, 21 (1993).
363. F. M. Ross, R. Kilaas, E. Snoeck, M. Hytch, A. Thorel, and L. Normand, in "Atomic Resolution Microscopy of Surfaces and Interfaces" (D. J. Smith and R. J. Hamers, Eds.), Materials Research Society Symposium Proceedings, Vol. 466, p. 245. Materials Research Society, Warrendale, PA, 1996.
364. H. Lichte, *Crystal Res. Technol.* 35, 887 (2000).
365. H. Lichte, M. Reibold, K. Brand, and M. Lehmann, *Ultramicroscopy* 93, 199 (2003).
366. P. A. Midgley, in "Proceedings of EUREM11," Vol. 2, p. 7. Committee of European Societies of Microscopy, Brussels, 1998.
367. W. Cao and C. Randall, *Solid State Comm.* 86, 435 (1993).
368. J. M. Cowley, *Ultramicroscopy* 41, 335 (1992).
369. R. A. Herring, G. Pozzi, T. Tanji, and A. Tonomura, *Ultramicroscopy* 50, 94 (1993).
370. R. A. Herring, G. Pozzi, T. Tanji, and A. Tonomura, *Ultramicroscopy* 60, 153 (1995).
371. H. Banzhof and K.-H. Herrmann, *Ultramicroscopy* 48, 475 (1993).
372. W. J. Vine, R. Vincent, P. Spellward, and J. W. Steeds, *Ultramicroscopy* 41, 423 (1992).
373. J. W. Steeds, R. Vincent, W. J. Vine, P. Spellward, and D. Cherns, *Acta Microsc.* 1, 1 (1992).
374. N. H. Dekkers and H. de Lang, *Optik* 41, 452 (1974).
375. H. Rose, *Utramicroscopy* 2, 251 (1977).
376. J. N. Chapman, P. E. Batson, E. M. Waddell, and R. P. Ferrier, *Ultramicroscopy* 3, 203 (1978).
377. M. Mankos, M. R. Scheinfein, and J. M. Cowley, *J. Appl. Phys.* 75, 7418 (1994).
378. M. R. McCartney, P. Kruit, A. H. Buist, and M. R. Scheinfein, *Ultramicroscopy* 65, 179 (1996).
379. G. Pozzi, *Optik* 66, 91 (1983).
380. G. Matteucci, G. F. Missiroli, and G. Pozzi, *Ultramicroscopy* 8, 403 (1982).
381. Q. Ru, N. Osakabe, J. Endo, and A. Tonomura, *Ultramicroscopy* 53, 1 (1994).
382. Q. Ru, *J. Appl. Phys.* 77, 1421 (1995).
383. Z. L. Wang, *Optik* 101, 24 (1995).
384. Q. Ru, J. Endo, T. Tanji, and A. Tonomura, *Appl. Phys. Lett.* 59, 2372 (1992).
385. Q. Ru, J. Endo, T. Tanji, and A. Tonomura, *Optik* 92, 51 (1992).
386. T. Kawasaki, J. Endo, T. Matsuda, N. Osakabe, and A. Tonomura, *J. Electron Microsc.* 35, 211 (1986).
387. K. Yamamoto, I. Kawajiri, T. Tanji, M. Hibino, and T. Hirayama, *J. Electron Microsc.* 49, 31 (2000).
388. J. Chen, T. Hirayama, G. Lai, T. Tanji, K. Ishizuka, and A. Tonomura, *Opt. Lett.* 18, 1887 (1993).
389. J. Chen, T. Hirayama, T. Tanji, K. Ishizuka, and A. Tonomura, *Opt. Comm.* 110, 33 (1994).
390. T. Hirayama, J. Chen, T. Tanji, and A. Tonomura, *Ultramicroscopy* 54, 9 (1994).
391. J. Chen, G. Lai, K. Ishizuka, and A. Tonomura, *Appl. Opt.* 33, 1187 (1994).
392. E. Völkl, *Microsc. Microanal.* 6, 211 (2000).
393. B. J. Thomson, G. B. Parrent, J. H. Ward, and B. Justh, *J. Appl. Meteorol.* 5, 343 (1966).
394. A. Tonomura, A. Fukuhara, H. Watanabe, and T. Komoda, *Japan. J. Appl. Phys.* 7, 295 (1968).
395. J. B. DeVelis, G. B. Parrent, and B. J. Thompson, *J. Opt. Soc. Am.* 56, 423 (1966).
396. T. Matsumoto, T. Tanji, and A. Tonomura, *Ultramicroscopy* 54, 317 (1994).
397. T. Matsumoto, T. Tanji, and A. Tonomura, *Optik* 97, 169 (1994).
398. L. A. Bursill, J. C. Barry, and P. R. W. Hudson, *Philos. Mag. A* 37, 789 (1978).
399. J. N. Ness, W. M. Stobbs, and T. F. Page, *Philos. Mag. A* 54, 679 (1986).
400. C. S. Baxter and W. M. Stobbs, *Ultramicroscopy* 16, 213 (1986).
401. C. B. Boothroyd, A. P. Crawley, and W. M. Stobbs, *Philos. Mag. A* 54, 663 (1986).
402. W. C. Shih and W. M. Stobbs, *Ultramicroscopy* 32, 219 (1990).
403. D. Özkaya and W. M. Stobbs, *Inst. Phys. Conf. Ser.* 119, 217 (1991).
404. F. M. Ross and W. M. Stobbs, *Philos. Mag. A* 63, 1 (1991).
405. P. Nallet, E. Chassaing, M. G. Walls, and M. J. Hÿtch, *J. Appl. Phys.* 79, 1 (1996).
406. C. P. Liu, R. E. Dunin-Borkowski, C. B. Boothroyd, P. D. Brown, and C. J. Humphreys, *Microsc. Microanal.* 3, 352 (1997).
407. S. J. Lloyd and R. E. Dunin-Borkowski, *Phys. Rev. B* 59, 2352 (1999).
408. R. E. Dunin-Borkowski, *Ultramicroscopy* 83, 193 (2000).
409. S. J. Lloyd, D. M. Tricker, Z. H. Barber, and M. G. Blamire, *Philos. Mag. A* 81, 2317 (2001).

410. H. Gong and J. N. Chapman, *J. Magn. Magn. Mater.* 67, 4 (1987).
411. S. J. Lloyd, J. C. Loudon, and P. A. Midgley, *J. Microsc.* 207, 118 (2002).
412. J. M. Cowley, *Ultramicroscopy* 34, 293 (1990).
413. M. Mankos, Z. J. Yang, M. R. Scheinfein, and J. M. Cowley, *IEEE Trans. Magn.* 30, 4497 (1994).
414. M. Mankos, J. M. Cowley, R. V. Chamberlin, M. R. Scheinfein, and M. B. Stearns, *IEEE Trans. Magn.* 30, 720 (1994).
415. M. Mankos, A. A. Higgs, M. R. Scheinfein, and J. M. Cowley, *Ultramicroscopy* 58, 87 (1995).
416. M. Mankos, J. M. Cowley, and M. R. Scheinfein, *Mater. Res. Soc. Bull.* 20, 45 (1995).
417. M. Mankos, M. R. Scheinfein, and J. M. Cowley, *IEEE Trans. Magn.* 31, 3796 (1995).
418. M. Mankos, M. R. Scheinfein, and J. M. Cowley, *IEEE Trans. Magn.* 32, 4150 (1996).
419. M. Mankos, J. M. Cowley, and M. R. Scheinfein, *Phys. Status Solidi A* 154, 469 (1996).
420. J. M. Cowley, M. Mankos, and M. R. Scheinfein, *Ultramicroscopy* 63, 133 (1996).
421. G. Matteucci and M. Muccini, *Ultramicroscopy* 53, 19 (1994).
422. T. Hirayama, T. Tanji, and A. Tonomura, *Appl. Phys. Lett.* 67, 1185 (1995).
423. T. Hirayama, G. Lai, T. Tanji, N. Tanaka, and A. Tonomura, *J. Appl. Phys.* 82, 522 (1997).
424. T. Hirayama, *Mater. Characterization* 42, 193 (1999).
425. K. Ogai, Y. Kimura, R. Shimizu, J. Fujita, and S. Matsui, *Appl. Phys. Lett.* 66, 1560 (1995).
426. T. Tanji, Q. Ru, and A. Tonomura, *Appl. Phys. Lett.* 69, 2623 (1996).
427. T. Tanji, S. Manabe, K. Yamamoto, and T. Hirayama, *Ultramicroscopy* 75, 197 (1999).
428. G. Lai, T. Hirayama, A. Fukuhara, K. Ishizuka, T. Tanji, and A. Tonomura, *J. Appl. Phys.* 75, 4593 (1994).
429. C. Jönsson, H. Hoffmann, and G. Möllenstedt, *Phys. Kondense. Mater.* 3, 193 (1965).
430. M. Keller, *Z. Phys.* 164, 274 (1961).
431. R. Buhl, *Z. Phys.* 155, 395 (1959).
432. H. Hoffmann and C. Jönsson, *Z. Phys.* 182, 360 (1965).
433. K. H. Gaukler and R. Schwarzer, *Optik* 2, 215 (1971).
434. F. Sonier, *J. Microsc.* 12, 17 (1971).
435. L. Sturkey, *Phys. Rev.* 73, 183 (1948).
436. G. Hoonjo and K. Mihama, *J. Phys. Soc. Japan* 9, 184 (1954).
437. K. Moliere and H. Niehrs, *Z. Phys.* 140, 581 (1955).
438. S. Miyake, *J. Phys. Soc. Japan* 17, 124 (1962).
439. H. Tomita and M. Savelli, *C. R. Acad. Sci. Paris B* 267, 580 (1968).
440. J. N. Chapman and M. J. Drummond, in "Proceedings of the II International Symposium on Electron Microscopy and Biophysics," Chandigarh, 1986, p. 193.
441. N. Yamamoto and J. C. H. Spence, *Thin Solid Films* 104, 43 (1983).
442. S. Miyake, K. Fujiwara, and K. Suzuki, *J. Phys. Soc. Japan* 18, 1306 (1963).
443. G. Möllenstedt and M. Keller, *Z. Phys.* 148, 34 (1957).

Encyclopedia of Nanoscience and Nanotechnology
www.aspbs.com/enn

Electron Raman Scattering in Nanostructures

R. Betancourt-Riera, R. Riera, J. L. Marín, R. Rosas
Universidad de Sonora, Hermosillo, Sonora, México

CONTENTS

1. INTRODUCTION

The study of the dispersion of light by the condensed matter constitutes a means of investigation of the substance. The theoretical investigations of Brillouin were the beginning of investigation of the inelastic dispersion of light, with the publication of a study about the interaction of light with liquids in 1922. One year later the work of Smekal continued, in atoms of two electron levels, foreseeing that the elementary excitements should produce a displacement in the frequency of the dispersed light. In 1928, Raman observed experimentally this effect, which is called the Smekal–Raman effect, now known as Raman scattering. In that same year, Landsberg and Mandelstam reported the same effect in quartz. The type of Raman scattering depends on the quasiparticles involved. In particular, if it is an electron, it is named electron Raman scattering.

Raman scattering allows the study of the vibrations and the structure of the molecules, which it makes possible the analysis of molecular complex mixtures, whose study by chemical means would be very complex and in some cases impossible to carry out. Also, it allows the investigation of the optical properties of the substance taking advantage of the growth of liquid crystals, fluorescent screens, optical memory, lasers, and other optical and electronic devices whose applications offer wide perspectives in the industry.

The study of Raman scattering is one of the most known methods of investigation of solids, with a very fast development. The employment of conventional sources of light, due to their low spectral density and their low coherence, has allowed the study of this phenomenon only in the region of small absorption, where the resonant effects are scarce, due to excitonic and impurities states. In 1962 Porto and Wood showed that the employment of a laser, as a substitute source of the solar light and lamps of gas discharges, could increase notably the experimental possibilities; this led to the discovery of phenomena whose detailed explanation is impossible in the frame of the traditional concepts of Raman scattering or luminescence in solids. The monochromatic, coherent and high spectral density character of the light laser allows one to explore the region of fundamental interest in the solids, that is, the region of strong absorption, which has allowed one to observe the phenomenon of multiphonon resonance Raman scattering.

The interest in the study of the nanostructured semiconductors is due to the possibility to observe phenomena such as the negative differential conductivity, the amplification of electromagnetic waves, the variation of the frequencies and intensities of the absorption and emission lines, and other nonlinear optical properties, which allows new applications in different branches of the science. In some cases these phenomena are impossible to observe in bulk materials. Thus,

ISBN: 1-58883-059-4/$35.00

such nanostructures might constitute the basis, in the near future, of an alternative class of solid-state lasers, transport, and optoelectronic devices.

The possibilities of fabrication of new electro-optical devices based in low dimensional systems have led to enormous interest in the investigation of semiconductor nanostructures such as quantum wells, superlattices, quantum wires, and quantum dots which are usually made from weakly ionic materials (for instance GaAs/AlAs). The nanostructured semiconductors can be synthesized by several methods, namely, porous silicon, zeolite cavities, molecular beam epitaxy with ionic attack, holographic nanolithography, etc. Such nanostructures might constitute the basis of an alternative class of solid-state lasers, transport, and optoelectronic devices, in the near future.

One of the most important processes in the physics of the condensed matter is the electron–phonon interaction, because it is involved in numerous phenomena such as the lifetime of the excitons, transport phenomena, etc. On the other hand, this theory is primordial for the analysis of the results of several experimental techniques like Raman scattering.

Polar optical oscillations in such systems play an important role in many physical processes, especially in the long-wavelength limit. One of the principal scattering mechanisms in these structures is through optical–phonon emission or absorption, which has led to many studies of the effects of the reduced dimensionality on the phonon modes. A continuum theory of optical phonons in quantum dot (QD), quantum well wire (QWW), and freestanding wire (FSW) was developed in [1–5], where the dispersion was illustrated for phonon modes with $l = 0$.

Raman scattering experiments are well known to provide a powerful tool for the investigation of different physical properties of semiconductor nanostructures [6–8]. In particular the electronic structure of semiconductor materials and nanostructures can be thoroughly investigated considering different polarizations for the incident and emitted radiation [6, 9]. In connection with this kind of experiment the calculation of the differential cross-section (DCS) for electron Raman scattering (ERS) remains a rather interesting and fundamental issue to achieve a better understanding of the manmade semiconductor nanostructures characterized by their mesoscopic dimensions [10–16].

The experiments of electron Raman scattering and resonant Raman scattering in nanostructures began to be carried out starting in the last five years. Some of the most recent works reported in the international literature are the following. Kohno et al. [17] studied chains of crystalline-Si nanospheres by means of Raman scattering spectroscopy. They found that the one-phonon Raman scattering peak from the chains was asymmetric and broader than that from bulk Si. This phenomenon can be attributed to phonon confinement in the silicon nanospheres. The phonon confinement became more obvious by decreasing the size of the silicon nanospheres in the chains. They also found that the Si nanospheres in the chains were under compressive stress by the covering oxide layers through the analysis of the Raman shift. Kaschner et al. [18] investigated ZnCdSe/ZnSe quantum dot structures which include planar and coherently strained three-dimensional islands with different sizes. They reported Raman scattering and photoluminescence investigations of ZnCdSe/ZnSe structures. Their results show that resonant Raman scattering can be used to investigate the stability of such structures under conditions of resonant excitation of the QDs. It was reported that resonant Raman scattering can be applied to the ZnSe material system to distinguish between the phonons ZnSe and the ZnSe-like longitudinal-optical (LO) phonons of $Zn_{0.8}Cd_{0.2}Se/ZnSe$ quantum wires presented in [19]. Artús et al. [20] provided direct experimental evidence of the existence of phonons arising from the QDs. To demonstrate this point they present a Raman scattering study of InAs/GaAs QDs for several coverage thicknesses L within the range of coherent three-dimensional island growth. For all the samples they detect a Raman peak at a frequency higher than the InAs LO frequency exhibits a dependence on L. Cazayous et al. [21] reported resonant Raman scattering by acoustical phonons in self-assembled QD structures. Most of the published Raman studies on self-assembled QDs were limited to the optical phonon frequency range [22–26]. It was demonstrated that valuable information about residual strain and chemical composition in the QD can be derived from the optical phonon Raman spectra [22, 24, 25]. Raman scattering by acoustic phonons in self-assembled QDs was reported very recently [27–29]. References [27, 28] deal with structures containing many QD layers, and the corresponding low frequency Raman spectra display features similar to the ones observed in SLs. Huntzinger et al. reported acoustic phonon Raman scattering in structures containing a single layer of InAs/InP QDs. Sobolev et al. [30] investigated the quenching of photoluminescence in InAs/GaAs quantum dots and quantum wells by 2 MeV electron irradiation. They demonstrate a much higher radiation hardness of the quantum dots. Possible mechanisms of this phenomenon are discussed in terms of the wavefunction localization and defect reactions. Heitz et al. [31] presented a time-integrated and time-resolved study of resonant Raman scattering for self-organized InAs/GaAs QDs yielding detailed information on the interaction of excitons with local phonon modes of the InAs QD layer. Kasuya et al. [32] presented experimental evidence on critical point structures in the electronic states found in resonant Raman scattering and optical absorption spectra. The recent Raman scattering measurements on single-wall nanotubes show evidence for the zone-folding effect in the phonon system [33, 34]. Jorio et al. [35] measured the resonant window for the resonant Raman scattering of one isolated single-wall carbon nanotube by using a frequency tunable laser E_l. They here report an asymmetric shape for the E_l dependent resonant window, showing the importance of including the detailed experimentally determined joint density of electronic states line shape in the resonant Raman theory for one-dimensional (1D) systems and considering incoherent scattering. Therefore, they show that resonant Raman scattering can be used to characterize the joint density of electronic states of a single-wall carbon nanotube with much better resolution than scanning tunneling spectroscopy measurements. In [36] it is shown that the Raman scattering technique can give complete structural information for one-dimensional systems, such as carbon nanotubes. Resonant confocal micro-Raman spectroscopy of an (n, m) individual single-wall nanotube makes it possible to assign its chirality uniquely by measuring one radial breathing mode frequency ω_{RBM} and using the theory of resonant transitions. A unique chirality assignment can be made for both metallic and semiconductor nanotubes of diameter d, using the parameters $\gamma_0 = 2.9$ eV and $\omega_{RBM} = 248/d$. Kneipp et al. [37] studied polarization effects in surface-enhanced resonant Raman spectroscopy from single-wall carbon nanotubes on silver colloidal clusters. The authors use single-wall carbon nanotubes as a model molecule to understand the increase of the depolarization ratios [38] measured in many surface-enhanced resonant Raman spectroscopy experiments compared to "normal" (resonant) Raman scattering [39–42]. Benoit et al. [43] presented a detailed low frequency Raman study of arc discharge multiwalled carbon nanotubes, which are characterized by transmission electron microscopy. They clearly identify low frequency modes in purified samples. They presented a model to interpret these modes based on the coupling of the radial-breathing mode of each individual wall of the multiwall nanotube. Intensity calculations have been carried out as well. The low frequency spectrum of the multiwall nanotube is shown to be very sensitive to the internal diameter of multiwall nanotube. At the first time, a comprehensive simulation of the low frequency Raman spectrum of multiwall nanotube is given. Preliminary results have already been presented elsewhere [44]. Maciel et al. [45] have used Raman spectroscopy to measure phonons in GaAs v-groove quantum wire structures containing (001) and (111) GaAs/AlAs superlattice barrier regions. Resonance enhancement permits the identification of modes in different regions of the structure, and the measured phonon frequencies provide structural information, which shows clear evidence of GaAs migration during growth from (001) surfaces into the grooves. Confined and interface phonons with large in-(111) plane wavevectors are observed. In [46] resonant Raman scattering for one single-wall carbon nanotube spectroscopy is overviewed. First order resonance Raman spectra of the radial breathing mode of one carbon nanotube is of importance for assigning (n, m) values to the nanotube. The assignment of (n, m) values is confirmed by the chirality dependence of the other phonon modes. Second order, one phonon emission, and the intervalley scattering processes of two-dimensional graphite and of single-wall carbon nanotubes are relevant to disorder-induced D-band Raman spectra. The dispersive nature of the D-band Raman spectra is explained by double resonance processes. Many weak Raman spectra appearing in the intermediate frequency range, which have been observed for a long time but were never assigned, have recently been assigned as double resonance Raman peaks. The second order Raman phonon frequencies can be used as a new fundamental tool for determining the phonon energy dispersion relations, especially for disordered materials and for zone boundary phonons. Husberg et al. [47] reported further investigations of the luminescence of the phase separated In-rich structures in c-$In_xGa_{1-x}N$/GaN DHs. The observation of luminescence line narrowing in resonant excitation experiments supports the QD-like structure of the In-rich phase. Annealing experiments were performed to gain insight into the process of formation of these inclusions. Toda et al. [48] have investigated the carrier relaxation mechanism in InGaAs/GaAs quantum dots by photoluminescence excitation spectroscopy. Near-field scanning optical microscopy successfully shows that a photoluminescence excitation resonance at a relaxation energy of 36 meV can be seen in all single-dot luminescence spectra and thus can be attributed to resonant Raman scattering by a GaAs LO phonon to the excitonic ground state. In addition, a number of sharp resonances observed in single-dot photoluminescence excitation spectra can be identified as resonant Raman features due to localized phonons, which are observed in the conventional Raman spectrum. The results reveal the mechanism for the efficient relaxation of carriers observed in self-assembled quantum dots: the carriers can relax within the continuum states and make transitions to the excitonic ground state by phonon emission.

On the other hand, recently it has also been reported in the international literature some theoretical works related with the electron Raman scattering and resonant Raman scattering; between them, we could mention the following. Wang and Sarma [49] provide (within the resonant RPA scheme) a compellingly generic theory for resonant Raman scattering theory for resonant Raman scattering experiments by including the valence-band electrons during the scattering processes for one-, two-, and three-dimensional semiconductor systems, following their earlier short paper [50] on 1D systems. Scamarcio et al. [51] unambiguously demonstrate that the strength of the electron–LO-phonon coupling increases with increasing confinement (decreasing size) in the strong confinement limit. This conclusion results from the size dependence of the ratio between two-phonon and one-phonon Raman cross-sections in a prototype system: CdS_xSe_{1-x} nanocrystals embedded in a glass matrix. It is important to stress that, in their experiments, the resonance condition has been kept constant (within <1 meV) for each investigated nanocrystal size. This has been achieved by fully exploiting the advantages of a Raman microprobe apparatus and a suitable sample in which the mean nanoparticle size increases monotonically along one direction. Due to the strong energy dependence of the Raman cross-section, keeping constant the resonance condition is mandatory for a meaningful comparison of spectra associated with nanocrystals having different sizes and hence different electronic transitions. Their approach represents an improvement with respect to previous resonant Raman scattering investigations in which the challenging experimental problem of keeping constant the resonance condition has not been properly solved [52–54]. Also, they report a polarization analysis of the Raman spectra in nanocrystals. Menéndez-Proupin and Cabo-Bisset [55] revised the adiabatic theory of resonance one-phonon Raman scattering in semiconductor nanocrystals and extended with perturbative nonadiabatic corrections, given by Albrecht's B term. This theory is confronted with the time-dependent perturbation approach, pointing at their differences and similarities. It is shown that both theories are equivalent in the limit of weak electron–phonon coupling and nondegenerate or uncoupled resonant states. Evaluations of the A and B terms for the confined LO phonon in CdSe and CdS nanocrystals are reported. These evaluations show that the B term can usually be neglected. Gonzalez and Menéndez-Proupin [56] studied interband absorption and luminescence small quantum dots under conditions similar to the experiments reported in [57, 58] (i.e., quasi-two-dimensional motion, magnetic fields in the interval 8 T $\leq B \leq$ 60 T, and temperatures well below 2 K).

Very recent experimental works [59–86], related to the electron Raman scattering and resonant Raman scattering phenomena in nanostructured materials, can be used in order to verify the several theoretical models considered in the interpretation of these processes. In the same way we could find other recent theoretical works in [87–105].

Among the various Raman scattering processes involved in this kind of research, ERS seems to be a useful technique providing direct information about the energy band structure and the optical properties of the investigated systems. ERS is qualitatively explained as a three-step process: in the first step the system absorbs a photon from the incident radiation and an electron–hole pair (EHP) is created in a virtual state (after an interband electron transition); in the second step the electron and the hole move independently of each other and emit one optical phonon performing intraband transitions. In the last step the electron and the hole move independently of each other and emit photons of secondary radiation performing intraband or interband transitions [14–16]. In the final step an excited phonon of frequency ω and a photon of frequency ω_s appear as a real state of the system and an EHP is annihilated or created in another excited state. Moreover, the DCS for ERS, in the general case, usually shows singularities related to interband and intraband transitions. The latter result strongly depends on the scattering configurations; namely, the structure of the singularities is varied when the photon polarization changes [106]. This peculiar feature of ERS allows one to determine the subband structure of the system by direct inspection of the singularity positions in the spectra.

For bulk semiconductors ERS has been studied in the presence of the external applied magnetic and electric fields [7, 107, 108]. In this work we study ERS in QW, QWW, and FS Wire of cylindrical shapes and QDs of spherical shape, considering transitions assisted by photons and within or without phonons. We assume electron confinement within the system, as well as parabolic bands at $T = 0$ K [109]. The assumption of $T = 0$ K eliminates the possibility of absorption of phonons in the electron–phonon interaction Hamiltonian, but it does not change the optical properties. The parabolic band approximation is commonly used in the II–VI and III–V semiconductors when one works near the center of Brillouin's zone in the reduced zone scheme. Under these dynamical conditions, since the potential barriers do not imply interband transitions, the nanostructures can be treated within the envelope function and effective mass approximations.

Regarding the complex valence band of III–V and II–VI semiconductor compounds, the contribution of band mixing is small in the confinement regime assumed in this work and is consistent with previous treatments [110, 111]. When the excitation energy is close to the bandgap energy, the valence band structure is very important in the strong confinement regime and only a conduction band can be assumed since the electronic contribution to the process is negligible, as has been pointed out recently by Fomin et al. [112]; however, this is not the case of the present work because the excitation energy considered herein is much higher than the bandgap energy.

2. RAMAN CROSS-SECTION

The DCS for ERS in a volume V per unit solid angle $d\Omega$ for incoming light of frequency ω_l and scattered light of frequency ω_s is given by [109]

$$\frac{d^2\sigma}{d\omega_s\, d\Omega} = \frac{V^2\omega_s^2\eta(\omega_s)}{8\pi^3 c^4 \eta(\omega_l)} W(\omega_s, \hat{\mathbf{e}}_s) \tag{1}$$

where $\eta(\omega)$ is the refraction index as a function of the radiation frequency, $\hat{\mathbf{e}}_s$ is the unit polarization vector for the emitted secondary radiation, c is the light velocity in vacuum, and $W(\omega_s, \hat{\mathbf{e}}_s)$ is the transition rate for the emission of secondary radiation (with frequency ω_s and polarization $\hat{\mathbf{e}}_s$) calculated according to the Fermi golden rule,

$$W(\omega_s, \hat{\mathbf{e}}_s) = \frac{2\pi}{\hbar} \sum_f |M_1 + M_2|^2\, \delta(E_f - E_i) \tag{2}$$

where, for an electron Raman scattering process without phonon-assisted transitions,

$$M_j = \sum_a \frac{\langle f|\widehat{H}_{js}|a\rangle\langle a|\widehat{H}_{jl}|i\rangle}{(E_i - E_a + i\Gamma_{a_j})} + \sum_b \frac{\langle f|\widehat{H}_{jl}|b\rangle\langle b|\widehat{H}_{js}|i\rangle}{(E_i - E_b + i\Gamma_{bj})} \tag{3}$$

and $j = 1, 2$ denotes electron or hole contributions, respectively.

In Eq. (3) $|i\rangle$ and $|f\rangle$ denote initial and final states of the system with their corresponding energies E_i and E_f. $|a\rangle$ and $|b\rangle$ are intermediate states with energies E_a and E_b. Γ_{aj} and Γ_{bj} are the corresponding lifetimes. In Figure 1 we show Feynman diagrams describing the considered processes.

The operator $\widehat{H}_{jl}$ is given in the form

$$\widehat{H}_{jl} = \frac{|e|}{\mu_0}\sqrt{\frac{2\pi\hbar}{V\omega_l}}\,\hat{\mathbf{e}}_l \cdot \hat{\mathbf{p}} \qquad \hat{\mathbf{p}} = -i\hbar\nabla \qquad j = 1, 2 \tag{4}$$

where μ_0 is the free electron mass. This operator describes the interaction with the incident radiation field in the dipole

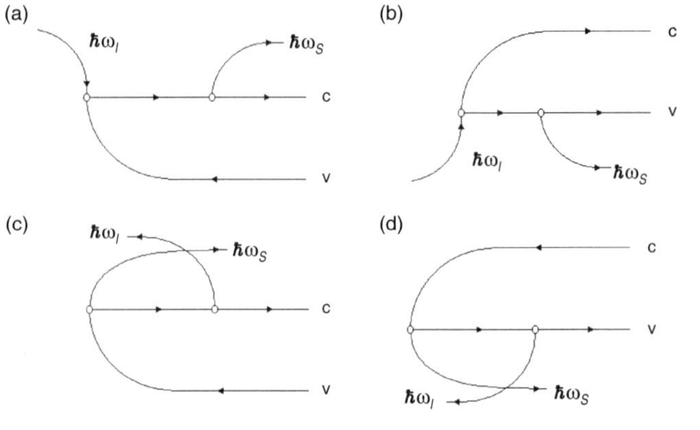

Figure 1. Diagrams contributing to electron Raman scattering for intersubband transitions. (a) and (b) show contributions from the electron or the hole, respectively. (c) and (d) show interference processes.

approximation. The interaction with the secondary radiation field is described by the operator:

$$\widehat{H}_{js} = \frac{|e|}{\mu_j}\sqrt{\frac{2\pi\hbar}{V\omega_s}}\hat{\mathbf{e}}_s \cdot \hat{\mathbf{p}} \tag{5}$$

This operator describes the photon emission by the electron (hole) after transitions between conduction (valence) subbands of the system. In Eq. (3) the intermediate state $|a\rangle$ represents an EHP virtual (after absorption of the incident photon), while the states $|b\rangle$ and $|c\rangle$ represent an EHP real (after the emission of a phonon and a secondary photon, respectively), and μ_j is the effective mass of the electron (hole).

In the initial state $|i\rangle$ we have a completely occupied valence band, an unoccupied conduction band, and an incident photon of energy $\hbar\omega_l$. Thus

$$E_i = \hbar\omega_l \tag{6}$$

The final state $|f\rangle$ involves an EHP excited in a real state and a secondary radiation emitted photon of energy $\hbar\omega_s$. Hence

$$E_f = E_1 + E_2 + E_g + \hbar\omega_s \tag{7}$$

For the intermediate state $|a\rangle$ the energies E_a are easily obtained from the previous discussion and using energy and momentum conservation laws we can evaluate the denominators in Eq. (3),

$$E_i - E_a = E_j - E_j' + \hbar\omega_s \tag{8}$$

$$E_i - E_b = E_j - E_j' - \hbar\omega_l \tag{9}$$

From the latter results it can be deduced that for semiconductors with large enough energy gap E_g (as is the case, for instance, with GaAs, CdTe, etc.) we have not considered the first term because $E_j - E_j' < E_g$ and considering that $\hbar\omega_l > E_g$, this corresponds to the "interference diagrams."

For an electron Raman scattering process with phonon-assisted transitions,

$$M_j = \sum_{a,b} \frac{\langle f|\widehat{H}_{js}|b\rangle\langle b|\widehat{H}_{jph}|a\rangle\langle a|\widehat{H}_{jl}|i\rangle}{(E_i - E_a + i\Gamma_{aj})(E_i - E_b + i\Gamma_{bj})} + \sum_{a,c} \frac{\langle f|\widehat{H}_{jph}|c\rangle\langle c|\widehat{H}_{js}|a\rangle\langle a|\widehat{H}_{jl}|i\rangle}{(E_i - E_a + i\Gamma_{aj})(E_i - E_c + i\Gamma_{cj})} \tag{10}$$

$|a\rangle$, $|b\rangle$, and $|c\rangle$ are intermediate states with energies E_a, E_b, and E_c while Γ_{aj}, Γ_{bj}, and Γ_{cj} are the corresponding lifetime widths. The "interference diagrams" can be neglected. In Figure 2 we show Feynman diagrams describing the considered processes.

In the initial state $|i\rangle$ we have a completely occupied valence band, an unoccupied conduction band, and an incident photon of energy $\hbar\omega_l$. Thus

$$E_i = \hbar\omega_l \tag{11}$$

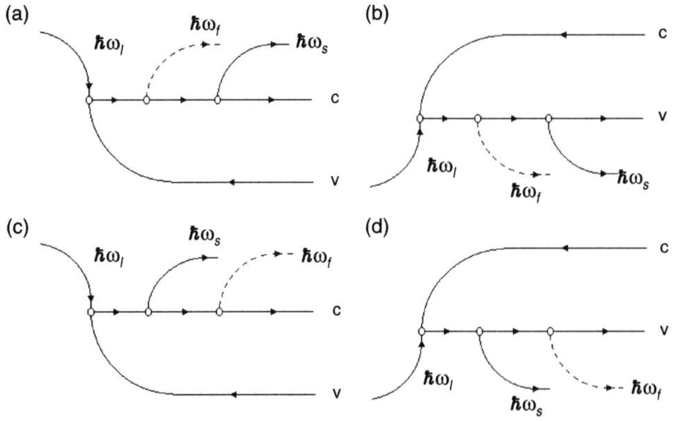

Figure 2. Diagrams contributing to one phonon-assisted electron Raman scattering for intersubband transitions. (a) and (c) show contributions from the electron. (b) and (d) show contributions from the hole.

The final state $|f\rangle$ involves an EHP excited in a real state, a secondary radiation emitted photon of energy $\hbar\omega_s$, and one phonon of energy $\hbar\omega_q$; hence

$$E_f = E_1 + E_2 + E_g + \hbar\omega_s + \hbar\omega_q \tag{12}$$

For the intermediate states $|a\rangle$, $|b\rangle$, and $|c\rangle$ the energies E_a, E_b, and E_c are easily obtained from the previous discussion and using energy and momentum conservation laws we can evaluate the denominators in Eq. (10):

$$E_i - E_a = E_j - E_j'' + \hbar\omega_s + \hbar\omega_q \tag{13}$$

$$E_i - E_b = E_j - E_j' + \hbar\omega_s \tag{14}$$

$$E_i - E_c = E_j - E_j' + \hbar\omega_q \tag{15}$$

Considering that the energy of the $\hbar\omega_l$ radiation incident is such that upon the electron absorbing the same it could reach the first subbands, and from the latter results, the contribution of the second term of Eq. (10) can be neglected if it is compared with the contribution of the first one.

We thus should consider, for the determination of the DCS, the contribution of just the first term in the right-hand side of Eq. (10) during the calculation of M_j.

3. MODEL DESCRIPTION OF ELECTRON–PHONON INTERACTION IN NANOSTRUCTURED SEMICONDUCTORS

Let us briefly summarize some details concerning the optical phonons and electron–phonon interaction Hamiltonian in QDs, QWWs, and FSWs. In [1, 2], within the framework of a macroscopic continuum coupling model, the mechanical phonon displacement vector **u**, the electrostatic potential ϕ, and the optical vibrational modes were obtained. These results are in close agreement with both experiments and microscopic calculations [113–115]. The main points of this treatment are: (i) we should solve a system of coupled differential equations for the displacement field **u** and the electric potential ϕ; (ii) we should apply matching conditions at the interfaces in close consistency with both the differential equations of the treatment and the involved physical principles; (iii) we are in general led to coupled oscillation modes involving a mixed character. For further details [113–115] should be consulted. The boundary conditions that need to be satisfied at the interface are the continuity of: (1) all the components of the displacement, **u**; (2) the electrostatic potential, ϕ; (3) the normal component of the mechanical stress tensor, σ_N; and (4) the normal component of the electric displacement, D_N. For a GaAs/AlAs system the zone center optical phonon frequencies in the two bulk materials are separated by 100 cm^{-1}, which leads to virtually complete confinement of the GaAs vibrations in GaAs and of the AlAs vibrations in AlAs (in fact, the displacements of the modes of GaAs frequency are zero at the first Al atom after the interface). In order to simplify the algebra, the boundary condition (1) is modified from **u** continuous to $\mathbf{u} = 0$ at the interface. With this modification, the stress boundary condition (3) becomes redundant.

Other cases also considered in the literature are free oscillations for FSW. In such cases the active medium is surrounded by vacuum (or quartz), thus allowing free oscillations of the surface. In this case the matching boundary conditions are $\boldsymbol{\sigma} \cdot \mathbf{N}|_{r\epsilon S} = 0$ and **u** is defined for just one side of the structure.

We will use a model for the optical oscillations in ionic crystals where a Lagrangian density is postulated and the solid is considered like a continuous distribution with density being the mass reduced of the ions and the volume of the primitive cell. So the Lagrangian density is expressed through the expression

$$L = \frac{1}{2}\rho\dot{\mathbf{u}} \cdot \dot{\mathbf{u}} - \frac{1}{2}\mathbf{u} \cdot \overleftrightarrow{\gamma} \cdot \mathbf{u} + \frac{1}{2}\mathbf{E} \cdot \overleftrightarrow{\beta} \cdot \mathbf{E} + (-\mathbf{u}) \cdot \overleftrightarrow{\alpha} \cdot \mathbf{E} + \frac{1}{2}u_{ik}\lambda_{iklm}u_{lm} \tag{16}$$

where **u** is the relative displacement between the ions and u_{ik} is the deformation tensor:

$$u_{ik} = \frac{1}{2}(\nabla_k u_i + \nabla_i u_k) \tag{17}$$

The first term of Eq. (16) is the kinetic energy density; the second describes the elastic interaction between the ions. The third term is the electric interaction. The fourth term is the coupling of electric field with the field of the displacements. The fifth and last term is related to the internal efforts and is the one that leads to the dispersion.

Considering that the fields vary harmonically with the time, the system of equations for **u** and ϕ is obtained as

$$\rho(\omega^2 - \omega_T^2)\mathbf{u} = \alpha\nabla\phi + \rho B_L^2\nabla\nabla \cdot \mathbf{u} - \rho B_L^2 \nabla \times \nabla \times \mathbf{u} \tag{18}$$

$$\nabla \cdot \nabla\phi = \frac{4\pi\alpha}{\epsilon_\infty}\nabla \cdot \mathbf{u} \tag{19}$$

If the displacement is written as

$$\mathbf{u} = \mathbf{u}_T + \mathbf{u}_L \quad \text{where} \quad \nabla \cdot \mathbf{u}_T = 0 \quad \text{and} \quad \nabla \times \mathbf{u}_L = 0 \tag{20}$$

then

$$\rho(\omega^2 - \omega_T^2)\mathbf{u}_T = -\rho B_T^2 \nabla \times \nabla \times \mathbf{u}_T \tag{21}$$

$$\rho(\omega^2 - \omega_L^2)\mathbf{u}_L = \rho B_L^2 \nabla\nabla \cdot \mathbf{u}_L \tag{22}$$

$$\nabla\phi = \frac{4\pi\alpha}{\epsilon_\infty}\mathbf{u}_L \tag{23}$$

$\omega_L = \sqrt{(\epsilon_0/\epsilon_\infty)}\omega_T$ is the Lydane–Sachs–Teller relation.

From this system of equations is easy to observe that the electric field originating from the media polarization is coupled only with the longitudinal oscillations. This field is also longitudinal and opposite to the displacement.

The Hamiltonian of the electron–phonon interaction is obtained considering that both functions $\mathbf{u}(r)$ and $\phi(r)$ form an orthogonal and complete system. The operators of creation and annihilation that satisfy the relation of commutation of the Bose statistical are introduced in order to obtain the quantized fields.

3.1. Electron–Phonon Interaction Hamiltonian for Quantum Well

The electron–LO-phonon interaction Hamiltonian is deduced in a systematic way from a model of dispersive polar optical vibrations for a double semiconductor heterostructure (DHS) as a QW. LO-phonon confinement effects are included in this model, which can be imagined as a jellium slab disposed between two rigid walls. The obtained Hamiltonian includes screening in the spirit of the static 3D Thomas–Fermi approximation and contains the 3D Fröhlich Hamiltonian as a limiting case. The case of a single heterostructure is also obtained as another limit.

3.1.1. Polarization Field

We are only concerned with the long wavelength LO-phonon and, therefore, a continuum approach along the lines of Born and Huang for the classical vibrational field is appropriated, which, afterward, must be properly quantized.

In the frame of this model the dynamical equations for the lattice are written in the form

$$\ddot{\mathbf{u}} = -\omega_T^2\mathbf{u} + \sqrt{\frac{\varepsilon_0 - \varepsilon_\infty}{4\pi}}\omega_T\mathbf{E} - v_a^2\nabla(\nabla \cdot \mathbf{u}) - v_b^2\nabla^2\mathbf{u} \tag{24}$$

$$\mathbf{P} = \sqrt{\frac{\varepsilon_0 - \varepsilon_\infty}{4\pi}}\omega_T\mathbf{u} + \frac{\varepsilon_\infty - 1}{4\pi}\mathbf{E} \tag{25}$$

where $\mathbf{u}(\mathbf{r}, t)$ is the effective relative ionic displacement (the units are such that $\mathbf{u} = \sqrt{\rho}\mathbf{w}$, where ρ is the mass density and **w** is the real displacement field with length units), **E** is the electric field, **P** is the polarization field associated with the vibrations, and $\mathbf{v}_a$, $\mathbf{v}_b$ are vector parameters with dimensions of velocity. Equations (24) and (25) are analogous to those of Born and Huang for an isotropic continuum but they include dispersion effects through the terms proportional to $\mathbf{v}_a$ and $\mathbf{v}_b$; this dispersive modification of Born and Huang equations is especially appropriate to account for the boundary conditions at the interfaces of the double heterostructure.

For a charge-free (and current-free) system Eqs. (24) and (25) are supplemented by the Maxwell equations,

$$\nabla\cdot\mathbf{D}=0\quad\nabla\cdot\mathbf{H}=0\quad\nabla\times\mathbf{H}=\frac{1}{c}\frac{\partial\mathbf{D}}{\partial t}\quad\nabla\times\mathbf{E}=-\frac{1}{c}\frac{\partial\mathbf{H}}{\partial t}\tag{26}$$

where the medium is assumed nonmagnetic and $\mathbf{D}=\mathbf{E}+4\pi\mathbf{P}$.

Assuming $\mathbf{u}(\mathbf{r},t)=\mathbf{u}(\mathbf{r})e^{-i\omega t}$ we obtain

$$\mathbf{E}=\left[\frac{4\pi}{\omega_T^2}(\varepsilon_0-\varepsilon_\infty)^{-1}\right]^{\frac{1}{2}}\left[\omega_T^2-\omega^2+v_a^2\nabla(\nabla\cdot)+v_b^2\nabla^2\right]\mathbf{u}\tag{27}$$

$$\mathbf{P}=\left[4\pi\omega_T^2(\varepsilon_0-\varepsilon_\infty)\right]^{-\frac{1}{2}}\left\{(\varepsilon_0-1)\omega_T^2-(\varepsilon_\infty-1)\omega^2+(\varepsilon_\infty-1)\left[v_a^2\nabla(\nabla\cdot)+v_b^2\nabla^2\right]\right\}\mathbf{u}\tag{28}$$

and

$$\mathbf{D}=\left[\frac{4\pi}{\omega_T^2}(\varepsilon_0-\varepsilon_\infty)^{-1}\right]^{\frac{1}{2}}\times\left\{\varepsilon_0\omega_T^2-\varepsilon_\infty\omega^2+\varepsilon_\infty\cdot\left[v_a^2\nabla(\nabla\cdot)+v_b^2\nabla^2\right]\right\}\mathbf{u}\tag{29}$$

Equations (24), (25), and (26) are satisfied by a displacement field decomposed in the form $\mathbf{u}=\mathbf{u}_T+\mathbf{u}_L$ where

$$\nabla\cdot\mathbf{u}_T=0\quad\nabla\times\mathbf{u}_T\neq0\quad\nabla\cdot\mathbf{u}_L\neq0\quad\nabla\times\mathbf{u}_L=0\tag{30}$$

For $\mathbf{u}_T$ the equation $\nabla\cdot\mathbf{D}=0$ is satisfied with $\mathbf{D}\neq0$ and the corresponding (dispersive) polariton modes are obtained. However, we are only interested in the longitudinal displacements $\mathbf{u}_L$ where equation $\nabla\cdot\mathbf{D}=0$ is satisfied automatically because $\mathbf{D}=0$. For $\mathbf{u}_L(\mathbf{r},\omega)$ we obtain

$$\left(\nabla^2+\frac{(\omega_{\rm LO}^2-\omega^2)}{v^2}\right)\mathbf{u}_L(\mathbf{r},\omega)=0\tag{31}$$

where $v^2=v_a^2+v_b^2$. In the bulk case Eq. (31) leads us to a quadratic dispersion law

$$\omega_q^2=\omega_{\rm LO}^2-v^2q^2\tag{32}$$

For the longitudinal electric and polarization fields we obtain

$$\mathbf{E}_L=-\left[\frac{4\pi\omega_{\rm LO}^2}{\varepsilon^*}\right]_L^{\frac{1}{2}}\mathbf{u}_L\quad\text{and}\quad\mathbf{P}_L=\left[\frac{\omega_{\rm LO}^2}{4\pi\varepsilon^*}\right]_L^{\frac{1}{2}}\mathbf{u}_L\tag{33}$$

where $(\varepsilon^*)^{-1}=\varepsilon_\infty^{-1}-\varepsilon_0^{-1}$ (and therefore $\mathbf{D}_L=0$).

Let us now consider a DHS with interfaces at $z=0$ and $z=d$ (the "z" axis is perpendicular to the interfaces). For $0<z<d$ we have a given material (say, GaAs) and for $z<0$ and $z>d$ we have another material (say, AlGaAs). The boundary conditions of continuous $\rho^{-1/2}\dot{u}_z$ and $v^2\rho^{1/2}\nabla\cdot\mathbf{u}$ at the interfaces are imposed in the spirit of a hydrodynamic (or jellium) model; the first quantity is the "z" component of the velocity field and the second quantity is the pressure. We will not discuss the more general solution of (31), but the one entailing complete confinement of the LO vibrations. This particular form of vibrations should provide an appropriate description for the observed highly confined LO phonon. Therefore, we solve Eq. (31) under the requirements

$$u_z(\mathbf{r},\omega)=0\qquad\text{for } z=0\quad\text{and}\quad z=d\tag{34}$$

From now the subscript "L" is avoided and $\mathbf{u}$ will describe purely longitudinal vibrations. Conditions (34) involve a "jellium" slab disposed between rigid walls at $z=0$ and $z=d$ as a macroscopic model for completely confined LO vibrations, that is, a model excluding penetration of the polar optical vibrations from a given layer into the adjacent ones. We must emphasize that this kind of model is in good agreement with experimental facts and numerical simulations and also provides the possibility of working out analytical expressions for the displacement and polarization fields.

After substitution of $\mathbf{u}(\mathbf{r})=\boldsymbol{\zeta}(z)e^{i\mathbf{q}_\perp\cdot\mathbf{r}_\perp}$ in (31) we obtain

$$\left(\frac{d^2}{dz^2}+q_z^2\right)\boldsymbol{\zeta}(z)=0\tag{35}$$

where

$$q_z^2=\frac{1}{v^2}(\omega_{\rm LO}^2-\omega^2)-q_\perp^2\geq0\quad\text{and}\quad\zeta_z(0)=\zeta_z(d)=0\tag{36}$$

For $q_z^2>0$ the solution of (35) is

$$\boldsymbol{\zeta}(z)=A\left[(\mathbf{q}_\perp+q_z\hat{\mathbf{e}}_z)e^{iq_zz}+(\mathbf{q}_\perp-q_z\hat{\mathbf{e}}_z)e^{-iq_zz}\right]\tag{37}$$

with $q_z=\frac{n\pi}{d}$, and $n=1,2,3,\ldots$

Let us remark that (37) satisfies Eq. (34) with the given boundary conditions $\nabla\times\mathbf{u}=0$ for the longitudinal displacements. It is easy to include the case $q_z=0$ just allowing n to be zero in (37). For $q_z^2<0$ no solutions exist satisfying the given boundary conditions.

From (37) we obtain the general solution for $\mathbf{u}$,

$$\mathbf{u}(\mathbf{r})=\sum_{\mathbf{q}_\perp}\sum_{n\geq0}A_{\mathbf{q}_\perp,n}\left[(\mathbf{q}_\perp+q_z\hat{\mathbf{e}}_z)e^{iq_zz}+(\mathbf{q}_\perp-q_z\hat{\mathbf{e}}_z)e^{-iq_zz}\right]e^{i\mathbf{q}_\perp\cdot\mathbf{r}_\perp}\tag{38}$$

where q_z and $A_{\mathbf{q}_\perp,n}$ are constant parameters. Let us note that the sums over $\mathbf{q}_\perp$ and n can be infinite. We must require $\omega^2=\omega_{\rm LO}^2-v^2q^2>0$ (only if $v\to0$ we obtain infinite sums). We should agree to sum in the following ways: n will be summed in the interval $0<n<N(q_\perp)$ where $N(q_\perp)$ is the higher integer less than $((\omega_{\rm LO}^2/v^2)-q_\perp^2)^{1/2}$ for a fixed $q_\perp$; after that $\mathbf{q}_\perp$ will be summed for all values inside the circle limited by $q_\perp^2=q_x^2+q_y^2=(\omega_{\rm LO}/v)^2$. Notice that the change $\mathbf{q}_\perp$ by $-\mathbf{q}_\perp$ does not alter the latter summation. Expression (38) can be transformed into

$$\mathbf{u}(\mathbf{r})=\sum_{\mathbf{q}_\perp}\sum_{n=-N(\mathbf{q}_\perp)}^{N(\mathbf{q}_\perp)}\mathbf{q}\left[A_{\mathbf{q}_\perp,|n|}e^{iq_zz}+A^*_{\mathbf{q}_\perp,|n|}e^{-iq_zz}\right]\tag{39}$$

with $\mathbf{q}=\mathbf{q}_\perp+q_z\hat{\mathbf{e}}_z$ and $A^*_{\mathbf{q}_\perp,|n|}=-A_{-\mathbf{q}_\perp,|n|}$, the latter condition ensuring that $\mathbf{u}$ is a real quantity.

Obviously, (38) and (39) are completely identical expressions, but (39) is more convenient in the following. Substitution of (39) in (33) yields $\mathbf{P}(\mathbf{r},t)$, the polarization field, while the corresponding canonical conjugate momentum is given by

$$\Pi(\mathbf{r}, t) = \frac{4\pi\varepsilon^*}{\omega_{\text{LO}}^2}\frac{\partial}{\partial t}\mathbf{P}(\mathbf{r}, t) \tag{40}$$

The parameters $A^*_{\mathbf{q}_\perp, |n|}$ must be chosen appropriately in order that the corresponding quantized fields $\mathbf{P}$ and Π could satisfy adequate commutation relations.

We just report the final expression for $\mathbf{P}(\mathbf{r})$,

$$\mathbf{P}(\mathbf{r}) = \sum_{\mathbf{q}_\perp}\sum_{n=-N(\mathbf{q}_\perp)}^{N(\mathbf{q}_\perp)}\left[\frac{\hbar\omega_{\text{LO}}^2}{8\pi\varepsilon^*\omega_q V}\right]^{1/2}\frac{\mathbf{q}}{|\mathbf{q}|}\left[e^{i\mathbf{q}\cdot\mathbf{r}}\hat{b}_{\mathbf{q}} + e^{-i\mathbf{q}\cdot\mathbf{r}}\hat{b}^+_{\mathbf{q}}\right] \tag{41}$$

where $\hat{b}_{\mathbf{q}}, \hat{b}^+_{\mathbf{q}}$ are second-quantization Bose operators such that

$$[\hat{b}_{\mathbf{q}}, \hat{b}_{\mathbf{q}'}] = [\hat{b}^+_{\mathbf{q}}, \hat{b}^+_{\mathbf{q}'}] = 0 \qquad [\hat{b}_{\mathbf{q}}, \hat{b}^+_{\mathbf{q}'}] = \delta_{\mathbf{q},\mathbf{q}'} \tag{42}$$

Expression (41) is the fundamental result of this section, which must be used to derive the electron–LO-phonon Hamiltonian for a DHS. The main difference with respect to the usual (bulk) polarization field is the discrete nature of $q_z = \frac{n\pi}{d}$ and the fact that $\mathbf{P}(\mathbf{r})$ is defined just for the layer $0 < z < d$.

3.1.2. Interaction Hamiltonian

The interaction Hamiltonian can be deduced from

$$\widehat{H}_I(\mathbf{r}) = \int \frac{e[-\nabla'\cdot\mathbf{P}(\mathbf{r}')]}{|\mathbf{r}-\mathbf{r}'|}e^{-\frac{|\mathbf{r}-\mathbf{r}'|}{\lambda}}d^3r' \tag{43}$$

where, for the sake of generality, we have included a screening λ in the spirit of the (bulk) Thomas–Fermi static approximation. Let us remark, however, that the obtained results are perfectly valid in the limit $\lambda \to \infty$.

Substitution of (41) in (43) yields, after straightforward integrations, the Hamiltonian for the DHS

$$\widehat{H} = \sum_{\mathbf{q}_\perp}\sum_{n=-N(\mathbf{q}_\perp)}^{N(\mathbf{q}_\perp)}\left[\Phi^{\text{DHS}}_{\mathbf{q},\lambda}(z)C_{\mathbf{q},\lambda}e^{i\mathbf{q}\cdot\mathbf{r}}\hat{b}_{\mathbf{q}} + HC\right] \tag{44}$$

where

$$C_{\mathbf{q},\lambda} = i\left[\frac{\omega_L}{\omega_q}\right]^{\frac{1}{2}}\frac{qC_F}{q^2+1/\lambda^2} \quad\text{and}\quad C_F = -\sqrt{\frac{2\pi e^2\hbar\omega_L}{V}}(\epsilon_{a\infty}^{-1}-\epsilon_{a0}^{-1}) \tag{45}$$

is the usual (screened) coupling factor of the bulk theory and

$$\Phi^{\text{DHS}}_{\mathbf{q},\lambda}(z) = \begin{cases}\frac{1}{2}e^{(Q-iq_z)z}[1-(-1)^n e^{-Qd}] & \text{for } z<0\\ 1-\frac{1}{2}e^{-iq_z z}[e^{-Qz}+(-1)^n e^{Q(z-d)}] & \text{for } 0<z<d\\ \frac{1}{2}e^{-(Q+iq_z)z}[(-1)^n e^{Qd}-1] & \text{for } z>d\end{cases} \tag{46}$$

where $Q^2 = q_\perp^2 + 1/\lambda^2$. $\epsilon_{a\infty}(\epsilon_{a0})$ is the high frequency (static) dielectric constant; the index a refers to the interior of the nanostructure.

The structure of $\widehat{H}$ is very clear from the physical point of view. Due to LO-phonon confinement the "z" component of the phonon wavevector $\mathbf{q}$ bears a discrete nature. The coupling factor $C_{\mathbf{q},\lambda}$ is the same as in the bulk (screened) Fröhlich Hamiltonian, but now a certain modulation function $\Phi^{\text{DHS}}_{\mathbf{q},\lambda}(z)$ is present describing the effect of confinement. In fact, we must notice that $\Phi^{\text{DHS}}_{\mathbf{q},\lambda}(z)$ is evanescent for $z < 0$ or $z > d$; for $0 < z < d$, $\Phi^{\text{DHS}}_{\mathbf{q},\lambda}(z)$ has two terms: the first term (equal to unity) is just the bulklike contribution, while the second term involves two contributions coming from the interface at $z = 0$ and $z = d$. The unscreened limit ($\lambda \to \infty$) is obtained without difficulties just taking $Q \to q_\perp$ in (46) and the corresponding expression for (45). A more realistic screening theory can be obtained from the unscreened $\widehat{H}$ just dividing $C_{\mathbf{q}}$ by a proper dielectric function $\varepsilon(\mathbf{q}, \omega)$ (see, for instance, [116]).

One important feature of the obtained expressions (44), (45), and (46) is that the 3D (bulk) limiting can be achieved if we first shift the coordinate origin by means of $z = z' + d/2$ and afterward take the limit $d \to \infty$. Hence we obtain the well-known screened Fröhlich Hamiltonian. Another important limiting case can be deduced from the expressions if we directly take $d \to \infty$ in (44), (45), and (46). This is the case of the single heterostructure (SHS) where

$$\Phi^{\text{SHS}}_{\mathbf{q},\lambda}(z) = \begin{cases}\frac{1}{2}e^{(Q-iq_z)z} & \text{for } z<0\\ 1-\frac{1}{2}e^{-(Q+iq_z)z} & \text{for } z>0\end{cases} \tag{47}$$

Substitution of (47) in place of $\Phi^{\text{DHS}}_{\mathbf{q},\lambda}$ in (44) provides the given Hamiltonian for SHS.

3.2. Electron–Phonon Interaction Hamiltonian for QWW and FSW

In the frame of the present treatment we shall restrict ourselves to the case of oscillations perpendicular to the wire axis (i.e., $q_z = 0$ and hence $U_z = 0$). We are thus considering a particular case, which, however, entails a direct interest for the study of certain physical processes (one phonon Raman scattering configuration [6], for instance) and gives us insight into the nature of the oscillations. It has been shown that the corresponding eigensolutions constitute a complete orthonormal basis of eigenvectors $\mathbf{U}_{nm}$ and the electron–phonon interaction Hamiltonian is derived using the second-quantization formalism and can be written as [2]

$$\widehat{H} = \sum_{n,m}C_{nm}\left[F_{nm}(r)e^{in\theta}\hat{b}_{nm} + HC\right] \tag{48}$$

where

$$C_{nm} = \left[\frac{\pi\omega_L\rho}{\omega_q}\right]^{\frac{1}{2}}r_0^2 C_F \tag{49}$$

where ρ is the reduced mass density, ω_q is the frequency of the nm mode, and the functions $F_{nm}(r)$ have different forms for the QWW and the FSW. For a QWW, we have that

$$F_{nm}^{\text{QWW}} = B_{nm} \begin{cases} \dfrac{1}{x} f_n\left(\mu \dfrac{r}{r_0}\right) - S_n(\mu)\left(\dfrac{r}{r_0}\right)^n & \text{for } r \leq r_0 \\ \left[\dfrac{1}{x} f_n(\mu) - S_n(\mu)\right]\left(\dfrac{r_0}{r}\right)^n & \text{for } r > r_0 \end{cases} \tag{50}$$

being

$$S_n(\mu) = \frac{\epsilon_{a\infty}}{\epsilon_{a\infty} + \epsilon_{b\infty}} \left[\frac{1}{n} f_{n-1}(\mu) + \frac{1}{x}\left(\frac{\epsilon_{b\infty} - \epsilon_{a\infty}}{\epsilon_{a\infty}}\right) f_n(\mu)\right] \tag{51}$$

while for a FSW

$$F_{nm}^{\text{FSW}} = B_{nm} \begin{cases} f_n\left(\mu \dfrac{r}{r_0}\right) - t_n(\mu, \nu)\left(\dfrac{r}{r_0}\right)^n & \text{for } r \leq r_0 \\ [f_n(\mu) - t_n(\mu, \nu)]\left(\dfrac{r_0}{r}\right)^n & \text{for } r > r_0 \end{cases} \tag{52}$$

where

$$t_n(\mu, \nu) = \left[y^2 f_n(\mu) f_{n+2}(\nu) + \frac{\epsilon_{a\infty}}{\epsilon_{b\infty}} g_n(\mu) f_n(\nu)\right] \times \left[\left(\frac{\epsilon_{a\infty}}{\epsilon_{b\infty}}\left(\frac{\beta_L}{\beta_T}\right)^2 \mu^2 + \nu^2\right) f_{n+2}(\nu)\right]^{-1} \tag{53}$$

and

$$g_n(\mu) = \frac{\beta_L^2}{\beta_T^2} \mu^2 f_n(\mu) - 2(n+1)\mu f_{n+1}(\mu) \tag{54}$$

where $\epsilon_{b\infty}(\epsilon_{b0})$ is the high frequency (static) dielectric constant, the index b refers to the exterior of the nanostructure, and B_{nm} is a normalization constant that can be determined from Eq. (34) of [2]. In this constant we may include the term ρ from Eq. (16) of same reference and this term will not appear later. $\beta_L(\beta_T)$ is a parameter describing the dispersion of the oscillations longitudinal (transverse).

On the other hand

$$\mu = q r_0 \quad \nu = Q r_0 \quad \left(\frac{r_0}{\beta_L}\right)^2 (\omega_L^2 - \omega_T^2) = R = \mu^2 - \left(\frac{\beta_T}{\beta_L}\right)^2 \nu^2 \tag{55}$$

$$q^2 = \frac{\omega_L^2 - \omega_q^2}{\beta_L^2} \quad Q^2 = \frac{\omega_T^2 - \omega_q^2}{\beta_T^2} \tag{56}$$

and $f_n(x)$ represents a solution of the Bessel equation of order n (this function should be bounded in its domain of definition) and m is the order of the zero, obtained in the secular equations Eq. (57), for the QWW, and (58) for the FSW. It must be noticed that $\omega_q < \omega_L$ for all the frequencies involved; thus, q_z is always a real quantity. On the contrary, Q is real for $\omega_q < \omega_T$, but a pure complex quantity for $\omega_T < \omega_q < \omega_L$. The eigenfrequencies of the oscillation modes for this case are reported in [4]; they can be obtained by solving the secular equation

$$\frac{\epsilon_{a\infty} + \epsilon_{b\infty}}{2} [f_{n-1}(\mu) f_{n+1}(\nu) + f_{n-1}(\nu) f_{n+1}(\mu)] + \left(\frac{\beta_L}{\beta_T}\right)^2 \frac{\epsilon_{a\infty}}{\nu^2} R^2 \left[f_{n-1}(\mu) + \frac{n}{\mu} f_n(\mu)\left(\frac{\epsilon_{b\infty}}{\epsilon_{a\infty}} - 1\right)\right] f_{n+1}(\nu) = 0 \tag{57}$$

and for the FSW case are given by solving the secular equation

$$2n(n-1)\left(\frac{\beta_L}{\beta_T}\right)^2 R^2 t_n(\mu, \nu) f_{n+2}(\nu) + 2n\nu^2 f_{n+2}(\nu)\,[\mu f_n'(\mu) - f_n(\mu)] + g_n(\mu) \times [2\nu f_n'(\nu) + (\nu^2 - 2n^2) f_n(\nu)] = 0 \tag{58}$$

for what $\omega_q = \omega_{nm}$.

3.3. Electron–Phonon Interaction Hamiltonian for QD

It has been shown that the corresponding eigensolutions constitute a complete orthonormal basis of eigenvectors $\mathbf{U}_{lnm}$ and the electron–phonon interaction Hamiltonian is derived using the second-quantization formalism and can be written as in Ref. [1],

$$\hat{H} = \sum_{l=0} \sum_{n} \sum_{m=-l}^{l} r_0 C_F \sqrt{\frac{4\pi}{3}} \Phi_{l,n} Y_l^m(\theta, \varphi)[\hat{b}_{lnm} + \hat{b}_{lnm}^+] \tag{59}$$

where the functions $\Phi_{l,n}(r)$ are given as

$$\Phi_{l,n}(r) = \frac{\sqrt{r_0}}{|\mathbf{u}|} \begin{cases} j_l\left(\mu_n \dfrac{r}{r_0}\right) - \dfrac{\mu_n j_l'(\mu_n) + (l+1)\frac{\epsilon_{b\infty}}{\epsilon_{a\infty}} j_l(\mu_n)}{l + (l+1)\frac{\epsilon_{b\infty}}{\epsilon_{a\infty}}} \left(\dfrac{r}{r_0}\right)^l \\ \text{for } r \leq r_0 \\ -\dfrac{\mu_n j_l'(\mu_n) + l\frac{\epsilon_{b\infty}}{\epsilon_{a\infty}} j_l(\mu_n)}{l + (l+1)\frac{\epsilon_{b\infty}}{\epsilon_{a\infty}}} \left(\dfrac{r}{r_0}\right)^l \\ \text{for } r > r_0 \end{cases} \tag{60}$$

where j_l is the spherical Bessel function, and μ and ν can be determined using Eqs. (55) and (56). The eigenfrequencies of the oscillation modes are reported in [1]; they can be obtained by solving the secular equation

$$\nu j_l'(\mu) F_l(\nu) = l(l+1) j_l(\mu) G_l(\nu) \tag{61}$$

where

$$F_l(\nu) = \frac{\omega_L^2 - \omega_T^2}{\beta_T^2}\left(\frac{r_0}{\nu}\right)^2 l[\nu g_l'(\nu) - l g_l(\nu)] + \left(l + (l+1)\frac{\epsilon_{b\infty}}{\epsilon_{a\infty}}\right)[\nu g_l'(\nu) + g_l(\nu)] \tag{62}$$

$$G_l(\nu) = \frac{\omega_L^2 - \omega_T^2}{\beta_T^2}\left(\frac{r_0}{\nu}\right)^2 \frac{\epsilon_{b\infty}}{\epsilon_{a\infty}}[l g_l(\nu) - \nu g_l'(\nu)] + \left(l + (l+1)\frac{\epsilon_{b\infty}}{\epsilon_{a\infty}}\right) g_l(\nu) \tag{63}$$

Here l represents the order of the spherical Bessel function and n is the order of the zero obtained in the secular equations (61). It can be appreciated that the energy of the modes, $\hbar\omega_q = \hbar\omega_{ln}$, does not depend on m.

In Eq. (60) the term $|\mathbf{u}|$ represents the norm of the vibrational amplitude and is equal to

$$|\mathbf{u}|^2 = \int_0^{r_0} \left\{ \left[-\frac{d}{dr} j_l\left(\mu \frac{r}{r_0}\right) + \frac{l(l+1)}{r} p_l g_l\left(\nu \frac{r}{r_0}\right) - \frac{t_l}{r_0} l \left(\frac{r}{r_0}\right)^{l-1} \right]^2 + \frac{l(l+1)}{r^2} \times \left[-j_l\left(\mu \frac{r}{r_0}\right) + \frac{p_l}{l}\frac{d}{dr}\left[r g_l\left(\nu \frac{r}{r_0}\right)\right] - t_l \left(\frac{r}{r_0}\right)^{l} \right]^2 \right\} r^2 dr \qquad (64)$$

with

$$p_l = \frac{\mu_k j_l'(\mu) - l j_l(\mu)}{l g_l(\nu) - \nu g_l'(\nu)} \qquad t_l = \frac{\gamma_0}{\nu_k^2} \frac{\left[\mu_k j_l'(\mu) + (l+1)\frac{\epsilon_{b\infty}}{\epsilon_{a\infty}} j_l(\mu)\right]}{l + \frac{\epsilon_{b\infty}}{\epsilon_{a\infty}}(l+1)} \qquad (65)$$

$$\gamma_0 = \frac{\omega_L^2 - \omega_T^2}{\beta_T^2} r_0^2 \qquad (66)$$

and the function $g_l(z)$ is defined as

$$g_l(z) = \begin{cases} j_l(z) & \text{for } z^2 > 0 \\ i_l(z) & \text{for } z^2 < 0 \end{cases} \qquad (67)$$

$i_l(z)$ is the modified spherical Bessel function.

Different modes of oscillation, with frequencies ω, contribute to the solution of the system of equations. These modes have several features; some of them are purely transverse and they are uncoupled to the electric field, and this is the reason because they do not interact via Fröhlich. The other oscillation modes are coupled to the electric field; these can have a longitudinal part and a transverse part. However, in this framework, the oscillation modes are purely longitudinal or purely transverse, for the QW. Also $n = 1$ for the QWW and the FSW, and $l = 0$ for the QD, where there are purely longitudinal modes. This can be observed in the Figure 3 where coupled modes $l = 0$ and $l = 1$ for the QD are shown.

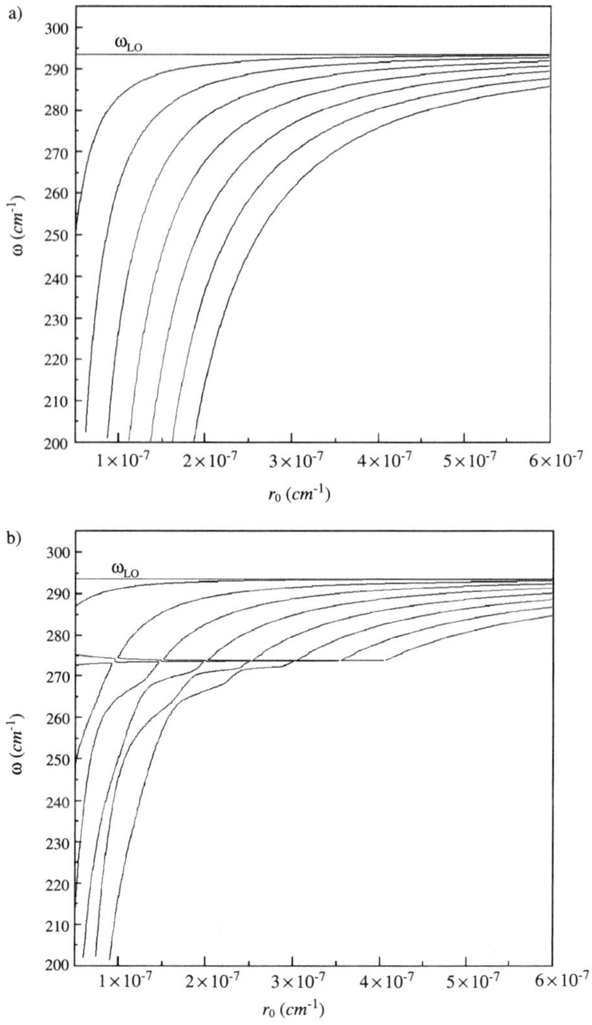

Figure 3. Coupled modes for: (a) $l = 0$, (b) $l = 1$.

4. DIFFERENTIAL CROSS-SECTION FOR QUANTUM WELL (QW)

Let us briefly describe the model and fundamental theory applied in our calculations. The problem of determining the stationary states of motion of a particle of mass μ in an external potential field leads to the solution of the equation

$$\left\{ \nabla^2 + \frac{2\mu}{\hbar^2}[E - V(r)] \right\} \Psi = 0 \qquad (68)$$

In our model we assume a z-direction grown semiconductor QW with interfaces located at $z = -\frac{d}{2}, z = \frac{d}{2}$. We also assume the quantum well with finite potential barriers at the interfaces. The solution of Schrödinger's equation, in the envelope function approximation, leads to

$$\varphi_{n_j}(r) = \left(\frac{2}{V}\right)^{\frac{1}{2}} \exp[-(i\mathbf{k}_{j\perp} \cdot \mathbf{r}_\perp)] \times A_j \begin{cases} \cos(k_{n_j} z) & n_j \text{ odd} \quad \text{for } |z| \le \frac{d}{2} \\ \sin(k_{n_j} z) & n_j \text{ even} \\ B_j e^{-\gamma_j |z|} & \quad \text{for } |z| > \frac{d}{2} \end{cases} \qquad (69)$$

and

$$\Psi_{n_j}(r) = \varphi_{n_j}(r) u_j(r) \qquad (70)$$

where

$$A_j = \frac{1}{\sqrt{1 - \frac{2}{d}\cos\left(\frac{k_{n_j} d}{2}\right)\left[\frac{1}{\gamma_{n_j}} + (-1)^{n_j} \frac{\sin\left(\frac{k_{n_j} d}{2}\right)}{k_j}\right]}} \qquad (71)$$

and

$$B_j = \exp\left(\frac{\gamma_{n_j} d}{2}\right) \begin{cases} \cos\left(\dfrac{k_{n_j} d}{2}\right) & \text{for } n_j \text{ odd} \\ \sin\left(\dfrac{k_{n_j} d}{2}\right) & \text{for } n_j \text{ even} \end{cases} \tag{72}$$

with

$$k_{n_j} d = n_j \pi - 2 \arcsin\left(\frac{1}{\sqrt{\frac{2\mu_{jo} V_j}{\hbar^2 k_{n_j}^2} - \frac{\mu_{jo}}{\mu_{ji}} + 1}}\right) \quad \text{and}$$

$$\gamma_{n_j} = \sqrt{\frac{2\mu_{jo} V_j}{\hbar^2} - \frac{\mu_{jo}}{\mu_{ji}} k_{n_j}^2} \tag{73}$$

The energies are given by

$$E_j = \frac{\hbar^2}{2\mu_{ji}}[k_{j\perp}^2 + k_{n_j}^2] \qquad n_j = 1, 2, \ldots \tag{74}$$

where $j = 1, 2$ denotes electron or hole quantities, V_j represents the band offset, μ_{ji} is the electron or hole effective mass in the nanostructures, V is the volume, and $u_j(r)$ is the Bloch function taken at $\mathbf{k} = 0$, where (by assumption) the band extrema are located. Label "$\perp$" means "perpendicular to the z axis." The energy bands are unfolded in a system of subbands, where n_j is the order of the zero of Eq. (73).

4.1. Raman Scattering Intensities

From Eqs. (69) and (4) we can obtain the matrix element

$$\langle a|\widehat{H}_{jl}|i\rangle = \frac{|e|}{\mu_0}\sqrt{\frac{2\pi\hbar}{V\omega_l}}\hat{\mathbf{e}}_l \cdot \langle a|\hat{\mathbf{p}}|i\rangle \tag{75}$$

where $\hat{\mathbf{p}} = -i\hbar\nabla$ for the electron contribution and

$$\langle a_1|\hat{\mathbf{p}}|i\rangle = \int \bar{\varphi}_{n_1'}(r)\bar{u}_1'(r)u_2(r)\hat{\mathbf{p}}\varphi_{n_2}(r)\,dV + \int \bar{\varphi}_{n_1'}(r)\bar{u}_1'(r)\varphi_{n_2}(r)\hat{\mathbf{p}}u_2(r)\,dV \tag{76}$$

Consider that:

(1) The crystal possesses a volume $V = N_0\Omega_0$, in which N_0 and Ω_0 are the number of unitary cells and their volume, respectively.
(2) The function $u_j(r)(\varphi_{n_j}(r))$ varies quickly (slowly) in the volume.

Then Eq. (76) comes in,

$$\langle a_1|\hat{\mathbf{p}}|i\rangle = \sum_{\text{cells}} \bar{\varphi}_{n_1'}(r)\hat{\mathbf{p}}\varphi_{n_2}(r)\int_{\Omega_0} \bar{u}_1'(r)u_2(r)\,dV + \sum_{\text{cells}} \bar{\varphi}_{n_1'}(r)\varphi_{n_2}(r)\int_{\Omega_0} \bar{u}_1'(r)\hat{\mathbf{p}}u_2(r)\,dV \tag{77}$$

upon considering the orthogonal character of the functions of Bloch. The first sum of expression (77) is annulled. The aforementioned and the definition of the moment between the bands of valence and conduction leads to the expression

$$\mathbf{p}_{cv}(0) = \int_{\Omega_0} \bar{u}_1'(r)\hat{\mathbf{p}}u_2(r)\,dV \tag{78}$$

where $\mathbf{p}_{cv}(0)$ is the interband momentum matrix element (evaluate at $\mathbf{k} = 0$). This allows us to write Eq. (75) as

$$\langle a|\widehat{H}_{jl}|i\rangle = \frac{|e|}{\mu_0}\sqrt{\frac{2\pi\hbar}{V\omega_l}}\hat{\mathbf{e}}_l \cdot \mathbf{p}_{cv}(0)\int \bar{\varphi}_{n_1'}(r)\varphi_{n_2}(r)\,dV \tag{79}$$

obtaining finally the expression for the matrix element

$$\langle a|\widehat{H}_{jl}|i\rangle = \frac{|e|}{\mu_0}\sqrt{\frac{2\pi\hbar}{V\omega_l}}\hat{\mathbf{e}}_l \cdot \mathbf{p}_{cv}(0)\begin{cases} T_{n_1', n_2}\delta_{\vec{k}_{1\perp}', -\vec{k}_{2\perp}} & \text{for } j = 1 \\ T_{n_2', n_1}\delta_{-\vec{k}_{2\perp}', \vec{k}_{1\perp}} & \text{for } j = 2 \end{cases} \tag{80}$$

where

$$T_{n_a, n_b} = A_a A_b \left\{\frac{[1 + (-1)^{n_a + n_b}]}{d} \times \left[\frac{\sin[(k_{n_a} - k_{n_b})\frac{d}{2}]}{k_{n_a} - k_{n_b}} - (-1)^{n_a}\frac{\sin[(k_{n_a} + k_{n_b})\frac{d}{2}]}{k_{n_a} + k_{n_b}}\right] + \frac{4B_a B_b}{(\gamma_{n_a} + \gamma_{n_b})d}\exp\left[-(\gamma_{n_a} + \gamma_{n_b})\frac{d}{2}\right]\right\} \tag{81}$$

where a (b) indicates the electron (hole) state. T_{n_a, n_b} represents the creation of an electron–hole pair due to a transition interband, where a photon of energy $\hbar\omega_l$ of incident radiation is annihilated.

If we applied the same treatment to the calculation of the other matrix elements of the expression (3), we obtain

$$\langle f|\widehat{H}_{jl}|a\rangle = (-1)^{j+1}\frac{|e|}{\mu_j}\sqrt{\frac{2\pi\hbar}{V\omega_s}}\hat{\mathbf{e}}_s \cdot \langle f|\hat{\mathbf{p}}|a\rangle \tag{82}$$

In a similar way we obtain that

$$\langle f|\hat{\mathbf{p}}|a\rangle = \int \bar{\varphi}_{n_j}(r)\bar{u}_j(r)\hat{\mathbf{p}}\varphi_{n_j'}(r)u_j'(r)\,dV + \int \bar{\varphi}_{n_j}(r)\bar{u}_j(r)\varphi_{n_j'}(r)\hat{\mathbf{p}}u_j'(r)\,dV \tag{83}$$

Then

$$\langle f|\hat{\mathbf{p}}|a\rangle = \sum_{\text{cells}} \bar{\varphi}_{n_j}(r)\hat{\mathbf{p}}\varphi_{n_j'}(r)\int_{\Omega_0} \bar{u}_j(r)u_j'(r)\,dV + \sum_{\text{cells}} \bar{\varphi}_{n_j}(r)\varphi_{n_j'}(r)\int_{\Omega_0} \bar{u}_j(r)\hat{\mathbf{p}}u_j'(r)\,dV \tag{84}$$

In this expression the second sum is equal to zero because the integral is calculated between symmetrical limits of an odd operator between functions of equal parity. Upon keeping in mind that the Bloch's functions are orthonormal, the integral of the first sum is equal to the unit. Then

$$\langle f|\widehat{H}_{js}|a\rangle = (-1)^{j+1}\frac{|e|}{\mu_j}\sqrt{\frac{2\pi\hbar}{V\omega_s}}\hat{\mathbf{e}}_s \cdot \int \bar{\varphi}_{n_j}(r)\hat{\mathbf{p}}\varphi_{n_j'}(r)\,dV \tag{85}$$

From here we obtain that

$$\langle f|\widehat{H}_{js}|a\rangle = (-1)^{j+1}\frac{|e|}{\mu_j}\sqrt{\frac{2\pi\hbar^3}{V\omega_s}} \times\left[\hat{\mathbf{e}}_s\cdot\mathbf{k}_{j\perp}I_{n_j,n'_j}+2i\frac{\hat{\mathbf{e}}_s\cdot\hat{\mathbf{e}}_z}{d}II_{n_j,n'_j}\right]\delta_{\mathbf{k}_{j\perp},\mathbf{k}'_{j\perp}} \quad (86)$$

being

$$I_{n_j,n'_j}=\begin{cases} A_jA'_j\left\{\frac{\left[1+(-1)^{n_j+n'_j}\right]}{d}\times\left[\frac{\sin\left[(k_{n_j}-k_{n'_j})\frac{d}{2}\right]}{k_{n_j}-k_{n'_j}}-(-1)^{n_j}\frac{\sin\left[(k_{n_j}+k_{n'_j})\frac{d}{2}\right]}{k_{n_j}+k_{n'_j}}\right]+\frac{4B_{n_j}B_{n'_j}}{(\gamma_{n_j}+\gamma_{n'_j})d}\frac{\mu_{jo}}{\mu_{ji}}\exp\left[-(\gamma_{n_j}+\gamma_{n'_j})\frac{d}{2}\right]\right\} & \text{for } n_j\neq n'_j \\ A_j^2\left\{\left[2-(-1)^{n_j}\frac{\sin(k_{n_j}d)}{k_{n_j}d}\right]+\frac{2B_{n_j}^2}{\gamma_{n_j}d}\frac{\mu_{jo}}{\mu_{ji}}\exp(-\gamma_{n_j}d)\right\} & \text{for } n_j=n'_j \end{cases} \quad (87)$$

$$II_{n_j,n'_j}=(-1)^{n_j}A_jA'_j(1-(-1)^{n_j+n'_j})k_{n'_j}\times\left[\frac{\sin\left[(k_{n_j}-k_{n'_j})\frac{d}{2}\right]}{k_{n_j}-k_{n'_j}}-(-1)^{n_j}\frac{\sin\left[(k_{n_j}+k_{n'_j})\frac{d}{2}\right]}{k_{n_j}+k_{n'_j}}\right] \quad (88)$$

This expression represents a intersubband transition, where a photon of secondary radiation was created, with energy $\hbar\omega_s$. In this calculation the photon wavevector in comparison with the electron wavevector has been neglected. Hence, in the final state we have $\mathbf{k}_e+\mathbf{k}_h=0$.

Considering that

$$|M_1+M_2|^2=|M_1|^2+|M_2|^2+2\,\mathrm{Re}(M_1)\mathrm{Re}(M_2)+2\,\mathrm{Im}(M_1)\mathrm{Im}(M_2) \quad (89)$$

we have three different contributions.

After substitution of Eqs. (80) and (86), in Eq. (3) we have

$$M_j=(-1)^{j+1}\beta_j\frac{A_0}{\mu_rE_0d}\sum_{n'_j}\frac{T_{n'_j,\,n_2\delta_{j1}+n_1\delta_{j2}}}{\frac{\hbar\omega_s}{E_0}+\beta_j\left[(k_{n_j}d)^2-(k_{n'_j}d)^2\right]+i\delta_{aj}}\times[d(\hat{\mathbf{e}}_s\cdot\mathbf{k}_{j\perp})I_{n_j,n'_j}+2i(\hat{\mathbf{e}}_s\cdot\hat{\mathbf{e}}_z)II_{n_j,n'_j}] \quad (90)$$

where

$$A_0=\frac{e^2}{\mu_0}\frac{2\pi\hbar^2}{V}\sqrt{\frac{1}{\omega_s\omega_l}}(\hat{\mathbf{e}}_l\cdot\mathbf{p}_{cv}(0)) \quad (91)$$

and

$$E_0=\frac{\hbar^2}{2\mu_rd^2}\quad \mu_r^{-1}=\mu_{1i}^{-1}+\mu_{2i}^{-1}\quad \beta_j=\frac{\mu_r}{\mu_{ji}}\quad \delta_{aj}=\frac{\Gamma_{aj}}{E_0} \quad (92)$$

After the substitution of Eq. (90) in Eqs. (2) and (1) we obtain that

$$\frac{d^2\sigma}{d\Omega\,d\omega_s}=\frac{V^2\omega_s^2\eta(\omega_s)}{8\pi^3c^4\eta(\omega_l)}\left|\frac{A_0}{\mu E_0d}\right|^2\frac{2\pi}{\hbar}\frac{1}{E_0}\sum_{k_{1\perp}}\sum_{n_1n_2}|\overline{M}_1+\overline{M}_2|\times\delta\left(\frac{\hbar\omega_l-E_g}{E_0}-\frac{\hbar\omega_s}{E_0}-\beta_1(k_{n_1}d)^2-\beta_2(k_{n_2}d)^2-(k_{1\perp}d)^2\right) \quad (93)$$

where

$$\overline{M}_j=(-1)^{j+1}\beta_j\sum_{n'_j}\frac{T_{n'_j,\,n_2\delta_{j1}+n_1\delta_{j2}}}{\frac{\hbar\omega_s}{E_0}+\beta_j\left[(k_{n_j}d)^2-(k_{n'_j}d)^2\right]+i\delta_{aj}}\times[d(\hat{\mathbf{e}}_s\cdot\mathbf{k}_{j\perp})I_{n_j,n'_j}+2i(\hat{\mathbf{e}}_s\cdot\hat{\mathbf{e}}_z)II_{n_j,n'_j}] \quad (94)$$

and considering that

$$\sum_{\mathbf{k}_\perp}\cdots\cdots\implies\frac{L_xL_y}{(2\pi)^2}\int_0^\infty\cdots\cdots\mathbf{k}_\perp\,d\mathbf{k}_\perp \quad (95)$$

we obtain

$$\frac{d^2\sigma}{d\Omega\,d\omega_s}=\left[\frac{d^2\sigma}{d\Omega\,d\omega_s}\right]_0+\left[\frac{d^2\sigma}{d\Omega\,d\omega_s}\right]_\perp+\left[\frac{d^2\sigma}{d\Omega\,d\omega_s}\right]_z \quad (96)$$

where

$$\left[\frac{d^2\sigma}{d\Omega\,d\omega_s}\right]_0=\sigma_0\frac{9}{2\sqrt{2}}\left(\frac{\hbar\omega_s}{E_0}\right)^2\sum_{n_1n_2}[\mathrm{Re}(M_{1p})\mathrm{Re}(M_{2p})+\mathrm{Im}(M_{1p})\mathrm{Im}(M_{2p})]S_0 \quad (97)$$

$$\left[\frac{d^2\sigma}{d\Omega\,d\omega_s}\right]_\perp=\sigma_0\frac{9}{4\sqrt{2}}\left(\frac{\hbar\omega_s}{E_0}\right)^2\sum_{n_1n_2}[|M_{1\perp}|^2+|M_{2\perp}|^2]\times\left(\frac{\hbar\omega_l-E_g}{E_0}-\frac{\hbar\omega_s}{E_0}-\beta_1(k_{n_1}d)^2-\beta_2(k_{n_2}d)^2\right)S_\perp \quad (98)$$

and

$$\left[\frac{d^2\sigma}{d\Omega\,d\omega_s}\right]_z=\sigma_0\frac{9}{4\sqrt{2}}\left(\frac{\hbar\omega_s}{E_0}\right)^2\sum_{n_1n_2}[|M_{1z}|^2+|M_{2z}|^2]S_z \quad (99)$$

Here

$$M_{jp}=M_{j\perp}\left(\frac{\hbar\omega_l-E_g}{E_0}-\frac{\hbar\omega_s}{E_0}-\beta_1(k_{n_1}d)^2-\beta_2(k_{n_2}d)^2\right)^{\frac{1}{2}}+iM_{jz} \quad (100)$$

and

$$M_{j\perp} = (-1)^{j+1}\beta_j \sum_{n'_j} \frac{T_{n'_j,\, n_2\delta_{j1}+n_1\delta_{j2}}}{\frac{\hbar\omega_s}{E_0} + \beta_j\left[(k_{n_j}d)^2 - (k_{n'_j}d)^2\right] + i\delta_{aj}} I_{n_j,\, n'_j} \tag{101}$$

$$M_{jz} = (-1)^{j+1}\beta_j \sum_{n'_j} \frac{2T_{n'_j,\, n_2\delta_{j1}+n_1\delta_{j2}}}{\frac{\hbar\omega_s}{E_0} + \beta_j\left[(k_{n_j}d)^2 - (k_{n'_j}d)^2\right] + i\delta_{aj}} II_{n_j,\, n'_j} \tag{102}$$

besides

$$S_0 = |\hat{\mathbf{e}}_s \cdot \hat{\mathbf{e}}_\perp||\hat{\mathbf{e}}_s \cdot \hat{\mathbf{e}}_z| \qquad S_\perp = |\hat{\mathbf{e}}_s \cdot \hat{\mathbf{e}}_\perp|^2 \qquad S_z = |\hat{\mathbf{e}}_s \cdot \hat{\mathbf{e}}_z|^2 \tag{103}$$

and

$$\sigma_0 = \frac{4\sqrt{2}Ve^4|\hat{\mathbf{e}}_l \cdot \mathbf{p}_{cv}(0)|^2 \eta(\omega_s)\mu_r^{\frac{1}{2}} E_0^{\frac{3}{2}}}{9\pi^2\mu_0^2\hbar^4c^4\eta(\omega_l)\omega_l\omega_s} \tag{104}$$

The contribution of the polarization S_0 can be neglected if it is compared with the contribution $S_\perp$ or S_z, if the electron effective mass is different from the hole effective mass.

Summation over the subband label must be done under the following requirements:

$$\frac{\hbar\omega_l - E_g}{E_0} - \frac{\hbar\omega_s}{E_0} - \beta_1(k_{n_1}d)^2 - \beta_2(k_{n_2}d)^2 \geq 0 \tag{105}$$

The differential cross-section for a semiconductor QW presents singular peaks for secondary radiation frequency ω_s such that

$$(s,e)(n_1, n'_1) = \beta_1\left[(k_{n'_1}d)^2 - (k_{n_1}d)^2\right] \tag{106}$$

$$(s,h)(n_2, n'_2) = \beta_2\left[(k_{n'_2}d)^2 - (k_{n_2}d)^2\right] \tag{107}$$

$$(n_1, n_2) = \frac{\hbar\omega_l - E_g}{E_0} - \beta_1(k_{n_1}d)^2 - \beta_2(k_{n_2}d)^2 \tag{108}$$

The intersubband transitions obey a selection rule requiring $n_1 + n_2$ to be odd for the polarization S_z, but for $S_\perp$ this is not a rule.

We have computed the scattering efficiency

$$\frac{1}{\sigma_0}\frac{d^2\sigma}{d\Omega\, d\omega_s} \quad \text{vs} \quad \frac{\hbar\omega_s}{E_0}$$

the so-called "emission spectra" of the electron raman scattering process for a given polarization $\hat{\mathbf{e}}_s$ of the emitted radiation. The physical parameters entering in our formulas were taken for the GaAs case: $E_g = 1.5177$ eV, $\varepsilon_{a\infty} = 10.90$, $\mu_{1i} = 0.0665\mu_0$, $\mu_{2i} = 0.45\mu_0$ (the heavy-hole band) and $\varepsilon_{b\infty} = 8.16$. We have set $\Gamma_f = 3$ meV, $\Gamma_{aj} = \Gamma_{bj} = 1$ meV, $V_1 = 0.968$ eV, and $V_2 = 0.6453$ eV.

In Figure 4 are shown two emission spectra for two wide $d = 3.5$ nm and $d = 5.0$ nm, and the polarization $\overline{Z}(\hat{\mathbf{e}}_l, S_z)Z$ of a QW. The incident radiation energy is $\hbar\omega_l = 2.4$ eV and several singularities are observed. The peaks position depends on the width in such a way that with the increase of width more peaks appear, because more energy levels

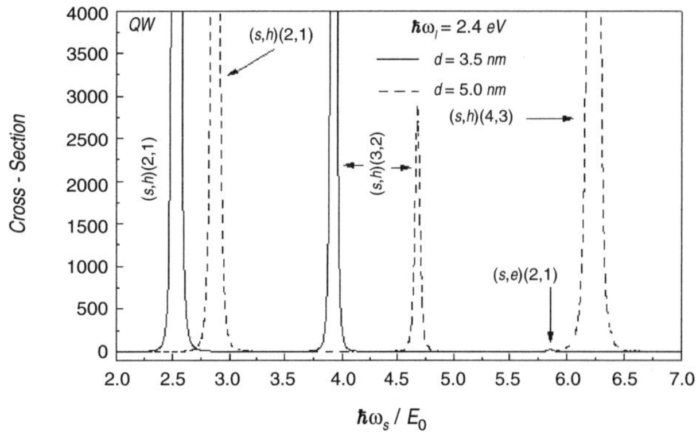

Figure 4. The emission spectra of the electron Raman scattering in QW for the polarization $\overline{Z}(\mathbf{e}_l, S_z)Z$ and for several widths.

exit from the interior of the QW. The intensity of the different singularities depends on the subbands involved in the transition.

In Figure 5 emission spectra for a QW with width $d = 5.0$ nm, incident radiation $\hbar\omega_l = 2.8$ eV, and polarization $\overline{Z}(\hat{\mathbf{e}}_l, S_z)Z$ are shown. The increasing of the width of QW allows the appearance of new singularities, but the previous ones are kept. This is due to the fact that with the increase of the width the energy levels descend, making it possible that the EHP can reach new subbands. If we increase the energy of the incident radiation, new peaks appear, and the previous ones are kept. This behavior is due to the fact that new subbands can be reached by the EHP. This can also be observed in the dispersion assisted or not by phonons in QDs and in QWWs. The intensity of the different singularities does not depend on the incident radiation for the polarization $\overline{Z}(\hat{\mathbf{e}}_l, S_z)Z$.

4.2. Raman Scattering Intensities with One Phonon-Assisted Transition in QWs

In this case we begin from Eqs. (69) and (44), taking into account the same previous treatment. The calculation of the matrix element is given by

$$\langle b|\widehat{H}_{jph}|a\rangle = \int \bar{\varphi}_{n'_j}(r)\bar{u}'_j(r)\widehat{H}_{jph}\varphi_{n''_j}(r)u''_j(r)\, dV \tag{109}$$

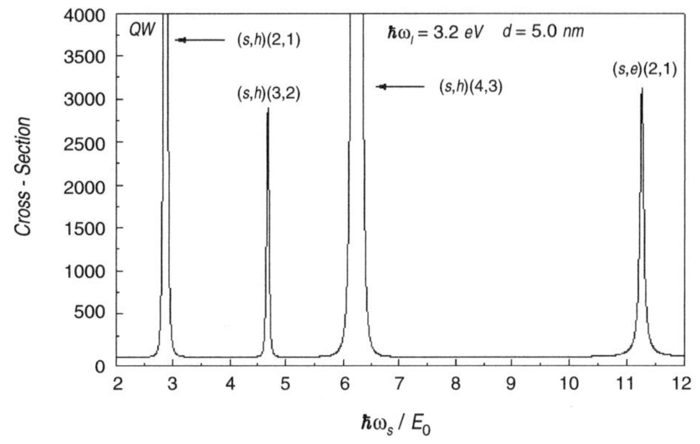

Figure 5. The emission spectra of the electron Raman scattering in QW for the polarization $\overline{Z}(\mathbf{e}_l, S_z)Z$.

From here we obtain that

$$\langle b|\widehat{H}_{jph}|a\rangle = \sum_{\text{cells}} \bar{\varphi}_{n_j'}(r)\widehat{H}_{jph}\varphi_{n_j''}(r)\int_{\Omega_0} \bar{u}_j'(r)u_j''(r)\,dV \tag{110}$$

considering the orthonormality of the functions of Bloch, it can be obtained that

$$\langle b|\widehat{H}_{jph}|a\rangle = \int \bar{\varphi}_{n_j'}(r)\widehat{H}_{jph}\varphi_{n_j''}(r), \tag{111}$$

which is given by

$$\langle b|\widehat{H}_{jph}|a\rangle = (-1)^{j+1}C_q\Xi_{n,n_j',n_j''}(q_\perp)\delta(k_{j\perp}' - k_{j\perp}'' - q_\perp) \tag{112}$$

$$\Xi_{n,n_j',n_j''}(q_\perp) = A_j'A_j''[\Pi_1(n,q_\perp) + \Pi_2(q_\perp)] \tag{113}$$

with

$$\begin{aligned}\Pi_1(n, q_\perp) = {} & \frac{(1+(-1)^{n_j'+n_j''})}{d} \\ & \times\left[\exp\left(iq_z\frac{d}{2}\right)[I_-'(iq_z) - (-1)^{n_j'}I_+'(iq_z)]\right. \\ & - \exp\left(q_\perp\frac{d}{2}\right)\{[I_-'(-q_\perp) + I_-'(q_\perp)] - (-1)^{n_j'} \\ & \left.\times[I_+'(-q_\perp) + I_+'(q_\perp)]\}\right] + \frac{(1-(-1)^{n_j'+n_j''})}{d} \\ & \times\left[\exp\left(iq_z\frac{d}{2}\right)[I_-''(iq_z) - (-1)^{n_j'}I_+''(iq_z)]\right. \\ & - \exp\left(q_\perp\frac{d}{2}\right)\{[I_-''(-q_\perp) + I_-''(q_\perp)] \\ & \left. - (-1)^{n_j'}[I_+''(-q_\perp) + I_+''(q_\perp)]\}\right]\end{aligned} \tag{114}$$

where

$$\begin{aligned}I_\pm'(q) = {} & \cosh\left(q\frac{d}{2}\right)\frac{(k_{n_j'} \pm k_{n_j''})\sin[(k_{n_j'} \pm k_{n_j''})\frac{d}{2}]}{(k_{n_j'} \pm k_{n_j''})^2 + q^2} \\ & + \sinh\left(q\frac{d}{2}\right)\frac{q\cos[(k_{n_j'} \pm k_{n_j''})\frac{d}{2}]}{(k_{n_j'} \pm k_{n_j''})^2 + q^2}\end{aligned} \tag{115}$$

$$\begin{aligned}I_\pm''(q) = {} & \cosh\left(q\frac{d}{2}\right)\frac{q\sin[(k_{n_j'} \pm k_{n_j''})\frac{d}{2}]}{(k_{n_j'} \pm k_{n_j''})^2 + q^2} \\ & - \sinh\left(q\frac{d}{2}\right)\frac{(k_{n_j'} \pm k_{n_j''})\cos[(k_{n_j'} \pm k_{n_j''})\frac{d}{2}]}{(k_{n_j'} \pm k_{n_j''})^2 + q^2}\end{aligned} \tag{116}$$

$$\Pi_2(q_\perp) = 4B_j'B_j''\sinh\left(q_\perp\frac{d}{2}\right)\frac{\exp[-(\gamma_{n_j'} + \gamma_{n_j''} + q_\perp)\frac{d}{2}]}{(\gamma_{n_j'} + \gamma_{n_j''} + q_\perp)d} \tag{117}$$

Here a quantum transition is realized and the hole or the electron emits a phonon of energy $\hbar\omega_q$. As can be observed, there exists a mixture between the confined part and the free part of the phonon $\Pi_1(n, q_\perp)$, although the contribution of this diminishes upon increasing $q_\perp$.

The first intermediate state, a, is indicated by the quantum numbers n_j'' and $k_{j\perp}''$, and for the second intermediate state, b, n_j', and $k_{j\perp}'$ will be used. The notation of a $''$ for the state a and a $'$ for the second intermediate state b will be used for the one phonon-assisted dispersion.

Substituting Eqs. (80), (86), and (112) in Eq. (3) the following expression is obtained:

$$\begin{aligned}M_j = {} & \beta_j\frac{A_0C_q}{\mu_rE_0d}\sum_{n_j',n_j''}\frac{(\delta_{k_{2\perp},k_{1\perp}'}\delta_{j1} + \delta_{k_{1\perp},k_{2\perp}'}\delta_{j2})}{\frac{\hbar\omega_s}{E_0} + \beta_j[(k_{n_j}d)^2 - (k_{n_j''}d)^2] + i\delta_{aj}} \\ & \times\frac{T_{n_j',n_2\delta_{j1}+n_1\delta_{j2}}\Xi_{n,n_j'',n_j'}(q_\perp)\delta(k_{j\perp} - k_{j\perp}' - q_\perp)}{[\frac{\hbar\omega_s}{E_0} + \frac{\hbar\omega_q}{E_0} + \beta_j[(k_{n_j}d)^2 - (k_{n_j'}d)^2 + (k_{j\perp}d)^2 - (k_{j\perp}'d)^2] + i\delta_{aj}]} \\ & \times[d(\hat{\mathbf{e}}_s\cdot\mathbf{k}_{j\perp})I_{n_j,n_j''} + 2i(\hat{\mathbf{e}}_s\cdot\hat{\mathbf{e}}_z)II_{n_j,n_j''}]\end{aligned} \tag{118}$$

If we substitute Eq. (118) into Eqs. (2) and (1), we obtain

$$\frac{d^2\sigma}{d\Omega d\omega_s} = \left[\frac{d^2\sigma}{d\Omega d\omega_s}\right]_0 + \left[\frac{d^2\sigma}{d\Omega d\omega_s}\right]_\perp + \left[\frac{d^2\sigma}{d\Omega d\omega_s}\right]_z \tag{119}$$

with

$$\begin{aligned}\left[\frac{d^2\sigma}{d\Omega\,d\omega_s}\right]_0 = {} & \sigma_{ph}\frac{1}{2\pi}\left(\frac{V}{d^3}\right)^2\left(\frac{\hbar\omega_s}{E_0}\right)^2\sum_n\sum_{n_1n_2}\int_0^\infty\frac{q_\perp\,dq_\perp}{q_\perp^2 + (\frac{n\pi}{d})^2} \\ & \times[\mathrm{Re}(M_{1p}^{ph})\mathrm{Re}(m_{2p}^{ph}) + \mathrm{Im}(M_{1p}^{ph})\mathrm{Im}(M_{2p}^{ph})]S_0\end{aligned} \tag{120}$$

$$\begin{aligned}\left[\frac{d^2\sigma}{d\Omega\,d\omega_s}\right]_\perp = {} & \sigma_{ph}\frac{1}{4\pi}\left(\frac{V}{d^3}\right)^2\left(\frac{\hbar\omega_s}{E_0}\right)^2\sum_n\sum_{n_1n_2}\int_0^\infty\frac{q_\perp\,dq_\perp}{q_\perp^2 + (\frac{n\pi}{d})^2} \\ & \times[|M_{1\perp}^{ph}|^2 + |M_{2\perp}^{ph}|^2]GS_\perp\end{aligned} \tag{121}$$

and

$$\begin{aligned}\left[\frac{d^2\sigma}{d\Omega\,d\omega_s}\right]_z = {} & \sigma_{ph}\frac{1}{4\pi}\left(\frac{V}{d^3}\right)^2\left(\frac{\hbar\omega_s}{E_0}\right)^2\sum_n\sum_{n_1n_2}\int_0^\infty\frac{q_\perp\,dq_\perp}{q_\perp^2 + (\frac{n\pi}{d})^2} \\ & \times[|M_{1z}^{ph}|^2 + |M_{2z}^{ph}|^2]S_z\end{aligned} \tag{122}$$

where

$$\begin{aligned}M_{jz}^{ph} = {} & \beta_j\sum_{n_j',n_j''}II_{n_j,n_j'}\Xi_{n,n_j',n_j''}\frac{T_{n_j'',n_2\delta_{j1}+n_1\delta_{j2}}}{[\frac{\hbar\omega_s}{E_0} + \beta_j[(k_{n_j}d)^2 - (k_{n_j''}d)^2] + i\delta_{bj}]} \\ & \times\frac{2}{[\frac{\hbar\omega_s}{E_0} + \frac{\hbar\omega_q}{E_0} + \beta_j\{[(k_{n_j}d)^2 - (k_{n_j'}d)^2] - (-1)^j[(q_\perp d)^2 - 2k_{1\perp}q_\perp d^2]\} + i\delta_{aj}]}\end{aligned} \tag{123}$$

$$\begin{aligned}M_{j\perp}^{ph} = {} & \beta_j\sum_{n_j',n_j''}I_{n_j,n_j'}\Xi_{n,n_j',n_j''}\frac{T_{n_j'',n_2\delta_{j1}+n_1\delta_{j2}}}{[\frac{\hbar\omega_s}{E_0} + \beta_j[(k_{n_j}d)^2 - (k_{n_j''}d)^2] + i\delta_{bj}]} \\ & \times\frac{(k_{1\perp}d - \delta_{j2}q_\perp d)}{[\frac{\hbar\omega_s}{E_0} + \frac{\hbar\omega_q}{E_0} + \beta_j\{[(k_{n_j}d)^2 - (k_{n_j'}d)^2] - (-1)^j[(q_\perp d)^2 - 2k_{1\perp}q_\perp d^2]\} + i\delta_{aj}]}\end{aligned} \tag{124}$$

$$M_{jp}^{ph} = M_{j\perp}^{ph} + iM_{jz}^{ph} \quad \text{and}$$

$$k_{1\perp}d = \beta_2q_\perp d + \sqrt{G^2 - \beta_1\beta_2(q_\perp d)^2} \tag{125}$$

with

$$\delta_{bj} = \frac{\Gamma_{bj}}{E_0} \qquad \sigma_{ph} = \frac{e^4\hbar|\hat{\mathbf{e}}_l \cdot \mathbf{p}_{cv}(0)|^2\eta(\omega_s)}{\pi\mu_0^2\mu_r^2c^4\eta(\omega_l)E_0^3\omega_l\omega_s}|C_F|^2 \quad (126)$$

and

$$G^2 = \frac{\hbar\omega_l - E_g}{E_0} - \frac{\hbar\omega_q}{E_0} - \frac{\hbar\omega_s}{E_0} - \beta_1(k_{n_1}d)^2 - \beta_2(k_{n_2}d)^2 \quad (127)$$

We have computed the scattering efficiency

$$\left(\frac{d^3}{V}\right)^2 \frac{1}{\sigma_{ph}}\left(\frac{d^2\sigma}{d\Omega\, d\omega_s}\right)_z \quad \text{vs} \quad \frac{\hbar\omega_s}{E_0}$$

As can be observed in Figure 6, in this case the one phonon-assisted dispersion does not produce new peaks associated with the emission of the phonon. This is due to the fact that the position and the intensity of these peaks, for the QW, depend on the phonon momentum, which takes continuous values [see Eq. (123)].

5. DIFFERENTIAL CROSS-SECTION FOR QUANTUM WELL WIRES (QWWs) AND FREESTANDING WIRES (FSWs)

The QWW geometry is cylindrical with circular cross-section of radius r_0 and length L. As explained, we consider a single conduction (valence) band, which is split into a subband system due to electron confinement within the structure:

$$\Psi_{l_jn_j} = (2\pi L)^{-\frac{1}{2}} \exp(-i(l_j\phi + k_jz))u_j \times \begin{cases} A_jJ_{l_j}\left(x_{l_jn_j}\dfrac{r}{r_0}\right) & \text{for } r \le r_0 \\ B_jK_{l_j}\left(y_{l_jn_j}\dfrac{r}{r_0}\right) & \text{for } r > r_0 \end{cases} \quad (128)$$

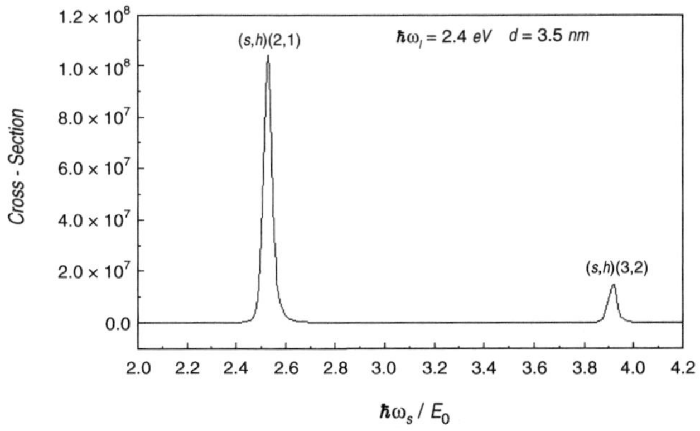

Figure 6. The emission spectra of the electron Raman scattering in QW for the polarization $\overline{Z}(\mathbf{e}_l, S_z)Z$. The transitions assisted by one phonon are considered.

with

$$B_j = \frac{J_{l_j}(x_{l_jn_j})}{K_{l_j}(y_{l_jn_j})}A_j \quad \text{and}$$

$$A_j = \left\{\left\|J_{l_j}(x_{l_jn_j})\right\|^2 + \left[\frac{J_{l_j}(x_{l_jn_j})}{K_{l_j}(y_{l_jn_j})}\right]^2 \left\|K_{l_j}(y_{l_jn_j})\right\|^2\right\}^{-\frac{1}{2}} \quad (129)$$

where

$$y_{l_jn_j} = \sqrt{\frac{2\mu_{jo}r_0^2V_j}{\hbar^2} - \frac{\mu_{jo}}{\mu_{ji}}x_{l_jn_j}}$$

J_{l_j} and K_{l_j} are the Bessel and modified Bessel functions, and μ_{jo} is the effective mass in the surrounding medium. The states are described by the quantum numbers: $l_j = 0, 1, \ldots;\ n_j = 1, 2, \ldots;\ k_j$. Taking as matching boundary conditions the continuity of the function Ψ and the current $(1/\mu_i)\frac{\partial\Psi}{\partial r}$ at the interface, the confinement energies are determined by the secular equation

$$\mu_{ji}y_{l_jn_j}K'_{l_j}(y_{l_jn_j})J_{l_j}(x_{l_jn_j}) = \mu_{jo}x_{l_jn_j}J'_{l_j}(x_{l_jn_j})K_{n_j}(y_{l_jn_j}) \quad (130)$$

The total energies are

$$E_j = \frac{\hbar^2}{2\mu_{ji}}\left[k_j^2 + \left(\frac{x_{l_jn_j}}{r_0}\right)^2\right] \quad (131)$$

The FSW can be modeled as a QWW with a infinitely high potential barrier; the wave function becomes

$$\Psi_j = \left[\sqrt{\pi L}r_0J'_{l_j}(x_{l_jn_j})\right]^{-1} J_{l_j}\left(x_{l_jn_j}\frac{r}{r_0}\right)\exp(-i(l_j\phi + k_jz))u_j \quad (132)$$

Here, $x_{l_jn_j}$ denotes the zeros of Bessel functions, $J_{l_j}(x_{l_jn_j}) = 0$. Thus, there will be an infinite number of states completely confined in the interior of the FSW. As we can observe the bands are unfolded in a system of subbands, l_j, which themselves are also unfolded in a system of subbands, n_j, and all of them are parabolic.

5.1. Raman Scattering Intensities

From Eqs. (128), (4), and (5), applying the same procedure, it can be obtained that

$$\langle a|\widehat{H}_{jl}|i\rangle = \frac{|e|}{\mu_0}\sqrt{\frac{2\pi\hbar}{V\omega_l}}\hat{\mathbf{e}}_l \cdot \mathbf{p}_{cv}(0) \times \begin{cases} T^0_{l'_1, n'_1, n_2}\delta_{l'_1, l_2}\delta_{k'_1, -k_2} & \text{for } j = 1 \\ T^0_{l'_2, n'_2, n_1}\delta_{l'_2, l_1}\delta_{-k'_2, k_1} & \text{for } j = 2 \end{cases} \quad (133)$$

where

$$T^v_{l, n_a, n_b} = A_aA_b\int_0^{r_0} J_{l+v}\left(x_{ln_a}\frac{r}{r_0}\right)J_{l+v}\left(x_{ln_b}\frac{r}{r_0}\right)r\,dr + B_aB_b\int_{r_0}^{\infty} K_{l+v}\left(y_{ln_a}\frac{r}{r_0}\right)K_{l+v}\left(y_{ln_b}\frac{r}{r_0}\right)r\,dr \quad (134)$$

with a (b) indicating the electron (hole) state. This term is similar to the obtained for the QW and it represents a interband transition, due to the emission of a photon and the creation of an EHP; but, in contrast with the QW, here the transition takes place between subbands with the same quantum number l_j.

In the solution of the second matrix element is convenient with the introduction of a new base,

$$\hat{\mathbf{e}}_\pm = \hat{\mathbf{e}}_x \pm i\hat{\mathbf{e}}_y \quad \hat{\mathbf{e}}_r \pm i\hat{\mathbf{e}}_\phi = e^{\mp i\phi}(\hat{\mathbf{e}}_x \pm i\hat{\mathbf{e}}_y) = e^{\mp i\phi}\hat{\mathbf{e}}_\pm \tag{135}$$

Thus, considering

$$\hat{\mathbf{p}} = -i\hbar\nabla \qquad \nabla = \hat{\mathbf{e}}_r\frac{\partial}{\partial r} + \frac{\hat{\mathbf{e}}_\phi}{r}\frac{\partial}{\partial\phi} + \hat{\mathbf{e}}_z\frac{\partial}{\partial z} \tag{136}$$

and after several operations and calculations we obtain that for $r < r_0$

$$\begin{aligned}\nabla J_l\left(x_{ln}\frac{r}{r_0}\right)e^{-i(l\phi+kz)} &= e^{-i(l\phi+kz)}\left\{-ikJ_l\left(x_{ln}\frac{r}{r_0}\right)\hat{\mathbf{e}}_z - \frac{x_{ln}}{2r_0}J_{l+1}\left(x_{ln}\frac{r}{r_0}\right)(\hat{\mathbf{e}}_x + i\hat{\mathbf{e}}_y)\right.\\ &\quad\left.+\frac{x_{ln}}{2r_0}J_{l-1}\left(x_{ln}\frac{r}{r_0}\right)(\hat{\mathbf{e}}_x - i\hat{\mathbf{e}}_y)\right\}\end{aligned} \tag{137}$$

and for $r > r_0$

$$\begin{aligned}\nabla K_l\left(y_{ln}\frac{r}{r_0}\right)e^{-i(l_j\phi+k_jz)} &= e^{-i(l_j\phi+k_jz)}\left\{-ik_zK_l\left(y_{ln}\frac{r}{r_0}\right)\hat{\mathbf{e}}_z - \frac{y_{ln}}{2r_0}K_{l+1}\left(y_{ln}\frac{r}{r_0}\right)(\hat{\mathbf{e}}_x + i\hat{\mathbf{e}}_y)\right.\\ &\quad\left.-\frac{y_{ln}}{2r_0}K_{l-1}\left(y_{ln}\frac{r}{r_0}\right)(\hat{\mathbf{e}}_x - i\hat{\mathbf{e}}_y)\right\}\end{aligned} \tag{138}$$

With these results it is obtained that

$$\begin{aligned}\langle f|\widehat{H}_{js}|b\rangle &= i\hbar(-1)^j\frac{|e|}{\mu_j}\sqrt{\frac{2\pi\hbar}{V\omega_s}}\\ &\quad\times\left\{\frac{(\hat{\mathbf{e}}_s\cdot\hat{\mathbf{e}}_+)}{r_0}Y_{j1}\delta_{n'_j,n_j-1} + \frac{(\hat{\mathbf{e}}_s\cdot\hat{\mathbf{e}}_-)}{r_0}Y_{j2}\delta_{n'_j,n_j+1}\right.\\ &\quad\left.-ik'_jY_{j3}(\hat{\mathbf{e}}_s\cdot\hat{\mathbf{e}}_z)\delta_{l_jl'_j}\right\}\delta_{k_jk'_j}\end{aligned} \tag{139}$$

where

$$\begin{aligned}Y_{jp} &= \frac{1}{2}\left[(-1)^pA_aA_b\int_0^{r_0}J_{l_a}\left(x_{l_an_a}\frac{r}{r_0}\right)J_{l_b}\left(x_{l_bn_b}\frac{r}{r_0}\right)r\,dr\right.\\ &\quad\left.-B_aB_b\int_{r_0}^{\infty}K_{l_a}\left(y_{l_an_a}\frac{r}{r_0}\right)K_{l_b}\left(y_{l_bn_b}\frac{r}{r_0}\right)r\,dr\right]\\ &\quad\text{for } p = 1,\ 2\end{aligned} \tag{140}$$

and

$$\begin{aligned}Y_{j3} &= A_aA_b\int_0^{r_0}J_{l_a}\left(x_{l_an_a}\frac{r}{r_0}\right)J_{l_b}\left(x_{l_bn_b}\frac{r}{r_0}\right)r\,dr\\ &\quad+B_aB_b\int_{r_0}^{\infty}K_{l_a}\left(y_{l_an_a}\frac{r}{r_0}\right)K_{l_b}\left(y_{l_bn_b}\frac{r}{r_0}\right)r\,dr\end{aligned} \tag{141}$$

Here we obtain three independent polarizations.

After substitution of Eqs. (133) and (139) into Eqs. (3) it can be obtained that

$$\begin{aligned}M_1 &= -A_0\beta_1\delta_{k_1k'_2}\sum_{l'_1,n'_1}\frac{T^0_{l'_1,n'_1,n_2}\delta_{l'_1,l_2}}{\frac{\hbar\omega_s}{E_0}+\beta_1\left[(x_{l_1n_1})^2-(x_{l'_1n'_1})^2\right]+i\delta_{a1}}\\ &\quad\times\left[\frac{(\hat{\mathbf{e}}_s\cdot\hat{\mathbf{e}}_+)}{r_0}Y_{11}\delta_{l'_1,l_1-1}+\frac{(\hat{\mathbf{e}}_s\cdot\hat{\mathbf{e}}_-)}{r_0}Y_{12}\delta_{l'_1,l_1+1}\right.\\ &\quad\left.+ik'_1Y_{13}(\hat{\mathbf{e}}_s\cdot\hat{\mathbf{e}}_z)\delta_{l'_1l_1}\right]\end{aligned} \tag{142}$$

$$\begin{aligned}M_2 &= A_0\beta_2\delta_{k_1k'_2}\sum_{l'_2,n'_2}\frac{T^0_{l'_2,n'_2,n_1}\delta_{l'_2,l_1}}{\frac{\hbar\omega_s}{E_0}+\beta_2\left[(x_{l_2n_2})^2-(x_{l'_2n'_2})^2\right]+i\delta_{a2}}\\ &\quad\times\left[\frac{(\hat{\mathbf{e}}_s\cdot\hat{\mathbf{e}}_+)}{r_0}Y_{21}\delta_{l'_2,l_2-1}+\frac{(\hat{\mathbf{e}}_s\cdot\hat{\mathbf{e}}_-)}{r_0}Y_{22}\delta_{l'_2,l_2+1}\right.\\ &\quad\left.+ik'_2Y_{23}(\hat{\mathbf{e}}_s\cdot\hat{\mathbf{e}}_z)\delta_{l'_2l_2}\right]\end{aligned} \tag{143}$$

After substitution of Eqs. (142) and (143) into Eqs. (2) and (1) we obtain

$$\frac{d^2\sigma}{d\Omega\,d\omega_s} = \sum_{p=0}^{3}\left[\frac{d^2\sigma}{d\Omega\,d\omega_s}\right]_p \tag{144}$$

$$\begin{aligned}\left[\frac{d^2\sigma}{d\Omega\,d\omega_s}\right]_0 &= \frac{9}{4\sqrt{2}}\sigma_0\left(\frac{\hbar\omega_s}{E_0}\right)^2\sum_{l_1,n_1,l_2,n_2}\{\mathrm{Re}(M_{12})\mathrm{Re}(M_{21})\\ &\quad+\mathrm{Re}(M_{22})\mathrm{Re}(M_{11})+\mathrm{Im}(M_{12})\mathrm{Im}(M_{21})\\ &\quad+\mathrm{Im}(M_{22})\mathrm{Im}(M_{11})\}\sqrt{\frac{g^2+\sqrt{g^4+\delta_f^2}}{\sqrt{g^4+\delta_f^2}}}S_0\end{aligned} \tag{145}$$

$$\begin{aligned}\left[\frac{d^2\sigma}{d\Omega\,d\omega_s}\right]_p &= \frac{9}{4\sqrt{2}}\sigma_0\left(\frac{\hbar\omega_s}{E_0}\right)^2\sum_{l_1,n_1,l_2,n_2}|M_{1p}+M_{2p}|^2\\ &\quad\times S_p\begin{cases}\sqrt{\dfrac{g^2+\sqrt{g^4+\delta_f^2}}{\sqrt{g^4+\delta_f^2}}} & \text{for } p\neq 3\\ \sqrt{g^2+\sqrt{g^4+\delta_f^2}} & \text{for } p = 3\\ p = 1,\ 2,\ 3\end{cases}\end{aligned} \tag{146}$$

where

$$\begin{aligned}M_{1p} &= -\beta_1\sum_{l'_1,n'_1}\frac{T^0_{l'_1,n'_1,n_2}\delta_{l'_1,l_2}}{\frac{\hbar\omega_s}{E_0}+\beta_1\left[(x_{l_1n_1})^2-(x_{l'_1n'_1})^2\right]+i\delta_{a1}}\\ &\quad\times Y_{1p}\begin{cases}\delta_{l'_1,l_1-1} & \text{for } p = 1\\ \delta_{l'_1,l_1+1} & \text{for } p = 2\\ \delta_{l'_1,l_1} & \text{for } p = 3\end{cases}\end{aligned} \tag{147}$$

$$M_{2p} = \beta_2 \sum_{l'_2, n'_2} \frac{T^0_{l'_2, n'_2, n_1} \delta_{l'_2, l_1}}{\frac{\hbar\omega_s}{E_0} + \beta_2[(x_{l_2 n_2})^2 - (x_{l'_2 n'_2})^2] + i\delta_{a2}} \times Y_{2p} \begin{cases} \delta_{l'_2, l_2-1} & \text{for } p = 1 \\ \delta_{l'_2, l_2+1} & \text{for } p = 2 \\ \delta_{l'_2, l_2} & \text{for } p = 3 \end{cases} \quad (148)$$

and

$$S_0 = |\hat{\mathbf{e}}_s \cdot \hat{\mathbf{e}}_+||\hat{\mathbf{e}}_s \cdot \hat{\mathbf{e}}_-| \qquad S_p \begin{cases} |\hat{\mathbf{e}}_s \cdot \hat{\mathbf{e}}_+|^2 & p = 1 \\ |\hat{\mathbf{e}}_s \cdot \hat{\mathbf{e}}_-|^2 & p = 2 \\ |\hat{\mathbf{e}}_s \cdot \hat{\mathbf{e}}_z|^2 & p = 3 \end{cases} \quad (149)$$

Besides

$$g^2 = \frac{\hbar\omega_l - E_g}{E_0} - \frac{\hbar\omega_s}{E_0} - \beta_1(x_{l_1 n_1})^2 - \beta_2(x_{l_2 n_2})^2 \quad (150)$$

for the QWW, FSW, and QD

$$E_0 = \frac{\hbar^2}{2\mu_r r_0^2} \quad (151)$$

In the deduction of Eqs. (146) and (145) we have taken

$$\delta(E_f - E_i) \to \frac{1}{\pi} \frac{\Gamma_f}{(E_f - E_i)^2 + \Gamma_f^2} \quad \text{and} \quad \delta_f = \frac{\Gamma_f}{E_0} \quad (152)$$

assuming a finite lifetime, Γ_f, for the EHP in the final state. We have four contributions; three of them (S_1, S_2, S_3) are independent of each other, but the contribution of the polarization S_0 can be neglected if it is compared with the contribution S_1 or S_2.

The differential cross-section for a semiconductor QWW presents singular peaks for secondary radiation frequency ω_s such that

$$(s, e)(l_1, n_1, l'_1, n'_1) = \beta_1[(x_{l'_1 n'_1})^2 - (x_{l_1 n_1})^2] \quad (153)$$

$$(s, h)(l_2, n_2, l'_2, n'_2) = \beta_2[(x_{l'_2 n'_2})^2 - (x_{l_2 n_2})^2] \quad (154)$$

$$(l_1, n_1, l_2, n_2) = \frac{\hbar\omega_l - E_g}{E_0} - \beta_1(x_{l_1 n_1})^2 - \beta_2(x_{l_2 n_2})^2 \quad (155)$$

As can be seen from Eqs. (153) or (154), these frequencies correspond to electron transitions connecting the subbands edges for a process involving just the conduction or just the valence band (i.e., intraband transitions). The values of frequencies reported in Eqs. (153) and (154) are associated with the emission of a photon by the electron or hole in intersubband transitions. The selection rules are $l_1 = l_2 \pm 1$ for the electron transitions and $l_2 = l_1 \pm 1$ for the hole transitions; this selection rule corresponds to $(\hat{\mathbf{e}}_s \cdot \hat{\mathbf{e}}_\mp)$ polarization and $l_2 = l_1$ to $(\hat{\mathbf{e}}_s \cdot \hat{\mathbf{e}}_z)$ polarization. The singularities involved in Eq. (155) are dependent on ω_l and correspond to interband transitions.

We have computed the scattering efficiency

$$\frac{1}{\sigma_0} \frac{d^2\sigma}{d\Omega\, d\omega_s} \quad \text{vs} \quad \frac{\hbar\omega_s}{E_0}$$

In Figure 7 we show emission spectra for the QWW and the QD with the polarization $\overline{Z}(\hat{\mathbf{e}}_l, S_2)Z$. If we compare the spectrum obtained for the QWW with $r_0 = 2.0$ nm with the one obtained for $r_0 = 2.5$ nm a displacement is observed to the right of peaks $(s, h)(1, 1, 0, 1)$ and $(s, e)(1, 1, 0, 1)$, which is of bigger magnitude for the peaks related to the electron, because for the finite potential barriers at the interface the states depend on r_0 and the effective mass for the QWW and QD [see Eqs. (130), (131), (187), and (188)]. The decrease of the intensity of the peaks can also be appreciated. This is due to the fact that with the increase of the radius the difference of energy of the subbands increases too.

This figure also shows that the peaks related to the electron transitions are bigger than those related to the hole transitions, because the hole possesses a bigger effective mass than the electron and also a lesser band offset. This can also be observed in the dispersion assisted or not assisted by phonons in QDs and in QWWs.

5.2. Raman Scattering Intensities with One Phonon-Assisted Transition for QWWs

From Eqs. (128) and (50) it can be obtained that

$$\langle b|\widehat{H}_{jph}|a\rangle = C_{nm} J^{l'_j n'_j}_{l''_j n''_j} \delta_{l'_j, l''_j - n} \delta_{k'_j, k''_j} \quad (156)$$

being

$$J^{l'_j n'_j}_{l''_j n''_j} = \sqrt{\frac{V\rho}{r_0}} \left\{ A'_j A''_j \int_0^{r_0} J_{l'_j}\left(x_{l'_j n'_j} \frac{r}{r_0}\right) F^{\text{QWW}}_{nm}(r) \times J_{l''_j}\left(x_{l''_j n''_j} \frac{r}{r_0}\right) r\, dr + B'_j B''_j \int_{r_0}^{\infty} K_{l'_j}\left(y_{l'_j n'_j} \frac{r}{r_0}\right) \times F^{\text{QWW}}_{nm}(r) K_{l''_j}\left(y_{l''_j n''_j} \frac{r}{r_0}\right) r\, dr \right\} \quad (157)$$

As can be observed the emission of the phonon of energy $\hbar\omega_q$ follows the rule of selection $l'_j - l''_j = n$. This means that the transition takes place between subbands with quantum numbers separated by n, which is a phonon quantum number.

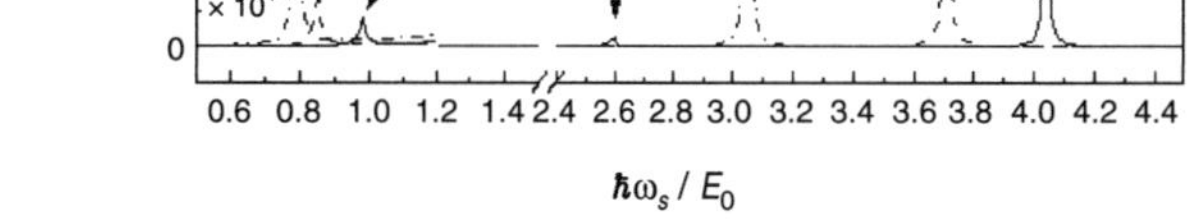

Figure 7. The emission spectra of the electron Raman scattering in QWWs and QDs for the polarization $\overline{Z}(\mathbf{e}_l, S_2)Z$. The left side of the figure was multiplied by 10^4.

After substitution of Eqs. (133), (156), and (139) into Eq. (10) it can be obtained that

$$M_1 = A_0 C_{nm}\beta_1 \sum_{l_1'',n_1'',l_1',n_1'} \frac{T^0_{l_1'',n_1'',n_2}\delta_{l_1'',l_2}\delta_{k_1'',-k_2}\delta_{k_1,k_1''}\delta_{k_1',k_1''}}{\left[\frac{\hbar\omega_s}{E_0}+\beta_1\left[(x_{l_1n_1})^2-(x_{l_1'n_1'})^2\right]+i\delta_{b1}\right]}$$
$$\times \frac{J^{l_1'n_1'}_{l_1''n_1''}\delta_{l_1',l_1''-n}}{\left[\frac{\hbar\omega_s}{E_0}+\frac{\hbar\omega_q}{E_0}+\beta_1\left[(x_{l_1n_1})^2-(x_{l_1''n_1''})^2\right]+i\delta_{a1}\right]}$$
$$\times\left[\frac{(\hat{\mathbf{e}}_s\cdot\hat{\mathbf{e}}_+)}{r_0}Y_{11}\delta_{l_1',l_1-1}+\frac{(\hat{\mathbf{e}}_s\cdot\hat{\mathbf{e}}_-)}{r_0}Y_{12}\delta_{l_1',l_1+1}\right.$$
$$\left.+ik_1'Y_{13}(\hat{\mathbf{e}}_s\cdot\hat{\mathbf{e}}_z)\delta_{l_1',l_1}\right] \tag{158}$$

$$M_2 = A_0 C_{nm}\beta_2 \sum_{l_2'',n_2'',l_2',n_2'} \frac{T^0_{l_2'',n_2'',n_1}\delta_{l_2'',l_1}\delta_{-k_2'',k_1}\delta_{k_2,k_2''}\delta_{k_2'k_2''}}{\frac{\hbar\omega_s}{E_0}+\beta_2\left[(x_{l_2n_2})^2-(x_{l_2'n_2'})^2\right]+i\delta_{b2}}$$
$$\times \frac{J^{l_2'n_2'}_{l_2''n_2''}\delta_{l_2',l_2''-n}}{\left[\frac{\hbar\omega_s}{E_0}+\frac{\hbar\omega_q}{E_0}+\beta_2\left[(x_{l_2n_2})^2-(x_{l_2''n_2''})^2\right]+i\delta_{a2}\right]}$$
$$\times\left[\frac{(\hat{\mathbf{e}}_s\cdot\hat{\mathbf{e}}_+)}{r_0}Y_{21}\delta_{l_2',l_2-1}+\frac{(\hat{\mathbf{e}}_s\cdot\hat{\mathbf{e}}_-)}{r_0}Y_{22}\delta_{l_2',l_2+1}\right.$$
$$\left.+ik_2'Y_{23}(\hat{\mathbf{e}}_s\cdot\hat{\mathbf{e}}_z)\delta_{l_2',l2}\right] \tag{159}$$

If we substitute Eqs. (158) and (159) into Eqs. (2) and (1) we obtain

$$\frac{d^2\sigma}{d\Omega\, d\omega_s} = \sum_{p=0}^{3}\left[\frac{d^2\sigma}{d\Omega\, d\omega_s}\right]_p \tag{160}$$

$$\left[\frac{d^2\sigma}{d\Omega d\omega_s}\right]_0 = \frac{\sqrt{2}}{4}\sigma_{ph}\left(\frac{\hbar\omega_s}{E_0}\right)^2 \sum_{nm}\sum_{l_1,n_1,l_2,n_2}\{\mathrm{Re}(M_{12}^{ph})\mathrm{Re}(M_{21}^{ph})$$
$$+\mathrm{Re}(M_{22}^{ph})\mathrm{Re}(M_{11}^{ph})+\mathrm{Im}(M_{12}^{ph})\mathrm{Im}(M_{21}^{ph})$$
$$+\mathrm{Im}(M_{22}^{ph})\mathrm{Im}(M_{11}^{ph})\}\sqrt{\frac{G^2+\sqrt{G^4+\delta_f^2}}{\sqrt{G^4+\delta_f^2}}}S_0\delta_{n,0} \tag{161}$$

$$\left[\frac{d^2\sigma}{d\Omega\, d\omega_s}\right]_p = \frac{\sqrt{2}}{4}\sigma_{ph}\left(\frac{\hbar\omega_s}{E_0}\right)^2 \sum_{nm}\sum_{l_1,n_1,l_2,n_2}|M_{1p}^{ph}+M_{2p}^{ph}|^2$$
$$\times S_p\begin{cases}\sqrt{\dfrac{G^2+\sqrt{G^4+\delta_f^2}}{\sqrt{G^4+\delta_f^2}}} & \text{for } p\neq 3\\ \sqrt{G^2+\sqrt{G^4+\delta_f^2}} & \text{for } p=3\\ p=1,\,2,\,3 & \end{cases} \tag{162}$$

where

$$M_{1p}^{ph} = \beta_1\frac{\omega_L}{\omega_q}\sum_{l_1'',n_1'',l_1',n_1'}Y_{1p}\frac{J^{l_1'n_1'}_{l_1''n_1''}}{\left[\frac{\hbar\omega_s}{E_0}+\frac{\hbar\omega_q}{E_0}+\beta_1\left[(x_{l_1n_1})^2-(x_{l_1''n_1''})^2\right]+i\delta_{a1}\right]}$$
$$\times\frac{T^0_{l_1'',n_1'',n_2}\delta_{l_1'',l_2}\delta_{l_1',l_1''-n}}{\left[\frac{\hbar\omega_s}{E_0}+\beta_1\left[(x_{l_1n_1})^2-(x_{l_1'n_1'})^2\right]+i\delta_{b1}\right]}\begin{cases}\delta_{l_1',l_1-1} & \text{for } p=1\\ \delta_{l_1',l_1+1} & \text{for } p=2\\ \delta_{l_1',l_1} & \text{for } p=3\end{cases} \tag{163}$$

$$M_{2p}^{ph} = \beta_2\frac{\omega_L}{\omega_q}\sum_{l_2'',n_2'',l_2',n_2'}Y_{2p}\frac{J^{l_2'n_2'}_{l_2''n_2''}}{\left[\frac{\hbar\omega_s}{E_0}+\frac{\hbar\omega_q}{E_0}+\beta_2\left[(x_{l_2n_2})^2-(x_{l_2''n_2''})^2\right]+i\delta_{a2}\right]}$$
$$\times\frac{T^0_{l_2'',n_2'',n_1}\delta_{l_2'',l_1}\delta_{l_2',l_2''-n}}{\left[\frac{\hbar\omega_s}{E_0}+\beta_2\left[(x_{l_2n_2})^2-(x_{l_2'n_2'})^2\right]+i\delta_{b2}\right]}\begin{cases}\delta_{l_2',l_2-1} & \text{for } p=1\\ \delta_{l_2',l_2+1} & \text{for } p=2\\ \delta_{l_2',l_2} & \text{for } p=3\end{cases} \tag{164}$$

being

$$\delta_{aj} = \frac{\Gamma_{aj}}{E_0} \qquad \delta_{bj} = \frac{\Gamma_{bj}}{E_0} \tag{165}$$

and

$$G^2 = \frac{\hbar\omega_l - E_g}{E_0} - \frac{\hbar\omega_q}{E_0} - \frac{\hbar\omega_s}{E_0} - \beta_1(x_{l_1n_1})^2 - \beta_2(x_{l_2n_2})^2 \tag{166}$$

The differential cross-section for a semiconductor QWW presents singular peaks for secondary radiation frequency ω_s:

$$(s,e)(l_1,n_1,l_1',n_1') = \beta_1\left[(x_{l_1'n_1'})^2-(x_{l_1n_1})^2\right] \tag{167}$$
$$(s,h)(l_2,n_2,l_2',n_2') = \beta_2\left[(x_{l_2'n_2'})^2-(x_{l_2n_2})^2\right] \tag{168}$$
$$(f,e)(l_1,n_1,l_1'',n_1'') = \beta_1\left[(x_{l_1''n_1''})^2-(x_{l_1n_1})^2\right]-\frac{\hbar\omega_q}{E_0} \tag{169}$$
$$(f,h)(l_2,n_2,l_2'',n_2'') = \beta_2\left[(x_{l_2''n_2''})^2-(x_{l_2n_2})^2\right]-\frac{\hbar\omega_q}{E_0} \tag{170}$$
$$(l_1,n_1,l_2,n_2) = \frac{\hbar\omega_l - E_g}{E_0} - \frac{\hbar\omega_q}{E_0} - \beta_1(x_{l_1n_1})^2 - \beta_2(x_{l_2n_2})^2 \tag{171}$$

We can see that the Eqs. (167) and (168) represent the same frequencies as (153) and (154) and have the same selection rule $l_j'' = l_j \pm 1$. As can be seen from (169) or (170), these frequencies correspond to electron transitions connecting the subband edges for a process involving just the conduction or just the valence band (i.e., intraband transitions). The values of frequencies reported in (169) and (170) are associated with the emission of a phonon by the electron or hole in intersubband transitions. The selection rules are $l_1' = l_1 \pm n + 1$ for the electron transitions and $l_2' = l_2 \pm n + 1$ for the hole transitions; this selection rule corresponds to $(\hat{\mathbf{e}}_s\cdot\hat{\mathbf{e}}_\mp)$ polarization and $l_j' = l_j + n$ to $(\hat{\mathbf{e}}_s\cdot\hat{\mathbf{e}}_z)$ polarization. The singularities involved in (171) depend on ω_l and correspond to interband transitions.

In Figure 8 the emission spectra for the QWW and the FSW for different phonon modes with $n = 0$ and a polarization $\overline{Z}(\hat{\mathbf{e}}_l, S_2)Z$ are presented. As is shown, the intensity of the peaks in the mode $m = 1$ is bigger than $m = 2$ and this is bigger than $m = 3$; this means that we only need to consider the contribution of the first phonon modes in the differential cross-section. On the other hand, the magnitude of the

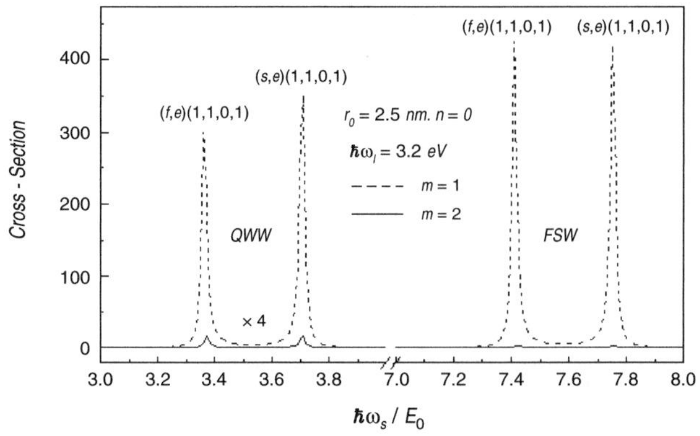

Figure 8. The emission spectra of the electron Raman scattering in QWW and FSW for the polarization $\overline{Z}(\mathbf{e}_l, S_2)Z$ for $n = 0$, varying the phonons mode of the system.

differential cross-section is bigger for the FSW than for the QWW, which is due to the fact that the boundary conditions make the form factor for the FSW bigger than for the QWW. A displacement to the right of peaks related to the emission of the phonon is also observed, in both spectra, due to the fact that $\hbar\omega_q$ diminishes with the increase of m [2].

In Figure 9 the emission spectra for a QWW with $n = 0$, $r_0 = 2.5$ nm, and $\hbar\omega_l = 3.2$ eV for the polarization $\overline{Z}(\hat{\mathbf{e}}_l, S_1)Z$ are presented. As can be observed if we compare with the polarization $\overline{Z}(\hat{\mathbf{e}}_l, S_2)Z$ (see Fig. 8) the change of the polarization affects the position of the peaks. This is due to the fact that the polarization change implies a change in the selection rules; that is to say, they change the subbands involved in the transitions, which also produces a decrease of the intensity of the DCS for the polarization $\overline{Z}(\hat{\mathbf{e}}_l, S_1)Z$. In this figure only singularities produced by holes transitions are shown, which is due to the fact that the number of electron states is very small (the electron tunneling is small), but its contribution increases with the increase of the radius. The decrease of the intensity of DCS for the polarization $\overline{Z}(\hat{\mathbf{e}}_l, S_1)Z$ is due to the fact that it is more probable that the electron or the hole makes a transition to a subband with a lesser l_j when emitting a photon.

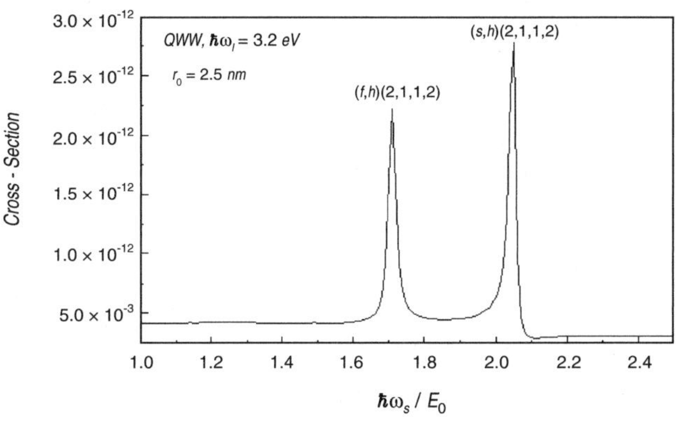

Figure 9. The emission spectra of the electron Raman scattering in QWW for the polarization $\overline{Z}(\mathbf{e}_l, S_1)Z$ for $n = 0$.

5.3. Raman Scattering Intensities with One Phonon-Assisted Transition for FSWs

From Eqs. (132) and (52) it can be obtained that

$$T^0_{l,\,n_a,\,n_b} = \delta_{n_a,\,n_b} \tag{172}$$

The aforementioned implies that electron and hole have the same quantum numbers,

$$Y_{j3} = \delta_{n_a,\,n_b},\ Y_{jp} = (-1)^p \frac{x_{l_j n_j} x_{l'_j n'_j}}{(x_{l'_j n'_j})^2 - (x_{l_j n_j})^2} \tag{173}$$

Then, by using Eqs. (52) and (157),

$$J^{l'_j n'_j}_{l''_j n''_j} = \sqrt{\frac{V\rho}{r_0}} \frac{1}{r_0^2 J'_{l'_j}(x_{l'_j n'_j}) J'_{l''_j}(x_{l''_j n''_j})} \times \int_0^{r_0} J_{l'_j}\left(x_{l'_j n'_j}\frac{r}{r_0}\right) F^{\mathrm{FSW}}_{nm}(r) J_{l'_j}\left(x_{l''_j n''_j}\frac{r}{r_0}\right) r\, dr \tag{174}$$

This means that in this case the creation of EHP is direct.

With these results Eqs. (158) and (159) remain in the form

$$M_1 = A_0 C_{nm} \beta_1 \sum_{l''_1, n''_1, l'_1, n'_1} \frac{\delta_{n''_1, n_2}\delta_{l''_1, l_2}\delta_{k''_1, -k_2}\delta_{k_1, k''_1}\delta_{k'_1, k''_1}}{\left[\frac{\hbar\omega_s}{E_0} + \beta_1[(x_{l_1 n_1})^2 - (x_{l'_1 n'_1})^2] + i\delta_{b1}\right]} \times \frac{J^{l'_1 n'_1}_{l''_1 n''_1}\delta_{l'_1, l''_1 - n}}{\left[\frac{\hbar\omega_s}{E_0} + \frac{\hbar\omega_q}{E_0} + \beta_1[(x_{l_1 n_1})^2 - (x_{l''_1 n''_1})^2] + i\delta_{a1}\right]} \times \left[\frac{(\hat{\mathbf{e}}_s\cdot\hat{\mathbf{e}}_+)}{r_0} Y_{11}\delta_{l'_1, l_1 - 1} + \frac{(\hat{\mathbf{e}}_s\cdot\hat{\mathbf{e}}_-)}{r_0} Y_{12}\delta_{l'_1, l_1 + 1} + ik'_1 Y_{13}(\hat{\mathbf{e}}_s\cdot\hat{\mathbf{e}}_z)\delta_{l'_1, l_1}\right] \tag{175}$$

$$M_2 = A_0 C_{nm} \beta_2 \sum_{l''_2, n''_2, l'_2, n'_2} \frac{\delta_{n''_2, n_1}\delta_{l''_2, l_1}\delta_{-k''_2, k_1}\delta_{k_2, k''_2}\delta_{k'_2 k''_2}}{\frac{\hbar\omega_s}{E_0} + \beta_2[(x_{l_2 n_2})^2 - (x_{l'_2 n'_2})^2] + i\delta_{b2}} \times \frac{J^{l'_2 n'_2}_{l''_2 n''_2}\delta_{l'_2, l''_2 - n}}{\left[\frac{\hbar\omega_s}{E_0} + \frac{\hbar\omega_q}{E_0} + \beta_2[(x_{l_2 n_2})^2 - (x_{l''_2 n''_2})^2] + i\delta_{a2}\right]} \times \left[\frac{(\hat{\mathbf{e}}_s\cdot\hat{\mathbf{e}}_+)}{r_0} Y_{21}\delta_{l'_2, l_2 - 1} + \frac{(\hat{\mathbf{e}}_s\cdot\hat{\mathbf{e}}_-)}{r_0} Y_{22}\delta_{l'_2, l_2 + 1} + ik'_2 Y_{23}(\hat{\mathbf{e}}_s\cdot\hat{\mathbf{e}}_z)\delta_{l'_2, l2}\right] \tag{176}$$

We just write the final results,

$$\frac{d^2\sigma}{d\Omega\, d\omega_s} = \sum_{p=0}^{3}\left[\frac{d^2\sigma}{d\Omega\, d\omega_s}\right]_p \tag{177}$$

$$\left[\frac{d^2\sigma}{d\Omega\, d\omega_s}\right]_0 = \frac{\sqrt{2}}{4}\sigma_{ph}\left(\frac{\hbar\omega_s}{E_0}\right)^2 \sum_{nm} \sum_{l_1, n_1, l_2, n_2} \{\mathrm{Re}(M^{ph}_{12})\mathrm{Re}(M^{ph}_{21}) + \mathrm{Re}(M^{ph}_{22})\mathrm{Re}(M^{ph}_{11}) + \mathrm{Im}(M^{ph}_{12})\mathrm{Im}(M^{ph}_{21}) + \mathrm{Im}(M^{ph}_{22})\mathrm{Im}(M^{ph}_{11})\} \sqrt{\frac{G^2 + \sqrt{G^4 + \delta_f^2}}{\sqrt{G^4 + \delta_f^2}}} S_0 \delta_{n,0} \tag{178}$$

$$\left[\frac{d^2\sigma}{d\Omega\, d\omega_s}\right]_p = \frac{\sqrt{2}}{4}\sigma_{ph}\left(\frac{\hbar\omega_s}{E_0}\right)^2 \sum_{nm}\sum_{l_1,n_1,l_2,n_2}\left|M_{1p}^{ph}+M_{2p}^{ph}\right|^2$$
$$\times S_p \begin{cases} \sqrt{\dfrac{G^2+\sqrt{G^4+\delta_f^2}}{\sqrt{G^4+\delta_f^2}}} & \text{for } p \neq 3 \\ \sqrt{G^2+\sqrt{G^4+\delta_f^2}} & \text{for } p = 3 \\ p = 1,\ 2,\ 3 \end{cases} \tag{179}$$

where

$$M_{1p}^{ph} = \beta_1 \frac{\omega_L}{\omega_q} \sum_{l_1'',n_1'',l_1',n_1'} Y_{1p} \frac{J_{l_1''n_1''}^{l_1'n_1'}}{\left[\frac{\hbar\omega_s}{E_0}+\frac{\hbar\omega_q}{E_0}+\beta_1[(x_{l_1n_1})^2-(x_{l_1''n_1''})^2]+i\delta_{a1}\right]}$$
$$\times \frac{\delta_{l_1'',l_2}\,\delta_{n_1'',n_2}\,\delta_{l_1',l_1''-n}}{\left[\frac{\hbar\omega_s}{E_0}+\beta_1[(x_{l_1n_1})^2-(x_{l_1'n_1'})^2]+i\delta_{b1}\right]} \begin{cases} \delta_{l_1',l_1-1} & \text{for } p=1 \\ \delta_{l_1',l_1+1} & \text{for } p=2 \end{cases} \tag{180}$$

$$M_{2p}^{ph} = \beta_2 \frac{\omega_L}{\omega_q} \sum_{l_2'',n_2'',l_2',n_2'} Y_{2p} \frac{J_{l_2''n_2''}^{l_2'n_2'}}{\left[\frac{\hbar\omega_s}{E_0}+\frac{\hbar\omega_q}{E_0}+\beta_2[(x_{l_2n_2})^2-(x_{l_2''n_2''})^2]+i\delta_{a2}\right]}$$
$$\times \frac{\delta_{l_2'',l_1}\,\delta_{n_2'',n_1}\,\delta_{l_2',l_2''-n}}{\left[\frac{\hbar\omega_s}{E_0}+\beta_2[(x_{l_2n_2})^2-(x_{l_2'n_2'})^2]+i\delta_{b2}\right]} \begin{cases} \delta_{l_2',l_2-1} & \text{for } p=1 \\ \delta_{l_2',l_2+1} & \text{for } p=2 \end{cases} \tag{181}$$

$$M_{13}^{ph} = \beta_1 \frac{\omega_L}{\omega_q} \frac{J_{l_2n_2}^{l_1n_1}\,\delta_{l_1,l_2-n},\,\delta_{n_1',n_2}}{\left[\frac{\hbar\omega_s}{E_0}+i\delta_{b1}\right]\left[\frac{\hbar\omega_s}{E_0}+\frac{\hbar\omega_q}{E_0}+\beta_1[(x_{l_1n_1})^2-(x_{l_2n_2})^2]+i\delta_{a1}\right]} \tag{182}$$

$$M_{23}^{ph} = \beta_2 \frac{\omega_L}{\omega_q} \frac{J_{l_1n_1}^{l_2n_2}\,\delta_{l_2,l_1-n}}{\left[\frac{\hbar\omega_s}{E_0}+i\delta_{b2}\right]\left[\frac{\hbar\omega_s}{E_0}+\frac{\hbar\omega_q}{E_0}+\beta_2[(x_{l_2n_2})^2-(x_{l_1n_1})^2]+i\delta_{a2}\right]} \tag{183}$$

The differential cross-section for a semiconductor FSW presents singular peaks for the secondary radiation frequency ω_s. This could be calculated with Eqs. (167), (168), (169), (170), and (171), and it has the same properties.

In Figure 8 also observed are a displacement to the right and more intensity of peaks for the FSW in comparison with the QWW, which is due to the different conditions of confining. The increment of the potential barrier is the cause of this displacement and also the increasing number of energy levels [117]. It also can be observed that the differences of intensity of DCS for $n = 0$ and the other phonons made in the FSW are bigger than in the QWW, because the FSW and the QWW have different dispersion laws for the phonon modes.

In Figure 10 the emission spectra for a FSW considering only $n = 0$ and for a polarization $\overline{Z}(\hat{\mathbf{e}}_l, S_2)Z$ are presented. The emission spectra for $r_0 = 2.5$ nm and $r_0 = 3.0$ nm can be observed. Then, a comparison between both radii illustrates that an increase of the radius causes a displacement of the peaks (associated to the emission of a phonon) toward the left. This fact is justified when the dispersion law and Eqs. (169) and (170) are analyzed; the displacement is influenced fundamentally by the energy of formation of the electron–hole pair, E_0. Also, the increase of the radius produces a decrease of the intensity of the DCS (fundamentally the transitions that involve the holes) due to the fact that the energy levels diminish. This also is observed in the one phonon-assisted dispersion in any nanostructures.

In Figure 11 the emission spectra of a FSW with $n = 0$, $r_0 = 2.5$ nm, and $\hbar\omega_l = 10.0$ eV for the polarization

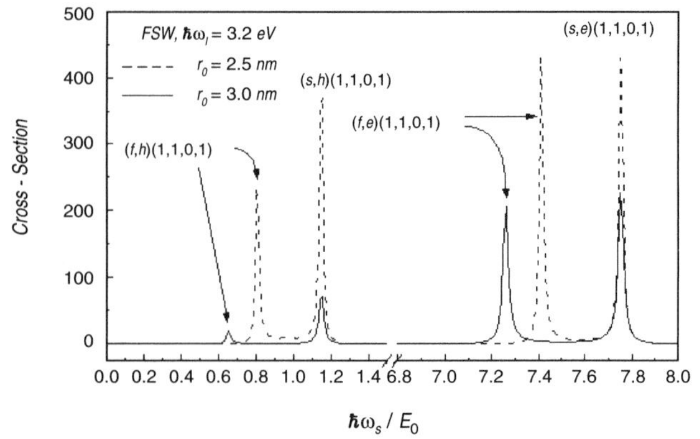

Figure 10. The emission spectra of the one phonon-assisted electron Raman scattering in FSW for the polarization $\overline{Z}(\mathbf{e}_l, S_2)Z$ for $n = 0$ and several radii.

$\overline{Z}(\hat{\mathbf{e}}_l, S_1)Z$ are presented. As can be observed if we compare with the polarization $\overline{Z}(\hat{\mathbf{e}}_l, S_2)Z$ (see Fig. 10), also in this case the change of polarization affects the position of the peaks; however, a bigger increase of the energy of incident radiation is required in order for the bands involved in transitions to be reached by the EHP. This is due to the fact that a change in the polarization changes the selection rules, which it means a change in the involved subbands in the transitions. This is observed in the QW and QWW.

If we compare with Figure 9 we can observe that in the case of the QWW the increment of energy cannot be very large, because of the finite height of the barrier. Thus, is enough to consider that the energy of the incident photon be equal to the gap of the external material in order for all the energy levels to be reached by the EHP. Also, due to the fact that the number of energy levels is finite, the selection rules will not always be completed, so that the peaks will not appear. In the case of the FSW peaks always can be obtained, because in this case we consider that the potential barriers are infinite and there exists an infinite number of energy levels.

In Figure 12 the emission spectrum is also observed for the FSW with $r_0 = 2.5$ nm, $\hbar\omega_l = 10.0$ eV, with polarization $\overline{Z}(\hat{\mathbf{e}}_l, S_2)Z$ for $n = 0$ and $n = 1$. It can be observed that the peak associated with the emission of a photon does not vary

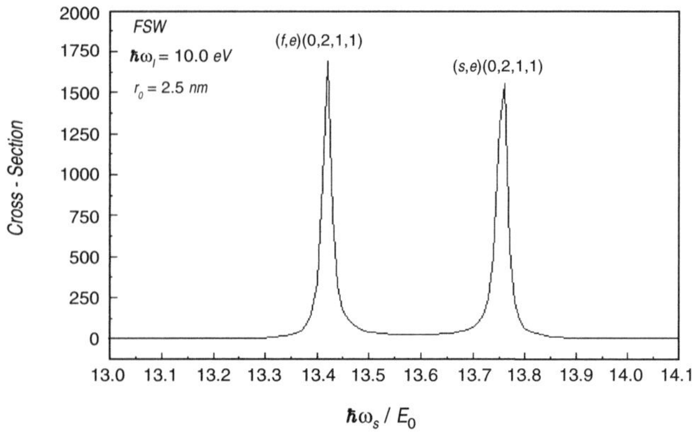

Figure 11. The emission spectra of the electron Raman scattering in FSW for the polarization $\overline{Z}(\mathbf{e}_l, S_1)Z$ for $n = 0$.

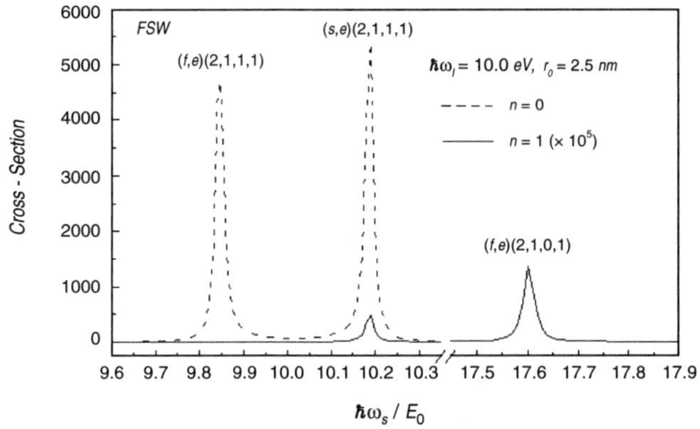

Figure 12. The emission spectra of the electron Raman scattering in FSW for the polarization $\overline{Z}(\mathbf{e}_l, S_2)Z$ for $n = 0$ and $n = 1$.

its position with the increase of the order of the phonon mode. The peak associated to the emission of a phonon (see Fig. 10) varies its position in accordance with the selection rule $l_2 = l_1 + n + 1$ [see Eqs. (180) and (181)]. It can also be appreciated that the intensity of the DCS diminishes with the increase of n. From the physical point of view this is due to the fact that for the emission of a phonon a transition intrasubband is less favorable for $n = 0$ than a transition serial intersubbands, for $n = 1$. This demonstrates that the fundamental contribution carries out the phonon mode $n = 0$.

6. DIFFERENTIAL CROSS-SECTION FOR QUANTUM DOT (QD)

We will consider a spherical QD of radius r_0 with single conduction (valence) band. This band is split in a system of subbands due to a complete confinement of the electrons in all directions. The solution of the Schrödinger equation, in the envelope function approximation, leads to

$$\Psi_{l_j m_j n_j} = Y_{n_j}^{m_j}(\theta, \varphi) u_j \begin{cases} \dfrac{A_j}{\sqrt{r}} J_{l_j+\frac{1}{2}}\left(x_{l_j n_j} \dfrac{r}{r_0}\right) & \text{for } r \le r_0 \\ \dfrac{B_j}{\sqrt{r}} K_{l_j+\frac{1}{2}}\left(y_{l_j n_j} \dfrac{r}{r_0}\right) & \text{for } r > r_0 \end{cases} \tag{184}$$

with

$$B_j = \frac{J_{l_j+\frac{1}{2}}(x_{l_j n_j})}{K_{l_j+\frac{1}{2}}(y_{l_j n_j})} A_j \tag{185}$$

and

$$A_j = \left\{ \| J_{l_j+\frac{1}{2}}(x_{l_j n_j}) \|^2 + \left[\frac{J_{l_j+\frac{1}{2}}(x_{l_j n_j})}{K_{l_j+\frac{1}{2}}(y_{l_j n_j})} \right]^2 \| K_{l_j+\frac{1}{2}}(y_{l_j n_j}) \|^2 \right\}^{-\frac{1}{2}} \tag{186}$$

In Eq. (184) $Y_{n_1}^{m_1}(\theta, \varphi)$ are the spherical harmonics. We use spherical coordinates r, θ, φ.

The confinement energies are determined by the secular equation

$$\mu_{ji} y_{l_j n_j} K'_{n_j+\frac{1}{2}}(y_{l_j n_j}) J_{l_j+\frac{1}{2}}(x_{l_j n_j}) = \mu_{jo} x_{l_j n_j} J'_{l_j+\frac{1}{2}}(x_{l_j n_j}) K_{n_j+\frac{1}{2}}(y_{l_j n_j}) \tag{187}$$

The energy is given by

$$E_j = \frac{\hbar^2}{2\mu_{ji} r_0^2} (x_{l_j n_j})^2 \tag{188}$$

where $x_{l_j n_j}$ denotes the zeros of Eq. (187). In this case the bands remain completely quantized, due to the total confinement to which they are subjected; however, due to the symmetry, this is $2l_j + 1$ times degenerate.

6.1. Raman Scattering Intensities

From Eqs. (184), (4), and (5) it can be obtained that

$$\langle a|\widehat{H}_{jl}|i\rangle = \frac{|e|}{\mu_0} \sqrt{\frac{2\pi\hbar}{V\omega_l}} \hat{\mathbf{e}}_l \cdot \mathbf{p}_{cv}(0) \begin{cases} T^{\frac{1}{2}}_{l'_1, n'_1, n_2} \delta_{l'_1, l_2} \delta_{m'_1, m_2} & \text{for } j=1 \\ T^{\frac{1}{2}}_{l'_2, n'_2, n_1} \delta_{l'_2, l_1} \delta_{m'_2, m_1} & \text{for } j=2 \end{cases} \tag{189}$$

This is the same as that obtained for the QWW. It represents an interband transition due to the emission of a photon and the creation of an EHP, but additionally the transition takes place between subbands with the same quantum number l_j and m_j.

In the solution of the second matrix element the introduction of a new base is convenient:

$$\hat{\mathbf{e}}_\pm = \frac{1}{\sqrt{2}} (\hat{\mathbf{e}}_x \pm i\hat{\mathbf{e}}_y) \quad \text{and} \quad \hat{\mathbf{e}}_z \tag{190}$$

$$\begin{aligned} \hat{\mathbf{e}}_r &= \sin\theta\cos\phi\hat{\mathbf{e}}_x + \sin\theta\sin\phi\hat{\mathbf{e}}_y + \cos\theta\hat{\mathbf{e}}_z \\ &= \frac{\sin\theta}{\sqrt{2}} (e^{-i\phi}\hat{\mathbf{e}}_+ + e^{i\phi}\hat{\mathbf{e}}_-) + \cos\theta\hat{\mathbf{e}}_z \end{aligned} \tag{191}$$

Then, by considering

$$\hat{\mathbf{p}} = -i\hbar\nabla \qquad \nabla = \hat{\mathbf{e}}_r \frac{\partial}{\partial r} + \frac{i}{\hbar r} \hat{\mathbf{e}}_r \times \widehat{\mathbf{L}} \tag{192}$$

applying the base given in Eq. (190) and introducing the operators

$$\widehat{L}_\pm = \frac{1}{\sqrt{2}} (\widehat{L}_x \pm \widehat{L}_y) \tag{193}$$

and considering that

$$\begin{aligned} \hat{\mathbf{e}}_r \times \widehat{\mathbf{L}} = {} & i\frac{\sin\theta}{\sqrt{2}} (e^{i\phi}\widehat{L}_- - e^{-i\phi}\widehat{L}_+) \hat{\mathbf{e}}_z \\ & + i\left(\cos\theta\widehat{L}_+ - \frac{\sin\theta}{\sqrt{2}} e^{-i\phi}\widehat{L}_z\right)\hat{\mathbf{e}}_- \\ & + i\left(\frac{\sin\theta}{\sqrt{2}} e^{-i\phi}\widehat{L}_z - \cos\theta\widehat{L}_-\right)\hat{\mathbf{e}}_+ \end{aligned} \tag{194}$$

we obtain for the matrix element

$$\langle f|\widehat{H}_{js}|b\rangle = \langle f|\widehat{H}_{js}|b\rangle_1 + \langle f|\widehat{H}_{js}|b\rangle_2 + \langle f|\widehat{H}_{js}|b\rangle_3 \quad (195)$$

where

$$\langle f|\widehat{H}_{js}|b\rangle_p = i\hbar(-1)^j \frac{|e|}{r_0\mu_j}\sqrt{\frac{2\pi\hbar}{V\omega_s}}\left[\Pi_{j1p}\delta_{l'_j,l_j+1} + \Pi_{j2p}\delta_{l'_j,l_j-1}\right] \times I_{l_jn_j}^{l'_jn'_j}\begin{cases}\delta_{m'_j,m_j+1}(\hat{\mathbf{e}}_s\cdot\hat{\mathbf{e}}_+) & \text{for } p=1\\ \delta_{m'_j,m_j-1}(\hat{\mathbf{e}}_s\cdot\hat{\mathbf{e}}_-) & \text{for } p=2\\ \delta_{m'_j,m_j}(\hat{\mathbf{e}}_s\cdot\hat{\mathbf{e}}_z) & \text{for } p=2\end{cases} \quad (196)$$

and

$$\Pi_{j11} = -\sqrt{\frac{(l_j+m_j+2)(l_j+m_j+1)}{2(2l_j+1)(2l_j+3)}} \qquad \Pi_{j21} = \sqrt{\frac{(l_j-m_j-1)(l_j-m_j)}{2(2l_j-1)(2l_j+1)}}$$
$$\Pi_{j12} = \sqrt{\frac{(l_j-m_j+2)(l_j-m_j+1)}{2(2l_j+1)(2l_j+3)}} \qquad \Pi_{j22} = -\sqrt{\frac{(l_j+m_j-1)(l_j+m_j)}{2(2l_j-1)(2l_j+1)}}$$
$$\Pi_{j13} = \sqrt{\frac{(l_j-m_j+1)(l_j+m_j+1)}{(2l_j+1)(2l_j+3)}} \qquad \Pi_{j23} = \sqrt{\frac{(l_j-m_j)(l_j+m_j)}{(2l_j-1)(2l_j+1)}} \quad (197)$$

$$I_{l_jn_j}^{l'_jn'_j} = A_jA'_j\int_0^{r_0} J_{l_j+\frac{1}{2}}\left(x_{l_jn_j}\frac{r}{r_0}\right)J_{l'_j+\frac{1}{2}}\left(x_{l'_jn'_j}\frac{r}{r_0}\right)r\,dr + B_jB'_j\int_{r_0}^{\infty} K_{l_j+\frac{1}{2}}\left(y_{l_jn_j}\frac{r}{r_0}\right)K_{l'_j+\frac{1}{2}}\left(y_{l'_jn'_j}\frac{r}{r_0}\right)r\,dr \quad (198)$$

This shows that in the emission of a photon, $\hbar\omega_s$, three independent polarizations appear in which the transitions occur between states that fulfill that $l'_j, l_j \pm 1$.

After substitution of Eqs. (189) and (195) into Eqs. (3) we will have that

$$M_1 = -A_0\beta_1\sum_{l'_1,m'_1,n'_1}\frac{\left[\Pi_{11p}\delta_{l'_1,l_1+1}+\Pi_{12p}\delta_{l'_1,l_1-1}\right]\delta_{l'_1,l_2}}{\frac{\hbar\omega_s}{E_0}+\beta_1\left[(x_{l_1n_1})^2-(x_{l'_1n'_1})^2\right]+i\delta_{a1}} \times I_{l_1n_1}^{l'_1n'_1}T_{l'_1n'_1,n_2}^{\frac{1}{2}}\delta_{m'_1,m_2}\begin{cases}(\hat{\mathbf{e}}_s\cdot\hat{\mathbf{e}}_+)\delta_{m'_1,m_1+1} & \text{for } p=1\\ (\hat{\mathbf{e}}_s\cdot\hat{\mathbf{e}}_-)\delta_{m'_1,m_1-1} & \text{for } p=2\\ (\hat{\mathbf{e}}_s\cdot\hat{\mathbf{e}}_z)\delta_{m'_1,m_1} & \text{for } p=3\end{cases} \quad (199)$$

$$M_2 = A_0\beta_2\sum_{l'_2,m'_2,n'_2}\frac{\left[\Pi_{21p}\delta_{l'_2,l_2+1}+\Pi_{22p}\delta_{l'_2,l_2-1}\right]\delta_{l'_2,l_1}}{\frac{\hbar\omega_s}{E_0}+\beta_2\left[(x_{l_2n_2})^2-(x_{l'_2n'_2})^2\right]+i\delta_{a2}} \times I_{l_2n_2}^{l'_2n'_2}T_{l'_2n'_2,n_1}^{\frac{1}{2}}\delta_{m'_2,m_1}\begin{cases}(\hat{\mathbf{e}}_s\cdot\hat{\mathbf{e}}_+)\delta_{m'_2,m_2+1} & \text{for } p=1\\ (\hat{\mathbf{e}}_s\cdot\hat{\mathbf{e}}_-)\delta_{m'_2,m_2-1} & \text{for } p=2\\ (\hat{\mathbf{e}}_s\cdot\hat{\mathbf{e}}_z)\delta_{m'_2,m_2} & \text{for } p=3\end{cases} \quad (200)$$

After substitution of Eqs. (199) and (200) into Eqs. (2) and (1) we obtain

$$\frac{d^2\sigma}{d\Omega\,d\omega_s} = \sum_{p=0}^{3}\left[\frac{d^2\sigma}{d\Omega\,d\omega_s}\right]_p \quad (201)$$

$$\left[\frac{d^2\sigma}{d\Omega\,d\omega_s}\right]_0 = \frac{27}{4}\sigma_0\left(\frac{\hbar\omega_s}{E_0}\right)^2\delta_f \times \sum_{l_1,n_1,m_1}\sum_{l_2,n_2,m_2}\{\mathrm{Re}(M_{11})\mathrm{Re}(M_{22}) + \mathrm{Re}(M_{12})\mathrm{Re}(M_{21}) + \mathrm{Im}(M_{11})\mathrm{Im}(M_{22}) + \mathrm{Im}(M_{12})\mathrm{Im}(M_{21})\}\frac{S_o}{g^4+\delta_f^2} \quad (202)$$

$$\left[\frac{d^2\sigma}{d\Omega\,d\omega_s}\right]_p = \frac{27}{4}\sigma_0\left(\frac{\hbar\omega_s}{E_0}\right)^2\delta_f\sum_{l_1,n_1,m_1}\sum_{l_2,n_2,m_2}\left|M_{1p}+M_{2p}\right|^2 \times \frac{S_p}{g^4+\delta_f^2} \qquad p = 1,\ 2,\ 3 \quad (203)$$

where

$$M_{1p} = -\beta_1\sum_{l'_1,m'_1,n'_1}\frac{\left[\Pi_{11p}\delta_{l'_1,l_1+1}+\Pi_{12p}\delta_{l'_1,l_1-1}\right]\delta_{l'_1,l_2}}{\frac{\hbar\omega_s}{E_0}+\beta_1\left[(x_{l_1n_1})^2-(x_{l'_1n'_1})^2\right]+i\delta_{a1}} \times I_{l_1n_1}^{l'_1n'_1}T_{l'_1n'_1,n_2}^{\frac{1}{2}}\delta_{m'_1,m_2}\begin{cases}\delta_{m'_1,m_1+1} & \text{for } p=1\\ \delta_{m'_1,m_1-1} & \text{for } p=2\\ \delta_{m'_1,m_1} & \text{for } p=3\end{cases} \quad (204)$$

$$M_{2p} = \beta_2\sum_{l'_2,m'_2,n'_2}\frac{\left[\Pi_{21p}\delta_{l'_2,l_2+1}+\Pi_{22p}\delta_{l'_2,l_2-1}\right]\delta_{l'_2,l_1}}{\frac{\hbar\omega_s}{E_0}+\beta_2\left[(x_{l_2n_2})^2-(x_{l'_2n'_2})^2\right]+i\delta_{a2}} \times I_{l_2n_2}^{l'_2n'_2}T_{l'_2n'_2,n_1}^{\frac{1}{2}}\delta_{m'_2,m_1}\begin{cases}\delta_{m'_2,m_2+1} & \text{for } p=1\\ \delta_{m'_2,m_2-1} & \text{for } p=2\\ \delta_{m'_2,m_2} & \text{for } p=3\end{cases} \quad (205)$$

The contribution of the polarization S_0 can be neglected if it is compared with the contribution S_1 or S_2.

The differential cross-section for a semiconductor QD presents singular peaks for secondary radiation frequency ω_s such that

$$(s,e)(l_1,n_1,l'_1,n'_1) = \beta_1\left[(x_{l'_1n'_1})^2-(x_{l_1n_1})^2\right] \quad (206)$$

$$(s,h)(l_2,n_2,l'_2,n'_2) = \beta_1\left[(x_{l'_2n'_2})^2-(x_{l_2n_2})^2\right] \quad (207)$$

$$(l_1,n_1,l_2,n_2) = \frac{\hbar\omega_l-E_g}{E_0} - \beta_1(x_{l_1n_1})^2 - \beta_2(x_{l_2n_2})^2 \quad (208)$$

As can be seen from Eqs. (206) or (207) these frequencies correspond to electron transitions connecting the subbands edges for a process involving just the conduction or just the valence band (i.e., intraband transitions). The values of frequencies in Eqs. (206) and (207) are associated with the emission of a photon by the electron or hole intersubband transitions. The selection rules are $l_1 = l_2 \pm 1$ for the electron transitions and $l_2 = l_1 \pm 1$ for the hole transitions. The singularities involved in Eq. (208) depend on ω_l and correspond to interband transitions.

If the spectra of the QWW are compared with those obtained for the QD with $r_0 = 2.5$ nm (see Fig. 7), it can be observed that the peaks are displaced to the right. This is due to the fact that the subbands of the QD are more separated that those of the QWW; the intensity of the peaks of QD is also smaller, since the QD has a smaller state density.

6.2. Raman Scattering Intensities with One Phonon-Assisted Transition

From Eqs. (184) and (59) it can be obtained that

$$\langle b|\widehat{H}_{jph}|a\rangle = r_0 C_F \sqrt{\frac{4\pi}{3}} II^{l'_j n'_j}_{l''_j n''_j} \Lambda^{l'_j n'_j}_{l''_j n''_j} \tag{209}$$

being

$$\begin{aligned} II^{l'_j n'_j}_{l''_j n''_j} &= A'_j A''_j \int_0^{r_0} J_{l'_j+\frac{1}{2}}\left(x_{l'_j n'_j}\frac{r}{r_0}\right)\Phi_{l,n}(r) J_{l''_j+\frac{1}{2}}\left(x_{l''_j n''_j}\frac{r}{r_0}\right) r\,dr \\ &\quad + B'_j B''_j \int_{r_0}^{0} K_{l'_j+\frac{1}{2}}\left(x_{l'_j n'_j}\frac{r}{r_0}\right)\Phi_{l,n}(r) \\ &\quad \times K_{l''_j+\frac{1}{2}}\left(x_{l''_j n''_j}\frac{r}{r_0}\right) r\,dr \end{aligned} \tag{210}$$

$$\begin{aligned} \Lambda^{l'_j n'_j}_{l''_j n''_j} &= \int Y^{m'_j}_{l'_j} Y^{m}_{l} Y^{m''_j}_{l''_j}\, d\Omega \\ &= (-1)^{l'_j}\left[\frac{(2l'_j+1)(2l+1)(2l''_j+1)}{4\pi}\right]^{\frac{1}{2}} \Sigma_1 \Sigma_2 \end{aligned} \tag{211}$$

and Σ are the $3j$ symbols given by

$$\Sigma_1 = \begin{pmatrix} l'_j & l & l''_j \\ -m'_j & m & m''_j \end{pmatrix} \quad \text{and} \quad \Sigma_2 = \begin{pmatrix} l'_j & l & l''_j \\ 0 & 0 & 0 \end{pmatrix} \tag{212}$$

which are different from zero if they fulfill the following selection rules:

1. $l'_j + l + l''_j$ is even,
2. $-m'_j + m_p + m''_j = 0$.

By applying the properties of the $3j$ symbols we can demonstrate that for $l = 0$ the transitions obey the selection rule $l'_j = l''_j$; that is to say, there will be intrasubband transitions. For $l = 1$ we have that $l'_j = l''_j \pm 1$; that is to say, there will be intersubband transitions. As it happens for the QWW and the FSW, the increase of l implies that the differential cross-section diminishes. This shows that the fundamental contribution is carried out by the phonon modes $l = 0$.

After substitution in Eq. (10) we obtain

$$\begin{aligned} M_1 &= \frac{A_0 r_0 C_F}{E_0}\sqrt{\frac{4\pi}{3}}\beta_1 \\ &\quad \times \sum_{l'_2, m'_2, n'''_{22}} \sum_{l''_2, m''_2, n''_2} \frac{I^{l'_1 n'_1}_{l_1 n_1} T^{\frac{1}{2}}_{l''_1 n''_1, n_2}}{\frac{\hbar\omega_s}{E_0} + \beta_1[(x_{l_1 n_1})^2 - (x_{l'_1 n'_1})^2] + i\delta_{b1}} \\ &\quad \times \frac{[\Pi_{11p}\delta_{l'_1, l_1+1} + \Pi_{12p}\delta_{l'_1, l_1-1}]\delta_{l''_1, l_2}}{\frac{\hbar\omega_s}{E_0} + \frac{\hbar\omega_q}{E_0} + \beta_1[(x_{l_1 n_1})^2 - (x_{l''_1 n''_1})^2] + i\delta_{a1}} II^{l'_1 n'_1}_{l''_1 n''_1} \Lambda^{l'_1 n'_1}_{l''_1 n''_1} \\ &\quad \times [(\hat{\mathbf{e}}_s \cdot \hat{\mathbf{e}}_+)\delta_{m'_1, m_1+1} + (\hat{\mathbf{e}}_s \cdot \hat{\mathbf{e}}_-)\delta_{m'_1, m_1-1} \\ &\quad + (\hat{\mathbf{e}}_s \cdot \hat{\mathbf{e}}_z)\delta_{m'_1, m_1}]\delta_{m''_1, m_2} \end{aligned} \tag{213}$$

$$\begin{aligned} M_2 &= \frac{A_0 r_0 C_F}{E_0}\sqrt{\frac{4\pi}{3}}\beta_2 \\ &\quad \times \sum_{l'_2, m'_2, n'''_{22}} \sum_{l''_2, m''_2, n''_2} \frac{I^{l'_2 n'_2}_{l_2 n_2} T^{\frac{1}{2}}_{l''_2 n''_2, n_1}}{\frac{\hbar\omega_s}{E_0} + \beta_2[(x_{l_2 n_2})^2 - (x_{l'_2 n'_2})^2] + i\delta_{b2}} \\ &\quad \times \frac{[\Pi_{21p}\delta_{l'_2, l_2+1} + \Pi_{22p}\delta_{l'_2, l_2-1}]\delta_{l''_2, l_1}}{\frac{\hbar\omega_s}{E_0} + \frac{\hbar\omega_q}{E_0} + \beta_2[(x_{l_2 n_2})^2 - (x_{l''_2 n''_2})^2] + i\delta_{a2}} II^{l'_2 n'_2}_{l''_2 n''_2} \Lambda^{l'_2 n'_2}_{l''_2 n''_2} \\ &\quad \times [(\hat{\mathbf{e}}_s \cdot \hat{\mathbf{e}}_+)\delta_{m'_2, m_2+1} + (\hat{\mathbf{e}}_s \cdot \hat{\mathbf{e}}_-)\delta_{m'_2, m_2-1} \\ &\quad + (\hat{\mathbf{e}}_s \cdot \hat{\mathbf{e}}_z)\delta_{m'_2, m_2}]\delta_{m''_2, m_1} \end{aligned} \tag{214}$$

Finally

$$\frac{d^2\sigma}{d\Omega\, d\omega_s} = \sum_{p=0}^{3}\left[\frac{d^2\sigma}{d\Omega\, d\omega_s}\right]_p \tag{215}$$

$$\begin{aligned} \left[\frac{d^2\sigma}{d\Omega\, d\omega_s}\right]_0 &= \frac{8}{3}\sigma_{ph}\left(\frac{\hbar\omega_s}{E_0}\right)^2 \delta_f \\ &\quad \times \sum_{l,n,m} \sum_{l_1, n_1, m_1} \sum_{l_2, n_2, m_2} \{\mathrm{Re}(M^{ph}_{11})\mathrm{Re}(M^{ph}_{22}) \\ &\quad + \mathrm{Re}(M^{ph}_{12})\mathrm{Re}(M^{ph}_{21}) + \mathrm{Im}(M^{ph}_{11})\mathrm{Im}(M^{ph}_{11}) \\ &\quad + \mathrm{Im}(M^{ph}_{12})\mathrm{Im}(M^{ph}_{21})\} \frac{S_0}{G^4 + \delta_f^2}\delta_{m,0} \end{aligned} \tag{216}$$

$$\begin{aligned} \left[\frac{d^2\sigma}{d\Omega\, d\omega_s}\right]_p &= \frac{8}{3}\sigma_{ph}\left(\frac{\hbar\omega_s}{E_0}\right)^2 \delta_f \\ &\quad \times \sum_{l,n,m} \sum_{l_1, n_1, m_1} \sum_{l_2, n_2, m_2} |M^{ph}_{1p} + M^{ph}_{2p}|^2 \\ &\quad \times \frac{S_p}{G^4 + \delta_f^2} \quad p = 1, 2, 3 \end{aligned} \tag{217}$$

where

$$\begin{aligned} M^{ph}_{1p} &= \beta_1 \sum_{l'_2, m'_2, n'''_{22}} \sum_{l''_2, m''_2, n''_2} \frac{[\Pi_{11p}\delta_{l'_1, l_1+1} + \Pi_{12p}\delta_{l'_1, l_1-1}]\delta_{l''_1, l_2}}{\frac{\hbar\omega_s}{E_0} + \frac{\hbar\omega_q}{E_0} + \beta_1[(x_{l_1 n_1})^2 - (x_{l''_1 n''_1})^2] + i\delta_{a1}} \\ &\quad \times \frac{II^{l'_1 n'_1}_{l''_1 n''_1}\Lambda^{l'_1 n'_1}_{l''_1 n''_1} I^{l'_1 n'_1}_{l_1 n_1} T^{\frac{1}{2}}_{l''_1 n''_1, n_2}}{\frac{\hbar\omega_s}{E_0} + \beta_1[(x_{l_1 n_1})^2 - (x_{l'_1 n'_1})^2] + i\delta_{b1}} \begin{cases} \delta_{m'_1, m_1+1} & \text{for } p = 1 \\ \delta_{m'_1, m_1-1} & \text{for } p = 2 \\ \delta_{m'_1, m_1} & \text{for } p = 3 \end{cases} \end{aligned} \tag{218}$$

$$\begin{aligned} M^{ph}_{2p} &= \beta_2 \sum_{l'_2, m'_2, n'''_{22}} \sum_{l''_2, m''_2, n''_2} \frac{[\Pi_{21p}\delta_{l'_2, l_2+1} + \Pi_{22p}\delta_{l'_2, l_2-1}]\delta_{l''_2, l_1}}{\frac{\hbar\omega_s}{E_0} + \frac{\hbar\omega_q}{E_0} + \beta_2[(x_{l_2 n_2})^2 - (x_{l''_2 n''_2})^2] + i\delta_{a2}} \\ &\quad \times \frac{II^{l'_2 n'_2}_{l''_2 n''_2}\Lambda^{l'_2 n'_2}_{l''_2 n''_2} I^{l'_2 n'_2}_{l_2 n_2} T^{\frac{1}{2}}_{l''_2 n''_2, n_1}}{\frac{\hbar\omega_s}{E_0} + \beta_2[(x_{l_2 n_2})^2 - (x_{l'_2 n'_2})^2] + i\delta_{b2}} \begin{cases} \delta_{m'_2, m_2+1} & \text{for } p = 1 \\ \delta_{m'_2, m_2-1} & \text{for } p = 2 \\ \delta_{m'_2, m_2} & \text{for } p = 3 \end{cases} \end{aligned} \tag{219}$$

The differential cross-section for a semiconductor QWW and FSW presents singular peaks for secondary radiation frequency ω_s:

$$(s, e)(l_1, n_1, l'_1, n'_1) = \beta_1[(x_{l'_1 n'_1})^2 - (x_{l_1 n_1})^2] \tag{220}$$

$$(s,h)(l_2,n_2,l_2',n_2') = \beta_1\left[(x_{l_2'n_2'})^2 - (x_{l_2n_2})^2\right] \quad (221)$$

$$(f,e)(l_1,n_1,l_1'',n_1'') = \beta_1\left[(x_{l_1''n_1''})^2 - (x_{l_1n_1})^2\right] - \frac{\hbar\omega_q}{E_0} \quad (222)$$

$$(f,h)(l_2,n_2,l_2'',n_2'') = \beta_1\left[(x_{l_2''n_2''})^2 - (x_{l_2n_2})^2\right] - \frac{\hbar\omega_q}{E_0} \quad (223)$$

$$(l_1,n_1,l_2,n_2) = \frac{\hbar\omega_l - E_g}{E_0} - \frac{\hbar\omega_q}{E_0} - \beta_1(x_{l_1n_1})^2 - \beta_2(x_{l_2n_2})^2 \quad (224)$$

We can see that Eqs. (220) and (221) represent the same frequencies as Eqs. (206) and (207). As can be seen from Eq. (222) or (223) these frequencies correspond to electron transitions connecting the subbands edges for a process involving just the conduction or just the valence band (i.e., intraband transitions). The values of frequencies in Eqs. (222) and (223) are associated with the emission of a phonon by the electron or hole intersubband transitions. The selection rules depend on the phonon oscillation modes. The singularities involved in (171) depend on ω_l and correspond to interband transitions.

In Figure 13 the emission spectra for a QD when varying the phonon mode of the system, for $n = 0$ and a polarization $\overline{Z}(\hat{\mathbf{e}}_l, S_2)Z$, are presented. It can be observed that the peaks associated with the emission of a phonon are lightly displaced to the right. The aforementioned is justified when analyzing the law of dispersion of the photon mode (see [1, 2]). However, when the peaks are due to the emission of photons, this situation does not occur; that is to say, the position of peaks is invariable under any oscillation modes [see Eqs. (220) and (222)]. On the other hand, the intensity of the effective differential cross-section diminishes with the increase of the order of the phonon modes. This is due to the fact that the intensity of the oscillator diminishes with the increase of the order of the phonon mode. If we observe the total contribution (see Fig. 14), the sum of the contribution of all the modes, we cannot appreciate the peaks corresponding to each phonon mode. This is due to the fact that the difference of energy between the phonon modes is very small.

In Figure 15 we can observe the emission spectra for one phonon-assisted electron Raman scattering in QDs and

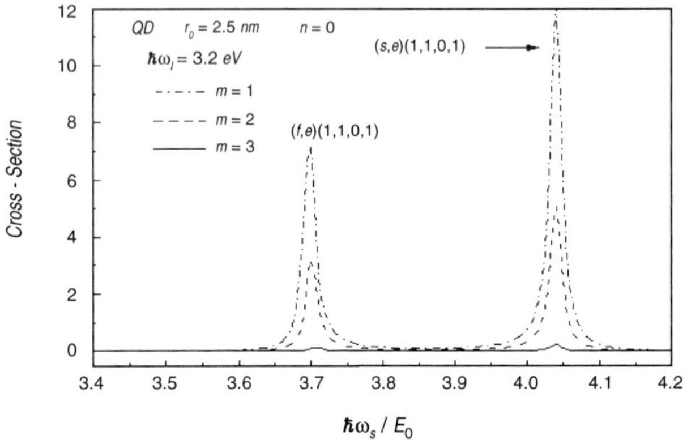

Figure 13. The emission spectra of the electron Raman scattering in QDs for the polarization $\overline{Z}(\mathbf{e}_l, S_2)Z$ for $n = 0$, varying the phonons mode of the system.

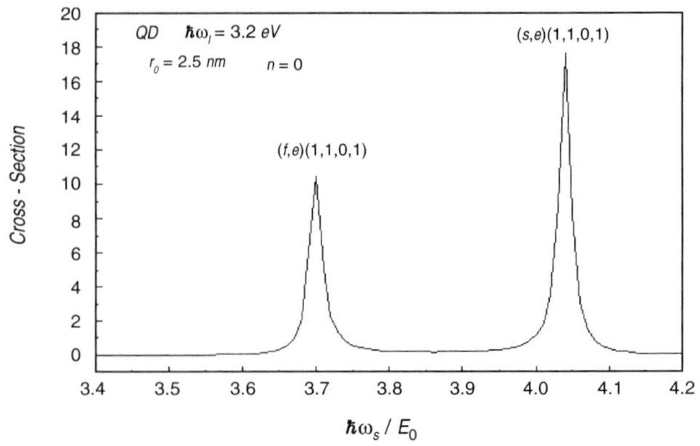

Figure 14. The emission spectra of the electron Raman scattering in QDs for the polarization $\overline{Z}(\mathbf{e}_l, S_2)Z$ for $n = 0$.

QWWs considering only the modes $n = 0$, for a polarization $\overline{Z}(\hat{\mathbf{e}}_l, S_2)Z$. If we compare these with those of Figure 7 we can appreciate the difference of intensity of the singularities; from here we can deduce that the electron–phonon interaction for this kind of dispersion is weak. It can be appreciated that the position of singularities does not change and peaks related to the emission of a phonon appear. If we compare with Figure 14 we can observe that the increase of the energy of the incident radiation produces a decrease of the intensity of the DCS.

In Figure 16 the emission spectra for a QD with $n = 0$, $r_0 = 2.5$ nm, and $\hbar\omega_l = 3.2$ eV for the polarizations $\overline{Z}(\hat{\mathbf{e}}_l, S_1)Z$ and $\overline{Z}(\hat{\mathbf{e}}_l, S_3)Z$ are presented. It can be observed that the polarization change does not affect the position of the peaks, but it only causes a decrease of the intensity of the DCS for $\overline{Z}(\hat{\mathbf{e}}_l, S_3)Z$. If we compare the spectrum for $\overline{Z}(\hat{\mathbf{e}}_l, S_1)Z$ with $\overline{Z}(\hat{\mathbf{e}}_l, S_2)Z$ we can observe that small changes take place in the intensity of the DCS. The aforementioned is due to the fact that only the polarization change implies changes in the degenerated states involved in the transitions.

If we compare the one phonon-assisted electron Raman scattering with the electron Raman scattering without the phonon participation, we can observe that, for the QD,

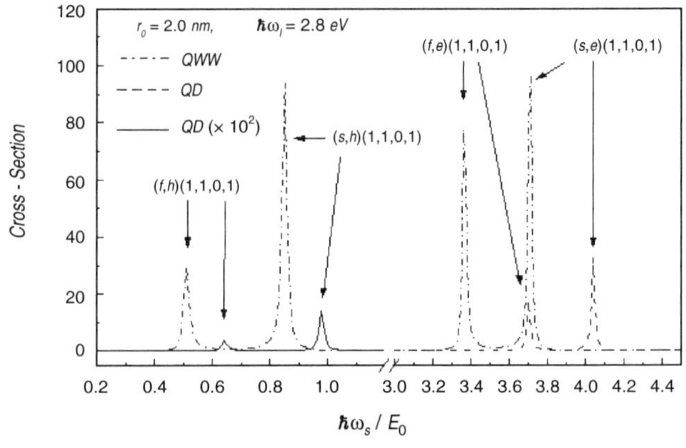

Figure 15. The emission spectra of the one phonon-assisted electron Raman scattering in QWWs and QDs for the polarization $\overline{Z}(\mathbf{e}_l, S_2)Z$ for $n = 0$. The left part of the curves that corresponding to the QD are multiplied by 10^2.

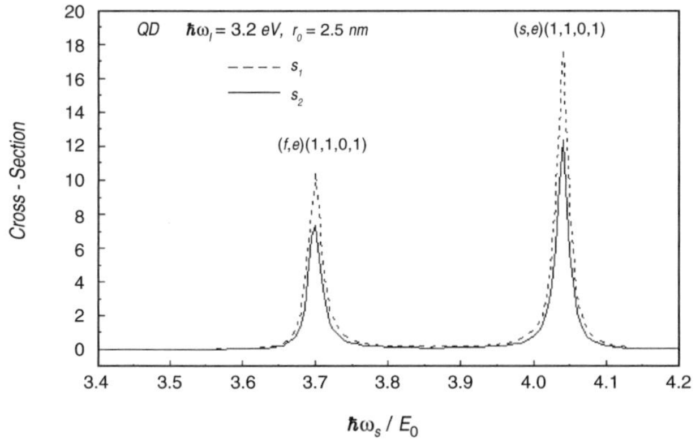

Figure 16. The emission spectra of the electron Raman scattering in QDs for the polarizations $\overline{Z}(\mathbf{e}_l, S_1)Z$ and $\overline{Z}(\mathbf{e}_l, S_3)Z$ for $n = 0$.

the quantum efficiency $\left(\frac{1}{V}\frac{d\sigma}{d\Omega}\right)$ of the system it is much less in the first one, but it is similar for the other nanostructures. From here we can conclude that the electron–phonon interaction is weak for this case. Therefore, the fundamental contributions to the spectrum are the singularities given by the emission of a photon of secondary radiation, which coincide in both dispersion types. The singularities described here correspond to intersubband transitions with smaller energy values than the gap. In this work we have showed that the EHP can be dispersed by any phonon mode, except the purely "transverse" [1, 2], because they are uncoupled from the electric field, because they do not interact via Fröhlich.

When we compare the dispersion assisted by one phonon in the three systems we can notice that for the QW the peaks corresponding to the emission of the phonon cannot be appreciated, which is due to the fact that their positions depend on the phonon and electron momenta, which vary in a very large range, so that the peaks cannot be observed. This same occurs with the QWW and the FSW because their contribution is very small in comparison with the QW; however, we cannot observe it because we worked with particles of $q_z = 0$.

The number of singularities in the case of QD and QWW will be determined by the radius and the energy of the incident radiation. We know that radius determines the number of subbands appearing, but we have supposed that QWs and FSWs have potential barriers of infinite height; therefore the number of subbands in this case is infinite. The energy of the incident radiation determines the number of singularities that should complete the law of conservation of the energy.

Finally, in the present work we have applied a simplified model for the electronic structure of system. In a more realistic case we should consider coupled band structure using a calculation model like that of Luttinger–Kohn or the Kane model. We also assumed an infinite potential barrier for the electron at the FSW interface. A calculation assuming a finite barrier would be better, but it is also possible to introduce a certain redefined effective mass for the infinite barrier case leading to the correct energy levels for the electrons and holes. The aforementioned assumptions would lead to better results but entail more complicated calculations. However, within the limits of our simple model we are able to account for the essential physical properties of the discussed problem. The fundamental features of the DCS, as described here, should not change very much in the real QWW case. It can be easily proved that the singular peaks in the DCS will be present irrespective of the model used for the subband structure and may be determined for the values of $\hbar\omega_s$ equal to the energy difference between two subbands: $\hbar\omega_s^{e(h)} = E_\alpha^{e(h)} - E_\beta^{e(h)}$, where $E_\alpha^{e(h)}(E_\beta^{e(h)})$ are the respective electron (hole) energies in the subbands. Similarly, we shall have a steplike dependence in the DCS for $\hbar\omega_l = \hbar\omega_s + \hbar\omega + E_g + E_\alpha + E_\beta$. This type of dispersion gives information about the contribution of the electrons and holes of separate form, and of the electron–phonon interaction. It also allows one to determine the structure of phonon subbands.

7. ONE PHONON RESONANT RAMAN SCATTERING IN QUANTUM WIRE, FREESTANDING WIRE, AND QUANTUM DOTS

Nanometer-scale confinements of the band electrons (band holes) in semiconductor materials provide varieties of quantum phenomena, such as low-dimensional electron states (hole states), dynamics of carriers in the systems, and increased exciton binding energies.

In previous sections the one phonon-assisted electron Raman scattering via Fröhlich interaction has been presented. The investigations for quantum well wires, freestanding wires, and quantum dots appear in [14, 118].

Within the strong-confinement regime and disregarding the exciton contribution, the electron intermediate states were considered as uncorrelated electron–hole pairs.

The present section is devoted to a model of the first order resonant Raman scattering in QWW, FSW, and QDs, in which electron–hole correction effects via Coulomb interaction are not taken into account.

7.1. One Phonon Resonant Raman Scattering in Quantum Wires and Freestanding Wires

7.1.1. Model and Applied Theory

In these calculations, the QWW geometry is cylindrical with circular cross-section of radius r_0 and length L. As was explained, we consider again a single conduction (valence) band, which is split into a subband system due to electron confinement within the structure. The solution of the Schrödinger equation in the envelope function approximation has been obtained in the previous sections where the envelope function and the energy states were presented.

7.1.2. Raman Cross-Section

The differential Raman cross-section $d^2\sigma/d\omega_s\,d\Omega$ of a volume V per unit solid angle $d\Omega$ for incoming light of frequency ω_l and scattered light of frequency ω_s is given by Eq. (1).

In this case

$$W(\omega_s, \hat{\mathbf{e}}_s) = \frac{2\pi}{\hbar}\sum_f |M_{fi}|^2 \delta(\hbar\omega_l - \hbar\omega_s - \hbar\omega_{ph}) \tag{225}$$

being

$$M_{fi} = \sum_{a,b} \frac{\langle f|\widehat{H}_r^+|b\rangle\langle b|\widehat{H}_{ph}|a\rangle\langle a|\widehat{H}_r^-|i\rangle}{(\hbar\omega_l - E_a + i\Gamma_a)(\hbar\omega_s - E_b + i\Gamma_b)} \tag{226}$$

In Eq. (226) $|a\rangle$ and $|b\rangle$ are intermediate states with energies E_a and E_b, Γ_a and Γ_b are the corresponding lifetime broadening, and $\widehat{H}_r^{\pm}$ and $\widehat{H}_{ph}$ are the electron–photon and electron–phonon interaction Hamiltonians, respectively. The sign $+(-)$ corresponds to the emission (absorption) of a photon.

The initial state $|i\rangle$ of the system consists of an incident photon of frequency ω_l and polarization $\hat{\mathbf{e}}_l$. In this case we have a completely occupied valence band and unoccupied conduction band. The final state $|f\rangle$ of the system consists of a photon of frequency ω_s and an excited phonon of frequency ω_p. Following Eqs. (1) and (225) the Raman scattering efficiency per unit solid angle can be obtained as

$$\frac{1}{V}\frac{d\sigma}{d\Omega} = \frac{V(\omega_l - \omega_{ph})^2\eta(\omega_s)}{8\pi^3 c^4 \eta(\omega_l)}\frac{2\pi}{\hbar^2}\sum_f |M_{fi}(\omega_l - \omega_{ph}, \hat{\mathbf{e}}_s, \omega_l, \hat{\mathbf{e}}_l)|^2 \tag{227}$$

The operator $\widehat{H}_r^+$ is given by Eqs. (4) and (5). This operator describes the interaction of the EHP with the incident and secondary radiation field in the dipole approximation.

In previous sections we summarized some details concerning the optical phonons and electron–phonon interaction Hamiltonian for QWW and FSW.

7.1.3. Calculation of Raman Scattering Intensities

If we consider direct allowed electron transitions between conduction (c) and valence (v) bands, Eqs. (128), (132), and (227), then the matrix element $\langle a|\widehat{H}_r^-|i\rangle$ (or $\langle f|\widehat{H}_r^+|b\rangle$), in the envelope function approximation, can be written as

$$\langle a|\widehat{H}_r^-|i\rangle = \frac{|e|}{\mu_0}\sqrt{\frac{2\pi\hbar}{V\omega_l}}\hat{\mathbf{e}}_l\cdot\hat{\mathbf{p}}_{cv}(0)T_{n_1, m_1', m_2}\delta_{n_1', n_2}\delta_{k_1', -k_2} \tag{228}$$

$$\langle f|\widehat{H}_r^+|b\rangle = \frac{|e|}{\mu_0}\sqrt{\frac{2\pi\hbar}{V\omega_s}}\hat{\mathbf{e}}_s\cdot\hat{\mathbf{p}}_{vc}(0)T_{n_1, m_1, m_2}\delta_{n_1, n_2}\delta_{k_1, -k_2} \tag{229}$$

where

$$T_{n, m_a, m_b} = A_{nm_a}^{(e)}A_{nm_b}^{(h)}\int_0^{r_0} J_n\left(x_{nm_a}^{(e)}\frac{r}{r_0}\right)J_n\left(x_{nm_b}^{(h)}\frac{r}{r_0}\right)r\,dr + B_{nm_a}^{(e)}B_{nm_b}^{(h)}\int_{r_0}^{\infty} K_n\left(y_{nm_a}^{(e)}\frac{r}{r_0}\right)K_n\left(y_{nm_b}^{(h)}\frac{r}{r_0}\right)r\,dr$$

and for the FSW

$$T_{n, m_a, m_b} = \delta_{m_a, m_b}$$

where $\hat{\mathbf{p}}_{cv}$ and $\hat{\mathbf{p}}_{vc}$ are the momenta matrix elements between the v and c bands (evaluated at $\mathbf{k} = \mathbf{0}$).

For the electron–phonon interaction matrix elements, we have

$$\langle b|\widehat{H}_{ph}|a\rangle = C_{nm}\big[\delta_{a_h, b_h}\langle c; n_1, m_1'|F_{nm}(r)e^{in\theta}|n_1, m_1; c\rangle - \delta_{a_e, b_e}\langle v; n_2, m_2'|F_{nm}(r)e^{in\theta}|n_2, m_2; v\rangle\big]$$

where

$$\langle i; n_j, m_j'|F_{nm}(r)e^{in\theta}|n_j, m_j; i\rangle = I_{n_j m_j m_j'}\delta_{n,0}\delta_{k_j, k_j'} \quad i = c, v \tag{230}$$

being

$$I_{n_j m_j m_j'} = \sqrt{\frac{V\rho}{r_0}}\left\{A_{n_j m_j'}A_{n_j m_j}\int_0^{r_0} J_n\left(x_{n_j m_j'}\frac{r}{r_0}\right) \times F_{0m}^{\mathrm{QWW}}(r)J_n\left(x_{n_j m_j}\frac{r}{r_0}\right)r\,dr + B_{n_j m_j'}B_{n_j m_j}\int_{r_0}^{\infty} K_n\left(y_{n_j m_j'}\frac{r}{r_0}\right) \times F_{0m}^{\mathrm{QWW}}(r)K_n\left(y_{n_j m_j}\frac{r}{r_0}\right)r\,dr\right\} \tag{231}$$

and for the FSW

$$I_{n_j m_j m_j'} = \delta_{m_j m_j'}\sqrt{\frac{V\rho}{r_0}}\left[r_0 J_{l_j}'(x_{n_j m_j})\right]^{-2} \times \int_0^{r_0} J_n^2\left(x_{n_j m_j'}\frac{r}{r_0}\right)F_{0m}^{\mathrm{FSW}}(r)r\,dr \tag{232}$$

Here, the phonon modes are restricted to $n = 0$ when the dipole approximation is used [7, 15, 110–112, 115, 119, 120].

The probability amplitude is given by

$$M_{fi} = M_1 - M_2 \tag{233}$$

$$M_j = \sum_{n_j m_j m_j' m_j'}\sum_{k_j}\left(\frac{e}{\mu_0}\right)^2\frac{2\pi\hbar}{\sqrt{\omega_s\omega_l}}\frac{\hat{\mathbf{e}}_s\cdot\hat{\mathbf{p}}_{vc}(0)\hat{\mathbf{e}}_l\cdot\hat{\mathbf{p}}_{cv}(0)}{V}C_{0m} \times \frac{(T_{n_1, m_1, m_2}I_{n_1 m_1 m_1'}T_{n_1, m_1', m_2}\delta_{j,1} + T_{n_2, m_1, m_2}I_{n_2 m_2 m_2'}T_{n_2, m_1, m_2'}\delta_{j,2})}{(\hbar\omega_l - E_a + i\Gamma_a)(\hbar\omega_l - \hbar\omega_{0m} - E_b + i\Gamma_b)} \tag{234}$$

After substitution of Eqs. (231) and (232) in Eq. (1) we obtain

$$\Phi_1(\hbar\omega_l, \hbar\omega_s) = \beta_1\sum_{n_1 m_1 m_2} T_{n_1, m_1, m_2} \times \sum_{m_1'}\frac{I_{n_1 m_1 m_1'}T_{n_1, m_1', m_2}}{[\hbar\omega_{0m} + E_{n_1, m_1, m_2} - E_{n_1, m_1', m_2} + i(\Gamma_a - \Gamma_b)]} \times [Z_1(n_1, m_1', m_2) - Z_2(n_1, m_1, m_2)] \tag{235}$$

$$\Phi_2(\hbar\omega_l, \hbar\omega_s) = \beta_2 \sum_{n_2 m_1 m_2} T_{n_2, m_1, m_2} \times \sum_{m_2'} \frac{I_{n_2 m_2 m_2'} T_{n_2, m_1, m_2'}}{[\hbar\omega_{0m} + E_{n_2, m_1, m_2} - E_{n_2, m_1, m_2'} + i(\Gamma_a - \Gamma_b)]} \times [Z_1(n_2, m_1, m_2') - Z_2(n_2, m_1, m_2)] \quad (236)$$

where

$$Z_1(n, m_a, m_b) = \frac{1}{[\hbar\omega_l - E_{n, m_a, m_b} + i\Gamma_a]^{\frac{1}{2}}} \quad (237)$$

$$Z_2(n, m_a, m_b) = \frac{1}{[\hbar\omega_s - E_{n, m_a, m_b} + i\Gamma_b]^{\frac{1}{2}}} \quad (238)$$

$$\beta_j = \sqrt{\frac{\mu_{ji}}{\mu_r}} \quad \text{and} \quad \frac{1}{\mu_r} = \frac{1}{\mu_{1i}} + \frac{1}{\mu_{2i}}$$

with

$$E_{n, m_a, m_b} = \frac{\hbar^2}{2\mu_{1i} r_0^2} (x_{nm_a}^{(e)})^2 + \frac{\hbar^2}{2\mu_{2i} r_0^2} (x_{nm_b}^{(h)})^2 + E_g \quad (239)$$

where the expressions

$$\frac{\hbar^2}{2\mu_{1i} r_0^2} (x_{nm_a}^{(e)})^2 \quad \text{and} \quad \frac{\hbar^2}{2\mu_{2i} r_0^2} (x_{nm_b}^{(h)})^2$$

are the "confined" part of the electron and hole energies, respectively [see Eqs. (131)].

Finally, substitution of Eqs. (235) and (236) into Eq. (1) gives

$$\frac{d^2\sigma}{d\Omega\, d\omega_s} = \frac{V\mu_r P}{2\pi^2} \left(\frac{e}{\mu_0 c}\right)^4 \frac{\eta(\omega_s)}{\eta(\omega_l)} \frac{\omega_s}{\omega_l} C_F^2 \times \sum_m \frac{\omega_L}{\omega_{0m}} |\Phi_1(\hbar\omega_l, \hbar\omega_s) - \Phi_2(\hbar\omega_l, \hbar\omega_s)|^2 \times \frac{\Gamma_p}{(\hbar\omega_l - \hbar\omega_s - \hbar\omega_{0m})^2 + \Gamma_p^2} \quad (240)$$

with $P = |(\hat{\mathbf{e}}_l \cdot \hat{\mathbf{p}}_{cv})(\hat{\mathbf{e}}_s \cdot \hat{\mathbf{p}}_{vc})|^2$. In Eq. (240), the substitution

$$\delta(E_f - E_i) \rightarrow \frac{1}{\pi} \frac{\Gamma_p}{(E_f - E_i)^2 + \Gamma_p^2}$$

has been used in order to take into account the phonon linewidth Γ_p in the Raman shift spectrum. The Raman scattering efficiency is given by

$$\frac{1}{V}\frac{d\sigma}{d\Omega} = \frac{\mu_r P}{2\pi^2} \left(\frac{e}{\mu_0 c}\right)^4 \frac{\eta(\omega_s)}{\eta(\omega_l)} C_F^2 \sum_m \frac{\omega_l - \omega_{0m}}{\omega_l} \frac{\omega_L}{\omega_{0m}} \times |\Phi_1(\hbar\omega_l, \hbar\omega_l - \hbar\omega_{0m}) - \Phi_2(\hbar\omega_l, \hbar\omega_l - \hbar\omega_{0m})|^2 \quad (241)$$

where the incoming and outgoing resonances are due to intersubband electron or hole transitions and are located at energies [see Eqs. (235)–(239)]

$$\hbar\omega_l = E_{n_1, m_a, m_b} \quad \text{and} \quad \hbar\omega_s = E_{n_1, m_a, m_b}$$

Figure 17 illustrates the behavior of ω_{om} frequency with radius r_0 and m optical phonon modes of the QWW (Fig. 17a) and FSW (Fig. 17b) using the parameters given in Table 1. These ω_{om} frequencies for $n = 0$ modes are simple confined modes of a purely longitudinal nature and they contain no surface mode contribution. In Figure 17a and b we can observe that the frequency decreases with the increase of the optical phonon mode for a fixed radius value and increases with the enlargement of the radius. Moreover, the ω_{om} frequency values tend to the ω_L value limit ($r_0 > 50$ Å, although, for clarity, it has not been shown). In all cases the frequency of the FSW is larger than the QWW. These results are explained with the formalism given in Section 7.1.

In Figure 18 the Raman scattering efficiency as a function of the incident energy for the FSW (Fig. 18a) and QWW (Fig. 18b) cases are shown. In both figures the solid (dashed) line corresponds to $r_0 = 15$ Å (20 Å). In order to indicate which subbands are included in the intersubband transitions, the subindex n_1, m_a, m_b is used. $m_a(m_b)$ belongs to the conduction (valence) subband. In all cases, when the radius is increased, the energy values associated with intersubbands transitions will be shifted to the left [see Eqs. (237) and (238)]. From the free boundary conditions we can see that the scattering intensity of the FSW is larger than the QWW. In each energy level, E_{n_1, m_a, m_b}, two peaks appear, which correspond to incoming and outgoing resonance separated by the energy of one phonon.

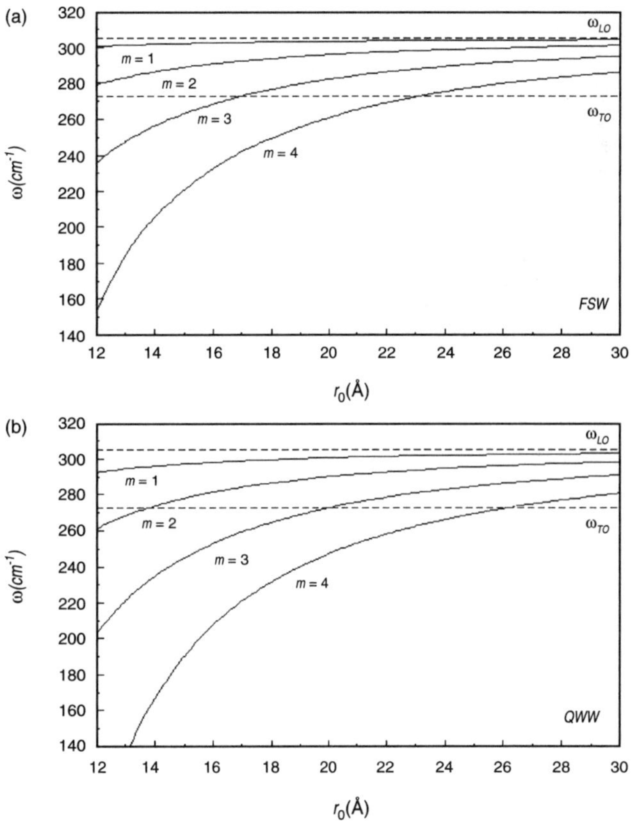

Figure 17. Oscillation modes of (a) freestanding wire and (b) quantum well wire.

Table 1. Values of the material parameters used for the numerical calculations illustrated in the figures [110].

	E_g (eV)	ω_L (cm^{-1})	ω_T (cm^{-1})	$\frac{\mu_e}{\mu_0}$	$\frac{\mu_h}{\mu_0}$	ε_∞	β_L (ms^{-1})	β_T (ms^{-1})	V_0^e (eV)	V_0^h (eV)
CdS	2.6	305	238	0.18	0.51	5.3	5.04 × 10^3	1.58 × 10^3	2.5	1.9
Glass	7	—	—	1	1	4.64	—	—	—	—

In Figure 19 the Raman cross-section as a function of frequency shift for different energies is shown. The $r_0 = 15$ Å radius has been chosen for FSW and QWW, respectively. The spectrum for $\hbar\omega_l = 3.34$ eV (Fig. 19a) corresponds to an incoming resonance with $n_1 = 0, m_1 = 1$ electron and $n_1 = 0$, $m_2 = 1$ hole states. There is a big contribution due to the $n = 0$, $m = 1$ phonon mode at 301 cm^{-1}. In this case the laser frequency fulfills the resonance condition. As in Figure 18, the Raman cross-section of the FSW is bigger than the QWW (see Fig. 19b). Figure 19b shows the spectra corresponding to QWW for different laser energies. The spectrum for $\hbar\omega_l = 2.99 + \hbar\omega_{01}$ eV belongs to an outgoing resonance with $n_1 = 0, m_1 = 1$ electron and $n_1 = 0$, $m_2 = 1$ hole states. There is a big contribution due to the $n = 0, m = 1$ phonon mode at 297 cm^{-1}. On the other hand, we can see that when the $\hbar\omega_l$ value increases the Raman intensity decreases, which can be justified from Eqs. (237) and (238) where the cross-section is inversely proportional to ω_l. Figure 19c shows the spectra for different energies of the QWW. We can see contributions from the three phonon modes $n = 0$, $m = 1$ to 3 with the biggest contribution from the $m = 2$ mode at 279 cm^{-1}, when the energy is $3.57 + \hbar\omega_{02}$ eV. This is mainly because the laser frequency has been chosen to be in outgoing resonance with the emission of a phonon of the $n = 0$, $m = 2$ mode frequency. The other contributions come from the $m = 1$ mode at 297 cm^{-1} and $m = 3$ at 245 cm^{-1}. Moreover, when the $3.57 + \hbar\omega_{01}$ eV energy is chosen, the biggest contribution corresponds to

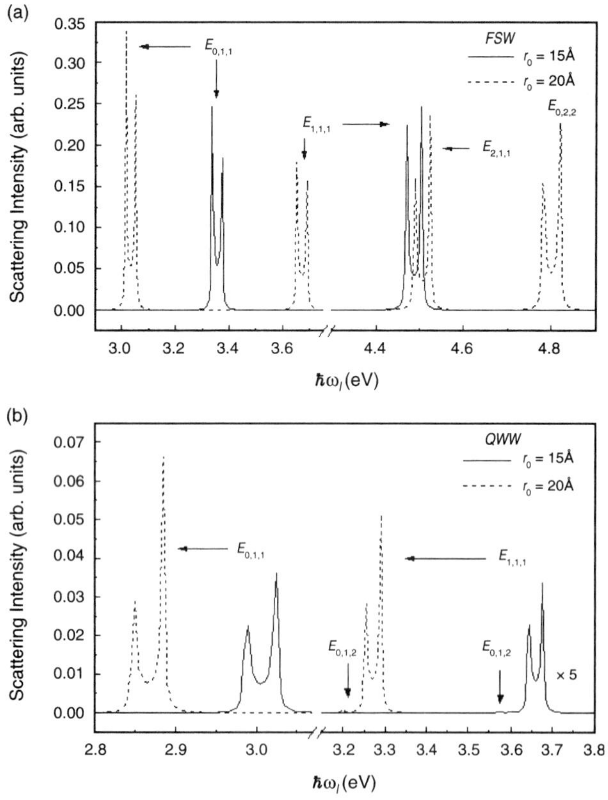

Figure 18. (a) Raman spectra for 20 and 15 Å radius CdS freestanding wire at two different laser energies. (b) Raman spectra for 20 and 15 Å radius CdS quantum wire embedded in glass at two different laser energies. The graph for 20 Å is multiplied by 5.

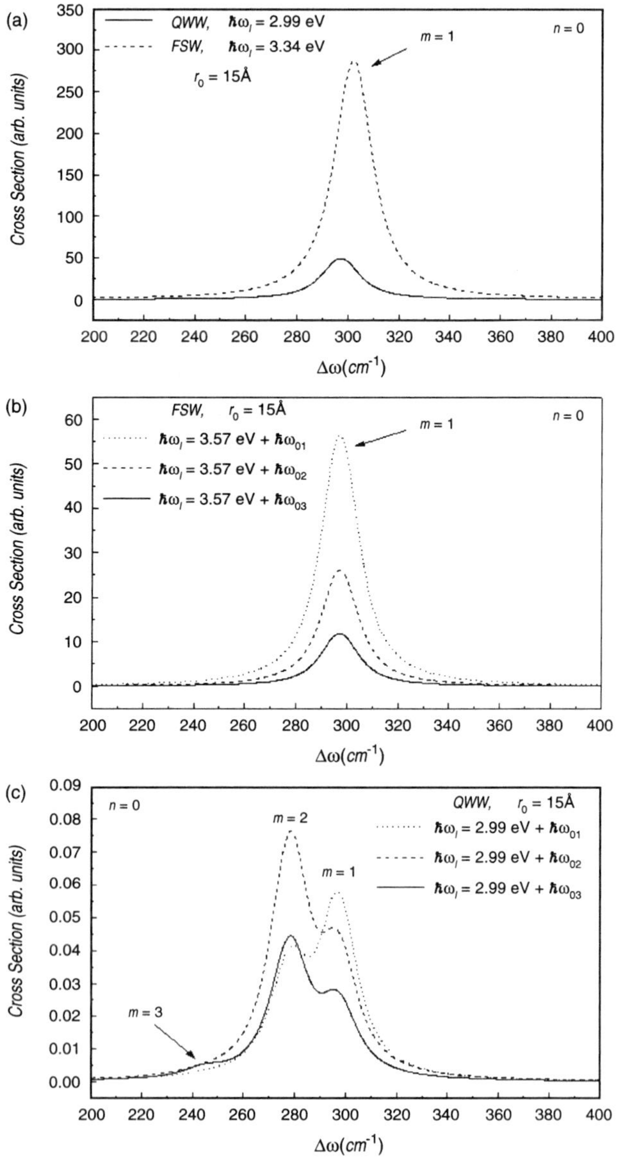

Figure 19. Differential cross section for (a) incoming resonance with $n_1 = 0, m_1 = 1$ electron $n_1 = 0, m_2 = 1$ hole states, (b) outgoing resonance with $n_1 = 0, m_1 = 1$ electron $n_1 = 0, m_2 = 1$ hole states for FWS, (c) outgoing resonance with $n_1 = 0, m_1 = 1$ electron $n_1 = 0, m_2 = 2$ hole states for QWW.

the $n = 0, m = 1$ mode frequency because in this case the laser frequency is in outgoing resonance with the emission of a phonon of this frequency. For $3.57 + \hbar\omega_{03}$ eV energy we can see situation similar to the $3.57 + \hbar\omega_{02}$ eV one, but the electron and hole subbands involved are $n_1 = 0, m_1 = 1$ and $n_1 = 0, m_2 = 2$, respectively. In the latter case, the laser frequency is in incoming resonance with the emission of a phonon of $n = 0, m = 1$ mode frequency.

A formalism for calculating Raman scattering intensities due only to the Fröhlich interaction from optical phonons in semiconductor QWW and FSW has been presented. A continuum model for the phonons which satisfies both electrostatic and mechanical boundary conditions at the interface has been used, in accordance with physical principles and differential equations. The case of oscillation perpendicular to the wire axis (i.e, $q_z = 0$ and hence $v_z = 0$) has been regarded. Here, only the phonon dispersions for $n = 0$ modes have been considered. In this case the modes of the systems are longitudinal and transverse in nature, but in other cases ($n \neq 0$) they are a mixture of LO, TO, and surface mode components.

When Raman intensities are being considered, it has been found that the only contribution comes from the dipole approximation term, which interacts only with $n = 0$ phonons, assuming cylindrical systems and neglecting valence band mixing, and is mainly responsible for the resonant intrasubband term. One band effective mass model with a finite potential for the conduction and valence bands has been considered. The phonon selection rules given early are strict for cylindrical system.

7.2. One Phonon Resonant Raman Scattering in Quantum Dots

In this section the analytical expressions for the dispersion of the $l_p = 0, 1, 2$ modes, with illustration for CdS spherical microcrystallites embedded in glass, are determined, following the work of [110]. The theory is somewhat more appropriate to the case of the microcrystallites since they can have sizes as low as 13 Å in radius, whereas quantum dots in the GaAs/AlAs system are normally fabricated by applying electric potentials to quantum wells or growth along terraces where the confining sizes in one or more directions are usually greater than 100 Å. A formalism for calculating the Raman scattering intensities from optical phonons in QDs via the Fröhlich interaction applied to CdS microcrystallite is also presented. In this section it is found that the dipole approximation interaction, which for perfectly spherical structures where valence band mixing is neglected couples only to $l_p = 0$ phonon modes, there also exists a contribution that is proportional to the wavevector of the light $\boldsymbol{\kappa}$ (electrical quadrupole approximation). This additional contribution takes place by $l_p = 1$ phonon mode emission. Although it is generally much smaller than the dipole approximation term, especially when in intrasubband resonance, it dominates when one is resonant with an intersubband transition.

In previous sections were presented the electron states, so as the envelope function for spherical QDs.

In [110] are presented some details concerning the optical phonons and electron–phonon interaction Hamiltonian where dispersion relation for optical phonon is given by

$$\tan\mu = \mu$$

for $l_p = 0$, while for $l_p = 1$ and phonon frequencies ω above the TO phonon frequency ω_T it is neccesary to solve the secular equation

$$\begin{aligned}
&[2\mu\cos\mu + (\mu^2 - 2)\sin\mu]\left\{\left[-\frac{\gamma_0}{\nu^2}\left(\frac{3}{\nu^2} - 1\right) + \left(1 + 2\frac{\epsilon_\infty^{(2)}}{\epsilon_\infty^{(1)}}\right)\right.\right.\\
&\quad \left.\times\left(1 - \frac{1}{\mu^2}\right)\right]\sin\nu + \left[-\frac{3\gamma_0}{\nu^3} + \frac{1}{\nu}\left(1 + 2\frac{\epsilon_\infty^{(2)}}{\epsilon_\infty^{(1)}}\right)\right]\cos\nu\Bigg\}\\
&= 2[\sin\mu - \mu\cos\mu]\left\{\frac{1}{\nu}\left[-1 - 2\frac{\epsilon_\infty^{(2)}}{\epsilon_\infty^{(1)}} + 3\frac{\gamma_0}{\nu^2}\frac{\epsilon_\infty^{(2)}}{\epsilon_\infty^{(1)}}\right]\cos\nu\right.\\
&\quad \left. + \frac{1}{\nu^2}\left[\frac{\epsilon_\infty^{(2)}}{\epsilon_\infty^{(1)}}\gamma_0\left(1 - \frac{3}{\nu^2}\right) + \left(1 + \frac{\epsilon_\infty^{(2)}}{\epsilon_\infty^{(1)}}\right)\right]\sin\nu\right\}
\end{aligned}$$

with

$$\gamma_0 = \frac{\omega_L^2 - \omega_T^2}{\beta_T^2}R_0^2 \quad \omega^2 = \omega_L^2 - \beta_L^2\left(\frac{\mu}{R_0}\right)^2 \quad \omega^2 = \omega_T^2 + \beta_T^2\left(\frac{\nu}{R_0}\right)^2$$

and $\beta_L(\beta_T)$ is a parameter describing the dispersion, assumed to be parabolic, of the LO (TO) phonon in the bulk.

7.2.1. Calculation of the Raman Scattering Intensities in Quantum Dots

Considering direct allowed electron transitions between conduction and valence bands the matrix element $\langle a|H_{E-R}^-|I\rangle$ (or $\langle F|H_{E-R}^-|b\rangle$) can be written as

$$\langle\mu|H_{E-R}^-|I\rangle = \frac{e}{\mu_0}\sqrt{\frac{2\pi\hbar}{\omega_l\eta_l^2}}\frac{\hat{\mathbf{e}}_l\cdot\hat{\mathbf{p}}_{cv}}{\sqrt{V}}\int\Psi_c^* e^{i\boldsymbol{\kappa}_l\cdot\mathbf{r}}\Psi_v\, d^3r$$

$\Psi_c(\Psi_b)$ being the electron (hole) wave function in the quantum dot. Since the photon wavevectors are small, the expansion of $e^{i\boldsymbol{\kappa}_l\cdot\mathbf{r}}$ up to terms proportional to $\boldsymbol{\kappa}$ can be made:

$$\begin{aligned}
\langle\mu|H_{E-R}^-|I\rangle &= \frac{e}{\mu_0}\sqrt{\frac{2\pi\hbar}{\omega_l\eta_l^2}}\frac{\hat{\mathbf{e}}_l\cdot\hat{\mathbf{p}}_{cv}}{\sqrt{V}}[\langle c; n_e, l_e, m_e|m_h, l_h, n_h; v\rangle\\
&\quad + i\boldsymbol{\kappa}_l\cdot\langle c; n_e, l_e, m_e|\mathbf{r}|m_h, l_h, n_h; v\rangle] \qquad (242)
\end{aligned}$$

For the electron–phonon interaction matrix elements they have

$$\begin{aligned}
\langle\mu_2|H_{E-P}|\mu_1\rangle &= \frac{C_F}{\sqrt{R_0}}\{\delta_{\mu'_h,\mu_h}\langle c; n'_e, l'_e|\Phi_{\mu,l_p}|l_e, n_e; c\rangle\\
&\quad \times\langle Y_{l'_e,m'_e}|Y_{l_p,m_p}|Y_{l_e,m_e}\rangle\\
&\quad - \delta_{e'_h,\mu_e}\langle v; n'_h, l'_h|\Phi_{\mu,l_p}|l_h, n_h; v\rangle\}\\
&\quad \times\langle Y_{l'_h,m'_h}|Y_{l_p,m_p}|Y_{l_h,m_h}\rangle \qquad (243)
\end{aligned}$$

where $\mu_e(\mu_h) = n_e, l_e, m_e(n_h, l_h, m_h)$ represents the set of quantum numbers for the electron (hole) state. The amplitude probability was found as

$$M_{FI} = \left(\frac{e}{m_0}\right)^2 \frac{2\pi\hbar}{\sqrt{\omega_l\omega_s\eta_l^2\eta_s^2}} \frac{\hat{\mathbf{e}}_s^* \cdot \hat{\mathbf{p}}_{vc}\hat{\mathbf{e}}_l \cdot \hat{\mathbf{p}}_{cv}}{V} \frac{C_F}{\sqrt{R_0}}[M_{FI}^{(0)} + M_{FI}^{(1)}] \tag{244}$$

where

$$\begin{aligned} M_{FI}^{(0)} = \sum_{n_e, n_h}\sum_{l'=0}^{\infty}\sum_{m'=-l'}^{l'} \langle Y_{l',m'}|Y_{l_p,m_p}|Y_{l',m'}\rangle \\ \times \frac{F_0(n_e, n_h)}{[\hbar\omega_l - E_g - E_e(n_e, l') - E_h(n_h, l') + i(\Gamma_e + \Gamma_h)]} \\ \times \left\{\sum_{n'} \frac{F_0(n_e', n_h)\langle c; n', l'|\Phi_{\mu,l_p}|l_e, n_e; c\rangle}{[\hbar\omega_s - E_g - E_e(n_e', l') - E_h(n_h, l') + i(\Gamma_e' + \Gamma_h)]}\right\} \\ \times \left\{\sum_{n'} \frac{F_0(n', n_e)\langle v; n', l'|\Phi_{\mu,l_p}|l, n_h; v\rangle}{[\hbar\omega_s - E_g - E_e(n_e, l') - E_h(n', l') + i(\Gamma_e' + \Gamma_h')]}\right\} \end{aligned} \tag{245}$$

Here, the incoming light with a wavevector $\boldsymbol{\kappa}_l \approx 0$ creates an electron–hole pair of energy $E_e(n_e, l') + E_h(n_h, l')$ via the dipole transition $F_0(n_e, n_h)$. Subsequently, as the electron (or hole) emits an LO phonon of frequency $\omega(\mu, l_p)$ an electron transition from (n_e, l', m') to (n', l', m') takes place [or from (n_h, l', m') to (n', l', m') hole states]. Finally, a scattered photon of frequency ω_s and wavevector $\boldsymbol{\kappa}_l \approx 0$ is created due to the electron–hole recombination with a probability given by the overlap integral $F_0(n_h, n')$ [or $F_0(n', n_e)$ if the phonon is emitted by a hole]. From the previous equation, a set of incoming and outgoing resonances follows at the frequencies:

$$\hbar\omega_l = E_g + E_e(n_e, l') + E_h(n_h, l')$$
$$\hbar\omega_s = E_g + E_e(n', l') + E_h(n_h, l') \quad l' = 0, 1, 2, \ldots$$

The second term in Eq. (244) is given by

$$\begin{aligned} M_{FI}^{(1)} = i\sum_{n,l'}\sum_{m'=-l'}^{l'} a_{l'-\frac{1}{2}\pm\frac{1}{2}}\langle Y_{l'\pm1,m'}|Y_{l_p,m_p}|Y_{l',m'}\rangle \\ \times\sum_{n'} F_1(n', l\pm1, n, l')\langle n', l'\pm1|\Phi_{\mu,l_p}|l', n\rangle \\ \times\left\{\frac{\kappa_l}{[\hbar\omega_l - E_g - E_e(n,l') - E_h(n', l'\pm1) + i(\Gamma_e + \Gamma_h')]}\right. \\ \times\left(\frac{1}{[\hbar\omega_s - E_g - E_e(n', l'\pm1) - E_h(n', l'\pm1) + i(\Gamma_e' + \Gamma_h')]}\right. \\ \left. - \frac{1}{[\hbar\omega_s - E_g - E_e(n,l') - E_h(n,l') + i(\Gamma_e + \Gamma_h)]}\right) \\ + \frac{\kappa_s}{[\hbar\omega_l - E_g - E_e(n,l') - E_h(n,l') + i(\Gamma_e + \Gamma_h)]} \\ \times\left(\frac{1}{[\hbar\omega_s - E_g - E_e(n', l'\pm1) - E_h(n,l') + i(\Gamma_e' + \Gamma_h)]}\right. \\ \left.\left. - \frac{1}{[\hbar\omega_s - E_g - E_e(n,l') - E_h(n', l'\pm1) + i(\Gamma_e + \Gamma_h')]}\right)\right\} \end{aligned} \tag{246}$$

where the wave function has been used to evaluate the matrix elements (242) and (243). The functions F_1 and a_l are given in an appendix of [110]. Here, the physical picture is similar to that given above for $M_{FI}^{(0)}$ but now the incoming and outgoing resonances are due to intersubband electron or hole transitions and are located at energies

$$\hbar\omega_s = E_g + E_e(n, l') + E_h(n', l' \pm 1)$$
$$\hbar\omega_s = E_g + E_e(n, l') + E_h(n, l')$$
$$\hbar\omega_l = E_g + E_e(n, l') + E_h(n', l' \pm 1)$$
$$\hbar\omega_l = E_g + E_e(n, l') + E_h(n, l') \quad l' = 0, 1, \ldots$$

7.3. Selection Rules

The Fröhlich interaction coupling [Eq. (245)] of the dipole approximation (the wavevector of the light $\boldsymbol{\kappa} = 0$) conserves the electron angular momentum $\Delta l' = l_e' - l_h' = 0$ and obviously $\Delta m' = m_e' - m_h' = 0$. A simple inspection of the angular matrix element $\langle Y_{l',m'}|Y_{l_p,m_p}|Y_{l',m'}\rangle$ determines that $l_p = 0, 2, 4, \ldots$ and $m_p = 0$; the function $Y_{l_p,0}$ needs to be even for a nonzero Raman scattering cross-section in Eq. (245) because the electron–phonon coupling must conserve the parity of the quantum dot functions. The degeneracy of the electron state due to the l_Z' component, where z is the quantization axis, has been considered in $M_{FI}^{(0)}$ through the indicated m' sum in Eq. (245). Now, this sum can be obtained using the addition theorem

$$\frac{2l'+1}{4\pi}P_{l'}(\cos\alpha) = \sum_{m'=-l'}^{l'} Y_{l',m'}^*(\theta_1, \phi_1)Y_{l',m'}(\theta_2, \phi_2) \tag{247}$$

where α is the angle between directions (θ_1, ϕ_1) and (θ_2, ϕ_2). Taking $\theta_1 = \theta_2 = \theta$ and $\phi_1 = \phi_2 = \phi$ Eq. (247) is transformed into

$$\frac{2l'+1}{4\pi} = \sum_{m'=-l'}^{l'} |Y_{l',m'}(\theta, \phi)|^2 \tag{248}$$

Thus, the term

$$I_\Omega = \sum_{m'=-l'}^{l'} \int_0^{2\pi}\int_{-\pi}^{\pi} |Y_{l',m'}(\theta, \phi)|^2 Y_{l_p,m_p}(\theta, \phi)\, d\Omega$$

of Eq. (245) is reduced to

$$I_\Omega = \frac{2l'+1}{4\pi}\int_0^{2\pi}\int_{-\pi}^{\pi} Y_{l_p,m_p}(\theta, \phi)\, d\Omega$$

Using the orthonormalization condition for the spherical harmonics,

$$I_\Omega = \frac{2l'+1}{4\pi}\delta_{p,0}\delta_{m_p,0}$$

Hence, the phonon modes are restricted to $l_p = 0$ for a spherical quantum dot when one uses the dipole approximation $\boldsymbol{\kappa} = 0$. One could have anticipated this selection rule since in the dipole approximation where $\boldsymbol{\kappa}_l, \boldsymbol{\kappa}_s = 0$, $m_p = 0$ for any axis of quantization. This implies for this most significant term in the Raman scattering cross-section, $M_{FI}^{(0)}$, that none of the higher order l_p phonons contribute and so no interaction with modes containing surface components is seen.

For the matrix element $M_{FI}^{(1)}$, which is proportional to $\boldsymbol{\kappa}$, the quantization axis z is chosen to be parallel to the light wavevector $\boldsymbol{\kappa}_l$. The selection rules are

$$\Delta l' = \pm 1 \quad \Delta m' = 0$$

$$m_p = 0 \quad \text{and} \quad l_p = 1, 3, 5, \ldots$$

The restriction $\Delta l' = \pm 1$ comes from the first order dipole transition which connects electron states of different parity implying that only phonons of odd parity give nonzero matrix elements. For the coefficient,

$$I_\Omega^{(1)}(l'+1, l'; l_p) = \sum_{m'=-l'}^{l'} a_{l'} \langle Y_{l'+1, m'} | Y_{l_p, m_p} | Y_{l', m'} \rangle$$

the following functional relation is obtained according to Eqs. (B3) and (B5) of [110]:

$$I_\Omega^{(1)}(l'+1, l'; l_p) = \sum_{m'=-l'}^{l'} \int_0^{2\pi} \int_{-\pi}^{\pi} |Y_{l', m'}(\theta, \phi)|^2 \times Y_{l_p, m_p}(\theta, \phi)\, d\Omega - I_\Omega^{(1)}(l'-1, l'; l_p)$$

Using Eq. (248) this expression is transformed to

$$I_\Omega^{(1)}(l'+1, l'; l_p) = \frac{2l'+3}{\sqrt{12\pi}} \delta_{l_p, 1} \delta_{m_p, 0} - I_\Omega^{(1)}(l'-1, l'; l_p)$$

where it can be shown that

$$I_\Omega^{(1)}(0, 1; l_p) = \frac{\delta_{l_p, 1} \delta_{m_p, 0}}{\sqrt{12\pi}}$$

From the two last relations it immediately follows the phonon selection rule $l_p = 1$. The other odd modes ($l_p = 3, 5, \ldots$) are forbidden in the scattering matrix element $M_{FI}^{(1)}$. This $l_p = 1$ selection rule breaks down when one considers configurations other than backscattering (i.e., when there is an angle between the incoming and scattered light directions). According to these selection rules the Raman cross-section for the Fröhlich interaction contribution in a spherical quantum dot or microcrystallite can be written as

$$\frac{\partial^2 \sigma}{\partial \Omega\, \partial \omega_s} = S_0 \sum_\mu \sum_{l_p} \{ |M_{FI}^{(0)}|^2 \delta_{l_p, 0} + |M_{FI}^{(1)}|^2 \delta_{l_p, 1} \} \times \frac{\Gamma_p}{[\hbar\omega_l - \hbar\omega_s - \hbar\omega_p(\mu, l_p)]^2 + \Gamma_p^2}$$

where the δ function in the calculation of the probability has been substituted by a Lorentzian in order to take into account the phonon linewidth Γ_p in the Raman shift spectrum. The coefficient S_0 is equal to

$$S_0 = \left(\frac{e}{m_0}\right)^4 \left(\frac{2P^2}{3c^2}\right) \hbar \left(\frac{\omega_s}{\omega_l}\right)^2 \frac{\eta_s}{\eta_l} \frac{|C_F|^2}{R_0}$$

with $P = \langle X | P_x | S \rangle$ [106].

The dispersion relations and scattering intensities for electron, hole, and lattice for a CdS microcrystallite embedded in glass in Figures 1, 2, and 3 of [110] can be appreciated.

In this section the one phonon Raman scattering cross-section using the band structure and phonon dispersions described previously has been calculated. It is important to remember the phonon selection rules already derived for the electron–phonon Fröhlich interaction. For the dipole approximation, with the wavevector of the light considered to be zero, there exist only intrasubband transitions. Due to the symmetry of the initial and final carrier states the phonons must be even in l_p and, furthermore, due to the degeneracy in energy for the carrier states of different m the only nonzero matrix element is for $l_p = 0$ phonon modes. The interaction term which is proportional to the carrier states and a nonzero matrix element, when summing over the m of the electrons or holes, only arises when $l_p = 1$ for the phonon modes.

In Figure 20 Raman scattering intensities as a function of frequency shift for different laser energies are shown.

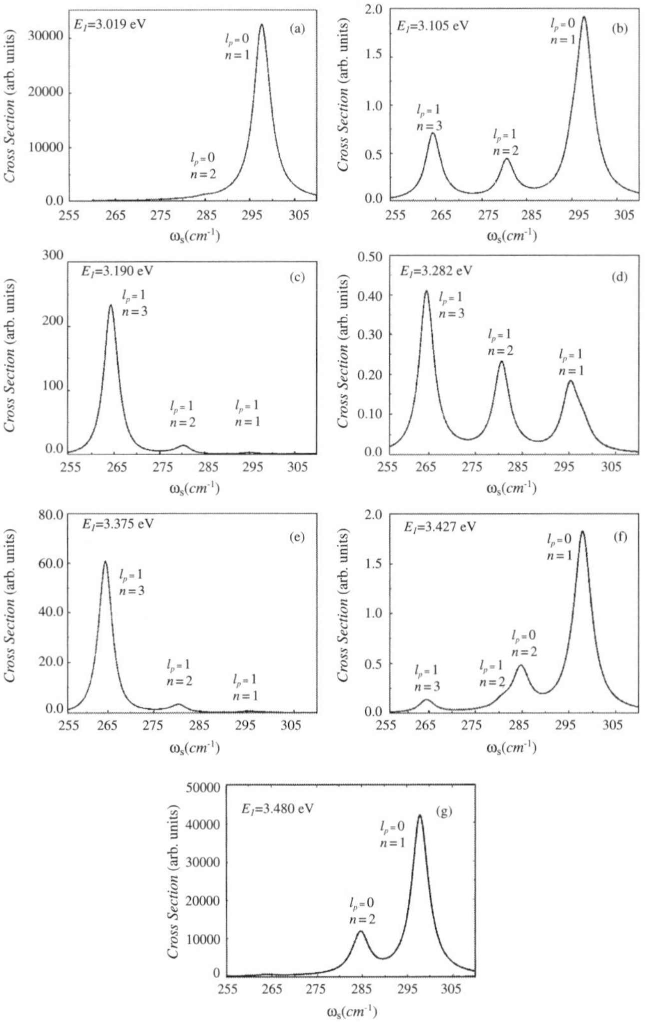

Figure 20. Raman spectra for a 20 Å radius CdS microcrystallite embedded in glass at different laser energies E_l. Reprinted with permission from [110], M. P. Chamberlain et al., *Phys. Rev. B* 51, 1680 (1995). © 1995, American Physical Society.

For the dipole approximation contribution to the matrix elements, $M_{FI}^{(0)}$ Eq. (245), the intrasubband contributions due to the $l = 0$, 1, and 2 electron and hole subbands have been considered. The contributions from higher electron and hole subbands have been neglected, but for laser energies away from the respective resonances these are effectively zero. For the matrix element proportional to the light wavevector, $M_{FI}^{(1)}$ Eq. (246), intersubband hole and electron transitions between the $l = 0$ and 1 carrier states (and vice versa) were considered. All the spectra have been broadened to have a full width at half maximum of 4 cm^{-1}. Since the TO component of the phonon modes does not couple to the electrons and holes via the Fröhlich interaction no interesting features are seen in the Raman spectra due to the LO and TO mode anticrossing between 238 and 255 cm^{-1} of Figures 2 and 3 of [110].

The Raman intensities in this region for a 20 Å radius crystallite are due to the higher order modes of the LO component and are much smaller than those due to the $n = 1$ to 3 modes shown in Figure 20. The spectrum for $E_l =$ 3.0193 eV (Fig. 20a) corresponds to an incoming resonance with the $l = 0$ electron and $l = 0$ hole states. It is almost exclusively due to the $l = 0$ electron and hole intrasubband transition contributions arising from interaction with the $l_p = 0$ phonons, although intrasubband transitions for $l = 1$ and 2 carrier states and intersubband transitions are also included in the spectra. The latter are several orders of magnitude smaller and cannot be seen in the scale of Figure 20a. There is a big contribution due to the $l_p = 0, n = 1$ phonon mode at 298 cm^{-1} with a very small contribution due to the $l_p = 0, n = 2$ phonon at 284 cm^{-1} since the n dependent part of the matrix element drops off rapidly as n increases. On the other hand, the spectrum for $E_l = 3.1900$ eV (Fig. 20c) corresponds to outgoing resonance with the $l = 0$ electron state and the $l = 1$ hole state. This is the resonance case for the $l = 0$ to 1 hole intersubband transition which takes place via $l_p = 1$ phonon emission. One can see contributions from the three phonon modes $l_p = 1, n = 1$ to 3 with the biggest contribution from the $n = 3$ mode at 264 cm^{-1}. This is mainly because the laser frequencies are chosen to be in outgoing resonance with the emission of a phonon of the $l_p = 1, n = 3$ mode frequency. However, it should be noted that if the outgoing resonance corresponding to the emission of a phonon with a frequency of the $l_p = 1, n = 1$, or $n = 2$ modes is taken, then the biggest contribution is still from the $n = 3$ mode. If one looks back at the dispersion relation for the $l_p = 1$ modes in Figure 2 of [110], it can be seen that the $n = 3$ mode is very close to the Fröhlich mode frequency for a 20 Å radius microcrystallite and so has a strong electrostatic contribution. The electrostatic contribution is evident from the bending in the phonon dispersion around the Fröhlich frequency and thus the interaction with electrons and holes will always be dominant for a mode around this frequency. Once again, the intrasubband spectra due to the $l = 0, 1$, and 2 electron and hole states have been included, but since the region considered is far from resonance for these transitions their contributions do not show up in Figure 20c.

It should also be pointed out that the intensity units for each of the spectra in Figure 20 are the same and so the intensity for $E_l = 3.1900$ eV is much smaller than for $E_l = 3.0193$ eV. The spectrum at the higher laser energy comes from the term which is proportional to $\boldsymbol{\kappa}_l$ or $\boldsymbol{\kappa}_s$ explaining the reduction in intensity by two orders of magnitudes. The laser energy $E_l = 3.1047$ eV of the spectrum in Figure 20b is halfway between that in Figure 20a and c. The spectrum shows a peak at 298 cm^{-1} due to the $l_p = 0, n = 1$ phonon mode and then two more peaks at 280 and 264 cm^{-1} due to the $l_p = 1$ phonon modes with $n = 2$ and 3, respectively. The highest frequency peak comes from the $\boldsymbol{\kappa} = 0$ contribution to the matrix element $M_{FI}^{(0)}$ and the other two from the $\boldsymbol{\kappa}$ dependent term $M_{FI}^{(1)}$. With the laser energy being out of resonance with the intersubband and intrasubband transitions both are visible in the spectrum. One can see that the $l_p = 1$, $n = 3$ phonon mode peak is bigger than the $l_p = 1$, $n = 2$ peak as it contains a larger electrostatic interaction due to the Fröhlich mode component at this frequency. The overall intensities are also smaller than the two previous cases because the laser frequency does not fulfill any resonance condition.

The spectra for $E_l = 3.3745$ eV (Fig. 20e) are for outgoing resonance with the $l = 0$ hole state and the $l = 1$ electron state. It is virtually identical to that for $E_l = 3.1900$ eV (Fig. 20c) since we are in resonance with an intersubband transition when only the $\boldsymbol{\kappa}$ dependent term involving $l_p = 1$ phonons has a significant contribution. At the lower energy of $E_l = 3.2823$ eV (Fig. 20d), we are not in outgoing resonance with the $l_p = 1, n = 3$ phonon mode so the $l_p = 1$, $n = 1$ and 2 mode peaks are of similar intensity, but the $n = 3$ phonon mode scattering is still the strongest because it contains a significant electrostatic Fröhlich component. We are further away from an intrasubband resonance than from an intersubband one so the $\boldsymbol{\kappa}$ dependent term is larger. However, a shoulder appears at the right-hand side of the $l_p = 1, n = 1$ mode peak at 298 cm^{-1} which is due to the contribution of the $l_p = 0, n = 1$ phonon from the intrasubband term. For $E_l = 3.4796$ eV (Fig. 20g) there exists incoming resonance for the $l = 1$ electron and hole states. Thus, only the intrasubband terms appear in the spectra. One should note that the $l_p = 0, n = 2$ mode at 285 cm^{-1} is much stronger in relation to the $l_p = 0, n = 1$ mode than was the case for a laser energy in resonance with the $l = 0$ electron and hole states (Fig. 20a). For the higher l carrier state the n-dependent part of the matrix element is not as for the $l = 0$ case and the drop off with increasing n is not as great. One might also expect that the overall intensities for the $l = 1$ intrasubband resonance would be smaller than for the $l = 0$ intrasubband resonance. The electron–phonon interaction for an $l = 1$ intrasubband is about 40% of that for the $l = 0$ intrasubband transition. However, the $l = 1$ electron (or hole) has three energy degeneracy m states which then implies a 120% increase for the $l = 1$ transition.

Hence the intensities for $E_l = 3.4796$ eV are bigger than for $E_l = 3.0193$ eV. Finally, the spectrum for $E_l = 3.4271$ eV (Fig. 20f) is similar to that for $E_l = 3.4796$ eV. The intensities are much reduced as we are away from intrasubband resonance. The $l = 1$ intersubband resonance at 3.3745 eV has also been moved and so the $l_p = 1$ modes from the intersubband transitions begin to appear. This is clearly visible for the $l_p = 1, n = 3$ mode at 264 cm^{-1} and also the $l_p = 1, n = 2$ mode appears as a shoulder at 280 cm^{-1}.

Klein et al. [52] have investigated electron–phonon coupling in CdSe microcrystallites both theoretically and experimentally. In considering the phonon modes of the nanospheres they have neglected the dispersion of the LO modes and calculated an independent set of internal modes (degenerate at the LO frequency due to lack of dispersion) and surface modes. These have been derived by applying electrostatic boundary conditions with the formalism of Mori and Ando [121] that was applied to quantum wells. Neglecting the mechanical boundary conditions and the LO dispersion means they were unable to describe the mixing of the confined and surface modes along with the mixed transverse longitudinal nature of the modes in general. When treating the Raman scattering they have only considered the dipole approximation which allows coupling to just $l_p = 0$ phonons. In the experiments, shoulders are seen on the spectra at the positions of the surface phonon frequency, which are disallowed for perfect spheres in the dipole approximation. The observation of the surface modes was explained by the relaxation of the $l_p = 0$ selection rule for monocrystals with a nonspherical shape. Although this is a possible explanation, the surface or Fröhlich mode shoulder could be due to $l_p = 1$ mode scattering from the $\boldsymbol{\kappa}$ dependent matrix element $M_{FI}^{(1)}$ from the intersubband electron (hole) transitions.

Finally in this section we have presented, following [110], a formalism for calculating Raman scattering intensities, due to the Fröhlich interaction, from optical phonons in semiconductor quantum dots and microcrystallites. In this treatment, the authors do not consider Raman scattering via the deformation potential interaction, which is important for laser energies that are far from the electron hole resonances. In particular, they used a continuum model for the phonons which satisfies both electrostatic and mechanical boundary conditions at the dot interface in accordance with the differential equations and physical principles involved. They also considered the CdS microcrystallite system and illustrated the phonon dispersions for $l_p = 1$ and 2 modes. It is seen that the modes of the system are generally both longitudinal and transverse in nature, being a mixture of LO, TO, and surface mode components as opposed to separate sets of confined and surface modes which have previously been assumed. Also they discussed the phonon properties of the CdS dots in glass, which is particularly interesting because of the positive phonon dispersion of the bulk TO mode which intermixes with LO and surface modes in the dot (for the GaAs/AlAs system both the TO and the LO mode frequencies decrease with increasing wavevector). When considering the Raman intensities, it has been found that the strongest contribution comes from the dipole approximation term which interacts only with $l_p = 0$ phonons, assuming perfectly spherical microcrystals and neglecting valence band mixing, and is mainly responsible for the resonant intrasubband term. They have shown that although the quadrupole contribution, proportional to the light wavevector $\boldsymbol{\kappa}$, is generally much smaller than the dipole interaction term it dominates for intersubband scattering. This intersubband interaction occurs only via $l_p = 1$ phonons.

In treating the electron–phonon interaction, they considered a one band effective mass model with a finite potential for the conduction and valence bands. Using this approach, the phonon selection rules given are strict for perfect spherical microcrystals. However, experiments performed by a number of authors give evidence for interaction with $l_p = 2$ phonons, implying a breakdown of these selection rules. It has been proposed elsewhere that this is either due to the nonspherical shape of the crystals [52] or mixing of the valence bands [122]. They believe that this is probably due to the hole subband mixing and propose to develop the formalism described here to calculate the magnitude of these effects using the same phonon model but increasing the complexity of the band structure description.

8. APPLICATIONS OF THE ELECTRON RAMAN SCATTERING AND RESONANT RAMAN SCATTERING PROCESSES IN NANOSTRUCTURES

A tunable intersubband Raman laser in multiple nanostructures can be designed theoretically as a new tunable source of infrared radiation based on intersubband electron Raman scattering in semiconductor nanostructures. The structure consists of three levels in two coupled quantum wells, quantum well wires, or quantum dots. A theoretical work on tunable intersubband Raman lasers in GaAs/AlGaAs multiple quantum well appears in [123].

8.1. Introduction

Following the work of Sun et al. [123] we can see there is a need for efficient sources of coherent mid-infrared radiation for applications such as remote sensing, laser radar, and wide-band communications. Rapidly evolving epitaxial growth techniques now allow the development of fundamentally new types of semiconductor lasers utilizing transitions between subbands in quantum wells and superlattices. With bandgap engineering, it is possible to adjust the intersubband transition energies over a wide range. The successful demonstration of the injection-pumped intersubband quantum-cascade laser operating in the 5-μm range has opened up the possibility of achieving coherent mid-to-far infrared light sources with high efficiency and compactness [124]. Although the quantum-cascade laser has been refined since its first demonstration nearly two years ago [125, 126], room-temperature continuous wave (cw) operation has not been achieved. This is primarily because intersubband lasers rely on elaborate pumping and relaxation schemes with multiple quantum wells to reach population inversion. The lifetime difference between the upper and the lower laser state in intersubband lasers can be artificially engineered but comes at the expense of the stimulated emission cross-section, as suggested in the optimization process of a four-level optically pumped intersubband laser [127]. An alternative scheme aimed at providing independent control of the intersubband lasing lifetimes has been proposed by use of the technique of electron Raman scattering [128]. Such a technique has previously been used in the investigation of transitions in confined systems [129]. In fact, a number of infrared sources based on stimulated Raman scattering in liquids [130] and semiconductors [131–135] have been successfully operated. It is therefore reasonable to consider a stimulated intersubband electron Raman oscillator (SIERO) based on intersubband transitions.

The proposed stimulated intersubband electron Raman oscillator has several advantages over the conventional stimulated Raman lasers made of bulk semiconductors. It allows the independent control of the lifetimes of laser states and the stimulated emission cross-section. It offers a wide tuning range in lasing wavelengths by the design of proper quantum well parameters, whereas the tuning of electron Raman lasers of bulk semiconductors is mainly achieved by variance of pump wavelength. For a given quantum well Raman laser structure, the lasing wavelength can be further tuned by the application of an electric field, whereas that of conventional Raman lasers requires a magnetic field.

8.2. Analysis

Any optically pumped intersubband laser (OPIL) can be practically represented by a three-level model, as shown in Figure 21a in which three subbands are confined within two coupled quantum wells, incorporating the ground state $|g\rangle$, the upper laser state $|u\rangle$, and the lower laser state $|l\rangle$. The optically pumped intersubband laser operating cycle consists of the absorption of one photon of pump frequency $\hbar\omega_p$, resonant with the energy separation between the upper laser and the ground states, and the subsequent emission of a photon of lasing frequency $\hbar\omega_{\text{out}}$, determined by the energy separation between the upper and the lower laser states. The necessary condition for cw lasing operation in this optically pumped intersubband laser structure is that the lifetime of the upper laser state τ_u be longer than that of the lower laser state τ_l. In comparison the stimulated intersubband electron Raman oscillator using the electron Raman scattering process can be operated between two levels as shown in Figure 21b. The Stokes intersubband electron Raman scattering can be described as an electron in the ground state $|g\rangle$ absorbing a photon of pump frequency $\hbar\omega_p$, re-emitting a photon of Stokes frequency $\hbar\omega_{\text{out}}$, and making a transition directly from the ground state $|g\rangle$ to the lower state $|l\rangle$. In fact, the transition is not entirely direct: The electron can be considered to arrive at its final destination via a virtual state $|u'\rangle$. In other words, it spends an extremely short time in the state $|u\rangle$, determined by the uncertainty principle associated with the upper state $|u\rangle$.

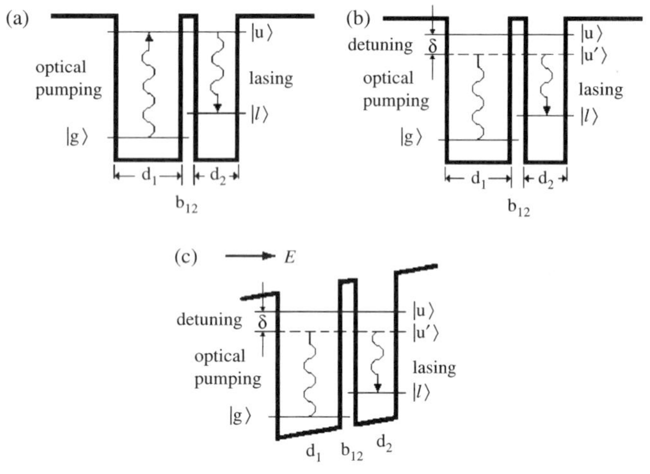

Figure 21. Conduction-band diagram of the laser states and the transitions in two coupled GaAs/AlGaAs quantum wells for (a) OPIL, (b) SIERO under zero bias, and (c) SIERO under nonzero bias. Indicated are well and barrier widths, subband levels, pump detuning, and pumping and lasing transitions.

A complete operating cycle of the stimulated intersubband electron Raman oscillator can therefore be described by the following processes: (a) an electron at the ground state $|g\rangle$ absorbs one photon with energy $\hbar\omega_p$, detuned from the energy separation $E_u - E_g$, and reaches the virtual state $|u'\rangle$; (b) it emits one photon of Stokes frequency ω_{out} and arrives at the lower laser state $|l\rangle$; and (c) it then relaxes back to the ground state $|g\rangle$ through nonradiative intersubband processes such as phonon scattering. There is clearly a similarity between the stimulated intersubband electron Raman oscillator and the optically pumped intersubband laser, with the only difference being that instead of the virtual state $|u'\rangle$, a real state $|u\rangle$ is used as the upper laser state in the optically pumped intersubband laser. Obviously, resonant pumping between the ground $|g\rangle$ and the upper laser state $|u\rangle$ is necessary in the optically pumped intersubband laser, while in the stimulated intersubband electron Raman oscillator, the pumping is below resonance with a detuning of $\delta = E_u - E_g - \hbar\omega_p$. The Stokes frequency $\hbar\omega_{\text{out}}$ is detuned by the same amount from the $|u\rangle$ to the $|l\rangle$ transition: $E_u - E_l = \hbar\omega_{\text{out}} + \delta$. Thus the electron Raman scattering energy conservation $E_l - E_g = \hbar\omega_p - \hbar\omega_{\text{out}}$ is satisfied.

The nonlinear susceptibility that is due to the electron Raman scattering is calculated in [123, 136]. The coefficient of linear off-resonant absorption by the detuned $|u\rangle$ to $|l\rangle$ transition also is obtained in [123].

To focus on the issue of tunability in [123] is considered a minimal configuration consisting of two coupled quantum wells as shown in Figure 21b and c, with the GaAs wells being confined by $Al_{0.3}Ga_{0.7}As$ barriers. The wide well is designed to have two confined states, whereas the narrow well has only one state when isolated. The two quantum wells form a three-level system when coupled. The energy levels of the confined states are primarily determined by the well widths. The dipole matrix elements, however, are controlled by the barrier width separating the two wells.

GLOSSARY

Carrier An electron, a hole or an exciton.

Conduction band Energy band in a crystalline solid partially filled.

Effective mass The mass of a particle which is moving through a crystalline solid.

Electron–phonon interaction The coupling between electrons and the oscillations of ions.

Exciton Electron-hole pair correlated by a coulomb-like interaction.

Free standing wire An isolated quantum well wire.

Gap Forbidden region for the electrons, which it is determined by the difference of energies between conduction and valence bands.

Hole Its the absence of an electron in the valence band.

Nanostructure Material medium whose size is of the order of few nanometers.

Phonon Collective elemental excitation which describes the oscillations of ions in solids.

Quantum dot A quantum system, in which the carriers are confined in all directions.

Quantum well A quantum system, in which the carriers are confined in only one direction, the other dimensions remain free.

Quantum well wire A quantum system, in which the carriers are confined in two directions, the other remains free.

Raman cross-section The probability of Raman scattering occurs.

Raman scattering A process in which the light is inelastic scattered by a material.

Selection rules The rules that describe the allowed and forbidden transitions for a given process.

Valence band Energy band in a crystalline solid completely filled.

REFERENCES

1. E. Roca, C. Trallero-Giner, and M. Cardona, *Phys. Rev. B* 49, 13704 (1994).
2. F. Comas, A. Cantarero, C. Trallero-Giner, and M. Moshinsky, *J. Phys.: Condens. Matter* 7, 1789 (1995).
3. C. Trallero-Giner, *Physica Scripta T* 55, 50 (1994).
4. C. Trallero-Giner and F. Comas, *Philos. Mag. B* 70, 583 (1994).
5. F. Comas, C. Trallero-Giner, and A. Cantarero, *Phys. Rev. B* 47, 7602 (1993).
6. "Light Scattering in Solids V" (M. Cardona and G. Güntherodt, Eds.), Topics in Applied Physics 66. Springer, Heidelberg, 1989.
7. M. Cardona, *Superlatt. Microstruct.* 7, 183 (1990).
8. M. V. Klein, *IEEE J. Quantum Electron.* QE-22, 1760 (1986).
9. A. Pinczuk and E. Burstein, in "Light Scattering in Solids I" (M. Cardona, Ed.), Springer Topics in Applied Physics 8. Springer, Heidelberg, 1983.
10. C. Colvard, T. A. Gant, M. V. Klein, R. Merlin, R. Fisher, H. Morkoc, and A. G. Gossard, *Phys. Rev. B* 31, 2080 (1985).
11. A. Cros, A. Cantarero, C. Trallero-Giner, and M. Cardona, *Phys. Rev. B* 46, 12627 (1992).
12. A. J. Shield, C. Trallero-Giner, M. Cardona, H. T. Grahn, K. Ploog, J. A. Hoisler, D. A. Tenne, N. T. Moshegor, and A. Toropov, *Phys. Rev. B* 46, 6990 (1992).
13. A. K. Sood, J. Menéndez, M. Cardona, and K. Ploog, *Phys. Rev. Lett.* 54, 2111 (1985).
14. J. M. Bergues, R. Riera, F. Comas, and C. Trallero-Giner, *J. Phys.: Condens. Matter* 7, 7273 (1995).
15. R. Riera, J. L. Marin, J. M. Bergues, R. Betancourt-Riera, and M. Fernández, *Rev. Mex.* 44, 167 (1998).
16. J. M. Bergues, R. Betancourt-Riera, J. L. Marin, and R. Riera, *Phys. Low-Dim. Struct.* 7/8, 81 (1996).
17. H. Kohno, T, Iwasaki, Y. Mita, and S. Takeda, *J. Appl. Phys.* 91, 3232 (2002).
18. A. Kaschner, M. Strassburg, A. Hoffmann, C. Thomsen, M. Bartels, K. Lischka, and D. Schikora, *J. Appl. Phys.* 76, 2662 (2000).
19. G. Lermann, T. Bischof, A. Materney, W. Kiefer, T. Kümmell, G. Bacher, A. Forchel, and G. Landwehr, *J. Appl. Phys.* 81, 1446 (1997).
20. L. Artús, R. Cuscó, S. Hernández, A. Pantané, A. Polimeni, M. Henini, and L. Eaves, *J. Appl. Phys.* 77, 3556 (2000).
21. M. Cazayous, J. R. Huntzinger, J. Groenen, A. Mlayah, S. Christiansen, H. P. Strunk, O. G. Schmidt, and K. Eberl, *Phys. Rev. B* 62, 7243 (2000).
22. J. Groenen, R. Carles, S. Christiansen, M. Albrecht, W. Dorsch, H. P. Strunk, H. Wawra, and G. Wagner, *Appl. Phys. Lett.* 71, 3856 (1997).
23. P. D. Persans, P. W. Deelman, K. L. Stokes, L. Schowalter, A. Byrne, and T. Thundat, *Appl. Phys. Lett.* 70, 472 (1997).
24. S. H. Kwok, P. Y. Yu, C. H. Tung, Y. H. Zhang, M. F. Li, C. S. Peng, and J. M. Zhou, *Phys. Rev. B* 59, 4980 (1999).
25. J. Groenen, C. Priester, and R. Carles, *Phys. Rev. B* 60, 16013 (1999).
26. A. A. Sirenko, M. K. Zundel, T. Ruf, K. Eberl, and M. Cardona, *Phys. Rev. B* 59, 4980 (1999).
27. J. L. Liu, G. Jim, Y. S. Tang, Y. H. Lou, K. L. Wang, and D. P. Yu, *Appl. Phys Lett.* 76, 586 (1999).
28. D. A. Tenne, V. A. Haisler, A. I. Toropov, A. K. Bakarov, A. K. Gutakovsky, D. T. A. Zahn, and A. P. Shebanin, *Phys. Rev. B* 61, 13785 (2000).
29. J. R. Huntzinger, J. Groenen, M. Cazayous, A. Mlayah, N. Bertru, C. Paranthoen, O. Dehaese, H. Carrére, E. Bedel, and G. Armelles, *Phys. Rev. B* 61, R10547 (2000).
30. N. A. Sobolev, A. Cavaco, M. C. Carmo, M. Grundmann, F. Heinrichsdorff, and D. Bimberg, Phys. *Status Solidi B* 224, 93 (2001).
31. R. Heitz, H. Born, A. Hoffmann, D. Bimberg, I. Mukhametzhanov, and A. Madhukar, *Appl. Phys. Lett.* 77, 3746 (2000).
32. A. Kasuya, M. Sugano, T. Maeda, Y. Saito, K. Tohji, H. Takahashi, Y. Sasaki, M. Fukushima, Y. Nishina, and C. Horie, *Phys. Rev. B* 57, 4999 (1998).
33. A. Kasuya, Y. Sasaki, Y. Saito, K. Tohji, and Y. Nishina, *Phys. Rev. Lett.* 78, 4434 (1997).
34. A. M. Rao, E. Richter, S. Bandow, B. Chase, P. C. Eklund, K. A. Williams, S. Fang, K. R. Subbaswamy, M. Menon, A. Thess, R. E. Smalley, G. Dresselhaus, and M. S. Dresselhaus, *Science* 275, 187 (1997).
35. A. Jorio, A. G. Souza Filho, G. Dresselhaus, M. S. Dresselhaus, R. Saito, J. H. Hafner, C. M. Lieber, F. M. Matinaga, M. S. S. Dantas, and A. M. Pimienta, *Phys. Rev. B* 63, 245416 (2001).
36. A. Jorio, R. Saito, J. H. Hafner, C. M. Lieber, T. McClure, G. Dresselhaus, and M. S. Dresselhaus, *Phys. Rev. Lett.* 86, 1118 (2001).
37. K. Kneipp, A. Jorio, H. Kneipp, R. Saito, S. D. Brown, K. Shafer, J. Motz, G. Dresselhaus, and M. S. Dresselhaus, *Phys. Rev. B* 63, 081401 (2001).
38. R. Saito, T. Takeya, T. Kimura, G. Dresselhaus, and M. S. Dresselhaus, *Phys. Rev. B* 57, 4145 (1998).
39. J. A. Creigthton, *Surf. Sci.* 124, 209 (1983).
40. T. J. Dines and R. D. Peacock, *J. Chem. Soc., Faraday Trans. 1* 84, 3445 (1988).
41. P. Jiang, C. Zhang, and G. Zhang, *Surf. Sci.* 171, L470 (1986)
42. A. Campion and P. Kambhampati, *Chem. Soc. Rev.* 27, 241 (1998).
43. J. M. Benoit, J. P. Buisson, O. Chauvet, C. Godon, and S. Lefrant, *Phys. Rev. B* 66, 073417 (2002).
44. J. P. Buisson, O. Chauvet, S. Lefrant, C. Stephan, and J. M. Benoit, *Mater. Res. Soc. Symp. Proc.* 633, A14.12.1 (2001).
45. A. C. Maciel, J. M. Freyland, L. Rota, C. Kiener, J. F. Ryan, U. Marti, D. Martin, F. Morier-Gemoud, and F. K. Reinhart, *Appl. Phys. Lett.* 68, 1519 (1996).
46. R. Saito, A. Jorio, A. G. Souza Filho, G. Dresselhaus, M. S. Dresselhaus, A. Grüneis, L. G. Cancado, and A. M. Pimienta, *Jpn. J. Appl. Phys.* 41, 4878 (2002).
47. O. Husberg, A. Khartchenko, H. Vogelsang, D. J. As, K. Lischka, O. C. Noriega, A. Tabata, L. M. R. Scolfaro, and J. R. Leite, *Physica E* 13, 1090 (2002).
48. Y. Toda, O. Moriwaki, M. Nishioka, and Y. Arakawa, *Physica E* 8, 328 (2000).

49. D.-W. Wang and S. Das Sarma, *Phys. Rev. B* 65, 125322 (2002).
50. S. Das Sarma and D.-W. Wang, *Phys. Rev. Lett.* 83, 816 (1999).
51. G. Scamarcio, V. Spagnolo, G. Ventruti, and M. Lugará, *Phys. Rev. B* 53, R10489 (1996).
52. M. C. Klein, F. Hache, D. Ricard, and C. Flytzanis, *Phys. Rev. B* 42, 11123 (1990).
53. A. P. Alivisatos, T. D. Harris, P. T. Carrol, M. L. Steigerwald, and L. E. Brus, *J. Chem. Phys.* 90, 3463 (1990).
54. J. J. Shiang, A. N. Golstein, and A. P. Alivisatos, *J. Chem. Phys.* 92, 3232 (1990).
55. E. Menéndez-Proupin and N. Cabo-Bisset, *Phys. Rev. B* 66, 085317 (2002).
56. A. Gonzalez and E. Menéndez-Proupin, *Physica E* 8, 333 (2000).
57. M. Hayne, C. L. Jones, R. Bogaerts, C. Riva, A. Usher, F. M. Peeters, F. Herlach, V. V. Moshchalkov, and M. Henini, *Phys. Rev. B* 59, 2927 (1999).
58. Y. Kim, F. M. Muntcanu, C. H. Perry, D. G Rickel, J. A. Simmons, and J. L. Reno, *Phys. Rev. B* 61, 4492 (2000).
59. H. Kohno, T. Iwasaki, Y. Mita, and S. Takeda, *J. Appl. Phys.* 91, 3232 (2002).
60. Y. Lin, J. E. Eldridge, J. Sichelschmidt, S. W. Cheong, and T. Wahlbrink, *Solid State Commun.* 112, 315 (1999).
61. A. Tanaka, S. Onari, and T. Arai, *J. Phys. Soc. Jpn.* 61, 4222 (1992).
62. O. Husberg, A. Khartchenko, D. J. As, H. Vogelsang, T. Frey, D. Schikora, K. Lischka, O. C. Noriega, A. Tabata, and J. R. Leite, *Appl. Phys. Lett.* 70, 1243 (2001).
63. N. N. Melnik, Y. G. Sadofyev, T. N. Zavaritskaya, and L. K. Vodop'yanov, *Nanotechnology* 11, 252 (2000).
64. L. M. Hernandez and I. Hernandez-Calderon, *Phys. Status Solidi B* 220, 205 (2000).
65. L. M. Hernandez and I. Hernandez-Calderon, *J. Vac. Sci. Technol. B* 18, 1542 (2000).
66. J. Gleize, F. Demangeot, J. Frandon, M. A. Renucci, M. Kuball, N. Grandjean, and J. Massies, *Thin Solid Films* 364, 156 (2000).
67. F. Demangeot, J. Gleize, J. Frandon, M. A. Renucci, M. Kuball, N. Grandjean, and J. Massies, *Phys. Status Solidi B* 216, 799 (1999).
68. A. Sayari, A. Mlayah, F. F. Charfi, R. Carles, and R. Plantel, *Phys. Status Solidi B* 207, 377 (1998).
69. A. Fainstein, Jusserand, and R. Andre, *Phys. Rev. B* 57, R9439 (1998).
70. M. Cazayous, J. Groenen, A. Zwick, A. Mlayah, R. Carles, J. L. Bischoff, and D. Dentel, *Phys. Rev. B* 66, 195320 (2002).
71. C. Rudamas, J. Martinez-Pastor, A. Garcia-Cristobal, P. Roussignol, J. M. Garcia, and L. Gonzalez, *Surface Sci.* 507, 624 (2002).
72. Y. Toda and Y. Arakawa, *IEEE J. Sel. Top. Quantum Electr.* 6, 528 (2000).
73. F. Findeis, A. Zrenner, M. Markmann, G. Bohm, and G. Abstreiter, *Physica E* 7, 354 (2000).
74. A. A. Sirenko, M. K. Zundel, T. Ruf, K. Eberl, and M. Cardona, *Phys. Rev. B* 58, 12633 (1998).
75. L. Saviot, B. Champagnon, E. Duval, and A. I. Ekimov, *Phys. Rev. B* 57, 341 (1998).
76. T. D. Krauss and F. W. Wise, *Phys. Rev. B* 55, 9860 (1998).
77. A. M. de Paula, L. C. Barbosa, C. H. B. Cruz, O. L. Alves, J. A. Sanjurjo, and C. L. Cesar, *Appl. Phys. Lett.* 69, 357 (1996).
78. P. Corio, M. L. A. Temperin, P. S. Santos, J. V. Romero, J. G. Huber, C. A. Luengo, S. D. M. Brown, G. Dresselhaus, M. S. Dresselhaus, M. S. S. Dantas, C. F. Leite, F. M. Matinaga, J. C. Gonzalez, and A. M. Pimienta, *Chem. Phys. Lett.* 350, 373 (2001).
79. H. C. Lee, *Int. J. Mod. Phys. B* 15, 3031 (2001).
80. L. J. Ci, J. Q. Wei, B. Q. Wei, J. Liang, C. L. Xu, and D. H. Wu, *Carbon* 39, 329 (2001).
81. C. Thomsem, P. M. Rafailov, H. Jantoljak, and S. Reich, *Phys. Status Solidi B* 220, 561 (2000).
82. A. Mlayah, A. Sayari, R. GracR, A. Zwick, R. Carles, M. A. Maaref, and R. Plantel, *Phys. Rev. B* 56, 1486 (1997).
83. K. Tohji, M. Sugano, A. Kasuya, Y. Nishima, Y. Saito, and H. Takahashi, *Appl. Surf. Sci.* 145, 657 (1999).
84. A. M. Rao, E. Richter, S. Bandow, B. Chase, P. C. Eklund, K. A. Williams, S. Fang, K. R. Subbaswamy, M. Menon, A. Thess, R. E. Smalley, G. Dresselhaus, and M. S. Dresselhaus, *Science* 275, 187 (1997).
85. R. Rinaldi, R. Cingolani, M. Ferrara, A. C. Maciel, J. Ryan, U. Marti, D. Martin, F. Moriergemoud, and F. K. Reinhartt, *Appl. Phys. Lett.* 64, 3587 (1994).
86. T. Egeler, *Festkorperprob-Adv. Solid State Phys.* 31, 315 (1991).
87. E. F. Bezerra, A. G. Souza, N. V. Freire, J. Mendes, and V. Lemos, *Phys. Rev. B* 64, 201306 (2001).
88. M. H. Zhang, Q. Huang, and J. M. Zhou, *J. Phys.: Condens. Matter* 9, 6257 (1997).
89. A. Cros, A. Cantarero, and I. V. Belitsky, *Solid State Electron.* 40, 715(1996).
90. P. D. Wang, Y. P. Song, C. M. S. Torres, M. C. Holland, D. J. Lockwood, P. Hawrylak, J. J. Palacios, P. C. M. Chistianen, J. C. Maan, and J. A. A. Perenboom, *Superlatt. Microstruct.* 7, 23 (1994).
91. K. J. Jin, S. H. Pan, and G. Z. Yang, *Phys. Rev. B* 50, 8584 (1994).
92. L. I. Korovin, S. T. Pavlov, and B. E. Eshpulatov, *Solid State Commun.* 89, 855 (1994).
93. G. Weber, *J. Phys.-Condens. Matter* 4, 9831 (1992).
94. E. Mariani, M. Sassetti, and B. Kramer, *Europhys. Lett.* 49, 224 (2000).
95. M. Sassetti, F. Napoli, and B. Kramer, *Microelectron. Eng.* 47, 301 (1999).
96. M. Sassetti and B. Kramer, *Ann. Phys. Berlin* 7, 508 (1998).
97. C. H. Liu, B. K. Ma, and C. Y. Chen, *Chin. Phys.* 11, 730 (2002).
98. M. I. Vasilevskiy, *Phys. Rev. B* 66, 195326 (2002).
99. E. Menéndez-Proupin, J. L. Penna, and C. Trallero-Giner, *Semicond. Sci. Technol.* 13, 871 (1998).
100. A. V. Fedorov, A. V. Baranov, and K. Inoue, *Phys. Rev. B* 56, 7491 (1997).
101. E. Menéndez, C. Trallero-Giner, and M. Cardona, *Phys. Status Solidi B* 199, 81 (1997).
102. A. Garcia-Cristobal, A. Cantarero, C. Trallero-Giner, and M. Cardona, *Physica B* 263, 809 (1999).
103. A. Garcia-Cristobal, A. Cantarero, C. Trallero-Giner, and M. Cardona, *Phys. Rev. B* 58, 10443 (1998).
104. V. Lopez, G. E. Marques, J. Drake, and C. Trallero-Giner, *Phys. Rev. B* 56, 15691 (1997).
105. M. I. Vasielevskiy, A. G. Rolo, and M. J. M. Gomes, *Solid State Commun.* 104, 381 (1997).
106. "Light Scattering in Solids II" (M. Cardona and G. Güntherodt, Eds.), Vol. 50 in Topics in Applied Physics. Springer, Berlin, 1982.
107. F. Bechstedt, R. Enderlein, and K. Peuker, *Phys. Status Solidi B* 68, 43 (1975).
108. R. F. Wallis and D. L. Mills, *Phys. Rev. B* 2, 3312 (1970).
109. R. Riera, F. Comas, C. Trallero-Giner, and S. T. Pavlov, *Phys. Status Solidi B* 148, 533 (1988).
110. M. P. Chamberlain, C. Trallero-Giner, and M. Cardona, *Phys. Rev. B* 51, 1680 (1995).
111. P. C. Sercel and K. J. Vahala, *Phys. Rev. B* 42, 3690 (1990).
112. V. M. Fomin, V. N. Gladilin, J. T. Devreese, E. P. Pokatilov, S. N. Balaban, and S. N. Klimin, *Phys. Rev. B* 57, 2415 (1998).
113. F. Comas, R. Pérez, C. Trallero-Giner, and M. Cardona, *Superlatt. Microstruct.* 14, 95 (1994).
114. R Pérez Alvarez, F. García Moliner, V. R. Velasco, and C. Trallero-Giner, *J. Phys. C* 5, 5389 (1993).
115. M. P. Chamberlain, M. Cardona, and B. K. Ridley, *Phys. Rev. B* 48, 14356 (1993).
116. T. Ando, A. Fowler, and F. Stern, *Rev. Mod. Phys.* 54, 437 (1982).
117. J. M. Bergues, R. Betancourt-Riera, R. Riera, and J. L. Marin, *J. Phys.: Condens. Matter* 7, 7983 (2000).
118. R. Betancourt-Riera, J. M. Bergues, R. Riera, and J. L. Marin, *Physica E* 5, 204 (2000).

119. C. Trallero-Giner, F. García Moliner, V. R. Velasco, and M. Cardona, *Phys. Rev. B* 45, 11944 (1992).
120. E. Menéndez, C. Trallero-Giner, and M. Cardona, *Phys. Status Solidi B* 199, 81 (1997).
121. N. Mori and T. Ando, *Phys. Rev. B* 40, 6175 (1989).
122. Al. L. Efros, A. I. Ekimov, F. Kozlowski, V. Petrova-Koch, H. Schmidbaur, and S. Shumilov, *Solid State Commun.* 78, 853 (1991).
123. G. Sun, J. B. Khurgin, L. Friedman, and R. A. Soref, *J. Opt. Soc. Am. B* 15, 648 (1998).
124. J. Faist, F. Capasso, D. L. Sivco, A. L. Hutchison, C. Sirtori, and A. Y. Cho, *Science* 264, 553 (1994).
125. J. Faist, F. Capasso, C. Sirtori, D. L. Sivco, A. L. Hutchison, M. S. Hybertsen, and A. Y. Cho, *Appl. Phys. Lett.* 67, 3057 (1995).
126. J. Faist, F. Capasso, C. Sirtori, D. L. Sivco, A. L. Hutchison, M. S. Hybertsen, and A. Y. Cho, *Phys. Rev. Lett.* 67, 411 (1996).
127. G. Sum and J. B. Khurgin, *IEEE J. Quantum Electron.* QE-29, 1104 (1993).
128. J. B. Khurgin, G. Sum, L. Friedman, and R. A. Soref, *J. Appl. Phys.* 78, 7398 (1995).
129. G. Abstreiter, R. Merlin, and A. Pineczuk, *IEEE J. Quantum Electron.* QE-22, 1771 (1986).
130. R. Y. Shen and N. Blombergen, *Phys. Rev. A* 137, 1786 (1965).
131. C. K. N. Patel and E. D. Shaw, *Phys. Rev. Lett.* 24, 451 (1970).
132. C. K. N. Patel, *Phys. Rev. Lett.* 28, 451 (1972).
133. A. Mooradian, S. R. J. Brueck, and F. A. Blum, *Appl. Phys. Lett.* 17, 481 (1970).
134. K. Suto, T. Kimura, and J. Nishizawa, *Appl. Phys. Lett.* 51, 1457 (1987).
135. K. Suto and S. Ogasawara, *J. Appl. Phys.* 66, 5151 (1989).
136. Y. R. Shen, "The Principles of Nonlinear Optics." Wiley, New York, 1984.

Encyclopedia of Nanoscience and Nanotechnology
www.aspbs.com/enn

Electronic Properties and Applications of Carbon Nanotubes

Markus Ahlskog
Helsinki University of Technology, Espoo, Finland

Christophe Laurent
Universite Paul-Sabatier, Toulouse, France

Mark Baxendale
Queen Mary, University of London, London, United Kingdom

Maria Huhtala
Helsinki University of Technology, Espoo, Finland

CONTENTS

1. INTRODUCTION

1.1. General Properties

The interest in organic, that is, carbon based, compounds as electronically and optically active materials has increased with an accelerating pace since the 1970s. With the emergence of the nanotechnology frontier in research, single organic molecules were often the focus of interest. With the discovery of the fullerene (C_{60}) in 1986 and especially the carbon nanotubes, all-carbon molecules are receiving a larger portion of the interest. Carbon nanotubes were discovered in 1991 by Sumio Iijima at the NEC laboratories in Japan rather accidentally during his TEM studies of fullerene material synthesized in an arc-discharge between graphite electrodes [1]. Although the interest in the fullerene molecule had been substantial, there were no widespread expectations of tubular extensions of the molecule. The carbon nanotube consists of one or multiple concentric shells of seamless graphite sheets; see Figure 1. A single wall nanotube (SWNT) has a diameter of around 1 nm, while the multiwalled nanotube (MWNT) may have any diameter in the range of 2–100 nm, with 10–20 nm being typical. Single nanotubes, even SWNTs, with lengths approaching 1 mm have been observed, giving astonishing aspect ratios (length/diameter). The typical length, however, is about 1 μm. As will be apparent from our brief discussion on nanotube synthesis in Section 2, there are various methods to produce nanotubes, some of which are related to the production of carbon fibers, an activity that well predates the discovery of the carbon nanotube. After the discovery, old research material on carbon fibers was given a second look, and indeed some old TEM images do show nanotubes [2]. Some methods of synthesis are especially developed with applications of molecular-level electronics in mind.

ISBN: 1-58883-059-4/$35.00

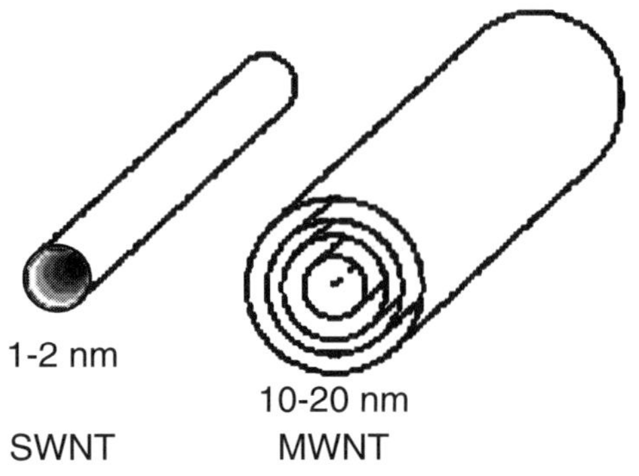

Figure 1. Carbon nanotubes can be either SWNT or MWNT.

Carbon nanotubes have several interesting physical and chemical properties that have attained much attention from both the basic and applied science communities. Depending on the wrapping angle of the graphite sheet, an individual nanotube can be either metallic or semiconducting. A fact of fundamental importance for this review is that the charge carriers in nanotubes of all sizes are truly delocalized. Ballistic conduction over μm-sized distances have been measured in carbon nanotubes. The axial Young's modulus of an individual nanotube sheet is approximated by the in-plane modulus of graphite ($\cong$1 TPa), which makes the carbon nanotube one of the stiffest materials that exists. A single carbon nanotube is an excellent freestanding electrical conductor. Chemically, the nanotube is perfectly stable in ambient conditions. In a vacuum, it can withstand temperatures far above 1000 °C without structural changes. These properties make it the ideal one-dimensional (1D) system. For technological applications, the major drawback is perhaps the insufficient control over the synthesis process. Presently the wrapping angle of the graphitic wall of the nanotube, which determines the electrical properties, can not be predetermined by the synthesicist. Nevertheless, the number of possible applications that have been forwarded within just ten years from the surprising discovery of the first nanotubes is substantial. Some of these already have been taken quite far toward realization (e.g., field emission displays), which is certainly promising for the future of the carbon nanotube.

In this chapter we will use the term nanotube to mean exclusively the carbon nanotube. However, nanotubes based on other elements besides carbon have also recently been synthesized. Most notable among these are boronitride (BN) tubes [3], which have a large energygap, and various metal chalcogenide tubes such as WS_2 [4] and MoS_2 [5]. In addition to nanotubes that are hollow and consist of concentric tubular shells, nanowires exist with diameters of about 10 nm. These nanowires are neither hollow nor are they derived from anisotropic layered materials. They have recently been synthesized from traditional semiconductor materials (Si, GaAs, InP) [6] as well as from certain metals (e.g., W [7]). Interesting electronic applications already have been demonstrated with semiconductor nanowires. It is possible that in the future the study and applications of nanotubes and nanowires will expand far beyond those made of carbon.

1.2. The Basic Geometry of Carbon Nanotubes

In the following section, we will define the basic geometries of the carbon nanotube. We will follow the presentation in Saito et al. [2]. The electronic properties in particular of a carbon nanotube are dependent on the geometry of the tube. A sheet of graphite can be wrapped in many different ways to make the wall of a carbon nanotube. Figure 2 illustrates the established nomenclature of three different types of nanotubes: (a) the armchair, (b) the zigzag, and (c) the chiral nanotube. The chiral nanotube obviously incorporates in principle an unlimited number of types with different wrapping angles relative to the tube axis. To define more precisely the lattice configuration of a single shell nanotube, we take a closer look on the honeycomb lattice of graphite in Figure 3. Vectors **C** and **T** are the chiral and translational vectors of a nanotube, respectively. They are defined by the unit vectors $\mathbf{a}_1$ and $\mathbf{a}_2$. Thus, for example, $\mathbf{C} = n\mathbf{a}_1 + m\mathbf{a}_2$. **T** points in the direction of the nanotube axis. To construct the nanotube, the graphite sheet is rolled so that the beginning and the end of **C** coincide. Thus the rectangle spanned by **C** and **T** is the unit cell of the nanotube lattice. The wrapping angle of the nanotube, or chiral angle θ, is given by the angle between $\mathbf{a}_1$ and **C**. Because of the hexagonal symmetry of the lattice, θ is limited to the range $0° \leq \theta \leq 30°$. The special cases are $\theta = 0°$, the zigzag nanotube, and $\theta = 30°$, the armchair nanotube. The chiral vector **C**, or in other words, the integer pair (n, m) defines the single shell nanotube. The tube diameter d_t and θ are given as

$$d_t = |\mathbf{C}|/\pi = a\sqrt{}(n^2 + m^2 + nm) \quad (1)$$

$$\cos(\theta) = \mathbf{C}\cdot\mathbf{a}_1/|\mathbf{C}||\mathbf{a}_1| = 2n + m/(2\sqrt{}(n^2 + m^2 + nm)) \quad (2)$$

where a is the length of the unit vectors. The armchair nanotubes are of the form (n, n) while the zigzag tubes are

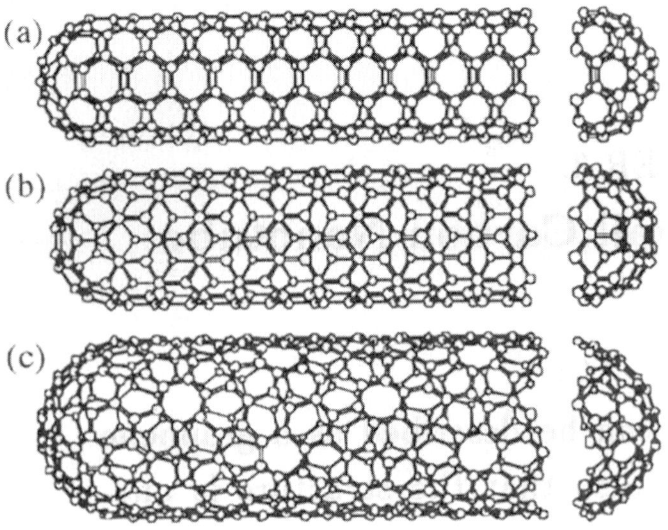

Figure 2. The different ways in which the graphitic wall of an individual carbon nanotube shell can be wrapped are generally presented as follows: (a) the armchair, (b) the zig-zag, and (c) the chiral nanotube. Reprinted with permission from [2], R. Saito et al., "Physical Properties of Carbon Nanotubes," Imperial College Press, London, 1998. © 1998, Imperial College Press.

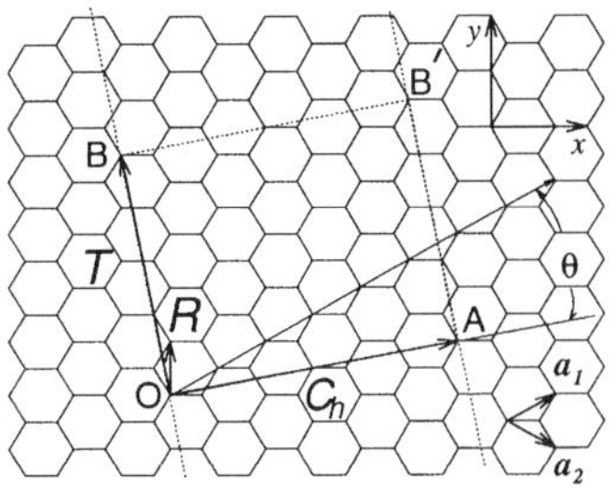

Figure 3. The definition of the vectors **C** and **T**, shown on the hexagonal lattice of a graphite sheet, that define the nanotube symmetry. Reprinted with permission from [2], R. Saito et al., "Physical Properties of Carbon Nanotubes," Imperial College Press, London, 1998. © 1998, Imperial College Press.

of the form $(n, 0)$. Furthermore, it follows from symmetry considerations that the restriction $0 < |m| < n$ can be imposed.

1.3. Various Forms of Carbon Nanotubes

Next we will discuss the different sizes and shapes a carbon nanotube can have. The variety that has already been observed is most impressive. As already mentioned, the carbon nanotube is a generic term that includes both single- and multiwalled nanotubes (SWNTs and MWNTs). The smallest nanotube reported to date has a diameter of only 4 Å [8]. Since increasing the curvature of the graphite wall increases the strain energy, this is also predicted to be the smallest possible diameter a nanotube can have. Typical diameters for SWNTs are between 1 and 2 nm, corresponding to n and m numbers of around 10. In MWNTs, the distance between the adjacent layers is about 0.34 nm, which is much more than the atomic separation (1.42 Å) within the individual layers. While typical values for the outer diameter are 10–20 nm, the inner diameter is typically 3–5 nm. Thus an ordinary MWNT has 10–30 layers, each of which are SWNTs of different diameters. The shell structure of MWNTs has not been studied thoroughly yet, but from geometrical considerations it is concluded that adjacent layers will generally be incommensurate [2]. Also, the larger the diameter of a MWNT is, the more likely the outer layers are disordered and noncircular.

The ends of nanotubes are generally closed in the growth process. While the wall of a carbon nanotube is made up of an exclusively hexagonal pattern, pentagons are needed to cap the tubes, as a close inspection of Figure 2 reveals [9]. More generally, pentagonal and heptagonal (5- and 7-cornered polygons, respectively) defects enable the graphitic sheet to take up more complicated structures than a simple tube, as we will discuss below. Unless special setups are used for the growth of SWNTs, they are usually assembled into ropes by their mutual van der Waals attraction. For example, a rope of a typical diameter of about 10 nm contains ~100 SWNTs. Since multiwalled nanotubes have a much higher bending stiffness, they do not arrange similarly.

Nanotubes grown under suitable conditions have a very low concentration of defects over μm-distances, that is, over hundreds or even thousands of interatomic spacings. However, if one were able to control the occurrence of defects, very useful and interesting nanotube structures would emerge. We will next show what kind of modified SWNTs and MWNTs have already been observed due to the occurrence of defects within the nanotubes. A single, stable defect structure in the hexagonal lattice of a graphite sheet is made up of a pentagon-heptagon pair, that is 5- and 7-sided polygons. One such defect can cause a sharp bend in a SWNT [10]. Figure 4 shows a SWNT with a sharp bend that is most likely caused by one or a few such simple defect structures. Defective nanotubes are especially interesting for electronic applications where the defect site may act as a tunneling barrier. In the case of MWNTs, highly defective tubes may be obtained with certain synthesis conditions, for example with relatively low growth temperatures. While high-quality MWNTs are very straight and stiff, very defective ones have a continuous and smooth curvature as seen in Figure 5. On the other hand, the curvature can be highly regular and so result in helices, such as is shown in Figure 6 [11]. An SWNT, or a single shell of a MWNT, can have a second nanotube branching out. The beginning point of the branch has similarly to the end sections pentagons dispersed among the hexagons to accommodate the necessary curvature. Such branching has been observed in a few cases to date [12].

As mentioned earlier, the ends of carbon nanotubes are usually closed. It is, however, possible to open nanotubes by a suitable chemical treatment. This is due to the stronger chemical reactivity of the end section with its larger curvature and pentagon structures. The opened nanotubes can be utilized for creating yet another side branch in the science of nanotubes, namely filled tubes [13]. Both SWNTs and MWNTs have been filled with various materials, such as fullerenes, simple metals, and molecular compounds. Nanotubes with fullerenes inside are called peapods, and are presently intensively investigated [14]. The study of the electronic properties of filled nanotubes is still in its infancy, and will not be considered in this review.

Finally, we will mention the μm-sized rings that were observed both in SWNT- and MWNT-based material rather early on. Figure 7 shows rings observed in MWNT material

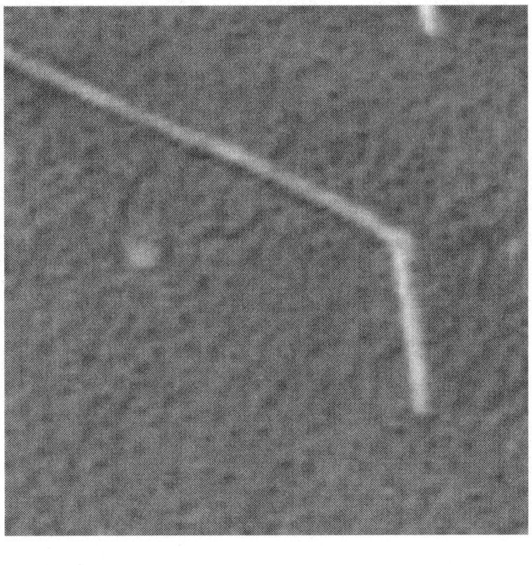

Figure 4. An AFM image of a SWNT with a sharp bend caused by a single defect site.

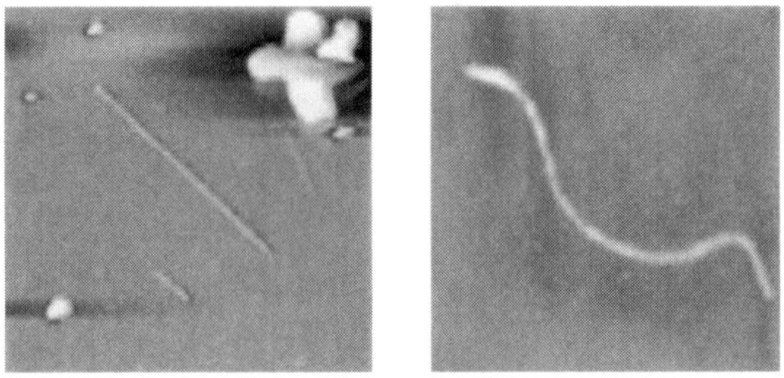

(a) (b)

Figure 5. AFM images of MWNTs produced by different synthesis conditions: (a) by the arc-discharge method and (b) by the CVD method. The curved appearance of the latter is seen to be due to a higher density of defects.

by SEM. Some claims were made for the SWNT rings to be genuine toroids, that is, seamless ring structures. Later research has, however, shown that ring structures are readily formed by the van der Waals force–mediated attraction between the two ends of a nanotube [15]. The curvature is in this case therefore not caused by defects but is determined by the competition between the strain energy of a bent nanotube and the van der Waals attraction energy. In fact, especially in the case of SWNTs, the ring may be composed of several turns of the nanotube (or a nanotube rope). Recently, it was claimed that a well defined single-turn ring can be made by the attraction between chemical groups attached to both ends of a nanotube [16].

1.4. Concluding Remarks

The prospects for widespread applications for carbon nanotubes in electronics (and elsewhere) are good for several reasons: As already discussed, the carbon nanotube is stable and durable both chemically and mechanically, and in addition is an excellent conductor. Furthermore, the previous discussion has made it clear that an impressive variability

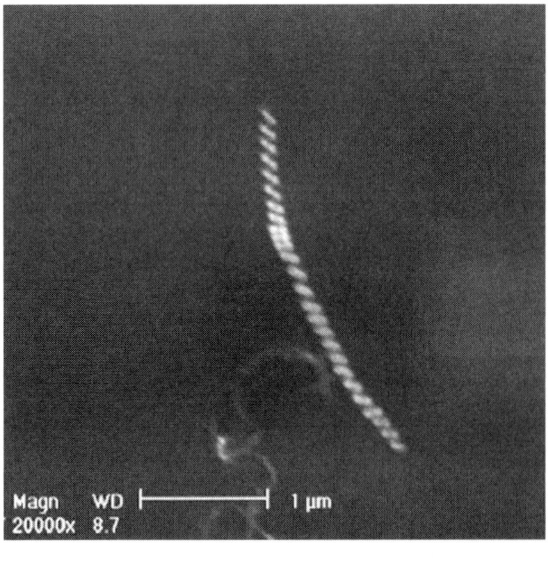

Figure 6. A SEM image of a coiled MWNT produced by the CVD method.

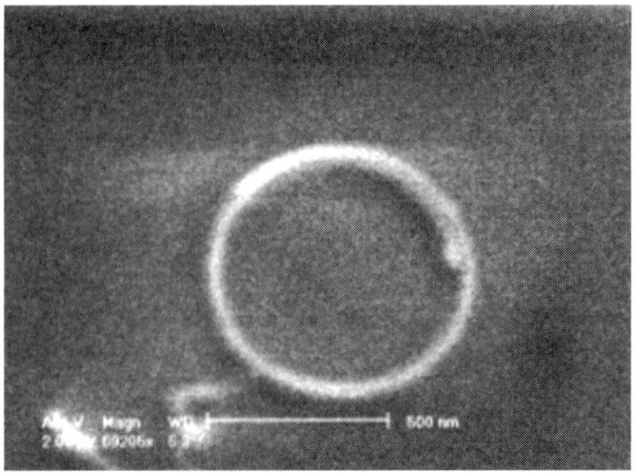

Figure 7. A SEM image of a carbon nanotube ring obtained from CVD grown MWNT material. The scale bar is 0.5 μm.

exists in the possible different nanotube structures, ranging from the single SWNT with its different chiralities to complex branched and filled nanotubes. To date the control over the synthesis process is very poor but there are no known fundamental obstacles to efficient and controlled production of nanotubes for technological applications. Meanwhile, the impact of carbon and noncarbon nanotubes in the physical sciences has already been important and continues to enrichen scientific research.

2. SYNTHESIS AND CHEMICAL PROPERTIES

2.1. Introduction

The formation of filamentous carbon from the decomposition of carbonaceous gases was reported in the 19th century [17]. There has been quite a lot of work since then on the topic, notably since the 1960s, and the formation mechanisms have been particularly discussed (see Rodriguez [18] for a review). Most authors observe the presence of a catalyst particle at the tip of each carbon filament and therefore propose a formation mechanism close to the vapor-liquid-solid (VLS) mechanism [19]. This involves four main steps.

1. Gas adsorption at the surface of the particle and decomposition producing surface carbon species.
2. Bulk and/or surface diffusion toward the other side (rear side) of the particle.
3. Carbon precipitation on the rear side and lift-off of the particle.
4. Migration of surface carbon species until the catalyst surface is totally covered and thus deactivated.

Transmission electron microscopy (TEM) images showing very small hollow nanofilaments were published in 1976 [20], but neither the nanometric dimension nor the notion of nanotube were stressed. However, looking back with today's knowledge reveals that these nanofilaments indeed are Iijima-type nanotubes. This shows that it is possible to prepare nanotubes with the relatively simple techniques used for the production of filamentous carbon, known as catalytic chemical vapor deposition (CCVD). It was proposed [21] that the crystallographical arrangement adopted by the metal particle will direct the precipitation of carbon toward the formation of filaments with different morphologies, that is, either with graphene planes perpendicular to the growth direction (nontubular fiber), with the so-called herringbone structure (hollow fiber), or as several concentric cylinders (thus forming what is now called a MWNT). For the latter case, a noteworthy growth model was proposed by Tibbetts [22].

However, it is the groundbreaking report by Iijima [1] on the obtaining of nanotubes with a diameter in the nanometer range and on their relation to the recently discovered fullerenes that triggered a worldwide research effort devoted to improving their synthesis, notably that of SWNTs. The most common synthesis methods (arc discharge, laser-vaporization, and CCVD) will be presented in the following section. First, the formation mechanisms will be briefly examined (see [23] for an early review).

Several authors using arc-discharge without a metal catalyst [24, 25] have proposed a growth mechanism for MWNTs based on the repeated incorporation of carbon dimers into an empty fullerene cage. Lengthening results from rearranging the carbon bonds to incorporate additional dimers near a pentagonal ring. Other researchers [26–28] have proposed that carbon atoms (C1, C2, C3) are adsorbed at the active dangling bond edge sites, thus always maintaining the tubes open during the growth. It was suggested [26, 27] that nanotube thickening occurs by the island growth of graphene basal planes on existing surfaces.

In the presence of metal catalyst, SWNTs are produced by arc-discharge, laser vaporization, and CCVD. Several nanotube formation mechanisms have been proposed but whatever the synthesis route, they basically fall into one of two categories: "one particle–one nanotube" or "one particle–many nanotubes," and they bear a certain relation with the VLS mechanism. In the case "one particle–one nanotube," observed when using CCVD, the diameter of the SWNTs is in the range 0.5–5 nm and is directly related to that of the catalyst particle. The critical diameter above which the formation of concentric carbon layers around the particle is favored was found to be about 3 nm [29, 30]. TEM images of both ends of isolated SWNTs support a base-growth mechanism [31]. In the "one particle–many nanotubes" mechanism, a large number of SWNTs bundles (or ropes) originate from the surface of a metal particle several tens of nm in diameter [32–34]. A narrow catalyst particle distribution centered on 15 nm appears to give the highest SWNT yields [35].

Whatever the synthesis route, research efforts on the synthesis and formation mechanisms have enabled the preparation of nanotubes samples of better as-produced purity and with an improved control on the diameter and number of walls. However, most importantly for the field of electronic properties, it is to be noted that an absolute control of the diameter and therefore of the helicity has so far not been found possible.

2.2. Most Common Synthesis Methods

In the following sections, we will only briefly describe the principal synthesis methods (arc-discharge, laser vaporization, and CCVD) and the characteristics of the so-prepared nanotubes. Each method could be the subject of a review on its own, which is beyond the scope of the present review. However, it is important to note some information that was not fully appreciated at the beginning of the development of the "nanotube" field, that is that each method produces different nanotube samples, that differences subsist depending on the experimental conditions for a given method, and that even different batches from a single source may not be exactly identical. A review, with 282 references, on the main synthesis methods and others (resistivity vaporization, electron- or ion-beam vaporization, sunlight-induced vaporization, flame synthesis, molten-salt electrolysis, chlorination of carbides and more) was published by Rakov [36].

2.2.1. Arc-Discharge

The arc-discharge method for the preparation of nanotubes is similar to that used for the synthesis of fullerenes. Following Iijima's report [1], modifications were proposed for the large-scale synthesis of nanotubes [37]. An inert gas atmosphere (typically He or Ar) is flown through a reaction vessel at a controlled pressure. Two graphite rods constitute the electrodes, between which a potential difference is applied. As the rods are brought close together, a discharge occurs, resulting in the formation of a plasma. A deposit, which may contain nanotubes under certain conditions, forms on the larger negative electrode (cathode) while the smaller positive electrode (anode) is consumed. Typically, the outer diameter of MWNTs prepared by the arc-discharge method ranges between 2 and 20 nm and the inner diameter ranges between 1 and 3 nm [38–41]. The length is of the order of one to several micrometers. MWNTs form regularly organized bundles, which may contain tens or hundreds of individual tubes. The bundles in turn are combined into fibers with a diameter of about 50 μm, which themselves form threads with a diameter of about 1 mm. It was furthermore noted that the arc-discharge MWNTs do contain many defects [42–44].

Using a metal catalyst was reported to produce SWNTs [45, 46]. A hole drilled in the anode is filled with a mixture of metal and graphite powder. The diameter distribution is very narrow: 0.7–1.6 nm [45] and 1.2 nm [46]. Ni-Y in particular emerged as a very good catalyst allowing a large-scale production of SWNTs [47]. Carbon deposits may be found in different places of the reaction chamber: as a soot deposited on the wall, as a hard shell at the surface of the cathode, as a soft belt (sometimes referred to as "collaret") around the cathode deposit, and as a web-like precipitate decorating the chamber walls. Various carbon species are found in the deposits including SWNTs, MWNTs, amorphous carbon, spherical metallic nanoparticles, empty or filled graphitic nanoparticles, and a few graphitic sheets, as well as round spherical metallic particles. The density of SWNTs is particularly high in the collaret. Purification procedures usually involving thermal and acidic treatments are required to eliminate the residual metal catalyst (the amount of which can reach 50 wt%) and the nontubular carbon species. The SWNT yield and diameter distribution depend on several other parameters including inert (He, Ar) or reactive (H_2, CH_4) gas nature, pressure and flow rate, electrodes nature, purity, diameter and position (vertical or horizontal), reactor geometry, arc-discharge variable (voltage, current plasma temperature), process duration, and configuration of cooling devices. DWNTs were also prepared by a hydrogen arc discharge method [48].

The SWNTs are mostly organized in bundles consisting of a few to a few tens of individual tubes forming a triangular lattice and more or less covered with some carbon soot. The diameter distribution of the SWNTs is very narrow (1.2–1.4 nm) and lengths reach up to several μm. The tube tips are generally seen closed, in agreement with observations for MWNTs, without any trace of catalyst species at the end. MWNTs filled with Sn, Te, Bi, Ge, Sb, Pb, Al, In, S, Se, Cd, Gd, or Hf have been observed [49–51].

2.2.2. Laser Vaporization

Guo et al. [52] proposed an original method in which a mixture of carbon and transition metal is vaporized by a laser impinging on a metal-graphite composite target. The reactor tube is placed inside a furnace (typical temperature 1200 °C, flowing Ar). It is claimed that in contrast to the arc method, direct vaporization allows far greater control over growth conditions, permits continuous production, and produces SWNTs in higher yield (70–90%) and of better quality. The SWNTs are about 1–1.6 nm in diameter, a very narrow distribution, and are arranged in large bundles (ropes) that may be 100 μm long, with a similar spacing between adjacent SWNTs. It is claimed that no amorphous coating is observed. Bimetallic catalysts (Co-Ni and Co-Pt mixtures) have a SWNT yield 10–100 times that for single metals alone. It was suggested that the principal effect of a mixture is to increase the mobility of carbon on and/or within the metal particle, resulting in a greater rate of SWNT precipitation from the particle.

Optimizations of the method using pulsed laser [53], separate carbon and metal-alloy targets [54], or without a furnace [55] have been reported. The diameter distribution can be slightly changed by varying the process temperature [56] and gas pressure or flow velocity [57] and by changing the chemical composition of the gas and of the target. The introduction of hydrogen (as admixture to Ar), sulfur (as a FeS additive to the carbon target), or, particularly, both leads to the efficient formation of SWNTs with large diameters from 2 to 5.6 nm [58]. In some conditions, laser vaporization SWNTs were also found [14] to contain C_{60} fullerenes (referred to as a "peapod"), which may coalesce, thus forming a second tube inside the first one [59].

2.2.3. CCVD

The CCVD methods basically consist of passing a gaseous flow containing a certain proportion of hydrocarbon (mostly CH_4, C_2H_2, C_2H_4, and C_6H_6 usually as a mixture with N_2 or H_2) or CO over small transition metal particles. Many parameters, including the temperature and duration of the treatment, the gas composition and flow rate, and the catalyst nature and size, will affect the nature of the carbon species in the resulting material. Indeed, as observed for studies on filamentous carbon and vapor-grown carbon fibers [18], several species may be obtained in addition to Iijima-type nanotubes, including amorphous carbon, carbon particles either connected or not to the metal or carbide particles and carbidic carbon particles. Here we will only consider a few post-1991 studies on the formation of MWNTs and SWNTs.

The decomposition of C_2H_2 and H_2-C_2H_2 on well-dispersed metal (Fe, Co, Ni, Cu) particles strongly adsorbed on a support (graphite flakes, SiO_2) produces MWNTs and graphitic filaments, sometimes covered by amorphous carbon [60, 61]. Co-SiO_2 was reported to be the best catalyst-support combination for the production of MWNTs. Straight and coiled tubes were obtained with inner and outer diameters of 3–7 and 15–20 nm, respectively, and up to 30 μm in length. The precipitation-ion-exchange method was found to provide a better dispersion of metals on SiO_2 than the impregnation technique [61]. The length of the MWNTs was found to depend on the duration of the catalytic process. However, the longest ones are also the thickest. The use of a zeolite as a support [62, 63] results in more finely dispersed metal particles (1–50 nm). Only on this Co-zeolite catalyst could thin tubes (4 nm, 2–3 walls) be observed, however in a very small amount. The influence of the addition of H_2 to CO on the MWNT or herringbone morphology was reported [64, 65]. The pyrolysis of metallocenes [66, 67], which act as a feedstock for both the transition metal and carbon, produces short MWNTs sometimes filled with the corresponding metal particles and coated with amorphous carbon.

For the preparation of SWNTs, it is of utmost importance that the metal particles are in the nanometer size range at the appropriate temperature. As previously noted, the critical size of the catalyst particles is ~3 nm. Achieving this is obviously related to experimental conditions such as reaction atmosphere, temperature, and duration, but also, most importantly, to the way the catalyst material is prepared. The first CCVD SWNTs were prepared by CO disproportionation catalyzed by preformed Mo particles a few nanometers in diameter [68] and by CH_4 decomposition on Fe nanoparticles formed *in-situ* by elective reduction of an oxide solid solution [69]. Several groups have reported progresses in the yield, control of the diameter, and of the number of walls. Indeed, detailed HREM observations often reveal the presence of DWNTs and small MWNTs in addition to SWNTs [29, 30]. As noted above, a mechanism such as the "yarmulke" [68] accounts for the formation of both SWNTs and DWNTs. The HiPCO method [70] involving CO and $Fe(CO)_5$ allows the production of large quantities of SWNTs. The diameter of the CCVD CNTs, in the range 1–5 nm, is closely correlated with the size of the catalytic particle. Co, Fe/Co, Fe/Mo, and Co/Mo appear to be the best catalysts for the formation of SWNTs, although the role of molybdenum has not yet been totally explained. Length is ranging between ~100 nm and several micrometers, and the surface is free of the amorphous carbon coating typically observed in catalytically produced MWNTs. The SWNTs are quite flexible and in general contain more defects than those produced by arc-discharge and laser vaporization.

CCVD methods are becoming increasingly attractive, owing to their great potential for the production of large quantities of nanotubes at a low cost, using standard technology. Moreover, one key advantage is the possibility to perform the oriented or localized growth of nanotubes.

2.3. Oriented or Localized Growth

The term "oriented" is used here to describe the formation of large quantities of CNTs aligned perpendicularly to the substrate surface, usually MWNTs of controlled dimensions (diameter 20–400 nm, length 0.1–60 μm). The key steps are substrate patterning, catalyst deposition, and the appropriate use of different CVD routes. The description of the numerous studies on this topic is well beyond the scope of this review. (see Huczko [71] and references therein).

The term "localized" describes the self-assembled nanointerconnections made by CNTs between islands on prepatterned substrate. The directed growth of SWNTs parallel to the plane of a silicon substrate was reported [72]. The SWNTs are suspended bridges grown from catalyst material placed on top of regularly patterned silicon tower structures. The key synthetic step in this approach involves developing a series of liquid-phase catalyst precursor materials, allowing for uniform film formation and large-scale catalyst patterning. Contact printing techniques were used to selectively deposit catalyst precursors on top of the silicon tower arrays. Calcination led to the formation of catalyst particles confined on the tower tops and subsequent CVD (CH_4, 900 °C, 20 min) growth yielded SWNTs emanating from the towers. Directed free-standing SWNT networks were formed by nanotubes growing to adjacent towers and becoming suspended above the surface. It is noted that a rational design of tower arrangements could lead to a variety of directed SWNT architectures. The liquid-phase catalyst precursors consist of three general components: inorganic chloride precursors, a removable triblock copolymer serving as the structure-directing agent for the chlorides, and an appropriate alcohol for dissolution of the inorganic and polymer compounds. The increase in yield was achieved through a so-called conditioning step that uses bulk amounts of catalysts placed near (upstream) the catalyst-patterned substrate in the CVD growth environment [73]. The influence of the CH_4 "wind" on the directional growth of the nanotubes was discussed [73, 74]. A hot-filament CVD technique was also reported [75]. SWNTs were grown onto more than 90% of 375 Si tips prefabricated on a wafer [76].

2.4. Chemistry of Carbon Nanotubes

The chemistry of carbon nanotubes, which encompasses the aspects of purification and functionalization, is a very important field for tailoring their structural and electronic properties and thus for the development of future applications. (see Rakov [77] and references therein). Purification aims at removing the nontubular material in raw nanotube samples. Several techniques, more or less complex, have been proposed, including solvent treatment followed by ultrafiltration [78], flocculation using aqueous surfactants [79, 80], oxidation and acid washing coupled with centrifugation, resuspension in surfactant solution, and cross-flow filtration steps [80–82]. These methods however lead to a mixture of nanotubes of differing lengths and still containing potential contamination with nontubular material. Polymer suspensions [83, 84] and chromatographic purification [85, 86] are useful to remove nontubular material. Field-flow fractionation on purified, shortened nanotubes [87], and size-exclusion chromatography on raw nanotubes [88, 89] have demonstrated some success at producing fractions separated by length. Capillary electrophoresis was reported to allow separation based on tube length [90].

An early paper by Ebbesen et al. [91] reported that a thermal oxidation treatment eliminates all nanoparticles and SWNTs caps so that open SWNTs can be obtained free of other carbon species. However, only ca. 1 wt.% of the initial material remained after oxidation. As noted by Chen et al. [92], the chemical reactions that selectively attack the sites of preferential reactivity in nanotubes (the caps where the pyramidalization and curvature is greatest) are not satisfactory for the production of processable materials, because this type of transformation is too localized to bring about the necessary change in bulk properties. In contrast, one needs to selectively attack a few of the π-bonds in a controlled manner without bringing about a total destruction of the graphene electronic structure. This presents a further challenge to experiment, because the reduced strain in the walls lessens the reactivity. Chemical functionalization was reviewed by Sinnott [93].

Several methods for attachment of oxygen-containing groups to the nanotube surfaces have been developed [94, 95]. Carboxylic, carbonyl, and hydroxyl groups were found covering the nanotube walls [96]. The concentration of acid groups on the surfaces of HNO_3-treated nanotubes was higher than that found for graphite treated under the same conditions [97]. Sonication of nanotubes prior to the functionalization in acids increases the concentration of acid groups. Oxidation of nanotubes with a H_2SO_4-HNO_3 mixture also leads to a higher concentration of functional groups on the surface [98]. Long-term acid treatment of nanotubes also induces modifications. Functional groups can be removed from the surface of nanotubes by heating. It should be noted that SWNTs obtained by different methods exhibit different behavior in the course of acid treatment and subsequent oxidation. The sensitivity of SWNTs to chemical processing was investigated by TEM [99].

To enhance the possibilities of atomic force microscopes with nanotube probe tips, the tip ends were functionalized using both solutions [87, 100, 101] and gases [102–104]. Modification of tips using a discharge in N_2 results in the formation of nitrogen-containing heterocycles at the tip ends. High degree of saturation of nanotube tips with functional groups is achieved by treatment with a low-pressure ammonia plasma followed by oxidation with a $NaClO_3$ solution [103].

Studies on fluorination of nanotubes were reviewed by Rakov [77] and Touhara and Okino [105]. Fluorination was found to have a strong effect on the electronic properties of nanotubes: untreated nanotubes were conductors while fluorinated nanotubes were insulators [106]. Fluorination below 325 °C appeared to be reversible. The interaction between fluorinated nanotubes with a composition close to C_2F and anhydrous hydrazine resulted in removal of the fluorine atoms and in recovery of the initial structure of the nanotubes. This was accompanied by recovery of many of their initial properties. Fluorinated nanotubes membranes are potentially interesting materials for electrochemical batteries or capacitors [105].

The reactions with silicon of individual nanotubes or nanotube bundles at 970 °C in ultrahigh vacuum resulted in the formation of Si-β-SiC(nanorod)-nanotube heterojunctions [107].

3. ELECTROMECHANICAL PROPERTIES

3.1. Introduction

Along with their electronic properties, carbon nanotubes have been praised for their mechanical properties. The praise is based on high strength, flexibility, and large aspect ratio. A carbon nanotube is a freestanding conductor and its mechanical properties are highly relevant for the electronic properties and the potential technological applications of the tube. For example, nanotube-based mechanical oscillators may have applications in high-frequency electronic circuits, and carbon nanotubes are already used in scanning probe microscopy both as STM and AFM tips. Thin, strong, and durable nanotubes come extremely close to an ideal scanning tip.

Mechanical properties of carbon nanotubes are closely related to the properties of a graphite sheet, but the tubular anisotropic form affects the mechanical behavior. The basis is the graphite sp^2 bond, which is the strongest of chemical bonds. In nanotubes the overall density of defects can be extremely low depending on the synthesizing method and prevailing synthesizing parametres. This has led to predictions of a very high axial strength.

3.2. Basic Mechanical Properties

Theoretical calculations of Young's modulus for individual SWNTs center around 1 TPa or slightly higher [108, 109], but values as high as 5.5 TPa [110] have been presented. The spread is due to different interaction models and, also to differing values of nanotube wall thickness—the wall thickness of an isolated nanotube is not a well-defined quantity. Most of the theoretical attention has been on SWNTs because modeling the interlayer interaction in MWNTs is a complicated matter. Lu [109] presents Young's modulus values for multiwalled tubes as well as SWNTs and obtains values from 0.97 TPa to 1.11 TPa with the value increasing slightly with the number of layers.

The small size of carbon nanotubes presents challenges also for experimental characterization. Nevertheless measurements have been performed. The first Young's modulus measurement [111] related thermal vibration amplitudes of MWNTs to their Young's modulus and obtained an average value of 1.8 TPa with a large spread. After that, with AFM techniques, values such as 1.28 ± 0.59 TPa for arc-discharge–produced tubes [112] have been obtained. The latest measuring technique by Poncharal et al., who induced vibrations on MWNTs by alternating potential and have measured the vibration frequencies [113]. Young's modulus values between 0.7 TPa and 1.3 TPa for tubes with a diameter less than 12 nm and between 0.1 TPa and 0.3 TPa for thicker tubes were reported. The drop is explained by an onset of a wavelike bending mode of the nanotube wall at the inner arc of the bending [113]. Measuring SWNTs is more complicated than measuring MWNTs due to their small diameters and the tendency to bundle. Krishnan et al. [114] report a measurement of individual SWNTs using the thermal vibrations method of Ref. [111], and they obtain an average value of 1.25 TPa. Although the current measurements suffer from inaccuracies due to vibration amplitude measurement and assumptions made on AFM tip characteristics, the current agreement is that defect-free nanotubes, both SWNTs and MWNTs, have a Young's modulus value around or slightly above 1 TPa, which is extremely high and sets nanotubes as the strongest known material albeit challenged by other nanotubular structures such as boronnitride tubes.

Theoretically, nanotube tensile strength is high, and this is supported by calculations in which SWNTs support as high as 30% of axial strain before brittle failure [115] and by more recent kinetic activation based calculations that give a yield strain of 17% with a chirality and temperature-dependent defect formation activation energy barriers [116]. Experimentally nanotube tensile strength has been measured for MWNTs by Yu et al. [117]. In the measurement, the contact is only to the outmost layer, and a sword-and-sheath type of failure of this outermost layer is reported. Tensile strength values ranging from 11 GPa to 63 GPa are reported [117]. The role of the inner layers is not, however, clear, and the failure process may be more complicated than has been proposed. For individual SWNTs, the experimental value of tensile strength is still an open question, but for bundles of SWNTs tensile strength values ranging between a few GPa and several tens of GPa depending on the bundle and measurement characteristics have been reported [118–120].

As a result of the high tensile strength, nanotubes can bear large strain before plastic deformation or brittle failure. The tubes buckle, flatten, form ripples, and generally deform under strain [110, 115, 121], but plastic deformation and relaxation occur only at elevated strain levels and temperatures [122–124]. An example of simulated nanotube behavior under strain is presented in Figure 8.

3.3. Defects in Carbon Nanotubes

As in any material, defects play an important role in nanotube properties. Structurally, defects make the tube less strong and thus in general defects are not desirable from the purely mechanical point of view. However, they alter the electronic properties locally, which can be utilized in the creation of single-tube devices. Defects are generated in the synthesizing process, and they can also be caused by mechanical manipulation, or, for example, by ion or electron beam bombardment of the tube. The most typical structural defects are fivefold (pentagon) and sevenfold (heptagon) rings in the sixfold (hexagonal) lattice. Other types of typical defects are vacancies and miscellaneous bonding configurations such as amorphous diamond. Noncarbon-based defects include substitutional atoms or atom groups. In addition to these, MWNTs exhibit diverse defects based on discontinuous inner layers. Defects may alter the tube form from a straight tube to a bulging, kinked, spiral, or even more miscellaneous form.

The effect of defects and deformations on nanotube conductivity has been studied in many reviews [125–131]. Defects and deformations induce scattering and localization,

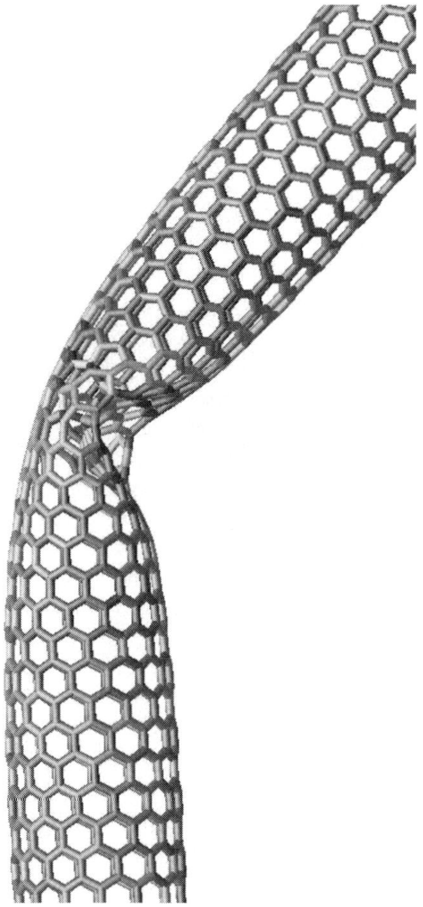

Figure 8. A molecular dynamics simulations based example of nanotube behavior under strain. The image shows a portion of a (14, 0) nanotube of 2632 atoms (tube length is 20 nm) bent 36 degrees with a constant bending rate in 300 K temperature and then, to enhance structural reconstruction, annealed in an elevated 3300 K temperature for a sequence of three 130 ps annealing and cooling simulation runs. In the bending process the tube buckled but remained hexagonally bonded. The defects visible in the image relax the strain and are a result of the third annealing [M. Huhtala, K. Mogami, A. Kuronen, and K. Kaski, to be published].

which result in conductivity drops, symmetry breaking based gap opening, and conductivity barriers. The scattering effects of isolated point defects are averaged over the circumference, and therefore their influence decreases with an increasing diameter and is small for all but the very thinnest tubes. However, if the defects or deformations accumulate locally, a tunneling barrier may form. AFM manipulation has been used to make sharp bends in SWNTs, which has resulted in the formation of tunnelling barriers [130–133]. The created changes are not, however, necessarily permanent. Another AFM based method for creating permanent tunneling barriers has been reported by Park et al. [134]. Nanotube defect manipulation offers interesting prospects both for devices with permanent characteristics and for devices with tunable properties, but a breakthrough requires more accurate and flexible mass manipulation of tube properties than is currently available.

3.4. Carbon Nanotubes as Mechanical Oscillators

Due to their small size, combined with a low mass density and high Young's modulus, nanotubes are predicted to have characteristic vibration frequencies in the gigahertz regime, which is extremely high for a mechanical vibrator [135, 136]. Such frequencies are not attainable with the current silicon-based micromechanical oscillator production techniques with which eigenfrequencies in the 10 MHz to 100 MHz regime can be attained [137]. A first approximation for the fundamental modes of a nanotube vibrator can be obtained from elasticity theory and is for a thin cylindrical transversally vibrating rod that is clamped from one end [138]

$$f_1 = 0.28\frac{d_t}{L^2}\sqrt{\frac{Y}{\rho}} \tag{3}$$

If values for tube diameter $d_t = 1$ nm, Young's modulus $Y = 1$ TPa, and density $\rho = 750$ kg/m^3 are used (The mass density of a SWNT is not a well-defined quantity; the value presented here is a rough estimate based on the nanotube atomic spacial density and the volume encompassed by a tubular cylinder depicting the tube of 1 nm radius, to compare with graphite mass density is 1900–2300 kg/m^3 [139]), a frequency of 10 MHz is obtained for a tube length of 1 μm and a frequency of 1 GHz for a length of 100 nm. For a rod clamped from both ends, the constant in front of the equation is altered according to the boundary conditions, but the overall dependency on d_t, L, Y, and ρ remains [138]. Based on the discussion above, nanotubes, and especially SWNTs, are capable of significantly smaller dimensions and higher frequencies than conventional silicon-based vibrators.

Currently, only a few studies of excited mechanical vibrations in nanotubes exist. Poncharal et al. apply alternating potentials to MWNTs and measure Young's modulus [113], and Reulet et al. [135] have performed a basic study of acoustoelectric coupling in ropes of SWNTs, and they report that depending on the excitation power, either electron heating or phase coherence breaking is observed.

To conclude, there are many things in the mechanical properties that have not been measured, and the measurements that exist suffer from a degree of inaccuracy. Most theory concentrates on SWNTs, and although the SWNTs often show more clearly the nanotube characteristic and desirable electronic properties, the theoretical aspects due to multiple layers still have many questions to answer, as do defectous-structure related matters. However, in nanoscale science and technology, electronic and mechanical will become more intimately connected, and therefore in the future the electromechanical properties of carbon nanotubes will play an important role.

4. ELECTRONIC STRUCTURE AND PROPERTIES

4.1. Tight Binding Model

The carbon atoms in nanotubes are sp^2-hybridized, thus each atom contributes one unpaired π-electron to the tube. Carbon nanotubes are electronically active materials due to these π-electrons. The basic electronic properties of the carbon nanotube are calculated by extending the tight-binding approach of 2D graphite to the geometrical configuration of the 1D nanotube system. The crystal momentum vector perpendicular to the nanotube axis is quantized due to the periodic boundary condition imposed by the finite circumference of the nanotube. This relatively simple model has been proven, mainly by low temperature STM spectroscopy, to explain well the basic electronic structure of the carbon nanotube. According to this model, the chirality of the carbon nanotube determines its division into metallic and semiconducting tubes. In the model, the curvature of graphitic wall of the nanotube is ignored. This has lately been shown to bring about important modifications to the model [140]. Still the model survives well enough to continue to serve as a basis for classification of the nanotubes into metallic and semiconducting ones.

In the following brief presentation of the basic tight-binding model of the single shell carbon nanotube, we use the nomenclature of Saito et al. [2]. The different geometric shapes that a carbon nanotube can have and the (n, m) indices used for their identification were introduced in Section 1. Due to the tubular 1D structure, the nanotube has N separate 1D energy bands. In a general form these are derived from the graphite energy levels by the general formula

$$E_\mu(k) = E_{g2D}\left(k\frac{\mathbf{K}_2}{|\mathbf{K}_2|} + \mu\mathbf{K}_1\right) \quad \left(\mu = 0, \ldots, N-1, -\frac{\pi}{T} < k < \frac{\pi}{T}\right) \tag{4}$$

E_{g2D} denotes the energy eigenvalues of a 2D graphite sheet, $\mathbf{K}_1$ and $\mathbf{K}_2$ are reciprocal lattice vectors corresponding to $\mathbf{C}$ and $\mathbf{T}$, respectively. Figure 9 shows how the carbon nanotube k-vectors are projected to the graphite momentum space lattice. In graphite the conduction and valence bands meet at the so called K points in the momentum space, which is responsible for its semimetallic behavior. In the tight-binding picture, the metallicity of a carbon nanotube is determined by whether the 1D lines of allowed k-vectors run through these points. The dispersion relations for the armchair tube (5, 5) are shown in Figure 10 as an example. In an energy diagram, whether the energybands will cross the Fermi energy or leave an energy gap depends on the (n, m), that is the chirality, as follows: The armchair nanotubes, with indices (n, n), have two bands that cross each other and the Fermi energy at the points $k = \pm 2\pi/3a$; they are thus always metallic, while only some of the zig-zag tubes are so. The general condition for the metallicity of a nanotube is that the factor $2n + m$ is a multiple of 3, a condition that indeed is fulfilled for the armchair tubes. As a further result, the theory predicts that one-third of all nanotubes are metallic while two-thirds are semiconducting. These rates of occurrence have not been confirmed by the presently performed experiments, which however is not surprising considering doping effects and the fact that almost nothing is known about the possible preference in the growth processes for certain nanotube symmetries over others.

As a general result, in all metallic 1D systems with half-filled bands, the so-called Peierls distortion will, by an atomic displacement, change the unit cell so that the band becomes filled and a gap opens up at the Fermi energy. The phenomenon is well known among conducting polymers, charge transfer salts, etc. [141]. In carbon nanotubes, however, it has been shown theoretically that the effect very quickly diminishes as a function of nanotube radius [2].

4.2. Experimental Studies: STM Imaging and Spectroscopy

The electronic structure of the SWNT that is predicted by the tight-binding calculations described above was put to test at first primarily by scanning tunneling spectroscopy, done with low temperature STM [142, 143]. In such experiments, it is possible to image the nanotube with atomic resolution along the length of the tube and with a width of a few unit cells. The width of the imaging is of course limited by the curvature of the nanotube. This imaging reveals (in the best cases) the structure and diameter of the nanotube, enabling one to determine the (n, m) indices of the nanotube with some confidence. Furthermore, the STM tip can be positioned at any point along the nanotube, separated by

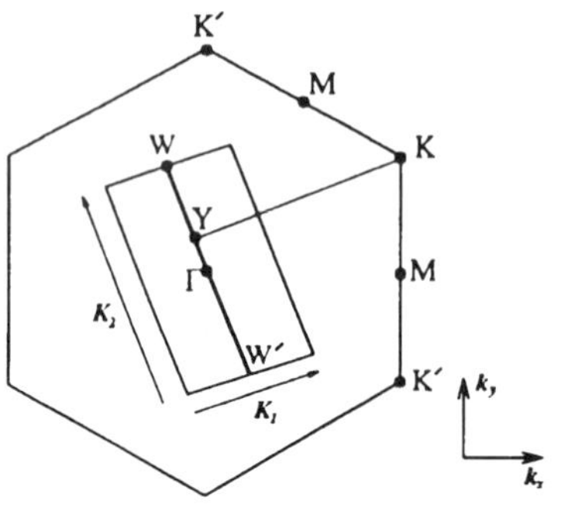

Figure 9. Allowed k-vectors of carbon nanotubes are formed by the vectors $\mathbf{K}_1$ and $\mathbf{K}_2$ in the reciprocal lattice space of graphite. The nanotube is metallic if the ratio of **YK** to $\mathbf{K}_1$ is an integer. Reprinted with permission from [2], R. Saito et al., "Physical Properties of Carbon Nanotubes," Imperial College Press, London, 1998. © 1998, Imperial College Press.

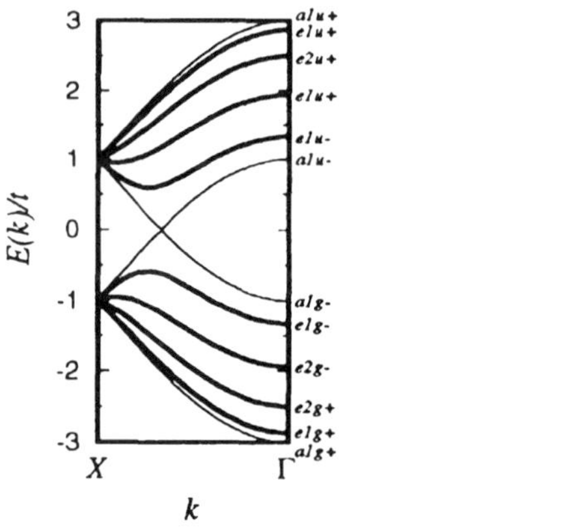

Figure 10. Energy dispersion relations of a (5,5) armchair tube. Reprinted with permission from [2], R. Saito et al., "Physical Properties of Carbon Nanotubes," Imperial College Press, London, 1998. © 1998, Imperial College Press.

a tunneling distance from the nanotube surface, and spectroscopic measurements be performed by taking the current-voltage characteristics. The I–V characteristics reveal the metallic or semiconducting character of the tube since the measured current is proportional to its density of states (DOS). This experimental technique thus addresses the fundamental question of the relation between the structure and the basic electronic properties of the individual nanotube. Figure 11 shows the principal differences between the DOS of a metallic and a semiconducting tube. As seen by STM spectroscopy, in the zero-bias region the metallic tube (Fig. 11a) has a finite conductance while the semiconducting tube has zero, corresponding to the basic Fermi surface properties. At higher voltages asymmetric peaks in the conductance appear. These are the Van Hove peaks due to the 1D DOS. The low-temperature STM experiments were carried out with SWNTs, and thus they cover only a very limited range of diameters, between 1 and 2 nm. For this range, the measured E_g for semiconducting nanotubes followed roughly the predicted $1/d_t$ dependence, being in the range 0.3–0.7 eV [142].

In the presentation above, the effects of nanotube curvature on the tight-binding model were not addressed. In more recent scanning tunneling spectroscopy measurements on SWNTs, a curvature-induced "extra" gap was found in zig-zag type metallic nanotubes, but not in tubes of armchair type [140]. This gap is also inversely dependent on the nanotube radius. The measured results were theoretically anticipated in refined calculations (see [140] and refs. therein). The curvature-induced gap has a size of a few 10s of meV for subnanometer tube diameters. From this energy scale we see that for all but the very smallest of possible tubes, the effect is washed out at room temperature.

Other scanning tunneling spectroscopy experiments have been performed that explore other topics such as individual quantum states in short SWNTs [144, 145], MWNTs [146],

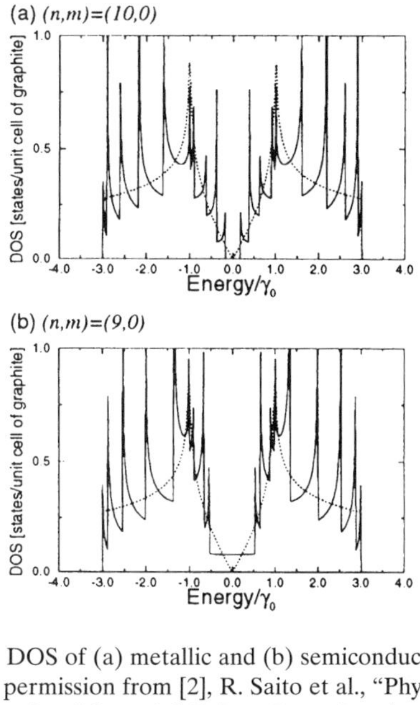

Figure 11. The DOS of (a) metallic and (b) semiconducting nanotubes. Reprinted with permission from [2], R. Saito et al., "Physical Properties of Carbon Nanotubes," Imperial College Press, London, 1998. © 1998, Imperial College Press.

doped nanotubes [147], crossing nanotubes [148], defects and junctions in SWNTs [149, 150], and peapods (SWNTs filled with C_{60}) [151].

4.3. Experimental Studies: Raman Spectroscopy

Another important tool for nanotube characterization is Raman spectroscopy. Particularly, the strong, so-called radial breathing mode is derived from the tubular structure of the nanotube and is absent from all other 2D graphitic forms of carbon. It is widely utilized to give a measure on the nanotube diameter [2]. Due to the singularities in the DOS of the 1D nanotubes and the electron-phonon coupling, very strong resonant Raman effects are observable in studies on carbon nanotubes. This enables the use of Raman spectroscopy to investigate the electronic structure of carbon nanotubes, particularly the occurrence of semiconducting and metallic properties. A possibly very important recent development is the ability to measure the Raman signal from individual nanotubes with the so-called micro-Raman technique. It was claimed to be possible to determine the (n, m) indices of the isolated nanotube with this technique [152]. This of course opens up the possibility of transport measurements on a nanotube of known symmetry.

4.4. Optical Properties

The optical physics of nanotubes has not moved forward in pace with the other developments in the field. This is due to unability to make samples consisting of nanotubes of a single symmetry. (A more recent exception to this statement was reported in [153].) Thus all optical signals are a mixture of signals from a very large variety of nanotubes. Furthermore, since nanotubes tend to aggregate together, especially SWNTs, any signal from semiconducting tubes is easily hidden by interaction with neighboring metallic tubes. Unlike transport measurements, optical measurements on single nanotubes are experimentally less feasible. On the other hand, however, the 1D singularities in the optical response of nanotubes give strong signals, just as in the case of Raman spectroscopy. Also, if aligned assemblies of nanotubes can be produced, the optical anisotropy of such a system is potentially huge. In a recent development, SWNTs were processed by sonication and surface-active chemicals to produce solutions of truly individually solvated tubes, coated with the surface-active hydrocarbon molecules. These SWNTs were rather short (due to the strong sonication), with an average length of only 130 nm. For perhaps the first time, photoexcited luminescence from nanotube solutions were observed in this case [154]. The luminescence spectrum was closely associated with the measured absorption spectrum, with only a minor red-shift. At the moment of writing, this work is intensively being pursued and is likely to significantly widen research in carbon nanotube systems. Photoconductivity measurements on single nanotubes have so far not been successful.

The research on the optical properties of carbon nanotube may in the near future bring forward a number of potential applications within optoelectronics, thus adding to the list that already exists within other fields of nanotube-based nanotechnology.

5. ELECTRONIC TRANSPORT PROPERTIES

5.1. Introduction

Within the realm of electronic transport physics, the carbon nanotube offers exciting possibilities both as a testing ground in basic research as well as in nanotechnology. In the case of the former, the near perfection in structure for lengths exceeding 1 μm makes the nanotube the best concrete example of a 1D quantum wire. On the other hand, in nanotechnology the prospect of devices made from a single nanotube, that is, intrananotube devices, offer the possibility of truly molecular-level electronics. Even if molecular electronics is eventually pursued with smaller molecules, carbon nanotubes may be used as connecting wires between metal structures of ~100 nm size and molecules of ~1 nm size.

Although the carbon nanotube is mainly "marketed" as a perfect 1D object, this is not always the case. In fact, depending on the circumstances one may have to treat the nanotube as anything from a 0D to a 3D object. Obviously, if a SWNT is short enough, it becomes a 0D object, or a quantum dot. If only the outer shell of a MWNT is considered, and the diameter of the tube is large enough with respect to some critical length scale of the particular situation, this outer shell becomes a 2D object. Finally, if we again consider a larger MWNT (in diameter), and the charge carrier coupling between the layers is strong enough, then we may have at hand an anisotropic 3D object. We can thus see that even without considering nanotubes of different chiralities and more elaborate topologies than the simple tubular one, the subject of electron transport in carbon nanotubes can be rather complicated.

As was discussed in the previous chapter, the energy gap in semiconducting SWNTs is ~0.5 eV, which is high enough with a good margin for the nanotube to function at room temperature as the active component for example, in field effect transistors. Other effects, such as the energy-level quantization in short nanotubes (L = 10–100 nm), require cryogenic temperatures for the effects to be observable. Roughly it can be said that in metallic nanotubes the interesting transport phenomena occur at low temperatures, while in semiconducting tubes much of the work is carried out at room temperature.

5.2. Bulk Transport

In this section we review work on the transport physics of single SWNTs, SWNT ropes, and MWNTs. This is clearly more significant than the subject of transport in macroscopic amounts of carbon nanotubes. However, the carbon nanotube may offer interesting applications as the conductive component in composites, when mixed together with an insulating host material. In order to have a composite conducting, the volume fraction of the conductive component has to exceed some critical value. Typically the conductive material consists of μm-sized particles of a more or less rounded shape. For conduction to occur, the particles have to touch each other frequently enough so that the conductive channels are formed over macroscopic distances. As prescribed by percolation theory, this occurs at a volume fraction of 16%, whereby the conductivity of the composite rises very sharply (as a function of filling percentage) with several orders of magnitude. However, it has already been demonstrated with conductive material consisting of 1D objects, for example fibrils of polyaniline [141], a conducting polymer, that there is no apparent sharp threshold for the onset of conduction, and more importantly, the composite becomes conducting at a much lower volume fraction of the conductive component. Obviously, this should also be achievable with nanotubes as the conductive components. In fact nanotube-based composite systems are already being considered for commercial applications.

5.3. Developments in Carbon Nanotube Transport

The theory of the basic electronic properties of carbon nanotubes was formed quickly after their discovery [155]. In the decade since then, the experimental investigations of carbon nanotube transport properties have been thorough but are still in a very dynamical phase. We will next review chronologically the development in experimental carbon nanotube transport physics. Larger amounts of carbon nanotubes were available two years after their discovery in 1991. The first experiments where electrodes were fabricated on single MWNTs were published in 1996 by Ebbesen et al. [156] and Langer et al. [157]. These groups used rather elaborate methods: The former deposited tungsten by focused ion beam deposition with direct writing of the electrodes, and the latter deposited a gold film on the MWNT and then removed it after patterning the excess gold with Ar ion-milling, leaving gold electrodes to the MWNT. The results in Ref. [156] were highly varying, showing two terminal resistances from ~100 Ω to ~100 MΩ. It is likely that the ion beams damaged excessively some of the nanotube sections between the electrodes. In the work by Langer et al. a rather low resistance (~10 kΩ) was achieved, and the group demonstrated quantum effects in the transport in the form of universal conductance fluctuations in the magnetoresistance.

Later, other groups undertook much more systematic investigations, concentrating either on SWNTs (often, in fact, SWNT ropes) or MWNTs. Basically, two methods have since been used to connect individual nanotubes to metal electrodes: Nanotubes can be deposited from a solution, such as by spinning, on top of prefabricated electrodes. Many electrode structures are needed due to the low success rate of getting just one tube on the electrodes. The alternative method is to deposit nanotubes on a substrate with alignment markers, to locate a tube with AFM or SEM, and then to fabricate electrodes on the tube with e-beam lithography. The main difference between the methods is thus that in the former the tube is on top of the electrodes, while in the latter the electrode covers the nanotube. The methods may result in different transport properties. It has, for example, been claimed that the metal evaporated on top of a tube effectively cuts it for charge carrier motion. More recently, the growth of SWNTs has been incorporated with the fabrication process. Single-walled nanotubes are grown from catalyst particles deposited on a wafer (see Section 2) with alignment markers. Thereafter metal electrodes are fabricated with e-beam lithography just as in the previous method. This latest method has enabled fabrication of samples in very large numbers. Some experimentation has already been undertaken where the principles of self-assembly are employed to deposit nanotubes onto electrodes in a more controlled and repeatable fashion [158]. Perhaps in the future it will also be possible to select between nanotubes of different chirality in the deposition process, if a single-chirality synthesis procedure does not arise. Despite all the experimental work done so far, the issue of contact resistance is not firmly understood [159].

With these methods, large numbers of carbon nanotubes have been investigated since the mid-90s. At first the contact resistances were typically in the range from several tens of kΩs to several MΩs (if conducting at all). The conductances were thus clearly smaller than the quantum of conductance ($G_0 = 1/13\ \mathrm{k}\Omega^{-1}$), and the nanotube was isolated by tunneling barriers from the electrodes. This led to a situation with the Coulomb blockade dominating the transport of metallic nanotubes at low temperatures. Both metallic and semiconducting tubes were found. With an effective gate electrode present, the latter could be driven on and off a conducting state at room temperature with a current modulation by a factor of 10^5 at best. Later contact resistances were improved to such a degree that the intrinsic resistance of the tube would dominate over the former. This has enabled the study of one-dimensional charge transport in both the ballistic and diffusive limits. Furthermore, a few reports have emerged about either proximity-driven or intrinsic superconductivity in SWNTs or SWNT ropes, although only two groups have made such claims so far [160, 161]. SWNTs were eventually made available almost to the same extent as MWNTs, so the studies on transport physics have emphasized these, since intershell interactions can be ignored. This has been the case in particular since the emergence of the individually grown SWNTs from catalytical particles, which evades the problem of coalescence of SWNTs into ropes. A major breakthrough that is pursued at this point is the complete physical characterization of SWNTs of known chirality. To date this has been possible only in STM studies with nanotubes dispersed on conducting surfaces.

In the last few years, a number of experiments relevant to transport properties have emerged in which individual nanotubes are manipulated with physical means for the purpose, for example, of making tunneling barriers within the tube. In the majority of these experiments, the AFM has been utilized for this goal. Single-walled tubes with tunneling barriers can also be obtained by searching for tubes with pre-existing kinks (sharp bends), but especially the manipulated kinks are important since it has been possible to place these very close to each other in a SWNT, separated by just a few tens of nanometers. This has enabled the fabrication of room-temperature SETs [132, 133]. In MWNTs on the other hand, it is less plausible to induce well-defined tunneling barriers at one point due to the complications from multiple shells. However, it is possible to remove the layers by different means one-by-one [162]. This may in the near future enable the study of interlayer transport properties. In another set of experiments, the conducting tip of an AFM is used as a local gate in close proximity with the tube to be investigated. This has enabled observation of individual scattering sites in a carbon nanotube [163].

5.4. Metallic Tubes

5.4.1. Ballistic Conduction

One of the most exciting aspects of transport in metallic carbon nanotubes is their ability for ballistic transport over relatively large distances, exceeding 1 μm. Ballistic transport implies a large mean free path τ_m for the electrons. This large τ_m was estimated on theoretical grounds by White et al. [164]. Again, the tubular structure plays a role in that the disorder is expected to be averaged over the circumference of the nanotube. A ballistic conductor has a conductance given by the Landauer formula:

$$G = N_{\mathrm{ch}} T_m G_0,\ G_0 = 2e^2/h \tag{5}$$

Where N_{ch} is the number of conducting channels, T_m is the transmission factor and G_0 the quantum of conductance in the noninteracting approximation. The factor 2 is due to spin degeneracy. In a much debated experiment by Frank et al. [165], an arc-discharge synthesized MWNT was sunk from the tip of a STM into a liquid metal (Hg,Ga) at room temperature. A conductance of $1G_0$ was measured repeatedly and independently of the distance that the tube penetrated into the liquid metal. However, a metallic carbon nanotube has two crossing 1D energy bands at the Fermi level, which yields a conductance of $2G_0$ for a single nanotube. Some of the more recent measurements on SWNTs have in fact, in accord with theory, approached but not exceeded this value, while all other transport experiments on MWNTs point toward a picture of more diffusive transport behavior. Therefore the controversy over the experiment by Frank et al. remains unsolved.

Two recent experiments [166, 167] in single SWNT samples grown by CVD have achieved the previously mentioned high conductances, a little below $2G_0$, in a fairly large number of samples. The temperature dependence of the conductance between 1 K and room temperature was weak, and even exhibited increasing conductance upon cooling. The ballisticity of the nanotubes, together with the fact that the contact resistances are large enough to cause reflections of the propagating modes at the nanotube-metal interfaces but weak enough to strongly suppress Coulomb blockade effects, results in a situation where the nanotube forms a resonant cavity for the charge carriers. If the wavelength of the conduction electrons is an integer fraction of the nanotube length, the wavefunctions form standing waves due to constructive interference between propagating and reflected electron waves. The wavelength can be tuned by changing the Fermi energy with a gate voltage (V_g). These effects can thus be seen as weak oscillations in I vs. V_g data at low temperatures. The work mentioned earlier demonstrated ballistic conduction over distances below 1 μm. Some recent work reports ballistic conduction for distances on the orders of tens of μms. The upper practical limit for the mean free path in SWNTs has thus not been established.

5.4.2. Luttinger Liquid

In the discussion above on ballistic conduction in carbon nanotubes we ignored e–e interaction effects. These do not necessarily affect the Landauer formula for ballistic conduction. The single-wall carbon nanotube in particular has been forwarded as a prime candidate for hosting the so-called Luttinger liquid (LL), which is formed by interacting electrons in a 1D system [168]. In a normal 3D metal, the effect of *e–e* interactions to the independent electron approximation is limited; the noninteracting electrons change to independent quasiparticles in a one-to-one correspondence, as described by the Fermi liquid theory. In a strictly 1D system, however, the Fermi liquid picture breaks down and is replaced by the LL. In the LL, the *e–e* interactions dominate, and the electrons are in a highly correlated state. The primary excitations are not single quasiparticles above the Fermi level, but plasmons, which are analogous to electromagnetic pulses in a transmission line. However, the LL requires a high degree of order, since disorder destroys the correlated ground state. Therefore, a clean, strictly 1D system is difficult to realize and to investigate. In a LL, the DOS at the Fermi level is suppressed in a power law fashion, and spin and charge excitations are separated. Dating the observations of LL behavior is limited to edge states in 2D electron gases, to certain compounds with 1D substructure, and to single-walled carbon nanotubes. In transport measurements, the LL behavior is seen in the tunneling conductance from a normal electrode over a tunneling barrier to the LL, which has a tunneling DOS that is suppressed at the Fermi level according to a power law: DOS $\sim (E - E_{\mathrm{F}})^{\alpha}$. The parameter α is determined by the strength of the interaction between the electrons. It follows from the energy dependence of the tunneling DOS, that, at small voltages the tunneling conductance becomes

$$eV \ll k_B T \quad G(T) \sim T^{\alpha} \tag{6}$$

and at large voltages

$$eV \gg k_B T \quad G(V) = dI/dV \sim V^{\alpha} \tag{7}$$

The T dependent conductance has to be observed at some intermediate temperature range, since Coulomb blockade always dominates at low temperatures if the contacts have a low transparency. Measurements of conductance with tunneling contacts have been undertaken with SWNTs both as a function of T and V [169, 170]. Scaling of the experimental data as a function of $eV/k_B T$ gave limiting behavior that supports the LL picture as given by Eqs. (6) and (7). In particular, different types of contacts give different values for α: Tunneling over a nanotube bend is said to correspond to an "end-contact," that is, the charge carriers tunnel to an end section of a carbon nanotube and have only one way to propagate, while a carbon nanotube laid onto a metal electrode gives a "bulk-contact" with two possible (opposite) propagation directions. The α differs theoretically by a factor of 2 between the two cases, which also was roughly observed in the experiments.

However, if restricted to $I(V, T)$ measurements, the differences to other models are usually very small. An experiment that could provide a strong proof would be, for example, one that would demonstrate the spin-charge separation inherent in a LL. Such work has not been reported to date.

5.4.3. Coulomb Blockade

Coulomb blockade effects are rather central in mesoscopic physics [171]. In the experimental studies on low temperature transport processes in single-carbon nanotubes, they have been almost always present. They occur most prominently in a situation where a mesoscopic scale conductor is isolated by tunneling barriers between two electrodes, and thus form an "island" between the electrodes. The island has a capacitance C, and if the charging energy of a single electron $E_c = e^2/C$, is larger than the thermal energy ($E_c \gg k_B T$), then the system will not conduct due to the *e–e* repulsion. The other fundamental requirement is the that the contact resistances have to be large enough to form tunneling barriers: $R_c \gg R_k = h/e^2$. Such a system forms a single electron transistor (SET). Coulomb blockade may also occur in a single junction system, which, however, requires a very high impedance in one of the conductors. The Coulomb blockade in a SET can be modulated by a gate electrode from full blocking to a conducting state limited only by the tunneling barriers.

When nanotube-based SETs are discussed, it should be noted that up to now the tunneling barriers from metal electrodes to the nanotube have been formed without the intentional deposition of an insulating layer between the metal and the tube, unlike in other SET systems. The tunneling barriers are said to form from unavoidable impurities or, especially in the case of SWNTs, from the bending of the tube on the electrodes. The capacitance between the electrode and the tube is formed from the self-capacitance of the tube and the barrier capacitance at the nanotube-electrode junctions. With a typical junction length of 200 nm and a nanotube length of 1 μm, the capacitance is $\sim 10^{-17}$ F. This gives an E_c of $\sim$10 meV, and thus the Coulomb blockade of a typical nanotube SET occurs at temperatures below 10 K ($k_B T$ should be $<0.1\ E_c$).

Nanotube SETs have been realized to date in different versions. In the simplest case, two metal electrodes are attached to the nanotube so that a tunneling barrier is formed between the tube and the metal (this easily happens unintentionally). The electrode spacing varies in the range 0.1–1 μm. This gives the spacing between the energy levels, ΔE, in the nanotube section to be measured, in the range 1–10 meV. That is, ΔE and E_c can be of the same order of magnitude. Energy-level quantization in the Coulomb blockade of SWNTs was for the first time measured in Refs. [172, 173].

A pure SET requires tunneling resistances much larger than R_K. As this condition is gradually relaxed, other effects are present alongside the Coulomb blockade, such as Kondo effects [174] and shell filling of individual quantum states [175]. As the tunneling resistance becomes small compared to R_K, and if the tube is ballistic, we recover the situation with ballistic conduction, already described.

More elaborate SETs can be built with nanotubes than the simple metal-nanotube systems described above. Figure 12 shows two crossing arc-discharge MWNTs; the lower one is contacted at both ends with gold electrodes and the upper tube with one electrode [176]. The lower tube acts as the island of a SET while the upper tube functions as a highly proximate gate electrode. Figure 13a shows the I–V

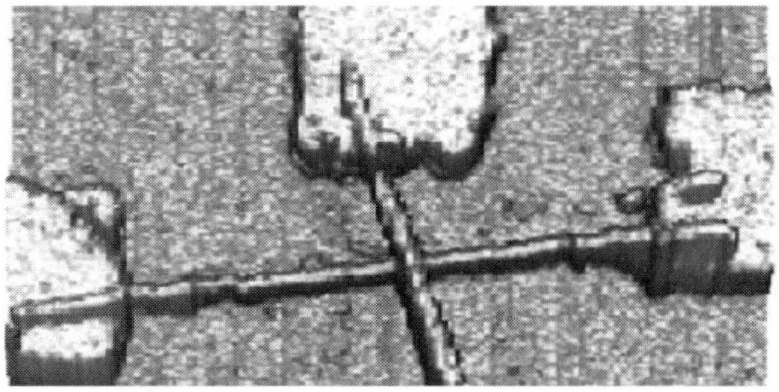

Figure 12. A SET made from two arc-discharge MWNTs. The horizontally directed (lower) tube forms the island, connected to source and drain electrodes, and the upper tube forms a highly proximate gate. Reprinted with permission from [176], M. Ahlskog et al., *Appl. Phys. Lett.* 77, 4037 (2000). © 2000, American Institute of Physics.

curves at different temperatures and Figure 13b shows current modulation using the nanotube gate. The close proximity of the nanotube gate translates into a small Coulomb oscillation period. It is of course also possible to place several tubes on the "island-tube" and thus create several local potential barriers on the tube. Two groups have used AFM manipulation to make two sharp bends very close to each other (<100 nm) on a single SWNT [132, 133]. In a SWNT, a sharp bend creates a tunneling barrier, thus two next to each other results in an intra-tube SET. The main advantage, besides the miniaturization aspect, is the very high Coulomb energy, due to the ultrasmall capacitances, resulting in a SET that works at temperatures close to room temperature. In another approach [177], single semiconducting nanotubes were covered with PMMA resist and small (~100 nm) openings developed into the resist, which allowed selective doping to be done on small sections of the tube, thus creating local islands separated by *p–n* junctions.

5.4.4. Superconductivity

Concerning the study of intrinsic transport phenomena in metallic carbon nanotubes, two other lines of development have appeared that we will briefly mention here, namely superconductivity and highly disordered tubes.

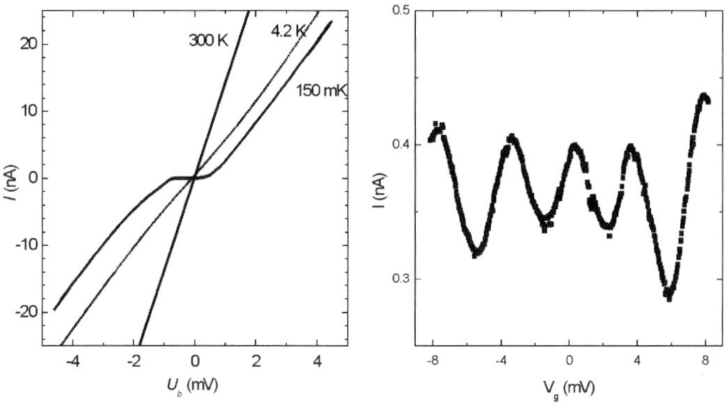

Figure 13. (a) The *I–V* curves at different temperatures of the SET of Figure 12 and (b) current modulation at 150 mK using the nanotube gate. Notice the small magnitude of the gate voltage (V_g) required to produce Coulomb oscillations. Reprinted with permission from [176], M. Ahlskog et al., *Appl. Phys. Lett.* 77, 4037 (2000). © 2000, American Institute of Physics.

Superconducting phenomena in connection with the single carbon nanotube were first observed as the proximity effect with SWNTs or SWNT ropes connecting two superconducting electrodes [178, 179]. Later, a smooth superconducting transition was observed at sub-Kelvin temperatures in a few SWNT rope samples that were connected to normal electrodes (Pt/Au) [160]. These SWNT ropes were suspended with a unique technique developed earlier. The observed transition was gradual rather than abrupt as is the case for 1D systems. The Meissner effect could not be checked in this case due to residual metallic impurities present in macroscopic amounts of the material. Superconductivity was also reported in bulk samples of ultrasmall diameter SWNTs grown inside zeolites [161]. A smooth transition at around 10–15 K was observed with a squid magnetometer, that is, the Meissner effect.

5.4.5. Transport in Disordered Tubes: MWNTs

One of the advantages of carbon nanotubes is the very high degree of structural perfection often seen in them. As was noted above, transport measurements often indicate ballistic behavior among SWNTs, while among MWNTs, with diameters in the 10–20 nm range, diffusive behavior is more the norm. The density of defects in MWNTs depends strongly on the synthesis conditions, and especially, it is believed, on the growth temperature. Arc-discharge tubes are grown at temperatures approaching 2000 °C while in the CVD growth the temperature is well below 1000 °C. The resulting structural difference is immediately apparent as can be seen in Figure 5. With present knowledge, the mean free path is rather small even in the arc-discharge MWNTs. Axial magnetoresistance measurements on such tubes revealed Aharonov–Bohm oscillations with a period that matched the nanotube circumference [180]. Transport measurements on highly disordered CVD grown tubes were undertaken recently [181]. These measurements probed the intrinsic behavior of the tubes since contact resistances were very low. A typical resistivity of 80 kΩ/μm was reached. At low temperatures (<10 K) a zero-bias anomaly developed that was not due to Coulomb blockade, since the contact resistances were low. Possible explanations are disorder-induced *e–e* interaction effects or raised electron gas temperature due to Joule heating in a diffusive conductor.

5.5. Semiconducting Tubes

Semiconducting nanotubes are especially important for device-oriented applications. To date, semiconducting behavior has been observed in single SWNTs. In MWNTs and SWNT ropes, there usually exists individual shells of both the metallic and semiconducting kinds, as has been demonstrated by the IBM group [162]. Therefore pure semiconducting behavior in MWNTs has rarely been mentioned. Semiconducting behavior in carbon nanotubes is demonstrated in a FET configuration. Typically with semiconducting SWNTs it is observed that the conducting state is attained with negative gate voltages, implying that the carbon nanotube forms a normally-off *p*-type conduction channel. Thus a semiconducting carbon nanotube is unintentionally *p*-doped, with oxygen as the likely dopant.

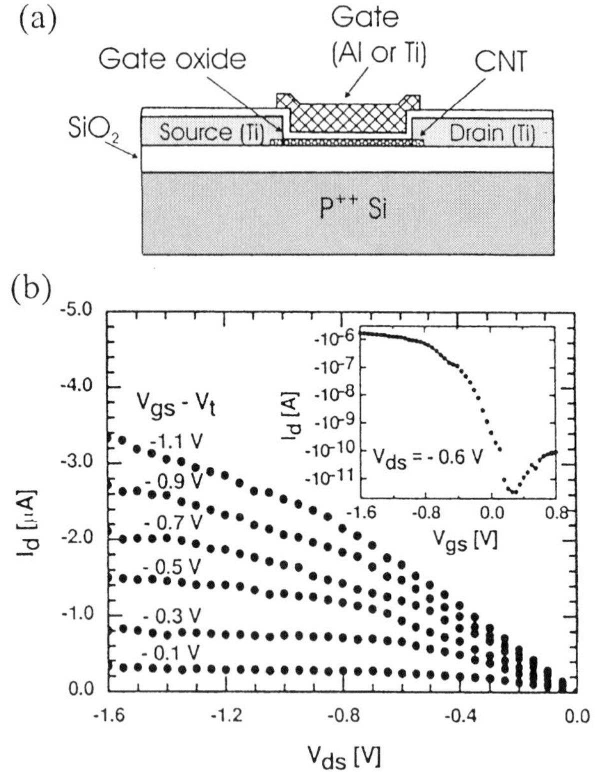

Figure 14. (a) Schematic picture of a FET made from an individual SWNT that is covered by a top gate. (b) The current vs. source-drain voltage. Inset: Current vs. gate-voltage. Reprinted with permission from [182], S. J. Wind et al., *Appl. Phys. Lett.* 80, 3817 (2002). © 2002, American Institute of Physics.

Figure 14 shows a schematic figure of such a device and its transistor characteristics [182]. The IBM group has shown that a higher transconductance can be achieved with SWNT-based FETs than with state-of-the-art silicon MOSFETs, which is encouraging, especially considering that the fabrication technique of nanotube-FETs is far from optimized. The resistance in the metallic state (ON state) is typically in the range 20 kΩ–1 MΩ. With the fabrication of gate electrodes that are strongly coupled to the nanotube, it is possible to reach ambipolar transistor action, achieving both *n*- and *p*-type behavior. Logic gates made from nanotube FETs have recently been demonstrated [183]. The nature of the Schottky barriers between bulk metal electrodes and the SWNT, a 1D object, is still being investigated [184].

6. ELECTRONIC APPLICATIONS

6.1. Introduction

Carbon nanotubes have attracted the label "building-blocks of nanotechnology" due to the range of realized and potential applications. Successful technology involving the nanoelectronic applications promises to be extremely lucrative, but the absence of reliable preparation methods for metallic or semiconducting nanotubes, either by selective synthesis or post-synthesis separation, is a major obstacle. This impediment has not been a factor in the development of nanotube-based electron sources for vacuum microelectronics. The recent establishment of pilot plants with gram

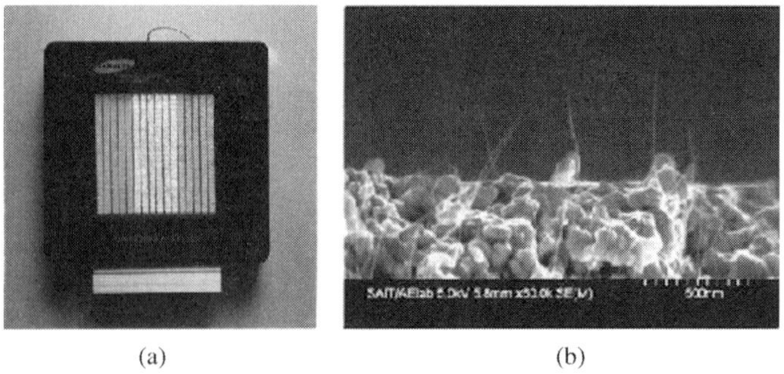

Figure 15. (a) 4.5 inch fully sealed FED display with red, green, and blue phosphor columns. (b) Cross-sectional SEM image SWNT-based cathode. SWNTs are aligned perpendicular to the substrate and firmly embedded into the metal electrode. Reprinted with permission from [206], W. B. Choi et al., *Appl. Phys. Lett.* 75, 3129 (1999). © 1999, American Institute of Physics.

quantities per day capacity has spurred those applications that require bulk quantities of carbon nanotubes. Mass production should drive the cost down from the present $500 per gram. Production of thousands of kilograms per week from a single plant by 2004 is predicted [185]. In the immediate term, controlled growth of aligned carbon nanotube patterns will be a significant factor in several emerging applications [186].

6.2. Prospects for Nanoelectronic Applications

6.2.1. Highly Integrated Circuits

Since discovery, exploitation of the electronic properties of carbon nanotubes outlined in the previous chapters has been the driving force behind much of the research efforts directed at applications. The prospect of device densities 10^4 times greater than present day microelectronics places semiconductor carbon nanotube electronics among a number of competing technologies poised to complement or replace CMOS in the <100 nm feature-size domain [187]. As outlined in Section 5, nanotube-based electronic devices have much to offer: conductivity, electromigration, and current carrying capacity of metallic nanotubes are as good as or better than the best metal nanowires, and semiconducting nanotubes exhibit carrier mobilities and transconductancies that meet or exceed those of the best conventional semiconductors. However, there are several major barriers to the development of highly integrated circuits: (i) Current synthesis techniques produce mixtures of metallic or semiconductor nanotubes, and these tend to form in ropes or bundles; (ii) the precise nature of nanotube-metal contacts is poorly understood, but the exact atomic arrangement at the interface and the gaseous environment appear to be significant [188]; and (iii) those nanotube SPM manipulation techniques used to demonstrate single devices are totally inappropriate for manufacture of highly integrated circuits. To date, no reliable method of producing contacted carbon nanotubes of predetermined electronic structure has been demonstrated. Complex arrays of nanotube-based devices are perhaps a distant prospect, but there appear to be no fundamental barriers to the development of the technology.

Nevertheless, the potential reward for successful technology is enormous and is likely to continue to be a motor of nanotube research.

6.2.2. Interconnect

The use of carbon nanotubes as wiring for interconnection of nanoscale circuit elements is being explored primarily because SWNTs can carry a current density of up to 10^9 Acm^{-2}, compared to 10^5 Acm^{-2} for normal metals [189]. Ballistic transport on the micrometer scale suggests that nanotubes are good candidates for high-bandwidth, dissipationless interconnect. Wet processing of nanotubes tends to degrade the electrical properties, so methods of guiding nanotubes to span the desired circuit elements during growth or deposition processes are being explored. Guiding through via holes [190], by electric field [191], or surface modification [192], as well as uncontrolled self-assembly methods have been demonstrated [193], with some degree of success [194]. However, a high-growth temperature (>900 °C) incompatible with standard silicon processing is required if CVD-grown carbon nanotubes are to exhibit truly ballistic transport, thus severely limiting the possibility of integration with CMOS circuits. The usefulness of nanotube ballistic interconnect is also limited by the fact that large contact resistances are necessarily involved. A circuit involving electrical contacts to a SWNT will have a resistance of at least $h/4e^2$, or 6.5 kΩ [195]. Moreover, although nanotube interconnect promises high frequency response, any conventionally wired external circuits will determine the response of a complete system.

6.3. Vacuum Microelectronics

Carbon nanotubes are ideal field emission electron sources for various vacuum microelectronic applications due to the high aspect ratio, current carrying capacity, and chemically passive tip. The essential physics of the field emission process is contained in the expression for the current density emitted from an ideal metal surface, the Fowler–Nordheim equation: $J = aF^2 \exp(-b\phi^{3/2}/\beta F)$ where J, F, ϕ, and β are the current density, applied macroscopic field, work function, and field enhancement factor (the ratio of the microscopic local field at a protrusion to the macroscopic applied field), respectively. The β factor calculated from the geometry of a single carbon nanotube is approximately equal to the aspect ratio, or ~1000. The average work function for an ensemble of nanotubes is ~5 eV, comparable to that of graphite and 0.4 eV greater than that of the most widely used field emission metal, molybdenum. Soon after discovery, it was assumed that nanotubes would be superior field emitters due to the high field enhancement at the tip; this was finally demonstrated in 1995 [196], and has since been the subject if intense investigation. Single SWNT [196], single MWNT [197], thin films of aligned nanotubes [198], and composite materials with nanotube inclusions [199] have all proved to be excellent field emission electron sources. The exact performance is highly dependent on the synthesis conditions, electronic screening effects, open- or closed-tip structure [200], and the role of surface absorbates [201]. Unlike ordinary metal surfaces, emission occurs from discrete energy states rather than continuous bands [202]. A low threshold field of typically 1–5 V/μm for a stable long-lifetime emitted current density of 10 mA/cm^2 is common, and current densities as high as 4 A/cm^2 have been obtained [203]. Nanotube-based field emission materials can be relatively easily manufactured by screen printing nanotube-loaded pastes [204] and require a moderate vacuum of 10^{-8} torr for acceptable performance. This has considerable advantage over microfabricated arrays of tungsten or molybdenum operating in 10^{-8} torr that are the bases of conventional vacuum microelectronics.

6.3.1. Field Emission Display

Field emission displays (FEDs) are the nanotube applications with the largest potential market and industrial interest. Conceptually the FED is simple; electrons are emitted from a cathode in vacuum, accelerated toward a parallel-plate anode by a directed electric field, where the kinetic energy is imparted to a phosphor producing emitted light. FEDs offer power consumption, brightness, response time, and operating temperature advantages over liquid crystal display (LCD) technology that currently dominates the flat panel display market. However, FEDs are technically complex, requiring not just a low threshold-voltage emission material, but also significant advances in phosphor technology, spacer elements capable of sustaining high vacuum and electric field, and pixel addressing circuitry [205]. Whether nanotube-based FEDs will overcome these obstacles to compete with low-cost flat panel LCD or emerging organic light emitting diode technologies is uncertain. Nevertheless, Samsung has produced several generations of prototype FED ranging from 4.5 inch (Fig. 15) with red-green-blue phosphor columns [206] to 9-inch moving-image display [207].

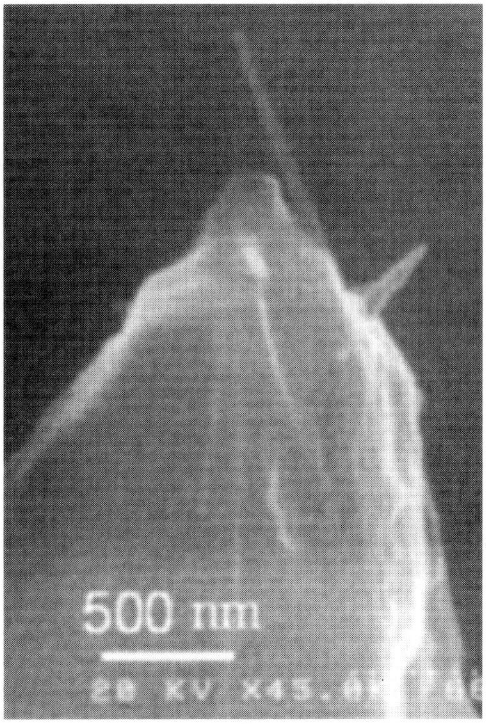

Figure 16. SEM image of the apex of a conventional micromachined Si AFM tip with protruding CVD-grown MWNT.

Lighting elements with nanotube-based field emission surfaces that work on essentially the same triode principles as the FED and suitable as pixels for very large area displays have been fabricated by Ise Electronics in Japan [208].

6.3.2. E-Beam Lithography

The present minimum feature size in the semiconductor industry is 130 nm, which is achieved using extreme ultraviolet optical lithography. This technology is currently limited by the wavelength of the light and the available UV optics used to project the pattern onto a silicon wafer. Alternative technologies are being explored for production of future <100 nm devices. Conventional electron beam lithography can already deliver 5 nm minimum feature size but is limited by the writing time as a single beam is used for writing the entire pattern. Thus arrays of microelectron guns operating in parallel are being considered as a route to drastically reduce the writing time while maintaining the high resolution of electron beam lithography. Carbon nanotubes are currently the most promising candidates for use as the emission source [209].

6.3.3. Microwave Devices

Electrons transported in vacuum have a higher mobility than in materials; consequently vacuum microelectronic devices find application when higher power and higher current levels than can be achieved in solid-state devices are required. One such application with a carbon nanotube field emitter is the microwave power amplifier tube. The outstanding technical challenge is to improve the lifetime of the emitter for operation >0.5 A/cm^2 [210].

6.3.4. X-Ray Source

The design of X-ray tubes has remained unchanged for a century, comprising thermionically generated electrons accelerated toward a metal target. Continuous and pulsed X-ray sources based on a carbon nanotube field emitter with a metal anode capable of producing medical diagnostic X-rays have been recently demonstrated [211, 212]. For biological samples, the quality of image produced by these sources is an improvement on those from conventional sources, probably because the energy distribution of emitted electrons is narrower than that of electrons thermionically emitted from metal wires. The compact geometry of these devices suggests that they may be further developed as portable sources and as probes for medical exploration.

6.4. Electrochemical Devices

The high electrochemically accessible surface area, high electrical conductivity, and desirable mechanical properties make porous nanotube arrays attractive materials for devices that use electrochemical double-layer charge injection.

6.4.1. Electrodes

High-power electrochemical capacitors (also called supercapacitors) based on carbon nanotube array electrodes are under development [213]. The capacitance of an electrochemical capacitor depends on the separation between the electrode charge and the counter charge in the surrounding electrolyte. With nanoscale separation and high surface area available in nanotube arrays, capacitances of typically 10–100 F/g can be achieved. This value is much greater than for conventional dielectric capacitors. A high degree of charge injection can be achieved with just a few applied volts and thus very efficient energy storage. Further developments are likely, with debate focussing on the suitability of MWNTs, SWNTs, and composite materials [214, 215].

6.4.2. Actuators

Electromechanical actuators based on SWNT sheets have been shown to generate higher stresses than natural muscle when operating in physiological conditions and higher strains than high-modulus ferroelectrics [216]. The actuation mechanism is a geometrical expansion of the carbon–carbon covalent bond caused by electrochemical double-layer charging; it requires low operating voltages to generate large strains. Work densities per cycle substantially higher than any previously known technology are predicted for SWNT sheets with mechanical properties close to those of individual SWNTs. This can be achieved by intertube binding and alignment optimization [217]. These promising results suggest that actuators based on a single (or a few hundred) nanotubes will lead to applications on the micro- or nanoscale in the biomedical and electronic fields.

6.5. Nanoprobes

The high stiffness, aspect ratio, and nanoscale tip make nanotubes excellent probes for sensing, testing, manipulation, and fabrication on the nanoscale.

6.5.1. Scanning Probe Microscopy

Both SWNTS and MWNTs have been extensively used as tips for scanning probe microscopy (SPM) in the life and physical sciences. The report by Dai et al. [218] on the use of MWNTs attached directly to silicon cantilevers signaled a major advance for SPM techniques. The demonstrated and potential advantages are such that nanotube probes are likely to replace conventional Si and Si_3N_4 probes in the near future. The dimensions and electronic and mechanical properties of nanotubes are ideal for SPM tips. The low tip-to-surface capacitance and contact area that is independent of clamping force give carbon nanotubes great advantages over micromachined SPM probes for electronic probing and contacting. Carbon nanotube tips have been applied to atomic force microscopy (AFM) in contact [218] and noncontact mode [219], spin-resolved scanning tunneling microscopy, chemical force microscopy, and scanning electrochemical microscopy [220]. This range of SPM techniques will be undoubtedly extended in the near future to those probes commonly used to give quantitative electronic information such as conducting AFM, scanning tunneling spectroscopy, electrostatic force microscopy, Kelvin probe microscopy, and scanned gate microscopy.

6.5.2. Sensors

The possibility of chemical sensor applications of nonmetallic nanotubes became apparent when thermoelectric power measurements revealed very high sensitivity to gaseous ambients that affect the amount of injected charge [221, 222]. MWNTs exhibit ultrahigh sensitivity at room temperature to O_2, N_2O, NH_3, CO, CO_2, and humidity [223]. Electronic tuning from *p*- to *n*-type by liquid electrolyte was demonstrated with a MWNT in a FET configuration [220]. The appeal of nanoscale sensing elements is the correspondingly low quantity of species required to stimulate a response. Nanotube sensors certainly have prospects for challenging conventional gas sensors for certain uses and becoming lab-on-a-chip sensing elements for biomedical applications. However major obstacles remain, the forward and reverse response of nanotube gas sensors is not symmetrical because the species that induce a response are often strongly physisorbed; also a response that can differentiate species in complex gas species is often required.

6.5.3. Manipulators

Nanoscale electromechanical manipulators, nanotweezers, based on carbon nanotubes have been developed for manipulation and electrical interrogation of nanostructures [224]. The tweezer action is driven by electrostatic attraction between two contacted nanotubes mounted on a tapered glass probe. Manipulators such as these are likely to be integrated with scanning probe microscopy and other measurement systems to provide the tools for the characterization and manipulation on the nanoscale [225].

6.5.4. Fabricators

Carbon nanotube AFM tips for fabricating oxide nanostructures in Si and Ti by anodization is a rapidly expanding field aimed at Tbit/cm^2 data storage [226]. The technique utilizes ambient moisture in the oxidation process with the tip biased negatively relative to the surface. Line widths of 5–10 nm suitable for antidots and tunnel junctions can be achieved.

6.6. Nanoelectromechanical Systems

Nanoeletromechanical systems (NEMS) is a term that embraces the actuator, manipulator, and sensor applications outlined above. The impact of NEMS is likely to be as significant as microelectromechanical systems [227]. Carbon nanotubes are promising for the design and development of NEMS, not only because of the excellent mechanical and electrical properties, but also because the significant progress in the fabrication of carbon nanostructures of the last few years points to possible implementation of recently proposed carbon nanotube-based NEMS devices such as a nonvolatile random access memory for molecular computing [228]. The predicted behavior of carbon nanotube nanoelectromechanical switches—the basis of many NEMS devices—is favorable [229], and electronic properties have been shown to be reversible with mechanical deformation by a local probe [230].

GLOSSARY

Density of states (DOS) The density of electronic states as a function of the energy.

Electron beam lithography A method for micro- and nanoscale fabrication where a pattern on a polymeric layer (resist) is exposed to electron irradiation. In research laboratories the electron exposure is performed with a modified scanning electron microscope (SEM).

Field effect transistor (FET) A transistor where the current between two electrodes (drain and source) is modulated by the electric field from a third electrode (gate).

Field emission The emission of electrons from a metal or semiconductor into vacuum or dielectric under the influence of a strong electric field.

Fullerene (C_{60}) A soccer-ball shaped spherical and hollow molecule consisting of 60 carbon atoms.

Magnetoresistance The electrical resistance of a sample or device as a function of the magnetic field.

Mesoscopic physics A new branch of condensed matter physics where physical phenomena are studied at length scales larger than the atomic level but smaller than macroscopic scales.

Molecular electronics A term for such devices that have single molecules as the active component.

Raman spectroscopy An optical characterization tool of matter that utilizes the Raman effect.

Scanning probe microscopy The generic term given to microscopy techniques that use a scanned micro- or nanoscale tip in immediate proximity to a surface to image its topology or other physical features with almost atomic resolution. The main types are scanning tunneling microscopy (STM) and atomic force microscopy (AFM).

Tunneling conduction A process for charge conduction where the charge carriers pass through an energy barrier by quantum mechanical tunneling.

Young's modulus A quantity that describes the stiffness of a material.

REFERENCES

1. S. Iijima, *Nature* 354, 56 (1991).
2. R. Saito, G. Dresselhaus, and M. S. Dresselhaus, "Physical Properties of Carbon Nanotubes," Imperial College Press, London, 1998.
3. A. Loiseau, F. Willaime, N. Demoncy, G. Hug, and H. Pascard, *Phys. Rev. Lett.* 76, 4737 (1996).
4. R. Tenne, L. Margulis, M. Genut, and G. Hodes, *Nature* 360, 444 (1992).
5. M. Remskar, A. Mrzel, Z. Skraba, A. Jesih, M. Ceh, J. Demsar, P. Stadelmann, F. Lévy, and D. Mihailovic, *Science* 292, 479 (2001).
6. Y. Huang, X. Duan, Y. Cui, L. J. Lauhon, K. Kim, and C. M. Lieber, *Science* 294, 1313 (2001).
7. Y. H. Lee, C. H. Choi, Y. T. Jang, and B. K. Ju, *Appl. Phys. Lett.*, in press.
8. N. Wang, Z. K. Tang, G. D. Li, and J. S. Chen, *Nature* 408, 50 (2000).
9. A. Krishnan, E. Dujardin, M. M. J. Treacy, J. Hugdahl, S. Lynum, and T. W. Ebbesen, *Nature* 388, 451 (1997).
10. J. Han, M. P. Anantram, R. L. Jaffe, J. Kong, and H. Dai, *Phys. Rev. B* 57, 14983 (1998).

11. A. Volodin, M. Ahlskog, E. Seynaeve, C. Van Haesendonck, A. Fonseca, and J. B. Nagy, *Phys. Rev. Lett.* 84, 3342 (2000).
12. W. Z. Li, J. G. Wen, and Z. F. Ren, *Appl. Phys. Lett.* 79, 1879 (2001).
13. S. C. Tsang, Y. K. Chen, P. J. F. Harris, and M. L. H. Green, *Nature* 372, 159 (1994).
14. B. W. Smith, M. Monthioux, and D. E. Luzzi, *Chem. Phys. Lett.* 315, 31 (1999).
15. M. Ahlskog, E. Seynaeve, R. J. M. Vullers, C. Van Haesendonck, A. Fonseca, K. Hernadi, and J. B. Nagy, *Chem. Phys. Lett.* 300, 202 (1999).
16. M. Sano, A. Kamino, J. Okamura, and S. Shinkai, *Science* 293, 1299 (2001).
17. P. Schützenberger and L. Schützenberger, *C. R. Séances Acad. Sci. Paris* 111, 774 (1890).
18. N. M. Rodriguez, *J. Mater. Res.* 8, 3233 (1993).
19. R. S. Wagner and W. C. Ellis, *Appl. Phys. Lett.* 4, 89 (1964).
20. A. Oberlin, M. Endo, and T. Koyama, *J. Cryst. Growth* 32, 335 (1976).
21. W. B. Downs and R. T. K. Baker, *J. Mater. Res.* 10, 625 (1995).
22. G. G. Tibbetts, *J. Cryst. Growth* 66, 632 (1984).
23. Ch. Laurent, E. Flahaut, A. Peigney, and A. Rousset, *New J. Chem.* 22, 1229 (1998).
24. R. Saito, G. Dresselhaus, and M. S. Dresselhaus, *Appl. Phys. Lett.* 60, 2204 (1992).
25. M. Endo and H. W. Kroto, *J. Phys. Chem.* 96, 6941 (1992).
26. S. Iijima, P. M. Ajayan, and T. Ichihashi, *Phys. Rev. Lett.* 69, 3100 (1992).
27. S. Iijima and T. Ichihashi, *Nature* 363, 603 (1993).
28. R. E. Smalley, *Mater. Sci. Eng. B* 19, 1 (1993).
29. J. H. Hafner, M. J. Bronikowski, B. R. Azamian, P. Nikolaev, A. G. Rinzler, D. T. Colbert, K. A. Smith, and R. E. Smalley, *Chem. Phys. Lett.* 296, 195 (1998).
30. A. Peigney, P. Coquay, E. Flahaut, R. E. Vandenberghe, E. De Grave, and Ch. Laurent, *J. Phys. Chem. B* 105, 9699 (2001).
31. Y. Li, W. Kim, Y. Zhang, M. Rolandi, D. Wang, and H. Dai, *J. Phys. Chem. B* 105, 11429 (2001).
32. Y. Saito, *Carbon* 33, 979 (1995).
33. Y. Saito, M. Okuda, N. Fujimoto, T. Yoshikawa, M. Tomita, and T. Hayashi, *Jpn. J. Appl. Phys.* 33, 526 (1994).
34. D. Zhou, S. Seraphin, and S. Wang, *Appl. Phys. Lett.* 65, 1593 (1994).
35. J. Gavillet, A. Loiseau, F. Ducastelle, S. Thair, P. Bernier, O. Stephan, J. Thibault, and J.-C. Charlier, *Carbon* 40, 1649 (2002).
36. E. G. Rakov, *Russ. Chem. Rev.* 69, 35 (2000).
37. T. W. Ebbesen and P. M. Ajayan, *Nature* 358, 220 (1992).
38. N. Hamada, S. Sawada, and A. Oshiyama, *Phys. Rev. Lett.* 68, 1579 (1992).
39. R. Saito, M. Fujita, G. Dresselhaus, and M. S. Dresselhaus, *Appl. Phys. Lett.* 60, 2204 (1992).
40. K. Tanaka, K. Okahara, M. Okada, and Y. Yamabe, *Chem. Phys. Lett.* 191, 469 (1992).
41. T. W. Ebbesen, H. Hiura, Y. Ochiai, S. Matsui, and K. Tanigaki, *Chem. Phys. Lett.* 209, 83 (1993).
42. S. Q. Feng, D. P. Yu, G. Hu, X. F. Zhang, and Z. Zhang, *J. Phys. Chem. Solids* 58, 1887 (1997).
43. O. Zhou, R. M. Fleming, D. W. Murphy, C. Chen, R. C. Haddon, A. P. Ramirez, and S. H. Glarum, *Science* 263, 1744 (1994).
44. V. Z. Mordkovich, M. Baxendale, S. Yoshimura, and R. P. H. Chang, *Carbon* 34, 1301 (1996).
45. S. Iijima and T. Ichihashi, *Nature* 363, 603 (1993).
46. D. S. Bethune, C. H. Kiang, M. S. De Vries, G. Gorman, R. Savoy, J. Vazquez, and R. Beyers, *Nature* 363, 605 (1993).
47. C. Journet and P. Bernier, *Appl. Phys. A* 67, 1 (1998).
48. J. L. Hutchison, N. A. Kiselev, E. P. Krinichnaya, A. V. Krestinin, R. O. Loutfy, A. P. Morawsky, V. E. Muradyan, E. D. Obraztsova, J. Sloan, S. V. Terekhov, and D. N. Zakharov, *Carbon* 39, 761 (2001).
49. C. Guerret-Piecourt, Y. Le Bouar, A. Loiseau, and H. Pascard, *Nature* 372, 761 (1994).
50. M. Ata, A. J. Hudson, K. Yamaura, and K. Kurihara, *J. Appl. Phys.* 34, 4207 (1995).
51. A. Loiseau and H. Pascard, *Chem. Phys. Lett.* 256, 246 (1996).
52. T. Guo, P. Nikolaev, A. Thess, D. T. Colbert, and R. E. Smalley, *Chem. Phys. Lett.* 243, 49, (1995).
53. A. Thess, R. Lee, P. Nikolaev, H. Dai, P. Petit, J. Robert, C. Xu, Y. H. Lee, S. G. Kim, A. G. Rinzler, D. T. Colbert, G. E. Scuseria, D. Tomanek, J. E. Fisher, and R. E. Smalley, *Science* 273, 483 (1996).
54. M. Yudasaka, T. Komatsu, T. Ichihashi, and S. Iijima, *Chem. Phys. Lett.* 278, 102 (1997).
55. W. K. Maser, E. Munoz, A. M. Benito, M. T. Martýnez, G. F. de la Fuente, Y. Maniette, E. Anglaret, and J.-L. Sauvajol, *Chem. Phys. Lett.* 292, 587 (1998).
56. F. Kokai, K. Takahashi, M. Yudasaka, R. Yamada, T. Ichihashi, and S. Iijima, *J. Phys. Chem. B* 103, 4346 (1999).
57. H. Kataura, Y. Kumazawa, Y. Maniwa, Y. Ohtsuka, R. Sen, S. Suzuki, B. Y. Achiba, *Carbon* 38, 1691 (2000).
58. S. Lebedkin, P. Schweiss, B. Renker, S. Malik, F. Hennrich, M. Neumaier, C. Stoermer, and M. M. Kappes, *Carbon* 40, 417 (2002).
59. S. Bandow, M. Takizawa, K. Hirahara, M. Yudasaka, and S. Iijima, *Chem. Phys. Lett.* 337, 48 (2001).
60. M. Jose-Yacaman, M. Miki-Yoshida, L. Rendon, and J. G. Santiesteban, *Appl. Phys. Lett.* 62, 657 (1993).
61. V. Ivanov, J. B. Nagy, Ph. Lambin, A. Lucas, X. B. Zhang, X. F. Zhang, D. Bernaerts, G. Van Tendeloo, S. Amelinckx, and J. Van Landuyt, *Chem. Phys. Lett.* 223, 329 (1994).
62. V. Ivanov, A. Fonseca, J. B. Nagy, A. Lucas, P. Lambin, D. Bernaerts, and X. B. Zhang, *Carbon* 33, 1727 (1995).
63. K. Hernadi, A. Fonseca, J. B. Nagy, D. Bernaerts, A. Fudala, and A. A. Lucas, *Zeolites* 17, 416 (1996).
64. S. Herreyre and P. Gadelle, *Carbon* 33, 234 (1995).
65. P. Pinheiro, M. C. Schouler, P. Gadelle, M. Mermoux, and E. Dooryhee, *Carbon* 38, 1469 (2000).
66. X. Y. Song, W. Cao, M. R. Ayers, and A. J. Hunt, *J. Mater. Res.* 10, 251 (1995).
67. R. Sen, A. Govindaraj, and C. N. R. Rao, *Chem. Phys. Lett.* 267, 276 (1997).
68. H. Dai, A. G. Rinzler, P. Nikolaev, A. Thess, D. T. Colbert, and R. E. Smalley, *Chem. Phys. Lett.* 260, 471 (1996).
69. A. Peigney, Ch. Laurent, F. Dobigeon, and A. Rousset, *J. Mater. Res.* 12, 613 (1997).
70. P. Nikolaev, M. J. Bronikowski, R. K. Bradley, F. Rohmund, D. T. Colbert, K. A. Smith, and R. E. Smalley, *Chem. Phys. Lett.* 313, 91 (1999).
71. A. Huczko, *Appl. Phys. A* 74, 617 (2002).
72. A. Cassell, N. Franklin, T. Tombler, E. Chan, J. Han, and H. Dai, *J. Am. Chem. Soc.* 1221, 7975 (1999).
73. N. R. Franklin and H. Dai, *Adv. Mater.* 12, 890 (2000).
74. Y. Homma, T. Yamashita, Y. Kobayashi, and T. Ogino, *Physica B*, in press.
75. L. Marty, V. Bouchiat, A. M. Bonnot, M. Chaumont, T. Fournier, S. Decossas, and S. Roche, *Microelec. Eng.* 61–62, 485 (2002).
76. E. Yenilmez, Q. Wang, R. J. Chen, D. Wang, and H. Dai, *Appl. Phys. Lett.* 80, 2225 (2002).
77. E. G. Rakov, *Russ. Chem. Rev.* 70, 827 (2001).
78. K. Tohji, H. Takahashi, Y. Shinoda, N. Shimizu, B. Jeyadevan, I. Matsuoka, Y. Sato, A. Kasuya, S. Ito, and Y. Nishina, *J. Phys. Chem. B* 101, 1974 (1997).
79. J. M. Bonard, T. Stora, J. P. Salvetat, F. Maier, T. Stockli, C. Duschl, L. Forro, W. A. De Heer, and A. Chatelain, *Adv. Mater.* 9, 827 (1997).

80. A. G. Rinzler, J. Liu, H. Dai, P. Nikolaev, C. B. Huffman, F. J. Rodriguez-Macias, P. J. Boul, A. H. Lu, D. Heymann, D. T. Colbert, R. S. Lee, J. E. Fisher, A. M. Rao, P. C. Eklund, and R. E. Smalley, *Appl. Phys. A* 67, 29 (1998).
81. A. C. Dillon, T. Genett, K. M. Jones, J. L. Alleman, P. A. Parilla, and M. J. Heben, *Adv. Mater.* 11, 1354 (1999).
82. F. Li, H. M. Cheng, Y. T. Xing, P. H. Tan, and G. Su, *Carbon* 38, 2041 (2000).
83. M. Yudasaka, M. Zhang, C. Jabs, and S. Iijima, *Appl. Phys. A* 71, 449 (2000).
84. J. N. Coleman, A. B. Dalton, S. Curran, A. Rubio, A. P. Davey, A. Drury, B. McCarthy, B. Lahr, P. M. Ajayan, S. Roth, R. C. Barklie, and W. J. Blau, *Adv. Mater.* 12, 213 (2000).
85. M. Holzinger, A. Hirsch, P. Bernier, G. S. Duesberg, and M. Burghard, *Appl. Phys. A* 70, 599 (2000).
86. S. Niyogi, H. Hu, M. A. Hamon, P. Bhomik, B. Zhao, S. M. Rozhenzak, J. Chen, M. E. Iktis, M. S. Meier, and R. C. Haddon, *J. Am. Chem. Soc.* 123, (2001).
87. J. Liu, A. G. Rinzler, H. Dai, J. H. Hafner, R. K. Bradley, P. J. Boul, A. Lu, T. Iverson, K. Shelimov, C. B. Huffman, F. J. Rodriguez-Macias, Y.-S. Shon, R. S. Lee, D. T. Colbert, and R. E. Smalley, *Science* 280, 1253 (1998).
88. G. S. Duesberg, J. Muster, V. Krstic, M. Burghard, and S. Roth, *Appl. Phys. A* 67, 117 (1998).
89. G. S. Duesberg, W. Blau, H. J. Byrne, J. Muster, M. Burghard, and S. Roth, *Synth. Met.* 103, 2484 (1999).
90. S. K. Doorn, R. E. Fields, H. Hu, M. A. Hamon, R. C. Haddon, J. P. Selegue, and V. Majidi, *J. Am. Chem. Soc.* 124, 3169 (2002).
91. T. W. Ebbesen, P. M. Ajayan, H. Hiura, and K. Tanigaki, *Nature* 367, 519 (1992).
92. Y. Chen, R. C. Haddon, S. Fang, A. M. Rao, P. C. Eklund, W. H. Lee, E. C. Dickey, E. A. Grulke, J. C. Pendergrass, A. Chavan, B. E. Haley, and R. E. Smalley, *J. Mater. Res.* 13, 2423 (1998).
93. S. B. Sinnott, *J. Nanosci. Nanotech.* 2, 113 (2002).
94. H. P. Boehm, *Carbon* 32, 759 (1994).
95. K. Esumi, M. Ishigami, A. Nakajima, K. Sawada, and H. Honda, *Carbon* 34, 279 (1996).
96. P. A. Cahill and C. M. Rohling, *Tetrahedron* 52, 5247 (1996).
97. R. M. Lago, S. C. Tsang, K. L. Liu, Y. K. Cheng, and M. L. H. Green, *Chem. Commun.* 1355 (1995).
98. R. Yu, L. Chen, Q. Lin, J. Lin, K.-L. Tan, S. C. Ng, H. S. O. Chan, G.-Q. Xu, and T. S. A. Hor, *Chem. Mater.* 10, 718 (1998).
99. M. Monthioux, B. W. Smith, B. Burteaux, A. Claye, J. E. Fischer, and D. E. Luzzi, *Carbon* 39, 1251 (2001).
100. S. S. Wong, E. Joselevich, A. T. Wooley, C. L. Cheung, and C. M. Lieber, *Nature* 394, 52 (1998).
101. S. S. Wong, A. T. Wooley, E. Joselevich, C. L. Cheung, and C. M. Lieber, *J. Am. Chem. Soc.* 120, 8557 (1998).
102. S. S. Wong, A. T. Wooley, E. Joselevich, and C. M. Lieber, *Chem. Phys. Lett.* 306, 219 (1999).
103. S. Delpeux, K. Metenier, R. Benoit, F. Vivet, L. Boufendi, S. Bonnamy, and F. Beguin, in AIP Conf. Proc., "Electronic Properties of Novel Materials—Science and Technology of Molecular Structures," 1999, p. 470.
104. B. Ni and S. B. Sinnott, *Phys. Rev. B* 61, R16343 (2000).
105. H. Touhara and F. Okino, *Carbon* 38, 241 (2000).
106. E. T. Mickelson, C. B. Huffman, A. G. Rinzler, R. E. Smalley, and J. L. Margrave, *Chem. Phys. Lett.* 296, 188 (1998).
107. Y. Zhang, T. Ichihashi, E. Landree, F. Niney, and S Iijima, *Science* 285, 1719 (1999).
108. E. Hernández, C. Goze, P. Bernier, and A. Rubio, *Phys. Rev. Lett.* 80, 4502 (1998).
109. J. P. Lu, *Phys. Rev. Lett.* 79, 1297 (1997).
110. B. I. Yakobson, C. J. Brabec, and J. Bernholc, *Phys. Rev. Lett.* 76, 2511 (1996).
111. M. M. J. Treacy, T. W. Ebbesen, and J. M. Gibson, *Nature* 381, 678 (1996).
112. E. W. Wong, P. E. Sheehan, and C. M. Lieber, *Science* 277, 1971 (1997).
113. P. Poncharal, Z. L. Wang, D. Ugarte, and W. A. de Heer, *Science* 283, 1513 (1999).
114. A. Krishnan, E. Dujardin, T. W. Ebbesen, P. N. Yianilos, and M. M. J. Treacy, *Phys. Rev. B* 58, 14013 (1998).
115. B. I. Yakobson, M. P. Campbell, C. J. Brabec, and J. Bernholc, *Comput. Mat. Sci.* 8, 341 (1997).
116. Ge. G. Samsonidze, Gu. G. Samsonidze, and B. I. Yakobson, *Phys. Rev. Lett.* 88, 65501 (2002).
117. M.-F. Yu, O. Lourie, M. J. Dyer, K. Moloni, T. F. Kelly, and R. S. Ruoff, *Science* 287, 637 (2000).
118. F. Li, H. M. Cheng, S. Bai, G. Su, and M. S. Dresselhaus, *Appl. Phys. Lett.* 77, 3161 (2000).
119. M.-F. Yu, B. S. Files, S. Arepalli, and R. S. Ruoff, *Phys. Rev. Lett.* 84, 5552 (2000).
120. D. A. Walters, L. M. Ericson, M. J. Casavant, J. Liu, D. T. Colbert, K. A. Smoth, and R. E. Smalley, *Appl. Phys. Lett.* 74, 3803 (1999).
121. M. R. Falvo, G. J. Clary, R. M. Taylor II, V. Chi, F. P. Brooks Jr., S. Washburn, and R. Superfine, *Nature* 389, 582 (1997).
122. M. Buongiorno Nardelli, B. I. Yakobson, and J. Bernholc, *Phys. Rev. Lett.* 81, 4656 (1998).
123. M. Buongiorno Nardelli, B. I. Yakobson, and J. Bernholc, *Phys. Rev. B* 57, R4277 (1998).
124. Q. Zhao, M. Buongiorno Nardelli, and J. Bernholc, *Phys. Rev. B* 65, 144105 (2002).
125. L. Chico, L. X. Benedict, S. G. Louie, and M. L. Cohen, *Phys. Rev. B* 54, 2600 (1996).
126. L. Chico, V. H. Crespi, L. X. Benedict, S. G. Louie, and M. L. Cohen, *Phys. Rev. Lett.* 76, 971 (1996).
127. M. Bockrath, W. Liang, D. Bozovix, J. H. Hafner, C. M. Lieber, M. Tinkham, and H. Park, *Science* 291, 283 (2001).
128. M. Buongiorno Nardelli and J. Bernholc, *Phys. Rev. B* 60, 16338 (1999).
129. A. Rochefort, D. R. Salahub, and P. Avouris, *Chem. Phys. Lett.* 297, 45 (1998).
130. L. Liu, C. S. Jayanthi, M. Tang, S. Y. Wu, T. W. Tombler, C. Zhou, L. Alexseyev, J. Kong, H. Dai, *Phys. Rev. Lett.* 84, 4950 (2000); T. W. Tombler, C. Zhou, L. Alekseyev, J. Kong, H. Dai, L. Liu, C. S. Jayanthi, M. Tang, and S.-Y. Wu, *Nature* 402, 769 (2000).
131. A. Maiti, A. Svizhenko, and M. P. Anantram, *Phys. Rev. Lett.* 88, 126805 (2002).
132. H. W. Postma, T. Teepen, Z. Yao, M. Grifoni, and C. Dekker, *Science* 293, 76 (2001).
133. D. Bozovic, M. Bockrath, J. H. Hafner, C. M. Lieber, H. Park, and M. Tinkham, *Appl. Phys. Lett.* 78, 3693 (2001).
134. J. Y. Park, Y. Yaish, M. Brink, S. Rosenblatt, and P. L. McEuen, *Appl. Phys. Lett.* 80, 4446 (2002).
135. B. Reulet, A. Yu. Kasumov, M. Koziak, R. Deblock, I. I. Khodos, Yu. B. Gorbatov, V. T. Volkov, C. Journet, and H. Bouchiat, *Phys. Rev. Lett.* 85, 2829 (2000).
136. Q. Zheng and Q. Jiang, *Phys. Rev. Lett.* 88, 045503 (2002).
137. A. Erbe, R. H. Blick, A. Tilke, A. Kriele, and J. P. Kotthaus, *Appl. Phys. Lett.* 73, 3751 (1998); A. Erbe, H. Krömmer, A. Kraus, R. H. Blick, G. Corso, and K. Richter, *Appl. Phys. Lett.* 77, 3102 (2000).
138. L. E. Kinsler, A. R. Frey, A. B. Coppers, and J. V. Sanders, "Fundamentals of Acoustics," 3rd Edition, John Wiley & Sons, New York, 1982.
139. "CRC Handbook of Chemistry and Physics," 75th edition, CRC Press, Inc., 1994.
140. M. Ouyang, J. L. Huang, C. L. Cheung, and C. M. Lieber, *Science* 292, 702 (2001).
141. M. Reghu, in "*Handbook of Conducting Polymers*," 2nd ed., (T. A. Skotheim, R. L. Elsenbaumer, and J. R. Reynolds, Eds.). Marcel Dekker, New York, 1998.
142. J. W. G. Wildöer, L. C. Venema, A. G. Rinzler, R. E. Smalley, and C. Dekker, *Nature* 391, 59 (1998).

143. T. W. Odom, J. L. Huang, P. Kim, and C. M. Lieber, *Nature* 391, 62 (1998).
144. L. C. Venema, J. W. G. Wildöer, J. W. Janssen, S. J. Tans, L. J. Hinne, T. Tuinstra, L. P. Kouwenhoven, and C. Dekker, *Science* 283, 52 (1999).
145. S. G. Lemay, J. W. Janssen, M. van den Hout, M. Mooij, M. J. Bronikowski, P. A. Willis, R. E. Smalley, L. P. Kouwenhoven, and C. Dekker, *Nature* 412, 617 (2001).
146. A. Hassanien, A. Mrzel, M. Tokumoto, and D. Tomanek, *Appl. Phys. Lett.* 79, 4210 (2001).
147. D. L. Carroll, Ph. Redlich, X. Blasé, J.-C. Charlier, S. Curran, P. M. Ajayan, S. Roth, and M. Rühle, *Phys. Rev. Lett.* 81, 2332 (1998).
148. J. W. Janssen, S. G. Lemay, L. P. Kouwenhoven, and C. Dekker, *Phys. Rev. B* 65, 115423 (2002).
149. M. Ouyang, J. L. Huang, C. L. Cheung, and C. M. Lieber, *Science* 291, 97 (2001).
150. M. Ouyang, J. L. Huang, and C. M. Lieber, *Phys. Rev. Lett.* 88, 66804 (2002).
151. D. J. Hornbaker, S.-J. Kahng, S. Misra, B. W. Smith, A. T. Johnson, E. J. Mele, D. E. Luzzi, and A. Yazdani, *Science* 295, 828 (2002).
152. A. Jorio, R. Saito, J. H. Hafner, C. M. Lieber, M. Hunter, T. McClure, G. Dresselhaus, and M. S. Dresselhaus, *Phys. Rev. Lett.* 86, 1118 (2001).
153. R. R. Schlittler, J. W. Seo, J. K. Gimzewski, C. Durkan, M. S. M. Saifullah, and M. E. Welland, *Science* 292, 1136 (2001).
154. M. J. O'Connell, S. M. Bachilo, C. B. Huffman, V. C. Moore, M. S. Strand, E. H. Haroz, K. L. Rialon, P.-J. Boul, W. H. Noon, C. Kittrell, J. Ma, R. H. Hauge, R. B. Weisman, and R. E. Smalley, *Science* 297, 593 (2002).
155. J. W. Mintmire, B. I. Dunlap, and C. T. White, *Phys. Rev. Lett.* 68, 631 (1992).
156. T. W. Ebbesen, H. J. Lezec, H. Hiura, J. W. Bennett, H. F. Ghaemi, and T. Thio, *Nature* 382, 54 (1996).
157. L. Langer, V. Bayot, E. Grivei, J. P. Issi, J. P. Heremans, C. H. Olk, L. Stockman, C. Van Haesendonck, and Y. Bruynseraede, *Phys. Rev. Lett.* 76, 479 (1996).
158. L. A. Nagahara, I. Amlani, J. Lewenstein, and R. K. Tsui, *Appl. Phys. Lett.* 80, 3826 (2002).
159. J. Tersoff, *Appl. Phys. Lett.* 74, 2122 (1999).
160. M. Kociak, A. Yu. Kasumov, S. Gueron, B. Reulet, L. Vaccarini, I. I. Khodos, Yu. B. Gorbatov, V. T. Volkov, and H. Bouchiat, *Phys. Rev. Lett.* 86, 2416 (2001).
161. Z. K. Tang, L. Zhang, N. Wang, X. X. Zhang, G. H. Wen, G. D. Li, J. N. Wang, C. T. Chan, and P. Sheng, *Science* 292, 2462 (2001).
162. P. G. Collins, M. S. Arnold, and Ph. Avouris, *Science* 292, 706 (2001).
163. M. T. Woodside and P. L. McEuen, *Science* 296, 1098 (2002).
164. C. T. White and T. N. Todorov, *Nature* 393, 240 (1998).
165. S. Frank, P. Poncharal, Z. L. Wang, and W. A. de Heer, *Science* 280, 1744 (1998).
166. W. Liang, M. Bockrath, D. Bozovic, J. H. Hafner, M. Tinkham, and H. Park, *Nature* 411, 665 (2001).
167. J. Kong, E. Yenilmez, T. W. Tombler, W. Kim, H. Dai, R. B. Laughlin, L. Liu, C. S. Jayanthi, and S. Y. Wu, *Phys. Rev. Lett.* 87, 106801 (2001).
168. F. D. M. Haldane, *Phys. Rev. Lett.* 47, 1840 (1981).
169. M. Bockrath, D. H. Cobden, J. Lu, A. G. Rinzler, R. E. Smalley, L. Balents, and P. L. McEuen, *Nature* 397, 598 (1999).
170. Z. Yao, H. W. Postma, L. Balents, and C. Dekker, *Nature* 402, 273 (1999).
171. "Single Charge Tunneling," H. Grabert and M. Devoret, Eds., Plenum Press, New York, 1992.
172. M. Bockrath, D. H. Cobden, P. L. McEuen, N. G. Chopra, A. Zettl, A. Thess, and R. E. Smalley, *Science* 275, 1922 (1997).
173. S. J. Tans, M. H. Devoret, H. Dai, A. Thess, R. E. Smalley, L. J. Geerligs, and C. Dekker, *Nature* 386, 474 (1997).
174. J. Nygård, D. H. Cobden, P. E. Lindelöf, *Nature* 408, 342 (2000).
175. W. Liang, M. Bockrath, and H. Park, *Phys. Rev. Lett.* 88, 126801 (2002).
176. M. Ahlskog, R. Tarkiainen, L. Roschier, and P. Hakonen, *Appl. Phys. Lett.* 77, 4037 (2000).
177. C. Zhou, J. Kong, E. Yenilmez, and H. Dai, *Science* 290, 1552 (2000).
178. A. F. Morpurgo, J. Kong, C. M. Marcus, and H. Dai, *Science* 286, 263 (1999).
179. A. Yu. Kasumov, R. Deblock, M. Kociak, B. Reulet, H. Bouchiat, I. I. Khodos, Yu. B. Gorbatov, V. T. Volkov, C. Journet, and M. Burghard, *Science* 284, 1508 (1999).
180. A. Bachtold, C. Strunk, J. P. Salvetat, J. M. Bonard, L. Forro, T. Nussbaumer, and C. Schönenberger, *Nature* 397, 673 (1999).
181. R. Tarkiainen, M. Ahlskog, P. Hakonen, and M. Paalanen, "Proceedings of Low Temperature" 23, in press.
182. S. J. Wind, J. Appenzaller, R. Martel, V. Derycke, and Ph. Avouris, *Appl. Phys. Lett.* 80, 3817 (2002).
183. A. Bachtold, P. Hadley, T. Nakanishi, and C. Dekker, *Science* 294, 1317 (2001).
184. S. Heinze, J. Tersoff, R. Martel, V. Derycke, J. Appenzaller, and Ph. Avouris, *Phys. Rev. Lett.* 89, 106801 (2002).
185. P. Ball, *Nature* 414, 142 (2001).
186. S. H. Dai, *Surf. Sci.* 500, 218 (2002).
187. L. Risch, *Mater. Sci. Eng. C* 19, 363 (2002).
188. P. Avouris, *Chem. Phys.* 281, 429 (2002).
189. Z. Yao, C. L. Kane, and C. Dekker, *Phys. Rev. Lett.* 84, 2941 (2000).
190. F. Kreupl, A. P. Graham, G. S. Duesberg, W. Steinhögl, M. Liebau, E. Unger, and W. Hönlein, *Microelectron. Eng.* 64, 399 (2002).
191. Y. Zhang, A. Chang, J. Cao, Q. Wang, W. Kim, Y. Li, N. Morris, E. Yenilmez, J. Kong, and H. Dai, *Appl. Phys. Lett.* 79, 3155 (2001).
192. J. Liu, M. J. Casavant, M. Cox, D. A. Walters, P. Boul, L. Wei, A. J. Rimberg, K. A. Smith, D. T. Colbert, and R. E. Smalley, *Chem. Phys. Lett.* 303, 125 (1999).
193. Y. Homma, Y. Kobayashi, T. Ogino, and T. Yamashita, *Appl. Phys. Lett.* 81, 2261 (2002).
194. N. R. Franklin, Y. Li, R. J. Chen, A. Javey, and H. Dai, *Appl. Phys. Lett.* 79, 4571 (2001).
195. Z. Yao, C. Dekker, and Ph. Avouris, *Top. Appl. Phys.* 80, 147 (2000).
196. A. G. Rinzler, J. H. Hafner, P. Nikoleav, L. Lou, S. G. Kim, D. Tomanek, D. Colbert, and R. E. Smalley, *Science* 269, 1550 (1995).
197. Y. Saito, K. Hamaguchi, T. Nishino, K. Hata, K. Tohji, A. Kasuya, and Y. Nishina, *Jpn. J. Appl. Phys.* 36, L1340 (1997).
198. Z. W. Pan, F. C. K. Au, H. L. Lai, W. Y. Zhou, L. F. Sun, Z. Q. Liu, D. S. Tang, C. S. Lee, S. T. Lee, and S. S. Xie, *J. Phys. Chem. B* 105, 1519 (2001).
199. P. G. Collins and A. Zettl, *Appl. Phys. Lett.* 69, 1969 (1996).
200. Y. Saito and S. Uehara, *Carbon* 38, 169 (2000).
201. R. Collazo, R. Schlesser, and Z. Sitar, *Appl. Phys. Lett.* 78, 2058 (2001).
202. J.-M. Bonard, T. Stöckli, F. Maier, W. A. de Heer, A. Châtelain, J.-P. Salvetat, and L. Forró, *Phys. Rev. Lett.* 81, 1441 (1998).
203. J.-M. Bonard, J.-P. Salvetat, T. Stöckli, L. Farró, and A. Châtelain, *Appl. Phys. A* 69, 245 (1999).
204. Q. H. Wang, A. A. Sethur, J. M. Lauerhaas, J. Y. Dai, E. W. Seeling, and R. H. Chang, *Appl. Phys. Lett.* 72, 2912 (1998).
205. A. P. Burden, *Materials World* 8, 22 (2000).
206. W. B. Coi, D. S. Chung, J. H. Kang, Y. W. Jin, I. T. Han, Y. H. Lee, J. E. Jung, N. S. Lee, G. S. Park, J. M. Kim, *Appl. Phys. Lett.* 75, 3129 (1999).
207. N. S. Lee, D. S. Chung, I. T. Han, J. H. Kang, Y. S. Choi, H. Y. Kim, S. H. Park, Y. W. Jin, W. K. Yi, and M. J. Yun, *Diamond Rel. Mater.* 10, 265 (2001).
208. Y. Saito, S. Uemura, and K. Hamaguchi, *Jpn. J. Appl. Phys.* 37, L346 (1998).

209. W. I. Milne, K. B. K. Teo, M. Chhowalla, G. A. J. Amaratunga, J. Yuan, J. Robertson, P. Legagneux, G. Pirio, K. Bouzehouane, D. Pribat, W. Bruenger, and C. Trautmann, *Current Appl. Phys.* 1, 317 (2001).
210. W. Zhu, C. Bowler, O. Zhou, G. Kochanski, and S. Jin, *Appl. Phys. Lett.* 75, 873 (1999).
211. H. Sugie, M. Tanemura, V. Filip, K. Iwata, K. Takahashi, and F. D. Okuyama, *Appl. Phys. Lett.* 78, 2578 (2001).
212. G. Z. Yue, Q. Qiu, B. Gao, Y. Cheng, J. Zhang, H. Shimoda, S. Chang, J. P. Lu, and O. Zhou, *Appl. Phys. Lett.* 81, 355 (2002).
213. C. Niu, E. K. Sichel, R. Hoch, D. Moy, and H. Tennent, *Appl. Phys. Lett.* 70, 1480 (1997).
214. E. Frackowiak and F. Béguin, *Carbon* 40, 1775 (2002).
215. K. Jurewicz, S. Delpeux, V. Bertagna, F. Béguin, and E. Frackowiak, *Chem. Phys. Lett.* 347, 36 (2001).
216. R. H. Baughman, C. Cui, A. A. Zakhidov, Z. Iqbal, J. N. Barisci, G. M. Spinks, G. G. Wallace, A. Mazzoldi, D. De Rossi, A. G. Rinzler, O. Jaschinski, S. Roth, and M. Kertesz, *Science* 284, 1340 (1999).
217. S. Roth and R. H. Baughman, *Current Appl. Phys.* 2, 211 (2002).
218. H. Dai, J. H. Hafner, A. G. Rinzler, D. T. Colbert, and R. E. Smalley, *Nature* 384, 147 (1996).
219. V. Barwich, M. Bammerlin, A. Baratoff, R. Bennewitz, M. Guggisberg, C. Loppacher, O. Pfeiffer, E. Meyer, H.-J. Güntherodt, J.-P. Salvetat, M. Bonard, and L. Forró, *Appl. Surf. Sci.* 157, 269 (2000).
220. M. Krüger, M. R. Buitelaar, T. Nussbaumer, C. Schönenberger, and L. Forró, *Appl. Phys. Lett.* 78, 1291 (2001).
221. P. G. Collins, K. Bradley, M. Ishigami, and A. Zettl, *Science* 287, 1801 (2000).
222. J. Kong, N. R. Franklin, C. Zhou, M. G. Chapline, S. Peng, K. Cho, and H. Dai, *Science* 287, 622 (2000).
223. O. K. Varghese, P. D. Kichambre, D. Gong, K. G. Ong, E. C. Dickey, and C. A. Grimes, *Sens. Actuators, B* 81, 32 (2001).
224. P. Kim and C. M. Lieber, *Science* 286, 2148 (1999).
225. S. Akita, Y. Nakayama, S. Mizooka, Y. Takano, T. Okawa, Y. Miyatake, S. Yamanaka, M. Tsuji, and T. Nosaka, *Appl. Phys. Lett.* 79, 1691 (2001).
226. H. Dai, N. Franklin, and J. Han, *Appl. Phys. Lett.* 73, 1508 (1998).
227. G. T. Kovacs, "*Micromachined Transducers Sourcebook*," McGraw-Hill, New York, 1998.
228. T. Rueckes, K. Kim, E. Joselevich, G. Y. Tseng, C.-L. Cheung, and C. M. Lieber, *Science* 289, 94 (2000).
229. M. Dequesnes, S. V. Rotkin, and N. R. Aluru, *Nanotechnology* 13, 120 (2002).
230. Y. Zhang, A. Chang, J. Cao, Q. Wang, W. Kim, Y. Li, N. Morris, E. Yenilmez, J. Kong, and H. Dai, *Appl. Phys. Lett.* 79, 3155 (2001).

Encyclopedia of Nanoscience and Nanotechnology
www.aspbs.com/enn

Electronic Properties of Carbon Nanotubes

J. González
Consejo Superior de Investigaciones Científicas, Madrid, Spain

CONTENTS

1. INTRODUCTION

Carbon nanotubes are tubular structures that can be thought of as the result of wrapping up the carbon honeycomb lattice of a graphene sheet. They were discovered in 1991 with the use of high-resolution transmission-electron microscopy, by observing the soot produced from an arc discharge between carbon rods [1]. Since then they have been a fascinating subject of research due to their remarkable mechanical, chemical, and electronic properties [2]. Carbon nanotubes have become very promising in the field of molecular electronics, in which atoms and molecules are envisaged as the building blocks in the fabrication of electronic devices. In this respect, the great expectations placed on the nanotubes are also shared by the fullerenes [3]. These also can be thought of as being assembled from the hexagonal carbon rings of a graphene sheet, by making a close cage by insertion of twelve pentagonal rings. The multiple forms and shapes in which the carbon-based materials may appear, with varying physical and chemical properties, are what make them so interesting for the purpose of designing and fabricating nanoscale devices.

The nanotubes as well as the fullerenes have a molecular structure that makes them particularly suitable to develop an alternative, at the nanometer scale, to the silicon-based integrated electronics. It is known, for instance, that nanotubes have a remarkable stiffness along the tubular direction, reinforcing the structure of the components based on them. On the other hand, the design of electronic devices at the molecular scale requires taking into account novel effects which stem from the reduced dimensionality of the systems under consideration. In the case of individual single-walled nanotubes at low temperatures, it has been shown that the electron waves may remain extended along the nanotubes over lengths of several microns [4, 5]. This means that the motion of the electrons cannot be understood in terms of classical diffusion, and that instead they show a genuine quantum-mechanical behavior, up to the point that interference patterns between the electron waves can be observed [6].

The behavior of the nanotubes as true quantum wires leads to remarkable predictions, like the ballistic transport along the longitudinal direction. This effect seems to have been observed in multiwalled nanotubes [7], which have an onion-like structure of concentric single-walled nanotubes. The very large current densities observed there (above 10^7 A cm^{-2}) seem to be only compatible with transport without dissipation in the nanotube structure. On the other hand, single-walled nanotubes are also commonly found forming close-packed ropes. Their behavior may give rise to interesting features, because the strong Coulomb repulsion between electrons existing in the isolated nanotubes is screened due to the intertube interactions between a large number of metallic nanotubes. A superconducting transition at temperatures below 1 K has been observed in ropes made of about three hundred nanotubes [8]. Superconductivity seems to be also a plausible phenomenon in small-diameter nanotubes, where the coupling of the electrons to lattice vibrations is largely enhanced [9].

The main goal in designing nanotechnology devices is to tailor the molecular structures to achieve a given functionality. In that respect, the various geometries that the nanotubes may adopt and the possibility of assembling them with other carbon structures open the way to the construction of common electronic devices such as diodes [10],

ISBN: 1-58883-059-4/$35.00

transistors [11–14], memory elements [15], and logic circuits [16, 17]. As explained in Section 2, the nanotubes may have metallic or semiconducting properties depending on the way the graphene sheet is wrapped [18–20]. The nanotube structure can change along the tubule from one class to the other by the presence of topological defects, that is, heptagonal and pentagonal rings in the hexagonal carbon lattice [21–23]. This kind of intramolecular junctions has shown diode-like rectifying properties [10]. Furthermore, transistors working with a single nanotube element have been produced, either by the field effect on semiconducting nanotubes [11–13] or by constraining single electrons to short islands between two buckles in a metallic nanotube [14]. An alternative to these structures can be the use of Y-junctions in which two single-walled nanotubes merge into one at a given angle [24, 25]. The response characteristics of such devices have shown interesting properties, including nonlinear transport behavior and current rectification [26–29].

One of the challenges of the carbon nanotube-based molecular electronics is the controlled high-yield production of the relevant structures in the design of the electronic devices. The progress in the development of new techniques and the synthesis of new structures is constant. Very promising is the formation of supramolecular assemblies in which fullerene cages are inserted in the hollow structure of the nanotube [30–32]. It has been shown that the presence of the fullerenes leads to a modulation of the gap in the case of the semiconducting nanotubes [33]. When the fullerenes are disposed forming an array inside the nanotube, the hybridization of the fullerene molecular orbitals with the states in the nanotube conduction band gives rise to a band with mixed fullerene-nanotube character [34]. Surely the interplay between the properties of the carbon nanotubes and the electronic features of the fullerenes has to give rise to new physical effects, opening the way to different routes in the design of components in molecular electronics.

2. METALLIC VERSUS SEMICONDUCTING NANOTUBES

2.1. Band Structure of Carbon Nanotubes

A remarkable feature of the single-walled carbon nanotubes is that their conduction properties depend on the helical arrangement of the hexagonal carbon rings on the tubular structure. Thus, carbon nanotubes can be metallic or semiconducting as a consequence of their particular geometry and, when a gap develops in the low-energy spectrum, it also depends on the diameter of the nanotube. These properties follow from the band structure of the tubular arrangements, which is composed of a certain number of one-dimensional subbands depending on the thickness of the nanotube. The conducting properties can be formally ascertained by addressing the question of whether the Fermi level crosses or not some of the subbands. The prediction of the metallic and semiconducting properties of the nanotubes depending on their geometry was actually made on theoretical grounds in 1992 [18–20]. Although the electronic properties of the nanotubes had been studied for some time, it was not until 1998 that their structure was atomically resolved by using scanning tunneling microscopy [35, 36]. In this way, it has been possible to establish experimentally the correspondence between the conducting properties and the geometric structure of the individual single-walled nanotubes.

Except in cases where the nanotube diameter is very small [37–39], the band structure of the individual nanotubes can be understood from the band dispersion of a graphene sheet, after applying the periodic boundary conditions arising from the winding of the hexagonal rings around the axis of the nanotube. The conduction properties can be obtained qualitatively from the hybridization of the carbon π orbitals, which gives rise to respective bonding and antibonding bands in the graphite sheet. The two-dimensional Brillouin zone is a hexagon and the electron energy-momentum dispersion $E(\mathbf{k})$ takes the following form in the tight-binding approximation [40]:

$$E(\mathbf{k}) = \pm t\sqrt{1 + 4\cos^2(\sqrt{3}k_x a/2) + 4\cos(\sqrt{3}k_x a/2)\cos(3k_y a/2)} \tag{1}$$

where a is the nearest-neighbor carbon–carbon distance ($\approx$0.14 nm) and t is the energy scale ($\approx$2.5 eV) given by the matrix element between neighboring π orbitals. The bonding π band of the dispersion relation (corresponding to the $-$ sign in front of the right-hand side of Eq. (1)) has the shape depicted in Figure 1. The cusps are attained at the corners of the hexagonal Brillouin zone, which correspond to the momenta

$$k_x = \pm\frac{4}{3\sqrt{3}}\frac{\pi}{a}, \qquad k_y = 0 \tag{2}$$

$$k_x = \pm\frac{2}{3\sqrt{3}}\frac{\pi}{a}, \qquad k_y = \pm\frac{2}{3}\frac{\pi}{a} \tag{3}$$

It has been shown that, for typical nanotubes with diameter above 1 nm, the carbon σ orbitals have a weak influence on the bonding π band, and that the bonding σ bands appear in the spectrum at energies below ~ -2 eV [18].

When the graphene sheet is not doped by impurities or other means, each carbon atom contributes with one electron to the above band structure which, taking into account the spin degeneracy, turns out to be half-filled. The Fermi level is found then at the cusps connecting the lower and upper branches of the band dispersion given by Eq. (1), so that the graphene sheet is a remarkable two-dimensional

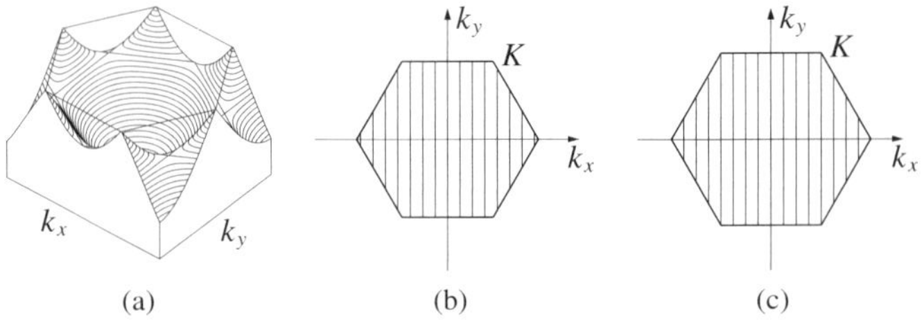

Figure 1. Representation of the first Brillouin zone and band structure of graphene and carbon nanotubes. (a) Dispersion of the bonding π band in graphene. (b) Plot of the allowed wavevectors (vertical lines) in the first Brillouin zone for a zigzag nanotube with $N = 11$. (c) Same plot as in (b) for a zigzag nanotube with $N = 12$.

system which has a set of isolated Fermi points (only two of them being inequivalent) when the carbon lattice is half-filled.

The passage from the electronic states in the graphene sheet to those allowed in the carbon nanotubes can be made by enforcing the conditions satisfied by the electron wavefunctions when winding around the waist of the tubule. This is the point where the geometry of the nanotube plays a crucial role, since many different possibilities exist characterized by the helicity of the carbon lattice on the tubular structure. A simple instance is given by the so-called *zigzag* nanotubes, in which the graphene lattice is wrapped in the direction depicted in Figure 2, and that receive their name from the /\/\/\/\ shape perpendicular to the tube axis. Another significant case corresponds to the *armchair* nanotubes, so called because of the pattern /‾‾__/‾‾__/‾‾\ that they display after forming the tubular arrangement with the axis as depicted in Figure 3.

The analysis of the zigzag and the armchair nanotubes gives insight about the general argument by which a single-walled nanotube can be catalogued as metallic or semiconducting. The different subbands of a nanotube can be obtained from the band dispersion in Eq. (1) by taking into account that the electron wavefunction $\Psi(\mathbf{r})$ has to be single-valued after taking a close path around the tubule. Its dependence on the position $\mathbf{r}$ (measured with the coordinates of the graphene sheet) is of the form

$$\Psi(\mathbf{r}) \sim \exp(i\mathbf{k} \cdot \mathbf{r}) \qquad (4)$$

This means that the argument of the exponential has to increase by a multiple of 2π times i after going once around the waist of the nanotube.

In the case of a zigzag nanotube with lattice vector $\mathbf{T}_1$ in the direction perpendicular to the axis, as shown in Figure 2, we must have

$$N\,\mathbf{T}_1 \cdot \mathbf{k} = 2\pi n \qquad (5)$$

where N is the number of hexagons found when going around the nanotube and n is an integer number. The preceding constraint translates into the quantization condition

$$k_x = \frac{2\pi}{\sqrt{3}a}\frac{n}{N} \qquad (6)$$

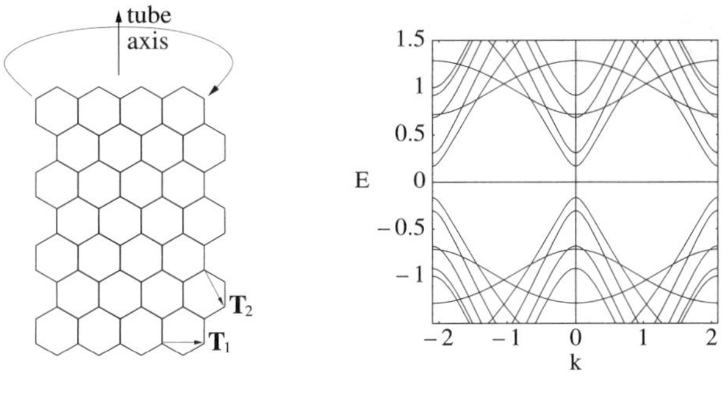

Figure 2. Left: Schematic representation of the wrapping action leading to a zigzag nanotube. Right: Band structure of a zigzag nanotube with $N = 11$. (The energy is measured in units of the overlap integral and the momentum in units of the inverse lattice spacing.)

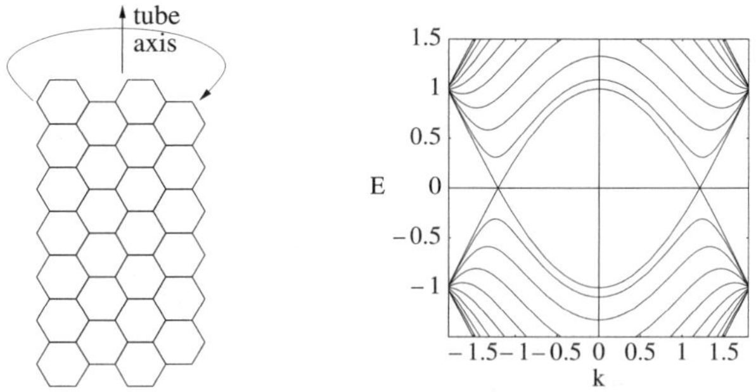

Figure 3. Left: Schematic representation of the wrapping that leads to an armchair nanotube. Right: Band structure of an armchair nanotube with $M = 10$ (in the same units as in Fig. 2).

which gives the allowed wavevectors in a zigzag nanotube. These correspond to a certain number of straight lines in the Brillouin zone, that have been shown for the particular cases of $N = 11$ and $N = 12$ in Figure 1. Each line corresponds to a one-dimensional subband for the tubular structure. In general, an undoped carbon nanotube can have metallic properties only when there is some subband passing by the points where the bonding and antibonding bands meet. According to Eqs. (2) and (3) and the quantization condition (6), that only happens in the case of a zigzag nanotube when the number N of hexagons around the circumference is a multiple of 3 [18, 19]. A narrow gap may open, however, due to the different strength of the electron transfer in the direction perpendicular to the tube axis, which leads to a shift of the Fermi points (in the k_x direction) from the corners of the Brillouin zone. If the number N is not a multiple of 3, a moderate gap Δ opens in the spectrum [18, 19], with a dependence on the diameter d of the form [2, 3]

$$\Delta = 2ta/d \qquad (7)$$

This sensitive dependence of the conducting properties of the nanotubes is illustrated by the band structure of the zigzag nanotube with $N = 11$ shown in Figure 2.

Moving now to the case of the armchair nanotubes, the periodic boundary conditions imply a similar quantization condition on the component k_y of the momentum, which is now in the direction transverse to the tubule axis. Calling M the number of lattice periods along the circumference of the nanotube, the requirement of having single-valued wavefunctions leads to the condition

$$k_y = \frac{2\pi}{3a}\frac{n}{M} \qquad (8)$$

n being again an integer number. The set of allowed wavevectors corresponds to a certain number of straight lines in the Brillouin zone, perpendicular to those found in the case of zigzag nanotubes. Now there is always a subband at the point where the bonding and antibonding π bands meet, irrespective of the value that M may take. This is shown in the band structure for $M = 10$ represented in Figure 3. It turns out, therefore, that all the armchair nanotubes have subbands crossing the Fermi level, and that they remain metallic even taking into account the effects due to the curvature of the tubule [18–20].

The geometric structures of the zigzag and the armchair nanotubes are special cases among all the possibilities by which the graphene sheet can be wrapped to form the nanotubes. The arrangement of the hexagonal carbon rings may show in general some helicity along the tubule. The construction of a helical arrangement can be visualized by starting from the lattice as depicted in Figure 2, and rolling up the tubule, not in the direction of the hexagon rows, but wrapping a row of hexagons onto the next above or below when completing the turn around the tubule. Obviously, that operation can be also done by joining a row of hexagons with the second row, the third, etc., above or below the starting hexagon.

The helicity can be measured then by the shift of a number m of lattice vectors $\mathbf{T}_2$ (defined in Fig. 2) when wrapping the graphene sheet. The degree of helicity m and the number n of hexagons in the row of the parent zigzag tubule characterize completely the geometric structure of the nanotube. These numbers are given usually in the notation (n, m). Thus, the zigzag nanotubes are represented by the $(n, 0)$ geometric structures, while it can be checked that the (n, n) tubules correspond to the armchair nanotubes. Carbon nanotubes which do not have zigzag or armchair structure are called *chiral* nanotubes. In the experiments, the (n, m) structure of the nanotube can be obtained from the measurements of the diameter of the tubule and the chiral angle ϕ formed by the tube axis and the rows of aligned hexagons.

The arguments allowing to understand the conduction properties of the zigzag and the armchair nanotubes can be applied also to the general case of the chiral nanotubes. It turns out that these have subbands crossing the K point of the Brillouin zone when the geometric structure given by (n, m) is such that $n - m$ is a multiple of 3 [18, 19]. In that case, a narrow gap may open at the Fermi level due to the same curvature effects invoked for the zigzag nanotubes. For the rest of chiral nanotubes, a moderate gap opens up in the spectrum, with a dependence on the nanotube diameter that follows the same law predicted for the zigzag nanotubes.

2.2. Experimental Measurements

The predictions regarding the sensitivity of the conducting properties on the geometric structure of the carbon nanotubes have been confronted in the experiments reported in [35] and [36]. In both experiments, individual single-walled nanotubes have been produced with the technique of laser vaporization, being deposited afterwards on a Au(111) substrate. In [36], measurements have been also reported on single-walled nanotubes at the surface of a rope. Topographic images of the individual nanotubes have been obtained at constant tunnel current in a scanning tunneling microscope (STM), leading to the resolution of the hexagonal structure of the carbon rings. This has made it possible to determine the degree of helicity which, together with the knowledge of the nanotube diameter, allows identification of the nanotube in the above catalogue given in terms of the (n, m) numbers.

Within the same experimental setting, scanning tunneling spectroscopy has been performed, in which the current I through the vacuum barrier between the STM tip and the nanotube is recorded as a function of the bias voltage V applied to the sample. This provides important information about the electronic structure, since the differential conductance dI/dV is proportional to the density of states in the nanotube. More precisely, the normalized differential conductance $(V/I)(dI/dV)$ seems to provide a good representation of the local density of electronic states [41]. From the measurements carried out in [35] and [36], it has been possible to determine the position of the peaks corresponding to the edges of the one-dimensional conduction and valence bands and the consequent gap in the semiconducting nanotubes.

Figure 4 contains one of the atomically resolved images of single-walled nanotubes reported in the experiment of [36]. As discussed there, the dark dots correspond to hexagonal carbon rings, and the degree of helical arrangement can be easily discerned in the images. The angle formed between the rows of hexagons and the tube axis determines the chiral angle. The measured values of the angle and the diameter are consistent with the structure of a $(14, -3)$ nanotube, which should be a moderate-gap semiconductor. The spectroscopy measurements show indeed a very small current for bias voltages between -300 and 400 mV. The peaks of the normalized differential conductance out of that range correspond to the edges of the one-dimensional subbands. As reported in [36], a gap can be estimated for the mentioned nanotube of the order of $\approx$750 meV. In general, in the measurements carried out in the semiconducting nanotubes, the current seems to be very small but nonvanishing within the gap, but this may be explained by the existence of tunneling through the nanotube to the gold substrate.

The majority of the measurements reported in [35] and [36] have determined the existence of moderate gaps, consistent with the geometric structure of semiconducting nanotubes. The experimental observations of the gap Δ fit very well to the dependence on the diameter given on theoretical grounds by Eq. (7). The parameter t which gives the best fit is very close to the nearest-neighbor overlap energy obtained from calculations in a single graphene sheet, $t = 2.5$ eV [42].

In the rest of the instances, the measurements of the differential conductance have not shown peaks in the spectra

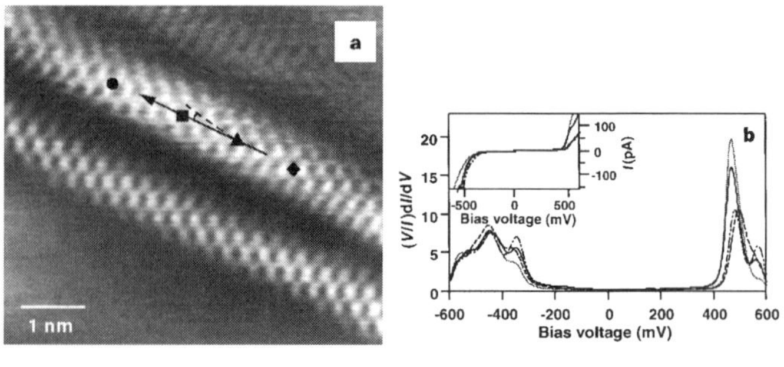

Figure 4. (a) Scanning tunneling microscope image of isolated single-walled nanotubes. As mentioned in [36], the black arrow highlights the tube axis, and the dashed line indicates the zigzag direction. (b) Plot of the normalized conductance and I–V values (inset) measured at the four locations marked in the carbon nanotube in (a). Reprinted with permission from [36], T. W. Odom et al., *Nature* 391, 62 (1998). © 1998, Macmillan Magazines Ltd.

over a wider range of energies, larger than 1.5 eV. The manifest nonvanishing value of the conductance, together with its smooth behavior, is the signature of the metallic character in the nanotubes. When this happens, the geometric structure of the nanotubes has been shown to correspond either to armchair nanotubes or to chiral nanotubes with the helicity suitable for metallic behavior. It has already been mentioned that, in the latter case, a small gap, of the order of $\sim$0.01 eV, should open at the Fermi level by effect of the curvature of the nanotube. Such a gap, however, has been only observed through high-resolution measurements of the conductance in zigzag nanotubes of $(3n, 0)$ type [43]. It is believed that, in the case of the chiral nanotubes of $(3n + m, m)$ type, the small-gap feature tends to be even tinier [44] and its observation may be then precluded by the instrumental noise.

The use of tunneling spectroscopy has given then important information about the band structure of the carbon nanotubes. It has shown that they can be considered as molecular wires. The proportion of the semiconducting nanotubes in the whole set considered in the experiments seems to agree with the ratio of two thirds predicted from pure theoretical arguments. Another important point is that the measures taken on metallic nanotubes in a rope are similar to those on isolated metallic nanotubes, as reported in [36], which seems to imply a weak intertube coupling of the nanotubes within a rope.

The fact that a slight variation in the geometric structure of a carbon nanotube may produce important changes in the conducting properties opens the possibility to build metal–semiconductor or semiconductor–semiconductor junctions in a single molecule. This requires changing the helicity within the same carbon nanotube, which can be achieved by introducing topological defects like combinations of heptagon and pentagon carbon rings in the nanotube lattice [21–23]. A pentagonal ring induces some local curvature in the hexagonal lattice, which can be superposed on a plane and has therefore no intrinsic curvature. The curvature of the pentagonal ring has to be counterbalanced with the opposite induced by an heptagonal ring in order to recover the tubular structure.

The combination of adjacent pentagonal and heptagonal rings has been proposed in [21] to produce changes in the chirality of a nanotube from an (n, m) structure to another of $(n \pm 1, m \mp 1)$ type. The particular case of a junction between $(8, 0)$ and $(7, 1)$ nanotubes has been analyzed, studying in detail the change in the band structure across the interface. This would be a typical instance of a semiconductor–metal junction, and the evolution of the gap along the nanotube has been established theoretically by computing the local density of states. The example of a semiconductor–semiconductor junction formed by $(8, 0)$ and $(5, 3)$ nanotubes has been also studied, stressing the appearance of interface states in the gap from the presence of three pentagon-heptagon defects [21].

Experimental observations of the carbon nanotube intramolecular junctions have been already carried out [10, 45]. In the experiments reported in [45], the atomic structure of the nanotube segments at each side of the junction has been resolved by using scanning tunneling microscopy. This has made it also possible to study the evolution through the junction of the different features in the local density of states. On the other hand, the experiments presented in [10] have investigated the peculiar transport properties of the intramolecular junctions.

In atomic force microscope images, the intramolecular junctions appear as kinks in individual carbon nanotubes, as illustrated in Figure 5 taken from [10]. The large angle formed by the two segments at each side of the junction requires that the heptagon and pentagon rings are located at opposite sides of the nanotube section. By employing the electrodes shown in the figure, it has been characterized that the nanotubes in Figure 5a give rise to a metal–semiconductor junction, while those in Figure 5b provide an example of a metal–metal junction [10].

The conductance through the kink is very different for the two samples shown in Figure 5. In the sample shown to the left, the resistance at zero bias voltage is extremely large, higher than 250 GΩ [10]. The current I measured as a function of the bias voltage V (I–V characteristics) displays a highly nonlinear behavior. With no voltage applied to the gate, the current shows a small increase when a negative bias voltage is applied to the electrodes, while it shows a sharp increase for a positive voltage above 2 V (applied to the upper electrode) [10]. The kink induces therefore a rectifying behavior in the current, which has led to the proposal that it could be used as a molecular device resembling a diode. It has been stressed in [10] that the I–V characteristics become more asymmetric as a gate voltage is applied to the substrate. This has been presented as evidence that the lower segment of the nanotube in Figure 5a is semiconducting. The upper segment is metallic in any event, since its resistance of 110 kΩ (measured at room temperature) does not show dependence on the gate voltage [10].

It has been established that the kink in Figure 5b produces a metal–metal intramolecular junction [10]. The behavior of its conductance, however, cannot be understood in the framework of the single-electron picture described previously, which does not take into account the effects of the Coulomb interaction. This leads to a strong correlation

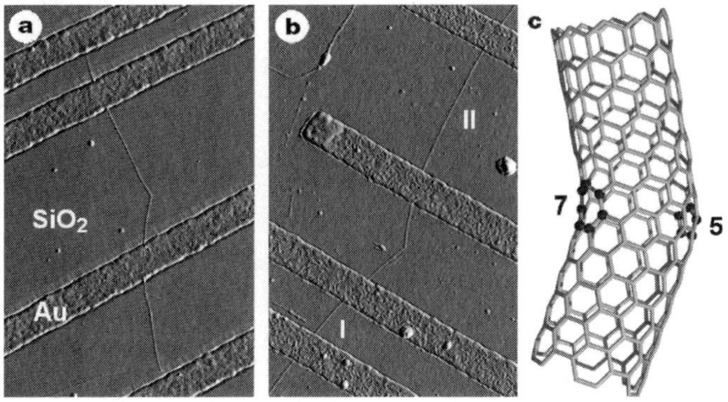

Figure 5. (a), (b) Atomic force microscope images of intramolecular junctions of carbon nanotubes. The pads seen in the figures correspond to titanium-gold electrodes embedded in the SiO_2 substrate, with the nanotubes placed on top of them. (c) Model of a kink created by the combination of the pentagonal and the heptagonal ring marked in the figure, and which gives rise to a junction between armchair and zigzag nanotube geometries. Reprinted with permission from [10], Z. Yao et al., *Nature* 402, 273 (1999). © 1999, Macmillan Magazines Ltd.

between the electrons in materials with reduced dimensionality. The features introduced by the electron-electron interaction in the behavior of the carbon nanotubes are reviewed in the following sections. See Table 1 for a summary of the physical constants reflecting the electronic properties of carbon nanotubes.

3. TRANSPORT PROPERTIES OF CARBON NANOTUBES

3.1. Luttinger Liquid Behavior

The metallic carbon nanotubes show transport properties that deviate remarkably from those of the conventional two- and three-dimensional metals. This is actually a consequence of their behavior as genuine one-dimensional conductors. Their transport properties cannot be understood from the standard Fermi liquid picture for interacting electrons, in which it is assumed that the elementary excitations are quasiparticles which behave similarly to free electrons, with just some characteristic parameters (like the effective mass) renormalized by the interaction. The failure to apply that Fermi liquid picture already anticipates the kind of exotic effects that may be found in molecular electronics, as well as it was anticipated on theoretical grounds many years ago that a new paradigm—the Luttinger liquid—should be used to describe the interacting electrons in one-dimensional systems [46, 47].

The observation of unconventional transport properties has been reported in [10], for instance, regarding the sample shown in Figure 5b. The two-terminal conductance G as a function of the temperature is represented in Figure 6, from the mentioned reference, where the behavior in the two different segments and through the metal–metal junction can be appreciated. The strong suppression of the conductance as the temperature decreases in the latter case is remarkable. In all the instances, the behavior can be fitted very well by a power-law dependence $G(T) \propto T^{\alpha}$ in a fairly large range of temperatures. The exponents corresponding to the upper and the lower segment of the sample are $\alpha = 0.34$ and 0.35, respectively, and the value obtained through the kink is $\alpha = 2.2$.

Table 1. Physical constants of carbon nanotubes.

Physical property	Constant [Ref.]
Minimum intrinsic resistance (individual metallic single-walled nanotubes)	≈6.5 kΩ
Bandgap (semiconducting nanotubes)	~0.5 eV [35, 36]
Energy gap ("metallic" zigzag nanotubes)	~0.05 eV [43]
Dielectric constant (single-walled nanotubes on Si/SiO_2 substrate)	≈1.4 [5, 55]
Charging energy (~3-μm-long single-walled nanotube)	~2.5 meV [5]
Charging energy (~2-μm-long multiwalled nanotube)	~0.5 meV [68]
Intertube resistance (nanotube ropes)	≳1 MΩ [54]
Superconducting transition temperature (nanotube ropes)	~0.5 K [8]
Critical supercurrent (nanotube rope between Au/Re contacts)	~1 μA [76]
Current density (multiwalled nanotubes)	>10^9 A/cm^2 [116]
Mobility (semiconducting nanotubes at room temperature)	>20,000 cm^2/Vs [117]

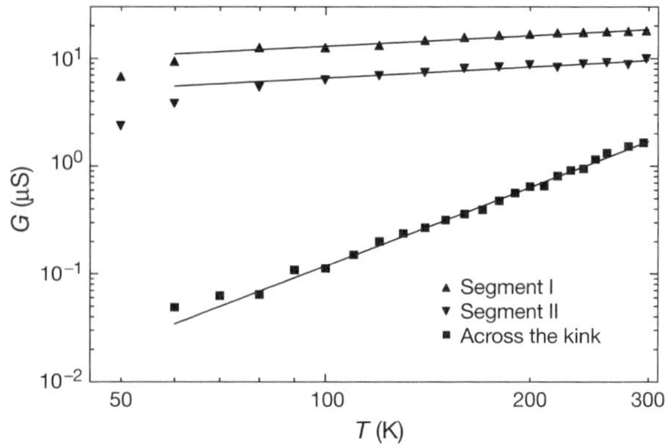

Figure 6. Log-log plot of the temperature dependence of the conductance, measured in the two nanotube segments and through the kink shown in Figure 5b. The straight lines correspond to the best fit to a power law for each set of data. Reprinted with permission from [10], Z. Yao et al., *Nature* 402, 273 (1999). © 1999, Macmillan Magazines Ltd.

As remarked by the authors of [10], the fact of producing contacts between the electrodes and the nanotube with sufficiently low resistance (about or smaller than 100 kΩ in their experiment) seems crucial to observe a definite behavior of the conductance. The suppression for decreasing temperature could be attributed, in principle, to activated transport over the tunnel barrier created by each contact. This would lead, however, to a dependence $G(T) \propto \exp(-\Delta/k_B T)$, in terms of the barrier height Δ, which cannot fit well the experimental data. The measurements of the conductance refer, therefore, to a property intrinsic to the carbon nanotube. This has to do with the effects of the electronic interaction, since the large suppression of the conductance through the junction cannot be accounted for by a model of noninteracting electrons with topological defects [48].

The behavior of the tunneling conductance is actually related to that of the density of states $\rho(E)$ at energies close to the Fermi level. It will be shown that the density of states has a power-law dependence $\rho(E) \propto E^{\alpha}$ in the framework of the Luttinger liquid theory, with the exponent α being a function of the interaction strength [46, 47]. The fact that $\rho(E)$ goes to zero at the Fermi level is a consequence of the absence of low-energy excitations with the properties of noninteracting electrons. The observation of the power-law dependence of the tunneling conductance can be considered then as another manifestation of the carbon nanotubes as genuine one-dimensional conductors.

The data in Figure 6 show that the Luttinger liquid behavior may extend over a wide range of temperatures, up to 300 K. At sufficiently low temperatures, the power-law behavior of the tunneling conductance is modified by the effect of the Coulomb blockade [49], which is described later in detail. That effect is observable in the experiments when the thermal energy (i.e., the temperature times the Boltzmann constant k_B) becomes low enough to be comparable to the discrete energy needed to put one more electron on the finite dimensions of the system. This quantity is the so-called charging energy E_c, which is given in terms of the electron charge e and the total capacitance C of the nanotube by the expression $E_c = e^2/2C$. The charging energy can be estimated as a few meV in the typical single-walled nanotubes (with lengths of the order of 1 μm) used in the experiments.

Evidence of power-law behavior has been also obtained from the measurements of the differential conductance dI/dV, in individual single-walled nanotubes [10] as well as in ropes of nanotubes [50]. Figure 7 illustrates the results of the experiments in ropes reported in [50]. For values of the bias voltage applied to the rope such that $eV \ll k_B T$, the effect of the temperature prevails and the differential conductance recovers the voltage-independent value given by $G(T)$. At higher bias voltage, the experimental values of dI/dV show a well-defined linear behavior in log-log scale as a function of the bias voltage at different temperatures. The exponent giving the best fit for the power-law dependence $dI/dV \propto V^\alpha$ is $\alpha = 0.36$ [50]. It is also remarkable that, upon scaling of the differential conductance by the values of T^α as dictated by the behavior of a tunnel junction [51, 52], the experimental data for the different temperatures fall into a unique universal curve when represented as a function of the scaled variable $eV/k_B T$ [50].

The exponents obtained for the transport through the tunnel junctions created by the contacts in the ropes and in the individual nanotubes are in very good agreement. From this fact, one can obtain significant information about the physical properties of the ropes. The agreement is consistent, for instance, with the results establishing the large suppression of the tunneling amplitude between the different nanotubes of a rope [53, 54]. This means that the only relevant intertube coupling arises from the Coulomb interaction.

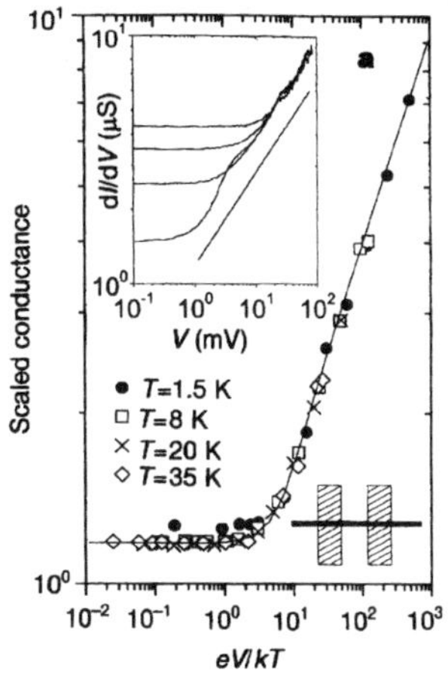

Figure 7. Log-log plot of the differential conductance in a rope, measured at different temperatures and scaled by T^α in each case. The inset shows the curves of differential conductance as a function of the bias voltage, for decreasing values of the temperature (from top to bottom). Reprinted with permission from [50], M. Bockrath et al., *Nature* 397, 598 (1999). © 1999, Macmillan Magazines Ltd.

Consequently, it can be concluded that the transport properties measured in individual nanotubes as well as in ropes are dominated by the two tunnel junctions in the circuit, which give rise to the observed power-law behavior. In the measurements through the kink reported in [10], the reason for the enhanced decrease of the conductance has to be found in the additional suppression of the tunneling of electrons between the two ends of the nanotube segments, according to the Luttinger liquid picture.

From a theoretical point of view, the clarification of the Luttinger liquid behavior of the metallic carbon nanotubes has been accomplished in [55] and [56]. The analyses carried out there assume that the repulsive Coulomb interaction between electrons is dominant in the single-walled carbon nanotubes. The discussion focuses on the low-energy regime in which only the four linear branches near the Fermi level (as observed in the band structure of Fig. 3, for instance) contribute to the electronic properties. When two electrons interact in the nanotube, there are a large number of possible processes, which can be classified depending on whether the electrons scatter near the same Fermi point or not, or whether the electrons shift from one Fermi point to the other. One of the main conclusions drawn in [55] and [56] is that the processes in which the electrons remain in their respective linear branches after scattering are largely dominant. The interaction strength of the rest of the processes (the so-called backscattering and Umklapp interactions) is suppressed in general by a factor inversely proportional to the number of subbands in the nanotube [55, 56].

The Hamiltonian H governing the low-energy electron dynamics can be approximated then by writing the interactions of the form density times density which arise from the Coulomb repulsion,

$$H = \frac{1}{2}\hbar v_F \int dk \sum_{r\sigma} \rho_{r\sigma}(k)\rho_{r\sigma}(-k) + \frac{1}{2}\int dk \sum_{r,s,\sigma,\sigma'} \rho_{r\sigma}(k)V(k)\rho_{s\sigma'}(-k) \tag{9}$$

$\rho_{r\sigma}(k)$ being the electronic density in the linear branch r for spin σ, and $V(k)$ being the Fourier transform of the Coulomb potential. The kinetic term in Eq. (9) gets the appropriate dimensions from the Fermi velocity v_F of the electrons and the Planck constant $\hbar(=h/2\pi)$.

In one spatial dimension, the density operators can be rescaled to satisfy canonical commutation relations characteristic of boson operators [46, 47]. The Hamiltonian (9) can be diagonalized in terms of these bosonic objects. For this reason, the states with well-defined energy do not correspond to the original electrons, but to collective excitations that represent physically wavelike modulations of charge or spin.

The transformation diagonalizing the Hamiltonian (9) is a pseudorotation determined by the parameter $\mu = 1/\sqrt{1+8V/\hbar v_F}$ [46, 47], where a suitable average value of the interaction strength V is assumed. All the properties of the Luttinger liquid are completely characterized by the parameter μ. Thus, the density of states $\rho(E)$ follows at low energies the power-law behavior

$$\rho(E) \propto E^\alpha \tag{10}$$

with an exponent $\alpha_{\text{bulk}} = (\mu + \mu^{-1} - 2)/8$ for the density of states measured in the bulk of the liquid [55, 56]. In the case of an open one-dimensional system, the density of states is substantially smaller at the ends of the liquid, having then a power-law behavior with an exponent $\alpha_{\text{end}} = (\mu^{-1} - 1)/4$ [56].

From the measurements of the conductance when tunneling into the bulk of the nanotube, it turns out that the best fit of the experimental data corresponds to a Luttinger liquid parameter μ around 0.22 [10]. This is the signature of a significant electron–electron repulsive interaction in the single-walled nanotubes, and indeed that value for μ agrees well with the theoretical estimate obtained by assuming that the Coulomb interaction is dominant in the individual nanotubes [55, 56]. An important consistency check is that the exponent that can be predicted for the tunneling into the end of the nanotube accounts for the large suppression of the conductance through the kink observed in [10]. Taking $\mu \approx 0.22$, the exponent α_{end} becomes ≈ 0.9. For the tunneling through the metal–metal junction, the conductance has to be proportional to the product of the tunneling density of states at each end of the nanotube segments. Thus, the corresponding exponent has to be twice the value of α_{end}, which is close to the estimate (≈ 2.1) from the fit of the experimental data [10]. Overall, there is therefore strong evidence that the Luttinger liquid picture applies to the tunneling processes in the single-walled nanotubes.

3.2. Ballistic Transport Properties

The experimental signatures of Luttinger liquid behavior are consistent with another remarkable observation in the carbon nanotubes, as it is the ballistic transport. This means that the charge can move along the nanotube in such a way that it is not disturbed by inelastic collisions. This behavior as a quantum conductor is opposite to the classical behavior in which the conduction takes place by diffusion of the electrons with a certain mean free path. One of the consequences of ballistic transport is that there cannot be dissipation of energy inside the ballistic conductor, and that the heat produced has to appear at the leads of the ballistic element. Moreover, another important property is that the conductance has to be quantized in units of $G_0 = 2e^2/h$, with each mode in the waveguide contributing with one of these quanta to the conductance [57–59].

The quantization of the conductance has been observed at room temperature in fibers of multiwalled nanotubes. The peculiarity of the experiment reported in [7] is that the fibers have been used in place of the tip of a scanning probe microscope. This has allowed the raising and lowering of the fiber inside a liquid metal, used as a second contact to close the circuit. By dipping the fiber into the metal, the current has been recorded as a function of the length of the fiber within the liquid. This has shown that the conductance increases by steps whose magnitude is very close to the quantum of conductance $G_0 = 2e^2/h$. The appearance of each step corresponds to the point at which one more multiwalled nanotube becomes in contact with the liquid metal [7].

The measurements reported in [7] provide a strong indication of ballistic transport in the multiwalled samples, since they have shown that the conductance does not decay over a variation of about 200 nm of the nanotube length submerged in the liquid. Supporting the same conclusion, there is also evidence that the heat produced by the current cannot be dissipated in the nanotubes. It has been estimated in [7] that the current densities produced in the experiment can be higher than 10^7 A cm^{-2}. These values are so large that, if the corresponding dissipation of energy had to take place within the fiber, it would give rise to a temperature well above that needed to burn the nanotubes [7].

The quantization of the conductance in the multiwalled nanotubes also sheds light on their internal structure. Each individual metallic nanotube has two modes contributing to the conduction properties, irrespective of the diameter of the tubule. This means that each metallic shell of a multiwalled nanotube could in principle contribute with two quantum units to the conductance, which is not observed in the experiment of [7]. This can be explained by the fact that only the outer shell participates in the conduction, which is plausible since, even in the event that the next layer were metallic, the resistivity in the direction perpendicular to the tubules has to be very large [3]. It still remains to be understood why the step observed in the conductance is given by G_0, instead of $2G_0$. This value accounts for the contribution of the two spin projections and the two propagating modes of the nanotubes, which suggests that the contribution of some of these degrees of freedom may be missing in the multiwalled samples [60].

The observation of ballistic propagation seems to require very pure nanotube samples and, indeed, there is evidence of the structural perfection of the multiwalled nanotubes used in the experiments of [7]. When the degree of purity is not so high, the ballistic propagation may be altered by the interactions with the impurities or defects in the samples. In the experiments presented in [61], for instance, the transport properties of several single-walled nanotubes with large intrinsic resistance have been measured, showing a highly nonlinear behavior upon variation of the length along the nanotube. For a separation between the contacts of the order of a few hundred nanometers, the resistance displays a very smooth dependence on the length of the nanotube segment. This has been interpreted as the signature of ballistic transport in the single-walled nanotubes [61]. Over larger distances, the resistance shows a steep increase, which is incompatible with Ohm's law in any event since this would imply a linear dependence of the resistance on the length of the wire. It has been argued that the overall behavior can be only consistent with a situation in which the transport is dominated by the elastic scattering with an increasing number of defects, which would give rise to the observed decrease of the probability amplitude over the length of the nanotube [61].

3.3. Low-Temperature Properties

In a different kind of experiments, it has been established that the electron wavefunctions can be extended over lengths of several microns in the single-walled nanotubes, when the conditions are such that the thermal energy is smaller than the single-particle level spacing in the tubes [4, 5]. These low-temperature experiments are measuring, therefore, a regime different from that probed by the experiments reporting the Luttinger liquid behavior, where the interaction between a manifold of single-particle levels is manifest.

The separation ΔE between the single-particle energy levels is dictated by the length L of the nanotube according to the expression

$$\Delta E = hv_F/2L \tag{11}$$

where v_F is the Fermi velocity ($\approx 8 \times 10^5$ ms^{-1}). For typical lengths of a few microns, the separation between the discrete energy levels is below 1 meV. When the thermal energy is below that value, single electrons can be added to the nanotube by resonant tunneling, that is, by adjusting the bias voltage applied to the leads so that the current increases by one step each time that a new molecular orbital becomes available [5]. Similarly, a series of sequential peaks can be observed in the conductance when the number of electrons is varied with the voltage applied to the gate [4]. The fact that the single-particle level splitting can be resolved by looking at the current steps working at suitably low temperatures is the signature that the electrons occupy molecular orbitals that correspond to delocalized electronic states.

Moreover, direct evidence of the extended character of the electron states has been obtained by imaging the electron wavefunctions in short nanotubes (with length $L \approx$ 30 nm) by scanning tunneling microscopy [62]. By measuring the tunneling conductance along the nanotube, the spatial modulation corresponding to the electron probability amplitude has been observed, finding agreement with the expected wavevectors of the quantized states in the nanotube. The electron wavefunctions of several discrete molecular levels have been discerned with this technique [62].

Another remarkable observation, related to the existence of extended electron wavefunctions, refers to the oscillatory behavior of the magnetoresistance of multiwalled nanotubes [63]. When these are aligned with the direction of the magnetic field, it has been shown that the resistance has modulations as a function of the enclosed flux. This is a consequence of the fact that the phase of the electron wavefunction is modified by the presence of the magnetic field (the so-called Aharonov–Bohm effect). When the electrons encircle the nanotubes in opposite directions, a phenomenon of quantum interference takes place, which reflects in the behavior of the resistance [63].

The absence of localization of the electronic states in the single-walled nanotubes may be surprising at first sight. On the one hand, there is the well-known fact that a single impurity has the ability to disrupt the conduction in a one-dimensional system [64, 65]. It has been shown, however, that the case of the nanotubes is special as the electrons feel the effect of impurities averaged over the circumference of the tube [66]. Thus, the metallic nanotubes may have very good conduction properties, with delocalized states over lengths of 10 μm or more for the nanotubes produced in the experiments. Furthermore, that localization length has to be an increasing function of the diameter of the nanotube [66].

The picture that emerges from the low-temperature experiments on transport is that the quality of the contacts used for the nanotubes determines the type of experimental observation. In some metallic nanotube devices, the resistance measured at room temperature is close to the theoretical lower limit of $\approx$6.5 kΩ, given by the inverse of the conductance $2G_0$ that would correspond to two modes propagating ballistically along a single-walled nanotube. In these cases, most part of the resistance has to be attributed to the nanotube, while the contacts between the metallic electrodes and the nanotube are nearly perfect. At low enough temperatures such that the thermal energy is below the single-particle level spacing within the nanotube, the electrons pass through the interface with little reflection and the experimental observations account for the propagation intrinsic to the nanotubes.

In the experiments reported in [6], a number of nanotube devices were measured with room-temperature resistances below 15 kΩ. In these samples the average value of the conductance has been found always between G_0 and $2G_0$, the departure from the latter value being attributed to the scattering of the electrons at the nanotube–electrode interface. Moreover, the two-dimensional plot of the differential conductance dI/dV as a function of the bias voltage V and the gate voltage V_g displays a quasiperiodic pattern of dips and peaks. It has been shown that the oscillation period scales appropriately with the inverse of the nanotube length, which supports that the observed pattern is due to the quantum interference of the electron waves after being scattered at the interfaces [6]. This shows the significance of taking into account the quantum coherence, represented by the extended character of the electron wavefunction, when designing low-temperature electronic devices at the molecular level.

Opposite to the cases of devices with almost transparent contacts, there are instances where the transmission between the nanotube and the metallic leads is dominated by much more suppressed tunneling processes. In these cases, the experimental observations give a measure of the difficulty that the current finds to go through the tunnel junction. In certain circumstances, it is possible to control the tunneling of single electrons into the nanotube, entering a regime with new physical properties which is described in the next section.

4. COULOMB BLOCKADE AND QUANTUM DOT BEHAVIOR

4.1. Carbon Nanotubes as Single-Electron Transistors

One of the main interests in the technological application of the carbon nanotubes arises from the possibility of developing electronic devices made of a single molecule. Semiconducting nanotubes have been proposed to act as field-effect transistors in [11] and [12]. In these devices, source and drain electrodes are attached to the semiconducting nanotube, while this is separated from the substrate (the gate electrode) by an oxide layer which acts as a dielectric. The capacitive coupling between the nanotube and the substrate is what makes it possible to change the density of charge carriers and the conduction properties in the nanotube by varying the voltage of the gate.

Unlike field-effect transistors, however, single-electron devices are based on the intrinsic quantum-mechanical character of the tunnel effect. In the case of metallic nanotubes, the development reported in [14] is that the electrons can be confined in short islands between two buckles of the tubule, so they can be added one by one by suitable variations of the voltage applied to the external gate.

The technique applied in [14] consists of using the tip of an atomic force microscope to press the nanotube against the substrate. The tip is then moved until the nanotube is strongly bent. In this way a buckle is created in the tubular structure, leading to a tunnel barrier for electron transport in a metallic nanotube. Two of these buckles can be arranged in series in an individual metallic nanotube, in order to study the transport properties through the two tunnel barriers [14]. One of the structures which have been produced with this technique can be observed in Figure 8. The short nanotube segment that appears there between the buckles has a length of the order of 25 nm.

When the tunneling of the electrons takes place into a short nanotube island, it costs a relatively large energy to add a single electron between the tunnel barriers. This is the well-known effect of Coulomb blockade [49]. On the one hand, there is the energy needed to overcome the electrostatic repulsion between the electrons which are confined in such a reduced dimension, the so-called charging energy E_c. As pointed out earlier, this is given by the expression $E_c = e^2/2C$, in terms of the total electrostatic capacitance C [49]. On the other hand, each new electron added is placed in the first unoccupied level, which requires spending an energy corresponding to the separation ΔE between quantized levels. That energy difference is inversely proportional to the length L of the island, according to the already quoted expression $\Delta E = hv_F/2L$ (or $\Delta E = hv_F/4L$ if the degeneracy between the two different low-energy modes in the metallic nanotubes has been lifted). Altogether, the energy needed to add an electron to the island is the sum of the two contributions, the so-called addition energy [67]

$$E_{\text{add}} = e^2/C + \Delta E \tag{12}$$

For the short nanotube segments created with the technique reported in [14], the charging energy and the level spacing ΔE have comparable magnitudes, and they give rise to an addition energy E_{add} of the order of 0.1 eV, which is considerably larger than the thermal energy k_BT at room temperature. For that reason, the transport properties resulting from the addition of single electrons to the island can be observed without resorting to the use of low temperatures. The number of electrons transferred between the two buckles is controlled by varying the gate voltage applied to one of the substrates, capacitively coupled to the nanotube, which allows the lowering of the successive empty energy levels of the nanotube island down to the Fermi level in the outer nanotube segments [14].

The single-electron transport properties are distinctively observed in the measurements of the differential conductance dI/dV through the tunnel barriers reported in [14]. The effect of Coulomb blockade gives rise to typical patterns in the intensity plot of the differential conductance as a function of the bias voltage V applied to the electrodes and the voltage V_g applied to the external gate. At fixed gate voltage, a gap can be generally observed in the measures of dI/dV as a function of the bias voltage, corresponding to the region where that is not large enough to reach the first unoccupied level within the island. At some values of V_g the gap closes, which marks the point at which the first empty level in the region between the tunnel barriers is aligned with the Fermi level outside the island. This produces a typical pattern of consecutive diamond-like regions with suppressed conductance in the intensity plot of the differential conductance in the (V_g, V) plane [14].

An illustration of the diamond-shaped regions (although corresponding to a different experiment reported in [68]) is given in Figure 9. The dark diamonds correspond to the regions where the differential conductance is suppressed. In each diamond the number of electrons in the nanotube segment is fixed, while it increases by one unit when shifting from a diamond to the next one. On the other hand, the addition energy E_{add} can be obtained from the height in bias voltage of the largest diamond, since that is the voltage required to establish the conduction through the nanotube segment. In the case of the short island of [14], the addition energy is $E_{\text{add}} \approx 120$ meV. The separation of the energy levels has been estimated as $\Delta E \approx 38$ meV, from which a value $E_c \approx 41$ meV has been obtained for the charging energy of the short island [14].

It turns out that, for the nanotube device described in [14], the level separation ΔE and the charging energy E_c have comparable magnitudes, and both of them are well above the thermal energy k_BT corresponding to room temperature. The behavior of the conductance at lower temperatures has been also studied in [14], reaching another

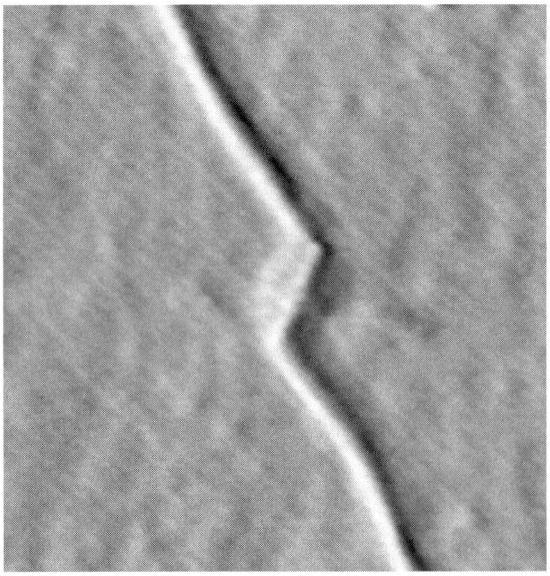

Figure 8. Atomic force microscope image of a short nanotube island between two buckles, formed by manipulation with the atomic force microscope tip. Image courtesy of Cees Dekker, Delft University of Technology.

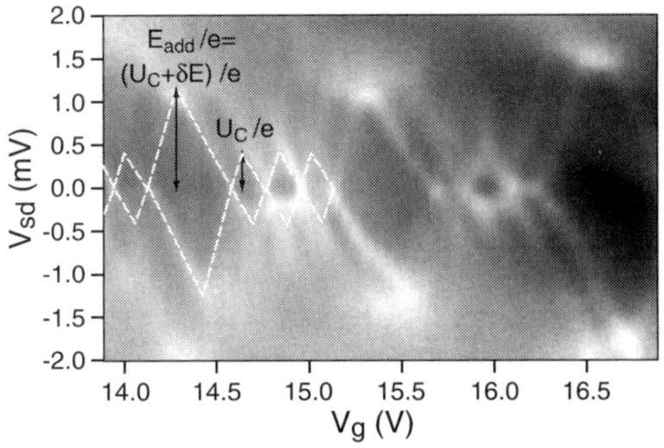

Figure 9. Intensity plot of the differential conductance of a multiwalled nanotube as a function of the bias voltage V_{sd} and the gate voltage V_g. The average conductance is of the order of $\sim 2e^2/h$. According to [68], ΔE_{add}, U_c, and δE denote the addition energy, the charging energy, and the single-electron level spacing, respectively. Reprinted with permission from [68], M. R. Buitelaar et al., *Phys. Rev. Lett.* 88, 156801 (2002). © 2002, American Physical Society.

important conclusion. The conductance of the nanotube follows a clear power-law dependence with decreasing temperature, pointing at a Luttinger liquid behavior of the kind already observed in [10] and [50]. It has been remarked that the exponents measured for that dependence do not correspond to a picture in which the transport proceeds with the sequential, independent tunneling through the two barriers that form the nanotube device. Instead, it has been shown that the data can be fitted with the exponent appropriate for correlated tunneling, in which the electrons propagate coherently through the nanotube island [14]. The corresponding value found for the parameter measuring the interaction strength, $\mu = 0.23$, is in very good agreement with the earlier estimates from Luttinger liquid behavior in carbon nanotubes [10, 50].

4.2. Quantum Dot Behavior

Effects related to the confinement of the electrons in a reduced dimension have been also observed in multiwalled nanotubes. In this case, the experiments reported in [68] have been realized at temperatures reaching values down to 280 mK. Such small temperatures are needed to discern charging energies and single-particle level spacings which range below 1 meV for multiwalled nanotubes with typical lengths of a few microns.

The experimental measurements of the differential conductance carried out in [68] are represented by the intensity plot of Figure 9. The dark regions correspond to lower values of the differential conductance, and it can be clearly seen that they form a sequence of diamonds as the gate voltage increases. There seems to be periodicity in the pattern formed by a large diamond and three consecutive smaller ones. Recalling that the number of electrons in the nanotube is increased by one unit from one diamond to the next and that the height of the diamond gives the energy required to add one electron, the observed pattern points at a fourfold degeneracy of the electronic levels in the multiwalled nanotube [68]. The height of the large diamond should correspond to the addition energy $E_{\text{add}} = e^2/C + \Delta E$, as in the measurement of the short island reported in [14]. The size of the smaller diamonds should give a measure of the charging energy alone, with the electrons being placed there in the same single-particle level [68].

The fourfold degeneracy is in agreement with the expected band structure of a metallic nanotube, in which two subbands cross at the two Fermi points of the one-dimensional structure. It may be surprising, however, to find that only two gapless modes contribute to the conduction properties in the multiwalled nanotubes. In general, these are significantly hole doped by the environment, and the Fermi level in the outermost tubule crosses several different subbands [69]. It may well happen to have an outer shell which is semiconducting, so that for appropriate choices of the gate voltage it may not contribute to the conductance of the multiwalled nanotube. As explained in [68], this seems to be the case of the experimental sample exhibiting the sequence of diamonds in the differential conductance. It has been observed that such a sequence appears for suitably large values of the gate voltage. According to the arguments given in [68], the features observed in the conductance should reflect the metallic properties of the tubule next to the outermost shell, as that nanotube would not be affected by the external doping.

As remarked earlier in this section, the height of the smaller diamonds in the intensity plot of the differential conductance gives a measure of the charging energy E_c. For the multiwalled nanotube sample considered in [68], the estimate is $E_c \approx 0.4$ meV. From the value of the addition energy, it turns out that $\Delta E \approx 0.8$ meV [68]. As explained in the preceding section, the level spacing arises from the quantization of the electron states in the finite length of the nanotube. It has been observed in [68] that the estimated value of ΔE corresponds to the propagation along the whole length of the nanotube ($\approx$2.3 μm), rather than to the distance between the electrodes ($\approx$300 nm). It has been argued that this further supports the idea that the outermost shell does not participate in the conduction of the multiwalled nanotube [68]. The quantization properties over the nanotube length show, as well as in the individual single-walled nanotubes, that the electron wavefunctions can be extended over large distances (above $\sim$1 μm) due to the weak influence of disorder with increasing diameter size [66].

Another physical effect reported in [68] refers to the behavior of the spin of the states with different electron number. By studying the behavior of the conductance under a magnetic field perpendicular to the tubules, it has been found that the spin follows the sequence $1/2, 0, 1/2, 0 \ldots$, upon adding one more electron each time to the nanotube. This can be explained if pairs of electrons have antiparallel spins, leading to the change from spin $1/2$ for an odd number of electrons to a vanishing value for an even number [68]. In general, the Coulomb repulsion between two electrons is minimized if they can be placed in degenerated orbitals, allowing them to have parallel spins [70]. This rule is not followed in the multiwalled nanotube considered in [68], which has been interpreted there as a signal that the perfect degeneracy between the modes of the two low-energy subbands in a metallic nanotube is spoiled in the experimental sample.

Moreover, it has been found in [68] that the differential conductance increases in the regions with an odd number of electrons as the temperature decreases below 1 K. This has been regarded as a manifestation of the Kondo effect, which deals in general with the properties of a static spin surrounded by delocalized electrons [71, 72]. In the Kondo system there is an energy scale, given by the so-called Kondo temperature, below which the itinerant electrons are able to screen the static spin, with a concomitant increase of the conductance. The Kondo temperature gives therefore a measure of the binding energy of the singlet state formed by the screening effect. According to the authors of [68], the coupling between the spin of the nanotube and the electrons in the leads is what gives rise to the formation of the singlet state and the increase of the conductance through the nanotube.

The carbon nanotubes appear then as ideal systems to study the properties of electrons confined in a very reduced spatial dimension. This confinement into so-called quantum dots can be caused by the presence of the own electrodes acting as tunnel barriers. It has been also proposed that the arrangement of two consecutive kinks made from topological defects (i.e., from the combination of pentagon and heptagon rings) can be used to lock the electrons in short nanotube segments [73]. It has been already shown experimentally that the resonant electron scattering between simple (nontopological) defects gives rise to the formation of intratube quantum dots, with conductance patterns similar to that shown in Figure 9 [74]. The charge states in the quantum dots can be actually imaged with scanned gate microscopy and electrostatic force microscopy, which may give interesting spatial information about the physical effects involved [75].

5. SUPERCONDUCTING CORRELATIONS IN CARBON NANOTUBES

5.1. Proximity-Induced Superconductivity

There have been several experiments revealing the existence of superconducting correlations in the carbon nanotubes. These observations have taken the form of a drastic drop in the resistance of the nanotube samples below certain temperature. In one of the most remarkable experiments, reported in [76], it has been shown that a rope of carbon nanotubes is able to carry an electric current with zero voltage drop, when embedded between superconducting contacts. The measurement of that so-called supercurrent implies therefore a vanishing resistance of the conductor. The experiment provides a realization of the proximity effect, by which the electronic properties of a normal metal change drastically when placed in contact with a superconductor [77, 78]. In the latter, there is no sign of electron-like particles at low energies and, instead, a condensate formed by pairs of bound electrons is found [79]. These so-called Cooper pairs may extend their propagation to the nearby normal metal, giving rise to the electric current without dissipation of energy.

The influence of the superconducting electrodes in the electronic properties of single-walled nanotubes has been investigated in the experiments reported in [76] and [80]. One of the main differences between these experiments is that, in the latter, the transport properties have been measured in a set of individual single-walled nanotubes. In the former experiment, the supercurrents have been observed in a massive rope made of about 200 nanotubes, and in a thin rope leading to a single nanotube at one of its ends. On the other hand, a common feature in both experiments is the low resistance attained for the samples that have shown the proximity effect. Typically the values measured at room temperature have been consistent with a resistance of the individual metallic nanotubes of the order or below the inverse of the conductance quantum, $h/e^2 \approx 25.8$ kΩ [76, 80]. Such values are comparable to the resistance $(2G_0)^{-1} = h/4e^2$ corresponding to the ballistic transport in individual nanotubes, which gives a measure of the high transparency of the contacts produced in the experiments.

In the experiments reported in [76], the ability to produce highly transparent junctions has been the result of using a remarkable technique allowing to suspend the nanotube ropes between the contacts. In the transport measurements, a drop to a vanishing resistance has been observed in the two nanotube samples mentioned earlier, below the temperature T_c of the transition of the electrodes to the superconducting state. The contacts were made of bilayer electrodes with respective temperatures $T_c \approx 1.1$ K for the Re/Au bilayer in the case of the thick rope, and $T_c \approx 0.4$ K for the Ta/Au bilayer in the case of the thin rope. By applying a magnetic field perpendicular to the nanotube axes, it has been possible to reduce the value of T_c as measured in the ropes, up to a point in which the transition disappears for a suitably large field [76]. This effect of the magnetic field is one of the genuine features of superconductivity, and it serves to corroborate the nature of the phenomenon observed in the experiment.

When increasing the current that flows along the rope, it can be supported without developing any resistance up to a maximum value, that is called the critical current. The behavior of the critical currents for the ropes studied in [76] has shown unconventional features, regarding their magnitude as well as their dependence with the temperature. The critical current should vanish, for instance, at the transition temperature of the contacts, but in the thick rope of [76] the behavior is very smooth instead near T_c. In the conventional picture of the proximity effect, the magnitude of the critical current should correspond to the expression $I_c = (\pi/2)\Delta/eR_N$, R_N being the normal state resistance and Δ the binding energy of the Cooper pairs in the superconducting condensate. As it has been pointed out in [76], the value of I_c estimated in that way is, however, 40 times smaller than what is actually measured in the thick rope. The thin rope shows a better agreement in the magnitude of the critical current, but this also displays a very unusual temperature dependence, with a flat behavior until the neighborhood of T_c is reached [76].

An explanation of the unconventional behavior of the critical currents in ropes has been presented in [81]. That work has shown the relevance of taking into account appropriately the interaction among the large number of metallic nanotubes that may be present in a rope. The Coulomb potential is not screened in an individual nanotube, but the interaction between the charges in different metallic nanotubes leads to a significant reduction of the effective interaction strength [81]. This can be understood by thinking that, instead of Eq. (9), the Hamiltonian appropriate for a rope with n metallic nanotubes is

$$H = \frac{1}{2}\hbar v_F \int dk \sum_{ar\sigma} \rho_{r\sigma}^{(a)}(k)\rho_{r\sigma}^{(a)}(-k) + \frac{1}{2}\int dk \sum_{a,\,b,\,r,\,s,\,\sigma,\,\sigma'} \rho_{r\sigma}^{(a)}(k)V_{(a,\,b)}(k)\rho_{s\sigma'}^{(b)}(-k) \quad (13)$$

the indices a, b labelling the electronic densities in the different metallic nanotubes. The Coulomb interaction is long-ranged and takes place therefore between all of them, so that it contributes equally to all the $V_{(a,b)}(k)$ terms. The Hamiltonian (13) can be diagonalized by passing to the total densities $\rho_{r\sigma}(k) = \sum_a \rho_{r\sigma}^{(a)}(k)$. It becomes clear that the Coulomb interaction is only felt in the channel of the total charge, while there are still $4n - 1$ noninteracting partial channels [81].

The preceding argument explains that the repulsive electron–electron interaction becomes less relevant as the number n of metallic nanotubes increases in the rope. The proximity effect for a Luttinger liquid in contact with a macroscopic superconductor has been studied in [82] and [83], showing that the Cooper pairs propagate along the one-dimensional metal but giving rise to a supercurrent I_c that decays with the length L as

$$I_c \propto 1/L^{1/\mu} \quad (14)$$

where μ is the Luttinger liquid parameter quoted above Eq. (10). Recalling that $\mu < 1$ in the case of a repulsive interaction, that kind of behavior can only account for the large critical current measured in the rope of [76] after the appropriate reduction in the strength of the Coulomb interaction is considered. It has been also shown in [81] that the temperature dependence of the critical currents can be reproduced by considering the one-dimensional propagation of the Cooper pairs, which gives further support to the picture of the single-walled nanotubes as genuine one-dimensional conductors.

In the experiment presented in [80], the resistance of individual single-walled nanotubes placed between Nb electrodes has been measured. The nanotubes have been capacitively coupled to the Si substrate, and changing the gate voltage V_g has allowed to increase the already high transparency of the contacts. Below the transition temperature of the Nb electrodes ($T_c \approx 9.2$ K) and for some interval of V_g, a dip has been observed in the broad peak of the resistance centered at zero bias voltage. That structure has disappeared by increasing the temperature above T_c, which shows its relation to the superconducting character of the electrodes [80]. Although the room-temperature resistances of the samples were comparable to those in the experiment of [76], no supercurrents have been found in this case. This can be attributed to the large strength of the repulsive electron–electron interaction in the individual nanotubes, supporting the point of view that the superconducting correlations are more likely to develop in ropes of nanotubes.

5.2. Superconductivity Inherent to Carbon Nanotubes

Superconducting properties have been also measured in nanotubes placed between metallic, nonsuperconducting contacts [8, 9]. These experimental observations open the way to use the carbon nanotubes as a testing ground to study superconductivity in molecular wires with a discrete number of conduction channels. It has been already shown that several interesting features arise, as a consequence of the finite number of metallic nanotubes involved and their small length as compared to the size of macroscopic conductors.

One of the main factors in the experiments reported in [8] is the good quality of the contacts produced for the ropes. This is essential to measure the transport properties intrinsic to the nanotubes since, under conditions of low transparency of the junctions, it is the effect of tunneling and Coulomb blockade which is measured at low temperatures, as described earlier. In the experiments of [8], the nanotube ropes have been suspended between Pt/Au bilayers by the same technique already used in the proximity-effect experiments [76]. This has allowed production of devices with a room-temperature resistance ranging from a few kiloohms down to a few hundred ohms. It has been checked that the bilayers used in the experiment do not become superconducting themselves at low temperatures. The fact that they are not the source of some kind of proximity-induced superconductivity in the nanotubes is supported by the absence of the effect in some of the shortest ropes [8].

The most clear observation of superconducting transition has been made in a rope with about 350 nanotubes and length $L \approx 1$ μm [8]. When measuring the resistance of the sample as a function of the temperature, a drop by two orders of magnitude has been observed below $\approx$0.5 K, as shown in Figure 10. The resistance does not completely vanish below the transition, but reaches a minimum value $R \approx 74$ Ω. As remarked in [8], this is a consequence of the fact that the resistance of a metallic nanotube has a minimum value, given in terms of the quantum of conductance by $(2G_0)^{-1} \approx 6.5$ kΩ. The value $R \approx 74$ Ω of the residual resistance is consistent then with the approximate number of metallic nanotubes in the rope contributing to the conduction in the superconducting state. The superconducting character of the observed transition is reinforced by the behavior under a magnetic field, which tends to decrease the transition temperature as shown in Figure 10 [8].

The finite length of the ropes can be also determinant in the development of the superconducting transition. In the experiments of [8], a rope with a room-temperature resistance of about one order of magnitude below that of the superconducting rope described in previous paragraphs has shown no sign of superconductivity at low temperatures (as it can be seen in Fig. 10). The absence of transition has been attributed to the comparatively small length of the rope ($L \approx 0.3$ μm) [8]. It has been argued that the superconducting coherence length has to be smaller than the rope length for the superconductivity to develop, that condition not being satisfied in the case of the smaller rope. Another sample with a resistance of approximately 1 kΩ and length $L \approx 1.6$ μm has shown a clear sign of transition in the drop of the resistance (although at a comparatively low temperature, below 0.15 K, as observed in Fig. 10 [8]).

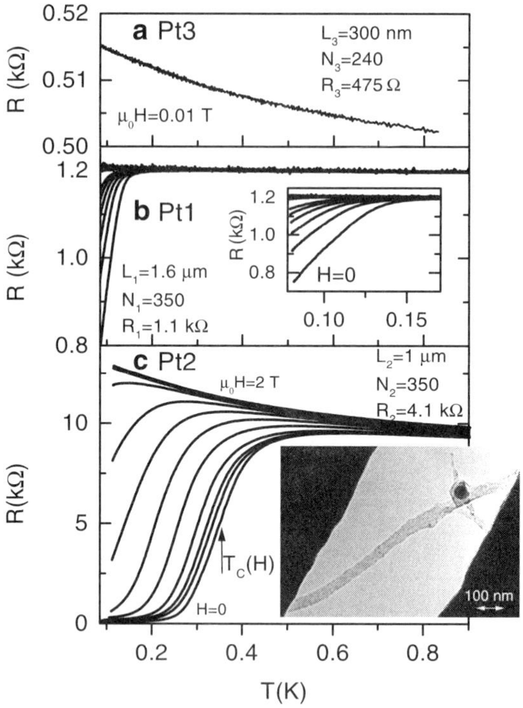

Figure 10. Behavior of the resistance at low temperatures for three different ropes. The length L, the estimated number of nanotubes N, and the room-temperature resistance R are indicated in each case. The inset of (b) shows the behavior of sample Pt1 in a magnetic field ranging from 0 up to 1 T. The plot in (c) displays also the dependence of the resistance of sample Pt2 in a magnetic field ranging from 0 up to 2 T (from bottom to top). The inset at the bottom shows a transmission electron microscope image of sample Pt2, with the dark spot corresponding to a Ni/Y catalyst particle. Reprinted with permission from [8], M. Kociak et al., *Phys. Rev. Lett.* 86, 2416 (2001). © 2001, American Physical Society.

A consistent explanation of the observed superconducting transitions requires taking into account the mechanism by which the repulsive Coulomb interaction can be overcome in the ropes of nanotubes. The superconducting transition signals the point at which bound electron pairs condense in the rope. This may happen if there is an effective attractive interaction in the electron liquid [79]. In the first place, it has to be considered that the strength of the Coulomb interaction is strongly reduced in a rope with a large number of metallic nanotubes, for the reasons already presented in the discussion of the proximity-induced superconductivity [81]. Furthermore, a source of attraction between the electrons is needed for the development of the superconductivity intrinsic to the ropes. That attraction arises in the scattering of two electrons by exchange of phonons (i.e., lattice vibrations), in similar manner as it happens in the case of macroscopic superconductors [79].

The effective attraction coming from the coupling of the electrons to the lattice vibrations can naturally overcome the repulsive Coulomb interaction [84, 85]. This is because the former takes place *within* each metallic nanotube of the rope while, as emphasized below Eq. (13), the Coulomb interaction operates in fact in a single interaction channel—that corresponding to the total charge density. Thus, although the strength of the bare Coulomb interaction is larger, the attraction between the electrons through phonon exchange prevails for sufficiently large number of metallic nanotubes in the rope [84, 85]. This argument makes clear that the relevant electron–phonon interactions come from the coupling to *intratube* lattice vibrations. The maximum energies that the optical phonons may reach in the nanotubes, of the order of ~0.2 eV [86–88], are also appropriate to give rise to the superconducting transitions observed experimentally.

Another important factor in the onset of the superconductivity is the small but finite electron tunneling amplitude that exists between different nanotubes in the rope. These have what has been called *compositional disorder*, meaning that they are made of a mixture of nanotubes with different helicities and diameters [53]. In these circumstances, neighboring nanotubes cannot have their lattices aligned, and this constitutes a great obstacle for conserving the longitudinal momentum when an electron hops from one nanotube to the other. Thus, in a compositionally disordered rope the intertube electron coherence is largely suppressed and the single-particle electron states have to be localized on individual nanotubes [53].

The coupling resistance between tubes in a rope has shown actually wide variations when measured in different experimental samples, with values ranging from 2 MΩ to 140 MΩ [54]. It has been argued that this can be only explained by assuming that transport in the transverse directions of the rope takes place by tunneling between metallic nanotubes of the same helicity. All this is in contrast to the case of an ideal crystalline rope, with perfect alignment of the nanotube lattices, where it has been estimated that the coupling between the nanotubes should give rise to a pseudogap in the density of states of about 0.1 eV [89–91].

It has been shown, however, that when the superconducting correlations develop in the individual nanotubes, the tunneling of Cooper pairs is the mechanism that reestablishes the intertube coherence in the rope [84]. The bound electron pairs are formed at zero total momentum, and they are not affected in the tunneling processes by the misalignment of neighboring nanotube lattices. The intertube coupling established by the tunneling of the electron pairs is a small quantity but, though small it may be, it is essential to account for the superconducting transition in the rope, since in a purely one-dimensional system the electron pairs cannot condense to form the superconducting state [92]. This becomes possible when there is a coherent propagation along the transverse directions of the rope [84]. The close packing of the bundle of nanotubes (forming a triangular lattice as viewed in a cross section of the rope) helps in this respect, since each nanotube may have on the average two metallic nearest neighbors, thus making possible the percolation of the Cooper pairs in the transverse directions of the rope.

The coupling of the electrons to the vibrations of the nanotube lattice and the intertube tunneling of the Cooper pairs are the two essential factors in the superconductivity of the ropes. The former can be actually enhanced by dealing with nanotubes of small diameter, since the electron–phonon coupling increases with the curvature of the nanotube [93]. In the experiment reported in [9], large superconducting correlations have been measured in nanotubes with a diameter of ≈ 4 Å inserted in a zeolite matrix. Such a transversal size is considerably smaller than that of typical nanotubes in a rope, which have a diameter of ≈ 1.4 nm. The measurements of the magnetic susceptibility presented in [9] show actually the tendency of the zeolite matrix with the nanotubes to expel the magnetic fields below a temperature of ≈ 10 K. This property corresponds to the usual Meissner effect in a macroscopic superconductor. The measurements of the conductance show also an unconventional behavior for a one-dimensional metallic system, in which that observable diverges as the temperature goes to zero [9]. A microscopic description of the large superconducting correlations observed in the nanotubes of short radius must take into account the enhanced electron–phonon coupling [93] as well as their particular band structure, in which more than two subbands cross the Fermi level of the nanotubes [37–39].

6. PERSPECTIVES

Carbon nanotubes have a great potential in the development of electronic devices with diverse functionality since their electronic properties are themselves diverse, depending on the geometry of the nanotube lattice, the contacts used in the devices, and the temperature. At room temperature the transport may be ballistic in samples with high structural perfection. In the cases where the contacts create tunnel junctions, one can expect nonlinear I–V characteristics which are the signature of the Luttinger liquid behavior. At much lower temperatures, quantum interference effects in the propagation of the electrons can be observed in samples with highly transparent contacts while, in the case of very thick ropes, the reduction in the strength of the Coulomb interaction may give rise to the superconductivity of the nanotubes.

The first step toward the use of the carbon nanotubes in molecular electronics requires the integration of several nanotube devices in order to produce the desired functionality. Important progress is already being made in this direction. In [15], the architecture of two perpendicularly crossed arrays of nanotubes has been proposed as a model for a nonvolatile random access memory. Each crossing point for two perpendicular nanotubes constitutes an addressable device element. The junctions have two stable positions, with a different separation between the crossed nanotubes that can be controlled electromechanically. This allows the definition of ON and OFF states at each crossing point, characterized by respective resistances which may differ in general by orders of magnitude. The feasibility of the proposal has been supported by the realization and investigation of junctions made of crossed ropes of nanotubes [15]. The measurement of the I–V characteristics of crossed nanotube junctions has been also accomplished in [94], with different combinations of individual single-walled nanotubes with metallic and semiconducting character.

Another experimental accomplishment which has opened the way for a nanotube-based electronics can be found in [16]. That work has reported the construction of the first circuit based on a single nanotube capable of performing a logic operation. The circuit represents what is called a voltage inverter, by which a logical 1 can be transformed into a logical 0 and vice versa. This is the realization of the NOT logic function, that is combined with the AND and OR logic operations to build the complex structure of modern microprocessors. A new experimental development has been needed in the construction of the logic circuit, since this requires to place in series two field-effect transistors being respectively of n-type (with excess of conduction electrons) and p-type (with conduction achieved by electron holes). While nanotubes are usually found with the latter character, the transformation to n-type has to be accomplished by doping the nanotubes with alkali metals or, in a simpler way, by heating the nanotubes in a vacuum as shown in [95]. An important feature of the circuit is the gain, which relates the strength of the output to that of the input signal and which, in this case, reaches the value of 1.6. This opens the possibility of assembling gates of the kind proposed into more complex circuits.

The operation of several small circuits built from the combination of nanotube field-effect transistors has been also shown in [17]. What is special in the integration of these devices is that each nanotube transistor has its own local gate, so that the effect of doping by varying the corresponding gate voltage can be controlled independently in each nanotube. A very large capacitive coupling has been achieved between the semiconducting nanotube and the nearby gate, making it possible to shift the Fermi level in the nanotube from the valence band (p-doped regime) to the conduction band (n-doped regime) under variations in the gate voltage. In this way, the integration of the nanotube transistors has allowed to realize several logic circuits, like an inverter, a NOR logic element, or a static random access memory element.

An interesting finding has been that the field-effect transistors made of single nanotubes can have better performance than the leading silicon transistor prototypes [96]. This has been realized in the process of building nanotube-based transistors with larger capacitive coupling between the nanotube and the gate electrode, that controls the density of charge carriers in the molecule. The advances in the design of the transistors have come from placing the gate electrode on top of the nanotube and using a thinner dielectric between them [96]. Thus, smaller variations in the voltage of the gate electrode can lead to significant changes in the resistance of the nanotube. This new kind of transistors has led to a high transconductance (the measure of the capability to carry electric current) at low voltages, outperforming in this respect the best silicon transistor prototypes [96].

The route toward the large-scale integration of nanotube devices presents great complexities, but carbon nanotubes have already shown the potential for more straightforward applications. One of them arises from the strong coupling between the electronic properties and mechanical deformations, that may include the twisting [97], bending [98], or stretching [99–102] of the carbon nanotubes. It has been shown that, in the case of semiconducting nanotubes, a semiconductor–metal transition can take place upon application of sufficient uniaxial strain [102]. The reverse trend has been also measured, by pushing a metallic carbon nanotube with the tip of an atomic force microscope to produce a decrease of nearly two orders of magnitude in the conductance [103–105]. These observations open the way to use carbon nanotubes as nanoscale mechanical sensors.

The carbon nanotubes have also shown the potential for piezoelectric applications. The injection of charge into the nanotubes can alter their structure, due to the fact that the carbon–carbon bonds modify their lengths according to the electrons or holes added [106]. These effects have been investigated in nanotube sheets, which are made of highly entangled mats of nanotube bundles. In the experiments reported in [107], the changes in the length of strips of such kind of nanotube paper have been measured as a function of the applied voltage, carrying the operation within a NaCl electrolyte. Thus, the expansion or contraction of the strips has been the result of the injection of electronic charge from the electrodes to the surface of the nanotube bundles, with the electrolyte ions forming layers of respective opposite charges to balance those in the nanotubes [107]. The electromechanical actuators thus designed have shown good performance, being able to generate higher stresses than those of natural muscles. An important advantage over conventional ferroelectric actuators is that the nanotube sheets can provide large strains with applied voltages of just a few volts. The mechanical performance should be enhanced in the case of nonbundled nanotubes, and it has been estimated that, for the sheets made of separate nanotubes, the actuator strain could be of the order of $\sim 1\%$ [107]. A number of possible uses of the nanotube actuators have been proposed, from biomedical applications to flow control at high temperatures.

The application of carbon nanotubes as chemical sensors has been also suggested. The nanotubes have the tendency to adsorb gas molecules in their surface. In the case of semiconducting nanotubes, this has been shown to lead to significant changes in the conduction properties [108, 109]. The gas molecules give rise to a transfer of charge that makes the nanotubes become *p*-doped semiconductors. The change in their conductivity can give then a measure of very small concentrations of particles in the chemical environment at room temperature, in a much more sensitive way than existing chemical sensors.

Carbon nanotubes can be used as tips in scanning probe microscopes, which provides several advantages over usual silicon tips [110, 111]. The ability that the nanotube tips have to buckle elastically reduces the damage that can be produced when crashing into the sample. They lead to an improvement of the resolution, as a consequence of their small diameter. Moreover, they can be modified at the ends to enable the manipulation of structures at the molecular scale [112]. The construction of nanoscale tweezers has been also possible by attaching a pair of carbon nanotubes to respective electrodes, and controlling the nanotube arms by the voltage applied between them [113]. Such a device has made possible the manipulation of different structures at the nanometer scale.

Finally, the technological applications of carbon nanotubes can also have a more direct impact in everyday life. They have been proposed for the construction of supercapacitors, which may take advantage of the large surface area accessible in nanotube arrays. These can give rise to capacitors with high power and storage capabilities. Anyhow, the carbon nanotubes may find the most interesting commercial application as electron sources in field-emission devices [114, 115]. These can be used in flat panel displays, as well as in lamps and X-ray sources. The emission is produced by applying a voltage between a surface with nanotube fibers, acting as a cathode, and a substrate with phosphor arrays. The high local fields created in the nanotube geometry make the electrons jump toward the anode, where the contact with the phosphor produces the spots of light in the display. The flat panel nanotube displays turn out to save more energy and to have higher brightness than liquid crystal displays. A similar field-emission effect can be applied to the generation of X-rays, when the anode is replaced by a metal surface, which can lead to interesting applications for medical purposes. All these developments stress once more the significance that the phenomena taking place in minute devices can have for the construction of useful engines, tailored for the needs of our time.

GLOSSARY

Ballistic transport Conduction property of a quantum wire by which the charge is carried without dissipation of energy from inelastic scattering with other quanta.

Electronic coherence Property of the electronic propagation in which the quantum state remains well-defined and continuous over a given spatial extension.

Luttinger liquid State of an electron system in which all the elementary excitations are given by wavelike modulations of either the charge or the spin texture.

Multiwalled nanotube Concentric arrangement of a large number of single-walled nanotubes in which these are disposed like shells in an onion-like structure.

Proximity effect Phenomenon by which the electronic properties of a normal metal change when it is placed in contact with a superconductor, and that is characterized by the development of superconducting correlations over a certain extension of the metal.

Quantum wire Structure in which the electrons are confined to move essentially in one dimension, in such a way that the electronic state keeps its coherence along the wire.

Single-walled carbon nanotube Macromolecule built from carbon bonds forming a hexagonal lattice that appears rolled up in a single tubular structure, with a diameter typically of nanometer size.

ACKNOWLEDGMENT

It is a pleasure to thank M. Bockrath, H. Bouchiat, C. Dekker, C. M. Lieber, and C. Schönenberger, for the help granted with the graphic material of this contribution.

REFERENCES

1. S. Iijima, *Nature* 354, 56 (1991).
2. R. Saito, G. Dresselhaus, and M. S. Dresselhaus, "Physical Properties of Carbon Nanotubes." Imperial College Press, London, 1998.

3. M. S. Dresselhaus, G. Dresselhaus, and P. C. Eklund, "Science of Fullerenes and Carbon Nanotubes." Academic Press, New York, 1996.
4. M. Bockrath, D. H. Cobden, P. L. McEuen, N. G. Chopra, A. Zettl, A. Thess, and R. E. Smalley, *Science* 275, 1922 (1997).
5. S. J. Tans, M. H. Devoret, H. Dai, A. Thess, R. E. Smalley, L. J. Geerligs, and C. Dekker, *Nature* 386, 474 (1997).
6. W. Liang, M. Bockrath, D. Bozovic, J. H. Hafner, M. Tinkham, and H. Park, *Nature* 411, 665 (2001).
7. S. Frank, P. Poncharal, Z. L. Wang, and W. A. de Heer, *Science* 280, 1744 (1998).
8. M. Kociak, A. Yu. Kasumov, S. Guéron, B. Reulet, I. I. Khodos, Yu. B. Gorbatov, V. T. Volkov, L. Vaccarini, and H. Bouchiat, *Phys. Rev. Lett.* 86, 2416 (2001).
9. Z. K. Tang, L. Zhang, N. Wang, X. X. Zhang, G. H. Wen, G. D. Li, J. N. Wang, C. T. Chan, and P. Sheng, *Science* 292, 2462 (2001).
10. Z. Yao, H. W. Ch. Postma, L. Balents, and C. Dekker, *Nature* 402, 273 (1999).
11. S. J. Tans, A. R. M. Verschueren, and C. Dekker, *Nature* 393, 49 (1998).
12. R. Martel, T. Schmidt, H. R. Shea, T. Hertel, and Ph. Avouris, *Appl. Phys. Lett.* 73, 2447 (1998).
13. C. Zhou, J. Kong, and H. Dai, *Appl. Phys. Lett.* 76, 1597 (1999).
14. H. W. Ch. Postma, T. Teepen, Z. Yao, M. Grifoni, and C. Dekker, *Science* 293, 76 (2001).
15. T. Rueckes, K. Kim, E. Joselevich, G. Y. Tseng, C.-L. Cheung, and C. M. Lieber, *Science* 289, 94 (2000).
16. V. Derycke, R. Martel, J. Appenzeller, and Ph. Avouris, *Nano Letters* 1, 453 (2001).
17. A. Bachtold, P. Hadley, T. Nakanishi, and C. Dekker, *Science* 294, 1317 (2001).
18. N. Hamada, S. Sawada, and A. Oshiyama, *Phys. Rev. Lett.* 68, 1579 (1992).
19. R. Saito, M. Fujita, G. Dresselhaus, and M. S. Dresselhaus, *Appl. Phys. Lett.* 60, 2204 (1992).
20. J. W. Mintmire, B. I. Dunlap, and C. T. White, *Phys. Rev. Lett.* 68, 631 (1992).
21. L. Chico, V. H. Crespi, L. X. Benedict, S. G. Louie, and M. L. Cohen, *Phys. Rev. Lett.* 76, 971 (1996).
22. R. Saito, G. Dresselhaus, and M. S. Dresselhaus, *Phys. Rev. B* 53, 2044 (1996).
23. Ph. Lambin, A. Fonseca, J. P. Vigneron, J. B. Nagy, and A. A. Lucas, *Chem. Phys. Lett.* 245, 85 (1995).
24. M. Menon and D. Srivastava, *Phys. Rev. Lett.* 79, 4453 (1997).
25. M. Menon and D. Srivastava, *J. Mat. Res.* 13, 2357 (1998).
26. C. Papadopoulos, A. Rakitin, J. Li, A. S. Vedeneev, and J. M. Xu, *Phys. Rev. Lett.* 85, 3476 (2000).
27. B. C. Satishkumar, P. J. Thomas, A. Govindaraj, and C. N. R. Rao, *Appl. Phys. Lett.* 77, 2530 (2000).
28. A. N. Andriotis, M. Menon, D. Srivastava, and L. Chernozatonskii, *Phys. Rev. Lett.* 87, 066802 (2001).
29. A. N. Andriotis, M. Menon, D. Srivastava, and L. Chernozatonskii, *Appl. Phys. Lett.* 79, 266 (2001).
30. B. W. Smith, M. Monthioux, and D. E. Luzzi, *Nature* 396, 323 (1998).
31. K. Hirahara, K. Suenaga, S. Bandow, H. Kato, T. Okazaki, H. Shinohara, and S. Iijima, *Phys. Rev. Lett.* 85, 5384 (2000).
32. S. Okada, S. Saito, and A. Oshiyama, *Phys. Rev. Lett.* 86, 3835 (2001).
33. J. Lee, H. Kim, S.-J. Kahng, G. Kim, Y.-W. Son, J. Ihm, H. Kato, Z. W. Wang, T. Okazaki, H. Shinohara, and Y. Kuk, *Nature* 415, 1005 (2002).
34. D. J. Hornbaker, S.-J. Kahng, S. Misra, B. W. Smith, A. T. Johnson, E. J. Mele, D. E. Luzzi, and A. Yazdani, *Science* 295, 828 (2002).
35. J. W. G. Wildöer, L. C. Venema, A. G. Rinzler, R. E. Smalley, and C. Dekker, *Nature* 391, 59 (1998).
36. T. W. Odom, J.-L. Huang, P. Kim, and C. M. Lieber, *Nature* 391, 62 (1998).
37. X. Blase, L. X. Benedict, E. L. Shirley, and S. G. Louie, *Phys. Rev. Lett.* 72, 1878 (1994).
38. O. Gülseren, T. Yildirim, and S. Ciraci, *Phys. Rev. B* 65, 153405 (2002).
39. K. Kanamitsu, *J. Phys. Soc. Japan* 71, 483 (2002).
40. J. González, F. Guinea, and M. A. H. Vozmediano, *Nucl. Phys. B* 406, 771 (1993).
41. J. A. Stroscio and R. M. Feenstra, in "Scanning Tunneling Microscopy" (J. A. Stroscio and W. J. Kaiser, Eds.), pp. 95–141. Academic Press, New York, 1993.
42. J. W. Mintmire, D. H. Robertson, and C. T. White, *J. Phys. Chem. Solids* 54, 1835 (1993).
43. M. Ouyang, J.-L. Huang, C.-L. Cheung, and C. M. Lieber, *Science* 292, 702 (2001).
44. C. T. White, D. H. Robertson, and J. W. Mintmire, in "Clusters and Nanostructured Materials" (P. Jena and S. Behera, Eds.), p. 231. Nova, New York, 1996.
45. M. Ouyang, J.-L. Huang, C.-L. Cheung, and C. M. Lieber, *Science* 291, 97 (2001).
46. V. J. Emery, in "Highly Conducting One-Dimensional Solids" (J. T. Devreese, R. P. Evrard, and V. E. Van Doren, Eds.). Plenum, New York, 1979.
47. J. Sólyom, *Adv. Phys.* 28, 201 (1979).
48. L. Chico, L. X. Benedict, S. G. Louie, and M. L. Cohen, *Phys. Rev. B* 54, 2600 (1996).
49. H. Grabert and M. H. Devoret, "Single Charge Tunneling." Plenum, New York, 1992.
50. M. Bockrath, D. H. Cobden, J. Lu, A. G. Rinzler, R. E. Smalley, L. Balents, and P. L. McEuen, *Nature* 397, 598 (1999).
51. M. P. A. Fisher and A. Dorsey, *Phys. Rev. Lett.* 54, 1609 (1985).
52. H. Grabert and U. Weiss, *Phys. Rev. Lett.* 54, 1605 (1985).
53. A. A. Maarouf, C. L. Kane, and E. J. Mele, *Phys. Rev. B* 61, 11156 (2000).
54. H. Stahl, J. Appenzeller, R. Martel, Ph. Avouris, and B. Lengeler, *Phys. Rev. Lett.* 85, 5186 (2000).
55. R. Egger and A. O. Gogolin, *Phys. Rev. Lett.* 79, 5082 (1997).
56. C. Kane, L. Balents, and M. P. A. Fisher, *Phys. Rev. Lett.* 79, 5086 (1997).
57. R. Landauer, *J. Phys. Cond. Matter* 1, 8099 (1989).
58. R. Landauer, *Philos. Mag.* 21, 863 (1970).
59. S. Datta, "Electronic Transport Properties in Mesoscopic Systems." Cambridge Univ. Press, Cambridge, 1995.
60. S. Sanvito, Y.-K. Kwon, D. Tománek, and C. J. Lambert, *Phys. Rev. Lett.* 84, 1974 (2000).
61. P. J. de Pablo, C. Gómez-Navarro, J. Colchero, P. A. Serena, J. Gómez-Herrero, and A. M. Baró, *Phys. Rev. Lett.* 88, 036804 (2002).
62. L. C. Venema, J. W. G. Wildöer, J. W. Janssen, S. J. Tans, H. L. J. Temminck Tuinstra, L. P. Kouwenhoven, and C. Dekker, *Science* 283, 52 (1999).
63. A. Bachtold, C. Strunk, J.-P. Salvetat, J.-M. Bonard, L. Forró, T. Nussbaumer, and C. Schönenberger, *Nature* 397, 673 (1999).
64. C. L. Kane and M. P. A. Fisher, *Phys. Rev. Lett.* 68, 1220 (1992).
65. C. L. Kane and M. P. A. Fisher, *Phys. Rev. B* 46, 15233 (1992).
66. C. T. White and T. N. Todorov, *Nature* 393, 240 (1998).
67. C. W. J. Beenakker, *Phys. Rev. B* 44, 1646 (1991).
68. M. R. Buitelaar, A. Bachtold, T. Nussbaumer, M. Iqbal, and C. Schönenberger, *Phys. Rev. Lett.* 88, 156801 (2002).
69. M. Krüger, M. R. Buitelaar, T. Nussbaumer, and C. Schönenberger, *Appl. Phys. Lett.* 78, 1291 (2001).
70. S. Tarucha, D. G. Austing, Y. Tokura, W. G. van der Wiel, and L. P. Kouwenhoven, *Phys. Rev. Lett.* 84, 2485 (2000).
71. J. Nygård, D. H. Cobden, and P. E. Lindelof, *Nature* 408, 342 (2000).
72. L. P. Kouwenhoven and L. Glazman, *Phys. World* 14, 33 (2001).

73. L. Chico, M. P. López Sancho, and M. C. Muñoz, *Phys. Rev. Lett.* 81, 1278 (1998).
74. M. Bockrath, W. Liang, D. Bozovic, J. H. Hafner, C. M. Lieber, M. Tinkham, and H. Park, *Science* 291, 283 (2001).
75. M. T. Woodside and P. L. McEuen, *Science* 296, 1098 (2002).
76. A. Yu. Kasumov, R. Deblock, M. Kociak, B. Reulet, H. Bouchiat, I. I. Khodos, Yu. B. Gorbatov, V. T. Volkov, C. Journet, and M. Burghard, *Science* 284, 1508 (1999).
77. D. Esteve, H. Pothier, S. Guéron, N. O. Birge, and M. H. Devoret, in "Mesoscopic Electron Transport" (L. L. Sohn, L. P. Kouwenhoven, and G. Schön, Eds.), p. 375. Kluwer, Dordrecht, 1997.
78. B. J. van Wees and H. Takayanagi, in "Mesoscopic Electron Transport" (L. L. Sohn, L. P. Kouwenhoven, and G. Schön, Eds.), p. 469. Kluwer, Dordrecht, 1997.
79. M. Tinkham, "Introduction to Superconductivity." McGraw-Hill, New York, Singapore, 1996.
80. A. F. Morpurgo, J. Kong, C. M. Marcus, and H. Dai, *Science* 286, 263 (1999).
81. J. González, *Phys. Rev. Lett.* 87, 136401 (2001).
82. D. L. Maslov, M. Stone, P. M. Goldbart, and D. Loss, *Phys. Rev. B* 53, 1548 (1996).
83. R. Fazio, F. W. J. Hekking, and A. A. Odintsov, *Phys. Rev. B* 53, 6653 (1996).
84. J. González, *Phys. Rev. Lett.* 88, 076403 (2002).
85. J. González, *Phys. Rev. B*, 67, 014528 (2003).
86. D. Sánchez-Portal, E. Artacho, J. M. Soler, A. Rubio, and P. Ordejón, *Phys. Rev. B* 59, 12678 (1999).
87. L. M. Woods and G. D. Mahan, *Phys. Rev. B* 61, 10651 (2000).
88. R. Saito, A. Jorio, A. G. Souza Filho, G. Dresselhaus, M. S. Dresselhaus, and M. A. Pimenta, *Phys. Rev. Lett.* 88, 027401 (2002).
89. P. Delaney, H. Joon Choi, J. Ihm, S. G. Louie, and M. L. Cohen, *Nature* 391, 466 (1998).
90. Y.-K. Kwon, S. Saito, and D. Tománek, *Phys. Rev. B* 58, R13314 (1998).
91. J.-C. Charlier, X. Gonze, and J.-P. Michenaud, *Europhys. Lett.* 29, 43 (1995).
92. H. J. Schulz, in "Correlated Electron Systems" (V. J. Emery, Ed.), Vol. 9. World Scientific, Singapore, 1993.
93. L. X. Benedict, V. H. Crespi, S. G. Louie, and M. L. Cohen, *Phys. Rev. B* 52, 14935 (1995).
94. M. S. Fuhrer, J. Nygård, L. Shih, M. Forero, Y.-G. Yoon, M. S. C. Mazzoni, H. J. Choi, J. Ihm, S. G. Louie, A. Zettl, and P. L. McEuen, *Science* 288, 494 (2000).
95. V. Derycke, R. Martel, J. Appenzeller, and Ph. Avouris, *Appl. Phys. Lett.* 80, 2773 (2002).
96. S. J. Wind, J. Appenzeller, R. Martel, V. Derycke, and Ph. Avouris, *Appl. Phys. Lett.* 80, 3817 (2002).
97. C. L. Kane and E. J. Mele, *Phys. Rev. Lett.* 78, 1932 (1997).
98. A. Rochefort, D. R. Salahub, and Ph. Avouris, *Chem. Phys. Lett.* 297, 45 (1998).
99. R. Heyd, A. Charlier, and E. McRae, *Phys. Rev. B* 55, 6820 (1997).
100. V. H. Crespi, M. L. Cohen, and A. Rubio, *Phys. Rev. Lett.* 79, 2093 (1997).
101. P. Zhang, P. E. Lammert, and V. H. Crespi, *Phys. Rev. Lett.* 81, 5346 (1998).
102. L. Yang and J. Han, *Phys. Rev. Lett.* 85, 154 (2000).
103. T. W. Tombler, C. Zhou, L. Alexseyev, J. Kong, H. Dai, L. Liu, C. S. Jayanthi, M. Tang, and S.-Y. Wu, *Nature* 405, 769 (2000).
104. L. Liu, C. S. Jayanthi, M. Tang, S.-Y. Wu, T. W. Tombler, C. Zhou, and L. Alexseyev, *Phys. Rev. Lett.* 84, 4950 (2000).
105. A. Maiti, A. Svizhenko, and M. P. Anantram, *Phys. Rev. Lett.* 88, 126805 (2002).
106. Yu. N. Gartstein, A. A. Zakhidov, and R. H. Baughman, *Phys. Rev. Lett.* 89, 045503 (2002).
107. R. H. Baughman, C. Cui, A. A. Zakhidov, Z. Iqbal, J. N. Barisci, G. M. Spinks, G. G. Wallace, A. Mazzoldi, D. De Rossi, A. G. Rinzler, O. Jaschinski, S. Roth, and M. Kertesz, *Science* 284, 1340 (1999).
108. J. Kong, N. R. Franklin, C. Zhou, M. G. Chapline, S. Peng, K. Cho, and H. Dai, *Science* 287, 622 (2000).
109. P. G. Collins, K. Bradley, M. Ishigami, and A. Zettl, *Science* 287, 1801 (2000).
110. H. Dai, J. H. Hafner, A. G. Rinzler, D. T. Colbert, and R. E. Smalley, *Nature* 384, 147 (1996).
111. J. H. Hafner, C. L. Cheung, and C. M. Lieber, *Nature* 398, 761 (1999).
112. S. S. Wong, E. Joselevich, A. T. Woolley, C. L. Cheung, and C. M. Lieber, *Nature* 394, 52 (1998).
113. P. Kim and C. M. Lieber, *Science* 286, 2148 (1999).
114. A. G. Rinzler, J. H. Hafner, P. Nikolaev, L. Lou, S. G. Kim, D. Tománek, P. Nordlander, D. T. Colbert, and R. E. Smalley, *Science* 269, 1550 (1995).
115. W. A. de Heer, A. Châtelain, and D. Ugarte, *Science* 270, 1179 (1995).
116. P. G. Collins, M. Hersam, M. Arnold, R. Martel, and Ph. Avouris, *Phys. Rev. Lett.* 86, 3128 (2001).
117. T. Dürkop, T. Brintlinger, and M. S. Fuhrer, in "Structural and Electronic Properties of Molecular Nanostructures" (H. Kuzmany, J. Fink, M. Mehring, and S. Roth, Eds.), pp. 242–246. AIP Conference Proceedings, New York, 2002.

Encyclopedia of Nanoscience and Nanotechnology
www.aspbs.com/enn

Electronic Structure of Semiconductor Nanoparticles

Sameer Sapra, J. Nanda, D. D. Sarma
Indian Institute of Science, Bangalore, India

CONTENTS

1. INTRODUCTION

The last two decades have witnessed a tremendous amount of research activity in the field of semiconductor nanoparticles [1–7]. Also known as nanocrystals, nanocrystallites, or quantum dots, these materials are viewed as promising candidates for technological applications in the near future for devices based on electronics and optoelectronics [8–10]. The electronic structure of these materials, which lie in the transition region between the molecular limit and the bulk limit, drastically changes as the size of the particle is varied. For example, the bandgap of a semiconductor such as CdS can be varied from 2.5 eV for bulk to about 4.5 eV for 20 Å nanoparticles [11, 12]. Optical properties, such as photoluminescence, can be tuned nearly continuously by merely changing the size of the nanomaterial and one can get emissions throughout the visible region of the electromagnetic spectrum from the same material, as in the case of CdSe, by choosing the particle size suitably [13, 14].

An understanding of these nanomaterials demands an accurate theoretical description. The first attempt to understand the bandgap variation with size in these low-dimensional systems was the use of the effective mass approximation (EMA) [15–26]. Though simple in approach, the infinite potential EMA led to overestimation of the gaps and to quantitatively wrong conclusions. Other approaches used were the effective bond orbital method (EBOM) [27, 28], the empirical pseudopotential method (EPM) [29–36], and the tight-binding approach (TB) [37–46].

Here we will discuss the EMA and the TB methods in detail as these two have been extensively used to correlate the size-gap variation, in conjunction with direct experimental investigation of the electronic structure of various nanoparticles.

We review the theoretical approaches in the first part of this chapter; we follow this up with a critical comparison of the theoretical results with experimental data in the latter part of this review.

2. BANDGAP VARIATION WITH NANOPARTICLE SIZE

2.1. Effective Mass Approximation

In crystalline materials, the energy-momentum relationship is defined by a complex set of equations. This relationship is known as the band dispersion of the crystal [47]. Semiconductors or insulators differ from metals in exhibiting an energy gap between the valence band and the conduction band. If the charge carrier has a small amount of kinetic energy, then the simplified description in Figure 1 is often applicable for a single-particle picture.

At low energies, near the band edges, delocalized electron (or hole) waves follow a quadratic equation describing the wave vector, k dependence of energy, E

$$E = \frac{\hbar^2 k^2}{2m^*}$$

where m^* is the effective mass of the charge carrier (electron or hole). The motion of single-charge carriers under the influence of external fields is described by this relation. Within the effective mass approximation, one can also describe the motion of a coupled electron-hole pair, the exciton. To use the effective mass approximation for describing the bandgap variation with size for nanoparticles, one

ISBN: 1-58883-059-4/$35.00

Encyclopedia of Nanoscience and Nanotechnology
Edited by H. S. Nalwa
Volume 3: Pages (181–192)

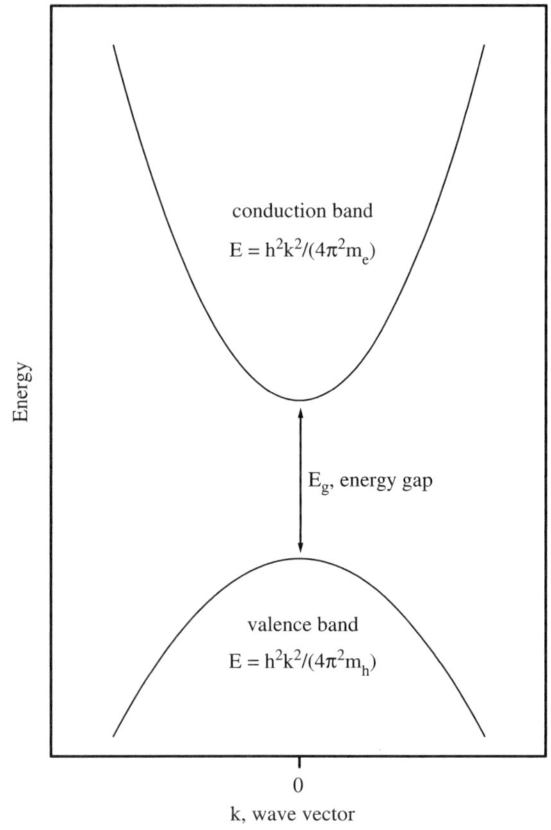

Figure 1. Schematic band dispersion: energy-momentum relation for a direct bandgap semiconductor.

needs to solve the Schrödinger equation for the envelope function ψ:

$$\left[-\frac{\hbar^2\nabla_e^2}{2m_e} - \frac{\hbar^2\nabla_h^2}{2m_h} - \frac{e^2}{4\pi\epsilon_0\epsilon r_{eh}} + V_0\right]\psi(r_e, r_h) = E\psi(r_e, r_h) \tag{1}$$

where the subscripts e and h refer to the electron and the hole with m and r being the mass and position vector, respectively, and $r_{eh} = |r_e - r_h|$. ϵ_0 and ϵ are the permittivity in vacuum and the relative dielectric constant of the material. The potential V_0 inside the nanoparticle is assumed to be zero and infinite outside for the EMA calculation corresponding to an infinite barrier height. Using a trial wave function, the above equation can be solved by approximate methods.

Brus [16–18] has considered a model for the particles that incorporates (a) the effective mass approximation for the kinetic energy of the electron and the hole, (b) an electrostatic potential term from classical continuum polarizability theory, and (c) tunneling of electrons and holes out of the crystallite at the surface in the case of a finite barrier height. For an infinite barrier height the model yields the following expression for the bandgap, E_R, of the quantum dot,

$$E_R = E_g + \frac{\hbar^2\pi^2}{2R^2}\left[\frac{1}{m_e} + \frac{1}{m_h}\right] - \frac{1.8e^2}{\epsilon R} - 0.248E_{Ry}^* \tag{2}$$

where E_g is the bulk bandgap, and R is the radius of the quantum dot. The effective Rydberg (in meV) is defined as

$$E_{Ry}^* = 13605.8\frac{1}{\epsilon^2}\left(\frac{m_0}{m_e} + \frac{m_0}{m_h}\right)^{-1} \tag{3}$$

The second term arises due to the Coulomb attraction, and the third term due to the spatial correlation between the electron and the hole. Brus used the EMA to calculate the bandgap variation with size for InSb, GaAs, CdS, and ZnO [17]. These results are shown in Figure 2.

Kayanuma [22] has identified two limiting cases depending upon the ratio of the radius of the quantum dot to the Bohr exciton radius, a_B of the bulk solid. For $R/a_B \gg 1$ the exciton can be pictured as a particle moving inside the quantum dot with only little increment in energy due to the confinement. This is the weak confinement regime. In the strong confinement regime where $R/a_B \ll 1$, confinement effects obviously dominate. It was pointed out that in this regime, the electron and hole should be viewed as individual particles in their respective single-particle ground states with little spatial correlation between them. Kayanuma further found that the strong confinement regime is not only limited to $R/a_B \ll 1$ but the effects are seen up to $R = 2a_B$. This is the regime where the effective mass approximation as given by Brus can be applied; however, we point out later that Eq. (2) overestimates the bandgap for all sizes, with the mismatch between the experiments and the calculated results increasing progressively for decreasing particle size.

Some improvement over the EMA with an infinite barrier was made when Shmidt and Weller [20] used a configuration interaction approach for the electron and the hole, treating

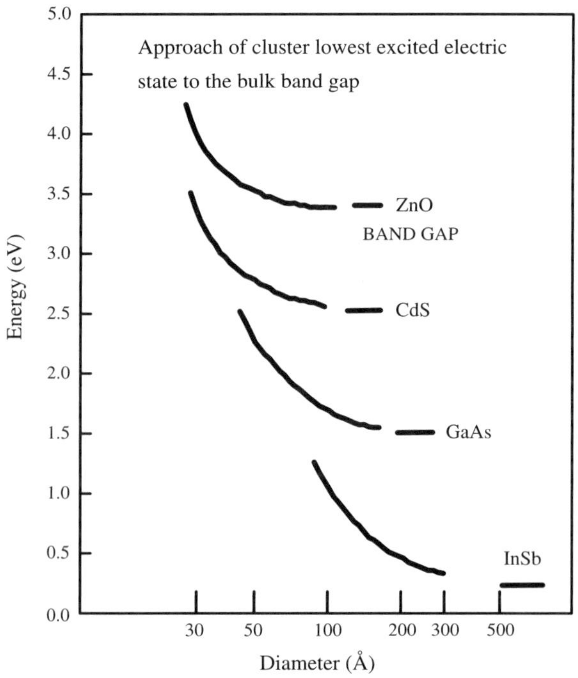

Figure 2. Calculated energy of the lowest excited electronic state of the cluster in relation to the bulk bandgap. Reprinted with permission from [18], L. E. Brus, *J. Phys. Chem.* 90, 2555 (1986). © 1986, American Chemical Society.

it like a two-electron atom. They also used the Hylleras functions and a perturbation expansion in R to obtain bandgap energies slightly better than those described by the single-particle EMA.

The effect of using infinite barrier potential is to neglect the presence of the surrounding medium which can be either a glassy matrix or a passivating agent to inhibit the growth of particle size and the ability of the electron and the hole to tunnel significantly beyond the confines of the nanoparticles. These shortcomings lead to an overestimation of bandgaps for very small crystallites. Finite barrier heights have therefore often been considered to account for the experimental data from small crystallites. Kayanuma and Momiji [23] have demonstrated an excellent agreement of the bandgap variation with size for CdS nanocrystals embedded in silicate glasses [48] by using a finite barrier potential at the interface. The confining potential $\overline{V}_e$, for the electron, and $\overline{V}_h$, for the hole, satisfies the relation $E_g + \overline{V}_e + \overline{V}_h = E_g^1$, where E_g^1 is the bandgap energy of the surrounding material. The resulting problem was solved variationally in the Hylleras coordinate system. The results obtained within this approach are shown in Figure 3 for the CdS system.

Further improvements to the EMA led to the emergence of the multiband EMA (MBEM) theory. In most of the semiconductors the conduction band is nondegenerate and the electron effective mass can be well described by the single-band theory. On the other hand, the top of the valence band is often degenerate. Hence, to describe the hole, one needs to consider various bands that contribute to the valence band edge. In CdSe nanocrystals, the size dependence of up to 10 excited states [24, 25] in the absorption spectra is successfully described by the uncoupled MBEM. This includes the valence band degeneracy but does not couple the valence and conduction bands. Banin et al. [26] have used the MBEM including the valence and conduction band coupling. They use an eight-band Luttinger Kohn Hamiltonian to calculate the quantum size levels in InAs [26].

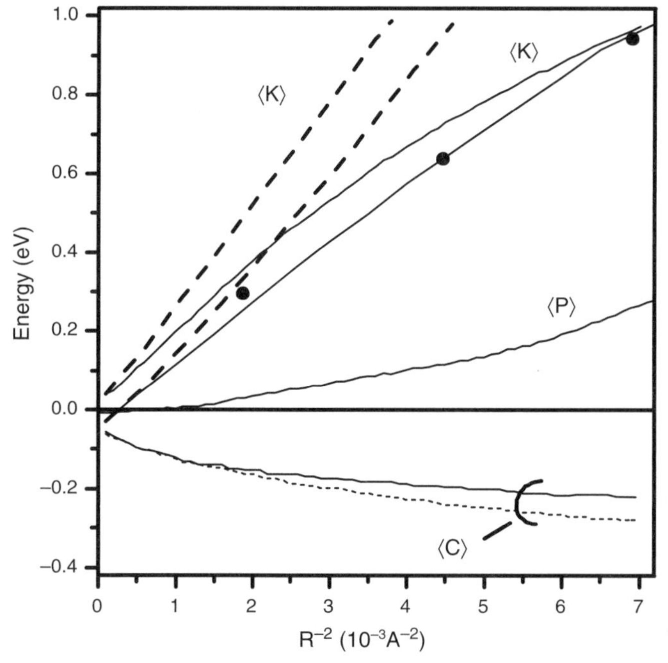

Figure 3. Calculated ground-state energy of the electron-hole system in CdS nanocrystals embedded in silicate glass. The bold line is the total energy. The expectation values of the kinetic energy ⟨K⟩, the Coulomb energy ⟨C⟩, and the potential energy ⟨P⟩ are also shown. The corresponding quantities calculated in the infinite barrier model are shown by the dashed lines. The origin of the energy is chosen at the bandgap energy of CdS. The experimental peak positions from [48] are plotted by small circles. Reprinted with permission from [23], Y. Kayanuma and H. Momiji, *Phys. Rev. B* 41, 10261 (1990). © 1990, American Physical Society.

2.2. Empirical Pseudopotential Method

The empirical pseudopotential method (EPM) has been used to investigate the electronic and optical properties of bulk semiconductor materials [49, 50]. The Schrödinger equation

$$H\Psi_{n,k}(r) = E_n(k)\Psi_{n,k}(r) \tag{4}$$

is solved at a set of k points in the Brillouin zone. The Hamiltonian H is given by

$$H = -\frac{\hbar^2}{2m}\nabla^2 + V_p(r) \tag{5}$$

where

$$V_p(r) = \sum_G [V_s(G)S_s(G) + iV_A(G)S_A(G)] \exp(iG \cdot r) \tag{6}$$

is an empirically determined pseudopotential. V_s and V_A are the symmetric and antisymmetric form factors, respectively. They are determined from the optical absorption data. S_s and S_A are the symmetric and antisymmetric structure factors, respectively, determined from the crystal structure of the semiconductor.

The wave functions $\Psi_{n,k}$ can be regarded as linear combinations of the plane waves

$$\psi_{n,k}(r) = \frac{1}{\sqrt{V}}\sum_G C_{G+k} \exp[i(G+k)\cdot r] \tag{7}$$

The Hamiltonian matrix consists of the elements

$$H(G'+k, G+k)$$
$$= \frac{1}{V}\int \mathrm{d}r \exp[-i(G'+k)\cdot r]H \exp[i(G+k)\cdot r] \tag{8}$$
$$= \frac{\hbar^2}{2m}(G+k)^2\delta_{G'G} + V(G'-G) \tag{9}$$

which is diagonalized at a grid of k-points in the Brillouin zone to obtain the band structure of the semiconductor.

For nanocrystals, the allowed wave vectors k are discrete. For a nanocrystal of radius R, a reasonable assumption is to use the bulk pseudopotential $V_p(r)$ inside the crystallite and terminate this potential at the edge of the cluster by an infinite potential. The wave vectors k_n of the allowed states are then given by $j_l(k_nR) = 0$. The HOMO and LUMO states at the band edges correspond to $l = 0$. The solution of $j_0(k_nR) = 0$ gives the allowed wave vectors k_n. Since k_n

is along the radial direction, it is projected onto each of the cartesian axes with equal magnitude to obtain the cartesian components of k,

$$k = \frac{\pi}{\sqrt{3}R}[n_x, n_y, n_z] \tag{10}$$

The energy levels at these k-states constitute the $l = 0$ band structure of the crystallites.

Figure 4a shows the bulk band structure of zinc-blende CdS. Ramakrishna and Friesner [29] calculated the band structure of CdS nanocrystals of $R = 15$ Å as shown in Figure 4b. In this figure, the energies of all the orbitals are specified relative to the energy of the HOMO orbital. As the cluster size is increased, all the orbitals corresponding to each band come closer in energy and the band structure of the clusters smoothly approaches that of the bulk.

From the calculated band structures of the clusters, Ramakrishna and Friesner calculated the variation of the bandgap with size for different sized CdS clusters. The bandgaps of the clusters were corrected for the electron-hole Coulomb attraction and the correlation energies by subtracting the last two terms given in Eq. (2) from the calculated bandgaps. The theoretical results for spherical and cubic shaped clusters crystallizing in the zinc-blende phase along with the experimental data [69] and the EMA calculations [18, 22] are shown in Figure 5a. Figure 5b shows the same results except that these results are for the wurtzite phase. In the case of GaAs, GaP, and Si nanocrystals of very small sizes, Ramakrishna and Friesner [29] also observed a red-shift in the bandgap. These red-shifts have not been predicted by any other theoretical technique.

Recently, Zunger and co-workers [30–36] employed the semiempirical pseudopotential method to calculate the electronic structure of Si, CdSe [33], and InP [32] quantum dots. Unlike EMA approaches, this method based on screened pseudopotentials allows the treatment of the atomistic character of the nanostructure as well as the surface effects, while permitting the multiband and intervalley coupling. The atomic pseudopotentials are extracted from first-principles LDA calculations on bulk solids. The single-particle LDA equation,

$$\left[-\frac{1}{2}\nabla^2 + V^{ps}_{\text{nonlocal}}(r) + V_{\text{LDA}}(r)\right]\Psi_i = \epsilon_i \Psi_i \tag{11}$$

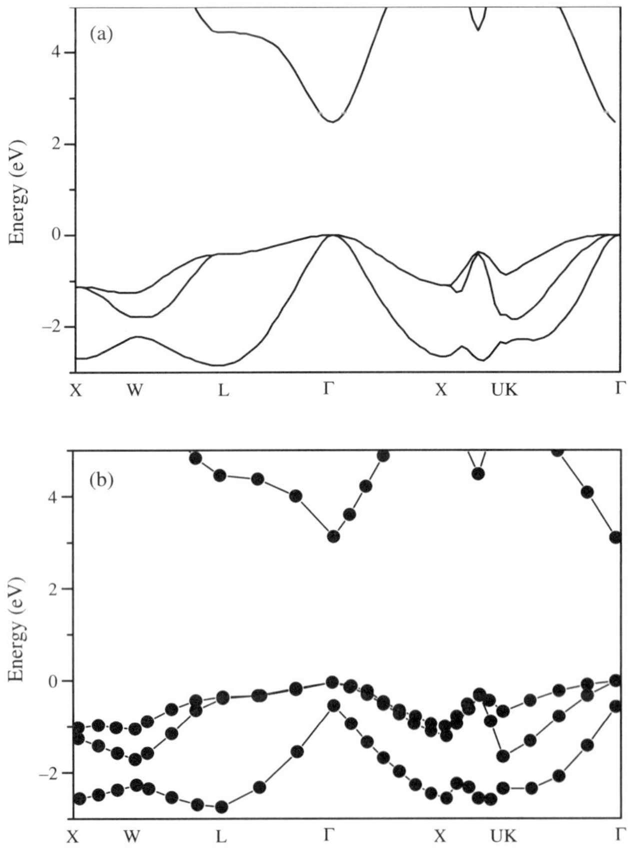

Figure 4. (a) A portion of the zinc-blende CdS crystal band structure near the top of the valence band. (b) The band structure of a 15 Å radius CdS cluster. Reprinted with permission from [29], M. V. Ramakrishna and R. A. Friesner, *Phys. Rev. Lett.* 67, 629 (1991). © 1991, American Physical Society.

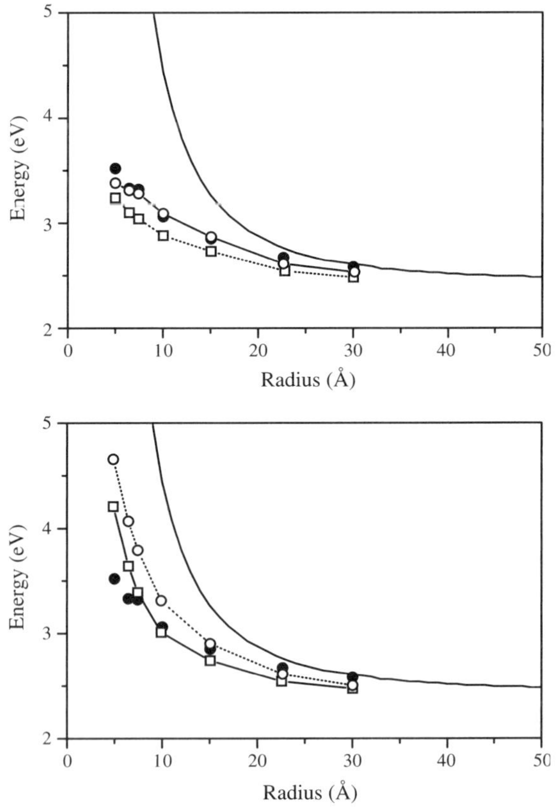

Figure 5. (a) Experimental and calculated exciton energies of CdS clusters. The solid circles are the experimental data [69] and the solid line is the result due to Bruss effective-mass model with $m_e = 0.19$, $m_h = 0.80$, $\epsilon = 5.5$, and $E_g = 2.5$ eV [18, 22]. (a) The open circles and squares are from our calculations of the zinc-blende CdS clusters with spherical and cubic shapes, respectively. (b) Same as (a), except that the open circles and the squares are calculated for the hexagonal CdS clusters. Reprinted with permission from [29], M. V. Ramakrishna and R. A. Friesner, *Phys. Rev. Lett.* 67, 629 (1991). © 1991, American Physical Society.

is solved to obtain the electronic structure of a solid. The ionic pseudopotential consists of the nonlocal part $V^{ps}_{\mathrm{nonlocal}}(r)$ and V_{LDA} is composed of the local ionic pseudopotential, $V^{ps}_{\mathrm{local}}(r)$, and the screening potential V_{HXC} made up of the interelectron Coulomb (Hartree) and the exchange and correlation (XC) parts. For a given crystal structure, the knowledge of the atomic pseudopotentials allows one to solve Eq. (11). The wave functions are LDA-like while the band structures, effective masses, and deformation potentials match experiments. These improvements over the empirical pseudopotential method provide a better description of the electronic structure of the quantum dots. However, large size of the Hamiltonian does not allow one to calculate the properties of large sized nanocrystals.

2.3. Tight-Binding Method

The tight-binding (TB) method or the LCAO approach was first described as an interpolation scheme to elucidate the electronic structure of solids by Slater and Koster [51]. Since then it has been used widely to study the electronic structure of various metals, semiconductors, and clusters. It has the advantage over the plane wave methods that it is computationally less expensive. Thus, it is possible to study very complex systems using the TB method and it has been used for complex alloys and doped compounds. In the late 1980s, Lippens and Lannoo [37] first used the tight-binding approach for quantum dots. Since then many groups have demonstrated the effectiveness and ease of use of this method.

For calculating the electronic properties of nanoparticles using the TB method, the first step is to calculate the band structure of the bulk material with the same lattice structure as observed in the nanoparticles. This bulk band structure can be calculated using any of the standard effectively single-particle, first-principles methods such as the LAPW, LMTO, or the pseudopotential methods. Then a TB model is selected to fit the bulk band structure. The specific choice of the TB model, including the basis set and the range of interactions, depends on the specific problem to be investigated. For describing the bandgap variation with the size in nanoparticles, one is obviously most interested in the energy region around the bandgap of the bulk material. The fitting of the band dispersions of the bulk material obtained within a first-principles approach by a suitable TB model yields the optimized set of parameters suitable to describe the material within that TB model. Assuming that these parameter strengths are transferable from the bulk to the nanoparticle case, one can use the same TB model to calculate the eigenspectrum, and consequently the electronic structure of any given size and shape. However, one important difference between the approach for the bulk and the nanoparticle arises from the finite size of nanoparticles and the resulting absence of periodicity in them. The Hamiltonian matrix for the nanoparticles needs to be set up in real space, with the dimension of the matrix being equal to the total number of orbitals in the cluster. This often makes it difficult to calculate the electronic structure of the nanoparticles with very large sizes. Direct diagonalization works well for small nanoparticles, but for larger particles, one needs to revert to other methods of diagonalization, such as the Lanczos technique [52]. In typical applications, by generating nanoparticles of different sizes, a curve for the bandgap variation with size is obtained. However, it is also possible to calculate the electron density of states as a function of size and correlate it to experimental photoemission and X-ray absorption spectra, as shall be discussed in the last section.

3. TIGHT-BINDING METHOD

Lippens and Lannoo [37] first used the TB method to describe the bandgap variation with size for CdS and ZnS nanocrystallites. They used the TB model with a sp^3s^* basis and the specific set of parameter values obtained earlier by Vogl et al. [53] for describing the bulk band structure. The calculations for CdS and ZnS crystallites were carried out for crystallites with different sizes, ranging between the smallest sized clusters with 17 atoms in 3 shells and the largest with 2563 atoms in 15 shells. The surface states or the dangling bonds were removed from the Hamiltonian, as they give rise to partially bonded electronic states in the nanoparticles and appear in the bandgap region, making the unambiguous determination of the bandgap difficult. These authors used direct diagonalization of the Hamiltonian for small crystallites with less than 83 atoms and the recursion technique for larger crystallites and obtained the variation of the valence and the conduction band edges as a function of size. An interesting feature in the variation of the band edges with size is the odd-even shell effect. Oscillations in the valence band (VB) or conduction band (CB) edge positions are observed for both CdS and ZnS. This is attributed to the manner in which the crystallites have been generated. The surface shell of the crystallites is either purely cationic or purely anionic in nature depending on whether the nanoparticle cluster considered has even or odd number of shells. The oscillations are a consequence of removing the dangling cationic or anionic orbitals. In order to compare the calculated results with experimental ones, it is necessary to take into account the changing contribution from the electron-hole interaction energy as a function of the size, besides the change in the bare bandgap energy calculated within the TB model. This is most conveniently achieved by assuming this energetic being given by the corresponding term in the EMA equation (Eq. (2)) also for the TB results. Thus, Lippens and Lannoo compared their results after the correction for the excitonic binding energy as discussed here with the existing experimental data [54, 55] as a function of particle size (Fig. 6). It is evident from the figure that this model agrees better with the experimental results compared to the single-band, infinite potential, effective mass approximation, especially for small crystallite sizes. However, the nearest-neighbor sp^3s^* tight-binding model is found to consistently underestimate the bandgap. Using the same sp^3s^* model, Lippens and Lannoo calculated the electronic structure and the bandgap variation as a function of size for various $A^{II}B^{VI}$ semiconductor nanocrystallites, where A = Zn, Cd and B = S, Se, Te [39]. These results are discussed later in this section, in comparison to results from other TB models.

Hill and Whaley employed a time-dependent tight-binding approach to determine the electronic structure of CdS and CdSe nanocrystals [40, 42]. This method has two

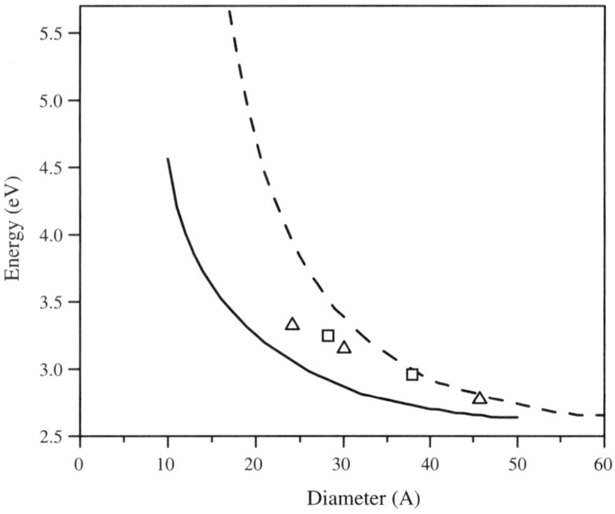

Figure 6. Comparison with experimental data for CdS. The tight-binding results are given by the solid curve and those of the effective-mass approximation (Eq. (2)) by the dashed curve. The experimental data are the exciton peak given in [54] (squares) and [55] (triangles). Reprinted with permission from [37], P. E. Lippens and M. Lannoo, *Phys. Rev. B* 39, 10935 (1989). © 1989, American Physical Society.

advantages. First, the size of the cluster is less restricted by computer memory limitations than in methods based on diagonalization of the tight-binding Hamiltonian directly. Second, the method can be extended to study excitonic states by propagating a two-particle (electron-hole) wave function and including the Coulomb interaction in the Hamiltonian, instead of introducing an *ad hoc* correction for the excitonic binding energy on the calculated single-particle bandgap. Furthermore, Hill and Whaley determined the surface atom contribution to the density of states of the nanocrystals. This is performed by calculating the density of states of the surface atoms and identifying the corresponding features in the total density of states. These results were further confirmed by two different approaches, namely (i) passivating the surface atoms with oxygen atoms and (ii) removing the dangling orbitals altogether to obtain the density of states of the unpassivated, but dangling orbital free, surfaces. Comparing these results with the density of states obtained from the usual unpassivated clusters, it is easy to identify the surface states. These authors have also studied the effect of terminating the surface with a cationic or an anionic atomic layer on the density of states. For both CdS and CdSe, a cationic surface gives rise to surface states near the conduction band edge, while the anionic surface shows states near the top of the valence band within the bandgap region. There are contributions to other parts of the density of states from the surface atoms too. These results are shown in Figure 1a and 1b in [42] for unpassivated Cd and Se. Figure 1c and 1d in [42] show the effect of passivating the Cd and Se surfaces, respectively, with oxygen. It is seen that oxygen passivation works well for the cationic surface, but fails to quench the surface states when there is an anionic surface. Removing the dangling bonds gets rid of the surface states completely, independent of the nature of the surface layer, as shown in Figure 1e and 1f in [42].

In Figure 9 in [40], a comparison between the theroretical results and the experimental photoemission data for CdS [56] for the variation of valence band maximum with size is shown. In the absence of any knowledge about the Fermi energy of the system, the experimental valence band maxima for the 38 Å crystallites were rigidly shifted in energy, such that this coincides with the calculated results. This gives a fair agreement between the experiment and the calculation for the 24 Å nanoparticles taking into account the large error bar. However, they could not compare the calculated density of states with the photoemission spectra of these samples, as Cd d levels that are a part of the bottom of the valence band and have a considerable contribution near the valence band edge, are absent in the sp^3s^* model. Thus, it suggests that the sp^3s^* model may be inadequate to account for all aspects of the experimental observations and further improvements are required in the calculational schemes.

Pérez Conde et al. [45] extended the standard sp^3s^* model by incorporating the spin-orbit coupling for CdSe and CdTe nanocrystals. The resulting bandgap variation with size for CdTe is shown in Figure 7 along with the experimental data from various groups [57–61]. Their result (open triangles) scales as a function of the nanoparticle diameter (D) as $1/D^{1.42}$ compared to the EMA (dotted line) which varies as $1/D^2$.

More recently, Allan et al. worked out the quantum confinement energies in zinc-blende III–V and group IV semiconductors [44]. They also used the tight-binding approximation including spin-orbit coupling, but expanded the basis to include the d orbitals ($sp^3d^5s^*$). Such a model gives conduction and valence bands for the bulk system in

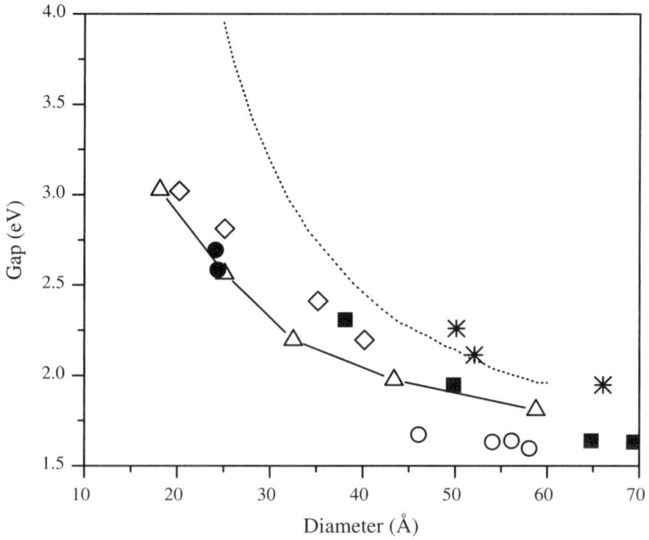

Figure 7. Absorption bandgap vs QD diameter. The experimental data from [57] (open diamonds) and [58] (closed squares) were measured at room temperature. The data from [59] (stars) were measured at 2 K. The values from [60] were taken at temperatures higher than room temperature (open circles). The sample studied in [61] (closed circles) was measured at 77 K (higher value) and room temperature (lower value). The theoretical TB values scale (open triangles) as a function of diameter as $1/D^{1.42}$, while the EMA gap (dotted line) varies as $1/D^2$. Reprinted with permission from [45], J. Pérez Conde et al., *Phys. Rev. B* 64, 113303 (2001). © 2001, American Physical Society.

good agreement with first-principles results. The calculated effective masses are also comparable to the experimental values. Using this basis, the variations of the HOMO and LUMO energies of nanoparticles of different diameters were obtained and related to the effective cluster diameter D using the following expressions:

$$E_{\text{HOMO}} = E_{BV} - \frac{1}{a_v D^2 + b_v D + c_v} \qquad (12)$$

$$E_{\text{LUMO}} = E_{BC} + \frac{1}{a_c D^2 + b_c D + c_c} \qquad (13)$$

where E_{BV} and E_{BC} are the bulk band edges and the a_v, b_v, c_v, a_c, b_c, and c_c are the fitting parameters for various III–V and IV semiconductors. Figure 3 in [44] shows a comparison of these calculated results with the $sp^3d^5s^*$ basis with those obtained from pseudopotential calculations [30] as well as with experimental results obtained from photoluminescence studies of InP [62–64]. It is clear from this figure that both the pseudopotential and $sp^3d^5s^*$ tight-binding results overestimate the experimental gap energy. While the same reference [44] also shows the comparison for InAs, the experimental results in this case from various groups are not in good agreement; hence, it is difficult to comment on the extent of agreement between the calculation and the experiment in the case of InAs.

Recently, our group has proposed [65] the sp^3d^5 tight-binding model with cation-anion nearest-neighbor and anion-anion next-nearest-neighbor interactions for the $A^{II}B^{VI}$ semiconductor compounds with A = Zn, Cd, Hg and B = S, Se, Te. After a careful analysis of the band structure of $A^{II}B^{VI}$ semiconductors, it has been shown that the tight-binding model with the sp^3d^5 orbital basis on both the cation and the anion with cation-anion and anion-anion interactions describes the *ab initio* band structure extremely well. These interactions are included after a careful analysis of the crystal orbital Hamiltonian overlap populations (COHP) which give the bonding and anti-bonding contributions to the total bonding for various types of interacting pairs. It was found that the d orbitals are very important for a correct description of the top of the valence band and the bottom of the conduction band. It was found that most of the features of the *ab initio* band dispersions are describable reasonably well by the nearest-neighbor model; however, certain subtle features that cannot be explained by the nearest-neighbor model are described by including the anion-anion interactions. The next-nearest-neighbor sp^3d^5 tight-binding model gives the correct bandgap and the band curvatures near the band extremal points. The accurate description achieved for the *ab initio* band structure within this sp^3d^5 next-nearest-neighbor TB model, suggests that this may be an appropriate starting point to describe the electronic structure of $A^{II}B^{VI}$ nanocrystallites. A comprehensive calculation has indeed been carried out for $A^{II}B^{VI}$ nanocrystallites (A = Zn, Cd and B = S, Se, Te) within this sp^3d^5 parametrization. Since such calculations not only provide an estimate of the bandgap as a function of the particle size, but also yield a reliable description of the density of states due to the use of a realistic basis set and interactions, photoemission spectra can be meaningfully compared with the calculated density of states for these clusters.

The nanocrystallites contain atoms on the surface which are comparable in number to those in the bulk of the crystal. Therefore, the dangling bonds at the surface give rise to substantial number of midgap states. There are several important consequences of this, such as a strong suppression of the bandgap emission and a red-shift of the photoluminescence spectra. At the level of calculations, these midgap states complicate the identification of the band edges and, consequently, the determination of the bandgap. Thus, it is often necessary to obtain reliable routes to quenching and removing these midgap states both experimentally and theoretically. One such route is to overcoat a nanoparticle with a thin layer of another material with a larger bandgap such as coating CdS with ZnS. This experimental approach was theoretically explored in [66] for CdS with a monolayer of ZnS coating. In Figure 8 we show the density of states of unpassivated and thus passivated CdS nanocrystallites of 46 Å diameter. The midgap states observed in the unpassivated calculation (see inset in the upper panel) are removed upon passivation of the cluster (inset, lower panel). This allows a correct determination of the bandgap.

In Figure 9 we show the bandgap variation with size for the various $A^{II}B^{VI}$ semiconductors calculated within the sp^3d^5 model [67], along with the results based on EMA results and those from the sp^3s^* model [39]. The Coloumb term (the third term in Eq. (2)) has been subtracted from the calculated bandgaps in order to facilitate comparison with the experimental data [11, 12, 55, 57–59, 68–80] that are also plotted in the same figure. The infinite potential

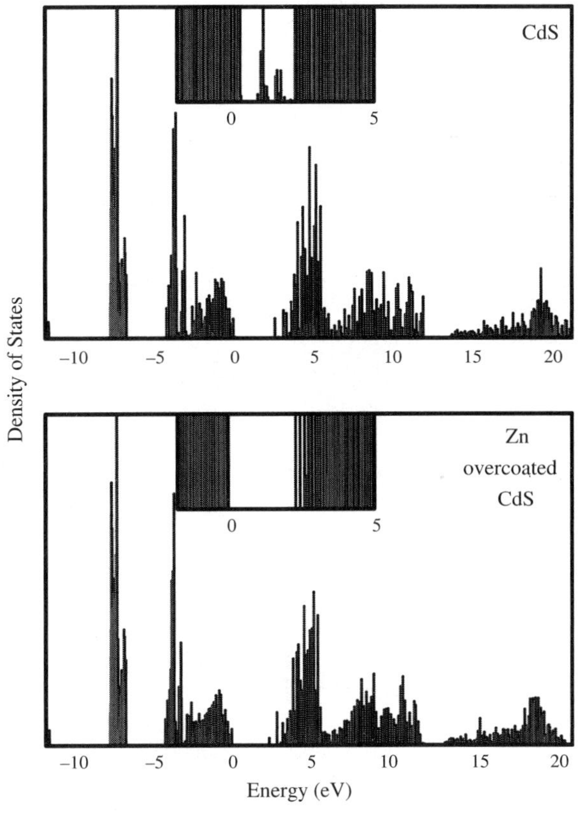

Figure 8. Unpassivated and Zn-passivated CdS nanocrystals. Passivation removes the midgap states as shown in the inset to the two panels. Reprinted from [66], S. Sapra et al., unpublished results.

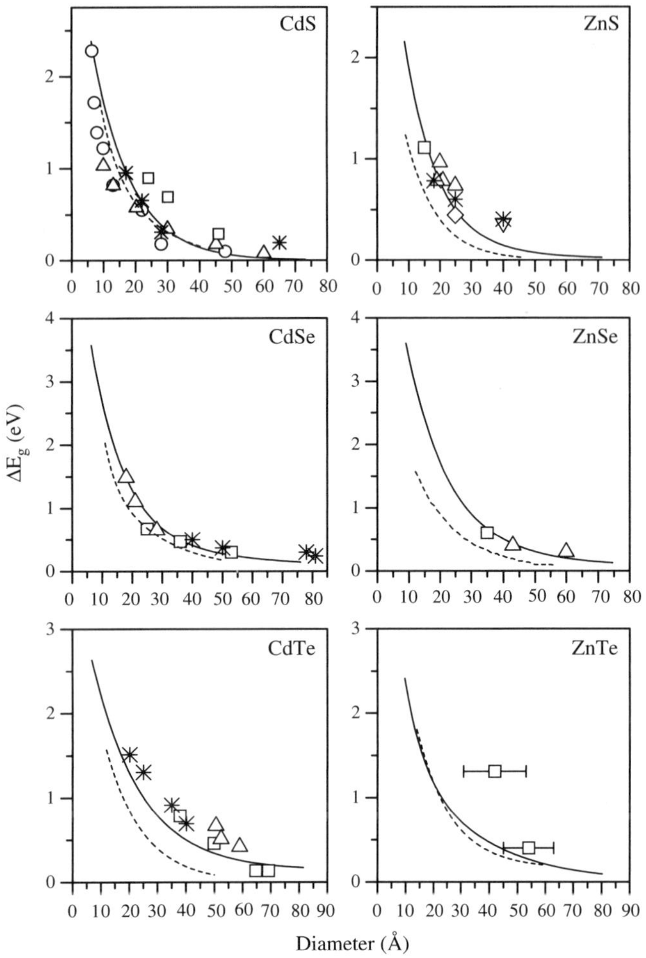

Figure 9. The sp^3d^5 TB model with the cation-anion and anion-anion interactions ([67], solid line) compared with the sp^3s^* TB nearest-neighbor model ([39], dashed line) and the experimental data points: CdS ∘ [68], △ [69], ∗ [70], squares [12]; CdSe squares [71], △ [72], ∗ [73]; CdTe squares [58], △ [59], ∗ [60]; ZnS squares [74], △ [55], ∇ [75], ◇ [76], ∗ [77]; ZnSe squares [78], △ [79]; ZnTe squares [80]. Reprinted from [67], S. Sapra and D. D. Sarma, unpublished results.

single-band EMA clearly overestimates the experimental bandgaps in every case. The results of [67] are found to describe the experimental data points considerably better than the nearest-neighbor sp^3s^* model.

4. VALENCE BAND SPECTROSCOPY

The first valence band photoemission study on nanocrystallites was reported by Colvin et al. [56] The system studied was a series of different sized CdS quantum dots on a metal surface such as gold (for 38 Å diameter cluster) or aluminum (for 24 Å and 70 Å diameter clusters) in order to avoid charging of the nanocrystallites due to electron removal from the sample. The nanoparticles were anchored to the metal surfaces using bifunctional organic groups as shown in Figure 10. The metal substrate did not contribute any observable spectral intensity possibly due to the short escape depth (10–30 Å) of the photoelectrons in the operational photon energy range. The experiments were carried

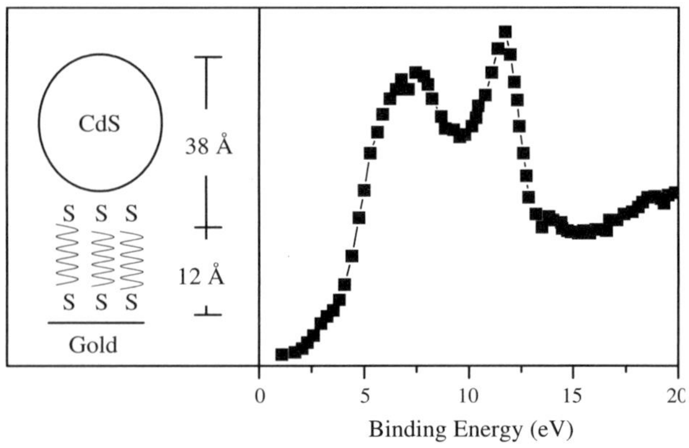

Figure 10. Photoemission spectra from CdS quantum dots bound to gold metal surface and schematic illustration of the sample. Reprinted with permission from [56], V. L. Colvin et al., *Phys. Rev. Lett.* 66, 2786 (1991). © 1991, American Physical Society.

out on a synchrotron source with variable photon energies in the range of 20–70 eV. Figure 10 shows a representative photoemission spectrum for the 38 Å CdS nanocrystallites. The data do not differ qualitatively from those of bulk CdS [81]. The valence band edges of the three differently sized nanoparticles were calculated from various spectra recorded with 50, 55, 60, and 70 eV photon energies. Extrapolation of the rising edge in each spectrum near the Fermi energy to the baseline gives the position of the valence band maximum. The experimentally determined valence band edges are plotted in Figure 11 as a function of the particle size. The solid line shows the infinite potential effective mass approximation with the effect of the dielectric material included. Due to the problem of fixing the reference of energy in absence of the corresponding bulk CdS spectrum, all the energies were rigidly shifted such that the data point for the 70 Å cluster coincides with the theoretical line. In this

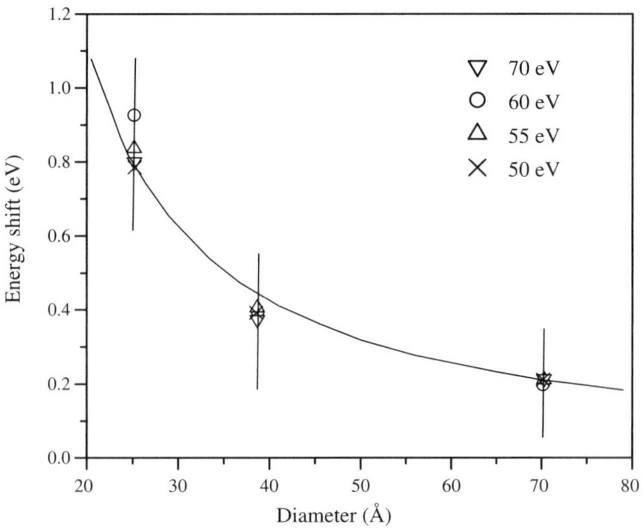

Figure 11. Energy shift of the valence band maximum vs cluster size. The shift for the largest size was fixed at 0.217 eV to match with the theoretical curve. The theory includes a dielectric polarization term as well as quantum confinement. Reprinted with permission from [56], V. L. Colvin et al., *Phys. Rev. Lett.* 66, 2786 (1991). © 1991, American Physical Society.

procedure, energy shifts for the smaller nanoparticles are essentially calculated with reference to the 70 Å cluster. Colvin et al. [56] did not observe any photoemission signal from the metal surfaces to which the nanoparticles were attached. Using this method, the chain length of the organic linker can be varied to study the changes in the photoemission spectra with the nanoparticle size.

Recently van Burren et al. carried out electronic structure measurements on nanocrystalline silicon [82]. The Si nanocrystallites were grown *in-situ* at 1700 °C in an argon buffer gas of 112 mTorr followed by hydrogen exposure to passivate the surface. The size of the crystallites was controlled by varying the temperature and Ar pressure. Photoemission measurements at a resolution of 0.25 eV and X-ray absorption studies with 0.05 eV resolution were carried out using synchrotron radiation in order to determine the valence band maximum and the bottom of the conduction band, respectively. The variations of valence band and conduction band edges are plotted in Figure 12. For Si, it is seen that the ratio of the shift of the valence band edge to the conduction band edge is always 2:1. This is in agreement with various electronic structure calculations reported earlier [83]. van Burren et al. [82] compared their experimental results with the calculated results based on pseudopotential method [84]. For all sizes of the nanoparticles, the theory was found to overestimate both the band edges and the bandgap.

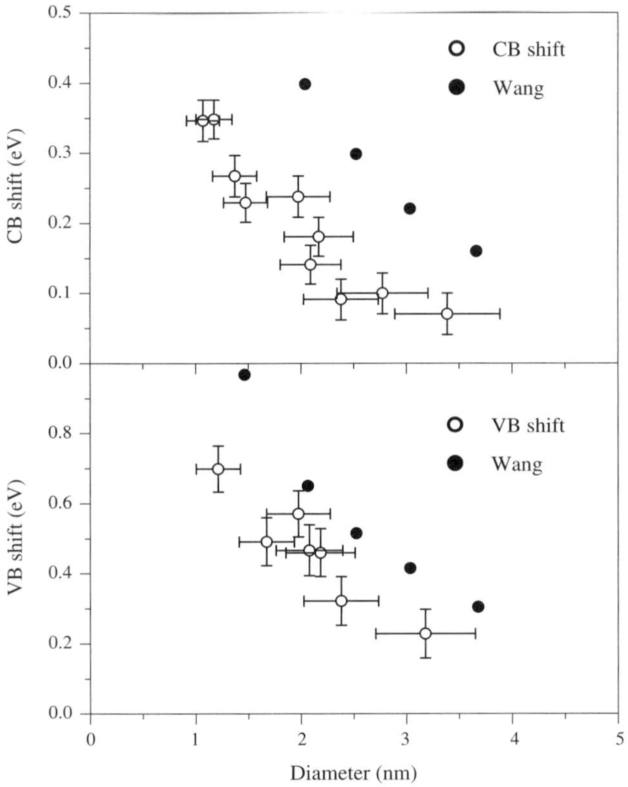

Figure 12. (a) The conduction band edge shift and (b) the valence band edge shift as a function of nanocrystal diameter. The solid circles represent the theoretically predicted VB and CB edge shifts [84] for a hydrogen terminated spherical silicon quantum dot. Reprinted with permission from [82], T. van Burren et al., *Phys. Rev. Lett.* 80, 3803 (1998). © 1998, American Physical Society.

Lüning et al. [85] studied four different sizes of CdS nanocrystallites, namely, cadmium thiolate $Cd_8(SR)_{16}$, $Cd_{17}S_4(SR)_{26}$ (13.5 Å), $Cd_{32}S_{14}(SR)_{36}$ (17.5 Å), and CdS nanoparticle of 40 Å diameter, using a combination of soft X-ray emission (SXE) spectroscopy for the valence band edge and soft X-ray absorption (SXA) spectroscopy for the conduction band edge. Figure 13 shows the SXE and SXA spectra for the various nanocrystallites along with those of bulk CdS. For small sized nanocrystallites one can easily observe a sharp feature in SXA spectra at about 163 eV; this has been attributed to a core excitonic peak near the conduction band edge. The exciton peaks are fitted with a Gaussian with a 700 meV width and subtracted from the SXA spectra to determine the conduction band edge. Apart from the bandgap variation, they also observed a systematic

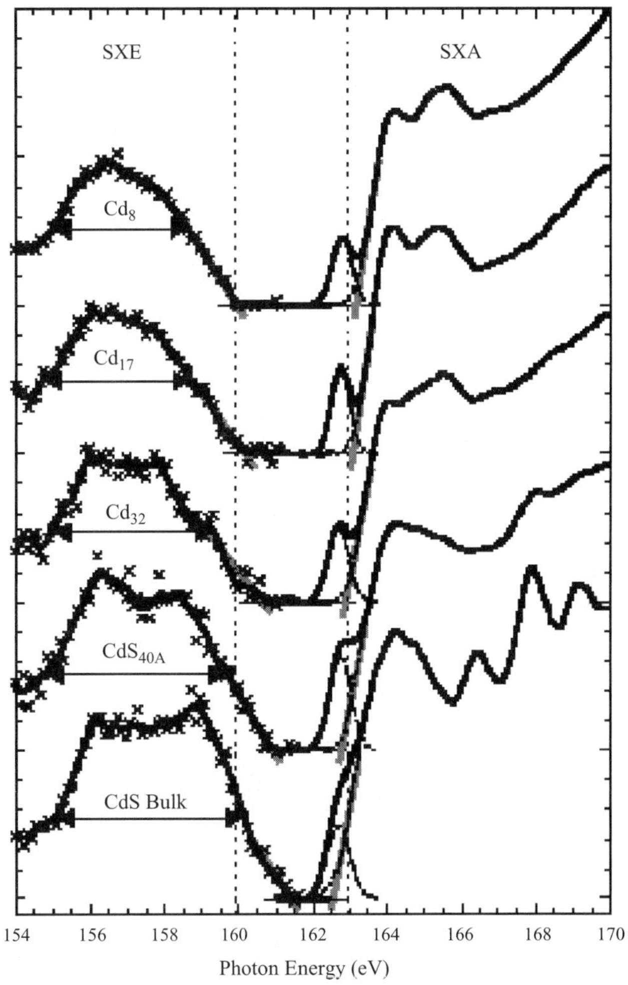

Figure 13. Right part: S $L_{2,3}$ SXA spectra (solid) of CdS bulk, a thiol-stabilized NC of 40 Å (CdS_{40}), the $Cd_{32}S_{14}(SR)_{36}$ (Cd_{32}), and the $Cd_{17}S_4(SR)_{26}$ (Cd_{17}) NC, and the Cd thiolate $Cd_8(SR)_{16}$ (Cd_8). The dotted lines show the separation into excitonic and LPDOS excitation. Left part: top part of the smoothed S L_3 SXE spectra (crosses show the non-smoothed data) mapping the local $S(s+d)$ symmetric DOS of the upper VB. The width of the upper VB is quantified by the arrows indicating the full width at half maximum of the upper VB. Gray bars indicate the linear extrapolation of the emission cutoffs and the LPDOS absorption onsets, respectively. Reprinted with permission from [85], J. Lüning et al., *Solid State Commun.* 112, 5 (1999). © 1999, Elsevier Science.

decrease in the band width of the upper part of the valence band (shown in Fig. 13), which is mainly composed of S $3p$ states; the width decreases from 4.8 eV for CdS bulk to 3.5 eV for Cd thiolate. However, the lower part of the valence band (not shown in the figure), which is derived from the more localized S $3s$ and Cd $4d$ states, is hardly affected by size quantization.

In order to separately determine the shifts in the valence band maximum (VBM) and conduction band minimum (CBM) for individual nanocrystallites, the S $2p$ core level photoemission spectra were aligned for all the spectra, serving as an internal energy reference, and the experimentally obtained shifts in the VBM and CBM are plotted in Figure 14. Also shown are the results of theoretical calculations using the tight-binding sp^3s^* model as well as using the infinite potential [15] (IP) and the finite potential [86] (FP) effective mass approximations (EMA). The FP-EMA agrees well with the experimental data whereas the TB model does not describe the data accurately. Figure 14 shows that the IP-EMA suggests a dominance of the conduction band shift in determining the variation in the total bandgap with size, with considerably lower contributions from the shifts in the valence band edge in contrast to the experimental results. It is interesting to note that EMA with a finite potential barrier changes the results more pronouncedly for the conduction band edge compared to those of IP-EMA, yielding a better agreement with the experimental results.

The efficacy of the realistic sp^3d^5 model [65, 67] has been established recently by comparing the valence band density of states of various sizes of ZnS nanocrystallites obtained from photoemission experiments with model calculations. Figure 15 shows the comparisons of experimental valence band spectra with calculated density of states for 1.8, 2.5, 3.5 nm ZnS nanoparticles and bulk ZnS. This figure shows an excellent agreement between the experimental density of states and that calculated using the sp^3d^5 model with next-nearest-neighbor interactions. It turns out that valence bands of the $A^{II}B^{VI}$ semiconductors have substantial admixtures from the cation d states. Therefore, it is essential to consider a model containing the d orbitals in order to compare the density of states to calculated results. Shifts in the valence band edges in the nanocrystallites compared to the bulk ZnS could be easily estimated from these valence band photoemission spectra and compared with the shifts calculated within the same model (Fig. 16). The shifts

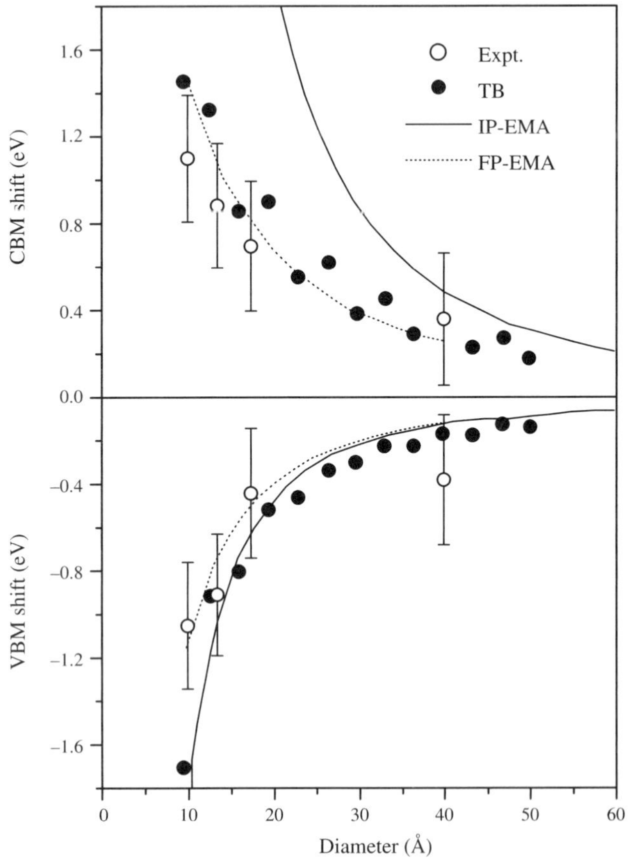

Figure 14. Experimentally determined contributions of the occupied and unoccupied states (black dots) to the total bandgap opening are compared to the EMA using infinitely high potential walls (solid lines, IP-EMA) [15], the EMA taking the finite height of the potential walls into account (dashed lines, FP-EMA) [86], and the tight-binding calculation of Lippens and Lannoo (gray diamonds, TB) [37]. Reprinted with permission from [85], J. Lüning et al., *Solid State Commun.* 112, 5 (1999). © 1999, Elsevier Science.

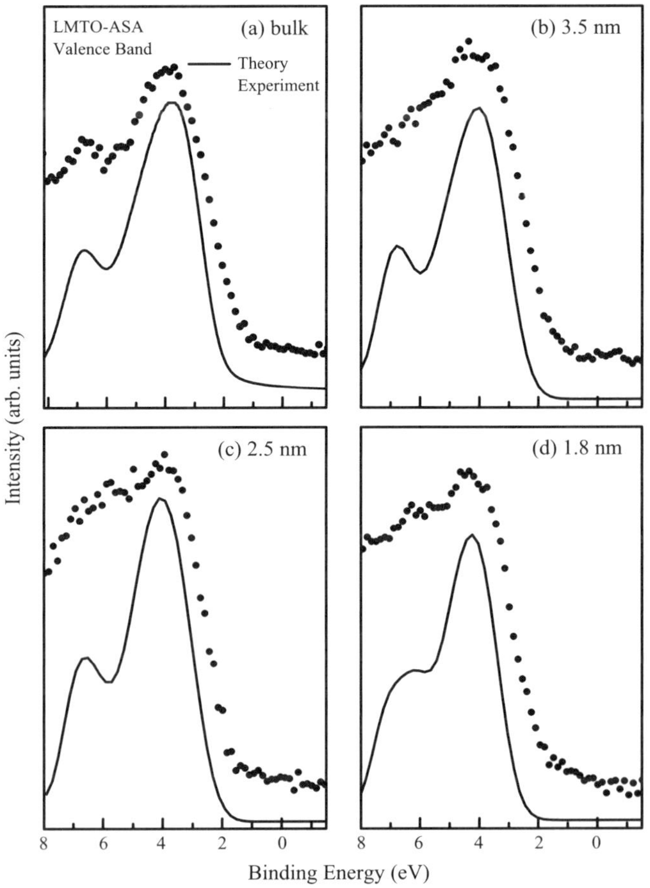

Figure 15. Valence band spectra: experimental (solid circles) and calculated spectra (solid line) for (a) bulk ZnS, (b) 3.5 nm nanocrystals, (c) 2.5 nm nanocrystals, and (d) 1.8 nm nanocrystals. The calculated spectra for bulk ZnS is obtained using the LMTO-ASA method and those for the nanocrystals are from the sp^3d^5 model with nearest-neighbor and anion-anion interactions.

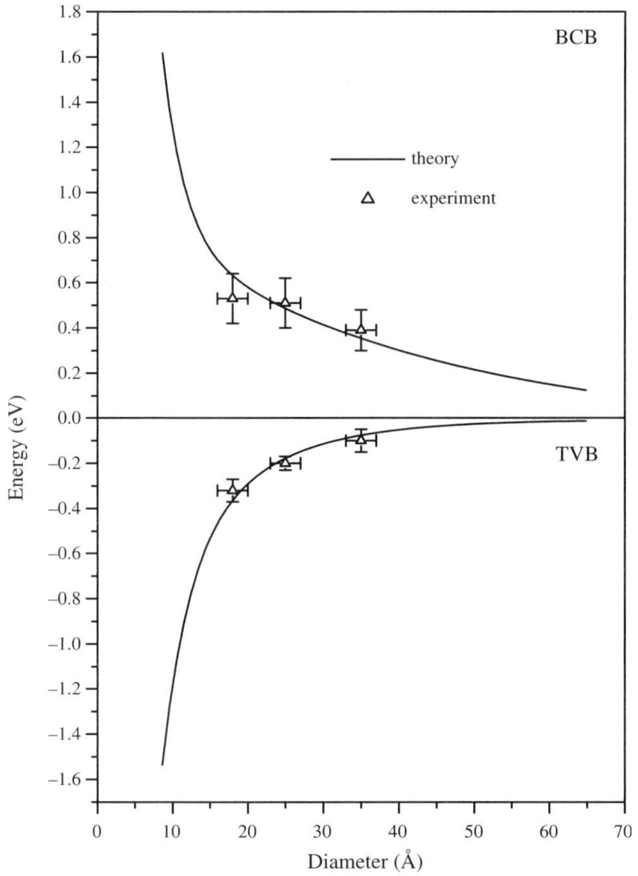

Figure 16. Variation of the top of the valence band (TVB) and bottom of the conduction band (BCB) with size as obtained using the sp^3d^5 model with nearest-neighbor and anion-anion interactions. The open-up-triangles on the TVB variation curve are experimental data points obtained from the photoemission experiments. The points on the BCB curve are obtained after subtraction of the TVB data points from the bandgap values obtained from the UV-Vis absorption.

in the conduction band edge could be extracted from the total shifts in bandgaps measured by optical absorption spectra and the shifts in the valence band edges from the photoemission spectra. Thus estimated shifts in the conduction band edges are also compared to the calculated ones as a function of size in Figure 16, indicating a good agreement.

5. CONCLUSIONS

We have reviewed here several effective and widely used techniques to describe the bandgap variation with size in semiconductor nanoparticles. The infinite potential EMA is a simple method to correlate the size of a nanoparticle with its bandgap; however, this approach overestimates the size. The finite potential EMA and the multiband EMA are improvements over the IP-EMA. The FP-EMA allows for the tunneling of charged electrons or holes outside the nanoparticles, while the MBEM approximation considers the degeneracy in the band structure at the top of the valence band. However, the EMA methods are incapable of providing a detailed electronic structure of the material.

The tight-binding method is a simple description that treats nanoparticle electronic states as a linear combination of the corresponding atomic orbitals. The initially used sp^3s^* orbital basis underestimates the bandgap and also lacks the d orbitals which are necessary to describe the density of states of the nanoparticles. The $sp^3d^5s^*$ orbital includes the d orbital in the basis and is an improvement over the sp^3s^* TB model. Inclusion of spin-orbit coupling further improves the description of the bandgap and the band edges of the nanoparticles. A good comparison with the experiment is provided by the sp^3d^5 model that includes the anion-anion interactions in addition to the usual cation-anion interactions.

In the last section, the work of various groups on valence band of the semiconductor nanoparticles has been reviewed. These experiments mainly describe the variation of the band edges and the evolution of the density of states as a function of size. Comparisons are made to the existing theoretical calculations both for the total bandgap variations as well as to shifts in the top of valence band and the bottom of the conduction band.

GLOSSARY

$A^{II}B^{VI}$ semiconductors Semiconductors with cation, A from Group IIb and the anion, B from Group VIa of the periodic table.

Band edge The top of the valence band or the bottom of the conduction band.

Bandgap The difference between the bottom of the conduction band and top of the valence band in case of semiconductors and insulators.

Band structure The relationship between the energy and the momentum vector.

Effective mass approximation (EMA) A theoretical method used to estimate the bandgap shift for nanocrystals. The conduction and the valence bands are assumed parabolic and are represented by the effective masses of the electron and the hole.

Mid gap states Electronic energy levels lying between the valence and the conduction band due to the presence of unsatisfied valencies at the surface of the nanocrystals.

Photoemission The emission of electrons from a material upon irradiation with X-rays or ultra-violet radiation.

Semiconductor nanocrystals Semiconductor nanoparticles with well defined crystalline arrangement of the atoms.

Semiconductor nanoparticles Semiconductor material particles with size in the regime 1–15 nm.

Tight binding (TB) method A calculation scheme for the electronic structure of materials based on the linear combination of atomic orbitals.

REFERENCES

1. A. D. Yoffe, *Adv. Phys.* 50, 1 (2001).
2. A. P. Alivisatos, *J. Phys. Chem.* 100, 13226 (1996).
3. A. D. Yoffe, *Adv. Phys.* 42, 173 (1993).
4. H. Weller, *Angew. Chem. Int. Ed. Engl.* 32, 41 (1993).
5. A. Hagfeldt and M. Gratzel, *Chem. Rev.* 95, 49 (1995).

6. Y. Wang and N. Herron, *J. Phys. Chem.* 95, 525 (1991).
7. C. B. Murray, C. R. Kagan, and M. G. Bawendi, *Ann. Rev. Mater. Sci.* 30, 545 (2000).
8. J. Nanda, K. S. Narayan, B. A. Kuruvilla, G. L. Murthy, and D. D. Sarma, *Appl. Phys. Lett.* 71, 1335 (1998).
9. K. S. Narayan, A. G. Manoj, J. Nanda, and D. D. Sarma, *Appl. Phys. Lett.* 74, 871 (1999).
10. W. Chen, D. Groquist, and J. Roark, *J. Nanosci. Nanotechnol.* 2, 47 (2002).
11. J. Nanda, B. A. Kuruvilla, K. V. P. Shafi, and D. D. Sarma, in "Physics of Semiconducting Nanostructures" (K. P. Jain, Ed.), p. 25. Narosa Publication, 1997.
12. J. Nanda, B. A. Kuruvilla, and D. D. Sarma, *Phys. Rev. B* 59, 7473 (1999).
13. B. O. Dabbousi, J. Rodriguez-Viejo, F. V. Mikulec, J. R. Heine, H. Mattoussi, R. Ober, K. F. Jensen, and M. G. Bawendi, *J. Phys. Chem. B* 101, 9463 (1997).
14. E. M. Chan, R. A. Mathies, and A. P. Alivisatos, *Nano Lett.* 3, 199 (2003).
15. Al. L. Efros and A. L. Efros, *Sov. Phys. Semicond.* 16, 772 (1982).
16. L. E. Brus, *J. Chem. Phys.* 79, 5566 (1983).
17. L. E. Brus, *J. Chem. Phys.* 80, 4403 (1984).
18. L. E. Brus, *J. Phys. Chem.* 90, 2555 (1986).
19. Y. Kayamma, *Solid State Commun.* 59, 405 (1986).
20. H. M. Shmidt and H. Weller, *Chem. Phys. Lett.* 129, 615 (1986).
21. S. V. Nair, S. Sinha, and K. C. Rustogi, *Phys. Rev. B* 35, 4098 (1987).
22. Y. Kayanuma, *Phys. Rev. B* 38, 9797 (1988).
23. Y. Kayanuma and H. Momiji, *Phys. Rev. B* 41, 10261 (1990).
24. D. J. Norris, A. Sacra, C. B. Murray, and M. G. Bawendi, *Phys. Rev. Lett.* 72, 2612 (1994).
25. D. J. Norris and M. G. Bawendi, *Phys. Rev. B* 53, 16338 (1996).
26. U. Banin, C. J. Lee, A. A. Guzelian, A. V. Kadavanich, A. P. Alivisatos, W. Jaskolski, G. W. Bryant, Al. L. Efros, and M. Rosen, *J. Chem. Phys.* 109, 2306 (1998).
27. G. T. Einevoll, *Phys. Rev. B* 45, 3410 (1992).
28. S. V. Nair, L. M. Ramaniah, and K. C. Rustogi, *Phys. Rev. B* 45, 5969 (1992).
29. M. V. Ramakrishna and R. A. Friesner, *Phys. Rev. Lett.* 67, 629 (1991).
30. H. Fu and A. Zunger, *Phys. Rev. B* 56, 1496 (1997).
31. A. Franceschetti and A. Zunger, *Phys. Rev. Lett.* 78, 915 (1997).
32. H. Fu and A. Zunger, *Phys. Rev. B* 55, 1642 (1997).
33. L. W. Wang and A. Zunger, *Phys. Rev. B* 51, 17398 (1995).
34. L. W. Wang and A. Zunger, *J. Chem. Phys.* 100, 2394 (1994).
35. L. W. Wang and A. Zunger, *J. Phys. Chem.* 94, 2158 (1994).
36. L. W. Wang and A. Zunger, *Phys. Rev. B* 53, 9579 (1996).
37. P. E. Lippens and M. Lannoo, *Phys. Rev. B* 39, 10935 (1989)
38. P. E. Lippens and M. Lannoo, *Phys. Rev. B* 41, 6079 (1990).
39. P. E. Lippens and M. Lannoo, *Mat. Sci. Eng. B* 9 485 (1991).
40. N. A. Hill and K. B. Whaley, *J. Chem. Phys.* 99, 3707 (1993)
41. L. M. Ramaniah and S. V. Nair, *Phys. Rev. B* 47, 7132 (1993).
42. N. A. Hill and K. B. Whaley, *J. Chem. Phys.* 100, 2831 (1994).
43. N. A. Hill and K. B. Whaley, *Phys. Rev. Lett.* 75, 1130 (1995).
44. G. Allan, Y. M. Niquet, and C. Delerue, *Appl. Phys. Lett.* 77, 639 (2000).
45. J. Pérez Conde, A. K. Bhattacharjee, M. Chamarro, P. Lavalland, V. D. Petrikov, and A. A. Lipovskii, *Phys. Rev. B* 64, 113303 (2001).
46. J. Pérez Conde and A. K. Bhattacharjee, *Phys. Rev. B* 63, 245318 (2001).
47. C. Kittel, "Introduction to Solid State Physics," 5th ed. Wiley, New York, 1976.
48. A. I. Ekimov, Al. L. Efros, and A. A. Onuschenko, *Solid State Commun.* 56, 921 (1985).
49. D. Brust, J. C. Phillips, and F. Bassani, *Phys. Rev. Lett.* 9, 94 (1962).
50. M. L. Cohen and J. R. Chelikowsky, "Electronic Structure and Optical Properties of Semiconductors." Springer, Berlin, 1989.
51. J. C. Slater and G. F. Koster, *Phys. Rev. B* 94, 1498 (1954).
52. E. Dagotto, *Rev. Mod. Phys.* 66, 763 (1994).
53. P. Vogl, H. P. Hjalmarson, and J. D. Dow, *J. Phys. Chem. Solids* 44, 365 (1983).
54. R. Rossetti, J. L. Ellison, J. M. Gibson, and L. E. Brus, *J. Chem. Phys.* 80 4464 (1984).
55. R. Rossetti, R. Hull, J. M. Gibson, and L. E. Brus, *J. Chem. Phys.* 82, 552 (1985).
56. V. L. Colvin, A. P. Alivisatos, and J. G. Tobin, *Phys. Rev. Lett.* 66, 2786 (1991).
57. T. Rajh, O. I. Micic, and A. J. Nozik, *J. Phys. Chem.* 97, 11999 (1993).
58. Y. Mastai and G. Hodes, *J. Phys. Chem. B* 101, 2685 (1997).
59. Y. Masumoto and K. Sonobe, *Phys. Rev. B* 56, 9734 (1997).
60. H. Arizpe-Chrez, R. Ramrez-Bon, F. J. Espinoza-Betrn, O. Zelaya-Angel, J. L. Marn, and R. Riera, *J. Phys. Chem. Solids* 61, 511 (2000).
61. A. M. Kapitonov, A. P. Stopak, S. V. Gaponenko, E. P. Petrov, A. L. Rogach, and A. Eychmuller, *J. Phys. Chem. B* 103, 10109 (1999).
62. M. Bruchez, Jr., M. Moronne, P. Gin, S. Weiss, and A. P. Alivisatos, *Science* 281, 2013 (1998).
63. O. I. Micic, J. Sprague, Z. Lu, and A. J. Nozik, *Appl. Phys. Lett.* 68, 3150 (1996).
64. O. I. Micic, H. M. Cheong, H. Fu, A. Zunger, J. R. Sprague, A. Mascarenhas, and A. J. Nozik, *J. Phys. Chem. B* 101, 4904 (1997).
65. S. Sapra, N. Shanthi, and D. D. Sarma, *Phys. Rev. B* 66, 205202 (2002).
66. S. Sapra, R. Viswanatha, and D. D. Sarma (unpublished results).
67. S. Sapra and D. D. Sarma (unpublished results); S. Sapra and D. D. Sarma, "Proceedings of the 3rd Japan-Korea Joint Workshop on First-Principles Electronic Structure Calculations," 2001, p. 105.
68. T. Vossmeyer, L. Katsikas, M. Giersig, I. G. Popovic, K. Diesner, A. Chemseddine, A. Eychmuller, and H. Weller, *J. Phys. Chem.* 98, 7665 (1994).
69. Y. Wang and N. Herron, *Phys. Rev. B* 42, 7253 (1990).
70. T. Torimoto, H. Kontani, Y. Shibutani, S. Kuwabata, T. Sakata, H. Mori, and H. Yoneyama, *J. Phys. Chem. B* 105, 6838 (2001).
71. B. O. Dabbousi, C. B. Murray, M. F. Rubner, and M. G. Bawendi, *Chem. Mater.* 6, 216 (1994).
72. A. L. Rogach, A. Kornowski, M. Gao, A. Eychmuller, and H. Weller, *J. Phys. Chem. B* 103, 3065 (1999).
73. S. Gorer and G. Hodes, *J. Phys. Chem.* 98, 5338 (1994).
74. Y. Nakaoka and Y. Nosaka, *Langmuir* 13, 708 (1997).
75. H. Inoue, N. Ichiroku, T. Torimoto, T. Sakata, H. Mori, and H. Yoneyama, *Langmuir* 10, 4517 (1994).
76. S. Yanagida, T. Yoshiya, T. Shiragami, and C. Pac, *J. Phys. Chem.* 94, 3104 (1990).
77. J. Nanda, S. Sapra, D. D. Sarma, N. Chandrasekharan, and G. Hodes, *Chem. Mater.* 12, 1018 (2000).
78. F. T. Quinlan, J. Kuther, W. Tremel, W. Knoll, S. Risbud, and P. Stroeve, *Langmuir* 16, 4049 (2000).
79. M. A. Hines and P. Guyot-Sionnest, *J. Phys. Chem. B* 102, 3655 (1998).
80. Y.-W. Jun, C.-S. Choi, and J. Cheon, *Chem. Commun.* 101 (2001).
81. L. Ley, R. A. Pollak, F. R. McFeely, S. P. Kowalczyk, and D. A. Shirley, *Phys. Rev. B* 9, 600 (1974).
82. T. van Burren, L. N. Dinh, L. L. Chase, W. J. Sickhans, and L. J. Terminello, *Phys. Rev. Lett.* 80, 3803 (1998).
83. T. van Burren, T. Tiedje, J. R. Daln, and B. M. Way, *Appl. Phys. Lett.* 63, 2911 (1993).
84. L. W. Wang and A. Zunger, in "Nanocrystalline Semiconductor Materials" (P. V. Kamat and D. Meisel, Eds.). Elsevier Science, New York, 1996.
85. J. Lüning, J. Rockenberger, S. Eisebitt, J.-E. Rubensson, A. Karl, A. Kornowski, H. Weller, and W. Eberhardt, *Solid State Commun.* 112, 5 (1999).
86. D. Schoos, A. Mews, A. Eychmuller, and H. Weller, *Phys. Rev. B* 49, 17072 (1994).

Encyclopedia of Nanoscience and Nanotechnology
www.aspbs.com/enn

Electronic Tunnelling Time in Nanostructures

V. Gasparian
California State University, Bakersfield, California, USA

M. Ortuño, O. del Barco
Universidad de Murcia, Spain

CONTENTS

1. INTRODUCTION

During the last few years key subjects and, consequently, terms in materials research have been continuously changing, indicating a tendency towards smaller and smaller scales. The physics of "low-dimensional structures" was replaced by the discipline of "submicron physics," emphasizing the effects due to a reduction of size. Then the term "mesoscopic systems" was introduced referring to typical length-scales ranging from a few nanometers up to a few micrometers. Mesoscopic systems are so small, that a complete quantum-mechanical treatment of electrons is required if one wants to describe their transport properties. On the other hand, they are so large, that an exact microscopic description, starting from precise location of impurities and sample boundaries, is not useful, since only the slightest change of the mesoscopic details will completely change the result. More recently still, the terms "nanophase" and "nanostructured materials" have become popular, indicating that scientists had learned to manipulate, synthesize, analyze, and observe objects approaching the molecular and atomic scales.

Often one distinguishes between "physical" and "chemical nanostructures." Under the term "physical nanostructures" are classified all artificially built up structures, as obtained, for example, by evaporation and subsequent deposition of materials. On the other hand, the term "chemical nanostructures" comprises all those nanophase materials that can be obtained by methods of chemical synthesis, such as the chemical compounds with chainlike or layer-type structures, as well as the cluster compounds. A review of the electronic properties of nanophase materials obtained from chemical synthesis was given by de Jongh [1]. A survey of chemically synthesized metal clusters was edited by Schmid [2]. The prospect of the applications of metal and semiconductor clusters in inorganic host structures was presented by Simon and Schön [3].

The time spent by a particle in a given region of space is not a new problem, although recently it has attracted a great deal of interest [4–16]. The problem has been approached from many different points of view, and there exists a huge literature on the tunnelling problem of electrons through a barrier, although tunnelling times have continued to be controversial even until now. As pointed out by Landauer and Martin [11], there is no clear consensus about simple expressions for the time in quantum mechanics (QM), where there is not a Hermitian operator associated with it. The problem of the tunnelling time of single electrons (SE) in nanostructures or in mesoscopic systems smaller than 10 nm becomes even more complicated, due to the Coulomb blockade effects [17] on small amounts of electrons and discreteness of electric charge.

In the present review we present the theoretical approaches to tunnelling times to illustrate the problems involved in nanostructures. But this plan proved to be more difficult than expected: although there exists an extended literature on tunnelling times (see [11] and [4] and references therein),

ISBN: 1-58883-059-4/$35.00

quantum-mechanical treatments mostly deal with propagating wavepackets in a more or less general way and they do not concentrate on tunnelling times in nanostructures. As far as we know, there is not yet a proper treatment about tunnelling times in very small nanostructures with single, localized electrons, where the radius of "localization" is in the same order of magnitude as the length of the barrier L. Thus it proved to be necessary not simply to present the existing models and theories on tunnelling time but to review them with respect to these necessities together with possible alternatives and to estimate future developments. In this context some original works will be analyzed from this point of view. We will be particularly concerned with the closed analytical treatment based on Green's function formalism.

In the main part of the chapter we study the Larmor clock approach to tunnelling time, based on measuring the spin rotation of an electron under a weak magnetic field acting on the region of interest. We also develop a Green's function formalism for the traversal, reflection, and dwell times based on the previous approach. The latter corresponds to the amount of time that a particle spends in a region independently of whether it later is transmitted or reflected. We review the rest of the existing major approaches to the time problem in Section 3. In Section 4, we present numerical results about the traversal time in rectangular barriers including finite size effects. We finally extract some conclusions and present open questions.

1.1. Tunnelling Time

Tunnelling refers to the classically impossible process of a particle to penetrate an energy barrier when its energy is smaller than the maximum of the potential of the barrier. The main magnitudes involved in the problem are the height V_0 and the length L of the potential barrier. If they are large, the probability to penetrate the barrier is very small and we say that it is an opaque barrier. Examples of tunnelling processes are α-decay, transmission of electromagnetic waves (EMW) in undersized waveguides, and tunnelling of electrons. A quantum particle usually is said to have an intrinsic "wave nature," often paraphrased by "wave functions" or "wavepackets." The probability to penetrate a barrier, which is quantified through the transmission coefficient (probability) T, strongly depends on the nature of the exponential decay of the wave function under the barrier.

Tunnelling of electrons has been of utmost importance for all fast effects in ME. The first device used as a fast switch was the semiconductor tunnel-diode which was commercially introduced in the late 1950s [18]. But again, only its total relaxation time was of interest and not the pure tunnelling time through the bounding barrier. Until relatively recently, little attention was paid to Hartman's theoretical work on tunnelling time of wavepackets in the 1960s [19]. His main striking result was that under certain circumstances (opaque barrier) the tunnelling time is independent of L and the traversal time can be less than the time that would be required to travel a distance equal to L in vacuum. Similar results were found by Rybachenko [20] for electrons in a rectangular barrier. Although these were excellent pioneer works, 30 years ago time was not ripe for a further evaluation with respect to practical consequences in ME or even to philosophical ones. Additionally, many physicists hesitated to deal with Hartman's results since a very fast tunnelling, or a zero tunnelling time, holds a serious consequence: the tunnelling velocity or the average velocity may become higher than the light velocity c. Thus superluminal speed can be expected [21, 22] or measured in some cases like in experiments where electromagnetic waves pass through a barrier [23–27] or through an optical gap [9, 10, 16].

Since SE tunnelling processes could be evaluated in many nanostructures [28–31], it provided a strong motive for advancing nanofabrication technologies and research on tunnelling, which become important even at room temperature, since the operating temperature of single electron devices is directly related to the geometrical size of the electron localization. In SE the discreteness of the electric charge becomes essential and a quantum-mechanical tunnelling of electrons in a system of rather opaque junctions can be much affected by Coulomb interactions. For ME purposes the electron interactions, the barrier height and shape, and thus the tunnelling probability can be varied at will by externally applied voltages or by injected charges.

Usually in QM we can only measure quantities for which we have introduced a Hermitian operator, for example, energy E, momentum p, coordinate y, and so on. For these quantities, expectation values can be calculated and checked experimentally. However, time appears in the standard quantum-mechanical approach only as a parameter and therefore its expectation value is not defined. Since the beginning of QM, people have been aware of the conceptual problem of how to introduce a time operator with an appropriate classical analog, and there have been different theoretical approaches to find a consistent description of this problem [4, 11, 32].

Moreover, according to QM a particle under a barrier, with energy E smaller than V_0, can only be observed with a strong inelastic in uence. If we fix its coordinate with an accuracy of Δy smaller than the length of the barrier L, it necessarily results in a variation of momentum, caused by the measurement, and correspondingly in a change of the kinetic energy of the particle. This change in energy must be greater than the energy difference between the barrier height V_0 and the energy of the particle E [33]. If such a measurement would be carried out by light quanta, then we would have $\hbar\omega \geq V_0 - E$. The latter result demonstrates impressively that it is practically impossible to measure the propagation time from one coordinate (position) to the next under a barrier. This means that in practice one must try to observe the particle outside the barrier, say left or right of the region of interest. For short wavepackets, where the length of the wavepacket approaches the barrier length L, this means "far" left and "far" right.

One can associate the traversal time with the time during which a transmitted particle interacts with the region of interest, as measured by some physical clock which can detect the particle's presence after leaving the region. For electrons, this approach can utilize the Larmor precession frequency of the spin produced by a weak magnetic field hypothetically acting within the barrier region [20, 34–36]. Similar procedures have been developed for electromagnetic waves in [37], where was proposed a clock based on the Faraday effect to measure their interaction time in a slab.

Another approach is to calculate the traversal time of a particle through a barrier by following the behavior of a wavepacket and determining the delay due to the structure of the region. In this approach one has to be careful with the interpretation of the results, since, for example, an emerging peak is not necessarily related to the incident peak in a causative way [38]. For more discussions on this problem see, for example, [11] and references therein. Martin and Landauer [39] studied the problem of the traversal time of classical evanescent electromagnetic waves by following the behavior of a wavepacket in a waveguide, and Ruiz and co-workers [15, 40] analyzed their behavior in the optical gap of a periodic structure. Japha and Kurizki [41] used the Faraday effect as a quantum clock for evanescent waves and studied its implications on two-photon correlations.

The problem of defining velocities is equally complicated as that of determining the time. One cannot use just one definition for the velocity both inside and outside the barrier at the same time. Usually, for a quantum particle when going from sub-barrier region to above-barrier region, one can do analytical continuation of the wave function. But in the first case there is an exponential decay of the wave function and in the second case we deal with a free propagation of the electron and so a wave function with oscillations. This analytical continuation is not correct for the velocity under the barrier, because one gets an imaginary velocity. So there is no definition of velocity for sub-barrier regions and as a consequence, in the limit of an opaque barrier or in the forbidden gap of a periodic system, there may be observed a "superluminal" speed.

1.2. Wavepacket Approach and Limits

The simplest model which illustrates the tunnelling problem for a quantum particle is a plane wave incident on a one-dimensional (1D) barrier. Part of the plane wave is reflected and part is transmitted. The above plane wave, which represents the electron in our model, is by nature infinitely large in space. The discussion of whether wavepackets with an infinite extension model the wave function of single photons or electrons and of whether they might be interpreted as signals is complicated [4, 11] and important for ME with nanostructures. Therefore it is better to consider a finite wavepacket and to look at its peak evolution in time (see Fig. 1). Thus the phase time is the time which elapses between the peak of the wavepacket entering the barrier and leaving it and can be defined as the energy derivative of the phase:

$$\tau_\varphi = \hbar \frac{d\varphi}{dE} \tag{1}$$

In some cases this time can be easily calculated, but as was mentioned before, it will lead us to Hartman's effect. We will see (Section 2) that, in general, more than one tunnelling time are involved in the problem: τ_y, τ_z, and the so-called Büttiker-Landauer $\tau^{\mathrm{BL}} = \sqrt{(\tau_y)^2 + (\tau_z)^2}$. Unfortunately this time is not additive in the sense, that when dividing the length of the barrier $L(y)$ arbitrarily into different parts, the total τ^{BL} tunnelling time is not the sum of the individual tunnelling times.

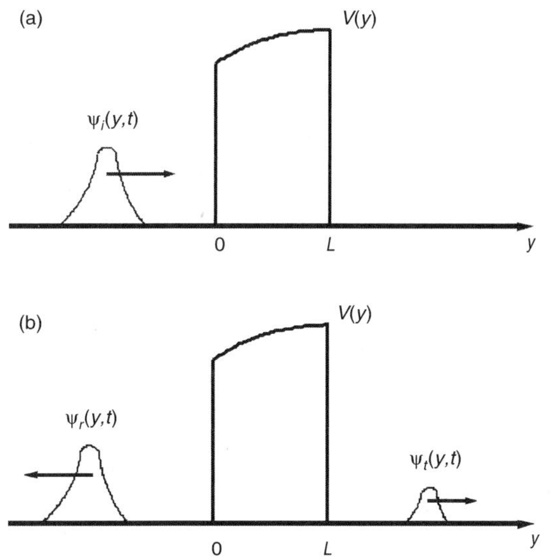

Figure 1. (a) The initial wavepacket enters from the left onto the barrier. (b) Transmitted and reflected wavepackets are moving away from the barrier in opposite directions.

To illustrate consequences from the above wavepacket model, let us consider a sharply peaked Gaussian wavepacket in space which starts to enter very far from the barrier to exclude any interaction. A wavepacket is an overlap of many plane waves with different wave numbers k. Hence, one may imagine the wavepacket as something like a group of electrons with different energies and velocities. The propagation will be dispersive and as a consequence the high-energy components of the packet will reach the barrier first. Due to the fact that higher energies can be transmitted more effectively than the low-energy components, the peak of the transmitted packet can leave the barrier long before the peak of the incident packet has arrived [11].

But what is really surprising is that even for the sub-barrier tunnelling, that is, when the wavepacket contains no energy components with energies above the rectangular barrier of height V_0, the transmitted packet will have a higher mean velocity than the free space propagation velocity. Numerical simulations show that one obtains very short tunnelling times when the spread of the Gaussian wavepacket is larger than the barrier width L. Within these restrictions even the simple rectangular barrier is an "electron accelerator." This is a manifestation of the aforementioned Hartman effect (see Section 2) which was treated by Rybachenko [20] for spin particles with analogous results. As we will show (Section 2), the tunnelling time component τ_y is independent of L and can be less than the time that would be required for a free particle to travel a distance equal to the barrier thickness L.

As a matter of principle, in ME for high information data rate, the spread of the Gaussian wavepacket must be small. Tunnelling time(s) then will depend to a higher degree on the size of the incident wavepacket and the shape of the barrier: tunnelling will become more sensitive to boundary effects at the barrier.

For the future SE logics in nanostructures, "pure" quantum-mechanical properties of monochromatic single electrons with an energy less than 1 eV will be predominant. For such particles the de Broglie wavelength $\lambda_B = \hbar/p$ will be in the same order of magnitude as the length of the barrier. At the same time, this wavelength will be comparable to the radius of localization (e.g., for electrons confined in the core of ligand-stabilized microclusters) in switches or mass memories (stores). Physically speaking, this means that the picture of a dispersive wavepacket is now failing. As far as we know, fast tunnelling for this case has not been treated theoretically until now.

1.3. Phase Time and Superluminal Velocity in Periodic Nanostructures

In ME the simultaneous transmission of electrons and of microwavepackets of selected optical signals between integrated microchips on wafers is of great importance and today much attention is paid to optoelectronics, since on the way to future ultimate miniaturization, the present generation of devices hopefully will be replaced by nanostructured systems.

While at the end of the preceding section we sketched the problems with localized electrons for the future SE logics in nanostructures, in the present section we first pay attention to microwaves in undersized waveguide barrier-systems and then to photons propagating in 1D periodic and quasi-periodic Fibonacci and Thue-Morse systems. We briefly report about the former first "superluminal" experiment, but we are mainly interested in the latter as there exists a considerable analogy between these periodic systems and, for example, chains of the above chemical nanostructures. So there also must exist forbidden bandgaps where electrons may propagate with "superluminal" speed. It must be noted that periodic structures can be easily built up in crystals of ligand-stabilized microclusters or likewise in chains or layers of supported cluster arrangements on structured wafers or other substrates or even in channels or layer spaces of porous chemical nanostructures (see [3]).

In order to avoid the problems involved with the dispersive nature of the electron's wavepacket and the invasive measuring process in QM, it was easier to look at a gaussian wavepacket of classical electromagnetic waves and to try to measure the delay time at a barrier. Indeed, in most of the past tunnelling experiments, instead of electrons, electromagnetic waves were used [8, 23], to exclude any electronic interaction with the tunnelling barrier. The analogy between the time-independent forms of the Schrödinger and the Maxwell equations confronts us again with Hartman's case: the possibility of achieving extremely high tunnelling velocities, even superluminal velocities.

Thus looking back, it was not so surprising that the actual discussion on "superluminal" speed started almost at the same time with the series of microwaves experiments by transmission through systems consisting of undersized waveguides [7, 8, 23–27]. Steinberg et al. [9] found "superluminal" velocities for electromagnetic waves in the photonic bandgap of multilayer dielectric mirrors. Spielman et al. [10] observed that the barrier traversal time of electromagnetic wavepackets tends to become independent of the barrier thickness for opaque barriers. This phenomenon is closely related to Hartman's theoretical prediction for electron tunnelling [19]. The theoretical explanation of this phenomenon can be found in the framework of classical Maxwell equations by following the time evolution of the wavepacket in time, as it was mentioned above (see, e.g., [22, 36, 40]). It was clear that parts of the microwavepacket were able to propagate with "superluminal" speed, proving the practical use of Hartman's effect.

The propagation of electromagnetic waves in 1D quasi-periodic Fibonacci and Thue-Morse systems was studied in [14]. It was shown that, under certain conditions, again the phase time becomes independent of the system size and so "superluminal" group velocities can be obtained for very narrow-frequency-band wavepackets.

2. LARMOR CLOCK APPROACH

Baz' [34, 35] proposed the use of the Larmor precession to measure the time spent by a spin-1/2 particle inside a sphere of radius $r = a$. He considered the effect of a weak homogeneous magnetic field **B** inside the sphere on an incident beam of particles of mass m and kinetic energy $E = \hbar^2k^2/2m$. Let us assume that the magnetic field is directed along the z axis and the incoming particles move along the y axis with their spin polarized along the x axis. As soon as a particle enters the sphere, its magnetic moment will start precessing about the field vector with the well-known Larmor frequency $\omega_L = 2\mu B/\hbar$. The precession will go on as long as the particle remains inside the sphere. The polarization of the transmitted (and reflected) particles is compared with the polarization of the incident particles. The angle $\theta_\perp$ in the plane xy, perpendicular to the magnetic field, between the initial and final polarizations is assumed to be given, in the lowest order in the field, by the Larmor frequency ω_L multiplied by the time τ_y spent by the particle in the sphere

$$\theta_\perp = \varphi_L \tau_y \tag{2}$$

The change in polarization thus constitutes a Larmor clock to measure the interaction time of the particles with the region of interest.

Rybachenko [20] considered the simpler problem of the interaction time of particles with a one-dimensional rectangular barrier of height V_0 and width L, for which everything can be calculated analytically. Rybachenko thought that the spin, in first order in the field, remains in the xy plane. For an opaque barrier, where there is a strong exponential decay of the wave function, he found a characteristic interaction time τ_y given by

$$\tau_y = \frac{\hbar k}{V_0 \xi} \tag{3}$$

where ξ is the inverse decay length in the rectangular barrier

$$\xi = (k_0^2 - k^2)^{1/2} \tag{4}$$

with $k_0 = (2mV_0)^{1/2}/\hbar$. This characteristic time τ_y is independent of the barrier thickness L. Instead of being proportional to the length, L is proportional to the decay length.

For an opaque barrier this decay length can become very short and so τ_y can be very small, in fact, smaller than the time that would be required for the incident particle to travel a distance L in the absence of the barrier. A similar result was found by Hartman [19] analyzing the tunnelling of a wavepacket through a rectangular potential barrier, which is known as Hartman's effect.

2.1. Büttiker Approach

Büttiker [36] argued that the main effect of the magnetic field is to tend to align the spin parallel to it in order to minimize the energy. It means that a particle tunnelling through a barrier in a magnetic field does not only perform a Larmor precession, but also a spin rotation produced by the Zeeman effect, which necessarily has to be included in the formalism. The idea behind this Zeeman rotation is the following. A beam of particles polarized in the x direction can be represented as a mixture with equal probabilities of particles with their z component equal to $\hbar/2$ and to $-\hbar/2$. In the barrier the kinetic energy differs by the Zeeman contribution $\pm\hbar\omega_L/2$, giving rise to a different exponential decay of the wave function depending on its spin component along the direction of the magnetic field. In the limit of small fields we have

$$\xi_\pm = \left(k_0^2 - k^2 \mp \frac{m\omega_L}{\hbar}\right)^{1/2} \cong \xi \mp \frac{m\omega_L}{2\hbar\xi} \tag{5}$$

where the sign indicates whether the z component of the spin is parallel (+) or antiparallel (−) to the field. The particles with spin $\hbar/2$ will penetrate the barrier more easily than the particles with spin $-\hbar/2$, and so the transmitted particles will have a net z component of the spin. This net component of the spin along the direction of the field defines a second characteristic time τ_z of the particle in the barrier.

For each of the spin components, one can define a characteristic time describing the interaction of the tunnelling particle with the barrier:

$$\lim_{\omega_L\to 0}\langle S_x\rangle = \frac{\hbar}{2}\left[1 - \frac{\omega_L^2(\tau_x^{BL})^2}{2}\right] \tag{6}$$

$$\lim_{\omega_L\to 0}\langle S_y\rangle = -\frac{\hbar}{2}\omega_L\tau_y^{BL} \tag{7}$$

$$\lim_{\omega_L\to 0}\langle S_z\rangle = \frac{\hbar}{2}\omega_L\tau_z^{BL} \tag{8}$$

Only two of these characteristic times are independent. The spin expectation values $\langle S_x\rangle$, $\langle S_y\rangle$, and $\langle S_z\rangle$ can be obtained in terms of the transmission amplitude for particles with $S_z = \pm\hbar/2$

$$t_\pm \equiv \sqrt{T_\pm}e^{i\varphi_\pm} \tag{9}$$

where $T_\pm$ and $\varphi_\pm$ are the corresponding transmission coefficient and phase, respectively.

For the special case of a 1D rectangular barrier, it is possible to find exact analytical expressions for the time. For energies smaller than the height of the barrier, $E < V_0$, Büttiker [36] obtained the following expression for the characteristic time associated with the direction parallel to the field τ_z^{BL}:

$$\tau_z^{BL} = -\frac{m}{\hbar\xi}\frac{\partial \ln T^{1/2}}{\partial\xi} \tag{10}$$

For the time τ_y^{BL} associated with the direction of propagation, perpendicular to the field, he found

$$\tau_y^{BL} = -\frac{m}{\hbar\xi}\frac{\partial\varphi}{\partial\xi} \tag{11}$$

Here T and φ are, respectively, the transmission coefficient (probability) and the phase accumulated by transmitted particles due to the rectangular barrier in the absence of the magnetic field. In Figure 2 we represent τ_y^{BL} and τ_z^{BL} for a rectangular barrier.

Büttiker assumed that the relevant interaction time depends on the times associated to both the Larmor precession and the Zeeman splitting, and is given by

$$\tau^{BL} = \left\{\left(\tau_y^{BL}\right)^2 + \left(\tau_z^{BL}\right)^2\right\}^{1/2} \tag{12}$$

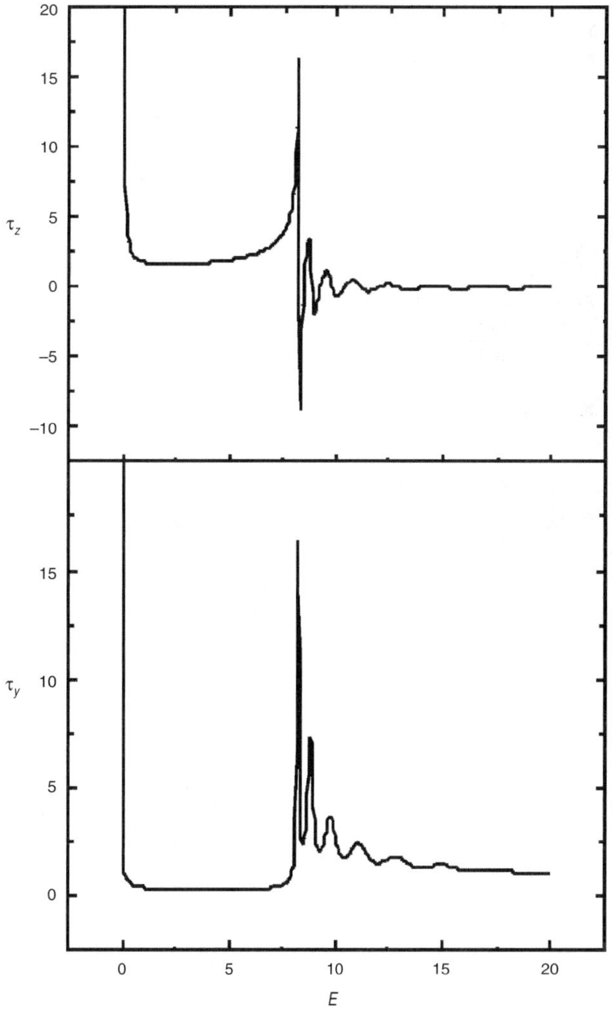

Figure 2. Components τ_y^{BL} and τ_z^{BL} of the Büttiker-Landauer time for a rectangular barrier. For energies higher than the barrier potential, both components oscillate with energy.

This traversal time is the so-called Büttiker-Landauer (BL) time for transmitted particles. Although it was obtained in the context of tunnelling, it is a general definition which applies for the traversal time of a particle or an electromagnetic wave through any given region of space. For a rectangular barrier, it is then given by

$$\tau^{BL} = \frac{m}{\hbar\xi}\left\{\left(\frac{\partial \ln T^{1/2}}{\partial \xi}\right)^2 + \left(\frac{\partial \varphi}{\partial \xi}\right)^2\right\}^{1/2} \quad (13)$$

When the energy E of an incident particle is well below the barrier height V_0 of an opaque rectangular barrier, Büttiker's result (13) is approximately equal to

$$\tau^{BL} \simeq \frac{mL}{\hbar\xi} \quad (14)$$

which is very different from the result of Rybachenko, Eq. (3). It is, however, in agreement with the traversal time obtained by Büttiker and Landauer [42] based on the transition from adiabatic to sudden limits for a time-modulated rectangular opaque barrier.

2.2. Green's Functions Method

We now derive a general expression for the traversal time using the Green's Function (GF) method [43, 44]. We will consider a 1D system with an arbitrary potential $V(y)$ confined to a finite segment $0 < y < L$, which we will call "the barrier." As for a rectangular barrier, we apply a weak magnetic field $\mathbf{B}$ in the z direction and confined to the barrier (see Fig. 3).

Our electron is incident on the barrier from the left with an energy E and with its spin polarized along the x direction. In the presence of the magnetic field, the Schrödinger equation takes the form

$$\left(-\frac{\hbar^2}{2m}\frac{d^2}{dy^2} + V(y) - E\right)\widehat{\Psi}(y) = -\mu B \begin{pmatrix} 1 & 0 \\ 0 & -1 \end{pmatrix} \widehat{\Psi}(y) \quad (15)$$

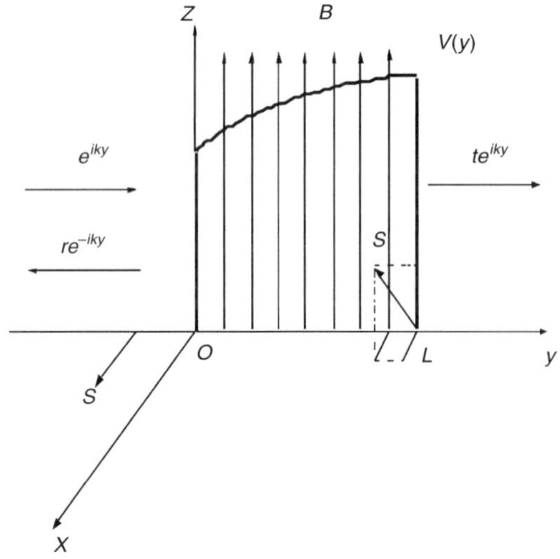

Figure 3. General potential barrier restricted to the interval $0 < y < L$ with a magnetic field applied.

where the column wavevector $\widehat{\Psi}(y)$ represents compactly both spin states.

The problem is solved by perturbation theory. In the lowest order in B, the spinor $\widehat{\Psi}(L)$ of the electron on the right end of the barrier is given by [43]

$$\widehat{\Psi}(L) = \begin{pmatrix} 1 \\ 1 \end{pmatrix}\psi(L) + \frac{e\hbar B}{2mc}\begin{pmatrix} 1 \\ -1 \end{pmatrix}\int_0^L \psi(y)G(y, L)\,dy \quad (16)$$

Here $\psi(y)$ is the solution of the spatial part of the Schrödinger equation in the absence of the magnetic field, which can be written in terms of the GF of the system as

$$\psi(y) = \exp(iky) - \int_0^L G(y, y')V(y')\exp(iky')\,dy' \quad (17)$$

where $G(y, y')$ is the retarded GF, whose energy dependence is not written explicitly. It should satisfy Dyson's equation:

$$G(y, y') = G_0(y, y') + \int_0^L G_0(y, y'')V(y'')G(y'', y')\,dy'' \quad (18)$$

where $G_0(y, y') = i(m/k\hbar^2)\exp(ik|y - y'|)$ is the free-electron GF. We can obtain all the relevant properties of the problem in terms of the GF, solution of the previous equation.

We first concentrate on the calculation of the traversal time. The expectation value of the component of the spin along the direction of the magnetic field of the transmitted electron is, up to second order in B:

$$\langle S_z\rangle = \frac{\hbar}{2}\langle\widehat{\Psi}(L)|\sigma_z|\widehat{\Psi}(L)\rangle = -\frac{e\hbar^2 B}{mc}\,\mathrm{Re}\left[\psi^*(L)\int_0^L \psi(y)G(L, y)\,dy\right] \quad (19)$$

We want to express the wave function $\psi(y)$ appearing inside the integral in the previous equation in terms of the GF. In order to do so, we take into account the following relationship between the wave function and the GF of a 1D system:

$$\psi(y) = -\frac{i\hbar^2 k}{m}G(0, y) \quad (20)$$

For one-dimensional systems also, we can further simplify the problem by writing the general expression of the GF, $G(y, y')$, in terms of its own expression at coinciding coordinates $y = y'$ [45]. One finds that the spin component along the direction of the magnetic field is given by

$$\langle S_z\rangle = \frac{e\hbar^2 B}{mc}|\psi(L)|^2\,\mathrm{Re}\int_0^L G(y, y)\,dy \quad (21)$$

Similarly, the spin component along the y direction is equal to

$$\langle S_y\rangle = -\frac{e\hbar^2 B}{mc}|\psi(L)|^2\,\mathrm{Im}\int_0^L G(y, y)\,dy \quad (22)$$

Büttiker-Landauer characteristic traversal times for the z and y directions are proportional to the corresponding spin components, Eqs. (8) and (7), and we finally arrive at

$$\tau^{\mathrm{BL}} = \hbar\left|\int_0^L G(y, y)\, dy\right| \tag{23}$$

Instead of defining the modulus of τ_z^{BL} and τ_z^{BL} as the central magnitude of the problem, we prefer to define a complex traversal time τ as

$$\tau = \tau_z^{\mathrm{BL}} + i\tau_y^{\mathrm{BL}} = \hbar \int_0^L G(y, y)\, dy \tag{24}$$

As we will see, other approaches also get a complex time. The two characteristic times of the problem can be written in a compact form as the real and imaginary parts of a single well-defined magnitude. Besides, these two time components may be separately relevant to different experimental results, and do not have to necessarily enter into the problem through the modulus, Eq. (23). We will come back to this question.

The final result, Eq. (24), only depends on the integral of the GF at coinciding coordinates. For practical purposes and in order to compare this result with those of other approaches, it is interesting to rewrite it in terms of the transmission t and reflection from the left r and from the right r' amplitudes [44, 45]. The spatial integral, over the length of the barrier, of the GF at coinciding coordinates can be expressed in terms of partial derivatives with respect to energy E:

$$\tau = \hbar \int_0^L G(y, y)\, dy = \hbar\left\{\frac{\partial \ln t}{\partial E} + \frac{1}{4E}(r + r')\right\} \tag{25}$$

This is a general expression, independent of the model considered.

The first term in the RHS of Eq. (25) mainly contains information about the region of the barrier. Most of the information about the boundary is provided by the term including the reflection amplitudes r and r'. This term is on the order of the wavelength λ over the length of the system L, and it becomes important for low energies and/or short systems. It can be neglected in the semiclassical WKB case and when r is negligible, for example, in the resonant case, when the influence of the boundaries is negligible. Certain approaches only obtain the contribution to the time proportional to an energy derivative, missing the terms proportional to the reflection amplitudes. The same type of problem arises when calculating densities of states or partial densities of states [46].

The components of the traversal time can be related to the density of states and the resistance. The imaginary part of $G(y, y)$ is proportional to the local density of states at the corresponding energy. So, τ_y^{BL} can also be written as $\tau_y^{\mathrm{BL}} = \pi\hbar L \nu_L(E)$, where $\nu_L(E)$ is the average density of states per unit energy and per unit length.

Thouless has shown [47] the existence of a dispersion relation between the localization length and the density of states. This relationship can be expressed [48] in the form of a linear dispersion relation between the real part, $\mathrm{Re}\ln t$, and the imaginary part, $\mathrm{Im}\ln t$, of the transmission amplitude. The self-averaging property of τ_z^{BL} and of τ_y^{BL} is therefore an immediate consequence of self-averaging of the localization length and of the density of states [48]. While τ_z and τ_y are additive, the total tunnelling time τ^{BL}, given by Eq. (23), is not the sum of the individual transmission times. This property has also been pointed out by Leavens and Aers [49]. It is a consequence of the fact that, for an infinitesimal B, the interference between the effects of the magnetic field in the separate regions $[0; y]$ and $[y; L]$ is of higher order than linear and does not contribute to the local times [49]. Mathematically speaking, we say that the BL time, Eq. (23), adds as the absolute value of complex additive numbers, and so it is not additive.

2.3. Reflection Time

For reflected particles we can proceed in the same way as we did for transmitted particles. We will use the subindex R to indicate that the magnitude corresponds to reflection, and we understand that similar magnitudes related to transmission will have no subindex. Proceeding as above, we find for the expectation values of the spin components of the reflected wave: and

$$\begin{aligned}\bullet\ \langle S_z\rangle_{\mathrm{R}} &= \frac{\hbar}{2}\langle(\widehat{\Psi}(0) - 1)\,|\sigma_z|\,(\widehat{\Psi}(0) - 1)\rangle \\ &= \frac{e\hbar^2 B}{mc}|\psi^*(0) - 1|^2\, \mathrm{Re}\int_0^L \psi(y) G(0, y)\, dy\end{aligned} \tag{26}$$

and

$$\langle S_y\rangle_{\mathrm{R}} = -\frac{e\hbar^2 B}{mc}|\psi^*(0) - 1|^2\, \mathrm{Im}\int_0^L \psi(y) G(0, y)\, dy \tag{27}$$

We can again define three new characteristic times, $\tau_{z,\mathrm{R}}^{\mathrm{BL}}$, $\tau_{y,\mathrm{R}}^{\mathrm{BL}}$, and $\tau_{x,\mathrm{R}}^{\mathrm{BL}}$, each of them associated with a component of the spin. Only two of these times are independent. Invoking Eqs. (26) and (27) and the relationship (20) between the wave function and the GF of a one-dimensional system, we arrive at

$$\tau_{y,\mathrm{R}}^{\mathrm{BL}} = \hbar\, \mathrm{Im}\, \frac{1 + r}{r} e^{-i2\theta(0)} \int_0^L G(y, y) e^{i2\theta(y)}\, dy \tag{28}$$

$$\tau_{z,\mathrm{R}}^{\mathrm{BL}} = \hbar\, \mathrm{Re}\, \frac{1 + r}{r} e^{-i2\theta(0)} \int_0^L G(y, y) e^{i2\theta(y)}\, dy \tag{29}$$

where $\theta(y)$ is a phase function given by

$$\theta(y) = \int_0^y \frac{im}{\hbar^2} \frac{dy'}{G(y', y')} \tag{30}$$

The characteristic times $\tau_{y,\mathrm{R}}^{\mathrm{BL}}$ and $\tau_{z,\mathrm{R}}^{\mathrm{BL}}$ are the real and imaginary components, respectively, of a complex quantity. This quantity is proportional to a new integral of the GF at coinciding coordinates, which in this case involves the phase function also. The previous integral can be written in terms of the transmission and reflection amplitudes. We arrive at the following expression for the complex reflection time [50]:

$$\tau_{\mathrm{R}} = \tau_{z,\mathrm{R}}^{\mathrm{BL}} + i\tau_{y,\mathrm{R}}^{\mathrm{BL}} \equiv \hbar\left\{\frac{\partial \ln r}{\partial E} - \frac{1}{4Er}(1 - r^2 - t^2)\right\} \tag{31}$$

This is again a general equation, independent of the model used.

For an arbitrary symmetric potential, $V((L/2)+y) = V((L/2)-y)$, the total phases accumulated in a transmission and in a reflection event are the same, as can be deduced from the form of the scattering matrix elements, and so the characteristic times for transmission and reflection corresponding to the direction of propagation are equal:

$$\tau_y^{\mathrm{BL}} = \tau_{y,\mathrm{R}}^{\mathrm{BL}} \tag{32}$$

as it immediately follows from Eqs. (25) and (31) (see also the review article by Hauge and Støvneng [4]). For the special case of a rectangular barrier, Eq. (32) was first found by Büttiker [36]. For an asymmetric barrier, Eq. (32) breaks down as discussed by Leavens and Aers [51].

2.4. Dwell Time

There is also another important characteristic time called the dwell time, about which there exists a vast literature (see, e.g., [4, 11] and references therein). This time was first introduced by Büttiker [36] region divided by the average number entering (or leaving) the barrier per unit time. It corresponds to the average time spent by a particle within the barrier irrespective of whether it is finally reflected or transmitted.

The dwell time in a neighborhood of y is defined as the ratio between the particle number in the interval $[y, y+dy]$ and the incoming current [36]:

$$d\tau^{(\mathrm{D})}(y) = \frac{|\psi(y)|^2}{J}\,dy \tag{33}$$

where $\psi(y)$ is the steady-state scattering solution of the time-independent Schrödinger equation. Obviously, Eq. (33) describes a balance equation: in the stationary case the injected current equals the decay rate of the probability in $[y, y+dy]$. The dwell time $\tau^{(\mathrm{D})}$ of a finite region within the context of a stationary-state scattering problem is obtained via a spatial integration of Eq. (33). So the dwell time $\tau^{(\mathrm{D})}$ is given by [36]

$$\tau^{(\mathrm{D})} \equiv \frac{m}{\hbar k}\int_0^L |\psi(y)|^2\,dy \tag{34}$$

Here the integral extends over the barrier, and $\hbar k/m$ is the incident flux. Again we want to calculate this time in terms of the transmission and reflection amplitudes.

Let us consider again a particle moving along the y direction in the presence of an arbitrary potential barrier $V(y)$ in the interval $[0, L]$. Taking explicitly into account that the wave function appearing in Eq. (34) is a solution of the Schrödinger equation, we arrive at [52]

$$\tau^{(\mathrm{D})} = -\frac{\hbar}{4k}\left[\psi^{*2}(y)\frac{\partial}{\partial E}\left(\frac{\psi'(y)\psi(y)}{|\psi(y)|^2}\right) + \psi^2(y)\frac{\partial}{\partial E}\left(\frac{\psi^{*\prime}(y)\psi^*(y)}{|\psi(y)|^2}\right)\right]_0^L \tag{35}$$

This expression is formally the same for particles incident from the left or from the right, but we have to remember that the corresponding wave functions will not be the same. García-Calderón and Rubio [53] arrived at the same result by a completely different method.

We can now rewrite Eq. (35) in terms of the retarded GF $G(y, y')$ of the system, as we have been doing for the other times. The dwell time is given by

$$\tau^{(\mathrm{D})} = \left[i\frac{\partial}{\partial E}\theta(y) - G(y,y)\frac{\partial}{\partial E}\left(\frac{G'(y,y)}{G(y,y)}\right)\right]_0^L \tag{36}$$

As it occurs for the wave function, the GF $G(y, y')$ depends on whether the particle arrives to the barrier from the left or from the right. After some cumbersome algebra, we can express the dwell time in terms of the transmission and reflection amplitudes:

$$\tau_-^{(\mathrm{D})} = \hbar\,\mathrm{Im}\left\{\left[\frac{\partial \ln t}{\partial E} + \frac{1}{4E}(r+r')\right] + \frac{1}{2}\left[\sqrt{R}\frac{\partial}{\partial E}\ln\frac{r}{r'} + \frac{1}{2E}(r-r')\right]\right\} \tag{37}$$

The subindex-indicates that the particle is coming from the left. R is the modulus square of the reflection amplitudes $R = |r|^2 = |r'|^2$. When the particle is coming from the right, the dwell time is given by an expression similar to Eq. (37), but interchanging r and r'. We will refer to this case with the subindex $+$.

Gasparian et al. [44] showed that the first term on the RHS of Eq. (37) is proportional to the density of states. Then, we finally arrive at the following expression for the dwell time:

$$\tau_\pm^{(\mathrm{D})} = \pi\hbar L\nu(E) \pm \frac{\hbar}{2}\,\mathrm{Im}\left[\sqrt{R}\frac{\partial}{\partial E}\ln\frac{r}{r'} + \frac{1}{2E}(r-r')\right] \tag{38}$$

For a symmetric potential we have that the reflection coefficients from the right and from the left are equal, $r = r'$, and we obtain $\tau_-^{(\mathrm{D})} = \tau_+^{(\mathrm{D})} = \pi\hbar L\nu(E)$, in agreement with the result of Gasparian and Pollak [43].

For an asymmetric barrier, it is easy to check that the contribution from the asymmetry is the opposite for particles coming from the left and from the right. Then we find that

$$\nu(E) = \frac{1}{2\pi\hbar L}\left(\tau_-^{(\mathrm{D})} + \tau_+^{(\mathrm{D})}\right) \tag{39}$$

This result was obtained in a much wider context by Iannaccone [54], which considered the relation between the dwell time and the density of states for a three-dimensional region Ω of arbitrary shape with an arbitrary number of incoming channels. He arrived at

$$\nu_\Omega(E) = \frac{1}{2\pi\hbar}\sum_{n=1}^{N}\tau_n^{(\mathrm{D})} \tag{40}$$

where $\nu_\Omega(E)$ is the density of states per unit volume, and $\tau_n^{(\mathrm{D})}$ is the dwell time for particles coming from the n-channel. This result shows that the density of states in Ω is proportional to the sum of the dwell times in Ω for all the incoming channels.

A controversial question concerning the dwell time is whether it satisfies or not the relation (see [49, 55, 56])

$$\tau^{(\mathrm{D})} = R\tau^{\mathrm{BL}}_{y,\mathrm{R}} + T\tau^{\mathrm{BL}}_{y} \tag{41}$$

This result is trivial for classical particles, for which the traversal time coincides with the y component of our complex traversal time and for which there is no interference between the reflected and the transmitted particles. For the quantum coherent case, this result is not so clear. We can prove this relation, which we believe must hold because a particle incident on the barrier is either transmitted or reflected. Reflection and transmission of a particle are mutually exclusive events in the sense of Feynman and Hibbs [57]; that is, a measurement can determine, without interfering with the scattering event, whether a particle has been transmitted or reflected. Our results for the y component of the transmission and reflection times, Eqs. (25) and (31), respectively, and for the dwell time, Eq. (37), allow us to prove exactly the previous relation between these times. On the other hand, our results also prove that a similar relation involving the full BL times does not hold. This relation has been claimed very often in the literature, and has been strongly criticized by other authors [11].

3. OTHER APPROACHES

In this section, we review other approaches to the problem of the traversal and reflection times. We would like to show that most results, obtained from very different points of view, are almost compatible and coincide with Eq. (25) for the traversal time and with Eq. (31) for the reflection time. Often, these approaches only obtain the contributions to the time proportional to the energy derivative of the logarithm of the transmission amplitude.

We start with the oscillatory incident amplitude and with the time-modulated barrier approaches. Then we review the Feynman path-integral approach, where the idea of a complex time arises more naturally. We finish with the kinetic approach, which is very convenient to study finite size effects and so the standard errors inherent to the problem.

3.1. Oscillatory Incident Amplitude

We assume an incident wave of oscillatory amplitude interacting with a time-independent potential, and study the shape distortion of the transmitted wave by the barrier. This method was proposed by Büttiker and Landauer [58, 59] and analyzed by Leavens and Aers [51] and Martin and Landauer [60]. The incident wave consists of two interfering plane waves:

$$\begin{aligned}\Psi(y,t) &= \exp\left\{i\left[ky - \frac{Et}{\hbar}\right]\right\} \\ &\quad + \exp\left\{i\left[(k+\Delta k)y - \frac{(E+\Delta E)t}{\hbar}\right]\right\} \\ &= 2\exp\left\{i\left[(k+\Delta k/2)y - \frac{(E+\Delta E/2)t}{\hbar}\right]\right\} \\ &\quad \times \cos\left(\frac{\Delta k y}{2} - \frac{\Delta E t}{2\hbar}\right)\end{aligned} \tag{42}$$

The energy difference between the two plane waves characterizes the oscillations in amplitude of the incident wave. In the region to the right of the barrier, we have the sum of two transmitted plane waves which can be written in the form

$$\begin{aligned}\Psi(y,t) &= t(E)\left\{i\left[ky - \frac{Et}{\hbar}\right]\right\} + t(E+\Delta E) \\ &\quad \times \exp\left\{i\left[(k+\Delta k)y - \frac{(E+\Delta E)t}{\hbar}\right]\right\}\end{aligned} \tag{43}$$

The shape distortion produced by the barrier on the transmitted wave will strongly depend on ΔE. If ΔE is small, the incident wave is modulated very slowly and in that case the transmitted wave (43) will reproduce the incident wave (42), in the sense that the destructive and constructive interferences will occur at the same time for both of them. As we increase ΔE, $t(E)$ and $t(E+\Delta E)$ will increasingly differ and the transmitted wave (43) will no longer reproduce the incident wave. We can assume that appreciable shape distortion will take place when a characteristic time delay, or dispersion in transit time, becomes comparable to or larger than the modulation period [58, 59]. Thus we define a new traversal time τ as $\hbar/\Delta E$, where ΔE is the energy difference which establishes the onset of significant distortion of the transmitted wave, that is, the energy such that $\Delta E|d\alpha(E)/dE| \approx 1$.

The analysis of this approach based on the WKB approximation led Büttiker and Landauer [58] to the following results. For $E < V(y)$, the phase of the transmission amplitude is of secondary importance as compared with the exponential decay of the modulus of $t(E)$. We can write the transmission amplitude in the form

$$t_{\mathrm{WKB}}(E) = \exp\left[-\int_{y_1}^{y_2} \xi(y)\,dy\right] \tag{44}$$

where ξ is the inverse decay length, given by Eq. (4), and y_1 and y_2 are the classical turning points. From this expression of the transmission amplitude, Büttiker and Landauer obtained for the traversal time for tunnelling

$$\tau_{\mathrm{WKB}}(E) = \frac{m}{\hbar}\left[-\int_{y_1}^{y_2} \frac{dy}{\xi(y)}\right] \tag{45}$$

When $E > V(y)$, the energy dependence of $t(E)$ comes primarily from the dependence of the phase ($|t(E)| = 1$), and then we can assume that $t(E)$ is of the form

$$t_{\mathrm{WKB}}(E) = \exp\left[-i\int_{0}^{L} K(y)\,dy\right] \tag{46}$$

with $K(y) = i\xi(y)$. For this case, in which the phase dominates, we have

$$\tau_{\mathrm{WKB}}(E) = \frac{m}{\hbar}\left[\int_{0}^{L} \frac{dy}{K(y)}\right] \tag{47}$$

It is easy to check that, for a rectangular barrier, the traversal time τ_{WKB} is equal to $mL/\hbar\xi$ for energies below the barrier height and equal to $mL/\hbar K$ for energies above the barrier. As it was shown by Martin and Landauer [60], the general analysis of this two-interfering-incident-waves approach yields characteristic times that depend on energy derivatives of the transmission coefficient

$$\tau = \hbar |t_E^{-1}| \left|\frac{dt_E}{dE}\right| = \hbar \left\{ \left(\frac{d\varphi}{dE}\right)^2 + \left(\frac{d\ln T}{dE}\right)^2 \right\}^{1/2} \quad (48)$$

As in the Büttiker and Landauer approach to the Larmor clock, the time is equal to the square root of the sum of the squares of two characteristic times, one involving energy derivatives of the phase and the other energy derivatives of the logarithm of the modulus of the transmission amplitude. The same result for the traversal time is also obtained in the modulated barrier approach [60].

It is interesting to note that this oscillatory amplitude approach without resort to the WKB approximation led Leavens and Aers [51] to complex times. Let us write the transmission amplitude as

$$t(E) = \exp[i\beta(E)] \quad (49)$$

where $\beta(E)$ is in general complex. For sufficiently small ΔE, we may expand $t(E+\Delta E)$ to lowest order in ΔE:

$$\begin{aligned} t(E+\Delta E) &\cong \exp\left[i\left(\beta(E) + \Delta E\frac{d\beta(E)}{dE}\right)\right] \\ &= t(E)\exp\left[i\Delta E\frac{d\beta(E)}{dE}\right] \end{aligned} \quad (50)$$

This expression should be substituted in Eq. (43) for the transmitted packet. For sufficiently small ΔE, the difference in exponents of the two components of the transmitted wave at $y = L$ and $t = \Delta t$ is greater than that of the two components of the incident wave at $y = 0$ and $t = 0$ by an amount

$$\begin{aligned} & i\left[\Delta kL - \frac{\Delta E}{\hbar\Delta k}\left(\Delta t - \hbar\frac{d\beta(E)}{dE}\right)\right] \\ &\qquad \cong i\Delta k\left[L - v(k)\left(\Delta t - \hbar\frac{d\beta(E)}{dE}\right)\right] \end{aligned} \quad (51)$$

with $v(k) \equiv \hbar^{-1}dE/dk = \hbar k/m$ being the group velocity. In the absence of the potential barrier, the traversal time associated with the propagation of the wavepacket from $y = 0$ to $y = L$ is the value of Δt for which $L - v(k)\Delta t = 0$, that is, $\tau = L/v(k)$. Formally, in the presence of the potential, Leavens and Aers [51] obtained from Eq. (51) the complex barrier interaction "time"

$$\tau^E = \frac{L}{v(k)} + \hbar\frac{d\beta(E)}{dE} \equiv -i\hbar\frac{\partial \ln t}{\partial E} \quad (52)$$

This final answer for the time is just proportional to $\partial \ln t/\partial E$, and so is correct for infinitely large systems only ($L \gg \lambda$). The difference between this expression for the traversal time and our general expression (25) is the term proportional to the reflection amplitude, which cannot be obtained with this type of approach.

We can deduce explicit expressions for all these times, and so see clearly the difference between Eq. (52) and Eq. (25), obtained with the GF formalism, for the special case of a rectangular barrier. Let us associate the real and imaginary components of this complex time, Eq. (52), with the previous characteristic times for the y and z components, and let us denote them as τ_y^E and τ_z^E. The explicit expressions for the two components of the traversal time τ_y^E and τ_z^E for this special case of a constant potential can be written in the form [51]

$$\begin{aligned} \tau_z^E &= -\hbar\frac{\partial \ln T^{1/2}}{\partial E} \\ &= \frac{mk_0^4}{2\hbar\xi^2k^2}\frac{2(\xi^2-k^2)\sinh^2(\xi L) + k^2\xi L\sinh(2\xi L)}{4k^2\xi^2 + k_0^2\sinh^2(\xi L)} \end{aligned} \quad (53)$$

$$\tau_y^E = \hbar\frac{\partial\varphi}{\partial E} = \frac{m}{\hbar k\xi}\frac{2\xi Lk^2(\xi^2-k^2) + k_0^4\sinh(2\xi L)}{4k^2\xi^2 + k_0^2\sinh^2(\xi L)} \quad (54)$$

After an obvious change of notation, it is easy to check that the times τ_y^E and τ_z^E are related to the exact results τ_y^{BL} and τ_z^{BL} through

$$\begin{aligned} \tau_y^E &= \tau_y^{\mathrm{BL}} + \frac{1}{2E}\operatorname{Im} r \\ -\tau_z^E &= \tau_z^{\mathrm{BL}} + \frac{1}{2E}\operatorname{Re} r \end{aligned} \quad (55)$$

where τ_z^{BL} and τ_y^{BL} are given by Eqs. (10) and (11). Figure 4 compares τ_y^E with τ_y^{BL} for a rectangular barrier. It can be

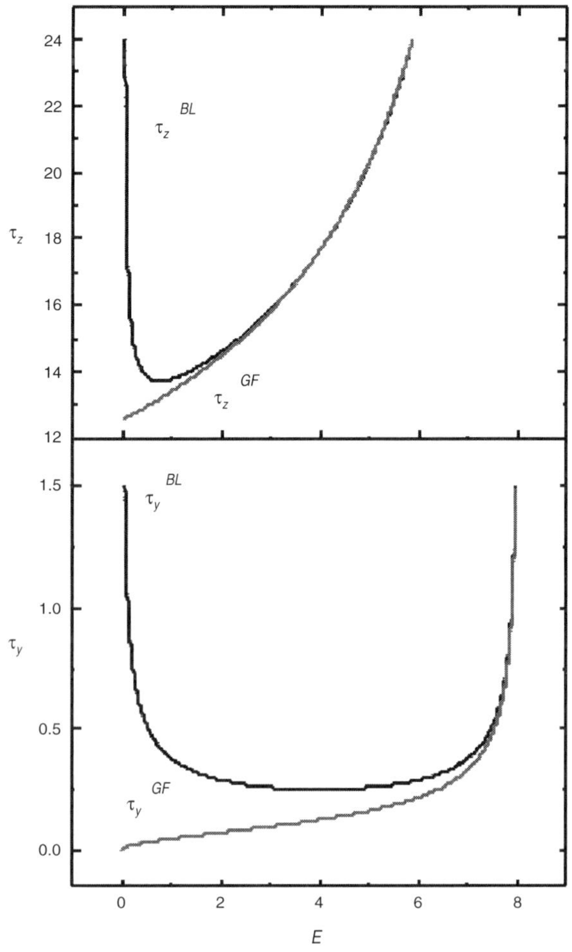

Figure 4. Components of the traversal time for a rectangular barrier according to Eqs. (25) and (54).

seen that, even for an opaque barrier ($\xi L \gg 1$), the differences between these times can be very significant. This is particularly so at very small energy E where τ_y^{BL} goes to zero as $E^{1/2}$, while τ_y^{E} diverges as $E^{-1/2}$, and, at the same time, τ_z^{BL} is approximately equal to 0, while τ_z^{E} diverges as E^{-1}. The oscillatory amplitude approach in general does not give the same answer as the GF formalism, based on the Larmor clock approach. As the difference between the corresponding tunnelling times is proportional to the amplitude of reflection, we concluded that it arises from boundary effects [44].

3.2. Time-Modulated Barrier

The time-modulated barrier approach to the traversal time was introduced by Büttiker and Landauer [42] (see also [58, 61]). Its basic idea is simple and can be explained as follows. Let us add to the static barrier potential which we discussed before a time-dependent potential which is zero everywhere except in the region of interest. So the 1D potential can now be written in the form

$$V(y,t) = V(y) + V_1 \cos(\omega t) \tag{56}$$

where V_1 is the amplitude of the small modulation added, and ω is its corresponding frequency. For the sake of simplicity, it is more convenient in this approach to consider the barrier restricted to the region $-L/2 < y < L/2$.

Suppose that there is a characteristic time τ during which the particle interacts with the barrier. If the period of the modulation $T = 2\pi/\omega$ is long compared to the time τ, then the particle sees an effectively static barrier during its traversal. In the opposite extreme, that is, for slowly tunnelling electrons, for which $\omega\tau > 1$, the barrier oscillates many times during the period of traversal of the electron. There is thus a crossover from a low-frequency behavior to a high-frequency behavior, and we expect to occur two distinct types of electron-barrier interactions, depending on the value of $\omega\tau$ as compared with unity.

We will use a rectangular barrier extensively, for illustrative purposes, but in principle all the results can be generalized to an arbitrary potential barrier by considering the adiabatic limit, $\omega \to 0$, of this inelastic scattering process [62]. The Hamiltonian for the time-modulated rectangular barrier in the scattering region is

$$H = -\frac{\hbar}{2m}\frac{d^2}{dy^2} + V_0 + V_1\cos(\omega t) \equiv H_0 + V_1 \cos(\omega t) \tag{57}$$

As it is well known from the time-dependent perturbation theory [63], incident particles with energy E, interacting with the perturbation $V_1\cos(\omega t)$, will emit or absorb modulation quanta $\hbar\omega$. In first-order corrections to the time-independent case, this means that inside the barrier, for $|y| > L/2$, the reflected and transmitted waves, used to represent the tunnelling electrons, we will now have a main feature at the initial energy E and also sidebands at the energies $E + \hbar\omega$ and $E - \hbar\omega$, as it is schematically represented in Figure 5. Taking V_1 as a perturbation, the two independent eigensolutions of the corresponding

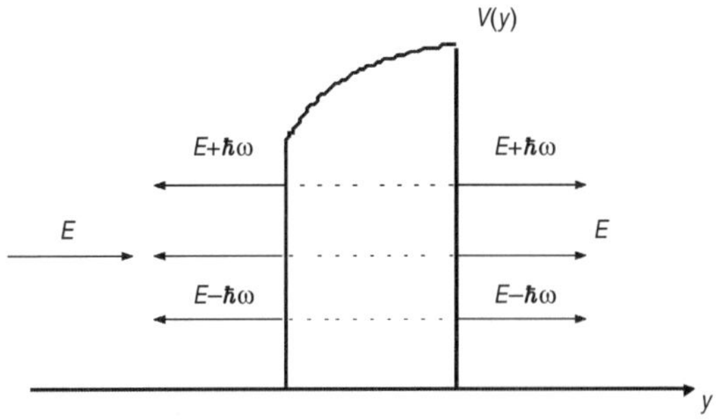

Figure 5. For an oscillating barrier, besides a main transmission and reflection component at the initial energy E, there are two lateral components at energies $E \pm \hbar\omega$. Reprinted with permission from [86], V. Gasparian et al., "Handbook of Nanostructured Materials and Nanotechnology" (H. S. Nalwa, Ed.), Vol. 2, Chapter 11, 1999. © 1999, Elsevier Science.

time-dependent Schrödinger equation, within the rectangular barrier, can be written as [63]

$$\Psi_{\mathrm{bar}}(y,t;E) = \varphi_E(y)\exp\left\{-\frac{iEt}{h}\right\}\exp\left\{-\frac{iV_1}{\hbar\omega}\sin\omega t\right\} \tag{58}$$

Here $\varphi_E = e^{\pm\xi y}$ is a wave function solution of the time-independent problem $H_0\varphi_E = E\varphi_E$, with the simpler Hamiltonian H_0.

As it was shown by Büttiker and Landauer [42], the next stage to find the solution for the oscillating rectangular barrier is to match Eq. (58) with the corresponding solutions at the same energy outside the scattering region. For an electron of energy E impinging on the scattering region, there will be reflected and transmitted waves at the three energies E, $E + \hbar\omega$, and $E - \hbar\omega$. So, if the electron is coming from the left, its wave function in the region to the left of the barrier, $y < -L/2$, will be of the form

$$\begin{aligned}\Psi_{\mathrm{inc+ref}} = {} & (e^{ik(y+L/2)} + re^{-ik(y+L/2)})e^{-iEt/\hbar} + r_+ e^{-ik_+(y+L/2)} \\ & \times e^{-i(E+\hbar\omega)t/\hbar} + r_- e^{-ik_-(y+L/2)} \\ & \times e^{-i(E-\hbar\omega)t/\hbar}\end{aligned} \tag{59}$$

where $k_\pm$ are the wavevectors corresponding to the side energy bands, defined as $k_\pm = (2m/\hbar^2)^{1/2}(E \pm \hbar\omega)^{1/2}$. Equation (59) represents an incident plane wave of unit amplitude and three reflected waves, one of amplitude r at the incident energy and two of amplitudes $r_\pm$ at energies $E \pm \hbar\omega$. To the right of the barrier ($y > L/2$), we have for the transmitted wave

$$\begin{aligned}\Psi_{\mathrm{tra}} = {} & te^{ik(y-L/2)}e^{-iEt/\hbar} + t_+ e^{ik_+(y-L/2)}e^{-i(E+\hbar\omega)t/\hbar} \\ & + t_- e^{ik_-(y-L/2)}e^{-i(E-\hbar\omega)t/\hbar}\end{aligned} \tag{60}$$

where t is the transmission amplitude at the energy of the incident wave and $t_\pm$ are the transmission amplitudes of the sidebands.

In the barrier, for an infinitesimal amplitude of the time-dependent potential, $V_1 \ll \hbar\omega$, we can expand Eq. (58) to lowest order in V_1 and represent the wave function in the form

$$\Psi_{\text{bar}} = [Be^{\xi y} + Ce^{-\xi y}]e^{-iEt/\hbar}\left[1 + \frac{V_1}{2\hbar\omega}e^{-i\omega t} - \frac{V_1}{2\hbar\omega}e^{i\omega t}\right] + [B_+e^{\xi+y} + C_+e^{-\xi+y}]e^{-i(E+\hbar\omega)t/\hbar} + [B_-e^{\xi-y} + C_-e^{-\xi-y}]e^{-i(E-\hbar\omega)t/\hbar} \tag{61}$$

$\xi_\pm$ are the inverse decay lengths for the sidebands, defined as $\xi_\pm = (2m/\hbar^2)^{1/2}(V_0 - E \mp h\omega)^{1/2}$. The coefficients r, $r_\pm$, B, $B_\pm$, C, $C_\pm$, t, and $t_\pm$ are determined by matching the wave functions and their derivatives at $y = -L/2$ and at $y = L/2$ in the usual manner. Note that the matching conditions must hold for all times; therefore, we have to match each time Fourier component separately. r and t play the role of the static reflection and transmission amplitudes, respectively. Using the standard matching relations, it is straightforward to show that for an almost completely reflecting barrier in the opaque limit, $\xi L \gg 1$, the coefficient t of the static barrier is given by the standard expression [63]

$$t = \frac{4k\xi}{k_0^2}e^{-\xi L}\exp\left\{-i\arctan\left[\frac{\xi^2 - k^2}{2k\xi}\right]\right\} \times \exp\left\{i\left[y - kL - \frac{Et}{\hbar}\right]\right\} \tag{62}$$

For the transmitted waves at the frequencies $(E/\hbar) \pm \omega$, Büttiker and Landauer found that their transmission coefficients are

$$t_\pm = \mp t\frac{V_1}{2\hbar\omega}(e^{\pm\omega\tau} - 1) \times \exp\left\{i\left[k_\pm \mp \frac{m\omega L}{2\hbar} - \frac{(E \pm \hbar\omega)t}{\hbar}\right]\right\} \tag{63}$$

$\tau = mL/\hbar\xi$ is the time it would take a particle with the velocity $v = \hbar\xi/m$ to traverse the opaque rectangular barrier. To obtain Eq. (63) it was additionally assumed that $\hbar\omega \ll E$, so that the wavevectors of the sidebands are approximately equal to $k_\pm \cong k \pm m\omega/\hbar k$, and also that $\hbar\omega \ll V_0 - E$, so that the decay lengths satisfy $\xi_\pm = \xi \mp m\omega/\hbar\xi$.

Note that for opaque barriers the traversal time τ^{BL} obtained in the Larmor clock approach, Eq. (14), coincides with the expression considered in the previous equation, $\tau = mL/\hbar\xi$. The classical time that one would obtain in the WKB limit at energies below the peak of the barrier is given by the integral

$$\tau = \int_{y_1}^{y_2} \frac{m}{\hbar\xi(y)}\,dy = \int_{y_1}^{y_2}\left\{\frac{m}{2(V_0(y) - E)}\right\}^{1/2} dy \tag{64}$$

where y_1 and y_2 are the classical turning points. This result also reduces to the value appearing in Eq. (63) for the case of a rectangular barrier, when $V_0(y)$ is constant.

The probability of transmission at the sideband energies, determined from Eq. (63), is

$$T_\pm = |t_\pm|^2 = \left(\frac{V_1}{2\hbar\omega}\right)^2(e^{\pm\omega\tau} - 1)^2 T \tag{65}$$

where T is the transmission coefficient for the static barrier. For small frequencies, so that $\omega\tau \ll 1$, the probabilities of transmission for the upper and lower sidebands obtained from Eq. (65) are the same and equal to

$$T_\pm = \left(\frac{V_1\tau}{2\hbar}\right)^2 T \tag{66}$$

Remember that τ is the approximate expression for the Büttiker–Landauer time for an opaque barrier, given by Eq. (14).

At high frequencies, the upper sideband is exponentially enhanced, while the lower sideband is exponentially suppresseds. So for an opaque barrier we do indeed have a rather well-defined crossover between tunnelling at high frequencies and tunnelling at low frequencies, with the characteristic time corresponding to the value given by Eq. (14). This characteristic crossover time is the same one appearing in the expression of the transmission coefficients of the sidebands in the adiabatic limit.

3.2.1. General Barrier

Let us briefly discuss the results of the general oscillating barrier problem following the papers of Hauge and Støvneng [4] and [62]. It was shown that in the adiabatic limit, $\omega \to 0$, the expression for the transmission coefficients for the sidebands, Eq. (65), can be generalized in the form

$$T\pm = |t_\pm|^2 \to \left(\frac{V_1|\tau^{\overline{V}}|}{2\hbar}\right)^2 |t(E, \overline{V})|^2 \tag{67}$$

where $|t|^2 = T$ and we have written explicitly the E and $\overline{V}$ dependence of the transmission amplitude t. $\overline{V}$ is the average value of the barrier potential in the scattering region, that is,

$$\overline{V} \equiv \frac{1}{L}\int_{-L/2}^{L/2} V(y)\,dy. \tag{68}$$

$\tau^{\overline{V}}$ is a complex quantity, with the dimensions of time, defined as

$$\tau^{\overline{V}} = i\hbar\frac{\partial \ln t(E, \overline{V})}{\partial\overline{V}} \tag{69}$$

This quantity characterizes the crossover from the adiabatic to the high-frequency limits, and we define it as the traversal time in the time-modulated barrier approach.

The corresponding definition of the reflection time appeals to the adiabatic limit of the reflected sidebands. Their reflection coefficients $R_\pm = |r_\pm|^2$ tend in the adiabatic limit to an expression that can be written as

$$|r_\pm|^2 \to \left(\frac{V_1|\tau_R^{\overline{V}}|}{2\hbar}\right)^2 |r(E, \overline{V})|^2 \tag{70}$$

where $|r|^2 = R$ is the static reflection coefficient. Again, we have explicitly written the E and $\overline{V}$ dependence of the reflection amplitude r. $\tau_{\rm R}^{\overline{V}}$ is a new complex quantity, playing the role of a reflection time, defined as

$$\tau_{\rm R}^{\overline{V}} = i\hbar \frac{\partial \ln r(E, \overline{V})}{\partial \overline{V}} \tag{71}$$

The complex times $\tau^{\overline{V}}$ and $\tau_{\rm R}^{\overline{V}}$ are related to the real quantities $\tau_z^{\rm BL}$, $\tau_y^{\rm BL}$, $\tau_{y,\rm R}^{\rm BL}$ and $\tau_{z,\rm R}^{\rm BL}$ and therefore the BL traversal time can be formally written in the form

$$\tau^{\rm BL} \equiv \hbar \left| \frac{\partial \ln t(E, \overline{V})}{\partial \overline{V}} \right| \tag{72}$$

It is still not so clear how this time, which was obtained from an analysis of the time-modulated barrier and which is valid for an arbitrary shaped potential $V(y)$, whose average is $\overline{V}$, can be justified as a traversal time for a general barrier [64]. In any case, note that Eq. (72) for an opaque barrier leads us to Eq. (14), which was obtained by Büttiker's analysis of the Larmor clock [36]. The BL reflection time can be defined as in Eq. (72), but replacing $t(E, \overline{V})$ by $r(E, \overline{V})$.

3.3. Complex Time

Although common sense dictates that the tunnelling time must be a real time and that there are no clocks that measure a complex time, nevertheless the concept of complex time in the theory of the traversal time problem of electrons appeared in many approaches (see [65], and references therein). The optical analog of the Larmor clock for classical electromagnetic waves based on Faraday effect lead us also to a complex time [37].

We saw, with the help of the GF formalism, that the two characteristic times appearing in the Larmor clock approach correspond to the real and imaginary components of a single quantity, which we define as a complex traversal (or reflection) time. In the subsection on the oscillatory incident amplitude, we also discussed that Leavens and Aers [51] arrived at a complex barrier interaction time, Eq. (52), by studying the shape distortion of the transmitted wave by the barrier.

It is in the Feynman path-integral approach where the concept of a complex time arises more naturally. Sokolovski and Baskin [55], using this kinematic approach to quantum mechanics, showed that a formal generalization of the classical time concept to the traversal time led to a complex quantity. The starting point for the Feynman path-integral approach [57] to the traversal time problem is the classical expression for the time that the particle spends in an arbitrary region [0, L], which can be calculated through the expression

$$\tau_{0L}^{\rm cl} = \int_0^T \theta(y(t'))\, \theta(L - y(t'))\, dt' \tag{73}$$

where θ is here the step function, equal to 1 when its argument is positive and zero otherwise. The two θ functions ensure that we only count the time while the particle is in the barrier region. To use Eq. (73) in the quantum regime one has to generalize the expression for the classical time by replacing the classical trajectory $y(t)$ in the previous expression by a Feynman path and average Eq. (73) over all possible paths that start at position 0 on the left side of the barrier and end at position L at time t. Each path is weighted by the quantity $\exp(iS\{y\})$, where

$$S\{y(t)\} = \int_0^t \left(\frac{m}{2} \left(\frac{dy}{dt'}^2 \right) - V(y(t')) \right) dt' \tag{74}$$

is the action associated with the path $y(t)$. As we are weighting each trajectory with a complex factor, it is natural to obtain a complex result for the average value. This weighting assumption has generated some controversies [4, 11, 66, 67]. Sokolovski and Baskin [55] arrived at the following complex time:

$$\tau_{0L} = i\hbar \int_0^L \frac{\delta \ln t}{\delta V(y)}\, dy \tag{75}$$

where $\delta/\delta V(y)$ represents the functional derivative with respect to the barrier potential.

This result, Eq. (75), is strictly equivalent to our expression of the traversal time. We would like to emphasize that this coincidence is quite natural, because in the tunnelling time problem we always deal with an *open* and *finite* system. The functional derivative with respect to the potential appearing in Eq. (75) is equivalent to a derivative with respect to energy plus a correction term proportional to the reflection coefficient (see Eq. (25)).

The modulus of this expression, Eq. (75), is the time that Büttiker [36] obtained for the tunnelling time in a square potential barrier and related to the Larmor clock times via

$$\operatorname{Re} \tau_{0L} = \tau_y \tag{76}$$

$$-\operatorname{Im} \tau_{0L} = \tau_z \tag{77}$$

Sokolovski and Connor [68] extended the Feynman path-integral approach to include the treatment of wavepackets. In their method the complex tunnelling formally appears as a transition element $\tau_{0L} = \langle \Psi_{\rm F} | \tau^{\rm cl} | \Psi_{\rm I} \rangle$ between the initial wavepacket $\Psi_{\rm I}$ and the final one $\Psi_{\rm F}$. Nevertheless, we have to note that Feynman and Hibbs [57] themselves do not associate any physical significance to transition elements.

Fertig [69, 70] avoided the problem of having to use wavepackets by considering restricted operators, for a fixed energy or for a fixed time. In this way, he was able to evaluate exactly the amplitude distribution for the traversal time for a rectangular barrier. He assumed that the weight of each path is proportional to $\exp(iS\{y\})$, where the action S is given by Eq. (74). He obtained the following amplitude distribution for the traversal time [70]:

$$F(\tau) = \frac{1}{2\pi t(E, V_0)} \int_{-\infty}^{\infty} e^{-i\omega t}\, t(E, V_0 - \omega)\, d\omega \tag{78}$$

where $t(E, V_0)$ is the transmission amplitude at energy E through a barrier of height V_0. With this probability amplitude distribution for the average traversal time for the square potential barrier, he arrives at

$$\langle \tau \rangle = -\frac{d\varphi}{dV_0} + \frac{i}{2} \frac{d \ln T}{dV_0}$$

which is the result of Sokolovski and Baskin [55].

The Wigner path distribution provides another approach to compute the traversal time. Jensen and Buot [71] used it to calculate the time for stationary waves, and Muga et al. [72] for wavepackets.

Yamada [73] claimed that the probability distribution of the tunnelling time cannot be properly defined. He arrived at this conclusion from the fact that Gell-Mann and Hartle's weak decoherence condition [74] does not hold for the tunnelling time due to the absence of a classical, dominant Feynman path. Yamada defined a range of values of the time and concluded that it is the only "speakable" magnitude.

Sokolovski [75] criticized the previous results on the grounds that the weak decoherence criterion is too restrictive. He showed that the probabilities for the outcomes of tunnelling time measurements can be described in terms of positive-operator-valued measurements, related to the interaction between the system and its environment.

3.3.1. Complex Time and Kramers-Kronig Relations

As it has just been shown, the concept of a complex time in the theory of the traversal time problem of electrons and electromagnetic waves (EMWs) has arisen in many approaches [4, 11, 37, 65]. In [37] was obtained, with the Faraday rotation scheme, a very similar result to Eq. (25) for the characteristic interaction time τ of an EMW. The Faraday rotation in the finite system, which is our magnetic clock, plays for light the same role as the Larmor precession for electrons [34, 36]. The emerging EMW is elliptically polarized and the major axis of the ellipse is rotated with respect to the original direction of polarization. All relevant information about both the angle of rotation and the degree of ellipticity is contained in a complex angle whose real part corresponds to the Faraday rotation, and whose imaginary part corresponds to the degree of ellipticity. This motivated us to associate a complex interaction time of the light in the region with magnetic field which can be written in terms of derivatives with respect to frequency as [37]

$$\tau(\omega) = -i\left[\frac{\partial \ln t}{\partial \omega} - \frac{r + r'}{4\omega}\right] = \tau_1(\omega) - i\tau_2(\omega) \qquad (79)$$

As was shown by Ruiz et al. [15], this is a general expression for the interaction time of an EMW with a one-dimensional region with an arbitrary index of refraction distribution, independent of the model considered. It can be rewritten in terms of the GF for photons analogously to Eq. (25), because all the general properties of the GF formalism for electrons which lead us to Eq. (25) are valid for any wave (sound or electromagnetic), whenever its propagation through a medium is described by a differential equation of second order [45].

Thus, most approaches indicate that the characteristic time associated with any wave (classical or quantum-mechanical) is a complex magnitude; which of the two components of this complex time is the most relevant depends on the experiment. It was shown in [15] that the real component $\tau_1(\omega)$ corresponds to the traversal time, and the main effect of the imaginary component $\tau_2(\omega)$ is to change the size of the wavepacket. Balcou and Dutriaux [16] experimentally investigated the tunnelling times associated with frustrated total internal reflection of light. They have shown that the real and imaginary parts of the complex tunnelling time correspond, respectively, to the spatial and angular shifts of the beam. Note that in most tunnelling experiments, instead of electrons, electromagnetic waves were used to exclude interaction effects (see, e.g., [10–23]).

It is known that the frequency dependence of the real and imaginary parts of certain complex physical quantities are interrelated by the Kramers-Kronig relations, for example, the real (dispersive) part of the complex dielectric function $\epsilon(\omega)$ to its imaginary (dissipative) part, the frequency-dependent real and imaginary parts of an electrical impedance, etc. [76]. The derivation of these relations is based on the fulfillment of four general conditions of the system: causality, linearity, stability, and that the value of the physical quantity considered is assumed to be finite at all frequencies, including $\omega \to 0$ and $\omega \to \infty$. If these four conditions are satisfied, the derivation of Kramers-Kronig relations is purely a mathematical operation which does not reflect any other physical properties or conditions of the system. These integral relations are very general and have been used in the theory of classical electrodynamics, particle physics, and solid-state physics as well as in the analysis of electrical circuits and electrochemical systems.

It is straightforward to show that the complex interaction time, $\tau(\omega)$, Eq. (79), is an analytical function of frequency in the upper half of the complex ω-plane (see, e.g., [76]). In other words, the four conditions mentioned above are fulfilled for the complex time (79) and the following relationship between the $\tau(\omega)$ and its complex conjugate $\tau^*(\omega)$ holds on the real axis (see Eq. (79)):

$$\tau(\omega) = \tau^*(-\omega) \qquad (80)$$

which means that the complex interaction time $\tau(\omega)$ has the following properties:

$$\tau_1(\omega) = \tau_1(-\omega), \quad \tau_2(\omega) = -\tau_2(-\omega) \qquad (81)$$

Therefore, the real part $\tau_1(\omega)$ is an even function of frequency and can have a finite value at zero frequency (for the slab we have $\tau_1^{\rm sl}(0) = L/vA$). As for the imaginary part $\tau_2(\omega)$, it is an odd function and must vanish in the limit of zero frequency: $\tau_2(0) = 0$. These conditions imply that the real and imaginary components of the time likewise obey Kramers–Kronig integral relations, and so we may write

$$\tau_1(\omega) - \tau_0 = \frac{2}{\pi}\mathbf{P}\int_0^\infty \frac{y\tau_2(y)}{y^2 - \omega^2}\,dy \qquad (82)$$

$$\tau_2(\omega) = -\frac{2\omega}{\pi}\mathbf{P}\int_0^\infty \frac{\tau_1(y) - \tau_0}{y^2 - \omega^2}\,dy \qquad (83)$$

where $\mathbf{P}$ means principal part and $\tau_0 = Ln/c$, that is, the crossing time in the dielectric system, without any boundary. In particular, if we make $\omega = 0$ in Eq. (83), we arrive at the so-called "macroscopic sum rule" for the complex interaction time

$$\tau_1(0) - \tau_0 = \frac{2}{\pi}\int_0^\infty \frac{\tau_2(y)}{y}\,dy \qquad (84)$$

Thus we see from Eq. (84) that in general if no imaginary component $\tau_2(\omega)$ exists at any frequency, then $\tau_1 = \tau_0$ always holds. In the case of the interaction time in the dielectric slab, the integral relations (82)–(84) can be verified, using the explicit expressions for the components of time (see Eqs. (85) and (86)).

The validity of the Kramers-Kronig relations for the complex interaction time has a rather deep significance because it may be demonstrated that these conditions are a direct result of the causal nature of physical systems by which the response to a stimulus never precedes the stimulus. It can serve also as a starting point to understand the origin of the complex time, and state that the interaction time for any classical or quantum-mechanical wave will always have two components: the real part $\tau_1(\omega)$ and the imaginary part $\tau_2(\omega)$. At this point it is worth mentioning that the experiments with, for example, undersized waveguides [23, 24] or periodic dielectric heterostructures [9, 10], where the so-called "superluminal velocities" have been observed for the barrier tunnelling time, need to be interpreted carefully.

For the dielectric slab, Eq. (79) leads us to the following expressions for the two time components [37]:

$$\tau_1^{sl}(\omega) = \frac{T\tau_0^{sl}}{2A}\left\{(1+A^2) + (1-A^2)\frac{\sin 2\Delta}{2\Delta}\right\} \qquad (85)$$

and

$$\tau_2^{sl}(\omega) = \frac{T\tau_0^{sl}}{2A}\frac{1-A^2}{2A}\left\{(1-A^2)\frac{\sin 2\Delta}{2} + (1+A^2)\frac{\sin^2\Delta}{\Delta}\right\} \qquad (86)$$

where $\tau_0^{sl} = L/v$ is the time that light with velocity $v = c/n_0$ would take to cross the slab, when reflection in the boundaries is not important, $\Delta = \omega\tau_0^{sl}$, $A = n_1/n_0$, n_0 is the refraction index of the slab, and n_1 is the refraction index of the two semi-infinite media outside the slab. T is the transmission amplitude for the slab in the absence of a magnetic field and is given by [63]

$$T = \left\{1 + \left(\frac{1-A^2}{2A}\sin\Delta\right)^2\right\}^{-1} \qquad (87)$$

The complex time components $\tau_1^{sl}(\omega)$, Eq. (85), and $\tau_2^{sl}(\omega)$, Eq. (86), are plotted against one another in the complex plane (see Fig. 6). We see that for small frequencies we have a skewed arc. With increasing frequency, the influence of the second terms in Eqs. (85) and (86), due to boundary effects, becomes less important and the curve, in the limit $\omega \to \infty$, approximates to an ideal circle.

Note that in the case of the Debye dispersion relations for the complex dielectric function $\epsilon(\omega)$, an ideal semicircle in the complex plane means that we deal with a single relaxation time. In our case it means that for high frequency or short wavelength, we deal with the classical crossing time, taking into account multiple reflection in the slab [15].

It is not difficult to show that in the limit $\omega \to \infty$ we have

$$(\tau_2^{sl})^2 + \left\{\tau_1^{sl} - \left[\frac{\tau_0^{sl}}{2A}(1+A^2) - r\right]\right\}^2 = r^2 \qquad (88)$$

which is the equation of a circle in the complex plane of $-\tau_2^{sl}$ and τ_1^{sl} with the center $\{\tau_0^{sl}/2A(1+A^2) - r, 0\}$ and with a radius given by

$$r = \frac{\tau_0^{sl}}{4A}\frac{(1-A^2)^2}{1+A^2}$$

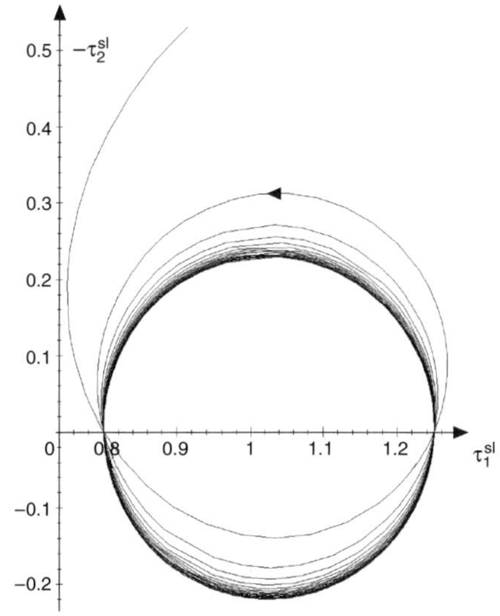

Figure 6. Complex plane interaction time diagram for a dielectric slab. The arrow indicates the direction of increasing frequency. Reprinted with permission from [86], V. Gasparian et al., "Handbook of Nanostructured Materials and Nanotechnology" (H. S. Nalwa, Ed.), Vol. 2, Chapter 11, 1999. © 1999, Elsevier Science.

Thus the two components, the real part $\tau_1(\omega)$ and the imaginary part $\tau_2(\omega)$, of the complex barrier interaction time for EMW are not entirely independent quantities, but connected by Kramers-Kronig relations. This means that the response to a stimulus never precedes the stimulus and thus the experiments [9, 10, 23, 24], where the so-called "superluminal velocities" have been observed need to be interpreted with respect to the fact that any classical or quantum-mechanical wave will always have two components; which of the two components is the most relevant depends on the experiment.

The validity of the Kramers-Kronig relations was only checked analytically for EMW, but in general this implies that they are also valid for all quantum particles represented by a differential equation of second order as indicated by the numerical calculations for the complex tunnelling time for electrons.

3.4. Escape Time of Electrons from Localized States

As we have seen in most approaches of the tunnelling time problem, only the scattering configuration in which the free electron (or wavepacket) is coming from the left (right) on a 1D arbitrary potential barrier has been considered. More than one tunnelling time is involved in this time problem, no matter whether we deal with the Büttiker-Landauer τ^{BL} or complex $\tau = \tau_1 - i\tau_2$ characteristic times (see Eqs. (12) and (24)). Furthermore, this seems not to be a peculiarity of a quantum-mechanical wave, but a general result, as the characteristic time associated with a classical wave is also a complex magnitude [37]. Which of the two components of this complex time is the most relevant depends on the experiment. For photons [37], the real part is proportional to the Faraday rotation or the density of optical modes, while the imaginary part gives the degree of ellipticity. In [16] the tunnelling times associated with frustrated total internal reflection of light were experimentally investigated, and the real and imaginary parts of the complex tunnelling time were shown to correspond to the spatial and angular shifts of the beam, respectively.

On the other hand, it is clear that there are other similar situations of practical interest which should be discussed in the same context of two time components as was done in the scattering problem. One of these situations is the escape of an electron from localized states in a quantum well with one or several surrounding barriers. The escape of an electron from a localized state in the quantum well connected to a continuum by a small barrier only by one side (Fig. 7) can be found, for instance, in miniaturized metal-oxide-semiconductor transistors, in which electrons arrive to the quantized accumulation or inversion layers after scattering and subsequently they can tunnel to the metal gate through a very narrow oxide layer which acts as the barrier [77]. Physically, this implies that the particles entering the quantum well region remain there for some time before being allowed to escape outside. Tunnelling escaping time has also been studied by transient-capacitance spectroscopy [78], where electrically injected electrons undergo an escaping process out of the quantum well, for example, in a three-barrier, two-well heterostructure [79, 80].

Let us consider a potential shape that includes a well and one surrounding barrier (see Fig. 7). A hard wall condition at $x = -w$ reduces the problem to escape to only one open channel, that is, transmission to the right.

To calculate the escape time $\tau^{\text{esc}}(k)$ of an electron from a quantum well when boundary effects can be neglected, we closely followed [81] and introduced the following complex quantity ($k = \sqrt{E}$):

$$\tau^{\text{esc}}(k) = -i\frac{d \ln t}{2k\, dk} \tag{89}$$

where $t = T^{1/2}e^{i\phi}$ is the complex amplitude of transmission of the electron through only the right barrier taking into account the hard wall condition at $x = -w$.

Using standard methods of quantum mechanics, it is easy to show that the Re and Im parts of the complex escape time

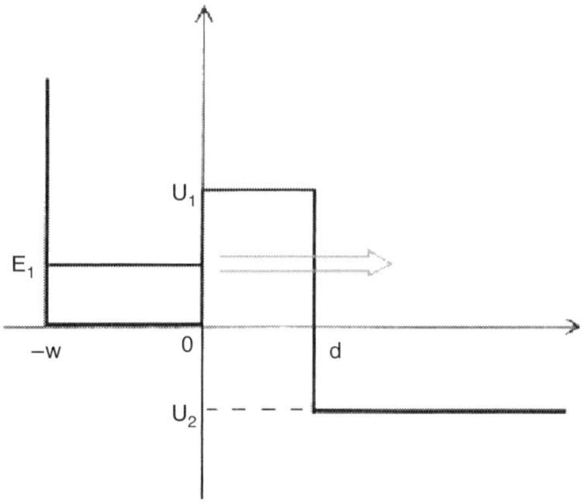

Figure 7. Schematic representation of the simplified potential profile, with a hard wall condition at $-w$.

$\tau^{\text{esc}}(k) = \tau_1^{\text{esc}}(k) - i\tau_2^{\text{esc}}(k)$ of an electron from a quantum well when boundary effects can be neglected and near the resonance energy E_1 and in the limit of an opaque barrier ($\gamma d \gg 1$) are given by [81]

$$\tau_1^{\text{esc}}(k_1) = \frac{\xi(1+\gamma w)(1+\beta^2)}{2k^2(1+\xi^2)}\exp(2\gamma d) \tag{90}$$

$$\tau_2^{\text{esc}}(k_1) = \frac{(1+\gamma w)(1+\beta^2)(1-\xi^2)}{4k^2(1+\xi^2)}\exp(2\gamma d) \tag{91}$$

where $\xi = \gamma/k_3$, $\gamma = \sqrt{E_1 - U_2}$, $\beta = k_1/\gamma$, and w, d, U_1, U_2 are defined in Figure 7.

Let us examine the relationship between the two components we have defined above and the lifetime expression

$$\tau_{\text{LT}}(k_1) = \frac{(1+\gamma w)(1+\beta^2)(1+\xi^2)}{16 \cdot k_1^2 \cdot \xi}\exp(2\gamma d) \tag{92}$$

which follows from an approximate perturbative approach based on Bardeen's perturbation Hamiltonian [82].

Despite the similarity between Eqs. (90), (91), and (92), they only allow a qualitative comparison at the bound level E_1. Since components τ_1^{esc} and τ_2^{esc} show a sharp variation around this energy, a comparison with Eq. (92) as a function of energy is interesting in order to study their behavior at the quasi-bound level, which is shifted with respect to E_1. Such comparison is shown in Figure 8 for the following parameter values: $U_1 = 1$, $U_2 = -1$, $w = 3$, for two d values ($d = 5$ in Fig. 8a, and $d = 2$ in Fig. 8b). An opaque barrier has been

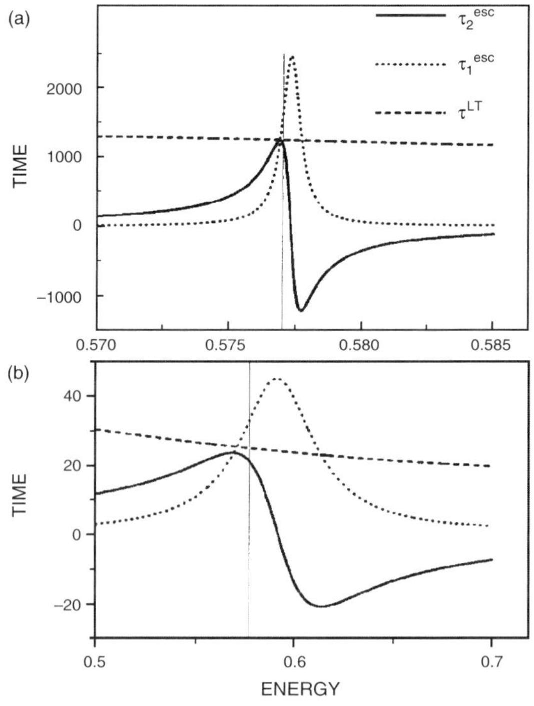

Figure 8. The time components as a function of energy for the potential profile depicted in Figure 7, with $U_1 = 1$, $U_2 = -1$, and $w = 3$, for two different d values: (a) $d = 5$, and (b) $d = 2$. The vertical line corresponds to the ground energy level when the barrier width is infinite. Reprinted with permission from [81], J. A. Lopez Villanueva and V. Gasparian, *Phys. Lett. A* 260, 286 (1999). © 1999, Elsevier Science.

chosen in Figure 8a to ensure the accuracy of the lifetime expression, but in Fig. 8b the estimated relative error for the lifetime expression is about 16%. The vertical line corresponds to the ground energy level when the barrier width is infinite. It is seen that the maximum of component τ_2^{esc} coincides with the τ_{LT} lifetime expression (92) at an energy very close to the bound level in the opaque barrier case (Fig. 8a). With the narrow barrier (Fig. 8b) both results deviate, but in this case the lifetime expression overestimates the lifetime by about 16% while the maximum of component τ_2^{esc} still provides an accurate value. Furthermore, this maximum is produced at an energy lower than the bound level, as predicted by the first-order perturbation theory.

Let us represent the complex time components $\tau_1^{esc}(k)$ and $\tau_2^{esc}(k)$ in the complex E plane. They are plotted against one another in Figure 9 and as it is seen, provide ellipse. This is what we expected, as the maxima of $\tau_1^{esc}(k)$ and $\tau_2^{esc}(k)$ are not the same. Nevertheless, there is a property of an ellipse that could be interesting: it is symmetric with respect to its main axis. Therefore, the maximum (and the minimum) of $\tau_2^{esc}(k)$ are found at the points where $\tau_1^{esc}(k)$ has a value of half its maximum. As this value is used to compute lifetime (width of the $\tau_1^{esc}(k)$ peak at half height), this width must be exactly the difference in energies between the maximum and minimum of $\tau_2^{esc}(k)$. It is easy to check that we have the following condition:

$$(E_{min} - E_{max})\tau_2^{esc} = 1$$

which confirms our previous conclusion concerning the fact that τ_2^{esc} at an energy close to the bound level in the well, coincides with the lifetime expression.

3.5. Wavepacket Approach

The delay time of a particle through a region can be directly calculated by following the behavior of its wavepacket. This approach has been criticized from different points of view, mainly due to the lack of causal relationship between the peaks or the centroids of the incident and transmitted wavepackets, and also because of the difficulties of an experimental setup to measure delay times. The dispersive character of electron propagation has been claimed as responsible for the acceleration of wavepackets under appropriate circumstances. High-energy components of the packet travel faster and are transmitted more effectively than the other components, and so the transmitted packet comes almost entirely from the front of the incident packet. However, similar results were also obtained for dispersiveness electromagnetic waves [15]. The fact that the transmitted wavepacket comes from the beginning of the incident wavepacket is mainly a consequence of interference effects. The results obtained from the wavepacket approach are similar to the results obtained with other approaches, and can be easily generalized to include finite size effects. In this section we review how to obtain delay times from the transmission coefficients.

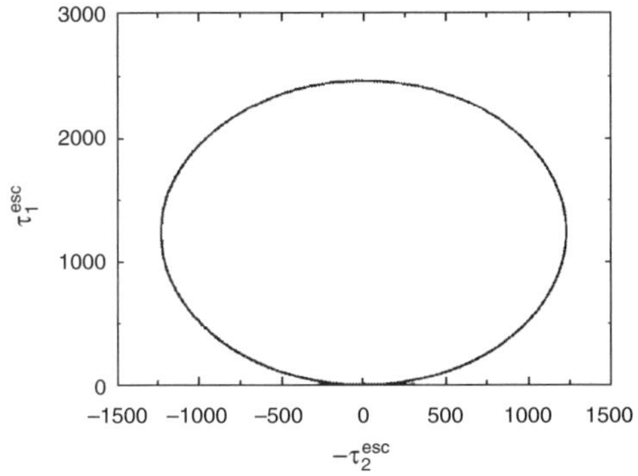

Figure 9. Complex plane escape time diagram for the quantum well used in this chapter.

Let us assume a region of interest, which in principle can be of any dimensionality, coupled to the outside by two 1D leads with a constant potential that we will assume equal to zero. We choose a coordinate system such that the incident lead extends from $-\infty$ to 0, and the other lead from L to $+\infty$. A gaussian wavepacket of spatial width σ_1 is incident from the incoming lead on the region of interest. This packet is characterized by a wave function of the form

$$\Psi(y,t) = \int_{-\infty}^{\infty} C \exp[-(k-k_0)^2/2(\Delta k)^2] \times \exp[iky - i\omega t]\, dk \qquad (93)$$

where C is a normalization constant, k_0 is the central wavenumber, $\omega = E/\hbar$, and $\Delta k = 1/\sqrt{2}\sigma_1$ is the spread of the packet in the wavenumber domain. The time evolution of this wavepacket is governed by the Schrödinger equation, although the results are directly applicable to any other type of wave, including classical electromagnetic waves. Part of the packet is transmitted and continues travelling outward along the second lead. Its wave function is given by

$$\Psi_t(y,t) = \int_{-\infty}^{\infty} C|t(k)|e^{i\varphi(k)} \exp[-(k-k_0)^2/2(\Delta k)^2] \times \exp[iky - i\omega t]\, dk \qquad (94)$$

$t(k)$ is the amplitude of transmission and $\varphi(k)$ is its phase, which here we prefer to write as functions of the wavenumber k. The functions $t(k)$ and $\varphi(k)$ contain all the relevant information to calculate the delay time of the electronic wave function due to the region of interest.

The general solution for finite size wavepackets has to be obtained numerically, but we can get close expressions for the delay time in the limit of very long wavepackets from a series expansion along the central wavenumber k_0. Let us assume that the wavepacket is so long (and so Δk so small) that $t(k)$, $\varphi(k)$, and $\omega(k)$ only change smoothly on the scale of Δk. Then, in evaluating $|\Psi_t(y,t)|^2$, where $\Psi_t(y,t)$ is given by Eq. (94), we can expand $t(k)$, $\varphi(k)$, and $\omega(k)$ to second order in $k - k_0$. We write the phase of the transmission amplitude as

$$\varphi(k) = \varphi(k_0) + \tau_1(\omega - \omega_0) + \frac{1}{2}\tilde{\tau}_1(\omega - \omega_0)^2 \qquad (95)$$

where ω_0 is the frequency corresponding to the central wavenumber, τ_1 is the first derivative of the phase $\tau_1 = d\varphi(\omega)/d\omega$, which roughly corresponds to the component τ_y of the Büttiker-Landauer time, and $\tilde{\tau}_1$ is its second derivative $\tilde{\tau} = d^2\varphi(\omega)/d\omega^2$. Analogously, we write the modulus of the transmission amplitude as

$$|\log t(k)| = |\log t(k_0)| + \tau_2(\omega - \omega_0) + \frac{1}{2}\tilde{\tau}_2(\omega - \omega_0)^2 \quad (96)$$

where $\tau_2 = d|\log t(\omega)|/d\omega$, and $\tilde{\tau}_2 = d^2|\log t(\omega)|/d\omega^2$. We can also expand the frequency ω in terms of k:

$$\omega = \omega_0 + v_g(k - k_0) + \frac{1}{2}a_g(k - k_0)^2 \quad (97)$$

v_g is the group velocity $v_g = d\omega/dk$, and a_g is its derivative with respect to k, $a_g = d^2\omega/dk^2$. For electrons, the group velocity is equal to $v_g = \hbar k/m$, while the group acceleration is $a_g = \hbar/m$.

If we keep terms up to second order in $k - k_0$, we can do analytically all the integrals appearing in the expression of the average value and the variance of y. After some trivial calculations we obtain that the average position $\langle y \rangle$ of the transmitted wavepacket as a function of time is given by

$$\langle y \rangle = (t - \tau_1)v_g\left[1 + \frac{\tau_2 a_g}{b}\right] - \frac{\tau_2\tilde{\tau}_1 v_g^3}{b} \quad (98)$$

where b is equal to

$$b = \frac{1}{(\Delta k)^2} - \tau_2 a_g - \tilde{\tau}_2 v_g^2 \quad (99)$$

We have chosen the phase of the incident wavepacket so that its peak is at the origin of coordinates in $t = 0$, in the absence of perturbations due to the presence of the barrier, that is, $\langle y_i \rangle(t = 0) = 0$. The peak of the transmitted wavepacket would be at the same position at a time τ implicitly defined by

$$\tau = t(\langle y_t \rangle = 0) \quad (100)$$

The barrier delays the gaussian wavepacket by an amount of time τ, which according to Eqs. (100) and (98) is equal to

$$\tau = \tau_1 + \tau_2\tilde{\tau}_1 v_g^2(\Delta k)^2 \quad (101)$$

This equation constitutes a useful tool for calculating the tunnelling time of an electron that has transversed a (rectangular) barrier as a function of the width of the incident wavepacket, Δk.

We can express Eq. (101) in terms of Büttiker–Landauer's times and their first derivatives. If we neglect the terms proportional to the reflection amplitudes in the expressions for Büttiker–Landauer's times, we have $\tau_1 = \tau_y$ and $\tau_2 = \tau_z$. Taking into account the dispersion relation for electrons, we also have

$$\tilde{\tau}_1 = \frac{1}{v_g}\tilde{\tau}_y \quad (102)$$

and a similar expression for τ_2 and τ_z. Thus Eq. (101) can be rewritten in the form

$$\tau = \tau_y + \tau_z\tilde{\tau}_y v_g(\Delta k)^2 \quad (103)$$

This is the main result of the long wavepacket approximation up to second order in Δk. We will check the limit of validity of this expression in the section on numerical results.

3.6. Time-of-Arrival Operator Approach

A direct attempt to obtain an observable for the time runs into the problem pointed out by Pauli [83] that the existence of such an operator would imply an unbounded energy spectrum, given the uncertainty relation between time and energy. The time-of-arrival is the natural candidate to become a property of the system, rather than an external parameter, and it plays a role similar to our traversal time.

Leon et al. [84, 85] have recently developed an interesting time-of arrival formalism, which we briefly describe. Classically, the time-of-arrival of a particle with coordinate q and momentum p at a point x is given by the expression

$$t_x(q, p) = \text{sgn}(p)\sqrt{\frac{m}{2}}\int_q^x \frac{dq'}{\sqrt{H(q', p) - V(q')}} \quad (104)$$

where $H(q, p)$ is the classical hamiltonian. Leon et al. (1996) showed that $t_x(q, p)$ is canonically conjugate to the hamiltonian $\{t_x(q, p), H(q, p)\} = -1$, where $\{\}$ are the Poisson brackets, but were unable to quantize this magnitude directly. Instead they considered first the time-of arrival for a free particle $t_{x0}(q, p) = m(x - q)/p$, which after symmetrization is quantized as

$$t_{x0}(q, p) = -e^{-ipx}\sqrt{\frac{m}{p}}q\sqrt{\frac{m}{p}}e^{ipx} \quad (105)$$

This operator is not self-adjoint, a difficulty related to the Pauli theorem. The measurement problem associated to the time-of-flight operator can be solved by means of a positive-operator-valued measure.

As a second step, Leon et al. [84] considered the canonical transformation that connects the free particle system with the general system including a potential $V(q)$. The Möller operators perform this canonical transformation. They are normally used to connect the free particle wavefunctions with the scattering and bound states of a given potential. In their formalism, Leon et al. applied these unitary operators to the time-of-arrival operator of the free particle in order to produce the corresponding operator in the presence of an arbitrary potential. Given this operator, one can then obtain the full distribution for the time-of-arrival of a particle at a position.

We believe that an especially interesting result of their approach is the average value of the time-of-arrival for a particle originally in state $\psi(k)$ which traverses a barrier with amplitude $\hat{t}(k)$. The expected average value is equal to

$$\tau = \frac{\int_{-\infty}^{\infty} dk|\hat{t}(k)\tilde{\psi}(k)|^2\tau_y(k)}{\int_{-\infty}^{\infty} dk|\hat{t}(k)\tilde{\psi}(k)|^2} \quad (106)$$

where $\tilde{\psi}(k)$ is the Fourier transform of $\psi(k)$. This expression has a very intuitive interpretation. The time corresponds to an average of the characteristic time τ_y over the transmitted components of the wavepacket. It is very easy to calculate this expression and can be applied to packets of any size. In the section on numerical results we will estimate the validity of this expression.

4. NUMERICAL RESULTS

The long wavepacket limit, when the spread of the wave function is longer than the size of the system, is well understood [86]. In all cases studied, the traversal and the reflection times correspond to the real component of time, τ_y. In this limit, the numerical problem reduces to the evaluation of the transmission and reflection amplitudes and their energy derivatives, which can be conveniently achieved through the use of the characteristic determinant method, introduced by Aronov et al. [45] and explained in Appendix A. Different similar mathematical methods, allowing us to take into account multiple interfaces consistently and exactly without the use of perturbation theory, have been proposed [87–91].

Numerical simulations are needed to clarify finite size effects. In this case we have to consider a specific wavepacket and evaluate its probability amplitude at different values of the time in order to calculate the amount of time taken to cross the system. In this section, we first review the results for periodic structures which, due to their complexity, are treated as nondispersive media, and so the conclusions are directly applicable to electromagnetic waves. Second, we consider the case of resonant tunnelling and we finish by presenting the results for finite size effects in electron tunnelling.

4.1. Periodic Structure

Let us consider a periodic arrangement of layers with potential V_1 and thickness d_1 alternating with layers with potential V_2 and thickness d_2. We assume that the energy is higher than max $\{V_1, V_2\}$, and so the wavenumber in the layers of the first and second type is $k_i = [2m(E - V_i)]^{1/2}/\hbar (i = 1, 2)$. In this case, the results for long wavepackets apply equally well to electromagnetic waves considering $k_i = \omega n_i/c$, where n_i is the index of refraction of the two types of layers. We concentrate on the simplest periodic case, which corresponds to the choice $k_1 d_1 = k_2 d_2$. This case contains most of the physics of the problem and is also used in most experimental setups [9]. Let us call a the spatial period, so $a = d_1 + d_2$. The periodicity of the system allows us to obtain analytically the transmission amplitude using the characteristic determinant method [40]:

$$t = e^{-ik_1 d_1} \left\{ \cos(N\beta a/2) - i\frac{\sin(N\beta a/2)}{\sin\beta a} \times \sqrt{\sin^2\beta a + \left[\frac{k_1^2 - k_2^2}{2k_1 k_2}\sin k_2 d_2\right]^2} \right\}^{-1} \tag{107}$$

where β plays the role of quasi-momentum of the system, and is defined by

$$\cos\beta a = \cos k_1 d_1 \cos k_2 d_2 - \frac{k_1^2 + k_2^2}{2k_1 k_2}\sin k_1 d_1 \sin k_2 d_2 \tag{108}$$

When the modulus of the RHS of Eq. (108) is greater than 1, β has to be taken as imaginary. This situation corresponds to a forbidden energy band. The term within brackets in Eq. (107) only depends on the properties of one barrier, while the quotient of the sine functions contains the information about the interference between different barriers. The transmission coefficient is equal to 1 when $\sin(N\beta a/2) = 0$ and β is different from 0. This condition occurs for

$$\beta a = \frac{2\pi n}{N} \qquad (n = 1, \ldots, N/2 - 1) \tag{109}$$

and we say that it corresponds to a resonant frequency.

For the reflection amplitude we have

$$r = te^{-ik_1 d_1}\frac{k_1^2 - k_2^2}{2k_1 k_2}\sin k_2 d_2 \frac{\sin(N\beta a/2)}{\sin\beta a} \tag{110}$$

With these expressions for the transmission amplitude, Eq. (107), and for the reflection amplitude, Eq. (110), we can calculate the traversal time through Eq. (25) and the reflection time via Eq. (31). From Eqs. (107) and (110), Ruiz et al. [15] calculated numerically the traversal time for electromagnetic waves considering a system of 19 layers ($N = 20$) with alternating indices of refraction of 2 and 1, and widths of 0.6 and 1.2, respectively. Their main conclusions are also applicable to the problem of an electron in a periodic potential. In Figure 10 we represent τ_1 and τ_2 for electromagnetic waves in a periodic system as a function of k_1. In the energy gaps, the traversal times are significantly smaller than the crossing time at the vacuum speed of light (horizontal line). The average of τ_1 with respect to wavenumber is equal to 22.8, and coincides with the classical crossing time, that is, for very short wavepackets, without including multiple reflections. It corresponds to the horizontal straight line in Figure 10.

The kinetic approach is suitable to study numerically the evolution of wavepackets with sizes of the order of the width of the region of interest. We will describe the numerical simulations of the time evolution of finite size wavepackets that cross the region of interest and measure the delay of the peak of the transmitted wave as a function of the size of the

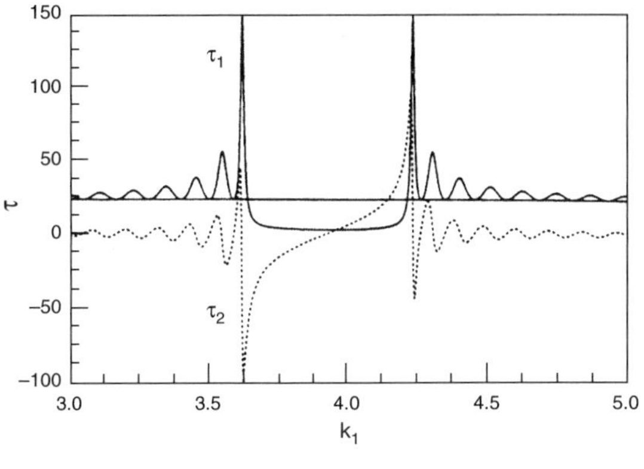

Figure 10. Traversal times versus the size of the wavepacket for a periodic system. The solid line corresponds to τ_1, and the dashed line to τ_2. The values of the parameters are $N = 20$, $n_1 = 2$, $n_2 = 1$, $d_1 = 0.6$, and $d_2 = 1.2$.

original packet. The simulations also calculate the change in size of the packets. First we consider the results for periodic structures without including dispersion effects, so we will use a nomenclature most appropriate for electromagnetic waves, although the results are equally valid for electrons, in the absence of dispersion, provided that we translate indices of refraction into their corresponding potentials. Later we will also present results about finite size effects in electron tunnelling.

Let us consider a three-dimensional layered system with translational symmetry in the Y-Z plane, consisting of N layers labelled $i = 1, \ldots, N$ between two equal semi-infinite media with a uniform dielectric constant n_0. The boundaries of the ith layer are given by y_i and y_{i+1}, with $y_1 = 0$ and $y_{N+1} = L$, so that the region of interest corresponds to the interval $0 \leq y \leq L$. Each layer is characterized by an index of refraction n_i. In the case of electrons, we assume that the energy E of the electron is higher than the potentials of the different layers and that the wavenumbers are inversely proportional to the indices of refraction; so the potential V_i in layer y is equal to $V_i = E(1 - (n_0/n_i)^2)$.

One calculates the position of the packet at different times and from this information one extracts the time taken by the packet to cross the region of interest. In particular, neglecting dispersion, one can measure the average positions $\bar{y}_1$ and $\bar{y}_2$ of the square of the modulus of the wavepacket at two values of t, t_1 and t_2, such that the packet is very far to the right of the structure at t_1 and very far to the left at t_2. These average positions are defined as

$$\bar{y}(t) = \int_{-\infty}^{\infty} y|\Psi(y, t)|^2 \, dy \tag{111}$$

The traversal time of the wavepacket through the region of interest is given by

$$\tau = t_2 - t_1 - \frac{(\bar{y}_2 - \bar{y}_1 - L)n_0}{c} \tag{112}$$

Although we refer to this time as a traversal time, it is learned that, strictly speaking, it is a delay time. Part of the interest of this type of simulations is to study how delay times relate to the previously obtained expressions for the traversal time.

Figure 11 represents the delay time versus the size of the wavepacket for two values of the central wavenumber, $k_0 = 3.927$ and $k_0 = 4.306$, which correspond to the center of the gap and to a resonance, respectively. There is again a strong similarity in the behavior of the traversal time and of the transmission coefficient [15]. The long wavepacket limit of the traversal time coincides with the characteristic time τ_1, while the short wavepacket limit is independent of wavenumber and equal to 29. The speed of the wave is greater than in vacuum for a wide range of sizes. The minimum size of the packets that travel faster than in vacuum is about 9, so that the corresponding width $2\sigma_1$ is very much the same as the size of the system. Velocities larger than in vacuum occur when the transmission coefficient is very small. In regions with a very small density of states, the traversal time is very short and, at the same time, transmission is very difficult due to the lack of states at the corresponding energies.

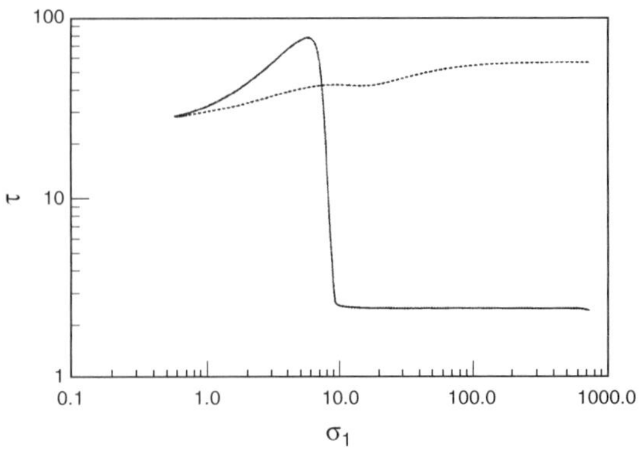

Figure 11. Traversal time versus the size of the wavepacket for a periodic system. The solid line corresponds to a central wavenumber $k = 3.927$, and the dashed line to $k = 4.3$.

The width of the transmitted packet σ_T is slightly smaller than the width of the incident packet σ_I. According to the results in the subsection on the wavepacket approach, we obtain that, in the absence of dispersion and up to second order in perturbation theory, this change in width depends on the derivatives with respect to frequency of τ_1 and τ_2. As the first of these derivatives is equal to zero in the center of the gap, one arrives at

$$\sigma_T^2 = \sigma_I^2 - \frac{v_g^2}{2} \frac{d\tau_2}{d\omega} \tag{113}$$

In order to check up to which sizes second-order perturbation theory is valid, Ruiz et al. [15] plotted $\sigma_I^2 - \sigma_T^2$ as a function of the size of the packet and compared it with the value of $(1/2)(d\tau_2/d\omega)$ obtained from the characteristic determinant. Second-order perturbation theory works adequately for a wide range of sizes and, in particular, for the sizes for which one obtains velocities larger than in vacuum. The error in the measurement of the traversal time of a single wavepacket is its width divided by its velocity. All the packets that travel faster than in vacuum are so wide that their uncertainty in the traversal time is larger than the traversal time itself and even larger than the time it would take a wave to cross the structure travelling at the same speed as in the vacuum.

4.2. Resonant Tunnelling

A double-barrier structure is a special case of a periodic system consisting of $N = 4$ interfaces with two evanescent regions separated by a propagating one. In the evanescent layers, the potential energy V_2 is larger than the energy of the electron E. The results of the previous part also applied to this case where one type of layers are evanescent. We merely have to replace k_2 by $-i\kappa$ where $\kappa = [2m(V_2 - E)]^{1/2}/\hbar$. (Correspondingly, $\sin k_2 d_2$ becomes $\sinh \kappa d_2$.) Double-barrier potential structures present resonant tunnelling, which has been studied for electrons since the early days of quantum mechanics [56]. Resonant tunnelling for electromagnetic waves is easier to carry out than corresponding experiments on electrons [24].

The traversal time τ for electromagnetic waves through a double-barrier structure was calculated by Cuevas et al. [40]

using the previous equations for the transmission and reflection amplitudes, Eqs. (107) and (110) with $N = 4$. They considered the finite resolution of the experimental devices by convoluting these expressions with a gaussian distribution function with a standard deviation of 6 MHz, which reproduces the same average height of the peak of the transmission coefficient as the corresponding experiments [24]. They found that at each resonance the Büttiker-Landauer traversal time is basically double the lifetime.

The dependence of the traversal time with frequency at a resonance is fairly universal. The phase of the transmission amplitude changes by an angle of π at each resonance, as predicted by Friedel's sum rule. Its frequency dependence can be fitted quite accurately by an arc tangent function. The time, proportional to the derivative of this phase, is a Lorentzian with the same central frequency and width as the Lorentzian corresponding to the transmission coefficient. As the lifetime τ_l of the resonant state is the inverse of the width of the transmission coefficient at half maximum, we conclude that it must be equal to half the traversal time at the maximum of the resonant peak:

$$\tau_l = \frac{1}{2}\tau_{\text{res}} \tag{114}$$

This result was obtained by Gasparian and Pollak [43] by considering the traversal time for an electron tunnelling through a barrier with loses, that is, with a decay time.

4.3. Finite Size Effects in Tunnelling

We have simulated the evolution of a finite wavepacket tunnelling across a rectangular barrier of height V_0 and length L and we have measured the delay of the peak of the transmitted wave as a function of the size of the original packet. The results are used to establish the limit of validity of the long wavepacket approach up to second order and of the time-of-arrival operator approach, previously analyzed.

As we have explained at the beginning of this section, one calculates the position of the packet at different times and from this information one extracts the time taken by the packet to cross the region of interest. Part of the interest of this type of simulations is to study how this time relates to the previously obtained expressions for the traversal time. In Figure 12 we represent the tunnelling time, τ, versus the width of the wavepacket in k-space, Δk, for an incident electron with a momentum $p_0 = 2.6$. The units are set by the choice $\hbar = m_e = 1$. The barrier parameters are $V_0 = 6$ and $L = 15$ (squares), 30 (circles). The numerical results are represented by geometrical figures, each one corresponding to a different barrier length L. The continuous curves represent the results obtained in the large wavepacket approach up to second order, Eq. (103). In the limit $\Delta k \to 0$ the tunnelling time tends to τ_y and is independent of L, in agreement with the so-called Hartman effect [19]. The continuous curves fit the numerical results relatively well up to values of Δk of the order of $3/L$. This limit of validity is quite general and has been checked for many different values of p_0, V_0, and L.

The width of the transmitted packet σ_T is slightly smaller than the width of the incident packet σ_I. According to the results in the subsection on the wavepacket approach,

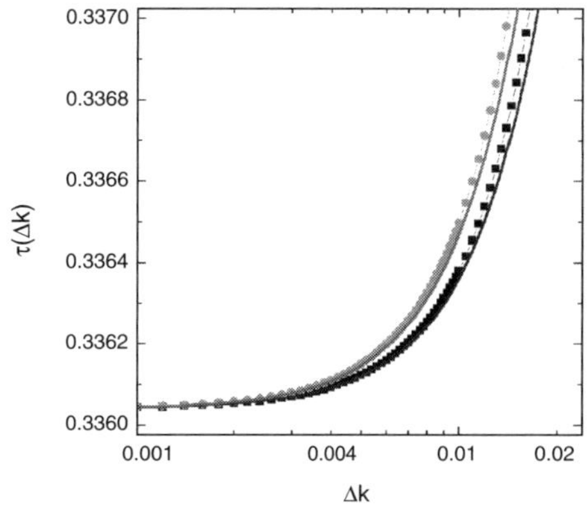

Figure 12. Tunnelling time τ versus the size of the wavepacket Δk for an incident electron with momentum $p = 2.6$ and barrier parameters $V_0 = 6$ and $L = 15$ (squares), 30 (circles). The curves represent the results for the long wavepacket approach up to second order.

we obtain that, up to second order in perturbation theory, this change in width depends on the derivatives with respect to frequency of τ_2. Second-order perturbation theory works adequately for a wide range of sizes.

We now want to analyze the applicability of the time-of-arrival operator approach previously studied. The average traversal time in this method is given by Eq. (106). Up to second order of approximation, this method coincides with the long wavepacket approach except for a term which depends on the second derivative of the Büttiker time τ_y with respect to the wavenumber k:

$$\tilde{\tilde{\tau}}_y \equiv \frac{\partial^2 \tau_y}{\partial k^2} \tag{115}$$

This term is negligibly small in most practical situations. Then it is feasible that the time calculated via Eq. (106) (to all orders in Δk) constitutes a very good approximation to the exact results.

In Figure 13 we show the tunnelling time τ versus Δk for an incident electron with momentum $p_0 = 2.5$ and rectangular barrier parameters $V_0 = 5$ and $L = 85$. The numerical results are represented by circles and the continuous

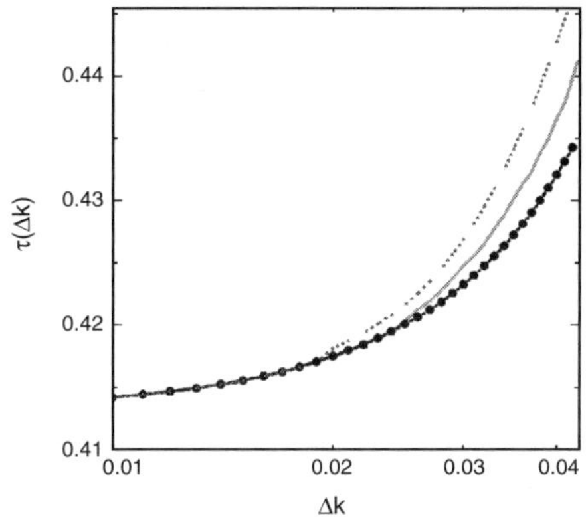

Figure 13. Tunnelling time τ versus Δk for an incident electron with momentum $p = 2.5$ and barrier parameters $V_0 = 5$ and $L = 85$. The numerical simulations (circles) and the results of the time-of-arrival approach (continuous curve) are similar up to sizes $\Delta k = 3.5/L$.

curve corresponds to Eq. (106), while the dashed curve corresponds to the wavepacket approach up to second order. We note that the weighting method fits the numerical results for a larger range of sizes. We estimate that it can be considered an excellent approximation for values of Δk smaller than $3.5/L$. The weighting method is also very practical since it only requires the evaluation of an integral in k and it can be applied to all kind of wavepackets.

5. CONCLUSIONS

In this review we have discussed the topic of tunnelling time in mesoscopic systems including nanostructures, particularly in 1D systems with arbitrary shaped potential. But the treatment of tunnelling time in "nanostructured materials" approaching the molecular and atomic scales is still open. In the field of tunnelling time, there are problems in any of the existing approaches, and we do not have a clear answer for the general question "How much time does tunnelling take?". Unfortunately no one of these approaches is completely adequate for the definition of the time in QM. Nevertheless, we note that all these different approaches can be consistently formulated in terms of Green's function, and their main differences can be fairly well understood.

For 1D systems we obtained closed expressions for the traversal and reflection times, Eqs. (25) and (31), in terms of partial derivatives of the transmission and reflection amplitudes with respect to energy. Results of other approaches can be related to these expressions and the main differences can be grouped into two categories: the complex nature of time and finite size effects.

Our conclusion about the complex nature of time is the following. It is clear that there are two characteristic times to describe the tunnelling of particles through an arbitrarily shaped barrier. (Similar conclusions can be reached for reflecting particles.) These two times correspond to the real and imaginary components of an entity, which we can choose as the central object of the theory. They are not entirely independent quantities, but are connected by Kramers-Kronig relations. Different experiments or simulations will correspond to one of their components or to a mixture of both. Büttiker and Landauer argue that these two times always enter into any physically meaningful experiment through the square root of the sum of their squares, and so claim that the relevant quantity is the modulus of the complex time.

As regards finite size effects, we believe that Eqs. (25) and (31) are exact, and adequately incorporate finite size effects. These effects correspond to the terms which are not proportional to derivatives with respect to energy. They are important at low energies and whenever reflection is important (as compared to changes in the transmission amplitude). Several approaches do not include finite size terms, since they implicitly consider very large wave functions. The WKB approximation, the oscillatory incident amplitude approach, and the wavepacket analysis, for example, do not properly obtain finite size effects. On the other hand, our GF treatment, based on the Larmor clock, the generalization of the time-modulated barrier approach, and the Feynman path-integral treatments arrived at exact expressions. In order to see that these expressions are all equivalent, one has to transform the derivative with respect to the average barrier potential, appearing in the time-modulated barrier approach, into an energy derivative plus finite size terms. The same has to be done with the functional derivative with respect to the potential appearing in the Feynman path-integral techniques.

Finite size effects can be very important in mesoscopic systems with real leads with several transmitting modes per current path. The energy appearing in the denominator of the finite size terms, Eqs. (25) and (31), corresponds in this case to the "longitudinal" energy of each mode, and so there is a divergence whenever a new channel is open. In the exact expressions there are no divergences; the problematic contributions of the finite size terms is cancelled out by the terms with energy derivatives.

GLOSSARY

WKB approximation This approximation takes advantage of the fact that the wavelength is changing slowly, by assuming that the wave function is not changed much from it would take if the height of the potential barrier were constant.

ACKNOWLEDGMENT

M. O. would like to acknowledge the Spanish Direction General de Investigatión Científica y Técnica for financial support, Project Number BFM2000-1059.

REFERENCES

1. de Jongh, in "Nanophase Materials" (C. Hadjipanayis and R. Siegel, Eds.), p. 349. Kluwer Academic, Dordrecht, 1994.
2. G. Schmid (Ed.), "Clusters and Colloids." VCH, Weinheim, 1994.
3. U. Simon and G. Schön, in "Handbook of Nanostructured Materials and Nanotechnology" (H. S. Nalwa, Ed.), Vol. 3, pp. 131–175. Academic Press, San Diego, 1999.
4. E. H. Hauge and J. A. Støvneng, *Rev. Mod. Phys.* 61, 917 (1989).
5. V. S. Olkhovsky and E. Recami, *Phys. Rep.* 214, 339 (1992).
6. C. R. Leavens and G. C. Aers, in "Scanning Tunnelling Microscopy III" (R. Wiesendanger and H. J. Güntherodt, Eds.), p. 105. Springer-Verlag, Berlin, 1993.
7. A. Enders and G. Nimtz, *Phys. Rev. B* 47, 9605 (1993).
8. A. Ranfagni, P. Fabeni, G. P. Pazzi, and D. Mugnai, *Phys. Rev. E* 48, 1453 (1993).
9. A. M. Steinberg, P. G. Kwiat, and R. Y. Chiao, *Phys. Rev. Lett.* 71, 708 (1993).
10. Ch. Spielman, R. Szipöcs, A. Stingl, and F. Krausz, *Phys. Rev. Lett.* 73, 708 (1994).
11. R. Landauer and Th. Martin, *Rev. Mod. Phys.* 66, 217 (1994).
12. Y.-P. Wang and D.-L. Zhang, *Phys. Rev. A* 52, 2597 (1995).
13. Y. V. Fyodorov and H.-J. Sommers, *J. Math. Phys.* 38, 1918 (1996).
14. R. Pelster, V. Gasparian, and G. Nimtz, *Phys. Rev. E* 55, 7645 (1997).
15. J. Ruiz, M. Ortuño, E. Cuevas, and V. Gasparian, *J. Phys. I France* 7, 653 (1997).
16. Ph. Balcou and L. Dutriaux, *Phys. Rev. Lett.* 78, 851 (1997).
17. G. L. Ingold and Y. V. Nazarov, in "Single Charge Tunnelling, Coulomb Blockade Phenomena in Nanostructures" (H. Grabert and M. Devoret, Eds.), p. 21. Plenum, New York, 1992).
18. L. Esaki, *Phys. Rev.* 109, 603 (1958).

19. T. E. Hartman, *J. Appl. Phys.* 33, 3427 (1962).
20. V. F. Rybachenko, *Sov. J. Nucl. Phys.* 5 (1967).
21. S. Bosanac, *Phys. Rev. A* 28, 577 (1983).
22. V. S. Olkhovski, E. Recami, F. Raciti, and A. K. Zaichenko, *J. Phys. I France* 5, 1351 (1995); A. M. Steinberg and R. Y. Chiao, *Phys. Rev. A* 49, 3283 (1994).
23. A. Enders and G. Nimtz, *J. Phys. I France* 2, 1693 (1992).
24. A. Enders and G. Nimtz, *Phys. Rev. E* 48, 632 (1993).
25. A. Enders and G. Nimtz, *J. Phys. I France* 3, 1089 (1993).
26. D. Mugnai, A. Ranfagni, R. Ruggeri, and A. Agresti, *Phys. Rev. E* 49, 1771 (1994).
27. A. Ranfagni, D. Mugnai, and A. Agresti, *Phys. Lett. A* 175, 334 (1993).
28. H. Grabert and M. H. Devoret, Eds., "Single Charge Tunnelling, Coulomb Blockade Phenomena in Nanostructures, p. 1. Plenum, New York, 1992.
29. D. V. Averin and K. K. Likharev, in "Single Charge Tunnelling, Coulomb Blockade Phenomena in Nanostructures" (H. Grabert and M. Devoret, Eds.), p. 311. Plenum, New York, 1992.
30. G. Schön and U. Simon, *Colloid Polym. Sci.* 273, 202 (1995).
31. J. G. A. Dubois, J. W. Gerristen, S. E. Shafranjuk, E. J. G. Boon, G. Schmid, and H. Kempen, *Europhys. Lett.* 33, 279 (1996).
32. Mandelstam and Tamm, *Izv. Akad. Nauk SSSR* 9, 122 (1945).
33. D. I. Blokhintsev, "Grundlagen der Quantenmechanik," p. 388 Harri Deutsch, Frankfurt/Main, 1963.
34. I. A. Baz', *Sov. J. Nucl. Phys.* 4, 182 (1967).
35. I. A. Baz', *Sov. J. Nucl. Phys.* 5, 161 (1967).
36. M. Büttiker, *Phys. Rev. B* 27, 6178 (1983).
37. V. Gasparian, M. Ortuño, J. Ruiz, and E. Cuevas, *Phys. Rev. Lett.* 75, 2312 (1995).
38. R. Landauer, *Nature* 365, 692 (1993).
39. Th. Martin and R. Landauer, *Phys. Rev. A* 45, 2611 (1992).
40. E. Cuevas, V. Gasparian, M. Ortuño, and J. Ruiz, *Z. Phys. B* 100, 595 (1996).
41. Y. Japha and G. Kurizki, *Phys. Rev. A* 60, 1811 (1999).
42. M. Büttiker and R. Landauer, *Phys. Rev. Lett.* 49, 1739 (1982).
43. V. Gasparian and M. Pollak, *Phys. Rev. B* 47, 2038 (1993).
44. V. Gasparian, M. Ortuño, J. Ruiz, E. Cuevas, and M. Pollak, *Phys. Rev. B* 51, 6743 (1995).
45. A. G. Aronov, V. Gasparian, and U. Gummich, *J. Phys.: Condens. Matter* 3, 3023 (1991).
46. V. Gasparian, T. Christen, and M. Büttiker, *Phys. Rev. A* 54, 4022 (1996).
47. D. J. Thouless, *Phys. Rep.* 136, 94 (1974).
48. V. Gasparian, B. L. Altshuler, A. G. Aronov, and Z. H. Kasamanian, *Phys. Lett. A* 132, 201 (1988).
49. C. R. Leavens and G. C. Aers, *Solid State Commun.* 67, 1135 (1988).
50. V. Gasparian, "Superlattices and Microstructures," to be published.
51. C. R. Leavens and G. C. Aers, *Solid State Commun.* 63, 1101 (1987); C. R. Leavens and G. C. Aers, in "Scanning Tunnelling Microscopy and Related Methods" (R. J. Behm, N. García, and H. Rohrer, Eds.), p. 59. Kluwer, Dordrecht, 1990.
52. V. Gasparian, M. Ortuño, J. Ruiz, and E. Cuevas, *Solid State Commun.* 97, 791 (1996).
53. G. García-Calderón and A. Rubio, *Solid State Commun.* 71, 237 (1989).
54. G. Iannaccone, *Phys. Rev. B* 51, 4727 (1995).
55. D. Sokolovski and L. M. Baskin, *Phys. Rev. A* 36, 4604 (1987).
56. J. A. Støvneng and E. H. Hauge, *Phys. Rev. B* 44, 1358 (1991).
57. R. P. Feynman and A. R. Hibbs, "Quantum Mechanics and Path Integrals." McGraw-Hill, New York, 1965.
58. M. Büttiker and R. Landauer, *Phys. Scr.* 32, 429 (1985).
59. M. Büttiker and R. Landauer, *IBM J. Res. Develop.* 30, 451 (1986).
60. Th. Martin and R. Landauer, *Phys. Rev. B* 47, 2023 (1993).
61. M. Büttiker, in "Electronic Properties of Multilayers and Low Dimensional Semiconductors" (L. E. J. M. Chamberlain and J. C. Portal, Eds.), p. 297. Plenum, New York, 1990.
62. M. Jonson, in "Quantum Transport in Semiconductors" (D. K. Ferry and C. Jacoboni, Eds.), p. 203. Plenum, New York, 1991.
63. L. D. Landau and E. M. Lifshitz, "Quantum Mechanics." Pergamon, New York, 1979.
64. A. P. Jauho and M. Jonson, *J. Phys.: Condens. Matter* 1, 9027 (1989).
65. Th. Martin, *Int. J. Mod. Phys. B* 10, 3747 (1996).
66. R. Landauer, *Ber. Bunsenges. Phys. Chem.* 95, 404 (1991).
67. A. P. Jauho, in "Hot Carriers in Semiconductor Nanostructures: Physics and Applications" (J. Shah, Ed.), p. 121. Academic Press, Boston, 1992.
68. D. Sokolovski and J. N. L. Connor, *Phys. Rev. A* 42, 6512 (1990).
69. H. A. Fertig, *Phys. Rev. Lett.* 65, 2321 (1990).
70. H. A. Fertig, *Phys. Rev. B* 47, 1346 (1993).
71. K. L. Jensen and F. Buot, *Appl. Phys. Lett.* 55, 669 (1989).
72. J. G. Muga, S. Brouard, and R. Sala, *Phys. Lett. A* 167, 24 (1992); S. Brouard, R. Sala, and J. G. Muga, *Europhys. Lett.* 22, 159 (1993); S. Brouard, R. Sala, and J. G. Muga, *Phys. Rev. A* 49, 4312 (1994).
73. N. Yamada, *Phys. Rev. Lett.* 82, 3350 (1999).
74. M. Gell-Mann and J. B. Hartle, in "Proceedings of the 3rd International Symposium on the Foundations of Quantum Mechanics in the Light of New Technology" (S. Kobayashi, H. Ezawa, Y. Murayama and S. Nomura, Eds.). Physical Society of Japan, Tokyo, 1990.
75. D. Sokolovski, *Phys. Rev. A* 66, 032107 (2002).
76. L. D. Landau and E. M. Lifshitz, "Electrodynamics of Continuous Media." Pergamon, New York, 1982.
77. F. Rana, S. Tiwary, and D. A. Buchanan, *Appl. Phys. Lett.* 69, 1104 (1996).
78. E. Martinet, E. Rosencher, F. Chevoir, J. Nagle, and P. Bois, *Phys. Rev. B* 44, 3157 (1991).
79. H. Z. Zheng, H. F. Li, Y. M. Zhang, Y. X. Li, X. Yang, P. Zhang, W. Zhang, and J. F. Tian, *Phys. Rev. B* 51, 11128 (1995).
80. J. Zinn-Justin, "Quantum Field Theory and Critical Phenomena." Oxford University Press, London, 2002.
81. J. A. Lopez Villanueva and V. Gasparian, *Phys. Lett. A* 260, 286 (1999).
82. J. Bardeen, *Phys. Rev. Lett.* 6, 57 (1961).
83. W. Pauli, "Handbuch der Physik" (S. Flugge, Ed.), Vol. V/1. Springer-Verlag, Berlin, 1926.
84. J. Leon, J. Julve, P. Pitanga, and F. J. de Urres, *Phys. Rev. D* 61, 062101 (2000).
85. J. Leon, *J. Phys. A* 30, 4791 (1997).
86. V. Gasparian, M. Ortuño, G. Schön and U. Simon, in "Handbook of Nanostructured Materials and Nanotechnology" (H. S. Nalwa, Ed.), p. 513. Academic Press, San Diego, 1999.
87. F. García-Moliner and J. Rubio, *J. Phys. C: Solid State Phys.* 2, 1789 (1969).
88. B. Velicky and I. Bartoŝ, *J. Phys. C: Solid State Phys.* 4, L104 (1971).
89. F. García-Moliner, *Ann. Phys.* 2, 179 (1977).
90. H. Ueba and S. G. Davison, *J. Phys. C: Solid State Phys.* 13, 1175 (1980).
91. E. Louis and M. Elices, *Phys. Rev. B* 12, 618 (1975).

Encyclopedia of Nanoscience and Nanotechnology
www.aspbs.com/enn

Embedded Metal and Alloy Nanoparticles

K. Chattopadhyay, V. Bhattacharya
Indian Institute of Science, Bangalore, India

CONTENTS

1. BASIC CONCEPTS

1.1. Introduction

Nanostructured materials can be defined as those materials which exhibit microstructure having the characteristic length scale of the order of a few nanometers in any dimensions. Based on dimensions, the nanostructured materials may be classified into many categories [1]. These may be of zero dimension (e.g., clusters), one dimension (e.g., single crystalline multilayers), two dimension (e.g., filamentary rods), and three dimension (e.g., embedded nanoparticles). In this article, only the embedded nanoparticles will be discussed.

Gleiter [2, 3] had first drawn attention to the fact that a higher volume percentage of the atoms (about 30% for very fine sizes) at the interfaces in nanocrystalline materials makes these materials different from the coarse-grained crystalline materials or glasses. This is because nanocrystalline materials have a significant contribution from the atoms configured at the grain boundaries. This is in addition to the effects arising from the fine-sized crystalline grains. Fine-grained materials share the following features: spatial confinement of atomic domain to less than 100 nm; a significant fraction of atoms associated with the interface and a higher probability of interactions between the constituent domains [4]. With such reduced length-scale, the physical and chemical behavior exhibited by these materials is different from free atoms or a bulk solid of same chemical composition [5].

Some of the examples of the size-related effects manifested in these nanostructured materials are "quantum-size effects" because of the spatial confinement of the delocalized valence electrons, alteration in the "many-body" atomic phenomena like melting (by lattice vibrations), suppression of the lattice-defect mechanisms like dislocation generation, and migration and alloying in immiscible systems. Exciting properties like giant magnetoresistance, superplasticity, optical transparency, etc. appear in materials below a critical (nanometric) length scale. This set of unique and improved properties has enhanced the drive for the study of these materials from both technological and fundamental viewpoints in the recent past. Advances in the high performance materials depend on the control of size, distribution, morphology of the constituent grains, and phases present at the nanoscale, nature of the interface between the constituent phases, and the interaction across the interface. Consequently, a significant amount of research is devoted to the controlled synthesis and property evaluation. The materials with embedded nanostructures are a special class of nanomaterials which are very promising because of the opportunities for materials development. This comes from the observations that the behavior of embedded nanoparticles is often different from both the free nanoparticles and the bulk. The combination of nanometric dimension and the type of constraining matrices can produce materials with new behavior. The properties of the embedded nanoparticles are intimately associated with the morphology and crystallography of these particles. However, several issues like the morphology of these small particles and their transformation behavior, are still not fully understood and are the subject of vigorous activities. The research on embedded nanomaterials covers all kinds of materials. However, in this article, we shall primarily focus on embedded metals and alloy particles, which have attracted significant attention in recent times.

1.2. Effects Controlling the Properties of Nanoembedded Materials

The properties of nanoscaled materials are different from that of bulk materials due to the changes in size, atomic configuration, and chemical structure. We will briefly discuss the different factors that lead to these changes.

ISBN: 1-58883-059-4/$35.00

Encyclopedia of Nanoscience and Nanotechnology
Edited by H. S. Nalwa
Volume 3: Pages (217–232)

1.2.1. Size Effect

The most obvious effect of reducing the particle size is the increase in surface and interface areas. The atoms lying at the surfaces or interfaces have a different environment both in terms of coordination and chemistry, since the structure of the grain boundaries is significantly different from that of the bulk crystal. The structure depends on orientation of the neighboring grains and interaction of the atomic potentials across the boundaries. The end result is a more open structure with local variation. Such a structure is also subjected to a relaxation process as the atoms try to arrive at a lower energy configuration. This difference in structure and chemistry of the boundary phase becomes important at nanoscale, since the boundary volume increases significantly with the reduction in the size of the particles. The above reasoning is equally applicable to free particles as well as for nanoembedded particles. One can estimate the volume fraction of the embedded particle interface regions by initially assuming a spherical shape and a reasonable thickness of the boundary where the effect of interactions of the potentials of different grains is significant. For the embedded case, the amount of this boundary phase will also depend on the volume fraction of these materials. Figure 1 gives a typical plot of the volume fraction of the boundaries as a function of size for three different volume fractions. The material, therefore, behaves like a two-phase composite and depending on the property of the boundary phase, can yield novel properties. Besides the effect of surface energy, one can anticipate additional effects. The presence of lattice defects near the interface represents a constrained state, which is structurally different from the constrained state of crystals or glasses. When the volume fraction of the defects becomes comparable to the volume fraction of the crystalline regions, this has a tremendous effect on the change in properties.

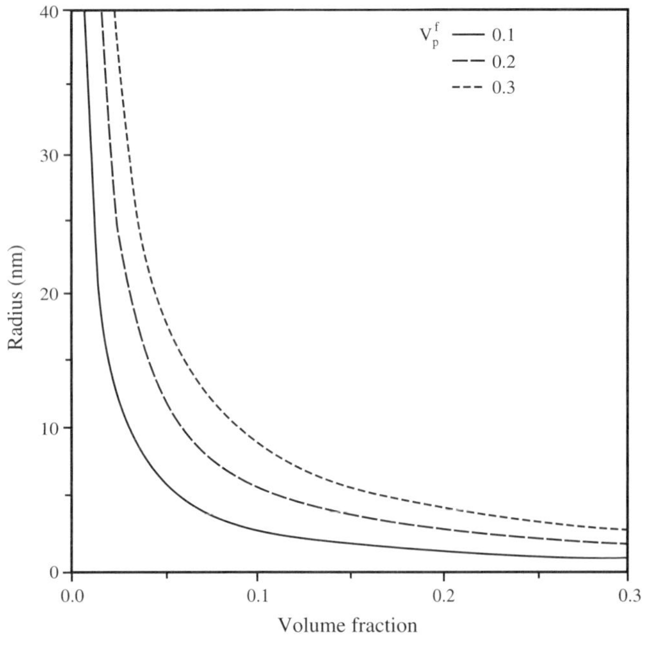

Figure 1. Typical plot of the volume fraction of the boundaries as a function of size for three different volume fractions.

In case the size of the particles is reduced such that the critical length scale of physical phenomena becomes comparable (viz. mean-free path of electrons, coherence length) with the characteristic microstructural length scale, size seems to play a key role in dictating the properties of such materials. For example, the band spacing is expected to expand with small confinement, and magnetocrystalline anisotropy can get evened out leading to a more isotropic behavior. With small size, one can observe significant changes in the alloying behavior. This leads not only to a change in the chemistry of the phases present but also to their structures. We will present in brief some of the recent results highlighting this scenario in the subsequent sections.

2. NANOEMBEDDED METALLIC PARTICLES

2.1. Synthesis

There exist several synthesis routes by which nanoembedded particles of metals and alloys can be manufactured. In the following sections, we briefly summarize some of the more popular techniques used in recent times.

2.1.1. Rapid Solidification

One of the popular techniques to synthesize nanoembedded metallic particles is by rapid solidification. This processing technique involves the achievement of a high cooling rate in the range 10^4–10^7 K/s, and significant undercooling of the melt that leads to a large departure from equilibrium conditions. This process involves the ejection of a stream of molten metal/alloy into a thin layer or fine droplets, which is brought in intimate contact with a cooling medium (solid substrate). The heat transfer is primarily by conduction through a solid substrate. The primary requirement for the synthesis of nanomaterials from liquid involves a high nucleation rate and a low growth rate of the solid phases involved in the liquid-solid phase change. The high nucleation rate can be achieved by increasing the undercooling and kinetically avoiding the heterogeneous nucleation sites. The principle of rapid solidification processing of embedded nanomaterials incorporates two major steps. In the first step, the dispersed phase should nucleate in the melt with a high nucleation rate and a low growth rate. In the second step, the matrix phase should nucleate and grow at a very high growth velocity, enabling trapping of the dispersed solid phase coexisting in the melt [6–7]. With a judicial choice of systems and processing conditions, we can observe that both the above processes are achievable in the rapid solidification process. Johns [8] and Cahn [9] have reviewed the rapid solidification process technology. There are several processes that can be classified under rapid solidification processing. A brief description of some of these is given below.

Atomization The process of gas atomization involves formation of fine droplets from a molten metal when it is impacted with an inert gas. The transfer of kinetic energy from the atomizing fluid to the molten metal results in the process of atomization. The cooling rates of the particles are controlled by several different parameters. These include thermal properties of the gas, and the parameters controlling the heat and fluid flow like viscosity and conductivity of the cooling media and the size of the droplets. Generally, with all the other parameters remaining constant, the cooling rate increases with a decreasing size of the droplets. Thus, finer sizes experience larger cooling rates and the melt experiences larger deviation from the equilibrium conditions.

Chill Block Melt Spinning In chill block melt spinning [10], which emerged as the most popular technique of rapid solidification, the molten material is forced through a nozzle onto a rotating wheel. The wheel is generally made of a good thermal conductor. A schematic of the melt spinning set up is shown in Figure 2. The heat transfer from the melt is primarily by conduction. The process parameters are optimized to obtain a mass flow through the nozzle in such a way that a dynamically stable pool of molten metal or alloy forms on the wheel surface from which the tape is spun out due to the rapid solidification at the pool/wheel interface. Thus, the rate of solidification is directly related to the surface velocity of the wheel. All the process parameters involved in melt spinning viz. nozzle diameter, nozzle to substrate distance, melt ejection pressure, velocity of the wheel, are critical and they control the length of puddle of the molten metal. The thickness of the ribbon is generally of the order of 10–100 μm and width of the sample varies from 1–3 mm. The thickness of the ribbon "t" is given by the expression [11]

$$t = (\alpha_T \cdot \Delta T / V_R \cdot \Delta T) \quad (1)$$

where α_T is the empirical heat transfer coefficient, V_R is the linear velocity of the substrate (that is, the speed of the rotating wheel), ΔT is the temperature difference between the opposite sides of the ribbon, ΔH is the latent heat per unit volume of the liquid metal, and l is the contact zone between the liquid metal and substrate.

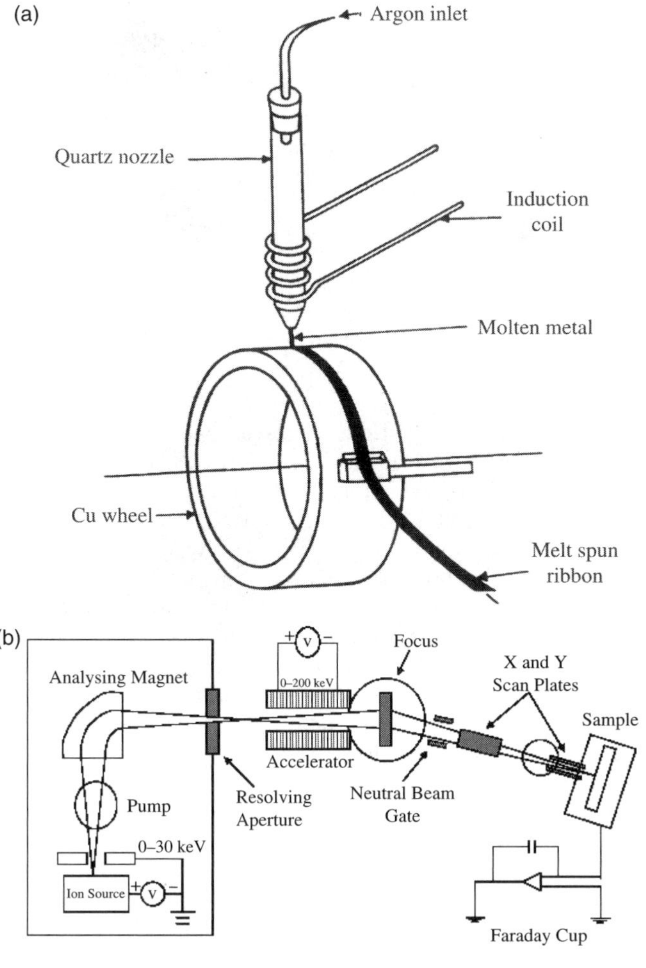

Figure 2. (a) A schematic of the melt spinning set-up is shown here. (b) A schematic showing the ion implantation set-up.

2.1.2. Mechanical Alloying

Mechanical alloying is a solid-state powder processing technique where mechanical energy is used to drive the system away from the equilibrium. This results in the development of highly nonequilibrium structures and microstructures. One of the ways the mechanical energy that is imparted on the sample can be stored is by creating a very large number of interfaces. This results in the reduction of the scale of the microstructures. The physical mechanisms responsible in case of ductile materials are repeated fracture, welding and rewelding of powder particles. For brittle materials, the process of fracture dominates the process although the actual mechanism is less clearly understood. Benjamin and his colleagues at Paul D. Merica Research Laboratory of the International Nickel Company (INCO) developed the process around 1976 [12]. Originally developed to produce oxide dispersion strengthened (ODS) nickel and iron-based superalloys for aerospace industry, mechanical alloying has been used to synthesize a variety of equilibrium and nonequilibrium alloy phases. The nonequilibrium phases synthesized include supersaturated solid solutions, amorphous alloys, metastable crystalline, and quasi-crystalline phases and nanostructured materials [13–15]. The amorphization of alloys by this technique was first obtained by grinding a Y-Co intermetallic compound in 1981 [16] and the Ni and Nb powders in the Ni-Nb system in 1983 [17]. Furthermore, it was realized that powder mixtures can be mechanically activated to induce chemical reactions. Known as mechanochemical reactions, this can produce at room temperature nanocomposites and a variety of commercially important materials having ultrafine microstructure [18]. Efforts are also underway since the early 1990's to understand the process fundamentals of mechanical alloying through modelling studies [19]. Maurice and Courtney [19] have developed a simple model for collision events involving ductile elements. Murty and Ranganathan [15] have prepared milling maps, taking into account the various parameters involved during milling like milling intensity, ball-to-powder weight ratio and the number of balls. Murty and Ranganathan [15] converted these factors into two energy parameters, that is, (i) energy of the balls, and (ii) total energy. The milling maps give the milling conditions required for amorphization. Recently, mechanical alloying has been used extensively to synthesize nanocrystals of pure metals and alloys [20, 21]. The process is also promising to synthesize embedded nanoparticles.

2.1.3. Physical Vapor Deposition

The physical vapor deposition (PVD) process involves the creation of a vapor phase species via evaporation, sputtering, or ablation of a target material. During transportation, the vapor phase species may undergo collision and ionization. Subsequently, it condenses onto a substrate where nucleation and growth occurs leading to the formation of films or droplets. The physical vapor evaporation process takes place in a vacuum, and therefore vacuum plays an important role on the deposition and growth of films. It is possible to introduce a reactive species in the chamber to initiate a chemical reaction with the vapor species prior to deposition. Known as reactive sputtering/ablation, this opens up the new possibility to synthesize nanoscaled materials including embedded nanoparticles. A detailed description of the PVD process is available in literature [22–24].

2.1.4. Sol–Gel

The evolution of colloidal solutions with rigidity (gel) from a liquid-like mixture of alkoxides and inorganic salts in a common solvent, which does not have the ability to resist shear stress (sol), is the essence of the sol–gel route [25]. The technique gradually became known as the sol–gel process because of the sudden characteristic change of viscosity at the onset of gel transition [25–27]. Sol–gel processing of nanocomposite materials has attracted much attention, because the process is simple and the conversions can occur readily with a variety of precursors available and reactions take place at room temperature. The gel generally contains a dispersed component in a dispersion medium and the dispersed component has three-dimensional networks [28]. The conversion of sol to gel and vice versa is brought about by a change in temperature, agitation, and peptization. Diphasic xerogels (gels dried conventionally in air) can be prepared by controlling the homogeneity of the sol. Two phases develop when the material is heterogeneous (*i.e.*, when the liquid has a considerable fraction of metal ions) or by the later addition of the metal phase precursor [29]. The derived gel is then a mixture of ceramic oxide and metal oxide. Under a suitable reduction environment, an extreme finely dispersed metallic phase in ceramic matrices can be obtained. Bhattacharya et al. [30, 31] has synthesized dispersion of lead in silica and alumina matrix. In the case of silica, the matrix is glassy while in the case of alumina, they report that the matrix is composed of ultrafine γ alumina. In both cases, the dispersoids of lead is of 10 to 20 nm in size. A typical micrograph of lead in alumina is shown in Figure 3. Iron nanoparticles are produced in a alumina matrix using Al (Osec-Bu)$_3$ as a sol–gel precursor [32]. The Fe/alumina nanocomposite exhibits weak saturation magnetization. Recently, Co nanocomposites have been synthesized using zirconia-modified silica aerogels [33]. Thus, sol–gel [34] has formed an excellent technique in recent times for synthesis of nanocomposites and embedded nanomaterials.

2.1.5. Ion Implantation

Ion implantation is a high technology approach used to modify surface properties of materials. The process is similar to a coating process, but it does not involve the addition of a layer on the surface. Originally developed for use in semiconductor applications, and, in fact, still used extensively in that capacity today, ion implantation utilizes highly energetic beams of ions (positively charged atoms) to modify surface structure and chemistry of materials at low temperature. The ion implantation process is conducted in a vacuum chamber at very low pressure (10^{-4}–10^{-5} torr). Large numbers of ions (typically 10^{16}–10^{17} ions/cm^2) bombard and penetrate a surface, interacting with the substrate atoms immediately beneath the surface. The typical depth of ion penetration is a fraction of a micron (or a few millionths of an inch). The interactions of the energetic ions with the material modify the surface, providing it with significantly different properties than the remainder of the material. Specific property changes depend on the selected ion beam treatment parameters—for instance, the particular ion species, energy, and the total number of ions that impact the surface. Ions are produced via a multistep process in a system. Ions are initially formed by stripping electrons from the source atoms in plasma. The ions are then extracted and passed through a mass-analyzing magnet, which selects only those ions of

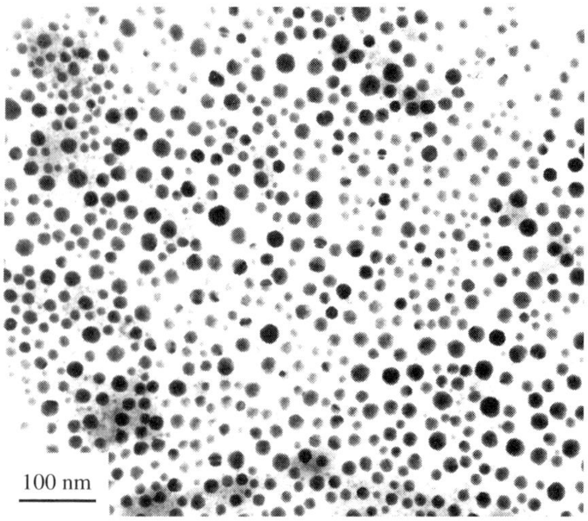

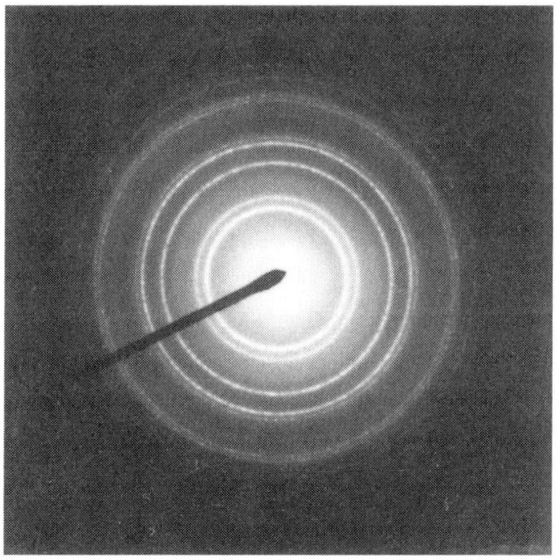

Figure 3. (Top) Typical bright field transmission electron micrograph showing Pb dispersed in γAl_2O_3 matrix, (Bottom) corresponding electron diffraction pattern. Reprinted with permission from [60], P. Bhattacharya, 2000.

a desired species, isotope, and charge state. The beam of ions is then accelerated using a potential gradient column. Typical ion energies are 10–200 keV. A series of electrostatic and magnetic lens elements shape the resulting ion beam, and scans it over an area in an end station containing the parts to be treated. Ion implantation offers numerous advantages for treating component surfaces. A primary benefit is the ability to selectively modify the surface without detrimentally affecting the bulk properties, largely because the process is carried out at low substrate temperatures. The process is also extremely controllable and reproducible and can be tailored to modify different surfaces in desired ways. In recent times, a significant amount of work on nanoembedded materials has been carried out on samples prepared by ion implantation [35–37]. A schematic of the ion implantation set-up is shown in Figure 2b. The implantation of ions by high-energy ion beam onto a substrate yield embedded particles with fairly uniform distribution and ultrafine sizes. The sizes of the particles are generally less than 20 nm and their density can be controlled by irradiation time. Johnson et al. at Lawrence Berkeley National Laboratory (LBNL) [38] have shown in recent times that it is possible to follow the deposition process by real-time electron microscopy/Rutherford backscattering.

2.2. Embedded Nanoparticles of Metals and Alloys in Inorganic Matrix

The most popular techniques for the synthesis of nanoembedded metallic particles in metallic matrix are via rapid solidification and ion implantation. Nanoembedded metallic particles can be synthesized by physical vapor deposition of alloys. Such films are often termed granular thin films and are the subject of great interest due to the advent of giant magnetoresistance and spintronics. Recently, dispersions of low melting metallic particles like lead, indium, tin, and bismuth in high melting metallic matrices have attracted major interests in the scientific community. In the following sections, we shall summarize some of the current work.

2.2.1. Synthesis of "Soft Nanoembedded" Particles by Rapid Solidification in Systems Exhibiting Miscibility Gap

Soft metallic phases of elements like Pb, Bi, In, and Sn can be dispersed in a higher melting point matrices of metals and alloys provided they form an immiscible alloy system. These soft, dispersed nanoembedded particles can be produced by rapid solidification processing (RSP) from the liquid phase in monotectic systems. The discussion below is restricted to those in the low-volume fraction regime of the dispersed phase because of the interest in developing a microstructure with fine dispersions of low-melting metals and alloys in a high-melting matrix. Figure 4 shows schematic of a typical binary monotectic phase diagram exhibiting a miscibility gap. The figure shows the possible metastable extensions mentioned above [39]. In order to evaluate the possible pathways of the microstructural evolution, one needs to determine, using thermodynamic calculations, the metastable phase diagram of the system. These

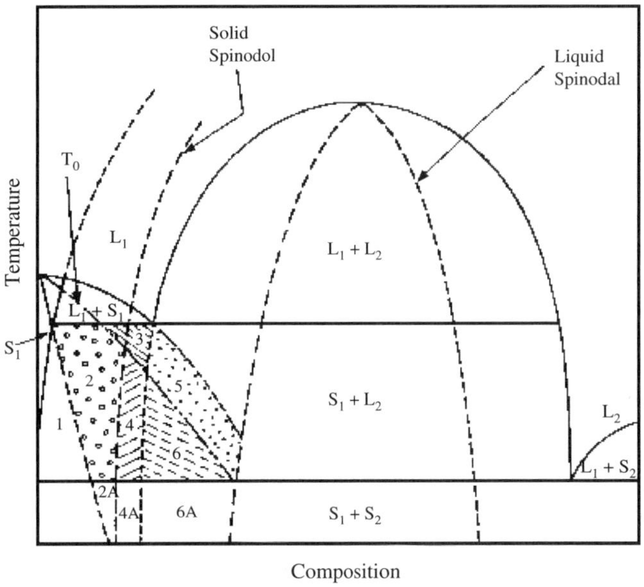

Figure 4. A typical monotectic phase diagram showing all the metastable extensions.

include the metastable extrapolation of the binodal and spinodal lines for liquid and a determination of equal, free-energy curve (T_o) between solid and liquid. In addition, one needs to determine the limit set by homogenous nucleation in case we are evaluating the stability of the liquid during quenching.

During solidification, the process of the development of the microstructure will be different, depending on the composition of alloy melt, that is, hypomonotectic or hypermonotectic. When an alloy melt is cooled through the immiscible domain (hypermonotectic composition), the liquid (L) phase separates into two liquids ($L_1 + L_2$). Depending on the melt composition, two possibilities exist that govern the kinetics of this phase separation. Spontaneous decomposition of liquid ($L \rightarrow L_1 + L_2$) occurs if the melt is cooled through the spinodal domain. However, if the melt is in binodal domain, the phenomena is one of nucleation and growth of L_2 from the melt. As the temperature decreases continuously, the primary solid phase (S_1) nucleates below the monotectic temperature (T_m) and the S/L interface moves at a very high velocity thereby trapping liquid L_2 within it. The size of the phase-separated liquid (L_2) droplets in the melt is dictated by the growth and coarsening rate and the time available before the L_2 droplets are trapped by the growing solid (S_1). In addition to the purely diffusional growth, the collisions of the different particles also play a major role in the coarsening process. As pointed out by Ulhmann et al. [40], the critical velocity required for the growing solid to trap the droplets is inversely proportional to the size of the droplet. As the growth velocity of the solid depends on the amount of the undercooling achieved during rapid solidification, a higher cooling rate is required to synthesize finer dispersions. The particles that are not trapped ultimately get engulfed at the grain boundaries creating a network of particles decorating the grain boundaries. This is more prominent at a lower cooling rate.

With a very high cooling rate, it is also possible to suppress the decomposition of liquid within the immiscible liquid domain, if the liquid does not enter the limit given by the spinodal regime. The liquid in such a case can kinetically avoid the process of nucleation and may enter the domain below the equal, free-energy curve T_o between the melt and the solid. In this domain, it is kinetically possible to achieve solidification without partitioning, leading to the formation of transient metastable solid solution. This solid is metastable and often decomposes spontaneously to yield a nanodispersed microstructure or may follow a typical precipitation process depending upon the composition in the G versus x curve. The size distribution in the latter case is narrower (region 5) compared to the liquid-phase separation route due to the large driving force for nucleation from the highly metastable solid solution and relatively slower diffusion in the solid state. In many cases, both liquid trapping and decomposition of the metastable solid solution can occur simultaneously. The microstructure under such a processing condition is characterized by a bimodal distribution of very fine dispersoids due to the solid-state decomposition and a coarser distribution of particles through the trapping of the phase-separating second liquid (L_2) by the moving interface.

The shapes of the particles depend on the evolutionary route it followed during the processing. Since the size is in nanodimension, the time the particles require to equilibrate with the surrounding matrix is very small and the equilibrium can easily be reached during solidification. The equilibrium shape of the particles in such a situation is dictated by the point group symmetries of the matrix and the particles. The symmetry of the entrapped liquid particles is (∞, ∞, ∞). The intersection group of the liquid particles in solid matrix will, therefore, be the same as the point group of the matrix. Upon solidification, the solid may conform to the same symmetry, if it does not have enough time in solid state to equilibrate. In case when the liquid transforms to solid in a massive manner without any solid partition and yield-dispersed microstructure through a subsequent solid-state decomposition, the symmetry of the particle will be dictated by the intersection symmetry of the point group of the solid particle and the matrix. Thus, a crystallographic shape analysis is a powerful tool to determine the evolutionary pathways of the embedded nanoparticles. This is further discussed in subsequent sections.

2.2.2. Nanoembedded Metallic and Intermetallic Particles by Nucleation and Devitrification

Even in the system where there is no miscibility gap, it is possible to synthesize nanoembedded particles via rapid solidification. This can be achieved in two ways. Nanoembedded particles can be obtained by controlling nucleation and growth. The basic principle is to promote a high nucleation rate of the embedding phase in the melt and restrict the growth rate. This is essentially followed by the rapid growth of the matrix phase, such that the matrix can entrap the fine particles formed earlier in the melt resulting in the formation of a desired nanocomposite [7]. This can be achieved by carefully selecting the alloy composition and controlling the undercooling of the melt, by selectively altering specific parameters during the process of rapid solidification. The addition of trace glass-forming elements can help achieve the above goals by controlling the viscosity of the melt and thereby promoting undercooling and altering the nucleation and growth processes. The synthesis of intermetallic Ti_2Ni nanoparticles embedded in a TiNi matrix via rapid solidification is an example of such an approach [6]. Another powerful way of synthesizing nanoembedded particles is to synthesis glassy alloy by choosing the correct chemistry and utilizing rapid solidification processing. This is followed by controlled devitrification. During the process of devitrification, many of these glasses yield nanocrystalline phases thus resulting in a composite of glassy matrix in which these particles are embedded [41]. The magnetic alloy "FINEMET" is the best-known example of this class of materials where the Fe_3Si particles are embedded in the glassy iron-rich matrix.

2.2.3. Structure of the Particle-Matrix Interface and Orientation Relationship

As discussed in the previous section, the embedded nanoparticles often exhibit well-defined morphology. The shape depends on the crystallography of the particles as well as the matrix. As pointed out in the previous section, the symmetries of the liquid particles are dictated by the matrix symmetry [42]. For example, the liquid will have a cuboctahedral shape (a regular octahedron bounded by eight $\{111\}$ and truncated on six $\{100\}$ facets) if the matrix is cubic, while it will adopt the shape of hexagonal prism or bipyramid if the matrix belongs to a hexagonal system [43]. Alternately, in the case of a glassy matrix, the liquid or glassy nanoparticles will possess spherical symmetry. In case the symmetry of the matrix and the particles are the same, the situation is relatively simpler. A large number of systems on which extensive investigations have been carried out falls into this category. The experimental results available for the dispersion of lead in different fcc matrices provide important insight. Both in aluminium and copper matrices, the lead particles show well-developed cubeoctahedral shape [44]. The orientation observed was cube-on-cube, despite the fact that the lattice parameter differences are large between Pb and the two matrices. The high-resolution imaging of the interface indicate that the interface is semicoherent with very little strain.

The shape in this case is clearly dictated by the intersection symmetry. In the case of Ni as matrix, the situation is different. The orientation of the solid Pb particles with Ni does not follow a cube-on-cube orientation relationship. The observed relationship is $(100)_{Ni} \parallel (100)_{Pb}$ and $[011]_{Ni} \parallel [001]_{Pb}$. The embedded particles in this case exhibit a distorted cubeoctahedral shape. Obviously, the liquid lead particle has a perfect cubeoctahedral shape dictated by the symmetry of the matrix. During solidification, the solid lead nucleates with a distinct orientation. The problem then is to fill up a cuboctahedral cavity with solid lead growing with an orientation, which does not conform to this symmetry. This is achieved by nucleating other variants such that the shape is closer to the preexisting cavity. This results in a distorted shape and a multigrained nanoembedded particle. The same principle seems to dictate the shapes and the nature of the interfaces of the nanoembedded particles having structure different from the matrix. The shapes of Pb, In, Bi, Cd, and Sn viewed along the [011] zone axis of the Al matrix are shown in Figure 5. The corresponding diffraction pattern is also shown in the same figure. From the projected images along prominent zone axes, one can reconstruct the actual three-dimensional (3d) shapes of the dispersoids. It has been experimentally observed that like Pb, the In particles conform to a cuboctahedral symmetry. A careful diffraction analysis shows that these In particles exhibit a cubic structure. However, in the case of Bi and Sn [45–47], one observes particles with a spectrum of shapes. These particles do not exhibit a strict orientation relationship with the matrix. In contrast, the Pb particles embedded in zinc matrices obtained by rapid solidification possess an epitaxial relationship with the matrix given by (0001) Zn‖(111)Pb and [11 $\bar{2}$0]Zn‖[1 $\bar{1}$0]Pb [43]. The shape of the Pb particles can be obtained by systematic tilting to many different zone axes and from the series of images we conclude that the shape is truncated hexagonal bipyramid consistent with the symmetry of the matrix. Larger Bi particles embedded in Zn matrix obtained by rapid solidification also corresponds to a truncated hexagonal bipyramid. However, fine Bi particles obtained from the decomposition of the solid solution matrix formed by diffusionless solidification show only hexagonal plates with rounded edges and corners [48]. Table 1 summarizes the crystallography of lead particles embedded in different matrices, the intersection group, and the expected shape based on the detailed study carried out by Chattopadhyay et al. [49]. Even if the intersection point group symmetry is the same, the actual shape in different matrices are different, since the anisotropy of surface energies in the solid state may be very different.

In case of ion-implanted samples, the shape of Pb on Al varies with size and shape, occasionally showing truncations on {100} planes and to a smaller extent on {111} facets. Thus, the equilibrium shapes of smaller inclusions include polyhedra mainly bounded by eight {111} facets and to a lesser extent by {100} facets [36, 37].

2.2.4. Shapes of Alloy Particles

Studies on alloy particles are limited, and only in recent times attempts have been made to synthesize such particles. Pb-In alloy particles embedded in aluminium have been

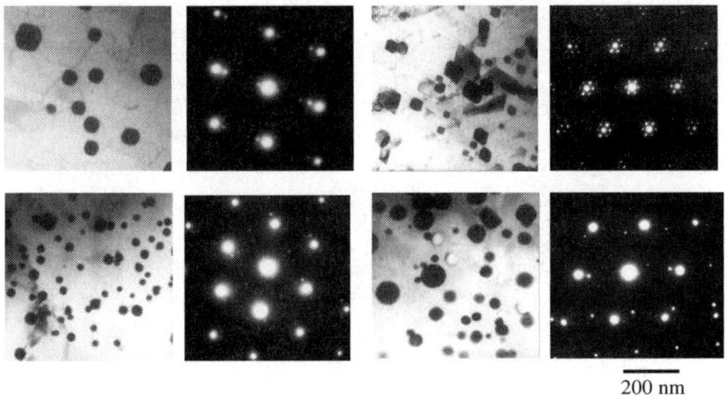

Figure 5. Electron micrograph and corresponding selected area diffraction pattern showing the orientation between the nanoscaled inclusion and aluminium is shown for (a) Pb, (b) In, (c) Bi, and (d) Sn.

Table 1. Crystallography of lead and equilibrium shape in different matrices [49].

System	Orientation relation	Matrix point group	Intersection group	Shape
Zn-Pb hcp-fcc	$(0001)_{Zn} \parallel (111)_{Pb}$ $[11\bar{2}0]_{Zn} \parallel [1\bar{1}0]_{Pb}$	6/mmm	$\bar{3}$m	Truncated hexagonal biprism
Cu-Pb fcc-fcc	$(111)_{Cu} \parallel (111)_{Pb}$ $[110]_{Cu} \parallel [110]_{Pb}$	m3m	m3m	Truncated cubeoctahedron
Al-Pb fcc-fcc	$(111)_{Al} \parallel (111)_{Pb}$ $[110]_{Al} \parallel [110]_{Pb}$	m3m	m3m	Truncated cubeoctahedron
Ni-Pb fcc-fcc	(100)Ni ‖ (100)Pb [011]Ni ‖ [001]Pb	m3m	4/mmm	Truncated cubeoctahedron

investigated by Bhattacharya et al. [50]. Lead and indium form a complete solid solution. These particles are cuboctahedral in shape and bounded by {111} and {100} facets and exhibits a cube-on-cube orientation relationship with aluminium matrix. In these particles, there is a preference for the formation of the 111 interfaces leading to the occasional occurrence of the octahedral shape.

Bi-crystalline Pb-Cd inclusions in an aluminium matrix consist mostly of an fcc Pb part forming the segment of a cuboctahedron and a hcp Cd slab attached to one of {111} Pb facets [51]. The close-packed planes and directions in the three structures are parallel to each other yielding a simple orientation relationship. Pb-Sn alloy particles embedded in aluminium have been recently studied by Johnson et al. [52] and Bhattacharya et al. [53]. Pb in most cases possess a cube-on-cube orientation relationship with the aluminium matrix. Sn shows different orientation relations. For ion-implanted samples, Johnson et al. [52] have reported two orientation relations for Sn particles with the matrix:

$$(111)\mathrm{Al} \parallel (100)\mathrm{Sn} \text{ and } [1\bar{1}0]\mathrm{Al} \parallel [001]\mathrm{Sn}$$

$$(111)\mathrm{Al} \parallel (100)\mathrm{Sn} \text{ and } [1\bar{1}0]\mathrm{Al} \parallel [0\bar{1}1]\mathrm{Sn}$$

Bhattacharya et al. [53] have reported an alternate orientation relationship of $(2\bar{2}0)$Al ‖ (101)Sn and [220]Al ‖ [020]Sn between Al and Sn. In this case, the particles were synthesized by rapid solidification. In most of the cases, the overall shape of these embedded nanoparticles are near cuboctahedral (Fig. 6) and the relationship among Pb, Sn, and Al can be written as

$$(2\bar{2}0)\mathrm{Al} \parallel (2\bar{2}0)\mathrm{Pb} \parallel (101)\mathrm{Sn} \text{ and}$$
$$[220]\mathrm{Al} \parallel [220]\mathrm{Pb} \parallel [020]\mathrm{Sn}.$$

In-Sn nanoalloy inclusions in Al have been recently investigated [54] and they show the presence of two phases, $\beta(In_3Sn)$ and $\gamma(InSn_4)$, at nanoscale. It has been shown that while β possesses an orientation relationship with Al, the γ phase only possesses a planar relationship. The indium-tin alloy particles exhibit a distorted cuboctahedral symmetry as shown in Figure 7. The research in alloy particles is still at a very early stage. However, already they show promise of extending our understandings.

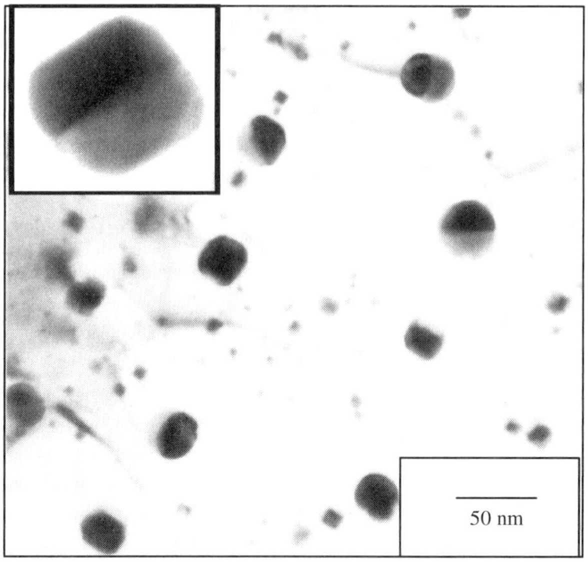

Figure 6. Micrograph shows Pb-Sn alloy particles. Majority of the partices correspond to a cuboctahedral symmetry. Reprinted with permission from [53], P. Bhattacharya et al., *J. Mater. Res.* 17, 2875 (2002). © 2002, Materials Research Society.

2.2.5. Defects in the Embedded Particles and the Matrix

There exist a limited number of studies to unravel defects at the interfaces and inside the nanoembedded particles. The possible defects that can arise from the geometrical effect of a constraining cavity have been discussed by Pond and Vlachavas [55]. In case the particles have the same intersection symmetry as that of the matrix, one does not expect any geometrical defect to occur. However, if they are different, defects can exist either at the interface as steps or in the interior as additional interfaces among variants. In the latter case, the particle does not remain a single crystal anymore. The twin boundaries observed in germanium crystal embedded in an aluminium matrix by Gouthama are a classical example of such a defect [56]. This is shown in Figure 8. In case of Pb particles embedded in an alloyed matrix of Cu and Zn with lower zinc concentration, the particles are cuboctahedron in shape bounded by 111 and 100 facets as shown in Figure 9. High-resolution electron microscopy reveals that the particle-matrix interface in the as-spun sample contains steps of height 0.42–1.9 nm on the {111} planes [57]. Repeated thermal recycling from room temperature to a temperature above the melting point leads to a complete removal of these steps indicating that these stepped structures are of kinetic origin and not geometrical. Recently, a model for the origin of these stepped interfaces has been proposed [58]. At small sizes, the embedded particles sometime exhibit twin or faults particularly if the matrix is rigid. As an example, stacking fault-like defects could be observed in lead embedded in an amorphous silica matrix [30].

In case of Pb-Cd eutectic inclusions in aluminium reported by Dahmen et al. [51, 58, 59], the hcp Cd parts of the inclusions are oriented with the close-packed planes and directions parallel to close-packed planes and directions of both Pb inclusion and an aluminium matrix. The different moire patterns from the Pb and Cd parts of the particles help identify the Pb/Cd interface containing close-packed planes of Pb(111) and Cd(0001) parallel to each other which is shown in Figure 10. The Pb/Cd interfaces are not always planar and slight curves and internal kinks are occasionally seen. The melt-quenched aluminium samples containing embedded lead, indium, bismuth, tin, and cadmium nanoparticles contain quenched in vacancies. In case of Pb-In alloy particles embedded in aluminium, the presence of dislocation loops in the particles [50] has been reported. In the case of Al-1.8 at%($Pb_{56}In_{44}$), the particles retain faceted shape even at 723 K, while some of the particles show slow atomic step activity. This is shown

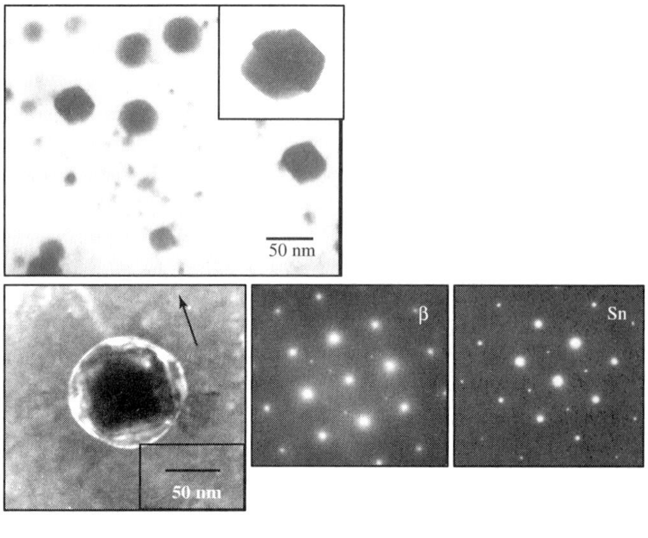

Figure 7. Micrograph shown In-Sn alloy particle corresponding to eutectic composition. The particle shows near-cuboctahedral symmetry and shows the presence of both β and Sn phases. Reprinted with permission from [62], V. Bhattacharya and K. Chattopadhyay.

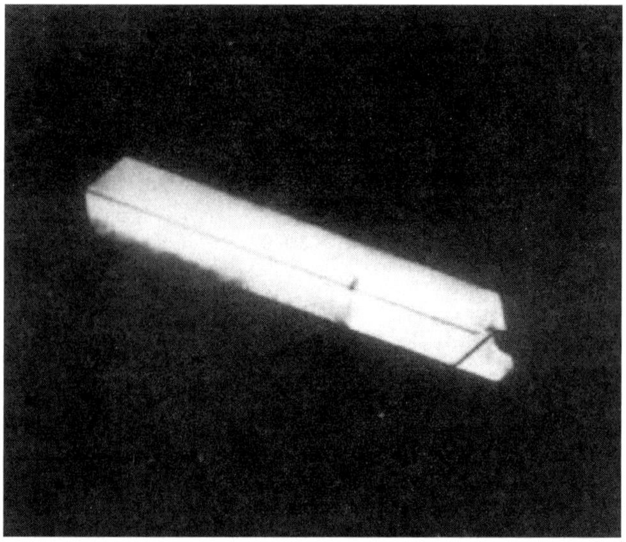

Figure 8. Twin plane shown in lath shaped Ge precipitate in Al-Ge alloy. Reprinted with permission from [133], Gouthama, 1989.

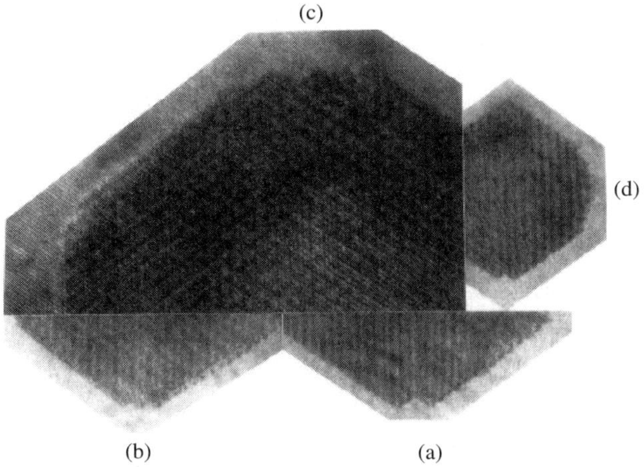

Figure 9. A set of high-resolution electron micrographs showing the particle-matrix interfaces. Steps of 111 type are observed on 011 and 001 facets. The development of steps on the 011 facet have been shown in (a) containing two steps, (b) containing three steps, (c) containing four steps, and (d) the saw teeth kind of appearance of 111 steps on the 001 facet. Reproduced with permission from [57], R. Goswami et al., *Acta Mater.* 46, 4257 (1998). © 1998, Elsevier Science.

in Figure 11. Bhattacharya [60] reported coarsening by mass transport occurring through dislocations, and provided microscopic evidence of particles interconnected by dislocations in ternary alloy of Al-2 at%($Pb_{31}In_{69}$).

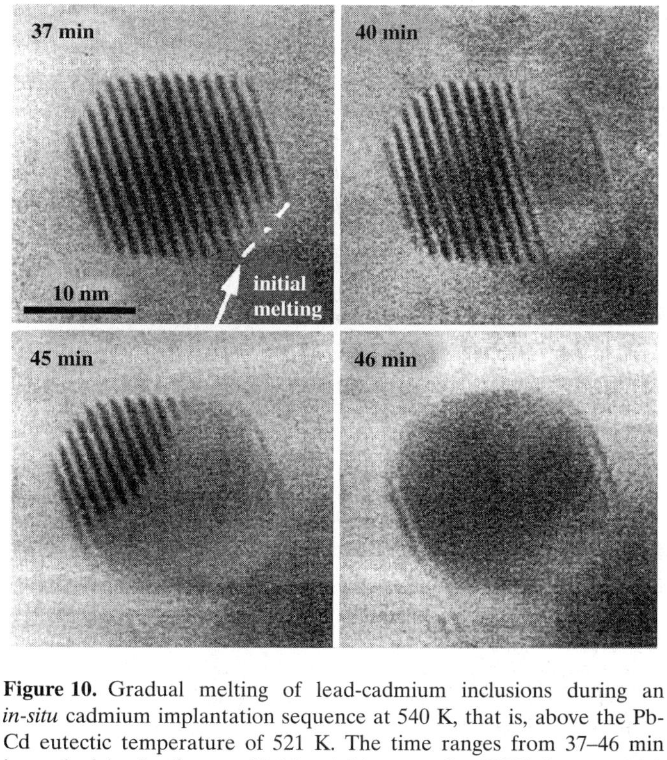

Figure 10. Gradual melting of lead-cadmium inclusions during an *in-situ* cadmium implantation sequence at 540 K, that is, above the Pb-Cd eutectic temperature of 521 K. The time ranges from 37–46 min is marked in the figures. Melting initiates at the {100} facets, which change into meniscus shape (37 min). At a later stage (40 min), the liquid menisci merge into a liquid droplet at a {111} facet. Melting then occurs in intermittent steps (45 min) until the entire inclusion is liquid (46 min). Reprinted with permission from [59], E. Johnson and U. Dahmen, *Micro. Microanal.* 3, 409 (1997). © 1997, Springer-Verlag.

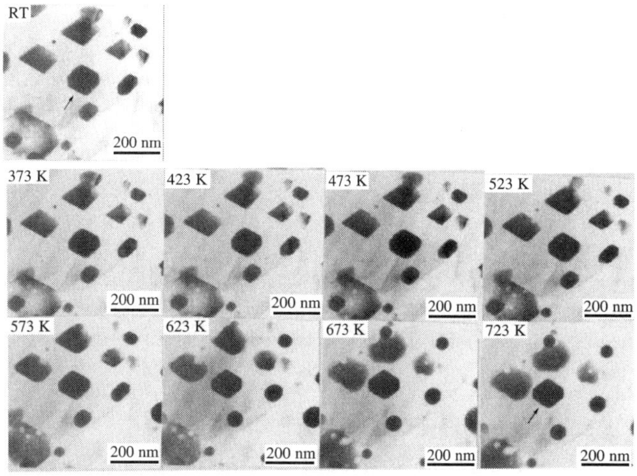

Figure 11. Hotstage sequences in Al-1.8 at % ($Pb_{70}In_{30}$) sample shows melting of Pb In nanoparticles. Interestingly, one particle marked in figure does not show roughening even at 723 K. Reprinted with permission from [60], P. Bhattacharya, 2000.

Johnson et al. [52, 59] have studied bi-phase alloy inclusions in aluminium matrix synthesized by ion implantation and observed the presence of voids. The presence of voids was also observed in many cases in particles in the rapidly solidified melt-spun samples containing lead-tin and lead bismuth alloy inclusions [61]. The presence of voids are also observed in case of alloy particles viz. Pb-Sn [53] and In-Sn [54, 62] in rapidly solidified samples.

2.3. Transformations in Embedded Particles

2.3.1. Shape Changes in Embedded Particles

The embedded nanoparticles bounded by sharp interfaces often exhibit superheating during melting [57]. Careful analysis of the ratio of the distance between 111 and 100 plane for nanoparticles of Pb embedded in Al matrix indicates that the values are widespread in the melt-spun state with the peak at around 1.18 [63]. This increases steadily with recycling and the ratio shifts to 1.3–1.5 indicating the dominance of 111 planes. This is so because along with the annihilation of surface defects, which act as heterogeneous nucleating site for phase change, the facets sharpen and surface energy ratio of the planes are altered. There is a tendency of increasing the surface energy for 111 facets with lower energy and decreasing defects leading to the emergence of superheating. Kjeldgaard and Johnson et al. [64, 65] have experimentally studied the morphological changes of liquid Pb inclusions embedded in an aluminium matrix. Their reports show a hysteresis in the morphology of liquid Pb inclusions. Immediately after melting, the {100} planes bulge outwards and become rounded, whereas the {111} planes remain flat until the temperature reaches around 500 °C when the inclusion rapidly becomes spherical. In a significant contribution, Gabrisch et al. [65] have focused on the characteristics of the equilibrium shape for small Pb inclusions in Al matrix. They observed a relation between the size and the location in an Al matrix. Using high-resolution electron microscopy, they established that unlike the equilibrium shape of free particles, the shape of inclusion changes with size and certain "magic sizes" are preferred [66]. This behavior can be ascribed to the oscillatory nature of strain energy that dominates at small sizes.

Bhattacharya et al. [50, 60] have conducted *in-situ* heating experiments on Pb-In nanoparticles embedded in an aluminium matrix in order to confirm the kinetic origin of shape of the particles. The results indicate that the behavior can be different from particle to particle [50]. Melting reports show that while some particles roughened to give spherical shape, there are other particles that do not exhibit roughening of both the 100 and 111 planes as shown in Figure 11. In a few cases, they report that while few adjacent sets of planes still remain faceted, the planes opposite to them have undergone roughening transition. It suggests that a major role is being played by interfacial segregation during roughening transitions. Since segregation can be different for different particles, the roughening behavior can also be different.

2.3.2. Liquid to Solid Transformation

Considerable interest in studying phase transformation in embedded nanoparticles exists with particular emphasis in liquid to solid transformation [67–69]. The primary issue is the nucleation behavior of the solid in the entrapped liquid. It has been shown that if the melt is divided into small metal clusters with no connectivity among them, large supercooling is achievable [70]. From the classical nucleation theory [71], energy due to the creation of interface is associated with the nucleation of solidification from the melt. The cause of the barrier is the energy expended for the creation of interface associated with the new phase that appears during solidification. Therefore, solidification always occurs below the thermodynamic freezing temperature. Classical nucleation theory predicts that the transformation of super-cooled liquid to solid phase can occur homogeneously. However, even in samples quoted to be very pure, certain amounts of impurities are present, and they act as a catalyst for heterogeneous nucleation [72] at temperatures above the homogeneous nucleation temperature. Turnbull [67–70, 73–75] have systematically studied undercooling of liquid metals and obtained a maximum undercooling of $\sim 0.18\ T_m$. It was then inferred that by entraining the droplets in emulsifying liquids, homogeneous nucleation could be achieved. But the later experiments by Southin and Chadwick [76] and Perepezko [77] succeeded in obtaining a much higher undercooling. Thus, it became clear that most of the Turnbull's results probably correspond to the heterogeneous solidification.

Heterogeneous Nucleation of Solidification Nucleation of solidification of individual droplets has been studied in the presence of clean substrates [73, 75, 78–80] and in the absence of substrate [81–83]. Irreproducible results in supercooling measurements led to the development of entrained droplet technique [84] and its adoption by Southin and Chadwick [76] to the study of solidification of low-melting metal in a higher melting matrix. Alloys from a eutectic or monotectic system are thermally treated to produce micron-sized dispersions in a matrix.

More recently, rapid solidification of monotectic alloys and other systems has been employed to produce a finely dispersed, nanoscaled embedded particle of a low-melting element in a high-melting matrix [43–47]. This provide an excellent opportunity to study the kinetics of heterogeneous nucleation of a solid in the entrapped liquid particles in contact with the matrix. Calorimetry, dilatometry, and microscopy can be employed to monitor the solidification behavior of these particles. Solidification of such fine droplets generally occurs by a single nucleation event with insignificant growth. In case the calculated critical nuclei size exceeds the size of these embedded liquid droplets, solidification might be bypassed. Droplet solidification is catalyzed by epitaxial nucleation by the surrounding matrix, and hence is controlled by the matrix and the crystallographic shape of the cavity inside the matrix. Doping of the matrix with trace alloying elements [85, 86] has enabled investigations of the influence of chemistry of droplet-matrix interface on heterogeneous nucleation.

The thermodynamic limit to which a liquid can be undercooled (0.5–0.75 T_m) was predicted by Kauzmann [87] from the entropic instability [88]. Experimentally, it is found that if the liquid can be cooled by avoiding crystallization, it forms glass between 0.3–0.75 T_m. Singh and Holz [89] gave the estimate of the limit of maximum supercooling that can be obtained from classical nucleation theory for homogeneous nucleation. The reasonable value that comes up from their results is $\sim 0.56\ T_m$. However, this does not take into account the variation of surface energy with temperature. In certain embedded nanoparticles [30, 46], a wide temperature range for solidification could be observed. Undercooling achievable during nucleation of solidification is not directly related to size. However, since the smaller sized particles are expected to contain less impurity, they will have a lesser concentration of heterogeneous nucleation sites. Therefore, the smaller particles would solidify later, closer to the homogeneous nucleation temperature [90]. A wide range of particle size distribution—an outcome of the processing route (like melt spinning)—is expected to yield a broad range of solidification event. However, broad solidification event is not true for all samples produced by melt spinning. The contact angle during heterogeneous nucleation of solidification is a very important parameter that plays a crucial role. There are two important variables that influence the heterogeneous nucleation. These are potency and concentration of nucleation sites. The wetting angle is a measure of the potency of the nucleation site in the framework of classical heterogeneous nucleation theory. A related question is correctness of the assumption of classical heterogeneous nucleation event. Kim and Cantor [91], and Goswami et al. [43] have shown that the number of catalytic nucleation sites N_c and the contact angle θ can be estimated for nanodispersed particles from controlled differential scanning calorimetric (DSC) experiments. Unreasonable low N_c values are obtained for a large number of cases. These results have suggested an apparent breakdown of the classical heterogeneous nucleation theory with a hemispherical cap model. However, the later work suggests that the microscopic details of the interface plays an important role in the nucleation process and may have biased the earlier results. The interface in the presence of defects cannot be treated as having a single contact angle. Instead, it can be modeled with a spread in the contact angle. This gives a reasonable reproduction of the DSC curve using the classical heterogeneous theory as shown by Goswami and Chattopadhyay [46]. In cases where the anisotropy of liquid-solid surface energy is very large, the nanodispersed particles bounded predominantly by low-surface energy interfaces can be used to test the classical heterogeneous nucleation theory. Such a study employing dispersion of lead in a zinc matrix strongly suggests the validity of the classical theory of heterogeneous nucleation [43].

The depression of the freezing point as a function of size has been studied extensively. Buffiat and Borel [92] have developed the thermodynamic criterion to estimate this depression. Recently, Sheng et al. [93] and Li et al. [94] have shown a size-dependent freezing behavior for Pb nanoparticles embedded in an Al matrix synthesized by mechanical milling. The freezing temperature decreased with an increase in milling time for Al-Pb and a subsequent reduction in the particle size of incoherent Pb nanoparticles. They have also investigated In, Sn, Bi, and Cd nanoparticles in ball milled samples. However, such an effect has not been established for embedded nanoparticles in an Al matrix obtained via rapid solidification. In this case, the particles have an orientation relation with a matrix and no apparent size-dependent effect can be detected.

Analysis by a classical nucleation theory gives us the free energy for homogeneous and heterogeneous nucleation. For bulk liquid freezing below equilibrium freezing point T_o, this is given by $\Delta G_v = cL_o \Delta T_f / T_o$, where $\Delta T_f = T_o - T_f$.

However, for particles with radius r, freezing starts at a different temperature T_r, and ΔG_v is modified to give the driving force for solidification as

$$\Delta G_V = \frac{\rho L_o}{T_o}\Delta T_f + \frac{3\alpha(\sigma_{lm} - \sigma_{sm})}{r} \tag{1}$$

where $0 < \alpha < 1$ because only a fraction of the liquid/matrix interface is replaced by a solid matrix interface upon nucleation.

Maximum ΔG is given by the condition,

$$\left(\frac{\partial \Delta G}{\partial R}\right) = 0 \tag{2}$$

$$R^* = \frac{2\sigma_{ls}}{\Delta G_v} \tag{3}$$

$$I = N_C \frac{kT}{h} \exp\left(-\frac{\Delta G^*}{kT}\right) \exp\left(-\frac{Q}{kT}\right) \tag{4}$$

where N_C = number of nucleation sites per unit volume and Q = activation energy for transferring atom across liquid-solid interface.

The nucleation frequency I is determined from ΔG^*. The ΔG^* contains ΔG_v and undercooling, ΔT_r for a given particle-matrix system. Both of these are size dependent for small particles system.

It can be shown that

$$\Delta T_r = \Delta T_o + \frac{T_o}{\rho L_o}\frac{3\alpha(\sigma_{sm} - \sigma_{lm})}{r} \tag{5}$$

$\Delta T_r = \Delta T_o + \alpha \Delta T_m$ where $\Delta T_m = T_o - T_m$ reflects the melting point depression. Thus, the undercooling of nanoparticles is larger compared to the "bulk materials" or corresponding micrometer counterparts.

Nanoparticles with more pronounced melting-point depressions show larger undercooling on freezing as is clear from Eq. (5). Indeed, such an analysis is supported by interrupted DSC thermograms of In nanoparticles embedded in Al produced by mechanical milling. Smaller particles, which melt first on heating, solidify at an undercooling that is larger than the undercooling of the larger particles. Thus, undercooling increases with a decrease in particle size, and apparent large undercooling is interpreted as a combined consequence of dependence on the need for nucleation of solid and the size-induced melting-point depression.

2.3.3. Melting

Melting of small particles has received considerable attention over the last few decades [95]. Results show that the melting points of fine particles have a strong dependence on their sizes. Ubbellohde [96] has reported the importance of Lindemann's criterion of melting. Lindemann [97] proposed melting as a vibrational instability released when the root mean square amplitude of vibration reaches a critical fraction of atomic distances. According to Lindemann, the melting temperature is given by $T_m(r)/T_m(\infty) = \exp[-(\alpha - 1)(r/3h - 1)^{-1}]$, where $T_m(\infty)$ is the bulk melting temperature, $T_m(r)$ is the melting temperature of a spherical particle with radius "r" and "h" corresponds to the height of a monolayer of atoms on the bulk crystals; α can be obtained directly and is given by σ_s^2/σ_v^2. σ_s^2, and σ_v^2 represents the mean-square displacements for the surface and the interior atoms of the corresponding crystals. Sheng et al. [98] have shown that when α is less than 1, the surface (interface) atoms are constrained to smaller vibrations than the interior atoms and superheating can be observed. However, if the interface atoms thermally vibrate with higher intensities, melting point depression takes place. There have been attempts to calculate the critical temperature for different planes where the instability will appear. However, only recently, a conclusive relationship between the melting temperature of surface planes and Debye temperature has been experimentally observed by Gråback and Bohr [99]. The change in the melting temperature with particle size was reported earlier. Buffiat and Borel [92] showed that the small particles of Au, Pt, and other the metals suffer a depression in the melting point as much as 30% for particles of 20–30 Å in diameter. Hoshino and Shimamura [100] have explained the change in the melting point in terms of the theory based on Lindemann's criterion. The mean-square amplitude of a solid surface is 1.5–2 times higher than the bulk, and the Lindemann's criteria can explain the depression in the melting point. Rossouw and Donnelly [101] have injected Ar ions into aluminium at room temperature. The melting temperature of argon at atmospheric pressure was 84 K. However, from precise measurements of diffraction intensities a metastable superheating of 480 K was reported [101]. The amplitude of vibration of embedded Ar particles was smaller than that for free solid argon. The Debye temperature, which was a measure of cohesion deduced from the intensities of the diffraction peaks of the embedded particles, was 139 K compared to 70 K for bulk Ar at atmospheric pressure. A high Debye temperature implies a higher cohesion, smaller vibration amplitudes, and generally a higher melting temperature. Since the bubbles are small, the amplitude of surface (interface) vibration is depressed due to the constraining effect of the Al matrix. Hence, superheating is possible because the surface (interface) atoms are constrained to smaller vibrations than the interior atoms. Clear evidence of superheating metallic particles due to embedding was first reported by Dages et al. [102] in an experiment where small (0.15 nm) silver nanocrystals were coated with gold. Following this, several investigations on the melting behavior of embedded nanoparticles has been carried out [99].

Allen et al. [103] first discussed the effect of embedding on the melting point. Allen et al. [103] and Malhotra and Van Arken [104] reported that micron-sized indium particles embedded in aluminium melt at a temperature higher than the bulk melting point of indium. They tried to explain superheating in terms of strain energy effect on melting. Sasaki and Saka [105] explained the elevation based on the interfacial energy difference between the liquid particle-matrix and the solid particle-matrix interfaces. Allen et al. [106] have shown that the melting temperature in most cases is inversely proportional to their particle size.

A large number of investigations on the melting of small particles of Pb, In, Bi, and Cd have been reported. Zhang and Cantor [107] observed that indium and lead particles embedded in aluminium matrix melt at a temperature 0–40 K above the bulk melting temperature. According to Zhang and Cantor [107], the elevation in the melting temperature was due to "superheating" phenomena caused by a kinetic difficulty in melt nucleation. Recent experiments on In and Pb particles embedded in Al and Cu matrices prepared by melt quenching [108] and ion implantation [35] shows that the particles superheat much above ($\geq$100 K) their bulk melting temperature.

According to Couchman and Jesser [109], the heterogeneous nucleation of melting at the crystalline surfaces played the most important role. The melting point of thin films or fine particles may be enhanced or depressed depending on the nature of the interfaces. The size-dependent change in melting point is given by the thermodynamic equation:

$$\frac{T_{ml}}{T_o} = 1 - \frac{3\left(\frac{\sigma_{sm}}{\rho_s} - \frac{\sigma_{lm}}{\rho_l}\right)}{rL_o}$$

where T_{ml} = melting point of the embedded particle; T_o = bulk melting point; σ_{sm} = solid particle-matrix interfacial energy; σ_{lm} = liquid particle-matrix interfacial energy; ρ_s = density of solid; ρ_l = density of liquid; L_o = latent heat per unit mass, and r = radius of the metal particle. Spaepen and Turnbull [110] have given an expression for the pressure difference developed due to the difference in thermal expansion of the matrix and the embedded particle

$$\Delta P = \frac{[12(\alpha_p - \alpha_m)\Delta T \mu_m k_p]}{(3k_p + 4\mu_m)}$$

where α_p = linear thermal expansion coefficient of the particle; α_m = linear thermal expansion coefficient of the matrix; k_p = bulk modulus of the precipitate; μ_m = shear modulus of the matrix. A change in the melting point can be obtained from the Claussius–Clapeyron equation

$$\Delta T = \frac{\Delta P T_o \Delta V}{L_v}$$

The effect of changes in volume during solidification of embedded particles has been discussed by Uhlmann [111] and Malhotra and Van Aken [104]. The volume change leads to an additional strain energy, which results in a reduction in the driving force for solidification. The change in free energy due to strain is given by

$$\Delta G_{st} = \frac{(2\mu_s k_l \Delta V^2)}{(3k_l + 4\mu_s)}$$

where μ_s = shear modulus of the embedded solid; k_l = bulk modulus of entrained liquid; ΔV = relative volume change. The reduction in the driving force has to be compensated by superheating. Efforts were undertaken to rationalize the maximum, observed superheating of Pb particles embedded in Al and Cu matrices [108] in terms of all the above pressure effects. Goswami and Chattopadhyay [108] have shown that the explanation for observed superheating for Pb embedded in Al, Cu, Ni, and Zn matrices [43, 44] cannot be explained in terms of all the pressure effects (Table 2).

The explanation for an increase in superheating is controversial due to strain energy effects [103, 104], interface energy effects [109, 112, 113], and a kinetic barrier to nucleation [102, 107, 114]. Sheng et al. [98] have tried to rationalize the difference in behavior of indium nano particles in different matrices. They attempted to rationalize the observed superheating for In particles in an aluminium matrix and melting point depression in an iron matrix. It was suggested that different types of Al/In and Fe/In interfaces [115] cause the different melting behaviors of the embedded indium particles. Cahn [116] suggested that the observed superheating originates from the epitaxy between the embedded particles and the matrix, and no superheating is expected from the incoherent interfaces. The melting behavior of the small particles may be different in an isotropic matrix. Experimental evidence of the influence of crystallographic shape on the melting point exist. Allen et al. [103] have shown that the plate-shaped particles do not show any size-dependent melting point depression. However, Metois and Heyraud [117] have found that the triangularly shaped Pb particles melt 3 K above that for the surrounding irregular particles, whose melting point is assumed to be the equilibrium melting temperature. Spiller [114] has reported evidence of superheating of 2 K for entrained Pb particles, condensed on either

Table 2. Calculations showing a change in temperature obtained due to different contributions effecting the melting point of Pb in different matrices [108].

	Volume	Thermal	Size	Total ΔT (calc.)	ΔT (expt.)
Al-Pb	17	4	54 (10)	75	103
Cu-Pb	17	14	48 (25)	79	125
Ni-Pb	17	26	53 (30)	96	−45
Zn-Pb	17	−47	40 (10)	10	(−7)
		+20		77	

Table 3. Superheating of lead in different matrices [108].

Size (nm)	Goswami and Chattopadhyay [44]	Kofman et al. (1990) (in Al_2O_3)	Peppiat and Sambles [118] (free particles)
10(Al-Pb)	103	−90	<−40
25(Cu-Pb)	125	−40	−30
30(Ni-Pb)	−5	−40	−20 to −30

graphite or amorphous carbon. In general, little research has been devoted to exploring the effect of the interface structure on the entrained particles on the melting behavior. There exist results for embedded Pb and Bi [114, 118] that suggest shape-dependent superheating ability. It has been found that on heat treatment, though the bounding planes do not change, superheating is observed. For Pb in Zn [43], the lead superheats only when the sample is thermally cycled several times to room temperature from above the melting point. Goswami et al. [57] attempted to correlate the shape with melting behavior. Table 3 summarizes the melting results of Pb in different matrices as reported by different investigators and Table 4 summarizes melting results of Pb in a Zn matrix obtained by careful DSC experiments.

Electron microscopy studies indicate an increase in sharpness of facets after recycling in a fraction of particles. It has been suggested that this is responsible for the observed superheating. The most significant of all the results observed is that of Pb dispersed in Cu-10% Zn matrix. The lead particles in as-spun particles show steps or ledges [57]. Repeated thermal recycling from room temperature to a temperature above the melting point leads to the complete removal of these steps. The particles are now bounded by sharp interfaces and the sample exhibits superheating during melting [57]. Careful analysis of the ratio of the distance between 111 and 100 planes indicates that the values are widespread in the melt-spun state with the peak at around 1.18. This increases steadily in recycling and the ratio shifts to 1.3–1.5 indicating the dominance of 111 planes. In addition to faceting, the recycling leads to the annihilation of surface defects, which act as heterogeneous nucleating sites. In the absence of nucleating agents (defects) and also with shortened 100, superheating is expected.

In order to understand the changes in the interface upon heating, *in-situ* X-ray diffraction studies and *in-situ* transmission electron microscopic studies were carried out. Sheng et al. [119] have monitored the melting of embedded lead nanoparticles in aluminium matrix using a high temperature *in-situ* X-ray diffraction technique. The analysis of X-ray diffraction data shows that melting took place when the root mean-square displacement of Pb atoms reached about 10%

Table 4. Summary of the DSC results on the melting of nanodispersed Pb in Zn matrix [43].

Sample	Melting peaks K	ΔT K
As cast	591 ± 0.6	Eutectic melting point
Melt spun	584 ± 0.2	−7
Melt spun and heat	589 ± 0.25	−2.5
treated 30 h at 550 K	651 ± 2	62 superheating

of their nearest neighbor atomic distances. Their investigation [119] suggests that the Lindemann's criterion [97] for melting is applicable to nanocrystals.

One of the issues that are addressed during *in-situ* experiments are how and where melting nucleates. Sasaki and Saka [105] have observed the melting processes for indium nanoparticles embedded in an aluminium matrix. Their observations suggest that melting initiates in the cavity at one of the {100} facets and then proceeds into the interior of the In particles. At the beginning of the melting process, the liquid droplets nucleated at a {100} facets and assume two different configurations alternatively before the liquid reaches the next {100} facets.

Johnson et al. [120] have reported results of *in-situ* studies for Pb in Al. Their results show that inclusions located in bulk Al are free of defects and sharply faceted along {111}, and to a lesser extent on {100} planes and assume dimensions of magic sizes. For larger inclusions, interfaces become more important and they assume a cuboctahedral shape. These shape changes affect the melting behavior of the particles. Gabrisch et al. [121] have first proposed that the difference in melting behavior of lead nanoparticles in different matrices is not only a function of inclusion size but also of their shape and interface structure.

Ohashi et al. [115] have investigated by *in-situ* microscopic studies the melting and solidification of indium nanoparticles embedded in an iron matrix. The melting temperature of small indium particles was found lower than that of the bulk material and decreases with decreasing radius "*r*." The difference in melting behavior of In nanoparticles embedded in an Fe matrix are in sharp contrast with that of particles embedded in an Al matrix. Saka et al. [105, 115] tried to rationalize the discrepancy in terms of interface energies between Al or Fe matrices and liquid and solid In.

Hagege and Dahmen [51] have investigated eutectic Pb-Cd inclusions in an aluminium matrix by *in-situ* microscopic studies. They have observed that the bimetallic inclusions were found to melt at the eutectic bulk melting point, irrespective of size. Melting was found to initiate at the triple interface junction of Pb, Cd, and Al [37, 51, 59].

2.4. Phase Selection in Small Particles

The questions of phase selection and phase stability of embedded pure metals are restricted to the issue of melting and solidification. These were discussed in previous sections. Multiphase isolated nanoparticles, however, provide newer challenges. Both the stability of the phases, as well as phase composition, can be altered as a function of size in these particles. The research in this direction is at a preliminary level and the salient findings are discussed in the next section.

2.4.1. Transformation in Multiphase Isolated Nanoparticles

As mentioned at the onset, the phase diagram that delineates the region of stability of different phases needs to be modified for nanosized particles. Recently, such a modification has been attempted for a Bi-Sn system [122]. As pointed out by Allen et al. there is a dramatic change in the phase coexistence with a significant depression of eutectic point [103]. This was verified by the observation of a molten layer on submicron Bi-Sn particles. Lee et al. observed crystalline to amorphous transition [122]. Recently, a series of experiments have been carried out by a group led by Mori [123–125] at an electron microscopy laboratory at Osaka University. They used a novel technique of *in-situ* metal deposition within the electron microscope to study the alloying behavior of nanosized particles. It was shown that when Bi is deposited on Sn nanoparticles, a spontaneous alloying takes place driving a crystalline to amorphous transition. Similar results were also observed in the case of Sn deposition on In particles. On the other hand, in systems like Au-Zn [126], spontaneous alloying leads to the formation of solid solution and intermetallic compounds viz. Au_3Zn and AuZn. These investigations highlight the size-dependent phase transformations that are possible in alloy nanoparticles.

2.5. Magnetic and Superconducting Transition in Small Particles

The embedded particles at small length scale provide exciting opportunities in tuning the magnetic properties. At very small sizes, they undergo superparamagnetic transitions. If the sizes of the particles are of the order of exchange interactions, the magnetic domains interact with each other. This leads to a significant decrease in magnetocrystalline anisotropy. Thus, the materials are expected to have enhanced soft magnetic properties. A classical example to such a class of material is FINEMET, where nanoscaled Fe-Si particles are embedded in an iron-based glassy matrix [127]. The exchange interaction is also employed to design cheaper, hard magnetic materials by diluting the expensive rare earth magnets with cheaper, soft ferromagnetic materials like iron. Because of the small sizes, the domains of the softer phase get coupled with the domains of the hard phase. Thus, the material behaves as a hard magnet although it contains a soft magnetic phase. The magnetic behavior of the nanoscaled materials is discussed in details elsewhere.

2.5.1. Magnetic Properties of Nanoembedded Particles

During the last two decades, amorphous and nanocrystalline materials are being investigated for applications in magnetic devices requiring either magnetically hard or soft materials. Nanomagnetic materials and nanocomposite magnetic materials are potential candidates for a variety of new applications in the fields of image sensors, catalysis, biotechnology, drug delivery systems, medical diagnostics, magnetic refrigeration, magnetically assisted chemical separation of radioactive and hazardous wastes, besides enhancing the recording density in information storage and reducing core losses for high frequency and microwave applications.

In particular, amorphous and nanocrystalline materials have been investigated for various soft magnetic applications. Due to the rapidly increasing working frequencies in several power electronics applications, such as power conversion banks, uninterruptible power supplies, magnetic filters, etc., there is a corresponding increasing interest in the use of special magnetic cores for transformers, inductive devices, switch mode power supplies, etc. In these materials, it has been shown that an averaging of the magnetocrystalline anisotropy over many grains coupled with an exchange length is the root cause of the magnetic softness of these materials. The $Fe_{73.5}Cu_1Nb_3Si_{13.5}B_9$, FINEMET alloy is one among these materials [128]. The nanostructured FINEMET alloy powders and ribbons were prepared by a mechanical alloying technique and a melt spinning technique and their soft magnetic properties were studied. Because of a weak coupling between the α-Fe(Si) nanograins, due to the amorphous ferromagnetic components, the FINEMET powders are inferior to those of melt-spun ribbons. But the Al addition in FINEMET alloy improves their soft magnetic property and it gives the better frequency responsibility.

2.5.2. Superconductivity in Small Particles

In 1911, while examining the electrical resistivity of pure mercury at a low temperature, Professor Kamerling Ohnes discovered that the resistivity vanished abruptly and completely at 4.15 K instead of decreasing continuously as expected upon cooling. Ohnes continued work on this topic till 1913, confirmed this unexpected result and the phenomena of disappearance of resistivity below a certain temperature (known as transition temperature) is known as superconductivity. Gradually, it was found that most metals exhibit superconductivity. In 1933, Meißner and Ochsenfeld performed an experiment to test whether superconductors behave the same way as Maxwell's law predicts a perfect conductor to behave. Maxwell's law predicts that a perfect conductor would not allow any change in flux. However, their experiments proved that rather than resisting changes in magnetic flux, a superconductor will expel the magnetic flux and acts as a perfect diamagnet. The description of a superconductor as a perfect diamagnet is known as Meißner effect.

Arising from the theoretical and experimental investigations of superconductivity, the concepts of two characteristic lengths, that is, the "London Penetration Depth" and "Coherence Length" emerge. The depth to which the magnetic flux penetrates before it exponentially decays to zero is known as the London Penetration Depth (λ). Pippard first formulated the concept of coherence length in 1953. When a superconductor is cooled below the transition temperature, some extra form of order sets in among the conduction electrons. So, coherence length (ξ) is the minimum length over which pair correlation takes place. The ratio of the penetration depth λ and the coherence length (ξ) is represented by a parameter κ known as the Gindsburg–Landau constant. When $\kappa < 0.71$, surface energy is positive and when $\kappa > 0.71$, surface energy is negative. If the electron mean-free path is very short, and λ increases, κ increases. Since the two characteristic length scales play a vital role in determining the type of transition taking place in superconductors, it can be intuitively felt that if the length scale of the sample becomes comparable to that of these two characteristic length scales, some new phenomena could be observed. The depression of a critical point for a superconducting transition occurs when the particle size is of the order of coherence length. In recent times, this has been quantitatively evaluated for nanoembedded lead particles in an aluminium matrix [129]. The results till date suggest that matrices do have influence over the nature of the depression in transition temperature. Inoue et al. [130, 131] have reported few results, which aimed at exploring the superconducting behavior of the composites containing a granular superconducting phase. It is well known that in some cases, the isolated particles may get coupled if the particle separation is of the order of London penetration depth. A percolation threshold exists when the entire composition behaves like a superconductor [132]. This depends upon size distribution, volume fraction, and interparticle separation of the second-phase particles. The size-dependent behavior of Type II superconductors has not been explored rigorously for the nanoembedded particles and of interest in recent times.

3. CONCLUSIONS

In this chapter, we have tried to give an overview of our current understanding of nanoembedded particles of metals and alloys with an emphasis on synthesis, structure, and stability. These types of materials have very wide applications, which include catalysis, spintronics, giant magnetoresistance, and tribological applications. There is a potential for these materials to develop additional functional properties making them prime candidates as small materials. Readers will find details of these in companion articles.

GLOSSARY

Intersection symmetry The common symmetry operations between the matrix and the particle.

Melting Solid to liquid phase transformation.

Monotectic system Binary system in which the two components are immiscible in the liquid state.

Nanoembedded Nanoembedded particles are a special class of nanomaterials, where the nanoscaled particles are constrained in a matrix.

Orientation relationship A set of planes and directions, which describe the particle and matrix orientation with respect to each other in space.

Rapid solidification The phenomena of rapid extraction of heat from the melt leading to high cooling rates.

Solidification Phase transformation from liquid to solid.

Superheating The elevation in melting temperature above the bulk equilibrium temperature.

REFERENCES

1. H. Gleiter, *Acta Mater.* (*Millenium-Issue*) 48, 1 (2000).
2. H. Gleiter, *Prog. Mater. Sci.* 33, 223 (1989).
3. H. Gleiter, *Nano. Mater.* 6, 3 (1995).
4. R. W. Siegel, *Nano. Mater.* 3, 1 (1993).
5. D. Chakraborty, *Bull. Mater. Sci.* 15, 411 (1992).
6. R. Nagrajan and K. Chattopadhyay, *Acta Metal. Mater.* 42, 947 (1994).
7. A. L. Greer, *Nature* 368, 688 (1994).
8. H. Johns, in "Rapid Solidification of Metals and Alloys," Monograph No. 8, Institute of Metallurgists, London, 1982.
9. R. W. Cahn, in "Materials Science and Technology" (R. W. Cahn, P. Hassen, and E. J. Kramer, Eds.), Vol. 9, p. 493. VCH Publishers, Weinheim, Germany, 1991.
10. S. Overshotl, *Electron. Power* 25, 347 (1979).
11. T. R. Anthony and H. E. Cline, *J. Appl. Phys.* 50, 245 (1979).
12. J. S. Benjamin, *Sci. Am.* 234(5), 40 (1976).
13. C. Suryanarayana, *Metals and Materials* 2, 195 (1996).
14. M. O. Lai and L. Lu, "Mechanical Alloying." Kluwer Academic Publishers, Boston, MA, 1998.
15. B. S. Murty and S. Ranganathan, *Intern. Mater. Rev.* 43, 101 (1998).
16. A. E. Ermakov, E. E. Yurchikov, V. A. Barinov, *Phys. Met. Metallogr.* 52(6), 50 (1981).
17. C. C. Koch, O. B. Cavin, C. G. McKamey, Scarbrough, *J. Appl. Phys. Lett.* 43, 1013 (1983).
18. P. G. McCormick, *Mater. Trans. JIM.* 36, 161 (1995).
19. D. R. Maurice and T. H. Courtney, *Metall. Trans.* 21, 289 (1990).
20. C. C. Koch and Y. S. Cho, *Nano. Mater.* 1, 207 (1992).
21. C. Suryanarayana, *Prog. Mat. Sci.* 46, 1 (2001).
22. D. A. Glocker and S. I. Shah, Eds., "Handbook of Thin Film Process Technology." Institute of Physics Publishing, Bristol, UK, 1995.
23. J. L. Vosen and W. Kern, Eds., "Thin Film Processing." Academic Press, Boston, 1991.
24. L. I. Maissel and M. H. Francombe, Eds., "An Introduction to Thin Film." Gordon and Breach, NewYork, 1973.
25. C. J. Brinker and G. W. Scherer, "Sol–Gel Science." Academic Press, Boston, 1990.
26. L. L. Herch and J. K. West, *Chem. Rev.* 90, 3 (1990).
27. C. N. R. Rao, *Mater. Sci. Eng. B* 18, 1 (1993).
28. J. Zarzycki, *J. Sol Gel Sci. Technol.* 8, 17 (1997).
29. R. A. Roy and R. Roy, *Mater. Res. Bull.* 19, 169 (1984).
30. P. Bhattacharya and K. Chattopadhyay, *Mater. Trans. JIM* 39, 75 (1998).
31. P. Bhattacharya and K. Chattopadhyay, "Proc. International Conference on Recent Advances in Metallurgical Processes II," Bangalore, India 1997.
32. V. Vendange and P. Colombus, *Mater. Sci. Eng. B* 168, 199 (1993).
33. A. Basumallick, K. Biswas, S. Mukherjee, and G. C. Das, *Mater. Lett.* 30, 363 (1997).
34. A. Chatterjee and D. Chakraborty, *J. Appl. Phys.* 23, 1097 (1990).
35. E. Johnson, S. Hinderberger, S. Q. Xiao, U. Dahmen, and A. Johansen, *Interface Sci.* 3, 279 (1996).
36. U. Dahmen, E. Johnson, S. Q. Xiao, and A. Johansen, *J. Surf. Anal.* 3, 185 (1997).
37. E. Johnson, V. S. Touboltsev, A. Johansen, U. Dahmen, and S. Hagege, *Nucl. Instr. Meth. Phys. Res. B* 127, 727 (1997).
38. E. Johnson, A. Johansen, L. Sarholt, and U. Dahmen, *J. Met. Nano. Cryst. Mater.* 10, 267 (2001).
39. R. Goswami, Ph.D. thesis, Indian Institute of Science, 1993.
40. D. R. Uhlmann, B. Chalmer, and K. A. Jackson, *J. Appl. Phys.* 35, 2986 (1964).
41. A. Inoue, *Acta Mater.* (*Millenium-Issue*) 48, 279 (2000).
42. J. W. Cahn and G. Kalonji, "Proc. International Conference on Solid → Solid Phase Transformations," The Metallurgical Society of AIME, 3 1982.
43. R. Goswami, W. T. Kim, K. Chattopadhyay, and B. Cantor, *Metal. Trans.* 23A, 3207 (1991).
44. R. Goswami and K. Chattopadhyay, *Acta Metal. Mater.* 43, 2837 (1995).
45. W. T. Kim and B. Cantor, *J. Mater. Sci.* 26, 2868 (1991).
46. R. Goswami and K. Chattopadhyay, *Acta Mater.* 44, 6, 2421 (1996).
47. V. Bhattacharya and K. Chattopadhyay (unpublished).
48. R. Goswami and K. Chattopadhyay, *Phil. Mag. Lett.* 72, 411 (1995).
49. K. Chattopadhyay and R. Goswami, *Prog. Mater. Sci.* 42, 287 (1997).
50. P. Bhattacharya, B. Mazumdar, and K. Chattopadhyay, *J. Elec. Micro.* 48 (suppl.), 1047 (1999).

51. S. Hagege and U. Dahmen, *Phil. Mag. Lett.* 74, 259 (1996).
52. E. Johnson, A. Johansen, L. Sarholt, and U. Dahmen, *J. Met. Nano. Cryst. Mat.* 10, 267 (2001).
53. P. Bhattacharya, V. Bhattacharya, and K. Chattopadhyay, *J. Mater. Res.* 17, 11, 2875 (2002).
54. V. Bhattacharya and K. Chattopadhyay, *Mater. Sci. Eng.* (in press).
55. R. C. Pond and D. S. Vlachavas, *Proc. Roy. Soc. Lond. A* 386, 95 (1983).
56. U. Dahmen, *MRS Bull.* 22 (1994).
57. R. Goswami, K. Chattopadhyay, and P. Ryder, *Acta Mater.* 46, 4257 (1998).
58. A. Landa, P. Wynblatt, E. Johnson, and U. Dahmen, *Acta Mater.* 48, 2557 (2000).
59. E. Johnson and U. Dahmen, *Micro. Microanal.* 3, 409 (1997).
60. P. Bhattacharya, Ph.D. thesis, Indian Institute of Science, 2000.
61. E. Johnson, A. Johansen, C. Nelson, and U. Dahmen, *J. Elec. Micro.* 51 (suppl.) 201 (2002).
62. V. Bhattacharya and K. Chattopadhyay (unpublished).
63. K. I. Moore, K. Chattopadhyay, and B. Cantor, *Proc. Roy. Soc. Lond. A* 414, 499 (1987).
64. L. Kjeldgaard, Ph.D. thesis, University of Copenhagen, 1999.
65. H. Gabrisch, L. Kjeldgaard, E. Johnson, and U. Dahmen, *Acta Mater.* 49, 4259 (2001).
66. U. Dahmen, E. Johnson, S. Q. Xiao, and A. Johansen, *J. Surf. Analysis* 3, 2 185 (1997).
67. D. Turnbull and J. C. Fisher, *J. Chem. Phys.* 17, 71 (1949a).
68. D. Turnbull, *J. Appl. Phys.* 20, 817 (1949b).
69. D. Turnbull, *J. Chem. Phys.* 18, 198 (1950a).
70. D. Turnbull, *J. Appl. Phys.* 21, 1022 (1950b).
71. J. W. Christian, "The Theory of Phase Transformation in Metals and Alloys." Pergamon Press, Oxford, 1975.
72. D. L. Zhang, K. Chattopadhyay, and B. Cantor, *J. Mater. Sci.* 26, 1531 (1991).
73. D. Turnbull and R. E. Cech, *J. Appl. Phys.* 21, 804 (1950).
74. D. Turnbull, *J. Met. Trans. AIME* 188, 1144 (1950c).
75. D. Turnbull, *Phys. Rev.* 83, 880 (1951).
76. R. T. Southin and G. A. Chadwick, *Acta Metal.* 26, 223 (1978).
77. J. H. Perepezko, *Mater. Sci. Eng.* 65, 125 (1984).
78. J. H. Holloman and D. Turnbull, *J. Metals Trans. AIME* 803 (1951).
79. B. E. Sundquist and L. F. Mondolfo, *Trans. Metal. Soc. AIME* 221, 157 (1961).
80. M. Stowell, T. Law, and J. Smart, *Proc. Roy. Soc. London* 318, 231 (1970).
81. L. Lacy, M. B. Robinson, and T. Z. Rathz, *J. Cryst. Growth* 51, 47 (1981).
82. A. J. Drehman and D. Turnbull, *Scripta Metal.* 15, 543 (1981).
83. A. J. Drehman and A. L. Greer, *Acta Metal.* 32, 323 (1984).
84. C. C. Wang and C. S. Smith, *Trans. Metal. Soc. AIME* 188, 156 (1950).
85. D. L. Zhang and B. Cantor, *Sci. Metal. Mater.* 24, 751 (1990).
86. C. R. Ho and B. Cantor, *Phil. Mag.* A66, 141 (1992).
87. W. Kauzmann, *Chem. Rev.* 43, 219 (1948).
88. A. T. Dinsdale, *CALPHAD* 15, 317 (1991).
89. H. B. Singh and A. Holz, *Sol. State Comm.* 45, 985 (1983).
90. K. I. Moore, D. L. Zhang, and B. Cantor, *Acta Metal.* 38, 1327 (1990).
91. W. T. Kim and B. Cantor, *Acta Metal. Mater.* 40, 3339, (1992).
92. Ph. Buffiat and J. P. Borel, *Phys. Rev. A* 13, 2287 (1976).
93. H. W. Sheng, K. Lu, and E. Ma, *Acta Mater.* 46, 14, 5195 (1998).
94. Y. Li, S. C. Ng, Z. P. Lu, Y. P. Feng, and K. Lu, *Phil. Mag. Lett.* 78, 1, 37 (1998).
95. R. W. Cahn, *Nature* 323, 23 (1986).
96. A. R. Ubbellohde, "The Molten State of Matter: Melting and Crystal Structure." Wiley, Chichester, 1978.
97. F. A. Lindemann, *Phys. Z.* 11, 609 (1910).
98. H. W. Sheng, G. Ren, L. M. Peng, Z. Q. Hu, and K. Lu, *J. Mater. Res.* 12, 1, 119 (1997).
99. L. Gråbaeck and J. Bohr, *Phys. Rev. Lett.* 64, 934 (1990).
100. K. Hoshino and K. Shimamura, *Phil. Mag. A* 40, 137 (1979).
101. C. J. Rossouw and S. E. Donnelly, *Phys. Rev. Lett.* 55, 2960 (1985).
102. J. Dages, H. Gleiter, and J. H. Perepezko, *Phys. Lett. A* 119, 79 (1986).
103. G. L. Allen, W. W. Gile, and W. A. Jesser, *Acta Metal.* 28, 1695 (1980).
104. A. K. Malhotra and D. C. Van Aken, *Phil. Mag. A* 71, 949 (1995).
105. K. Sasaki and H. Saka, *Phil. Mag. A* 63, 6, 1207 (1991).
106. G. L. Allen, R. A. Bayles, W. W. Gile, and W. A. Jesser, *Thin Solid Films* 144, 297 (1986).
107. D. L. Zhang and B. Cantor, *Acta Metal. Mater.* 39, 1595 (1991).
108. R. Goswami and K. Chattopadhyay, *Acta Metal. Mater.* 42, 383 (1994).
109. P. R. Couchman and W. A. Jesser, *Phil. Mag.* 35, 787 (1977).
110. F. Spaepen and D. Turnbull, *Scripta Metal.* 13, 149 (1979).
111. D. R. Uhlmann, *J. Non-Crys. Solids* 41, 347 (1980).
112. H. Saka, Y. Nishikawa, and T. Imura, *Phil. Mag. A* 57, 895 (1998).
113. H. W. Sheng, G. Ren, L. M. Peng, Z. Q. Hu, and K. Lu, *Phil. Mag. Lett.* 73, 179 (1996).
114. G. D. T. Spiller, *Phil. Mag.* 46, 535 (1982).
115. T. Ohashi, K. Kuroda, and H. Saka, *Phil. Mag. B* 65, 5, 1041 (1992).
116. R. W. Cahn, *Nature* 342, 630 (1989).
117. J. J. Metois and J. C. Heyraud, *J. Phys. France* 50, 3175 (1983).
118. S. J. Peppiat and J. R. Sambles, *Proc. Roy. Soc. A* 345, 387 (1975).
119. H. W. Sheng, K. Lu, and E. Ma, *Nano. Mater.* 10, 5, 865 (1998).
120. E. Johnson, A. Johansen, U. Dahmen, S. Chen, and T. Fujii, *Mater. Sci. Eng. A* 304–306, 187 (2001).
121. H. Gabrisch, L. Kjeldgaard, E. Johnson, and U. Dahmen. *Acta Mater.* 49, 4259 (2001).
122. J. G. Lee, H. Mori, and H. Yasuda, *Phys. Rev. B* 66, 012105 (2002).
123. J. G. Lee, H. Mori, and H. Yasuda, *Phys. Rev. B* 65, 132106 (2002).
124. H. Mori and H. Yasuda, *Mater. Sci. Eng. A* 312, 99 (2001).
125. H. Yasuda, K. Mitsuishi, and H. Mori, *Phys. Rev. B* 64, 094101 (2001).
126. H. Yasuda and H. Mori, *Phys. Rev. Lett.* 69, 3747 (1992).
127. K. Raviprasad and K. Chattopadhyay, *Scripta Metal. Mater.* 32, 1049 (1995).
128. M. Manivel Raja, K. Chattopadhyay, B. Majumder, and A. Narayanasamy, *J. Alloys Compounds* 297, 199 (2000).
129. K. Chattopadhyay, V. Bhattacharya, and A. P. Tsai, *Mater. Res. Proc.* 703, V13.22.1–6 (2002).
130. A. Inoue, N. Yano, K. Matsuzaki, and T. Masumoto, *J. Mater. Sci.* 22, 123 (1987).
131. A. Inoue, M. Oguchi, K. Masuzaki, and T. Masumoto, *Internat J. Rapid Solidification* 1, 273 (1984).
132. R. Goswami, S. Banerjee, A. K. Roychoudhuri, and K. Chattopadhyay, *J. Appl. Phys.* 73 (1993).
133. Gouthama, Ph.D. thesis, Department of Metallurgy, Indian Institute of Science, 1989.

Encyclopedia of Nanoscience and Nanotechnology
www.aspbs.com/enn

Epitaxy of Ion-Irradiated Quartz

Klaus-Peter Lieb
Universität Göttingen, Göttingen, Germany

CONTENTS

1. INTRODUCTION

Covalent Si-based materials like silicon carbide, nitride, and oxide, in both their various crystalline phases and amorphous states, are of high technological interest since they uniquely combine important material properties: high mechanical strength, high thermal and chemical stability, low mass density, wide bandgap, and other interesting electronic and optical properties. Quartz (SiO_2) and related silicate glasses have applications in optical devices such as wave guides, optical switches, and amplifiers. Because of the high melting points of these materials, ion-beam processing and implantation are indispensable tools for electrical and optical dopings. These materials are very sensitive to ion bombardment and amorphize even at very low fluences. Therefore, damage accumulation and (radiation-induced) impurity diffusion and their relevance for materials modifications are important issues. Quartz in its challenging behavior shows fascinating similarities under high pressure and during ion implantation concerning amorphization, melting, and phase transformations but, despite numerous attempts, remains one of the most difficult materials to describe theoretically.

The long-range ordered crystalline state is usually the lowest free-energy form of elemental and compound solids. Irradiation-induced atomic rearrangements may occur along the trajectories of swift heavy ions or in collision cascades generated by low-energy ion beams, and can result in the loss of translational or orientational order ("amorphization"), at temperatures below which thermally activated diffusion processes may lead to epitaxial recrystallization. Over the past two decades, many attempts have been made to achieve epitaxial recovery of the crystalline state in Si-based compounds, disordered or amorphized by ion irradiation, and to model, on the basis of topological and bonding arguments, optimal conditions for epitaxy. Solid phase epitaxial regrowth (SPEG) of thin amorphous SiO_2 films, which the present chapter focuses on, has therefore been studied in considerable detail.

SiO_2, like Si, SiC, and Si_3N_4 in their various crystalline polymorphous and amorphous phases, can be considered as networks of corner-sharing $[SiX_4]$ tetrahedra formed by four X = C, N, O, or Si atoms around each Si atom. In α-quartz, each O atom is bound to two Si atoms. The Si—O bond length is 0.161 nm and the O—O tetrahedral distance 0.260 nm; the Si—O—Si intra-tedrahedral angle ranges between $\theta = 108.74°$ and 110.54°. This tetrahedral structure unit is very robust and preserved intact upon irradiation. The slight structural changes induced by irradiations or amorphization concern the Si—Si and O—O distances (by about 5 pm) and the Si—O bond length (at most 2 pm). The angle θ shows more pronounced changes of 7.5–12° [1–4]; Figure 1 illustrates the atomic configuration and topology of α-quartz.

The topology and connectivity of such polyhedral networks can be characterized by the two numbers $\{V, C\}$, where V denotes the number of vertices in the polytope and C the average number of polytopes sharing each vertex; for SiO_2 $\{V, C\} = \{4, 2\}$. According to Gupta and Cooper [5], the *structural freedom* f of each vertex is defined as the difference between the number d of degrees of freedom and the number of constraints h, originating from the surrounding structure:

$$f = d - h = d - C\,\delta[1 - (\delta + 1)/2V] \qquad (1)$$

where δ is the dimensionality of the network-forming polytope. For SiO_2 we have $d = \delta = 3$ and therefore, like for Si, the structural freedom vanishes: $f = 0$. According to Hobbs and co-workers [6], this is important for any transformation of an ordered network into a disordered network and vice versa. SiO_2, crystalline and amorphous, represents a fully connected network and is therefore most likely to become disordered by ion implantation (and rather unlikely to be thermally recrystallized). Indeed, SiO_2 and Si become disordered at the rather low energy of 7 eV/atom and

ISBN: 1-58883-059-4/$35.00

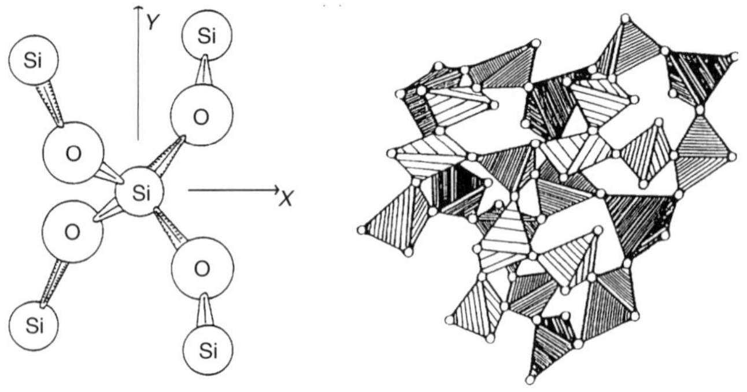

Figure 1. Atomic configuration and topology of SiO_2.

11 eV/atom, as compared to SiC (26 eV/atom) and Si_3N_4 (>700 eV/atom) [7–11].

This chapter is organized as follows: after the present short introduction into the phases and topology of the SiO_2 network and the ion analytical methods used, in Section 2 we briefly summarize recent results on ion-beam induced damage and the kinetics of damage accumulation. Recrystallization of such amorphized or damaged nm layers, via dynamic, chemically guided, or laser processes, will be covered in Section 3. Finally, in Section 4 we touch upon luminescence properties of nanocrystalline particles in silica, and discuss possible future doping and processing techniques and applications in optoelectronics. For previous review papers on ion-induced amorphization of quartz, we refer to the work of Mazzoldi and Miotello [12], Hobbs and co-workers [6], Götz [7], Douillard and Duraud [13], Bolse [14], and Piao et al. [15]. As alkali ion diffusion in silicate is of importance in the context of this chapter, we also mention the corresponding report by Frischat [16] and briefly summarize some important results in Section 4. Optically active defect centers and nanoparticles in silica have been reviewed, for example, by Haglund [17], Polman [18], and Townsend et al. [19]. For theoretical treatments of the of SiO_2 phases and dynamic instabilities under pressure, we mention [20–23] and references given therein.

For studying the microstructure in SiO_2 layers, tens to hundreds of nm thick, various experimental techniques have been employed, such as transmission electron microscopy (TEM), X-ray diffraction (XRD), electron spin resonances (ESR), nuclear magnetic resonance (NMR), profilometry and ellipsometry, Raman and other optical spectroscopy. Ion-irradiation-induced damage and epitaxy of thin quartz films have also been studied with ion analytical methods, which offer unsurpassed advantages as will be shown. The crystallinity and/or the damage profile of the matrix can be measured by means of Rutherford backscattering channeling spectroscopy (RBS-C) for the Si-sublattice using a 1- to 2-MeV α-particle beam [24, 25]. This technique also provides the depth profiles (and in favorable cases even the lattice locations) of heavier implanted ions. For depth-profiling light elements or isotopes, either after implantation or as constituents of the SiO_2 matrix, one may use resonant nuclear reaction analysis with 100- to 500-keV proton beams (RNRA) or elastic recoil detection analysis with energetic heavy ions (ERDA) [26, 27]. The latter technique allows one to scan simultaneously the depth profiles of all light- and medium-weight elements in the near-surface region of the samples, while RNRA offers the advantage of being sensitive to an individual isotope distribution only. By combining these methods, a full picture of important transport and microstructural properties of the matrix and the implants can be gained. There are a number of recent articles and books surveying the various ion-beam analytical methods and their combined use in thin-film analysis [24–27]. A brief description is given in Box 1.

2. ION-BEAM-INDUCED AMORPHIZATION OF QUARTZ

2.1. Damage Accumulation and Amorphization

Before describing processing techniques, which were developed to epitaxially regrow amorphous nm surface layers in α-quartz, it appears advantageous to characterize the type and degree of radiation damage in this matrix introduced by energetic ion impact. For a general discussion of radiation defect production see, for example, [40]. The effects of irradiating α-quartz or silicon dioxide have been extensively studied both experimentally and theoretically. According to Douillard and Duraud [13] and Piao et al. [15], these studies have shown that upon various particle irradiations α-quartz gradually becomes disordered and evolves towards a single amorphous state, which slightly differs from ordinary vitreous silica and which is classified by geologists as *metamict*. This amorphous end product is an "optically isotropic, glass-like material with virtually identical density, thermal expansion and elastic properties" [15]. It is

(i) almost independent of the irradiation process, that is, whether neutrons, ions, electrons, or photons are used for the irradiation;
(ii) nearly indistinguishable from the end product of an ordinary vitreous silica; but
(iii) has a slightly higher density (by 3–4%) as compared to vitreous silica.

Early irradiation experiments on quartz [28–30] by fast reactor neutrons (>1 MeV) at low neutron fluences ($\Phi_n < 10^{16}$ neutrons/cm^2) indicated that the neutrons create point defects, mainly B'-centers and oxygen hole centers. Their concentration increases as a power function with the neutron fluence. For medium fluences ($1 \times 10^{16} < \Phi_n < 1 \times 10^{20}$ neutrons/cm^2), α-quartz under-goes an $\alpha \rightarrow \beta$ phase transition, followed by full amorphization at $\Phi_n > 1 \times 10^{20}$ neutrons/cm^2. These studies suggest that amorphization proceeds by progressive disordering of a highly defective β-matrix [31, 32]. Measurements of the density and refractive index after high-fluence neutron irradiations of α-quartz showed no difference relative to vitreous silica.

There also exist extensive irradiation studies of quartz with protons and α-particles (for which electronic stopping dominates) and with heavy ions (in the regimes of both electronic and nuclear stopping) [28, 33–39]. Macaulay-Newcombe, Thompson, and collaborators [35, 36] investigated the effects of implanting heavy ions in quartz. Fischer et al. [37] studied defect production and annealing up to 1100 °C in quartz after 150-keV Ar and 35-keV

Box 1. Ion-beam analysis.

Ion-beam analytical methods are particularly suited for depth profiling the various components in ion-beam irradiated quartz (matrix, dopants) and to monitor the damage profiles and/or crystallinity of the matrix. For that reason we briefly explain the essence of these techniques. More detailed descriptions are given in [24–27].

Rutherford backscattering spectroscopy (RBS) mainly measures the depth profiles of medium-mass (or heavy) elements in a matrix composed of light (or medium-mass) elements. The primary beam (element number Z_1, mass number M_1, well-defined energy E_0) hits the target. The energy E_f and yield N_f of the backscattered particles are analyzed at the scattering angle θ in a Si surface barrier detector or a magnetic spectrometer having the solid angle $d\Omega$. The backscattering energy E_f depends on the mass M_2 of the backscattering nucleus (isotope sensitivity) and its location x relative to the surface (depth sensitivity). The beam energy E_0 is chosen so that the scattering is purely Coulombian. Hence the differential scattering cross section scales approximately as $d\sigma/d\Omega \propto (Z_1Z_2e^2/E)^2 \sin^{-4}(\theta/2)$, where E denotes the incident energy just before the collision and thus is corrected for the energy loss $\Delta E_1^{in}(x)$ of the beam between the surface and the target nucleus: $E = E_0 - \Delta E_1^{in}$.

In the present application of ion-beam irradiated quartz, RBS with 1- to 3-MeV α-particles is very well suited to measure the profiles of the implants such as Rb, Cs, Xe, and Ba, whose signals are well separated in energy from those of the SiO_2 matrix and enhanced in intensity due to their relatively large Z_2 values. RBS is less suited to differentiate between the oxygen isotopes ^{16}O and ^{18}O, due to their smaller charge $Z_2 = 8$ and mass $M_2 = 16$ and 18 u, relative to silicon ($Z_2 = 14$, $M_2 = 28$, 29, 30 u). The depth resolution depends on the energy definition of the beam, the straggling on its way in and out, the surface roughness, and the energy resolution of the detector. To very good approximation, RBS can be considered nondestructive.

Extensions to ion beams of higher masses, ion energies above the Coulomb barrier, and focused beams of micrometer diameter (microbeam) are known.

RBS channeling is related to RBS, but additionally uses the fact that the stopping power for the incoming beam along a crystalline channel direction is much smaller than that outside the channel (random direction). When travelling along a channel, the ions thus experience a reduced hit probability and consequently give a smaller backscattering yield (virgin curve).

RBS channeling provides the main information concerning the damage profiles and/or amorphous and recrystallized zones of the irradiated quartz samples. In most of the work a 1.0-MeV α-particle beam and a three-axis goniometer were used.

Elastic recoil detection analysis is complementary to RBS in the sense that a beam of heavy projectiles (Z_p, M_p) is directed onto a matrix composed of lighter elements/isotopes (Z_T, M_T). Again a Coulombian elastic collision takes place, which changes both the energy E_p and direction of the projectile and accordingly ejects the hit atom from its site and possibly out of the sample. Measuring the ejectile's energy E_T (in a silicon surface barrier detector) and mass M_T (by a time-of-flight equipment) allows one to measure the depth distributions of the matrix and possible light dopants.

In the epitaxy of ion-irradiated quartz, TOF-ERDA was applied to determine the concentration profiles of ^{16}O and ^{18}O in the annealed matrix and that of light implanted ions, such as ^{1}H, ^{7}Li, ^{20}Ne, and ^{23}Na. In the present work, beams of 30- to 55-MeV ^{107}Ag or ^{113}I and the TOF-ERDA equipment installed at the University of Helsinki were used. The element sensitivity is comparable to that quoted above for RBS, while the depth sensitivity is somewhat lower. Care should be taken concerning the amorphization of the samples due to the high projectile mass and energy loss. Some sputtering is inevitable. Otherwise ERDA can be considered nondestructive. It is a very versatile ion beam method, which scans a whole range of elements in a single measurement.

Nuclear reaction analysis (NRA) is a depth- and isotope-sensitive profiling method, which is mostly used for particular light isotopes in the target, such as ^{1}H, ^{15}N, or ^{18}O, for which nuclear resonance reactions with narrow resonances and sufficient cross sections are known. Sometimes also reactions with smoothly varying cross sections are useful. In the case of a narrow resonance, the sensitive depth x is defined by the energy loss of the beam, $\Delta E_1^{in}(x)$, which equals the difference between the initial beam energy E_0 and the resonance energy E_R. Measuring the yield of ejectiles or γ-rays for constant projectile number as function of the beam energy, scanned in small steps, provides the depth profile of the component searched for. For instance, the ^{18}O content can be determined by the $^{18}O(p,\alpha)^{15}N$ resonance reaction at $E_R = 198$ keV.

The depth resolution depends again on the energy precision of the beam from the accelerator, its straggling inside the target, and the roughness of the surface. Although NRA is in principle nondestructive, modifications of the samples by the rather intense analyzing beam cannot always be avoided (amorphization, heating, diffusion, even disintegration). NRA is most suitable for strong and narrow resonances in proton-induced reactions, but often suffers from long measuring times and the need to cover different resonance energies for different isotopes/elements.

He implantation, that is, in the regime of nuclear stopping power. These authors observed three defect annealing stages depending on the energy density deposited by nuclear stopping:

(i) simple defects at energy densities of <0.1 keV/nm^3, which can be removed by thermal annealing at 500 °C;

(ii) amorphized microregions generated between 0.1 and 0.25 keV/nm^3, which anneal out at temperatures up to 1000 °C;

(iii) amorphous layers (as seen by RBS-C) produced at energy densities larger than 0.25 keV/nm^3, which cannot be annealed below 1100 °C.

For low- and medium-mass ions (H, N, Ne, Na) in the 10–100 keV range, Harbsmeier and Bolse [38, 39] recently carried out detailed analyses, whose results we shall now describe in more detail. Typically, the 1-mm-thick single-crystalline (0001) quartz samples were irradiated at liquid nitrogen temperature with ions in the fluence range from 10^{13} to 10^{17} ions/cm^2. The 1-μA beam homogeneously covered the sample by means of a sweeping system. The crystallinity of the matrix was analyzed with RBS channeling. As an example, Figure 2 displays spectra in the (0001) channel obtained after implanting 50-keV $^{23}Na^+$ ions at fluences of $\Phi = 2 \times 10^{13}$–6×10^{15} ions/cm^2, together with spectra taken in *random* and in-channel (*virgin*) directions from a nonirradiated sample. Below 5×10^{13} ions/cm^2, the backscattering spectra clearly indicated accumulation of radiation damage in the surface layer, giving rise to bell-shaped damage profiles (see Fig. 2). However, above a fluence of 5×10^{13} ions/cm^2, the backscattering signal reached the random yield, which we take as the *criterion of full amorphization* of this layer. Note that the thickness of the amorphous layer increased when the ion fluence was further increased. These two trends, first the rising damage profile and afterwards the increasing thickness of the amorphous layer, are a common feature in ion-irradiated Si and Si-based ceramics, for example, in SiC, Si_3N_4, SiO_2 [41, 42]. When-ever the mass of the implanted ions was much higher than that of Si and O, RBS channeling also allowed one to determine the depth profiles of the implanted ions. In the case of low-mass implants, TOF-ERDA or nuclear resonance reactions were used to measure their profiles.

The measured implantation profile of 50-keV Na ions was compared with the one calculated with the Monte Carlo code TRIM95 [50]. The measured projected range, $R = 7.8 \times 10^{17}$ at/cm^2, and distribution width, $\Delta R = 5.9 \times 10^{17}$ at/cm^2, agree rather well with the predictions of TRIM95, $R = 6.9 \times 10^{17}$ at/cm^2 and $\Delta R = 5.3 \times 10^{17}$ at/cm^2.

Irradiations of quartz with swift heavy ions at energies of up to 10 MeV/u were carried out by Douillard et al. [43] and by Wilson et al. [44] for 730-MeV Pb and 540-MeV Ni ions, by Ramos and co-workers [45, 46] for 0.5- to 10-MeV Au ions, and by Trautmann and collaborators for 90-MeV Ar ions [47] and 820-MeV Pb ions [48]. The ion-induced modifications in the near-surface regions were again analyzed by means of RBS channeling, TEM, profilometry, and AFM. These experiments provided information on amorphization, swelling or compaction, surface roughening, and sputtering as well as single-ion track formation and surface craters. Most important, these investigations demonstrated that very similar damages are produced via electronic stopping as via nuclear stopping.

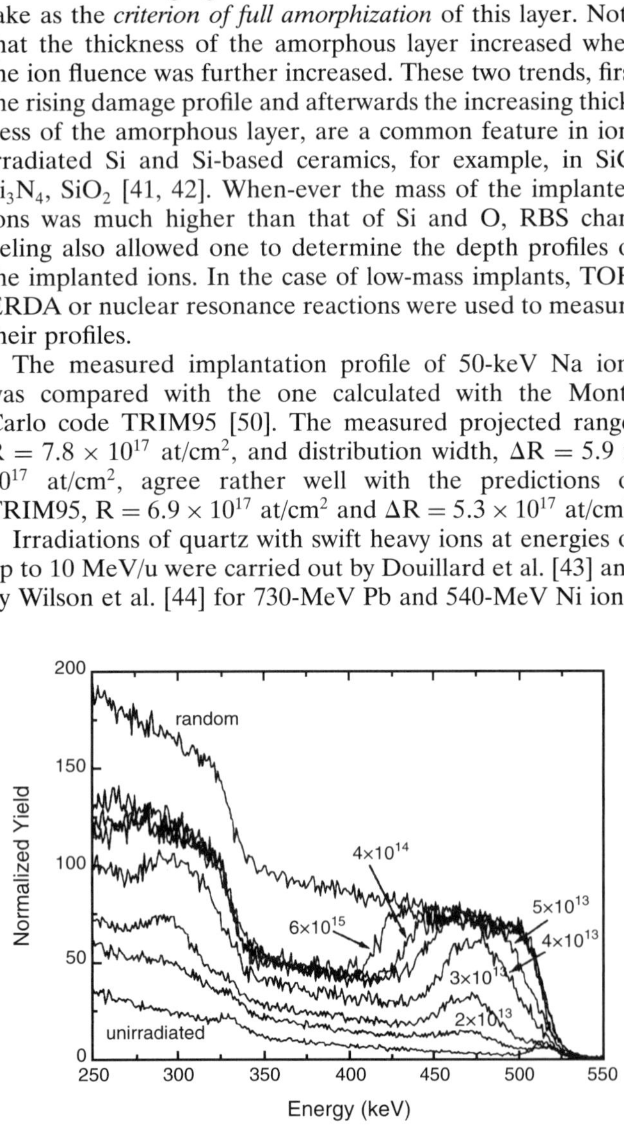

Figure 2. RBS channeling spectra in the (0001) channel obtained after implanting 50-keV $^{23}Na^+$ ions at fluences of $\Phi = 2 \times 10^{13}$–6×10^{15} ions/cm^2, together with spectra taken in *random* and *virgin* directions from a nonirradiated sample [39].

2.2. Model Calculations

For modeling the damage accumulation, in relation to the ion fluence Φ and the depth distribution of the deposited energy density, $E_D(\Phi, z)$, we define the damage depth profile $\chi(z)$ and the integral damage $I_\chi(\Phi) = \int \chi(z)\,dz$, summed over the full implantation profile. Evidently, $\chi(z) \leq \chi^{max} \leq 1$, where χ^{max} denotes the maximum damage of the distribution. The limit $\chi^{max} = 1$ indicates that the backscattering yield in the channeling direction reaches the random level. The experimental functions $\chi(z)$ and $I_\chi(\Phi)$ were derived from the RBS-C spectra by means of the program DAMAGE [49] and compared with the calculated deposited energy density $E_D(\Phi,z) = \Phi F_D(z)$. The quantity $F_D(z)$ was estimated with the Monte Carlo code TRIM95 [50]:

$$F_D(z) = E_d[2n_V(z) + n_D(z)] \quad (2)$$

Here $n_D(z)$ and $n_V(z)$ denote the number of displacement and replacement collisions, respectively, and $E_d = 25$ eV is the displacement threshold energy. In Figure 3, two of the bell-shaped damage profiles $\chi(z,\Phi)$ after 20-keV H_2^+ and 50-keV Na^+ implantations are compared with the calculated energy density $F_D(z)$. Obviously, the damage profile $\chi(z,\Phi)$ does not directly scale with $F_D(z)$, but follows the power-law function $[F_D(z)]^{n_1}$. The exponent n_1 is of the order of 3–5 (see Table 1). The experimentally deduced fluence dependences of the maximum damage $\chi^{max}(\Phi)$ and integrated damage $I_\chi(\Phi)$ can be parametrized again with the power law [39]:

$$\chi^{max}(\Phi) = (\Phi/\Phi_C)^\alpha \text{ and } I_\chi(\Phi)/I_\chi(\Phi_C) = (\Phi/\Phi_C)^\beta \quad (3)$$

where the exponents α and β are given in Table 1. It is at the critical ion fluence Φ_C that the backscattering yield first reaches the random level and an amorphous layer starts growing.

Various models have been developed to describe the damage accumulation and transition to full amorphization. In Figure 4, the 50-keV Na data are compared with the predictions of the direct impact model by Dennis and Hale [51], the damage overlap model by Gibbons [52], and the nucleation and growth model by Avrami [53], as discussed by Campisano et al. [54] and Harbsmeier and Bolse [38, 39]. Clearly, the latter model fits the data best. Indeed, the maximum damage follows the function

$$\chi^{max}(\Phi) = 1 - \exp[(-K\Phi)^n], \text{ where } K = R_a V_a^{n-1} \quad (4)$$

Here R_a denotes the nucleation rate and V_a the growth rate of the disordered phase. It was concluded [39] that for all

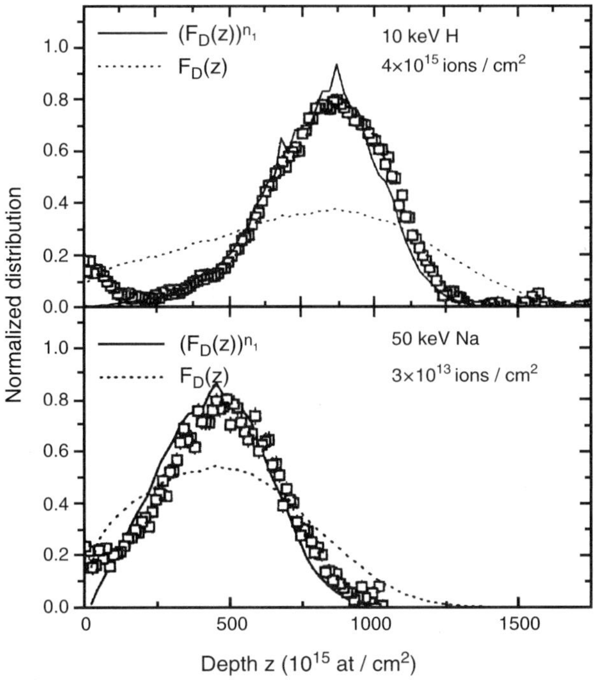

Figure 3. Comparison of the measured and calculated damage profiles after 20-keV H_2^+ ($\Phi = 4 \times 10^{15}/cm^2$) and 50-keV $^{23}Na^+$ ion implantation ($\Phi = 3 \times 10^{13}/cm^2$). The measured damage profiles do not linearly scale with the deposited energy density $F_D(z)$, but with the power $n_1 = 5.7 \pm 1.0$ in the case of $^2H_2^+$ and $n_1 = 3.2 \pm 0.1$ in the case of Na [39].

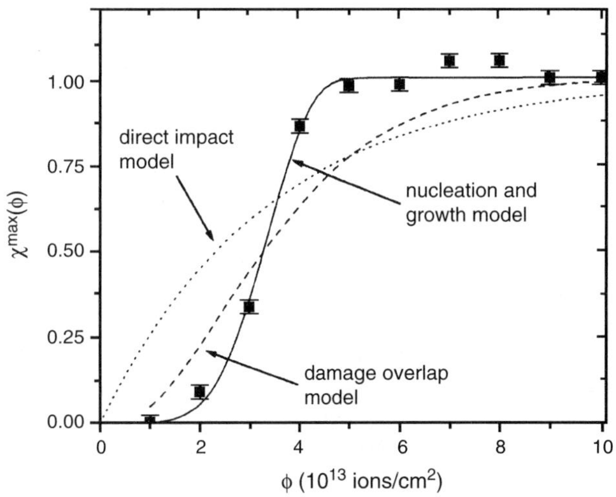

Figure 4. Comparison of the maximum damage $\chi^{max}(\Phi)$ with the predictions of the three models discussed in the text. Evidently the nucleation and growth model by Avrami–Johnson–Mehl gives the best fit to the data [39].

the cases listed in Table 1, the exponent n lies between 3 and 6, indicating three-dimensional defect nucleation. Furthermore, the parameter K is proportional to the maximum deposited energy density F_D^{max}, and the critical fluence Φ_c scales with $1/F_D^{max}$. The authors summarize their conclusions: "Amorphization of quartz occurs by three-dimensional nucleation and growth of defect agglomerates or small spatially separated amorphous zones in a still crystalline matrix, until, after a deposition of a certain critical density $E_C = 1.92 \pm 0.07$ eV/atom (=0.04 displacements per atom), a continuous amorphous layer has formed."

2.3. Build-up and Relaxation of Stresses During Ion Implantation

The accumulation of defects and the excess volume created during ion im-plantation are connected to the build-up of a huge strain field in the damaged region, which, however, is relaxed in the high-fluence regime. Density changes in the irradiated volume may also cause shear forces relative to the nonirradiated parts of the sample (below and beneath the irradiated volume) resulting in a bending of the whole sample. Furthermore, the possible local melting of small volumes as a consequence of ion-induced thermal spikes may exhibit similarities to the processes of pressure-melting or thermal quenching of SiO_2. It is therefore appropriate to briefly discuss similarities between ion-irradiation and pressure-induced modifications of quartz and silicon dioxide.

According to Polman and collaborators [55], three different radiation-induced deformation phenomena are known in SiO_2: densification, radiation-enhanced viscous flow, and anisotropic deformation. Densification (compaction) in amorphous SiO_2 is caused by the arrangement of the SiO_2 network in smaller, more compact rings as a consequence of atomic collisions. Densification eventually saturates with fluence and results in a tensile stress if the irradiated region is constrained by a substrate. The radiation-induced viscosity has been found to be inversely proportional to the nuclear energy loss over several orders of magnitude. Concerning the anisotropic deformation, Snoeks et al. [55] proposed a model which considers the competition between bulk and surface deformation effects. Depending on the ion energy and ratio between nuclear and electronic stopping power, these effects can give rise to either compressive stress (for >3.6-MeV Xe ions) or tensile stress (for low-energy Xe ions).

Table 1. Irradiation conditions and values of the critical fluence Φ_c, the maximum of the deposited energy density F_d^{max}, the exponents of the fitted power laws (n_1, α, β), and the parameters n and K used in the nucleation and growth model [39].

Ion	Energy (keV)	Φ_c^a ($10^{15}/cm^2$)	$F_d^{max\,a}$	n_1	α	β	n	K (10^{-16} cm^2)
H_2	20	530 (20)	0.04	5.7 (10)	8.6 (16)	0.32 (7)	5.7 (4)	2.08 (2)
N	50	15 (1)	1.8	2.4 (3)	2.3 (1)		3.2 (4)	81 (2)
N_2	100	13 (1)	1.8	2.1 (3)	2.6 (4)		3.5 (6)	100 (3)
Ne	50	5.5 (3)	3.0	2.4 (3)	3.3 (4)	0.3 (1)	3.9 (6)	230 (20)
Na	50	4.2 (3)	3.5	3.2 (4)	3.2 (4)	0.10 (1)	5.1 (4)	283 (4)
Na	100	5.2 (3)	2.9	2.2 (3)	2.8 (1)	0.6 (1)	3.6 (1)	246 (2)

[a] In units of 10^{-14} eV cm^2/at ion.

The high-pressure behavior of α-quartz has been a subject of extensive experimental and theoretical studies, but is far from being understood. The stable phases of SiO_2 under pressure are α-quartz, coesite, and stishovite (in this order). At room temperature, the phase transformation to coesite (at $\approx$3 Gpa) and to stishovite (at $\approx$7 Gpa) are inhibited and quartz remains metastable up to about 15–25 Gpa [56, 57]. This seems to be a common property of systems that amorphize under pressure: the parent phase is driven far beyond its stability range before it amorphizes. The crystalline-to-amorphous transition is accompanied by an increase in the silicon coordination. Kingman et al. [57] have shown that the pressure-induced amorphization is preceded by another reversible phase transition at 21 Gpa. This new phase persists up to 30 Gpa, beyond which irreversible amorphization sets in. Hemley et al. [56] proposed that pressure-induced amorphization in SiO_2 represents the metastable extension of the melting curve. In parallel argumentation to a similar behavior of ice at room temperature, it has been argued that the melting of quartz first leads to an extremely unrelaxed, high-density amorphous state, which then relaxes towards an energetically more favorable glass-like state. At the end of this process, pressure glass and thermal glass are almost indistinguishable.

The build-up and relaxation of stress related to changes of the atomic density can be monitored via X-ray diffraction, transmission electron microscopy, Raman spectroscopy, or simply by measuring, with a profilometer, the step height and curvature between the implanted and virgin parts of the sample. In the case of 50-keV Na ion implantations at fluences of more than 10^{15} ions/cm^2, which we discussed before [39], the step size Δh was found to be proportional to the size of the amorphized layer. The ion-induced decrease of the density in quartz due to radiation damage can be as large as 20%. In the case of 50-keV Na ions mentioned above, the maximum pressure amounted to 1.6 Gpa at 10^{14} ions/cm^2. It decreased by about one order of magnitude when the fluence was increased to 10^{15} ions/cm^2. The step size rose from 5 nm to 22 nm in this fluence range, as illustrated in Figure 5.

The swelling of quartz has also been investigated with fast heavy ions using energies of several MeV/u, where the energy loss is due to electronic stopping. For 4-MeV/u Pb ions, the step height Δh was found to increase linearly up to a fluence of 6×10^{11} ions/cm^2, at a rate of $\Delta h/\Phi \approx 1\ \mu m/10^{12}$ ions/cm^2 $= 10^{-16}$ cm^3/ion. For higher fluences the step height reached saturation. It has been customary to normalize the step height Δh to the total ion range R. In quartz the ratio Δh/R also reached a saturation value. For the case of 2.3-MeV/u Ar ions, the saturation step height of $\Delta h_{sat} = 2.9\ \mu m$, which corresponds to $\Delta h_{sat}/R = 0.16$, was reached at a fluence of 2×10^{14} ions/cm^2. Trautmann et al. [48] pointed out that this value of $\Delta h_{sat}/R$ just equals the relative difference in density between the amorphous state ($\rho_a = 2.22$ g/cm^3) and the crystalline state ($\rho_X = 2.65$ g/cm^3). Like in the case of nuclear stopping, it is important to note that no swelling occurs below a critical (electronic) energy loss of $S_e = 1.8(5)$ keV/nm. Above this threshold value, the ratio Δh/R increases linearly with S_e, the slope being 1×10^{13} cm^2 nm/keV.

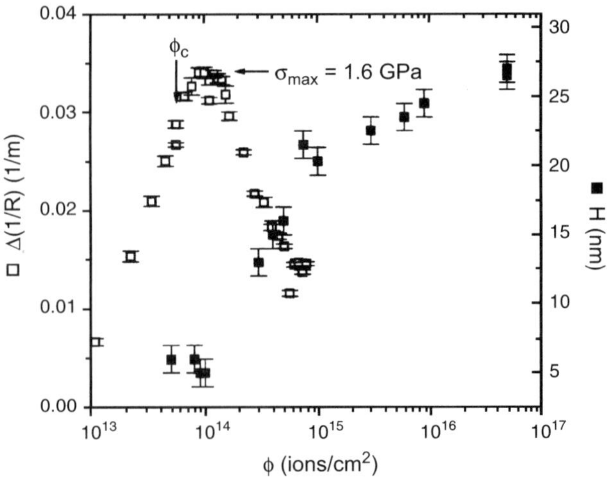

Figure 5. Amorphization of quartz after 50-keV Na implantation [39]. The fluence dependence of the change of curvature, $\Delta(1/R)$, and the step height $H \equiv \Delta h$ are shown.

Ion irradiation of amorphous solids produces plastic deformation in the form of stress-induced creep and anisotropic growth consisting of an expansion perpendicular to the ion-beam direction, together with a shrinkage parallel to it [58]. Both phenomena have been treated by Trinkaus [59] in terms of shear stress relaxation in thermal spikes, followed by the freezing-in of strain during rapid cooling. The viscosities measured over a wide range of ion masses (Ne–Au) and energies (0.15–8 MeV) [55] covering the stopping power range between 0.2 and 4 keV/nm indicate a local spike temperature between 2000 K and 4000 K.

2.4. Surface Effects on SiO_2 Induced by Ion Bombardment

Defect production and sputtering at the surface provide further possibilities to study the energy transfer of incident ions to silicon dioxide. Changes of the surface morphology via single-ion impact and high-fluence sputter erosion were investigated by means of energy-dispersive X-ray reflectivity and/or atomic force microscopy [60, 61]. Mayer et al. [60] reported on $\approx$1-keV proton, He^+, and Xe^+ irradiations of SiO_2 layers (35 nm–2 μm) on Si wafers. For the Xe^+ irradiations, these authors observed roughening and ripple formation at the surface and explained them in terms of the Bradley–Harper theory [62]. This approach considers stochastic sputtering (atom removal) as well as surface diffusion and viscous flow. In the case of proton and He^+ implantations, sputtering was found to be negligible and therefore surface smoothing occurs as a consequence of diffusion. The Bradley–Harper theory turned out to account for both phenomena. The measured relaxation rates of the rms roughness during light-ion bombardment were interpreted by a radiation-enhanced relaxation mechanism of short-lived defects. The extracted viscosity is comparable to the thermal viscosity at about 1000 °C [60].

Sputter erosion and the evolution of surface ripples ("waves") in fused silica as function of the ion fluence and angle of incidence were recently studied by Flamm et al. [61], who irradiated the samples with 0.6- to 1.5-keV Ar ions at fluences of 200–600 μA/cm^2 and inclined incidence to the surface normal. Patterns of surface ripples with wavelengths of 30–250 nm were established. The orientation of these patterns depends on the angle of incidence: up to about 70° the ripples were oriented perpendicular to the projection of the ion beam on the surface, while at 80° the ripple orientation had changed parallel to the beam direction. This rotation of the ripple pattern as function of the angle of incidence was also observed in the case of Xe irradiation of graphite [63]. Figure 6 shows AFM pictures of the ripple patterns obtained for 0.8-keV Ar ions on SiO_2 as function of the tilt angle, showing the rotation of the pattern. The measured fluence dependences of the rms roughness σ and ripple wavelength λ at this energy are plotted in Figure 7. The ion flux density was 400 μA/cm^2 corresponding to about 2.5×10^{15} ions/cm^{2}s. The lines are fits with the scaling functions $\sigma(t) \propto t^{\beta}$ and $\lambda(t) \propto t^{\gamma}$, with the scaling exponents $\beta = 0.46 \pm 0.03$ and $\gamma = 0.15 \pm 0.01$. The authors pointed out that these power-law scalings do not agree with the predictions of the Bradley–Harper theory in the linear approximation and that nonlinear effects have to be included.

Gebeshuber and collaborators [68] have recently reported on nanodefects on α-quartz (0001) surfaces produced by 1-keV Ar^+ single-ion impact at normal incidence and scanned via AFM. The AFM images exhibit hillocks, few nm high, but no craters.

3. EPITAXIAL RECRYSTALLIZATION OF QUARTZ

3.1. Dynamic Epitaxy

The high energy density of some 1–10 keV/nm, which an energetic heavy ion beam deposits in the irradiated material along its track or in the collision cascade, may not only destroy the crystalline order, but rather induce self-organizing processes of the microstructure, such as epitaxy, texturing, grain growth, surface ripples, and even

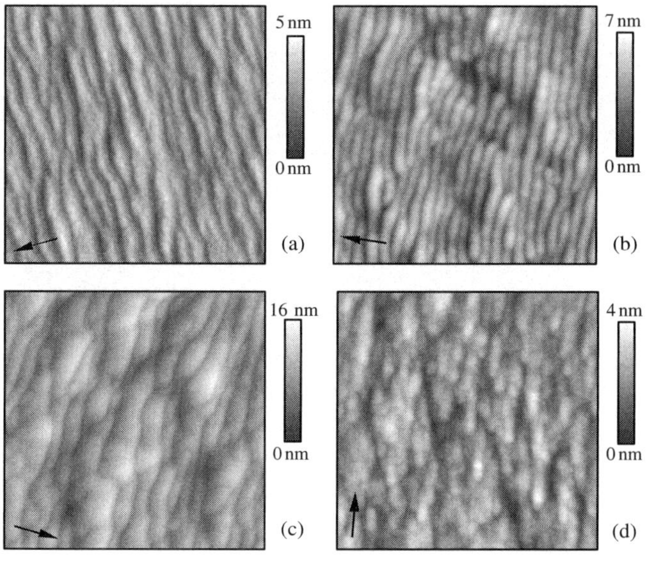

Figure 6. Surface ripples on SiO_2 after 0.8-keV Ar ion irradiation at various angles of incidence [61]. The direction of the ion beam is indicated by the arrow. (a) 50°; (b) 60°; (c) 70°; (d) 80°.

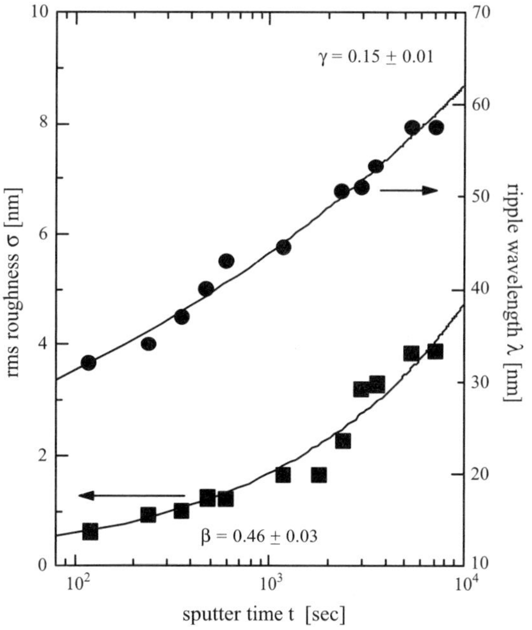

Figure 7. Evolution of the rms roughness σ and ripple wavelength λ with the sputtering time t, in the case of 0.8-keV Ar^+ ions incident at an angle of 60° to the surface normal and at a flux density of 400 μA/cm^2 [61]. The two quantities scale with the exponents $\beta = 0.46 \pm 0.03$ and $\gamma = 0.25 \pm 0.01$, respectively.

magnetic textures. "Dynamic" epitaxial regrowth, which is often called ion-beam-induced epitaxial growth (IBIEG), is a process in which the radiation-induced damages in the crystalline matrix anneal out during the implantation process itself, due to vacancy and self-interstitial diffusion in the cascade or spike zone, restoring the long-range order of the matrix. Concerning future ion-beam doping of quartz for optoelectronic purposes, this process appears to be very promising to retain the crystalline structure, depending of course on the clustering and/or solving properties of the dopants. Dynamic epitaxy of quartz has been reported by Townsend and collaborators [64] after Cu, Ga, and Ag implantations at 55 keV beam energy and at a fluence of 6.4×10^{16} ions/cm^2. Above 225 °C, the implantation profiles of these metals were found to deviate from the predictions of the TRIM code [50], showing bimodal distributions and indicating (radiation-enhanced) diffusion processes during the implantation. Evidence of recrystallization of the amorphized quartz layer at 600 °C was gained for Ag ions and also for Ga ions. These authors recently reported on the formation of Cu nanoclusters in SiO_2 at an implantation fluence of 8×10^{16} ions/cm^2 and also studied their behavior during excimer laser annealing [65].

Dhar et al. [66] investigated 50-keV ^{20}Ne irradiations of quartz at fluences between 1×10^{13} and 4×10^{15} ions/cm^2 and substrate temperatures ranging from liquid nitrogen temperature to 973 K. The aim of this study was to determine, for each ion fluence Φ, the temperature at which no damage was left after the implantation. Figure 8 illustrates RBS channeling spectra obtained at a fixed Ne fluence of 1×10^{15} ions/cm^2, over the full temperature range up to 973 K.

The damage profiles deduced from these spectra are given in Figure 9. Fully amorphized surface layers were found

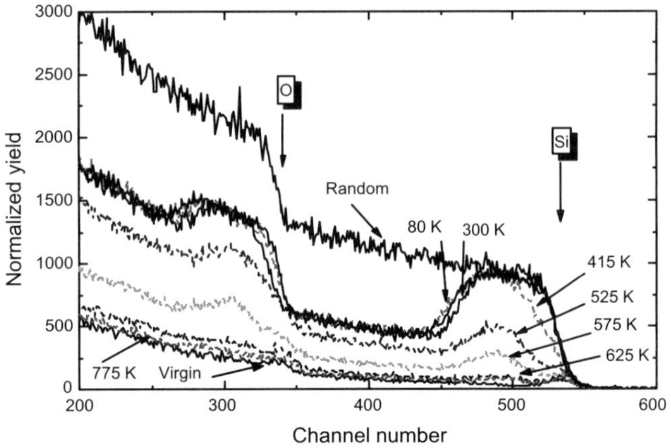

Figure 8. RBS channeling spectra of α-quartz samples irradiated with 50-keV Ne ions at a fluence of 1×10^{15} ions/cm^2 at the temperatures indicated. Above 415 K, the amorphous layer starts to transform into a bell-shaped damage profile, which disappears at 625 K. For this and higher sample temperatures, the ion-beam-induced damages are annealed out during the implantation process, leaving a single-crystalline matrix [66].

up to 500 °C implantation temperature. The thickness of the amorphized layer decreased from 11×10^{17} at/cm^2 for the as-implanted sample to 8.5×10^{17} at/cm^2 at 773 K. For higher temperatures, the damage profiles had the bell shape described in Section 2, their maximum damage χ^{max} decreasing with increasing temperature, until above 900 K no damage was left in the sample anymore.

Such series of irradiations and RBS-C analyses were performed over the full fluence and temperature range indicated above. In Figure 10, the deduced critical fluence Φ_c for epitaxy is plotted versus the temperature. One notes a slow increase of Φ_c with the temperature up to about 800 K, followed by a sudden increase above 800 K. The data closely follow a model proposed by Morehead and Crowder [67],

$$\Phi_c(T) \propto [1 - (L^0/R^0)\exp(-E_{ax}/2k_BT)]^{-2} \quad (5)$$

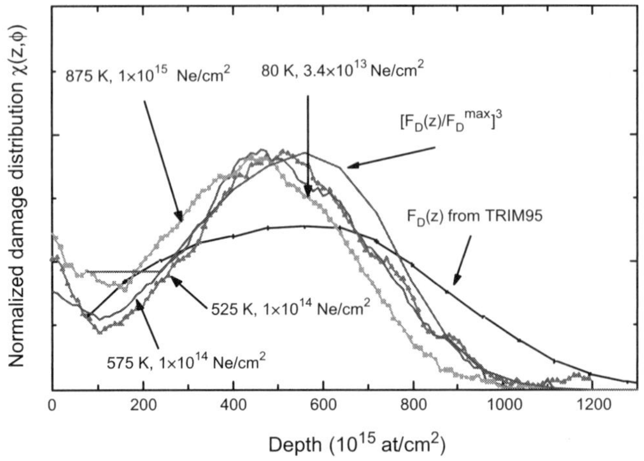

Figure 9. Damage profiles obtained at various sample temperatures and Ne ion fluences. Also shown are the calculated damage profiles $F_D(z)$ in linear and third power [66].

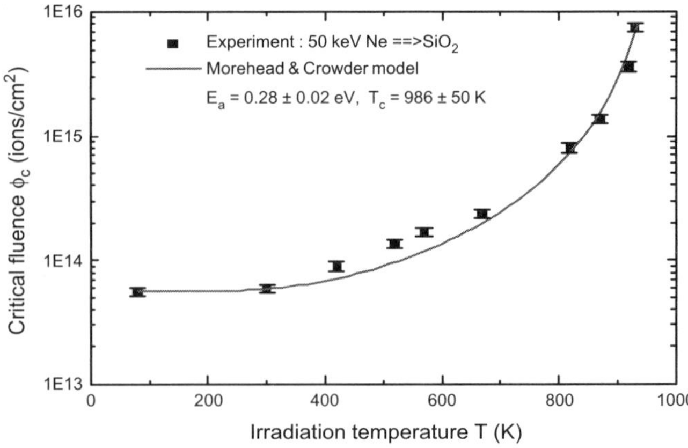

Figure 10. Temperature dependence of the critical fluence Φ_c for 50-keV Ne ion irradiations in quartz. The solid line corresponds to a fit according to the model by Moorehead and Crowder with the parameters $T_c = 986 \pm 50$ K and $E_a = 0.28 \pm 0.02$ eV [66].

The fit gives the activation energy $E_{ax} = 0.26 \pm 0.02$ eV and the critical temperature $T_c = (E_{ax}/k_B)\ln(L^0/R^0) = 986 \pm 50$ K. The value of T_c indicates that temperature at which epitaxy becomes independent of the ion fluence Φ. The parameter L^0 is the diffusion length at infinite temperature and R^0 the radius of the radiation-damaged zone (for each impact ion), in which no thermal diffusion of radiation-induced defects takes place at low temperatures. The ratio L^0/R^0 does not depend on the temperature [67].

The deduced activation energy $E_{ax} = 0.26$ eV agrees well with the known activation energy $E_a = 0.28$ eV for defect diffusion in quartz and suggests that the same annealing process is active below and above Φ_c. The values of E_{ax} and T_c depend on the deposited energy density F_D and consequently on the ion species and energy. For 1.5-MeV Xe irradiations of SiO_2 at 15–1023 K, Wang et al. [69], from *in-situ* transmission electron microscopy, found $T_c = 1446$ K and $E_a = 0.17$ eV. For ⟨100⟩ InP crystals irradiated with 100–600 keV boron, silicon, argon, and selenium ions, Wendler et al. [70] combined RBS-channeling, TEM, and optical spectroscopy and deduced values of T_c =350–420 K and $E_{ax} = 0.35$ eV.

Recently, Dhar and collaborators [71] extended the work on dynamic epitaxy of quartz to the implantation of 170-keV Ba ions by measuring damage profiles in the range from room temperature up to 1175 K. Figure 11 illustrates RBS-channeling spectra for some of these samples. Evidently, the Ba profile shown in the upper part of the spectra is essentially stable, while the matrix partially recrystallizes around 1100–1125 K, but never reaches the virgin status in the top layer. Possibly Ba stabilizes in nanocrystals at this temperature, which disturbs the progression of the crystalline/amorphous (c/a) interface towards the surface; TEM analyses are required to check this conjecture.

3.2. Chemically Guided Epitaxy of Silicon Dioxide

3.2.1. First Attempts

While thermal recrystallization of ion-irradiated Si and Si-based compounds *in vacuo* is a common process [72–74], previous attempts at the epitaxy of ion-irradiated synthetic

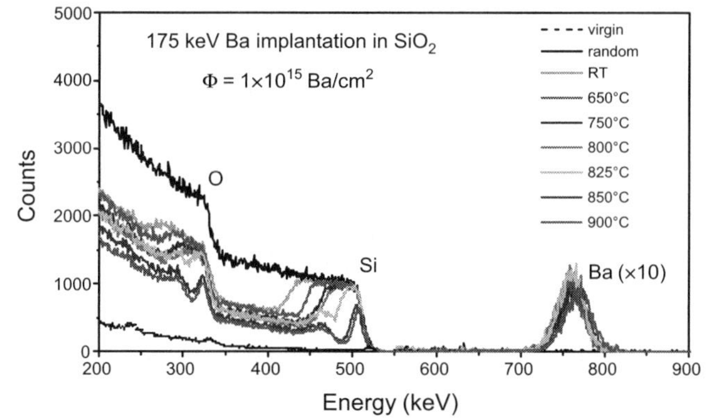

Figure 11. RBS channeling spectra of Ba-irradiated quartz [71]. The Ba ion fluence was $1 \times 10^{15}/cm^2$ and the sample temperature varied from RT up to 900 °C. One notes that the SiO_2 matrix gets amorphized during implantation at room temperature and stays damaged near the surface up to and including 900 °C, while the thickness of the amorphous layer decreases for rising sample temperature. The Ba depth profile is essential immobile at all temperatures.

quartz were generally not successful up to about 1270 K and led to partial epitaxy at most [75–78]. This is particularly true concerning thermal annealing of the irradiated samples *in vacuo*, in which case the results for different ion species, energies, and fluences are somewhat conflicting, depending on the type and concentration of impurities in the matrix before ion implantation. For instance, when irradiating natural Brazilian quartz with Si and O ions at a fluence of 1×10^{16} ions/cm^2 and annealing the samples in air, Devaud et al. [77] observed SPEG of the matrix to be completed at 1120 K and concluded that impurity OH groups were responsible for recrystallization. As early as 1980, Arnold and Peercy [75] in their implantation studies of noble gas ions in $Li_2O \cdot 2SiO_2$ and of Li ions in quartz had found evidence of the special role of alkali ions for epitaxy in silicates. These authors obtained crystalline $Li_2O \cdot 2SiO_2$ when annealing the irradiated sample at 775 K, and recrystallization of Li-doped SiO_2 at 970–1070 K, respectively. After the implantation of 3-MeV Ar ions in quartz, Wang et al. found recrystallization to occur below half the melting temperature [76]. More recently, Harbsmeier et al. [78] carried out similar experiments on α-quartz after 50-keV carbon and nitrogen ion implantations at a fluence of $10^{16}/cm^2$ and annealing in air up to 1670 K. In these experiments, SPEG was found to set in at 820 K (after N-doping) and at 1470 K (after C-doping), visible as a reduced thickness of the amorphous layer and a movement of the amorphous/crystalline (a/c) interface towards the surface. However, no full recrystallization was achieved below 1670 K, the highest annealing temperature used in these experiments.

3.2.2. Epitaxy after Cs Implantation and Annealing in Oxygen

The breakthrough in solving this problem emerged from systematic outdiffusion studies of Na ions implanted into various metals [79–84], Si [74], and SiO_2 [85] as well as into Ni/Si and Ni/SiO_2 bilayers [86, 87], which were performed at the University of Göttingen. Roccaforte et al. [88–92], when investigating Cs-ion implantations in synthetic α-quartz, achieved full epitaxy below 1170 K when annealing the samples in air or oxygen. Similar observations were subsequently made for other alkali ions, for example, by Gustafs-son et al. for Li ions [93], by Dhar et al. for Na ions [94], and by Gasiorek et al. for Rb ions [95]. Finally, Bolse et al. [96] reported on the (so far unsuccessful) attempt to recrystallize quartz after proton implantation. This so-called chemically guided epitaxy via alkali-ion irradiation turned out to be extremely efficient, since after the annealing process the implanted alkali atoms were found to have diffused to the surface, from where they evaporated. Figure 25 illustrates a typical set of RBS channeling spectra for the case of 250-keV Cs^+-ion implantations and annealing in oxygen up to 1120 K [88]. One notes that up to 970 K the a/c interface remains at the as-deposited location, then moves towards the surface, and finally disappears at 1150 K, leaving a fully recrystallized matrix. The recrystallization speed follows an Arrhenius dependence with an activation energy of $E_{ax} = 2.83 \pm 0.20$ eV.

Figure 12 also illustrates various steps of the diffusion of the implanted Cs: the Gaussian implantation profile is stable

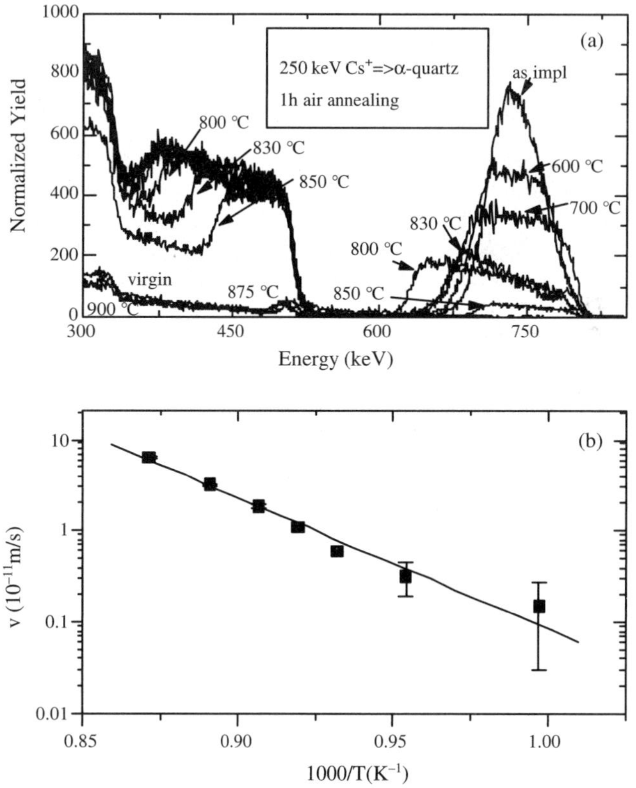

Figure 12. Epitaxy of α-quartz after Cs implantation and annealing in air [88]. (a) RBS channeling spectra at 900 keV α-particle energy obtained after implanting 2.5×10^{16} Cs^+ ions/cm^2 at 250 keV and annealing for 1 h in air at the temperatures given. The Si signal indicates epitaxy between 850 and 875 °C. For the evolution of the Cs profile, see text. (b) Arrhenius plot of the recrystallization speed v corresponding to the activation energy $E_{ax} = 2.83 \pm 0.20$ eV and the pre-exponential factor $v_0 = 152 \pm 10$ m/s.

up to about 825 K, but starts broadening at 870 K, until at 1075 K it extends over the full amorphous layer and reaches the c/a interface. During this process more and more Cs leaves the sample. At 1150 K, that is, when epitaxy of SiO_2 has been completed, no more Cs can be identified in the sample. Chemically guided epitaxy therefore is a catalytic process. Figure 13 displays the retained Cs fraction after a 1-h annealing as function of the temperature. The loss of implanted Cs follows an Arrhenius dependence with an activation energy of $E_a^{Cs} = 0.98 \pm 0.10$ eV.

The third partner in this process is the oxygen, whose role was highlighted by using an isotopically enriched ^{18}O atmosphere during the annealing. In Figure 14, the ^{18}O and Cs depth profiles in the recrystallizing layer are compared with each other. [In these first experiments, the ^{18}O concentration profiles were monitored via the nuclear reaction $^{18}O(p,\alpha)^{15}N$ at 151 keV resonance energy [97]. In the subsequent Cs experiments [91, 98], time-of-flight elastic recoil detection analysis (TOF-ERDA [26]) was used to depth profile ^{18}O. TOF-ERDA was also most useful after Li and Na ion implantation for a simultaneous profiling of all light elements, such as ^{7}Li, $^{16,18}O$, and ^{23}Na, while hydrogen profiling was done with the resonant reaction $^{1}H(^{15}N,\alpha\gamma)^{12}C$ (see below).] Returning to Figure 14, the ^{18}O and Cs profiles barely overlap at 775 K, but starting at 875 K ^{18}O moves into the amorphized layer and extends over its full width at 1075 K. It turned out that up to half the oxygen content in the surface layer was replaced by ^{18}O, indicating a pronounced $^{16}O \Leftrightarrow ^{18}O$ exchange between the silica matrix and the annealing gas during the recrystallization process.

These findings were substantiated by a number of additional experiments [92, 95, 98]. First, it was verified that neither annealing *in vacuo* (after Cs implantation) nor annealing in oxygen (after Si, O, or noble gas ion implantations) led to epitaxy, which is in good agreement with previous results. Furthermore, the recrystallization temperature T_x was measured for various ion fluences and annealing times. Figure 15 shows the recrystallization speed at the fixed temperature of 1070 K strongly increasing with the Cs fluence, while Figure 16a–d shows the corresponding channeling spectra as a function of the implanted Cs fluence (in the range from 5×10^{15} to 1×10^{17} Cs^+/cm^2) and the annealing time (1–5 h at T = 1070 K). Evidently, full epitaxy was achieved at the highest ion fluence within a 2-h annealing, whereas for the lowest fluence the layer was still fully amorphous after a 5-h annealing.

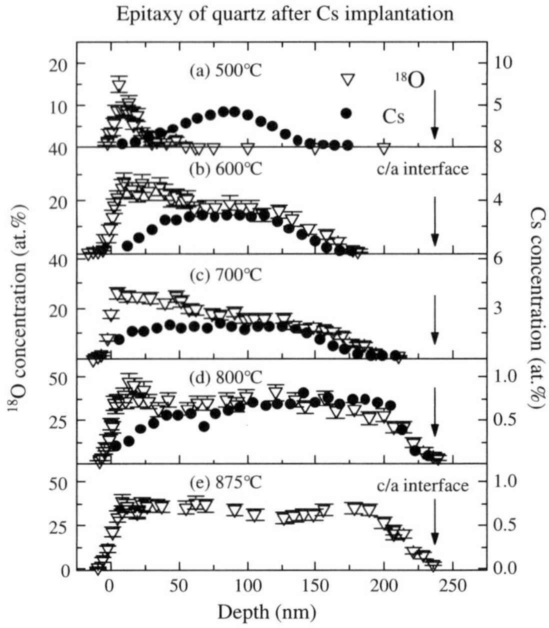

Figure 14. Epitaxy of α-quartz after Cs implantation and annealing in $^{18}O_2$ [97]. For increasing annealing temperature, one first notes in-diffusion of ^{18}O, then broadening of the Cs profile, and finally complete loss of Cs.

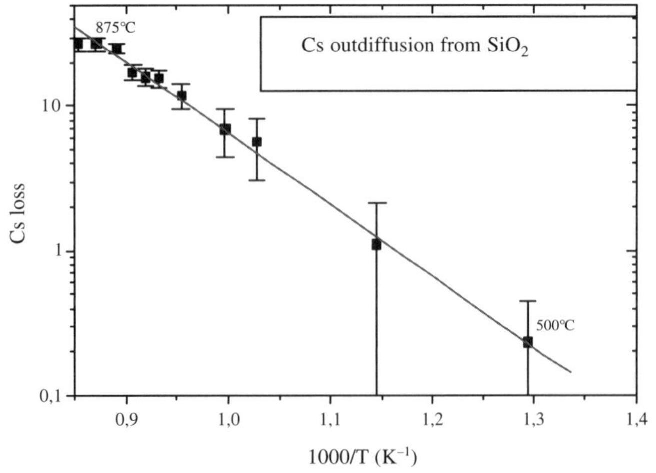

Figure 13. Epitaxy of α-quartz after Cs implantation and annealing in $^{18}O_2$. The retained Cs fraction plotted as function of the annealing temperature displays Arrhenius behavior with an activation energy of 0.98 ± 0.10 eV.

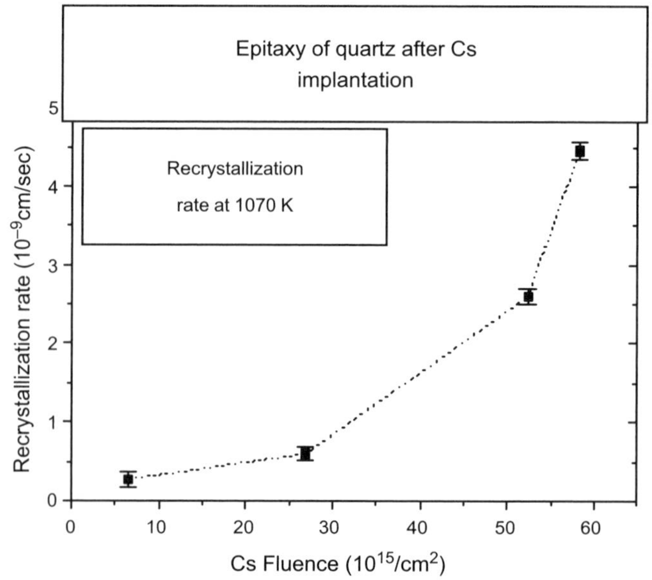

Figure 15. Recrystallization speed after Cs implantation and annealing at 1070 K as function of the Cs fluence.

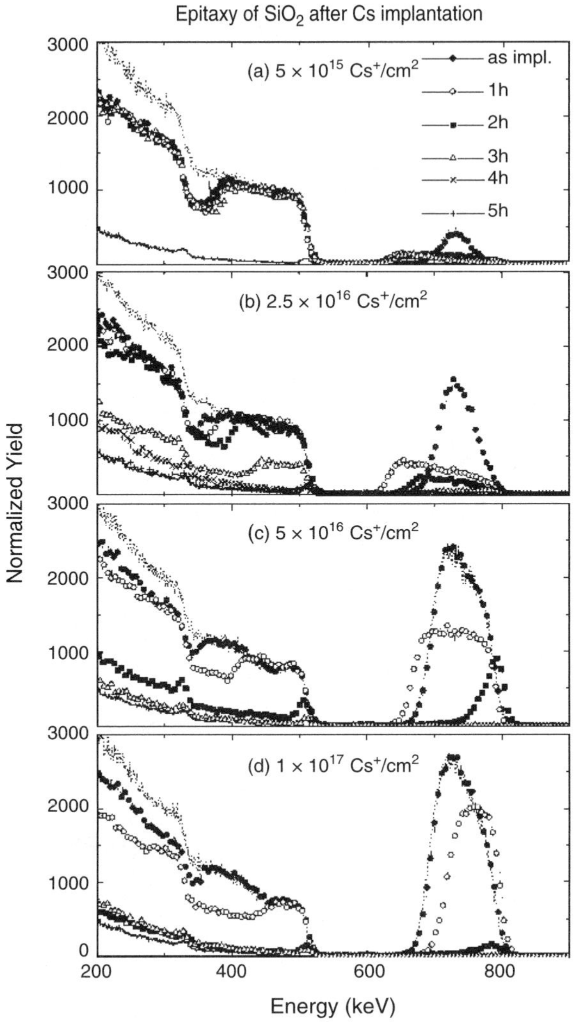

Figure 16. Epitaxy of α-quartz after Cs implantation and annealing in $^{18}O_2$ [92]. (a)–(d) RBS channeling spectra obtained for various Cs fluences and annealing times.

3.2.3. Experiments with H, Li, Na, and Rb Ion Implantations

The studies on chemically guided epitaxy of SiO_2 layers after Cs implantation were extended in two directions:

(a) epitaxy after the implantation of other alkali ions or protons, such as Li, Na, and Rb, and

(b) epitaxy with Cs ions of samples amorphized by other methods.

For details concerning the proton and alkali ion implantations, see Table 2. Most irradiations were carried out at LN temperature and the annealing was done in air or in $^{18}O_2$ atmosphere. The same ion analytical methods as described above were used to depth profile the various species. For all elements with $Z < 20$ (except H), the evolution of the profiles of the implanted species (Li, Na) and the $^{16}O \Leftrightarrow {}^{18}O$ exchange between the annealing gas and the matrix were monitored via TOF-ERDA. As an example, Figure 17 compares the temperature variation of the retained Na and Cs fractions (relative to their implanted doses and integrated over the full profiles in the damaged region) and that of the integrated ^{18}O content [94]. The major difference lies in the temperature shift between the two ion species. The temperature T^R, at which half of the implanted ions have left the sample, is clearly lower for Na ($\approx$830 K) than for Cs ($\approx$1065 K). On the other hand, the quality of the epitaxy of the quartz matrix is about equally good for both ions and much better than after Li implantation, where full epitaxy was never achieved [93]. Finally, in the case of proton irradiations, no epitaxy could be found and the samples stayed amorphous up to 1225 K. The hydrogen implantation profile was stable up to 725 K, but then diffused out at around 875 K [96].

We finally discuss the case of chemically guided epitaxy after Rb^+ implantations in quartz recently studied in detail by Gasiorek and collaborators [95, 99]. Figure 18 shows RBS spectra, damage profiles, and Rb depth distributions after 175-keV implantation at LN temperature and 1-h annealing in $^{18}O_2$ atmosphere at the temperatures given. For increasing temperature, one notes a gradual motion of the a/c interface to the surface and a loss of Rb. The quality of epitaxy achieved is as good as for Na and Cs (and better than for Li), and the values of T^{Rb} and T_x are very similar to the ones for Cs. What is different is that the epitaxy occurs in two steps, as shown in Figure 19. The recrystallization speed follows a double Arrhenius dependence with activation energies of $E^L_{ax} = 0.58 \pm 0.12$ eV below and $E^H_{ax} = 3.55 \pm 0.32$ eV above 1175 K, respectively. The origin of this behavior is not understood yet.

In this system, a systematic comparison of annealing in air and ^{18}O was performed and the influences of the gas pressure and Rb fluence were investigated as well. Besides RBS channeling and TOF-ERDA analyses, detailed atomic

Table 2. Processing parameters of chemically guided SPEG in quartz after alkali ion implantation.

Implanted ion			Processing parameters				
Type	Energy (keV)	Fluence ($10^{16}/cm^2$)	Annealing gas	T^R (K)	T_X (K)	E_{ax} (eV)	Ref.
^1H	20	5.0	air	830 (50)	—[a]	—	[96]
^7Li	15	2.5	^{18}O	870 (59)	925[b]		[93]
^{23}Na	50	2.5	air	835 (40)	1070		[94]
Rb	175	2.5, 5.0	air	1035	1170	0.58 (12), 3.55 (32)	[99]
			^{18}O			0.54 (9), 1.60 (36)	
Cs	250	2.5	air	1065 (48)	1150	2.83 (20)	[88]

[a] No epitaxy.
[b] Incomplete epitaxy.

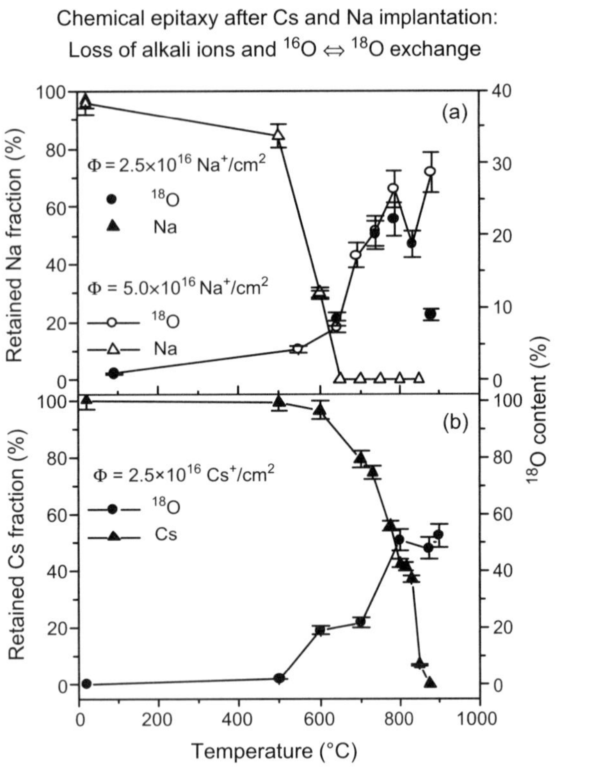

Figure 17. (a) Average fraction of implanted Na retained in the quartz samples (relative to the total implanted Na content at fluences of 2.5 and 5.0×10^{16} Na^+ ions/cm^2) and of ^{18}O having entered the samples during annealing in $^{18}O_2$ [94]. (b) A comparison is made with the results on Cs-doped samples [94].

force microscopy was done for these samples in order to gain information on compaction or swelling during ion bombardment and epitaxy. Figure 20 displays the temperature dependence of the step height at the boundary between irradiated and virgin parts of the sample during annealing in $^{18}O_2$, for 2.5×10^{16} Rb/cm^2 implanted at 250 keV. At 1123 K, when recrystallization has been completed, the step height has reached zero.

3.2.4. Separation of Amorphization and Epitaxy

In the studies presented so far in this section, the implanted alkali ions induced both the radiation damage and amorphization of the matrix during implantation and, under proper annealing conditions, its recrystallization. Of course, this correlation can be relaxed and attempts have been made to separate amorphization (via ion implantation or amorphous silica deposition via evaporation) and chemically guided epitaxy.

Figure 21 illustrates channeling backscattering spectra obtained for three such cases [98]. The spectra plotted in Figure 21a refer to a 90-nm-thick amorphous SiO_2 layer deposited via electron-gun evaporation on (0001) crystalline α-quartz. The process is sketched in Figure 22. The sample was then post-irradiated with 250-keV Cs ions at a fluence of 2.5×10^{16}/cm^2 and annealed for 1 h in air at 840 °C or 875 °C. The Cs post-implantation increased the mass

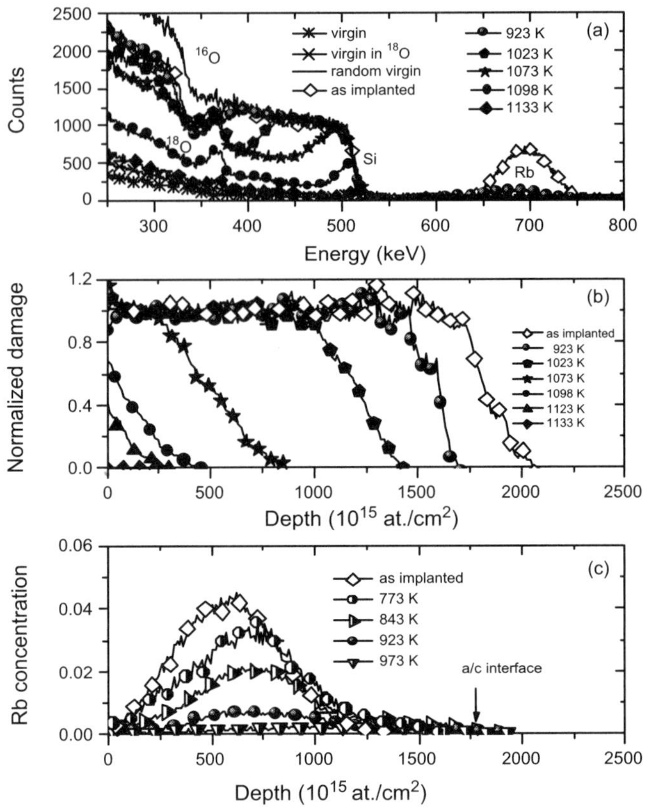

Figure 18. Epitaxy of quartz after 275-keV Rb implantation and annealing in ^{18}O. (a) RBS-C spectra of samples, annealed in $^{18}O_2$ atmosphere for 1 h between 673 and 1133 K. (b) Depth distributions of the irradiation damage as function of the annealing temperature. (c) Rb concentration profiles extracted from the RBS spectra.

density of the amorphous layer by about 6%. Annealing in air at 1120 K led to full epitaxy and outdiffusion of Cs. This experiment confirmed that chemically guided epitaxy of silica is successful, inrrespective of the amorphization process.

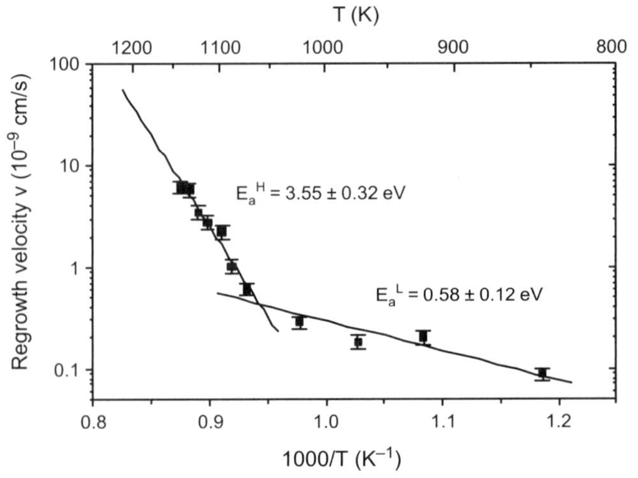

Figure 19. Epitaxy of α-quartz after Rb implantation and annealing in $^{18}O_2$. Recrystallization speed $v(T)$ deduced from the Si signal. The two straight lines correspond to the activation energies indicated.

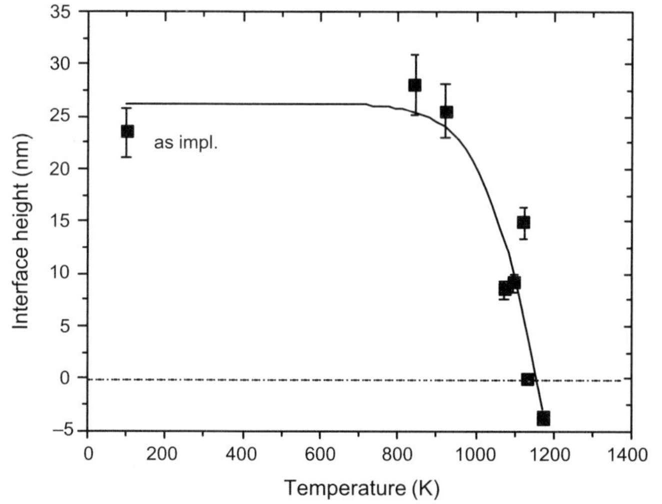

Figure 20. Height difference between implanted and unimplanted regions for quartz irradiated with 275-keV Rb ions at $2.5 \times 10^{16}/cm^2$ and annealed in ^{18}O at the temperature given.

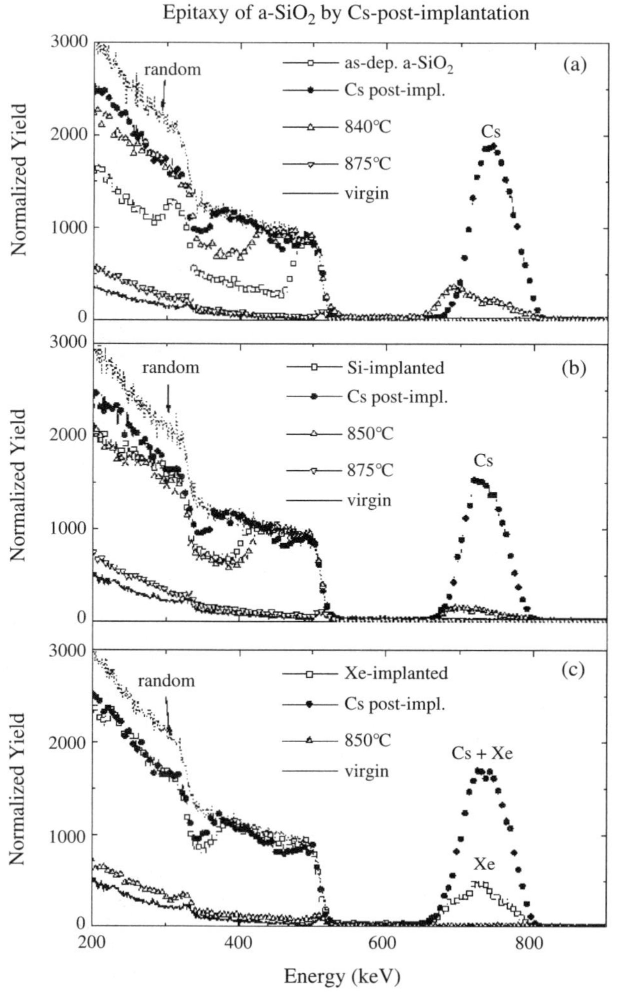

Figure 21. Epitaxy of a-SiO_2 by Cs post-implantation and annealing in air. (a) A 90-nm-thick amorphous SiO_2 layer evaporated onto quartz is post-irradiated with 250-keV Cs ions and annealed at 840 °C or 875 °C in $^{18}O_2$. (b), (c) Double implantation with Si (or Xe) and Cs ions and subsequent annealing in $^{18}O_2$ at 850 °C and/or 875 °C.

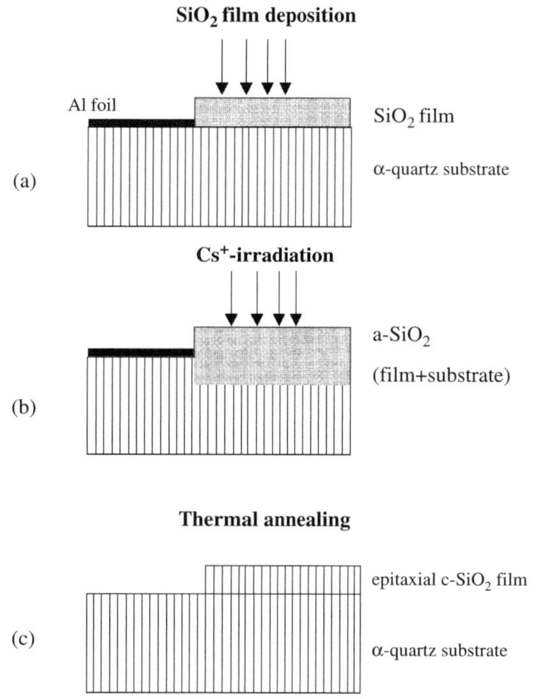

Figure 22. Epitaxy of a thin a-SiO_2 layer via Cs post-implantation.

In the other two experiments illustrated in Figure 21, the quartz samples were first surface-amorphized via ion implantation by either 70-keV Si or 100-keV Xe ions and afterwards epitaxially regrown by chemical SPEG with 250-keV Cs^+ ions under air [93]. Indeed, these double irradiations and subsequent annealing at 1150 K or 1125 K again led to full SPEG. The corresponding channeling spectra are displayed in Figure 21b and c. This series of experiments might be important in the context of ion doping quartz with luminescent species and subsequent epitaxy via chemical SPEG, as will be discussed later.

3.2.5. A Possible Scenario

The obvious correlation of epitaxy of the SiO_2 matrix, outdiffusion of the implanted alkali ions, and $^{16}O \Leftrightarrow {}^{18}O$ exchange between the matrix and the annealing gas was explained qualitatively by Roccaforte et al. [97, 98] who considered changes of the tetrahedral $[SiO_4]$ network and the structural freedom parameter f introduced before. Extended X-ray absorption fine structure (EXAFS) measurements [14] indicated that the $[SiO_4]$ units remain almost preserved, even during high-fluence irradiations. This means that the network connectivity parameter $C = 2$ does not change during amorphization (see Fig. 23). The elemental interhedral connection of SiO_2 can be represented by the basic chain unit

$$(O_3)\equiv Si—O—Si\equiv(O_3) \tag{6}$$

where $(O_3)\equiv Si$ indicates the continuation of the network. Introducing *network modifiers* like alkali-oxide breaks the continuity of the network by creating oxygen atoms, which do not bridge adjacent tetrahedra. In the case of alkali ion implantation only a marginal fraction of nonbridging oxygen atoms has formed, which arises by the "scavenging" of oxygen atoms by the implanted ions located in weakly bound

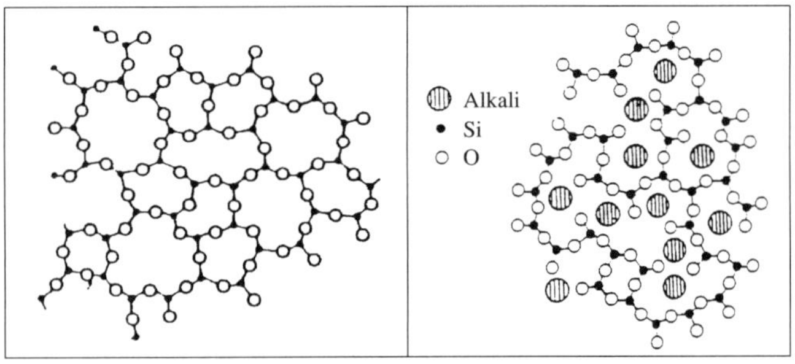

Figure 23. Left: Disturbed [SiO_4] tetrahedra network. Right: Incorporation of alkali ions in SiO_2.

interstitial positions. When oxygen diffuses into the alkali-doped matrix, alkali oxide (i.e., $Cs_2^+O^{-2}$) is dissolved in the matrix by forming an alkali-glass silicate, in which bridging oxygen atoms are replaced with nonbridging ones. The effect of Cs implantation and annealing thus may be described as

$$\begin{aligned}&(O_3)\equiv Si—O—Si\equiv(O_3) + 2Cs + O^*\\&\to (O_3)\equiv Si—O—Cs\,Cs—O^*—Si\equiv(O_3)\\&\to (O_3)\equiv Si—O^*—Si\equiv(O_3) + (Cs_2^+O^{*^{-2}})_{jump} \quad (7)\end{aligned}$$

where O* denotes external ^{18}O. Those oxygen atoms (O or O*) with a strong bond to Si and a weaker one to Cs can easily migrate through the opened glassy structure. The pronounced oxygen migration is thought to arise from preferential diffusion paths, which are opened at broken tetrahedral linkages.

The correlation between diffusion of Cs and ^{18}O suggests that each oxygen isotope exchange is accompanied by a "jump" of Cs from one linkage to another, as indicated above. When at 800 °C, Cs and ^{18}O reach the c/a interface, the high flexibility of the structure caused by the formation of nonbridging oxygen atoms favors epitaxial regrowth.

On the basis of this scenario, Roccaforte et al. [97] also estimated the amount of oxygen diffusing in the amorphous layer. Considering the lattice parameter of quartz along the c-axis (0.54 nm) as the average jump distance, the number of diffusion steps N_{step} of each Cs atom can be taken from the typical distance which Cs needs to reach the surface or the c/a interface, $\Lambda = 71 \pm 5$ nm at 800 °C and for 1 h. This corresponds to $N_{step} = 131 \pm 9$ diffusion steps. The diffusion of Cs implanted into an a-SiO_2/Si bilayer has been studied by Fishbein and Plummer [100] by means of RBS. At 970–1270 K, the diffusion constant (under nitrogen and in the constraint geometry due to the presence of the SiO_2/Si interface) is given as $D = 0.50\exp(-2.9\ eV/k_BT)\ cm^2/s$. The estimated diffusion length at 1075 K during 1 h, $\Lambda = 81$ nm, is in good agreement with this number.

Since according to this interpretation the isotope exchange of one oxygen atom is accompanied by the movement of two Cs ions, the total number of oxygen atoms N(O*) is given by $N(O^*) = N_{Cs}\ N_{step}/2$. Using the experimentally determined Cs-content, $N_{Cs} = 11.4 \times 10^{15}$ at/cm^2, the authors arrived at $N(O^*) = 7.5 \times 10^{17}$ at/cm^2, which is rather close to the experimental ^{18}O content of $(6.2 \pm 0.4) \times 10^{17}$ at/cm^2. This rough estimate evidently neglects the anisotropic mobility of alkali ions in "preferential pathways" or "channels and pockets" in SiO_2 [101–106] to be discussed in the next subsection.

3.2.6. Alkali Ion Diffusion in SiO_2 and Silica

Diffusion of alkali ions in quartz, a-SiO_2, and various silicate glasses under different environments and the role of defect structures related to the impurities in the samples have been studied in much detail over several decades and have been reviewed extensively by, among others, Day [105], Frischat [16], Ingram [101], and Jund et al. [103]. Although the role of different types of defects, and in particular those associated with alkali ions, in the diffusion kinetics (and the optical and electrical properties of the SiO_2 matrix, see Section 4) is still not fully understood, we would like to briefly allude here to some concepts and results dealing with the diffusion of alkali ions in SiO_2. The main body of the experimental information derives from Na diffusion [107–116] and some from diffusion studies of Rb tracers in the context of mineral dating in geosciences. Si and O diffusion studies in quartz have recently been carried out by Bejina and Jaoul [117] and by Hamann [118]. Concerning the role of Al-associated defects in alkali ion diffusion in quartz, we refer to the work by Breton, Giradet, and collaborators [119].

The non-Debye behavior observed in the electric response has led to local ion motion based on a "forward-backward hopping mechanism" or hopping processes over random potential barriers [103]. Ingram introduced the notion of "preferential pathways" governing the dynamics on large scales. Recent molecular dynamics (MD) calculations of Na^+ diffusion in NSn silicate glasses of the structure formula $(Na_2O)n(SiO_2)$ ($n = 2, 3, 4$) and neutron scattering measurements [107] appear to confirm this Ingram concept and gave evidence of the following details [102–104]:

1. Na diffusion is much faster than Si or O diffusion, its diffusion constant being two orders of magnitude larger. Up to about 2500 K, the SiO_2 matrix can thus be considered as frozen.
2. At temperatures $T < 3000$ K, Na diffusion proceeds along "preferential pathways" or "channels with pockets" [101]. This means that even after diffusion times of several ns, large fractions of the SiO_2 matrix have not been "visited" by migrating Na ions and the diffusion trajectories do not fill the space uniformly. The sodium-free regions form a percolating structure around a channel network ("Swiss cheese" structure). The characteristic distance between these channels is of the order of 0.66 nm [104].
3. The MD calculations indicated uncorrelated Na diffusion within the channels and did not support the cooperative motion proposed by Greaves [106]. There are preferred sites for the sodium ions inside the channels associated with dangling bonds. These bonds do not disappear until the network has rearranged itself.
4. According to Horbach et al. [104], the fast Na diffusion within the channels and the rearrangement of the channel network within SiO_2—although proceeding on very different time scales—are closely related to each other.

As mentioned before, the diffusion of implanted Cs ions in a-SiO_2 at temperatures relevant for chemically guided SPEG was of Arrhenius type with an activation energy of E_a = 2.9 eV [100].

3.3. Epitaxy of Quartz via Laser Annealing

Epitaxy in ion-irradiated quartz by means of dynamic or chemically guided SPEG has the advantage that both methods may provide full epitaxial regrowth of the a-SiO_2 layers. But these methods may have severe drawbacks in optoelectronic device fabrication. In the case of chemically guided SPEG, annealing in oxygen at elevated temperatures may influence the dopants via oxidation and/or diffusion. With regard to oxidation, dynamic SPEG is more favorable (since it proceeds in high vacuum), but it requires implantation into heated samples, and the layer structures and implantation profiles may suffer from unwanted diffusion of the dopants and the matrix and from ripening precipitates. For that reason, attempts have been made to overcome these obvious deficiencies by laser-annealing ion-doped SiO_2 samples.

Stepanov et al. [139] reported on laser annealing experiments of silica after Ag or Cu ion implantations, at a typical energy of 50 keV and fluences of 4–8 × 10^{16} ions/cm^2. Pulsed laser treatment with a KrF excimer laser was applied (wavelength $\lambda = 248$ nm, average energy density = 0.25 J/cm^2, pulse length = 25 ns FWHM, up to 250 pulses per irradiation spot). Properties of the metal nanoparticles were studied via RBS and optical reflectivity and transmittance. The changes in cluster size and diffusion of the implants were found to depend on the energy absorption by the clusters themselves on the one hand and the repeated melting and long-term warming up of the matrix on the other hand. Dissolution of dopant clusters, long-range atomic diffusion of dopant atoms, and formation of new dopant clusters are the consequence of this complicated mechanism, which is far from being fully understood. Electroluminescence (EL) of these samples was recorded as functions of the sample temperature and pulse structure of excitation [141], giving evidence of a stabilization of the defect centers in silica.

First attempts of laser-induced epitaxy in α-quartz after Cs, Ba, and Xe ion implantation have been undertaken by Schaaf et al. [146], using a pulsed XeCl excimer laser (λ = 208 nm, pulse length = 55 ns, up to 150 pulses, laser energy density $\leq$4.7 J/cm^2). A typical RBS-C spectrum obtained after implantation of 250-keV Cs ions at a fluence of 2.5 × 10^{16}/cm^2 and laser annealing is illustrated in Figure 24. At energy densities between 3.6 and 4.7 J/cm^2, one notes the diffusion of the implanted Cs towards the surface (and some loss of Cs) and a reduction of the thickness of the amorphous layer, but both processes are incomplete and suggest that a higher laser power is required to achieve full epitaxy. Preliminary results on the laser annealing after the implantation of 3.5 × 10^{16} Ba ions/cm^2 at 250 keV into silica have been obtained for laser powers of up to 5.0 J/cm^2.

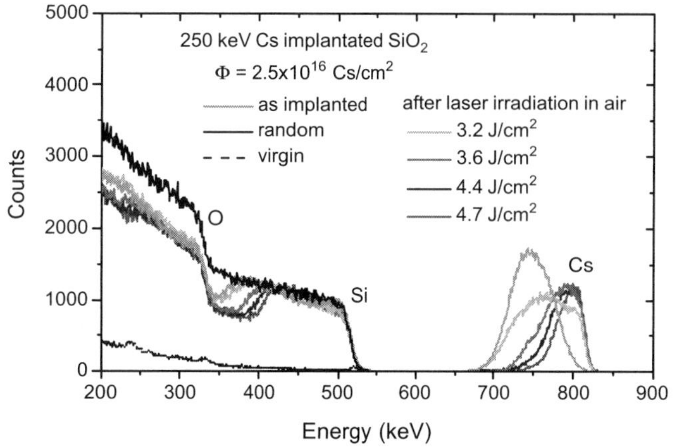

Figure 24. RBS-C spectra taken after implantation of 2.5 × 10^{16} Cs ions/cm^2 at 250 keV into α-quartz and laser annealing with a pulsed excimer laser of the energy densities indicated.

4. LUMINESCENCE OF IMPLANTED NANOCLUSTERS IN SiO_2

One of the possible future applications of (ion-beam) doped quartz layers appears to be in photonic device design and production by making use of luminescence effects. It is well known that amorphous SiO_2 containing silicon, germanium, or metal nanocrystals shows intense photoluminescence (PL), electroluminescence (EL), and even ion-beam luminescence (IBL) in the visible range [120–143], but little is known about such luminescence phenomena in ion-beam doped, crystalline quartz. For that reason, we briefly review some information on PL and EL in a-SiO_2 and then describe first attempts of laser annealing to grow optically active crystalline SiO_2 elements.

Si and Ge nanoclusters in a-SiO_2 can be produced by several methods including ion implantation (see [122–134]), chemical vapor deposition [121, 144, 145], sputtering, and laser ablation/deposition. The generation of strong PL from Si nanocrystals after Si^+ ion implantation into a-SiO_2 films or a-SiO_2/Si bilayers is rather well understood and many attempts have been undertaken to optimize the light output in different regions of the visible spectrum [122–128, 130, 134]. The obtained PL spectra have some common features: the PL peak is quite broad, due to the size distribution of the nanocrystals, and its position shifts towards smaller wavelengths when reducing the cluster size. Both findings are a consequence of quantum confinement. An additional and less well-understood component of the PL spectrum arises from color centers in the a-SiO_2 matrix. The PL spectrum thus contains contributions from both the recombination of quantum defined excitons in the nanocrystals and ion-irradiation defects. By etching layer by layer of the photoactive SiO_2 film, Bongersma et al. [136] have been able to locally identify the various sources of PL. Defect luminescence originates from near-surface nanocrystals, while large nanocrystals luminescing at long wavelengths ($\lambda \approx 900$ nm) are mainly located in the center of the film, where the Si concentration is highest. A large density of small nanocrystals luminescing at $\lambda \approx 700$ nm, which escape detection by electron microscopy, was found near the SiO_2 surface and SiO_2/Si interface. Luterova et al. [142] recently reported on blue PL in SiO_2 films doped with Si nanocrystals after picosecond pulsed UV excitation. These authors found that the PL signal increases more than linearly with the exciting photon intensity and decays in time via a superposition of exponential functions. Space- and time-resolved investigations have considerably contributed to the microscopic description and interpretation of luminescence.

Electroluminescence and EL light amplification in Si nanoclusters embedded in substoichiometric SiO_x ($x < 2$) have recently been reviewed by the Catania group [121, 144, 145]. The excitation occurred by impact of hot electrons with the Si nanoclusters and subsequent electron-hole generation and recombination. The EL signal was found to strongly depend on the amount and distribution of Si in the samples, via both the stoichiometry of the SiO_x matrix and the size of the Si nanocrystals. Efficient EL requires a compromise between the parameter x (increased injection current and excitation probability due to increased Si content) and the nanocrystal size (increasing light output in the visible spectrum for decreasing nanocrystal diameter). Pacifici et al. [144] also studied defect production and annealing of these Si nanocrystals under 2-MeV ion bombardment and the influence of these defects on the PL output.

Violet PL of implanted Ge nanocrystals in SiO_2 was first observed by Bao and co-workers [129, 130] and a Caltec–Amsterdam collaboration [135]. The Rossendorf group studied PL and EL in Ge-doped SiO_2/Si bilayers in detail [131–134] and, based on their findings, developed the first integrated Si-based optocoupler containing Ge nanocrystals and luminescing in the blue [131]. When using two subsequent Ge implantations and optimizing the ion energies and fluences, these authors achieved a nearly homogeneous distribution of 2- to 6.5-nm diameter clusters throughout the SiO_2 layer, after annealing the samples at 500–1200 °C in a dry N_2 atmosphere [132]. Electron microscopy combined with RBS indicated a cluster-free zone at the surface and the accumulation of small clusters near the SiO_2/Si interface. These optocoupling devices hold great promise for integrated optoelectronic devices, especially in the field of sensors and biotechnology.

Optical properties of metallic clusters or nanocrystals in a-SiO_2 have been reviewed, by for example, Townsend, Chandler, Zhang, and collaborators [19, 138, 140], and by Polman, Atwater, and collaborators [18, 135–137]. Rebohle et al. [132] reported on strong PL in the violet after doping a-SiO_2 with Sn ions, 2–20 times higher in intensity than after Si or Ge implantation. A further line of research for optimizing the light output in certain spectral regions is doping the nanocrystals with impurities, such as Er [137]. Furthermore, double ion irradiations of a-SiO_2 samples, first with low-energy C ions at high fluences of the order of 5×10^{16} to $1 \times 10^{18}/cm^2$ and subsequently with high-energy Ar, Kr, Xe, or U ions at low fluences of some 10^{12} ions/cm^2 have been shown to produce intense blue-violet PL [143].

The question whether one of the methods, which have been developed for epitaxial regrowth of SiO_2 after ion irradiation and discussed in Section 3, leads to PL or EL of doped quartz, is still open. First experiments in this direction by Dhar et al. [71] after 175-keV Ba ion implantation have recently been successful. Going back to Figure 11, partial dynamic epitaxy was achieved, when implanting 1×10^{15} ions/cm^2 at temperatures up to 900 °C into α-quartz: for increasing temperature, the c/a interface has moved towards the surface, leaving, however, a very thin nonrecrystallized subsurface layer. Figure 25 displays PL and EL spectra taken at 12 K and room temperature for this series of samples.

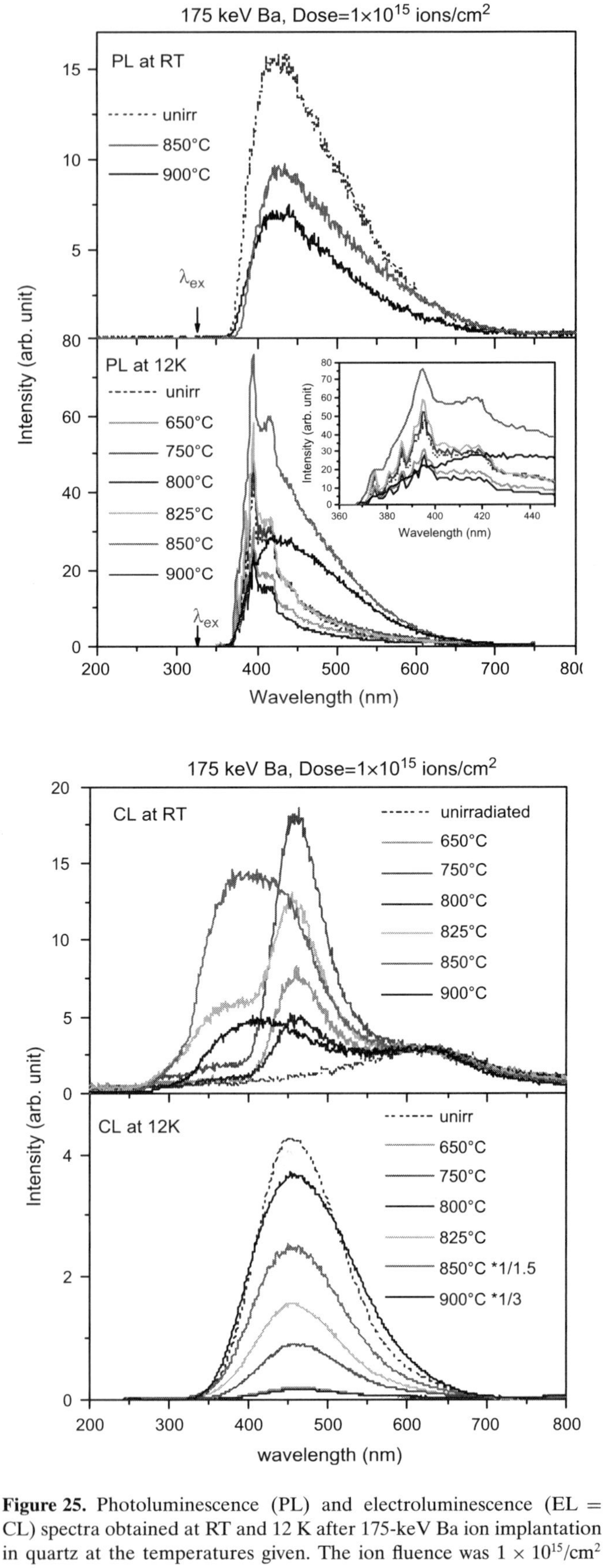

Figure 25. Photoluminescence (PL) and electroluminescence (EL = CL) spectra obtained at RT and 12 K after 175-keV Ba ion implantation in quartz at the temperatures given. The ion fluence was $1 \times 10^{15}/cm^2$ in all the cases.

Clearly luminescence can be stimulated by both 325-nm photon and 5-keV electron excitation. An interpretation of these spectra, some of which strongly depend on the sample temperature, is in progress. Now already we emphasize the strong dependence of the CL spectrum both in shape and intensity, on the implantation temperature (see Fig. 25 upper right corner). A strong blue and violet peak appears at an irradiation temperature, where the a/c interface has started moving to the surface indicating the onset of epitaxy. It is interesting to note that the violet peak survives the epitaxy and may be attributed to the implanted Ba ions.

This study has been recently extended by Sahoo and collaborators [147] to Ge ion-implantation of quartz at elevelated temperatures. The CL yield was measured as function of the sample temperature during ion irradiation (300–1223 K) and CL analysis (30–300 K). In this case, only partial dynamic epitaxy was achieved at rates of 0.1 nm/s (at 920 K) to 0.3 nm/s (at 1250 K) for a fluence of 7×10^{14} Ge-ions/cm^2. Above 920 K, two intense violet bands at 381 and 418 nm appeared in the CL spectra, which were associated with properties of the incorporated Ge ions (Ge—Si bonds, Ge nanoclusters, Ge oxide) and not with the damaged SiO_2 matrix.

GLOSSARY

Chemically guided epitaxy of ion-irradiated quartz The radiation defects and ion-beam amorphized surface layer can be transformed into a quartz single crystal, if alkali ions are implanted near the amorphous/crystalline interface and the sample is annealed in air or oxygen. Amorphization may occur by the alkali ion implantation itself or by implantation of other ions.

Dynamic epitaxy Ion irradiation of a single crystal at elevated temperature leaving the long-range order of the sample intact. The unavoidable radiation defects anneal out due to the thermal mobility of the implanted ions and the matrix atoms.

Elastic recoil detection analysis Ion-beam analytical method in which a beam of heavy projectiles is directed onto a matrix composed of lighter elements/isotopes.

Laser epitaxy Quartz, which has been surface-amorphized by ion bombardment, can be recrystallized by intense laser irradiation.

Luminescence of nanoclusters in silica Photo- and/or electroluminescence in silica arises after the implantation of semiconductor (Si, Ge), rare earth, or other ions. The spectral shape and intensity of the luminescence light after UV photon or keV electron irradiation depend on the ripening conditions of the nanoparticles as well as on the implanted ion species and the defects created during ion implantation in the silica.

Nuclear reaction analysis A depth- and isotope-sensitive profiling method used mostly for particular light isotopes in the target for which nuclear resonance reactions with narrow resonances and sufficient cross sections are known.

Rutherford backscattering spectroscopy Ion-beam analytical method that mainly measures the depth profiles of medium-mass (or heavy) elements in a matrix composed of light (or medium-mass) elements.

Sputter erosion Ion beam impact produces atomic transport near the sample surface leading to removal of surface atoms and self organization (ripple formation, smoothing).

ACKNOWLEDGMENTS

The author gratefully acknowledges discussions with Michael Aziz, Wolfgang Bolse, Sankar Dhar, Stanislava Gasiorek, Hans Hofsäß, Malcolm D. Ingram, Juhani Keinonen, Fabrizio Roccaforte, Wolfgang Skorupa, and Ulrich Vetter. It is a pleasure to thank Lucie Hamdi for helping to correct the manuscript. This work was funded by Deutsche Forschungsgemeinschaft (SFB 345, Li-325/8).

REFERENCES

1. L. Merill, *J. Phys. Chem. Ref. Data* 11, 1043 (1982).
2. S. R. Eliot, "Physics of Amorphous Materials." Longman, New York, 1990.
3. K. J. Kingma, R. J. Hemley, H.-K. Mao, and D. R. Veblen, *Phys. Rev. Lett.* 70, 3927 (1993); *Phys. Rev. Lett.* 72, 1302 (1994).
4. V. Mnitriev, V. Torgashev, P. Tolédano, and E. K. H. Salje, *Europhys. Lett.* 37, 553 (1997).
5. A. R. Cooper, *Phys. Chem. Glasses* 19, 60 (1978); P. K. Gupta and A. R. Cooper, *J. Non-Cryst. Solids* 123, 14 (1990); P. K. Gupta, *J. Non-Cryst. Solids* 192, 79 (1995).
6. L. W. Hobbs, *Nucl. Instr. Meth. B* 91, 30 (1994); *J. Non-Cryst. Solids* 192/193, 79 (1995); L. W. Hobbs, A. N. Sreeram, C. E. Jesurum, and B. A. Berger, *Nucl. Instr. Meth. B* 116, 18 (1996).
7. G. Götz, in "Ion Beam Modifications of Insulators" (P. Mazzoldi and G. W. Arnold, Eds.), pp. 412–448. (Elsevier, Amsterdam, 1987.)
8. S. J. Zinkle, *Nucl. Instr. Meth. B* 91, 234 (1994).
9. D. F. Mullica, H. O. Perkins, D. A. Grossie, L. A. Boatner, and B. C. Sales, *J. Solid State Chem.* 62, 371 (1986).
10. J. K. N. Lindner, R. Zuschlag, and E. H. te Kaat, *Nucl. Instr. Meth. B* 62, 314 (1992).
11. R. K. Eby, R. C. Ewing, and R. C. Birtcher, *J. Mater. Res.* 7, 3080 (1992).
12. P. Mazzoldi and A. Miotello, *Rad. Eff.* 90, 39 (1986).
13. L. Douillard and J. P. Duraud, *J. Phys. III (France)* 6, 1677 (1996).
14. W. Bolse, *Nucl. Instr. Meth. B* 141, 133 (1998); 148, 83 (1999).
15. F. Piao, W. G. Oldham, and E. E. Haller, *J. Non-Cryst. Solids* 276, 61 (2000).
16. G. H. Frischat, *Phys. Chem. Glasses* 25, 110 (1984); W. Beier and G. H. Frischat, *J. Non-Cryst. Solids* 73, 113 (1985).
17. R. F. Haglund, *Mat. Sci. Eng. A* 253, 275 (1998).
18. A. Polman, *J. Appl. Phys.* 82, 1 (1997).
19. P. D. Townsend, P. J. Chandler, and L. Zhang, "Optical Effects of Ion Implantation." Cambridge Univ. Press, 1994; P. D. Townsend and D. E. Hole, *Vacuum* 63, 641 (2001).
20. J. S. Tse and D. H. Klug, *Phys. Rev. Lett.* 67, 3559 (1991).
21. M. S. Somayazulu, S. M. Sharma, and S. K. Sikka, *Phys. Rev. Lett.* 73, 98 (1994).
22. J. Badro, J. L. Barrat, and P. Gillet, *Phys. Rev. Lett.* 76, 772 (1996); *Europhys. Lett.* 42, 643 (1998).
23. J. R. Chelikowsky, E. H. King, N. Troullier, J. L. Martins, and J. Glinnemann, *Phys. Rev. Lett.* 65, 3309 (1990); R. M. Wentzcovitch, C. da Silva, J. R. Chelikowsky, and N. Binggeli, *Phys. Rev. Lett.* 80, 2149 (1998).
24. J. R. Tesmer and M. Nastasi, "Handbook of Modern Ion Beam Material Analysis." Material Science Society, Pittsburgh, 1995.
25. M. Nastasi, J. W. Mayer, and J. K. Hirvonen, "Ion Solid Interactions—Fundamentals and Applications." Cambridge Univ. Press, 1996.

26. J. Tirira, Y. Serruys, and P. Trocellier, Eds. "Forward Recoil Spectrometry." Plenum, New York, 1996.
27. K. P. Lieb, *Contemp. Phys.* 40, 385 (1999).
28. E. Lell, N. J. Kreidl, and J. R. Hensler, "Radiation Effects in Quartz, Silica and Glasses." Pergamon, New York, 1966.
29. W. Primak, "Compacted States in Vitreous Silica." Gordon and Breach, New York, 1975.
30. E. P. Ernisse and C. B. Norris, *J. Appl. Phys.* 45, 5169 (1974).
31. S. L. Chan, L. F. Gladden, and S. R. Elliott, *J. Non-Cryst. Solids* 106, 413 (1988); in "The Physics and Technology of Amorphous SiO_2" (R. A. B. Devine, Ed.), p. 83. Plenum, New York, 1988.
32. M. Guzzi, F. Piot, G. Spinolo, A. Vedda, and A. Paleari, *J. Phys.: Condens. Matter* 4, 8635 (1992).
33. A. L. Odom and W. J. Rink, *Science* 246, 107 (1989).
34. R. R. Gulamova, E. M. Gasenov, and R. Alimov, *Nucl. Instr. Meth. B* 127/128, 497 (1997).
35. R. G. Macaulay-Newcombe and D. A. Thompson, *Nucl. Instr. Meth. B* 1, 176 (1984).
36. R. G. Macaulay-Newcombe, D. A. Thompson, J. Davies, and D. V. Stefanovic, *Nucl. Instr. Meth. B* 46, 180 (1990).
37. H. Fischer, G. Götz, and H. Karge, *Phys. Stat. Sol. A* 76, 249 (1983); *Phys. Stat. Sol. A* 76, K171 (1983).
38. F. Harbsmeier and W. Bolse, *Mater. Sci. Forum* 248–249, 279 (1997).
39. F. Harbsmeier and W. Bolse, *J. Appl. Phys.* 83, 4049 (1998).
40. R. S. Averback and D. N. Seidman, *Mat. Sci. Forum* 15–18, 963 (1987); A. M. Stoneham, *Nucl. Instr. Meth. B* 91, 1 (1994); W. Bolse, *Mat. Sci. Eng.* R12, 40 (1994).
41. W. Bolse, J. Conrad, T. Rödle, and T. Weber, *Surf. Coat. Techn.* 74/75, 927 (1995).
42. W. Bolse, J. Conrad, F. Harbsmeier, M. Borowski, and T. Rödle, *Mater. Sci. Forum* 248–249, 319 (1997).
43. L. Douillard, J. Jollet, J. P. Duraud, R. A. B. Devine, and E. Doorhyee, *Rad. Eff. Def. Sol.* 124, 351 (1994); L. Douillard, Thesis, Université Paris VI, 1992.
44. L. H. Wilson, J. B. Xu, R. A. B. Devine, and R. P. Webb, *Nucl. Instr. Meth. B* 118, 473 (1996).
45. S. M. M. Ramos, C. Clerc, B. Canut, J. Chaumont, M. Toulemonde, and H. Bernas, *Nucl. Instr. Meth. B* 166, 31 (2000).
46. M. Toulemonde, S. M. M. Ramos, H. Bernas, C. Clerc, B. Canut, J. Chaumont, and C. Trautmann, *Nucl. Instr. Meth. B* 178, 331 (2001).
47. C. Trautmann, J. M. Costantini, A. Meftah, K. Schwartz, J. P. Stoquert, and M. Toulemonde, *Mat. Res. Soc. Symp. Proc.* 50, 133 (1999).
48. C. Trautmann, M. Boccanfuso, A. Benyagoub, S. Klaumünzer, K. Schwartz, and M. Toulemonde, *Nucl. Instr. Meth. B* 191, 144 (2002).
49. J. Conrad, Doctoral Thesis, Göttingen, Cuvillier, 1997.
50. J. F. Biersack and J. M. Monoyan, *Nucl. Instr. Meth. B* 35, 215 (1988).
51. J. R. Dennis and E. B. Hale, *J. Appl. Phys.* 49, 1119 (1978).
52. J. F. Gibbons, *Proc. IEEE* 1062 (1972).
53. M. Avrami, *J. Chem. Phys.* 7, 1103 (1939); 8, 212 (1940); 9, 177 (1941).
54. S. U. Campisano, S. Coffa, V. Raineri, F. Priolo, and E. Rimini, *Nucl. Instr. Meth. B* 80/91, 514 (1993).
55. E. Snoeks, T. Weber, A. Caciato, and A. Polman, *J. Appl. Phys.* 78, 4723 (1995); M. L. Brogersma, E. Snoek, and A. Polman, *Appl. Phys. Lett.* 71, 1628 (1997).
56. R. J. Hemley, A. P. Jepcoat, H. K. Mao, L. C. Ming, and H. Manghani, *Nature (London)* 334, 52 (1988); R. M. Hazen, L. W. Finger, R. J. Hemley, and H. K. Mao, *Solid State Commun.* 72, 507 (1989).
57. K. J. Kingman, C. Meade, R. J. Hemley, H. Mao, and D. R. Veblen, *Science* 259, 666 (1993); K. J. Kingman, R. J. Hemley, H. Mao, and D. R. Veblen, *Phys. Rev. Lett.* 70, 25 (1993).
58. S. Klaumünzer, *Rad. Eff. Def. Solids* 110, 79 (1989); M.-D. Hou, S. Klaumünzer, and G. Schumacher, *Phys. Rev. B* 41, 1144 (1990).
59. H. Trinkaus, *Nucl. Instr. Meth. B* 107, 155 (1996); *J. Nucl. Mat.* 246, 244 (1977); H. Trinkaus and A. I. Ryazanov, *Phys. Rev. Lett.* 74, 5072 (1995).
60. T. M. Mayer, E. Chason, and A. J. Howard, *Appl. Phys.* 76, 1633 (1994).
61. D. Flamm, F. Frost, and D. Hirsch, *Appl. Surf. Sci.* 179, 95 (2001).
62. R. M. Bradley and J. M. E. Harper, *J. Vac. Sci. Techn. A* 6, 2390 (1988).
63. S. Habenicht, W. Bolse, K. P. Lieb, K. Reimann, and U. Geyer, *Phys. Rev. B* 60, R2200 (1999); S. Habenicht, *Phys. Rev. B* 63, 125419 (2001).
64. N. D. Skelland and P. D. Townsend, *J. Phys. D: Appl. Phys.* 27, 1672 (1994).
65. A. L. Stepanov, D. E. Hoyle, and P. D. Townsend, *Nucl. Instr. Meth. B* 191, 468 (2002).
66. S. Dhar, W. Bolse, and K. P. Lieb, *Nucl. Instr. Meth. B* 148, 683 (1999); *J. Appl. Phys.* 85, 3120 (1999).
67. F. F. Morehead and B. L. Crowder, *Radiat. Eff.* 6, 27 (1970).
68. I. C. Gebeshuber, S. Cernusca, F. Aumayr, and H. P. Winter, in press.
69. S. X. Wang, L. M. Wang, R. C. Ewing, and R. H. Doremus, *J. Appl. Phys.* 81, 587 (1997).
70. E. Wendler, T. Opfermann, and P. I. Gaiduk, *J. Appl. Phys.* 82, 5965 (1997).
71. S. Dhar, S. Gasiorek, P. Sahoo, U. Vetter, H. Hofsäß, S. N. Kulkarni, and K. P. Lieb, submitted.
72. J. W. Rabalais, A. H. Al-Bayati, K. J. Boyd, D. Marton, J. Kulik, Z. Zhang, and W. K. Chu, *Phys. Rev. B* 53, 10781 (1996); D. Marton, K. J. Boyd, and J. W. Rabalais, *Chem. Phys. Lett.* 283, 215 (1998).
73. E. Kasper, I. Eisele, and H. C. Evan, Eds., *Mat. Sci. Eng. B* 89 (2002).
74. W. Bolse, W. H. Wang, C. Illgner, K. P. Lieb, J. Keinonen, and J. C. Ewerth, *Thin Solid Films* 295, 169 (1997).
75. G. W. Arnold and P. S. Peercy, *J. Non-Cryst. Solids* 41, 359 (1980); G. W. Arnold, P. S. Peercy, and B. L. Doyle, *Nucl. Instr. Meth.* 182/183, 733 (1981).
76. S. X. Wang, L. M. Wang, and R. C. Ewing, *J. Appl. Phys.* 81, 587 (1997).
77. G. Devaud, C. Hayzelden, M. J. Aziz, and D. Turnbull, *J. Non-Cryst. Solids* 134, 129 (1991).
78. F. Harbsmeier, W. Bolse, M. R. da Silva, M. F. da Silva, and J. C. Soares, *Nucl. Instr. Meth. B* 136–138, 363 (1998).
79. N. Scapellato, M. Uhrmacher, and K. P. Lieb, *J. Phys. F: Metal Phys.* 18, 677 (1988).
80. A. Kehrel, K. P. Lieb, and M. Uhrmacher, *Mat. Sci. Eng. A* 115, 43 (1989).
81. A. Kehrel, K. P. Lieb, N. Scapellato, and M. Uhrmacher, *J. Nucl. Mat.* 173, 284 (1990).
82. A. Kehrel, J. Keinonen, P. Haussalo, K. P. Lieb, and M. Uhrmacher, *Rad. Eff. Def. Solids* 118, 297 (1991).
83. J. Keinonen, A. Kehrel, K. P. Lieb, and M. Uhrmacher, *Nucl. Instr. Meth. B* 66, 209 (1992).
84. M. Uhrmacher and K. P. Lieb, *Nucl. Instr. Meth. B* 68, 175 (1992).
85. M. Schwickert, K. P. Lieb, W. Bolse, M. Gustafsson, and J. Keinonen, *Nucl. Instr. Meth. B* 147, 238 (1999).
86. P. Haussalo, K. P. Lieb, W. Bolse, C. Illgner, and J. Keinonen, *Nucl. Instr. Meth. B* 120, 266 (1996).
87. K. P. Lieb, P. Haussalo, W. Bolse, C. Illgner, J. Keinonen, and M. Niederdrenk, *Thin Solid Films* 302, 162 (1997).
88. F. Roccaforte, W. Bolse, and K. P. Lieb, *Appl. Phys. Lett.* 73, 1349 (1998); *Addendum: Appl. Phys. Lett.* 74, 1922 (1999).
89. F. Roccaforte, W. Bolse, and K. P. Lieb, *Nucl. Instr. Meth. B* 148, 692 (1999).

90. F. Roccaforte, S. Dhar, F. Harbsmeier, and K. P. Lieb, *Appl. Phys. Lett.* 75, 2509 (1999).
91. F. Roccaforte, M. J. Gustafsson, W. Bolse, J. Keinonen, and K. P. Lieb, *Nucl. Instr. Meth.* 166/167, 148 (2000).
92. F. Roccaforte, W. Bolse, and K. P. Lieb, *J. Appl. Phys.* 89, 3611 (2001).
93. M. Gustafsson, F. Roccaforte, J. Keinonen, W. Bolse, L. Ziegeler, and K. P. Lieb, *Phys. Rev. B* 61, 3327 (2000).
94. S. Dhar, S. Gasiorek, M. Lang, K. P. Lieb, J. Keinonen, and T. Sajavaara, *Nucl. Instr. Meth. B* 193, 283 (2002).
95. S. Gasiorek, S. Dhar, and K. P. Lieb, *Nucl. Instr. Meth. B* 193, 283 (2002).
96. W. Bolse, M. Gustafsson, F. Harbsmeier, and F. Roccaforte, *Nucl. Instr. Meth. B* 161–163, 644 (2000).
97. F. Roccaforte, F. Harbsmeier, S. Dhar, W. Bolse, and K. P. Lieb, *Appl. Phys. Lett.* 76, 3709 (2000).
98. F. Roccaforte, Doctoral Thesis, Göttingen, Cuvillier, 1999.
99. S. Gasiorek, S. Dhar, T. Sajavaara, J. Keinonen, and K. P. Lieb, *J. Appl. Phys.*, submitted. S. Gasiorek, Doctoral Thesis, Göttingen, in preparation.
100. B. J. Fishbein and J. D. Plummer, *Appl. Phys. Lett.* 50, 1200 (1987).
101. M. D. Ingram, *Phys. Chem. Glasses* 28, 215 (1987); *Philos. Mag. B* 60, 739 (1989); P. Maass, A. Bunde, and M. D. Ingram, *Phys. Rev. Lett.* 68, 3064 (1992).
102. W. Smith, G. N. Greaves, and M. J. Gillan, *J. Chem. Phys.* 103, 3091 (1995); J. Oviedo and J. F. Sanz, *Phys. Rev. B* 58, 9047 (1998).
103. P. Jund, W. Kob, and R. Julien, *Phys. Rev. B* 64, 134303 (2001); P. Jund and R. Julien, *Philos Mag. A* 79, 223 (1999).
104. J. Horbach, W. Kob, and K. Binder, *Philos Mag. B* 79, 1981 (1999); *Phys. Rev. Lett.* 88, 125502 (2002).
105. D. E. Day, *J. Non-Cryst. Solids* 21, 343 (1976).
106. G. N. Greaves, *J. Non-Cryst. Solids* 71, 203 (1985); *Miner. Mag.* 64, 441 (2000).
107. A. Meyer, H. Schober, and D. B. Dingwell, to be published.
108. K. L. Dunning, G. K. Hubler, J. Comas, W. H. Lucke, and H. L. Hughes, *Thin Solid Films* 19, 145 (1973).
109. R. Terai and R. Hayami, *J. Non-Cryst. Mat.* 18, 217 (1975).
110. L. Tian, R. Dieckmann, C.-Y. Hui, Y.-Y. Lin, and J. G. Couillard, *J. Non-Cryst. Sol.* 286, 146 (2001).
111. V. F. Antyushin, B. I. Sysoev, and V. F. Syronov, *Phys. Stat. Sol. A* 56, K91 (1979).
112. V. S. Gerasimenko and A. Yu. Posudievskii, *Ukrain-Fizychnyi-Zhurnal* 26, 2030 (1981).
113. H. Jain, *Phys. Stat. Sol. A* 80, K15 (1983).
114. N. G. Stenina, L. Sh. Bazarov, M. Ya. Shcherbakova, and R. I. Mashkovtsev, *Phys. Chem. Min.* 10, 180 (1984).
115. K. B. Hitt and J. J. Martin, *J. Appl. Phys.* 54, 5030 (1983).
116. P. Campone, M. Magliocco, G. Spinolo, and A. Vedda, *Phys. Rev. B* 52, 15903 (1995).
117. F. Bejina and O. Jaoul, *Phys. Earth Planet. Interior* 97, 145 (1996).
118. D. R. Hamann, *Phys. Rev. Lett.* 81, 3447 (1998).
119. J. Breton and C. Girardet, *Phys. Rev. B* 33, 8748 (1986); J. Plata, J. Breton, and C. Girardet, *Phys. Rev. B* 38, 3482 (1988); C. Girardet, J. Plata, J. Breton, and A. Hardisson, *Phys. Rev. B* 38, 5648 (1988); V. Delgado, J. Breton, J. Plata, E. Alvira, and C. Girardet, *Phys. Rev. B* 43, 5046 (1991).
120. S. S. Iyer and Y.-H. Xie, *Science* 260, 40 (1993).
121. G. Franzó, A. Irrera, E. C. Moreira, M. Miritello, F. Iacona, D. Sanfilippo, G. Di Stefano, P. G. Fallica, and F. Priolo, *Appl. Phys. A* 74, 1 (2002).
122. T. Shimizu-Iwayama, K. Fujita, S. Nakao, K. Saitoh, T. Fujita, and N. Itoh, *J. Appl. Phys.* 75, 7770 (1994).
123. J. G. Zhu, C. W. White, J. D. Budai, S. P. Withrow, and Y. Chen, *J. Appl. Phys.* 78, 4386 (1995).
124. P. Mutti, G. Ghisletti, S. Bertoni, L. Bonaldi, C. F. Cerafolini, L. Meda, E. Grilli, and M. Guzzi, *Appl. Phys. Lett.* 66, 851 (1995).
125. W. Skorupa, R. A. Yankov, I. E. Tyschenko, H. Fröb, T. Böhm, and K. Leo, *Appl. Phys. Lett.* 68, 2410 (1996).
126. K. S. Min, K. V. Sheheglov, C. M. Yang, H. A. Atwater, M. L. Brongersma, and A. Polman, *Appl. Phys. Lett.* 69, 2033 (1996).
127. L. Rebohle, J. von Borany, R. A. Yankov, W. Skorupa, I. E. Tyschenko, H. Fröb, and K. Leo, *Appl. Phys. Lett.* 71, 2809 (1997).
128. H. Kozlowski, H. E. Porkann, V. Petrova-Koch, and F. Koch, *Mat. Res. Soc. Symp. Proc.* 452, 657 (1997).
129. X. M. Bao, T. Gao, F. Yan, and S. Tong, *Mat. Res. Soc. Conf. Proc.* 438, 477 (1997).
130. L. S. Liao, X. M. Bao, X. Q. Zheng, N. S. Li, and N. B. Miu, *Appl. Phys. Lett.* 68, 850 (1996).
131. L. Rebohle, J. von Borany, D. Borchert, H. Fröb, T. Gebel, M. Helm, W. Möller, and W. Skorupa, *Electrochem. Sol. State Lett.* 4, G57 (2001).
132. L. Rebohle, J. von Borany, W. Skorupa, H. Fröb, and S. Niedermeier, *Appl. Phys. Lett.* 77, 969 (2000).
133. A. Markowitz, L. Rebohle, H. Hofmeister, and W. Skorupa, *Nucl. Instr. Meth. B* 147, 361 (1999).
134. L. Rebohle, T. Gebel, J. von Borany, W. Skorupa, M. Helm, D. Pacifici, G. Franzó, and F. Priolo, *Appl. Phys. B* 74, 53 (2002).
135. K. S. Miu, K. V. Shcheglov, C. M. Yang, H. A. Atwater, M. L. Brongersma, and A. Polman, *Appl. Phys. Lett.* 68, 2511 (1996).
136. M. L. Brongersma, A. Polman, K. S. Min, and H. A. Atwater, *J. Appl. Phys.* 86, 759 (1999).
137. A. Polman, *Physica B* 300, 79 (2001).
138. P. D. Townsend and D. E. Hole, *Vacuum* 63, 641 (2001).
139. A. L. Stepanov, D. E. Hole, and P. D. Townsend, *Nucl. Instr. Meth. B* 166/167, 882 (2000); *Nucl. Instr. Meth. B* 191, 468 (2002).
140. R. J. Brooks, D. E. Hole, P. D. Townsend, Z. Wu, J. Gonzalo, A. Suarez-Garcia, and P. Knott, *Nucl. Instr. Meth. B* 190, 709 (2002).
141. Z. Wu, A. Türkler, R. Brooks, D. E. Hole, P. D. Townsend, S. F. Köster, K. Kurt, J. Gonzalo, and A. Suarez-Garcia, *Nucl. Instr. Meth. B* (2002).
142. K. Luterova, I. Pelant, I. Mikulska, R. Tomasiunas, D. Muller, J. J. Grob, J. L. Rehspringer, and B. Honerlage, *J. Appl. Phys.* 91, 2896 (2002).
143. Z. G. Wang, Y. F. Yin, E. Q. Zie, Z. Y. Zhu, M. D. Hou, X. X. Chen, Y. M. Sun, and Q. X. Zhang, *Nucl. Instr. Meth. B* 193, 685 (2002).
144. D. Pacifici, E. C. Moreira, G. Franzó, V. Martorino, F. Priolo, and F. Iacona, *Phys. Rev. B* 65, 144109 (2002).
145. L. Pavesi, L. Dal Negro, C. Mazzoleni, G. Franzó, and F. Priolo, *Nature* 408, 440 (2000).
146. P. Schaaf, S. Dhar, S. Gasiorek, and K. P. Lieb, in preparation.
147. P. K. Sahoo, S. Dhar, S. Gasiorek, and K. P. Lieb, *Nucl. Instr. Meth. B*, in press, *J. Appl. Phys.*, submitted.

Encyclopedia of Nanoscience and Nanotechnology
www.aspbs.com/enn

Epoxy/Clay Nanocomposites

Nandika Anne D'Souza
University of North Texas, Denton, Texas, USA

CONTENTS

1. INTRODUCTION

When two or more phases are mixed together to make a composite, a combination of properties that are not available in any of the individual components is possible. In particulate-filled plastics, the role of clay as a filler has been clearly established [1]. When clays were previously used as fillers, they were primarily utilized for increasing the modulus of the matrix resin they are dispersed in. The scale of dispersion is macroscale. Separately, clays have been used in water dispersions aimed at enhanced oil recovery. In the early 1990s, a merger of the two technologies occurred. Researchers at Toyota determined that, when montmorillonite clay was treated with amino acids, the individual layers of the clay could be separated [2–10]. The decreased van der Waals attraction was exploited to swell resins such as polyamides into the intergallery spacing and separate out the platelets. The resulting system then consists of the host resin and individual platelets whose approximate diameter would be around 2 μm, and whose thickness would be 1 nm. Thus, a new classification of composites was born: nanocomposites. The word "nanocomposite" refers to composites where the reinforcement has at least one dimension in the nanometer scale. The nanoscale filler belongs to a class of expandable smectic clays since, for nanoscale reinforcement, key requirements are that the individual layers of the clay can be separately dispersed into the host resin. Because the building blocks of a nanocomposite are nanoscale, they have an enormous surface area ($\sim$760 m^2/g), leading to a high interfacial area between the filler and the matrix. The unique properties of the nanocomposite arise from the interactions of its phases at the interfaces [11, 12]. In contrast, in a conventional composite based on micron-sized fillers such as carbon fibers and glass fibers, the interfaces between the filler and matrix constitute a much smaller volume fraction of the bulk material, and therefore influence its properties to a much smaller extent.

Nanocomposites have found widespread applications in the automobile industry (exterior and interior body parts), packaging industry (bottles, containers, plastic films, and fuel tanks), electronic industry (packaging materials and exterior parts of electronic devices), coating industry (paints, wire enamel coatings, etc), and aerospace industry (body parts of airplane and exterior surface coatings) [11–13]. A key to the burgeoning interest is that the reinforcement aspect ratio and scale require much lower volume fractions compared to micro- and macroreinforced polymers Where filled plastics have filler volume fractions of around 40–50% and fiber-reinforced materials have fractions around 60–70% to meet effectiveness based on scaling laws, nanocomposites often contain reinforcements below 7%. Since the basis of improvement is based on expansion of the smectic layers of the clay and subsequent increased particle wetting, the influence of clay treatment on dispersion has driven much of the earlier research [14–17]. Scientists have also tried to correlate and explain the behavior of nanocomposites through thermodynamic models to understand clay treatments and the formation of nanocomposites [18–24]. Improved barrier properties [25–29], higher mechanical properties [30–34], improved flame retardance [35, 11–13], and increased dimensional stability [36–40] have been determined in polymer nanocomposites. All of these benefits are obtained without significantly raising the density of the compound or reducing light transmission [18].

The multifunctional properties are dependent on the ability to disperse the inorganic material throughout the polymer and increase the interfacial area. First, at the nanometer level, if the individual smectic layers retain some range of order, an intercalated dispersion is obtained. If the layers are completely separated with an interlayer distance

ISBN: 1-58883-059-4/$35.00

Encyclopedia of Nanoscience and Nanotechnology
Edited by H. S. Nalwa
Volume 3: Pages (253–265)

greater than 100 nm, a delaminated or exfoliated dispersion results. On the macroscopic level, it is possible for the clay to self-aggregate so that clay-rich and polymer-rich domains result, giving rise to a morphology similar to that of particle-reinforced polymers, where individual self-aggregated regions can range in size from 10 to 50 μm [16, 18]. Factors limiting the successful dispersion of the layered silicates are the hydrophilic nature of the silicates and the largely hydrophobic nature of most engineering polymers. To produce an intercalated nanocomposite, the polymer has to wet the clay particles to some extent, so that the polymer chains are intercalated between the clay galleries. To make delaminated or exfoliated nanocomposites, a higher degree of wetting is required [16, 34]. To counter the problem of particle agglomeration, the clay interlayer surfaces of the silicate are chemically treated to make the silicate less hydrophilic, and therefore more wettable by the polymer. This is accomplished by a cation exchange process where hydrophilic cations such as Na^+, K^+, and Ca^{2+} are exchanged by alkyl ammonium cations. The role of alkyl ammonium cations in the organosilicates is to lower the surface energy of the inorganic host, and improve the wetting characteristics with the polymer. In general, the length of the alkyl ammonium cations determines the hydrophobicity of silicate layers.

Epoxy nanocomposites, in particular, have been actively investigated, although the number of critical parameters is far larger than for thermoplastics. Organic treatment of the clay surface, curing agent, epoxy type, temperature of curing, and method of premixing are all interactive variables that must be considered. Thus, epoxy nanocomposites pose an interesting scientific and technological challenge where the state of current investigations point as much to the understanding of the present as to where the research needs to be headed. Keeping that in mind, this review of epoxy montmorillonite is divided into separate sections. A review of montmorillonite and key factors affecting nano- as opposed to microcomposite dispersions is covered in Section 2. Efforts to develop epoxy nanocomposites and the influence of different curing agents is covered in Section 3. The role of clay concentration in dispersion is covered in Section 4. The final section, Section 5, is geared toward unique results that stimulate the curiosity for future research.

2. MONTMORILLONITE

Clay is distinguished from sand and silt on the basis of size [41]. Sand has a grain size between 2–0.050 mm, silt is between 0.050–0.002 mm, and clay is below 0.002 mm. On the basis of the SiO_4 tetrahedra, six soil silicates are generally identified: cyclosilicates, inosilicates, neosilicates, phyllosilicates, sorosilicates, and tectosilicates. Clay minerals belong to the phyllosilicates (Table 1) [42]. Phyllosilicates are two-dimensional arrays of silicon–oxygen tetrahedra and two-dimensional arrays of aluminum or magnesium–oxygen–hydroxyl octahedra. In the case of silicon–oxygen sheets, silicon atoms are coordinated with four oxygen atoms. The oxygen atoms are located on the four corners of a regular tetrahedron with the silicon atom in the center. This is shown in Figure 1 [43]. In the sheet, three neighboring tetrahedra share three of the four oxygen atoms of each tetrahedron, and the fourth oxygen atom of each tetrahedron is pointed downward, as shown in Figure 2 [43]. The structure of clay, however, is not a three-dimensional network of simple linkages of silicon–oxygen units. By linking three oxygens of each tetrahedron with adjacent tetrahedra, a tetrahedral sheet is formed. The silicon–oxygen sheet is called the tetrahedral sheet or the silica sheet. In this network of silica tetrahedra, one oxygen in each tetrahedron remains electrically imbalanced, and is therefore linked to Al or Mg in octahedral coordination (Figs. 3 and 4). In the tetrahedral sheet, Si^{4+} replaces Al^{3+}. In the octahedral sheet, Al^{3+} replaces Mg^{2+}. The substitution of higher valence atoms results in a deficit positive charge or an excess of negative charge. The excess of negative charge within the layers is balanced by adsorption of cations. These clays, which are hydrophilic in nature, have Na^+, K^+, or Ca^{2+} cations in the layer gallery. The imbalance leads to the possibility of cation exchange which varies between the different clay minerals. For example, kaolinite has a cation exchange capacity (CEC) of 1–10 meq/g, while montmorillonite is in the range of 70–100 meq/g.

Table 1. Major phyllosilicate minerals in soils [41].

Layer type	Group name	Charge per unit formula	Common minerals
1:1	Kaolinite–serpentine	~0	Kaolinite, halloysite, chyrsotile, lizardite, antigorite
2:1	Pyrophyllite–talc	~0	Pyrophyllite, talc
	smectite or montmorillonite	~0.25–0.6	Montmorillonite (smectite) bedellite, notronite, saponite, hectorite, sauconite
	Mica	~1	Muscovite, paragonite, biotite, phlogopite
	Brittle mica	~2	Margarite, clintonite
	Illite	2	Illite
	Vermiculite	0.6–1.9	Vermiculite
2:1:1	Chlorite	Variable	Chlorite

The packing of a silicate tetrahedron and aluminum octahedron sheets results in a layered structure. While several layers can be stacked, each layer is an independent crystal unit. Within each layer, there is a repetition of structure, and therefore it is referred to as a unit cell. The distance between a certain plane in the layer and the corresponding plane in the next layer is called the basal or d spacing. The basal

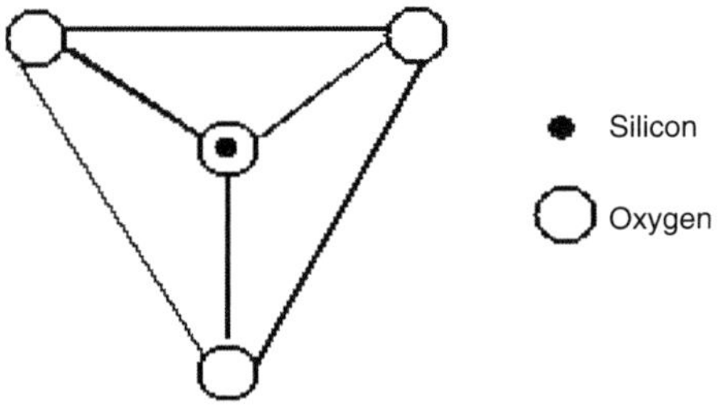

Figure 1. Tetrahedral arrangement of Si and O.

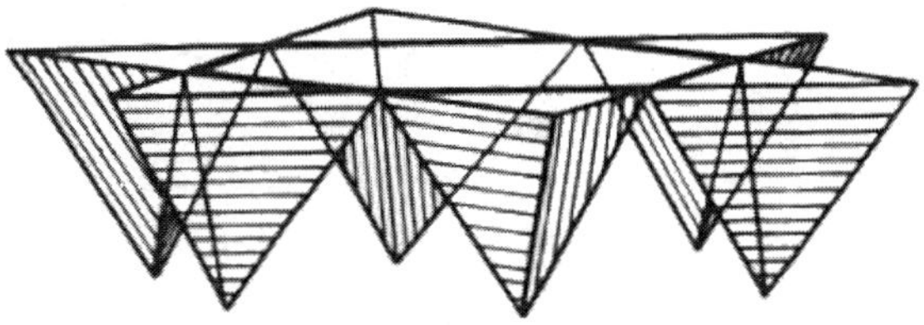

Figure 2. Perspective sketch of tetrahedral linking.

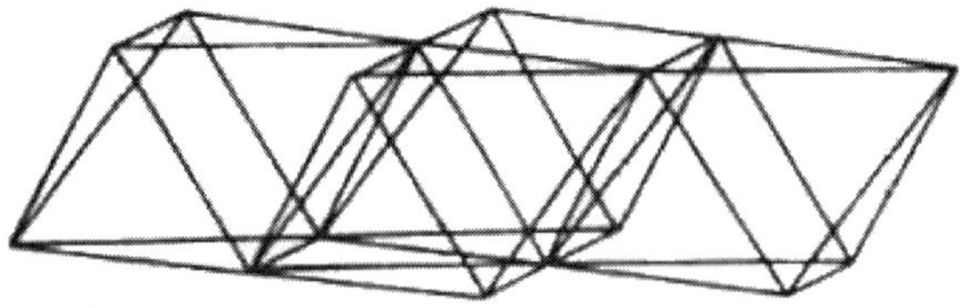

Figure 4. Perspective sketch of octahedral linking.

spacing can be determined by XRD. The basal spacing is around 7.2 Å in kaolinite and 9.2 Å in 2:1 layer clays. The bonding between the layers is weak (van der Waals type), but a strong covalent bond exists between the atoms of the same layer. The suggested crystallographic structure for a layered silicate is shown in Figure 5 [43]. This structure is derived from pyrophyllite and talc by substitution of certain atoms for other atoms.

The bonds between the layers can be strong, as in the case of kaolinite, or weak, as in montmorillonite. On the basis of the number of tetrahedral to octahedral sheets in one layer, dimorphic (1:1), trimorphic (2:1), or tetramorphic (2:2) types are identified. Kaolinite, for instance, has a 1:1 structure where one tetrahedral sheet has one octahedral sheet. The smectite group, however, is a 2:1 since two tectahedral sheets are related to 1 octahedral sheet.

Montmorillonite-layered silicates (MLSs) are distinguished by the location and type of cations in the layer. Due to a high concentration of negative charge, the surface oxygens of tetrahedrally substituted MLS are stronger electron donors than the surface oxygens of octahedrally substituted MLS. The high CEC is responsible for interest in combining MLS with polymers as the potential for swelling and separation of the individual sheets is high. Alternative clays such as magadite and mica have been investigated, but to a smaller extent [44]. An exfoliated nanocomposite was synthesized [45], but the high cost of synthetic magadite over the low cost of natural montmorillonite has limited extensive investigation.

2.1. Surface Treatment of Clay Minerals

Clays are hydrophilic by nature, while polymers are hydrophobic and organophilic. To encourage interaction between the clay and the polymer, the surface of the clay is treated to make it hydrophobic and more organophilic. This is a cation exchange reaction where inorganic cations such as Ca^{2+}, K^+, and Na^+ are replaced by organically modified cations, such as alkyl ammonium ($R\text{–}NH_3^+$) groups. R is a long aliphatic residue. The cationic head of the aliphatic group is found close to the surface layer due to the presence of negative charge on the layer, and the tail without charge is found away from the layer. Four types of structures are possible, depending on the packing density, temperature, and chain length of the aliphatic groups [43]. When the alkyl ammonium chains lie parallel to the host layers, a lateral monolayer or lateral bilayer is formed. If alkyl ammonium chains radiate away from the host layers, an extended paraffin-type monolayer or paraffin-type bilayer structure is formed [46, 47]. These are shown in Figures 6–9.

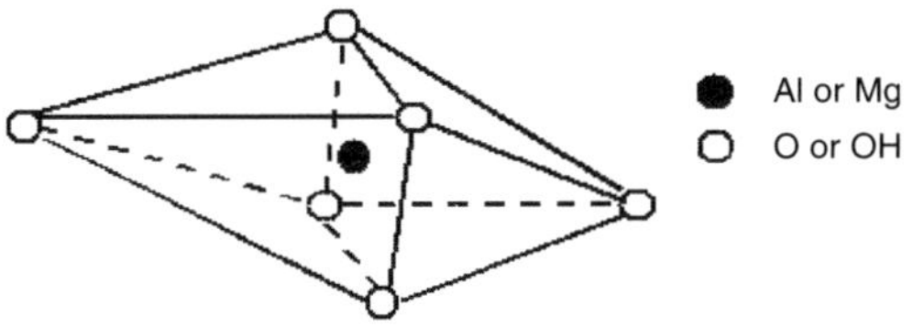

Figure 3. Octahedral arrangement of Al or Mg with O or OH.

The hydrophilic clays are typically intercalated by alkyl ammonium ion (onium) using an ion-exchange reaction. Based on the alkyl chain length, X-ray diffraction and FTIR are typically use to determine which structure is formed in the clay prior to introduction into the polymer. When alkyl ammonium ions with 12–18 carbons are ion exchanged for the inorganic cations in smectic clays with a CEC around 100–120 meq/100 g, a lateral bilayer (18 Å results) or an inclined paraffin-like structure (20–23 Å) results. At higher concentrations of onium ions, a lipid-like bilayer is also formed. For epoxies in particular, a detailed study was conducted to understand the influence of alkyl chain length on the basal spacing of the clay. Lan et al. [32] calculated the final structure by first assuming that the gallery cations would orient vertically from their initial monolayer or bilayer orientation. Basal spacings were expected to follow the relation $d_{001} = 1.27(n-1) + d_A + r_M$, where $(n-1)$ is the number of methylene groups in the onium ion chain, d_A is the basal spacing for the NH_4^- montmorillonite (12.8 Å), r_M is the van der Waals radius of the methyl end group (3 Å), and 1.17 Å is the contribution due to the CH_2 segments when the chain adopts an all-trans configuration. Table 2 shows the results. The difference in the epoxy-solvated value and the calculated value for $x = 1$ was caused by the onium ion taking an inclined orientaion as opposed to a vertical one.

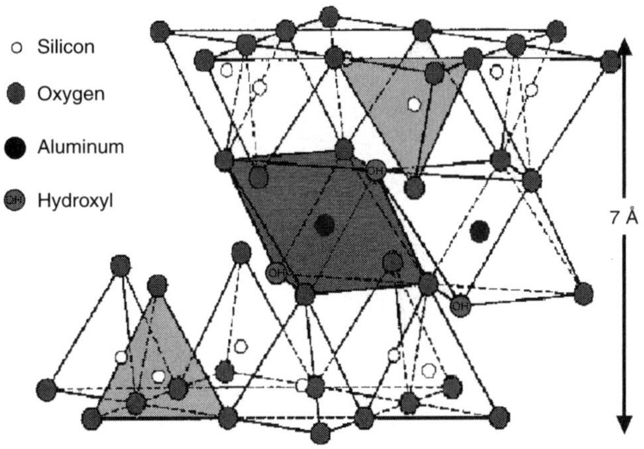

Figure 5. Suggested layered structure of mica-type layered silicates.

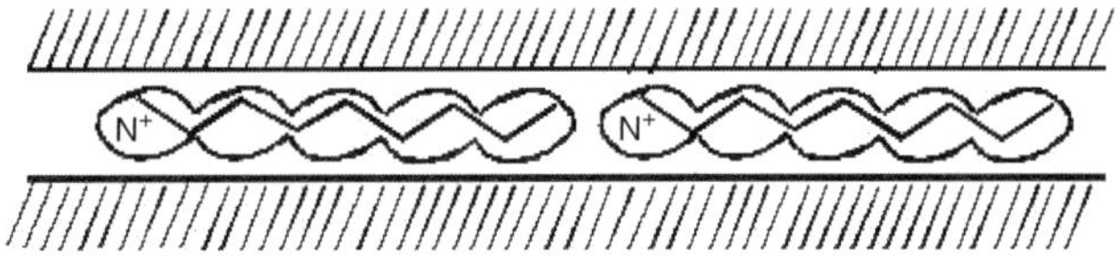

Figure 6. Lateral monolayer.

The similar values between the measured and predicted values for $x > 8$ implied that, for higher chain lengths, a vertical orientation was obtained, and multiple epoxy monomers can be accommodated in the galleries. Vaia developed a schematic structure model for organically modified fluorohectorite (FH–C_n) [43]. Fourier transform infrared spectroscopy (FTIR) along with X-ray diffraction were used. By monitoring the frequency shifts of the asymmetric CH_2 stretching and bending vibrations, it was found that the alkyl ammonium chains exist in states with varying degrees of order. Depending upon the number of carbon atoms present in an alkyl ammonium chain, the model adopts a more ordered structure. For the least number of carbon atoms, that is, when the chain is short, the molecules are isolated from each other. For intermediate chain lengths, a structure with varying degrees of in-plane disorder is formed. At higher chain lengths, an increase in interlayer order was determined. Thus, as the interlayer packing density increases, the alkyl ammonium chains adopt a more disordered structure.

2.2. Dispersion of Montmorillonite

As stated before, the alkyl ammonium chains in the galleries make the clay organophilic. The surface energy of the layered silicates is reduced due to the presence of alkyl ammonium chains, thus allowing organic species with varying degrees of polarities to intercalate between the layers. In considering particulate-reinforced polymer composites, an important parameter is the dispersion of reinforcement. Uniform dispersion is key to better mechanical and thermal properties. The addition of clay in a host matrix produces three different types of dispersion:

- intercalated dispersion
- exfoliated dispersion
- immiscible dispersion.

Figure 10a shows an intercalated dispersion. As it indicates, a finite penetration of polymer chains into silicate galleries results in the finite expansion of silicate galleries. A well-ordered multilayered structure with retention of clay structure is obtained. Extensive penetration of polymer chains into silicate galleries, which results in disorder

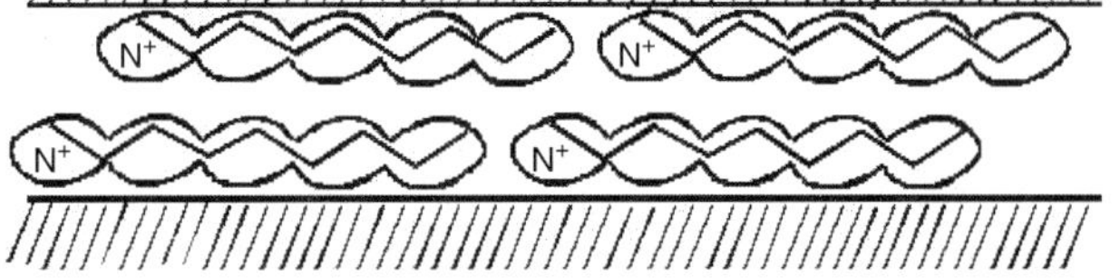

Figure 7. Lateral bilayer.

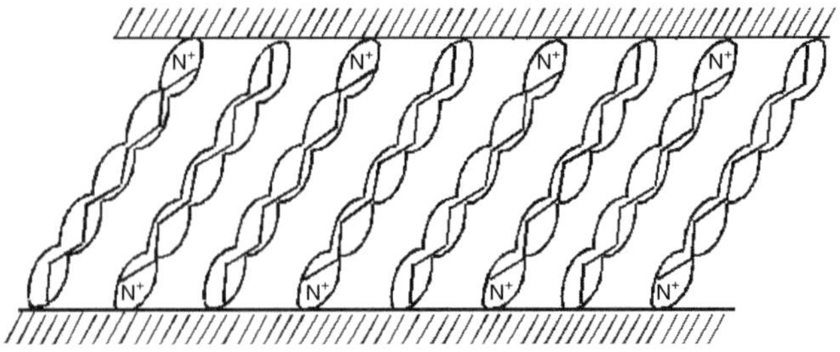

Figure 8. Paraffin-type monolayer.

and eventual delamination with a complete disruption of clay structure, is called an exfoliated dispersion. The exfoliated dispersion is shown in Figure 10b. An immiscible or macrosystem represents complete immiscibility of clay and polymer on the molecular level. Both clay and polymer retain their individual identities without affecting their structure. This effect is shown in Figure 10c. The factors which affect the dispersion and miscibility in a polymer clay nanocomposite are as follows:

- packing density of clays
- chain length of organic cations
- charge on the layer
- processing temperature
- shear during nanocomposite formation
- type of bonding at polymer/silicate interface
- polymer–polymer, polymer–silicate, and silicate–silicate interaction.

2.3. Influence of Epoxy Resin, Surfactant Treatment, and Clay on Dispersion

Since treatment of the clay is vital to generating nanocomposites and epoxy polymerization, is a critical contributor to the generation of nanocomposites, it is important to consider if the alkyl ammonium surfactant itself contributes to the polymerization. Lan et al. [32, 33] argued that the acidity of the alkyl ammonium ions may catalyze the homopolymerization of the digycidyl either of bisphenol A molecules inside the clay galleries. Kornmann et al. [37] investigated the effect of the cation exchange capacity of the base clay on epoxy polymerization. If the CEC is

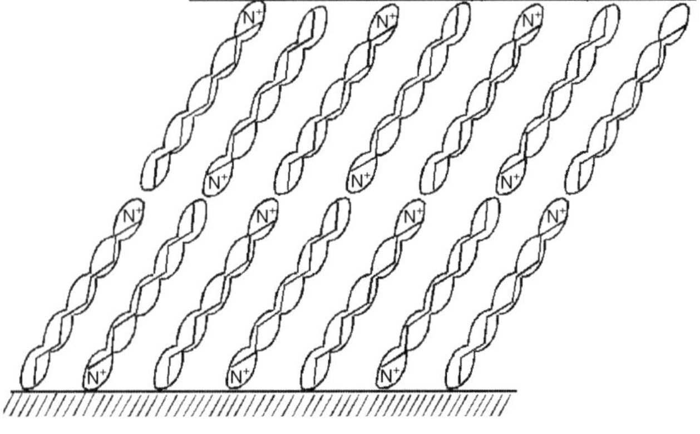

Figure 9. Paraffin-type bilayer.

Table 2. Cation chain-length influence on *d* spacing in montmorillonite.

$CH_3(CH_2)_xNH_3^+$	Air dried	Epoxy solvated	Calculated distance between plates	Initial cation orientation
3	13.5	16.5	19.6	Monolayer
7	13.8	27.2	24.7	Monolayer
9	13.8	30	27.2	Monolayer
11	15.6	31.9	29.8	Bilayer
15	17.6	34.1	34.9	Bilayer
17	18	36.7	37.4	Bilayer

high, more alkyl ammonium surfactant would lie next to the clay, and therefore the effect of homopolymerizaiton would be clarified. First, in treating the low CEC clay, an increase in (001) spacing from 9.7 to 17.2 Å was obtained for the organoclay, whereas the low CEC clay swelled from 12.4 to 21.4 Å. The organoclays, when swollen in DGEBA resin at 75 °C, quickly increased in *d* spacing to 30 Å. This was associated with the development of a lateral bilayer of perpendicular orientation. As the time for swelling reached 24 h, the interlamellar spacing of the higher CEC clay was constant around 34 Å, while the low CEC clay was completely exfoliated. An increase in viscosity of the low CEC clay was also apparent, which the authors explained as resulting from self-polymerization. However, they also acknowledged that the exfoliated state with increased effective clay particle size would lead to increased viscosity as well. Thus, Kornmann et al. showed that a low CEC permits higher diffusion of alkyl ammonium ions. Butzloff et al. [48] also investigated the competing homopolymerization versus epoxy + curing agent polymerization. For a given clay, they showed that the kinetics of the two-part system given by the epoxy + organoclay are substantially lower than epoxy + hardener; therefore, the homopolymerization of the epoxy and clay had little influence on the final nanocomposite. By increasing the swelling temperature to 80 °C and decreasing the time for epoxy + clay swelling to 1 h, they found that the DSC trace showed only a single hump as opposed to the two-stage hump shown by Lan et al. [32] at 75°. They investigated the influence of changing clay concentration, and found that all of the kinetics of the three part (epoxy + clay + curing agent) system dominated the homopolymerization. However, there were definite signs that the alkyl ammonium ions were participatory in the reaction since changing the heating rate changed the area under the curve of the polymerization. One of the few attempts to study a high T_g epoxy resin was recently made by Becker et al. [49]. They investigated the tri- and tetrafunctional epoxies. They found that, while the DGEBA system had an exfoliated dispersion, the high-crosslink-density resins had intercalated dispersions. A general increase in modulus and fracture toughness was obtained. This was related to the intercalated dispersion. The comparative results of curing agent, epoxy, and clay on dispersion are shown in Table 3.

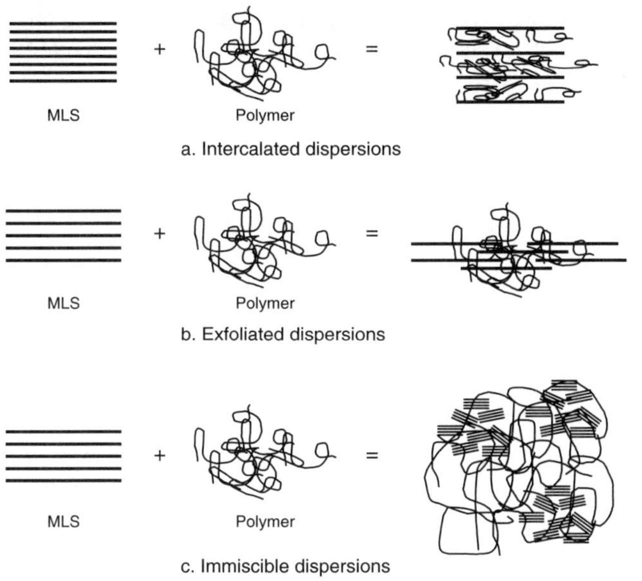

Figure 10. Dispersions of MLS in polymers.

3. PROCESSING OF POLYMER CLAY NANOCOMPOSITES

Techniques for polymerization include melt intercalation, solution polymerization, and *in-situ* polymerization. Melt intercalation is primarily applicable to thermoplastics, and is adaptable to conventional processing technologies such as injection and extrusion. It is briefly reviewed in Section 3.1. The most generally followed approach in epoxy nanocomposites is the *in-situ* polymerization covered in Section 3.2. Solution polymerization is followed when the polymer is mixed with an organic solvent, the clay swells, and the polymer chains diffuse between the clay platelets. When the solvent is evaporated, often an intercalated nanocomposite results. The use of a solvent with epoxy has been successful when high clay loadings are considered. This is elaborated on in Section 3.3.

3.1. Melt Intercalation

In 1993, Vaia and his colleagues were the first to report melt intercalation to make thermoplastic nanocomposites. The method is not applicable for thermosets because the process involves the melting of a polymer. The process involves blending of a molten thermoplastic with organoclay, and then annealing at a temperature above the glass transition temperature of the polymer. The mechanism is shown in Figure 11 [43]. The observation that polymer chains can undergo center of mass diffusion in essentially two dimensions is rather surprising because the unperturbed chain diameters (10 nm) are approximately an order of magnitude greater than the interlayer distance between the silicate sheets (1 nm). The proposed driving force for this mechanism is the important enthalpic contribution of the polymer–clay interactions during the blending and annealing steps. Since the process involves the blending of polymer with clay, melt intercalation is one of the most popular methods to prepare nanocomposites in industry. Different thermoplastics have been tried so far to make nanocomposites by this technique, for example, polypropylene, polyethylene, nylon, and polystyrene.

Table 3. Curing agent and clay treatment effects on intercalated and exfoliated dispersions in epoxy nanocomposites.

Epoxy resin	Curing agent	Clay treatment	Intercalated or exfoliated?
DGEBA	$NH_2CHCH_3CH_2(OCH_2CHCH_3)_xNH_2$, $x = 2.6$ Commercial name: JEFFAMINE 230	Clay treated with octadecylamine $C_{18}NH_3$	Exfoliated, $d > 88$ nm [48, 53]
DGEBA	JEFFAMINE 230	$C_{18}NH_3$ I.28E™(Nanocor)	Intercalated Mixed at 65 °C and degassed, cured at 75 °C
DGEBA	3,3′–dimethylmethylenedi(cyclohexylamine) (3DCM)	$C_{18}NH_3$	Intercalated $D \sim 40$ Å [37]
DGEBA	Amicure bisparaaminocyclohexylmethane (PACM)	$C_{18}NH_3$	Intercalated (d 37 nm) [52]
DGEBA	4,4′–diaminodiphenyl methane (DDM)	$C_{18}NH_3$	Exfoliated [53, 61]
DGEBA	m–phenylene diamine (mPDA)	$C_{18}NH_3$ I.30E™ (Nanocor)	Intercalated [33, 55]
DGEBA	4,4′–diaminodiphenylsulfone (DDS)	$C_{18}NH_3$	Exfoliated [68]
DGEBA	JEFFAMINE D400 ($N = 5.6$)	Bis(2–hydroxy–ethyl)methyl tallow ammonium MMT	Exfoliated [58]
Trifunctional epoxy (Araldite XVMY0505)	Aromatic hardener HY5200	Dimethyl benzyl hydrogenated tallow ammonium chloride	Intercalated [56]
DGEBA	Methyl tetra hydrophthalicanhydride (MeThPa)	$C_{18}NH_3$ and $C_{18}NM_3$	Exfoliated Exfoliated [53]
DGEBA	Nadic methyl anhydride	$C_{18}NH_3$	Intercalated [35]
DGEBA	Boron trifluoride monoethylamine (BTFA)	$C_{18}NH_3$	Intercalated [35]
DGEBA	Benzyldimethylamine	$C_{18}NH_3$	Exfoliated [35]
DGEBA	Methylinedianiline (MDA)	$C_{18}NH_3$	Intercalated [35]
TGAP (trifunctional triglycidyl p–amino phenol	Diethyltoluene diamine (DETDA)	$C_{18}NH_3$	Exfoliated [49]
Tetrafunctional tetraglycidyldiamino diphenylmethane (TGDDM)	Diethyltoluene diamine (DETDA)	$C_{18}NH_3$	Exfoliated [49]

3.2. *In-Situ* Polymerization

This was the first method used to synthesize polymer clay nanocomposites. The first polymer used was nylon (polyamide 6). However, it is the central processing technique for epoxy nanocomposites. Strategies with conventional epoxies follow two paths. In one case, the epoxy monomer swells the clay, and subsequently, the curing agent is added (two stage). In the second case, epoxy and the curing agent are mixed, and the clay is added (one stage). The two-stage method is generally followed. Figure 12(a) shows the flowchart of a single-stage reaction where epoxy + hardener + MLS are added together. Figure 12(b) is a flowchart for the two-stage polymerization route. The first step is to achieve the swelling of organoclay in monomer. This is a time-consuming step since the swelling depends on the polarity of monomer molecules, surface treatment of clay, and swelling temperature. Then, depending upon the type of polymer, the reaction is initiated. In the case of

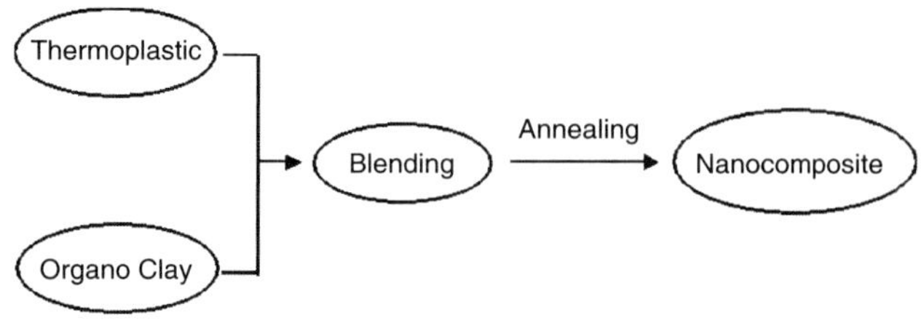

Figure 11. Flowchart of the "melt intercalation."

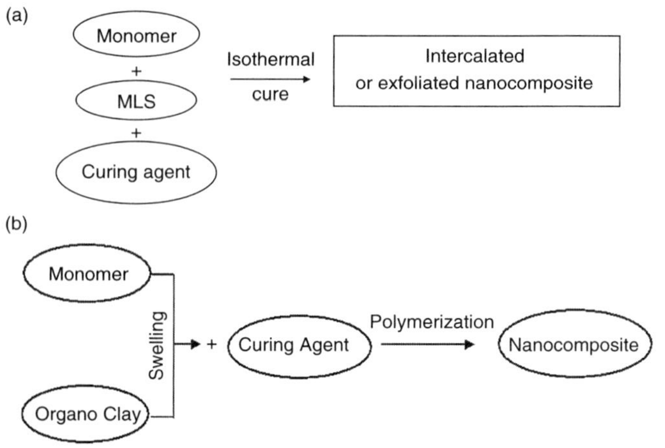

Figure 12. (a) One-stage epoxy *in-situ* polymerization. (b) Flowchart of the "*in-situ* polymerization" two-step process.

thermoplastics, the polymerization is achieved by the addition of free radicals or by an increase in temperature. In the case of thermosets, a curing agent is added to initiate polymerization. Figure 13 shows the interactions between the clay and the monomer at different stages of polymerization [43].

The driving force for this "*in-situ* polymerization" depends on the polarity of monomer molecules. Due to the high surface energy of the clay, monomer molecules get diffused between the clay layers. The diffusion is continued until equilibrium is reached. During polymerization, the monomer starts to react with the curing agent or free radical. This reaction lowers the overall polarity of the intercalated molecules so that more polar molecules are driven between the clay layers. The most important thing is to control the polymerization occurring between the layers (intragallery polymerization). If the cure kinetics is lower between the layers than outside the layers (extragallery polymerization), delamination of clay is possible. Therefore, greater importance is given to intragallery polymerization than to extragallery polymerization.

One of the first publications on developing epoxy nanocomposites was by Wang and Pinnavaia [26]. Prior to their paper, the only report was a filed patent [50]. The patent utilized a solvent intermediate. They introduced the organoclay into epoxy at 75°. After stirring for 30 min, they gradually increased the temperature at 20 °C/min, and obtained a powder. The reaction took place between 200–300 °C. The powder obtained was a delaminated nanocomposite. When no clay was present, the epoxy had an onset polymerization temperature of 384 °C, which went to 229 °C when 0.2% clay was present. The enthalpy of reaction was similar with and without clay. Comparing clays of different acid values, they determined that the interlayer accessibility and gallery cation acidity were important parameters guiding nanocomposite formation. Examining reaction kinetics, they determined that, while the epoxy activation energy was around 75 kJ/mol, the clay-filled epoxies had activation energies over 100 kJ/mol. The higher values were associated with the steric hindrance caused by intercalated clay layers. The first publication of epoxy nanocomposites, which was compatible with curing agents and coating-based processing

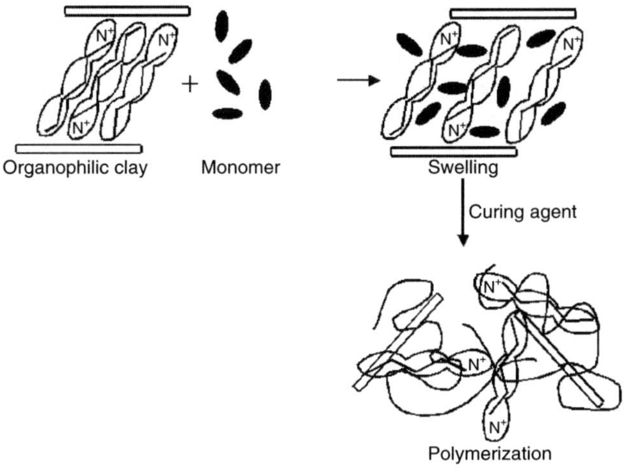

Figure 13. Chemical steps involved in the "*in-situ* polymerization."

techniques, was by Messersmith and Giannelis [35]. Nanocomposites using anhydride, BDMA and BTFA were used. TEM showed layer spacings of over 100 Å. Using a low T_g curing agent, Lan and Pinnavaia [30] established the possibility of increasing both the UTS and modulus (a generally unachievable combination in macrofilled composites.) Lan et al. [33] established the foundation for the formation of epoxy nanocomposites by setting up the need for epoxy swelling of the clay galleries prior to reaction initiation. Lee and Jang [51] investigated *in-situ* polymerization of epoxy with an emulsified epoxy. They indicated that an approximate 0.5 nm increase in *d* spacing over the based montmorillonite (001) peak was obtained. Kornmann et al. [37, 52] showed that the demarcation between exfoliated and intercalated dispersions is primarily associated with the nature of the curing agent. They postulated that, first, the clay's high surface energy attracts the polar monomer molecules. Once saturation is reached and the hardener is introduced, the polymerization serves to decrease the polarity of the system adjacent to the clay. To compensate for the decreased polarity, more polar species diffuse to the clay surfaces, pushing the platelets apart, and leading to an exfoliated system. This last aspect is governed by competition between intragallery polymerization and extragallery polymerization, which proceeds at two rates [33]. Kornmann et al. compared three systems of different reactivities. By comparing the slope of the curing peak in the DSC, they showed that the slower the reaction (the smaller the initial slope), the higher the likelihood for an exfoliated nanocomposite. However, decreasing the reaction rate of any curing agent did not increase the likelihood for exfoliation. It is clear that intragallery versus extragallery polymerization—is an active variable. One could decrease the temperature for curing of a highly reactive curing agent to slow the reaction rate; however, then, the migration of the curing agent between the clay galleries would be impeded. Thus, extragallery polymerization would dominate. The migration capability not due to polarity differences between the curing agents (all curing agents had similar polarities), but to the flexibility of the main chain. A supportive case for this argument is the work done by Kong and Park [64]. They associated the increased likelihood exfoliation in a slow reacting DDS agent to its low electronegativity ($pK_a = 2.15$) which provides a medium of low viscosity for the clay layers. PDA with $pK_a = 4.$ and MDA with pK_a intermediate between DDS and PDA lead to intercalated and partially exfoliated nanocomposites. Slow curing delays the extragallery gelation, and provides time for intragallery polymerization. Conducting real-time SAXS, they showed that, at room temperature, no change in the diffraction peak of the organoclay is evident even after 1 h. At 50° the DGEBA molecule has sufficient thermal energy, and diffuses into the intergallery of C_{18} clay from the edge to the inner region. The change in *d* spacing is sudden and not gradual, indicating that the C_{18} amine attached to the edge of the clay interacts with the DGEBA, and changes the configuration from mono- or bislanting to a bilipid perpendicular to the clay surface. The sudden increase in *d* spacing permits more DGEBA to diffuse into the galleries.

A second increase is associated with the clay layer folding, translating, and expanding from the self-polymerization of DGEBA with the catalytic effect of a protonated C_{18} amine cation. The third increase takes place from the cross linking of epoxy resin with the amine and the amine curing agent in the intergallery region. Jiankun et al. [53] investigated the relationship between the acidic nature of the clay and the generation of intercalated or exfoliated dispersions in epoxy clay nanocomposites. They found that, in the DDM curing agent, the more acidic clay resulted in exfoliated dispersions as opposed to the less acidic clay. Thus, Lan et al.'s prediction was confirmed independently. *In a more acidic clay, the clay is catalytic to the amine curing reaction of the epoxy. The intragallery epoxy polymerization is accelerated, and becomes comparable to that of the extragallery polymerization. The clay can then be exfoliated.*

The case for catalytic activity of the clay was not a contributor to the anhydride-cured completely exfoliated nanocomposites which resulted from both the low and highly acidic organoclays. The difference in the curing chemistry of amine over anhydride curing agents and its influence on intercalation over exfoliation were related more to the viscosity of the amine-curing agents. Thus, in the less acidic clays, the amine was unable to penetrate the galleries since both chemical and physical interactions were restricted. Brown et al. [58] investigated a hydroxyl-substituted quaternary ammonium modifier, and found a high level of exfoliation. The premise is that the catalytic activity of the acidic onium ions is more pronounced when the alkyl chains of the onium ions provide an intragallery hydrophobic environment compatible with both the epoxy polymer precursors and the diamine curing agent [54]. Triantafillidis et al. [54] very recently investigated the use of a difunctional amine. Here, the onium ions act concomitantly as a clay surface modifier, intragallery polymerization catalyst, and curing agent.

Tolle and Anderson [55] investigated the migration of the curing agent into the epoxy-swelled clays using SAXS. They conducted the mixing using the two-stage method at 60°, and then monitored the cure temperature effect on exfoliation and intercalation. They showed that, for the same curing temperature, the purity of the organoclay was critical to developing an exfoliated structure.

3.3. Polymer Solution Intercalation

This method is different from the "*in-situ* polymerization" in that it makes use of a preformed polymer instead of monomer. Polymer solution intercalation is based on a solvent system in which the polymer is soluble and the silicate layers are swellable. The first step is the swelling of clay in a solvent, usually a polar solvent (Fig. 14). Then the polymer, dissolved in the solvent, is added to the solution, and intercalates between the clay layers. The last step is the removal of the solvent by evaporation, normally under vacuum. After solvent removal, the intercalated structure remains.

The driving force for polymer solution intercalation is the entropy gained by desorption of solvent molecules, which compensates for the decrease in conformational entropy

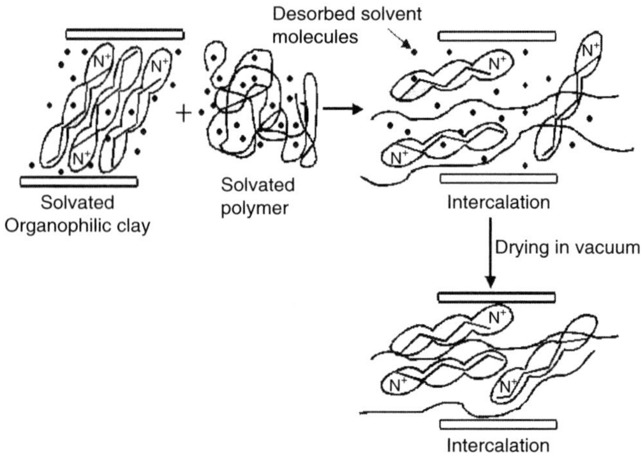

Figure 14. Intercalation of the polymer by the "solution."

of the intercalated polymer chains. Therefore, a relatively large number of solvent molecules need to be desorbed from the clay to accommodate the incoming polymer chains. The advantage of this method is that it offers the possibility of synthesizing intercalated nanocomposites based on polymers with low or no polarity.

The water-soluble polymers, which have been used to make nanocomposites by this technique, are poly(vinyl alcohol), poly(vinyl pyrrolidione), poly(ethylene oxide), poly (ethylene vinyl alcohol), and linear poly(ethylenimine) using water as the solvent. Salahuddin et al. [56] report on the use of acetone in conjunction with a dimethyl benzyl hydrogenated tallow ammonium-chloride-treated montmorillonite. To the suspension of the organoclay and acetone, epoxy and hardener were added. Based on the dispersion scaling that stacks of parallel layers with an average of ten lamellae form a primary clay particle (10 nm), the association of a few particles would lead to a few hundred nanometers, and large-scale aggregates of primary particles form microaggregates of 0.1–10 μm, they determined that the use of acetone resulted in uniform platelet distribution [57]. Even though the dispersion was intercalated, transparency equal to the unmodified epoxy was obtained. The resulting composites containing over 60% MLS had a 100% maximum increase in Vicker's hardness. Brown et al. [58] also used acetone, but their suspension was based on using acetone, organoclay, and epoxy. The concept explored by them was to extend the ability of the epoxy to enter between the platelets in order to enhance the intragallery reaction during the curing stage. Comparing a range of surfactants, they found that bis(2–hydroxy–ethyl) methyl tallow surfactant compatibilizes the diamine/epoxy mixture, resulting in a highly exfoliated nanocomposite, and it also serves to catalyze the epoxy/diamine reaction. A clay-concentration-dependent processing scheme was followed. Epoxy with acetone was mixed with the layered silicates at room temperature. Room-temperature drying and vacuum drying at 90 °C were followed to remove remnant acetone. The curing agent was added at 75 °C, and the mixture was cured. For clay concentrations greater than 10%, 100 psi of vacuum was applied during the cure.

4. CLAY CONCENTRATION EFFECTS

Butzloff et al. investigated the clay concentration effect in epoxies [48]. First, by comparing the reaction between just the montmorillonite and the epoxy to the inclusion of hardener, they determined that the migration of the epoxy between the galleries of the clay prior to introduction of the curing agent was clay concentration dependent. Without hardener, they found that

1. increased clay concentration decreased the epoxy homopolymerization peak substantially (Fig. 15)
2. the slope of the exotherm was clay concentration dependent (Fig. 16).

They also found that the epoxy + clay system showed only one peak compared to two peaks shown in the work of Lan et al. [27]. Lan et al. had attributed this to a difference in the concentration of reactive acidic onium protons inside the nanoclay gallery as compared with the chemical environment outside the gallery. Therefore, Butzloff et al. determined that the single peak observed indicated that the acid protons from inside or outside the nanoclay galleries behaved no differently in the attack of the epoxide ring. It was therefore concluded that the gallery environment in the premix was not sufficiently different to cause the peak-splitting effect observed in the work of Lan et al. This brought up an interesting premix temperature effect on intercalation versus exfoliation since the preparation used by Butzloff et al. was 80 °C with a 1 h shear, whereas the preparation reported by Lan et al was 75 °C with a 24 h shear. No significant change in the area under the curve or enthalpy as a function of composition for the resin was obtained. When hardener was introduced into the epoxy + montmorillonite mixture, there was a significant change in the concentration effect of the montmorillonite on the epoxy. Introduction of 1% montmorillonite by weight decreased the reaction enthalpy, but after that, there is little change in enthalpy, except for a peak value at 2.5% montmorillonite by weight. It was ascertained that at 2.5 wt% montmorillonite, a critical concentration is established, analogous to the critical micelle concentration in surfactant sols. Activation energy and the Arrenhius frequency factor for the nonisothermal kinetic runs both peaked at this concentration. TEM results of the epoxy nanocomposite are shown in Figure 17. The assertion of a 2.5% critical concentration related to charge aggregation was based on observation of the 45° edge–edge interaction seen in the TEM micrographs. Charge nonuniformity due to different layer spacings of adjacent platelet stacks, or tactoids, was indicated as a mechanism for aggregate formation. An interesting change in dispersion was evident at higher clay concentrations. The TEM of the 5% MLS nanocomposite is shown in Figure 18. This micrograph indicates a change to a much more compact parallel arrangement of the intercalated regions of nanoclay. Little or no space between these was left for individual exfoliated platelets. This structure is short range, not long range. It is clearly illustrated in this figure that some regions of parallel cards or intercalated montmorillonite plates have their grains oriented at right angles to others. No flocs or micelle-like structures were observed. All compositions were completely exfoliated. X-ray diffraction patterns for the MLS and composites are shown in Figure 19. The X-ray pattern of the treated montmorillonite clay showed a strong 001 reflection at $2\theta = 72°$, corresponding to the interlayer spacing of the regular stacked nanoclay structure, and the 002 reflection at $2\theta = 8.02°$. Another point made on epoxy nanocomposites in their work was that, since the chemical formula for the nanoclay is not that for an ideal montmorillonite, the 003 ($2\theta = 12.01°$) and 004 ($2\theta = 16.25°$)

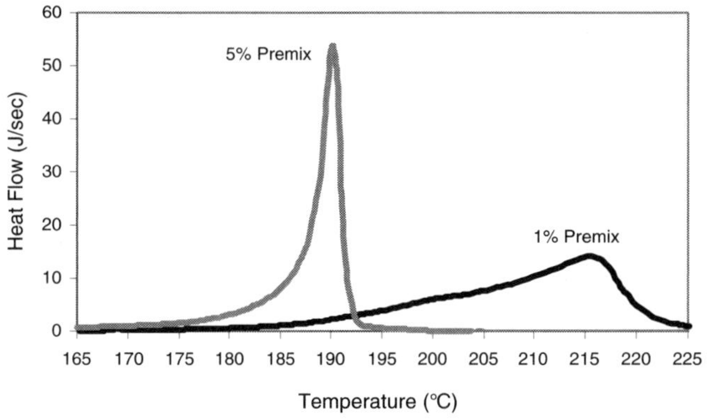

Figure 15. Epoxy + clay homopolymerization.

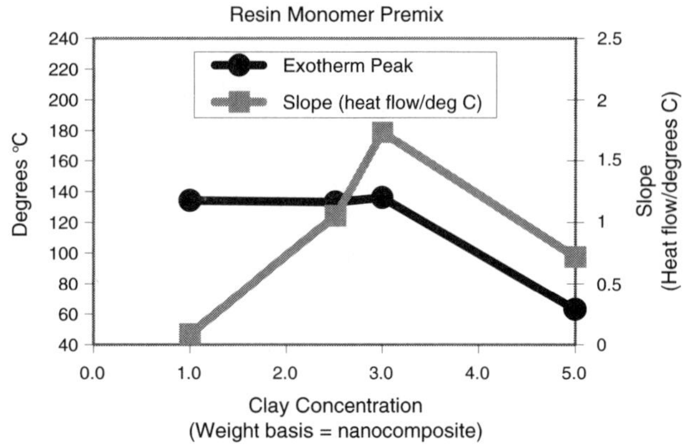

Figure 16. Exotherm peak on curing of epoxy + curing agent + clay mix as a function of clay concentration and slope of the exothermal peak.

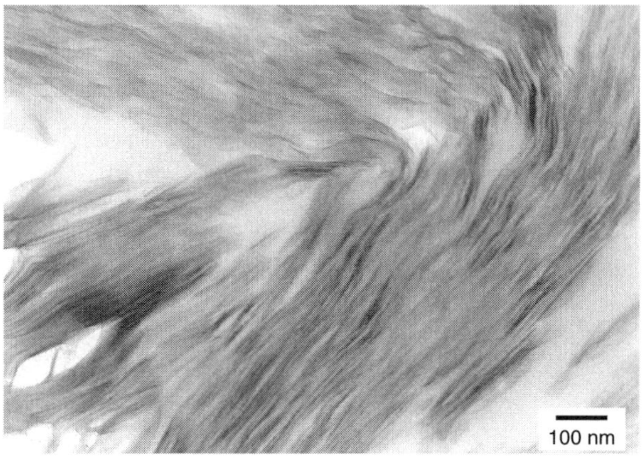

Figure 17. TEM of 45° platelet edge association for 2.5% montmorillonite in epoxy nanocomposite.

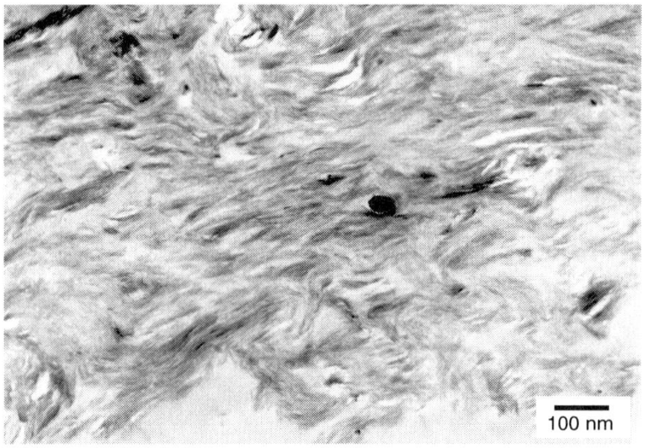

Figure 18. TEM of platelet edge to face association at 5% montmorillonite.

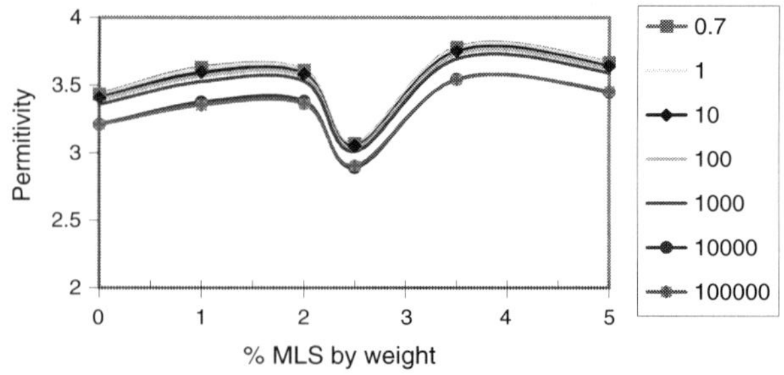

Figure 20. Dielectric permittivity as a function of MLS concentration showing minima at 2.5% montmorillonite by weight.

reflections are weak for the treated nanoclay [59]. Thus, they performed slower scan speeds, and determined that the 005 ($2\theta = 20.16°$) reflection overlaps a 02;11 band at $2\theta = 19.72°$. This band at $2\theta = 19.72°$ corresponds to a turbostratic stacking disorder common in smectic minerals, and is confirmed by a 20;13 asymmetrical band at $2\theta = 34.84°$ [60]. The critical concentration determined by Butzloff et al. has also demonstrated other phenomena. For instance, the dielectric permittivity–concentration (Fig. 20) and thermal conductivity–concentration (Fig. 21) results show a minima and maxima at the same concentration. The presence of a peak in the property-concentration profile led to a hypothesis that the behavior of the clay within the epoxy resin is similar to a colloid suspension. This was investigated separately by Park et al. [61]. Using an octadodecyl amine montmorillonite, they determined that the organoclays had an effect of decreasing the pH, increasing the Boehm's acid value, or decreasing the base value. Using another analysis by Panzer et al. [62], Park et al. determined that modification of the clay results in an acid value higher than the unmodified clay due to an increase in cation groups on the clay surface or reaction of the –OH groups at the clay surface. Their experimental schedule showed complete exfoliation of the clay in epoxy. A result similar to that was shown by Butzloff and D'Souza [48]. Park et al. analyzed the FTIR spectra of the nanocomposites. Peaks at 1027 and 793 cm^{-1} of MLS associated with the Si–O vibrations were retained. Comparing pristine and organoclay, a new peak at 3000–2900 cm^{-1} due to the CH_2 mode of the epoxy and the disappearance of the 918 cm^{-1} peak of the epoxy ring indicated that the clay interlayer was modified by the octadodecyl amine, and that the epoxy ring was opened by homopolymerization [63]. It has been generally presumed that the influence of the surface treatment of clay is to decrease the hydrophilic nature of clay in favor of hydrophobic. Park et al. compared the contact angle measurements of diiodomethane, ethylene glycol, and water on the treated and untreated clay. They showed that the hydrophilic nature of the clay is retained even after organo treatment. The greater response to diiodomethane indicated that the decreased pH of the untreated clay increased the acid values of the organoclay.

The curing agent effect on exfoliation has been associated with its electronegativity [64]. Kong and Park found that, when DDS was the curing agent, a completely exfoliated system was obtained. One approach to minimizing the amount of organic presence in the epoxy nanocomposite was recently investigated by Triantafillidis et al. [65]. They substituted some organic cation exchange sites with inorganic cations within the same galleries [66, 67]. Such a mixed exchange ion clay was considered "homostructured" because each gallery in the stacked tactoid would be equivalent in composition and intercalation properties. Homostructured mixed-ion clays were distinguished from "heterostructured" clays by the fact that, in the latter, the intercalation properties of each gallery would be different. Further, the state of homostructure versus heterostructure was dependent on

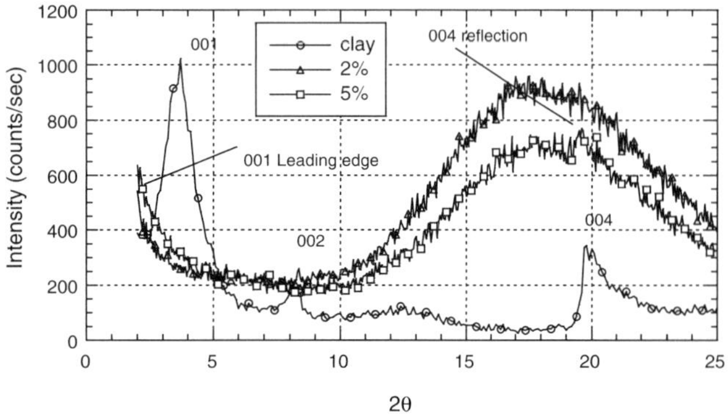

Figure 19. X-ray of the 001, 002, and 004 reflections in the clay and nanocomposite.

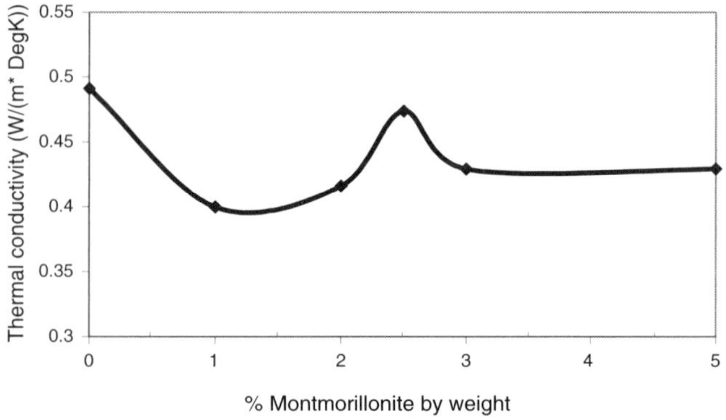

Figure 21. Thermal conductivity as a function of MLS concentration showing maxima at 2.5% montmorillonite by weight.

the nature of the inorganic and organic exchange ions, the charge density of the clay layers, and the method used to form the mixed-ion composition. While the thermodynamic driving force would direct the reaction toward segregation into different galleries, trapping of the ions kinetically would permit presence within the same gallery. The key solution presented in the mixed ion homostructure was the ability to develop uniform gallery spacing. Triantafillidis et al. synthesized low-onium-content ions using Jeffamine D2000, which is a long-chain primary diamine. The basal spacing of the clays was independent of the amount of onium ion substitution, indicating that an increased onium ion presence only served to increase chain coiling. Onium ion substitution of both montmorillonite and hectorite was carried out. It was found that only high onium substitutions (above 50% in MLS and 65% in fluorohectite) led to exfoliated nanocomposites.

5. EPOXY NANOCOMPOSITE PROPERTIES

5.1. Flammability

Thermooxidative degradation comparison of epoxy and its nanocomposite have shown that the nanocomposites are inferior to the epoxy at lower temperatures, but at high temperatures, above 400 °C, the nanocomposites fared better [68]. In a departure from the conventional nanocomposite synthesis, Kim et al. [69] investigated intercalating triphenyl phosphate (TPP) into the galleries of a mica-type silicate. A significant improvement in flame retardance in the corresponding epoxy nanocomposite was obtained.

5.2. Glass Transition and Beta Transition Effects

The influence of the clay on the glass transition was investigated substantially, especially in the context of confinement of polymers [70–73]. For controlled pore geometries, they determined two glass transitions related to the liquid plug in the core and an interacting surface layer. A hump in the specific heat plots was indicative of material not confined in the pores. The influence of solvent elution and its influence on the generation of two glass transitions were also reviewed [74]. In epoxy nanocomposites, Butzloff et al. [48] reported no significant influence of the MLS on the glass transition at concentrations below 5%. A substantial drop in glass transition was evident at higher concentrations. No change in the beta relaxation temperature was determined for the same systems. However, recently, Becker et al. [49] investigated the beta transition, and found that a decrease in both the glass transition and beta relaxation (associated with the crankshaft rotation of the hydroxy ether segments of the cross-linked network) decreased with increased clay presence. It is to be noted that their study was on high T_g epoxy resins. The decreased T_g was associated with a decreased density in the epoxy resin. Using organoclay density of 1.6748 g/cm^3 as determined by gas-displacement pycnometry, they found that the actual densities were slightly lower than that of the theoretically predicted. They inferred that the lower density was related to poorer packing of the polymer molecules near the silicate layers.

Acetone solution polymerized epoxy nanocomposites have displayed no change in the sub-T_g relaxations and minimal changes in the glass transition. A decreased tan δ has often resulted from the introduction of MLS into epoxy [58].

5.3. Thermal Degradation

A critical limitation in using the organically treated clays is considered to be the thermal stability of the organic treatment. For montmorillonite, Xie et al. [75] conducted a detailed analysis of evolution of volatiles in organically modified montmorillonite. For temperatures between 200 and 500 °C organics, CO_2 long-chain alkyl fragments are detected. The surfactant degradation is supposed to present an upper limit to processing temperatures of the polymers they are introduced in. It is to be noted that Ranade et al. [76], investigating another thermosetting nanocomposite processed using the solution polymerization route, determined that, even when the surfactant degradation takes place, the small fraction that it represents need not do any damage to the host polymer. DSC and TGA scans of the as-processed sample showed significant exotherms in the first heating run (taken from 30 to 400 °C). The second scans, however, showed a single glass transition temperature. To determine if the PAI was affected by the surfactant degradation, FTIR was conducted on samples before and after being subjected to 1 h at 400°. There was complete overlap of the spectra, showing no degradation. Thus, they surmise that interacting polymer chains between the clay layers serve not just to shield the polymer as typically inferred, but rather are influenced by the level of intercalation. For a highly intercalated material, the increased solvation between the chains and the end-tethered alkyl ammonium group ensures protection from the alkyl ammonium degradation with the clay acting as barriers for thermooxidative stability.

5.4. Mechanical Properties

The general results for mechanical properties reported are of increased modulus, decreased or retention of ultimate tensile strength, and decreased strain to failure. The compressive strength of epoxies is a significant issue to be considered, while exfoliated and intercalated nanocomposites have demonstrated no significant change in compression [77, 78]. However, the mechanisms of yielding changes from large-scale yielding in unfilled epoxies with no void formation to yielding with shear bands and gradual expansion of these shear bands to cover the entire sample [78]. Fracture toughness increased with clay concentration up to 5%, which was correlated to decreasing distance between particle distance (particle referred to self-aggregated clumps of groups of intercalated clays in the epoxy). The crack propagation front in the unfilled epoxies was straight, but the filled nanocomposites displayed significant crack branching along the length. The creation of new surfaces rather than changes in host matrix deformation was determined to be the leading contributor to the increased fracture toughness. They attributed the intercalated morphology to be a significant contributor to the increased toughness. Based on a prediction that effective toughening required a particle size not smaller than 0.1 μm [79], they inferred that an exfoliated structure was ineffective in toughening the matrix.

The mechanical properties of nanocomposites prepared by using hectorite, montmorillonite, and fluorohectorite were examined in early work [33]. Relative to the unmodified epoxy, the failure strength and modulus of the hectorite reinforced epoxies improved. Fluorohectorite showed a decrease in failure strength, but an improvement of tensile strength. Montmorillonite, relative to hectorite, showed a higher increase in tensile strength, but a low modulus. No values of failure strain were given.

6. SUMMARY

Epoxy nanocomposites are significantly influenced by the method of processing. Exfoliated and intercalated dispersions are obtained by changing the chemical environment of the clay + epoxy interaction. The strong dependence of the pH of the clay and the critical concentration of the clay + epoxy interaction point to a strong likelihood that the anisotropic charge distribution of the clay is operative in the epoxy. Significant improvements in modulus of the epoxy can be obtained, but the decreased strain to failure is of concern. Ongoing research is now concentrated on translating the nanocomposite technology to composite technology in the form of ternary composites.

GLOSSARY

Clay A family of 2:1 layered- or phyllosilicate materials. The layers made up of two silica tetrahedral fused to an edge-shared octahedral sheet of either aluminum or magnesium hydroxide.

Epoxy A class of thermosetting synthetic polymers containing epoxy groups. Epoxies are used in surface coatings, laminates, and adhesives.

Montmorillonite One of the most commonly used layered silicates having a chemical structure: $(M_{1/3}(Al_{5/3}Mg_{1/3})Si_4O_{10}(OH)_2)$. Montmorillonite is miscible with hydrophilic polymers, such as poly (ethylene oxide) and poly (vinyl alcohol).

Nan Phase mixing of atleast two dissimilar materials occurring on a nanometer scale.

ACKNOWLEDGMENTS

The author thanks the Army Research Laboratory and NASA for support of nanocomposite research in her laboratory.

REFERENCES

1. J. A. Manson and L. H. Sperling, "Polymer Blends and Composites." Plenum, New York, 1976.
2. Y. Kojima, K Fukumori, A. Usuki, A. Okada, and T. Kurauchi, *J. Mater. Sci. Lett.* 12, 889 (1993).
3. Y. Kojima, A. Usuki, M. Kawasumi, A. Okada, Y. Fukushima, T. Kurauchi, and O. Kamigaito, *J. Mater. Res.* 8, 1185 (1993).
4. Y. Kurokawa, H. Yasuda, M. Kashiwagi, and A. Oyo, *J. Mater. Sci. Lett.* 16, 1670 (1997).
5. K. Yano, A. Usuki, A. Okada, T. Kurauchi, and O. Kamigaito, *J. Polymer Sci.* 31, 249 (1993).
6. A. Usuki, Y. Kojima, M. Kawasumi, A. Okada, Y. Fukushima, T. Kurauchi, and O. Kamigaito, *J. Mater. Res.* 8, 1179 (1993).
7. A. Usuki, Y. Kojima, M. Kawasumi, A. Okada, Y. Fukushima, T. Kurauchi, and O. Kamigaito, *J. Mater. Res.* 8, 1174 (1993).
8. Y. Kojima, A. Usuki, M. Kawasumi, A. Okada, T. Kurauchi, and O. Kamigaito, *J. Polymer Sci.* 31, 98 (1993).
9. Y. Kojima, A. Usuki, M. Kawasumi, A. Okada, O. Kurauchi, O. Kamigaito, and K. Kaji, *J. Polymer Sci.* 32, 625 (1994).
10. Y. Kojima, T. Matsuoka, H. Iakahashi, and T. Kurauchi, *J. Appl. Polymer Sci.* 51, 68 (1994).
11. R. Dagani, *Chem. Eng. News*, 77, 25 (1999).
12. B. Miller, *Plastics Formulating Compounding* 30 (May/June 1997).
13. L. M. Sherman, *Plastics Technol.* 52 (June 1999).
14. H. Bidadi, P. A. Schroeder, and T. J. Pinnavaia, *J. Phys. Chem. Solids* 49, 1435 (1998).
15. R. Krishnamoorti and E. P. Giannelis, *Macromolecules* 30, 4097 (1997).
16. E. P. Giannelis, *Adv. Mater.* 8, 29 (1996).
17. S. D. Burnside and E. P. Giannelis, *Chem. Mater.* 7, 1597 (1995).
18. R. A. Vaia, Polymer Melt Intercalation in Mica-Type Layered Silicates, Ph.D. Dissertation, Cornell University, May 1995.
19. R. Krishnamoorti, R. A. Vaia, and E. P. Giannelis, *Chem. Mater.* 8, 1728 (1996).
20. R. A. Vaia, B. B. Sauer, O. K. Tse, and E. P. Giannelis, *J. Polymer Sci.* 35, 59 (1997).
21. R. A. Vaia, S. Vasudevan, W. Krawiec, L. G. Scanlan, and E. P. Giannelis, *Adv. Mater.* 7, 154 (1995).
22. R. A. Vaia, K. D. Jandt, E. J. Kramer, and E. P. Giannelis, *Chem. Mater.* 8, 2628 (1996).
23. R. A. Vaia and E. P. Giannelis, *Macromolecules* 30, 8000 (1997).
24. R. A. Vaia and E. P. Giannelis, *Macromolecules* 30, 7990 (1997).
25. Z. Wang and T. J. Pinnavaia, *Chem. Mater.* 10, 1820 (1998).
26. M. S. Wang and T. J. Pinnavaia, *Chem. Mater.* 6, 468 (1994).
27. T. Lan, P. D. Kaviratna, and T. J. Pinnavaia, *J. Phys. Chem. Solids* 57, 1005 (1996).
28. P. D. Kaviratna, T. J. Pinnavaia, and P. A. Schroeder, *J. Phys. Chem. Solids* 57, 1897 (1996).
29. P. B. Messersmith and E. P. Giannelis, *J. Polymer Sci.* 33, 1047 (1995).
30. T. Lan and T. J. Pinnavaia, *Chem. Mater.* 6, 2216 (1994).
31. Z. Wang and T. J. Pinnavaia, *Chem. Mater.* 10, 3769 (1998).
32. T. Lan, P. D. Kaviratna, and T. J. Pinnavaia, *Chem. Mater.* 6, 57 (1994).
33. T. Lan, P. D. Kaviratna, and T. J. Pinnavaia, *Chem. Mater.* 7, 2144 (1995).
34. E. P. Giannelis, *J. Minerals* 44, 28 (1992).
35. P. B. Messersmith and E. P. Giannelis, *Chem. Mater.* 6, 1719 (1994).
36. Z. Shen, Nanocomposites of Polymers and Layered Silicates, Ph.D. Dissertation, Monash University, Australia, Oct. 2000.
37. X. Kornmann, H. Lindberg, and L. A. Berglund, *Polymer* 42, 1303 (2001).
38. S. D. Burnside, H. C. Wang, and E. P. Giannelis, *Chem. Mater.* 11, 1055 (1999).
39. E. P. Giannelis, *Chem. Mater.* 2, 627 (1990).
40. P. B. Messersmith and E. P. Giannelis, *Chem. Mater.* 5, 1064 (1993).
41. K. H. Tan, "Principles of Soil Chemistry." Marcel Dekker, 1998.
42. H. van Olphen, "An Introduction to Clay Colloid Chemistry," pp. 57–76. Wiley, 1977.
43. A. Ranade, M.S. Thesis, University of North Texas, 2001. http://www.library.unt.edu/theses/open/20012/ranade_ajit/thesis.pdf.
44. Z. Wang, T. Lan, and T. J. Pinnavaia, *Chem. Mater.* 8, 2200 (1996).
45. Z. Wang and T. J. Pinnavaia, *Chem. Mater.* 10, 1820 (1998).
46. G. Lagaly, *Solid State Ion* 22, 43 (1986).

47. E. Hackett, E. Manias, and E. P. Giannelis, *J. Chem. Phys.* 108, 7410 (1998).
48. P. Butzloff, N. A. D'Souza, D. Garrett, and T. D. Golden, *Polymer Eng. Sci.* 41, 1794 (2001).
49. O. Becker, R. Varley, and G. Simon, *Polymer* 43, 4365 (2002).
50. A. Usuki, T. Mizutani, Y. Fukushima, M. Fujimoto, K. Fukumori, Y. Kojima, N. Satao, T. Kurauchi, and O. Kamigaito, U.S. Patent 4,889,885, 1989.
51. D. C. Lee and L. W. Jang, *J. Appl. Polymer Sci.* 68, 1997 (1998).
52. X. Kornmann, H. Lindberg, and L. A. Berglund, *Polymer* 42, 4493 (2001).
53. L. Jiankun, K. Yucai, Q. Zongneng, and Y. Xiaosu, *J. Polymer Sci., Part B, Polymer Phys.* 39, 115 (2001).
54. C. S. Triantafillidis, P. C. LeBaron, and T. J. Pinnavaia, *J. Solid State Chem.* 167, 354 (2002).
55. T. B. Tolle and D. P. Anderson, *Composites Sci. Technol.* 62, 1033 (2002).
56. N. Salahuddin, A. Moet, A. Hiltner, and E. Baer, *Eur. Polymer J.* 38, 1477 (2002).
57. J. Mering, *Trans. Faraday. Soc.* 42, 205 (1946).
58. J. M. Brown, D. Curliss, and R. A. Vaia, *Chem. Mater.* 12, 3376 (2000).
59. G. W. Brindley and G. Brown, Eds., "Crystal Structures of Clay Minerals and Their X-Ray Identification," Monograph 5, pp. 125–195. Mineralogical Society, London, 1980.
60. D. M. Moore and R. C. Reynolds, Jr., "X-Ray Diffraction and the Identification and Analysis of Clay Minerals," 2nd ed., pp. 335–341. Oxford University Press, 1997.
61. S. J. Park, D. I. Seo, and J. R. Lee, *J. Colloid Interface Sci.* 251, 160 (2002).
62. J. Panzer, *J. Colloid Interface Sci.* 44, 142 (1973).
63. P. LeBaron, Z. Wang, and T. J. Pinnavaia, *Appl. Clay Sci.* 15, 11 (1999).
64. D. Kong and C. E. Park, *Chem Mater.* 15, 419 (2003).
65. C. S. Triantafillidis, P. C. leBaron, and T. J. Pinnavaia, *Chem. Mater.* 14, 4088 (2002).
66. T. J. Pinnavai, H. Z. Shi, and T. Lan, U.S. Patent 5, 866,645, 1999.
67. T. J. Pinnavai, H. Z. Shi, and T. Lan, U.S. Patent 6,261,640, 2001.
68. C. L. Chiang, C. C. M. Ma, F. Y. Wang, and H. C. Kuan, *Eur. Polymer J.* (2003).
69. J. Kim, K. Lee, K. Lee, J. Bae, J. Yang, and S. Hong, *Polymer Degradation Stability* 79, 201 (2003).
70. J. Y. Park and G. B. McKenna, *Phys. Rev. B* 61, 6667 (2000).
71. G. B. McKenna, "Conference Proceedings ANTEC, 2000," Vol. II, 2004.
72. G. B. McKenna, *J. Phys. IV France* 10, 7 (2000).
73. C. L. Jackson and G. B. McKenna, *Chem. Mater.* 8, 2128 (1996).
74. O. K. C. Tsui, T. P. Russsell, and C. J. Hawker, *Macromolecules* 34, 5535 (2001).
75. W. Xie, Z. Gao, W. Pan, R. A. Vaia, D. Hunter, and A. Singh, *Thermochimica Acta.* 367, 229 (2001).
76. A. Ranade, N. A. D'Souza, and B. Gnade, *Polymer* 43, 3759 (2002).
77. T. J. Pinnavaia, J, Massam, Z. Wang, T. Lan, and G. Beall, *Abstr. Papers, Am. Chem. Soc.* 215, 216 (1998).
78. A. S. Zerda and A. J. Lesser, *J. Polymer Sci., Part B, Polymer Phys.* 39, 1137 (2001).
79. C. B. Bucknall, A. Karpodinis, and X. C. J. Zhang, *J. Mater. Sci.* 29, 3377 (1994).

Encyclopedia of Nanoscience and Nanotechnology
www.aspbs.com/enn

Equilibrium Properties of Mesoscopic Quantum Conductors

L. Saminadayar

Université Joseph Fourier, Grenoble, France and Centre National de la Recherche Scientifique, Grenoble, France

C. Bäuerle

Centre National de la Recherche Scientifique Grenoble, France

D. Mailly

Centre National de la Recherche Scientifique, Marcoussis, France

CONTENTS

1. INTRODUCTION

The emergence of mesoscopic physics has led to the discovery of many striking new phenomena in solid-state physics in the last two decades [1]. This field is also intimately related to the progress in fabrication techniques: the possibility of creating objects of submicron size has allowed us to fabricate and manipulate conductors which are fully coherent.

In solid-state physics, one usually considers *macroscopic* systems. This term often refers to the notion of the thermodynamic limit: the number of particles N and the volume of the system Ω both tend to infinity, whereas the ratio $n = N/\Omega$ is kept constant [2, 3]. This idea is also closely related to the physical size of the system: the sample is considered as macroscopic as soon as its size is larger than some characteristic length, for example, the typical distance between two particles, $n^{-1/3}$. Below this size, the system is said to be *microscopic*.

It is well known from our daily experience that macroscopic objects obey classical mechanics, whereas microscopic ones are governed by quantum mechanics. This dichotomy between microscopic and macroscopic behavior is quite familiar: small particles exhibit wave-like attributes, and they must be described by quantum mechanics that allows for wave behavior like diffraction or interference. Electrons have been observed to interfere in many experiments in vacuum. However, if one considers a large number of electrons in a disordered medium, like a macroscopic piece of metal at room temperature, the conductivity is described in a classical way *via* the Boltzmann equation, which leads to the Drude formula.

The question is then: is it possible to observe the wave-like behavior of the electrons in a solid? Actually, the characteristic length which is relevant is the length over which the electronic wave keeps a well-defined phase, namely, the *phase-coherence length* l_ϕ. This phenomenon is well known in optics: incoherent light cannot give rise to interference patterns.

At room temperature, the phase-coherence length of an electron in a metal is on the order of a nanometer, roughly the $n^{-1/3}$ factor mentioned above. However, at low temperature, let us say below 1 K, this phase-coherence length increases, and may reach several micrometers in metals or

ISBN: 1-58883-059-4/$35.00

even more than 10 μm in the highest quality semiconductor heterojunctions. Combined with the progress of fabrication techniques, this allows us to observe the quantum behavior of the electrons in solids. It should be stressed that a micron size sample is, in a sense, really macroscopic when compared to microscopic scales (for example, the interatomic distances): if a mesoscopic sample behaves like a large molecule to some extent, it still contains a rather large number of atoms and electrons (more than 10^{20}). However, as the electronic wavefunction is fully coherent over the whole sample, this sample is really a *quantum conductor* [4, 5].

It is important to point out that the physics of such a system is completely different from the physics of free electrons in vacuum. First, as we said, the sample is macroscopic, and some notions, reminiscent of the standard solid-state physics, are still relevant: the Fermi wavelength, Fermi level, Fermi velocity, or chemical potential still make sense; more importantly the energy spectrum is discrete or, at least, the interlevel spacing $\Delta \approx L^{-d}$, where L is the size of the system and d is the dimensionality, may become comparable or even smaller than the temperature T. Secondly, disorder is still present in solids at the micron scale: thus, the electronic wave propagates in a random medium, and its electronic motion is still diffusive.

It is often tempting to make analogies between interference phenomena in mesoscopic physics and optics. Such a comparison may sometimes be appropriate: for example, the Young slit experiment and the Sagnac effect are very similar to some transport experiments on mesoscopic rings like the Aharonov–Bohm conductance oscillations [6, 7] or the quantization of the conductance, which may be understood in the light of the theory of waveguides [8]. There are, however, two important differences: first, electrons are fermions, and this obviously strongly affects the energy spectra of mesoscopic samples, and consequently their transport and thermodynamic properties; second, electrons are charged particles, and couple to the vector potential of the electromagnetic field. This provides a powerful tool to control interference effects simply by applying a magnetic flux.

This chapter is organized as follows: in the first part, we give the fundamental length scales which are important in mesoscopic physics. We then give an overview of the different materials commonly used in this field. In the second part, we present a comprehensive overview of the different fabrication techniques. Finally, the third part is devoted to the thermodynamics of theses mesoscopic systems.

2. MESOSCOPIC SYSTEMS

2.1. Mesoscopic Samples

2.1.1. Characteristic Lengths

Mean-free path A mesoscopic sample is a disordered sample: even at zero temperature, electrons are scattered by static defects like impurities, grain boundaries, or the edge of the sample. Such events are elastic scattering in the sense that their energy is conserved during the collision. The disorder just acts like a static, random potential which adds to the lattice potential. In such a system, Bloch states are no more eigenstates, but the system is still Hamiltonian. It should be pointed out that the translational invariance of the crystal lattice is destroyed by such defects, but this usually does not affect the electronic properties of the system. The typical length associated with these scattering processes is l_e, often called the mean-free path. The time associated with these collisions is τ_e, and they are related via the relation $l_e = v_F \tau_e$, v_F being the Fermi velocity.

On the contrary, other collisions are inelastic in the sense that the energy of an electron is not conserved. Such processes are irreversible, and are related to the coupling of the electrons with their environment, that is, other electrons, phonons, or photons. The inelastic length is given by $l_{\text{in}} = \min\{l_{e-e}, l_{e-\text{photon}}, l_{e-\text{phonon}}\}$. At high temperature (typically above 1 K), the dominant mechanism is electron–phonon scattering. At low temperature, however, the dominant process is electron–electron scattering [2, 3].

Another important source of decoherence is electron–photon scattering. This is especially the case in micrometer size samples, where small dissipation (10^{-15} W) is sufficient to heat the conduction electrons at very low temperatures. Extreme care should therefore be taken for external radiofrequency filtering [9–11] when working in the millikelvin temperature range.

Phase-coherence time After an inelastic scattering event, the energy of the electron changes, and the phase of the wavefunction is randomly distributed between 0 and 2π; thus, the quantum coherence is lost, and the phase-coherence time is mainly limited by the inelastic time $\tau_\phi \approx \tau_{\text{in}}$. It is important to note, however, that elastic scattering also leads to dephasing: the wavevector $\vec{k}$ changes to $\vec{k'}$ after such a diffusion; elastic scattering implies only that $|\vec{k}| = |\vec{k'}|$, but there is a priori no condition on their respective directions. The point is that this dephasing is perfectly deterministic and reproducible: two successive electrons with the same wave vector $\vec{k}$ will be scattered and dephased in exactly the same way, which could be calculated if the scattering potential were known. The phase coherence is thus preserved, and interference effects are not destroyed. On the contrary, inelastic scattering depends on the state of the environment the electron interacts with at the time of the interaction. In this case, the dephasing is random, and the phase coherence is lost. This is why, at room temperature, the dominant scattering process is the electron–phonon scattering, and the phase-coherence length is very short, typically $l_\phi \approx$ 1–10 nm. In the frawework of Fermi liquid theory, the available phase space at low temperaturc tends to zero. As a consequence, electron–electron, electron–photon, and electron–phonon couplings all tend to zero, and hence, the phase-coherence length should diverge [12]. Recent experiments, however, seem to show that this is not the case. Presently, there is still an ongoing debate concerning this point, and we will not address this issue in this chapter [13].

Finally, it is important to mention magnetic impurities. As they are static defects at low temperature, scattering by magnetic impurities is elastic as the energy of the electron is conserved. However, the electronic spin is flipped in such a collision, and the phase coherence may be lost. The exact effect of magnetic scattering on the phase coherence time, especially when entering the Kondo regime, is far from being understood [14]. In this chapter, we will not elaborate on this point, and will consider only systems containing no magnetic impurities.

Thermal diffusion length and the Thouless energy At distances beyond the elastic mean-free path l_e, electrons propagate in a random medium. This diffusive nature of the movement is characterized by the diffusion coefficient $D = (1/d)v_F^2\tau_e = (1/d)v_F l_e$, where d is the dimensionality of the sample. To propagate over a distance L, an electron then needs a diffusion time $\tau_d = L^2/D$. In a semiclassical picture, each diffusion path l is characterized by a probability $\Psi_l = |\Psi_l| \exp(iS_l/\hbar)$, where $S_l = \int_l \vec{k}d\vec{l} - Et_l$, with E being the energy of the electron and t_l the diffusion time along the path l. Over the whole sample of size L, this diffusion time is then simply τ_D. If one considers an energy range larger than $2\pi\hbar/\tau_D$, the phase of the electrons in this energy range will be distributed between 0 and 2π, and interference effects will not be observable anymore. This defines the Thouless energy (or correlation energy) $E_c = h/\tau_D = hD/L^2$. When the energy range involved is larger than E_c (e.g., when $k_BT \geq E_C$), interference effects do *not* disappear; they are simply no longer observable [15]. If the size of the sample is smaller than l_e, the time for an electron to travel across the sample becomes simply L/v_F, and the Thouless energy simply expresses as $E_c = hv_F/L$.

2.1.2. Disorder Configurations

In a macroscopic sample, one usually characterizes the disorder by some characteristic length, say the elastic mean-free path l_e. Such a parameter is relevant when considering the disorder from a "global" point of view. From a mesoscopic point of view, things may be quite different: the electronic wavefunctions are fully coherent over the whole sample, and the acquired phase depends on the precise path one electron follows. Thus, the interference pattern depends on the *microscopic* disorder configuration of the sample. Moving even a single impurity drastically affects the electronic properties of the sample. That is why two samples identical from a macroscopic point of view may behave in a completely different way due to their microscopic individuality (their *fingerprints*). This phenomenon is equivalent to the speckles observed when a coherent light beam diffracts in a random medium.

2.1.3. Quantum Coherence and the Effect of Aharonov–Bohm Flux

The most important parameter that physicists can use to probe a mesoscopic sample is the magnetic flux. As an electron is a charged particle, it couples to the vector potential $\vec{A}$ (the momentum changes as $\vec{p} \to \vec{p} + e\vec{A}$ in the Hamiltonian, with e the charge of the electron) *even if the magnetic field $\vec{B}$ is zero* ($\vec{B} = \vec{\nabla} \times \vec{A} = \vec{0}$). Note, however, that as the field is zero, or at least very weak in all of the experiments,[1] the effect of the magnetic field on the trajectories of the electrons is negligible. When propagating along a path i, the wavefunction Ψ acquires a phase simply given by $S = \int_i (\vec{k}(\vec{r}) + e\vec{A}(\vec{r}))d\vec{r}$. The first term is simply the equivalent of the optical path, whereas the second one characterizes the quantum coupling of the charge with the magnetic flux. This shows how applying a small magnetic field can indeed control the interference pattern of a mesoscopic sample [16]. There is no equivalent of such a possibility in optics: this is a powerful way to play with the quantum, wave-like nature of the electrons.

2.2. Materials

2.2.1. Ballistic Versus Diffusive Versus Localized

The different length scales for a mesoscopic sample are the Fermi wavelength λ_F, the elastic mean-free path l_e, and the size of the sample L. The ratio between λ_F and l_e characterizes the strength of the disorder: for $\lambda_F \ll l_e$ (or, equivalently, $k_F l_e \gg 1$ or $h/\tau_e \ll E_F$), the disorder is said to be "weak," whereas for $k_F l_e \ll 1$, the disorder is said to be "strong."

Considering the ratio between these different length scales, one can distinguish different regimes for a mesoscopic sample.

Ballistic regime ($\lambda_F \ll l_e$ and $L \leq l_e$) In this regime, the disorder is very weak, and the elastic mean-free path is on the order of the size of the sample. In this case, the phase coherence length is mainly limited by electron–electron collisions. The trajectories of the electrons is mainly governed by the shape of the sample, implying that the reflections at the edges of the samples are specular. In this case, transport properties as well as equilibrium properties depend on the shape of the sample. Such systems are powerful tools to probe the energy spectra of quantum billiards.

Diffusive regime ($\lambda_F \ll l_e \ll L$) In such systems, electrons experience a large number of collisions during the traversal of the sample. Their movement is rather a Brownian motion, a random walk between impurities. The phase coherence length is then given by $l_\phi = \sqrt{D\tau_\phi}$. In this regime, the exact shape of the sample does not affect its electronic properties; only its size is relevant.

Localized regime In the case of a strong disorder, Anderson has suggested that each electron is confined in a part of the sample, and cannot travel through it: its wavefunction is exponentially decreasing on a length scale ξ, and the electron is localized in a domain of size ξ^d, with d the dimensionality of the sample, and the sample becomes an insulator [17]. For $d = 3$, there is a critical value for the disorder below which the sample becomes insulating, and one observes a metal-to-insulator transition. For $d = 1$ and $d = 2$, on the other hand, electrons are localized for an arbitrary small disorder [18]. Recent experiments, however, show that there is indeed a metal-to-insulator in some two-dimensional electron gas. As both the experimental and theoretical situations are at least unclear, we will not address this topic in this chapter [19].

Two limits are then to be considered[2]: when $\xi \prec L$, electrons are confined in some regions of the sample, and

[1] Except in the case of the quantum Hall effect, which we will not address in this chapter.

[2] Note that one always has $\xi \geq l_e$.

conduction occurs by hopping from domain to domain. On the other hand, when $\xi \succ L$, localization domains are larger than the sample: electrons are indeed localized, but they can still explore the whole sample.

2.2.2. Dimensionality

One defines the dimensionality of a sample by comparing its size with the intrinsic characteristic lengths [20]. Usually, the most relevant length scale is the Fermi wavelength. Considering a rectangular sample of sizes L_x, L_y, and L_z, with $L_x \prec L_y \prec L_z$, one has

$$\lambda_F \ll L_x \prec L_y \prec L_z\text{: 3D (bulk samples)}$$

$$L_x \leq \lambda_F \ll L_y \prec L_z\text{: 2D (films)}$$

$$L_x \prec L_y \leq \lambda_F \ll L_z\text{: 1D (quantum wires)}$$

$$L_x \prec L_y \prec L_z \leq \lambda_F\text{: 0D (quantum dots)}$$

Such a definition is certainly the most relevant from a microscopic point of view. Note, however, that when considering transport properties, and due to the quantum nature of a mesoscopic conductor, one can also define the dimensionality of a sample by comparison with the phase-coherence length:

$$l_\phi \ll L_x \prec L_y \prec L_z\text{: 3D}$$

$$L_x \leq l_\phi \ll L_y \prec L_z\text{: 2D}$$

$$L_x \prec L_y \leq l_\phi \ll L_z\text{: 1D}$$

2.2.3. Metals

Metals have a high charge carrier density of about 10^{22} cm^{-3}. Because of this high carrier density, the Fermi wavelength is very short, in the range of the angström. Moreover, it is impossible to use gates to modulate this electron density (a too important voltage would be necessary in the case of metals). Another consequence is that the Coulomb interaction is very efficiently screened on the scale of the Thomas–Fermi vector $q_{TF} = 2\pi e^2/\rho_0$, with ρ_0 the carrier density at the Fermi level. Even if metals can be very pure from a chemical point of view, the intrinsic disorder usually makes them diffusive conductors. The elastic mean-free path l_e is on the order of 1–100 nm, and the phase coherence length l_ϕ is on the order of a micrometer.[3]

At low temperature, some metals become superconductors. This provides a new degree of freedom and a wide variety of mesoscopic effects. In particular, the superconducting state is quite different on a mesoscopic scale as compared to its macroscopic equivalent.

2.2.4. Semiconductors

Bulk semiconductors In semiconductors, the carrier densities can range practically between 10^{14} and 10^{19} cm^{-3}. Moreover, this density can be controlled using metallic gates deposited at the surface of the sample or simply by varying the doping concentration.

In the case of very pure semiconductors, for example, those obtained by molecular beam epitaxy, the elastic mean-free path is basically limited by the distance between two doping impurities. This leads easily to l_e of ≈100 nm, whereas l_ϕ is on the order of several micrometers. Finally, another important difference between metals and semiconductors is that, in the latter, the effective mass of the electrons, which is related to the band structure, can be very small.

Heterojunctions To reduce the dimensionality of a conductor, one may reduce the thickness of the film itself. However, it is quite difficult to obtain real two-dimensional conductors on the scale of the Fermi wavelength λ_F. An alternative way consists of playing with the band structure of two different semiconductors. Using the impressive control of growth offered by the molecular beam epitaxy, it is possible to grow two different semiconductors on top of each other, especially if their lattice parameters are matched [21]. The most common example is GaAs and GaAlAs (III–V heterostructures), but there also exist II–VI heterostructures (CdTe/HgCdTe) or even IV–IV heterostructures (SiGe).

The different band structure, mainly the energy gap and the work function,[4] causes changes in the charge transfer between the two adjacent materials in order to equalize the electrochemical potentials. Electrons are attracted to the remaining holes, and the dipole layer formed at the interface leads to the band bending at the vicinity of the interface. True two-dimensional electron (or hole) gas at the scale of λ_F can be formed using this technique [22].

The spatial separation between charge carriers and doping impurities leads to very high-mobility materials.[5] The electronic density is typically in the range of 10^{11} cm^{-2}, leading to a relatively large Fermi wavelength, on the order of 300 Å. This large Fermi wavelength allows us to create easily true 1D or 0D structures. Moreover, the use of electrostatic gates on the top of the sample allows us to deplete the 2D electron gas underneath. Using this technique, one can modulate *in-situ* and in a reversible way the shape of the 2D electron gas, allowing us to create a wide variety of quantum devices, like quantum wires or quantum dots [20]. Moreover, the edges defined by electrostatic gates are by far less rough than those produced by etching techniques.

3. SAMPLE FABRICATION TECHNIQUES

3.1. The Size to Reach

Typically, one wants to be able to tailor samples with a size smaller than 1 μm. But the smallest size is not the ultimate goal since the roughness of the edge may play an important role. In metals, the Fermi wavelength λ_F is very short, and the roughness is always much larger than the Fermi wavelength. But in semiconductor samples, where λ_F can be several tens of nanometers, the roughness can be on

[3] In extremely clean metals, obtained by molecular beam epitaxy (MBE), the phase-coherence length can reach ≈20 μm at best.

[4] For example, in GaAlAs [23, 24], the gap E_g varies linearly with the concentration of aluminum as $E_g = 1.424 + 1.247x$ at 300 K, x being the concentration in aluminum. Moreover, the mismatch in the lattice parameter does not exceed 0.3%.

[5] The highest mobility achieved in GaAs-GaAlAs heterostructures [25] is $14 \cdot 10^6$ cm$^2 \cdot$ V$^{-1} \cdot$ s^{-1}. Mobilities of ≈10^6 cm$^2 \cdot$ V$^{-1} \cdot$ s^{-1} are currently achieved in this material.

the same order. In this case, one wants the edges to be defined with a precision much smaller than λ_F. As discussed in the previous paragraph, the dimensionality of the sample depends strongly on the physics involved. For interference effects, the phase-coherence length is the characteristic length which is on the order of 1 nm at low temperatures for a good metal and more than 10 μm for high-quality 2DEG. So typically, one wants to be able to fabricate samples with a width smaller than 100 nm for metal structures and a few hundred nanometers for semiconducting ones.

3.2. Nanofabrication Technique

First of all, let us recall a standard process flow. Figure 1 resumes the main steps one must follow. The starting material is the substrate, which can be the system one wants to pattern or just a flat and neutral surface used as a support. By spinning, the substrate is coated with a layer of resist.

The resist is a material sensitive to irradiation. After exposure, the resist is developed, and the exposed (nonexposed) area will be cleared off for the case of positive (negative) resist. A rich variety of processes can be done after the lithography. A commonly used process is the lift-off technique. In this case, one covers the whole patterned substrate with a metal, for instance. The resist is then completely removed by rinsing it with a strong solvant. Only the part which had been previously patterned will be covered by the metal, so one may say that we have replaced the design on the resist by a solid pattern made of metal. This metal can be just the structure wanted or can be used as a mask for a subsequent etching process. Other processes can also be used, such as ion implantation, electrochemical growth, and so on.

3.3. Optical Lithography

Optical lithography is the dominant lithography in industry. With this technique, UV light is shed through a mask, which contains the drawing information, on a resist. The

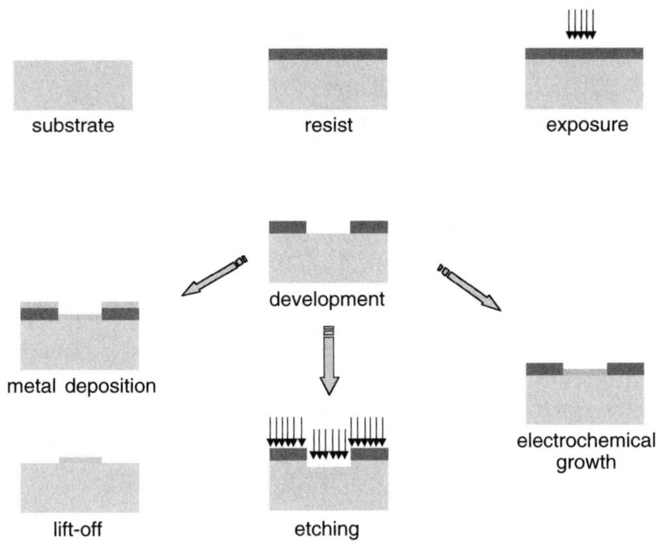

Figure 1. Typical flowchart for a fabrication process.

resolution is mostly limited by the diffraction, and hence depends on the wavelength of the light. This explains why short wavelengths are employed. Optical lithography, which started with UV (400–366 nm) is now in the DUV range (248–193 nm), and EUV (13 nm) is the next predicted range. DUV lithography can reach the sub-100 nm range, but with a complexity and costs which are too high for any scientific laboratory. For instance, the complexity of masks which use phase-shift techniques to overcome diffraction makes them difficult, and hence very expensive, to produce. Only mass production can afford such high costs. Refractive optics are presently not available in the EUV range. The fabrication of reflective optics at this wavelength is also very delicate, and it is hopeless that this technique will be inexpensive for laboratory use. Classical optical lithography, on the other hand, which uses a quartz plate mask directly pressed onto the resist with a standard DUV light, is not able to produce samples with a sub-100 nm resolution.

3.4. Electron-Beam Lithography

The possibility to finely focus an electron beam has been exploited in electron microscopy for a long time. Starting in the 1960s, focused electron beams have been used to expose a resist, and a 0.1 μm resolution was readily obtained. Ten years later, a 10 nm wide line was demonstrated using an inorganic resist. Unfortunately, this technique is essentially sequential: the electron beam is scanned pixel by pixel on the resist to draw the entire design; hence, the process is too slow to be included in an industrial processes. It is, on the other hand, the perfect technique for the laboratory. An advantage of this maskless technique is its versatility. The drawing can be easily changed on a computer with no additional cost.

In the following, we detail the electron-beam lithography to explain the resolution and limitations of this technique.

3.4.1. Resolution and Proximity Effect

Most of the resists employed in nanotechnology are polymers. The effect of the electron is to break the chain, hence leaving a polymer with a small chain, giving a better solubility. This resist is then sensitive to a very small energy compared to the one of the electron beam. Typically, one needs 10 eV to break a polymer chain, whereas for technical reasons, the focused electron beam is accelerated at several tens of kilovolts. It is then important to know how the electrons diffuse into the resist and lose their energy in order to understand how the resist is affected.

An analytical treatment is quite complex, especially in three dimensions. Monte Carlo simulations are widely used to follow the electron trajectory. Figure 2 shows electron trajectories obtained for electrons with energies of 10 and 20 kV in a silicon substrate covered with 400 nm of PMMA resist. The effects of electron diffusion are twofold. First, a forward diffusion which enlarges the spot in the resist is observed. Second, a backscattering diffusion, mainly from electrons diffused in the substrate back into the resist, but far from the initial impact of the electron, takes place. This latter effect, known as the proximity effect, has important consequences, as we will see later. The energy of the beam is quite important, as can be seen in Figure 2: higher energy

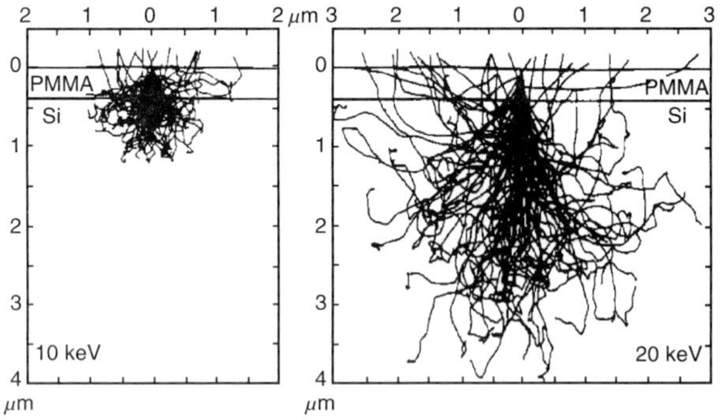

Figure 2. Monte Carlo simulation of electron trajectories in silicon substrate covered with PMMA at 10 and 20 keV. Adapted with permission from [28], D. F. Kyser and N. S. Viswanathan, *J. Vac. Sci. Technol.* 12, 1305 (1975). © 1975, Vacuum Society.

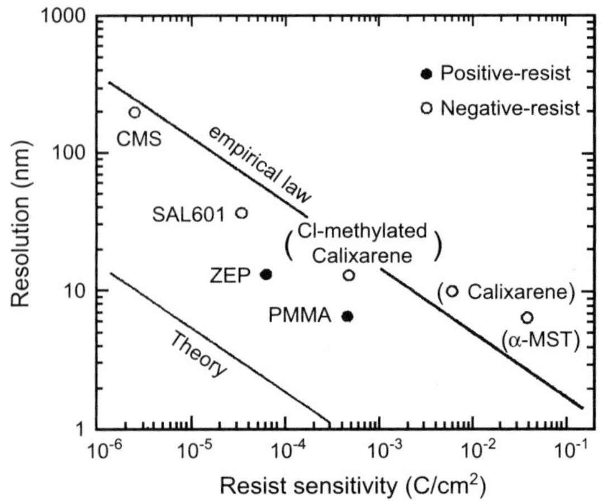

Figure 3. Resolution and sensitivity of various organic resists positive and negative for electron-beam lithography. Polymethylmetacrylate (PMMA) is presently the best organic resist, and is most widely used in scientific laboratories around the world. An example of sub-10 nm lines of PMMA with a 50 nm pitch is shown in Figure 4.

decreases the forward scattering angle and shrinks the effective beam spot. On the other hand, the electrons penetrate more deeply into the substrate as their energy increases, and are backscattered at larger distances from the impact. In other words, higher energy dilutes the proximity effect. This is the reason why recent electron-beam machines use a 100 kV source.

The total dose received by the resist at one point depends on the exposed dose at that point, but also on the vicinity around this point. Hence, a large square uniformly exposed, for instance, will be more dosed in the center than on the edge. It is also very difficult to expose two large patterns close to each other. The gap between these two patterns being exposed by proximity may result in an unwanted connection between them. Arrays of lines with a very small pitch are also very difficult. It is possible, however, to correct the dose at each point by calculating the proximity effect of the overall pattern. Softwares have been developed for that purpose, but cannot completely cure the effect of the diffusion since it may require a negative dose at certain points!

The problem of proximity effect arises from the sensitivity of organic resists to small energies. It is thus natural to try to use a resist which needs higher energy to be exposed. This is the case of an inorganic materials, for example, NaCl, AgF_2, or Al_2O_3. Such inorganic resists have been used to demonstrate the finest lines obtained by e-beam lithography, around 1 nm. The beam energy in that case gives rise to the partial or total sublimation of the resist. For instance, on AlF_3, the electron energy evaporates fluor, leaving a layer of aluminum. Hydrocarbon films have also been used where, under irradiation, a polymerization takes place. In most cases, the dose necessary to expose these types of resists is orders of magnitude higher than with a conventional resist. The total time to expose the pattern can reach nonreasonable values. Furthermore, this type of resist can be used only with thin layers which enable any lift-off process. Another possibility to avoid proximity effects is to use a very small energy. In this case, however, it is very difficult to focus the beam in conventional electron optics due to chromatic aberration. Another drawback is the forward scattering which rapidly enlarges the beam in the resist.

Figure 3 shows the best resolution obtained with different organic resists and the dose needed with e-beam lithography. One should keep in mind that the maximum current available in an e-beam system with a small spot size (less than 10 nm) is about 100 pA with a field effect gun source. This means that exposing an area of 100 μm $\times$ 100 μm at a nominal dose of 10^{-1} C/s takes more than 27 h!

An electron microscope with a computer-assisted deflection system is the basic tool for e-beam lithography. It is enough to make simple patterns in a single field. The available field size depends on the desired resolution. Lens aberrations induce severe distortions at the edge of the field which depend on the field size. With a conventional microscope, 50 μm $\times$ 50 μm is usually the maximum size one can afford to produce sub-100 nm structures. The nanostructure then has to stand within a single field since there is no possibility to displace the sample holder with enough accuracy to stitch with the previous writing field. The stability of the electron column is also a problem for long time exposure. Dedicated machines have been built to overcome the difficulties mentioned above with conventional electron microscopes. They include a laser-interferometry-controlled stage with an accuracy better than 1 nm to measure mechanical displacement. A feedback to the electron deflection is usually chosen for the field alignment. The overall field stitching accuracy is on the order of 20 nm. Using patterned marks on the sample, it is also possible to align several layers of lithography. The mark detection system combined with the laser interferometry also allows us to calibrate the deflection amplifier and to correct field distortion. It is simply done by moving a mark at different positions in the field. The exact position of the mark is known using the laser interferometry, and is compared to the position of the mark obtained by deflecting the beam. All of these essential features explain that there is at least one order of magnitude in the price of such a machine compared to a standard electron microscope.

3.5. Other Charged Particle Lithography

Focused ion beam lithography arose rapidly after electron-beam lithography as a good candidate for nanofabrication. Ions offer several advantages compared to electrons. First,

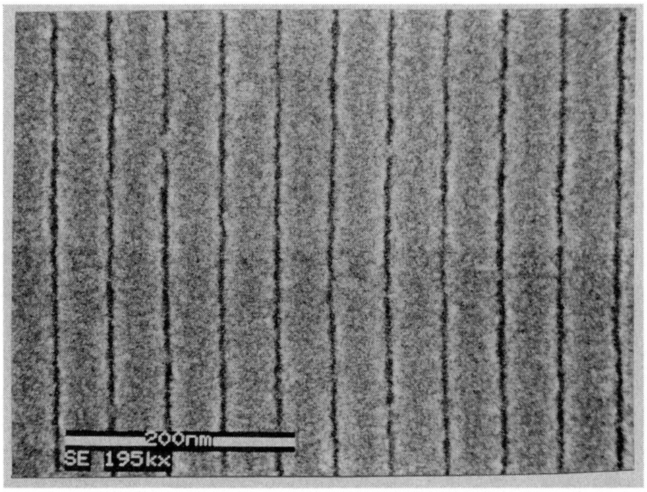

Figure 4. The image shows an example of sub-10 nm lines with a 50 nm pitch. Reprinted with permission from [26], C. Vieu et al., *Microwave Elect. Eng.* 35, 253 (1997). © 1997, American Institute of Physics.

they very quickly deliver their energy, and consequently, a much smaller dose is necessary. Second, the throughput is much better, and the proximity effect is much smaller. In addition, ions can directly erode the material, and is a resist-free process which can be very interesting for materials which are sensitive to pollution by organic materials.

In this etching mode, by varying the dose, it is also possible to produce three-dimensional structures: the paradigm of nanofabrication. Finally, at higher energy, one can locally implant atoms. On the other hand, ion lithography did not take an important place among the nanofabrication techniques. The major reason is the difficulty in producing fine spots with enough current and good stability. Recent progress in ion optics and ion source technology succeed in producing sub-10 nm spot sizes with a particle density sufficient for etching. Figure 5 shows an 8 nm line produced by such a high-resolution ion system [27].

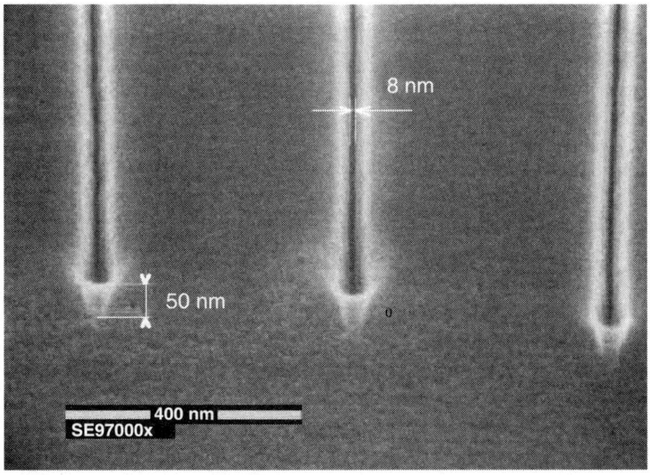

Figure 5. 8 nm line in GaAs produced using a 20 kV gallium focused ion beam. Adapted with permission from [27], J. Gierak et al., *Microwave Elect. Eng.* 57–58, 865 (2001). © 2001, American Institute of Physics.

3.6. Near-Field Techniques

Shortly after their discovery, near-field techniques have been used to produced nanostructures. The ultimate resolution has been obtained by the IBM group, who wrote the acronym of their company using Xenon atoms with a scanning tunneling microscope (STM). But such nanostructures are unfortunately very volatile. An STM can also be used in a more conventional way as a source of focused electrons. Indeed, the size of the electron spot of an STM is approximately on the order of the sample-to-tip distance, and a 10 nm spot size can readily be obtained.

Exposure of the resist is complicated by the fact that most of the resists are nonconducting. Another problem due to the proximity of the tip to the resist is the swelling of the resist under irradiation that can damage the tip. STM lithography is used more without resist by electrochemical process. For instance, it is possible to remove a group of atoms by applying a pulse on a gold surface [31]. Local oxidation is also an electrochemical process that is widely used with atomic force microscopy (AFM) lithography. The native water film on the surface of a sample at room temperature is the medium for this anodization process. In GaAlAs/GaAs samples, it is possible, for instance, to locally oxidize the surface, and the oxide formation destroys the two-dimensional electron gas beneath. Several mesoscopic structures have been produced with this technique [32]. Another example is the use of niobium, which can be anodized [33]. One of the advantages of this technique is that, using a small voltage on the tip, one can visualize the structures obtained at a higher voltage. Usually, one is limited to a small writing field because of the hysteresis of the piezodisplacement. In most cases, this technique is combined with other techniques like optical lithography. The possibility of visualization, with the AFM in the nonwriting mode, allows for the alignment of the two steps.

4. PERSISTENT CURRENTS: THEORETICAL ASPECTS

Usually, one considers *transport* properties of quantum conductors, measured by connecting voltage and current probes to the sample. In this case, however, two important properties of such a measurement must be pointed out.

- First, the strong coupling between these voltage and current probes certainly affects the quantum properties of the sample, and thus the measurement itself.
- Second, in a transport experiment, one only probes an energy range eV around the Fermi energy, with e being the electron charge and V the applied voltage. That is why one cannot access the entire energy spectrum

It is therefore very interesting to deal with the *equilibrium* properties of mesoscopic samples. It has to be stressed that such experiments in the field of mesoscopic physics are by far much more difficult than transport experiments. This is why there is only a very small number of experimental data available.

The existence of persistent currents was first suggested by London in 1937 [34], in his studies on the diamagnetism of aromatic rings (benzene rings). In 1938, Hund suggested that such an effect could be present in clean, metallic samples at low temperature [35]. The amplitude of the persistent currents was first calculated by Bloch and Kulik in the case of a clean, 1D ring [36, 37], but their existence in a real, diffusive 3D metallic ring was only been predicted by Büttiker et al. [38] in 1983.

It is important to note that the persistent current we are considering here is a *nondissipative* current flowing in a *nonsuperconductor* ring. Another interesting point is that persistent currents and orbital magnetism are two phenomena completely equivalent from a physical point of view. Only the geometry of the sample makes one term or the other more "intuitive."

4.1. A Simple Picture: The 1D Ballistic Ring

The simplest model for the persistent currents is the case of a pure, 1D metallic ring, without disorder. Although somewhat "academic," this example allows us to present the main idea of the problem. Let us consider a ring of perimeter L pierced by a magnetic flux Φ. We take the ring to be smaller than the phase-coherence length l_ϕ, and we neglect its self-induction. The Hamiltonian for the electrons of the rings is then simply given by

$$\mathcal{H} = \frac{1}{2m}[\vec{p} - e\vec{A}]^2 + V(\vec{r}) \tag{1}$$

where $\vec{p}$ is the momentum of the electron, e is its charge, $\vec{A}$ is the vector potential, and $V(\vec{r})$ is the periodic potential of the lattice. A simple gauge transformation $\vec{A} \Rightarrow \vec{A} + \vec{\nabla}(\int \vec{A} \cdot d\vec{l})$ leads to the Hamiltonian of free electrons $\mathcal{H}_0 = p^2/2m + V$, whereas a phase is added to the wavefunction Ψ: $\Psi(x) \Rightarrow \Psi(x)\exp(ie/\hbar \int \vec{A} \cdot d\vec{l})$. This wavefunction then obeys the new boundary conditions [39]

$$\Psi(x+L) = \Psi(x)\exp\left(i\frac{e}{\hbar}\oint \vec{A}\cdot d\vec{l}\right) = \Psi(x)\exp\left(2i\pi\frac{\Phi}{\Phi_0}\right) \tag{2}$$

where $\Phi_0 = h/e$ is the flux quantum. These boundary conditions also lead to a new quantization for the wave vector: $k = 2\pi/L(n + \Phi/\Phi_0)$.

It should be noted that, in this case, the boundary conditions can be controlled simply by varying the magnetic flux. Moreover, such boundary conditions show that wavefunctions, eigenenergies, as well as *any thermodynamic property of the system are periodic with magnetic flux [40], with periodicity* $\Phi_0 = h/e$.

In analogy with known results on Bloch's states, one can define a velocity for each energy level [38–41]:

$$v_n = \frac{1}{\hbar}\frac{\partial \varepsilon_n}{\partial k} = \frac{L}{e}\frac{\partial \varepsilon_n}{\partial \phi} \tag{3}$$

This velocity is equivalent to a current which is given by

$$i_n = -\frac{ev_n}{L} = -\frac{\partial \varepsilon_n}{\partial \phi} \tag{4}$$

At zero temperature, the net current is then simply the sum of the currents carried by the N levels:

$$I_N = \sum_{n=0}^{N} i_n = \sum_{n=0}^{N} -\frac{\partial \varepsilon_n}{\partial \phi} = \frac{\partial E(N,\Phi)}{\partial \phi} \tag{5}$$

where E is the total energy of the N electrons of the ring. However, as can be seen in Figure 6, two consecutive levels carry two currents of the same amplitude, but of opposite sign: the net current is then simply given by the last occupied level [47], that is, the Fermi level. We thus obtain for the amplitude of the persistent current

$$I_0 = \frac{ev_F}{L} \tag{6}$$

In this expression, v_F/L is simply the time needed for an electron to perform one turn around the ring. It should be noted that this expression can be rewritten as a function of the Thouless energy hv_F/L:

$$I_0 = \frac{ev_F}{L} = \frac{hv_F}{L}\frac{e}{h} = \frac{E_c}{\phi_0} \tag{7}$$

It should be stressed that the persistent current depends strongly on the number of electrons in the ring and on its parity, both in amplitude as well as in sign: for N

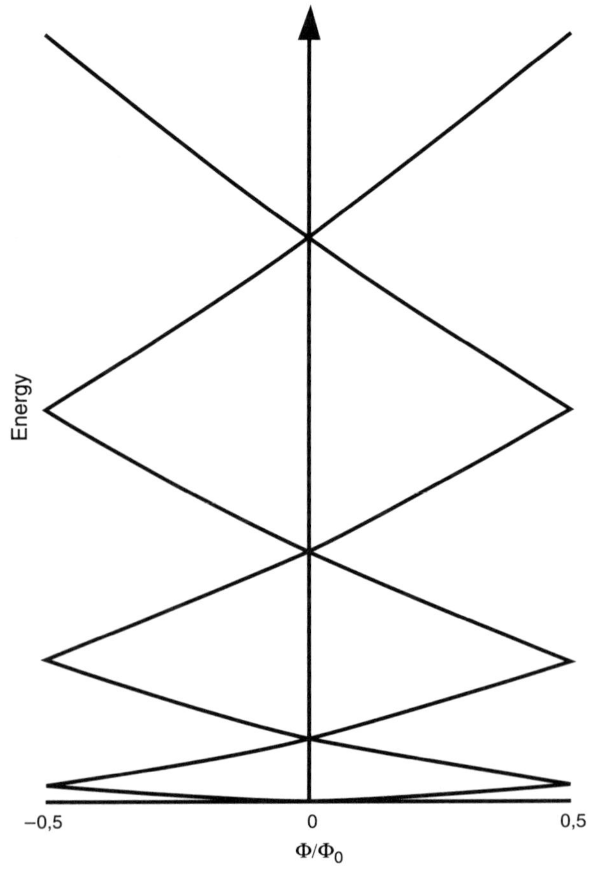

Figure 6. Energy spectrum of a pure, ballistic one-dimensional ring as a function of the magnetic flux.

even, the current is paramagnetic, whereas for N odd, it is diamagnetic.[6]

This very simple approach for the pure 1D ring allows us to give a good estimate for the order of magnitude of the persistent current. Moreover, the main features, such as the dependence on the parity of the number of electrons, remain true, even in the more realistic 3D, disordered ring.

4.2. Realistic Ring

4.2.1. Introduction

In this section, we will consider the case of diffusive, 3D rings. As we have stressed above, each sample is unique due to its specific disorder configuration. To take into account this unicity, we will consider a large number of rings, which is equivalent to averaging over disorder configurations: one obtains the *average current*. As we will see, this average current is measured in many-ring experiments. Fluctuations from this average value are also of interest, as they are accessible experimentally: this is called the *typical current*. This typical current is a good approximation of the current measured in single-ring experiments.

Another important property of the 3D case is the *spectral rigidity*. In 1D, we have seen that two successive energy levels have opposite slopes. In 3D, two successive levels repel each other: this correlation between energy levels leads to a correlation in the slopes of the energy levels [43], as can be seen in Figure 7. The slopes of two successive levels are almost identical, and this correlation extends over an energy range corresponding to the Thouless (or correlation) energy E_c. In other words, it is necessary to explore an energy range E_c to find a level of opposite slope [44, 45]. An important consequence of that is that the sign of the persistent current still depends on the number of electrons, but one has to add E_c/Δ electrons, with Δ being the mean level spacing, to reverse the sign of the persistent current [46].

4.2.2. Noninteracting Electrons

Average current The calculation of the average current [47] raises an interesting problem of statistical physics. In an experiment on many rings, the number of electrons in each ring is fixed, whereas the chemical potential μ is not. We are thus dealing with the *canonical* ensemble [48, 49]. This point is very important, as it has been shown that the persistent current calculated in the grand canonical ensemble (μ fixed) is exponentially small, $I \approx \exp(-L/2l_e)$.

On the other hand, the calculation in the canonical ensemble can be related to the calculation in the grand canonical ensemble, which is much easier to perform [1]. The canonical average persistent current is given by [50]

$$\langle I_N \rangle = -\frac{\partial \langle \mathcal{F} \rangle}{\partial \Phi}\bigg|_N = -\frac{\partial \langle \Omega \rangle}{\partial \Phi}\bigg|_\mu \tag{8}$$

where N is the number of electrons, $\mathcal{F}$ is the free energy, Ω is the grand potential, and Φ is the magnetic flux. With μ being the sample and flux-dependent chemical potential,

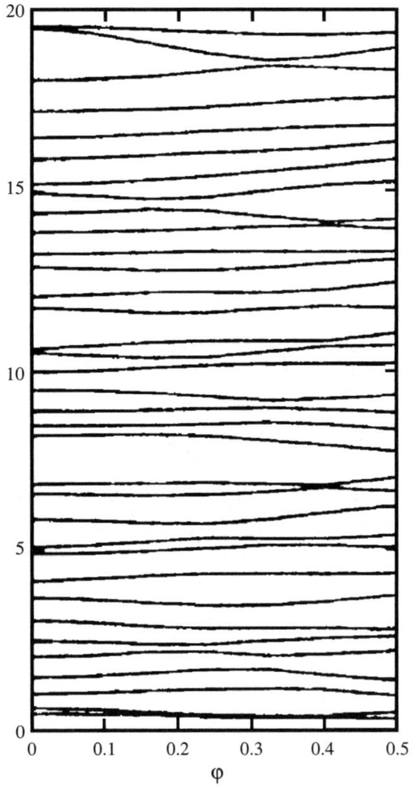

Figure 7. Energy spectrum of a real, three-dimensional diffusive ring as a function of the magnetic flux. Adapted with permission from [50], G. Montambaux, in "Quantum Fluctuations," 1996. © 1996, Elsevier Science.

one can then expand the expression (8) as a function of $\delta\mu(\phi) = \mu(\phi) - \langle \mu \rangle$, where $\langle \mu \rangle$ is flux independent:

$$-\frac{\partial \langle \Omega \rangle}{\partial \phi}\bigg|_\mu = -\frac{\partial \langle \Omega \rangle}{\partial \phi}\bigg|_{\langle \mu \rangle} - \delta\mu(\phi)\frac{\partial}{\partial \mu}\frac{\partial \langle \Omega \rangle}{\partial \phi}\bigg|_{\langle \mu \rangle} \tag{9}$$

$$= -\frac{\partial \langle \Omega \rangle}{\partial \phi}\bigg|_{\langle \mu \rangle} - \delta\mu(\phi)\frac{\partial}{\partial \phi}\frac{\partial \langle \Omega \rangle}{\partial \mu}\bigg|_{\langle \mu \rangle} \tag{10}$$

The first term is simply the grand canonical current, which is exponentially small, and will be neglected. The term $\partial\langle \Omega \rangle/\partial\mu$ corresponds to the number of electrons. Using the relation $\delta\mu = -\delta N \partial\mu/\partial N|_\phi$, we obtain

$$\langle I_N \rangle = -\frac{\partial \mu}{\partial N}\bigg|_\phi \left\langle \delta N \frac{\partial N}{\partial \phi}\bigg|_{\langle \mu \rangle} \right\rangle \tag{11}$$

where $\partial\mu/\partial N|_\phi$ is the level spacing. Finally, one obtains [52–55]

$$\langle I_N \rangle = -\frac{\Delta}{2}\frac{\partial}{\partial \phi}\langle \delta N_\mu^2 \rangle \tag{12}$$

The number of electrons is simply given by $N = \int_{-\varepsilon_F}^{0} \rho(\varepsilon)d\varepsilon$. The fluctuation in the number of electrons is

[6] This is the case for aromatic rings: for benzene ($N = 3$), for example, the persistent current is diamagnetic.

given by

$$\begin{aligned}\langle \delta N_\mu^2 \rangle &= \langle (N - \langle N \rangle)^2 \rangle \\ &= \left\langle \int_{-\varepsilon_F}^{0} (\rho(\varepsilon) - \rho_0) d\varepsilon \int_{-\varepsilon_F}^{0} (\rho(\varepsilon') - \rho_0) d\varepsilon' \right\rangle \\ &= \int_{-\varepsilon_F}^{0} \int_{-\varepsilon_F}^{0} (\langle \rho(\varepsilon)\rho(\varepsilon') \rangle - \rho_0^2) d\varepsilon d\varepsilon' \\ &= \int_{-\varepsilon_F}^{0} \int_{-\varepsilon_F}^{0} K(\varepsilon, \varepsilon') d\varepsilon d\varepsilon'\end{aligned}$$

where $K(\varepsilon, \varepsilon')$ is the two-point correlation function of the density of states. The average current is then given by

$$\langle I_N \rangle = -\frac{\Delta}{2} \frac{\partial}{\partial \phi} \int_{-\varepsilon_F}^{0} \int_{-\varepsilon_F}^{0} K(\varepsilon, \varepsilon') d\varepsilon d\varepsilon' \tag{13}$$

It has been shown [56] that the spectral form factor $\widetilde{K}(t)$ can be related to the return probability to the origin $P(t) = P(\vec{r}, \vec{r}, t)$:

$$\tilde{K}(t) = \frac{1}{4\pi^2} t P(t) \tag{14}$$

which leads to

$$\langle I_N \rangle = -\frac{\Delta}{4\pi^2} \frac{\partial}{\partial \phi} \int_0^\infty \frac{P(t)}{t} dt \tag{15}$$

This return probability contains two terms: the first one is flux independent, and will be ignored. The interference term $P_{\text{int}}(t)$ can be expressed as a function of the winding number of the different trajectories m:

$$P_{\text{int}}(t) = \sum_{m=-\infty}^{\infty} P_m(t) \cos\left(4\pi m \frac{\phi}{\phi_0}\right) \tag{16}$$

Inserting this into Eq. (15), one obtains

$$\langle I_N \rangle = \frac{2}{\pi} \frac{\Delta}{\phi_0} \sum_{m=1}^{\infty} m \sin\left(4\pi m \frac{\phi}{\phi_0}\right) \int_0^\infty \frac{P_m(t)}{t} dt \tag{17}$$

Knowing the expression for $P_m(t)$: $P_m(t) = 1/\sqrt{4\pi D t} \times \exp(-(m^2 L^2/4Dt))$, one finally obtains [57]

$$\langle I_N \rangle = \frac{2}{\pi} \frac{\Delta}{\phi_0} \sum_{m=1}^{+\infty} \sin\left(4\pi m \frac{\phi}{\phi_0}\right) \exp\left(-m \frac{L}{l_\phi}\right) \tag{18}$$

This current has a periodicity of $\Phi_0/2$, and is paramagnetic for small magnetic flux. It should be stressed, however, that the amplitude of this current is on the order of $\langle I_N \rangle \approx \Delta/\phi_0$. Taking a level spacing of 100 μK (for a metallic ring with a typical radius of 1 μm), one obtains a current in the range of the pA. Such a current would certainly not be measurable, and is by far much lower than the experimentally observed value.

Typical current The typical current I_{typ} is defined as the fluctuations around the average current [47]:

$$I_{\text{typ}}^2 = \langle I^2 \rangle - \langle I \rangle^2 \approx \sqrt{\langle I^2 \rangle} \tag{19}$$

Starting from the expression for the current

$$I = -\frac{\partial \mathcal{F}}{\partial \Phi} = \frac{\partial}{\partial \phi} \int_{-E_F}^{0} \varepsilon \rho(\varepsilon, \Phi) d\varepsilon \tag{20}$$

one obtains

$$\begin{aligned}I_{\text{typ}}^2 \approx \langle I^2 \rangle &= \frac{\partial}{\partial \Phi} \frac{\partial}{\partial \Phi'} \int_{-E_F}^{0} \varepsilon \varepsilon' \langle \rho(\varepsilon, \Phi) \rho(\varepsilon', \Phi') \rangle \, d\varepsilon d\varepsilon' &(21) \\ &= \frac{\partial}{\partial \Phi} \frac{\partial}{\partial \Phi'} \int_{-E_F}^{0} \varepsilon \varepsilon' K(\varepsilon - \varepsilon', \Phi, \Phi') \, d\varepsilon d\varepsilon' &(22)\end{aligned}$$

Performing a Fourier transform[7] and using again the relation (14), one obtains [50, 56]

$$I_{\text{typ}}^2 = \frac{1}{8\pi^2} \frac{1}{\phi_0^2} \int_0^\infty \frac{(t, \phi)}{t^3} dt \tag{23}$$

where P''_{int} denotes the second derivative of P_{int} with respect to Φ. Using Eq. (16), one finally obtains

$$\begin{aligned}I_{\text{typ}}^2 = \frac{96}{(2\pi)^2} \left(\frac{E_c}{\phi_0}\right)^2 \sum_{m=1}^{\infty} \frac{1}{m^3} \left[1 + m \frac{L}{l_\phi} + \frac{1}{3} m^2 \left(\frac{L}{l_\phi}\right)^2\right] \\ \times \sin^2\left(2\pi m \frac{\phi}{\phi_0}\right) \exp\left(-m \frac{L}{l_\phi}\right)\end{aligned} \tag{24}$$

Keeping only the first harmonic, and assuming $l_\phi \ll L$, we find for the typical current [57]

$$I_{\text{typ}} \approx \frac{\sqrt{96}}{2\pi} \frac{E_c}{\phi_0} \approx 1.56 \frac{E_c}{\phi_0} \tag{25}$$

This current is Φ_0 periodic. It should be noted that the amplitude is again of order E_c/Φ_0. This result can be rewritten as $I_{\text{typ}} \propto E_c/\Phi_0 \propto e/\tau_D \propto ev_F/L \cdot l_e/L$, where τ_D is the diffusion time. As derived in Section 4.1, the typical current is hence simply given by the time needed for an electron to perform one turn around the ring.

Extensivity One important property of the typical current is the fact that its amplitude increases only as $\sqrt{N_R}$, where N_R is the number of rings, since the typical current is given by the *fluctuations* around the average value.

On the contrary, the average current, such as any average value, grows simply as N_R. This has been extensively studied in the case of conductance oscillations [51], but is also true for thermodynamics properties.

4.2.3. Interacting Electrons

Motivated by the first experimental observations, where a much larger amplitude of the persistent current has been obtained than theoretically predicted, electron–electron interaction has been recognized as an important contribution to the persistent current [58]. The calculation is made in the Hartree–Fock approximation, and one assumes a screened Coulomb interaction [3?], $U(\vec{r} - \vec{r}') = U_{2D}\delta(\vec{r} - \vec{r}')$, with $U_{2D} = 2\pi e^2/q_{TF}$, q_{TF} being the Thomas–Fermi wavevector. In the Hartree–Fock approximation, the total energy E reads [59]

$$E = E^0 - \frac{U}{4} \frac{\partial}{\partial \Phi} \int n^2(\vec{r}) \, d\vec{r} \tag{26}$$

[7] We have omitted the classical part of P which does not depend on the flux.

where E^0 is the total energy for the noninteracting electrons. Given that[8] $n(\vec{r}) = 2\int_0^{\mu} \rho(\vec{r}, \omega)\, d\omega$, one finds for the interaction contribution to the average current

$$\langle I_{ee} \rangle = \left\langle \frac{\partial E}{\partial \Phi} \right\rangle = -U \frac{\partial}{\partial \Phi} \int \rho(\vec{r}, \omega_1)\rho(\vec{r}, \omega_2)\, d\vec{r}\, d\omega_1\, d\omega_2 \tag{27}$$

Again, this integral can be expressed as a function of $P(t)$:

$$\langle I_{ee} \rangle = -\frac{U\phi_0}{\pi} \frac{\partial}{\partial \Phi} \int_0^{\infty} \frac{P(t, \Phi)}{t^2}\, dt \tag{28}$$

Indexing by m the winding number of the trajectories, one obtains finally [60]

$$\langle I_{ee} \rangle = 16 \frac{U\rho_0}{2\pi} \lambda_0 \frac{E_c}{\phi_0} \times \sum_{m=1}^{+\infty} \frac{1}{m^2} \left[1 + m \frac{L}{l_\phi} \right] \times \sin\left(4\pi m \frac{\phi}{\phi_0} \right) \exp\left(-m \frac{L}{l_\phi} \right) \tag{29}$$

with ρ_0 being the average density of states at the Fermi level and $\lambda = U\rho_0$ the interaction coupling constant. In the simple limit $l_\phi \ll L$, and considering only the first harmonic, one finds an average current of the order E_c/Φ_0, a result obtained in the simple model of Section 4.1. It should be noted that this current is much larger than the noninteracting current calculated in Section 4.2.2. Another interesting point is that the prefactor is proportional to the interaction parameter U: this implies that the sign of the average current depends on the attractive or repulsive nature of the interaction. Finally, it should be stressed that this result is independent of the statistical ensemble: coulombian interactions locally fix the electron density [53, 58], leading to this insensitivity to the statistical ensemble. Calculations including exact coulombian interactions lead to somewhat unclear results [61–65].

We should also mention that the fluctuations of the persistent current (the typical current) are much larger than its average value (even when including the interaction term): $\sqrt{\langle I^2 \rangle} \gg \langle I \rangle$. However, the typical current for N_R rings varies as $\sqrt{N_R}$, whereas the average current varies as N_R: for few (or single)-ring experiments, the Φ_0 periodic typical current dominates, whereas for a large number of rings, the signal is dominated by the $\Phi_0/2$ periodic average current.

Finally, it should be stressed that the calculation of the typical current is made *only* for noninteracting electrons. Attempts to include coulombian interactions [66–68] are more difficult to perform and interpret.

5. EXPERIMENTAL RESULTS

5.1. Orders of Magnitude

Due to the experimental difficulty, only a few experimental studies on persistent currents are available. In the following, we give a review of these experiments. There are two distinct sets of experiments: first, the *many-ring experiments* which have been carried out on a very large number of rings, and second, the *single-ring experiments*. Both kind of experiments have been performed on metals and semiconductor heterojunctions. More recently, experiments have been carried out on a small number of rings.

Let us recall briefly the order of magnitude for the typical and average current. The typical current, for N_R rings, is given by

$$I_{\text{typ}} = \sqrt{\langle I^2 \rangle} = \frac{\sqrt{96}}{2\pi} \frac{ev_F}{L} \frac{l_e}{L} \sqrt{N_R} \approx 1.56 \frac{ev_F}{L} \frac{l_e}{L} \sqrt{N_R} \tag{30}$$

whereas the average current is given by

$$\langle I \rangle = \frac{16}{2\pi} \lambda \frac{E_c}{\phi_0} N_R \tag{31}$$

The coupling constant λ, when taking into account all of the orders of the interactions, is typically of the order 10^{-1}. This gives, for the average current,

$$\langle I \rangle \approx 0.25 \frac{ev_F}{L} \frac{l_e}{L} N_R \tag{32}$$

The average current, even when taking into account coulombian interactions, is one order of magnitude lower than the typical current.

In a metal, the Fermi velocity is on the order of $10^7\ \text{m}\cdot\text{s}^{-1}$, whereas for semiconductors, it is typically $10^5\ \text{m}\cdot\text{s}^{-1}$. The elastic mean-free path in a metal is typically 20 nm, and about 10 μm in a heterojunction. For N_R rings of radius 2 μm, one obtains the following.

- For metals:

$$I_{\text{typ}} \approx 0.3\sqrt{N_R}\,[nA] \approx 430\sqrt{N_R}\,[\mu_B] \tag{33}$$

$$\langle I \rangle \approx 0.05\, N_R\,[nA] \approx 70\, N_R\,[\mu_B] \tag{34}$$

$$E_c \approx 60\,\text{mK} \tag{35}$$

- For semiconductor heterojunctions:

$$I_{\text{typ}} \approx 2\sqrt{N_R}\,[nA] \approx 2700\sqrt{N_R}\,[\mu_B] \tag{36}$$

$$\langle I \rangle \approx 0.5\, N_R\,[nA] \approx 670\, N_R\,[\mu_B] \tag{37}$$

$$E_c \approx 380\,\text{mK} \tag{38}$$

For single- or few-ring experiments, the signal is dominated by the typical current, whereas for many-ring experiments, it is the average current which is measured. However, in all cases, the signal to be measured is rather small, and such experiments are always an experimental challenge.

In all of the experiments performed up to now, the signal detected is the magnetic flux generated by the persistent currents. Basically, two different techniques have been employed. First is the dc SQUID, either a macroscopic (standard) one [69] or an on-chip micro-SQUID [70]. The second one involves an RF resonator that allows us to detect, at the same time, both the magnetic flux generated and the "conductivity" of isolated rings.

[8] The factor 2 takes into account the spin.

5.2. Many-Ring Experiments

In many-ring experiments, at least when the number of rings is very large, the measured physical quantity is the average current, as it grows like the number of rings. These experiments are also easier to perform, as one deals with a "macroscopic" object, and hence the detector is simpler to design.

5.2.1. Metallic Rings

The first experimental observation of the existence of persistent currents was performed by Lévy and co-workers on a network of 10^7 copper rings [71], as shown in Figure 8. In this experiment, the rings were squares of perimeter 2.2 μm, which gives $\Phi_0 \approx 130$ G,[9] and the phase-coherence length was much larger than the perimeter of the rings.

The signal is detected using a commercial dc SQUID. It is crucial to eliminate the contribution due to magnetic impurities from the signal of the rings. For that purpose, the authors used the nonlinearity of the signal coming from the persistent currents: the magnetic field is modulated at low frequency, and the signal is detected as the second and third harmonic of the magnetic response. The procedure is repeated at several values of the magnetic field.

The experimental data are reported in Figure 9. The signal displays clear oscillations as a function of the magnetic field, with periodicity $\Phi_0/2$. The amplitude of the persistent current, deduced from the magnetic response, is 0.4 nA per ring, corresponding to $3 \cdot 10^{-3} ev_F/L$ per ring. This result, although somehow larger than predicted, is in relatively good agreement with theory, taking into account electron–electron interactions.

In this experiment, the determination of the sign of the magnetic response relies on some assumptions for the data processing. In the paper, the authors stated a *diamagnetic* response at zero field. This result is quite surprising as it would correspond to an *attractive* interaction, which is quite unlikely in a metal-like copper. On the other hand, this sign has been confirmed by a recent experiment on silver rings [72].

5.2.2. Semiconductor Rings

Another experiment has been performed on a large number of rings in a semiconductor heterojunction [73] (Fig. 10). The rings were 10^5 squares of mean perimeter 8 μm (corresponding to $\Phi_0 \approx 10$ G).[10] In such semiconductor rings, the level spacing Δ is on the order of 25 mK, much higher than in metallic rings, where it is on the order of $\approx$10 μK. In this experiment, the experimental technique to detect the persistent current is somewhat different from the technique used in the experiment by Lévy et al. Instead of measuring the dc magnetic response of the rings, the authors study the ac response of the rings to an RF excitation. Using this technique, they measure the ac complex conductance of the rings, from which they deduce the persistent currents.

[9] The sample-specific parameters in this experiment were: $l_e = 20$ nm, $E_c = 80$ mK.

[10] The sample-specific parameters in this experiment were: $l_e = 3$ μm, $E_c = 200$ mK, $l_\phi = 8$ μm.

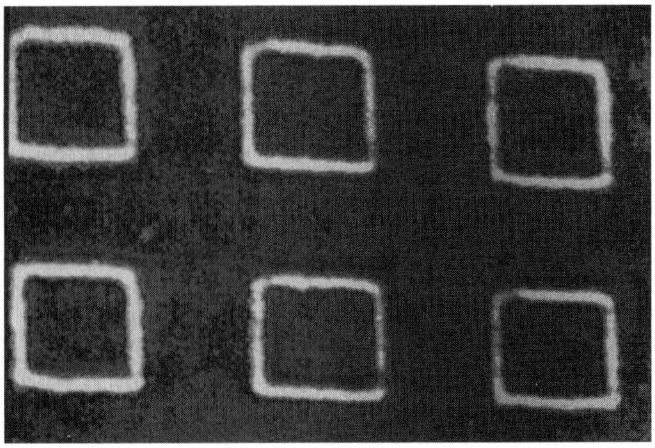

Figure 8. Picture of a part of the sample used in the experiment of [71]. It consists of an array of 10^7 copper squares, of perimeter 2.2 μm. Adapted with permission from [71], L. P. Lévy et al., *Phys. Rev. Lett.* 64, 2074 (1990). © 1990, American Physical Society.

The quantity measured in this experiment is the magnetic susceptibility of the rings, $\chi(\omega) = \chi'(\omega) + \imath\chi''(\omega)$. The complex ac conductance of the rings is then deduced by $\chi(\omega) \propto \imath\omega G(\omega)$. At low frequency,[11] the imaginary part of $G(\omega)$ is just proportional to the derivative of the persistent current with respect to the flux.

The magnetic susceptibility is measured using a resonating technique. The resonator consists of a meander stripline on top of which the rings are deposited. The meander, open at both ends, is made of 20 cm of superconducting niobium. The fundamental frequency of the resonator is 380 MHz. The shift in the resonance frequency and the variations of the quality factor are proportional to the imaginary and real parts of the ac complex conductance of the rings.

The experimental conductance (see Fig. 11) shows $h/2e$ oscillations, as expected, for experiments on many rings. However, the amplitude found for the persistent current, on the order of 1.5 nA per ring, is almost an order of magnitude larger than predicted. More surprising, the measured signal implies a *diamagnetic* zero field persistent current, that is, an *attractive* interaction between the electrons. Again, such an attractive interaction is very unlikely in this two-dimensional electron gas.

It should be noted, however, that in such an experiment, the frequency is quite close to the level spacing. This may affect the response of the rings, and makes a direct comparison with the experiment by Lévy et al. somewhat difficult.

5.3. Single-Ring Experiments

Single-ring experiments are a true experimental challenge. In such experiments, it is the *typical* current which is detected, as it is roughly one order of magnitude larger than the average current for one ring. It should be stressed, however, that as the average current is an extensive quantity in contrast to the typical current that varies like $\sqrt{N_R}$, the signal to be detected in a many-ring experiment is orders of

[11] In this experiment, the characteristic frequency is given by the inverse of the inelastic mean free time τ_{in}^{-1}.

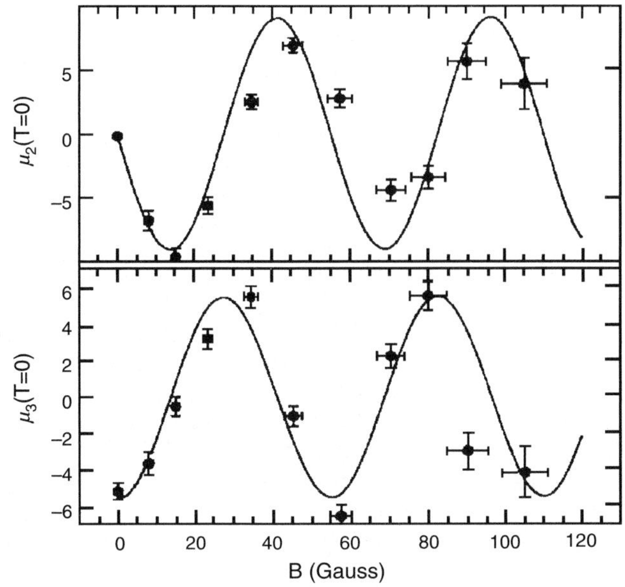

Figure 9. Dependence of the second and third harmonic of the response of the SQUID as a function of the magnetic field. In this experiment [71], Φ_0 corresponds to 130 G. Both harmonics show clear oscillations as a function of the magnetic field, with a periodicity $\Phi_0/2$. Adapted with permission from [71], L. P. Lévy et al., *Phys. Rev. Lett.* 64, 2074 (1990). © 1990, American Physical Society.

magnitude larger than the signal to be detected in a single-ring experiment.

5.3.1. Metallic Rings

The first single-ring experiment was performed by Chandrasekhar et al. [74] on a single gold ring. In this experiment, three different samples were measured: two were rings of diameter 2.4 and 4.0 μm, and the third one was a rectangle of dimensions[12] 1.4 μm × 2.6 μm.

The experimental setup consists of a homemade miniature dc SQUID. The SQUID itself has a sensitivity of $6 \cdot 10^{-8}\Phi_0$. The pick-up loop consists of a counterwound niobium loop in order to minimize the sensitivity to the static background field. To maximize the coupling between the pick-up coil and the sample, both were fabricated on the same chip, and the coil was deposited around the gold ring (see Fig. 12). Moreover, the field coil consists of a niobium line deposited around the ring.

In this experiment, the authors detect the modulation of the flux measured by the SQUID as a function of the magnetic field, which is swept over a few Φ_0. The magnetic field is modulated at low frequency (≈4 Hz), and the signal is detected at f and $2f$. The background signal is subtracted numerically using a quadratic form, and the amplitude is extracted from the Fourier transform (power spectrum) of the data as a function of the magnetic field (see Fig. 13).

As a result, the authors found a persistent current with Φ_0 periodicity and an amplitude of 3 ± 2 nA, 30 ± 15 nA, and 6 ± 2 nA for the three samples investigated, whereas

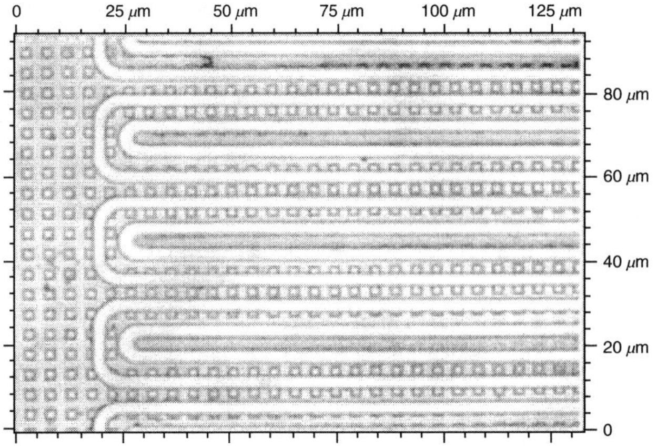

Figure 10. Optical photograph of the sample used in the experiment of [73]. It consists of an array of 10^5 GaAs/AlGaAs rings. On the top of them, one can see the niobium meander stripline used as the resonator. Reprinted with permission from [73], B. Reulet et al., *Phys. Rev. Lett.* 75, 124 (1995). © 1995, American Physical Society.

the theoretical values are 0.09, 0.27, and 0.25 nA, respectively. Obviously, the measured signal is 30–150 times larger than expected. Different arguments have been invoked to explain this discrepancy. It should be noted, however, that the observed signal is on the order of ev_F/L, that is, the signal one should find for a ballistic ring ($l_e \approx L$). On the other hand, it is very unlikely that gold rings behave as ballistic rings, and the theoretical explanation of this experimental observation remains an open question.

The sign of the persistent current is quite difficult to determine in such a single-ring experiment. The authors claim that

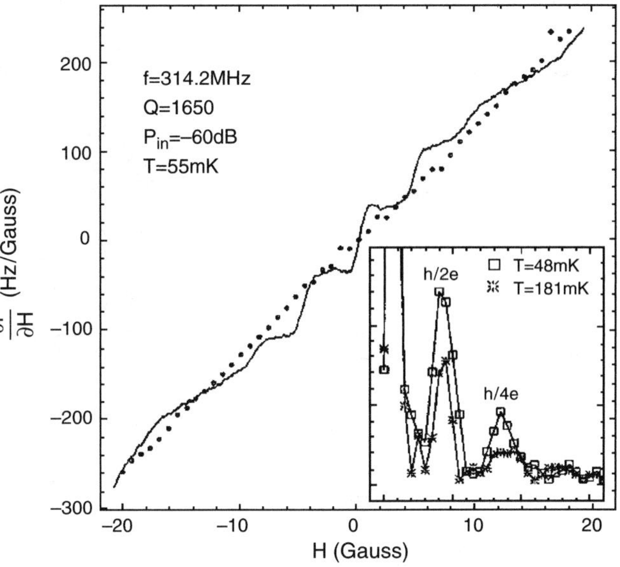

Figure 11. Derivative of the resonance frequency of the resonator used in [73]. The linear background (dotted line) is due to the diamagnetism of niobium. Superimposed on this, one clearly sees the $h/2e$ periodic signal due to the persistent currents in the rings(solid line). The inset shows the Fourier transform of the signal at two different temperatures. Reprinted with permission from [73], B. Reulet et al., *Phys. Rev. Lett.* 75, 124 (1995). © 1995, American Physical Society.

[12] $l_\phi = 12$ μm; $l_e \approx 70$ nm in this experiment.

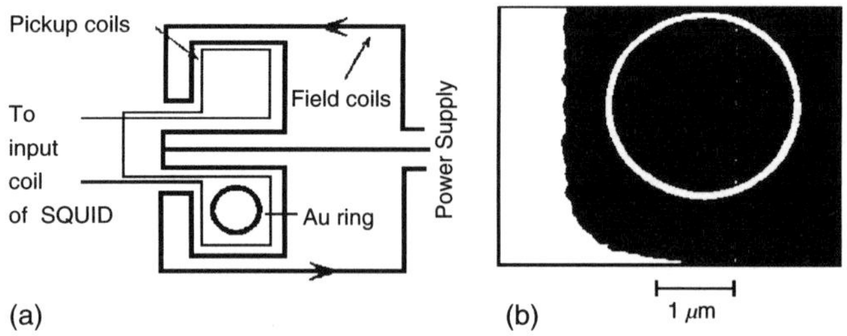

Figure 12. (a) Schematic picture of the sample used for the experiment of [74], displaying the counterwound niobium pick-up loop, the field coil, and the gold ring. (b) Picture of the ring. The white part is a corner of the pick-up loop. Reprinted with permission from [74], V. Chandrasekhar et al., *Phys. Rev. Lett.* 67, 3578 (1991). © 1991, American Physical Society.

the samples studied showed a paramagnetic signal. A clear statement, however, as stressed by the authors, is difficult due to the few samples measured, and due to the extreme experimental difficulty.

5.3.2. Semiconductor Rings

Another experiment was performed by Mailly and coworkers on a single, isolated ring, etched into a semiconductor heterojunction [75]. In this case, the signal is expected to be larger than in metallic rings as the elastic mean-free path is much larger compared to the latter case.

The ring was etched into a two dimensional electron gas at the interface of a GaAs–GaAlAs heterojunction. The mean

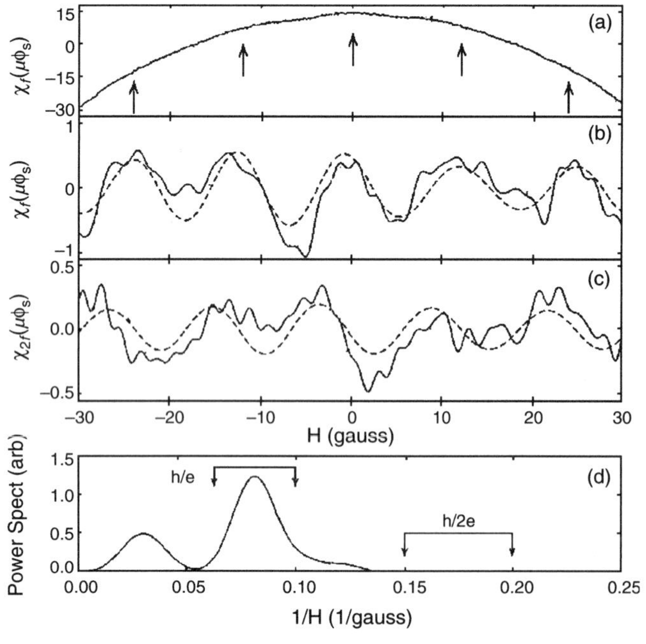

Figure 13. (a) First harmonic of the response of the dc SQUID as a function of the magnetic field. (b) Same data after subtraction of a quadratic background. (c) Second $(2f)$ harmonic of the response of the dc SQUID after subtraction of a quadratic background. (d) Power spectrum of the data displayed in (b). One clearly observes a peak at h/e frequency. Reprinted with permission from [74], V. Chandrasekhar et al., *Phys. Rev. Lett.* 67, 3578 (1991). © 1991, American Physical Society.

perimeter in this sample is on the order of 6 μm, which corresponds to $\Phi_0 \approx 10$ G).[13]

An important advantage of semiconductors is the possibility of using gates on the sample. This allows us to modify *in-situ* the geometry of the sample simply by applying a dc voltage to the gates. In this experiment, the authors used two different gates (see Fig. 14). The first is used to separate the ring from the reservoirs: the presence of these ohmic contacts allows us to measure at the same time the conductance and the persistent current oscillations, and hence to check the electronic temperature and the coherence of the electrons in the ring. The second gate is evaporated on top of one arm of the ring. By polarizing this gate ("open" ring), one can suppress all of the interference effects in the ring, both the Aharonov–Bohm oscillations and the persistent currents. This allows us to perform a "zero" measurement, equivalent to measuring the SQUID with no ring. The advantage is that this can be made on the same sample. Moreover, the subtraction of the signal obtained with the ring "closed" and "open" allows us to experimentally suppress the background signal of the detector.

In this experiment, a sophisticated on-chip micro-SQUID technique was employed. With such a design, no pick-up coil is needed: the SQUID itself is deposited exactly on the top of the ring. This has two major advantages. First, the absence of a pick-up coil reduces the inductance of the setup. Second, and most important, in such a geometry, the coupling between the ring and the SQUID is basically optimal, as the SQUID has exactly the same shape as the ring. The SQUID is actually designed as a gradiometer, consisting of two counterwound loops in order to compensate the externally applied static magnetic field (see Fig. 14). The two Josephson junctions are made using Dayem microbridges, evaporated at the same time as the second level of the gradiometer. For a detailed description of the miro-SQUID gradiometer technique, we refer the reader to [76].

The measurement consists of sweeping the magnetic field over several Φ_0, and recording the critical current of the SQUID. This is made successively for the "closed" and "open" rings. The signal is then obtained by taking the Fourier transform of the difference between the two measurements. The noise is evaluated at the same time by taking the Fourier transform of the difference between two "closed" or "open" ring measurements.

In the Fourier spectrum (see Fig. 15), a clear peak is observed at the Φ_0 frequency corresponding to a value of 4 ± 2 nA for the persistent current amplitude, in good agreement with the theoretical prediction ev_F/L. No measurable signal was observed at the $\Phi_0/2$ frequency, as expected when comparing the theoretical signal and the noise level of the experiment. The sign of the persistent current, on the other hand, was impossible to determine in a reliable way.

This experiment proves that, in the case of very weak disorder and a small number of channels, standard theory gives a correct description of the persistent current amplitude. Moreover, in such samples, electron–electron interactions are much enhanced due to the low electron density. This suggests that these interactions are unlikely to strongly enhance the amplitude of the persistent currents.

[13] $l_e \approx 10\,\mu$m, $l_\Phi \approx 25\ \mu$m, $E_c \succ 1$ K in this experiment.

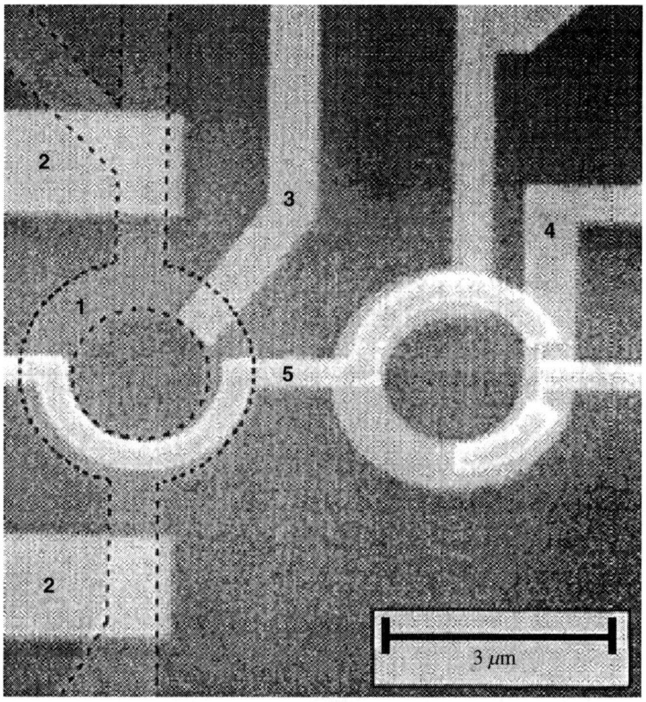

Figure 14. Picture of the sample used in [75]. 1) GaAs-GaAlAs ring (dashed line). 2), 3) gold gate used to isolate the sample from ohmic contacts and to suppress the signal. 4) Gold calibration loop. 5) First level of the micro-SQUID gradiometer containing the two Dayem micro-bridges on the right. The picture was taken before the evaporation of the second level of the micro-SQUID gradiometer. Reprinted with permission from [75], D. Mailly et al., *Phys. Rev. Lett.* 70, 2020 (1993). © 1993, American Physical Society.

5.4. Few-Rings Experiments

More recently, two experiments were also performed on an ensemble of a few rings, either metallic or semiconductor. In this case, the small number of rings (typically ten rings) allows us to check the theory concerning the ensemble

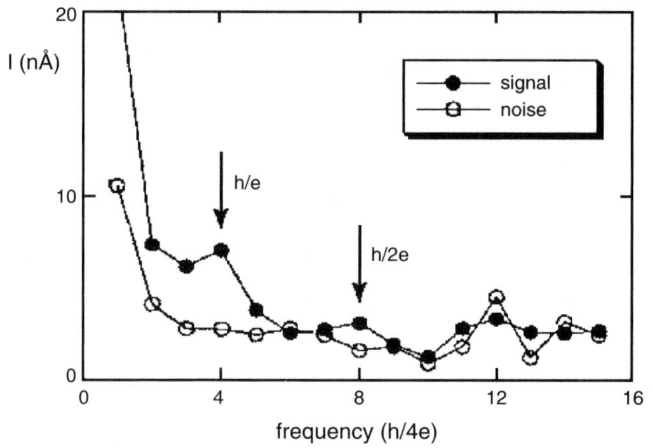

Figure 15. Fourier spectrum of the magnetization of the ring of [75] in units of nanoamperes. The arrows indicate the h/e and $h/2e$ frequency. Open dot is the experimental noise. One clearly observes the signal at the h/e frequency corresponding to a persistent current of 4 ± 2 nA in the ring. Reprinted with permission from [75], D. Mailly et al., *Phys. Rev. Lett.* 70, 2020 (1993). © 1993,

averaging, and should allow us to observe both the h/e and $h/2e$ components of the persistent current. Moreover, in the experiment on semiconductor rings, the authors were able to check the effect of a connection (ohmic contacts) between the rings.

5.4.1. Metallic Rings

In this experiment, Jariwala et al. [77] used a similar experimental setup as for the experiment on the single gold ring. The sample (see Fig. 16) consists of a line of 30 isolated gold rings of radius 1.3 μm (perimeter 8 μm) corresponding to a flux period[14] of $\Phi_0 \approx 8$ G.

To extract the persistent current signal from the background signal, the magnetic field is modulated at low frequency (typically $\approx$2 Hz), and detected at the first, second, and third harmonics of the response of the SQUID.

In this experiment, both the h/e and $h/2e$ components were detected (see Fig. 17). For the h/e component, the authors found a current of $I_{\text{typ}} = 0.35$ nA $= 2.3\, E_c/\Phi_0$ per ring, in good agreement with theoretical predictions, taking into account e–e interactions. To obtain this result, the authors divided the total signal by $\sqrt{N_R}$ to account for the random sign of the persistent current. This suggests that the amplitude of the persistent current measured in the single-ring experiment [74] is somewhat overestimated.

The $h/2e$ component was found to be $\langle I_N \rangle = 0.06$ nA per ring, corresponding to $0.44\, E_c/\Phi_0$. This result is in line with the results found in the previous experiment of Lévy et al. on copper rings, and only a factor of 2 larger than the theoretical predictions when taking into account electron–electron interactions.

The sign of the persistent current, on the other hand, is much more surprising. In this experiment, the sign of the average current is *diamagnetic*. Although this has been seen in previous experiments on many rings [71, 72], in this work, the determination of the sign is unambiguous. As we have seen, such a diamagnetic response is quite unlikely as it corresponds to an *attractive* interaction between the electrons. Clearly, such a discrepancy between experiment and theory may be attributed to an unexplored physical phenomenon that modifies the ground state of the electron gas. The authors of this experiment explain their result in light of a recent theory on zero temperature dephasing in metals [84]. However, as the status of such theories is still quite controversial, we will not go further into this point.

Finally, it should be noted that, in such an experiment on 30 rings, both the average current, that grows like N_R, and the typical current, that grows like $\sqrt{N_R}$, have the same amplitude. This proves that 30 rings are not enough for ensemble averaging, and the exact variation of I_{typ} and $\langle I_N \rangle$ with the number of rings remains experimentally an open question.

5.4.2. Semiconductor Rings

Another experiment was performed on a small number of semiconductor rings by Rabaud and co-workers [78]. This experiment was performed on two arrays of 4 and 16 rings (actually squares) etched into a two-dimensional electron

[14] $l_e \approx 87$ nm, $l_\Phi \approx 16$ μm, $E_c \approx 7$ mK in this experiment.

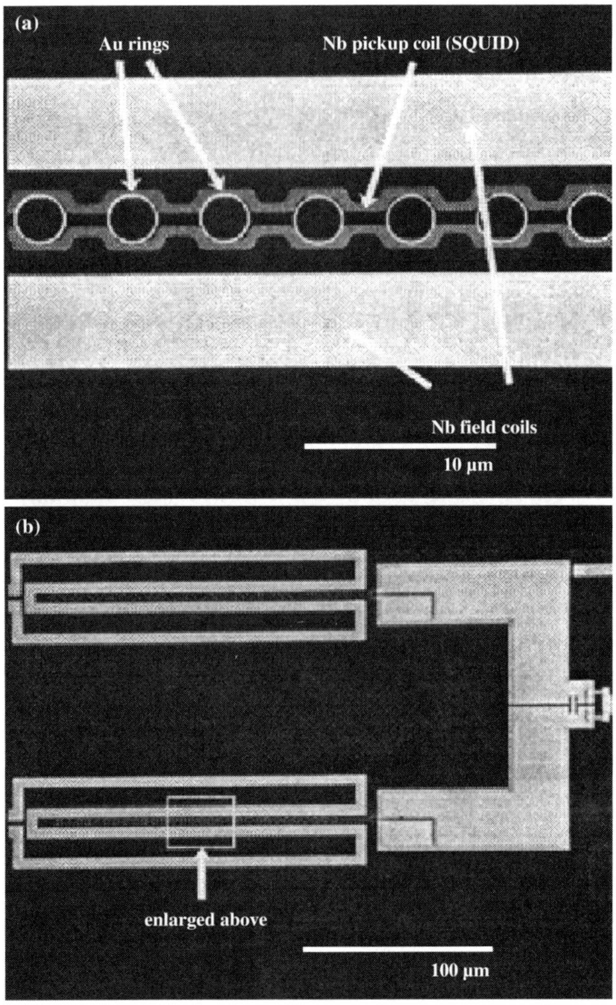

Figure 16. Picture of the sample used in the experiment of [77]. (a) Close-up view showing the gold rings, the niobium pick-up coil, and the niobium field coil. (b) Larger view showing the entire gradiometer. Reprinted with permission from [77], E. M. Q. Jariwala et al., *Phys. Rev. Lett.* 86, 1594, (2000). © 2000, American Physical Society.

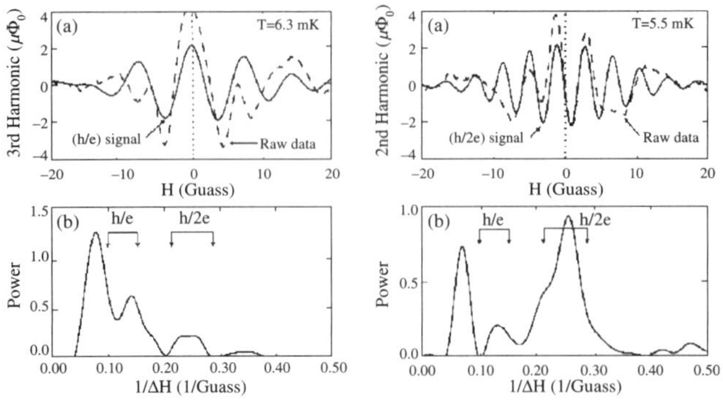

Figure 17. Magnetic response of the array of 30 gold rings of the experiment of [77]. Left panel: (a) raw data (dashed line) and h/e contribution (solid line) extracted from the Fourier spectrum displayed in (b). Right panel: (a) raw data (dashed line) and $h/2e$ contribution (solid line) extracted from the Fourier spectrum displayed in (b). Reprinted with permission from [77], E. M. Q. Jariwala et al., *Phys. Rev. Lett.* 86, 1594 (2000). © 2000, American Physical Society.

gas at the interface of a GaAs GaAlAs heterojunction. The squares were 3 μm $\times$ 3 μm, perimeter 12 μm, corresponding to $\Phi_0 \approx 5$ G).[15] Using an original setup containing three different metallic gates (see Fig. 18), Rabaud et al. were able to measure in the same experiment (same cooling down run) both the signal of connected and isolated rings. This original setup also permits us to suppress the signal via a gate, and check the "zero" of the detector. This allows us to perform an *in-situ* subtraction of the background. Such a technique has the advantage of measuring at the same time both the signal and the noise in order to have an unambiguous determination of the magnetic signal.

In this experiment, the authors measured a clear h/e periodic signal, of amplitude 2.0 ± 0.3 nA per ring for the 4-ring sample, and 0.35 ± 0.07 nA per ring for the 16-ring sample (see Fig. 19), to be compared with the theoretical values 2.18 nA per ring for the 4-ring sample and 1.09 nA per ring for the 16-ring sample. The experimental results are in relatively good agreement with the theoretical values. The discrepancy observed for the 16-ring sample may be attributed to an overestimation of the elastic mean-free path, which is determined on wires fabricated from the same wafer of the GaAs/AlGaAs heterojunction. However, the complete lithographic process, quite complicated in this experiment, may affect l_e, mainly because of the roughness of the edges after etching.

In this work, the authors were also able to measure the persistent currents in the same array of rings, but this time with an ohmic connection between the rings. Measurements on both isolated and connected rings can be made basically at the same time, by simply applying a dc voltage on the gates on the top of the arms connecting the rings. The purpose of this experiment was to measure the persistent current in a sample much larger than l_ϕ, that is, a macroscopic sample from the quantum physic point of view. This work was stimulated by theoretical models that calculated persistent currents in arrays of rings, showing that they do not vanish, but are only reduced by some geometrical factor [79].

The lines of 4 and 16 rings used in the experiment were, respectively, $\approx$60 and $\approx$250 μm, both much larger than the phase-coherence length. In that sense, these line of rings are macroscopic objects. The authors found a current of amplitude 1.7 ± 0.3 nA per ring for the 4-ring sample, and 0.40 ± 0.08 nA per ring for the 16-ring sample, whereas the theoretical values, calculated in [79], were respectively, 1.25 nA per ring and 0.62 nA per ring. There is obviously a discrepancy between experimental and theoretical values. However, it should be noted that the theoretical model was developed for diffusive (metallic) rings, which is certainly not the case in heterojunction rings. Moreover, coulomb interactions are not taken into account for the typical current; in heterojunctions, the low electronic density strongly enhances the interactions.

The key result of this experiment is the fact that the ratio between the amplitude of the persistent currents observed in connected and isolated rings is on the order of 1 for both samples. This shows that persistent currents are basically

[15] $l_e \approx 10$ μm; $l_\Phi \approx 25$ μm; $E_c \approx 500$ mK in this experiment

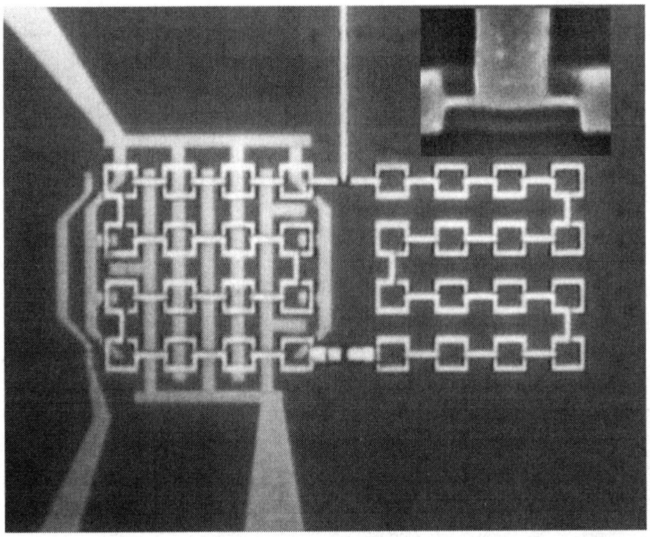

Figure 18. Optical photograph of the sample used for the experiment of [78]. The three metallic gates and the aluminum micro-SQUID gradiometer are clearly visible. Inset shows a SEM picture of the two Dayem microbridges used as Josephson junctions for the SQUID. Adapted with permission from [78], W. Rabaud et al., *Phys. Rev. Lett.* 86, 3124 (2001). © 2001, American Physical Society.

unaffected by the connection between the rings. This suggests that, even in a macroscopic sample, there should be a reminiscence of the quantum nature of electrons.

Finally, it is interesting to compare this experiment with the experiment performed on 30 metallic rings. In the experiment on semiconductor rings, no observable signal was detected at the $h/2e$ frequency for both samples. At least for the 16-ring sample, this result is quite surprising, as the average current grows linearly with the number of rings. As

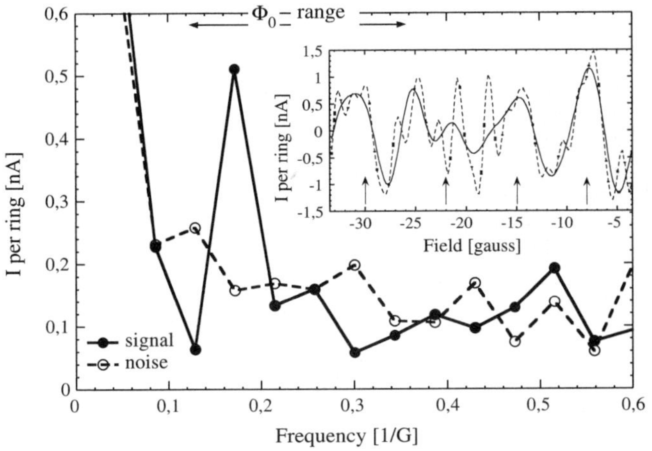

Figure 19. Power spectrum of the magnetization due to the persistent current in a line of 16 connected rings in units of nanoamperes per ring. The arrow indicates the h/e frequency window. Open symbols are the experimental noise. One clearly sees a peak in in the "signal" curve, absent from the "noise" curve. Inset shows the raw data (dashed line) after substraction of the background and after bandpassing the signal over the h/e frequency range (solid line). Adapted with permission from [78], W. Rabaud et al. *Phys. Rev. Lett.* 86, 3124 (2001). © 2001, American Physical Society.

a comparison, in the experiment on metallic rings, the signals at h/e and $h/2e$ for 30 rings were of similar amplitude. Again, this shows that the ensemble averaging, in the case of persistent currents, is still not fully understood.

6. CONCLUSION

Persistent currents are certainly one of the most spectacular manifestations of the quantum coherence of the electrons in a mesoscopic system: it manifests as a permanent, nondissipative current flowing around a normal, nonsuperconducting ring. The amplitude of this current is on the order of a nanoampere, whereas the resistance of the ring can be on the order of a kilohm, for example, for the case of semiconductor rings.

Although heavily controversial at the beginning, the existence of such currents is well established, from both a theoretical as well as an experimental point of view. However, many questions remain open, and experimental results point out the lack of a deep understanding of this phenomenon.

First, the amplitude experimentally observed seems different from the theoretical predictions. Most of the experimental results are about an order of magnitude larger than the theoretical predictions. However, it must be stressed that all of these experiments are very difficult, as they deal with the measurement of very small magnetic signals. From this point of view, and taking into account the different approximations in the theoretical models, it seems difficult in the absence of new experimental results to draw a definitive conclusion concerning the validity of the theoretical predictions on the amplitude of the persistent currents.

More surprising is the sign observed in the many-ring experiments. As we have seen, the sign of the zero-field magnetic response due to the average persistent currents are directly related to the sign of the interaction between electrons. In at least two experiments on many rings, both metallic or semiconductor, the sign was found to be diamagnetic, whereas in the first experiment on copper rings, there were indications that it was also diamagnetic. This result is quite intriguing, as it should correspond to an attractive interaction between electrons. Such an attractive interaction is very unlikely in "standard" metals like copper or gold, or even in GaAs-GaAlAs heteronjunctions. Clearly, there are many open questions in the description of the average persistent current of interacting electrons.

Another interesting point is the change of persistent currents when the sample evolves from a true mesoscopic sample to a macroscopic sample. Only one such experiment has been carried out up to now, and the result found is that persistent currents are not significantly modified when the size of the sample increases. These results suggest that persistent currents should be observable in a macroscopic objet; by extension, in the spirit of the evolution from Aharonov–Bohm oscillations to weak localization, one may think about observing the zero-field magnetic response of a standard two-dimensional metal.

Finally, there are natural extensions of this problem of equilibrium properties of mesoscopic conductors that have been largely unexplored. An interesting problem is a ballistic dot of different shape, that is, quantum billiards. In this case, the properties of the energy spectrum are no longer given by the impurity configuration, but by the specular scattering at the boundaries of the sample [80–82]. The orbital magnetism of these systems should be controlled by the regular or chaotic nature of the billiard, and must be understood in light of the quantum chaos theory. Another point is the connection between the persistent currents and the zero temperature decoherence: it has been proposed that the anomalously high amplitude of the average current may be related to the decoherence of the electron due to the RF environment [83, 84]. It should also be interesting to study thermodynamic properties different from the persistent current. One example is the specific heat of mesoscopic samples. In an equivalent way to persistent currents, the specific heat should oscillate with the magnetic flux. However, such a measurement is certainly very difficult, as the energy involved in such a phenomenon is again on the order of the Thouless energy. Such an experiment would imply strong improvements in the sensitivity of present detectors. Finally, one subject of major interest at present is the possibility of observing the Kondo effect in artificial nanostructures [85]. Coupling this with a persistent current measurement should allow us to directly probe the reality and the extension of the Kondo cloud [86–88].

GLOSSARY

Aharonov–Bohm flux Additional dephasing of an electronic wavefunction due to the coupling between the electric charge of the electron and the vector potential.

Ballistic conductor Conductor pure enough to allow the electrons to be scattered only by the edge of the sample.

Equilibrium (thermodynamic) property Property related to the ground state of a system.

Mesoscopic (quantum) conductor Conductor whose size is between the microscopic scale (atoms and molecules) and the macroscopic scale (daily objects).

Molecular beam epitaxy Technique used to grow semiconductors with a very high purity. This technique allows the control of atomic layer growth.

Nanolithography Technique used to fabricate samples at the nanometer scale. This includes, UV lithography, e-beam lithography, X-rays lithography etc.

Quantum coherence length Length over which the associated wave of an electron maintains a well defined phase. Beyond this length scale, the electron looses its quantum, wave-like nature.

Transport property Property related to the response of a conductor to an external excitation, e.g. the current in response to a given voltage.

REFERENCES

1. Y. Imry, "Introduction to Mesoscopic Physics." Oxford University Press, 1997.
2. N. W. Ashcroft and N. D. Mermin, "Solid State Physics." Saunders College, 1976.
3. C. Kittel, "Introduction to Solid State Physics, 6th ed." Wiley, 1986.
4. R. Landauer, *IBM J. Res. Dev.* 1, 223 (1957).
5. R. Landauer, *Philos. Mag.* 21, 863 (1970).
6. D. Y. Sharvin and Y. V. Sharvin, 34, 272 (1981).
7. R. A. Webb, S. Washburn, C. P. Umbach, and R. B. Laibowitz, *Phys. Rev. Lett.* 54, 2696 (1985).
8. B. J. van Wees, H. van Houten, C. W. J. Beenakker, J. G. Williamson, L. P. Kouwenhouven, D. van der Marel, and C. T. Foxon, *Phys. Rev. Lett.* 60, 848 (1988).
9. A. B. Zorin, *Rev. Sci. Instrum.* 66, 4296 (1995).
10. D. Vion, P. F. Orfila, P. Joyez, D. Esteve, and M. H. Devoret, *J. Appl. Phys.* 77, 2519 (1995).
11. D. C. Glattli, P. Jacques, A. Kumar, P. Pari, and L. Saminadayar, *J. Appl. Phys.* 81, 7350 (1997).
12. B. L. Altshuler and A. G. Aronov, in "Electron-Electron Interactions in Disordered Conductors" (A. L. Efros and M. Pollak, Eds.). North Holland, Amsterdam, 1985.
13. P. Mohanty, E. M. Q. Jariwala, and R. A. Webb, *Phys. Rev. Lett.* 78, 3366 (1997).
14. F. Schopfer, C. Bäuerle, W. Rabaud, and L. Saminadayar, *Phys. Rev. Lett.* 90, 056801 (2003).
15. J. Edwards and D. Thouless, *J. Phys. C* 5, 807 (1972).
16. S. Chakravarty and A. Schmid, *Phys. Rep.* 140, 193 (1986).
17. P. W. Anderson, 109, 1492 (1958).
18. E. Abrahams, P. W. Anderson, D. C. Licciardello, and T. V. Ramakrishna, *Phys. Rev.* 42, 673 (1979).
19. V. M. Pudalov, M. D'Iorio, S. V. Kravchenko, and J. W. Campbell, *Phys. Rev.* 70, 1866 (1993).
20. C. W. J. Beenaker and H. van Houten, 44, 1 (1991).
21. L. Esaki and R. Tsu, Internal Report RC 2418, IBM Research, March 1969.
22. T. Ando, A. B. Fowler, and F. Stern, *Rev. Mod. Phys.* 54, 437 (1982).
23. J. Blackmore, *J. Appl. Phys.* 53, R123 (1982).
24. S. Adachi, *J. Appl. Phys.* 58, R1 (1985).
25. V. Umansky, R. de Picciotto, and M. Heiblum, *Appl. Phys. Lett.* 71, 683 (1997).
26. C. Vieu, M. Meijas, F. Carcenac, G. Faini, and H. Launois, *Microwave Elect. Eng.* 35, 253 (1997).
27. J. Gierak, D. Mailly, G. Faini, J. L. Pelouard, P. Denk, J. Y. Marzin, A. Septier, G. Schmid, J. Ferré, R. Hydman, C. Chappert, J. Flicstein, B. Gayral, and J. M. Gerard, *Microwave Elect. Eng.* 57–58, 865 (2001).
28. D. F. Kyser and N. S. Viswanathan, *J. Vac. Sci. Technol.* 12, 1305 (1975).
29. C. Vieu, F. Carcenac, A. Pépin, Y. Chen, M. Méjias, A. Leibib, L. Ferlazzo, L. Couraud, and H. Launois, *Appl. Surf. Sci.* 164, 111 (2000).
30. D. M. Eigler and A. K. Schweizer, *Nature* 344, 524 (1990).
31. C. Lebreton and Z. Wang, *Surf. Sci.* 382, 193 (1997).
32. S. Lüscher, T. Heinzel, K. Ensslin, W. Wegscheider, and M. Bichler, *Phys. Rev. Lett.* 86, 2118 (2001).
33. V. Bouchiat, M. Faucher, T. Fournier, B. Pannetier, C. Thirion, W. Wernsdorfer, N. Clement, D. Tonneau, H. Dallaporta, S. Safarov, J. C. Villégier, D. Fraboulet, D. Mariolle, and J. Gauthier, "Micro- and Nano-engineering 2001, MNE: Proceedings of the 27th International Conference on Micro- and Nano-Engineering," Grenoble, France, Sept. 2001.
34. F. London, *J. Phys. (France)* 8, 379 (1937).
35. F. Hund, *Ann. Phys. (Leipzig)* 32, 102 (1938).
36. F. Bloch, *Phys. Rev.* 137, A 787 (1965).
37. I. O. Kulik, *JETP* 31, 1172 (1970).
38. M. Büttiker, Y. Imry, and R. Landauer, *Phys. Lett.* 96A, 365 (1983).
39. N. Byers and C. N. Yang, *Phys. Rev. Lett.* 7, 46 (1961).
40. H. F. Cheung, Y. Gefen, E. K. Riedel, and W.-H. Shih, *Phys. Rev. B* 37, 6050 (1988).
41. R. Landauer and M. Büttiker, *Phys. Rev. Lett.* 54, 2049 (1985).
42. H. F. Cheung and E. K. Riedel, *Phys. Rev. B* 40, 9498 (1989).
43. T. Brody, J. Flores, J. French, P. Mello, A. Pandey, and S. Wong, *Rev. Mod. Phys.* 53, 385 (1981).
44. H. Bouchiat, G. Montambaux, and D. Sigeti, *Phys. Rev. B* 44, 1682 (1991).

45. Y. Gefen, B. Reulet, and H. Bouchiat, *Phys. Rev. B* 46, 15922 (1992).
46. E. K. Riedel and F. von Oppen, *Phys. Rev. B* 47, 15449 (1993).
47. H. F. Cheung, E. K. Riedel, and Y. Gefen, 62, 587 (1989).
48. H. Bouchiat and G. Montambaux, *J. Phys. France* 50, 2695 (1989).
49. G. Montambaux, H. Bouchiat, D. Sigeti, and R. Friesner, *Phys. Rev. B* 42, 7647 (1990).
50. G. Montambaux, in "Quantum Fluctuations" (S. Reynaud, E. Giacobino, and J. Zinn-Justin, Eds.). Elsevier, Amsterdam, 1996.
51. C. P. Umbach, C. Van Haesendonck, R. B. Laibowitz, S. Washburn, and R. A. Webb, *Phys. Rev. Lett.* 56, 386 (1986).
52. B. Altshuler, Y. Gefen, and Y. Imry, *Phys. Rev. Lett.* 66, 88 (1991).
53. A. Schmid, *Phys. Rev. Lett.* 66, 80 (1991).
54. F. von Oppen and E. K. Riedel, *Phys. Rev. Lett.* 66, 84 (1991).
55. E. Akkermans, *Europhys. Lett.* 15, 709 (1991).
56. N. Argaman, Y. Imry, and U. Smilansky, *Phys. Rev. B* 47, 4440 (1993).
57. I. S. Gradshteyn and I. M. Ryzhik, "Table of Integrals, Series and Products." Academic, New York, 1965.
58. V. Ambegaokar and U. Eckern, *Phys. Rev. Lett.* 65, 381 (1990).
59. B. L. Altshuler, A. G. Aronov, and P. A. Lee, *Phys. Rev. Lett.* 44, 1288 (1980).
60. G. Montambaux, *J. Phys. I (France)* 6, 1 (1996).
61. A. Müller-Groeling, H. A. Weindenmüller, and C. Lewenkopf, *Europhys. Lett.* 22, 193 (1993).
62. G. Bouzerar, D. Poilblanc, and G. Montambaux, *Phys. Rev. B* 49, 8258 (1994).
63. M. Abraham and R. Berkovits, *Phys. Rev. Lett.* 70, 1509 (1993).
64. H. Kato and D. Yoshioka, *Phys. Rev. B* 50, 4943 (1994).
65. G. Bouzerar and D. Poilblanc, *Phys. Rev. B* 52, 10772 (1995).
66. M. Ramin, B. Reulet, and H. Bouchiat, *Phys. Rev. B* 50, 4943 (1994).
67. U. Eckern and A. Schmid, *Europhys. Lett.* 18, 457 (1992).
68. R. A. Smith and V. Ambegaokar, *Europhys. Lett.* 20, 161 (1992).
69. J. Clarke, W. M. Goubeau, and M. B. Ketchen, *J. Low Temp. Phys.* 25, 99 (1976).
70. C. Chapelier, M. E. Khatib, P. Perrier, A. Benoît, and D. Mailly, in "SQUID 91: Superconducting Devices and Their Applications" (H. Koch and H. Lubbig, Eds.). Springer, 1991.
71. L. P. Lévy, G. Dolan, J. Dunsmuir, and H. Bouchiat, *Phys. Rev. Lett.* 64, 2074 (1990).
72. R. Deblock, R. Bel, B. Reulet, H. Bouchiat, and D. Mailly, *Phys. Rev. Lett.* 89, 206803 (2002).
73. B. Reulet, M. Ramin, H. Bouchiat, and D. Mailly, *Phys. Rev. Lett.* 75, 124 (1995).
74. V. Chandrasekhar, R. A. Webb, M. J. Brandy, M. B. Ketchen, W. J. Gallagher, and A. Kleinsasser, *Phys. Rev. Lett.* 67, 3578 (1991).
75. D. Mailly, C. Chapelier, and A. Benoît, *Phys. Rev. Lett.* 70, 2020 (1993).
76. W. Rabaud, L. Saminadayar, C. Bäuerle, K. Hasselbach, A. Benoît, D. Mailly, and B. Etienne, in "Toward the Controllable Quantum States" (H. Takayanagi and J. Nitta, Eds.). 2003.
77. E. M. Q. Jariwala, P. Mohanty, M. B. Ketchen, and R. A. Webb, *Phys. Rev. Lett.* 86, 1594 (2000).
78. W. Rabaud, L. Saminadayar, D. Mailly, K. Hasselbach, A. Benoît, and B. Etienne, *Phys. Rev. Lett.* 86, 3124 (2001).
79. M. Pascaud and G. Montambaux, *Phys. Rev. Lett.* 82, 4512 (1999).
80. L. P. Levy, D. H. Reich, L. Pfeiffer, and K. West, 189, 204 (1993).
81. D. Ullmo, K. Richter, and R. A. Jalabert, *Phys. Rev. Lett.* 80, 4955 (1993).
82. Y. Noat, H. Bouchiat, B. Reulet, and D. Mailly, *Phys. Rev. Lett.* 80, 4955 (1998).
83. V. E. Kravtsov and V. I. Yudson, *Phys. Rev. Lett.* 70, 210 (1993).
84. V. E. Kravtsov and B. L. Altshuler, *Phys. Rev. Lett.* 84, 3394 (2000).
85. D. Goldhaber-Gordon, H. Shtrikman, D. Mahalu, D. Abusch-Magder, U. Meirav, and M. A. Kastner, *Nature* 391, 156 (1998).
86. V. Ferrari, G. Chiappe, E. V. Anda, and M. A. Davidovich, *Phys. Rev. Lett.* 82, 5088 (1999).
87. K. Kang and S. C. Shin, *Phys. Rev. Lett.* 85, 5619 (2000).
88. I. Affleck and P. Simon, *Phys. Rev. Lett.* 86, 2854 (2001).

Encyclopedia of Nanoscience and Nanotechnology
www.aspbs.com/enn

Evanescent Wave Spectroscopy

V. G. Bordo
Russian Academy of Sciences, Moscow, Russia

H.-G. Rubahn
Syddansk Universitet, Odense M, Denmark

CONTENTS

1. INTRODUCTION

Suppose an electromagnetic wave hits the interface between two transparent media having indices of refraction n_1 and n_2. If this occurs at an angle close to normal incidence, then part of the wave is transmitted into the new medium and part is reflected back into the original medium. However, if the wave arrives from the medium with greater index of refraction ($n_1 > n_2$) at an angle of incidence θ_i greater than the critical angle

$$\theta_c = \arcsin\left(\frac{n_2}{n_1}\right) \tag{1}$$

only a vanishing (or *evanescent*) tail of the electromagnetic field penetrates into the optically less dense medium (Fig. 1) [1]. The wave vector of this *evanescent wave* (EW) possesses a component along the interface

$$k_{\parallel} = \frac{2\pi}{\lambda} n_1 \sin\theta_i \tag{2}$$

and a component perpendicular to it

$$k_{\perp} = i\frac{2\pi}{\lambda}\sqrt{n_1^2 \sin^2\theta_i - n_2^2} \tag{3}$$

where λ is the wavelength of light in vacuum. The imaginary component $k_{\perp}$ implies that the EW amplitude decays exponentially with increasing the distance from the surface into the optically less dense medium. Usually a penetration depth δ is defined

$$\delta = 1/|k_{\perp}| \tag{4}$$

at which the EW amplitude decreases by a factor of e. Depending on the angle of incidence δ varies from an formally infinite value at the critical angle to values of the order of the wavelength for more grazing incidence. This way the EW penetration depth spans a scale of only a few hundred nanometers in the visible spectral range (i.e., the light is confined to the interface).

Hence optical methods relying on the evanescent wave should provide access to surface and interface processes. This advantage of *evanescent wave spectroscopy* (EWS) has been widely used in the past for the construction of surface-sensitive atom and molecule sensors. EWs in such sensors are implemented for example via excitation of waveguide modes in planar waveguides or in optical fibers. If the EW frequency is resonant to the optical transition frequency of molecules in the media attaching the interface, the fluorescence signal from the molecules or the dielectric losses at the resonant frequencies can be used to obtain the concentration of molecules at the interface. More detailed discussions of evanescent wave sensors can be found for example in [2, 3].

In this chapter we will focus on more recent and more sophisticated spectroscopic approaches in the evanescent wave regime of a surface. If one investigates not only peak intensities of spectroscopic lines but the *optical line shapes* at an interface between a solid and a gaseous medium much more detailed physical information can be obtained. Since the gas atoms or molecules striking the surface in general change their quantum states, polarization state, or velocity distribution the resonant optical response of a gas in the close vicinity of a surface differs from that inside a gas volume. However, departing further into the gas interior the gas particles "forget" the way they have been scattered by the surface since they undergo spontaneous relaxation or

ISBN: 1-58883-059-4/$35.00

Encyclopedia of Nanoscience and Nanotechnology
Edited by H. S. Nalwa
Volume 3: Pages (287–295)

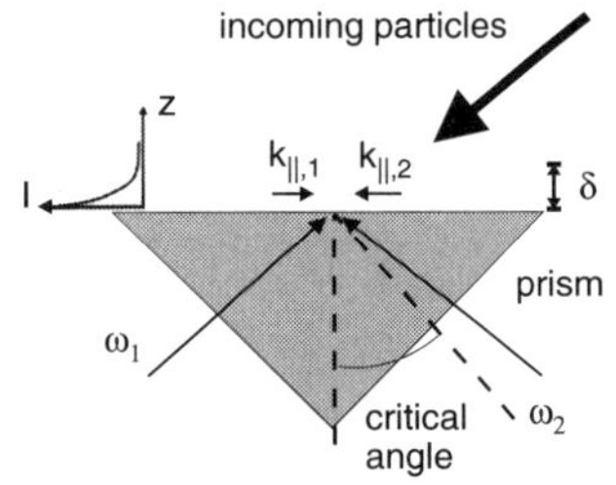

Figure 1. Scheme for multiphoton evanescent wave particle excitation.

experience collisions with other particles. Consequently we define that gas layer bordering a surface for which the optical response differs from that in the bulk gas as the *gas boundary layer* (GBL).

Let us inspect a nonsaturated optical transition. Then the GBL can be characterized by the following *memory lengths* determining the scales at which gas particles "remember" the quantum state which they had immediately after the bounce from the surface (Fig. 2):

(i) The population memory length within which most of the gas particles departing from the surface do not change the population of their quantum state; this length is of order $v_T/\gamma_{||}$ with v_T the most probable velocity and $\gamma_{||}$ the rate of the longitudinal (energy) relaxation of the quantum state [4].
(ii) The polarization memory length within which the particles do not change their polarization; this length can be estimated as $v_T/\gamma_\perp$ with $\gamma_\perp$ the rate of the transverse (phase) relaxation of the transition [4].
(iii) The velocity distribution memory length within which the particles do not change their velocities; this length is of the order of the mean free path.

All memory lengths decrease if the gas pressure increases. In the case of saturation of the optical transitions by the excitation light the population and the polarization memory lengths are determined not only by the corresponding relaxation rates but also by the Rabi frequencies of the transitions [4].

Particles departing from the surface within the GBL possess transient (non-steady-state) populations and/or induced dipole moments. If the latter is the case then the induced polarization of the gas is not proportional to the electric field of the wave at a given point. Hence the optical response of the GBL is *nonlocal*.

Note that the spatial scale within which the gas atoms or molecules contribute to the signal is set by the way the optical spectra are obtained. This *detection length* in the case of fluorescence spectra is determined by whatever is larger,

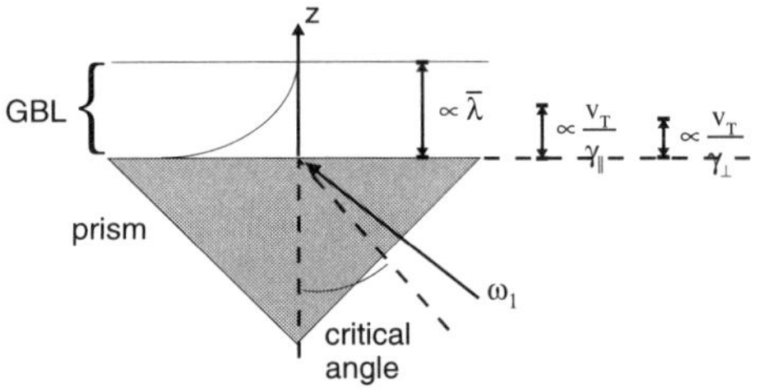

Figure 2. Important length scales for evanescent wave spectroscopy.

the EW penetration depth or the population memory length. For absorption and reflection spectra the detection length coincides with the EW penetration depth.

In general, the gas optical response depends on the relation between the detection length and the memory lengths. When the detection length is less than or comparable to the memory length of some quantity, the EW spectra of the gas are sensitive to the behavior of that quantity during the gas-surface scattering.

Evanescent wave spectroscopy of the gas boundary layer is related to *selective reflection spectroscopy* (SRS) [5]. In selective reflection spectroscopy the frequency-dependent reflectivity of a monochromatic optical beam is monitored which illuminates under near-normal incidence the interface between a dielectric window and a resonant vapor. This technique has already been used at the beginning of last century to deduce atomic dispersion spectra [6]. Although the illuminated gas volume in SRS is proportional to the absorption length, only the gas layer of thickness of the order of a wavelength contributes to the reflection spectrum.

The first evidence of the influence of gas–surface interactions on the observed spectra was given by Cojan in 1954 [7]. He studied the wavelength dependence of the reflection coefficient of light normally incident at an interface between a glass and a mercury vapor near its resonance line. The spectrum was observed to be narrower than the Doppler width when the homogeneous linewidth, $\gamma_\perp$, was less than the inhomogeneous broadening, $2\pi v_T/\lambda$. Cojan [7] made the first attempt to develop a theoretical explanation of SRS taking into account wall collisions. A more appropriate theory providing good qualitative agreement with the experimental results was developed later by Schuurmans [8].

More recent SRS work allowed one to determine the strengths of the attractive [9] as well as resonant repulsive van der Waals interaction [10] between alkali atoms and dielectrics, supported by developed theoretical approaches [11–13]. In terms of the memory lengths the situation corresponds to the case where the polarization memory length is greater than the detection length in SRS, which has the value $\lambda/2\pi$.

2. EVANESCENT WAVE SPECTROSCOPY OF ATOMIC VAPORS

This section deals with applications of evanescent waves to the detailed spectroscopy of atomic vapors near dielectric surfaces.

2.1. Evanescent Wave Absorption Spectroscopy

Boissel and Kerherve reported in 1981 the absorption of the sodium D_2 line when the sodium vapor was excited by an EW at a sapphire prism surface [14]. This can be thought of as the first experiment on *linear* EWS of atomic vapors. The authors calculated—based on the theory of optical response of atoms in the vicinity of a bounding surface—the absorption of the EW in sodium vapor via the reflection coefficient of light from the prism–vapor interface. The boundary conditions for the equation describing the atomic polarization were set separately for the atoms arriving at the prism surface and for the atoms departing from it. For the former ones it was assumed that their polarization equals zero at infinite times before the collision with the surface, whereas the latter ones were assumed to be diffusively scattered by the surface with a zero polarization at the moment of desorption. The number density of sodium atoms was of the order of 10^{16} cm^{-3}; therefore both Lorentz–Lorenz correction and collisional broadening effects had to be incorporated into the theory. Despite these theoretical efforts, a clear discrepancy between the calculations and the experimental results was found. The authors suggested as a possible reason the effect of attraction of sodium vapor by the sapphire surface.

2.2. Nonlinear Doppler-Free Evanescent Wave Spectroscopy

The first experiments on *nonlinear* EWS have been carried out by S. Moneau and co-authors in 1986 at the interface between sodium vapor and a glass prism surface [15] in a Doppler-free pump–probe scheme. Here, two counterpropagating beams from a single-mode continuous wave (cw) tunable dye laser irradiated the prism at an angle of incidence in the vicinity of the critical angle (cf. Fig. 1). One of the beams (pump) was amplitude-modulated, whereas the other one (probe) monitored the reflectivity when scanning across the sodium D_1 and D_2 lines. As the wave vectors of the two EWs were opposite, the rccorded reflection spectra exhibited Doppler-free saturated absorption lineshapes for $\theta_i > \theta_c$. With decreasing θ_i and passing across θ_c, the absorption lineshapes suddenly turned into saturated dispersion lineshapes (Fig. 3). This transition took place within an angular interval of less than 1 milliradian.

The reflectivity change δR can be described in terms of a change of the complex refractive index $\delta n = \delta n_R + i\delta n_I$ and in terms of the angular detuning $\psi = \theta_i - \theta_c$,

$$\delta R \propto \delta n_I/\sqrt{\psi} \qquad \text{for } \psi > 0 \tag{5}$$

$$\propto \delta n_R/\sqrt{|\psi|} \qquad \text{for } \psi < 0 \tag{6}$$

provided that $|\delta n| \ll |\psi| \ll 1$. The saturated absorption lineshapes are dominated by the transit time broadening

$$\Delta \nu_T \approx (2\pi v_T/\lambda)(2\psi/\tan\theta_c)^{1/2} \tag{7}$$

which in general exceeds the natural linewidth of 10 MHz.

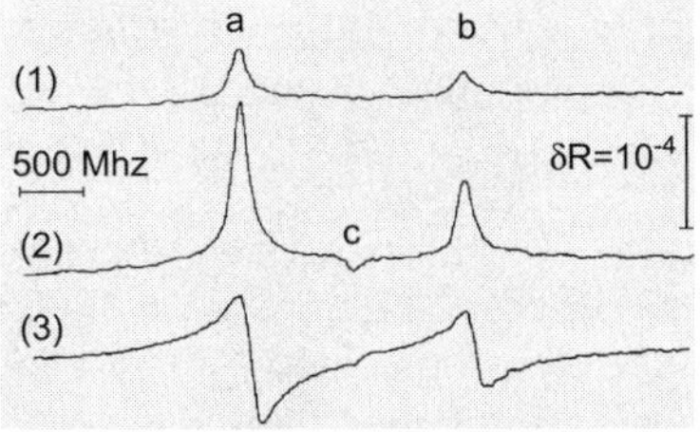

Figure 3. Probe reflectivity change around the sodium D_2 line. The three lines correspond to angles of incidence of $\theta_i - \theta_c = 1.4$ (1), 0.4 (2), and −0.6 mrad (3). Peaks a and b are due to the two hyperfine levels of the sodium ground state, whereas c is a crossover resonance. Reprinted with permission from [15], P. Simoneau et al., *Opt. Commun.* 59, 103 (1986). © 1986, Elsevier Science.

By using well collimated beams, the authors were able to resolve completely the hyperfine structure of the D_1 line and to observe both the main and crossover resonances.

2.3. Evanescent Wave Spectroscopy of Sublevel Resonances

EWS can be used to probe atomic sublevel coherence in the close vicinity of a surface. Such an approach has first been realized for the Zeeman sublevels of the $3\,^2S_{1/2}$ ground state of sodium atoms near a glass prism surface in 1991 [16]. A circularly polarized pump beam resonant to the D_1 line created magnetization in the atomic vapor via optical pumping, whereas the *p*-polarized probe beam excited an EW at the prism surface to monitor the reflectivity of the interface. A magnetic field was applied perpendicular to the plane of incidence of the probe beam and caused a precession of the magnetization of the vapor. This precession corresponds to oscillations between the ground state sublevels $m_z = \pm 1/2$ and therefore to a modulation of the index of refraction of the vapor

$$n_{2\pm} = 1 + \chi_0(1 \mp m_z)/2 \tag{8}$$

where χ_0 is the susceptibility of the unpolarized vapor and the opposite signs refer to the two eigenpolarizations. The angle of incidence of the probe beam was set near the critical angle for total internal reflection and the reflected light beam was analyzed at two different angles (±45°) with respect to the plane of incidence. In the linear regime ($|\chi_0| \ll 1$ with the angular detuning $|\psi| \ll 1$) the difference between the two signals is given by

$$\Delta S \approx -I_0\sqrt{2}n_1(n_1^2 - 1)^{-3/4}\,\mathrm{Re}(\chi_0/\sqrt{\psi})m_z \tag{9}$$

with I_0 the probe beam intensity and n_1 the index of refraction of the prism (the divergence at $\psi = 0$ is induced by invoking the linear approximation). The signal is therefore directly proportional to the magnetization component along the evanescent wave vector.

In order to obtain the highest sensitivity, the pump beam should be modulated with a frequency near the Larmor frequency. Then ΔS is expected to oscillate sinusoidally at the modulation frequency which can be determined via lock-in detection with the modulation frequency as reference.

2.4. Detection of Evanescent Waves by Means of Atomic Spectroscopy

The components along and perpendicular to the interface of an evanescent wave vector determine the Doppler and transit-time broadenings, respectively. They can be deduced by means of analysis of the atomic spectral lineshapes in spectra excited by an EW. This was carried out in 1997 for the D_2 line of a cesium atomic vapor [17].

The cesium vapor cell consisted of a glass cylinder with a truncated glass prism attached to one end of the cylinder. Such a setup allows one to excite the vapor in the cell by both a laser beam propagating perpendicular to the glass–vapor interface (i.e., a volume wave) and an EW travelling along the prism surface. The volume wave resonant to the $6\,^2S_{1/2}(F=4) \leftarrow 6\,^2P_{3/2}(F')$ transition produced velocity-selective hyperfine pumping to the $6\,^2S_{1/2}(F=3)$ state. The change in the population due to optical pumping was monitored by measuring the absorption of the EW at the $6\,^2S_{1/2}(F=3) \leftarrow 6\,^2P_{3/2}(F')$ transition as a function of the pump wave frequency. The resulting absorption spectrum was analyzed in terms of the EW penetration depth [Eq. (4)]. Agreement between theoretical and experimental red wings of the absorption spectra was obtained.

In another experiment the pump laser was switched off and the intensity change of the reflected probe beam due to absorption of the EW at the $6\,^2S_{1/2}(F=3) \leftarrow 6\,^2P_{3/2}(F')$ transition was registered. The width $\Delta\omega$ of the Doppler-broadened absorption line was measured as a function of the angle of incidence for angular values larger than the critical angle. The obtained dependence was in good agreement with the relation $\Delta\omega = k_{\|} v_T \propto \sin\theta_i$ which follows from Eq. (2). The quantity $k_{\|}$ was associated with a "pseudo momentum" $\hbar k_{\|}$, which is conserved during the interaction between an atom and a quantum of an EW.

Saturated absorption spectroscopy utilizing two EWs was used by the same group to demonstrate that the Doppler shift in an EW is determined by the quantity $k_{\|}$ [18]. For that purpose, counterpropagating EWs were generated at the prism surface. One of them (pump) was set at a fixed angle of incidence θ_{pump} slightly larger than the critical angle, whereas the other (probe) was set at various angles θ_{probe} larger than the critical angle. The laser frequency was scanned across the cesium $6\,^2S_{1/2}(F=3) \leftarrow 6\,^2P_{3/2}(F')$ transitions. The observed saturated absorption spectra exhibited, besides Lamb dips at the transition frequencies ω_i and ω_j, also split crossover resonances around the frequency $\omega_{\text{cr}} = (\omega_i + \omega_j)/2$ (Fig. 4). The splitting between the crossover resonance components is equal to

$$\Delta = |\omega_i - \omega_j| \frac{|\sin\theta_{\text{pump}} - \sin\theta_{\text{probe}}|}{\sin\theta_{\text{pump}} + \sin\theta_{\text{probe}}} \qquad (10)$$

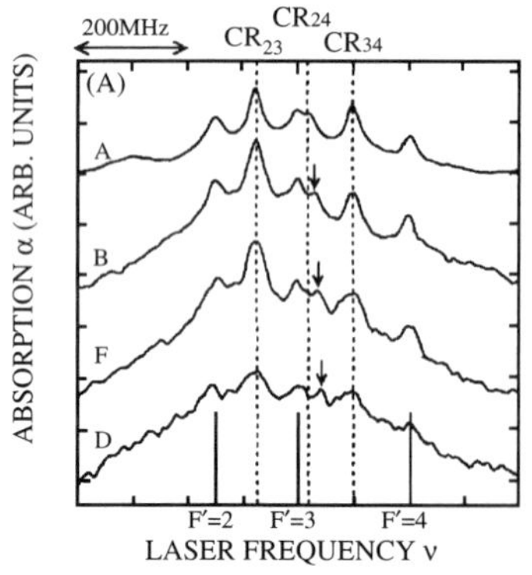

Figure 4. Saturated absorption spectra, induced by counterpropagating evanescent waves from the same laser in a cesium cell with truncated prism. The different scans have been obtained at probe angles of 44, 50.1, 55.5, and 59.4 degrees, respectively (from top to down). "CR" denotes crossover resonances of hyperfine states $F' = 2, 3$, and 4 in the upper state. Reprinted with permission from [18], T. Matsudo et al., *Opt. Commun.* 145, 64 (1998). © 1998, Elsevier Science.

At $\theta_{\text{pump}} = \theta_{\text{probe}}$, the crossover resonance becomes a single peak at ω_{cr} as in the typical scheme of saturated absorption spectroscopy. The measurements carried out at different θ_{probe} thus confirm Eq. (10) and therefore the dependence of the Doppler shift on the angle of incidence.

3. EVANESCENT WAVE SPECTROSCOPY FOR SURFACE SCIENCE

In this section we consider how the strong spatial localization of an EW field near a surface can be utilized for studying the processes of adsorption, desorption, and atom–surface scattering (i.e., typical surface science problems).

3.1. One-Photon Evanescent Wave Spectroscopy

The optical response of atoms desorbing from a surface is essentially transient and their contribution to the gas spectrum differs from that of the atoms arriving at the surface. This phenomenon has a classical analogy. An atom with a transition frequency ω_0 can be represented as an oscillating dipole with an eigenfrequency ω_0. The EW with the frequency ω causes forced oscillations with two components: free oscillations with the frequency ω_0 and oscillations with the frequency of the external force, ω. The former decay exponentially and hence cannot be observed in the steady-state regime. However, in the transient regime the spectrum of the dipole radiation consists of two components, ω and ω_0.

In the case of atomic excitation by an evanescent field quantitative results for that spectrum can be obtained in the density matrix formalism [19, 20]. Decay of the atomic polarization occurs when an atom moves away from the surface. The overall fluorescence intensity is given by an integral over the gas volume, whereas the relative contribution of the free decaying term with respect to that of the forced term can be estimated as $R = (\gamma_\perp \tau)^{-1}$ with τ the time-of-flight across the EW. The atoms moving to the surface are in the steady-state regime and give contributions only to the component at the frequency ω. For typical atomic parameters the quantity R can greatly exceed unity. This means that the dominant contribution to the fluorescence spectrum arises from the desorbing atoms and is centered at the atomic transition frequency ω_0.

These conclusions were confirmed experimentally in fluorescence spectra of sodium atoms excited by an EW resonant to the $3S_{1/2} \leftarrow 3P_{3/2}$ transition [21]. A flux of sodium atoms was directed onto a glass prism surface situated in a vacuum apparatus. The laser exciting the EW was set at a fixed frequency. The resulting fluorescent light was spectrally analyzed with the help of a Fabry–Perot etalon. The fluorescence line of Na atoms was Doppler-broadened with the linewidth determined by the prism surface temperature and not dependent on the atomic flux. This confirmed the theoretical prediction that the fluorescence signal arises mainly from the desorbed atoms. Thus this technique enables one to enhance the spectral contribution of the atoms leaving the surface over that for the atoms impinging on the surface.

3.2. Two-Photon Evanescent Wave Spectroscopy

A rigorous comparison of experimentally determined fluorescence lineshapes with theoretical predictions asks for a good signal-to-noise ratio. This can readily be obtained by applying a two-photon excitation scheme. For example, sodium atoms could be excited by two counterpropagating EWs at a glass prism surface (Fig. 1) [22]. In the experiment, one of the exciting lasers was set in resonance with the $3P_{3/2} \leftarrow 5S_{1/2}$ transition whereas the second laser was scanned across the adjacent $3S_{1/2} \leftarrow 3P_{3/2}$ transition. The resulting fluorescence light from the $3S_{1/2} \leftarrow 4P_{1/2,3/2}$ transitions was observed as a function of detuning of the second laser (Fig. 5). Sodium atoms were directed in the plane of incidence of the laser beams toward the prism surface at an angle of about 60° with respect to the surface normal. Due to the Doppler shift, the exciting EWs were out of resonance with the atoms arriving at the surface whereas they were in resonance with the atoms desorbing from the surface. For this reason, the contribution to the fluorescence spectrum of the atoms from the source itself was negligible.

The mean free path of the sodium atoms greatly exceeded the EW penetration depth. Due to this fact the fluorescence lineshapes were described by a Maxwell distribution modified by a "flux correction" determined by the atomic velocity component normal to the surface, v_z,

$$f_K = \frac{2}{\pi v_T^4} v_z \exp\left(-\frac{v^2}{v_T^2}\right) \quad (11)$$

This correction appears in problems relevant to the statistical properties of gas atoms which strike a surface and results in the well-known cosine law of Knudsen [23].

A rigorous theory of the two-photon evanescent wave spectra [22] gave results that agreed very well with the experimental data. Note that the main difference between two-photon excitation spectra in the nonevanescent and in the evanescent wave regime is that the relaxation rates are affected by transit time broadening and that the Rabi frequency is modified in accordance with the evanescent character of the wave pumping the lower transition. A comparison between the measured and the calculated lineshapes clearly demonstrated that the velocity distribution function of the atoms leaving the surface obeys Knudsen's cosine law.

The gas spectra excited by EWs can also be influenced by the atoms *adsorbed* on the surface. Especially in the case of nonlinear optical spectra this influence is expected since the EW amplitude is determined by the boundary conditions

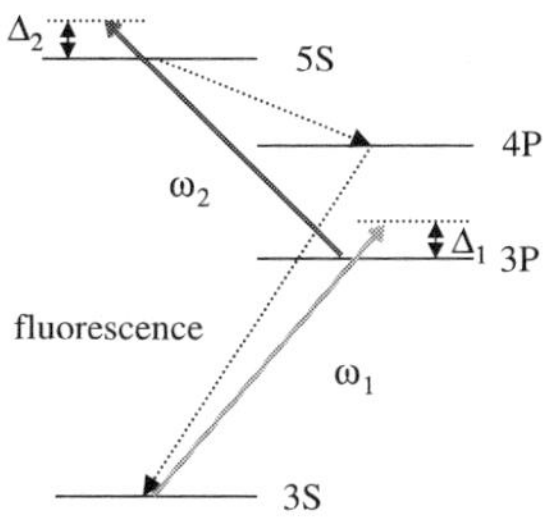

Figure 5. Sodium term scheme for two-photon evanescent wave fluorescence spectroscopy.

at the surface and hence by the adsorbate properties. In particular, the atoms adsorbed at the surface are polarized by the exciting field and thus create a surface displacement current

$$\mathbf{j}(\omega) = -i\omega N_s \alpha_s(\omega) \mathbf{E}_t(\omega) \quad (12)$$

where ω is the EW frequency, N_s is the surface number density, $\alpha_s(\omega)$ is the linear polarizability of an adsorbed atom, $\mathbf{E}_t(\omega)$ is the tangential component of the electric field amplitude at the interface, and the local-field correction has been neglected. Accounting for this current in the boundary conditions for Maxwell's equations one ends up with a dependence of the EW amplitude on the parameters N_s and $\alpha_s(\omega)$. Specifically, these quantities determine the Autler–Townes splitting [24] in the fluorescence spectrum of the gas excited by two EWs resonant to adjacent transitions and thus they become measurable via this splitting (Fig. 6).

This idea was realized for adsorption of sodium atoms at a glass prism surface in the setup described in the preceding section [25, 26]. Both waves were excited near the critical angle for total internal reflection in order to decrease the time-of-flight broadening which would wash out the Autler–Townes splitting. The splitting was then measured as a function of the prism surface temperature. Assuming the Langmuir model of adsorption, it was possible to extract from this dependence the adsorption energy, the pre-exponential factor for the rate of desorption, and the linear polarizability of the adsorbed atoms. The deduced values of the parameters were then used to deduce the surface coverage as a function of the surface temperature. This method obviously can be used for highly accurate *in-situ* monitoring of the surface coverage in a gas atmosphere.

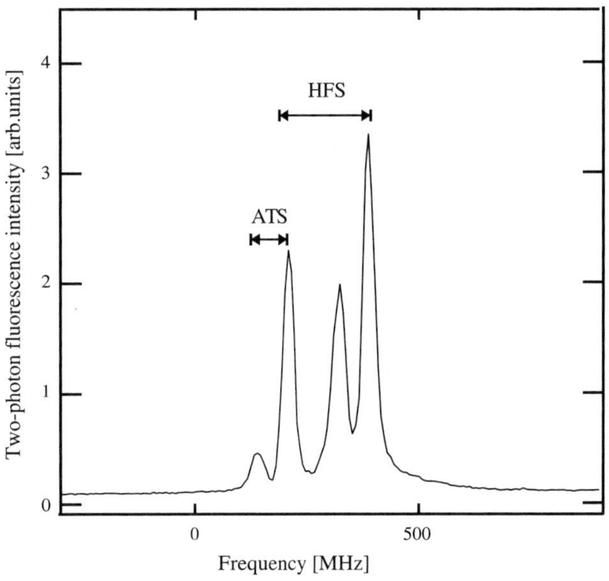

Figure 6. Two-photon sodium spectra in the Autler–Townes-split (ATS) domain. For comparison, the hyperfine structure splitting (HFS) is also noted.

3.3. Two-Photon Evanescent-Volume Wave Spectroscopy

An extremely small EW penetration depth into the gas allows one to excite gas atoms only within the so-called "Knudsen layer" having the thickness of the order of the mean free path [27]. This has been exploited in the previous sections.

Additional spectroscopic possibilities open up if one excites the boundary gas layer by a volume electromagnetic wave propagating normally to the surface in addition to the EW wave. Because of different Doppler shifts for the atoms moving to the surface and away from it, one can distinguish between their contributions. Therefore the combination of EW and volume waves in a two-photon excitation scheme enables one to separate spectrally the contributions of the atoms immediately before the collision with the surface from those immediately after the scattering by it [28, 29]. The Doppler-broadened two-photon fluorescence lineshape is then determined by the velocity distribution functions of the atoms approaching the surface and departing from it (i.e., contains information about the gas–surface scattering dynamics).

The experimental setup for studying sodium atoms scattering at a glass prism surface contained a truncated prism to allow normal incidence of one of the beams onto the glass–sodium vapor interface. The cw lasers excited two adjacent transitions in sodium atoms. Two excitation geometries were investigated: (i) the "normal configuration," where laser 1 resonant to the $3S_{1/2} \leftarrow 3P_{3/2}$ transition excited a volume wave and laser 2 resonant to the $3P_{3/2} \leftarrow 5S_{1/2}$ transition excited an EW (Fig. 7) and (ii) the "inverse configuration" (Fig. 8), where the character of the waves was exchanged.

In both cases laser 1 was set at a fixed frequency whereas laser 2 was scanned across the upper transition. The flux of sodium atoms reached the prism surface at an angle $\theta_0 = 55°$ with respect to the surface normal. The flux was directed in the plane of incidence of the laser beam exciting the EW along its propagation direction.

In the normal configuration and for negative detuning of laser 1, Δ_1, a narrow peak was observed besides the Doppler-broadened line, whose position changed with Δ_1. For positive Δ_1, the intensity of the narrow peak decreased. In the inverse configuration the narrow peak could also be observed, but this time at the left-wing side of the Doppler-broadened line. For negative Δ_1, the narrow peak disappeared. In this case, in contrast to the normal configuration, a clear two-peak shape of the broad fluorescence line was

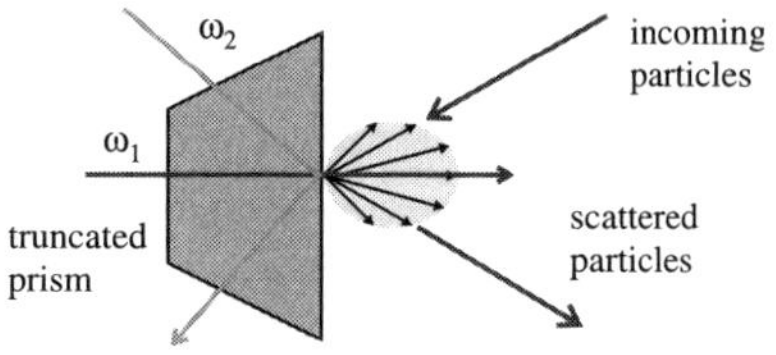

Figure 7. "Normal configuration" for two-photon evanescent-volume wave spectroscopy of sodium atoms, where the pump laser provides Doppler-broadened bulk excitation, whereas the probe laser uses evanescent wave excitation.

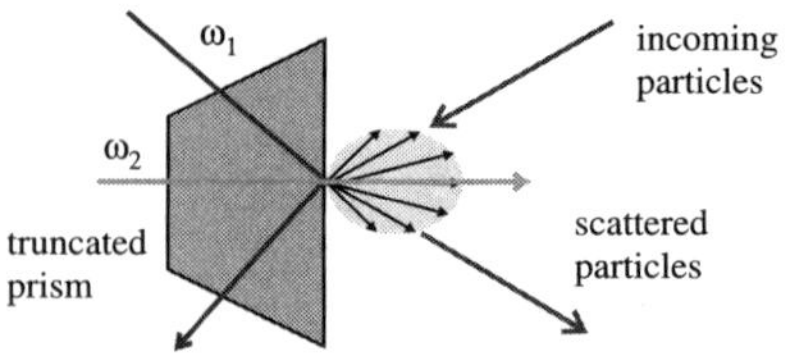

Figure 8. Same as Figure 7, but "inverse configuration," exchanging the character of pump and probe waves.

observed. The separation between the two maxima was independent of either the intensity of laser 1 or its detuning.

The observed features can be explained by the conditions of two-photon resonance in crossed fields taking into account the Doppler shifts for both transitions. For negative detunings of laser 1 only atoms arriving at the surface give contribution to the spectrum. The narrow peak at positive detunings of laser 2 arises from the atoms approaching the surface inside a small angular interval around θ_0 (i.e., emanating directly from the sodium source). The broader line also originates from atoms arriving at the surface but which had been scattered after mutual collisions in front of the surface. Its shape is asymmetric and is determined by a nonequilibrium velocity distribution function of such atoms. For positive detunings of laser 1 only atoms departing from the surface contribute to the spectrum.

In the inverse configuration, the left wing of the spectrum arises from the atoms arriving at the surface whereas the right wing arises from the ones departing from it. The atoms flying directly from the source contribute to the left wing only when Δ_1 is positive. The dip at the line center that corresponds to $\Delta_2 = 0$ arises from the fact that, as follows from Knudsen's cosine law [see Eq. (11)], there are no atoms leaving the surface with $v_z = 0$. This dip is washed out because of the transit-time and power broadenings.

Thus combining measurements in the normal and inverse configurations one can obtain comprehensive information on the *two-dimensional* velocity distribution functions of different groups of atoms in the immediate vicinity of the surface.

The relation between the atoms arriving at the surface and the atoms departing from it is given by the gas–surface scattering kernel, a fundamental quantity in gas–surface interactions.

The scattering kernel $R(\mathbf{v} \to \mathbf{v}')$ is the probability density that an atom striking the surface with velocity between $\mathbf{v}$ and $\mathbf{v} + d\mathbf{v}$ will reemerge at practically the same point with velocity between $\mathbf{v}'$ and $\mathbf{v}' + d\mathbf{v}'$ [30]. It determines the boundary condition for the velocity distribution function $f(\mathbf{v})$ of the gas atoms

$$f^+(\mathbf{v}')v'_z = \int R(\mathbf{v} \to \mathbf{v}')f^-(\mathbf{v})|v_z|\, d\mathbf{v} \qquad (13)$$

where the superscripts $-$ and $+$ denote the atoms arriving at the surface for which $v_z < 0$ and the atoms scattered by the surface for which $v_z > 0$, respectively.

In order to comply with different scattering mechanisms the scattering kernel can be divided into two parts:

$$R(\mathbf{v} \to \mathbf{v}') = R_{\text{des}}(\mathbf{v} \to \mathbf{v}') + R_{\text{dir}}(\mathbf{v} \to \mathbf{v}') \qquad (14)$$

Here, the kernel

$$R_{des}(\mathbf{v} \rightarrow \mathbf{v}') = Sf_K(\mathbf{v}') \tag{15}$$

corresponds to desorption following trapping at the surface with S the sticking probability and f_K determined by Eq. (11). The kernel $R_{dir}(\mathbf{v} \rightarrow \mathbf{v}')$ describes direct scattering. The latter quantity can be expressed in terms of the probability $N(\epsilon, \mathbf{Q})$ that the state of the substrate phonon system after scattering differs from its initial state by energy ϵ and momentum $\mathbf{Q}$ related to the velocities $\mathbf{v}$ and $\mathbf{v}'$ by the conservation laws.

The experiment was carried out in the normal configuration described previously [31]. As noted, the two-photon fluorescence spectrum obtained for positive laser 1 detuning arises from the atoms departing from the surface. At the right wing of such a spectrum the atoms which were scattered without being trapped at the surface (i.e., scattered via a direct channel) contribute to two-photon intensity.

Using an expression for the probability $N(\epsilon, \mathbf{Q})$ derived in the trajectory approximation [32] one obtains by comparing calculated and measured spectra a value of the mean energy transfer to surface phonons. Then from the fitting of the total two-photon signal the mean probability of direct scattering $\eta = 1 - S$ can also be obtained. The measurements carried out for different surface temperatures showed that this quantity decreases on lowering the temperature. This result is consistent with the fact that atoms stick better if the surface is colder.

4. PROSPECTS FOR NANOSCIENCE

Evanescent wave spectroscopy probes the dynamics of an interface region of thickness less than the wavelength of the exciting radiation. In the visible optical range typical detection lengths approach the range of 100 nm—hence this is a dynamic *nanospectroscopy* of the interface. The spectra of gaseous particles at an interface region at such a scale not only display features connected with the interface structure but can also reflect the nonlocality of its optical response. This implies that EW spectra are very sensitive to the interactions between particles of the contacting media at the boundary. A most important prospect of the possibility to distinguish spectrally between the contributions of gas particles arriving at the surface and departing from it is that it should allow us to gain deeper insight into chemical reactions on the surface [33]. Under steady-state conditions a measurement of the decrease of the flux of molecules arriving at the surface is directly related to the surface reaction rate. Such an approach does not require the study of the transient, nonequilibrium stage of the reaction and can be utilized equally well under low and high pressures of reactant gases. It could help to solve the so-called pressure gap problem in catalysis [34]: one needs to study catalytic reactions in combined experiments at low pressure and at high pressure in order to provide a more direct basis for comparison of ideal and technical catalysts. Other surface specific optical techniques such as second harmonic generation, sum-frequency generation, or reflectance anisotropy spectroscopy are based on monitoring signal intensities from adsorbed species. In such a case—in contrast to EWS—the kinetics of a reaction can be followed only during the transient stage.

The EWS method is restricted to gas pressures at which the EW penetration depth into the gas is comparable with one of the memory lengths. For a penetration depth of 100 nm and for a gas at room temperature the upper gas pressure limit is of the order of 1 atm. If the pressure is further increased the interface specific signal disappears due to collisions between the particles. In such a regime the evanescent wave variant of second-order nonlinear optics from the gas such as second harmonic generation or sum-frequency generation could become a viable tool [35]. This is based on the broken inversion symmetry of the gas atoms or molecules in close vicinity to the surface. The distances from the surface where the lowering of the symmetry takes place are determined by the range of the surface potential and are much smaller than any other optical probe scales. Thus the relevant signals are formed very close to the surface and they reflect the details of the gas–surface interaction even at atmospheric or higher gas pressures.

We envision another development of resonant optics at a gas–solid interface that is related to evanescent wave spectroscopy. In the visible spectral range the EW depth of penetration into the gas is less than 1 μm. An atom with a mean thermal velocity passes this distance within a few nanoseconds. Hence the EW profile can be thought of as a "pulse" for an atom moving across it. At gas pressures well below 1 atm typical atomic relaxation times are larger than the "pulse duration" (i.e., the atom interacts coherently with the EW field). Due to this fact, nonstationary coherent optical effects can be observed in the EW spectra [36, 37]. For systems where elastic atom–surface scattering is predominant, such effects will be determined by a phase shift of the atomic wavefunction when the atom is moving across the atom–surface potential.

The methods described could be extended to nanostructured surfaces by means of an additional spatial localization of the fluorescence light in *lateral* directions at the surface. This could possibly be achieved by the tip of a *scanning near-field optical microscope* (SNOM) [38] for collecting the fluorescence photons. In such a modification of a SNOM, called a *photon scanning tunneling microscope* (PSTM) [39], photons from an internally reflected light beam tunnel through the region between a sample surface and a probe tip and are collected by a detection system. The presence of a sample modulates the waveform of the evanescent field, and this modulation manifests itself as a spatial variation in the field intensity at a given height above the sample surface (Fig. 9).

Instead of using evanescent wave excitation via the surface one could also imagine exciting fluorescence light within nanoscaled objects, which then are probed via near-field detection. As demonstrated in Figure 10 via measurements on light-emitting nanofibers on a mica surface, the changes in intensity picked up by the fiber probe provide topographical information about the sample surface as well as information about the optical properties of the sample. Such a scheme can be used in a spectroscopic mode by detecting the signal as a function of the wavelength of the exciting radiation.

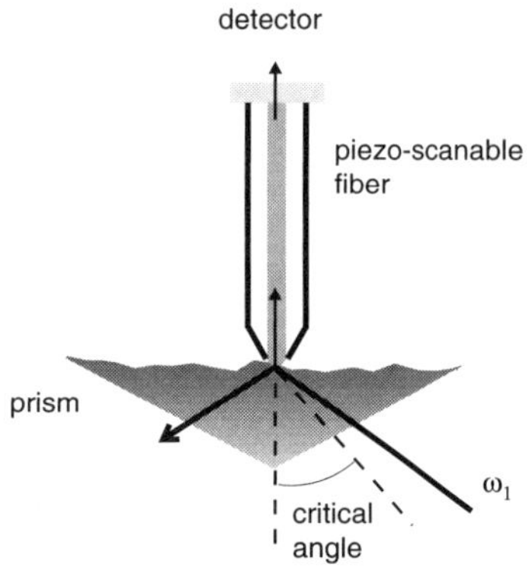

Figure 9. Principle of photon scanning tunneling microscopy, a SNOM variant that explicitly uses evanescent waves.

One of the crucial points for the success of such a spectroscopic application is the sensitivity of the detection system to a weak signal arising from a small number of molecules below the PSTM tip. Suppose one investigates the spectra of a gas in the volume beneath the PSTM tip with the dimensions $100 \times 100 \times 100\ \text{nm}^3$. At gas pressures of the order of 1 atm such a volume contains about $N = 10^4$ molecules. Suppose now that the intensity of the EW resonantly exciting the gas molecules is high enough to saturate the molecular transition. Then the number of photons emitted per second equals $\gamma_{||} \times N/2$. With typically $\gamma_{||} = 10^7\ \text{s}^{-1}$ one obtains a fluorescence intensity of 5×10^{10} photons per second. Transmissivities of fiber-optic scanning tunneling probes are of the order of 10^{-5} [38], and so we conclude that such a fluorescence signal can be detected in the PSTM scheme.

Finally, let us note that although EWS has been demonstrated to be a powerful tool for the analysis of structure and composition of various interfaces, its full potential for solving problems of surface science, heterogeneous catalysis, and rarefied gas dynamics has not yet been realized. In particular, the possibility to identify spectrally the gas fluxes to the surface and apart from it in EWS combined with the PSTM scheme could provide a tool for monitoring the formation process of nanostructures on a surface (*nanofabrication*) and for studying catalysis on nanostructured surfaces (*nanocatalysis*).

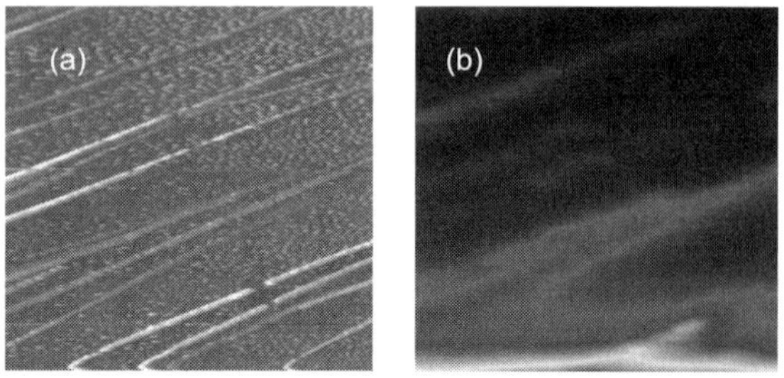

Figure 10. Photon scanning tunneling microscopy on nanoscaled optical fibers ("nanofibers") made of fluorescing organic molecules [40]. The nanofibers are excited in the far field via focused ultraviolet light from the lower side of the image and probed in the near field. Image size $50 \times 50\ \mu\text{m}^2$. (a) Topographic image. (b) Optical image. Waveguiding along selected fibers is clearly visible.

GLOSSARY

Evanescent wave (EW) An electromagnetic wave having an amplitude decreasing in the direction perpendicular to its wave vector at a scale of the wavelength.

Evanescent wave spectroscopy (EWS) Spectroscopy using EWs for excitation of optical spectra.

Gas boundary layer (GBL) The gas layer bordering a surface for which the optical response differs from that in the bulk gas.

Photon scanning tunneling microscopy (PSTM) The kind of scanning microscopy where photons from an internally reflected light beam tunnel through the region between a sample surface and a probe tip and are collected by a detection system.

Scanning near-field optical microscope (SNOM) A microscope where a subwavelength-sized light source or detector (viz., the optical probe) is placed at subwavelength distances from the sample surface, in the region called "near field." The probe can scan over the surface thus providing its imaging.

ACKNOWLEDGMENTS

It is a pleasure to acknowledge the experimental help of our colleagues F. Balzer, C. Henkel, L. Jozefowski, A. Lindinger, and J. Loerke in the course of the cited work of the authors. We also thank the Danish Research Agency SNF for supporting this work.

REFERENCES

1. M. Born and E. Wolf, "Principles of Optics." Pergamon Press, Oxford, 1975.
2. R. A. Potyrailo, S. E. Hobbs, and G. M. Hieftje, *Fresenius J. Anal. Chem.* 362, 349 (1998).
3. F. de Fornel, "Evanescent Waves." Springer, Berlin, 2000.
4. L. Allen and J. H. Eberly, "Optical Resonance and Two-Level Atoms." Wiley, New York, 1975.
5. P. R. Berman, M. Ducloy, and D. Bloch, *Phys. Rev. A* 63, 043410 (2001), and references therein.
6. R. Wood, *Philos. Mag.* 18, 187 (1909).
7. J. L. Cojan, *Ann. Phys. (Paris)* 9, 385 (1954).
8. M. F. H. Schuurmans, *J. Phys. (Paris)* 37, 469 (1976).
9. M. Oria, M. Chevrollier, D. Bloch, M. Fichet, and M. Ducloy, *Europhys. Lett.* 14, 527 (1991).
10. H. Failache, S. Saltiel, M. Fichet, D. Bloch, and M. Ducloy, *Phys. Rev. Lett.* 83, 5467 (1999).
11. M. Fichet, F. Schuller, D. Bloch, and M. Ducloy, *Phys. Rev. A* 51, 1553 (1995).
12. J. Guo, J. Cooper, and A. Gallagher, *Phys. Rev. A* 53, 1130 (1996).
13. T. A. Vartanyan and A. Weis, *Phys. Rev. A* 63, 063813 (2001).
14. P. Boissel and F. Kerherve, *Opt. Commun.* 37, 397 (1981).
15. P. Simoneau, S. Le Boiteaux, Cid B. De Araujo, D. Bloch, J. R. Rios Leite, and M. Ducloy, *Opt. Commun.* 59, 103 (1986).
16. D. Suter, J. Äbersold, and J. Mlynek, *Opt. Commun.* 84, 269 (1991).
17. T. Matsudo, H. Hori, T. Inoue, H. Iwata, Y. Inoue, and T. Sakurai, *Phys. Rev. A* 55, 2406 (1997).
18. T. Matsudo, Y. Takahara, H. Hori, and T. Sakurai, *Opt. Commun.* 145, 64 (1998).
19. V. G. Bordo, *Sov. Phys. JETP* 74, 137 (1992).
20. V. G. Bordo, *Surf. Sci.* 269/270, 169 (1992).

21. V. G. Bordo, C. Henkel, A. Lindinger, and H.-G. Rubahn, *Opt. Commun.* 137, 249 (1997).
22. V. G. Bordo and H.-G. Rubahn, *Phys. Rev. A* 60, 1538 (1999).
23. F. O. Goodman and H. Y. Wachman, "Dynamics of Gas-Surface Scattering." Academic Press, New York, 1976.
24. C. Delsart and J.-C. Keller, *J. Phys.* (*Paris*) 39, 350 (1978).
25. V. G. Bordo and H.-G. Rubahn, *Opt. Express* 4, 59 (1999).
26. V. G. Bordo and H.-G. Rubahn, *Phys. Status Solidi A* 175, 265 (1999).
27. C. Cercignani, "Rarefied Gas Dynamics. From Basic Concepts to Actual Calculations." Cambridge Univ. Press, Cambridge, UK, 2000.
28. V. G. Bordo, J. Loerke, and H.-G. Rubahn, *Phys. Rev. Lett.* 86, 1490 (2001).
29. V. G. Bordo, J. Loerke, L. Jozefowski, and H.-G. Rubahn, *Phys. Rev. A* 64, 012903 (2001).
30. E. M. Lifshitz and L. P. Pitaevskii, "Physical Kinetics." Pergamon Press, Oxford, 1986.
31. V. G. Bordo and H.-G. Rubahn, *Phys. Rev. A* 67, 012901 (2003).
32. R. Brako and D. M. Newns, *Phys. Rev. Lett.* 48, 1859 (1982); *Surf. Sci.* 117, 42 (1982).
33. V. G. Bordo, *Phys. Status Solidi A* 175, 271 (1999).
34. H. P. Bonzel, *Surf. Sci.* 68, 236 (1977).
35. V. G. Bordo, *Opt. Commun.* 132, 67 (1996).
36. V. G. Bordo, *Opt. Commun.* 101, 37 (1993).
37. V. G. Bordo, *Opt. Commun.* 111, 61 (1994).
38. H. Heinzelmann and D. W. Pohl, *Appl. Phys. A* 59, 89 (1994).
39. R. C. Reddick, R. J. Warmack, and T. L. Ferrel, *Phys. Rev. B* 39, 767 (1989).
40. V. S. Volkov, S. I. Bozhevolnyi, and H.-G. Rubahn, in preparation.

Encyclopedia of Nanoscience and Nanotechnology
www.aspbs.com/enn

Extreme Ultraviolet Nanolithography

Paul B. Mirkarimi
Lawrence Livermore National Laboratory, Livermore, California, USA

CONTENTS

1. INTRODUCTION

Moore's law predicts that the number of transistors or memory bits on integrated circuits (ICs) will double about every 1.5–2.0 years [1, 2]. Adherence to Moore's law is believed to be necessary to continue to decrease costs per transistor or memory bit; one important practical consequence of this is the production of much faster computers at constant or decreasing prices. For Moore's law to continue, critical features such as transistor gate lines will need to shrink to several tens of nanometers in width, bringing the process within the realm of nanotechnology. Accompanying this change will be a transition from conventional microlithography techniques to nanolithography techiques. The leading nanolithography candidate for producing feature sizes of below 45 nm in the high-volume manufacturing of integrated circuits is extreme ultraviolet lithography (EUVL) [3–5]. Because of the potential impact of the technology, it has been well covered in the popular literature [1, 6–10] as well as in scientific journals.

The highest resolution lithography used in high-volume integrated circuit manufacturing employs ultraviolet (UV) light with a wavelength of 248 nm, although the industry is in the process of implementing lithography tools/processes that employ light with a wavelength of 193 nm [2, 5]. Before the insertion of EUV lithography, it is also possible that lithography technology utilizing light with a wavelength of 157 nm will be used (likely for manufacturing 65 nm feature sizes) [2, 11]. The particular numbers of 248 nm, 193 nm, and 157 nm for the wavelength are chosen because there are lasers, which serve as the source of light for these lithographic technologies, that operate at these wavelengths. EUV lithography will use light with a wavelength of approximately 13 nm and it should be noted that there are no suitable lasers operating near this wavelength. The 13-nm wavelength was chosen based on the available source technology and on the wavelength region for optimal EUV reflectivity for the Mo/Si multilayer-coated optics, which will be discussed later. One advantage of EUV lithography is that the relatively low wavelength implies that it would enable the semiconductor industry to adhere to Moore's law for several generations of ICs.

In lithography light is shone on a partially transmitting mask with slits in the shape of the features to be transferred onto a chip. The light pattern emanating from the mask is reduced in size by a factor of 4 and projected onto an area of a silicon wafer. The silicon wafer is coated with a polymer film called a resist. The resists are designed such that a small exposure to the UV light changes the structure of the material. For negative resists, the UV light exposure may cause the polymeric chains to crosslink, making them more resistant to etching. For positive resists, the UV light exposure may break bonds, making the resist more easily etched. Etchants can then be used which are designed to preferentially etch either the exposed areas of the resist (for a negative resist) or the unexposed areas (for a positive resist). The developed resist can then act as a template for subsequent processes. For example, exposed Si could be doped with impurities by ion irradiation or etched and filled in with metals and/or insulators, to make a circuit pattern. It is also possible to have coatings between the Si wafer and resist, and the patterns can then be formed in these coatings. Real-world integrated circuits are produced with processes that have many different coating and etch steps. Leading edge lithography techniques, such as EUV lithography, will be first used for the etch steps that require the smallest features. The primary difference between EUVL and previous lithographic technologies is that EUVL uses *reflective* optics and a *reflective* mask. This is due to the fact that sufficiently thick and robust materials with sufficient transmission at EUV wavelengths do not exist. The development of suitable Mo/Si multilayer coatings with high reflectivities in the EUV wavelength range of ~13 nm have enabled this technology to move forward [12–24]. The development of this nanolithographic technique is driving other areas

ISBN: 1-58883-059-4/$35.00

Encyclopedia of Nanoscience and Nanotechnology
Edited by H. S. Nalwa
Volume 3: Pages (297–306)

further into the nanoscale regime. For example, the aspheric optic substrates require a figure (shape) and a finish of a few tenths of a nanometer be achieved simultaneously [19, 25]. Interferometry has had to be advanced to enable one to accurately measure the figure to a few tenths of a nanometer on an aspheric surface [26, 27]. The multilayer coatings require thickness control to a few tenths of a nanometer and the interfaces must be controlled with sub-nanometer accuracy to maintain a high EUV reflectivity [16, 24, 28]. In the mask, defects in the multilayer that are as small as several nanometer, in height and several tens of nanometer, in width are unacceptable, and defects on top of the multilayer that are only several tens of nanometer, in size are unacceptable [29–31].

There have been a few reviews of EUVL technology in the past several years [32–35]. In this chapter, the primary aspects of EUVL technology are covered with an emphasis on the most recent results; in particular, an emphasis is placed on the multilayer-based mask blank, which is one of the highest risk areas for the technology and where new nanoscience and nanotechnology is being developed to reduce that risk.

2. EXTREME ULTRAVIOLET LITHOGRAPHY (EUVL) SYSTEM OVERVIEW

A schematic of the components of an all-reflective EUVL system [36–38] is shown in Figure 1, and in Figure 2 is an illustration of the components of a prototype EUVL system called the Engineering Test Stand [36, 39]. The *source* sends EUV light to condenser optics which serve to collect and shape the light. Condenser mirrors which reflect light at near-grazing incidence can be made with metallic-coated substrates. Condenser mirrors needed to reflect EUV at other angles need to be multilayer-coated to achieve a sufficient reflectivity. The EUV light is then used to illuminate a mask. The mask consists of a multilayer-coated substrate that has absorbing lines deposited on it in a particular pattern. The EUV light impinging on the multilayer is reflected and the light impinging on the absorber lines is not reflected. Thus the EUV light emanating from the mask is representative of the pattern on the mask. The EUV light from the mask is then sent to the projection optics, which serve to

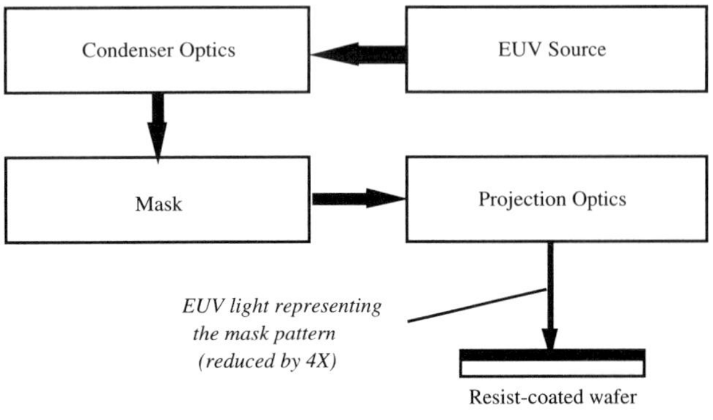

Figure 1. Schematic illustration of the components of an EUV lithography tool.

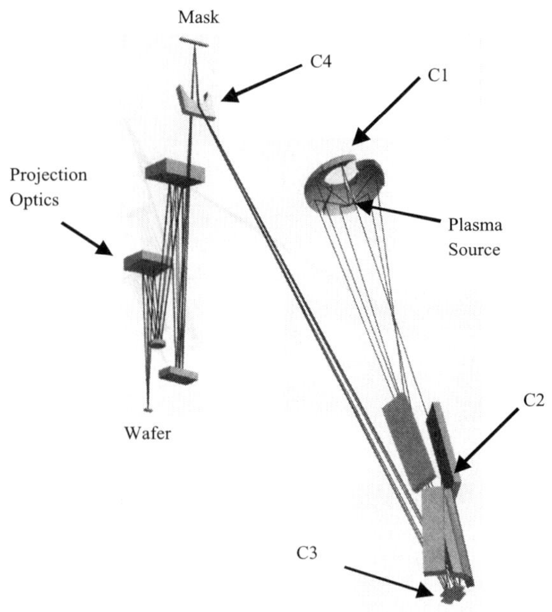

Figure 2. Schematic illustration of a prototype EUV lithography tool called the EUV Test Stand. The optics with "C" in front of them are the condenser, or illuminator optics.

reduce the pattern size by a factor of 4. This reduced pattern of EUV light impinges on a resist-coated wafer, where the resist is designed to be sensitive to EUV light. A pattern is formed on the resist-coated wafer as described in Section 1.

Wafer throughput is an important consideration in designing and building an EUV lithography tool since it has a strong impact on the cost of ownership, which essentially represents the economic viability of the lithography process. The current reported target for commercial EUV lithography tools is 80 wafers per hour [40], although that number is likely to increase [41]. The main driver for throughput is the intensity of EUV light arriving at the wafer surface and the (energy) sensitivity of the resist material. In terms of EUV light intensity, the power of the EUV source and the reflectivity of the multilayer coatings have the most dramatic effect. As is discussed in Section 3.3, while there have been improvements in the sources, significant advances are still required for the technology to be viable. The multilayer coatings have EUV reflectivities approaching 70% [23, 24] and do not represent a significant obstacle in the advancement of EUVL technology. For the resist, significant advances have been reported in synthesizing resist materials requiring as little as ~4 mJ/cm, [2, 42].

Another important consideration is wafer yield, which is impacted significantly by the quality of the mask. The pattern of the mask is copied to thousands of wafers in a high-volume production environment, and imperfections in the mask can lead to imperfections in the integrated circuits being synthesized on the wafers. Fewer acceptable integrated circuits means a lower yield and a larger cost of ownership. Like the EUV source, improvements in the mask have been made but advances are still required to make the technology commercially viable. The reflective nature of the mask makes it more susceptible to imperfections, as will be discussed in Section 3.2.

3. PRIMARY EUVL SYSTEM COMPONENTS

3.1. Optics

Unlike conventional optical microlithography techniques which employ transmissive optics, EUV lithography employs reflective optics. The first set of optics, commonly referred to as the *illuminator* or *condenser* optics, serve to collect and shape the EUV light from the source to illuminate the mask. For illuminator optics designed to reflect at grazing incidence, simple metallic films can have sufficient EUV reflectivity at those angles. For illuminator optics designed to reflect EUV light at angles closer to normal incidence, multilayer films are required. To reflect near 13.4 nm at near-normal incidence, Mo/Si multilayer films must have a bilayer period thickness (the thickness of one Mo layer plus one Si layer) of about 7 nm. Typically 40–60 bilayer periods are deposited. A schematic illustration of a reflective Mo/Si multilayer film is shown along with a cross-sectional transmission electron micrograph of a real Mo/Si multilayer in Figure 3.

Light reflected from the mask passes to the *projection* optics. Several of the projection optics have an aspherical shape, and require *simultaneous* control of the aspheric figure (shape) and surface finish (roughness) to a few tenths of a nanometer, levels not required previously. The finish requirement is driven primarily by the fact that the EUV reflectivity of the multilayer coatings, which can have a significant impact on the throughput of the EUVL system because of the number of optics, is a function of the substrate finish. The reflective coatings also cannot increase the figure error by more than a tenth of a nanometer or so, which requires that the coating thickness profile across the optic be controlled with extreme precision [22, 28, 43]. Coated optics have been successfully produced for the EUVL prototype tool called the Engineering Test Stand [28, 43] (ETS) and for other, smaller EUV tools [44]. The coatings met the specifications for commercial-grade EUVL optics but the substrate figure/finish needs further improvement [43].

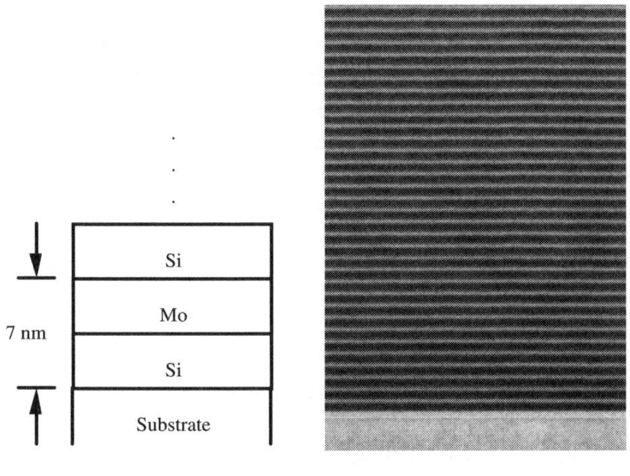

Figure 3. Schematic illustration of a reflective multilayer film along with a cross-sectional transmission electron micrograph of a Mo/Si multilayer film.

It is important that the spatial position of the projection optics not shift in the EUVL tool due to temperature fluctuations. This has led to the requirement that the substrate be made of low thermal expansion coefficient materials (LTEM). Zerodur and ULE are two of the leading LTEM candidates, with near-zero coefficients of thermal expansion (CTE). Zerodur has two phases (crystalline and amorphous), approximately twelve components, and consists mainly of SiO_2 and Al_2O_3 [45, 46]. The crystalline and amorphous phases have CTEs with opposite signs which enables the near-zero CTE to be achieved. ULE is a single-phase amorphous material consisting of SiO_2 and TiO_2 [47, 48].

The primary challenge with the optic substrates is to develop a process to manufacture sufficient numbers of optic substrates meeting the stringent figure and finish specifications with acceptable costs. For the coatings, there are several areas where work is needed. One is to develop multilayer films capable of withstanding exposure to EUV light over years in a high (but not ultrahigh) vacuum environment, where oxidation and carbon contamination could degrade the reflectivity (and hence wafer throughput) of the EUVL tool over time. Ru-capped multilayer films currently show promise for this application [24, 49]. More on environmental issues can be found in Section 4.2. Another area is to develop multilayer coatings in which the reflectance is not degraded significantly for substrates with roughness levels of a few tenths of a nanometer. EUV reflectivities approaching 70% near $\lambda = 13.4$ nm have been achieved for Mo/Si on ultrasmooth Si substrates [23, 24], and a goal is to achieve similar results on somewhat rougher LTEM for the projection optics. Conventional magnetron sputtering deposition techniques have yielded Mo/Si coatings with EUV reflectivity and thickness profile control values that meet or exceed the values from any competing deposition techniques; however, the EUV reflectivity is very sensitive to substrate roughness [50, 51] and it may be difficult to manufacture relatively large quantities of projection optic substrates that meet the ~0.1 nm desired values for the surface finish. Recent results with ion beam sputter deposition [52] show promise in this area and there is also an indication of work on this topic using ion-assisted electron beam deposition [23].

Mo/Si films typically have a stress of roughly 400–450 MPa [53, 54] and coating stress is an issue in that it has the potential to deform the precisely figured optics. One challenge is to mitigate stress effects without degrading the coating reflectivity. The use of compensating buffer layers [53], annealing [55–57], and other techniques [54, 58] have shown promise. There has been theoretical work suggesting that coating stress is not a significant issue in that most of the deformation due to stress will be spherical, and spherical deformation is expected to be compensable in the lithographic process [59, 60]. This theoretical work focused on the requirements of a prototype EUVL tool and it is possible that coating stress could become an issue for more advanced EUV lithographic tools. The arsenal of tools developed to combat stress should make this a tractable problem if it does arise.

3.2. Mask

EUV lithography employs a reflective mask. A schematic illustrating the difference between reflective and transmissive masks is shown in Figure 4. The EUVL mask blank consists of a square format LTEM substrate such as ULE or zerodur coated with a reflective Mo/Si multilayer thin film. LTEM is used so that thermal distortion of the masks will be reduced, since thermal distortion can shift and blur the light pattern coming from the mask [61]. The multilayer film is then coated with a buffer layer and an absorber layer, and processed with an electron beam lithographic tool to form a patterned EUVL mask.

A key challenge in making EUV lithography commercially viable is the ability to produce nearly defect-free mask blanks. *Critical* defects are imperfections that will cause some form of error in the pattern of light reflected from the mask, resulting in imperfections in the integrated circuit pattern formed on the wafer. Recent estimates suggest that a mask blank defect density of ~0.0025 critical defects per cm^2 will be needed [62]. This means that there can be an average of only ~1 of these defects per two 6" × 6" square mask blank, and defects as small as several tens of nanometers in diameter can be critical defects, as will be discussed later in this section.

There are two primary reasons why the defect requirements are more stringent for EUVL masks. The first reason, which is not inherent to EUV lithography, is the fact that the patterns of the mask need to be made smaller because the feature sizes to be printed are getting smaller as Moore's law is followed. This would mean that smaller defect sizes will become more of an issue for any mask technology. Note that the implementation of techniques such as phase shifting with the 248 and 193-nm lithographies is also likely to make smaller defect sizes more of an issue with transmissive masks.

The second reason for more stringent defect requirements, which is inherent to EUV lithography, is the reflective nature of the mask. This reflective mask gives rise to a new class of mask defects, referred to as multilayer phase defects. Particles as small as a few tens of nanometers in diameter on the substrate disturb the multilayer as it is deposited, resulting in the layers over the particle being shifted in height relative to adjacent regions. As illustrated in Figure 5a, EUV light reflected from the multilayer in the affected region will be out of phase with the light reflected from adjacent regions. If the size of the defective multilayer is large enough, it will result in a defect in the lithographically processed wafer. It is most likely to result in what is called a proximity defect, where a lithographically printed feature such as a line has an unacceptable variation in width.

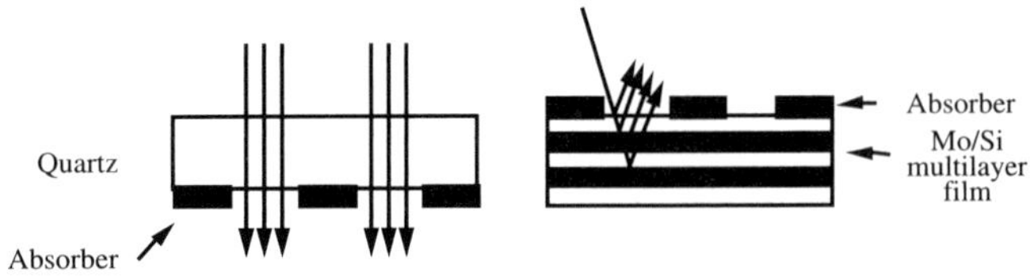

Figure 4. Schematic illustration of a conventional transmission mask and an EUVL reflecting mask.

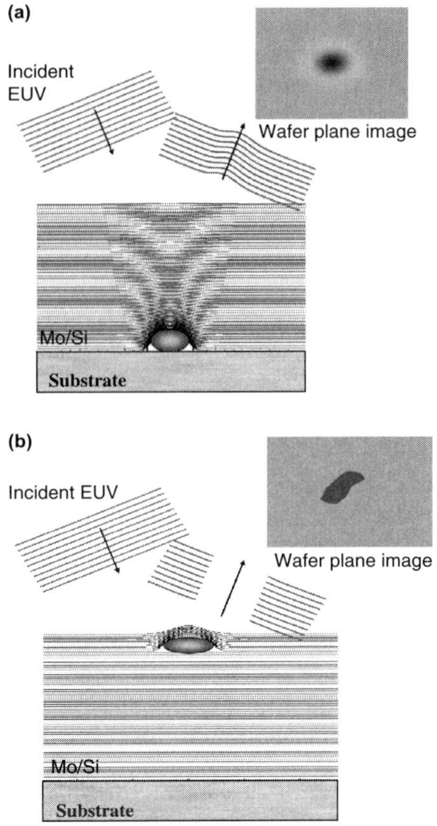

Figure 5. Schematic illustration of (a) a phase defect and (b) an amplitude defect in an EUVL mask blank.

This linewidth variation is often referred to as a CD (critical dimension) variation and it can affect functions such as timing in an integrated circuit. It should be noted that multilayer phase defects can result from particles arising early in the deposition process as well as from particles on the substrate.

The other type of mask blank defect is commonly referred to as an amplitude defect. This is illustrated in Figure 5b. This defect reduces the reflectivity of the multilayer in an area such that it results in an alteration of the lithographic features being formed on the wafer. Prime examples would be an absorbing particle sitting on top of the multilayer or a particle embedded in the top layers of the multilayer film. This defect is analogous to absorbing particles sitting in the transmissive area of a conventional transmissive mask.

In order to investigate the response of multilayer coatings to nanoscale substrate particles, a technique was developed to deposit monodisperse Au nanoparticles on Si substrates [63]. It was observed that the angle of incidence of the deposition flux can have a dramatic effect on the evolution of the multilayer defects [64, 65]. This is shown by the XTEM images in Figure 6. For near-normal incidence deposition, the multilayer tends to smooth over the substrate particle, mitigating its effect; however, for off-normal incidence deposition, this does not occur. It has also been suggested that the resputtering of Mo and Si during the ion beam sputter deposition process can enhance the smoothing properties of the coating [66].

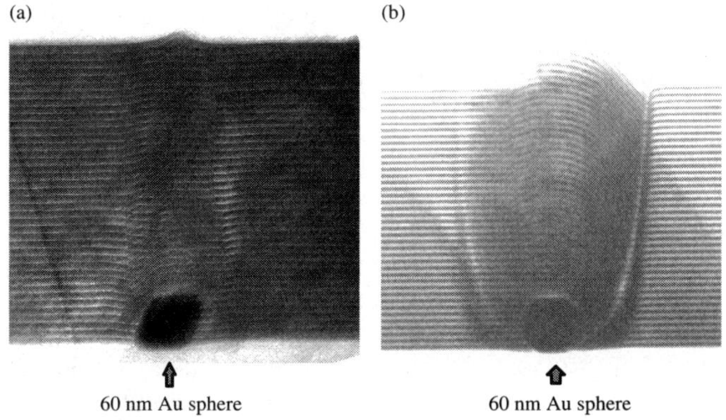

Figure 6. Cross-sectional transmission electron micrographs of Mo/Si films deposited on 60-nm-diameter substrate particles with the deposition flux at (a) near-normal incidence (~8°) and (b) off-normal incidence (~48°) with respect to the substrate normal.

Substrate particles <50 nm in diameter are a particular concern since it is unknown if metrology will be available to reliably detect these very small particles on square format LTEM substrates. To attempt to clean the substrates of these particles, one would need a reliable metrology tool, and even then it is unknown if the particles can be removed without roughening or otherwise damaging the substrate. A strategy was suggested to enhance the smoothing properties of the Mo/Si coatings to planarize the substrates, making it as if the small particles never existed [67]. Progress has been made in this area [68], and substrate particles as large as 50 nm diameter can be smoothed to yield a surface with a defect height of ~1 nm [69], making the defect benign according to the EUV printability theory [29]. The key process step involves etching of the Si layers in the Mo/Si multilayer [69]. This smoothing process also yields excellent film thickness/reflectance uniformity, good EUV reflectivity, and reasonable film stress, and also smooths substrate roughness [52, 69]. One outstanding issue for the smoothing technique is that it must be compatible with ultraclean deposition processes, that is, the introduction of the secondary ion source for etching cannot introduce significant amounts of particles. Another issue is that the effect of smoothing on substrate pits has not been quantified yet, although it has been reported using a different Mo/Si process that no significant difference was observed in the smoothing of substrate bumps versus divots when the substrate defects were very small (3–5 nm in height) [66].

Methods are also being developed to clean the substrates of these particles. A supersonic hydrocleaning method [70] is being developed which mixes deionized water with a carrier gas; the liquid breaks into droplets which are accelerated to a high velocity by the carrier gas and directed towards the substrate surface to remove particles. The technique shows promise and a relatively high particle removal efficiency has been reported [70]; however, the effect of this technique on the surface finish of the substrate was not discussed, which can be an important factor in evaluating potential cleaning techniques.

When there are errors in the metal patterns on a conventional transmissive mask, they can usually be repaired by removing or adding material with a focused ion beam (FIB) tool. This technique can also be used with the patterned EUVL mask to repair absorber features [71, 72]. It had been said that one of the drawbacks to EUV lithography was that it was not possible to repair the multilayer-based mask blank, since this FIB-technique would not work for multilayer defects. In response to this, several techniques have been developed to address this issue [73–75]. One technique developed to repair multilayer phase defects entails irradiating the defective area with a 10–20 kV, high current electron beam [73, 76]. This heats the multilayer in the defective region and activates silicide formation between the layers, causing them to contract. A temperature approaching 800 °C is believed to be needed to achieve the necessary contraction; such a temperature would severely degrade the multilayer structure [77–79] except for the fact that the annealing is to be done over a very short period of time (milliseconds). It was experimentally demonstrated that an electron beam could be used to controllably produce depressions in Mo/Si multilayer films with nanometer-scale depths without any significant degradation in the EUV reflectivity [73]. Due to tool limitations, this demonstration occurred over lateral sizes of hundreds of microns. To be viable, the lateral size of the beam footprint will need to be reduced to the order of 100 nm, which, considering the energy and current requirements, may require further optimization of current electron beam column design. This is an example of EUV lithography driving other technologies further into the nanoscale regime.

Another technique designed to repair amplitude defects in or on the multilayer film entails milling out a small crater in the defective area [74, 80]. According to theory, it is possible to put a crater in the multilayer which does not become a defect if (i) there is a sufficient number of unaffected multilayer bilayers under the crater to reflect the EUV light, and (ii) the slope of the sidewalls of the crater is shallow enough. Condition (ii) puts the minimum lateral size of the crater at several microns and condition (i) limits the depth to <20 bilayers if a 60-bilayer multilayer is used. It will also be necessary to put a capping layer on the surface of the crater [74] since otherwise there would be exposed Mo layers within the crater and Mo oxidation has been shown to result in a significant decrease in EUV reflectivity [81]. Craters have been produced in multilayer films with the appropriate depths of several tens of nanometers using an ion beam, but the lateral size was on the order of millimeters and needs to be reduced to <5 μm eventually. Thin Si was used to coat the crater surface and the reflectivity loss inside the crater was reduced to several percent; further improvement will be needed [74, 82].

A key challenge is to produce multilayer coatings with nearly no defects added during the coating process. Significant progress was made in the mid-1990s, taking a process yielding roughly thousands of defects/cm^2 to a process yielding ~0.10 defects/cm^2 for particles 130 nm in diameter and larger [83, 84]. In addition to improving the cleanliness of the handling processes, one of the changes believed to be responsible for this advance was a move to ion beam sputtering from conventional magnetron sputtering [84, 85]. Most mask blank development work is now done with ion beam sputtering [86–88]. Two reasons possibly differentiating ion beam and magnetron sputter deposition processes with respect to particles are (i) conventional magnetron sputtering can produce particle sources/traps in the immediate vicinity of the magnetron sources [89–91], and (ii) Mo/Si multilayers tend to amplify the effect of substrate particles, likely due to the off-normal flux components inherent to conventional magnetron sputtering [64]. It should be noted that by employing rotating magnets in the magnetron sources and by collimating the deposition flux, it should be possible to mitigate the above effects and make the magnetron deposition process more competitive for this application. Note that magnetron sputtering is competitive for the coating of optics because small coating defects are not currently considered to be a significant concern for the optics.

Subsequent progress in defect reduction during the coating process has been made since 1997, but at a slower rate, where the best repeatable defect levels are ~0.05 defects/cm^2 for defects 90–100 nm and larger [92, 93]. The difficulty of achieving sufficiently low defect densities during the coating process is one of the most significant challenges facing EUV lithography. Development of the repair techniques discussed above could relax the defect specification during the coating process [94].

As mentioned earlier, the multilayer-based mask blank must be processed to produce a pattern of absorbing lines and open spaces on the multilayer surface. A buffer layer is first deposited to facilitate the patterning of the absorber layer while protecting the multilayer from damage. A leading candidate for the buffer layer is SiO_2 and for the absorber layer it is Cr or TaN [95–97]. It should be noted that Ru is being considered not only as a capping layer for the optics, but also for the mask, where it has the potential to replace SiO_2 as leading candidate for the buffer layer [98]. For additional information on the topic of EUVL mask patterning, the reader is referred to papers by Hector [94] and Hector and Mangat [99].

There are two techniques to mitigate the effects of multilayer defects during and after the mask blank is patterned. The first technique, which we will call *defect obscuration*, entails putting the pattern on the mask blank in such a way as to cover multilayer defects with the absorber lines [94, 100]. While this technique has potential, it is unknown if it would be able to mitigate more than a few defects on a mask at a time; this would depend somewhat on where the mask defects are located, which will likely be a random process. The second technique, termed *defect compensation*, entails removing or adding material of an absorber line that is near a multilayer phase defect such that the effect of the multilayer defect is negated during the lithographic process [75]. An advantage that the defect obscuration and defect compensation techniques have over the multilayer repair techniques described earlier is that a smaller advancement, and hence a smaller investment, would be needed to make these techniques viable. Two disadvantages that these techniques have are that they cannot be used by the mask blank manufacturer to qualify the blanks before shipping, and some investment in terms of patterning is required before these techniques can be attempted (i.e., an unsuccessful repair attempt on a mask blank is less expensive than a failed repair attempt on a patterned mask). With the challenges posed by the EUVL stringent mask defect specifications, there can be a strong argument that all of these techniques should be vigorously pursued.

After the mask is fabricated, protecting it from particle contamination is also an issue; this is discussed in Section 4.2.

3.3. Source

The source of EUV radiation must provide sufficient power (>47 watts) [101, 102] at EUV wavelengths and must do so for a reasonable length of time before maintenance is required. This is needed to achieve a sufficient wafer throughput to make the technology commercially viable. An additional consideration is that a minimal amount of debris be emitted from the source, since this debris could deposit on the illuminator optics, reducing the reflectivity and significantly limiting their lifetime. The technical challenges in the source area have been formidable and the source is currently one of the highest risk areas for EUVL. One positive change over the past several years is that there are now a number of source concepts being developed [101, 103–111] at companies and institutions across the globe. One of the leading EUV source candidates is a laser-based plasma source employing a Xe target, where there is a slight preference for a liquid-based Xe target at this time [104, 112]. A high-powered pulsed laser is focused onto the xenon and the laser energy is used to raise the xenon atoms to an excited state. When the xenon deexcites it give off photons with wavelengths in the EUV part of the electromagnetic spectrum. Previously, metal targets were used instead of xenon; however, xenon has lower debris [113]. One of the reasons why it is challenging to obtain sufficient EUV power is that the conversion efficiencies of the sources are typically <1%. This means that less than 1% of the power input to the Xe-based source from something like a Nd:YAG laser is converted into usable EUV power [112].

Electric capillary discharge sources are promising in that if successful, their costs should be considerably less than the laser-based plasma source, since very large, high-powered lasers should not be needed. Debris and power scale up are the two most significant challenges facing the development of the capillary discharge sources [105, 107].

Despite the greater worldwide participation and progress that has been made over the past several years, significant progress in the EUV source area is still required for EUV lithography to be commercially viable.

4. SUPPORTING TECHNOLOGIES

4.1. Resist

Standard resist materials attenuate EUV light too readily, making it necessary to develop new resist processes. A key metric for the resist is the dose needed to process the resist. A lower dose requirement means that fewer EUV photons are needed to process a given wafer, which translates into increased wafer throughput. The dose requirement must be met while satisfying other criteria, primarily maintaining a sufficiently low line edge roughness (LER) for the features formed during the lithographic printing process [42]. An impressive dose level below 4 mJ/cm^2 has been recently met; however, some improvement of the LER is needed

[42]. Recent lithographic results from the Engineering Test Stand indicate that the resist performance will need to be improved further [114].

4.2. Environment

Because EUV light is attenuated in air, it will be necessary to keep the beam path in vacuum ($P > 10^{-8}$ torr), although it is probably not practical to maintain ultrahigh vacuum conditions ($P < 10^{-9}$ torr) inside of the EUVL tool. Long-term exposure of the reflective coatings to EUV light, and in particular under conditions where gaseous species in the ambient may react with, or condense on, the coating, is a serious concern. One mechanism for degradation of the coating reflectance is enhanced oxidation of the Si-terminated Mo/Si surface [115]. It has been suggested that low-energy secondary electrons created by exposure of the Si-terminated Mo/Si to EUV light could interact with water molecules from the environment to enhance oxidation of the surface [24, 116]. Terminating the Mo/Si with a thin Ru film on the surface shows promise for hindering surface oxidation with a minimal impact of the EUV reflectivity from the Ru layer [24].

Another mechanism for degradation of the coating reflectance is the deposition of hydrocarbons on the surface, believed to be due to the cracking of hydrocarbons from electrons emitted from the Mo/Si during EUV exposure [115, 117–119]; carbon, like most materials, absorbs EUV light. This mechanism is not surprising considering that it has been known for many years that carbon deposition can occur on surfaces exposed to high-energy light in synchrotron beamlines [120–122]. It has been observed that optimizing the thickness of the Si layer terminating the Mo/Si multilayer can reduce the carbon buildup on the multilayer [118].

It has also been observed that ethanol can be added to the environment to significantly hinder oxidation, but it does allow carbon deposition [117, 123]. It has been suggested that the carbon could be removed via *in-situ* molecular oxygen cleaning [124]; one challenge with this approach is to remove the carbon without inducing further oxidation. Ongoing work using atomic hydrogen to clean the surface of carbon contamination shows promise [115].

Work on high-reflectance Mo/Si coatings on thin test substrates exposed to ambient conditions suggested that there is some stress relaxation that occurs in the first several weeks, but otherwise the coatings are stable over time under ambient conditions [22, 53, 125].

As noted earlier, the mask must be nearly defect-free and must remain so during the operation of the lithographic tool. Current lithographic technologies encase the transmissive mask in a *pellicle*, which essentially is a transparent cover that allows light through. If a few small particles land on the pellicle, it is not as significant as if these particles land directly on the mask surface, because the particles are offset from the mask plane by a distance equal to the pellicle thickness. For EUVL masks it is not plausible to use a pellicle during EUV exposure because suitable materials strongly attenuate EUV light. A removable pellicle can be used for the EUVL mask for all of the time except for times when it is in the lithographic tool, and this is planned [94, 102].

In order to protect the mask from particles during operation of the lithographic tool, a thermophoretic protection technique has been developed [126, 127]. The mask is maintained at a slightly elevated (<10 °C) temperature to the surroundings, providing a force opposing particle motion to the mask surface. This force arises because the gas molecules closer to the slightly warmer mask will have a higher kinetic energy and will tend to move away from the mask to the cooler region. In moving away from the mask, these molecules will collide with particles moving towards the mask, providing a net force on the particles away from the mask. Experiments to date have shown that the basic technique works [127]. One drawback to this technique is that it works better for higher pressures, and higher pressures can entail more EUV attenuation and possibly contamination. Instituting differential pumping to enable a higher pressure in the vicinity of the mask relative to the rest of the lithographic tool is a potential remedy.

4.3. Nanometrology

Defect inspection tools are currently relied on to measure defect levels on masks as well as processed wafers. The most commonly used technique scans the sample with light in the visible spectral range and detectors look at light reflected/scattered off of the surface. If the surface is relatively smooth, it is possible for defects on the surface to provide an optical signature in the scattered light that can be measured.

EUV lithography will require advances in the capability of these defect detection tools [128–132]. For wafers the extension is straightforward in that smaller defects would need to be detected as feature sizes shrink regardless of whether EUVL or some other lithographic technology is used to synthesize the features. For the mask there are EUVL-specific issues pertaining to advancing the defect detection techniques [128, 133].

Interferometry is needed to determine if the optics meet the figure (shape) specifications [26, 27] and if the masks meet the flatness specifications [134]. It is also needed to align the optics in the EUVL tool. As noted earlier, the figure specifications for the projection optics are very stringent, and a figure specification does not mean much if there is no way to measure it. The fact that some of the optics are aspheres serves to complicate the measurements even further. This has and will continue to drive the resolution capability of interferometry further into the nanoscale regime. For EUV lithography there is work employing interferometers that operate with visible [26, 27, 135, 136] and EUV light [136, 137]. The visible light interferometers currently have an absolute accuracy of ~0.2 nm rms [27, 135] and will need to drive that to ~0.10 nm rms for commercial-grade EUVL tools [27]; it has been suggested that one way to achieve this is to move to a lensless interferometry [27].

Miscellaneous metrology tools will be needed to characterize nanoscale features, whether it be in the development of EUVL technology or in the examination of the features of the integrated circuit patterns formed on the wafer during the lithographic process. Atomic force microscopy (AFM) is a very valuable nanoscale metrology tool for characterizing surface parameters such as high spatial frequency roughness and defect topology; but existing AFM technology is likely to be sufficient for most applications relating to EUVL. Scanning electron microscopy (SEM) is also useful for characterizing surfaces, and advancement in the capabilities of CD-SEMs, which measure parameters such as line edge roughness in critical dimension features in integrated circuits, will be needed [138].

5. CONCLUSIONS

EUV lithography is expected to be used to manufacture integrated circuits with feature sizes below 45 nm, which will enable extremely powerful integrated circuits to be made for microprocessor and memory applications. The very low wavelength of $\lambda = 13$ nm should enable EUVL technology to be used for several generations of integrated circuit manufacturing. The two key elements differentiating EUVL from other optical lithography technologies operating at $\lambda \geq$ 157 nm are (i) the use of reflective optics and a reflective mask, and (ii) a source that is not a laser.

A prototype EUV lithography tool has been successfully constructed and work is underway to build more advanced EUV lithography tools. In addition to system integration, there has been a significant amount of work in component development. While significant progress has been made, component development issues represent the primary challenge to the successful implementation of the technology. In particular, power/stability/debris issues with the source and defect issues in the mask blank are the two greatest challenges. In order to combat mask defect issues, new nanoscience and nanotechnology has been developed. It is likely that additional advances in nanoscience and nanotechnology will occur in the development of EUVL, and if EUVL is successful it could enable widespread advances in nanotechnology by enabling nanoscale features to be formed in a well-controlled, high-volume manufacturing environment.

GLOSSARY

EUV printability theory Theoretical models used to determine if a defect in the mask, such as in the multilayer coating in the mask, will become a printed feature (or printed variation in a feature) in the wafer when exposed in an extreme ultraviolet lithography system.

EUV resist Material which when exposed to EUV light will become more or less resistant to etching.

Microlithography Branch of lithography applied to integrated circuit fabrication.

Nanolithography A more up-to-date term for "microlithography," reflecting the fact that printed features are approaching the nanometer regime in integrated circuit fabrication.

Substrate The bulk material upon which a film or coating is deposited. Commonly used materials are silicon and quartz.

ACKNOWLEDGMENTS

The author acknowledges valuable conversations with EUVL stakeholders from across the globe. In particular, he wishes to thank colleagues at the Virtual National Laboratory, which consists of Lawrence Livermore, Lawrence Berkeley, and Sandia National Laboratories, and the EUV LLC, which consists of Intel, Motorola, AMD, IBM, Infineon, and Micron, for their help and support. Most of the work performed to prepare this manuscript was supported under the auspices of the USDOE by the University of California Lawrence Livermore National Laboratory and funded by the EUV LLC through a cooperative research and development agreement.

REFERENCES

1. G. Stix, *Sci. Am.* April, 2001.
2. P. Gargini, *Proc. SPIE* 4688, 25 (2002).
3. Next Generation Lithography Workshop, International SEMATECH, Colorado Springs, CO, 2000.
4. G. Dao, R. S. Mackay, and P. Seidel, *Proc. SPIE* 4688, 29 (2002).
5. B. Fay, *Microelec. Eng.* 61–62, 11 (2002).
6. G. Zorpette, *Red Herring*, p. 134, Dec. 19, 2000.
7. J. Markoff, *New York Times*, p. 1, Dec. 12, 2000.
8. J. Carey and M. Kunii, *Business Week*, p. 74, April 16, 2001.
9. W. J. Holstein, *U.S. News and World Report*, p. 47, Oct. 6, 1997.
10. T. Beardsley, *Sci. Am.*, p. 15, Dec. 1997.
11. Photomask: *SPIE Bacus Newsletter* 18, 12 (2002).
12. E. Spiller, *Appl. Opt.* 15, 2333 (1976).
13. T. W. Barbee, Jr., S. Mrowka, and M. C. Hettrick, *Appl. Opt.* 24, 883 (1985).
14. A. E. Rosenbluth, *Rev. Phys. Appl.* 23, 1599 (1988).
15. D. G. Stearns, R. S. Rosen, and S. P. Vernon, *Appl. Opt.* 32, 6952 (1993).
16. D. L. Windt and W. K. Waskiewicz, *J. Vac. Sci. Technol. B* 12, 3826 (1994).
17. R. L. Schlatmann, C. Liu, J. Verhoeven, E. J. Puik, and M. J. V. Wiel, *Appl. Surf. Sci.* 78, 147 (1994).
18. E. Spiller, "Soft X-Ray Optics." SPIE Optical Engineering Press, Bellingham, 1994.
19. See *Appl. Opt.* 32, Dec. 1, 1993.
20. J. H. Underwood, in "OSA Trends in Optics and Photonics, Extreme Ultraviolet Lithography" (G. D. Kubiak and D. R. Kania, Eds.), Vol. 4, p. 162. Optical Society of America, Washington, DC, 1996.
21. C. Montcalm, S. Bajt, P. B. Mirkarimi, E. Spiller, F. J. Weber, and J. A. Folta, *Proc. SPIE* 3331, 42 (1998).
22. J. A. Folta, S. Bajt, T. W. Barbee, R. F. Grabner, P. B. Mirkarimi, T. Nguyen, M. A. Schmidt, E. Spiller, C. C. Walton, M. Wedowski, and C. Montcalm, *Proc. SPIE* 3676, 702 (1999).
23. E. Louis, A. E. Yashin, P. Gorts, S. Oestreich, R. Stuik, E. L. G. Maas, M. J. H. Kessels, F. Bijkerk, M. Haidl, S. Mullender, and M. Mertin, *Proc. SPIE* 3997, 406 (2000).
24. S. Bajt, J. B. Alameda, J. T. W. Barbee, W. M. Clift, J. A. Folta, B. Kaufmann, and E. A. Spiller, *Opt. Eng.* 41, 1797 (2002).
25. D. W. Sweeney, H. N. Chapman, R. M. Hudyma, and D. R. Shafer, *Proc. SPIE* 3331, 2 (1998).
26. G. E. Sommargren, in "Phase Shifting Diffraction Interferometry for Measuring Extreme Ultraviolet Optics," p. 108. Optical Society of America, Washington, DC, 1996.
27. G. E. Sommargren, D. W. Phillion, M. A. Johnson, N. Q. Nguyen, A. Barty, F. J. Snell, D. R. Dillon, and L. S. Bradsher, *Proc. SPIE* 4688, 316 (2002).

28. C. Montcalm, R. F. Grabner, R. M. Hudyma, M. A. Schmidt, E. Spiller, C. C. Walton, M. Wedowski, and J. A. Folta, *Appl. Opt.* 41, 3262 (2002).
29. E. M. Gullikson, C. Cejan, D. G. Stearns, P. B. Mirkarimi, and D. W. Sweeney, *J. Vac. Sci. Technol. B* 20, 81 (2001).
30. T. Pistor, Y. Deng, and A. Neureuther, *J. Vac. Sci. Technol. B* 18, 2926 (2000).
31. S. D. Hector, E. M. Gullikson, P. Mirkarimi, E. Spiller, P. Kearney, and J. Folta, *Proc. SPIE* 4562 (2002).
32. A. M. Hawryluk, N. M. Ceglio, and D. A. Markle, *Microlithog. World*, p. 17, Summer 1997.
33. C. W. Gwyn, R. Stulen, D. Sweeney, and D. Attwood, *J. Vac. Sci. Technol. B* 16, 3142 (1998).
34. R. H. Stulen and D. W. Sweeney, *IEEE J. Quantum Electron.* 35, 694 (1999).
35. D. Attwood, "Soft X-rays and Extreme Ultraviolet Radiation." Cambridge University Press, Cambridge, 2000.
36. D. A. Tichenor et al., *Proc. SPIE* 3997, 48 (2000).
37. H. Meiling, J. Benschop, U. Dinger, and P. Kurz, *Proc. SPIE* 4343, 38 (2001).
38. T. Watanabe et al., *J. Vac. Sci. Technol. B* 18, 2905 (2000).
39. D. A. Tichenor et al., *Proc. SPIE* 4688, 72 (2002).
40. P. J. Silverman, *Proc. SPIE* 4343, 12 (2001).
41. P. J. Silverman, "1st International Symposium of EUV Lithography," 2002.
42. J. Cobb, P. Detinger, L. Hunter, D. O'Connell, G. Gallatin, B. Hinsberg, F. Houle, M. Sanchez, W.-D. Domke, S. Wurm, U. Okoroyanawu and S. H. Lee, *Proc. SPIE* 4688, 412 (2002).
43. R. Soufli, E. Spiller, M. A. Schmidt, J. C. Davidson, R. F. Grabner, E. M. Gullikson, B. B. Kaufmann, S. Mrowka, S. L. Baker, H. N. Chapman, R. M. Hudyma, J. S. Taylor, C. C. Walton, C. Montcalm, and J. A. Folta, *Proc. SPIE* 4343, 51 (2001).
44. E. Spiller, F. J. Weber, C. Montcalm, S. L. Baker, E. M. Gullikson, and J. H. Underwood, *Proc. SPIE* 3331, 62 (1998).
45. H. Bach, Ed., "Low Thermal Expansion Glass Ceramics." Springer-Verlag, Heidelberg, 1995.
46. I. Mitra et al., *Proc. SPIE* 4688, 462 (2002).
47. S. T. Gulati, in "Optical Waveguide Materials" (M. M. Broer, G. H. Siegel, R. T. Kersten, and H. Kawazoe, Eds.), Vol. 244, p. 67. Materials Research Society, Pittsburg, PA, 1992.
48. K. E. Hrdina, B. Z. Hanson, P. M. Fenn, and R. Sabia, *Proc. SPIE* 4688, 454 (2002).
49. M. Singh and J. Braat, *Proc. SPIE* 3997, 412 (2000).
50. P. B. Mirkarimi, S. Bajt, and M. Wall, *Appl. Opt.* 39, 1617 (2000).
51. D. G. Stearns, D. P. Gaines, D. W. Sweeney, and E. M. Gullikson, *J. Appl. Phys.* 84, 1003 (1998).
52. E. Spiller, S. Baker, P. B. Mirkarimi, V. Sperry, E. Gullikson, and D. Stearns, *Appl. Opt.*, submitted.
53. P. B. Mirkarimi, *Opt. Eng.* 38, 1246 (1999).
54. M. Shiraishi, W. Ishiyama, T. Oshino, and K. Murakami, *Jpn. J. Appl. Phys.* 39, 6810 (2000).
55. C. Montcalm, *Opt. Eng.* 40, 469 (2001).
56. R. R. Kola, D. L. Windt, W. K. Waskiewicz, B. E. Weir, R. Hull, G. K. Celler, and C. A. Volkert, *Appl. Phys. Lett.* 60, 3120 (1992).
57. P. B. Mirkarimi and C. Montcalm, *Proc. SPIE* 3331, 133 (1998).
58. T. Feigl, S. Yulin, T. Kuhlmann, and N. Kaiser, *Jpn. J. Appl. Phys.* 41, 4082 (2002).
59. A. K. Ray-Chaudhuri, S. E. Gianoulakis, P. A. Spence, M. P. Kanouff, and C. D. Moen, *Proc. SPIE* 3331, 124 (1998).
60. P. A. Spence, M. P. Kanouff, and A. K. Ray-Chaudhuri, *Proc. SPIE* 3676, 724 (1999).
61. S. E. Gianoulakis and A. V. Ray-Chaudhuri, *J. Vac. Sci. Technol. B* 16, 3440 (1998).
62. S. D. Hector, P. Kearney, C. Montcalm, J. Folta, C. Walton, W. Tong, J. Taylor, P. Y. Yan, and C. Gwyn, *Proc. SPIE* 4186, 733 (2000).
63. P. B. Mirkarimi, S. L. Baker, and D. G. Stearns, *J. Vac. Sci. Technol. B* 19, 628 (2001).
64. P. B. Mirkarimi, S. L. Baker, M. A. Wall, P. A. Kearney, and D. G. Stearns, *Solid State Technol.* 43, 95 (2000).
65. P. B. Mirkarimi and D. G. Stearns, *Appl. Phys. Lett.* 77, 2243 (2000).
66. T. Ogawa, M. Ito, M. Takahashi, H. Hoko, H. Yamanashi, E. Hoshino, S. Okazaki, K. Sekine, and I. Kataoka, *Jpn. J. Appl. Phys.* 41, 2285 (2002).
67. P. B. Mirkarimi, S. Bajt, and D. G. Stearns, U.S. Patent 6, 319, 635 B1, 2001.
68. P. B. Mirkarimi, E. A. Spiller, D. G. Stearns, V. Sperry, and S. L. Baker, *IEEE J. Quantum Electron*. 37, 1514 (2001).
69. P. B. Mirkarimi, E. A. Spiller, D. G. Stearns, S. L. Baker, and V. Sperry, unpublished.
70. T. Yoneda, H. Hoko, E. Hoshino, T. Ogawa, S. Okazaki, Y. Isobe, T. Matsumoto, and T. Mizoguchi, *Microelec. Eng.* 61-62, 213 (2002).
71. T. Liang, A. Stivers, R. Livengood, P.-Y. Yang, G. Zhang, and F.-C. Lo, *J. Vac. Sci. Technol. B* 18, 3216 (2000).
72. J. R. Wasson, K. Smith, P. J. S. Mangat, and S. D. Hector, *Proc. SPIE* 4343 (2001).
73. P. B. Mirkarimi, D. G. Stearns, S. L. Baker, J. W. Elmer, D. W. Sweeney, and E. M. Gullikson, *J. Appl. Phys.* 91, 81 (2002).
74. A. Barty, P. B. Mirkarimi, D. G. Stearns, D. Sweeney, H. N. Chapman, M. Clift, S. Hector, and M. Yi, *Proc. SPIE* 4688, 385 (2002).
75. A. K. Ray-Chaudhuri, G. Cardinale, A. Fisher, P.-Y. Yan, and D. W. Sweeney, *J. Vac. Sci. Technol. B* 17, 3024 (1999).
76. D. G. Stearns, D. W. Sweeney, and P. B. Mirkarimi, patent pending.
77. R. S. Rosen, D. G. Stearns, M. A. Villiardos, M. E. Kassner, S. P. Vernon, and Y. Cheng, *Appl. Opt.* 32, 6975 (1993).
78. H. Takenaka and T. Kawamura, *J. Electron Spectrosc. Relat. Phenom.* 80, 381 (1996).
79. J. M. Liang and L. J. Chen, *J. Appl. Phys.* 79, 4072 (1996).
80. D. G. Stearns, D. W. Sweeney, and P. B. Mirkarimi, patent pending.
81. J. H. Underwood, E. M. Gullikson, and K. Nguyen, *Appl. Opt.* 32, 6985 (1993).
82. S. HauRiege, unpublished.
83. S. P. Vernon, D. R. Kania, P. A. Kearney, R. A. Levesque, A. V. Hayes, B. Druz, E. Osten, R. Rajan, and H. Hedge, in "OSA Trends in Optics and Photonics, Extreme Ultraviolet Lithography" (G. D. Kubiak and D. R. Kania, Eds.), Vol. 4, p. 44. Optical Society of America, Washington, DC, 1996.
84. P. A. Kearney, C. E. Moore, S. I. Tan, S. P. Vernon, and R. A. Leavesque, *J. Vac. Sci. Technol. B* 15, 2452 (1997).
85. P. A. Kearney, personal communication, 2002.
86. H. W. Becker, L. Aschke, B. Schubert, J. Krieger, F. Lenzen, S. A. Yulin, T. Feigl, T. Kuhlmann, and N. Kaiser, *Proc. SPIE* 4688, 503 (2002).
87. H. Yamananashi, T. Ogawa, H. Hoko, B.-T. Lee, E. Hoshino, M. Takahashi, and S. Okazaki, *Proc. SPIE* 4688, 710 (2002).
88. J. Hue, E. Quesnel, V. Muffato, C. Pelle, D. Granier, S. Favier, and P. Besson, *Microelec. Eng.* 61–62, 203 (2002).
89. G. S. Selwyn, C. A. Weiss, F. Sequeda, and C. Huang, *J. Vac. Sci. Technol. A* 15, 2023 (1997).
90. G. S. Selwyn, C. A. Weiss, F. Sequeda, and C. Huang, *Thin Solid Films* 317, 85 (1998).
91. T. Hattori, Ed., "Ultraclean Surface Processing of Silicon Wafers." Springer-Verlag, Berlin, 1998.
92. C. C. Walton, P. A. Kearney, P. B. Mirkarimi, J. M. Bowers, C. Cerjan, A. L. Warrick, K. Wilhelmsen, E. Fought, C. Larson, S. Baker, S. C. Burkhart, and S. D. Hector, *Proc. SPIE* 3997, 496 (2000).

93. J. A. Folta, J. C. Davidson, C. C. Larson, C. C. Walton, and P. A. Kearney, *Proc. SPIE* 4688, 173 (2002).
94. S. D. Hector, *Proc. SPIE* 4688, 134 (2002).
95. P. Y. Yan, *Proc. SPIE* 4688, 150 (2001).
96. P. J. S. Mangat, S. Hector, S. Rose, G. Cardinale, E. Tejnil, and A. Stivers, *Proc. SPIE* 3997, 76 (2000).
97. M. Takahashi, T. Ogawa, E. Hoshino, H. Hoko, B. T. Lee, A. Chiba, H. Yamanashi, and S. Okazaki, *Proc. SPIE* 4343 (2001).
98. B. T. Lee, E. Hoshino, M. Takahashi, T. Yoneda, H. Yamananashi, H. Hoko, A. Chiba, M. Ito, T. Ogawa, and S. Okazaki, *Microelec. Eng.* 61–62, 233 (2002).
99. S. Hector and P. Mangat, *J. Vac. Sci. Technol. B* 19, 2612 (2001).
100. A. Stivers, T. Liang, B. Lieberman, P. Y. Yan, and F.-C. Lo, Presentation at 3rd Int. Workshop on EUVL Lithography, ASET and Int. SEMATECH, 2002.
101. B. A. M. Hansson, L. Rymell, M. Berglund, O. Hemberg, E. Janin, J. Thoresen, S. Mosesson, J. Wallin, and H. M. Hertz, *Proc. SPIE* 4688, 102 (2002).
102. H. Meiling, J. Benschop, R. Hartman, P. Kurz, P. Hoghoj, R. Geyl, and N. Harned, *Proc. SPIE* 4688, 52 (2002).
103. J. Pankert et al., *Proc. SPIE* 4688, 87 (2002).
104. W. P. Ballard, L. J. Bernardez, R. E. Lafon, R. J. Anderson, Y. Perras, A. Leung, H. Shields, M. Petach, R. J. St. Pierre, and R. Bristol, *Proc. SPIE* 4688, 302 (2002).
105. N. R. Fornaciari, H. Bender, D. Buchenauer, J. Dimkoff, M. Kanouff, S. Karim, C. Romeo, G. Shimkaveg, W. T. Silfast, and K. D. Stewart, *Proc. SPIE* 4688, 110 (2002).
106. U. Stamm, I. Ahmad, V. Borisov, F. Flohrer, K. Gabel, S. Gotze, A. Ivanov, O. Khristoforov, D. Klopfel, P. Kohler, J. Kleinschmidt, V. Korobotchko, J. Ringling, G. Schriever, and A. Y. Vinokhodov, *Proc. SPIE* 4688, 122 (2002).
107. E. Robert, T. Gonthiez, O. Sarroukh, A. L. Thomann, R. Viladrosa, C. Fleurier, J. M. Pouvesle, and C. Cachoninlle, *Proc. SPIE* 4688, 672 (2002).
108. V. Banine and R. Moors, *Proc. SPIE* 4343, 203 (2001).
109. R. Lebert, K. Bergmann, L. Juschkin, O. Rosier, and W. Neff, *Proc. SPIE* 4343, 215 (2001).
110. R. H. Moyer, H. Shields, A. Martos, S. W. Foranaca, R. J. S. Pierre, and M. B. Petach, *Proc. SPIE* 4343, 249 (2001).
111. G. D. Kubiak, L. J. Bernardez, and K. Krenz, *Proc. SPIE* 3331, 81 (1998).
112. H. Shields, S. W. Fornaca, M. B. Petach, M. Michaelian, R. D. McGregor, R. H. Moyer, and R. St. Pierre, *Proc. SPIE* 4688, 94 (2002).
113. G. D. Kubiak, L. J. Bernardez, K. D. Krenz, D. J. O'Connell, R. Gutowski, and A. M. Todd, in "OSA Trends in Optics and Photonics, Extreme Ultraviolet Lithography" (G. D. Kubiak and D. R. Kania, Eds.), Vol. 4, p. 66. Optical Society of America, Washington, DC, 1996.
114. S. H. Lee et al., *Proc. SPIE* 4688, 266 (2002).
115. S. Graham, M. E. Malinowski, C. E. Steinhaus, P. A. Grunow, and L. E. Klebanoff, *Proc. SPIE* 4688, 431 (2002).
116. N. Koster et al., *Microelec. Eng.* 61–62, 65 (2002).
117. H. Meiling, B. Mertens, F. Stietz, M. Wedowski, R. Klein, R. Kurt, E. Louis, and A. Yakshin, *Proc. SPIE* 4506, 93 (2001).
118. M. Malinowski, C. Steinhaus, M. Clift, L. E. Klebanoff, S. Mrowka, and R. Soufli, *Proc. SPIE* 4688, 442 (2002).
119. R. Kurt, M. v. Beek, C. Crombeen, P. Zalm, and Y. Tamminga, *Proc. SPIE* 4688, 702 (2002).
120. K. Boller, R.-P. Haelbich, H. Hogrefe, W. Jark, and C. Kunz, *Nucl. Instrum. Meth.* 208, 273 (1983).
121. T. Koide et al., *Nucl. Instrum. Meth. in Phys. Res.* A246, 215 (1986).
122. R. A. Rosenberg and D. C. Mancini, *Nucl. Instrum. Meth. in Phys. Res.* A291, 101 (1990).
123. L. E. Klebanoff, M. E. Malinowski, P. Grunow, W. M. Clift, C. Steinhaus, A. H. Leung, and S. J. Haney, *Proc. SPIE* 4343, 342 (2001).
124. M. Malinowski, P. Grunow, C. Steinhaus, M. Clift, and L. Klebanoff, *Proc. SPIE* 4343, 347 (2001).
125. T. W. Barbee, Jr., J. C. Rife, W. R. Hunter, M. P. Kowalski, R. G. Cruddace, and J. F. Seely, *Appl. Surf. Sci.* 78, 147 (1993).
126. L. E. Klebanoff and D. J. Rader, U.S. Patents 6, 153, 044, 2000, and 6, 253, 464, 2001.
127. D. J. Rader, D. E. Dedrick, E. W. Beyer, A. H. Leung, and L. E. Klebanoff, *Proc. SPIE* 4688, 182 (2002).
128. A. Stivers, T. Liang, M. Penn, B. Lieberman, G. Sheldon, J. Folta, C. Larson, P. Mirkarimi, C. Walton, E. Gullikson, and M. Yi, *Proc. SPIE* 4889 (2002).
129. D. Pettibone, A. Veldman, T. Liang, A. R. Stivers, P. Mangat, B. Lu, S. Hector, J. Wasson, K. Blaedel, E. Fisch, and D. M. Walker, *Proc. SPIE* 4688, 363 (2002).
130. M. Yi, M. Park, P. Mirkarimi, C. Larson, and J. Bokor, *Proc. SPIE* 4688, 395 (2002).
131. T. Liang, A. Stivers, P. Y. Yan, E. Tejnil, and G. Zhang, *Proc. SPIE* 4562, 288 (2001).
132. M. Yi, T. Haga, C. Walton, and J. Bokor, *J. Vac. Sci. Technol. B* 19, 2401 (2001).
133. M. Yi, T. Haga, C. Walton, C. Larson, and J. Bokor, *Jpn. J. Appl. Phys.* 41, 4101 (2002).
134. K. L. Blaedel, J. S. Taylor, S. D. Hector, P. Y. Yan, A. Ramamoorthy, and P. D. Brooker, *Proc. SPIE* 4688, 767 (2002).
135. K. Otaki, T. Yamamoto, Y. Fukuda, K. Ota, I. Nishiyama, and S. Okazaki, *J. Vac. Sci. Technol. B* 20, 295 (2002).
136. K. A. Goldberg, P. Naulleau, J. Bokor, and H. N. Chapman, *Proc. SPIE* 4688, 329 (2002).
137. P. P. Naulleau, K. A. Goldberg, S. H. Lee, C. Chang, D. Attwood, and J. Bokor, *Appl. Opt.* 38, 7252 (1999).
138. J. Roberts, Presentation at First Int. Symp. on EUV Lithography, 2002.

ENN Encyclopedia of Nanoscience and Nanotechnology
www.aspbs.com/enn

Fast Atomic Force Microscopy

Fredy R. Zypman
Yeshiva University, New York, USA

CONTENTS

1. INTRODUCTION

The atomic force microscope (AFM) was invented in the mid-1980s by Binning et al. [1]. It belongs to the family of scanning probe microscopes (SPMs), which, by different mechanisms, obtain local information of surfaces at scales down to the nanometer. It is a natural extension of its two SPM precursors, the scanning tunneling microscope (STM) and the stylus profilometer (SP), created few years before. The STM [2] consists of a metallic tip placed above a sample under study. Upon the application of a bias voltage between tip and sample, electrons move from tip to sample (or vice versa) via quantum mechanical tunneling. The flow of electrons is sensed by an external circuit as a current that, in turn, is used as the characteristic parameter of the system. This current is then processed *ex-situ* to obtain relevant information of the sample surface like topography and local electron energy level distribution. The stylus profilometer was introduced by J. B. P. Williamson. It also consists of a tip that interrogates the surface, in this case only for topographic features. Its principle of operation is extremely simple—although not necessarily its actual realization. A hard and relatively narrow tip is placed in contact with the sample under study and then moved across the surface, always remaining in contact. The vertical motion of the tip is, then, a direct measurement of surface protuberances.

The AFM is an extension of the SP and a complement to the STM. The AFM, similarly to the SP, has a tip that is scanned over a surface. However, the AFM tip (and other relevant mechanical parts) is lighter than its corresponding SP counterpart. In addition, the AFM's tip, at its apex, has a much smaller radius of curvature than that of the SP, allowing for the detection of minute morphological features of the sample's surface. The light mass of the AFM's tip allows for its quick response in time to sample variations or to variations in time due to placement of the tip at various locations. This is central to the operation of the AFM and will be discussed further. On the other hand, the STM has a spatial resolution similar to that of the AFM, but while the former is appropriate for studying conductors, the latter is also good for studying insulating materials. While the STM is suited to measure the energies of electrons in the (conducting) material, the AFM is suited to extract forces between tip and sample.

The operation of the AFM can be characterized according to its functionality. When the user seeks to obtain the shape of the surface under study, the AFM is said to operate in *topographic mode*. In this mode of operation, the tip is scanned across the sample to obtain the height of the surface at different locations. The AFM can also be used in *spectroscopic* mode, in which case we seek to measure interaction forces between tip and sample.

Figure 1 shows a schematic representation of the tip–sample system. There, the whole microscope assembly has been conceptually lumped into a support (left of figure), which comprises the link of the cantilever to the glass substrate from which it was sculpted, the link of the substrate to a larger massive holder, and, finally, the link of the holder to the external frame of the microscope. The tip, the sensing piece of the AFM is, as mentioned before, the central component. The tip is a protrusion manufactured at the end of a cantilever. The figure has not been drawn to scale so that we can show both the cantilever and the tip. In reality, typical cantilevers are about 100 μm long, while the tips are a hundred times smaller. Given that AFM measurements rely on the motion of the tip, any theory of AFM must take into consideration the cantilever, since its inertia dominates that of the tip.

ISBN: 1-58883-059-4/$35.00

Encyclopedia of Nanoscience and Nanotechnology
Edited by H. S. Nalwa
Volume 3: Pages (307–319)

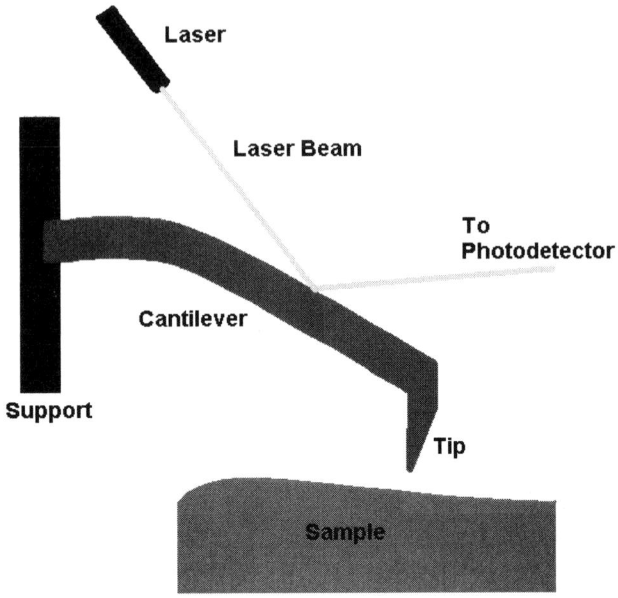

Figure 1. Standard AFM setting.

2. TOPOGRAPHIC MODE

To understand the operation of the AFM in topographic mode, let us refer to Figure 1. There, the surface shows a flat region on the right and a bump on its left. Suppose that, initially, the tip dangles over the rightmost flat area of the sample and the whole assembly is at a nominal rest—in practice, perfect rest is not achieved due to the ever-present uncontrolled vibrations such as thermal vibrations of microscopic origin or simply crass street traffic. If we center our attention on the tip, it will be in static equilibrium when the total forces acting on it vanish. Thus

$$F_{\text{interaction}} = F_{\text{elastic}} \tag{1}$$

where $F_{\text{interaction}}$ is the force that the sample exerts on the tip and F_{elastic} is the force that the flexed cantilever applies on the tip. In a simplified picture, we can view the cantilever as a spring that can be extended or contracted in accordance with $F_{\text{interaction}}$.

For data collection, a laser beam impinges on the top face of the cantilever. The beam reflected off the cantilever is directed toward a system of photodetectors, which, in turn, outputs a voltage proportional to the angle of deflection of the beam. By simple geometry, this angle can be related to the vertical position of the tip. Thus, the voltage output from the photodetectors represents a measure of the vertical position of the tip. In the static geometrical configuration just described, the reflected beam lies along a constant line, and the voltage output from the photodetectors does not change, which is a statement of the fact that the tip stays put at a constant height.

Although laser detection, almost as originally introduced, is still widely and successfully used, other techniques have been introduced lately [3].

At some instant in time, the tip is set to move toward the left, relative to the sample. Sometime later, when the tip starts touching the base of the bump, the tip–sample gap will begin to narrow. Almost simultaneously, the photodetector's voltage changes, indicating that the relative tip–sample height has changed. This happens as a result of a loss of static equilibrium: at the base of the bump, the tip–sample force changes all of a sudden as a result of the narrowing of the gap, while the elastic force remains the same. Equation (1) does not hold, the tip is not in equilibrium, and it accelerates, trying to change its position. However, the AFM, in topographic mode, is designed so that the photodetector's voltage remains fixed. To achieve this, an electromechanical transducer (a piezoelectric material) is attached to the cantilever to move it vertically when it senses a change in voltage from the photodetectors. Via a feedback loop, the piezoelectric moves the cantilever in such a way that the photodetector output regains its original voltage value. However, in order to achieve this, the piezoelectric must now have an applied voltage different from its initial value in order to keep the cantilever higher. This piezoelectric voltage permits the final image reconstruction. The idea is that as the tip visits a variety of horizontal positions on the sample surface, its height relative to the sample is kept constant at the expense of applying a bias voltage to the piezoelectric. This bias voltage defines a two-variable function $V_P(x, y)$, where V_P is the voltage applied to the piezoelectric and x and y are the Cartesian coordinates of the tip on a horizontal sample grid. In principle, a three-dimensional, graphic rendering of $V_P(x, y)$ corresponds to the topography of the sample's surface.

The description just given to unveil surface information with the AFM is, to a good extent, reliable, but it lacks in accuracy exactitude. We are not referring here to the experimental errors intrinsic with any measuring apparatus, but rather to the tacit limitations imposed by an algorithm of reconstruction based on the model given before. First, the model supposes that there is a one-to-one relationship between the tip–sample separation and the piezoelectric voltage. That can be seen immediately not to be true in general. As a counterexample, consider a perfectly flat surface. According to our simple model, as the tip moves horizontally above the sample, the piezoelectric voltage remains fixed since it never receives a signal to change from the photodetector system. Let us suppose that the flat surface comprises two regions of different materials A and B, separated by a horizontal line boundary on the surface. In general, as the horizontal projection of the tip's position crosses the boundary, the forces on the tip will change with a concomitant change in the reflected laser beam direction, the output voltage from the photodetectors, and, finally, the new voltage applied to the piezoelectric to bring "everything" back to where it was before. Thus, we see that a reliable algorithm needs to take into consideration not only morphology but also chemical information of the sample. Another problem with the simple algorithm is that it assumes the cantilever to be appropriately described by a spring—with its associated mass. For static measurements, that is fine. Nevertheless, how do we know when the static regime has been overcome? Many of the applications of AFM deal with samples in motion, which implies that the horizontal scanning has to be performed fast enough so that the sample does not move significantly during the time the picture is taken. Under fast motion, the spring fails to be a good paradigm for the cantilever. It is similar to what would happen to a rubber band with a mass attached at its end: at low speeds, the rubber band acts as a spring, but, at higher speeds, other vibrations appear that cannot be taken into account by a single-degree-of-freedom system.

3. SPECTROSCOPIC MODE

In spectroscopic mode, we inquire about tip height changes while the tip–sample separation is varied. In most contemporary AFM systems, the sample is moved up or down, while we look for tip deflections via the laser–photodetector arrangement. The idea is that from the known motion (experimentally measured) of the tip we can extract—that is, reconstruct—interaction forces between tip and sample. In the pure spectroscopic mode, we are not interested in the topography of the sample but rather in the tip–sample force as a function of their separation.

In short, the aim of this mode of operation is as follows:

- Vary and monitor $z_S(t)$, the absolute vertical position of the sample's surface versus time.
- Measure directly $z_T(t)$, the absolute vertical position of the tip as a function of time.
- Reconstruct, via appropriate algorithms, $F_{\text{interaction}}(z)$, where $F_{\text{interaction}}$ has been defined previously and $z = |z_T - z_S|$ is the relative tip–sample separation for all times.

To obtain an idea of this mode of operation, we once more call on the simple—albeit somewhat incomplete—spring model to describe the mechanical response of the cantilever. Again, this picture is sufficient to describe a motion in which all parts move very slowly. A few practical applications—but by no means the majority—fall in this category.

Figure 2 depicts a representative configuration. The sample is z_S above a fixed vertical reference, which in the picture is taken to be vt, where v is the speed of the platform that holds the sample and t is the time elapsed since the beginning of the experiment. The tip is z_T above the same reference and varies in time in an unknown way a priori. The tip–sample distance is $z(t)$. The cantilever has been removed from the picture and replaced instead by a single mass–spring arrangement, with mass m and spring constant k.

Figure 3 illustrates a typical force versus separation curve. In the figure, the pale horizontal line represents the line of null force. Points below the line (negative) correspond to attractive forces, and points above the line correspond to repulsive forces. Thus, we see that the force–separation curve is negligible at large distances, corresponding to the fact that tip and sample do not interact when they are far away. Likewise, at medium-range separations (on the order of nanometers), the interaction is attractive. Finally, at very small separations, core–core atomic repulsive forces dominate, and the overall force becomes large and repulsive.

Let us use Figures 1–3 to gain insight into the kind of signal we expect to measure. In what follows, we will assume a common measuring setup in which the feedback mechanism to maintain the tip–sample separation fixed is not engaged,

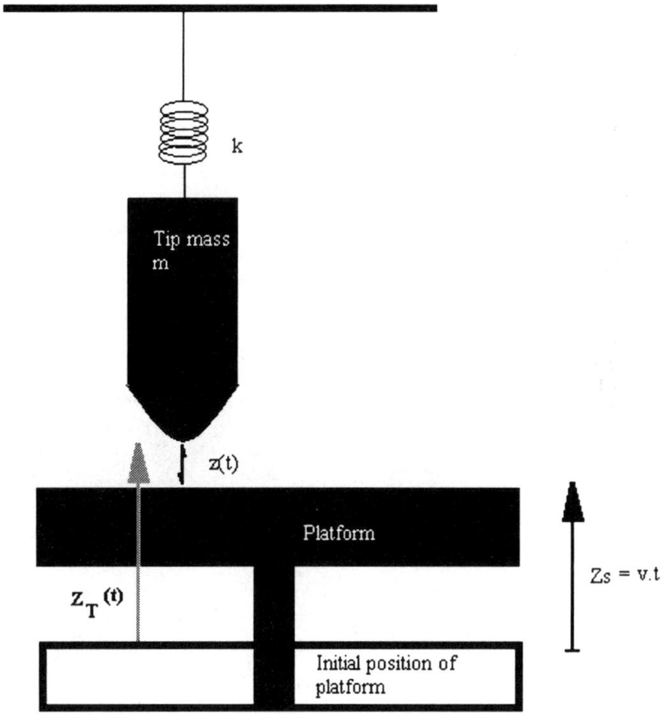

Figure 2. Detailed tip–sample configuration. The tip–cantilever system is modeled as a lumped mass–spring.

in which case the output signal is the voltage from the photodetectors. The initial condition of the system has the following properties (Fig. 2):

- The tip is at its natural equilibrium point, corresponding to the balance between the spring force (upward) and the gravitational force (downward).
- The platform that holds the sample is in its lowermost position.
- Concomitantly, the tip–sample separation lies on the right-hand side of the horizontal scale in Figure 3, resulting in a nil force.

Subsequently, the platform starts moving up with a slow constant speed v (Fig. 2). During the initial period, the tip and sample are relatively far from each other, and therefore the tip stays put at its natural equilibrium point. At some

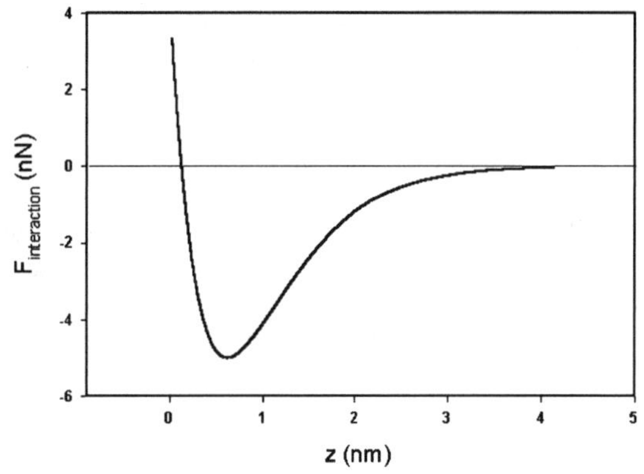

Figure 3. Typical force–separation curve.

moment, the platform moves up enough to decrease the tip–sample distance to the point at which slight interactive forces become appreciable. At that time, the tip will start to move down since the total net force now points downward, due to $F_{\text{interaction}}$. As the tip moves down, the angle that the cantilever makes with a horizontal line increases.

Consequently, the reflected beam off the cantilever will decrease its angle θ_R from the horizontal and thus induce a change (consider an increase) in the voltage of the photodetectors. As time progresses, and the tip–sample separation shrinks, the photodetector voltage, $V_{PH}(t)$, increases as a nonlinear function of time. Since $\Delta V_{PH} \propto \Delta\theta_R$ (the proportionality constant known by calibration in practical situations), and $\Delta\theta_R$ determines, from simple geometry, the position of the tip $z_T(t)$, $z(t)$ can be obtained from

$$z(t) = z_T(t) - Vt \tag{2}$$

The force can be found by using Newton's second law:

$$m\frac{d^2 z_T}{dt^2} = F_{\text{interaction}} + kz_T(t) - mg \tag{3}$$

where $g \cong 9.8\ \text{m/s}^2$. Equation (3) can be rewritten as

$$F_{\text{interaction}} = m\frac{d^2 z_T}{dt^2} - kz_T(t) + mg \tag{4}$$

In Eqs. (2)–(4), the function $z_T(t)$ is completely known from the experiment described previously. Therefore, the function d^2z_T/dt^2 is also completely known. In this form, Eq. (4), in combination with Eq. (2), provides a parametric form for the function $F(z)$.

Thus, we have shown in this section how to obtain $F(z)$ from $V_{PH}(t)$. Notice that the central theoretical assumption is that the cantilever can be conceptually lumped into a mass–spring system. The reconstruction algorithm (which we just described) is then simply based on using Newton's law to the model system.

4. MODULATION TECHNIQUES

Methods that are more sensitive (in the sense that they increase the signal-to-noise ratio) have been designed to extract force–separation curves. They are based on the concept of modulation—in which some input parameter is forced to oscillate at a prescribed frequency. The simplest improvement consists of forcing the tip–cantilever structure to oscillate at a constant frequency, for example, the natural frequency ω_0 of free oscillation with which it would freely vibrate when no sample is nearby. To analyze the problem and understand what information can be obtained, let us consider Figure 4. There, again, the cantilever–tip assembly is well represented by a spring (of constant k_t) and a mass (m_t). In addition, we now introduce k_i, a spring constant that characterizes the interaction between the tip and the sample. This can be understood as follows: a small-amplitude oscillatory tip in the presence of a force field (Fig. 3) will sense a small interval of forces. In this small interval, the force–separation curve can be well approximated by a straight line with slope $dF(z)/dz$. Thus, in this interval, the effect

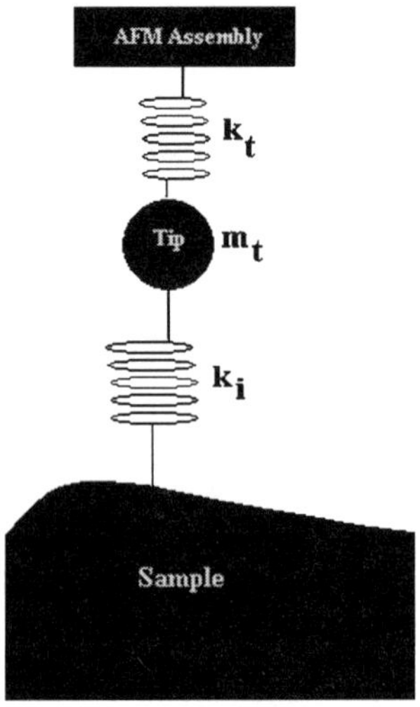

Figure 4. As in Figure 2, but also taking into consideration the elastic properties of the sample.

of the interaction force is the same as that of a spring—"interaction spring"—with equivalent spring constant $k_i = dF(z)/dz$. In this derivation, we are assuming that the sample is very heavy and stiff, which is a very good starting point to isolate the effect of the interaction force. Later, when showing how to extract vibrational properties of the sample, we will remove these assumptions.

The equation of motion for the tip is

$$m_t\frac{d^2 z(t)}{dt^2} = \left[A_{\text{in}}e^{i\omega_{\text{in}}t} - z(t)\right]k_t - k_i z(t) - b\frac{dz(t)}{dt} \tag{5}$$

where $z(t)$ is the position of the tip—the sample is at rest. The first term inside the parentheses corresponds to a forced vibration of the cantilever, with input amplitude A_{in} and frequency ω_{in}. The two terms with the spring constants represent the restitution forces of the springs, and the last term accounts for a friction term.

In steady state, Eq. (5) accepts a solution of the form

$$z(t) = A_{\text{out}}e^{i\omega_{\text{in}}t} \tag{6}$$

Upon substitution of Eq. (6) into Eq. (5)

$$A_{\text{out}} = \frac{\omega_{\text{tt}}^2}{(\omega_{\text{tt}}^2 + \omega_{\text{it}}^2 - \omega_{\text{in}}^2) + i\gamma\omega_{\text{in}}}A_{\text{in}} \tag{7}$$

where we have used the shorthand notation

$$\omega_{\text{tt}} \equiv \sqrt{\frac{k_t}{m_t}} \qquad \omega_{\text{it}} \equiv \sqrt{\frac{k_i}{m_t}} \qquad \gamma \equiv \frac{b}{m_t}$$

Next, as mentioned previously, we require that $\omega_{\text{in}} = \omega_0$, the natural frequency in the absence of interaction—the one it would have if ω_{it} were 0. Then Eq. (7) takes the form

$$A_{\text{out}} = \frac{\omega_0^2}{\omega_{\text{it}}^2 + i\gamma\omega_0}A_{\text{in}} \tag{8}$$

Equation (8) immediately provides the physical quantities' root mean square,

$$a^{\text{rms}} \equiv \sqrt{\int_0^{2\pi/\omega_0} [\text{Re}(Ae^{i\omega_0 t})]^2\, dt}$$

and phase Φ with respect to the input signal:

$$\frac{a_{\text{out}}^{\text{rms}}}{a_{\text{in}}^{\text{rms}}} = \frac{\omega_0}{\gamma} \frac{1}{\sqrt{1 + (\omega_{\text{it}}^2/\gamma\omega_0)^2}} \tag{9}$$

$$\Phi = -\arctan\left(\frac{\omega_0 \gamma}{\omega_{\text{it}}^2}\right) \tag{10}$$

Equation (9) or Eq. (10) forms the basis for the classification of the modulation techniques in AFM, according to the quantity being detected. Thus, among many others, measurements can be made in *amplitude detection mode* or *phase detection mode*.

4.1. Amplitude Detection Mode

In this case, Eq. (9) forms the basis of the algorithm to convert experimental data into force information. Concretely, $a_{\text{in}}^{\text{rms}}$ is known since it is the input amplitude. The quantity $a_{\text{out}}^{\text{rms}}$ is also known since it corresponds to the rms output voltage from the photodetector system. Finally, γ and ω_0 are known from calibration of the cantilever–tip system. We will use Q, the quality factor instead of γ, which, for small damping—as is usually the case—are related by $\gamma = \omega_0/Q$. Therefore, Eq. (9) permits the calculation of ω_{it}. By using the definition of ω_{it} and the relationship between k_{i} and force, we finally arrive at the very handy relationship:

$$F(z) = \frac{m_{\text{t}}\omega_0^2}{Q} \int_{+\infty}^{z} dz \sqrt{\left[Q \frac{a_{\text{in}}^{\text{rms}}}{a_{\text{out}}^{\text{rms}}(z)}\right]^2 - 1} \tag{11}$$

where we have explicitly written the z dependence of $a_{\text{out}}^{\text{rms}}$—which varies as a result of the tip being embedded in a different force field. The constant of integration was chosen so that there is no interaction force at large distances, $F(+\infty) = 0$—recall that $a_{\text{out}}^{\text{rms}}/a_{\text{in}}^{\text{rms}} = Q$ when no interaction force is present.

4.2. Phase Detection Mode

This mode is conceptually the same as the previous one, but we use Eq. (10) instead of Eq. (9) as the reconstruction algorithm. In this case, the force is

$$F(z) = -\frac{m_{\text{t}}\omega_0^2}{Q} \int_{+\infty}^{z} \frac{dz}{\tan[\Phi(z)]} \tag{12}$$

where the limits of integration were chosen recalling that, at resonance, $\Phi(+\infty) = -\pi/2$.

Strictly speaking, the equations derived previously are valid under the hypothesis that dF/dz is smaller than the cantilever's effective spring constant. If we tried to set the tip to vibrate around the point in space for which $dF/dz = k_{\text{t}}$, then any small deflection of the tip closer to the sample will make the interaction force increase, pulling the tip toward the sample since the opposite elastic forces cannot balance the attractive interaction force. Once this process of approach has started, there is no natural return: the interaction force keeps increasing and the tip is pulled closer to the sample's surface. Eventually, when the tip reaches the repulsive region of $F_{\text{interaction}}$, it will stop. This process of approach of the tip to the sample is called *snap to contact*. Since this approach is fast, even for modern standard electronic instrumentation, the measurement of the force curve in this spatial range presents a challenge. Clearly, a simple solution to the problem would be the use of stiffer cantilevers, that is, cantilevers with larger k_{t}. The drawback of this solution is that the system becomes less sensitive to force detection, since a stiff cantilever requires larger forces to produce the same deflections as that of a soft one. Solutions to this problem using a soft cantilever exist, which involve hardware and software modifications. We will discuss here Tapping imaging and continuous modeling of the cantilever.

4.3. Tapping

Tapping mode [4] was originally invented to minimize the problems associated with shear forces when operating the microscope in contact mode. As explained before, the signal from standard AFM is obtained from a multi-photodetector system. We explained how the vertical position of the tip could be attained from the output voltage of the photodetectors, V_{PH}. It should be clear, however, that, when scanning the tip in contact across a surface, the reflected laser beam will make an angle that depends not only on the height of the tip, but also on any additional slight tilt that the top of the cantilever might have. This tilt is normally due to small-angle tip rotations produced by the torque of shear tip–sample forces. In this general situation, the reflected beam angle does not provide enough information to determine the static state of the cantilever–tip system. Tapping mode solves this problem by lightly tapping the sample with a vibrating tip while horizontally scanning. The tip moves horizontally when far from the surface, thus eliminating the shear interactions and rendering the conversion from laser angle to vertical position simple. In addition, the reduction in lateral forces is imperative when characterizing soft samples such as biological matter. In this case, simply dragging the tip across the sample would destroy the sample. However, Tapping mode is used beyond its original topographic application to obtain information about interaction forces. Given that the tip quickly traverses large intervals of the force–separation curve, it is necessary to take full account of the curve in any analysis. This makes the problem highly nonlinear and not amenable to analytical solutions. Numerical solutions to the problem have been obtained for various parameters that define the system. A particularly interesting solution has been devised by Sarid [5]. He studied numerically the dynamics of a tip, modeled as a sphere, in the presence of a planar sample, that interact via a Jones seven–one potential. He found the tip motion to be almost characterized by a single frequency. The force, on the other hand, is a train of pulses corresponding to repulsive wall–tip interactions. Viewed as a function of the stiffness of the material, the indentation of the sample is inversely proportional to the stiffness and the indentation force becomes smaller in magnitude.

5. HIGH-FREQUENCY SPECTROSCOPY

Force curves can also be reliably obtained from simple deflection measurements by considering the cantilever as an extended beam—a realistic assumption. The experimenter can just measure the deflection of the cantilever versus the separation in order to obtain force curves, even at high tip speeds, beyond the snap-to-contact point. Only minor hardware changes are needed, for example, changes in the parameters defining the light source, such as controlling the laser beam width. It turns out that one needs to take as many measurements as frequencies one believes are important in the motion of the cantilever. Let us say one decides that only two frequencies are important to describe the motion of the cantilever. Then one could obtain a deflection curve for a given laser beam width and a second one for a different laser beam width. Experimentally, it is better to take the two curves at the same time by, for example, modulating the beam width. A more mathematically elegant way to collect the data is to take a deflection–separation curve for a given laser width and then get a second curve for a small change in the width, so that one gets the signal and its derivative with respect to the beam width at constant separation. This second prescription allows for a generalization of the measurement process in terms of *moments* of the signal–width curves—one such curve for each tip–sample separation.

6. DYNAMICS OF THE CANTILEVER

Cantilever dynamics have been studied in the past [6]. Here, we describe solutions relevant to the new generation of fast AFM. In particular, we show how to solve the problem of high-frequency response and, *at the same time*, explicitly include the tip–sample interaction force.

The cantilever is modeled as a beam governed by the following equation [5]:

$$\frac{\partial^2}{\partial x^2}\left(EI\frac{\partial^2 u}{\partial x^2}\right) + \rho A\frac{\partial^2 u}{\partial t^2} = 0 \tag{13}$$

where $u(x, t)$ is the deflection of the cantilever with respect to the horizontal at position x along the cantilever and at time t, E is Young's modulus, I is the (area) moment of inertia through the middle axis perpendicular to the plane of motion, ρ is the density of the cantilever, and A is the cross-sectional area.

To fix ideas, we consider the cross-sectional area to be a constant. In this case, the parameters do not vary with x. Thus, we can define the constant

$$V^2 \equiv \frac{EI}{\rho A^2} \tag{14}$$

Now we can rewrite (13) as

$$AV^2\frac{\partial^4 u}{\partial x^4} + \frac{\partial^2 u}{\partial t^2} = 0 \tag{15}$$

To solve Eq. (15), we need to set appropriate boundary conditions (BCs). These conditions will also introduce the tip–sample interaction force that one wants to measure. At the fixed end of the cantilever, $x = 0$, we impose no deflection

$$u(x = 0, t) \equiv 0 \tag{16a}$$

Also, since the lever is not free to rotate about that point,

$$\frac{\partial u}{\partial x}(x = 0, t) \equiv 0 \tag{16b}$$

At the free end, $x = L$, the cantilever experiences no torque if we consider vertical interaction forces

$$\frac{\partial^2 u}{\partial x^2}(x = L, t) \equiv 0 \tag{16c}$$

Finally, at the free end, the internal force on that layer of the cantilever must equal the external force

$$-EI\frac{\partial^3}{\partial x^3}u(x = L, t) = F[u(x = L, t)] \tag{16d}$$

6.1. Static Solution

It is common to use the solution to the previous problem for which the interaction force F is slowly varying with time. This hypothesis typically breaks down once the tip reaches the snap-to-contact point because then the magnitude of the force changes its value very quickly. However, if one insists on a quasistatic solution, one could build the whole force–separation curve as a patch of a series of solutions for which $F = F_S$, with F_S a constant force.

In this static case, Eq. (15) reduces to

$$\frac{\partial^4 u_S(x, t)}{\partial x^4} = 0 \tag{17}$$

where we have added the subindex S to u to indicate *static solution*. This equation, in conjunction with the four BCs, provides a unique solution

$$u_S(x) = \frac{(3F_S L)x^2 - (F_S)x^3}{6EI} \tag{18}$$

If the AFM's laser beam monitors the deflection of one point of the cantilever, for example, $x = L$, then the measured deflection, z_S, would be

$$z_S(t) = \frac{F_S(t)L^3}{3EI} \tag{19}$$

As the force varies slowly with time, so does the deflection [we have explicitly written this time dependence in Eq. (19) to stress that measurements are made in time steps]. This corresponds to a one-degree-of-freedom simple harmonic oscillator (SHO) approximation of the cantilever with spring constant $3EI/L^3$. As the model makes strong approximations, it must render a solution with strong limitations. In particular, it will fail when the tip–sample interaction force does not vary slowly. Under these circumstances, the cantilever will support high-frequency vibrational modes and, consequently, a one-degree-of-freedom model will not be enough to determine the state of the system. In the quasistatic regime, on the other hand, one degree of freedom is enough because if one measures $z(t)$ by using Eqs. (18) and (19), one can get the whole shape of the cantilever.

6.2. Generalizing the SHO Solution

In order for the SFM to be of use in fast imaging, where a force–distance curve must be obtained for each pixel and as quickly as possible, as, for example, in the study of biomolecular motion, the quasistatic solution must be called into question.

With that in mind, we now generalize the previous problem by allowing for the cantilever to sense not only a constant force, but also a gradient of the force–distance curve as the tip moves away or toward the surface. This will enrich the solution for the deflection $u(x, t)$ by allowing the possibility of vibrations superimposed on the SHO solution.

Therefore, we write the force as

$$F = F_S + G[u(L, t) - u_S(L)] \tag{20}$$

In writing Eq. (20), we assumed, as in the BC, that the force on the cantilever is applied on the tip at $x = L$. Here, F_S is the average force the tip senses at one location above the sample and G is the derivative of the force–distance curve at that same position.

We write the solution $u(x, t)$ to Eq. (15) as the sum of the static solution and a superposition of normal modes

$$u(x, t) = u_S(x) + \sum_n f_n(x)\cos(\omega_n t + \varphi_n) \tag{21}$$

where $f_n(x)$ are the normal modes, ω_n are the angular frequencies corresponding to the $f_n(x)$, and φ_n is the phase of mode n.

Substituting expression Eq. (21) into Eq. (15) and noticing that the $f_n(x)$ are linearly independent, we obtain

$$\frac{d^4 f_n(x)}{dx^4} - \frac{\omega_n^2}{AV^2} f_n(x) = 0 \tag{22}$$

with solutions

$$f_n(x) = f_n^1 \cos kx + f_n^2 \sin kx + f_n^3 \cosh kx + f_n^4 \sinh kx \tag{23}$$

where f_n^1, f_n^2, f_n^3, and f_n^4 are constants and k is the wavenumber.

Next, we apply the BCs Eq. [(16a)–(16d)] and end up with BCs for each individual mode $f_n(x)$

$$f_n^1 + f_n^3 = 0 \tag{24a}$$

$$f_n^2 + f_n^4 = 0 \tag{24b}$$

$$-f_n^1 \cos kL - f_n^2 \sin kL + f_n^3 \cosh kL + f_n^4 \sinh kL = 0 \tag{24c}$$

$$\begin{aligned} k^3(-f_n^1 \sin kL + f_n^2 \cos kL - f_n^3 \sinh kL - f_n^4 \cosh kL) \\ = \gamma(f_n^1 \cos kL + f_n^2 \sin kL + f_n^3 \cosh kL \\ + f_n^4 \sinh kL) \end{aligned} \tag{24d}$$

where $\gamma = G/EI$.

In order for the system of equations (24) not to have trivially null solutions, we must require that the determinant of the system be 0. Thus

$$\begin{aligned} k^3(1 + \cos kL \cdot \cosh kL) \\ = \gamma(\sin kL \cdot \cosh kL - \cos kL \cdot \sinh kL) \end{aligned} \tag{25}$$

With the definitions $\beta = \gamma L^3$ and $\xi_n = k_n L$, one obtains a transcendental equation for the frequencies ω_n for a given G

$$\frac{1 + \cos \xi_n \cdot \cosh \xi_n}{\sin \xi_n \cdot \cosh \xi_n - \cos \xi_n \cdot \sinh \xi_n} \xi_n^3 = \beta \tag{26}$$

We can use these frequencies to find, for each n, the normal modes f_n^i $(i = 1, 2, 3, 4)$ (in fact, we will obtain three ratios between the f_n^i since the system is, by construction, indeterminate). After some tedious but easy algebra, one obtains

$$\frac{f_n^1}{f_n^4} = -\frac{\xi_n^3}{\beta} \frac{\cos \xi_n + \cosh \xi_n + \sinh \xi_n - \sin \xi_n}{\cos \xi_n - \cosh \xi_n + \sin \xi_n - \sinh \xi_n} \equiv R_n \tag{27a}$$

$$\frac{f_n^2}{f_n^4} = -1 \tag{27b}$$

$$\frac{f_n^3}{f_n^4} = \frac{\xi_n^3}{\beta} \frac{\cos \xi_n + \cosh \xi_n + \sinh \xi_n - \sin \xi_n}{\cos \xi_n - \cosh \xi_n + \sin \xi_n - \sinh \xi_n} = -R_n \tag{27c}$$

Finally, using (21), one arrives at the complete solution to the problem

$$\begin{aligned} u(x, t) = u_S(x) + \sum_n f_n^4 \Big[R_n \cos\Big(\frac{\xi}{L}x\Big) - \sin\Big(\frac{\xi}{L}x\Big) \\ - R_n \cosh\Big(\frac{\xi}{L}x\Big) + \sinh\Big(\frac{\xi}{L}x\Big)\Big] \\ \times \cos(\omega_n t + \varphi_n) \end{aligned} \tag{28}$$

In the next section, we will derive an expression for the signal that is actually measured. That signal is a consequence of the motion $u(x, t)$ but is not it.

6.3. Detailed Study of the Laser Width and Its Use in Force Measurements

In optical detection, a light beam intensity reflected from the top of the cantilever is measured at a photodetector system. If the beam has a distribution of intensities $g(x)$ (Fig. 5), each of these intensities contributes to the total signal.

The signal originating at position x at the cantilever is proportional to the deflection $u(x, t)$ and to the light intensity at that point. If $u(x, t)$ were 0, there would not be any signal because the two photodiodes would be balanced; if the intensity were 0, then the signal would also be 0 since there would be no light to measure. Therefore, up to instrument constants, the signal at the photodiode is

$$S(t) = \int_0^L u(x, t) \cdot g(x)\, dx \tag{29}$$

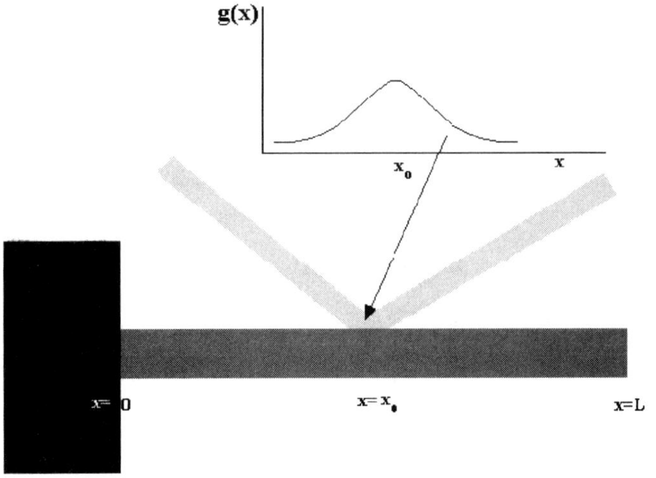

Figure 5. A laser beam of Gaussian width impinges on the cantilever.

where $g(x)$ is the light beam intensity profile. We assume it has a Gaussian shape

$$g(x) = \frac{1}{\sigma\sqrt{\pi}} \exp\left(-\left(\frac{x-x_0}{\sigma}\right)^2\right) \tag{30}$$

By using Eqs. (27), (28), (30), and (18), the integral in Eq. (29) can be carried out explicitly. The result is

$$\begin{aligned} S(t) = S_{\text{static}} + \sum_n f_n^4 \Bigg[& R_n \exp\left(-\left(\frac{k\sigma}{2}\right)^2\right) \cos kx_0 \\ & - R_n \exp\left(\left(\frac{k\sigma}{2}\right)^2\right) \cosh kx_0 \\ & - \exp\left(-\left(\frac{k\sigma}{2}\right)^2\right) \sin kx_0 \\ & + \exp\left(\left(\frac{k\sigma}{2}\right)^2\right) \sinh kx_0 \Bigg] \\ \times \cos(\omega_n t + \varphi_n) & \end{aligned} \tag{31}$$

where S_{static} is the following constant:

$$S_{\text{static}} = \frac{F_s}{3EI}\left[(3L - x_0)x_0^2 + \frac{3}{2}\sigma^2(L - x_0)\right] \tag{32}$$

The constants f_n^4 are obtained from the initial conditions.

6.4. Further Generalization to the SHO Solution

Although the assumption in Eq. (20) for the force is a generalization over what is commonly used, it is natural to ask whether the problem could be solved for an arbitrary force shape. In this section, we present the theory to obtain such a solution. Regrettably, no closed-form solution can be obtained for the general case. However, by writing an algorithm following these steps, it will be easy to write computer code to obtain numerical solutions.

In Figure 6, a realistic force–separation curve is shown to be made up of the concatenation of eight linear segments. Each segment is characterized by a tip–sample separation, an average force, and a slope. For example, segment 2 is characterized by s_2, F_2, and G_2. For that same segment, the tip stays there during the period of time from $t = 2\Delta t$ to $t = 3\Delta t$. In general, we shall consider M intervals labeled m, that is, $m = 0, 1, 2, 3, \ldots, M-2, M-1$. For negative values of m (and time), we assume that the force is 0 (far-away region) and the cantilever is in its horizontal equilibrium position. Zero volts are detected at the photodetector.

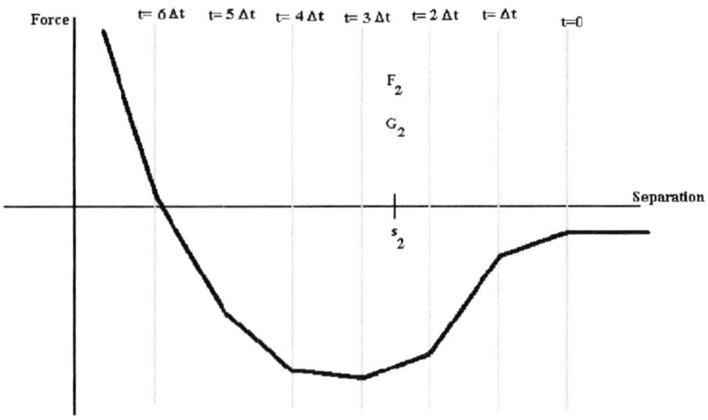

Figure 6. Reconstruction of the overall force from linear pieces.

Experimental measurements provide $S(t)$ at thousands or even millions of points, so it is necessary to account for situations where the number of data points is larger than M. Let us then divide each of the previously mentioned segments into P smaller segments. Each of these smaller segments will comprise a time interval δt, so that $\Delta t = P\delta t$. There will be a total number of points $D = PM$. We choose P so that D is the number of data points.

The signal $S(t)$ in Eq. (31) depends on the light beam width σ and the point x_0 at which the light beam touches the cantilever. Therefore, it is possible to collect many signals, $S_q(t)$ $(q = 1, 2, 3, \ldots, Q)$, simultaneously. Each of these signals could consist of a measurement of $S(t)$ for a given σ. In this case, σ can be written as $\sigma = \sigma_0 + [q - (Q+1)/2]\Delta\sigma$. However, σ is not the only parameter available. The position of the light beam could also be changed. However, whatever one decides to choose, the following general derivation applies.

We choose Q to be equal to the number of frequencies the cantilever is assumed to support (it may be larger, but never smaller than that). This can be determined before the tip interacts with the surface, that is, when the tip and the surface are far away. It is true that the frequency spectrum will change as the tip approaches the sample. However, the number of frequencies involved should not change since the cantilever is driven more by inertia than by the tip–sample force. In a typical experiment, one should excite the tip with a broad array of frequencies and study the cantilever's response. If three frequencies were excited, then one would choose $Q = 3$. For safety, one may want to take $Q = 4$ and check if one gets null coefficients for modes above the third.

Equation (31) can now be rewritten to simplify the analysis

$$S_p^{qm} = S_{\text{static}}^{qm} + \sum_{n=1}^{n=Q} f_n^{4m} \psi_n^q \cos(\omega_n p\delta t + \varphi_n) \tag{33}$$

where ψ_{nq} is shorthand for the expression in brackets in (31).

The left-hand side of Eq. (33) must be compared to the $QPM = QD$ data points. The right-hand side can then be obtained by fitting this expression to experiments. To complete the process, one needs to choose M, the number of regions into which we want to divide the whole "separation" range. It seems that $M = 10$ should be a good guess. By doing this fitting, we are solving an inverse problem, namely, that of obtaining the force–separation curve from the measured signals.

7. HISTORICAL NOTES

As mentioned previously, the details of the first AFM were published in the mid-1980s by Binning et al. Their implementation of the AFM was an actual merging between the STM and the stylus profilometer. In that original setup, they used an SP tip and monitored its vertical position with the STM. Their main claims at the time were that they could measure, in air, extremely small interaction forces (10^{-18} N), that the technique provided spatial atomic resolution, and that it was relevant for the study of nonconducting surfaces, where STM is inappropriate.

In general terms, their promises became a reality. Operating in air as opposed to high vacuum has opened many new research lines. Studies in biomolecules have been spurred by not having to operate in a vacuum. Traditional high-resolution microscopes in biology, such as the transmission electron microscope, the scanning electron microscope, and lately the scanning tunneling microscope, use dissected samples. By operating in air, the AFM can study biomolecules *in vivo*. Often, it is more desirable to use the AFM not in air, but embedded in a liquid (water, serum, buffer solution) so that biological molecules can be studied in their natural environment. In addition, by operation in liquid one avoids the creation of adhesion forces due to water meniscus formation between tip and sample in humid air. These types of forces are so large that they may mask the smaller molecular interactions between the tip and the sample. Some of the applications of AFM in biology and medicine are structural studies of DNA and RNA, ligand–receptor interactions, protein–nucleic acid systems, chromosomes, cellular membranes, peptides, molecular crystals, and polymeric biomaterials [7, 8].

Nevertheless, the impact of AFM has also made its way into the studies of materials in general [9]. The AFM is used to solve processing and materials problems in a wide range of technologies affecting the electronics, telecommunications, biomedical, chemical, automotive, aerospace, and energy industries. The materials under investigation include thin- and thick-film coatings, ceramics, composites, glasses, synthetic and biological membranes, metals, polymers, and semiconductors. The AFM is also used to study phenomena such as lubrication, plating, polishing, abrasion, adhesion, cleaning, corrosion, etching, and friction [10]. Although trying to name all of the applications would be a major task, we mention a few of the most prominent ones. Some applications in the microelectronics industry include measurements of surface roughness, depth of nanoridges in thin-film epitaxial deposition, detection of defects in growth of layers on substrates, efficiency of chemomechanical polishing, statistical classification of grain sizes in contacts with deposits, detection and sorting of defects in photolitography masks, identification of residual photoresist, and analysis of bits in compact disc media. In basic science studies, AFM is heavily used in the characterization of the atomic structure of pure surfaces or surfaces with adsorbed atomic clusters. AFM has also provided an immense boost to the fundamental understanding of sample elastic properties and of adhesion and friction. Although friction is not yet understood from the point of view of microscopic theories, its is very clear that such an understanding will ultimately come from an exchange of ideas between atomistic computational calculations and direct observations with AFM.

Forces were initially measured by looking at force-induced changes in the amplitude or phase of oscillating cantilevers near resonance [11]. In these first attempts, there was a tradeoff between increasing the signal-to-noise ratio and increasing the available bandwidth of the system. Clearly, in principle, one would like to maximize both parameters to be able to obtain highly precise measurements in as quick a time interval as possible. To improve these tasks, the basic method was shortly refined by using frequency modulation (FM) detection [12]. Instead of measuring amplitude or phase, this approach measures (via an FM demodulator) instantaneous frequency changes in the vibrating cantilever induced by $\partial F/\partial z$. This new procedure increased the sensitivity tenfold compared with previous techniques as demonstrated by studying magnetic transitions on CoPtCr thin-film magnetic disks.

Additional progress in improving the accuracy of the AFM was restricted for a while by reconstruction algorithms based on very simplified models of the AFM system. In the beginning, to advance the technique, it was enough to use a reconstruction algorithm based on a single mass attached to a spring to model all the inertia and elastic properties of the cantilever–tip system. Soon, models that were more realistic were put forward. First, Sarid [6] extended the simple model to a rectangular cantilever and studied its modes of vibration. Other theoretical variations on the same idea dealt with more realistic modeling of the oft-used V-shaped cantilevers [13] via variational techniques.

In the mid-1990s, people realized that thermal vibrations of the tip in close proximity to the sample could be used as a technique to extract intermolecular forces in aqueous solutions [14]. The researches were able to detect forces acting in the length scale of a few nanometers, which is the relevant range for adhesion, friction, and colloidal interactions. They base their analysis on the fact that the tip deflection is (in a fluid) distributed according to the probability density

$$p(s) = C \exp\left(-\frac{V(s)}{k_B T}\right)$$

where $V(s)$ is the tip–sample interaction energy versus separation, K_B is Boltzmann's constant, T is the absolute temperature, and C is a normalization constant. By measuring s with a high-speed acquisition card at many intervals (50,000 points at 100 KHz), they were able to experimentally construct $p(s)$, from which $V(s)$ can be evaluated. There exist theoretical calculations based on multimode analysis of the cantilever dynamics [15] that agree with the experimental results regarding modal frequencies. The power spectrum curves, however, do not match exactly. The authors justify this disagreement in terms of poorly characterized manufacturers' parameters.

Further progress in the interpretation of the experimental results was put forward [16] by comparison of the frequency spectra between theoretical results from beam models and optical interferometric measurements. This work showed that, for the case of free tip–cantilever systems, experiments and theory agree, lending credence to the beam model as a data processing algorithm. However, when the tip interacts significantly with the sample, there is a disagreement between the resonant frequencies. The researchers argued that the disagreement is related to the cantilever–tip geometry and nonlinearities in the force curve. They also suggest taking into account tip deformation and material transfer between tip and sample in more realistic models as that would be more consistent with their observations.

Studies were performed comparing theoretically and experimentally the dynamics of the cantilever for the specific case of the cantilever immersed in water or gas [17]. For this purpose, the researchers solved the coupled electromechanic–hydrodynamic problem. In addition, they performed experimental tests for the cantilever in air, water, and bromoform. They found an excellent match between the two sets of resonance frequencies for any of the three embedding media.

Numerical integration techniques to study the dynamics of the cantilever in noncontact mode were known to provide poor results when the snap-to-contact mechanical instability was reached. An interesting theoretical result was found to overcome this problem [18]. The researchers found analytical solutions based on the Krilov–Bogoliubov–Mitropolsky method for the motion of a tip in noncontact but in the possible presence of instabilities. They applied this result to the case of a van der Waals interaction and found good agreement with numerical integration for distances larger than the snap-to-contact point. For closer distances, this analytical method provides a qualitatively acceptable result for the frequency shifts induced by the interaction, while the numerical integration fails even qualitatively.

Theoretical studies for cantilevers in viscous fluids presented a formidable challenge. However, successful results were obtained with the goal of extracting interaction forces from thermal vibrations [19]. The researchers were able to obtain analytical results for the motion under the hypothesis that the cantilever's length is much larger than its width, the fluid is incompressible, and the viscosity is small. For larger viscosities, they presented numerical techniques to find the kinematics of the cantilever.

AFM has been used to obtain elastic information on the sample. This capability was soon recognized after the inception of the technique. However, quantitative measurements were hindered by the irreproducibility of the vibrational power spectrum of the cantilever. This irreproducibility was largely due to the impossibility to fully characterize the shape of thin tips. This problem was partially overcome [20] by the introduction of micromachined, geometrically reproducible, tips. These studies show that the power spectrum depends on the forcing amplitude. However, for small forcing amplitudes, the power spectrum remains unchanged, only varying linearly with the tip–sample separation. This permits the problem to be inverted, in other words, to explicitly obtain elastic constants of the sample from the power spectrum of the cantilever's vibration.

Although physically different, conservative and dissipative force fields could not be separated with reconstruction algorithms until three years ago. Then, theoretical analysis [21] of shifts in frequency *and* shifts in quality factor showed that the two types of forces could be separated. As expected, frequency shifts are related to conservative forces, and quality factor shifts are related to dissipation. The researchers found that, under typical experimental conditions, phonon drag in elastic nanoindentation is negligible and thus most energy loss is due to internal viscous damping forces or intrinsic irreversibilities of the tip–sample contact. However, the calculation also shows that soft materials might lose energy through phonon drag.

Particularly important in the extraction of elastic properties from samples is the realization that common force reconstruction algorithms are incomplete since they require motion information on the sample, which is not readily available in commercial microscopes. To overcome this difficulty, a new algorithm was designed, in which the tip–sample interaction damping and Young's modulus can be derived from contact resonance frequencies [22, 23]. The beauty of this result is that it is independent of the spatial location of the driver. The only constraint is that the excitation be as frequency independent as feasible in the frequency range of interest.

Experimental work using AFM in fluids [24] has shown that water is compressed between tip and sample and contributes to dissipation. The main result of these studies is that, when using AFM dynamically in water, the resulting image is, in fact, a map of contours of constant stiffness. More sophisticated force reconstruction algorithms were developed that explicitly account for force nonlinearities, based on variational principles and studying the degree on anharmonicity of the tips orbit, when the tip is driven at high amplitudes and thus is able to probe the whole range of the interaction force [25]. These developments show that, under these conditions, the frequency shift versus tip–sample separation is a function that, in appropriate units, lies between the force and the energy curves.

AFM in dynamic mode has also been used to study the elastic properties of anisotropic crystals [26]. In this case, the cantilever was excited at frequencies higher than the fundamental one. This method is particularly relevant for the detection of different metallurgical phases in an otherwise homogeneous sample.

In the search for high information-gathering speeds, new AFM designs, with modified cantilevers, were devised [27]. Specifically, the cantilever and its attached vertical piezoelectric control were substituted by a cantilever made purely of ZnO, a piezoelectric material. In this form, the frequency bandwidth content and the quality factor are increased. In typical Tapping mode imaging, one can acquire a 1-megapixel^2 image in a few minutes. With this new technique, the same image can be obtained in seconds. Damping is reduced by an active feedback circuit, in which the position of the tip, $z(t)$, is monitored and, via a phase shifter, converted to a velocity, $\dot{z}(t)$, which is fed back into the input circuit that controls the cantilever motion so that it reduces the speed and thus the friction.

There exist studies that compare experimental dynamical properties (frequency spectra) with theoretical results for cantilevers immersed in incompressible fluids [28]. These studies find that if only information about the lowest frequency is of interest, then a simple harmonic oscillator model appropriately describes the experimental results for quality factors $Q \geq 1$ as long as viscosity is not neglected. Neglecting viscosity renders large disagreements between theory and experiments.

The AFM has become a standard tool for biological studies. Among outstanding open problems in molecular biology is the understanding of the pathways by which biomolecules react. Due to temporal and spatial resolution limitations, typically only average values of the forces involved can be measured. The solution to this technical issue is of high priority, as it would clarify specific binding in ligand–receptor and antigen–antibody complexes, protein unfolding, and the mechanical properties of DNA [29]. To solve this problem, Heymann and Grubmüller [29] propose obtaining force profiles around molecules from force spectra obtained at various loads. This allows for augmented spatial resolution and thus for more site specificity regarding intermolecular forces. Other works have reported on the direct measurement of chemical bond formation [30]. This group worked at low temperature to avoid creep and drift and used nonlinear algorithms (as opposed to frequency shifts) to extract forces. In addition, they separated long-range forces from local chemical interactions.

During the past few years, a number of results have refined what was put forward during the 1990s. Experimentally, lateral forces have been measured in noncontact mode [31]; the collection of mechanical properties data from samples has been increased to a few megahertz by the use of stroboscopic techniques [32]; very fast acquisition rates, up to 100 MHz, have been achieved [33]; and other sensing techniques have been developed based on piezoresistive cantilevers [34]. Theoretical studies [35] have also further improved previous models in terms of speed and spatial resolution.

Hamilton's principle is used to obtain repulsive and van der Waals forces from samples by modeling and computing the acquisition via piezoelectric actuators (excited at their bases) and sensors, in contact and noncontact modes, respectively [36]. There have been attempts to extend steady-state treatments in noncontact spectroscopy by introducing transients via nonautonomous treatments [37]. The authors of this work argue that such a treatment is unavoidable since steady-state theories predict instabilities not yet observed experimentally. In addition, in the presence of flow, chaos can be generated, which cannot at all be studied within a steady-state framework.

Sometimes, the surfaces under study are so stiff that studying their elastic properties from spectral changes in the first mode of the cantilever (the softer one) is not possible. However, higher modes have been successfully used in such applications [38]. In particular, this work provides explicit results for the sensitivity of the various modes of flexural and torsional vibrations. A theoretical solution, in combination with the fast acquisition rates of the tip's position (megahertz), exists that allows for the extraction of force–separation curves, including the snap-to-contact instability, even for nominally flimsy cantilevers [39–45]. In this work, the cantilever is modeled as a continuous beam, and an ill-posed inverse problem (from kinematics to force) is solved by regularization techniques. Results agree with observational data.

8. FUTURE PERSPECTIVES

We have presented the status of fast atomic force microscopy to the present time. In retrospect, it is now clear that AFM has played, during the last decade, an enormous role as an imaging tool in the elucidation of new structures in organic as well as inorganic materials. It is also apparent what the role of AFM will be during the next 10 years. With the boom in nanoscience and nanotechnology, it is imperative that SPMs become more quantitative to provide detailed information on the transport, elastic, and electromagnetic properties with nanometric resolution. In addition, they must be quantitative as manipulation tools, to provide accurate positioning of small elements within relatively larger, synthetic architectures.

Since the invention of the transistor, electronics has been driven by technology toward smaller dimensions, higher information densities, and higher processing speeds. To that end, many ingenious solutions have been developed, some dealing with the intrinsic properties of the materials used and some with devices. Thus, there was, in the 1960s, the inception of silicon as a better alternative to germanium in the semiconductor industry. In the 1970s, there appeared quantum wells and superlattices made of alternate nanolayers of gallium arsenide and its alloys. On one hand, electrons in gallium arsenide and its alloys have larger mobilities and coherence lengths than those in silicon, so electronics could be made to respond faster, taking advantage of quantum effects—such as discrete quantum states—for information storage. On the other hand, gallium arsenide crystals can be grown, for example, on gallium aluminum arsenide with excellent adhesion. This is because the two crystalline structures are in registry. This allows for the building of multilayer structures with up to hundreds of layers, an impossible task to achieve at that time with silicon and its derivate materials. In addition, gallium arsenide is an ideal material for optoelectronics, that is, the integration of electronics and optics in a device. Due to its direct bandgap, gallium arsenide can efficiently emit and absorb light, that is, without the need for phonon exchange.

The development toward atomic control of nanometric devices has exploded in the last decade. With the introduction of SPMs, it is not only possible to study matter at the nanoscale but also to modify it. The most notable advances in nanoscience and nanotechnology, since the beginning of the 21st century, are, arguably, the inception of nanowires and nanodots and the better understanding and use of biomolecular self-organization. Nanowires are wires of a few to a few hundred atoms long and a few atoms thick, with excellent tensile properties. They can be made into conductors or semiconductors and are thus likely to be used as contacts or active optoelectronic devices (detectors, emitters, etc). Today, the most common materials studied for nanowires use are silicon [46], carbon [47], and metals [48]. It has been possible to fabricate and characterize these wires under laboratory conditions, but, thus far, it has not been possible to fully tame them. For example, a large number of nanotubes can be grown on substrates and viewed with an SPM or other techniques. However, their actual position cannot be perfectly controlled. Thus, it will be by about 2010 that these nanotubes will be wired industrially.

At this stage, AFM is used to structurally and electrically characterize these novel nanotubes. For electrical characterization, the AFM is used in the new surface potential imaging mode in which a bias voltage is externally applied between the tip and the sample to cancel the effect of all the other interaction forces. In this form, charge distribution along the wire can be obtained. Nanodots are atomic-size clusters of atoms. Arrays of metallic nanodots are important to guide light in optoelectronics applications below the diffraction limit [49].

Noncontact and Tapping AFM studies provide quantitative information on the charge injection and retention properties of silicon nanodots. At frequencies close to plasmon resonances, optical extinction through the nanodots is dominated by absorption, as opposed to the typical scattering mechanism. Sub-diffraction-limit wavelength light propagates coherently through the structure via near-field nearest-neighbor light-tail interactions. AFM has been used to both fabricate the array and characterize it both geometrically and electromagnetically.

The ability of biological molecules to arrange themselves into large, useful arrays will be exploited in the next generation of integrated electronics. Integrated electronics will require the ability to order nanowires (the elemental building blocks) into desired architectures. For example, DNA naturally self-assembles by molecular recognition, and it is compatible with biological technologies (polymerase chain reaction, enzymatic cutting). Peptide nucleic acids, which mimic DNA, have been attached to single-wall nanotubes [50]. Thus, nanotube connectivity to electrodes and among themselves can be controlled, with the possibility of obtaining self-assembled architectures in the nanoscale.

Metallization of nanowires can be obtained naturally or by human design. For example, silicon nanowires can be made conducting by wrapping them in Au films [51]. Carbon nanotubes are naturally conductors or semiconductors, depending on the way their surface is wrapped around the main axis. These properties have been investigated by STM and by AFM.

Nanometric micromachines that operate as voltage gate pumps and valves (membrane channels and pore proteins) have also been studied by AFM [52]. AFM will be used to obtain, with industrial quality, morphological and chemical information on self-assembled nanodevices. It will also be used to study conduction channels within nanowires as well as charge effects at the interface of nanowires used for one-dimensional quantum wells and superlattices.

Perhaps the major leap in the advance of AFM in the next decade will be the development of more robust quantitative algorithms to gain information on the tip–sample interaction. Since molecules are routinely attached to AFM tips, these developments will allow one to obtain a large database of molecule–molecule force–separation curves. A technician could then gather force information from the sample and, by comparing it with the database, he or she will be able to produce a high-quality chemical map of the sample, with the high degree of confidence with which today a technician uses nuclear magnetic resonance as an analytical chemical technique.

GLOSSARY

AFM attractive mode Operation of the AFM under the condition that the probe tip and the sample attract each other. This happens for relatively large tip–sample separation. At closer distances, atomic core–core repulsion dominates and the condition of attractiveness does not hold. This mode of operation has the advantage of being gentle on the sample, thus avoiding potential permanent damage such as scratches. One disadvantage is that the magnitude of the forces is small and thus their measurement is more delicate.

AFM reconstruction algorithm A mathematical prescription to process the experimentally available data (tip's deflection vs. nominal tip–sample separation) and convert it into useful information, namely surface topography and force–separation curves (i.e., chemical specificity).

AFM repulsive mode The tip and sample remain relatively close to each other. Under these conditions, core–core repulsion dominates the interaction. This mode of operation gives a strong signal because nominally it follows the sample's profile. However, for soft samples, as is the case in all biological applications and in many inorganic cases, the tip will likely move the sample material, thus changing the sample one wishes to study.

Atomic force microscope A modern microscope in which information is gathered via interaction forces between a probe and the sample under study.

Atomic force spectroscopy Use of an atomic force microscope to extract probe–sample forces as a function of separation.

REFERENCES

1. G. Binning, C. F. Quate, and Ch. Gerber, *Phys. Rev. Lett.* 56, 930 (1986).
2. G. Binning and H. Rohrer, *Sci. Am.* 253, 40 (1985).
3. P. Kielczyński, W. Pajewski, and M. Szalewski, *IEEE Trans. Ultrasonics* 46, 233 (1999).
4. www.di.com.
5. D. Sarid, "Exploring Scanning Probe Microscopy with Mathematica," Chap. 9. Wiley, New York, 1997.
6. D. Sarid, "Scanning Force Microscopy with Applications to Electric, Magnetic and Atomic Forces." Oxford Univ. Press, New York, 1991.
7. P. G. Arscott and V. A. Bloomfield, *Trends Biotechnol.* 8, 151 (1990).
8. C. Bai, P. Zhang, Y. Fang, E. Cao, and C. Wang, *J. Korean Phys. Soc.* 31, S47 (1997).
9. H. J. Guntherodt and R. Wiesendanger, Eds., "Scanning Tunneling Microscopy II." Springer-Verlag, Berlin, 1993.
10. H.-Q. Li, A. Chen, S. Roscoe, and J. Lipkowski, *J. Electroanal. Chem.* 500, 299 (2001).
11. Y. Martin, D. W. Abraham, and H. K. Wickramasinghe, *Appl. Phys. Lett.* 52, 1103 (1988).
12. T. R. Albretch, P. Grüter, D. Horne, and D. Rugar, *J. Appl. Phys.* 69, 668 (1991).

13. G. Y. Chen, R. J. Warmack, T. Thundat, D. P. Allison, and A. Huang, *Rev. Sci. Instrum.* 65, 2532 (1994).
14. J. P. Cleveland, T. E. Schäfer, and P. K. Hansma, *Phys. Rev. B* 52, R8692 (1995).
15. M. V. Salapaka, H. S. Bergh, J. Lai, A. Majumdar, and E. McFarland, *J. Appl. Phys.* 81, 2480 (1997).
16. U. Rabe, K. Janser, and W. Arnold, *Rev. Sci. Instrum.* 67, 3281 (1996).
17. F.-J. Elmer and M. Dreier, *J. Appl. Phys.* 81, 7709 (1997).
18. N. Sasaki and M. Tsukada, *Jpn. J. Appl. Phys.* 37, L533 (1998).
19. J. E. Sader, *J. Appl. Phys.* 84, 64 (1998).
20. K. Yamanaka, A. Noguchi, T. Tsuji, T. Koike, and T. Goto, *Surf. Interface Anal.* 27, 600 (1999).
21. U. Dürig, *Surf. Interface Anal.* 27, 467 (1999).
22. U. Rabe, E. Kester, and W. Arnold, *Surf. Interface Anal.* 27, 386 (1999).
23. U. Rabe, K. Janser, and W. Arnold, *Rev. Sci. Instrum.* 67, 3281 (1996).
24. M. Lantz, Y. Z. Liu, X. D. Cui, H. Tokumoto, and S. M. Lindsay, *Surf. Interface Anal.* 27, 354 (1999).
25. U. Dürig, *Appl. Phys. Lett.* 75, 433 (1999).
26. T. Drobek, R. W. Stark, M. Gräber, and W. M. Heckl, *New J. Phys.* 1, 15.1 (1999).
27. T. Sulchek, R. Hsieh, J. D. Adams, G. G. Yaralioglu, S. C. Minne, C. F. Quate, J. P. Cleveland, A. Atalar, and D. M. Adderton, *Appl. Phys. Lett.* 76, 1473 (2000).
28. J. W. M. Chon, P. Mulvaney, and J. E. Sader, *J. Appl. Phys.* 87, 3978 (2000).
29. B. Heymann and H. Grubmüller, *Phys. Rev. Lett.* 26, 6126 (2000).
30. M. A. Lantz, H. J. Hug, R. Hoffmann, P. J. A. van Schendel, P. Kappenberger, S. Martin, A. Baratoff, and H.-J. Güntherodt, *Science* 291, 2580 (2001).
31. O. Pfeiffer, R. Bennewitz, A. Baratoff, and E. Meyer, *Phys. Rev. B* 65, 161403(R) (2002).
32. E. Dupas, G. Gremaud, and A. Kulik, *Rev. Sci. Instrum.* 72, 3891 (2001).
33. H. Kawakatsu, S. Kawai, D. Saya, M. Nagashio, D. Kobayashi, H. Toshiyoshi, and H. Fujita, *Rev. Sci. Instrum.* 73, 2317 (2002).
34. T. L. Porter, M. P. Eastman, D. L. Pace, and M. Bradley, *Sens. Actuators, A* 88, 47 (2000).
35. D. Drakova, *Rep. Prog. Phys.* 64, 205 (2001).
36. R.-F. Fung and S.-C. Huang, *J. Vibrations Acoustics* 123, 502 (2001).
37. M. Gauthier, N. Sasaki, and M. Tsukada, *Phys. Rev. B* 64, 085409 (2001).
38. J. A. Turner and J. S. Wiehn, *Nanotechnology* 12, 322 (2001).
39. F. R. Zypman and S. J. Eppell, *J. Vac. Sci. Technol., B* 16, 2099 (1998).
40. B. A. Todd, S. J. Eppell, and F. R. Zypman, *J. Appl. Phys.* 88, 7321 (2000).
41. S. J. Eppell, B. A. Todd, and F. R. Zypman, *Materials Issues and Modeling for Device Nanofabrication* 584, 189 (2000).
42. F. R. Zypman and S. J. Eppell, U.S. Patent 6,145,374, 2000.
43. F. R. Zypman and J. S. Eppell, U.S. Patent 6,452,170, 2002.
44. B. A. Todd, S. J. Eppell, and F. R. Zypman, *J. Appl. Phys.* 88, 6973 (2000).
45. B. A. Todd, S. J. Eppell, and F. R. Zypman, *Appl. Phys. Lett.* 79, 1888 (2001).
46. R. Barsotti, J. E. Fischer, C. H. Lee, and J. Mahmood, "Proceedings of the Materials Research Society Fall Meeting," p. 112. www.mrs.org.
47. A. T. Johnson and M. Freitag, "Proceedings of the Materials Research Society Fall Meeting," p. 121. www.mrs.org.
48. G. Rubio-Bollinger, S. R. Bahn, N. Agrait, K. W. Jacobsen, and S. Vieira, *Phys. Rev. Lett.* 87, 026101 (2001).
49. E. A. Boer, L. D. Bell, M. L. Brongersma, and H. A. Atwater, *J. Appl. Phys.* 90, 2764 (2001).
50. F. Rosei, *Science* 296, 328 (2002).
51. C. P. Li, "Proceedings of the Materials Research Society Fall Meeting," p. 109. www.mrs.org.
52. J. Schmidt, D. Ho, and C. Montemagno, "Proceedings of the Materials Research Society Fall Meeting," p. 93. www.mrs.org.

ENN Encyclopedia of Nanoscience and Nanotechnology
www.aspbs.com/enn

Fe–Cu Nanoalloying

J. Z. Jiang

Zhejiang University, Hangzhou, People's Republic of China; and Technical University of Denmark, Lyngby, Denmark

X. J. Liu

Tohoku University, Sendai, Japan; and National Huaqiao University, Quanzhou, People's Republic of China

CONTENTS

1. INTRODUCTION

Nanostructured materials, consisting of small crystallites of diameter 1–100 nm, often have novel physical and chemical properties, differing from those of the corresponding bulk materials, for example, solid solubility enhancement in nanoalloys [1]. The solubility of a solute A in a solvent B, is controlled by the chemical potential of A in B. Obviously, if the atomic structure of B is changed, the chemical potential and hence the solubility of A in B may be enhanced (or reduced). Thus, the solute solubility of nanostructured materials is expected to be different from the one of single crystals with the same chemical composition. Under thermodynamic equilibrium, the solid solubilities of iron in copper and copper in iron are negligible at low temperatures [2]. However, nanoalloying (i.e., alloying when the elemental components in a system have a size ranging 1–100 nm at least in one dimension) in the iron–copper system has been reported using nonequilibrium processing techniques (e.g., liquid quenching [3], thermal evaporation [4, 5], sputtering [6, 7], ion-beam mixing [8], and mechanical milling). The nanoalloying process has been intensively investigated in the system prepared by mechanical milling as compared to other nonequilibrium processing techniques. This chapter, therefore, focuses on recent developments of nanoalloying in the iron–copper system, as a model system for those with positive heats of mixing, by a mechanical milling process.

More than 30 years ago mechanical alloying was developed as a way to circumvent the limitations of conventional alloying [9]. In this metallurgical process, powder particles are subjected to severe mechanical deformation from collisions with balls and vial and are repeatedly deformed, cold welded, and fractured. In recent years, mechanical alloying has emerged as a versatile technique for producing materials far from equilibrium, for example, nanostructured alloys and amorphous alloys [10–12]. By comparison with other novel techniques [e.g., liquid quenching (or melt-spinning), thermal evaporation, and sputtering, etc.], it has the advantages of relatively low-temperature processing, easy control of compositions, relatively inexpensive equipments needed, and the possibility for easily scaling up to tonnage quantities of material. These advantages greatly facilitate the characterization and application of the resulting metastable alloys.

In a variety of systems it has been demonstrated that mechanical alloying of element powder blends allows metastable alloys to be synthesized. In particular, much work has concentrated on the formation of metastable phases in alloy systems which exhibit a negative heat of mixing of the alloy components and the phase formation has been studied from both the thermodynamic and the kinetic viewpoints [10–12]. The mechanism of phase formation has been explained by an interdiffusion reaction of the components

ISBN: 1-58883-059-4/$35.00

Encyclopedia of Nanoscience and Nanotechnology
Edited by H. S. Nalwa
Volume 3: Pages (321–333)

occurring during the milling process. Metastable phase formation is observed if the formation of a stable phase is suppressed under the specific milling conditions. In particular, intermetallic compounds can be energetically destabilized due to chemical disorder introduced by the deformation during milling [13]. This often results in the formation of extended solid solutions or amorphous phases. On the other hand, phase formation in alloy systems with positive heats of mixing is far from being understood, as in these systems a diffusional reaction generally results in decomposition of the alloy. Nevertheless, the formation of solid solutions prepared by mechanical alloying has been observed in several systems with positive heats of mixing, such as Cu–M (M: Ag [14], Cr [15], V [16, 17], Ta [18–21], W [22], Co [23], and Fe). Among them the Fe–Cu system has been extensively studied [24–81]. Therefore, the Fe–Cu system is selected as a model system to study phase formation during mechanical alloying of elemental powder blends with a positive heat of mixing. By using the CALPHAD (*Cal*culation of *Ph*ase *D*iagrams) method [48, 55, 68, 82], it was predicted that the positive free energy of mixing at 673 K in the Fe–Cu system would be a maximum of about 7.5 kJ/g-atom for the composition of $Fe_{50}Cu_{50}$. For the discussion of the alloying process in the Fe–Cu system, we therefore mainly concentrated on this composition.

This chapter is organized in the following way. The typical mechanical milling process is mentioned in Section 2. Then, several problems on the phase formation of the alloy system with positive heats of mixing prepared by mechanical alloying will be discussed, for example, (1) whether alloying occurs on an atomic level, (2) the solid solubility in the Fe–Cu system; (3) where the positive energy is stored in the alloy; (4) the decomposition process of the alloy, (5) the phase diagram of nanometer-sized Fe–Cu grains, and (6) the kinds of the magnetic properties the new materials have. Finally, summaries will be given in Section 9.

2. MECHANICAL MILLING

Mechanical alloying begins by blending individual powder constituents, for the Fe–Cu system, which are usually powders of iron (99.9%) and Cu (99.9%) with particle sizes smaller than 100 μm. The powder blend is mixed to a desired composition and loaded in Ar atmosphere into hardened steel vials and balls in a planetary ball mill (e.g., Fritsch Pulverisette 5) or a vibrational ball mill (e.g., SPEX 8000 mixer/mill). The vial is sealed under an Ar atmosphere to prevent oxidation of both the Fe and Cu powders. In the final stage, the morphology of the sample varies from powder to pellet depending on the composition. For example, for $Fe_{90}Cu_{10}$, the sample is powder (less than 100 μm), and for $Fe_{50}Cu_{50}$, a pellet shape of about 0.7 mm in diameter and about 0.3 mm in thickness is observed. In many laboratories a small amount of lubricant is added into the vials to prepare fine powders. But, on the other hand, it is found that the milled samples contain contamination from the lubricant as well as from the abrasion of the vials and balls [83].

3. ALLOYING PROCESS ON AN ATOMIC LEVEL

To monitor the alloying process, the samples collected after various milling times are investigated by various techniques [e.g., X-ray diffraction, neutron diffraction, Mössbauer spectroscopy, and extended X-ray absorption fine structure (EXAFS)]. Typical X-ray diffraction patterns for the $Fe_{50}Cu_{50}$ samples after different milling times are shown in Figure 1. The elemental lines of body-centered cubic (bcc)-Fe lose intensity and broaden significantly during milling. After 100 h the bcc-Fe peaks completely disappear and only face-centered cubic (fcc) peaks remain. From the X-ray diffraction patterns the lattice constants and the average grain sizes of different phases can be determined. In Figure 2 are shown the grain size and the lattice constant of the fcc phase, together with the grain size of the bcc phase, as a function of milling time. Initially, the milling process reduces the grain sizes of both elemental powders. Upon milling, the grain size of the fcc phase saturates around 15 nm. On the lattice constant of the fcc phase, a large increase was observed from 10 to 50 h milling, suggesting that alloying between Fe and Cu may take place. The increase of the lattice parameter has been attributed to the magnetovolume effects [5]. Similar results have been reported by electron and neutron diffraction measurements [43, 44]. However, the X-ray, neutron, and electron diffraction results from the Fe–Cu system cannot absolutely confirm that in the final stage the samples have been alloyed on an atomic level when the element components form coherent composites with elemental spacing <10 nm [84]. In an attempt to examine whether the alloying on an atomic level occurs, Mössbauer spectroscopy [24, 25, 32, 33, 36, 38, 44, 49, 50, 56] and EXAFS [41, 51, 54, 57, 60, 77] measurements were performed.

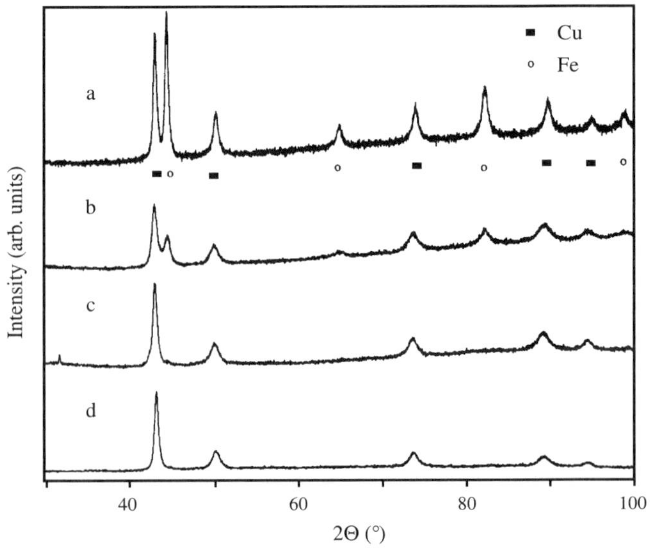

Figure 1. Typical X-ray diffraction patterns for $Fe_{50}Cu_{50}$ samples after different milling times: (a) 10, (b) 30, (c) 50, and (d) 100 h. Reprinted with permission from [32], J. Z. Jiang et al., *Appl. Phys. Lett.* 63, 2768 (1993). © 1993, American Institute of Physics.

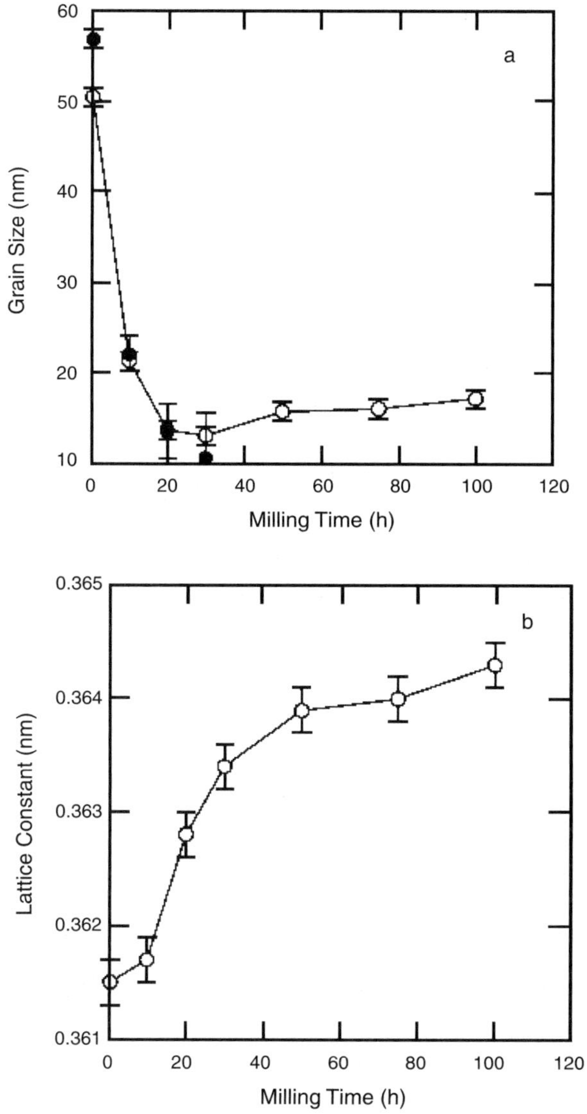

Figure 2. (a) Grain size vs milling time for fcc (open circles) and bcc (filled circles) phases. (b) Lattice parameter vs milling time for fcc phases for $Fe_{50}Cu_{50}$ samples [65].

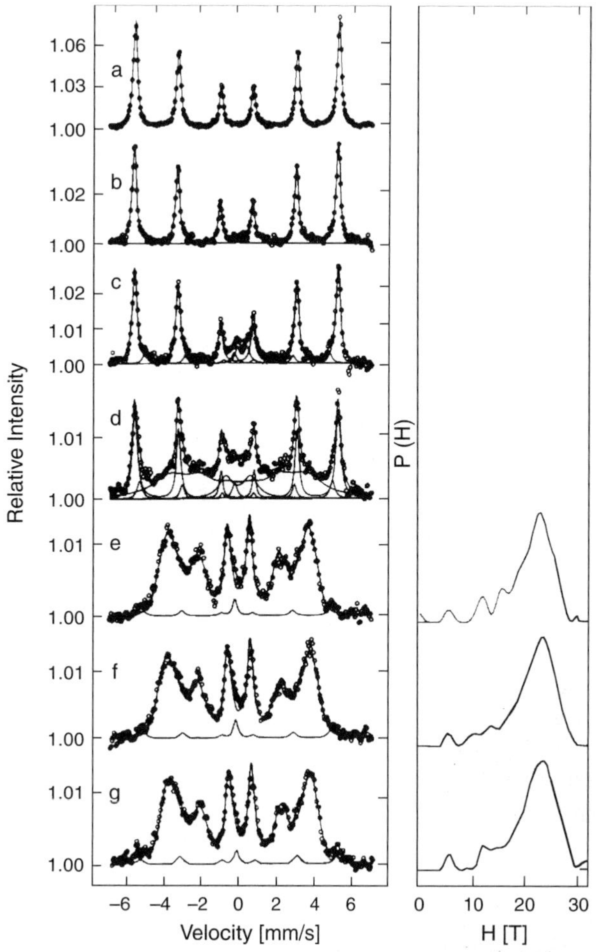

Figure 3. Room-temperature CXMS spectra and the corresponding hyperfine field distributions for $Fe_{50}Cu_{50}$ samples after different milling times [30]; (a) 0.5, (b) 10, (c) 20, (d) 30, (e) 50, (f) 75, and (g) 100 h. Reprinted with permission from [32], J. Z. Jiang et al., *Appl. Phys. Lett.* 63, 2768 (1993). © 1993, American Institute of Physics.

Mössbauer spectroscopy is sensitive to the local atomic surroundings of the iron atoms and can therefore be used to determine whether alloying occurs on an atomic level during milling. For thick samples (>100 μm) resonant lines in transmission Mössbauer measurements could become broadened. (Note that for the $Fe_{50}Cu_{50}$ composition, samples consist of small pellets with an average thickness of about 300 μm [32, 33].) Therefore, Mössbauer measurements in a backscattering geometry [i.e., conversion X-ray Mössbauer spectroscopy (CXMS) or conversion electron Mössbauer spectroscopy] are required to avoid the thickness effect. One example is given in Figure 3, where CXMS spectra and the corresponding hyperfine field distributions for the $Fe_{50}Cu_{50}$ samples after different milling times are shown. It should be emphasized that the samples used for CXMS measurements are the as-milled sample, prepared without any lubricant and without any post-treatments (e.g., cold rolling) which could influence the structure of the metastable Fe–Cu alloys. For milling times of 0.5 and 10 h, the Mössbauer spectra are identical to those of bcc-Fe. A significant change is observed after 20 h with the appearance of two nonmagnetic spectra and one new magnetic spectrum. The two nonmagnetic spectra are attributed to Fe in fcc Cu solid solution and to a fcc Fe phase coherently attached to the fcc-Cu matrix, respectively. These results indicate that a transformation of bcc-Fe into fcc-Fe structure occurs. The new magnetic sextet with a hyperfine field of about 300 kOe observed in Figuer 3c can be attributed to bcc-Fe atoms surrounded by Cu atoms. Based on the relative area of the sextet, the Cu concentration in the bcc-Fe phase is estimated to be less than 3 at% by assuming Fe atoms surrounded by one or two Cu atoms. For a milling time of 30 h, mixing of Fe and Cu is observed. This results in an increase in the relative area of the paramagnetic doublet and the formation of a broadened magnetic sextet with an average hyperfine field of 220 kOe. However, the relative area of the component with a hyperfine field of about 300 kOe does not change much. On further increasing the milling time (>50 h), three different components can be distinguished in the Mössbauer spectrum. The first is a broadened sextet caused by a hyperfine field distribution, which is attributed to the fcc-FeCu solid solution. The hyperfine field can be analyzed in terms of different environments for the iron atoms arising from near neighbor interactions. Since interactions influencing the hyperfine field are of a rather short-range character, it is valid to consider only the nearest neighbor interactions in the fcc structure [85]. By fitting, as shown in Figure 4, it is found that the distribution of nearest neighbor configurations in the fcc-FeCu phase follows a binomial distribution, indicating a random solid solution. This infers that many different environments of iron atoms exist in the fcc-FeCu solid solution and demonstrates that Fe and Cu are alloyed on an atomic level. The second and third components are a tiny amount of the residual bcc-Fe phase and the fcc-Fe phase, respectively. The tiny amount of the fcc-Fe phase was found in all samples for milling time longer or equal to 10 h. It should be noted that the bcc- and fcc-Fe phases are not detected by X-ray diffraction measurement. This can be explained by the very small amount (~2.6 vol%), the small crystallite sizes, and similarity of the lattice constants between fcc-FeCu and fcc-Fe phases. By comparing the Mössbauer spectra (Fig. 3e–g) with those for the small bcc-Fe particles having an average particle size of 2.4 nm [86], it is obvious that the $Fe_{50}Cu_{50}$ sample after a 100 h milling does not consist of a mixture of small Fe and Cu particles.

This conclusion is further supported by EXAFS measurements [41, 51, 54, 57, 60, 77], differential scanning calorimetry measurements [26–31, 34, 54], vacuum ultraviolet reflectance measurements [71], and magnetic measurements [37, 39] (see Section 8). For example, EXAFS studies [41, 51, 54, 57, 60, 77] on the as-milled $Fe_{50}Cu_{50}$ sample indicated that the first coordination shell around both the Fe and Cu atoms consists of a mixture of Fe and Cu atoms. In agreement with the Mössbauer spectroscopy measurements, the EXAFS measurements demonstrate that atomic level mixing has occurred around Fe and Cu sites during ball milling. The change of the copper Fermi level in the milled samples also indicated occurrence of alloying at the atomic level during milling [71].

In addition, transmission electron microscopy (TEM) and high resolution transmission electron microscopy (HRTEM) studies [38, 63, 64] of the mechanical alloying process have also been performed in Fe–Cu specimens. It is found that in the early stage of milling shear bands and grain boundary structures have been formed. Small bcc grains with sizes less than 5 nm exist in the specimens. In the intermediate stage [64] a bcc-to-fcc phase transformation occurs by simultaneous shearing process when the copper content reaches 20 at% in bcc-Fe_{rich}Cu grains with a grain size of about 20 nm, as shown in Figure 5. The interface is clearly illustrated

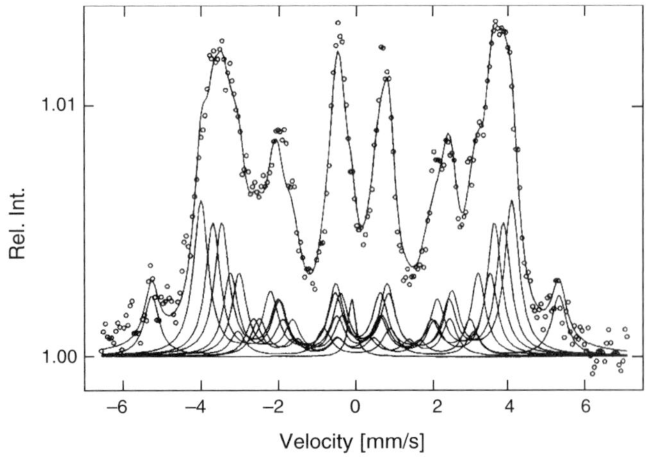

Figure 4. Mössbauer spectrum for an as-milled fcc-$Fe_{50}Cu_{50}$ sample fitted as a superposition of nine subspectra. Reprinted with permission from [38], J. Z. Jiang and F. T. Chen, *J. Phys. Condens. Matter* 6, L343 (1994). © 1994, Institute of Physics.

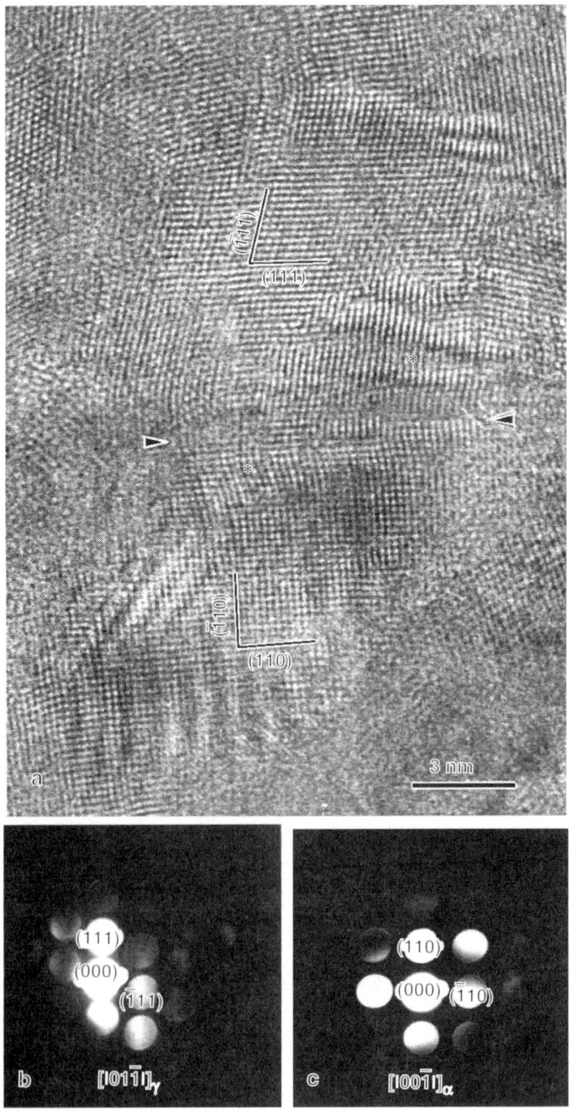

Figure 5. An HRTEM image of an interface between bcc- and fcc-Fe_{rich}Cu in the sample milled for 30 h. (a) The NW orientation relationships, that is, $[001]_\alpha//[011]_\gamma$ and $(110)_\alpha//(111)_\gamma$, exist between the two phases. The arrows indicate the interfaces and asterisks mark lattice distortion regions. The EDPs and EDS from fcc-Fe_{rich}Cu (top) and bcc-Fe_{rich}Cu (bottom) are shown in (b), (d), (c), and (e), respectively [64].

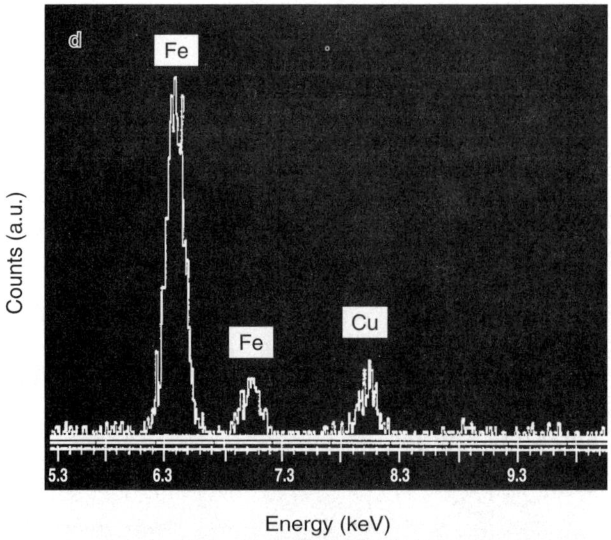

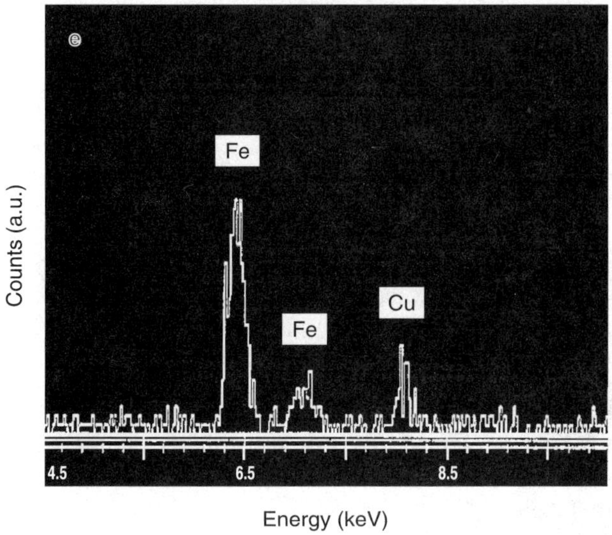

Figure 5. Continued.

in Figure 5a together with the electron diffraction patterns (EDP) from both the upper grain (Fig. 5b) and the lower grain (Fig. 5c). Both the lattice image and the EDP of the lower grain can be identified as $[001]_{bcc}$, while those of the upper grain can be identified as $[011]_{fcc}$. Nanoscale composition analyses (Fig. 5d and e) revealed that both grains have a similar composition to $Fe_{80}Cu_{20}$. The upper grain was unambiguously identified as fcc-$Fe_{80}Cu_{20}$. The Nishiyama–Wasserman (NW) orientation relationships, $[001]_{\alpha}//[011]_{\gamma}$ and $(110)_{\alpha}//(111)_{\gamma}$, were found, which are typical orientation relationships held in a Martensitic transformation. After prolong milling, the structure is fine and homogeneous with a distribution of grain size in a range of a few nm to a few tens of nm. Deformation twins are often observed in the mechanically alloyed specimen. The formation of deformation twins could result from the high strain rates induced by mechanical milling. A nanoscale composition analysis [63] (with an electron beam size around 1 nm) of the $Fe_{16}Cu_{84}$ specimen after the longest milling time shows that the average iron content in both the interior of grains and interfaces between grains are close to the designed composition. Their results support the earlier conclusion that supersaturated Fe–Cu solid solutions are formed.

According to the experimental results reported, it can be concluded that true alloying occurs in the Fe–Cu system during milling. Initially, the milling process reduces the crystallite sizes of both elemental powder blends and additional interfaces between Cu and Fe regions are formed. Further, refinement of the microstructure increases the density of interfaces making interdiffusion of Cu and Fe energetically possible. Finally a random fcc-FeCu solid solution is formed, in which Fe and Cu are alloyed on an atomic level.

4. SOLID SOLUBILITY IN THE Fe–Cu SYSTEM

From X-ray diffraction measurements of $Fe_{100-x}Cu_x$ samples in the final stage [19, 21–23, 26, 39] it is found that for $x \geq 40$ the bcc peaks disappear completely and only fcc peaks remain. When $x \leq 20$, only bcc peaks remain and the solutions have the bcc structure. For $20 < x < 40$, both fcc and bcc diffraction lines coexist even after extended milling. These findings have been confirmed by TEM investigations [25]. The obtained values of lattice constant for the Fe–Cu alloys agree well with data reported in these alloys prepared by evaporation techniques [87, 88]. A schematic illustration of the solid solutions prepared by various techniques in the Fe–Cu system is shown in Figure 6. It is clear that the single fcc phase field is extended significantly by mechanical alloying. The observed fcc–bcc transition can be predicted by the calculated free energy curves determined by the CALPHAD method [48, 55, 68]. When $x > 40$, the free energy of mixing for the fcc structure is smaller than that for

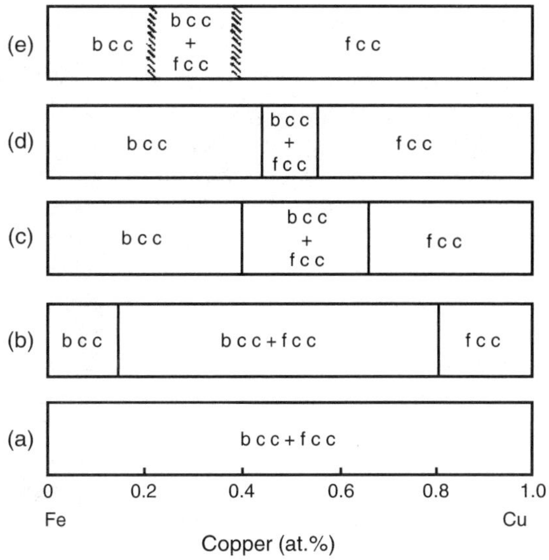

Figure 6. Schematic diagram of the equilibrium phase boundary at room temperature (a), and the phase boundaries obtained by liquid quenching (b), thermal evaporation (c), sputtering (d), and mechanical alloying (e) [65].

the bcc structure, for $x < 20$ the bcc structure is more stable, and in between, differences of the free energy between both structures are small, resulting in the existence of both structures in the samples as observed. However, from the calculated free energy curves only one crossover ($\Delta G = 0$) point exists; one expects the fcc (or bcc) phase does not immediately transform into the bcc (or fcc) phase once the solute content is beyond the crossover concentration. This question has been further investigated. Hong and Fultz [68], based on the results obtained from Mössbauer spectroscopy measurements and modifying CALPHAD-type free energy calculations to consider defect enthalpies induced by mechanical milling, about 1.5 kJ/mol for the bcc phase and 0.5 kJ/mol for the fcc phase, suggested the two-phase coexistence links in part to spatial heterogeneity in the defect distribution induced by milling. Once crossover in the free energy curves is reached in a region, the bcc-to-fcc (or fcc-to-bcc) transformation will immediately proceed to completion. The heterogeneity of stored defect energy results in different crossover points in different local regions, and this leads to the two-phase region observed. On the other hand, Schilling et al. [70], based on the results obtained from X-ray absorption near-edge structure spectroscopy measurements, found that the compositions of the two phases in coexistence are nearly the same (i.e., they are practically polymorphs with the same composition as the overall mixture). This rules out the possibilities of any two-phase stable or metastable equilibrium under a common-tangent construction that would dictate two fixed end compositions with varying phase fractions across the two-phase region. Furthermore, they [70] proposed that the transformation of one phase to another is not only governed by free energy, but only when there is a sufficient driving force (ΔG) to overcome a transformation barrier. This is contrast to the model suggested by Hong and Fultz [68], in which no such barrier is present. A single fcc (or bcc) phase becomes unstable as the composition passes the free energy crossover. Phase transformation (e.g., via Martensitic transformation observed in [64]) occurs, which requires a driving force ($\Delta G \neq 0$) to overcome the transformation barrier arising most likely from the strain energy associated with nucleation and growth. At compositions outside the two-phase region, sufficient driving force is available to accomplish the transformation.

Hong and Fultz [68] further studied the milling intensity effect on the composition range for the two-phase coexistence. They found that with higher milling intensity, there was a narrowing of the range of compositions for fcc and bcc coexistence, and a shift toward Fe-rich compositions. However, Schilling et al. [70] found that low-temperature milling at around 77 K gives a similar two-phase coexistence composition as compared to that using room temperature milling, although it was thought that low-temperature milling would influence the two-phase coexistence composition range because low-temperature milling effectively reduced the powder agglomeration and cold welding. The formation of a fcc-$Fe_{50}Cu_{50}$ solid solution by mechanical alloying requires the free energy to be raised by about 7.5 kJ/mol. However, there is a basic question (i.e., where the energy is stored in the as-milled solid solutions). The following section will deal with this problem.

5. DRIVING FORCE FOR THE ALLOYING

The mechanism by which a solid solution is formed by mechanical alloying in systems exhibiting positive free energies of mixing has been extensively studied in the last few years. First, Veltl et al. [21] suggested that the energy stored in the grain boundaries serves as a driving force for the formation of a solid solution. Their assumption is based on the observation that alloying occurs when the grain sizes of elemental components decrease down to the nanometer range and that a substantial amount of enthalpy can be stored in nanocrystalline metals due to the large grain boundary area [89]. In the Fe–Cu system, mechanical alloying leads to the formation of single-phase solid solutions of up to 60 at% Fe in Cu and 20 at% Cu in Fe. X-ray diffraction investigations reveal that the average grain size of the elemental components prior to alloying is not smaller than 5 nm. (Note that X-ray line broadening is determined by the volume-weighted average grain size). An increase of the grain size from 5 to 15 nm during alloying corresponds to a maximum decrease in the grain boundary enthalpy of the alloy of $3\sigma_{gb}V_m(1/d_1 - 1/d_2) \approx 1.4$ kJ/g-atom, where V_m is the average molar volume and $\sigma_{gb} = 0.5$ J/m^2 is the grain-boundary enthalpy of Cu as measured on nanocrystalline samples [90]. Compared to the free energy of mixing this value may not be sufficient to explain the alloying behavior in the Fe–Cu system.

Second, Yavari and Desré [91] suspected that alloying of immiscible elements in the systems with high solubilities for oxygen originates from a high oxygen content thus resulting in a negative heat of mixing of the multicomponent mixture. This suggestion becomes less possible in Fe–Cu samples due to low oxygen contents (e.g., less than 0.5 wt% in a fcc-$Fe_{50}Cu_{50}$ solid solution milled for a 100 h) [32]. Eckert et al. [27–31] suggested another mechanism, which explains alloy formation by an enhanced solubility due to the high dislocation density during the milling of nanocrystalline powders. The effect of point defects formed during plastic deformation on the alloying process has also been stressed [52]. However, Ma et al. [34, 48] estimated the magnitude of the contribution of the strain relief in the Fe–Cu solid solutions. They found that for a root-mean-square (rms) strain of 0.3%, usually observed in the Fe–Cu system, only an enthalpy of less than 0.5 kJ/g-atom is associated, which indicated the strain energy induced by mechanical deformation is insignificant in the Fe–Cu system for the alloying process. (Note that strain energy will be on the order of 1 kJ/g-atom in other systems in which a rms strain of a few percent is observed.)

Later, Yavari et al. [26, 40, 47] proposed alternative alloying mechanisms, based on the assumption that upon deformation of the composite, elemental fragments with small tip radii are formed and a small fraction of materials is expected to be in crystallite diameters of 2 nm or less, corresponding to the tail of the particle size distribution. In these cases the capillary pressure forces the atoms at the tips of the fragments to dissolve. This process will continue to full dissolution, due to generation of such small particles by necking at tips of larger ones with continued milling. They further estimated the chemical energy of the Fe–Cu interface as a function of grain size. They found that for grain sizes less than 2 nm, the formation of solid solutions can be energetically possible. A similar mechanism was also suggested earlier by Gente et al. [23] in a study of the immiscible Co–Cu system. They suggested that the chemical enthalpy associated with the interface between the elemental components can enhance the free energy of a composite above that of the related solid solution thus providing a driving force for alloying in systems with a positive free energy of mixing. Gente et al. [23] and Yavari et al. [40, 47] both claimed that the dimensions of the element components before alloying may be significantly smaller than the average grain size, which has indeed been observed in recent high-resolution TEM studies of Fe–Cu [43, 63, 64]. The chemical energy stored in the Fe/Cu interfaces is very high, for example, up to 6.3 kJ/g-atom for elemental domains of 1 nm size and a chemical interfacial energy of 0.3 J/m^2 [23]. This value is comparable to the free energy of mixing for a $Fe_{50}Cu_{50}$ alloy. Hence, the driving force for the formation of the single-phase solid solution could originate mainly from the chemical enthalpy of the Fe/Cu interfaces formed during ball milling. The main difference between the models proposed by Yavari et al. [26, 40, 47] and Gente et al. [23] is that Yavari et al. assume incoherent interfaces between the elemental components, whereas Gente et al. suggest the formation of composites with coherent elemental domains before alloying occurs. So far, microstructural information obtained might not be sufficient to decide between the two models and further investigation is needed.

The kinetic process of the nanoalloying in the Fe–Cu system has been investigated by means of Monte Carlo simulation [46] and HRTEM measurements [64]. Both investigations suggested that mechanical deformation-induced diffusion is an important factor in controlling the alloying process. Monte Carlo simulation [46] showed that with an increasing shearing frequency, the interfaces become increasingly rough, leading to refinement of the microstructure and eventually to homogeneous alloying of the system. The HRTEM study [64] of a sample milled in the intermediate stage revealed that a bcc-to-fcc phase transformation occurs by a simultaneous shearing process when the copper content reaches 20 at% in bcc-Fe_{rich}Cu grains with the grain size of about 20 nm. Thus, the kinetic process of the nanoalloying was suggested as follows: Initially, the milling process reduces the crystallite sizes of both elemental Fe and Cu powder blends to a few tens of nanometers due to the mobility of dislocations. Many defects, (e.g., lattice distortion, dislocation, and low- and high-angle grain boundaries) are formed in these nanostructured materials. The solubilities of Fe in Cu and Cu in Fe are then enhanced according to the Gibbs–Thomson equation [92]. This promotes the defect-induced interdiffusion processes in the system, which has indeed been experimentally confirmed by Mössbauer measurements [32] and the nanoscale composition analyses. Note that a diffusional reaction in the systems with a positive heat of mixing generally results in decomposition of the alloy under equilibrium conditions. However, by decreasing the grain size, the solubility, especially iron in fcc Cu matrix, can be large even under equilibrium condition (details given in Section 7). The bcc-Fe_{rich}Cu grains which have the Martensitic transformation temperature lower than the local temperature at collisions with balls and vial become unstable and then the bcc → fcc phase transformation occurs by simultaneous shearing of the whole grain. On further increasing the milling time, although the average crystallite size of the fcc phases remains unchanged, interdiffusion between fcc-Fe_{rich}Cu and fcc-$FeCu_{rich}$ is favored because of the similarity of the lattice structures. The last stage of milling is a homogenization process until a complete fcc-FeCu solid solution is formed. It is concluded that the alloying process in the Fe–Cu system during milling is kinetically governed by the atomic shear effect and the defect-induced interdiffusion process.

6. THERMAL STABILITY OF THE ALLOYS

Using a high energy ball milling technique, many new materials have been produced, which may have potential industrial applications. Structural stability of the new materials prepared should be studied.

The thermal stability of the Fe–Cu solid solutions has been investigated extensively [24–31, 33, 34, 36, 41, 42, 44, 47, 56, 69]. In bcc Fe-rich solid solutions, segregation of Cu to the grain boundaries was observed [56]. Figure 7 shows typical X-ray diffraction patterns for the $Fe_{50}Cu_{50}$ samples after milling and for those subsequently annealed at different temperatures for one hour in vacuum. With annealing at 473 K, the (110) peak of the bcc-Fe phase appears. However, within the accuracy of the measurements, the grain size of the fcc phase does not change with respect to the as-milled sample. The lattice constant of the fcc phase decreases, corresponding to a reduction of the iron content in the fcc phase due to Fe precipitation [33]. By increasing the annealing temperature from 473 to 873 K the amount of the bcc-Fe phase increases at the expense of that of the fcc-FeCu phase. Thereby the lattice constant of the fcc-FeCu phase is further reduced and the grain sizes of the fcc and bcc phases increase up to about 50 nm. However, due to the similarity of the lattice constants between fcc-Fe, fcc-Fe_{rich}Cu, and fcc-Cu_{rich}Fe phases, it is impossible to detect the early coherent stage of the decomposition by diffraction techniques [84]. On the basis of a simulation of the saturation magnetization of the fcc-$Fe_{50}Cu_{50}$ solid solution as a function of annealing time at a given temperature, Drbohlav and Yavari [47] suggested that fcc-Fe_{rich}Cu and fcc-Cu_{rich}Fe zones are formed in the early stage by the spinodal decomposition process. Furthermore, in the Mössbauer measurements [33], it is found that at the beginning, fcc iron precipitates coherently in the fcc-FeCu matrix. Upon further annealing, the Fe particles grow and transform into the stable bcc-Fe phase. The average hyperfine field of the fcc-FeCu component decreases gradually as the annealing temperature increases, which indicates a continuous reduction of the Fe content in the fcc-FeCu phase. The decomposition process of the fcc-$Fe_{50}Cu_{50}$ solid solution prepared

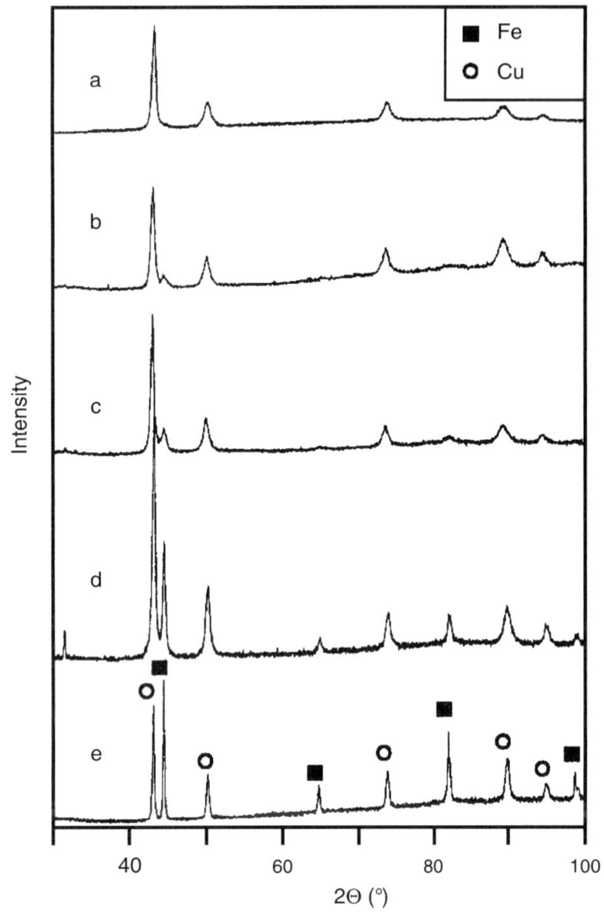

Figure 7. Typical X-ray diffraction patterns for various $Fe_{50}Cu_{50}$ samples in the as-milled state (a) and annealed at 473 (b), 573 (c), 673 (d), and 873 K (e) [33]. Reprinted with permission from [33], J. Z. Jiang et al., *Appl. Phys. Lett.* 63, 1056 (1993). © 1993, American Institute of Physics.

by mechanical alloying can be summarized as follows: Initially, composition fluctuations within grains occurs (i.e., formation of coherent fcc-Fe_{rich}Cu and fcc-Cu_{rich}Fe domains). Upon further annealing decomposition proceeds while the fcc-Fe particles grow and transform into the stable bcc-Fe phase. Finally, separated bcc-Fe and fcc-Cu phases are formed.

Xu et al. [69] further studied the low-temperature ball milling effect at elevated temperatures (373–523 K) on the decomposition of the unstable fcc-$Fe_{50}Cu_{50}$ solid solution prepared by high-energy ball milling. They found that despite the reduced driving force due to the homogenizing effect of the deformation, milling-induced deformation significantly enhanced the initial decomposition rate as compared to the isothermal annealing at the same temperatures used. At 423 K, the initial milling (10 h) led to the formation of bcc phase. Increasing the milling time, the relative amount of the bcc phase reached a maximum at about 20 h and then decreased. After 80 h, the bcc phase disappeared. By further milling, only the fcc phase was detected. A similar reversal fcc-to-bcc-to-fcc phase transformation process during milling was observed in samples milled at 473 K $\geq T \geq$ 373 K. On the other hand, at 523 K, after 2 h, bcc phase appeared, and its fraction increased with milling time up to about 30 h. Further milling did not change the relative fraction ratio of fcc to bcc phase. The observed results were explained by an effect-temperature model, which was originally proposed by Martin [93]. Under competition between externally driven mixing and thermal back-diffusion toward equilibrium, the model revealed that the configuration of a regular binary solution would be the same as that in thermal equilibrium at the effect temperature, T_{eff}. Based on this model, Xu et al. [69] constructed a dynamic Fe–Cu phase diagram in T_{eff}. T_{eff} decreased with increasing milling temperature. At 423 K, T_{eff} still favors a single fcc phase while at 523 K, T_{eff} drops from the fcc single-phase region (e.g., at T^1_{eff}) to lower values (e.g., T^2_{eff}) corresponding to the two-phase region, where coexistence of fcc and bcc phases is expected. It should be noted that T_{eff}, estimated from the model for the Fe–Cu system, was significantly higher than their melting points.

7. Fe–Cu BINARY PHASE DIAGRAM WITH NANOMETER-SIZED PARTICLES

It is mentioned in Section 4 that the solid solubilities of Fe in the fcc phase and Cu in the bcc phase are enhanced by mechanical alloying, as shown in Figure 6. The enhancement of solid solubility could be linked with small grain sizes induced by milling. Although some attempts [21, 26–31, 91] tried to explain the extension of the solid solubility during the mechanical milling process, the subject is still not well understood. Recently, a method to calculate phase diagram with nanometer-sized grains was proposed [94], which is given as follows.

Using the CALPHAD method [95, 96], the Gibbs energy of the solid solution phase in the bulk state is described by the regular solution model. For instance, the Gibbs energy of the fcc and bcc phases in the Fe–Cu binary system is expressed as

$$G^{\phi} = {}^0G^{\phi}_{Cu}x^{\phi}_{Cu} + {}^0G^{\phi}_{Fe}x^{\phi}_{Fe} + RT(x^{\phi}_{Cu}\ln x^{\phi}_{Cu} + x^{\phi}_{Fe}\ln x^{\phi}_{Fe}) + L^{\phi}_{FeCu}x^{\phi}_{Cu}x^{\phi}_{Fe} + G^{mag} \quad (\phi = \text{bcc and fcc phases}) \quad (1)$$

where ${}^0G^{\phi}_{Cu}$ and ${}^0G^{\phi}_{Fe}$ are the Gibbs energy of pure components Cu and Fe in the respective reference state, x^{ϕ}_{Cu} and x^{ϕ}_{Fe} is the mole fraction of components Cu and Fe, respectively, and L^{ϕ}_{CuFe} is the temperature and composition dependent interaction energy in binary system.

The Gibbs energy of pure components Cu and Fe in its different phase states is taken from the SGTE database [97]. The thermodynamic assessment of the Fe–Cu phase diagram has been carried out by several groups [98–103]. The latest version was performed by Chen and Jin [103], whose parameters are used in the present work. G^{mag} is the magnetic contribution, which is described by the Hillert and Jarl model [104]. The magnetic term to the Gibbs energy is given by

$$G^{mag} = RT\ln(B+1)g(\tau) \quad (2)$$

where $\tau = T/Tc$, T_c is the Curie temperature, and B is the average magnetic moment per atom. $g(\tau)$ is given by

$$g(\tau) = 1 - \left[\frac{79\tau^{-1}}{140} + \frac{474}{497}\left(\frac{1}{p} - 1\right)\left(\frac{\tau^3}{6} + \frac{\tau^9}{135} + \frac{\tau^{15}}{600}\right)\right]\Big/ D \qquad \tau < 1$$

$$g(\tau) = -\left[\frac{\tau^{-5}}{10} + \frac{\tau^{-15}}{315} + \frac{\tau^{-25}}{1500}\right]\Big/ D \qquad \tau > 1$$

where

$$D = \frac{518}{1125} + \frac{11692}{15975}\left(\frac{1}{p} - 1\right)$$

The value of p, which is the fraction of the magnetic enthalpy absorbed above the critical temperature, depends on the structure; for the bcc phase $p = 0.40$ and for the fcc phase $p = 0.28$.

In the system with nanometer-sized grains, the pressure and volume contributions should be considered. x_0 is the concentration of Fe in the bulk state at certain T_0 and P_0. Thus, the chemical potential corresponding to the system with nanometer-sized grains can be written as the following series:

$$\mu(T, x, P) = \mu(T_0, x_0, P_0) + \left.\frac{\partial \mu}{\partial x}\right|_{x=x_0}(x - x_0) + \left.\frac{\partial \mu}{\partial P}\right|_{P=P_0} \times (P - P_0) + \left.\frac{\partial^2 \mu}{\partial x^2}\right|_{x=x_0}(x - x_0)^2 + \left.\frac{\partial^2 \mu}{\partial P^2}\right|_{P=P_0} \times (P - P_0)^2 + \left.\frac{\partial^2 \mu}{\partial x\,\partial P}\right|_{\substack{x=x_0\\P=P_0}} \times (x - x_0)(P - P_0) + \cdots \quad (3)$$

In this treatment, we will restrict our considerations to second-order terms. The Gibbs–Duhem relationship is given by the expression

$$-V\,dP + S\,dT + x_{Cu}\,d\mu_{Cu} + x_{Fe}\,d\mu_{Fe} = 0 \quad (4)$$

Using the Gibbs–Duhem relationship, we may obtain the equations

$$\frac{\partial \mu_{Cu}}{\partial P} = -\frac{V}{x_{Cu}} \qquad \frac{\partial \mu_{Fe}}{\partial P} = -\frac{V}{x_{Fe}}$$

$$\frac{\partial^2 \mu_{Cu}}{\partial P^2} = \frac{\kappa V}{x_{Cu}} \qquad \frac{\partial^2 \mu_{Fe}}{\partial P^2} = \frac{\kappa V}{x_{Fe}}$$

$$\frac{\partial^2 \mu_{Cu}}{\partial P\,\partial x} = -\frac{V_{Fe}}{x_{Cu}^2} \qquad \frac{\partial^2 \mu_{Fe}}{\partial P\,\partial x} = -\frac{V_{Cu}}{x_{Fe}^2}$$

where V_i and V are the molar volumes of an i element and alloy, respectively, and κ is the isothermal compressibility coefficient. $\partial\mu/\partial x$ and $\partial^2\mu/\partial x^2$ can be calculated from Eq. (1).

If we consider the phase equilibrium of the bcc and fcc phases in the Cu–Fe binary system, the equilibrium equation $\mu_{Cu}^{bcc}(T_0, x^{bcc}, P^{bcc}) = \mu_{Cu}^{fcc}(T_0, x^{fcc}, P^{fcc})$, $\mu_{Fe}^{bcc}(T_0, x^{bcc}, P^{bcc}) = \mu_{Fe}^{fcc}(T_0, x^{fcc}, P^{fcc})$ should hold, and the equation may be written as

$$\mu_i^{bcc}(T_0, x_0^{bcc}, P_0^{bcc}) - \mu_i^{fcc}(T_0, x_0^{fcc}, P_0^{fcc}) + \frac{\partial \mu_i^{bcc}}{\partial x}(x_{Fe}^{bcc} - x_0^{bcc}) - \frac{\partial \mu_i^{fcc}}{\partial x}(x_{Fe}^{fcc} - x_0^{fcc}) - \frac{V^{bcc}}{x_i^{bcc}}(P^{bcc} - P_0^{bcc}) + \frac{V^{fcc}}{x_i^{fcc}}(P^{fcc} - P_0^{fcc}) + \frac{\partial^2 \mu_i^{bcc}}{\partial x^2}(x_{Fe}^{bcc} - x_0^{bcc})^2 - \frac{\partial^2 \mu_i^{bcc}}{\partial x^2}(x_{Fe}^{fcc} - x_0^{fcc})^2 + \frac{\kappa^{bcc}V^{bcc}}{x_i^{bcc}}(P^{bcc} - P_0^{bcc})^2 - \frac{\kappa^{fcc}V^{fcc}}{x_i^{fcc}}(P^{bcc} - P_0^{fcc})^2 + \frac{V_j^{bcc}}{x_i^2}(x_{Fe}^{bcc} - x_0^{bcc}) \times (P^{bcc} - P_0^{bcc}) - \frac{V_j^{fcc}}{x_i^2}(x_{Fe}^{fcc} - x_0^{fcc})(P^{fcc} - P_0^{fcc}) = 0 \quad (5)$$

$i, j =$ Cu and Fe, respectively.

The molar volume and the isothermal compressibility coefficient of alloy are described as $V^\phi = (1 - x_0^\phi)V_{Cu}^\phi + x_0^\phi V_{Fe}^\phi$ and $\kappa^\phi = (1 - x_0^\phi)\kappa_{Cu}^\phi + x_0^\phi \kappa_{Fe}^\phi$, respectively. Both X_0 and P_0 can be chosen to be the concentration and pressure associated with the bulk phases. Therefore, in the bulk state, there are the equilibrium equations as follows:

$$\mu_{Cu}^{bcc}(T_0, x_0^{bcc}, P_0^{bcc}) = \mu_{Cu}^{fcc}(T_0, x_0^{fcc}, P_0^{fcc}) \qquad \text{and}$$

$$\mu_{Fe}^{bcc}(T_0, x_0^{bcc}, P_0^{bcc}) = \mu_{Fe}^{fcc}(T_0, x_0^{fcc}, P_0^{fcc})$$

The pressure P^ϕ in the ϕ phase is given by the Laplace law,

$$P^\phi = P_0^\phi + \frac{2\sigma^\phi}{r^\phi} \quad (6)$$

where r is the radius of the particle, P_0^ϕ is the pressure in the gaseous phase, and σ^ϕ is the surface tension in the ϕ phase.

Experimental data for surface tension of solids are limited, and the existing theoretical attempts did not give the dependence of structure on surface tension. For example, the different structures such as the bcc-Fe and fcc-Fe phases are not distinguished in the literature [105–107]. A method to predict the surface tension of solids for pure metals and binary alloys in various structures was suggested [108]. The calculated surface tension results for the bcc and fcc phases in the Cu–Fe system at 400 K are given in Figure 8, which correspond to the metastable bcc and fcc phases because the miscibility gap of the fcc phase exists at 400 K. It is seen that the surface tensions of both bcc- and fcc-Fe phases, 2894 and 2807 mN/m, respectively, sharply decrease due to a small addition of copper. There exist minima at about 4 at% copper. Then, they increase with increasing copper content. For pure bcc- and fcc-Cu, the surface tensions are 1622 and 1730 mN/m, respectively.

Two equations can be obtained from Eq. (5) with respect to Cu and Fe, respectively, and x_{Fe}^{bcc} and x_{Fe}^{bcc} can be calculated by simultaneously solving these two equations, which correspond to the concentrations of Fe with the particle radius (r). On the basis of this consideration, in principle phase equilibria in binary systems with nanometer-sized

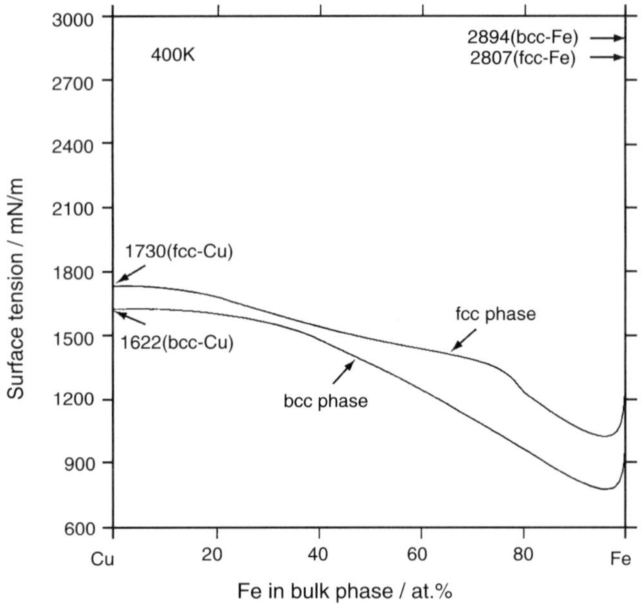

Figure 8. Calculated surface tensions of metastable fcc and bcc phases at 400 K [94].

grains can be calculated. The fcc/bcc phase equilibria in the Fe–Cu system with various grain sizes are shown in Figure 9. Two main characteristics for the phase diagram with decreasing grain size can be deduced: (1) the solubilities of Fe in the fcc phase and Cu in the bcc phase are enhanced by decreasing grain size, and (2) a negative temperature dependence of solid solubility is found in both fcc and bcc phases. The calculated results indicate that the solid solubility of Fe in the fcc phase is larger than that of Cu in the bcc phase for the same grain size at low temperatures, in agreement with the experimental ones reported previously

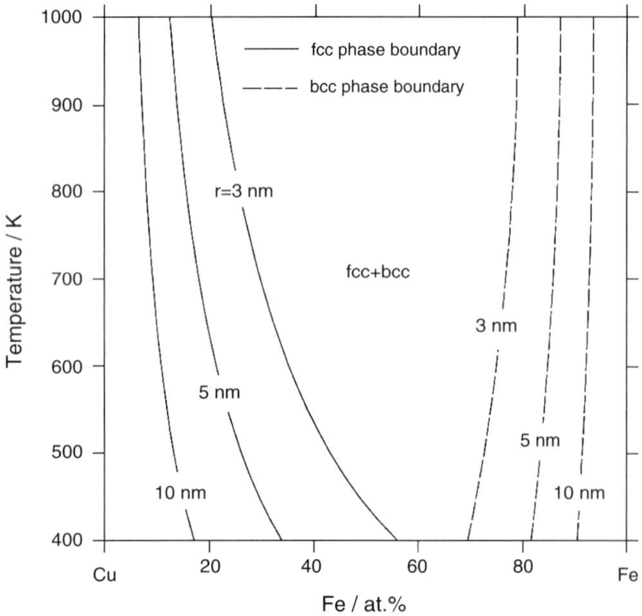

Figure 9. Calculated fcc/bcc phase equilibria with various nanometer-sized particles [108].

in Figure 6. Some quantitative difference of solubility may originate from other factors such as high dislocation density and internal strains, etc., which were not considered in the present calculation.

8. MAGNETIC PROPERTIES OF THE ALLOYS

Like the studies of the thermal stability of the Fe–Cu solid solutions, the magnetic properties of these alloys have been also studied by several groups [25, 34–37, 39, 42, 47, 49, 75, 81]. The saturation magnetization ($M(T)$) of the as-milled fcc-$Fe_{50}Cu_{50}$ alloy as a function of temperature [39], measured in a 1 T field, is shown in Figure 10. The magnetization decreases with increasing temperature up to about 510 K. With a further increase in temperature, the magnetization increases. By extrapolation of the curve below this minimum to zero magnetization, the Curie temperature for the metastable fcc-$Fe_{50}Cu_{50}$ solid solution is estimated to be $T_c = 505 \pm 5$ K. The temperature dependence of saturation magnetization in the temperature range above 500 K can be explained by the decomposition of the fcc-$Fe_{50}Cu_{50}$ alloy. When the annealing temperature reaches 473 K, the fcc-$Fe_{50}Cu_{50}$ phase decomposes into fcc-Fe, fcc-Fe_{rich}Cu, fcc-Cu_{rich}Fe, and bcc-Fe phases. With higher annealing temperatures the amount of the bcc-Fe phase rapidly increases at the expense of the Fe content in the fcc-FeCu and fcc-Fe phases. Although the saturation magnetization for the fcc-FeCu solid solution is zero when the temperature is above 500 K, the bcc-Fe phase can still maintain high saturation magnetization values (e.g., about 208 emu/g at 500 K and about 185 emu/g at 750 K) [109]. This accounts for the increase

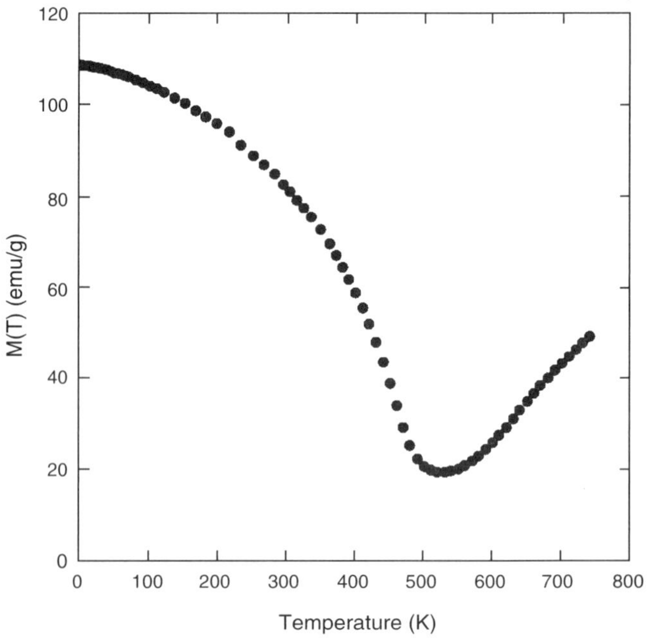

Figure 10. The saturation magnetization of an as-milled fcc-$Fe_{50}Cu_{50}$ sample as a function of temperature in 1 T field. Reprinted with permission from [39], J. Z. Jiang et al., *J. Phys. Condens. Matter* 6, L227 (1994). © 1994, Institute of Physics.

Table 1. Summary of the saturation magnetic moment, M_s, values per atom versus the Cu concentration or number of electrons per atom at 4 K of the metastable alloys in the Fe–Cu system [81].

Cu content (%)	10	20	25	30	35	40	45	50	60	65	70	80
Electrons per atom	26.3	26.6	26.75	26.9	27.05	27.2	27.35	27.5	27.8	27.95	28.1	28.4
M_s (μ_B)	2.26	1.89	1.69	1.47	1.30	1.30	1.21	1.09	0.83	0.73	0.62	0.25

of the saturation magnetization in the sample with temperatures above 510 K. The saturation magnetization [37, 39] for the as-milled fcc-$Fe_{50}Cu_{50}$ alloy at temperatures lower than 200 K ($0.4T_c$) follows Bloch's law, $M(T)/M(0) = 1 - BT^{3/2}$, or $(1 - M(T)/M(0)) = B_{3/2}(T/T_c)^{3/2}$, with $M(0) =$ 108.7 emu/g and $B = 0.00457$ (emu/g) $K^{-3/2}$. The values of the parameters B and $B_{3/2}$ for the as-milled fcc-$Fe_{50}Cu_{50}$ alloy are many times larger than those for crystalline Fe and Ni but are similar to those for the amorphous alloys [39]. It is known that the disordered structure in amorphous alloys leads to the high values of B and $B_{3/2}$ [110]. Consequently, the larger values of B and $B_{3/2}$ in the fcc-$Fe_{50}Cu_{50}$ solid solution imply that the atomic structures are very disordered. This is in good agreement with results obtained in Section 3.

The microscopic arrangement of the magnetic moment in the fcc-$Fe_{100-x}Cu_x$ alloys has been investigated [25, 39, 49, 75]. The bulk magnetic moments data at 5 K were reported [75]. The system is nonmagnetic for concentration below about 5 at% Fe. Above this concentration the magnetic moment increases at a rate of about 2.85 μ_B/Fe atom with a pronounced tendency to level off at higher Fe concentration, similar to what is observed in the case of Fe–Ni [111]. The onset and linear dependence of the bulk magnetization on the Fe concentration for less than 30% can be interpreted by means of a simple model: charge transfer from Cu to Fe bands. Many theoretical investigations on the fcc phase of Fe reveal a high-spin state, stable at high atomic volume and favoring strong ferromagnetism (i.e. an almost full spin-up band). When Cu concentration in the system increases, a transfer of electrons takes place from the delocalized ($s + p$)-like bands to the spin-down $3d$-like band of Fe, which is weak ferromagnetism. Consequently, a decrease of magnetic moment is expected as observed in [75]. The nonlinear trend of the bulk magnetization as the concentration of Fe is increased over 30% needs a mechanism different from the simple charge transfer model, which might be explained by the two-magnetic-state model of Fe suggested by band theory of ferromagnetism for fcc Fe alloys [112–115]. It has been suggested that an Fe atom with a larger atomic volume in the high-spin state has a larger magnetic moment. The second state, a low-spin state, with a low atomic volume is associated with a lower magnetic moment, and it favors the antiferromagnetic coupling. The existence of both states may result in the nonlinear trend of the bulk magnetization as the concentration of copper. Further investigation is required to clarify the two-magnetic-state mechanism. Low-temperature magnetization measurements of $Fe_{100-x}Cu_x$ ($100 > x > 0$) alloys prepared by mechanically milling were performed by Mashimo et al. [81]. The saturation magnetization moment per atom versus the number of electrons per atom at 4 K of the metastable alloys in the Fe–Cu system was obtained together with other binary systems. The values are summarized in Table 1. They found that the saturation magnetization moment reaches a local maximum at about 26.2 electrons per atom, which is similar to the cases in the Fe–Co and Fe–Ni systems. It decreases with increasing Cu concentration and approaches zero at about 28.6 electrons per atom. Their results were consistent with other binary systems in the Slater–Pauling curve.

The coercive field of the fcc-$Fe_{50}Cu_{50}$ at 300 K is only a few tens of Oe [35, 37, 42, 47, 49], which means a negligible structural anisotropy. The result reflects the disordered atomic configurations in the alloy. The temperature dependencies of the coercive field of some Fe–Cu solid solutions have been studied by Crespo et al. [49]. They found that for the $Fe_{20}Cu_{80}$ sample, an anomalous temperature dependence of the coercive field is observed, which has also been observed during the decomposition of some alloys (i.e., a high coercive field of about 150 Oe at 300 K, a minimum value at 50 K, and a further increase to 260 Oe at 10 K). This behavior could be explained by the existence of a distribution in composition, that is, different Curie temperatures, which would give rise to isolated ferromagnetic grains as the temperature continuously increases above the Curie temperatures of such distributed phases. However, for temperatures below 50 K, the coercive field increases with decreasing temperature. At the moment, no plausible explanation for this experimental result has been found. The room-temperature Mössbauer spectrum [32, 69] of the as-milled $Fe_{50}Cu_{50}$ sample (see Fig. 3g) shows that the intensity ratio of the second (or fifth) line to the first (or sixth) line is about 0.4. This means that the spins in the inside of a pellet prefer the normal direction of the pellet, which could be caused, via the magnetoelastic effect, by strain stress induced by the ball milling in the sample. Low-energy ball milling at 423 K for 240 h of the unstable fcc-$Fe_{50}Cu_{50}$ solid solution led to a random orientation of the spins [69].

9. SUMMARY

The studies of mechanical alloying on the Fe–Cu system, as a model system for those with positive heats of mixing, are reviewed. Several problems involved in the mechanical alloying process are discussed. The results demonstrate that alloying, on the atomic level, in systems with positive heats of mixing occurs by the high-energy ball milling. The driving force for the formation of the alloy is mainly derived from the chemical enthalpy associated with Fe/Cu interfaces formed during the ball milling. The alloying process in the Fe–Cu system during milling is kinetically governed by the atomic shear effect and the defect-induced interdiffusion process. The solubilities in the Fe–Cu system were largely extended, especially for the single fcc structure, which is linked with grain-size enhanced solid solubilities. The magnetic parameters of the fcc $Fe_{50}Cu_{50}$ alloy were investigated in detail (e.g., the Curie temperature, coercive field, orientation of spins, and the saturation magnetization). Furthermore, the thermal stability of the alloy was investigated and shows that iron particles already precipitate at temperatures around 473 K.

GLOSSARY

Driving force The difference in Gibbs free energy between parent phase and daughter phase.

Equilibrium system The system has the lowest Gibbs free energy for all possible configurations.

Grain boundary The boundary between at least two grains.

Immiscible system System with a positive heat of mixing.

Interface energy The free energy of an interface with a unit of J/m^2.

Mechanical alloying Alloying occurs by means of mechanical treatment.

Metastable alloys Alloys have configurations, which lie at a local minimum in Gibbs free energy.

Nanoalloying At least two components, having a nanometer size (1–100 nm) at least in one direction, mix on an atomic level.

Nanostructured material Material consists of components which have a nanometer size (1–100 nm) at least in one direction.

Solid solubility The concentration limit of one component in a solid solution.

Solid solution Solid material with at least two elements which mix on an atomic level.

Unstable alloys Alloys have configurations, which do not lie at a minimum in Gibbs free energy.

ACKNOWLEDGMENTS

J. Z. J. acknowledges all of the members of his research group at the Technical University of Denmark and Zhejiang University who have contributed to the research presented here. Financial support from (Zhejiang University) the Danish Technical Research Council, the Danish Natural Sciences Research Council, and the Danish Research Academy is gratefully acknowledged.

REFERENCES

1. H. Gleiter, *Acta Mater.* 48, 1 (2000).
2. T. B. Massalski, H. Okamoto, P. R. Subramanian, and L. Kacprzak, "Binary Alloy Phase Diagrams," 2nd ed., Vol. 2, p. 1736. ASM International, Materials Park, OH, 1990.
3. W. Klement, Jr., *Trans. Metall. Soc. AIME* 233, 1180 (1965).
4. E. F. Kneller, *J. Appl. Phys.* 35, 2210 (1964).
5. C. L. Chien, S. H. Liou, D. Kofalt, W. Yu, T. Egami, and T. R. McGuire, *Phys. Rev. B* 33, 3247 (1986).
6. K. Sumiyama, T. Yoshitake, and Y. Nakamura, *Trans. Jpn. Inst. Metals* 26, 217 (1985).
7. K. Sumiyama and Y. Nakamura, *Hyperfine Interactions* 53, 143 (1990).
8. L. J. Huang and B. X. Liu, *Appl. Phys. Lett.* 57, 1401 (1990).
9. J. S. Benjamin, *Metall. Trans.* 1, 2943 (1970).
10. W. L. Johnson, *Progr. Mater. Sci.* 30, 81 (1986).
11. C. C. Koch, in "Materials Science and Technology" (R. W. Cahn, P. Haasen, and E. J. Kramer, Eds.), Vol. 15, p. 193. VCH, Weinheim, 1991.
12. M. Oehring, Z. H. Yan, T. Klassen, and R. Bormann, *Phys. Status Solidi A* 131, 671 (1992).
13. R. Bormann, in "International Conference on Materials by Powder Technology," Dresden, Germany, 23–26 March 1993, p. 247. DGM Informationsgesellschaft Verlag, Germany, 1993.
14. K. Uenishi, K. F. Kobayashi, K. N. Ishihara, and P. H. Shingu, *Mater. Sci. Eng. A* 134, 1342 (1991).
15. Y. Ogino, T. Yamasaki, S. Murayama, and R. Sakai, *J. Non-Cryst. Solids* 117–118, 737 (1990).
16. T. Fukunaga, M. Mori, K. Inou, and U. Mizutani, *Mater. Sci. Eng. A* 134, 863 (1991).
17. K. Sakurai, M. Mori, and U. Mizutani, *Phys. Rev. B* 46, 5711 (1992).
18. T. Fukunaga, K. Nakamura, K. Suzuki, and U. Mizutani, *J. Non-Cryst. Solids* 117–118, 700 (1990).
19. K. Sakurai, Y. Yamada, M. Ito, C. H. Lee, T. Fukunaga, and U. Mizutani, *Appl. Phys. Lett.* 57, 2660 (1990).
20. K. Sakurai, Y. Yamada, C. H. Lee, T. Fukunaga, and U. Mizutani, *Mater. Sci. Eng. A* 134, 1414 (1991).
21. G. Veltl, B. Scholz, and H. D. Kunze, *Mater. Sci. Eng. A* 134, 1410 (1991).
22. E. Gaffet, C. Louison, M. Harmelin, and F. Foudot, *Mater. Sci. Eng. A* 134, 1380 (1991).
23. C. Gente, M. Oehring, and R. Bormann, *Phys. Rev. B* 48, 13244 (1993).
24. P. H. Shingu, K. N. Ishihara, K. Uenishi, J. Kuyama, B. Huang, and S. Nasu, in "Solid State Powder Processing" (A. H. Clauer and J. J. de Barbadillo, Eds.), p. 21. The Minerals, Metals and Materials Society, Warrendale, PA, 1990.
25. K. Uenishi, K. F. Kobayashi, S. Nasu, H. Hatano, K. N. Ishihara, and P. H. Shingu, *Z. Metall.* 83, 132 (1992).
26. A. R. Yavari, P. J. Desre, and T. Benameur, *Phys. Rev. Lett.* 68, 2235 (1992).
27. J. Eckert, R. Birringer, J. C. Holzer, C. E. Krill III, and W. L. Johnson, *Mater. Res. Soc. Symp. Proc.* 238, 739 (1992).
28. J. Eckert, J. C. Holzer, and W. L. Johnson, *Scripta Metall. Mater.* 27, 1105 (1992).
29. J. Eckert, J. C. Holzer, C. E. Krill III, and W. L. Johnson, *J. Mater. Res.* 7, 1980 (1992).
30. J. Eckert, J. C. Holzer, and W. L. Johnson, *J. Appl. Phys.* 73, 131 (1993).
31. J. Eckert, J. C. Holzer, C. E. Krill III, and W. L. Johnson, *J. Appl. Phys.* 73, 2794 (1993).
32. J. Z. Jiang, U. Gonser, C. Gente, and R. Bormann, *Appl. Phys. Lett.* 63, 2768 (1993).
33. J. Z. Jiang, U. Gonser, C. Gente, and R. Bormann, *Appl. Phys. Lett.* 63, 1056 (1993).
34. E. Ma, M. Atzmon, and F. E. Pinkerton, *J. Appl. Phys.* 74, 955 (1993).
35. T. Ambrose, A. Gavrin, and C. L. Chien, *J. Magn. Magn. Mater.* 124, 15 (1993).
36. P. Crespo, A. Hernando, R. Yavari, O. Drbohlav, A. G. Escorial, J. M. Barandiaran, and I. Orue, *Phys. Rev. B* 48, 7134 (1993).
37. A. Hernando, P. Crespo, J. M. Barandiaran, A. G. Escorial, and R. Yavari, *J. Magn. Magn. Mater.* 124, 5 (1993).
38. J. Z. Jiang and F. T. Chen, *J. Phys.: Condens. Matter* 6, L343 (1994).
39. J. Z. Jiang, Q. A. Pankhurst, C. E. Johnson, C. Gente, and R. Bormann, J. *Phys.: Condens. Matter* 6, L227 (1994).
40. A. R. Yavari, *Mater. Sci. Eng. A* 179–180, 20 (1994).

41. P. Crespo, A. Hernando, A. G. Escorial, K. M. Kemner, and V. G. Harris, *J. Appl. Phys.* 76, 6322 (1994).
42. O. Drbohlav and A. R. Yavari, *J. Magn. Magn. Mater.* 137, 243 (1994).
43. J. Y. Huang, A. Q. He, and Y. K. Wu, *Nanostruct. Mater.* 4, 1 (1994).
44. P. P. Macri, P. Rose, R. Frattini, S. Enzo, G. Principi, W. X. Hu, and N. Cowlam, *J. Appl. Phys.* 76, 4061 (1994).
45. M. Qi, M. Zhu, and D. Z. Yang, *J. Mater. Sci.* 13, 966 (1994).
46. P. Bellon and R. S. Averback, *Phys. Rev. Lett.* 74, 1819 (1995).
47. O. Drbohlav and A. R. Yavari, *Acta Mater.* 43, 1799 (1995).
48. E. Ma and M. Atzmon, *Mater. Chem. Phys.* 39, 249 (1995).
49. P. Crespo, M. J. Barro, I. Navarro, M. Vazquez, and A. Hernando, *J. Magn. Magn. Mater.* 140–144, 85 (1995).
50. T. Li, Y. Z. Li, Y. H. Zhang, C. Gao, S. Q. Wei, and W. H. Liu, *Phys. Rev. B* 52, 1120 (1995).
51. M. Angioloni, F. Cardellini, M. Krasnowski, G. Mazzone, and M. V. Antisari, *Microsc. Microanal. Microstruct.* 6, 601 (1995).
52. M. Angioloni, F. Cardellini, M. Krasnowski, G. Mazzone, M. Urchulutegui, and M. V. Antisari, *Mater. Sci. Forum* 195, 13 (1995).
53. A. Hernando, P. Crespo, A. G. Escorial, J. M. Barandiaran, M. Urchulutegui, and M. V. Antisari, *Europhys. Lett.* 32, 585 (1995).
54. P. J. Schilling, J.-H. He, J. Cheng, and E. Ma, *Appl. Phys. Lett.* 68, 767 (1996).
55. G. Mazzone and M. V. Antisari, *Phys. Rev. B* 54, 441 (1996).
56. B. Fultz, C. C. Ahn, S. Spooner, L. B. Hong, J. Eckert, and W. L. Johnson, *Metall. Mater. Trans. A* 27, 2934 (1996).
57. V. G. Harris, K. M. Kemner, B. N. Das, N. C. Koon, A. E. Ehrlich, J. P. Kirkland, J. C. Woicik, P. Crespo, A. Hernando, and A. Garcia Escorial, *Phys. Rev. B* 54, 6929 (1996).
58. J. Y. Huang, Y. D. Yu, Y. K. Wu, and H. Q. Ye, and Z. F. Dong, *J. Mater. Res.* 11, 2717 (1996).
59. J. Y. Huang, A. Q. He, Y. K. Wu, H. Q. Ye, and D. X. Li, *J. Mater. Sci.* 31, 4165 (1996).
60. E. Ma, J. H. He, and P. J. Schilling, *Phys. Rev. B* 55, 5542 (1997).
61. M. Angioloni, G. Mazzone, A. Montone, and M. V. Antisari, *Mater. Sci. Forum* 235–238, 175 (1997).
62. M. J. Barro, E. Navarro, P. Agudo, A. Hernando, P. Crespo, and A. Garci Escorial, *Mater. Sci. Forum* 235–238, 553 (1997).
63. J. Y. Huang, Y. D. Yu, Y. K. Wu, and H. Q. Ye, *Acta Mater.* 45, 113 (1997).
64. J. Y. Huang, J. Z. Jiang, H. Yasuda, and H. Mori, *Phys. Rev. B* 58, R11817 (1998).
65. J. Z. Jiang, C. Gente, and R. Bormann, *Mater. Sci. Eng. A* 242, 268 (1998).
66. F. Cardellini, V. Contini, G. D'Agostino, and A. Filipponi, *Mater. Sci. Forum* 269–272, 473 (1998).
67. K. Tokumitsu, *Mater. Sci. Forum* 269–272, 467 (1998).
68. L. B. Hong and B. Fultz, *Acta Mater.* 46, 2937 (1998).
69. J. Xu, G. S. Collins, L. S. J. Peng, and M. Atzmon, *Acta Mater.* 47, 1241 (1999).
70. P. J. Schilling, J. H. He, R. C. Tittsworth, and E. Ma, *Acta Mater.* 47, 2525 (1999).
71. R. N. Suave and A. R. B. de Castro, *J. Electron Spectrosc. Related Phenom.* 101–103, 653 (1999).
72. K. Tokumitsu, *Mater. Sci. Forum* 312–314, 405 (1999).
73. J. Z. Jiang, J. S. Olsen, L. Gerward, and S. Mørup, *Nanostruct. Mater.* 12, 847 (1999).
74. I. F. Vasconcelos and R. S. de Figueiredo, *Nanostruct. Mater.* 11, 935 (1999).
75. L. E. Bove, C. Petrillo, F. Sacchetti, and G. Mazzone, *Phys. Rev. B* 61, 9457 (2000).
76. S. H. Mahmood, G. M. Al-Zoubi, and A. F. Lehlooh, in "Proc. First Regional Conference on Magnetic and Superconducting Materials," Vol. 2, p. 825. World Scientific, Singapore, 2000.
77. S. Wei, W. Yan, Y. Li, W. Liu, J. Fan, and X. Zhang, *Physica B* 305, 135 (2001).
78. G. Principi, *Hyperfine Interactions* 134, 53 (2001).
79. Y. Todaka, P. G. McCormick, K. Tsuchiya, and M. Umemoto, *Mater. Trans.* 43, 667 (2002).
80. F. M. Lucas, B. Trindafe, B. F. D. Costa, and G. Le Caer, *Key Eng. Mater.* 230–232, 631 (2002).
81. T. Mashimo, X. S. Huang, X. Fan, K. Koyama, and M. Motokawa, *Phys. Rev. B* 66, 132407 (2002).
82. A. Jansson, Trita Mac 0340, Mater. Res. Center KTH, Stockholm, Sweden, 1987.
83. F. Faudot, E. Gaffet, and M. Harmelin, *J. Mater. Sci.* 28, 2669 (1993).
84. C. Michaelsen, *Philos. Mag. A* 72, 813 (1995).
85. F. E. Fujita, in "Application of Nuclear Techniques to the Studies of Amorphous Metals" (U. Gonser, Ed.), p. 173. International Atomic Energy Agency, Vienna, 1981.
86. F. Bødker, S. Mørup, and S. Linderoth, *Phys. Rev. Lett.* 72, 282 (1994).
87. K. Sumiyama, T. Yoshitaka, and Y. Nakamura, *Acta Metall.* 33, 1791 (1985).
88. E. Kneller, *J. Appl. Phys.* 35, 2210 (1964).
89. H. J. Fecht, E. Hellstern, Z. Fu, and W. L. Johnson, *Metall. Trans. A* 21, 2333 (1990).
90. B. Günther, A. Kumpmann, and H.-D. Kunze, *Scripta Metall.* 27, 833 (1992).
91. A. R. Yavari and P. J. Desré, *Mater. Sci. Forum* 88–90, 43 (1992).
92. P. G. Shewmon, "Transformations in Mctals," p. 300. McGraw–Hill, New York, 1969.
93. G. Martin, *Phys. Rev. B* 30, 1424 (1984).
94. X. J. Liu, C. P. Wang, J. Z. Jiang, I. Ohnuma, R. Kainuma, and K. Ishida, submitted for publication.
95. L. Kaufman and H. Beinstein, "Computer Calculations of Phase Diagrams." Academic, New York, 1970.
96. B. Sundman, B. Jansson, and J. O. Andersson, *CALPHAD* 9, 153 (1985).
97. A. T. Dinsdale, *CALPHAD* 15, 317 (1991).
98. L. Kaufman and H. Nesor, *CALPHAD* 2, 117 (1978).
99. M. Hasebe and T. Nishizawa, *CALPHAD* 4, 83 (1980)
100. P. A. Linquist and B. Uhrenius, *CALPHAD* 4, 193 (1980).
101. Y. Y. Chuand, R. Schmid, and Y. A. Chang, *Metall. Mater. Trans. A* 15, 1921 (1984).
102. M. Hasebe and T. Nishizawa, *CALPHAD* 5, 105 (1981).
103. Q. Chen and Z. P. Jin, *Metall. Mater. Trans. A* 26, 417 (1995).
104. M. Hillert and M. Jarl, *CALPHAD* 2, 227 (1978).
105. A. S. Skapski, *Acta Metall.* 4, 576 (1956).
106. W. R. Tyson and W. A. Miller, *Surface Sci.* 62, 267 (1977).
107. C. L. Reynolds, P. R. Couchman, and F. E. Karasz, *Philos. Mag.* 34, 659 (1976).
108. X. J. Liu, C. P. Wang, J. Z. Jiang, I. Ohnuma, R. Kainuma, and K. Ishida, submitted for publication.
109. B. D. Cullity, "Introduction to Magnetic Materials," p. 617. Addison–Wesley, Reading, MA, 1972.
110. C. L. Chien and R. Hasegawa, *Phys. Rev. B* 16, 2115 (1977).
111. J. Crangle and G. C. Hallam, *Proc. Roy. Soc. London, Ser. A* 272, 119 (1963).
112. R. J. Weiss, *Proc. Phys. Soc.* 82, 281 (1963).
113. A. Katsuki and K. Terao, *J. Phys. Soc. Japan* 26, 1109 (1969).
114. M. Shiga, *Solid State Commun.* 10, 1233 (1972).
115. O. K. Anderson, J. Madsen, U. K. Poulsen, O. Jepsen, and J. Kollar, *Physica B* 86–88, 249 (1977).

Encyclopedia of Nanoscience and Nanotechnology
www.aspbs.com/enn

Ferroelectric Nanocrystal Dispersed Oxide Glasses

B. K. Chaudhuri
Indian Association for the Cultivation of Science, Kolkata, India; and National Sun-Yat Sen University, Kaohsiung, Taiwan, Republic of China

H. Sakata
Tokai University, Kitakaname, Hiratsuka, Kanagawa, Japan

S. Mollah
National Sun-Yat Sen University, Kaohsiung, Taiwan, Republic of China; and Aligarh Muslim University, Aligarh, India

H. D. Yang
National Sun-Yat Sen University, Kaohsiung, Taiwan, Republic of China

CONTENTS

1. INTRODUCTION

During the last two decades, semiconducting properties of many multicomponent and transition metal oxide (TMO) and other metal oxide glasses have been studied because of their probable technological applications [1–4] in optical and electrical memory switches, oxygen sensors, etc. The high temperature ($T > \theta_D/2$, θ_D being the Debye temperature) conductivity data in many of these glasses, for example, V_2O_5–TeO_2 [5], V_2O_5–Bi_2O_3 [6], V_2O_5–P_2O_5 [7], V_2O_5–MnO–TeO_2 [8], V_2O_5–SnO–TeO_2 [9], V_2O_5–P_2O_5–TiO_2 [10], V_2O_5–CoO–TeO_2 [11], etc., were explained by using the "small polaron" hopping (SPH) model [12–15] based on strong electron–phonon interaction. On the other hand, in the low temperature phase ($T < \theta_D/2$), a variable-range hopping (VRH) conduction mechanism [15, 16] with $T^{-1/4}$ dependence was used to explain the conductivity data. The dielectric constants of these glasses are, in general, very low and follow Debye-type dielectric relaxation behavior. Another class of semiconducting glasses like CdS nanoparticles dispersed in sol–gel derived polymers [17], NiO–SiO_2 nanocomposites [18], ZnS nanocrystals stabilized in silica [19], etc. have been studied. Recently Itoigawa et al. [20] also observed the formation of nanosize silver and bismuth particles, respectively, in the Na_2O–B_2O_3–Ag_2O and Na_2O–B_2O_3–BiO_3 glasses. Lately Sakuri and Yamaki [2] reported the presence of such nanocrystalline phases in the V_2O_5–CoO–TeO_2 glass following multiphonon tunneling conduction mechanism. In the high temperature range (above $\theta_D/2$), these glasses also followed the SPH conduction mechanism but the VRH mechanism was found to be

ISBN: 1-58883-059-4/$35.00

not strictly followed [2] as these glasses showed extremely large values of the density of states (DOS) at the Fermi level $N(E_F)$, which is not possible. Similarly, Ag_2O doped Bi–Sr–Ca–Cu–O glasses, precursors for high T_c superconductors, also dispersed with nanocrystalline silver particles showed a high value [21] of $N(E_F)$ estimated from the VRH model. But none of these nanocrystal-dispersed glasses are ferroelectric (FE) or they do not show high dielectric permittivity. However, some of the TMO glasses containing $BaTiO_3$, $PbTiO_3$, TeO_2, etc. are found to show FE behavior due to the presence of nanocrystalline phases.

The aforementioned ferroelectric behavior in nanocrystals of pure FE material like $PbTiO_3$ is, however, not new [22]. Glass ceramics containing ferroelectric phases have been studied for long time to obtain useful dielectric and electro-optic properties [23–26]. Nanophase materials and nanocomposites, characterized by an ultrafine grain size (<50 nm), have created high interest in recent years by virtue of their unusual mechanical, electrical, optical, and magnetic properties. Pure nanocrystalline ferroelectric $PbTiO_3$ studied earlier showed a critical size of the nanocrystals [27] (different for different systems) at which the dielectric constant showed a maximum value. There is, however, a large difference in the ferroelectric behavior (Curie temperature, dielectric constant, polarization, etc.) of the bulk and its nanocrystalline counterpart [27]. But still today, ferroelectricity in pure glass has not been discovered. Ferroelectricity in liquid crystalline phase has, however, been observed [28].

Our major interest in this chapter is to report the synthesis, characterization, and study of transport and other properties of such FE glass–nanocomposite (hereafter referred to as FGNC) materials. It has also been shown how different theoretical models are used to explain the electrical transport and dielectric relaxation behavior of these FGNCs. This chapter will focus on this new class of materials for further theoretical and experimental studies and also find new applications of the glass–nanocrystal composites showing giant dielectric permittivity.

2. FERROELECTRIC NANOCRYSTAL DISPERSED OXIDE GLASSES

Recently it has been observed that some multicomponent oxide glasses [29–39] dispersed with ferroelectric nanocrystalline phases are formed for suitable choice of the glass compositions with TiO_2, $BaTiO_3$, $PbTiO_3$, $SrTiO_3$, $PbZrO_3$, $KTiOPO_4$, $Bi_2VO_{5.5}$, $Li_2Ge_7O_{15}$, PbS, etc. Some of these nanocrystal-dispersed glasses (referred to as FGNC) are found to show ferroelectric or relaxor type ferroelectric behavior [29–30, 40–43]. Figure 1 schematically represents the nanocrystals dispersed in the glassy matrix. Some important parameters of studied FGNC materials (both opaque and transparent) are shown in Table 1 for comparison.

Many low melting TMO glasses like vanadium phosphate glasses (V_2O_5–P_2O_5, for example), containing different concentrations of TiO_2, $BaTiO_3$, $PbTiO_3$ etc., have been prepared in recent years, which contain nanocrystalline ferroelectric phases embedded in the glass matrix [29, 30]. All these FGNCs are, however, opaque. The alkali metal doped Bi–Sr–Ca–Cu–A–O (A = Li, K, Na) glasses are also found to be ferroelectric for typical concentrations [42, 44]. As mentioned, the dielectric constants of these opaque TMO glasses are found to be extremely large (more than $BaTiO_3$) [30, 44]. Such a giant dielectric constant has also been obtained very recently in crystalline $CaCu_3Ti_4O_{12}$ oxide [45, 46] and La containing $PbTiO_3$ [47]. These materials of high dielectric constant are very important for their applications in devices. However, the origin of such a high dielectric constant is not yet clearly elucidated. In the nanoparticle dispersed glassy systems, the presence of nanocrystalline TiO_2, $BaTiO_3$, $PbTiO_3$, $SrTiO_3$, $PbZrO_3$, $KTiOPO_4$, $Bi_2VO_{5.5}$, $Li_2Ge_7O_{15}$, PbS, etc. phases are considered to be responsible for giant dielectric constant. As stated, this is a new class of material of immense technological as well as fundamental importance. By varying the glass compositions or the quenching rate, the size as well as the concentration of the FE nanoparticles/clusters can be varied.

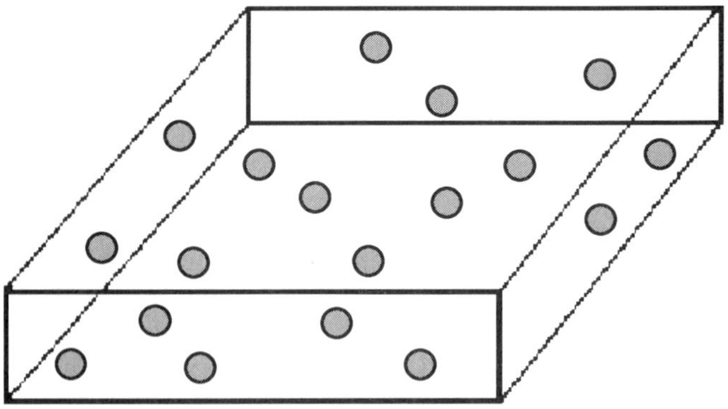

Figure 1. Schematic diagram of a ferroelectric nanocomposite glass. The dispersed nanocrystals are shown by dark black spots in the glass matrix.

The amorphous characters in the aforementioned oxide glasses and FGNCs are, in general, examined by an X-ray powder diffraction (XRD) technique. Nevertheless, depending on the smaller size and concentrations of the embedded nanocrystals, XRD is not sufficient to detect the presence of the nanocrystals in these glasses and FGNCs. XRD simply indicates amorphous character of these systems. Transmission electron microscopic (TEM) study can clearly reveal the presence of nanocrystalline phases in these FGNC materials. Some physical properties (like high dielectric constant) also indicate the presence of such nanocrystalline phase in some typical TMO glasses. It is also noticed from the TEM studies that many of the semiconducting TMO glasses, earlier reported to be pure homogeneous glasses, are found to contain nanocrystalline particles or clusters of different sizes (5–100 nm) and concentrations embedded in the corresponding glass matrices. Using the TEM technique, very little microstructural study of the semiconducting glassy system was, however, accomplished earlier [6, 11]. It is found that the size and concentration of the nanocrystals affect the transport, optical, and other properties of these ferroelectric nanocrystal-dispersed glasses. This behavior is similar to that observed in ferroelectric materials, for example, $PbTiO_3$ where dielectric permittivity and other properties strongly depend on the grain sizes [22].

Table 1. Some important physical parameters of different ferroelectric glass nanocrystal composites.

Glass matrix/FE nanoparticle	Cluster size (nm)	T_g (K)	T_c (K)	ε' at 1 kHz and 300 K	Ref.
$70(80V_2O_5–20PbO)/30BaTiO_3$	70–90	300	393	3.5×10^4	[30]
$50SrB_4O_7/50Bi_2VO_{5.5}$	15	730	763	3.2×10^3	[31, 32]
$50.5PbO–25.5ZrO_224TiO_2/PbS$	15.3	—	—	3.5×10^3	[34]
$80SiO_2/20KTiOPO_4$(KTP)	10–50	600	—	—	[35]
$(12Li_2O–88GeO_2)/Li_2Ge_7O_{15}$	10	—	283	—	[36–38]
$10(90V_2O_5–10P_2O_5)/90BaTiO_3$	~100	250	290	—	[50]

Note: T_g, T_c, and ε' represent, respectively, the glass transition temperature, Curie temperatures, and dielectric constants.

It was reported earlier [48] that the TMO glasses prepared with $BaTiO_3$, etc. contain nanocrystalline clusters and these FGNC samples, for example, $V_2O_5–Bi_2O_3 + xBaTiO_3$, showed giant dielectric constants because of the presence of nanocrystal/clusters of $BaTiO_3$ embedded in the glass matrix. The transport and dielectric properties of the pure host glasses, viz. $80V_2O_5–20P_2O_5$, $80V_2O_5–10Bi_2O_3$, $80V_2O_5–20PbO$, etc., where FE nanocrystals are precipitated (hereafter referred to as VP glass) by adding TiO_2, $BaTiO_3$ etc., have been well studied and reported in the literature [6, 7, 49]. Dielectric constants of these VP type base glasses are, on the other hand, very small and none of them exhibited ferroelectric behavior as mentioned earlier. The TiO_2, $PbTiO_3$, $BaTiO_3$, etc. with high melting oxide content in the VP glasses or similar other mostly binary TMO glasses, in spite of this, show extremely high dielectric constant [48] (compared to the pure base glass, VP, or similar other TMO glasses). Very little increase of conductivity observed in these systems is assigned to the appearance of nanocrystalline phases. Interestingly, the general behavior of all the ferroelectric FGNC composites, studied so far, is found to be similar.

Another class of nanocrystal-dispersed glasses [35, 39], for example, $KTiOPO_4$ dispersed oxide glasses, is found to be transparent. The transparent ferroelectric glass nanocomposites (TFGNCs) are suitable for second harmonic generation. More about these glass nanocomposites will be discussed separately in Section 8. Some of the important parameters of these TFGNCs are shown in Table 1 for comparison.

3. SYNTHESIS

3.1. Melt Quenching Technique

The melt quenching technique is widely used to prepare glasses and glass–ceramics. It needs higher temperature compared to that required for the sol–gel technique, but it takes less time. The TiO_2, $BaTiO_3$, $SrTiO_3$, $PbTiO_3$, etc. doped vanadate glasses (or the VP glasses) or other low melting TMO glasses (VP type base glasses), in general, can be prepared by a fast quenching method [48, 50]. For example, VP glass can be used to prepare a typical $(80V_2O_5–20P_2O_5) + xTiO_2$ type FGNC with different concentrations of TiO_2 ($x = 0.5, 5, 10, 20$, and 30 wt% TiO_2). Here $80V_2O_5–20P_2O_5$ (or VP glass) acts as the base glass. In brief, V_2O_5, $NH_4H_2PO_4$, and TiO_2 (or $BaTiO_3$, $SrTiO_3$, $PbTiO_3$, etc. can be used), each of purity 99.99% or better, were well mixed in appropriate proportions and preheated in air at about 500 °C for five hours with occasional grinding [48, 50]. The preheated mixed oxide samples were melted in air taking in a platinum crucible at about 1100 to 1250 °C (depending on the TiO_2, $BaTiO_3$, $SrTiO_3$, etc. concentrations) for about 30 minutes. The homogeneous melt was then quickly quenched between two cooled copper or steel blocks resulting in thin (0.6–0.7 mm thickness) pellets of the glassy FGNC samples. Ferroelectric glass nanocomposites in the pseudobinary systems like $(100 - x)SrB_4O_7–xBi_2VO_{5.5}$ ($0 \leq x \leq 70$) were prepared by the splat quenching technique [31, 32]. Strontium borate and bismuth vanadate were melted in a platinum crucible at 1100 °C. The melt was then quenched between two flat stainless steel blocks placed at a temperature around 100 °C to prevent cracking at the time of glass formation. The prepared glasses were then annealed well below the glass transition temperature (T_g). The Cr^{3+} doped lithium germanate glasses were also obtained [36–38] by similar fast quenching of the $12\%Li_2O–88\%GeO_2{:}0.5\%Cr_2O_3$ melt from 1350 °C. Subsequent annealing of the glass at different (480–520 °C) temperatures resulted in the precipitation of ferroelectric $Li_2Ge_7O_{15}$ nanocrystals in the glass matrix [36–38]. More details about the preparation of some other nanocrystal–glass composite samples by the melt quenching technique are available in the literature [51–53].

3.2. Sol–Gel Technique

The sol–gel technique is one of the most important methods to prepare glasses and ceramics of uniform grain sizes. The advantage of this technique is the mixing of the raw materials in the atomic scale during preparation. The disadvantage of this technique is that it is time consuming and the glass/ceramic may get through unnecessary water molecules. Glass–ceramics containing micrometer sized lead zirconate titanate (PZT) was synthesized by the sol–gel method and subsequently nanometer-size lead sulfide (PbS) phase was grown by treatment in hydrogen sulphide gas [33, 34]. A typical sol–gel method is as follows. One first needs to synthesize a gel of the target composition, viz. $50.5PbO–25.5ZrO_2–24TiO_2$. Here a solution of ethyl alcohol and acetic acid in the volume ratio of 75:25 was used. An appropriate amount of tetra-isopropyl orthotitanate ($C_{12}H_{28}O_4Ti$) was added to the solution and stirred for one hour. Subsequently, a suitable amount of lead acetate–acetic acid solution was added and stirred for more than one hour. An adequate amount of $Zr[O(CH_2)_3CH_3]_4$ was then added and stirred for two hours. The sol was dried at 70 °C for 72 hours. Then it was subjected to several heat treatments [33, 34] to get the crystalline PZT phase in the glass matrix. This glass–ceramic was cold pressed and sintered at 900 °C for 2 hours in PbO atmosphere. The sintered pellet was polished with alumina powder and reduced in H_2S at different temperatures. Nanometer size particles of PbS were precipitated in the glass ceramic matrix. The detailed preparation procedure of transparent glass–nanocomposites, like xKTiOPO$_4$(KTP)–$(1 - x)$SiO$_2$, by the sol–gel method has been described elsewhere [35, 39]. Multicomponent K_2O–TiO_2–P_2O_5–SiO_2 glass was prepared by the sol–gel method first [35, 39]. When this glass was heated above 650 °C, it was transformed into a dense nanocomposite of glass and KTP nanocrystals [35, 39]. Preparation of nanocomposites in some other techniques are also found in the literature [54–58].

4. DENSITY

The density of the FGNCs is generally measured by using the Archimedies method. Concentration dependent density of the FGNCs showed nonlinear variation [42]. This is an important property of nanocrystal-dispersed glasses. Since both size and concentrations of the nanocrystals depend on the compositions and there is also a critical size of the nanocrystals above which physical properties change [42] abruptly; the density of the FGNC system showed nonlinear variation with concentration. Both density and the glass transition temperature (T_g) showed little increase with increasing TiO_2 or $BaTiO_3$ concentration in these glass–nanocrystal composites [59]. But the corresponding increase of the average intersite separation (R) between two TMO (vanadium, for example) ions (V–O–V) with large increase of TiO_2 content (5–30 wt%) was relatively small (about 5%). R was calculated from the relation $(R = 1/N)^{1/3}$, where N is the vanadium ion concentrations. A similar small variation of R was also observed for the $BaTiO_3$ containing V_2O_5–Bi_2O_3–xSrTiO$_3$ glass–nanocrystal composites [29, 41] that followed the nonadiabatic hopping conduction mechanism.

The relationship between the density and the composition of nanocrystal-dispersed glasses is frequently expressed by the effective volume occupied by 1 g. wt of oxygen (V_o^*). The values of V_o^* can be determined from the theoretical formula given by Drake et al. [60]. Both density and V_o^* change in a similar manner with composition indicating that the topology of the glass–nanocrystal composites do not change appreciably for higher concentration of $SrTiO_3$ (or $BaTiO_3$). Variation of density (ρ) of the $(1 - x)(90V_2O_5$–$10P_2O_5) + x$BaTiO$_3$ glass–nanocomposites [50] with concentrations (x) of $BaTiO_3$ is shown in Figure 2. Initially, the density decreases with x and then shows a maximum for the highest concentration of $BaTiO_3$ (Fig. 2). The particle diameters of a typical composite also show a nonlinear variation with the change of concentration of composite components [34]. Density of the 20KTP–80SiO$_2$ type FGNC system increased with the increase of heat treatment (annealing) time and temperature (being 2.34 g/cm^3 at 850 °C with 0.2% porosity) [35]. The density and V_o^* values of a typical TiO_2 containing FGNC for different TiO_2 concentrations (along with average size of the precipitated nanoparticles/clusters) are shown in Table 2.

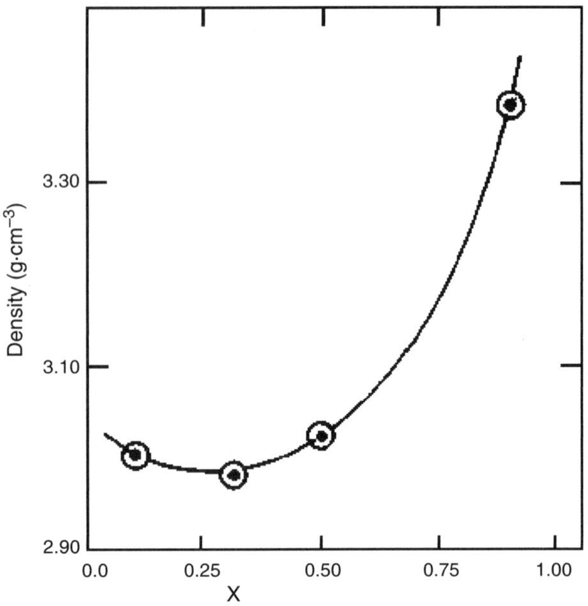

Figure 2. Variation of density of a typical glassy ferroelectric nanocomposite, viz. $(1 - x)(90V_2O_5$–$10P_2O_5) + x$BaTiO$_3$ with concentrations (x) of $BaTiO_3$ Reprinted with permission from [50], M. Sadhukhan et al., *J. Chem. Phys.* 105, 11326 (1996). © 1996, American Institute of Physics.

5. DIFFERENTIAL THERMAL ANALYSIS STUDY

The glass transition temperature (T_g) is being measured from the differential thermal analysis (DTA) curve with heating rate 10 °C per minute (say). Figure 3 shows the DTA traces for typical as-quenched and heat-treated $50SrB_4O_7$–$50Bi_2VO_{5.5}$ ferroelectric glass nanocomposites [31]. In Figure 3a, the first exotherm was attributed to the crystallization of $Bi_2VO_{5.5}$. The second exothermic peak was associated with the crystallization of $Bi_{12}V_2O_{23}$ and $BiVO_4$ impurity phases. The third peak was attributed to the reformation of $Bi_2VO_{5.5}$ due to reaction of $Bi_{12}V_2O_{23}$ and $BiVO_4$ impurity phases. The fourth peak was due to the crystallization of the host glass matrix SrB_4O_7. All these findings were consistent with the XRD study of the sample [31]. The heat-treated samples did not show the first exotherm peak as shown in Figure 3b and c. Elaborate DTA study of $50SrB_4O_7$–$50Bi_2VO_{5.5}$ glass-nanocomposites has also been made [32]. The DTA curve of the $(80V_2O_5$–$20PbO) + x$ wt% $BaTiO_3$ type glass-nanocomposites showed [30] an increase of T_g with the corresponding increase of $BaTiO_3$ concentration (x). DTA traces of several glasses [60–63] indicated a clear correlation of T_g with the change of coordination number of the network former and the formation of nonbridging oxygen (NBO) atoms which means a destruction or depolymerization of the network structure. The increase of T_g, in general, indicates the increase of coordination number of the network former. In contradiction to this, formation of NBO causes a decrease of T_g as observed in lead vanadate glasses [15]. Thus the continuous increase of T_g with

Table 2. Certain significant physical structures of a typical ferroelectric glass nanocrystal composite system, viz. ($80V_2O_5$–$20P_2O_5$) glass embedded with x wt% TiO_2 nanocrystalline particles/clusters [64].

Values of x (wt%)	Grain/cluster size (nm)[a]	ρ^b (gm/c.c.)	γ_p	V_o^* (c.c.)	Q (μV K^{-1})	η	T_g (°C)
5	10	3.063	25.25	11.38	−146	0.734	270 ± 5
10	20	3.065	24.47		−142	0.780	278 ± 5
20	40	3.082	17.38	11.99	−138	0.827	284 ± 5
30	80	3.216	12.25	12.03	−135	0.861	298 ± 5

Note: Seebeck coefficients (Q) were obtained from the best fitting of the experimental thermoelectric power data with Eq. (1).
[a] Indicates average size of the grains/cluters present in the FGNC.
[b] Error in the estimation of density was within ±0.005%.

x indicated strengthening of the bond [30]. The difference between T_g and crystallization temperature (T_{cr}) was less than 30 °C and it increased with the increase of x. Accordingly, the stability of the glasses increased with the increase of $BaTiO_3$. The values of T_g for TiO_2 containing FGNCs are given in Table 2.

6. SPECTROSCOPIC CHARACTERIZATION

6.1. X-Ray Diffraction

The amorphous character of all the ($80V_2O_5$–20PbO) + x wt% T (where T = $BaTiO_3$) type nanocrystal dispersed glassy system is demonstrated from the study of XRD patterns (with Cu K_α radiation) as shown in Figure 4a and b [30]. No crystalline peak was seen from the XRD patterns of this system. The corresponding annealed samples, however, showed the presence of $BaTiO_3$ as demonstrated in Figure 4c and d. The nanocrystals precipitated in these glasses (referred to as VPTI glasses) during glass formation are mostly $BaTiO_3$ or TiO_2 as indicated from the XRD patterns of the corresponding glass–ceramic samples obtained by annealing the as-quenched glass samples at 300 °C for a few hours in air (Fig. 4c and d). The very large dielectric permittivity of these multicomponent glasses compared to that of the base glasses V_2O_5–PbO or many other TMO glasses [12, 43, 48] also confirmed the presence of nanocrystalline $BaTiO_3$ or TiO_2 in the glass matrix. This result supported the earlier observation [48] that the dielectric constant of nanocrystalline TiO_2 or $BaTiO_3$ was much higher than that of bulk TiO_2 or $BaTiO_3$. The infrared (IR) spectra of all the glass samples exhibited almost identical

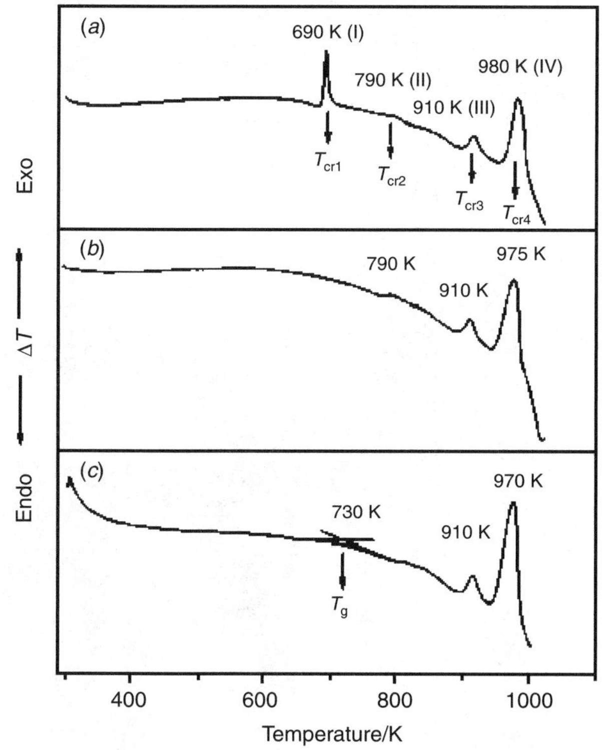

Figure 3. Differential thermal analysis traces for (a) as-quenched, (b) 447 °C, and (c) 547 °C heat-treated $50SrB_4O_7$–$50Bi_2VO_{5.5}$ glass nanocomposites. T_g and T_{cr} stand, respectively, for glass-transition and crystallization temperatures. Reprinted with permission from [31], N. S. Prasad and K. B. R. Varma, *J. Mater. Chem.* 11, 1912 (2001). © 2001, The Royal Society of Chemistry.

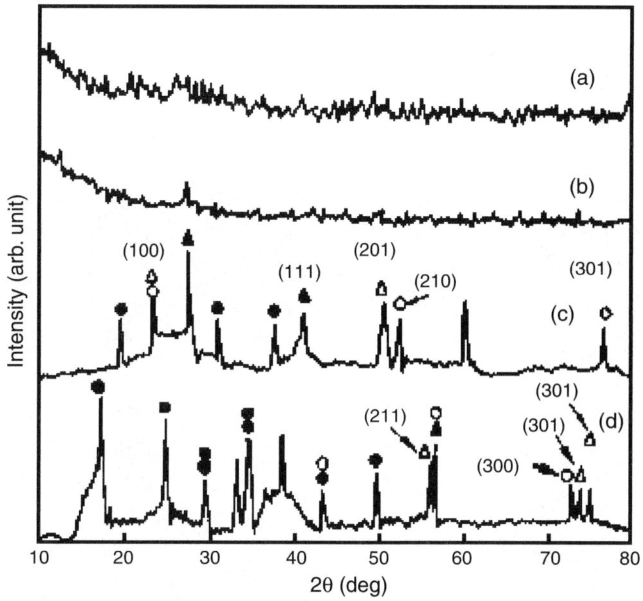

Figure 4. X-ray diffraction pattern of the as-quenched ($80V_2O_5$–20PbO) + x wt% $BaTiO_3$ glassy nanocomposite for (a) x = 15 and (b) x = 30 wt%. XRD of the corresponding annealed samples (annealed at 300 °C for 9 h) for (c) x = 15 and (d) x = 30 wt% with some of the identified crystalline peaks: PbV_2O_6(●), $PbTiO_3$ (○), PbV_2O_7 (♦), $BaTiO_3$ (△), and TiO_2(▲). Reprinted with permission from [30], M. Sadhukhan et al., *J. Appl. Phys.* 85, 3477 (1999). © 1999, American Institute of Physics.

features [30, 43] as will be discussed. The observed fine $Bi_2VO_{5.5}$ crystallites present in the SrB_4O_7 glass matrix were also confirmed by XRD studies [31, 32]. Systematic XRD studies corroborated the crystallization of $Bi_2VO_{5.5}$ phase along with minor impurity phases like $Bi_{12}V_2O_{23}$ and $BiVO_4$, depending on the temperature range of heating of the as-quenched samples [31, 32]. The XRD spectra well confirmed the presence of PZT crystals embedded in the glass matrix and the diffraction lines corresponding to PbS phase were broadened due to the small size of these crystallites [34].

6.2. Scanning Electron Microscopy

The scanning electron microscopic (SEM) study of the nanocrystal dispersed glassy composite gives clear evidence of the presence of crystallites in the glass matrix. The glassy character of the as-quenched FGNC samples was confirmed by SEM study as shown in Figure 5a and b for the two samples $(80V_2O_5\text{–}20PbO) + x$ wt% A (where A = $BaTiO_3$, etc.) with $x = 15$ and 30 wt% respectively [30]. The SEM micrographs of a typical sample (with $x = 30$ wt%) annealed at two different temperatures (290 and 300 °C) are shown, respectively, in Figure 5c and d. It is concluded from the SEM photographs (Fig. 5) that no crystallite was present before annealing the glasses. The presence of nanocrystals was, however, observed in the corresponding annealed samples consistent with the XRD pattern (Fig. 4). The SEM micrographs of ferroelectric PZT in glass–nanocrystal composite samples also showed identical results [34].

6.3. Transmission Electron Microscopy

Transmission electron microscopy is very important for the microstructural study of nanocomposites. The TEM studies of fine glass powder are made on carbon grids [21], which clearly reports the presence of nanocrystalline particles/clusters present in all the annealed glasses [48, 50]. The nanocrystals of TiO_2 or $BaTiO_3$ are formed in the glass matrices because of the large difference between the melting point of TiO_2 or $BaTiO_3$ than those of the base glasses (VP glasses). TEM indicated the presence of nanocrystalline $BaTiO_3$ or TiO_2 particles/clusters in these glasses (with maximum grain/cluster sizes varying from 5 to 80 nm with change of $BaTiO_3$, TiO_2, etc. content from 5 to 30 wt%). For lower concentration ($x = 5$–10 wt%), the size as well as concentration of the nanoparticles was very low. The TEM micrographs of $(80V_2O_5\text{–}20PbO) + 30$ wt% $BaTiO_3$ glass and the corresponding glass ceramic (annealed at 300 °C for 9 h) samples are shown, respectively, in Figure 6a and b. The selective area electron diffraction (SAED) pattern of the same glass and glass ceramics sample are shown correspondingly in Figure 6c and d. The TEM as well as the SAED patterns of the glass showed their amorphous nature (Fig. 6a and c) and those of the glass ceramic sample (Fig. 6b and d) indicated the distributed nanocrystalline particles/clusters in the glass matrix. These findings were in agreement with the XRD and SEM patterns of these samples [30]. Similar nanocrystalline particles were also observed in the V_2O_5–CoO–TeO_2 glasses [11] and also in the $(Bi_3Pb_1)Sr_3Ca_{4-m}(M_m)O_x$ glasses [44, 64] with M = Li, K, Cr, Mn, Zn, Ti, Ag. Prasad and Varma [31] found the presence of $Bi_2VO_{5.5}$ crystallites in SrB_4O_7 glass matrix by TEM and SAED study. Li et al. [35] also located the presence of KTP nanocrystals in SiO_2 transparent glass matrix.

6.4. Infrared Characterization

The IR spectra of the ferroelectric nanocrystals dispersed oxide glasses are generally studied in KBr matrix. Figure 7 shows the IR spectra of the $(80V_2O_5\text{–}20PbO) + x$ wt%

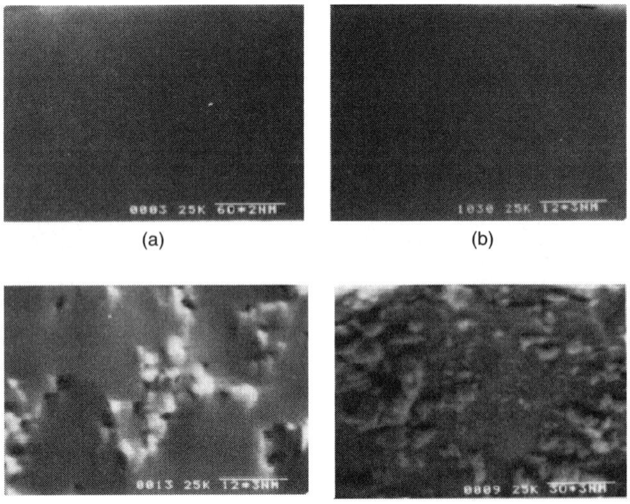

Figure 5. Scanning electron micrographs of the $(80V_2O_5\text{–}20PbO) + x$ wt% $BaTiO_3$ glasses for (a) $x = 10$ and (b) $x = 30$ wt% and those of the glass ceramics with $x = 30$ wt% annealed at different temperatures: (c) 290 °C and (d) 300 °C for 9 h each. Reprinted with permission from [30], M. Sadhukhan et al., *J. Appl. Phys.* 85, 3477 (1999). © 1999, American Institute of Physics.

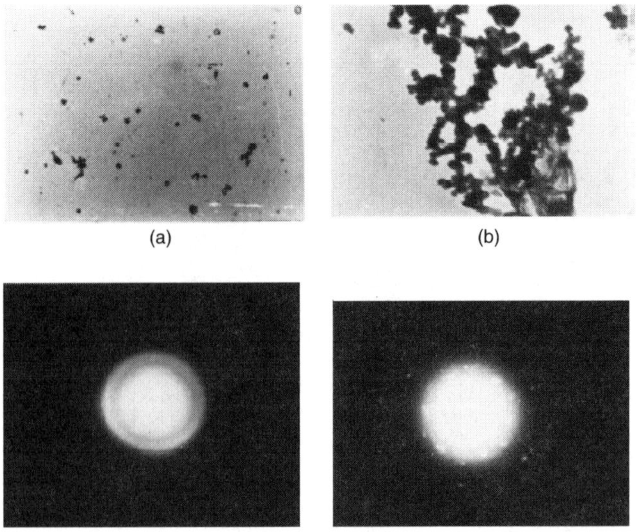

Figure 6. Tunneling electron micrographs of the $(80V_2O_5\text{–}20PbO) + x$ wt% $BaTiO_3$ glass-nanocomposite with $x = 30$ wt% (a) and the corresponding glass ceramic sample (obtained by annealed at 300 °C for 9 h) (b). The selective area electron diffraction pattern of the same glass (c) and the corresponding glass ceramic sample (d). Reprinted with permission from [30], M. Sadhukhan et al., *J. Appl. Phys.* 85, 3477 (1999). © 1999, American Institute of Physics.

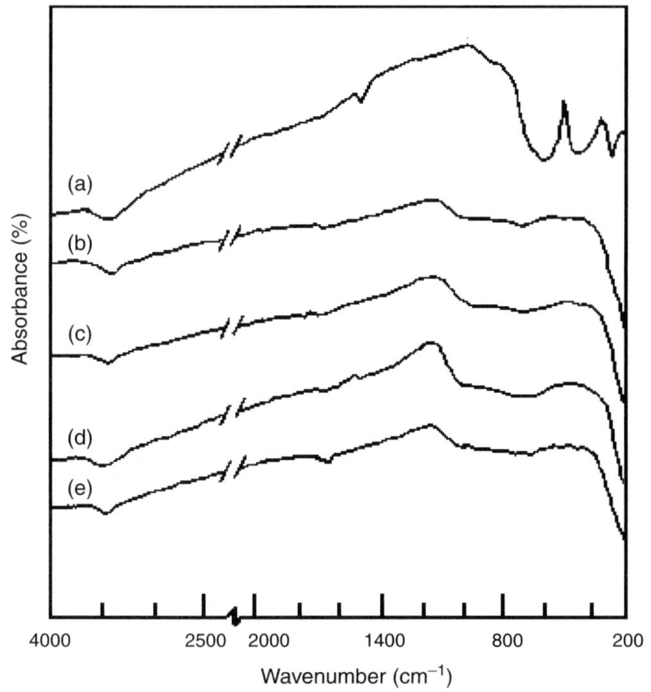

Figure 7. Infrared absorption spectra of (a) pure $BaTiO_3$, and $(80V_2O_5–20PbO) + x$ wt% $BaTiO_3$ glass-nanocomposites, respectively, for $x = 10$ (b), 15 (c), 20 (d), and 30 (e) wt% of $BaTiO_3$. Reprinted with permission from [30], M. Sadhukhan et al., *J. Appl. Phys.* 85, 3477 (1999). © 1999, American Institute of Physics.

$BaTiO_3$ ($x = 10, 15, 20$, and 30) glass–nanocomposites along with the pure $BaTiO_3$. The characteristic features of the IR spectra of all the glasses were found to be identical. The vibrational frequencies corresponding to the $Pb_2V_2O_7$ and $Pb_3V_2O_8$ structural units as reported for the lead vanadate glasses [65, 66] are not clearly observed from the IR spectra of these $(80V_2O_5–20PbO) + x$ wt% $BaTiO_3$ glasses [30]. Sadhukhan et al. [30] found a water band around 3400 cm^{-1} and an –OH stretching peak at 2920 cm^{-1} in all the glasses and crystalline $BaTiO_3$ due to the hygroscopic nature of the powder sample [67, 68]. In all the glass composites, signatures of the bonds and peaks of $BaTiO_3$ were not well visualized. However, instead of a band around 1500 cm^{-1} present in crystalline $BaTiO_3$, a new band around 1600 cm^{-1} was present in all the glasses. This might be due to the glassy environment of $BaTiO_3$ nanoparticles. The characteristic phonon frequency (ν_{ph}) estimated from the IR spectra was of the order of 10^{13} Hz corresponding to an IR band around 1000 and 1600 cm^{-1} for different glasses. This value of the phonon frequency agreed with that obtained from the conductivity data [30] of the sample.

7. PROPERTIES

7.1. Thermal Properties and Thermoelectric Power

The study of thermal property of glassy materials is difficult due to the higher thermal resistivity of the materials. However, that of glass–ceramics becomes much easier due to the crystalline nature of the glass–ceramics. The Seebeck coefficient (Q) of the FGNC sample is higher than that of the pure base glass as it contains crystalline nanoparticles of relatively higher conductivity [29, 41]. The Q values of FGNCs are determined by measuring the thermoelectric power (TEP) of the samples with a temperature difference around ~8 K between the two electrodes [29, 41]. For TEP measurements, samples of higher thickness (~1 mm or more) were generally used. All the FGNC systems showed negative values of thermoelectric power between 300 and 450 K. Table 2 shows the Q values of a typical TiO_2 containing FGNC system indicating it as an n-type semiconductor. Like many ternary vanadate glasses [49, 69–73], no appreciable temperature dependence of the TEP was observed between 300 and 400 K for the FGNCs. However, the magnitude of Q decreased with an increase in TiO_2 or $BaTiO_3$, etc. concentrations in the FGNCs (that is, with increase of grain size and their concentrations). The Q values were fitted with Heike's formula [73], viz.

$$Q = k_B/e[\ln\{C/(1 - C)\} + \eta] \qquad (1)$$

where e is the electronic charge and η is a constant of proportionality between the heat transfer and the kinetic energy of carrier. Values of $\eta < 1$ indicated a small polaron hopping transport mechanism [5, 29, 72–77] while for $\eta > 2$, the conduction was due to large polaron hopping [15, 60]. Equation (1) implies no temperature dependence of Q. Though it was reported earlier that the values of Seebeck coefficients of the glasses followed [5, 72–74] Heike's law, this model was also found to be applicable to FGNCs and the estimated values of η were almost constant and less than one (Table 2) indicating the validity of the small polaron hopping conduction mechanism in these FGNCs. Similar values of η (=0.44–0.98 depending on the concentrations) were also reported in other multicomponent vanadate glasses [49, 70] dispersed with nanocrystals. Interestingly, the values of η for these FGNCs were also positive and <1. For many TMO glasses η values are large and negative [42, 44].

7.2. Electrical Conductivity

7.2.1. dc Conductivity

For the temperature dependent dc conductivity (σ_{dc}) measurement, the thin pellets of the composites (diameters ~8 mm and thickness 0.3–0.4 mm) are generally used. The two polished faces of the pellets are gold plated by sputtering. Interestingly, like density, concentration dependent conductivity of the FGNC also showed an anomaly (change of slope) around the same concentration where density illustrated an anomaly between 5 and 10 wt% TiO_2 as shown in Figure 8 for typical $80V_2O_5–20P_2O_5 + xTiO_2$ glass-nanocomposites [64]. The dielectric constant of the same sample (discussed in a separate section) also decreased for TiO_2 content higher than 5 wt% as shown in Figure 8 for 0.1 kHz. This indicates that the dielectric constant is smaller (or larger) for the FGNC with grain/cluster size greater (or smaller) than a critical size (between 5 and 10 nm) [64]. This new result also supported the outcome of Zhang and co-workers [78]. It was shown by Wang et al. [22] that there was a critical grain size (~10 nm) around which the

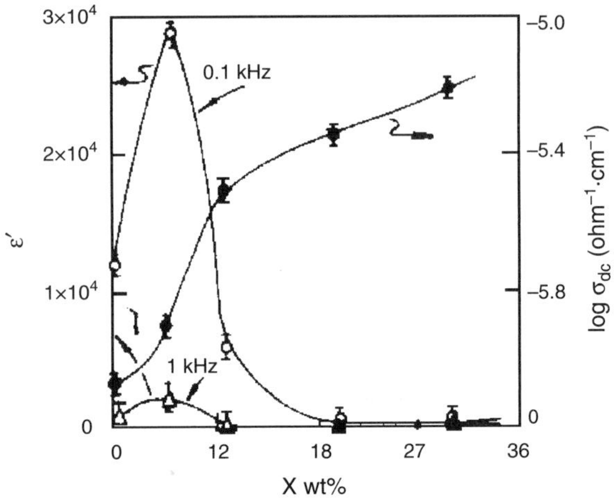

Figure 8. TiO_2 concentration (x) dependent logarithmic dc conductivity (log σ_{dc}) of the ($80V_2O_5$–$20P_2O_5$) + $xTiO_2$ glass–nanocrystal composites at a fixed temperature (300 K). The dielectric constant (ε') at two different frequencies (0.1 and 1 kHz) and at a fixed temperature (300 K) of the same glass-nanocomposite. Reprinted with permission from [64], D. K. Modak et al., *Sol. State Phys.* 43, 368 (2000).

dielectric constant of nanocrystalline ferroelectric $PbTiO_3$ was maximum. Most of the FGNC systems also behave in a similar way.

Thermal variations of the dc electrical conductivity (σ_{dc}) of the ferroelectric nanoparticle dispersed glassy system have been shown by different groups [29, 30, 50]. The logarithm of the dc conductivity (log σ_{dc}) vs $10^3/T$ curve (Fig. 9) for a typical FGNC ($80V_2O_5$–$20PbO + x$ wt% $BaTiO_3$ with $x = 10$–30) indicated a change of slope around $\theta_D/2$ (Debye temperature θ_D being different for different samples) [30]. All the other FGNCs also showed an analogous nature of variation of conductivity with change of temperature. As mentioned, density increased with increasing TiO_2 or $BaTiO_3$ concentration (x) exhibiting a break around $x =$ 5 wt% (Fig. 2) which was also reflected in the dc conductivity vs TiO_2 concentration (x) curves shown earlier (Fig. 8). For lower concentrations ($x < 10$ wt%), σ_{dc} at a fixed temperature (300 K) increased slowly with increase of TiO_2 content in the FGNC (Fig. 8). This behavior was in sharp contrast to those of other pure vanadate glasses (base glasses), for example, V_2O_5–MnO_2–TeO_2 [8] and V_2O_5–CoO–TeO_2 [11], where a decrease in conductivity with increase of the second transition metal ion, viz. Mn or Co ions, was observed. The increase of σ_{dc} in these VPTI type FGNCs with increasing TiO_2 content (say) was comparatively very small [6] (the values of log σ_{dc} at room temperature are −5.90, −5.5, −5.35, and −5.20 $ohm^{-1} cm^{-1}$, respectively, for 5, 10, 20, and 30 wt% TiO_2). The corresponding increase of R (from 0.498 to 0.526 nm for $x = 5$–30 wt% TiO_2) was also little. As the V ion concentration in all the glasses was fixed (since the glass composition $80V_2O_5$–$20P_2O_5$ or $80V_2O_5$–20PbO was fixed for all the VPTI type FGNCs), this small increase of conductivity was not supposed to be due to an increase of V ion concentration. Again, little increase in the estimated values of average intersite separation R (separation between the transition metal ions) with increasing TiO_2 concentration in the VPTI type FGNC samples should rather cause a decrease of conductivity. The corresponding decrease of activation energy (W) and hence the observed small increase of conductivity with increase of TiO_2 (for example) content was, therefore, considered to be due to the presence of precipitated nanocrystalline particles/clusters uniformly dispersed in the glass–nanocrystal composites. This conclusion was also supported from the results of earlier observation [73] that the conductivity in nanocrystalline phase increased in glasses and in ceramic semiconductors while it decreased in the metallic systems. This is an interesting property of the nanocrystalline grains. No theoretical model has yet been developed explaining this increase of conductivity in the glassy system with the appearance of nanocrystalline phase. Therefore, dc conductivity of the FGNCs is not solely due to the hopping between the TMO ions. In fact, the precipitated nanoclusters also contribute to the total conductivity. The magnitude of this contribution also depends on the size and concentration of the nanoclusters as in other FGNCs. With increase of TiO_2 content (from $x = 5$ to 30 wt%) in the FGNC system, like $80V_2O_5$–$20P_2O_5 + xTiO_2$ FGNCs, the rise of conductivity

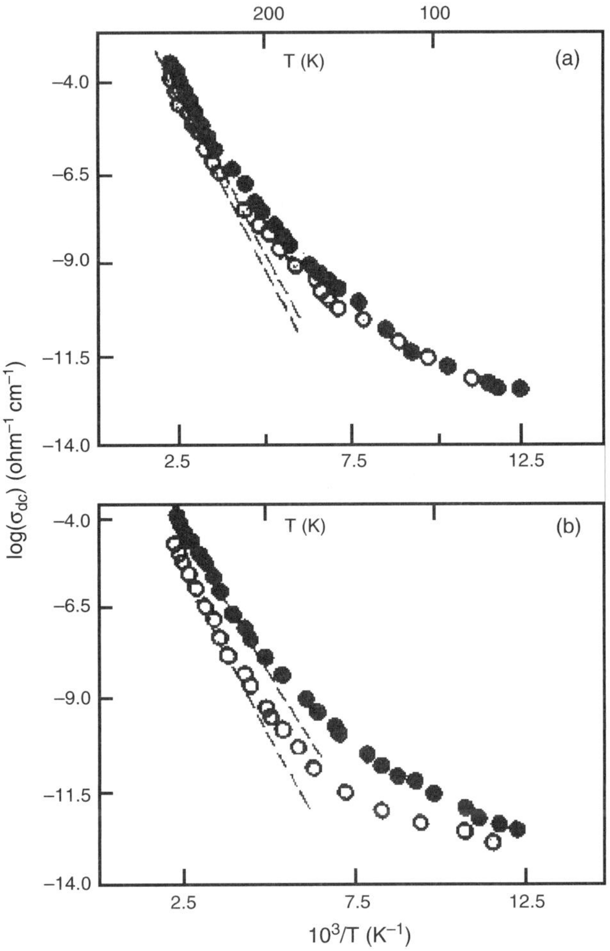

Figure 9. The logarithmic dc conductivity (log σ_{dc}) as a function of inverse temperature ($10^3/T$) of the ($80V_2O_5$–20PbO) + x wt% $BaTiO_3$ glass-nanocomposites with (a) $x = 10$ (●), 20 (○), (b) 15 (●), and 30 wt% (○). The labels on the top x axis give temperature (T). Reprinted with permission from [30], M. Sadhukhan et al., *J. Appl. Phys.* 85, 3477 (1999). © 1999, American Institute of Physics.

was quite sharp between 5 and 10 wt% TiO_2 (where nanoparticle size varying between 5 and 20 nm) and for the values of x between 10 to 30 wt% (maximum grain/cluster size varying from 20 to 80 nm), the corresponding rise in conductivity with concentration was slow (Fig. 8). These two regions of conductivity are separated by an anomaly (change of slope) occurring between $x = 5$ and 10 wt% TiO_2 (Fig. 8). Similar types of concentration (x) dependent anomalies in the conductivity and other properties, above a particular (or critical) size and concentration of the nanoparticles, were also present in many other nontransition metal semiconducting oxide glasses. The importance of the contribution of such nanoclusters in the transport and other properties of glasses is, in general, overlooked as the overall behavior of the nanocrystal dispersed glasses and pure nanocrystal free glasses appears identical.

The relation between $\log \sigma_{dc}$ and inverse temperature (T^{-1}) is almost linear (Fig. 9) indicating the temperature dependent activation energy (W). This behavior of dc conductivity was described by the relation $\sigma_{dc} = (\sigma_o/T)\exp(-W/k_B T)$, where σ_o is the pre-exponential factor. The values of W were obtained from the plot of $\log \sigma_{dc}$ vs $1000/T$ (Fig. 9). At temperatures lower than $\theta_D/2$, the linearity in the $\log \sigma_{dc}$ vs T^{-1} curve deviated as seen from Figure 9. Similar thermal variation of dc conductivity (σ_{dc}) was also exhibited in V_2O_5–P_2O_5 [79] and many other TMO glasses [4, 69, 70] and followed a polaron hopping conduction mechanism. The high temperature ($T > \theta_D/2$) conductivity data of the VPTI type FGNCs thick pellets were interpreted in terms of the phonon-assisted hopping model given by [15]

$$\sigma_{dc} = (\sigma_o/T)\exp(-W/k_B T) \quad (2)$$

where $W = W_H + W_D/2$ (for $T > \theta_D/2$), $W = W_H + W_D/4$ (for $T < \theta_D/4$), and

$$\sigma_o = \nu_{ph} N e^2 R^2 C(1-C)\exp(-2\alpha R)/k_B \quad (3)$$

ν_{ph} is the optical phonon frequency ($\sim 10^{13}$ Hz), α is the decay constant of the localized wave function, W_H is the polaron hopping energy, W_D is the disorder energy, and other parameters have usual meanings as mentioned earlier. The importance of the tunneling term $\exp(-2\alpha R)$ in Eq. (3) for the FGNCs could be understood [79] by plotting $\log \sigma_{dc}$ against W at a chosen temperature for all the VPTI type FGNCs. The temperature T_e (say), estimated from the slope of such a plot, would be close to the experimental temperature when the hopping is considered in the adiabatic regime [5, 79]. On the other hand, T_e would be very different from the experimental temperature when the hopping is considered in the nonadiabatic regime. The FGNC system was found to follow a nonadiabatic hopping conduction mechanism [64]. It is interesting to mention that Sayer and Mansingh [79] reported an adiabatic hopping conduction mechanism for the TiO_2 free base glass $80V_2O_5$–$20P_2O_5$. Therefore, a change of conduction mechanism occurred in the TiO_2 containing VPTI air quenched FGNCs. It is, however, rather difficult to uniquely identify the type of small polaron hopping conduction mechanism, adiabatic or nonadiabatic, only if the temperature dependence of conductivity is used, because a hopping process of nonadiabatic small polarons requires several restrictions on the electron transfer integral between neighboring hopping sites. So we have also attempted to confirm the nature of hopping conduction using other theoretical methods discussed later.

Dc conductivity data of FGNCs were well fitted with Eq. (2) above $\theta/2$. The estimated phonon frequencies agreed well with those estimated from the infrared spectra as mentioned (shown in Table 3). The polaron radius r_p was estimated from the formula [75], viz.

$$r_p = (1/2)(\pi/6N)^{1/3} \quad (4)$$

The estimated values of r_p ($\sim$0.2 Å) suggested strong localization in these FGNCs. The polaron hopping energy W_H was calculated from the relation [10]

$$W_H = e^2/4\varepsilon_p(1/r_p - 1/R) \quad (5)$$

where $1/\varepsilon_p$ ($=1/\varepsilon_\alpha - 1/\varepsilon_s$, ε_s and ε_α are the static and high frequency dielectric constants of the FGNCs, respectively). W_H can be calculated from Eqs. (4) and (5) under the approximation $\varepsilon_p = \varepsilon_\alpha = n^2$, where n is the refractive index of the FGNC. The values of W_H were obtained from fitting of the conductivity data with Schnakenberg's model [76] [Eq. (6)] and were found to be of the order of $\sim$0.53, 0.42, 0.35, and 0.26 eV, respectively, for TiO_2 = 5, 10, 20, and 30 wt% in VPTI glass-nanocomposites [72]. These values of W_H were then used to estimate ε_p (=2.047, 2.486, 2.847, 3.932, respectively, for 5, 10, 20, and 30 wt% TiO_2 in VPTI glass-nanocomposites) [72]. The polaron binding energy (W_p) can be obtained from the formula [24] $W_p = e^2/\varepsilon_p r_p$. The calculated values of W_p are shown in Table 3. Using the values of ε_p, the calculated values of

Table 3. Selected valuable parameters of the TiO_2 nanocrystals dispersed $(80V_2O_5$–$20P_2O_5) + xTiO_2$ FGNCs [64] (FGNC in Table 2) obtained by the fitting of their high and low temperature dc conductivity data.

Values of x (wt%)	θ_D (K)	R (nm)	$\nu_{ph} \times 10^{13}$ (Hz)[a]	W^b (eV) at 450 K	W_H^b (eV)	W_p^b (eV)	W_D^b (eV)
5	664 ± 2	0.498	1.384 (3.00)	0.677	0.53	1.751	0.15 ± 0.01
10	625 ± 2	0.506	1.302 (3.03)	0.553	0.42	1.420	0.13 ± 0.01
20	583 ± 2	0.519	1.215 (3.03)	0.461	0.35	1.206	0.10 ± 0.01
30	555 ± 2	0.526	1.158 (3.03)	0.425	0.26	0.863	0.18 ± 0.1

[a] Values of ν_{ph} are obtained from the fitting of the high temperature conductivity data with Eq. (2) and the corresponding values within the parenthesis are obtained from the infrared spectra of the samples with KBr.
[b] W, W_H, W_p, and W_D stand, respectively, for activation energy, polaron hopping energy, polaron binding energy, and disorder energy.

refractive index (n) agree quite well with those determined from the Berwster angles [64]. The values of polaron binding energy W_p (Table 3) were found to decrease a little with increase of TiO_2 concentration (or with increase of nanocrystalline grain/cluster size) in the FGNCs. The disorder energy W_D can be estimated independently from the generalized polaron hopping model of Schnakenberg [72]. In this model, optical multiphonon and acoustical single-phonon processes determine the conductivity in the high temperature region. According to this model, the expression for the dc conductivity can be written as

$$\sigma_{\mathrm{dc}} T \approx \sin h(h\nu_{\mathrm{ph}}\beta)^{1/2} \exp[4W_H/h\nu_{\mathrm{ph}} \tan h(h\nu_{\mathrm{ph}}\beta/4)] \times \exp(-W_D\beta) \quad (6)$$

where $\beta = 1/k_BT$. Equation (6) is fitted to the experimental high temperature dc conductivity data by a least squares method. The values of the parameters W_H and W_D obtained from fitting the dc conductivity data with Eq. (6) are shown in Table 3. These values of W_H and W_D appeared to be a little higher than those of many other vanadate glasses referred in this chapter. This is considered to be due to the presence of dispersed nanocrystalline phases in the FGNCs. Precipitation of this phase caused some modifications of the glass network structure and hence some changes in the values of W_H and W_D (compared to those of the base glass without TiO_2) were observed. According to Eq. (2), the difference existing between W and W_H (Table 3) arose from the disordering energy term ($W_D/2$). This discrepancy had been explained as the effect of the partial charge of the cations of the glass forming oxides on activation energy for hopping conduction in transition metal oxides [4].

It has already been mentioned that Eq. (2) can only fit the high temperature (above $\theta_D/2$) dc conductivity data. For the low temperature regime ($T < \theta_D/2$) where the polaron binding energy is smaller [15, 16], σ_{dc} was explained, as usual, by the VRH models of Mott and Davis [15]. The expression for the conductivity in the three-dimensional VRH model is given by

$$\sigma_{\mathrm{dc}} = A \exp[-(B/T)^{1/4}] \quad (7)$$

where A and B are constants, $B = 19.4\alpha^3/k_BN(E_F)$ with the $N(E_F)$ as the density of states at the Fermi level. Figure 10 shows the plot of $\log \sigma_{\mathrm{dc}}$ against $T^{-1/4}$ for $(1-x)(90V_2O_5$–$10P_2O_5) + xBaTiO_3$ FGNCs. These curves indicate straight-line behavior with slightly different slopes. The values of

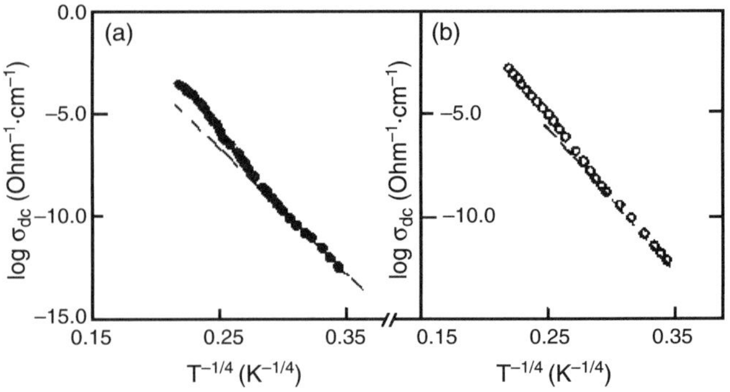

Figure 10. Plot of the logarithm of dc conductivity ($\log \sigma_{\mathrm{dc}}$) as a function of $T^{-1/4}$ for different concentrations $x = 0.9$ (a) and $x = 0.3$ (b) of $BaTiO_3$ in $(1-x)(90V_2O_5$–$10P_2O_5) + xBaTiO_3$ glass nanocomposites. Reprinted with permission from [50], M. Sadhukhan et al., *J. Chem. Phys.* 105, 11326 (1996). © 1996, American Institute of Physics.

$N(E_F)$ were estimated from the best fitting of Eq. (7). The estimated values of $N(E_F)$ are shown in Table 4 for a typical FGNC ($80V_2O_5$–$20P_2O_5 + xTiO_2$) [64]. Knowing $N(E_F)$ and other parameters, the nature of hopping conduction (adiabatic or nonadiabatic) can also be decided from the theory of Emin and Hilstein [74]. According to this theory, the hopping conduction mechanism is adiabatic for $J > \varphi$ and nonadiabatic for $J < \varphi$ where

$$\varphi = (2k_BTW_H/\pi)^{1/4}(h\nu_{ph}/\pi)^{1/2} \quad (8)$$

J is the polaron bandwidth related to the electron wave function overlap on the adjacent sites. For the present VPTI type glassy–nanocomposite system, the limiting values of φ estimated from the right hand side of Eq. (8) are 4.12×10^{-2}, 3.78×10^{-2}, 3.49×10^{-2}, and 3.54×10^{-2} eV, respectively, with 5, 10, 20, and 30 wt% TiO_2. The values of J were also estimated independently from the relation [15] $J = e^3[N(E_F)/\varepsilon_p^3]^{1/2}$. Using the values of $N(E_F)$ and ε_p one finds $J = 1.89 \times 10^{-3}$, 4.24×10^{-3}, 3.66×10^{-3}, and 2.99×10^{-3} eV for the VPTI type FGNCs with $x = 5$, 10, 20, and 30 wt% TiO_2, respectively. Comparing the estimated values of J and φ, it is observed that J is about one order of magnitude less than that of φ for all the glass compositions. This indicates nonadiabatic hopping conduction in all the VPTI type FGNCs, which is in agreement with the results obtained earlier from the plot of $\log \sigma_{\mathrm{dc}}$ vs activation energy (W) curves [64]. So the TiO_2 containing VPTI

Table 4. Various model parameters obtained from the best fitting of the ac and dc conductivity data of the nanoparticle dispersed $(80V_2O_5$–$20P_2O_5) + xTiO_2$ (FGNC) [64] sample with different TiO_2 concentrations (x).

Values of x (wt%)	W_{HO}^{a} (eV)	r_p' (Å)	$N(E_F)$ (eV^{-1})[b]	Exponent (s) at 300 K
5	0.8759	0.365	2.65×10^{17} (7.26×10^{18})	0.085
10	0.7102	0.408	1.01×10^{17} (2.46×10^{19})	0.095
20	0.6030	0.588	1.04×10^{17} (3.06×10^{19})	0.031
30	0.4315	0.806	1.82×10^{17} (5.04×10^{19})	0.028

[a] W_{HO} stands for polaron hopping energy between two sites at an infinite distance.
[b] Values of $N(E_F)$ were obtained from the fitting of the dc conductivity data with the VRH model [Eq. (7)]. The corresponding values within the brackets are obtained from the ac conductivity data of these FGNCs.

type FGNCs followed the nonadiabatic hopping conduction mechanism. A similar nonadiabatic hopping conduction mechanism was also reported for the $BaTiO_3$ doped V_2O_5–Bi_2O_2 FGNCs [29, 41, 50]

Attempts were made to fit the conductivity data of these nanoparticle dispersed glasses (FGNCs) with the percolation theoretical model of Triberis and Friedman [77] who applied percolation theory to the small polaron hopping regime. Considering correlation due to energy of a common site in a percolation cluster, they [77] obtained $\sigma_{dc} = \sigma_o' \exp[-(T_o'/T)^{1/4}]$, where σ_o' and $T_o'(= C'\alpha^3/k_B N_o)$ are constant. N_o is the constant DOS in the mobility gap, and the constant C' has two values, 12.5 and 17.8, in the high and low temperature regimes, respectively. It may further be noted that this equation is similar to that of Mott's VRH model [Eq. (7)] with slightly different values of the constant T_o'. In Figure 10, two different slopes (high and low temperatures) are not clearly observed in the $\log \sigma_{dc}$ vs $T^{-1/4}$ plot for the FGNC system. The difference between the low and high temperature slopes might, however, be larger at higher temperatures beyond the range of measurements. So it was concluded that the percolation theory of Triberis and Friedman [77] was not suitable for application to these FGNCs in the measured temperature range.

An estimation of the small polaron coupling constant (γ_p), which is a measure of the electron–phonon interaction in the VPTI type nanocrystal dispersed glasses, was made from the relation [12] $\gamma_p = W_H/h\nu_{ph}$. Using the value of W_H from Table 3, the authors [64] calculated γ_p which varied from 25.25 to 12.44 for different VPTI type FGNCs. These values of γ_p were much larger compared to those of $BaTiO_3$-free V_2O_5–Bi_2O_3 glasses [60] or other TMO glasses where γ_p varied from 4 to 9. Such high values of γ_p were also obtained by Sega et al. [9] for the multicomponent V_2O_5–NiO–TeO_2 (γ_p = 19–23) and V_2O_5–MnO–TeO_2 (γ_p = 21–25) FGNCs of different concentrations. All these FGNCs, in addition, followed the nonadiabatic small polaron hopping conduction mechanism. According to Austin and Mott [12], higher values of γ_p (>4) indicate a strong electron–lattice (phonon) coupling. So it was found that the electron–lattice interaction decreased with the increase of TiO_2 content in the VPTI type FGNCs. The decrease of γ_p with decreasing TiO_2 concentrations indicated a decrease of electron–phonon interaction with an increase of concentration and size of the nanoclusters. Considering the diffusion of electrons in small polaron hopping, the hopping carrier mobility [75, 77] was estimated from the relation $\mu = 2\pi e R^2 \beta / h(\pi\beta/4W_H)^{1/2} J^2 \exp(-W\beta)$ (for nonadiabatic hopping). Using the aforementioned values of R and W_H, the values of μ estimated at 300 K are 4.24×10^{-8}, 5.17×10^{-7}, 5.33×10^{-6}, and 2.26×10^{-5} $cm^{-2}\,V^{-1}\,s^{-1}$, respectively, for x = 5, 10, 20, and 30 wt% TiO_2 FGNCs. These values of μ were comparable to those of other multicomponent vanadate glasses, viz. V_2O_5–Bi_2O_3–$SrTiO_3$ [29], V_2O_5–Sb_2O_3–TeO_3 [70], etc. The observed increase in mobility with concentration (x) was again attributed to the change of nanocrystalline grain sizes with increase of TiO_2 content in the VPTI type FGNCs. The localization condition for hopping electrons, viz. $\mu \ll 0.01$ $cm^{-2}\,V^{-1}\,s^{-1}$ [80], was satisfied for the VPTI type FGNCs. This result indicates that electrons in these FGNCs are localized mainly at the V ion site which is similar to that observed in the V_2O_5–Sb_2O_3–TeO_2 type of FGNCs [70]. Therefore, conduction in these glassy systems was primarily due to a polaron-hopping mechanism between the vanadium ions of different valence states though the authors [64] considered there to be little increase of conductivity due to the precipitation on nanocrystalline particles. The carrier concentrations N_c calculated from the relation [49, 70] $\sigma_{dc} = eN_c\mu$ were 5.56×10^{21}, 1.14×10^{22}, 1.57×10^{21}, and 5.24×10^{19} eV^{-1} for the VPBT type FGNCs, respectively, with TiO_2 concentration of 5, 10, 20, and 30 wt%. Concentration (x) dependent N_c showed little discontinuity (increase and then decrease with increasing x) around x = 10 wt% which might also be due to the phase separation in the FGNCs.

7.2.2. Ac Conductivity

For the frequency and temperature dependent dielectric constants (ε) and ac conductivity $[(\sigma_{ac}(\omega))]$ measurements, the two polished faces of each of the samples (thickness 0.25–0.30 mm) were gold plated by sputtering and then annealed at about 150 °C for half an hour before conductivity measurement. This was done for better electrical contact with the sample surface. Conductivity measurements were made in the ohmic region as determined from the study of current–voltage (I–V) method.

It is well known that the frequency dependent ac conductivity data of amorphous or powdered semiconductors follow the relation [13, 14]

$$\sigma_{ac}(\omega) = A\omega^S \tag{9a}$$

where A is a constant weakly dependent on temperature and s is the frequency exponent, generally less than unity. The ac conductivity was calculated from the total conductivity $\sigma_t(\omega)$ measured at frequency ω and at a fixed temperature. Both dc conductivity (σ_{dc}) and total conductivity, $\sigma_t(\omega)$, are measured independently and then ac conductivity $[\sigma_{ac}(\omega)]$ is estimated from the relation [13–14]

$$\sigma_t(\omega) = \sigma_{dc} + \sigma_{ac}(\omega) \tag{9b}$$

In the glassy FGNC system, the ac and dc conductivities arise due to completely different processes [81]. In other words, ac conductivity represents the dc conductivity in the limit [81] $\omega \to 0$. Analyses of the ac conductivity $[\sigma_{ac}(\omega) = \sigma_t(\omega) - \sigma_{dc}]$ data of the semiconducting TMO glasses are generally made in the framework of quantum mechanical tunneling (QMT) [82], correlated barrier hopping (CBH) [12], and overlapping large polaron tunneling (OLPT) [83] models. In the following section, the temperature dependent ac conductivity data and the frequency exponent (s) of the nanoparticle dispersed glasses have been analyzed in terms of the OLPT model which is found to be the most appropriate for these nanocrystal–glass composite systems. Long [82] proposed the polaron tunneling model where the potential wells of two sites overlap thereby reducing the value of polaron hopping energy [84, 85] due to the long-range nature of the dominant Coulomb interaction. The polaron hopping energy has the form [81] $W_H = W_{HO}(1 - r_p/R)$ where W_{HO} is the polaron hopping between two sites at infinite distance. The ac conductivity for the OLPT model [83] is given by

$$\sigma_{ac}(\omega) = (\pi^4/12)e^2(k_BT)^2[N(E_F)]^2 \times[\omega R_\omega^4/(2\alpha k_BT + W_{HO}r_p/R_\omega^2)] \quad (10)$$

where $R_\omega = (1/2\alpha)[\ln(1/\ln(\omega\tau))]$ is the optimum hopping length at a frequency ω calculated by the quadratic equation $R_\omega'^2 + [\beta W_{HO} + \ln(\omega\tau_o)]R'_\omega - \beta W_{HO}r'_p = 0$ (where $R'_\omega = 2\alpha R_\omega$, $r'_p = 2\alpha r_p$, and $\beta = 1/k_BT$). The frequency exponent (s) of $\sigma_{ac}(\omega)$ in this model is calculated from the relation

$$s = 1 - (4 + 6\beta W_{HO}r'_p/R'_\omega)/(1 + \beta W_{HO}r'_p/R'_\omega)^2/R'_\omega \quad (11)$$

Thus, the OLPT model [Eq. (11)] predicts that the exponent(s) [in Eq. (9a)] should be both temperature and frequency dependent.

At high temperatures (above $\theta_D/2$), the temperature dependence of both σ_{dc} and $\sigma_t(\omega)$ were strong and consequently the measured $\sigma_{ac}(\omega)$, at all frequencies, coincided with σ_{dc} in the high temperature region. Figure 11 represents the plots of $\log\sigma_{ac}$ of a typical glass nanocomposite [50] as a function of inverse temperature and frequency, viz. $\log\omega$. The solid lines are obtained by a least-square fitting procedure. It is evident that $\sigma_{ac}(\omega)$ obeys the universal relation $\sigma_{ac}(\omega) = A\omega^S$, suggesting that the loss mechanism should have a distribution of relaxation times. The plots of exponents s (calculated from the slopes of the curves) as a function of temperature are shown in Figure 12 (for two typical FGNC samples with $x = 10$ and 20 wt%) which indicated that the exponent (s) decreased with increasing temperature and then exhibited a minimum at a temperature around 300 K and subsequently increased. This typical behavior of s suggested that the OLPT model [Eq. (11)] was appropriate for these FGNC systems. In the intermediate temperature range (i.e., below 300 K), the values of s resided around the theoretical curves (solid lines in Fig. 12) for various values of the normalized polaron radius r'_p (shown in Table 4). In the high-temperature regime (above 300 K), an increase in s with increasing temperature was observed which was consistent with the behavior of the OLPT model. The best fit to the experimental points has been observed for the values of W_{HO} and r'_p as shown in Table 4. Both W_{HO} and s decrease with increasing TiO_2 content in the VPTI type FGNCs (i.e., with increase of nanocrystal size and concentration).

Ac conductivity of the FGNC system increased linearly with increasing temperature [i.e., $\sigma_{ac}(\omega)\alpha T^n$ with $n = 1$] over a limited range of temperature below 175 K as predicted by the QMT model [82]. Hence the experimental values of σ_{ac} could also be fitted with the QMT model that predicted a linear temperature dependence of $\sigma_{ac}(\omega)$ due to weak ac conductivity and the corresponding expression for the $\sigma_{ac}(\omega)$ can be written as

$$\sigma_{ac}(\omega) = \{\kappa Ke^2k_BT[N(E_F)]^2\omega R_\omega^4\}/\alpha \quad (12)$$

where κ is a constant factor and varies slightly between different treatments, and R_ω [$=(1/2\alpha)\ln(1/\omega\tau_o)$] is the characteristic tunneling distance and the corresponding frequency exponent has the form $s = 1 + 4/\ln(\omega\tau_o)$. Therefore, according to the QMT model, the exponent s was

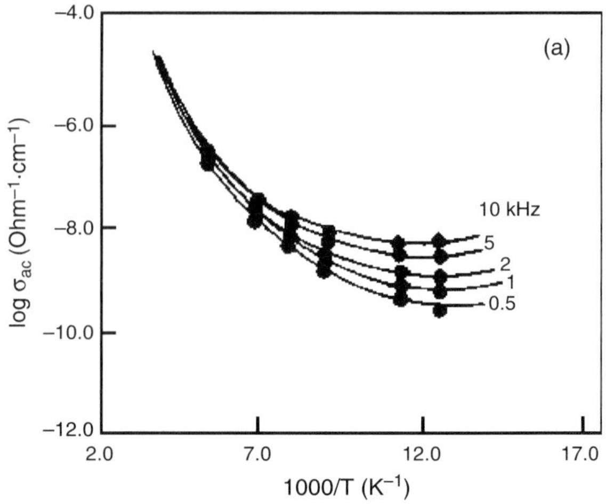

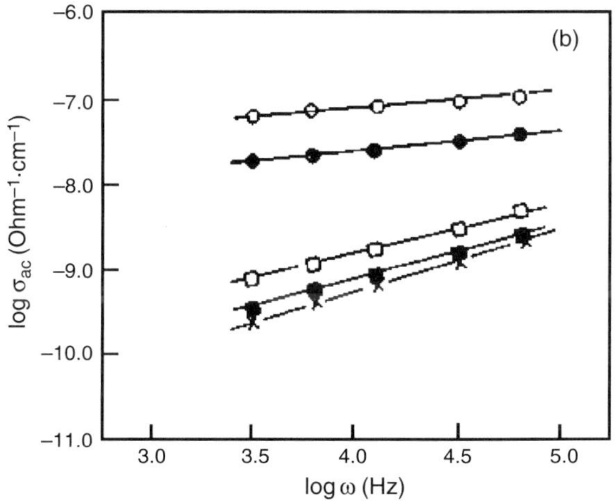

Figure 11. (a) Inverse temperature dependent logarithmic ac conductivity ($\log\sigma_{ac}$) of a typical $(1 - x)(90V_2O_5\text{–}10P_2O_5) + xBaTiO_3$ glass-nanocomposite with $x = 0.3$ at different frequencies (0.5–10 kHz). (b) Frequency dependent ac conductivity of the same glass-nanocomposite sample with $x = 0.1$ at different temperatures [182 (○), 176 (●), 120 (□), 96 (♦), 88 K (x)]. From the slope of these curves, the values of s (frequency exponent) were estimated. Similar curves were obtained for other glass concentrations. (Solid lines are guides for the eyes.) Reprinted with permission from [50], M. Sadhukhan et al., *J. Chem. Phys.* 105, 11326 (1996). © 1996, American Institute of Physics.

independent of temperature that appeared to be valid for the FGNCs only in the low temperature region. The values of the DOS at the Fermi level, $N(E_F)$, were obtained by least-square fitting of the linear region in the low temperature phase for the same composition. The values of $N(E_F)$ presented in Table 4 seem to be reasonable and comparable to those obtained from the dc conductivity data (Table 4) of the same nanoparticle dispersed glasses. However, although the QMT mechanism appears to be consistent with the thermal variation in $\sigma_{ac}(\omega)$ of these VPTI nanocomposites in the low temperature regime, it completely failed to interpret the observed temperature dependence of the exponent s. The QMT model as discussed predicted a value of $s = 0.81$ (assuming $\tau_o = 10^{-13}$ sec and $\omega/2\pi = 10^4$ sec^{-1}), independent of temperature. It has already been

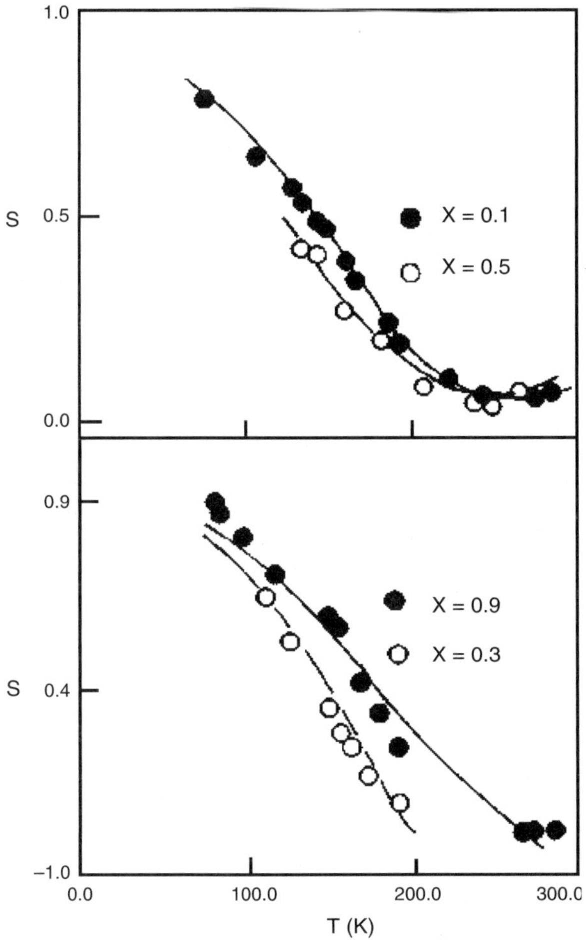

Figure 12. Thermal variation of frequency exponent (S) for $(1-x)\cdot(90V_2O_5–10P_2O_5) + xBaTiO_3$ ($x = 0.1$, 0.3, 0.5, and 0.9) glass-nanocomposites fitted with the OLPT model [Eq. (10)]. The continuous lines are the best-fit curves. Reprinted with permission from [50], M. Sadhukhan et al., *J. Chem. Phys.* 105, 11326 (1996). © 1996, American Institute of Physics.

shown in Figure 12 that s decreases with increasing temperature and then it increases after reaching a certain minimum value, which is also in complete contradiction to the prediction of the simple QMT model. However, at higher temperatures, the ac conductivity starts to deviate from linearity and the temperature (around $\theta_D/2$) at which deviation from linearity starts increases with increasing frequency. Thus the QMT model can explain only the experimental ac conductivity data at low temperature of these VPTI type FGNCs.

It is to be mentioned here that the OLPT model [Eq. (10)] is also applicable in the low temperature range if it is approximated to low temperature condition. One can show that at low temperature this equation reduces to the form

$$\sigma_{ac}(\omega) = (\pi^4/12)e^2(k_BT)^2[N(E_F)]^2[\omega R_\omega^4/2\alpha k_BT]$$
$$= (\pi^4/24)e^2(k_BT)[N(E_F)]^2\omega R_\omega^4/\alpha \quad (13)$$

Equation (13) can explain the data of Figure 11 and the estimated $N(E_F)$ values fairly agree (within ±2%) with the corresponding values estimated from the QMT model. The OLPT model is, therefore, applicable for a wide range of temperature (both low and high) for these glass-nanocomposites. And the failure in temperature dependence of the exponent s as shown by the QMT model can also be avoided.

It was also observed that the CBH model [12] which correlates the barrier height W with the intersite separation R for single-electron hopping [84] was not applicable for the ac conductivity in these VPTI type nanoparticle dispersed glasses, even though this model was found to hold good for the base glass [69] V_2O_5–P_2O_5 (with 80–90 mol% V_2O_5) in the temperature range 80–400 K. A similar feature was observed as well for the $SrTiO_3$ doped V_2O_5–Bi_2O_3 FGNC system [41, 48]. The presence of TiO_2 in the V_2O_5–P_2O_5 glass, therefore, not only favors the formation of nanocrystals but also modifies the network structure, which in turn was responsible for the change of physical behavior of the nanocrystal dispersed glass. The nonlinear behavior of the TiO_2 doped VPTI type FGNCs, as discussed, was in addition a special property of these FGNCs which might be responsible for the observed differences in the conductivity mechanism in the TiO_2 doped FGNCs from that of the pure base glass, viz. vanadium phosphate (VP) glasses. The thermal variation of dielectric constant (both real and imaginary) of these glasses to be discussed is also interesting, indicating nonlinear behavior of these glasses with $x > 5$ wt% TiO_2 [64].

7.3. Dielectric Properties

Dielectric constants of the FGNC composites are, in general, measured by a GR-1615A Bridge and also by a HP 4291 impedance analyzer [48, 50]. Figures 13 and 14 show, respectively, the thermal variations of the real part (ε') of the dielectric constants of $80V_2O_5$–$20PbO + 30$ wt% $BaTiO_3$ and $80V_2O_5$–$20P_2O_5 + 10$ wt% TiO_2 nanoparticle dispersed FGNCs for different frequencies (0.1–100 kHz) [30, 64]. Interestingly, the values of dielectric constants of these FGNCs were several orders (three to four) of magnitude higher than those of the undoped $80V_2O_5$–$20P_2O_5$ glasses [5]. It was also observed from the T vs ε' curve (Fig. 13) that this relaxation peak above 360 K was shifted to a higher temperature region with the increase of frequency. For higher concentrations ($x > 5$ wt% TiO_2, or $BaTiO_3$) at a fixed frequency [30, 64], where grain size became larger, or at higher frequencies (for a fixed concentration), the dielectric constant decreased (Figs. 13 and 14). A maximum dielectric constant for the VPTI type FGNC was observed with x around 5 wt% (having smallest grain/cluster size ~10 nm) at 100 kHz [64]. This behavior was also observed in the V_2O_5–P_2O_5–$KTiO_3$ glasses [86].

The Debye type relaxation behavior [87] can be characterized by a complex dielectric constant of the form $\varepsilon^* = \varepsilon_\alpha + (\varepsilon_s - \varepsilon_\alpha)/(1 + j\omega\tau)$ where ε_s and ε_α are the static and high frequency dielectric constants, respectively. ω is the angular frequency and τ is the dielectric relaxation time. Using the relation $\varepsilon^* = \varepsilon' + i\varepsilon''$, we get both real ($\varepsilon'$) and imaginary ($\varepsilon''$) parts of dielectric constants expressed as $\varepsilon' = \varepsilon_\infty + (\varepsilon_s - \varepsilon_\infty)/(1 + \omega^2\tau^2)$ and $\varepsilon'' = (\varepsilon_s - \varepsilon_\infty)\omega\tau/(1 + \omega^2\tau^2)$. Debye type dielectric relaxation

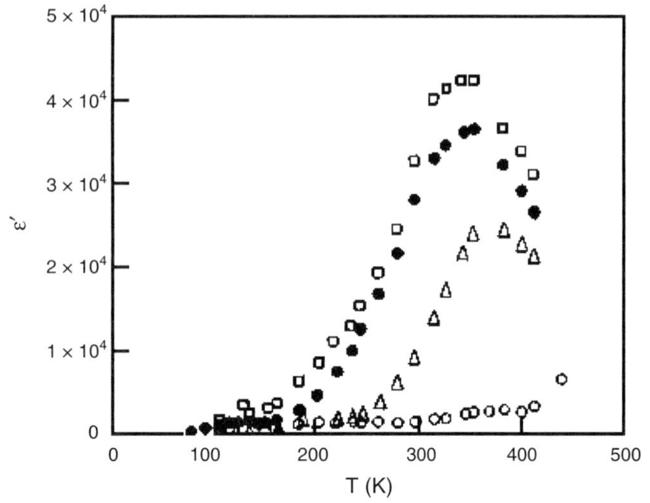

Figure 13. The thermal distinction real part of dielectric constant (ε') of the glass-nanocomposite sample ($(80V_2O_5$–$20PbO) + 30$ wt% $BaTiO_3$ (annealed at 300 °C for 9 h in air) showing one broad peak around 390 K at four different frequencies, 100 Hz (□), 1 kHz (●), 10 kHz (△), and 100 kHz (○), corresponding to ferroelectric transition of $BaTiO_3$ at 393 K. Reprinted with permission from [30], M. Sadhukhan et al., *J. Appl. Phys.* 85, 3477 (1999). © 1999, American Institute of Physics.

behavior is also characterized by relaxation frequency $f_c = 1/2\pi\tau = \nu_{\mathrm{ph}} \exp(-W/k_B T)$, where ν_{ph} is the characteristic phonon frequency. This relaxation process can be more clearly resolved [88] from the calculation of dielectric modulus M^* $(=M' + iM'') = \varepsilon^{*-1} = \omega\varepsilon_0 Z^*$, where ε_0 is the vacuum permittivity and Z^* is the complex impedance. The real (M') and imaginary (M'') parts of the complex dielectric modulus M^* are related to the dielectric constants as

$$M' = \varepsilon'/(\varepsilon'^2 + \varepsilon''^2) \quad \text{and} \quad M'' = \varepsilon''/(\varepsilon'^2 + \varepsilon''^2) \qquad (14)$$

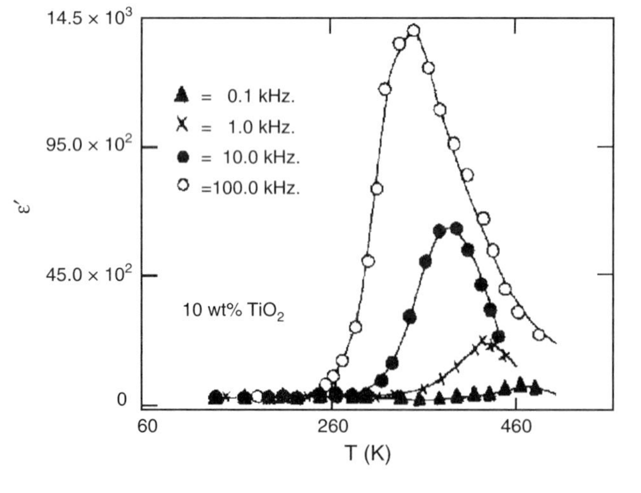

Figure 14. Real part of dielectric constant (ε') of $(80V_2O_5$–$20P_2O_5) + xTiO_2$ glass–nanocrystal composites for a fixed ($x = 10$ wt%) TiO_2 content at different frequencies (0.1 to 100 kHz) as a function of temperature. Reprinted with permission from [64], D. K. Modak et al., *Sol. State Phys.* 43, 368 (2000).

Again according to Cole and Cole [89], the complex dielectric constant can be expressed as

$$\varepsilon^* = \varepsilon_\infty + (\varepsilon_s - \varepsilon_\infty)/(1 + i\omega\tau_m)^{1-\phi} \qquad (15)$$

where τ_m is the mean relaxation time and ϕ is an empirical parameter lying between 0 and 1. The slope of the Cole–Cole plot [89] at $\varepsilon'' = 0$ gives the value of $(1 - \phi)\pi/2$. A value of $\phi = 0$ leads to a single relaxation time and $\phi = 1$ gives an infinite distribution of relaxation time which is very common in the glassy system.

The thermal variations of loss tangent $\tan\delta$ $(=\varepsilon''/\varepsilon')$ for TiO_2 doped nanoparticle dispersed FGNCs were large (since $\varepsilon'' > \varepsilon'$) indicating a very high loss factor in these FGNCs [64, 78]. The high loss of these glasses might be due to the more conducting nature of the precipitated nanocrystals embedded in the glass matrix, or the effects of space-charge polarization [90] of nanocrystalline TiO_2. In the high temperature region, a sharp increase of the loss tangent was observed with increasing frequency. Interestingly, at lower concentrations ($x \leq 10$ wt% which was almost below the critical concentration x above which "phase transition" occurred), loss tangent of the VPTI type glass–nanocrystal composites showed peaks, which shifted to a higher temperature region for higher concentration for a constant frequency of 10 kHz. The $\tan\delta$ vs temperature curves of higher concentrations did not show any peak. It should be mentioned that Kundu et al. [34] did not observe any peak in temperature variation of $\tan\delta$ curves and the loss was also very low. Because of the absence of a well defined loss peak and the uncertainties involved in the subtraction of the dc conductivity from the total conductivity, it was difficult to use conventional methods of estimating the relaxation frequency f_c and the nature of the dispersion (i.e., whether it is characterized by a single relaxation time or by a distribution of relaxation times). However, the following two approaches may be used [79] for the analysis of the data. The first involves the dielectric modulus, defined by Macedo et al. [88], and the second approach is to extract information about the dielectric relaxation time from the temperature or frequency variation of the dielectric constant alone assuming a symmetric distribution of relaxation times. Although no peak was observed in the temperature dependent ε'' curves of $80V_2O_5$–$20P_2O_5 + 5$ wt% TiO_2 nanoparticle dispersed FGNC (Fig. 15) for a fixed frequency up to the highest temperature limit measured (460 K), it exhibited Debye type dielectric relaxation behavior characterized by relaxation frequency f_c $(=1/2\pi\tau)$ [64]. It is interesting to note that Prasad et al. [31, 32], however, found peak in temperature variation of dielectric loss (ε'') data of $50SrB_4O_7$–$50Bi_2VO_{5.5}$ type FGNCs. The relaxation process in these FGNCs can be more clearly resolved by using Eq. (14). It was seen from the thermal variation of M' for a fixed (20 wt%) TiO_2 doped VPTI type FGNC (Fig. 16) that M' varies slowly with temperature for fixed measured frequency (10 kHz) above 260 K [64]. This type of temperature dependence of M' indicates that the dielectric constants of the samples are thermally activated. Similarly M'' vs T curves for a sample with 30 wt% TiO_2 (Fig. 17) at different fixed frequencies (1–100 kHz) showed peaks within the measured temperature limit around 110 K. The peak position of M''

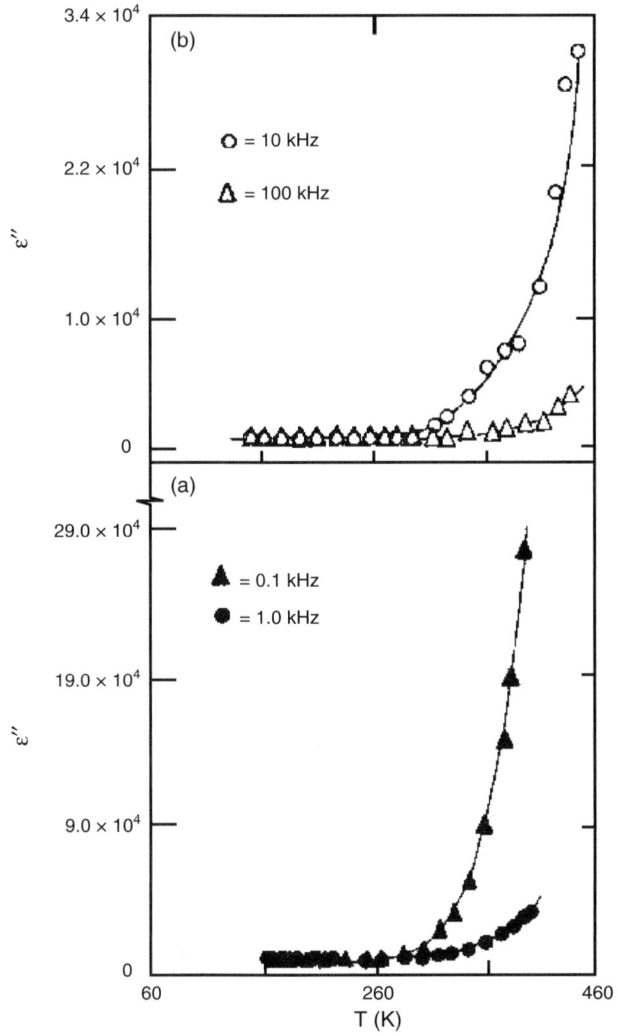

Figure 15. Imaginary part of dielectric constant (ε'') as a function of temperature for $(80V_2O_5–20P_2O_5) + xTiO_2$ glass–nanocrystal composites (with $x = 5$ wt% TiO_2) at frequencies 0.1 and 1.0 kHz (a) and 10 and 100 kHz (b). Reprinted with permission from [64], D. K. Modak et al., *Sol. State Phys.* 43, 368 (2000).

gave the temperature at which the measuring frequency was given by $f_c = \nu_{ph} \exp(-W/k_BT)$. Attempts were also made to find the nature of the dielectric relaxation behavior of these FGNCs from the Cole–Cole plot [89] of Eq. (15) at different fixed temperatures with different frequencies [64]. The semicircular nature of the curves as predicted from the Cole–Cole plot [89] for the TiO_2 or $BaTiO_3$ nanocrystal doped VPTI type FGNCs was not prominent [30, 50, 64]. From the Cole–Cole plot [89], it was also evident that ϕ resided between 0 and 1, which suggested that all these glass–nanocrystal composite samples had a distribution of relaxation times.

Grain size effects on ferro- and antiferroelectric oxides have been well investigated theoretically and experimentally [90, 91]. For ferroelectric $PbTiO_3$, the estimated critical size of the nanoparticle clusters [40] was between 8 and 12 nm. The high dielectric constant of these nanoparticle dispersed glasses might be related to the space charge polarization or ionic relaxation polarization occurring at the

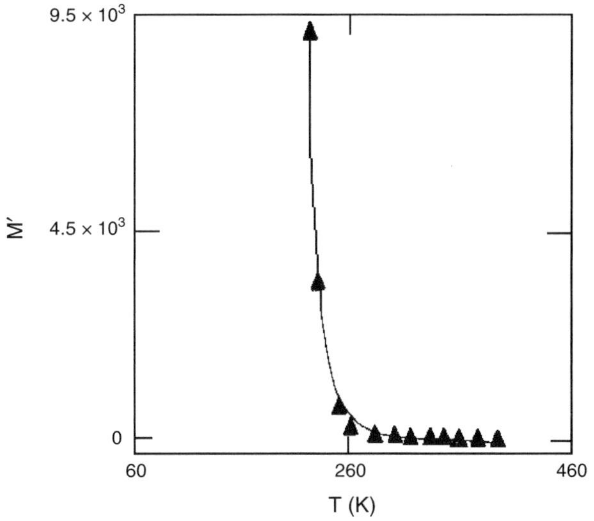

Figure 16. Thermal variation of the real part of dielectric modulus (M') for $(80V_2O_5–20P_2O_5) + xTiO_2$ glass–nanocrystal composites for a fixed TiO_2 concentration ($x = 20$ wt%) at 10 kHz. Reprinted with permission from [64], D. K. Modak et al., *Sol. State Phys.* 43, 368 (2000).

interface as suggested by Zhong et al. [90, 91]. The dielectric loss of the VPTI type nanoparticle dispersed FGNC was also very high which might be related to the more conducting character of the precipitated nanoclusters.

7.4. Ferroelectric Properties

For a typical FGNC system, viz. $Bi_{2-x}Pb_xSr_2Ca_2Cu_{3-y}K_y$ ($x = 0.1–0.3$, $y = 0.2–0.3$), the ferroelectric property is discussed. The T_g values of this system were found to be (340, 350, 360, and 400 K) depending on the values of x and y [44]. For a typical FGNC, viz. ($x = 0.2$, $y = 0.2$), its

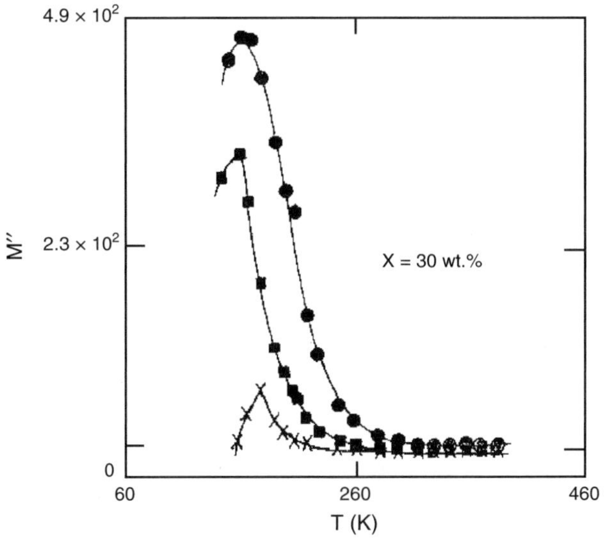

Figure 17. Thermal variation of imaginary part of dielectric modulus (M'') of $(80V_2O_5–20P_2O_5) + xTiO_2$ glass–nanocrystal composites for a fixed TiO_2 concentration ($x = 30$ wt%) at 1 (x), 10 (♦), and 100 (○) kHz. Reprinted with permission from [64], D. K. Modak et al., *Sol. State Phys.* 43, 368 (2000).

TEM micrograph clearly indicated the presence of microcrystals/clusters (10–50 nm size). The concentration of the nanoparticles slightly increased with the increase of Pb concentration. As mentioned earlier, these nanoparticles are considered to be responsible for the ferroelectric behavior of K containing FGNCs since nanoparticle free glasses are not ferroelectric. Interestingly, the K free glasses are reported to become also superconductors [59] by annealing around 840 °C. Hereafter these K free glasses are referred to as SG glasses. But due to the presence of alkali metal (partial substitution of Cu by K) in these glasses, they show FE behavior and do not become superconductors by annealing [44] around 840 °C. The K doped ferroelectric glasses are termed FG glasses.

The structure of Bi–Sr–Ca–Cu–O glasses depends to a great extent on the Cu content and its valence state [21]. So the addition of alkali metal (K, say) in the Cu site distorted the average structure of the K free SG type glass to a great extent [44]. Though both Bi_2O_3 and CuO are not glass-forming oxides, it is well established that both Bi_2O_3 and CuO are necessary to form a glassy state in the Bi–Sr–Ca–Cu–O system as reported by several authors [78, 79]. Bi_2O_3 enhances the glass-forming ability in the Bi-based FGNC systems; Bi_2O_3 acts as "glass former" in the presence of CuO. It has been reported [92] that Bi^{3+} ions are six coordinated (distorted BiO_6 octahedra) in the Bi-based glasses. The oxygen coordination number of Cu^+ in Cu_2O is 2 and that of Cu^{2+} in CuO is 4. The density of CuO is more than Cu_2O, concluding that CuO structure is more compact than that of Cu_2O. Recently Sato et al. [93] proposed a glass network structure unit of $Bi_4Sr_3Ca_3Cu_4O_x$ glasses as shown in Figure 18. In this model, BiO_6 octahedra are connected to other octahedra through two-coordinated Cu^+ ions and particularly in the K free SG type glasses Cu^+ ions take part mainly in the formation of networks. The other cations of Cu^{2+}, Ca^{2+}, and Sr^{2+} may be distributed randomly in the surrounding network. Partial replacement of Cu sites by K changed some ionic states of Cu and distorted the glass network structure in favor of dipolar ordering and hence appearance of ferroelectricity as observed from the dielectric properties discussed. The difference between the glass structures of superconducting precursor glass (SG type glass) can be compared in terms of the oxygen molar volume [94, 95] V_o^*. The value of V_o^* of the present FGNC was calculated using the

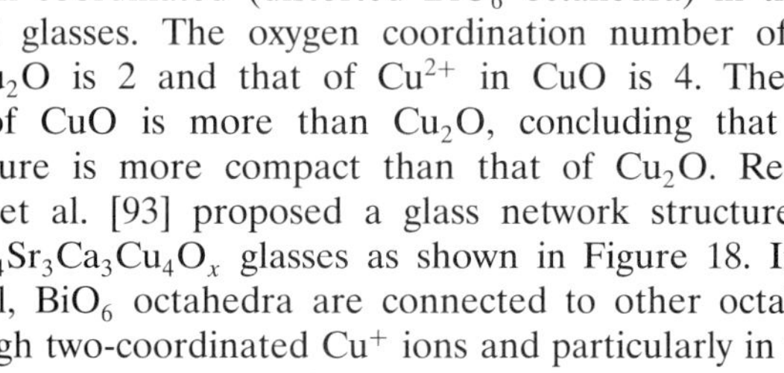

Figure 18. Network structure model of $Bi_4Sr_3Ca_3Cu_4O_x$ glasses. In K doped glasses, Cu ions are partially replaced by K ions. Reprinted with permission from [93], R. Sato et al., *J. Non-Cryst. Solids* 160, 180 (1993). © 1993, Elsevier Science.

equation

$$V_o^* = \frac{(M_{CuO} - 16C)X_{CuO} + M_K X_K + M_{Ca} X_{Ca} + M_{Sr} X_{Sr} + M_{Pb} X_{Pb} + M_{Bi} X_{Bi}}{d[(5-C)X_{Cuo} + 3X_K + 3X_{Ca} + 3X_{Sr} + 1X_{Pb} + 3X_{Bi}]} \quad (16)$$

where M_{CuO}, M_K, M_{Ca}, M_{Sr}, M_{Pb}, and M_{Bi} are the molecular weights of CuO, K_2CO_3, $CaCO_3$, $SrCO_3$ PbO, and Bi_2O_3, respectively. X_{CuO}, X_K, X_{Ca}, X_{Sr}, X_{Pb}, and X_{Bi} are the mole fractions of CuO, K_2CO_3, $CaCO_3$, $SrCO_3$, PbO, and Bi_2O_3, respectively. The parameters d and C are the density of the glass and fraction of reduced Cu ions, respectively. Both T_g (400 °C) and V_o^* (14.52) of the K containing FGNCs were a little higher than those of the K free SG glasses ($T_g = 340$ °C, $V_o^* = 11.23$) [44]. This indicated that K doped glasses had a more stable glass network structure than that of the K free SG glasses. However, the network structure also depended on the K and Pb concentrations.

The dielectric constant of the Bi-based glassy precursor ($Bi_4Sr_3Ca_3Cu_4O_x$) for high T_c superconductors (SG glasses) did not show any ferroelectric behavior [96–100]. Figure 19a shows the variation of the dielectric constant (ε') of a typical K-doped FGNC, viz. (with $x = 0.2$, $y = 0.2$) as a function of temperature for different frequencies 0.1, 0.12, 1.0, 10, 100 kHz. The appearance of small broad peak in the (ε'–T) curve (Fig. 19a) around 600 K along with the large peak for every frequency indicated the presence of the relaxation mode in the glass. The ε' values of the FGNCs are of the same order as those of the Bi-based multicomponent SG glasses (Fig. 19b) [96–100]. It has already been shown [78] that the dielectric constant of nanocrystalline TiO_2 is much larger than that of the corresponding bulk sample. As mentioned earlier [30–32, 34, 50, 64], the $Bi_2VO_{5.5}$, PbS, $BaTiO_3$, and TiO_2 ferroelectric nanocrystals containing FGNCs also showed high dielectric constants as shown in Table 1.

The dielectric constant data of the K doped FGNC followed the Curie–Weiss relation $\varepsilon' = C/(T - T_0)$, where C is the Curie constant. T_0 is the extrapolated intersection of the high temperature part of the plot with the temperature axis as shown in Figure 20a [44]. The order of transition was identified by finding the ratio of the slopes $\frac{\partial(1/\varepsilon)}{\partial T}$ below and above T_{cf} [44]. The value of this ratio (=−2.63) indicated that the transition was of first order type. Prasad et al. [31, 32], also analyzed the dielectric constant data of the SrB_4O_7–$Bi_2VO_{5.5}$ type FGNCs in the same way. In case of second order phase transition, T_0 is practically the same as the transition temperature or the Curie point T_{cf}, while in case of a first order transition T_0 is lower than the Curie temperature T_{cf}. T_{cf} of this FGNC was found to vary from 500 to 530 K depending on K concentrations. It is more interesting to mention here that for any of the K-free superconducting precursor glasses (SG glasses), becoming superconductors by annealing, the plot of $1/\varepsilon'$ vs T (Fig. 20b) did not at all match with that of K doped FGNC showing ferroelectric behavior (Fig. 20a). Detailed analysis of the dielectric properties of other SG glasses without showing ferroelectric behavior can be found in the text [96–100].

Here we like to mention that frequency dependent conductivity $\sigma_{ac}(\omega) \propto \omega^s$ (where $s < 1$ is frequency exponent [13, 14]) of these FGNCs also supported the small polaron hopping conduction mechanism [86]. Murawski

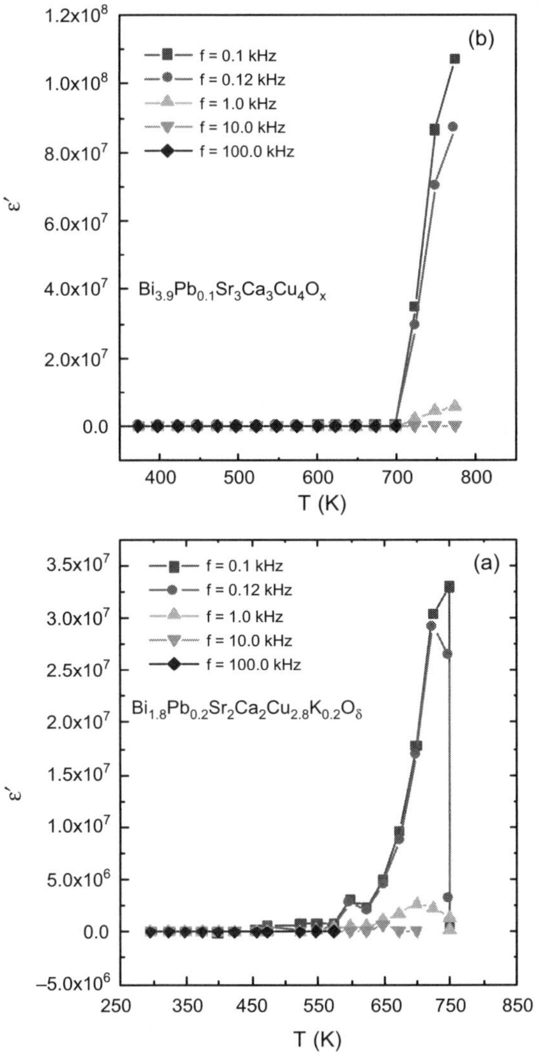

Figure 19. Thermal variation of real part of dielectric constant (ε') of (a) $Bi_{1.8}Pb_{0.2}Sr_2Ca_2Cu_{2.8}K_{0.2}O\delta$ and (b) $Bi_{3.9}Pb_{0.1}Sr_3Ca_3Cu_4O_x$ samples at different frequencies, 0.1–100 kHz. Reprinted with permission from [44], S. Mukherjee et al., *J. Appl. Phys.* 94, 1211 (2003), © 2003, American Institute of Physics.

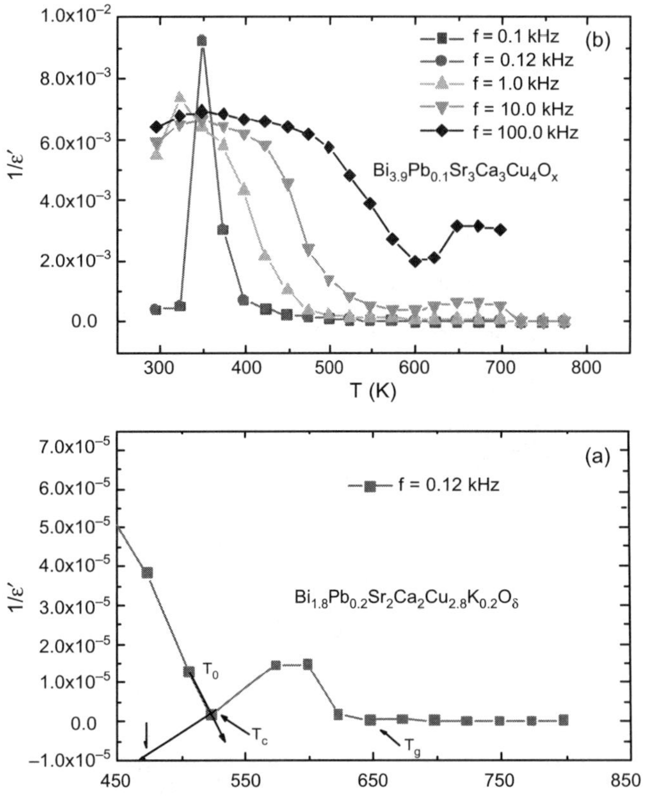

Figure 20. Variation of inverse dielectric constant ($1/\varepsilon'$) with temperature of (a) $Bi_{1.8}Pb_{0.2}Sr_2Ca_2Cu_{2.8}O_\delta$ at a fixed frequency (0.12 kHz) and (b) $Bi_{3.9}Pb_{0.1}Sr_3Ca_3Cu_4O_x$ at different frequencies (0.1–100 kHz). Reprinted with permission from [44], S. Mukherjee et al., *J. Appl. Phys.* 94, 1211 (2003), © 2003, American Institute of Physics.

and Barczynski [101] showed that a fractal structure was responsible for relaxation currents and the total conductivity $\sigma_t(\omega) = \sigma_{dc} + \sigma_{ac}(\omega)$ can be expressed as $\sigma_t = \sigma_{dc}[1 + K(d)(\omega/\omega_c)^r]$, where $r = 1 + d - d_f > 1$, d is the dimensionality of the space containing relevant clusters, and d_f is the fractal dimensionality of such clusters. For frequency $\omega < \omega_c = 2\pi f_c$, where f_c is the peak frequency at which dielectric constant data showed peaks and $K(d)$ is a dimensionally dependent constant related to the statistics of the contributing clusters ($K = 0.001$–0.0025) for many TMO glasses [90]. For a typical glass sample with $x = 30$ wt% of TiO_2, Mukherjee et al. [44] estimated d_f from the knowledge of ac conductivity and dielectric constant data of this sample. The estimated value of d_f was ~2.71 meaning that in the clusters the percolation paths also have three-dimensional character in the nanocrystal dispersed FGNCs.

Prasad et al. [31] observed that $50SrB_4O_7$–$50Bi_2VO_{5.5}$ glass nanocomposites heat-treated at 820 K/h did not show a polarization (P) versus electric field (E) hysteresis loop at room temperature (300 K) indicating that the coercive field required to switch on the polarization at this temperature was much higher. However, the sample exhibited a hysteresis loop in the vicinity of the ferroelectric to paraelectric transition temperature (720 K) associated with a remnant polarization (P_r) of 5.63×10^{-9} C cm^{-2} and coercive field (E_c) of 1250 V/cm (Fig. 21a). However, the values of P_r and E_c were, respectively [102], 2.25×10^{-8} C cm^{-2} and 650 V/cm for micrometer size crystallites containing $Bi_2VO_{5.5}$ ceramic. The loops disappeared and the plots became linear (Fig. 21b) at 780 K, which was beyond the phase transition temperature of this glass-nanocomposite. On cooling, the loops reappeared around 730 K, which was similar to that of the ferroelectric behavior of $Bi_2VO_{5.5}$ ceramics. Similar behavior was also exhibited by the K doped Bi–Sr–Ca–Cu–O FGNCs as discussed.

7.5. Piezoelectric Properties

Though transport and other properties of the ferroelectric nanocrystal dispersed oxide glasses have been studied widely, the piezoelectric nanocrystal–glass composites have not been so extensively investigated. However, composites having perovskite dispersed polymer and glasses have been studied as electronic–ceramics [103–110].

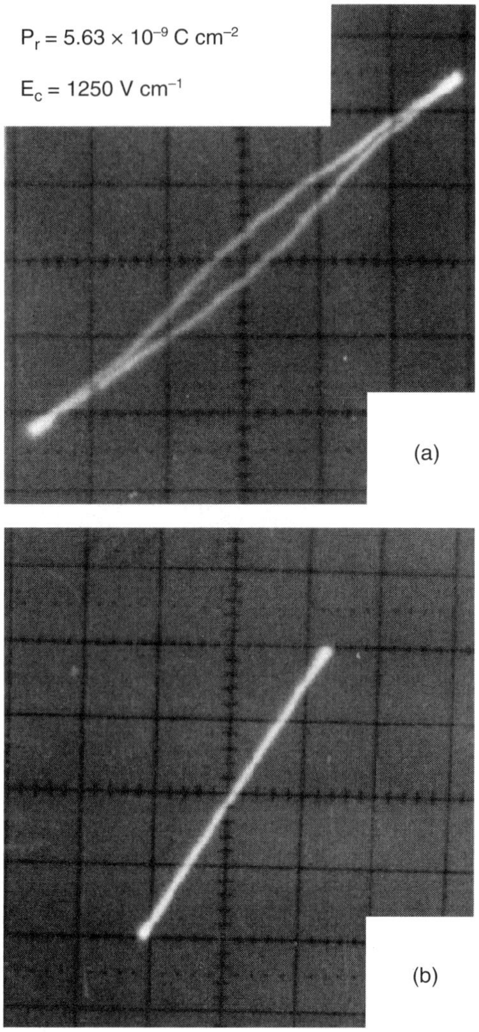

Figure 21. Hysteresis loops recorded on 820 K heat-treated $50SrB_4O_7$–$50Bi_2VO_{5.5}$ glass nanocomposites at two different temperatures, (a) 720 and (b) 780 K. P_r and E_c stand respectively for remnant polarization and coercive field. Reprinted with permission from [31], N. S. Prasad and K. B. R. Varma, *J. Mater. Chem.* 11, 1912 (2001). © 2001, The Royal Society of Chemistry.

The piezoelectric particulate reinforced nanocomposite ceramics have been broadly reported [111–119]. Various piezoelectric–nanocomposite systems like PZT/Ag, PZT/Al_2O_3, MgO/PZT, $BaTiO_3$/Ni, $BaTiO_3$/SiC, etc. have been explored by several groups [111–119]. The $PbZr_{0.5}Ti_{0.5}O_3$ type system can be used to prepare piezoelectric–glass nanocomposites [120]. The relative permittivity (ε') of these piezoelectric–nanocomposite systems is in the range of 1000–3000 at a temperature of 300 K and frequency of 1 kHz [119]. The dielectric constant and dielectric loss of the piezoelectric–nanocomposite system generally have an abrupt increase above the critical field as shown in Figure 22. This effect has been confirmed by different research groups [121]. Hagemann [122] proposed that this change is due to hysteretic domain wall motion and the resultant reorientation by spontaneous polarization. Similar to piezoelectric nanoparticle–glass composites,

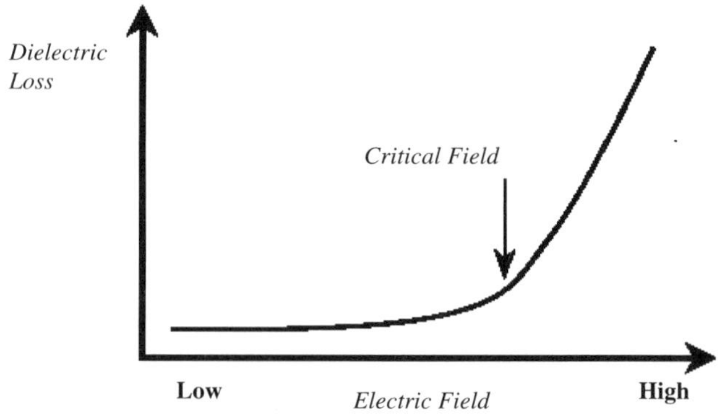

Figure 22. General trend of dielectric loss with applied electric field of piezoelectric glass-nanocomposites. Hysteretic domain wall motion above a critical field (indicated by arrow) leads to an increased dielectric loss. Reprinted with permission from [119a], S. R. Panteny et al., preprint (2002).

the pyroelectric nanoparticle–glass composites are not widely studied. Some composite films of lead titanate, PZT, lanthanum–lead titanate, and calsium–lanthanum–lead titanate into copolymer matrices for pyroelectric sensor application are already available in the literature [123–128].

7.6. Mechanical Properties

The mechanical behavior of a material is determined by its type of bonding and defect structure. Nanophase materials show increasing hardness with decreasing grain size. When the grain size of the material is reduced to the nanoscale, its mechanical property becomes strongly influenced due to the fact that a large fraction of the atoms in the material lie in the grain boundaries. So the material becomes superhard possessing hardness rivaling diamond [129, 130]. Glasses are generally very brittle. Dispersing nanocrystals inside the glass matrix can decrease its brittleness. Toughening is an extremely important property of glass for their practical use. A piezoelectric secondary phase (e.g., PZT particles) can increase the toughening when incorporated into the glass matrix. The works on PZT-containing ceramic matrix composites suggest that energy dissipation due to the piezoelectric effect as well as the motion of the domain wall should contribute to toughening. The sintering process of the mixtures containing different concentrations of PZT particles should be optimized in order to obtain a pore-free material. Transparent glass fiber reinforced silicate glass matrix composites with improved mechanical and optical properties are important for technical applications. The studies of mechanical properties of these products, viz. fracture strength, toughness, and thermal shock resistance, and the optical properties, viz. composite effective refractive index and light transmittance, are also very important for future investigation and analysis. Hot pressing is generally used to obtain dense glass-nanocomposite materials. The hardness of ferroelectric glass-nanocomposites are generally done by heat treatment of the material to produce a fine dispersion of particles within the microstructure which increases the resistance to domain wall motion [131].

8. TRANSPARENT FERROELECTRIC GLASS NANOCOMPOSITES

Some of the glass-nanocomposites are transparent since these have a big bandgap for light absorption for electrons, and so light absorption is very low and thus transmitted light fraction is very high. The nanoparticles in the glassy matrix are so small that light is unable to interact with them (i.e., the object is smaller than light wavelength) and thus the transparency of the glass-nanocomposite is not hampered by the inclusion of nanoparticles in the glass matrix. On the other hand, if the refractive index of the dispersed nanoparticles is the same as that of the base glass, it will also not affect the transparency of the glass. These materials have several advantages like easy formability, optical transparency, and low cost. Examples of some TFGNCs are $LiNbO_3$ nanocrystal embedded Li_2O–SiO_2–Al_2O_3 glass matrix and $KTiOPO_4$ (KTP) nanocrystals dispersed in SiO_2 glass matrix [35, 39].

In $KTiOPO_4$ (KTP)/SiO_2 TFGNC, the ferroelectric KTP nanocrystals were directly precipitated from K_2O–TiO_2–P_2O_5–SiO_2 parent glasses by heat treatment [35, 39]. The SAED, SEM, and TEM of these TFGNCs showed the distribution of the nanoparticles in the glass matrix beautifully before and after heat treatment. Particle sizes were measured from TEM photographs [35, 39]. The ultraviolet (UV) absorption band in these TFGNCs originated from the Ti–O band. There was also a strong absorption near 2800 nm, which corresponded to the well-known IR absorption band of the OH-band stretching mode. The transparency of these TFGNCs decreased with the increase of heat-treating temperature as shown in Figure 23. With the increase of heat-treatment temperature, the particle size increased and hence became able to interact with the light and as a result the transmittance decreased. However, these TFGNCs had a transmittance cutoff edge in the UV range at ~350 nm [35, 39].

The second order nonlinear susceptibility is generally induced only in a noncentrosymmetric crystal structure or anisotropic materials. In other words, second harmonic generation (SHG) in glasses should be forbidden as glasses have macroscopic inversion symmetry. Therefore, SHG and other electro-optical effects are usually unexpected in glassy systems. In fact, no SHG signal was observed in different SiO_2-based glasses [35] like $20P_2O_5$–$80SiO_2$, $10K_2O$–$10TiO_2$–$80SiO_2$, $10TiO_2$–$10P_2O_5$–$80SiO_2$, and $20TiO_2$–$80SiO_2$. But the SHG signal has been reported in some poled glasses [132] and photoinduced glasses [133–135] though the origin of the SHG signal was completely unclear. SHG was also observed in GeO_2-, CeO_2-, and Eu_2O_3-containing SiO_2 glass fibers which was due to the defects and additives and may be the origin of the SHG centers in the glasses [135–138]. The multicomponent K_2O–TiO_2–P_2O_5–SiO_2 parent glasses, when annealed at 550 °C, did not contain resolvable crystallites [39] as analyzed by XRD and hence did not exhibit any SHG signal (Fig. 24a). Nevertheless, when annealed at 750 °C, it contained KTP nanocrystals dispersed in SiO_2 glass matrix and exhibited the SHG signal with a narrow bandwidth at 532 nm signifying a true SHG process (Fig. 24b). These results suggested that the SHG in KTP/SiO_2 TFGNC originated from KTP nanocrystals embedded in SiO_2 matrix [35]. It is to be noted here that a third order nonlinear optical susceptibility or third harmonic generation in silver nanocrystal dispersed BaO–B_2O_3–P_2O_5 glass was observed around the surface plasmon resonance band [51].

9. APPLICATIONS

Nanophase engineering with organic and inorganic materials is expanding enormously to manipulate optical and electronic functions. Semiconducting nanoparticles have been the subject of increasing interest since the demonstration in 1975 of the enhancement of their third order nonlinear optical properties [139]. Very small semiconductor (<10 nm) or metal particles in glass composites and semiconductor/polymer composites show interesting quantum effects and nonlinear electrical and optical properties. Nanocomposite materials consisting of small metal or oxide particles dispersed in glass matrices are attracting much

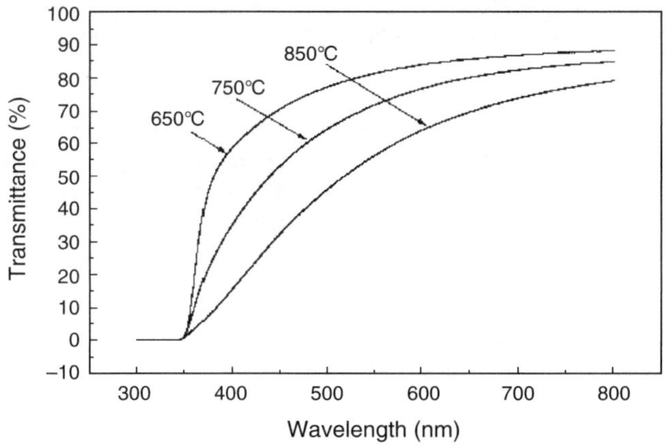

Figure 23. Wavelength variation of optical transmittance (%) of $20(K_2O$–TiO_2–$P_2O_5)$–$80SiO_2$ system heat-treated at various temperatures (650, 750, and 850 °C) for 4 hours. Reprinted with permission from [35], D. L. Li et al., *J. Non-Cryst. Solids* 271, 45 (2000). © 2000, Elsevier Science.

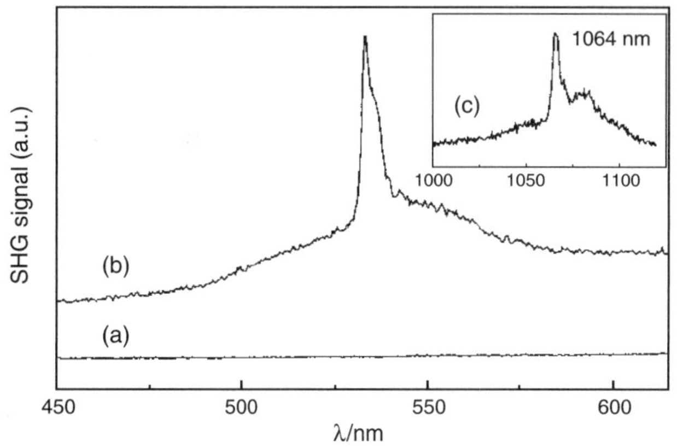

Figure 24. The SHG (532 nm) spectra for (a) $30(K_2O$–TiO_2–$P_2O_5)$–$70SiO_2$ multicomponent glass, (b) 30KTP–$70SiO_2$ nanocomposites prepared by heating at 750 °C, and (c) fundamental beam (1064 nm). Reprinted with permission from [39], D. L. Li et al., *J. Non-Cryst. Solids* 261, 273 (2000). © 2000, Elsevier Science.

attention because of their potential use in a variety of fields, such as optics, magnetism, and electronics [140–146]. Semiconducting nanoparticles are particularly interesting due to their possible applications in diverse areas like photocatalysis, solar cells, display panels, and new devices, viz. single electron transistors and so on [147–149]. A high third order optical nonlinearity of semiconducting nanocrystal composites allow a wide range of useful function, such as high speed all-optical switching and other signal processing operations. Ferroelectric nanoparticles can be treated as built in electrical dipoles, which can produce electro-optic effects, photorefraction, and harmonic generation. The FGNC systems are both semiconducting and ferroelectric with a giant dielectric constant. The FGNCs with high permitivitty may be attractive for applications in microelectronics. Because of the ferroelectric behavior of glass-nanocomposites, they have additional applications as sensor materials [150]. When an ordinary glass is doped with quantized semiconductor colloids, it becomes a high-performance optical medium with potential applications in optical computing. Very small particles have special atomic structures with discrete electronic states that give rise to an extraordinary property like superparamagnetism. Magnetic nanocomposites have been used for mechanical force transfer (ferrofluids), high-density information storage, and magnetic refrigeration. Piezoelectric nanocrystal–glass composites are more sensitive materials for high frequency hydrophones.

Transparent FGNCs having nonlinear optical properties are used in transducers, sensors, and actuators in the fields of communication, energy, and health. These new totally inorganic materials will be attractive for various optical applications (switching and memory devices) where currently standard monolithic glasses cannot be used due to poor mechanical properties. More conventional applications of wide commercial interest such as architectural and building materials are also possible including applications where high fire and thermal shock resistance are required and where standard laminated glasses are not suitable due to their organic components. The superhardness of the glass-nanocomposites can be used as a material requiring friction and wear resistance. The superhard nanostructured films promise revolutionary improvements for wear protection for applications in high speed machining as well as in the emerging field of miniaturized moving parts in microelectromechanical systems [129, 130].

10. CONCLUSIONS

The ferroelectric glass-nanocomposites (FGNCs) are formed by uniformly dispersing nanocrystalline particles/clusters of TiO_2, $BaTiO_3$, $PbTiO_3$, $SrTiO_3$, $PbZrO_3$, $KTiOPO_4$, $Bi_2VO_{5.5}$, $Li_2Ge_7O_{15}$, PbS, etc. in the glass matrices. These FGNCs show giant dielectric constant, ferroelectric, and piezoelectric behavior depending on their composition. The electrical and the dielectric properties of these FGNCs are found to depend on the size and concentration of the dispersed nanoclusters in the glass matrices. For lower concentrations of FE nanocrystals ($\leq$5 wt% with average grain or cluster size $<$10 nm), the increase of conductivity was found to be small. A sharp rise of conductivity occurred for the FGNCs with 5 and 10 wt% TiO_2 embedded in $80V_2O_5$–$20P_2O_5$ glass. For higher concentrations ($>$10 wt% with average grain or cluster size $>$20 nm), the conductivity increased slowly with increase of concentration. The increase of electrical conductivity compared to the base glass was considered to be due to the precipitation of nanocrystalline phase in the base glass matrix. A phase separation appeared in the FGNC between 5 and 10 wt% TiO_2 showing nonlinearity in the concentration dependent conductivity and dielectric permittivity. The dielectric constant also showed the maximum value for the FGNC between 5 and 10 wt %. The high dielectric constant of these FGNCs might be related to the space charge polarization or ionic relaxation polarization occurring at the interface as suggested by Zhang et al. [78]. Very strong polaronic couplings (γ varies from 12 to 25, generally for TMO glasses; $\gamma > 4$ indicates very strong electron–phonon interaction constant) observed for these FGNCs play an important role for their high dielectric constant. It appears that there is an average critical grain/cluster size ($\sim$10 nm) above which dielectric constant of these FGNCs tends to decrease. In the literature, grain size effects on ferro- and antiferroelectric oxide materials have been well established both theoretically and experimentally [29, 77]. For ferroelectric $PbTiO_3$, the estimated critical size resides between 8.2 and 13.8 nm [29]. Recently the high dielectric constant in $CaCu_3Ti_4O_{12}$ [45, 46] was also explained by considering semiconducting grains with insulating grain boundaries. Further study of dielectric and conductivity relaxation behavior in these FGNCs would be interesting. Problems related to the uniform size and distribution of the nanoparticles and preparation of the samples in the form of wire, tapes, thin films, etc. from these nanoparticle-dispersed glasses are yet to be explored. More elaborate theoretical analysis is necessary to understand the origin of the giant dielectric constant in these systems. Moreover, our understanding of higher harmonic generations in nanocrystal dispersed glassy materials is no way complete. Thorough experimental and theoretical investigations on such materials are highly encouraging. SHG signal in TFGNCs was found to originate from the KTP nanocrystals embedded in SiO_2 glassy matrix [35, 39].

The ferroelectric nanocrystal dispersed oxide glasses will most probably gain rapidly increasing importance in the near future. In general, properties, production, and characterization methods and their interrelations are, however, not yet satisfactorily understood. Hence, efforts need to be made to enable the directed tailoring of nanophase, nanoscopic, and nanocomposite materials for future technological and industrial applications. The main disadvantage of ferroelectric crystals grown within a glass medium is the difficulty in poling them because of the low dielectric permittivity of the glass matrix. However, the precipitating of nanocrystalline FE metallic phase in glass matrix is easy and it enhances the dielectric constant of the glass enormously [33].

GLOSSARY

AC conductivity Frequency dependent conductivity. The conductivity of the material changes both with temperature and frequency. From the study of electric conductivity of solid materials (composites etc.), mechanism of electron or ion transport, responsible for the electrical conductivity, can be explained.

Correlated barrier hopping (CBH) Mechanism of electrical conductivity. The charge carrier is assumed to hop between site pairs over the potential barrier (W) separating them. For the neighboring sites at a distance R the effective potential barrier will be reduced to a value Wo due to overlap of Coulomb potential of the two sites. Thus the barrier height is correlated with the site separation R (for references see text).

Curie temperature The ferroelectric property of a crystal disappears above a critical temperature TC. This temperature is called Curie temperature. Above the Curie temperature, the ferroelectric phase become paraelectric, like ferromagnetic to paramagnetic transition.

Ferroelectricity It is a special property of some crystalline materials showing spontaneous polarization. A necessary, but not sufficient, condition for a solid to be ferroelectric is the absence of a canter of symmetry. In total there are 21 classes of crystals, which lack a center of symmetry. The classes are based on the rotational symmetry of crystals.

Ferroelectric glass nanocrystal composite (FGNC) When some transition metal oxides are mixed with ferroelectric oxides PZT etc. nanocrystal of the ferroelectric oxides are precipitated in the glass matrix. These nanocrystals embedded glassy phase is called the glass-nanocrystal composite. Depending on the grain size and concentration of these nanoparticles, the properties of the FGNC change.

Melt-quenching This technique is widely used for preparing disordered glassy and amorphous materials. The materials (metallic oxides, in particular) are melted to the liquid phase and quickly quenched to room temperature.

Overlapping large polaron tunneling (OLPT) If there is appreciable overlap of the lattice distortion in the electron wave function, this model is used to explain ac conductivity of semiconductors in disordered materials. The electronic wave function in this case extends over many atomic sites, hence the polaron is termed as large polaron. (See text for details).

Phonon The vibrations of the atoms or molecules are considered to be quantized and phonon is the quantized lattice vibration.

Piezoelectric Of the 21 classes of feroelectric crystals, 20 are piezoelectric, i.e. these crystals become polarized under the influence of external stress.

Polaron Localized charged carriers (formed due to strong interaction between the phonon and the carriers) which strongly interacts with the lattice. If the electron wave function extends over many atomic sites, the polarons are called large polarons, otherwise polarons are called small polarons.

Pyroelectric Ten out of the 20 piezoelectric classes exhibit pyroelectric effects. These pyroelectric crystals are spontaneously polarized.

Quantum mechanical tunneling (QMT) In this process variable range phonon assisted quantum mechanical tunneling between sites close to Fermi level is considered for explaining electrical conductivity in disordered materials like composites, glass etc.

Scanning electron microscope (SEM) This is a technique for studying the surface structure of the materials.

Transition metal oxide (TMO) glasses The transition metal (Fe, Co, Ni, Cu, Mn etc) oxides when melted at higher temperature and quickly quenched to room temperature, forms oxide glasses. These glasses are semiconductors because of the the hopping of electrons/holes from higher valence to lower valence states of the transition metal ions. Strong electron-phonon interaction is responsible for the polaron formation in these glasses.

Transmission electron microscope (TEM) This is a technique for studying the surface structure of the materials.

ACKNOWLEDGMENT

B. K. Chaudhuri is grateful to the National Science Council (NSC), ROC, for partial financial assistance during his visit to the Department of Physics, NSYSU, Kaohsiung, Taiwan, ROC.

REFERENCES

1. H. Hirashima, Y. Watanabe, and T. Yoshida, *J. Non. Cryst. Solids* 95 & 96, 825 (1987).
2. Y. Sakuri and J. Yamaki, *J. Electrochem. Soc.* 132, 512 (1985).
3. S. Chakraborty, H. Satou, and H. Sakata, *J. Appl. Phys.* 82, 5520 (1997).
4. C. H. Chung and J. D. Mackenzie, *J. Non. Cryst. Solids* 42, 357 (1980).
5. V. K. Dhawan, A. Mansingh, and M. Sayer, *J. Non-Cryst. Solids* 51, 87 (1982).
6. A. Ghosh and B. K. Chaudhuri, *J. Non-Cryst. Solids* 83, 151 (1986).
7. G. N. Greaves, *J. Non. Cryst. Solids* 11, 427 (1973).
8. H. Mori, T. Kitami, and H. Sakata, *J. Non-Cryst. Solids* 168, 157 (1994).
9. K. Sega, Y. Kuroda, and H. Sakata, *J. Mater. Sci.* 33, 1303 (1998).
10. H. Hirashima, K. Nishii, and T. Yoshida, *J. Am. Ceram. Soc.* 66, 704 (1983).
11. H. Sakata, K. Sega, and B. K. Chaudhuri, *Phys. Rev. B* 60, 3230 (1999).
12. I. G. Austin and N. F. Mott, *Adv. Phys.* 18, 41 (1969).
13. N. F. Mott, *Adv. Phys.* 16, 49 (1967).
14. A. Miller and E. Abrahams, *Phys. Rev.* 120, 745 (1960).
15. N. F. Mott and E. A. Davis, in "Electronic Processes in Non-Crystalline Materials," 2nd ed. Clarendon, Oxford, 1979.
16. N. F. Mott, *Philos. Mag.* 19, 835 (1969).
17. M. Casu, A. Lai, A. Musinu, G. Piccaluga, S. S. Solinas, S. Bruni, F. Cariati, and E. Beretta, *J. Mater. Sci.* 36, 3731 (2001).
18. B. Capeon, T. Gacoin, J. M. Nedelec, S. Turrel, and M. Bouazaoui, *J. Mater. Sci.* 36, 2565 (2001).
19. N. Hebalkar, A. Cobo, S. R. Sainkar, S. D. Pradhan, W. Vogel, J. Urcan, and S. K. Kulkarni, *J. Mater. Sci.* 36, 4377 (2001).
20. H. Itoigawa, T. Kamiyama, and Y. Nakamura, *J. Non-Cryst. Solids* 210, 95 (1997); *ibid.* 220, 210 (1997).
21. S. Chatterjee, S. Banerjee, and B. K. Chaudhuri, *J. Appl. Phys.* 81, 7320 (1997).
22. Y. G. Wang, W. L. Zhong, and P. L. Zhang, *Phys. Rev. B* 51, 11439 (1996).
23. A. M. Glass, K. Nassau, and J. W. Shiever, *Appl. Phys. Lett.* 31, 249 (1977).
24. M. Takashinge, T. Mitsui, T. Nakamura, Y. Akiawa, and M. Jang, *Jpn. J. Appl. Phys.* 20, L159 (1981).

25. A. M. Glass, K. Nassau, and J. W. Shiever, *J. Appl. Phys.* 48, 5213 (1977).
26. M. Takashinge, T. Nakamura, H. Ozawa, R. Uno, N. Tsuyo, and K. Ari, *Jpn. J. Appl. Phys.* 19, L255 (1980).
27. G. Burns and B. A. Scott, *Phys. Rev. B* 7, 3088 (1993).
28. T. Matsumoto, A. Fukuda, M. Johno, Y. Motoyama, T. Yui, S. S. Seomun, and M. Yamashita, *J. Chem. Phys.* 9, 2051 (1999).
29. M. Sadhukhan, S. Chakraborty, D. K. Modak, and B. K. Chaudhuri, *Philos. Mag. B* 72, 139 (1996).
30. M. Sadhukhan, D. K. Modak, and B. K. Chaudhuri, *J. Appl. Phys.* 85, 3477 (1999).
31. N. S. Prasad and K. B. R. Varma, *J. Mater. Chem.* 11, 1912 (2001).
32. N. S. Prasad, K. V. R. Varma, and S. B. Lang, *J. Phys. Chem. Solids* 62, 1299 (2001).
33. T. K. Kundu and D. Chakravorty, *Appl. Phys. Lett.* 67, 2732 (1995).
34. T. K. Kundu, M. Mukherjee, D. Chakravorty, and L. E. Cross, *J. Appl. Phys.* 83, 4380 (1998).
35. D. L. Li, L. B. Kong, L. Y. Zhang, and X. Yao, *J. Non-Cryst. Solids* 271, 45 (2000).
36. R. S. Meltzer, W. M. Yen, H. Zheng, S. P. Feofilov, M. J. Dejneka, B. M. Tissue, and H. B. Yuan, 64, 100201 (2001).
37. S. A. Basun, A. A. Kaplyanskii, and S. P. Feofilov, *Sov. Phys. Solid State* 34, 1807 (1992).
38. S. A. Basun, A. A. Kaplyanskii, and S. P. Feofilov, *Ferroelectrics* 143, 163 (1993).
39. D. Li, Y. Lin, L. Zhang, and X. Yao, *J. Non-Cryst. Solids* 261, 273 (2000).
40. S. Chottopadhyay, A. Pushan, V. R. Palkar, and M. Manu, *Phys. Rev. B* 52, 13177 (1995).
41. S. Chakraborty, M. Sadhukhan, D. K. Modak, K. K. Som, H. S. Maiti, and B. K. Chaudhuri, *Philos. Mag. B* 71, 1125 (1995).
42. A. A. Bhagat and T. M. Kamel, *Phys. Rev. B* 63, 012101 (2001).
43. S. Bhattacharya, H. Sakata, and B. K. Chaudhuri, *Phys. Rev. B*, 68, 16201 (2003).
44. S. Mukherjee, H. Sakata, B. K. Chaudhuri, S. Mollah, and H. D. Yang, *J. Appl. Phys.* 94, 1211 (2003).
45. D. C. Sinclair, T. B. Adams, F. D. Morrison, and A. R. West, *Appl. Phys. Lett.* 80, 2153 (2002).
46. Y. Lin, Y. B. Chen, T. Garret, S. W. Liu, C. L. Chen, L. Chen, R. P. Bontchev, A. Jacobson, J. C. Jiang, E. I. Meletis, J. Horwitz, and H.-D. Wu, *Appl. Phys. Lett.* 81, 631 (2002).
47. B. G. Kim, S. M. Cho, T. Y. Kim, and H. M. Jang, *Phys. Rev. Lett.* 86, 3404 (2002).
48. S. Chakraborty, M. Sadhukhan, D. K. Modak, and B. K. Chaudhuri, *J. Mater. Res.* 30, 5139 (1995).
49. H. Mori and H. Sakata, *J. Mater. Sci.* 31, 1621 (1996).
50. M. Sadhukhan, D. K. Modak, and B. K. Chaudhuri, *J. Chem. Phys.* 105, 11326 (1996).
51. Y. Hamanaka, N. Hayashi, A. Nakamura, and S. Omi, *J. Lumin.* 87–89, 859 (2000).
52. K. Uchida, S. Kaneko, S. Omi, C. Hata, H. Tanji, Y. Asahara, A. J. Ikushima, T. Tokizaki, and A. Nakamura, *J. Opt. Soc. Amer. B* 11, 1236 (1994).
53. F. Hache, D. Ricard, and C. Flytzanis, *J. Opt. Soc. Amer. B* 3, 1647 (1986).
54. M. Fujii, A. Mimura, S. Hayashi, Y. Yamamoto, and K. Murakami, *Phys. Rev. Lett.* 89, 206805 (2002).
55. A. Mimura, M. Fujii, S. Hayashi, D. Kovalev, and F. Koch, *Phys. Rev. B* 62, 12625 (2000).
56. M. Fujii, A. Mimura, S. Hayashi, and K. Yamamoto, *Appl. Phys. Lett.* 75, 184 (1999).
57. M. Fujii, A. Mimura, S. Hayashi, K. Yamamoto, C. Urakawa, and H. Ohta, *J. Appl. Phys.* 87, 1855 (2000).
58. H. Hofmann, P. Bowen, N. Jongen, and J. Lemaitre, *Scripta Mater.* 44, 2197 (2001).
59. B. K. Chaudhuri, *Chinese J. Phys.* 38, 211 (2000).
60. C. F. Drake, J. A. Stephens, and B. Yates, *J. Non-Cryst. Solids* 28, 61 (1986).
61. J. E. Shelpy, *J. Amer. Ceram. Soc.* 57, 436 (1975).
62. J. E. Shelpy, *J. Appl. Phys.* 46, 193 (1975).
63. T. Nishida, M. Ogatta, and Y. Takashina, *Bull. Chem. Soc. Jpn.* 59, 2204 (1986).
64. D. K. Modak, U. K. Mondal, H. Sakata, and B. K. Chaudhuri, *Sol. State Phys.* 43, 368 (2000).
65. S. Mondal and A. Ghosh, *Phys. Rev. B* 48, 9388 (1993).
66. V. Dimitrov and Y. Dimitrov, *J. Non-Cryst. Solids* 122, 133 (1990).
67. N. F. Borrelle, B. D. Mc Swain, and G. Su, *Phys. Chem. Glasses* 4, 119 (1963).
68. T. Quan and C. F. Adams, *J. Phys. Chem.* 70, 331 (1966).
69. A. Mansing and V. K. Dhawan, *J. Phys. C* 11, 3439 (1978).
70. H. Mori and H. Sakata, *J. Ceram. Soc. Jpn.* 102, 562 (1994).
71. J. Appel, "Solid State Physics," Vol. 21, p. 193. Academic Press, New York, 1968.
72. N. N. Ritten and T. Abraham, *Am. Cerm. Bull.* 76, 51 (1997).
73. R. R. Heikes, A. A. Maradudin, and R. C. Miller, *Ann. Phys. (N.Y.)* 8, 733 (1963).
74. D. Emin and T. Hilstein, *Ann. Phys. (N.Y.)* 53, 439 (1969).
75. D. Emin, *Phys. Rev. Lett.* 32, 303 (1975).
76. J. Schnakenberg, *Phys. Status Solidi* 28, 623 (1968).
77. G. P. Triberis and L. Friedman, *J. Phys. C* 18, 2281 (1985); G. P. Triberis, *J. Non-Cryst. Solids* 74, 1 (1979).
78. L. D. Zhang, H. F. Zhang, G. Z. Wang, C. M. Mo, and Y. Zhang, *Phys. Status Solidi A* 157, 483 (1996).
79. M. Sayer and A. Mansingh, *Phys. Rev. B* 6, 4629 (1972).
80. M. H. Cohen, *J. Non-Cryst. Solids* 4, 391 (1970).
81. S. R. Elliott, *Adv. Phys.* 36, 135 (1987).
82. A. R. Long, *Adv. Phys.* 31, 553 (1982).
83. M. Polak, *Phys. Rev.* 138, 1822 (1965); *Philos Mag.* 23, 517 (1971).
84. A. L. Efros, *Philos. Mag. B* 43, 829 (1981).
85. G. E. Pike, *Phys. Rev. B* 6, 1572 (1972).
86. D. K. Modak, Ph.D. Thesis, Jadavpur University, Kolkata, India, 2000.
87. H. Fröhlich, "Theory of Dielectrics." Oxford Universsity Press, 1958.
88. P. B. Macedo, C. T. Moynighan, and R. Bose, *Phys. Chem. Glasses* 13, 171 (1972).
89. K. Cole and R. H. Cole, *J. Chem. Phys.* 9, 341 (1941).
90. W. L. Zhong, Y. G. Wang, P. L. Zhang, and B. D. Qu, *Phys. Rev. B* 50, 689 (1994).
91. W. L. Zhong, B. Jiang, P. L. Ma, H. M. Cheng, Z. H. Yang, and L. X. Li, *J. Phys.: Condens. Matter* 5, 2619 (1993).
92. W. H. Dumbaugh, *Phys. Chem. Glasses* 27, 119 (1986).
93. R. Sato, T. Komatsu, and K. Matusita, *J. Non-Cryst. Solids* 160, 180 (1993).
94. H. Mori, H. Matsumo, and H. Sakata, *J. Non-Cryst. Solids* 276, 78 (2000).
95. T. Szorenyi, K. Bali, and I. Hevesi, *J. Non-Cryst. Solids* 70, 297 (1985).
96. K. K. Som and B. K. Chaudhuri, *Phys. Rev. B* 41, 1581 (1990).
97. S. Chatterjee, S. Banerjee, S. Mollah, and B. K. Chaudhuri, *Phys. Rev. B* 53, 5942 (1996).
98. T. Komatsu, R. Sato, C. Hirose, K. Matusita, and Y. Matushita, *Jpn. J. Appl. Phys.* 27, L2293 (1988).
99. Y. Akamatsu, M. Tatsumisago, N. Tohge, S. Tsuboi, and T. Minami, *Jpn. J. Appl. Phys.* 27, L1696 (1988).
100. S. Mollah, K. K. Som, K. Bose, and B. K. Chaudhuri, *Phys. Rev. B* 46, 11075 (1992).
101. L. Murawski and R. J. Barczynski, *J. Non-Cryst. Solids* 185, 84 (1995).
102. K. Shantha and K. B. R. Varma, *J. Amer. Ceram. Soc.* 83, 1122 (2000).

103. A. Herczog, *J. Amer. Ceram. Soc.* 47, 107 (1964).
104. V. F. Janas and A. Safari, *J. Amer. Ceram. Soc.* 78, 2345 (1995).
105. H. I. Hsiang and F. S. Yen, *Jpn. J. Appl. Phys.* 33, 3991 (1994).
106. T. Yamamoto, K. Urabe, and H. Banno, *Jpn. J. Appl. Phys.* 32, 4272 (1993).
107. M. H. Lee, A. Halliyal, and R. E. Newnham, *J. Amer. Ceram. Soc.* 72, 986 (1989).
108. G. S. Gong, A. Safari, S. J. Jang, and R. E. Newnham, *Ferroelectr. Lett.* 5, 131 (1986).
109. F. Duan, C. P. Fang, Z. Y. Ding, and H. Q. Zhu, *Mater. Lett.* 34, 184 (1998).
110. B. Houng, C. Y. Kim, and M. J. Haun, *IEEE Trans. Ultra. Ferro. Freq. Contl.* 47, 808 (2000).
111. K. Nihara, *J. Ceram. Soc. Jpn.* 99, 974 (1991).
112. T. Nagai, H. J. Hwang, M. Yasuoka, M. Sando, and K. Nihara, *J. Kor. Phys. Soc.* 32, S1271 (1998).
113. C. E. Borsa, S. Jiao, R. I. Todd, and R. J. Brook, *J. Microsc.* 77, 305 (1995).
114. L. C. Stearns, J. Zhao, and M. P. Harmer, *J. Euro. Ceram. Soc.* 10, 473 (1992).
115. H. J. Hwang, S. Ueda, and K. Nihara, *Ceram. Trans.* 44, 399 (1994).
116. T. Nagai, H. J. Hwang, M. Yasuoka, M. Sando, and K. Nihara, *Jpn. J. Appl. Phys.* 37, 3377 (1998).
117. H. Hyuga, Y. Hayashi, T. Sekino, and K. Nihara, *Nanostruct. Mater.* 9, 547 (1997).
118. T. Ohji, Y.-K. Jeong, Y.-H. Chao, and K. Nihara, *J. Amer. Ceram. Soc.* 81, 1453 (1998).
119. (a) S. R. Panteny, R. Stevens, and C. R. Bowen, preprint (2002). (b) S. R. Panteny, C. R. Bowen, and R. Stevens, *Key Eng. Mater.* 206, 1313 (2002).
120. A. M. Morales, M. Gonzales, and J. M. Hruby, in "Proc. 7th Foresight Conference on Molecular Nanotechnology," 1999.
121. K. Wu and W. Schulze, *J. Amer. Ceram. Soc.* 75, 3385 (1992).
122. H. J. Hagemann, *J. Phys. C: Solid State Phys.* 11, 3333 (1978).
123. Q. Zhang, H. L. Chan, Q. Zhou, and C. L. Choy, *Mater. Res. Innovat.* 2, 283 (1999).
124. H. L. W. Chan, Y. Chen, C. L. Choy, *J. Amer. Ceram. Soc.* 81, 1231 (1998).
125. Q. F. Zhou, H. L. W. Chan, and C. L. Choy, *J. Mater. Proc. Techn.* 63, 281 (1997).
126. Q. F. Zhou, L. S. Yin, J. X. Zhang, H. L. W. Chan, and C. L. Choy, *J. Kor. Phys. Soc.* 32, S1380 (1998).
127. N. M. Shorrocks, A. Patel, M. J. Walker, and A. D. Parasons, *Microelectron. Eng.* 29, 59 (1995).
128. B. Willing, M. Kohli, K. Brooks, P. Muralt, and N. Setter, *Ferroelectrics* 201, 147 (1997).
129. S. L. Grishick, W. W. Gerberich, J. V. R. Heberlein, and P. H. McMurry, NSF Foundation Nanostructure grant 9871863.
130. F. Di Fonzo, A. Gidwani, M. H. Fan, D. Neumann, D. I. Iordanoglou, J. V. R. Heberlein, P. H. McMurry, S. L. Girshick, N. Tymiak, W. W. Gerberich, and N. P. Rao, *Appl. Phys. Lett.* 77, 910 (2000).
131. I. Harris and T. Bailey, in "Nd–Fe Permanent Magnets: Their Present and Future Applications" (I. Mitchell, Ed.). Elsevier, England, 1985.
132. R. A. Mayers, N. Mukherjee, and S. R. J. Bruek, *Opt. Lett.* 16, 1732 (1991).
133. J. Wasylak, J. Kucharski, and I. V. Kityk, *J. Appl. Phys.* 85, 425 (1999).
134. I. V. Kityk, E. Golis, J. Fihpecki, J. Wasylak, and V. M. Zacharko, *J. Mater. Sci.* 14, 1292 (1995).
135. T. E. Tasi, C. G. Askins, and E. J. Friebeble, *Appl. Phys. Lett.* 61, 390 (1992).
136. U. Obsterberg and W. Margulis, *Opt. Lett.* 11, 516 (1986).
137. Y. Fuji, B. S. Kawasaki, K. O. Hill, and D. C. Johnson, *Opt. Lett.* 5, 48 (1980).
138. Y. Sasaki and Y. Ohmori, *Appl. Phys. Lett.* 39, 466 (1981).
139. C. Flystanis, in "Quantum Electronics: A Treatise," Vol. 1a. Academic Press, New York, 1995.
140. S. Komaren, *J. Mater. Chem.* 2, 1719 (1992).
141. W. F. Smith, in "Principles of Materials Science and Engineering." McGraw–Hill, Singapore, 1986.
142. R. W. Siegel, *Nanostruct. Mater.* 3, 1 (1993).
143. G. Piccaiuga, A. Carrias, G. Ennas, and A. Musinu, *J. Mater. Res.* 13, 1 (2000) and references therein.
144. R. Uyeda, *Progr. Mater. Sci.* 35, 1 (1991).
145. L. E. Brus, *J. Chem. Phys.* 79, 5566 (1983).
146. A. P. Alivisatos, *Science* 271, 933 (1996).
147. D. L. Klein, R. Roth, A. K. Lim, A. P. Alivisatos, and P. Meeuen, *Nature* 389, 699 (1997).
148. V. L. Colvin, M. C. Schlamp, and A. P. Alivisatos, *Nature* 370, 354 (1994).
149. R. N. Bhargava, *J. Lumin.* 70, 85 (1996).
150. M. E. Lines and A. M. Glass, "Principles and Applications of Ferroelectricity and Related Materials." Clarendon, Oxford, 1979.

Encyclopedia of Nanoscience and Nanotechnology
www.aspbs.com/enn

Ferroelectric Nanodomains

A. Gruverman
North Carolina State University, Raleigh, North Carolina, USA

CONTENTS

1. INTRODUCTION

Ferroelectrics are a class of polar dielectrics characterized by a reversible spontaneous polarization [1]. Ferroelectrics possess a unique set of physical properties, that makes them attractive for application in a variety of electronic devices, such as nonvolatile memories and microelectromechanical systems. Their piezoelectric, electrooptic, and nonlinear optical properties are largely determined by arrangement of ferroelectric domains. Domains form during phase transition to relieve mechanical stress associated with transformation strains and to minimize the depolarization energy associated with the spontaneous polarization. The presence of the domain structure is one of the fundamental properties of ferroelectrics. Domain arrangements reflect the effects of sample conductivity, mechanical strains, and imperfections such as dislocations, vacancies, and impurities. Polarization reversal under an external electric field proceeds via domain nucleation and motion of domain walls. Direct observation of ferroelectric domains provides a wealth of information on static and dynamic properties of ferroelectrics. A number of techniques, such as polarized optical microscopy, electron microscopy, surface decoration, pyroelectric emission, and etching, have been employed for visualization of domain structures [1]. The main limitations of these methods have been either low spatial resolution or the destructive nature of the visualization techniques.

Rapid development of electronic devices based on ferroelectric thin films [2] generated a strong need for studies of ferroelectric properties at the nanoscale. Fortunately, this need appeared at the same time as new techniques for nanoscale characterization of materials became available. Specifically, scanning force microscopy (SFM) has emerged as a powerful tool for high-resolution characterization of virtually all types of materials, such as metals, semiconductors, dielectrics, polymers, and biomolecules [3–5]. In the field of ferroelectricity, the application of the SFM technique resulted in a real breakthrough providing an opportunity for nondestructive nanoscale visualization of domain structures in ferroelectric thin films. The employment of SFM makes possible nanoscale mapping of the surface potential, evaluation of local electromechanical properties, and dielectric constant measurements. In other words, characterization by means of SFM provides crucial information on the dielectric properties of ferroelectrics with unprecedented spatial resolution. Another promising development related to SFM is modification of the ferroelectric properties at the nanoscale via local polarization reversal induced by a conductive SFM tip.

Figure 1 illustrates the pace with which publications on SFM studies of ferroelectrics have been evolving over the last 10 years. An increasing number of papers indicates the growing importance of SFM in the field of ferroelectricity. Research groups in the United States, Europe, and Asia are actively using SFM for high-resolution characterization of ferroelectric materials in both bulk and thin-layer forms.

The purpose of this chapter is to review the application of SFM to nanoscale studies of ferroelectric domains. The wide variety of physical properties of ferroelectric materials allows the use of various SFM modes for domain imaging, depending on which physical mechanism provides the higher domain contrast. In this paper, experimental results related to different mechanisms of domain contrast are presented along with consideration of limitations and advantages of each SFM-based approach. Particular emphasis is given to one specific imaging method, namely, piezoresponse force microscopy (PFM), which proved to be the most effective approach for nanoscale studies of ferroelectric domains in bulk crystals and thin films. Future directions in nanoscale studies of ferroelectric materials are discussed.

ISBN: 1-58883-059-4/$35.00

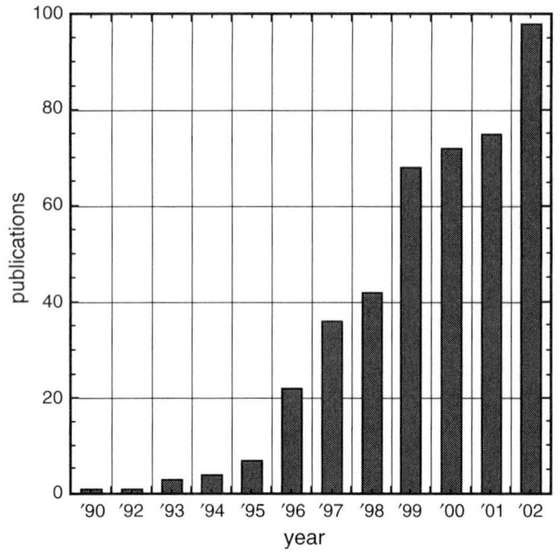

Figure 1. Number of publications on SFM studies of ferroelectrics per year.

2. FERROELECTRIC DOMAIN IMAGING IN SCANNING FORCE MICROSCOPY (SFM)

Scanning force microscopy can be considered as a combination of a surface force apparatus and a surface profilometer, as it is based on local monitoring of the interaction forces between a probing tip and a sample [4]. The forces acting on the tip after it has approached the sample surface cause a deflection of the cantilever according to Hooke's law. This deflection can be detected optically or electrically with subangstrom accuracy and is controlled by a feedback device, which regulates the vertical position of the tip over the surface. By keeping the deflection constant while scanning the sample, a three-dimensional map of the surface topography can be obtained. Besides this constant force mode, many other modes have been developed. The response of the cantilever to the externally modulated force (for example, due to an applied ac bias) can be used to map such physical properties as mechanical stiffness, friction, electric fields, density of electronic states, and so forth.

Depending on the type of tip-sample force interaction (attracting or repelling), the SFM can operate in two different regimes: noncontact or contact, respectively. In the noncontact regime, the tip is scanned over the surface at a distance of 10–100 nm, which is controlled, for example, by monitoring the resonant frequency of the cantilever [6]. The tip-sample interaction in this regime is dominated by the long-range polarization and electrostatic forces. Because of this feature, noncontact SFM can be used for ferroelectric domain imaging by detecting the electrostatic field of the surface polarization charges. This mode of SFM is called electrostatic force microscopy (EFM) [7]. Quantitative information on local surface potential related to spontaneous polarization can be obtained by means of scanning surface potential microscopy (SSPM), or Kelvin probe force microscopy (KPFM) [8, 9], a technique complementary to EFM. Scanning capacitance microscopy (SCM) can be used to generate an image of trapped charges by measuring local tip-sample capacitance [8, 10]. SCM is not necessarily a noncontact mode of operation, as the probing tip can be placed in direct contact with the sample surface to measure the sample capacitance. General disadvantages of noncontact methods include susceptibility to screening effects, sensitivity to sample surface conditions, and low resolution in ambient air.

In the contact regime, the probing tip is in mechanical contact with the sample surface and senses repulsive short-range forces. The difference in mechanical, structural, electrochemical, dielectric, and piezoelectric properties of opposite ferroelectric domains can provide domain contrast in the SFM contact regime. Contact SFM methods of domain imaging include a topographic mode of atomic force microscopy (AFM), lateral (friction) force microscopy (LFM), piezoresponse force microscopy (PFM), and scanning nonlinear dielectric microscopy (SNDM). SFM-based methods of ferroelectric domain imaging are summarized in Table 1 and are discussed in the following sections.

3. NONCONTACT DOMAIN IMAGING

3.1. Electrostatic Force Microscopy

Imaging of ferroelectric domains in the noncontact mode is based on the detection of the modulated electrostatic interaction force between the probing tip and polarization charges. There are two modulation methods in the noncontact imaging.

In the conventional approach, the cantilever is made to oscillate near its resonant frequency by use of a piezoelectric bimorph. When the tip is brought close to the surface, the attractive force gradient acting on the tip alters the force constant k_0 of the cantilever as $k' = k_0 - \partial F/\partial z$. This in turn leads to a change in the resonant frequency and in the vibration amplitude. A feedback loop adjusts the tip-sample distance to keep the amplitude of oscillation constant [6]. Obviously, in the case of the ferroelectric sample, there is an electrostatic contribution to the attractive force due to the Coulomb interaction between a surface polarization charge and an image charge Q_t in the probing tip. As the tip crosses the wall, it experiences a change in the force gradient, and the feedback loop alters the tip-sample distance to keep the gradient and the vibration amplitude constant. This produces a variation of contrast in the feedback signal image, which can be interpreted as an image of the domain wall (Fig. 2) [11]. Since the Coulomb force is

Table 1.

Ferroelectric domain imaging by SFM	
Non-contact imaging	
EFM	Electrostatic Force Microscopy
SSPM/KPFM	Scanning Surface Potential Microscopy/ Kelvin Probe Force Microscopy
SCM	Scanning Capacitance Microscopy
Contact imaging	
LFM	Lateral (friction) Force Microscopy
PFM	Piezoresponse Force Microscopy
AFM	Atomic Force Microscopy
SNDM	Scanning Nonlinear Dielectric Microscopy

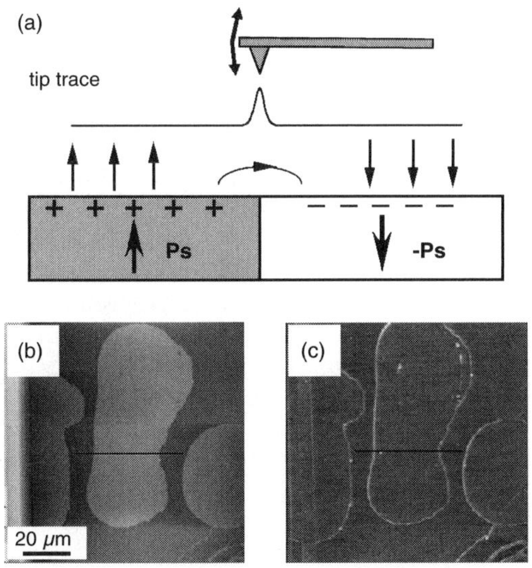

Figure 2. (a) Sketch illustrating a mechanism of domain wall image formation in EFM. (b) Topographic and (c) EFM images of the etched lead titanate crystal. The domain structure consists of antiparallel *c*-domains, which appear with the same contrast in the EFM image, and domain walls appear as bright lines due to the surface charge gradient at the 180° domain walls. Topographic contrast is due to the differently etched surfaces of positive and negative domains Reprinted with permission from [11], P. Lehnen et al., *J. Phys. D: Appl. Phys.* 33, 1932 (2000). © 2000, Institute of Physics.

proportional to the product of the polarization and image charges, the force gradient signal provides information only on the polarization magnitude and not the sign. This implies that the contrast of opposite 180° domains will be the same and that only domain walls will be visible because of the spatial variation of the charge density in the vicinity of a 180° domain boundary.

Using this approach, the pioneering work on SFM domain imaging has been performed by Saurenbach and Terris in a single crystal of gadolinium molybdate [12]. A rather wide image of the domain wall (about 10 μm, while the actual wall thickness is on the order of several lattice constants) had been attributed to the tip-sample separation and the finite size of the tip, which broadened any sharp changes in the force gradients. Later works of Luthi et al. [13–16] and Eng et al. [17–19] demonstrated that lateral resolution in EFM can be significantly improved: in single crystals with cleaved polar surfaces, such as GASH and TGS, the width of the walls measured in EFM was in the range of 8 to 80 nm.

However, this method of domain imaging may suffer from the cross-talk with other sources of the force gradient, such as van der Waals forces. As a result, the force gradient image is usually a superposition of domain and surface topographic features. In the case of domains of irregular shape and complex surface topography, the interpretation of the EFM images could be quite difficult. One of the ways to alleviate this problem is to use a liftmode technique, which combines the contact and noncontact modes. In this approach, the tip scans each line twice: first, recording the topography in the contact regime, and second, retracing the topographic line at the predetermined height while detecting the variations in the vibration amplitude. In this case, since the tip-sample distance is kept constant during the second scan, the force gradient is related to the surface charge.

Another method of circumventing the cross-talk effect is by using a dual-modulation scheme, developed for the detection of static surface charges by two IBM groups [6, 7, 20, 21]. In this approach, also used by Saurenbach and Terris, the cantilever is additionally modulated by an ac voltage $V_t = V_{ac} \cos \omega t$ applied between the probing tip and the bottom electrode. The frequency of the electrostatic modulation is chosen so that it is well below the frequency of mechanical modulation to avoid any resonance effect. In this case, an additional capacitance component will be introduced to the force acting on the tip, which can be written as

$$F = F_{cap} + F_{coul} = \frac{1}{2}\frac{\partial C}{\partial z}V_t^2 + \frac{Q_s Q_t}{4\pi\varepsilon_0 z^2} \qquad (1)$$

where C is the tip-surface capacitance and z is the tip-surface separation. The total charge induced in the tip will be $Q_t = -(Q_s + Q_e) = -(Q_s + CV_t)$. The force gradient can be expressed as

$$\frac{\partial F}{\partial z} = \left[\frac{V_{ac}^2}{4}\frac{\partial^2 C}{\partial z^2} + \frac{Q_s}{2\pi\varepsilon_0 z^2}\right] + \frac{Q_s V_{ac}\cos \omega t}{2\pi\varepsilon_0 z^2}\left[\frac{C}{z} - \frac{1}{2}\frac{\partial C}{\partial z}\right] + \frac{V_{ac}^2}{4}\frac{\partial^2 C}{\partial z^2}\cos 2\,\omega t \qquad (2)$$

Three terms in Eq. (2) represent a dc component and first and second harmonics of the force gradient. In the absence of net polarization ($Q_s = 0$), the only oscillating signal that can be measured is the signal oscillating at 2ω. For electrically polarized samples ($Q_s \neq 0$), the first harmonic is nonzero and can be used to obtain information on the surface charge distribution. This method of domain imaging, which has been used by a number of groups [22–31], is sensitive to the charge only and allows determination of its sign by monitoring the phase of the first harmonic signal with a lock-in.

Figure 3 illustrates an application of this technique to domain imaging in a tetragonal $Pb(Zr,Ti)O_3$(PZT) film. Prior to the imaging, a small part of the film was polarized by scanning with a positively biased tip, and two lines were written across this area with a tip under a negative bias. The positively and negatively polarized domains appear as bright and dark areas in Figure 3, respectively, because of uncompensated polarization charges of newly switched domains. At the same time, this image illustrates one of the limitations of the EFM method, namely, an unwritten area shows only slight variation of the contrast, although it contains as-grown domains with polarization normal to the film surface, which was confirmed by another SFM method. Furthermore, the contrast of the written structure gradually fades and almost disappears within several hours. This behavior is due to the accumulation of surface charge on the film surface, which neutralizes polarization charges and causes a uniform contrast over the surface due to zero net charge. Therefore, although the EFM charge detection mode has the advantage of distinguishing between topographic features and the electrostatic signal, the domain contrast in this mode can easily be obscured. In addition, a surface contamination layer

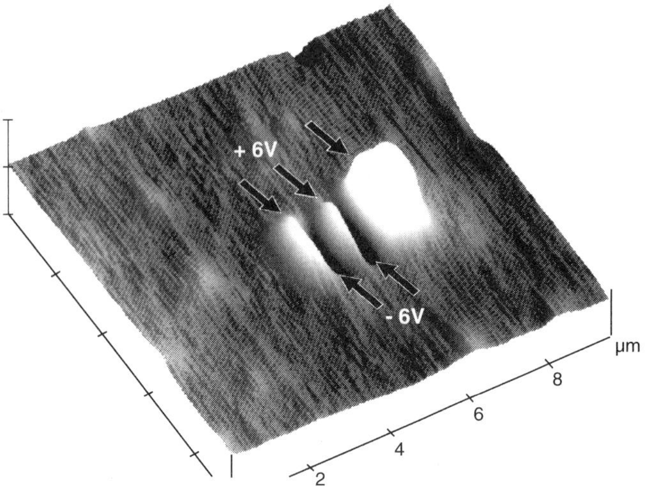

Figure 3. EFM charge image of a PZT film. Bright and dark areas correspond to positively and negatively poled regions, respectively. The scanning area is $10 \times 10\ \mu m^2$. Image courtesy of H. Yokoyama, National Institute of Advanced Industrial Science and Technology, Japan.

always present on the sample surface under ambient conditions can change or even conceal the image of real domain structure. Conducting experiments in a vacuum or in an inert atmosphere can eliminate these detrimental effects and make possible detailed investigation of the spatial distribution of polarization charges and stray electric fields at ferroelectric surfaces. Another problem is that quantification of the EFM signal is problematic since the tip-sample capacitance is difficult to measure accurately.

3.2. Scanning Surface Potential Microscopy

A method closely related to EFM is the SSPM method of domain imaging based on detection of a surface potential associated with spontaneous polarization. In the presence of a surface potential, the electrostatic force acting on the tip depends on the tip-sample potential difference V_s:

$$F = \frac{1}{2}(V_t - V_s)^2 \frac{\partial C}{\partial z} + \frac{Q_s Q_t}{4\pi\varepsilon_0 z^2} \quad (3)$$

The absolute value of the surface potential can be measured with the so-called nulling method [28]. In this approach, the probing tip is additionally biased by a dc voltage, so that $V_t = V_{dc} + V_{ac} \cos \omega t$. In the case of complete screening of polarization charges by adsorbed surface charges ($Q_s = 0$), the three components of the electrostatic force can be written as

$$F_{dc} = \frac{1}{2}\left((V_{dc} - V_s)^2 + \frac{1}{2}V_{ac}^2\right)\frac{\partial C}{\partial z} \quad (4)$$

$$F_{1\omega} = (V_{dc} - V_s)V_{ac}\frac{\partial C}{\partial z} \cos\ \omega t \quad (5)$$

$$F_{2\omega} = \frac{1}{4}V_{ac}^2 \frac{\partial C}{\partial z} \cos\ 2\omega t \quad (6)$$

The first harmonic of the electrostatic force is then nullified by adjusting the constant bias on the tip so that $V_{dc} = V_s$. By detecting the nullifying V_{dc} value during scanning, a surface potential image can be obtained. This approach has been extensively used by Kalinin and Bonnell to study polarization screening processes in ferroelectrics [32–34]. Figure 4 shows schematic diagrams of domain structure, surface topography, and surface potential in a single crystal of barium titanate. Corrugated surface topography is an indication of *a*- and *c*-domain regions (with in-plane and out-of-plane spontaneous polarization, respectively). An SSPM image provides additional information on the domain structure: inverted potential contrast within the *c*-domain region indicates the presence of antiparallel *c*-domains. The low value of the surface potential change across the 180° domain boundaries (on the order of 150 mV) is due to almost complete screening of polarization by surface charges. The SSPM approach has also been used for nanometer detection of two-dimensional potential profiles at the semiconductor interfaces [35].

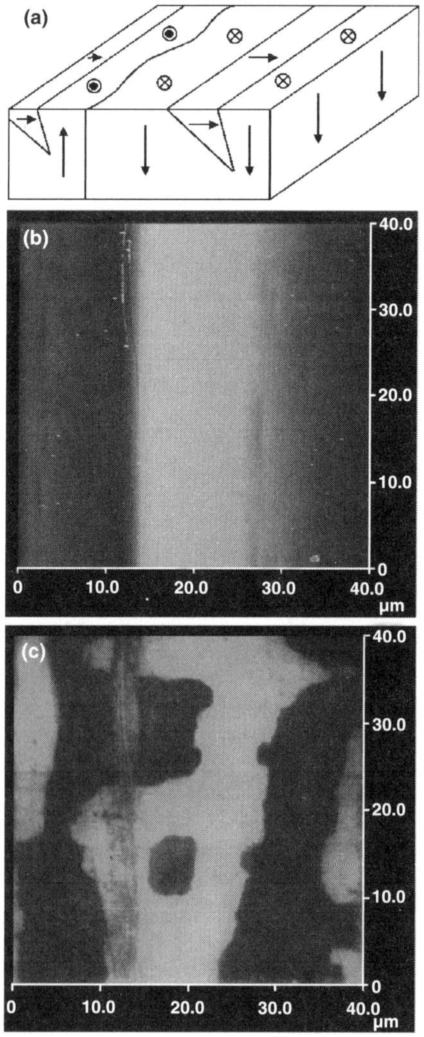

Figure 4. SSPM imaging of the barium titanate surface. (a) Schematic diagrams of domain structure. (b) Surface topography. (c) *a*-domain region with *c*-domain wedges. Reprinted with permission from [32], S. V. Kalinin and D. A. Bonnell, *Phys. Rev. B* 63, 125411 (2001). © 2001, American Institute of Physics.

3.3. Scanning Capacitance Microscopy

Additional information on local electronic properties of ferroelectrics can be obtained by monitoring the second harmonic component of the electrostatic force signal. Since it is proportional to the capacitance gradient (Eq. (6)), this method is often referred to as scanning capacitance microscopy (SCM). This technique has been used to map carrier distribution in gate oxides and dopant profiles in nonuniformly doped semiconductors [36, 37]. In an original SCM setup, where the tip is held at some distance from the surface [8, 38], the constant capacitance contour contains information on the thickness of the sample, its dielectric constant, and charge carrier concentration underneath the tip. Alternatively, surface topography and capacitive force imaging can be obtained sequentially, that is, the capacitive force image can be obtained with the tip scanned at a predetermined distance from the surface following the previously measured topography profile. In this case, the measured capacitance signal includes information only on the dielectric constant and charge concentration.

In a more widely used SCM configuration, the probing tip is placed in direct contact with the sample surface, which allows independent recording of the surface topography. The tip-sample capacitance is measured by a capacitance sensor electrically connected to the cantilever. This approach has a very strong potential for charge storage and detection of trapped charge in semiconductor heterostructures [10, 39, 40]. Application of a variable dc bias to the probing tip results in the shift of the free carriers in a semiconductor and formation of the depletion or accumulation layer underneath the tip, which is equivalent to the capacitance change. A modulation ac voltage is used to detect the capacitance derivative $\partial C/\partial V$ as a function of the tip dc bias.

With respect to ferroelectrics, the SCM method can be considered as a complementary approach to SSPM in the sense that it allows estimation of the carrier concentration in the subsurface regions of ferroelectric samples and evaluation of effect of the free carriers on polarization screening. Figure 5 shows the SCM image of the $SrBi_2Ta_2O_9$ film acquired at a dc bias of -1 V. Prior to the imaging, a central part of the film was polarized by a positive voltage applied to the tip. The observed contrast is due to the different effect of the imaging dc bias on a screening charge layer in oppositely poled regions. A higher concentration of the electrons in the positively poled area yields a smaller layer thickness and, therefore, larger capacitance, which leads to higher SCM contrast. Measurements of the $\partial C/\partial V$–V curves yield a voltage shift between oppositely poled regions:

$$\Delta V = P_s \frac{\varepsilon_i W + \varepsilon_f d_i}{\varepsilon_0 \varepsilon_i \varepsilon_f} \qquad (7)$$

where ε_i and ε_f are the dielectric constants of a surface dielectric layer and a ferroelectric sample, respectively, and d_i and W are the thicknesses of the dielectric layer and a depletion (accumulation) layer, respectively.

Still, quantitative interpretation of the SCM data is difficult since it requires an accurate account of the parameters of the dielectric layer and nonlocal cantilever contribution to the capacitance signal. One attractive possibility would be to use SCM and SSPM on cross-sectional samples of ferroelectric capacitors, which would allow characterization of the depletion layers in the vicinity of the film/electrode interfaces [37]. A likely problem of relatively low spatial resolution of SCM can be overcome by using a wedge-like sample, which was recently suggested by Lu et al. [41]. In this case, the depth profiling of the carrier concentration can be performed by changing the lateral position of the probing tip at the slope of the wedge sample.

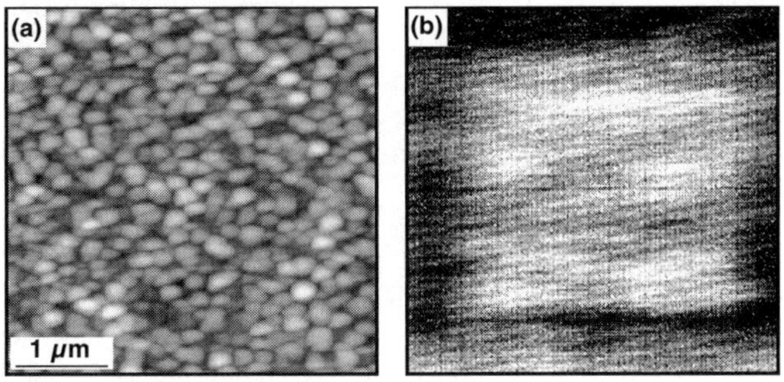

Figure 5. SCM image of the $SrBi_2Ta_2O_9$ film acquired at a dc bias of $+1$ V. Prior to the imaging a central part of the film was polarized by a positive voltage. The higher concentration of the electrons in the positively poled area yields a smaller depletion layer thickness and, therefore, a larger capacitive force. Image courtesy of Y. Haga, Sony Corporation Research Center, Japan.

4. CONTACT DOMAIN IMAGING

Contact domain imaging can be divided into static and dynamic, or voltage-modulated, methods (Table 2). Static imaging methods, such as a topographic mode of AFM and lateral force microscopy, make use of the surface domain-dependent properties of ferroelectrics, such as surface corrugations associated with the presence of different domains, difference in structure of polar faces of opposite domains, variations in friction forces, and so on. Dynamic methods, which include PFM and SNDM, are based on voltage modulation and detection of the electrical and mechanical responses of opposite ferroelectric domains to the applied ac voltage. The contact SFM imaging methods provide significant advantages, such as high lateral resolution (well below 10 nm), a possibility of the three-dimensional reconstruction of domain structure, and effective control of nanodomains. However, interpretation of the domain images

Table 2.

SFM Contact imaging	
Static imaging	
LFM	Lateral (friction) Force Microscopy
AFM	Atomic Force Microscopy
Voltage modulated imaging	
PFM	Piezoresponse Force Microscopy
SNDM	Scanning Nonlinear Dielectric Microscopy

could be complicated by cross-talk between different mechanisms involved in the domain contrast formation.

4.1. Domain Imaging via Polarization-Dependent Friction

The first imaging of antiparallel domains via polarization-dependent friction was performed by Luthi et al. [13, 14]. With this approach, domain structure has been revealed on freshly cleaved surfaces of single crystals of GASH and TGS [17, 19, 42–44]. In most of the papers, the imaging mechanism is attributed to the permanent charging of the probing tip by a ferroelectric surface. Electrostatic tip-sample interaction causes an additional contribution to the lateral force acting on the tip and results in different torsion of the cantilever when the tip is scanning surfaces of opposite 180° domains. The lateral resolution has been reported to be less than 8 nm. The image contrast depends on the scanning direction and can be reversed by switching from forward to backward scan, which is an indication of the tribological effect rather than of the surface morphology (Fig. 6). A complementary mechanism of the domain contrast in LFM can be the difference in surface structure of opposite domains, which gives rise to different friction coefficients of the regions occupied by these domains [45, 46]. One of the greatest limitations of this method is that it is extremely sensitive to the surface conditions affecting sample tribological properties: adhesion layers, interfacial wetting, contamination, and roughness. As a result, its application is mainly limited to crystals with atomically flat surfaces of cleavage planes, such as GASH and TGS. On atomically flat terraces of the freshly cleaved surfaces, even small variations in the friction forces can easily be detected. However, even in these crystals, friction images exhibit a wide diversity and should be interpreted with caution. For example, because of the different orientation of molecules on the chemically homogeneous terraces making up the surface of individual domains, frictional contrast can occur not only between opposite domains, but also inside individual domains [46]. Long exposure to the ambient environment could lead to deterioration of surface quality and to degradation of the domain contrast.

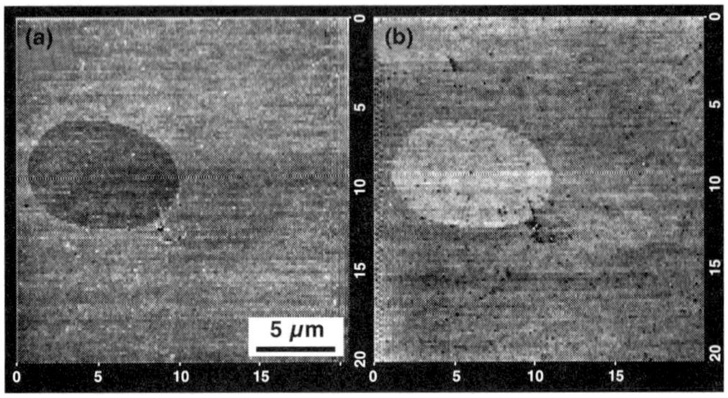

Figure 6. Friction force micrograph of a GASH cleavage surface. An imbedded domain exhibits opposite contrast compared with the surroundings because of the difference in tip-surface friction forces. Scan direction: (a) left to right; (b) right to left.

4.2. Domain Imaging via Surface Topography

The conventional topography mode of SFM has been used for domain studies via investigation of the domain-related surface morphology of ferroelectrics. There are several mechanisms that can provide morphological contrast of ferroelectric domains: (1) topographic steps at domain boundaries due to the structural difference between positive and negative ends of domains [47]; (2) inclination of the *cleaved* surfaces according to the polarity of domains and the direction of cleavage propagation [47–49]; (3) surface corrugation at the junction of *a*- and *c*-domains in perovskite ferroelectrics [11, 17, 50–58].

Topographic steps several angstroms in height have been observed at the 180° domain boundaries on the cleaved surface of TGS crystals by Bluhm et al. [45] and Eng et al. [25, 59]. This effect was explained by the relative shift of atom positions in opposite domains. An additional factor, which can affect the surface topography and reveal domain structure, is the different etching behaviors of positive and negative domains. Selective etching with subsequent topographic imaging has been used to reveal nanoscale domains in $LiNbO_3$ crystals [29, 60]. For hydrophilic materials such as TGS and GASH, exposing a sample to a humid atmosphere can reveal domains due to selective surface etching by water vapor. Topographic imaging of the etched surface can be used for identification of domain polarity [15, 43, 45]. At the same time, this feature of TGS could be a complicating factor: fine morphological structures of ferroelectric domains on opposite cleavage faces of TGS vary strongly, even for domains of equal polarity. Etching of positive domains can result in both etch hole formation and recrystallization of islands from the saturated solution at the surface, depending on which molecular layer is exposed to ambient air after the cleavage. Etch patterns can easily be confused with domain structure.

Another mechanism, which can lead to domain topographic contrast, is surface corrugation at the 90° domain walls separating domains with in-plane polarization (*a*-domains) and out-of-plane polarization (*c*-domains). With this approach, *a*-*c*-domain structure has been observed in $BaTiO_3$ and $PbTiO_3$ crystals and PZT thin films [11, 51–58]. A difference between *a* and *c* lattice constants of the tetragonal cell produces a lattice distortion at the junction of *a*- and *c*-domains and surface inclination with a characteristic angle determined as $\theta = \pi/2 - 2\arctan(a/c)$ (Fig. 7). Domain structures consisting of alternating *a*- and *c*-domains arise to reduce elastic energy in mechanically strained ferroelectric samples, such as epitaxial thin films [61–65]. SFM can provide a simple and nondestructive method for studying domain patterns in epitaxial ferroelectric films by topographic imaging of their surfaces. Figure 7 shows a topographic image of a $Pb(Zr_{0.20},Ti_{0.80})O_3$ film deposited by laser ablation on a $LaAlO_3$ substrate. The *a*-*c*-domain arrangement appears as a rectangular structure with height variations in the range of 1.5–3.5 nm, occurring as a result of twinning between *a*- and *c*-domains. A value of surface tilting of approximately 2° was measured at the 90° domain boundaries, which is consistent with the *c*/*a* ratio of the unit cell of the film at room temperature. Considering

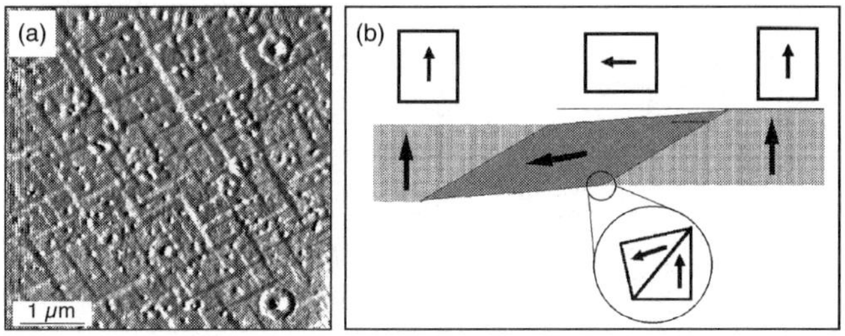

Figure 7. (a) Topographic image of a PZT film on a $LaAlO_3$ substrate showing a rectangular structure of *a*- and *c*-domains. (b) Lattice matching at the 90° domain wall through the formation of a strained unit cell. Twinning occurs on the (101) plane. Arrows indicate the polarization vectors [86]. Reprinted with permission from [86], A. Gruverman et al., *Integrated Ferroelectrics* 19, 49 (1998). © 1998, Gordon and Breach Science.

that the tetragonality of a PZT film with a Zr/Ti ratio of 20/80 is about 4%, the 90° domain wall should stand at an angle of 43° 44′ to the polar direction. The relative tilt between the surfaces of two adjacent *a*- and *c*-domains, therefore, will be 2° 32′, which fits fairly well with the value obtained from the SFM measurements.

There are obvious limitations to the applicability of the SFM topographic mode for domain imaging. Any treatment of the surface during sample preparation inevitably eliminates the fine structure of morphological steps associated with domain patterns. Therefore, only crystals with cleavage planes, like TGS and GASH, are suitable for SFM topographic studies. Also, since formation of 180° domain walls does not influence the surface topography, this method is not applicable for imaging of domain structure consisting of antiparallel *c*-domains, which is of direct interest for investigation of the polarization reversal processes in ferroelectrics. In some limited cases, the topographic mode can be used for visualization of opposite 180° domains via detection of the static piezoelectric sample deformation induced by an external dc bias [66], as will be discussed in one of the following sections.

4.3. Domain Imaging via Nonlinear Dielectric Response

Recently, Cho et al. have developed a pure electrical dynamic method of domain delineation [67–70]. This method, termed scanning nonlinear dielectric microscopy

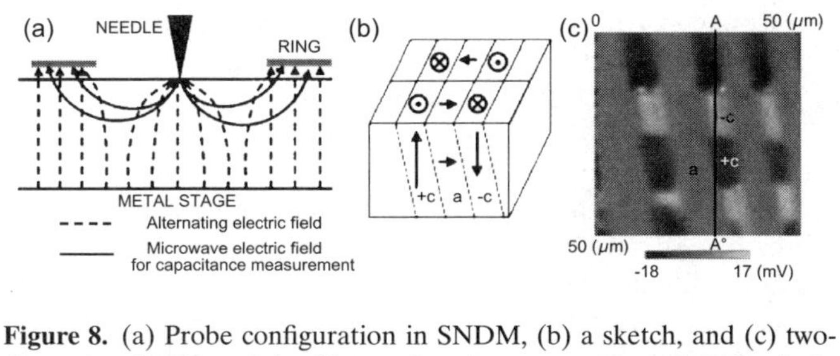

Figure 8. (a) Probe configuration in SNDM, (b) a sketch, and (c) two-dimensional image of the 90° *a*-*c*-domain structure in a $BaTiO_3$ single crystal. Reprinted with permission from [68], Y. Cho et al., *Appl. Phys. Lett.* 75, 2833 (1999). © 1999, American Institute of Physics.

(SNDM), is based on the detection of the capacitance variation with an alternating electric field. To measure the capacitance variation, Cho et al. developed a special lumped constant resonator probe by using an electrolytically polished tungsten needle and a LC resonance circuit operating in the microwave frequency range. Application of the modulation voltage E (in the range of 100 Hz to 1000 Hz) across the sample leads to an oscillating change ΔC_s in the capacitance between the needle and the bottom electrode due to the nonlinear dielectric response of the sample with the first harmonic component proportional to the nonlinear dielectric constant ε_{333}:

$$\frac{\Delta C_s}{C_s} = \frac{\varepsilon_{333}}{\varepsilon_{33}} E \cos \omega t \tag{8}$$

where C_s and ε_{33} are a static capacitance and a linear dielectric constant, respectively. The change in the capacitance is measured by detecting the modulated high-frequency signal (around 1.3 GHz) of the oscillator with a demodulator and a lock-in amplifier. The sign of an even rank tensor, such as the linear dielectric constant, does not depend on the polarization direction. On the other hand, the lowest order of the nonlinear dielectric constant ε_{333} is a third-rank tensor, similar to the piezoelectric constant, so the sign of ε_{333} changes with inversion of the spontaneous polarization. Therefore, a polarization map can be obtained by point-to-point detection of the field-induced changes in the nonlinear dielectric constant. This method, as it is designed, allows nanoscale detection of antiparallel 180° domains in the surface layer with a thickness much smaller than the probe size (<10 nm). According to [71], subnanometer lateral resolution can be obtained by detecting the higher order nonlinear dielectric constants. However, in this case the imaged region will become even shallower. Figure 8 shows a two-dimensional SNDM image of *a*-*c*-domain structure in single-crystal barium titanate. The sign of ε_{333} in the $+c$ domain is negative, whereas it is positive in the $-c$ domain. Furthermore, the magnitude of ε_{111} is zero in the *a*-domain. It is possible to measure ferroelectric polarization parallel to the sample surface by detecting ε_{311} constant with the use of different configuration of electrodes [70]. Therefore, three-dimensional domain structure can be determined by SNDM.

Since in SNDM the probing tip is in contact with the sample surface, nanoscale domain dots can be switched by applying a relatively low dc bias to the probe, which in combination with high spatial resolution can be used for ultra-high-density data storage [69].

Closely related near-field scanning microwave techniques have been used for domain imaging and dielectric constant measurements in single crystals of $LiNbO_3$, $BaTiO_3$, and deuterated triglycine sulfate and thin films of $Ba_{0.6}Sr_{0.4}TiO_3$ [72–75]. However, the lateral resolution has been just below 1 μm because of the size of the inner probe of the resonator.

4.4. Domain Imaging via Static Piezoresponse

The next domain imaging method makes use of the piezoelectric properties of ferroelectrics and therefore is often referred to as piezoresponse. It is based on the detection of local piezoelectric deformation of the ferroelectric sample induced by an external electric field. Since all ferroelectrics exhibit piezoelectric properties, application of an external voltage results in the deformation of a ferroelectric sample. Depending on the relative orientations of the applied field and the polarization vector, sample deformation can be in the form of elongation, contraction, or shear. For the converse piezoelectric effect, the field-induced strain S_j can be expressed as [76]

$$S_j = d_{ij} E_i \tag{9}$$

where d_{ij} is the piezoelectric coefficient and E_i is the applied field.

On the other hand, with the thermodynamic approach it can be shown that the piezoelectric coefficient relates to the spontaneous polarization P_s via the following expression [77, 78]:

$$d_{ij} = \varepsilon_{im} Q_{jmk} P_{sk} \tag{10}$$

where ε_{im} is the dielectric constant and Q_{jmk} is the electrostriction coefficient. For a single domain ferroelectric of tetragonal symmetry, the relation between the piezoelectric coefficients and the polarization in the reduced matrix notation can be written as

$$d_{33} = 2\varepsilon_{33} Q_{33} P_{s3} \tag{11}$$

The longitudinal piezoelectric coefficient, d_{33}, represents an expansion or contraction of the sample along the polar direction when the applied field is parallel to it. The transverse piezoelectric coefficient, d_{31}, represents an expansion or contraction of the sample in the direction perpendicular to the applied field. The d_{15} coefficient describes shear deformation of the ferroelectric sample.

The linear coupling between the piezoelectric and ferroelectric constants implies that the domain polarity can be determined from the sign of the field-induced strain. This is illustrated in Figure 9, which shows a simple case of a ferroelectric sample consisting of two antiparallel *c*-domains. In the absence of the external field and under zero stress within the sample, the two domains have equal size in polar direction since spontaneous strain is proportional to the square of the polarization and under ideal conditions should be the same in antiparallel domains [50]. Application of the electric field along the polar direction results in the elongation of the domain with polarization parallel to the applied field and in the contraction of the domain with opposite polarization. The field-induced strain in this case can be

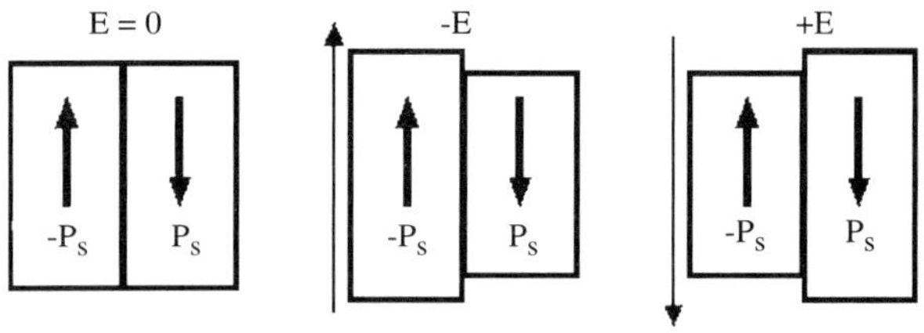

Figure 9. Deformation of *c*-domains as a result of the converse piezoelectric effect.

written as

$$S = \frac{\Delta L}{L} = \pm d_{33} E \tag{12}$$

where ΔL is the sample deformation and L is the sample thickness. Equation (4) can be rewritten as

$$\Delta L = \pm d_{33} V \tag{13}$$

where V is an applied voltage. The ± sign reflects the opposite piezoelectric coefficients for antiparallel domains. Therefore, opposite domains can be visualized by monitoring their voltage-induced surface displacement.

Because of its extremely high vertical sensitivity, nanoscale topography variations can be routinely measured in SFM. However, domain imaging based on detecting the static piezoelectric deformation is difficult to implement unless a sample has a very smooth surface. In a sample with an average surface roughness of several nanometers per square micrometer, the static cantilever deflection due to the piezoelectric deformation (typically on the order of several angstroms) will be superimposed on the much larger deflection signal due to the surface roughness, which will make domain imaging very problematical.

A seemingly obvious conclusion from Eq. (13) is that the contrast between opposite *c*-domains can be infinitely enhanced by increasing the imaging voltage. However, there is a strict limitation imposed on this parameter: to perform *nondestructive* visualization of domain structure, the imaging voltage should be kept below the coercive voltage of the ferroelectric sample. In addition, a high imaging voltage will lead to an increased contribution of the electrostatic signal to the tip-sample interaction, which in some cases can obscure the domain image. Given that a typical value of the coercive field in a 200-nm-thick $Pb(Zr,Ti)O_3$ ferroelectric film is approximately 50 kV/cm, the imaging voltage should not exceed 1 V, otherwise the imaging process will change the domain structure by inducing the polarization reversal. In a PZT film with the d_{33} constant of about 200 pm/V the surface displacement induced by an external voltage of 1 V will be only 0.2 nm. Obviously, such a displacement could not be reliably detected in ferroelectric films, where topographic features can be on the order of several nanometers. The static approach can be applied in some limited cases, for example, to ferroelectric samples with relatively high values of piezoelectric constants and coercive fields. Wang et al. [66] used this approach to delineate domains in a doped crystal of $Sr_{0.61}Ba_{0.39}Nb_2O_6$ (SBN), known for a high concentration of pinning centers, which gives rise to an increase in local coercive fields. Because of this feature of the SBN sample, even under an applied voltage of 200 V, there exist nonswitched *c*-domains antiparallel to the external field. At the same time, this voltage is high enough to produce a surface indentation of about 2 nm due to contraction and elongation of opposite domains (Fig. 10), which makes them discernible in the topographic mode. Another interesting observation reported in [66] includes the photostimulated nanoscale topographic changes due to the piezoelectric effect by an internal electric field.

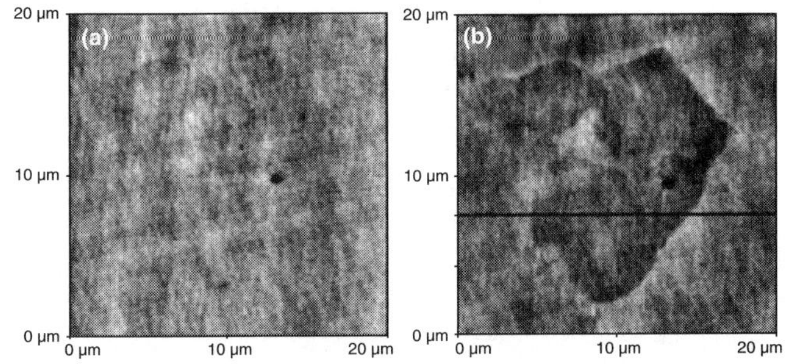

Figure 10. Topographic images of a SBN crystal under normal external electric fields of (a) $E = 0$, and (b) $E = 200$ V/mm. Reprinted from [66], Y. G. Wang et al., *Phys. Rev. B* 61, 3333 (2000). © 2000, American Physical Society.

4.5. Domain Imaging via Dynamic Piezoresponse: Piezoresponse Force Microscopy

A problem of low sensitivity of a static piezoresponse mode can be circumvented by employing a dynamic piezoresponse imaging method based on the voltage-modulation approach, which increases sensitivity by three orders of magnitude. In this approach, known as piezoresponse force microscopy (PFM), an ac modulation (imaging) voltage $V = V_0 \cos \omega t$ is applied to the ferroelectric sample, and surface displacement is measured by a standard lock-in technique by detecting the vertical vibration of the cantilever, which follows sample surface oscillation. A domain map can be obtained by scanning the surface while detecting the first harmonic component of the normal surface vibration (vertical piezoresponse):

$$\Delta L = \Delta L_0 \cos(\omega t + \varphi) \tag{14}$$

where $\Delta L_0 = d_{33} V_0$ is a vibration amplitude and φ is a phase difference between the imaging voltage and piezoresponse, which provides information on the polarization direction. With the modulation voltage applied to the probing tip, positive domains (polarization vector oriented downward) will vibrate in phase with the applied voltage so that $\varphi(+) = 0°$, whereas vibration of negative domains (polarization vector oriented upward) will occur in counter-phase: $\varphi(-) = 180°$.

The SFM piezoresponse mode, after being developed for detection of polarized regions in ferroelectric copolymer films of vinylidene fluoride and trifluoroethylene [79], had been successfully applied for visualization of domain structure in PZT thin films and quickly became one of the most widely used methods for nanoscale characterization of ferroelectrics [80–82].

One of the significant advantages of the PFM method is that it also allows delineation of domains with polarization parallel to the sample surface (*a*-domains) [57, 83, 84]. In the lateral PFM approach, developed by Eng [83], *a*-domains are visualized by detecting the torsional vibration of the cantilever. Application of the modulation voltage across the sample generates sample vibration in the direction parallel to its surface due to the piezoelectric shear deformation (Fig. 11). This surface vibration, translated via

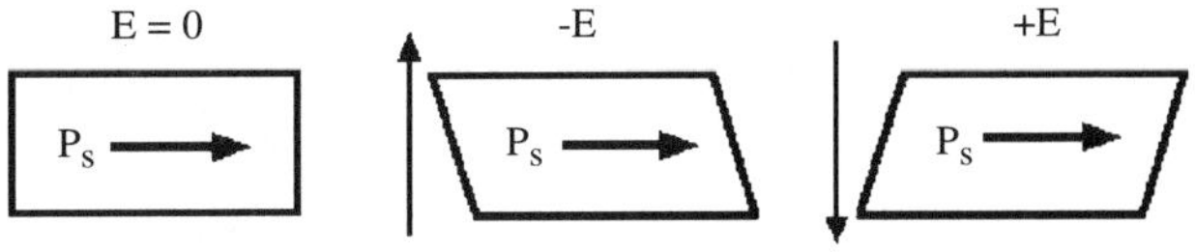

Figure 11. Deformation of *a*-domains as a result of the converse piezoelectric effect.

the friction forces to the torsional movement of the cantilever, can be detected in the same way as the normal cantilever oscillation in vertical PFM. The amplitude of the in-plane oscillation (lateral piezoresponse) is given by

$$\Delta X_0 = d_{15} V_0 \tag{15}$$

whereas the polarization direction can be determined from the phase signal since oscillation phases of opposite *a*-domains differ by 180°. It should be noted that Eq. (15) could be used only when the in-plane polarization vector is perpendicular to the physical axis of the cantilever. However, in the general case the in-plane polarization vector can be oriented arbitrarily with respect to the cantilever. Therefore, to obtain a complete picture of the in-plane distribution of polarization, X and Y components of the lateral piezoresponse image should be recorded by physically rotating the sample by 90°. Although quantitative analysis of the lateral piezoresponse signal is rather difficult because of the complexity of the friction mechanism involved, it can readily provide valuable information on in-plane distribution of polarization facilitating the three-dimensional reconstruction of the nanoscale domain arrangement [83].

Figure 12 presents experimental results on simultaneous acquisition of vertical and lateral piezoresponse images of a $SrBi_2Ta_2O_9$ film. A grain in the upper part of the image shows a strong vertical piezoresponse signal (dark contrast in Fig. 12b), whereas its contribution to the lateral piezoresponse signal is rather weak (gray contrast in Fig. 12c), suggesting predominantly out-of-plane orientation of the polarization vector. On the other hand, grains in the lower part of the image exhibit gray contrast in Figure 12b and black-white contrast in Figure 12c, which is an indication of the in-plane polarization.

The PFM experimental setup is shown in Figure 13. During PFM imaging the modulation frequency is usually kept far above the cutoff frequency of the SFM feedback to avoid

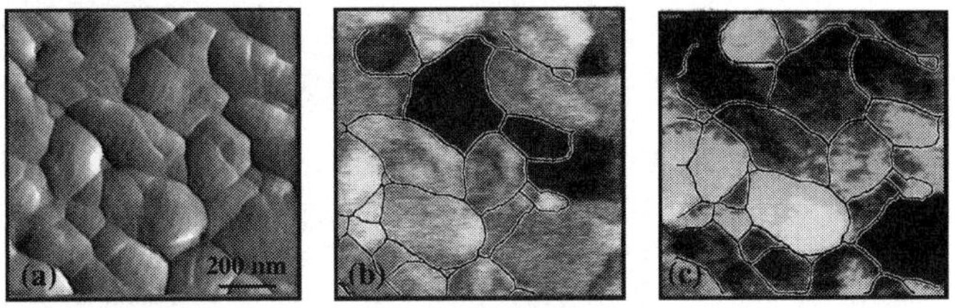

Figure 12. Topographic (a), normal piezoresponse (b), and lateral piezoresponse (c) images of the SBT film. A large grain in the upper part of the images has black contrast in (b) and gray contrast in (c), suggesting out-of-plane polarization. Grains in the lower part of the images exhibit gray contrast in (b) and black-white contrast in (c), which is an indication of the in-plane polarization. The scanning area is $1 \times 1\ \mu m^2$. Image courtesy of S. V. Kalinin, Oak Ridge National Laboratory.

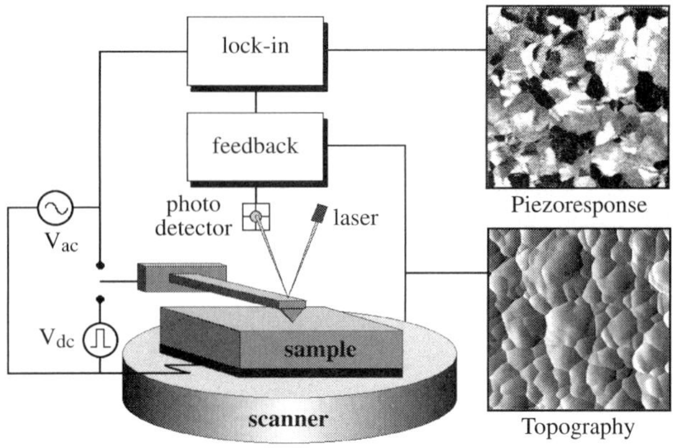

Figure 13. Experimental setup of piezoresponse force microscopy.

cross-talk between the static cantilever deflection due to the surface topographic features and modulated deflection due to the piezoelectric effect. With this setup it is possible to perform simultaneous acquisition of the piezoresponse and topographic images.

In terms of exciting the piezoelectric vibration of the sample, there are two main approaches in PFM. In one approach, the vibration is generated locally by applying a modulation voltage between the bottom electrode and the conductive SFM tip, which scans the bare surface of the film without a deposited top electrode. A great advantage of this approach is the possibility of establishing a correlation between domain configurations and film microstructure. In addition, this method can be used for nanoscale domain writing and direct investigation of domain wall interaction with microstructural features, such as defects and grain boundaries, for local spectroscopy measurements and investigation of electrical and mechanical coupling between adjacent grains. Furthermore, this approach offers extremely high resolution, potentially allowing investigation of the microscopic mechanism of the domain wall motion.

However, the electric field generated by the SFM tip in this configuration is highly inhomogeneous, which makes quantitative analysis of the field-dependent parameters difficult. This problem is exacerbated by the likely presence of a contamination layer at the film surface, which increases the resistance of the tip-sample electric contact. As a result, an increased time constant of the electric circuitry makes it difficult to extend the experiments on switching behavior to the micro- and nanosecond range, which is of direct application interest. In an alternative PFM approach, domain structure can be visualized through the top electrode of a ferroelectric capacitor. In this case the piezoelectric vibration is generated in a film region underneath the deposited top electrode, which is much larger than the tip-sample contact area. The modulation voltage can be applied either by using an external wire attached to the top electrode or, in the case of a micrometer-size electrode, directly through the conductive SFM tip. In both cases the piezoelectric displacement is probed locally by the SFM tip. In such a configuration, a homogeneous electric field is generated throughout the ferroelectric film, which allows quantitative treatment of domain wall dynamics and investigation of the polarization reversal mechanism in ferroelectric capacitors. Because of the reduced time constant, fast pulse switching and transient current measurements can be accomplished in submicrometer capacitors, thus making PFM suitable for memory device testing. Combined PFM studies of the ferroelectric films and capacitors bring about complementary information and provide direct assessment of the effect of the electrical and mechanical boundary conditions on domain stability and polarization reversal.

5. APPLICABILITY OF SFM TO DOMAIN IMAGING

Versatility of SFM can provide critical information on the dielectric properties of ferroelectrics at the nanoscale. Each of the SFM-based methods of domain imaging has its own advantages and limitations. Noncontact methods (EFM, KPFM) are the most universal methods, which can be applied to all ferroelectrics; however, they are susceptible to screening effects and have low spatial resolution in air. Static imaging methods (AFM, LFM) have limited applicability because of the specific domain contrast mechanisms. The SNDM approach is characterized by high imaging resolution and universality but is technically challenging and, like most of the SFM methods, provides information on the surface properties of ferroelectrics. The PFM method is the only method based on analysis of bulk properties. Therefore, combined study of ferroelectric samples by various SFM methods seems to be the best way to obtain a complete picture of nanoscale properties of ferroelectrics. A summary of different SFM methods of domain imaging is presented in Table 3.

It is generally accepted that the PFM method is the most suitable technique for investigation of the domain structure in ferroelectric materials, particularly in thin films [85]. This view is based on several key features of PFM:

- **High lateral resolution in domain imaging.** In the PFM approach, the electromechanical response of the sample is detected locally by the probing tip, which is in contact with the sample surface. In this case, the lateral resolution is determined by the size of the tip-sample contact area, which is a function of the tip radius, the contact force, and the relative hardnesses of the tip and sample materials [86, 87]. It has been demonstrated that in typical PFM experimental conditions the 180° domains can be imaged with a resolution better than 10 nm [88, 89].
- **Easy implementation** of the PFM imaging method stems from the fact that it can yield data by analyzing samples in the ambient environment and without special sample preparation. In addition, PFM detects the modulated surface displacement due to the converse piezoelectric effect, which minimizes cross-talk between the piezoresponse and topographic signals. As a result, topographic features have a negligible effect on domain image.
- **Three-dimensional reconstruction of domain structure.** The PFM approach can be used to determine the orientation of out-of-plane and in-plane components of local polarization, which opens the possibility for three-dimensional reconstruction of the nanoscale domain structures [57, 83, 84, 90, 91].

Table 3.

Method	Imaging mechanism	Features	Materials studied
EFM, KPFM	Electrostatic tip-sample interaction	• Measures surface potential and charge distribution • Low resolution in air • Susceptible to screening effects	Films: PZT, SBT, Crystals: $PbTiO_3$, $BaTiO_3$, $LiNbO_3$, GASH, TGS
AFM	Static surface deformation	• High lateral resolution • Limited applicability	Films: PZT Crystals: $PbTiO_3$, $BaTiO_3$, GASH, TGS
LFM	Polarization dependent surface friction	• High lateral resolution • Sensitive to surface conditions • Limited applicability	Crystals: GASH, TGS
SNDM	Nonlinear dielectric response	• High resolution • Measures in-plane and out-of-plane polarization	Films: PZT Crystals: $BaTiO_3$, $LiTaO_3$, $LiNbO_3$
PFM	Piezoelectric effect	• High resolution • Measures in-plane and out-of-plane polarization and piezocoefficients • Detects domains through top electrodes	Films: PZT, SBT, $PbTiO_3$ Crystals: $PbTiO_3$, $BaTiO_3$, $LiTaO_3$, $LiNbO_3$, TGS, $Sr_{0.61}$ $Ba_{0.39}Nb_2O_6$

- **Effective manipulation and control of nanosize domains**. In PFM, a probing tip can be used not only for domain imaging but also for modification of the original domain structure by inducing local polarization reversal. Application of a voltage pulse to the probing tip generates a strong and sufficiently localized electric field, which makes possible writing of ferroelectric domains of nanoscale dimensions and their subsequent readout [80–82, 85, 92–95].
- **Nanoscale hysteresis loop measurements** can be accomplished by monitoring a piezoresponse signal as a function of a poling dc voltage. These measurements are closely related to investigation of the switching behavior and provide information on coercive voltage, switching time, and local polarization at the nanoscale level [81, 85, 96–102].
- **Insensitivity to charge screening effects.** Since PFM is based on detecting the bulk properties of the ferroelectrics, it is less susceptible to the effect of the surface charges. As such, PFM allows reproducible imaging of a true polarization distribution.
- **Possibility of domain imaging in capacitor structures.** PFM allows detection of the piezoelectric displacement of a ferroelectric layer through the top electrode, which is one of the most significant advantages of this method [103–109]. This allows one to circumvent the problems of inhomogeneous tip-generated field distribution and possible contribution of Maxwell stress to the piezoresponse signal [87, 110], which are the major drawbacks of the PFM approach when applied to the samples with a bare surface. Applicability of PFM to nanoscale characterization of capacitor structures is of direct application relevance since it can be used for testing of ferroelectric memory devices.

In spite of the apparent simplicity of the PFM method, quantification of the PFM data, particularly in the case of thin films, is nontrivial because of the complexity of the tip-sample interaction. Experimental conditions, such as driving voltage, frequency, loading force, cantilever force constant, tip apex radius, and ambient environment, as well as physical properties of the samples (thickness, dielectric constants, orientation, defect structure, crystallinity, electrode material), should be taken into account to avoid misinterpretation of the PFM results [87]. Despite this difficulty, PFM has become one of the main tools for high-resolution characterization of ferroelectric crystals and thin films. PFM can provide crucial information on the correlation between domain configurations and microstructural features, basic mechanisms of polarization reversal, the scaling effect, and the intrinsic variability of properties at the nanoscale. One of the examples of the recent breakthrough results made possible because of PFM is characterization of the switching behavior of 0.04 μm^2 ferroelectric capacitors [111].

6. SFM OF FERROELECTRICS

Application of SFM to ferroelectric materials opened new possibilities for their characterization, control of ferroelectric properties at the nanoscale, and fabrication of regular nanodomain patterns for use in functional devices.

6.1. Characterization

Ferroelectric thin films are currently attracting significant interest because of the recent achievements in their integration into nonvolatile random access memory devices characterized by high-speed access, low power consumption, almost unlimited endurance, and extreme radiation hardness [112]. However, for ferroelectric memories to become high-volume, mainstream devices competitive with other types of nonvolatile memories, two major requirements must be fulfilled. The first is the ability to manufacture at high device density in the gigabit range, where the constituent ferroelectric capacitors are reduced to submicrometer dimensions. The second is the necessity to ensure that ferroelectric memory devices of these dimensions can achieve the necessary performance, reliability, and lifetime requirements. These two requirements demand intensive studies of ferroelectric films to achieve a major breakthrough in understanding their nanoscale properties. For high-density memories, the ferroelectric capacitor must be scaled to area and thickness dimensions, which may approach the stability limits for the ferroelectric phase. In addition, the intrinsic capacitor-to-capacitor variability becomes an issue as the number of grains per capacitor becomes small. Despite significant progress in the development of ferroelectric films and devices over the last several years, there remains a great deal that is not known regarding the domain structure of thin films and its role in determining device switching speed and lifetime. The major limitation in acquiring this crucial information has been the lack of experimental methods to characterize ferroelectric switching behavior at the microscopic level. Most of such experiments have been performed by destructive methods, such as transmission electron microscopy.

In this respect, SFM has a significant advantage over other high-resolution techniques, as it provides the possibility of direct nondestructive high-resolution characterization of ferroelectric materials at the nanoscale level. Over the last 10 years SFM has proved to be an indispensable tool capable of dielectric, surface potentiometric, and electromechanical characterization of ferroelectrics at the micro- and nanoscale levels. Application of SFM has rendered direct information on local surface potential and charge distribution, nanoscale domain arrangement, and domain kinetics during phase transition [17, 23, 32, 34, 43, 54, 113]. SFM has been successfully applied to high-resolution studies of static and dynamic properties of domains in ferroelectric thin films and capacitors [80–111]. Recent achievements facilitated by the SFM approach include manipulation of domains as small as 20 nm in diameter [69, 85], direct study of domain nucleation and growth during polarization reversal [104], investigation of the mechanisms of polarization decay and ferroelectric fatigue at the subgrain level [82, 88, 94, 97], evaluation of the switching behavior of individual nanocapacitors via hysteresis loop measurements [105–110], and investigation of the dielectric breakdown mechanism [114]. Figure 14 illustrates several examples of SFM capabilities in the nanoscale characterization of ferroelectrics. These results demonstrate that SFM is well positioned to become a major characterization and testing tool for ferroelectric devices. Nanoscale SFM characterization in combination with phenomenological theory for ferroelectric films will lead to better understanding of their basic properties and significant improvement in the performance of nanoscale ferroelectric devices.

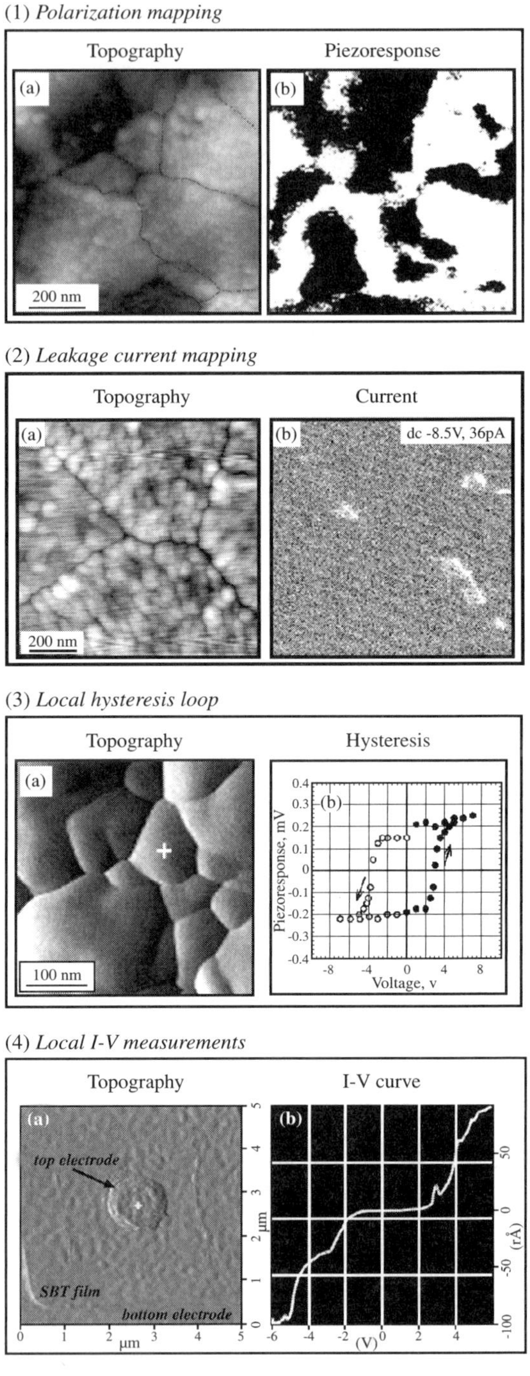

Figure 14. Examples of SFM capabilities in high-resolution characterization of ferroelectrics. (1) Simultaneous acquisition of surface topography and polarization distribution. Reprinted with permission from [85], A. Gruverman et al., *Annu. Rev. Mater. Sci.* 28, 101 (1998). © 1998, Material Research Society. (2) Mapping of leakage current sites. Reprinted with permission from [114], A. Gruverman et al., *Mater. Res. Soc. Symp. Proc.* 655, CC8.5 (2001). © 2001, Material Research Society. (3) Measurement of local (within 20-nm size region) coercive voltage and piezoelectric coefficient. Reprinted with permission from [85]. (4) Measurement of current-voltage characteristics of a micrometer-size ferroelectric capacitor. Reprinted with permission from [114].

6.2. Control

SFM provides a unique opportunity for controlling ferroelectric properties at the nanoscale and direct studies of the domain structure evolution under an external electric field, which cannot be matched by previously available techniques. A conductive probing tip can be used not only for domain visualization but also for modification of the initial domain structure. Application of a small dc voltage between the tip and bottom electrode generates an electric field of several hundred kilovolts per centimeter, which is higher than the coercive voltage of most ferroelectrics, thus inducing local polarization reversal.

When an electric field is applied opposite to the polarization direction of a single-domain ferroelectric, the switching mechanism involves several steps [115]. First, new domains with the reverse polarization direction nucleate, mainly at the surface, and then grow through the sample thickness (forward growth). Second, the grown domains expand sideways as new domains continue to form. Finally, the growing domains coalesce to complete the polarization reversal of a ferroelectric. The contribution of the forward and sideways growth mechanisms is a function of the applied field and the electrodes and to a large extent determines the switching time. Until recently, observation of domain dynamics during switching in thin films has not been performed because of the absence of an appropriate experimental technique. It has become possible with the help of the PFM method. This approach allows direct studies of the mechanism of polarization reversal and measurements of the key parameters of domain dynamics, such as nucleation rate, domain wall velocity, spatial distribution of nucleation sites, etc.

However, a poor time resolution, which is determined by the time required for image acquisition (at a scan rate of 2 Hz and an image resolution of 256 × 256 pixels, it takes about 2 min to acquire an image), makes the *in-situ* measurements of domain dynamics during fast switching processes difficult. Although SFM can readily be used to investigate slow polarization relaxation processes with characteristic times on the order of minutes and above, it is a challenge to deduce the mechanism of domain transformation when polarization reversal occurs in a matter of microseconds and faster.

This problem is usually circumvented by studying of domain structure dynamics in a quasi-static regime using *step-by-step switching*. This method has previously been used at the macroscopic level in classical switching experiments on correlating domain structure evolution and transient current in ferroelectric crystals [115] and later was applied to thin films [116, 117]. In this approach, partial reversal of polarization is generated by applying a voltage pulse shorter than the total switching time with subsequent piezoresponse imaging of the resulting domain pattern. By applying a sequence of voltage pulses of successively increasing duration and acquiring piezoresponse images after each pulse, a consistent picture of time-dependent behavior of domain structure can be obtained, providing information on the domain wall velocity, its spatial anisotropy, and its field dependence. To avoid data misinterpretation due to spontaneous backswitching between the pulses, stability of the produced intermediate patterns should be checked by acquiring domain images at different time intervals after single pulse application.

An example of this approach is illustrated in Figure 15, which shows a sequence of piezoresponse images of a growing domain in the PZT film. Based on the contrast of the domain, it is assumed that it fully extended through the film thickness after the first pulse and further growth occurs via the sideways expansion only. The diameter of the smallest stable domain produced by the first voltage pulse of 5 V varied from 20 nm to 40 nm. Since the electric field generated by a probing tip quickly decreases with distance, the rate of domain growth, after fast initial expansion, declines until the domain stops growing. To describe the sideways expansion of the domain it is necessary to take into account the field dependence of the domain wall velocity and the spatial distribution of the electric field generated by the probing tip [118].

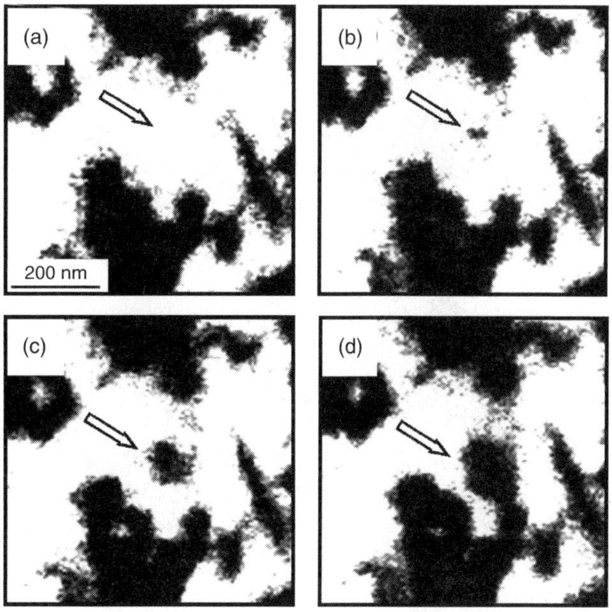

Figure 15. Sequence of PFM images of PZT film showing a sideways growth of a domain in the center of the image. The growth is induced by application of 5-V voltage pulses of increasing duration to the same spot by a fixed probing tip. (a) Original domain structure. (b–d) Domain images after application of voltage pulses. Pulse duration: (b) 0.25 s, (c) 0.5 s, (d) 1 s.

6.3. Fabrication

SFM capability of domain control creates a possibility of developing ferroelectrics with specifically designed nanoscale domain patterns, which can find application in novel electronic devices. For example, Hidaka et al. [81] and Cho et al. [69] proposed to use SFM as a basis for high-density data storage with a ferroelectric layer as a recording medium and nanoscale domains as data bits. Recently, a pioneering work by Kalinin et al. [119] has demonstrated the feasibility of practical application of ordered domain structures for the fabrication of nanostructures. Manipulation of polarization of ferroelectric substrates in SFM opens a new way to assemble nanostructures consisting of oxide substrates, metal nanoparticles, and organic/biological molecules. The idea of using ferroelectric templates for nanofabrication is based on utilizing the local surface electronic properties and surface chemical reactivity, which can be controlled by switching the direction of spontaneous polarization. On the surface of a ferroelectric, an abrupt change in the normal component of the spontaneous polarization results in bound polarization charges and in the appearance of a depolarizing field [120]. Energy consideration requires that this field must vanish outside the ferroelectric, that is, it must be compensated for. In a ferroelectric capacitor, this field can be completely compensated (screened) by the charges on the electrodes.

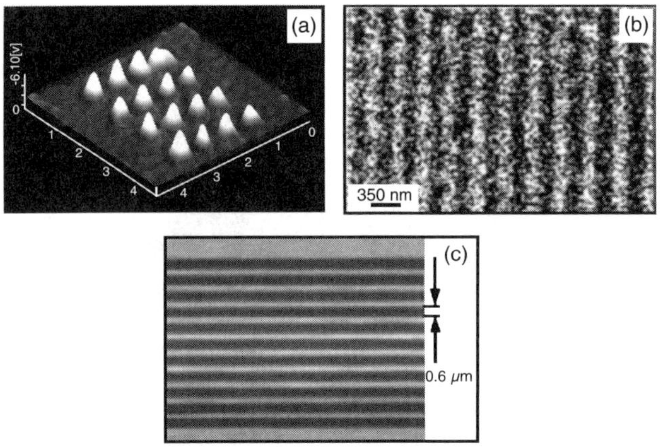

Figure 16. SFM-generated nanoscale domain patterns in (a) $BaTiO_3$, (b) $Sr_{0.61}Ba_{0.39}Nb_2O_6$ (Reprinted with permission from [122], K. Terabe et al., *Appl. Phys. Lett.* 81, 2044 (2002). © 2002, American Institute of Physics), and (c) $RbTiOAsO_4$ single crystal. (Reprinted with permission from [121], G. Rosenman et al., *Appl. Phys. Lett.* 82, 103 (2003). © 2003, American Institute of Physics.

In addition, screening by charge absorption at the surface can be accompanied by accumulation of the free carriers in the ferroelectric sample just near its surface. Surface potential, determined by the relative contribution of the external (surface charges) and internal (nonequilibrium carriers) screening mechanisms, is a crucial parameter for assembling molecular structures. Therefore, investigation of the interface properties and the deconvolution of the external and internal screening in ferroelectrics is critical for developing new nanofabrication methods using domain patterned ferroelectric templates.

Several groups reported generation of regular domain patterns with nanoscale periodicity by SFM [42, 69, 81, 117, 121, 122]. Nanodomain patterns can be successfully fabricated not only in thin films but also in bulk crystals, in spite of the fact that the tip-generated electric field quickly decreases with distance to the bulk of the sample [58, 121]. Examples of SFM-fabricated nanodomain patterns in $BaTiO_3$, $Sr_{0.61}Ba_{0.39}Nb_2O_6$, and $RbTiOAsO_4$ single crystals are shown in Figure 16.

For successful application of this domain-based approach to fabrication of novel electronic devices, several key issues related to the scale of domain engineering should be addressed.

First, reproducibility of high-resolution domain writing is necessary for the fabrication of reliable, large-scale ferroelectric templates. This requirement critically depends on the uniformity of the properties of the ferroelectrics at the nanoscale. This problem can be addressed based on statistical analysis of the PFM data of written domains in thin films [98, 123]. With this approach the effect of grain size and orientation on spatial homogeneity and reproducibility of the domain writing in thin films has been determined. The same approach can be used for bulk crystals and thin films to investigate the statistical variations in domain sizes as a function of the microstructure of the samples, their dielectric properties, and domain writing conditions.

Second to be considered is the AFM domain writing resolution, that is, the ability to scale the size of the written domains below 50 nm. There are a number of factors affecting a process of polarization reversal in SFM: microstructure, interface electrical and mechanical conditions, spatial distribution of an SFM-generated electric field, etc. The role of these factors can be different for different types of materials and samples. For example, the domain writing conditions in single crystals and thin films will be quite different.

Third, long-term stability of the written domain patterns is a key requirement necessary for the fabrication of large-scale nanodomain patterns. This problem is closely related to the problem of thermodynamic stability of the ferroelectric phase at the nanoscale. One of the most important factors affecting domain stability is the screening of spontaneous polarization. EFM and SSPM studies can provide an insight into the process of polarization screening at the nanoscale and will open a way for local stabilization (and destabilization when necessary) of the domain patterns. This combined approach allows quantitative examination of screening processes as a function of time, type of interface, template microstructure, and ambient environment. Preliminary studies showed that domains as small as 20 nm in diameter could be written and detected by SFM [69]. In addition, it has been found recently that ferroelectric films as thin as 1.6 nm (two unit cells) still exhibit ferroelectric behavior [124]. SFM studies of domains of these dimensions can contribute to better understanding of the size effects and stability issues in ferroelectrics.

Fourth, and probably the most serious problem for the use of SFM in nanodomain fabrication, is throughput [125], which is limited by the tip scanning rate: whereas the write speed in ferroelectrics can be in the nanosecond range, the typical scanning speed of most scanning probe microscopes is in the range of tens of micrometers per second. The likely solution of this problem lies in the use of massive parallelism, as has recently been demonstrated by the IBM Millipede concept [126]. High data rates have been achieved through the use of large (32×32) two-dimensional arrays of SFM probes, which operate simultaneously.

7. CONCLUSIONS

Rapid development of ferroelectric-based devices generated a strong need for extensive investigation of the nanoscale properties of ferroelectric materials. Application of SFM to ferroelectrics opened new possibilities not only for their high-resolution characterization, but also for control of ferroelectric properties at the nanoscale and for fabrication of nanodomain patterns for use in functional devices. This paper presented a review of the recent advances in nanoscale investigation of ferroelectrics by means of SFM. Applicability of different SFM modes to domain imaging and mechanisms of domain contrast has been discussed. Comparative analysis of SFM-based imaging methods and problems that can be addressed by SFM has been presented. SFM has made a decisive contribution to solving such problems as investigation of the basic mechanism of polarization reversal in ferroelectric thin films, three-dimensional reconstruction of domain structure in ferroelectric films at the subgrain level, establishing a correlation between domain configurations and film microstructure, and mechanisms of degradation effects. Future trends in the characterization of ferroelectrics and in the fabrication of ferroelectric-based functional nanostructures have been discussed.

GLOSSARY

Ferroelectric Crystalline dielectric possessing spontaneous polarization the direction of which can be reversed by application of an electric field.

Ferroelectric domains Regions of uniform polarization in a ferroelectric.

Piezoelectricity Property of a crystal to exhibit electric polarity when subject to mechanical stress (direct piezoelectric effect) or mechanical deformation upon application of an electrical field (converse piezoelectric effect).

ACKNOWLEDGMENTS

The author expresses his gratitude to Prof. A. Kingon, B. J. Rodriguez, and Prof. R. J. Nemanich, NCSU, for their help and support. The author also expresses his sincere thanks to Prof. Y. Cho, Prof. W. Kleemann, Dr. S. V. Kalinin, Dr. Y. Haga, Dr. K. Terabe, Prof. G. Rosenman, and Dr. H. Yokoyama for kindly allowing the use of their data in this review.

REFERENCES

1. M. E. Lines and A. M. Glass, "Principles and Applications of Ferroelectric and Related Materials." Clarendon, Oxford, 1977.
2. "Nanoelectronics and Information Technology, Advanced Electronic Materials and Novel Devices" (R. Waser Ed.). Wiley-VCH Verlag GmbH, Weinheim, 2003.
3. D. Sarid, "Scanning Force Microscopy with Applications to Electric, Magnetic and Atomic Forces," Oxford Series in Optical and Imaging Sciences. University Press, Oxford, 1991.
4. R. Wiesendanger, "Scanning Probe Microscopy and Spectroscopy: Methods and Applications." Cambridge University Press, Cambridge, UK, 1994.
5. G. Binnig, C. F. Quate, and Ch. Gerber, *Phys. Rev. Lett.* 56, 930 (1986).
6. Y. Martin, C. C. Williams, and H. K. Wickramasinghe, *J. Appl. Phys.* 61, 4723 (1987).
7. J. E. Stern, B. D. Terris, H. J. Mamin, and D. Rugar, *Appl. Phys. Lett.* 53, 2717 (1988).
8. Y. Martin, D. W. Abraham, and H. K. Wickramasinghe, *Appl. Phys. Lett.* 52, 1103 (1988).
9. M. Nonnenmacher, M. P. O'Boyle, and H. K. Wickramasinghe, *Appl. Phys. Lett.* 58, 2921 (1991).
10. R. C. Barrett and C. F. Quate, *J. Appl. Phys.* 70, 2725 (1991).
11. P. Lehnen, J. Dec, and W. Kleemann, *J. Phys. D: Appl. Phys.* 33, 1932 (2000).
12. F. Saurenbach and B. D. Terris, *Appl. Phys. Lett.* 56, 1703 (1990).
13. R. Luthi, H. Haefke, K.-P. Meyer, E. Meyer, L. Howald, and H.-J. Guntherodt, *J. Appl. Phys.* 74, 7461 (1993).
14. R. Luthi, H. Haefke, K.-P. Meyer, E. Meyer, L. Howald, and H.-J. Guntherodt, *Surf. Sci.* 285, L498 (1993).
15. R. Luthi, H. Haefke, W. Gutmannsbauer, E. Meyer, L. Howald, and H.-J. Guntherodt, *J. Vac. Sci. Technol., B* 12, 2451 (1994).
16. R. Luthi, E. Meyer, L. Howald, H. Haefke, D. Anselmetti, M. Dreier, M. Ruetschi, T. Bonner, R. M. Overney, J. Frommer, and H.-J. Guntherodt, *J. Vac. Sci. Technol., B* 12, 1673 (1994).
17. L. M. Eng, M. Friedrich, J. Fousek, and P. Gunter, *J. Vac. Sci. Technol., B* 14, 1191 (1996).
18. L. M. Eng, J. Fousek, and P. Gunter, *Ferroelectrics* 191, 419 (1997).
19. L. M. Eng, M. Bammerlin, C. Loppacher, M. Guggisberg, R. Bennewitz, E. Meyer, and H.- J. Güntherodt, *Surf. Interface Anal.* 27, 422 (1999).
20. B. D. Terris, J. E. Stern, D. Rugar, and H. J. Mamin, *Phys. Rev. Lett.* 63, 2669 (1989).
21. B. D. Terris, J. E. Stern, D. Rugar, and H. J. Mamin, *J. Vac. Sci. Technol. A* 8, 374 (1990).
22. L. M. Blinov, R. Barberi, S. P. Palto, M. P. De Santo, and S. G. Yudin, *J. Appl. Phys.* 89, 3960 (2001).
23. E. Z. Luo, Z. Xie, J. B. Xu, and I. H. Wilson, *Phys. Rev. B* 61, 203 (2000).
24. J. W. Hong, S.-I. Park, Z. G. Khim, *Rev. Sci. Instrum.* 70, 1735 (1999).
25. L. M. Eng, M. Bammerlin, C. Loppacher, M. Guggisberg, R. Bennewitz, R. Lüthi, E. Meyer, and H. -J. Güntherodt, *Appl. Surf. Sci.* 140, 253 (1999).
26. S. Tsunekawa, J. Ichikawa, H. Nagata, and T. Fukuda, *Appl. Surf. Sci.* 137, 61 (1999).
27. J. W. Hong, D. S. Kahng, J. C. Shin, H. J. Kim, and Z. G. Khim, *J. Vac. Sci. Technol., B* 16, 2942 (1998).
28. J. W. Hong, K. H. Noh, and S. Park, et al., *Phys. Rev. B* 58, 5078 (1998).
29. H. Bluhm, A. Wadas, R. Wiesendanger, A. Roshko, J. A. Aust, and D. Nam, *Appl. Phys. Lett.* 71, 146 (1997).
30. C. H. Ahn, T. Tybell, L. Antognazza, K. Char, R. H. Hammond, M. R. Beasly, Ø. Fischer, and J. M. Triscone, *Science* 276, 1100 (1997).
31. G. Zavala, J. H. Fendler, and S. Trolier-McKinstry, *J. Appl. Phys.* 81, 7480 (1997).
32. S. V. Kalinin and D. A. Bonnell, *Phys. Rev. B* 63, 125411 (2001).
33. S. V. Kalinin, C. Y. Johnson, and D. A. Bonnell, *J. Appl Phys.* 91, 3816 (2002).
34. S. V. Kalinin and D. A. Bonnell, *Appl. Phys. Lett.* 78, 1116 (2001).
35. R. Shikler, N. Fried, T. Meoded, and Y. Rosenwaks, *Phys. Rev. B* 61, 11041 (2000); R. Shikler, T. Meoded, N. Fried, and Y. Rosenwaks, *Appl. Phys. Lett.* 74, 2972 (1999).
36. C. C. Williams, J. Slinkman, W. P. Hough, and H. K. Wickramasinghe, *Appl. Phys. Lett.* 55, 1662 (1989).
37. P. A. Rosenthal, E. T. Yu, R. L. Pierson, and P. J. Zampardi, *J. Appl. Phys.* 87, 1937 (2000).
38. D. W. Abraham, C. C. Williams, J. Slinkman, and H. K. Wickramasinghe, *J. Vac. Sci. Technol., B* 9, 703 (1991).
39. J. W. Hong, S. M. Shin, C. J. Kang, Y. Kuk, Z. G. Khim, and S.-I. Park, *Appl. Phys. Lett.* 75, 1760 (1999).
40. J. T. Jones, P. M. Bridger, O. J. Marsh, and T. C. McGill, *Appl. Phys. Lett.* 75, 1326 (1999).
41. X. M. Lu, F. Schlaphof, S. Grafstrom, C. Loppacher, L. M. Eng, G. Suchaneck, and G. Gerlach, *Appl. Phys. Lett.* 81, 3215 (2002).
42. A. Gruverman, O. Auciello, and H. Tokumoto, *J. Vac. Sci. Technol., B* 14, 602 (1996).
43. A. Correia, J. Massanell, N. Garcia, A. P. Levanyuk, A. Zlatkin, and J. Przeslawski, *Appl. Phys. Lett.* 68, 2796 (1996).
44. H. Bluhm, U. D. Schwarz, and R. Wiesendanger, *Phys. Rev. B* 57, 161 (1998).
45. H. Bluhm, R. Wiesendanger, and K.-P. Meyer, *J. Vac. Sci. Technol., B* 14, 1180 (1996).
46. H. Bluhm, U. D. Schwarz, K.-P. Meyer, and R. Wiesendanger, *Appl. Phys. A* 61, 525 (1995).
47. N. Nakatani, *Jpn. J. Appl. Phys.* 18, 491 (1979).
48. V. Ya. Shur, A. L. Gruverman, N. Yu. Ponomarev, and N. A. Tonkachyova, *Ferroelectrics* 126, 371 (1992).
49. V. Ya. Shur, A. L. Subbotin, and V. P. Kuminov, *Ferroelectrics* 140, 101 (1993).

50. F. Jona and G. Shirane, "Ferroelectric Crystals." Pergamon Press, Oxford, 1962.
51. A. Seifert, F. F. Lange, and J. S. Speck, *J. Mater. Res.* 10, 680 (1995).
52. J. Munoz-Saldana, G. A. Schneider, and L. M. Eng, *Surf. Sci.* 480, L402 (2001).
53. L. M. Eng and H.-J. Guntherodt, *Ferroelectrics* 236, 35 (2000).
54. M. Takashige, S. Hamazaki, Y. Takahashi, et al., *Ferroelectrics* 240, 1359 (2000).
55. M. Takashige, S. Hamazaki, N. Fukurai, F. Shimizu, and S. Kojima, *Ferroelectrics* 203, 221 (1997).
56. Y. G. Wang, J. Dec, and W. Kleemann, *J. Appl. Phys.* 84, 6795 (1998).
57. C. S. Ganpule, V. Nagarajan, B. K. Hill, A. L. Roytburd, E. D. Williams, R. Ramesh, S. P. Alpay, A. Roelofs, R. Waser, and L. M. Eng, *J. Appl. Phys.* 91, 1477 (2002).
58. A. Gruverman, J. Hatano, and H. Tokumoto, *Jpn. J. Appl. Phys.* 36, Part 1, 2207 (1997).
59. L. M. Eng, M. Bammerlin, C. Loppacher, et al., *Ferroelectrics* 222, 411 (1999).
60. V. Ya. Shur, E. L. Rumyantsev, E. V. Nikolaeva, E. I. Shishkin, D. V. Fursov, R. G. Batchko, L. A. Eyres, M. M. Fejer, and R. L. Byer, *Appl. Phys. Lett.* 76, 143 (2000).
61. W. Pompe, X. Gong, Z. Suo, and J. S. Speck, *J. Appl. Phys.* 74, 6012 (1993).
62. B. S. Kwak, A. Erbil, J. D. Budai, M. F. Chisholm, L. A. Boatner, and B. J. Wilkens, *Phys. Rev. B* 49, 14865 (1994).
63. S. K. Streiffer, C. B. Parker, A. E. Romanov, L. Zhao, J. S. Speck, W. Pompe, and C. F. Foster, *J. Appl. Phys.* 83, 2742 (1998).
64. N. A. Pertsev, A. G. Zembilgotov, and A. K. Tagantsev, *Phys. Rev. Lett.* 80, 1988 (1998).
65. V. G. Koukhar, N. A. Pertsev, and R. Waser, *Phys. Rev. B* 64, 214103 (2001).
66. Y. G. Wang, W. Kleemann, T. Woike, and R. Pankrath, *Phys. Rev. B* 61, 3333 (2000).
67. Y. Cho, A. Kirihara, and T. Saeki, *Jpn. J. Appl. Phys.* 36, Part 1, 360 (1997).
68. Y. Cho, S. Kazuta, and K. Matsuura, *Appl. Phys. Lett.* 75, 2833 (1999).
69. Y. Cho, K. Fujimoto, Y. Hiranaga, Y. Wagatsuma, A. Onoe, K. Terabe, and K. Kitamura, *Appl. Phys. Lett.* 81, 4401 (2002).
70. H. Odagawa and Y. Cho, *Appl. Phys. Lett.* 80, 2159 (2002).
71. Y. Cho and K. Ohara, *Appl. Phys. Lett.* 79, 3842 (2001).
72. C. Gao, F. Duewer, Y. Lu, and X.-D. Xiang, *Appl. Phys. Lett.* 73, 1146 (1998).
73. Y. Lu, T. Wei, F. Duewer, Y. Lu, N.-B. Ming, P. G. Schultz, and X.-D. Xiang, *Science* 276, 2004 (1997).
74. D. E. Steinhauer and S. M. Anlage, *J. Appl. Phys.* 89, 2314 (2001).
75. D. E. Steinhauer, C. P. Vlahacos, F. C. Wellstood, Steven M. Anlage, C. Canedy, R. Ramesh, A. Stanishevsky, and J. Melngailis, *Appl. Phys. Lett.* 75, 3180 (1999).
76. W. G. Cady, "Piezoelectricity: An Introduction to the Theory and Applications of Electromechanical Phenomena in Crystals." Dover Publications, New York, 1964.
77. A. F. Devonshire, *Philos. Mag.* 40, 1040 (1949).
78. A. F. Devonshire, *Adv. Phys.* 3, 85 (1954).
79. P. Guthner and K. Dransfeld, *Appl. Phys. Lett.* 61, 1137 (1992).
80. K. Franke, J. Besold, W. Haessle, and C. Seegebarth, *Surf. Sci. Lett.* 302, L283 (1994).
81. T. Hidaka, T. Maruyama, I. Sakai, M. Saitoh, L. A. Wills, R. Hiskes, S. A. Dicarolis, and J. Amano, *Integrated Ferroelectrics* 17, 319 (1997).
82. A. Gruverman, H. Tokumoto, S. A. Prakash, S. Aggarwal, B. Yang, M. Wuttig, R. Ramesh, O. Auciello, and V. Venkatesan, *Appl. Phys. Lett.* 71, 3492 (1997).
83. L. M. Eng, H.-J. Guntherodt, G. A. Schneider, U. Kopke, and J. M. Saldana, *Appl. Phys. Lett.* 74, 233 (1999).
84. A. Roelofs, U. Böttger, R. Waser, F. Schlaphof, S. Trogisch, and L. M. Eng, *Appl. Phys. Lett.* 77, 3444 (2000).
85. A. Gruverman, O. Auciello, and H. Tokumoto, *Annu. Rev. Mater. Sci.* 28, 101 (1998).
86. A. Gruverman, O. Auciello, and H. Tokumoto, *Integrated Ferroelectrics* 19, 49 (1998).
87. S. V. Kalinin and D. A. Bonnell, *Phys. Rev. B* 65, 125408 (2002).
88. C. S. Ganpule, A. L. Roytburd, V. Nagarajan, B. K. Hill, S. B. Ogale, E. D. Williams, R. Ramesh, and J. F. Scott, *Phys. Rev. B* 65, 014101 (2002).
89. O. Auciello, A. Gruverman, H. Tokumoto, S. A. Prakash, S. Aggarwal, and R. Ramesh, *MRS Bull.* 23, 33 (1998).
90. C. Harnagea, A. Pignolet, M. Alexe, and D. Hesse, *Integrated Ferroelectrics* 38, 667 (2001).
91. M. Abplanalp, L. M. Eng, and P. Gunter, *Appl. Phys. A* 66, Part 1, Suppl., S231 (1998).
92. P. Paruch, T. Tybell, and J. M. Triscone, *Appl. Phys. Lett.* 79, 530 (2001).
93. T. Tybell, C. H. Ahn, and J. M. Triscone, *Appl. Phys. Lett.* 75, 856 (1999).
94. J. Woo, S. Hong, D.-K. Min, H. Shin, and K. No, *Appl. Phys. Lett.* 80, 4000 (2002).
95. S. Hong, J. Woo, H. Shin, J. U. Jeon, Y. E. Pak, E. L. Colla, N. Setter, E. Kim, and K. No, *J. Appl. Phys.* 89, 1377 (2001).
96. J. Hong, H. W. Song, S. Hong, H. Shin, and K. No, *J. Appl. Phys.* 92, 7434 (2002).
97. H. Y. Guo, J. B. Xu, I. H. Wilson, Z. Xie, E. Z. Luo, S. Hong, and H. Yan, *Appl. Phys. Lett.* 81, 715 (2002).
98. J. A. Christman, S.-H. Kim, H. Maiwa, J.-P. Maria, B. J. Rodriguez, A. I. Kingon, and R. J. Nemanich, *J. Appl. Phys.* 87, 8031 (2000).
99. M. Abplanalp, J. Fousek, and P. Gunter, *Phys. Rev. Lett.* 86, 5799 (2001).
100. C. Harnagea, A. Pignolet, M. Alexe, D. Hesse, and U. Gösele, *Appl. Phys. A* 70, 261 (2000).
101. A. Pignolet, C. Schafer, K. M. Satyalakshmi, C. Harnagea, D. Hesse, and U. Gösele, *Appl. Phys. A* 70, 283 (2000).
102. C. Harnagea, A. Pignolet, M. Alexe, K. M. Satyalakshmi, D. Hesse, and U. Gösele, *Jpn. J. Appl. Phys.* 38, Part 2, L1255 (1999).
103. S. Hong and N. Setter, *Appl. Phys. Lett.* 81, 3437 (2002).
104. S. Hong, E. L. Colla, E. Kim, D. V. Taylor, A. K. Tagantsev, P. Muralt, K. No, and N. Setter, *J. Appl. Phys.* 86, 607 (1999).
105. M. Alexe, C. Harnagea, D. Hesse, and U. Goesele, *Appl. Phys. Lett.* 75, 1793 (1999).
106. M. Alexe, A. Gruverman, C. Harnagea, N. D. Zakharov, A. Pignolet, D. Hesse, and J. F. Scott, *Appl. Phys. Lett.* 75, 1158 (1999).
107. C. S. Ganpule, A. Stanishevsky, S. Aggarwal, J. Melngailis, E. Williams, R. Ramesh, V. Joshi, and C. A. Paz de Araujo, *Appl. Phys. Lett.* 75, 3874 (1999).
108. O. Auciello, A. Gruverman, and H. Tokumoto, *Integrated Ferroelectrics* 15, 107 (1997).
109. J. A. Christman, R. Woolcott, Jr., A. Kingon, and R. J. Nemanich, *Appl. Phys. Lett.* 73, 3851.
110. S. Hong, H. Shin, J. Woo, and K. No, *Appl. Phys. Lett.* 80, 1453 (2002).
111. S. Tiedke, T. Schmitz, K. Prume, A. Roelofs, T. Schneller, U. Kall, R. Waser, C. S. Ganpule, V. Nagarajan, A. Stanishevsky, and R. Ramesh, *Appl. Phys. Lett.* 79, 3678 (2001).
112. O. Auciello, J. F. Scott, and R. Ramesh, *Phys. Today* 51, 22 (1998).
113. V. Likodimos, M. Labardi, and M. Allegrini, *Phys. Rev. B* 66, 024104 (2002).
114. A. Gruverman, C. Isobe, and M. Tanaka, *Mater. Res. Soc. Symp. Proc.* 655, CC8.5 (2001).
115. E. Fatuzzo and W. J. Merz, "Ferroelectricity." North-Holland, Amsterdam, 1967.
116. O. Lohse, S. Tiedke, M. Grossmann, and R. Waser, *Integrated Ferroelectrics* 22, 123 (1998).

117. T. Tybell, P. Paruch, T. Giamarchi, and J.-M. Triscone, *Phys. Rev. Lett.* 89, 097601 (2002).
118. E. J. Mele, *Am. J. Phys.* 69, 557 (2001).
119. S. V. Kalinin, D. A. Bonnell, T. Alvarez, X. Lei, Z. Hu, J. H. Ferris, Q. Zhang, and S. Dunn, *NanoLetters* 2, 589 (2002).
120. V. M. Fridkin, "Photoferroelectrics." Springer-Verlag, Berlin, 1979.
121. G. Rosenman, P. Urenski, A. Agronin, Y. Rosenwaks, and M. Molotski, *Appl. Phys. Lett.* 82, 103 (2003).
122. K. Terabe, S. Takekawa, M. Nakamura, K. Kitamura, S. Higuchi, Y. Gotoh, and A. Gruverman, *Appl. Phys. Lett.* 81, 2044 (2002); K. Terabe, M. Nakamura, S. Takekawa, K. Kitamura, S. Higuchi, Y. Gotoh, and Y. Cho, *Appl. Phys. Lett.* 82, 433 (2003).
123. A. Gruverman, *Appl. Phys. Lett.* 75, 1452 (1999).
124. S. K. Streiffer, J. A. Eastman, D. D. Fong, C. Thompson, A. Munkholm, M. V. Ramana Murty, O. Auciello, G. R. Bai, and G. B. Stephenson, *Phys. Rev. Lett.* 89, 67601 (2002).
125. C. F. Quate, *Surf. Sci.* 386, 259 (1997).
126. P. Vettiger, J. Brugger, M. Despont, U. Drechsler, U. Durig, W. Haberle, M. Lutwyche, H. Rothuizen, R. Stutz, R. Widmer, and G. Binnig, *Microelectron. Eng.* 46, 11 (1999).

ENN Encyclopedia of Nanoscience and Nanotechnology
www.aspbs.com/enn

Ferroelectric Nanomaterials

Pushan Ayyub

Tata Institute of Fundamental Research, Mumbai, India

CONTENTS

1. INTRODUCTION

1.1. Basic Concepts

1.1.1. Ferroelectrics

A ferroelectric is a solid that exhibits a *spontaneous* dielectric polarization, P_S (when $T < T_C$, the Curie temperature), that can be *reversed* by the external application of an electric field, E. The spontaneous (remanent) polarization at $E = 0$ can have one of two values, $\pm|P_R|$, depending on the sign of E just before it was switched off. We define these quantities with respect to the dielectric hysteresis curve (Fig. 1). The spontaneous polarization is defined as the dipole moment per unit volume and its axis is usually a high symmetry direction of the crystal structure. Of the 20 noncentrosymmetric, piezoelectric crystal classes, only 10 (those with C_n, C_{nv}, and C_{1h} point symmetries) have a unique, "polar" axis because their primitive cell has a nonzero dipole moment. These are the 10 pyroelectric crystal classes. A crystal belonging to any of these classes *may* show ferroelectricity provided the polarization is reversible. The usual signatures of a ferroelectric system are: (a) a peak in the temperature dependence of the low frequency dielectric response function (ε) at $T \approx T_C$, and (b) a hysteresis in the P–E curve for $T < T_C$. In most ferroelectrics, the temperature dependence of ε_0 above the Curie point (i.e., in the paraelectric phase) can be described fairly accurately by the Curie–Weiss law,

$$\varepsilon = \varepsilon_0 + C/(T - T_0) \qquad \text{for } T > T_0 \tag{1}$$

where C is the Curie–Weiss constant and T_0 is the Curie–Weiss temperature. In general, a spatially coherent alignment of electric dipoles occurs only in limited regions ("domains") of the crystal, such that a ferroelectric crystal normally has a polydomain structure reminiscent of a ferromagnet. However, such domains may be switched in the same direction ("poled") by the application of a suitable external field. In an antiferroelectric, on the other hand, the ground state consists of an antiparallelly ordered array of local dipoles, which results in the net polarization being zero.

The paraelectric to ferroelectric phase transition may be of the second order ($P_S \to 0$ continuously as $T \to T_C$, although dP_S/dT is discontinuous) or of the first order, in which case P_S has a discontinuity at T_C. Similar continuous or discontinuous behaviors are also exhibited by ε in the two cases. Conventionally, a ferroelectric is termed "displacive" when the elementary dipoles strictly vanish in the paraelectric phase, and "order–disorder" when these dipoles are nonvanishing but thermally average out to zero in the paraelectric phase. In a displacive transition, the displacements of the atoms in the low-temperature ferroelectric phase from their corresponding positions in the high-temperature paraelectric phase are small compared to the internuclear distances. Just below T_C such displacements can indeed be infinitesimally small in the case of a second order displacive phase transition. However, in an order–disorder transition, the displacements across T_C are comparable to the internuclear distances. Though it is often possible to rigorously distinguish between the two types of systems on the basis of the dynamics of their phase transition and the nature of the soft mode involved (whether propagating or diffusive), it is difficult to classify certain ferroelectrics and antiferroelectrics strictly as "displacive" or "order–disorder." The basic physics and phenomenological aspects of ferroelectricity are covered by a number of excellent textbooks [1–4]. Some of the more common ferroelectrics and antiferroelectrics (out of more than a thousand known ones) are listed in Table 1.

ISBN: 1-58883-059-4/$35.00

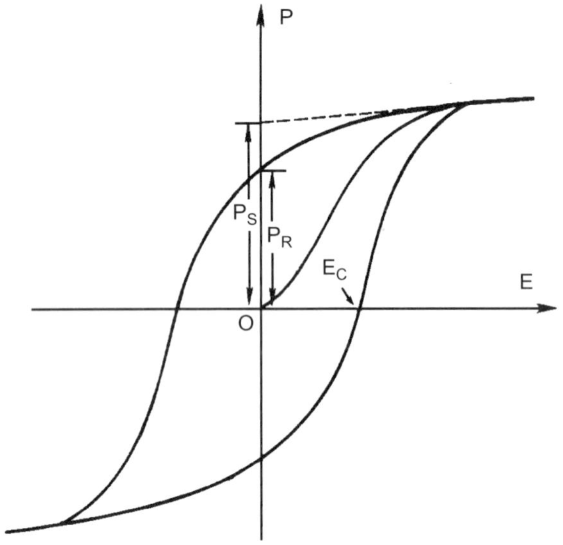

Figure 1. The dielectric hysteresis curve (polarization vs applied electric field) for a typical ferroelectric.

1.1.2. Size Effects in Ferroelectrics

The study of finite size effects in ferroelectric nanoparticles and thin films is an important contemporary problem. Though ferroelectric materials (which are also, by definition, piezoelectric) have been widely used as bulk capacitors and transducers, there are a number of interesting current and futuristic applications that involve these materials in microscopic form. These include sensors of various types and nanorobotic and microelectromechanical devices. In addition, interest in ferroelectrics is growing rapidly because its high dielectric constant can be utilized in dynamic random access memories (D-RAM), while its capacity for being polarized in opposite directions makes it an attractive candidate for nonvolatile random access memories (NV-RAM) and electrically erasable programmable read-only memories (EEPROM). An important motivation for the study of size effects in ferroelectrics is to determine the ultimate level to which a device based on such systems can be miniaturized, since certain properties may get degraded while others are enhanced with a decrease in the characteristic size of the ferroelectric element. From the point of view of basic physics, it is no less interesting to investigate model ferroelectric systems in which the particle size is smaller than the correlation length corresponding to the long range dipolar ordering. For example, the existence of the superparaelectric state (analogous to superparamagnetism in magnetic fine particles) in nanocrystalline ferroelectrics is yet to be established unambiguously.

Table 1. List of some common ferroelectrics (F) and antiferroelectrics (A).

Chemical formula	Structural family	F/A	T_C (°C)
$BaTiO_3$	perovskite	F	120, 5, −90
$PbTiO_3$	perovskite	F	490
$PbZrO_3$	perovskite	A	235
$NaNbO_3$	perovskite	A	354
$KNbO_3$	perovskite	F	435, 225, −10
$BiFeO_3$	Perovskite	F	850
$LiNbO_3$	$LiNbO_3$ type	F	1210
$LiTaO_3$	$LiNbO_3$ type	F	665
$PbNb_2O_6$	tungsten bronze	F	570
$PbTa_2O_6$	tungsten bronze	F	260
$Sr_{0.6}Ba_{0.4}Nb_2O_6$	tungsten bronze	F	75, −213
$Ba_2NaNb_5O_{15}$	tungsten bronze	F	560, 300
KH_2PO_4	KDP	F	−150
$(NH_4)H_2PO_4$	KDP	A	−125
$(NH_2CH_3CO_2H)_3{\cdot}H_2SO_4$	TGS	F	49
$NaNO_2$	nitrite	F	163

Note: The third column lists the ferroelectric/antiferroelectric transition temperature as well as lower transitions.

1.1.3. Experimental Aspects

It is pertinent to make a few brief comments about the experimental techniques involved in these studies. Other than direct observations using electron microscopes and atomic force microscopes, the most popular method for characterizing the particle size is based on X-ray diffraction (XRD) line broadening. Assuming that the small crystallite size is the only cause of line broadening, the coherently diffracting domain size (d_{XRD}) can be related to the full width at half maximum (FWHM) of a particular diffraction peak by the Scherrer equation: $d_{XRD} = K\lambda/\beta(\theta)\cos\theta$, where λ is the X-ray wavelength, $\beta(\theta)$ is the instrument-corrected FWHM of the diffraction line, θ is the diffraction angle, and $K \approx 1$ for equiaxed particles. Different aspects of ferroelectric phenomena are governed by different length scales, such as particle size, grain size, ferroelectric domain size, film thickness, and the lateral dimensions of the relevant devices. The important "intrinsic" length scales are the ferroelectric correlation length and the electron mean free path.

The important characteristics of a ferroelectric material are manifested mainly through its dielectric, piezoelectric, pyroelectric, acoustic, optical, electro-optic, and nonlinear optic properties, and a reduction in the crystallite size may affect many of these. Clearly, the ferroelectric T_C of a system may be measured by studying the temperature dependence of any one of a number of properties, such as the dielectric response, tetragonal distortion, mode softening, etc., though there are subtle differences in the information obtained from these distinct properties. It is, therefore, possible for the nature of the "size effect" to actually depend on the method used to observe it. Thus, structural measurements may show that a ferroelectric system has become pseudo-cubic at a particular critical size, while vibrational spectra may still show evidence for a low symmetry structure.

1.2. Ferroelectricity in Perovskites

Most of the technologically important ferroelectrics and antiferroelectrics are oxides that possess the well-known perovskite structure. These include $BaTiO_3$, $PbTiO_3$, $PbZrO_3$, $KNbO_3$, and mixed oxides such as $Pb(Zr_{1-x}Ti_x)O_3$ and $(Pb_{1-y}La_y)(Zr_{1-x}Ti_x)O_3$. Since much of the work on finite sized ferroelectrics also relates to this class of compounds, we briefly recapitulate the mechanism of ferroelectric ordering in them. Most perovskites are oxide ceramics with the general formula ABO_3. In their cubic unit cell, the cube corners are occupied by the "A" cations, which may be monovalent (e.g., K^+, Na^+, Ag^+) or bivalent (e.g., Ba^{2+}, Sr^{2+}, Pb^{2+}, Ca^{2+}, Cd^{2+}), while the body center is occupied by the smaller B cations, which are pentavalent (e.g., Nb^{5+} or Ta^{5+}) or tetravalent (e.g., Ti^{4+} or Zr^{4+}). The O^{2-} ions are situated at the face centers (Figure 2a). The perovskite lattice is basically a three-dimensional network of BO_6 octahedra, as shown in Figure 2b. The phenomenon of ferroelectricity is controlled by a delicate interplay between the long-range Coulomb forces that favor the ferroelectrically ordered state and short-range repulsive forces that favor the paraelectric state.

The cubic perovskite structure shown in Figure 2a actually represents the high-temperature ($T > T_C$), high-symmetry (cubic) paraelectric phase. It is the off-centering of the B cation in the oxygen octahedron that leads—through a lowering of the symmetry—to the creation of a local dipole moment, and ultimately to ferroelectric or antiferroelectric ordering. The structural distortion from the high-symmetry (paraelectric) phase is therefore of fundamental importance to the stabilization of ferroelectric order in this class of systems. This concept is central to the understanding of finite size effects in displacive ferroelectrics.

In the case of second order transitions of the displacive type, or of relatively "gentle" first order transitions, the static displacements involved in the transition from the high symmetry to the low symmetry structure are similar to the lattice vibrational amplitudes. The link between lattice dynamics and ferroelectricity is the Lyddane–Sachs–Teller equation:

$$\frac{\varepsilon(0)}{\varepsilon(\infty)} = \prod_i \frac{[\omega_{LO}(\vec{k}=0)]_i^2}{[\omega_{TO}(\vec{k}=0)]_i^2} \qquad (2)$$

Here $\varepsilon(0)$ and $\varepsilon(\infty)$ are the static and optical dielectric constants, respectively, while ω_{LO} and ω_{TO} are the longitudinal and transverse optical modes, the product being over all optic branches. If any one of the TO modes (the "soft mode") exhibits an anomalous temperature dependence of the type $\omega_{TO}^2 = K(T - T_0)$, a behavior that is experimentally verified, then $\varepsilon(0)$ would diverge as $T \rightarrow T_0$, as expected across the ferroelectric transition.

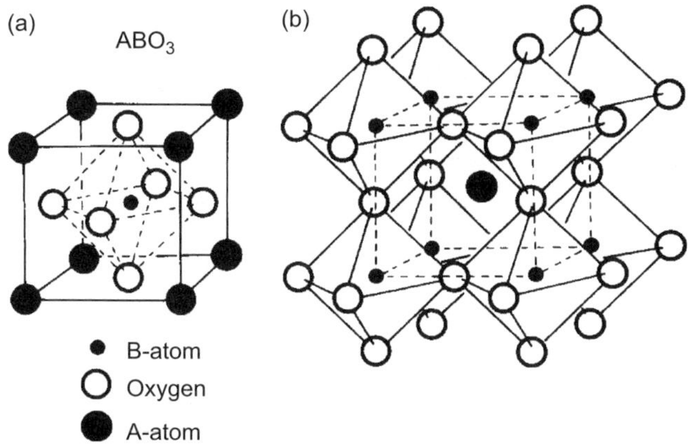

Figure 2. (a) A unit cell of the cubic perovskite (ABO_3) structure, (b) The cubic perovskite lattice consisting of a three-dimensional network of BO_6 octahedra.

The nature of the soft mode can be visualized rather simply in the case of a typical perovskite (ABO_3) compound. The allowed normal modes of vibration (optical modes) are shown in Figure 3. The first three modes are infrared active (i.e., they involve an oscillating dipole moment) while the fourth is "silent" (i.e., neither infrared nor Raman active). Generally, the lowest frequency TO mode is the soft mode (Fig. 3a). As $T \rightarrow T_0$, the frequency of only this mode decreases continuously and at T_0, the restoring force for this mode vanishes due to a cancellation of the long-range (dipole) attractive forces and the short-range repulsive forces. The structure of the low-symmetry phase ($T < T_0$) is determined by the frozen-in soft mode displacements. Referring to Figure 3a, it is clear (since the mode-softening takes place at $k \approx 0$) that a dipole moment would be produced along the z-direction in all the unit cells. Further, the lattice would undergo a *tetragonal distortion*, getting slightly elongated in the direction of the dipole moment and contracted perpendicular to it.

1.3. Size Effects in Ferroelectrics: A Historical Perspective

Though nanoparticle physics, in general, picked up momentum only in the 1980s, research on finite size effects in ferroelectric materials started surprisingly early (in the early 1950s) by Jaccard and co-workers. The first systematic study [5] involved potassium dihydrogen phosphate, KH_2PO_4 (KDP), which is a rather complicated system in that it has characteristics of both displacive and order–disorder mechanisms. In this study of KDP fine particles embedded in an insulating medium with a low dielectric constant, it was found that the spontaneous polarization (which can be considered as the order parameter in ferroelectric

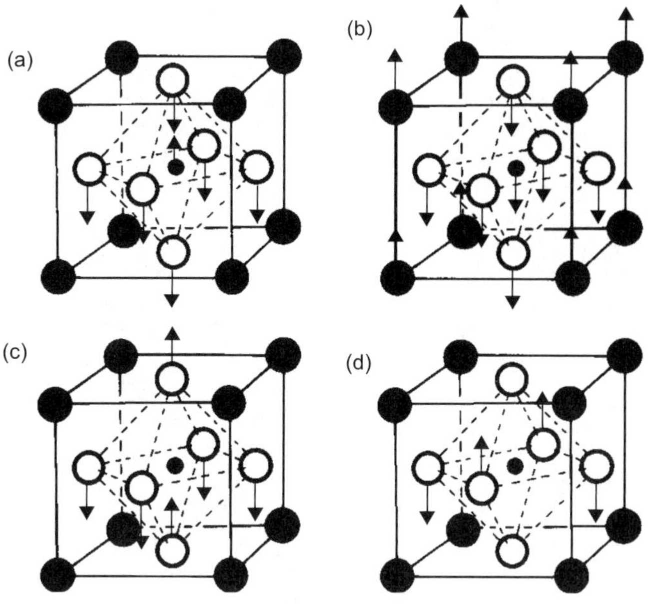

Figure 3. The atomic displacements in the perovskite unit cell (see Fig. 2) for the four normal optical modes of vibration. Though only the vibrations along the z-direction are shown, each of these modes is actually triply degenerate.

systems) was hindered when the particle size was less than 150 nm. Below a critical particle size that depends upon the embedding medium, the depolarization field was thought to prevent polar, ferroelectric structures. Bulk ferroelectric behavior was exhibited only when the particle size was above 0.5 μm.

A contrasting result was obtained in $BaTiO_3$, in which the paraelectric phase (with a cubic perovskite structure) undergoes a tetragonal distortion of about 1% below $T_C \approx 125$ °C to a polar phase. Ball milled fine particles of $BaTiO_3$ in the size range of 0.03–2.3 μm showed a small *increase* in the T_C and a smearing of the transition with decreasing particle size [6, 7]. No critical size (for the destabilization of ferroelectricity) was observed in this size range. The authors concluded from their electron diffraction studies that the deviation from normal ferroelectric behavior was due to the existence of a 10–15 nm thick surface layer that was under a higher tetragonal strain than the bulk and had a *higher* T_C. This layer appeared to remain ferroelectric up to a few hundred degrees higher than the bulk T_C. They suggested that the misfit between the surface layer and the interior caused the broadening of the transition region and the lattice distortions.

The inconsistencies in the early results can be ascribed to the fact that some of the samples were prepared by mechanical grinding, in which case the observations often did not reflect intrinsic size effects but were influenced by residual surface stresses, inhomogeneities, and imperfect crystallization. In spite of such problems, Jaccard et al. were able to make the key observation that the spontaneous tetragonal strain (c/a) in the ferroelectric phase decreases with decreasing particle size.

During the 1960s and 1970s there was considerable progress in the study of grain size effects in ferroelectric *ceramics* or sintered compacts. Such systems, and the results pertaining to them, differ in certain essential aspects from the quasi-free nanoparticles—which would appear to be more suitable for the study of "intrinsic" size effects. However, since ferroelectric ceramics are technologically important, one cannot ignore the possibility of enhancing some of their properties by controlling the size of the polycrystalline grains. The salient observations on ceramic samples of $(Pb_{1-y}La_y)(Zr_{1-x}Ti_x)O_3$ or PLZT [8], $Pb(Zr_{1-x}Ti_x)O_3$ or PZT [9], and other systems can be summarized as follows. With a decrease in the grain size—generally in the 1–5 μm range—(i) there is a decrease in the peak dielectric constant ε_{max}, (ii) the peak of the dielectric response near T_C becomes broader, and (iii) the ferroelectric T_C goes up. According to Martrena and Burfoot [9], these observations can be explained in terms of an intergranular effect (involving the size dependence of the domain wall mobility) and an intragranular effect, where each grain is assumed to be a monodomain and one considers the effect of different-sized grains having different values of T_C. The dielectric response of the system was simulated using a distribution model [10], which assumes that the T_C for different grains obeys a Gaussian distribution. Below each individual T_C, the grains are assumed to have a constant dielectric constant, while the Curie–Weiss law is obeyed above T_C.

Martrena and Burfoot also made use of the concept of a diffuse phase transition (DPT). They considered the crystal or ceramic to consist of many noninteracting microregions that can independently switch from a ferroelectric to a paraelectric state. Several physical situations that could result in a DPT were proposed. In the case of solid solutions such as PZT and PLZT, microscopic variations in the local chemical composition could give rise to a T_C distribution and hence a DPT. The DPT effectively leads to a Curie *region* rather than the conventional Curie *peak*. We will later introduce a physical situation that can lead to a DPT even in *pure* compounds where there is no compositional variation.

2. SIZE EFFECTS IN PEROVSKITE FERROELECTRICS

During the period described, the study of ferroelectric fine particles was mainly an academic exercise, except in the technologically important case of piezoelectric ceramics. After a period of relative inactivity (1975–1987), there was a rather sudden renewal of interest in ferroelectric nanomaterials due to a revival of the idea of a ferroelectric memory device. The possible applications of ferroelectrics and piezoelectrics as microelectromechanical devices, sensors, etc. also provided a strong motivation for a spurt in both theoretical as well as experimental activities in this area. By this time a lot of progress had been made in the development and improvement of synthesis techniques for nanomaterials. These included wet chemical techniques such as sol–gel, co-precipitation, spray-pyrolysis, and microemulsion-mediated reactions (Section 8), and physical vapor deposition techniques such as sputtering, evaporation–condensation, and laser ablation (Section 9). This made it possible to study different compounds in the form of phase-pure, ultrafine particles with a relatively narrow size distribution, as also high-quality ultrathin films on desired substrates.

2.1. $BaTiO_3$

Barium titanate has a cubic perovskite structure in its high-symmetry, paraelectric phase (above $T_C \approx 125$ °C). With a decrease in the temperature, it transforms successively to three structurally distinct ferroelectric phases: a tetragonal phase ($T_C < T < 5$ °C), an orthorhombic phase (5 °C $< T < -90$ °C), and a rhombohedral phase ($T < -90$ °C). All three transitions are of the first order.

Early measurements involved sintered samples with grain size in the submicrometer to few micrometer range and relatively broad size distributions. In samples with grain size in the 1–50 μm range, the dielectric function (ε) was found to depend strongly on size in the ferroclectric phase but not so in the paraelectric phase [11]. The strong dependence of ε on the grain size was explained from the observation that the width of the 90° domains is proportional to the square root of the grain diameter and assuming that the value of ε is determined mainly by the 90° domain walls [12]. In the absence of ferroelectric domains above T_C, the value of ε in the high temperature phase is almost size-independent. For sintered (but not hot-pressed) samples with 20 μm $< d <$ 0.2 μm, the tetragonal to orthorhombic transition is relatively insensitive to the particle size, d [13]. There is, however, an increase in the thermal hysteresis with a decrease in d, indicating that the latent heat of transition is strongly size-dependent. Uchino et al. [14] studied the variation of the tetragonal distortion (c/a) in $BaTiO_3$ fine particles (0.1–1.0 μm) prepared by co-precipitation and hydrothermal routes. Identifying T_C as the temperature at which $c/a \to 1$, they determined the critical size for the existence of ferroelectricity (d_{crit}) in $BaTiO_3$ to be 120 nm. This implies that the $T_C < T_R$ (room temperature) when $d < d_{\text{crit}}$.

Detailed X-ray diffraction (XRD) measurements on smaller-sized (0.01–1.0 μm) $BaTiO_3$ nanocrystals indicate [15] that the structural phase change associated with the ferroelectric transition does not occur when $d < 50$ nm. This was supported by second harmonic generation (SHG) measurements and differential scanning calorimetry (DSC). Interestingly, *no change* was observed in the Raman spectra of the tetragonal structure, with a decrease in the particle size. Note that while XRD and SHG probe the average crystal structure over several unit cells, Raman spectral response is determined by the symmetry of individual unit cells. These results appear to imply that the polar distortion of the TiO_6 octahedra continues to be maintained irrespective of the particle size, though the *average* structure over many unit cells does become cubic. The authors suggest that such local distortions (in apparently "cubic" nanoparticles) exist only at the time scales of light scattering ($\approx 10^{-13}$ s). These results are supported by EXAFS and XANES measurements of the local structure around Ti atoms [16], which show that the magnitude of the Ti atom off-center displacement *does not* depend on the particle size. A careful study of size effects in well-characterized sol–gel derived nanocrystalline $BaTiO_3$ showed that with decreasing size, the cubic–tetragonal transformation temperature shifts downward *only slightly* before getting substantially suppressed, while the tetragonal–orthorhombic transformation temperature shifts slightly upward before getting suppressed [17]. These authors also point out that though hydroxyl defects are clearly observed by infrared (IR) spectroscopy in nanocrystalline samples synthesized at low temperature, the Raman activity observed in XRD-cubic material cannot be ascribed to their presence only.

A more precise understanding of the locally acentric structure in $BaTiO_3$ nanoparticles (15–155 nm) has been recently obtained from electron paramagnetic resonance measurements on Mn^{2+}-doped samples [18]. Comparatively larger crystallites consist of a regular ferroelectric core with a tetragonality gradient toward the outer surface, surrounded by a peripheral layer with strongly distorted symmetry. In particles smaller than $\approx$40 nm the regular core no longer exists at room temperature. Thus, the regular core undergoes a size-driven transition into the paraelectric phase, but the distorted surface layer does not participate in this transition. The size-driven transition is considerably smeared out by the grain-size distribution. The application of Landau theory to individual $BaTiO_3$ particles suggests a critical size of $\approx$50 nm at room temperature whereas in the size-distributed nanopowder a considerably lower mean size ($\approx$25 nm) represents the effective threshold for the size-driven transition into the paraelectric phase.

The value of d_{crit} obtained in these experiments ($\approx$50 nm) contradicts an earlier study of a surfactant-stabilized *suspension* of $BaTiO_3$ particles [19]. Probing the spontaneous polarization through the electro-optic Kerr effect, the authors claim to have detected a permanent dipole moment even in 10 nm particles. However, such a claim may be contested because: (a) the nanoparticles were prepared by a relatively crude ball-milling technique and some relatively large particles may have remained, (b) such a mechanical attrition process may create surface damage and induce defects that show a "polar" response, and (c) the size determination process with arbitrary correction factors appears questionable. Studies on $BaTiO_3$-based glass ceramic composites also show that the T_C, $\varepsilon_{\max}$, and dielectric peak broadening all scale systematically with the mean $BaTiO_3$ crystallite size, but the value of d_{crit} was found to be only 17 nm [20].

2.2. $PbTiO_3$

Lead titanate is a typical displacive ferroelectric with very good dielectric, pyroelectric, and piezoelectric properties. It has a cubic perovskite structure above $T_C = 490$ °C and transforms to a tetragonal ferroelectric phase below it. The room temperature tetragonal distortion, $(c - a)/a$, is as large as 6%, as compared to only 1% for $BaTiO_3$. Due to its obvious application potential, detailed size effect studies have been carried out in this system during the past decade.

Ishikawa et al. [21] prepared $PbTiO_3$ fine particles ($22 < d_{\text{XRD}} < 52$ nm) by an alkoxide route and made a Raman study of samples with different average size as a function of temperature. They measured T_C indirectly using the principle that in displacive ferroelectrics, the frequency (ω_s) of the ferroelectric soft mode (which is usually the lowest TO mode) tends to vanish as $T \to T_C$ [22]. For each sample, ω_s was plotted as a function of T and the T_C was identified as the temperature at which ω_s extrapolates to zero. They found that T_C decreases monotonically when the average particle size, $d < 50$ nm, and its deviation from the bulk value, $T_C(\infty)$, can be fitted to the equation

$$T_C(d) = T_C(\infty) - C/(d - d_{\text{crit}}) \tag{3}$$

with $d_{\text{crit}} = 12.6$ nm and $C = 588.5$ °C. Note that d_{crit} is the size at which $T_C \to 0$ K. Zhong et al. [23] also made Raman-scattering measurements (at room temperature) on sol–gel-derived $PbTiO_3$ nanoparticles (20–200 nm) and found that the ω_s decreases with decreasing particle size, implying a lowering of T_C with size. Even when T_C is identified as the peak of the heat flow curve in DSC, it is found to vary with particle size following Eq. (3) but with $d_{\text{crit}} = 9.1$ nm. However, results obtained from such indirect measurements can be influenced by surface depolarization fields that may affect the TO mode frequency.

The nanostructure of ultrafine ferroelectric $PbTiO_3$ particles (20–2000 nm) has been imaged by high-resolution transmission electron microscopy (HR-TEM) [24]. High-resolution images and selected area diffraction patterns showed that all the particles had tetragonal structure, and that the c/a ratio and domain size both decreased with decreasing particle size. The particles became monodomain when $d < 20$ nm. A domain wall width of 1.4 nm was deduced from strain contrast shown by 90° domain walls, but there was no evidence of amorphous surface layers.

Because it has a large spontaneous polarization, a high T_C, only one structural transition (at T_C), and there is a satisfactory conceptual understanding of its transition mechanism, $PbTiO_3$ is certainly the candidate of choice for studying finite size effects in ferroelectric systems. However, the studies reported suffer from certain drawbacks: (i) the transition temperature is measured either indirectly, or using only a single method, (ii) there are no proper studies of the size dependence of the dielectric response, which is the single most important property of ferroelectrics, and (iii) no clear correlation is made between the dielectric properties and the structural parameters (particularly the ferroelectric distortion). The following study [25] fills up some of these lacunae and provides data that could lead to a clearer conceptual understanding.

Ultrafine $PbTiO_3$ particles ($20 < d_{XRD} < 200$ nm) were synthesized by co-precipitation. Though the average particle size was obtained from X-ray line broadening (d_{XRD}), specific surface area measurements (gas adsorption), and scanning electron micrographs (see Fig. 4), the X-ray domain size (d_{XRD}) is expected to be more relevant to the problem since ferroelectric phenomena in displacive systems are controlled by lattice vibrational mechanisms. The T_C was obtained from three complementary measurements: (i) dielectric response vs temperature, (ii) temperature dependence of the tetragonal distortion (c/a), which scales with the spontaneous polarization, and (iii) differential scanning calorimetry. XRD measurements were also made down to 15 K to confirm that the large deviation observed in c/a at room temperature is an inherent particle size effect.

Figure 5 shows the variation of the ferroelectric distortion (c/a) with particle size at room temperature. The unit cell dimensions do not change down to $\approx$100 nm, below which c decreases and a increases with decreasing size. The resulting reduction in c/a is especially rapid below $\approx$50 nm. To estimate the critical size at which $PbTiO_3$ transforms to the cubic (paraelectric) phase *at room temperature*, the c/a vs d_{XRD} data were fitted to an equation of the form

$$y = y_\infty - C\exp[C(d_{crit} - x)] \tag{4}$$

with $x \equiv d_{XRD}$, $y \equiv c/a$, $y_\infty \equiv 1.065$, and $C \equiv y_\infty - 1$. Using the critical size (d_{crit}) as a fitting parameter, one finds—by extrapolation—that $c/a \to 1$ when $d_{crit} \to 7.0$ nm (ferroelectric order cannot be sustained below this size). The dielectric, thermal, and structural (variable temperature XRD) data also indicated that the T_C decreases monotonically with decreasing particle size.

A study of the temperature dependence of the low frequency dielectric response function, ε (Fig. 6, top), shows that with a decrease in the particle size, T_C decreases (Fig. 7, inset), ε_{max} decreases, and the peaks become increasingly broader. Samples with $d_{XRD} \leq 26$ nm exhibited no peaks in the ε vs T curves. In larger particles of $PbTiO_3$ ($d_{XRD} \geq 60$ nm), a decrease in frequency leads to an increase in ε_{max}, but the ferroelectric T_C remains constant at $\cong$500 °C. In smaller particles of $PbTiO_3$ (e.g., with $d_{XRD} = 31$ nm), however, the T_C moves up with an increase in frequency from 476 °C at 500 kHz to 488 °C at 1 MHz. Also, for a particular frequency, the temperature interval between the maxima in $\tan\delta$ (dielectric loss tangent) and ε is larger in

Figure 4. Scanning electron micrographs (×50,000) of chemically synthesized samples of nanocrystalline $PbTiO_3$ with an average size of about 80 nm (top) and 40 nm (bottom).

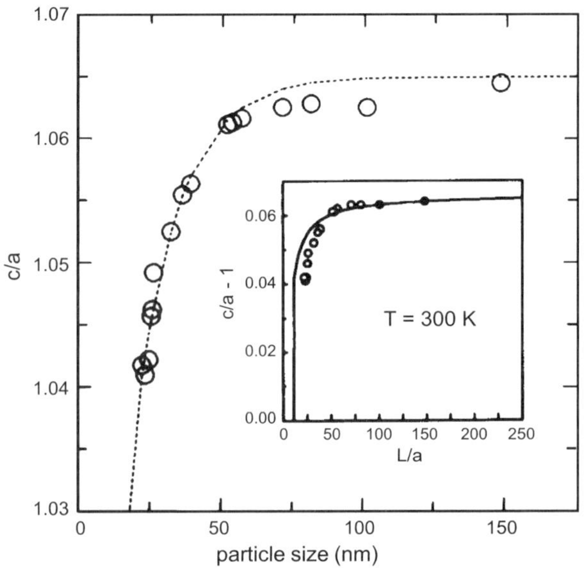

Figure 5. Variation of the ferroelectric distortion (measured at room temperature) with the particle size (d_{XRD}) in nanocrystalline $PbTiO_3$. The dashed line represents a fit to Eq. (4). The inset shows ($c/a - 1$) as a function of the reduced particle size, where $L \equiv d_{XRD}$, and a is the lattice constant. The solid line represents a fit to the model described in Section 4.3.

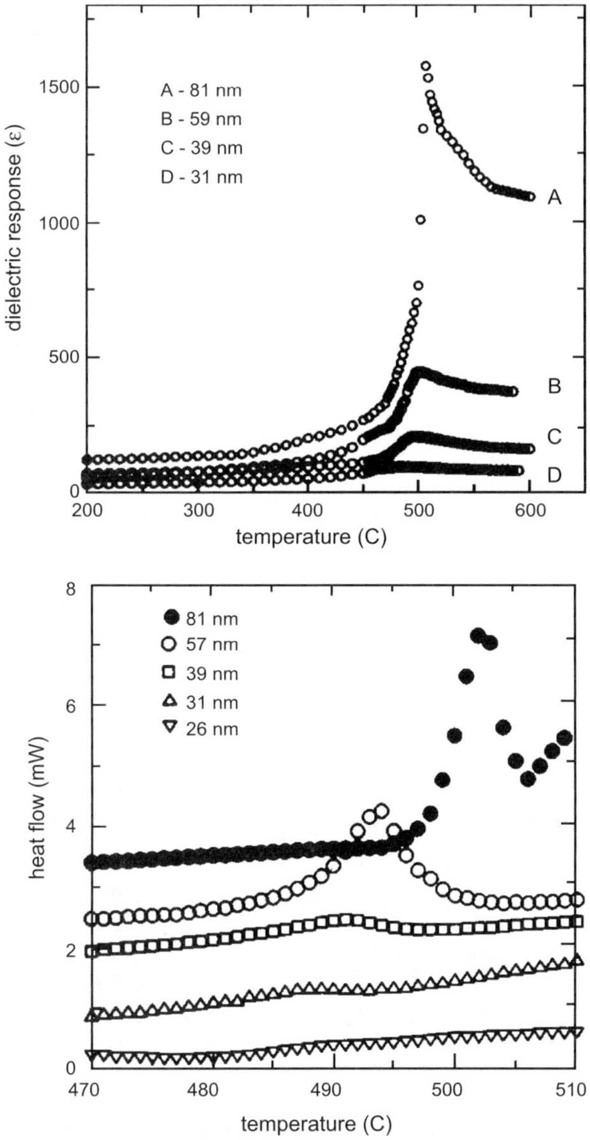

Figure 6. Top: Temperature dependence of the dielectric response function (ε) for $PbTiO_3$ samples with different average particle size (d_{XRD}) measured at 1 MHz. Bottom: Differential scanning calorimetric study of the effect of particle size on the thermal transition at T_C in nanocrystalline $PbTiO_3$, scanned at 5 °C/min.

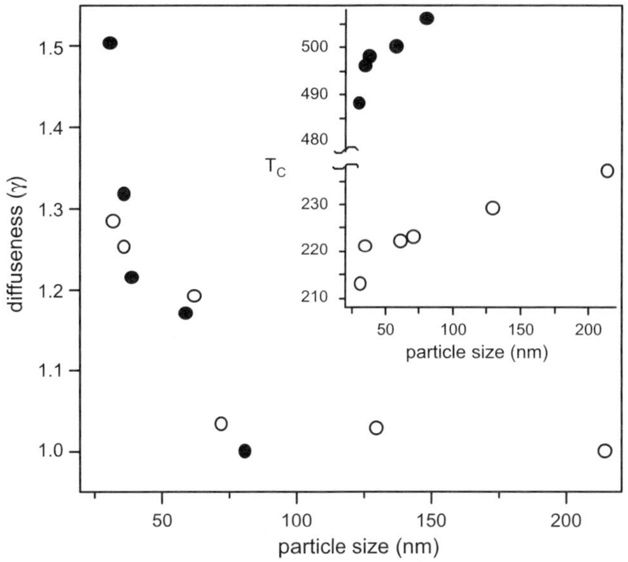

Figure 7. Particle size dependence of the diffuseness coefficient (γ) in $PbTiO_3$ (solid circles) and $PbZrO_3$ (open circles). The variation in the "dielectric" T_C with particle size for the same two systems is shown in the inset.

finer particles. These properties are typical of materials with a diffused phase transition and will be discussed later (Section 4.4). With a reduction in the particle size, the nature of the changes in the "dielectric" and the "thermal" transitions (observed by DSC) were found to be qualitatively similar. Thus, the peak in the heat flow shifted toward lower temperatures and became broader and flatter with decreasing size in $PbTiO_3$ (Fig. 6, bottom).

A quantitative estimate of the "diffuseness" of the ferroelectric phase transition in nanoparticles can be obtained by fitting the ε–T data ($T > T_{max}$, where T_{max} is corresponds to the peak in ε) to thc equation

$$\frac{1}{\varepsilon} - \frac{1}{\varepsilon_{max}} = \frac{1}{C}\left(T - T_{max}\right)^{\gamma} \quad (5)$$

where C is a constant and the critical exponent $\gamma = 1$ for a perfect Curie–Weiss ferroelectric. According to the prediction of the local compositional fluctuation model, $\gamma = 2$ for a system with a completely diffuse transition. For systems exhibiting intermediate degrees of diffuseness we expect γ to have values between 1 and 2. Figure 7 shows that the ferroelectric phase transition becomes increasingly diffuse with a reduction in the particle size (see Section 4.4 for a discussion).

2.3. $PbZrO_3$

Though size effects have been studied for over 35 years in ferroelectrics, there were no systematic studies in nanocrystalline antiferroelectrics prior to the study of $PbZrO_3$ by Chattopadhyay et al. [26] in 1997. At 300 K, $PbZrO_3$ has an orthorhombic structure with eight formula units per crystallographic unit cell. This structure is derived from a cubic perovskite prototype (the high-temperature paraelectric phase) by *antiparallel* displacements of the Pb ions along one of the original [110] directions—which becomes the *a*-axis of the orthorhombic phase. Ignoring the displacements of the Pb ions (so as to compare the paraelectric and ferroelectric unit cells), the ordered phase has a pseudo-tetragonal unit cell whose lattice constants (a_T, c_T) are related to the real orthorhombic unit cell (a, b, c) through $a_T = a/\sqrt{2}$ and $c_T = c/2$. At $T_C \approx 500$ K, $PbZrO_3$ shows a strong dielectric anomaly.

Figure 8 shows the dependence of the pseudo-tetragonal distortion (c_T/a_T) on the particle size in sol–gel-derived ultrafine particles of $PbZrO_3$ ($d_{XRD} \geq 30$ nm). The crystallographic unit cell remains relatively undistorted (from the bulk) down to $\approx$100 nm but thereafter shows an increasing tendency to become cubic. From a downward extrapolation of these data, one may conclude that the crystal structure would become perfectly cubic below a size of 23 nm. With

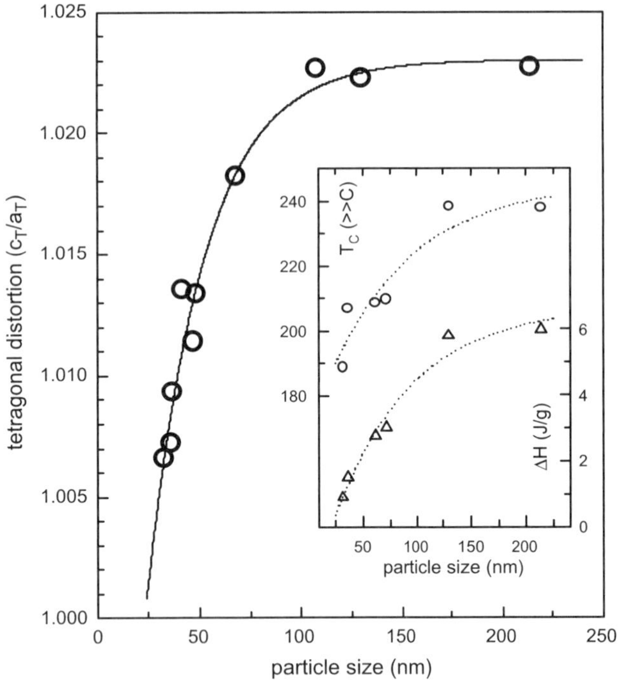

Figure 8. The particle size dependence of the pseudo-tetragonal distortion of the $PbZrO_3$ unit cell, measured at room temperature. The solid line represents a least-squares fit to an empirical function of the form $(1 - e^{-x})$. The inset shows the variation in the "thermal" T_C (obtained from DSC) and the heat of transition with particle size.

decreasing particle size, there is a monotonic reduction in the T_C as observed by both dielectric (Fig. 7, inset) and thermal (Fig. 8, inset) measurements. The ferroelectric phase transition also becomes increasingly diffuse with a reduction in the particle size (Fig. 7). The broadening of the transition is also reflected by the fact that the heat of transition (measured by DSC) decreases with decreasing size (Fig. 8, inset). In general, the nature of size effects in antiferroelectric $PbZrO_3$ is qualitatively similar to those in the ferroelectrics $PbTiO_3$ and $BaTiO_3$.

2.4. Pb(Zr,Ti)O$_3$

Binder has shown that when only intrinsic effects are considered (i.e., contributions from domain walls, defects, and impurities are disregarded) a decrease in particle size should, in general, cause a reduction in the T_C, a reduction in ε_{max}, and a broadening of the dielectric response peak [27]. However, in solid solutions, a broadening in the dielectric response (and other critical parameters) may also take place because of the disordering of the two ions occupying equivalent crystallographic positions. A number of ferroelectric mixed oxides or solid solutions such as PZT, barium strontium titanate (BST), and strontium bismuth tantalate (SBT) are technologically important and studies of particle size effects have been carried out in a few of these compounds. Since it has a number of important applications, it is not surprising that the majority of such studies involve PZT.

PZT or $Pb(Zr_{1-x}Ti_x)O_3$ is obtained by a random replacement of the Ti^{4+} ions in $PbTiO_3$ by Zr^{4+} ions. The ferroelectric T_C increases monotonically from 235 °C ($x = 0$) to 490 °C ($x = 1$). With decreasing x, the structure transforms from tetragonal (as in $PbTiO_3$) to rhombohedral at $x \approx 0.5$ to orthorhombic (as in $PbZrO_3$) at $x \approx 0.05$. The tetragonal to rhombohedral phase boundary is almost temperature-independent (morphotropic), though its exact position and width depends crucially on the purity and local homogeneity of the sample. The piezoelectric properties of PZT are particularly good in the neighborhood of the morphotropic phase boundary and may be further improved by suitable doping. Both the cubic–rhombohedral as well as the cubic–tetragonal transitions get progressively suppressed below ≈ 0.6 μm. Though the ε–T curves get flattened with decreasing size there no discernible change in T_{max} in the rather narrow size range (0.5–0.8 μm) investigated by the authors [28]. They also point out that the values of ε_{max} are higher in samples sintered at higher temperature (and have higher density) even though the X-ray domain size does not differ much between them. This is ascribed to the higher porosity and the discontinuity of electric flux between grains in low-density samples.

Long ago, Kittel [29] had obtained a relation between the grain size (t) and the ferroelectric domain size (d) from a simple model involving the competing effects of the depolarization energy and the domain wall energy:

$$d = \left[\left(\frac{\sigma}{\varepsilon^* P_0^2} \right) t \right]^{1/2} \tag{6}$$

In PZT ceramics, Cao and Randall [30] found that though this relation was valid in the grain size range of 3–10 μm, there was a deviation at lower grain sizes, probably due to the clamping of domain walls at the grain boundaries. Randall et al. [31] have attempted to separate out the role of extrinsic (i.e., due to domain wall, lattice defects, and impurities) and intrinsic contributions to particle size effects in Nb-doped PZT. They suggest that extrinsic contributions to the dielectric and piezoelectric properties get drastically suppressed when the grain size falls below about 2 μm while intrinsic properties get affected below ≈ 0.8 μm, mainly through a progressive reduction in the spontaneous strain.

The dielectric property of *amorphous* nanoparticulate PZT (synthesized by the sol–gel process) was studied by Srinivasan et al. [32]. The results were explained using a model that assumes that the glassy phase contains frozen-in dipoles with random magnitudes and orientations formed by a distortion of the dielectrically soft prototype units (the BO_6 octahedra), which are expected to retain their identity even in the glassy phase.

2.5. Other Solid Solutions

Size effects have also been investigated in a number of other mixed ferroelectric or antiferroelectric oxides. The effect of grain size reduction on the piezoelectric properties in *sintered* ceramics of $(Pb_{1-y}La_y)(Zr_{1-x}Ti_x)O_3$ or PLZT was studied by Okazaki et al. [33]. The authors measured the piezoelectric coefficients (d_{31} and g_{31}) and the planar coupling factor (k_p) in hot-pressed, low-porosity, electrically polarized PLZT samples ($y = 0.2$, $x = 0.35$), with grain sizes (G) in the 1–14 μm range. The size dependence of each property (X) could be expressed by a general equation of the form $X = A + B/G^{0.5}$. They also observed an increase in the coercive field with decreasing size. This paper provided an early guide to the optimization of piezoelectric properties of fine-grained ceramics.

Zhang et al. [34] have carried out a detailed study of size effects for different values of x in BST. Deviations from "bulk" properties in solid solutions typically occur at much larger particle sizes than in pure compounds. Thus, in the case of BST, $d_{crit} = 317$ nm (for $x = 0.3$), 246 nm (for $x = 0.5$), and 176 nm (for $x = 0.7$). The diffuseness coefficient (γ) again increases with decreasing size. However (unlike in $PbTiO_3$ or $PbZrO_3$) the transition becomes *completely* diffuse ($\gamma = 2$) below 300 nm (for $x = 0.3$). We emphasize that in solid solutions such as BST it is difficult to obtain a homogeneous cation distribution, which may be the main reason for the extremely diffuse transition and reduced values of ε_{max}.

In $(Pb_{1-x}La_x)TiO_3$, Raman studies of the soft mode indicate that both the phase transition temperature *and* pressure get reduced with a reduction in particle size [35]. A detailed study [36] of size effects in $Pb(Sc_{0.5}Ta_{0.5})O_3$ in the 10–160 nm range revealed that c/a reduces to unity (at 0 °C) when $d_{XRD} \leq 53$ nm. The T_C decreases gradually and the phase transition becomes more diffuse with decreasing size. Representative studies on size effects in ferroelectric solid solutions (mostly based on the perovskite structure) are summarized in Table 2.

3. SIZE EFFECTS IN OTHER TYPES OF FERROELECTRICS

The preceding part of this chapter shows that nanoparticles of *displacive* systems such as $BaTiO_3$, $PbTiO_3$, and $PbZrO_3$ exhibit certain basic similarities such as (i) a monotonic decrease in the T_C with particle size below $\approx$100 nm, (ii) existence of a critical particle size below which ferroelectricity cannot be observed, (iii) a broadening of the ferroelectric transition peak and behavior reminiscent of a diffuse phase transition at low sizes, and (iv) an interrelation between the T_C, the ferroelectric distortion, and the particle size.

The behavior of a typical *order–disorder* system such as $NaNO_2$, however, is somewhat different. Bulk $NaNO_2$ undergoes a first order ferroelectric–paraelectric transition at 165 °C, which is associated with a relatively large latent heat. Marquardt and Gleiter [46] prepared nanocrystalline $NaNO_2$ (down to 5 nm) by three different methods and studied the ferroelectric phase transition by differential thermal analysis. Very little change in T_C was observed down to 5 nm, whereas displacive systems start showing deviations in the T_C at sizes as large as 100 nm. This clearly emphasizes the importance of the size-induced *structural distortions* in nanoparticles of displacive systems. The other obvious difference in the two types of system is that thermal as well as dielectric phase transitions are usually not observed at all in nanoparticles of displacive systems that are smaller than $\approx$25 nm, while even 5 nm $NaNO_2$ nanoparticles show a clear transition peak. Size effects, therefore, appear to be much weaker in order–disorder ferroelectrics.

Again, no change in T_C ($\approx$440 K) was found in a study of 40–50 nm $NaNO_2$ particles impregnated in the free spaces between the 200 nm SiO_2 spheres that form the three-dimensional (3D) lattice in synthetic opal [47]. However, the authors reported a giant enhancement in the dielectric constant of the $NaNO_2$-impregnated opal (at 100 Hz) above 550 K by about 10^7 times! They ascribed this to the melting of the dispersed $NaNO_2$ in the host lattice and the formation of electrolytic drops. This is similar to the large dielectric constants observed in electrolyte-saturated porous ceramics. More recently, the temperature evolution of the structure of nanodispersed $NaNO_2$ confined within a porous glass with an average pore size of $\approx$7 nm has been studied across T_C with the help of neutron diffraction [48]. This study suggests that there is a lattice softening above T_C, manifested by

Table 2. Studies of size effects in ferroelectric solid solutions (other than PZT) based on the perovskite structure.

System	Sample specs	Nature of study	Ref.
(Pb, La)(Zr, Ti)O_3 or PLZT	(1) sintered ceramics	piezoelectric properties	[33]
	(2) sintered ceramics	$\varepsilon - T$, T_C	[37]
	(3) sintered ceramics	electro-optic properties	[38]
$(Pb_{1-x}La_x)TiO_3$ or PLT	(1) 0.6–11 μm, sinter	$\varepsilon - T$, T_C	[39]
	(2) $0 \leq x \leq 0.2$, sol–gel	Raman at high press. & temp.	[35]
	(3) sol–gel	Raman, soft mode	[40]
$(Pb_{1-x}Sr_x)TiO_3$	sintered ceramic	S/TEM, AFM, DPT	[41]
$Pb(Sc_{1/2}Ta_{1/2})O_3$	10–160 nm	$\varepsilon - T$, T_C, DPT, XRD-T, DSC, $d_{crit} = 10$–20 nm	[36]
$Pb(Zn_{1/3}Nb_{2/3})O_3$–$BaTiO_3$/$PbTiO_3$	model	compositional fluctuation at nanoscale	[42]
$Pb(Mg_{1/3}Nb_{2/3})O_3$–$PbTiO_3$	granular thin film	(1) local piezoelectric properties using SFM	[150]
(Pb, Ba)TiO_3	$0 \leq x \leq 1$, sol–gel	micro-Raman, soft mode	[43]
$(Ba_{1-x}Sr_x)TiO_3$ or BST	$x = 0.3, 0.5, 0.7$, sol–gel	(1) $\varepsilon - T$, T_C, DPT	[34]
		(2) barrier height, relaxation frequency	[44]
$(Sr_xBa_{1-x})Nb_2O_6$ or SBN	sol–gel	$\varepsilon - T$, T_C, DSC, Raman	[45]

a pronounced growth in the thermal vibration parameters. The appearance of an ionic current due to oxygen diffusion in the lattice above T_C is presumed to lead to the giant increase in the dielectric response.

KNO_3 is also believed to be an order–disorder ferroelectric. On heating it transforms from an orthorhombic ferroelectric phase (II) to a trigonal paraelectric phase (I) at 130 °C. On cooling, phase I transforms to a metastable trigonal phase (III) at 125 °C, which goes to phase II at 115 °C. Westphal [49] has studied the effect of particle size on the II–I, I–III, and III–II phase transitions and found that the III–II transition shifted to lower temperature, while the II–I transition temperature increased with decreasing size. However, these measurements were made on powdered single crystals, in which surface stress effects and deformations could play a major role. It is also essential to extend these results to the submicrometer size range and use samples with a narrower size distribution to in order to obtain more reliable results.

4. CONCEPTUAL APPROACHES

4.1. Size Induced Structural Phase Transition

All the perovskite-type oxide ferroelectrics appear to exhibit a decrease in the tetragonal (ferroelectric) distortion with a reduction in particle size. It is important to note that this effect is a manifestation of a much more general phenomenon and is *not* restricted to ferroelectrics. Detailed studies show that in a large number of partially covalent oxides, the crystal lattice tends to transform into structures of higher symmetry when the particle size is reduced [50]. In systems such as Fe_2O_3 and Al_2O_3, the size-induced lattice distortion is large enough to actually produce a crystallographic transition to a high-symmetry structure [51]. In most other cases, there is a gradual reduction in an asymmetry parameter—such as $(c/a - 1)$—with decreasing size.

A decrease in the particle size may cause a change in the bulk lattice parameters via surface stresses [52, 53]. The *anisotropic* nature of the lattice expansion in the nanoparticles of most oxides can be understood in terms of a tendency of such systems to become increasingly *ionic* with decreasing size [54]. As a result, the interatomic bonds lose their directional character and the crystal tends to assume a structure with a comparatively higher symmetry. Interestingly, the size dependence of many important physical properties (such as superconductivity and magnetism, in addition to ferroelectricity) is often directly or indirectly connected to the anisotropic lattice distortion that accompanies a reduction in the particle size [50].

4.2. Overview of Theoretical Models

Experiments on perovskite-type ferroelectrics and antiferroelectrics indicate that a reduction in the average particle size leads to a decrease in the Curie temperature, the ferroelectric distortion (c/a), the heat of transition, and the soft mode frequency. Various attempts have been made to obtain a conceptual understanding of these observations. It is beyond the scope of this chapter to provide a complete review of the theoretical studies of size effects in ferroelectric nanoparticles. We will merely attempt to present a flavor of the area and briefly discuss some simple, general models. Attempts to explain the basic experimental features using the Landau–Devonshire phenomenological theory have been reasonably successful. For a finite-sized, inhomogeneous ferroelectric with a second order phase transition, Zhong et al. [55] write the total free energy in the form

$$\int \left[\frac{1}{2} A(T - T_{C\infty})P^2 + \frac{1}{4} BP^4 + \frac{1}{2} D(\nabla P)^2 \right] dv + \int \frac{1}{2} D\delta^{-1} P^2 \, dS \tag{7}$$

where P is the polarization, and $T_{C\infty}$ refers to the "bulk" value of T_C. D is connected to the correlation length (ξ): $D = \xi^2 \, |A\,(T - T_{C\infty})|$. The volume and surface integrals represent, respectively, the free energies of the interior and the surface. Compared to the analogous expression for an infinite, homogeneous ferroelectric, Eq. (7) contains additional contributions due to the gradient term and the surface term. The "extrapolation length," δ, turns out be particle size dependent. The authors carry out numerical calculations to obtain the size dependence of the polarization as well as the T_C and show that both decrease with decreasing size and vanish as $d \rightarrow d_{\text{crit}}$. A few years later, Rychetský and Hudák analytically solved the phenomenological model for the second order phase transition in spherical ferroelectric particles and obtained the temperature and size dependence of both the static and dynamical dielectric susceptibilities [56].

In these treatments, the surface charge is assumed to be fully compensated, such that the depolarization fields can be ignored. Shih et al. [57] have considered the effect of the depolarization field and a space charge layer—which tend to break up the particles into domains of different polarization. For simplicity, they assume cubic particles with alternating domains separated by 180° domain walls and incorporate the depolarization energy as well as the domain wall energy into the Landau free energy density. Since the multidomain ferroelectric state disappears in small enough particles, the presence of a depolarization field is found to substantially lower the ferroelectric T_C.

More recently, Jiang and Bursill [58] have extended the macroscopic phenomenological approach by calculating the size dependence of the Landau–Ginzburg–Devonshire coefficients [A, B, and D in Eq. (7)] themselves. Other than predicting the size driven phase transition to a paraelectric state for $d < d_{\text{crit}}$, the model can explain the changes in the other physical parameters and appears to give a relatively better description of the changes in the latent heat than previous models. It also yields a simpler free energy expression, which allows some of the physical properties to be discussed analytically. Other than the macroscopic Landau theories, there have also been reasonably successful attempts to investigate size effects with the help of a microscopic pseudo-spin theory based on the Ising model in a transverse field [59, 60].

4.3. Microscopic Model for Nanoferroelectrics

The nature of size-induced changes in displacive ferroelectrics such as $PbTiO_3$ can be understood with the help of the simple model [61, 62] that we now describe. Interestingly, the model also allows one to make certain predictions about order–disorder ferroelectrics which appear to be consistent with the available data. The model is based on two simplifying assumptions: (a) the finite sized ferroelectric system is quasi-free (i.e., it consists of loosely aggregated, unclamped nanoparticles) and (b) the ferroelectric nanoparticles are not electrically isolated. Under these conditions, we can ignore the effects of external strain and depolarization.

Experiments show that a reduction in the size of ABO_3-type displacive ferroelectrics and antiferroelectrics is accompanied by a monotonic decrease in the ferroelectric T_C, which is the temperature at which the high symmetry paraelectric structure transforms to a low symmetry phase in which the B ions no longer occupy the centrosymmetric positions in the BO_6 octahedra. When the particle size is made sufficiently small ($d < d_{\text{crit}}$), the ferroelectric system reverts to the paraelectric phase, *however low the temperature*. In other words, the ferroelectric to paraelectric transition may be made to occur as a function of increasing temperature as well as decreasing size. The following microscopic model naturally leads to a size-induced structural phase transition of the first order.

The displacement of the centrosymmetric B-ion to an off-centric position at the ferroelectric T_C is accompanied by a distortion in the cubic (paraelectric) unit cell to one of lower symmetry (e.g., in $PbTiO_3$, a tetragonal one with sides $a \times a \times c$. The tetragonal distortion ($c/a - 1$) is experimentally found to scale with the order parameter. The displacement (d) of each such ion clearly requires some energy, and the system would undergo a ferrodistortive transition only when it gains a *larger* amount of energy from the interaction of the resulting dipoles, which are aligned in some particular fashion.

The dipolar interaction energy of the stable ordered phase, ($\sim Jzd^2$), should be comparable to kT, where z is the coordination number. Thus, dipolar alignment may take place provided $d \geq \sqrt{kT/Jz}$. Note that this argument itself implies the possibility of driving the transition by varying either the temperature or the system size (which controls the average value of z). So the model describes the temperature-driven transition in the bulk material, as well as the size-driven transition in finite systems.

Phenomenologically, we can describe the interaction energy between the nearest neighbor dipoles by $Jd_id_j\alpha_i\alpha_j$, where $\alpha = \pm 1$ is an Ising variable, which accounts for the fact that the central ion can be displaced in two opposite directions. J is positive for ferroelectrics and negative for antiferroelectrics. The effective Hamiltonian for the problem is thus the sum of an elastic part and an Ising part. From symmetry considerations, we obtain

$$H = \sum_i \left(\frac{1}{2}\lambda_2 d_i^2 + \frac{1}{4}\lambda_4 d_i^4\right) - J\sum_{i,j} d_i d_j \alpha_i \alpha_j \tag{8}$$

This Hamiltonian is solved by treating the Ising variable via an inhomogeneous mean-field theory and the displacement variable within a saddle point approximation. The free energy density is obtained in the form $F(d) = E(d) + I(d)$, where the elastic part, $E(d)$, is given by

$$E(d) = \frac{1}{2}\lambda_2 d^2 + \frac{1}{4}\lambda_4 d^4 \tag{9}$$

and the Ising part, $I(d)$, is given by

$$I(d) = \frac{1}{2}Jzm^2d^2 - T\ln 2\cosh\left(\frac{Jzmd^2}{kT}\right) \tag{10}$$

where $m \equiv \langle\alpha\rangle = \tanh(Jzmd^2/T)$. With increasing d, $E(d)$ increases monotonically and rapidly, while $I(d)$ decreases more slowly. The saddle point values of d, obtained by minimizing $F(d)$, are $d = 0$ and $d = d_0 = \sqrt{(Jzm^2 - \lambda_2)/\lambda_4}$. Clearly, it is this two-minima structure of the free energy that is responsible for the first order ferroelectric ($d = d_0$) to paraelectric ($d = 0$) transition.

Taking the model parameters (λ_2, λ_4, and J) for $PbTiO_3$ from experimental data [25], it is possible to plot the phase diagram in the temperature–particle size plane (Fig. 9). The phase boundary for displacive systems ($\lambda_2 > 0$, solid line) represents first order transitions for all $T > 0$. Thus, the theory predicts a depression in T_C with reducing size. In systems with $\lambda_2 < 0$ (which usually describes order–disorder ferroelectrics), size effects are greatly suppressed, as shown by the dotted curve in Figure 9. One would need to study very small particles at temperatures very close to T_C to observe such effects. This is also in qualitative agreement with experimental observations [46]. Thus, in general and to the first order, particle size effects in ferroelectrics appear to arise mainly from a lower average coordination number for the atoms close to the surface of a nanoparticle.

This model has the important advantage of simplicity—having only three fitting parameters—while the phenomenological Landau theories [55, 57] have about twice as many. Also, by making J negative, the model can be easily extended to describe displacive-type *anti*ferroelectrics.

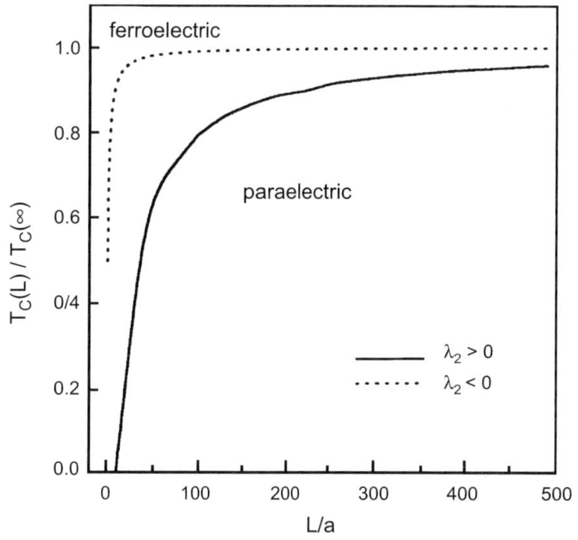

Figure 9. The phase diagram of nanocrystalline ferroelectrics (reduced temperature against reduced particle size) obtained from the model described in Section 4.3. The solid line separates the ferroelectric phase from the paraelectric phase in typical displacive systems, while the dotted line does the same for typical order–disorder systems ($L \equiv d_{\text{XRD}}$, a = lattice constant).

4.4. Diffused Phase Transition

We now discuss another important experimentally observed feature in ferroelectric and antiferroelectric nanoparticles. The dielectric behavior (ε vs T) of nanoparticles of ferroelectric $PbTiO_3$ [25] and antiferroelectric $PbZrO_3$ [26] can be summarized as follows. With a decrease in the particle size: (i) the peaks in the ε–T curve become broader, (ii) there is an increase in the separation between the peaks in the $\tan\delta$ (dielectric loss) vs T and ε vs T curves, (iii) there is an increasing deviation from the Curie–Weiss law (increase in γ), (iv) both the ε vs T and $\tan\delta$ vs T curves show marked frequency dispersion, (v) the decrease in the tetragonal distortion (c/a) as $T \rightarrow T_C$ becomes more gradual in samples with smaller particle size, (vi) values of T_C obtained from different measurements (dielectric, thermal, and structural) do not necessarily coincide, and (vii) with a decrease in frequency, T_C decreases while $\varepsilon_{\max}$ increases.

The broadening of the dielectric and thermal peaks at the ferroelectric transition and most of the other observed features are known characteristics of the so-called "relaxor" ferroelectrics which exhibit a diffuse phase transition (DPT) [63–66]. A quantitative measure of the degree of diffuseness can be obtained by fitting the dielectric data to Eq. (5). The fitted value of the critical exponent (γ) lies between 1 (for an ideal Curie–Weiss ferroelectric) and 2 (for a ferroelectric with a totally diffused transition). Other than nanocrystalline ferroelectrics, a DPT is also exhibited by many complex perovskite ferroelectrics, random dipole ferroelectrics, and ferroelectrics with a graded composition. The near-T_C dielectric response of such relaxor ferroelectrics is often substantially higher than what might be expected from the LST relation [Eq. (2)], making them technologically important. The feature common to all relaxor ferroelectrics is that they possess structural inhomogeneities or compositional variations at the submicrometer scale.

Applying the Landau–Ginzburg free energy formulation to an inhomogeneous medium, Li et al. [67] have provided a thermodynamic approach to DPT in mesoscopically inhomogeneous ferroelectrics. They find that a coherent coupling can take place across the interface between two chemically different dipole-ordered regions. They assume each localized cluster to have a mean-field character, such that the DPT can be understood as an inhomogeneous condensation of localized soft modes.

DPTs were discovered in perovskite solid solutions of the type $A(B_{1-x}C_x)O_3$, in which, generally, there is no long range crystallographic ordering of the B and C cations. Local compositional fluctuations are therefore present throughout the crystal, so that it could consist of microscopic domains of ABO_3 and ACO_3 as well as intermediate compositions (Fig. 10, top). Since ABO_3 and ACO_3 would—in general—have different values of T_C, this would lead to a microlevel T_C distribution about a central value corresponding to the T_C for $A(B_{1-x}C_x)O_3$. There could also be an intermixture of ferroelectric and paraelectric regions if only one of the constituents of the solid solution is ordered at the observation temperature. The scale of heterogeneity is believed to be ~10–20 nm [65].

One can now understand the conceptual similarity between a solid solution and a nanocrystalline system in a simple manner [25]. A nanocrystalline ferroelectric with an average size in the range of 10–50 nm would consist of a collection of crystallites with varying sizes (due to the inherent size distribution) centered around the mean size (Fig. 10, bottom). Recall that in the size range being discussed, particles of different sizes also have different values of T_C. Since the variation of T_C with size is more conspicuous in smaller particles, the latter should exhibit a more diffused phase transition. Many workers have invoked the T_C-distribution model [9] to explain the dielectric behavior of sintered ceramics, in which a complicated stress distribution occurs due to the high compaction. In the comparatively loosely aggregated nanocrystalline material studied by us, the observed strain and the T_C appear to depend directly on the particle size. However, in this case, the strain dependence is *inherently* a size effect and is not an artifact of sintering or pressure compaction.

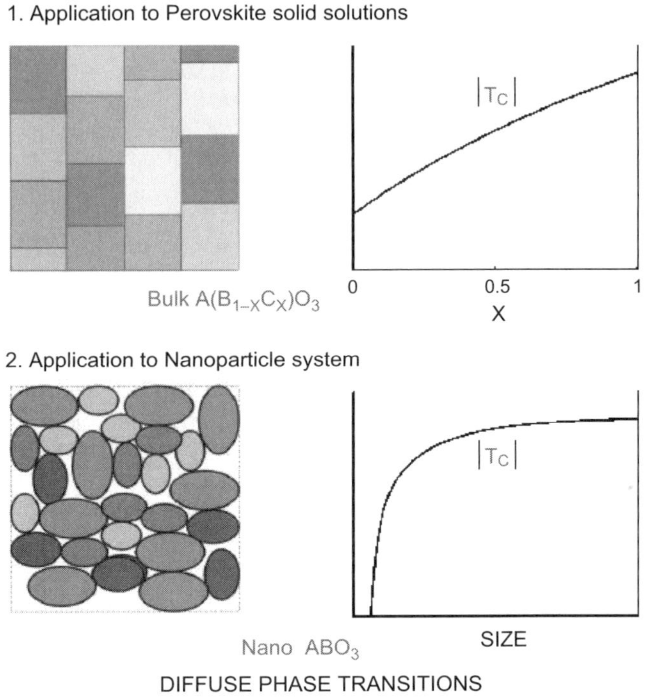

Figure 10. A schematic equivalence of the phenomenon of diffuse phase transition in a typical ferroelectric solid solution (top) with a variation in the local stoichiometry and a nanocrystalline ferroelectric system (bottom) with a particle size distribution. In both situations, different shades represent local regions with different values of T_C.

5. FERROELECTRIC THIN FILMS

5.1. Size Effects in Thin Films

During the past decade, a very large amount of developmental work has been carried out on the synthesis and characterization of ferroelectric thin films—as evidenced by the publication of nine large volumes of the Materials Research Society (MRS) Symposium Proceedings titled *Ferroelectric Thin Films*. This has been fueled mainly by the intense interest in the development of ferroelectric memory devices. A detailed discussion of the technological aspects and the related synthesis problems are not within the scope of this chapter. Readers interested in applications aspects of ferroelectric thin films may refer to the recent and excellent reviews by Scott [68] and Shaw et al. [69], in addition to the aforementioned MRS Symposium Proceedings. Here, we attempt to cover some of the important aspects of inherent size effects in ferroelectric thin films. We will also present an interesting effect in antiferroelectric thin films which appear to show ferroelectric behavior below a critical thickness.

The important length scales in thin films are the film thickness, the grain size (in granular films), and the lateral device size (in thin film devices), and it is interesting to observe the consequences of any of these quantities approaching the ferroelectric critical size (d_{crit}). We also point out that in thin films, it is often difficult to distinguish genuine size effects from extraneous effects related to crystalline quality, microstructural heterogeneities, and mechanical stresses induced by the substrate.

5.2. Intrinsic Effects

A large body of experimental data indicates that size/thickness effects in ferroelectric thin films are manifested mainly in the following five ways [70, 71]. A decrease in the thickness (t) usually leads to (i) a larger coercive field, (ii) a lower T_C, (iii) a smaller remanent polarization, (iv) a lower dielectric constant, and (v) a broadening of the dielectric peak at the ferroelectric transition.

In bulk ferroelectrics, the measured value of the coercive field E_C (the external field required to reverse the macroscopic polarization, see Fig. 1) is almost always much smaller than the theoretical or intrinsic value predicted by the Landau–Ginzburg mean field theory. This is due to domain nucleation and domain wall motion. The energy required to switch the polarization of such small nucleation volumes is much smaller than that required for the collective polarization reversal of the entire crystal. Thus, the obvious way to increase E_C would be to use particles that are so small or films that are so thin that domain nucleation is inhibited. Recently, Ducharme et al. [72] have been able achieve the intrinsic value of E_C in Langmuir–Blodgett films of a polyvinylidene fluoride (PVDF)-based polymeric ferroelectric with $t \approx 1$ nm. The measured E_C is in good agreement with theoretical predictions and is *independent* of t in the range 1–15 nm.

Considering only intrinsic effects, the shift in the T_C and the jump in the spontaneous polarization (P_S) at the T_C as a function of the film thickness (t) are given, respectively, by

$$\Delta T_C = At^{-1} \quad \text{and} \quad \Delta P_S(T_C) = Bt^n \tag{11}$$

where A and B are constants independent of temperature and thickness, while the exponent n is of the order unity. In addition, the film thickness may affect other types of phase transitions. For example, the morphotropic phase boundary (see Section 2.4) that separates the tetragonal phase from the rhombohedral phase in $Pb(Zr_{1-x}Ti_x)O_3$ (PZT) shifts from $\approx$0.5 in the bulk to about 0.2 in thin films [73]. There have been numerous attempts to calculate the intrinsic, thickness dependent electric susceptibility, χ_f. Starting from the continuum, phenomenological free energy, Qu et al. [74] obtain the expression:

$$\chi_f = \chi_b\left[1 + \frac{C_0}{(L - L_0)}\right] \tag{12}$$

where C_0 is independent of L, the film thickness.

The study of ultrathin ferroelectric films also gives us an opportunity to investigate the possible existence of the superparaelectric (SPE) state. This is analogous to the phenomenon of superparamagnetism observed in ultrafine magnetic particles. The latter occurs when the particle size is small enough and the temperature is low enough such that the magnetic anisotropy energy of individual particles is smaller than the thermal energy. Under this condition the magnetization vector of individual particles continues to flip spontaneously by a coherent reorientation of atomic spins. Similarly, the SPE state would occur when the energy required to coherently flip the polarization vector of individual ferroelectric particles is smaller than kT. A simple calculation based on the *intrinsic* coercive field and remanent polarization of typical ferroelectrics, tells us that for the switching energy to approach kT, one requires a particle having a volume $\approx 1\ \text{nm}^3$ [69].

5.3. Extrinsic Effects

Changes in ferroelectric properties usually occur when the film thickness is below ~200 nm, but the actual value of this "critical thickness" often depends sensitively on the nature of the microstructure (and hence on the synthesis conditions), interfacial stresses, crystallographic orientation, and epitaxy. As mentioned earlier, it may be difficult to distinguish between particle size effects and thickness effects since the microstructure often changes continuously with t—thinner films being usually composed of smaller particles. Ren et al. [75] provide an example of this in their study of free standing polycrystalline $PbTiO_3$ films. They find that the domain structure and dynamic response change drastically when the thickness is reduced below 200 nm, because the thinner films are composed of nanosized, single-domain grains. From a study of the effect of granularity on the dielectric response of apparently dense radio-frequency (rf)-sputtered thin films of $BaTiO_3$, Dudkevich et al. [76] found that ε decreases considerably and the ferroelectric state becomes unstable when the coherently scattering X-ray domain size (d_{XRD}) ≤ 30 nm. Since the thinner films are also usually composed of smaller particles, this phenomenon is apt to be misinterpreted as a "thickness" effect.

The issue of thickness induced phase transitions in ferroelectrics is interesting and important. Batra et al. [77] had suggested long ago that a ferroelectric exhibiting a second order transition in the bulk should exhibit a new type of first order transition. This was ascribed to the presence of thickness-dependent depolarization fields in a ferroelectric thin film deposited on a semiconductor surface.

Another type of thickness-dependent phase transition has recently been reported [78]. The study involves the two antiferroelectric compounds $BiNbO_4$ (which is antiferroelectric up to 360 °C, ferroelectric between 360 and 570 °C, and paraelectric above 570 °C) and $PbZrO_3$ (which has a

$T_C \approx 230°C$. In high-purity $PbZrO_3$ samples, it is possible to stabilize a rhombohedral *ferroelectric* phase in a temperature range of 220–230 °C. Polycrystalline $BiNbO_4$ and c-axis oriented $PbZrO_3$ films (100 nm $\leq t \leq$ 900 nm) were pulsed laser deposited on Si (100) wafers. Atomic force micrographs showed that the thinner films were (expectedly) composed of smaller grains in both systems. Dielectric hysteresis (P vs E) measurements showed that in both $PbZrO_3$ and $BiNbO_4$ thin films, the thicker films are antiferroelectric, while—below a characteristic thickness—they show ferroelectric behavior. Note that the critical thickness (t_C) below which ferroelectricity is found to be stabilized is quite large—about 500 nm. The observed stabilization of ferroelectricity in thin films of normally antiferroelectric materials may be ascribed to the effect of the electric field produced at the semiconductor–dielectric interface. For a typical semiconductor–insulator interface, the interfacial field $E_{\text{if}} \approx 100$ kV cm^{-1}, which is high enough to induce ferroelectricity in many antiferroelectric crystals. The nucleation of $PbZrO_3$ and $BiNbO_4$ grains on the substrate takes place at 700 °C (which is above their respective T_C's) in the presence of a strong electric field (E_{if}) that tends to orient the local dipoles along the field, leading to ferroelectric ordering. In thicker films ($t > t_C$), the grains are larger and have multiple domains, which stops this sequential polarization process and the system shows its original antiferroelectric property.

In general, the piezoelectric coefficients of thin films are appreciably lower than those of the corresponding bulk crystals or ceramics. This can be mainly ascribed to (a) the clamping of the film by the substrate, which retards field-induced motion and (b) a reduction in the extrinsic contribution to the piezoelectric response—such as a reduction in the mobility of certain types of domain walls [69].

6. DEVICE APPLICATIONS

The high level of interest in the synthesis, structure, and properties of nanoferroelectrics is obviously related to their current and potential use in devices of various types. Applications are envisaged in memory devices [79] such as D-RAM, NV-RAM, and EEPROM. Important nonmemory applications are in nanorobotic and microelectromechanical (MEMS) devices [80], and in sensors and detectors of various types.

Size effects have a direct bearing on applications. One of the main problems involving ferroelectric thin film capacitors of finite area has been pointed out by Scott [68]. This arises from a reduction in the total quantity of switched charge due to the effect of fringing fields. Thus, in a circular disc of diameter 2 μm and thickness 200 nm (aspect ratio = 10:1) the fringe fields produce a 7% reduction in the switched charge. But this effect increases nonlinearly with decreasing aspect ratio and is about 23% for an aspect ratio of 5:1.

The ferroelectric NV-RAM works on the principle that at zero applied field, a ferroelectric system has two thermodynamically stable states described by $\pm P_R$ (see Fig. 1), each direction of polarization encoding either "0" or "1." Since no external field is required to maintain these states, the memory device is nonvolatile. The polarization can be switched between the two states by the application of an external electric field, $E > \pm E_C$. The criteria for an ideal NV-RAM are fast read/write speed, high recording density, high endurance and retention, radiation hardness, compatibility with current integrated circuit (IC) technology, nondestructive readout capability, low power operation, and commercial viability. Property for property, ferroelectric-based memories are expected to beat most competing technologies. The recent discovery of a series of titanate-based layered perovskite compounds with very high values of spontaneous polarization might turn out to be very useful for NV-RAM applications. For example, a c-axis oriented $Bi_{4-x}Nd_xTi_3O_{12}$-based capacitor was found to have a switchable remanent polarization ($2P_r$) of over 100 μC/cm^2 and fatigue-free behavior [81].

Each memory cell consists of a ferroelectric capacitor sandwiched between metal electrodes and integrated with an access transistor. Thus the chip layout is composed of a matrix of capacitors and transistors. In order to achieve 1 Gbit NV memories, one needs to fabricate ferroelectric capacitors with lateral dimensions of 100 nm or less. The fundamental question to be answered is: what is the smallest size of the capacitor that will still behave like a ferroelectric? One is particularly concerned with the possibilities of depolarizing fields that might become large enough to inhibit switching. According to a recent review [82], the smallest switchable ferroelectric memory cells up to now have been fabricated by Mitsubishi–Symetrix (1 μm × 1 μm) [83], NEC (0.7 μm × 0.7 μm) [84], U. Maryland (100 nm × 100 nm) [85], and MPI Halle (75 nm × 75 nm) [86]. The fabrication process involved is either e-beam lithography, focused ion-beam milling, or self-assembly. Materials and process parameters for ferroelectric D-RAM applications have been reviewed by Scott [87]. As an example, Figure 11 shows a sample of PZT nanodots patterned by e-beam lithography on a $SrTiO_3$ substrate by the group at MPI. See also Table 3.

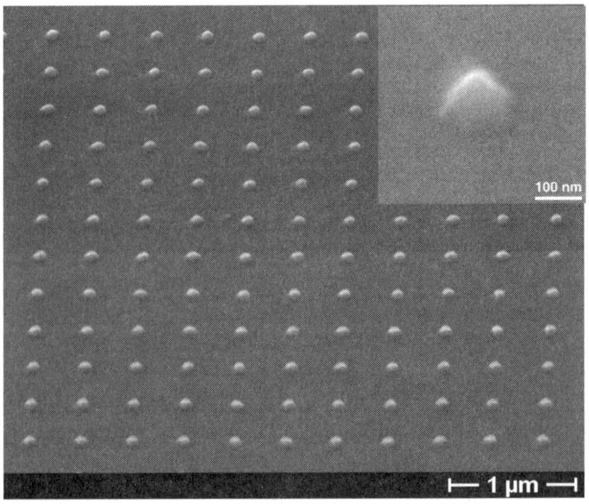

Figure 11. Scanning electron micrograph of large-scale-ordering of PZT (30/70) nanoparticles on $SrTiO_3$ (111) substrate, patterned by electron beam lithography and crystallized at 650 °C. Inset shows magnified view of single dot. Courtesy of Dr. M. Alexe, Max Planck Institute of Microstructure Physics, Halle.

Table 3. Approximate size of smallest switching single ferroelectric cells obtained by various research groups.

Research	Element dimension	Material used	Process	Ref.
Mitsubishi–Symetrix	1000 nm × 1000 nm	(Ba, Sr)TiO_3	e-beam lithography	[83]
NEC	700 nm × 700 nm	$SrBi_2Ta_2O_9$	e-beam lithography	[84]
University of Maryland	100 nm × 100 nm	Pb(Nb,Zr,Ti)O_3	focused ion beam milling	[85]
MPI, Halle	100 nm × 100 nm	$SrBi_2Ta_2O_9$	e-beam lithography	[86]
	75 nm × 75 nm	Pb(Zr,Ti)O_3		

To be useful as a memory element, a ferroelectric should ideally have a low dielectric constant, moderate spontaneous polarization, T_C higher than the operating temperature range, switching speed in the nanosecond range, high chemical and environmental stability, and low fatigue properties. The need for low voltage operation (<5 V) demands that film thickness be in the submicrometer range and the coercive field be low. Two classes of materials, the perovskites (mostly PZT, doped PZT, and BST) and the layered perovskites SBT and $SrBi_2Nb_2O_9$ (SBN) have turned out to be particularly useful for this purpose. For many integrated ferroelectric thin film applications, SBT has now become the material of choice because: (a) it exhibits very little fatigue, (b) it has low leakage currents, and most importantly, (c) bulk properties are retained even in extremely thin films. In recent years, some success has also been achieved in depositing $PbTiO_3$ films on Si [88].

The simpler form of the ferroelectric memory is the destructive readout (DRO) consisting basically of a transistor and a capacitor. Each read operation consists of switching the capacitor, comparing with the linear response of an unswitched reference capacitor, and finally resetting the capacitor. The signal is measured and the state of the bit identified by means of a sense amplifier at the end of each bit line. The reset operation of the DRO device obviously enhances the fatigue. A more sophisticated version is the nondestructive readout (NDRO) where the information can be read over and over again, without the need to switch until the next write operation. The NDRO involves a field effect transistor (FET) with a ferroelectric thin film replacing the metal gate in a conventional FET. Here the polarization of the ferroelectric element (0 or 1) can be sensed without reversal by measuring the source–drain current. The magnitude of the current would depend on whether the gate is biased positively or negatively. Though such a device has many attractive features and some progress has been made in its development [89], a true ferroelectric memory FET has not yet been realized. The main problem is short memory retention and this is ascribed to the depolarizing field and the gate leakage current [90].

Ferroelectric thin films and nanoscale systems are also very useful in several nonmemory applications. These include IR detectors, gas sensors, surface acoustic wave devices, electro-optic switches and modulators, optical FETs, and waveguides. A lot of interest has been generated due to the futuristic applications of MEMS (which include transducers, actuators, accelerometers, etc.) that incorporate piezoelectric thin films such as PZT. Recent developments in this field are now covered by a new journal (*Journal of MEMS*). Other applications of ferroelectric thin films include multilayer capacitors, boundary layer capacitors, varistors, etc.

7. COMPOSITE STRUCTURES

The formation of micro- or nanodispersed ferroelectric phases in ceramics, polymers or glasses with different well-defined geometric structures has opened up a new dimension in the engineering and optimization of desired physical properties. This affords, for example, the interesting possibility of combining the flexibility of a polymer with the piezoelectric or pyroelectric property of an oxide. In addition to the specific properties of its two phases, a biphasic composite is characterized by the scale of the dispersion and the nature of connectivity. In the Newnham convention, the connectivity of a biphasic composite is specified by n–m ($0 \leq n, m \leq 3$) since each phase may be self-connected in zero, one, two, or three dimensions in an orthogonal coordinate system. Thus, a "1–2" connectivity pattern implies that the first phase is self-connected in the form of 1D chains or fibers, while the second phase is self-connected in the form of 2D layers. The convention does not specify whether the fibers of the first phase are parallel or perpendicular to the layers of the second. By convention, the first index (n) usually denotes the ferroelectric phase while the second (m) denotes the matrix (polymer, glass, other ceramic, etc.). Three of the more common and practically realizable connectivity patterns relevant to ferroelectric-polymer systems are 0–3, 1–3 and 3–3 (see Fig. 12). A 0–3 composite consists of dispersed ferroelectric particles in a matrix, a 1–3 composite represents isolated ferroelectric fibers in a matrix, while a 3–3 composite consists of two interpenetrating networks. Similarly, a 3–1 composite may consist of a ferroelectric matrix with parallel cavities filled with a polymer. Some of the earlier work involving ferroelectric composites has been reviewed by Mendiola and Jimenez [91].

7.1. Piezoelectric Properties

Early activity in this area mostly involved piezoelectric–polymer composites [92] and, in view of the strong application potential of such materials, this trend has persisted. PZT, the most studied piezoelectric, has a high piezoelectric strain coefficient (d_{33}) but low voltage coefficient (g_{ij}),[1] which is a drawback in many transducer applications. On the other hand, piezoelectric polymers such as PVDF have comparatively low d values, but their g values are high due to a low dielectric constant. Thus, one expects a PZT/PVDF

[1] The third rank tensors d (piezoelectric strain coefficient) and g (piezoelectric voltage coefficient) are conventionally defined as $d_{ij} = dP_i/d\sigma_j$ and $g_{ij} = dE_i/d\sigma_j$ ($i = 1, 2, 3$ and $j = 1, 2, 3, 4, 5, 6$) where P_i is the polarization, E_i is the electric field, and σ is the stress tensor (of rank 2). The two coefficients are related by $g_{ij} = d_{ij}/\varepsilon_0\varepsilon$ where ε is the dielectric constant and ε_0 is the permittivity of free space.

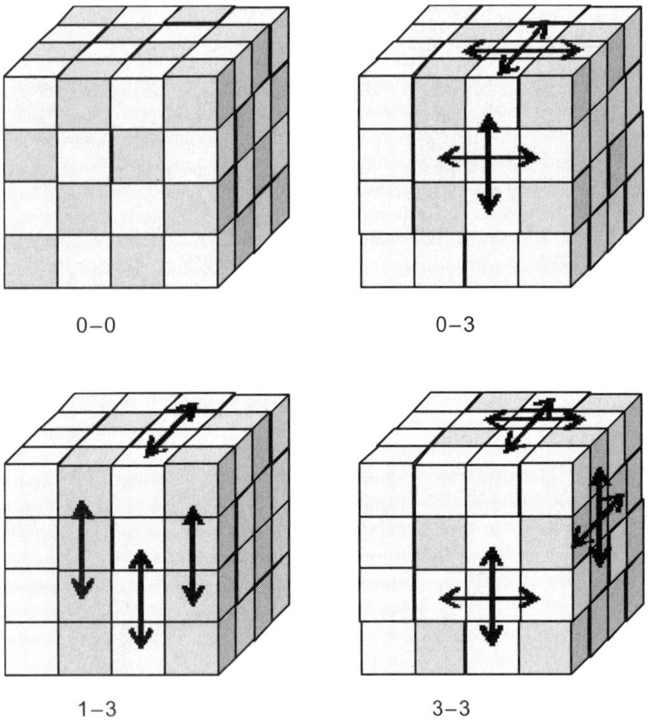

Figure 12. Four of the ten possible connectivity patterns in a biphasic composite system. The pink cells represent the piezoelectric phase, while the yellow cells represent the polymer.

composite to have a combination of desirable piezoelectric properties in addition to having lower density and higher flexibility than sintered PZT. Pioneered by Robert Newnham's group at Penn State University, PZT–polymer composites with 0–3, 1–3, and 3–3 connectivities have been intensely studied and are widely used in ultrasonic imaging and hydrophones. The superior properties of such composites over their monolithic counterparts arise from the macroconnectivity of the stiff, high-permittivity PZT elements in the compliant, low-permittivity polymer matrix. The polymeric matrices used include PVDF, PVA, PVC, and epoxy resins. The presence of the polymer reduces the dielectric constant and the density, leading respectively to a larger voltage output and lower acoustic impedance. Recently, hot-pressed 0–3 PZT–PVDF (50/50 volume fraction) has been found to show reasonably high d_{33} and good mechanical properties [93]. There are several studies of ceramic–ceramic nanocomposites as well. For example, Cheung et al. [94] prepared a dispersion of $BaTiO_3$ nanoparticles in a sol–gel matrix, also of $BaTiO_3$. Films having this composite structure were made by spin coating. Similar 0–3 ceramic–ceramic nanocomposites were also formed using PZT [95] and some of these may find application as ultrasonic transducers. Explicit formulas are available for calculating the effective piezoelectric coefficients of 0–3 composites of ferroelectric particles in a ferroelectric matrix in terms of the composition and the constituents' piezoelectric, dielectric, and elastic properties [96].

Even though the largest hydrostatic figures of merit for biphasic PZT–polymer composites have actually been obtained from 3–*n* types of structure, such as 3–3 [97], 3–2, and 3–1 [98], most of the older techniques used for the synthesis of these architectures are laborious and result in poorly controlled structures. Recently Smay et al. [99] have made use of a robotic direct-write deposition technique ("robocasting") to fabricate a range of such 3–*n* structures and have presented a systematic study of the piezoelectric and dielectric properties as a function of connectivity and PZT/polymer ratio.

7.2. Dielectric Properties

A proper choice of composite architecture can lead to a substantial enhancement in the dielectric constant. Recently, a dielectric constant of over 130 has been observed [100] in a 0–3 $BaTiO_3$–cyanoresin composite (51/49 volume fraction), while the dielectric constant of the pure polymer is only 21 (at 1 kHz). This is one of the highest dielectric constants observed so far in $BaTiO_3$-based polymer composites. Bai et al. [101] have recently reported large enhancements in the dielectric constant of a 0–3 composite where the matrix was an *irradiated* copolymer. High-energy irradiation converted the polymer from a normal ferroelectric to a relaxor ferroelectric, leading to a reduction in the P–E hysteresis at high fields. The dispersed ferroelectric phase was $Pb(Mg,Nb)O_3$–$PbTiO_3$, which is itself a relaxor ferroelectric with a high room temperature dielectric constant. Such a composite showed a weakly temperature dependent dielectric constant of over 250. At optimal composition, the breakdown strength too was very high (120 MV/m), leading to a very large energy storage density of 15 J/cm^3. However, the value of the ceramic content in this case is so high that the material probably loses its flexibility.

Dang et al. [102] have recently reported a three-phase composite with a dielectric constant above 500. The composite consists of an insulating polymer matrix and a homogeneous distribution of small particles of $BaTiO_3$ and Ni. Such a three-phase polymer–matrix composite is flexible and can be easily fabricated into various shapes at low processing temperatures. The electrical properties of such composites can be analyzed by percolation theory when the metallic filler concentration is close to the percolation threshold. Similarly, the incorporation of metallic Ag particles into SBN [103] and Pt particles in PZT [104] leads to significant improvements in the dielectric properties and also to lower processing temperatures. Kundu et al. [105] observed a drastic increase in the dielectric constant of a PZT/glass composite on introducing 4–15 nm PbS particles. In samples with PbS volume fraction varying between 0.07 and 0.14, the dielectric constant ranges between 900 and 3500 (at 1 kHz).

The glass ceramic composite has also been used as a vehicle for basic studies of size effects in ferroelectrics. Thus, McCauley et al. [106] have grown dispersed nanometer-sized $BaTiO_3$ particles in a residual glass matrix. They find that the T_C, ε_{max}, and the width of the ε–T curve all scale systematically with the mean size of the $BaTiO_3$ crystallites. The extrapolated critical size for the destabilization of ferroelectric order was found to be 17 nm.

7.3. Pyroelectric Properties

Ferroelectric–polymer composites also show excellent pyroelectric properties such as the figure of merit p/ε, where p is the pyroelectric coefficient and ε is the dielectric constant.

For example, the value of p/ε for a PZT/polymer composite has been reported to be up to six times higher than in monolithic PZT [107]. Similar results have also been reported for polymer composites involving $PbTiO_3$, $BaTiO_3$, and TGS. Taking the depolarization of the ferroelectric particles into consideration, Wang et al. [108] have derived a general formula for the pyroelectric coefficient of a composite in terms of the dielectric constant of the two components, the volume fraction and the depolarization coefficient. Ploss et al. [109] studied the pyroelectric coefficient of a 0–3 $PbTiO_3$/PVDF–TRFE composite as a function of the relative polarizations of the ceramic and the copolymer. They find that the composite with the ceramic and the copolymer polarized in the same direction exhibits the strongest pyroelectricity but relatively weak piezoelectricity.

7.4. Optical Properties

Several ferroelectric oxides such as $BaTiO_3$, $(Pb, La) \cdot TiO_3$, $LiNbO_3$, and $KNbO_3$ have useful nonlinear optical (NLO) properties because of their polar crystallographic structure. Transparent TeO_2-based glasses containing dispersed ferroelectric crystallites are promising NLO glass candidates. The advantage of using a TeO_2-based glass as matrix is that its refractive index is close to that of several ferroelectric crystals, which leads to low scattering losses at the glass-ceramic interface. Other types of glasses have also proved useful. Transparent glass–ceramic composites in which ≈40 nm $SrBi_2Ta_2O_9$ particles were dispersed in lithium borate glass appears to be a promising candidate for NLO applications [110]. At room temperature, the system shows a pronounced second harmonic generation (at 532 nm) and a ferroelectric hysteresis with a low dielectric loss.

From a calculation of the electro-optic Kerr constant of a composite containing nonlinear, centrosymmetric dielectric microcrystals, it has been shown that the nonsphericity of the microcrystals may lead to an enhancement of the Kerr constant provided the permittivity of the glass ceramic phase is much smaller than that of the inclusions [111].

7.5. Magnetoelectric Properties

The magnetoelectric (ME) effect is characterized by the appearance of an electrical polarization (magnetization) on applying a magnetic (electric) field. A material with a high ME coefficient should find application as magnetoelectric sensors in radioelectronics, microwave electronics, and optoelectronics. Other than the "intrinsic" ME effect shown by many single phase materials (Cr_2O_3, yttrium iron garnets, rare earth ferrites, etc.), an "extrinsic" ME effect may be exhibited, for example, by a ferrite/piezoelectric (ceramic/ceramic) composite, so-called since neither of the two components is individually a ME system. Thus, when a magnetic field is applied to such a composite, the ferrite particles change shape due to magnetostriction, and the strain is passed on to the piezoelectric phase, which results in an electric polarization. The ME effect has, in fact, been observed in composites such as $CoFe_2O_4/BaTiO_3$, (Ni–Co–Mn)ferrite/PZT, etc. Theoretical calculations of the ME effect in composites have been made using Green's function techniques and considering linear and nonlinear coupling between piezoelectric and piezomagnetic particles. Interestingly, a giant linear ME effect is predicted in, for example, a composite of ferromagnetic rare earth iron alloys filled with a ferroelectric polymer such as PVDF [112]. Nan et al. [113] have studied a three-phase multiferroic composite consisting of particles of Tb–Fe–Dy alloy (Terfenol-D), dispersed (with a volume fraction f) in a mixture of PZT and PVDF. The ME response of such a composite is due to the large magnetostriction of Terfenol-D and the large piezoelectricity of PZT, and the coupling between the two through a flexible matrix. The percolation threshold for such a system (as measured from dielectric, piezoelectric, and ME properties) occurs at $f_T \approx 0.12$. For $f < f_T$, the system exhibits increasing piezoelectric and ME response. But these quantities drop sharply as f_T is approached and vanish at f_T, above which volume fraction the composite only shows magnetostriction. The expected giant ME effect remains unrealized as yet.

8. SYNTHESIS OF NANOCRYSTALLINE FERROELECTRICS

A wide variety of chemical techniques have been successfully used for synthesizing nanocrystalline ferroelectrics with controlled mean size and size distribution (see [114] for a recent review). These include sol–gel, co-precipitation, spray-pyrolysis, freeze-drying, microemulsion-mediated reactions, hybrid dry–wet processes, etc. Conventional "dry" powder-mixing processes are usually unsuitable because the temperatures involved are often too high—resulting in grain growth and a broad size distribution. Some of the more useful wet techniques are briefly described.

8.1. Sol–Gel Process

A *sol* is a dispersion of solid particles (less than ≈100 nm in size) in a liquid, while a *gel* is a rigid (or semisolid), porous interconnected network. The sol–gel process [115, 116] consists of the following generic steps (also described schematically in Fig. 13):

(i) preparation of a homogeneous solution of the precursors (e.g., metal alkoxides) in an organic solvent miscible with water [or with the reagent used in step (ii)]
(ii) conversion of this solution into a sol by treatment with suitable reagents
(iii) gelation of the sol via polycondensation
(iv) shaping the gel (or viscous sol) into the desired form (granular, spherical, thin film, fiber, etc.)
(v) low temperature sintering (usually <500 °C) of the shaped gel into the final nanocrystalline product

The precursor metal alkoxides, $M(OR)_n$ (where M is the metal and R is an alkyl group) can be synthesized by reacting the corresponding metal chloride with an alcohol in the presence of a base such as ammonia. The general reaction is

$$MCl_n + nROH + nNH_3 = M(OR)_n + nNH_4Cl$$

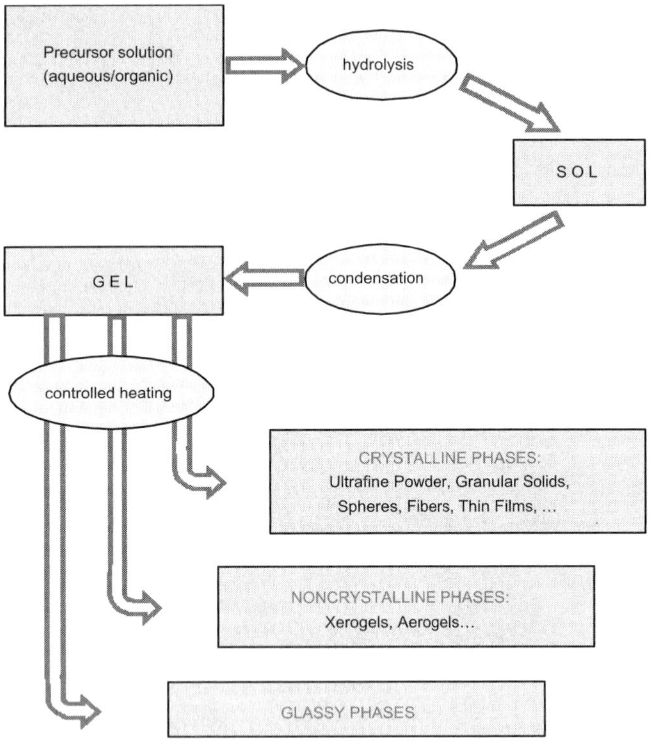

Figure 13. A schematic flow chart of the sol–gel process for the synthesis of low-dimensional oxide/ceramic materials.

The metal alkoxides can then be hydrolyzed to yield oxides, hydrated oxides, and hydroxides in the amorphous, crystalline, or nanocrystalline form. A number of ferroelectric oxides have been successfully synthesized using the basic sol–gel process as well as its many variations. Ozaki [117] has given a detailed discussion of the preparation of a variety of ultrafine Ba and Pb based perovskite compounds and their solid solutions. Jo and Yoon [118] have provided a comparative study of the advantages of the sol–gel technique over the conventional oxide–mixing route in the case of ferroelectric $(Ba_{0.2}Pb_{0.8})TiO_3$. Fukui et al. [119] have used complex alkoxides to prepare submicrometer $Pb(Zr_{1-x}Ti_x)O_3$. Zhong et al. [23] have prepared $PbTiO_3$ fine particles by the basic sol–gel technique. They dissolved lead acetate in 2-ethylhexanol at 80 °C and added a stoichiometric amount of titanium *n*-butoxide to the solution, which was then hydrolyzed by adding water. The resulting precipitate was washed, filtered, and dried. $PbTiO_3$ fine particles of different average size in the 20–200 nm range were obtained by calcining the dried precipitate at different temperatures between 550 and 700 °C. Earlier, Ishikawa et al. [21] had achieved similar results starting from a stoichiometric mixture of titanium tetra-isopropoxide and lead isopropoxide, both dissolved in isopropyl alcohol. Several hybrid techniques based on the basic sol–gel have also been successfully used. Chattopadhyay et al. [26] synthesized nanocrystalline $PbZrO_3$ by using NH_4OH to co-precipitate Pb^{2+} and Zr^{4+} ions from an aqueous solution of $Pb(NO_3)_3$ and $ZrOCl_2{\cdot}8H_2O$. The mixed precipitate was converted to a stable sol by ultrasonication and gelated using 2-ethyl hexanol in presence of a surfactant. The gel was dried and calcined to obtain nano-$PbZrO_3$.

8.2. Co-Precipitation and Complexation

In this process, it is essential to achieve conditions that allow simultaneous and not sequential or selective precipitation of the relevant cations. This would ideally allow a molecular level mixing of the components and may be expected to lead to a totally homogeneous product. However, even in this case the resulting precipitate would be a mixture of different phases, probably with different particle sizes. The formation of the final product would then proceed via solid-state atomic diffusion among these particles, resulting in particle growth and a relatively broad size distribution. Wherever possible, it is preferable to form a *complex* of all the cations that would *directly* lead to the desired phase on thermal decomposition. The decomposition temperature should obviously be optimized to minimize particle growth.

As a simple example, we consider the synthesis of nanocrystalline $PbTiO_3$ by coprecipitation from $Pb(NO_3)_2$ and $TiCl_4$ [25]. $TiCl_4$ (dissolved in HCl) was precipitated as $Ti(OH)_4$ by adding NH_4OH before being converted to soluble $Ti(NO_3)_4$ by adding con. HNO_3. The presence of NH_4NO_3 in the solution prevents the formation of insoluble $PbCl_2$ during the subsequent step. An aqueous solution of $Pb(NO_3)_2$ was added to the $Ti(NO_3)_4$ solution in stoichiometric proportion. The cations were co-precipitated as hydroxides by adding the solution to NH_4OH while keeping the pH controlled at 9–10. The mixed hydroxide precipitate was heated to various temperatures between 460 and 850 °C to produce $PbTiO_3$ nanoparticles in the size range 20–200 nm.

Co-precipitation from organometallic precursors has been used to synthesize ultrafine $BaTiO_3$ [120]. Potassium titanyl oxalate solution was slowly added to a barium acetate solution, being continuously stirred. The precipitate was dried and calcined at 550 °C to produce $BaTiO_3$. A similar technique (using strontium nitrate instead of barium acetate) can be employed to produce $SrTiO_3$.

8.3. Microemulsion-Mediated Reactions

A microemulsion is a thermodynamically stable, optically isotropic dispersion of two immiscible liquids (usually water and oil) consisting of nanosized domains of one liquid in the other, stabilized by an interfacial layer of surface active molecules (surfactant). A water-in-oil microemulsion consists of essentially monodispersed droplets (10–100 nm in size) of an aqueous phase in a continuous hydrocarbon phase. The dispersed aqueous phase can be used as nanoreactors for the precipitation of uniform-sized nanoparticles of a variety of metals, oxides, halides, etc. [121, 122]. Typically, the two reactants are confined within the aqueous cores of two different suitably chosen microemulsion systems. On adding the two, the reactants come into contact in a controlled manner since the aqueous droplets continually collide, coalesce, and exchange contents. The resulting precipitated species remains confined within the core of the microemulsion and is subsequently separated, dried, and heat-treated—if necessary—to obtain the nanoparticulate solid (Fig. 14). The droplet size, and hence the particle size of the final product, is controlled essentially by the water/surfactant ratio, in addition to other microemulsion parameters.

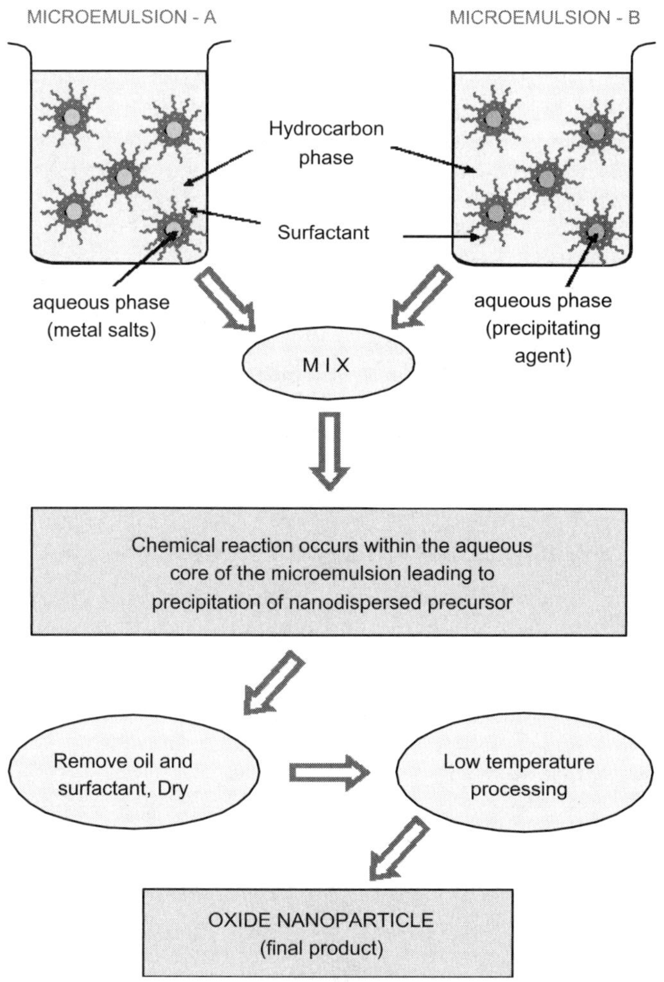

Figure 14. Schematic diagram of the conventional scheme used for microemulsion-mediated precipitation of oxide nanoparticles.

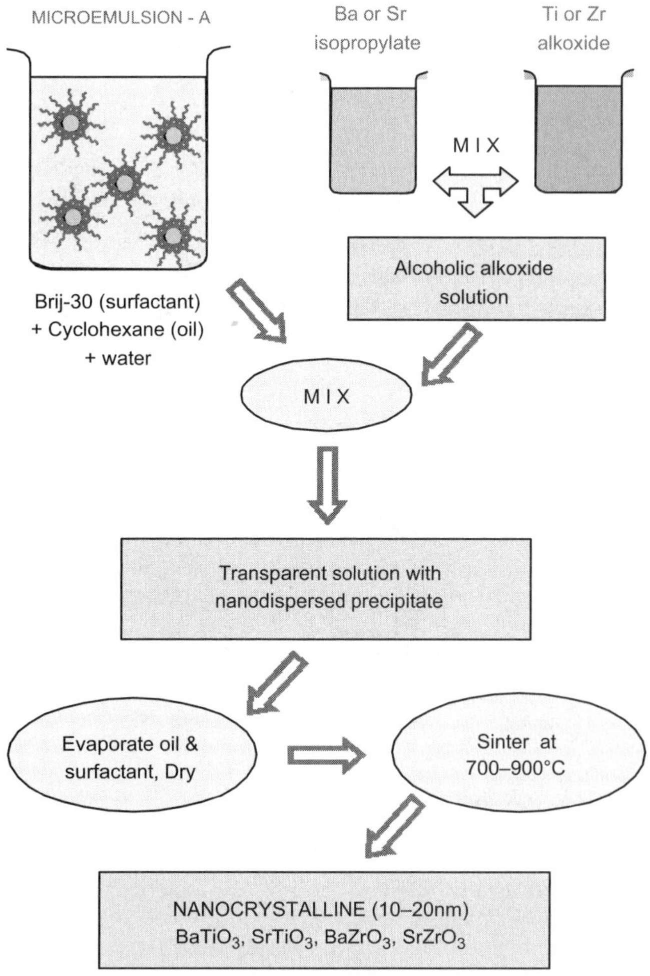

Figure 15. Schematic diagram of the microemulsion-mediated sol–gel reaction for producing oxide nanoparticles.

Herrig and Hemplemann [123] have reported a microemulsion-mediated sol–gel process for the synthesis of perovskite oxides such as $BaTiO_3$, $BaZrO_3$, $SrTiO_3$, and $SrZrO_3$ in the form of ultrafine particles stabilized as colloids in an organic solvent. The metal alkoxide solution was prepared by refluxing Ba or Sr in isopropanol at 125 °C under dry N_2. The Ba/Sr isopropylate was cooled to room temperature and equimolar quantities of $Ti(nprop)_4$ or $Zr(nprop)_4$ added to it. Immediately thereafter, a water-in-cyclohexane microemulsion with Brij-30 (polyoxyethylene-4-laurylether) as surfactant was poured into the solution. This results in a transparent solution consisting of nanodispersed metal precipitates. The alcohol, cyclohexane, and water were removed from the solution by vacuum evaporation and the residue was washed in acetone using a centrifuge. The dried precipitates were sintered at various temperatures to obtain 9 nm $BaTiO_3$ (at room temp.), 18 nm $SrTiO_3$ (700 °C, 1 h), 11 nm $BaZrO_3$ (800 °C, 1 h), or 19 nm $SrZrO_3$ (900 °C, 1 h). A flow chart of the process is shown in Figure 15.

Lu and Wu [124] used a water–oil emulsion to synthesize 100 nm (and larger) $PbTiO_3$ microparticles. The aqueous phase contained a solution of equimolar lead nitrate and titanium oxynitrate (produced by hydroxide precipitation from $TiCl_4$ using NH_4OH, and dissolving the precipitate in HNO_3) dispersed in *n*-octane with 1 wt% SPAN-80 as surfactant. Due to the nature of the surfactant as well as the low surfactant/water ratio used, the resulting mixture is an *emulsion* (which phase-separates spontaneously with time) rather than a microemulsion. The advantage of this technique is that an additional precipitation step is unnecessary and the $PbTiO_3$ fine particles are obtained directly by removing the solvents and subsequent thermal treatment. The disadvantage of not using a microemulsion is that the resulting particle size was rather large: 100–200 nm for samples calcined at 700 °C.

8.4. Hydrothermal Reactions

Hydrothermal synthesis generally involves water as a catalyst as well as a component and is used to describe reactions at temperatures above 100 °C and pressures above 1 atm. This route offers certain important advantages such as ease of use, high product purity, good control over particle size, and low process cost. For example, $BaTiO_3$ powder has been synthesized by a low temperature, low pressure hydrothermal route [125] based on the reaction of TiO_2 (anatase) powder in an alkaline aqueous solution of

$Ba(OH)_2$ at 90 °C. Hydrothermal routes generally involve the processing of an aqueous solution or suspension of the precursor gel powder at somewhat elevated temperatures and pressures. Such precursor gel powders are usually nm-sized gels containing precipitated metal salts (such as chlorides, sulfates, and nitrates) or metal alkoxides. However, spherical gel powders have also been used in hydrothermal synthesis. Thus Choi et al. [126] have synthesized spherical nanoparticles of $PbTiO_3$, $BaTiO_3$, and $SrTiO_3$ from spherical TiO_2 gel powder, while spherical nanoparticles of $PbZrO_3$ and $Pb(Zr,Ti)O_3$ where prepared from spherical gels of ZrO_2 and $ZrTiO_4$, respectively. The TiO_2 and ZrO_2 spherical gel powders were first synthesized by the thermal hydrolysis of $TiCl_4$ and $ZrOCl_2$, respectively, while the spherical $ZrTiO_4$ gel powder was obtained from thermal hydrolysis of a mixture of $TiCl_4$ and $ZrOCl_2$ in an alcohol–water solution. The sources for the *A*-site cations of the perovskite compounds were $Pb(CH_3COO)_2{\cdot}3H_2O$, $Ba(OH)_2{\cdot}8H_2O$, and $Sr(OH)_2{\cdot}8H_2O$. The most interesting point about this process is that the spherical precursor gel powders retain their shape during the hydrothermal treatment to produce rather large (400–800 nm) but perfectly spherical particles of the perovskite oxide product.

Since the hydrothermal technique leads to considerable energy savings due to relatively low processing temperatures, several modifications of the basic process have been investigated. Komarneni et al. [127] have used a combination of microwave and hydrothermal techniques to synthesize a number of ultrafine perovskite such as $BaTiO_3$, $SrTiO_3$, $(Ba,Sr)TiO_3$, $PbTiO_3$, $Pb(Zr,Ti)O_3$, $Pb(Mg,Nb)O_3$, etc. The use of microwaves enhances the reaction kinetics by one or two orders of magnitude over the conventional hydrothermal process. Since the reaction vessel is sealed during the process, the synthesis of Pb- and Bi-based systems, for example, is environmentally safer. The microwave-hydrothermal process typically uses a microwave digestion system operating at 2.45 GHz and 600 W and at a maximum pressure of 200 psi. As a typical example, the authors prepared $BaTiO_3$ from a solution of $Ba(NO_3)_2$, $TiCl_4$, and KOH at 20 psi. The product consisted of spherical, unagglomerated particles having a size in the range of 100–200 nm.

9. DEPOSITION OF FERROELECTRIC THIN FILMS

During the past couple of decades, major advances have taken place in the techniques for the deposition of uniform, homogeneous, device-quality thin films. Several different techniques have been successfully employed for the synthesis of high quality ferroelectric thin films suitable for basic studies as well as device applications. A process for depositing ferroelectric thin film based devices should generally possess the following characteristics [128]:

(1) possibility of integration with metallic or conductive oxide electrode layers with physicochemical properties different from the ferroelectric material,
(2) compatibility with integrated device processing with specific microstructure on specific substrates,
(3) production of highly oriented as well as polycrystalline films and heterostructures with specific properties,
(4) ability to produce patterned structures, superlattices, and layered heterostructures,
(5) perfect reproducibility of the deposition process,
(6) economical with high deposition rates.

The three most successful techniques that satisfy most of these criteria are pulsed laser ablation deposition (PLAD), metal-organic chemical vapor deposition (MOCVD), and plasma/ion-beam sputter deposition (PSD/IBSD). Here we provide brief descriptions of these processes. It is relevant to note that these techniques can also be suitably modified to produce nanocrystalline thin films of various materials (metals, oxides, nitrides, sulfides, etc.) with controlled grain size in the 5–100 nm range [129]. Some of the other techniques that have been used with varying degrees of success for producing ferroelectric films are chemical vapor deposition, metal-organic deposition, liquid source mist deposition, and sol–gel.

9.1. Pulsed Laser Ablation Deposition

The main advantage of PLAD is that it allows rapid and *stoichiometric* deposition of multication as well as multicomponent oxide films. A typical PLAD system requires an excimer laser (such as a KrF laser operated at 248 nm, with an output power density of $\sim$1 J cm^{-2}, and a repetition rate of 1–10 Hz), a rotating target holder, and a substrate holder that can be heated to about 800 °C. The process allows considerable control over the composition, microstructure, and properties. The process parameters that mainly determine these quantities are the ambient gas (usually O_2) pressure, the substrate temperature, the target-to-substrate distance, the laser energy, and wavelength.

9.1.1. Recent Applications

SBT films were grown by PLAD as a part of a metal–ferroelectric–insulator–semiconductor FET structure (SBT/SiON/Si) by Sugiyama et al. [130]. Nakaiso et al. [131] deposited ferroelectric $Sr_2(Ta_{1-x}Nb_x)_2O_7$ films on Pt/Ti/SiO_2/Si and SiO_2/Si(100) substrates. Excellent ferroelectric properties were observed [132] in $(Pb,La)(Zr,Ti)O_3$ deposited directly on metal surfaces. PZT thin films deposited on different substrates (Au coated Si, Al_2O_3, Corning glass, etc.) by reactive pulsed laser (Nd-YAG) deposition show good piezoelectric and ferroelectric properties [133]. Cillessen et al. [134] have shown that the coercive field in PLAD-deposited thin-film ferroelectric capacitors consisting of PZT sandwiched between $La_{0.5}Sr_{0.5}CoO_3$ electrodes is comparable to bulk ceramic values. Textured thin-film capacitors with a columnar microstructure were found to show lower switching voltages than epitaxial films. High quality $BaTiO_3$ thin films were epitaxially grown on $SrTiO_3$ (100) substrates from a single crystal target [135]. The authors also grew $BaTiO_3/YBa_2Cu_3O_7$ bilayer structures on $LaAlO_3$ (100) substrates. Jia et al. [136] deposited $Ba_{0.5}Sr_{0.5}TiO_3$ thin films on conducting RuO_2 [grown epitaxially on (100) $LaAlO_3$], which showed pure (111) orientation normal to the substrate surface and exhibited dielectric constant above 600 and dielectric loss of 0.03–0.05 at 100 kHz under zero dc bias. Palkar et al. [137] were able to deposit perfectly *c*-axis oriented, single phase, ferroelectric $PbTiO_3$ films directly on (100) Si.

9.2. Metal-organic Chemical Vapor Deposition

MOCVD outstrips all other techniques as far as compositional control, large-area film uniformity, and compatibility to ULSI fabrication technology are concerned. Since it is extensively used in commercial IC fabrication, the process parameters and instrumentation are understood extremely well. However, the major disadvantage still faced by this technique in the processing of ferroelectric oxides is the nonavailability of some of the metal-organic precursors with acceptable purity, vapor pressure, and chemical stability. A typical experimental setup (used for the synthesis of $PbTiO_3$ and PZT thin films) [138] consists of a cold-wall reactor with horizontal flow design using a resistively heated inconel susceptor. The system is equipped with three temperature and pressure controlled liquid-source stainless steel bubblers containing the organometallic precursors (for Pb, Ti, and Zr). In the case of low vapor pressure solid precursors, one directly sublimates the solid or evaporates the melt into the carrier gas, which is usually N_2, He, or Ar. For most materials, better thickness and compositional uniformity can be achieved when the processing occurs at pressures below about 100 Torr.

9.2.1. Recent Applications

The application of MOCVD to the synthesis of ferroelectric thin films has been reviewed very recently [139]. Many common ferroelectrics such as SBT, BST, PZT, etc. have been successfully grown by MOCVD and there is substantial volume of literature on the subject. In MOCVD-grown $BaTiO_3$ films on (100)Pt/(100)MgO substrates, the grain size was found to be strongly affected by $p(O_2)$ and showed a maximum when $p(O_2)$ was between 66 and 93 Pa. The dielectric constant increased from 93 to 640 with increasing grain size from 20 to 130 nm, showing a broad peak at 350–380 K [140]. Leedham et al. [141] have addressed the important issue of new, single source MOCVD precursors for SBT, SBN, and SBTN. It is clear that MOCVD remains the most promising technique for ferroelectric thin film based device fabrication and integration.

9.3. Plasma Sputter Deposition

There is extensive work on the PSD of ferroelectric thin films using multiple element targets as well as multicomponent oxide targets. The target (cathode) is connected to a dc or rf power supply (100 V to few kV) while the substrate (anode) may be grounded, electrically floating, or at small dc bias and is typically heated to a few hundred °C. After evacuating the chamber, a gas (typically a few millitorr of Ar) is introduced and serves as the medium in which the plasma discharge is initiated and sustained. It is advantageous to confine the plasma magnetically using a magnetron system. When a visible discharge is maintained between the electrodes, there is a flow of current accompanied by a deposition on the substrate. The process parameters that control the composition, microstructure, and film quality include the gas pressure, the substrate temperature, the target-substrate geometry, and the applied dc/rf voltage. Thus, oriented and epitaxial films can usually be grown when conditions are such that the mean kinetic energy of the sputtered atoms is rather low when they arrive at the substrate. The use of elemental metallic targets in the deposition of PZT films with controlled stoichiometry and properties has been demonstrated [142]. The authors used a disc-shaped target with several unequal sectors of Pb, Zr, and Ti. They suggest that such a process is governed by three factors: (a) the formation of a reproducible oxide layer on the target surface, (b) the stability of the oxide species during transport through the plasma, and (c) the nucleation and growth of the film on the substrate.

9.3.1. Recent Applications

Remiens et al. [143] recently deposited PZT (and Nb substituted PZT) films on Si by rf-magnetron sputtering and studied the effect of process parameters on dielectric, ferroelectric, and piezoelectric properties. Kikuchi et al. [144] deposited ferroelectric PLZT thin films on 200 mm diameter $Pt/Ti/SiO_2/Si$ substrates in a multichamber production system. Using rf-magnetron sputtering from elemental targets, highly oriented (111) $PbTiO_3$ films were deposited on amorphous sapphire for optical studies [145]. SBT films have been deposited by rf-magnetron sputtering on a $Pt/Ti/SiO_2/Si(100)$ structure between 500 and 600 °C. The maximum remnant polarization value ($2P_r$) of the SBT film was 15 $\mu C/cm^2$ [146]. BST films have been grown on $SrTiO_3$, $LaAlO_3$, MgO, α-Al_2O_3, and $Y_3Fe_5O_{12}$ by rf sputtering [147]. In recent times, many other ferroelectric thin films have also been grown using this technique.

The IBSD technique typically uses an Ar^+ beam directed at 45° to the target normal. In a more successful modification, multiple ion beams have been used, where each beam is directed at different elemental targets such as Pb, Ba, Zr, and Ti [148].

10. LOCAL PROBE MEASUREMENTS

Clearly, it is essential to understand the basic phenomena governing size effects in ferroelectric nanomaterials if we want to apply them to the next generation of high density electronic devices and microelectromechanical systems. In addition to some of the important studies described in the preceding sections, a number of new and interesting results have started arriving from various types of high-resolution, local probe measurements such as scanning force microscopy (SFM). These techniques allow us to directly visualize the ferroelectric domain structure and dynamics. SFM-based methods for ferroelectric domain imaging make use of either the piezoelectric property of ferroelectrics or the presence of surface charge associated with the presence of the spontaneous polarization [149]. The two basic modes of the SFM can be utilized in the following ways. (a) *Noncontact Mode*: In the noncontact (attractive) mode, the SFM can be employed as an electrostatic force microscope (EFM), which can be used to detect the static surface charge—proportional to the normal component of the electrical polarization. The surface charge induces an opposite image charge on the probe tip, such that the resulting force gradient depends only on the magnitude but not the sign of the surface polarization. It follows that oppositely oriented

180° domains would present the same contrast in EFM and only domain walls would be clearly seen due to a variation in the charge density in their vicinity. A contrast between opposite domains can be introduced with the help of an external bias voltage. (b) *Contact* (*Piezoresponse*) *Mode*: When the probe tip is in contact with the ferroelectric sample, the domain structure can be observed in the SFM piezoresponse mode by monitoring the local electromechanical oscillation of the sample caused by an ac voltage applied through the probe tip. The sample vibration due to an ac signal of frequency ω also has a second harmonic component at 2ω due to the electrostrictive effect and the dielectric response.

Studying the piezoelectric properties of individual grains in relaxor ferroelectric films [solid solution of $PbTiO_3$ and $Pb(Mg,Nb)O_3$] by SFM, Shvartsman et al. [150] observe a clear correlation between the values of the effective piezoelectric coefficients and the grain size. Larger grains show stronger piezoelectric behavior, spontaneous polarization, and coercivity. A contrasting result was reported by Buhlmann et al. [151], who deposited PZT films by reactive sputtering on conductive Nb:$SrTiO_3$ and patterned features down to 50 nm by e-beam lithography. Piezoelectric sensitive SFM in the contact mode revealed a strong *increase* of the piezoelectric response for feature sizes with lateral dimensions below 300 nm (in 200 nm films). This behavior is believed to be due to vanishing *a*-domains.

Higgins et al. [152] have used scanning piezoresponse microscopy to show that in a ferroelectric thin film that has been depoled using the shrinking hysteresis loop method (similar to degaussing a ferromagnet) the zero-field, zero-polarization state actually consists of an array of up and down polarized domains, and each domain is roughly the size of the grains in the sample. Gruverman et al. [153] have studied the microscopic correlation between the crystallinity, domain arrangements, and switching behavior of SBT thin films deposited by PLAD and flash MOCVD. They show that the domain contrast of an individual grain and its switchability are strongly dependent on the grain orientation.

Scanning probe techniques have also been used to demonstrate that single, 10 nm wide nanowires of $BaTiO_3$ are ferroelectric [154]. These nanowires, synthesized by a solution-based method, are essentially single crystalline, defect-free ferroelectrics. Due to their one-dimensional structure, they should serve as the building blocks for fabricating a variety of nanoscale functional devices. Electrical polarization can be reproducibly induced on such nanowires but is retained only for a few days.

It appears safe to anticipate that in the next few years, many of the fundamental problems in nanoferroelectrics would be addressed and resolved using scanning probe techniques.

GLOSSARY

Antiferroelectric A solid with an antiparallelly ordered array of local dipoles, resulting in a net zero polarization.

Coercive field (E_C) The external electric field required to reverse the macroscopic polarization of a ferroelectric.

Displacive ferroelectric A type of ferroelectric in which the elementary dipoles strictly vanish in the paraelectric phase. The atomic displacements in the low-temperature (ferroelectric) phase with respect to their corresponding positions in the high-temperature (paraelectric) phase are small compared to the inter-nuclear distances.

Ferroelectric A solid that exhibits a spontaneous, field-reversible dielectric polarization, for $T < T_C$, the Curie temperature.

Order–disorder ferroelectric A type of ferroelectric in which the elementary dipoles are non-vanishing but thermally average out to zero in the paraelectric phase. The atomic displacements on both sides of T_C are comparable to the inter-nuclear distances.

Piezoelectric effect Development of electrical polarity on the application of stress. This effect and its converse (the development of strain on applying an electric field) are both linear in nature, a reversal in the stimulus resulting a reversal of the response.

Pyroelectric effect Change in the spontaneous polarization with temperature.

Relaxor ferroelectric In such systems, the temperature dependence of the dielectric constant (ε) displays a broadened peak. Unlike a normal ferroelectric, it obeys a modified Curie-Weiss law in which the inverse dielectric constant varies as some power (between 1 and 2) of the temperature.

REFERENCES

1. M. E. Lines and A. M. Glass, "Principles and Applications of Ferroelectrics and Related Materials." Clarendon Press, Oxford, 1977.
2. T. Mitsui, I. Tatsuzaki, and E. Nakamura, "An Introduction to the Physics of Ferroelectrics." Gordon and Breach, London, 1976.
3. R. Blinc and B. Zeks, "Soft Modes in Ferroelectrics and Antiferroelectrics." Elsevier, New York, 1974.
4. Y. Xu, "Ferroelectric Materials and Their Applications." North-Holland, Amsterdam, 1991.
5. C. Jaccard, W. Känzig, and M. Peter, *Helv. Phys. Acta* 26, 521 (1953).
6. M. Anliker, H. R. Brugger, and W. Känzig, *Helv. Phys. Acta* 27, 99 (1954).
7. W. Känzig, *Phys. Rev.* 98, 549 (1955).
8. K. Okazaki and K. Nagata, *J. Am. Ceram. Soc.* 56, 82 (1973).
9. H. T. Martrena and J. C. Burfoot, *J. Phys. C* 7, 3182 (1974).
10. H. Diamond, *J. Appl. Phys.* 32, 909 (1961).
11. K. Kinoshita and A. Yamaji, *J. Appl. Phys.* 47, 371 (1976).
12. G. Arlt, D. Hennings, and G. de With, *J. Appl. Phys.* 58, 1619 (1985).
13. K. Kanata, T. Yoshikawa, and K. Kubota, *Solid State Commun.* 62, 765 (1987).
14. K. Uchino, E. Sadanaga, and T. Hirose, *J. Am Ceram. Soc.* 72, 1555 (1989).
15. S. Schlag, H. F. Eicke, and W. B. Stern, *Ferroelectrics* 173, 351 (1995).
16. A. I. Frenkel, M. H. Frey, and D. A. Payne, *J. Synchrotron Radiation* 6, 515 (1999).
17. M. H. Frey and D. A. Payne, *Phys. Rev. B* 54, 3158 (1996).
18. R. Bottcher, C. Klimm, D. Michel, H. C. Semmelhack, G. Volkel, H. J. Glasel, and E. Hartmann, *Phys. Rev. B* 62, 2085 (2000).
19. R. Bachmann and K. Bärner, *Solid State Commun.* 68, 865 (1988).
20. C. A. Randall, D. E. McCauley, and D. P. Cann, *Ferroelectrics* 206, 325 (1998).

21. K. Ishikawa, K. Yoshikawa, and N. Okada, *Phys. Rev. B* 37, 5852 (1988).
22. G. Burns and B. A. Scott, *Phys. Rev. Lett.* 25, 167 (1970); *Phys. Rev. B* 7, 3088 (1973).
23. W. L. Zhong, B. Jiang, P. L. Zhang, J. M. Ma, H. M. Cheng, Z. H. Yang, and L. X. Li, *J. Phys.: Cond. Mat.* 5, 2619 (1993).
24. B. Jiang, J. L. Peng, L. A. Bursill, and W. L. Zhong, *J. Appl. Phys.* 87, 3462 (2000).
25. S. Chattopadhyay, P. Ayyub, V. R. Palkar, and M. S. Multani, *Phys. Rev. B* 52, 13177 (1995).
26. S. Chattopadhyay, P. Ayyub, V. R. Palkar, A. V. Gurjar, R. M. Wankar, and M. S. Multani, *J. Phys.: Cond. Mat.* 9, 8135 (1997).
27. K. Binder, *Ferroelectrics* 73, 43 (1987).
28. S. K. Mishra and D. Pandey, *J. Phys.: Cond. Mat.* 7, 9287 (1995).
29. C. Kittel, *Phys. Rev.* 70, 965 (1946).
30. W. Cao and C. A. Randall, *J. Phys. Chem. Solids* 57, 1499 (1996).
31. C. A. Randall, N. Kim, J. P. Kucera, W. Cao, and T. R. Shrout, *J. Am. Ceram. Soc.* 81, 677 (1998).
32. M. R. Srinivasan, P. Ayyub, M. S. Multani, V. R. Palkar, and R. Vijayaraghavan, *Phys. Lett. A* 101, 435 (1984).
33. K. Okazaki, H. Igarashi, K. Nagata, and A. Hasegaka, *Ferroelectrics* 7, 153 (1974).
34. L. Zhang, W.-L. Zhong, C.-L. Wang, P.-L. Zhang, and Y.-G. Wang, *J. Phys. D* 32, 546 (1999).
35. J. Meng, G. Zou, J. Li, Q. Cui, X. Wang, Z. Wang, and M. Zhao, *Solid State Commun.* 90, 643 (1994).
36. Y. Park, K. M. Knowles, and K. Cho, *J. Appl. Phys.* 83, 5702 (1998).
37. A. J. Burggraaf and K. Keizer, *Matter. Res. Bull.* 10, 521 (1975).
38. P. C. Osbond, R. W. Whatmore, V. A. Griffiths, and F. W. Ainger, *Ferroelectrics* 54, 203 (1984).
39. K. Keizer and A. J. Burggraaf, *Phys. Status Solidi A* 26, 561 (1974).
40. Q. F. Zhou, H. L. W. Chan, Q. Q. Zhang, and C. L. Choy, *J. Appl. Phys.* 89, 8121 (2001).
41. X. H. Wang, T. L. Ren, Z. L. Gui, and L. T. Li, *Ferroelectrics* 262, 1219 (2001).
42. Z. X. Yue, X. L. Wang, L. Y. Zhang, and X. Yao, *Acta Phys. Sinica* 6, 913 (1995).
43. J. F. Meng, R. S. Katiyar, and G. T. Zou, *J. Phys. Chem. Solids* 59, 1161 (1998).
44. L. Zhang, W.-L. Zhong, C.-L. Wang, Y.-P. Peng, and Y.-G. Wang, *Europhys. J. B* 11, 565 (1999).
45. S. G. Lu, C. L. Mak, and K. H. Wong, *J. Am. Ceram. Soc.* 84, 79 (2001).
46. P. Marquardt and H. Gleiter, *Phys. Rev. Lett.* 48, 1423 (1982).
47. S. V. Pan'kova, V. V. Poborchii, and V. G. Solov'ev, *J. Phys.: Cond. Mat.* 8, L203 (1996).
48. A. V. Fokin, Yu. A. Kumzerov, N. M. Okuneva, A. A. Naberezhnov, S. B. Vakhrushev, I. V. Golosovsky, and A. I. Kurbakov, *Phys. Rev. Lett.* 89, 175503 (2002).
49. M. J. Westphal, *J. Appl. Phys.* 74, 6107 (1993).
50. P. Ayyub, V. R. Palkar, S. Chattopadhyay, and M. S. Multani, *Phys. Rev. B* 51, 6135 (1995).
51. P. Ayyub, M. S. Multani, M. Barma, V. R. Palkar, and R. Vijayaraghavan, *J. Phys. C* 21, 2229 (1988).
52. M. Ya. Gamarnik, *Phys. Status Solidi* 178, 59 (1993).
53. P. Ayyub, in "Frontiers in Materials Modelling and Design" (V. Kumar, S. Sengupta, and B. Raj, Eds.), p. 228. Springer-Verlag, Heidelberg, 1997.
54. V. R. Palkar, P. Ayyub, S. Chattopadhyay, and M. S. Multani, *Phys. Rev. B* 53, 2167 (1996).
55. W. L. Zhong, Y. G. Wang, P. L. Zhang, and B. D. Qu, *Phys. Rev. B* 50, 698 (1994).
56. I. Rychetský and O. Hudák, *J. Phys.: Cond. Mat.* 9, 4955 (1997).
57. W. Y. Shih, W.-H. Shih, and I. A. Aksay, *Phys. Rev. B* 50, 15575 (1994).
58. B. Jiang and L. A. Bursill, *Phys. Rev. B* 60, 9978 (1999).
59. M. G. Cottam, D. R. Tilley, and B. Zeks, *J. Phys. C* 17, 1793 (1984).
60. C. L. Wang, Y. Xin, X. S. Wang, and W. L. Zhong, *Phys. Rev. B* 62, 11432 (2000).
61. K. Sheshadri, R. Lahiri, P. Ayyub, and S. Bhattacharya, *J. Phys.: Cond. Mat.* 11, 2459 (1999).
62. P. Ayyub, S. Chattopadhyay, K. Sheshadri, and R. Lahiri, *Nanostruct. Mater.* 12, 713 (1999).
63. G. Smolenskii, *J. Phys. Soc. Jpn. Suppl.* 28, 26 (1970).
64. K. Uchino and S. Nomura, *Ferroelectrics Lett.* 44, 55 (1982).
65. L. E. Cross, *Ferroelectrics* 76, 241 (1987).
66. V. S. Tiwari and D. Pandey, *J. Am. Ceram. Soc.* 77, 1819 (1994).
67. S. Li, J. A. Eastman, R. E. Newnham, and L. E. Cross, *Phys. Rev. B* 55, 12067 (1997).
68. J. F Scott, *Ferroelectrics Rev.* 1, 1 (1998).
69. T. M. Shaw, S. Trolier-McKinstry, and P. C. McIntyre, *Annu. Rev. Mater. Sci.* 30, 265 (2000).
70. M. H. Francombe, "Physics of Thin Films: Mechanic and Dielectric Properties." Academic, San Diego, 1993.
71. S. Trolier-McKinstry, P. Chindaudom, K. Vedam, and R. E. Newnham, in "Optical Characterization of Real Surfaces and Films" (M. H. Francombe and J. L. Vossen, Eds.). Academic, New York, 1994.
72. S. Ducharme, V. M. Fridkin, A. V. Bune, S. P. Palto, L. M. Blinov, N. N. Petukhova, and S. G. Yudin, *Phys. Rev. Lett.* 84, 175 (2000).
73. J. F. Scott, *Phase Transitions* 30, 107 (1991).
74. B. Qu, W. Zhong, K. Wang, and Z. Wang, *Integ. Ferroelectrics* 3, 7 (1993).
75. S. B. Ren, C. J. Lu, H. M. Shen, and Y. N. Wang, *Phys. Rev. B* 55, 3485 (1997).
76. V. P. Dudkevich, V. A. Bukreev, Vl. M. Mukhortov, Yu. I. Golovko, Yu. G. Sindeev, V. M. Mukhortov, and E. G. Fesenko, *Phys. Status Solidi A* 65, 463 (1981).
77. I. P. Batra, P. Wurfel, and B. D. Silverman, *Phys. Rev. Lett.* 30, 384 (1973).
78. P. Ayyub, S. Chattopadhyay, R. Pinto, and M. S. Multani, *Phys. Rev. B* 57, R5559 (1998).
79. J. F. Scott, in "Ferroelectric Memories." Springer-Verlag, New York, 2000.
80. W. S. Trimmer, "Micromechanics and MEMS: Classic and Seminal Papers to 1990." IEEE Press, Piscataway, NJ, 1997.
81. U. Chon, H. M. Jang, M. G. Kim, and C. H. Chang, *Phys. Rev. Lett.* 89, 087601 (2002).
82. M. Alexe, C. Harnagea, A. Visinoiu, A. Pignolet, D. Hesse, and U. Gösele, *Scripta Mater.* 44, 1175 (2001).
83. H. Uchida, N. Soyama, K. Kageyama, K. Ogi, and C. A. Pax de Araujo, *Integ. Ferroelectrics* 16, 41 (1997).
84. K. Ammanuma and T. Kunio, *Jpn. J. Appl. Phys. Part 1* 35, 1400 (1996).
85. C. S. Ganpule, A. Stanishevsky, Q. Su, S. Aggarwal, J. Melngailis, E. Williams, and R. Ramesh, *Appl. Phys. Lett.* 75, 409 (1999).
86. M. Alexe, C. Harnagea, D. Hesse, and U. Gösele, *Appl. Phys. Lett.* 75, 1793 (1999).
87. J. F. Scott, *Annu. Rev. Mater. Sci.* 28, 79 (1998).
88. V. R. Palkar, S. C. Purandare, and R. Pinto, *J. Appl. Phys.* 32, R1 (1999).
89. H. Ishiwara, *J. Semicond. Technol. Sci.* 1, 1 (2001).
90. T. P. Ma and J. P. Han, *IEEE Electron Device Lett.* 23, 386 (2002).
91. J. Mendiola and B. Jimenez, *Ferroelectrics* 53, 159 (1984).
92. R. E. Newnham, D. P. Skinner, and L. E. Cross, *Mater. Res. Bull.* 13, 525 (1978).
93. E. Venkatragavaraj, B. Satish, P. R. Vinod, and M. S. Vijaya, *J. Phys. D* 34, 487 (2001).
94. M. C. Cheung, H. L. W. Chan, Q. F. Zhou, and C. L. Choy, *Nanostruct. Mater.* 11, 837 (1999).
95. Q. F. Zhou, H. L. W. Chan, and C. L. Choy, *Thin Solid Films* 375, 95 (2000).
96. C. K. Wong, Y. M. Poon, and F. G. Shin, *J. Appl. Phys.* 90, 4690 (2001).

97. K. Rittenmyer, T. Shrout, W. A. Schulze, and R. E. Newnham, *Ferroelectrics* 41, 189 (1982).
98. A. Safari, R. E. Newnham, L. E. Cross, and W. A. Schulze, *Ferroelectrics* 41, 197 (1982).
99. J. E. Smay, J. Cesarano III, B. A. Tuttle, and J. A. Lewis, *J. Appl. Phys.* 92, 6119 (2002).
100. C. K. Chiang and R. Popielarz, *Ferroelectrics* 275, 1 (2002).
101. Y. Bai, Z. Y. Cheng, V. Bharti, H. S. Xu, and Q. M. Zhang, *Appl. Phys. Lett.* 76, 3804 (2000).
102. Z.-M. Dang, Y. Shen, and C.-W. Nan, *Appl. Phys. Lett.* 81, 4814 (2002).
103. C. Wang, Q. F. Fang, and Z. G. Zhu, *Appl. Phys. Lett.* 80, 3578 (2002).
104. N. Duan, J. E. ten Elshof, H. Verweij, G. Greuel, and O. Dannapple, *Appl. Phys. Lett.* 77, 3263 (2000).
105. T. K. Kundu, M. Mukherjee, D. Chakravorty, and L. E. Cross, *J. Appl. Phys.* 83, 4380 (1998).
106. D. McCauley, R. E. Newnham, and C. A. Randall, *J. Am. Ceram. Soc.* 81, 979 (1998).
107. A. S. Bhalla, R. E. Newnham, L. E. Cross, W. A. Schulze, J. P. Dougherty, and W. A. Smith, *Ferroelectrics* 33, 139 (1981).
108. Y. Wang, W. Zhong, and P. Zhang, *J. Appl. Phys.* 74, 521 (1993).
109. B. Ploss, F. G. Shin, H. L. W. Chan, and C. L. Choy, *IEEE Trans. Dielectric Electrical Insulation* 7, 517 (2000).
110. G. S. Murugan, K. B. R. Varma, Y. Takahashi, and T. Komatsu, *Appl. Phys. Lett.* 78, 4019 (2001).
111. A. A. Berezhnoi, M. I. Vasilev, A. O. Volchek, A. V. Dotsenko, and V. A. Tsekhomskii, *Opt. Spectrosc.* 92, 230 (2002).
112. C. W. Nan, M. Li, and J. H. Huang, *Phys. Rev. B* 63, 144415 (2001).
113. C. W. Nan, L. Liu, N. Cai, J. Zhai, Y. Ye, Y. H. Lin, L. J. Dong, and C. X. Xiong, *Appl. Phys. Lett.* 81, 3831 (2002).
114. V. R. Palkar and S. C. Purandare, *Ferroelectrics Rev.* 2, 169 (2000).
115. L. L. Hench and J. K. West, *Chem. Rev.* 90, 33 (1990).
116. R. C. Mehrotra, "Structure and Bonding," Vol. 77. Springer-Verlag, Berlin, 1992.
117. Y. Ozaki, *Ferroelectrics* 49, 285 (1983).
118. K. H. Jo and K. H. Yoon, *Matter. Res. Bull.* 24, 1 (1989).
119. T. Fukui, C. Sakurai, and M. Okuyama, *J. Mater. Res.* 7, 791 (1992).
120. H. S. Potdar, P. Singh, S. B. Deshpande, P. D. Godbole, and S. K. Date, *Mater. Lett.* 7, 791 (1992).
121. V. Pillai, P. Kumar, M. J. Hou, P. Ayyub, and D. O. Shah, *Adv. Colloid Interface Sci.* 55, 241 (1995).
122. J. Fendler and I. Dekany, "Nanoparticles in Solids and Solutions." Kluwer, Dordrecht, 1996.
123. H. Herrig and R. Hemplemann, *Mater. Lett.* 27, 287 (1996).
124. C.-H. Lu and Y.-P. Wu, *Mater. Lett.* 27, 13 (1996).
125. A. T. Chien, J. S. Speck, F. F. Lange, A. C. Daykin, and C. G. Levi, *J. Mater. Res.* 10, 1784 (1995).
126. J. Y. Choi, C. H. Kim, and D. K. Kim, *J. Am. Ceram. Soc.* 81, 1353 (1998).
127. S. Komarneni, Q. Li, K. M. Stefansson, and R. Roy, *J. Mater. Res.* 8, 3176 (1993).
128. O. Auciello, C. M. Foster, and R. Ramesh, *Annu. Rev. Mater. Sci.* 28, 501 (1998).
129. P. Ayyub, R. Chandra, P. Taneja, A. K. Sharma, and R. Pinto, *Appl. Phys. A* 73, 67 (2001).
130. H. Sugiyama, K. Kodama, T. Nakaiso, M. Noda, and M. Okuyama, *Integr. Ferroelectrics* 34, 1521 (2001).
131. T. Nakaiso, H. Sugiyama, M. Noda, and M. Okuyama, *Jpn. J. Appl. Phys. Pt.1* 39, 5517 (2000).
132. R. Klarmann, J. Hemberger, W. Biegel, and B. Stritzker, *Ferroelectrics* 241, 1777 (2000).
133. F. Craciun, P. Verardi, M. Dinescu, and G. Guidarelli, *Thin Solid Films* 344, 90 (1999).
134. J. F. M. Cillessen, M. W. J. Prins, and R. M. Wolf, *J. Appl. Phys.* 81, 2777 (1997).
135. Z. H. Chen, D. F. Cui, J. Zhang, H. B. Lu, Y. L. Zhou, L. Li, and G. Z. Yang, *Ferroelectrics* 196, 343 (1997).
136. Q. X. Jia, A. T. Findikoglu, R. Zhou, S. R. Foltyn, and X. D. Wu, *Integr. Ferroelectrics* 14, 167 (1997).
137. V. R. Palkar, S. C. Purandare, S. P. Pai, S. Chattopadhyay, P. R. Apte, R. Pinto, and M. S. Multani, *Appl. Phys. Lett.* 68, 1582 (1996).
138. C. M. Foster, Z. Li, M. Buckett, D. Miller, P. M. Baldo, L. E. Rehn, G. R. Bai, D. Guo, H. You, and K. L. Merkle, *J. Appl. Phys.* 78, 2607 (1995).
139. A. C. Jones and P. R. Chalker, *J. Phys. D* 36, R80 (2003).
140. T. Tohma, H. Masumoto, and T. Goto, *Mater. Trans.* 43, 2880 (2002).
141. T. J. Leedham, A. C. Jones, H. O. Davies, N. L. Tobin, P. A. Williams, J. F. Bickley, and A. Steiner, *Integr. Ferroelectrics* 36, 111 (2001).
142. K. Sreenivas, M. Sayer, and P. Garrett, *Thin Solid Films* 172, 251 (1989).
143. D. Remiens, E. Cattan, C. Soyer, and T. Haccart, *Mater. Sci. Semicond. Proc.* 5, 123 (2002).
144. S. Kikuchi, Y. Miyaguchi, T. Jimbo, I. Kimura, M. Tanimura, K. Suu, and M. Ishikawa, *Integr. Ferroelectrics* 46, 105 (2002).
145. Q. Zhao, Z. X. Fan, Z. S. Tang, X. J. Meng, J. L. Song, G. S. Wang, and J. H. Chu, *Surface Coatings Tech.* 160, 173 (2002).
146. Y. Nishioka and H. Ishiwara, *Integr. Ferroelectrics* 46, 265 (2002).
147. S. F. Karmanenko, A. I. Dedyk, N. N. Isakov, Y. J. Oh, V. I. Sakharov, and I. T. Serenkov, *Integr. Ferroelectrics* 46, 265 (2002).
148. S. B. Krupanidhi, H. Hu, and V. Kumar, *J. Appl. Phys.* 71, 376 (1992).
149. A. Gruverman, O. Auciello, and H. Tokumoto, *Annu. Rev. Mater. Sci.* 28, 101 (1998).
150. V. V. Shvartsman, A. Yu. Emelyanov, A. L. Kholkin, and A. Safari, *Appl. Phys. Lett.* 81, 117 (2002).
151. S. Buhlmann, B. Dwir, J. Baborowski, and P. Muralt, *Integr. Ferroelectrics* 50, 261 (2002).
152. M. J. Higgins, A. Krishnan, M. M. J. Treacy, and S. Bhattacharya, *Appl. Phys. Lett.* 80, 3373 (2002).
153. A. Gruverman, K. Hironaka, Y. Ikeda, K. M. Satyalakshmi, A. Pignolet, M. Alexe, N. D. Zakharov, and D. Hesse, *Integr. Ferroelectrics* 27, 1203 (1999).
154. W. S. Yun, J. J. Urban, Q. Gu, and H. Park, *Nano Lett.* 2, 589 (2002).

Encyclopedia of Nanoscience and Nanotechnology
www.aspbs.com/enn

Field Emission of Carbon Nanotubes

Z. P. Huang
NanoLab, Inc., Brighton, Massachusetts, USA

Y. Tu
Boston College, Chestnut Hill, Massachusetts, USA

D. L. Carnahan
NanoLab, Inc., Brighton, Massachusetts, USA

Z. F. Ren
Boston College, Chestnut Hill, Massachusetts, USA

CONTENTS

1. FIELD EMISSION AND FIELD-EMISSION MATERIALS

Field emission—also called Fowler–Nordheim tunneling [1]—is a process where electrons tunnel through the surface potential barrier of a metal–vacuum interface and float in the vacuum region during the presence of an extremely high electric field (Fig. 1). This quantum-mechanical tunneling of electrons is an important mechanism for thin surface potential barriers, just like those in heavily doped semiconductor–metal junctions. Different from thermionic emission, electron field emission occurs without emitter heating (cold cathode). As a result, a field-emission electron source has several prominent advantages: less power consumption, low thermionic noise, high current density (typically 100 A/cm^2 for field emission, 0.5 A/cm^2 for thermionic emission), high coherence, low energy spread, miniature volume, and so on [2].

The Fowler–Nordheim model describes the tunneling current I as a function of the voltage between the cathode and anode V:

$$I/V^2 = a\exp(-b/V) \quad (1)$$

where $a = (\alpha A\beta^2/1.1\phi)\exp[B(1.44 \times 10^{-7})/\phi^{1/2}]$, $b = 0.95B\phi^{3/2}/\beta$, $\beta = E/V$, $A = 1.54\times10^{-6}$, and $B = 6.87\times10^7$. α is the apparent emission area, β is the surface local field enhancement factor (geometric field enhancement factor), E is the electric field intensity at the surface, and ϕ is the electron work function of the emitter. All these parameters characterize electron field emission.

Metals, semiconductors, and graphite have electron work functions of a few electron volts (Table 1), and therefore field emission will not occur until the electric field exceeds 5×10^7 V/cm [2]. In practice, however, it is impossible to build up such an electric field uniformly over any area larger than 10^{-8} cm^2. Field emission always occurs in some tiny protruding areas (10^{-15}–10^{-8} cm^2) associated with a reduced electric field ($<10^6$ V/cm). The reason for this simply is a local enhancement of the electric field in those tiny protruding areas, which dramatically decreases the field intensity needed for field emission. For large-scale, uniform field emission, an array comprising many molybdenum microtips was proposed and developed by Spindt in 1968 [3] (Spindt cathode, TFFEC). The Spindt cathode realized a 1–10 μA/tip field-emission current at 300 V gate bias (60 nm spacing between the gate and tip) [4]. Subsequent improvements in process have enabled the Spindt cathode to produce a current over 100 μA/tip at the gate voltage below

ISBN: 1-58883-059-4/$35.00

Encyclopedia of Nanoscience and Nanotechnology
Edited by H. S. Nalwa
Volume 3: Pages (401–416)

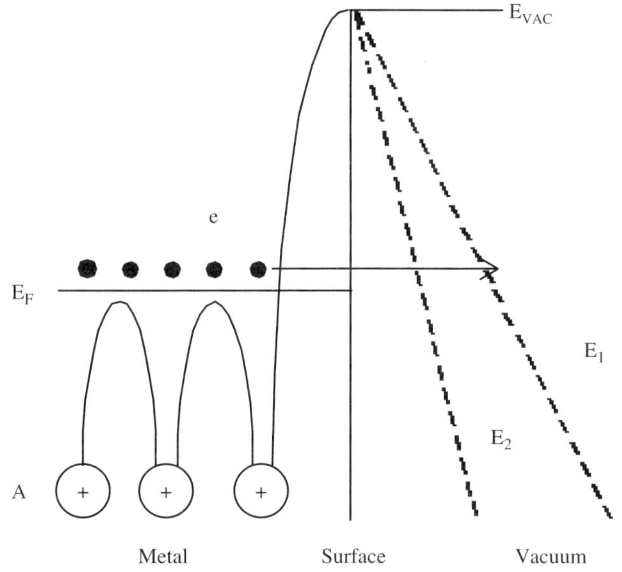

Figure 1. Energy band diagram model for metal. E_F: metal Fermi level, E_{VAC}: vacuum level at metal surface, e: electrons in conduction band, A: metal atoms. Dashed line represents the band diagrams in the electric fields applied to metal, $E_1 < E_2$.

100 V [4]. Following the Spindt cathode, various configurations of emitters in pure and compound materials have been studied in the past three decades, and a continuous effort is still being devoted to achieving cathodes of high performance.

An ideal material for the field emitter should have a high mechanical strength and melting point, excellent electrical conductivity and thermal conductivity, a low work function and stable chemical properties, and be inexpensive and easy to process. From the engineering point of view, field emitters should be able to deliver a high current density at a low electric field. Molybdenum shows not only a high hardness, but also a high melting point and good electrical conductivity (Table 2). Head and shoulders above other materials, molybdenum microtips (Spindt cathode) exhibited an availability of 100–1000 A/cm^2 emission current [2]. However, the inherent features of double evaporation (a technique for fabricating molybdenum microtips) restrains the microtip array from being scaled up. Silicon and polysilicon are prevalent materials in the semiconductor industry.

Table 1. Work functions of metals, silicon, and graphite.

Material	Work function	Material	Work function
Mg	3.61	In	3.97
Al	4.19	Sn	4.43
Ti	3.83	Sb	4.56
Cr	4.44	Ba	2.48
Mn	4.08	Ta	4.20
Fe	4.65	W	4.55
Co	4.79	Pt	5.40
Ni	5.15	Au	5.21
Cu	4.59	Pb	4.04
Mo	4.21	Si (P)	4.30
Pd	5.17	Si (N)	3.30
Ag	4.42	Graphite	4.6–5.4

Table 2. Melting point and resistivity of materials commonly used for field emission.

Material	Melting point (°C)	Resistivity ($\mu\Omega \cdot$ cm)
Mo	2617	5.17
W	3410	5.6
Si	1410	100–10^{11}
C	3550	1375

Arrays of silicon and polysilicon microtips can be potentially scaled up, and the fabrication is also compatible with up-to-date integrated-circuit techniques. Despite a great effort with silicon and polysilicon microtips through the 1980s and 1990s [5–11], the uniformity issue with array scale up and the high cost involved remained. In addition, silicon and molybdenum microtips require a working condition of ultrahigh vacuum and high voltage. Therefore, their current density and stability are limited [12–14, 72–74]. As a potential field-emission material, diamond drew great attention to its negative electron affinity (NEA), and excellent performance in high-speed, high-temperature, and high-power applications [14–23, 68–71]. Negative electron affinity can contribute an intensive electron beam at a low electric field. Unfortunately, diamond fails to benefit from this virtue due to a small concentration of electrons in its conduction band (wide bandgap material). It has been proven that shallow-level n-type doping of wide-bandgap materials is extremely difficult. The shallow donors such as P, Li, and Na showed a very low solubility and activity in diamond [24]. Moreover, the chemical vapor deposition process for diamond is not completely compatible with the fabrication techniques of integrated circuits.

2. FIELD EMISSION OF CARBON NANOTUBES

Since the discovery of carbon nanotubes in 1991 [25], they have been regarded as a new generation of materials for field emission. The carbon nanotube, a seamless small carbon tube formed by rolling up a single or several pieces of graphene sheets, is predicted to have numerous advantages as a field emitter: (1) high chemical stability and high mechanical strength, (2) high melting point and reasonable conductivity, (3) high aspect ratios (10–100) with very small tip radius (5–50 nm) to greatly enhance the local electric field, (4) large-scale production at low cost, (5) an extremely low electron work function, and so on (Tables 2, 3) [13, 26–30].

2.1. Electron Work Function

The electron work function is an energy threshold where electrons start to be emitted into vacuum from a metal conduction band, that is, the difference of energy between vacuum and the Fermi levels (Fig. 1). The work function varies with different metals, and ranges from 2 to 6 eV (Table 1). The modified Fowler–Nordheim equation $(I/V^2) = (a\beta/\phi)\exp(-b\phi^{3/2}/\beta V)$ shows clearly that the field-emission current I is reversely proportional to the electron work function ϕ. For practical applications, therefore, the emitter with a small work function is preferred. Since the emission current I and the voltage V can be measured experimentally, the work function ϕ is determined by the Fowler–Nordheim equation if the local field enhancement factor β is known. Nevertheless, ϕ usually is not well identified due to an ambiguity in both the effective emission area (smaller than apparent emission area) and the local field enhancement factor.

Sinitsyn et al. [26] experimentally estimated the electron work function of carbon nanotubes. Two kinds of carbon nanotubes were used: 10 nm thick bundles comprising 8–11 Å diameter single-wall carbon nanotubes (SWNTs), and 10 nm diameter multiwall carbon nanotubes (MWNTs). They calculated the β using the Fowler–Nordheim equation (FN equation) for field emission and the Schottky equation for thermal emission, respectively. By the I–V characteristics measured (FN curves), they found that both SWNTs and MWNTs exhibited extremely low work functions, 1.1 eV for bundled SWNTs and 1.4 eV for MWNTs. Obraztsov et al. [31] assessed MWNT ϕ using a simplified approximation of $\beta \approx H/R$ (H is the nanotube height, and R is the radius of the nanotube tip). They calculated a ϕ in 0.2–1.0 eV for MWNTs. These values are much smaller than the work functions of metal, silicon, and graphite (Table 2). No theoretical model has explained such a low work function for pure carbon nanotubes so far, but some possible explanations have been proposed in different aspects. One explanation highlights the role of imperfections in the tubular structure [26]. The theoretical study reveals that imperfections in the nanotube structure may cause dramatic reductions in the ionization potential and work function compared to those defect-free tubules. For instance, taking away one carbon atom from a C_{60} molecule will result in a 0.4 eV decrease in its ionization potential. On the other hand, numerous experimental observations verified that imperfect and uncapped tubular structures do exist in SWCNTs and MWCNTs. In agreement with this explanation, MWNTs generally include more impurities and defects, so they show a lower work function than SWNTs. Another explanation is associated with the curvature of graphene sheets that form a nanotube [32, 33]. When the graphene sheets are bent, the atomic bonding shows more sp^3 character than sp^2. The character change from sp^2- to sp^3-like hybridization will produce some sp^3-like atom lines along the nanotube surface, and consequently lead to a decrease in the work function and effective surface barrier. A similar case to carbon nanotubes is diamond, which totally consists of sp^3-hybridized carbon. The sp^3 hybridization endows the diamond surface with negative electron affinity, which dramatically reduces the height of the surface potential barrier.

In other studies, the work function of carbon nanotubes was measured by photoelectron emission (PEE) and field-emission electronic energy distribution (FEED). Shiraishi et al. [34] used PEE to investigate SWNT and MWNT work functions, and obtained 5.05 eV for SWNT and 4.95 eV for MWNT. Fransen et al. [35] measured the work function of an individual MWNT mounted on a tungsten tip by FEED associated with the I–V characteristic curve. Their measurement showed a 7.3 ± 0.5 eV work function for MWNTs. Obviously, these values are inconsistent with those calculations from the FN curve slope. Some are inclined to believe FEED and FEE measurements rather than the calculation by the FN curve slope due to the indefinite field-enhancement factor and effective field-emission area in the FN equation. Clearly, the electron work function for carbon nanotubes is more complicated than for metal, diamond, and graphite. The structure, defect, and surface state of carbon nanotubes (specifically, cap nature, the edge structure of graphene sheet in an open end, etc.) strongly affect the value of the work function. Therefore, they should be taken into consideration together. Our current knowledge about the work function of carbon nanotubes is limited. A comprehensive understanding still awaits more detailed studies.

2.2. Field-Enhancement Factor

The local field-enhancement factor β (also called the geometric enhancement factor) reflects the enhancement of an electric field at the emitter surface by an electrode configuration. It is represented as $E_{\text{local}} = \beta\, dE_{\text{macro}}$, where E_{local} is the local field intensity, E_{macro} is the macroscopic field intensity, and d is the distance between the cathode and anode. A perfectly planar surface should have a unitary field-enhancement factor.

The field-enhancement factor β is obtained by either the slope of the FN curve or the aspect ratio of the emitter. The FN equation is modified as $I/V^2 = (a\beta/\phi)\exp(-b\phi^{3/2}/\beta V)$. From the slope $(-b\phi^{3/2}/\beta)$ of the logarithmic NF equation ($\log(I/V^2)$ versus $1/V$), the field enhancement factor β is easily worked out if the work function is known. As the effective emission area and the work function for carbon nanotubes usually are difficult to be determined, β only can be estimated by the slope. For typical vertically aligned carbon nanotubes, they can be treated as whisker-shaped emitters, and thus the local field-enhancement factor β is approximated as a simple function of the nanotube aspect ratio:

$$\beta \approx H/R \tag{2}$$

where H and R are the height and radius of the carbon nanotube tip, respectively [28]. It should be noted that Eq. (2) excludes the screening effect (see Section 2.5), so it is suitable for single and low site density carbon nanotubes. On the other hand, the radius R in Eq. (2) for field emission could be much smaller than the actual radius of the nanotube tip because field emission may come from a few spots instead of the whole tip. As a result, the β from Eq. (2) could be much smaller than that from the slope of the FN curve.

There have been numerous experimental studies to assess MWNT and SWNT local field-enhancement factors [13, 35–39]. The values reported range from 500 to 7,600,000, which is 10–10^4 times larger than the field-enhancement factor of allotropic diamond. The maximum value is comparable to the field-enhancement factor of molybdenum and silicon (Table 3). Clearly, such a strong field enhancement directly benefits from the extra high aspect ratio and the tiny tip size of carbon nanotubes. Field enhancement is also related to carbon nanotube alignment and site density. Bonard et al. [37] estimated the field-enhancement factors of SWNTs and MWNTs by the slope of the FN

Table 3. Comparison of the field-emission characteristic of silicon, carbon nanotubes, diamond, and molybdenum from the cited references.

Material	Turn-on field intensity (V/μm)	Emission current	Field-enhancement factor (1/cm^{-1})	Work function (eV)
Silicon tips	40–400 [90–92]	10–100 μA (~100 V/μm) (100 tips) [90–92]	100–400,000 [90–92]	4.30 Si (P) 3.30 Si (N)
Carbon nanotubes	<15 V/μm (SW) [26] 1.5 V/μm [31]	100 mA/cm^2 (30 V/μm) [26]	1300 (max), 500–800 [13]	1.1 (MW) 1.3 (SW) [26] 5.05 (SW) 4.95 (MW) [34]
			7,600,000 [35]	7.3 ± 0.5 (MW) [35]
	1.0 V/μm (SW) [36]	90 μA/cm^2 (3 V/μm) [36]	17,000–33,000 [36]	0.2–1.0 (SW) [36]
	1.5–4.5 V/μm (SW) [37]	10 mA/cm^2 (SW, 3.9–7.8 V/μm) [37]	1000–3000 (MW) 2500–10,000 (SW) [37]	
	0.7 V/μm (MW) [38]	1 mA/cm^2 (3.0 V/μm) [38] 0.1–1.0 × 10^6 A/cm^2 (capped) 10–100 A/cm^2 (uncapped) [39]	80,000–190,000 [38] 20,000–40,000 (capped) 60,000–300,000 (uncapped) [39]	
	1.34–1.45 V/μm (65% MW) [40]	3–7 mA/cm^2 [40]		
Diamond	15 (0.01 A · m/cm^2, B-doped) [14] 0.2–0.3 (N$^+$-doped polycrystal) [87]	10 mA/cm^2 (30 V/μm, B-doped) [14]		5.0–5.3
	20–50 (B-doped polycrystal), 5–11 (undoped nanocrystal) [88]	1 mA/cm^2 (20–80 V/μm) [88]	70–203 (B-doped polycrystal) 223–745 (undoped nanocrystal) [88]	
		0.5 mA/cm^2 (5–8 V/μm) [89]		
	15 (N-type T-amorphous) [90]			
Metal tips (Mo, Spindt cathode)	Above 10–100 V/μm [2]	100–1000 A/cm^2 [2]	100,000–500,000 [2]	4.21

curve in the low-current regime ($d = 125\ \mu$m, $\phi = 5$ eV). The β ranged from 2500 to 10,000 for SWNTs, and 1000–3000 for MWNTs. It is easy to understand that, for the smaller tip size, SWNTs normally showed a higher local field-enhancement factor than MWNTs. Kim et al. and Saito and Uemura [36, 39] reported the even higher field-enhancement factors of 17,000–33,000 (SWNTs) and 60,000–300,000 (MWNTs), respectively. It is noted that the field-enhancement factors from different reports vary within a wide range. Besides the obscure effective emission area and work function, the variation may also be attributed to the different experimental conditions, such as electrode configuration, carbon nanotube morphology, structure, site density, and so on. For example, capped and uncapped MWNTs showed a substantial difference in the field-enhancement behavior [39]. The β for uncapped MWNTs could be several times larger than it is for capped MWNTs (Table 3).

2.3. Turn-On Field Intensity

Turn-on voltage is an important parameter for field-emission devices. It represents the ability to work at low voltage. To avoid taking an electrode configuration into consideration, the turn-on electric field intensity is used instead of the turn-on voltage. It is defined as a macroscopic field needed to draw a specific current density (usually 10 μA/cm^2) [37]. Experimentally, the turn-on field intensity is also identified as a voltage where the I–V characteristic ($\log(I/V^2)$–$1/V$ curve) begins to deviate from the Fowler–Nordheim model [40]. For a material, the turn-on field strongly depends on its work function and field-enhancement factor.

Turn-on field intensity is also a function of the structure, morphology, and geometry of carbon nanotubes, particularly the alignment, length, diameter, cap, defects, and so on. With a low work function and an extremely high field enhancement factor, carbon nanotubes demonstrate a turn-on field intensity between 0.1–5 V/μm, 10–100 times smaller than the turn-on field intensities of metal, silicon, and diamond (Table 3). The carbon nanotube therefore is a material especially suited for those field-emission devices that are expected to work at low voltage. Theoretically, SWNTs should have a lower turn-on field intensity compared to MWNTs due to the smaller size and uncapped end. In experiments, conversely, MWNTs show a lower turn-on field intensity than SWNTs because the impurities and defects in MWNTs lead to a change of work function, and then a decrease of the turn-on field intensity. On the other hand, the effect of carbon nanotube caps on the turn-on field has not been completely clarified yet. Saito and Uemura [39] found that uncapped carbon nanotubes not only emitted at a lower voltage, but also kept a higher current in comparison with those capped nanotubes, whereas Bonard et al. [41] claimed that uncapped nanotubes emitted at twice the voltages needed for capped nanotubes. Apparently, these totally different results of Saito and de Pablo's experiments also include a conflict of the work functions and field-enhancement factors between capped and uncapped carbon nanotubes.

2.4. Field-Emission Mechanism and Characteristic

Carbon nanotube behavior strongly depends on its structural properties. It cannot be simply treated as a metallic tip in field-emission behavior and mechanism. Metal tips, for instance, have a monotonic work function of ~5.0 eV, whereas carbon nanotubes show a wide range of work function (0.2–7 ± 0.5 eV) due to the impurity in their structure, carbon bonding sp^3 hybridization, and so on. Therefore, the field-emission behavior and mechanism for carbon nanotubes are also more complicated.

Field-emission electron energy distribution is efficient tool for the study of field-emission characteristics and mechanisms. For metal tips, the emitted electrons originate from the conduction band, that is, around the Fermi level. The work function simply is the difference between the vacuum level and metal Fermi level (~5.0 eV) (Fig. 1), and the electronic density of state (DOS) in the energy band follows Fermi–Dirac statistics [Fig. 2(c)]. Correspondingly, a single peak in FEED spectra appears around the Fermi level, and the typical full width at half maximum (FWHM) of FEED is fixed at 0.45 eV [42]. For carbon nanotubes, a multiple-peak FEED spectra was observed [35, 43, 44]. The multiple peaks in FEED reflected that field-emission electrons originate from several states of energy levels, that is, nonmetallic electronic states. Fransen et al. [35] studied the multiple-peak spectra from an individual MWNT mounted on a tungsten tip, and found that some peaks with a larger FHWM (0.21–0.71 eV) were shifting linearly with the applied electric field. The shifted peaks were distinguished as arising from the field penetration into nanotubes or a voltage drop in the ohmic resistance of nanotubes and the contacting resistance between the nanotubes and tungsten tip. The other peaks with a smaller FWHW (0.17 eV), corresponding to the resonant states from a nanotube tip, did not shift with applied field [Fig. 2(a)]. From an individual SWNT at room temperature, Dean et al. [43] observed a peak corresponding to adsorbate states (3.0 eV above Fermi level), and no peak shift was shown in FEED spectra with applied electric field. When the temperature increased to 900 K, the emission current decreased an order of magnitude from its initial value, which indicated adsorbate shifting. As the temperature increased further, additional peaks appeared in the FEED spectra (above the Fermi level), and the adsorbates and adsorbate tunneling states were removed, leading to an abrupt drop in emission current [Fig. 2(b)]. In another study [45] of a single SWNT, Dean and Chalamala found different regimes associated with temperature and applied electric field. The first regime corresponds to the adsorption (at low fields and low temperatures) and desorption (at either high fields and high currents or temperatures above 400 °C) through resonant tunneling (adsorbates). The other regimes correspond to intrinsic field emission from a clean nanotube, and have a far lower emission current at comparable voltages. All of the studies show that nonmetallic electronic states play a key role in the emission mechanism

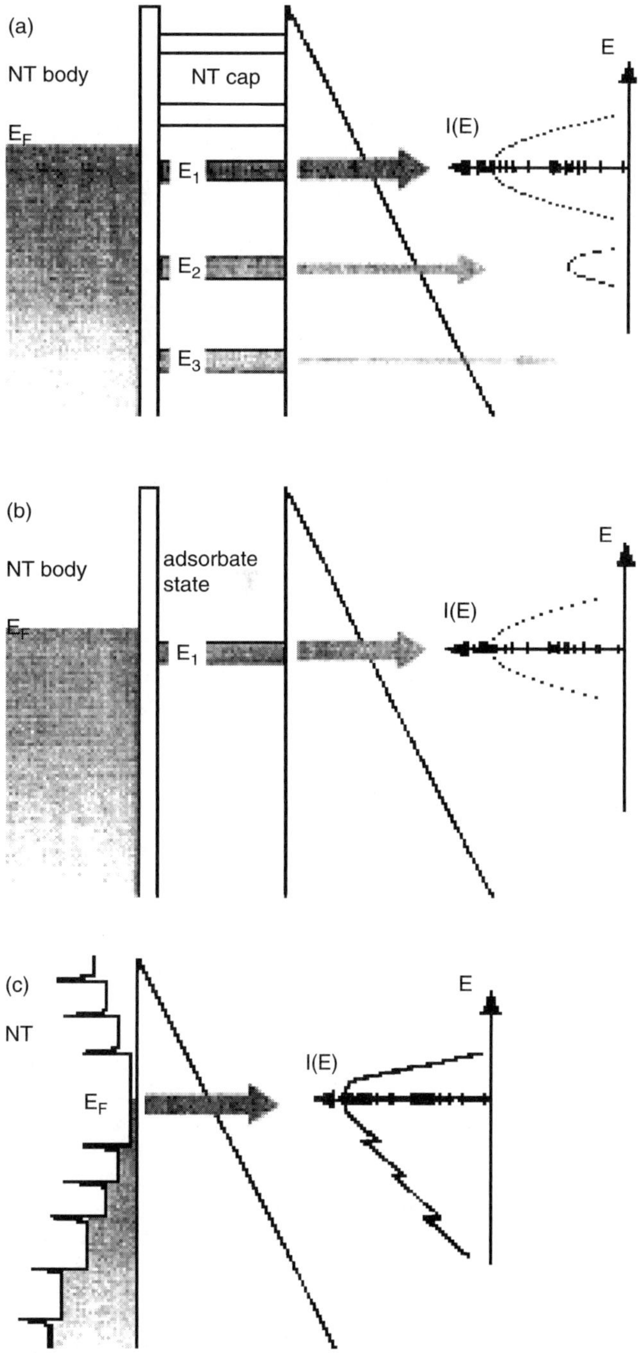

Figure 2. Energy band models of the field emission from the carbon nanotubes with (a) cap, (b) adsorbate resonant tunneling, (c) metallic behavior. The *E–I* curves show the current distribution with electronic energy. Reprinted with permission from [41], J. M. Bonard et al., *Solid-State Electron.* 45, 893 (2001). © 2001, Elsevier Science.

of carbon nanotubes, and strongly affect the field-emission characteristic.

Regarding the geometric position of carbon nanotubes for field emission, field-emission microscope patterns reveal that the emitted electrons originate from the edge rings of graphene layers for uncapped MWNTs and from the caps for both capped MWNTs and SWNTs [39, 41, 44, 46–50].

Rinzler et al. [49] previously proposed that emitted electrons came from an unraveled linear carbon chain at the graphene edge of uncapped MWNTs. Saito and Uemura proved Rinzler's assumption by the field-emission microscopy (FEM) patterns taken from uncapped MWNTs [39]. The FEM patterns show a "doughnut-like" annular bright ring and a black spot in the central region, which corresponds to the cavity of an uncapped nanotube. Through another two patterns taken from capped MWNTs with clean surface and absobates, Takakura et al. [44] further verified that emitted electrons arose from the whole cap of capped nanotubes. Apparently, those patterns contain six bright spots, and each spot exhibits the shape of a pentagonal ring. Moreover, the six spots are also fivefold symmetrized [Fig. 3(a) and (b)]. All of the symmetries in the patterns completely agree with the structure of a nanotube cap. According to the theory of nanotube capping, any nanotube cap must include six pentagonal carbon rings, and they are isolated from each other [51–53].

Figure 4 shows a typical field-emission characteristic of carbon nanotubes (I–V and $\ln(I/V^2)$–$1/V$). The I–V curve in Figure 4 can be divided into three sections with the voltage from low to high, that is, nonlinear–linear–nonlinear distribution. In the low-voltage region, the current increases slowly, and then switches to a sharp linear rise following the turn-on voltage. In the high-voltage region, the current begins to increase slowly again until saturation. The measurement of the field-emission electron energy distribution associated with the I–V characteristic curves [35, 54] revealed that, at low voltages and currents, the emitted electrons came from the energy levels close to the Fermi level, which corresponds to the linear section in the $\ln(I/V^2)$–$1/V$ curve. The I–V linearity exhibited an intrinsically metallic field-emission characteristic as predicted by the Fowler–Nordheim theory. As the peak position in the V-FEED spectra shifted forward to lower kinetic energies (E–E_F) at high voltages and current densities, the I–V curve correspondingly became linear. This indicated that the energy of electrons had changed. Electron energy levels were moving away from the Fermi level. Simultaneously, the I–V characteristic showed an ohmic behavior.

Saturation is an important aspect in the limit of field-emission current. Several mechanisms leading to current saturation have been proposed [35, 45, 54–56]. Vacuum space charges, one of the mechanisms, occurs when an extra high electric field (over 7 V/nm) is applied. Considerable emitted electrons distribute in the space between two electrodes, and create an additional electric field opposite the applied field. Overlapping the additional field, the intensity of the applied field will be cut down, and thus the field-emission current will also be limited. Since the intensity of the additional field is directly proportional to the applied field, the emission current eventually shows saturation with a gradual increase of applied field. Another mechanism for current saturation is solid-state transport, that is, the serial resistance of the emitter and the contacting resistance of the emitter and substrate. The ohmic resistances play the key role of limiting the current at high voltage and current density. These two mechanisms of space charges and solid-state transport are well understood through studying metal emitters.

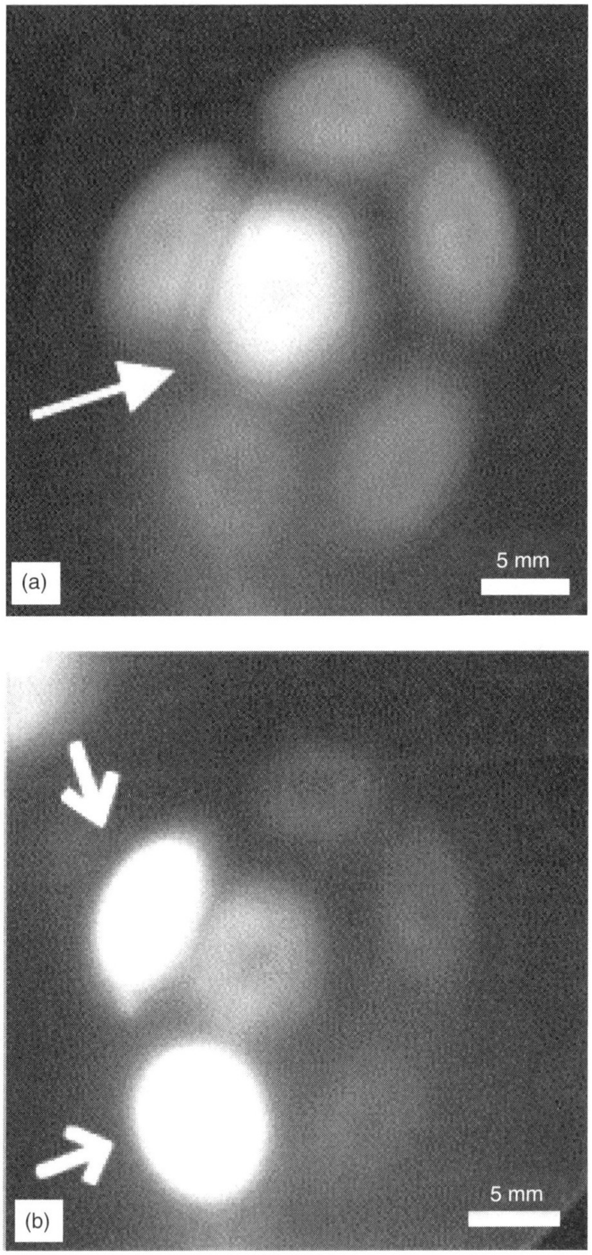

Figure 3. Field-emission patterns from an MWNT with (a) clean surface obtained just after heat cleaning at about 1300 K, and (b) two adsorbates onto pentagons. Reprinted with permission from [44], A. Takakura et al., *Ultramicroscopy* 95, 139 (2003). © 2003, Elsevier Science.

Adsorbates and adsorbate tunneling enhancement, one kind of current mechanism found in semiconductors, is another mechanism to limit the emission current of carbon nanotubes. Dean and Chalamala [45] introduced adsorbates to a clean individual SWNT by exposing them to H_2O at a partial pressure of 10^{-7} torr for 5 min. The adsorbates caused an increase in field-emission current by a factor of 200 (at 1400 V), and the I–V curve with adsorbates showed a strong current limit between 100 and 300 nA, in sharp contrast to the behavior of an SWNT without adsorbate (clean carbon nanotubes). An individual carbon nanotube

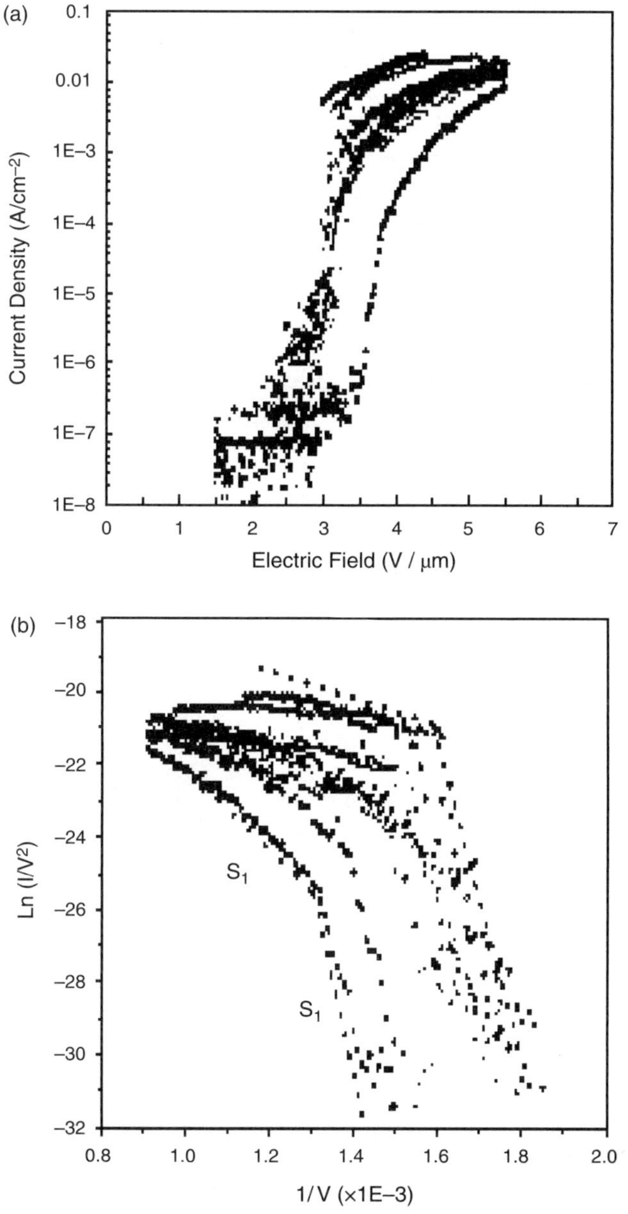

Figure 4. Group of typical field-emission characteristics from nine different regions of a carbon-nanotube-based panel display. (a) *I–V* characteristics. (b) Fowler–Nordheim plots. Reprinted with permission from [36], J. M. Kim et al., *Diamond Rel. Mater.* 9, 1184 (2000). © 2000, Elsevier Science.

was selected for the study to highlight the current-limiting effect and avoid any interaction with the neighboring nanotubes. The emission current in this experiment was also kept at a comparable low level (2 μA) so as to rule out vacuum space charges. This study clearly demonstrates that, similar to the case of semiconductors, the current saturation for carbon nanotubes can arise from adsorbates. The adsorbates are removed from adsorbate tunnels with increasing of applied electric field and emission current, which decreases the adsorbate-tunneling enhancement, and finally induce the saturation of the field-emission current. The adsorbates can also return to the nanotube surface (adsorbate tunnels) when the applied electric field is reduced. Correspondingly, the *I–V* characteristic shows a reversible behavior between the up and down sweeps of the applied field. Besides vacuum space charges and solid-state transport, adsorbate and adsorbate tunneling should be taken into consideration for the current-limiting mechanisms of carbon nanotubes.

2.5. Site Density of Carbon Nanotubes—Field-Screening Effect

The field-screening effect was observed by Nilsson et al. [57] on a carbon nanotube film. That is, the electric field on each nanotube is shielded by nearby neighboring nanotubes, and thus it restrains the field penetration. Nilsson et al. simulated the screening effect, as shown in Figure 5. In their scanning field-emission measurement, they clearly showed that a decreased site density of carbon nanotubes would reduce the screening effect and increase the emission site density and emission current, as in Figure 6. The authors predicted that an intertube distance of about two times the height of nanotubes would optimize the emitted current per unit area. Several other groups also reported similar results [58–62].

To control the site density of carbon nanotube film, several methods have been proposed. Nilsson et al. [57] used microcontact printing to deliver an Fe-containing catalyst solution to a substrate. By changing the Fe concentration, the site density of carbon nanotubes in film can be changed. However, the carbon nanotubes are randomly oriented. Ren et al. [63–67] used electron-beam lithography to make a catalyst array of Ni nanoparticles, and then grew well-aligned nanotubes on the nickel nanoparticles by plasma-enhanced chemical vapor deposition (PE–CVD).

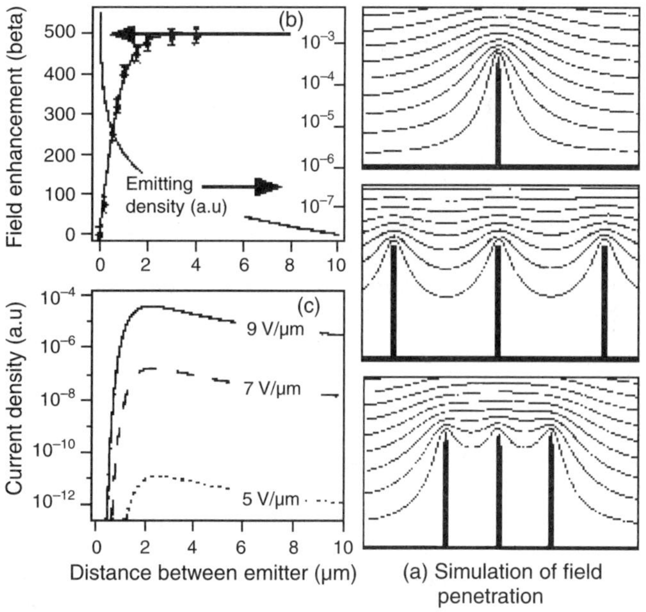

Figure 5. (a) Simulation of the equipotential lines of the electrostatic field for tubes of 1 μm height and 2 nm radius, for distances between tubes of 4, 1, and 0.5 μm, along with the corresponding changes of the field-enhancement factor β and emitter density (b), and current density (c) as a function of the distance. Reprinted with permission from [57], L. Nilsson et al., *Appl. Phys. Lett.* 76, 2071 (2000). © 2000, American Institute of Physics.

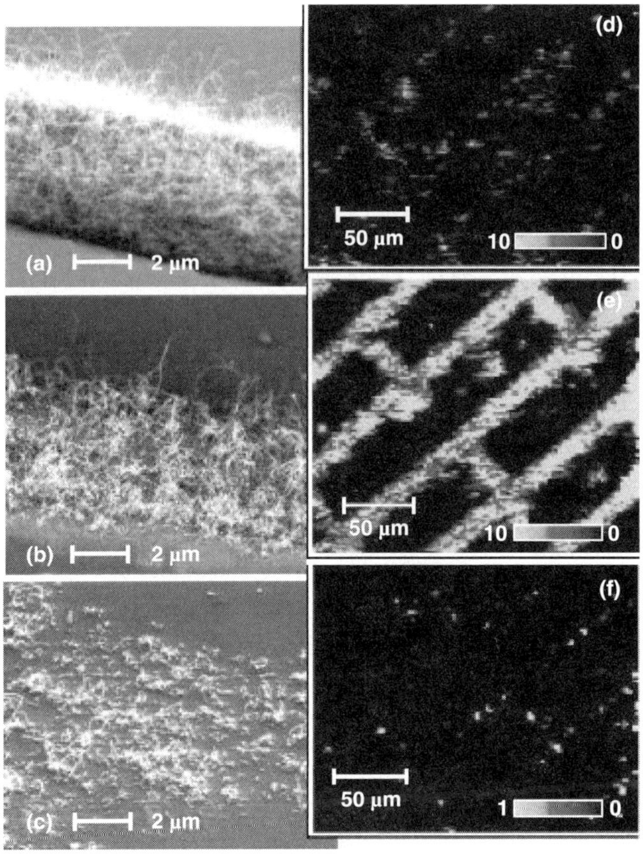

Figure 6. SEM images of patterned carbon nanotube films showing regions of (a) high, (b) medium, and (c) low density, with the corresponding FE maps of current density (d)–(f). The films were produced with ethanolic inks of 10(a), 40(b), and 60 mM $Fe(NO_3)_3 \cdot 9H_2O$. The FE maps were taken under identical conditions using 100 V in constant voltage mode. The color scale corresponds to 0–10 μA/pixel for images (d), (e), and to 0–1 μA/pixel in image (f). Reprinted with permission from [57], L. Nilsson et al., *Appl. Phys. Lett.* 76, 2071 (2000). © 2000, American Institute of Physics.

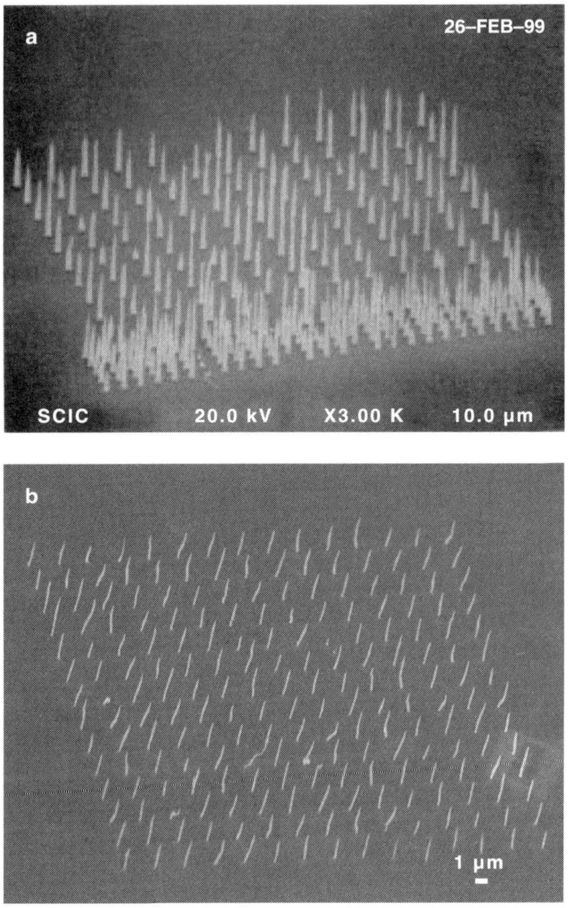

Figure 7. SEM images of aligned carbon nanotubes grown from Ni nanoparticle arrays of *e*-beam photolithography with controllable diameter, site density, and site position. (a) 200 nm thick nanotubes in 1 and 3 μm spacings. (b) 100 nm thick nanotubes in 1.5 μm spacing.

Furthermore, the carbon nanotubes were precisely positioned on a substrate (Fig. 7). Due to the expensive equipment used and the extensive labor requirement, this method cannot be used for mass production.

Recently, Tu et al. [68] applied pulse-current electrochemical deposition to prepare different site densities of Ni nanoparticles by changing the deposition current and duration time. Aligned carbon nanotube arrays with a site density from 10^6 to 10^9 /cm^2 were grown from the Ni nanoparticles using PE–CVD, as shown in Figure 8. Field-emission properties were improved for the CNT array with low site density and long length [61].

Another method that used nanosphere lithography to fabricate Ni nanoparticles in a periodic array was developed by Huang et al. [69]. A monolayer of self-assembled polystyrene spheres was coated on a silicon substrate as a mask (Fig. 9). By Ni deposition, a certain thickness of nickel went through the interspaces of polystyrene spheres to form Ni nanoparticles in a honeycomb array. The polystyrene layer was washed away, and only the Ni nanoparticles remained on the silicon substrate. Aligned carbon nanotubes were grown from the honeycomb array of Ni nanoparticles by PE–CVD (Fig. 10). The carbon nanotubes can have an intertube distance between 0.1 and 5 μm, depending on the size of the polystyrene spheres. Correspondingly, the site density is varied around 10^7–10^8 nanotubes/cm^2.

2.6. Field-Emission Stability

A stable field emission is imperative for practical devices. The residual gases in vacuum influence the field-emission stability dramatically. It has been found that metal and silicon emitters are very sensitive to residual gases. Various residual gases originate from the components that constitute a device, such as a cathode, anode, sealing frit, spacer, and so on. Although pumped out in outgassing process and absorbed by getters after sealing, the residual gases are difficult to remove completely. They can evolve and further increase the pressure during device operation. The residual gases not only react with emitters, changing the surface properties, but also ionize under an electric field to bombard the emitters and modify the tip radius, which leads to field-emission instability and even device failure. It was found that unstable field emission and irreversible degradation occurred with the Mo and Si emitters exposed to O_2,

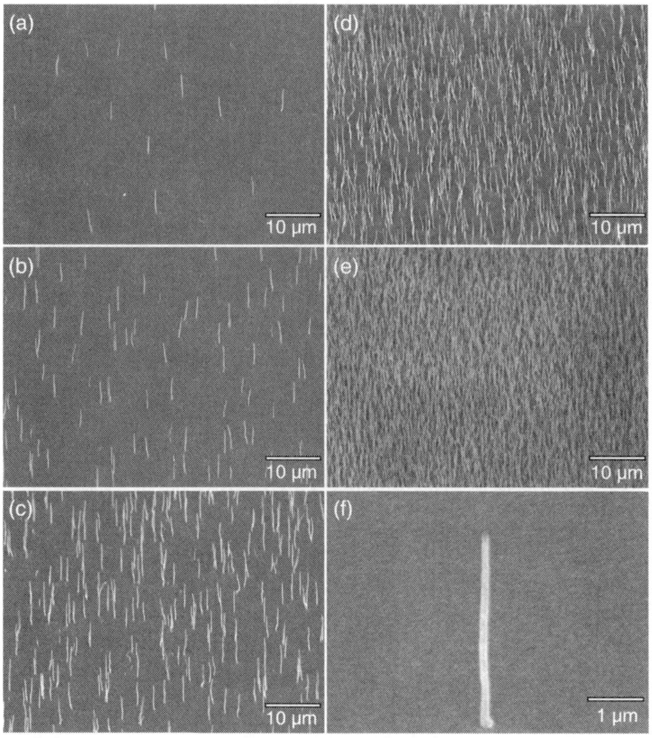

Figure 8. SEM images of aligned carbon nanotubes with site density of (a) 7.5×10^5 cm^{-2}, (b) 2×10^6 cm^{-2}, (c) 6×10^6 cm^{-2}, (d) 2×10^7 cm^{-2}, (e) 3×10^8, and (f) a single standing carbon nanotube. Reprinted with permission from [68], Y. Tu et al., *Appl. Phys. Lett.* 80, 4018 (2002). © 2002, American Institute of Physics.

CO_2, and H_2O during operation. The reason is that the formation of metal oxides and silicon dioxides changes the surface barrier width, surface conductivity, and tip radius [70, 71]. For a stable field emission, Mo and Si emitters have to work at a pressure below 10^{-9} torr, which is an impractical pressure for any static vacuum system.

Figure 9. SEM image of a monolayer of self-assembled polystyrene nanospheres on silicon (scale bar: 1 μm). Reprinted with permission from [69], Z. Huang et al., *Appl. Phys. Lett.* 82, 460 (2003). © 2003, American Institute of Physics.

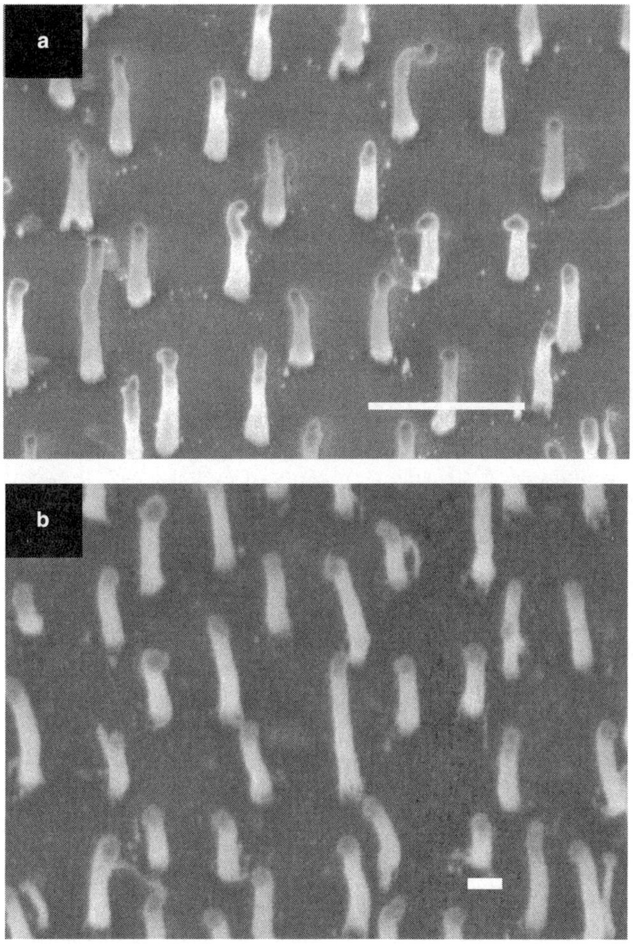

Figure 10. As-grown periodic carbon nanotube arrays (a) from 1040 nm polystyrene sphere mask (scale bar: 1 μm), (b) from 496 nm polystyrene sphere mask (scale bar: 100 nm). Reprinted with permission from [69], Z. Huang et al., *Appl. Phys. Lett.* 82, 460 (2003). © 2003, American Institute of Physics.

Carbon nanotubes have exhibited a chemical inertness and a tolerance to ion sputtering, so they are stable over a wide range of pressure. Sinitsyn et al. [26] investigated the pressure influence on the field-emission characteristics of SWNTs and MWNTs at 291 K. Field emission was obtainable from 10^{-7} to 10^{-2} torr for both SWNTs and MWNTs, and a stable emission remained until up to 10^{-5} torr pressure. Moreover, pressure-induced current degradation was also restorable with the pressure decrease. Other experiments [36, 37, 39, 48, 72, 73] showed similar results in the field-emission stabilities of SWNT and MWNT. The stable emission currents from 0.01 to 20 mA/cm^2 remained for 1 h to thousands of hours. Noticeably, Saito and Uemura [39] used MWNTs with a cathode ray tube (CRT) to replace a conventional thermionic cathode. In their CRT dc-driving lifetime test, a constant field-emission current of 210 μA at 12 kV anode voltage and 10^{-8} torr pressure was sustained over 8000 h without degradation.

An individual carbon nanotube and a group of carbon nanotubes showed different behaviors in current stability. Saito and Uemura [39] traced a probe current emitted from an individual MWNT. Some stepwise fluctuations were recorded at low currents, and large steps appeared in the high-current region. The stepwise fluctuation was attributed to intrinsic field-emission characteristics. In contrast, the total current from a group of comparable MWNTs showed no stepwise fluctuation. The stable current without fluctuation was interpreted as a macroeffect from a group of MWNTs, that is, individual fluctuations were averaged in the macroeffect.

Emission current stability is also correlated to the current density [37, 74]. Bonard et al. [37] studied the dependence of current fluctuation with time in different current regimes. In a low-current regime ($\leq$10–25 μA/cm^2), the frequency and amplitude of the fluctuation increased with the current increase, and the maximum amplitude was 1–10 μA/cm^2. The emission current became stable gradually as the current increased to near saturation (above 10–25 μA/cm^2) (Fig. 11).

Residual gases are also responsible for the field-emission instability of carbon nanotubes. Exposed to H_2, H_2O, Ar, CO, N_2, and O_2, respectively, carbon nanotubes showed different behaviors. The experimental result and discussion are given in Section 2.7.

2.7. Field-Emission Degradation and Failure

Field-emission degradation is a sustained decrease of emission current while the applied voltage remains constant. It includes reversible and irreversible degradations. Field-emission failure occurs after an evolutional degradation of field emission or disastrous damage to the field emitter. Field-emission degradation and failure generally arise from physical and chemical modifications, such as surface absorption and oxidation, Joule heating, gas ion bombardment, defects, and so on.

Reversible degradation normally involves a reversible V–I characteristic of field emission. Dean and Chalamala [45] proposed that the reversible V–I characteristic of carbon nanotubes resulted from surface adsorption. They investigated a reversible V–I characteristic from individual SWNTs using a field-emission microscope, and found an evident hysteresis between the V–I curves of the "up" voltage sweep and the "down" voltage sweep in a clean 10^{-9} torr vacuum, as shown in Figure 12. In a 10^{-7} torr unbaked vacuum (unclean), however, the hysteresis disappeared, that is, the I–V curves became reversible (Fig. 13). On the other hand, FEM images revealed that adsorbates shifted from enhanced tunneling states to induce a current decrease over time between voltage steps during the up sweep (behind the saturation point), and reoccupied the enhanced tunneling states with a higher tunneling enhancement leading to a current increase over time between voltage steps during the down sweep (Fig. 14). The study has solidly proven that adsorbate-tunneling enhancement results in a reversible V–I characteristic and reversible degradation.

The observation of SWNT and MWNT degradation has been reported in a number of papers [13, 26, 36, 37, 39, 48, 72–76]. Kim et al. [36] observed the field-emission stability of an SWNT-based flat panel display at 1.0 mA current (1.5 mA corresponded to 90 μA/cm^2) for 10 h. In the initial stages of field emission, no degradation occurred, but only <10% fluctuations were found over a 4.5 in panel. The <10% fluctuations were interpreted as a field stress effect because the field-emission current showed a high stability in the extended stage.

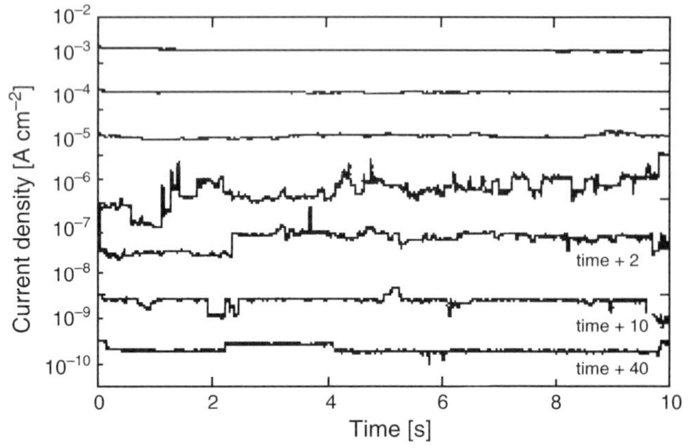

Figure 11. Short-term current stability at constant voltage on MWNTs at 10^{-7} mbar. Reprinted with permission from [37], J. M. Bonard et al., *Appl. Phys. Lett.* 73, 918 (1998). © 1998, American Institute of Physics.

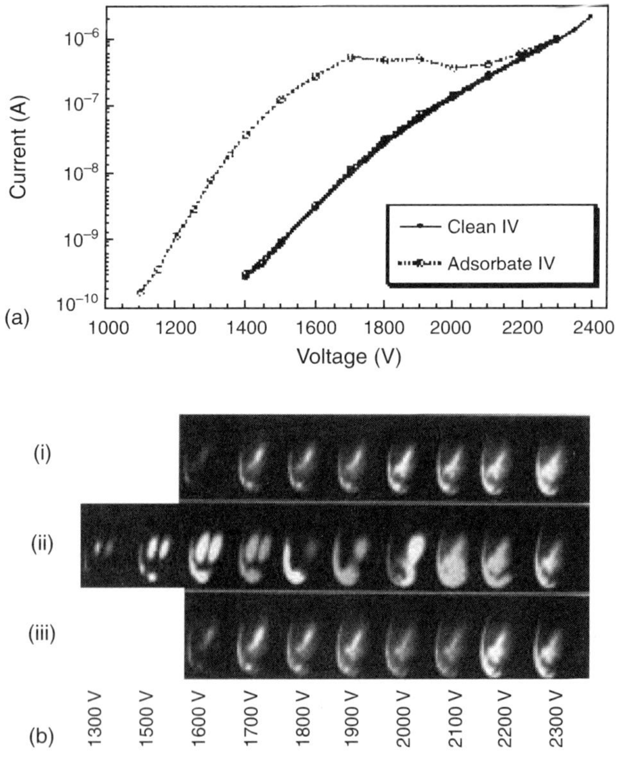

Figure 12. (a) Field-emission characteristics of a single SWNT with and without adsorbates. [(b)(i)] The FEM pattern of a clean SWNT is stable over the voltage sweep, but [(b)(ii)] the FEM pattern of the same nanotube with an adsorbate changes during the I–V sweep concurrent with the onset of current saturation. [(b)(iii)] At 2300 V, the adsorbate effects disappear, resulting in a clean nanotube field-emission pattern. During the downward I–V sweep, the images match those of the clean SWNT shown in [(b)(i)]. Reprinted with permission from [45], K. A. Dean et al., *Appl. Phys. Lett.* 76, 375 (2000). © 2000, American Institute of Physics.

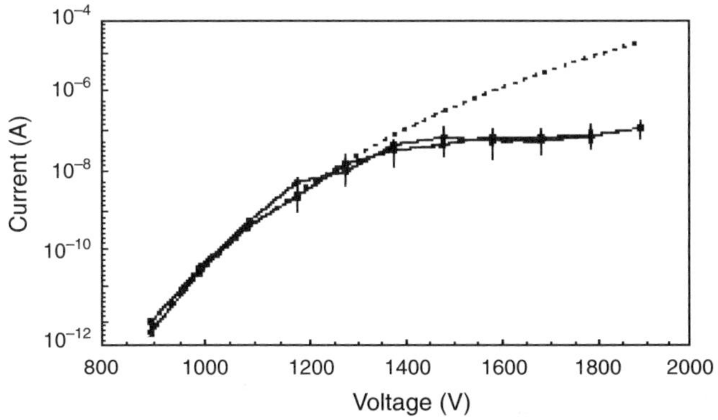

Figure 13. Current saturation behavior of an individual SWNT. The dashed line is an NF equation fit to the low-voltage data. Reprinted with permission from [45], K. A. Dean et al., *Appl. Phys. Lett.* 76, 375 (2000). © 2000, American Institute of Physics.

Further, Zhu et al. [48] assessed the field-emission durability of an SWNT with different current densities. The durability was tested by the measurement of an electric field needed to maintain the current density as a constant until the SWNTs were destroyed. One test at 20 mA/cm^2 current density supported by a field of 5.5 V/μm (10 μA current, 250 μm anode-emitter distance) was sustained for 75 h. The other test at a higher current density of 500 mA/cm^2 (100 μA current, 100 μm anode-emitter distance) ran for 25 h, and the field needed for the current density increased from 16.5 to 20 V/μm. This result clearly shows that faster degradation and failure usually occurs at the higher current density where more Joule heating has been produced. More discussions on Joule heating are given in [76].

Nilsson et al. [76] observed the field-dependent and current-dependent degradations of a carbon nanotube thin film in different current regions. For field-dependent degradation, it typically occurred at the first scan of the field and low current levels. The electrostatic force removed loosely adhered carbon nanotubes, and the nanometer-sized amorphous carbon allotropes from the nanotube tip apex to lower the emission site density and lead to field-emission degradation. The field-induced modifications on carbon nanotubes have been experimentally verified in a number of reports [37, 77, 78] by scanning electron microscope (SEM) and transmission electron microscope (TEM) observations. For current-dependent degradation, it occurred at high current levels. The current for Nilsson's study (>0.3 μA), however, is not considered to be at the level (>3 mA) that can cause a substantial heating dissipation in the solid transport through a nanotube body. Therefore, Nilsson et al. attributed the current-dependent degradation to a poor contact between the carbon nanotubes and the substrate, that is, in those poor contacting areas, the electron transport from the substrate through nanotubes to vacuum can produce enough heat to burn the nanotubes and substrate away. SEM examination has clearly shown some melted regions in the silicon interface without the coverage of nanotubes. For a stable emission and high current density, it is necessary to improve the contact of carbon nanotubes and substrate.

Bonard et al. [37] compared SWNT and MWNT degradations at 150 μA/cm^2 current density and 10^{-7} mbar pressure. First, a gradual degradation and a monotonic decrease in the field-enhancement factor were observed for both SWNTs and MWNTs through the experiment. Second, a fast degradation occurred to the SWNTs, and the current density decreased more than 50% from its initial value within 10 h. By SEM observations, it was found that the SWNT site density was substantially reduced after the experiment. Compared with SWNTs, MWNT degradation was much slower, only a 5–10% decrease of current density from its initial value in the same period (Fig. 15). It is easy to understand that the single shell of SWNTs is more sensitive to ion bombardment and residual gas. SWNTs thereby are easily modified and deformed, whereas the multishells of MWNTs are more tolerant to those influences.

In other studies [13, 36, 39, 72], no degradation occurred to MWNTs except for some small fluctuations. de Heer [13]

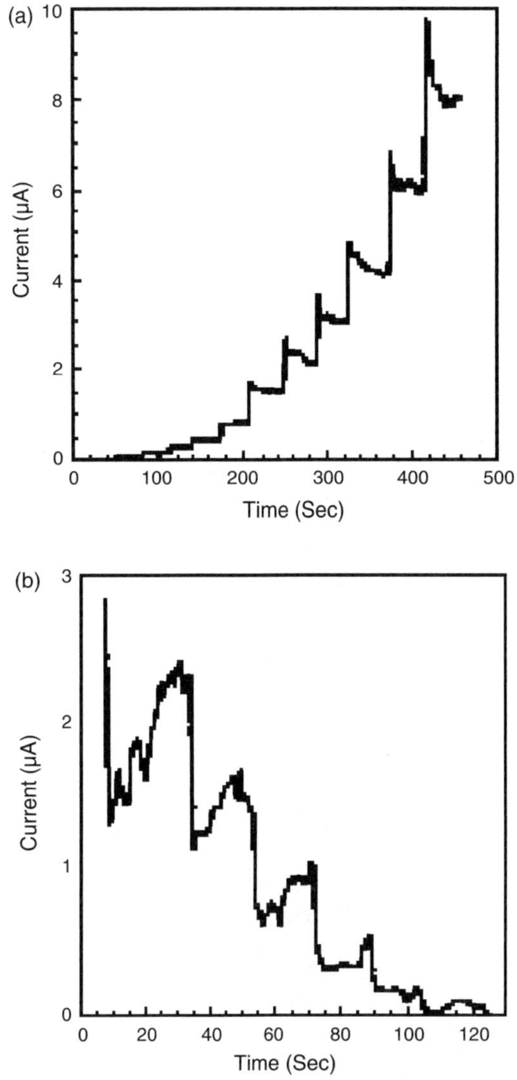

Figure 14. Changes in current with time during an I–V measurement. Step changes in current correspond to the steps in applied voltage. (a) The up sweep shows consistent drops in current over time between the voltage steps above the saturation onset. (b) The down sweep shows an increase in current over time when measured in nonideal vacuum (10^{-7} torr, unbaked phosphor). Reprinted with permission from [45], K. A. Dean et al., *Appl. Phys. Lett.* 76, 375 (2000). © 2000, American Institute of Physics.

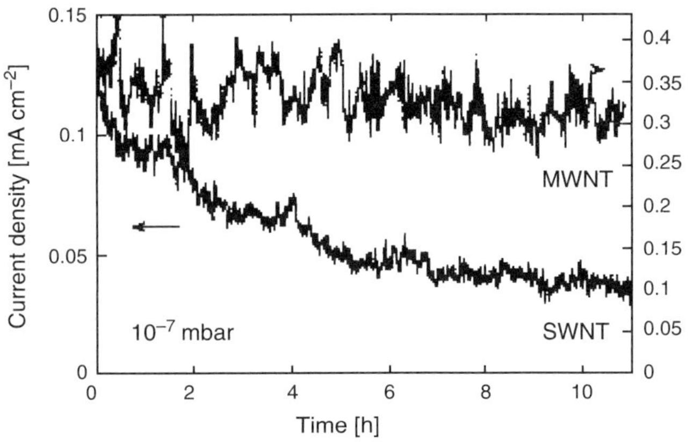

Figure 15. Long-term field-emission stabilities of SWNTs (left scale) and MWNTs (right scale) at constant voltage and 10^{-7} mbar pressure. Reprinted with permission from [37], J. M. Bonard et al., *Appl. Phys. Lett.* 73, 918 (1998). © 1998, American Institute of Physics.

et al. tested an MWNT-based electron gun at a 30 μA/mm^2 current density for 48 h. Only 10% current fluctuations were measured, and the fluctuations could be decreased to 2% by a feedback system. Wang et al. [72] observed a similar result on their matrix-addressable diode flat panel display. A stable current of 76 μA/mm^2 (fluctuation <8%) was tested for 12 h. Saito and Uemura [39] recorded a time trace of MWNT at an average current of 110 μA over 8000 h, and no degradation was measured (4% fluctuation).

For the field-emission failure of carbon nanotubes, de Pablo et al. [79] proposed an electrical failure model induced by defects. They used a nanocontact scanning force microscope to examine an MWNT rope where both ends were connected to a voltage ramp. While a voltage was repeatedly applied to this MWNT rope, every I–V curve and corresponding change in rope morphology were recorded in turn until an electrical failure occurred. Eventually, de Pablo et al. located the electrical failure at a small region of a single nanotube where a 4 nm protrusion had originally stuck out from surface before the experiment. The diameter of apparent protrusion was 62 ± 5 nm. At such a scale, the irregular region can be identified as a defect. Accordingly, they further proposed that defects could be a mechanism of emission current limit for carbon nanotubes, especially for SWNTs.

Dean and Chalamala, Wadhawan et al., and Hata et al. [73, 80, 81] investigated H_2, HO_2, Ar, CO, N_2, and O_2 influences on the field-emission characteristic and degradation of carbon nanotubes. Dean and Chalamala [73] found that SWNTs operated at 3–4 μA current and 10^{-9} torr pressure for 350 h without substantial degradation. In addition, no degradation (some small fluctuations) was found when SWNTs operated for 80–100 h under the following conditions: (1) 1 μA current and 10^{-7} torr pressure with HO_2 exposure (HO_2 partial pressure 0.036 torr), and (2) 3–4 μA current and 10^{-7} torr pressure with H_2 and Ar exposures, respectively (H_2 partial pressure 0.15 torr, Ar partial pressure 0.015 torr). However, an irreversible degradation occurred when SWNTs operated at 3–4 μA current and 10^{-7} torr pressure with $H_2O_2$and O_2 exposures (Fig. 16). After 48 h, the currents reduced by 25% for H_2O and 75% for O_2. These experimental results demonstrated clearly: (1) that SWNT degradation depends on current level, and (2) the fast, inreversible degradation in oxygen and the different percents of current drops between H_2O and O_2 highlighted carbon nanotube sensitivity to oxygen and oxygen concentration. As a result, Dean and Chalamala proposed that, in poor vacuum, SWNT emitters were more stable than unballasted metal emitters because SWNTs never showed any protrusion growth, sputter-induced mobile atoms, and an arcing process that metal emitters did. Although SWNTs are susceptible to damage in oxygen, they can operate in a high-vacuum water environment and comparably low current. Therefore, carbon nanotubes are among the best materials for field-emission devices. Similar to Dean and Chalamala's observation, Wadhawan et al. [80] found no significant change in the field-emission characteristics for both the SWNTs and MWNTs exposed to Ar and H_2 gases. O_2 exposure temporarily increased the turn-on voltage of SWNTs by 22%, and decreased the field-emission current by two orders of magnitude. The degradation was completely reversible by operating in ultrahigh vacuum for 40 h. For MWNTs, the operation in O_2 at a higher voltage (compared to SWNTs) caused a 43% increase in turn-on voltage and a decrease of three orders of magnitude in emission current. The degradation on MWNTs was only partially reversible. Recently Hata et al. [81] studied the field emission from MWNTs in the ambient gases of H_2, CO, N_2, and O_2. The experimental result obtained is also in agreement with Dean and Chalamala's conclusion on SWNTs.

From those investigations in this section, one can conclude that the field-emission degradation and failure for carbon nanotubes arise from high current density (Joule heating), the nature of carbon nanotubes, structure defects, adsorbate state and gas absorption, oxygen concentration, ion bombardment, working pressure, and so on.

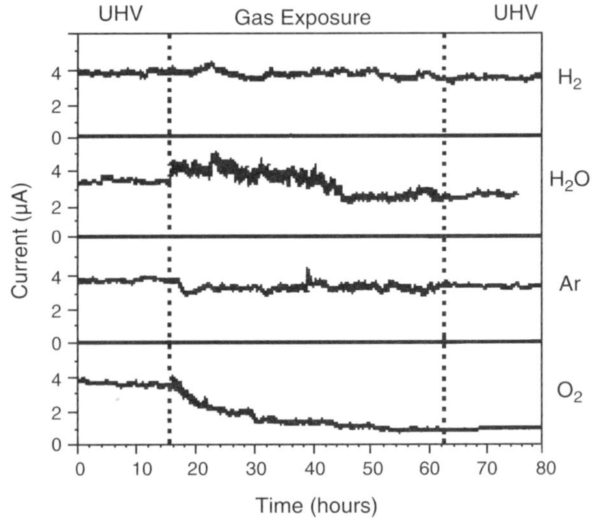

Figure 16. Operation of a single SWNT emitter under several gas ambiences. Total exposures were 0.15 torr (150,000 L) for H_2 and 0.015 torr (15,000 L) each for H_2O, Ar, and O_2. Reprinted with permission from [45], K. A. Dean et al., *Appl. Phys. Lett.* 75, 3017 (1999). © 1999, American Institute of Physics.

3. FIELD-EMISSION APPLICATION OF CARBON NANOTUBES AND FLAT PANEL DISPLAY

Practical applications require different field-emission current densities. Flat panel displays call for a uniform current density of 1–10 mA/cm^2 over a large area, whereas microwave amplifiers demand a current density above 500 mA/cm^2 for a high power gain at 1–10 GHz. Presently, the available current density for carbon nanotubes approximates 0.1–100 mA/cm^2 (individual reports claimed 1–4 A/cm^2), much larger than silicon and diamond. Carbon nanotubes are suitable for those applications in modest current density.

Carbon nanotubes are applied to field emission in two ways: (1) an individual nanotube for a point electron source, and (2) multiple carbon nanotubes for a parallel electron-beam source.

An individual nanotube on a probe for a scanning tunnel microscope and an atomic force microscope is the typical application of a point electron source. Schmid and Fink [82] demonstrated that individual carbon nanotubes can behave as a coherent electron source. They mounted a single nanotube onto the tungsten tip of a low-energy electron point source microscope, and recorded the in-line holograms for a 7 nm wide carbon fiber that was used as a scattering object. This single nanotube electron source operated at low voltage, and could show an interference pattern with the same order of fringes as those shown by a conventional tungsten atomic point source. In addition, carbon nanotubes are relatively–chemically inert and tolerant to ion sputtering. These properties associated with the high coherence of emitted electrons and the low operating voltage make individual carbon nanotubes a unique point electron source.

In parallel electron-beam applications, most of the attention has been given to field-emission flat panel displays (FEDs) because flat panel displays have a giant market and demand a moderate current density [36, 40, 72, 83–86]. Moreover, carbon nanotubes can be produced in large scale, and can easily be built into composite materials for FED purposes. Rapid progress has been made in the past decade. In 1998, Saito et al. demonstrated an MWNT lighting element structured inside a cathode ray tube (CRT) [83]. The MWNTs were grown on graphite disks by dc arc discharge. Baked at 450–500 °C in air to treat a silver paste and remove a cylindrical mold, these graphite disks attached with carbon nanotubes were mounted onto a cathode ray tube to replace a conventional filament cathode. By applying a voltage to the lighting elements, the green, blue, and red fluorescent screens mounted on three CRTs were lighted, respectively. Right after Saito, Wang et al. [72] built a 32×32 matrix-addressable flat panel display by patterning carbon nanotube-epoxy (on a cathode) and phosphor-coated ITO film (on an anode, i.e., screen) into 32 straight lines, respectively. The nanotube-epoxy lines and the phosphor-coated ITO lines cross over each other perpendicularly. Addressable pixels were formed at every intersection of nanotube-epoxy lines with phosphor-coated ITO lines. When the voltage addresses a nanotube-epoxy line and an ITO line, the pixel at the intersection will be excited. In 2000, Kim et al. [36] successfully assembled and characterized a 4.5 in, 128-line matrix-addressable flat panel display. This prototype showed an unusually high brightness (1800 cd/m^2) at a lower voltage (740 V, 0.25 duration, 15.7 kHz frequency) compared to a Spindt-type FED (300 cd/m^2 at 6000 V). Further, a fully sealed 128-line addressable FED in the color mode also was experimentally fabricated. Although these diode-type flat panel displays are only able to display letters and stationary images, they have clearly demonstrated carbon nanotube feasibility for flat panel displays. Recently, a great breakthrough in 5 in triode-type FEDs was made by Jung et al. [84]. The triode-type display for the first time realized a full-color display in the moving-image mode [Fig. 17(c)], and the brilliant image was displayed at a resolution of 240 × 120 lines [Fig. 17(d)]. A matrix-addressable circuit was used to drive this FED. To form a triode structure, backside photolithography associated with photosensitive SWNT paste was initially introduced to protect the surface properties of SWNT paste from the interruption of a subsequent process [Fig. 17(a)]. The adoption of a thick gate wall suppressed the diode effect and color cross in subpixels [Fig. 17(b)]. With the decrease of spacer thickness and gate hole size, the turn-on voltage and anode voltage of the triode-type display reduced to 40–45 and 400–800 V, respectively. The breakthrough demonstrates that carbon nanotubes are compatible with current FED fabrication. The remarkable achievements outline the prospect of carbon nanotubes in field-emission applications.

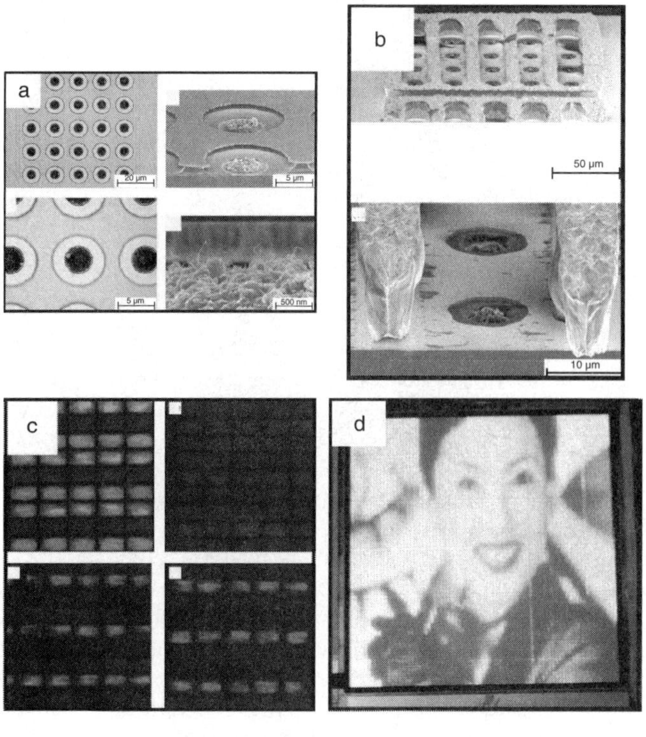

Figure 17. (a) Triode-type carbon nanotube emitter array. (b) Thick gate wall structure. (c) Magnified color pixels. (d) Video—running, full color, moving image of 5 in diagonal carbon nanotube flat panel display. Reprinted with permission from [84], J. E. Jung et al., *Phys. B: Cond. Matter* 323, 71 (2002). © 2002, Elsevier Science.

4. CONCLUSION

A constant topic for vacuum microelectronics is how to achieve a cathode of high performance because the cathode is the core of a field-emission device. A cathode of high performance should exhibit not only a high current density at low operating voltage, but also a stable field emission in harsh environments. Accomplishing this purpose depends on the developments of theory, technology, and material. It has been theoretically and experimentally proven that carbon nanotubes are an inherent good material for field emission. The low electron work function and unique geometry endow the cathode with the features of low turn-on voltage and high local field enhancement, which make the field-emission device able to work at high current density and low voltage. Their chemical inertness, strong carbon bonding, and sputter etching-free feature lead to field emission with reduced environmental dependence.

Carbon nanotubes have shown great potential to fulfill the demands for field-emission applications. In the past decade, rapid progress had been made in carbon nanotube synthesis, field-emission mechanism, and field-emission characterization. The control over the carbon nanotube diameter, alignment, and site density have improved field-emission properties. The initial experiments and analysis in field-emission stability, degradation, and failure also have been carried out. The present barriers to commercialization include field-emission uniformity, reliability, and lifetime. The imperative task is establishing a comprehensive theory and model to control carbon nanotubes in both morphology and structure, developing the ability to manipulate one single nanotube, and better understanding the field-emission mechanism, including various defects and surface-state properties. More detailed studies should be devoted to the electron work function, the interface property of nanotubes and substrate, the degradation and failure mechanism, and so on. With the development of theory and technology, one can expect that various field-emission devices of carbon nanotubes will be commercially available in the near future.

GLOSSARY

Absorbate states Absorbates on semiconductor surface bring in electronic states in energy band and impact on the electrical behavior of semiconductor. These electronic states are called absorbate states. Absorbrate states can be removed or replaced with the status change of semiconductor surface, leading to an unstable and deviant operation of devices.

Band gap In a semiconductor material, the minimum energy necessary for an electron to transfer from the valence band into the conduction band, where it moves more freely.

Conduction band A vacant or partially occupied set of many closely spaced electronic levels resulting from an array of a large number of atoms forming a system in which the electrons can move freely or nearly so. In reference to the conductivity in semiconductors, it is the band that accepts the electrons from the valence band.

Donor An atom that is likely to give off one or more electrons when placed in a crystal.

Electron Particle with spin 1/2 and carrying a single negative charge (1.6×10^{-19} Coulomb).

Fermi–Dirac statistics Fermi–Dirac statistics provides the probability that an electron in thermal equilibrium with a large system occupies energy level E. Fermi-Dirac distribution function is given by:

$$f(E) = \frac{1}{1 + e^{(E-E_F)/kT}}$$

The system in thermal equilibrium is characterized by its temperature T and its Fermi energy E_F.

Fermi level For metals, the highest occupied molecular orbital in the valence band at absolute zero temperature ($T = 0$ K). In insulators and semiconductors, the valence and conduction bands are separated. Therefore the Fermi-Level is located in the band gap. Based on Fermi-Dirac distribution function, the probability for an electron to occupy Fermi level is one half.

Gate Electrode of a field emission transistor, which controls the current from the emitter.

Impurity A foreign atom in a crystal.

Interface Boundary between two materials.

Ionization The process of adding or removing an electron to/from an atom thereby creating a charged atom (i.e. ion).

***I–V* curve** Current versus voltage characteristics.

***I–V* measurement** Current versus voltage measurement.

Negative electron affinity Electron affinity is defined as the energy required removing an electron from the bottom of conduction band to a distance far from the material surface, i.e. to vacuum level. Hence no energy barrier prevents low-energy electrons from escaping into the vacuum if vacuum level lies below the conduction band minimum. In this case the surface has a negative electron affinity.

Quantum mechanics Theory that describes particles as a wave function.

sp^3 hybridization The phenomenon of mixing up of different orbitals of same energy level of an atom to produce equal number of hybrid-orbitals of same energy and identical properties is known as hybridization. A hybrid orbital contains maximum two electrons with opposite spin. The process of hybridization in which one *s*-orbital and three *p*-orbitals overlap to produce four hybrid-orbital is known as sp^3 hybridization. These hybrid-orbitals are not only identical in shape and energy but also at an angle of 109.5° from each other (tetrahedral fashion).

Static vacuum system A closed vacuum system without exhaust.

Substrate The material in which a device is embedded or on to.

Surface barrier Increased potential at the surface of a material.

Thermionic emission Process where electrons are emitted across a confining barrier. The driving force of this process is the thermal energy that provides a non-zero density of electrons at those energies larger than the barrier.

Tunneling Quantum mechanical process where a particle can pass through a barrier rather than having to go over the barrier.

Valence band Made up of occupied molecular orbitals and is lower in energy than the so-called conduction band. It is generally completely full in semiconductors. When heated, the electrons from this band jump out of the band across band gap and into the conduction band, making the material conductive.

ACKNOWLEDGMENTS

The work at NanoLab Inc. was supported by the U.S. Army Natick Soldier Systems Center under Grant DAAD16-03-C-0020, and the U.S. Army Research Laboratory under Grant DAAD17-02-C-0004. The work in Boston College was supported partly by DOE under Grant DE-FG02-00ER45805, partly by the U.S. Army Natick Soldier Systems Center under Grants DAAD16-00-C-9227 and DAAD16-02-C-0037, partly by the NIH under Grant CA-97945-01, and partly by the NSF under Grant ECS-0103012.

REFERENCES

1. R. H. Fowler and L. W. Nordheim, *Proc. R. Soc. London A* 119, 173 (1928).
2. I. Brodie and P. R. Schwoebel, *Proc. IEEE* 82, 1006 (1994).
3. C. A. Spindt, *J. Appl. Phys.* 39, 3504 (1968).
4. C. A. Spindt, I. Brodie, L. Humphrey, and E. R. Westerberg, *J. Appl. Phys.* 47, 5248 (1976).
5. H. F. Gray, "Proceedings of the 29th International. Field Emission Symposium," Stockholm, Sweden, 1982, p. 111.
6. G. J. Campisi, H. F. Gray, R. Greene, H. Gray, and G. Campisi, "Technical Digest IEDM," 1985, p. 175.
7. J. W. Faust and E. D. Palick, *J. Electrochem. Soc.* 130, 1413 (1983).
8. E. D. Palick, H. F. Gray, and P. B. Klein, *J. Electrochem. Soc.* 130, 956 (1983).
9. T. Sakai, T. Ono, M. Nakamoto, and N. Sakuma, *J. Vac. Sci. Technol. B* 16, 770 (1998).
10. R. B. Marcus, T. S. Ravi, T. Gmitter, K. Chin, D. Liu, W. J. Orvis, D. R. Ciarlo, C. E. Hunt, and J. Trujillo, *Appl. Phys. Lett.* 56, 236 (1990).
11. J. Itoh, Y. Tohma, K. Morikawa, S. Kanemaru, and K. Shimizu, *J. Vac. Sci. Technol. B* 13, 1968 (1995).
12. W. Zhu, "Vacuum Microelectronics." Wiley, 2001.
13. W. A. de Heer, A. Chatelain, and D. Ugarte, *Science* 270, 1179 (1995).
14. W. Zhu, G. P. Kochanski, S. Jin, and L. Seibles, *J. Vac. Sci. Technol. B* 14, 2011 (1996).
15. F. J. Himpsel, J. A. Knapp, J. A. Van Vechten, and D. E. Eastman, *Phys. Rev. B* 20, 624 (1979).
16. M. W. Geis, J. A. Gregory, and B. B. Pate, *IEEE Trans. Electron Devices* 38, 619 (1991).
17. M. W. Geis, N. N. Efremow, J. D. Woodhouse, M. D. McAleese, M. Marchywka, D. G. Socker, and J. F. Hochedez, *IEEE Electron Device Lett.* 12, 456 (1991).
18. C. Wang, A. Garcia, D. C. Ingram, M. Lake, and M. E. Kordesch, *Electron. Lett.* 27, 1459 (1991).
19. N. S. Xu, R. V. Latham, and Y. Tzeng, *Electron. Lett.* 29, 1596 (1993).
20. K. Okano, K. Hoshina, M. Iida, S. Koizumi, and T. Inuzuka, *Appl. Phys. Lett.* 64, 2742 (1994).
21. M. W. Geis, J. C. Twichell, and T. M. Lyszczarz, *J. Vac. Sci. Technol. B* 14, 2060 (1996).
22. K. Okano, S. Koizumi, S. R. P. Silva, and G. A. J. Amaratunga, *Nature* 381, 140 (1996).
23. N. A. Fox, W. N. Wang, T. J. Davis, J. W. Steeds, and P. W. May, *Appl. Phys. Lett.* 71, 2337 (1997).
24. S. A. Kajihara, A. Antonelli, J. Bernholc, and R. Car, *Phys. Rev. Lett.* 66, 2010 (1991).
25. S. Iijima, *Nature* 354, 56 (1991).
26. N. I. Sinitsyn, Yu. V. Gulyaev, G. V. Torgashov, L. A. Chernozatonskii, Z. Ya. Kosakovskaya, Yu. F. Zakharchenko, N. A. Kiselev, A. L. Musatov, A. I. Zhbanov, Sh. T. Mevlyut, and O. E. Glukhova, *Appl. Surf. Sci.* 111, 145 (1997).
27. P. G. Collins and A. Zettl, *Appl. Phys. Lett.* 69, 1969 (1996).
28. P. J. F. Harris, "Carbon Nanotubes and Related Structures." Cambridge University Press, 1999.
29. E. W. Wong, P. E. Sheehan, and C. M. Lieber, *Science* 277, 1971 (1997).
30. J. W. Mintmire, B. I. Dunlap, and C. T. White, *Phys. Rev. Lett.* 68, 631 (1992).
31. A. N. Obraztsov, A. P. Volkov, and I. Pavlovsky, *Diamond Rel. Mater.* 9, 1190 (2000).
32. H. Hiura, T. W. Ebessen, J. Fujita, K. Tanigaki, and T. Takada, *Nature* 367, 148 (1994).
33. J. Robertson, *Diamond Rel. Mater.* 5, 797 (1996).
34. M. Shiraishi and M. Ata, *Carbon* 39, 1913 (2001).
35. M. J. Fransen, Th. L. van Rooy, and P. Kruit, *Appl. Surf. Sci.* 146, 312 (1999).
36. J. M. Kim, W. B. Choi, N. S. Lee, and J. E. Jung, *Diamond Rel. Mater.* 9, 1184 (2000).
37. J. M. Bonard, J. P. Salvetat, T. Stockli, W. A. de Heer, L. Forro, and A. Chatelain, *Appl. Phys. Lett.* 73, 918 (1998).
38. M. Hirakawa, S. Sonoda, C. Tanaka, H. Murakami, and H. Yamakawa, *Appl. Surf. Sci.* 169, 662 (2001).
39. Y. Saito and S. Uemura, *Carbon* 38, 169 (2000).
40. J. L. Kwo, M. Yokoyama, W. C. Wang, F. Y. Chuang, and I. N. Lin, *Diamond Rel. Mater.* 9, 1270 (2000).
41. J. M. Bonard, H. Kind, T. Stockli, and L. O. Nilsson, *Solid-State Electron.* 45, 893 (2001).
42. J. W. Gadzuk and E. W. Plummer, *Rev. Mod. Phys.* 45, 487 (1973).
43. K. A. Dean, O. Groening, O. M. Kuttel, and L. Schlapbach, *Appl. Phys. Lett.* 75, 2773 (1999).
44. A. Takakura, K. Hata, Y. Saito, K, Matsuda, T. Kona, and C. Oshima, *Ultramicroscopy* 95, 139 (2003).
45. K. A. Dean and B. R. Chalamala, *Appl. Phys. Lett.* 76, 375 (2000).
46. Y. Saito, K, Hamaguchi, K. Hata, K. Uchida, Y. Tasaka, F. Ikasaki, M. Yumura, A. Kasuya, and Y. Nishina, *Nature* 389, 554 (1997).
47. Y. Saito, K. Hamaguchi, K. Hata, T. Nishina, K. Iiata, K. Tohji, A. Kasuya, and Y. Nishina, *Jpn. J. Appl. Phys.* 36, L1340 (1997).
48. W. Zhu, C. Bower, O. Zhou, G. P. Kochanski, and S. Jin, *Mater. Res. Soc. Symp. Proc.* 558, 75 (2000).
49. A. G. Rinzler, J. H. Hafner, P. Nikolaev, L. Luo, S. G. Kim, D. Tomanek, P. Nordlander, D. T. Colbert, and R. E. Smalley, *Science* 269, 1550 (1995).
50. W. A. de Heer, J. M. Bonard, K. Fauth, A. Chatelain, L. Forro, and D. Ugarte, *Adv. Mater.* 8, 87 (1997).
51. M. S. Dresselhaus, G. Dresselhaus, and P. C. Eklund, "Science of Fullerenes and Carbon Nanotubes." Academic, San Diego, 1996.
52. M. Fujita, R. Saito, G. Dresselhaus, and M. S. Dresselhaus, *Phys. Rev. B* 45, 13834 (1992).
53. M. S. Dresselhaus, G. Dresslhaus, and P. C. Eklund, *J. Mater. Res.* 8, 2054 (1993).
54. R. Schlesser, R. Collazo, C. Bower, O. Zhou, and Z. Sitar, *Diamond Rel. Mater.* 9, 1201 (2000).
55. J. P. Barbour, W. W. Dolan, J. K. Trolan, E. E. Martin, and W. P. Dyke, *Phys. Rev.* 92, 45 (1953).
56. W. A. Anderson, *J. Vac. Sci. Technol. B* 11, 383 (1993).
57. L. Nilsson, O. Groening, C. Emmenegger, O. Kuettel, E. Schaller, L. Schlapbach, H. Kind, J. M. Bonard, and K. Kern, *Appl. Phys. Lett.* 76, 2071 (2000).
58. S. H. Jeong, H. Y. Hwang, K. H. Lee, and Y. S. Jeong, *Appl. Phys. Lett.* 78, 2052 (2001).

59. M. Chhowalla, C. Ducati, N. L. Rupesinghe, K. B. K. Teo, and G. A. Amaratunga, *Appl. Phys. Lett.* 79, 2079 (2001).
60. K. B. K. Teo, M. Chhowalla, G. A J. Amaratunga, W. I. Milne, G. Pirio, P. Legagneux, F. Wyczisk, D. Pribat, and D. G. Hasko, *Appl. Phys. Lett.* 80, 2011 (2002).
61. S. H. Jo, Y. Tu, Z. P. Huang, D. L. Carnahan, D. Z. Wang, and Z. F. Ren, *Appl. Phys. Lett.* (in press).
62. Y. T. Feng, S. Z. Deng, J. Chen, and N. S. Xu, *Ultramicroscopy* 95, 93 (2003).
63. Z. F. Ren, Z. P. Huang, D. Z. Wang, J. G. Wen, J. W. Xu, J. H. Wang, L. E. Calvet, J. Chen, J. F. Klemic, and M. A. Reed, *Appl. Phys. Lett.* 75, 1086 (1999).
64. Z. P. Huang, J. Moser, M. Sennett, H. Gibson, M. J. Naughton, J. G. Wen, and Z. F. Ren, *Mater. Res. Soc. Symp. Proc.* 633, A13.22.1 (2001).
65. J. G. Wen, Z. P. Huang, D. Z. Wang, J. H. Chen, S. X. Yang, Z. F. Ren, J. H. Wang, L. E. Calvet, J. Chen, J. F. Klemic, and M. A. Reed, *J. Mater. Res.* 16, 3246 (2001).
66. K. B. K. Teo, M. Chhowalla, G. A. J. Amaratunga, W. I. Milne, D. G. Hasko, G. Pirio, P. Legagneux, F. Wyczisk, and D. Pribat, *Appl. Phys. Lett.* 79, 1534 (2001).
67. V. I. Merkulov, D. H. Lowndes, Y. Y. Wei, G. Eres, and E. Voelkl, *Appl. Phys. Lett.* 76, 3555 (2000).
68. Y. Tu, Z. P. Huang, D. Z. Wang, J. G. Wen, and Z. F. Ren, *Appl. Phys. Lett.* 80, 4018 (2002).
69. Z. P. Huang, D. L. Carnahan, J. Rybczynski, M. Giersig, M. Sennett, D. Z. Wang, J. G. Wen, K. Kampa, and Z. F. Ren, *Appl. Phys. Lett.* 82, 460 (2003).
70. S. Itoh, T. Niiyama, and M. Yokoyama, *J. Vac. Sci. Technol. B* 11, 647 (1993).
71. B. R. Chalamala, R. M. Wallace, and B. E. Gnade, *J. Vac. Sci. Technol. B* 16, 859 (1998).
72. Q. H. Wang, A. A. Setlur, J. M. Lauerhaas, J. Y. Dai, E. W. Seelig, and R. P. H. Chang, *Appl. Phys. Lett.* 72, 2912 (1998).
73. K. A. Dean and B. R. Chalamala, *Appl. Phys. Lett.* 75, 3017 (1999).
74. D. H. Kim, H. S. Yang, H. D. Kang, and H. R. Lee, *Chem. Phys. Lett.* 368, 439 (2003).
75. L. R. Baylor, V. I. Merkulov, E. D. Ellis, M. A. Guillorn, D. H. Lowndes, V. Melechko, M. L. Simpson, and J. H. Whealton, *J. Appl. Phys.* 91, 4602 (2002).
76. L. Nilsson, O. Groening, P. Groening, and L. Schlapbach, *Appl. Phys. Lett.* 79, 1036 (2001).
77. Z. L. Wang, *Adv. Mater.* 12, 1295 (2000).
78. A. Cumings and A. Zettl, *Science* 289, 602 (2000).
79. P. J. de Pablo, S. Howell, S. Crittenden, B. Walsh, E. Graugnard, and R. Reifenberger, *Appl. Phys. Lett.* 75, 3941 (1999).
80. A. Wadhawan, R. E. Stallcup, II, K. F. Stephens, II, J. M. Perez, and I. A. Akwani, *Appl. Phys. Lett.* 79, 1867 (2001).
81. K. Hata, A. Takakura, and Y. Saito, *Ultramicroscopy* 95, 107 (2003).
82. H. Schmid and H. W. Fink, *Appl. Phys. Lett.* 70, 2679 (1997).
83. Y. Saito, S. Uemura, and K. Hamaguchi, *Jpn. J. Appl. Phys.* 37, 346 (1998).
84. J. E. Jung, Y. W. Jin, J. H. Choi, Y. J. Park, T. Y. Ko, D. S. Chung, J. W. Kim, J. E. Jang, S. N. Cha, W. K. Yi, S. H. Cho, M. J. Yoon, C. G. Lee, J. H. You, N. S. Lee, J. B. Yoo, and J. M. Kim, *Phys. B: Cond. Matter* 323, 71 (2002).
85. C. J. Lee, J. H. Park, S, Y. Kang, and J. H. Lee, *Chem. Phys. Lett.* 326, 175 (2000).
86. J. Yu, Q. Zhang, J. Ahn, S. F. Yoon, Rusli, Y. J. Li, B. Gan, K. Chew, and K. H. Tan, *Diamond Rel. Mater.* 10, 2157 (2001).
87. K. Okano, T. Yamada, H. Ishihara, S. Koizumi, and J. Itoh, *Appl. Phys. Lett.* 70, 2201 (1997).
88. S. Jou, H. J. Doerr, and R. F. Bunshah, *Thin Solid Films* 280, 256 (1996).
89. B. L. Weiss, A. Badzian, L. Pilione, T. Badzian, and W. Drawl, *Appl. Phys. Lett.* 71, 794 (1997).
90. K. C. Park, J. H. Moon, S. J. Chung, J. Jang, M. H. Oh, and W. I. Milne, *Appl. Phys. Lett.* 70, 1381 (1997).
91. P. A. Lewis, B. W. Alphenaar, and H. Ahmed, *Appl. Phys. Lett.* 79, 1348 (2001).
92. K. Ehara, S. Kanemaru, T. Matsukawa, and J. Itoh, *Appl. Surf. Sci.* 146, 172 (1999).
93. H. Busta, D. Furst, A. T. Rakhimov, V. A. Samorodov, B. V. Seleznev, N. V. Suetin, and A. Silzars, *Appl. Phys. Lett.* 78, 3418 (2001).
94. S. M. Sie, "VLSI Technology." McGraw-Hill, New York, 1983.
95. Editorial Board of Technical Handbook of Electronic Industry, "Technical Handbook of Electronic Industry," Vol. 7, p. 582. National Defence Industry Press, China, 1991.

ENN Encyclopedia of Nanoscience and Nanotechnology
www.aspbs.com/enn

Filling of Carbon Nanotubes

Pawan K. Tyagi, Manoj K. Singh, D. S. Misra

Indian Institute of Technology, Mumbai, India

CONTENTS

1. INTRODUCTION

The discovery of carbon nanotubes by Ijima [1] in 1991 started the onset of activity in this novel class of materials. Carbon nanotubes are tubules of very narrow (1.4–50 nm) diameter. It might be justified to say that the nanotubes are another allotrope of carbon, diamond, graphite, and buckminster fullerenes (C_{60}, C_{70}, etc.) being the others. Out of these allotropes only graphite has planar structure and graphite planes truly are the building blocks of the carbon nanotubes. Nanotubes have strong potential for applications due to the interesting electrical, mechanical, and chemical properties they possess. In addition, the interesting possibility of filling of the tubes with different materials enables us to make nanostructures in a controlled fashion. The filling of nanotubes has recently been discussed in a few excellently written reviews [2–5]. The confined existence of the material inside the tubes might introduce interesting new properties in these materials and show physical phenomena which have not yet been seen.

Initially nanotubes were synthesized by the arc discharge method [6–9]. Later on, however, a host of other techniques [10–13] were exploited for the synthesis of nanotubes. Currently plasma arc jet [13], laser ablation, chemical vapor deposition (CVD) [10, 11] pyrolysis, electrolytic conversion [12], and arc discharge are the most popular methods for the synthesis of the nanotubes. The filling of the carbon nanotubes received immediate attention because of the known existence of the metal particle entrapped by the fullerene [11, 12]. Nanotubes are either single walled or multiwalled as shown in Figure 1a and b. Single walled tubes are divided into three categories depending on their structure as shown in Figure 2a–c. In multiwalled tubes the concentric shells of the graphite planes are held together by weak van der waals forces. Single walled tubes mostly exist in bundles.

1.1. Structure of Carbon Nanotubes

It is convenient to describe the structure of a single walled carbon nanotube in terms of a honeycomb lattice of a single plane of graphite (a graphene layer). The basic unit cell of a honeycomb lattice is a two-dimensional (2D) parallelogram with a two-atom basis. The unit vectors of a graphene layer are shown in Figure 3b. In general a tube is defined in terms of its chirality. The axes of the graphene layer are along $\mathbf{a}_1$ and $\mathbf{a}_2$, respectively (i.e., the two directions termed as zigzag). The bisector of these two axes is called the "armchair" axis. Chiral vector is defined as a vector C_h drawn from its origin to a point on a honeycomb lattice defining chirality of the tube. As the coordinates of points on a zigzag axis are $(m, 0)$ the chiral angle of zigzag type tubes is zero. Likewise the chiral angle (θ) of a tube along the armchair direction is 30°. Any other type of tube falls in between these categories and the chiral angle lies between $0 < \theta \leq 30°$. The structure of various tubes is described in Figure 2a–c. Various structure parameters of the tubes are listed in Table 1.

2. SYNTHESIS

2.1. Multiwalled Carbon Nanotubes

Carbon nanotubes are readily prepared by striking an arc between graphite electrodes in ~500 torr of helium, which is considerably higher than the pressure of helium used for the production of fullerene soot. A current of 60–100 A across the plasma with potential drop of about 25 V gives high yields of carbon nanotubes. The arcing process can be optimized such that the major portion of the carbon anode is deposited on the cathode in the form of carbon nanotubes and graphitic nanoparticles [6]. Carbon nanotubes have been produced in large quantities by using plasma arc jets [13] by optimizing the quenching process in the arc between

ISBN: 1-58883-059-4/$35.00

Encyclopedia of Nanoscience and Nanotechnology
Edited by H. S. Nalwa
Volume 3: Pages (417–430)

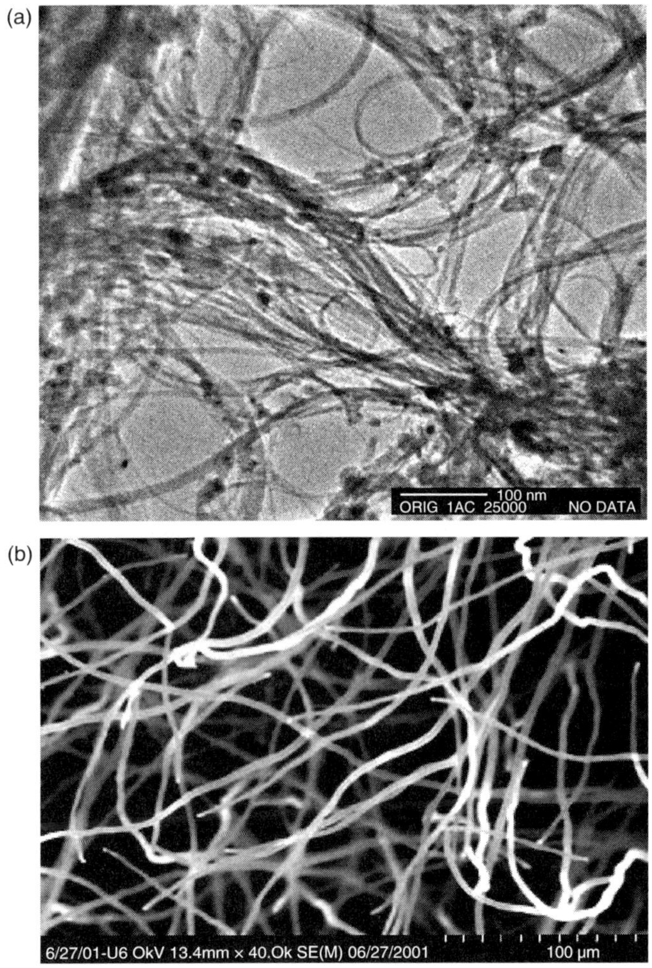

Figure 1. (a) Bundles of single walled carbon nanotubes. (b) Multi-walled carbon nanotubes.

a graphite anode and a cooled Cu electrode [14]. Studies with scanning tunneling microscopy reveal that the deposition of carbon vapor on cooled substrates of highly oriented pyrolytic graphite gives rise to tubelike structures [15]. Electrolysis in molten halide salts with carbon electrodes under Ar atmosphere also yields carbon nanotubes with high concentration [16]. Besides the conventional arc-evaporation technique, carbon nanotubes are produced by the decomposition of hydrocarbons, such as acetylene, under inert conditions around 700 °C over iron/graphite [17], cobalt/graphite [18], or iron/silica [19] catalysts. The presence of transition metal particles is essential for the formation of nanotubes by the pyrolysis process and the diameter of the nanotubes is determined by the size of the metal particles [20]. Sen et al. [21] prepared carbon nanotubes and metal-filled onionlike structures by the pyrolysis of ferrocene, cobaltocene, and nickelocene under reductive conditions, where the precursors act as a source of the metal catalyst as well as carbon. It was also shown that the pyrolysis of benzene in the presence of ferrocene or $[Fe(CO)_5]$ gives high yields of nanotubes. The diameter of the nanotubes depends on the concentration of the carbon source and size of metal precursors [22].

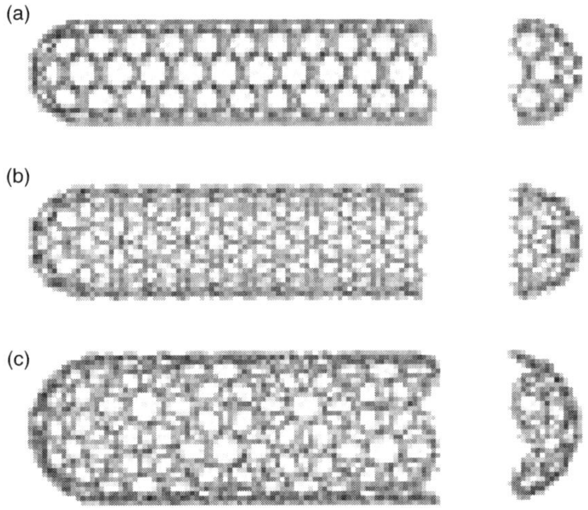

Figure 2. Structural models of (a) armchair, (b) zigzag, and (c) a general tube. The chiral angle (θ) lies between 0° and 30° for a general tube. Reprinted with permission from [3b], M. S. Dresselhaus et al., *Carbon* 33, 883 (1995). © 1995, Elsevier Science.

2.2. Aligned Carbon Nanotube Bundles

Carbon nanotubes are potential candidates for use as field emitters [23]. The synthesis of aligned nanotubes bundles is of particular relevance to this application. Aligned nanotube

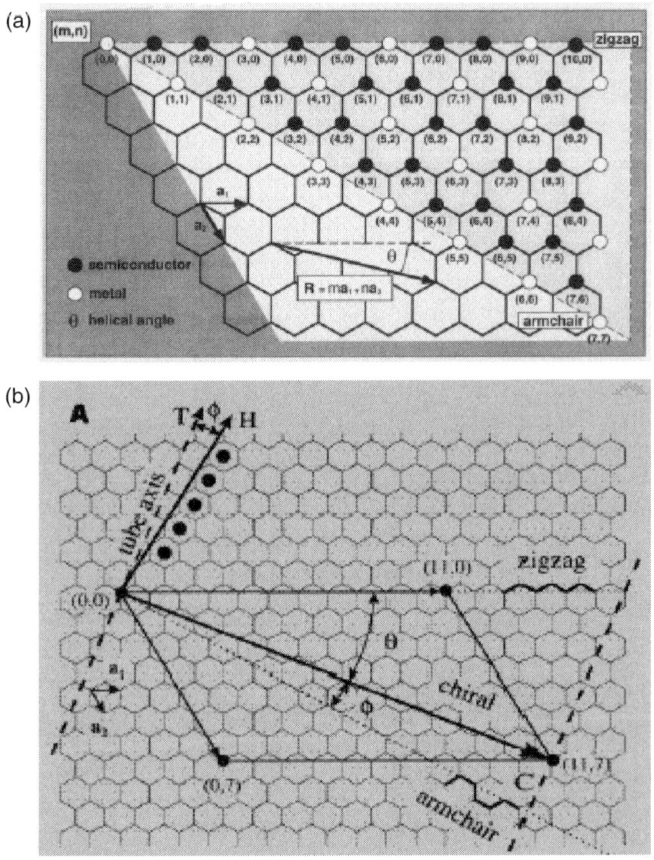

Figure 3. (a) Graphene layer and unit vectors of a single walled carbon nanotube and (b) chiral vector and chiral angle of a (11,7) tube. Reprinted with permission from [103], J. W. G. Wildoer et al., *Nature* 391, 59 (1998). © 1998, Macmillan Magazines Ltd.

Table 1. Parameters of carbon nanotubes.

Symbol	Name	Formula	Value
a_{C-C}	carbon–carbon distance		1.421Å
a	length of unit vector	$\sqrt{3}a_{c-c}$	2.46 Å
$\mathbf{a}_1$, $\mathbf{a}_2$	unit vectors	$(\frac{\sqrt{3}}{2}, \frac{1}{2})a$, $(\frac{\sqrt{3}}{2}, -\frac{1}{2})a$ in (x, y)	
coordinates			
$\mathbf{b}_1$, $\mathbf{b}_2$	reciprocal lattice vector	$(\frac{1}{\sqrt{3}}, 1)\frac{2\pi}{a}$, $(\frac{1}{\sqrt{3}}, -1)\frac{2\pi}{a}$ in (x, y) coordinates	
C_h	chiral vector	$C_h = na_1 + ma_2 = (n, m)$ n, m: integers	
L	circumference of nanotube	$L = \|C_h\| = a\sqrt{n^2 + m^2 + nm}$ $0 \leq \|m\| \leq n$	
d_t	diameter of nanotube	$d_t = \frac{L}{\pi} = \frac{\sqrt{n^2+m^2+nm}}{\pi}a$	
θ	chiral angle	$\sin\theta = \frac{\sqrt{3}m}{2\sqrt{n^2+m^2+nm}}$ $\cos\theta = \frac{2n+m}{2\sqrt{n^2+m^2+nm}}$ $0 \leq \|\theta\| \leq 30°$ $\tan\theta = \frac{\sqrt{3}m}{2n+m}$	
SWNTs of chirality (n, m) are (i) semiconducting if $(n - m)$ is not a multiple of 3 (ii) metallic otherwise			
d_R	the highest common divisor of $(2n + m, 2m + n)$		
$\mathbf{T}$	translational vector of 1D unit cell	$T = t_1a_1 + t_2a_2 \equiv (t_1, t_2)$ t_1, t_2: integers	
T	length of T	$T = \frac{\sqrt{3}L}{d_R}$	
N	number of hexagons per 1D unit cell	$N = \frac{2(n^2+m^2+nm)}{d_R}$ $2N \equiv n_C$/unit cell	
$\mathbf{R}$	symmetry vector[a]	$\mathbf{R} = pa_1 + qa_2 \equiv (p, q)p, q$: integers[a] $d = mp - nq, 0 \leq p \leq n/d, 0 \leq q \leq m/d$	
$\mathbf{M}$	number of 2π revolutions	$\mathrm{M} = \frac{[(2n+m)p+(2m+n)q]}{d_R}$ M: integer $N\mathbf{R} = MC_h + dT$	
R	basic symmetry operation	$R = (\Psi\|\tau)$	
Ψ	rotation operation	$\Psi = 2\pi\frac{M}{N}$, $(\chi = \frac{\psi L}{2\pi})$ Ψ: radians	
τ	translation operation	$\tau = \frac{dT}{N}$ τ, χ : length	

[a] **R** and R refer to same symmetry operation.
Source: Reprinted with permission from [3b], M. S. Dresselhaus et al., *Carbon* 33, 883 (1995). © 1995, Elsevier Science.

bundles have been obtained by chemical vapor deposition over transition metal catalysts embedded in the pores of mesoporus silica or the channels of alumina membranes [24, 25]. Terrones et al. [26, 27] prepared aligned nanotubes over silica substrates, laser patterned with cobalt. Ren et al. [28] employed plasma-enhanced CVD on nickel-coated glass with acetylene and ammonia mixtures for this purpose. The mechanism of the growth of aligned nanotubes by this method and the exact role played by metal particles in alignment are not clear, although a nucleation process involving the metal particles is considered to be important. Fan et al. [29] have obtained aligned nanotubes by employing CVD on porous silicon and plain silicon substrates patterned with iron films. Here again, the exact roll of iron particles in alignment of the tubes is not known. The work of Pan et al. [30] shows that the tip growth mechanism might be responsible for alignment.

Aligned nanotubes can be obtained by pyrolysis of mixtures of organometallic precursors and hydrocarbons [21, 22]. One would expect that the transition metal nanoparticles, produced *in-situ* in the pyrolysis, may not only nucleate the formation of carbon nanotubes but also align them. This aspect has been examined by carrying out the pyrolysis of metallocenes along with additional hydrocarbon sources in a suitably designed apparatus [31–33]. A ferrocene–acetylene mixture appears to be ideal to prepare large quantities of compact bundles of aligned carbon nanotubes. Andrews et al. [34] have carried out the pyrolysis of ferrocene–acetylene mixtures to obtain aligned carbon nanotubes. The advantage of this procedure is that the aligned bundles of nanotubes can be produced in one step at relatively low cost without prior preparation of the substrates. Pyrolysis of iron(ii)phthalocyanine also yields aligned nanotubes [35]. Hexagonally ordered arrays of the tubes are produced using alumina templates with ordered pores [36]. By employing catalytic chemical vapor deposition, Mukhopadhyay et al. [37] obtained large quantities of quasi-aligned carbon nanotubes using metal-impregnated zeolite templates.

2.3. Single Walled Carbon Nanotubes

The nanotubes generally obtained by the arc method or hydrocarbon pyrolysis are multiwalled and have several graphitic sheets or layers. Single walled nanotubes (SWNTs) were first prepared by metal-catalyzed dc arcing of graphite rods [7] under helium atmosphere. The graphite anode was filled with metal powder (Fe, Co, Ni) and the cathode was of pure graphite. SWNTs are generally formed in the material deposited behind the cathode. Various metal catalysts have been used to make SWNTs by this route. Dai et al. [38] prepared SWNTs by the disproportionation of CO at 1200 °C over molybdenum particles of a few nanometers diameter dispersed in a fumed alumina matrix. Saito et al. [39] compared SWNTs produced by using different catalysts and found that a cobalt or an iron/nickel bimetallic catalyst gives rise to tubes forming a highway-junction pattern. Nickel catalyst yields long and thin tubes radially growing from the metal particles. SWNTs are also prepared by using various oxides (Y_2O_3, La_2O_3, and CeO_2) as catalyst [40]. The arc-discharge technique, although cheap and easy to implement, leads to low yields of SWNTs. Journet et al. [41] obtained an ~80% yield of SWNTs in the arc by using 1 atomic yttrium and 4.2 atomic nickel as catalyst.

Arc evaporation of graphite rods filled with nickel and yttria under helium atmosphere (660 Torr) gives rise to weblike deposits of SWNT bundles [33] on the chamber walls near the cathode. SWNTs have been produced in more than 70% yields by the condensation of a laser-vaporized carbon–nickel–cobalt mixture at 1200 °C [42]. These SWNTs are nearly uniform in diameter and self-assemble into ropes consisting of 100 to 500 tubes in a 2D triangular lattice. Under controlled conditions of pyrolysis, dilute hydrocarbon–organometallic mixtures yield SWNTs [33, 43–45]. Pyrolysis of a nickelocene–acetylene mixture at 1100 °C yields SWNTs [43]. Pyrolysis of ferrocene–thiophene mixtures also yields SWNTs but the yield appears to be somewhat low. Laplaze et al. [46] have demonstrated that concentrated solar energy can be employed to vaporize graphite to synthesize SWNTs. Nikloaev et al. [47] obtained SWNTs using a gas-phase catalytic method involving the pyrolysis of [$Fe(CO)_5$] and CO. The decomposition of CO on a silica-supported Co/Mo catalyst also yields SWNTs [48]. Colomer et al. [49] obtained SWNTs in high yield by passing a H_2/CH_4 mixture over a transition-metal deposited MgO substrate. Flahaut et al. [50] have synthesized SWNTs by passing a H_2/CH_4 mixture over transition-metal oxide spinels, obtained by the combustion route. Zeolites containing one-dimensional channels have also been investigated toward the synthesis of monosized SWNTs [51].

3. VARIOUS ROUTES FOR FILLING THE TUBES

3.1. Filling MWNT

3.1.1. Capillary Filling by Chemical Routes [52]

The idea that the carbon nanotubes could be filled with a variety of metals to prepare the nanowires of metals was very appealing. One has to, however, make sure that the tubes are open ended. This is because in general the tubes are closed at both ends. The ends are also the weakest spot in the structure as they contain pentagons of carbon atoms. The opening of the tube ends can be achieved by boiling the suspensions of nanotubes [53] in aqueous nitric acid for several hours at 150 °C. This generally results in the opening of the ends of about 80–90% tubes. We describe the methodology of filling the tubes using the **wet chemical route**.

One Step Method for Filling the Tubes [54] A sample of closed nanotubes (0.5 g), azeotropic nitric acid (68% ca., 100 cc), and soluble nitrate (0.5–1 g) was heated to reflux for 4.5–12 h. The nitric acid was decanted off subsequently. The filtered residue was dried at 60 °C. When the resulting powder is calcined at 450 °C in Ar flow, about 60% of the nanotubes contain oxide nanocrystals of 3–5 nm diameter and 10–30 nm length. Using this technique it is possible to fill the oxides of **Ni, Co, Fe, U** [53], **Nd, Eu, La, Ce, Pr, Y, Zr, and Cd** [54], **Mo** [55], and **Sn** [56, 57].

Two Step Method for Filling the Tubes [54] In the two step process, the nanotubes are opened by refluxing in azeotropic HNO_3 for 8–24 h. The filtered residue is heated to 900 °C to remove the acidic groups attached to the surface of the tubes. Subsequently the sample of the tubes is mixed with the metal complexes. The filtered residue is dried and later on calcined at 450 °C in Ar flow. The two step method is useful for systems that do not dissolve in nitric acid. In particular this is found to be useful for filling the tubes with **Rh, Ru** [58], **Pd** [59], **Co, Au** [60, 61], **and Ag** [61, 62]. For filling of the tubes with **Au** and **Ag** the nanotube powder is mixed with solutions of $AuCl_3$ and $AgNO_3$, respectively. This results in filling of tubes with $AuCl_3$ and $AgNO_3$. Further reduction of the mixture to 300–400 °C in H_2 flow leads to filling of the tubes with Au and Ag [54].

SnO [56] The filling of the tubes with **SnO** is achieved by heating 100 mg of nanotube powder with 0.8 g of $SnCl_2$ in 2 ml of hot concentrated hydrochloric acid. The pH of the solution has to be raised slowly to 10.2 by dropwise addition of Na_2CO_3. The resulting mixture was dried at 160 °C for 3 h. Multiwalled tubes contained **SnO** in the core region.

3.1.2. Filling with Molten Solid Route

The filling of the carbon nanotubes by heating them in the presence of molten metal was immediately explored after the discovery of the tubes when Ajayan and Ijima [52] attempted to fill the tubes with **Pb** in a sealed quartz ampoule. Various elements that have been filled using these techniques are discussed.

Eutectic Mixture of UCl_4 and KCl [63] **UCl_4** and **KCl** are mixed in different molar ratios and are mixed with opened multiwalled tubes. The mixture is then heated in a sealed quartz ampoule at a temperature 150 °C greater than the liquid temperature of each eutectic [63]. Using this technique the surface tension of the composite material is suitably adjusted to achieve the filling.

Filling the MWNTs with Pb, Se, Sb [52] Lead (**Pb**) was evaporated on carbon nanotube samples using e-beam evaporation techniques. When the tubes were heated to 400 °C, the melting temperature of lead, the filling took place. Similar results were obtained for **Sb** and **Se**.

As discussed the capillary filling of the MWNTs has been achieved using two dominant routes. In the first case, the tubes are mixed with chemical solutions containing nitrates and halides salts (wet chemical route) of the filler material. In the second case the powder of the tubes is mixed and heated with the filler itself. In the majority of the cases the compound of the filler material has to be reduced in the flow of either Ar or H_2 gas at sufficiently high temperature. That raises the question whether the filling is of pure intended solid or a mixed phase of the two. So far the evidence shows that the complete reduction of the salt has occurred. One should, however, not rule out the presence of the localized nonhomogeneity of the starting compound. The limitation of the wet chemical route is the limited availability of the salts, less or no control of the composition of the resulting filler, and process based impurities.

In the filling with molten route capillary action, surface tension, and wetting play important roles in the process [64, 65]. Generally the filling of the nanotubes by capillary action is restricted to low surface tension metals such as **Bi**, **Cs**, **S**, and **Se** [65, 66]. When filling the nanotubes two competing phenomena, the rise of the molten liquid in the capillary and the wetting of the walls accompany each other. If the diameter of the tube is too small, generally the wetting of the tube surface results. This is one of the main reasons for the poor efficiency of the filling of SWNTs. The capillary action to fill the nanotubes by the forces of surface tension is governed by the contact angle (θ) between the liquid and solid interface [67]. If the θ is more than 90°, the filling of a liquid inside a capillary can only be achieved by applying the pressure difference. On the other hand if the value of θ is less than 90° the liquid gains energy by rising in the capillary. Figure 4 shows the cases of different contact angles.

As discussed, the open-ended multiwalled nanotubes (MWNTs) are filled with **Pb**, **Bi**, **Cs**, **S**, and **Se** [52, 65–67] by heating with molten metals. It has been concluded that the filling in this case is limited to only the low surface tension substances (with $T < 180$ mN/m). This, however, requires much more serious attention because the surface tension decreases steeply with temperature. One should also not discount the carbonaceous impurities affecting the surface tension of the molten metal. Table 2 shows the surface tension of various elements popular for filling the tubes. The data are represented at the melting point of the materials. To circumvent the threshold value in a few cases the mixtures of the compounds have also been used for filling the tubes [68–73].

Another important parameter for filling the tubes with heating in the molten state of the materials could be the viscosity of the material. It may be difficult to fill the tubes with a highly viscous metal.

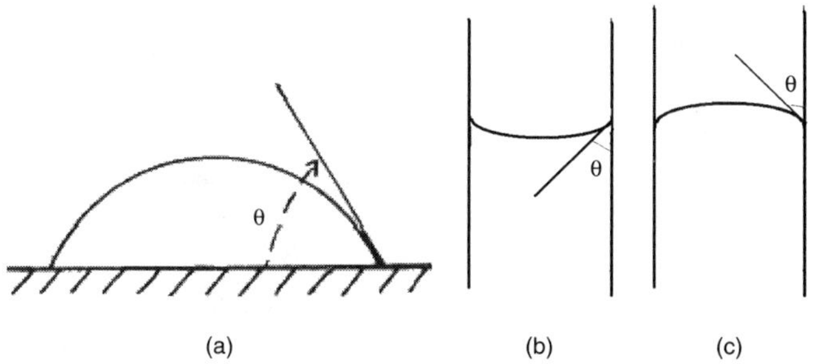

Figure 4. (a) Description of the contact angle of liquid and wall of a capillary tube. (b) and (c) The contact angle for concave and convex surface, respectively.

Table 2. Surface tension at melting point of some of the materials used for filling the tubes.

Elements/compounds	Surface tension (mN/m)
HNO_3	43
Sulphur	61
Cesium	67
Selenium	97
Vanadium oxide	80
Lead oxide (PbO)	132
Bismuth oxide (Bi_2O_3)	200
Tellurium	190
Lead	470
Uranium oxide	513
Nickel	1856
Indium	568
Uranium	1653

3.1.3. Filling by Arc Discharge

Arc discharge experiments have also been used for filling the materials inside the MWNTs [74]. In this method the holes are drilled in the graphite anodes that are packed with the metal to be filled inside the tubes. With this arrangement, the filled MWNTs are deposited on cathodes. The important work in filling of the nanotubes using this route was perfected by Guerret-Piecourt et al. [74]. Figure 5 shows the filling of the tubes by the arc discharge technique. A correlation is observed between the efficiency of a filling material (filler) and the incomplete shell structure of some of the elements. It is speculated that the elements with incomplete shells are efficient fillers. Clearly a lot more experimentation is essential to confirm this result. It has been reported that in all 41 elements are inserted inside the MWNTs: some partially and some fully. A variety of elements such as transition elements (**Cr**, **Ni**, **Re**, **Au**), rare earths (**Sm**, **Gd**, **Dy**, **Yb**), and covalents (**S**, **Ge**, **Se**, **Sb**) are filled nearly continuously and **Mn**, **Co**, **Fe**, **Pd**, **Nb**, **Hf**, **Os**, **B**, **Te**, and **Bi** are filled partially with small gaps in between [74–78]. The procedure of filling the tubes is discussed.

***Filling the MWNTs with Ge, S, Se, and Sb* [*74–78*]** For filling of the multiwalled tubes with these elements the arc discharge method was used. The anode and cathode were graphite rods of diameter 9 mm and lengths 38 and 72 mm, respectively. A 6 mm diameter hole was drilled in the anode to a depth 38 mm and was filled with graphite and element powder. The ratio of the graphite and the element in the mixture was 70% and 30%. The electric arc conditions were 100–110 A, 20–30 V and He pressure was 0.5 bar. The same technique is also used for filling the tubes with **Cr**, **Ni**, **Re**, **Au**, **Sm**, **Gd**, **Dy**, and **Yb**.

***Filling TaC and CeC_2 inside MWNTs* [*79, 80*]** Tantalum powder filled graphite rods are used in an arc discharge apparatus as the positive electrode (anode). The negative electrode (cathode) is pure graphite. The powder filled rod is initially heated to 2000 °C to form carbide and is then used for deposition. The analyses of the tubes show that there are no oxide impurities and the fillings are either pure carbide or pure metal.

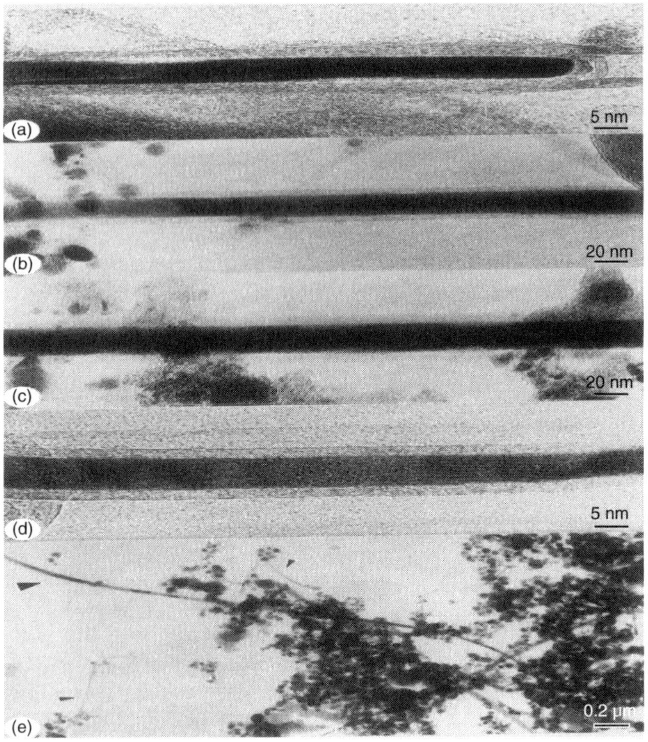

Figure 5. Filling of the core region of multiwalled tubes using arc discharge with (a) Cr, (b) Ni, (c) Dy, (d) Yb, and (e) Cr. Reprinted with permission from [74], G. Guerret-Piecourt et al., *Nature* 372, 761 (1994). © 1994, Macmillan Magazines Ltd.

It was concluded by these studies that the elements with incomplete outer electronic shells perhaps have the best chance of filling the tubes fully. However, only one such study has been reported and more data are required before it can be used as a rule of thumb. It has also been reported that a small concentration of S impurity helps considerably in filling the tubes. Using this method carbides such as **LaC_2** [79, 80], **YC_2** [81, 82], **CeC_2** [79, 83], **Gd_2C_3** [74, 84], **TiC** [74, 85], **ZrC** [86], **TaC** [80, 83, 87], **HfC** [88], **MoC** [87, 89], and others [90] have been inserted in the MWNTs. This technique is disadvantageous because in the high temperature environment of the arc discharge it is rather easier for elements to convert to the carbide form and only a few elements are inserted without carbide formation.

3.1.4. Catalytic Filling of MWNTs [10, 11]

Another technique for generating filled MWNTs is to use the catalyst itself to fill the tubes at the time of synthesis. In this process an electroplated **Ni** layer is used as a catalyst for the growth of nanotubes by a CVD technique [10]. The copper substrates are subjected to NH_3 treatment for the production of the nanocatalyst and used as templates for nanotube growth. The method has been adopted to fill Ni and NiPt [11] inside the tubes and can be easily used for filling **Cr**, **Co**, **Fe**, and **Pt** inside the MWNTs. The transmission electron microscope (TEM) image of the nanorods of diameter ~40 nm and length ~0.5 μm nanorod and therefore the structure of the **Ni** rod may have interesting features that were not seen earlier. In Figure 6 one can clearly see the uniform and dense **Ni** filling inside the tube. The diameter of the **Ni** nanorod is approximately 40 nm and its length is 500 nm. It is also evident that the symmetry of the end cap of the tube is enforced on the end of the tube.

To understand the microstructure of the Ni nanorods inside carbon nanotubes, a nanodiffraction study was performed at various locations of the tubes. Typically the size of the area chosen for nanodiffraction was 15 nm. The locations were (a) inside the Ni nanorod, (b) near the tip of the tube, (c) near the lower tip of the tube, and (d) on the walls. The nanoarea diffraction patterns at all four locations are shown in Figure 7. Narrow diffraction spots observed from face-centered cubic, Ni nanorods in the case of Figure 7 indicate that the structure of the rods is perfectly crystalline in this region. No evidence of another phase, such as Ni_3C, is found at this location. Closer to the tip (Fig. 7b), the diffraction pattern becomes intensely blurred with thicker

(a)

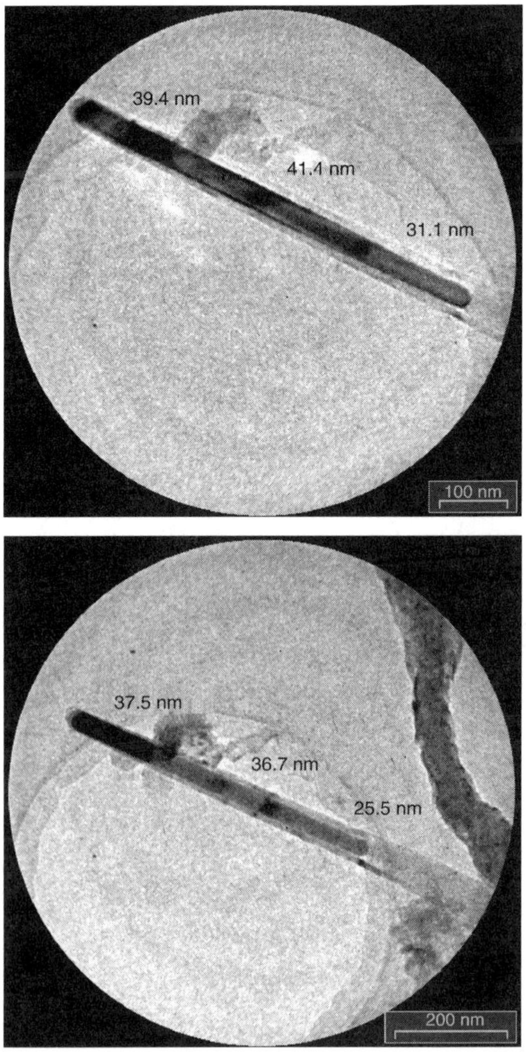

(b)

Figure 6. TEM images of the Ni nanorods inside the multiwalled carbon nanotube. Reprinted with permission from [11], M. K. Singh et al., *J. Nanosci. Nanotech.* 3, 165 (2003). © 2003, American Scientific Publishers.

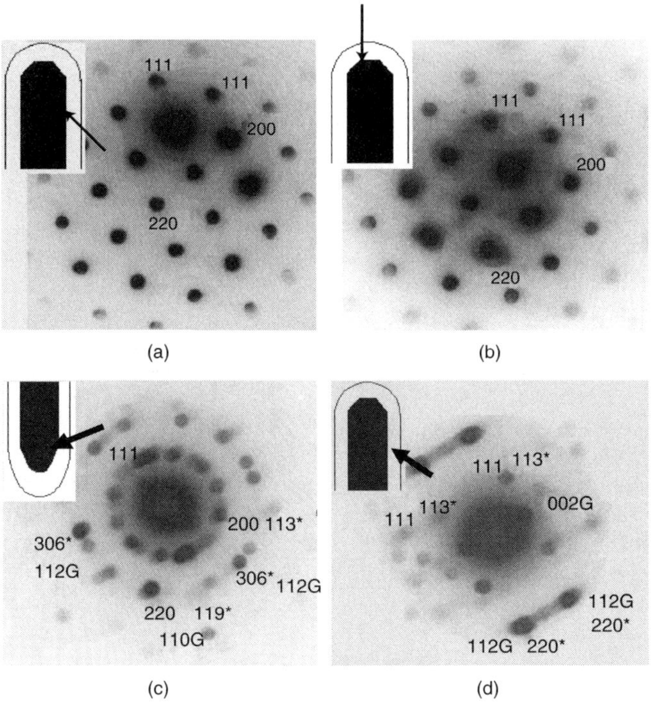

Figure 7. Electron diffraction patterns of the Ni nanorods inside the multiwalled carbon nanotubes prepared in our laboratory. The diffraction pattern was recorded at different locations of the tubes. It is observed that the nanorods are crystalline and of pure phase in the core region. Near the walls graphite and Ni_3C phases are seen. Reprinted with permission from [11], M. K. Singh et al., *J. Nanosci. Nanotech.* 3, 165 (2003). © 2003, American Scientific Publishers.

spots, implying that the structure now may not be perfectly crystalline. Figure 7c shows the nanodiffraction observed near the lower tip of the tubes. It shows diffraction spots corresponding to superposition of several phases. The dominant spots belong to the (110) and (112) planes of graphite, (110) and (220) planes of Ni, and (306) and (113) planes of Ni_3C. Similarly the nanoarea diffraction (Fig. 7d) within the walls of the tubes indicates the presence of (112), (002) planes of graphite structure and some evanescent patterns of (113), (220) planes of the Ni_3C phase that are possibly present at the interface between the graphite walls and the Ni filling.

The disadvantage of the previous technique may be that it is restricted to filling by only those elements that can be used as the catalyst for the growth of the nanotubes. Similarly **Fe** filling of the multiwalled tubes has been accomplished using catalytic decomposition of cyclohexane [91]. This technique also relies on *in-situ* filling of the tubes at the time of synthesis.

3.2. Filling SWNTs

Filling of SWNTs has been attempted using the previous techniques also. However, the additional constraint in case of SWNTs is their small diameter and as a result the capillary action may not be as efficient. An excellent review on filling of the SWNTs was published by Monthioux [3a]. The filling of SWNTs is all the more important because the dimensions are much smaller and the chances of seeing quantum effects due to the reduction of the size will be high in this case. Following the work of David Luzzi, several studies have been published on the filling of the SWNTs with fullerenes [92]. This structure is known as a "peapod" [93]. Remarkable behavior is reported for C_{60} molecules when encapsulated in ~1.4-nm-diameter SWNTs during *in-situ* electron-beam irradiation within an electron microscope, including dimerization [94, 95], coalescence [95–97], and diffusion [94]. The latter is probably more spectacular because it may lead to applications where motion and transport are required, such as the electronics industry (e.g., nanodevices) or biochemistry and medicine (e.g., nanopipes or nanosyringes). It is also possible to attach a foreign atom to the fullerene cage and then insert it in SWNTs. Research in this direction is being carried out vigorously by various groups [94, 97–100]. Actually, the synthesis of "endotubular metallofullerenes" such as $\mathbf{GdC_{82}}$ [98–100], $\mathbf{LaC_{82}}$, $\mathbf{La_2C_{82}}$, $\mathbf{SmC_{82}}$ [100], $\mathbf{La_2C_{80}}$ [101, 102], and $\mathbf{Er_xSc_{3-x}NC_{80}}$ SWNTs [102] is already achieved. Some of these are described.

3.2.1. Filling of SWNTs with Halides of Lanthanides (La, Nd, Sm, Eu, Gd, Tb, and Yb)

As-prepared SWNT powder (30 mg) was mixed with the anhydrous halides of these elements in dry box conditions. The samples of the mixture were sealed in a quartz ampoule under vacuum and placed in a tube furnace. The temperature of the furnace was raised to 10–50 °C above the melting points of the respective lanthanides (Table 3) at the rate 3–5 °C min^{-1}. The sample was held at a high temperature for one hour and then furnace cooled to room temperature.

The seminal work on filling of SWNTs was carried out by Meyer et al. [72] at Oxford who showed the actual 2D nature of the KI crystals inside the SWNT. In this work the image of single atoms arranged in a truly 1D crystal lattice inside the nanotube of 1.6 nm diameter is produced. A clear arrangement of alternate K and I atoms is evident in the TEM image. The spectacular result (Fig. 8) is that the coordination number of the atoms of the 1D crystal is different from the bulk and the lattice appears to be distorted. The normally square lattice is distorted more along the ⟨100⟩ direction than along ⟨110⟩ resulting in a "pseudo square" lattice having the circular symmetry of a nanotube. The distortion corresponds to a "Poisson ratio" of 0.293. The Oxford team has also succeeded in filling halides such as $\mathbf{(KCl)_x}$ $\mathbf{(UCl_4)_y}$ and $\mathbf{AgCl_xBr_y}$ [104], $\mathbf{CdI_2}$ and $\mathbf{ThCl_4}$ [104],

Table 3. The melting point and the filling temperature of lanthanide [71].

Material	Melting temperature	Filling temp.
$LaCl_3$	860	910
$NdCl_3$	784	834
$SmCl_3$	686	706
$EuCl_3$	850	860
$GdCl_3$	609	659
$TbCl_3$	588	638
$YbCl_3$	854	904

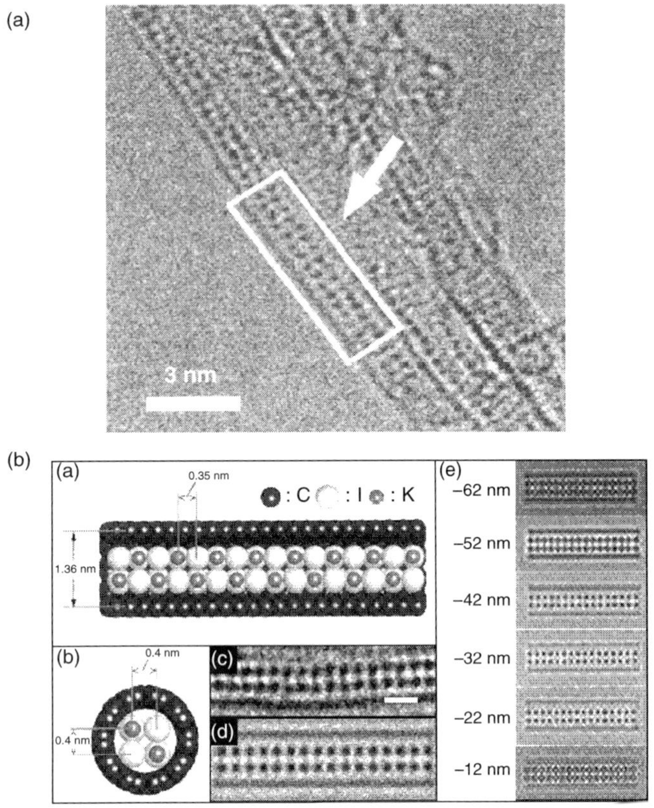

Figure 8. (a) Filling of potassium iodide crystals in a 1.6 nm diameter carbon nanotube. (b) Simulated model of filling. Reprinted with permission from [73], J. Sloan et al., *Chem. Phys. Lett.* 329, 61 (2000). © 2000, Elsevier Science.

$\mathbf{CdCl_2}$ [69, 104], $\mathbf{TbCl_3}$ [69], **TiCl** and $\mathbf{PbI_2}$ [70], **KI** [69, 72, 105, 106], $\mathbf{ZrCl_4}$ [96], and $\mathbf{AgCl_xI_{1-x}}$ [107].

3.2.2. Filling with AgCl, AgBr, and $AgBr_{0.2}Cl_{0.8}$

Filling with these halides was achieved by the previously discussed route. The halides of Ag and Br were mixed and ground with SWNT powder. The mixture was sealed in a silica tube under vacuum and then heated to temperatures higher than the melting temperature of the halide. The heating temperatures, respectively, for **AgCl**, **AgBr**, and $\mathbf{AgBr_{0.2}Cl_{0.8}}$ were 833, 800, and 783 K.

Other halides such as $\mathbf{MoCl_5}$ and $\mathbf{FeCl_3}$ [108] have also been filled inside SWNTs. Recently the insertion of chromium oxide in SWNTs was achieved by Mittal et al. [109–111] using a simple chemical process to be described. Another example of filling $\mathbf{Sb_2O_3}$ was reported recently [63, 112, 113]. Generally speaking, filling SWNTs with oxides has rarely been attempted so far.

3.2.3. Filling with Sb_2O_3 [113]

As-prepared powder (0.03 g) of SWNTs is mixed and ground with 0.15 g of $\mathbf{Sb_2O_3}$ powder. The mixture is sealed in a quartz ampoule and heated at a rate 1 K min^{-1} to 1033 K [113] and slowly cooled to room temperature.

3.2.4. Filling with Pt, Au, and Pd [59, 61, 62]

Platinum and gold are filled inside the single walled carbon nanotubes by mixing a purified powder (5 mg) of SWNTs with approx. 10 mg of $AuCl_4 \cdot xH_2O$. The mixture was dried at 373 K for 2 h and then heated to 643 K in vacuum (10^{-3} Torr). For the preparation of **Pt** filled tubes $H_2PtCl_6 \cdot 6H_2O$ was mixed with purified powder of single walled tubes. The mixture was heated to 773 K in vacuum in a quartz ampoule. The **Pd** filling of the tubes is achieved by heating the purified nanotube powder with $PdCl_2$ in a sealed quartz ampoule at 873 K [59, 62].

3.2.5. Filling Single Chain Compound of $ZrCl_4$ in SWNT [96]

$\mathbf{ZrCl_4}$ powder was mixed with SWNTs in a 1:1 mass ratio under dry box conditions. The mixture was ground thoroughly and then the sample was sealed in a quartz ampoule and heated at 2 K/min to 350 °C and kept at that temperature for one hour. The samples were furnace cooled to room temperature.

3.3. Fullerene Cages Inside SWNTs

The most interesting supermolecules which can be filled inside SWNTs are $\mathbf{C_{60}}$, $\mathbf{C_{70}}$, $\mathbf{C_{84}}$, etc. Fullerene cages appear to fill the SWNTs almost naturally. The efficiency of fullerene filling can easily be termed highest. It is very likely that the van der Waals force between the fullerene cages and tubes is responsible for this. Because of the attraction force between the wall and cages the fullerene molecule is in an energetically favorable position and remain stable.

3.3.1. Making "Peapod" Structure with C_{60} Inside SWNTs

For making peapod ($\mathbf{C_{60}}$ and $\mathbf{C_{70}}$ inside SWNTs) structures pulse laser deposited SWNTs are used. The sample of purified SWNTs is heated with fullerene powder at 650 °C for two hours inside an evacuated quartz ampoule. It is observed that the fullerenes fill the SWNTs with high density. Filling of the SWNTS with the fullerenes occurs spontaneously when the tubes are heated with the fullerene powder. Arrangement of fullerene cages inside the SWNTs is shown in Figure 9.

3.3.2. Filling of $Gd@C_{82}$ Inside the SWNTs [98–100]

$\mathbf{Gd@C_{82}}$ doped with a range of metallic elements such as **Gd**, **La**, **Sm** etc. is being filled inside the SWNTs as shown in Figure 10. For this purpose carbon soot containing

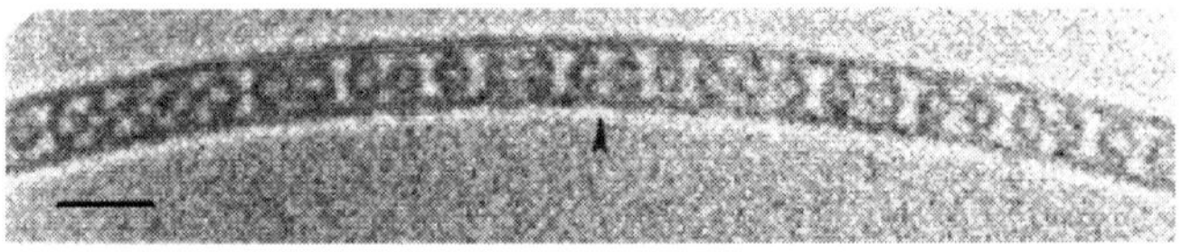

Figure 9. Filling C_{60} inside SWNT "peapod" structure. Reprinted with permission from [94], B. W. Smith et al., *Chem. Phys. Lett.* 315, 31 (1999). © 1999, Elsevier Science.

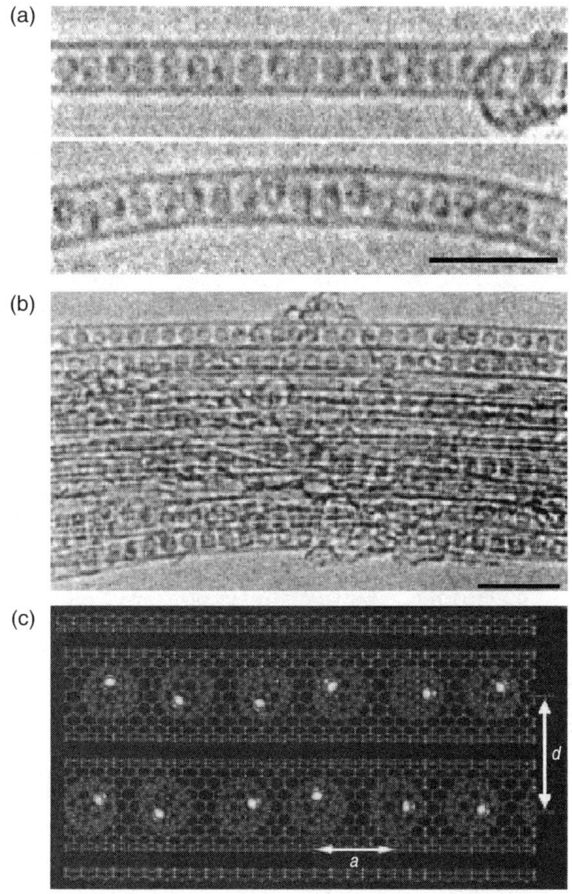

Figure 10. (a) and (b) are high resolution TEM (HRTEM) images of the isolated and bundles SWNTs containing the Gd@C_{82} fullerenes. Dark spots seen on the fullerene cages correspond to encapsulated Gd atoms that are oriented randomly with respect to tube axis (bar = 5 nm). (c) The computer model of Gd@C_{82} in the nanotube. Reprinted with permission from [98], K. Hirahara et al., *Phys. Rev. Lett.* 85, 5384 (2000). © 2000, American Physical Society.

Gd@C_{82} and other metallofullerenes was generated with gadolinium containing graphite rods in a dc arc deposition apparatus. Subsequent to the understanding that the fullerenes fill the nanotubes spontaneously, the fullerene cage **C_{82}** was carried out in a multistage high performance liquid chromatography apparatus. The purified SWNT powder was refluxed in HNO_3 for 12 hours at 140 °C prior to the experiment. The filling of the **Gd@C_{82}** was achieved by heating the mixture of **Gd@C_{82}** and SWNT powder at 500 °C for 12 hours. A similar procedure was adopted for filling the inner core of SWNTS with other metallofullerenes such as **La@C_{82}**SWNT, **La_2@C_{82}**SWNT, **Sm@C_{82}**SWNT [100], **La_2@C_{80}**SWNT [101, 102], and **Er_xSc_{3-x}N@C_{80}**SWNT [102].

3.3.3. Filling Sc_2@C_{84} Inside SWNTs [92]

Sc_2@C_{84} powder is mixed with SWNT powder and heated at 650 °C in ampoule. This sample procedure results in an extremely high density of fullerene molecule inside SWNTs, confirming our earlier statement that **C_{60}** is a natural filler for SWNTs. As shown in Figure 11 the density of **Sc_2@C_{84}** inside SWNTs is very high. The interatomic distance between atomic positions of Sc inside C_{84} in SWNTs is 0.35 nm.

3.3.4. Chemical Solution Route for Filling CrO_x in SWNTs

Recently Monthioux et al. proposed two solution phase chemistry-based methods for filling SWNTs with **CrO_x**. The first method was originally developed to tentatively intercalate SWNT ropes with **CrO_3**. As-prepared SWNTs from a Y/Ni-anode electric arc method were soaked into HCl together with CrO_3 in large excess. The mixture was left at room conditions (temperature, light) in open air at various times ranging from 3 h to 1 month. Evidence for filling was obtained via TEM after 3 h, with an optimum in yield (~30%) obtained after 2 days of soaking time. From microprobe EDX analysis and nanoprobe electron energy loss spectroscopy, it was concluded that CrO_x fills inside SWNTs, while some Cl (in addition to C, Cr, and O) was found in SWNTs gathered into ropes.

4. STRUCTURE AND PROPERTIES OF THE NANOMATERIALS INSIDE THE TUBE

In spite of the tremendous progress made in fabrication of the single walled as well as multiwalled carbon nanotubes, the properties of the materials filled inside the tubes have not received as much attention. This is probably mainly due to the technical difficulties related to the complete filling of the tubes and the uncertainties about the purity of the tubes. However, there have been some results on the characterization of the filled material. HRTEM and the electron diffraction patterns of nickel filling inside the nanotubes have been studied by Singh et al. [10] in some detail. X-ray diffraction

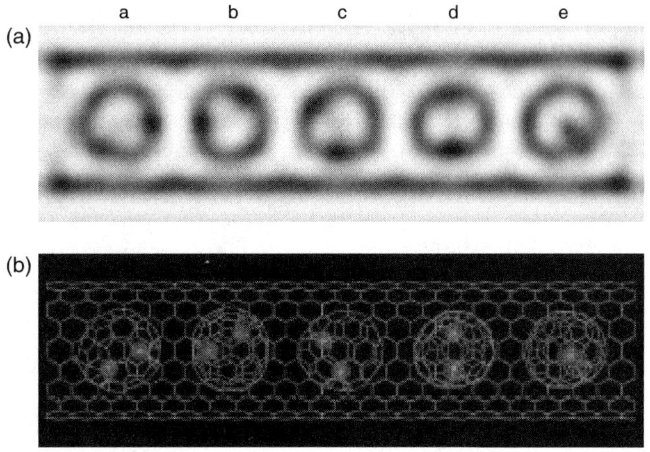

Figure 11. (a) HRTEM image of various orientations of Sc_2@C_{84} molecules inside SWNTs. (b) Simulation models for molecules observed in SWNTs. Reprinted with permission from [92], K. Suenaga et al., *Phys. Rev. Lett.* 90, 055506-01 (2003). © 2003, American Physical Society.

(XRD) studies show that the **Ni** nanorods inside the multiwalled tubes are under tensile stress and all XRD peaks are downshifted toward lower 2θ values as compared to the bulk **Ni**. The compression of the lattice inside the nanotubes has also been studied by computer simulation and the results are interesting [89]. It is also observed by Singh et al. that the **Ni** nanorods inside the tubes are highly cylindrically symmetric. The cylindrical cap of the **Ni** nanorod may have an entirely different crystalline structure than otherwise obtained. The results of encapsulation of $\mathbf{La_2C_3}$ inside the nanotubes, on the other hand, show that the nanotube end cap follows the symmetry of the carbide itself, while Mn_3C encapsulation inside the nanotubes follows the cylindrical symmetry of the nanotubes. It is also reported that the Ni_3C phase is mostly present at the interface of the filling and the carbon nanotube walls [11].

For the multiwalled carbon tubes filled with Fe there are reports by Grobert et al. [114] that the coercivity of the filling inside the tubes is much higher than the bulk iron and even higher than the **Ni** and **Co** [114]. Similarly γ Fe filled tubes show interesting hysteresis shifts in magnetization studies [115]. A large body of work is available by Loiseau et al. in studying the scanning tunneling microscopy (STM) of **Cr** filled nanotubes. They suggest that there is a strong interaction between the **Cr** filling and carbon nanotubes. Interesting features are observed in STM studies as a function of biasing voltage. Theoretical estimates of the Raman signal enhancement by **Ag** filling of the carbon nanotubes are also reported.

The observation of the quantum effects due to the confinement of the materials inside the tube has not received sufficient attention. This is mainly because the filling of the SWNTs where the confinement effects are most likely to be present has been achieved only in a limited fashion. To see the confinement effects one has to achieve efficient filling of material inside the single walled tubes. At least complete filling of the material inside the tube length of 100 nm and a uniform high density of such filled tubes must be achieved to effectively see the influence of confinement effects. The size of the tubes has to match the parameters corresponding to the confinement one is envisaging. For instance the diameter of the tubes must be of the order of the electron mean path if the confinement effect is to be seen in resistivity of the filled material. Similarly, to observe the confinement effects on the bandgap of the materials one has to use the size of the tubes of the order of a few lattice constants. For silicon the diameter of the tube required may be 1.5 nm. Interesting magnetic confinement effects might be possible to observe if the number of atoms confined in the tube is really small (50–100). This has not been achieved so far and controlled filling of the tubes is still a bit distant.

On the other hand, very interesting structural properties and characteristics of the filled tubes are already evident. For instance, structural modifications of filled material are reported for $\mathbf{ZrCl_4}$ filled tubes. It is observed that under the electron beam of 300 kV of a transmission electron microscope the continuous filling converts to clusters of $\mathbf{ZrCl_4}$. Clusters as small 1 nm can be easily formed inside the nanotube. In addition it is observed that the walls of the nanotubes are not affected by the interaction with the beam. The image contrast reveals clear images of Zr atomic sites and it is speculated that the Cl content of the clusters decreases continuously with the beam interaction as a function of time. Similar observations are also reported for the Bi filled inside the SWNTs. In that case, however, the electron irradiation of the tubes results in Bi loss through the defects in the walls. The "peapod" and "endofullerenes" also show interesting properties as function of electron irradiation. It is reported that these materials have a chain structure inside the tubes and can move or diffuse in the tube freely. This remarkable behavior was reported for $\mathbf{C_{60}}$ molecules when encapsulated in 1.4-nm-diameter SWNTs during *in-situ* electron-beam irradiation within an electron microscope, including dimerization [94, 95, 116], coalescence [95–97], and diffusion [94].

A truly one-dimensional lattice of $\mathbf{Gd@C_{82}}$ is observed inside SWNT using reflection high energy electron diffraction analysis [98]. $\mathbf{C_{82}}$ molecules with entrapped Gd atoms are located in a one-dimensional arrangement with interatomic distances 0.105 ± 0.05 nm. It would be of huge interest to perform magnetic measurements on such one-dimensional arrangements of atomic moments.

5. SUMMARY OF FILLING OF TUBES USING VARIOUS TECHNIQUES

Table 4. Multiwalled carbon nanotube.

Elements/material	Technique	Ref.
MoO_3	heating	[117]
Ni, NiS	arc discharge	[77, 118, 157]
Cr, CrS	arc discharge	[77, 157]
Ni	laser-assisted production	[119]
Ni	CVD	[120, 123, 148]
Co	CVD	[34, 121, 122, 148]
Fe	pyrolysis	[124, 138, 168]
B	arc discharge	[125]
B	heating	[126]
V_2O_5	heating	[127]
Pb	heating	[52]
SnO	pH-controlled precipitation process	[56]
Cu	hydrogen arc	[128, 159]
Mn	arc discharge	[129]
Co	arc discharge	[129, 159]
Ceramic powders	arc discharge	[130]
CuO	chemical method	[131]
Si/Pd	microwave plasma-enhanced CVD (MPE-CVD)	[132]
Ru	chemical method	[58, 161]
BN	a substitution chemical reaction	[133]
Pd/Si	MPE-CVD	[132]
Fe	MPE-CVD	[134]
Ni	catalytic decomposition	[135, 143]
Co	catalytic decomposition	[136]
N_2	arc discharge	[137]

continued

Table 4. Continued

Elements/material	Technique	Ref.
Li	electrochemical method	[139, 145, 154, 155, 158]
Co	catalytic decomposition	[140]
Ru	catalytic decomposition of propene	[141]
Co	catalytic decomposition of acetylene	[142]
Si	Radio Frequency-plasma hot filament CVD	[147]
Se, S, Sb, and Ge	arc discharge	[75]
Pt	thermal chemical vapor deposition	[148]
Pd	thermal chemical vapor deposition	[148]
KCl	chemical method	[153]
Ge	arc discharge	[157, 159]
S	arc discharge	[157]
Pt	a template carbonization method	[160]
Mn_3C	arc discharge	[163]
α-LaC_2	arc discharge	[164]
Fe	Laser-assisted thermal chemical vapor deposition	[165]
LiCl	constant voltage electrolysis	[167]
Sn	constant voltage electrolysis	[167]
Pd	MPE-CVD	[169]
TiC	thermal	[171]
Ga	furnace heating with graphite	[172]

Table 5. Single walled carbon nanotube.

Elements/material	Technique	Ref.
Li	electrochemical doping	[144, 146]
K	electrochemical doping	[144]
Bi	by heating solid bismuth	[149]
Ni	arc discharge	[150, 157]
Co	arc discharge	[150, 157, 162, 174]
Fe	arc discharge	[150, 157]
C_{60}	heating	[151]
$LnCl_3$	capillary filling method	[152]
C_{60}	heating	[155]
Ge	arc discharge	[157, 162]
Cr	arc discharge	[157]
S	arc discharge	[157]
Ag	capillary filling method	[68]
Au	chemical method heating	[158]
Ag	chemical method heating	[158]
Pt	chemical method heating	[158]
Pd	chemical method heating	[158]
Fe	laser-assisted thermal chemical vapor deposition	[165]
Ni	laser vaporization	[166]
CrO_2	room temperature filling	[175]
C_{60}	alkali–fullerene plasma method	[170, 173]
Bi	arc discharge	[174]
Glass (SiO_2)	chemical method	[176]

GLOSSARY

C_{60} A structure consisting of 60 carbon atoms connected in a network of closed hexagons and pentagons.

Capillary A tube with very fine diameter.

Carbonaceous impurities The impurities such as amorphous carbon particles, graphite particles and other carbon based particle.

Catalyst chemical vapor deposition (CCVD) In which the gas precursors are decomposed over a catalyst.

Chemical vapor deposition (CVD) process In this process the environment of gas precursors is used for the deposition of the films.

Confinement The phenomena of restricting or limiting the motion or otherwise of the atoms or molecules in an enclosure.

Electron energy loss spectroscopy A technique based on electron energy loss very sensitive for characterizing the graphite structure.

High-resolution transmission electron microscopy (HRTEM) Used for deciphering the structure of the materials. The technique is extremely useful for studying the nanomaterials.

Nanodiffraction When electron diffraction pattern is studied from a nanometer size area of a sample it is termed as nanodiffraction.

Peapods The structure containing fullerenes inside the carbon nanotubes is called peapods.

Pyrolysis A high temperature process in which the gas precursors are decomposed in a high temperature zone.

REFERENCES

1. S. Iijima, *Nature* 354, 56 (1991).
2. M. Terrones, N. Grobert, W. K. Hsu, Y. Q. Zhu, W. B. Hou, H. Terrones, J. P. Hare, H. W. Kroto, and D. R. M. Walton, *Mater. Res. Soc. Bull.* 43 (1999).
3. (a) M. Monthioux, *Carbon* 40, 1809 (2002). (b) M. S. Dresselhaus, G. Dresselhaus, and R. Saito, *Carbon* 33, 883 (1995).
4. P. M. Ajayan, *Prog. Crystal Growth Characterization Mater.* 34, 31 (1997).
5. C. N. R. Rao, R. Sen, and A. Govindaraj, *Current Opinion Solid State Mater. Sci.* 1, 279 (1996).
6. T. W. Ebbesen and P. M. Ajayan, *Nature* 358, 220 (1992).
7. S. Iijima and T. Ichihashi, *Nature* 363, 603 (1993); D. S. Bethune et al. *Nature* 363, 605 (1993).
8. R. S. Ruoff, D. C. Lorents, B. Chan, R. Malhotra, and S. Subramoney, *Science* 259, 346 (1993).
9. Y. Murakami et al., *J. Phys. Chem. Solids* 54, 1861 (1993).
10. M. K. Singh, P. P. Singh, E. Titus, D. S. Misra, and F. LeNormand, *Chem. Phys. Lett.* 354, 331 (2002).
11. M. K. Singh, E. Titus, P. K. Tyagi, K. N. Unni, D. S. Misra, M. Roy, A. K. Dua, C. S. Cojocaru, and F. LeNormand, *J. Nanosci. Nanotech.* 3, 765 (2003).
12. J. B. Bai, A.-L. Hamon, A. Marraud, B. Jouffrey, and V. Zymla, *Chem. Phys. Lett.* 365, 184 (2002).
13. N. Hatta and K. Murata, *Chem. Phys. Lett.* 217, 398 (1994).
14. D. T. Colbert, J. Zhang, S. M. McClure, P. Nikolaev, Z. Chen, J. H. Hafner, D. W. Oweens, P. G. Kotula, C. B. Carter, J. H. Weaver, and R. E. Smalley, *Science* 266, 1218 (1994).
15. M. Ge and K. Sattler, *Science* 260, 515 (1993).
16. W. K. Hsu, M. Terrones, J. P. Hare, H. Terrones, H. W. Kroto, and D. R. M. Walton, *Chem. Phys. Lett.* 262, 161 (1996).

17. M. José-Yacamán, M. Miki-Yoshida, L. Rendón, and J. G. Santiesteban, *Appl. Phys. Lett.* 62, 202 (1993).
18. V. Ivanov, J. B. Nagy, Ph. Lambin, A. Lucas, X. B. Zhang, X. F. Zhang, D. Bernaerts, G. Van Tendeloo, S. Amelinckx, and J. Van Landuyt, *Chem. Phys. Lett.* 223, 329 (1994).
19. K. Hernadi, A. Fonseca, J. B. Nagy, J. Riga, A. Lucas, and D. Bernaerts, *Synthetic Metals* 77, 31 (1996).
20. N. M. Rodriguez, *J. Mater. Res.* 8, 3233 (1993).
21. R. Sen, A. Govindaraj, and C. N. R. Rao, *Chem. Phys. Lett.* 267, 276 (1997).
22. R. Sen, A. Govindaraj, and C. N. R. Rao, *Chem. Mater.* 9, 2078 (1997).
23. W. A. deHeer, J. M. Bonard, K. Fauth, A. Chatelain, L. Forro, and D. Ugrate, *Adv. Mater.* 9, 87 (1997).
24. W. Z. Li, S. S. Xie, L. X. Qian, B. H. Chang, B. S. Zou, W. Y. Zhou, R. A. Zhao, and G. Wang, *Science* 274, 1701 (1996).
25. G. Che, B. B. Laxmi, C. R. Martin, E. R. Fisher, and R. S. Ruoff, *Chem. Mater.* 10, 260 (1998).
26. M. Terrones, N. Grobert, J. Olivares, J. P. Zhang, H. Terrones, K. Kordatos, H. K. Hsu, J. P. Hare, P. D. Townsend, K. Prassides, A. K. Cheetham, H. W. Kroto, and D. R. M. Walton, *Nature* 388, 52 (1997).
27. M. Terrones, N. Grobert, J. P. Zhang, H. Terrones, J. Olivares, W. K. Hsu, J. P. Hare, A. K. Cheetham, H. W. Kroto, and D. R. M. Walton, *Chem. Phys. Lett.* 285, 299 (1998).
28. Z. F. Ren, Z. P. Huang, J. W. Xu, J. H. Wang, P. Bush, M. P. Seigal, and P. N. Provencio, *Science* 282, 1105 (1998).
29. S. Fan, M. C. Chapline, N. R. Franklin, T. W. Tombler, A. M. Cassel, and H. Dai, *Science* 283, 512 (1999).
30. Z. W. Pan, S. S. Xie, B. H. Chang, L. F. Sun, W. Y. Zhou, and G. Wang, *Chem. Phys. Lett.* 299, 97 (1999).
31. B. C. Satishkumar, A. Govindaraj, and C. N. R. Rao, *Chem. Phys. Lett.* 307, 158 (1999).
32. C. N. R. Rao, R. Sen, B. C. Satishkumar, and A. Govindaraj, *Chem.Commun.* 1525 (1998).
33. C. N. R. Rao, A. Govindaraj, R. Sen, and B. C. Satishkumar, *Mater. Res. Innov.* 2, 128 (1998).
34. R. Andrews, D. Jacques, A. M. Rao, F. Derbyshire, D. Qian, X. Fan, E. C. Dickey, and J. Chen, *Chem. Phys. Lett.* 303, 467 (1999).
35. S. Huang, A. W. H. Mau, T. W. Turney, P. A. White, and L. Dai, *J. Phys. Chem. B* 104, 2193 (2000).
36. J. Li, C. Papadopoulos, J. M. Xu, and M. Moskovits, *Appl. Phys. Lett.* 75, 367 (1999).
37. K. Mukhopadhyay, A. Koshio, T. Sugai, N. Tanaka, H. Shinohara, Z. Konya, and J. B. Nagy, *Chem. Phys. Lett.* 303, 117 (1999).
38. H. Dai, A. G. Rinzler, P. Nikolaev, A. Thess, D. T. Colbert, and R. E. Smalley, *Chem. Phys. Lett.* 260, 471 (1996).
39. Y. Saito, M. Okuda, and T. Koyama, *Surf. Rev. Lett.* 3, 863 (1996).
40. Y. Saito, K. Kawabata, and M. Okuda, *J. Phys. Chem.* 99, 16076 (1995).
41. C. Journet, W. K. Maser, P. Bernier, A. Loiseau, M. Lamydela Chapelle, S. Lefrant, P. Deniard, R. Lee, and J. E. Fischer, *Nature* 388, 756 (1997).
42. A. Thess, R. Lee, P. Nikolaev, H. Dai, P. Petit, J. Robert, C. Xu, Y. H. Lee, S. G. Kim, A. G. Rinzler, D. T. Colbert, G. E. Scuseria, D. Tomanek, J. E. Fischer, and R. E. Smalley, *Science* 273, 483 (1996).
43. B. C. Satishkumar, A. Govindaraj, R. Sen, and C. N. R. Rao, *Chem. Phys. Lett.* 293, 47 (1998).
44. S. Seraphin and D. Zhou, *Appl. Phys. Lett.* 64, 2087 (1994).
45. H. M. Cheng, F. Li, G. Su, H. Y. Pan, L. L. He, X. Sun, and M. S. Dresselhaus, *Appl. Phys. Lett.* 72, 3282 (1998).
46. D. Laplaze, P. Bernier, W. K. Maser, G. Flamant, T. Guillard, and A. Loiseau, *Carbon* 36, 685 (1998).
47. P. Nikolaev, M. J. Bronikowski, R. K. Bradley, F. Rohmund, D. T. Colbert, K. A. Smith, and R. E. Smalley, *Chem. Phys. Lett.* 313, 91 (1999).
48. B. Kitiyanan, W. E. Alvarez, J. H. Harwell, and D. E. Resasco, *Chem. Phys. Lett.* 317, 497 (2000).
49. J. F. Colomer, C. Stephan, S. Lefrant, G. Van Tendeloo, I. Willems, Z. Kónya, A. Fonseca, Ch. Laurent, and J. B. Nagy, *Chem. Phys. Lett.* 317, 83 (2000).
50. E. Flahaut, A. Govindaraj, A. Peigney, Ch. Laurent, A. Rousset, and C. N. R. Rao, *Chem. Phys. Lett.* 300, 236 (1999).
51. H. D. Sun, Z. K. Tang, J. Chen, and G. Li, *Appl. Phys. A* 69, 381 (1999).
52. P. M. Ajayan and S. Ijima, *Nature* 361, 333 (1993).
53. S. C. Tsang, Y. K. Chen, P. J. F. Harris, and M. L. H. Green, *Nature* 372, 159 (1994).
54. Y. K. Chen, A. Chu, J. Cook, M. L. H. Green, P. J. F. Harris, R. Heesom, M. Humphries, J. Sloan, S. C. Tsang, and J. F. C. Turner, *J. Mater. Chem.* 7, 545 (1997).
55. Y. K. Chen, M. L. H. Green, and S. C. Tsang, *J. Chem. Soc. Chem. Commun.* 21, 2489 (1996).
56. J. Sloan, J. Cook, J. R. Heeson, M. L. H. Green, and J. P. Hutchison, *J. Cryst. Growth* 173, 81 (1997).
57. J. Sloan, J. Cook, J. R. Heeson, M. L. H. Green, J. P. Hutchison, and R. Tenne, *J. Mater. Chem.* 7, 1089 (1997).
58. J. Sloan, J. Hammer, M. Zweifka-Sibley, and M. L. H. Green, *J. Chem. Soc. Chem. Commun.* 3, 347 (1998).
59. R. M. Lago, S. C. Tsang, K. L. Lu, Y. K. Chen, and M. L. H. Green, *J. Chem. Soc. Chem. Commun.* 13, 1335 (1995).
60. S. C. Tsang, J. J. Davis, M. L. H. Green, H. A. O. Hill, Y. C. Leung, and P. J. Sadler, *J. Chem. Soc. Chem. Commun.* 17, 1803 (1995).
61. A. Govindaraj, B. C. Satishkumar, M. Nath, and C. N. R. Rao, *Chem. Mater.* 12, 202 (2000).
62. A. Chu, J. Cook, J. R. Heeson, and J. Sloan, *Chem. Mater.* 8, 2751 (1996).
63. J. Sloan et al., *J. Solid State Chem.* 140, 83 (1998).
64. M. R. Pederson and J. Q. Broughton, *Phys. Rev. Lett.* 69, 2689 (1992).
65. P. M. Ajayan, T. W. Ebbesen, T. Ichihashi, S. Iijima, K. Tanigaki, and H. Hiura, *Nature* 362, 522 (1993).
66. E. Dujardin, T. W. Ebbesen, H. Hiura, and K. Taginaki, *Science* 265, 1850 (1994).
67. T. W. Ebbessen, *J. Phys. Chem. Solids* 57, 951 (1996).
68. J. Sloan, D. M. Wright, H. G. Woo, S. Bailey, G. Brown et al., *Chem. Comm.* 8, 699 (1999).
69. J. Sloan, S. Friedrichs, E. Flahaut, G. Brown, S. R. Bailey et al., *Amer. Inst. Phys. Conf. Proc.* 591, 277 (2001).
70. E. Flahaut, J. Sloan, K. S. Coleman, and M. L. H. Green, *Amer. Inst. Phys. Conf. Proc.* 591, 283 (2001).
71. C. Xu, J. Sloan, G. Brown, S. Bailey, CV. Williams, et al., *Chem. Comm.* 24, 2427 (2000).
72. R. R. Meyer, J. Sloan, R. E. Dunin-Borkowski, A. Kirkland, M. C. Novotny et al., *Science* 289, 1324 (2000).
73. J. Sloan, M. C. Novotny, S. R. Bailey, G. Brown, C. Xu, et al., *Chem. Phys. Lett.* 329, 61 (2000).
74. G. Guerret-Piecourt, Y. LeBouar, A. Loiseau, and H. Pascard, *Nature* 372, 761 (1994).
75. A. Loiseau and H. Pascard, *Chem. Phys. Lett.* 256, 246 (1996).
76. A. Loiseau, *Full. Sci. Tech.* 4, 1263 (1996).
77. A. Loiseau and F. Willaime, *Appl. Surf. Sci.* 164, 227 (2000).
78. M. Demoncy, O. Stephan, N. Brun, C. Colliex, A. Loiseau, and H. Pascard, *Eur. Phys. J.* 4, 147 (1998).
79. Y. Murakami, T. Shibita, K. Okuyama, T. Arai, H. Suematsu, and Y. Yoshida, *J. Phys. Chem. Solids* 54, 1861 (1994).
80. M. Tomita, Y. Saito, and T. Hayashi, *Jpn. J. Appl. Phys.* 32, L280 (1993).

81. R. S. Ruoff, D. C. Lorents, B. Chan, R. Malhotra, and S. Subramoney, *Science* 259, 346 (1993).
82. Y. Saito, M. Yoshikawa, M. Inagaki, M. Tomita, and T. Hayashi, *Chem. Phys. Lett.* 204, 277 (1993).
83. Y. Yosida, *Appl. Phys. Lett.* 62, 3447 (1993).
84. S. A. Majetich, J. O. Artman, M. E. McHenry, N. T. Nuhfer, and S. W. Staley, *Phys. Rev. B* 48, 16845 (1993).
85. S. Seraphin, D. Zhou, J. Jiao, J. C. Withers, and R. Loufty, *Appl. Phys. Lett.* 63, 2073 (1993).
86. S. Bandow and Y. Saito, *Jpn. J. Appl. Phys.* 32, L1677 (1994).
87. M. Terrones, W. K. Hsu, A. Schilder, H. Terrones, N. Grobert, J. P. Hare, Y. Q. Zhu, M. Schwoerer, K. Prassieds, H. W. Kroto, and D. R. M. Walton, *Appl. Phys. A* 66, 307 (1998).
88. M. Ata, K. Yamaura, and J. P. Hudson, *Adv. Mater.* 7, 286 (1995).
89. J. P. Hare, W. K. Hsu, H. W. Kroto, A. Lappas, K. Praissdes, M. Terrones, and D. R. M. Walton, *Chem. Mater.* 8, 6 (1996).
90. S. Subramoney, *Adv. Mater.* 10, 1157 (1998).
91. Z.-J. Liu, R. Che, Z. Xu, and L.-M. Peng, *Synthetic Metals* 128, 191 (2002).
92. K. Suenaga, T. Okazaki, C.-R. Wang, S. Bandow, H. Shinohara, and S. Iijima, *Phys. Rev. Lett.* 90, 055506-1 (2003).
93. S. Okada, S. Saito, and A. Oshiyama, *Phys. Rev. Lett.* 86, 3835 (2001).
94. B. W. Smith, M. Monthioux, and D. E. Luzzi, *Chem. Phys. Lett.* 315, 31 (1999).
95. D. E. Luzzi and B. W. Smith, *Carbon* 38, 1751 (2000).
96. G. Brown, S. Bailey, J. Sloan, C. Xu, S. Friedrichs, et al., *Chem. Comm.* 9, 845 (2001).
97. B. W. Smith, M. Monthioux, and D. E. Luzzi, *Nature* 296, 323 (1998).
98. K. Hirahara, K. Suenaga, S. Bandow, H. Kato, T. Okazaki, et al., *Phys. Rev. Lett.* 85, 5384 (2000).
99. K. Suenaga, M. Tence, C. Mory, C. Colliex, H. Kato, et al., *Science* 290, 2280 (2000).
100. K. Suenaga, K. Hirahara, S. Bandow, S. Iijima, T. Okazaki, et al., *Amer. Inst. Phys. Conf. Proc.* 591, 256 (2001).
101. B. W. Smith, D. E. Luzzi, and Y. Achiba, *Chem. Phys. Lett.* 331, 137 (2000).
102. D. E. Luzzi, B. W. Smith, R. Russo, B. C. Satishkumar, F. Stercel, and N. R. C. Nemes, *Amer. Inst. Phys. Conf. Proc.* 591, 622 (2001).
103. W. G. Wildoer, L. C. Venema, A. G. Rinzler, R. E. Smalley, and C. Decker, *Nature* 391, 59 (1998).
104. J. L. Hutchison, R. C. Doole, R. E. Dunin-Borkowski, J. Sloan, and L. H. Green, *Jeol News* 34E, 10 (1999).
105. J. Sloan, M. C. Novotny, S. R. Bailey, G. Brown, C. Xu, et al., *Chem. Phys. Lett.* 329, 61 (2000).
106. M. Wilson and P. A. Madden, *J. Am. Chem. Soc.* 123, 2101 (2001).
107. J. Sloan, M. Terrones, S. Nufer, S. Friedrichs, S. R. Bailey, et al., *Chem. J. Am. Chem. Soc.* 124, 2116 (2002).
108. J. Mittal, M. Monthioux, and H. Allouche, In: (2001).
109. J. Mittal, H. Allouche, and M. Monthioux, In: (2000).
110. J. Mittal, M. Monthioux, H. Allouche, and O. Stephan, *Chem. Phys. Lett.* 339, 311 (2001).
111. J. Mittal, M. Monthioux, H. Allouche, O. Stephan, and W. Bacsa, *Amer. Inst. Phys. Conf. Proc.* 591, 273 (2001).
112. S. Friedrichs, R. R. Meyer, J. Sloan, A. I. Kirkland, J. L. Hutchison, and M. L. Green, *Chem. Comm.* 11, 929 (2001).
113. S. Friedrichs, J. Sloan, J. L. Hutchison, M. L. H. Green, R. R. Meyer, and A. I. Kirkland, *Phys. Rev. B* 64,045406/1 (2001).
114. N. Gobert, W. K. Hsu, Y. Q. Zhu, J. P. Hare, and H. W. Kroto, *Appl. Phys. Lett.* 75, 3363 (1999).
115. C. Prados, P. Crespo, J. M. Gonzalez, A. Hernando, J. F. Marco, R. Gancedo, M. Terrones, R. M. Walton, and H. W. Kroto, *Phys. Rev. B* 65, 113405 (2002).
116. B. W. Smith and D. E. Luzzi, *Chem. Phys. Lett.* 321, 169 (2000).
117. Y. K. Chen, M. L. H. Green, and S. C. Tsang, *Chem. Commun.* 21, 2489 (1996).
118. G. Guerret-Piecourt, Y. LeBouar, A. Loiseau, and H. Pascard, *Nature* 372, 761 (1994).
119. H. Kanzow, A. Schmalz, and A. Ding, *Chem. Phys. Lett.* 295, 525 (1998).
120. Q. Liang, L. Z. Gao, Q. Li, S. H. Tang, B. C. Liu, and Z. L. Yu, *Carbon* 39, 897 (2001).
121. W. Z. Li, J. G. Wen, M. Sennet, and Z. F. Ren, *Chem. Phys. Lett.* 368, 299 (2003).
122. C. Singh, M. Shaffer, I. Kinloch, and A. Windle, *Physica B* 323, 339 (2002).
123. G. W. Ho, A. T. S. Wee, J. Lin, and W. C. Tjiu, *Thin Solid Films* 388, 73 (2001).
124. N. Mayne, M. Grobert, R. Terrones, M. Kamalakaran, H. Rühle, W. Kroto, and D. R. M. Walton, *Chem. Phys. Lett.* 338, 101 (2001).
125. Ph. Redlich, J. Loeffler, P. M. Ajayan, J. Bill, F. Aldinger, and M. Rühle, *Chem. Phys. Lett.* 260, 465 (1996).
126. W. Han, Y. Bando, K. Kurashima, and T. Sato, *Chem. Phys. Lett.* 299, 368 (1999).
127. P. M. Ajayan, O. Stéphan, Ph. Redlich, and C. Colliex, *Nature* 375, 564 (1995).
128. A. A. Setlur, J. M. Lauerhaas, J .Y. Dai, and R. P. H. Chang, *Appl. Phys. Lett.* 69, 354 (1996).
129. J. M. Lambert, P. Bernier, and P. M. Ajayan, *Synthetic Metals* 70, 1475 (1995).
130. Y. Q. Zhu, H. W. Kroto, D. R. M. Walton, H. Lange, and A. Huczko, *Chem. Phys. Lett.* 365, 457 (2002).
131. H.-Q. Wu, X.-W. Wei, M.-W. Shao, J.-S. Gu, and M.-Z. Qu, *Chem. Phys. Lett.* 364, 152 (2002).
132. S.-H. Tsai, C.-T. Shiu, S.-H. Lai, and H.-C. Shih, *Carbon* 40, 1597 (2002).
133. D. Golberg, Y. Bando, L. Bourgeois, K. Kurashima, and T. Sato, *J. Appl. Phys.* 86, 2364 (1999).
134. H. L. Chang, C. H. Lin, and C. T. Kuo, *Thin Solid Films* 420, 219 (2002).
135. Z. Li, J. Chen, X. Zhang, Y. Li, and K. Fung, *Carbon* 40, 409 (2002).
136. X. Ma, Y. Cai, N. Lun, Q. Ao, S. Li, F. Li, S. Wen, *Mater. Lett.* in press.
137. R. Droppa, Jr., P. Hammer, A. C. M. Carvalho, M. C. Dos Santos, and F. Alvarez, *J. Non-Cryst. Solids* 299, 874 (2002).
138. S. Liu, X. Tang, Y. Mastai, I. Felner, and A. Gedanken, *J. Mater. Chem.* 10, 2502 (2000).
139. Z. Yang and H. Wu, *Solid State Ionics* 143, 173 (2001).
140. J. S. Lee, G. H. Gu, H. Kim, J. S. Suh, I. Han, N. S. Lee, J. M. Kim, and G.-S. Park, *Synthetic Metals* 124, 307 (2001).
141. M. L. Mabudafhasi, R. Bodkin, C. P. Nicolaides, X.-Y. Liu, M. J. Witcomb, and N. J. Coville, *Carbon* 40, 2737 (2002).
142. A. K. Sinha, D. W. Hwang, and L.-P. Hwang, *Chem. Phys. Lett.* 332, 455 (2000).
143. C. Liang, G. W. Meng, L. D. Zhang, N. F. Shen, and X. Y. Zhang, *J. Crystal.* 218, 136 (2000).
144. A. Claye, S. Rahman, J. E. Fische, A. Sirenko, G. U. Sumanasekera, and P. C. Eklund, *Chem. Phys. Lett.* 333, 16 (2001).
145. G. Maurin, Ch. Bousquet, F. Henn, P. Bernier, R. Almairac, and B. Simon, *Solid State Ionics* 136, 1295 (2000).
146. A. Claye, J. E. Fischer, and A. Métrot, *Chem. Phys. Lett.* 330, 61 (2000).
147. Y. F. Zhang, Y. H. Tang, Y. Zhang, C. S. Lee, I. Bello, and S. T. Lee, *Chem. Phys. Lett.* 330, 48 (2000).
148. C. J. Lee, J. Park, J. M. Kim, Y. H. Jeong Y. Lee, and K. No, *Chem. Phys. Lett.* 327, 277 (2000).
149. C. Kiang, *Carbon* 38, 1699 (2000).
150. J. Jiao and S. Seraphin, *J. Phys. Chem. Solids* 61, 1055 (2000).
151. B. W. Smith and D. E. Luzzi, *Chem. Phys. Lett.* 321, 169 (2000).
152. C. Xu, J. Sloan, G. Brown, S. Bailey, V. C. Williams, S. Friedrichs, K. S. Coleman, E. Flahaut, J. L. Hutchison, and R. E. Dunin-Borkowski, *Chem. Commun.* 24, 2427 (2000).

153. W. K. Hsu, W. Z. Li, Y. Q. Zhu, N. Grobert, M. Terrones, H. Terrones, N. Yao, J. P. Zhang, S. Firth, R. J. H. Clark, A. K. Cheetham, J. P. Hare, H. W. Kroto, and D. R. M. Walton, *Chem. Phys. Lett.* 317, 77 (2000).
154. F. Leroux, K. Méténier, S. Gautier, E. Frackowiak, S. Bonnamy, and F. Béguin, *J. Power Sources* 81, 317 (1999).
155. B. Burteaux, A. Claye, B. W. Smith, M. Monthioux, D. E. Luzzi, and J. E. Fischer, *Chem. Phys. Lett.* 310, 21 (1999).
156. G. Maurin, Ch. Bousquet, F. Henn, P. Bernier, R. Almairac, and B. Simon, *Chem. Phys. Lett.* 312, 14 (1999).
157. N. Demoncy, H. Pascard, A. Loiseau, N. Brun, C. Colliex, and O. Stéphan, *Synthetic Metals* 103, 2380 (1999).
158. G. T. Wu, C. S. Wang, X. B. Zhang, H. S. Yang, Z. F. Qi, and W. Z. Li, *J. Power Sources* 75, 175 (1998).
159. A. A. Setlur, J. Y. Dai, J. M. Lauerhaas, and R. P. H. Chang, *Carbon* 36, 721 (1998).
160. T. Kyotani, L.-F. Tsai, and A. Tomita, *Chem. Commun.* 7, 701 (1997).
161. J. Cook, J. Sloan, R. J. R. Heesom, J. Hammer, and M. L. H. Green, *J. Chem. Soc. Chem. Commun.* (1996).
162. Y. Dai, J. M. Lauerhaas, A. A. Setlur, and P. H. Chang, *Chem. Phys. Lett.* 258, 547 (1996).
163. M. Liu and J. M. Cowley, *Carbon* 33, 749 (1995).
164. M. Liu and J. M. Cowley, *Carbon* 33, 225 (1995).
165. R. Alexandrescu, A. Crunteanu, R.-E. Morjan, I. Morjan, F. Rohmund, L. K. L. Falk, G. Ledoux, and F. Huisken, *Infrared Phys. Technol.* 44, 43 (2003).
166. T. Guo, P. Nikolaev, A. Thess, D. T. Colbert, and R. E. Smalley, *Chem. Phys. Lett.* 243, 49 (1995).
167. A. T. Dimitrov, G. Z. Chen, I. A. Kinloch, and D. J. Fray, *Electrochim. Acta* 48, 91 (2002).
168. P. C. P. Watts, W. K. Hsu, V. Kotzeva, and G. Z. Chen, *Chem. Phys. Lett.* 366, 42 (2002).
169. H. Chan, K. H. Hong, S. H. Lai, X. W. Liu, and H. C. Shih, *Thin Solid Films* 423, 27 (2003).
170. G.-H. Jeong, T. Hirata, R. Hatakeyama, K. Tohji, and K. Motomiya, *Carbon* 40, 2247 (2002).
171. Y. Gao, J. Liu, M. Shi, S. H. Elder, and J. W. Virden, *Appl. Phys. Lett.* 74, 3642 (1999).
172. Z. Liu, Y. Gao, and Y. Bando, *Appl. Phys. Lett.* 81, 4844 (2002).
173. G.-H. Jeong, R. Hatakeyama, T. Hirata, K. Tohji, K. Motomiya, and N. Sato, *Appl. Phys. Lett.* 79, 4213 (2001).
174. C. H. Kiang, J. S. Choi, T. T. Tran, and A. D. Bacher, *J. Phys. Chem. B* 103, 7449 (1999).
175. J. Mittal, M. Monthioux, H. Allouche, and O. Stephan, *Chem. Phys. Lett.* 339, 311 (2001).
176. J. DiMaio, S. Rhyne, Z. Yang, K. Fu, R. Czerw, J. Xu, S. Webster, Y.-P. Sun, D. L. Carroll, and J. Ballato, *Information Sci.* 149, 69 (2003).

Encyclopedia of Nanoscience and Nanotechnology
www.aspbs.com/enn

Fivefold Twinned Nanoparticles

H. Hofmeister
Max Planck Institute of Microstructure Physics, Halle, Germany

CONTENTS

1. INTRODUCTION

Twinning is widespread in crystalline materials of various origin and nature. Basic concepts and definitions of twinning are treated in many textbooks and review papers [1–3]. A description of the crystallographic fundamentals of twinning can be found in the *International Tables for Crystallography*, Volume D: "Physical Properties of Crystals" [4]. Twins may form as a result of erroneously attaching atoms or molecules to a growing crystal such that two crystals appear to be growing out of or into each other. Character and rule of twinning can be understood by considering the sequence of atomic layers added to a crystal during growth. Stacking of close-packed planes in face-centered cubic crystals is possible in three different positions denoted **A**, **B**, and **C** leading to a regular growth sequence **ABCABCABCABCA**. If, for example, the central **A** layer of this sequence is followed by a layer of misplaced atoms assuming the wrong position **C**, upon which a regular stacking appears again, then the following sequence will form: **ABCABCACBACBA**. In this way the crystal lattice is mirrored at the central layer **A**, which is easier to see if the central letter **A** is replaced by a vertical line | representing a mirror or twin plane: **ABCABC|CBACBA**. There are two general types of twin style: contact and penetration [5, 6]. The one considered here is contact twins that have a composition plane, the twin plane, that forms a boundary between the twinned subunits.

Twinning often has a serious effect on the outward shape and symmetry of a crystal, in particular in the case of repeated twinning. Two types of repeated twinning are known: lamellar and cyclic. Lamellar twinning forms from parallel contact twins repeating continuously, one after another. Cyclic twinning requires nonparallel coplanar composition planes. If these twin planes enclose an angle being an integer part of 360°, then a complete circle can be formed by cyclic twinning. Some classic minerals like cassiterite (SnO_2), wurtzite (ZnS), and rutile (TiO_2) form cyclic twins called "trilling," "fourling," "sixling," or "eightling" quite according to their twin angles of 120°, 90°, 60°, or 45°, respectively. Cyclic twinning is also found in minerals of the spinel ($MgAl_2O_4$) group whose specific rule of twinning bears its name, the spinel twin law. Here a twin plane is parallel to one of the octahedral habit planes enclosing an angle of 70.53°, which is close to $2\pi/5$. Repeated cyclic twinning according to this twin law does not form a complete circle, but leaves a small gap. Nevertheless, it enables the formation of "fivelings" of a number of crystals.

Actually, repeated twinning, also called polysynthetic twinning, or multiple twinning, is rather common in natural minerals and crystalline materials. There are also known examples of twin compounds composed of cyclicly arranged twin pairs [7]. However, it is the type of fivefold twinning on alternate coplanar twin planes in small particles, creating "fivelings" of unique morphology, for which the term "multiply twinned particles" (usually abbreviated as MTPs) was applied. The term MTP will herewith be used for such particles, otherwise, for example, for fivefold twinning in thin films, the term "fivefold twinned structures" is used. The unique morphology of MTPs and the unusual symmetry of the arrangement of building units are essential structural features, mostly at dimensions of a few nanometers, but fairly often also up to micrometer or even millimeter dimensions.

Fivefold twinning in thin films and nanoparticles of nanometer dimensions is in itself a whole class of materials, the origin of widespread structures resulting from a great variety of substances and fabrication processes involved. These are introduced in this chapter together with the issues of synthesis, formation mechanisms, and stability and lattice defects. Various illustrative examples are aimed at emphasizing the importance of this phenomenon in the area of nanostructured materials. It always has attracted the attention not only of crystal growth and crystallography research,

ISBN: 1-58883-059-4/$35.00

but also of cluster physics, physical chemistry, surface science, thin film growth, and materials research. The occurrence of quintuples of twins and local fivefold structures includes also quite different materials systems ranging from biological materials to minerals, such as proteins [8], polyoxometalates [9], viruses [10], surfactant bilayers [11], natural diamond [12], and self-assembled metal nanoparticle superlattices [13]. Within the minerals, even particular diamond species of extraterrestrial origin, which were contained in meteorites, are found [14]. Appropriate examples will be mentioned when discussing various aspects of fivefold twinning. This chapter is accompanied by an almost complete list of references that makes available results and experiences of previous work in a greater context.

1.1. Crystallographic Characteristics

The characteristics of materials that favor MTP formation are: (i) face-centered cubic (fcc) or diamond cubic (dc) crystals, (ii) low twin boundary energy, and (iii) a surface energy anisotropy with, for example, $\sigma\{111\} < \sigma\{100\}$, where {111} and {100} are the indices of surfaces of lowest energy for cubic crystals. However, any other crystal allowing repeated cyclic twinning with twin angles of about $2\pi/5$ would fit as well, if the twin boundary energy is not exceedingly large. The structural peculiarities of such particles comprise the following characteristics: (i) They are composed of equisized subunits of tetrahedral shape, (ii) the subunits join together on adjacent bounding faces (twin planes), (iii) the subunits enclose an angle of $\sim 2\pi/5$, and (iv) the involved tetrahedra share common axes of fivefold symmetry. Shape and composition of particles formed according to the above construction scheme are (i) the decahedron (pentagonal bipyramid), consisting of 5 tetrahedra with 1 fivefold axis, bounded by 10 triangular faces, and (ii) the icosahedron, consisting of 20 tetrahedra that share 6 fivefold axes and one common point at the center, bounded by 20 triangular faces. Composition and shape of the decahedron (point group symmetry D_{5h}) and the icosahedron (point group symmetry I_h) are schematically shown in Figure 1. As tetrahedral subunits of regular fcc or dc lattice, respectively, cannot form a complete space-filling structure, there remains an angular misfit (resulting in a gap of 7.35° for the decahedron), which is not considered in the drawings. Strictly speaking, the fivefold axes in such materials are only of pseudo-fivefold symmetry, unless there is some rearrangement of the lattice.

For single crystalline particles of most of the materials (cubic crystals) considered here, the common growth form is that of a cuboctahedron that is bounded by triangular octahedron faces, or {111}, and square cube faces, or {100}. This semiregular or Archimedian solid is drawn as a hard sphere model in Figure 2. Different from that, MTPs such as the icosahedron (Platonic solid), also drawn in Figure 2, are bounded by triangular faces of equal type, or {111}, only. This octahedron face is energetically favored for fcc and dc crystals because of the surface energy anisotropy of most of these materials, according to the above mentioned condition (iii) of MTP formation. That is how MTPs minimize their surface energy by approaching a spherical shape, which is most effectively achieved with the icosahedron.

1.2. Modes of Appearance

The mode of appearance of fivefold twinned particles depends on both their orientation with respect to a planar substrate (or a matrix) and the evolution of their surface morphology as influenced by the growth conditions. With respect to a planar substrate, there are four possible high symmetry orientations for both types of MTPs. Decahedra may be situated (i) with their fivefold axis perpendicular to the substrate plane, as in "fivefold" orientation or (011); (ii) with the fivefold axis parallel to the substrate plane, as in "parallel" or (001); (iii) with one tetrahedral bounding face resting on the substrate, as in "face" orientation or (111); and (iv) with the common edge of two tetrahedra resting on the substrate, as in "edge" orientation or (112). As can be seen from Figure 3, this gives as projection on the substrate (or imaging) plane a regular pentagon, a rhombic, a shortened pentagon, or a slightly less shortened pentagon, respectively. Icosahedra may be situated (i) with the common edge of two tetrahedra resting on the substrate and two fivefold axes parallel to it, as in "edge" orientation or (112); (ii) with one corner resting on the substrate and one fivefold axis parallel to it, as in "parallel" orientation or (001); (iii) with one fivefold axis perpendicular to the substrate, as in "fivefold" orientation, or (011); and (iv) with one tetrahedral bounding face resting on the substrate, as in "face" orientation or (111). Figure 4 shows the corresponding projections on the imaging plane giving a hexagon shortened along a diagonal, a hexagon elongated along a diagonal, a regular decagon, or a regular hexagon, respectively. The assignment of orientations in terms of crystal axes indices (in braces)

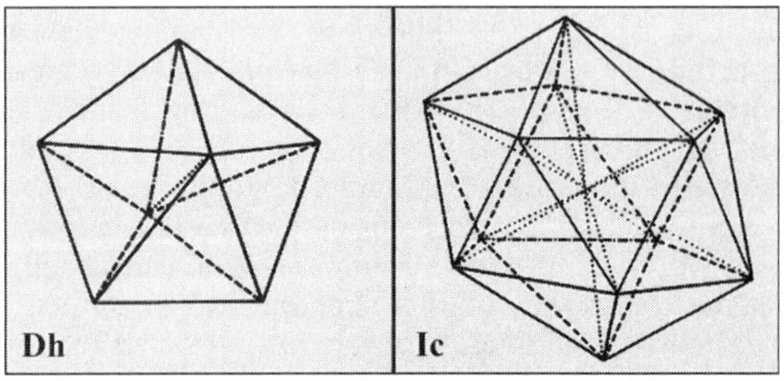

Figure 1. Shape of MTPs and their composition of tetrahedra: (Dh) decahedron and (Ic) icosahedron. Adapted with permission from [220], H. Hofmeister, *Cryst. Res. Technol.* 33, 3 (1998). © 1998, Wiley-VCH.

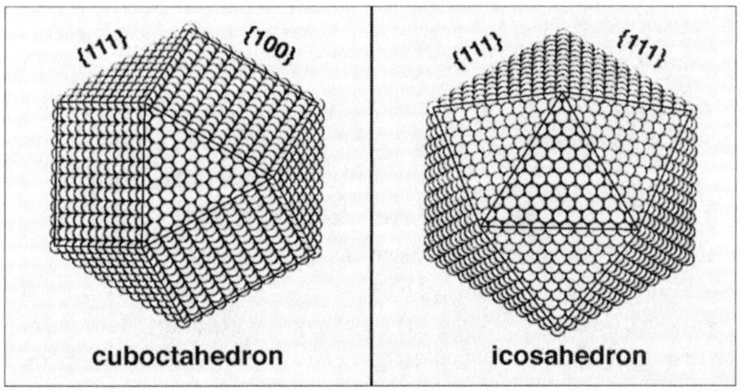

Figure 2. Hard sphere models of the surface morphology of cuboctahedron and icosahedron.

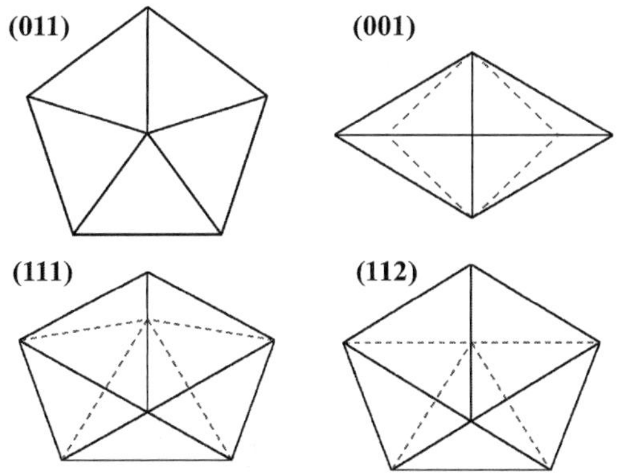

Figure 3. Orientation of decahedra on a substrate: "fivefold" (011), "parallel" (001), "face" (111), and "edge" (112).

concerns one, two, or five tetrahedra situated in the corresponding orientation.

The growth conditions are usually described by a growth parameter α that relates the rates of growth along different crystal directions. For fcc or dc materials, it is given by $\alpha = \sqrt{3}v_{100}/v_{111}$, where v_{100} is the growth velocity of cube faces and v_{111} is that of octahedron faces. The effect of α on the crystal morphology is a continuous variation starting, for example, with a perfect cube shape for $\alpha = 1$, via the cuboctahedron shape shown in Figure 2 for $\alpha = 1.5$, which corresponds to thermal equilibrium growth, to the octahedron shape for $\alpha = 3$. Under thermal equilibrium, the growth morphology may also be described by a parameter β relating the surface free energies ε_{100} and ε_{111} of the lowest energy surfaces. However, crystal growth usually is far from thermal equilibrium; thus the shape evolution is not characterized by minimizing the surface energy, but rather the growth rate of each face as determined by the kinetics. For MTPs this may lead to deviations from the ideal shapes introduced above, which range, for example, for the decahedron from a star-shaped (A) to a strongly faceted (B) and a prism-shaped specimen (C) corresponding to a variation of growth parameters from 3 to 2 and 1.5, respectively, as shown in Figure 5. Shapes B and C are named

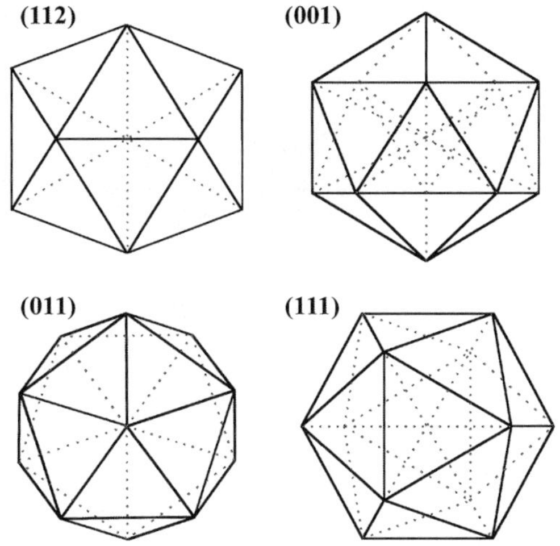

Figure 4. Orientation of icosahedra on a substrate: "edge" (112), "parallel" (001), "fivefold" (011), and "face" (111).

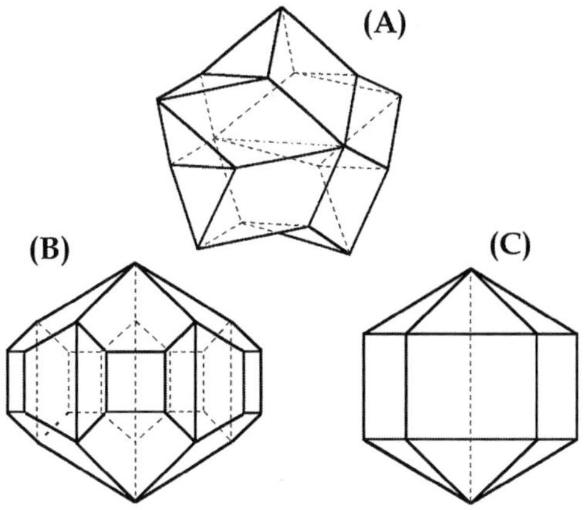

Figure 5. Deviations from the ideal shape of decahedra: (A) star-shaped, (B) faceted (Marks), and (C) prism-shaped (Ino).

"Marks decahedron" [15, 16] and "Ino decahedron" [17] by those who first introduced the corresponding models. An example of the formation of re-entrant edges, where twin boundaries emerge to the surface, as well as faceted dimples at the emergence points of fivefold axes, represents the Cu decahedron in Figure 6. Accordingly, re-entrant edges, faceted dimples, and pyramidal capping of triangular faces may occur at icosahedra [18].

1.3. Features of Fivefold Twinning

1.3.1. Natural Origin

Fivefold twinning, being a widespread habit of nanoparticles and nanostructured materials, actually is not only found in synthetic materials, but also in structures with a natural origin. As a most striking example, one may mention the polyhedral forms of certain viruses and their pentagonal aggregation. This was predicted in 1956 by Crick and Watson [19] and was confirmed in 1958 and later, mainly by electron microscopy means [10, 20–23]. Another example from biology is the hollow icosahedron configuration *in vivo* of

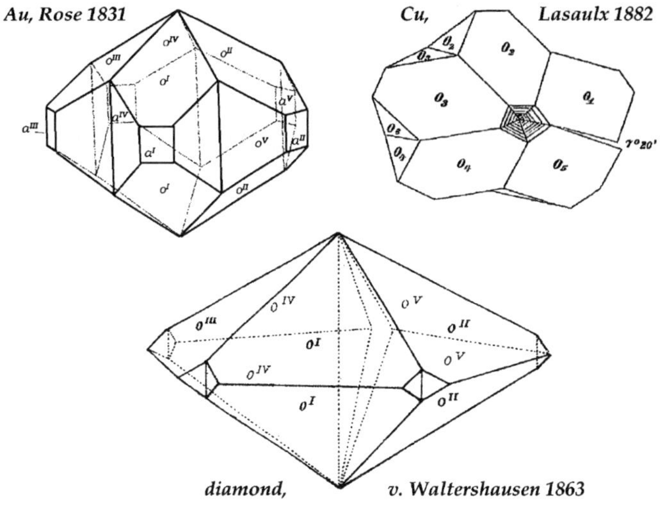

Figure 6. Early findings of fivefold twinning (fivelings) of natural occurrence.

certain protein supermolecules [8]. In the course of studying the presolar history of matter, fivefold twinned diamonds of extraterrestrial origin have been found in meteorites [14, 24]. By far, the most numerous and earliest examples of natural occurrence of fivefold twinning are known from the field of mineralogy. In former centuries, an essential part of the contemporary materials science was developed by mineralogists, mining engineers, and metallurgists. Thus it is understandable that in the first half of the 19th century the natural formation of "fivelings" in some minerals was known and therefore reported subsequently in textbooks and shape catalogues [25–29].

It was as early as 1831 that Rose [30] reported the observation of a strongly faceted decahedron of gold, the schematic drawing of which, shown in Figure 6, is an astonishing precursor of the Marks decahedron given in Figure 5B. Next to gold, fivefold twinning was also frequently observed in diamond of natural origin. This was reported for the first time by von Waltershausen [31]. Besides slightly faceting at the twin boundary tips, the original drawing, shown in Figure 6, contains a gap between subunits o^{IV} and o^{V}, corresponding to a defect caused by the lack in space filling with five tetrahedra. The first finding of a copper fiveling was reported 1882 by von Lasaulx [32]. This multiply twinned crystallite, also shown in Figure 6, is characterized by a nearly star-like shape and pentagonal dimples at the emergence of the fivefold axis. The indication of a 7°20′ gap between subunits O_1 and O_5 is based on theoretical considerations rather than on experimental observation. The MTPs of natural origin usually exhibit sizes around 1 to 2 mm. Further findings of natural fivefold twinned crystallites including, besides the already mentioned Au, diamond, and Cu, also Ag, sphalerite (ZnS), marcasite (FeS_2), magnetite (Fe_3O_4), and spinel ($MgAl_2O_4$), are presented together with the year of first mention of the corresponding mineral and related references in Table 1. One particular observation was reported by von Rath in 1877 [33] concerning an approximately 2 mm long pentagonal needle of gold whose shape corresponds to an elongated form of the prism-shaped decahedron in Figure 5C. This way, all essential shape variations of decahedra as introduced in Figure 5 were known by the end of the 19th century. The formation of fivefold twin junctions by cyclic twinning in naturally occurring substances, as confirmed by continued observations in the first half of the 20th century, was supported from a theoretical crystallography point of view by Herrmann [34] who introduced the noncrystallographic point-groups D_{5h} of the decahedron and I_h of the icosahedron. A description of simple forms of these noncrystallographic classes was given by Niggli [35].

Table 1. Natural occurrence of fivefold twinned structures.

Matter	(First mention) Refs.
Au	(1831) [26, 28, 30, 33, 397, 398, 423]
Ag	(1944) [29]
Cu	(1882) [29, 32]
C (dc)	(1863) [7, 12, 14, 24, 25, 27, 29, 31, 424, 425]
ZnS	(1882) [7, 27, 32]
FeS_2	(1977) [426]
$MgAl_2O_4$	(1877) [427]
Fe_3O_4	(1984) [428]
virus	(1958) [10, 20–23]
protein	(1997) [8]

1.3.2. Synthetic Origin

The investigation of fivefold twinned structures in synthetic nanoparticles and thin films started in the second half of the 20th century by Segall [36] with the observation of pentagonal grains of pyramidal shape in cold rolled Cu upon thermal etching in 1957. This was followed in 1959 by the observation of pentagonal whiskers (i.e., rod-like shape) of Ni, Fe, and Pt grown from the vapor phase on W substrates by Melmed and Hayward [37] who also explained the peculiar shape by assuming five twinned fcc subunits with only slight lattice distortions. Mackay [38] presented in 1962 a hard sphere model of icosahedra, described them as being made up of 20 tetrahedra, discussed their characteristics, calculated the density of closed shell icosahedra, and demonstrated a mechanism of transition to the fcc structure. In the same year Schlötterer [39–41] reported on fivefold twinned pyramidal grains of Ni grown by electrodeposition, and one year later Wentorf [42] described fivefold twinned crystallites of synthetic diamond with indications of a small-angle grain boundary accommodating the angular misfit. In 1964 Faust and John [43] reported on Si and Ge fivefold twinned grains grown from the melt. Skillman and Berry [44] found fivefold twinned particles of AgBr grown from solution. Ogburn et al. [45, 46] communicated the observation of pentagonal dendrites of Cu grown from the vapor phase. Schwoebel [47, 48] reported pentagonal pyramids of Au grown in fivefold orientation on Au(110) and Au(100) surfaces, and Gedwill et al. [49] obtained fivefold twinned grains of pyramidal shape in the deposition of Co by hydrogen reduction of $CoBr_2$, respectively. Similarly, in 1965 De Blois [50, 51] found the formation of pentagonal shaped whiskers of Ni by hydrogen reduction of $NiBr_2$. In the same year Bagley [52–54] proposed a model of pentagonal decahedra made up of five twinned tetrahedra whose orthorhombic lattice only slightly deviates from the fcc crystal lattice. In 1966 Downs and Braun [55] found fivefold twinned grains in the plating of Ni by thermal decomposition of nickel carbonyl.

The discovery of decahedral and icosahedral particles of Au and Ag formed in the early stage of thin film growth on alkali halide and mica substrates as well as by evaporation in inert-gas atmosphere in 1966 [56–64] is connected with extended availability and improved capabilities of electron microscopes at this time, which favored focusing on fivefold twinned structures of nanometer dimensions. This way, already in the first ten years of exploration, broad experimental evidence of the phenomenon, correct nomenclature, clear models, and reasonable insight in formation mechanisms was achieved. Since then a continuous and even increasing interest in fivefold twinned structures in nanoparticles and thin films produced a more than linear increase (from 1 in 1957 to 25 in 2001) of publications per year and was more and more devoted to technologically important materials like diamond, semiconductors, and Ni. The fivefold twin structure could be made visible, in particular, by

HREM as it is shown in Figure 7 by the example of a decahedron of Rh [65]. The HREM image (left) of the MTP situated in fivefold orientation (twin boundaries marked by arrow heads), that means with the fivefold twin junction perpendicular to the image plane, together with the corresponding diffractogram (right) give a clear representation of its symmetry as well as spacings (e.g., of {111} and {200} planes) and angular relations of the lattice of the tetrahedral subunits involved. Utilization of dedicated experimental techniques like cluster source equipped molecular beam devices for synthesis [66] and real-time video recording equipped electron microscopes for characterization [67–71] enabled elucidating new models and mechanisms of MTP formation as well as uncovering a rich variety of new materials having such structures. The appeal of fivefold symmetry was tremendously encouraged with the disclosure of icosahedral quasicrystals and related phases [72–75] and with the invention of the quasilattice concept to describe these structures basing on local icosahedral packing of atoms contained in tetrahedrally close-packed and related phases of intermetallic compounds [76–79]. Between both fields there are certain relations from a structural point of view, such as via decagonal twinned crystalline approximant phases [79–82].

2. MATERIALS SYNTHESIS AND FORMATION MECHANISMS

2.1. Materials Overview

Fivefold twinned structures may be found in any crystalline material that allows twinning on alternate coplanar twin planes enclosing an angle of about $2\pi/5$. Favorite materials throughout the periodic table of elements are the transition metals Fe, Co, Ni, Cu, Ru, Rh, Pd, Ag, Ir, Pt, Au; the lanthanide's Sm and Yb; as well as the group II element Mg, group III elements Al and In, and group IV element Pb, that have fcc crystal lattice, at least for the modification present in multiply twinned particles. Additionally, the group IV elements C, Si, and Ge with dc crystal lattice contribute to the MTPs. Further, multiply twinned structures are known from a number of alloys like Au-(Fe, Co, Ni, Cu, Pd), Al-(Li, Cr, Mn, Fe, Cu, Zr), Ni-(Zr, Ti), Pt-(Fe, Rh), and Si-Ge. There exists also a considerable list of binary and ternary compounds from which MTPs have been reported, including AgBr; the nitrides and carbides BN, TiN, TiCN and BC, Cr_2C_{2-x}, SiC; the oxides Fe_2O_3, Fe_3O_4, SnO_2, $BaTiO_3$, and B_6O; and further the compound semiconductors GaP, $CuInSe_2$, CdSe, and CdTe. The list of materials, to which even the molecular crystals fullerite C(60) and C(76) must be added, as well as supramolecular polyoxometalates and surfactants, is still increasing.

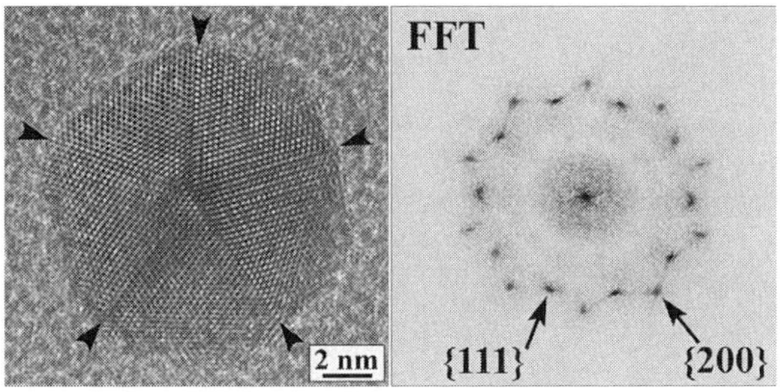

Figure 7. Decahedral particle of Rh in 5-fold orientation grown by vapor deposition on NaCl. Adapted with permission from [97], H. Hofmeister, *Mater. Sci. Forum* 312–314, 325 (1999). © 1999, Trans. Tech. Publications.

A summary of these materials together with their main characteristics and the routes of synthesis applied is given in Table 2, where elements, alloys, compounds, and composite materials with fivefold twinned structures are listed. For each entry, the table contains the year of first mention and a number of representative references. This summary must be completed by composite materials consisting of MTPs embedded in a matrix like Ge, Si, or Si-Ge precipitates in Al alloys [83–88]; Cu precipitates in Ni-Zn-Cu alloy [89]; and Au, Ag, or Co precipitates in polymer or glass matrix [90–96]. As an example of matrix-embedded MTPs, Figure 8 shows a decahedral particle of Ag grown by precipitation in glass [97]. The HREM image (left) of the particle (twin boundaries marked by short lines) clearly shows its nearly spherical shape determined by the metal-matrix interface energy. The accompanying diffractogram (right) reveals symmetry, lattice plane spacings, and angular relations according to the approximately fivefold orientation. Another class of materials should be finally mentioned, namely colloidal crystals or self-assembled superlattices consisting of two- or three-dimensional arrangements of metal nanoparticles that form fivefold twinned structures quite according to those previously described [13].

2.2. Routes of Synthesis

2.2.1. Vapor Phase Techniques

The synthesis of fivefold twinned nanoparticles and thin films may be proceeded by a large number of various processes and specific techniques. Generally, they differ by the state of the material applied in the synthesis. We distinguish synthesis (i) from the vapor phase, (ii) from the liquid phase, and (iii) from the solid phase. Vapor phase synthesis (i) includes (a) heterogeneous nucleation and growth of particles and thin films by various methods of either physical or chemical vapor deposition on substrates, and (b) homogeneous nucleation and growth of particles by aggregation within an inert-gas atmosphere. Most of the early work on metal MTPs has been done according to the process scheme given in (a) by thermal evaporation of the metal within an evacuated chamber and condensation of the metal vapor on appropriate substrates [57, 60, 98–111]. Physical vapor deposition was also applied in the formation of multiply twinned nanoparticles and thin films of Ge [112–121], SnO_2 [122], Fe_2O_3 [123], and C(60) or C(76) [124–128]. Chemical vapor deposition has been used for formation of multiply twinned nanoparticles and thin films, for example, of diamond [18, 129–132] (particles) and [133–146] (thin films), or Si and Si-Ge [147–149], TiN [150–152], TiCN [153, 154], SiC [152], GaP [155], and BN [74, 132] from precursor molecules, the decomposition or reaction of which provides the species deposited. The inert-gas aggregation technique (b) was successful in producing MTPs of most of the metals [64, 101, 156–167] and Si and Ge [168, 169], as well as alloys of

Table 2. Materials with fivefold twinned structures: (A) elements, (B) alloys, (C) compounds, and (D) composites.

Materials	Synthesis	Characteristics	Refs.
A. Elements			
Mg (1981)	inert-gas aggregation, cluster beam expansion	decahedra, closed shell icosahedra	[160, 242]
Fe (1959)	PVD, inert-gas aggregation	decahedral whiskers, decahedra	[37, 101, 158, 159, 293]
Co (1964)	solid-phase reduction, inert-gas aggregation	decahedra, icosahedra	[49, 64, 293, 430]
Ni (1959)	PVD, electrodeposition, inert-gas aggregation, colloidal synthesis	hollow whiskers, icosahedra, decahedra, thin films, rod-shaped decahedra	[37, 39, 157, 159, 165, 190, 200, 202, 204, 206, 208, 431]
Cu (1957)	electrodeposition, inert-gas aggregation, colloidal synthesis, PVD, e-beam	thin films, rod-shaped decahedra, icosahedra	[36, 39, 45, 98, 100, 157, 159, 178, 193, 231, 363, 388]
Ru (1988)	colloidal synthesis	structural fluctuations	[71]
Rh (1981)	solid phase reduction, PVD, colloidal synthesis, electrodeposition	decahedra, icosahedra	[65, 108, 226, 371, 374, 410, 435, 436]
Pd (1966)	PVD, inert-gas aggregation, electrodeposition, colloidal synthesis	decahedra, icosahedra, double icosahedra	[56, 63, 64, 99, 101, 107, 159, 161, 174, 198, 249, 438, 439]
Ag (1966)	PVD, inert-gas aggregation, electrodeposition, colloidal synthesis, e-beam	decahedra, icosahedra, rod-shaped decahedra	[60, 63, 98, 105, 157–159, 166, 174, 205, 207, 211, 217, 230, 255, 258, 291, 292, 349, 354, 440, 441]
Ir (1997)	electrodeposition	decahedra, icosahedra, shape variations	[210, 216]
Pt (1959)	PVD, inert-gas aggregation, colloidal synthesis, electrodeposition	decahedral whiskers, shape variations, decahedra, icosahedra	[37, 104, 161, 191, 210, 376, 411, 444]
Au (1964)	PVD, inert-gas aggregation, colloidal synthesis, electrodeposition, solid-phase reduction	decahedra, icosahedra, rod-shaped decahedra, double icosahedra; self-assembled nanoparticles superlattice	[13, 40, 47, 57–59, 61, 62, 64, 105, 106, 157–159, 161, 173, 182, 189, 210, 212, 237, 239, 240, 254, 259, 324–327, 335, 365, 366a, 378, 382, 387, 400, 407, 411, 447–450]
Sm (1992)	inert-gas aggregation	decahedra	[66]
Yb (1993)	inert-gas aggregation	decahedra, icosahedra	[162, 452]
Al (1997)	e-beam irradiation	decahedra, icosahedra	[232]
In (1999)	physical vapor deposition	decahedra, tetragonal lattice sub-units	[109, 111, 340]
C (dc) (1963)	high-pressure melt synthesis, CVD, dynamic-shock synthesis	decahedra, misfit faults, icosahedra, thin films, star-shaped & rod-shaped decahedra	[18, 42, 129–132, 138, 139, 141–143, 145, 350–352, 456, 457]
Si (1964)	melt growth, matrix precipitation, CVD, PVD, inert-gas aggregation	decahedra, misfit faults, thin films, star-shaped decahedra	[43, 147, 149, 168, 169, 185, 186, 234, 266, 270]
Ge (1964)	melt growth, matrix precipitation, PVD, inert-gas aggregation, solid phase crystallization, ion implantation	star-shaped & rod-shaped decahedra, thin films, decahedra	[43, 96, 112–116, 120]
Pb (2000)	inert-gas aggregation, PVD	decahedra, tetragonal lattice sub-units	[110, 164]
B. Alloys			
Au-Fe (1999)	PVD, electrodeposition	decahedra, icosahedra	[213, 385]
Au-Co (2002)	electrodeposition	decahedra, icosahedra	[213]
Au-Ni (2002)	electrodeposition	decahedra, icosahedra	[213]
Au-Cu* (1997)	inert-gas aggregation, electrodeposition	decahedra, icosahedra, rod-shaped decahedra	[170, 210, 215, 432]
Au-Pd (1990)	colloidal synthesis	decahedra of rounded shape	[428]
Fe-Pt (2002)	colloidal synthesis, inert-gas aggregation	decahedra, icosahedra, ordered phase transition	[171, 434]

continued

Table 2. Continued

Materials	Synthesis	Characteristics	Refs.
Ni-Zr (1985)	rapid solidification	thin films, quasicrystal approximants	[80, 437]
Ni-Ti (1986)	rapid solidification	thin films, quasicrystal approximants	[437]
Rh-Pt (1990)	colloidal synthesis	decahedra	[371, 433]
Al-Li* (1985)	matrix precipitation, rapid solidification	thin films, icosahedra, quasicrystal approximants	[219, 442, 443]
Al-Cr* (1993)	rapid solidification	star-shaped decahedra	[82, 187]
Al-Mn* (1985)	rapid solidification	thin films, icosahedra, star-shaped decahedra	[81, 82, 339, 451]
*Al-Fe** (1987)	rapid solidification	thin films, quasicrystal approximants	[79, 188]
*Al-Cu** (1988)	rapid solidification	thin films, icosahedra, quasicrystal approximants	[187, 442]
Al-Zr* (1985)	rapid solidification	thin films, tetrahedrally closed-packed structure	[453]
steel (1983)	steel processing	decahedra, tetrahedrally closed-packed structure	[454, 455] 2
Si-Ge (2001)	CVD	decahedra, thin films	[149]
C (60) (1992)	PVD, aerosol synthesis	decahedra, icosahedral clusters, star-shaped decahedra	[124–126, 128, 405]
C (76) (1995)	PVD	decagonal twinning	[127]
C. Compounds			
AgBr (1964)	solution growth	decahedra	[44, 429]
BN (1985)	CVD, e-beam irradiation	decahedra, thin films, star-shaped decahedra	[132, 228, 403]
TiN (1988)	CVD	rod-shaped & star-shaped decahedra, thin films	[150–152]
TiCN (1985)	CVD	decahedra	[153, 154]
BC (2002)	arc evaporation	icosahedra	[167]
Cr_3C_{2-x} (1991)	reactive sputtering	decagonal twinning	[183]
SiC (1996)	CVD	star-shaped decahedra	[152]
Fe_2O_3 (1983)	PVD	decahedra	[123]
SnO_2 (1996)	PVD	thin films, multiple twin junctions	[122]
B_6O (1998)	high-pressure melt synthesis, pulsed laser deposition	icosahedra, hierarchic structure, decahedra	[179, 180, 184, 241, 341, 343]
$BaTiO_3$ (1998)	solid phase chemical reaction	thin films, multiple twin junctions	[445, 446]
GaP (1988)	CVD	thin films, multiple twin junctions	[155]
CdTe (1993)	PVD	hollow whiskers	[406]
$CuInSe_2$ (1994)	molecular beam epitaxy	thin films, multiple twin junctions	[353]
CdSe (2002)	colloidal synthesis	icosahedra	[418]
polyoxometalate (2001)	self-assembly	icosahedra, hierarchic structure	[9]
surfactant (2001)	self-assembly	hollow icosahedra	[11]
D. Composites			
Co/polymer (2000)	colloidal synthesis	polytetrahedral packing	[93]
Cu/Ni-Zn-Cu (1993)	matrix precipitation	decahedra	[89]
Ag/glass (1991)	ion exchange, ion implantation	decahedra, icosahedra	[90, 94]
Au/polymer (1998)	colloidal synthesis, PVD	decahedra, icosahedra	[91, 92]
Ge/Al (1986)	matrix precipitation	decahedra, rod-shaped decahedra	[83, 84, 85, 86]
Si/Al (2001)	matrix precipitation	decahedra	[88]
Si-Ge/Al (2001)	matrix precipitation	decahedra	[87]
Ge/silica (2001)	ion implantation, co-sputtering	decahedra	[95, 96]

Note: Entries in italics refer to tenfold twinning, asterisk signs indicate possible changes of order or stoichiometry in alloys.

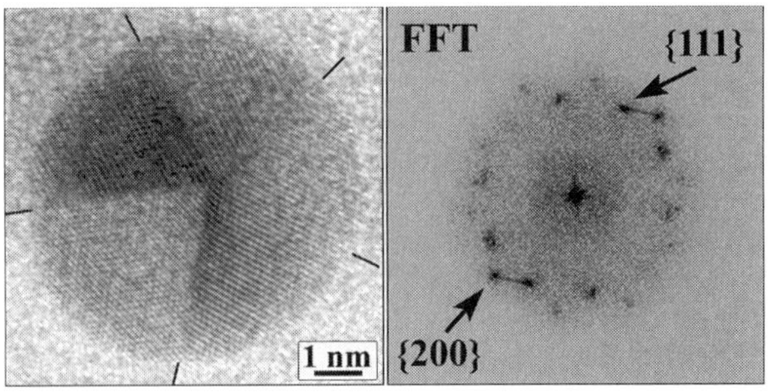

Figure 8. Decahedral particle of Ag in fivefold orientation grown by precipitation in glass. Adapted with permission from [97], H. Hofmeister, *Mater. Sci. Forum* 312–314, 325 (1999). © 1999, Trans. Tech. Publications.

Au-Cu and Fe-Pt [170, 171]. The process scheme given in (b) mainly enables particulate mass production of the corresponding material [159, 169, 172–174] and investigation of unsupported particles [164, 175–178]. The vapor phase route also includes modern techniques of materials synthesis such as pulsed laser deposition [179–181] and sputtering [95, 182, 183]. A typical example of growth from the vapor phase is given in Figure 9. The rhombic profile of the decahedral Rh particle [108] shown in the HREM image (left) is due to the (001) orientation of the base tetrahedral subunit relative to the electron beam. The accompanying diffractogram exhibits spots originating from {111}, {200}, and {220} lattice plane fringes of the (001) oriented base and two (112) oriented top tetrahedra marked by squares and circles, respectively. Additional spots marked by open arrows result from Moiré type contrast features in the particle center due to superposition of subunits of both orientations [166].

2.2.2. Liquid Phase Techniques

Liquid phase synthesis (ii) includes (a) growth from the melt and (b) growth from solution via precipitation by chemical means (b1) or by electrodeposition (b2). High-pressure melt growth was used for diamond [42] and B_6O [184] synthesis. Precipitation from alloy melt was utilized for the growth of Si and Ge MTPs [43, 185, 186]. Most of the quasicrystalline phases and their approximants have been produced by rapid cooling of Al-Mn and similar alloy melts [74, 187, 188]. The solution route (b1), or colloidal synthesis, has been frequently used for wet chemical formation of multiply twinned AgBr [44] and metal particles of Au, Ag, Cu, Pt, Pd, Ni, and Ru, as well as Pt- alloys [91, 189–198]. The solution route (b2), or electrodeposition, is widely utilized for fabricating protective coatings, mostly of Ni, whose appearance depends on grain size and texture, which may be controlled by the electrode potential. Fivefold twinned grains are the main constituents of films with ⟨110⟩ texture. Multiply twinned structures have been reported for Au, Ag, Cu, Pt, Pd, Ni, Ir, Rh, and Cu- alloys deposited as films [40, 199–204] or particles [45, 205–213]. As it was mainly revealed by the work of Da-ling Lu [209, 210, 214–216], the crystal habit of fcc metal particles is controlled by the electrode potential in solution, that means icosahedra and decahedra are formed at lower potential, whereas at higher potential less or no MTPs occur. Typical examples of MTPs grown from solution can be found in Figure 18 where HREM image contrast features of icosahedral Ag particles formed by hydrolysis of a mixed solution of tetraethoxy orthosilicate and silver nitrate in ethanol plus water [217] are presented.

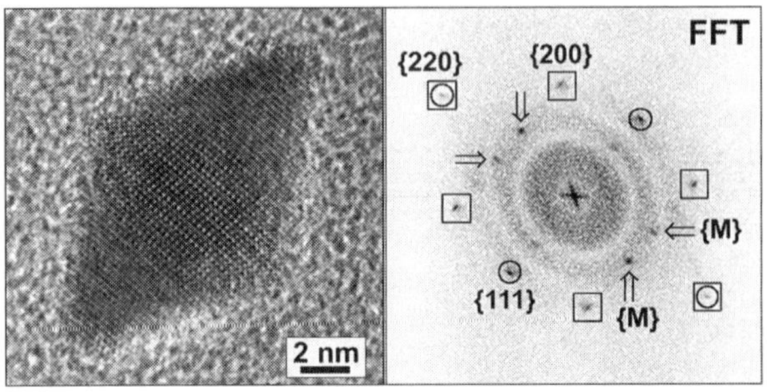

Figure 9. Decahedral particle of Rh in parallel orientation grown by vapor deposition on NaCl.

2.2.3. Solid Phase Techniques

Solid phase synthesis (iii) includes several subroutes: (a) precipitation from solid solutions in crystalline or glassy hosts, (b) solid phase crystallization from the amorphous phase, (c) solid phase reduction by reactive gases like H_2 or CO of highly disperse metal compounds, and (d) irradiation-assisted processing by electron beam or ion beam impact applied to induce particle formation in a matrix or on a substrate. Precipitation of fivefold twinned nanoparticles in crystalline matrix was reported for Au, Ag, Cu, Ni, Al-Li, Si, Ge, and Si-Ge [83–89, 181, 182, 186, 218, 219] where shape deviations, such as rod-like particle shapes, depending on orientation relations between precipitate and matrix as well as on the respective interface energy, were frequently observed. Precipitation of MTPs in glassy hosts was observed for Ag in soda lime glass, doped by ion exchange, upon thermal processing [90, 94, 220–222] and for Si in SiO_x by thermal decomposition of SiO [223]. Ag decahedra formed in glass matrix keep their structural peculiarities even when stretching the glass at elevated temperatures results in elongation of the previously spherical particles [220]. The amorphous-to-crystalline phase transition according to process scheme (b) has been studied intensively for thin films of Ge [113–117, 119, 224] and powder particles of Si [147, 148], which exhibit a distinct tendency to fivefold twinned structure formation. Solid phase reduction, or process scheme (c), has been used for decades to fabricate highly dispersed, supported metal particles, such as for application in heterogeneous catalysis. MTPs of Au, Ag, Cu, Rh, Co, and Ni are reported for this subroute [49, 50, 55, 217, 225–227] to occur on appropriate carriers. In recent years, a number of techniques for irradiation-assisted processing was developed to produce new nanoparticles and nanoparticulate composites. These techniques include electron beam irradiation to induce particle formation by reduction and aggregation of precursors where MTPs have been observed for BN, Ag, Cu, Al, Si, and Ge [95, 96, 228–234], as well as ion beam irradiation to introduce, that means to implant, dopants like Ag or Ge in a matrix so as to enable particle formation. MTPs originating from the latter technique were found for Ag and Ge in glassy hosts [94–96].

Finally, as a further route to fivefold twinned nanostructures, one should mention here the self-assembly of ligand-stabilized metal nanoparticles into superlattices of two or even three dimensions [13] that exhibit fivefold twin junctions similar to the pentagonal aggregations of virus particles [22]. When using a micelle route to preferentially create multiply twinned metal particles [235, 236], one could compose in this way a fivefold twinned superlattice of MTPs.

2.3. Formation Mechanisms

2.3.1. Nucleation-Based Formation

The formation of MTPs and fivefold twinned structures is an important issue, since understanding of the relevant mechanisms may help to control conditions for preferred formation or prevention of such structures. The great variety of materials and processes involved cannot be attributed to only one mechanism of formation. In general, we distinguish (i) nucleation-based and (ii) growth-mediated formation of fivefold twins. The nucleation (i) or noncrystallographic packing of atoms, is complemented by layer-by-layer growth in the course of which the noncrystallographic arrangements transform to quintuples of twins. The growth-mediated formation (ii) may proceed by cyclic twinning operations due to (a) misstacking of atoms (growth twinning) or (b) mismatch of lattices (deformation twinning) during growth.

MTPs are observed in the nucleation stage (a) of thin film growth on substrates via physical [237–240] and chemical vapor deposition [18, 129, 130], as well as in inert-gas aggregation [159, 161], melt growth [241], solution growth [189], electrodeposition [201], and solid phase crystallization [116]. Sometimes the noncrystallographic nature of the nuclei formed is emphasized by the name "paracrystalline nuclei" [201]. The preferred formation of closed shell structures with icosahedral arrangement is confirmed by the observation of magic numbers in the mass spectra of transition metal clusters [242–248]. Fcc metal clusters obey the building plan of Mackay icosahedra [38] as it has been shown for five-shell Pd clusters of 561 atoms by direct imaging [249]. The first steps of evolution for such clusters, starting from a 1 shell nucleus, by shape maintaining layer-by-layer growth, contain 13, 55, 147, 309, 561, ... atoms as schematically drawn in Figure 10. Likewise, pentagonal decahedra may evolve from a nucleus of decahedral shape whose initial growth sequence contains 7, 23, 54, 105, 181, ... atoms. During growth the noncrystallographic packing of atoms is transformed to a fivefold twinned arrangement of translationally ordered subunits whose small size enables compensation of the angular misfit [92].

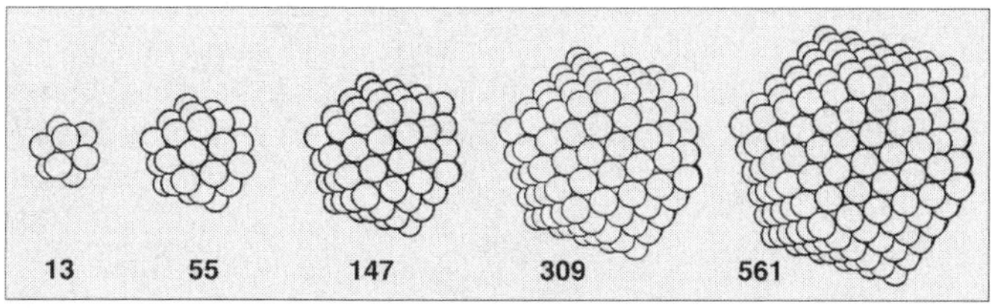

Figure 10. First steps of fcc closed shell cluster evolution of icosahedral shape.

The nucleation of fivefold symmetry in dc materials proceeds, according to their bonding characteristics, via cage rather than closed shell structures having pentagonal dodecahedron (20 atoms) and truncated pentagonal bipyramid (15 atoms) shapes, which are analogues of icosahedron and decahedron, respectively. The first steps of a layer-like growth sequence of decahedral shape, containing 15, 60, 140, 265, 490, ... atoms are illustrated in Figure 11. With the dodecahedron nucleus, a growth sequence of 20, 100, 292, 568, 994, ... is obtained when its 12 pentagonal faces are decorated by truncated pentagonal bipyramids. Always three of these attached cages of the first layer share one atom, at which formation of a tetrahedral subunit of a dc lattice may start in the course of further layer-like growth [250, 251]. In the above cluster models, the tetrahedral bond is preserved with bond angles and bond lengths only slightly differing from that of bulk dc crystals. The outer atoms have dangling bonds that may be saturated, for example, by hydrogen. 15 atoms and 20 atoms hydrogenated carbon cage clusters correspond to the hydrocarbon molecules hexacyclopentadecane and dodecahedrane [252], respectively, which are assumed to be effective in the nucleation of diamond MTPs by methane decomposition [129]. The formation of fivefold twinned structures of Ge was proposed to originate from a 15 atoms nucleus formed in the amorphous phase [116]. A 100 atoms cluster first has been proposed to explain defect structures in the heteroepitaxial growth of Si on spinel [250].

2.3.2. Growth-Mediated Formation

If not nucleated from the beginning, MTPs also may form during growth by repeated cyclic twinning. The main source of growth twinning is misstacking of atoms at faces of low growth rate so as to produce reentrant edge configurations, which enable accelerated growth along a twin boundary [43, 117, 120, 186, 253]. In the particle stage of growth, twinning may proceed by the formation of primary, secondary, and tertiary twins on pre-existing tetrahedra as shown schematically in Figure 12. This process is found to operate not only to form decahedra, but also icosahedra by successive stacking of tetrahedra. Rather soon after the nucleation mechanism has been introduced, alternatively the tetrahedra stacking mechanism began to be discussed [57, 93, 169, 205, 254–257]. First it was observed during *in-situ* investigation of the epitaxial growth of Au on MgO [258] and was later confirmed by an *ex-situ* study on the growth of Au on

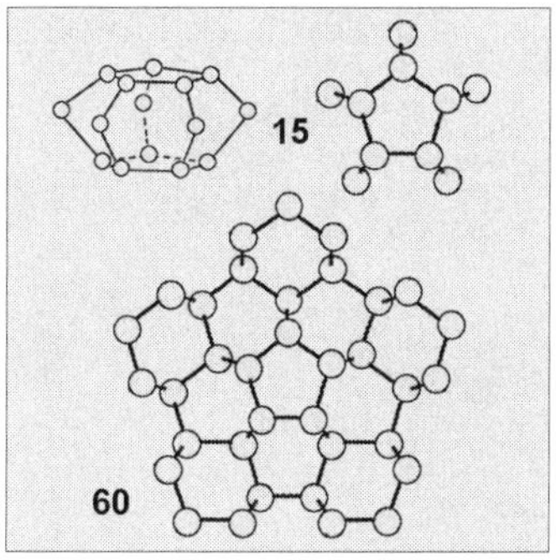

Figure 11. First steps of dc cluster evolution of decahedral shape.

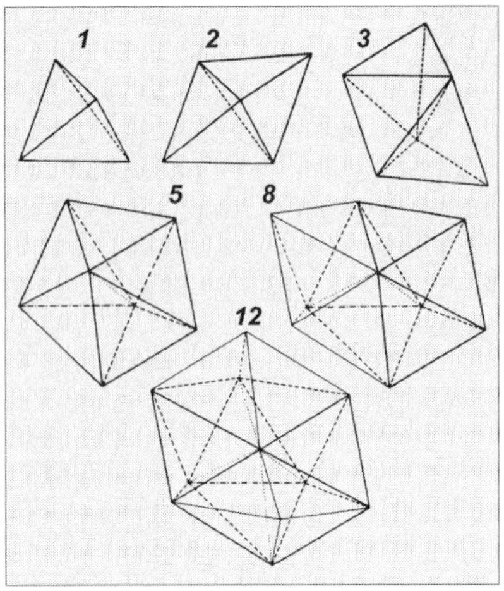

Figure 12. Successive stacking of tetrahedra in twin position resulting in MTP formation.

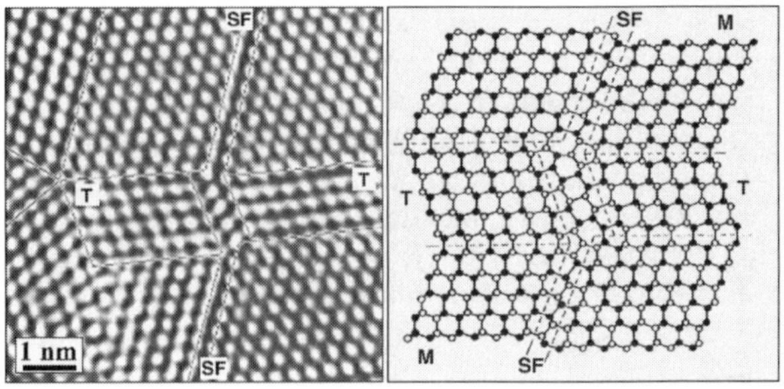

Figure 13. Secondary twin formation upon intersection of a twin band **T** by a stacking fault **SF** during solid phase crystallization of Ge. Adapted with permission from [97], H. Hofmeister, *Mater. Sci.* Forum 312–314, 325 (1999). © 1999, Trans. Tech. Publication.

AgBr [259]. The formation of multiply twinned structures by successive twinning on alternate cozonal twin planes also has been found in thin film growth of Ni, Ge, and SnO_2 [117, 122, 134, 202, 260]. In all cases, the formation of triple twin junctions is a decisive step in favor of fivefold twinning. Frequently, networks of interlinked threefold and fivefold twin junctions are observed in these films. Local fivefold twinned structures have been considered as essential growth stimulating constituents of preferred fcc growth of van der Waals crystals [261–263].

In addition to misstacking of atoms during growth as one possible origin of repeated twinning, the intersection of stacking faults and twin lamellae, introduced into the lattice of growing thin films because of plane strain deformations, must be considered. Deformation twinning may serve as a means of relaxing plane strains. The ability of fivefold twinned structures to accommodate large interfacial strains due to lattice misfit and thermal expansivity differences is known from the heteroepitaxial growth of semiconductors on insulating substrates [155, 264–266]. Strain-induced twin formation starts with the introduction of 90° Shockley partial dislocations passing through the strained lattice [267–270]. Successive penetration of a strained lattice by dislocations on alternate twin planes consequently will lead to the crossing of twins. A simple case of twin intersection, the penetration of a stacking fault **SF** through a twin **T**, observed in the solid phase crystallization of Ge thin films [117, 220], is shown in Figure 13. In the crossing region, a secondary twin is formed. At the intersection of stacking fault and twin boundary, five-membered rings occur resulting from dislocation reactions [270]. The latter will act as seeds of prospective fivefold twin junctions upon propagation of additional dislocations on adjacent planes and will also lead to an extension of the secondary twin.

Sometimes, the origin of a certain fivefold twinned structure cannot be attributed to only one of the above discussed formation mechanisms, but may result from an interplay of growth twinning and deformation twinning. At various stages of thin film growth, extended structures containing several individual multiple twins may occur. During thin film growth of Au and Ag, decahedral and icosahedral MTPs have been observed to form polyparticles via coalescence preserving almost completely their previous structures [271, 272]. Networks of interlinked fivefold and threefold twin junctions have been found in electrodeposited Ni films having ⟨110⟩ texture [202, 204, 260]. Similar networks occur at advanced stages of the solid phase crystallization of amorphous Ge [114, 119, 120, 224, 253, 273] as well as in the chemical vapor deposition of diamond [134, 136, 139] and Si [149]. In the applied range of temperature and film thickness, it is generally assumed that kinetic factors dominate the growth that has been described as "solid-like growth" by Marks [272]. However, the formation of MTPs has also been discussed as being due to transformation to a higher symmetrical arrangement [274].

3. STABILITY AND PHASE TRANSITIONS

3.1. Stability of Fivefold Twins

Frequently, MTPs and fivefold twinned structures do not exist separately, but in coexistence or even competition with structures that exhibit regular crystal lattice without twins. Their stability is an intriguing issue mainly because of the discrepancy between noncrystallographic packing of atoms and its extension in three-dimensional space. Experimental and theoretical investigations of clusters, a few atoms up to several hundreds or even a few thousands of atoms in size, aimed at determining stable forms and their size limits, have been mostly done on rare-gas [262, 275–281] and transition metal [243, 246, 282–289] clusters produced in supersonic beams during gas expansion. Atomistic studies, using data from electron diffraction [164, 176–178, 276–278, 290–292] and mass spectrometry [242–244, 293, 294], by means of molecular dynamics and Monte Carlo simulations employing various pair interaction potentials [245, 247, 279, 280, 287, 295–297], revealed a wealth of knowledge on magic numbers and growth sequences [242, 244–248], thermal stability, shape and structure [246, 282, 284, 296, 298–307], phase transitions [248, 262, 289, 297, 308–316], and melting behavior [288, 310, 317] of clusters of icosahedral, decahedral, fcc crystalline, or disordered structure. However, there has been predicted not a global minimum of potential energy for a multitude of structural motifs and cluster configurations, but very small energy differences such that clusters do not necessarily have a single stable structure at realistic temperatures [282, 309, 317, 318]. Moreover, there has been found not a single sequence of phase transitions like icosahedral to decahedral to single crystalline fcc and its dependence on size and temperature, but also a reversal of this sequence [311] as well as a gradual instead of immediate transition [297, 308]. In addition, from molecular dynamics simulations there has been predicted, besides the layer-by-layer growth, also a certain probability of misstacking of atoms leading to island growth in twin position, which enables transition of decahedral to icosahedral shape by growth as it was experimentally observed on a much larger size scale [259]. Finally, experimental magic numbers associated with structures based on Mackay icosahedra have been classified by atomistic simulation to be of kinetic origin [319]. Even if the intermolecular potential disfavors the icosahedral structure, it occurs frequently due to potential characteristics that enhance kinetic trapping effects. The existence of such kinetic effects suggests that it will be possible to control structures of clusters and nanoparticles by tuning external parameters to enable design of nanomaterials properties.

The above findings are analogous to the configurational instabilities inherent to particles of sizes smaller than 8 nm [320, 321]. Real-time video recording of HREM investigations on very small metal particles revealed fast changes between a number of structures including cuboctahedra (single crystal), single twinned cuboctahedra, fivefold twinned decahedra, and icosahedra [67–71, 89, 322–327]. Some of these structures may be understood as result of a fivefold twin junction (also described as wedge disclination [328] or line disclination [329]) entering into and moving through a particle [330, 331]. Steps of this movement will include also asymmetric decahedra like the one of Ag shown as example in Figure 14. An eccentric position of the fivefold twin junction can be observed more often the smaller the particles are. The structural transformations observed along with a much higher rate of particle rotations in the presence of an electron beam may be understood in terms of statistical fluctuations with the probability of a particular configuration depending on size and temperature [320, 321].

MTPs consisting of regular fcc or dc subunits contain spatial discontinuities that introduce inhomogeneous strains. Additional strain and twin energy resulting from the specific composition of MTPs may be balanced by a reduction of surface energy up to a certain size above which transformation to single crystalline particles of cuboctahedral shape was expected. Strain relief by structural modifications such as homogeneous lattice distortions or the introduction of lattice defects as inhomogeneous lattice distortions may extend the range of stability. Energy balance considerations including cohesive, surface, adhesive (i.e., concerning particle/substrate interaction), elastic strain, and twin boundary energy aimed at calculation of stable size regions for MTPs of transition metals in comparison to their single crystalline counterparts [16, 17, 332] provide stable and quasistable size limits around 30 and 300 nm for icosahedral and decahedral MTPs of Ag, respectively. Transitions from multiply twinned structures to single crystalline fcc have been observed for very small metal particles in gas expansion experiments by electron diffraction techniques from which crossover sizes of 3.8 nm have been derived for Cu [333], whereas in comparable experiments a size-independent transition was found for Ag [178] and a dependence on the type of inert gas was found for Pb [164]. On the other hand, experimental studies on MTPs gave evidence of their extension to sizes far above the size limits derived from stability considerations (see Section 4.2.). One of the reasons for this behavior is that they may undergo lattice transformations and in many cases exhibit lattice defects.

3.2. Lattice Transformations and Defects

The lack in space filling that results when composing MTPs of regular fcc tetrahedral subunits raises the question of whether the lattice of some or all of these subunits may adopt a slightly changed state of uniform distortion. To allow for the absence of spatial discontinuities in MTPs, some kind of structural modification or lattice defect is needed. This may be brought about by elastic strains acting on the tetrahedral subunits as first described by S. Ino to calculate their stability [17]. A slight, uniform distortion, for example, transforms the tetrahedral subunit fcc lattice into one having body-centered orthorhombic (bco) point group symmetry so as to enable a Bagley decahedron with twin angle 72° [52–54]. Figure 15 shows the fcc unit cell of lattice parameter **a** inside which a bco unit cell of lattice parameters **a**, **b**, **c** is drawn. As long as $\mathbf{a} = \sqrt{2}\,\mathbf{b}$ and $\mathbf{b} = \mathbf{c}$ (i.e., the nearest neighbor distance), the inscribed tetrahedron has fcc characteristics. The required uniform distortion is achieved

Figure 14. Asymmetric decahedral MTP of Ag grown by physical vapor deposition on alumina with eccentric position of the fivefold junction.

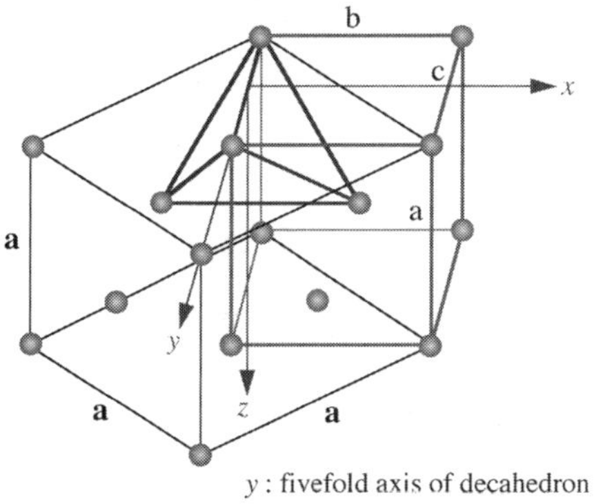

Figure 15. Tetrahedral twin subunit of a decahedron transformed from fcc to bco lattice.

by applying a biaxial stress to elongate **c** (**c** = 1.0515 **b**) and to shorten **a** (**a** = 1.3764 **b**) [334]. This transformation preserves the close packing, but widens the angle between the triangular faces meeting at the *y*-axis from 70.52° to 72°. Another uniform distortion transforms the tetrahedral subunit fcc lattice into one having rhombohedral (rho) point group symmetry so as to enable a Mackay icosahedron with twin angle 72° [38]. Figure 16 shows the fcc unit cell of lattice parameter **a** inside which a rho unit cell of lattice parameters **b** is drawn. As long as the rhombohedral cell angle is $\alpha = 60°$, the inscribed tetrahedron has fcc characteristics. By applying an uniaxial stress along the cube diagonal direction close packing is preserved in the icosahedron with a rhombohedral structure, but α is enhanced to 63.43° [334]. The nearest neighbor distance however is different now for interplane atoms (**b** = OA, OB, OC) intraplane atoms (**c** = AB, AC, BC) with **c** = 1.0515 **b**. Consequences of these model considerations for lattice characteristics, diffraction patterns, and image contrast features have been demonstrated by crystallographic and electron microscopy studies on Au particles [334–338].

Accommodation of the angular misfit by transformation to the rho lattice has been reported also for Al-Mn multiple twins [339]. Contrary to the above examples where lattice distortions are assumed uniformly throughout all tetrahedral subunits, there are also reports about tetragonal lattice distortions in only one or two subunits while the remaining tetrahedra exhibit fcc lattice. This behavior has been observed for decahedral MTPs of In [109, 111, 340] and Pb [110]. The lattice of In bulk metal usually has base-centered tetragonal (bct) point group symmetry and adopts fcc structure only in multiply twinned nanoparticles, whereas the lattice of Pb bulk metal usually has fcc point group symmetry and adopts bct structure not only in MTPs, but also in single crystalline and single twinned nanoparticles. Icosahedral MTPs of the Fe-Pt intermetallic phase have been assumed to adopt the $L1_0$ superstructure, which was not found in untwinned nanoparticles of this material [171]. Interestingly, the $L1_0$ superstructure has not only promising magnetic properties, but also enables formation of perfect decahedra without any need of distortion. Based on this structural characteristic, an icosahedron model has been proposed that consists of two such $L1_0$ decahedra, having one common vertex and their fivefold axes in line, completed by a "belt" of ten slightly distorted tetrahedra [171]. It should be noted here that, similar to chemically ordered $L1_0$ decahedra of Fe-Pt, icosahedra of a few materials have been found that do not require elastic straining to close the angular gap, because their lattice characteristics already fit to the condition that the tetrahedron angle α amounts to 63.43°. This has been reported for MTPs of C(76) having monoclinic lattice [127], for B_6O where oxygen atoms are three-coordinated to icosahedral B(12) clusters in a rho lattice [179, 180, 184, 241, 341], as well as for BC with rho lattice [167]. Although clusters having fivefold symmetry are well known as entities in crystal structures [342–344], up to now only the above mentioned B(12) have been found to be arranged in hierarchical packing from which icosahedral MTPs may form.

Elastic strains in fivefold twinned structures of fcc and dc materials determine not only the general structural characteristics [16, 17, 275, 328, 332, 345, 346], but also that of the twin boundaries involved [347]. At sizes distinctly above 10 nm, inhomogeneous elastic strains [348] allow rather large reductions of the strain energy stored in MTPs such that stress relief processes may occur involving the formation of lattice defects [332]. Typically, planar defects such as stacking faults and secondary twin boundaries are observed [115, 321, 349]. A particular stress-relieving configuration observed in fivefold twinned structures of Si and Ge [115, 117, 169] is shown in Figure 17. It consists of regular arrays of tetrahedrally arranged stacking faults emerging at stair-rod dislocations. Such stacking of fault arrays results in an angular lattice dilatation in the respective twin subunit, while the neighboring subunits remain undistorted. Two pairs of stacking faults are sufficient to accommodate the angular gap at the length scale of the particle shown here. More extended arrays in combination with small angle boundaries have been observed at Si particles of larger dimensions. Localized strains, defects, and misfit faults, which often simply consist of a small angle boundary, are by far the most reported inhomogeneities in MTPs and fivefold twinned structures of diamond [27, 129, 350–352], Si [169], TiN [151], and $CuInSe_2$ [353]. Individual dislocations

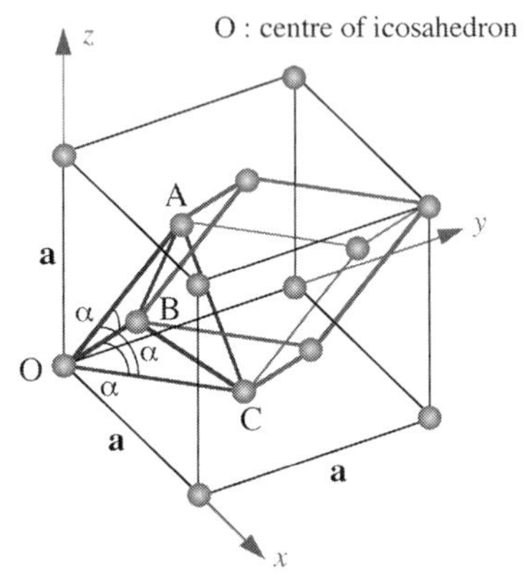

Figure 16. Tetrahedral twin subunit of an icosahedron transformed from fcc to rho lattice.

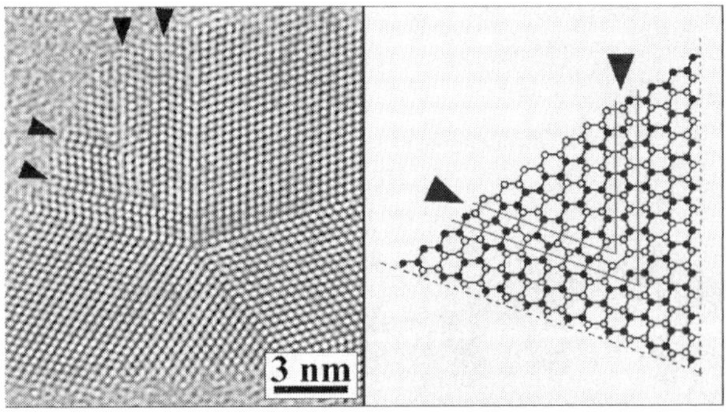

Figure 17. Array of two pairs of stacking faults (marked by arrow heads) emerging from stair- rod dislocations in one tetrahedral twin subunit of a Ge MTP (left) and the corresponding model representation of a twin sub-unit with one pair of stacking faults (right). Adapted with permission from [220], H. Hofmeister, *Cryst. Res. Technol.* 33, 3 (1998). © 1998, Wiley-VCH.

[105, 349, 354] and point defect agglomerations [107, 348] are rather scarcely observed.

4. STRUCTURAL CHARACTERIZATION

4.1. Characterization Methods

4.1.1. Electron Microscopy and Diffraction Methods

Electron diffraction as employed to the study of cluster beams using refined methods of diffraction peak analysis [164, 176–178, 277, 278, 290–292], similar methods have been applied in XRD studies [305, 355], enabled one to distinguish between MTPs and single crystal structures on a scale of only a few nanometers or even less. The importance of electron microscopy for structural characterization of MTPs and fivefold twinned structures in synthetic materials from the very beginning has already been pointed out in Section 1.3.2. This essential role results from the submicrometer size scale at which the phenomenon of multiple twinning mostly was found, thus being the actual domain of electron microscopy structural characterization. Utilization of a considerable number of methods and techniques ranging from simple shadow casting [10, 40] to state-of-the-art investigations devoted to, for example, observation under ultrahigh vacuum and at low temperature conditions [111], revealed many of structural characteristics that otherwise could not have been elucidated. Within the continuously increasing number and quality of electron microscopy studies, there have been employed electron diffraction pattern recording of individual MTPs and calculation of such patterns [337, 356–362], *in-situ* experiments to follow growth and transformation processes inside the electron microscope [234, 255, 258, 355, 363], weak-beam dark-field and related imaging modes for visualizing the internal structure of MTPs [259, 335, 336, 364, 365], HREM [105, 109, 191, 217, 273, 325, 326, 349, 354, 366–373] and corresponding image contrast calculation [121, 174, 175, 347, 362, 369, 372, 374–391], tilt series to study how external shape and internal structure of MTPs change with their orientation to the electron beam [166, 190, 192, 362, 392], real time observation of fast processes such as structural fluctuations or particle coalescence [67, 70, 322, 323, 325, 326, 381, 393], and combination of XRD or X-ray absorption spectroscopy investigations with TEM or HREM studies [194, 355, 394].

4.1.2. Selected Area Electron Diffraction

Special attention is devoted to selected area electron diffraction (SAED), from which fivefold symmetry may be recognized in a direct manner, and HREM, from which unique "fingerprints" may be obtained. Actually, before and besides HREM imaging, it is the SAED pattern of an individual decahedron in fivefold orientation, one of the first was published 1964 by Schlötterer [40] and another striking example 1972 by Ino et al. [395], which is directly convincing and allows one to examine with high accuracy the symmetry as well as spacings and angular relationships of multiply twinned particles. To illustrate the capabilities of this method, Figure 18 shows as an example the SAED pattern of a decahedral Ni grain in fivefold orientation within an electrodeposited thin film having ⟨110⟩ texture. This grain of about 400 nm extension in a plane perpendicular to its fivefold axis exhibits secondary twin boundaries in two of the tetrahedral units. Accordingly, in the diffractogram a slight splitting of related spots of {111} and {222} type can be seen, which indicates an inhomogeneous relaxation of elastic strains due to the space filling gap. For the sake of clarity, no assignment of spots has been added to the SAED pattern, but two circles are drawn enclosing the innermost spots of {111} and {200} type. From this rather complex electron diffraction pattern, it can be clearly seen that not one single crystal, but a grain consisting of five subunits in well-defined orientation relationship are transmitted by the electron beam. Likewise, diffraction patterns from regions of 1 nm size of multiply twinned Au nanoparticles obtained by means of a microdiffraction equipment operated in the scanning transmission mode [396] confirm the particle composition of twinned subunits.

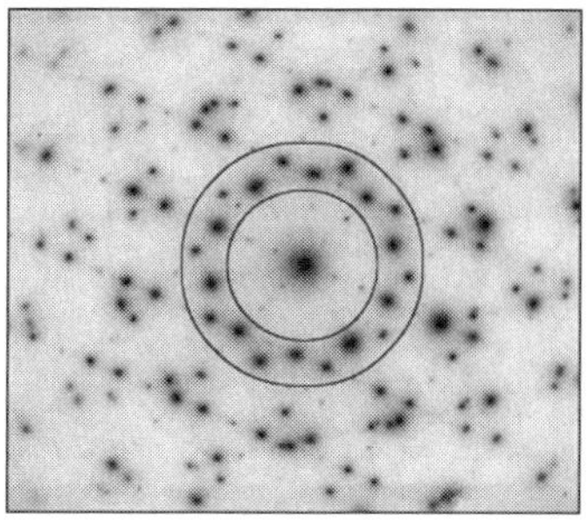

Figure 18. Selected area electron diffraction pattern of a decahedral Ni grain (fivefold orientation) within an electrodeposited thin film.

4.1.3. High Resolution Electron Microscopy

Imaging of lattice plane fringes by high resolution electron microscopy of MTPs frequently reveals, in combination with diffractogram analysis and image contrast calculation, a clear signature of particle shape and internal structure [166, 382, 384, 385, 390]. This is demonstrated in Figure 19 for Ag icosahedra arranged in various orientations with respect to the electron beam. It comprises HREM image, diffractogram, and particle model of the particles in "face" orientation, "edge" orientation, "fivefold" orientation, and one tilted around 10° out of "edge" toward "fivefold" orientation (from top to bottom). The edge length of the HREM images corresponds to 4.8 nm. These image contrast features depend on the configuration of tetrahedral subunits that are oriented such as to give rise to lattice plane contrasts. In the "face" oriented icosahedron, for example, there are six twin planes parallel to the electron beam, or the axis of observation, leading to six sets of {111} lattice plane fringes. However, there must be considered superposition of lattice plane fringes where tetrahedral units are stacked one above another. That is why the image details cannot be straightforwardly interpreted in terms of lattice planes and increasingly become more complicated the more superposition occurs. The highly complex contrast patterns in HREM images of icosahedral particles due to superposition of various lattice segments cause corresponding complex spot patterns in the diffractogram. However, the frequently observed

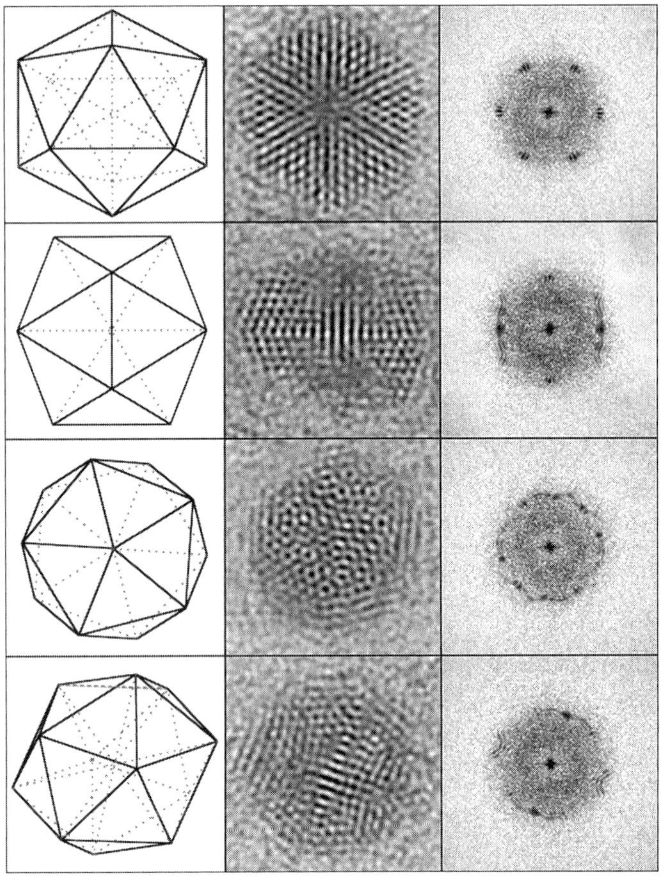

Figure 19. HREM "fingerprints" (edge length 4.8 nm) of icosahedral Ag particles in various orientations: "face," "edge," "fivefold," and tilted around 10° out of "edge" toward "fivefold," from top to bottom, together with diffractogram (right) and model (left).

diffractogram spot splitting, as shown in Figure 19, is no direct evidence of angular misalignment or spatial mismatch of the lattice of twinned subunits. Actually, the shape of image regions of equal lattice plane fringe arrangement is reflected in the diffractogram fine structure [390], which is related in a certain way to the electron diffraction spot fine structure of polyhedral crystallites observed earlier [397]. The fine structure of diffractogram spots also is found for calculated HREM images of icosahedra assuming a rho pointgroup symmetry without any lattice defects. By Fourier transform processing of various projections of tetrahedral subunit model images, the interference nature of the phenomenon has convincingly been demonstrated [390].

4.2. Size and Shape

The additional strain and twin energy associated with the formation of MTPs may be balanced by a reduction of surface energy up to a certain size (see Section 3.1), above which transformation to single crystalline structures is expected. Experimentally observed fivefold twinned structures however, not only frequently exceed the size limits based on thermodynamic considerations, but also exhibit distinct deviations from the nearly spherical shape into various types of rod-like or even star-like particle shapes. One reason for this behavior is the accommodation of angular misfit by the introduction of lattice transformations or lattice defects (see Section 3.2). Another reason is that obviously certain growth conditions not only favor deviation from the ideal MTP shape (see Section 1.2), but also enable exceedingly large particle size. Besides the two examples of Figures 20 and 21, which show a large decahedral MTP of Pd in "fivefold" orientation and a large icosahedral MTP in "parallel" orientation, the most impressive examples of extremely large MTPs of various materials (i.e., those having micrometer size and above) are compiled in Table 3. These include decahedral particles of the molecular C(60) crystal fullerite [125] exceeding the millimeter scale of size, and icosahedral particles of boron suboxide B_6O [341] with sizes around 40 μm. While most of the MTPs on the micrometer scale are of decahedral shape, the above-mentioned extraordinary large icosahedra forming material exhibits a rhombohedral structure with a rhombohedral unit cell angle of $\alpha = 63.1°$ being very close to the one for ideal icosahedral twinning.

Multiply twinned rod-like particles may form from decahedral nuclei by preferential growth along the fivefold axis. First observations were made within Au crystals of natural occurrence [33, 398–400]. Different from regular decahedra, these particles exhibit extended prism faces of {001} type. Their multiply twinned nature is revealed most easily from tilting experiments in the electron microscope, as has been shown recently for rod-like silver particles grown by inertgas evaporation technique [166]; this reference also sums up the literature about MTPs of rod-like shape in synthetic materials. As may be concluded from the model shown in Figure 5C, rotation around the long axis of the particle, situated perpendicular to the electron beam, is found to produce two characteristic image contrast patterns, separated from one another by 18° rotation, both having rotational periodicity of 36°. According to a rather recent publication, decahedral nanorods have also been fabricated via a bioreduction route [401]. Elongation of icosahedral MTPs toward rod-like shape may be achieved not simply by growth, but by successive growth twinning, this way reaching beyond the shape of complete icosahedra. As shown by the model in Figure 22, particles of elongated shape can be formed by stacking two icosahedra into each other such that they share five tetrahedra grouped around a common fivefold axis, as in a decahedron. Characteristic image contrasts of triple rhombic shape result from positioning one of the fivefold axes

Figure 20. Decahedral particle of Pd in fivefold orientation grown on KI substrate by vapor deposition.

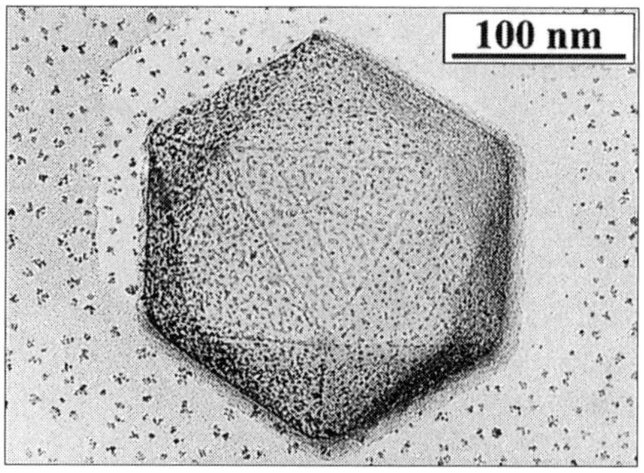

Figure 21. Pt-C shadow casting of an icosahedral particle of Ag grown on AgBr substrate by vapor deposition. Adapted with permission from [220], H. Hofmeister, *Cryst. Res. Technol.* 33, 3 (1998). © 1998, Wiley-VCH.

of these particles, being their long axis, parallel to the substrate, or perpendicular to the electron beam [387]. Hence the three decahedral regions involved show {111} lattice plane fringes within rhombic areas (shaded in Figure 22). In addition, these decahedra in "edge" orientation exhibit {220} lattice plane fringes within square areas (hatched in Figure 22). From atomic-scale simulations of copper polyhedral nanorods [391, 402], both types of rod-like MTPs have been found as stable geometrical structures. Multiply twinned particles of star-like shape may form, as deviation from the decahedron shape, by reduced growth rate along the five twin boundaries of the tetrahedral subunits. Hence,

Table 3. Size extrema found in fivefold twinned materials.

Material	Approx. size	Type	(Year) Ref.
Cu	100 μm	Dh	(1957) [36]
diamond	100 μm	Dh	(1963) [42]
Ni	3 μm	Dh	(1964) [40]
Si	500 μm	Dh	(1964) [43]
Co	40 μm	Dh	(1964) [49]
Ni	8 μm	Dh	(1966) [55]
Ni	~2 mm	Dh (rod)	(1966) [51]
Ag	100 μm	Dh	(1968) [46]
Cu	300 μm	Dh	(1969) [225]
diamond (natural)	1 mm	Dh	(1972) [12]
Ni	50 μm	Dh	(1976) [207]
Au (natural)	800 μm	Dh (rod)	(1978) [398]
diamond	600 μm	Dh	(1979) [350]
TiN	5 μm	Dh (rod)	(1988) [150]
C (60)	2 mm	Dh	(1993) [125]
Yb	1.5 μm	Dh	(1993) [162]
TiN	10 μm	Dh (star)	(1996) [152]
SiC	50 μm	Dh (star)	(1996) [152]
Au	60 μm	Dh	(1996) [399]
B_6O	40 μm	Ic	(1998) [341]
Au	4 μm	Dh (star)	(2001) [212]
Si	40 μm	Dh (star)	(2001) [88]
Cu	1 μm	Dh (rod)	(2000) [392]
surfactant bilayer	1 μm	Ic (hollow)	(2001) [11]
BC	10 μm	Ic	(2002) [167]
Pd	1 μm	Dh, Ic	(2002) [198]

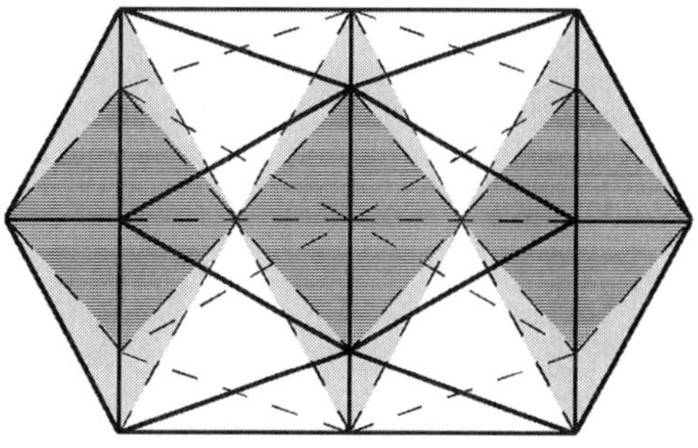

Figure 22. Schematic drawing of a "twinned icosahedron" particle consisting of 35 tetrahedral subunits.

these tetrahedra exhibit {111} truncations at their peripheral corners resulting in a star decagon projection when viewed in "fivefold" orientation. Star-like MTPs first have been reported for Cu of natural occurrence [32] and later for synthetic materials such as diamond [18, 42], Ge and Si [43, 88, 185], BN [132, 403], colloidal gold [212, 404], TiN and SiC [151, 152], Al-Cr-Si alloy [82], and also C(60) [405]. Finally it should be mentioned that a number of multiply twinned structures are hollow, that means they exhibit an external shape of fivefold symmetry but an internal void of variable extension. Hollow fivefold twinned structures are mainly found in whiskers of pentagonal cross-section [50, 51, 406] and in organic materials such as proteins or surfactant bilayers [8, 11].

5. PROPERTIES AND APPLICATIONS

5.1. Structure-Sensitive Properties

Physical and chemical properties of materials assembled of fivefold twinned nanoparticles may differ from materials consisting of untwinned nanoparticles in a variety of aspects according to their respective structural characteristics. These differences concern properties sensitive to the surface energy, the lattice symmetry, the internal structure, and the surface structure, and they may cause changes, such as of the melting point, magnetic moment, electronic transition, and chemical reactivity, respectively. For MTPs embedded in a matrix of foreign material instead of the surface structure, the interface structure has to be considered, which via particle-matrix interaction may influence the elastic properties of the composite. In studies devoted to the properties of multiply twinned nanoparticles mostly the influence of their real structure on heterogeneous catalysis is stressed [65, 108, 156, 163, 217, 226, 363, 371, 374, 407–412] since adsorption and reactivity are highly structure-sensitive properties. In a very recent investigation, Au MTPs have been found to lower selectivity and activity in the partial hydrogenation of unsaturated aldehydes with respect to the desired product allyl alcohol [227]. That means MTPs of Au are not useful for this reaction path. For separated Au(55) clusters of closed shell composition, an extraordinary high resistance against oxidation has recently been reported [413] that most probably is due to their icosahedral morphology. Tetrahedral subunits with the rho lattice of B_6O perfectly fit together at a common vertex without dislocations needed to accommodate an angular gap, thus enabling the growth of rather large icosahedra of boron suboxide [179, 180, 184, 241, 341]. Consequently, glide planes are locked in these particles that may result in a low density material of extraordinary hardness. Quite similar is the situation with massive icosahedral crystals of boron carbide, for which the well known hardness of this compound could be further improved because of being multiply twinned with icosahedral symmetry [167]. Precipitation hardening in structural alloys of AL-Si-Ge is dependent on the precipitate morphology, which is largely determined by twinning [87]. Multiply twinning completely changes the interface with the matrix, and consequently the strengthening effect of these precipitates in the metal matrix is reduced. Since the formation of multiple twin junctions apparently promotes the growth of Si nanowires in the oxide-assisted route [414], it will be interesting to see to which extent this structure may influence the optoelectronic properties of the material.

5.2. Symmetry-Dependent Properties

The appearance of spontaneous ferromagnetic order in Pd nanoparticles of about 6.8 nm size has been explained by a transition from single crystalline to multiply twinned structure with decreasing size [313]. The icosahedral symmetry is considered to contribute to the onset of ferromagnetic ordering or to the increase of an already existing magnetic moment. The main driving force in this transition has been shown to be the strong surface anisotropy of fcc single crystals being replaced by the energetically more stable icosahedral arrangement below the above size [313]. Stoichiometric Fe-Pt nanoparticles of 3 to 6 nm in size are found to preferentially exhibit icosahedral structure upon appropriate thermal processing [171]. Icosahedral MTPs of this alloy are assumed to be stabilized by transition to the $L1_0$ ordered phase, which exhibits large magnetocrystalline anisotropy [171]. This may be the basis for future magnetic materials with nanometer dimensions. For studying their physical properties the Raman, Brillouin, and elastic tensors of materials that exhibit fivefold point group symmetry have been calculated [415–417]. Concerning the optoelectronic properties of nanoparticles there is a very recent report on an excellent combination of fluorescence spectroscopy and HREM of isolated semiconductor nanoparticles allowing both methods to be applied to the same specific particle [418]. This way changes not only in size, but also in structural as well as morphological characteristics can be correlated to fluorescence properties of isolated nanoparticles. First results of the investigation of CdSe nanoparticles on transparent Si_3N_4 substrate indicate that the emission of strong fluorescence is not restricted to single crystalline particles of about 8 nm size, since icosahedral MTPs of slightly smaller size also show such emission [418]. More and systematic studies are needed to ascertain the role of the respective structural characteristic in this behavior.

6. SUMMARY

The aim of this chapter is to emphasize by illustrative examples and comprehensive references the importance of the widespread habit of fivefold twinning in nanostructured materials and to shed some light on the multitude of its facets. In particular, it shall enable us to link synthesis and processing of technologically promising or even important materials, their fivefold twinning characteristics, and their physical and chemical properties. This also includes the issue of comparing nanoparticulate materials, which preferentially have fivefold twinned structure to those being mainly in the untwinned state (see, e.g., [227]). For more detailed reading about this fascinating and rather complex phenomenon, some review articles concerning experimental as well as theoretical work in this field may be recommended. These are "Structure of Small Metallic Particles," by M. Gillet [419]; "Noble Metal Clusters" by R. Monot [420], "Comparison Between Icosahedral, Decahedral, and Crystalline Lennard–Jones Models Containing 500 to 6000 Atoms," by B. Raoult et al. [279]; "Phase Instabilities in Small Particles," by P. M. Ajayan and L. D. Marks [320]; "The Energetics and Structure of Nickel Clusters: Size Dependence," by C. L. Cleveland and U. Landman [284]; "Experimental Studies of Small Particle Structures," by L. D. Marks [421]; "Growth and Structure of Supported Metal Catalysts," by P. J. F. Harris [411]; "Preferred Structures in Small Particles," by N. Doraiswamy and L. D. Marks [321]; "Shells of Atoms," by T. P. Martin [294]; "Crystallography of Clusters," by J. Urban [384]; "Pentagonal Symmetry and Disclinations in Small Particles," by V. G. Gryaznov et al. [328]; and "Structure, Shape, and Stability of Nanometric Sized Particles" by M. J. Yacaman et al. [422]. The review "Forty Years Study of Fivefold Twinned Structures in Small Particles and Thin Films," by H. Hofmeister [220], gives a comprehensive record of four decades work (1957–1997) on fivefold twinned structures in small particles and thin films. The present chapter shall not only make available models and experimental findings of previous investigations in a greater context, but also stimulate future studies on this phenomenon.

GLOSSARY

Dislocation A line defect in a crystal, along which the lattice is displaced by a certain amount perpendicular or parallel to the dislocation line.

Ferromagnetic order Chemical order in a crystal that exhibits interaction at the atomic level, causing the unpaired electron spins to line up parallel with each other in a domain where a magnetic moment results.

Fluorescence The emission of light by a substance immediately after the absorption of energy from light of usually shorter wavelength.

Glide plane A low index crystal plane along which translation of one part of a crystal relative to the other part may proceed by the movement of dislocations.

Pair interaction potential Intermolecular potential describing the interaction between pairs of atoms derived from empirical models of interatomic bonding, used for computer simulation of bond energy and atomic structure of clusters.

Point group symmetry A method of denoting the combination of symmetry elements that a crystal contains.

Stacking fault A planar defect in a crystal where one part is displaced relative to the other part, such that the displacement does not correspond to a translational symmetry operation.

ACKNOWLEDGMENTS

This work is dedicated to Hans-Ude Nissen on the occasion of his 70th birthday. The author is greatly indebted to Christopher H. Gammons who drew his attention to the natural occurrence of gold fivelings that were known to mineralogists long ago. Special thanks are due to Wolfgang Neumann for many suggestions and critical discussions.

REFERENCES

1. R. Cahn, *Adv. Phys.* 3, 202 (1954).
2. P. Hartmann, *Z. Kristallogr.* 107, 225 (1956).
3. I. Kostov and R. I. Kostov, in "Crystal Habits of Minerals" (M. Drinov, Ed.), p. 34. Academic Publishing House, Sofia, 1999.
4. T. Hahn and H. Klapper, in "International Tables for Crystallography, D: Physical Properties of Crystals" (A. Authier, Ed.), ch. 3.3. Kluwer, Dordrecht, 2003, in print.
5. M. L. Senechal, *N. Jb. Miner. Mh.* 518 (1976).
6. M. Senechal, *Sov. Phys. Crystallogr.* 25, 520 (1980).
7. C. Palache, *Amer. Mineral.* 17, 360 (1932).
8. J. Walz, T. Tamura, N. Tamura, R. Grimm, W. Baumeister, and A. J. Koster, *Mol. Cells* 1, 59 (1997).
9. A. Müller, P. Körgerler, and A. W. M. Dress, *Coord. Chem. Rev.* 222, 193 (2001).
10. R. C. Williams and K. Smith, *Biophysica Acta* 28, 464 (1958).
11. M. Dubois, B. Demé, T. Gulik-Krzywicki, J.-C. Dedieu, C. Vautrin, S. Désert, E. Perez, and T. Zemb, *Nature* 411, 672 (2001).
12. R. Casanova, B. Simon, and G. Turgo, *Amer. Mineral.* 57, 1871 (1972).
13. Z. L. Wang, *Adv. Mater.* 10, 1 (1998).
14. T. L. Daulton, D. D. Eisenhour, T. J. Bernatowicz, R. S. Lewis, and P. R. Buseck, *Geochim. Cosmochim. Acta* 60, 4853 (1996).
15. L. D. Marks, *J. Cryst. Growth* 61, 556 (1983).
16. L. D. Marks, *Philos. Mag. A* 49, 81 (1984).
17. S. Ino, *J. Phys. Soc. Jpn.* 27, 941 (1969).
18. J. Bühler and Y. Prior, *J. Cryst. Growth* 209, 779 (2000).
19. F. H. C. Crick and J. D. Watson, *Nature* 177, 473 (1956).
20. A. Klug, R. E. Franklin, and S. P. F. Humphreys-Owen, *Biochim. Biophys. Acta* 32, 203 (1959).
21. J. T. Finch and A. Klug, *Acta Crystallogr.* 13, 1051 (1960).
22. G. Millman, B. G. Uzman, A. Mitchell, and R. Langridge, *Science* 152, 1381 (1966).
23. J. L. Melnick, E. R. Rabin and A. B. Jenson, *J. Virol.* 2, 78 (1968).
24. T. L. Daulton, D. D. Eisenhour, P. R. Buseck, R. S. Lewis, and T. J. Bernatowicz, in "Abstracts of the 25th Lunar & Planetary Science Conference," p. 313. Houston, 1994.
25. C. Hintze, in "Handbuch der Mineralogie. I. Elemente und Sulfide" p. 3. Verlag von Veit und Co., Leipzig, 1904.
26. F. Wallerant, in "Cristallographie" (C. Beranger, Ed.), p. 117. Librairie Polytechnique, Paris, 1909.
27. A. von Fersmann and V. Goldschmidt, in "Der Diamant" p. 204. Carl Winters Universitätsbuchhandlung, Heidelberg, 1911.
28. V. M. Goldschmidt, in "Atlas der Kristallformen" p. Pl–48. Carl Winters Universitätsbuchhandlung, Heidelberg, 1918.
29. C. Palache, H. Berman, and C. Frondel, in "The System of Mineralogy of J. D. Dana and E. S. Dana; I. Elements, Sulfides, Sulfosalts, Oxides" p. 88. John Wiley and Sons, New York, 1944.
30. G. Rose, *Pogg. Ann.* 23, 196 (1831).
31. S. von Waltershausen, *Nachr. Königl. Wissensch. Ges. Göttingen* 135 (1863).
32. A. von Lasaulx, *Sitzungsber. Niederrhein. Gesellsch.* 39, 95 (1882).
33. G. von Rath, *Z. Kristallogr.* 1, 1 (1877).
34. C. Hermann, *Z. Kristallogr.* 79, 186 (1931).
35. P. Niggli, in "Lehrbuch der Mineralogie und Kristallchemie," p. 123. Verlag Gebrüder Borntraeger, Berlin, 1941.
36. R. L. Segall, *J. Metals* 9, 50 (1957).
37. A. J. Melmed and D. O. Hayward, *J. Chem. Phys.* 31, 545 (1959).
38. A. Mackay, *Acta Crystallogr.* 15, 916 (1962).
39. H. Schlötterer, in "Proc. 5th Int. Congr. on Electron Microscopy" (S. S. Breese Jr. Ed.), p. DD6. Academic Press, New York, 1962.
40. H. Schlötterer, *Z. Kristallogr.* 119, 321 (1964).
41. H. Schlötterer, *Metalloberfläche* 18, 33 (1964).
42. R. H. Wentorf, in "The Art and Science of Growing Crystals" (J. Gilman, Ed.), p. 176. J. Wiley & Sons, New York, 1963.
43. J. W. Faust and H. F. John, *J. Phys. Chem. Solids* 25, 1407 (1964).
44. D. C. Skillman and C. R. Berry, *Photogr. Sci. Eng.* 1964, 65 (1964).
45. F. Ogburn, B. Paretzkin, and H. S. Peiser, *Acta Crystallogr.* 17, 774 (1964).
46. J. Smit, F. Ogburn, and C. J. Bechtold, *J. Electrochem. Soc.* 115, 371 (1968).
47. R. L. Schwoebel, *Surf. Sci.* 2, 356 (1964).
48. R. L. Schwoebel, *J. Appl. Phys.* 37, 2515 (1966).
49. M. A. Gedwill, C. J. Altstetter and C. M. Wayman, *J. Appl. Phys.* 35, 2266 (1964).
50. R. W. DeBlois, *J. Appl. Phys.* 36, 1647 (1965).
51. R. W. DeBlois, *J. Vac. Sci. Technol.* 3, 146 (1966).
52. B. G. Bagley, *Nature* 208, 674 (1965).
53. J. A. R. Clarke and J. D. Bernal, *Nature* 211, 280 (1966).
54. B. G. Bagley, *J. Cryst. Growth* 6, 323 (1970).
55. G. L. Downs and J. D. Braun, *Science* 154, 1443 (1966).
56. J. G. Allpress, H. Jaeger, P. D. Mercer, and J. V. Sanders, in "Proc. 6. Int. Congr. on Electron Microscopy" (R. Uyeda, Ed.), p. 489. Maruzen Co. Ltd., Tokyo, 1966.
57. J. G. Allpress and J. V. Sanders, *Surf. Sci.* 7, 1 (1967).
58. M. Gillet and E. Gillet, in "Proc. 6. Int Congr. on Electron Microscopy" (R. Uyeda, Ed.), p. 633. Maruzen Co. Ltd., 1966.
59. S. Ino, *J. Phys. Soc. Jpn.* 21, 346 (1966).
60. S. Ino and S. Ogawa, in "Proc. 6. Int Congr. on Electron Microscopy" (R. Uyeda, Ed.), p. 521. Maruzen Co. Ltd., Tokyo, 1966.
61. S. Ino and S. Ogawa, *J. Phys. Soc. Jpn.* 22, 1365 (1967).
62. S. Ogawa, D. Watanabe, S. Ino, T. Kato, and H. Ota, *Sci. Rep. Res. Inst. Tohoku Univ. A* 18 Suppl., 171 (1966).
63. S. Ogawa, S. Ino, T. Kato, and H. Ota, *J. Phys. Soc. Jpn.* 21, 1963 (1966).
64. K. Kimoto and I. Nishida, *Jpn. J. Appl. Phys.* 6, 1047 (1967).
65. G. Rupprechter, K. Hayek, and H. Hofmeister, *J. Catal.* 173, 409–422 (1998).
66. P. Melinon, G. Fuchs and M. Treilleux, *J. Phys. I* (France) 2, 1263 (1992).
67. L. R. Wallenberg, J. O. Bovin, and G. Schmid, *Surf. Sci.* 156, 256 (1985).
68. S. Iijima and T. Ishihashi, *Phys. Rev. Lett.* 56, 616 (1986).
69. D. J. Smith, A. K. Petford-Long, R. Wallenberg, and J. O. Bovin, *Science* 233, 872 (1986).
70. L. R. Wallenberg, J. O. Bovin, A. K. Petford-Long, and D. J. Smith, *Ultramicroscopy* 20, 71 (1986).
71. J. O. Malm, J. O. Bovin, A. Petford-Long, D. J. Smith, G. Schmid, and N. Klein, *Angew. Chem., Int. Ed.* 27, 555 (1988).
72. D. Shechtman, I. Blech, D. Gratias, and J. W. Cahn, *Phys. Rev. Lett.* 53, 1951 (1984).
73. D. Levine and P. J. Steinhardt, *Phys. Rev. Lett.* 53, 2477 (1984).
74. K. Hiraga, M. Hirabayashi, A. Inoue, and T. Masumoto, *Sci. Rep. Res. Inst. Tohoku Univ. A* 32, 309 (1985).
75. D. R. Nelson, *Sci. Am.* 254, 42 (1986).
76. D. R. Nelson and B. I. Halperin, *Science* 229, 233 (1985).
77. K. H. Kuo, *J. de Physique Colloque* C3 47, 425 (1986).
78. K. H. Kuo, *J. Electron Microsc. Techn.* 7, 277 (1987).
79. K. F. Fung, X. D. Zuo, and C. Y. Yang, Phil. Mag. 55, 27 (1987).
80. W. J. Jiang, Z. K. Hei, Y. X. Guo, and K. H. Kuo, *Philos. Mag.* 52, L53 (1985).
81. T. R. Anantharaman, *Current Science* 58, 1067 (1989).
82. A. K. Srivastava and S. Ranganathan, *Acta Mater.* 44, 2935 (1996).

83. U. Dahmen and K. H. Westmacott, *Science* 233, 875 (1986).
84. J. Douin, U. Dahmen, and K. H. Westmacott, *Coll. Phys. C1* 51, 809 (1990).
85. J. Douin, U. Dahmen, and K. H. Westmacott, *Philos. Mag. B* 63, 867 (1991).
86. S. Q. Xiao, S. Hinderberger, K. H. Westmacott, and U. Dahmen, *Philos. Mag. A* 73, 1261 (1996).
87. D. Mitlin, U. Dahmen, V. Radmilovic, and J. W. Morris Jr., Mater. Sci. *Eng., A* 301, 231 (2001).
88. Y. T. Pei and J. T. M. de Hosson, *Acta Mater.* 49, 561 (2001).
89. M. Fujimoto, K. Hoshi, M. Nakazawa, and S. Sekiguchi, *Jpn. J. Appl. Phys.* 32, 5532 (1993).
90. M. Dubiel, H. Hofmeister, and J. Hopfe, *Beitr. Elektronenmikroskop. Direktabb. Oberfl.* 24, 49 (1991).
91. S. T. Selvan, Y. Ono, and M. Nogami, *Mater. Lett.* 37, 156 (1998).
92. W. Vogel, J. Bradley, O. Vollmer, and I. Abraham, *J. Phys. Chem. B* 102, 10853 (1998).
93. F. Dassenoy, M. J. Casanove, P. Lecante, M. Verelst, E. Snoeck, A. Mosset, T. O. Ely, C. Amiens, and B. Chaudret, *J. Chem. Phys.* 112, 8137 (2000).
94. G. L. Tan, H. Hofmeister, and M. Dubiel, in "Proceedings of Autumn School on Materials Sciences and Electron Microscopy 2000" (D. S. Su and S. Wrabetz, Eds.), p. 67. FHI Berlin, Berlin-Dahlem, 2000.
95. W. K. Choi, Y. W. Ho, S. P. Ng, and V. Ng, *J. Appl. Phys.* 89, 2168 (2001).
96. M. Klimenkov, W. Matz, S. A. Nepijko, and M. Lehmann, *Nucl. Instr. Methods Phys. Res., Sect. B* 179, 209 (2001).
97. H. Hofmeister, *Mater. Sci. Forum* 312–314, 325 (1999).
98. A. Nohara, S. Ino and S. Ogawa, *Jpn. J. Appl. Phys.* 7, 1144 (1968).
99. S. Ino, *J. Electron Microsc.* 18, 237 (1969).
100. K. Reichelt and S. Schreiber, *Surf. Sci.* 43, 644 (1974).
101. Y. Fukano, *Jpn. J. Appl. Phys.* 13, 1001 (1974).
102. M. Takahashi, T. Suzuki, H. Kushima, and S. Ogasawara, *Jpn. J. Appl. Phys.* 17, 1499 (1978).
103. A. Renou, PhD Thesis, University Aix–Marseille III (1979).
104. M. Gillet, A. Renou, and J. M. Miquel, in "Growth and Properties of Metal Clusters" (J. Bourdon, Ed.), p. 185. Elsevier, New York, 1980.
105. L. D. Marks and D. J. Smith, *J. Microsc.* 130, 249 (1983).
106. Y. Ohtsuka, *Acta Crystallogr. Sect. A* 40, C (1984).
107. H. Hofmeister, *Z. Phys. D* 19, 307 (1991).
108. G. Rupprechter, K. Hayek, and H. Hofmeister, *Vacuum* 46, 1035 (1995).
109. M. Tanaka, M. Takeguchi, and K. Furuya, *Surf. Sci.* 433–435, 491 (1999).
110. Y. Wu, Q. Chen, Ma. Takeguchi, and K. Furuya, *Surf. Sci.* 462, 203 (2000).
111. Y. Oshima, T. Nannguo, H. Hirayama, and K. Takayanagi, *Surf. Sci.* 476, 107 (2001).
112. S. Mader, *J. Vac. Sci. Technol.* 8, 247 (1971).
113. N. G. Nakhodkin, Y. A. Barabanenkov, A. F. Bardamid, A. I. Novoselskaya, and K. I. Yakimov, *Ukrainian Fiz. Zh.* 34, 1355 (1989).
114. V. Bykov, H. Hofmeister, T. Junghanns, and S. Nepijko, in "Proc 7. Oxford Conf. on Microscopy of Semiconducting Materials" (A. G. Cullis and N. J. Long, Eds.), p. 51. IOP Publishing Ltd., Bristol, 1991.
115. H. Hofmeister, A. F. Bardamid, T. Junghanns, and S. A. Nepijko, *Thin Solid Films* 205, 20 (1991).
116. T. Okabe, Y. Kagawa, and S. Takai, *Philos. Mag. Lett.* 63, 233 (1991).
117. H. Hofmeister and T. Junghanns, *Mater. Sci. Forum* 113–115, 631 (1993).
118. H. Hofmeister and T. Junghanns, *Nanostruct. Mater.* 3, 137 (1993).
119. H. Hofmeister and T. Junghanns, *Trans. Mater. Res. Soc. Japan* 16B, 1581 (1994).
120. H. Hofmeister and T. Junghanns, *J. Non-Crystalline Solids* 192 & 193, 550 (1995).
121. W. Neumann, H. Hofmeister, D. Conrad, K. Scheerschmidt, and S. Ruvimov, *Z. Kristallogr.* 211, 147 (1996).
122. J. G. Zheng, X. Q. Pan, M. Schweizer, F. Zhou, U. Weimar, W. Göpel, and M. Rühle, *J. Appl. Phys.* 79, 7688 (1996).
123. C. Leclercq, H. Batis, and M. Boudeulle, *J. Microsc. Spectrosc. Electron.* 8, 243 (1983).
124. Y. Saito, Y. Ishikawa, A. Oshita, H. Shinohara, and H. Nagashima, *Phys. Rev. B* 46, 1846 (1992).
125. M. Haluska, H. Kuzmany, M. Vybornov, P. Rogl, and P. Fejdi, *Appl. Phys. A* 56, 161 (1993).
126. W. L. Zhou, W. Zhao, K. K. Fung, L. Q. Chen, and Z. B. Zhang, *Physica C* 214, 19 (1993).
127. H. Kawada, Y. Fujii, H. Nakao, Y. Murakami, T. Watanuki, H. Suematsu, K. Kikuchi, Y. Achiba, and I. Ikemoto, *Phys. Rev. B* 51, 8723 (1995).
128. Y. Kim, L. Jiang, T. Iyoda, K. Hashimoto, and A. Fujishima, *Appl. Surf. Sci.* 130–132, 602 (1998).
129. S. Matsumoto and Y. Matsui, *J. Mater. Sci.* 18, 1785 (1983).
130. Y. H. Lee, P. D. Richards, K. J. Bachmann, and J. T. Glass, *Appl. Phys. Lett.* 56, 620 (1990).
131. D. Dorignac, S. Delclos, and F. Phillipp, *Philos. Mag. B* 81, 1879 (2001).
132. T. Oku and K. Hiraga, Diamond *Relat. Mater.* 10, 1398 (2001).
133. J. Narayan, A. R. Srivatsa, M. Peters, S. Yokota. and K. V. Ravi, *Appl. Phys. Lett.* 53, 1823 (1988).
134. J. Narayan, *J. Mater. Res.* 5, 2414 (1990).
135. C. Wild, N. Herres, and P. Koidl, *J. Appl. Phys.* 68, 973 (1990).
136. B. E. Williams, J. T. Glass, R. F. Davis, and K. J. Kobashi, *J. Cryst. Growth* 99, 1168 (1990).
137. J. F. DeNatale, A. B. Harker, and J. F. Flintoff, *J. Appl. Phys.* 69, 6456 (1991).
138. J. L. Hutchison and D. Shechtman, in "Proc. EUREM92" (A. Lopéz-Galindo and M. Rodríguez-Garcia, Eds.), p. 713. Secretariado de Publicaciones de la Universidad de Granada, Granada, 1992.
139. D. Shechtman, J. L. Hutchison, L. H. Robins, E. N. Farabaugh, and A. Feldman, *J. Mater. Res.* 8, 473 (1993).
140. P. Wurzinger, M. Joksch, and P. Pongratz, in "Proc. Microsc. Semicond. Mater. Conf." (A. G. Cullis, A. E. Staton-Bevan, and J. L. Hutchison, Eds.), p. 157. IOP Publ., Bristol, 1993.
141. C. H. Chu and M. H. Hon, *Mater. Chem. Phys.* 38, 131 (1994).
142. S. Barrat and E. Bauer-Grosse, *Diamond Relat. Mater.* 4, 419 (1995).
143. J.-H. Choi, S.-H. Lee, and J.-W. Park, *Mater. Chem. Phys.* 45, 176 (1996).
144. W. N. Wang, N. A. Fox, T. J. Davis, D. Richardson, G. M. Lynch, J. W. Steeds, and J. S. Lee, *Appl. Phys. Lett.* 69, 2825 (1996).
145. C.-S. Yan and Y. K. Vohra, *Diamond Relat. Mater.* 8, 2022 (1999).
146. S. Delclos, D. Dorignac, F. Phillipp, F. Silva, and A. Gicquel, *Diamond Relat. Mater.* 9, 346 (2000).
147. H. Hofmeister, J. Dutta, and H. Hofmann, *Phys. Rev. B* 54, 2856 (1996).
148. J. Dutta, R. Houriet, H. Hofmann, and H. Hofmeister, *Nanostruct. Mater.* 9, 359 (1997).
149. W. Qin, D. G. Ast, and T. I. Kamins, *Jpn. J. Appl. Phys.* 40, 4806 (2001).
150. T. N. Millers and A. A. Kuzjukévics, *Prog. Cryst. Growth & Charact. Mat.* 16, 367 (1988).
151. H. E. Cheng and H. M. Hon, *J. Cryst. Growth* 142, 117 (1994).
152. H. E. Cheng, T. T. Lin, and M. H. Hon, *Scripta Mater.* 36, 113 (1996).
153. W. P. Sun, D. J. Cheng, and M. H. Hon, *J. Cryst. Growth* 71, 787 (1985).
154. D. J. Cheng, W. P. Sun, and M. H. Hon, *Thin Solid Films* 146, 45 (1987).
155. F. Ernst and P. Pirouz, *J. Appl. Phys.* 64, 4526 (1988).

156. G. Turner and E. Bauer, in "Proc. 5. Int. Cong. Electron Microsc." (S. S. Breese Jr., Ed.), p. DD3. Academic Press, New York, 1962.
157. N. Wada, *Jpn. J. Appl. Phys.* 7, 1287 (1968).
158. R. Uyeda, *J. Cryst. Growth* 24/25, 69 (1974).
159. T. Hayashi, T. Ohno, S. Yatsuya, and R. Uyeda, *Jpn. J. Appl. Phys.* 16, 705 (1977).
160. T. Ohno and K. Yamauchi, *Jpn. J. Appl. Phys.* 20, 1385 (1981).
161. A. Renou and M. Gillet, *Surf. Sci.* 106, 27 (1981).
162. M. Arita, N. Suzuki, and I. Nishida, *Nagoya University Research Bulletin B* 37, 39 (1993).
163. J. Urban, H. Sack-Kongehl, and K. Weiss, *Catal. Lett.* 49, 101 (1997).
164. M. Hyslop, A. Wurl, S. A. Brown, B. D. Hall, and R. Monot, *Eur. Phys. J. D* 16, 233 (2001).
165. B. Rellinghaus, S. Stappert, E. F. Wassermann, H. Sauer, and B. Spliethoff, *Eur. Phys. J. D* 16, 249 (2001).
166. H. Hofmeister, S. A. Nepijko, D. N. Ievlev, W. Schulze, and G. Ertl, *J. Cryst. Growth* 234, 773 (2002).
167. B. Q. Wei, R. Vajtai, Y. J. Jung, F. Banhart, G. Ramanath, and P. M. Ajayan, *J. Phys. Chem. B* 106, 5807 (2002).
168. Y. Saito, *J. Cryst. Growth* 47, 61 (1979).
169. S. Iijima, *Jpn. J. Appl. Phys.* 26, 365 (1987).
170. D. K. Saha, K. Koga, and H. Takeo, *Nanostruct. Mater.* 8, 1139 (1997).
171. S. Stappert, B. Rellinghaus, M. Acet, and E. Wassermann, in "Nanoparticulate Materials" (R. K. Singh, R. Partch, M. Muhammed, M. Senna, and H. Hofmann, Eds.), p. 73. MRS, Warrendale, PA, 2002.
172. Y. Saito, S. Yatsuya, K. Mihama, and R. Uyeda, *Jpn. J. Appl. Phys.* 17, 1149 (1978).
173. P. Gao, *Z. Phys. D* 15, 175 (1990).
174. C. Altenhein, S. Giorgio, J. Urban, and K. Weiss, *J. Phys. D* 19, 303 (1991).
175. M. Flüeli, R. Spycher, P. Stadelmann, Ph. Buffat, and J. P. Borel, *J. Microsc. Spectrosc. Electron.* 14, 351 (1989).
176. B. Hall, PhD Thesis, EPFL Lausanne, (1991).
177. D. Reinhard, B. D. Hall, D. Ugarte, and R. Monod, *Z. Phys. D* 26 Suppl., 76 (1993).
178. D. Reinhard, B. D. Hall, P. Berthoud, S. Valkealathi, and R. Mono, *Phys. Rev. B* 58, 4917 (1998).
179. S. W. Yu, G. H. Wang, S. Y. Yin, Y. X. Zhang, and Z. G. Liu, *Phys. Lett. A* 268, 442 (2000).
180. F. Ding, G. Wang, S. Yu, J. Wang, W. Shen, and H. Li, *Eur. Phys. J. D* 16, 245 (2001).
181. J.-P. Barnes, A. K. Petford-Long, R. C. Doole, R. Serna, J. Gonzalo, A. Suárez-García, C. N. Afonso, and D. Hole, *Nanotechnology* 13, 465 (2002).
182. R. A. Roy, R. Messier, and J. M. Cowley, *Thin Solid Films* 79, 207 (1981).
183. E. Bouzy, G. Le Caer, and E. Bauer-Grosse, *Philos. Mag. Lett.* 64, 1 (1991).
184. H. Hubert, B. Devouard, L. A. J. Garvie, M. O'Keefe, P. R. Buseck, W. T. Petuskey, and P. F. McMillan, *Nature* 391, 376 (1998).
185. H. Fredriksson, M. Hillert, and N. Lange, *J. Inst. Metals* 101, 285 (1973).
186. K. Kobayashi and L. M. Hogan, *Philos. Mag. A* 40, 399 (1979).
187. J. J. Hu and P. L. Ryder, *Philos Mag.* 68, 389 (1993).
188. M. Ellner and U. Burkhardt, *J. Alloys Compd.* 198, 91 (1993).
189. N. Uyeda, M. Nishino, and E. Suito, *J. Colloid Interface Sci.* 43, 264 (1973).
190. M. Brieu and M. Gillet, *Thin Solid Films* 100, 53 (1983).
191. N. J. Long, R. F. Marzke, M. McKelvy, and W. S. Glausinger, *Ultramicroscopy* 20, 15 (1986).
192. M. Brieu and M. Gillet, *Thin Solid Films* 167, 149 (1988).
193. A. C. Curtis, D. G. Duff, P. P. Edwards, D. A. Jefferson, B. F. G. Johnson, A. I. Kirkland, and A. S. Wallace, *J. Phys. Chem. B* 92, 2270 (1988).
194. D. G. Duff, P. P. Edwards, J. Evans, J. T. Gauntlett, D. A. Jefferson, B. F. G. Johnson, A. I. Kirkland, and D. J. Smith, *Ang. Chem., Int. Ed.* 28, 590 (1989).
195. W. Vogel, D. G. Duff, and A. Baiker, *Langmuir* 11, 401 (1995).
196. G. M. Chow, M. A. Markowitz, R. Rayne, D. N. Dunn, and A. Singh, *J. Colloid Interface Sci.* 183, 135 (1996).
197. C. Greffié, M. F. Benedetti, C. Parron, and M. Amouric, *Geochim. Cosmochim. Acta* 60, 1531 (1996).
198. Q. Li, M. Shao, S. Zhang, X. Liu, G. Li, K. Jiang, and Y. Qian, *J. Cryst. Growth* 243, 327–330 (2002).
199. R. Breckpot, *Anales Real Soc. Esp. Fisica y Quimica B* 51, 31 (1965).
200. I. Epelboin, M. Froment, and G. Maurin, *Plating* 56, 1356 (1969).
201. I. Epelboin, M. Froment, and G. Maurin, in "Electrocystallization; Proc. 28. Meeting Int. Soc. Electrochem" p. 371. Varna, 1977.
202. C. R. Hall and S. A. H. Fawzi, *Philos. Mag. A* 54, 805 (1986).
203. M. Froment and G. Maurin, *J. Microsc. Spectrosc. Electron.* 12, 379 (1987).
204. H. Hofmeister and N. Atanassov, in "Proceedings of the 11th European Conference on Electron Microscopy, Dublin 1996" (Committee of Europ. Soc. of Microsc., Ed.), p. 333. Brussels, 1998.
205. N. Pangarov and V. Velinov, *Electrochim. Acta* 13, 1641 (1968).
206. M. Froment and J. Thevenin, *Metaux Corros. Ind.* 54, 43 (1975).
207. C. Digard, M. Maurin, and J. Robert, *Metaux Corros. Ind.* 51, 255 (1976).
208. J. Thevenin and M. Froment, *J. Microsc. Spectrosc. Electron.* 1, 7 (1976).
209. D.-L. Lu, Y. Okawa, K. Suzuki, and K.-I. Tanaka, *Surf. Sci.* 325, L397 (1995).
210. D.-L. Lu and K.-I. Tanaka, *J. Solid State Electrochem* 1, 187 (1997).
211. V. Radmilovic, K. I. Popov, M. G. Pavlovic, A. Dimitrov, and S. H. Jordanov, *J. Sol. State Electrochem.* 2, 162 (1998).
212. B. Bozzine, A. Fanigliulo, and M. Serra, *J. Cryst. Growth* 231, 589 (2001).
213. D.-L. Lu, K. Domen, and K.-I. Tanaka, *Langmuir* 18, 3226 (2002).
214. D.-L. Lu, Y. Okawa, M. Ichihara, A. Aramata, and K. Tanaka, *J. Electroanal. Chem.* 406, 101 (1996).
215. D.-L. Lu and K. Tanaka, *Phys. Rev. B* 55, 13865 (1997).
216. D.-L. Lu and K.-I. Tanaka, *J. Cryst. Growth* 181, 395 (1997).
217. P. Claus and H. Hofmeister, *J. Phys. Chem. B* 103, 2766 (1999).
218. K. Chattopadhay and P. Ramachandrarao, *J. Cryst. Growth* 36, 355 (1976).
219. M. D. Ball and D. J. Lloyd, *Scr. Met.* 19, 1065 (1985).
220. H. Hofmeister, *Cryst. Res. Technol.* 33, 3 (1998).
221. H. Hofmeister, W. G. Drost, and A. Berger, *Nanostruct. Mater.* 12, 207 (1999).
222. C. Mohr, M. Dubiel and H. Hofmeister, *J. Phys.: Condens. Matter* 13, 525 (2001).
223. H. Hofmeister and U. Kahler, in "Silicon Chemistry: From Molecules to Extended Systems" (P. Jutzi and U. Schubert, Eds.), Wiley-VCH, Weinheim (2003), in print.
224. H. Hofmeister, P. Werner, and T. Junghanns, in "Physics and Chemistry of Finite Systems: From Clusters to Crystals" (P. Jena, S. N. Khanna, and B. K. Rao, Eds.), p. 1251. Kluwer Academic Publ., Dordrecht, 1992.
225. A. Nohara and T. Imura, *J. Phys. Soc. Jpn.* 27, 793 (1969).
226. G. Rupprechter, G. Seeber, K. Hayek, and H. Hofmeister, *Phys. Status Solidi A* 146, 449 (1994).
227. C. Mohr, H. Hofmeister, and P. Claus, *J. Catal.* 213, 86 (2003).
228. Y. Matsui, *J. Cryst. Growth* 66, 243 (1984).
229. J. O. Malm, G. Schmid, and B. Morun, *Philos. Mag.* 63, 487 (1991).
230. P. J. Herley and W. Jones, *J. Chem. Soc. Faraday Trans.* 88, 3213 (1992).
231. P. J. Herley, N. P. Fitzsimons, and W. Jones, *J. Chem. Soc. Faraday Trans.* 91, 719 (1995).
232. B. S. Xu and S.-I. Tanaka, *Nanostruct. Mater.* 8, 1131 (1997).

233. S. Thiel, M. Dubiel, S. Schurig, and H. Hofmeister, in "Proceedings of the 11th European Conference on Electron Microscopy, Dublin 1996" Committee of Europ. Soc. of Microsc., Brussels, 1998, p. 445.
234. M. Takeguchi, M. Tanaka, H. Yasuda, and K. Furuya, *Surf. Sci.* 493, 414 (2001).
235. T. Takami, K.-I. Sugiura, Y. Sakata, T. Takeuchi, and S. Ino, *Appl. Surf. Sci.* 130–132, 834 (1998).
236. T. Takami, M. Brause, D. Ochs, W. Maus-Friedrichs, V. Kempter, and S. Ino, *Surf. Sci.* 407, 140 (1998).
237. S. Ogawa and S. Ino, *J. Vac. Sci. Technol.* 6, 527 (1969).
238. H. Sato and S. S. Shinozaki, *J. Appl. Phys.* 41, 3165 (1970).
239. S. Ogawa and S. Ino, *J. Cryst. Growth* 13/14, 48 (1972).
240. E. Gillet and M. Gillet, *J. Cryst. Growth* 13/14, 212 (1972).
241. P. F. McMillan, H. Hubert, A. Chizmeshya, W. T. Petuskey, L. A. J. Garvie, and B. Devouard, *J. Solid State Chem.* 147, 281 (1999).
242. T. P. Martin, T. Bergmann, H. Göhlich, and T. Lange, *Chem. Phys. Lett.* 176, 343 (1991).
243. T. P. Martin, T. Bergmann, H. Göhlich, and T. Lange, *Z. Phys. D* 19, 25 (1991).
244. T. P. Martin, T. Bergmann, H. Göhlich, and T. Lange, *J. Phys. Chem. B* 95, 6421 (1991).
245. J. Uppenbrink and D. J. Wales, *J. Chem. Soc. Faraday Trans.* 87, 215 (1991).
246. J. Uppenbrink and D. J. Wales, *J. Chem. Phys.* 96, 8520 (1992).
247. J. P. K. Doye and D. J. Wales, *Chem. Phys. Lett.* 247, 339 (1995).
248. J. P. K. Doye, D. J. Wales, and R. S. Berry, *J. Chem. Phys.* 103, 4234 (1995).
249. V. V. Volkov, G. van Tendeloo, G. A. Tsirkov, N. V. Cherkashina, M. N. Vargaftik, I. I. Moiseev, V. M. Novotortsev, A. V. Kvit, and A. L. Chuvilin, *J. Cryst. Growth* 163, 377 (1996).
250. C. Gerstengarbe and W. Neumann, in "Publications of the 12th Electron Microscopy Conference" (J. Heydenreich and H. Luppa, Eds.), p. 481. Dresden, 1988.
251. H. Hofmeister, *Phys. Bl.* 53, 37 (1997).
252. L. A. Paquette, D. W. Balogh, R. Usha, D. Kountz, and G. G. Christoph, *Science* 211, 575 (1981).
253. H. Hofmeister, in "Nanophase Materials, NATO ASI Series E" (G. C. Hadjipanayis and R. W. Siegel, Eds.), p. 209. Kluwer Academic Publ., Dordrecht, 1994.
254. E. Gillet and M. Gillet, *Thin Solid Films* 4, 171 (1969).
255. G. Honjo and Yagi K, *J. Vac. Sci. Technol.* 6, 576 (1969).
256. J. G. Allpress and J. V. Sanders, *Austral. J. Physics* 23, 23 (1970).
257. S. A. Nepijko, V. I. Styopkin, and R. Scholz, *Poverchnostj* 4, 116 (1984).
258. K. Yagi, K. Takayanagi, K. Kobayashi, and G. Honjo, *J. Cryst. Growth* 28, 117 (1975).
259. H. Hofmeister, *Thin Solid Films* 116, 151 (1984).
260. S. A. H. Fawzi. PhD Thesis, University of Warwick (1984).
261. B. W. van de Waal, *J. Cryst. Growth* 158, 153 (1996).
262. B. W. van de Waal, *Phys. Rev. Lett.* 76, 1083 (1996).
263. B. W. van de Waal, PhD Thesis, University of Twente (1997).
264. M. S. Abrahams, J. L. Hutchison, and G. K. Booker, *Phys. Status Solidi A* 63, K3 (1981).
265. C. Gerstengarbe and W. Neumann, in "Publications of the 11th Electron Microscopy Conference" (J. Heydenreich and H. Luppa, Eds.), p. 253. Dresden, 1984.
266. K. C. Paus, J. C. Barry, G. R. Brooker, T. B. Peters, and M. G. Pitt, in "Proc. Micros. Semicond. Mater. Conf." (A. G. Cullis and D. B. Holt, Eds.), p. 35. Adam Hilger, Bristol, 1985.
267. W. Wegscheider, K. Eberl, G. Abstreiter, H. Cerva, and H. Oppolzer, *Appl. Phys. Lett.* 57, 1496 (1990).
268. G. Wagner and P. Paufler, *Z. Kristallogr.* 195, 17 (1991).
269. D. M. Hwang, S. A. Schwarz, T. S. Ravi, R. Bhat, and C. Y. Chen, *Phys. Rev. Lett.* 66, 739 (1991).
270. W. Wegscheider, K. Eberl, G. Abstreiter, H. Cerva and H. Oppolzer, in "Proc. 7th Oxford Conf. Microsc. Semicond. Mater." (N. J. Long and A. G. Cullis, Eds.), p. 21. IOP Publishing Ltd., Bristol, 1991.
271. D. J. Smith and L. D. Marks, *J. Cryst. Growth* 54, 433 (1981).
272. L. D. Marks, *Thin Solid Films* 136, 309 (1986).
273. H. Hofmeister and T. Junghanns, in "Proc. Autumn School1991 of the Int. Centre of Electron Microsc." (J. Heydenreich and W. Neumann, Eds.), p. 245. MPI Halle, Halle, 1992.
274. S. Krafczyk, H. Jacobi, and H. Follner, *Cryst. Res. Technol.* 32, 163 (1997).
275. J. Farges, M. F. de Feraudey, B. Raoult, and G. Torchet, *Acta Crystallogr. A* 38, 656 (1982).
276. J. Farges, M. F. de Feraudy, B. Raoult, and G. Torchet, *J. Chem. Phys.* 78, 5067 (1983).
277. J. Farges, M. F. de Feraudy, B. Raoult, and G. Torchet, *J. Chem. Phys.* 84, 3491 (1986).
278. J. Farges, M. F. de Feraudy, B. Raoult, and G. Torchet, *Adv. Chem. Phys.* 70, 45 (1988).
279. B. Raoult, J. Farges, M. F. de Feraudy, and G. Torchet, *Philos. Mag. B* 60, 881 (1989).
280. B. W. van de Waal, *J. Chem. Phys.* 90, 3407 (1989).
281. B. W. van de Waal, *J. Chem. Phys.* 98, 4909 (1993).
282. M. B. Gordon, F. Cyrot-Lackmann, and M. C. Desjonqueres, *Surf. Sci.* 80, 159 (1979).
283. R. Mosserri and J. F. Sadoc, *Z. Phys. D* 12, 89 (1989).
284. C. L. Cleveland and U. Landman, *J. Chem. Phys.* 94, 7376 (1991).
285. J.-Y. Yi, D. J. Oh and J. Bernholc, *Phys. Rev. Lett.* 67, 1594 (1991).
286. Q. Wang, M. D. Glossmann, M. P. Iniguez, and J. A. Alonso, *Philos. Mag. B* 69, 1045 (1994).
287. D. J. Wales, L. J. Munro, and J. P. K. Doye, *J. Chem. Soc., Dalton Trans.* 5, 611 (1996).
288. Y. J. Lee, J. Y. Maeng, E. K. Lee, B. Kim, S. Kim, and K. K. Han, *J. Comput. Chem.* 21, 380 (2000).
289. F. Baletto, R. Ferrando, A. Fortunelli, F. Montaleni, and C. Mottet, *J. Chem. Phys.* 116, 3856 (2002).
290. B. D. Hall, M. Flueli, R. Monot, and J.-P. Borel, *Z. Phys. D* 12, 97 (1989).
291. B. D. Hall, M. Flueli, R. Monot, and J. P. Borel, *Phys. Rev. B* 43, 3906 (1991).
292. D. Reinhard, R. D. Hall, D. Ugarte, and R. Monod, *Phys. Rev. B* 55, 7868 (1997).
293. M. Pellarin, B. Baguenard, J. L. Valle, J. Lerme, M. Broyer, J. Miller, and A. Perez, *Chem. Phys. Lett.* 217, 349 (1994).
294. T. P. Martin, *Phys. Rep.* 273, 199 (1996).
295. B. M. Smirnov, *Chem. Phys. Lett.* 232, 395 (1995).
296. J. P. K. Doye and D. J. Wales, *J. Chem. Soc. Faraday Trans.* 93, 4233 (1997).
297. T. Ikeshoji, G. Torchet, M.-F. de Feraudy, and K. Koga, *Phys. Rev. B* 63, 031101 (2001).
298. M. R. Hoare and P. Pal, *Adv. Phys.* 20, 161 (1971).
299. J. J. Burton, in "Materials Science Research" (G. C. Kuczynski, Ed.), p. 17. Plenum Press, New York, 1975.
300. J. Xie, J. A. Northby, D. L. Freeman, and J. D. Doll, *J. Chem. Phys.* 91, 612 (1989).
301. A. Sachdev and R. I. Masel, *J. Mater. Res.* 8, 455 (1993).
302. A. Sachdev, R. I. Masel, and J. B. Adams, *Z. Phys. D* 26, 310 (1993).
303. I. G. Garzón and A. Posada-Amarillas, *Phys. Rev. B* 54, 11796 (1996).
304. C. L. Cleveland, U. Landman, M. N. Shafigullin, P. W. Stephens, and R. L. Whetten, *Z. Phys. D* 40, 503 (1997).
305. C. L. Cleveland, U. Landman, T. G. Schaaff, M. N. Shafigullin, P. W. Stephens, and R. L. Whetten, *Phys. Rev. Lett.* 79, 1873 (1997).
306. I. L. Garzón, K. Michaelian, M. R. Beltrán, A. Posada-Amarillas, P. Ordejón, E. Artacho, D. Sánchez-Portal, and J. M. Soler, *Eur. Phys. J. D* 9, 211 (1999).
307. L. G. Gonzalez and J. M. Montejano-Carrizales, *Phys. Status Solidi B* 220, 357 (2000).

308. J. L. Aragón, *Chem. Phys. Lett.* 226, 263 (1994).
309. R. S. Berry, B. M. Smyrnov, and A. Y. Strizhev, *J. Exp. Theor. Phys.* 85, 588 (1997).
310. C. L. Cleveland, W. D. Luedtke, and U. Landman, *Phys. Rev. Lett.* 81, 2036 (1998).
311. F. Baletto, C. Mottet, and R. Ferrando, *Phys. Rev. Lett.* 84, 5544 (2000).
312. C. Barreteau, M. C. Desjonquères, and D. Spanjaard, *Eur. Phys. J. D* 11, 395 (2000).
313. L. Vitos, B. Johansson, and J. Kollar, *Phys. Rev. B* 62, R11957 (2000).
314. W. H. Zhang, L. Liu, J. Zhuang, and Y. F. Li, *Phys. Rev. B* 62, 8276 (2000).
315. F. Baletto and R. Ferrando, *Surf. Sci.* 490, 361 (2001).
316. F. Baletto, C. Mottet, and R. Ferrando, *Eur. Phys. J. D* 16, 25 (2001).
317. C. L. Cleveland, W. D. Luedtke, and U. Landman, *Phys. Rev. B* 60, 5065 (1999).
318. I. G. Garzón, K. Michaelian, M. R. Beltrán, A. Posada-Amarillas, P. Ordejón, E. Artacho, D. Sánchez-Portal, and J. M. Soler, *Phys. Rev. Lett.* 81, 1600 (1998).
319. F. Baletto, J. P. K. Doye, and R. Ferrando, *Phys. Rev. Lett.* 88, 075503 (2002).
320. P. M. Ajayan and L. D. Marks, *Phase Transitions* 24–26, 229 (1990).
321. N. Doraiswamy and L. D. Marks, *Philos. Mag. B* 71, 291 (1995).
322. S. Iijima, in "Proc. XI. Int. Cong. on Electron Microsc." (T. Imura, S. Maruse, and T. Suzuki, Eds.), p. 87. The Japanese Society of Electron Microscopy, Tokyo, 1986.
323. K. Harada, H. Endoh, and R. Shimizu, *Technol. Reports Osaka Univ.* 37, 221 (1987).
324. S. Iijima, in "Microclusters" (Y. Nishina, S. Ohnishi, and S. Sugano, Eds.), p. 186. Springer-Verlag, Berlin, 1987.
325. L. R. Wallenberg, PhD Thesis, University of Lund (1987).
326. M. Mitome, Y. Tanishiro, and K. Takayanagi, *Z. Phys. D* 12, 45 (1989).
327. T. Kizuka, T. Kachi, and N. Tanaka, *Z. Phys. D* 26 Suppl., 58 (1993).
328. V. G. Gryaznov, J. Heydenreich, A. M. Kaprelov, S. A. Nepijko, A. E. Romanov, and J. Urban, *Cryst. Res. Technol.* 34, 1091 (1999).
329. G. P. Dimitrakopulos, P. Komninou, T. Karakostas, and R. C. Pond, *Interface Science* 7, 217 (1999).
330. P. M. Ajayan and L. D. Marks, *Phys. Rev. Lett.* 60, 585 (1988).
331. J. Dundurs, L. D. Marks, and P. M. Ajayan, *Philos. Mag. A* 57, 605 (1988).
332. A. Howie and L. D. Marks, *Philos. Mag. A* 49, 95 (1984).
333. D. Reinhard, B. D. Hall, P. Berthoud, S. Valkealathi, and R. Monod, *Phys. Rev. Lett.* 79, 1459 (1997).
334. C. Y. Yang, *J. Cryst. Growth* 47, 274 (1979).
335. K. Heinemann, M. J. Yacaman, C. Y. Yang, and H. Poppa, *J. Cryst. Growth* 47, 177 (1979).
336. M. J. Yacaman, K. Heinemann, C. Y. Yang, and H. Poppa, *J. Cryst. Growth* 187 (1979).
337. C. Y. Yang, M. J. Yacáman, and K. Heinemann, *J. Cryst. Growth* 47, 283 (1979).
338. C. Y. Yang, K. Heinemann, M. J. Yacáman, and H. Poppa, *Thin Solid Films* 58, 163 (1979).
339. M. J. Carr, *J. Appl. Phys.* 59, 1063 (1986).
340. Q. Chen, M. Tanaka, and K. Furuya, *Surf. Sci.* 440, 398 (1999).
341. H. Hubert, L. A. J. Garvie, B. Devouard, P. R. Buseck, W. T. Petuskey, and P. F. McMillan, *Chem. Mater.* 10, 1530 (1998).
342. R. I. Kostov and I. Kostov, *Cryst. Res. Technol.* 23, 973 (1988).
343. A. Mackay, *Naturwissenschaften* 391, 334 (1998).
344. I. Kostov and R. I. Kostov, in "Crystal Habits of Minerals" (M. Drinov, Ed.), p. 45. Academic Publishing House, Sofia, 1999.
345. P. M. Ayajan, L. D. Marks, and J. Dundurs, in "Mater. Res. Soc. Symp. Proc.," p. 469. Materials Research Society, 1987.
346. V. G. Gryaznov, A. M. Kaprelov, A. E. Romanov, and I. A. Polonski, *Phys. Status Solidi B* 167, 441 (1991).
347. O. A. Shenderova and D. W. Brenner, *Phys. Rev. B* 60, 7053 (1999).
348. L. D. Marks, *Surf. Sci.* 150, 302 (1985).
349. L. D. Marks and D. J. Smith, *J. Cryst. Growth* 54, 425 (1981).
350. N. J. Pipkin and D. J. Davies, *Philos. Mag. A* 40, 435 (1979).
351. J. Narayan, A. R. Srivatsa, and K. V. Ravi, *Appl. Phys. Lett.* 54, 1659 (1989).
352. B. E. Williams, H. S. Kong, and J. T. Glass, *J. Mater. Res.* 5, 801 (1990).
353. T. Wada, T. Negami, and M. Nishitani, *Appl. Phys. Lett.* 64, 333 (1994).
354. D. J. Smith and L. D. Marks, *Philos. Mag. A* 44, 735 (1981).
355. K. Koga, H. Takeo, T. Ikeda, and K. Oshima, *Phys. Rev. B* 57, 4053 (1998).
356. K. Kimoto and I. Nishida, *J. Phys. Soc. Jpn.* 22, 940 (1967).
357. A. Renou and M. Gillet, *Thin Solid Films* 44, 75 (1977).
358. A. Gomez, P. Schabes-Retchkiman, and M. J. Yacaman, *Thin Solid Films* 98, L95 (1982).
359. W. Neumann and C. Gerstengarbe, in "Publications of the 11th Electron Microscopy Conference" (J. Heydenreich and H. Luppa, Eds.), p. 148. Dresden, 1984.
360. P. Schabes-Retchkiman, A. Gomez, G. Vazquez-Polo, and M. J. Yacaman, *J. Vac. Sci. Technol. A* 2, 22 (1984).
361. A. Renou and A. Rudra, *Surf. Sci.* 156, 69 (1985).
362. A. I. Kirkland, D. A. Jefferson, D. Tang, and P. P. Edwards, *Proc. R. Soc. London. Ser. A* 434, 279 (1991).
363. B. C. Smith and P. L. Gai, in "Proc. 8th Europ. Cong. Electron Microscopy" (À. Csanády, P. Röhlich, and P. Szabo, Eds.), p. 1151. Progr. Committee 8. Europ. Cong. Electron Microscopy, Budapest, 1984.
364. M. J. Yacamán and T. Z. Ocana, *Phys. Status Solidi A* 42, 571 (1977).
365. H. Hofmeister, H. Haefke, and M. Krohn, *J. Cryst. Growth* 58, 507 (1982).
366. T. Komoda, *Jpn. J. Appl. Phys.* 7, 27 (1968).
367. T. Komoda, *Bull. Jap. Inst. Metals* 7, 661 (1968).
368. L. D. Marks, *Ultramicroscopy* 18, 445 (1985).
369. J. M. Penisson and A. Renou, in "Proc. Europ. Electron Microscopy Meeting," IOP Publ., Bristol, 1988.
370. S. Giorgio, J. Urban, and W. Kunath, *Philos. Mag. A* 60, 553 (1989).
371. J.-O. Bovin and J.-O. Malm, *Z. Phys. D* 19, 293 (1991).
372. D. J. Wales, A. I. Kirkland, and D. A. Jefferson, *J. Chem. Phys.* 91, 603 (1989).
373. M. J. Yacaman, R. Herrera, A. Gomez, S. Tehuacanero, and P. Schabes-Retchkiman, *Surf. Sci.* 237, 248 (1990).
374. M. J. Yacaman, D. Romeu, S. Fuentes, and J. M. Domingues, *J. Chim. Phys.* 78, 861 (1981).
375. J. C. Barry, L. A. Bursill, and J. V. Sanders, *Australian J. Phys.* 38, 437 (1985).
376. P. L. Gai, M. J. Goringe, and J. C. Barry, *J. Microsc.* 142, 9 (1986).
377. D. G. Duff, A. C. Curtis, P. P. Edwards, D. A. Jefferson, B. F. G. Johnson, A. I. Kirkland, and D. E. Logan, *Angew. Chem., Int. Ed.* 26, 676 (1987).
378. M. Flüeli, R. Spycher, P. A. Stadelmann, P. A. Buffat, and J.-P. Borel, *Europhys. Lett.* 6, 349 (1988).
379. M. Flueli, R. Spycher, P. A. Stadelmann, P. A. Buffat, and J. P. Borel, IOP Publishing Ltd., Bristol (1988), p. 309.
380. D. A. Jefferson and A. I. Kirkland, in "Electron Beam Imaging of Non-Crystalline Materials" (K. Knowles, Ed.), p. 71. The Institute of Physics, Bristol, 1988.
381. M. Flueli, PhD Thesis, EPFL, Lausanne (1989).
382. P.-A. Buffat, M. Flüeli, R. Spycher, P. Stadelmann, and J.-P. Borel, *Faraday Discuss.* 92, 173 (1991).
383. W. Neumann and H. Hofmeister, in "Proc. Autumn School 1992 of the Int. Centre of Electron Microscopy" (J. Heydenreich and W. Neumann, Eds.), p. 183. MPI Halle, Halle, 1993.
384. J. Urban, *Cryst. Res. Technol.* 33, 1009 (1998).

385. D. K. Saha, K. Koga, and H. Takeo, *Eur. Phys. J. D* 9, 539 (1999).
386. J. A. Ascencio, M. Pérez, and M. J. Yacamán, *Surf. Sci.* 447, 73–80 (2000).
387. S. A. Nepijko, H. Hofmeister, H. Sack-Kongehl, and R. Schlögl, *J. Cryst. Growth* 213, 129 (2000).
388. J. Urban, H. Sack-Kongehl, K. Weiss, I. Lisiecki, and M.-P. Pileni, *Cryst. Res. Technol.* 35, 731 (2000).
389. J. A. Ascencio, M. Pérez-Alvarez, S. Tehuacanero, and M. José-Yacamán, *Appl. Phys. A* 73, 295 (2001).
390. H. Sauer and H. Sack-Kongehl, in "Abstracts Dreiländertagung Elektronenmikroskopie," p. 87. Innsbruck, 2001.
391. J. W. Kang and H. J. Hwang, *Nanotechnology* 13, 524 (2002).
392. I. Lisiecki, A. Filankembo, H. Sack-Kongehl, K. Weiss, M.-P. Pileni, and J. Urban, *Phys. Rev. B* 61, 4968 (2000).
393. A. Renou, J. M. Penisson, and M. F. Gillet, *Z. Phys. D* 12, 139 (1989).
394. W. Vogel, B. Rosner, and B. Tesche, *J. Phys. Chem. B* 97, 11611 (1993).
395. S. Ino, S. Ogawa, T. Taoka, and H. Akahori, *Jpn. J. Appl. Phys.* 11, 1859 (1972).
396. A. G. Dhere, R. J. de Angelis, P. J. Reucroft, and J. Bentley, *Ultramicroscopy* 18, 415 (1985).
397. W. Neumannn, J. Komrska, H. Hofmeister, and J. Heydenreich, *Acta Crystallogr. A* 4, 890 (1988).
398. C. Hintze, in "Handbuch der Mineralogie. I. Elemente und Sulfide," p. 236. Verlag von Veit und Co., Leipzig, 1904.
399. V. M. Kvasnytsya, *Dokl. Acad. Nauk* 7, 587 (1978).
400. C. H. Gammons, *Can. Mineral.* 34, 1 (1996).
401. G. Canizal, J. A. Ascencio, J. Gardea-Torresday, and M. J. Yacaman, *J. Nanoparticle Res.* 3, 475 (2001).
402. J. W. Kang and H. J. Hwang, *J. Phys. C* 14, 2629 (2002).
403. K. Hiraga, T. Oku, M. Hirabayashi, and T. Matsuda, *J. Mater. Sci. Lett.* 8, 130 (1989).
404. R. Hernandez, G. Diaz, A. Vazquez, Y. Reyesgasga, and M. J. Yacaman, *Langmuir* 7, 1546 (1991).
405. B. Pauwels, D. Bernaerts, S. Amelinckx, G. VanTendeloo, J. Joutsensaari, and E. I. Kauppinen, *J. Cryst. Growth* 200, 126 (1999).
406. A. E. Romanov, I. A. Polonsky, V. G. Gryaznov, S. A. Nepijko, T. Junghanns. and N. I. Vitrykhovski, *J. Cryst. Growth* 129, 691 (1993).
407. N. R. Avery and J. V. Sanders, *J. Catal.* 18, 129 (1970).
408. L. D. Marks and A. Howie, *Nature* 282, 196 (1979).
409. B. Moraweck and A. J. Renouprez, *Surf. Sci.* 106, 35 (1981).
410. M. J. Yacaman, S. Fuentes, and J. M. Domingues, *Surf. Sci.* 106, 472 (1981).
411. P. J. F. Harris, *Int. Materials Review* 40, 97 (1995).
412. M. J. Yacaman, M. Marin-Almazo, and J. A. Ascencio, *J. Mol. Catal. A* 173, 61 (2001).
413. H.-G. Boyen, G. Kästle, F. Weigl, B. Koslowski, C. Dietrich, P. Ziemann, J. P. Spatz, S. Riethmüller, C. Hartmann, M. Möller, G. Schmid, M. G. Garnier, and P. Oelhafen, *Science* 297, 1533 (2002).
414. S. T. Lee, N. Wang, Y. F. Zhang, and Y. H. Tang, *MRS Bull.* 24, 36 (1999).
415. J. Brandmüller and R. Claus, *Indian J. Pure Appl. Phys.* 26, 60 (1988).
416. J. Brandmüller and R. Claus, *Croat. Chem. Acta* 61, 267 (1988).
417. Y.-J. Jiang, L.-J. Liao, G. Chen, and P.-X. Zhang, *Acta Crystallogr. A* 46, 772 (1990).
418. F. Koberling, A. Mews, U. Kolb, I. Potapova, M. Burghard, and T. Basché, *Appl. Phys. Lett.* 81, 1116 (2002).
419. M. Gillet, *Surf. Sci.* 67, 139 (1977).
420. R. Monot, in "Proc. Int. Symp. on the Physics of Latent Image Formation in Silver Halides" (A. Baldereschi, Ed.), p. 175. World Scientific Publ., Singapore, 1984.
421. L. D. Marks, *Rep. Prog. Phys.* 57, 603 (1994).
422. M. J. Yacaman, J. A. Ascencio, H. B. Liu, and J. Gardea-Torresdey, *J. Vac. Sci. Technol. B* 19, 1091 (2001).
423. V. M. Kvasnytsya, *Miner. Zh.* 78, 45 (1997).
424. G. Rose and A. Sadebeck, in "Physikalische Abhandlungen der Königlichen Akademie der Wissenschaften zu Berlin," p. 85. F. Dümmler's Verlagsbuchhandlung, Berlin, 1876.
425. P. Groth, in "Die Mineraliensammlung der Universität Strassburg. Ein Supplement zu den vorhandenen mineralogischen Handbüchern" (Karl J. Tübner, Ed.), p. 4. Strassbourg, 1878.
426. J. Kourimsky and F. Tvrz, in "Bunte Welt der Minerale," p. 59. Artia-Verlag, Prague, 1977.
427. J. Strüver, *Atti della Reale Accademia dei Lincei* 275, 109; plate 1 (1877).
428. A. M. Dimkin and A. A. Permyakov, "Ontogenesis of Magnetite" Ural Naucn. Centr Akad. Nauk USSR, Sverdlovsk (1984), p. 188.
429. G. Bögels, J. G. Buijnsters, S. A. C. Verhaegen, H. Meekes, P. Bennema, and D. Bollen, *J. Cryst. Growth* 203, 554 (1999).
430. O. Kitakami, H. Sato, Y. Shimada, F. Sato, and M. Tanaka, *Phys. Rev. B* 56, 13849 (1997).
431. M. Gillet and M. Brieu, *Z. Phys. D* 12, 107 (1989).
432. D.-L. Lu and K.-I. Tanaka, *Surf. Sci.* 409, 283 (1998).
433. M. J. Yacaman, M. Avalos-Borja, A. Vazquez, S. Tehuacanero, P. Schabes, and R. Herrera, in "Mat. Res. Symp. Proc." (R. D. Bringans, R. M. Feenstra, and J. M. Gibson, Eds.), p. 371. Materials Research Society, Pittsburgh, 1990.
434. Z. R. Dai, S. H. Sun, and Z. L. Wang, *Surf. Sci.* 505, 325 (2002).
435. G. Rupprechter, PhD Thesis, Leopold-Franzens-Universität Innsbruck (1995).
436. G. Rupprechter, K. Hayek, and H. Hofmeister, *Nanostruct. Mater.* 9, 311 (1997).
437. K. H. Kuo, in "Proc. XI. Int. Cong. on Electron Microscopy," The Japanese Society of Electron Microscopy, Tokyo, 1986, p. 159.
438. F. Robinson and M. Gillet, *Thin Solid Films* 98, 179 (1982).
439. A. Renou and J. M. Penisson, *J. Cryst. Growth* 78, 357 (1986).
440. S. Giorgio and J. Urban, *Appl. Phys. Lett.* 52, 1467 (1988).
441. G. Bögels, H. Meekes, P. Bennema, and D. Bollen, *J. Phys. Chem. B* 103, 7577 (1999).
442. M. Audier and P. Guyot, *Acta Metall. Mater.* 36, 1321 (1988).
443. K. S. Vecchio and .B. Williams D, *Metall. Trans. A* 19, 2875 (1988).
444. A. Rodriguez, C. Amiens, B. Chaudret, M.-J. Casanove, P. Lecante, and J. S. Bradley, *Chem. Mater.* 8, 1978 (1996).
445. A. Recnik and D. Kolar, in "Proceedings of the 11th European Conference on Electron Microscopy, Dublin 1996" (Committee of Europ. Soc. of Microsc., Ed.), p. 710. Brussels, 1998.
446. A. Recnik and D. Kolar, in "Proceedings of the 11th European Conference on Electron Microscopy, Dublin 1996" (Committee of Europ. Soc. of Microsc., Ed.), p. 712. Brussels, 1998.
447. C. Solliard, P. Buffat, and F. Faes, *J. Cryst. Growth* 32, 123 (1976).
448. J. G. Pérez-Ramírez, M. J. Yacamán, A. Díaz-Pérez, and L. R. Berriel-Valdos, *Superlattices Microstruct.* 1, 485 (1985).
449. A. R. Thölén, *Phase Transitions* 24–26, 375 (1990).
450. M. Shimoda, T. J. Sato, A. P. Tsai, and J. Q. Guo, *Phys. Rev. B* 62, 11288 (2000).
451. R. D. Field and H. L. Fraser, *Mater. Sci. Eng.* 68, L17 (1984–1985).
452. M. Arita, N. Suzuki, and I. Nishida, *J. Cryst. Growth* 132, 71 (1993).
453. H. Q. Ye, D. N. Wang, and K. H. Kuo, *Ultramicroscopy* 16, 273 (1985).
454. G. D. Sukhomlin and A. V. Andreeva, *Phys. Status Solidi A* 78, 333 (1983).
455. D. Carron and R. Portier, in "Proc. 14th Int. Cong. Electron Microscopy" (H. A. Calderón Benavides and M. José Yacaman, Eds.), p. 13. IOP Publ., Bristol, 1998.
456. W. Luyten, G. van Tendeloo, S. Amelinckx, and J. L. Collins, *Philos. Mag. A* 66, 899 (1992).
457. M. Mikiyoshida, L. Rendon, S. Tehuacanero, M. J. Yacaman, *Surf. Sci.* 284, L444 (1993).

Encyclopedia of Nanoscience and Nanotechnology
www.aspbs.com/enn

Fluoride-Based Magnetic Heterostructures

D. H. Mosca, N. Mattoso, W. H. Schreiner

Departamento de Física, Universidade Federal do Paraná, 81531-990 Curitiba PR, Brazil

M. Eddrief, V. H. Etgens

Laboratoire de Minéralogie et Cristallographie de Paris, Université Paris VI et Paris VII, F-75252 Paris, France

CONTENTS

1. INTRODUCTION

The growth of epitaxial insulators shows a potential for applications in microelectronics devices such as semiconductor-on-insulator (SOI) structures, monolithic integrated circuits, improved gate insulators in metal–insulator–semiconductor field effect transistors (MISFETs), and passivation layers of compound semiconductor devices [1].

Fluorides have attracted much attention particularly by their ability to grow epitaxially on semiconductor crystalline substrates in opposition to other dielectrics, such as SiO_2 and SiN_x, which are mainly amorphous. In particular, group IIA fluorides such as CaF_2, SrF_2, and BaF_2 show many desirable properties such as low vapor pressure, easy molecular sublimation with the right stoichiometry, and good insulating behavior at room temperature.

The small lattice mismatch between CaF_2 and Si (+0.6%) at room temperature allowed various applications for CaF_2 thin films, the most important being on SOI devices. The integration of metals on CaF_2/Si heterostructures is equally well used in MIS devices, resonant tunneling diodes, and resonant tunneling transistors. We note that the availability of devices based on CaF_2/Si structures has sped up research in this interesting field in the past few years.

Progress in thin films on insulators and semiconductors is demanded by large scientific and business-related interests. The most significant support for these investigations of magnetism in thin films arises from the magnetic recording industry.

This review describes not only one specific class of metallic films integrated on silicon owing to CaF_2 epilayers, but also prototypical ferromagnetic thin films widely used in many applications. More general subjects are also discussed in this review such as magnetism, magnetotransport, and the mechanical behavior of thin films, thin-film technology, and surface-sensitive measurements.

This review is divided into four main topics, with three of them focused on CaF_2-based heterostructures, that is, CaF_2/Si(111), Fe/CaF_2 heterostructures, and metals/CaF_2/Si(111). An original contribution about the initial growth of Fe on CaF_2(111) has been included in Section 3. The focus on CaF_2(111) epitaxially grown on Si(111) is due to experimental results that clearly pointed to the (111) plane as the best experimental choice. An overview of the results obtained for different metals epitaxially grown on CaF_2 is discussed in Section 4. In Section 5, we address the reports involving Fe and metallic alloys grown on fluorides in general as well as multilayered and granular thin films. Finally, the role of physical properties on the overall behavior of the systems and the potential areas of technological application are discussed.

1.1. CaF_2

Calcium fluoride, or fluorite, is a very important industrial mineral with many different applications. It is used as a flux in steelmaking, as a source of fluorine in fluorinated water, as a precursor to hydrofluoric acid industrialization, and as a prime material for special optical lenses.

Fluorite is a popular mineral and is mined worldwide. It can be found in a variety of colors, depending on the hydrocarbon impurities. CaF_2 belongs to the halide mineral group,

ISBN: 1-58883-059-4/$35.00

is brittle with a hardness of 4 on the Mohs scale, and cleaves easily along [111] planes.

Fluorite is an isometric mineral with a face-centered cubic structure, *Fm3m* symmetry, and lattice parameter of $a_0 = 0.5462$ nm. Its density is 3.18 g/cm^3 at 20 °C, and the refractive index is between 1.43 and 1.44 for the visible spectrum. The crystal structure of CaF_2 is shown in Figure 1.

The electronic structure of fluorite is dominated by a wide gap of 12.1 eV at room temperature, making it a very good insulator. Calcium fluoride is clearly ionic with the strongly electronegative fluorine capturing two electrons from the cationic calcium valence shell. Its valence band is dominated by fluorine p states, whereas, on the contrary, the conduction band is formed mostly by strongly mixed calcium s and d states [2].

CaF_2 is considered the ideal material by researchers working on semiconductor-on-insulator (SOI) devices due to both its crystal structure and its insulating characteristics [3].

1.2. CaF_2/Si

The use of naturally grown silicon oxide on silicon for insulating purposes has been a natural choice employed over and over. The problem with silicon oxide is its amorphous structure, which precludes postgrowth of heteroepitaxial films. In fact, in several areas of scientific and technological interest, the growth of crystalline epitaxial insulators is preferred: SOI structures, gate insulators, three-dimensional structures, metal–insulator–semiconductor (MIS) structures, and the integration of magnetic films onto silicon are among those that should be mentioned. For the demanding optoelectronics, electrical transport, and magnetic devices, high-quality epitaxial insulators are needed in combination with sequential metal, semiconductor, or magnetic epitaxial layers.

Surveying the potential epitaxial insulators on silicon, the group IIA fluorides arise as natural candidates [1, 3], and, among these, fluorite is by far the most studied material.

The lattice match of face-centered cubic fluorite with silicon (cubic *Fd3m* diamond structure) is very good, but for growth by molecular beam epitaxy (MBE), epilayer strain is unavoidable in the process. This is, in fact, produced in the cooling cycle after CaF_2 growth, due to the large difference in thermal expansion coefficients between CaF_2 and Si [1], which eventually induces cracking in thick epilayers [3].

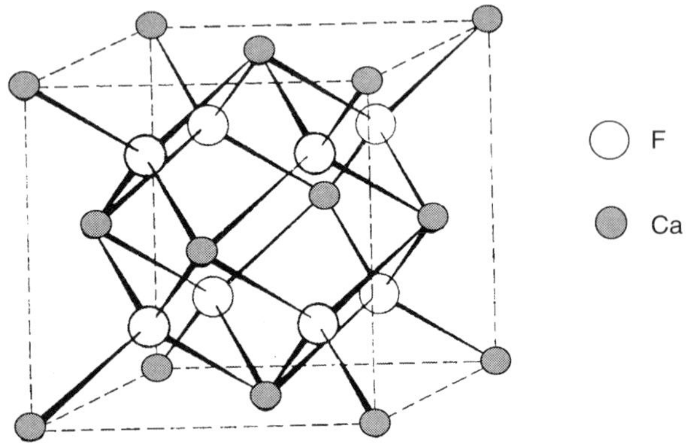

Figure 1. Crystal structure of CaF_2.

Furthermore, the growth of good-quality heteroepitaxial films of CaF_2 on Si(111) substrates has been demonstrated, leading to structures grown layer by layer with adequate smoothness. This aspect and the optimal growth and thickness conditions of epitaxial fluorite will be reviewed in Section 2.

1.3. Fe/CaF_2/Si

Several different metals such as Cu, Pt, Al, Co, and Fe, among others, have been deposited with reasonable epitaxial results. Here we stress the magnetic Fe heteroepitaxial epilayers deposited on CaF_2/Si(111).

The driving force behind such studies of MIS structures concerns the production of hybrid ferromagnetic–semiconductor devices. The last decade has seen tremendous development of solid-state electronics and optoelectronics, while the circuit elements involving magnetic materials have been under-researched. This trend is changing rapidly due to the discovery and application of the giant magnetic resistance (GMR) effect, together with the upcoming spintronics field, where the conduction electron spin can be manipulated [4, 5].

The direct deposition of ferromagnetic transition metals on silicon produces unwanted silicide formation even at room temperature [6]. The operation of this hybrid structure at higher temperatures could lead to irreversible structural and phase changes. An alternative solution would be the growth of a diffusion blocking layer between the silicon and the magnetic metallic films, with CaF_2 displaying the required properties for moderate temperature growth and operation.

In the last decade, several works were published on the structural, thermal, electrical, magnetic, and mechanical aspects of epitaxial Fe deposited on CaF_2/Si. These efforts will be reviewed in Section 3.

An overview on the other metals successfully grown on CaF_2 are described and discussed in Section 4.

1.4. Fe on Other Fluorides

Although the bulk of this review is selective with respect to the more common research of Fe epitaxially grown on CaF_2/Si(111), we include a discussion of other fluoride substrates with characterizations that can be found in the literature.

The deposition of Fe thin films on LiF crystals was investigated in the 1970s without the tools actually required for interface characterization that are currently employed in surface physics and materials science. More recently, the initial growth stage of Fe on group IA fluorides (LiF, KF, RbF, ...) has been investigated by numerous methods, revealing rather poorly defined interfaces together with a strong interfacial chemistry. The growth of Fe films and clusters on group IIA fluorides (MgF_2, CaF_2, BaF_2, ...) reveals, on the other hand, well-defined interfaces and, due to this, they have received more attention.

Chemical sensitivity, particularly to atmospheric exposure, precludes the investigation of some fluoride compounds such as CdF_2 as suitable substrate materials.

Certainly the most interesting fluoride compounds are the transition metal fluorides of the form MF_2 (M = Mn, Fe, Co,

Zn, ...), which have been used as model systems to study finite-size and other phase transition phenomena because of their well-known Ising-like magnetic interaction and well-defined magnetic and crystalline structures. Perhaps the most interesting system is Fe/FeF_2, which has received much attention due to a number of novel and surprising magnetic properties. Another very interesting system consists of magnetic clusters embedded in a MgF_2 matrix, which has been pointed to as the granular system with the largest tunneling magnetoresistance. All of these topics, together with their detailed literature references, will be described and discussed in Section 5.

1.5. Technological Perspectives

The growth of heteroepitaxial $Fe/CaF_2/Si$ structures has been shown to be feasible. Less successful seems the tailoring of Fe/CaF_2 multilayers where the growth of CaF_2 on Fe represents the major difficulty. This alternate deposition of Fe and CaF_2 multilayers leads to granular films of iron and calcium fluoride, even at room temperature, with intermixing at the interfaces. This process is followed by the formation of several FeF_3 compounds involving Fe^{3+} ions.

The basic research aspects of heteroepitaxial $Fe/CaF_2/Si$ structures have been advanced in the last decade, but so far no device has been produced to our knowledge. The technological perspectives nevertheless look bright for this type of MIS system in the future.

Many other fluorides such as LiF, MnF_2, and FeF_2 have been successfully used as substrates for the deposition of high-quality Fe thin films. Each can provide useful information, but the best candidates are MnF_2 and FeF_2 because of their high crystal quality in combination with a promising avenue for prototypical spin-based device exploration.

2. EPITAXIAL CaF_2 ON Si(111)

In 1989, Schowalter and Fathauer [3] published a complete review of the growth of CaF_2 on Si(111) where various aspects of the epitaxial growth of films were analyzed, such as the crystalline quality of epitaxial layers of CaF_2, the surface energy, the cracking of epitaxial layers, and the CaF_2/Si interface.

With the democratization of surface study techniques and spectroscopic and structural techniques based on synchrotron radiation in the 1990s, it was rapidly possible to obtain a deeper understanding of the initial growth stages of the epitaxial films of CaF_2 on Si single-crystal substrates. However, the most studied subsystem to date is $CaF_2/Si(111)$. Two main reasons can be pointed out for this choice among others. First, the (111) plane of both CaF_2 and Si is the densest plane, facilitating epitaxial growth. Second, the surface energy of the (111) CaF_2 plane is lower, in comparison with planes of other directions [3], thus helping the condensation. CaF_2 films grown on the Si(111) surface have smoother surfaces and a better crystalline quality than those grown on the Si(100) surface.

In virtue of this scenario, the initial growth stages and structural quality of epitaxial films of CaF_2 on Si(111) have been extensively studied, while only a few reports are found for Si surfaces with different orientation.

Sugiyama and Oshima [1] have published a review on the MBE growth of fluorides where the initial growth stages, the growth mode, and the type of epitaxy are extensively discussed in the light of techniques such as X-ray photoelectron spectroscopy (XPS), X-ray photoelectron diffraction (XPD), near-edge X-ray absorption fine structure (NEXAFS), medium-energy ion scattering (MEIS), X-ray standing waves (XSW), scanning tunneling microscopy (STM), atomic force microscopy (AFM), and low-energy electron microscopy (LEEM). The interfacial structures of $CaF_2/Si(111)$ have been studied by high-resolution transmission electron microscopy (HRTEM) and X-ray diffraction.

Next, we will briefly describe the initial nucleation stage of CaF_2 on Si(111) for two of the most important growth conditions.

2.1. Initial Growth at Low Temperature

Two types of epitaxies are reported for CaF_2 films grown on Si(111), depending on the substrate temperature [7]: type A below 500 °C and type B above 500 °C. Type A epitaxy is characterized by the coincidence in the surface plane between the $[1\bar{1}0]$ and $[\bar{2}11]$ axes of the CaF_2 and Si, whereas, for type B, these axes are exchanged between film and substrate, which corresponds to a 180° film rotation with respect to type A epitaxy. The two distinct epitaxial relations are shown in Figure 2.

Denlinger et al. [8] have shown by *in-situ* XPS, XPD, and low-energy electron diffraction (LEED) that, for CaF_2 films grown at 450 °C, a fraction of the CaF_2 molecules are dissociated and can consequently induce reactions between Ca—Si and Si—F at the interface. It has also been observed that the Si—F bonds inhibit the lateral growth of the first monolayer, resulting in nucleation centers for the second monolayer. The second monolayer grows preferentially on the first one, even if it remains discontinuous, promoting a three-dimensional (3D) growth that results in worse epitaxial quality as reported by Wollschläger and Meier [9].

More recently, Wang et al. [10] have shown by AFM that the deposition at 410 °C leads to the coexistence of type A and type B epitaxies. In summary, the growth of CaF_2 films on Si(111) at substrate temperatures below 500 °C is initially

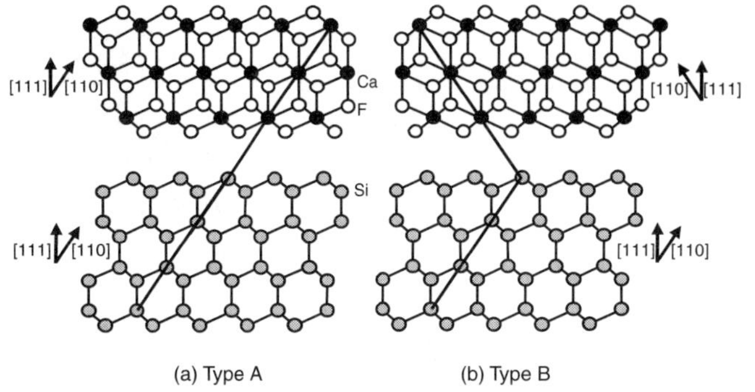

Figure 2. Schematic drawings of the two distinct epitaxial relations of the CaF_2 on Si(111): (a) type A orientation in which the orientation of the film is identical to that of the Si; (b) type B orientation in which the orientation of the film is rotated by 180° about the surface normal of the substrate.

two dimensional (2D), with the increase in the thickness becoming 3D-like with a predominant type A epitaxy.

2.2. Initial Growth Stage at High Temperature

Denlinger et al. [8] have also studied the initial growth of CaF_2 films on Si(111) at higher substrate temperatures and concluded that, for substrate temperatures above 600 °C, Ca—Si bonds and CaF occur at the interface as a consequence of the CaF_2 dissociation. Another interesting result is the nucleation of the second monolayer, which only starts after the Si surface is totally covered. In contrast with low-temperature experimental findings, no Si—F bonds were observed, corroborating the hypothesis that Si—F bonds reduce the diffusion length at the surface. Additionally, X-ray photoemission diffraction results show that the type B epitaxy of the successive layers is associated with Ca—F bonds.

Huang et al. [11] elaborated a structural model for the CaF_2/Si(111) interface with an intermediate monolayer of $CaSi_2$ between the substrate and the CaF_2, owing to high-resolution grazing incidence X-ray diffraction and X-ray reflectivity measurements performed at synchrotron radiation. This completely agrees with the results of Denlinger et al. [8] for high substrate temperatures; that is, the formation of Ca—Si bonds and CaF occurs at the interface as a consequence of the CaF_2 dissociation.

LEED, XPS, and STM measurements were recently employed by Sumiya [12] to study the phase evolution of the CaF_2 surface reconstructions as a function of the temperature of the substrate during growth and thermal annealing with STM images showing these surface reconstructions. XPS also indicates that the surface at high temperatures consists of CaF. The STM images are clear and disclose the formation of a structure of CaF rows at 700 °C. Surface analyses show that the surface of the islands at 640 °C possesses a (1×1) reconstruction and the rows are proceeding from a (3×1) reconstruction.

Thus, a type B epitaxy of CaF_2 on Si(111) is found under low CaF_2 growth rates and at high deposition temperatures (>600 °C) due to the dissociation of the CaF_2 to CaF at the Si surface. This promotes the formation of a CaF/Si interface with Ca directly bonded to Si to form a $CaSi_2$ compound. For CaF_2 films grown above 700 °C, the row structure is observed, resulting in better structural quality. These results show the influence of the morphology of the surface on the quality of the films [1, 12].

2.3. CaF_2/Si(111) Interface

The initial stage of epitaxial growth of CaF_2 on Si(111) has been studied by several techniques. Although the reported results are partly inconsistent with each other, it is clear that the chemical reaction at the initial coverage of CaF_2 on Si(111) and the interface structure depend strongly on the deposition conditions (see [1] and references therein).

There is a consensus that both Ca and F bond to Si at the interface at high deposition temperatures (>600 °C). The proposed structural model of a stoichiometric CaF_2/Si(111) interface is based on the existence of Si—Ga and Si—F bonds. However, it is thought that the dissociation of CaF_2 molecules to CaF at the interface may promote a predominance of Si—Ca bonds and the resulting depletion of F atoms. The Ca atoms are mainly situated at the T4 bonding sites; that is, Ca is located on the threefold site on top of a second layer of Si atoms [1]. The CaF_2/Si interface is atomically abrupt since a single layer of film forms on the Si surface until the atomic terraces have been filled. The nucleation of the second layer only starts after the completion of the first layer. The epitaxial film has a predominant type B orientation. When the growth is performed below 500 °C, there is both more F and more disorder at the interface. It is also generally stated that at a lower substrate temperature the critical size of the one-layer CaF_2 island is rather small, so that the Si substrate is not fully covered when nucleation of the second layer starts.

There is no conclusive report on the structural changes occurring at the interfaces during postdeposition annealing of the CaF_2/Si heterostructures.

3. Fe/CaF_2 HETEROSTRUCTURES

The number of publications on Fe/CaF_2 heterostructures grown on Si(111) in the last years attests to the growing interest in this system. In this section, the growth process, with special emphasis on the importance of the deposition temperature on microstructure, on chemical stability, and on magnetic behavior, will be drawn.

3.1. Growth and Epitaxy

Epitaxial α-Fe films grown by the electron beam evaporation technique on a CaF_2 epilayer on Si(111) substrates show the [110] orientation along the growth direction. The structure of nanometer-thick Fe films has been studied using several X-ray diffraction techniques and scanning and transmission electron microscopy as well [13].

Grazing incidence diffraction χ-scans confirm that the growth of α-Fe films on CaF_2(111) proceeds along the $[1\bar{1}0]$ direction with an in-plane relation such that the α-Fe$[1\bar{1}0]$ direction is parallel to the $CaF_2[1\bar{2}1]$ direction, meaning a 30° rotation of α-Fe(110) planes with respect to CaF_2(220) ones. Two main defects could be identified in the epilayers. First, twins are clearly apparent for samples deposited above room temperature. Second, regular dislocation lines appear on some sample regions. Scanning electron microscopy micrographs of the shiny metallic surface of the samples reveals a regular geometric pattern, which constitutes isosceles triangles interpreted as dislocation lines emerging at the surface. The planes normal to the α-Fe(110) plane, which have a mutual angular relationship, were identified as being due to the following Miller indices: $[11\bar{2}]$, $[1\bar{1}2]$, and $[1\bar{1}0]$. Thus, the regularity of dislocation lines emerging at the film surface strongly suggests the structural quality of the epitaxial Fe films.

By using geometrical arguments, the $[110]_{\alpha\text{-Fe}} \| [111]_{CaF_2}$ and $[1\bar{1}0]_{\alpha\text{-Fe}} \| [1\bar{2}1]_{CaF_2}$ epitaxial relations could be explained as a particular case of the Nishiyama–Wassermann epitaxy [14]. Taking into account only the lattice parameter, one readily would expect the same orientation for Fe grown on Si and on CaF_2. This is not verified, and despite the lattice parameter of CaF_2 and Si differing by only 0.6% at room temperature, α-Fe grows on Si(111) with the [111] axis along the growth direction [15].

After Schowalter and Fathauer [3], the difference in surface free energy, estimated at 0.45 J m^{-2} for $CaF_2(111)$ and at 1.24 J m^{-2} for Si(111), can be used to explain these two distinct orientations of Fe films on both materials. Another argument that can be equally used to explain this difference is the formation of a silicide interlayer between Fe and Si by a solid-state interdiffusion [4, 16], even at room temperature (RT). This interfacial chemistry can help to accommodate the large lattice misfit.

Over the last few years, the initial growth of metallic Fe on CaF_2, notably over thin $CaF_2(111)$ epilayers on Si(111), has been extensively studied [17, 18]. According to ultrahigh vacuum scanning transmission electron microscopy with nanometer resolution, the initial growth of Fe proceeds by three-dimensional islanding on CaF_2 substrates. A very high nucleation density has been observed on both bulk and thin CaF_2 epilayers and this for a wide range of substrate temperatures, from 20 to 300 °C. This indicates a limited surface diffusion with nucleation occurring at defect sites (notably fluorine sites) with a high trapping energy ($\sim$1 eV). Although the Fe island distribution is correlated to chemical or defect dominating growth, Fe is not expected to form compounds with CaF_2. The analyses of particle size revealed an almost monodisperse distribution of Fe islands on $CaF_2(111)$ surfaces with no statistical differences between diameters and spatial distribution of islands for growth from RT up to 300 °C. More than 85% of the population of Fe islands have a diameter of 2.0 nm. The number of Fe islands per unit area was found to be 7×10^{12} islands/cm^2 with an average separation between Fe island centers around 3.7 nm. The stage of Fe island nucleation on CaF_2/Si(111) surfaces held at room temperature occurs at least for coverage between 0.4 and 1.3 ML Fe (1 ML = 7.7×10^{14} atoms/cm^2). Another interesting result concerns the coalescence and/or coarsening that have a weak influence on secondary nucleation which seems to occur via surface diffusion and incorporation in the sides of preexisting islands. A more in depth analysis of the Fe initial growth stage will be given later with recent unpublished results.

A set of samples has been prepared in an MBE multigrowth chamber. First, a 100-nm-thick CaF_2 epilayer has been grown on a hydrogen-terminated Si(111) surface. These $CaF_2(111)$/Si substrates were introduced in the growth chamber and annealed at 250 °C for 4 h to promote structural ordering. The substrate temperature was decreased to 150 °C, and once the temperature was stabilized, Fe was deposited at a rate of around 0.1 nm/min. In order to check the chemical stability, Fe films were also prepared with substrate temperatures of 450 °C, which is higher than the decomposition temperature of CaF_2.

In-situ X-ray photoemission spectroscopy (XPS) has been used to probe the core levels of the chemical species (Mac II spectrometer). At several stages of deposition, the samples were cooled to room temperature and transferred to an XPS chamber under ultrahigh vacuum (UHV). As expected, the slow increase in the Fe/Ca and Fe/F intensity ratio indicates that the initial growth of Fe proceeds by islanding on the CaF_2 surface.

In-situ reflection high-energy electron diffraction (RHEED) measurements were performed using a variable-incidence Staib RHEED system operating at 12 keV under grazing angle. For quantitative analyses, the diffraction pattern was recorded and analyzed using a high-sensitivity charge-coupled device (CCD) camera linked to a computer image-processing system. Figure 3a and c shows the RHEED diagrams of the $CaF_2(111)$ surface taken along the $[2\bar{1}\bar{1}]_{CaF_2}$ and $[1\bar{1}0]_{CaF_2}$ azimuths, respectively, immediately prior to the start of Fe growth. These two directions, which differ from each other by 30°, reveal a flat and well-structured $CaF_2(111)$ surface. Figure 3b and d shows the RHEED pattern diagrams along the $[2\bar{1}\bar{1}]_{CaF_2}$ and $[1\bar{1}0]_{CaF_2}$ azimuths, respectively, after the growth of 1.8 nm equivalent thickness of Fe deposited at 150 °C. The beginning of the Fe growth is characterized by a two-dimensional diffraction that fades into an increasing background. Once the Fe coverage increases, the Fe spots become visible and more elongated with thickness, within a complex RHEED pattern that is associated with small fractions of aligned epitaxial regions.

Partial indexing of the Fe diffraction spots of the RHEED diagram, taken along the $[2\bar{1}\bar{1}]_{CaF_2}$ and $[1\bar{1}0]_{CaF_2}$ azimuths, is given in Figure 3b and d. According to the RHEED diagram taken along the $[2\bar{1}\bar{1}]_{CaF_2}$ azimuth shown in Figure 3b, the initial nucleation of Fe on $CaF_2(111)$ exhibits two domains with the same epitaxial relationship along the growth direction $[110]_{\alpha\text{-Fe}} \| [111]_{CaF_2}$, but with two distinct in-plane matching planes: $[1\bar{1}0]_{\alpha\text{-Fe}} \| [2\bar{1}\bar{1}]_{CaF_2}$ and $[1\bar{1}2]_{\alpha\text{-Fe}} \| [2\bar{1}\bar{1}]_{CaF_2}$.

The RHEED diagram taken along the $[1\bar{1}0]_{CaF_2}$ azimuth shown in Figure 3d reveals twinning domains with a $[111]_{\alpha\text{-Fe}} \| [111]_{CaF_2}$ epitaxial relationship along the growth direction and two distinct in-plane matching planes: $[1\bar{1}0]_{\alpha\text{-Fe}} \| [1\bar{1}0]_{CaF_2}$ and $[\bar{1}10]_{\alpha\text{-Fe}} \| [1\bar{1}0]_{CaF_2}$. As usual, in the case of Fe epitaxied directly on a Si(111) substrate [19], it can adopt two different stacking sequences along the (111) direction during growth: one aligned (type A) and the other rotated by 180° with respect to the surface normal of the substrate (type B). The symmetry of the diffraction of each crystal domain is clearly shown in Figure 3d. These two α-Fe domain orientations have a twinning epitaxial relationship.

Therefore, the RHEED pattern diagrams reflect a complex situation, including multidomain epitaxy. There are at least three epitaxial modes of the α-Fe on $CaF_2(111)$. We summarize the domain epitaxial relationships at low Fe coverage as follows [20]:

Type A epitaxy:

$[111]_{\alpha\text{-Fe}} \| [111]_{CaF_2}$ along the growth direction
$[1\bar{1}0]_{\alpha\text{-Fe}} \| [1\bar{1}0]_{CaF_2}$ in-plane matching of planes

Type B epitaxy:

$[111]_{\alpha\text{-Fe}} \| [111]_{CaF_2}$ along the growth direction
$[\bar{1}10]_{\alpha\text{-Fe}} \| [1\bar{1}0]_{CaF_2}$ in-plane matching of planes

Type C epitaxy:

$[110]_{\alpha\text{-Fe}} \| [111]_{CaF_2}$ along the growth direction
$[1\bar{1}2]_{\alpha\text{-Fe}} \| [2\bar{1}\bar{1}]_{CaF_2}$, $[1\bar{1}2]_{\alpha\text{-Fe}} \| [1\bar{1}0]_{CaF_2}$, and $[1\bar{1}2]_{\alpha\text{-Fe}} \| [2\bar{1}\bar{1}]_{CaF_2}$ with in-plane matching of planes

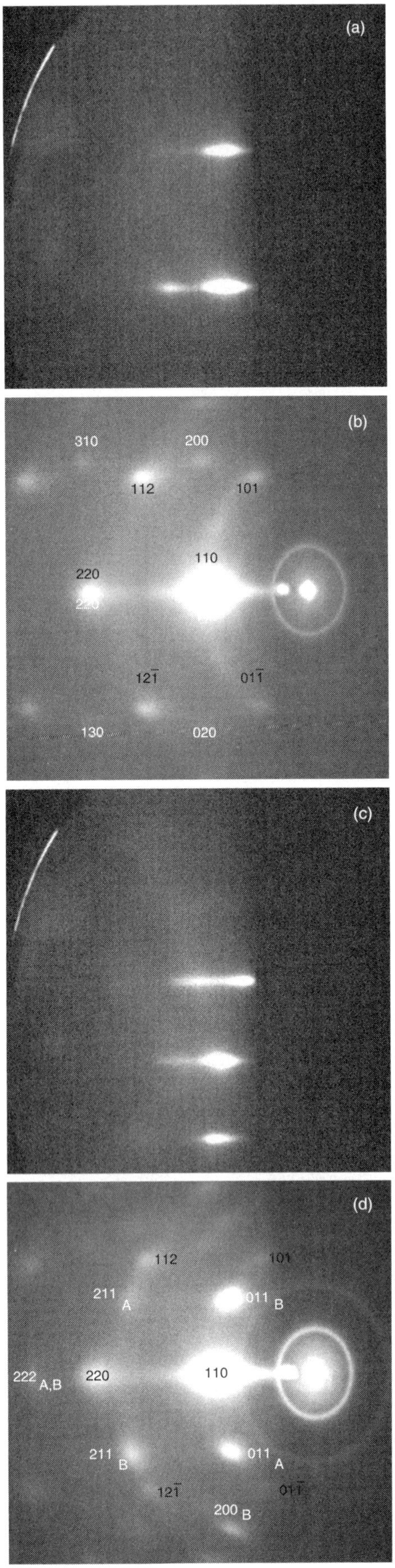

It is worth noting that the RHEED analyses along both $[2\bar{1}\bar{1}]_{CaF_2}$ and $[1\bar{1}0]_{CaF_2}$ directions reveal a so-called type C epitaxy with three in-plane matching planes rotated 30° from each other. This type C epitaxy becomes predominant when the Fe coverage increases, as will be shown later.

The in-plane lattice parameter evolution of the growing Fe film has been equally monitored through the RHEED profiles using a high-sensitivity CCD camera. Streak positions and distances have been measured as a function of time by making intensity profiles along a predefined region. The in-plane lattice parameter evolution along the $[1\bar{1}0]_{CaF_2}$ azimuth is presented in Figure 4 as a function of the equivalent Fe coverage (deposition time is given in the top axis). The streak spacing between $CaF_2(0,\bar{1})$ and $(0,1)$ has been used as reference lattice parameter to extract the misfit for deposition at 150 °C. From the beginning of the growth, a slow evolution from $(1/\sqrt{2})a_{CaF_2} = 0.3863$ nm to a value of $a_{interface} = 0.4129$ nm is observed, followed by a smooth decrease to a value of $a_{bulk} = 0.4084$ nm, which is 0.8% greater than the Fe bulk value $\sqrt{2}a_{\alpha\text{-Fe}} = 0.4053$ nm.

When the Fe coverage increases, the domains formed by the stacking of planes with higher atomic density overcome the twinned domains, leading to large crystallites (mosaic crystals) with the $[110]_{\alpha\text{-Fe}}$ orientation axis aligned along the $[111]_{CaF_2}$ growth axis. Figure 5 shows a conventional θ–2θ scan in Bragg–Brentano geometry of the (110) Bragg reflection of the samples protected with a 10-nm-thick amorphous ZnSe cap layer to prevent reaction with air.

In summary, structural and XPS analyses do not indicate the existence of a relevant interfacial chemistry. Indeed, the multidomain epitaxial regions are still observed. The characteristic growth process and structural relaxation of the Fe on $CaF_2(111)$ seems to be controlled not simply by Fe film thickness, but instead by the balance between the intrafilm interaction strength and the Fe–CaF_2 interaction strength. These features may strongly affect the magnetic response at low coverage of Fe. A more detailed study is necessary to clarify these questions.

3.2. Thermochemical Stability

Chemical reaction between Fe and CaF_2 is commonly rejected based on thermochemical considerations [18]. The formation of iron silicides promoted by pinholes in the thin CaF_2 epilayer has been suggested to occur rather than a chemical reaction between Fe and CaF_2. However, recent results on multilayered and granular Fe thin films and fluorides (CaF_2 and LiF), grown at room temperature, have

Figure 3. RHEED patterns of the 100-nm-thick $CaF_2(111)$ surface taken along (a) $[2\bar{1}\bar{1}]_{CaF_2}$ and (c) $[1\bar{1}0]_{CaF_2}$ azimuths, respectively. The RHEED patterns taken along the same azimuths as (a) and (c) but after the deposition 1.8 nm equivalent Fe coverage are shown in (b) and (d), respectively. Single-crystal spot diffraction patterns of the epitaxial domains were indexed by following the in-plane matching planes according to the notation: $[1\bar{1}2]_{\alpha\text{-Fe}}\|[2\bar{1}\bar{1}]_{CaF_2}$ (white) and $[1\bar{1}0]_{\alpha\text{-Fe}}\|[2\bar{1}\bar{1}]_{CaF_2}$ (black) for type C epitaxy with $[2\bar{1}\bar{1}]_{CaF_2}$ azimuth, and $[2\bar{1}\bar{1}]_{\alpha\text{-Fe}}\|[1\bar{1}0]_{CaF_2}$ (black) for type C epitaxy, $[1\bar{1}0]_{\alpha\text{-Fe}}\|[1\bar{1}0]_{CaF_2}$ (white with subindex A), and $[\bar{1}10]_{\alpha\text{-Fe}}\|[1\bar{1}0]_{CaF_2}$ (white with subindex B) for the epitaxics A and B, respectively, with $[0\bar{1}1]_{CaF_2}$ azimuth.

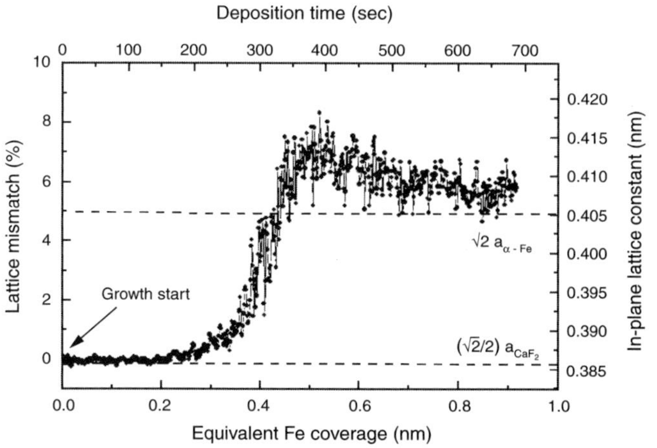

Figure 4. Lattice parameter evolution during growth of a small amount of Fe at 150 °C as visualized by RHEED. The reference was taken along the $[1\bar{1}0]_{CaF_2}$ azimuth on in-plane lattice parameter of CaF_2.

shown the intermixing at the interfaces and the formation of imperfect compounds involving Fe^{3+} ions [21, 22].

The chemical mechanism of compound formation is still elusive. However, it seems that CaF_2 films in contact with a moist atmosphere with subsequent adsorption of small molecules (O_2, H_2O, CO_2, N_2, etc.) are important agents for causing a degenerative effect on CaF_2 epitaxial films and subsequently promoting the chemical reaction between Fe and CaF_2. Recent experiments on $CaF_2(111)$ surfaces damaged by radiation with low-energy electrons have shown that the damaged surface is highly reactive with O_2 and CO_2 [23]. In general, the structure of the fluorine-terminated surface is almost free of defects. Exposing this surface to an oxygen dose of some hundred langmuir, however, produces defect formation, each one occupying an ionic site that has been interpreted as hydroxyl groups substituting surface fluorine ions [24]. Therefore, these studies seem to indicate that a perfect fluoride surface is chemically inert and will not be degraded by exposure to oxygen or water, but preexisting surface defects in the form of vacancies or F centers can play an important role in surface reactivity.

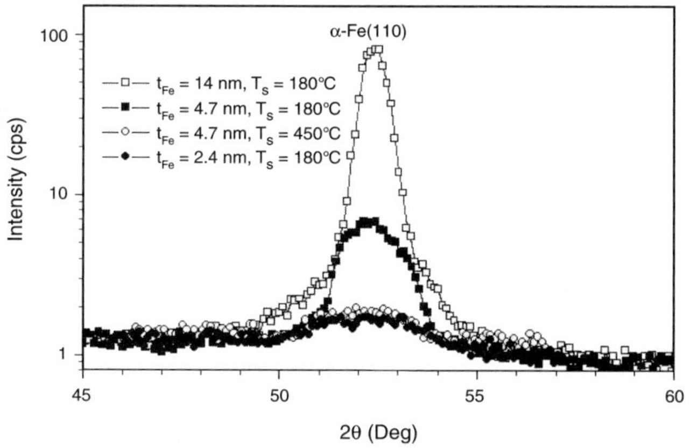

Figure 5. X-ray diffraction pattern in θ–2θ scan of the (011) Bragg reflection of the samples protected with a 10-nm-thick amorphous ZnSe cap layer.

Now, let us address the dependence of the crystal quality on the Fe deposition temperature. According to θ–2θ diffraction experiments performed for 30-nm-thick Fe films, the α-Fe(110) reflection increases in intensity steadily with deposition temperature up to 450 °C, whereas the α-Fe(200) signal becomes weaker. Concomitantly, the full width at half-maximum associated with the Fe(110) reflection decreases from 5.9° to 2.1° (or 64%) when the deposition temperature increases from 150 to 450 °C, respectively. This clearly indicates that the crystal quality increases with temperature. The best crystalline quality concerning grain size and stress release is found for samples deposited at 450 °C [25].

The thermochemical stability of the material obtained by sequential deposition of discontinuous Fe and CaF_2 thin layers at room temperature definitely plays an important role. The reactivity of Fe and F is well documented [26–30] and the intermixing of Fe and CaF_2 at the interfaces is also strongly demonstrated [21, 22, 31].

Composite films were deposited as multilayers onto 40-nm $CaF_2(111)$ epilayers previously grown on HF-etched Si(111) wafers in BALZERS UMS 500P electron beam equipment with a base pressure of 2×10^{-8} mbar. For comparative analyses, 30-nm-thick Fe films were grown at 450 °C on $CaF_2(111)$/Si(111). The samples discussed here follow the convention where the subindices placed after the parentheses indicate the repetition number of Fe and CaF_2 layers in each sample. Such samples are denoted as follows:

Sample A. $(CaF_2$ 3.4 nm/Fe 1.5 nm$)_{20}$/$CaF_2(111)$ 40nm/Si(111)
Sample B. $(CaF_2$ 3.4 nm/Fe 3.2 nm$)_{10}$/$CaF_2(111)$ 40 nm/Si(111)
Sample C. Fe 30 nm/$CaF_2(111)$ 40 nm/Si(111)

Transmission electron microscopy (TEM) reveals that sample A is composed almost entirely of isolated Fe-rich granules with an average size of 20 nm, dispersed in a disordered CaF_2 matrix. For sample B, the plan-view micrographs show Fe-rich granules forming a percolation network where magnetic closure structures are readily facilitated. Selected-area electron diffraction (SAED) of samples A and B exhibits diffraction patterns consisting of rings that correspond to ferric trifluoride (FeF_3) reflections together with a $Ca(OH)_2$–CaF_2 mixture. Several crystalline forms of FeF_3 are present in these samples: rhombohedral FeF_3 (r-FeF_3) and hexagonal tungsten bronze FeF_3 (HTB-FeF_3) species, namely, anhydrous HTB-FeF_3 and HTB-$FeF_3(H_2O)_{0.33}$ [28–30]. TEM analyses of sample C reveal an excellent structural quality with respect to the first two samples with predominant reflections of α-Fe(110) planes stacking along the growth direction.

In this way, the complex chemical reactions involving ferric trifluoride by-products must be taken into account when considering future applications of Fe/CaF_2 structures.

3.3. Magnetic Properties

3.3.1. Epitaxial Fe on CaF_2

The magnetic properties of Fe films on $CaF_2(111)$ were investigated by both *in-situ* and *ex-situ* methods. It has been demonstrated, for example, that *in-situ* surface magnetooptic

Kerr effect (SMOKE) measurements are unsuitable for the detection of a ferromagnetic signal in the island array. The ferromagnetic response cannot be observed for a maximum applied field of 1 kOe if the island diameter is smaller than 2.8 nm. Besides, the SMOKE analyses cannot distinguish between superparamagnetic and nonmagnetic behavior. An alternative approach was to cover the Fe island array with a protective silver capping for *ex-situ* magnetic measurements. In these measurements, a superparamagnetic response was observed [32].

To clarify the magnetic response in a better way, we have performed magnetic measurements on the set of samples discussed in Section 3.1, which were cooled to room temperature after deposition and capped with a 10-nm-thick amorphous ZnSe layer to prevent reaction with air. This procedure has been tested as being efficient to protect samples for several weeks in air. As an additional precaution, samples were stored in a N_2-saturated atmosphere and quickly prepared to the measurements in an alternating gradient force magnetometer as well as in a SQUID magnetometer. This amorphous cap layer is less constrained to the Fe islands than the Ag cap layer from the point of view of mechanical relaxation and strain-induced effects.

Figure 6 shows the saturation magnetization moment per square centimeter (m_s) versus the Fe thickness for a set of three samples grown at a deposition temperature of 180 °C as well as for one sample grown at 450 °C. The same deposition rate of about 0.1 nm/min has been used for all samples. Figure 6 shows that m_s increases linearly with thickness and one can find a slope ($M_s = 1700$ G) that is in very good agreement with bulk Fe magnetization at room temperature ($M_s = 1710$ G). The extrapolation of this line intercepts the thickness axis at a value of 0.48 nm. A crude interpretation will lead to the existence at the interface of a magnetically dead layer, which includes a number of Fe atoms corresponding to a 0.48-nm-thick layer. The deposition performed at the higher temperature of 450 °C also suggests the presence of an equivalent dead layer. Since the slope extracted from the data is quite close to the bulk Fe value, the dead layer may result from a less accurate thickness calibration, from a surface chemistry with an interfacial phase, or from a superparamagnetic behavior induced by islanding at the CaF_2 surface. Another possibility is that magnetization is being lost by strain induced in the Fe epilayer by the lattice mismatch between Fe and CaF_2 at the interface.

Transmission electron microscopy has been equally used to confirm the thickness calibration. In fact, while the effective Fe thickness remains a difficult parameter to determine, the equivalent Fe coverage depends only on the shutter time control for a given deposition rate. The thickness of 0.48 nm is very close, however, to the thickness registered by RHEED ($\sim$0.5 nm) for the Fe structural relaxation process (see Fig. 4).

Figure 7 shows the magnetization versus the field measured with the applied field in plane for the sample set of Figure 6. No appreciable magnetic anisotropy was observed for magnetic fields applied along different directions in the film plane. By increasing the Fe layer thickness, both the remanence and the coercivity increase. The higher deposition temperature strongly reduces the coercive field, H_c, whereas the saturation magnetization is only slightly affected. This suggests a structural relaxation in the sample. Also, the magnetic results point to the existence of misaligned epitaxial regions and multiphase domains at low Fe coverage more than to the existence of an interfacial compound. It is worth noting, however, that both assumptions lead to a loss of magnetization as inferred from the magnetic measurements.

By remembering that the Fe nucleation proceeds by three-dimensional islanding on CaF_2, a superparamagnetic response at low Fe coverage is expected. However, the structural results concerning the formation of misoriented epitaxial domains presented in Section 3.1 require a much more careful analysis. The evidence of different orientations of α-Fe at the interface imply a complex magnetic response, strongly dependent on the relaxation of the in-plane lattice parameter, which persists up to 1-nm-thick Fe coverage. Our present results demonstrate the importance of an

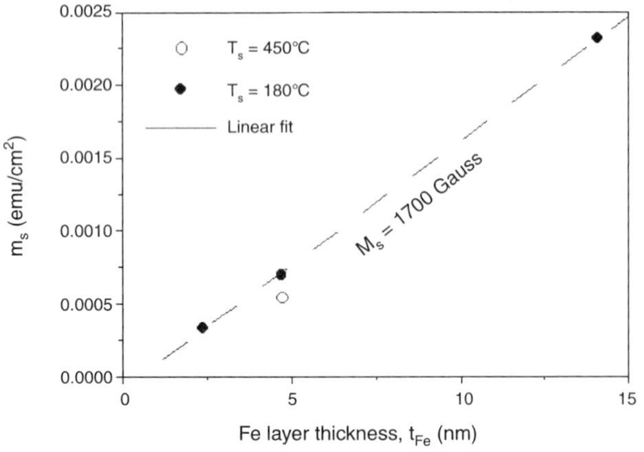

Figure 6. Saturation magnetization moment per square centimeter (m_s) versus Fe thickness for three samples deposited at 180 °C. Result of one sample deposited at 450 °C is also shown. A linear fit (dashed line) of the samples deposited at 180 °C leads to $M_s = 1700$ and a magnetic dead layer thickness of about 0.48 nm.

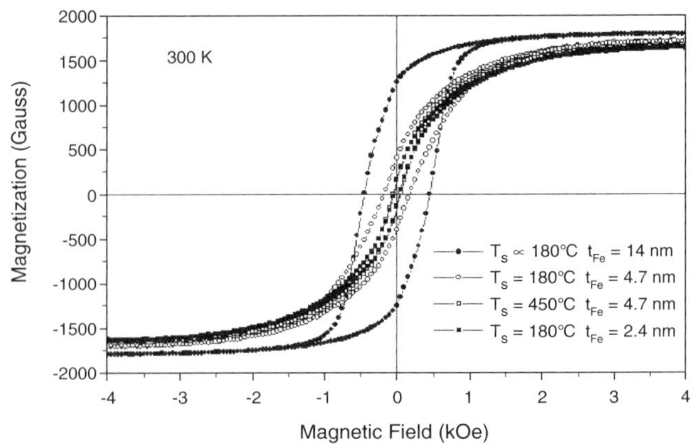

Figure 7. Magnetization versus magnetic field curves for samples with $T_s = 180$ °C; $t_{Fe} = 14$ nm, $H_c = 451$ Oe (solid circles); $T_s = 180$ °C, $t_{Fe} = 4.7$ nm, $H_c = 172$ Oe (open circles); $T_s = 450$ °C, $t_{Fe} = 4.7$ nm, $H_c = 49$ Oe (open squares); and $T_s = 180$ °C, $t_{Fe} = 2.4$ nm, $H_c = 40$ Oe (solid squares).

in-situ characterization of the early stages of Fe growth on CaF_2 to understand the magnetic properties of Fe/CaF_2 heterostructures.

3.3.2. Fe/CaF_2 Multilayers

The structural and magnetic properties of Fe/CaF_2 multilayers have been investigated by Mosca et al. [21] and Kita et al. [22]. Both reports indicate that discontinuous multilayered structures are formed. Discontinuous multilayers are an intermediary structure between an ideal multilayer thin film and a granular system as shown in Figure 8.

Discontinuous multilayered structures consisting of Fe/CaF_2 present interesting magnetic behavior of a cluster-glass phase with the presence of short-range antiferromagnetic ordering within Fe-rich granules. Magnetic irreversibility occurs mainly due to frustration of the antiferromagnetic interactions in the magnetic structure of Fe^{3+} ions. Sample A, described in Section 3.2, presents the irreversibility (branching of field cooling (FC) and zero field cooling (ZFC) curves) commonly observed in spin-glass systems. The irreversibility temperature, T_{irr}, associated with the bifurcation in the ZFC/FC curves, can be attributed either to magnetic clusters or to frustration of antiferromagnetic interactions. FC magnetization curves exhibit a broad maximum around 100 K that is progressively attenuated by increasing the applied field. The magnetic response of sample B differs markedly from that of sample A. The direct-current (dc) data suggest a ferromagnetic behavior coexisting with frustrated antiferromagnetic interactions in the percolated network of magnetic particles. According to Hansen and Morup [33], T_{irr} is dependent on the size distributions of the assembly of magnetic particles. Thus, the large value of T_{irr} indicates an increase in the size of the magnetic particles in sample B.

Mössbauer spectra of samples A and B measured at room temperature and zero field reveal that ferromagnetic order is not present in sample A, as expected for a composite sample consisting of granules, whereas the characteristic α-Fe sextets are observed for sample B, which depicts ferromagnetic behavior at room temperature. For sample A, three different Fe sites could be identified: HTB-$FeF_3(H_2O)_{0.33}$, FeF_2, and a singlet that can be ascribed to small superparamagnetic Fe clusters dispersed in a CaF_2 matrix. For sample B, the same Fe sites plus a sextet corresponding to α-Fe were identified. It is clear that the contributions due to HTB-$FeF_3(H_2O)_{0.33}$, FeF_2, and the small Fe clusters come from the reaction between Fe and CaF_2 at the interfaces.

A possible explanation for these magnetic results is that frustrated random antiferromagnetic interactions exist in the assembly of granules consisting of antiferromagnetic ferric trifluorides FeF_3 [28–30]. According to Ferey et al. [28], the magnetic structure of the crystalline forms of FeF_3 consist of a tetrahedral subnetwork of Fe^{3+} cations where the number of magnetic interactions involved in the square and triangular platelets of the cationic subnetwork gives rise to the frustrating topology of the structure.

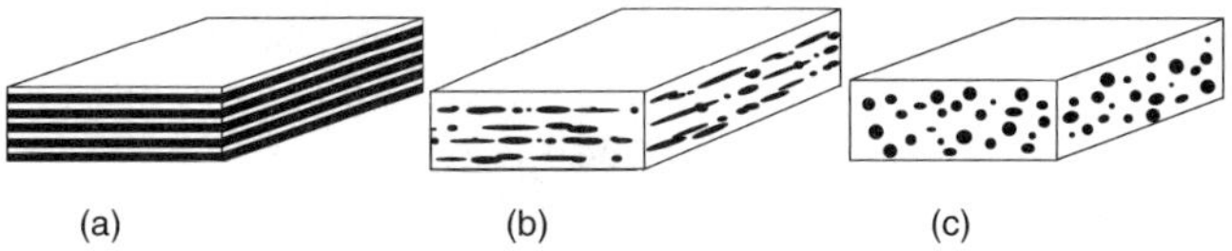

Figure 8. Schematic drawings of (a) multilayered film, (b) discontinuous multilayered film, and (c) granular film.

We believe that our aforementioned results are consistent with a chemical reaction between Fe and CaF_2, implying Fe^{3+} antiferromagnetic compounds rather than a frustrated random superparamagnetic phase. The magnetic behavior of sample A is consistent with an almost complete reaction of Fe with CaF_2 at room temperature, forming FeF_3 compounds. Sample B shows a ferromagnetic behavior that implies that nonreacted Fe still exists in this film where three-dimensional, rather than bidimensional, Fe islanding occurs during the growth stage. Sample C, consisting of a continuous and homogeneous Fe film grown at 450 °C, also exhibits a small magnetic irreversibility at moderate high temperatures.

Magnetic measurements of samples A and B annealed at 450 °C for 1 h in vacuum [31] reveal a strong magnetic irreversibility, which can be explained by the spatial distribution of antiferromagnetic aggregates consisting of crystalline forms of FeF_3 with a Néel temperature around 140 K.

Magnetoresistance measurements [34] using the standard alternating-current (ac) four-point method of samples A and B before and after annealing reveal a small negative magnetoresistive effect ($<0.2\%$) at room temperature. Up to 10 kOe, the resistivity systematically decreases as a function of the applied field without evidence of saturation.

As discussed in the last paragraph of Section 3.2, experimental efforts are still needed to clarify the magnetic and magnetotransport properties of Fe/CaF_2 structures.

3.4. Self-Assembled Fe Nanowires on CaF_2

Linear arrays of Fe stripes regularly spaced 15 nm can be obtained by photolysis of ferrocene that is selectively adsorbed between CaF_2 stripes that were self-assembled on a stepped Si(111) surface [35]. These self-assembled Fe nanowires represent a very interesting type of magnetic nanostructure for patterned storage media and related applications. To date, there is no work on the magnetism for the samples fabricated by this method.

3.5. Mechanical Properties

Nanomechanical tests of CaF_2/Si(111) and Fe/CaF_2/Si(111) samples were made with a Nanoindenter IIs (Nano-Instruments Co.) using a Berkovich diamond indenter tip [25, 36]. Nanoscratch tests were performed using the available scratch option for the nanoindenter to measure the tangential forces during the scan of the lateral face of the inverted triangular pyramid of the Berkovich tip. Typical parameters of the nanoscratch tests used were relative speed of 5 μm/s, scanning distance of 400 μm, and load in the range of 20–4000 μN.

The nanoscratch tests with 40-nm-thick CaF_2 films on Si(111) clearly show a pronounced plateau at the friction coefficient curves. For the macroscopic roughness of the 20-nm-thick sample and depth profiles during the scratch tests limited to 40 nm, the friction coefficient varies from 0.15 to 0.25. A similar behavior has been verified on samples as grown at 300 °C (polycrystalline) and subsequently rapidly thermal annealed (RTA) at 450 °C for 1 min (mosaic crystal), with the exception of a secondary plateau, present in the RTA sample, which can be related to the Ca silicide layer formed at the interface between CaF_2 and Si [36].

Nanomechanical properties of epitaxial 30-nm-thick α-Fe films grown on CaF_2/Si(111) at deposition temperatures ranging from 30 to 450 °C were also reported in the literature [25]. The results show dependence for in-plane residual strain with hardness and nanoscratch measurements. For example, the nanoindentation results show an increase in the pop-in events in the load versus penetration curve upon increasing deposition temperature, indicating a correlation of film crystal quality with mechanical behavior. The biggest nanoscratch resistance was achieved for a sample deposited at 150 °C, which has larger hardness and better adhesion.

Delamination of the Fe film from the CaF_2 support was observed at a load of about 1200 μN. The load value where the delamination occurs for a tip penetration smaller than the thickness of the bilayer Fe/CaF_2 is defined as the critical load for adhesion, and the ratio of the critical load with the contact area is defined as the critical stress [37]. By using these definitions, a two-order-of-magnitude increase in adhesion has been observed for the sample deposited at 150 °C with respect to room temperature deposition. The samples deposited at 150 °C were also qualitatively harder, showing a smaller plastic deformation. It is worth noting that the crystalline quality of the Fe films is enhanced by increasing the deposition temperature from 20 to 450 °C, whereas the relative strain of the Fe plane spacing parallel to the surface displays a minimum at a deposition temperature of 150 °C [25]. These results indicate that an Fe deposition temperature of about 150 °C minimizes the internal strain relaxation of the Fe films and, consequently, favors better adhesion of the Fe films to the CaF_2 substrate.

4. OTHER METALS ON CaF_2/Si(111)

Besides ferromagnetic iron, several metals have been deposited on epitaxial CaF_2, leading to interesting three-dimensional device structures.

Although the lattice parameter of Al is approximately three-quarters the lattice parameter of CaF_2, Cho et al. [7] have succeeded in growing epitaxial CaF_2 films on Al and vice versa, thus fabricating a technically sound epitaxial metal/insulator/metal structure on Si.

Cuberes et al. [38] demonstrate the successful deposition of Au/CaF_2 structures on the reconstructed (7 × 7) Si(111) surface. These authors characterized the interfaces using the ballistic electron emission microscopy (BEEM) technique, showing that the spectral features of the density of states depend on the CaF_2 epitaxial thin film.

Mattoso et al. [39] have shown that Cu grows epitaxially on CaF_2/Si(111) with the following epitaxial relationships: Cu(111)$\|$$CaF_2(111)\|$Si(111) along the growth direction and Cu[1$\bar{1}$0]$\|$$CaF_2$[1$\bar{1}$0]$\|$Si[1$\bar{1}$0] for the in-plane matching. Despite a mismatch of about 34% between Cu ($a_0 = 0.36150$ nm) and CaF_2 ($a_0 = 0.54626$ nm), a geometric coincidence between three Cu lattices and two lattices of CaF_2 brings the misfit to only 0.74%. This geometric argument was used to explain the heteroepitaxy of Cu on CaF_2.

The most promising structure of all reported up to now seems to be the $CoSi_2$/CaF_2 heterostructure on silicon. In a series of papers, the group at the Tokyo Institute of Technology demonstrated the feasibility of the growth of metal-like cobalt silicide on a CaF_2 insulator. Since 1992, Saitoh et al. [40] have shown devices presenting quantum interference of so-called hot electrons. This group reported on the first hot-electron transistor, the resonant tunneling diode, and the resonant tunneling transistor, establishing a very successful line of research followed by many scientists worldwide thereafter.

In 1998, LaBella et al. [41] demonstrated the hot-electron transport properties in ultrathin Pt/CaF_2/Si(111) MIS heteroepitaxial structures.

More recently, Shusterman et al. [42] have reported on successful epitaxial growth of Al on CaF_2/Si(111), as well as Cu/Al(3 nm)/CaF_2/Si(111) structures, indicating the same relationships described by Mattoso et al. [39] for Cu/CaF_2/Si(111) structures.

In summary, beyond magnetic Fe a few other metals such as Au, Pt, Al, and Cu as well as Co silicides can be epitaxially grown on CaF_2/Si heterostructures. Geometric relationships have been applied to explain the epitaxy of metals on CaF_2, but very few studies exist on the initial growth stages of these metals and their dependence on the deposition parameters. Such studies can extend the possibility of future applications beyond those already reported.

5. Fe AND THEIR ALLOYS ON FLUORIDES

Several fluorides grow epitaxially on semiconductor substrates, opening many possibilities for new electrical and optical devices. Further progress in fabricating novel and attractive devices using fluorides could be achieved by the integration of epitaxial metals on fluoride/semiconductor heterostructures. In this section, we address several kinds of fluorides that were epitaxially grown on semiconductors, especially on Si substrates.

Next, we describe the epitaxial growth, structural characterization, and magnetic properties of Fe and their alloys on fluorides. However, there have been very few reports on this subject, except for FeF_2 and LiF substrates.

Finally, nanostructured multilayer films and nanogranular thin films consisting of ferromagnetic 3D metals and fluorides are described and discussed.

5.1. Fluorides on Semiconductors

The epitaxial growth of alkali metal fluorides, such as KF and RbF, is a very difficult task since they are dissociatively adsorbed due to the reaction of halogen atoms with Si dangling bonds [43]. Concomitant with this fact, the fluorination of the Si surface exposed to fluorine leads to spontaneous surface etching at room temperature [44, 45].

LiF on Si is another interesting system because LiF is the material with the most extreme ultraviolet (UV) transmission of all, and therefore it is largely used for special UV and X-ray optics.

Cosset et al. [46] have reported on the deposition of textured crystalline LiF film on a monocrystalline Si(100) substrate. They found that LiF is almost site equivalent for adsorption on the Si surface and tends to form clusters for all growth conditions. From 300 to 550 °C, reactions between the LiF adsorbate and the substrate occur as a consequence of the dissociation of LiF and fluorination of the surface [47]. Similar results were obtained for LiF on Si(111) substrates with evidence of the formation of SiF_x and Li—Si complexes at temperatures higher than 300 °C [48]. Thus, good heteroepitaxial growth of LiF on Si(100) or Si(111) can be ruled out.

Prior work on epitaxial CaF_2 on Si can be extended to MgF_2 and CdF_2, which both have a smaller lattice parameter than that of Si. Although a CdF_2 crystal has the same fluoride structure as CaF_2, CdF_2 possesses a rather lower chemical stability [49]. An appropriate combination of MgF_2 and CdF_2 fluorides with CaF_2 could allow better lattice matching with Si. The optical coating applications of MgF_2 as infrared optical components of, for example, windows, mirrors, lenses, and prisms, make this an attractive alternative.

The most useful studies of the epitaxial growth of fluorides using an MBE apparatus have been done mainly with respect to group IIA fluorides such as CaF_2, SrF_2, and BaF_2 [50–52]. These cubic fluoride structures are closely related to the diamond structure of Si, thus giving high-quality crystalline thick films. Although the thermal expansion coefficients close to the room temperature of the fluorides ($10–30 \times 10^{-6}$ K^{-1}) are comparatively larger than those of Si (2.33×10^{-6} K^{-1}), suggesting that the cubic fluorides do not exhibit completely crack-free epitaxial growth on Si, the cubic fluorides are ionic solids with many desirable properties, such as large free energy of formation, low vapor pressure, simple molecular sublimation in UHV, and ease of growth as stoichiometric compounds [52].

Some rare earth lanthanide trifluorides, such as LaF_3, CeF_3, and NdF_3, can also be grown epitaxially on Si(111) substrates [53]. These trifluorides have a smaller linear thermal expansion coefficient compared to those of the group IIA fluorides. The trifluorides have a water-insoluble nature and possess a hexagonal crystal structure, which was successfully used as a template for epitaxial growth of rare earth metals [54]. A quasi-stoichiometric GaF_3, which has a rhombohedral crystal structure, can also be obtained by sublimation under UHV [55]. Yttrium fluoride, YF_3, produces low-index film layers on glass that exhibit good transparency in the UV through infrared (10 μm) regions.

The properties of some fluorides are briefly reviewed in Table 1. More data on the MBE growth of fluorides on semiconductors such as GaAs, Ge, and InP can be found in [1].

5.2. Fe on FeF_2

It is easily recognized that the growth of Fe on FeF_2 is currently an active area of research with goals that range from the fundamental understanding of the exchange anisotropy phenomena [62, 63] of such a system to its potential application to the magnetic domain stabilization

Table 1. Fluoride properties.

Compound	Crystalline structure	Lattice parameter (nm)	Lattice mismatch with Si at room temperature (%)	Energy bandgap (eV)	Antiferromagnetic Néel temperature T_N (K)	Ref.
LiF	Cubic NaCl	0.4026	−26.0	12	—	[56]
KF	Cubic NaCl	0.5348	−1.5	11	—	[1, 3]
RbF	Cubic NaCl	0.5652	+4.1	10	—	[1, 3]
MgF_2	Tetragonal	$a = 0.4640$ $c = 0.3060$	−15.0	10.8	—	
CdF_2	Cubic fluoride	0.5388	−0.8	8	—	[1, 3]
CaF_2	Cubic fluoride	0.54629	+0.6	12.1	—	[1, 3, 52]
SrF_2	Cubic fluoride	0.57996	+6.8	11.3	—	[1, 3, 52]
BaF_2	Cubic fluoride	0.62001	+14.2	11.0	—	[1, 3, 52]
MnF_2	Tetragonal	$a = 0.5112$ $c = 0.5256$	—		67	[57]
FeF_2	Tetragonal	$a = 0.4690$ $c = 0.3309$	—		78.3	[58]
CoF_2	Tetragonal	$a = 0.4710$ $c = 0.3169$	—		37.7	[59]
NiF_2	Tetragonal	$a = 0.4651$ $c = 0.3084$	—		73	[60]
ZnF_2	Tetragonal	$a = 0.4703$ $c = 0.3133$	—		—	
GaF_3	Rhombohedral	$a = 0.5200$ $\alpha = 57.5°$	—	9.6	—	[55, 61]
LaF_3	Hexagonal	a = 0.7187 $c = 0.7350$	+7.4	—	—	[52]
CeF_3	Hexagonal	$a = 0.7112$ $c = 0.7279$	+6.7	—	—	[52]
NdF_3	Hexagonal	$a = 0.7030$ $c = 0.7199$	+5.4	—	—	[52]

in magnetoresistive (MR) heads [64] and spin-valve-based devices [65]. Next, we address the research activity on the magnetic fluorides.

FeF_2 is a magnetic insulator with a body-centered tetragonal crystal structure where Fe^{2+} ions sit at the unit cell center ordering antiferromagnetically with the ions at the corners. The interaction between the Fe^{2+} ion spins is the superexchange interaction via F^- ions and is sensitive to a change in distance. When the distance between Fe^{2+} ions decreases, the exchange integral increases, resulting in a concomitant increase in the Néel temperature, T_N. FeF_2 is a uniaxial antiferromagnet material with a large uniaxial magnetocrystalline anisotropy along the *c* axis [66]. Therefore, the $FeF_2(110)$ surface has equal numbers of spins from the two antiparallel sublattices and is called compensated, whereas the $FeF_2(100)$ surface is uncompensated since its surface plane contains spins that point in a single direction.

Nogués et al. [66, 67], using conventional electron beam evaporation, have successfully grown (110)- and (100)-oriented Fe thin films on FeF_2 previously grown on MgO and Al_2O_3 substrates. Highly textured (110)- and (100)-oriented Fe thin films were also grown by electron beam evaporation on the (100) and (110) crystal faces of FeF_2 using the Bridgman–Stockbarger method [58].

The magnetic characterization of these Fe/FeF_2 bilayer structures has shown a number of novel and interesting properties such as a large exchange bias [67], a positive exchange bias [66, 68], a perpendicular ferromagnetic–antiferromagnetic coupling between Fe and FeF_2 [58], and an asymmetric magnetization reversal in the Fe exchange-biased layer [69, 70].

Exchange bias (EB) is a phenomenon that occurs in systems where a ferromagnet (FM) is in direct contact with an antiferromagnet (AFM). When a sample with an FM/AFM interface is cooled below T_N (assuming that the Curie temperature is larger than T_N) in the presence of a magnetic field, the center of the magnetization loop of the FM/AFM bilayer shifts away from the zero field. The magnitude of this shift is known as the exchange bias field H_{EB}. Since the discovery of this phenomenon by Meiklejohn and Bean [62, 63], extensive efforts have been devoted to its full understanding because of both its fundamental interest and its importance in technological applications [71].

Currently, technological interest is focused on the unidirectional anisotropy, or exchange bias field produced in an FM film that is coupled to an appropriate AFM film. Berkowitz and Takano [72] reviewed the experimental characterization and theoretical models for a variety of metallic and oxide film couples. Almost all the reported investigations with insulating AFMs involve the monoxides NiO, CoO, and $(Ni_xCo_{1-x})O$. Typically, the exchange couples with metallic AFM films involve Mn-based AFM films such as FeMn (30–55 at% Mn), which is currently in use in most commercial exchange-biased MR read heads. However, the crystal structure of metallic AFMs often deviate more strongly from cubic symmetry than do the structures of the insulating AFMs.

The simple crystal and spin structure of FeF_2 (as well as MnF_2) make it a prototype system for understanding the EB phenomenon and for investigating practical problems such as the unwinding of the magnetic domain state and the annihilation of domains with field cycling.

There have been reports on the growth of Fe on the transition metal fluorides such as FeF_2, MnF_2, and ZnF_2, which have similar crystal and spin structures but very different anisotropies, a crucial parameter to investigate and control the exchange bias phenomenon in an FM/AFM interface [73].

Furthermore, the bilayers consisting of Fe deposited on the (110) crystal face of FeF_2 and MnF_2 exhibit an unconventional EB phenomenon [66]. The main feature of normal EB is the displacement of the hysteresis loop to fields that point opposite to the applied cooling field. Nogués et al. [66] applied strong cooling fields to the same samples where normal (or negative) exchange bias has been measured in weak cooling fields and discovered positive exchange bias in Fe/FeF_2 and Fe/MnF_2 systems, that is, a shift to the right of the hysteresis loop ($H_{EB} > 0$). The role of magnetic domains in the ferromagnet in positive exchange-biased bilayers was recently investigated by Kiwi et al. [74, 75]. Indeed, the fundamental physics of exchange bias coupling continue to excite considerable research [76, 77]. The synthetic layout of FM/AFM exchange-biased bilayers and magnetization loops with negative and positive exchange bias is shown in Figure 9.

Perpendicular coupling between the Fe spins and the ordered FeF_2 spins was discovered by studying the freezing field dependence of the exchange bias in Fe films on (110) and (100) FeF_2 surfaces [58]. Similar perpendicular anisotropy was also found in uniaxial FeF_2–CoPt heterosystems [78, 79].

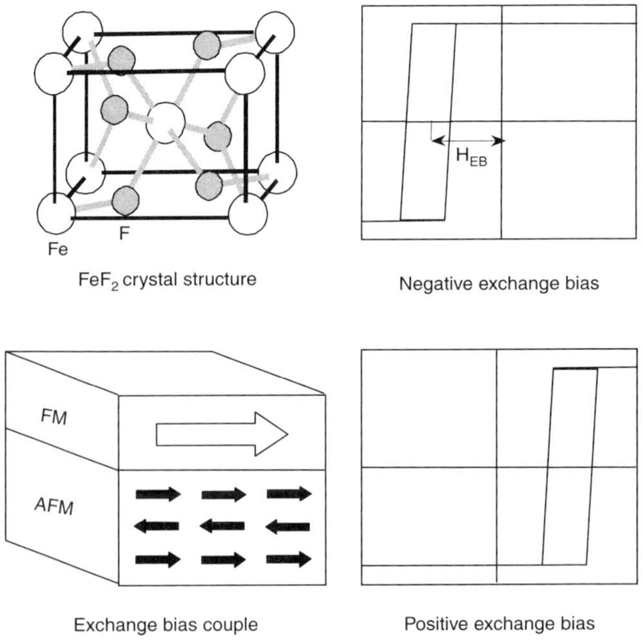

Figure 9. Schematic drawings of the crystal structure of FeF_2 and FM/AFM exchange-biased bilayer structure where the crystalline orientation of the AFM layer is reflected in the periodic alternating pattern of three rows. Scheme of the FM magnetization loops with negative and positive exchange bias observed in the exchange-biased structures with AFM fluoride.

Although the perpendicular anisotropy is strongly related to the domain configuration at the FM/AFM interface, the spin wave excitation also plays a crucial role. For instance, Mössbauer effect studies of Fe(110) surfaces covered by MnF_2 show that surface magnetic effects persist for at least 10 monolayers of the (110) Fe sample below T_N, indicating that the classical framework of spin wave models must be taken into account [80].

5.3. Fe and their Alloys on Other Fluorides

Several studies have demonstrated that crystalline thin films of ferromagnetic metals and alloys can be grown on LiF crystal substrates.

At the beginning of the 1970s, different groups in the former Soviet Union worked on the structural, magnetic, and magnetooptical properties of Fe—Co and Fe—Ni alloy thin films, as well as on double-layer ferromagnetic films, deposited on LiF crystal substrates, but many of these references are difficult to access [81–83]. According to these reports, the ferromagnetic thin films have good crystal quality and exhibit magnetic properties similar to those of the bulk material. Detailed information on the interfacial chemistry is not available in these studies since all the thin films that were described have a thickness of several tens of nanometers.

More recently, X-ray rocking curves and transmission electron microscopy analyses have been used to show that the epitaxial growth of some cubic metals such as Ag, Al, Cu, Cr, Fe, V, and Mo on rock salt substrates is improved markedly by covering the air-cleaved rock salt with a LiF intermediate layer [84]. Also, the structural properties of ferromagnetic Fe—Pd and Fe—Pt alloy thin films on LiF crystals are improved significantly in comparison to MgO crystals and glass substrates [85].

In a recent work, epitaxial (100) Fe films grown on the cleaved LiF substrates present unidirectional anisotropy and displaced hysteresis loops explained by a possible interfacial antiferromagnetic layer between the film and the substrate [86].

Surprisingly, there have been only a few reports on the growth of Fe and its alloys on other fluorides, except $(Fe_xZn_{1-x})F_2$, which is an Ising antiferromagnet commonly obtained using coevaporation of FeF_2 and ZnF_2 sources.

McChesney et al. [87] have studied the epitaxial thin films of $(Fe_xZn_{1-x})F_2$ grown on $MgF_2(100)$ substrates by coevaporation of FeF_2 and ZnF_2 sources. Such $(Fe_xZn_{1-x})F_2$ single-crystal thin films with fixed Zn concentration exhibit high quality and were used to study the magnetic phase diagram of dilute magnets. Shi et al. [88] have used $(Fe_xZn_{1-x})F_2$/Co bilayers grown on MgO(100) substrates to demonstrate that antiferromagnetic–ferromagnetic interface exchange bias can be controlled by diluting antiferromagnetic FeF_2 with nonmagnetic ZnF_2. In particular, a neutron scattering study of the random-field Ising film $Fe_{0.5}Zn_{0.5}F_2$ near the percolation threshold has been performed [89, 90] and modeled [91, 92].

It is worth noting that during the last 20 years dilute antiferromagnetic systems such as $(Fe_xZn_{1-x})F_2$ and $(Fe_xMn_{1-x})F_2$ have been extensively used as model systems to study the spin dynamics and phonon relaxation associated with local magnon modes [93–99].

5.3.1. Fe/Fluoride Multilayered Films

The interfacial chemistry between Fe and the fluorides was recently investigated in Fe/MgF_2, Fe/CaF_2, and Fe/LiF multilayered thin films and granular systems. In contrast to the interface between Fe and MgF_2 and CaF_2, which is markedly abrupt at relatively high temperatures under UHV, the Fe/LiF structures exhibit interface mixing and compound formation.

Ferromagnetic 3D metal/fluoride structures are commonly obtained by alternating deposition in conventional UHV evaporation systems as well as by codeposition using UHV evaporation or radiofrequency (RF) sputtering.

The interfacial phenomena of ferromagnetic Fe and nonmagnetic fluoride insulators (LiF, CaF_2, and MgF_2) have been investigated by means of Mössbauer spectroscopy, ferromagnetic resonance, and magnetization measurements [22, 100–102].

Kita et al. [22] investigated multilayered thin films of Fe and fluorides (CaF_2, LiF) produced by UHV electron beam evaporation. They found that mixing occurs at the interfaces, forming imperfect compounds. Mössbauer studies show that Fe and the fluorides are mixed at their interface and confirm the presence of ionic Fe atoms at the interfaces with hyperfine field contributions similar to that of FeF_3. The chemical mixing is more pronounced at the Fe-on-LiF interface than at the LiF-on-Fe interface. Unidirectional anisotropy due to an exchange anisotropy in both Fe/CaF_2 and Fe/LiF is also observed in both multilayered thin films. It is suggested that exchange anisotropy is due to magnetic exchange coupling between ferromagnetic and antiferromagnetic parts in the films, that is, the presence of an antiferromagnetic layer. The thickness of the antiferromagnetic layer was estimated to be less than 4 nm, which is the thickness of a fluoride layer.

Differently from the Fe/LiF, the Fe/MgF_2 interface is chemically stable and enhancement of the hyperfine field was observed in Fe/MgF_2 multilayered films deposited on Kapton and Si(111) wafer substrates [102]. The enhancement of the hyperfine field indicates that the fluoride causes a large electronic and magnetic modulation at the interfaces. Multilayer structures are clearly observed in films with Fe layer thickness greater than 3 nm, whereas with Fe layer thickness less than 2 nm a granular-like structure appears to be present in these films. The Fe atoms form clusters small enough to exhibit superparamagnetic characteristics. It is worth noting that this island structure in the Fe/MgF_2 is responsible for the tunneling magnetoresistance found in the same system [103].

5.3.2. Fine Magnetic Particles in Fluoride Films

Large spin-dependent tunneling magnetoresistance (TMR) in insulating nanogranular systems consisting of metallic ferromagnetic particles embedded in an insulating host was first found in Co—(Al—O) by Fujimori et al. [104, 105] and since then TMR has been investigated for many kinds of insulating nanogranular systems [106]. These granular films are easily obtained in systems consisting of a combination of a ferromagnetic 3D metal (such as Fe, Co, and Ni) and an oxide that has a large heat of formation. Fluorides such as MgF_2, BaF_2, and CaF_2 have also a large heat of formation and are chemically stable. Some nanogranular structures have already been obtained such as Fe—(Mg—F) [107, 108], Co—(Mg—F) [108], and (Fe—Co)–(Mg—F) [109] films.

Granular thin films composed of Fe or Co metal or Fe_xCo_{1-x} alloy and MgF_2 have been investigated by XRD and transmission electron microscopy (TEM) which confirm the nanogranular structure consisting of bcc Fe or Fe—Co alloy and network-like crystalline Mg—F with MgF_2 intergranules. The largest value ever reported in insulating nanogranular TMR thin films has been obtained in the FeCo–(MgF) system [109]. This large TMR effect is attributed to spin-dependent tunneling associated with superparamagnetic Fe or FeCo nanogranules separated by tunnel barriers consisting of MgF insulating intergranules. More recently, large tunneling magnetoresistance effects of Co clusters embedded in a MgF_2 matrix have been reported [110]. A possible explanation for the enhanced spin polarization of the tunneling electrons is the dominated d-electron tunneling due to Co(3d)–F(2p) hybridization.

6. FINAL REMARKS

The objective of this review was to resume the state of the art of research on the metal/CaF_2/Si heteroepitaxial growth with a short incursion into other fluoride systems. The feasibility of several MIS structures shows that this area of research has gained maturity. The heteroepitaxial growth of CaF_2 thin films on Si is clearly dominated as shown in a number of publications. Although still not completely understood, it has been shown that the electrical properties and epitaxial relations of CaF_2/Si(111) depend on the growth conditions. Many research groups have used CaF_2/Si(111) as a prototypical system for studying the atomic and electronic structure at the ionic/covalent interface. The problem now is found centered in the successful deposition of metal films on CaF_2. Using the newly available deposition and characterization techniques, an effort has to be made to study the different metal/CaF_2 interfaces. These interfaces clearly influence the quality of the heteroepitaxial structures and films as well as their transport, magnetic, or optical behavior and quality. Useful and attractive devices using the properties of CaF_2 will be realized when the interfacial atomic structure and chemical reactions can be controlled on the atomic level. The future for these types of MIS and multilayer structures seems as interesting as ever.

GLOSSARY

Epitaxial growth Deposit of a crystalline layer onto a single-crystal seed layer, maintaining the spatial coherence at the interface between crystal lattices.

Epitaxial relationship Preferential parallelism of the lattice planes of two crystal structures spaced by an interface.

Exchange bias Exchange interaction across the interface between antiferromagnets and ferromagnets with pinning of the ferromagnet magnetization.

Field cooling Magnetization procedure in which a magnetic field is applied when the sample is cooling down.

Giant magnetoresistance Phenomenon associated with an appreciable decrease in the resistance when a applied magnetic field, strong enough to magnetically saturate the system, brings the magnetization of all magnetic entities into alignment.

Magnetic anisotropy Preferential alignment of the magnetization along a certain direction of a material under a given magnetizing field.

Superexchange interaction Long-range indirect magnetic coupling between localized magnetic moments mediated by atoms with pairing of electrons.

Superparamagnetism Magnetic behavior observed above a certain blocking temperature in the systems consisting of non-interacting single-domain magnetic particles with diameters below a certain critical size.

Zero-field cooling Magnetization procedure in which the magnetic field is switched off when the sample is cooling down.

REFERENCES

1. M. Sugiyama and M. Oshima, *Microelectron. J.* 27, 361 (1996).
2. J. Kudrnovský, N. E. Christensen, and J. Masek, *Phys. Rev. B* 43, 12597 (1991).
3. L. J. Schowalter and R. W. Fathauer, *Crit. Rev. Solid State Mater. Sci.* 15, 367 (1989).
4. G. A. Prinz, *Science* 283, 330 (1999).
5. S. A. Wolf, D. D. Awschalom, R. A. Buhrman, J. M. Daughton, S. von Molnar, M. L. Roukes, A. Y. Chtchelkanova, and D. M. Treger, *Science* 294, 1488 (2001).
6. E. G. Michel, *Appl. Surf. Sci.* 117, 294 (1997).
7. C. C. Cho, H. Y. Liu, B. E. Gnade, T. S. Kim, and Y. Nishioka, *J. Vac. Sci. Technol., A* 10, 769 (1992).
8. J. D. Denlinger, E. Rotenberg, U. Hessinger, M. Leskovar, and M. A. Olmstead, *Phys. Rev. B* 51, 5352 (1995).
9. J. Wollschläger and A. Meier, *J. Appl. Phys.* 79, 7373 (1996).
10. C. R. Wang, B. H. Müller, and K. R. Hofmann, *Thin Solid Films* 410, 72 (2002).
11. K. G. Huang, J. Zegenhagen, J. M. Phillips, and J. R. Patel, *Physica B* 221, 192 (1996).
12. T. Sumiya, *Appl. Surf. Sci.* 156, 85 (2000).
13. N. Mattoso, D. H. Mosca, W. H. Schreiner, I. Mazzaro, and S. R. Teixeira, *Thin Solid Films* 272, 83 (1996).
14. F. Grey and J. Bohr, *Appl. Surf. Sci.* 65–66, 35 (1993).
15. S. M. Rezende, J. A. S. Moura, F. M. de Aguiar, and W. H. Schreiner, *Phys. Rev. B* 49, 15105 (1994).
16. H. J. Kim, D. Y. Noh, J. H. Je, and Y. Hwu, *Phys. Rev. B* 59, 4650 (1999).
17. K. R. Heim, S. T. Coyle, G. G. Hembree, J. A. Venables, and M. M. R. Scheinfein, *J. Appl. Phys.* 80, 1161 (1996).
18. K. R. Heim, G. G. Hembree, and M. R. Scheinfein, *J. Appl. Phys.* 76, 8105 (1994).
19. Y. T. Cheng, Y. L. Chen, W. J. Meng, and Y. Li, *Phys. Rev. B* 48, 14729 (1993).
20. J. W. Edington, in "Practical Electron Microscopy in Materials Science Series," p. 98. MacMillan Philips, Eindhoven, 1975.
21. D. H. Mosca, N. Mattoso, W. H. Schreiner, A. J. A. de Oliveira, W. A. Ortiz, W. H. Flores, and S. R. Teixeira, *J. Magn. Magn. Mater.* 231, 337 (2001).
22. E. Kita, M. Ochi, T. Erata, and A. Tasaki, *J. Magn. Magn. Mater.* 117, 294 (1992).
23. V. M. Bermudez, *Appl. Surf. Sci.* 161, 227 (2000).
24. M. Reichling and C. Barth, *Phys. Rev. Lett.* 83, 768 (1999).

25. N. Mattoso, D. H. Mosca, W. H. Schreiner, C. M. Lepienski, I. Mazzaro, and S. R. Teixeira, *Thin Solid Films* 323, 178 (1998).
26. G. Ferey, F. Varret, and J. M. D. Coey, *J. Phys. C: Solid State Phys.* 12, L531 (1979).
27. M. Leblanc, G. Ferey, P. Chevallier, and R. de Pape, *J. Solid State Chem.* 47, 53 (1983).
28. G. Ferey, R. de Pape, M. Leblanc, and J. Pannetier, *Rev. Chim. Miner.* 23, 427 (1986).
29. M. Leblanc, R. Pape, G. Ferey, and J. Pannetier, *Solid State Commun.* 58, 171 (1986).
30. J. M. Greneche, A. Le Bail, M. Leblanc, A. Mosset, F. Verret, J. Galy, and G. Ferey, *J. Phys. C: Solid State Phys.* 21, 1351 (1988).
31. J. Varalda, W. A. Ortiz, A. J. A. de Oliveira, D. H. Mosca, N. Mattoso, W. H. Schreiner, and S. R. Teixeira, *J. Magn. Magn. Mater.* 226–230, 1738 (2001).
32. K. R. Heim, G. G. Hembree, K. E. Schmidt, and M. R. Scheinfein, *Appl. Phys. Lett.* 67, 2878 (1995).
33. M. F. Hansen and S. Morup, *J. Magn. Magn. Mater.* 203, 214 (1999).
34. J. Varalda, Caracterização magnética de filmes de ligas e multicamadas magnéticas, M.S. thesis, Universidade Federal de São Carlos, São Carlos, Brazil, 2000 (in Portuguese).
35. J.-L. Lin, D. Y. Petrovykh, A. Kirakosian, H. Rauscher, F. J. Himpsel, and P. A. Dowben, *Appl. Phys. Lett.* 78, 829 (2001).
36. C. M. Lepienski, N. Mattoso, D. H. Mosca, W. H. Schreiner, I. Mazzaro, S. R. Teixeira, W. A. A. Macedo, and M. D. Martins, *Mater. Res. Soc. Symp. Proc.* 522, 457 (1998).
37. B. Bhushan and B. K. Gupta, in "Handbook of Tribology" p. 15. McGraw–Hill, New York, 1991.
38. M. T. Cuberes, A. Bauer, H. J. Wan, M. Prietsch, and G. Kaindl, *Appl. Phys. Lett.* 64, 2300 (1994).
39. N. Mattoso, D. H. Mosca, I. Mazzaro, S. R. Teixeira, and W. H. Schreiner, *J. Appl. Phys.* 77, 2831 (1995).
40. W. Saitoh, T. Suemasu, Y. Kohno, M. Watanabe, and M. Asada, *Jpn. J. Appl. Phys.* 34, 4481 (1992).
41. V. P. LaBella, Y. V. Shusterman, L. J. Schowalter, and C. A. Ventrice, Jr., *J. Vac. Sci. Technol., A* 16, 1692 (1998).
42. Y. V. Shusterman, N. L. Yakovlev, and L. J. Schowalter, *Appl. Surf. Sci.* 175–176, 27 (2001).
43. R. Souda, W. Hayami, T. Aizawa, S. Otani, and Y. Ishizawa, *Phys. Rev. B* 47, 4092 (1993).
44. D. L. Flamm, V. M. Donnely, and J. A. Mucha, *J. Appl. Phys.* 52, 3633 (1981).
45. H. F. Winters and F. A. Houle, *J. Appl. Phys.* 54, 1218 (1983).
46. F. Cosset, A. Celerier, B. Barelaud, and J. C. Vareille, *Thin Solid Films* 303, 191 (1997).
47. H. Guo, H. Kawanowa, and R. Souda, *Surf. Sci.* 445, 394 (2000).
48. H. Guo, H. Kawanowa, and R. Souda, *Mater. Sci. Eng., B* 72, 160 (2000).
49. N. S. Sokolov, S. V. Gastev, N. L. Yakovlev, A. Izumi, and S. Furukawa, *Appl. Phys. Lett.* 64, 2964 (1994).
50. R. F. C. Fallow, P. W. Sullivan, G. M. Williams, G. R. Jones, and D. G. Camerson, *J. Vac. Sci. Technol.* 19, 415 (1981).
51. H. Ishiwara and T. Asano, *Appl. Phys. Lett.* 40, 66 (1982); T. Asano, H. Ishiwara, and N. Kaifu, *Jpn. J. Appl. Phys.* 22, 1474 (1983).
52. R. F. C. Fallow, S. Sinharoy, R. A. Hoffman, J. H. Rieger, W. J. Takei, J. C. Greggi, Jr., S. Wood, and T. A. Temofonte, *Mater. Res. Soc. Symp. Proc.* 37, 181 (1985).
53. S. Sinharoy, R. A. Hoffman, J. H. Rieger, W. J. Takei, and R. F. C. Fallow, *J. Vac. Sci. Technol., B* 3, 722 (1985).
54. R. F. C. Fallow and C. H. Lee, *Mater. Res. Soc. Symp. Proc.* 187, 211 (1991).
55. A. S. Barriere, G. Couturier, G. Gevers, H. Guegan, T. Seguelond, A. Thabti, and D. Bertault, *Thin Solid Films* 173, 243 (1989).
56. B. Kunz, *Phys. Rev. B* 26, 2056 (1981).
57. T. Yagi, *J. Geophys. Res.* 84, 113 (1979).
58. T. J. Moran, J. Nogués, D. Lederman, and I. K. Schuller, *Appl. Phys. Lett.* 72, 617 (1998).
59. S. Reed, *J. Am. Chem. Soc.* 76, 5279 (1954).
60. J. Austin, *J. Phys. Chem. Solids* 30, 1282 (1969).
61. K. Aizawa and H. Ishiwara, *Jpn. J. Appl. Phys.* 33, 5178 (1994).
62. W. P. Meiklejohn and C. P. Bean, *Phys. Rev.* 102, 1413 (1956).
63. W. P. Meiklejohn and C. P. Bean, *Phys. Rev.* 105, 904 (1957).
64. C. Tsang and R. Fontana, *IEEE Trans. Magn.* 18, 1149 (1982).
65. B. Dieny, V. S. Speriosu, S. S. P. Parkin, J. C. Scott, B. A. Gurney, D. R. Wilhoit, and D. Mauri, *Phys. Rev. B* 43, 1297 (1991).
66. J. Nogués, D. Lederman, T. J. Moran, and I. K. Schuller, *Phys. Rev. Lett.* 76, 4624 (1996).
67. J. Nogués, D. Lederman, T. J. Moran, I. K. Schuller, and K. V. Rao, *Appl. Phys. Lett.* 68, 3186 (1996).
68. J. Nogués et al., *Phys. Rev.* 59, 6984 (1999).
69. M. Fitzsimmons et al., *Phys. Rev. Lett.* 84, 3986 (2000).
70. C. Leighton et al., *Phys. Rev. Lett.* 86, 4394 (2001).
71. J. Nogués and I. K. Schuller, *J. Magn. Magn. Mater.* 192, 203 (1999).
72. A. E. Berkowitz and K. Takano, *J. Magn. Magn. Mater.* 200, 552 (1999).
73. M. S. Lund, W. A. A. Macedo, K. Liu, J. Nogués, I. K. Schuller, and C. Leighton, *Phys. Rev. B* 66, 054422 (2002).
74. M. Kiwi, J. Mejía-López, R. D. Portugal, and R. Ramírez, *Solid State Commun.* 116, 315 (2000).
75. M. Kiwi, J. Mejía-López, R. D. Portugal, and R. Ramírez, *Europhys. Lett.* 48, 573 (1999).
76. J. L. Almeida and S. M. Rezende, *Phys. Rev. B* 65, 92412 (2002).
77. S. M. Rezende, A. Azevedo, F. Aguiar, J. Fermin, W. Egelhoff, and S. S. P. Parking, *Phys. Rev.* 66, 64109 (2002).
78. B. Kagerer, Ch. Binek, and W. Kleemann, *J. Magn. Magn. Mater.* 217, 139 (2000).
79. Ch. Binek, B. Kagerer, S. Kainz, and W. Kleemann, *J. Magn. Magn. Mater.* 226–230, 1814 (2001).
80. H. Tang, Z. Q. Qiu, Y. W. Du, G. P. Stern, and J. C. Walker, *J. Appl. Phys.* 63, 3659 (1988).
81. A. A. Potseluiko and V. G. Pyn'ko, *Sov. Phys. Solid State* 15, 435 (1973).
82. E. S. Mushailov, *Sov. Phys. Solid State* 16, 1013 (1974).
83. A. B. Vaganov and V. V. Vasilevskii, *Sov. Phys. Solid State* 16, 1328 (1977).
84. S. Koster, N. Herres, M. Rey, and K. Reichelt, *J. Appl. Phys.* 59, 278 (1986).
85. V. G. Pynko, A. S. Komalov, and L. V. Ivaeva, *Phys. Status Solidi A* 63, K127 (1981).
86. A. A. Potseluiko and V. G. Pyn'ko, *Crystallogr. Rep.* 46, 848 (2001).
87. J. McChesney, M. Hetzer, H. Shi, T. Charlton, and D. Lederman, *J. Mater. Res.* 16, 1769 (2001).
88. H. Shi, D. Lederman, and E. E. Fullerton, *J. Appl. Phys.* 91, 7763 (2002).
89. D. P. Belanger and H. Yoshizawa, *Phys. Rev. B* 47, 5051 (1993).
90. D. P. Belanger, J. Wang, Z. Slanic, S.-J. Han, R. M. Nicklow, M. Lui, C. A. Ramos, and D. Lederman, *J. Magn. Magn. Mater.* 140–144, 1549 (1995).
91. Z. Slanic, D. P. Belanger, and J. A. Fernandez-Baca, *Phys. Rev. Lett.* 82, 426 (1999).
92. K. Jonason, C. Djurberg, P. Nordblad, and D. P. Belanger, *Phys. Rev. B* 56, 5404 (1997).
93. E. Montarroyos, C. B. Araujo, and S. M. Rezende, *J. Appl. Phys.* 50, 2033 (1979).
94. E. Montarroyos, S. S. Vianna, C. B. de Araujo, S. M. Rezende, and A. R. King, *J. Magn. Magn. Mater.* 31–34, 557 (1983).
95. S. S. Vianna, C. B. Araujo, and S. M. Rezende, *Phys. Rev. B* 30, 3516 (1984).
96. S. M. Rezende and E. F. Silva, Jr., *Phys. Rev. B* 31, 570 (1985).
97. S. S. Viana, S. M. Rezende, and C. B. Araujo, *Phys. Rev. B* 32, 428 (1985).

98. F. C. Montenegro, S. M. Rezende, and M. D. Coutinho Filho, *J. Appl. Phys.* 8, 3755 (1988).
99. F. C. Montenegro, J. C. O. Jesus, F. L. A. Machado, E. Montarroyos, and S. M. Rezende, *J. Magn. Magn. Mater.* 104–110, 277 (1992).
100. M. Ochi, E. Kita, N. Saegusa, T. Erata, and A. Tasaki, *J. Phys. Soc. Jpn.* 61, 35 (1991).
101. H. Ohta, M. Hayashi, M. Motokawa, and E. Kita, *J. Phys. Soc. Jpn.* 66, 3272 (1997).
102. M. Nishikawa, O. Kazunaga, E. Kita, H. Yanagihara, T. Erata, and A. Tasaki, *J. Magn. Magn. Mater.* 238, 91 (2002).
103. K. Ono, C. Yasui, H. Yanagihara, T. Koyano, K. Ota, and E. Kita, *Jpn. J. Appl. Phys.* 41, 97 (2002).
104. H. Fujimori, S. Mitani, and S. Ohnuma, *Mater. Sci. Eng., B* 31, 219 (1995).
105. H. Fujimori, S. Mitani, and S. Ohnuma, *J. Magn. Magn. Mater.* 156, 311 (1996).
106. H. L. Bai and E. Y. Jiang, *Chin. Sci. Bull.* 46, 529 (2001).
107. T. Furubayashi and I. Nakatani, *Mater. Sci. Eng., A* 217–218, 307 (1996).
108. N. Kobayashi, S. Ohnuma, T. Masumoto, and H. Fujimori, *J. Magn. Soc. Jpn.* 24, 571 (2000).
109. N. Kobayashi, S. Ohnuma, T. Masumoto, and H. Fujimori, *J. Appl. Phys.* 90, 4159 (2001).
110. B. Hackenbroich, H. Zare-Kolsaraki, and H. Micklitz, *Appl. Phys. Lett.* 81, 514 (2002).

Encyclopedia of Nanoscience and Nanotechnology
www.aspbs.com/enn

Focused Ion Beam Nanofabrication

Andrei V. Stanishevsky

University of Alabama at Birmingham, Birmingham, Alabama, USA

CONTENTS

1. INTRODUCTION

The development of fabrication processes of materials and devices at nanoscale is one of the major challenges of nanotechnology. Among the nanofabrication techniques available, focused ion beams (FIB) have become a popular tool for surface modification of materials and functional structure prototyping at the micro- and nanoscale. Focused ion beams have a diameter less than 1 μm and are produced by using electrostatic lenses to focus the image of a point source, usually a liquid metal ion source, onto the substrate, and to deflect it in a precise fashion [1]. A schematic representation of a FIB apparatus is given in Figure 1.

FIB systems producing a few picoamperes ion beam probes with a diameter down to 100 nm were reported as early as the 1960s. An imaging resolution of 50 nm was demonstrated in 1974 with the ion systems based on gas phase field ionization sources. The liquid metal ion source was invented in 1975. Hitachi pioneered the use of FIBs for integrated circuit cross-sectioning for failure analysis in 1985 [1]. According to Melngailis [2], there were about 35 FIB machines in the world by 1986, and now the number is several hundred.

The past decade has seen many advances in FIB tools including improvements in both ion optics and tool automation. Ion beam optics has improved rapidly to the point where the beam size and profile are no longer the limiting factors in many applications of FIBs ranging from failure analysis to prototype nanodevice manufacturing. Most FIB equipment on the market today uses gallium liquid metal ion sources (LMIS) with the service life of up to 2000 μA-h and accelerating voltages in the range of 10–50 kV. Other LMISs have typically much shorter service lives, and they include AuPdAs, Si, Co, $Au_{77}Ge_{14}Si_9$, $Co_{36}Nd_{64}$, etc.

Over the last two decades, focused ion beams have emerged as one of the important tools for both scientific research and industrial applications. Commercially available FIB systems have reached resolutions close to electron beam systems (about 5 nm), but micro- and nanoengineering with FIBs is a sequential process. It is therefore much slower than the parallel processes of optical or electron beam lithography and it should not be expected to provide a technology for mass production of nanosize structures and devices any time soon. FIB technology has already found applications in the semiconductor industry as a complement to other processes for certain operations on individual chips on the wafer such as failure analysis, rewiring, and defect repair. FIB will find, probably, its niche in the fabrication of small series of specific nanoscale devices—for example in a research environment—or in customization of unique devices. Examples of current FIB applications include micro- and nano-optical components, samples for transmission electron microscopy, scanning microscopy probes, microtools, semiconductor devices, surface nanodevices, and three-dimensional (3D) nanostructures and components of microelectromechanical systems.

Principles of focused ion beam equipment design, generation of ion beams, and the method applications are thoroughly discussed in numerous reviews [1–10]. Several other reviews outline specific applications of FIBs for microfabrication [11], nanomachining [12], two-dimensional and three-dimensional shapes, ion lithography, and the preparation of site-specific specimens [13].

FIBs are used mostly for defect repair of optical and X-ray masks, circuit modification, device modification, failure analysis of circuits and devices, circuit and device cross-sectioning, transmission electron microscopy (TEM) sample preparation, ion microscopy, secondary ion mass spectroscopy, and experiments on direct maskless implantation of devices. Since the incident ions produce both

ISBN: 1-58883-059-4/$35.00

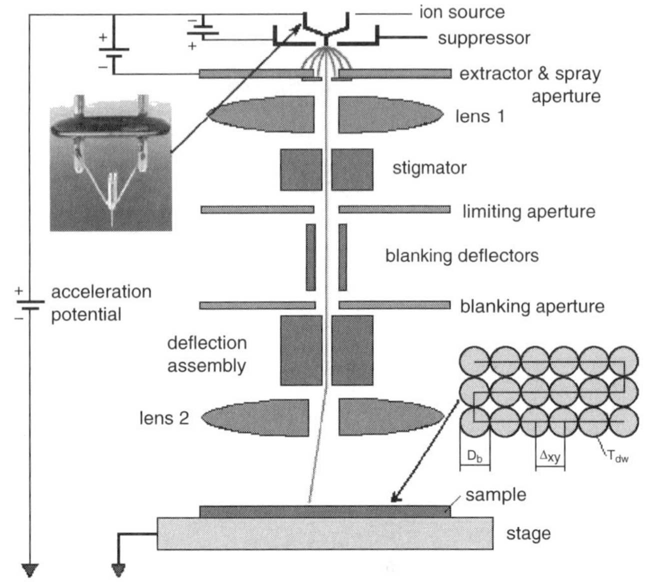

Figure 1. A schematic of a focused ion beam column. A photograph of a gallium liquid metal ion source and an example of beam raster are shown in inserts.

secondary electrons and ions, images formed from either are possible. FIB provides results superior to other techniques in fabrication of cross-sections in materials and devices for failure analysis. Figure 2 gives an example of the image of the FIB polished cross-section in a Cu-metallized test structure in comparison with the same area after mechanical polishing. Grain structure of Cu layer is clearly visible in FIB polished cross-section. An example of FIB fabricated lamellae for TEM studies is shown in Figure 3. The thickness of the lamellae of less than 40 nm can be achieved.

The main applications of FIB processes today are mostly concerned with local ion sputtering or modification of the material. The sputtering yield is usually 1–3 atom/ion. A number of factors determine the micromachining results, including ion beam diameter, current and mass, angular effects caused by edges or depth variations across the

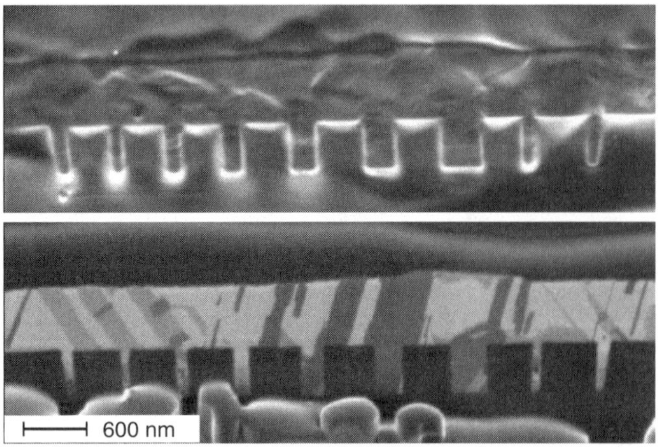

Figure 2. FIB secondary electron images of the cross-sections of copper plated electron beam patterned PMMA. Top: cleaved and mechanically polished. Bottom: the same area after FIB polishing.

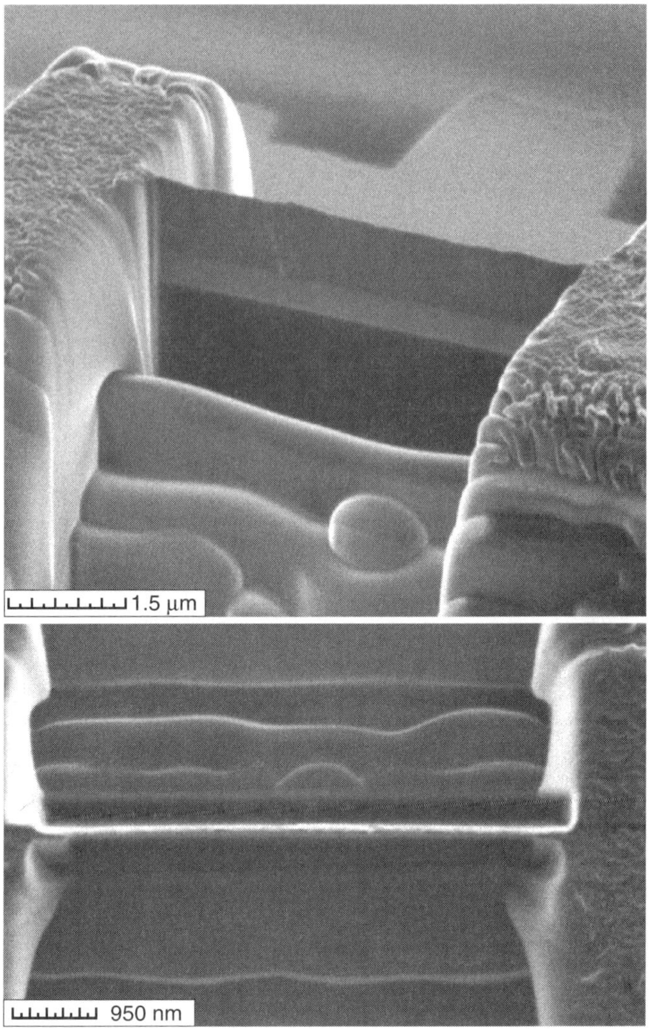

Figure 3. FIB fabricated lamella for TEM studies at 60° tilt (top) and top views (bottom). The lamella's thickness is between 60 and 100 nm.

pattern, and redeposition of sputtered material. The ion beam diameter determines the resolution of the ion-milled structure. To enhance the sputter yield by a factor between 2 and 30 and to suppress redeposition of sputtered material, an approach called FIB gas assisted etching (GAE) is often used.

FIB is a very useful, ultra-high-vacuum compatible tool for a maskless, direct, and local implantation of dopant ion species in semiconductors. Due to the small lateral straggling, the still increasing resolution in the further development of the instruments and the flexibility of the writing schedule programming for a step and repeat mode FIB can gain importance in the field of nanoscale semiconductor devices.

This chapter is not intended to describe FIB principles and equipment in detail but rather to show current results and potential opportunities for FIB fabrication at the sub-100 nm scale. The FIB fabricated structures presented in photographs in this chapter were prepared by the author using a Micrion FIB-2500 machine capable of operating at up to 50 kV accelerating voltage and with a Ga^+ ion beam diameter down to 5 nm.

2. FOCUSED ION BEAM (FIB) SPUTTERING AND IMPLANTATION AT NANOSCALE

2.1. FIB Implantation

FIB implantation takes place in conditions of much higher local ion currents than conventional implantation. Compared to the fundamental meaning of the damage production by ion irradiation, however, there exist only a few systematic investigations concerning high current density irradiations with a focused ion beam. For nanoscale patterning the appropriate level of interaction between ions and solids is translated toward low dose effects. In this case local defect injection now plays a major role due to the small thickness of active layers used, and to the high sensitivity of many materials and structures to ion bombardment.

The dramatic changes of magnetic, ferroelectric, optical, electric, and mechanical properties of FIB implanted materials have been observed. The effects of a high ion current during the FIB process on the implant distribution profiles, defect formation and annealing, chemical composition modification, and phase transformations are not studied in detail, especially at the sub-100 nm scale. It is obvious that the lateral straggling effect during implantation is a limiting factor in nanoscale applications. Some approaches to fabricate functional nanoscale FIB implanted structures will be given in Section 4, and this technique has yet to be optimized.

Shinada et al. [14] utilized FIB optics in single-ion implantation, which enables one to implant dopant ions one-by-one into a fine semiconductor region until the necessary number is reached. The authors have optimized the process in order to improve the controllability of dopant atom position to within 100 nm. The average aiming precision of 60 nm was achieved in one-by-one ion implantation of 60 keV Si ions into CR-39 (a fission track detector). Mitan et al. [15] have demonstrated formation of $CoSi_2$ lines using FIB implantation of 200 keV As^{2+} ions through the surface $Si/Co/SiO_2$ layers into the underlying Si substrate followed by the removal of the surface layers and annealing. The authors believe that the linewidth below 100 nm can be achieved.

Most FIB machines operate with Ga^+ ions at accelerating voltages of 30–50 kV, and this results in implantation and damage depths of less than 100 nm. This surface layer becomes structurally and chemically modified. For example, the exposure of a ferroelectric $Pb(Zr,Ti)O_3$ (PZT) material to the 50 kV Ga^+ ion beam dramatically changes the composition of a surface layer [16]. Auger electron spectroscopy reveals ~26–57 at% Ga in the 5 nm thick surface layer depending on ion dose. Oxygen concentration is significantly reduced when compared with a nondamaged film, while lead is practically lost. The composition of the surface layer at the dose of 6.24×10^{16} ion/cm^2 was found as $Ga_{0.53}Zr_{0.13}Ti_{0.21}O_{0.13}$. The annealing increases the amount of oxygen to 50–60 at%. The distribution of gallium in the damaged layer is not uniform. It was observed that gallium has a strong tendency to segregate in nanosize droplets (<50 nm) on the PZT surface during FIB milling. The FIB exposed 100 nm thick PZT films were found to lose the ferroelectric properties becoming dielectric after thermal annealing.

When a 50 kV Ga^+ FIB was applied to bulk diamond and so-called "tetrahedral amorphous carbon" (ta-C) films, the formation of Ga-rich amorphous graphitic layer was observed, as well as strong surface swelling due to the Ga implantation and substrate phase transformations [17]. Surface swelling is a common effect during conventional ion implantation. Huey and Langford [18] observed such an effect during low dose FIB irradiation of silicon surface. Surface protrusions with a height as small as 1 nm can be fabricated in an ordered manner. Dotsch and Wieck [19] have shown that the insulating lines in GaAs can be created using 100 kV Ga^+ FIB implantation. Roshchin et al. [20] noticed that thin magnetic films are extremely sensitive to ion irradiation even at 10^{12} cm^{-2} ion doses although Xiong et al. [21] found no effect of ion-induced damage on the magnetic properties of the magnetic nanoelements.

2.2. Sputtering

The majority of FIB applications employ physical sputtering of a substrate material to form either trench or protrusion patterns with various outlines. The precise shape of the sputtered region produced by FIB depends on the size of the region, the ion dose, the profile of the beam, and the beam scan parameters (scan strategy, time the beam is present in a spot or dwell time) [2, 4, 7]. The slope of the sidewalls of a sputtered region depends on the inverse gaussian profile of the ion beam current density. Both the speed and direction of scan are found to play a role in determining the results of material removal. The redeposition of a material sputtered from the bottom of the milled region causes a distortion of the sidewall profile. In a case of a narrow sub-100 nm groove or hole the redeposition effects will be the most important for the sidewalls of the structure. Furthermore, the reflected ions on the sidewalls (glancing incidence angle) will also have an effect on the ion-milled structure. At the base of the nearly vertical sidewalls a deep trench is formed (self-focusing effect) [2, 7]. Other observed effects such as an enhancement of the sputtering yield with increasing ion incident angle, precipitation of implanted species, and surface morphology changes can also play a significant role at nanoscale.

Campbell et al. [12] produced 60 nm wide and 120 nm deep trenches with nearly vertical sidewalls in Si using a 30 kV Ga^+ FIB process. Damage of silicon consisted of amorphization up to the depth of 80 nm, and possible gallium implantation up to depth of 50 nm according to TEM studies. Gallium precipitation was suggested at higher ion doses. Gierak et al. [5, 22] designed a 30 kV Ga^+ FIB machine and demonstrated a ~8–10 nm trench width in AlF_3/GaAs multiplayer samples and a hole diameter down to 10 nm in 10 nm thick SiC membranes. Sub-100 nm FIB sputtered patterns in both diamond and ta-C films were prepared by Stanishevsky et al. [17, 23, 24]. Using the lowest available beam current (2–5 pA) and an optimized beam scanning procedure, submicrometer tips with heights of ~800 nm and radii of ~40 nm at the top were milled in both a 1 μm thick ta-C film and chemical vapor deposition (CVD) diamond microcrystals. A minimum size of the milled through trench of ~30 nm has been achieved at these ion currents in 1 μm CVD diamond crystallites. A beam diameter of ~10 nm was estimated for the beam current of 3 pA. It was found that the effects of material swelling and redeposition hinder the trench formation at short dwell times. Similar trench sizes in ta-C film have been reported. An example of a narrow trench FIB milled in a ta-C film is shown in Figure 4. The depth of this 20 nm wide trench was estimated at ~400 nm.

Taniguchi et al. [25] and Russell et al. [26] have demonstrated the FIB fabrication of field emitter tips with a radius of less than 100 nm in a diamond crystal by sputtering the area around the central mesa. Olbrich et al. [27] proposed the FIB milling procedure to fabricate high aspect ratio diamond tips for atomic force microscopy. Kuball et al. [28] used FIB sputtering for the fabrication of GaN nanopillars. GaN pillars as small as 20 to 30 nm in diameter were fabricated. Using secondary electron imaging, the quality of the GaN pillars was investigated as a function of the gallium ion beam current used for the sputtering. Ion beam currents as low as 1 to 4 pA are needed to fabricate such a nanometer-size GaN structure, which makes the process very long.

Besides the grooves and pillars, it is also possible to fabricate metal wires with a nanosize width by the direct sputtering of a metal layer as shown in Figure 5. The 300 nm thick Ni wires with widths of ~80 nm and lengths of several hundred micrometers have been made by sputtering the Ni film on SiO_2 substrate using an optimized FIB scan procedure. The gap separating the nanowire from the remaining Ni film was found to be insulating despite the ion induced damage.

The ability of FIB to form narrow grooves in a thin metal layer was investigated as a key approach in the fabrication of novel structures for observation of such phenomena as single electron tunneling and giant magnetoresistance. Using a finely focused beam, narrow line patterns with narrow spacings were delineated on a Ge/nitrocellulose double layer resist [29]. It was shown that the line spacing and the length are limited mainly by the process of peeling off the Ge layer during the FIB sputtering. FIB was applied to form narrow grooves in magnetic and nonmagnetic metal films using Ge/nitrocellulose double layers as self-development resist and nanopatterns were formed to reveal the usefulness of FIB direct write processes for the fabrication of magnetic microdevices. To show the effectiveness of the present method for fabricating nanoscale grooves using a finely focused beam, 40 or 60 nm wide and 7.8 μm long grooves with a spacing of 500 nm were fabricated. Of course, much smaller structures can be currently fabricated as demonstrated in Figure 6. Nanogaps and nanobridges with widths down to 20 nm have been fabricated in Ni/Cu two-layer thin film heterostructures on SiO_2. The electrical

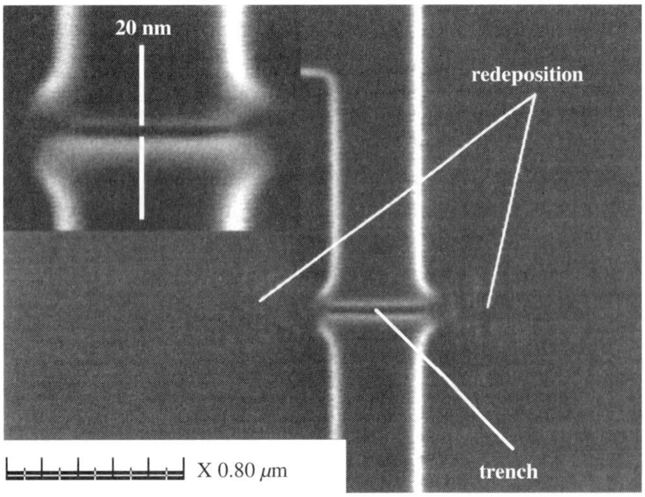

Figure 4. FIB milled 20 nm trench in ta-C film. Two areas of 200 nm depth were originally milled, and the trench connects these areas. The ion dose used for the trench formation was a priori found to provide a depth of ~400 nm. The spread of the redeposited material is visible at the both ends of the trench.

Figure 5. Direct FIB patterning of nanowires with a width down to 80 nm in 300 nm thick Ni film on quartz.

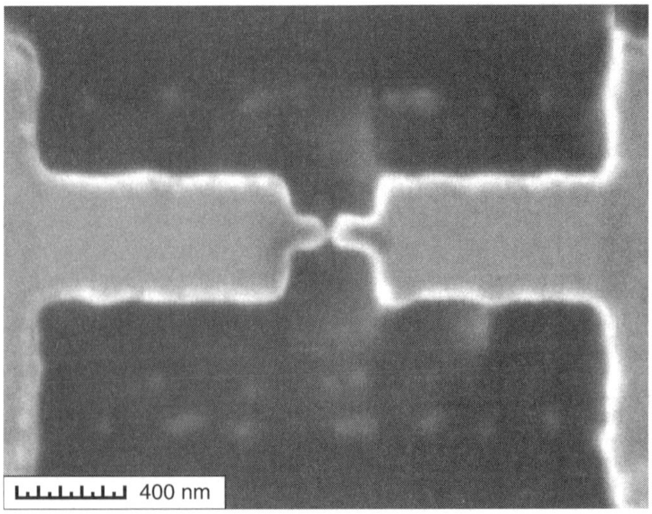

Figure 6. Nanobridges in Ni(10 nm)/Cu(50 nm) film on SiO_2/Si substrate. The bridge width is ~20 nm.

properties of such structures have not yet been investigated to date. It is unclear if there are effects of ion implantation, damage, and precipitation of implanted atoms on the electric properties of such nanostructures.

The observations of Ga precipitation in the FIB fabricated structures have been made [16, 30]. This effect may have serious consequences in FIB prototyping of nanostructures. In recent experiments [30], a focused ion beam was used to implant an array of spots in a silicon host. The FIB was programmed to inject a 1 mm^2 area with ion doses of 2×10^{10} ions/spot at a spot spacing of 10 μm. It was shown that the regions around the edges of the drilled holes do contain fine Ga precipitates with a size of ~5 nm. This proof-of-principle experiment conclusively demonstrated that the FIB can produce ordered spots on a substrate and that it is even possible to produce precipitates around the edges of the ion-drilled regions. Another observation of Ga precipitation during the FIB milling of a doughnut-like trench in quartz is shown in Figure 7. Relatively large (70–100 nm) droplets of Ga are formed along the edges of the sputtered region in some cases.

The effect of the ripple formation during the FIB sputtering is observed on tilted surfaces [24, 31]. This effect may become important when prototyping three-dimensional nanostructures, for example, optical lenses and emitter tips. Figure 8 shows the surface morphology changes of a ta-C film FIB sputtered at 60° incident angle. The height (30–40 nm) and period (150–400 nm) of the ripples are in the nanoscale range, and they depend little on the ion beam scan procedure. Understanding ripple formation on surfaces created by ion bombardment is of both fundamental and technological interest. Although there is a general consensus about the evolution of ripple structure as being due to competition between surface roughening introduced by sputtering and smoothing created by various processes, all the ripple formation mechanisms are yet to be identified. Candidates include pure thermal diffusion, viscous flow,

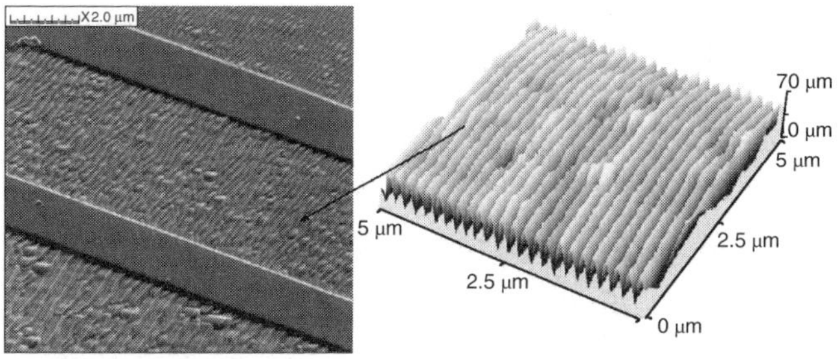

Figure 8. FIB secondary electron (left) and atomic force microscopy (right) images of ripples formed on the surface of ta-C film during 60° tilted FIB irradiation. The beam scan parameters were the same for all three areas in left image. The ion dose increases from the top to bottom area.

ion-enhanced or -inhibited diffusion, and preferential sputtering without actual mass movement.

The redeposition of sputtered material during the FIB patterning process seems to be the most important when dealing with nanosize structures. One problem associated with this effect is the need for much higher ion doses to fabricate, for example, a sub-100 nm diameter hole with certain depth when compared to a few micrometer diameter hole of the same depth. This is demonstrated in Figure 9. In this case, square 1-μm wide through apertures were FIB milled in a 2 μm thick silicon membrane with increasing ion doses. The image contrast variations at the edges of the apertures appear when the dose increases indicating the formation of a redeposited layer on the sidewalls. When the ion beam breaks through the membrane, the aperture width

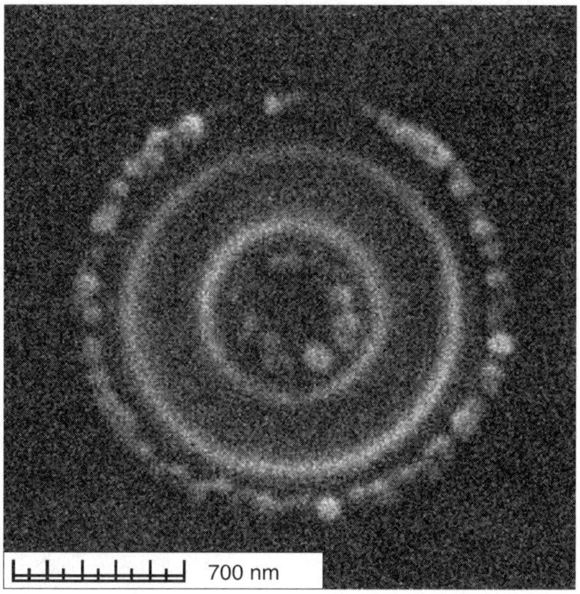

Figure 7. Gallium nanoparticles formed around a 50 keV Ga^+ FIB milled 100 nm deep doughnut-like trench in quartz.

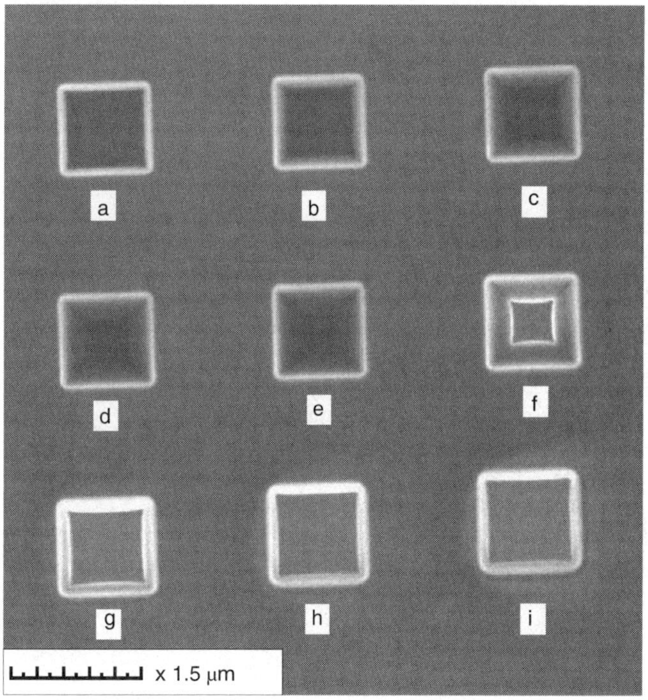

Figure 9. Redeposition effect of sputtered material onto the sidewalls during the membrane fabrication in 2 μm thick Si substrate. The ion dose increases from (a) to (i). The ion beams breaks through the substrate in the area (f).

is much smaller than expected (Fig. 9f). It requires an ion dose twice as high to remove the redeposited layer and to reach the size close to preset one (Fig. 9i). The fabrication of smaller through apertures requires even larger ion doses and, respectively, a very long process time. The redeposited material is much more difficult, if possible at all, to remove in the case of deep nanowide holes and trenches in a bulk substrate.

The diameter of the through apertures can be controlled by using a controlled endpoint termination technique. In this technique a Faraday cup is placed underneath the membrane and the transmitted ion current is measured *in-situ* during FIB milling. The milling can then be terminated, by blanking the ion beam, once a defined ion current is passing through the emerging hole in the membrane. The ordered arrays of through apertures with diameters down to <20 nm in 500 nm thick gold membranes have been successfully fabricated [32]. The size of such apertures can be further reduced employing the interesting effect recently described in [33, 34]. According to the authors, when a whole membrane containing 60 nm nanoholes is additionally irradiated by 3 keV Ar^+ ions, the diameter of a nanohole reduces to nanometer size until it collapses. Figures 10 and 11 demonstrate this effect for a case of additional irradiation of the area containing the arrays of FIB fabricated nanoholes in Si_3N_4 coated gold membrane with the same 50 kV Ga^+ FIB. One can see the shrinkage of the 100 nm holes down to 30 nm diameter after low dose irradiation (Fig. 10). Surprisingly, much larger holes shrink in the same way (Fig. 11). This cannot be currently explained by the theory proposed in [33, 34], and further investigations of this effect are of great interest. The shape of the initially cylindrical cross-section of the hole can be changed during the hole shrinking, and this can affect the nanoscale optical properties of such structures [32].

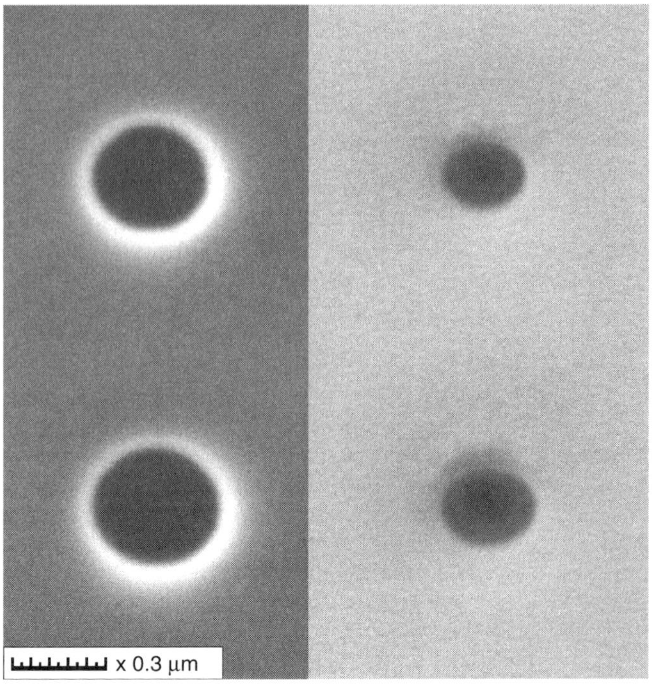

Figure 11. The effect of FIB low-dose irradiation of a 30 μm × 30 μm area of a 400 nm thick Si_3N_4 coated gold membrane on the changes of a 300 nm diameter through holes. Left image—as-milled holes, Right—after the low-dose exposure. Note the contrast change inside and around the holes.

2.3. FIB Gas Assisted Etching

The introduction of a reactive gas can enhance or reduce the sputtering rate of a particular material and provide higher selectivity between different materials. The redeposition effect can also be substantially reduced. The principle of ion beam gas assisted etching (GAE) is to adsorb suitable precursor gas molecules on a substrate and expose them to an ion beam. The ion beam causes a reaction between the adsorbed gas molecules and the substrate atoms, forming volatile compounds and thus an enhanced removal of material. The gases used in GAE applications should not react spontaneously with the substrate. Reactive gases commonly used in a FIB GAE processes are XeF_2 for SiO_2, Si_3N_4, and W etching, Cl_2 and I_2 for aluminum and silicon, O_2 for photoresists, H_2O for diamond, and Br_2 for chromium. Other gases or gas mixtures are under investigation [35, 36]. There are currently no reports on the application of FIB GAE to the fabrication of sub-100 nm structures. One reason for that is the requirement of a very short time period (dwell time) of the beam interaction with the adsorbed gas molecules to produce a greater enhancement of the sputtering rate. It has been shown that the yield of the reaction depends linearly on the surface coverage with gas molecules. When the focused ion beam moves along the surface, it consumes the adsorbed molecules, and a certain time (often called the "refresh time") is needed for the adsorbed layer to recover. If the size of the FIB exposed area is very small, for example, for making a nanohole, the beam is present in this area for too a long time or returns to the same spot quickly. This prevents formation of the adsorbed layer of the gas molecules, therefore diminishing the effect of GAE. Possibly only when the pattern size is large (or when a relatively big number of nanostructures is patterned simultaneously) can one benefit from the FIB GAE.

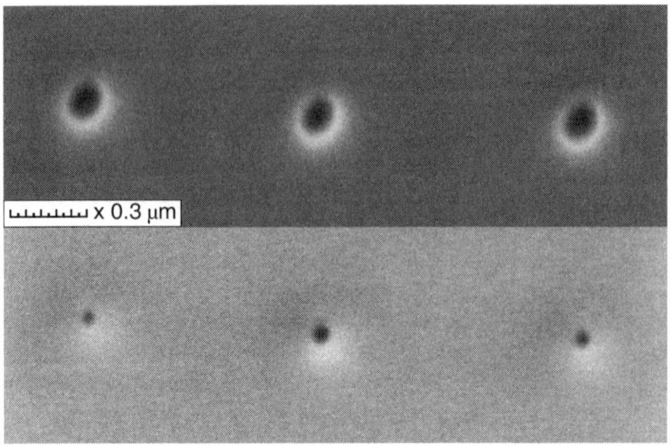

Figure 10. Shrinking of a nanohole diameter in patterned 400 nm thick Si_3N_4 coated gold membrane after additional FIB low dose exposure of a larger 20 μm × 20 μm surface area around the holes. Top—original as-milled diameter. Bottom—after the low dose exposure.

3. THREE-DIMENSIONAL NANOSTRUCTURES

The patterns of nanosize trenches or protrusions produced by the FIB sputtering give one example of possible three-dimensional nanostructure fabrication in a FIB process. Two other approaches to fabricate nanoscale structures using FIB induced deposition and FIB irradiation combined with chemical etching are described.

3.1. FIB Induced Deposition

In the case of FIB induced deposition, the incident energetic ions cause a fragmentation of the adsorbed molecules of a suitable gaseous compound leaving behind a deposit. FIB induced deposition enables the fabrication of complex three-dimensional shapes with overhanging features. A variety of gaseous precursors can be used, although few were utilized to date. FIB induced deposition has been used to fabricate layers of W, Au, Pt, Cu, C, and SiO_2. An example of the SiO_2 pillars formed by FIB deposition using TEOS and O_2 as the precursor gases is shown in Figure 12. Even at the smallest beam diameters, the size of the deposited structures is relatively large. The size of the deposited structures can be reduced by the FIB trimming using physical sputtering process as shown in Figure 13, but this is not always convenient.

The most successful work in truly nanoscale 3D direct deposition has been done by Matsui et al. [37]. This group fabricated 80 nm spirals of up to several micrometers tall using 30 kV Ga^+ FIB induced carbon deposition from an aromatic hydrocarbon precursor. Microcoil, drill, and bellows with 0.1 μm dimensions were also fabricated. That three-dimensional nanostructure fabrication process is illustrated in Figure 14. In this process, the beam is scanned at digital mode. First, a pillar is formed on the substrate by fixing a beam position. After that, the beam position is moved within a diameter of a pillar and then fixed until the deposited terrace thickness exceeds an ion range, which is a few tens of nanometers. This process is repeated to make three-dimensional structures. The key point to make three-dimensional structures in this approach is to adjust a beam-scan speed to avoid the sputtering of the deposited terrace, while providing the conditions for the growth of the overhanging part of the terrace. The growth conditions in x and y directions are controlled by ion beam deflectors. The growth of the z direction is determined by a deposition rate; that is, the height of structure is proportional to the irradiation time when the deposition rate is constant.

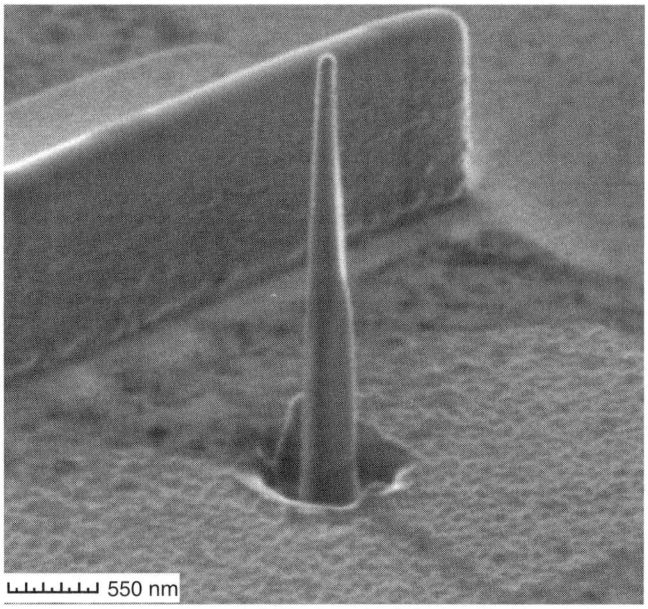

Figure 13. SiO_2 pillar with the tip diameter less than 100 nm after FIB trimming.

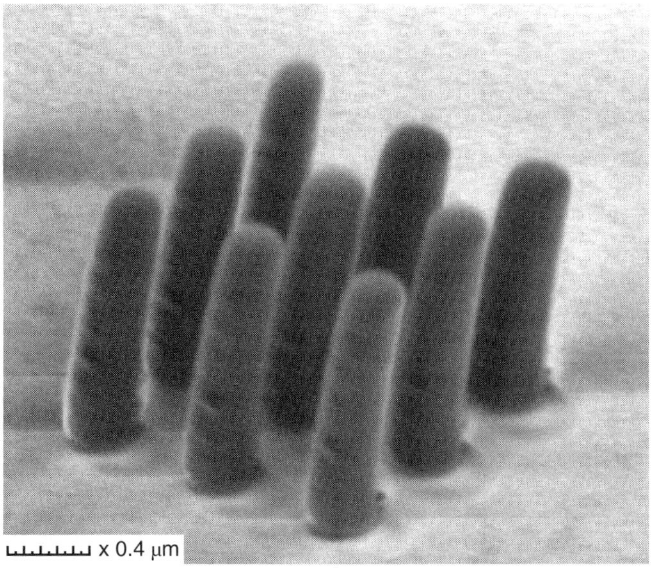

Figure 12. FIB induced deposition of SiO_2 nanopillars using tetratetraethylorthosilane (TEOS) and oxygen.

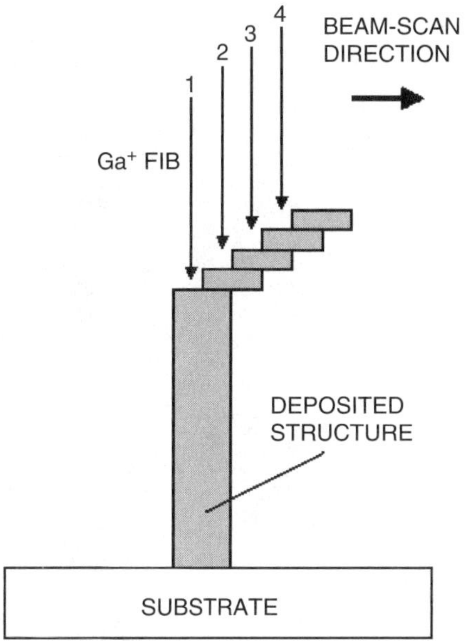

Figure 14. Fabrication process for three-dimensional nanostructure by FIB induced deposition proposed by Matsui et al. [37].

It was concluded from the experimental results that FIB direct write deposition processes may become interesting tools for the fabrication of micro- and nano-systems in the field of electronics, mechanics, optics, and biology.

FIB deposition of other materials, particularly W and Pt, has not yet been realized to fabricate sub-100 linewidths [38]. Direct write metallization is an important approach for circuit modification and prototyping. It was proposed that the FIB induced deposition starts with the nucleation of nanoscale metal deposits scattered over the substrate surface. Despite local impacts of the ion beam within the irradiated area of the substrate the localization of the nucleation spots is not correlated to the scan path of the ion beam. The nanoscale tungsten particles preserve their position and typical shape during further deposition. Only after merging of the particles into a contiguous tungsten layer does the second regime of growth characterized by deposition of tungsten on a tungsten surface set in. In this regime the deposition process is determined by the total ion dose and the average current density the sample was subjected to.

There is a need in more theoretical studies of FIB induced deposition for nanoscale structures. The effect of much faster growth of a deposit inside the trenches than on a flat surface is commonly observed in FIB induced deposition, but the comparison of the composition and the properties of the deposits in these two situation has not yet been reported. This effect is demonstrated in Figure 15. The substrate containing FIB sputtered narrow grooves was first coated with a ~500 nm thick SiO_2 layer, which also filled the whole depth of the grooves. After that, the FIB sputtering was used again to mill narrower grooves into the FIB-deposited SiO_2 layer. The ~150 nm thick layer of tungsten was finally deposited. The tungsten deposit filled also the whole depth of the grooves, resulting in the ~50 nm feature size at the groove tip. Although this structure outlines the possible nanoscale design of FIB fabricated nanovias and interconnects for nanodevice prototyping, the properties and stability of these nanodeposits have yet to be investigated.

Figure 15. FIB induced deposition of SiO_2 dielectric layer and tungsten-plug in nanovias.

Another major problem in FIB induced deposition is the presence of the ion beam tail. This leads to a substantial deposition around complex patterns as shown in Figure 16. Although this tungsten microheater is not a nanostructure, it indicates the common problem of excessive deposition in unwanted areas in the FIB induced deposition process. One approach to overcome this problem is to use FIB GAE with suitable reactive gases to clean the fabricated structure as shown in the bottom picture in Figure 16. Other precursor compounds and beam scan strategies also have to be investigated.

3.2. Etching of FIB Exposed Surface

It is well known that a high concentration of p^+ doping in Si drastically reduces its etch rate in certain chemical etchants such as KOH and hydrazine. The FIB irradiation of selected regions on Si substrate leads to the same result. The exposure to Ga^+ FIB with ion doses $>10^{15}$ cm^{-2} of selected

Figure 16. Ion beam tail effect on the deposited pattern. Top: as-deposited tungsten meander microheater; Bottom: the same structure after GAE cleaning.

areas reduces the etching rate of such areas to a negligible value when compared to bulk Si (100). This approach has been utilized to fabricate various micro- and nanoscale three-dimensional structures. Koh et al. [39] demonstrated formation of ordered arrays of nanopyramids by wet etching of a 60 kV Si^+ FIB irradiated Si (100) surface in a hydrazine solution for 15 s at 115 °C. An example of similar arrays initially patterned by 50 kV Ga^+ FIB with an ion dose of $\sim 1 \times 10^{15}$ cm^{-2} and subsequently etched in 10% KOH solution at 80 °C is shown in Figure 17. The thickness of the insoluble layer in Si is ~100 nm at these FIB parameters. When a much larger ion dose is used to irradiate the substrate this results in a pattern similar to that shown in Figure 17; the deep holes are formed due to sputtering of the substrate material. The KOH etching removes unexposed material leaving funny three-dimensional structures because the ion-damaged sidewalls of these holes are stable to the chemical etching (see, for example, Fig. 18). The thickness of the walls in such structures depends on the ion species and their energy, the substrate material and the beam scan strategy. Using the described approach many other complex 3D micro- and nanostructures can be fabricated. Figure 19 shows an array of FIB delineated bridges with lengths of ~30 μm. The narrowest bridge has a width of ~100 nm. The bridge with the preset width of 50 nm did not survive the etching/rinsing process. The minimum width of nanocomponents produced in this approach has yet to be determined. Complex shape cantilevers with a width of ~100 nm produced by FIB and KOH etching were also demonstrated

Figure 18. There-dimensional glasslike structures on KOH wet-etched Si, roughly representing damage profile produced by FIB at large ion doses.

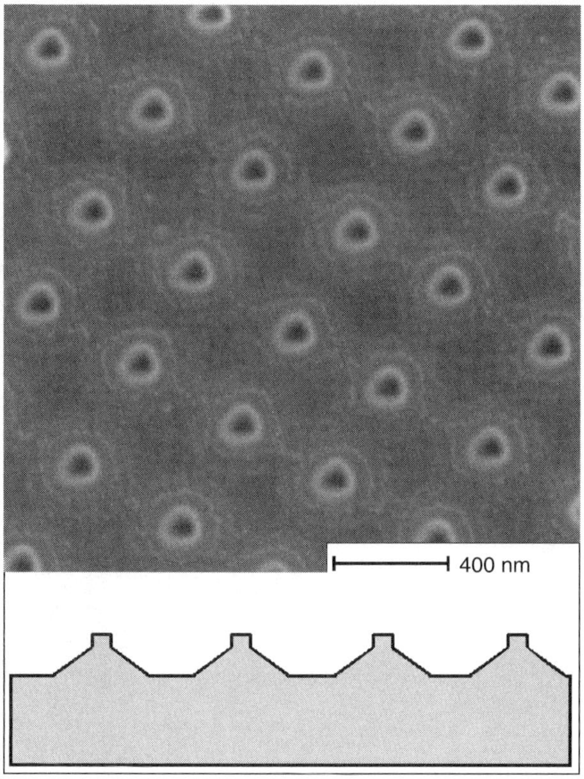

Figure 17. Pyramid array formed by FIB implanted dots after wet etching of unexposed Si (100) surface. The surface profile is graphically shown.

[40]. It is possible to create other quite complex structures in such an approach, like gridlike membranes with ~100 nm component width as shown in Figure 20. These structures may have attractive mechanical properties for prototyping micro- and nanomechanical components. However, the contamination of the fabricated structures with Ga or other ion species can be a limiting factor.

FIB induced surface modification can also have the opposite effect and enhance the rates of chemical or reactive ion etching processes depending on the reactive compound. Enhanced etching of FIB exposed InP in HF was observed [41]. The FIB fabricated patterns were applied

Figure 19. Si nanobridges fabricated by KOH etching of a FIB patterned Si (100) surface. The minimum width of 80 nm for 30 μm long wire was achieved.

Figure 20. A segment of a gridlike nanomembrane fabricated by KOH wet etching of Si (100) substrate with a pattern defined by FIB induced damage.

to create index coupled feedback gratings in GAInAsP/InP laser structures. The width of the pattern elements was ~100 nm. Koshida et al. [42] fabricated 40 nm wide refractory Mo and W metal lines on Si substrates by FIB exposure of MoO_3 and WO_3 inorganic resists, development, and subsequent reduction in dry hydrogen gas.

Titanium and TiO_2 nanowires for potential chemical sensor applications with lengths up to 100 mm were produced by etching FIB exposed Ti and TiO_2 films on SiO_2 with appropriate chemical solutions. An example of TiO_2 100 nm single-grain wide nanowire is shown in Figure 21.

An interesting approach to produce nanosize channels in alumina membranes was proposed by Liu et al. [43]. The

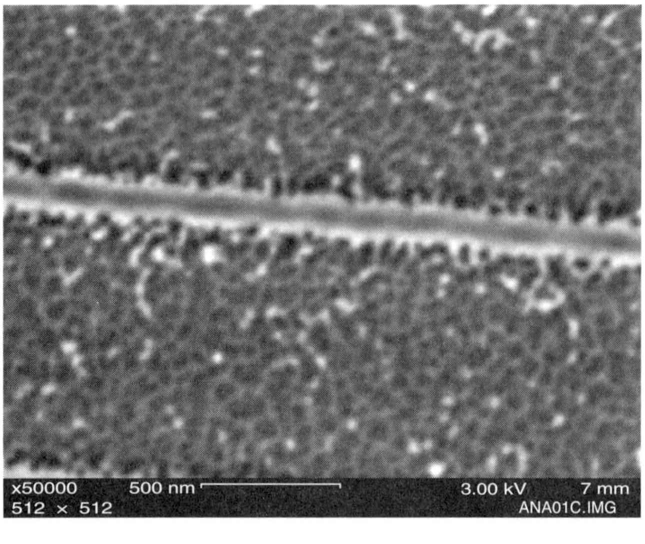

Figure 21. FIB delineated metal oxide (TiO_2) nanowires on SiO_2.

surface of polished Al was irradiated by FIB followed by anodization of the surface and removal of the remaining Al substrate. It was found that concaves as shallow as a few nanometers can effectively guide growth of the nanochannels to an aspect ration of ~100. A pore size of 20–50 nm seems to be achieved.

4. FUNCTIONAL SUB-100 nm STRUCTURES

A variety of working nanostructures has been produced using FIB sputtering and implantation processes. Formation of narrow grooves is a key process in fabrication of novel structures for observation of such phenomena as single electron and photon tunneling, Coulomb blockade, and giant magnetoresistance. The ability to fabricate nanometer-wide trenches and holes was used to make novel Josephson junction devices [44–47], nanosize shadow masks in silicon membranes [48] to subsequently deposit InGaN quantum dots and quantum wires by metal organic CVD. Ochiai et al. [49] demonstrated wedge emitter fabrication using focused ion beam physical sputtering. The emitter–collector gaps of various widths ranging from 10 to 1000 nm were produced as shown in Figure 22. The formation by FIB sputtering of a nanohole in a metal layer was followed by the chemical etching of underlying SiO_2 to remove the ion induced damage in the emitter structure.

The FIB implantation is increasingly used to fabricate numerous magnetic nanostructures [20]. It was found that thin magnetic films are extremely sensitive to ion irradiation even at 10^{12} cm^{-2} ion doses. Reducing the lateral dimensions of patterns in magnetic materials to the level of certain critical parameters (exchange length, domain wall width, Barkhausen jump length) can lead to new insights into nanomagnetic applications such as ultrahigh density recording.

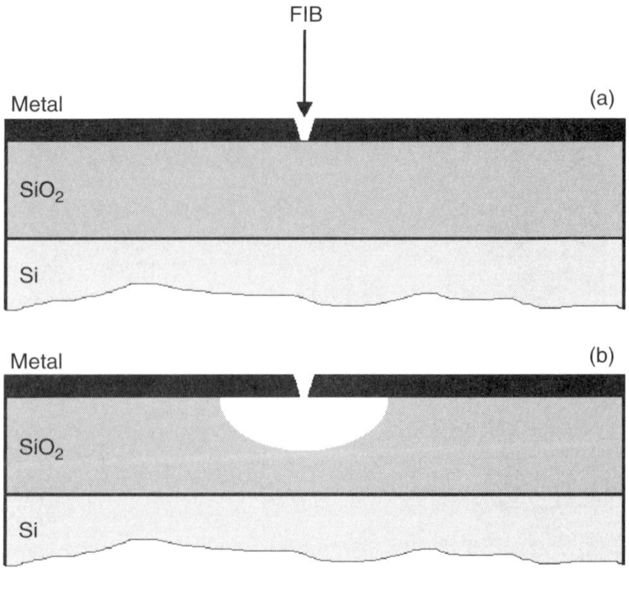

Figure 22. Fabrication process for a Mo field emitter with a lateral wedge. (a) Sample cross-section of a Mo/SiO_2/Si structure with FIB sputtering for the production of an emitter–collector gap. (b) Subsequent wet etching for the suppression of the structural damage.

Anders et al. [50] have fabricated magnetic nanostructures using focused ion beams to pattern perpendicular media. The perpendicular granular media consisted of a 20 nm $Co_{70}Cr_{20}Pt_{10}$ magnetic film protected by 5 nm of CN sputtered onto a Cr–Ta (1 nm)/Ni–Al (6 nm) underlayer structure grown on glass substrates. Arrays of square islands with sizes ranging from 50 to 730 nm and periods from 70 to 750 nm were fabricated and studied using atomic force microscopy (AFM) and magnetic force microscopy. Small islands with periods of 130 nm are single domain while larger islands show multidomain states similar to the unpatterned film. The smallest islands exhibit a thermal stability far superior to the continuous media (about three orders of magnitude higher). The high thermal stability of the nanoislands can be explained, in part, by a substantial reduction in the demagnetizing fields caused by the FIB patterning.

Similar experiments by Xiong et al. [21] have demonstrated that very high quality magnetic nanostructures can be fabricated by FIB milling even in thin films of soft magnetic materials such as permalloy. No evidence was found for ion-induced damage to the magnetic properties of the nanoelements. The length scales of 100 nm achieved using the FIB fabrication method are suitable for single-domain magnetic devices. This finding allows a high-throughput rate of device prototyping and could even be significant for the future commercialization of certain niches of magnetic nanotechnology and magnetoelectronics.

FIB implantation of B^+ ions into *n*-type doped silicon yields lateral *npn* junctions [51] producing white luminescence. It was found that the actual ion beam size plays only a small role because optical radiation originates from nanoclusters only in depleted regions of about 100 nm in width. These junctions emit light from their nanoscale depletion regions if operated in reverse biased breakdown mode, depending on voltage polarity on either side of the implanted area. Bacher et al. [52] realized luminescent structures based on CdTe/CdMnTe quantum dots using 100 kV, 30 nm Ga^+ ion beam diameter. The maximum of photoluminescence shifts from 1.69 to 1.77 eV with the ion dose increase from 10^{13} to 5×10^{13} cm^{-2}. Kuball et al. [53] fabricated pillars with a lateral size down to 30 nm in GaN/AlGaN heterostructures. Although the larger (~300 nm) GaN/AlGaN HFETs fabricated by FIB exhibit photoluminescence despite the FIB induced damage, the results for smaller structures have not been reported.

Kondo et al. [54] fabricated Si MOSFETs with a nanoscale channel by changing the effective channel lengths from 27 to 82 nm using FIB implantation and SiO_2 implantation masks with very narrow widths (see Fig. 23), and they have investigated Coulomb blockade phenomena in such structures. Coulomb blockade oscillations were observed in the temperature range from 4.2 to 13 K. Kim et al. [55] used FIB techniques utilizing both lithographic and nanoparticle formation processes to fabricate a single-electron transistor that can operate at room temperature. The results for the drain current as a function of the gate voltage at different source voltages at room temperature clearly showed Coulomb oscillations indicative of Coulomb-blockade effects. These results indicated that FIB fabricated single electron transistor structures operating at room temperature hold promise for potential applications in ultra-high-density memory devices.

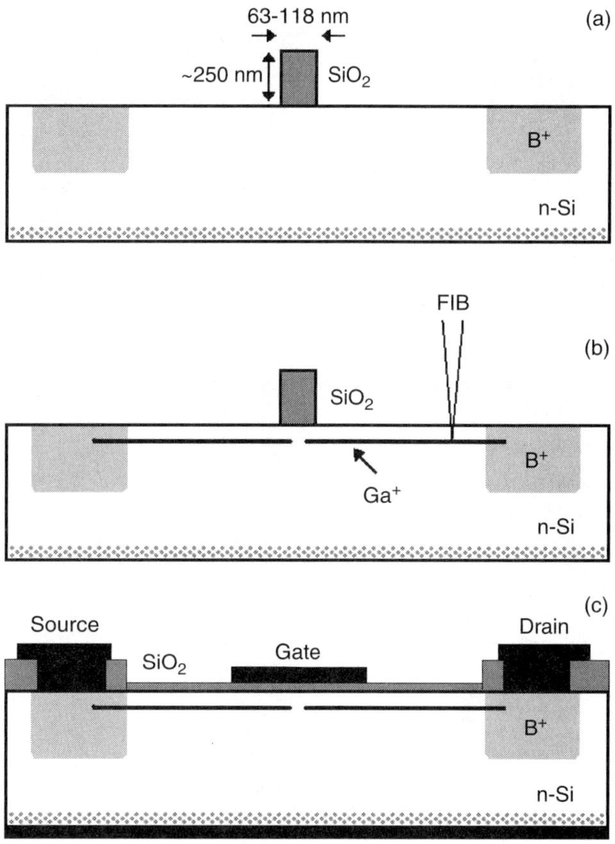

Figure 23. Schematic diagram showing the fabrication sequence of a transistor structure. (a) A base structure with SiO_2 implantation masks processed using e-beam lithography and electron cyclotron resonance plasma etching, and p^+ regions formed by conventional ion implantations of boron. (b) By scanning a focused Ga^+ ion beam across the SiO_2 masks, the *p*-type source/drain regions were formed across the nonimplanted gap. (c) A schematic diagram of the final device structure.

Another kind of a nanotransistor structure was proposed by Dotsch and Wieck [19] who defined insulating lines in two-dimensional electronic layers in GaAs using a 100 kV Ga^+ ion beam with a diameter of 30 nm. A lateral so-called in-plane-gate field-effect transistor-like structure was demonstrated.

FIB patterning of thin film multilayer stacks to form ferroelectric PZT capacitor heterostructures was conducted [16, 56] to investigate the effect of scaling on such structures. It was found that the thermally annealed FIB patterned working nanocapacitor (Fig. 24) consists of a dielectric (sidewall layer) and ferroelectric (bulk) parts. The thickness of the sidewall damaged layer is ~5–10 nm. The contribution of this sidewall layer is significant in sub-100 nm structures, and it will affect the results of the measurements of the capacitor characteristics. The volume of the dielectric part in a 70 nm capacitor reaches ~50%, and the actual lateral size of the ferroelectric part is ~50 nm.

Nanofabrication of charge density wave (nano-CDW) sub-100 nm structures by applying a hydrogen focused ion beam to the one-dimensional conductor, molybdenum blue bronze ($K_{0.3}MoO_3$) crystal [57, 58], was reported. The nano-CDW

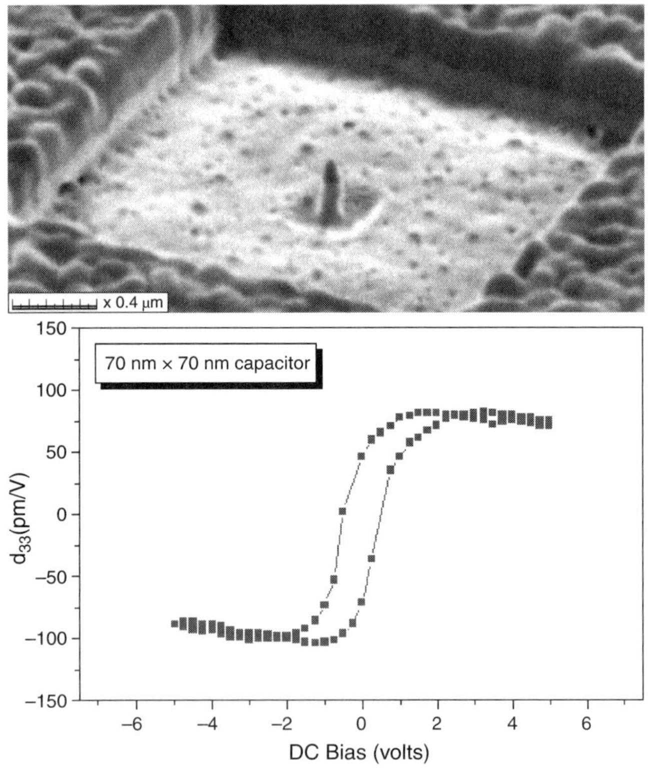

Figure 24. FIB fabricated $Pt/La(Sr_{0.5}Co_{0.5})O_3/Pb(Nb_{0.04}Zr_{0.28}Ti_{0.68})O_3/La(Sr_{0.5}Co_{0.5})O_3/Pt/SiO_2$ ferroelectric nanocapacitor structure with a lateral size of 70 nm (top) and its piezoelectric hysteresis loop (bottom).

structures show negative resistance above the threshold fields, which monotonously increased with decreasing temperature. The one-dimensionality of electrons has been confirmed by negative magnetic resistance, whose magnitude reduces with decreasing dimensionality.

Several functional sensor nanostructures have been also proposed. Sandhu et al. [59] fabricated bismuth Hall-effect nanoprobes by FIB trimming with dimensions of 120 × 120 nm for direct imaging of magnetic materials using scanning Hall probe microscopy. These nanoprobes were used for room temperature imaging of crystalline garnet films and strontium ferrite permanent magnets. The Hall coefficient and magnetic field sensitivity of the Bi nanoprobes were 3.3×10^{-4} $\Omega\,G^{-1}$ and 7.2 $G/Hz^{1/2}$, respectively, at driving current of 40 μA. Puers et al. [60] fabricated a miniature "Pirani" pressure sensor. Although the minimum lateral dimension of that sensor is ~1 μm, it is worth mentioning this device here. The sensor's working principle is that of Pirani-type vacuum sensors. However, unlike most Pirani sensors, the working range of the device is around atmospheric pressure. The device active area was fabricated by FIB induced deposition of W, followed by FIB trimming to fabricate a suspended W bridge of ~10 × 1 μm size. The structure was encapsulated in a FIB deposited SiO_2 cage (protective cap). The electric resistivity of W was 180 $\mu\Omega$ cm, and it was able to withstand a current density of 7.5×10^9 A/m^2.

AFM cantilevers with integrated nanoelectrodes were proposed by Lugstein et al. [61]. Such a scanning probe is capable of simultaneously measuring topography and local electrochemistry at a sample surface. Nanosize electrodes were patterned directly on cantilevers for scanning probe microscopy by FIB trimming.

The emerging area where the FIB finds numerous applications is micro- and nano-optics. Naghski et al. [62] have proposed the modification of optical structures by FIB patterned grooves. Optical channel waveguides formed by FIB implantation-induced mixing of AlGaAs multiple quantum well structures and subsequent oxidation of the mixed regions have the potential of significantly reducing the size of integrated photonic waveguide structures. The calculations presented for the submicrometer channel waveguides show that reductions in size by at least an order of magnitude are possible for directional couplers and other structures involving curved channel waveguide sections. Such size reductions would allow the realization of significantly higher levels of device integration than those currently possible.

The FIB capability to produce nanoholes with a high aspect ratio is useful in fabrication of the apertures in near-field scanning optical microscopy (NFSOM) tips [63, 64]. Danzebrink et al. [64] used FIBs to fabricate slit-shaped apertures that show a polarization dependent transmission efficiency and a lateral resolution of <100 nm at a wavelength of 1064 nm. These slits were fabricated in metallized silicon cantilevers. Lehrer et al. [65] realized the NFSOM aperture probes with the actual hole diameter down to 60 nm. These aperture probes are based on metal-coated AFM sensors. Apertures with circular and rectangular shape with dimensions below 100 nm were fabricated. Optical near-field measurements (wavelength of light 1064 nm) were used to demonstrate the functionality of these probes with a resolution down to 60 nm (1/17 of wavelength). The properties of NFSOM aperture probes can be further improved by realizing the subwavelength surface plasmon full-eye structures as shown in Figure 25.

Observation of photon tunneling gated by light at a different wavelength in the FIB created array of nanometer scale cylindrical channels in a thick gold film has been demonstrated [32]. In these experiments the rectangular arrays of nanometric cylindrical holes have been patterned into a freestanding 400 nm thick gold membrane using focused gallium ion beam milling. To avoid stress in the gold membrane, the gold film is supported by 50 nm thick Si_3N_4 membrane with a 5 nm intermediate Cr layer. Since there is no glass/metal interface, milling artifacts such as redeposited material can be avoided leading to better-defined structure geometry. Relatively large areas (50 μm × 50 μm) of the gold membranes were patterned, creating nanosized holes with dimensions down to 20 nm. The diameter of the hole can be monitored by using a controlled endpoint termination technique. A drop of poly-3-butoxy-carbonylmethyl-urethane (3BCMU) polydiacetylene solution in chloroform was deposited onto the gold membrane surface. After solvent evaporation polydiacetylene film was left on both the top and the bottom surfaces of the perforated gold membrane. Such 3BCMU polydiacetylene films hold the current record for the largest fast nonresonant optical $x^{(3)}$ nonlinearity. Polarization properties of gated light provided strong proof of the enhanced nonlinear optical mixing in nanometric channels involved in the process. This suggests the

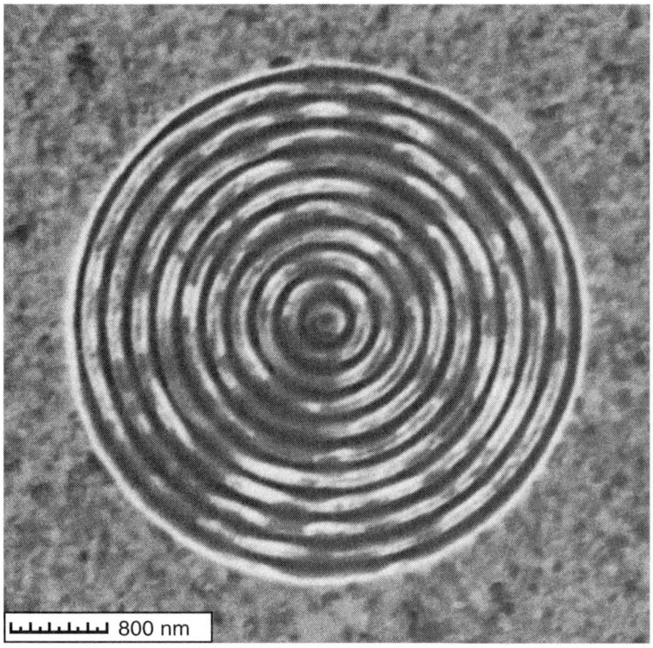

Figure 25. FIB milled optical ring structure with ~80 nm linewidth in 300 nm thick Ag thin film on quartz.

possibility of building a new class of "gated" photon tunneling devices for massive parallel all-optical signal and image processing. More complex nanostructures (see, for example, Fig. 26) for nanophotonic applications are currently under investigation.

The formation of nanocrystals in FIB irradiated materials has been observed. Tarumi et al. [66] found this effect for amorphous Ni–P alloy where the crystallographically oriented nanocrystals were formed. The FIB induced formation of arranged nanocrystal regions in optical materials may have a great impact on design of new optical nanodevices.

An interesting approach to locally electrochemically deposit metal structures was shown by Spiegel et al. [67]. In this study the Cu conductive lines with a width down to 300 nm were grown on a FIB irradiated Si surface. Although

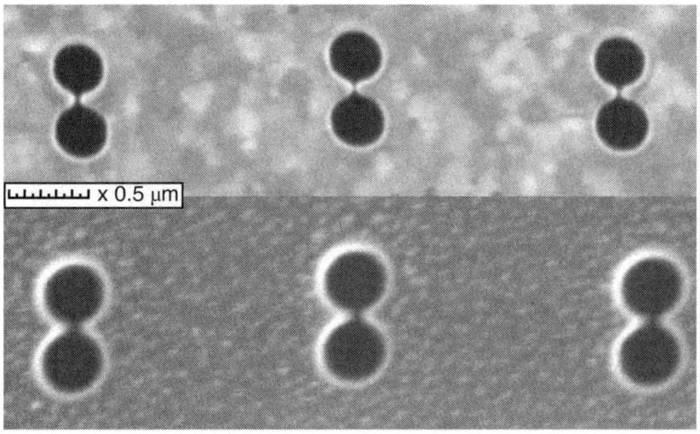

Figure 26. Fabrication of nanogaps for nano-optical structures in a Si_3N_4 coated gold membrane. Bottom image shows the surface facing the FIB. Top—the opposite view. Note the nanoscale gap formed in each case.

this is much larger than 100 nm, this approach is interesting to explore. The copper nanostructures were deposited on *p*-type silicon (*p*-Si) by means of a selective electrochemical reaction. *P*-type Si was implanted with Ga^+ ions at ion energy of 30 keV with different ion fluences, 1.0×10^{12} to 1.0×10^{16} cm^{-2}. In a second step Cu was deposited selectively at the implantation sites at different cathodic potentials and for different exposure times. Deposition time, as well as potential, strongly affects the morphology of the Cu deposit. Extended polarization leads to overgrowth of the initial defect patterns resulting in a loss of lateral growth control. By adjusting the implant and electrochemical conditions it was possible to obtain deposits as fine as 300 nm (width), which is also the lateral resolution of the FIB process used (i.e., corresponds to the ion beam distribution).

5. SUMMARY

Each technique that can be utilized for nanofabrication, such as laser NFSOM, scanning probe microscopy, electron beam lithography, and FIB patterning, has its own excellent features.

The focused ion beam is a powerful tool for engineering at nanoscale. The main benefits are the high flexibility of the nanostructure designs that can be realized. The slow processing is the main drawback of the FIB. This technique is best suited for prototype structures and nanodevice fabrication.

Future tasks for researchers dealing with FIBs for nanoengineering are to explore novel applications of the high-resolution ion beams thereby opening new niche markets for this technology. A breakthrough in FIB technology applications will be the demonstration of potential mass production capacity.

GLOSSARY

FIB induced deposition Based on the dissociation of adsorbed molecules of a metal organic or another gaseous precursor by incident ions, which results in a deposited layer. Growth rate is proportional to the reaction yield and the ion flux.

Focused ion beam (FIB) A method for a high-vacuum micro and nano-machining based on a liquid metal ion source (LMIS), which produces an optically bright beam of ions that are accelerated focused to sub-100 beam diameter, and rastered over a surface in a desired pattern, causing implantation and sputtering of the target.

Gas assisted etching (GAE) The enhancement of removal of material from a certain area in FIB processing. When a gas is introduced near the surface of the sample during FIB milling, the sputtering yield can selectively increase depending on the chemistry between the gas and the material. The etchant is selected based on its ability to react with the sputtered material to form volatile molecules that do not bond to the sample material. GAE ion beam induced reaction kinetics is based on an assumption that the yield of the reaction depends linearly on the density of the adsorbed gas molecules.

Liquid metal ions source (LMIS) Uses the quantum mechanical field emission/field evaporation mechanism, appearing at high electric field strengths, for ionization of metal ions. The molten metal covers a sharp emitter tip, which has several kV potential with respect to an extraction electrode aperture. The high field strength at the emitter tip enables the metal atoms to evaporate and ionize. Ions are extracted through the extraction aperture and can be further accelerated.

REFERENCES

1. M. Ultaut, Focused ion beams, in "Handbook of Charged Particle Optics," pp. 429–465. CRC Press, Boca Raton, FL, 1997.
2. J. Melngailis, *J .Vac. Sci. Technol. B* 5, 469 (1987).
3. K. Nikawa, *J. Vac. Sci. Technol. B* 9, 2566 (1991).
4. L. R.Harriott, *Nucl. Instrum. Methods B* 55, 802 (1991).
5. J. Gierak, D. Mailly, G. Faini, J. L. Pelouard, P. Denk, F. Pardo, J. Y. Marzin, A. Septier, G. Schmid, J. Ferre, R. Hydman, C. Chappert, J. Flicstein, B. Gayral, and J. M. Gerard, *Microel. Eng.* 57, 865 (2001).
6. P. D. Prewett and G. L. R. Mair, "Focused Ion Beams from Liquid Metal Ion Sources." Wiley, New York, 1991.
7. W. Driesel, *Phys. Status Solidi A* 146, 523 (1994).
8. J. Melngailis, A. A. Mondelli, I. L. Berry III, and R. Mohondro, *J. Vac. Sci. Technol. B* 16, 927 (1998).
9. K. Edinger, Focused ion beams for direct writing, in "Direct Write Technologies for Rapid Prototyping" (A. Pique and D. Chrisey, Eds.), pp. 347–383. Academic Press, San Diego, 2002.
10. S. Reyntjens and R. Pueis, *J. Micromech. Microeng.* 11, 287 (2001).
11. D. F. Moore, J. H. Daniel, and J. F. Walker, *Microel. J.* 28, 465 (1997).
12. L. C. I. Campbell, D. T. Foord, and C. J. Humphreys, *Electron Microsc. Anal.* 153, 657 (1997).
13. R. M. Langford, A. K. Petford-Long, M. Rommeswinkle, and S. Egelkamp, *Mater. Sci. Technol.* 18, 743 (2002).
14. T. Shinada, H. Koyama, C. Hinoshita, K. Imamura, and I. Ohdomari, *Jpn. J. Appl. Phys.* 41, L287 (2002).
15. M. M. Mitan, D. P. Pivin, T. L. Alford, and J. W. Mayer, *Thin Solid Films* 411, 219 (2002).
16. A. Stanishevsky, B. Nagaraj, J. Melngailis, R. Ramesh, L. Khriachtchev, and E. McDaniel, *J. Appl. Phys.* 92, 3275 (2002).
17. A. Stanishevsky and L. Khriachtchev, *J. Appl. Phys.* 86, 7052 (1999).
18. B. D. Huey and R. M. Langford, *Nanotechnology* 14, 409 (2003).
19. U. Dotsch and A. D. Wieck, *Nucl. Instrum. Methods B* 139, 12 (1998).
20. I. V. Roshchin, J. Yu, A. D. Kent, G. W. Stupian, and M. S. Leung, *IEEE Trans. Magn.* 37 2101 (2001).
21. G. Xiong, D. A. Allwood, M. D. Cooke, and R. P. Cowburn, *Appl. Phys. Lett.* 79, 3461 (2001).
22. J. Gierak, A. Septier, and C. Vieu, *Nucl. Instrum. Methods A* 427, 91 (1999).
23. A. Stanishevsky, *Diamond Relat. Mater.* 8, 1246 (1999).
24. A. Stanishevsky, *Thin Solid Films* 398–399, 560 (2001).
25. J. Taniguchi, N. Ohno, S. Takeda, I. Miyamoto, and M. Komuro, *J. Vac. Sci. Technol. B* 16, 2506 (1998).
26. P. E. Russell, T. J. Stark, D. P. Griffis, J. R. Phillips, and K. F. Jarausch, *J. Vac. Sci. Technol.* B 16, 2494 (1998).
27. A. Olbrich, B. Ebersberger, C. Boit, Ph. Niedermann, W. Hanni, J. Vancea, and H. Hoffmann, *J. Vac. Sci. Technol. A* 17, 1570 (1999).
28. M. Kuball, F. H. Morrissey, M. Benyoucef, I. Harrison, D. Korakakis, and C. T. Foxon, *Phys. Status Solidi A* 176, 355 (1999).
29. M. Yoshida, S. Murakami, M. Nakayama, J. Yanagisawa, F. Wakaya, T. Kaito, and K. Gamo, *Microelectron. Eng.* 57–58, 877 (2001).
30. R. A. Zuhr, J. D. Budai, P. G. Datskos, C. M. Egert, A. Meldrum, K. A. Thomas, C. W. White, L. C. Feldman, M. Strobel, and K-H. Heinig, *Mater. Res. Soc. Symp. Proc.* 536, 251 (1999).
31. A. Datta, Y.-R. Wu, and Y. L. Wang, *Phys. Rev. B* 63, 125407 (2001).
32. I. I. Smolyaninov, A. V. Zayats, A. Stanishevsky, and C. C. Davis, *Phys. Rev. B* 66, 205414 (2002).
33. J. Li, D. Stein, C. McMullan, D. Branton, M. J. Aziz, and J. A. Golovchenko, *Nature* 412, 166 (2001).
34. D. Stein, J. L. Li, and J. A. Golovchenko, *Phys. Rev. Lett.* 89, 276106 (2002).
35. K. Edinger and T. Kraus, *J. Vac. Sci. Technol. B* 18, 3190 (2000).
36. K. Edinger, *J. Vac. Sci. Technol. B* 17, 3058 (1999).
37. S. Matsui, T. Kaito, J. Fujita, M. Komuro, K. Kanda, and Y. Haruyama, *J. Vac. Sci. Technol. B* 18, 3181 (2000).
38. H. Langfischer, B. Basnar, H. Hutter, and E. Bertagnolli, *J. Vac. Sci. Technol. A* 20, 1408 (2002).
39. M. Koh, S. Sawara, T. Goto, Y. Ando, T. Shinada, and I. Ohdomari, *Jpn. J. Appl. Phys.* 39, 2186 (2000).
40. J. Brugger, G. Beljakovic, M. Despont, N. F. de Rooij, and P. Vettiger, *Microelectron. Eng.* 35, 401 (1997).
41. S. Rennon, L. Bach, H. Konig, J. P. Reithmaier, A. Forchel, J. L. Gentner, and L. Goldstein, *Microelectron. Eng.* 57, 891 (2001).
42. N. Koshida, S. Watanuki, K. Yoshida, K. Endo, M. Komuro, and N. Atoda, *Jpn. J. Appl. Phys.* 31, 4483 (1992).
43. C. Y. Liu, A. Datta, and Y. L. Wang, *Appl. Phys. Lett.* 78, 120 (2001).
44. H. W. Seo, Q. Y. Chen, C. Wang, W. K. Chu, T. M. Chuang, S. F. Lee, and Y. Liou, *Int. J. Modern Phys. B* 15, 3359 (2001).
45. H. Y. Zhai, Q. Y. Chen, X. W. Xu, M. Strikovski, J. Miller, and W. K. Chu, *Physica C* 341, 1587 (2000).
46. D. H. A. Blank, W. Booij, H. Hilgenkamp, B. Vulink, D. Veldhuis, and H. Rogalla, *IEEE Trans. Appl. Super.* 5, 2786 (1995).
47. M. Nakayama, J. Yanagisawa, F. Wakaya, and K. Gamo, *Jpn. J. Appl. Phys.* 38, 7151 (1999).
48. M. Lachab, M. Nozaki, J. Wang, Y. Ishikawa, Q. Fareed, T. Wang, T. Nishikawa, K. Nishino, and S. Sakai, *J. Appl. Phys.* 87, 1374 (2000).
49. C. Ochiai, A. Sawada, H. Noriyasu, M. Takaia, A. Hosono, and S. Okuda, *J. Vac. Sci. Technol. B* 19, 904 (2001).
50. S. Anders, S. Sun, C. B. Murray, C. T. Rettner, M. E. Best, T. Thomson, M. Albrecht, J. U. Thiele, E. E. Fullerton, and B. D. Terris, *Microelectron. Eng.* 61, 569 (2002).
51. H. Rocken, J. Meijer, A. Stephan, U. Weidenmuller, H. H. Bukow, and C. Rolfs, *Nucl. Instrum. Methods* 181, 274 (2001).
52. C. Bacher, T. Kummell, D. Eisert, A. Forchel, B. Konig, W. Ossau, C. R. Becker, and G. Landwehr, *Appl. Phys. Lett.* 75, 956 (1999).
53. M. Kuball, M. Benyoucef, F. H. Morrissey, and C. T. Foxon, *MRS Internet J. Nitride Sem. Res.* 5, U816 (2000).
54. H. Kondo, K. Izumikawa, M. Sakurai, S. Baba, H. Iwano, S. Zaima, and Y. Yasuda, *Jpn. J. Appl. Phys.* 38, 7222 (1999).
55. T. W. Kim, D. C. Choo, J. H. Shim, and S. O. Kang, *Appl. Phys. Lett.* 80, 2168 (2002).
56. C. S. Ganpule, A. Stanishevsky, S. Aggarwal, J. Melngailis, E. Williams, R. Ramesh, V. Joshi, and C. Paz de Araujo, *Appl. Phys. Lett.* 75, 3874 (1999).
57. H. Kubota, T. Sumita, S. Takami, T. Shinada, and I. Ohdomari, *J. Phys. IV* 9, 175 (1999).
58. T. Sumita, T. Nagai, H. Kubota, T. Matsukawa, and I. Ohdomari, *Synth. Met.* 103, 2234 (1999).
59. A. Sandhu, H. Masuda, K. Kurosawa, A. Oral, and S. J. Bending, *Electron. Lett.* 37, 1335 (2001).

60. R. Puers, S. Reyntjens, and D. De Bruyker, *Sensors Actuators A* 97, 208 (2002).
61. A. Lugstein, E. Bertagnolli, C. Kranz, A. Kueng, and B. Mizaikoff, *Appl. Phys. Lett.* 81, 349 (2002).
62. D. H. Naghski, J. T. Boyd, H. E. Jackson, and A. J. Steckl, *Opt. Comm.* 150, 97 (1998).
63. J. Voigt, F. Shi, K. Edinger, P. Guthner, and I. W. Rangelow, *Microelectron. Eng.* 57–58, 1035 (2001).
64. H. U. Danzebrink, T. Dziomba, T. Sulzbach, O. Ohlsson, C. Lehrer, and L. Frey, *J. Microsc.* 194, 335 (1999).
65. C. Lehrer, L. Frey, S. Petersen, T. Sulzbach, O. Ohlsson, T. Dziomba, H. U. Danzebrink, and H. Ryssel, *Microelectron. Eng.* 57, 721 (2001).
66. R. Tarumi, K. Takashima, and Y. Higo, *Appl. Phys. Lett.* 81, 4610 (2002).
67. A. Spiegel, L. Staemmler, M. Dobeli, and P. Schmuki, *J. Electrochem. Soc.* 149, C432 (2002).

Encyclopedia of Nanoscience and Nanotechnology
www.aspbs.com/enn

Formation of Bulk Nanostructured Alloys

H. W. Kui

The Chinese University of Hong Kong, Shatin, N.T., Hong Kong, People's Republic of China

CONTENTS

1. INTRODUCTION

The notion of generating systems with plenty of interfacial areas was first perceived and suggested by Turnbull [1] and Gleiter [2]. By "plenty," it means that the ratio, interfacial-area/volume $\sim 10^7$ cm^{-1}. Physically, there would be just about as many atoms residing at the grain boundaries as there are in the bulk. These interfacial atoms would manifest themselves in novel physical and chemical behaviors, as described in a number of reviews [3–6], which are not found in conventional polycrystalline materials of coarse grain size (grain size > 10 μm), of which the ratio, interfacial-area/volume, is negligibly small. For instance, the Hall-Petch (H-P) relation says that the yield strength of a material is inversely proportional to the square root of its grain size, r. But the H-P relation is explained on the basis of dislocation pile-ups at grain boundaries [7]. On reducing the grain size of a grain to a few nanometers, confining a dislocation in it would become more and more difficult. It is apparent that the conventional model may have to be revised, providing opportunities in both fundamental research and applications. Magnetic nanostructured material provides another example. A well-known synthesis technique involves annealing an amorphous material at an elevated temperature [8]. Nanocrystals then emerge with the amorphous phase serving as the background. The system exhibits attractive, soft magnetic properties. There are other preparation methods besides devitrifying a glassy specimen, but the interesting magnetic properties at very small grain size, $r \sim 15$ nm, remain [9].

The attractive mechanical and magnetic properties of nanostructures render them important functional materials in industrial applications. In the electronic industry, thin nanostructured specimens can very often serve these purposes. However, nanostructures only in their bulk form are applicable in many other areas. There is therefore a pressing need for the synthesis of bulk nanostructures. By "bulk," we mean that the dimension of a specimen in any direction is >1 mm. In this review, the focus is on the synthesis of bulk nanostructures.

2. TECHNIQUES OF SYNTHESIZING NANOSTRUCTURED ALLOYS

2.1. Gas Condensation and Powder Compaction

The first attempt to produce bulk nanostructured materials is by a method that involves condensation and compaction [10, 11]. In the experiment, metals are evaporated onto a cold finger (cooled by liquid nitrogen) to form very fine powders in an ultra-high vacuum (UHV) chamber. The cold finger acts as a collector for the fine powders, which are subsequently scrapped off by a device made of Teflon®. The scrapped powders are led into a compaction device, also located inside the UHV chamber, for a fusion/bonding process. The *in-situ* compaction process provides a clean environment for the nanometer-sized powders to avoid contamination. Although the synthesis process is done using great precaution, an actual nanostructured specimen unavoidably contains flaws [5], which include micropores; trapped gas and the size of the constituent grains is not uniform. In fact, an as-pressed specimen has a density about 90% of its bulk density or theoretical value.

The flaws can be, and in many cases, proved fatal to the promising properties of the nanostructured materials. For example, a pore behaves like a crack, causing premature failure under tensile stress. Sanders et al. [12, 13] introduced warm compaction, which is very effective in driving out absorbed and trapped gases, bringing the density of a compact to $\geq$98% of the theoretical value, albeit some micropores are still there. Since the compaction takes place at an

ISBN: 1-58883-059-4/$35.00

Encyclopedia of Nanoscience and Nanotechnology
Edited by H. S. Nalwa
Volume 3: Pages (485–495)

elevated temperature, grain growth is expected, but not serious. Most importantly, it turns out that the mechanical properties of these heat-treated compacts improve significantly.

The processing of nanostructured materials by means of a UHV method is costly and slow, but the powders so produced are clean. Driven by economic considerations, other modified techniques of producing nanometer-sized powders were introduced. Mechanical alloying or ball milling [14–16] is a notable one in which a large amount of fine powders can be produced in a relatively short period. Besides, it is capable of manufacturing exotic powders that are not accessible by other means. During ball milling, the materials are more open to contamination. This problem is also common to the other methods for the mass production of ultra-fine powders, such as sol–gel synthesis [17] and spray conversion processing [18]. Consequently, even though these modified methods are less costly, the defective nature of the nanostructured compacts can only be worse than before.

2.2. Severe Plastic Deformation

In this method [19], metallic materials are subjected to a very large degree of plastic deformation. There are two major experimental arrangements. First, two channels made of special die are joined to each other at an angle ϕ. A metallic billet is then repeatedly forced through the device to bring about plastic deformation. This process is called equal channel angular (ECA) pressing. An as-fabricated specimen can be rod shaped with a diameter of $\leq$20 mm and a length in the range of 70–100 mm. Second, an anvil with a cylindrical hole (flat base) on its surface is used as a support. A metallic ingot is put into the hole. A second anvil or a cylindrical plunger that just fits into the hole of the first anvil is placed above the ingot. The two anvils are then pressed against each other very hard. Meanwhile, the plunger is allowed to rotate. The surface friction induces surface shear on the ingot, causing serious straining. This process is termed severe torsion straining. The specimens prepared by this method are usually disk shaped with a diameter of 10 to 20 mm and a thickness of 0.2 to 0.5 mm.

The grain size, r, of an as-treated specimen can be of nanometer dimension, with $r \sim 20$ nm in the case of high carbon steel. The grain boundaries are high angled, but wavy as observed under TEM. There is huge stress in the grains, arising partly from a large number of dislocations located near the grain boundaries. TEM does not detect any cracks or pores, an improvement over those nanostructures synthesized by powder compaction.

Cold rolling can also bring about severe plastic deformation. Recently, Wang et al. [20] were able to produce nanocrystalline Cu by severely cold rolling commercially available pure Cu. It is interesting that an as-prepared specimen exhibits high tensile ductility in sharp contrast to its coarse-grained counterpart.

2.3. Electrodeposition

Electrodeposition techniques can be employed to produce bulk nanocrystalline materials [21, 22] with a thickness that can be as large as ~2 mm. It turned out that an as-prepared specimen contains negligible porosity. The constituent grains can be either textured as illustrated by the commercially available nanocrystalline Ni or equiaxed [23]. If the experimental conditions are carefully adjusted, the grain-size distribution can also be narrow.

By an electrodeposition technique, Lu et al. [24] were able to synthesize nanocrystalline Cu, which displays superplasticity in a rolling experiment. A conventional coarse-grained Cu specimen on cold rolling breaks when its thickness is reduced by about a factor of 8. The factor on the other hand increases to 50 for nanostructures synthesized by the electrodeposition technique.

3. UNDERCOOLED LIQUIDS AND BULK AMORPHOUS ALLOYS

3.1. By Nucleation and Growth

3.1.1. Introduction

A few terminologies are first introduced to facilitate the discussions below. The thermodynamic melting temperature (or liquidus) of an alloy is denoted by T_1. As an alloy melt is cooled to a temperature T below its T_1, it is called an undercooled melt with an undercooling, ΔT, defined as $\Delta T = T_1 - T$. On crystallization or solidification, the undercooled molten ingot turns into a solid called undercooled specimen, which can be a crystal, an amorphous solid, a quasi-crystal, or a mixture of them.

Physically, as an alloy melt is undercooled below its T_1, hardening occurs, that is, the liquid becomes more and more difficult to flow. The hardening takes place homogeneously throughout the entire undercooled melt. A quantitative measure of the hardening is its viscosity, η. It increases only slowly when ΔT is small. However, at large ΔT, the hardening process picks up speed and η increases by some orders of magnitude to 1×10^{12} poise in a narrow temperature range. At $\eta \sim 1 \times 10^{12}$ poise, the hardened molten alloy is termed an amorphous solid. The temperature at which $\eta \sim 1 \times 10^{12}$ poise is called the glass transition temperature of the alloy system. Therefore, an amorphous solid is a frozen liquid. Please refer to [25, 26] for an excellent review on undercooled liquid.

Most often, the continuous hardening process of an undercooled melt can be interrupted by crystallization, transforming it into a crystalline solid(s). In the transformation process, a nucleus of the stable crystal forms in the undercooled liquid, creating a liquid-solid interface. This is called nucleation. Next, the liquid-solid interface advances into and consumes the undercooled melt, completely transforming it into a solid. This step is called growth.

3.1.2. Formation of Bulk Glasses

It is clear that in order for a melt to become an amorphous solid on continuous cooling, the nucleation and growth process must be bypassed. Klement et al. [27] demonstrated that Au-Si could be quenched into an amorphous solid with a cooling rate estimated to be on the order of 10^6 K s^{-1}. It is, however, not certain whether the glass formation is primarily due to the suppression of nucleation, that of growth or both. On the more fundamental aspects, Chen and Turnbull [28] demonstrated the kinetic glass transition by calorimetry.

Turnbull and Cohen [29–31] introduced a free volume model to explain the glass transition.

Since the synthesis involves rapid quenching or rapid solidification, one dimension of the as-prepared amorphous alloys is thin to facilitate heat removal [27]. This limits the potential applications of amorphous alloys, which exhibit attractive mechanical and magnetic properties. The search for bulk glass is therefore an urgent matter.

Nucleation can be brought about by two different origins. First, the surfaces of impurities provide a periodic potential for the attachment of the atoms from the metastable phase, helping nucleation. This type of nucleation is called heterogeneous nucleation. Second, in the absence of impurities, a pure undercooled melt can form a nucleus by structural fluctuation. This kind of nucleation is termed homogeneous nucleation. Experience [32] indicates that homogeneous nucleation rarely occurs.

To form bulk glass, one approach is to remove potent nucleation impurities from a melt and this is demonstrated in the case of Pd-Ni-P [33], in which a wholly amorphous ingot of diameter ~1 cm was prepared by a fluxing technique. The technique was extended to Pd-Cu-Ni-P [34], leading to the formation of an amorphous rod specimen of diameter 72 mm.

The second approach to bulk glass formation is to search for systems that resist nucleation or growth or both when quenching through the undercooling regime at slow cooling rates, typically 1 K s. The earlier bulk glasses formed by this method, to name a few, are La-Al-Ni [34], Zr-Cu-Ni-Al [35, 36], and Zr-Ti-Cu-Ni-Be [37]. To summarize, Inoue [38] introduced three empirical rules for the search of bulk glass formation systems (extracted from [38]):

1. multicomponent system consisting of more than three elements;
2. significant difference in atomic-size ratios above 12% among the main constituent elements;
3. negative heats of mixing among their elements.

3.1.3. Formation of Nanostructured Bulk Amorphous Alloys

When bulk glass is annealed at an elevated temperature, crystallization proceeds by the advancement of the crystal-amorphous or crystal-liquid interface [39]. After a volume that is amorphous is swept by the crystal-amorphous interface, it is turned into a crystalline solid. Very often, even if the dimension of the microstructures of the crystallized solid is of nanometer size [40], it becomes extremely brittle, rendering it useless for applications. When such an amorphous phase is crystallized in a differential scanning calorimeter (DSC), the thermogram would display one or two closely overlapped crystallization peaks [40].

On the other hand, if an amorphous specimen is only partially crystallized into nanocrystals that disperse themselves uniformly in the remaining amorphous phase, a process called nanocrystallization by Inoue [38] occurs. The physical properties are enhanced. In particular, the ductility and high strength associated with the amorphous phase is further improved. When such a glass is subjected to thermal scan in a DSC, there would be two crystallization peaks, the first one corresponding to the emergence of the nanocrystals while the second one corresponds to the crystallization of the amorphous matrix.

To bring about nanocrystallization, Inoue et al. [41] added certain impurities into a bulk glass former. They are chosen in such a way that the three empirical rules for bulk glass formation stated above are somewhat violated. For instance, it is suggested that to form bulk glass, the heats of mixing are negative among the constituent elements. The added impurities are then purposely chosen such that when mixed, there would be negative heats of mixing, zero heats of mixing, and positive heats of mixing. This alteration generates microscopic homogeneity that eventually leads to a primary crystallization and then a crystallization event for the remaining amorphous matrix. The following four criteria are proposed by Inoue [42] to achieve nanocrystallization (extracted from [42]):

1. multistage crystallization processes;
2. existence of homogeneous nucleation sites in an amorphous phase;
3. suppression of growth reaction caused by segregation of a solute element with low atomic diffusion at the nanocrystal/amorphous interface; and
4. high thermal stability of the remaining amorphous phase by the enrichment of solute elements from the primary crystalline phase.

To illustrate it, Inoue added Ti to $Zr_{58}Al_{12}Ni_{10}Cu_{20}$ to change its composition to $Zr_{53}Al_{12}Ti_5Ni_{10}Cu_{20}$. Ti has a zero heat of mixing against Zr. The single stage crystallization then becomes a two-stage crystallization process (criteria 1 and 2). Since the reaction during the first peak is not polymorphic, that is, the composition of the nanocrystals is different from that of the amorphous phase, in this case Al atoms are rejected to the crystal/amorphous interface, stabilizing the amorphous phase and slowing down the diffusion motion or growth velocity (criterion 3). Finally, the crystal/amorphous interface stop moving when more Al builds up in the rest of the amorphous phase (criterion 4). The nanocrystals emerged during nanocrystallization appear to be spherical (homogeneous nucleation in criterion 1) and have a grain size of 10–15 nm.

3.2. Formation of Bulk Metallic Nanostructured Alloys Involving Spinodal Mechanism

3.2.1. Introduction

A typical phase diagram of a binary eutectic alloy of elements A and B is shown in Figure 1. Let ε_{AA}, ε_{BB}, and ε_{CC} be the bond energies between A-A atoms, B-B atoms, and A-B atoms, respectively. Then $1/2[\varepsilon_{AA} + \varepsilon_{BB}] < \varepsilon_{AB}$. There are two interesting considerations. First, if the crystal structures of A and B are different, the free energy curves of this system at a temperature above its eutectic temperature T_E is shown in Figure 1b(i), in which the free energy curves for the crystalline phases do not join to each other smoothly. On the other hand, the liquid phase, which does not have to meet any long-range order, can be represented by a single, smooth, free-energy curve. Moreover, owing to the large entropic term in $G_1 = H_1 - TS_1$, where G_1 is the Gibbs

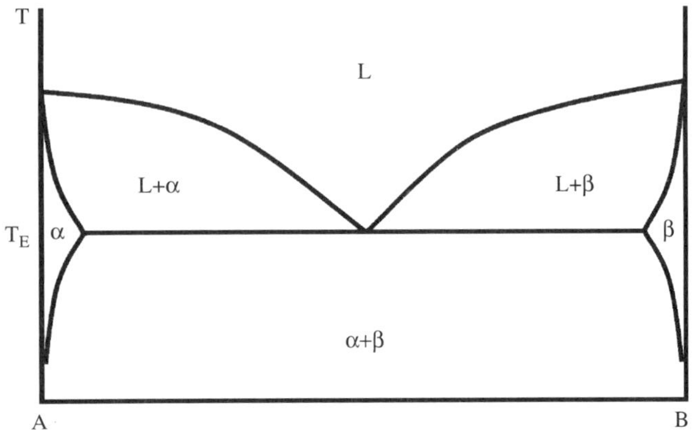

Figure 1. A typical phase diagram of a binary alloy of A-B.

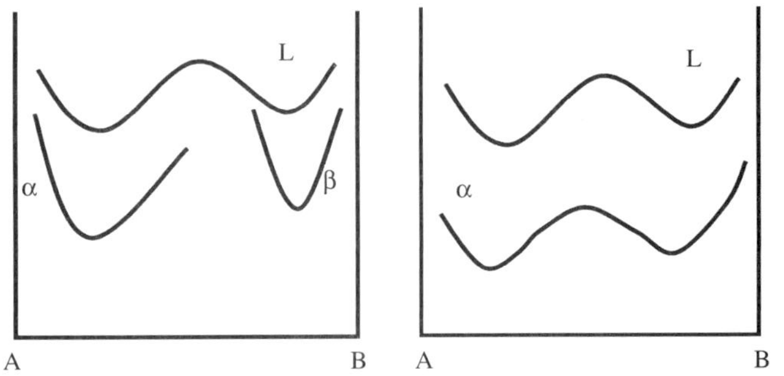

Figure 3. (a) The development of a hump in the liquid free-energy curve shown in Figure 2a at a temperature well below T_E. (b) The development of a hump in the liquid free-energy curve shown in Figure 2b at a temperature substantially below T_E.

free energy, H_1 is the enthalpy, T is the temperature in K, S_1 is the entropy, and the subscript 1 stands for the liquid phase, it concaves upward. Second, if A and B have the same crystal structure, the corresponding solid free-energy curves, again at a temperature above T_E, may be depicted as the one shown in Figure 2b. The one for the liquid phase is also a single, smooth curve and concaves upward, again for the same reasons. From now on, we would focus only on the free-energy curves of the liquid phase in both cases.

We assume that the condition $1/2[\varepsilon_{AA} + \varepsilon_{BB}] < \varepsilon_{AB}$ persists in the liquid state at all ΔT. Then, as a binary eutectic melt of composition, A_xB_y is cooled from above, its liquidus, T_1, to a temperature T, which is below T_E; it is metastable with an undercooling, $\Delta T = T_1 - T$. On further cooling or at low values of T, the entropic term in $G_1 = H_1 - TS_1$ becomes less dominant and H_1 begins to take charge, which brings about metastable liquid phase separation. So far, the discussion has been limited to A_xB_y, a fixed composition of the A-B alloy. On relaxing this condition to include the entire composition range, we will see a hump emerging on the liquid free energy curve as the temperature is lowered well below T_E as shown in Figure 3a and 3b. A free-energy diagram describes the behavior of various phases at a fixed temperature. When the free-energy curves are considered as a function of temperature, it culminates in a metastable liquid phase miscibility gap in a phase diagram (please refer to Fig. 4).

In general, a liquid miscibility gap does not have to lie below all the solid phases, for example, the one in a monotectic phase diagram [43]. On the other hand, consider a eutectic alloy in which the solid and the liquid phase share the same composition range. Since $H_s < H_l$ and $S_s < S_l$, where the subscript s stands for the solid phase, the liquid miscibility gap as described above would lie below the eutectic temperature, T_E.

There are two decomposition reactions in the metastable liquid miscibility gap, namely, by liquid nucleation and growth (metastable), and by liquid spinodal decomposition (metastable). In the former, on quenching, a homogeneous undercooled melt decomposes by first forming a nucleus of a more stable liquid phase. The transformation continues by the advancement of the liquid-nucleus interface. The morphology in the absence of coarsening is island-like. A typical example is shown in Figure 4 of a liquid of composition x_{NG}, which on reaching T decomposes into two liquids of compositions x_1 and x_2. In the latter case [44–46], an original homogeneous undercooled melt undergoes local composition fluctuation, characterized by uphill diffusion.

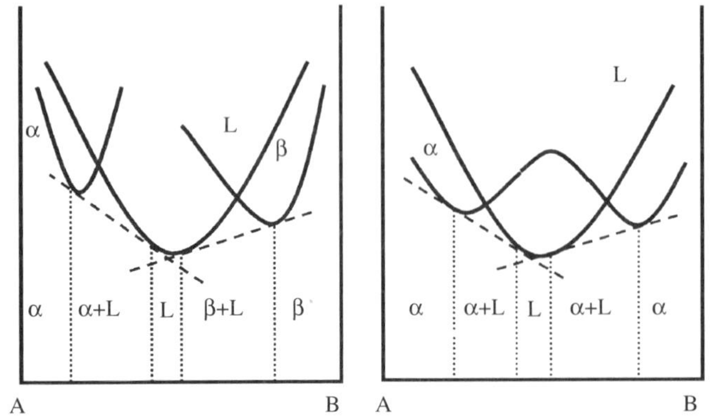

Figure 2. (a) Free-energy curves of the binary A-B alloy at a temperature $T > T_E$ (eutectic temperature). Assume that A and B have different crystal structures. (b) Free-energy curves of the binary system A-B alloy again at a temperature $T > T_E$. Assume that A and B have the same crystal structure.

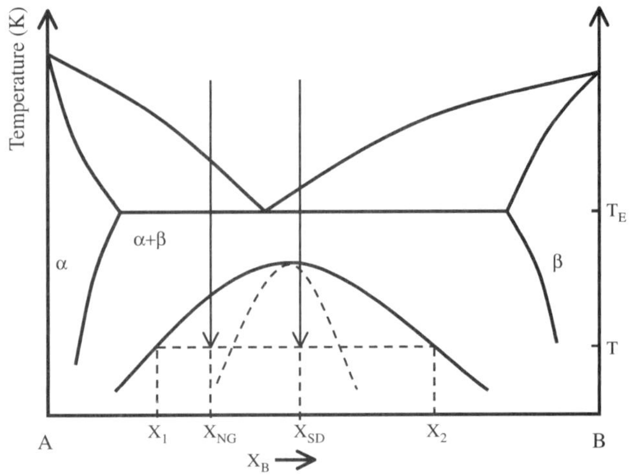

Figure 4. The metastable liquid miscibility gap below T_E for the A-B binary alloy.

There is no sharp interface during the incipient stage of the decomposition. But as the uphill diffusion process continues, the composition gradient sharpens and at the end it turns into a steep interface. In Figure 4, on reaching the spinodal regime, a homogeneous melt of composition x_{SD} continues to build up the concentration gradient until the compositions of its split components—both liquids—reach x_1 and x_2, leaving behind sharp interfaces. The transformation does not start out from certain points in the melt, as in the case of the decomposition by nucleation and growth, but occurs simultaneously throughout the entire specimen. Consequently, the decomposed melt is filled with interpenetrating networks of characteristic wavelength, λ. Cahn [43] and Hilliard [46] predicted that the system would evolve to a characteristic wavelength λ_m. As a result, the microstructure is quite uniform. A diagram showing the network morphology of undercooled $Pd_{80}Si_{20}$ is shown in Figure 5. Cahn [43] and Hilliard [46] also predicted that $\lambda \propto [(T_{CS} - T)/T_{CS}]^{-1/2}$, where T_{CS} is the temperature of the chemical spinodal of the liquid. Since $(T_{CS} - T)$ scales with ΔT, if we can undercool the melt to a very large ΔT, forcing λ into the nanometer regime, it becomes a nanostructure on crystallization.

The decomposition, either by nucleation and growth or by spinodal decomposition, results in the formation of multiple metastable/undercooled liquids. The structures can be described as island-like liquid droplets embedded in another liquid matrix or liquid networks interpenetrating each other. The crystallization kinetics, in the presence of plenty of interfaces, can be different from that of conventional systems.

3.2.2. Nanostructures from Bulk Glassy Alloys I

Chen and Turnbull [47] found that when amorphous $Pd_{78}Au_6Si_{16}$ is rapidly heated to a temperature just above its glass transition temperature, T_g, for a few minutes, two interconnected amorphous phases of different electron density emerge. There are only tiny traces of nanometer crystalline inclusions in one of the decomposed amorphous phases. They suggested that the transformation follows the course:

$$\text{Glass} \rightarrow \text{melt} \rightarrow \text{(phase separation) 2 melt phases} \rightarrow \text{2 crystalline phases.}$$

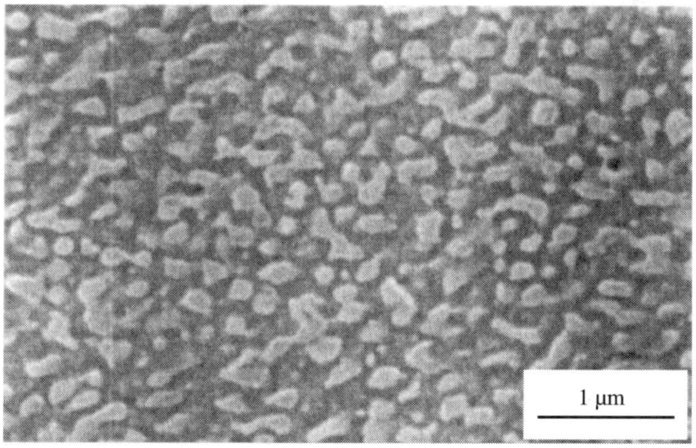

Figure 5. Typical morphologies of a system that has undergone a spinodal mechanism.

Chen and Turnbull [47] studied the crystallization kinetics of amorphous $Pd_{80}Si_{20}$ in a similar manner. The experimental results suggested that the transformation follows the course:

$$\text{Glass} \rightarrow \text{melt} \rightarrow \text{(phase separation) crystalline phase} + \text{melt} \rightarrow \text{crystalline phase} + \text{second crystalline phase.}$$

The heats of mixing the pure liquid metals are negative, leading the authors to suggest that the two liquids formed from the homogeneous liquid have a high degree of unique local order. Hence, the overall energy is lowered.

Since the interconnected amorphous microstructure observed in thermally annealed $Pd_{78}Au_6Si_{16}$ alloy is similar to that exhibited by phase-separated silicate glasses, Chou and Tumbull [48] proposed that it is the spinodal mechanism that leads to the interconnected morphology in amorphous $Pd_{78}Au_6Si_{16}$ upon annealing. They went further to conduct the small-angle X-ray scattering (SAXS) experiments. together with large-angle X-ray scattering and calorimetry on various alloys. The SAXS is a suitable tool for the study of the composition segregation in a material, thereby on the evolution of spinodal decomposition, which exhibits continuous composition variation. Experimentally, the behavior of the incipient decomposition was generally consistent with the predictions of Cahn's spinodal theory. The authors then suggested that the following transformation courses for various amorphous systems were rapidly heated to just above their respective T_g and annealed for a short period:

$Pd_{82}Si_{18}$

$$\text{Glass} \rightarrow (T_g)\ \text{melt} \rightarrow \text{homogeneous fcc solution}$$

$Pd_{80}Au_{3.5}Si_{16.5}$

$$\text{Glass} \rightarrow (T_g)\ \text{melt} \rightarrow \text{phase separation} \rightarrow \text{crystalline phase} + \text{melt} \rightarrow \text{second crystalline phase } Pd_{74}Au_8Si_{18}$$

$$\text{Glass} \rightarrow (T_g)\ \text{melt} \rightarrow \text{(phase separation) 2 melt phases} \rightarrow \text{2 crystalline phases.}$$

Most of the phase-separated specimens studied by Chen and Turnbull [47], and Chou and Turnbull [48] were annealed at temperatures just above their respective T_g. Tanner and Ray [49] were able to prepare amorphous phase-separated specimens in the Zr-Ti-Be system by quenching the melts at cooling rates of 10^5–10^7 K s^{-1}. The two amorphous phases, which are of nanometer size, mix up in two different ways. In the first, discrete, separated particles embed themselves in an amorphous matrix. In the second, a characteristic interconnected structure appears in the system of $Zr_{36}Ti_{24}Be_{40}$. Calorimetry confirms that there are two glass transition temperatures However, the nature of the phase-separation process is not known. The experimental results indicate that at cooling rates of 10^5–10^7 K s^{-1}, there is enough time for liquid-phase separation to take place and complete, but too short even for crystallization to start. The specimens prepared by Tanner and Ray [49] were not bulk nanostructured alloys due to the necessity of fast quenching rates.

3.2.3. Nanostructures from Bulk Glassy Alloys II

Peker and Johnson [37] found that $Zr_{41.2}Ti_{13.8}Cu_{12.5}Ni_{10.0} \cdot Be_{22.5}$ is an excellent bulk glass former. To understand why it forms glass so readily. Johnson's group studied in detail the crystallization kinetics of Zr-Ti-Cu-Ni-Be by using a high vacuum electrostatic levitation (HVESL) method [50], which enables the heating, melting, and cooling/undercooling of roughly a 3 mm-diameter drop. The themogram is obtained by means of a pyrometer. Undercooled specimens of various degrees of undercooling were prepared this way.

In the experiment, an undercooled $Zr_{41.2}Ti_{13.8}Ni_{10.0}Be_{22.5}$ melt is cooled to a temperature T, which is just below 800 K. As soon as it reaches T, the heat treatment process is changed to an isothermal condition. A small hump will first appear on the thermogram, to be followed by two overlapping, but distinct crystallization peaks. Both peaks are very large compared to the small hump preceding them. On the contrary, on skipping the prolonged isothermal run and allowing the melt to cool continuously to a temperature close to or below T_g, the molten drop can become an amorphous solid.

When subjected to field ion microscopy and atom probe techniques [51], an as-prepared amorphous specimen displays composition inhomogeneities of nanometer size. Some of the regions are Be rich while the others are Be poor, but both are amorphous. There is an amorphous phase separation. Compared to the themogram recorded earlier [50], the authors suggested that the small hump is due to the heat released during phase separation. Since the hump occurs at a temperature way above T_g, it is a liquid state phase separation. The undercooled liquid components eventually are frozen to amorphous state on further cooling.

The two phases in the amorphous phase separated specimen have different compositions or different T_g. In a reheating experiment that brings the temperature of the specimen to a value that is intermediate between the two glass transition temperatures, the phase with lower T_g crystallizes, leaving the other one in the amorphous state. The final morphology then becomes a bulk specimen with nanocrystals embedded in an amorphous matrix. Busch and Johnson [52] found that $Zr_{46.2}Ti_{8.8}Cu_{7.5}Ni_{10}Be_{27.5}$, on the other hand, does not exhibit any decomposition reaction, even though its composition is very close to $Zr_{41.2}Ti_{13.8}Ni_{10}Cu_{12.5}Be_{22.5}$. The peculiar behavior arises from the fact that its chemical spinodal is below its T_g. This is not surprising for, in general, the chemical spinodal line of a miscibility gap changes steeply with composition.

The introduction of a metastable miscibility gap also helps explain another dilemma. $Zr_{41.2}Ti_{13.8}Ni_{10.0}Cu_{12.5}Be_{22.5}$ is a bulk glass former. Accordingly, during its cooling from a melt to become a bulk glass and even if the as-formed glass is reheated to a temperature near the glass transition temperature, nucleation events should be sparse. It [53] was found, to the contrary, that as the bulk glass is reheated to a temperature somewhat above T_g, plenty of nanocrystals emerge, embedding themselves in the amorphous matrix. Additionally, they are very small, indicating that the glass-forming ability of $Zr_{41.2}Ti_{13.8}Ni_{10.0}Cu_{12.5}Be_{22.5}$ is likely to be crystal growth limited. The amorphous phase separation in undercooled $Zr_{41.2}Ti_{13.8}Ni_{10.0}Cu_{12.5}Be_{22.5}$ and the behavior of $Zr_{46.2}Ti_{8.8}Cu_{7.5}Ni_{10.0}Be_{27.5}$ in the undercooling regime illustrate that the glass-forming ability of Zr-Ti-Ni-Cu-Be is not necessarily crystal growth limited.

Schneider et al. [53, 54] attempted to unveil the nature of phase-separation process by employing small-angle neutron scattering (SANA). The experimental data is generally consistent with the predictions of Cahn's theory on spinodal decomposition.

Liquid $Mg_{62}Li_3Cu_{25}Y_{10}$ [55] specimens were cast into strip form with a thickness of 1 mm by means of copper molds. The as-quenched specimens were then studied by high-resolution transmission electron microscopy (HRTEM). By phase contrast, there are two apparently amorphous regions in a HRTEM image. Crystallization has started to take place in one of the phase-separated regions. The authors then suggested that a phase separation occurs in the liquid state during quenching. The one that is Mg-Li-rich one undergoes incipient crystallization.

3.2.4. Nanostructures from Bulk Undercooled Molten Alloys

It was found that when an alloy melt is immersed in an oxide flux at an elevated temperature, preferably about 200 K above the liquidus, T_l, of the alloy, impurities can be effectively removed from the melt into the flux. When free of potent heterophase impurities, the alloy melt can be undercooled substantially below its T_l. There are now two unique experimental procedures that this molten alloy can offer. First, it can be cooled down from its T_l to a temperature deep in the undercooling regime with very slow cooling rates, sometimes less than 1 K s^{-1}, until it is interrupted by crystallization. This enables the formation of a large piece of undercooled specimen with various degrees of undercoolings. For example, molten $Pd_{40}Ni_{40}P_{20}$ [33] can be directly quenched to its glassy state bypassing crystallization with a cooling rate of $\sim$1 K s^{-1}. The as-formed amorphous ingot can have a diameter $>$1 cm. Second, a continuous cooling cycle can be modified to include intermediate isothermal annealing periods [56]. If crystallization events take place during isothermal runs, the initial bulk undercooling of an undercooled specimen just before crystallization is well defined, simplifying microstructural analysis.

It is not clear why the fluxing technique is so effective in obtaining large undercooling for an undercooled melt. Presumably the flux, itself an oxide, helps absorb heterophase impurities in the molten alloy, which are mostly oxides, into itself. For those homophase impurities, the high temperature fluxing forces them to homogenize with the bulk melt. Physically, the large undercooling of a melt below its T_l is attributed to its very different short-range order from a crystalline phase [32]. Phenomenally, the interfacial free energy, γ_{lc}, between the liquid and crystalline phases assumes a large value, leading to large ΔT or large, free-energy change between the liquid and crystalline phases for transformation.

In subsequent experiments, Lee and Kui [57, 58], and Yuen et al. [59, 60] employed a fluxing technique with a slow cooling rate of 8 K s^{-1} to study the crystallization kinetics of undercooled molten alloys that crystallized at different ΔT. AS a result, the microstructures of undercooled $Pd_{80}Si_{20}$

[57, 58] and $Pd_{40.5}Ni_{40.5}P_{19}$ [59, 60], both eutectic alloys, were recorded as a function of ΔT. Since a slow cooling rate was used, the initial bulk undercooling just before crystallization could be measured with precision, that is, the established relationship between microstructure and ΔT is relevant and reliable. Based on the experimental results, it can be concluded that there is a metastable liquid miscibility gap for each of them. In fact, the entire metastable liquid miscibility gap of Pd-Si [61] was determined experimentally. In the following, undercooled $Pd_{80}Ni_{20}$ will be the leading example with occasional references to undercooled $Pd_{40.5}Ni_{40.5}P_{19}$.

AT $0 < \Delta T < 190$ K, the microstructures consist of long dendrites (pure Pd_3Si without any inclusions) embedded in a eutectic matrix as shown in Figure 6a, similar to that of a typical solidified metal. At $190 < \Delta T < 220$ K, there is a sudden microstructural change. The long dendrites disappear and are replaced by island-like structures (Pd_3Si). Pd particles are often found, dispersing themselves either inside or at the periphery of the islands as shown in Figure 6b. The background matrix is Pd_9Si_2. At $\Delta T > 220$ K, it is an interconnected network, consisting of two subnetworks. The one with a lighter color is Pd_3Si. Again, there are Pd inclusions but the population of Pd particles at the periphery of the Pd_3Si subnetwork increases over the one with $190 < \Delta T < 220$ K. The dark subnetwork has a composition of Pd_9Si_2 (Fig. 6c).

The microstructures revealed in undercooled specimens in the undercooling regimes of $190 < \Delta T < 220$ K and $\Delta T > 220$ K provide a clue to the transformation mechanisms in undercooled $Pd_{80}Si_{20}$. The microstructures have three characteristics. First, the microstructures shown in $190 < \Delta T < 220$ K are island-like, whereas those in $\Delta T > 220$ K are intertwining networks. Second, the phases in both undercooling regimes remain unchanged, but there is a sudden refinement in grain size by as much as 30 times, as one crosses over from $\Delta T < 220$ K to $\Delta T > 220$ K. Third, the island-like structures in $190 < \Delta T < 220$ K become connected after a long period of annealing. The microstructures versus ΔT are therefore generally consistent with the predictions of a miscibility gap (Fig. 4). In the nucleation and growth regime, the characteristic microstructures should be island-like, while those inside the chemical spinodal are network-like. Seward et al. [62] showed that a network-like morphology is not a sufficient condition for spinodal mechanism as they demonstrated that a system of spherical particles becomes interconnected on annealing. To remove any doubt about the origin of the network structure, an undercooled melt with $\Delta T = 210$ K at an undercooling just above the chemical spinodal, was annealed for a prolonged period (>30 min). The spherical structures coarsen, building up a somewhat connected morphology, which is distinctively different, however, from that in Figure 6c. Therefore, the network structure shown in Figure 6c, which takes place at a larger undercooling, results from a spinodal mechanism. There is additional evidence indicating that the microstructure shown in Figure 6c indeed comes from spinodal decomposition. The sudden grain refinement transition observed boosts a spinodal mechanism for undercooled molten Pd-Si. There are a number of models [63–70] proposed to explain the phenomenon of grain refinement in metals at large ΔT.

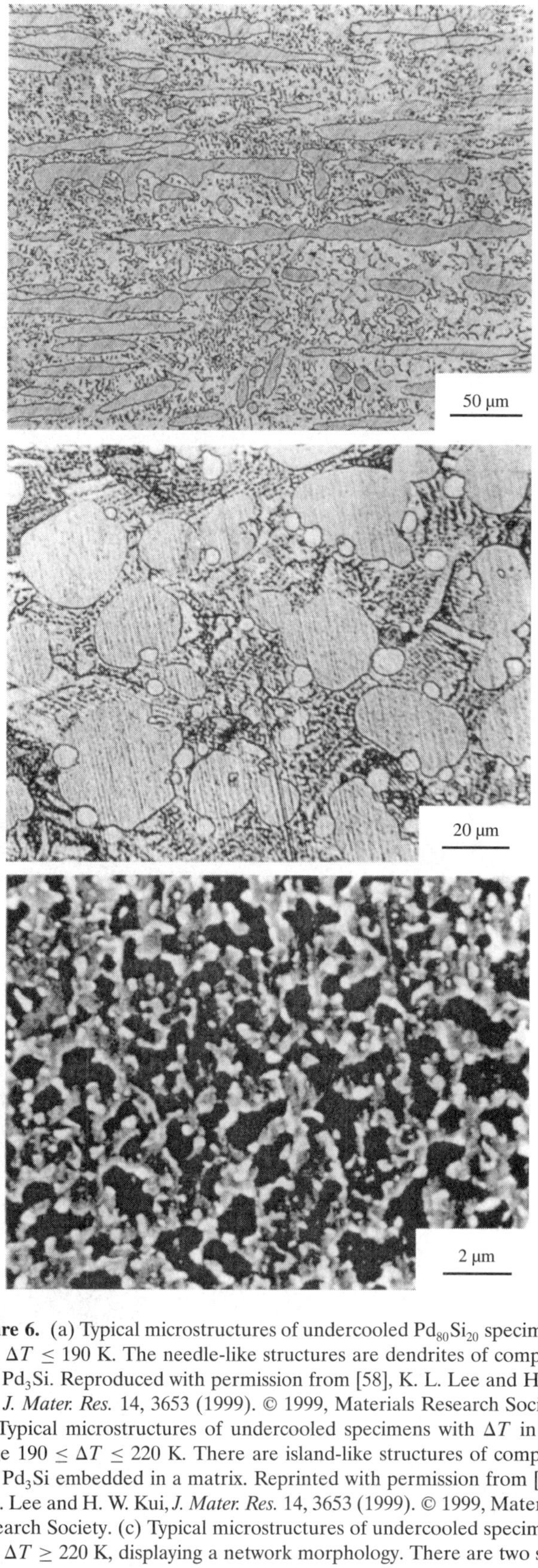

Figure 6. (a) Typical microstructures of undercooled $Pd_{80}Si_{20}$ specimens with $\Delta T \leq 190$ K. The needle-like structures are dendrites of composition Pd_3Si. Reproduced with permission from [58], K. L. Lee and H. W. Kui, *J. Mater. Res.* 14, 3653 (1999). © 1999, Materials Research Society. (b) Typical microstructures of undercooled specimens with ΔT in the range $190 \leq \Delta T \leq 220$ K. There are island-like structures of composition Pd_3Si embedded in a matrix. Reprinted with permission from [58], K. L. Lee and H. W. Kui, *J. Mater. Res.* 14, 3653 (1999). © 1999, Materials Research Society. (c) Typical microstructures of undercooled specimens with $\Delta T \geq 220$ K, displaying a network morphology. There are two subnetworks with Pd percipitations (small white dots). The composition of the lighter subnetwork is Pd_3Si, while that of the dark subnetwork is Pd_9Si_2. Reprinted with permission from [58], K. L. Lee and H. W. Kui, *J. Mater. Res.* 14, 3653 (1999.) © 1999, Materials Research Society.

It generally requires that the crystal growth velocity be very large when a grain refinement process takes effect. In undercooled $Pd_{80}Si_{20}$ melt, the crystal growth velocity is slow at $\Delta T \approx 200$ K, a condition that is only favorable to a spinodal mechanism.

There is an important question left. Does the metastable liquid miscibility gap occur in the solid state or in the liquid state? A thermocouple was used to record the thermal history of a molten Pd-Si ingot as it was cooled below its T_1 and crystallized. There were only crystallization peaks, excluding any solid-state reaction that took place in the undercooled specimen. When a spinodal decomposition reaction occurs in a binary solid, the two emerging solids should have the same crystal structure. The Pd_3Si subnetwork and the Pd_9Si_2 subnetwork found in undercooled Pd-Si have very different crystal structures. They do not form the right couple for spinodal decomposition. On the other hand, as a liquid undergoes spinodal decomposition, only short-range order has to be matched, which is readily met in the liquid. Therefore, it is concluded that there is a metastable liquid miscibility gap in Pd-Si alloy.

The crystallization kinetics of two undercooled spinodals intermixing with each other are of interest. It was found that their respective crystallization processes could be quite independent of each other. As crystallization proceeds in one, the other spinodal can stay in the liquid state.

The characteristic wavelength, λ, of undercooled Pd-Si near but below its chemical spinodal is $\sim$1 μm. According to Cahn [43], it decreases rapidly as the temperature of the undercooled liquid moves away from the chemical spinodal. The formation of bulk nanostructured alloy is then one step away. Two different experimental approaches were employed for this goal. In the first, undercooled molten specimens were crystallized at well-defined ΔT for quantitative results. Accordingly, Guo and Kui [71] prepared, again by a fluxing technique, a series of undercooled $Pd_{82}Si_{18}$ specimens. The largest ΔT achieved was $\sim$360 K. The measured λ versus ΔT is shown in Figure 7a, which depicts a rapid decrease of grain size against ΔT at small ΔT, but leveling off at $\Delta T \geq 300$ K. The smallest λ is 50 nm, occurring at $\Delta T = 360$ K. A morphological transition appears and is marked in Figure 7a. The microstructures obtained in undercooled specimens with $\Delta T < 320$ K are interconnected (Fig. 7b), but changes to more or less equiaxed grains for $\Delta T > 320$ K (Fig. 7c). It appears that the driving force for this transition is due to surface tension, as demonstrated in the case of undercooled Pd-Ni-P ingots [72]. For specimens with $\lambda \geq 180$ nm, the networks are interconnected. On the other hand, for $\lambda \leq 180$ nm, the networks are prone to break up into droplets to reduce surface areas.

An as-prepared undercooled ingot, taking the one with $\Delta T = 360$ K as an example, enjoys three privileges. First, its diameter can be as large as 1 cm, so it is a bulk nanostructured alloy. Second, micropores are absent. Third, the microstructures are uniform, which is a result of the spinodal mechanism.

The second approach to bulk nanostructured alloys formation involves rapid quenching. In the experiment, a fluxed molten specimen inside an evacuated fused silica tube was directly quenched in water. The undercooled specimen was

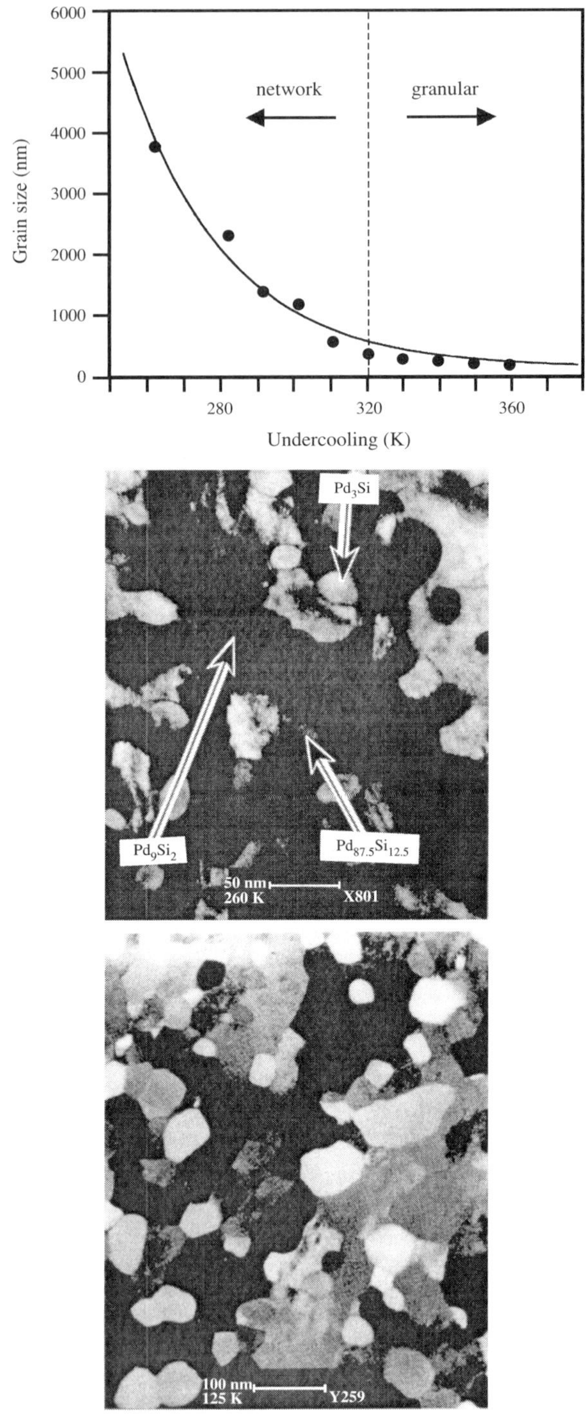

Figure 7. (a) A plot showing grain size versus ΔT. There is a morphological transition at $\Delta T > 320$ K, at which the network morphology ($\Delta T < 320$ K) changes to granular structure ($\Delta T > 320$ K). Reprinted with permission from [71], W. H. Guo and H. W. Kui, *Acta Mater.* 48, 2117 (2000). © 2000, Elsevier Science. (b) A TEM micrograph showing the microstructures of an undercooled Pd-Si specimen with $\Delta T =$ 320 K. Connectivity of the subnetwork is still quite obvious. Reprinted with permission from [71], W. H. Guo and H. W. Kui, *Acta Mater.* 48, 2117 (2000). © 2000, Elsevier Science. (c) A TEM micrograph displaying the microstructures of an undercooled specimen with $\Delta T = 350$ K. Compared with the micrograph shown in Figure 7b, there are more equiaxed grains. Reprinted with permission from [71], W. H. Guo and H. W. Kui, *Acta Mater.* 48, 2117 (2000). © 2000, Elsevier Science.

therefore in rod shape, but still a bulk sample for its diameter is ≥2 mm. It is estimated that the quenching rate is a few hundred degrees per second. The maximum ΔT achieved in this situation is certainly larger than that under isothermal conditions. As a result of the rapid solidification, λ versus ΔT is not available and the structure is even farther away from equilibrium. The microstructures can basically be divided into three categories. In the case [73] where the relative cooling rate is the least but larger than that in [71], there are still two crystalline networks intermixing with each other, but λ can be as small as 5–6 nm (Fig. 8a). The grains are equiaxed indicating that during water quenching, the liquid spinodals with very small λ still found time for network breakup (please see that the electron diffraction in the insert of Fig. 8a is ring, made up of diffraction spots). Also, the grain size distribution is very sharp, consistent with the predictions of Cahn's spinodal theory (Fig. 8b). In the case where [73] the relative cooling rate is somewhat faster, there are again two networks, but one is amorphous while the other one remain crystalline (Fig. 8c). The diffraction pattern, shown as an insert in Figure 8c, reveals an interesting feature. It resembles that taken from a single crystal. But the pattern was actually taken from an area with a diameter >100 nm. Inside that region, there are plenty of the crystalline spinodals all contributing to the diffraction spots and yet the final pattern is that of a single crystal. A plausible explanation is that during quenching. $L \rightarrow L_1 + L_2$ by spinodal decomposition. Although L_1 crystallizes, the quenching rate is able to freeze L_2. It is apparent that the quenching rate is slow enough for the spinodal mechanism to proceed, but was not long enough for the liquid spinodals to break up. As crystallization starts out from one place (heterophase impurities), it grows epitaxially along the L_1 network until it completely crystallizes. L_2 is immune to what happens in L_1. In the case that the relative quenching rate is the fastest, it is expected that both spinodal would be amorphous. This does not occur in undercooled Pd-Si, but takes place in undercooled Pd-Ni-P [74]. Unfortunately, transmission electron microscopy (TEM) cannot resolve the two amorphous phases using the phase contrast method. Presumably, the composition difference between the two amorphous phases is not large enough or the wavelength is too small to be resolved by the TEM.

The rapid quenching rate does not destroy the three privileges as previously mentioned. An as-prepared specimen is still in bulk form, free of micropores and uniform grain size.

The bulk nanostructured alloys so far prepared by spinodal mechanism are still not large enough for precision mechanical testing and direct industrial applications. In a recent attempt, again by a fluxing technique, Ng and Kui [75] were able to prepare a bulk Pd-Si rod of diameter 8 mm and length 80 mm. The grain size is not as uniform as the one in [71], ranging from 50 nm to 100 nm. Micropores are absent, but there is a groove or cavity running along the length of the undercooled specimen, which originates from the volume change as the molten rod crystallizes at very large undercooling.

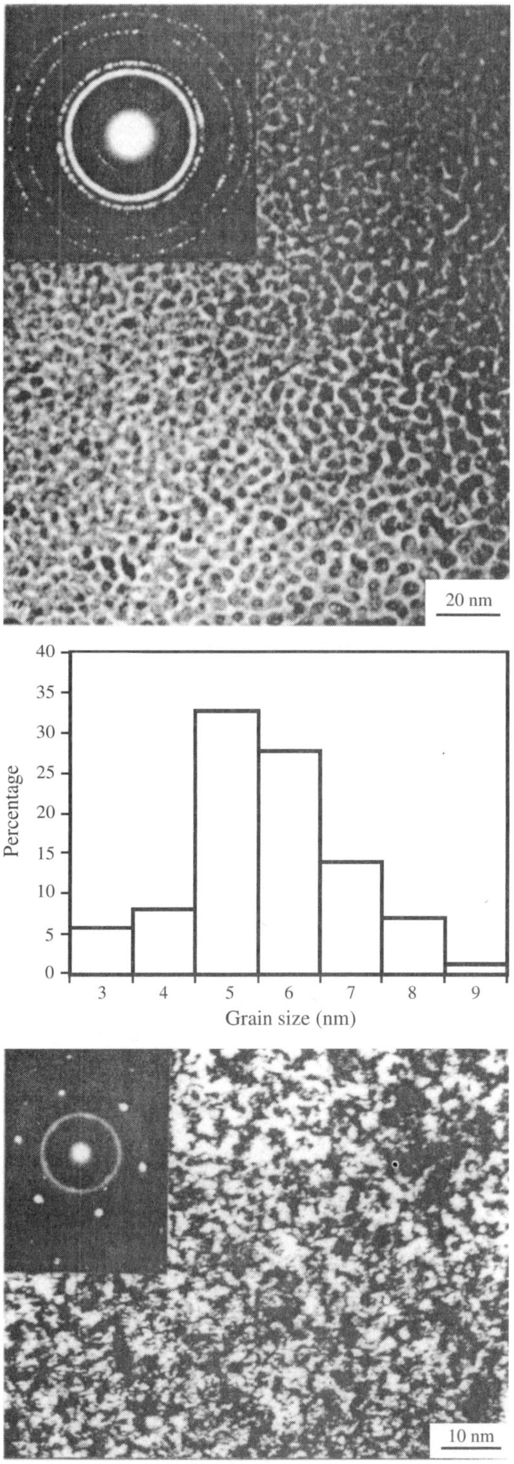

Figure 8. (a) A TEM micrograph showing a network morphology. There are two phases present—the lighter phase a solid solution Pd while the dark one is Pd_4Si Reprinted with permission from [72], W. H. Guo et al., *J. Mater. Res.* 15, 1605 (2000). © 2000, Materials Research Society. (b) The grain size distribution of the nanostructured specimen shown in Figure 8a. Reprinted with permission from [72], W. H. Guo et al., *J. Mater. Res.* 15, 1605 (2000). © 2000, Materials Research Society. (c) A TEM micrograph depicting two subnetworks—the white one is crystalline while the dark region is amorphous. Reprinted with permission from [72], W. H. Guo et al., *J. Mater. Res.* 15, 1605 (2000). © 2000, Materials Research Society.

GLOSSARY

Bulk metallic glass An amorphous solid with a physical dimension ≥ 1 mm in any direction.

Bulk nanostructure A 3-dimensional nanostructure with a physical size ≥ 1 mm in any direction.

Dendrite Branched crystal with the appearance of trees.

Dislocation A line defect in a crystal, distorting it from being a lattice of regular periodicity.

Hall–Petch equation It states that the yield stress σ_y is inversely proportional to the square root of grain size D. Mathematically, $\sigma_y = \sigma_o + k(D)^{-1/2}$, where σ_o is a frictional stress required to move dislocations and k is the Hall–Petch slope.

Heterogeneous nucleation The nucleus (or nuclei) forms with the help of foreign particles in a nucleation and growth process.

Homogeneous nucleation The nucleus (or nuclei) forms by an intrinsic structural fluctuation in a nucleation and growth process.

Liquid phase separation When undercooled into a liquid miscibility gap, a homogeneous liquid splits into liquid components.

Liquidus The thermodynamic melting temperature of an alloy, denoted by T_1.

Metastable liquid miscibility gap A miscibility gap for liquids, which are at a temperature below their respective T_1.

Miscibility gap A phase boundary separating single-phase field from that of mixtures.

Nucleation and growth One type of phase transformation in which a nucleus of the stable phase forms first in a less stable phase. Then crystal growth starts out from the nucleus, consuming the less stable phase.

Phase diagram A diagram describing the distribution of the thermodynamic equilibrium phases of an alloy at different temperatures and compositions.

Polymorphic transformation It occurs in single component systems when different crystal structures are stable over different temperature ranges.

Spinodal decomposition One type of phase transformation in which nucleation barrier is absent (no need for the formation of nucleus). It is best described by uphill diffusion of the like species.

Undercooled molten alloy An alloy melt at a temperature below its T_1.

Undercooled specimen An alloy that had solidified at a temperature T, which below its T_1. However, T is not specified.

Undercooled specimen with ΔT An alloy that had solidified at an undercooling of ΔT.

Undercooling The notation for undercooling is ΔT. Mathematically, $\Delta T = T_1 - T$, where T_1 and T are the liquidus and the temperature of an alloy melt under the condition that $T_1 > T$.

Viscosity The degree of resistance to flow when a liquid or solid is under a shear stress.

ACKNOWLEDGMENT

The work described in this paper was partially supported by two grants from the Research Grants Council of the Hong Kong Special Administration Region (Project No. CUHK 4170/00E and Project No. CUHK 4184/01E).

REFERENCES

1. D. Turnbull, *Metall. Trans.* 12A, 695–708 (1981).
2. H. Gleiter, in "Deformation of Polycrystals: Mechanisms and Microstructures" (N. Hansen, A. Horsewell, T. Leffers, and H. Lilholt, Eds.), p. 15. Riso National Laboratory, Roskilde, 1981.
3. C. Suryanarayana, *Internat. Mater. Rev.* 40, 41 (1995).
4. D. G. Morris, Mechanical Behavior of Nanostructured Materials, in "Materials Science Foundations, Series 2" (M. Magini and F. H. Wohlbier, Eds.). Trans. Tech. Publications, Uetikon-Zurich, 1998.
5. J. R. Weertman, D. Farkas, K. Hemker, H. Kung, M. Mayo, R. Mitra, and H. van Swygenhoven, *MRS Bull.* 24(2), 44 (1999).
6. C. C. Koch, D. G. Morris, K. Lu, and A. Inoue, *MRS Bull.* 24(2), 54 (1999).
7. G. E. Dieter, "Mechanical Metallurgy," p. 189. McGraw-Hill Book, New York, 1986.
8. Y. Yoshizawa, S. Oguma, and K. Yamauchi, *J. Appl. Phys.* 64, 6044 (1988).
9. G. Herzer, in "Nanomagnetism" (A. Hernando, Ed.), p. 111. Kluwer, Dordrecht, 1993.
10. R. Birringer, H. Gleiter, H. P. Klein, and P. Marquardt, *Phys. Lett.* A102, 365 (1984).
11. R. W. Siegel, *MRS Bull.* 15(10), 60 (1990).
12. P. G. Sanders, J. A. Eastman, and J. R. Weertman, *Acta Mater.* 45, 4019 (1997).
13. P. G. Sanders, J. A. Eastman, and J. R. Weertman, *Acta Mater.* 46, 4195 (1998).
14. P. H. Shingu, B. Huang, and S. R. Nishitani, *Nasu S. Suppl. Trans. Japan Inst. Metals* 29, 3 (1988).
15. J. R. Thompson and C. Politis, *Europhys. Lett.* 3, 199 (1987).
16. C. C. Koch, *Nanostructured Mater.* 9, 13 (1997).
17. P. Colomban, *J. Mater. Res.* 13, 803 (1998).
18. B. H. Kear and L. E. McCandlish, *Nanostructured Mater.* 3, 19 (1993).
19. R. Z. Valiev, R. K. Islamgaliev, and I. V. Alexandrov, *Prog. Mater. Sci.* 45, 103 (2000).
20. Y. Wang, M. Chen, F. Zhou, and E. Ma, *Nature* 419, 912 (2002).
21. G. McMahon and U. Erb, *Microstru. Sci.* 17, 447 (1989).
22. G. Palumbo, S. J. Thorpe, and K. T. Aust, *Scr. Metall. Mater.* 24, 1347 (1990).
23. F. Dalla Torre, H. Van Swygenhoven, and M. Victorial, *Acta Mater.* 50, 3957 (2002).
24. L. Lu, M. L. Sui, and K. Lu, *Acta Mater.* 49, 4127 (2001).
25. D. Turnbull, *Contemporary Phys.* 10, 473 (1969).
26. D. Turnbull and M. H. Cohen, *J. Chem. Phys.* 29, 1049 (1958).
27. W. Klement, R. H. Willens, and P. Duwez, *Nature* 187, 869 (1960).
28. H. S. Chen and D. Turnbull, *Apply. Phys. Lett.* 10, 284 (1967).
29. M. H. Cohen and D. Turnbull, *J. Chem. Phys.* 31, 1164 (1959).
30. D. Turnbull and M. H. Cohen, *J. Chem. Phys.* 34, 120 (1961).
31. D. Turnbull and M. H. Cohen, *J. Chem. Phys.* 52, 3038 (1970).
32. D. Turnbull, *J. Chem. Phys.* 20, 411 (1952).
33. H. W. Kui and D. Turnbull, *Appl. Phys. Lett.* 45, 615 (1984).
34. A. Inoue, N. Nishiyama, and H. M. Kimura, *Mater. Trans.,* JIM 38, 179 (1997).
35. A. Inoue, T. Zhang, and T. Masumoto, *Mater. Trans.,* JIM 30, 965 (1990).
36. A. Inoue, T. Zhang, and T. Masumoto, *Mater. Trans.,* JIM 31, 177 (1990).
37. A. Peker and W. L. Johnson, *Appl. Phys. Lett.* 63, 2342 (1993).

38. A. Inoue, *Acta Mater*. 48, 279 (2000).
39. A. J. Drehman and A. L. Greer, *Acta Met.* 32, 323 (1983).
40. C.-P. Chou and F. Spaepen, *Acta Met.* 23, 609 (1975).
41. A. Inoue, T. Zhang, and Y. H. Kim, *Mater. Trans.* JIM 38, 749 (1997).
42. A. Inoue, *Nanostruct. Mater.* 6, 53 (1995).
43. J. W. Cahn, *Trans. Met. Soc.* AIME 242, 166 (1968).
44. J. W. Cahn, *Acta Met.* 9, 795 (1961).
45. J. W. Cahn, *Acta Met*. 10, 179 (1962).
46. J. E. Hilliard, "Phase Transformation," p. 497. American Society for Metals, Metals Park, OH, 1970.
47. H. S. Chen and D. Turnbull, *Acta Met.* 17, 1021 (1969)
48. C.-P. Chou and D. Turnbull, *J. Non-Cryst. Solids* 17, 169 (1975).
49. L. E. Tanner and R. Ray, *Scripta Met.* 14, 657 (1980).
50. Y. J. Kim, R. Busch, W. L. Johnson, A. J. Rulison, and W. K. Rhim, *Appl. Phys. Lett.* 68, 1057 (1996).
51. R. Busch, S. Schneider, A. Peker, and W. L. Johnson, *Appl. Phys. Lett.* 67, 1544 (1995).
52. R. Busch and W. L. Johnson, *Appl. Phys. Lett.* 72, 2695 (1998).
53. S. Schneider, P. Thiyagarajan, and W. L. Johnson, *Appl. Phys. Lett.* 68, 493 (1996).
54. S. Schneider, P. Thiyagarajan, U. Geyer, and W. L. Johnson, *Physica B* 241–243, 918 (1998).
55. W. Liu and W. L. Johnson, *J. Mater. Res.* 11, 2388 (1996).
56. W. H. Guo, C. C. Leung, and H. W. Kui, "Bulk Metallic Glass, Materials Research Society Symposium Proceedings" (W. L. Johnson, A. Inoue, and C. T. Liu, Eds.), Vol. 554, p. 211. Materials Research Society, Warrendale, PA, 1998.
57. K. L. Lee, "Phase Separation and Crystallization in Undercooled Pd-Si Melts," Ph.D. Thesis, The Chinese University of Hong Kong, 1996.
58. K. L. Lee and H. W. Kui, *J. Mater. Res.* 14, 3653 (1999); K. L. Lee and H. W. Kui, *J. Mater. Res.* 14, 3663 (1999).
59. C. W. Yuen, "Liquid Phase Separation in Molten Pd-Ni-P Alloy," M. Phil. Thesis, The Chinese University of Hong Kong, 1996.
60. C. W. Yuen, K. L. Lee, and H. W. Kui, *J. Mater. Res.* 12, 314 (1997); C. W. Yuen and H. W. Kui, *J. Mater. Res.* 13, 3034 (1998); C. W. Yuen and H. W. Kui, *J. Mater. Res.* 13, 3043 (1998).
61. S. Y. Hong, W. H. Guo, and H. W. Kui, *J. Mater. Res.* 14, 3668 (1999).
62. T. P. Seward III, D. R. Uhlmann, and D. Turnbull, *J. Am. Ceram. Soc.* 51, 634 (1968).
63. J. L. Walker, "Principles of Solidification" (B. Chalmers, Ed.), p. 112. Wiley, New York, 1964.
64. K. A. Jackson, J. D. Hunt, D. R. Uhlmann, and T. P. Seward III, *Trans. Metall. Soc. A.I.M.E.* 236, 149 (1966).
65. C. F. Lau and H. W. Kui, *Acta Metall.* 39, 323 (1991).
66. G. Horvay, *Internat J. Heat Mass Transfer* 8, 195 (1965).
67. K. K. Leung, C. P. Chiu, and H. W. Kui, *Scr. Metall. Mater.* 32, 1559 (1995).
68. J. Z. Xiao, K. K. Leung, and H. W. Kui, *Appl. Phys. Lett.* 67, 3111 (1995).
69. J. Z. Xiao and H. W. Kui, *Scr. Mater.* 37, 1017 (1997).
70. J. Z. Xiao, H. Yang, and H. W. Kui, *Mat. Res. Soc. Symp. Proc.* 481, 1998. Materials Research Society. p. 15.
71. W. H. Guo and H. W. Kui, *Acta Mater*. 48, 2117 (2000).
72. W. H. Guo, L. F. Chua, C. C. Leung, and H. W. Kui, *J. Mater. Res.* 15, 1605 (2000).
73. W. H. Guo and H. W. Kui, "Supercooled Liquid, Bulk Glassy and Nanocrystalline States of Alloys," Materials Research Symposium Proceddings (A. Inoue, A. R. Yavari, W. L. Johnson, and R. H. Dauskardt, Eds.), Vol. 644. Warrendale, PA: Materials Research Society, L5.4.1., 2000.
74. C. C. Leung, W. H. Guo, and H. W. Kui, *Appl. Phys. Lett.* 77, 64 (2000).
75. K. L. Ng and H. W. Kui, *Phil. Mag. B* 82, 1777 (2002).

Encyclopedia of Nanoscience and Nanotechnology
www.aspbs.com/enn

Formation of Nanostructured Polymers

Beinn V. O. Muir, Christine K. Luscombe, Wilhelm T. S. Huck
University of Cambridge, Cambridge, United Kingdom

CONTENTS

1. INTRODUCTION

Nanotechnology has received a tremendous amount of interest over the last decade, not only from the scientific community, but also from the business sector and the general public [1, 2]. Although nanotechnology is still a relatively new science [3], it has already shown great potential for exciting technological innovations, with an enormous impact on areas as diverse as information technology, medicine, and energy supply [4, 5]. The miniaturization of components, and consequently the devices themselves, would increase the speed of devices and information storage capacity. More importantly, nanotechnology should lead to completely new functional devices since nanostructures are small enough to have fundamentally different physical properties that are governed by quantum effects. If nanometer-sized features are fabricated in materials that are currently used in electronic, magnetic, and optical applications, quantum behavior should lead to unique properties. The interaction of nanostructures with biological materials is also unexplored, and this is certain to be another important area of research in the future [6].

Polymers have played a vital role in the development of man from prehistoric times. Since ancient times, biopolymers such as wood, wool, and starch have been used as a construction material, clothing, and thickening agent, respectively. In the early 20th century, polymers, including plastics, synthetic rubbers, and elastomers, were synthesized for use in many applications. Recently, electrically conductive and light-emitting polymers have been discovered [7], paving the way for visual display units constructed from a flexible polymer film. Polymers are showing increasing versatility in a large number of applications, including electronic devices [8], medicine [9], and in the rapidly advancing field of nanotechnology.

A polymer is a macromolecule constructed from a collection of covalently connected monomers. For example, poly(styrene) (PS) $(\text{–}CH_2\text{–}CH(C_6H_5)\text{–})_n$ is built from the combination of styrene monomers $CH_2{=}CH(C_6H_5)$ with the monomeric unit $CH_2\text{–}CH(C_6H_5)$. The simplest polymers are linear polymers, and there are other possible forms, such as branched or star shaped (see Fig. 1). In solution, polymers can form spheres, random coils, and helices. There are two main methodologies used for polymer synthesis. One method is chain polymerization, where compounds with multiple bonds act as monomers. The most common functionality used for polymerization is the alkene group, with carbonyl and alkyne groups also being utilized occasionally. The activated intermediate (a radical, carbocation, or carboanion) adds across the multiple bond to form a new, and larger, intermediate which then adds across a new monomer. Another method is step (condensation) polymerization. This uses bifunctional monomers which undergo standard condensation reactions to form long chains. Suitable reactions include esterification, amidation, and etherification, as well as many organometallic coupling reactions [10].

Protein engineering, which allows the design of new proteins with novel functions, is a very powerful tool that can be used to fuse the biological and synthetic world. However, as it is currently impossible to predict the folding of unknown protein sequences, and even harder to design, express, and harvest *de novo* proteins, it is clear that other building blocks are required. Inspired by these biological examples, it is clear that polymers could be ideal components for man-made nanostructures because of their natural 5–100 nm length scale, ease of synthesis, and rich chemistry. Unfortunately, it is presently impossible to design and fabricate arbitrary polymers with a well-defined shape and functionality. If we are to bridge the gap between the biological world and nanofabrication, major advances in the control over polymeric materials at the nanometer level are necessary.

In this chapter, we will look at a number of different strategies to synthesize or fabricate polymers with well-defined nanoscale dimensions. The text is structured around

ISBN: 1-58883-059-4/$35.00

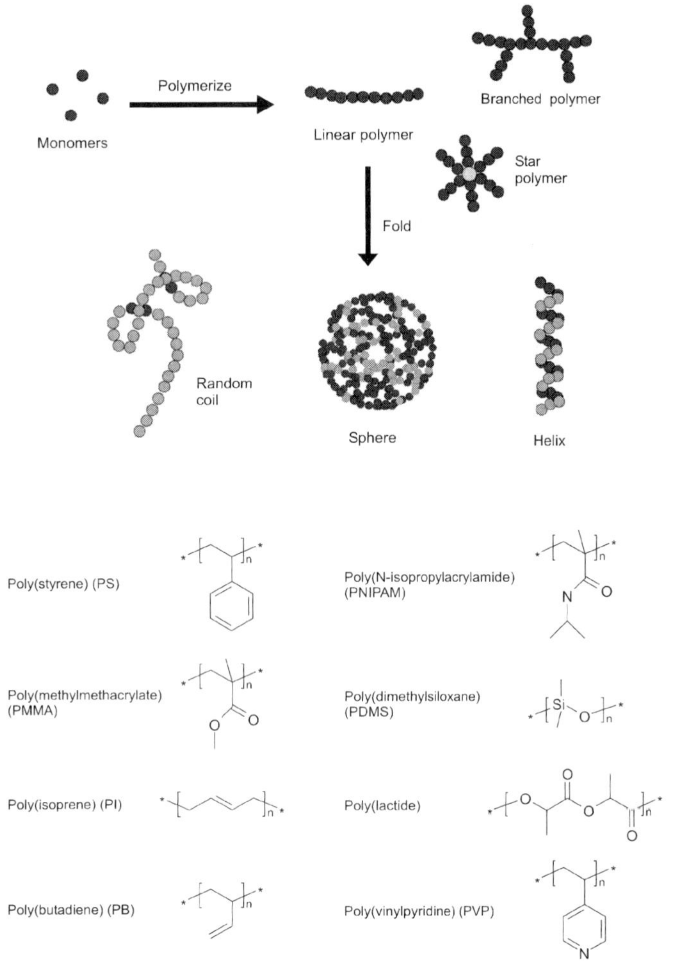

Figure 1. The upper section is a schematic drawing of linear, branched, and star polymers, including their possible shapes in solution. The lower section shows the structures of the polymers discussed in this chapter.

four main themes: top-down techniques, molecular self-assembly, chemical synthesis, and structures using biological materials. The top-down approach involves subtractive or lithographic techniques, where bulk material is structured into small features. In contrast, a bottom-up approach can be achieved via self-assembly or self-organization, processes which rely on the formation of noncovalent linkages between the building blocks (see Fig. 2). This alternative to conventional lithography uses molecular assembly, starting with molecular (Ångström-sized) building blocks, and linking them together into larger and larger structures. Nanosized structures can also be built up via the chemical synthesis of new covalent bonds between atoms. This leads to a diverse array of functionalized nanoparticles. Finally, polymers form the basis of biological function, and great inspiration can be obtained through the study of polymeric structures in biology. Research into the structure of such biological nanomachines will have important consequences on the fabrication of future nanosized devices. There is a growing interest in the use of biological polymers as building blocks for nonnatural structures. DNA is an obvious candidate, but protein engineering is beginning to make a sufficiently large impact, allowing the use of proteins as building

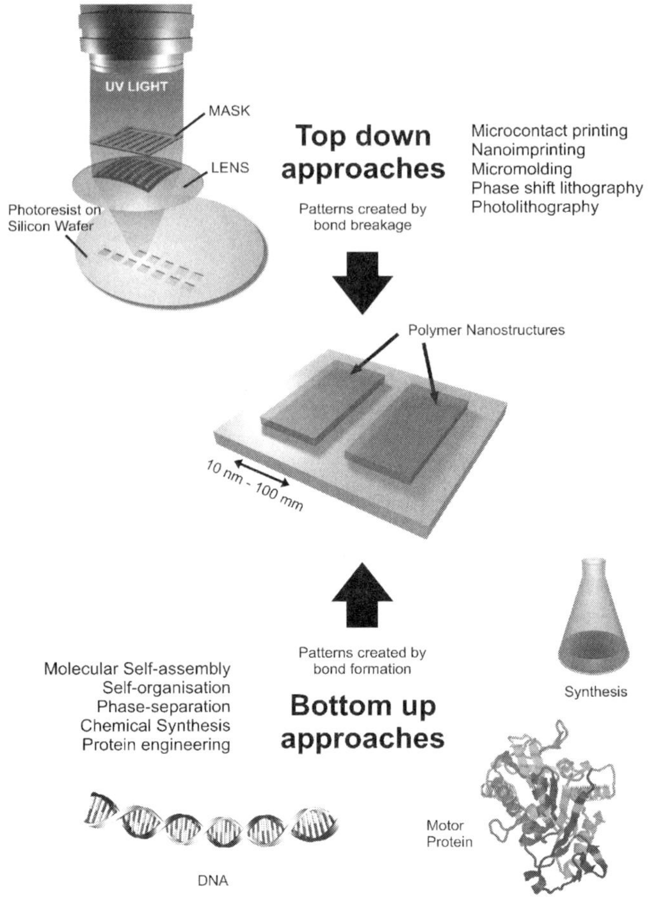

Figure 2. Overview of nanostructure fabrication technologies, illustrating the top-down and bottom-up approaches.

blocks. These developments will be discussed at the end of this chapter.

2. TOP-DOWN TECHNIQUES

The continuing drive in the semiconductor industry to fabricate smaller structures using top-down techniques will soon require dimensional control at sub-100 nm length scales, comparable to the size of polymers. State-of-the art photolithographic techniques can now more or less routinely fabricate features with dimensions below 100 nm, and at best can produce structures as small as 10 nm. Although patterned polymer films are used as sacrificial etch resist layers which are not present in the finished structure, it is clear that polymers are an integral part of these techniques. In a number of other lithographic methods, classified as soft-lithographic techniques, polymers are shaped into submicron structures, and form a more integral part of the final structure or patterning procedure. These alternative methods have the common feature of not using light or other short-wavelength energy sources to generate patterns, but rely on printing, imprinting, molding, or other mechanical processes instead. Before discussing nanoimprint lithography and a number of soft-lithographic techniques, we would like to illustrate the different uses of polymers in techniques such as *e*-beam and extreme ultraviolet (EUV) lithography.

2.1. Chemically Amplified Resist Materials

Over the last 15 years, silicon-based microelectronic devices have seen a dramatic increase in performance, due to the capability to produce smaller and smaller features, and to increase their surface density. Chemically amplified resist (CAR) materials, developed by Fréchet et al. at IBM, have been the main driving force behind this miniaturization [11, 12]. In particular, research has focused on positive tone photoresists consisting of polymers with acid-labile pendant protective groups, photoacid generator molecules, and additives [13]. Upon exposure to radiation, the photoacid generator molecules undergo a chemical reaction that generates a low amount of acid. During the postexposure bake, this acid diffuses and cleaves off the acid-labile protective groups. This deprotection step converts the insoluble polymers into soluble polymers, which can be washed away in the developing step. Their sensitivity relies on the chemistry of the photoacid catalysis, whereby a single radiation event (which generates the acid) can induce the chemical transformation of hundreds of molecules. The ultimate resolution of these resists is dependent on the interplay between a number of complex functions, including acid mobility and deprotection rate kinetics. Recent work uses X-ray and neutron reflectivity measurements to study the reaction front of the acid deprotection step into the polymer film [14]. The next generation of CAR materials needs to be capable of imaging feature sizes well below 100 nm, using deep-UV (DUV—248 nm), extreme UV (EUV—13.4 nm), or *e*-beam radiation sources. The development of these resists requires an enormous combined effort of all of the major resist manufacturers and end users [15]. EUV resists need to be applied in very thin films (<100 nm), as every organic material absorbs high amounts of short-wavelength radiation. Langmuir–Blodgett (LB) techniques have been proposed to prepare ultrathin, defect-free resist layers of poly(N–alkylmethacrylamide), although their use as EUV resists has not been shown [16]. Current resists are based on linear polymers, but it is conceivable that chain entanglement might eventually limit the minimum resolution. Recently, Fréchet and co-workers synthesized a range of dendrimers bearing thermally labile t-Boc end groups which could be used for sub-100 nm lithography [17]. Using both DUV and *e*-beam lithography, features as small as 50 nm with aspect ratios of 2.5:1 were achieved. Dendrimers combine the advantages of high molecular weights (and hence good processing properties) with low entanglement. In an attempt to make the lithographic process more environmentally friendly, Fréchet also reported the synthesis of water-soluble photoresists based on the photo-cross-linking of poly(methacrylates) containing sugar groups [18]. CAR resists do not have to rely on photoacids; it is also possible to use photogenerated bases and styrenic copolymers which are susceptible to base-catalyzed β elimination [19] or decarboxylation [20].

The high sensitivity of chemically amplified resists makes them ideal candidates for overcoming the low throughput in *e*-beam lithography, and to make fast *e*-beam writing possible. Negative *e*-beam resists are generally based on the epoxy-cross-linking chemistry of novolac polymers (glycidyl ethers) [21], with added phenylcarbinols [22] or calixarenes [23] for acid-hardening strategies.

The molecular weight of the polymers used as resists is extremely important. Poly(methylmethacrylate) (PMMA) has been the standard high-resolution positive tone resist for *e*-beam lithography since it was first introduced in the early 1980s [24]. The polymer undergoes chain scission, and shows a reduction in M_w upon exposure. The developer is strongly selective toward the dissolution of the low-molecular-weight components [25]. When high M_w PMMA is used (>500 kD), the fabrication of high-density nanostructures becomes very difficult as the irradiated polymer chains form a gel in the developer solution [26]. The development can be aided by sonication and a careful choice of developing solution, which contracts the radius of gyration of the PMMA inside the nanosized features. The highest resolution achieved with *e*-beam lithography is now about 3–4 nm [27, 28]. There is also some concern that high M_w polymers are too large compared to the structures to be resolved, which might lead to irregular features. Studies have found 20–40 nm nodes in both high and low PMMA [29]. It is therefore likely that these are caused by polymer aggregates, which will adhere to the sidewalls of the polymer features formed after developing [30].

In a fairly recent development, supercritical CO_2 ($scCO_2$) has been introduced as an alternative to developing in aqueous base [31]. It was shown that both silicon-containing [32] and a novel class of fluoropolymers containing acid cleavable tetrahydropyranyl groups [33] were good photoresists in combination with $scCO_2$ developing, facilitating the formation of 200 nm features [34].

2.2. Nanoimprinting

One of the major problems of *e*-beam lithography is its low throughput, which makes it a very costly technique. A viable methodology would involve the creation of nanomasters using lithographic techniques such as *e*-beam, focused ion beam, or extreme UV, which can then be replicated into other materials using inexpensive techniques. Over the last years, nanoimprinting has developed into a very powerful technique to generate nanometer scale features in both polymeric [35] and inorganic substrates [36]. Imprinting or embossing is a well-known technique used to introduce microstructures into hard polymers by pressing a master with surface relief features into a thermoplastic polymer film that is heated close to [37] or above its T_g [38]. This process is the basis for many commercial products, perhaps the best known of which are compact disks [39]. Chou and co-workers have demonstrated that features as small as 25 nm, extending over remarkably large areas, can be generated [40]. Nanoimprint lithography (NIL) has the potential of high throughput due to parallel processing. It does not require sophisticated tools, and allows nanoscale replication for data storage [41, 42]. Furthermore, nonplanar surfaces can be imprinted via the use of planarization layers. Multiple layers of imprinted features can then be prepared [43]. Large areas can be patterned by rolling a cylindrical mold over a flat polymer film; however, this process has not been demonstrated with sub-100 nm features [44]. A prototype "nano-CD" with 10 nm features and 400 Gbits/in^2 has been fabricated as a demonstration of the power of the nanoimprint technique

[45]. NIL is also compatible with conventional device processing techniques [46], and has been used to prepare a number of nanoelectronic devices, including GaAs metal–semiconductor–metal photodetectors [47]. When polymer films doped with dye molecules were imprinted in the submicron range, polymer films with anisotropic optical properties due to molecular alignment of chromophores [48] were formed. Similarly, organic light-emitting devices were imprinted with 200 nm gratings, which is a first step toward the fabrication of photonic crystals or lasers [49].

The quality of the nanoimprinting process depends on a number of experimental parameters, such as T_g, viscosity in the melt, and adhesion of the polymer to the mold [50]. PMMA has been most widely used as the imprintable material, but a range of thermoplastic and thermosetting polymers is under investigation to optimize the imprinting and subsequent etching steps [51, 52]. As a first step toward the integration of biological systems and nanotechnology, the interaction of cells with nanoembossed polymers has been studied [53]. Surprisingly, it was found that cells can detect nanosized "spikes," and do not adhere to these surfaces. Surfaces created in this way could have important applications in tissue engineering [54].

2.3. Soft Lithography

Soft lithography is a term coined by Xia and Whitesides [55] to describe techniques which allow the patterning of surfaces without the use of the expensive "traditional" lithographic techniques described in the previous section. These techniques are based on self-assembly and molding as low-cost alternatives for micro- and nanopatterning.

The most widely used soft-lithographic technique is microcontact printing (μCP) [56]. In this technique, a patterned elastomeric stamp formed from poly(dimethyl siloxane) (PDMS) is coated with a chemical ink (see Fig. 3). A monolayer is then transferred when the stamp is brought into conformal contact with the substrate. Submicron surface relief structures can easily be introduced in PDMS by curing the polymers against a lithographically prepared master. The advantage of μCP is the ability to pattern surfaces *chemically* at the submicron level. The fabrication of sub-100 nm surface relief features has proven to be much more difficult as nanoscale structures easily collapse [57, 58].

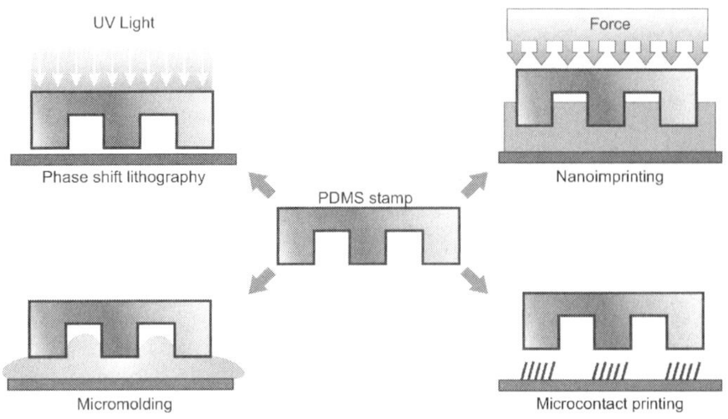

Figure 3. Soft-lithographic techniques using PDMS stamps.

Researchers at IBM have made considerable progress by systematically changing the cross-linking density and molecular weights of the PDMS, in addition to using a variety of nanosized filler materials to improve the stiffness of the polymer materials [59, 60]. It must be ensured that the Young's modulus of the PDMS does not exceed a certain value; otherwise, the stamps formed will be too stiff to come into intimate contact with the surface. The specific shape of the features on the stamp is also very important. Recently, Huck and co-workers replicated PDMS stamps against 200 nm gratings in InP [61]. On the stamps, 135 nm wide and approximately 100 nm high protruding lines were separated by 65 nm wide gaps. It was found that even hard PDMS with additional amounts of cross linking did not always accurately reproduce these gratings. The attractive (van der Waals) forces between the two polymer surfaces at either side of the 65 nm gaps are sufficiently large to make the lines collapse. Mere scanning with an AFM tip was sufficient to cause this collapse. An interesting way to achieve stamps with nanometer-sized features is via anisotropic etching of Si(100) substrates. The fast etching along the preferred crystal planes generates extremely sharp etch pits or grooves. Replicating these pits or grooves into soft PDMS immediately yields stamps with a feature size in the sub-50 nm range without the need for expensive lithographic equipment [62]. Commercially available Sylgard 184 PDMS (which is most commonly used for microcontact printing) is not stiff enough to replicate these sharp features accurately. The minimum feature sizes that can be obtained with this soft elastomer are in the 100–200 nm size range. By exploiting a multilayer strategy of hard and soft PDMS, Whitesides and co-workers recently succeeded in producing composite stamps with sub-50 nm features, which were still sufficiently deformable to enable printing over large areas [63, 64]. The elastomeric nature of the PDMS can be exploited using mechanical compression of the stamp to generate smaller features. New masters can be fabricated through stamp compression, and subsequent replication of the compressed features into a hard poly(urethane) polymer [65]. After a number of cycles, PDMS stamps with sub-100 nm features were obtained [66].

A major problem which becomes significant at the 100 nm level is the rapid diffusion of the small molecules that are commonly used as inks. This limits the resolution and reliable reproduction of features using μCP to approximately 100 nm [67]. The spreading of the ink has been exploited in the fabrication of very narrow lines. By leaving the stamp on the surface for an extended period of time, the gap between two adjacent printed lines eventually narrows to approximately 100 nm [68]. Although lateral feature sizes below 100 nm are very difficult to achieve, it is possible to combine μCP and the assembly of ultrathin polymer films with nanometer precision [69, 70]. Such films can be used to amplify the initial printed pattern, and greatly enhance the etch resistance of the monolayer underneath [71]. It has been shown that polymers can be used as inks, and present a solution toward diffusion-limited feature resolution [72]. In recent work, dendrimers were used as inks to pattern silicon wafers into 135 nm wide dendrimer lines on a 200 nm pitch [61].

Replica molding (REM) is a soft-lithography technique initially used to fabricate polymer microstructures with controlled structures on planar or curved surfaces. The process involves molding a prepolymer onto a previously fabricated master, followed by polymer curing and replica removal. The technique was subsequently extended to produce nanostructures using a PDMS stamp as the mold [65]. The soft elastomeric stamp reduces the stresses involved in the separation of the master and the replica, and therefore reduces any potential damage to the master and replica. Whitesides and co-workers have demonstrated the replication from a PDMS master of an array of 60 nm wide features into polyurethane (PU). Multiple copies were made without any noticeable reduction in the PU replica pattern fidelity [73].

The next soft-lithography technique to be discussed here is micromolding in capillaries (MIMIC), which was also developed by Whitesides and co-workers [74]. When a patterned PDMS stamp is brought into contact with a substrate, which may also be structured, an empty network of capillaries is formed. When a sufficiently fluid prepolymer is placed in contact with network openings, the liquid fills the cavities via capillary action. The prepolymer can then be cured and the PDMS stamp removed to yield polymeric patterned structures with sub-100 nm feature dimensions [75]. Through careful selection of the capillary surface energy, the exact structure of the final polymer film can be fine tuned [76]. An important advantage of MIMIC over standard lithographic techniques is that pattern formation can be carried out in a single step, a valuable characteristic for any microfabrication technique.

Phase-shift lithography is a technique which combines photolithography with soft lithography to form features at or below 100 nm [77]. The process involves the exposure of a thin layer of photoresist to polychromatic ultraviolet light passing through an elastomeric stamp which is in conformal contact. This allows standard photolithography to be carried out in the near field, with relatively simple and convenient apparatus. Whitesides and co-workers have used composite PDMS stamps to further reduce the minimum feature size possible with phase-shift lithography [64]. This work describes the fabrication of high-density arrays of ring-like structures with edge lengths down to 30 nm. These nanowells can be used as containers for the growth of inorganic nanocrystals

3. SELF-ASSEMBLY

A self-assembly, or bottom-up, approach to nanostructured polymer films is fundamentally different from the top-down lithographic techniques described in the previous section. Instead of locally destroying or deforming material to generate features, bottom-up approaches rely on the formation of covalent or noncovalent bonds between basic building blocks to generate larger structures [78, 79]. The main advantages of this approach are the ease of reaching the nanometer domain and the low capital expenditure on equipment. One of the main drawbacks of this method is the difficulty in making arbitrary features, as one relies heavily on thermodynamically defined structures. These limitations, which will become apparent below, will need to be taken into account in any device design.

3.1. Phase Separation of Block Copolymers

Block copolymers comprised of two (or more) flexible, chemically incompatible, and dissimilar blocks, for example, poly(styrene) and poly(isoprene), can microphase separate into a variety of morphologies with nanometer-scale dimensions [80, 81]. This self-assembly process is driven by an unfavorable mixing enthalpy and a small mixing entropy, while the covalent bond between the two blocks prevents macrophase separation. The microphase separation can form various morphologies, including spheres, lamellae, inverse spheres, and several more complex shapes, depending on the polymers used and on their volume fractions [82, 83]. If the morphology can be controlled [84, 85] and turned into a useful structure, phase separation of block copolymers could be a powerful tool for fabricating nanostructures without additional lithography and processing steps [86].

Phase separation of polymers into lamellar structures has been used to generate 50–100 nm thick periodic layers with different refractive indexes, which can be used as photonic crystals. Solution cast films of symmetrical poly(styrene)–poly(isoprene) (PS–PI) diblock copolymer films showed single, well-defined reflectivity peaks in the visible wavelength region [87]. The peak reflective wavelength could be tuned from 350 to 600 nm by adding homopolymers to the block copolymers, which increased the thickness of the lamellar layers [88]. By blending a triblock copolymer such as poly(styrene)–b–poly(butadiene)–b–poly(t–butyl methacrylate) (PS–PB–PBMA) with a PS–PBMA diblock copolymer, noncentrosymmetric lamellar phases were obtained with three different alternating layers [89]. Other morphologies are easily accessible by varying the two polymers in the diblocks. Dense periodic arrays of holes and dots have been fabricated by exploiting the microphase separation of poly(styrene)–poly(butadiene) (PS–PB) or PS–PMMA diblock copolymer films [90]. After microphase separation, one of the polymer components can be selectively etched and rinsed away. A typical example shows 20 nm holes, separated by 40 nm and extending over very large areas. The phase-separated structures were transferred into an underlying silicon nitride film via selective etching of one of the polymer blocks, and subsequent reactive ion etching (RIE) into the inorganic substrate. Alternatively, the pattern can be transferred into a thin poly(imide) film on a silicon substrate. Subsequent *e*-beam evaporation of metals and lift off of the poly(imide) mask results in dense arrays of metal dots [91].

Porous membranes have been formed from the diblock copolymer poly(t–butyl acrylate)–b–poly(2–cinnamoylethyl methacrylate) (PBA–PCEMA). The PCEMA block can be cross linked photochemically, while the PBA block contains the t–butyl group which may be cleaved by hydrolysis. The PBA block undergoes cylindrical phase formation, which can then be used to create thin films with regularly packed and uniformly sized nanochannels with diameters less than 20 nm. Films prepared in this way may serve as the skin layers for membranes with channels tunable both in size and in the functional groups within the cavity [92]. The long-range lateral order of the microdomains can be improved using chemically [93] or topographically patterned substrates [94–96], the interplay between crystallization and microphase separation [97], and epitaxial growth on crystalline substrates [98].

A prerequisite for the successful use of block copolymers as nanoscopic templates is the control over the orientation of the microdomains. In particular, for cylindrical microdomains, an orientation normal to the substrate is required. This thermodynamically stable orientation can be achieved in thin polymer films by controlling the surface energy of the substrate, choosing the right solvent for drop or spin casting of the block copolymer and thermal annealing. Electrical poling can also be used to orient the microdomains [99]. The alignment of cylindrical domains and the subsequent removal of one of the components of the diblock copolymers leads to a nanoporous structure. The porous film can be filled via electrodeposition, which leads to the formation of nanowires ($\sim$15 nm in diameter) in a polymer matrix [100]. The unfilled, nanoporous film yields a nanoelectrode array, which at low scan rates behaves like a macroelectrode, but at high scan rates, the nanoelectrodes act independently [101].

When the surface is patterned into areas of different surface tension, phase separation can be driven to follow the underlying pattern [102, 103] (see Fig. 4). A thin film of a poly(styrene) and poly(vinylpyridine) (PS/PVP) blend was spun cast on top of alternating hydrophobic and hydrophilic stripes. As the two polymers are strongly incompatible, demixing takes place during spin coating. After the solvent has completely evaporated, the topography of the polymer film mimics the underlying surface energy pattern. Rinsing the films with cyclohexane selectively removes one of the polymers (PS), leaving patterns of PVP behind. Recently, results have been published which indicate that control over phase separation can be extended to the sub-100 nm size regime [104]. Further developments in this direction could lead to the generation of functional patterns of various designs in microphase-separated structures. However, the kinetic pathways through which the equilibrium-phase-separated states are reached delicately depend on the particular pattern, its size, and overall film thickness [105], and in practice, this means that it is quite difficult to obtain a well-defined phase-separated structure that follows the underlying pattern.

3.2. Polymer Nanotubes

Polymeric nanotubes have received considerable attention recently as potential "molecular wires" and replacements for carbon nanotubes which are difficult to functionalize. Polymers with suitable electronic and optoelectronic properties have been structured into nanofibers as it was envisaged that they would be used in nanoscale devices. Polymer structures were produced within the pores of

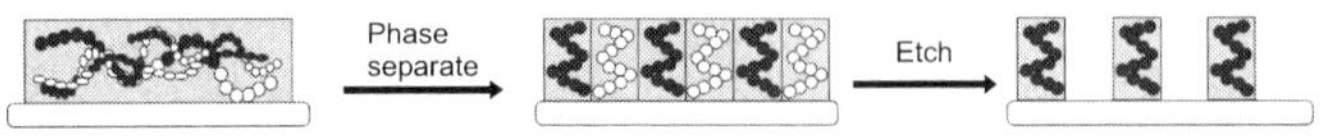

Figure 4. Phase separation of block copolymers leading to well-organized nanoscale surface features.

nanoporous membranes [106] or zeolites [107]. Specifically, well-defined poly(p–phenylene vinylene) (PPV) fibers have been produced inside hexagonally ordered templates formed from liquid crystalline materials [108]. The formation of nanowires without the use of templates is also possible, although there is only limited control over the shape, dimension, and orientation of the fibers [109]. Nonetheless, impressively regular polypeptide nanotubes show the potential of this kind of approach [110]. The formation of individual nanosized objects from block copolymers is also possible, although the design parameters are less well understood. A number of general concepts have been described. Amphiphilic block copolymers, for example, poly(stryrene)–block–poly(acrylic acid), form so-called crew-cut aggregates [111] when they are first dissolved in a good solvent and water is slowly added, leading to controlled aggregation of the hydrophobic blocks [112]. Similar nanostructures can be formed based on the thermoreversible dissolution of block copolymers in single solvents [113]. The amphiphilic properties of the blocks can be tailored via hydrogen bonding with small surfactant molecules. These occupy polar or charged sites on one of the blocks, leading to a controlled, shielded assembly of individual crew-cut aggregates [114]. A disadvantage of these methods is the sensitivity toward the dissolution–precipitation process and the instability of the objects formed. "Freezing" the self-assembled nanostructures can be achieved by incorporating photochemically cross-linkable groups or blocks [115]. The cross-linked fibers can subsequently be "sculptured" by selectively dissolving or degrading one of the blocks, and are stable enough to be suspended in other solvents [116]. An inverse approach can also be used to form organic–inorganic mesoporous materials, where self-organized surfactant phases form templates for the mineralization of a variety of inorganic minerals [117]. Instead, the block copolymers aid in the controlled, nanostructured mineralization of aluminosilicates [118], which are subsequently removed, to render individual polymeric "hairy objects" [119]. An elegant approach toward polymeric nanostructures has been explored by Steinhart et al. [120]. By simply placing a polymer melt or solution on a nanoporous surface, monodisperse polymer nanotubes form after the polymers wet the inside of the pores. Wall thickness, composition, length, orientation, and ordering of the polymeric nanotubes can be easily controlled by varying the polymers used, the surface energy, and the pattern of the mold. Instead of wetting the inside of capillaries to create hollow nanofibers, one can also coat the outside of ultrathin fibers, and subsequently remove this template [121]. The templates are formed via electrospinning [122] of easily degradable polymers, and fibers with diameters as small as 100 nm can be prepared [123]. Interestingly, fibers with surprisingly regular elongated pores in the 200 nm region were obtained by electrospinning dichloromethane solutions of poly(L–lactide) [124]. It is thought that spinodal phase separation as a result of very rapid solvent evaporation, followed by solidification of the polymer fibers, leads to the formation of these structures. Using a sol–gel coating technique, these fibers can also be used as templates for nanoporous TiO_2 tubes, which could have important catalytic properties and applications in solar cells [125].

4. POLYMER NANOPARTICLES

Thus far, we have discussed numerous examples where polymers are structured at the nanometer level via lithographic methods or by the exploitation of self-organization and phase separation. Nanoparticles have been fabricated using a variety of synthetic techniques, for example, via miniemulsion polymerization, which has been optimized to yield particles ranging from 20 to 50 nm [126]. Landfester and co-workers have exploited the miniemulsion process to prepare semiconducting polymer nanospheres, which can be deposited via spin coating from aqueous dispersions [127, 128]. Miniemulsions are understood as stable emulsions consisting of stable droplets with a size of 20–500 nm created by shearing a system containing oil, water, surfactant, and a highly water-insoluble compound [129, 130]. Polymer particles are formed by polymerizing monomers that are dissolved in the oil phase [131], or by evaporating the solvent, leaving polymer residue behind [132]. Miniemulsions also allow the formation of hollow particles that can be used as nanocapsules [133]. It has also been possible to form inorganic microcapsules filled with organic liquids or polymers [134], chitosan-grafted biopolymer nanocapsules [135], molecularly imprinted polymer nanospheres for use as affinity receptors [136], and semiconducting polymer nanospheres [137].

Polymeric nanostructures are more commonly prepared via a bottom-up approach by synthesizing polymers with well-defined shapes (see Fig. 5). Synthesis can be carried out on assemblies that are spatially constrained to form novel structures [138]. Functionalized polymeric nanoparticles have a huge potential in a range of nanotechnological applications, ranging from vectors for drug-delivery systems [6] to templating agents for nanoporous microelectronic materials [139].

4.1. Dendrimers

Dendrimers, with their perfectly branched architectures, are the best known examples of spherical nanoparticles consisting of single molecules [140, 141]. Their highly branched structure gives rise to interesting properties which are different from those exhibited by linear polymers with similar molecular weights [142–147]. Dendrimers can be synthesized via divergent [148–155] or convergent [156–158] pathways, leading to full control over the placement of every monomer in the ever-growing molecule. There is no need to discuss dendrimers in great detail here, as they have been the topic of numerous books and review articles over the last few years [159–170]. As a result of their branched architecture, the shell of the dendrimers can be made denser than the inside (core). This has been exploited to make molecular containers [171, 172] large enough to encapsulate up to five large dye molecules [173, 174]. It is possible to produce dyes with modified properties via the synthesis of dendritic chromophores which consist of a fluorescent core attached to a dendrimer [175–181]. Polyamidoamine (PAMAM) dendrimers have shown promising results as delivery agents in DNA transfection studies [182]. The full control over the synthesis enables the placement of functional groups at any position. For example, redox-active metal porphyrins in the center of large dendrimers can become completely shielded from the environment, altering the redox potential [183]. A number of other redox-active moieties have been investigated, including tris(bipyridine) ruthenium (II) complexes [184], fullerines [185], ferrocenes [186–189], oligothiophenes [190], oligonaphthalenes [191], and 4,4′-bipyridinium [192]. Research has shown that the dendritic structures have an insulating effect on the redox sites [193]. Even more elaborate structures contain light-harvesting groups at the periphery of the dendrimers, combined with electroactive cores, as a first step toward molecules that are able to transform solar power into electrical energy [194–199]. Conjugated dendrimers have also been synthesized for use in light-emitting devices [200–203]. The shielding of the dendrimer core from the environment has been used to introduce simple catalytic functionality [204–214]. Photochemically responsive cores have been incorporated into dendritic structures so that they may potentially be used as molecular switches [215]. Azo linkers which can undergo a photochemical cis-trans isomeration have also been investigated [216–222]. Dendrimers with hydrophilic outer rims and hydrophobic interiors provide an ideal microenvironment for reactions that proceed via apolar transition states. The reactants (which do not react significantly in the polar solvent in which the dendrimer is dissolved) diffuse toward the interior, with its more apolar "climate." The reaction then proceeds to yield polar products, which are subsequently expelled from the interior and diffuse back into the solvent [223]. One of the major problems associated with dendrimer use is the daunting synthesis of these large (M_w over 10,000) molecules with numerous functional groups. Instead of making molecularly perfect structures, considerable effort has been devoted to the facile synthesis of structures that have a similar construction, but are far easier to synthesize.

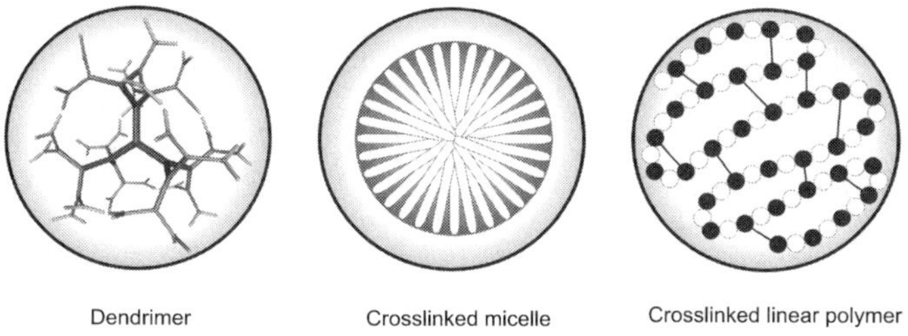

Figure 5. Selection of polymer nanoparticles with different molecular architectures.

4.2. Cross-Linked Micelles

Discrete nanoparticles may also be synthesized via new methodologies such as intramolecular chain collapse, as demonstrated in a recent elegant example by Hawker and co-workers [224]. First, linear chains with benzocyclobutene (BCB) coupling groups were synthesized using a controlled radical polymerization. The monodisperse polymers that were obtained assumed a spherically shaped random coil in solution. Collapse of these chains, followed by intramolecular cross linking, leads to the formation of discrete nanoparticles [225]. This strategy allows the synthesis of polymeric nanoparticles on a multigram scale with minimal difficulty.

Instead of intramolecular cross linking of single chains, a number of groups have utilized intermolecular cross linking of preassembled polymer chains, to synthesize well-defined spherical nanoparticles, known as polymer micelles. These polymer micelles are prepared through the self-assembly of amphiphilic block copolymers in aqueous solutions [226, 227]. The particles have diameters ranging from 20 to 200 nm, and are generally spherical. Other shapes have also been observed, including fractal structures consisting of toroidal networks, worm-like micelles [228, 229], and vesicles [230]. Polymer micelles only exist above critical micelle concentrations, and show a strong dependency on solvent polarity and salt concentration.

Covalent cross linking of the chains stabilizes the structures, forming cross-linked dendrimer-like nanospheres [231, 232], polymer brushes [233], and shell cross-linked knedel-like (SCK) structures. SCKs combine the features of a hard shell with a fluid cavity, block copolymer micelles, and dendrimers. These amphiphilic copolymer micelles consist of a rigid, but permeable cross-linked outer layer and a mobile uncross-linked interior. It is believed that these types of structures will find applications in phase-transfer reactions, drug delivery, and industrial support catalysts [234]. The resulting materials, consisting of nanoparticles embedded in an insulating matrix, have semiconducting properties which are independent of the type of DNA used [235].

A large range of different structures is possible through the careful design of each block within the diblock copolymer. Nanofibers may be created, where one block is the cross-linked core and the second block forms the concentric shell [236, 237]. Hollow nanospheres [238, 239] can be formed by varying the segments of the block copolymers used [240], and by postfunctionalization of the cross-linked nanoparticles [241]. Pore generation within nanospheres was carried out via the introduction of a porogen (a short chain polymer) into the micelle core. The core structure was locked in shape through cross linking, and the porogen could then be removed. These cavities were then used to estimate the thymine–adenine binding constant within the pores [242].

5. BIOLOGICAL POLYMERIC ASSEMBLIES

This chapter has thus far focused on synthetic polymers that have well-defined nanoscale dimensions or polymers that self-assemble and self-organize into nanostructures. As explained in the Introduction, polymers are ideal nanoscale building blocks, and are widely used in the natural world as part of the cellular machinery. Much of the inspiration for the research reviewed in this chapter has come from biological macromolecules and macromolecular assemblies. There are many hundreds of examples of biological polymers and their nanoscale functionality, and as a result, we would like to focus on the only real biopolymer that can be designed and synthesized from molecular building blocks: DNA. DNA has found its way into nanotechnology because of its unique self-assembly and recognition properties, in addition to its ease of synthesis. Here, we illustrate the wide range of research that has exploited DNA, including the organization of nanoparticles and the direct assembly into nanostructures. This section will conclude with the very few examples of protein engineering which give a hint of future developments involving complex proteins in well-defined nonbiological devices.

Nature uses polymers such as DNA, proteins, and a range of biopolymers as building blocks to form working nanoscale devices, including viruses, bacteria, and ultimately, the cells of higher organisms. The ribosome contained within cells is constructed from two subunits of RNA and protein. Their function is to utilize the information coded within DNA to fabricate polypeptides within the cytoplasm. This is a perfect example of a polymeric device used to manufacture new polymeric structures. Muscle contraction is another example where nanostructures feature in biological function. The globular protein actin can polymerize into long fibers which are widely distributed within cells. In muscle, these actin fibers are interdigitated with myosin aggregates, and it is the relative movement of these two sets of fibers which gives the molecular basis of muscle contraction.

5.1. DNA Structures

The bottom-up approach toward structurally well-defined nanoarchitectures needs a molecular toolkit that allows the facile assembly of molecular building blocks into two- and three-dimensional structures [243]. DNA is potentially one of the most versatile and well-understood "glues" available to the biologist, chemist, and materials scientist [244, 245]. For this reason, a number of groups have utilized DNA to generate nanostructured materials [246–248] or to arrange nanoparticles [249, 250]. The main reason for the attractiveness of DNA as a nanoscopic building block is the hybridization of complementary single strands into a double helix. This hybridization is based on the formation of multiple hydrogen bonds between the matching bases in each strand. The process is very robust, and leads to stable structures, but more importantly, it is a very selective process: only complementary strands are able to form the double helix. This allows the "programming" of all assembling processes, which is an almost unique property for any "assembler" in the nanometer-size scale [251]. Nanoparticles coated with different strands of DNA (via a thiol linker) will recognize each other in solution and aggregate [252]. The aggregation leads to a color shift of the nanoparticle solution [253], and can be used as a colorimetric detection of specific DNA sequences [254, 255]. This strategy can also be used to induce the formation of binary or ternary nanoparticle networks, combining large and small particles [256]. The aggregation is not limited to gold or silver nanoparticles, but can be extended to incorporate semiconductor quantum dots such as CdSe [257]. The aggregation has almost exclusively been studied in solution, but is also very effective in the layer-by-layer build up of DNA and gold nanoparticle composites [258]. An elegant example of the programmable hybridization is the use of DNA as a molecular ruler that governs the exact distance between two metal nanoparticles [259]. A single strand of DNA with a number of recognition sites has been decorated with 1.4 nm nanoparticles containing different short complementary sequences. The length of the DNA that interconnects the particles can simply be tailored by synthesizing longer sequences. An even more complex experiment used the combined molecular recognition powers of DNA/RNA and of streptavidin–biotin to organize 1.4 nm Au nanoparticles in yet more complicated patterns [260]. DNA has also been used as a template to grow nanometer-scale conducting silver wires. Instead of linking two nanoparticles together, a 16 μm long single strand of λ DNA was used to bridge the gap between two electrodes, both functionalized with short DNA strands. Silver ions were then deposited onto the organic DNA double-helix bridge, which were reduced to form metallic silver. Overall, 100 nm wide, 12 μm long metallic wires were fabricated [261]. Although these are only crude proof-of-principle experiments, they nonetheless demonstrate the power of self-assembly of macromolecules in nanofabrication.

All of these strategies use DNA as a programmable assembler to connect inorganic building blocks. Perhaps the most stunning structures that have been formed from DNA have been synthesized by Seeman and [262, 263]. In that work, DNA is the material from which complicated 2-D and 3-D molecules are built, entirely relying on the hybridization of complementary strands [264, 265]. The first innovation was the use of three- or four-arm junctions through appropriate sequence design, hybridization conditions, and annealing procedures [266, 267]. These junctions later served as corners in more complex structures, by combining the "sticky ends" of the junctions with short rigid pieces of double-stranded DNA. Three-arm junctions were combined to form triangular [268] or square grids [269], or even cubes [270]. It is important to note that this DNA cube does not actually resemble a cube when in solution; its structure is based on the topology, or interconnectivity, that is equivalent to a cube. The synthesis of these molecules is extremely complicated, and a solid-phase strategy was adopted to increase both yields and the amounts of nucleotides that can be introduced [271] (see Fig. 6).

To improve the structural rigidity of the nanostructures, double crossovers (DX) were introduced, which fuse two double helices together [272, 273]. These DNA structures do yield recognizable structures in solution compared to the DNA structures described above. Amazingly, triangles can be hooked up to a longer chain, leading to a necklace pattern that can be visualized with AFM [274]. The spacing between the triangles can easily be tailored via the supporting strand. By designing DX building blocks with "sticky ends" that were linked to form 2-D crystalline networks, periodic arrays as large as 2×8 μm were formed, and the network repeat units could be characterized by AFM [275]. The stiff, essentially linear DX double double helices form dense mats or porous networks, completely following the self-assembly designs that are programmed into the polymers [276].

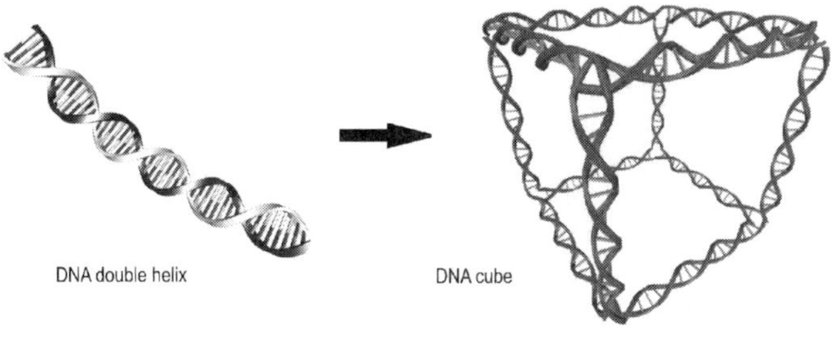

Figure 6. Self-assembled DNA structure.

5.2. Biological Nanomachines

So far, we have discussed a wide range of different nanoscale structures that were assembled from polymeric building blocks. However, almost all of these structures are very far from actual working devices; in fact, the only process where polymers are used to fabricate devices is in the (conceptually) old-fashioned process of lithography. Yet, as we have briefly discussed in the Introduction, polymers are the building blocks for devices in the biological world. For example, the ribosome forms one of the most impressive pieces of cellular "machinery," and consists of two large subunits. Each subunit, in turn, consists of RNA and over 20 proteins. These proteins unwrap DNA, read the information it contains, assemble proteins from amino acids, and rewrap the DNA again [277, 278].

These are the kind of truly astonishing devices that will need to be developed if nanotechnology is going to make a huge impact on fields such as medicine, chemistry, and materials science. Over the last decade, considerable progress in this direction has been made. One of the most impressive examples involves harnessing motor proteins into powering inorganic devices. The rotary nature of the ATPase enzyme was conclusively demonstrated by Noji et al., with the attachment of actin filaments to the motor protein. They showed that the work performed by the enzyme was almost 100% efficient [279]. Montemagno and co-workers [280] fabricated tiny biomolecular motors constructed from three elements: lithographically produced nickel posts with a height of 200 nm and a width of 80 nm, a recombinant F_1–ATPase enzyme anchored to the nickel posts via histidine tags, and an individual silicon nanopropellor attached to each enzyme. These silicon rotors were functionalized with a biotinylated histidine-rich peptide to facilitate attachment to the F_1–ATPase enzyme. The addition of Na_2ATP caused the rotors to rotate in an anticlockwise direction at speeds of up to 8.3 r/s. The motion of the propellers could be stopped by the addition of sodium azide, a potent ATPase inhibitor. This conclusively demonstrated that the movement was due to the F_1–ATPase motors. Research has also focused on harnessing the movement of kinesin, another motor protein powered by the hydrolysis of ATP. Kinesin acts by moving along tubular protein filaments, known as microtubules, and the drive of this research has been to restrict the movement of microtubules along predefined pathways. In recent work by Uyeda and co-workers, lithographic techniques were used to create embedded glass tracks which confine the movement of the microtubules [281]. Their novel approach overcame the common drawbacks of systems involving motor proteins, including the loss of filaments over track boundaries [282, 283] and bidirectional movement [284]. The system developed by Uyeda and co-workers displayed linear unidirectional movement through the incorporation of arrowheads along the tracks. This allowed the fabrication of so-called microtubule circulators, and enabled the active transport of tubules from a randomly distributed configuration to one in which only one of two sides is populated. It is possible that future devices will be constructed from a combination of microtubules and conventionally fabricated silicon structures.

6. SUMMARY

The range of examples reviewed in this chapter illustrate how control over polymers and polymeric materials can aid the fabrication of nanoscale devices. Synthetic routes designed to control polydispersities and molecular weights of polymers were essential in the development of photoresists that can now be used in extreme UV lithography to create patterns in the 150 nm, and ultimately in the sub-100 nm, regime. Alternative techniques exploiting contact printing, molding, and embossing or imprinting all rely on the nanostructuring of polymers to replicate nanoscale features in an economically viable way. At the same time, synthetic capabilities have expanded to give perfect control over the phase separation of block copolymers, which is another rapidly expanding field of research. Nanoparticles can also be obtained by synthesizing polymer molecules with increasingly large dimensions, while maintaining control over their three-dimensional shape and functionality. Examples include dendrimers, hyperbranched polymers, cross-linked micelles, and core–shell particles. These have great potential as nanoreactors, delivery vehicles, transfection agents, and rheology modifiers. Natural systems are unmatched in their controlled three-dimensional shape and their ability to perform work at the molecular level, as the examples of the motor proteins surely show. Slowly but surely, however, synthetic polymers are approaching a level of complexity that can finally result in the mimicking of biological polymers. The field of polymers in nanotechnology will continue to show a diverse array of developments, leading to increasingly complex structures and the incorporation of these synthetic polymers into nanoscale devices.

GLOSSARY

Biological polymers Polymers which exist in nature e.g. DNA, proteins, polysaccarides, gums, and resins.

Block copolymers A copolymer where each type of monomer is arranged in substantial sequences or blocks. ~AAAAABBBBBAAAAA~

Copolymer A polymer which is composed of two or more types of monomer unit.

Cross-linked micelles Discrete nanoparticles formed by the cross-linking of polymer to achieve spherical shapes.

Dendrimer A unique type of polymer which is a highly branched three-dimensional macromolecule with highly controlled structures, a single molecular weight, and the flexibility for tailored functionalization.

Deoxyribonucleic acid (DNA) A double-stranded helix of nucleotides which carries the genetic information of a cell. It encodes the information for proteins and is able self-replicate.

Microcontact printing (μCP) A type of soft-lithographic technique which involves the use of a patterned elastomeric stamp formed from poly(dimethyl siloxane) (PDMS) with a chemical ink to transfer monolayers on to a substrate when the stamp and the substrate are brought into contact.

Molecular weight (M) The weight average molar mass of the molecules present in the polymeric mixture.

Molecular wires Molecular wires are single chain conducting molecules which have their ends attached to electrodes.

Monomer A small molecule which can be joined together to form larger molecules such as dimers, trimers or polymers.

Phase separation When there are two or more components in a system which are immiscible then the components separate into their respective phases.

Photolithography A process of transferring geometric shapes on a mask to a surface using radiation commonly UV, extreme UV, *e*-beam, or X-ray. A layer of photoresist is placed on the surface using spin-coating, and parts of it are removed during the lithographic process to transfer the shapes.

Photoresist Organic substances (monomeric or polymeric) which are sensitive to light. There are two types of photoresists: positive and negative. Positive photoresists are often polymeric material which have photolabile groups which cause the material to become soluble in the developer on exposure to light. Negative photoresists are often monomeric material which polymerize on exposure to light.

Polymer A large molecule constructed from many smaller strucutral units called monomers.

Self-assembly Self-assembly is a thermodynamic process by which molecules come together to build desired supramolecular structures due to specific intermolecular interactions. Examples of self-assembly are often seen in nature.

Soft lithography A lithographic technique which allows the patterning of surfaces with the use of expensive traditional photolithographic processes. Different types of soft lithographic techniques include phase shift lithography, microcontact printing, nanoimprinting, and micromolding.

Spin coating A process to form thin films of material by placing a solution of the material on the center of a disc and spinning very rapidly allowing the material to spread out evenly and thereby forming a thin film.

Synthetic polymers Man-made polymers which were synthesized either by chain polymerization or step (condensation) polymerization.

REFERENCES

1. J. C. Love and G. M. Whitesides, *Sci. Amer.* 285, 38 (2001).
2. G. Stix, *Sci. Amer.* 285, 32 (2001).
3. P. Ball, *Nature* 408, 904 (2000).
4. G. M. Whitesides, *Sci. Amer.* 285, 78 (2001).
5. K. E. Drexler, *Sci. Amer.* 285, 74 (2001).
6. C. M. Niemeyer, *Angew. Chem., Int. Ed.* 40, 4128 (2001).
7. J. H. Burroughes, D. D. C. Bradley, A. R. Brown, R. N. Marks, K. Mackay, R. H. Friend, P. L. Burns, and A. B. Holmes, *Nature* 347, 539 (1990).
8. R. H. Friend, *Pure Appl. Chem.* 73, 425 (2001).
9. M. J. Liu and J. M. J. Fréchet, *Pharm. Sci. Technol. Today* 2, 393 (1999).
10. H.-G. Elias, "An Introduction to Polymer Science." VCH, Weinheim, 1997.
11. J. M. J. Fréchet, H. Ito, and C. G. Willson, "Proceedings of Microcircuit Engineering," Colloque Internationale sur la Microlithographie, 1982, Vol. 82, p. 260.
12. S. A. MacDonald, C. G. Willson, and J. M. J. Fréchet, *Acc. Chem. Res.* 27, 151 (1994).
13. G. M. Walraff and W. D. Hinsberg, *Chem. Rev.* 99, 1801 (1999).

14. E. K. Lin, C. L. Soles, D. L. Goldfarb, B. C. Trinque, S. D. Burns, R. L. Jones, J. L. Lenhart, M. Angelopoulos, C. G. Willson, S. K. Saija, and W.-L. Wu, *Science* 297, 372 (2002).
15. R. L. Brainard, C. Henderson, J. Cobb, V. Pao, J. F. Mackevich, U. Okoroanyanwu, S. Gunn, J. Chambers, and S. Connolly, *J. Vac. Sci. Technol. B* 117, 3384 (1999).
16. Y. Guo, F. Feng, and T. Miyashita, *Macromolecules* 32, 1115 (1999).
17. D. C. Tully, A. R. Trimble, and J. M. J. Fréchet, *Adv. Mater.* 12, 1118 (2000).
18. J. M. Havard, N. Vladimirov, J. M. J. Fréchet, S. Yamada, C. G. Willson, and J. D. Byers, *Macromolecules* 32, 86 (1999).
19. E. J. Urankar and J. M. J. Fréchet, *J. Polym. Sci A; Polym. Chem.* 35, 3543 (1997).
20. J. M. J. Fréchet, M.-K. Leung, E. J. Urankar, C. G. Willson, J. F. Cameron, S. A. MacDonald, and C. P. Niesert, *Chem. Mater.* 9, 2887 (1997).
21. P. Argitis, I. Raptis, C. J. Aidinis, N. Glezos, M. Baciocchi, J. Everett, and M. Hatzakis, *J. Vac. Sci. Technol. B* 13, 3030 (1995).
22. S. Uchino, J. Yamamoto, S. Migitaka, K. Kojima, M. Hashimoto, and H. Shiraishi, *J. Vac. Sci. Technol. B* 16, 3684 (1998).
23. J. Fujita, Y. Ohnishi, S. Manako, Y. Ochiai, E. Nomura, T. Sakamoto, and S. Matsui, *Jpn. J. Appl. Phys.* 36, 7769 (1997).
24. A. N. Broers, *J. Electrochem. Soc.* 128, 166 (1981).
25. D. G. Hasko, S. Yasin, and A. J. Mumtaz, *J. Vac. Sci. Technol. B* 18, 3441 (2000).
26. E. A. Dobisz, S. L. Brandow, R. Bass, and J. Mitterender, *J. Vac. Sci. Technol. B* 18, 107 (2000).
27. S. Yasin, D. G. Hasko, and H. Ahmed, *Appl. Phys. Lett.* 78, 2760 (2001).
28. W. Chen and H. Ahmed, *J. Vac. Sci. Technol. B* 11, 2519 (1993).
29. E. A. Dobisz, S. L. Brandow, E. Snow, and R. Bass, *J. Vac. Sci. Technol. B* 15, 2318 (1997).
30. H. Namatsu, M. Nagase, T. Yamguchi, K. Yamazaki, and K. Kurihara, *J. Vac. Sci. Technol. B* 16, 3315 (1998).
31. R. D. Allen, K. J. Chen Rex, and P. M. Gallagher-Wetmore, *Proc. SPIE* 2438, 250 (1995).
32. C. K. Ober, A. H. Gabor, P. M. Gallagher-Wetmore, and R. D. Allen, *Adv. Mater.* 9, 1039 (1997).
33. S. Yang, J. Wang, K. Ogino, S. Valiyaveettil, and C. K. Ober, *Chem. Mater.* 12, 33 (2000).
34. N. Sundararajan, S. Yang, K. Ogino, S. Valiyaveettil, J. Wang, X. Zhou, C. K. Ober, S. K. Obendorf, and R. D. Allen, *Chem. Mater.* 12, 41 (2000).
35. S. Y. Chou, P. R. Krauss, and P. J. Renstrom, *Science* 272, 85 (1996).
36. S. Y. Chou, C. Keimel, and J. Gu, *Nature* 417, 835 (2002).
37. N. Stutzmann, T. A. Tervoort, C. W. M. Bastiaansen, K. Feldmann, and P. Smith, *Adv. Mater.* 12, 557 (2000).
38. E. P. Herzig, "Micro-Optics." Taylor & Francis, London, 1997.
39. M. Madou, "Fundamentals of Microfabrication." CRC Press, Boca Raton, FL, 1997.
40. S. Y. Chou, P. R. Krauss, and P. J. Renstrom, *Appl. Phys. Lett.* 67, 3114 (1995).
41. B. D. Terris, H. J. Mamin, M. E. Best, J. A. Logan, D. Rugar, and S. A. Rishton, *Appl. Phys. Lett.* 69, 4262 (1996).
42. M. Li, J. Wang, L. Zhuang, and S. Y. Chou, *Appl. Phys. Lett.* 76, 673 (2000).
43. X. Sun, L. Zhuang, W. Zhang, and S. Y. Chou, *J. Vac. Sci. Technol. B* 16, 3922 (1998).
44. H. Tan, A. Gilbertson, and S. Y. Chou, *J. Vac. Sci. Technol. B* 16, 3926 (1998).
45. S. Y. Chou, P. R. Krauss, W. Zhang, L. Guo, and L. Zhuang, *J. Vac. Sci. Technol. B* 15, 2897 (1997).
46. I. Martini, D. Eisert, M. Kamp, L. Worschech, and F. Forchel, *Appl. Phys. Lett.* 77, 2237 (2000).
47. Z. Yu, S. Schlabitsky, and S. Y. Chou, *Appl. Phys. Lett.* 74, 2381 (1999).
48. J. Wang, X. Sun, L. Chen, L. Zhuang, and S. Y. Chou, *Appl. Phys. Lett.* 77, 166 (2000).
49. J. Wang, X. Sun, L. Chen, and S. Y. Chou, *J. Vac. Sci. Technol. B* 17, 2957 (1999).
50. B. Faircloth, H. Rohrs, R. Tiberio, R. Ruolff, and R. R. Krchnavek, *J. Vac. Sci. Technol. B* 18, 1866 (2000).
51. H. Schulz, H. C. Scheer, T. Hoffmann, C. M. Sotomayor Torres, K. Pfeiffer, G. Bleidiessel, G. Grützner, C. H. Cardinaud, F. Gaboriau, M. C. Peignon, J. Ahopelto, and B. Heidari, *J. Vac. Sci. Technol. B* 18, 1861 (2000).
52. H. Schulz, D. Lyebyedyev, H. C. Scheer, K. Pfeiffer, G. Bleidiessel, G. Grützner, and J. Ahopelto, *J. Vac. Sci. Technol. B* 18, 3582 (2000).
53. C. D W. Wilkinson, A. S. G. Curtis, and J. Crossan, *J. Vac. Sci. Technol. B* 16, 3132 (1998).
54. A. Curtis and C. Wilkinson, *Trends Biotechnol.* 19, 97 (2001).
55. Y. Xia and G. M. Whitesides, *Angew. Chem. Int. Ed.* 37, 550 (1998).
56. A. Kumar and G. M. Whitesides, *Science* 263, 60 (1994).
57. E. Delamarche, H. Schmid, B. Michel, and H. Biebuyck, *Adv. Mater.* 9, 741 (1997).
58. A. Bietsch and B. Michel, *J. Appl. Phys.* 88, 4310 (2000).
59. H. Schmid and B. Michel, *Macromolecules* 33, 3042 (2000).
60. B. Michel, A. Bernard, A. Bietsch, E. Delamarche, M. Geissler, D. Juncker, H. Kind, J. P. Renault, H. Rothuizen, H. Schmid, P. Schmidt-Winkel, R. Stutz, and H. Wolf, *IBM J. Res. Dev.* 45, 697 (2001).
61. H.-W. Li, D.-J. Kang, M. G. Blamire, and W. T. S. Huck, *Nano Lett.* 2, 347 (2002).
62. J. L. Wilbur, E. Kim, Y. Xia, and G. M. Whitesides, *Adv. Mater.* 7, 649 (1995).
63. T. W. Odom, J. C. Love, D. B. Wolfe, K. E. Paul, and G. M. Whitesides, *Langmuir* 18, 5314 (2002).
64. T. W. Odom, V. R. Thalladi, J. C. Love, and G. M. Whitesides, *J. Am. Chem. Soc.* 124, 12112 (2002).
65. Y. Xia, E. Kim, X.-M. Zhao, J. A. Rogers, M. Prentiss, and G. M. Whitesides, *Science* 273, 347 (1996).
66. Y. Xia, J. J. McClelland, R. Gupta, D. Qin, X.-M. Zhao, L. L. Sohn, R. J. Celotta, and G. M. Whitesides, *Adv. Mater.* 9, 147 (1997).
67. E. Delamarche, H. Schmid, A. Bietsch, N. B. Larsen, H. Rothuizen, B. Michel, and H. A. Biebuyck, *J. Phys. Chem. B* 102, 3324 (1998).
68. Y. Xia and G. M. Whitesides, *J. Am. Chem. Soc.* 117, 3274 (1995).
69. S. L. Clark and P. T. Hammond, *Adv. Mater.* 10, 1515 (1998).
70. W. T. S. Huck, A. D. Stroock, and G. M. Whitesides, *Angew. Chem. Int. Ed.* 39, 1058 (2000).
71. W. T. S. Huck, L. Yan, A. Stroock, R. Haag, and G. M. Whitesides, *Langmuir* 15, 6862 (1999).
72. L. Yan, W. T. S. Huck, X. M. Zhao, and G. M. Whitesides, *Langmuir* 15, 1208 (1999).
73. X.-M. Zhao, Y. Xia, and G. M. Whitesides, *J. Mater. Chem.* 7, 1069 (1997).
74. E. Kim, Y. Xia, and G. M. Whitesides, *Nature* 376, 581 (1995).
75. Y. Xia, E. Kim, and G. M. Whitesides, *Chem. Mater.* 8, 1558 (1996).
76. E. Kim and G. M. Whitesides, *J. Phys. Chem. B* 101, 855 (1997).
77. J. A. Rogers, K. E. Paul, R. J. Jackman, and G. M. Whitesides, *Appl. Phys. Lett.* **70**, 2658 (1997).
78. J. P. Spatz, *Angew. Chem. Int. Ed.* 41, 3359 (2002).
79. J. P. Spatz, A. Roescher, and M. Moller, *Adv. Mater.* 8, 337 (1996).
80. F. S. Bates and G. H. Frederickson, *Annu. Rev. Phys. Chem.* 41, 525 (1990).
81. G. Krausch and R. Magerle, *Adv. Mater.* 14, 1579 (2002).

82. M. Bockstaller, R. Kolb, and E. L. Thomas, *Adv. Mater.* 13, 1783 (2001).
83. H. A. Klok and S. Lecommandoux, *Adv. Mater.* 13, 1217 (2001).
84. P. Mansky, T. P. Russell, C. J. Hawker, M. Pitsikalis, and J. Mays, *Macromolecules* 30, 6810 (1997).
85. T. Thurn-Albrecht, J. De Rouchy, T. P. Russell, and H. M. Jaeger, *Macromolecules* 33, 3250 (2000).
86. P. Mansky, P. Chaikin, and E. L. Thomas, *J. Mater. Sci.* 30, 1987 (1995).
87. A. Urbas, Y. Fink, and E. L. Thomas, *Macromolecules* 32, 4748 (1999).
88. A. Urbas, R. Sharp, Y. Fink, E. L. Thomas, M. Xenidou, and L. J. Fetters, *Adv. Mater.* 12, 812 (2000).
89. T. Goldacker, V. Abetz, R. Stadler, I. Erukhimovich, and L. Leibler, *Nature* 398, 137 (1999).
90. M. Park, P. M. Chaikin, R. A. Register, and D. H. Adamson, *Appl. Phys. Lett.* 79, 257 (2001).
91. M. Park, C. Harrison, P. M. Chaikin, R. A. Register, and D. H. Adamson, *Science* 276, 1401 (1997).
92. G. Liu, J. Ding, A. Guo, M. Herfort, and D. Bazett-Jones, *Macromolecules* 30, 1851 (1997).
93. L. Rockford, Y. Liu, P. Mansky, and T. Russell, *Phys. Rev. Lett.* 82, 2602 (1999).
94. M. J. Fasolka, D. J. Harris, A. M. Mayes, M. Yonn, and S. G. J. Mochrie, *Phys. Rev. Lett.* 79, 3018 (1997).
95. C. Park, J. Y. Chen, M. J. Fasolka, A. M. Mayes, C. A. Ross, E. L. Thomas, and C. De Rosa, *Appl. Phys. Rev. Lett.* 79, 848 (2001).
96. R. A. Segalman, H. Yokoyama, and E. J. Kramer, *Adv. Mater.* 13, 1152 (2001).
97. G. Reiter, G. Castelein, P. Hoerner, R. Giess, A. Blumen, and J. Sommer, *Phys. Rev. Lett.* 83, 4625 (1999).
98. D. E. Rosa, C. Park, B. Lotz, J. C. Wittmann, L. J. Fetters, and E. L. Thomas, *Macromolecules* 33, 4871 (2000).
99. T. Thurn-Albrecht, R. Steiner, J. DeRouchey, C. M. Stafford, E. Huang, M. Bal, M. Tuominen, C. J. Hawker, and T. P. Russell, *Adv. Mater.* 12, 787 (2000).
100. T. Thurn-Albrecht, J. Schotter, G. A. Kästle, N. Emley, T. Shibauchi, L. Krusin-Elbaum, K. Guarini, C. T. Black, M. T. Tuominen, and T. P. Russell, *Science* 290, 2126 (2000).
101. E. Jeoung, T. H. Galow, J. Schotter, M. Bal, A. Ursache, M. T. Tuominen, C. M. Stafford, T. P. Russell, and V. M. Rotello, *Langmuir* 17, 6396 (2001).
102. M. Böltau, S. Walheim, J. Mlynek, G. Krausch, and U. Steiner, *Nature* 391, 877 (1998).
103. A. Karim, J. F. Douglas, B. P. Lee, S. C. Glotzer, J. A. Rogers, R. Jackman, E. J. Amis, and G. M. Whitesides, *Phys. Rev. E* 57, R6273 (1998).
104. P. D. Peters, X. M. Yang, Q. Wang, J. J. de Pablo, and P. F. Nealey, *J. Vac. Sci. Technol. B* 18, 3530 (2000).
105. L. Kielhorn and M. Muthukumar, *J. Chem. Phys.* 111, 2259 (1999).
106. C. R. Martin, *Acc. Chem. Res.* 259, 957 (1995).
107. C. G. Wu and T. Bein, *Science* 264, 1757 (1994).
108. R. C. Smith, W. M. Fischer, and D. L. Gin, *J. Am. Chem. Soc.* 119, 4092 (1997).
109. H. Qiu, M. Wan, B. Matthews, and L. Dai, *Macromolecules* 34, 675 (2001).
110. M. R. Ghadiri, J. R. Granja, and L. K. Buehler, *Nature* 369, 301 (1994).
111. Y. Yu, L. Zhang, and A. Eisenberg, *Langmuir* 12, 5980 (1996).
112. Y. Yu and A. Eisenberg, *J. Am. Chem. Soc.* 119, 8383 (1997).
113. L. Desbaumes and A. Eisenberg, *Langmuir* 12, 36 (1996).
114. K. de Moell, G. O. R Alberda van Ekenstein, H. Nijland, E. Polushkin, G. ten Brinke, R. Maeki-Ontto, and O. Ikkala, *Chem. Mater.* 13, 4580 (2001).
115. S. Stewart and G. Liu, *Angew. Chem. Int. Ed.* 39, 340 (2000).
116. G. Liu, J. Ding, L. Qiao, A. Guo, B. P. Dymov, J. T. Gleeson, T. Hashimoto, and K. Saijo, *Chem. Eur. J.* 5, 2740 (1999).
117. C. T. Kresge, M. E. Leonowics, W. J. Roth, J. C. Vartuli, and J. S. Beck, *Nature* 359, 710 (1992).
118. R. Ulrich, A. Du Chesne, M. Templin, and U. Wiesner, *Adv. Mater.* 11, 141 (1999).
119. M. Templin, A. Franck, A. Du Chesne, H. Leist, Y. Zhang, R. Ulrich, V. Schaedler, and U. Wiesner, *Science* 278, 1795 (1997).
120. M. Steinhart, Wendorff, A. Greiner, R. B. Wehrspohn, K. Nielsch, J. Schilling, J. Choi, and U. Goesele, *Science* 296, 1997 (2002).
121. H. Hou, Z. Jun, A. Reuning, A. Schaper, J. H. Wendorff, and A. Greiner, *Macromolecules* 35, 2429 (2002).
122. D. H. Reneker, A. L. Yarin, H. Fong, and S. Koombhongse, *J. Appl. Phys.* 87, 4531 (2000).
123. M. Bognitzki, H. Hou, M. Ishaque, T. Frese, M. Hellwig, C. Schwarte, A. Schaper, J. H. Wendorf, and A. Greiner, *Adv. Mater.* 12, 637 (2000).
124. M. Bognitzki, W. Czado, T. Frese, A. Schaper, M. Hellwig, M. Steinhart, A. Greiner, and J. H. Wendorff, *Adv. Mater.* 13, 70 (2001).
125. R. A. Cauros, J. H. Schattka, and A. Greiner, *Adv. Mater.* 13, 1577 (2001).
126. G. Zhang, A. Niu, S. Peng, M. Jiang, Y. Tu, M. Li, and C. Wu, *Acc. Chem. Res.* 34, 249 (2001).
127. K. Landfester, R. Montenegro, U. Scherf, R. Günter, U. Asawapirom, S. Patil, D. Neher, and T. Kietzke, *Adv. Mater.* 14, 651 (2002).
128. M. Antonietti and K. Landfester, *Chem. Phys. Chem.* 2, 207 (2001).
129. K. Landfester, *Macromol. Rapid. Commun.* 22, 896 (2001).
130. K. Landfester, *Adv. Mater.* 13, 765 (2001).
131. K. Landfester, F. Tiarks, M. Willert, H.-P. Hentze, and M. Antonietti, *Macromol. Chem. Phys.* 201, 1 (2000).
132. K. Landfester, *J. Disper. Sci. Technol.* 23, 167 (2002).
133. F. Tiarks, K. Landfester, and M. Antonietti, *Langmuir* 17, 908 (2001).
134. B. Z. Putlitz, K. Landfester, H. Fischer, and M. Antonietti, *Adv. Mater.* 13, 500 (2001).
135. E. Marie, K. Landfester, and M. Antonietti, *Biomacromolecules* 3, 475 (2002).
136. D. Vaihinger, K. Landfester, I. Krauter, H. Brunner, and G. E. M. Tovar, *Macromol. Chem. Phys.* 203, 1965 (2002).
137. K. Landfester, R. Montenegro, and U. Scherf, *Adv. Mater.* 14, 651 (2002).
138. K. Tajima and T. Aida, *Chem. Commun.* 24, 2399 (2000).
139. J. L. Hedrick, R. D. Miller, C. J. Hawker, K. R. Carter, W. Volksen, D. Y. Yoon, and M. Trollsas, *Adv. Mater.* 10, 1049 (1998).
140. P. J. Flory, *J. Am. Chem. Soc.* 74, 2718 (1952); E. Buhleier, W. Wehner, and F. Vögtle, *Synthesis* 155 (1978).
141. G. R. Newkome, Z. Yao, G. R. Baker, and V. K. Gupta, *J. Org. Chem.* 50, 2003 (1985).
142. K. L. Wooley, J. M. J. Fréchet, and C. J. Hawker, *Polymer* 35, 4489 (1994).
143. C. J. Hawker, E. Malmström, C. W. Frank, and J. P. Kampf, *J. Am. Chem. Soc.* 119, 9903 (1997).
144. T. H. Mourey, S. R. Turner, M. Rubinstein, J. M. J. Fréchet, C. J. Hawker, and K. L. Wooley, *Macromolecules* 25, 2401 (1992).
145. K. L. Wooley, C. J. Hawker, J. M. Pochan, and J. M. J. Fréchet, *Macromolecules* 26, 1514 (1993).
146. C. J. Hawker, P. J. Farrington, M. E. Mackay, K. L. Wooley, and J. M. J. Fréchet, *J. Am. Chem. Soc.* 117, 4409 (1995).
147. S. M. Grayson and J. M. J. Fréchet, *J. Am. Chem. Soc.* 122, 10335 (2000).
148. M. Liu and J. M. J. Fréchet, *Polym. Bull.* 43, 379 (1999).
149. L. Bu, W. K. Nonidez, J. W. Mays, and N. B. Tan, *Macromolecules* 33, 4445 (2000).
150. R. Haag, A. Sunder, and J. F. Stumbe, *J. Am. Chem. Soc.* 122, 2954 (2000).

151. D. A. Tomalia, H. Baker, J. Dewald, M. Hall, G. Kallos, S. Martin, J. Roeck, J. Ryder, and P. Smith, *Macromolecules* 19, 2466 (1986).
152. D. A. O'Sullivan, *Chem. Eng. News* 20 (1993).
153. G. R. Newkome, C. N. Moorefield, G. R. Baker, A. L. Johnson, and R. K. Behera, *Angew. Chem., Int. Ed. Engl.* 30, 1176 (1991).
154. G. R. Newkome, C. N. Moorefield, G. R. Baker, M. J. Saunders, and S. H. Grossman, *Angew. Chem., Int. Ed. Engl.* 30, 1178 (1991).
155. N. Launay, A. M. Caminade, R. Lahana, and J. P. Majoral, *Angew. Chem.* 33, 1589 (1994).
156. C. J. Hawker and J. M. J. Fréchet, *J. Chem. Soc., Chem. Commun.* 1010 (1990).
157. H. Sahota, P. M. Lloyd, S. G. Yeates, P. J. Derrick, P. C. Taylor, and D. M. Haddleton, *J. Chem. Soc., Chem. Commun.* 2445 (1994).
158. J. W. Leon and J. M. J. Fréchet, *Polym. Bull.* 35, 449 (1995).
159. S. M. Grayson and J. M. J. Fréchet, *Chem. Rev.* 101, 3819 (2001).
160. D. A. Tomalia, A. M. Naylor, and W. A. Goddard, III, *Angew. Chem., Int. Ed. Engl.* 29, 138 (1990).
161. H. B. Mekelburger, W. Jaworek, and F. Vögtle, *Angew. Chem.* 104, 1609 (1992).
162. J. M. J. Fréchet, *Science* 263, 1710 (1994).
163. J. Issberner, R. Moors, and F. Vögtle, *Angew. Chem.* 106, 2507 (1994).
164. G. R. Newkome, C. N. Moorefield, and F. Vögtle, "Dendritic Molecules: Concepts, Syntheses, Perspectives." VCH, Weinheim, 1996.
165. J. M. J. Fréchet and C. J. Hawker, in "Comprehensive Polymer Science, 2nd Supplement" (S. L. Aggarwal and S. Russo, Eds.), p. 140. Pergamon, Oxford, 1996.
166. O. A. Matthews, A. N. Shipway, and J. F. Stoddart, *Prog. Polym. Sci.* 23, 1 (1998).
167. D. K. Smith and F. Diederich, *Chem.-Eur. J.* 4, 1353 (1998).
168. M. Fischer and F. Vögtle, *Angew. Chem., Int. Ed. Engl.* 38, 885 (1999).
169. A. W. Bosman, H. M. Janssen, and E. W. Meijer, *Chem. Rev.* 99, 1665 (1999).
170. S. Hecht and J. M. J. Fréchet, *Angew. Chem., Int. Ed.* 40, 74 (2001).
171. C. J. Hawker, K. L. Wooley, and J. M. J. Fréchet, *J. Chem. Soc., Perkin Trans.* 1, 1287 (1993).
172. J. F. G. A. Jansen, E. de Brabander van den Berg, and E. W. Meijer, *Science* 266, 1226 (1994).
173. S. Stevelmans, J. C. M. van Hest, J. F. G. A. Jansen, D. A. F. J. van Boxtel, E. M. M. de Brabander van den Berg, and E. W. Meijer, *J. Am. Chem. Soc.* 118, 7398 (1996).
174. D. M. Watkins, Y. SayedSweet, J. W. Klimash, N. J. Turro, and D. A. Tomalia, *Langmuir* 13, 3136 (1997).
175. P. R. L. Malenfant, L. Groenendaal, and J. M. J. Fréchet, *J. Am. Chem. Soc.* 120, 10990 (1998).
176. J. J. Apperloo, R. A. J. Janssen, P. R. L. Malenfant, L. Groenendaal, and J. M. J. Fréchet, *J. Am. Chem. Soc.* 122, 7042 (2000).
177. J. J. Apperloo, P. R. L. Malenfant, J. M. J. Fréchet, and R. A. J. Janssen, *Synth. Met.* 121, 1259 (2001).
178. J. J. Apperloo, R. A. J. Janssen, P. R. L. Malenfant, and J. M. J. Fréchet, *J. Am. Chem. Soc.* 123, 6916 (2001).
179. O. Karthaus, K. Ijiro, M. Shimomura, J. Hellmann, and M. Irie, *Langmuir* 12, 6714 (1996).
180. J. Hofkens, W. Verheijen, R. Shukla, W. Dehaen, and F. C. De Schryver, *Macromolecules* 31, 4493 (1998).
181. M. Kawa and J. M. J. Fréchet, *Thin Solid Films* 331, 259 (1998).
182. J. F. Kukowska-Latallo, A. U. Bielinska, J. Johnson, R. Spindler, D. A. Tomalia, and J. R. Baker, Jr., *Proc. Nat. Acad. Sci. U.S.A.* 93, 4897 (1996).
183. P. J. Dandliker, F. Diederich, A. Zingg, J. P. Gisselbrecht, M. Gross, A. Louati, and E. Sanford, *Helv. Chim. Acta* 80, 1773 (1997).
184. F. Vögtle, M. Plevoets, M. Nieger, G. C. Azzellini, A. Credi, L. D. Cola, V. De Marchis, M. Venturi, and V. Balzani, *J. Am. Chem. Soc.* 121, 6290 (1999).
185. I. Jestin, E. Levillain, and J. Roncali, *Chem. Commun.* 2655 (1998).
186. C. M. Cardona and A. E. Kaifer, *J. Am. Chem. Soc.* 120, 4023 (1998).
187. D. K. Smith, *J. Chem. Soc., Perkin Trans.* 2, 1563 (1999).
188. Y. Wang, C. M. Cardona, and A. E. Kaifer, *J. Am. Chem. Soc.* 121, 9756 (1999).
189. C. M. Cardona, T. D. McCarley, and A. E. Kaifer, *J. Org. Chem.* 65, 1857 (2000).
190. L. L. Miller, B. Zinger, and J. S. Schlechte, *Chem. Mater.* 11, 2313 (1999).
191. R. Toba, J. Maria Quintela, C. Peinador, E. Roman, and A. E. Kaifer, *Chem. Commun.* 857 (2001).
192. C. B. Gorman and J. C. Smith, *Acc. Chem. Res.* 34, 60 (2001).
193. C. B. Gorman, J. C. Smith, M. W. Hager, B. L. Parkhurst, H. Sierzputowska-Gracz, and C. A. Haney, *J. Am. Chem. Soc.* 121, 9958 (1999).
194. A. Adronov, S. L. Gilat, J. M. J. Fréchet, K. Ohta, F. V. R. Neuwahl, and G. R. Fleming, *J. Am. Chem. Soc.* 122, 1175 (2000).
195. A. Adronov and J. M. J. Fréchet, *Chem. Commun.* 1701 (2000).
196. S. L. Gilat, A. Adronov, and J. M. J. Fréchet, *Polym. Mater. Sci. Eng.* 77, 91 (1997).
197. S. L. Gilat, A. Adronov, and J. M. J. Fréchet, *J. Org. Chem.* 64, 7474 (1999).
198. S. L. Gilat, A. Adronov, and J. M. J. Fréchet, *Angew. Chem., Int. Ed.* 38, 1422 (1999).
199. A. Adronov, P. R. L. Malenfant, and J. M. J. Fréchet, *Chem. Mater.* 12, 1463 (2000).
200. M. Halim, I. D. W. Samuel, J. N. G. Pillow, A. P. Monkman, and P. L. Burn, *Synth. Met.* 102, 1571 (1999).
201. M. Halim, I. D. W. Samuel, J. N. G. Pillow, and P. L. Burn, *Synth. Met.* 102, 1113 (1999).
202. M. Halim, I. D. W. Samuel, J. N. G. Pillow, and P. L. Burn, *Adv. Mater.* 11, 371 (1999).
203. M. Halim, I. D. W. Samuel, J. N. G. Pillow, and P. L. Burn, *Synth. Met.* 102, 922 (1999).
204. P. R. L. Malenfant, M. Jayaraman, and J. M. J. Fréchet, *Chem. Mater.* 11, 3420 (1999).
205. P. R. L. Malenfant and J. M. J. Fréchet, *Macromolecules* 33, 3634 (2000).
206. H. F. Chow and C. C. Mak, *J. Org. Chem.* 62, 5116 (1997).
207. C. C. Mak and H. F. Chow, *Macromolecules* 30, 1228 (1997).
208. I. Morao and F. P. Cossio, *Tetrahedron Lett.* 38, 6461 (1997).
209. Q. S. Hu, V. Pugh, M. Sabat, and L. Pu, *J. Org. Chem.* 64, 7528 (1999).
210. P. B. Rheiner, H. Sellner, and D. Seebach, *Helv. Chim. Acta* 80, 2027 (1997).
211. T. Habicher, F. Diederich, and V. Gramlich, *Helv. Chim. Acta* 82, 1066 (1999).
212. F. Djojo, E. Ravanelli, O. Vostrowsky, and A. Hirsch, *Eur. J. Org. Chem.* 1051 (2000).
213. B. R. Cook, T. J. Reinert, and K. S. Suslick, *J. Am. Chem. Soc.* 108, 7281 (1986).
214. K. S. Suslick and B. R. Cook, *J. Chem. Soc., Chem. Commun.* 200 (1987).
215. M. Smet, L. X. Liao, W. Dehaen, and D. V. McGrath, *Org. Lett.* 2, 511 (2000).
216. D. L. Jiang and T. Aida, *Nature* 388, 454 (1997).
217. T. Aida, D. L. Jiang, E. Yashima, and Y. Okamoto, *Thin Solid Films* 331, 254 (1998).
218. Y. Wakabayashi, M. Tokeshi, D. L. Jiang, T. Aida, and T. Kitamori, *J. Lumin.* 83, 313 (1999).
219. D. M. Junge and D. V. McGrath, *Chem. Commun.* 847 (1997).
220. D. V. McGrath and D. M. Junge, *Macromol. Symp.* 137, 57 (1999).
221. D. M. Junge and D. V. McGrath, *J. Am. Chem. Soc.* 121, 4912 (1999).
222. S. Li and D. V. McGrath, *J. Am. Chem. Soc.* 122, 6795 (2000).

223. M. E. Piotti, F. Rivera, Jr., R. Bond, C. J. Hawker, and J. M. J. Fréchet, *J. Am. Chem. Soc.* 121 , 9471 (1999).
224. E. Harth, B. Van Horn, V. Y. Lee, D. S. Germack, C. P. Gonzalez, R. D. Miller, and C. J. Hawker, *J. Am. Chem. Soc.* 124, 8653 (2002).
225. D. Mecerreyes, V. Lee, C. J. Hawker, J. L. Hedrick, A. Wursch, W. Volksen, T. Magbitang, E. Huang, and R. D. Miller, *Adv. Mater.* 13, 204 (2001).
226. I. Astafieva, X. F. Zhong, and A. Eisenberg, *Macromolecules* 26, 7339 (1993).
227. A. Qin, M. Tian, C. Ramireddy, S. E. Webber, P. Munk, and Z. Tuzar, *Macromolecules* 27, 120 (1994).
228. L. Zhang and A. Eisenberg, *Science* 268, 1728 (1995).
229. S. Förster, N. Hemsdorf, W. Leube, H. Schnablegger, M. Regenbrecht, and S. Akari, *J. Phys. Chem.* 103, 6657 (1999).
230. D. E. Discher and A. Eisenberg, *Science* 297, 967 (2002).
231. A. Guo, G. Liu, and J. Tao, *Macromolecules* 29, 2487 (1996).
232. V. Butun, A. B. Lowe, N. C. Billingham, and S. P. Armes, *J. Am. Chem. Soc.* 121, 4288 (1999).
233. J. Ding and G. Liu, *Langmuir* 15, 1738 (1999).
234. K. B. Thurmond, II, T. Kowalewski, and K. L. Wooley, *J. Am. Chem. Soc.* 118, 7239 (1996).
235. S.-J. Park, A. A. Lazarides, C. A. Mirkin, P. W. Brazis, C. R. Kannewurf, and R. L. Letsinger, *Angew. Chem., Int. Ed.* 39, 3845 (2000).
236. G. Liu, L. Qiao, and A. Guo, *Macromolecules* 29, 5508 (1996).
237. Y.-Y. Won, H. T. Davis, and F. S. Bates, *Science* 283, 960 (1999).
238. S. Stewart and G. Liu, *Chem. Mater.* 11, 1048 (1999).
239. H. Huang, E. E. Remsen, T. Kowalewski, and K. L. Wooley, *J. Am. Chem. Soc.* 121, 3805 (1999).
240. Z. Zhang, G. Liu, and S. Bell, *Macromolecules* 33, 7877 (2000).
241. K. B. Thurmond, II, T. Kowalewski, and K. L. Wooley, *J. Am. Chem. Soc.* 119, 6656 (1997).
242. J. Zhou, Z. Li, and G. Liu, *Macromolecules* 35, 3690 (2002).
243. C. M. Niemeyer, *Chem. Eur. J.* 7, 3188 (2001).
244. J. J. Storhoff and C. A. Mirkin, *Chem. Rev.* 99, 1849 (1999).
245. C. M. Niemeyer, *Angew. Chem., Int. Ed.* 40, 4128 (2001).
246. A. M. Cassell, W. A. Scrivens, and J. M. Tour, *Angew. Chem., Int. Ed. Engl.* 37, 1528 (1998).
247. C. M. Niemeyer, M. Adler, S. Gao, and L. F. Chi, *Angew. Chem., Int. Ed.* 39, 3055 (2000).
248. C. M. Niemeyer, M. Adler, B. Pignataro, S. Lenhert, S. Gao, L. F. Chi, H. Fuchs, and D. Blohm, *Nucl. Acids Res.* 27, 4553 (1999).
249. R. C. Mucic, J. J. Storhoff, C. A. Mirkin, and R. L. Letsinger, *J. Am. Chem. Soc.* 120, 12674 (1998).
250. S.-J. Park, A. A. Lazarides, C. A. Mirkin, and R. L. Letsinger, *Angew. Chem., Int. Ed.* 40, 2909 (2001).
251. C. A. Mirkin, *Inorg. Chem.* 39, 2258 (2000).
252. C. A. Mirkin, R. L. Letsinger, R. C. Mucic, and J. J. Storhoff, *Nature* 382, 607 (1996).
253. J. J. Storhoff, A. A. Lazarides, R. C. Mucic, C. A. Mirkin, R. L. Letsinger, and G. C. Schatz, *J. Am. Chem. Soc.* 122, 4640 (2000).
254. R. Elghanian, J. J. Storhoff, R. C. Mucic, R. L. Letsinger, and C. A. Mirkin, *Science* 277, 1078 (1997).
255. R. A. Reynolds, III, C. A. Mirkin, and R. L. Letsinger, *J. Am. Chem. Soc.* 122, 3795 (2000).
256. R. C. Mucic, J. J. Storhoff, C. A. Mirkin, and R. L. Letsinger, *J. Am. Chem. Soc.* 120, 12674 (1998).
257. G. P. Mitchell, C. A. Mirkin, and R. L. Letsinger, *J. Am. Chem. Soc.* 121, 8122 (1999).
258. T. A. Taton, R. C. Mucic, C. A. Mirkin, and R. L. Letsinger, *J. Am. Chem. Soc.* 122, 6305 (2000).
259. A. P. Alivisatos, K. P. Johnsson, X. Peng, T. E. Wilson, C. J. Loweth, M. P. Bruchez, Jr., and P. G. Schultz, *Nature* 382, 609 (1996).
260. C. M. Niemeyer, W. Burger, and J. Peplies, *Angew. Chem., Int. Ed. Engl.* 37, 2265 (1998).
261. E. Braun, Y. Eichen, U. Sivan, and G. Ben-Yoseph, *Nature* 391, 775 (1998).
262. N. C. Seeman, *Annu. Rev. Biophys. Biomol. Struct.* 27, 225 (1998).
263. N. C. Seeman, *Acc. Chem. Rev.* 30, 357 (1997).
264. N. C. Seeman, Y. Zhang, and J. Chen, *J. Vac. Sci. Technol. A* 12, 1895 (1993).
265. N. C. Seeman, *Acc. Chem. Res.* 30, 357 (1997).
266. N. R. Kallenbach, R.-I. Ma, and N. C. Seeman, *Nature* 305, 829 (1983).
267. R.-I. Ma, N. R. Kallenbach, R. D. Sheardy, M. L. Petrillo, and N. C. Seeman, *Nucl. Acids Res.* 14, 9745 (1986).
268. J. Qi, X. Li, X. Yang, and N. C. Seeman, *J. Am. Chem. Soc.* 118, 6121 (1996).
269. J.-H. Chen, N. R. Kallenbach, and N. C. Seeman, *J. Am. Chem. Soc.* 111, 6402 (1989).
270. J. Chen and N. C. Seeman, *Nature* 350, 631 (1991).
271. Y. Zhang, and N. C. Seeman, *J. Am. Chem. Soc.* 116, 1661 (1992).
272. T.-J. Fu and N. C. Seeman, *Biochem.* 32, 3211 (1993).
273. X. Li, X. Yang, J. Qi, and N. C. Seeman, *J. Am. Chem. Soc.* 118, 6131 (1996).
274. X. Jiang, L. A. Wenzler, J. Qi, X. Li, and N. C. Seeman, *J. Am. Chem. Soc.* 120, 9779 (1998).
275. E. Winfree, F. Liu, L. A. Wenzler, and N. C. Seeman, *Nature* 394, 539 (1998).
276. C. Mao, W. Sun, and N. C. Seeman, *J. Am. Chem. Soc.* 121, 5437 (1999).
277. R. Garrett, *Nature* 400, 811 (1999).
278. P. B. Moore and T. A. Steitz, *Nature* 418, 229 (2002).
279. H. Noji, R. Yasuda, M. Yoshida, and K. Kinosita, *Nature* 386, 299 (1997).
280. R. K. Soong, G. D. Bachand, H. P. Neves, A. G. Olkhovets, H. G. Craighead, and C. D. Montemagno, *Science* 290, 1555 (2000).
281. Y. Hiratsuka, T. Tada, K. Oiwa, T. Kanayama, and T. Q. P. Uyeda, *Biophys. J.* 81, 1555 (2001).
282. H. Suzuki, A. Yamada, K. Oiwa, H. Nakayama, and H. Mashiko, *Biophys. J.* 72, 1997 (1997).
283. D. V. Nicolau, H. Suzuki, H. Mashiko, T. Taguchi, and S. Yoshikawa, *Biophys. J.* 77, 1126 (1999).
284. J. R. Dennis, J. Howard, and V. Vogel, *Nanotechnol.* 10, 232 (1999).

Encyclopedia of Nanoscience and Nanotechnology
www.aspbs.com/enn

Friction of Model Self-Assembled Monolayers

J. A. Harrison, G. T. Gao
U.S. Naval Academy, Annapolis, Maryland, USA

R. J. Harrison
Catawba Valley Community College, Hickory, North Carolina, USA

G. M. Chateauneuf, P. T. Mikulski
U.S. Naval Academy, Annapolis, Maryland, USA

CONTENTS

1. INTRODUCTION

The experimental preparation of supported, oriented monolayers of polyfunctional organic molecules with a variety of molecular structures was first achieved in the early 1980s [1–5]. This technique involved the adsorption of disulfides on zerovalent gold substrates, which spontaneously organized into films. These films have come to be known as self-assembled monolayers (SAMs). Several types of SAM systems have been developed and characterized, including organosilicon–oxide, carboxylic acid–metal oxide, and organosulfur–noble metal (Au or Ag). Because self-assembly is a viable means of controlling the physical and chemical properties of solid surfaces, many studies have been devoted to the development of thin films of functionalized molecules for optical, electronic, mechanical, or biochemical applications [6–8]. Of all of the monolayer systems, the thiol SAMs on Au(111) have received the most extensive experimental [9–21] and theoretical examination [22–39].

Technological applications, such as microelectromechanical systems (MEMS) [40–44] and magnetic storage devices [45–47], have piqued interest in the potential use of organic films for boundary layer lubricants. In these applications, the thickness of the lubricating film is on the order of a monolayer. At these thicknesses, adhesion of the molecules to the substrate is of paramount importance, particularly in applications that involve the relative motion of solid surfaces in contact. Thus, monolayers formed from self-assembly, which are covalently bonded to the substrate, such as alkanethiols and alkylsilanes, are good candidates for boundary layer lubricants. As a result, the mechanical and tribological properties of self-assembled monolayer materials have been studied a great deal using scanning probe microscopies [48–71].

The atomic force microscope (AFM) provides a unique and powerful means of directly investigating the tribology of atomic-scale contacts. However, interpretation of data obtained from AFM experiments is notoriously difficult for a number of reasons: absolute quantitative measurements are often not possible, the structure of the AFM tip may not be known, nor is its structure necessarily preserved during the course of an experiment, and the preparation and characterization of samples can be difficult, and involves many uncertainties [65].

ISBN: 1-58883-059-4/$35.00

Molecular dynamics (MD) simulation is a useful tool that can be used to model atomic-scale probes in sliding contact with SAMs, providing information that would otherwise be unavailable [72, 73]. Like AFM experiments, MD experiments are faced with a number of difficult challenges, most notably problems of scale. Despite the exponential growth in computing power, it will be quite some time before the time and length scales of laboratory experiments can be approached. Computational groups that have undertaken MD investigations of the tribology of SAMs have often taken very different approaches. Broadly, the differences can be addressed by a number of basic questions: How are atomic-interactions modeled? How are the initial SAMs and the initial probing tips designed? How are external forces used to control load and temperature? How are the challenges of length and time scales handled? These questions are not independent of one another. However, they do provide a useful framework within which the work in this field can be understood. This review focuses on work by groups that have undertaken computational studies aimed at complementing AFM or surface force apparatus (SFA) tribology experiments with SAMs. This review does not provide an exhaustive discussion of the modeling work that deals with the structure and dynamics of monolayer films. The reader is referred to the original papers for such a detailed discussion. In this work, the structure is discussed because it is a starting point for the MD studies of the tribology of SAMs. This work is intended to discuss MD simulations that focus on the tribology of SAMs, highlighting some significant issues associated with these studies (and MD simulations in general), as well as indicating some possible future directions.

2. THE UNITED-ATOM MODEL

The works of Hautman and Klein [22, 23] are among the earliest MD studies concerned with the structure and dynamics of long-chain molecules on metallic substrates. Although these are not tribological studies, they are particularly relevant because they employ a basic molecular model that has been subsequently adopted by most groups undertaking computational studies of alkanethiol SAMs. More recently, the structure [22–39], compression [74, 75], and the friction [76, 77, 79–92] of alkanethiol monolayers have been the subject of considerable theoretical examination. In addition, the particular system under investigation, $SH(CH_2)_{n-1}CH_3$ chains anchored onto a gold surface, represents a type of sample that has been extensively examined in AFM friction studies of SAMs.

The classical empirical model utilized by Hautman and Klein is the united-atom potential [93, 94]. This model has been used in MD studies of the structural and dynamical properties of a variety of complex molecular systems such as solid alkanes [95, 96], layers of long-chain molecules [97–100], and organic liquids, including thiols [101, 102]. Hautman and Klein adopted the basic model of these studies, and extended it to model the interaction of chain groups with a gold substrate.

In the basic united-atom model, the CH_2 groups, the terminal CH_3 group, and the anchoring sulfur atom are represented as spherical pseudoatoms. This allows for computational efficiency and fewer parameters at the expense of some realism. Essentially, the pseudoatom represents an average over the motions of hydrogen atoms attached to the carbon atom of interest. To eliminate the time scale issues associated with high-frequency vibration, bonds between pseudoatoms are held rigid [103]. By constraining the system to evolve only through the low-frequency angular deformations of chain backbones, the equations of motion can be evolved using relatively large time steps, for example, time steps on the order of 5 fs.

The intramolecular potential includes a harmonic angular-bending potential, where the angle is that formed by three consecutive Pseudoatoms. Also included is a torsional potential that is a function of the dihedral angle associated with four consecutive pseudoatoms (the middle two atoms define the axis perpendicular to the plane containing the dihedral angle). pseudoatoms in the same chain that are farther apart than fourth nearest neighbors, and pseudoatoms from different chains, interact through Lennard–Jones potentials. This qualitatively characterizes the model of atomic interactions utilized by most groups that have undertaken computational studies of the structural [22–39], mechanical [74, 75], and tribological [76, 77, 79–92] properties of alkanethiol SAMs.

To describe the interactions between pseudoatoms and the metallic substrate, a straightforward approach would be to integrate the 12–6 interactions over the half space describing the substrate. The resulting form, however, does not accurately describe basic properties, such as binding energy, distance above the surface, and dispersion coefficient [104, 105]. Instead, Hautman and Klein utilized a 12–3 form for the surface potential. Coefficients for the CH_2 and CH_3 pseudoatoms are fixed using a combination of scaling arguments to connect with Lennard–Jones parameters and existing fits to Kr on Ag(111) [105]. Parameters for the S–Au surface interaction were fixed using thermal desorption and electron energy loss spectroscopy measurements for dimethyl disulfide on Au(111).

In their initial study, Hautman and Klein equilibrated arrays of 90 $SH(CH_2)_{15}CH_3$ (hexadecanethiol) chains at room temperature, with periodic boundary conditions imposed to model infinite monolayer surfaces [22, 23]. The use of periodic boundary conditions is a standard tool utilized in all of the theoretical studies discussed in this review. Two model systems, referred to as models I and II, were investigated for comparison. Both models make use of the potential outlined above. Model I leaves the angle between the S–C bond and the surface normal unconstrained, while Model II imposes a harmonic bond-bending potential with a potential minimum at an angle of 100° with respect to the horizontal. The force constant is taken to be the same as for the bond-bending potential describing the hydrocarbon chains. These two models represent limiting behaviors.

After equilibration, Model I shows an ordered all-*trans* arrangement of chains uniformly tilted with respect to the surface normal. There were very few gauche defects, and they were confined primarily to the ends of chains. As expected, Model II shows very different characteristics near the head groups. Due to the extra constraint causing the

S–C bond to lie approximately parallel to the gold substrate, nearly every chain exhibits a gauche defect near the head group, causing the chains to take on a characteristic "L" shape. Far from the head groups, however, the chains take on an ordered uniformly tilted arrangement as in Model I. Although Model I is somewhat more ordered, the models are very similar in the region far from the head groups. This ordered uniformly tilted arrangement appears to be a general property at this packing density, and does not appear to be sensitive to the details of how the chains are anchored to the underlying substrate. This structure has been observed in many simulations of anchored long chains using a number of potential energy functions [106–113]. The similarity between Models I and II, in the regions far removed from the gold substrate and chain-head groups, suggests that the tribology of these systems may not be very sensitive to the details of the substrate–chain interaction. It is likely that the structure of the interface between sliding surfaces will be of paramount importance, and these limiting models investigated by Hautman and Klein show no significant difference in this region.

3. STRUCTURE OF ALKANETHIOL SAMs

There have been many theoretical studies of the structure of long-chain molecules on substrates [22–39, 106–115]. By far, the majority of the theoretical work has focused on the structure of model alkanethiols. Experimental studies of the structure of alkanethiol SAMs on Au(111) have shown a slightly distorted $(\sqrt{3} \times \sqrt{3})R30°$ triangular lattice structure corresponding to a chain–chain spacing of approximately 5 Å and an area of 21.6 Å^2 per chain [9–21]. The hydrocarbon tails self-organize into a well-ordered monolayer with chains tilted uniformly toward their next-nearest neighbors. Additional experimental studies of the structure of alkanethiol SAMs at low temperatures (50 K to room temperature) revealed a $c(4 \times 2)$ superlattice structure [116–118] (four chains per unit cell, with two types of chains whose carbon backbones are oriented differently from one another). Hautman and Klein pointed out that several phases might exist in SAMs over the accessible temperature range [22, 23]. In addition, multiple structures might exist for chains of different lengths at a given temperature [18]. Few of the theoretical studies on this system have been able to reproduce the superlattice structure. Only two papers to date have produced a superlattice structure [31, 37]. In what follows, we discuss a cross section of the theoretical papers dealing with the structure of alkanethiol monolayers.

In one of the earliest works [24] on the structure of alkanethiols, the collective tilt behavior of chains containing 20 psuedoatoms and 20 carbon atoms as a function of density was examined using the united-atom potential and an all-atom potential [93, 94]. Bond lengths were constrained in the simulations using both potentials, between pseudoatoms in the united-atom simulations, and C–C and C–H bonds in the all-atom simulations. Simulation cells contained 90 molecules in a triangular lattice with periodic boundary conditions imposed.

Conformational, or gauche, defects were most often present at the ends of the chains. This phenomenon has been observed in earlier simulations [22, 23] and subsequent simulations of alkylsilane monolayers [110], long-chain alcohol molecules [108], and monolayers composed of alkane chains [109, 122]. Karaborni [106, 107] utilized an anisotropic united-atom model to model the approach and separation of opposing monolayers. Simulations were performed by compressing 128 molecules, with 19 CH_2 segments and a COOH head group, evenly distributed on two plates (25 Å^2 per molecule) at 7 $m \cdot s^{-1}$. These simulations also showed that increasing the tilt angle of the molecules decreased the number of gauche defects in the midsections of both monolayers. This same link between tilt angle and gauche defects was observed in the simulation of long-chain alcohols [108].

Bareman and Klein determined that the average tilt angle increases with increasing area per chain (density), and that the molecules are in a rotator phase (molecules rotating about their molecular axis) at 300 K for both potential models [24]. However, the backbone orientation distribution functions and the reorientation relaxation times differed depending upon the potential used. Use of the proper all-atom parameters led to monolayers that exhibited the collective tilt angle and density behavior observed in experiments, while the united-atom potential did not. Lastly, the united-atom parameters, which were fit to the bulk properties of short alkane chains, were shown to significantly underestimate the effective diameter of the C_{20} chains studied. This study, and another by Klein and co-workers [25], showed that the inclusion of hydrogen atoms has a significant impact on the structure of the SAMs. In fact, stable next-nearest neighbor configurations could not be obtained using the united-atom potential. In simulations of crystalline n–alkanes, the explicit treatment of hydrogen atoms was essential to obtain the proper packing structure [28, 39]. Thus, there remain some unanswered questions regarding the validity of the united-atom potential for simulations of SAMs [22–24, 28, 30, 31].

Garrison and co-workers used MD and the united-atom potential to do an exhaustive search of the phase space of chain orientations and tilt directions to determine possible candidates for the $c(4 \times 2)$ superlattice [31] of alkanethiol SAMs. The simulation systems were composed of 64 chains placed over eight layers of Au to form the $(\sqrt{3} \times \sqrt{3})R30°$ structure. The top three layers of Au atoms were allowed to move according to Newton's equations of motion, the four intermediate layers were coupled to a thermostat, and the bottom-most layer was rigid. In contrast, in all previous simulations of alkanethiols on Au, the S–Au attachment to the surface was fixed [22–30, 32–39].

One structure for the $c(4 \times 2)$ superlattice for chains with 4–6 carbon atoms and three structures for chains with 12 or more carbon atoms were obtained. None of these structures contained chains that were in a perfectly perpendicular herringbone arrangement, as depicted by most theoretical papers [22–30, 32–39]. Rather, a skewed herringbone structure was obtained with the interplanar angle different from 90°. However, only two of the proposed structures were stable with respect to time at temperatures below 100 K. The remaining two structures (for long chains) only survived several hundred picoseconds before the chain backbones reoriented themselves. This process typically involved a small amount of untilting (1–5°). The authors speculated that this might be the preferred mechanism for monolayers to undergo structural changes, and that adding bulky groups to the chains might prevent this reorientation.

One of the most recent investigations of alkanethiol self-assembly on gold was conducted by Goddard and co-workers [37]. The goal of that work was to develop a force field that would be capable of predicting the $c(4 \times 2)$ superlattice structure which was stable over a wide temperature range with no disulfide bonds to the substrate. For a model to be able to predict the superlattice structure, it appears that an accurate force field for the S–Au interaction must be used. Early studies of alkanethiol structures on gold, where the sulfur–psuedoatom interaction within the united-atom potential and the surface was modeled as a continuum interacting with pseudoatoms via a structureless potential, were unable to predict the superlattice structure. In contrast, by using mobile S atoms and dynamic Au atoms, Bhatia and Garrison identified four possible $c(4 \times 2)$ superlattice structures for different length alkanethiols, which were all unstable [31].

Goddard and co-workers [37] performed quantum mechanical calculations on Au clusters of 15 atoms and periodic systems of two Au layers. Both of these systems showed that the *fcc* bonding site had the lowest energy. Then a force field was developed by using the results of the quantum calculations on the periodic system. Using the developed force field, MD simulations on monolayers with chain lengths from C_1 to C_{18} were conducted. The $c(4 \times 2)$ superstructure was reproduced at 300 K. Simulations on C_{12} chains between 50 and 400 K showed that the tilt angle decreased and the monolayer thickness increased as the temperature increased.

A small number of theoretical studies have also examined the structure of hydrocarbon chains attached to silica [29, 111]. For example, the structure of monolayers containing a hydrocarbon chain with azobenzene surrounded by other pure hydrocarbon chains was examined [29]. The simulation cell was an (8×8) square lattice that corresponded to a 21 Å^2/chain surface density. Two configurations, *cis* and *trans*, of the azobenzene chain were examined. The main difference in the monolayers studied was that the *cis* monolayers had looser packing, and the *cis*–azobenzene chains were shorter than the surrounding hydrocarbon chains. The *trans*–azobenzene chains were higher than the surrounding monolayers.

4. OPPOSING MONOLAYERS

The earliest computational studies of SAMs that examined tribology are those of Tupper and Brenner [75, 76] and Glosli and McClelland [77]. Both groups utilized the united-atom potential model [91, 92] to simulate the monolayer chains. Glosli and McClelland [77] focused solely on the friction between identical, ordered monolayers, which models the geometric setup found in SFA experiments [78]. Each monolayer consisted of 36 tightly packed $SH(CH_2)_5CH_3$ chains, physically pinned to the substrates by a force constant of 50 K/Å^2. The chains were arranged in a two-dimensional triangular lattice with a spacing of 4.44 Å. This resulted in an area of 17.1 Å^2/chain. The short length of the chains, six carbon units long, and the small spacing between chains resulted in the chains being perpendicular to the substrate rather than tilted (the spacing and the area per chain on gold is 4.99 and 21.9 Å^2, respectively). This eliminates gauche defects, and provides a symmetric computational cell. As mentioned previously, the united-atom method was utilized, with an LJ 6–12 potential for interactions between chains and a smoothing function for nonbonded interactions. The simulation temperature was ramped from 20 to 300 K, with weak and strong interfacial attractions applied to investigate the effect on shear stress. Although this system is not exactly the type of system examined in the AFM experiments of the friction of alkanethiols on gold, a number of features are nonetheless noteworthy. The commensurate geometry of the identical layers led to clear organized behavior during sliding. Layers responded in a synchronized fashion such that the instantaneous shear stress showed a very clear periodic variation, exhibiting steep drops to large negative shear stresses.

Chandross et al. [110] recently performed simulations that examined the adhesion and friction of identical alkylsilane SAMs attached to silicon dioxide substrates (Fig. 1). Systems with chain lengths of 6–8 carbon atoms with 100–1600 molecules on each surface were examined. One hundred chains per monolayer corresponds to a surface that is 43 Å × 50 Å. These simulations showed that large negative instantaneous shear stress results with system sizes up to hundreds of chains. However, in monolayers comprised of 1600 chains, all chains were not precisely synchronized. Thus, the instantaneous shear stress of the entire system exhibited much less severe variations as groups of chains slipped at slightly different times. While the 1600-chain system never exhibited a negative instantaneous shear stress, the local slipping behavior was the same as in much smaller systems of 400, and even 100 chains.

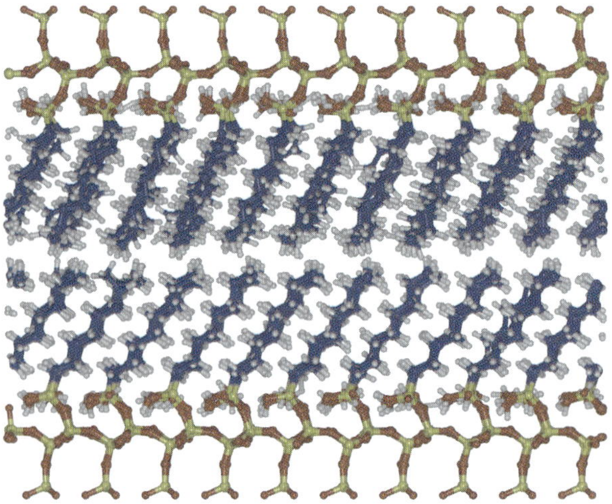

Figure 1. Two alkylsiloxane monolayers composed of C_8 chains attached to silicon dioxide (SiO_2) substrates, which are 4 oxygen atoms thick. The system is being sheared at a velocity of 1 m · s^{-1} under a load of 200 MPa. Oxygen, silicon, carbon, and hydrogen atoms are colored red, yellow, blue, and light gray, respectively. This figure is based on the work described in [110] by M. Chandross, G. S. Grest, and M. J. Stevens at Sandia National Labs.

The analysis by Glosli and McClelland of their slides, conducted under a variety of conditions, revealed two separate mechanisms for the dissipation of energy. The first mechanism was an abrupt, almost discontinuous, plucking mechanism, where the layers jump between potential maxima. This mechanism was most evident at low-temperature and high-head-group interaction strengths. Because of the strong interfacial attraction, the friction force was found to be independent of the velocity of the slide. Because bond lengths were held fixed, vibration in these studies refers to modes associated with angular variations along chain backbones. The second mechanism was also a thermal mechanism, where vibrational excitations helped the chains to slide over potential maxima. This latter mechanism was most evident at high-temperature and low-head-group interaction strengths, where the friction force decreased because of this gliding of the energetic chains over one another.

Similar investigations undertaken by Tupper and Brenner [75, 76] utilized identical alkanethiol monolayers consisting of longer chains, $SH(CH_2)_{15}CH_3$ and $SH(CH_2)_{20}CH_3$, chemisorbed via the sulfur end group to Au(111). The simulation cell contained 64 chains per monolayer attached to gold substrates. In these simulations, a Taylor expansion for the torsional terms was added to the harmonic potential, as well as a harmonic term to model bond bending. Interchain interactions were modeled with an LJ 6–12 term, and the rigid-bond constraint was replaced by a modified Hautman and Klein harmonic bond-stretching potential for modeling compression. Consequently, the time step was shortened to 0.25 fs, so that fast bond-length oscillations were accurately modeled. In turn, due to computational constraints, a sliding speed of $0.1\ nm \cdot ps^{-1}$ or $100\ m \cdot s^{-1}$ was utilized. This is order of magnitudes higher than experimental speeds, and alarmingly close to thermal speeds. However, simulations of the compression phase at speeds of $0.1\ nm \cdot ps^{-1}$ and $0.01\ nm \cdot ps^{-1}$ showed no noteworthy differences. The recent work of Chandross et al. [110] also confirms this with respect to slides conducted at speeds of 2.0 and $20\ m \cdot s^{-1}$ in a load regime of 0.2–2.0 GPa. This is not to say that the sliding speeds utilized are not a serious concern. Compression (or sliding) speed is one of the compromises that must be made to undertake simulations of this scale. These simulations were done at 300 K using Hoover's algorithm for computing NVT dynamics. The opposing monolayers did not entangle upon compression, but rather experienced a uniform increase in tilt angle. Sliding the upper monolayer relative to the lower caused a change in the tilt angle of 180° during 20 ps of sliding. The authors conclude that energy is dissipated primarily by oscillations in the tilt angle.

Also undertaken by Tupper and Brenner were slides of a single monolayer of $SH(CH_2)_{15}CH_3$ chains, or hexadecanethiol, against a single-layer gold surface [76]. These were the first attempts to investigate a system that resembled systems examined in AFM experiments. These simulations revealed a friction versus load response that remained linear (consistent with Amontons' law), despite the fact that the monolayer exhibited structural changes under compression [76]. It is possible that this is a consequence of the fact that these slides were conducted at fixed height with a flat probing gold surface. In other words, the friction versus load response may reflect more the interaction between the two ordered flat arrays comprised of gold atoms in the probe against the end groups of the chains, rather than the underlying structure of the monolayer.

The studies of Fujihira and co-workers [82–86] also made use of the united-atom potential energy function. In fact, their early studies [82–84] utilized the same computational setup employed by Glosli and McClelland [77], except that the LJ potentials between chains, and within a chain for neighbors more distant than three psuedoatoms, were altered. Bond lengths were also kept rigid during the simulations. Constraining bonds to be rigid may seem like a step backwards; however, the authors make an intriguing suggestion: unless energy fluctuations are strong enough to excite bond-stretching vibrations quantum mechanically, the rigid-bond constraint may be more appropriate in classical models. The implication is that these studies are restricted to weakly interacting systems or systems under low loads. With rigid bonds, it was practical to utilize a sliding speed of $2\ m \cdot s^{-1}$. Although this was still much faster than experimental speeds, it is clearly separated in scale from thermal speeds in this system. Furthermore, in their recent studies [85, 86], the properties of the cantilever assembly (discussed below) were carefully adjusted to account for the difference between modeled and experimental sliding speeds.

A recurring theme that characterizes the initial studies of Fujihira and co-workers was a clear separation between weak and strong energy modes and their relationship to the work done during sliding. Weak modes include the sum of LJ 6–12 energies between atoms separated by the interface (interfilm potential), the sum of LJ 6–12 potential energies between atoms within the monolayer (intrafilm potential), and the film–substrate potential. Strong modes are composed of the pinning of chains, bond bending, and torsional potentials. While the weak modes exhibited a dramatic build up and release of mechanical energy at slip, the strong modes showed only a weak correlation with stick-slip events. Dissipated energy eventually showed up in all available energy modes including strong modes, although it was the structured build up and release of mechanical energy associated with weak modes that governed frictional behavior. It is possible that the dramatic build up of energy is dependent upon several simulation details, such as load, sliding speed, and probe geometry (in addition, all of these things are intimately linked to the probe–monolayer potential). The strong separation between weak and strong modes is likely due to the fact that the loads probed were very low, and the tip geometry was not commensurate with the sample [72, 120].

5. PYRAMIDAL TIPS

The work of Bonner and Baratoff [79] extended the early studies by adopting a simple model of an AFM cantilever. They modeled the cantilever as a deformable, pyramidal cluster of 285 atoms initially arranged in (001) layers. The cantilever was connected to a rigid support by three orthogonal springs. Relatively soft springs were used for the sliding and loading directions, and a very stiff spring for the transverse direction. Damping was applied to reduce lever response to rapid SAM motions. The gold atoms comprising the tip interacted with each other through an embedded-atom-model potential. The top two layers of the tip were rigid, the velocities were rescaled at regular intervals to regulate temperature in the next two layers, and no constraints were applied to the remaining atoms. The SAM, comprised of 120 $S(CH_2)_{10}CH_3$ chains was anchored to the deformable gold substrate, that was also partitioned into rigid, thermal, and free atoms.

Due to computational limitations, the radii of curvature of experimental tips (no less than about 15 nm) are far too large to model atomically, although it is commonly suggested that the atomic information extracted from AFM measurements is a reflection of local tip asperities in contact with the sample. Consequently, the model system here may reasonably reflect the appropriate size of a tip asperity in contact with a SAM domain.

Even with imposing the rigid bond-length constraint, the system was at the limit of what was computationally feasible at the time. Consequently, only some general qualitative features were extracted from these simulations. For instance, at low temperature, the same SAM structure previously reported by Hautman and Klein [22, 23] was observed. Increasing the temperature caused the chains to rotate about their tilt axis, to precess about the substrate normal, and to eventually untilt as thermally activated gauche defects preferentially appeared near the methyl end groups. Lateral scans with the tip in contact with the SAM showed a fairly consistent sequence of stick-slip events. Each event involved a few atoms below the tip. These events were sufficient to eventually align the tilt of the SAM with the sliding direction. In its wake, the tip left behind a significant number of gauche defects and a vacancy island. Although the vacancy island quickly disappeared, the gauche defects annealed out only slightly over the duration of the slide.

In the studies of Koike and Yoneya [80, 81], all atoms were modeled explicitly, and comparative studies of the friction of a number of different kinds of monolayers were undertaken. The studies addressed Langmuir–Blodgett (LB) films composed of hydrogenated and fluorinated carboxylic acids. Room-temperature studies of completely hydrogenated or fluorinated chains of the same length, with different packing densities, were conducted. From the simulation perspective, the distinction between LB films and SAMs is minor. Both are modeled as organized long chains in sliding contact with a probing tip. To model individual atoms, including the terminating hydrogen and fluorine atoms, the consistent valence force-field (CVFF) potential, developed for protein structure analysis, was utilized. This potential requires the specification of various parameters that describe the stretching of bonds, bending of bonds, torsional angles, van der Waals interactions, and Coulomb interactions for all atoms.

Using a flat, multilayered SiO_2 tip and sliding at constant load, many of the general trends reported in previous studies were also seen in these studies. With regard to the particular systems studied, the highly fluorinated films exhibited higher friction than the purely hydrogenated films (the friction coefficient is about three times higher). The authors cited a number of experimental studies that reported this dramatic difference. The authors changed the hydrogen and fluorine torsion and bond angle potentials in an effort to change the stiffness of the films, and repeated the friction simulations. They concluded that these contributions to the potential energy could not account for the calculated differences in the friction. A similar analysis, interchanging the intermolecular hydrogen and fluorine potentials, was not carried out because the conformations of the films would have been altered. A particularly interesting analysis showed a correlation between root-mean-square fluctuations in the individual energy modes of the system and friction. The authors asserted that dramatically larger energy fluctuations were apparent in the highly fluorinated system in 1–4 LJ intramolecular interactions in the LB film (and to some extent, also in bond-stretching interactions), and that they were the origin of the difference in friction between the fluorinated and the hydrogenated systems. No strong differences in the root-mean-square energy fluctuations were observed with respect to the bond angles, torsional angles, interfacial, or Coulomb energies.

Fujihara and co-workers [85, 86] also examined a number of tips under positive, zero (but still in contact), and negative loads. The modeled sliders contained harmonic springs, damping, and feedback controls, with the specifications carefully tuned to model basic features of experimental tips, and to account for the mismatch between practical simulation and experimental sliding speeds. Variations in tip structure included incommensurate versus commensurate geometries with respect to the arrangement of chains in the sample, and in the size of molecular-sized tip asperities. The magnitude of the friction, and the frictional dynamics, were seen to be very sensitive to these variations. The authors observed that it was possible to achieve a structured periodic response, reflecting the periodic array of the SAM, by utilizing an incommensurate tip containing an asperity with the thickness of a single atom. Thus, the results of this setup seem qualitatively similar to an experimental AFM's ability to resolve a local sample structure.

A key problem encountered in MD simulation is the time-scale issue. Practical considerations limit atomic simulations to a few nanoseconds with sliding, or indentation, speeds of 1–100 $m \cdot s^{-1}$. Experiments are conducted on time scales that are orders of magnitude larger, and at sliding speeds that are orders of magnitude smaller, than simulation speeds. Therefore, physical pictures obtained from such simulations may be questionable when compared with experimental results. To overcome this problem, Leng and Jiang proposed a temporal–hybrid method to simulate the scanning process over SAMs in the time scale of AFM experiments. In this method, the thermal motion of the individual atoms in the AFM tip is ignored. The tip and cantilever assembly is driven by a support moving at a constant scanning velocity. Molecular dynamics is used to simulate the relaxation of the SAMs, but the motion of the tip is modeled with the harmonic equations plus the interaction between the tip and the SAMs. Two time scales are involved in the simulation: one describing the motion of the tip, and one corresponding to the relaxation of the SAMs. The time scale of the motion of the tip is approximately the same as it is in experiments ($\sim$0.5 μs). In contrast, the relaxation of the SAMs is related to all atomic-scale motion; thus, the time step of the integration has to be small ($\sim$1 fs). As the relaxation of the SAMs is supposed to be very fast, within several picoseconds, the dynamic evolution of the SAMs is a quasistatic process. With this kind of approach, the scanning speed can be as low as 10 $\mu m \cdot s^{-1}$, a practical speed in AFM experiments. All simulations utilizing this temporal–hybrid method were conducted with a rigid tip in contact with a deformable substrate. The relaxation of the atomic substrate leads to the inevitable contact damping that makes surface forces nonconservative.

This method was compared to conventional MD simulations, and results showed that when the resonant frequency of the cantilever was at least one order of magnitude less than that of the atomic substrate, the energy modes of the cantilever and the substrate could be separated. However, the hybrid method did not capture the higher vibrational modes of atoms in the contact area caused by the atomic-plucking motion, which arise from the motion of the slider against the SAM. This did not influence the friction force.

Using this hybrid approach, the indentation and sliding of a rigid (001) Au facet in contact with SAMs of alkylthiolate chains were investigated. The elastic modulus in the film thickness direction was chain-length independent, but the effective shear modulus was chain-length dependent. Longer chains had higher lateral stiffness than shorter chains, leading to lower friction, in agreement with AFM experimental results.

Recently, adhesion and friction in chemical force microscopy were studied using the same method. The scanning tip was modeled by attaching the alkanethiol chains $[-S(CH_2)_7CH_3]_{40}$ or $[-S(CH_2)_7OH]_{40}$ to an Au tip. Interactions between the tip and the SAMs were between the hydrophobic methyl (CH_3) groups or the hydrophilic hydroxyl (OH) terminal groups. Results obtained include force–distance curves, friction curves, friction–sliding-distance curves, interfacial molecular or hydrogen bonding structures, and load distributions among different molecular chains. Results of this study showed that adhesion and friction for the OH–OH contact pairs were much larger than those for the CH_3–CH_3 pairs, due to the formation of hydrogen bonds. The tip position versus cantilever support position curves showed that, during the retraction process of a chemical force microscope (CFM) tip from a surface, the CFM tip was away from the sample surface slightly, while the spring underwent enlongation in the normal direction, before the tip broke away from the surface. Single molecule forces were distributed unevenly at the contact area.

Diestler and co-workers [130, 131] showed that the atomic description of friction of the scanning surface probe device is greatly complicated by wide variations in the temporal regime of the measurement. At one extreme, the system remained in equilibrium for the duration of the measurement, and one can employ statistical thermodynamics to compute friction. However, at the other extreme, the system remained out of equilibrium, and the dynamic history of the system had a strong impact on friction. Obviously, the temporal–hybrid method proposed by Leng and Jiang is based upon treating the sliding as a quasistatic process. Therefore, this method cannot be applied to systems that are far from equilibrium; this includes systems at high loads, systems where there is significant penetration of the tip into the monolayer, and systems undergoing chemical restructuring.

6. ATOMICALLY SHARP TIPS AND INFINITELY FLAT TIPS

Harrison and co-workers developed and used [119–124] a hydrocarbon potential energy function that is one of the few that allows for chemical reactions (with the accompanying changes in atomic hybridization) and long-range interactions [125] to examine the tribology of hydrocarbon monolayers attached to diamond.

Modeling the breaking and forming of chemical bonds is not a straightforward task because it requires the potential energy function to go beyond simple pairwise interactions to include multibodied terms representing the chemical environment. The reactive empirical bond-order (REBO) potential [126, 127] was the first potential developed to model chemical reactions of hydrocarbons, such as those needed to model the chemical vapor deposition growth of diamond from gaseous reactants. In REBO potential, in addition to the terms representing the repulsive and attractive interactions of a bond, there exists a bond-order factor that depends on bond angles, bond lengths, and atomic coordination in the vicinity of the bond. This many-body nature of the bond-order term makes the bond energy dependent on the local environment surrounding the bond, and therefore allows the REBO potential to describe chemical reactions. Harrison and Brenner used this potential in their study of the wear of diamond surfaces in sliding contact, and reported the first case of simulated tribochemistry [128].

Recently, Harrison and co-workers examined the compression and friction of hydrocarbon chains attached to diamond using various probes. The REBO potential lacks the long-range intermolecular and torsional interactions needed to model the organized, tilted structure of SAMs. However, its ability to model chemical reactions is crucial to the simulation of tribochemistry. With that in mind, the REBO potential was modified, and the resulting potential is known as the adaptive intermolecular REBO (AIREBO) potential [125]. In particular, modifications contained in the AIREBO potential include long-range LJ interactions between pairs of atoms that are smoothly turned on and off based on the surrounding chemical environment. Since its creation, the AIREBO potential has been used to undertake a number of interesting mechanical [109] and tribological studies of SAMs [119–124]. In addition, this potential was also used to model the bending of empty and filled single-wall carbon nanotubes [129].

In these simulations equilibrated hydrocarbon monolayers with approximately the same packing density as alkanethiols on gold had an ordered arrangement of chains which were uniformly tilted with respect to the surface normal [109, 120, 122] (Fig. 2). Chains typically had few gauche defects [109], which were primarily confined to the ends of the chains. The position of the gauche defects within the chains was observed in previous simulations by Hautman and Klein [22, 23], and subsequent simulations of alkylsilane monolayers [110]. Karaborni [107] utilized an anisotropic united-atom potential to model the approach and separation of opposing monolayers. Simulations were performed by compressing 128 molecules, with 19 CH_2 segments and a COOH head group, evenly distributed on two plates (25 Å^2/molecule), at 7 $m \cdot s^{-1}$. These simulations showed that

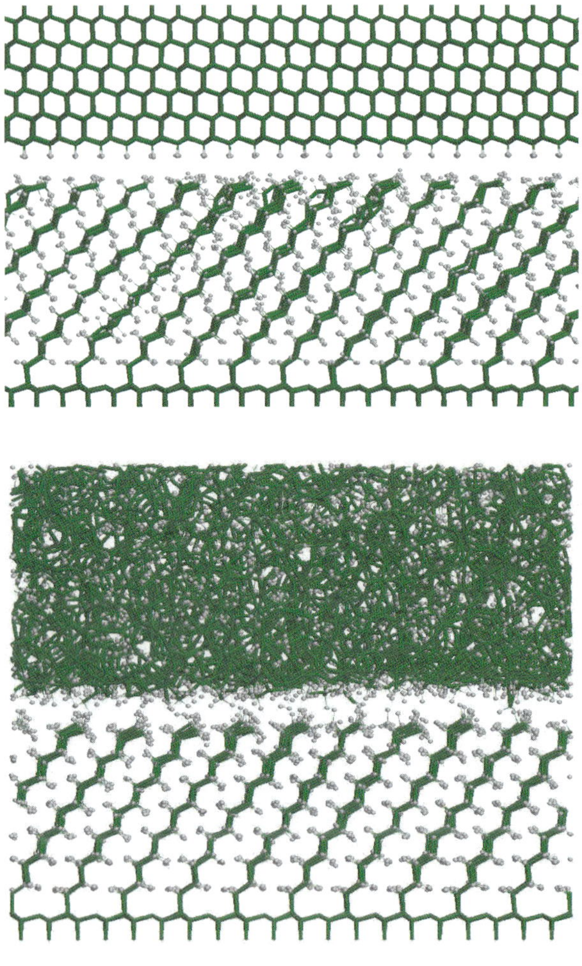

Figure 2. In the upper panel, a monolayer composed of C_{18} n–alkane chains in the (2 × 2) arrangement on diamond (111). This packing is referred to as a tightly packed system. The counterface (probe) is a hydrogen-terminated diamond (111) surface. In the lower panel, C_{14} chains are attached to diamond (111) in a tightly packed arrangement. This system is in sliding contact with a diamond-like carbon probe. This is the ending configuration for the highest load (around 30 nN) investigated for the data reported in Figure 15. In both panels, carbon–carbon bonds appear as green cylinders, and hydrogen atoms as white spheres.

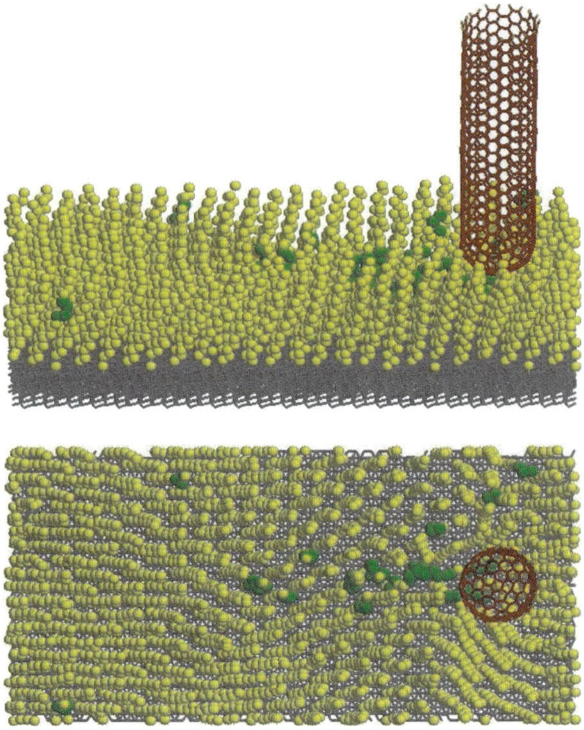

Figure 3. Monolayer of C_{13} n–alkane chains shown in yellow (hydrogen atoms are not shown for clarity). The alkane chains are attached to a diamond (111) substrate in the (2 × 2) arrangement (shown in gray). A capped [10, 10] SWNT is used as the probe (shown in red) [139]. Gauche defects in the C_{13} monolayer are colored green. The tube has slid for 38 Å in the direction of the chain tilt (left to right). In the bottom panel, the view is looking down the long axis of the tube onto the plane that contains the monolayer. In the top panel, the system is rotated away from the viewer, and viewed at an angle.

increasing the tilt angle of the molecules decreased the number of gauche defects in the midsections of both monolayers. This same link between tilt angle and gauche defects was observed in the simulation of long-chain alcohols [108].

The initial studies of Harrison and co-workers utilized tips of two extremes: infinitely flat, hydrogen-terminated diamond counterfaces (Fig. 2, upper panel), which emphasize the friction mechanisms associated with compression of the SAM, and capped carbon nanotubes (Fig. 3), which emphasize penetration into the SAM. Recent studies made use of amorphous carbon tips (Fig. 2, lower panel), which make the monolayer and the probe incommensurate. The compression (indentation) of monolayers composed of n–alkane chains of lengths ranging from C_8 to C_{18}, using hydrogen-terminated diamond (111) and carbon nanotube probes, were examined. The chains were attached to a diamond (111) substrate in the (2 × 2) arrangement because this yields approximately the same packing density (21.9 Å^2/chain) as alkanethiols on Au(111). These simulations showed that the number of defects formed, and their location, depended upon the contact geometry of the probe, the length of the alkane chains, and the compression speed. For instance, when a hydrogen-terminated diamond (111) probe was used to compress n–alkane monolayers [119–123], gauche defects formed primarily on the ends of the chains. Slower compression speeds led to fewer defects being formed. In contrast, when the probe was a capped, single-wall carbon nanotube (SWNT), the majority of the gauche defects formed were localized to the region of the monolayer adjacent to, and under, the nanotube (Fig. 3) [109, 121, 123, 124]. The small radius of the nanotube probe allowed it to "slice" into the monolayer, interacting with a small number of hydrocarbon chains and localizing defects near the tip, irrespective of the length of the n–alkane chains [123, 124]. Due to the collective tilt angle of the chains, some chains were pushed closer to the substrate (causing an increase in their tilt angle), some were pushed farther from the substrate (causing a decrease their tilt angle), and some were partially "pinned" under the tube as a result of the indentation. The distribution of defects around the indentation site was not uniform. That is, a larger number of defects were generated on the side of the tube where the tilt angle of the chains decreased.

The effect of probe stiffness on indentation and friction was examined by using rigid, flexible, and multiwall carbon nanotube probes [109, 121, 123, 124]. During indentation, the small contact area of the nanotubes resulted in the number of defects formed being a function of penetration depth into the monolayer. Because a flexible nanotube compresses slightly, and bends a bit to one side as it interacts with the C_{13} monolayer, it does not penetrate a given monolayer as deeply as a rigid nanotube. As a result, for probes in this size range, a flexible tube generates fewer defects than a rigid nanotube as it is pushed into a hydrocarbon monolayer [109]. One important difference between the rigid and the flexible nanotube was that the rigid tube was able to penetrate the monolayer and pin chains against the diamond substrate. Severe indents such as these sometimes led to broken bonds between the chains and the substrate [109, 121].

The elastic modulus of C_8, C_{13}, and C_{22} monolayers was extracted from compression studies of these monolayers using an infinitely flat surface [109]. Plots of stress σ versus strain ε for the compression of these monolayers showed that the elastic modulus of C_8 was approximately an order of magnitude larger than the value determined for C_9 alkanethiols on gold using a quartz crystal microbalance [132]. This is not surprising due to the fact that atomic-scale experiments sample much larger contact areas than those examined in these simulations. Domains and defects are present in the experimental samples which are not present in the simulated systems. The computed elastic modulus of C_8 was the largest, followed by C_{13} and C_{22} [109]. That is, the shorter chains were stiffer in the compression direction (which is perpendicular to the shearing direction). This phenomenon was recently observed during simulations of the compression of opposing alkylsilane monolayers [110].

Harrison and co-workers used various probes to examine the friction of monolayers composed of *n*–alkane chains with lengths ranging from C_8 to C_{22} [119–124]. The effects of chain length, packing density, sliding direction, probe rigidity, and contact area on friction and wear were all examined. The effect of probe rigidity on friction was examined by using flexible and rigid SWNT probes, and a double-wall nanotube (DWNT) probe, to investigate the friction of C_8, C_{13}, and C_{22} *n*–alkane monolayers. After penetration into the monolayers, all of these tubes "plowed" through the monolayers, creating gauche defects (Fig. 3) [123, 124]. These defects built up in the wake of the tube, and the number of defects depended upon sliding speed, penetration depth, and sliding direction. The number of defects increased as the penetration depth (and load) increased, as noted earlier. The effect of sliding direction on defect formation was also examined. The total number of gauche defects versus sliding distance is shown in Figure 4 for the case when a flexible DWNT "slices" through a C_{13} monolayer. These two sets of data correspond to different sliding directions. When sliding in the direction parallel to the tilt angle of the chains (left to right in Fig. 3), the number of defects increased with sliding distance (blue line, Fig. 4) until the periodic image of the simulation cell was reached. That is, when the tube passed through an area of the monolayer that had already been perturbed (approximately ten unit cells of sliding), a marked decrease in the number of defects occurred [124]. Continued sliding through the perturbed region caused the number of defects to decrease almost linearly with sliding distance. This severe decrease in the number of defects did not occur when the tube motion was stopped and the defects were allowed to anneal. In other words, the motion of the tube through an already perturbed region accelerated the rate of defect removal. Adjusting the sliding direction such that it made a small angle with the original sliding direction, and so the tube did not slide over its wake (a previously perturbed region), changed the number of defects created (red line, Fig. 4) during sliding. In this case, the number of defects continued to increase as sliding progressed.

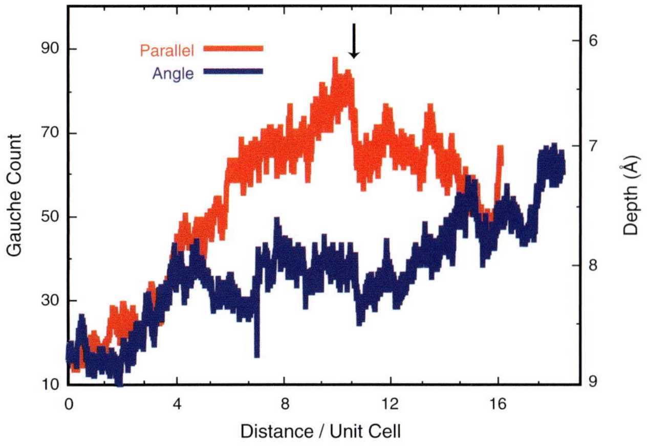

Figure 4. Number of gauche defects generated versus sliding distance when a flexible DWNT is in sliding contact with a C_{13} *n*–alkane monolayer. Data shown in red correspond to sliding along the direction of chain tilt. The black arrow corresponds to the point where the tube slides through an already perturbed area. (That is, the tube is sliding through its periodic image.) Data shown in blue correspond to sliding at a small angle away from the initial sliding direction. The unit cell in the tilt direction is 8.71 Å. The load on the nanotube is 13.8 nN. (See [139] for more details regarding the DWNT.)

While the motion of the end of the nanotube, or the depth of the tube in the monolayer, is intimately linked to the number of defects formed, the correlation does not appear to be a strong one [124]. Figure 5 shows the depth of the nanotube and the gauche defects generated during sliding with the chain tilt (blue line, Fig. 4). For approximately the first 12 ps (12 Å or ~1.38 unit cells) of sliding, the end of the DWNT (capped region) was "stuck" in the C_{13} monolayer. Because the top of the nanotube was moved, the capped region began to "pull out" of the monolayer (decreasing its depth). There was no appreciable change in the number of gauche defects formed during this time because the capped end of the tube moved little. After the first 12 ps of sliding, the capped end of the DWNT "slipped," and began moving through the monolayer, increasing the number of gauche defects as it moved. The capped end of the nanotube continued this type of weak "stick-slip" motion through the monolayer for the remainder of the simulation. This caused the number of defects to increase monotonically as sliding progressed. Removal of the tube at any time from the monolayer ultimately resulted in the annealing of the gauche defects.

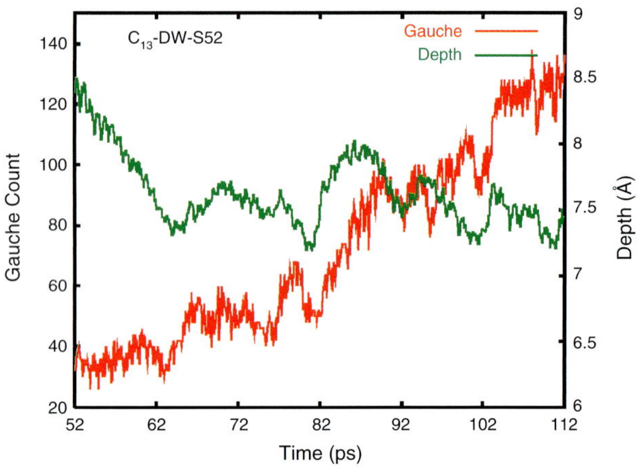

Figure 5. Number of gauche defects (red line) generated when sliding in the direction of chain tilt and the depth of the DWNT (green line) versus sliding time when a flexible DWNT is in sliding contact with a C_{13} *n*–alkane monolayer. The unit cell in the sliding (tilt) direction is 8.71 Å. The load on the nanotube is 13.8 nN. (See [139] for more details regarding the DWNT.)

Friction versus load data for a flexible and a rigid SWNT sliding through C_{13} monolayers are shown in Figure 6. These data show that, for nanotubes with small contact areas, the flexibility of the probe did not have a marked influence on the measured friction at a given load. The probe rigidity simply limits the loads (penetration depths) that can be examined with a given probe. Sliding through the monolayer caused the flexible nanotube to buckle [123, 124]. Using a DWNT as the probe increased the range of loads that could be examined prior to tube buckling to approximately 15 nN [124].

A number of scanning probe microscope (SPM) experiments reported that the friction of alkanethiols decreases with increasing chain length [58, 60, 63, 64, 67]. After some critical chain length is reached (which is believed to correspond to a well-packed film), all chain lengths exhibit similar friction versus load behavior. Single-wall carbon nanotubes were used to examine the friction of tightly packed, *n*–alkane monolayers composed of C_8, C_{13}, and C_{22} chains [123, 124]. Because the diameter of the nanotube probes was approximately 15 Å, the measured friction was a function of the penetration depth into the monolayer. In addition, the experimental trend of longer chains yielding lower friction than shorter chains was not reproduced. Single nanotubes have been used as AFM probes to image very small features on surfaces [133]. However, they have not been used to examine friction. The SPM experiments to date that have examined friction have utilized tips that are one–two orders of magnitude larger than the diameter of the nanotubes used in these simulations. Thus, comparison with experimental data is problematic.

To increase the contact area between the probe and the monolayers, a hydrogen-terminated diamond (111) surface was used as the probe [119, 120, 122, 123]. In this case, the behavior of the friction as a function of load for C_8, C_{13}, $C_{18,}$ and C_{22} chains more closely resembled the trends observed in experimental data. At most loads, the average friction of the shortest chains was generally higher, and that of the longest chains was generally lower. However, subsequent simulations on these systems using the infinitely flat probe have shown that it is very difficult to reduce the error bars sufficiently to separate these curves (Fig. 7). It is likely that the well-ordered nature of the monolayers and the commensurate probe are responsible for this phenomenon. Other, more realistic probes (discussed below) may be able to better reproduce the experimental data. In addition, the majority of the chains moved in a synchronized way during sliding. This resulted in approximate periodicities in a number of quantities associated with sliding, which are discussed below [120].

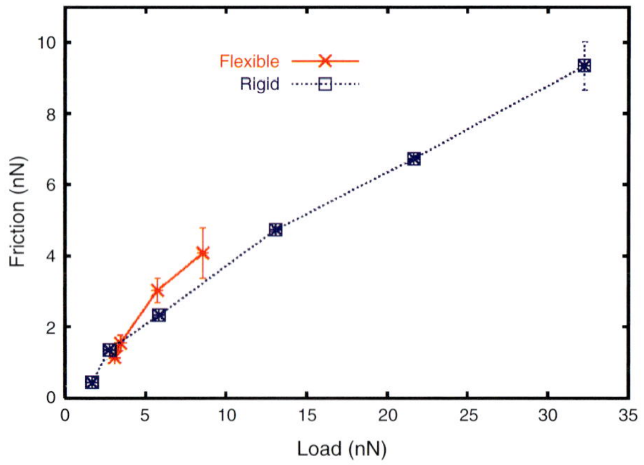

Figure 6. Friction versus load data when two carbon nanotube probes are in sliding contact with C_{13} *n*–alkane chains attached to a (111) diamond substrate in the (2 × 2) arrangement. The sliding direction corresponds to the direction of the chain-tilt angle. Data shown in red and blue correspond to a flexible and a rigid [10, 10] single-wall nanotube, respectively. Error bars correspond to one standard deviation. Each point was calculated from 40 ps slides.

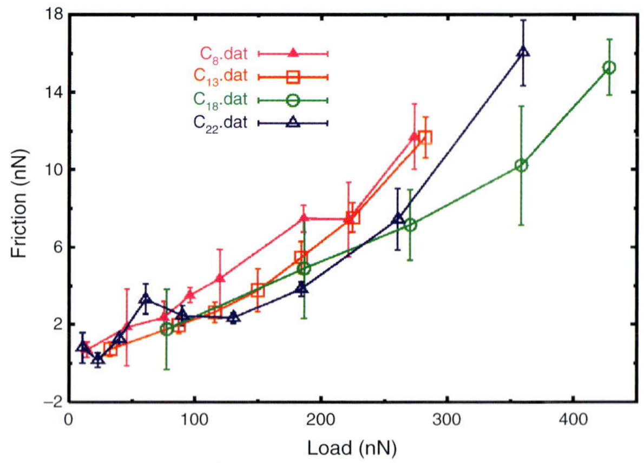

Figure 7. Friction versus load for a hydrogen-terminated diamond (111) surface in sliding contact with *n*–alkane monolayers. Pink, red, green, and blue lines correspond to data for C_8, C_{13}, C_{18}, and C_{22} carbon atom chains. Error bars indicate minimum and maximum friction values obtained by window averaging the data over unit-cell lengths of sliding (the unit-cell length is from the (2 × 2) arrangement of chains in the sliding direction). The sliding direction corresponds to the chain-tilt angle. Lines are draw to aid the eye. See [119] and [122] for more simulation details.

Simulations that examined the effects of packing density [119] on friction were motivated by the AFM experiments of Perry and co-workers [50, 51], who examined the friction of a series of spiroalkanedithiols. These simulations examined both tightly and a series of loosely packed monolayers with a hydrogen-terminated diamond probe. The tightly packed system consisted of C_{18} *n*–alkane chains attached to diamond (111) in the (2 × 2) arrangement. Loosely packed systems were created by randomly removing chains (20, 40, or 54) from the tightly packed system or by attaching the C_{18} chains to the substrate in a different arrangement, such as (4 × 4). These systems were equilibrated to 300 K, and compressed with a hydrogen-terminated diamond (111) counterface. The number of gauche defects per chain formed for each system during equilibration and compression is shown in Figure 8. The equilibration phase lasted for approximately 20–25 ps for all systems. It is apparent that removing chains from the tightly packed system, or increasing the space between chains, caused an increase in the disorder of the monolayers. Compressing all but the tightly packed monolayers further increased the disorder.

Friction versus load data for these systems are shown in Figure 9. At moderate to high loads, all of the loosely packed systems (more disordered) exhibited higher friction than the tightly packed (more ordered) system. This result agrees with the experimental findings of Perry and co-workers [50, 51]. A detailed analysis of the quantities associated with sliding was conducted. Those results, in conjunction with data from other studies [119, 120], revealed the following. For tightly packed systems explored with an infinitely flat probe, there are a number of properties that are correlated with the frictional force. Some properties, such as the generation of gauche defects, are correlated with a slight time lag [122].

A C_{13} chain isolated from a tightly packed monolayer is shown in Figure 10. The variation in a number of the

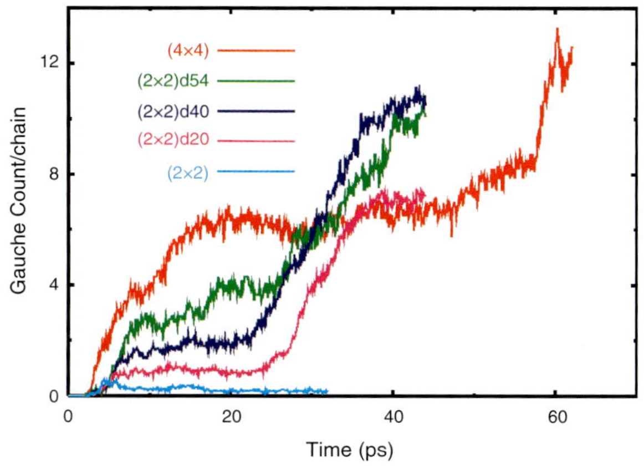

Figure 8. Number of gauche defects per chain for the equilibration and compression of C_{18} *n*–alkane monolayers attached to diamond (111). Pink and aqua lines correspond to the tightly and loosely packed systems of [119], respectively. Data represented by the red line is for C_{18} chains attached to diamond (111) in the (4 × 4) arrangement. All other systems had the chains attached in the (2 × 2) arrangement. The tightly packed system (aqua line) has 72 chains. 20, 40, and 54 chains have been randomly removed from the surfaces represented by the pink, blue, and green lines, respectively.

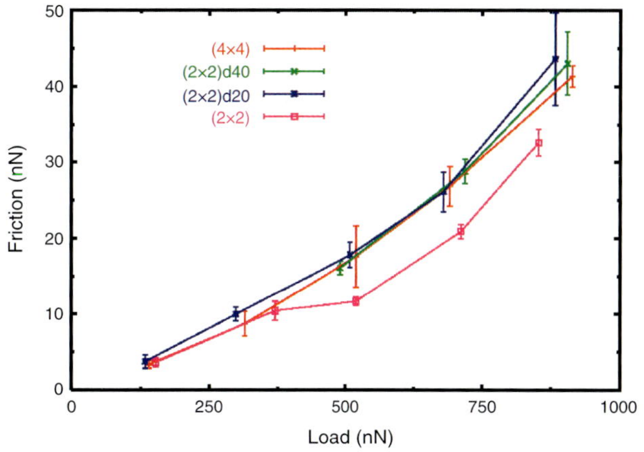

Figure 9. Friction versus load for a hydrogen-terminated diamond (111) surface in sliding contact with C_{18} *n*–alkane monolayers. Pink and blue lines correspond to the tightly and loosely packed systems of [119], respectively. The systems are the same as those described in the caption of Figure 8, with the omission of the data for the system with 54 chain removed. Error bars indicate minimum and maximum friction values obtained by window averaging the data over unit-cell lengths of sliding. The sliding direction corresponds to the chain-tilt angle. Lines are drawn to aid the eye. See [119] for more simulation details.

quantities defined in Figure 10 with time is shown in Figures 11 and 12. During equilibration and compression, the chains in tightly packed *n*–alkane monolayers exhibited a right-handed configuration, where the chain circulates counterclockwise starting from the attachment site, or a left-handed configuration. (See [120] for a picture of these chains.) The handedness of the chains may be correlated because groups of chains within the monolayer shared the same configuration. However, not all chains in the simulations shared the same handedness. The fact that two

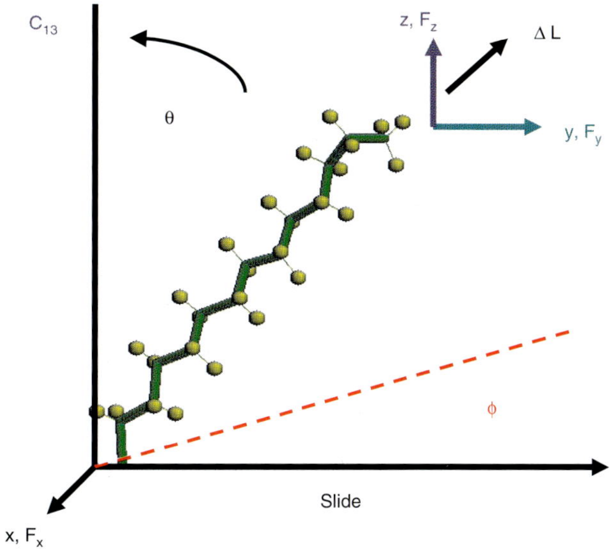

Figure 10. Single C_{13} *n*–alkane chain extracted from an equilibrated monolayer on diamond (111). Azimuthal angle φ (red), tilt angle θ (black), and change in length of the chain ΔL (black) are all defined. Sliding direction (with the tilt angle) is from left to right in the figure. See [120] for simulation details.

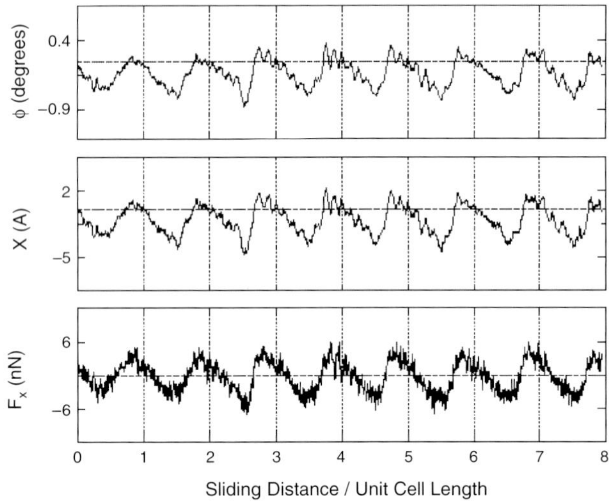

Figure 11. Average force on the counterface perpendicular to the sliding direction F_x, average x coordinate on the top carbon atoms of the C_{13} chains, and the azimuthal angle φ as a function of sliding distance. The sliding distance has been divided by the unit cell distance in the sliding direction for the (2 × 2) packing of the chains on diamond (111). The sliding direction is along the tilt of the chains (positive y in Fig. 10). See [120] for simulation details.

configurations existed may be related to the compression speed; chains became pinned before they could adjust to their neighbors. The fixing of the handedness of the chains by compression broke the symmetry of the system slightly. Analysis of the azimuthal angle φ as a function of sliding distance (Fig. 11) showed that the chains are pushed from side to side as each wave of hydrogen atoms from the countersurface passes over them. (It should be noted that the unit-cell distance of the countersurface was half that of the monolayer.) All of the chains underwent this synchronized motion. The motion was not symmetric about zero because the chains were preferentially shifted to one side upon compression. This periodicity is also apparent in the average lateral position of the top layer of carbon atoms in the chains (x) and the lateral force (F_x). The sliding direction is along the chain tilt direction, or the y direction (Fig. 11).

This approximate periodicity has also manifested in the trajectory of the carbon atoms at the ends of the chains. Figure 12 shows the motion of these carbon atoms in the sliding (y) and the compression (z) directions. Each wave of hydrogen atoms interacted with the top of the chains, slightly compressing and stretching them. When the countersurface hydrogen atoms passed by, the chains sprang back. Due to the ordered nature of the monolayer, the stretching of the chains ΔL extended throughout the entire chain. In fact, the change in bond lengths between carbon–carbon bonds in the alkane chains is the property that was most strongly correlated with friction [119]. Higher loads led to restricted movement of the chains in the tightly packed system compared to the loosely packed system (labeled *d*20 in Figs. 8 and 9) [119]. Twenty chains were randomly removed from the tightly packed system to form this loosely packed system (d20). In addition, sliding initiated larger bond-length fluctuations in the loosely packed system (d20), which ultimately led to more energy dissipation via vibration [119]. In the tightly packed system, the vibrational energy in the carbon–carbon bonds of the chains was a linear function of the work done during sliding (Fig. 13). Because the countersurface was rigid and sliding at constant velocity, the work was proportional to the friction. Thus, the vibrational energy imparted to the carbon–carbon bonds of the chains had the same behavior with load as friction. The approximate torsional energy, also shown in Figure 13, was calculated by mapping the torsional energy of each set of four consecutive carbon atoms in the chain backbones to the potential energy of an uncompressed butane molecule [119]. This method

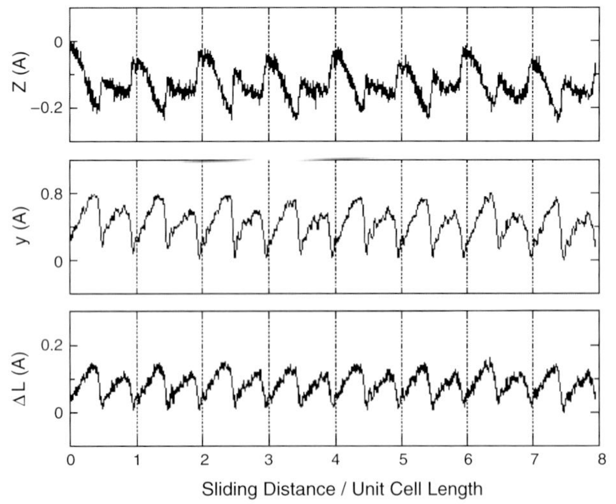

Figure 12. Average displacement of the carbon–carbon bonds in the alkane chains ΔL and the average y and z coordinates of the top carbon atoms in the alkane chains (C_{13}) as a function of sliding distance. The sliding distance has been divided by the unit cell distance in the sliding direction for the (2 × 2) packing of the chains on diamond (111). The sliding direction is along the tilt of the chains (positive y in Fig. 10). See [120] for simulation details.

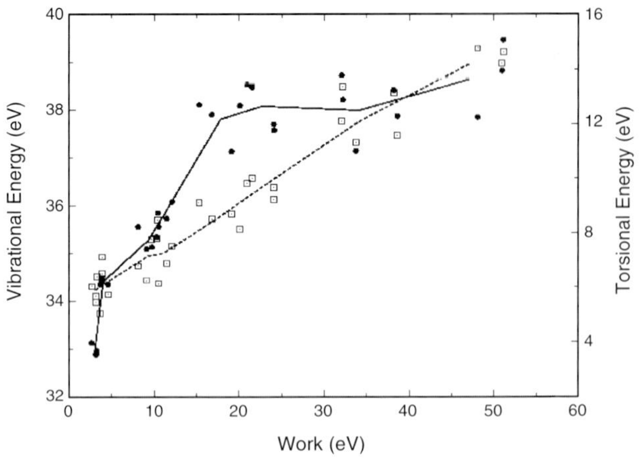

Figure 13. Vibrational (open squares) and approximate torsional (filled circles) energy as a function of work for a monolayer of tightly packed C_{13} alkane chains. The hydrogen-terminated diamond (111) counterface was held rigid in these simulations. The results of 8 sliding simulations at different loads are shown. Each simulation is divided up into 4 unit cells, and the average energy for each unit cell is plotted. Thus, there are 32 data points. Other simulation details can be found in [119].

of calculating the torsional energy, or the energy associated with gauche defects, implies that the contribution of the torsional energy to energy dissipation plateaus at a given load. Thus, the torsional energy and the friction did not respond the same way to the application of load.

While the creation of gauche defects dissipates energy, their generation may not be the main channel of energy dissipation in these tightly packed systems, as shown in the data presented here. However, there are additional possibilities. First, the torsional energy contribution was mapped to an uncompressed butane molecule. Because the conformation of an individual chain is altered under load, this may not be an appropriate mapping. Furthermore, the result of the formation of a defect with respect to the surrounding region is untreated (changes in the orientation of the chain tip can significantly affect the local order in this region). Finally, the flattening of the torsional angle contribution may be related to the sliding speed. It is possible that the characteristic time for a gauche defect to anneal is longer than the time between subsets of hydrogen atoms on the countersurface interacting with the ends of the chains. If that were the case, the number of defects would continue to increase, eventually reaching some constant level as it does here. In short, the flattening of the torsional energy contribution with load says more about the number of gauche defects than it does about their contribution to energy dissipation. What is clear is that, for these sliding conditions (geometry, sliding speed, etc.), energy dissipation via vibration was determined to be the main channel for energy dissipation. Any variation in simulation conditions, such as probe geometry, sliding speed, or packing density, could open up additional channels of energy dissipation or change the relative importance of existing channels.

When utilizing a commensurate tip in sliding contact with a perfectly ordered monolayer, it may be useful to look at how the friction and load forces are distributed over the set of monolayer chains. Are there trends that manifest themselves in a significant fraction of the individual chains or is there significant variation across the monolayer? In the former case, one could conduct a robust analysis in a comparative study that targets a particular property by way of varying system geometry or simulation parameters over a limited set of simulation runs. In the latter case, more extensive simulations involving independent runs of similarly prepared systems may be required to conduct a statistically significant analysis.

Figure 14 shows how friction and load are distributed across two similarly prepared systems. The systems are linear hydrocarbon chains anchored in a (2 × 2) arrangement on diamond (111), with a hydrogen-terminated diamond (111) surface serving as the counterface. Starting from the same initial configuration, the counterface was slid against the tilt of the monolayer chains in one system, with the tilt of the monolayer chains in the other. There are 52 chains in the model system. Thus, there are 52 points for each system showing the friction and load attributed to each chain. These forces were calculated by summing all forces between atoms in a single chain and all atoms in the probing counterface. There is clearly a great deal of overlap between the two systems, and also a wide range of variation within individual systems (some chains even have a net effect of pushing

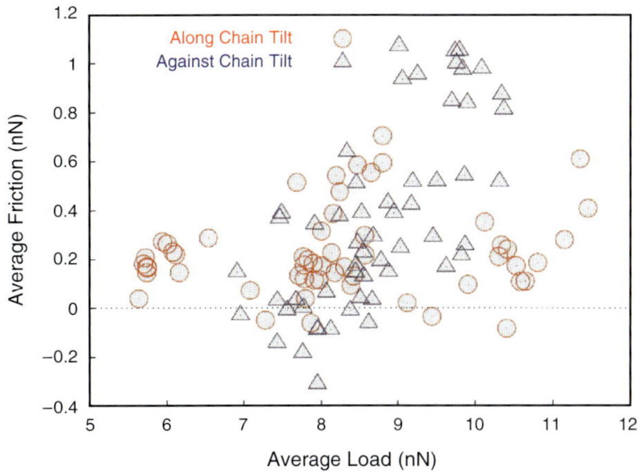

Figure 14. Average friction and load plotted for each chain in a monolayer composed of C_{16} hydrocarbon chains in a (2 × 2) arrangement on diamond (111). The counterface was a hydrogen-terminated diamond (111) surface. Sliding directions along (positive y) (red circle) and against (negative y) (blue triangle) the tilt angle are given in Figure 10. The monolayer was composed of 52 chains. Sliding speeds were 100 $m \cdot s^{-1}$, and the temperature was maintained at 300 K. Other simulations details are as in [119].

the probe along). It is clear that the overall friction was significantly higher when sliding against the chain tilt. However, the higher friction was associated with a subset of the chains (what is going on locally), and not with the monolayer as a whole (what is going on globally). This suggests that additional runs of independently prepared systems would be required to see if this is a consistent trend associated with sliding direction.

The simulations to date have utilized two types of extreme probes: a very small carbon nanotube, and an infinitely flat hydrogen-terminated diamond probe. In an effort to model more realistic probes, Harrison and co-workers have begun making use of amorphous carbon tips with surface roughness. These tips offer several advantages over the previously used probes. First, because their structure is amorphous, many of the issues associated with commensurate contacts will be removed [72]. Second, the irregular structure is likely to resemble the surface of an AFM tip, particularly one coated with a diamond film.

A recent experimental study reported measurements of the structural and friction properties of mixed and single component films of alkanethiols on Au(111) [48]. Two linear regimes were distinguishable in the friction versus load data of the mixed film. At low loads, the mixed film exhibited a very rapid rise in friction. At higher loads, the slope of the curve decreased to a level comparable with the single-component film. The friction versus load data of the single component film was linear over the entire range of loads investigated. With these results in mind, MD simulations of the friction of mixed and pure component SAMs were undertaken. A brief discussion of some preliminary findings follows.

Figures 2 (lower panel) and 15 depict model sliding systems where the probing tip is made from a diamond-like-carbon (DLC) mixture of hydrogen and carbon atoms. Within the bulk, this tip was consistent with some typical benchmark properties of DLC (density, percent hydrogen content) [134]. The surface structure of the tip exhibited the desired traits of moderate surface roughness and incommensurability with the diamond (111) substrate underlying the SAM. Two systems, referred to as a pure and mixed SAM, were examined. The pure SAM was comprised of 100 linear hydrocarbon chains, each containing 14 carbon atoms (C_{14}) (Fig. 2). The mixed SAM was comprised of an equal mixture of 12-carbon-atom (C_{12}) chains and 16-carbon-atom chains (C_{16}) (Fig. 15). Thus, each of the SAMs contained an equal number of carbon and hydrogen atoms. The chains were anchored to a thin diamond (111) substrate in a (2×2) arrangement.

After equilibrating the systems at 300 K, identical tips were used to indent both systems up to loads near 30 nN. From the indents, a number of system configurations were extracted and used as starting configurations for slides. Figure 16 shows a comparison of friction versus load for these pure and mixed SAM systems. Over the load range investigated, the response of both systems is approximately linear, and the slope of the data for the mixed system is over three times larger than the pure system. This was also observed in AFM experiments that examined alkanethiols of the same lengths [48]. It is likely that this behavior is related to the high degree of disorder in surface structure of the mixed SAM. The relationship between order and friction is well established [50, 51, 72, 119], and the sliding interfaces shown in Figures 2 (lower panel) and 15 are strikingly different in structure.

Initial analysis of energy modes revealed that the thermostats utilized at the top of the tip and the bottom of the sample, to maintain the system at 300 K, are too strong to establish a clear correlation between frictional properties and weak energy modes. Consequently, these systems are being restudied with a weakened thermostat that is applied to all atoms in the system in the dimensions transverse to the sliding direction. Although thermostats are typically applied only to atoms at the periphery of the system, it has been demonstrated that all nonrigid atoms can be coupled to the thermostat, provided the coupling is weak and not applied in the sliding direction [72].

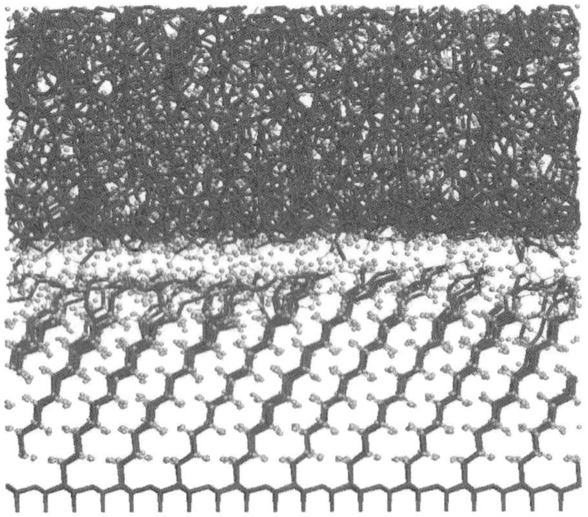

Figure 15. Mixed component SAM in sliding contact with a diamond-like carbon tip. The SAM is a random mixture of C_{12} and C_{16} *n*–alkane chains. This is the ending configuration for the highest load investigated in Figure 16 (around 30 nN). The tip density is 1.9 g/cm^3. The tip is composed of 38 and 61% sp^3- and sp^2-hybridized carbon atoms, and is 36% hydrogen. Color coding is as in Figure 2. The sliding direction is along the chain-tilt angle (left to right).

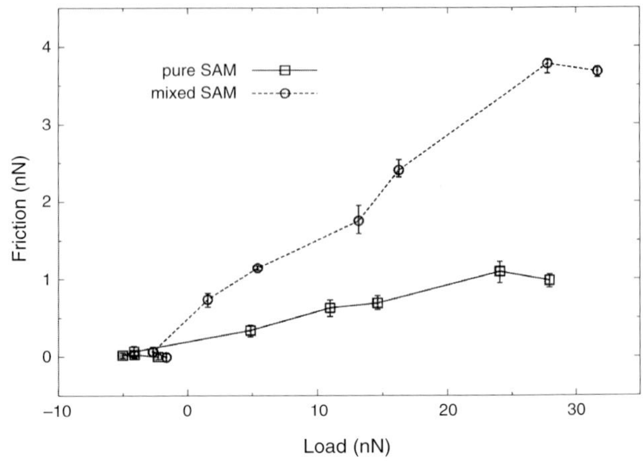

Figure 16. Comparison of the friction versus load for a single-component SAM composed of C_{14} chains versus a mixed SAM with equal numbers of C_{12} chains and C_{16} chains (Fig. 15). Error bars indicate minimum and maximum friction values obtained by window averaging the data over unit-cell lengths of sliding (the unit-cell length is from the (2×2) arrangement of chains in the sliding direction). The sliding direction is along the chain-tilt angle (left to right in Fig. 15). Lines are drawn to aid the eye.

Because the friction responses of these similar systems are clearly distinct from one another, these preliminary results suggest that further investigation will yield some insight into fundamental friction mechanisms. It will be interesting to extend these investigations to higher loads to see if the slope of the friction versus load curve of the mixed SAM ever matches that of the pure SAM, as is seen experimentally [48].

Recently, there has been much interest in monolayers composed of polydiacetylene. The friction of unpolymerized and polymerized diacetylene monolayers has been measured using the AFM [135]. These authors assert that the presence of the cross-linked molecules in polymerized films alters the structure of the friction versus load data, compared to unpolymerized films, by inhibiting lateral displacement within the monolayers. Moreover, the friction exhibited a directional dependence (anisotropy) in the polymerized films. This is thought to arise from the presence of domains associated with the linear conjugated polymer backbones. Harrison and co-workers have begun to examine the friction in monolayers composed of alkynes, or model unsaturated hydrocarbon layers, using the AIREBO potential. Particular interest was given to the extent to which polymerization, and its position in the chain, affects the lateral stiffness of the film. Careful attention was also given to the way in which these factors might influence the frictional response of the films.

Preliminary results revealed that the application of load initiates a small degree of polymerization, or carbon–carbon bond formation, between chains. Increasing the load generally increased the degree of polymerization. This process was irreversible. That is, when the load was removed, the polymerization remained. When these films were in sliding contact with an amorphous DLC probe, the polymerization continued with increasing sliding distance [136].

7. FUTURE DIRECTIONS

Matching simulation time scales with experimental time scales is one of the most challenging issues related to MD simulations. Experimental sliding speeds are many orders of magnitude smaller than what is currently feasible with MD. The methodology adopted by Leng and Jiang is successful in bridging this gap. However, its use is restricted to systems at very low loads. A more powerful technique is one that could be flexible with respect to the kinds of systems that can be modeled. If phenomena such as wear, chemical restructuring of the tip, or even less dramatic phenomena such as penetration of soft tip asperities into SAM samples, are to be investigated, using fast sliding speeds may be unavoidable in MD simulations. By altering properties of the modeled cantilever assembly, such as the damping of the harmonic springs used to control the motion, mass of the support to which the springs and the tip atoms are attached, and the feedback loop (for regulating the position of the rigid support), it may be possible to reasonably duplicate the essential features of experimental systems using relatively high sliding speeds [87, 88].

Rather than trying to directly model an AFM cantilever, simulations can be conducted in the simpler constant height and constant load modes. In these limits, there is no special tuning of cantilever properties that requires consideration of the sliding speed. It is sufficient to simply use a sliding speed that is separated in scale from thermal speeds. Furthermore, such a simplified approach may help develop fundamental insights into friction mechanisms, particularly if both external force modes (constant height and constant load) are studied together over a wide range of average loads. The wide variety of computational studies undertaken to date hint at a rich variety of physical mechanisms that are relevant to the understanding of friction. However, the large differences in potential models, constraints, and methodologies make it difficult to establish an integrated view of the friction of SAMs.

In addition to the modeling of external forces, care must also be given to the structure of the tip. Comparable experimental and theoretical length scales are thus another challenging problem. The radius of curvature of scanning probe microscope tips is currently no less than about 150 Å. Depending upon the type of potentials used, it is not currently computationally feasible to do MD at this scale. Moreover, it is also not clear that this is the relevant length scale. It is likely that localized asperities from the tip in contact with SAM samples are responsible for the phenomena observed with the AFM. Not much is presently known, theoretically or experimentally, about the surface structure of AFM tips over lengths on the order of angstroms. To what degree is the surface atomically rough? To what degree are there chemically unsaturated atoms at the surface? These are crucial questions that are just beginning to be explored.

This review clearly shows that work in this field is still in its early stages. There are a number of exciting possibilities yet to explore. It is likely that realistic modeling of a variety of systems will require a potential that can address a wide range of phenomena, including the breaking and forming of bonds and nonbonded, long-range interactions. The AIREBO potential discussed above is such a potential, although it is currently limited to the treatment of hydrogen and carbon atoms. Recently, the model was extended to include fluorine, and it is reasonable to expect that other atom types can similarly be added [137]. But there are many kinds of interactions that cannot be addressed within the basic structure of this potential model. For instance, what is the role of humidity and other gases? Most AFM experiments are not conducted in vacuum. What about the chain–substrate interaction? The structure of $SH(CH_2)_{n-1}CH_3$ chains on gold, the most commonly studied SAM, is still an issue of debate. Recent experimental evidence suggests a dimer arrangement as opposed to the adsorption of sulfur atoms at the threefold hollow sites of the gold substrate [138]. It would be a very challenging task to model such a variety of interactions in a consistent manner, although it is likely that this variety is necessary to capture all of the major facets involved with an experimental AFM–SAM system.

While directly connecting with experimental AFM–SAM systems is a long-term goal, there is much room for fundamental studies that elucidate friction mechanisms without necessarily directly modeling experimental systems. The investigations undertaken thus far indicate a rich variety of phenomena. As computational power continues to grow, these investigations will also grow in breadth and in depth.

GLOSSARY

Adaptive integrated reactive empirical bond-order (AIREBO) potential Based on the REBO potential, this potential energy function is used to model physical and chemical reactions in hydrocarbon systems of all phases.

Alkanethiol Saturated hydrocarbon chain with a sulfur head group [$-SH(CH_s)_nCH_3$].

Atomic force microscopy (AFM) Probing a molecular surface using a movable cantilever, and obtaining atomic-scale force and position information.

Langmuir–Blodgett (LB) The method of applying a single-monolayer thickness of self-assembled alkanethiol-type chains, in an amphiphillic environment, to a substrate using compression of the molecules on a liquid.

Lennard–Jones (LJ) interactions A model for the repulsive and attractive interactions that occur between nonbonded atoms.

Molecular dynamics (MD) Computational method that calculates the atomic-scale trajectory of atoms as a function of time that are allowed to move according to Newton's equations of motion.

Nanotube A nanometer-scale folded carbon sheet, capped or uncapped, used to probe surface structure.

Reactive empirical bond-order (REBO) potential A potential energy function used to model physical and chemical reactions in gas- and solid-phase hydrocarbon systems.

Self-assembled monolayer (SAM) A film of hydrocarbon chains that are chemisorbed to a substrate by their sulfur- or silica-containing head groups.

ACKNOWLEDGMENTS

The authors' work was supported by the U.S. Office of Naval Research and the Air Force Office of Scientific Research under Contracts N0001403WX20412 and NMIPR035203542, respectively. The authors thank M. J. Stevens, G. S. Grest, and M. Chandross for providing a picture of their work, and R. W. Carpick, K. J. Wahl, M. Salmeron, S. S. Perry, and M. O. Robbins for helpful discussions.

REFERENCES

1. R. G. Nuzzo and D. L. Allara, *J. Am. Chem. Soc.* 105, 4481 (1983).
2. C. D. Bain, E. B. Troughton, Y.-T. Tao, J. Evall, G. M. Whitesides, and R. G. Nuzzo, *J. Am. Chem. Soc.* 111, 321 (1989).
3. J. P. Folkers, P. E. Laibinis, and G. M. Whitesides, *Langmuir* 8, 1330 (1992).
4. A. Ulman, *Chem. Rev.* 96, 1533 (1996).
5. F. Schreiber, *Prog. Surf. Sci.* 65, 151 (2000).
6. C. E. D. Chidsey and D. N. Loiacono, *Langmuir* 6, 682 (1990).
7. N. L. Abbott, J. P. Folkers, and G. M. Whitesides, *Science* 257, 1380 (1992).
8. D. L. Allara, *Biosens. Bioelectron.* 10, 771 (1995).
9. L. H. Dubois and R. G. Nuzzo, *Ann. Rev. Phys. Chem.* 43, 437 (1992).
10. G. M. Whitesides and P. E. Laibinis, *Langmuir* 6, 87 (1990).
11. S. D. Evans, R. Sharma, and A. Ulman, *Langmuir* 7, 156 (1991).
12. J. T. Woodward, M. L. Walker, C. W. Meuse, D. J. Vanderah, G. E. Poirier, and A. L. Plant, *Langmuir* 16, 5347 (2000).
13. G. E. Poirier, *Langmuir* 15, 1167 (1999).
14. G. E. Poirier, M. J. Tarlov, and H. E. Rushmeirer, *Langmuir* 10, 3383 (1994).
15. L. H. Dubois, B. R. Zegarski, and R. G. Nuzzo, *J. Chem. Phys.* 98, 678 (1993).
16. P. Fenter, P. Eisenberger, and K. S. Liang, *Phys. Rev. Lett.* 70, 2447 (1993).
17. L. H. Dubois, B. R. Zegarski, and R. G. Nuzzo, *Ann. Rev. Phys. Chem.* 43, 437 (1992).
18. N. Camillone, C. E. D. Chidsey, G.-Y. Liu, and G. J. Scoles, *J. Chem. Phys.* 94, 8493 (1991).
19. A. Badia, R. B. Lennox, and L. Reven, *Acc. Chem. Res.* 33, 475 (2000).
20. H. S. Kato, J. Noh, M. Hara, and M. Kawai, *J. Phys. Chem. B* 106, 9655 (2002).
21. C. G. Zeng, B. Wang, H. Q. Wang, K. D. Wang, J. L. Yang, J. G. Hou, and Q. S. Zhu, *J. Chem. Phys.* 117, 851 (2002).
22. J. Hautman and M. L. Klein, *J. Chem. Phys.* 91, 4994 (1989).
23. J. Hautman and M. L. Klein, *J. Chem. Phys.* 93, 7483 (1990).
24. J. P. Bareman and M. L. Klein, *J. Phys. Chem.* 94, 5202 (1990).
25. J. Hautman, J. P. Bareman, W. Mar, and M. L. Klein, *J. Chem. Soc., Faraday Trans.* 87, 2031 (1991).
26. J. I. Siepmann and I. R. McDonald, *Mol. Phys.* 75, 255 (1992).
27. J. I. Siepmann and I. R. McDonald, *Mol. Phys.* 79, 457 (1993).
28. W. Mar and M. L. Klein, *Langmuir* 10, 188 (1994).
29. L. Xing and W. L. Mattice, *Langmuir* 12, 3024 (1996).
30. R. Bhatia and B. J. Garrison, *Langmuir* 13, 765 (1997).
31. R. Bhatia and B. J. Garrison, *Langmuir* 13, 4038 (1997).
32. W. D. Luedtke and U. Landman, *J. Phys. Chem.* 102, 6566 (1998).
33. Z. Zhang, T. L. Beck, J. T. Young, and F. J. Boerio, *Langmuir* 12, 1227 (1996).
34. R. D. Mountain, J. B. Hubbard, C. W. Meuse, and V. Simmons, *J. Phys. Chem. B* 105, 9503 (2001).
35. L. Zhang, K. Wesley, and S. Jiang, *Langmuir* 17, 2675 (2001).
36. J. J. Gerdy and W. A. Goddard, *J. Am. Chem. Soc.* 118, 3233 (1996).
37. L. Z. Zhang, W. A. Goddard, and S. Y. Jiang, *J. Chem. Phys.* 117, 7342 (2002).
38. A. Pertsin and M. Grunze, *Langmuir* 10, 3668 (1994).
39. J. P. Ryckaert, I. R. McDonald, and M. L. Klein, *Mol. Phys.* 67, 957 (1989).
40. C. H. Mastrangelo, in "Tribology Issues and Opportunities in MEMS" (B. Bhushan, Ed.), p. 367. Kluwer Academic, Norwell, MA, 1998.
41. U. Srinivasan, R. T. Howe, and R. Madoudian, in "Tribology Issues and Opportunities in MEMS" (B. Bhushan, Ed.), p. 597. Kluwer Academic, Norwell, MA, 1998.
42. M. Dugger, in "Nanotribology: Critical Assessment and Research Needs" (S. M. Hsu and Z. C. Ying, Eds.), p. 123. Kluwer Academic, Norwell, MA, 2002.
43. V. V. Tsukruk, in "Nanotribology: Critical Assessment and Research Needs" (S. M. Hsu and Z. C. Ying, Eds.), p. 347. Kluwer Academic, Norwell, MA, 2002.
44. R. Madoudian, *Mater. Res. Soc. Bull.* 23, 47 (1998).
45. T. E. Karis, in "Nanotribology: Critical Assessment and Research Needs" (S. M. Hsu and Z. C. Ying, Eds.), p. 291. Kluwer Academic, Norwell, MA, 2002.
46. S. S. Perry, G. A. Somorjai, C. M. Mate, and R. L. White, *Tribol. Lett.* 1, 47 (1995).
47. S. S. Perry, G. A. Somorjai, C. M. Mate, and R. L. White, *Tribol. Lett.* 1, 233 (1995).
48. E. Barrena, C. Ocal, and M. Salmeron, *Surf. Sci.* 482–485, 1216 (2001).
49. M. Salmeron, *Tribol. Lett.* 10, 69 (2001).
50. S. Lee, Y. S. Shon, R. L. Guenard, T. R. Lee, and S. S. Perry, *Langmuir* 16, 2220 (2000).
51. Y. S. Shon, S. Lee, R. Colorado, S. S. Perry, and T. R. Lee, *J. Am. Chem. Soc.* 122, 7556 (2000).
52. E. Barrena, C. Ocal, and M. Salmeron, *J. Chem. Phys.* 113, 2413 (2000).
53. B. D. Beake and G. J. Leggett, *Langmuir* 16, 735 (2000).
54. B. D. Beake and G. J. Leggett, *Phys. Chem. Chem. Phys.* 1, 3345 (1999).
55. J.D. Kiely and J.E. Houston, *Langmuir* 15, 4513 (1999).
56. E. Cooper and G. J. Leggett, *Langmuir* 15, 1024 (1999).
57. A. R. Burns, J. E. Houston, R. W. Carpick, and T. A. Michalske, *Phys. Rev. Lett.* 82, 1181 (1999).
58. E. Barrena, S. Kopta, D. F. Ogletree, D. H. Charych, and M. Salmeron, *Phys. Rev. Lett.* 82, 2880 (1999).
59. E. Barrena, C. Ocal, and M. Salmeron, *J. Chem. Phys.* 111, 9797 (1999).
60. H. I. Kim, M. Graupe, O. Oloba, T. Doini, S. Imaduddin, T. R. Lee, and S. S. Perry, *Langmuir* 15, 3179 (1999).
61. S. S. Wong, H. Takano, and M. D. Porter, *Anal. Chem.* 70, 5200 (1998).
62. J. A. Harrison and S. S. Perry, *MRS Bull.* 23, 27 (1998).
63. A. Lio, D. H. Charych, and M. Salmeron, *J. Phys. Chem. B* 101, 3800 (1997).
64. A. Lio, C. Morant, D. Ogletree, and M. Salmeron, *J. Phys. Chem. B* 101, 4767 (1997).
65. R. W. Carpick and M. Salmeron, *Chem. Rev.* 97, 1163 (1997) and references therein.
66. H. I. Kim, T. Koini, T. R. Lee, and S. S. Perry, *Langmuir* 13, 7192 (1997).
67. X. Xiao, J. Hu, D. H. Charych, and M. Salmeron, *Langmuir* 12, 235 (1996).

68. M. T. McDermott, J.-B. D. Green, and M. D. Porter, *Langmuir* 13, 2504 (1997).
69. Y. Liu, D. F. Evans, Q. Song, and D. W. Grainger, *Langmuir* 12, 1235 (1996).
70. S. S. Perry, S. Lee, Y. S. Shon, R. Colorado, and T. R. Lee, *Tribol. Lett.* 10, 81 (2001).
71. M. Garcia-Parajo, C. Longo, J. Servat, P. Gorostiza, and F. Sanz, *Langmuir* 13, 2333 (1997).
72. M. O. Robbins and M. H. Müser, in "Modern Tribology Handbook" (B. Bhushan, Ed.), Vol. 1, p. 717. CRC Press, Boca Raton, FL, 2001.
73. J. A. Harrison, S. S. Stuart, and D. W. Brenner, in "Handbook of Micro/Nano Tribology" 2nd ed. (B. Bhushan, Ed.), p. 525. CRC Press, Boca Raton, FL, 1998.
74. K. J. Tupper and D. W. Brenner, *Langmuir* 10, 2335 (1994).
75. K. J. Tupper, R. J. Colton, and D. W. Brenner, *Langmuir* 10, 2041 (1994).
76. K. J. Tupper and D. W. Brenner, *Thin Solid Films* 253, 185 (1994).
77. J. N. Glosli and G. M. McClelland, *Phys. Rev. Lett.* 70, 1960 (1993).
78. J. N. Israelachvili, A. M. Homola, and P. M. McGuiggan, *Science* 240, 189 (1988).
79. T. Bonner and A. Baratoff, *Surf. Sci.* 377–379, 1082 (1997).
80. A. Koike and M. Yoneya, *Langmuir* 13, 1718 (1997).
81. A. Koike and M. Yoneya, *J. Chem. Phys.* 105, 6060 (1996).
82. T. Ohzono, J. N. Glosli, and M. Fujihira, *Jpn. J. Appl. Phys.* 37, 6535 (1998).
83. M. Fujihira and T. Ohzono, *Jpn. J. Appl. Phys.* 38, 3918 (1999).
84. T. Ohzono, J. N. Glosli, and M. Fuguhira, *Jpn. J. Appl. Phys.* 38, L675 (1999).
85. T. Ohzono and M. Fujihira, *Tribol. Lett.* 9, 63 (2000).
86. T. Ohzono and M. Fujihira, *Phys. Rev. B* 62, 17055 (2000).
87. Y. S. Leng and S. Jiang, *J. Am. Chem. Soc.* 124, 11764 (2002).
88. Y. S. Leng and S. Jiang, *Phys. Rev. B* 63, 193406 (2001).
89. Y. S. Leng and S. Jiang, *Phys. Rev. B* 64, 115415 (2001).
90. Y. S. Leng and S. Jiang, *J. Chem. Phys.* 113, 8800 (2000).
91. Y. S. Leng and S. Jiang, *Tribol. Lett.* 11, 111 (2001).
92. L. Li, S. Chen, and S. Jiang, in "Interfacial Properties on the Submicron Scale" (J. E. Frommer and R. Overney, Eds.), p. 168. ACS Press, Washington, DC, 2001.
93. J.-P. Ryckaert and A. Bellemans, *J. Chem. Soc., Faraday Discuss.* 66, 95 (1978).
94. P. van der Ploeg and H. J. C. Berendsen, *Mol. Phys.* 49, 233 (1983).
95. J. P. Ryckaert and M. L. Klein, *J. Chem. Phys.* 85, 1613 (1986).
96. J. P. Ryckaert, M. L. Klein, and I. R. McDonald, *Phys. Rev. Lett.* 58, 698 (1987).
97. P. van der Ploeg and J. J. C. Berendsen, *J. Chem. Phys.* 76, 3271 (1982).
98. J. Harris and S. A. Rice, *J. Chem. Phys.* 89, 5898 (1988).
99. J. P. Bareman, G. Cardini, and M. L. Klein, in "Atomic Scale Calculations in Materials Science" (J. Tersoff, D. Vanderbilt, and V. Vitek, Eds.), MRS Symposium Series, Vol. 141, p. 411. Materials Research Society, Pittsburgh, 1989.
100. J. P. Bareman, G. Cardini, and M. L. Klein, *Phys. Rev. Lett.* 60, 2152 (1988).
101. W. L. Jorgenson, *J. Phys. Chem.* 90, 6379 (1986).
102. W. L. Jorgenson and J. Tirado-Rives, *J. Am. Chem. Soc.* 110, 1657 (1988).
103. J. P. Ryckaert, G. Ciccotti, and H. J. C. Berendsen, *J. Comput. Phys.* 23, 327 (1977).
104. W. Steele, "The Interaction of Gases with Solid Surfaces." Pergamon, New York, 1974.
105. G. Cardini, S. F. O'Shea, and M. L. Klein, *J. Chem. Soc., Faraday Discuss.* 80, 227 (1985).
106. S. Karaborni, *Langmuir* 9, 1334 (1993).
107. S. Karaborni, *Phys. Rev. Lett.* 73, 1668 (1994).
108. J. T. Buontempo, S. A. Rice, S. Karaborni, and J. I. Siepmann, *Langmuir* 9, 1604 (1993).
109. A. B. Tutein, S. J. Stuart, and J. A. Harrison, *J. Phys. Chem. B* 103, 11357 (1999).
110. M. Chandross, G. S. Grest, and M. J. Stevens, *Langmuir* 18, 8392 (2002).
111. M. J. Stevens, *Langmuir* 15, 2773 (1999).
112. A. B. Sieval, B. van den Hout, H. Zuilhof, and E. J. R. Sudhölter, *Langmuir* 16, 2987 (2000).
113. S. Rovillard, E. Perez, R. Ionov, M. Voue, and J. De Coninck, *Langmuir* 15, 2749 (1999).
114. T.-W. Li, I. Chao, and Y.-T. Tao, *J. Phys. Chem. B* 102, 2935 (1998).
115. K.-Q. Yu, Z.-S. Li, and J.-Z. Sun, *Langmuir* 18, 1419 (2002).
116. G. E. Poirier and M. J. Tarlov, *Langmuir* 10, 2853 (1994).
117. R. G. Nuzzo, E. M. Korenic, and L. H. Dubois, *J. Chem. Phys.* 93, 767 (1990).
118. P. Fenter, A. Eberhardt, and P. Eisenberger, *Science* 266, 1216 (1994).
119. P. T. Mikulski and J. A. Harrison, *J. Am. Chem. Soc.* 123, 6873 (2001).
120. P. T. Mikulski and J. A. Harrison, *Tribol. Lett.* 10, 29 (2001).
121. J. A. Harrison, S. J. Stuart, and A. B. Tutein, in "Interfacial Properties on the Submicron Scale" (J. E. Frommer and R. Overney, Eds.), p. 216. ACS Press, Washington, DC, 2001.
122. A. B. Tutein, S. J. Stuart, and J. A. Harrison, *Langmuir* 16, 291 (2000).
123. J. A. Harrison, P. T. Mikulski, S. J. Stuart, and A. B. Tutein, in "Nanotribology: Critical Assessment and Research Needs" (S. M. Hsu and Z. C. Ying, Eds.), p. 55. Kluwer Academic, Norwell, MA, 2002.
124. J. A. Harrison, P. T. Mikulski, S. J. Stuart, and A. B. Tutein (in preparation).
125. S. J. Stuart, A. B. Tutein, and J. A. Harrison, *J. Chem. Phys.* 112, 6472 (2000).
126. D. W. Brenner, *Phys. Rev. B* 42, 9458 (1990).
127. D. W. Brenner, O. A. Shenderova, J. A. Harrison, S. J. Stuart, B. Ni, and S. B. Sinnott, *J. Phys. C* 14, 783 (2002).
128. J. A. Harrison and D. W. Brenner, *J. Am. Chem. Soc.* 116, 10399 (1994).
129. B. Ni, S. B. Sinnott, P. T. Mikulski, and J. A. Harrison, *Phys. Rev. Lett.* 88, 2055051 (2002).
130. D. J. Diestler, G. T. Gao, and X. C. Zeng, *Phys. Chem. Chem. Phys.* 3, 1175 (2001).
131. D. J. Diestler, *J. Chem. Phys.* 117, 3411 (2002).
132. N. D. Shinn, T. M. Mayer, and T. A. Michalske, *Tribol. Lett.* 7, 67 (1999).
133. S. S. Wong, J. D. Harper, J. P. T. Lansbury, and C. M. Lieber, *J. Am. Chem. Soc.* 120, 603 (1998).
134. A. Erdemir and C. Donnet, in "Modern Tribology Handbook" (B. Bhushan, Ed.), Vol. 2, p. 871. CRC Press, Boca Raton, FL, 2001.
135. M. D. Mowery, S. Kopta, D. F. Ogletree, M. Salmeron, and C. E. Evans, *Langmuir* 15, 5118 (1999).
136. G. M. Chateauneuf, G. Gao, P. T. Mikulski, and J. A. Harrison (in preparation).
137. S. J. Stuart, L. Hu, and B. M. Dickson (in preparation).
138. G. J. Kluth, C. Carraro, and R. Maboudian, *Phys. Rev. B* 59, 10449 (1999).
139. For details regarding the nanotube probe, see: J. A. Harrison, S. J. Stuart, D. H. Robertson, and C. T. White, *J. Phys. Chem. B* 101, 9682 (1997).

Encyclopedia of Nanoscience and Nanotechnology
www.aspbs.com/enn

Fullerene Dimers

Hidekazu Shimotani
University of Tokyo, Kashiwa-shi, Chiba, Japan

Nita Dragoe
Université Paris-Sud, Orsay, France

Koichi Kitazawa
Japanese Science and Technology Corporation, Kawaguchi, Saitama, Japan

CONTENTS

1. INTRODUCTION

This chapter outlines the main synthesis routes, the structural features, and the physicochemical properties of fullerene dimers. This relatively simple class of materials proved to be rather difficult to study owing, in general, to their low solubility. The interest manifested in these materials is related to their promising electronic and optical properties, and in particular to the intramolecular charge transfer.

The discovery of fullerenes [1] by mass spectroscopy in the mid-1980s attracted wide interest, both from physicists and chemists. Soon thereafter, fullerenes in isolable amounts were obtained by a procedure developed by Krätschmer et al. [2]. This is the point when theoretical works, the prior predominant ones, could be thoroughly checked by experiments. After the initial frenzy when fullerenes were believed to be useful for just about anything, a concept encouraged by the unexpected properties observed for the new entry in the carbon–allotropes family, a more systematic work was started. The properties of fullerenes and derivatives, such as superconductivity [3], ferromagnetism [4], nonlinear optical properties [5], or even HIV protease blocking [6], are now generally understood as a result of their structure. The famous member of the fullerene family C_{60}, the soccer ball (**1**), was the most analyzed, and this is undoubtedly due to its relatively high abundance, its unique spherical shape, and high symmetry I_h. Fullerene dimers were reported soon after the synthesis of the new monomers. In the search for other properties of fullerene derivatives and an understanding of their structure–properties relationship, systematic synthetic organic chemistry work was done (there are fewer inorganic chemistry works in this field, although several works, for instance by Fagan et al. [7], are famous for proving the electronic nature of C_{60}). The chemistry of fullerenes is today a mature field; the interested reader can consult the works of Diederich and Thilgen [8], Hirsch [9], or Taylor [10, 11]. In spite of sustained research activity in the field of fullerenes, there have not been many fullerene dimers synthesized up to now. This apparent lack of interest in dimeric fullerenes is partly related to the general low solubility of fullerene dimers, and hence, to the difficulty of their synthesis and purification, as well as partly to the difficulty of separating pure samples from the multitude of the isomers obtained. In this sense, experimental methods rarely used by chemists for structural characterization ought to be used, for instance, scanning tunneling microscopy (STM). There is a large variety in the synthesis methods of these molecules, and these will be described along with their structures. We will classify these molecules into two main categories: with one or more direct connections between the fullerene cages, and without direct connections between the fullerene cages. Although this separation seems artificial, it points out the different electronic properties which can be expected from these materials: intercage electronic interaction in one case,

ISBN: 1-58883-059-4/$35.00

Encyclopedia of Nanoscience and Nanotechnology
Edited by H. S. Nalwa
Volume 3: Pages (529–544)

and donor/acceptor electronic transfer in the other. Aside from this classification, we will divide our discussion into other two categories according to the presence or absence of atoms other than the carbon. The latter category is relatively undeveloped, that of all-carbon fullerene dimers, new molecular carbon allotropes [12]. We will also include in our discussion molecules which are not, strictly speaking, dimers, such as C_{119} [13]. However, this is an outstanding example of the synthesis of new molecules starting from a fullerene dimer, and we will detail here all compounds reported in the literature having two fullerene-like cages, even nonequivalent. Most of the compounds synthesized are dimers of C_{60}; only a few examples are known for other fullerenes.

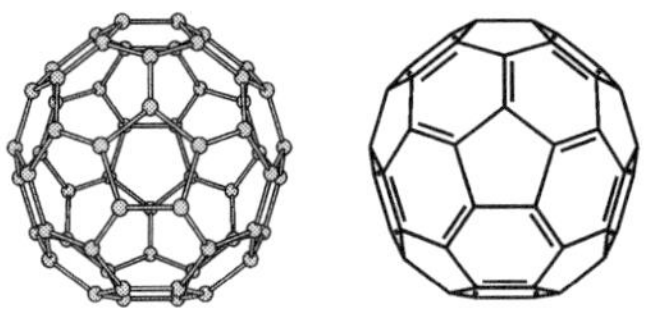

1

We will focus our attention on the all-carbon fullerene derivatives, not only because they might be considered as new allotropes, but also because they may be used for the synthesis of endohedral, higher fullerenes, or even nanotubes by a "coalescence" reaction; these possibilities will be discussed later. Another interesting feature of the dimers is that their electronic behavior should remain largely the same as that of fullerenes, while gaining other characteristics as a result of the intercage couplings. As an example, $C_{120}O$ and other dimers have been found to manifest an electronic communication between the cages [14] (see Section 2). C_{60} dimers also have been claimed to offer a route to the synthesis of endohedral fullerene compounds [15], and *ab initio* and density functional theory (DFT) calculations have been employed to find a reaction path toward the formation of endohedral fullerene compounds. The proposed mechanism involves a [2 + 2] cycloaddition yielding C_{120}, while it allows the formation of a stable "window" through which an atom can enter one of the cages. Finally, fullerene dimers have been sought as proofs for the structures of more complex molecules, for instance, fullerene polymers. In an exceptional example, C_{120} has been shown to be the prototype of a C_{60} polymer, synthesized, and whose structure has been correctly assigned by Rao et al. [16].

As one can see, these molecules have interesting structures, modulable properties, can be used as starting materials for the synthesis of other materials, and can serve as structural prototypes.

2. SYNTHESIS OF FULLERENE DIMERS

2.1. Dimers with Direct Connections Between Fullerene Cages

2.1.1. All-Carbon Fullerene Dimers

Special attention is paid to the all-carbon fullerene dimers because they may be used for the synthesis of higher fullerenes, for instance, by a coalescence reaction [17] observed to take place under mass spectroscopy (MS) conditions. This reaction was not experimentally proven on a macroscopic scale, but such an opportunity exists and may offer a "rational" approach to the synthesis of higher fullerenes.

In the family of all-carbon dimers with a direct connection between the fullerene cages, the most prominent member is C_{120} (**2**) [18]; synthesized by an unusual mechanochemical reaction, it was rather fully characterized. The synthesis involves milling C_{60} in the presence of a catalyst such as KCN. A possible reaction mechanism is shown in Scheme 1. The method used by Komatsu involved high-speed vibration milling (HSVM), but this dimer can also be obtained by manually grinding C_{60} and lithium under inert atmosphere [19, 20], high-temperature phototransformation [21], hydrostatic pressure [22], or by pressurization of a template compound [23]. By the latter method, with pressurization under 5 GPa at 200 °C of an $(ET)_2C_{60}$ organic molecular crystal, yields of about 80% have been obtained. A detailed analysis of the infrared (IR) signatures of (**2**), the corresponding ultraviolet (UV) polymer, and their assignments by molecular dynamics (MD) calculations have been published recently [24].

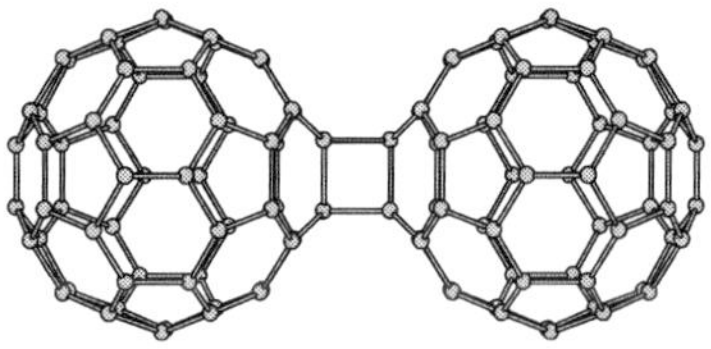

2

When the starting material is endohedral[60]fullerene, which is C_{60} encapsulating one or more atoms, the resulting C_{120} will contain one or more atoms in C_{60} cages. Because amounts of the available endohedral[60]fullerenes are very small, a mixture of hollow C_{60} and endohedral[60]fullerenes was used for the synthesis. Thus, only one of the two C_{60} cages of reported endohedral fullerene dimers contains an atom. The reported endohedral fullerene dimers are $^3He@C_{120}$ [25] and $N@C_{120}$ [26]. The encapsulated ^3He and N atoms can be used as labeling because the environment of the ^3He and N atoms in the dimers can be monitored by ^3He–NMR (nuclear magnetic resonance) and electron paramagnetic resonance (EPR), respectively.

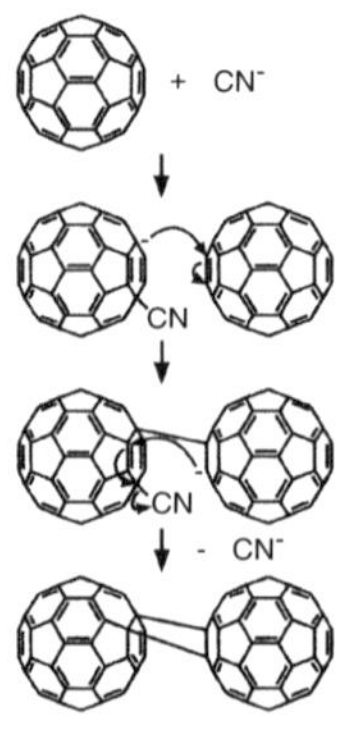

Scheme 1. Possible mechanism of formation of C_{120}.

There is another recent report on the synthesis of a different kind of dumbbell C_{120} dimer, in which the connection between two fullerenes is made face to face on two hexagons [27], but we believe that there is not enough evidence to prove this structure. This face-to-face dimer would be more likely to give a coalescence reaction, and the authors of this report discussed such a possibility.

Following pioneering work by McElvany et al. [28] who observed a C_{119} cluster by mass spectroscopy, Krätschmer and co-workers [29] isolated two isomers of a different molecule which could be assigned to C_{119} (**3**). Of these two isomers, subtle experimental analysis clearly identified one of the structures of these two compounds as (**3a**). This compound can best be described as a dimer of two 58-atoms clusters, connected through a three-carbon bridge. The synthesis method of (**3**) involved the solid-state heating at 550–600 °C of $C_{120}O$ under argon; hence, presumably, the mechanism involved the elimination of a CO molecule, followed by a reorganization of the C_{119} fragment. This is a nice example of overcoming the high energy barrier required to remove a carbon atom from a C_{60} cluster, and this method could allow the synthesis of other clusters (it has been attempted, unsuccessfully, for the decomposition of $C_{120}OS$) [30]. According to McElvany's work, C_{129} and C_{139} can also be obtained; it is likely that the same solid-state procedure will lead to a mixture of these compounds, but no attempts to obtain these compounds were made until now.

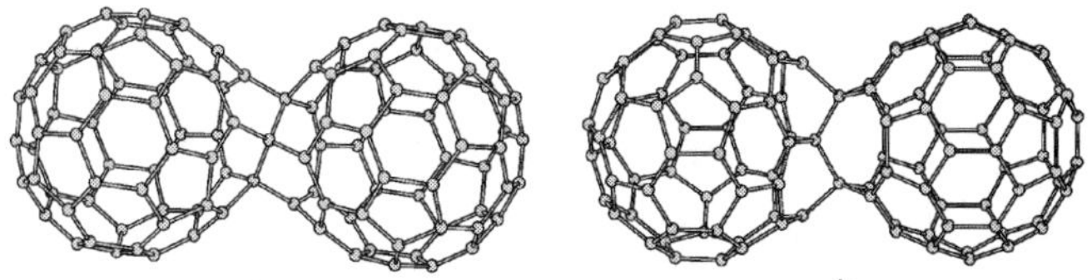

a b

3

Among other possible dimeric structures, theoretical calculations suggest that a pentagon-like linked structure for C_{121} should be stable [31].

2.1.2. Fullerene Dimers Containing Heterogeneous Atoms

As expected, in this category, many more dimers can be included. One of the most interesting compounds in this class is $C_{120}O$ (**4**); it has been used for the synthesis of other compounds, among them the intriguing C_{119}, and it was the first dimer to show an intercage electronic communication. On the other hand, as found by Taylor, this dimer is present as impurity in all samples of C_{60} exposed to light and air. This had major implications on the EPR results published so far, as was clearly demonstrated by Reed and co-workers [32].

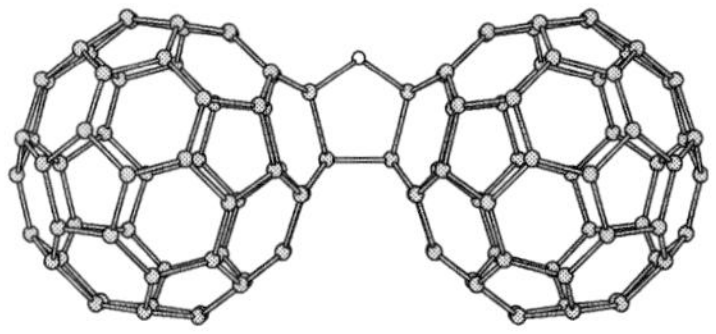

4

Independently, Krätschmer and co-workers [33] and Smith et al. [34] obtained $C_{120}O$ by a thermolysis of a mixture of $C_{60}O$ and C_{60} in solid state and in solution, respectively. The two fullerene cages are directly connected, forming a furan-like ring. It is noteworthy that $C_{120}O$ also has been isolated from the as-purchased C_{60} from different providers by Taylor et al. [35]. They also noted that C_{60} "degrades" to $C_{120}O$ under the influence of light, in solid state, in the presence of oxygen. It is likely that this involves the formation of the $C_{60}O$ which is unstable relative to $C_{120}O$.

Using the same procedure, by the thermolysis of $C_{61}H_2$ in the presence of C_{60}, Smith et al. synthesized an analog dimer, $C_{120}CH_2$ (**5**) [34], in which a cyclopentane connects two fullerene cages. It is likely that the same experimental procedure can lead to higher oligomers. A different approach for the synthesis of this dimer was reported by Fujiwara and Komatsu [36] by using a mechanochemical solid-state reaction. The dimer is bridged by a silicon atom (**6**). The mechanochemical reaction was performed on a mixture of C_{60}, dichlorodiphenylsilane, and lithium powder under HSVM conditions. This dimer is sensitive to moisture in solution, which transforms it into C_{60} and $C_{60}H_4$.

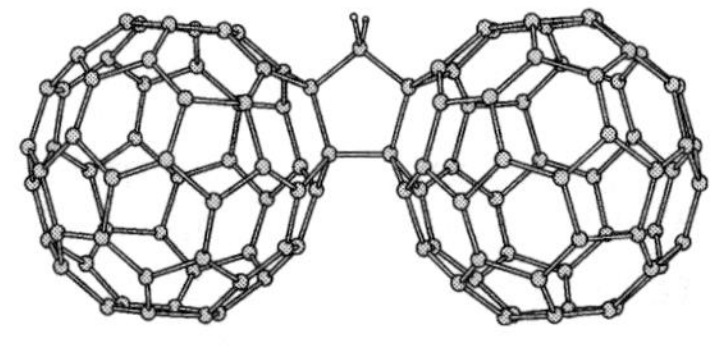

5

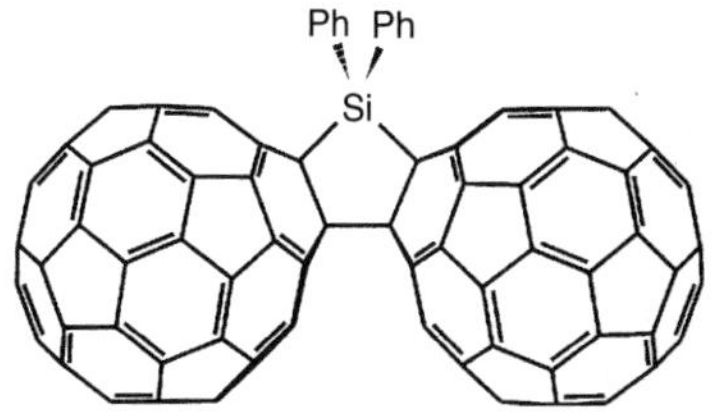

6

The $C_{60}/C_{60}O/C_{60}O_2$ system was used to synthesize $C_{120}O_2$ (**7**) by Krätschmer et al. [37, 38]. A highly symmetrical $C_{120}O_2$ ($C_{2\nu}$) has two furanoid links between the cages, the bridges connecting to 6:6 positions on the fullerene core, and similarly with the $C_{120}OS$ (**8**) structure synthesized by the same solid-state technique [39]. The bridging pattern involves four sp^3 carbons on each of the fullerene cages, and there are several possibilities for obtaining this. The most stable structure is the one having the highest number of double bonds at the 6:6 bond (see the discussion on $C_{122}H_4$ below). Lower symmetry isomers of $C_{120}O_2$ (C_1) have been synthesized by Gromov et al. [40]; a *cis*–1 *syn* compound is shown in Figure 1.

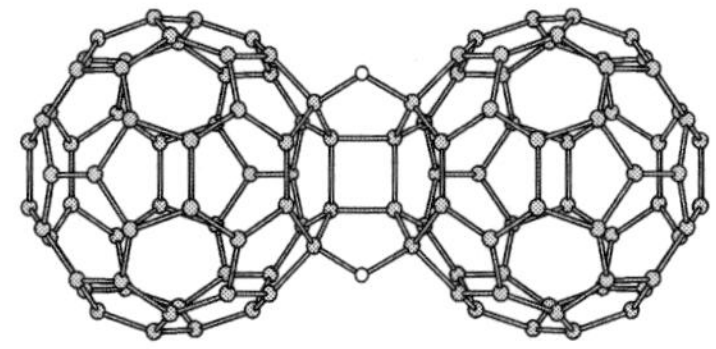

7

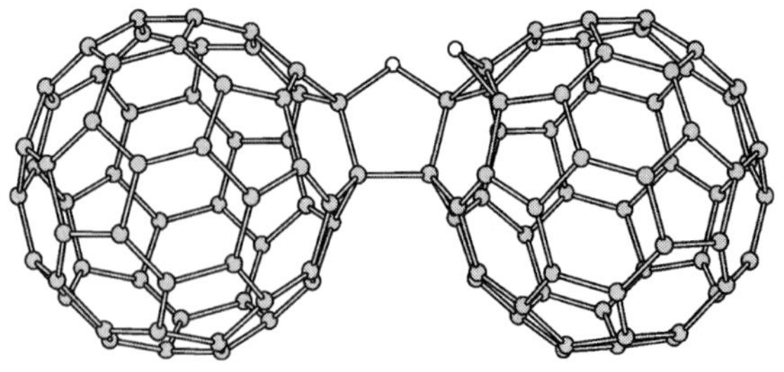

Figure 1. Molecular structure of *cis*–1 *syn* isomer of $C_{120}O_2$.

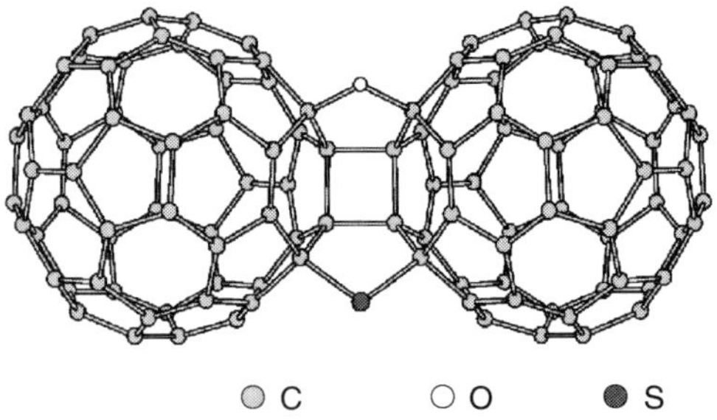

8

Similar in structure to $C_{120}O$ and $C_{120}CH_2$, a new dimer has been isolated from the product of thermolysis of cyclopropane derivatives. The thermolysis of methano[60]fullerenes containing hydrogen atoms such as ethoxycarbonylmethano[60]fullerene (**9**) under inert atmosphere yields $C_{122}H_4$ (**10**) [41]. This dimer has a similar structure to $C_{120}O_2$ and $C_{120}OS$, that is, two direct intercage bonds forming a part of a cyclobutane ring and two indirect intercage links. The cyclobutane edges belonging to C_{60} are located between a five-membered ring and a six-membered ring. The CH_2 bridgehead atoms are connected to the five-membered ring.

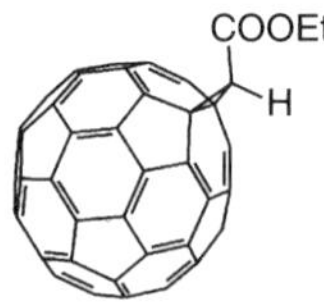

9

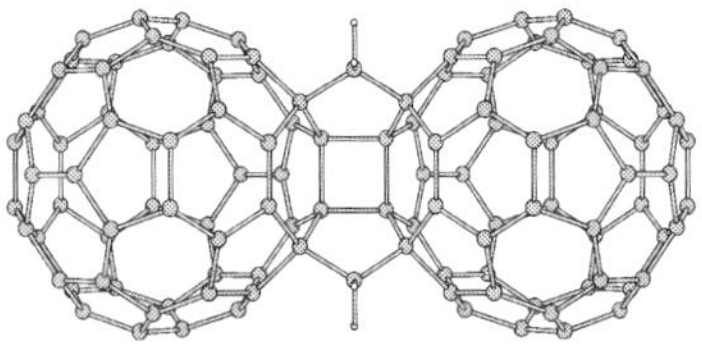

10

Another interesting dimer in this category is $(C_{59}N)_2$ (**11**), which has been obtained by Wudl and co-workers [42]. The dimer was synthesized by "organic chemical methods" (Scheme 2). The first step was the 1,3–dipolar cycloaddition of methoxyethoxymethylazide (MEM–N_3) to C_{60}, followed by the photooxygenation of the as-obtained 6:5–azahomo[60]fullerene; this yields a C_{60}–N–MEM–ketolactam. Finally, the C_{60}–N–MEM–ketolactam was treated with a large excess of *p*–toluenesulfonic acid monohydrate in *o*–dichlorobenzene. From this dimer, hydoroaza[60]fullerene ($C_{59}HN$, **12**) was obtained by irradiation through a Kapton filter in *o*–dichlorobenzene in the presence of tributyltin hydride in argon at 25–35 °C [43]. A recent discussion concerning this dimer is due to Reuther and Hirsch [44]. It is also notable that $(C_{59}N)_2$ can be obtained as a byproduct during the synthesis of a nitrogen endohedral fullerene [45].

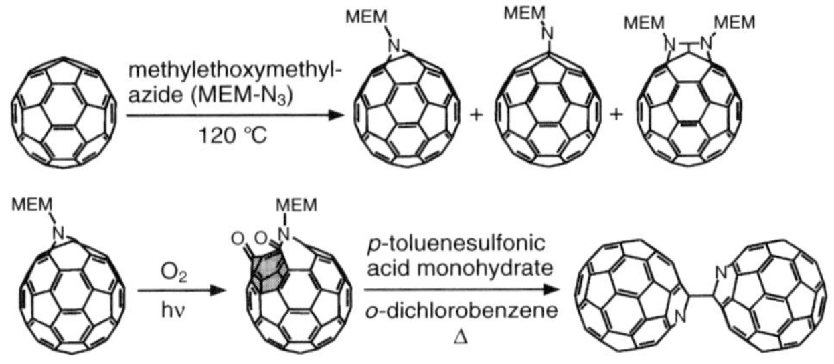

Scheme 2. Synthesis of $(C_{59}N)_2$.

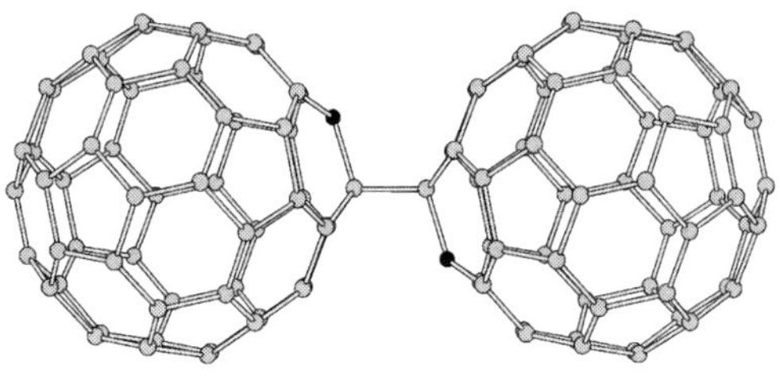

11

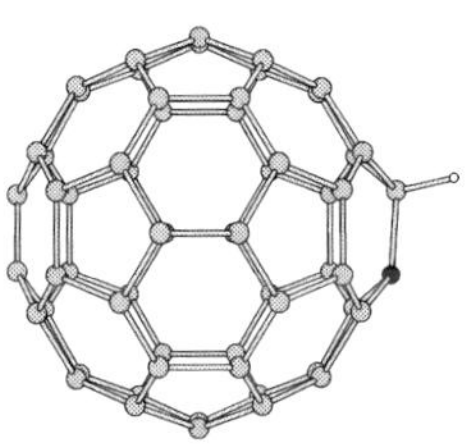

12

One of the earliest directly connected dimers was synthesized by Fagan et al. [46] during the study of a *t*–BuLi addition to C_{60} (Fig. 2). The synthesis of this class of compounds is quite straightforward, as a radicalic coupling of fullerene derivatives (Scheme 3). Several dimers have been obtained by this method [47], and it is notable that these dimers are in equilibrium with the C_{60} radicals [48]. The addition pattern to forming these dimers is, in general, in 1,4 on the C_{60} cage, regardless of the energetically more favorable position on 1,2; this is due to the strong steric interaction favoring the former addition pattern. An interesting topic for the study of these dimers is the energy barrier for the rotation function of the addends; several calculations have been performed on this [49].

Another single-bonded dimer was obtained by Schick et al. [50] by reacting C_{60} in air-saturated benzene with a large excess of morpholine. The obtained precipitate, 50% yield, is a dimer bis–[(4,4′–dimorpholino)–1,4–dihydro[60]fullerenyl] embodying two C_{60} cages directly connected and having two morpholine addends in the 1,4

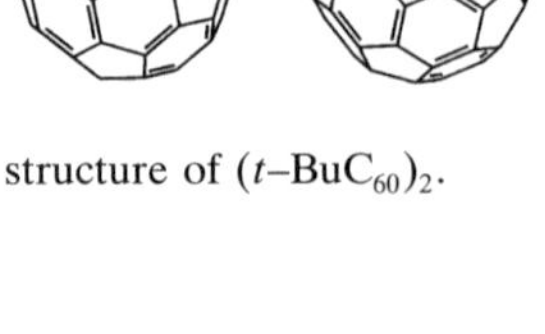

Figure 2. Molecular structure of $(t\text{–BuC}_{60})_2$.

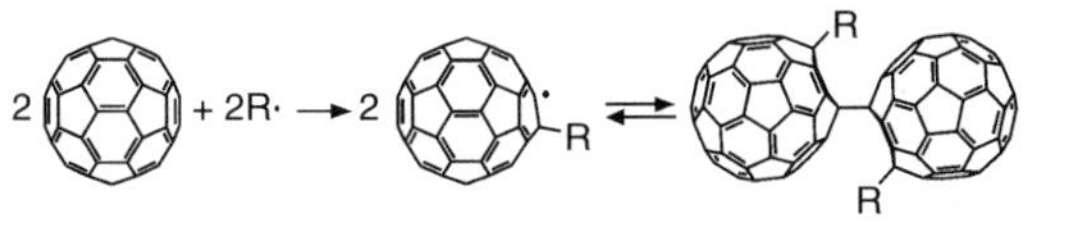

Scheme 3. Synthesis of $(RC_{60})_2$.

pattern. The formation of this structural type is also determined by the presence of aminated C_{60} radicals.

2.2. Dimers Without Direct Connections Between Fullerene Cages

2.2.1. All-Carbon Fullerene Dimers

In the family of all-carbon dimers without a direct connection between fullerene cages, we include C_{121} (**13a, 13b**), C_{122} [51] (**14**), and $C_{n(60+5)}$ [52] (**15**). The case of $C_{n(60+5)}$ is very interesting, and its synthesis is ingenious. The authors [52] were interested in the synthesis of molecular carbon allotropes of the fullerene–acetylene hybrid type (**16**). The synthesis involves oxidative coupling of diethynylmethano[60]fullerene. Limited by the solubility problems, the authors adopted a tether synthesis strategy in which, having attached functional groups to the fullerene core, the solubility problem was avoided.

C_{121} can be described as a dumbbell-like dimer or more appealing as a bis([60]fullerene)methane. A connection of two fullerenes through one carbon atom gives D_{2d} (**13a**) and C_s (**13b**) symmetry fullerene dimers. Similar in structure and formation is C_{122}, in which an ethylene group connects two fullerenes, giving a D_{2h} symmetry molecule. The latter was first reported by Strongin and co-workers [53] by the reaction of an "atomic carbon" with a fullerene. The ingenious idea of making all-carbon allotropes by the reaction of C_{60} with "carbon" is unfortunately impeded by the use of dangerous chemicals.

C_{121}, C_{122}, and other trimers [54] can be obtained by a solid-state decomposition of methano[60]fullerenes. The synthesis of **13a**, **13b**, and **14** is based on the formation of a methanofullerenecarbene as an intermediate, as proposed earlier by Osterodt and Vögtle [55] (Scheme 4). By using the same strategy, higher oligomers of the all-carbon type can also be obtained [54].

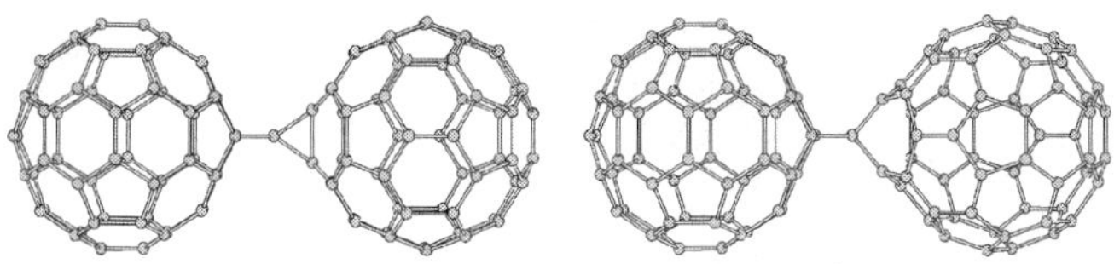

a b

13

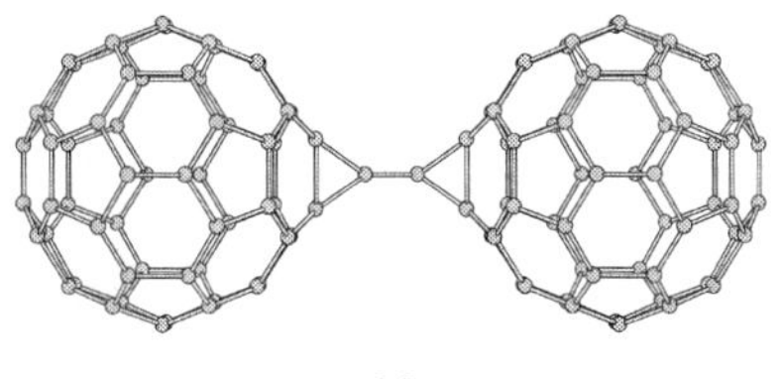

14

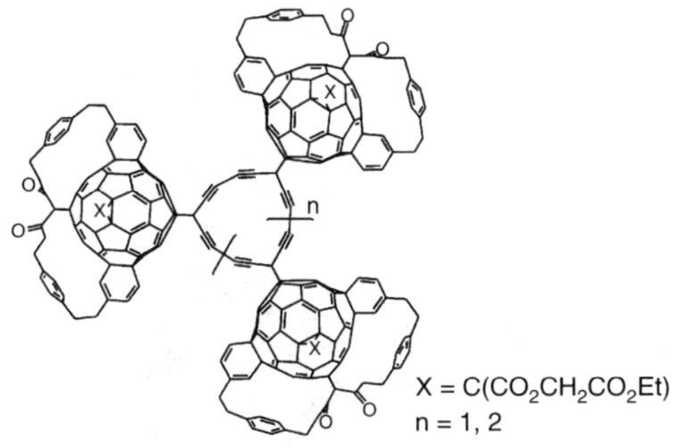

15

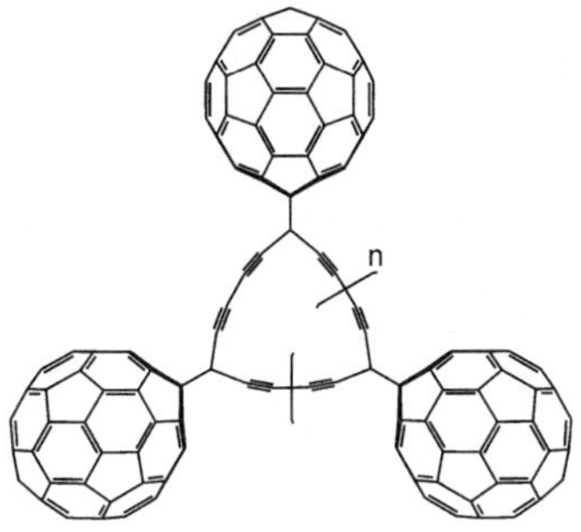

16

2.2.2. Fullerene Dimers Containing Heterogeneous Atoms

This kind of dimer is relatively easy to synthesize, compared with all previous ones, and several such dimers have been characterized. The ease of synthesis is also supported by a rather high solubility, depending on the flexibility of the linkage. According to Mullen and co-workers [56], about 1 g of such a compound can be synthesized and separated in one day on a laboratory scale. Comparatively, in our experience, the isolation of a few milligrams of pure C_{122} requires several weeks. This is one of the reasons for the relative abundance of dimers without a direct connection between fullerenes, similarly called *dumbbells*.

One fullerene dimer in this category, a dyad prepared by the dipolar cycloaddition of a fullerene with N–methylglycine

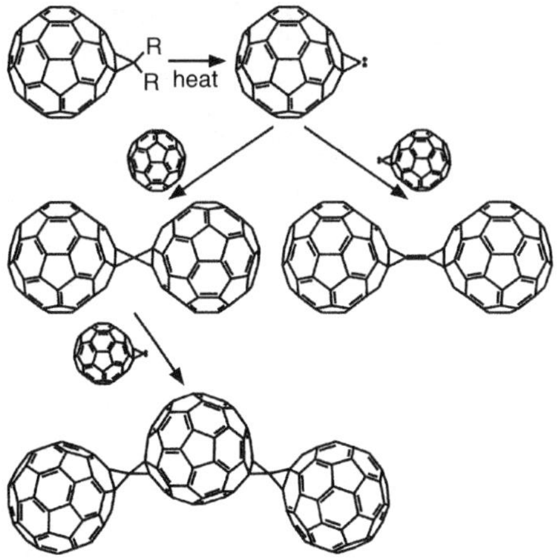

Scheme 4. Synthesis of C_{121} and C_{122} based on the formation of a methanofullerenecarbene as an intermediate.

(**17**), has been the object of study for electronic interactions [57]. 1,3–dipolar cycloaddition of N–methylglycine with C_{60} was also used for the synthesis of the fullerene dimer **18** [58]. Significant work also has been done in this area by Rubin and co-workers [59]. They have synthesized a series of "ball-and-chain" systems based on C_{60} in order to study the intramolecular electronic transfer. The synthetic procedures used by the above-mentioned authors involved a diene component in a Diels–Alder reaction with C_{60}. One of the reaction products which we would like to mention here is a fullerene dimer with a dumbbell-like structure (Fig. 3). This dimer has several possible conformations, and they have been compared with a pair of bolas (Fig. 4). Among the three possible conformations of this dimer, the folded–folded one is the most probable, taking into account molecular geometry optimizations. Such a conformation would be based on π–π attractive interactions between the fullerene cages, and probably imposes consequences in the electrochemical and photophysical properties of this dimer.

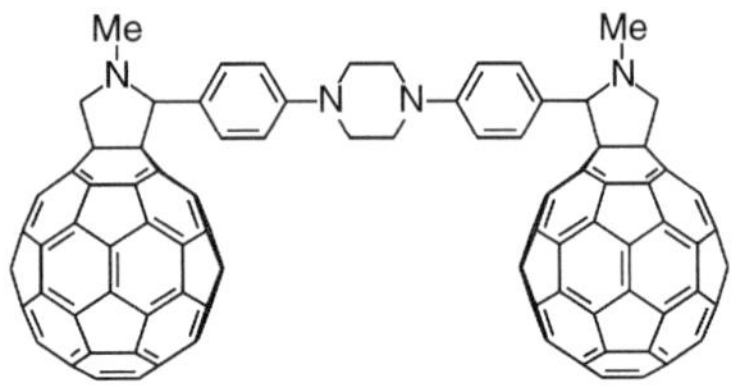

17

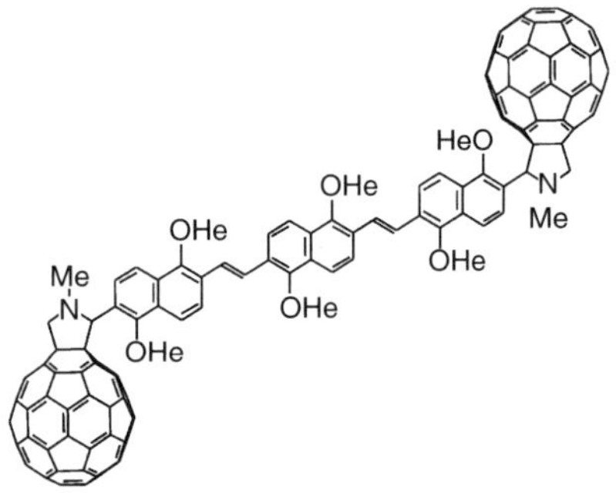

18

Another dimer formed by the Diels–Alder reaction is the "dumbbell"-like compound (**19**) [60]. The authors improved the solubility of **19** by the introduction of hexyloxyl groups, whose oligomers of up to seven [60]fullerenes were observed by matrix-assisted laser desorption ionization time-of-flight mass spectroscopy (MALDI–TOF MS) and size exclusion chromatography (**20**) [61]. The ^{1}H–NMR of the dimer of **20**

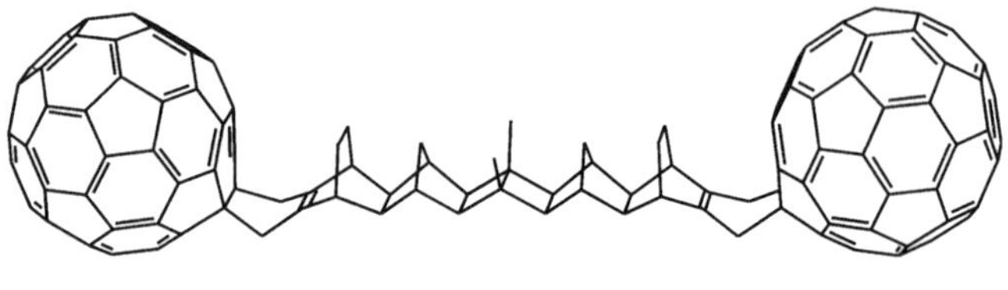

Figure 3. Molecular structure of dumbbell 14-bond bis–C_{60} adduct.

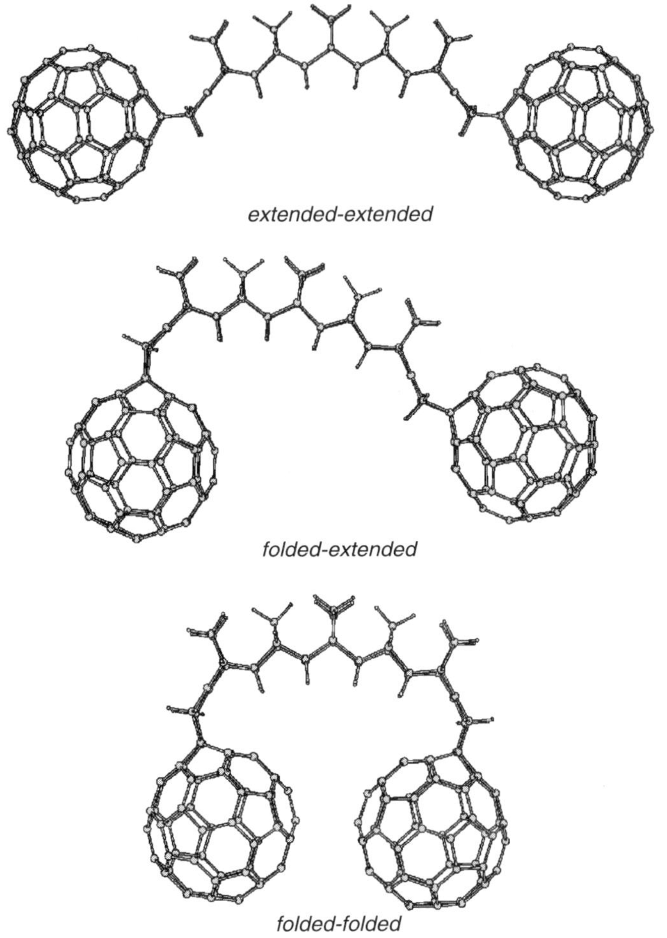

Figure 4. Three unique conformations for dumbbell system (Fig. 3).

($n = 0$) shows that it can exist either in the *cis* or in the *trans* conformation (Scheme 5).

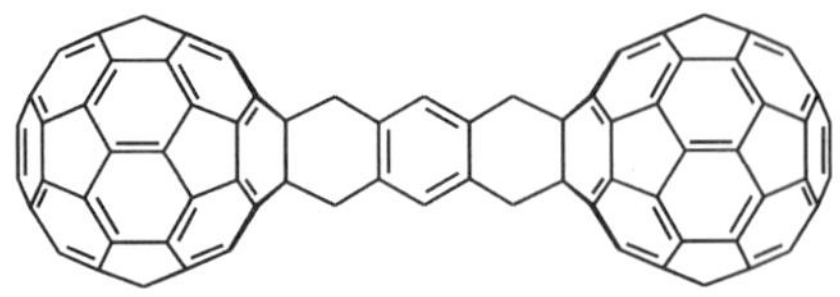

19

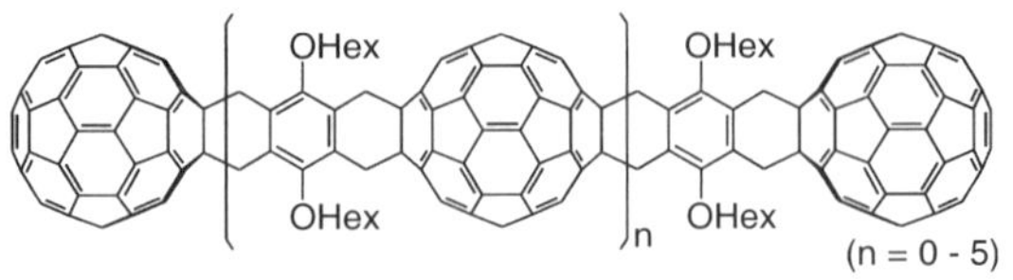

20

Diels–Alder reactions were also performed by solid-state reactions; thus *anti*-bis[60]fullerene adducts of pentacene (**21**) were synthesized by the HSVM reaction of C_{60} with pentacene [62]. The *syn* adduct of the dimer was not obtained by HSVM, while the *syn* derivative of the bis[60]fullerene adducts of 6,13-substituted pentacene (**22**)

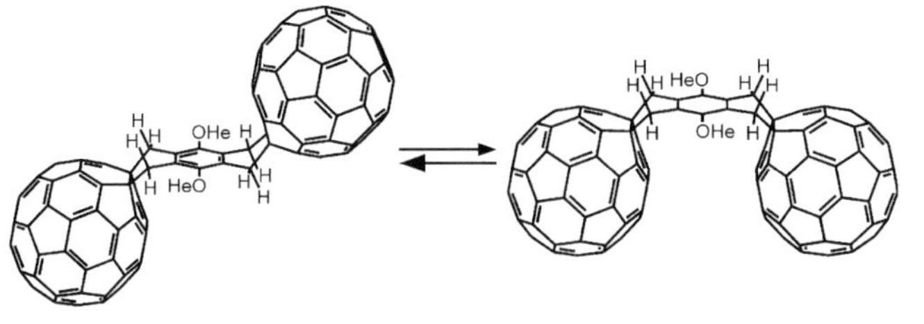

Scheme 5. *cis* and *trans* conformers of the bis–hexyloxy-substituted "dumbbell" compound **21** ($n = 0$).

was synthesized in solution [63] when only trace amounts of the *anti* compound were formed.

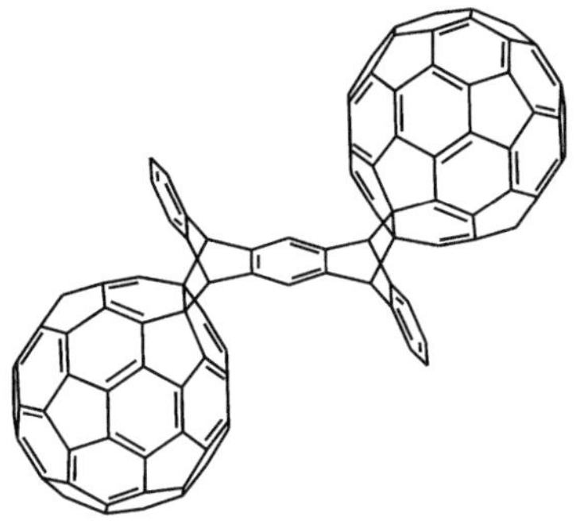

21

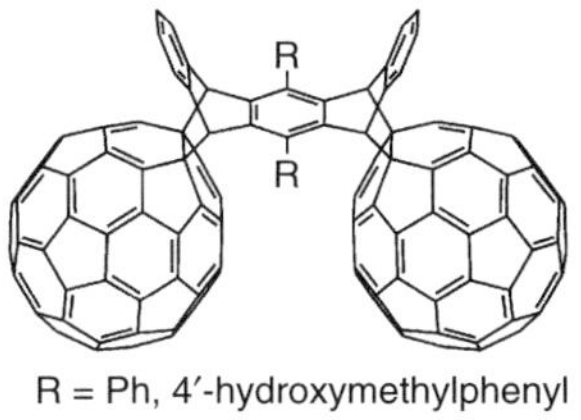

R = Ph, 4′-hydroxymethylphenyl

22

An unusually strong electronic communication between two C_{60} cages has been observed for a fullerene–metal sandwich complex formed by reacting $Rh_6(CO)_9(dppm)_2\cdot(\mu_3\text{–}\eta_2, \eta_2, \eta_2\text{–}C_{60})$ with C_{60}. The product obtained has a structure like that shown in Figure 5, in which two C_{60} are linked through an octahedral Rh_6 framework [64].

Another metal bridged fullerene dimer is C_{70}–Sm–C_{70}, which is formed in the fulleride $Sm_{2.78}C_{70}$ [65]. While the other reported fullerides are generally van der Waals or ionic crystals, two [70]fullerenes in this fulleride are bridged by an Sm atom with single covalent bonds, whose lengths are 2.49(1) and 2.47(1) Å.

Aside from those, bis–([60]fullereno[1,2–d]isoxazole–3–yl) (**23**) was synthesized by dipolar cycloaddition [66]. More specifically, (**23**) was generated by the double 1,3–dipolar cycloaddition of C_{60} to cyanogen di–*N*–oxide. The solubility of this dimer into CS_2, chlorobenzene, and *o*–dichlorobenzene is poor, but it is well soluble in 1–chloronaphthalene/2–chloronaphthalene 9/1(v/v).

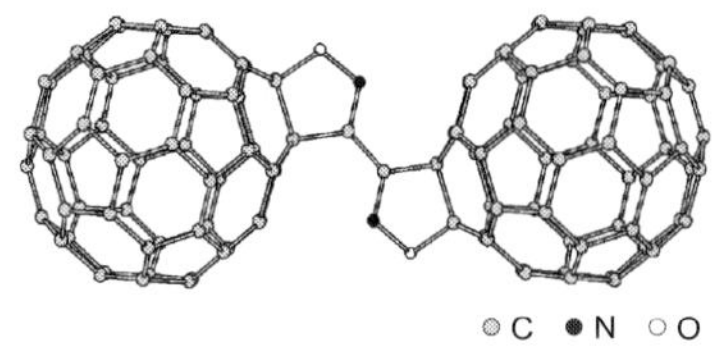

23

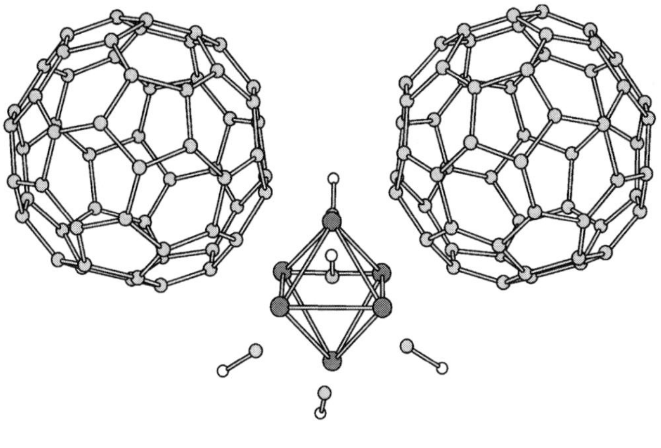

Figure 5. Molecular structure of $Rh_6(CO)_5(dppm)_2(CNR)(C_{60})_2$. RNC and dppm ligands are omitted for clarity. ($R = CH_2C_6H_5$.)

In the search for molecules having charge separation states, Hudhomme and co-workers [67] developed strategies for the synthesis of C_{60} TTF triads. The idea behind the interest in such a compound lies in the stabilization of the charge separation state $C_{60}^{\cdot-}$ $TTF^{\cdot+}$, as a result of the formation of a heteroromatic 1,3 dithiolium cation upon oxidation. The obtained C_{60}–TTF–C_{60} triad is formed by the reductive elimination of the tetrakis (bromine) TTF in the presence of C_{60}. The compound (**24**) is formed by rigid links of TTF to C_{60}; this approach allows the control of both the relative orientation and the distance between the donor/acceptor groups. As expected, the product is not very soluble, and its thermal instability impeded its purification. Its structure is inferred nonetheless from the synthesis of several soluble derivatives (**25**) in which the C_{60} is cyclopropanated, as well as from MS, UV–vis and electrochemical properties.

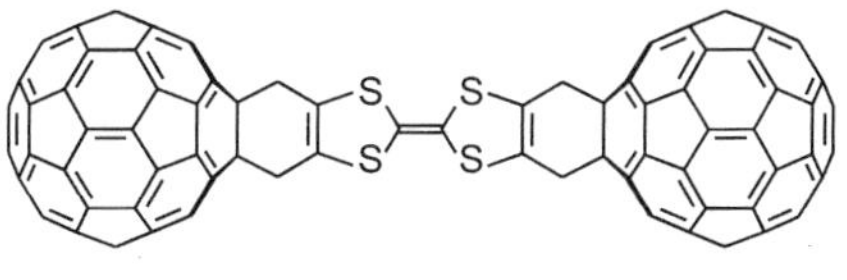

24

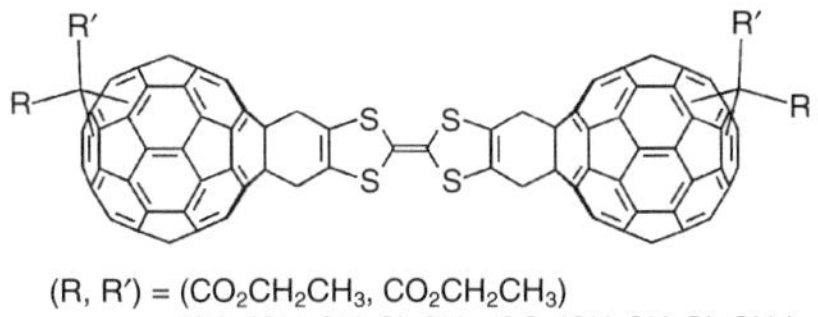

(R, R′) = ($CO_2CH_2CH_3$, $CO_2CH_2CH_3$)
($CO_2(CH_2CH_2O)_3CH_3$, $CO_2(CH_2CH_2O)_3CH_3$)
(CO_2CH_3, Ph)

25

A recent class of compounds in this category was opened by the synthesis of supramolecular dimers, initially reported by Diederich et al. [68]. Although the chemistry of fullerenes is well developed to date, little work has been done in the area of supramolecular assemblies bearing fullerene units.

The synthesis of a derivative of C_{60} with dibenzo–24–crown–8 ether was accomplished, as well as the one of C_{60} with dibenzylammonium hexafluorophosphate; these form a supramolecular C_{60} dimer (DG = −17.0 kJ in $CDCl_3$:CD_3CN 9:1) (**26**). The hydrogen bonds and ion–dipole interactions led to a structure having the dibenzylammonium component threaded through the cavity of the crown ether macrocycle. As in most of the dimers where there is a large distance between the cages, no electronic interactions between the cages were detected by means of electrochemistry.

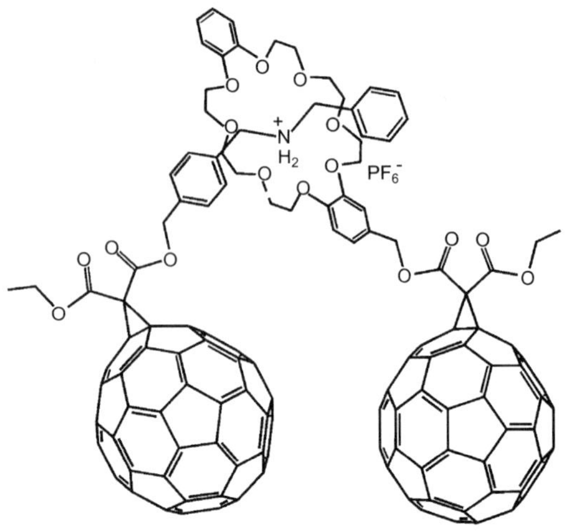

26

There is another type of supramolecular dimer, which is based on quadruple hydrogen bonding (**27**) [69]. The noncovalent bonding of this compound is different from that of **26**.

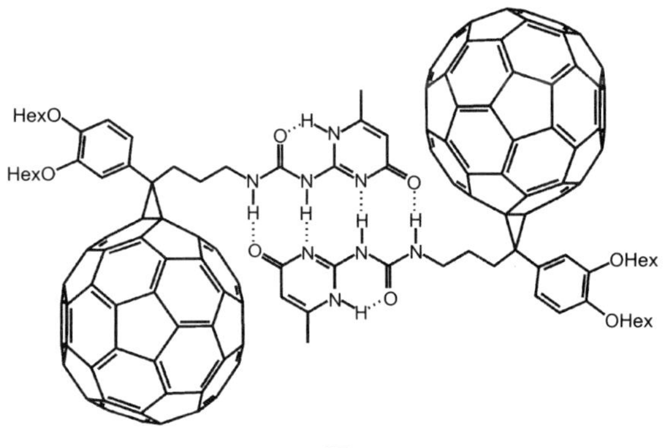

27

Murata et al. reported two C_{60} dimers (**28** and **29**) bridged by a bicyclic framework, both synthesized by solid-state reactions [70, 71]. The dimer **28** was synthesized by heating the solid mixture of C_{60} and 2,3–diazanaphthalene at 150 °C for 2 h. The dimer **29** was synthesized by processing the solid mixture of C_{60} and 4 equivalents of phtalazine using high-speed vibration milling. The interesting point about these compounds is that **28** and **29** were isomerized into **30** and **31**, respectively, by heating in solution.

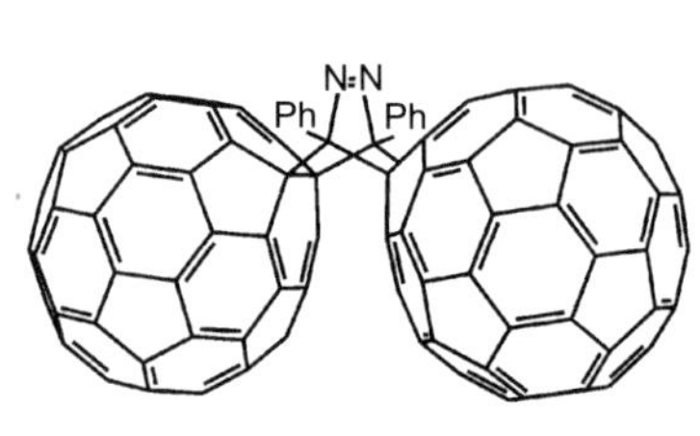

28

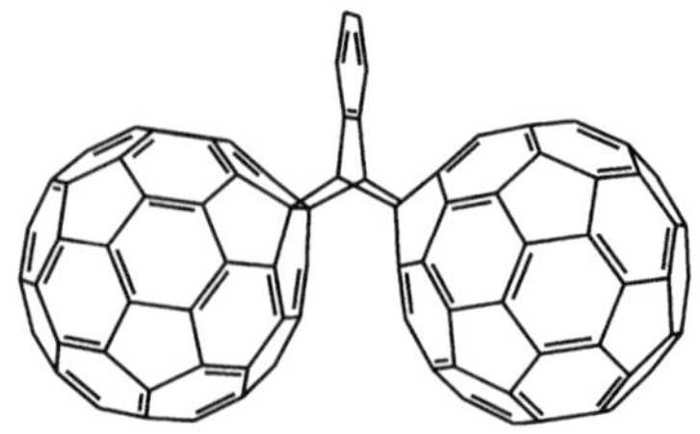

29

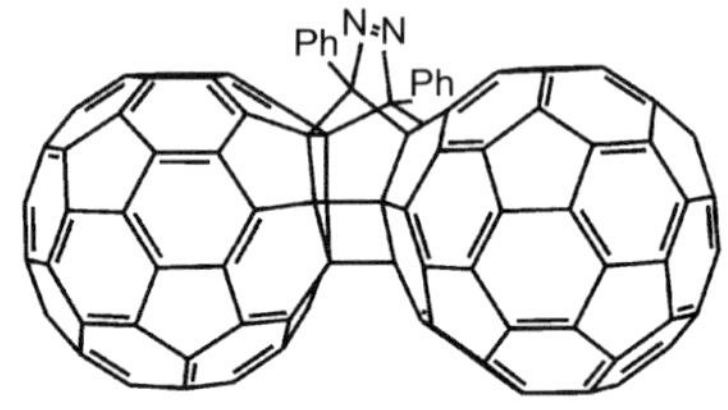

30

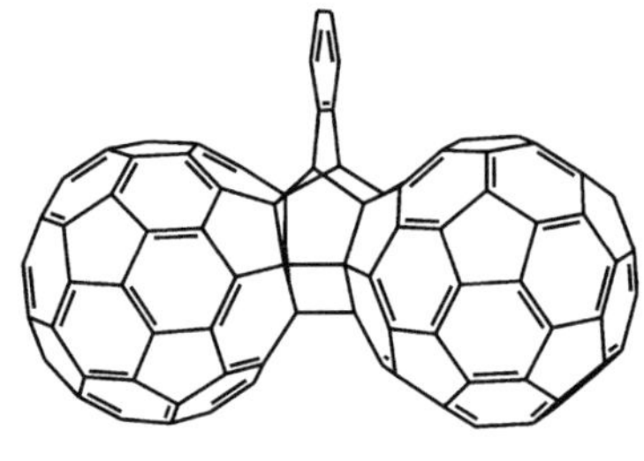

31

Another dimer that transforms into an isomer with direct intercage bonds, *m*–phenylenebis(arylmethanofullerene) (**32**), was synthesized by Knol et al. [72]. Irradiation in deoxygenated *o*–dichlorobenzene isomerizes about 60% of **32** into **33**. This process is significantly retarded by molecular oxygen. The dimer with direct intercage bonds (**33**) isomerizes completely into **32** by refluxing in *o*–dichlorobenzene. This is the first example of a controlled [2 + 2] cycloaddition process of fullerenes.

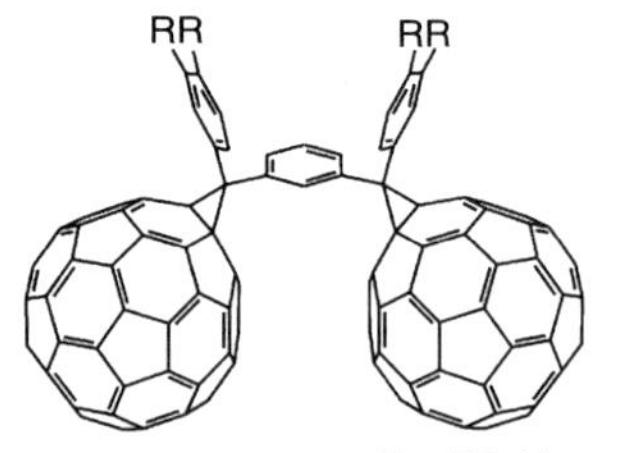

R = OC_6H_{13}

32

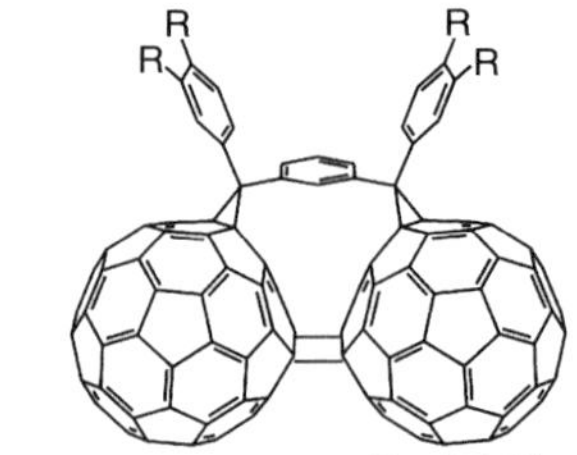

R = OC_6H_{13}

33

The cage-structured C_{60} dimer (**34**), which consist of four aliphatic chains and two C_{60} end caps, was synthesized by the coupling of two [60]fullerenes and four 1,3–diaminopropanes [73].

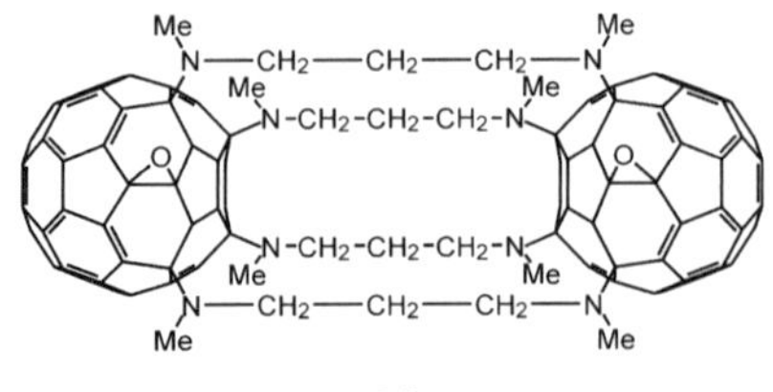

34

3. SPECTROSCOPIC CHARACTERIZATION

3.1. Mass Spectroscopy

Mass spectroscopy plays an important role in the characterization of fullerene dimers because this technique requires substantially fewer samples (and machine time) than other analysis methods such as NMR, X-ray diffraction, and so on. However, the fragmentation of fullerene dimers into monomers and the low solubility of fullerene dimers made most of the ionization methods inaccessible for these samples. Typically, only MALDI–TOF–MS gives sufficient mass accuracy and intensity for most of the fullerene dimers. The representative fullerene dimers whose mass spectrum was successfully obtained by MALDI–TOF–MS are $C_{120}OS$ [39], $C_{120}H_4$ [41], C_{121} [76], and C_{122} [76]. All of them were analyzed by using 9–nitroanthracene (9–NA) as a matrix and negative-ion mode. The mass spectra of $C_{120}H_4$, C_{121}, and C_{122} were also obtained by using dithranol as a matrix.

Several ionization techniques were tested to obtain the mass spectrum of C_{120} (**2**) [18]. A signal for C_{120} and C_{60} was obtained by MALDI–TOF–MS and by Fourier-transform ion cyclotron resonance (FTICR). The C_{60} signal was formed by fragmentation of the dimer, which is unavoidable for any of the fullerene dimers reported to date. Fast atom bombardment (FAB), atmospheric pressure chemical ionization (APCI), and electrospray ionization (ESI) in both the positive- and negative-ion modes showed only a peak of C_{60}.

The mass spectra of $C_{120}O$ (**4**) were obtained by MALDI–TOF–MS using 9–NA as a matrix [33] in the negative-ionization mode. In addition to the peaks of $C_{120}O$, the peaks of $C_{120}O_2$, $C_{120}O_3$, $C_{60}O$, and C_{60} were detected in the spectrum. The mass spectra of $C_{120}O$ were also obtained both by negative-ion APCI with N_2 as reagent gas and by negative-ion ESI with N,N,N′,N′–tetramethyl–1,4–phenylenediamine as the electron transfer reagent, but laser desorption ionization (LDI) gives only C_{60} and $C_{60}O$ fragment ions [75].

Not only the molecular ion peak, but also the fragmentation pattern provides information on the molecular structure. For instance, a different fragmentation pattern was observed in the MALDI–TOF–MS observed for the two isomers $C_{120}O_2$ [40].

FAB–MS is usually not suitable for fullerene dimers because the volatility of good solvents for fullerene dimers such as 1,2–dichlorobenzene and 1–chloronaphthalene are too high as a matrix in FAB–MS. In order to solve the problem, Gross et al. [74] applied negative-ion low-temperature FAB–MS to fullerene dimers. In this method, a mixture of 1–chloronaphthalene and 3–nitrobenzyl alcohol and the dissolved sample was frozen with liquid nitrogen. This procedure enabled the use of these solvents in the FAB–MS, and showed high-quality molecular ion peaks for C_{119}, C_{120}, $C_{120}O$, $C_{120}O_2$, and $C_{120}OS$.

3.2. Nuclear Magnetic Resonance

^{13}C–NMR is one of the most powerful tools for the structure analysis of fullerene dimers, and in general was the method of choice. Because few carbon atoms in fullerene cages are bound to protons, there are small differences of relaxation time and of the nuclear Overhauser effect. This makes it possible to quantitatively count the number of carbon atoms of ^{13}C–NMR spectra. Unfortunately, there are some difficulties in obtaining satisfactory ^{13}C–NMR spectra: low solubility of dimers, long relaxation time, as well as the concentration of resonances of fullerene cage atoms in a small domain. For these reasons, many ^{13}C–NMR experiments on dimers required a long acquisition time, relaxation reagent, and a relatively high field NMR. It is not uncommon in this field to have routine spectra acquired for a few days.

A typical example of the structure determined by ^{13}C–NMR is **14** [76]. The ^{13}C–NMR spectrum shown in Figure 6 was measured under the following conditions: instrument: JEOL LNM270 (^{1}H 270 MHz), solvent: 1–chloronaphthalene with 20% v/v 1–methylnaphthalene–d_{10}, relaxation reagent: $Cr(acac)_3$ 10 mM, flip angle: 90°, pulse repetition time: 10.000 s (acquisition time 3.146 s + pulse delay 6.854 s), number of scans: 34,464. There are other two peaks at 84.71 and 128.90 ppm out of the range of Figure 6. By using a partial ^{13}C-enriched sample, the resonance at 128.90 ppm was detected to originate in the bridging carbon atoms. There are three small resonances, 11 resonances of double intensity, and one resonance of quadruple intensity of the small peaks. The intensity of the resonance at 84.71 ppm is about same as the small resonances. Considering their number and intensity, the only reasonable explanation is that the quadruple intense peak is the overlap of two double intensity peaks, and the symmetry of the dimer is thus C_{2h}. The three small resonances originate in the carbon atoms located in the mirror plane of the molecule.

The structural as well as mechanistic aspects of synthesis of fullerene dimers can be elucidated by using ^{13}C–NMR on partially ^{13}C marked compounds. Thus, the formation of $C_{120}O$ [75] and C_{121} [76] was checked by using ^{13}C partially marked C_{60}. We also note that ^{3}He NMR studies were performed on the ^{3}He-incarcerated C_{120} molecules [25, 77].

The experimental conditions of acquisition for some dimers are summarized in Table 1.

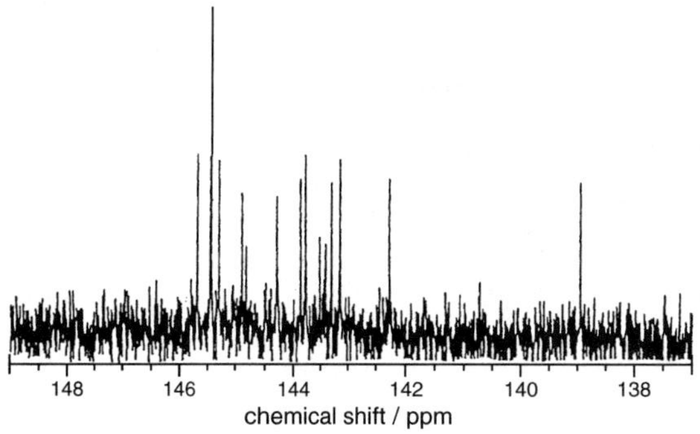

Figure 6. ^{13}C–NMR of C_{122} (**14**). Reprinted with permission from [76], N. Dragoe et al., *J. Am. Chem. Soc.* 123, 1294 (2001). © 2001, American Chemical Society.

Table 1. Experimental conditions of reported ^{13}C–NMR of fullerene dimers.

Molecule	Solvent	^{13}C abundance ratio	$Cr(acac)_3$ (mM)	Flip angle (°)	Pulse repetition time (s)	Number of scans	Ref.
C_{120}	ODCB–d_4	natural	0	N/A	N/A	N/A	[18]
C_{121}	1–chloronaphthalene/1–methylnaphthalene–d_{10} (4:1)	natural	10	60	5.146	44,735	[76]
C_{122}	1–chloronaphthalene/1–methylnaphthalene–d_{10} (4:1)	natural	10	90	10	34,464	[76]
$C_{120}O_2$	$CS_2/CDCl_3$ (98:2)	7%	35	N/A	6	36,000	[37]
$C_{120}O_2$ (minor)	ODCB acetone–d_6 (in capillary)	7%	17.5	90	2.5	92,160	[40]
$C_{120}OS$	ODCB–d_4	7%	17	70	2	151,200	[39]

3.3. FT–IR and Raman Spectroscopy

Because of its high symmetry, C_{60} shows only 4 and 11 peaks in IR and Raman spectra respectively, in spite of the large number of carbon atoms. For the case of fullerene dimers whose symmetries are lower than that of C_{60}, their IR and Raman spectra show hundreds of peaks. Thus, these spectra must be used in combination with other experimental and theoretical data, and in general have less impact on the structural identification of these compounds. Lebedkin et al. obtained strong evidence for the structure of C_{119} from a comparison of the experimental Raman spectrum and theoretical Raman spectra of several possible isomers [78].

The characteristic vibrations of fullerene dimers are six cage–cage vibrations. One of them is derived from the combination of translations of two C_{60} cages; the others are derived from the combination of rotations of two C_{60} cages. Lebedkin et al. reported the Fourier transform (FT)–Raman spectra of the vibrations of C_{120}, $C_{120}O$, and $C_{120}O_2$ [79]. The observed wavenumbers for the cage–cage vibrations of C_{120} agree well with the theoretical predictions made by Porezag et al. [80]. In general, the intercage vibrations ("diatomic like") are useful for structural analysis because their frequency can be directly related to the strength of the connection between the cages. The wavenumbers of these vibrations are lower than 200 cm^{-1}, and their intensities are very small. Thus, they are difficult to observe experimentally. In Figure 7, we show the dimer's vibrations modes, as calculated for C_{121} [**13(b)**] [86].

3.4. X-Ray Diffraction

Since the first synthesis of C_{60} by Krätschmer and co-workers [2], it has been observed that C_{60} is difficult to obtain in an ordered single crystalline form. This also has been the case for their derivatives. Luckily, C_{120} (**2**) can be obtained in an ordered single crystalline form; its structure was proven by single-crystal analysis, and consists of two fullerene cages bonded through a four-membered ring at the 6:6 position (i.e., at the connection between two hexagons in a fullerene framework). We note that the cage–cage distance is slightly longer, at 0.1575 nm, than a usual sp^3–sp^3 bond. The shape of C_{120} caused it to be dubbed a *dumbbell*, and this compound is thought to be the repetitive unit of a type of all-carbon fullerene polymer obtained previously by UV irradiation [16].

Another fullerene dimer analyzed by X-ray diffraction (XRD) was $C_{120}O$; however, due to the disorder, its structure could not be analyzed in detail [81].

4. THEORETICAL CALCULATIONS

In many cases, experimental IR, Raman, and NMR spectra of fullerene dimers were analyzed with the help of theoretical calculations. Theoretical calculations have been employed for the study of fullerenes; several possible structures of fullerene dimers were examined energetically and geometrically.

The three dimers, that is, $C_{120}O_2$, $C_{120}OS$, and $C_{122}H_4$, are structurally similar; from ^{13}C–NMR data alone, it is difficult to find the correct structure since several structures with the same symmetry are possible. As an example, the three possible structures of $C_{122}H_4$ are shown in Figure 8. The differences between them are the location of the two direct intercage bonds and the two methylene bridges. Among these, the most stable ones were determined by quantum chemical calculations, and it was found that their stability

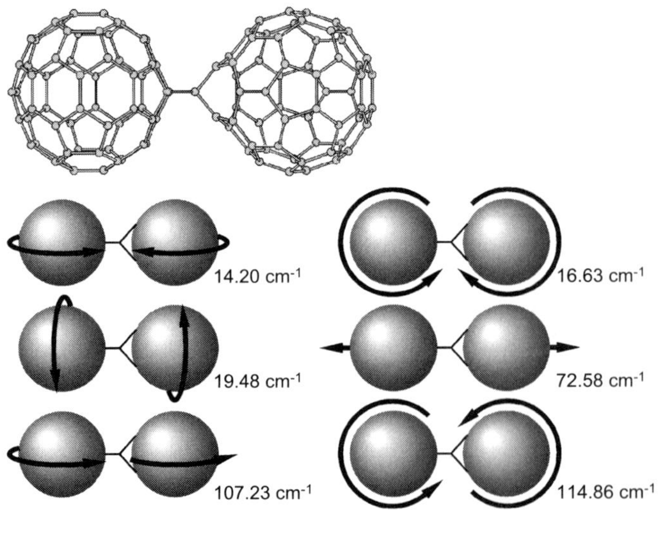

Figure 7. Cage–cage vibrational modes of C_{121} [**13(b)**].

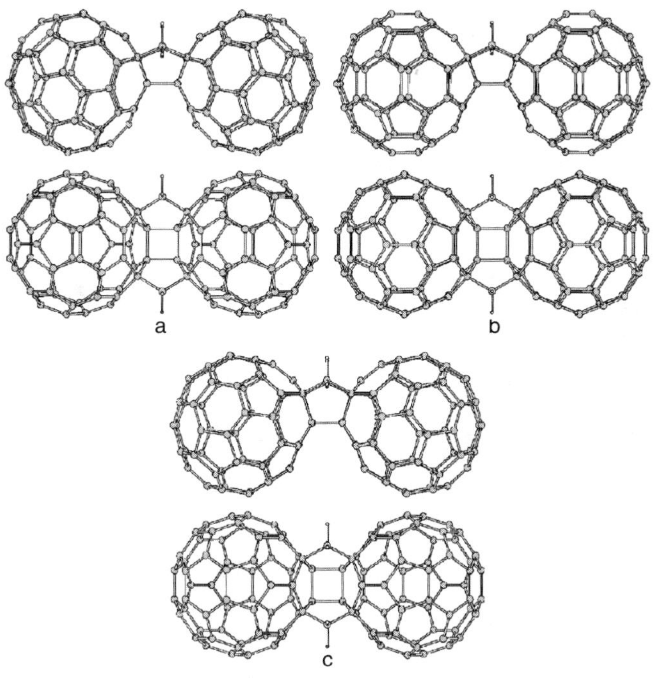

Figure 8. Three possible structures of $C_{122}H_4$.

follows the same rule. The rule is that the most stable structure has the maximum number of double bonds at 6:6 positions, that is, the minimum number of double bonds at 6:5 positions for the dominant Kekule structure [82, 83]. The dominant Kekule structures of three possible structures of $C_{122}H_4$ are shown in Figure 9. Because of mirror symmetry, only one of two C_{60} cages is shown for each structure. The most stable structure is shown in Figure 9(a), and the next is Figure 9(b). The stability is related to the number of double bonds between two six-membered rings: 28, 27, and 24 for (a), (b), and (c), respectively. The same behavior was found for other fullerene derivatives [84].

13(a) consists of two methano[60]fullerenes, while **13(b)** consists of a methano[60]fullerene and a homo[60]fullerene. Generally, a methano[60]fullerene structure is more stable than the corresponding homo[60]fullerene structure [85]. In the case of C_{121}, the computed strain energy of the spiro structure of the bridging atom of **13(a)** is larger than **13(b)**, and this difference compensates for the difference of stability between a methano[60]fullerene and a homo[60]fullerene. Consequently, according to theoretical investigations, Gibbs's free energy of **13(a)** is larger than

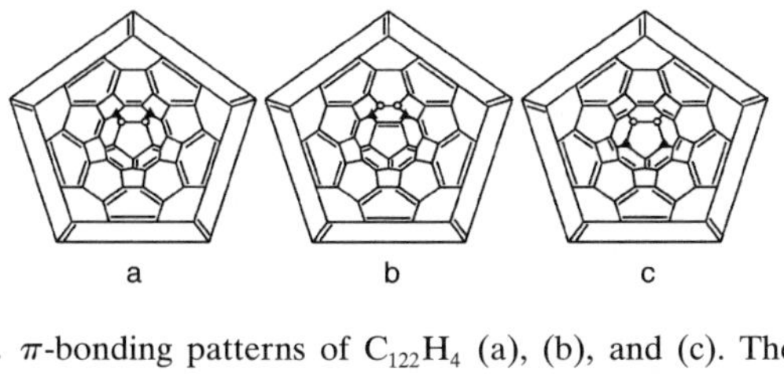

Figure 9. π-bonding patterns of $C_{122}H_4$ (a), (b), and (c). The carbon atoms that are connected to a carbon atom of another C_{60} cage and a methylene bridge are marked by an open circle and a closed triangle, respectively.

that of **13(b)** [86]. Actually, it was observed experimentally that **13(a)** and **13(b)** isomerize each other, and the yield for the synthesis of **13(b)** is larger than of **13(a)** [76].

Concerning the properties of **13(b)**, we note that this is the only molecule embodying both types of connection to fullerenes. It has also shown intercage electronic communication; the first reduction potential is close to that of C_{60}, with peak separations for the addition of further electrons. This asymmetry in the structure and electronic properties can explain the asymmetric images obtained by STM [54].

Theoretical calculations have been employed to explore the structure of several all-carbon derivatives [87, 88].

5. PROPERTIES

5.1. Electrochemical Properties

Electrochemical studies have been performed on C_{60} and derivatives since their discovery, mainly to study the abilities of these molecules to be electrochemically reduced. From these studies, it has been found that, in general, C_{60} derivatives are less prone to reduction compared to C_{60} by 50–60 meV. For the fullerene dimers, similar behavior should be observed, with the difference that the two fullerene moieties can interact, thus leading to reduced species having large electronic delocalization. For these systems, one of the most important characteristics was the detection of such electronic communication which would manifest by a splitting between the first and second reduction potentials, third and the fourth, and so on. For a dimeric system in which the two electroactive units are separated (i.e., noninteracting), the first and second reduction potentials should be degenerate. The degree of such "electronic communication" was not ascertained in detail, other than the splitting between the pairs of reduction potentials. One of the most interesting systems to be discussed is C_{120}, and apparently there is no electronic interaction between the fullerene cages in this dimer. Cyclic voltammetry measurements suggest the decomposition to C_{60} after the first electron reduction process [18].

Balch et al. [89] observed an "electronic communication" between the fullerene cages of $C_{120}O$, that is, a stepwise addition of electrons to the fullerene cores. Supported by EPR measurements, they found that the reversible reductions of $C_{120}O$ are sequential: the addition of one electron to one C_{60} cage in a first step changes the reduction potential for the second addition to the other C_{60} cage, that is, a communication between the C_{60} frameworks. In a similar way, although not discussed in detail, in the di–aza dimer $C_{118}N_2$, some degree of interaction occurs [42].

As well as for $C_{120}O$, an electronic communication between the C_{60} cages of silicon-bridged dimer **(6)** was observed by cyclic voltammetry [36].

An interesting system is the unsymmetrical dimer C_{121}. The differential pulse voltammogram of the unsymmetrical C_{121} (C_s) [**13(b)**] shows six distinct peaks located at −1044, −1124, −1436, −1532, −1988, and −2160 mV versus ferrocene/ferricium (Fig. 10) [76]. These six peaks in pairs correspond to each of the first three reductions of the C_{60} cage. The differences between the reduction potential pairs are about 80, 80, and 135 mV, which are larger than the differences between the reduction potentials of isomeric homofullerenes and methanofullerenes. Thus, the splitting arises not from the difference of the two C_{60} cages of C_{121} (C_s), but from intercage electronic interactions. It is interesting that the first reduction potential of C_{121} (C_s) is slightly more positive than that of C_{60}, while the reduction potentials of C_{60} adducts are expected to be shifted to more negative potentials. That may reflect the unusual electronic structure that involves the coupling of two dissimilar carbon cages: methano[60]fullerene and homo[60]fullerene.

Similarly, intercage electronic communication in $C_{122}H_4$ was detected by means of electrochemical measurements (Fig. 10) [41]. Each of the three electrochemical reduction peaks for C_{60} are split. It is likely that this electronic communication is related to the through-space orbital overlap deduced to exist in HOMO/LUMO for this dimer, by means of electronic calculations. In a similar way, it is likely that $C_{120}O_2$, having a structure similar to that of $C_{122}H_4$, should manifest such intercage interactions.

A remarkable effect was found for $Rh_6(CO)_5(dppm)_2 \cdot (CNR)(C_{60})_2$ ($R = CH_2C_6H_5$), whose cyclic voltammetry experiments showed six well-separated, reversible, one-electron redox waves [64]. The large peak separations for this compound (0.19, 0.24, and 0.29 V) reflect strong electronic couplings between the fullerene cages. In contrast, for other dimers showing electronic communication, the separations between the first two redox waves are about 0.05–0.08 V [41, 76, 89].

The six reduction peaks of a supramolecular C_{60} dimer with quadruple hydrogen bonding (**27**) measured by cyclic voltammetry in toluene/acetonitrile mixture (4:1 v/v) were not split [69]. These results indicate that there is no intercage interaction in the ground state. Like **28**, the split of reduction peaks in the cyclic voltammogram of the dimers

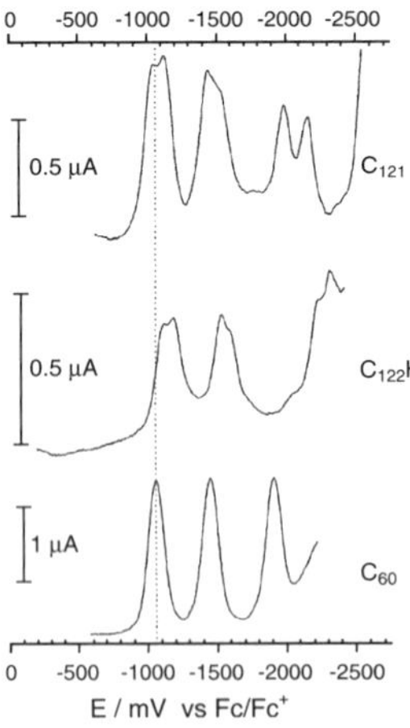

Figure 10. Differential-pulse voltammogram of C_{121} (top trace), square-wave voltammogram of $C_{122}H_4$ (middle trace), and C_{60} (bottom trace) in *o*-dichlorobenzene with 0.1 M TBAP as supporting electrolyte. For the DPV, the parameters were: scan rate, 20 mV; pulse amplitude, 50 mV; sample width, 17 ms; pulse width 50 ms; and pulse period, 200 ms. For the SWV, the parameters were: pulse step, 2 mV; square-wave frequency, 15 Hz; square-wave amplitude, 10 mV; 256 samples/point for $C_{122}H_4$; pulse step, 4 mV; square-wave amplitude, 25 mV; square-wave frequency, 15 Hz; and 256 samples per point for C_{60}. The first reduction potential of C_{60} is shown by the dotted line. Reprinted with permission from [76], N. Dragoe et al., *J. Am. Chem. Soc.* 123, 1294 (2001). © 2001, American Chemical Society.

without direct intercage bond, **18** and **22**, was not observed. The cyclic voltammetry of **18** shows four quasireversible reduction waves at potentials rather similar to all 1,2–dihydrofullerenes, that is, slightly more negative potentials than those found for fullerenes [90].

In summary, intercage electronic interactions were observed for the dimers with direct intercage bonds and C_{121}, and not for the dimers without direct intercage bonds. The metal sandwich dimer (Fig. 5), which has no intercage bonds, shows exceptionally strong intercage electronic interactions.

31 and **32** have direct intercage bonds, and thus they are appropriate for studying the influence of direct intercage bonds on intercage electronic communication. Indeed, electronic communication was investigated by electrochemical methods, and the splits between the first and second reduction peaks of **29**, **30**, **31**, and **32** measured by differential pulse voltammograms were 0.06, 0.06, 0.09, and 0.09 eV, respectively. These results showed that the formation of direct intercage bonds enhanced intercage electronic communication.

5.2. Photosensitive Properties

5.2.1. Photostability

One of the important problems is the stability of fullerene derivatives, and particularly their photostability. This problem has not yet been systematically addressed, as far as C_{120} is concerned. Some reports [91–93] concluded that C_{120} is stable under light illumination, while others [94–96] concluded that C_{120} dissociated into two C_{60} monomers in the presence of light. Lebedkin et al. [95] and Bachilo et al. [96], who support that C_{120} is unstable to light, indicated that photodissociation of C_{120} occurs via excited triplet states.

There are fewer reports of photostability of other dimers. Lebedkin et al. reported that $C_{120}O$ and $C_{120}O_2$ did not dissociate under the influence of light [95]. Fujitsuka et al. also reported that $C_{120}O$ is stable under light illumination [97]; this is in agreement with the observation of $C_{120}O$ presence in samples of C_{60} kept in air, as observed by Taylor et al. [98]. Gruss et al. studied the EPR signal of $C_{59}N\cdot$ by pulse laser irradiation (532 nm) of $(C_{59}N)_2$ [99]. The signal intensity did not decrease after more than 10^5 laser pulses, which indicates that the major decay channel of the $C_{59}N\cdot$ is redimerization.

From the available data, most of the dimers are quite stable under the influence of light.

5.2.2. Absorption and Fluorescence Properties

The optical properties of C_{120} have been most intensively studied among the fullerene dimers. The photophysical optical properties of this dimer were reported by Ma et al. [91], and were observed to be rather similar to that of C_{60}. Noticeable is a weaker triplet–triplet absorption in the case of C_{120} comparable to that of C_{60}. All values of C_{120} reported here were determined in toluene. C_{120} has absorption peaks at 434 and 698 nm; this feature is similar to those of other 1,2–dihydro[60]fullerene derivatives (**35**) which have 58π-electron systems. Ma et al. [91] compared the absorption spectra of C_{120} and several 1,2–dihydro[60]fullerene derivatives. These spectra showed that the molar absorptivity of C_{120} at the 0–0 absorption band is larger than that of other C_{60} derivatives by a factor of about 2, although they are similar. This indicates that C_{120} can be treated as behaving like a mix of two independent 1,2–dihydro[60]fullerene cages with negligible interactions with regard to electronic transitions. Concerning the absorption properties, fluorescence spectrum, lifetime, and quantum yield, the C_{120} dimer is rather similar to 1,2–dihydro[60]fullerene derivatives than to the parent C_{60}.

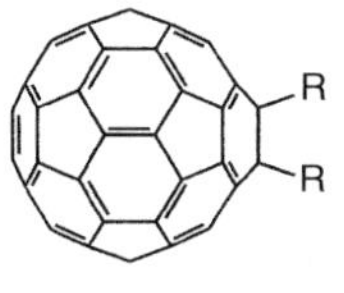

35

For the $C_{120}O$ dimer, the absorption peaks of $C_{120}O$ in toluene appeared at 327, 418(sh), 535(sh), and 691 (Fig. 11). This profile is also similar to 1,2–dihydro[60]fullerenes and C_{120}, but is different from C_{60}. The fluorescence spectrum and lifetime (1.6 ns) in toluene are also similar to those of 1,2–dihydro[60]fullerenes (1.3–1.5 ns) and C_{120} (1.5–1.6 ns), but different from that of C_{60} (1.2 ns) [97].

Fujitsuka et al. reported the optical properties of C_{121} [**13(b)**] [100]. The optical properties of C_{121} are interesting because this molecule has two different chromophores, namely, a methanofullerene moiety and a homofullerene moiety. Homofullerenes, which keep a 60π-electron system, have similar absorption spectral profiles as pristine fullerenes, while methanofullerenes, whose number of π electron is reduced to 58, have different absorption spectral features. C_{121} in *o*–dichlorobenzene shows absorption bands at 689, 425, and 331 nm, with shoulder bands around 600 and 530 nm (Fig. 11). Among them, 689 and 425 nm bands are typical features of methanofullerenes, and 600 and 530 nm bands are typical features of homofullerene, and both of them have a 331 nm band. Consequently, the absorption spectrum of C_{121} is explained by the simple superposition of the spectra of a methanofullerene and

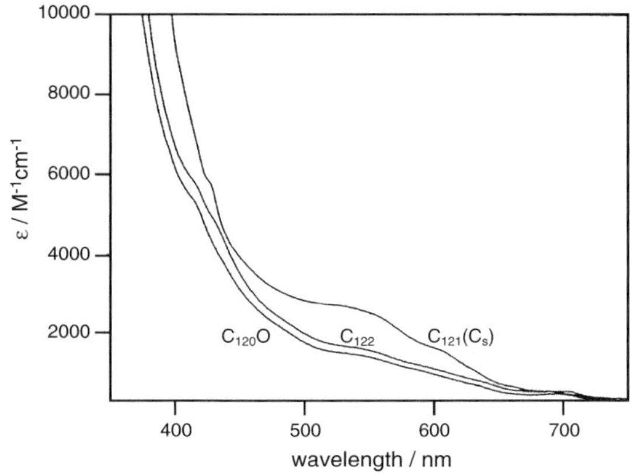

Figure 11. UV/vis spectra of $C_{121}(C_s)$, C_{122}, and $C_{120}O$ in ODCB, determined using solutions of concentrations of 0.1 mg/mL. Reprinted with permission from [76], N. Dragoe et al., *J. Am. Chem. Soc.* 123, 1294 (2001). © 2001, American Chemical Society.

a homofullerene, and no sign of intercage electronic interaction was observed in the absorption spectrum. The fluorescence quantum yield of C_{121} was estimated to be 8.4×10^{-4}. The value is similar to C_{120} (7.9×10^{-4}) and $C_{61}H_2$ (8.7×10^{-4}), and about three times of that of pristine C_{60} (3.2×10^{-4}).

In contrast to C_{121}, C_{122} consists of two identical methanofullerenes without a homofullerene moiety. The absorption spectrum of C_{122} shows similar features as other methanofullerenes like C_{120} and $C_{120}O$, that is, absorption bands at ~420 and ~700 nm [76]. As shown in Figure 11, C_{122} does not have absorption bands typical of homofullerenes (530 and 600 nm), as was the case for C_{121}. In the case of C_{122}, the absorption spectrum is explained by the simple superposition of two methanofullerene spectra.

The absorption spectrum of $(C_{59}N)_2$ is different from both pristine C_{60} and methano[60]fullerenes. The spectrum is similar to that of $C_{59}HN$, and has absorption peaks at ~420, ~600, ~720, and ~800 nm in toluene [101].

5.3. Thermal Properties

Few thermal property experiments have been performed on fullerene dimers. Their importance is related to the feasibility of growing thin films for device and/or transport property studies. Among these studies, it is known that the [2 + 2] photopolymerized polymer reverses to C_{60} upon heating at about 200 °C [102]. Similar in structure, the model molecule C_{120} dissociates quantitatively into C_{60} by heating its *o*–dichlorobenzene solution at 175 °C for 15 min [18] or by heating in solid state [20]. For other dimers, heating at higher than about 200 °C approaches isomerization or to decomposition, as is the case of C_{121} and C_{122}, respectively [103].

A special case was found for the thermal decomposition of fullerene oxides. While heating a mixture of C_{60} and $C_{60}O$ yields fullerene oligomers, mostly $C_{120}O$ and $C_{120}O_2$, the heating of these species was found to lead to C_{119} [29].

The thermal stability of fullerene dimers was less studied since this method is generally destructive, and requires a relatively large amount of sample (milligram range). However, we speculate that the use of the solid-state reactions will eventually lead to the synthesis of higher fullerenes or small well-defined carbon nanotubes.

6. CHEMISTRY OF FULLERENE DIMERS

The effect of dimerization on chemical reactivity was examined in the case of [2 + 1] cycloaddition to C_{120}. The mono [77] and decakis [104] cyclopropanation of C_{120} by diethyl bromomaronate (Scheme 6) was reported by Fujiwara et al.

The cycloaddition to a fullerene derivative that already has a substituent can take place in nine possible sites (Fig. 12). For the case of bis addition of diethylbromomaronate to C_{60}, which forms bis[di(ethoxycarbonyl)-methano][60]fullerene [$C_{62}(CO_2Et)_4$ (**36**)], Hirsch et al. found that the cis-1 isomer was not formed; the relative yields of the other isomers are summarized in Table 2 [105]. In the case of cyclopropanated C_{120} ($C_{121}(CO_2Et)_2$), besides the cis-1 isomer, neither cis-2 nor cis-3 isomers were formed. That is probably due to the larger steric effect of the C_{60} group than that for the diethoxycarbonylmethano group.

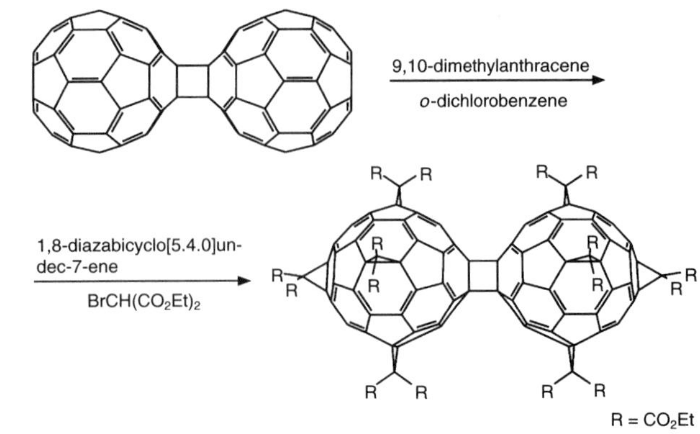

Scheme 6. Cyclopropanation of C_{120} by diethyl bromomaronate.

A trans-1 isomer also was not found, probably due to a low yield because of fewer reaction sites than for the other isomers. The trans-1 isomer is also the lowest yield isomer of $C_{62}(CO_2Et)_4$, except for the cis-1 isomer. Aside from the cis- and trans-1 isomers, the orders of yield of isomers of $C_{121}(CO_2Et)_2$ and $C_{62}(CO_2Et)_4$ are the same (Table 2), and hence are governed by the electronic effects on the C_{60} cage.

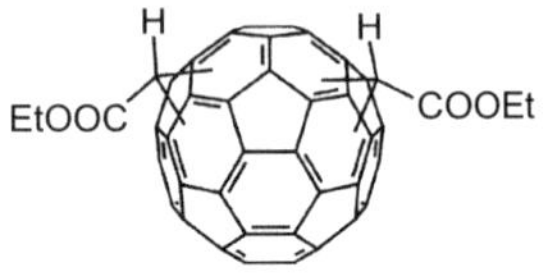

36

When an excess amount of diethyl bromomaronate (17 equivalents) was used for synthesis, surprisingly, a highly symmetrical decakis-cyclopropanated derivative of C_{120} (**37**) was synthesized in 26% yield. This product was cleaved by light irradiation or by heating, similarly to C_{120}, to give pentakis adducts of C_{60} (Scheme 7). The decakis adduct can be transformed into the eicosa–carboxylic acid derivative in 91% yield (Scheme 7).

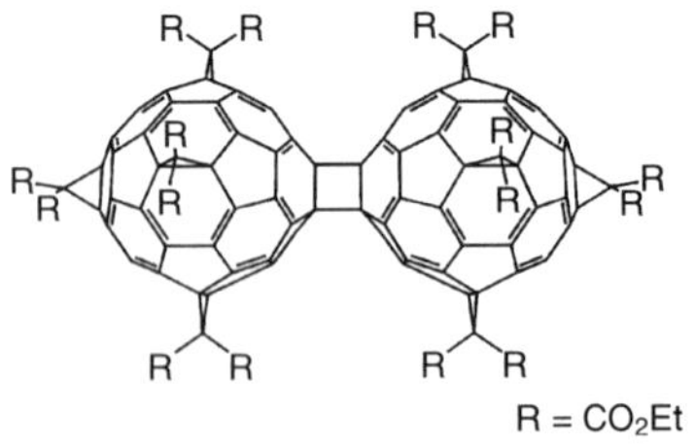

37

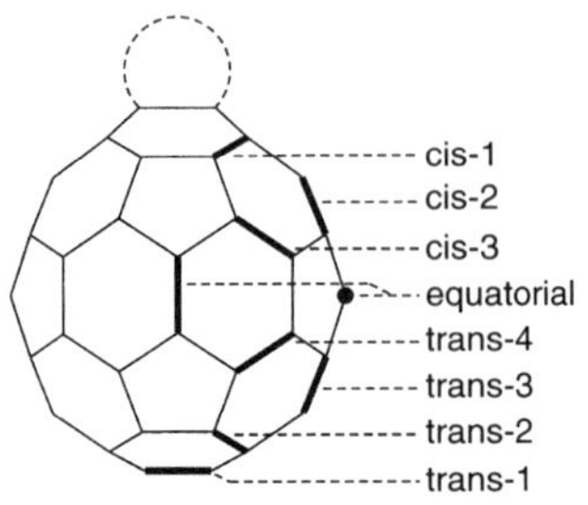

Figure 12. Nine possible reaction sites of monofunctionalized fullerenes by cycloaddition.

Table 2. Relative yield of the isomers of $C_{62}(CO_2Et)_4$ (**36**) and $C_{121}(CO_2Et)_2$.

	Yield (%)	
Isomer	$C_{62}(CO_2Et)_4$	$C_{121}(CO_2Et)_2$
cis-1	0	0
cis-2	2	0
cis-3	6	0
equatorial	38	58.9
trans-4	9	6.9
trans-3	30	26.9
trans-2	13	7.3
trans-1	2	0

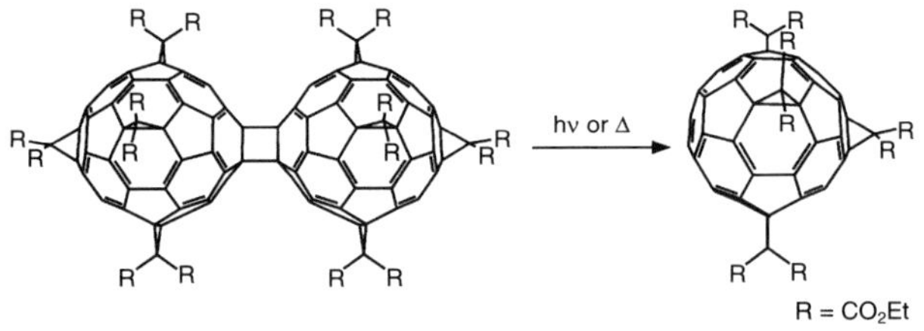

Scheme 7. Cleavage and hydrolysis of the decakis adduct of C_{120}.

GLOSSARY

Carbon nanotube Cylindrical carbon clusters. All carbon atoms in carbon nanotubes are trivalent. The major part of nanotubes consists of a graphene rolled into a cylinder.

Endohedral fullerene Fullerenes encapsulating one or some atoms in their cage.

Fullerene Hollow closed-cage molecules consist of even number of carbon atoms. The number of carbon atoms is equal or greater than 20. All carbon atoms of fullerenes are trivalent.

Higher fullerene Fullerenes that consist of more than 60 carbon atoms.

Homofullerene Fullerene with an methanoannulene moiety in its cage. The methanoannulene moiety is formed by fusion of a five-membered ring and a neighboring six-membered ring for all reported homofullerenes.

Methanofullerene Fullerene with a methanobridge. The methanobridge is at a bond between two six-membered rings for all reported methanofullerenes.

ACKNOWLEDGMENTS

We are grateful for the financial support of the Japan Science and Technology Corporation (JST) under the SORST program.

REFERENCES

1. H. W. Kroto, J. R. Heath, S. C. O'Brien, R. F. Curl, and R. E. Smalley, *Nature* 318, 162 (1985).
2. W. Krätschmer, L. D. Lamb, K. Fostiropoulos, and D. R. Huffman, *Nature* 347, 354 (1990).

3. A. F. Hebard, M. J. Rosseinsky, R. C. Haddon, D. W. Murphy, S. H. Glarum, T. T. M. Palstra, A. P. Ramirez, and A. R. Kortan, *Nature* 350, 600 (1991).
4. P. M. Allemand, K. C. Khemani, A. Koch, F. Wudl, K. Holczer, S. Donovan, G. Gruner, and J. D. Thompson, *Science* 253, 301 (1991).
5. Z. H. Kafafi, J. R. Lindle, R. G. S. Pong, F. J. Bartoli, L. J. Lingg, and J. Milliken, *Chem. Phys. Lett.* 188, 492 (1992).
6. S. H. Friedman, D. L. Decamp, R. P. Sijbesma, G. Sradanov, F. Wudl, and G. L. Kenyou, *J. Am. Chem. Soc.* 115, 6506 (1993).
7. P. J. Fagan, J. C. Calabrese, and B. Malone, *Acc. Chem. Res.* 25, 134 (1992).
8. F. Diederich and C. Thilgen, *Science* 271, 317 (1996).
9. A. Hirsch, "The Chemistry of the Fullerenes." Georg Thiem Verlag, New York, 1994.
10. R. Taylor, *Synlett* 776 (2000).
11. R. Taylor, "Lecture Note on Fullerene Chemistry." Imperial College Press, London, 1999.
12. In this chapter, we use the terminology "dimer," although for most of the cases, the proper term is "bisfullerene."
13. S. W. McElvany, J. H. Callahan, M. M. Ross, L. D. Lamb, and D. R. Huffman, *Science* 260, 1632 (1993).
14. A. L. Balch, D. A. Costa, W. R. Fawcett, and K. Winkler, *J. Phys. Chem.* 100, 4823 (1996).
15. S. Patchkovskii and W. Thiel, *J. Am. Chem. Soc.* 120, 556 (1998).
16. A. M. Rao, P. Zhou, K. A. Wang, G. T. Hager, J. M. Holden, Y. Wang, W. T. Lee, X.-X. Bi, P. C. Eklund, D. S. Cornett, M. A. Duncan, and I. J. Amster, *Science* 259, 955 (1993).
17. C. Yeretzian, K. Hansen, F. Diedrich, and R. L. Whetten, *Nature* 359, 44 (1992).
18. G.-W. Wang, K. Komatsu, Y. Murata, and M. Shiro, *Nature* 387, 583 (1997).
19. S. Lebedkin, A. Gromov, S. Giesa, R. Gleiter, B. Renker, H. Rietschel, and W. Krätschmer, *Chem. Phys. Lett.* 285, 210 (1998).
20. K. Komatsu, G.-W. Wang, Y. Murata, T. Tanaka, K. Fujiwara, K. Yamamoto, and M. Saunders, *J. Org. Chem.* 63, 9358 (1998).
21. B. Burger, J. Winter, and H. Kuzmany, *Synth. Met.* 86, 2329 (1997).
22. V. A. Davydov, L. S. Kashevarova, A. K. Rakhmanina, V. M. Senyavin, V. Agafonov, R. Ceolin, and H. Szwarc, *JETP Lett.* 68, 928 (1998).
23. Y. Iwasa, K. Tanoue, T. Mitani, A. Izuoka, T. Sugawara, and T. Yagi, *Chem. Commun.* 1411 (1998).
24. V. C. Long, J. L. Musfeld, K. Kamaras, G. B. Adams, J. B. Page, Y. Iwasa, and W. E. Mayo, *Phys. Rev. B* 61, 13191 (2000).
25. K. Komatsu, G.-W. Wang, Y. Murata, T. Tanaka, and K. Komatsu, *J. Org. Chem.* 63, 9358 (1998).
26. B. Goedde, M. Waiblinger, P. Jakes, N. Weiden, K.-P. Dinse, and A. Weidinger, *Chem. Phys. Lett.* 334, 12 (2001).
27. S. Osawa, J. Onoe, and K. Takeuchi, *Fullerene Sci. Technol.* 6, 301 (1998).
28. S. W. McElvany, J. H. Callahan, M. M. Ross, L. D. Lamb, and D. R. Huffman, *Science* 260, 1632 (1993).
29. A. Gromov, S. Ballenweg, S. Giesa, S. Lebedkin, W. E. Hull, and W. Krätschmer, *Chem. Phys. Lett.* 267, 460 (1997).
30. W. Krätschmer (personal communication).
31. D. K. Yu, R. Q. Zhang, and S. T. Lee, *Eur. Phys. J. D* 15, 57 (2001).
32. P. Paul, K.-C. Kim, D. Sun, P. D. W. Boyd, and C. A. Reed, *J. Am. Chem. Soc.* 124, 4394 (2002).
33. S. Lebedkin, S. Ballenweg, J. Gross, R. Taylor, and W. Krätschmer, *Tetrahedron Lett.* 36, 4971 (1995).
34. A. B. Smith, H. Tokuyama, R. M. Strongin, G. T. Furst, W. J. Romanow, B. T. Chait, U. A. Mirza, and I. Haller, *J. Am. Chem. Soc.* 117, 9359 (1995).
35. R. Taylor, M. P. Barrow, and T. Drewello, *Chem. Commun.* 2497 (1998).
36. K. Fujiwara and K. Komatsu, *Org. Lett.* 4, 1039 (2002).
37. A. Gromov, S. Lebedkin, S. Ballenweg, A. G. Avent, R. Taylor, and W. Krätschmer, *Chem. Commun.* 209 (1997).
38. A. Gromov, S. Ballenweg, W. Krätschmer, S. Giesa, and S. Lebedkin, in "Fullerenes and Fullerene Nanostructures, Proceedings of the International Winterschool on Electronic Properties of Novel Materials" (H. Kuzmany, J. Fink, M. Mehring, and S. Roth, Eds.), p. 96. World Scientific, Singapore, 1998.
39. S. Giesa, J. H. Gross, W. E. Hull, S. Lebedkin, A. Gromov, R. Gleiter, and W. Krätschmer, *Chem. Commun.* 465 (1999).
40. A. Gromov, S. Lebedkin, W. E. Hull, and W. Krätschmer, *J. Phys. Chem.* 102, 4997 (1998).
41. N. Dragoe, H. Shimotani, M. Hayashi, K. Saigo, A. de Bettencourt-Dias, A. L. Balch, Y. Miyake, Y. Achiba, and K. Kitazawa, *J. Org. Chem.* 65, 3269 (2000).
42. J. C. Hummelen, B. Knight, J. Pavlovich, R. Gonzalez, and F. Wudl, *Science* 269, 1554 (1995).
43. M. Keshavarz-K, R. González, R. G. Hicks, G. Srdanov, V. L. Srdanov, T. G. Collins, J. C. Hummelen, C. Bellavia-Lund, J. Pavlovich, F. Wudl, and K. Holczer, *Nature* 383, 147 (1996).
44. U. Reuther and A. Hirsch, *Carbon* 38, 1539 (2000).
45. F. Fulop, F. Simon, A. Rockenbauer, L. Korecz, and A. Janossy, "CP591 Electronic Properties of Molecular Nanostructures" (H. Kuzmany, Ed.), p. 11. American Institute of Physics, 2001.
46. P. J. Fagan, P. J. Krusic, D. H. Evans, S. A. Lerke, and E. Johnston, *J. Am. Chem. Soc.* 114, 9697 (1992).
47. K. Komatsu, N. Takimoto, Y. Murata, T. S. M. Wan, and T. Wong, *Tetrahedron Lett.* 37, 6153 (1996).
48. M. Yoshida, A. Morishima, Y. Morinaga, and M. Iyoda, *Tetrahedron Lett.* 35, 9045 (1994).
49. S. Osawa, E. Osawa, and M. Harada, *J. Org. Chem.* 61, 257 (1996).
50. G. Schick, K.-D. Kampe, and A. Hirsch, *J. Chem. Soc., Chem. Commun.* 2023 (1995).
51. N. Dragoe, S. Tanibayashi, K. Nakahara, S. Nakao, H. Shimotani, L. Xiao, K. Kitazawa, Y. Achiba, K. Kikuchi, and K. Nojima, *Chem. Commun.* 85 (1995).
52. H. L. Anderson, R. Faust, Y. Rubin, and F. Diederich, *Angew. Chem., Int. Ed. Engl.* 33, 1366 (1994).
53. T. S. Fabre, W. D. Treleaven, T. D. McCarley, C. L. Newton, R. M. Landry, M. C. Saraiva, and R. M. J. Strongin, *J. Org. Chem.* 63, 3522 (1998).
54. N. Dragoe, K. Nakahara, H. Shimotani, L. Xiao, S. Tanibayashi, T. Higuchi, and K. Kitazawa, *Fullerene Sci. Technol.* 8, 545 (2000).
55. J. Osterodt and F. Vögtle, *Chem. Commun.* 547 (1996).
56. P. Belik, A. Gugel, J. Spickerman, and K. Mullen, *Angew. Chem., Int. Ed. Engl.* 32, 78 (1993).
57. A. L. de Lucas, N. Martín, L. Sánchez, and C. Seoane, *Tetrahedron Lett.* 37, 9391 (1996).
58. J. L. Segura and N. Martín, *Tetrahedron Lett.* 40, 3239 (1999).
59. J. M. Lawson, A. M. Oliver, D. F. Rothenfluh, Y.-Z. An, G. A. Ellis, M. G. Ranasinghe, S. L. Khan, A. G. Franz, P. S. Ganapathi, M. J. Shephard, M. N. Paddon-Row, and Y. Rubin, *J. Org. Chem.* 61, 5032 (1996).
60. P. Belik, A. Gügel, J. Spickermann, and K. Müllen, *Angew. Chem.* 105, 9597 (1993); *Angew. Chem. Int. Ed. Engl.* 32, 78 (1993).
61. A. Gügel, P. Belik, M. Walter, A. Kraus, E. Harth, M. Wagner, J. Spickermann, and K. Müllen, *Tetrahedron* 52, 5007 (1996).
62. Y. Murata, N. Kato, K. Fujiwara, and K. Komatsu, *J. Org. Chem.* 64, 3438 (1999).
63. G. P. Miller and J. Mack, *Org. Lett.* 2, 3979 (2000).
64. K. Lee, H. Song, B. Kim, J. T. Park, S. Park, and M.-G. Choi, *J. Am. Chem. Soc.* 124, 2872 (2002).
65. D. H. Chi, Y. Iwasa, X. H. Chen, T. Takenobu, T. Ito, T. Mitani, E. Nishibori, M. Takata, M. Sakata, and Y. Kubozono, *Chem. Phys. Lett.* 359, 177 (2002).
66. H. Irngartinger and A. Weber, *Tetrahedron Lett.* 37, 4137 (1996).

67. D. Kreher, M. Cariou, S.-G. Liu, E. Levillain, J. Veciana, C. Rovira, A. Gorgues, and P. Hudhomme, *J. Mater. Chem.* 12, 2137 (2002).
68. F. Diederich, L. Echegoyen, M. Gomez-Lopez, R. Kessinger, and J. F. Stoddart, *J. Chem. Soc., Perkin Trans. 2* 1577 (1999).
69. J. J. González, S. González, E. M. Priego, C. Luo, D. M. Guldi, J. de Mendoza, and N. Martín, *Chem. Commun.* 163 (2001).
70. Y. Murata, M. Suzuki, and K. Komatsu, *Chem. Commun.* 2338 (2001).
71. Y. Murata, N. Kato, and K. Komatsu, *J. Org. Chem.* 66, 7235 (2001).
72. J. Knol and J. C. Hummelen, *J. Am. Chem. Soc.* 122, 3226 (2000).
73. H. Isobe, A. Ohbayashi, M. Sawamura, and E. Nakamura, *J. Am. Chem. Soc.* 122, 2669 (2000).
74. J. H. Gross, S. Giesa, and W. Krätschmer, *Rapid Commun. Mass Spectrom.* 13, 815 (1999).
75. A. B. Smith, III, H. Tokuyama, R. M. Strongin, G. T. Furst, W. J. Romanow, B. T. Chait, U. A. Mirza, and I. Haller, *J. Am. Chem. Soc.* 117, 9359 (1995).
76. N. Dragoe, H. Shimotani, J. Wang, M. Iwaya, A. de Bettencourt-Dias, A. L. Balch, and K. Kitazawa, *J. Am. Chem. Soc.* 123, 1294 (2001).
77. K. Fujiwara, K. Komatsu, G.-W. Wang, T. Tanaka, K. Hirata, K. Yamamoto, and M. Saunders, *J. Am. Chem. Soc.* 123, 10715 (2001).
78. S. Lebedkin, H. Rietschel, G. B. Adams, J. B. Page, W. E. Hull, F. H. Hennrich, H.-J. Eisler, M. M. Kappes, and W. Krätschmer, *J. Chem. Phys.* 110, 11768 (1999).
79. S. Lebedkin, A. Gromov, S. Giesa, R. Gleiter, B. Renker, H. Rietschel, and W. Krätschmer, *Chem. Phys. Lett.* 285, 210 (1998).
80. D. Porezag, M. R. Pederson, T. Frauenheim, and T. Köhler, *Phys. Rev. B* 50, 14963 (1995).
81. M. M. Olmstead, D. A. Costa, K. Maitra, B. C. Noll, S. L. Phillips, P. M. Van Calcar, and A. L. Balch, *J. Am. Chem. Soc.* 121, 7090 (1999).
82. P. W. Fowler, D. Mitchell, R. Taylor, and G. Seifert, *J. Chem. Soc. Perkin Trans. 2* 1901 (1997).
83. H. Shimotani, N. Dragoe, and K. Kitazawa, *Phys. Chem. Chem. Phys.* 4, 428 (2002).
84. N. Matsuzawa, D. A. Dixon, and T. Fukunaga, *J. Phys. Chem.* 96, 7594 (1992).
85. F. Diederich, L. Isaacs, and D. Philp, *Chem. Soc. Rev.* 23, 243 (1994).
86. H. Shimotani, N. Dragoe, and K. Kitazawa, *J. Phys. Chem. A* 105, 4980 (2001).
87. A. A. Shvartsburg, R. R. Hudgins, R. Gutierrez, G. Jungnickel, T. Frauenheim, K. A. Jackson, and M. F. Jarrold, *J. Phys. Chem. A* 103, 5275 (1999).
88. D. K. Yu, R. Q. Zhang, and S. T. Lee, *Eur. Phys. J. D* 15, 57 (2001).
89. A. L. Balch, D. A. Costa, W. R. Fawcett, and K. Winkler, *J. Phys. Chem.* 100, 4823 (1996).
90. C. Boudon, J. P. Gisselbrecht, M. Gross, L. Isaacs, H. L. Anderson, R. Faust, and F. Diederich, *Helv. Chim. Acta.* 78, 1334 (1995).
91. B. Ma, J. E. Riggs, and Y.-P. Sun, *J. Phys. Chem. B* 102, 5999 (1998).
92. M. Fujitsuka, C. Luo, O. Ito, Y. Murata, and K. Komatsu, *J. Phys. Chem. A* 103, 7155 (1999).
93. H. S. Cho, S. K. Kim, D. Kim, K. Fujiwara, and K. Komatsu, *J. Phys. Chem. A* 104, 9666 (2000).
94. K. Komatsu, G.-W. Wang, Y. Murata, T. Tanaka, K. Fujiwara, K. Yamamoto, and M. Saunders, *J. Org. Chem.* 63, 9358 (1998).
95. S. Lebedkin, W. E. Hull, A. Soldatov, B. Renker, and M. M. Kappes, *J. Phys. Chem. B* 104, 4101 (2000).
96. S. M. Bachilo, A. F. Benedetto, and R. B. Weisman, *J. Phys. Chem. A* 105, 9845 (2001).
97. M. Fujitsuka, H. Takahashi, T. Kudo, K. Tohji, A. Kasuya, and O. Ito, *J. Phys. Chem. A* 105, 675 (2001).
98. R. Taylor, M. P. Barrow, and T. Drewello, *Chem. Commun.* 2497 (1998).
99. A. Gruss, K.-P. Dinse, A. Hirsch, B. Nuber, and U. Reuther, *J. Am. Chem. Soc.* 119, 8727 (1997).
100. M. Fujitsuka, O. Ito, N. Dragoe, S. Ito, H. Shimotani, H. Takagi, and K. Kitazawa, *J. Phys. Chem. B* 106, 8562 (2002).
101. M. Keshavarz-K, R. González, R. G. Hicks, G. Srdanov, V. I. Srdanov, T. G. Collins, J. C. Hummelen, C. Bellavia-Lund, J. Pavlovich, F. Wudl, and K. Holczer, *Nature* 383, 147 (1996).
102. A. M. Rao, P. Zhou, K. A. Wang, G. T. Hager, J. M. Holden, Y. Wang, W. T. Lee, X. X. Bi, P. C. Eklund, D. S. Cornett, M. A. Duncan, and I. J. Amster, *Science* 259, 955 (1993).
103. N. Dragoe, H. Shimotani, and K. Kitazawa (unpublished).
104. K. Fujiwara and K. Komatsu, *Chem. Commun.* 1986 (2001).
105. A. Hirsch, I. Lamparth, and H. R. Karfunkel, *Angew. Chem., Int. Ed. Engl.* 33, 437 (1994).

Encyclopedia of Nanoscience and Nanotechnology
www.aspbs.com/enn

Fullerene Lipid Films

Naotoshi Nakashima
Nagasaki University, Bunkyo, Nagasaki, Japan

CONTENTS

1. INTRODUCTION

The chemistry and physics of fullerenes have received increasing attention by many research groups [1–6]. The solution electrochemistry of fullerenes and the electrochemistry of fullerene films have been widely studied, and are the subject of an intense research focus from both theoretical and experimental approaches [7–10]. Fullerenes with long chain derivatives form a variety of aggregated structures, such as monolayers, self-assembled monolayers (so-called SAM), lipid bilayers, Langmuir–Blodgett (LB) films, and liquid crystals (Fig. 1). The present author [11–27] and others [28–42] have been interested in combining fullerene chemistry and the chemistry of lipid bilayer membranes. Here, we mainly focus on fullerenes with an ordered structure in aqueous systems, and briefly review a lipid bilayer-based aqueous superstructure from fullerene amphiphiles, as well as the structure and electrochemistry of fullerene lipid films and fullerene–lipid composite films.

2. BILAYER-BASED SUPERSTRUCTURES OF FULLERENE–AMPHIPHILES IN WATER

A variety of artificial lipids are known to form liposomal lipid bilayer membranes in water [43–44]. Some of the bilayer aggregates show the growth to superstructures, such as helical bilayers and tubes [45–48]. In this section, molecular bilayer-based superstructure formations from water-soluble fullerenes are summarized. Cassell et al. described that C_{60}-*N*,*N*-dimethylpyrrolidinium iodide **1** (Fig. 2) forms supramolecular nanorods (Fig. 3) in dimethylsulfoxide with one part water and then adding one part benzene, and that an aqueous solution of the compound treated with ultrasonication and then filtered gives vesicles with diameters of 10–70 nm [49]. Sano and coworkers described the vesicle formation by a bola-amphiphilic fullerene **2** (Fig. 2), although the bilayer structure is not evident [37]. Hirsch et al. revealed by transmission electron microscopy (TEM) of a freeze fracture replica that a C_{60}-carrying globular amphiphile **3** (Fig. 2) forms a vesicular structure in water [34, 38]. Zhou et al. reported that an amphiphile **4** (Fig. 2), a potassium salt of pentaphenylfullerene with charged cyclopentadienide units, forms spherical vesicles in water (Fig. 4) [41]. The average hydrodynamic radius and aggregation number of the vesicle evaluated by laser light scattering were about 17 nm and 1.2×10^4, respectively.

We synthesized a C_{60}-carrying ammonium amphiphile **5** (Fig. 2), which is soluble in DMF, DMSO, hot chloroform, and hot alcohols, but was insoluble in THF, acetonitrile, ethylacetate, hexane, and toluene [17, 20]. Negative-stained transmission electron microscopy (TEM) revealed that the aqueous solution of **5** forms both fibrous aggregates (Fig. 5) and disk-like aggregates with 10–12 nm of thickness. This superstructure is formed via self-organization of **5** in aqueous solution. A possible molecular model for the disk-like aggregate is shown in Figure 6. The wavenumber of the asymmetric and symmetric methylene stretching vibrations in the FTIR spectra of cast films of **5** from both methanol and aqueous solutions appeared at 2922 ± 0.1 and 2850 ± 0.1 cm^{-1}, respectively, over a temperature range of 10–40 °C, indicating that the methylene chain in **5** contains a *gauche* conformation. The UV–visible absorption maximum for an aqueous solution of **5** appeared at 271 nm; this

ISBN: 1-58883-059-4/$35.00

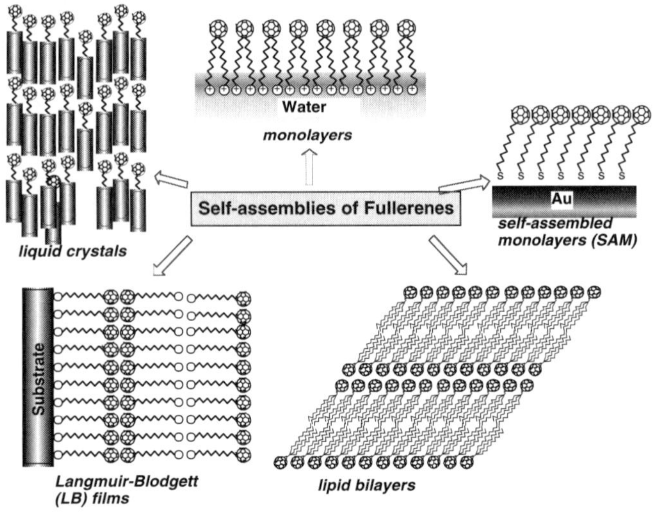

Figure 1. Schematic drawing of self-assembled fullerenes.

maximum is shifted to longer wavelength by 3 and 5 nm, respectively, compared to those in micellar solutions of hexadecyltrimethylammonium bromide or sodium dodecylsulfate. These shifts are due to the electronic interaction of the fullerene moieties in **5** in an aggregated state.

The phase transition between a crystal phase and a liquid-crystal phase is the most fundamental characteristic of lipid bilayer membranes [50]; however, aqueous aggregates of **1–5** do not possess a phase transition. The results described above indicate that the introduction of a hydrophilic moiety to fullerene enables the formation of bilayer-based assemblies in water. This kind of fullerene-based biomembrane-like carbon nanosuperstructures in aqueous solution is of interest from aspects of both fundamental and biological applications [51]. The construction of electroactive fullerene nanoarchitectures with an ordered structure based on self-assembled molecular organization would possess many chemical and biochemical applications.

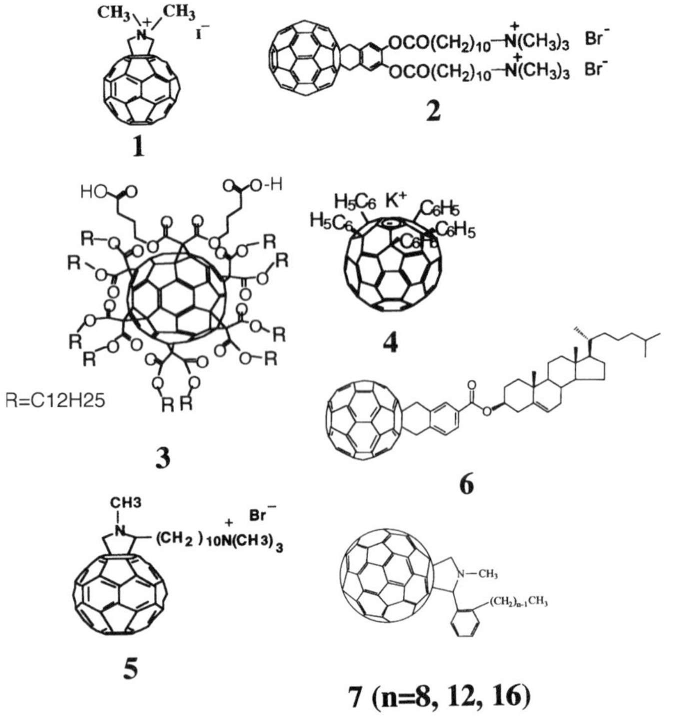

Figure 2. Chemical structure of compounds **1–7**.

Figure 3. TEM micrograph of the nanorod formed from compound **1**. Reprinted with permission from [49], M. Cassell et al., *Angew. Chem. Int. Ed.* 38, 2403 (1999). © 1999, Wiley-VCH.

3. FULLERENE GEL

Oishi et al. and Ishi-i et al. reported the formation of organic gel from fullerene derivatives in organic solvents [52–53]. They found that a methanol solution of the fullerene-carrying diammonium salt **2** (10–20 mM) transfers gradually

Figure 4. Bilayer vesicle model for aqueous solution of **4**. Reprinted with permission from [41], S. Zhou et al., *Science* 291, 1944 (2001). © 2001, American Association for the Advancement of Science.

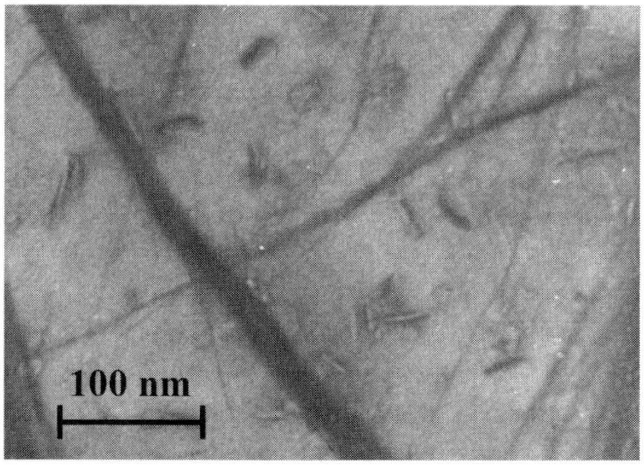

Figure 5. Typical TEM image of an aqueous solution of compound 5. Reprinted with permission from [20], N. Nakashima et al., *Chem. Eur. J.* 7, 1766 (2001). © 2001, Wiley-VCH.

into an organic gel [52]. The TEM image of the gel showed a number of fibrous aggregates with a 10–20 nm diameter. They also reported chirally ordered fullerene assemblies in organic gel systems of a cholesterol-appended C_{60} fullerene **6** (see Fig. 2) [53]. We have described that gel-like films of a tetraoctylphosphonium ion on electrodes provide suitable microenvironments for the electrochemistry of C_{60} [19].

4. STRUCTURE OF FULLERENE THIN FILMS

Cast films of fullerenes with long chains form multilayered structures. By means of X-ray study, cast films of **5** were found to show a molecular bilayer structure (Fig. 7) [21]. Cast films of **2** show sharp, periodical peaks up to order 13, indicating the formation of a well-ordered multibilayer structure [52]. Chikamatsu et al. synthesized compound **7** ($n = 8, 12, 16$) (Fig. 2), and suggested that cast films of **7** form bilayers with interdigitated structures [54]. No phase transition appears for cast films of these compounds.

We reported the synthesis and characterization of self-assembled triple-chain fullerene lipids **8–10** (Fig. 8), where the cross-sectional area of the triple alkyl chain is close to that of C_{60} moiety [11, 15–16, 24]. Compounds **8–10** are not soluble in water and methanol because of the lack of a high hydrophilic moiety in the chemical structure, whereas they are soluble in a variety of organic solvents such as DMF, DMSO, chloroform, benzene, and hexane. They are also soluble in ethanol and isopropanol, although the solubility in the latter two solvents is not high.

Cast films of **8–10** are readily prepared by solvent evaporation. The structure of fullerene lipid bilayer films was investigated by X-ray diffraction. Cast films of **8** form a multibilayer structure with the molecular layer tilting by 42.8° from the basal plane [11, 24]. Cast films of **9** and **10** gave diffraction peaks at $2\theta = 1.97°$ and $2\theta = 2.12°$, respectively, which can be assigned to the (001) plane of these films. These data indicate that the cast films of both **9** and **10** also form multibilayer structures. The X-ray diagrams of the films were temperature dependent. The diffraction peak of the **8**-film maintained at temperatures below 35 °C, but it almost disappeared at 50 °C, indicating that the **8**-film at the higher temperatures is rather disordered. This temperature dependence is reversible, that is, when the temperature is cycled back to 25 °C, the X-ray intensity of the films is recovered, although aging is required.

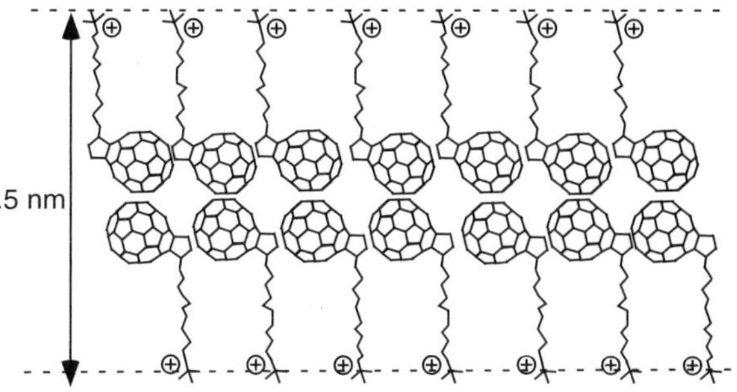

Figure 7. Drawing for the molecular-bilayer structure of 5. Reprinted with permission from [20], N. Nakashima et al., *Chem. Eur. J.* 7, 1766 (2001). © 2001, Wiley-VCH.

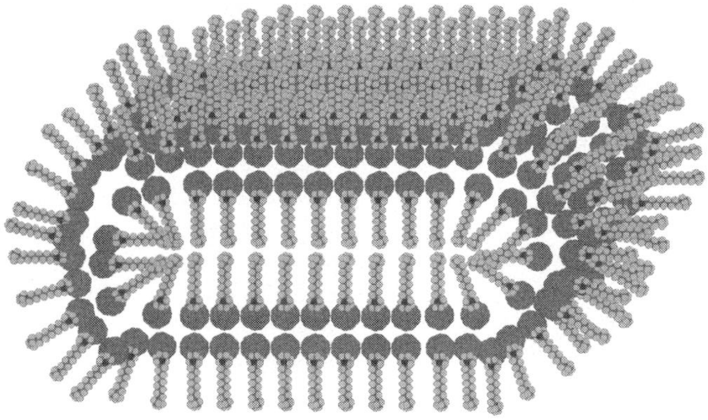

Figure 6. Cartoon model of the disk-like structure of 5. Reprinted with permission from [20], N. Nakashima et al., *Chem. Eur. J.* 7, 1766 (2001). © 2001, Wiley-VCH.

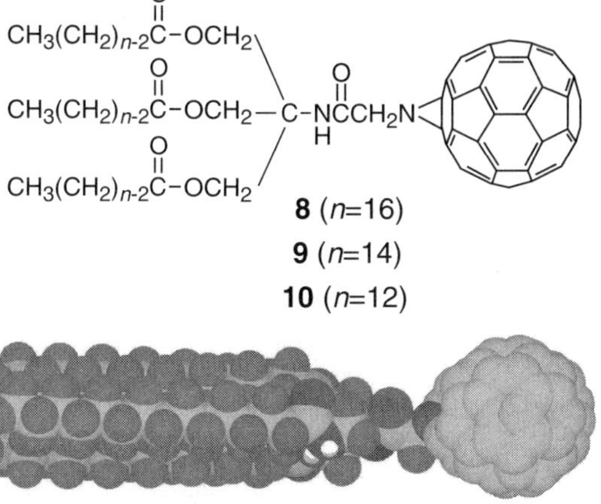

Figure 8. Chemical structure of fullerene lipids 8–10 and the CPK model of 8. Reprinted with permission from [24], T. Nakashima et al., *Chem. Eur. J.* 8, 1641 (2002). © 2002, Wiley-VCH.

5. PHASE TRANSITION OF FULLERENE LIPID FILMS

An interesting feature of cast films of **8–10** is that they show phase transitions [11, 24]. The phase transition of lipid bilayers between the crystalline and liquid crystalline phases is one of the most fundamental characteristics of lipid bilayer membranes, and the lipid bilayer properties depend on the fluidity change of the bilayers [50]. Differential scanning calorimetry (DSC) of cast films of **8** show two endothermic peaks at 35.2 °C (main peak) and 47.0 °C (subpeak) in air, 39.7 °C (main peak) and 47.6 °C (subpeak) in the presence of water, 33.8 °C (main peak) and 47.9 °C (subpeak) in the presence of 0.5 M aqueous tetraethylammonium chloride (TEAC), and 31.5 °C (main peak) and 43.8 °C (subpeak) in the presence of acetonitrile. Interestingly, the phase transition behavior of **8** in various media is not much different, that is, the lipid possesses a similar molecular orientation in the different environments. Retention of the phase transition of lipids in organic solvents is not very unusual [55].

6. SPECTRAL PROPERTIES OF FULLERENE LIPID FILMS

Cast films of fullerene lipids exhibit unique spectral properties [11, 24]. The wavenumber of the asymmetric and symmetric methylene stretching vibrations in the FT–IR spectra of the **8** films changed drastically near 35 °C (Fig. 9a). The shifts of $\nu_{as}(CH_2)$ from 2917.9 to 2922.0 cm^{-1} and $\nu_s(CH_2)$ from 2849.2 to 2851.0 cm^{-1} are ascribable to the *trans-gauche* conformational change of the long alkyl chain [56–57], indicating that the main peak in the DSC thermogram of the **8** films is attributable to the bilayer phase transition typically observed for liposomal and synthetic lipid bilayer membranes. In contrast, no temperature dependence is observed for cast films of **9** and **10** (Fig. 9b, c). From the observed $\nu_{as}(CH_2)$ and $\nu_s(CH_2)$ vibration, both **9** and **10** are found to form *gauche* conformation in the alkyl chains at the measured temperatures; therefore, we detect no main transition for films of **9** and **10**. The origin of the small endothermic peaks is revealed by UV–vis spectroscopy. The UV–vis spectra of films of **8–10** in air at temperatures below 35 °C showed three bands (labeled as bands *A*, *B*, and *C*) with absorption peaks near 215, 262, and 330 nm, respectively (Fig. 10); the peak maxima for bands *A*, *B*, and *C* shifted to longer wavelengths upon increasing the temperature above the small endothermic peaks. Thus, we know that the small endothermic peaks are due to the change in the orientation of the C_{60} moieties in the films. Interestingly, the phase transition of the films regulates an electronic interaction between the C_{60} moieties of lipids **8–10**.

7. A MODEL FOR FULLERENE LIPID BILAYERS

All data obtained from X-ray study, DSC, and spectral measurements suggest a possible schematic model for the structure and phase transition behavior for cast films of **8–10** (Fig. 11). Phase 1 consists of the lipid with a rigid crystalline state, where all alkyl chains form a *trans* conformation. Lipid

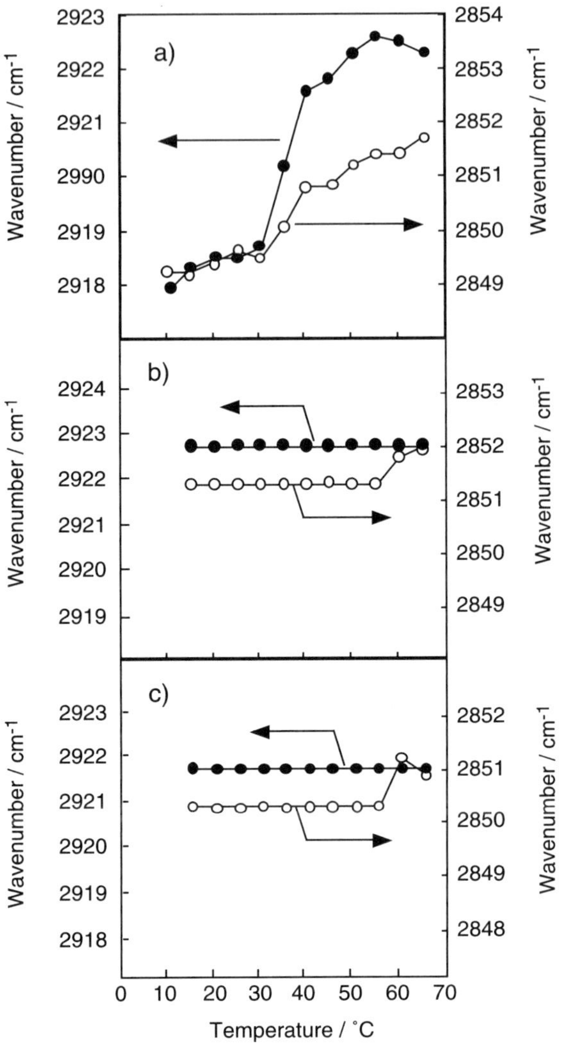

Figure 9. Temperature dependence of the wavenumber in the FTIR spectra of cast films of (a) **8**, (b) **9**, and (c) **10**. Reprinted with permission from [24], T. Nakanishi et al., *Chem. Eur. J.* 8, 1641 (2002). © 2002, Wiley-VCH.

8 only forms this state. Phase 2 is a fluid state in which the alkyl chains contain a *gauche* conformation, while the orientation of the fullerene moieties is almost the same as in phase 1. Phase 3 is a fluid state with a less ordered structure than phases 1 and 2 [24].

8. MONOLAYERS AND LANGMUIR–BLODGETT (LB) FILMS OF A FULLERENE LIPID

The derivatives of C_{60} and C_{70} form monolayers at the air–water interface, and the formed monolayers can be transferred onto solid substrates [58–70]. Single components of C_{60} and C_{70} form monolayers on water under limited experimental conditions [66–68]. The fullerene lipid **8** forms stable monolayers at the air–water interface [15, 71]. The surface pressure–area isotherm of **8** from the spreading solution of 1.0×10^{-4} M gives a limiting area of 0.27 nm^2/molecule, suggesting the formation of a multiplayer on water. The isotherms from the $\mathbf{8} = 1.0 \times 10^{-5}$ M

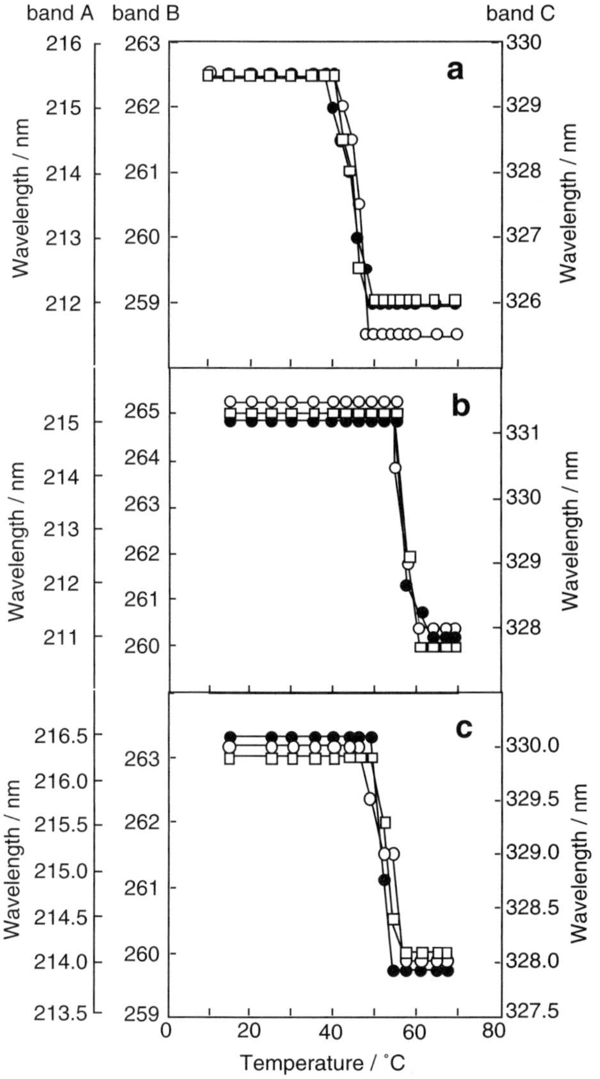

Figure 10. Temperature dependence of peak wavelengths in the UV–vis absorption spectra of cast films of (a) **8**, (b) **9**, and (c) **10**. Reprinted with permission from [24], T. Nakanishi et al., *Chem. Eur. J.* 8, 1641 (2002). © 2002, Wiley-VCH.

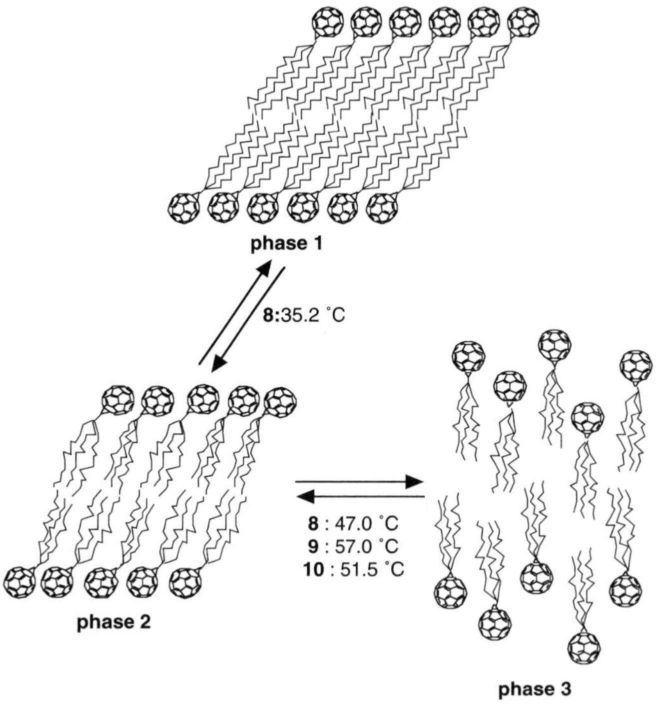

Figure 11. Possible model for the phase transitions of cast film of **8–10**. Reprinted with permission from [24], T. Nakanishi et al., *Chem. Eur. J.* 8, 1641 (2002). © 2002, Wiley-VCH.

solution give limiting areas of 0.78 and 0.98 nm^2/molecule, suggesting the formation of monolayers with hexagonal and the simple square packing, respectively. How is the molecule spatially arranged in the monolayer? An *in-situ* XR technique revealed the nanostructure of the fullerene lipid monolayer on water, that is, the fullerene moiety was in contact with the water surface, and the lipid tail extended to the air (Fig. 12) [71]. The isotherms obtained in the dark and in the light were almost identical; however, deposited LB films in the light or in the dark show different behavior. In the dark or in the light (ca. 190 lux), at a surface pressure of 30 mN/m, the monolayer of **8** was transferred onto a quartz plate.

The LB film fabricated in the light shows structureless electronic spectra with no temperature dependence, whereas the LB film prepared in the dark shows a temperature-dependent UV–vis absorption spectra (Fig. 13). At the lower temperatures, an electronic interaction between the fullerene moieties in the LB film prepared in the dark exists. Upon heating over 47 °C, which is identical to the subphase transition temperature of the cast films of **8** [11, 24], the electronic interaction is loosened. The temperature dependence is reversible, that is, the electronic interaction between the fullerene moieties is controllable by the phase change of the film. A light-driven aggregate structural change [72] in the LB film prepared in the light is suggested. Cast films of **8** did not show such a fluorescent light-driven change in the spectra. The molecular orientations of the fullerene moieties of the LB film and the cast film of **8** are not significantly different, that is, the fundamental property of the self-assembled bilayer membrane film is maintained in the LB film prepared in the dark. This is important from the viewpoint of the introduction of bilayer properties of the fullerene lipids to the LB system. This information is useful for designing and constructing new functional supramolecular ultrathin fullerene films at the air/water interface and on solid surfaces.

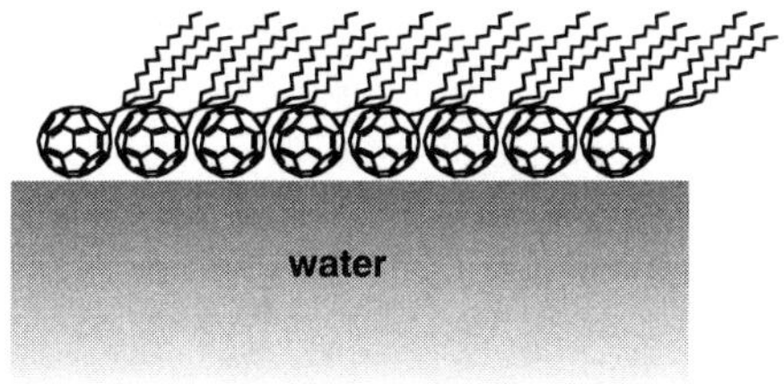

Figure 12. Model for the monolayer of fullerene lipids on water.

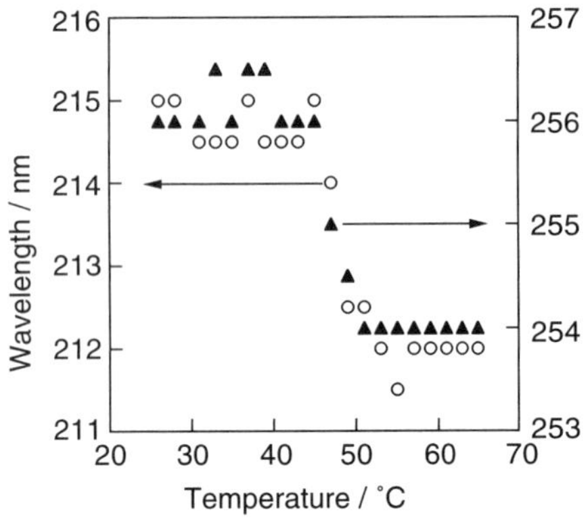

Figure 13. Temperature-dependent UV–vis absorption spectra of an LB film of **8**. Reprinted with permission from [15], T. Nakanishi et al., *Chem. Lett.* 1219 (1998). © 1998, The Chemical Society of Japan.

9. ELECTROCHEMISTRY OF FULLERENE THIN FILMS

The electrochemistry of fullerenes in solution phase and their films on electrode surfaces have been extensively studied, and are the subject of intense research focus. Chlistunoff and coworkers summarized half-wave potentials ($E_{1/2}$) of fullerenes in various media [7]. Supporting electrolyte cations, solvents, and temperature affect the reduction potentials of fullerenes. Dubois and coworkers investigated the correlation between $E_{1/2}$ values of the first three-consecutive one-electron transfer processes of C_{60} and the carbon chain length of tetraalkylammonium perchlorate supporting electrolytes in four different organic solvents [73]. They found a monotonous positive shift of $E_{1/2}$ for the first and second reductions with an increasing chain length. Fawcett and coworkers described the thermodynamics of the reduction of C_{60} in benzonitrile containing tetraalkylammonium salts [74]. They reported that $C_{60}^{\bullet-}$ and C_{60}^{2-} are associated with two or more tetraalkylammonium cations. S.-Guillous et al. quantitatively investigated the ion pairing of tetraalkylammonium cations with C_{60}^{n-} ($n = 1$–4) in four different aprotic organic solvents by cyclic voltammetry at a 10 mm diameter Pt disk microelectrode [75]. Chlistunoff et al. reviewed the electrochemistry of the pristine fullerenes and fullerene thin films in organic solution systems [7]. Echegoyen et al. summarized the redox properties of the pristine fullerenes and their large number of derivatives in an organic solution [9]. In this section, we review the aqueous electrochemistry for electrodes modified with films of: (1) C_{60}, (2) fullerene lipids, and (3) fullerenes (including higher fullerenes and metallofullerenes)–lipid composites (Fig. 14). The aqueous electrochemistry of fullerene films is of importance for both fundamentals and applications in understanding the science and nanotechnology of nanocarbons [16].

9.1. C_{60} Cast Thin Films

The aqueous electrochemistry of C_{60} cast films on electrodes is very unstable and complicated. Szucs et al. assumed that a C_{60} trianion is formed during the reduction in the presence

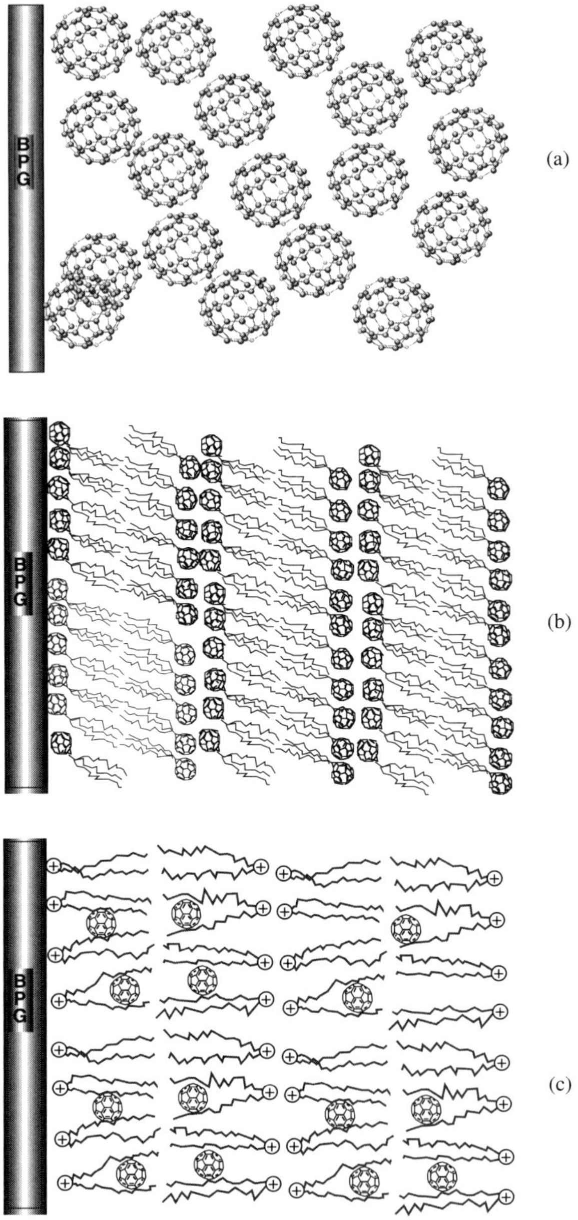

Figure 14. Schematic drawing for (a) C_{60}, (b) fullerene lipids, and (c) fullerene–lipid composites on electrodes.

of a sodium ion; however, the corresponding cathodic process was not detected [76]. Davis and coworkers reported that two reduction and reoxidation peaks appearing on the cyclic voltammograms (CVs) of C_{60} film-modified electrodes disappear after several cycles [77]. The results indicate that the stable aqueous electrochemistry of C_{60} films on electrodes is not possible. This is in sharp contrast to an organic solution system, in which C_{60} shows six reduction processes leading to the generation of a C_{60} hexaanion [7–10].

Szucs et al. examined the effect of cations, including Li^+, Na^+, K^+, Rb^+, Cs^+, Ca^+, and Ba^{2+}, on the first reduction potential for a cast film of C_{60} on an electrode surface in an aqueous solution, and found that the reduction peaks are shifted to less negative potentials with the decrease of the hydration energy of these cations [78]. This suggests that the binding of fullerene anions and electrolyte cations determines the formal potentials of the redox reaction. A similar electrochemical behavior is observed for C_{70} film-coated electrodes [79].

9.2. Fullerene Lipid Cast Films

Figure 15 shows a typical cyclic voltammogram of a cast film of **8** on a basal-plane pyrolytic graphite (BPG) electrode, where well-defined redox waves leading to the generation of a fullerene dianion of **8** in the films on the electrodes in water containing 0.5 M tetra-n-butylammonium chloride (TBAC) are evident [24]. The first reduction/oxidation process is stable in the potential cycling, but the second reduction process is unstable (almost disappears after ten scans) because of the formation an electroinactive film of the lipid on the electrode. Similar cyclic voltammetric behavior is observed for the **9**- and **10**-modified electrodes. The formal potentials for **9**- and **10**-modified electrodes in water are not much different; for example, in 0.5 M TBAC aqueous solution, they are −431, −435, and −440 mV versus SCE for **9**- and **10**-modified electrodes, respectively. Electroactive amounts of the fullerene moiety on **8–10** increase with a decrease in the alkyl-chain length. About 50% of the C_{60} moieties in the film of **10** on the electrode are electroactive. Interestingly, the fullerene moieties in the films are oriented in such a way as to undergo facile electrochemical communication with the electrodes.

The electrochemistry for **8–10** electrodes show thin-layer electrochemical behavior in the scan range of 10–30 $mV \cdot s^{-1}$ and a diffusion-controlled electron transfer in the range of 0.1–3.0 $V \cdot s^{-1}$.

From the analysis of thermodynamics for the binding of electrolyte cations and the electrogenerated radical monoanion and dianion of C_{60} moieties in films of **8–10** on electrodes, the electron transfer mechanism for the modified electrodes has been revealed [16]. A theoretical treatment for the binding was made with the following equations (1) and (2):

$$E_{1/2,1} = E_1^{0\prime} + \frac{RT}{F}\ln K_1 + \frac{pRT}{F}\ln c \qquad (1)$$

$$E_{1/2,2} = E_2^{0\prime} + \frac{RT}{F}\ln\frac{K_2}{K_1} + \frac{(q-p)RT}{F}\ln c \qquad (2)$$

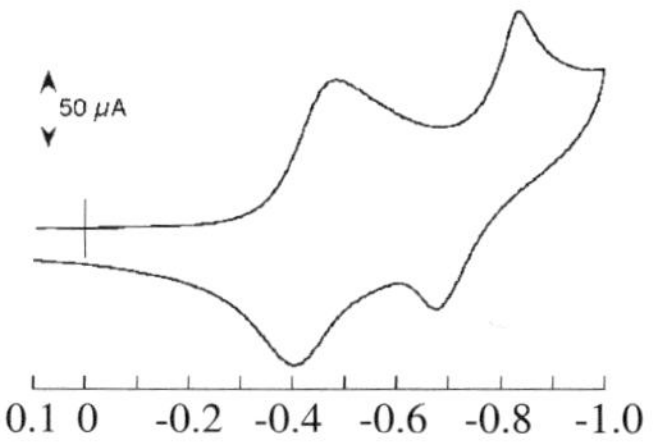

Figure 15. Typical cyclic voltammogram for a cast film of **8** on a BPG electrode in water containing 0.5 M TBAC at 55 °C. Scan rate, 100 mV/s.

where $E_{1/2,1}$ and $E_{1/2,2}$ are the half-wave potential for the first and second redox processes, $E_1^{0\prime}$ and $E_2^{0\prime}$ are the standard redox potentials for the two elemental redox processes, p and q are the numbers of electrons involved in the first and second reduction processes, respectively, K_1 and K_2 are the binding constants between the fullerene radical monoanions and the electrolyte cations, respectively, and c is the concentration of electrolytes. The electrolytes used are: tetramethylammonium chloride (TMAC), trimethybenzylammonium chloride (TMBAC), TEAC, tetraethylphosphonium chloride (TEPC), triethylbenzylammonium chloride (TEBAC), tetra-n-propylammonium chloride (TPAC), tri-n-butylbenzylammonium chloride (TBBAC), TBAC, and tetra-n-butylphosphonium chloride (TBPC). Details of the analysis of the binding are described in the literature [16]. Table 1 summarizes p, q, and K_1 values for an **8**-coated electrode. The interesting features are as follows.

1. For all cations used, p is almost 1 and q is almost 2.
2. K_1 shows a strong alkyl-chain-length dependence. This is in sharp contrast to the binding of electrochemically generated fullerene anions in organic solvents with tetraalkylammonium cations, where no alkyl-chain-length dependence has been reported [73, 74]. K_1 values for tetraalkylphosphonium cations are about one order of magnitude greater than those for tetraalkylammonium cations with the same alkyl chain length. The "softness" of tetraalkylphosphonium cations would explain this stronger complex formation since fullerene anions are "soft" anions.
3. For the binding with fullerene radical anions, the introduction of a benzyl moiety to the electrolyte shows a stronger effect than for methyl or ethyl groups, but a weaker effect than for the butyl group.

As shown above, the electrochemistry for electrodes modified with C_{60} and fullerene lipids shows an electrolyte cation dependence, and the formal potential of the fullerene moieties depends on the binding constants between the fullerene anion and electrolyte cations. In contrast, the electrochemistry for fullerene–lipid composite film-coated electrodes shows anion dependence, as shown in the following section.

Table 1. Parameters for the binding of the reduced forms of **8** and alkylammonium ions and phosphonium ions at the reduction processes of **8** films on BPG at 328 K.

Electrolyte cation	p	q	K_1/M^{-1}
TMAC	1.03	—	3.80×10
TMBAC	1.08	—	5.24×10^3
TEAC	1.05	—	7.93×10^3
TEPC	1.16	—	5.00×10^4
TEBAC	1.04	—	1.79×10^5
TPAC	1.10	2.15	0.96×10^6
TBBAC	0.97	1.99	1.01×10^8
TBAC	0.97	2.11	1.71×10^8
TBPC	1.06	1.98	1.70×10^9

9.3. Fullerene–Lipid Composite Thin Films

9.3.1. C_{60}–Lipid Composite Thin Films

In sharp contrast to the electrochemistry of C_{60}-coated electrodes [76–78], electrodes modified with films of C_{60}–cationic lipids exhibit stable electron transfer reactions. CVs for a C_{60} embedded in a cast film of an artificial poly(ion-complexed) lipid, ditetradecyldimethylammonium poly(styrene sulfonate) (**11**) on a BPG electrode, show two reversible electron transfers in an aqueous solution, and that the generated radical monoanion and the dianion are very stable (Fig. 16a) [13]. Both the first and second redox couples did not change, even after 50 cycles.

Reversible redox waves leading to the generation of a C_{60} trianion are observed in the cyclic voltammograms of a C_{60}/tridodecylmethylammonium bromide (**12**)-coated electrode (Fig. 16b) [12]. The observed formal potentials are $E_1^{0\prime} = -0.12$, $E_2^{0\prime} = -0.74$ and $E_3^{0\prime} = -1.21$ V versus SCE. Stronger binding between the fullerene anions and tridodecylmethylammonium cations enables the detection of three reduction peaks within the potential window in the aqueous system. The radical monoanion of C_{60} generated in this system is very stable, that is, virtually no change in the CVs for the first redox wave is observed, even after a 30 min hold at -0.5 V. The hold at -1.0 V, where the dianions of the fullerenes are generated, caused a gradual decrease in the current, suggesting the formation of electro-inactive films on the electrodes. The radical trianion generated in this system is not so stable.

The CVs for cast films of C_{60}/**12** on BPG are essentially the same for the measurements in water containing NaCl, KCl, $CaCl_2$, tetraethylammonium chloride, and tetrabutylammonium chloride as an electrolyte. The results suggest that these electrolyte cations are not major counter cations of the fullerene anions; instead, tridodecylmethylammonium acts as the counter cation during the reduction of C_{60}. The CVs for cast films of C_{60}/**12** on BPG show electrolyte anion dependence. The cathodic peak potentials of the electrodes are shifted to a positive direction with an increase of the hydration energy of anions (F^-, Cl^-, Br^-, NO_3^-). This is derived from a stronger complex formation of tridodecylmethylammonium and anions with smaller hydration energy, such as NO_3^-.

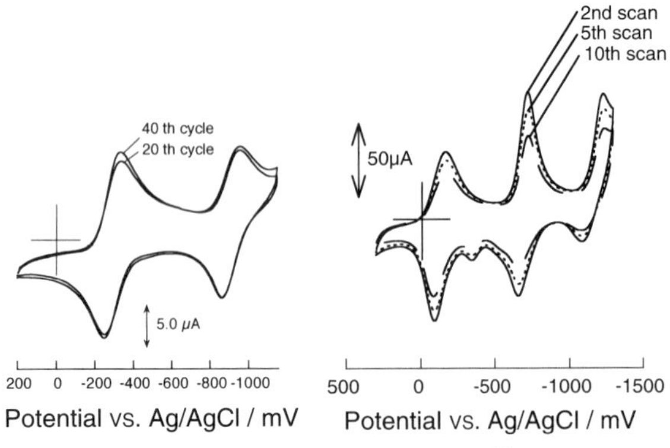

Figure 16. Cyclic voltammograms for C_{60} embedded in cast films of (a) ditetradecyldimethylammonium poly(styrene sulfonate) (**11**) and (b) tridodecylmethylammonium bromide **12** on BPG electrodes in water containing 0.5 M KCl at 25 °C. Scan rate, 100 mV/s. (a) Reprinted with permission from [13], N. Nakashima et al., *Angew. Chem. Int. Ed.* 37, 2671 (1998). © 1998, Wiley-VCH Verlag GmbH. (b) Reprinted with permission from [12], N. Nakashima et al., *Chem. Lett.* 633 (1998). © 1998, The Chemical Society of Japan.

9.3.2. The Higher Fullerene C_{84}–Lipid Composite Thin Films

Reports [10, 80] describing the electrochemical properties of higher fullerenes have been limited due to their low abundance compared with the more readily available C_{60} and C_{70}. From the isolated pentagon rule, 24 isomers of C_{84} are predicted to exist, and two major isomers of C_{84} with molecular symmetry D_2 and D_{2d} have been isolated and used for experiments. In an organic solution, C_{84} has been reported to show four or five reduction processes [81–84], of which the first three reduction processes are stable and the last two are unstable.

The CVs for a cast film of a single component of C_{84} on an electrode gave no well-defined redox current in an aqueous solution. In contrast, well-defined multiple electron transfer reactions for a composite film of C_{84}/tetaoctylammonium bromide (**13**) cast on an electrode occur (Fig. 17). The electrode gives multiple cathodic peaks, together with their corresponding anodic peaks. The formal potentials for the six redox couples estimated from their CVs at 5 mV/s are: +477, +204, −145, −254, −636, −900, and −1110 mV. Since C_{84} exists as two isomers, the assignment of the multiple redox process is complex. The large positive shift of the formal potentials of C_{84} at the modified electrode compared to those in an organic solution [84] suggests the strong binding of C_{84} anions with a tetraoctylammonium cation in the film [16]. Hydrophobic microenvironments afforded by the tetraoctylammonium matrix contribute to the generation and stability of C_{84} anions.

9.3.3. Metallofullerene–Lipid Composite Thin Films

Metallofullerenes are novel and attractive carbon nanomaterials that encapsulate one or more metal atoms inside the hollow of fullerene cages, and that have been reported to exhibit unique properties that are not seen in empty fullerenes [85–86]. Despite the low abundance compared with the empty-caged fullerenes, researchers have disclosed

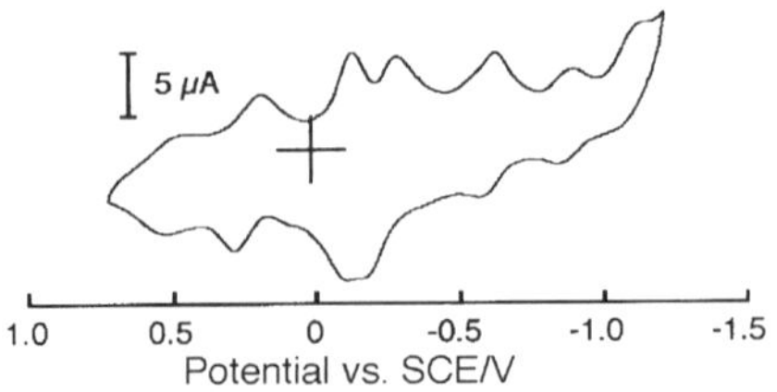

Figure 17. Typical cyclic voltammogram for C_{84} embedded in a cast film of tetraoctylammonium bromide (**13**) on a BPG electrode in water containing 0.5 M TEAC at 25 °C. Scan rate, 100 mV/s. Adapted with permission from [22], N. Nakashima et al., *Chem. Lett.* 748 (2001). © 2001, The Chemical Society of Japan.

their interesting properties and functions for solid metallofullerenes of and metallofullerenes in organic solutions by means of a variety of physical and physicochemical techniques. La@C_{82}-A is the first metallofullerene reported by Smalley and coworkers [87] that was identified to possess C_{2v} symmetry [88–89].

Almost no electrochemistry is observed for an La@C_{82}-A film-coated electrode. In sharp contrast, an La@C_{82}-A/**12**-coated electrode shows well-defined cyclic voltammograms with a major cathodic peak potential at 0.28, −0.69, and −1.0 V (versus SCE) which would correspond to the first, second, and third reductions of La@C_{82}-A, respectively (Fig. 18). The observed cyclic voltammograms were stable for potential cycling. The first reduction process was found to be stable for the negative-end potential hold, while the second and third reduction processes were unstable. The higher stability of the first reduction state of La@C_{82}-A would be due to the formation of a closed shell structure like La@C_{82}-A anion in o-dichlobenzene. La@C_{82} embedded in the lipid film on the electrode is rather stable for dioxygen.

The first, second, and third cathodic peak potentials for La@C_{82}-A in *o*-dichlorobenzene have been reported to be −124, −1103, and −1260 mV versus Ag/$AgNO_3$ [88]. Large positive shifts of the cathodic peak potentials were observed for La@C_{82}-A embedded in the cationic lipid film compared to those in an organic solution. This shift suggests the strong binding of La@C_{82}-A anions with alkyl ammonium cation in these films [16]. Hydrophobic microenvironments afforded by the alkyl ammonium matrices contribute to the generation and stability of the metallofullerene anions.

9.3.4. Electrochemistry of C_{60} in Gel-Like Films

The chemistry and physics of organic gels that have potential applications in the construction of "intelligent" material systems are also an exciting area. Organic gels are known to provide suitable microenvironments for the electrochemistry and regulated electrochemistry of some redox active molecules [90–91]. The CVs for a gel-like membrane of C_{60}/tetraoctylphosphonium bromide (**14**) on a BPG electrode showed three redox couples corresponding to C_{60}/C_{60}^{-}, C_{60}^{-}/C_{60}^{2-}, and C_{60}^{2-}/C_{60}^{3-} in the potential window in the aqueous medium [19]. This result shows that the

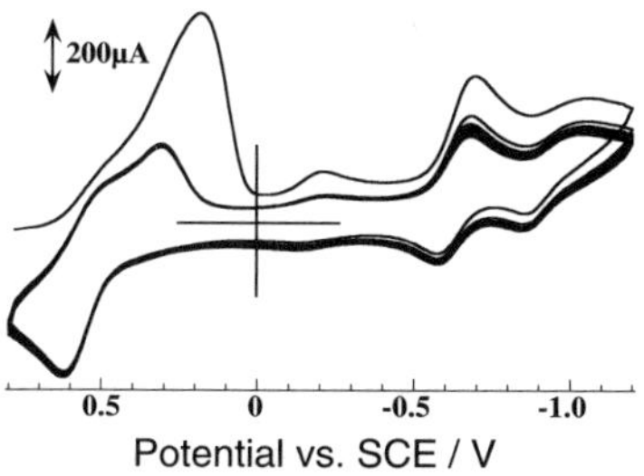

Figure 18. Typical cyclic voltammograms for La@C_{82}-A embedded in a cast film of tridodecylmethylammonium bromide (**12**) on a BPG electrode in water containing 0.5 M TEAC at 25 °C. Scan rate, 100 mV/s. Reprinted with permission from [27], N. Nakashima et al., *J. Phys. Chem. B* 106, 3523 (2002). © 2002, American Chemical Society.

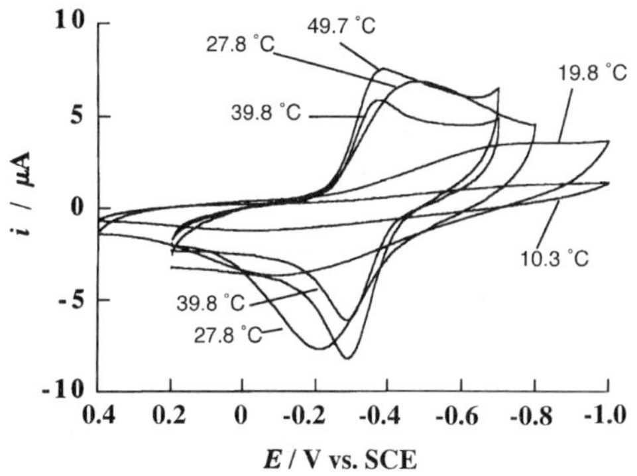

Figure 19. Temperature-dependent cyclic voltammograms for C_{60} embedded in a cast film of dihexadecyldimethylammonium poly(styrene sulfonate) (**15**) on a BPG electrode in water containing 0.5 M TEAC at 25 °C. Scan rate, 100 mV/s. Reprinted with permission from [14], N. Nakashima et al., *J. Phys. Chem. B* 102, 7328 (1998). © 1998, American Chemical Society.

electrochemistry of fullerene C_{60} incorporated in a gel-like membrane on an electrode is possible. The formal potentials for the electrochemistry of C_{60} in the membrane were $E_1^{0\prime} = -0.17$, $E_2^{0\prime} = -0.66$, and $E_3^{0\prime} = -1.23$ V, which are close to those for the C_{60}/**13**-coated electrode. The voltammograms were stable for potential cycling over the first two redox waves, leading to the generation of C_{60}^{2-}. These findings afford the opportunity to undergo electrochemical communication of fullerenes and related materials at gel-modified electrode systems that possess many possible applications in a variety of areas.

9.3.5. Aqueous Electrochemistry at a p-Type Semiconductive Diamond Electrode

A *p*-type semiconductive diamond has appeared as a unique electrode material [92, 93] due to the wide electrochemical potential window in aqueous media [94, 95], low background current [96], anticorrosiveness, and fast electron transfer kinetics for outer-sphere redox species [97–98]. Wu et al.

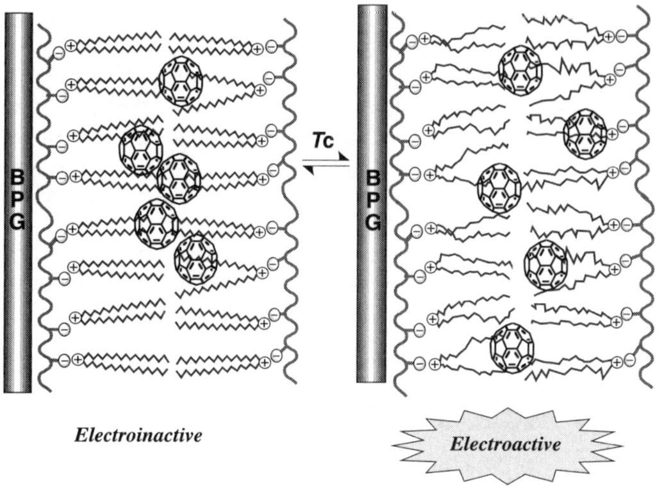

Figure 20. Schematic drawing of the redox switch of fullerenes by a lipid bilayer phase transition.

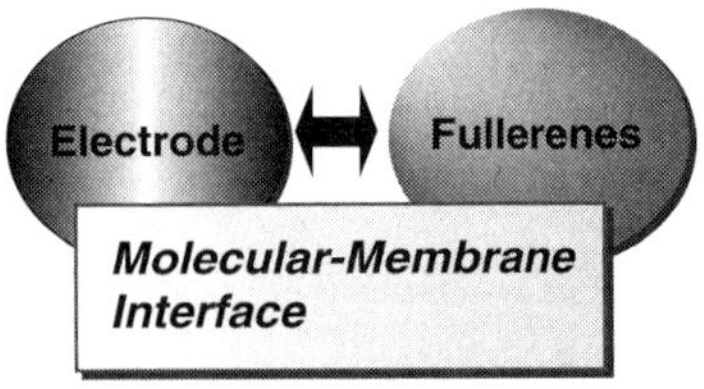

Figure 21. Schematic drawing of a molecular membrane interface.

observed five reduction steps of C_{60} in a nonaqueous solution using a diamond thin-film electrode [99]. The aqueous electrochemistry of C_{60} incorporated into a cast film of an artificial lipid **12** on a boron-doped *p*-type semiconductive diamond electrode has been investigated. The electrode shows one quasi-reversible and one sigmoidal wave in a cyclic voltammogram. This is quite different from those using a BPG electrode. The redox mechanism of C_{60} in the film on the diamond electrode is explained using an electron energy diagram. The surface states in the hole transfer processes from the diamond electrode to C_{60} in the lipid film are significant.

9.4. Fullerene Redox Switch

The redox switch is of importance in the design and construction of molecular-switchable devices [100–101]. Osteryoung square-wave voltammetry for BPG electrodes modified with films of fullerene lipids (**8–10**) show a strong temperature dependence. The plots of the reduction current for an **8**-modified electrode show breaks near 35 and 48 °C that are close to the respective main and subphase transitions of **8** films, thus suggesting that the phase transition of **8**-films affects the electrochemistry of the modified electrodes [25]. For **9–10** modified electrodes, some temperature-dependent structural changes of the films that are not detectable by DSC affect the electrochemistry.

The electrochemistry of C_{60} embedded in the artificial lipid film is tunable by the temperature-controlled phase transition of the lipid bilayer films on electrodes [14]. As shown in Figure 19, the electrochemistry for a BPG electrode modified with a film of C_{60}-dihexadecyldimethylammonium poly(styrene sulfonate) (**15**) shows a strong temperature dependence; at the lower temperatures, the electrode gives almost no faradaic current of C_{60} (OFF state). On the contrary, a redox current is evident at higher temperatures (ON state). The break observed near 28 °C is very close to the phase transition of the bilayer films of **15**. This temperature dependence is reversible, that is, the temperature-driven "on–off" switching of C_{60} electrochemistry is possible (Fig. 20).

The construction of fullerene/lipid electrical devices based on the nature of self-organized lipid bilayer membranes would be of interest for the application and utilization of fullerenes.

10. CONCLUSIONS

Fullerenes with an ordered structure in aqueous systems and a lipid bilayer-based aqueous superstructure from fullerene amphiphiles, as well as the structure and electrochemistry of fullerene lipid films and fullerene (higher fullerene, metallofullerene)–lipid composite films, were reviewed. Fullerene-based biomembrane-like carbon nanosuperstructures created in an aqueous solution may be adapted for biochemical and biological use. The discovery that a simple C_{60}-carrying ammonium amphiphile forms both fibrous and disk-like superstructures is of interest since the superstructure is formed through self-organization of the amphiphile in an aqueous solution. The fullerene lipids formed multibilayer structure films that possess phase transitions attributable to the lipid bilayer phase transition typically observed for liposomal and synthetic lipid bilayer membranes and/or to the change in orientation of the C_{60} moieties. We could schematically show the phase transitions of the fullerene lipid films. The aqueous electrochemistry for electrodes modified with films of: (1) C_{60}, (2) fullerene lipids, and (3) fullerenes (including higher fullerenes and metallofullerenes)–lipid composites was reviewed here, and the differences in the electron transfer mechanism are revealed. The aqueous electrochemistry of fullerene films is of importance for both fundamentals and applications in understanding the science and nanotechnology of nanocarbons. We have pointed a factor that plays an important role in the generation of multiply-charged fullerene anions at the electrode devices.

Finally, we would like to emphasize that multibilayer lipid films on electrodes provide hydrophobic microenvironments that are suitable for the facile electrochemistry of fullerene thin films, that is, the lipid films function as a molecular membrane interface that connects the fullerenes and metal electrodes (Fig. 21). Fullerene nanoarchtectures using self-assembled fullerenes are of interest for the construction of nanocarbon-based molecular devices.

GLOSSARY

Cyclic voltammograms Typical electrochemical method that measures electron transfer reactions of electroactive compounds.

Fullerene lipids Fullerene-carrying amphiphiles that form molecular-bilayer structure.

Langmuir–Blodgett films Deposited films on substrates from monolayers on water by the langmuir–Blodgett technique.

Lipid phase transition Phase transition between a crystalline phase and a liquid crystalline phase.

Metallofullerenes Novel carbon nanomaterials that encapsulate one or more metal atoms inside the hollow of fullerene cages.

REFERENCES

1. W. E. Billups and M. A. Ciufolini (Eds.), "Buckminsterfullerenes." VCH, New York, 1993.
2. K. Prassides (Ed.), "Physics and Chemistry of the Fullerenes." Kluwer Academic, Boston, 1994.
3. H. W. Kroto, "The Fullerenes; New Horizons for the Chemistry, Physics and Astrophysics of Carbon." Cambridge University Press, Cambridge, 1997.
4. A. Hirsch (Ed.), "Fullerenes and Related Structure." Springer, Berlin, 1999.

5. K. M. Kadish and R. S. Ruoff (Eds.), "Fullerenes: Chemistry, Physics and Technology." Wiley-Interscience, New York, 2000.
6. E. Osawa (Ed.), "Perspectives of Fullerene Nanotechnology." Kluwer Academic, Dordrecht, 2002.
7. J. Chlistunoff, D. Cliffel, and A. J. Bard, in "Handbook of Organic Conductive Molecules and Polymers" (H. S. Nalwa, Ed.), Vol. 1, pp. 333–412. Wiley, Chichester, 1997.
8. L. Echegoyen and L. E. Echegoyen, *Acc. Chem. Res.* 31, 593 (1998).
9. L. Echegoyen, F. Diederich, and L. Echegoyen, in "Fullerenes: Chemistry, Physics and Technology" (K. M. Kadish and R. S. Ruoff, Eds.), Chap. 1. Wiley-Interscience, New York, 2000.
10. C. L. Reed and R. D. Bolskar, *Chem. Rev.* 100, 1075 (2000).
11. H. Murakami, Y. Watanabe, and N. Nakashima, *J. Am. Chem. Soc.* 118, 4484 (1996).
12. N. Nakashima, T. Kuriyama, T. Tokunaga, H. Murakami, and T. Sagara, *Chem. Lett.* 633 (1998).
13. N. Nakashima, T. Tokunaga, T. Nakanishi, Y. Nonaka, H. Murakami, and T. Sagara, *Angew. Chem. Int. Ed.* 37, 2671 (1998).
14. N. Nakashima, Y. Nonaka, T. Nakanishi, T. Sagara, and H. Murakami, *J. Phys. Chem. B* 102, 7328 (1998).
15. T. Nakanishi, H. Murakami, and N. Nakashima, *Chem. Lett.* 1219 (1998).
16. T. Nakanishi, H. Murakami, T. Sagara, and N. Nakashima, *J. Phys. Chem. B* 103, 304 (1999).
17. H. Murakami, M. Shirakusa, T. Sagara, and N. Nakashima, *Chem. Lett.* 815 (1999).
18. N. Nakashima, "Recent Advances in the Chemistry and Physics of Fullerenes and Related Materials" (K. M. Kadish and R. F. Ruoff, Eds.), Vol. 8, p. 131. Electrochemical Society, Pennington, NJ, 2000.
19. T. Nakanishi, H. Murakami, T. Sagara, and N. Nakashima, *Chem. Lett.* 340 (2000).
20. N. Nakashima, T. Ishii, M. Shirakusa, T. Nakanishi, H. Murakami, and T. Sagara, *Chem. Eur. J.* 7, 1766 (2001).
21. N. Nakashima, N. Wahida, M. Mori, H. Murakami, and T. Sagara, "Fullerenes" (P. V. Kamat, D. M. Guldi, and K. M. Kadish, Eds.), Vol. 11, p. 4. Electrochemical Society, Pennington, NJ, 2001.
22. N. Nakashima, N. Wahida, M. Mori, H. Murakami, and T. Sagara, *Chem. Lett.* 748 (2001).
23. H. Ohwaki and N. Nakashima, *Trans. Mater. Res. Soc. Jpn.* 26, 933 (2001).
24. T. Nakanishi, M. Morita, H. Murakami, T. Sagara, and N. Nakashima, *Chem. Eur. J.* 8, 1641 (2002).
25. H. Murakami, R. Matsumoto, Y. Okusa, T. Sagara, M. Fujitsuka, O. Ito, and N. Nakashima, *J. Mater. Chem.* 12, 2026 (2002).
26. M. Komatsu, T. Sagara, and N. Nakashima, *J. Electrochem. Soc.* 149, E227 (2002).
27. N. Nakashima, M. Sakai, H. Murakami, T. Sagara, T. Wakahara, T. Akasaka, and T. Nakanishi, *J. Phys. Chem. B* 106, 3523 (2002).
28. H. Hungerbühler, D. M. Guldi, and K.-D. Asmus, *J. Am. Chem. Soc.* 115, 3386 (1993).
29. R. V. Bensasson, J.-L. Garaud, S. Leach, G. Miquel, and P. Seta, *Chem. Phys. Lett.* 210, 141 (1993).
30. J. L. Garaud, J. M. Janot, G. Miquel, and P. Seta, *J. Membr. Sci.* 91, 259 (1994).
31. S. Niu and D. Mauzerall, *J. Am. Chem. Soc.* 118, 5791 (1996).
32. J. M. Janot, P. Seta, R. V. Bensasson, and S. Leach, *Synth. Metals* 77, 103 (1996).
33. H. T. Tien, L.-G. Wang, X. Wang, and A. L. Ottova, *Bioelectrochem. Bioenerg.* 42, 161 (1997).
34. M. Hetzer, S. Bayerl, X. Camps, O. Vostrowsky, A. Hirsch, and T. M. Bayerl, *Adv. Mater.* 9, 913 (1997).
35. M. Cassell, C. L. Asplund, and J. M. Tour, *Angew. Chem. Int. Ed.* 38, 2403 (1999).
36. S. F. Marcel, K. J. Lie, and S. H. W. Cheung, *Lipids* 34, 1223 (1999).
37. M. Sano, K. Oishi, T. Ishii, and S. Shinkai, *Langmuir* 16, 3773 (2000).
38. M. Brettreich, S. Burghardt, C. Böttcher, T. Bayerl, S. Bayerl, and A. Hirsch, *Angew. Chem. Int. Ed.* 39, 1845 (2000).
39. M. Braun and A. Hirsch, *Carbon* 38, 1565 (2000).
40. S. F. Marcel, K. J. Lie, S. H. W. Cheung, and J. C. M. Ho, *Lipids* 36, 649 (2001).
41. S. Zhou, C. Burger, B. Chu, M. Sawamura, N. Nagahara, M. Toganoh, U. E. Hackler, H. Isobe, and E. Nakamura, *Science* 291, 1944 (2001).
42. M. Li, M. Xu, N. Li, Z. Gu, and X. Zhou, *J. Phys. Chem.* 106, 4197 (2002).
43. T. Kunitake, Kimizuka, N. Higashi, and N. Nakashima, *J. Am. Chem. Soc.* 106, 1978 (1984) and references therein.
44. N. Nakashima, Y. Yamaguchi, H. Eda, M. Kunitake, and O. Manabe, *J. Phys. Chem. B* 101, 215 (1997) and references therein.
45. N. Nakashima, S. Asakuma, and T. Kunitake, *Chem. Lett.* 1709 (1984).
46. N. Nakashima, S. Asakuma, and T. Kunitake, *J. Am. Chem. Soc.* 107, 509 (1985).
47. T. Simizu and M. Masuda, *J. Am. Chem. Soc.* 119, 2812 (1997).
48. I. Nakazawa, M. Masuda, Y. Okada, T. Hanada, K. Yase, M. Asai, and T. Shimizu, *Langmuir* 15, 4757 (1999).
49. M. Cassell, C. L. Asplund, and J. M. Tour, *Angew. Chem. Int. Ed.* 38, 2403 (1999).
50. D. Chapman, "Biomembrane and Functions." Verlag Chemie, Weinheim, 1984.
51. S. R. Wilson, in "Fullerenes: Chemistry, Physics and Technology," pp. 437–465. Wiley-Interscience, New York, 2000.
52. K. Oishi, T. Ishi-i, M. Sano, and S. Shinkai, *Chem. Lett.* 1089 (1999).
53. T. Ishi-i, Y. Ono, and S. Shinkai, *Chem. Lett.* 808 (2000).
54. M. Chikamatsu, T. Hanada, Y. Yoshuda, N. Tanigaki, K. Yase, H. Nishikawa, T. Kodama, I. Ikemoto, and K. Kikuchi, *Mol. Cryst. Liq. Cryst.* 316, 157 (1998).
55. N. Nakashima, K. Nakayama, M. Kunitake, and O. Manabe, *J. Chem. Soc., Chem. Commun.* 887 (1990).
56. N. Nakashima, N. Yamada, T. Kunitake, J. Umemura, and T. Takenaka, *J. Phys. Chem.* 90, 3374 (1986).
57. Y. Yamaguchi and N. Nakashima, *Anal. Sci.* 10, 863 (1994).
58. J. Chlistunoff, D. Cliffel, and A. J. Bard, *Thin Solid Films* 257, 166 (1995).
59. M. Matsumoto, H. Tachibana, R. Azumi, M. Tanaka, T. Nakamura, G. Yunome, M. Abe, S. Yamago, and E. Nakamura, *Langmuir* 11, 660 (1995).
60. D. M. Guldi, Y. Tian, J. H. Fendler, H. Hungerbühler, and K.-D. Asmus, *J. Phys. Chem.* 99, 17673 (1995).
61. S. E. Campbell, G. Luengo, V. I. Srdanov, F. Wudl, and J. N. Israelachvili, *Nature* 382, 520 (1996).
62. P. Wang, Y. Maruyama, and R. M. Metzger, *Langmuir* 12, 3932 (1996).
63. V. V. Tsukruk, M. P. Everson, L. M. Lander, and W. J. Brittain, *Langmuir* 12, 3905 (1996).
64. M. I. Sluch, I. D. W. Samuel, A. Beeby, and M. C. Petty, *Langmuir* 14, 3343 (1998).
65. F. Cardullo, F. Diederich, L. Echyegoyen, T. Habicher, N. Jayaraman, R. M. Leblanc, J. F. Stoddart, and S. Wang, *Langmuir* 14, 1955 (1998).
66. C. Jehoulet, Y. S. Obeng, Y.-T. Kim, F. Zhou, and A. J. Bard, *J. Am. Chem. Soc.* 114, 4237 (1992).
67. Y. Tomioka, M. Ishibashi, H. Kajiyama, and Y. Taniguchi, *Langmuir* 9, 32 (1993).
68. M. Yanagida, A. Takahara, and T. Kajiyama, *Bull. Chem. Soc. Jpn.* 73, 1429 (2000).

69. P. Tundo, A. Perosa, M. Selva, L. Valli, and C. Giannini, *Colloids Surf. A* 190, 295 (2001).
70. H. Tachubana, R. Azumi, A. Ouchi, and M. Matsumoto, *J. Phys. Chem. B* 105, 42 (2001).
71. E. Mouri, T. Nakanishi, N. Nakashima, and H. Matsuoka, *Langmuir* (in press).
72. S. Wang, R. M. Leblanc, F. Arias, and L. Echegoyen, *Langmuir* 13, 1672 (1997).
73. D. Dubois, G. Moninot, W. Kutner, M. T. Jones, and K. M. Kadish, *J. Phys. Chem.* 96, 7137 (1992).
74. W. R. Fawcett, M. Opallo, M. Fedurco, and J. W. Lee, *J. Am. Chem. Soc.* 115, 196 (1993).
75. B. S.-Guillous, W. Kutner, M. T. Jones, and K. M. Kadish, in "Fullerenes: Recent Advances in the Chemistry and Physics of Fullerenes and Related Materials" (K. M. Kadish and R. S. Ruoff, Eds.), p. 1020. Electrochemical Society, Pennington, NJ, 1994.
76. A. Szucs, A. Loix, J. B. Nagy, and L. J. Lamberts, *J. Electroanal. Chem.* 379, 189 (1995).
77. J. Davis, H. A. O. Hill, A. Kurz, A. D. Leighton, and A. Y. Safronov, *J. Electroanal. Chem.* 429, 7 (1997).
78. A. Szucs, A. Loix, J. B. Nagy, and L. Lamberts, *J. Electroanal. Chem.* 402, 137 (1996).
79. A. Szucs, M. Tolgyesi, M. Csiszar, J. B. Nagy, and M. Novok, *J. Electroanal. Chem.* 441, 287 (1998).
80. L. Echegoyen, F. Diederich, and L. E. Echegoyen, in "Fullerenes: Chemistry, Physics and Technology" (K. M. Kadish and R. S. Ruoff, Eds.), Chap. 1. Wiley-Interscience, New York, 2000.
81. M. S. Meier, J. F. Guarr, J. P. Selegue, and V. K. Vance, *J. Chem. Soc., Chem. Commun.* 63 (1993).
82. P. L. Boulas, M. T. Jones, K. M. Kadish, R. S. Ruoff, D. C. Lorents, and D. S. Tse, *J. Am. Chem. Soc.* 116, 9393 (1994).
83. J. P. Selegue, J. P. Show, T. F. Guarr, and M. S. Meier, in "Recent Advances in the Chemistry and Physics of Fullerenes and Related Materials" (K. M. Kadish and R. S. Ruoff, Eds.), p. 1274. Electrochemical Society, Proceeding Series, Pennington, NJ, 1994.
84. P. L. Boulas, M. T. Jones, R. S. Ruoff, D. C. Lorents, R. Malhotra, D. S. Tse, and K. M. Kadish, *J. Phys. Chem.* 100, 7573 (1996).
85. H. Shinohara, in "Fullerenes: Chemistry, Physics and Technology" (K. M. Kadish and R. S. Ruoff, Eds.), p. 357. Wiley-Interscience, New York, 2000.
86. S. Nagase, K. Kobayashi, T. Akasaka, and T. Wakahara, in "Fullerenes: Chemistry, Physics and Technology" (K. M. Kadish and R. S. Ruoff, Eds.), p. 395. Wiley-Interscience, New York, 2000.
87. Y. Chai, T. Guo, C. Jin, R. E. Haufler, L. P. F. Chibante, J. Fure, L. Wang, J. M. Alford, and R. E. Smalley, *J. Phys. Chem.* 95, 7564 (1991).
88. T. Akasaka, T. Wakahara, S. Nagase, K. Kobayashi, M. Waelchli, K. Yamamoto, M. Kondo, S. Shirakura, S. Okubo, Y. Maeda, T. Kato, M. Kako, Y. Nakadaira, R. Nagahata, X. Gao, E. V. Caemelbecke, and K. M. Kadish, *J. Am. Chem. Soc.* 122, 9316 (2000).
89. T. Akasaka, T. Wakahara, S. Nagase, K. Kobayashi, M. Waelchli, K. Yamamoto, M. Kondo, S. Shirakura, Okubo, Y. Maeda, T. Kato, M. Kako, Y. Nakadaira, X. Gao, E. V. Caemelbecke, and K. M. Kadish, *J. Phys. Chem. B* 105, 2971 (2001).
90. M. Watanabe, Y. Tadenuma, M. Ban, K. Sanui, and N. Ogata, *J. Intell. Mater. Syst. Struct.* 4, 216 (1993).
91. N. Oyama, T. Tatsuma, and K. Takahashi, *J. Phys. Chem.* 97, 10504 (1993).
92. K. Okano, H, Naruki, Y. Akiba, T. Kurosu, M. Iida, Y. Hirose, and T. Nakamura, *Jpn. J. Appl. Phys.* 28, 1066 (1989).
93. N. S. Lewis, *J. Phys. Chem. B* 102, 4843 (1998).
94. H. B. Martin, A. Argoitia, U. Landau, A. B. Anderson, and J. C. Angus, *J. Electrochem. Soc.* 143, L133 (1996).
95. F. Bouamrane, A. Tadjeddine, J. E. Butler, R. Tenne, and C. L-Clément, *J. Electroanal. Chem.* 405, 95 (1996).
96. S. Alehasham, F. Chambers, J. W. Strojek, G. M. Swain, and R. Ramesham, *Anal. Chem.* 67, 2812 (1995).
97. J. W. Strojek, M. C. Granger, G. M. Swain, T. Dallas, and M. W. Holtz, *Anal. Chem.* 68, 2031 (1996).
98. N. Vinokur, B. Miller, Y. Avyigal, and R. Kalish, *J. Electrochem. Soc.* 143, L238 (1996).
99. Z. Wu, T. Yano, D. A. Tryk, K. Hashimoto, and A. Fujishima, *Chem. Lett.* 503 (1998).
100. L. Feringa, "Molecular Switches." Wiley-VCH, Weinheim, 2000.
101. A. E. Kaifer and M. G-Kaifer, "Supramolecular Electrochemsitry," pp. 103–141. Wiley-VCH, Weinheim, 1999.

Encyclopedia of Nanoscience and Nanotechnology
www.aspbs.com/enn

Fullerene Mechanochemistry

Guan-Wu Wang
University of Science and Technology of China, Hefei, Anhui, China

CONTENTS

1. INTRODUCTION

According to the type of energy source supplied to the system, the different fields of chemistry are classified as thermochemistry, electrochemistry, photochemistry, etc. The term mechanochemistry was proposed by Ostwald as early as 1919 as an energy source for chemical reactions [1]. A narrow field, tribochemistry, was used for reactions initiated by friction during milling of solid reagents [2].

Mechanochemistry involves the transformation of mechanical energy into the driving force for the chemical reactions of solid reagents. Mechanochemistry is generally done in the absence of any solvent. The external dynamic force generated from the mechanical agitation can induce the solid to its vibrationally and electronically excited structures, thus destabilizing the electronic structure of the bonding and making the solid prone to chemical reaction. The activation of the local reaction sites by the mechanical energy caused by the stress, friction, shear deformation, etc., together with the high concentration and closest contact of the reacting species due to the lack of solvation, would be particularly advantageous for the chemical reaction in the solid state.

A recent overview on mechanochemistry by Fernández-Bertrán [3] and an excellent and lengthy review on mechanical alloying and milling by Suryanarayana [4] have appeared. Mechanochemistry was mainly applied to processes related to inorganic materials, such as alloy, ceramics, cermets, ferrites, semiconductors and superconductors, ferroelectrics, mineral fertilizers, catalysts, ceramic construction materials, etc. Fewer examples of mechanochemical reactions by ball milling involving organic compounds have been reported [5–14].

Fullerenes (mainly C_{60} and C_{70}) were first discovered in 1985 [15], and macroscopic amount of fullerenes was made available in 1990 [16]. The discovery and availability in large quantity of fullerenes have stimulated their extensive research in a wide variety of fields such as physics, chemistry, material science, biology, etc. Initially viewed as aromatic and quite chemically inert molecules, fullerenes can actually undergo various chemical reactions characteristic of electron-deficient alkenes. Several review articles [17–21] and an excellent monograph [22] regarding fullerene chemistry have been published.

The solubility of fullerenes in common organic solvents is so low that the use of a large amount of solvents is inevitable in their solution reactions. The reactions of fullerenes in the absence of any solvent have advantages over their solution reactions from the viewpoint of the solubility problem of fullerenes and benign environmental concern by eliminating the usage of harmful organic solvent. Therefore solvent-free reaction of fullerene is an attractive and appealing method to synthesize fullerene derivatives. Fullerene chemistry and mechanochemistry started to merge in the mid-1990s. A review article on the mechanochemical approach to fullerene chemistry at a very early stage of fullerene mechanochemistry was published by Braun in 1997 [23]. Since then, much more work on various solvent-free reactions of fullerenes has been done. The solid-state mechanochemical reaction of fullerene C_{60} based on the authors' own work was reviewed in 1999 [24]. Here a full coverage of fullerene mechanochemistry, including most recent research work in this field, is presented.

2. MECHANOCHEMICAL DEVICES

Different types of mechanochemical devices have been used to provide mechanical activation energy. They differ in their capacity and efficiency, and can be generally classified into two categories: grinding devices and milling devices.

ISBN: 1-58883-059-4/$35.00

2.1. Grinding Devices

The most convenient and easy method for a mechanochemical reaction is the use of a mortar and pestle, which promotes the reaction through grinding, mixing, and triturating. This method has been widely used by Toda and co-workers to investigate various solvent-free organic reactions [25–27]. A new version of mortar and pestle called Mortar Grinder RM 100 is available from Retsch GmbH & Co. KG; the rotary speed of the mortar is 90 rpm.

2.2. Milling Devices

2.2.1. Shaker/Mixer Mills

Shaker mills such as SPEX mills are most commonly used for laboratory investigations. A shaker mill contains a vial, consisting of the sample and grinding balls, secured in the clump and swung energetically back and forth several thousand times a minute. SPEX mills are manufactured by SPEX CertPrep, Metuchen, NJ. A SPEX shaker mill can process about 10–20 g of the sample at a time.

Most of the reactions described in this article were done with the so-called "high-speed vibration mill." This high-speed vibration mill has almost the same working mechanism as the SPEX shaker mill, and consists of a capsule and a milling ball, which are made of stainless steel (Fe-Cr-Ni with a composition of 74:18:8 wt%). The inner diameter and length of the capsule are 9.0 mm and 26 mm, respectively, and the diameter of the milling ball is 6.0 mm. The capsule containing the sample and milling ball was fixed in a home-built vibrating machine so that the capsule was shaken along its long axis horizontally with a slight fluctuation. The vibration frequency is 3500 cycles per minute [28]. Because of the small size of the presently used capsule, only a small amount (usually not more than 200 mg) of sample can be handled.

2.2.2. Planetary Ball Mills

The planetary ball mill owes its name to the planet-like movement of its vials. These vials are arranged on a rotating support disk and a special drive mechanism causes them to rotate around their own axes. The centrifugal force produced by the vials rotating around their own axes and that produced by the rotating support disk both act on the material and grinding balls in the vial. Because the vials and supporting disk rotate in opposite directions, the centrifugal forces alternately act in like and opposite directions. The material in the vial is ground due to the running down, lifting off, traveling, and colliding of the grinding balls. Planetary ball mills (Pulverisette) are available from Fritsch, GmbH in Germany. Compared to the SPEX shaker mills, Fritsch Pulverisettes are lower-energy mills. A few hundred grams of the sample can be milled in a planetary ball mill at a time.

2.2.3. Attritor Mills

A conventional ball mill consists of a rotating horizontal drum half-filled with small steel balls. In contrast, an attritor consists of a vertical drum with a series of impellers inside it. A powerful motor rotates the impellers. Set progressively at right angles to each other, the impellers energize the steel balls. The operation of an attritor is simple. The material to be ground and the grinding medium are placed in the stationary tank. The mixture is then agitated by a rotating shaft with arms. The laboratory attritor works up to ten times faster than the conventional ball mills. Commercial attritors are available from Union Process, Akron, OH. Attritors are the mills in which a large quantity of the sample from about 0.5 to 40 kg can be milled at a time.

3. MECHANOCHEMICAL REACTIONS OF FULLERENES

The simplest method of mechanochemistry is the use of mortar and pestle, while ball milling is suitable when a prolonged reaction time is required. Various organic reactions have been shown to take place in the solid state when a mortar and pestle is used without solvents [25–27]. In these examples, essentially the same reactions are reported to occur regardless of the presence or absence of solvent. For most of the reactions shown in the following sections, the so-called "high-speed vibration milling" (HSVM) technique is utilized. The stainless-steel capsule is vigorously shaken at a frequency of 3500 cycles per minute. The effective maximum local pressure in the stainless-steel capsule is not less than 15,000 bars during the high-speed milling process [29]. Some solid-state reactions can be totally different from those in the liquid phase under such extreme conditions. Indeed, some novel solvent-free reactions of fullerenes have been observed to occur only under HSVM conditions, and will be discussed in the following sections.

3.1. Nucleophilic Addition of Organozinc Reagent to C_{60}

The nucleophilic addition of amines, organolithiums, and Grignard reagents to C_{60} is one of the most common and widely used reactions to functionalize C_{60}. All these reactions were done in organic solvents such as benzene and toluene [19]. The first application of HSVM technique to fullerene chemistry was the Reformatsky-type reaction of C_{60}. The addition of organozinc reagents to C_{60}, which had not even been attempted in solution, was investigated under solvent-free HSVM conditions. The reactants of C_{60}, zinc powder, and ethyl bromoacetate in a molar ratio of 1:20:5 along with a stainless-steel milling ball were put into a stainless-steel capsule. This mixture was vigorously milled for 20 min by shaking the capsule at the speed of 3500 cycles per minute by the use of a high-speed vibration mill. The reaction mixture was acidified and separated to afford mainly the expected product **1** in 17.2% and some minor side products along with 72.5% recovered C_{60} (Fig. 1) [30]. It was found that longer reaction time increased the yield of compound **3** at the expense of the main product **1**.

3.2. Synthesis of Fullerene Dimers and Trimers

The formation, structure, and characteristics of all-carbon fullerene oligomers and polymers are of particular interest because of their potential application in molecular devices,

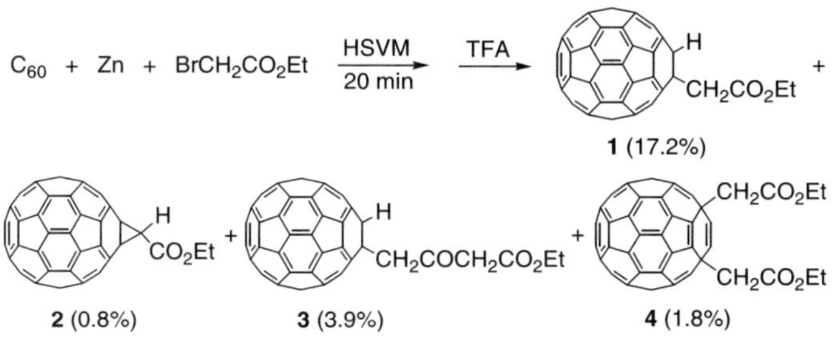

Figure 1. Reaction of C_{60} with zinc and ethyl bromoacetate under HSVM conditions [30].

optoelectronics, and nanotechnology. These materials have been prepared by photochemical irradiation [31–37], by alkali metal doping [38–40], and under high-pressure/high-temperature condition [41–45]. However, the insolubility of these materials in any organic solvent has precluded their characterization from the organic chemistry point of view; therefore, their detailed structures remain to be elucidated. It has been suggested that fullerene dimer with [2 + 2] structure is the essential subunit in the fullerene oligomers and polymers. Obviously, the preparation and characterization of fullerene dimers and trimers are of paramount importance for understanding the structure and properties and eventual application of the fullerene oligomers and polymers. Fortunately, fullerene dimers and trimers can be uniquely obtained by mechanochemical reactions.

3.2.1. Synthesis of C_{60} Dimer, C_{120}

The reaction of cyanide ion with C_{60} in a solution of *o*-dichlorobenzene (ODCB)-DMF gives a stable $(CN)C_{60}^-$ anion, which can be quenched with various electrophiles to afford cyanated C_{60} derivatives [46]. In sharp contrast to the reaction in liquid phase, the solid-state reaction of C_{60} with KCN under HSVM conditions for 30 min unexpectedly and selectively afforded the formal [2 + 2]-type dimer C_{120} (**6**) in 18% other than the cyanated C_{60} derivative **5**, along with 70% of unconsumed C_{60} (Fig. 2) [47]. The X-ray crystal structure unequivocally showed that the dumbbell-shaped C_{120} is connected by a cyclobutane ring and this four-membered ring is square rather than rectangular as predicted theoretically. Furthermore, the [2 + 2] structure of C_{120} is stable and does not rearrange to other C_{120} isomers such as those with peanuts-shaped structure.

The amount of KCN need not necessarily be equivalent to that of C_{60}; a small amount such as 0.2 molar equivalent

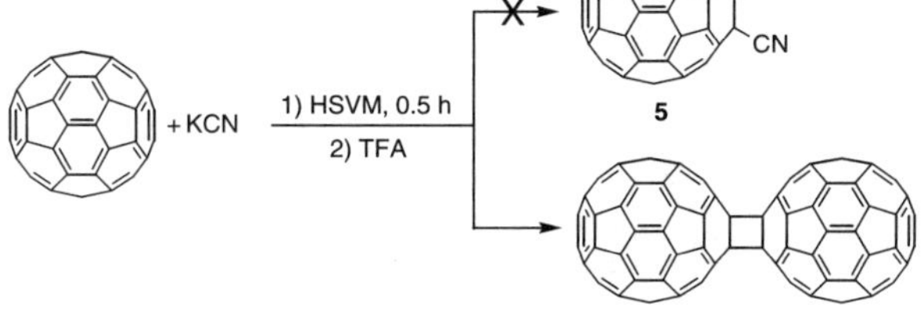

Figure 2. Synthesis of C_{120} from C_{60} and KCN under HSVM conditions [47].

is enough [28]. Apparently, the solid-state reaction of C_{60} with cyanide proceeds in a quite different reaction pathway from the solution reaction. Formation of C_{120} can also be obtained by the reaction of C_{60} with potassium salts such as K_2CO_3 and CH_3CO_2K, metals such as Li, Na, K, Mg, Al, and Zn, and organic bases such as 4-(dimethylamino)- and 4-aminopyridine [28]. This mechanochemical reaction appears to reach an equilibrium state with a ratio of C_{60} to C_{120} in about 7:3 either starting from a mixture of C_{60} and a chosen reagent or from pure C_{120}. For these reactions, the formation of C_{120} was most probably initiated by electron transfer reaction. Once the radical anion $C_{60}^{\cdot-}$ is formed, it would couple with a neutral C_{60} to give $C_{120}^{\cdot-}$, which would transfer one electron to another C_{60} to afford the neutral C_{120}. Helium-3 labeled C_{60} dimer, $^3He@C_{120}$ (**7**) (Fig. 3), was prepared by the above-mentioned HSVM reaction using a helium-3 labeled C_{60} sample in which the approximate ratio of $^3He@C_{60}$ to empty C_{60} is about 1:1000 [28]. $^3He@C_{120}$ and its derivatives can be monitored by helium-3 nuclear magnetic resonance (^{3}He NMR) [48]. Similarly, $N@C_{120}$ has been synthesized by the mechanochemical reaction of C_{60} containing approximately 0.01% of $N@C_{60}$ with additives such as KCN, $CuSO_4$, $CaCl_2$, and P_2O_5 [49]. $N@C_{120}$ has been measured by electron paramagnetic resonance (EPR).

[60] Fullerene dimer was also obtained as the main product from the HSVM reaction of C_{60} and C_{70} with 4-aminopyridine, which was designed to synthesize cross dimer C_{130} [50]. Dimerization of C_{60} was also found to occur without the presence of any catalyst. Treatment of C_{60} itself under HSVM conditions resulted in a small amount of C_{120} and a large amount of insoluble C_{60} polymers; the latter was not formed in the reaction of C_{60} and KCN [28]. A similar phenomenon was observed by prolonged mechanical milling in a conventional horizontal ball mill at a rotational speed of 95 ± 1 rpm [51, 52].

Shinohara and coworkers described the preparation of C_{120} in 15% overall yield by simply hand-grinding a mixture of C_{60} and K_2CO_3 [53]. The same research group also reported that C_{120} was obtained as the main product (8%) from a hand-ground reaction mixture of C_{60} and C_{70} with K_2CO_3 aiming to synthesize cross dimer C_{130}, which was isolated in 3% yield [54].

It should be mentioned that C_{120} has been synthesized in high yield (~80%) by squeezing the organic molecular crystal $(ET)_2C_{60}$ at 5 GPa and 200 °C using a wedge-type cubic anvil high-pressure apparatus [55]. While C_{60} powder itself was squeezed in a simple piston-cylinder pressure cell at 1 GPa and ~200 °C for a few minutes, only ca. 3% of C_{120} and mainly insoluble residue of C_{60} oligomers and polymers along with 15–20% unchanged C_{60} were obtained [56]. Longer treatment resulted in a decrease in the yield of C_{120} and an increase in the insoluble portion.

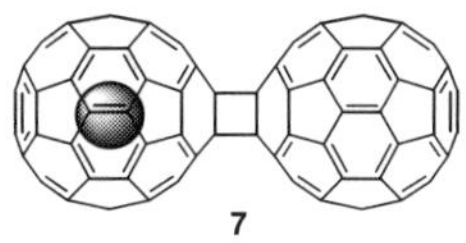

7

Figure 3. The structure of $^3He@C_{120}$ [48].

3.2.2. Synthesis of Cross Dimer C_{130}

Similar to the mechanochemical synthesis of C_{120}, the C_{60}-C_{70} cross dimer C_{130} (**8**) can be obtained by the HSVM reaction of C_{60} with C_{70} in the presence of a catalyst. Reaction of equal amounts of C_{60} and C_{70} with two equivalents of 4-aminopyridine for 30 min under HSVM conditions gave the cross dimer C_{130} (1.5%) in addition to C_{120} (6%), and unreacted C_{60} (46%) and C_{70} (44%) (Fig. 4) [50]. It should be noted that C_{70} dimer, C_{140}, could not be detected in the reaction mixture. The cross dimer C_{130} was found to be a single pure isomer, in which the C_{60} cage is attached to the C_{70} cage at the 1,2-bond in a [2 + 2] fashion. This cross dimerization can even occur by hand-grinding an equimolar mixture of C_{60} and C_{70} with four equivalents of K_2CO_3 in a mortar and pestle for 15 min [54]. Two isomers of the cross dimer C_{130} (3%) along with C_{120} (8%) were obtained (Fig. 4). Again, no detectable quantity of C_{70} dimer, C_{140}, was observed under the reaction condition. These results indicate that dimerization of the C_{60} and C_{70} mixture does not proceed statistically with equal probabilities between the two fullerene units. The formation of the two isomers of C_{130} involves the two most reactive bonds, that is, 1,2- and 5,6-bonds, of C_{70}. The presence of K_2CO_3 appears to be crucial to facilitate the cross dimerization between C_{60} and C_{70} under hand-grinding condition as no reaction was observed in its absence. Furthermore, other reagents that can act as electron donors such as 4-aminopyridine were not effective. The cross dimerization of C_{60} and C_{70} in the presence of K_2CO_3 was proposed to proceed via nucleophilic addition rather than electron transfer reaction [54].

3.2.3. Synthesis of C_{70} Dimer, C_{140}

Effort to obtain C_{70} dimer, C_{140}, by the reaction of C_{70} with various reagents under HSVM conditions was fruitless. However, the synthesis of C_{140} can be realized by hand-grinding a mixture of C_{70} and two equivalents of K_2CO_3 in a mortar and pestle [53]. The presence of K_2CO_3 is crucial for the dimerization. No reaction occurred in the absence of K_2CO_3, and moreover, other reagents such as 4-aminopyridine were ineffective. Five [2 + 2] structural isomers of C_{140} were obtained in a total yield of 8% (Fig. 5). This result is in sharp contrast with the dimerization reaction of C_{70} in high yield (20–30%) under high-pressure (1 GPa) and high-temperature (200 °C) conditions which gave only one single isomer of C_{140} (**10**), a [2 + 2] cycloaddition cap-to-cap C_{70} dimer with C_{2h} molecular symmetry (Fig. 5) [56]. The reason for not being able to obtain C_{140} under HSVM conditions is not clear. However, it is likely that dimer C_{140} is more labile and much more easily dissociated to monomer C_{70} when compared with C_{120} under the HSVM conditions.

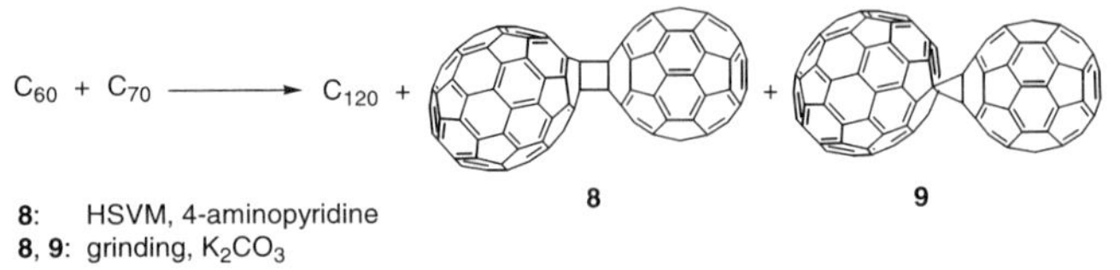

Figure 4. Synthesis of cross dimer C_{130} from C_{60} and C_{70} under HSVM and grinding conditions [50, 54].

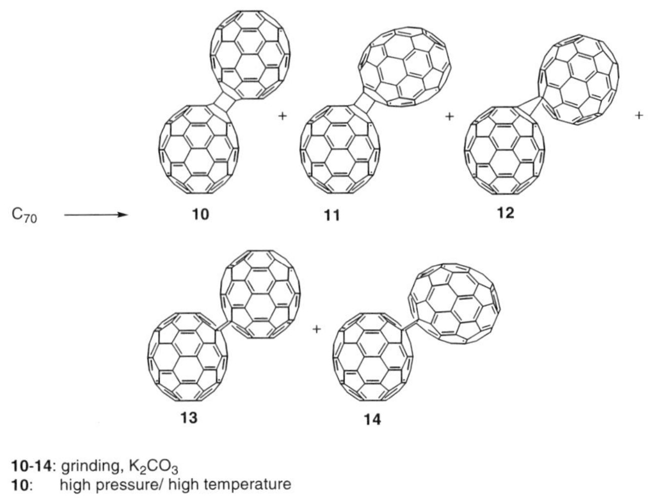

Figure 5. Synthesis of C_{140} by dimerization of C_{70} under grinding and high pressure/high temperature [53, 56].

3.2.4. Synthesis of C_{60} Trimer, C_{180}

The effects of various reagents including potassium salts, metals, and aromatic amines on the yield of C_{120} were surveyed to find the optimal reaction condition. It was found that fullerene trimer C_{180} was also formed in these reactions [57, 58]. The analysis of the reaction mixture of C_{60} with one equivalent of 4-aminopyridine for 30 min under HSVM conditions on a HPLC Cosmosil 5PBB column eluted with ODCB showed two small peaks (fractions A and B) of the trimer C_{180}, appearing after major peaks for C_{60} and the dimer C_{120}. The combined components in fractions A and B amounted to 4% yield as C_{180}. Further analysis on a Cosmosil Buckyprep column eluted with toluene revealed that the fraction A is a mixture of at least three structural isomers (**15–18**) while fraction B is a single component with a cyclic triangular structure (**19**) (Fig. 6), as verified by scanning tunneling microscopy (STM) [58].

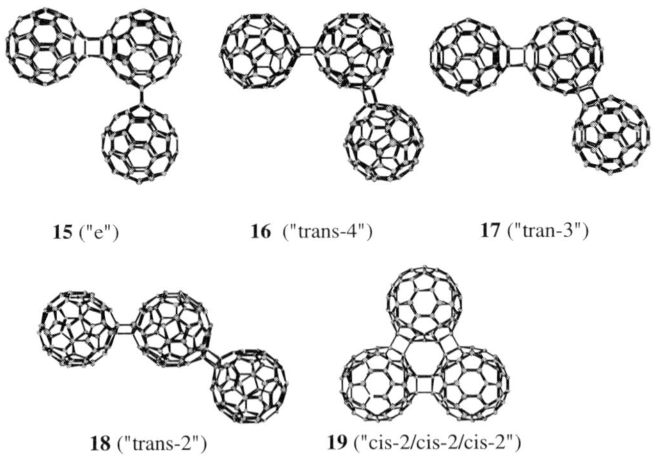

Figure 6. Isomeric structures of C_{60} trimer, C_{180} [57, 58].

3.3. [2 + 4] Cycloaddition Reactions of C_{60}

3.3.1. Diels–Alder Reaction of Condensed Aromatics to C_{60}

The [4 + 2] cycloaddition of C_{60} with anthracene is a well-studied reaction in solution [59–61]. The highest yield of the monoadduct is 39% in the conventional thermal conditions [59]. The Diels–Alder reaction of a 1:1.2 mixture of C_{60} and anthracene for 1 h under HSVM conditions was very efficient and gave 55% of the corresponding monoadduct and isomeric mixture of bisadducts in 19% [62]. The reaction appeared to reach an equilibrium state between the reactants and products after a reaction time of about 30 min. The reverse reaction (dissociation) of the monoadduct was found to give a similar component ratio. These results demonstrate that a chemical equilibrium is established under the present solid-state reaction system in spite of the heterogeneous reaction conditions. The HSVM reaction of C_{60} with one equivalent of pentacene afforded the double [4 + 2] adduct **21** (11%) in addition to the symmetrical monoadduct **20** (19%) and positional isomers of bisadducts **22** (15%) (Fig. 7) [62]. Apparently adduct **21** was formed by trapping the thermodynamically less favorable monoadduct **23** by another molecule of C_{60} (Fig. 8). In contrast, the reaction of C_{60} with pentacene in solution selectively gave the monoadduct **20** which has higher thermodynamic stability [63]. Adduct **21** could not be obtained from the solution reaction. Thus, in this particular case, the solid-state reaction appears to be advantageous in trapping the kinetic product. The HSVM treatment of C_{60} with 9,10-dimethylanthracene, tetracene, naphtho[2,3-α]pyrene afforded the corresponding [4 + 2] cycloadducts in fairly good yields.

3.3.2. Diels–Alder Reaction of Phthalazine to C_{60}

When the reaction of C_{60} with one equivalent of phthalazine was carried out in a solution of 1-chloronaphthalene at 255 °C for 1 h, the open-cage fullerene derivative **24** was obtained in 44% yield together with 23% of recovered C_{60} by a one-pot reaction. In contrast, the HSVM reaction of

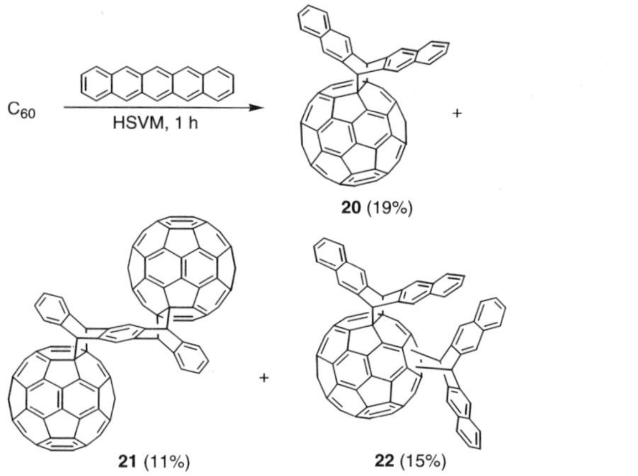

Figure 7. Reaction of C_{60} with pentacene under HVSM conditions affording the monoadduct, bisadducts, and C_{60} dimer derivative [62].

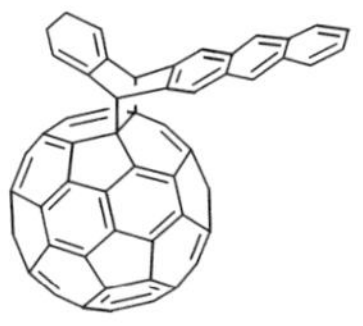

23

Figure 8. Structure of less stable [4 + 2] adduct from C_{60} and pentacene [62].

C_{60} with four equivalents of phthalazine for 1 h followed by heating this solid reaction mixture at 200 °C for 2 h afforded dimeric derivative **25** as the sole product in 14% along with 61% of unchanged C_{60} (Fig. 9) [64]. Compound **25** underwent intramolecular [2 + 2] cycloaddition between the two C_{60} cages located in proximity to give a new C_{60} dimer in quantitative yield either upon irradiation with visible light in chloroform or by heating in ODCB.

3.3.3. Diels–Alder Reaction of Di(2-pyridyl)-1,2,4,5-tetrazine to C_{60}

Upon HSVM treatment of a mixture of C_{60} and di(2-pyridyl)-1,2,4,5-tetrazine under HSVM conditions for 30 min, a highly efficient reaction took place and adduct **26** was obtained as a single product quantitatively [65]. The yield of compound **26** obtained by HSVM method is obviously higher than that carried out in refluxing toluene (50–60%) [66]. Compound **26** is highly sensitive to moisture and is transformed by the addition of water molecule followed by an unusual series of rearrangements to give the novel and unexpected 1,2,3,4-tetrahydro-C_{60} derivative **27** (Fig. 10). Other nucleophiles such as alcohols, thiols, and primary and secondary amines react with **26** similarly. The reaction is catalyzed by weak acid catalyst such as silica gel and requires ambient light [67].

The synthesis of dimeric compound **28** was achieved by the HSVM reaction of C_{60} with 0.5 equivalent of di(2-pyridyl)-1,2,4,5-tetrazine, followed by the [4 + 2] cycloaddition of the resultant **26** with C_{60} by heating the solid mixture at 150 °C for 2 h to afford 27% of **28** along with 40% of recovered C_{60} (Fig. 10) [65]. This dimeric compound **28** was obtained only by the solid-state reaction.

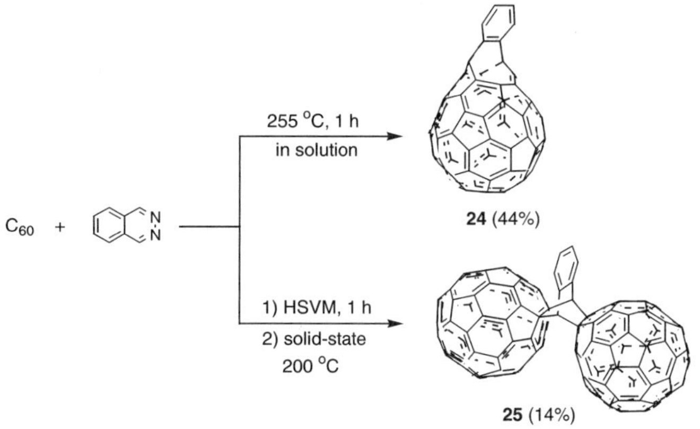

Figure 9. Diels–Alder reaction of C_{60} with phthalazine under HSVM conditions and in solution [64].

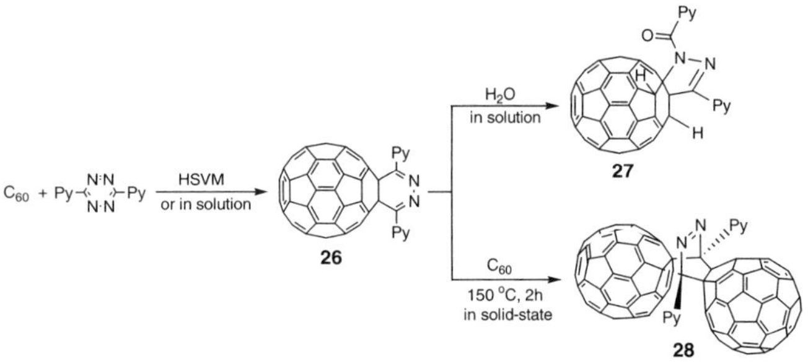

Figure 10. Reaction of C_{60} with di(2-pyridyl)-1,2,4,5-tetrazine under HSVM conditions and in solution [65–67].

The reaction in solution has been reported to result in only hydrogenation of **26** [66, 67]. A new novel C_{60} dimer with direct intercage bonds was formed quantitatively by the intramolecular [2 + 2] cycloaddition when a solution of **28** in ODCB was irradiated with room light for 3 h [65].

3.4. [2 + 3] Cycloaddition Reactions of C_{60}

3.4.1. 1,3-Dipolar Cycloaddition of Organic Azides to C_{60}

Upon the thermal reaction in solution, it is reported that organic azides and C_{60} give triazoline derivatives only at low temperature and the triazolines are converted to 5,6-open and 6,6-closed azafullerenes by heating the solution at 100 °C [68]. The HSVM reaction of C_{60} and azides for 30 min afforded triazoline derivatives **29** in 62–76%. The triazoline **29b** was then heated in the solid state at 120 °C for 2 h to give the 5,6-open and 6,6-closed azafullerenes **30b** and **31b** in higher yields than in solution (Fig. 11). The fact that triazoline derivative **29b** could not be converted to azafullerenes **30b** and **31b** under HSVM conditions indicates that the temperature in the current HSVM system does not rise up to 100 °C [24].

3.4.2. 1,3-Dipolar Cycloaddition of Diazo Compounds to C_{60}

The HSVM reaction of C_{60} with 9-diazofluorene did not allow the isolation of azo-compound **32** but gave 9,9-fluorenofullerene **33** produced by denitrogenation (Fig. 12) [24].

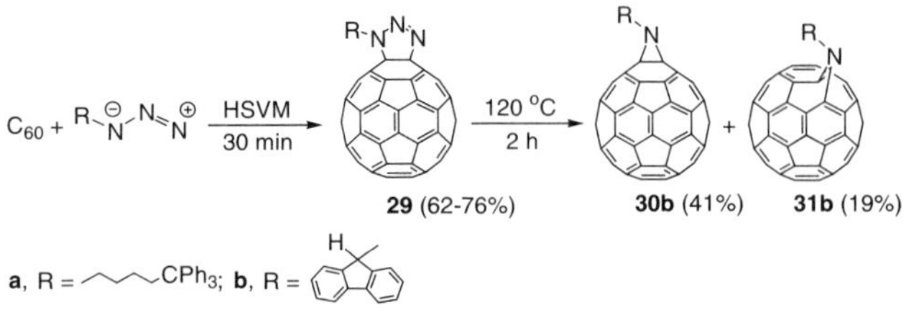

Figure 11. Reaction of C_{60} with azides under HSVM conditions followed by thermal solid-state reaction [24].

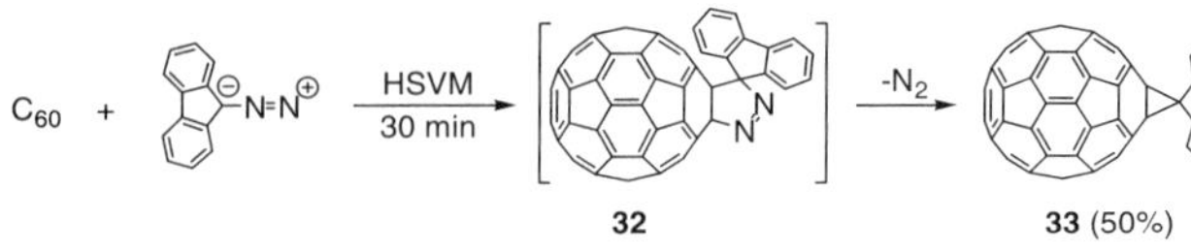

Figure 12. Reaction of C_{60} with 9-diazofluorene [24].

3.4.3. 1,3-Dipolar Cycloaddition of Azomethine Ylide to C_{60} and C_{70}

The 1,3-dipolar cycloaddition of azomethine ylide generated *in-situ* from *N*-methylglycine and paraformaldehyde to C_{60} to give fulleropyrrolidine was reported at a very early stage of fullerene chemistry by Prato and co-workers [69]. The HSVM technique is applied to the Prato reaction of C_{60} under solvent-free conditions. A mixture of C_{60}, one equivalent of *N*-methylglycine, and one equivalent of an aldehyde was vigorously shaken under HSVM conditions for 1 h; the expected fulleropyrrolidines **34a–e** were obtained in moderate yields (Fig. 13) [70].

Similarly, the HSVM reaction of C_{70} with one equivalent of *N*-methylglycine and one equivalent of paraformaldehyde for 1 h gave three monoadduct isomers **35**, **36**, and **37** in 41% yield with a ratio of 47 : 36 : 16 (Fig. 14).

3.4.4. Cycloaddition of Glycines to C_{60} and C_{70} [70]

While working on the Prato reaction of C_{60}, besides the corresponding fulleropyrrolidines, compound **34a** was isolated unexpectedly as a minor product in all reactions of C_{60} with sarcosine and aldehydes. Intrigued by this observation, the novel reactions between fullerenes and *N*-substituted glycines were found. The HSVM reaction of C_{60} with sarcosine (1:1) for 1 h gave **34a** and C_{120} in 19% and 23%, respectively. Furthermore, when sarcosine was substituted by *N*-ethylglycine, the same kind of reaction occurred, and **38** and C_{120} were obtained in 16% and 19%, respectively (Fig. 15).

When a mixture of C_{70} and *N*-methylglycine was treated with HSVM, the reaction gave monoadducts **35** and **36** in total yield of 23% with relative ratio of 1.5:1 (Fig. 16). The reason for the absence of **37** is not known. It should be noted that there is no evidence of the formation of C_{70} dimer, C_{140}, either from the Prato reaction of C_{70} or the direct reaction of C_{70} with *N*-methylglycine. This is consistent with the failure of formation of C_{140} from the reactions of C_{70} with potassium salts, metals, and amines under HSVM conditions.

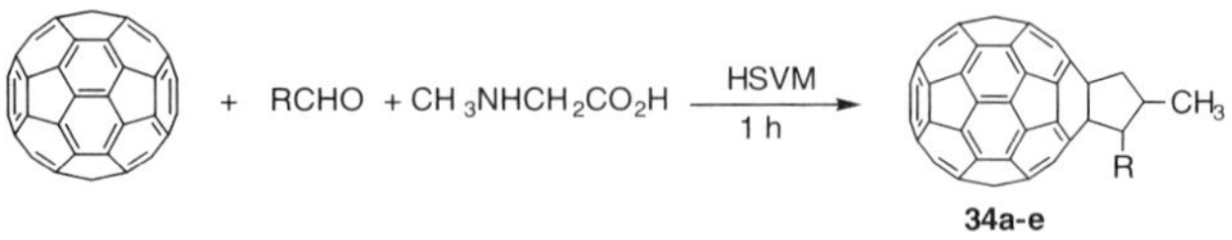

a, R = H **b**, R = C_6H_5; **c**, R = *p*-NO_2-C_6H_4; **d**, R = *p*-CH_3O-C_6H_4; **e**, R = *p*-$(CH_3)_2N$-C_6H_4

Figure 13. Prato reaction of C_{60} under HSVM conditions [70].

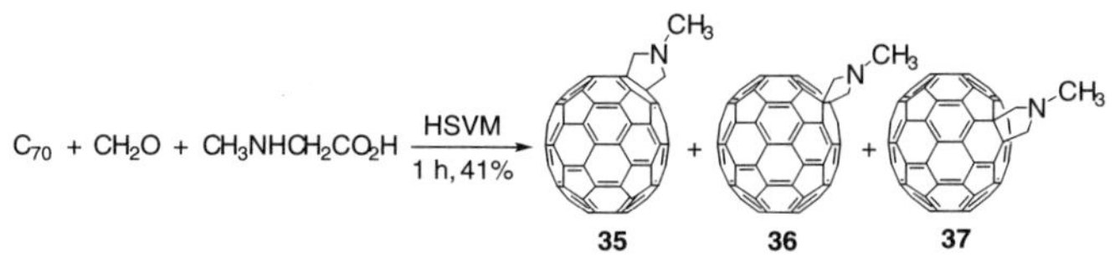

Figure 14. Prato reaction of C_{70} under HSVM conditions [70].

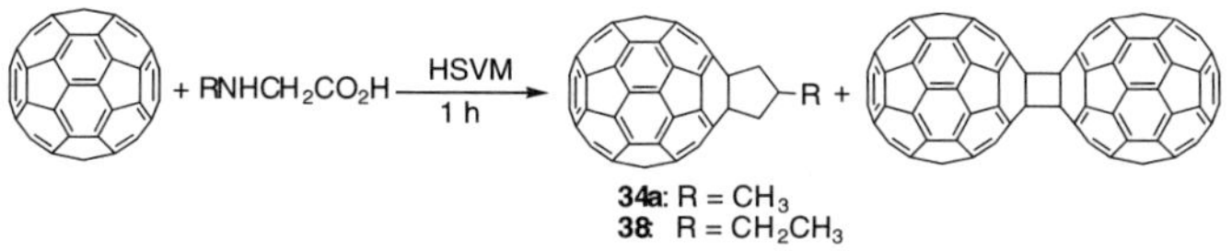

Figure 15. Reaction of C_{60} with *N*-alkylglycines under HSVM conditions [70].

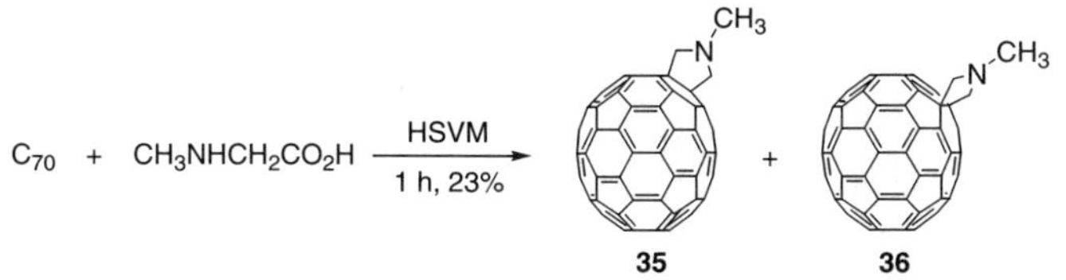

Figure 16. Reaction of C_{70} with *N*-methylglycine under HSVM conditions [70].

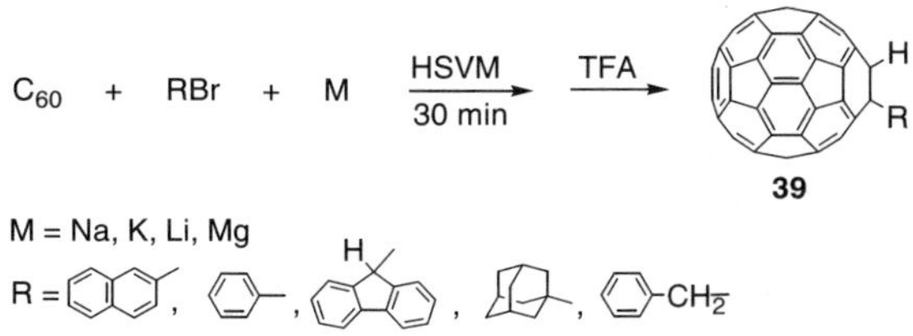

Figure 17. Reaction of C_{60} with alkali metals and bromides under HSVM conditions [71].

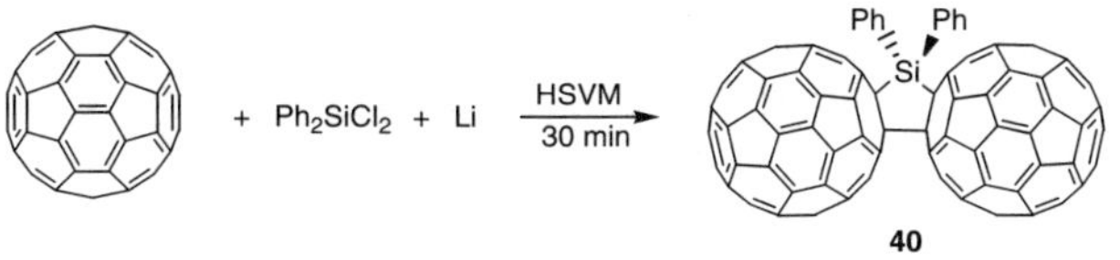

Figure 18. Reaction of C_{60} with dichlorodiphenylsilane and lithium under HSVM conditions [82].

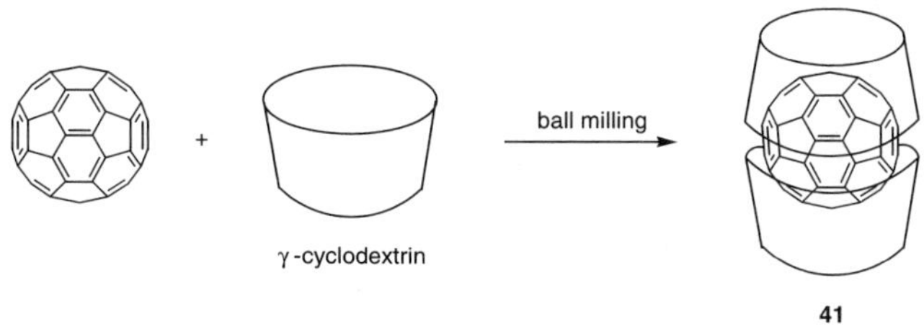

Figure 19. Formation of inclusion complex of C_{60} and γ-CD by ball milling [83].

3.5. Reaction of Organic Bromides and Alkali Metals with C_{60}

Reaction of C_{60} with organolithiums and Grignard reagents affording alkyl, aryl, or alkynyl C_{60} derivatives is one of the earliest studied reactions in solution [17–22]. A mixture of C_{60}, aryl or alkyl bromide, and metal (Li, Na, K, Mg) in a molar ratio of 1:2 ~ 3:4 ~ 6 was vigorously shaken for 30 min under HSVM conditions. The resulting black reaction mixture was acidified to give the corresponding aryl or alkyl C_{60} derivatives **39** in moderate yield and small amount of C_{120} (Fig. 17) [71]. The HSVM reaction of C_{60} with metals with higher reducing power is known to give a mixture of C_{60} and C_{120} in a 30:70 ratio. The formation of either the C_{60} dimer or the C_{60} polymer is suppressed by the presence of organic bromide. Alkali metals were found effective for these reactions, while the usage of magnesium gave poor result.

3.6. Reaction of Dichlorodiphenylsilane and Lithium with C_{60}

The silicon species such as silylene [72, 73], disilirane [74, 75], and silyllithium [76] have been reported to react with C_{60} to give mono- or disilyl derivatives of C_{60} [72–81]. Unlike the above-mentioned HSVM reaction of C_{60} with alkyl bromide and lithium which gave the alkyl C_{60} derivatives, the HSVM reaction of C_{60} with dichlorodiphenylsilane and lithium powder for 30 min gave 7.5% of adduct **40**, a C_{60} dimer fused with a silacyclopentane ring, together with 67.4% of recovered C_{60} (Fig. 18) [82]. The reaction mechanism is not quite clear yet, but it was assumed that the C_{60} radical anion formed by one-electron transfer from lithium was playing an important role in the reaction.

3.7. Formation of Fullerene Complexes

Braun and co-workers studied the mechanochemical synthesis of the supramolecular host-guest complex of C_{60} and gamma-cyclodextrin (γ-CD) [83]. The equipment used in the mechanochemical investigation was a 50 ml agate ball mill (diameter: 70 mm, height: 15 mm) (Pulverisette, Type 701, Fritsch, Germany). A homogenized mixture of 10 mg of C_{60} and 100 mg of γ-CD was ball-milled for about 20 h at room temperature. A 1:2 C_{60}-γ-CD complex **41** with a bicapped structure was supposed to be formed (Fig. 19). It has been proposed that the mechanochemical enhancement of the reactivity of C_{60} towards γ-CD proceeds via the amorphization of both the originally crystalline phases and guest molecules followed by a solid phase dissolution in amorphous media under vigorous mechanochemical treatment [83]. The C_{60}/δ-CD and C_{70}/δ-CD inclusion complexes have been prepared by the same method [84]. The solid-state inclusion complexation of C_{60} and C_{70} as well as some derivatives of C_{60} with γ-CD, and that of C_{60} and C_{120} with sulfocalix[8]arene, was attained under HSVM conditions more efficiently than classical ball milling [85]. The efficiency of complex formation under HSVM conditions was reflected in the very short reaction time (10 min) for the complexation of C_{60} with γ-CD as compared to about 20 h by classical ball milling [83] or more than 24 h in refluxing aqueous solution [86], and highest solubility of the formed complexes in water.

Another type of complex, that is, charge transfer complex, was reported to be obtained by grinding 1,4-diazabicyclooctane (DABCO) and C_{60} at room temperature in a dry atmosphere. A number of mixtures were prepared with different mole ratios of DABCO : C_{60}. The formed charge transfer complex was found to dissolve readily in water due to its charge separation and increase in dipole moment. However, the reaction of an aqueous solution of DABCO with C_{60} stirred vigorously for 2 h at 60 °C failed to afford the complex formed in the solid state [87].

4. CONCLUSIONS

Various solvent-free mechanochemical reactions of fullerenes have been investigated. Most of the studied mechanochemical reactions of fullerenes were done by high-energy HSVM technique, and many fewer reactions were done by low-energy ball milling and hand-grinding. It has been found that some mechanochemical reactions under HSVM conditions have the same reaction mechanisms as in the solution reactions, but some other mechanochemical reactions under HSVM conditions undergo quite different reaction pathways from those in the solution reactions and therefore give totally different products. Some fullerene derivatives can be uniquely obtained only under solvent-free and HVSM conditions. The fullerene mechanochemistry is still in its infancy. Even though knowledge of the fullerene mechanochemistry has been accumulated over the past years, many more experiments including more various reactions and the adjustment of variables, such as type of mill, milling speed, milling time, and ball to sample weight ratio, have to be done in order to fully understand the unique mechanochemical behavior and be able to propose a possible theoretical model for the fullerene chemistry under HSVM conditions. Furthermore, the extension of the HSVM technique to more general organic reactions (without the involvement of fullerenes) would be a promising research field to produce some unusual results.

GLOSSARY

Fullerene The third form of carbon, it is a family of caged compounds composed of only carbon element. A fullerene C_n consists of 12 pentagonal rings and any number of hexagonal ones, m, such that $m = (C_n - 20)/2$ (Euler's theorem). The most abundant fullerenes are C_{60} and C_{70}.

Fullerene chemistry The chemistry of fullerenes. Fullerenes undergo various kinds of reactions such as nucleophilic addition, Diels–Alder reaction, 1,3-dipolar cycloaddition, radical addition, oxidation, reduction, etc.

High-speed vibration milling (HSVM) It is a technique used to effectively activate the reaction system by vigorous milling of the reaction mixture with a vibration speed of several thousand cycles per minute.

Mechanochemistry Involves the transformation of mechanical energy into the driving force for the chemical reactions. The activation of the local reaction sites is generated by the stress, friction, shear deformation, etc. Mechanochemistry is generally done under solvent-free condition.

Solvent-free reaction Reaction conducted in the absence of any solvent.

ACKNOWLEDGMENT

This article was made possible by the financial support from "Hundred Talents Program" of Chinese Academy of Sciences and National Science Fund for Distinguished Young Scholars.

REFERENCES

1. W. Ostwald, "Handbuch der allgemeine Chemie. B. I," p. 70. Academische Verlagsgesellschaft, Leizpig, 1919.
2. G. Heinicke, "Tribochemistry." Academie-Verlag, Berlin, 1987.
3. J. F. Fernández-Bertrán, *Pure Appl. Chem.* 71, 581 (1999).
4. C. Suryanarayana, *Prog. Mater. Sci.* 46, 1 (2001).
5. S. A. Rowlands, A. K. Hall, P. G. McCormick, R. Street, R. J. Hart, G. F. Ebell, and P. Donecker, *Nature* 367, 223 (1994).
6. M. O. Rasmussen, O. Axelsson, and D. Tanner, *Synth. Commun.* 27, 4027 (1997).
7. L. D. Field, S. Sternhell, and H. V. Wilton, *Tetrahedron* 53, 4051 (1997).
8. L. D. Field, S. Sternhell, and H. V. Wilton, *Tetrahedron Lett.* 39, 115 (1998).
9. J. Fernández-Bertrán, J. C. Alvarez, and E. Reguera, *Solid State Ionics* 106, 129 (1998).
10. J. Fernández-Bertrán, L. Castellanos-Serra, H. Yee-Medeira, and E. Reguera, *J. Solid State Chem.* 147, 561 (1999).
11. V. D. Makhaev, A. P. Borisov, and L. A. Petrova, *J. Organometa. Chem.* 590, 222 (1999).
12. J. Fernández-Bertrán, L. M. Alfonso, J. C. Alvarez, and E. Reguera, *J. Incl. Phenom. Macro.* 37, 131 (2000).
13. M. Nüchter, B. Ondruschka, and R. Trotzki, *J. Prakt. Chem.* 342, 720 (2000).
14. V. P. Balema, J. W. Wiench, M. Pruski, and V. K. Pecharsky, *Chem. Commun.* 724 (2002).
15. H. W. Kroto, J. R. Heath, S. C. O'Brien, R. F. Curl, and R. E. Smalley, *Nature* 318, 162 (1985).
16. W. Krätschmer, L. D. Lamb, K. Fostiropoulos, and D. R. Huffman, *Nature* 347, 354 (1990).
17. R. Taylor and D. R. M. Walton, *Nature* 363, 685 (1993).
18. A. Hirsch, *Angew. Chem., Int. Ed. Engl.* 32, 1138 (1993).
19. A. Hirsch, *Synthesis* 895 (1995).
20. F. Diederich and C. Thilgen, *Science* 271, 317 (1996).
21. A. Hirsch, *Top. Curr. Chem.* 199, 1 (1999).
22. A. Hirsch, "The Chemistry of the Fullerenes." Thieme, Stuttgart, 1994.
23. T. Braun, *Fullerene Sci. Techn.* 5, 1291 (1997).
24. K. Komatsu, M. Murata, G.-W. Wang, T. Tanaka, N. Kato, and K. Fujiwara, *Fullerene Sci. Techn.* 7, 609 (1999).
25. F. Toda, *Synlett* 303 (1993).
26. F. Toda, *Acc. Chem. Res.* 28, 480 (1995).
27. K. Tanaka and F. Toda, *Chem. Rev.* 100, 1025 (2000).
28. K. Komatsu, G.-W. Wang, Y. Murata, T. Tanaka, K. Fujiwara, K. Yamamoto, and M. Saunders, *J. Org. Chem.* 63, 9358 (1998).
29. F. Dachille and R. Roy, *Nature* 186, 34 (1960).
30. G.-W. Wang, Y. Murata, K. Komatsu, and T. S. M. Wan, *Chem. Commun.* 2059 (1996).
31. A. M. Rao, P. Zhou, K.-A. Wang, G. T. Hager, J. M. Holden, Y. Wang, W.-T. Lee, X.-X. Bi, P. C. Eklund, D. S. Cornett, M. A. Duncan, and I. J. Amster, *Science* 259, 955 (1993).

32. Y. Wang, J. M. Holden, X.-X. Bi, and P. C. Eklund, *Chem. Phys. Lett.* 217, 413 (1994).
33. Y.-P. Sun, B. Ma, C. E. Bunker, and B. Liu, *J. Am. Chem. Soc.* 117, 12705 (1995).
34. B. Burger, J. Winter, and H. Kuzmany, *Z. Phys. B* 101, 227 (1996).
35. J. Onoe and K. Takeuchi, *Phys. Rev. B* 54, 6167 (1996).
36. J. Onoe, A. Nakao, and K. Takeuchi, *Phys. Rev. B* 55, 10051 (1997).
37. J. Onoe and K. Takeuchi, *Phys. Rev. Lett.* 79, 2987 (1997).
38. S. Pekker, A. Jánossy, L. Mihaly, O. Chauvet, M. Carrard, and L. Forró, *Science* 265, 1077 (1994).
39. P. W. Stephen, G. Bortel, G. Faigel, M. Tegze, A. Jánossy, S. Pekker, G. Oszlanyi, and L. Forró, *Nature* 370, 636 (1994).
40. M. C. Martin, D. Koller, A. Rosenberg, C. Kendziora, and L. Mahaly, *Phys. Rev. B* 51, 3210 (1995).
41. H. Yamawaki, M. Yoshida, Y. Kakudate, S. Usuba, H. Yokoi, S. Fujiwara, K. Aoki, R. Ruoff, R. Malhotra, and D. Lorents, *J. Phys. Chem.* 97, 11161 (1993).
42. Y. Isawa, T. Arima, R. M. Fleming, T. Siegrist, O. Zhou, R. C. Haddon, L. J. Rothberg, K. B. Lycons, H. L. Carter Jr., A. F. Hebard, R. Tycko, G. Dabbagh, J. J. Krajewski, G. A. Thomas, and T. Yagi, *Science* 264, 1570 (1994).
43. M. Núñez-Regueiro, L. Marques, J.-L. Hodeau, O. Béthoux, and M. Perroux, *Phys. Rev. Lett.* 74, 278 (1995).
44. C. Goze, F. Rachdi, L. Hajji, M. Núñez-Regueiro, L. Marques, J.-L. Hodeau, and M. Mehring, *Phys. Rev. B* 54, R3676 (1996).
45. P.-A. Persson, U. Edlund, P. Jacobsson, D. Johnels, A. Soldatov, and B. Sundqvist, *Chem. Phys. Lett.* 258, 540 (1996).
46. M. Keshavarz-K, B. Knight, G. Srdanov, and F. Wudl, *J. Am. Chem. Soc.* 117, 11371 (1995).
47. G.-W. Wang, K. Komatsu, Y. Murata, and M. Shiro, *Nature* 387, 583 (1997).
48. K. Fujiwara, K. Komatsu, G.-W. Wang, T. Tanaka, K. Hirata, K. Yamamoto, and M. Saunders, *J. Am. Chem. Soc.* 123, 10715 (2001).
49. B. Goedde, M. Waiblinger, P. Jakes, N. Weiden, K.-P. Dinse, and A. Weidinger, *Chem. Phys. Lett.* 334, 12 (2001).
50. K. Komatsu, K. Fujiwara, and Y. Murata, *Chem. Commun.* 1583 (2000).
51. Z. G. Liu, H. Ohi, K. Masuyama, K. Tsuchiya, and M. Umemoto, *J. Phys. Chem. Solid* 61, 1119 (2000).
52. Z. G. Liu, H. Ohi, K. Masuyama, K. Tsuchiya, and M. Umemoto, *J. Mater. Res.* 15, 1528 (2000).
53. G. S. Forman, N. Tagmatarchis, and H. Shinohara, *J. Am. Chem. Soc.* 124, 178 (2002).
54. N. Tagmatarchis, G. S. Forman, A. Taninaka, and H. Shinohara, *Synlett* 235 (2002).
55. Y. Iwasa, K. Tanoue, T. Mitani, A. Izuoka, T. Sugawara, and T. Yagi, *Chem. Commun.*, 1411 (1998).
56. S. Lebedkin, W. E. Hull, A. Soldatov, B. Renker, and M. M. Kappes, *J. Phys. Chem. B* 104, 4101 (2000).
57. K. Komatsu, K. Fujiwara, and Y. Murata, *Chem. Lett.* 1016 (2000).
58. M. Kunitake, S. Uemura, O. Ito, K. Fujiwara, Y. Murata, and K. Komatsu, *Angew. Chem. Int. Ed.* 41, 969 (2002).
59. K. Komatsu, Y. Murata, N. Sugita, K. Takeuchi, and T. S. M. Wan, *Tetrahedron Lett.* 34, 8473 (1993).
60. J. A. Schlueter, J. M. Seaman, S. Taha, H. Cohen, K. R. Lykke, H. H. Wang, and J. M. Williams, *J. Chem. Soc., Chem. Commun.* 972 (1993).
61. M. Tsuda, T. Ishida, T. Nogami, S. Kurono, and M. Ohashi, *J. Chem. Soc., Chem. Commun.* 1296 (1993).
62. Y. Murata, N. Kato, K. Fujiwara, and K. Komatsu, *J. Org. Chem.* 64, 3483 (1999).
63. J. Mack and G. P. Miller, *Fullerene Sci. Techn.* 5, 607 (1997).
64. Y. Murata, N. Kato, and K. Komatsu, *J. Org. Chem.* 66, 7235 (2001).
65. Y. Murata, M. Suzuki, and K. Komatsu, *Chem. Commun.* 2338 (2001).
66. G. P. Miller and M. C. Tetreau, *Org. Lett.* 2, 3091 (2000).
67. G. P. Miller, M. C. Tetreau, M. M. Olmstead, P. A. Lord, and A. L. Balch, *Chem. Commun.* 1758 (2001).
68. T. Grösser, M. Prato, V. Lucchini, A. Hirsch, and F. Wudl, *Angew. Chem., Int. Ed. Engl.* 34, 1343 (1995).
69. M. Maggini, G. Scorrano, and M. Prato, *J. Am. Chem. Soc.* 115, 9798 (1993).
70. G.-W. Wang, T.-H. Zhang, E.-H. Hao, L.-J. Jiao, Y. Murata, and K. Komatsu, *Tetrahedron*, 59, 55 (2003).
71. T. Tanaka and K. Komatsu, *Synth. Commun.* 29, 4397 (1999).
72. T. Akasaka, W. Ando, K. Kobayashi, and S. Nagase, *J. Am. Chem. Soc.* 115, 1605 (1993).
73. T. Akasaka, W. Ando, K. Kobayashi, and S. Nagase, *Fullerene Sci. Techn.* 3, 339 (1993).
74. T. Akasaka, W. Ando, K. Kobayashi, and S. Nagase, *J. Am. Chem. Soc.* 115, 10366 (1993).
75. T. Akasaka, T. Suzuki, Y. Maeda, M. Ara, T. Wakahara, K. Kobayashi, S. Nagase, M. Kato, Y. Nakadaira, M. Fujitsuka, and O. Ito, *J. Org. Chem.* 64, 566 (1999).
76. T. Kusukawa and W. Ando, *Angew. Chem., Int. Ed. Engl.* 35, 1315 (1996).
77. T. Kusukawa, Y. Kabe, T. Erada, B. Nestler, and W. Ando, *Organometallics* 13, 4186 (1994).
78. T. Kusukawa, Y. Kabe, and W. Ando, *Organometallics* 14, 2142 (1995).
79. T. Kusukawa, A. Shike, and W. Ando, *Tetrahedron* 52, 4995 (1996).
80. T. Kusukawa, K. Ohkubo, and W. Ando, *Organometallics* 16, 2746 (1997).
81. T. Kusukawa and W. Ando, *Organometallics* 16, 4027 (1997).
82. K. Fujiwara and K. Komatsu, *Org. Lett.* 4, 1039 (2002).
83. T. Braun, Á. Buvári-Barcza, L. Barcza, I. Konkoly-Thege, M. Fodor, and B. Migali, *Solid State Ionics* 74, 47 (1994).
84. K. Süvegh, K. Fujiwara, K. Komatsu, T. Marek, T. Ueda, A. Vértes, and T. Braun, *Chem. Phys. Lett.* 344, 263 (2001).
85. K. Komatsu, K. Fujiwara, Y. Murata, and T. Braun, *J. Chem. Soc., Perkin Trans.* 1, 2963 (1999).
86. T. Andersson, K. Nilsson, M. Sundahl, G. Westman, and O. Wennerströn, *J. Chem. Soc., Chem. Commun.* 604 (1992).
87. K. I. Priyadarsini, H. Mohan, A. K. Tyagi, and J. P. Mittal, *Chem. Phys. Lett.* 230, 317 (1994).

Encyclopedia of Nanoscience and Nanotechnology
www.aspbs.com/enn

Fullerene Nanomaterials

Shashadhar Samal
Kwangju Institute of Science and Technology, Kwangju, South Korea

CONTENTS

1. INTRODUCTION

During a series of experiments in 1985 to probe the nature and chemical reactivity of the species produced during the nucleation of carbon plasma, Kroto et al. discovered C_{60} as a stable molecule [1]. This stability is due to geodesic and electronic properties inherent in the truncated icosahedral cage structure. The molecule was named buckminsterfullerene after R. Buckminster Fuller, American architect, engineer, and constructor of geodesic domes. Back in 1966, the possible presence of hollow carbon cages in carbon vapors was indicated [2]. Osawa in 1970 and Bochvar and Gal'pern in 1973 offered unique theoretical studies and proposed the spherical icosahedral-symmetric structure for a C_{60} carbon cage [3]. Subsequent to the paper reporting the discovery of C_{60}, the novel proposal of a truncated icosahedral cage structure did not receive instant universal acceptance. In 1990 Krätschmer et al. confirmed the presence of C_{60} in arc-processed graphite [4]. They extracted a soluble material, which formed crystals. The X-ray analysis showed the material to consist of 10-Å diameter spheroidal molecules. Subsequent mass spectrometric and infrared data provided the first unequivocal evidence for C_{60} and a host of higher members of the fullerene family, the most prominent one being C_{70} [5]. The extracted C_{60} compound yielded a single ^{13}C nuclear magnetic resonance (NMR) line, which proved that all 60 carbon atoms are equivalent as is expected for the truncated icosahedral buckminsterfullerene structure. The early work on the experimental discovery, properties, and theoretical work on buckminsterfullerene was reviewed in 1991 by Kroto et al. [6].

According to the Euler's theorem, each molecule of the fullerene family (C_n) consists of 12 pentagons and m number of hexagons ($m \neq 1$), conforming to the relation $m = (n - 20)/2$ [i.e., a C_n can be made up of $2(m + 10)$ carbon atoms]. Therefore the 60 carbon atoms of C_{60} are comprised of 12 pentagons and 20 hexagons forming 32 faces. The bonds shared by two hexagons (1.40 Å) (6,6-bonds) (**1**) are slightly shorter than the pentagonal edges (6,5-bonds) (1.46 Å) (**2**) (Fig. 1). All the carbon atoms in C_{60} are equivalent. The carbon atoms are arranged at the 60 vertices of a truncated icosahedron (I_h). Molecules with I_h symmetry, of which C_{60} is the most prominent member (other examples being the smallest fullerene C_{20} stable in its fully hydrogenated form and C_{80}), have the highest degree of symmetry of any known molecule. The point group I_h has three symmetry operations involving the fivefold (C_s) (**3**), the threefold (C_3) (**4**), and the twofold (C_2) (**5**) axes of symmetry (Fig. 2). On the other hand, C_{70} consists of 12 pentagons and 25 hexagons. It has 5 types of carbon atoms (**6**) and 8 different C–C bonds (**7**, **8**) (Fig. 3). It has the shape of a rugby ball (**9**) and has D_{5h} symmetry (Fig. 4). The geometry of the C_{70} polar caps is very similar to that of C_{60}. In contrast to C_{60}, this fullerene has an equatorial belt consisting of 5 hexagons. Like C_{60}, the pentagonal edges are longer bonds, whereas the bond shared by 2 hexagons is shorter. The inner diameter of the C_{60} molecule is determined by NMR experiments to be 7.1 ± 0.07 Å [7]. The outer diameter of the C_{60} molecule is estimated by taking into account the size of the π-electron cloud associated with the carbon atoms. Considering that the interplanar distance between graphite layers is 3.35 Å the outer diameter of C_{60} is estimated as $7.09 \pm 3.35 = 10.34$ Å, where 3.35 Å accounts

ISBN: 1-58883-059-4/$35.00

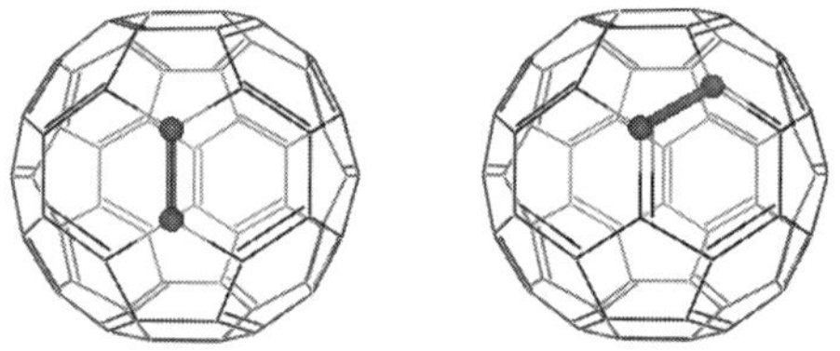

1 2

Figure 1. The buckminsterfullerene (C_{60}) showing a 6,6-ring junction double bond (**1**) and a 6,5-ring junction single bond (**2**).

for the thickness of the π-electron cloud surrounding the carbon atoms. The C_{70} long axis inner diameter is 7.96 Å and the short axis inner diameter is 7.12 Å [8]. There can be several giant fullerenes such as C_{180}, C_{240}, C_{540}, etc. They represent the higher members of the icosahedral fullerene family. Most studies have been done on C_{60} and C_{70} with few reports on C_{76}, C_{78}, C_{80}, C_{82}, and C_{84} [9]. Some of the important physical constants of C_{60} and C_{70} are listed in Table 1. In C_{60} the highest occupied molecular orbital/lowest unoccupied molecular orbital (HOMO–LUMO) gap is calculated to be 1.5–2.0 eV. The LUMO is triply degenerate and energetically low-lying [10], making C_{60} a good electron acceptor. For C_{70} the LUMO is doubly degenerate. However, the energy separation between the LUMO and the LUMO + 1 is small. From cyclovoltametry (CV) studies, it is found that both C_{60} and C_{70} are capable of having six negative oxidation states in solution. The six electrons completely fill the LUMO level of C_{60}. In the CV the potential separation between any two successive reductions is almost constant (~450 mV). This result is as expected if the LUMO is triply generate. The choice of the solvent and the supporting electrolyte in the electrochemical cell as well as temperature are important for generating a stable C_{60}^{n-} or C_{70}^{n-} species in solution. For C_{60} a mixture of acetonitrile and toluene is the preferred solvent. The formation and stability of C_{60}^{n-} and C_{70}^{n-} species suggest that the electron transfer reaction with atoms or molecules is possible, which include metals and metal complexes. Reduction potential of C_{60} is 0.99 eV versus saturated calomel electrode (SCE). It accepts six electrons electrochemically, generating up to hexa-anion [11, 12]. Oxidation of C_{60} is relatively difficult (IP + 1.76 eV) [11]. Like C_{60}, C_{70} also displays six easily accessible one-electron reductions.

1.1. Fullerene: Superaromatic or Superalkene?

In the fullerenes the π-bonds are conjugated. The question whether fullerenes have to be considered aromatic

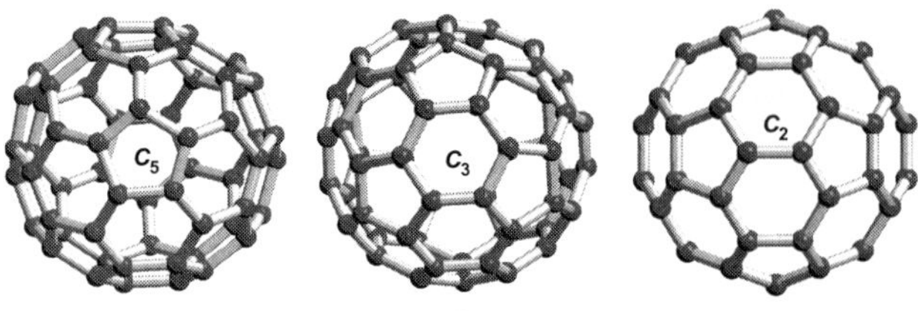

3 4 5

Figure 2. The symmetry axes of the regular truncated icosahedron.

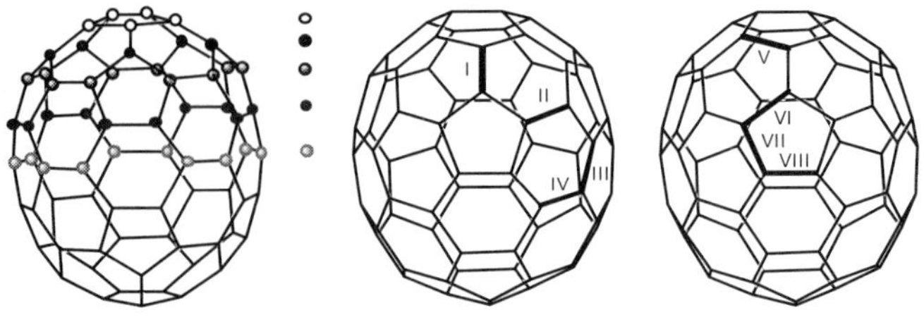

6 7 8

Figure 3. The C_{70} with five types of carbon atoms and eight different C–C bonds, four between 6,6-ring junctions (**I–IV**) and four between 5,6-ring junctions (**V–VIII**).

has remained a debate. A hexagon can be fused to two pentagons in three different ways (Fig. 5). Stable fullerenes have either entirely or predominantly the *meta* orientation of pentagones. The *para* orientation is only found in fullerenes made of 84 carbons or more. In such higher fullerenes strain is less significant due to the lower curvature of the cage surface. C_{60} is the smallest possible fullerene, which obeys the "isolated pentagon rule" [13]. All pentagons in C_{60} are isolated by hexagons. Thus, out of the possible arrangements of a hexagon fused to two pentagons, there is only one structure (**13**) in which there is no double bond on the five-membered rings. A double bond in a five-membered ring increases strain and hence decreases stability. Theoretical calculations indicate that in C_{60}, there can be 12,500 resonating structures [14]. There is only one resonating structure in which all the five bonds of the pentagons are single bonds. In C_{60}, all the double bonds are localized between 6,6-ring junctions (i.e., all the 6,6-ring junctions are double bonds and all the 5,6-ring junctions are single bonds). The hexagons, though seeming to be benzene-like, behave like cyclohexatrienes, and each pentagon has the [5]radialene character. Therefore C_{60} behaves in most of its reactions as a superalkene, rather than a superaromatic. Since there are no hydrogen atoms, substitution reactions are not possible, although substitution of boron for carbon has been observed [15]. Addition reactions are facilitated with considerable ease.

Even though the fullerenes undergo addition reactions easily, the π-bonds are conjugated. There is ample evidence from both experiment and theoretical calculations that fullerenes and their derivatives can sustain ring currents, just like the planar polycyclic aromatic hydrocarbons,

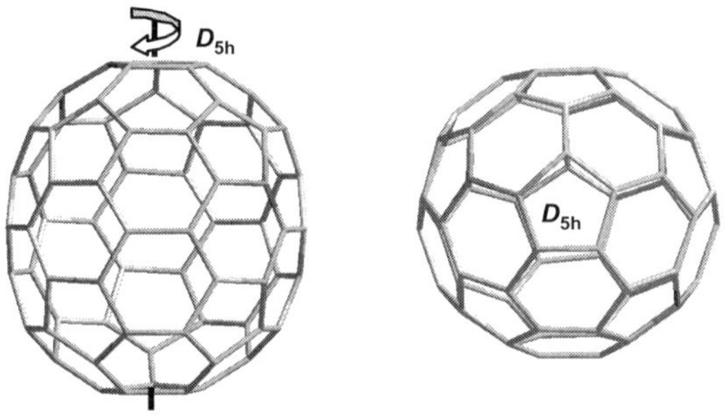

9 10

Figure 4. The pentagons superpose when viewed along the D_{5h} symmetry axis passing through the poles.

Table 1. Some important physical constants of C_{60} and C_{70}.

C_{60}		C_{70}	
Physical constant	Value	Physical constant	Value
5,6-Bond length (Å)	1.46	Long axis diameter (Å)	7.96
6,6-Bond length (Å)	1.40	Short axis diameter (Å)	7.12
Inner diameter (Å)	7.10	Number of carbon atom types	5
Outer diameter (Å)	10.34	Number of distinct C–C bonds	8
Binding energy per C atom (eV)	7.40	Binding energy per C atom (eV)	7.42
Heat of formation (kcal/g of C)	10.16	Heat of formation (kcal/g of C)	9.65
First ionization potential (eV)	7.58	First ionization potential (eV)	7.61
Electron affinity (eV)	2.65 ± 0.05	Electron affinity (eV)	2.72

Source: Selected data taken from M. S. Dresselhaus, G. Dresselhaus, and P. C. Eklund, in "Science of Fullerenes and Carbon Nanotubes, Ch. 3, pp. 63 and 67. Academic Press, San Diego, 1996.

as a consequence of the cyclic delocalization of π-electrons. The magnetic properties of fullerenes clearly reflect delocalized character of the conjugated π-electron system. Depending upon the number of π-electrons, within the loops of the hexagons and the pentagons there can occur diamagnetic and paramagnetic ring currents. For neutral C_{60}, for example, the hexagons are diamagnetic whereas the pentagons are paramagnetic. Such ring currents are taken as a measure of aromaticity in planar molecules. The same criterion can be applied to the three-dimensional fullerenes. Due to the presence of both the diamagnetic and paramagnetic ring currents, the fullerenes are described to possess "anomalous aromatic character." The center of each cage feels the combined effects of all ring currents. The best means of study of the overall degree of aromaticity or antiaromaticity is to study both experimentally and theoretically the endohedral fullerenes [16]. Theoretical estimates of heats of hydrogenation based on the reaction $C_{60} + H_2 \rightarrow C_{60}H_{60}$ indicate a stabilization per double bond in C_{60} about one-half that in benzene [17]. Based on several studies, including the heat of hydrogenation, it can be stated that C_{60} exhibits an aromatic stability, which is significantly inferior to that of benzene. The aromaticity of fullerenes is elaborately discussed in a review article [18].

1.2. Structure and Reactivity

Each carbon in fullerenes is sp^2-hybridized but not planar like the ideal sp^2 carbon in graphite. The curvature of the trigonal bonds in C_{60} confers some sp^3 character to the carbon atoms (i.e., each carbon atom is slightly pyramidized).

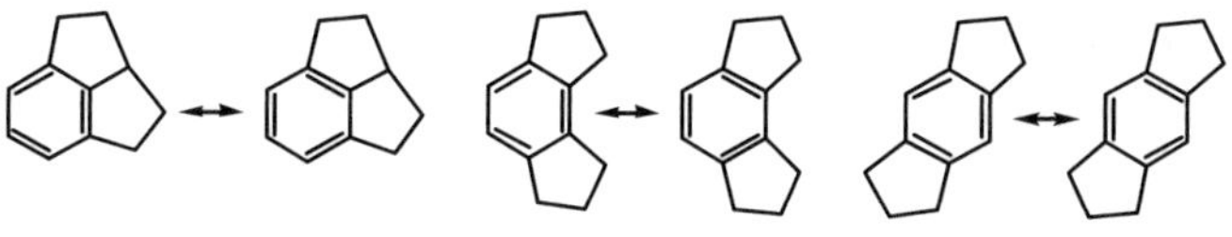

11 12 13 14 15 16

Figure 5. Three possible ways a hexagon can be fused to two pentagons. In structure **13** there are no double bonds on the pentagons.

The pyramidization angle ($\theta_{\sigma\pi} - 90°$) is 11.6° for C_{60} [19]. This angle is 0° for graphite. Hence the π-bonds on the C_{60} surface are bent. The pentagons are responsible for generating the curved surface of the fullerenes. When a pentagon is surrounded by five cyclohexatrienes sharing the bonds of the pentagon, such that all the double bonds are between 6,6-ring junctions, a curved surface is generated. The pentagon together with its five neighboring hexagons has the bowl shape of a corannulene molecule. However, the shape is not an exact replication of the fullerene polar caps, because corannulene has hydrogen atoms on the boundary carbons of its hexagons. The sp^2 networks of fullerenes have no boundaries, which can be saturated by hydrogen atoms. For C_{60}, a spherical shape distributes the strain as evenly as possible (i.e., the stain energy on a curved surface is minimum) [20]. As the number of carbon atoms in the cage increases, the curvature is reduced, and the chemical behavior should approach that of graphite, with the difference that graphite has dangling bonds at the edges. In C_{70} the polar curvature is same as that of C_{60}; the corannulene subunits have the same type of bond length alternation. The strain calculated per C atom of C_{60} is 10.16 kcal mol^{-1} [21]. Exohedral addition to C_{60} relieves this strain, while endohedral additions cause increase in strain. When isolated nitrogen atoms are inserted into C_{60}, the N atoms do not bond to C atoms. These atoms remain spectroscopically indistinguishable from nitrogen atoms *in vacuo* [22]. Similarly, F atom, methyl radicals are chemically inert when placed inside C_{60} [23].

The driving force for addition is for relieving of strain [24] and a change of the hybridization of carbon atoms involved in addition reaction from sp^2 to sp^3. The change in pyramidization consequent upon hybridization change is clearly seen in the X-ray structures that show distortion upon relieving strain. Addition to fullerene is exothermic due to relief of strain. It is therefore often not possible to stop the reaction at the monoadduct stage. Multiaddition is facilitated with considerable ease. A series of addition sequences ultimately lead to a product with the remaining unreacted six-membered rings, and then further addition is not possible. Since there are a number of double bonds, addition leads to a mixture of mono- and polyadducts. Furthermore, the polyadduct can form regioisomers; the regioselectivity of addition is governed by avoidance of products with a double bond in the 5,6-ring junction.

C_{60} can be viewed as a close cage molecule holding six nonoverlapping pyracylene molecules arranged in an octahedral fashion (Fig. 6). Four pyracylene units are arranged around the belt of the C_{60} molecule, and the remaining two centered on the top and bottom of this equatorial belt (**17**). The six pyracylene units join up in a manner so as to create eight hexagons thus accounting for the 20 hexagons. There is one reactive double bond located in the center of each unit. Many of the addition reactions of C_{60} (and C_{70}) involve the scission of this central double bond in the pyracylene unit. Addition reactions such as cycloaddition, addition of nucleophiles, radicals, including reactions like hydrogenation, halogenations, hydroboration, hydrometallation, and metal complex formations, etc., take place at the 6,6-bonds. Addition at the six 6,6-ring

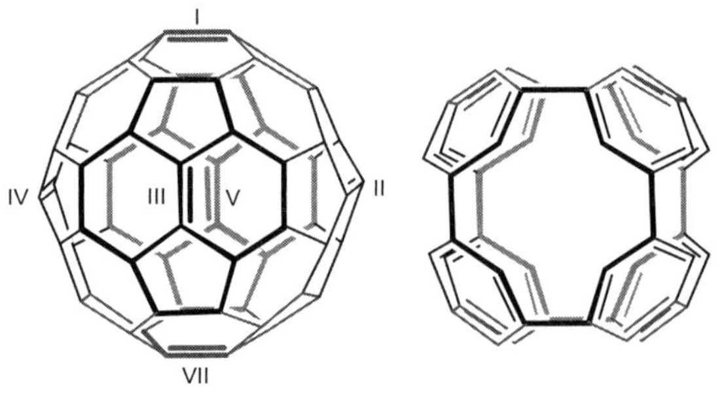

17 18

Figure 6. C_{60} showing six pyracylene units. The six double bonds at the 6,6-ring junctions (**17**, I–VI) are the most reactive ones for addition reactions. Other double bonds are not shown for reasons of clarity. When these six bonds are removed, an aromatic framework of eight benzene rings remains (**18**).

junctions of the six pyracylene units leaves behind an aromatic framework of eight benzene rings (**18**). This is the driving force for octahedral addition. Fullerenes are therefore electron-attracting and reaction with nucleophiles is preferred.

In this chapter, a concise account of the chemical reactions carried out on the curved surface of the fullerenes, specifically of C_{60} and C_{70}, is discussed. Since the literature on fullerenes has grown enormously [25], it is impractical to integrate all the reactions and properties of the products in the limited scope of this chapter. Nevertheless, an attempt is made to cover the important ones as much as possible, keeping the approach overtly simple and coherent. Though comprehensive articles integrating all the developments into one are very few [26], a vast number of up-to-date reviews and monographs on specialized topics are available. These are referred to in each section, which will provide a further link to the original works.

2. ENDOHEDRAL METALLOFULLERENES

Since the face-centered cubic packing structure of C_{60} has two vacant tetrahedral sites and one octahedral site per C_{60} molecule, spheres of radius 1.12 and 2.06 Å, respectively, can be accommodated in these sites. A number of exohedral metal complexes of fullerenes, such as MC_{60}^{+}, M = Li–Cs, Fe, Co, Ni, Cu, Rh, La, U, and VO, are formed by direct attachment of M^{+} with C_{60} by charge transfer [27]. For a C_{60} ternary fulleride like Ba_2CsC_{60}, the fulleride adopts a novel orientation ordering in which the Ba cations occupy the tetrahedral sites and Cs the octahedral sites. Some of the alkali metal exohedral complexes show superconductivity [28]. The C_{60} cavity is roomy enough to accommodate one or more metal ions leading to the formation of endohedral fullerene metal complexes, $M_m@C_n$. The metallofullerene in macroscopic amount is prepared by laser vaporization of a graphite–metal composite in a helium-filled tube-oven at high temperature [29]. Group 3 metals (M = Sc, Y, La), most of the lanthanide series elements, Group 2 metals (M = Ca, Sr, Ba), alkali metals (M = Li, Na, K, Cs), and some tetravalent metals (M = U, Zr, Hf) have been encapsulated into a fullerene cage to form mono-, di-, and trimetallofullerenes. Mass spectrometry plays a key role in their identification and characterization. The endohedral nature of metallofullerenes, the electronic structures, and properties are determined by synchrotron X-ray powder diffraction and various spectrometric means such as NMR, electron paramagnetic resonance (EPR), ultraviolet-visible near-infrared (UV-vis-NIR), X-ray photoelectron spectroscopy (XPS), and CV. One of the most distinct features of metallofullerenes is their superatom character. They can be viewed as a positively charged core metal surrounded by a negatively charged carbon cage. Several chemical reactions of endohedral metallofullerenes are reported. These novel forms of fullerene-based materials have attracted wide interest not only in physics and chemistry but also in such interdisciplinary areas as materials and biological sciences [30]. The materials have potential application as new types of superconductors, organic ferromagnets, nonlinear optical materials, functional molecular devices, magnetic resonance imaging agents, and biological tracing agents.

2.1. Synthesis

The synthetic method of choice is the arc-evaporation technique in which the arc burns a composite rod made from a metal oxide and graphite in a depressurized He gas. The soot produced contains a number of metallofullerenes and empty fullerenes, which are extracted with solvents such as pyridine, toluene, and carbon disulphide. Laser vaporization of a mixture of C_{60} and La_2O_3 is a convenient method to introduce lanthanum atom [31]. Yttrium atoms are also introduced by laser irradiation of a mixture of graphite and Y_2O_3 [32]. Production efficiency and the species of metallofullerenes increase if metal carbides are used as metal sources [33]. The size of the fullerene cage is also a determining factor. For example, when the cage is C_{82} fullerene, its metallofullerenes are formed and extracted efficiently. A number of methods including high pressure liquid chromatography (HPLC) have been used for isolation of endohedral metallofullerenes [34]. A silica-packed HPLC column in which tetraphenylporphyrin moicties are anchored to the silica surface is used to separate $M@C_{82}$ (M = La, Y) from empty fullerenes [35]. Metallofullerenes like $Ln@C_{80}$, $Ln@C_{82}$, $Ln_2@C_{80}$, $Gd@C_{82}$, and $La@C_{2n}$ are efficiently extracted from the insoluble part of the soot after toluene Soxhlet extraction by way of the high-temperature and high-pressure extraction by pyridine [36]. Aniline is a suitable extraction solvent for a number of metallofullerenes [37].

2.2. Structure

A number of studies by electron spin resonance (ESR), extended X-ray absorption fine structure (EXAFS), scanning tunneling microscopy, transmission electron microscopy, and synchrotron powder X-ray diffraction have strongly suggested that the metal atom(s) are indeed inside the fullerene cages. Since there are many isomers for endohedral metallofullerenes, the determination of structures and symmetries has long been of fundamental interest.

Some metallofullerenes such as $M@C_{82}$ (M = La, Y, Sc), $La_2@C_{80}$, $Sc_2@C_{84}$, and $Sc_3@C_{84}$ have been extensively investigated. For the C_{82} fullerene, there are nine isomers of $M@C_{82}$ (M = La, Y, Sc). For $La@C_{82}$ the observation of two octets in the ESR spectrum suggests two different conformations [38], existing under thermal equilibrium [39]. Like La, $M@C_{82}$ with other trivalent metal atoms like Y, Sc forms several isomers, two of which are dominant. With divalent metal ions (Tm, Ca) three isomers for $Tm@C_{82}$ and four isomers for $Ca@C_{82}$ are known [40]. Theoretical calculations, EPR, EXAFS, and synchrotron X-ray diffraction studies show that M is located off-center in $M@C_{82}$, and $M@C_{82}$ is most stable when M approaches the center of one hexagonal ring [41].

Two metal atoms can even be encapsulated inside the higher fullerene cages forming dimetallofullerenes, $M_2@C_{80}$ (M = La, Sc), of which $La_2@C_{80}$ is a representative example. C_{80} fullerene has seven isomers. The two La atoms are encapsulated inside the most stable isomers. The two La atoms are located equivalently on the C_2 axis facing the hexagon rings of C_{80}. To know whether encapsulated metal atom(s) are rigidly attached to cage carbons or move about ^{139}La-NMR study has been carried out, which confirmed that the metal atoms rotate inside the fullerene cage [42]. For the C_{84} fullerene, there are 24 distinct isomers. Three isomers of $Sc_2@C_{84}$ have been isolated [43]. For one isomer only one single ^{45}Sc-NMR signal observed confirms that the two Sc atoms are equivalent. For another isomer the ^{45}Sc-NMR spectra show two distinct signals of equal intensity, indicating that two Sc ions in C_{84} are at inequivalent positions. At about 383 K, however, the two Sc ions move and rapidly exchange their positions, resulting in an equivalent NMR line [44].

The tri-scandium fullerene, $Sc_3@C_{82}$, has been characterized by ESR [45]. The results of theoretical calculations show three scandium atoms form an equilateral Sc_3 trimer within the C_{82} cage retaining a threefold symmetry axis [46]. The three Sc atoms are in a dynamic state inside the cage. Introducing nitrogen gas into the arc generator during vaporization of the packed graphite rods containing metal oxides has produced high yields of trimetallic nitride endohedral metallofullerenes, $Sc_3N@C_{82}$ [47]. The complex can be viewed as a positively charged planar cluster of atoms inside a negatively charged icosahedral carbon cage, $[Sc_3N]^{+6}@[C_{82}]^{-6}$.

2.3. Electronic State

The incorporation of metal atom(s) inside the fullerene results in an electron transfer from the metal atom(s) to the carbon cage under the formation of a stable ion pair (e.g., $La^{3+}C_{82}^{3-}$ for $La@C_{82}$), confirmed from EPR and UV photoelectron spectroscopy (UPS) [48]. Similar measurements for $Y@C_{82}$ indicate the charge-transfer electronic state of $Y^{3+}C_{82}^{3-}$. In studies of Group-2-based samples (M = Ca, Sr, Ba), the resultant structures are $M^{2+}C_{82}^{2-}$. As a result of the transfer of two electrons from each Sc to C_{84} cage, $Sc_2@C_{84}$ is also described as $(Sc^{2+})_2C_{84}^{4-}$ concluded from the powder diffraction data and XPS studies [49]. C_{80} in $M_2@C_{80}$ (M = La, Pr, Ce) acquires the six valence electrons from the two M atoms, forming the same electronic structures, $(M^{3+})_2C_{80}^{6-}$. Cyclovoltametry studies indicate that fullerenolanthanides are stronger electron donors as well as stronger electron acceptors than empty fullerenes [50]. The calculated electrostatic potential of the inner surface of empty fullerenes shows that the carbon cage is positive at any position [51]. However, the electrostatic potentials become highly negative for metallofullerenes.

2.4. Reactivity

A large number of experimental studies have been performed on the chemical derivatization of empty fullerenes such as C_{60} and C_{70}. Up to now, however, the chemical properties of endohedral metallofullerenes have been studied scantily due to the low yield and separation difficulties. It is of great interest whether endohedral metallofullerenes can also be functionalized and how the endohedral doping of metal atom(s) modifies the reactivities of empty fullerenes. Photochemical cycloaddition of disilacyclopropane with $La@C_{82}$, $Gd@C_{82}$, and $La_2@C_{80}$ is one of the earliest chemical derivatizations reactions reported [52]. Under thermal conditions the reaction of these three endohedral fullerenes is observed, though the higher fullerenes, C_{76}, C_{78}, C_{82} and C_{84} as well as C_{60} and C_{70}, do not react thermally with disilacyclopropane. Since the electronegativities of germanium and silicon are similar, it is also found that photochemical and thermal reactions of $La@C_{82}$ with digermacyclopropane afford an exdohedral adduct [53]. The chemical reactions of $La@C_{82}$ and diphenyldiazomethane (Ph_2CN_2) have been reported and the products (several possible isomers) have been characterized by EPR [54]. Water soluble multihydroxyl endohedral metallofullerols [$Gd@C_{82}(OH)_m$ ($m = 20$) and $Pr@C_{82}O_m(OH)_n$ ($m = 10$, $n = 10$)] are reported [55]. These derivatives are very important and have applications in the field of biomedicine. A symmetric derivative of the trimetallic nitride endohedral metallofullerene, $Sc_3N@C_{80}$ with ^{13}C labeled 6,7-dimethoxyisochroman-3-one, is obtained by cycloaddition reaction of the *o*-quinodimethane intermediate with the endohedral complex [47].

2.5. Potential Applications

Endohedral metallofullerenes with radioactive atoms inside the fullerene cage have potential applications in materials science, biology, and medicine. Promising applications are demonstrated in radiochemical techniques [56]. Radioactive endohedral $^{7}Be@C_{60}$ can be detected using radiochemical and radiochromatographic methods [57]. Radioisotopically labelled metallofullerenes, such as ^{140}La-labelled $La@C_{82}$ and $La_2@C_{80}$, are administered to rats and the distribution of the metallofullerenes in organs is studied. Large portions of La fullerenes are found in liver and blood [58]. Endohedral metallofullerenes such as water-soluble $Gd@C_{82}(OH)_n$ can be used as a magnetic resonance imaging (MRI) contrast agent [59]. It is found to be much more effective than the commercially available Gd-MRI contrast agents. Endohedral holmium metallofullerenes have been utilized in a model metallofullerene radiotracer study. The ^{165}Ho metallofullerenes are chemically functionalized to impart water solubility and then neutron activated to ^{166}Ho in order to determine their biodistribution and metabolism properties. The results have been evaluated for potential applications of lanthanide metallofullerenes as new diagnostic or therapeutic radiopharmaceuticals [60].

3. ADDITION OF RADICALS

Alkyl and aryl radicals (R• = Me•, Et•, Ph•, PhS•, $PhCH_2^•$, $CBr_3^•$, $CF_3^•$, $Me_2CH^•$, $Me_3C^•$, $Me_3CO^•$, 1-adamantyl) react with C_{60} to form $R–C_{60}^•$ radical adducts. Single or multiple radical additions are possible depending on the type of radical involved [61]. Addition takes place at the 6,6-double bond of the pyracylene unit, with the creation of new radical center on the C_{60} carbon. Often delocalization of the radical is not possible due to an unfavorable double bond in the pentagonal ring. In such cases two $R–C_{60}^•$ radicals dimerize leading to $RC_{60}C_{60}R$. Dimerization does not take place in multiple additions.

A photochemically generated benzyl radical reaction with C_{60} leads to both radical and nonradical adducts R_nC_{60} ($R = PhCH_2$, $n = 1$ to at least 15) [62]. The radical adducts with $n = 3$ and 5 are stable above 50 °C and have been identified by ESR spectroscopy as the allylic $R_3C_{60}^•$ (**22**) and cyclopentadienyl $R_5C_{60}^•$ (**24**) radicals (Fig. 7). The unpaired electrons are highly localized on the C_{60} surface. The extraordinary stability of **22** and **24** is attributed to the steric protection of the surface radical sites by the surrounding benzyl substitutents. Mass spectrometric analyses of the product from addition of photochemically generated methyl radicals to C_{60} show formation of $(CH_3)_nC_{60}$ with $n = 1$ to at least 34.

In addition to the benzyl and methyl radicals, a variety of photochemically generated reactive, neutral radicals efficiently add to C_{60} to yield remarkably persistent radical adducts. The photolysis of C_{60} in the presence of donor molecules produces radical anion of C_{60} [63]. The radical anion $C_{60}^{•-}$ is obtained by photolysis of C_{60} solution containing benzylic trialkyltin derivatives. Reaction of a number of photolytically and thermally generated alkyl radicals with C_{60} leading to $R–C_{60}^•$ radical adducts (monoadducts) are identified by the proton and ^{13}C hyperfine interactions obtained from their electron spin resonance spectra [64]. From ^{13}C NMR spectra it is shown that the unpaired electron in $R–C_{60}^•$ is mostly confined to the two fused, six-membered rings without delocalization of the unpaired electron over the C_{60} sphere. The ESR spectrum of the $(CH_3)_3C–C_{60}^•$ is obtained by several photochemical and thermal methods including the UV irradiation of a saturated solution of C_{60} in benzene containing *tert*-butyl bromide. The spectrum is remarkably persistent after irradiation and shows a 10-line hyperfine manifold having the binomial intensity distribution appropriate for nine equivalent protons interacting with a single unpaired electron. The tri- and pentabenzyl adducts having allylic and pentadienylic structures found for C_{60}-benzyl radical reaction are not observed for *tert*-butyl radical.

ESR spectra of several simple monoalkyl radical adducts of C_{60}, studied in solution over a broad temperature range, revealed a substantial barrier to internal rotation about the bond connecting the alkyl substituent to the C_{60} [65]. An analysis of the temperature-dependent shape of the spectrum of *tert*-butyl-C_{60} affords activation energy for internal rotation of 8.2 kcal/mol. The rotation of the *tert*-butyl group is strongly hindered at low temperatures. Similar barriers exist for other alkyl adducts (except methyl). The alkyl substituents adopt a preferred equilibrium conformation relative to the carbon framework (Fig. 8). Alkoxy radicals RO• from the photolysis of ROSSOR add to C_{60} yielding $RO–C_{60}^•$ adducts (R = Me, Et, iso-Pr, *tert*-Bu, iso-$PrCH_2$, *tert*-$BuCH_2$) which are detected by ESR spectroscopy [66].

The ability of C_{60} to efficiently react with radicals leads to consideration of C_{60} as a "radical sponge" [67]. This property of fullerenes being a radical scavenger is useful for several practical applications. Antifriction, antiwear, and antiseize properties of lubricating oil are improved in the presence of C_{60}, because C_{60} inhibits unwanted polymerization. The use of a dissolved fullerene catalyst as inhibitor of coking reactions to thereby inhibit formation of coke on a catalyst or catalyst support by elimination of nucleating points or growth regions for such coke formation is reported. In the petrochemical and refining industries, the efficient production of many hydrocarbon products is often hindered by fouling resulting from unwitting polymerization of reactive olefins like 1,3-butadiene, isoprene, or styrene. Addition of fullerenes in ppm levels inhibits the thermal polymerization of activated olefins like styrene and butyl methacrylate. The extent of polymerization and the molecular weight of the recovered polymer are strongly dependent on fullerene concentration. Fullerenes apparently inhibit polymerization by scavenging monomer species that propagate chain growth.

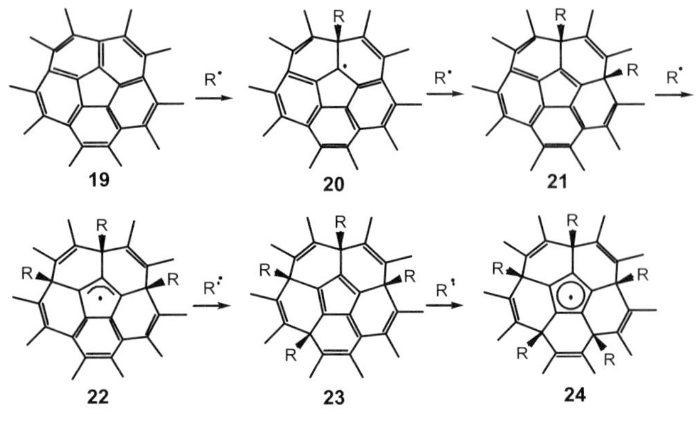

Figure 7. Multiple additions of free radicals to C_{60} leading to formation of persistent radical species (**20**, **22**, **24**). Adapted from [62].

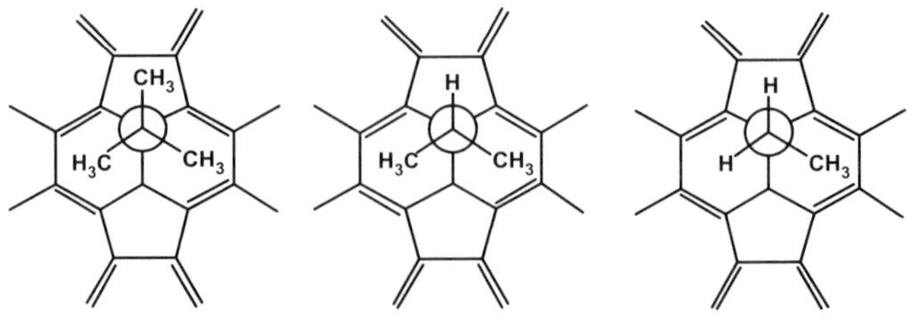

Figure 8. Preferred equilibrium conformations of the alkyl radicals. Adapted from [65].

4. SIMPLE FUNCTIONALIZATION REACTIONS

The π-bonds of fullerenes undergo addition reactions readily. Hydrogenation, halogenation, arylation, and alkylation of fullerenes have been extensively studied and comprehensively reviewed [68]. Since fullerenes contain a number of double bonds multiple addition takes place leading to mixture of isomers. The Schlegel diagrams for C_{60} (**25**) and C_{70} (**26**) with numbering of each carbon atoms is useful in identifying these isomers (Fig. 9). The 6,6-ring junction is the preferable position for 1,2-addition (**27**) (Fig. 10). 1,4-Addition (**28**) leads to introduction of unfavorable 5,6-double bonds and so does the 1,6-addition (**29**). Such additions are preferred only for sterically demanding addends [69].

4.1. Hydrogenation

Hydrogenation of a benzene solution of C_{60} is carried out using hydrogen at just above atmospheric pressure with Pd–C or Pt–C catalysts. However, the hydrofullerene obtained becomes rapidly oxidized to fullerenols. The reduction of the cage is not affected with Raney nickel as the catalyst. Hydroboration, hydrozirconation, or reduction by diimide leads to $C_{60}H_2$ [70]. Since all the 6,6-ring junction π-bonds are equivalent in C_{60}, only one dihydrogen product is obtained. With respect to a hydrogenated π-bond in C_{60}, a second molecule of H_2 can add to any one of the eight regiosymmetrically different double bonds, leading to eight possible tetrahydro derivatives. Indeed the reaction to tetrahydro derivative $C_{60}H_4$ leads to an isomeric mixture and all eight isomers are isolated [70b]. The tetrahydro[60]fullerene mixture consists of 30% of the 1,2,3,4-isomer **30**, shown as the Schlegel diagram (Fig. 11). Isolation of this product in good yields has fundamental significance in relation to the overall chemistry of fullerenes and understanding their aromaticity. The principal tetrahydro product is the one obtained by addition of the second hydrogen molecule to a π-bond in the same ring to which the first H_2 molecule was added. Subsequent addition could have occurred at any of the remaining eight different bonds of C_{60} with equal ease (except for the *trans*-1 double bond which is statistically disadvantaged by a factor of 4:1) had all the π-bonds

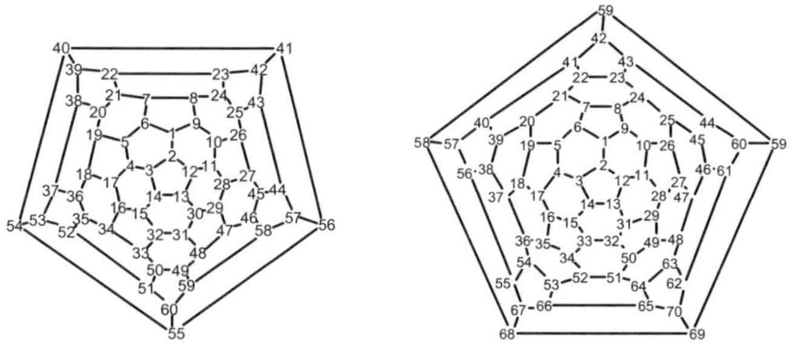

25 **26**

Figure 9. Schlegel diagrams of C_{60} and C_{70} with the carbon atoms numbered. The π-bonds are omitted for reasons of clarity. Adapted from [68].

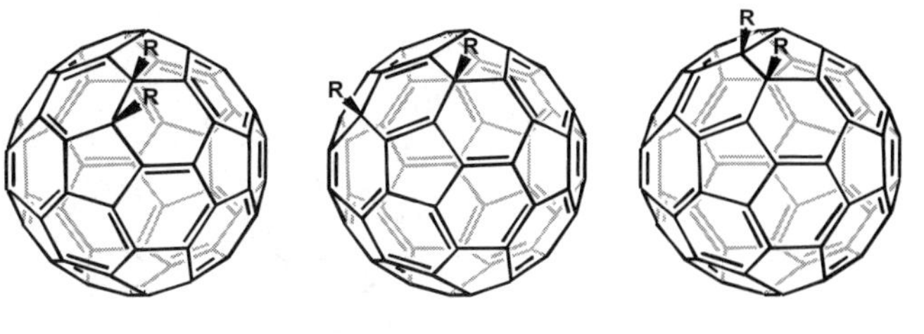

27 **28** **29**

Figure 10. Various modes of additions, showing unfavorable 6,5-ring junction double bonds in the 1,4- and 1,6-addition modes.

been localized. The preferential formation of the 1,2,3,4-tetrahydroderivative shows that there is some delocalization of the electron (i.e., the cage is partly aromatic in character). Addition in one hexagon leads the hexagon to lose its aromatic character and hence further addition is favored there.

Unlike C_{60}, all the double bonds of C_{70} are not equivalent. Hence a number of isomeric dihydro[70]fullerenes, in principle, are possible. The reduction of C_{70} by di-imide leads to 1,2- and 5,6-dihydro[70]fullerenes (**31**, **32**) in a 25:1 ratio [70b] indicating that the 1,2-bond is more reactive compared to the 5,6-bond (Fig. 12). The curvature is greater across the 1,2-bond (i.e., the extent of π-bond bending is more in this position and hence addition is favored there). Further reduction of C_{70} by di-imide gives six tetrahydro[70]fullerenes, 1,2,3,4- and 1,2,5,6-isomers (**33**, **34**) each comprising about 25% of the total. This again is due to loss of conjugation upon addition to one double bond in a six-membered ring. The remaining π-bonds of the ring become more reactive for subsequent additions compared to other π-bonds in the molecule.

Higher addition products of C_{60} are obtained by warming a stirred benzene solution of C_{60} with Zn/HCl for 10 minutes. Reduction by Zn–HCl gives $C_{60}H_{36}$ as a major product, which degrades slowly to $C_{60}H_{18}$ (**35**) (Fig. 13) on strong heating. C_{70} gives $C_{70}H_{36}$, $C_{70}H_{38}$, and $C_{70}H_{40}$ [71]. Reduction takes longer time since C_{70} is less reactive than C_{60}. Higher fullerenes such as C_{78} and C_{84} on reduction with Zn/HCl produce $C_{78}H_{36-48}$ and $C_{84}H_{48-52}$ as major components [72]. Formation of stable $C_{60}H_{36}$ (**36**) is a consequence of the four benzenoid rings generated upon hydrogen addition (see Fig. 13). These rings are substantially aromatic due to the removal of strain from the adjacent pentagons [73]. Both $C_{60}H_{18}$ and $C_{60}H_{36}$ can easily be converted into parent C_{60} by action of 2,3-dichloro-5,6-dicyanobenzoquinone (DDQ) in refluxing toluene [74].

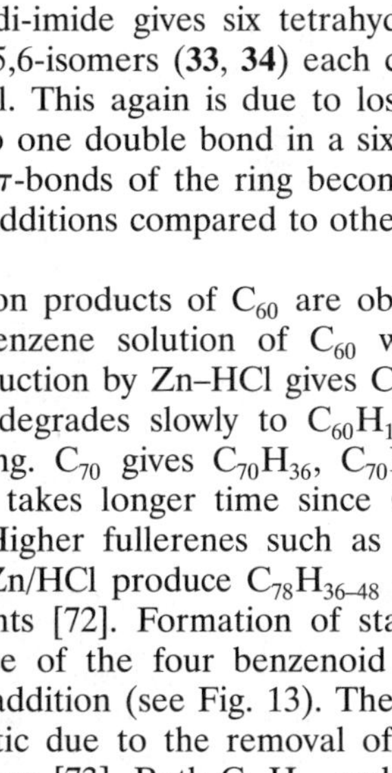

30

Figure 11. Schlegel diagram for the 1,2,3,4-tetrahydro[60]fullerene.

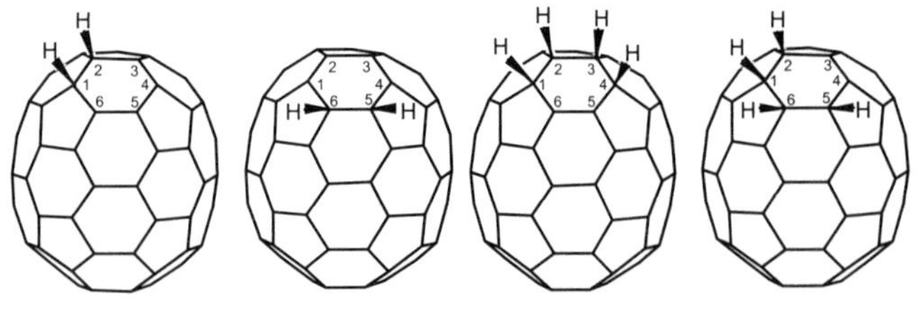

31 32 33 34

Figure 12. The di- and tetrahydro derivatives of C_{70}, the major products of diimide hydrogenation.

Reduction using Zn/DCl leads to $C_{60}D_{44}$ and $C_{70}D_{48}$. This higher hydrogenation level is attributed to the greater stabilities of C–D bonds in comparison to the C–H bonds.

4.2. Halogenation

4.2.1. Fluorination

The drive for fluorination of C_{60} was to get a material with properties analogous or superior to Teflon. However, this expectation had one point overlooked, the cage structure, which cannot flex significantly and hence high fluorination levels cannot be achieved. The principal features that dominate fluorination of fullerenes are the addition of fluorine at the 6,6-ring junctions and the tendency to generate aromatic rings in the fluorinated products, as is seen for hydrogenation. A number of fluorinated derivatives are now known [75].

Fluorination is initially carried out using either F_2 or XeF_2 as fluorinating agents at ca. 70 °C. Since these reagents react readily with the solvents used to dissolve the fullerenes (benzene, toluene, dichloromethane, etc.), fluorination is carried out under heterogeneous conditions. Fluorination of the solid fullerenes occurs layer by layer and can be seen by visual swelling of the samples [76]. In contrast to hydrogenation and many other addition reactions, fluorination of solid C_{70} is faster than C_{60}, in which efficient molecular packing restricts lattice penetration [77]. Heterogeneous reaction leads to inefficient fluorination and hence the product consists of a mixture of fluorinated and unreacted fullerenes [78]. Under these conditions, the only pure fluorinated fullerene isolated is $C_{60}F_{48}$. Fluorination of fullerenes by metal fluorides is carried out at high temperatures, ca. 400–500 °C, under vacuum. In this method a mixture of fluorinated products is obtained which needs extensive HPLC separation. $C_{60}F_{48}$ is a white solid. It exists in two chiral forms (**37**, **38**), the structures of which are determined by two-dimensional (2D)^{19}F NMR (Fig. 14) [79]. $C_{60}H_{48}$ is a concentrated source of fluorine and has been used for fluorination of aromatics [80]. Heating C_{60} with MnF_3 at 330 °C under vacuum and HPLC separation lead to $C_{60}F_{36}$, isolated in two isomeric forms, *T* and C_3. Addition of 36 fluorine atoms leaves behind benzenoid hexagonal rings, four for *T* (**39**) and three for C_3 isomer (**40**) (Fig. 15) [81].

Fullerene $C_{60}F_{18}$, a lemon yellow compound, and presumed to be an intermediate in $C_{60}F_{36}$ synthesis, is obtained by fluorination with K_2PtF_6 at 600 K [82]. It is isostructural with $C_{60}H_{18}$. It has an aromatic ring and the bonds in the aromatic rings are shorter than in benzene. The X-ray crystal structure shows the hexagonal aromatic ring at the center of the fluorinated face [83]. $C_{60}F_{16}$ is produced from fluorination of C_{60} with K_2PtF_6 at 465 °C and its structure is confirmed from ^{19}F NMR [84]. $C_{60}F_{20}$, an off-white solid, is formed by fluorinating C_{60} with either a mixture of MnF_3 and KF at 480 °C or Cs_2PbF_6 at 580 °C for 6 h. This compound has a unique structure with flattened poles and an expanded equator. $C_{60}F_2$ is formed in very low yield on fluorinating with K_2PtF_6 at 470 °C. A number of fluorinated derivatives of C_{70} are isolated, though in low yields. The prominent ones are one isomer each of $C_{70}H_{34}$, $C_{70}H_{42}$, $C_{70}F_{44}$, six isomers of $C_{70}H_{36}$, and four isomers of $C_{70}H_{40}$ [85]. Fullerenes $C_{76}F_{36}$, $C_{76}F_{38}$, $C_{76}F_{40}$, $C_{76}F_{42}$, and $C_{78}F_{42}$ are isolated in small quantities from fluorination of C_{76} and C_{78} with MnF_3 at 450–500 °C. Fluorination of C_{82} and C_{84} by MnF_3 at 470 °C leads to fluoro compounds $C_{84}F_{44}$, $C_{84}F_{40}$, and $C_{84}F_{44}$. In view of the low yields, preliminary characterizations of some of the compounds are available [86].

Fluorofullerenes are very reactive toward nucleophilic reactions. Although hydrophobic in the presence of a co-solvent such as acetone, an immediate exothermic reaction takes place with water. Such nucleophilic reaction means that fluorofullerenes would be unsuitable lubricants. Reaction with oxygen traces occurred during fluorination leading to formation of epoxides. For example, $C_{60}F_{18}O$ is isolated by HPLC from the product mixture from reaction leading to $C_{60}F_{18}$. The crystal structure of $C_{60}F_{18}O$ shows the oxygen atom being present in the form of a bridge at the 5,6-ring junction. Three isomers of the compound

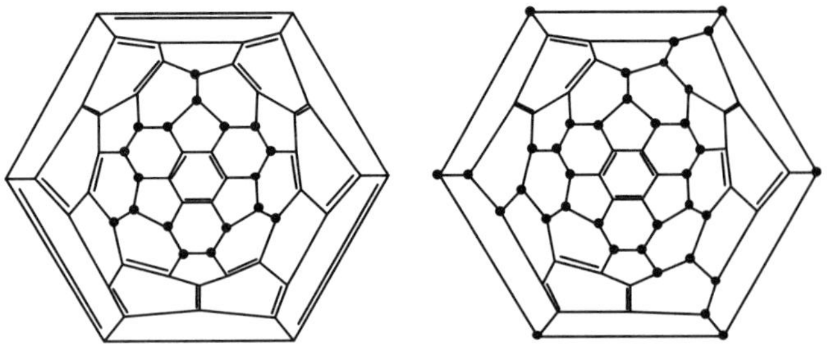

35 36

Figure 13. Schlegel diagrams for $C_{60}H_{18}$ and $C_{60}H_{36}$ (● = H). Adapted from [72].

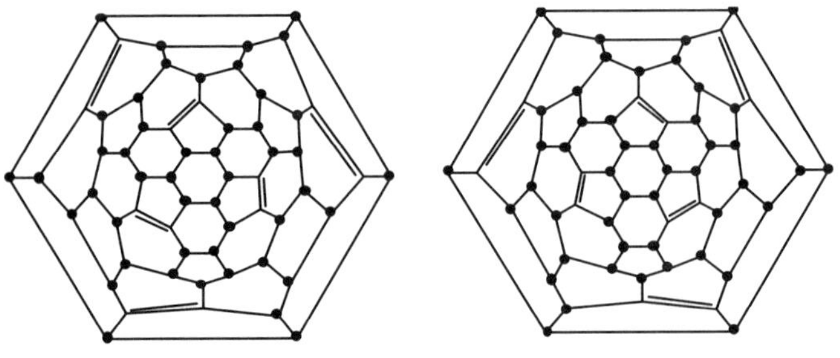

37 38

Figure 14. Two chiral forms of $C_{60}F_{48}$ (**37** *SS*, **38** *RR*) (● = F). Adapted from [79].

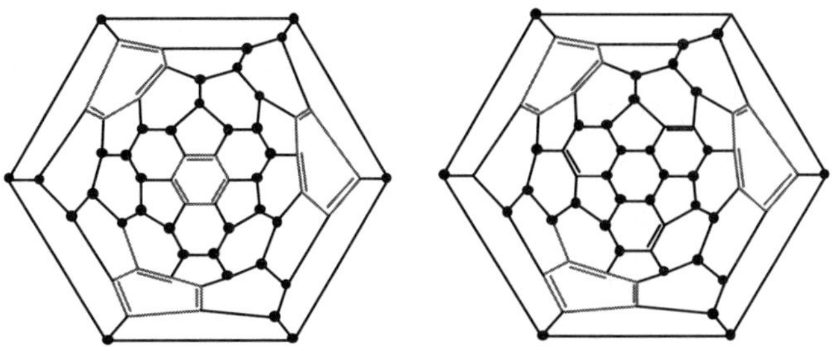

39 40

Figure 15. Two isomeric forms of $C_{60}F_{36}$, *T* (left) and C_3 (right). Adapted from [81].

are possible, which are isolated [87]. Seven isomers of bis-ethers $C_{60}F_{18}O_2$ are isolated and characterized by 1D and 2D ^{19}F NMR [88].

4.2.2. Chlorination

A reaction of a toluene solution of C_{60} with ICl leads to $C_{60}C_{16}$ (**41**) (Fig. 16), which is a very useful synthon for many phenylated fullerenes and derivatives [89]. Reaction of C_{70} with ICl furnishes $C_{70}Cl_{10}$ (**42**) [90]. Several of the chlorine atoms are in 1,4-positions, leading to two double bonds in the pentagon, which are also conjugated. In $C_{70}Cl_{10}$, pentagon in the center has five exocyclic double bonds. The structure is unique for a halogenated fullerene in having no double bonds in pentagons. The eclipsing adjacent chlorine atoms in both the compounds are regions of high reactivity. The compound $C_{70}Cl_{10}$ decomposes slowly in air, which is attributed to instability arising from the adjacent and eclipsed chlorines. Methoxy groups can replace all chlorine atoms from polychlorinated derivative to give polymethoxy derivatives.

4.2.3. Bromination and Iodination

Bromination of C_{60} in excess bromine gives a quantitative yield of yellow crystals of $C_{60}Br_{24}$ along with some $C_{60}Br_8$ and $C_{60}Br_6$ [91]. The bromine adds on to C_{60} by a 1,4-addition process. The bromo compounds decompose to C_{60} on heating and the stability order is $C_{60}Br_{24} > C_{60}Br_8 > C_{60}Br_6$. The instability of $C_{60}Br_8$ and $C_{60}Br_6$ may reflect the presence of four and two double bonds in pentagonal rings in their respective structures. Further, $C_{60}Br_6$, which is isostructural with $C_{60}Cl_6$, contains an eclipsing steric interaction [92]. C_{70} yielded a mixture of products of unknown composition. Iodination of fullerenes does not occur because steric hindrance requires the iodines to be well apart, which requires many double bonds to be on the pentagons.

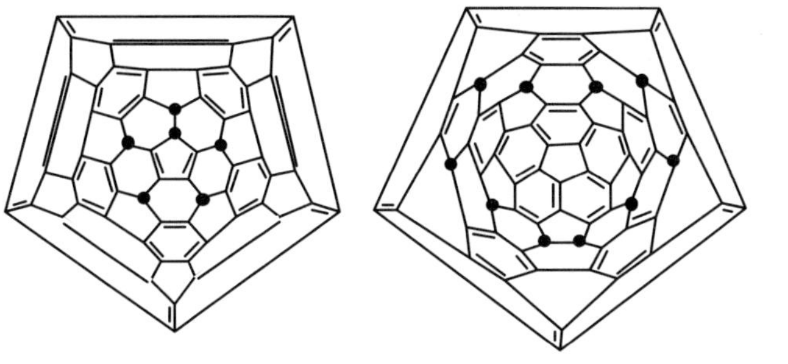

41 42

Figure 16. Structures of $C_{60}Cl_6$ and $C_{70}Cl_{10}$ (● = Cl). Adapted from [89].

4.3. Nitration, Sulphonation, and Hydroxylation

Nitration of C_{60} is carried out reacting nitronium ion generated from a mixture of H_2SO_4 and KNO_3 at 95 °C or with nitronium tetrafluoroborate in the presence of alkyl or aryl carboxylic acid in methylene chloride at ambient temperature under atmospheric pressure of N_2 [93]. In the latter reaction instantaneous formation of nitronium alkyl or aryl carboxylate salt is observed. Electrophilic attack of nitronium ion from $RCOO^-NO_2^+$ (R = C_6H_5, *p*-C_6H_4–Br) to C_{60} forms an intermediate $C_{60}(NO_2^+)_x$ and nucleophilic 1,3- or 1,5-substitution followed by hydrolysis finally gives a polyhydroxy organocarboxylate derivative consisting of an average of 13–15 hydroxyl substituents and 4–5 *p*-bromobenzoate moieties per C_{60} molecule. This upon hydrolysis gives the corresponding water-soluble fullerols consisting of 18–20 hydroxyl groups per C_{60} molecule. Nitro derivatives are also prepared by reacting C_{60} in the inert CCl_4 solution with conc. HNO_3 and NO_2 gas. The product has the composition $C_{60}(NO_2)_x(OH)_y$. Computation of structures, energetics, and vibrations of $C_{60}(OH)_x$ (x = 2–36) and $C_{60}(NO_2)_y$ has also been studied [94]. The computations offer a partial dissociation heat through which thermodynamic stability is measured. $C_{60}(OH)_6$ and $C_{60}(OH)_{18}$ exhibit the best value for partial dissociation heat. C_{60} is sulphonated in fuming sulphuric acid. The reaction involves initial oxidation of C_{60} followed by *in-situ* trapping of electrophilic C_{60} cation. Reactivity of C_{60} toward oxidative sulphonation reaction induced by H_2SO_4-SO_3 is highly enhanced by P_2O_5, leading to hexacyclosulphated fullerene intermediates, which are converted to dodecahydroxy-fullerenes upon hydrolysis [95]. Hydroxy groups can be added to C_{60} by reaction with BH_3–THF complex followed by hydrolysis with glacial acetic acid, NaOH/H_2O_2 or NaOH. The reaction of C_{60} in toluene with potassium and hydrolysis of the product is a convenient route to fullerols. A water-soluble endohedral metallofullerol, $Pr@C_{82}O_m(OH)_n$ ($m \sim 10$ and $n \sim 10$), is synthesized through the reaction of the endohedral metallofullerene, $Pr@C_{82}$, with concentrated HNO_3 and subsequent hydrolysis [96].

4.4. Amination

Neutral nucleophiles such as amines are added to C_{60} to form the corresponding adduct. In this process a maximum of 12 amine molecules can be added to C_{60} nucleus but the most stable one is the adduct which contains six amino groups. The addition takes place across the 6,6-ring junction of the pyracylene unit resulting in an octahedral array. Controlled reactions lead to both 1,2- and 1,4-addition products. The 1,2-addition product is most stable, but depends upon the steric demand of the addend. The reaction of C_{60} with an excess of morpholine or piperidine in benzene in the presence of oxygen leads to isolation of hydrogenated aminoadducts with a preferred 1,4-addition [97]. If the addend is very bulky, monoamine derivative of fullerene can be isolated. Addition of azacrown ethers in toluene to C_{60} at room temperature results in the monoamine derivative of fullerenes. Monoamination can be efficiently carried out following a titration technique in which addition of the amine to C_{60} solution is carried out in small portions until about 40% of the fullerene remains unreacted.

5. REACTIONS OF HALOFULLERENES

The halofullerenes undergo substitution reactions in the absence and presence of Lewis acid catalysts. When the nucleophile is an aromatic compound, the reaction can be considered an electrophilic aromatic substitution by the halogenofullerene (i.e., fullerenation of aromatics). Pure C_{60} and C_{60}/C_{70} mixtures undergo $AlCl_3$- or $FeCl_3$-catalyzed reaction with aromatics such as benzene and toluene to polyarene units by initial protonation of the cage by the $HAlCl_3OH$ super acid [98]. Ar and H are added across fullerene double bonds. In the products $C_{60}H_nAr_n$, the species with $n = 12$ and 16 are especially obtained when the aromatic group is Ph. The mechanism involves direct fullerenation of the aromatic ring by initial protonation of fullerene to the corresponding fullerene cation and subsequent Friedel–Crafts alkylation. Heating a solution of C_{60}, bromine and $FeCl_3$, in benzene leads to a C_{60} derivative with 12 phenyl groups on the cage [99]. HPLC separation of the reaction mixture leads to $C_{60}Ph_5H$ as the main product, along with $C_{60}Ph_n$ ($n = 4, 6, 8, 10, 12$), $C_{60}Ph_nO_2$ ($n = 4, 6, 8, 10, 12$), $C_{60}Ph_nOH$ ($n = 7, 9, 11$), and a number of other products from spontaneous oxidation of $C_{60}Ph_5H$ [100].

Products with well-defined structures are obtained starting from chlorofullerenes. Reaction of $C_{60}Cl_6$ with $FeCl_3$ in benzene produces $C_{60}Ph_5Cl$, which can be reduced to $C_{60}Ph_5H$ on reaction with PPh_3/H_2O [101]. This is a general reaction for a number of aromatics leading to products of the type $C_{60}Ar_5H$. The mechanism of substitution of the halogen involves an intermediate penta-arylfullerene cation, $C_{60}Ph_5^+$, isolated as the $AlCl_3$ catalyst [102]. Other by-products of the reaction are $C_{60}Ph_2$ and $C_{60}Ph_4$. On exposure to air, both $C_{60}Ph_5H$ and $C_{60}Ph_5Cl$ spontaneously oxidize with elimination of the hydrogen or chlorine attached to the cage and the ortho hydrogen of the adjacent phenyl group, giving a benzo[*b*]furanyl[60]fullerene; a second oxygen is located in the pentagon that is surrounded by phenyl groups [103]. $C_{60}Ph_5Cl$ reacts with CNBr in the presence of $FeCl_3$ producing a bromoisoquinolino[60]fullerene arising from nucleophilic replacement of the chorine by N≡C-Br, followed by electrophilic substitution into the ortho position of the adjacent phenyl ring. Reaction of bromoisoquinolino[60]fullerene with OH^- ion leads to isoquinolino[3′,4′:1,2][60]fullerene [104]. $C_{60}Cl_6$ reacts with allyltrimethylsilane in the presence of $TiCl_4$ to form a C_s symmetrical, hexaallyl[60]fullerene derivative, $C_{60}(CH_2CH = CH_2)_6$ [105]. The use of $FeCl_3$ as the Lewis acid catalyst in this reaction also produces the same product but in low yields.

Reaction of $C_{70}Cl_{10}$ with $FeCl_3$ in benzene produces $C_{70}Ph_8$ and $C_{70}Ph_{10}$ as the major phenylated derivatives, along with $C_{70}Ph_9OH$ as a minor component [106]. $C_{70}Ph_8$ gets converted to a bis-lactone, a fullerene with a hole (called a "holey fullerene"), which is a bis-lactone, $C_{70}Ph_8O_4$ [107]. Reaction is light catalyzed and proceeds via insertion of oxygen into [5,6]-bonds followed by oxidative cleavage of the adjacent double bond.

$C_{60}F_{18}$ reacts with benzene/$FeCl_3$ to give the triphenylated compound $C_{60}F_{15}Ph_3$. The structure of the compound is determined from ^1H and ^{19}F NMR studies, which shows that the phenyl groups replaced the three most accessible halogens of $C_{60}F_{18}$ [108].

6. η^5-FULLERENE METAL COMPLEXES

In the exohedral fullerene–transition metal complexes, the fullerenes act as an olefinic ligand and bind the metal center in η^2-fashion [109]. There are several examples of η^5-fullerene metal complexes [110]. For this purpose a cyclopentadiene moiety is generated on the fullerene surface from a pentaadduct, such as $C_{60}Ph_5H$. A number of cyclopentadienyl (Cp) metal complexes (η^5-$C_{60}Ph_5$)MLn (**43**) (MLn = Li, K, Tl, Cu · PEt_3) are synthesized (Fig. 17) using the cyclopentadienyl ring generated through deprotonation of the parent compound $C_{60}Ph_5H$ (**44**) (Fig. 18) [111]. In the process, the carbon atoms bearing the phenyl groups change the hybridization from sp^2 to sp^3. The cyclopentadienide moiety in the penta-adduct **44** appears to be electronically isolated from the bottom 50 sp^2 carbon atoms with the surrounding 5 sp^3 carbon atoms. Theoretical and electrochemical evidence, however, indicates that these two parts show a strong sign of mutual electronic interaction [112]. The η^5-fullerene metal complexes show promise as catalysts and materials of unique steric, electronic, and photonic properties.

Treatment of the pentaphenyl adduct $C_{60}Ph_5H$ with an alkali metal alkoxide such as LiO^tBu, KO^tBu, and TlOEt leads to Li(η^5-$C_{60}Ph_5$), K(η^5-$C_{60}Ph_5$), and Tl(η^5-$C_{60}Ph_5$), respectively. From the ^1H and ^{13}C NMR spectra it is observed that the reaction leads to conversion of the C_s symmetric $C_{60}Ph_5H$ to a new species possessing C_{5v} symmetry. The singlet peak due to the CpH proton (δ5.31 for $C_{60}Ph_5H$) disappears in the formation of the complexes. The synthesis of $[Cu(\eta^5\text{-}C_{60}Ph_5)(Ph)]^-$ involves the fivefold addition of an organocopper reagent to C_{60}. This is accomplished by the reaction of C_{60} with an excess amount of organocopper reagent prepared from PhMgBr and CuBr · SMe_2 followed by quenching with aqueous NH_4Cl. The reaction of $C_{60}Ph_5H$

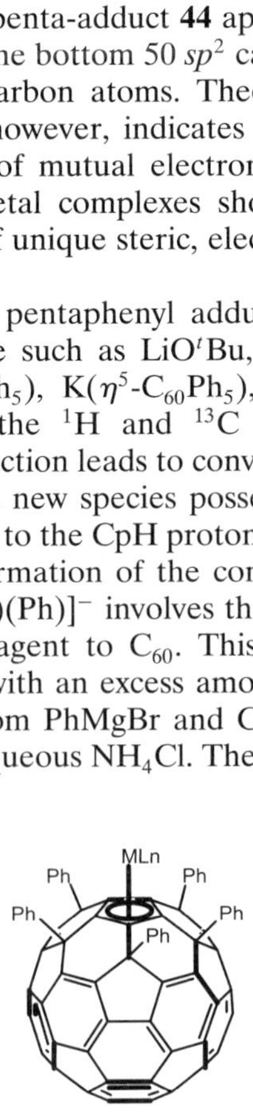

43

Figure 17. The pentahaptofullerene metal complexes. Adapted from [110].

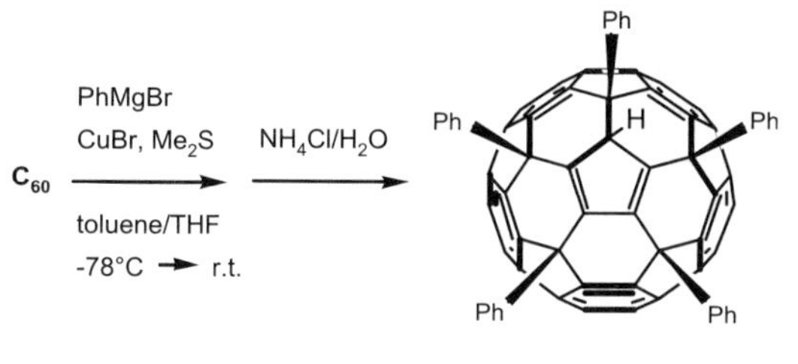

44

Figure 18. Synthesis of the pentaphenyl fullerene adduct $C_{60}Ph_5H$. Adapted from [111].

with $Cu(O^tBu)(PEt_3)$ gives a phosphine-coordinated transition metal complex $Cu(\eta^5\text{-}C_{60}Ph_5)(PEt_3)$. These complexes are sensitive to air and moisture but thermally stable in solution at room temperature.

The X-ray single crystal structure analysis for the thallium(I) complex shows the thallium atom as deeply buried in a cavity created by the five phenyl groups. The metal atom is equally bonded to the five Cp carbons with an average bond length of 2.87 Å. The thallium complex adopts approximately C_5 symmetric conformation with the phenyl groups arranged in a chiral propeller array. The ^1H and ^{13}C NMR spectra of the complex indicate free rotation of the phenyl groups at room temperature in solution. However, the solid phase structure implies that the enantiomer separation is possible by a suitable modification of the aromatic substitutents that can inhibit ring rotation.

The structural constraints of the fullerene core cause significant deformation of local structure around the Cp ring. Still the C_{60} core largely retains its original spherical form, with the average lengths of the C–C double bonds (1.38 Å) and the single bonds (1.44 Å) in the C_{50} moiety being similar to those of monofunctionalized C_{60} derivatives. The UV-vis spectra of both the cyclopentadiene derivative $C_{60}Ph_5H$ and the thallium complex show that the 50 π-electron C_{50} moiety is a chromophore of intensity comparable to that of the parent C_{60}. The absorption extends to fairly long wavelengths (670 nm for $C_{60}Ph_5H$, 720 nm for the complex) suggesting that these fullerene derivatives have electronic properties similar to those of the parent C_{60}.

The reaction is very general for various aryl and alkenyl groups as well as for methyl group [113] (Fig. 19). The synthesis of the alkenyl product is accomplished using a 1-alkenyl copper reagent prepared from the corresponding lithium reagent instead of the Grignard reagent employed for the synthesis of the arylated and the methylated analog (Fig. 20). The alkenyl substitutents can be *E* or

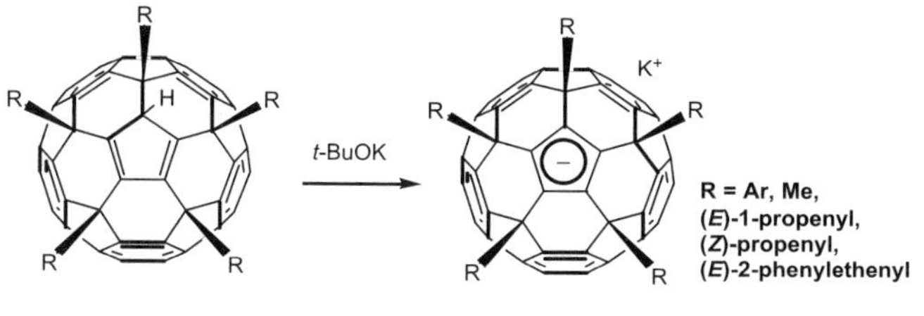

45 46

Figure 19. The versatility of penta-addition to C_{60}. Adapted from [113c].

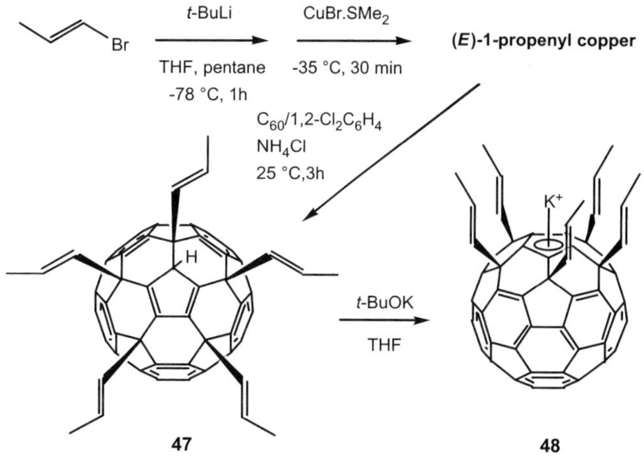

47 48

Figure 20. Regioselective penta-addition of 1-alkenyl copper reagent to C_{60}. Adapted from [113c].

Z depending upon the alkenyllithium reagent used. The reagents react with retention of stereochemistry. Thus, the product $C_{60}[(E)\text{-1-propenyl}]_5H$ (**47**) is found to be C_s symmetrical from NMR studies. Upon conversion to $C_{60}[(E)\text{-1-propenyl}]_5K$ (**48**), the symmetry changes to the expected C_{5v}. The alkenyl substituents extend upward from the fullerene core, which is the most stable structure obtained by Monte Carlo structural search with a molecular mechanics (MM2) force field.

The mechanism of addition of the five R groups is not clear, though a rationale for the mechanism is proposed [110] where the reaction goes in a stepwise manner, first putting two R groups in a 1,4-manner (**49**) [114], then the third and the fourth, and finally the fifth (**51**) (Fig. 21). There is a series of experimental evidence that supports this pathway. When 1,4-dibenzylated fullerene is treated with phenylcopper reagent, dibenzyltriphenyl compound is formed quantitatively, indicating that the 1,4-di-adduct lies on the pathway to the penta-adduct. Interestingly, the reaction stops at the tri-addition stage when a phenyl Grignard reagent is used instead of a phenyl copper reagent.

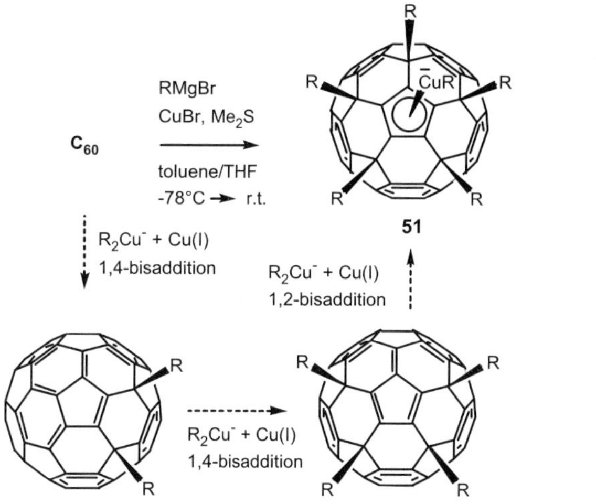

49 50

Figure 21. Proposed mechanism of penta-addition. Adapted from [110].

The reaction of the 1,4-dibenzylated fullerene derivative with PhMgBr cleanly affords the triaddition product.

Similar reactions are also investigated with C_{70} [115]. The reaction of C_{70} with the organocopper reagent does not lead to the expected pentakis adduct but a tris adduct $C_{70}(4\text{-}CF_3\text{–}C_6H_4)_3H$ (**52**) with an indene-like substructure flanked by the three sp^3 carbons (Fig. 22). The tris adduct **52** is deprotonated with metal alkoxides (KO^{t}Bu, TlOEt) to form the corresponding C_s-symmetric metal complexes $C_{70}(4\text{-}CF_3\text{–}C_6H_4)_3M$ [M = K, Tl] (**53**). The *ab initio* MO studies (HF/3-21G$^{(*)}$) suggest that the fullerene group $C_{70}(4\text{-}CF_3\text{–}C_6H_4)_3^-$ in $C_{70}(4\text{-}CF_3\text{–}C_6H_4)_3M$ coordinates to the metal as an η^5-indenylide rather than an η^3-ligand. The η^5-coordination is confirmed from the X-ray diffraction study of the thallium complex. In contrast to C_{60}, the higher fullerene C_{70} has two areas of different reactivity. One is the inert equator area, and the other is the more reactive C_{60}-like region near the poles. The latter possesses two different kinds of double bonds, 10 $C_a{=}C_b$ and 10 $C_c{=}C_c$. The first step of the tris addition is likely to be the regioselective carbocupration of the most reactive double bond $C_a{=}C_b$ with one molecule of $(4\text{-}CF_3\text{–}C_6H_4)_2Cu^-$ giving the metalated fullerene **55**, which may equilibrate with the 1,4-addition product **56**. This species decomposes to form the 1,4-bisarylation product **57**, Cu(0), and one electron, which is given to the Cu(I) species that exists in excess in the solution. The third aryl group adds to $C_c{=}C_c$ to form the stable copper(I) species with three aryl groups (**58**), which gives $C_{70}(4\text{-}CF_3\text{–}C_6H_4)_3H$ upon hydrolysis.

Contrary to the isolated six cyclopentadienyl anion in the pentakis-arylated fullerene ligand $C_{60}Ph_5^-$, the corresponding five-membered ring of the C_{70} analog is directly conjugated with the remaining polyene moiety of the C_{70} cage. Therefore, various structures are conceivable for the complexes $C_{70}(4\text{-}CF_3\text{–}C_6H_4)_3M$ as to the position of the metal binding site. Among those the most probable ones are η^3- and η^5-complexes, in which the metal atom is coordinated to the three or five sp^2 carbon atoms flanked by the three sp^3 carbon atoms. Such a consideration is a reasonable one since the pyramidization at the three sp^3 carbons makes the adjacent three sp^2 carbons [C(1), C(2), C(2′)] more planar, and this planarity should favor the interaction with the metal atom. The C_s symmetry observed in the NMR spectra of the complexes is consistent with both the η^3- and η^5-structures. The *ab initio* molecular orbital calculations (HF/3-21G$^{(*)}$) and X-ray crystallographic analysis show that the thallium atom is located equidistant from the three aryl groups and bonded to the pentagon flanked by the three sp^3 carbons.

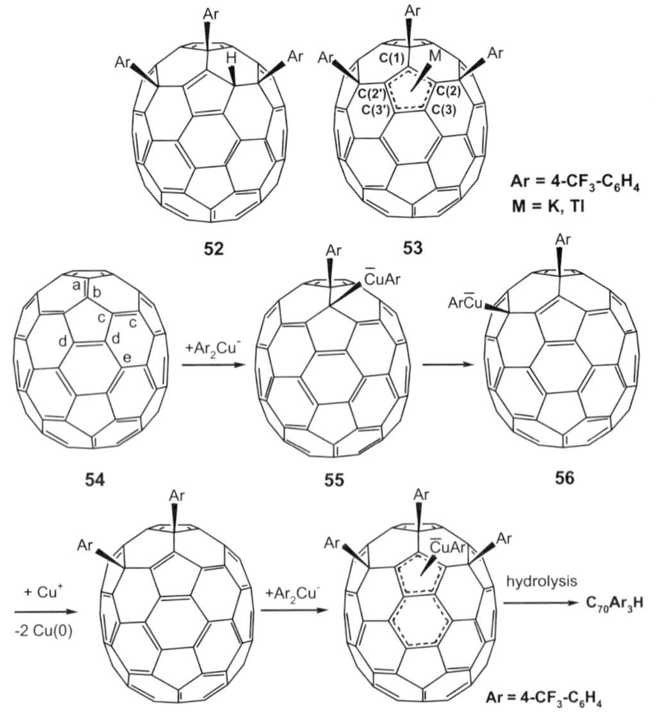

Figure 22. The indenelike fullerene ligand, its metal complexes via tris-arylation of C_{70}, proposed mechanism of trisarylation, and subsequent complexation. Adapted from [115].

7. FULLERENE–FERROCENE HYBRID

Ferrocene is composed of a pair of 6π-electron pentagonal carbon arrays and a 6*d*-electron iron(II) atom. Fusion of ferrocene and C_{60} through face-to-face resulting in a "bucky ferrocene" is intriguing, since such a molecule would manifest the properties typical of both ferrocenes and fullerenes. The ferrocene/fullerene hybrids of C_{60} (**60**) and C_{70} (**62**) by the use of penta- (**59**) and tri-adducts (**61**) of fullerenes, respectively, are reported (Fig. 23) [116].

The hybrid molecule $Fe(C_{60}Me_5)Cp$ is obtained in low yield by reaction between $FeBrCp(CO)_2$ and a thallium complex of pentamethylated [60]fullerene $Tl(C_{60}Me_5)$. A more expeditious route is heating $C_{60}Me_5H$ (**59**) and $[FeCp(CO)_2]_2$ together in benzonitrile at 180 °C for 8 h that was produced in 52% yield. The X-ray crystal structure shows the five methyl groups to protrude outward at an

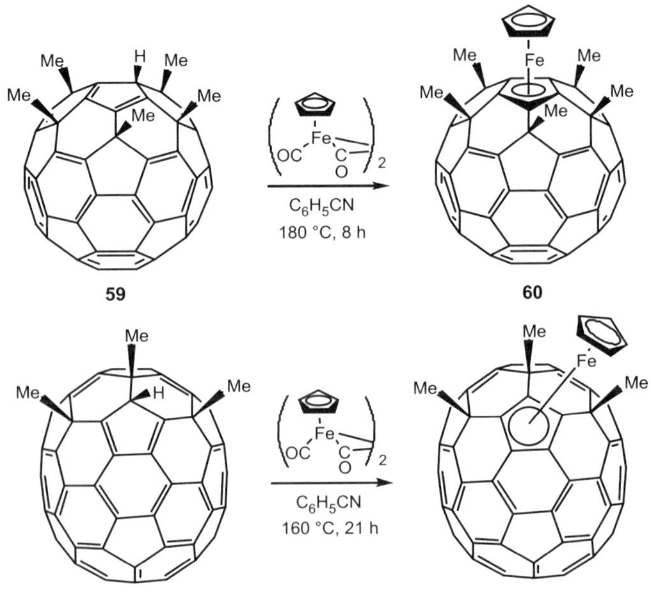

Figure 23. Synthesis of ferrocene fullerene hybrids. Adapted from [116].

angle of 42° relative to the symmetry axis of the molecule. The methyl and Cp hydrogen atoms are in van der Waals contact with each other. The Cp group and cyclopentadienide in the fullerene moiety are arranged in a staggered manner. All C–C bond lengths in each pentagon are equal to each other and hence both rings are aromatic. The distances between the pentagon carbon atoms and the iron are comparable to those in known ferrocene derivatives. ^1H and ^{13}C NMR spectra indicate that **60** has C_{5v} symmetry and that the methyl groups rotate freely (a single signal at δ2.51 ppm). The ^{13}C NMR signal of the Cp carbon atoms indicates strong electron-withdrawing effects of the fullerene moiety. The UV/visible spectrum is similar to those of C_{60}. The hybrid compound can be oxidized reversibly like ferrocene and reduced reversibly like fullerene. Such physicochemical properties of **60** are consistent with the presence of homoconjugation between the bottom C_{50} moiety and the Cp anionic moiety.

Treatment of $C_{70}Me_3H$ with $[FeCp(CO)_2]_2$ in benzonitrile at 160 °C for 21 h leads to **62** in 31% yield. ^1H and ^{13}C NMR spectra at 25 °C indicate that **62** is C_s symmetric and that the methyl groups rotate freely [singlet signals at δ2.81 (3H) and 2.93 ppm (6H)]. The Cp protons resonate also as a singlet at δ4.89 ppm, indicating free rotation of the Cp group. The X-ray crystal structure of **62** shows the distances between the pentagon carbon atoms and the iron as well as the C–C bond lengths in the Cp ligand as comparable to those of ferrocene. A small but significant difference between **60** and **62** is that the cyclopentadienide C–C bond connected to a hexagon in the "belt region" of the C_{70} core is slightly longer than the remaining four C–C bonds in the pentagon, a feature known for indenyl iron complexes.

8. RING EXPANSION REACTIONS

Fullerene ring expansions by a number of $[2+n]$-cycloaddition reactions, where $n = 1$–4, have been reported. The subject has been comprehensively reviewed [117], some with exhaustive details [118].

8.1. [2 + 1] Cycloadditions: Reactions Leading to Bridging

[2 + 1]-Cycloaddition reactions on fullerenes are convenient routes to exohedral functionalization via metal, oxygen, carbon, and nitrogen bridging. Addends suitably modified lead to introduction of a number of functional groups attached to the bridging position. Both single and multiple additions are possible. Multiple additions lead to regiosymmetric isomers.

8.1.1. Metal Bridging

Fullerenes are versatile multifunctional ligands. Through the π-bonds they can react with a variety of metal centers. Some of the metal complexes form cocrystals readily with the fullerenes. For example, mixing solutions of C_{60} or C_{70} and ferrocene results in crystallization of the solid adducts with compositions $C_{60} \cdot 2\{(\eta^5\text{-}C_5H_5)_2Fe\}$ [119] and $C_{70} \cdot 2\{(\eta^5\text{-}C_5H_5)_2Fe\}$ [120]. A benzene solution of C_{60} and $Fe_4(CO)_4(\eta^5\text{-}C_5H_5)_4$ produces crystals of $C_{60} \cdot \{Fe_4(CO)_4(\eta^5\text{-}C_5H_5)_4\} \cdot 3C_6H_6$ [121]. Extensive study involving a number of metal centers, such as, Ru, Re, Ir, Os, Pt, etc., covalently bound to fullerenes is known and the subject is reviewed periodically [122].

In reaction with metal derivatives, C_{60} behaves as an alkene (giving η^2-coordination) rather than an aromatic (η^6-coordination). Treatment of C_{60} with $(\eta^5\text{-}C_5H_5)_2ZrHCl$ in benzene yields a deep red solution which is believed to contain a mixture of adducts: $\{(\eta^5\text{-}C_5H_5)_2ZrCl\}_nC_{60}H_n$ with $n = 1, 2$, and 3. Hydrolysis of this mixture produces $C_{60}H_2$ along with $C_{60}H_4$ and $C_{60}H_6$ [123]. A number of stable η^2-C_{60} complexes of tungsten and molybdenum are known. Photolysis of $W(CO)_4(Ph_2PCH_2CH_2PPh_2)$ and C_{60} in 1,2-dichlorobenzene produces both $(\eta^2\text{-}C_{60})W(CO)_3(Ph_2PCH_2CH_2PPh_2)$ and $(C_{60})\{W(CO)_3(Ph_2 \cdot PCH_2CH_2PPh_2)\}_2$ [124]. Similarly, photolysis of $Mo(CO)_4(Ph_2PCH_2CH_2PPh_2)$ and C_{60} produces *mer*-$(\eta^2\text{-}C_{60})Mo(CO)_3(Ph_2PCH_2CH_2PPh_2)$ along with a mixture of $C_{60}\{Mo(CO)_3(Ph_2PCH_2CH_2PPh_2)\}_2$ and $C_{60}\{Mo(CO)_3 \cdot (Ph_2PCH_2CH_2PPh_2)\}_3$ [125]. The radical $Re(CO)_5$ obtained from $Re_2(CO)_{10}$ through photolysis or from $(\eta^3\text{-}Ph_3C) \cdot Re(CO)_4$ through thermolysis in the presence of CO adds to C_{60} to produce $C_{60}\{Re(CO)_5\}_2$ [126].

Treatment of C_{60} with osmium tetroxide in the presence of pyridine yields either the single addition product, $C_{60}O_2OsO_2(py)_2$ (**63**), and/or a mixture of five double addition products, $C_{60}\{O_2OsO_2(py)_2\}_2$ (**64**) (Fig. 24) [127]. The single addition product can be separated from the double addition products, since the single addition product has a higher solubility in toluene. Osmylation of C_{70} in pyridine produces two single addition products, $C_{70}\{O_2OsO_2(py)_2\}$ [128]. The osmyl groups added to the C_a–C_b and to the C_c–C_c bonds of the fullerene, respectively.

The carbonyls of Os, Ru, and Fe also react with C_{60}. Heating toluene solution of C_{60} and $Os_3(CO)_{11}(NCMe)$ or other lower valent compounds such as $Os_3(CO)_{11}(PPh_3)$ and $Os_3(CO)_{10}(PPh_3)_2$ results in the triangular Os_3 compounds (Fig. 25) [129]. $Ru_3(CO)_{12}$ reacts with C_{60} to form soluble $Ru_3(CO)_9(\mu_3\text{-}\eta^2, \eta^2, \eta^2\text{-}C_{60})$ (Fig. 26) [130] and insoluble Ru_3C_{60} [131]. $Ru_3(CO)_{12}$ also reacts with C_{70} to form two products, $Ru_3(CO)_9(\mu_3\text{-}\eta^2, \eta^2, \eta^2\text{-}C_{70})$ and $\{Ru_3(CO)_9\}_2(\mu_3\text{-}\eta^2, \eta^2, \eta^2\text{-}C_{70})$ [132]. The ruthenium carbide clusters, $Ru_5C(CO)_{15}$ and $Ru_6C(CO)_{17}$, react with C_{60} followed by treatment PPh_3 to form $Ru_5C(CO)_{11}(PPh_3)(\mu_3\text{-}\eta^2, \eta^2, \eta^2\text{-}C_{60})$ and $Ru_6C(CO)_{12}(Ph_2PCH_2PPh_2)(\mu_3\text{-}\eta^2, \eta^2, \eta^2\text{-}C_{60})$ [133]. In these complexes C_{60} reacts with a triangular face of the ruthenium cluster.

Iron carbonyl $Fe_3(CO)_{12}$ reacts with 1,2-(3,5-cyclohexadieno)C_{60} to form $Fe(CO)_3$ adduct (**68**) in which the metal is coordinated to the exterior cyclohexadiene portion rather than to the fullerene itself (Fig. 27) [134]. UV irradiation of toluene solutions of C_{60} and $Fe_2S_2(CO)_6$ results in the

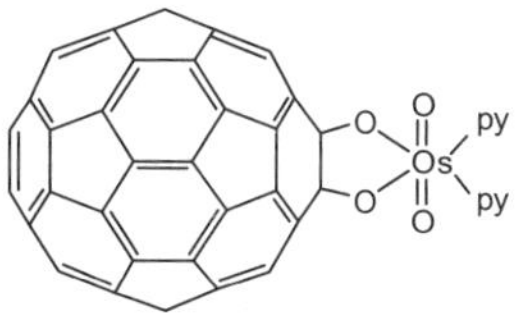

63

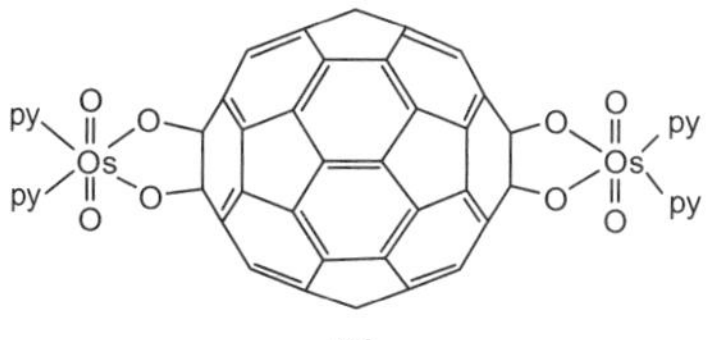

64

Figure 24. Structures of $C_{60}O_2OsO_2(py)_2$ and $C_{60}\{O_2OsO_2(py)_2\}_2$ (one of the five isomers shown). Adapted from [127].

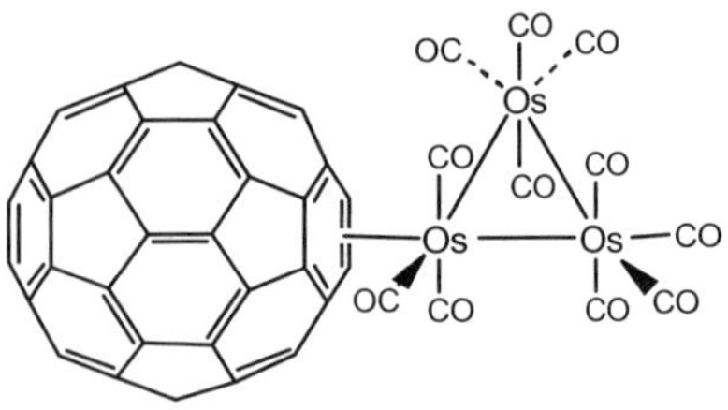

65

Figure 25. A triangular Os_3 complex of C_{60}. Adapted from [129].

formation of $C_{60}S_2Fe_2(CO)_6$, along with some multiple addition products [135].

$Ir(CO)Cl(PPh_3)_2$ and related complexes with different phosphine ligands react with C_{60} [136]. The reaction leads to formation of adduct of the type $(\eta^2\text{-}C_{60})Ir(CO)Cl(ER_3)_2$ (E = P or As, R = alkyl or aryl). Slow diffusion of a concentrated benzene solution of C_{60} into a benzene solution of $Ir(CO)Cl(PPh_3)_2$ produces the adduct $(\eta^2\text{-}C_{60})Ir(CO)Cl(PPh_3)_2 \cdot 5C_6H_6$ (Fig. 28) [136]. Like fullerenes, their epoxides also react very efficiently with iridium complexes. $C_{60}O$ reacts with $Ir(CO)Cl(PPh_3)_2$ to form crystals of $(\eta^2\text{-}C_{60}O)Ir(CO)Cl(PPh_3)_2 \cdot 5C_6H_6$ [137]. Reaction of a mixture of the two $C_{70}O$ isomers with $Ir(CO)Cl(PPh_3)_2$ yields $(\eta^2\text{-}C_{70}O)Ir(CO)Cl(PPh_3)_2 \cdot 5C_6H_6$ [138]. Heating a benzene solution of the indenyl iridium complex, $(\eta^5\text{-}C_9H_7)Ir(CO)(\eta^2\text{-}C_8H_{14})$, with C_{60} produces the complex $(\eta^2\text{-}C_{60})Ir(CO)(\eta^5\text{-}C_9H_7)$ [139], which dissolves in aromatic and chlorinated solvents to give stable solutions.

$RhH(CO)(PPh_3)_3$ is a catalyst used for the homogeneous hydrogenation of olefins. Efforts to catalyze the hydrogenation of C_{60} using this have not been successful. However, it reacts with C_{60} forming the complex $(\eta^2\text{-}C_{60})RhH(CO)(PPh_3)_2$ [140]. The reaction of C_{60} with $(\eta^5$-bicyclo[3.2.0]hepta-1,3-dienyl)$(\eta^4$-tetraphenylcyclobutadiene)cobalt(I) proceeds via a ring slippage reaction through an η^3-intermediate that undergoes a [4 + 2] cycloaddition to the fullerene [141]. Mixing solutions of C_{60} and $(Ph_3P)_2Pt(\eta^2\text{-}C_2H_4)$ produce $(\eta^2\text{-}C_{60})Pt(PPh_3)_2$ [142]. Related complexes with nickel, palladium, and platinum can also be obtained by treatment of C_{60} with $M(PR_3)_4$ [143]. Multiple addition to C_{60} has been examined through the reactions of $M(PEt_3)_4$ (M = Pt or Pd) with this fullerene. With an excess of $M(PEt_3)_4$, a stable hexaaddition

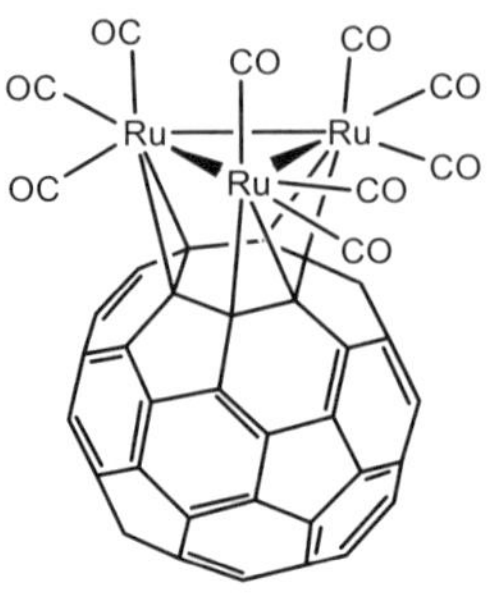

66

Figure 26. The structure of $Ru_3(CO)_9(\mu_3\text{-}\eta^2, \eta^2, \eta^2\text{-}C_{60})$. Adapted from [130].

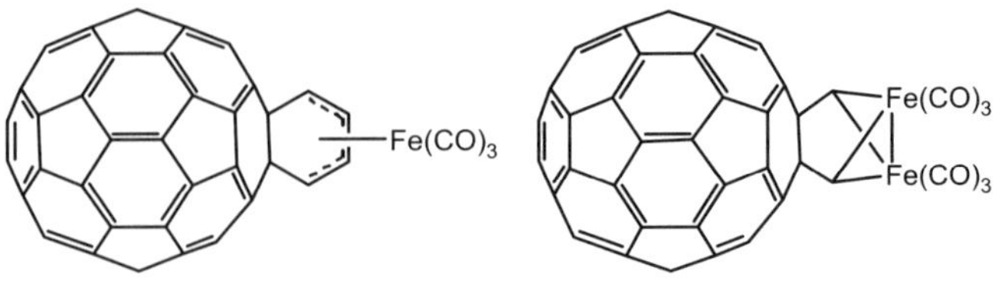

67 68

Figure 27. $Fe(CO)_3$ adducts of C_{60}. Adapted from [134].

product is obtained [144]. Addition of the labile complex, $Pd_2(dba)_3 \cdot CHCl_3$ (dba, dibenzylideneacetone), to a benzene solution of C_{60} yields a black, amorphous, air-stable solid [145]. This material is insoluble in common organic solvents and, consequently, is believed to have a polymeric structure.

8.1.2. Oxygen Bridging

The mass spectra of C_{60} and C_{70} are accompanied by satellite ions in low abundance. Changes in the relative ion abundances of these peaks with the method of sample preparation and as a function of evaporation into the mass spectrometer suggest that there are several different fullerene adducts corresponding to the addition of 14, 15, 16, and 17 daltons. The mass increases are assigned to CH_2, CH_3, O, and OH, respectively. The M + 16 adduct is interpreted as sequential cyclopropanation and epoxidation product of monoxides [146]. Ultraviolet irradiation of C_{60} in the presence of oxygen in hexane gives rise to the corresponding epoxides [147]. When these epoxides are heated the corresponding ethers are formed. The addition of oxygen takes place across the 6,6-junction of pyracylene unit of C_{60}. The epoxides are degraded to the carbonyl containing compounds upon extensive irradiation. The eight symmetry nonequivalent bonds in C_{70} yield as many as eight isomeric epoxides and an equal number of oxidoannulenes. The isolation of $C_{70}O$, a minor component of the fullerene mixture obtained by resistive heating of graphite, is reported [148]. The compound is an annulene with the oxygen atom bridging an unspecified 6,6-ring fusion. Addition of O_3 onto C_{60} leads to molozonide. The spherical nature of C_{60} places stringent steric constraints on the conversion of the molozonide intermediate to the ozonide. In the presence of toluene molozonide rearranges to 1,2-diketone. The oxidation occurs at the various double bonds of C_{60} generating a mixture of oxidized product having ketonic, ester, and epoxide functionalities [149]. Small amounts of C_{60} epoxides are detected, but the bulk of the product consists of highly oxidized derivative of the fullerene.

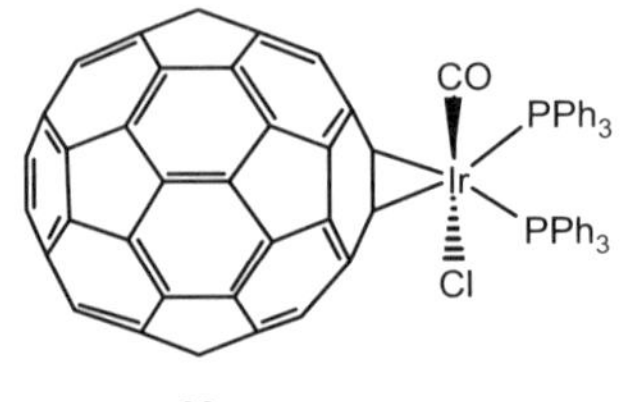

69

Figure 28. Structure of $(\eta^2\text{-}C_{60})Ir(CO)Cl(PPh_3)_2$. Adapted from [136].

The photooxidation of C_{60} in benzene furnishes a mono oxide $C_{60}O$ as the sole isolable product in 7% yield [150]. Addition of benzil significantly enhances both the yield (16%) and the rate of oxidation. Electrochemical oxidation of C_{60} in benzene generates $C_{60}O_n$ (n = 1–4) [151]. A 1,2-diepoxy[60]fullerene is isolated in the analysis of the metabolic transformation of C_{60} [152]. From the spectroscopic data, out of the three possible symmetrical structures the diepoxide is identified as the 1,2;3,4-diepoxy[60]fullerene. It is seen that the oxygen atom transferred from the active species is introduced at a double bond adjacent to an existing epoxide. In addition to the diepoxide, two triepoxides are also isolated and identified as 1,2;3,4;9,10-triepoxy[60]fullerene and 1,2;3,4;11,12-triepoxy[60]fullerene.

8.1.3. Carbon Bridging

The cyclopropanation reaction is synthesis of a C_{61} or C_{71} species, the extra carbon atom forming a bridge. Cyclopropanation is accomplished through (i) the reaction with stabilized carbanion, (ii) addition of a carbene, and (iii) the thermal addition of diazo compounds followed by thermolysis or photolysis of the resulting intermediates. Addition may take place across either a 6,6- or a 5,6-ring junction, resulting in either closed-fullerene or open-fulleroid structures. A total of four configurations are possible, that is, 6,6-open, 6,6-closed, 6,5-open, and 6,5-closed (Fig. 29). If the derivative of fullerene contains cyclopropane rings (6,5-closed and 6,6-closed), the product is called methanofullerene; otherwise it is called fulleroid (6,5,6-open and 6,6-open).

A bridging carbon atom can be introduced by several routes, such as the Bingel reaction, which is the reaction of the fullerenes with a stabilized α-halocarbanion [153]. Treatment of C_{60} with (i) 2-bromomalonic ester in the presence of NaH, (ii) reaction of the bromo compound with C_{60} in the presence of DBU (1,8-diazabicyclo[5.4.0]undec-7-ene), (iii) a compound with a reactive methylene group with I_2 or CBr_4 in the presence DBU (*in-situ* formation of 2-iodo- or 2-bromomalonates) are the three most used routes for Bingel reaction (Fig. 30) [154]. NaH is nonnucleophilic; primary and secondary amines cannot be used as they add on to C_{60}. The mechanism of the reaction is the initial formation of a carbanion, which attacks C_{60}; the resulting C_{60} carbanion intermediate eliminates Br^- resulting in the methanofullerene. Reaction of phosphonium ylides or sulphur ylides also leads to methanofullerenes via an addition–elimination mechanism (Fig. 31) [155].

Cyclopropanation of fullerenes using haloalkanes can be carried out using reactions other than the Bingel route. A fast and simple route for the selective synthesis of methanofullerenes that use the reaction between iodo- and bromo-reagents and the C_{60} dianion generated by electrochemical reduction has been reported (Fig. 32) [156]. Addition takes place at the 6,6-ring junction.

Reaction of C_{60} with dibromocarbene generated from $PhHgCBr_3$ leads to efficient cyclopropanation, though fullerene dimers are formed as the product. Methanofullerene-containing stable phosphorous ylide is obtained by the reaction of C_{60} with a mixture of triphenylphosphine and dimethyl acetylene carboxylate. A carbanion generated as a reaction intermediate adds on to C_{60} [157].

In a number of reactions diazo compounds are used as precursors of carbenes. Thermal addition of the diazo compound to C_{60} leads to the formation of 6,6-closed and 6,5-open cycloadducts depending upon the condition of the reaction. There can be two possible routes to cyclopropanation: (1) thermal decomposition of the diazo compound to carbenes and addition of carbenes to the 6,6-bond in

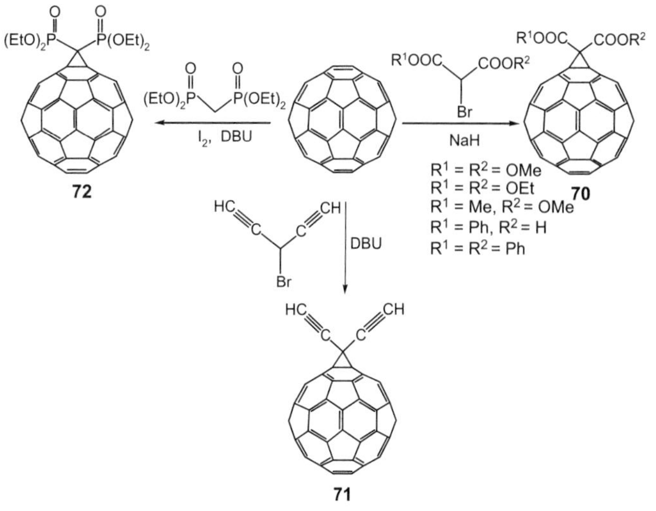

Figure 30. Three most used routes for Bingel reaction. The reactions shown are illustrative examples. Adapted from [153, 154].

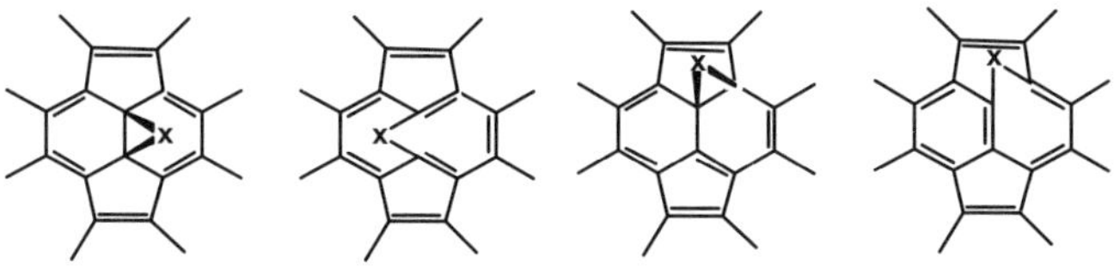

Figure 29. Four possible modes of bridging (X = C, N, O, metal atoms).

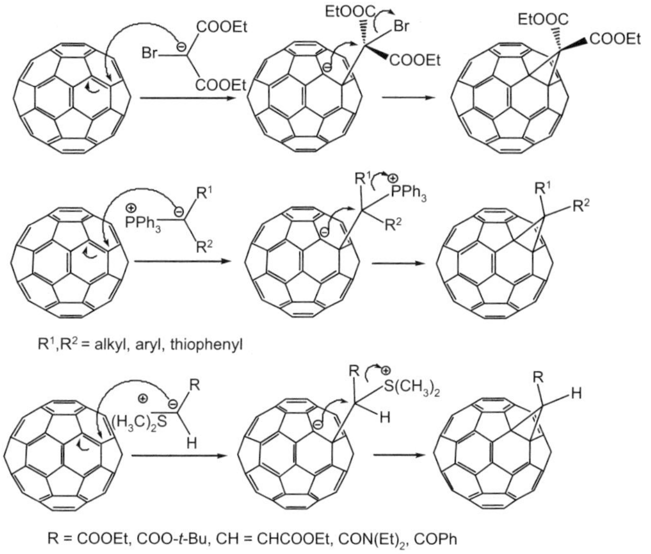

Figure 31. Mechanism of cyclopropanation by Bingel and related routes. Adapted from [155].

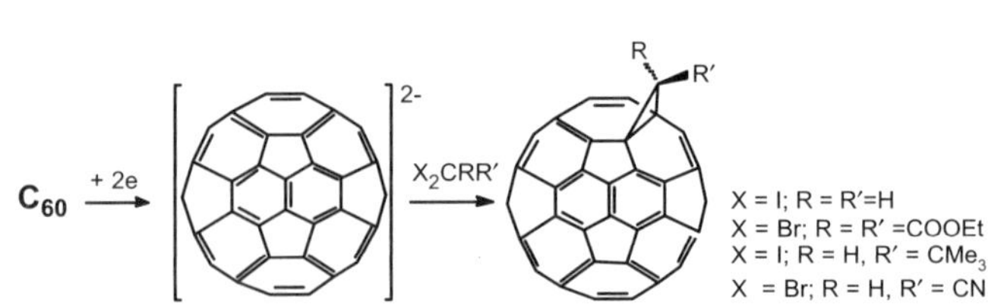

Figure 32. Electrochemical addition of halogen derivatives to C_{60} leading to methanofullerenes. Adapted from [156].

C_{60} and (2) elimination of nitrogen from the pyrazoline intermediate. Formation of the pyrazoline intermediate **73** (Fig. 33) and elimination of nitrogen can lead to both 6,5-open (**74**) and 6,6-closed (**75**) isomers. The 6,5-open isomers containing stabilizing substituents in the methane bridge rearrange to the thermodynamically more stable 6,6-closed isomers. The pyrazolone is prepared by addition of diazomethane to a 6,6-bond of C_{60} [158]. Thermolysis of the pyrazolone furnishes the rearranged fulleroid (6,5-open isomer). The photolysis of the pyrazolone leads to, in addition to the fulleroid, the methanofullerene (6,6-closed isomer) [159].

Addition of diazomethane to C_{70} is more complex because there are eight different types of bonds in C_{70}. Several studies [160] have revealed that the bonds **I** and **II** in C_{70} (**76**) (Fig. 34) are the shortest bonds suggesting that they should be most reactive. Further it is observed that bond **I** is more reactive than **II**, which is attributed to greater local curvature [128]. In fact addition of diazomethane takes place at position **I** leading to a mixture of isomeric pyrazolines, which upon photolysis gives two isomeric $C_{71}H_2$ cyclopropanes. Thermolysis of the pyrazolines (**77**, **78**) produces two isomeric 6,5-open isomers [162].

A number of diazo compounds such as dimethyldiazomethane, diethyl diazoacetate, and diphenyl diazomethane undergo 1,3-dipolar addition reaction with C_{60} followed by elimination of N_2 to introduce $(CH_3)_2C$, $(C_2H_5)_2C$, Ar_2C groups, respectively, into the carbon framework (Fig. 35). The reaction of dimethyldiazomethane with C_{60} leads to the annulene the 6,5-open isomer, which upon

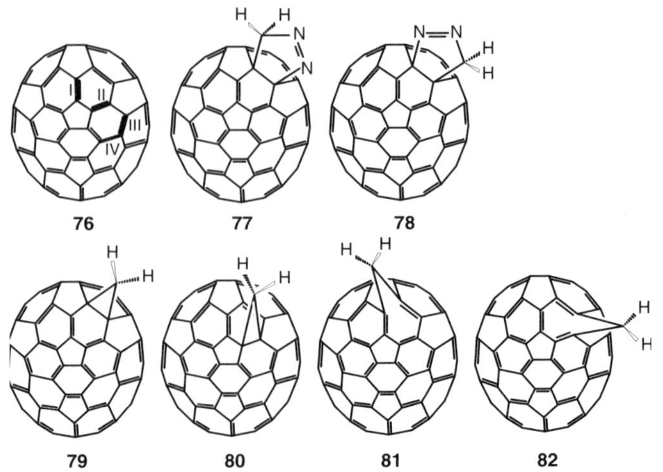

Figure 34. Addition of diazomethane to bond **I** of C_{70} leading to two isomeric pyrazolines, which furnish 6,6-closed or 6,5-open isomers on photochemical or thermal reaction conditions, respectively. Adapted from [162].

heating is converted to methanofullerene [163]. A similar conversion is observed for a fulleroid carrying phenyl and methyl butyrate substituents under photochemical conditions [164]. The mechanism of the conversion is stated to be di-π-methane rearrangement (Fig. 36). The diradical formed is stable due to the attachment of phenyl groups. For fulleroids with methyl substituent the thermal conversion to methanofullerene is a disrotatory migration of the bridging methylene group to a 6,5-ring junction followed by [1,5]-sigmatropic shift to the 6,6-ring junction. As stated previously there can be four possible configurations for a bridging group introduction to C_{60}. Among these the 6,5-open isomer is the lowest energy product. The energetic preference for this structure is due to maximizing the number of double bonded 6,6-ring junctions. The 6,5-open isomer is stable under kinetically controlled conditions while the 6,6-closed isomer is stable under thermodynamically controlled conditions.

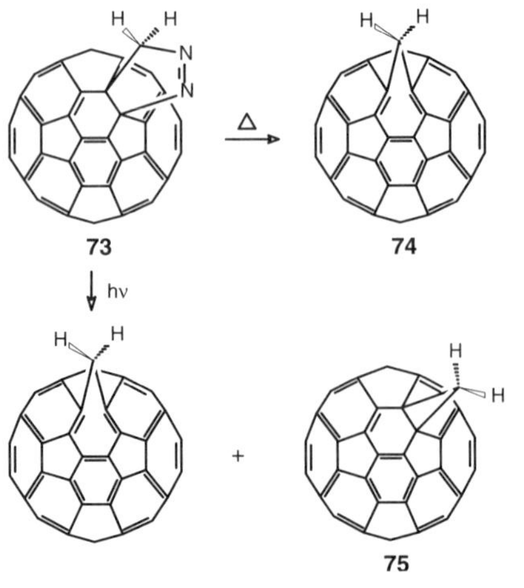

Figure 33. Thermal and photochemical decomposition of the pyrazolone intermediate obtained from addition of diazomethane to C_{60}. Adapted from [159].

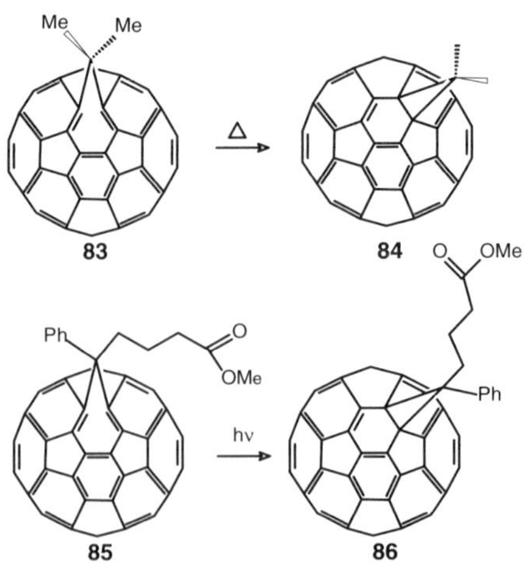

Figure 35. Conversion of the 6,5-open isomers (fulleroid) to 6,6-closed isomer (methanofullerene) under both thermal and photochemical conditions. Adapted from [163].

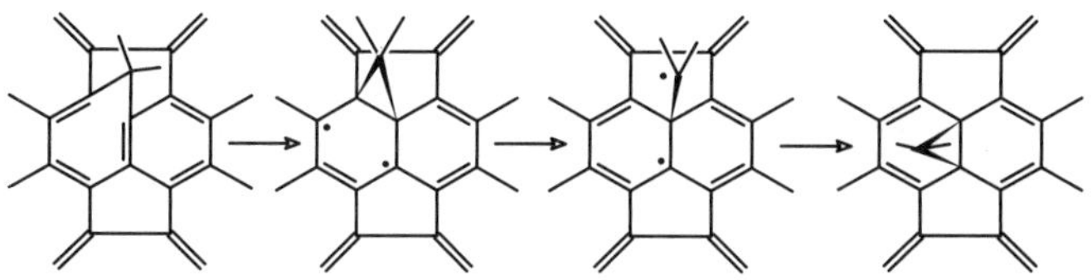

Figure 36. Di-π-methane reaction mechanism for photochemical transformation of the fulleroid to methanofullerene.

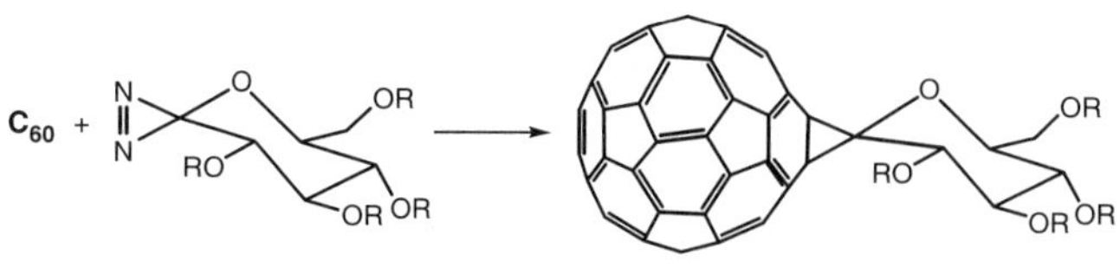

Figure 38. Spiro-linked *C*-glycosides of C_{60}. Adapted from [168].

In some cases of addition reactions involving diazo compounds a specific product may be formed exclusively. Thus the thermal or photochemical reaction of fullerene with quinoid-type diazo compounds with C_{60} in *o*-dichlorobenzene under an atmosphere of nitrogen affords exclusively 6,6-closed spiromethanofullerenes [165]. In some cases a reactive substituent on the methylene bridge can take part in further reaction. In the thermal addition of diazoketone, the 6,6-closed isomer further rearranges to the dihydrofuranofullerene derivative [166]. The production distribution from the reaction of carbenes with C_{60} can be influenced by catalysts, for example, $Rh_2(OAc)_4$ complex catalysis of the reaction of ethyl diazoacetate and ethyl diazomalonate with C_{60} [167]. The catalyst increases the selectivity and yield. Reaction carried out without catalyst gives a 6,5-open isomer as the predominant product, whereas in presence of the catalyst the 6,6-closed isomer is almost exclusively formed (Fig. 37).

Alkyl and acyl-protected 1-azi derivatives can be used as precursors of nucleophilic carbenes of monosaccharides. These derivatives react with acceptor- and donor-substituted alkenes to yield spiro-linked glucosylidene cyclopropanes. The *O*-benzyl protected diazirine reacts with C_{60} to afford the monoglycosylated adduct, **90** (Fig. 38) [168].

Stable hydrazones and their derivatives are also used for cyclopropanation. The reaction proceeds through the initial 1,3-dipolar cycloaddition reaction of the hydrazone to C_{60} followed by nitrogen elimination from the resulting pyrazoline intermediate, **91** (Fig. 39). The reaction is carried out in the presence of MnO_2–KOH–EtOH. Mixtures of 6,6-closed and 6,5-open derivatives are obtained [169]. Thermolysis of lithium or sodium salts of tosylhydrazones leads to mixtures of 6,6-closed and 6,5-open isomers. Reaction conditions are optimized to yield the 6,5-open isomer. On prolonged reaction, the kinetic product rearranges into the more stable 6,6-closed thermodynamic product [170]. Reaction of hydrazones containing *p*-diarylaminoaryl electron-donating groups affords the thermodynamically more favored 6,6-closed isomer as the only product even at low temperature [171].

The silyl enol ether under the influence of KF/18-crown-6 or tetrabutylammonium fluoride can generate nucleophilic species *in-situ*. The reaction of C_{60} and silyl enol ether of 4-methoxyacetophenone in the presence of the complex of KF and 18-crown-6 leads to efficient cyclopropanation (Fig. 40) [172].

Interaction between Substituents and Fullerene Cage

Substituents attached to the bridging carbon atom can exercise considerable influence on the fullerene sphere though not directly attached to it. Cyclovoltametry is the most direct method for studying the influence of addends on the electronic properties of fullerenes. CV data of a number of spiroannelated fullerene analogs show remarkable substituent effects. For a number of fluorenofullerenes isolated exclusively as adducts on 6,6-ring junctions [173], the CV spectra clearly demonstrate the striking influence of electron-donating and electron-withdrawing groups of the fluorenyl group on the C_{60} moiety (Fig. 41). This indicates that the substituents, even though separated from the fullerene cage by a spiro atom, can still strongly interact with the fullerene. Interestingly, the CV data of diphenylmethanofullerene with electron-donating and electron-withdrawing groups in the *p*,*p*′-positions of the phenyl rings are independent of the functional groups, indicating that similar interactions as seen for the spiroannelated fullerenes is not operating in the diphenylmethanofullerene.

This is explained by periconjugation, which is the interaction between the p_z orbitals of the addends and the "p_z" orbitals of the adjacent fullerene carbon atoms (Fig. 42). The mode of interaction of the p_z orbitals is geometrically completely different from spiroconjugation as seen from the relative orientations of the orbitals involved. The crucial difference between 9,9-fluorenofullerenes and diphenylmethanofullerene is that in the latter the phenyl rings are free to rotate but prefer to lie "parallel" to the fullerene surface, whereas the aromatic rings in fluorenofullerenes are held rigidly perpendicular to the surface of the spheroid.

Periconjugation is also possible when the substitutents attached to the bridging carbon atom carrying π-bonds are strongly electron withdrawing. Among a number

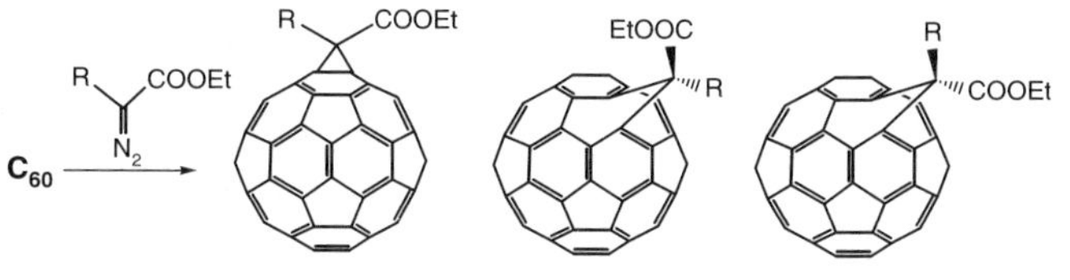

Figure 37. Mixture of closed and open adducts and their distribution is influenced by metal complex used as the catalyst. Adapted from [167].

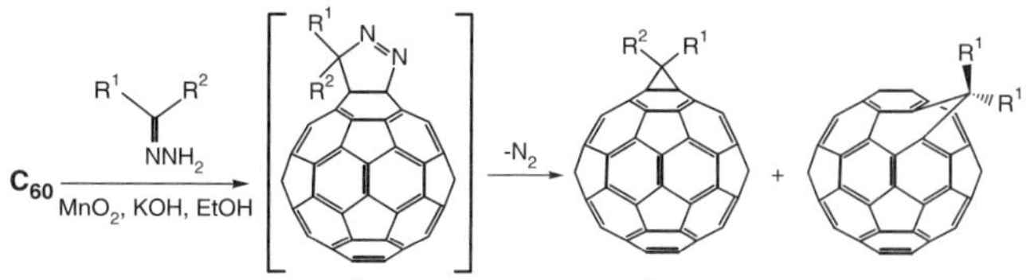

Figure 39. Addition of hydrazones and subsequent nitrogen loss leading to 6,6-closed and 6,5-open isomers. Adapted from [169].

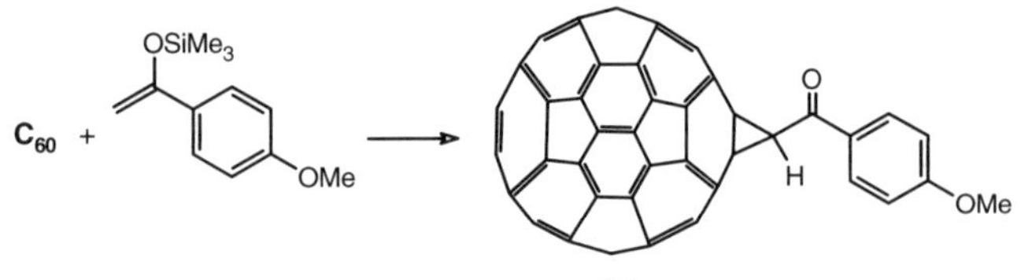

94

Figure 40. Cycloaddition of silyl enol ether to C_{60}. Adapted from [172].

of methanofullerenes with cyano, nitro, and carboethyl substituents on the bridging carbon (Fig. 43) [174], the bis(cyano)methanofullerene showed an irreversible reduction peak, which is 160 mV more positive than the corresponding step in C_{60}. A comparative study of methanofullerenes carrying electron-donating and electron-withdrawing substituents, using Hammett σ_m, shows that the cyano substituent appears to be more electron withdrawing than predicted by the Hammett relation. This is ascribed to periconjugation, as seen for the spiroannelated fullerenes. A number of electroactive substitutents such as ferrocenyl, tetrathiafulvalene, etc. are appended to the bridging carbon in C_{60} (Fig. 44) [175].

Introduction of Coordinating Ligands The Bingel reaction is versatile since many derivatives with active methylene groups are known. Functional groups with specific properties can be appended choosing the right kind of reactants, such as introduction of groups with coordination sites (Fig. 45). Reaction of a stabilized carbanion from bis(4-pyridyl)chloromethane with C_{60} under Bingel conditions leads to introducing pyridyl rings to the fullerene [176], which can act as a ligand for generation of metallo-organic frameworks.

A C_{60} derivative containing a 2,2′-bipyridine (bpy) moiety (**108**) is used for electrochemical, quantum chemical, and spectroscopic techniques to study the dynamics of the reduction of bpy-C_{61} (Fig. 46) [177]. Quantum chemical calculations showed that triply and quadruply reduced bpy-C_{61} undergoes facile conversion among three of its structural isomers.

Fullerene-substituted 2,2′:6′,6″-terpyridine (tpy) ligands (**109**) are synthesized [178] (Fig. 47). Polyethyleneoxy spacers are introduced between the tpy ligand and the fullerene to allow the synthesis of a series of Ru^{II} diad and triad

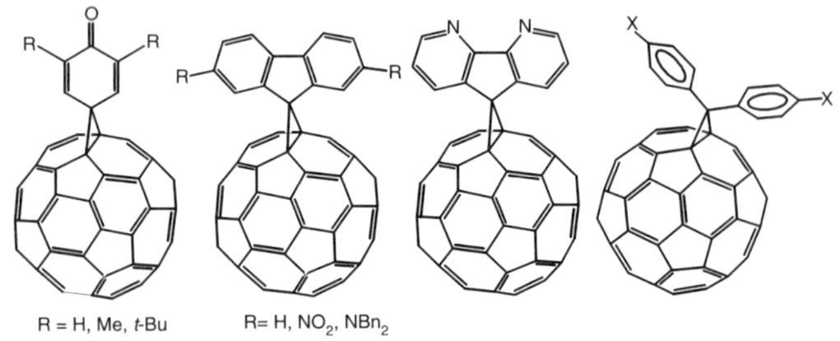

95 96 97 98

Figure 41. Spiroannelated fullerenes used for CV studies show a remarkable substituent effect, which is not observed for the diphenyl derivatives. Adapted from [173a].

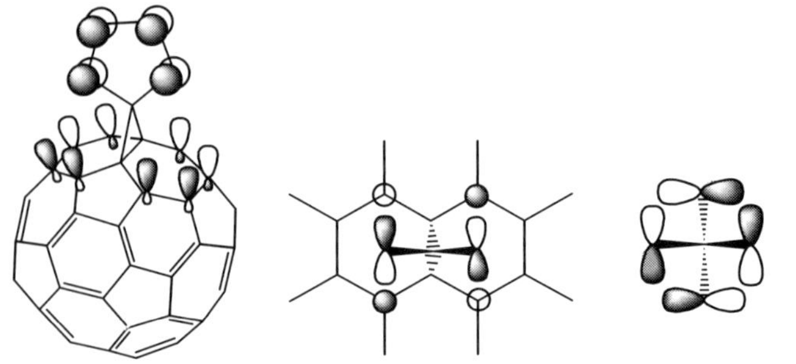

Figure 42. Orbital interaction in spiroannelated fullerenes by periconjugation, as different from interaction in spiroconjugation. Adapted from [173a].

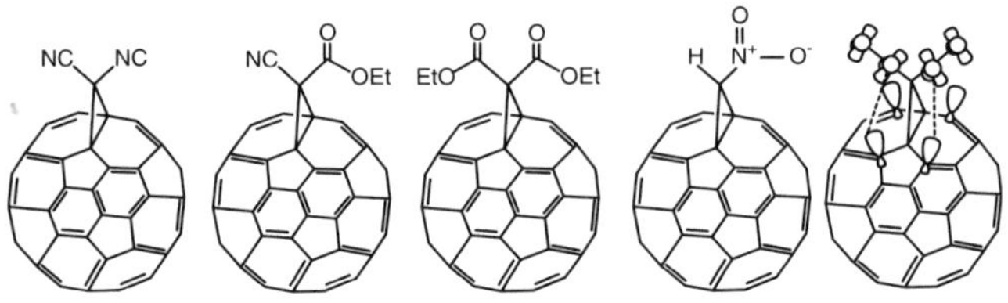

99 100 101 102 103

Figure 43. Methanofullerenes carrying electron-donating and electron-withdrawing substituents and periconjugation for fullerene derivative with cyanosubstitutents. Adapted from [174].

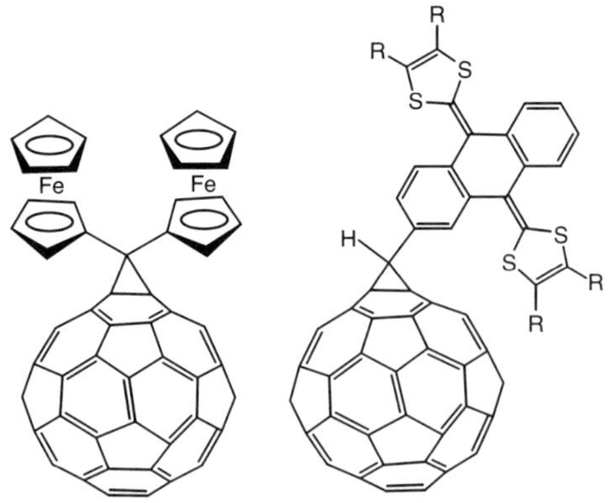

104 105

Figure 44. Attaching electroactive moiety to C_{60}. Adapted from [175].

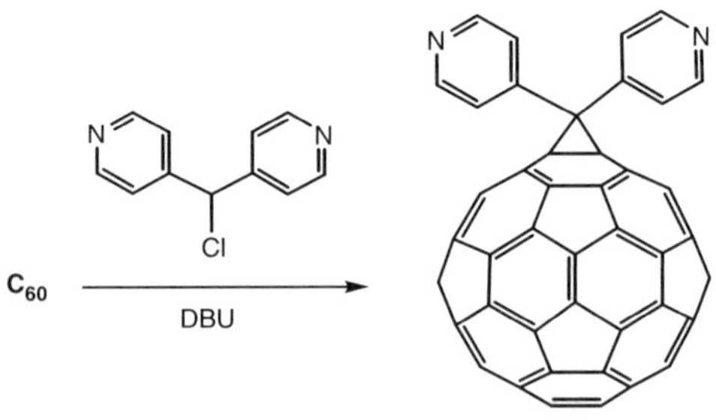

106

Figure 45. Pyridyl derivative of C_{60}. Adapted from [176].

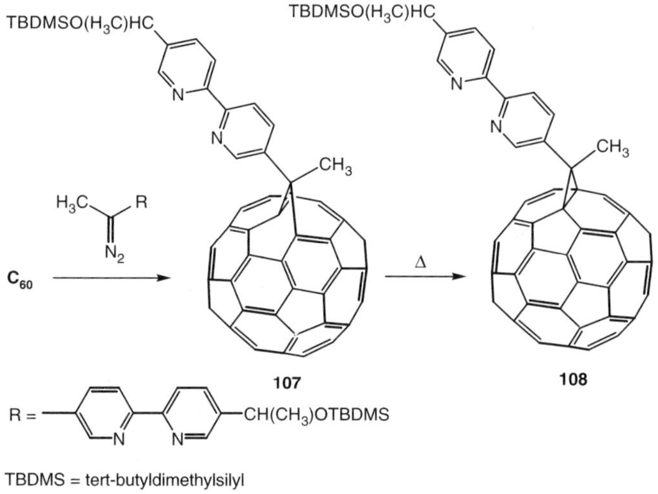

Figure 46. C_{60} derivative containing a 2,2′-bipyridine (bpy) moiety. Adapted from [177].

systems with variable metal–fullerene distances. The electrochemical interaction between the pendant fullerenes and metal-based photocenters are studied. Electron-releasing dimethylamino substituent is seen to have a dramatic effect upon the Ru^{II}–Ru^{III} redox processes for these complexes.

Multiple Additions Multiple additions on C_{60} lead to a number of regioisomeric products. Regioselective and stereoselective syntheses starting from bis-adducts to hexakis adducts have been reviewed [179]. The problem is best illustrated by the sequential cyclopropanation of C_{60} with diethyl 2-bromomalonate in the presence of base. With reference to a group already existing on a 6,6-ring junction, a second addend can occupy any one of the nine possible 6,6-bond positions. These are *cis*-1, -2, -3, *trans*-1, -2, -3, -4, e_{face}, and e_{edge} (**111**) (Fig. 48). Eight bisadducts are possible because the two positions e_{face} and e_{edge} are identical if the two addends of the bisadduct are identical. In case of addition of 2-bromomalonate to $C_{61}(COOEt)_2$, of the possible eight bisadducts, $C_{62}(COOEt)_4$, seven regioisomers have been isolated in various quantities [180], affording the *e*-$C_{62}(COOEt)_4$ in highest yield (~15%) followed by *trans*-3-$C_{62}(COOEt)_4$ (~12%).

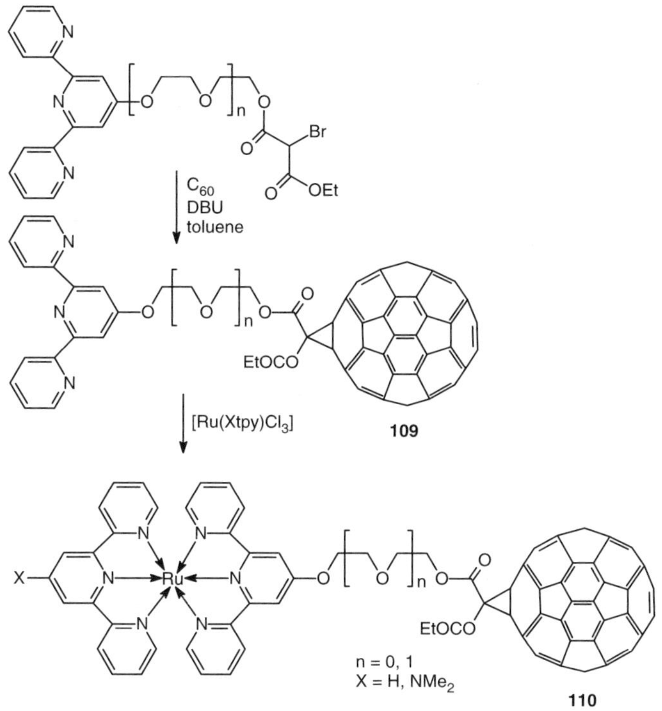

Figure 47. C_{60} with terpyridine ligands and their Ru^{II} complexes. Adapted from [178].

The cyclopropane rings fused to the carbon spheres have generally been found to be highly stable, though the monoanion of 1,2-methano[60]fullerene-61,61-dinitrile reverts back to C_{60} [181]. Exhaustive controlled potential electrolysis (CPE) of diethyl-1,2-methano[60]fullerene-61,61-dicarboxylate leads to removal of the cyclopropane ring to yield predominantly C_{60} [182]. This is called the retro-Bingel reaction, which has been employed to remove up to four substituted cyclopropane rings from fullerenes. If CPE is not exhaustive, the fullerene adduct undergoes isomerization by the migration of the cyclopropane ring on the C_{60} surface. This electrochemical "walk on the sphere" rearrangement has been observed for the six regioisomeric bis-malonate adducts, **112** (Fig. 49) [183]. Coulometrically controlled two electron per molecule electrolysis of six regioisomers leads to a mixture of regioisomeric bis(methano)adducts, but none of the mixtures contained *cis*-2 and *cis*-3 adducts. These derivatives, with two cyclopropane rings located in close proximity in one hemisphere, are the most strained of the isolable regioisomeric bis(methano)adducts.

The malonate bridges appear to be migrating around the fullerene sphere, but not in a random manner. The major isomer formed is always the *trans*-2 derivative (40–50%). About 10% of the *trans*-1 adduct is produced in every case, and the *trans*-1 and *trans*-2 isomers, therefore, appear roughly in their expected 1:4 statistical ratio. The *trans*-1 and *trans*-2 derivatives predominate over the *trans*-3 and *trans*-4 ones, while *e*-isomer is also formed in high percentage (~25%). The preferred formation of both *e*- and *trans*-isomers could be a consequence of their inherently

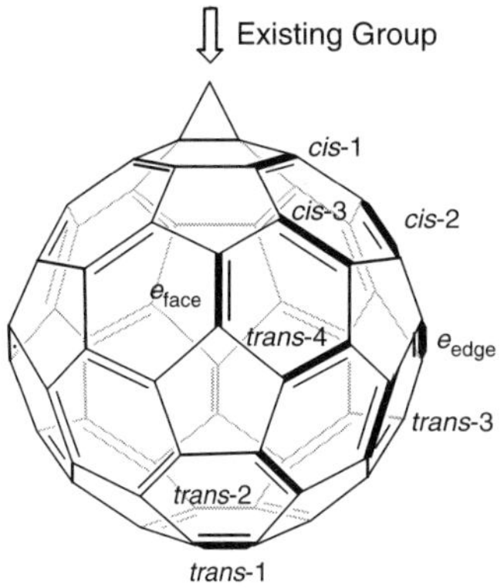

111

Figure 48. Eight 6,6-ring junction positions to which a second cyclopropane ring can be attached in reference to the existing cyclopropane ring. Adapted from [180].

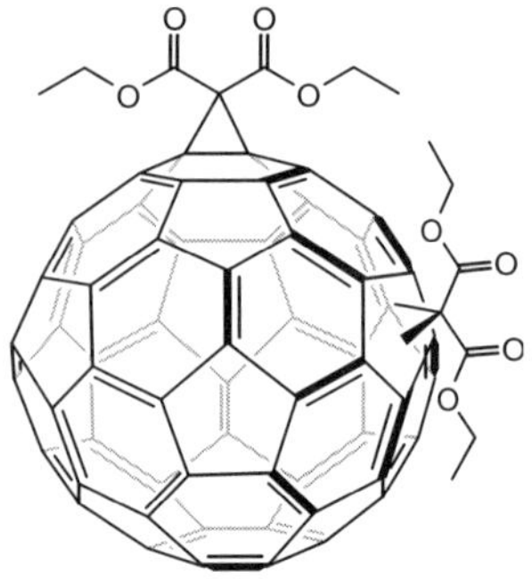

112

Figure 49. Migration of di(alkoxycarbonyl)methano bridges in C_{60} bisadducts. Adapted from [183].

higher stability. As a working hypothesis, it is proposed that, after introduction of two electrons into an isomer, a bond to one of the cyclopropane rings is broken, thus giving rise to a singly bound stabilized malonate anion, which might undergo "moon walking" around the carbon sphere. This would presumably delocalize the second electron. The migration could occur one foot at a time, involving sequential formation and breaking of methano bridges at 6,5- and 6,6-junctions, until the thermodynamic equilibrium of the dianionic species is reached.

It is shown that multiple additions to the fullerene core show a remarkable high regioselectivity [184], because addends already bound provide directing influence for further addition. Among the possible sites for a subsequent attack of a 6,6- double bond, *e* positions are significantly preferred followed by *trans*-3 positions [185]. The regioselectivity of attacks to the remaining *e* positions becomes even more pronounced as the number of addends already bound in *e* positions increases. Thus, upon moving from bis- to higher adducts, the regioselectivity of the additions becomes remarkably enhanced. The introduction of a third addend in the Bingel reaction occurs preferentially in the *e* position. Starting from the *e*,*e*,*e* trisadduct, stepwise *e* additions take place with amazing succession leading in high yield to hexakisadduct with a pseudooctahedral, all-*e* addition pattern [184a]. Starting from C_{60} and 8 equiv. of diethyl 2-bromomalonate/1,8-diazabicyclo[5.4.0]undec-7-ene (DBU), hexakisadduct is directly obtained [186].

Addition of 9,10-dimethylanthracene (DMA) to the reaction mixture enhances the overall yield of the hexakis adduct. The Diels–Alder reaction of C_{60} with DMA is reversible at room temperature. The reversible templated activation of C_{60} using 10-fold excess of DMA leads to $C_{60}(DMA)_2$ and $C_{60}(DMA)_3$ as the predominant products. Among the various regioisomers of $C_{60}(DMA)_3$, the C_3-symmetrical *e*,*e*,*e*-isomer with an incomplete octahedral addition pattern is by far the most abundant. The remaining unsaturated octahedral bonds are significantly preferred for subsequent attacks. After addition to the fullerene, the template directs diethyl malonate addends in the Bingel reaction regioselectively into *e* positions, ultimately yielding hexakisadduct **112** (Fig. 50) [187]. Monoadducts of fullerenes also undergo efficient templated activation of *e* 6,6-bonds. Thus, the bis(alkynyl)methanofullerene **113** reacts with diethyl 2-bromomalonate/DBU (8 equiv.) and

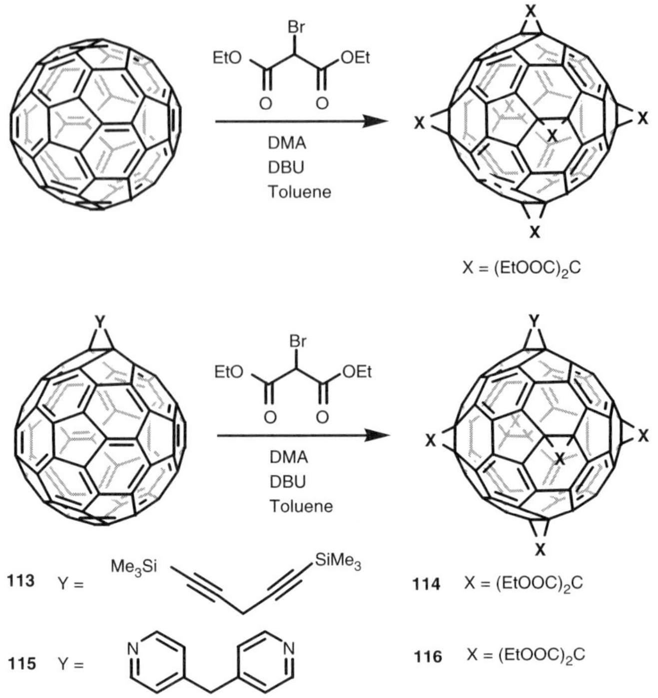

Figure 50. Synthesis of hexakisadducts of C_{60} with a pseudo-octahedral addition pattern. Adapted from [187].

DMA (12 equiv.) to provide hexakisadduct **114** in 28% yield [188]. Without template, the yield is only 12% and the targeted product is contaminated with an inseparable impurity. Also, monoadduct **115** is transformed in 35% yield into hexakisadduct **116** [189].

Template activation has been observed in several other cases too, such as the fulleropiperazine **117** and the triazoline **119** (Fig. 51) [190]. Starting from **117** and **119**, the mixed

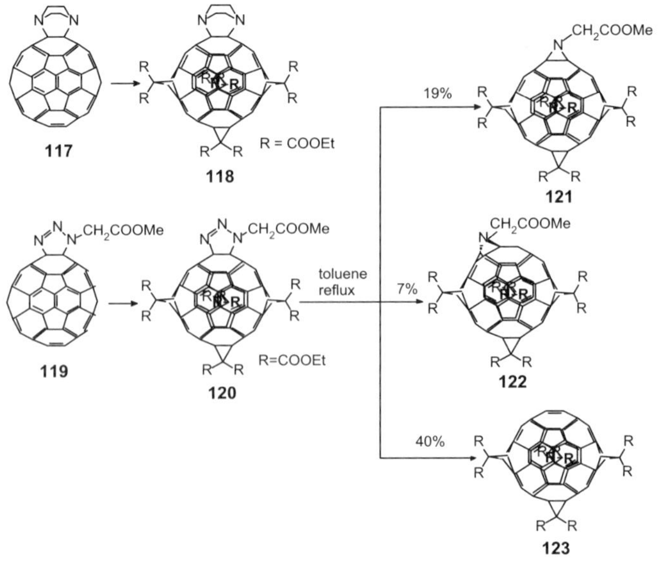

Figure 51. Synthesis of mixed hexadducts of fulleropiperazine and triazoline, and the thermal decomposition products of the triazoline. Adapted from [190].

hexaadducts **118** and **120** are formed, respectively. Extrusion of N_2 from the triazoline **119** by refluxing in toluene leads preferably to aziridine **121** than the fulleroid **122**. Further, **122** rearranges slowly at room temperature to **121**. Unfavorable steric interactions between the ester groups of the aza-bridge with those of a nearby methano bridge in **122** could be the possible reason for this rearrangement. In addition, in **121** a complete octahedral addition pattern is present, which is thermodynamically favorable.

Azides can be used as protecting groups for octahedral 6,6-bonds, since the triazoline group in **120** undergoes facile cycloreversion. For example, cycloreversion of the triazoline group in **120** leads to the synthesis of pentaadduct like **123** in comparatively high yields. Since a direct synthesis of pentaadducts of C_{60} with di(ethoxycarbonyl)methylene yields mixtures of regioisomers, this cycloreversion reaction is of synthetic importance. The pentakis adduct **123** can be considered as an excellent reagent in the sense that further reactions can be carried out almost exclusively at the remaining octahedral 6,6-bond. Oxidation of **123** by $KMnO_4$ in the presence of dicyclohexano-18-crown-6 followed by hydrolysis by acetic acid leads to the formation of the diol, which on further oxidation by $Pb(ac)_4$ gave a dioxetane. Addition to this vacant octahedral site is particularly favorable since it generates a residual fullerene π-chromophore consisting of an eight stable benzenoid ring substructure (see Fig. 6) [186].

Heating a sample of crystalline C_{60}-anthracene monoadduct **124** at 180 °C for 10 min affords 48% each of C_{60} and bisadduct **125**, resulting from an intermolecular anthracene-transfer reaction (Fig. 52) [191]. In this preparation of *trans*-1 bisanthracene adduct, the crystal packing provides the molding effect characteristic of a template. The bisadduct is used for the synthesis of the hexakis adduct **126** by Bingel reaction of the bromomalonate. The two anthracene units direct the four malonate addends regiospecifically into the *e* positions. Thermal removal of the anthracene templates provides tetrakisadduct **127** with all four addends aligned along the equatorial belt of C_{60} [192]. Like **123**, the tetrakis adduct **127** with two vacant reaction sites can add on with suitable reactant selectively in the octahedral positions.

Tether-Directed Functionalization Tether-directed remote functionalization of C_{60}, followed by addition to the vacant sites and subsequent removal of the tether, is an elegant approach for the synthesis of C_{60} regioisomers of defined structure. Such addition patterns are difficult to obtain by thermodynamically or kinetically controlled reactions with free untethered reagents. From semiempirical PM3 calculations the anchor-tether-reactive group conjugates such as **128** are designed (Fig. 53) [193]. The conjugate **128**, after attachment through a Bingel reaction to C_{60}, undergoes Diels–Alder additions at the two *e* positions on opposite sides, yielding trisadduct **129** in 60% yield with complete regioselectivity [194]. Reaction of **129** with a large excess of diethyl 2-bromomalonate and DBU in toluene gives by sequential *e* additions in 73% yield the yellow hexakisadduct **130**.

Addition reactions on **129** are also carried out in controlled steps, which provide tetrakis- and pentakisadducts. The pentakis adduct is left with only one vacant reactive site in the *e* 6,6-bond. However, repeated treatment of this pentakisadduct with diazomethane provides the heptakisadduct and octakisadducts with high regioselectivity and excellent yields, via hexakisadducts. The intermediate pyrazolines could only be isolated during the formation to hexakisadducts. These are not stable during the subsequent conversions in which N_2 extrusion occurs quickly even at low temperatures. The products obtained by both thermolysis and photolysis of pyrazoline are similar to those isolated in the addition of diazomethane to C_{60} followed by N_2 extrusion: the thermal process exclusively provides the 6,5-open derivative, whereas photolysis gives a 1:1 mixture of 6,5-open and 6,6-closed methanofullerenes.

The final step in the process is to remove the tether. Since the reactive ends of the tether have undergone [2 + 4] cycloaddition reaction, a cycloreversion reaction can lead to removal of the tether. However, this was found to be not successful for the hexakis- or the pentakisadducts. Eventually an elegant procedure is introduced, which could successfully remove the tether [195]. The $^1O_2$ ene reaction at the two cyclohexene rings of **130** yields a mixture of

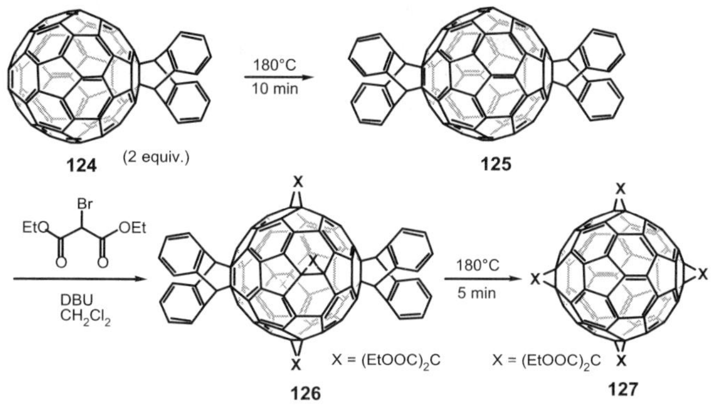

Figure 52. Templated synthesis of tetrakisadduct using anthracene as the reversible template. Adapted from [191].

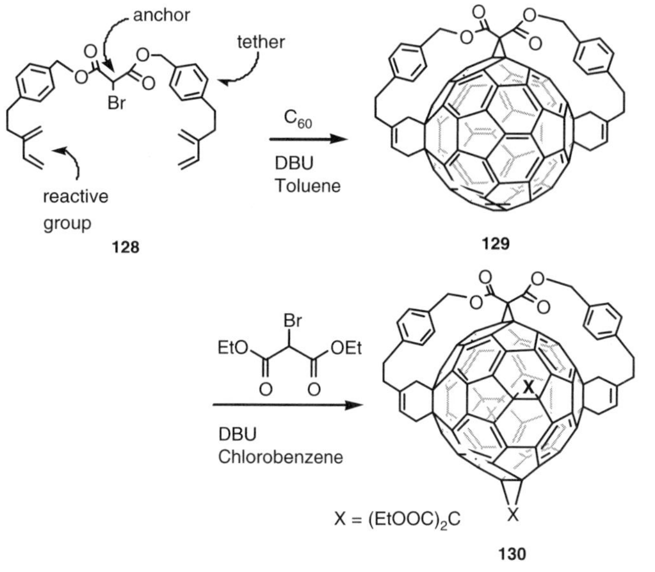

Figure 53. Tether-directed remote functionalization and subsequent *e*-attacks generating octahedral hexakisadduct. Adapted from [193].

isomeric allylic hydroperoxides **131** with endocyclic double bonds (Fig. 54). Reduction of **131** *in-situ* with PPh_3 gives a mixture of isomeric allylic alcohols. The allylic alcohol is heated in toluene together with toluene-4-sulfonic acid and dimethyl acetylene dicarboxylate (DMAD) [196]. This leads to the dehydration of the alcohol to the corresponding bis(cyclohexa-1,3-diene) derivative **132**. Further the diene **132** reacts with DMAD via a Diels–Alder reaction resulting in the tetrakisadduct **133**. Transesterification of **133** yields the octakis(ethyl ester) **134**, with all the addends in the equatorial belt. This tetrakis adduct has two reactive 6,6-bonds at the pole, activated by the four malonate addends in the *e* positions. Thus, addition of 2 equiv. of dialkynyl bromide affords hexakisadduct, which after deprotection, leads to the reactive compound tetrakisethynylated **135** with scope for further functionalization.

The fullerene properties change with increasing reduction of the conjugated chromophore resulting from increasing functionalization. Thus, the physical properties and chemical reactivity of the multiple adducts are comprehensively investigated as a function of degree and pattern of addition and the nature of the addends using UV-vis, IR, and NMR spectroscopy, voltammetry, calculations of LUMO energies and electron affinities, and chemical competition experiments. With increasing reduction in the conjugated fullerene π-chromophore, (i) the optical (HOMO–LUMO) gap in the UV-vis spectrum shifts to higher energy, (ii) the number of reversible one-electron reductions decreases, and the first reduction potential becomes increasingly negative, (iii) the computed LUMO energy increases and the electron affinity decreases, (iv) the reactivity of the fullerene toward nucleophiles and carbenes or dienophiles in cycloadditions decreases, and (v) the capacity for photosensitization of $^1O_2$ formation decreases [197]. The smallest perturbation of the fullerene π-chromophore by multiple additions occurs when all-methano addends are introduced along an equatorial belt [198]. Thus the properties of the parent fullerene in the highly functionalized fullerene derivatives are considerably retained if the addends are in an equatorial belt.

Macrocyclization A double Bingel reaction between C_{60} and bismalonate derivatives leads to macrocyclization. This is the most versatile and simple method for the preparation of covalent bisadducts of C_{60} with high regio- and diastereoselectivity. All possible bisaddition patterns (**136**), except *cis*-1, have been obtained by this tether-directed remote functionalization (Fig. 55).

Examples of regiosymmetric bisadducts are given in Figure 56 (**137–141**) [199]. Double Bingel reaction of bis(ethyl malonate), iodine, and DBU in toluene at 20 °C generates the macrocyclic bisadducts. Each macrocyclic regioisomer forms as a mixture of diastereoisomers, depending on how the EtOCO groups at the two methano bridge carbon atoms are oriented with respect to each other. Three diastereoisomers, in principle, are possible: *in–in*, *in–out*, and *out–out*. The *out–out* diastereoisomer is almost exclusively formed, except for **140**, which has the *in–out* geometry [200]. The tether in the bisadducts can be readily removed, making it a true template. Thus, transesterification of the fullerene crown ether conjugate **142** yields the symmetrical *trans*-1 bisadduct **143** (Fig. 57) [201].

Blocking the *trans*-1 positions by double Bingel macrocyclization leads to molecules that have three vacant octahedral sites for further reaction. Though the tether blocks the *trans*-1 positions (**144**), one of the *e*-positions is not accessible to cyclopropanation as the tether masks this position (Fig. 58) [202]. In targeting construction of highly organized three-dimensional scaffolds where all the addends in the octahedral sites are different (**145**), the *trans*-1 bisadduct **144** is used as the starting material.

The selective formation of *trans*-1 bisadduct has been challenging because this regio isomer is the least likely to form on statistical and electronic grounds. For successful formation of this bisadduct, the tether in the intermediates has to favor the second intramolecular [2 + 4] cycloaddition at the *trans*-1 position over attack at any nearby

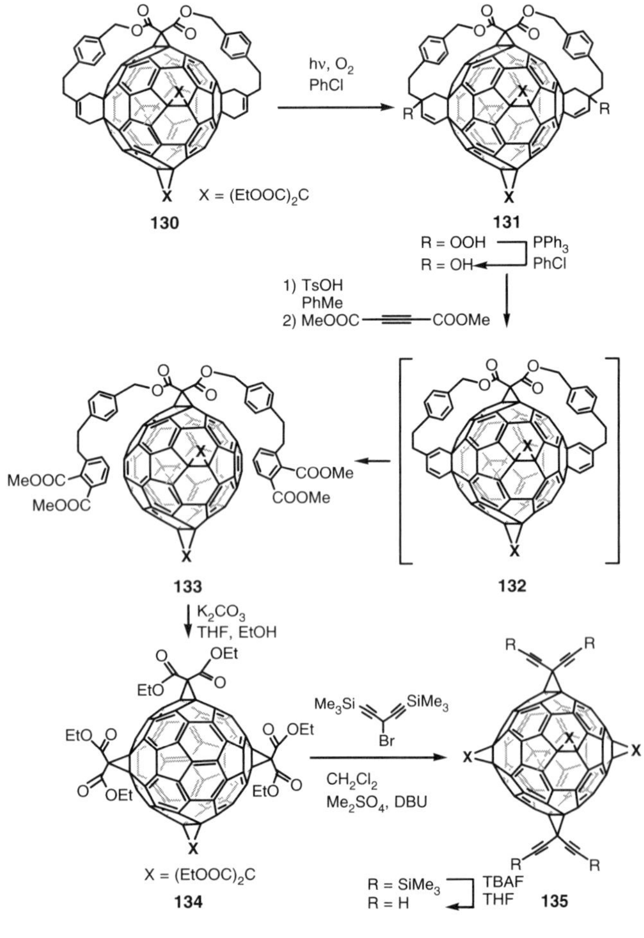

Figure 54. Removal of tether leading to tetrakisadduct with all addends in the equatorial belt. Adapted from [196].

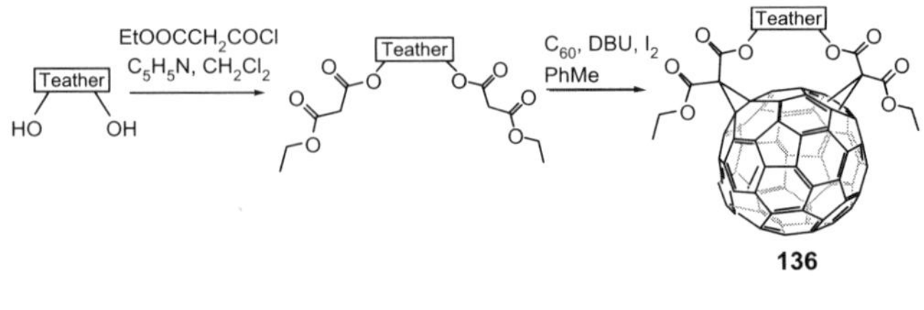

Figure 55. Tether-directed remote functionalization using a double Bingel reaction lead to bisadducts of high regioselectivity.

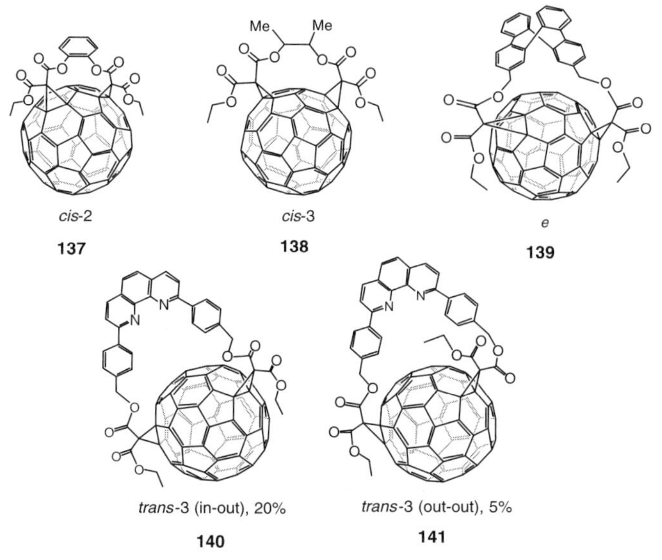

Figure 56. Examples of regiosymmetric bisadducts. Adapted from [199].

double bonds, including the four closely positioned *trans*-2 bonds (Fig. 59). A suitable tether is designed using AM1 semiempirical calculations. The tether length has the effect of significantly bending the alkyne and aromatic moieties for both the *trans*-1 and *trans*-2 approaches. The crude experimental mixture contained a 1:1.5–2 ratio of the *trans*-1 to *trans*-2 regioisomeric diastereomers. The *trans*-3, *trans*-4, and *e* regioisomers are not detected experimentally. The least polar *trans*-1 bisadduct (**148**) is isolated in 25% yield. The *trans*-1 bisadduct represents a strategically protected octahedral building block. The three available *e*-positions are strongly activated toward subsequent nucleophilic additions. Reaction of **148** with malonate under *in-situ* bromination conditions affords the mixed pentakis adducts **149**. Elimination of the tether by sequential Diels–Alder and retro-Diels–Alder steps leads to the trisadducts **150**. Further reaction leads to unprecedented heterohexaadducts by attaching different sets of addends to the vacant *e*-positions. The addends 4,5-diazafluorene and 2,7-dinitrofluorene added to these *e*-positions with exquisite selectivity under the *in-situ* bromination/nucleophilic addition conditions, and the mixed hexaadducts **151** are formed almost instantaneously at 25 °C. These hexaadducts are of interest, as the diazafluorene and dinitrofluorene addends are known

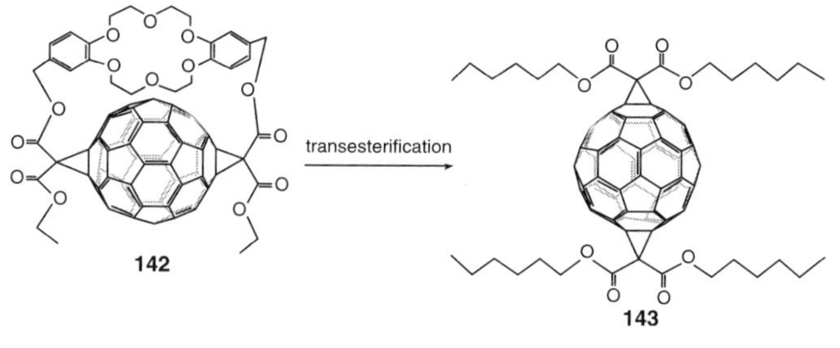

Figure 57. Synthesis of *trans*-1 bisadducts by removal of the tether. Adapted from [201].

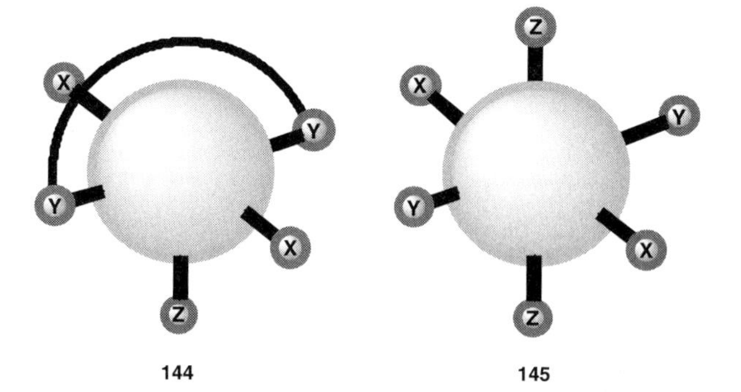

Figure 58. Double Bingel reaction at the *trans*-1 positions leaves three octahedral sites for further reaction. Adapted from [202].

for their metal complexing ability and electron-withdrawing homoconjugative effect on the fullerene π-system.

The high symmetry of C_{60} provides an opportunity to expand the diversity of molecular libraries. One can make a useful analogy between C_{60} and Werner's octahedral framework. Contrary to the freely rotating ligand-to-metal bonds in the octahedron, the 6,6-bonds in the octahedral sites to C_{60} are fixed in space, being included in the three planes of symmetry that are implicitly part of the T_h point group. In the octahedral model, there is only one possible diastereomer for each of the all-*trans* and all-*cis* configurations,

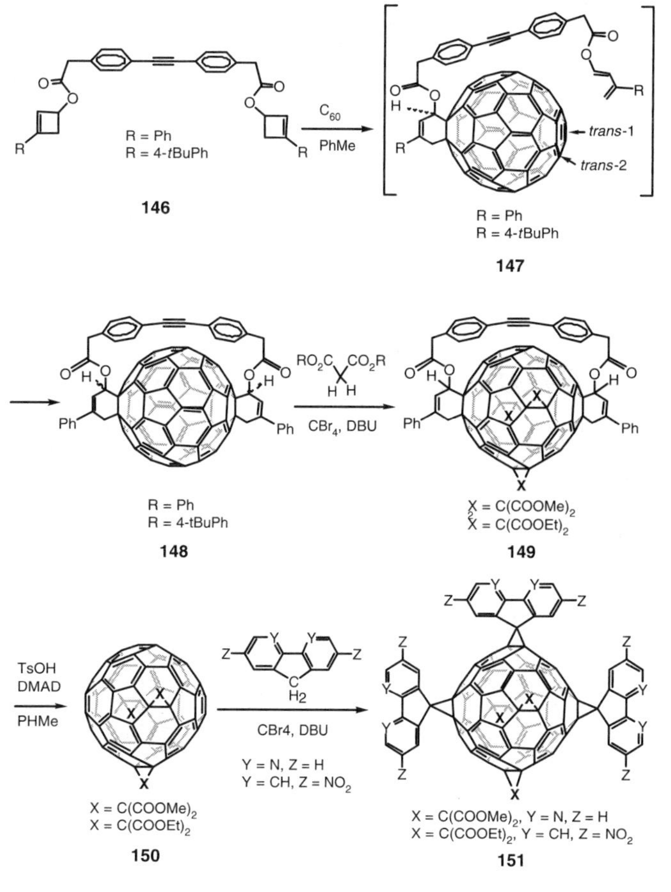

Figure 59. Synthesis of tris- and hexakisadducts of C_{60} by regiocontrolled additions. Adapted from [202].

while in the pseudo-octahedral (T_h) model, each of the all-*trans* and all-*cis* relationships gives rise to a pair of compounds that are regioisomers or *permutational* isomers.

The number of isomers that arise from $A_2 + B_2 + C_2$ functionalization at these bonds increases to 7 and goes up to 30 if all six adding groups are different. Using the "*mer*-3 + 3" stepwise regiocontrol strategy where the *trans*-1 bisadduct is the starting material, all seven isomers have been synthesized (Fig. 60) [203]. Following a similar strategy, complete control over addend permutation at all six pseudooctahedral positions of C_{60} has been achieved [204]. The C_{60} regioisomer with all six addends different (ABCDEF) is synthesized, which will eventually lead to all 30 possible regioisomers.

Hexakis adduct that features the location of all addends along an equatorial belt rather than evenly distributed over the entire carbon sphere is reported [205]. In the chiral D_2-symmetric structure A (Fig. 61), the addition sites are aligned in a distinct helical array whereas B, with a D_{3d} symmetric addition pattern, features a circumferential ("Saturn"-like) functionalization about the equator

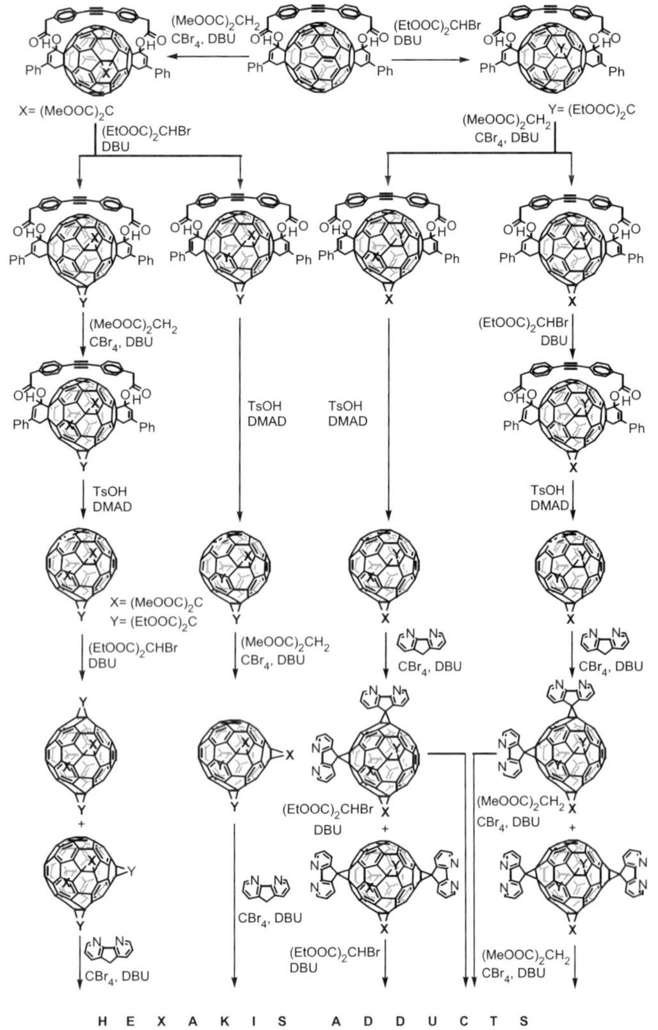

Figure 60. The *mer*-3 + 3 approach toward synthesis of pseudooctahedral regioisomers: synthesis of the hexakisadducts of the type $A_2 + B_2 + C_2$. Adapted from [203].

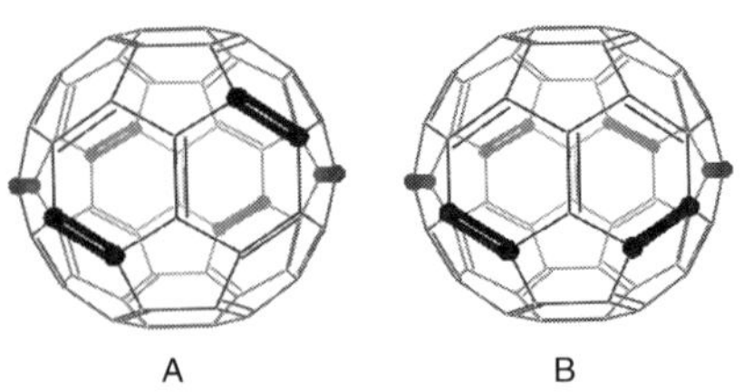

Figure 61. The D_2 (**A**) and D_{3d} (**B**) symmetric hexakis addition patterns that feature all addends in an equatorial belt. Adapted from [205].

which dissects the residual π-electron chromophore of the fullerene into two equal polar halves with no direct π-electron conjugation. The D_2-symmetric hexakisadduct reported features addition pattern A. The target compound is obtained in two sequential tether-directed remote fictionalization steps. The first consisted of the preparation of *trans*-1 bisadduct (±**152**) by Bingel macrocyclization of C_{60} with dibenzo[18]crown-6 appended with two bismalonate moieties. Cleavage of the crown ether template and *tert*-butyl ester groups affords tetraacid **153**, which is converted to **156** (Fig. 62).

Fourfold intramolecular Bingel addition of **156** gives the target compound (±**157**) as a single hexakisadduct in 10% yield. The molecular formula of the hexakisadduct is revealed by high-resolution matrix-assisted laser desorption ionization time-of-flight mass spectrometry (MALDI-TOF-MS). 1,3-Benzenedimethanol-tethered bismalonates are well known to yield *cis*-2 addition patterns on the fullerenes [206]. Fourfold addition of **156** with *cis*-2 selectivity can only give (with equal probability) two hexakis adducts, with addition patterns **A** or **B** (see Fig. 61). Single crystal X-ray structure determination nicely reveals the helical nature of the addition pattern **A** and allows an unambiguous assignment of structure ±**157**. π-Electron conjugation between the unsubstituted poles in ±**157** is maintained through two *trans*-stilbene-like bridges (Fig. 63). As a result of this extended conjugation the compound is red in solution (end absorption 600 nm). This contrasts with the light yellow color (end absorption ~450 nm) with pseudooctahedral addition patterns, in which the residual π-electron is reduced to a benzenoid "cubic cyclophane"-type structure (see Fig. 6). The novel hexakisadduct undergoes further functionalization at the central 6,6-bond of each pole. A Bingel reaction with diethyl 2-bromomalonate affords the heptakisadduct. Addition of 20-fold excess of the same 2-bromomalonate to ±**157** produces the octakisadduct.

The covalent cyclotriveratrylene-C_{60} adducts are prepared by the tether-directed Bingel reaction [207], which gives the two C_3-symmetrical *trans*-3,*trans*-3,*trans*-3 and *e*,*e*,*e* trisadducts with a high degree of regiospecificity. Cyclo-[*n*]-alkylmalonate is used in a one-step Bingel reaction with C_{60} that leads to products with remarkable stereoselectivity [208]. When cyclo-[2]-dodecylmalonate (**158**) is treated with C_{60} under Bingel conditions, the *trans*-3 bisadduct (**159**) is obtained almost exclusively (Fig. 64). If cyclo-[2]-octylmalonate is used under the same conditions, a mixture of several adducts as well as polymeric material is formed. In order to get the *trans*-1 bisadduct, cyclo-[2]-hexadecylmalonate is used. A mixture of *trans*-3 and *trans*-1 (55:45) is obtained. When C_{60} is treated

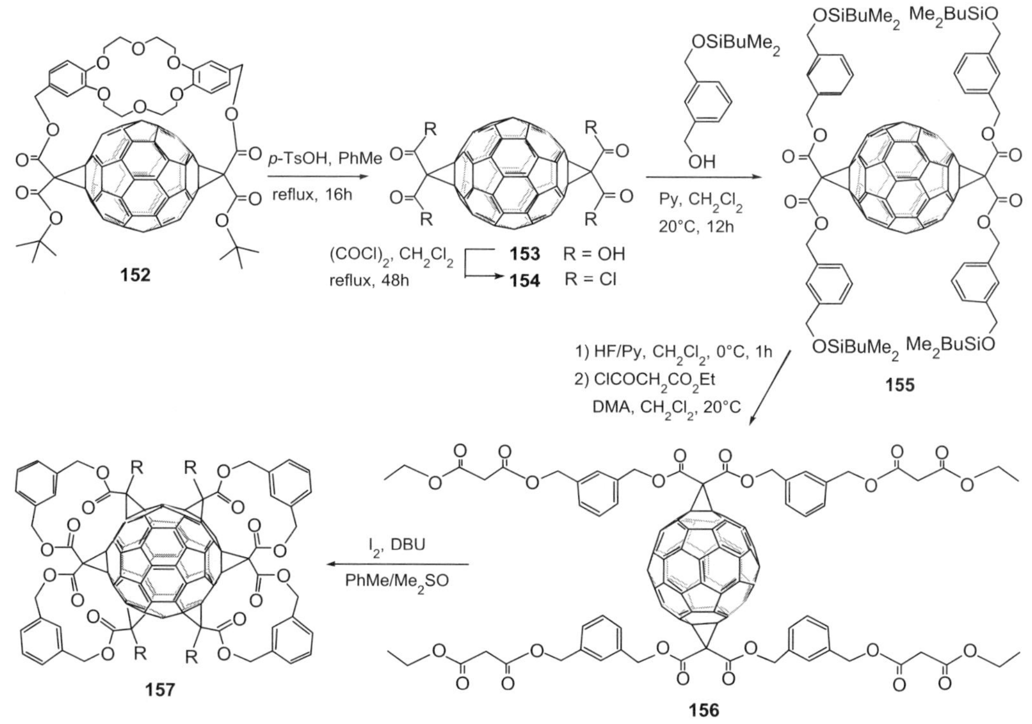

Figure 62. Synthesis of chiral hexakis cyclopropanated fullerene with all addends located along an equatorial belt. Adapted from [205].

with cyclo-[2]-butyl-octylmalonate or cyclo-[2]-octyl-tetradecylmalonate, the exclusive formation of adducts that have a C_s-symmetric bisaddition pattern is observed. In the former case, the *cis*-2 bisadduct is isolated and in the latter the *e*-bisadduct. The reaction of cyclo-[3]-octylmalonate (**160**) with C_{60} leads to *e,e,e* isomer (**159**) in 94% relative and 42% isolated yield (Fig. 65). A less polar by-product *trans*-4,*trans*-4,*trans*-4 trisadduct is formed. Reaction of the trisadduct **161** with another equivalent of cyclo-[3]-octylmalonate leads to the hexaadduct **162**. The trisadduct is also treated with an excess of didodecylmalonate under template activation conditions with 9,10-dimethylanthracene to produce the hexaadduct **163**. When cyclo-[4]-octylmalonate is used as the tether component in the reaction with C_{60}, the main product formed is *trans*-1,*e′*,*e″*-tetraadduct.

8.1.4. Nitrogen Bridging

Aliphatic and aromatic azides readily add on to fullerenes under thermal and photochemical reaction conditions. Since a wide range of aliphatic and aromatic azides are available, they offer considerable synthetic possibilities for functionalization of the fullerene sphere. Depending upon the reaction condition and the nature of the azide, nitrogen bridging takes place at the 6,6- or the 6,5-ring junctions. In some cases addition takes place by the azide decomposing first with the loss of a nitrogen molecule to nitrene (analog of carbene) intermediate, which adds to C_{60}. The azide can also add directly to C_{60} by a [2 + 3] cycloaddition reaction leading to isolation of a stable triazoline derivative, which then undergoes nitrogen extrusion by thermolysis or photolysis. In a number of reactions, one of the two products (6,6-closed and 6,5-open) constitutes the major component, whereas the other component may be present in very small amounts. Multiple additions may also take place leading to isolation of regiosymmetric products. Excellent collections of literature reviews are available in the subject [209].

Alkyl azides, such as [(trimethylsilyl)-ethoxy] methyl, and benzyl azides (RCH_2N_3) add on to C_{60} to give the corresponding azafulleroids (Fig. 66) [210]. Heating a mixture of the azide and C_{60} in chlorobenzene overnight produces a more polar product (A) stable at room temperature, along with the less polar azafulleroid. The more polar product is

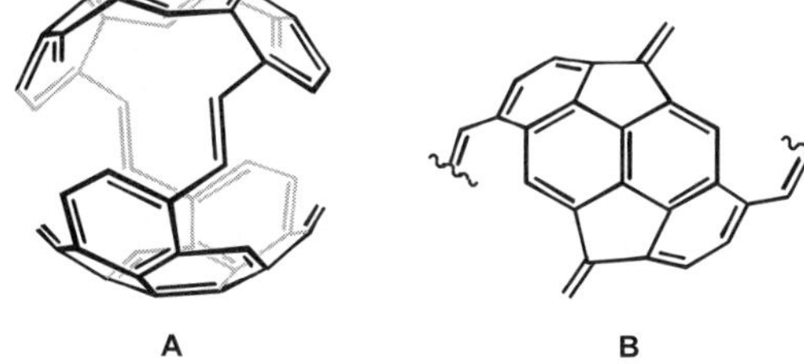

Figure 63. (A) Residual π-electron chormophore, (B) a view of the polar cap showing the reactive 6,6-bond. Adapted from [205].

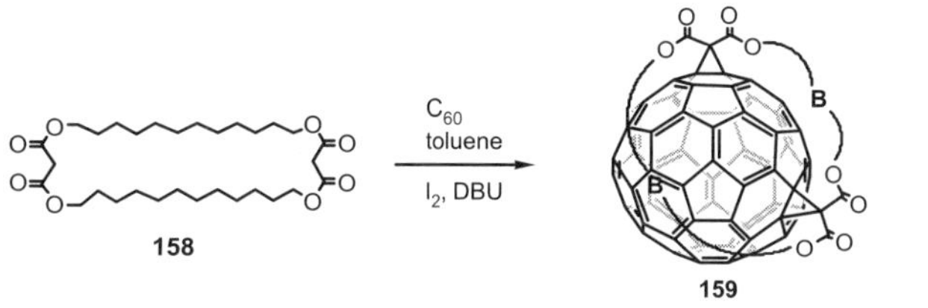

Figure 64. Bingel reaction of cyclo[2]dodecylmalonate with C_{60} leading to *trans*-3 bisadduct. Adapted from [208].

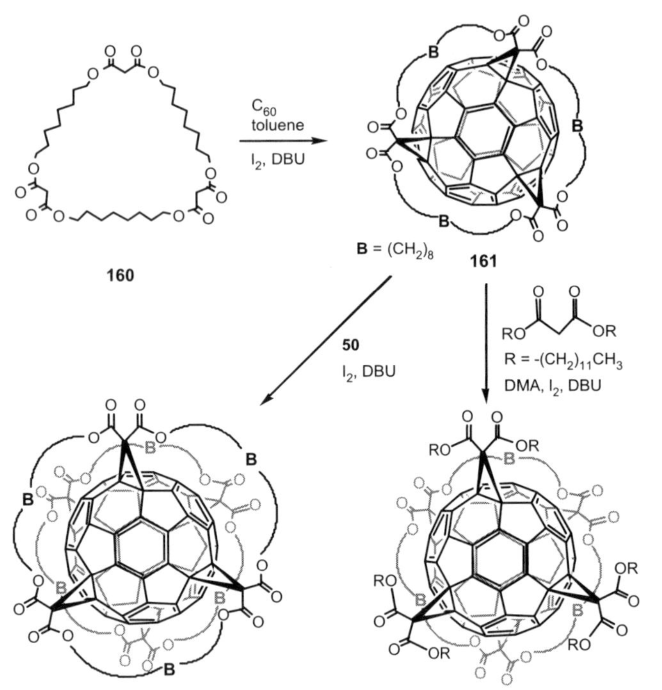

Figure 65. Synthesis of the *e*,*e*,*e* trisadduct and the two hexakisadducts. Adapted from [208].

evidently the triazoline derivative of C_{60}. When heated in refluxing chlorobenzene for a few hours or for a few minutes in solid state the triazoline is transformed to the less polar azafulleroid (**164**), the structure of which is confirmed by ^{13}C and ^{15}N NMR studies. The fast atom bombardment mass spectrometry (FAB-MS) of the final product shows the typical M^+ cluster with loss of the alkyl or benzyl moiety to give $[C_{60}N]^+$ and C_{60}^+. Addition across the 6,5-ring junction leads to polar triazoline, which under thermal conditions loses a molecule of nitrogen giving the less polar azafulleroid.

Reaction of isocyanurato-substituted azides with C_{60} leads to *N*-isocyanurato-substituted 6,5-open and 6,6-closed and isomers. The yield of the product depends upon the azide used. In some cases the azahomofullerene (6,5-open) is the major product, whereas in other cases the main product is the aziridinofullerenes (6,6-closed) (Fig. 67) [211]. Presumably, under the thermal reaction conditions (180 °C, *o*-dichlorobenzene) the azafulleroid rearranges to the aziridine.

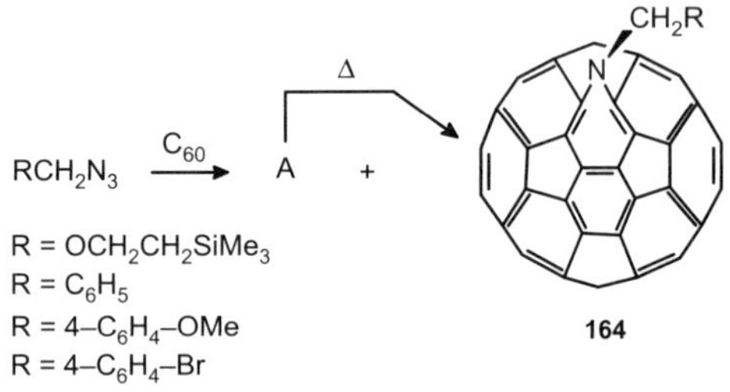

Figure 66. Azafulleroids from addition of alkyl azides to C_{60}. Adapted from [210].

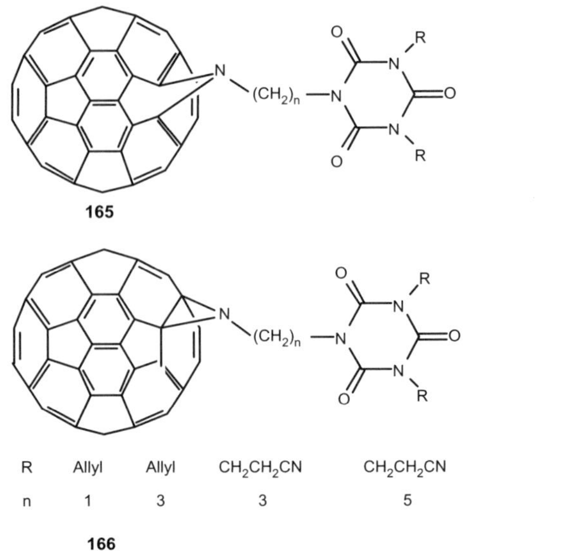

Figure 67. Reaction of isocyanourato-substituted azides with C_{60} gives both open and closed isomers. Adapted from [211].

The reaction of C_{60} with aryl azides (PhN_3, $4\text{-}NCC_6H_4N_3$) at room temperature leads to isolable triazoline derivatives, **167** (Fig. 68). Photolysis of this derivative selectively furnishes fulleroaziridines (**169**) by the loss of a molecule of nitrogen. Thermolysis on the other hand gives azafulleroids (**168**) as the main product. The rearrangement of the azafulleroid can be induced photochemically to the fulleroaziridine [212]. Besides aziridines, the photochemical reactions of azides with C_{60} predominantly yield C_s symmetrical bisadducts. The mechanism of 1,3-cycloaddition of methyl azide to C_{60} forming the triazoline and subsequent extrusion of a molecule of nitrogen is theoretically studied by semiempirical (AM1) and density functional quantum chemical calculations (B3LYP/6-31G*) [213]. Additions take place at the 6,6-double bond forming the triazoline derivative, the structure of which has been confirmed by X-ray crystallography [214]. Subsequent thermal extrusion of nitrogen molecule proceeds by a stepwise mechanism in which

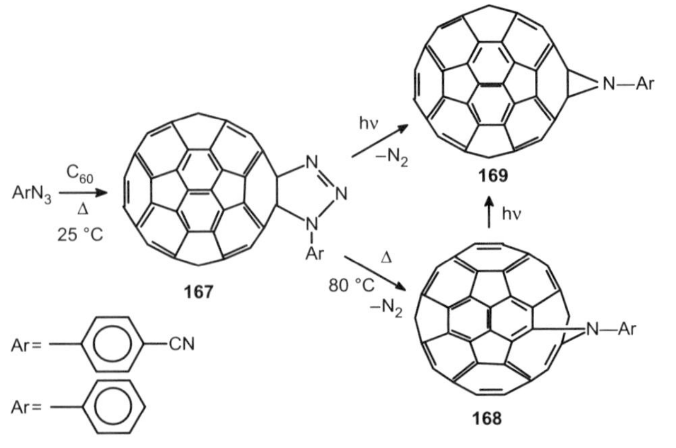

Figure 68. Addition of 2,3-diphenyl-2*H*-aziridien to C_{60}, synthesis of triazolinofullerne, and its conversion to aziridinofullerenes. Adapted from [212].

the cleavage of the N–N bond precedes the cleavage of the C–N bond. Under thermal reaction conditions the formation of the 6,5-open azafulleroid is explained on the basis of the steric effect of the leaving nitrogen molecule that prevents the attack of the nitrene nitrogen atom on the 6,6-bond and facilitates attack on the 6,5-bond.

The reaction of C_{60} to acyl nitrenes, generated by photolysis of acyl azides (N_3COR), leads to fulleroaziridine derivatives as the predominant products. These are formed in excellent yields by trapping of the nitrene at a 6,6-ring junction. Thus, the first step in the case of acyl azide addition to C_{60} is the formation of a resonance-stabilized nitrene intermediate, which then adds on to the 6,6-ring junction. The reaction of C_{60} with an azido formate leads to the isolation of the fulleroaziridine with a closed 6,6-structure. The fulleroaziridine formed, on heating, rearranges to an oxazole (**170**) (Fig. 69) [215]. Upon further heating it rearranges to fullerooxazoles bound to the 6,6-ring fusion sites.

A similar reaction of aroyl azides generates the corresponding acyl nitrenes (**171**) by photolysis of the azides and the nitrenes added on to C_{60} giving the fullerene adducts with the closed 6,6-ring fused fulleroaziridine structures, **172** (Fig. 70) [216]. The results confirm the nitrene addition to be the key step for the formation of fulleroaziridine derivatives. The fulleroaziridine on boiling in tetrachloroethane rearranged to yield the fulleooxazole, **173**.

From the foregoing results a generalization emerges. The addition of the azide can take place leading to the 6,5-open azafulleroid or 6,6-closed fulleroaziridine as the principal products, depending upon whether the reaction is thermal or photochemical, respectively. In cases of the fulleroaziridine formation, further rearrangement to fullerooxazole takes place on heating the 6,6-azafulleroid if a C=O group neighbors the nitrene. The azide can decompose to give the nitrene intermediate if it can be resonance stabilized, and then the nitrene intermediate can react with C_{60}. If the nitrene cannot be resonance stabilized, addition of the azide takes place as the first step leading to the formation of the triazoline intermediate which loses nitrogen, furnishing the desired product. This generalization is supported by many other examples. An N-hydroxysuccinimide functionalized perfluorophenyl azide reacts photochemically with C_{60} with the addition of the nitrene occurring at the 6,6-ring junction (Fig. 71) [217]. Further, the ester group of the aziridine is converted to the reactive functional group, which allows for the attachment of other organic or bioactive molecules to C_{60}.

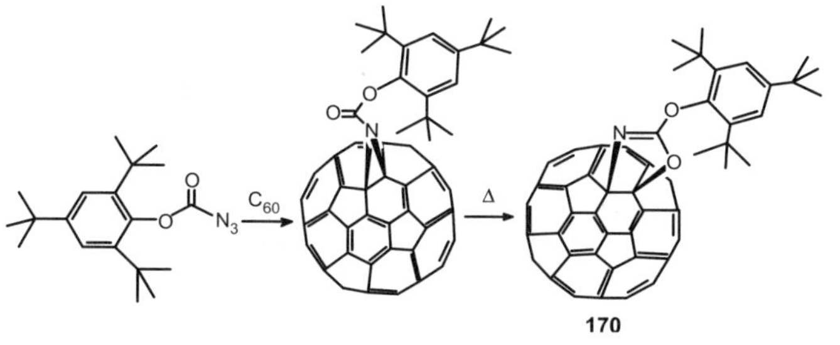

Figure 69. Acyl azide addition at the 6,6-ring junction of C_{60} and rearrangement of fulleroaziridine to fullerooxazole on heating. Adapted from [215].

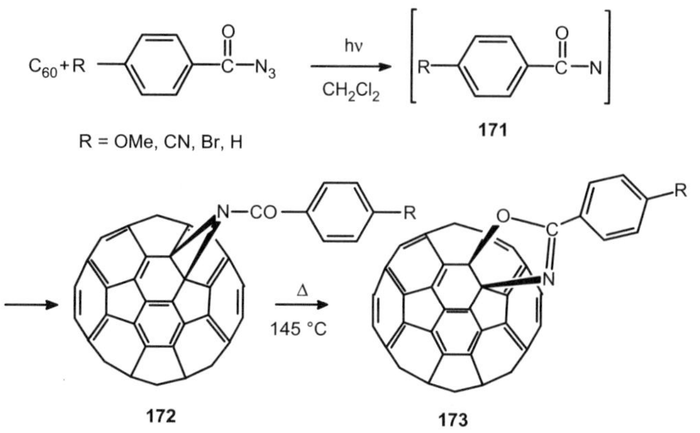

Figure 70. Photochemically generated nitrenes react with C_{60} to give the fulleroaziridine, which rearranges to fullerooxazole on heating. Adapted from [216].

Thermal addition of 2-azidobenzothiazole and 1-azido-4-(3′,5′-dimethyl-1′-pyrazalyl)tetrafluoro benzene to C_{60} is studied to discover whether the formation of aziridine or the fulleroid can be explained by the duality of azide decomposition mechanism. The addition of 2-azidobenzothiazole leads to the 6,5-open azafulleroid whereas 1-azido-4-(3′,5′-dimethyl-1′-pyrazalyl)tetrafluoro benzene furnishes 6,6-closed fulleroaziridine (Fig. 72) [218].

In addition to the azides, another source of nitrene for capture by C_{60} is the base-induced α-elimination from substituted O-4-nitrophenyl sulphonyl hydroxamic acid. Rearrangement of the resulting fulleroaziridine bearing N-ethoxycarbonyl group under the influence of chloromethylsilane results in the quantitative formation of a fullerooxazolidin-2-one, which can be cleaved to yield 1-hydroxy-2-N-methylamino[60]fullerene. Isomerization of the 6,6-bridged adducts to the corresponding oxazolofullerene derivatives occurs upon heating, with the exception of the *tert*-butoxycarbonyl adducts, which eliminates isobutene and carbon dioxide to yield the parent aziridinofullerene $C_{60}NH$. In this work it is reported that isolation of closed 6,6-bridged fulleroaziridines is accompanied by

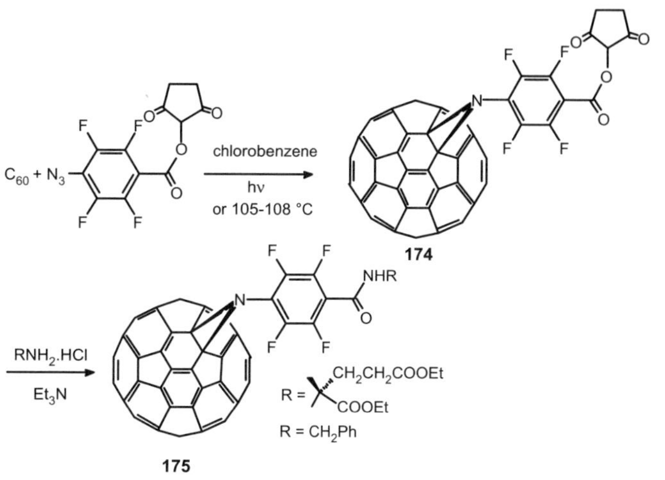

Figure 71. Reaction of azidofluorobenzoate C_{60}. Adapted from [217].

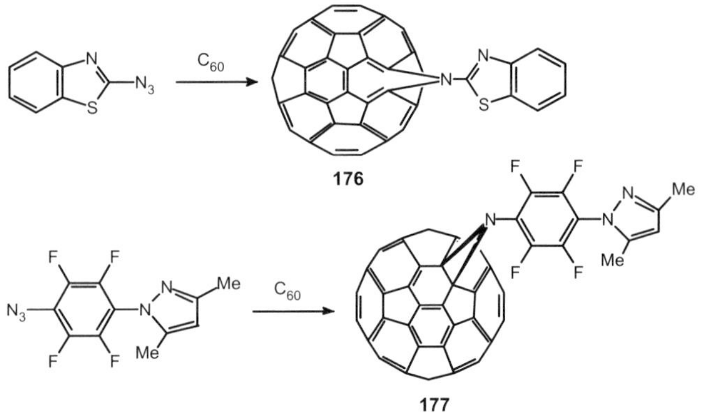

Figure 72. Formation of 6,6-closed and 6,5-open isomers depends upon the duality of azide decomposition mechanism. Adapted from [218].

a small amount of previously undetected closed 6,5-adducts [219]. The report of the closed 6,5-aziridine derivatives ($C_{60}NCOOR$, R = Et, *tert*-butyl, and 2,4,6-tri-*tert*-butylphenyl) [220] is reexamined [221] and it is concluded that the addition of these nitrenes to C_{60} furnishes the fullerene derivative, which are 6,5-open azaannulenes rather than the 6,5-closed aziridines.

An example of 6,6-open-bridged compounds by selective reduction of the C–C bond of 6,6-closed-bridged azirido-fullerene is reported. The fulleroaziridine is reduced in acid medium (Zn/glacial acetic acid) to give the dihydro derivative having a bridged 10-membered ring (Fig. 73) [222]. The removal of the *N*-protecting group leads to $C_{60}H_2NH$ (**181**) which is remarkably air stable, although under basic conditions it suffers an unusual oxidative ring contraction to the parent fulleroaziridine.

The regiochemistry of azide addition to C_{70} is more complicated since it has five sets of different carbon atoms. Treatment of C_{70} with (2-methoxyethoxy)methyl azide (MEM-N_3) or methyl azidoacetate at moderate temperatures gives three (two with C_s symmetry and one with C_1 symmetry) out of a possible six triazoline isomers [223]. The addition pattern shows chemoselectivity as well as regioselectivity. The major product arises from addition of azide to the double bond of C_{70} possessing the greatest local curvature (polar 6,6-bonds). Selective thermolysis of the least stable triazoline, the one resulting from addition to the least curved 6,6-double bond, allows solution of the structure of the three isomers. Thermolysis of the triazoline isomers produces mixtures of azafulleroids and fulleroaziridines (Fig. 74).

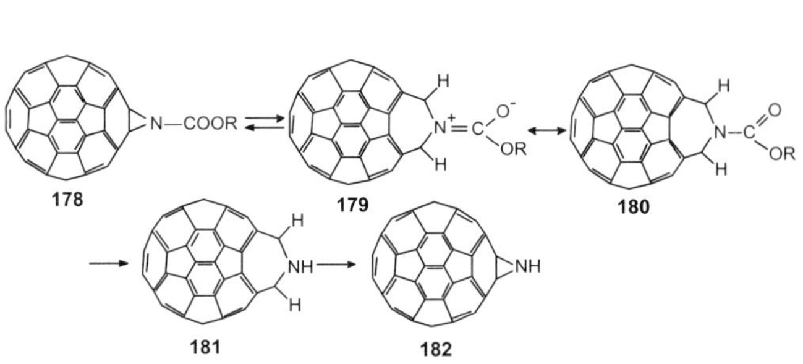

Figure 73. An example of 6,6-open nitrogen-bridged C_{60} derivative. Adapted from [222].

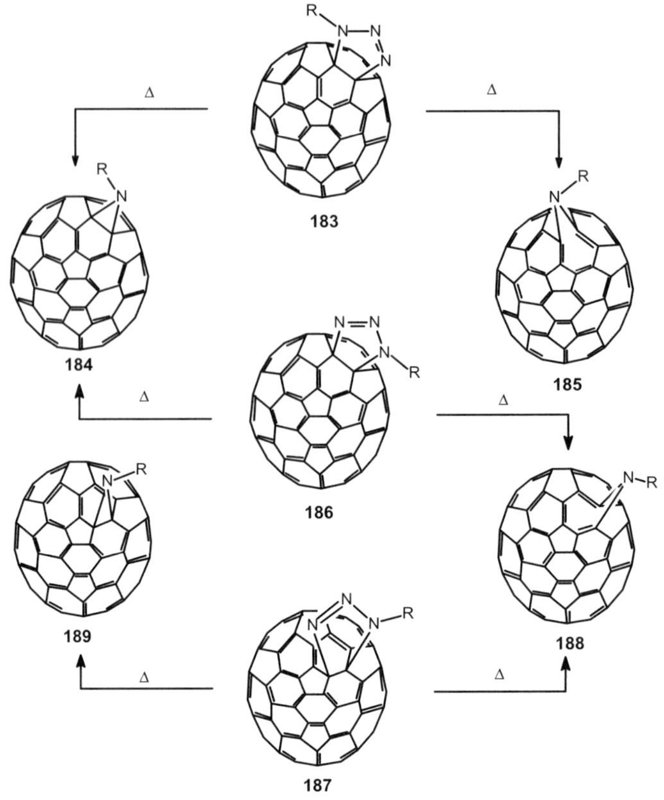

Figure 74. Thermolysis of three triazolines of C_{70} gives azafulleroids and fulleroaziridines. Adapted from [223].

The bridging nitrogen with useful functional groups can impart specific properties to the fullerene derivative. Aqueous C_{60} solution under photoirradiation expresses several biological activities due to singlet oxygen ($^1O_2$). A C_{60} derivative with an acridine group (**190**) (Fig. 75) is synthesized and used for the study of DNA-cleaving activity [224]. An acridine moiety is chosen because of its intercalating activity with DNA double strands. The DNA-cleaving activity of C_{60} and the acridine adduct, both solubilized in water with poly(vinylpyrrolidone), are tested under visible light irradiation using pBR322 supercoiled plasmid. Under dark

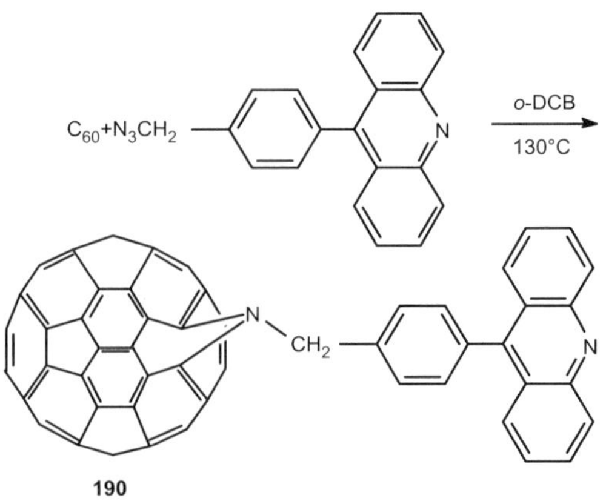

Figure 75. A C_{60} derivative with acridine group on the bridging nitrogen atom used for DNA cleaving studies. Adapted from [224].

conditions none of the chemicals show any DNA-cleaving activity. Under visible light irradiation, C_{60} itself shows weak but significant DNA-cleaving activity at 0–5 °C within 4 h of photoirradiation. C_{60} showed stronger DNA-cleaving activity at higher temperature. Within 1 h of visible light irradiation at 35–40 °C, about 25% of the supercoiled DNA is converted to the nicked DNA. In comparison, the fullerene–acridine adduct showed stronger activity even at 0–5 °C within 4 h of visible light irradiation. About 50% of the supercoiled DNA is cleaved in 2 h and more than 90% is converted to nicked DNA after 6 h of photoirradiation. The minimum dose of the acridine adduct is 17.5 μM at 0–5 °C for 4 h of visible light irradiation.

Synthetic lipid bilayer membranes possess fundamental physicochemical properties similar to those of biomembranes and can be immobilized as molecular lipid films. The synthesis and characterization of a C_{60}-bearing triple-chain lipid on the bridging nitrogen (**191**) is reported (Fig. 76) [225]. Spectral evidence suggests the lipid to exist as a closed aziridine structure at a 6,6-ring junction of C_{60}. The amphiphile is not soluble in water because of the lack of a suitable hydrophilic moiety in the molecule but it is soluble in organic solvents. The phase transition between the crystalline phase and liquid crystalline phase is one of the most fundamental characteristics of lipid bilayer membranes. The work demonstrates that the lipid forms multiplayer membrane films, which possess main and subphase transitions, and the subtransition regulates electronic properties of the fullerene.

8.1.5. Bisazafullerenes and Fullerides

As described in the previous section, addition of diazocompounds or azides to C_{60} leads to fulleroids ($C_{60}NR$) by extrusion of N_2 of the initially formed 6,6-bridged pyrazolines or triazolines. In $C_{60}NR$ the π-electron system of C_{60} remains intact. The opening of a 6,5-single bond means a fullerene core that contains a bridged nine-membered ring. The 6,5-azafulleroids are isolated in good yields. However, the C=C adjacent to the nitrogen atom bonds are very susceptible to singlet oxygen. The *N*-ethoxymethoxy ethyl (MEM)-substituted 6,5-azafulleroid (**192**) reacts with singlet oxygen to afford the ring-opened *N*-MEM keto lactam, **193** (Fig. 77) [226]. The structure of the holed fullerene is proved by FAB-MS and ^{13}C NMR studies. The resonances at 198.5 and 163.6 ppm are assigned to ketone and lactam carbon atoms, respectively.

Following this observation, multiple additions of azides to C_{60} are carried out [227]. Addition of one equivalent more

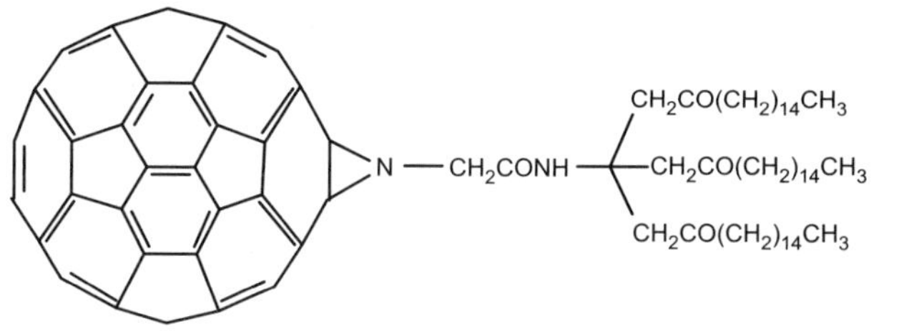

191

Figure 76. A C_{60} bearing lipid that forms self-organized multilayer films. Adapted from [225].

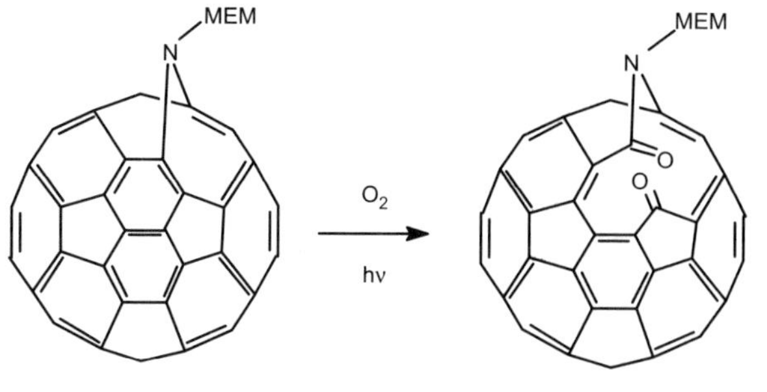

192 **193**

Figure 77. Reaction showing the C=C adjacent to nitrogen as susceptible to singlet oxygen, leading to the first fullerene with a hole. Adapted from [226].

of the azide N_3R to $C_{60}NR$ leads to the bisazafullerenes, **194** (Fig. 78). The second nitrogen atom also forms a bridge by breaking a neighboring 6,5-single bond. The resulting diazafulleroids, having three 7-membered and one 11-membered ring, are formed in an extraordinary regioselectivity pathway. Treatment of C_{60} with methyl azidoacetate furnishes triazoline, which is not isolated. In addition to a mixture of nitrogen-bridged monoadducts with 6,5-open and 6,6-closed structures, a bisadduct is formed in which two neighboring 6,5-open bonds are imino bridged. This addition pattern of methyl azidoacetate to C_{60} is proved to be a general case with a number of other azides, such as substituted benzyl azides RN_3 (R = benzyl, 4-methoxybenzyl, 4-bromobenzyl) reacting a similar way. The addition of trimethylsilylethoxymethyl azide ($SEMN_3$) to C_{60} is a special case, because of the formation of the bisadduct as the major product. When a dilute solution of C_{60} and excess $SEMN_3$ in chlorobenzene is heated to reflux for approximately 12 h, the diazafulleroid is formed as the main product (60%) along with the 6,5-open (30%) and 6,6-closed (10%) monoadducts. The bisadduct is not stable in

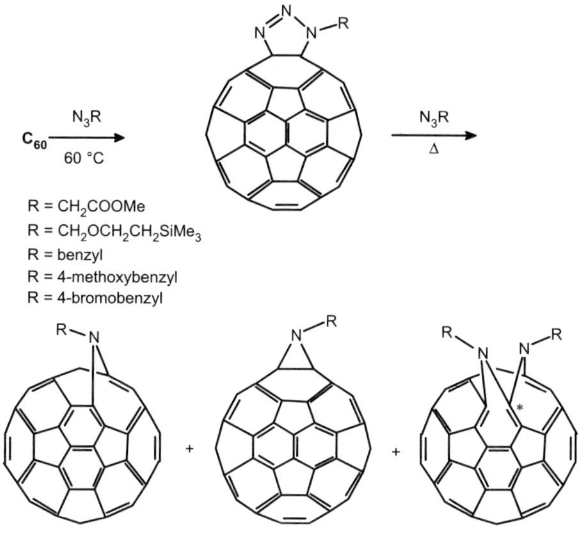

194

Figure 78. Regioselective formation of diazafulleroids. Adapted from [227].

refluxing chlorobenzene. On heating for about 12 h, the bisadduct is converted quantitatively to a 7:1 mixture of 6,5-open and 6,6-closed monoadducts. The bisazafulleroid has a unique structure in that it has a highly symmetrical ^{1}H and ^{13}C NMR signal pattern. The two R groups are magnetically equivalent. The compound has C_s symmetry with the mirror plane crossing the four carbon atoms of two 6,6-junctions. The distinct ^{13}C NMR signal at $\delta 160$ is attributed to the central bridgehead C atom (carbon in the bisazafulleroid marked with an asterisk, **194**). A twofold addition of azidoformates under thermal conditions leads to reaction of azides with C_{60} as nitrenes (rather than azides). This reaction leads to all conceivable eight regioisomeric bisadducts [228]. In contrast to nucleophilic cycloadditions of C_{60}, the *cis*-1 isomers are among the preferably formed regioisomers. The *cis*-1 adducts represent the first examples of open *trans*-annular 6,6-bonds. As in case of C_{60}, bisazafulleroids are formed when the reaction of C_{70} with alkyl azides, $N_3CH_2CO_2R$ (R = Me, Et), is carried out at 120 °C [229]. Two C_s-symmetric isomers (**195**, **196**) are formed as the only regioisomers (Fig. 79).

Reacting C_{60} with diethyldiazidomalonate leads to a doubly bridged fulleroid [230], which exhibits a base peak at m/z 906 (M^+) in agreement with the expected molecular mass of the doubly bridged fulleroid (Fig. 80). Three structures are considered for the compound, corresponding to a double 6,5-insertion, a double 6,6-insertion, and a 6,5/6,6-insertion. From NMR results and considering the nonequivalence of the two carboethoxy groups, a double 6,5-insertion (**197**) is proposed as the most plausible structure of the double-bridged fulleroid.

Treatment of C_{60} with an excess of $N_3(CH_2)_nN_3$ ($n = 2$ or 3) in refluxing chlorobenzene furnishes the corresponding bisazafulleroids [231]. The addition at the two 6,5-ring junctions of the same five-membered ring or at the two five-membered ring junctions of different five-membered rings is possible (Fig. 81). From ^{1}H NMR studies and characteristic absorption profiles in the UV-visible spectra, the bisazafulleroid is assigned the structure in which the addition is shown at the two 6,5-ring junctions of the same five-membered ring (**200**). This work is extended to the synthesis of chiral bisazafulleroids, expecting that the chirospectroscopic properties of the chiral bisazafulleroids would demonstrate some unique properties (Fig. 82) [232]. Two C_2 chiral diols are converted to their diazides, which furnish the chiral bisfulleroids on treatment with C_{60}. The results establish that tunable chiral substituent significantly enhances the chirospectroscopic response in these fullerene derivatives.

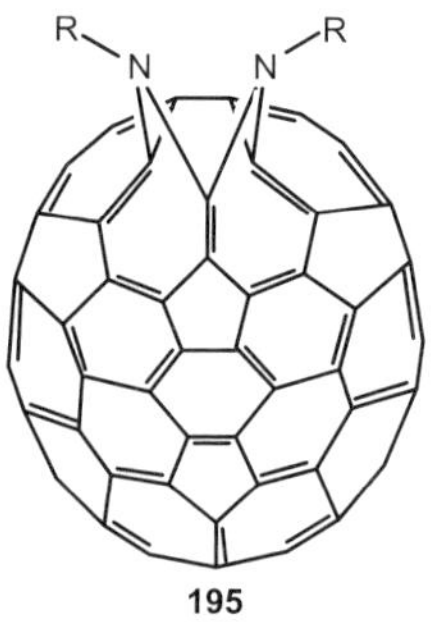

195

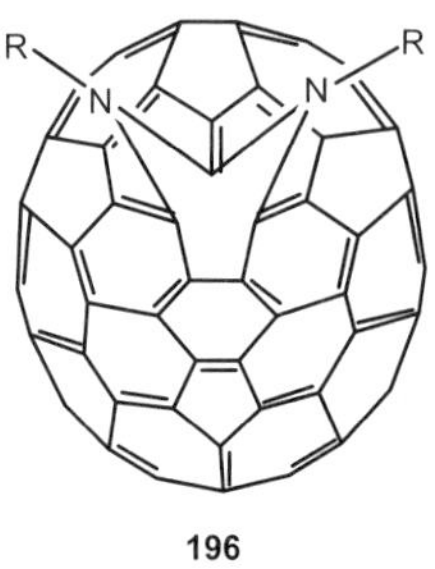

196

Figure 79. Regioisomeric bisazafulleroids of C_{70}. Adapted from [229].

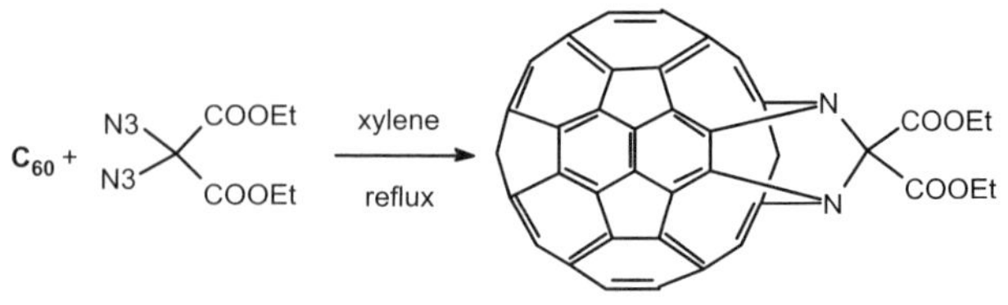

double 6,5-insertion

197

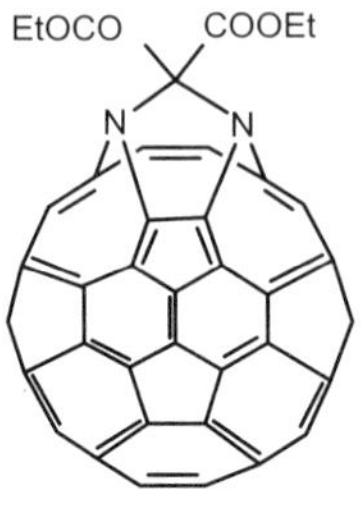

double 6,6-insertion

198

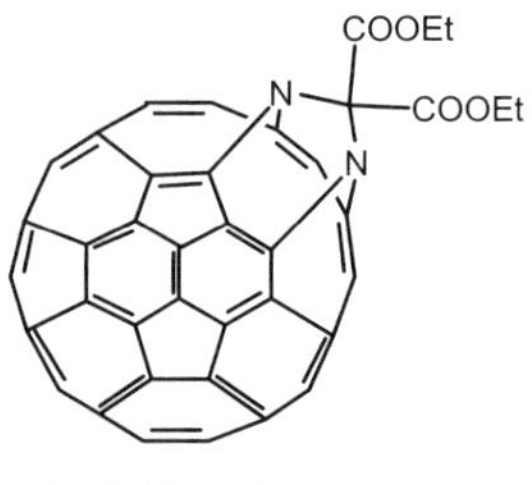

6,5/6,6-insertion

199

Figure 80. Double-bridged fulleroid, the double 6,5-insertion being the most plausible structure. Adapted from [230].

A bisazide has the structural constraints to allow double 6,5-open insertion in two different five-membered rings, as well as a Diels–Alder addition (Fig. 83) [233]. This reaction is pursued to effectively create an opening within the C_{60} core. The opening has been achieved in which four bonds of the fullerene cage are cleaved. The reaction of the diazidobutadiene with C_{60} followed by N_2 extrusion affords iminofullerene derivative. Subsequent reaction with singlet oxygen yields endo peroxide intermediate, which rearranges by spontaneous [2 + 2 + 2] ring opening. This is further treated with 5,6-dicyano-1,4-benzoquinone or oxygen to afford the desired ring-opened fullerene (**205**), which is an unprecedented class of ligand for supramolecular chemistry. This unique one-pot reaction provides efficient formation of the largest orifice created so far on the C_{60} molecule, thus giving the opportunity to test if small atoms or molecules could be made to pass through it. Incorporation experiments with helium and hydrogen are followed by ^{1}H and ^{3}He NMR studies. The results provide a fully convincing demonstration of a H_2 molecule or a He atom inside the ring-opened fullerene [234].

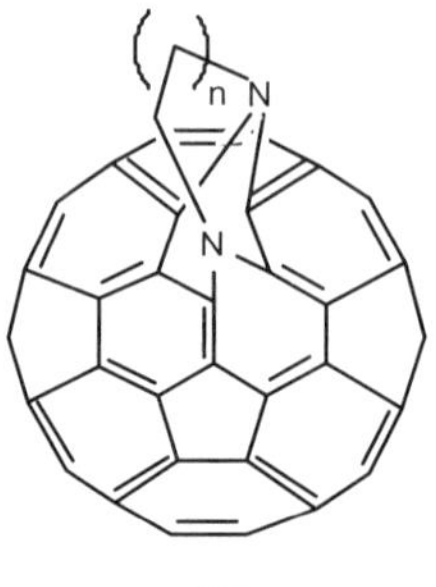

200

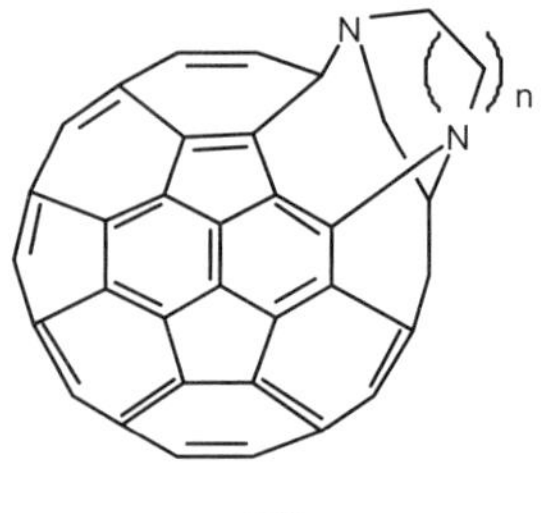

201

Figure 81. A bisazafulleroid: addition occurs at the two 6,5-ring junctions of the same five-membered ring. Adapted from [231].

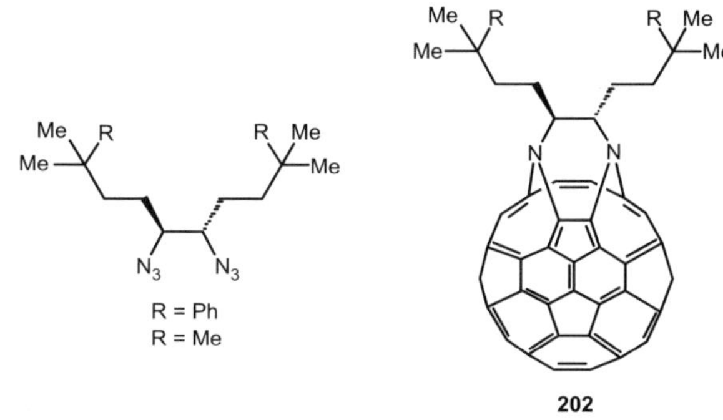

Figure 82. Chiral bisazafulleroid. Adapted from [232].

The open-cage fullerene derivative **205** can incorporate a He atom and a H_2 molecule in 1.5% and 5% yields, respectively. To achieve 100% encapsulation of these gas molecules, it is desirable to have a larger orifice on the fullerene surface. An open-cage fullerene derivative has been synthesized which has a 13-membered-ring orifice on a C_{60} cage [235]. The derivative has an orifice of 5.64 Å along the long axis and 3.75 Å along the short axis. Encapsulation of a H_2 molecule into it does take place to produce the endohedral complex in 100% yield [236], and that $H_2@C_{60}$ can be generated in the gas phase by restoration of the C_{60} cage from the endohedral complex upon laser irradiation.

8.2. [2 + 2] Cycloadditions

The curved geometry and lack of significant conjugation of the π-bonds in C_{60} endow the π-bonds with efficient ene character resulting in a number of [2 + 2] cycloaddition reactions. *o*-Benzyne typically reacts with polycyclic aromatic compounds by adding across the 1,4-positions within a ring, not between rings, thus participating in [2 + 4] cycloaddition reactions. However, with C_{60} and C_{70}, the addition of benzyne takes place by [2 + 2] cycloaddition leading to the attachment of a benzene ring via a cyclobutane ring [237]. Addition takes place at the 6,6-ring junction. The ^1H NMR spectra of the diadduct are complex resulting from the

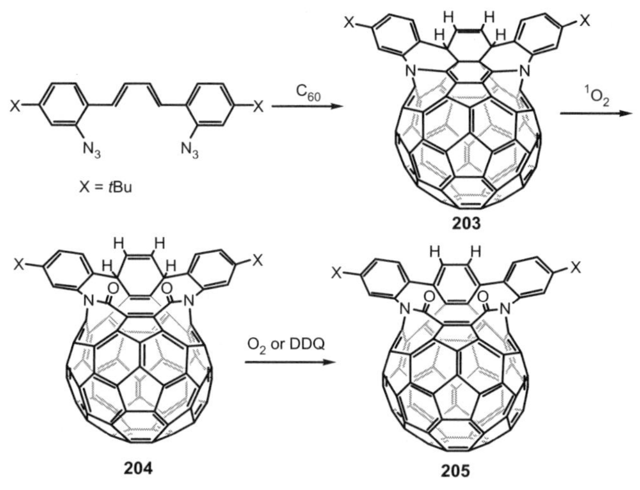

Figure 83. Ring-opened fullerene: ligands for supramolecular chemistry. Adapted from [233].

many regiochemical possibilities for adding a second benzyne molecule.

Under photochemical reaction conditions C_{60} reacts with a number of dienes. Again, instead of the [2 + 4] cycloadduct, the products are formed through [2 + 2] cycloaddition. In most of the cases both the double bonds of the diene undergo addition to give two different products with defined regioselectivity if appropriate substitutents are present on the dienes (Fig. 84). However, the stereochemistry of the double bond of the initial diene involved in the reaction is lost in all cases. The predominant products are the *trans*-cyclobutanes. The stereochemistry of [2 + 2] photocycloaddition of *trans,trans*-, *cis,cis*-, and *cis,trans*-hexa-2,4-diene to C_{60} [238] involves the addition of diene to C_{60} in the excited triplet state through electron transfer from diene followed by rapid conversion of the resulting ion pairs to [2 + 2] cycloadducts. The collapse of the ion pairs determines the diastereoselectivity. In the process the thermodynamically more stable *trans*-cyclobutane adducts predominate. The lifetime of the biradical intermediate is sufficiently large for rotation about the C_2–C_3 bond. If the diene is an unsymmetrical one, cycloaddition can involve either of the two double bonds. Photochemical cycloaddition of *trans*-hepta-3,5-dien-2-one to C_{60} is an example that utilizes both of its nonequivalent C=C bonds in giving different regioisomeric products [239]. Another example where the two C=C bonds are nonequivalent is *trans*-2-methylhepta-2,5-dien-4-one. In this compound one of the two electronically and sterically nonequivalent double bonds adds on photochemically to C_{60} to give a single product [239]. In contrast to the formation of this cycloadduct in good yields, photoreaction of C_{60} with mesityl oxide requires a 1000-fold excess of the mesityl oxide and the reaction is reversible.

Acetylenes can also undergo photochemical [2 + 2] cycloaddition with C_{60}. Thus C_{60} on reaction with diethylaminopropyne leads to a cycloadduct, which on hydrolysis cleaves the cyclobutene ring resulting in an amide (Fig. 85) [240]. Photoaddition of a silyl ketene acetal to C_{60} takes place at room temperature in wet toluene to afford a fullerene substituted carboxylic ester instead of the expected [2 + 2] cycloadduct. The mechanism is suggested to involve a zwitterion formed through a radical ion-pair [241]. An addend with both a double and a triple bond adds on to C_{60} through the triple bond, if the triple bond is electron rich. Thus an addend with *N*-diethylamine moiety attached to a triple bond under photocycloaddition with C_{60} leads to a fused cyclobuteneamine, which suffers photooxidation leading to a fullerenocarbamide [242].

Photochemical [2 + 2] addition of C_{60} and C_{70} with yndiamine leads to cyclobutene adducts, which suffers from self-sensitized photooxidation to furnish fullerene anhydrides $C_{62}O_3$ (**223**) and $C_{72}O_3$ (**225**) (Fig. 86) [243]. Reaction of C_{60} with *N,N,N′N′*-tetraethylenediamine gives the corresponding cyclobutene intermediate, which upon self-sensitized photooxygenation in toluene gives a diamide. The cyclobutanamine (R = CH_3) is unique in that it has a photosensitizer (the dihydrofullerene) and a photooxidizable group (the enamine) in the same molecule. On brief exposure to air at room light the enamine double bond gets cleaved producing the ketoamide. On treatment with *p*-toluene sulphonic acid the diamide hydrolyzes to a

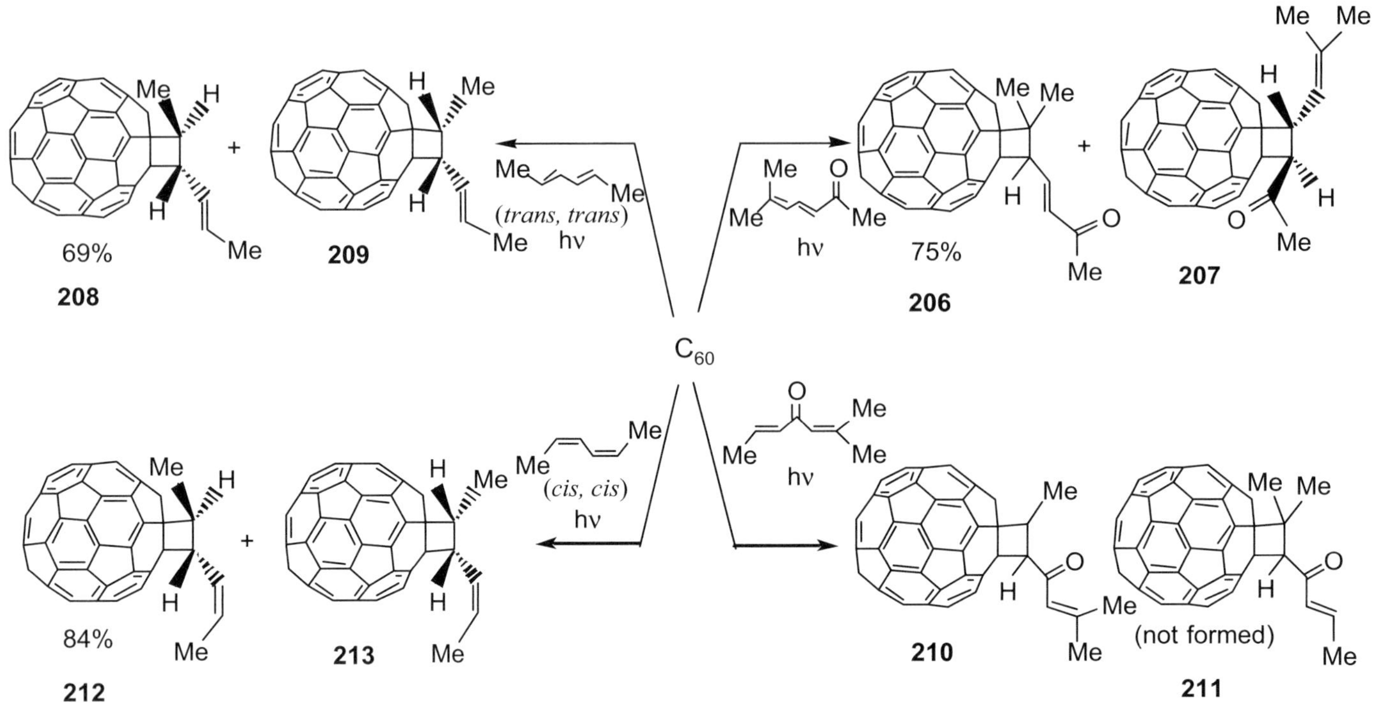

Figure 84. Photocycloaddition of dienes to C_{60} leading to cyclobutane adducts of defined regiochemistry. Adapted from [238].

fullerene dicarboxylic acid anhydride. Compared to C_{60}, C_{70} reacts ~4 times faster. Addition of a thioyndiamine takes place across the 1,9-bond of C_{70}.

A highly strained ketene acetal undergoes stepwise [2 + 2] cycloaddition to give a cycloadduct, the C–C bond of which is cleaved upon heating in aqueous H_2SO_4 to give an alcohol derivative, **227** (Fig. 87) [244]. A variety of functionally substituted derivatives of C_{60} can be prepared by [2 + 2] photocycloaddition to cyclic enones. The preparation and resolution of enantiomers of *cis*- and *trans*-fused [2 + 2] photoadduct of C_{60} and cyclic enones are reported [245]. Photocycloaddition of 3-methylcyclohexenone to C_{60} forms both *cis*- and *trans*-adducts in a 40:60 ratio. The course of reaction upon UV irradiation of crown ether fulleroid and 3-methylcyclohexen-2-one in benzene solution is followed by electrospray ionization mass spectrometry (ESI-MS) (Fig. 88). Photocycloaddition of a number of cyclic enones is studied [246]. Bulky substitutents in the enones hinder addition of the enones to C_{60}.

Reaction of C_{60} with quadricylene gives a stable adduct, a derivative of C_{60} bearing a reactive double bond in the side chain, **233** [247] (Fig. 89). The norbornene double bond

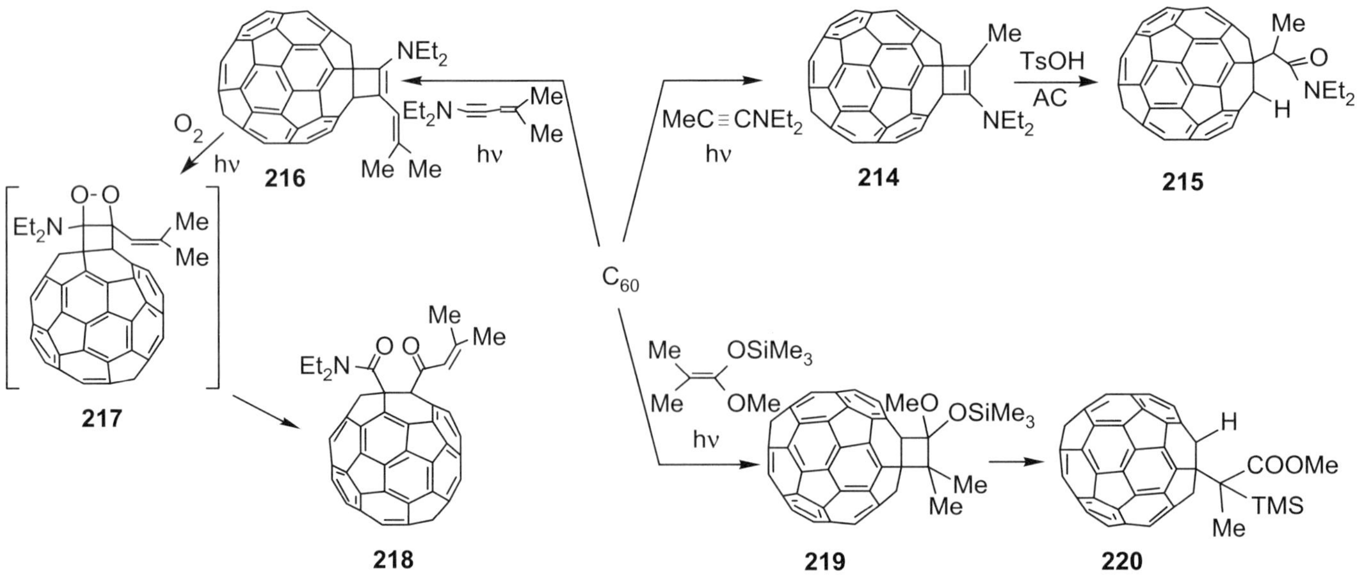

Figure 85. Photocycloaddition of alkenes, alkynes to C_{60}. Adapted from [240].

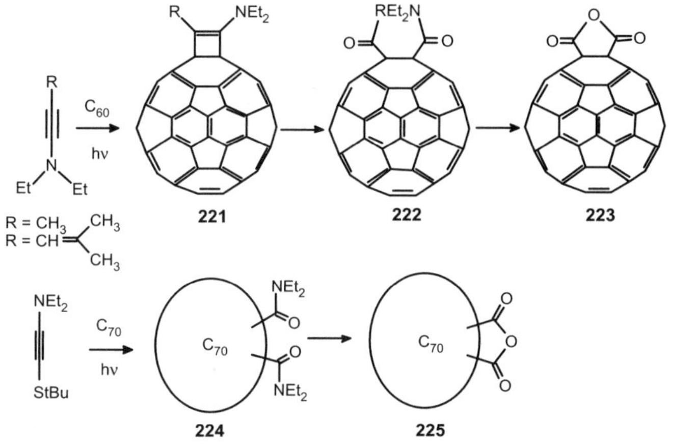

Figure 86. Photochemical [2 + 2] addition of C_{60} and C_{70} with yndiamine and thioyndiamine leading to synthesis of fullerene anahydrides. Adapted from [243].

is unhindered and reacts readily with electrophiles such as PhSCl to afford quantitatively the product of *anti* addition. [2+4] Cycloaddition of norbornene-containing C_{60} with 3,6-di(2-pyridyl)-*s*-tetrazine providing powerful chelating sites is reported [248] (Fig. 90). The diene 3,6-di(2-pyridyl)-*s*-tetrazine (**235**) reacts site specifically at the external site with norbornene-containing C_{60} molecules. The products are oxidized to the corresponding pyridazine adducts. Similarly, an acetylene derivative bearing ester group is added by an additional [2 + 2] cycloaddition to norbornene-containing C_{60} (**239**). These compounds represent the first example of a new class of polydentate nitrogen ligands incorporating an attached C_{60} moiety.

8.3. [2 + 3] Cycloaddition

8.3.1. TMM Adducts

A five-membered ring can be attached to C_{60} via [2 + 3] cycloaddition of trimethylenemethane (TMM). A nonpolar TMM is the one generated from a 7-alkylidene-2,3-diazabicycloheptene, whereas a polar TMM is generated by thermolysis of methylenecyclopropanes. These TMM precursors have provided simple access to five-membered ring compounds. The parent TMM-C_{60} adduct **240** is synthesized by palladium-catalyzed [2 + 3] cycloaddition by mixing C_{60} with a stoichiometric amount of $[Pd(PPh_3)_4]$ and $Ph_2PCH_2CH_2PPh_2$ in benzene which results in the palladium–C_{60} complex and then adding a benzene solution

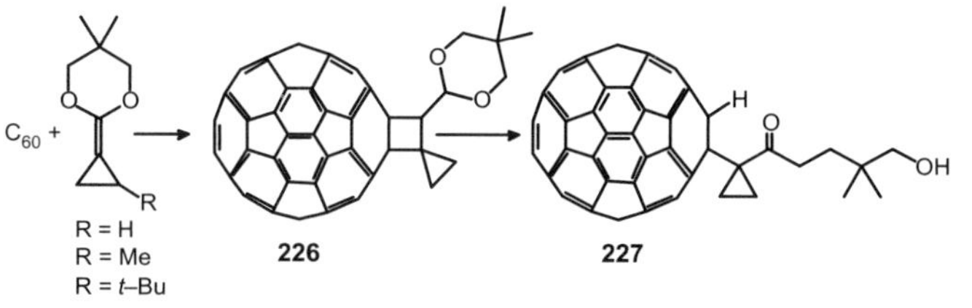

Figure 87. Cycloaddition of dimethyleneketene acetal to C_{60}. Adapted from [244].

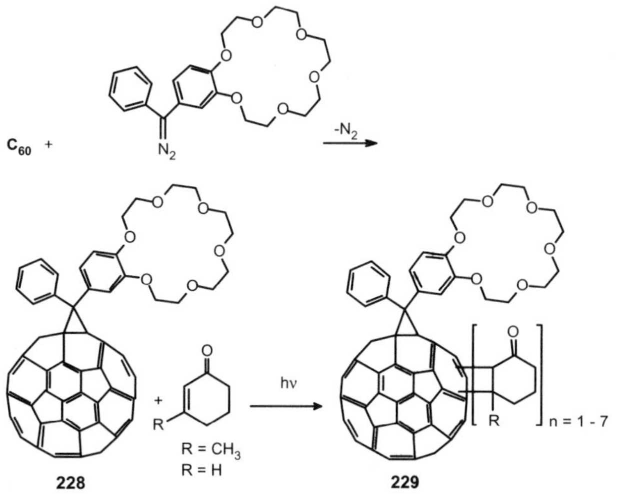

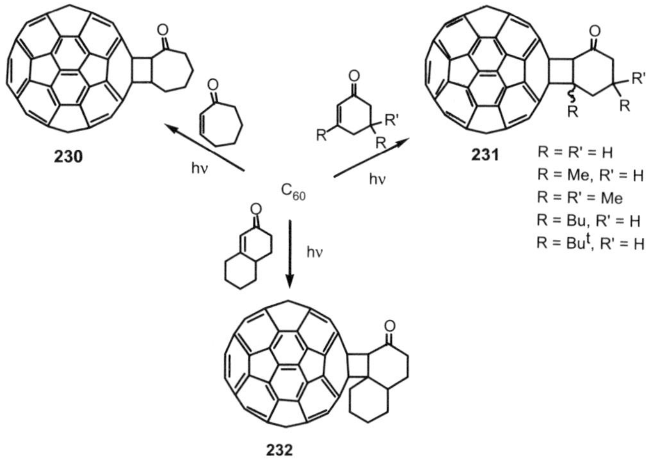

Figure 88. Photocycloaddition of cyclic enones. Adapted from [246].

of the TMM precursor (Fig. 91) [249]. Use of TMM precursors leads to further functionalization of C_{60} via the five-membered ring. The TMM precursor, which gives the best results in terms of isolation and characterization, is the methylene cyclopropanone ketal (Fig. 92) [250], which on treating with C_{60} results in the isolation of two products, **241** and **242**, the ester **242** being a possible silica gel catalyzed hydrolysis product of the ketene acetal **243**. The ester **242** is further functionalized through the alcoholic OH group to benzoate **244**, methacrylic ester **245** (Fig. 93) [251], and a number of such compounds varying R and R′.

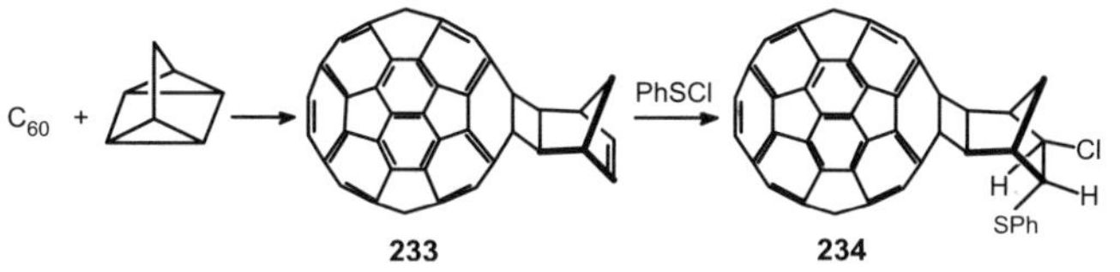

Figure 89. Addition of quadricyclane to C_{60} leading to norbornene-containing fullerene. Adapted from [247].

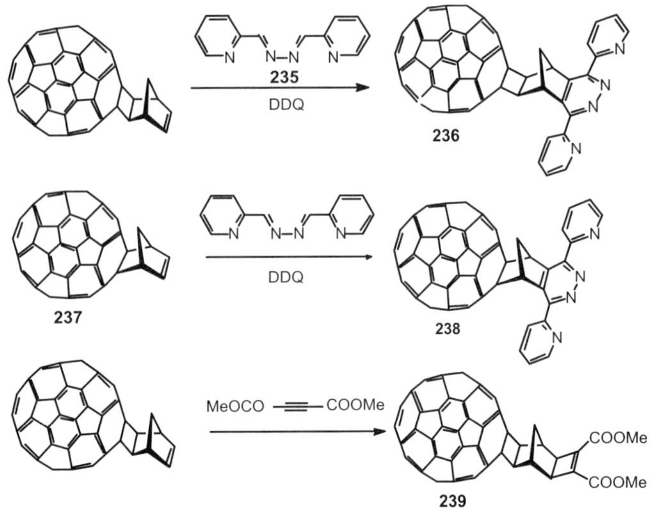

Figure 90. Attachment of chelating capacity to C_{60} through a [2 + 2] cycloadduct. Adapted from [248].

The [2 + 3] cycloadditions of buta-2,3-dienoates with C_{60} provide a very efficient route to the synthesis of cyclopentene attachment to fullerene (Fig. 94) [252]. In the presence of phosphine catalysts, ethyl buta-2,3-dienoate and ethyl but-2-ynoate undergo [2 + 3] dipolar cycloaddition to C_{60}. A number of derivatives are synthesized by varying the alkyl group.

8.3.2. Further Functionalization via Five-Membered Ring

Reaction leading to fulleropyrrolidines by 1,3-dipolar cycloaddition of azomethine ylides is proven to be a very facile route to functionalization of C_{60} [253], due to the many ways available to generate the reactive intermediates from a wide variety of easily accessible starting materials. Azomethine ylides, such as $H_2C{=}N^+(CH_3)$–CH_2^-, are expediently generated following the Prato reaction, which is the decarboxylation of immonium salts obtained from a mixture of *N*-methylglycine (sarcosine) and paraformaldehyde. The reaction affords exclusively the product of cycloaddition across a 6,6-ring junction of C_{60}. The scope of the azomethine ylide addition is very broad. A number of α-aminoacids and aldehydes can be used. Aldehydes other than formaldehyde lead to 2-substituted *N*-methyl-3,4-fulleropyrrolidines, whereas condensation of *N*-functionalized glycines with aldehydes leads to *N*-substituted fulleropyrrolidines (Fig. 95) [254]. Glycine

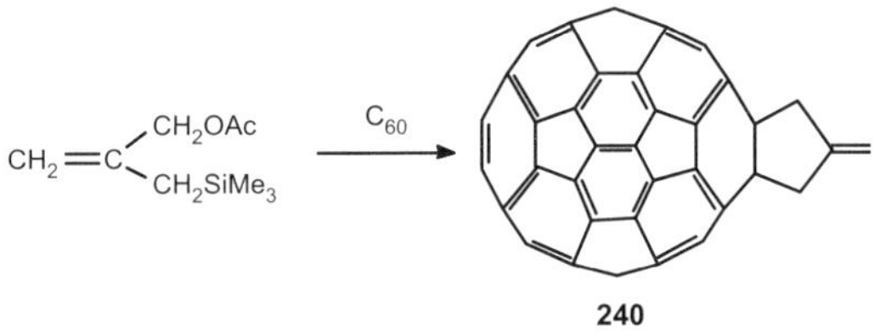

Figure 91. Synthesis of trimethylmethane cycloadduct of C_{60}. Adapted from [249].

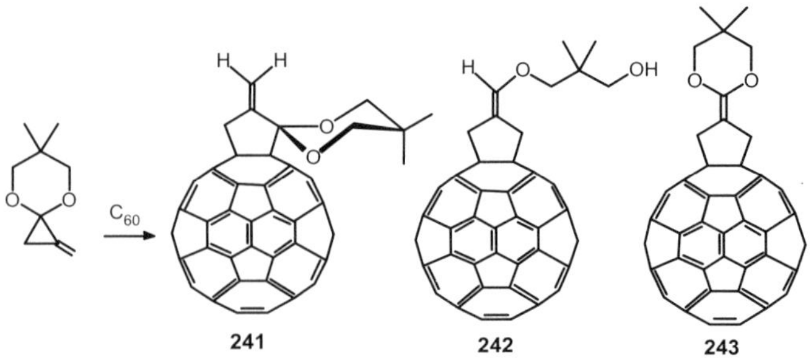

Figure 92. Trimethylmethane derivatives of C_{60}. Adapted from [250].

cannot be used, because the parent fulleropyrrolidine is unstable, probably because the generated secondary amine reacts with fullerene. Since the addend is symmetrical, there are eight possibilities for bis(*N*-methylpyrrolidine)-C_{60} formation, out of which six isomers have been isolated and characterized [255]. Unless there is a specific interest in multiple adducts, for the isolation of pure single addition products, the reaction is typically carried out until about 50–60% C_{60} is consumed, when the product mixture contains about 40–50% of the monoadduct.

Fulleroproline *tert*-butyl ester prepared by reacting glycine *tert*-butyl ester, paraformaldehyde, and C_{60} in refluxing toluene is unstable (Fig. 96) [256]. This is due to the reactivity of amines with fullerenes. Treating the reaction mixture directly with acetic anhydride results in stable acetamido derivative. When fulleroproline *tert*-butyl ester is hydrolyzed by trifluoromethanosulphonic acid, the protonated amino acid is obtained which reacts with acetic anhydride to give the *N*-acetyl derivative, **252**. The *N*-acetyl proline is a water-soluble compound and a potential candidate for synthesis of peptides containing C_{60}.

Synthesis of *N*-substituted fulleropyrrolidine without the carboxylic acid function on the fused five-membered ring is possible. The synthesis of C_{60}-fused *N*-benzylpyrrolidine

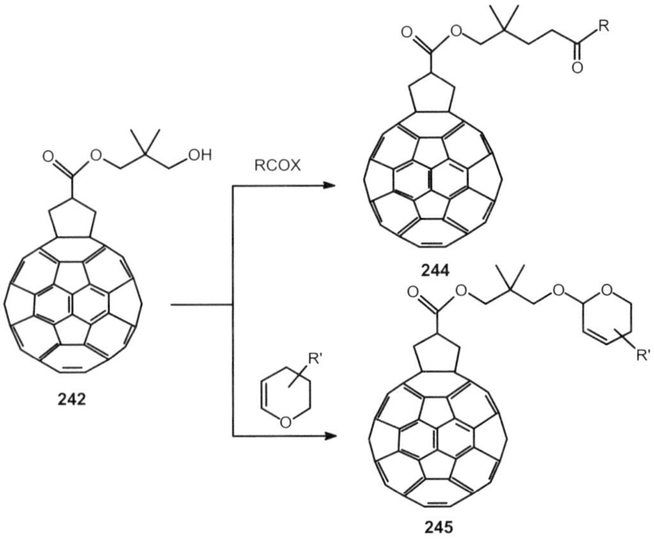

Figure 93. Further functionalization of trimethylmethane C_{60} cycloadduct. Adapted from [251].

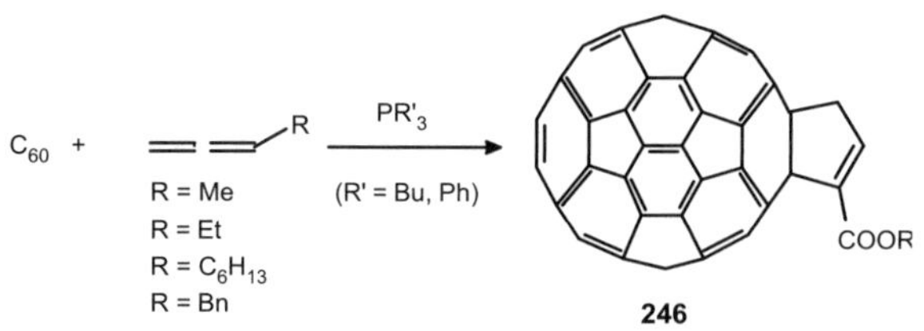

Figure 94. [2 + 3] Cycloaddition of buta-2,3-dienoates with electron-deficient alkenes. Adapted from [252].

is accomplished by generating *in-situ* *N*-benzyl azomethine ylide by desilylation of *N*-(methoxymethyl)-*N*-[(trimethylsilyl) methyl]amine catalyzed by trifluoroacetic acid (Fig. 97) [257].

A direct reaction between amino acid esters and C_{60} has also been reported [258]. Thermal reaction between glycine ethyl ester and C_{60} results in an unexpected product, **254** (Fig. 98), formed by an initial reaction of acetone present in solvent reacts with glycine ester generating a 1,3-dipolar intermediate, which adds on to C_{60}. Under photochemical conditions, the isolated product is **255**. Photochemical addition of sarcosine provides **256**; the reaction is twice as fast as the glycine ester reaction. Thus the methyl group on the nitrogen has a positive effect on the reaction rate. The *t*-butyl ester **257** is hydrolyzed by trifluoromethane sulphonic acid to yield the dicarboxylic acid **258**, which is useful for biological studies because of the appended hydrophilic function.

In addition to amino acids, under photochemical conditions, certain tertiary amines are found to add onto C_{60} leading to pyrrolidine derivatives. It is shown that C_{60} and triethylamine (TEA) undergo a characteristic photoinduced electron transfer–proton transfer reaction in a room temperature toluene solution leading to the formation of *N*-ethyl, 2′,5′-dimethylpyrrolidine derivative [259]. The C_{60}–TEA ion–pair intermediate in a nonpolar solvent is formed, which gets transformed to the cycloadduct. Similarly, photoinduced reaction of C_{60} with $RNMe_2$ (R = Ph and Me) gives adducts formed by the addition of a α-C–H bond of the amine to a 6,6-ring junction of C_{60}. On further irradiation in the presence of C_{60}, the adduct (R = Me) is converted to the pyrrolidine derivative [260]. Aziridines under thermal conditions undergo ring opening to furnish azomethine ylides and hence can be used for the synthesis of fulleropyrrolidines. Thus heating the aziridine with C_{60} in toluene leads to the pyrrolidine, **259** (Fig. 99) [254].

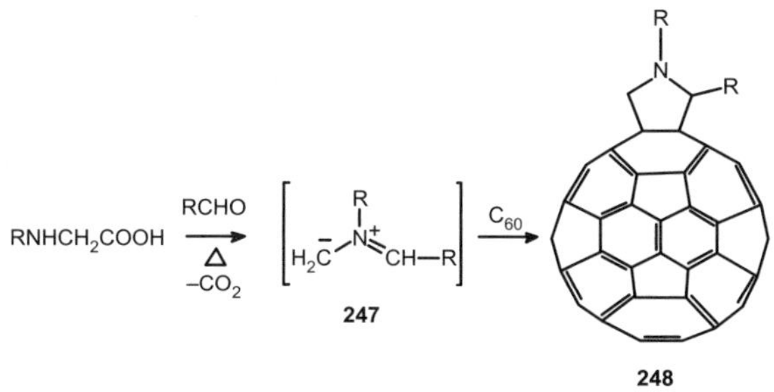

Figure 95. Addition of azomethine ylides to C_{60}. Adapted from [254].

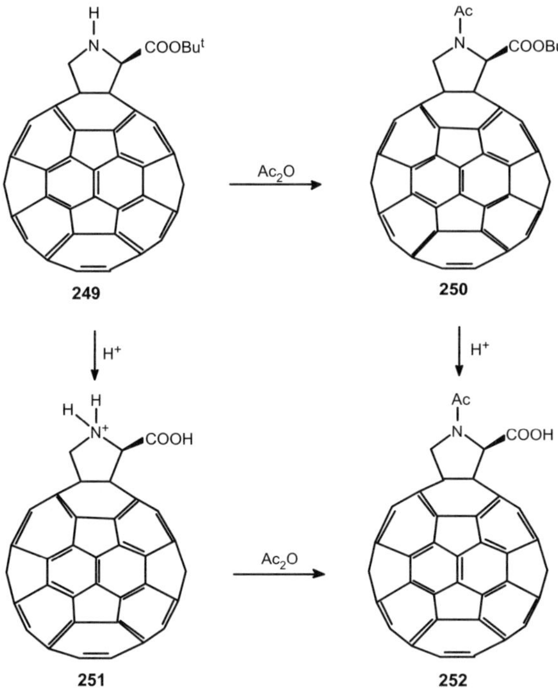

Figure 96. Synthesis of water-soluble *N*-acetylfulleroproline. Adapted from [256].

Yet another method for the synthesis of azomethine ylides involves heating 3-triphenylmethyl-5-oxazolidinone with C_{60}, which provides **260** (Fig. 100) [254]. Reacting **260** with trifluoromethanesulfonic acid gives the parent fulleropyrrolidine. Though reasonably stable at ambient temperature in dilute solutions, when the solution is concentrated it decomposed. However, addition of dansyl chloride to **261** brings about the isolation of dansyl pyrrolidine derivative **262**. The possibility of further reacting **261** through the secondary nitrogen, such as reaction with acid halides in presence of pyridine, provides an extremely versatile approach to many functionalized fullerenes, **263**.

A number of fulleropyrrolidine–bridge–ferrocene diads (**264–267**) have been synthesized by condensing the suitably functionalized ester of sarcosine and paraformaldehyde (Fig. 101) [261]. A systematic steady-state fluorescence and time-resolved flash photolysis investigation on these diads is carried out. The nature of the spacer has important consequences on the lifetime of the charge-separated states.

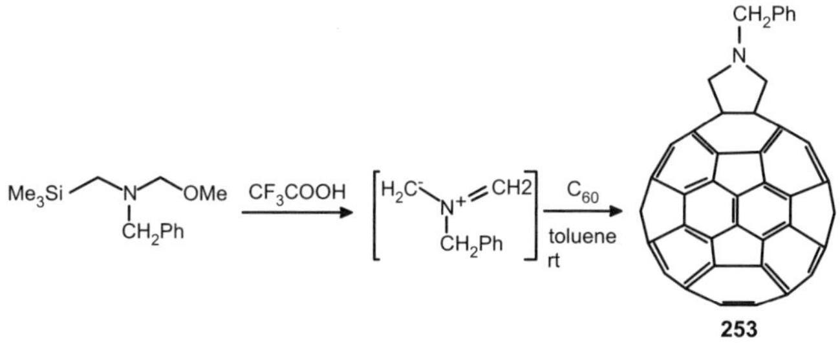

Figure 97. Cycloaddition of *N*-benzyl azomethine ylide to C_{60}. Adapted from [257].

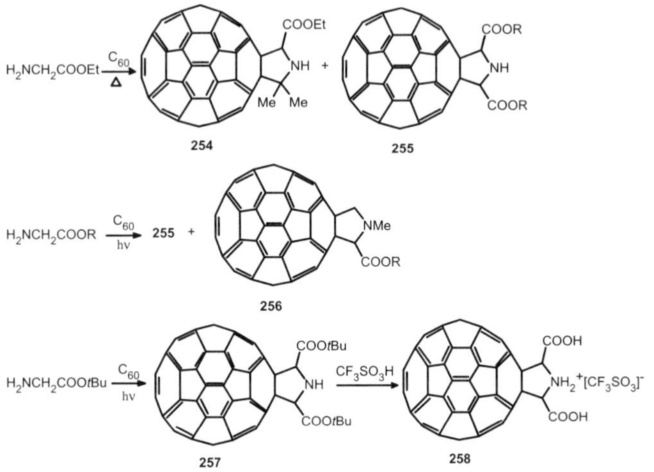

Figure 98. Reaction between amino acid esters and C_{60}. Adapted from [258].

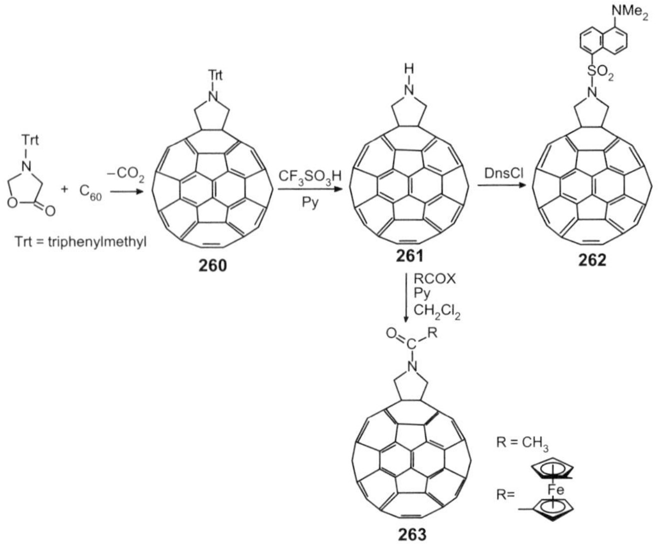

Figure 100. Addition of azomethine ylides by heating 3-triphenylmethyl-5-oxazolidinone: a versatile route to functionalized fullerenes. Adapted from [254].

Electron transfer occurs from the ferrocene donor to the fullerene acceptor upon irradiation at 532 nm.

In addition to attaching ferrocene, a number of fulleropyrrolidines are synthesized in which a strong electron donor such as tetrathiafulvalene is attached to the electronegative fullerene moiety. The substituents on the pyrrolidene are either a rigid array or a flexible arm. Such compounds are appealing as on–off components in molecular switches. The electrochemical properties of a number of these compounds are systematically studied [262]. A covalently functionalized fullerene comprising an electron donating aniline group coupled to the fullerene unit via the pyrrolidine unit is reported and its photophysical properties studied (Fig. 102) [263]. Photoinduced electron transfer to C_{60} across an extended 11-bond rigid hydrocarbon bridge leads to creation of a long-lived charge-separated state [264].

$Ru(bpy)_3^{2+}$ complex, a photoactive compound known to give energy transfers in the excited states, is coupled with C_{60}. Azomethine ylide cycloaddition to C_{60} leads to fulleropyrrolidine, which has a bipyridyl group. Reaction of the bipyridyl-functionalized derivative with $Ru(bpy)_2Cl_2 \cdot 2H_2O$ leads to a C_{60} derivative covalently linked to Ru(II) tris(bipyridine) complex (Fig. 103) [265].

The synthesis of compounds possessing stronger electron-withdrawing properties as compared to fullerene by itself is reported. The tetracyanoquinodimethane (TCNQ) and dicyanoquinonimine (DCNQ) fragments are attached to fullerene (Fig. 104) [266]. These fullerene derivatives are precursors suitable for the synthesis of intermolecular charge-transfer complexes.

Nicotine has a range of biological activities and is used as a pharmacological tool. To check the influence of C_{60} on the toxicity of nicotine a fullerene-fused natural nicotine is reported. However, the nicotine derivative is not water-soluble and hence a *N*-substituted glycine is used. It is reacted with several aldehydes leading to the formation of a number of nicotine analogs, **273** (Fig. 105) [267]. All the compounds exhibit moderate solubility in water–DMSO. One of the compounds tested is found to be active against a variety of microorganisms.

Cycloaddition of chiral azomethine ylides from chiral aldehydes or chiral amino acids is of interest by itself

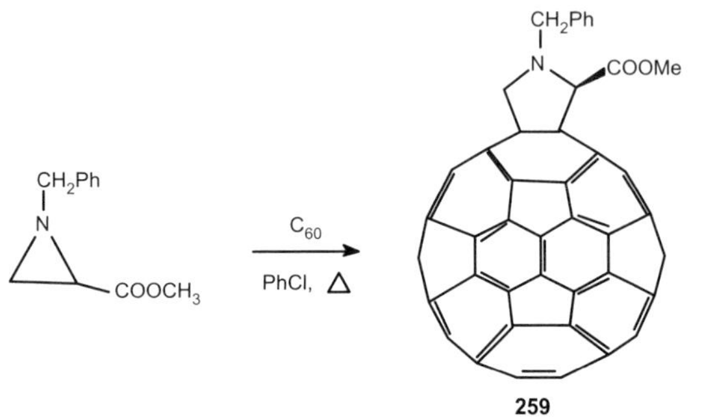

Figure 99. Addition of azomethine ylides obtained by thermal ring opening of aziridines. Adapted from [254].

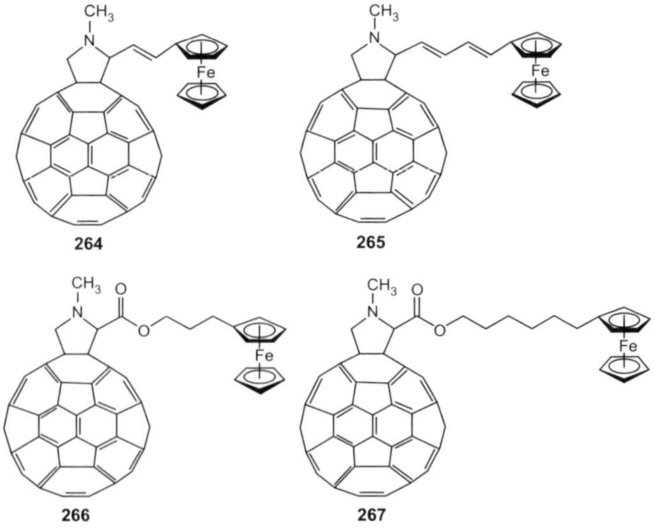

Figure 101. Fulleropyrrolidine–ferrocene diads. Adapted from [261].

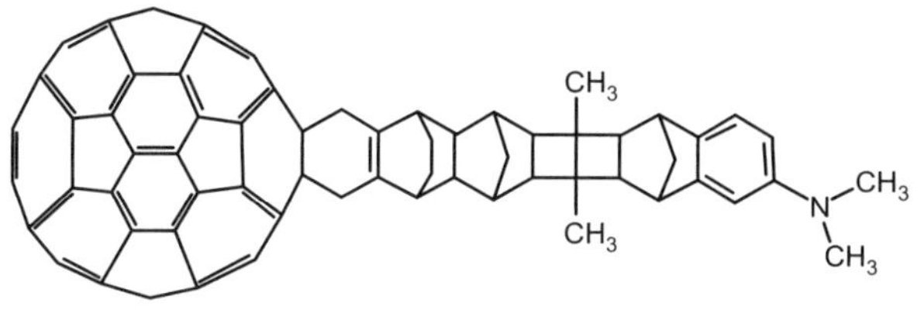

268

Figure 102. Photoinduced electron transfer to C_{60} across extended 11-bond hydrocarbon bridges. Adapted from [263].

because it allows one to prepare disymmetric compounds with unusual chirality of the fullerene derivatives. The reaction of *O*-substituted (2*S*,4*R*)-4-hydroxypyrrolidine-2-carboxylic acid, formaldehyde, and C_{60} in refluxing toluene affords two diastereomeric monoadducts (Fig. 106) [268]. *In-situ* condensation of the (+)-2,3-*O*-isopropylidene-D-glyceraldehyde with sarcosine in refluxing toluene affords chiral azomethine ylide. The addition of the latter to C_{60} also gives two diastereomeric fullerenopyrrolidines [269].

Through the functionality on the pyrrolidine ring, C_{60} is anchored to silica for the construction of a new support for HPLC. The thermal reaction of C_{60} with *N*-[3-(triethoxysilyl)propyl]-2-ethoxycarbonylaziridine affords the corresponding fullerenopyrrolidine derivative. It is then anchored to silica gel via the reaction of triethoxysilyl group to silica surface, **277** (Fig. 107) [270].

8.3.3. Miscellaneous Functionalizations via the Five-Membered Ring

A pyrrolidine ring with a C–N double bond can be attached to C_{60} leading to pyrroline-fused C_{60} by 1,3-dipolar cycloaddition of nitrile ylides. The addend ylide can be generated by the photochemical cleavage of azirine derivatives. The reaction of C_{60} with a 10-fold excess of 2,3-diphenylazirine affords the corresponding pyrrole [271]. The reaction is also useful for the synthesis of monoaryl-substituted products (Fig. 108) [272]. The 1,3-dipolar cycloaddition of 1,3-diphenyl nitrilimine generated *in-situ* gives a dinitro heterocycle-fused C_{60}. In the presence of triethylamine a solution of C_{60} and *N*-(α-chlorobenzylidine)-*N*′-phenyl hydrazine gives the desired adduct (Fig. 109) [273]. Intramolecular charge-transfer interaction between the benzene rings and the C_{60} sphere is studied.

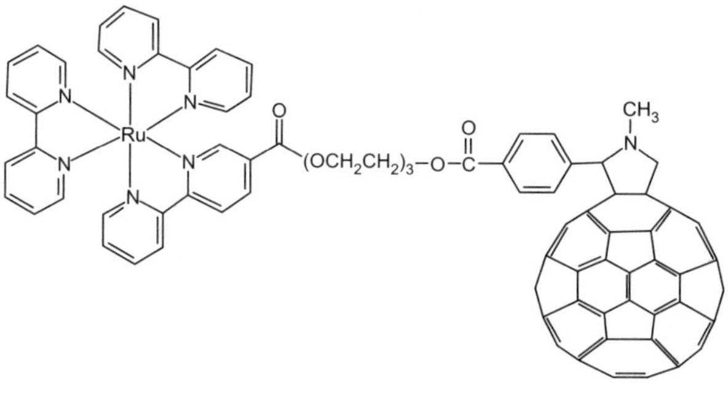

269

Figure 103. Attaching ruthenium complex to C_{60}. Adapted from [265].

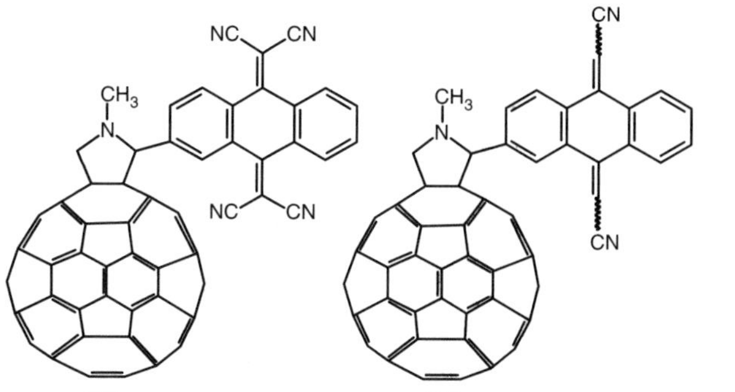

271 **272**

Figure 104. Attaching electron-withdrawing groups to C_{60} via the pyrrolidine ring. Adapted from [266].

Nitrile oxide generated *in-situ* adds on to C_{60} in a dipolar fashion leading to the formation of isoxazoline. To a solution of C_{60} in toluene addition of triethylamine leads to the formation of a number of isoxazolines (Fig. 110) [274]. Since a number of oximes are available with various substituents, the reaction offers considerable possibilities for functionalization of C_{60}. 1,3-Dipolar cycloaddition of pyrazolidinium ylides to C_{60} is reported [275]. The obtained bicyclic γ-lactam structures are not stable enough for further reaction on the side-chain. The synthesis of fullerene-fused 2-imino-1,3-thiazolidines is carried out. C_{60} reacts with the heterocyclic masked 1,3-dipoles, the 5-imino-1,2,4-thiadiazolidine-3-ones furnishing corresponding fullerene thiazolidines (Fig. 111).

8.4. [2 + 4] Cycloaddition Reactions

8.4.1. Addition of Simple Dienes

In the [2 + 4] cycloaddition reaction the fullerene nucleus acts as a reactive dienophile [276]. This is due to the fact that inter-five-membered ring bonds are electrophilic in nature and fulvenoid. Fullerene is substantially more reactive than typical dienophiles, the reactivity being comparable with those of maleimides [277]. Dienes with electron-donating substituents readily undergo Diels–Alder reaction with the fullerenes. Up to six molecules of diene can be added to both C_{60} and C_{70} across the six pyracylene units. A number of classical dienes such as anthracene, furan, cyclopentadiene, nitrile oxide, isobenzofuran, *o*-quinodimethanes, etc. undergo [2 + 4] cycloaddition reaction with C_{60} to give the corresponding cycloadducts. C_{60} addition takes place across the junction of two six-membered rings rather than

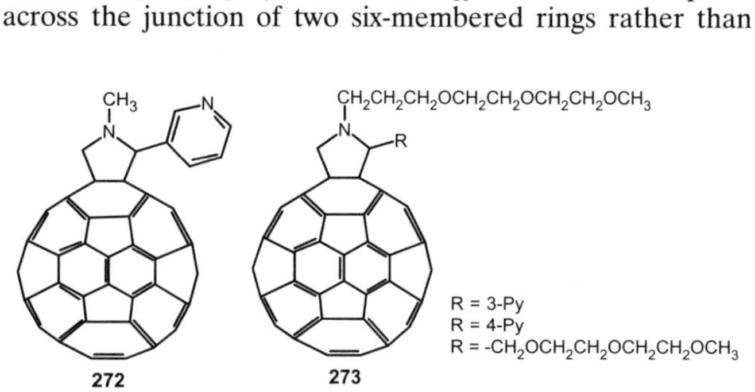

272 **273**

Figure 105. Synthesis of fullerene-fused nicotines. Adapted from [267].

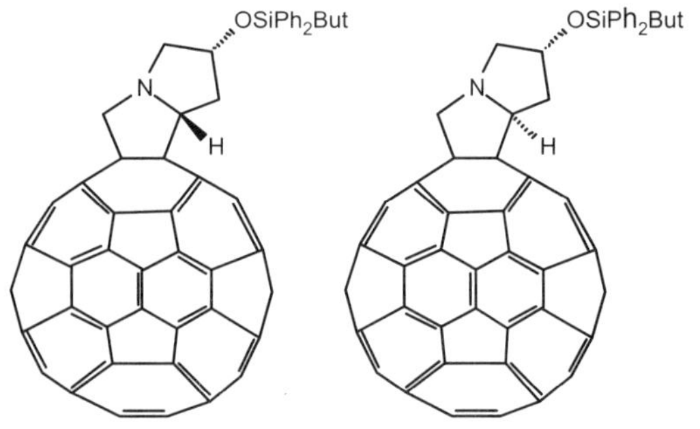

274 275

Figure 106. Attaching chiral azomethine ylides to C_{60}. Adapted from [268].

at the junction of five- and six-membered rings. In reaction of C_{60} with excess cyclopentadiene, it is not possible to isolate the cycloadduct. However, in using freshly distilled cyclopentadiene, a monoadduct is isolated in 74% yield [278]. The cycloadduct undergoes ready retro-Diels–Alder reaction. However, reaction with a pentamethylcyclopentadiene leads to a stable cycloadduct [279]. Hydrogenating the generated double bond on the diene moiety can prevent the reversal of the reaction. The hydrogenated product becomes thermally stable [280]. Cyclohexadiene exhibits somewhat lower reactivity than cyclopentadiene. Again, hydrogenation of the lone double bond on adduct prevents retro-Diels–Alder reaction [281]. Like reaction with cyclopentadiene the reaction between anthracene and C_{60} is facile but the adducts are not be isolated and characterized [282]. The multiple addition product formation by the addition of excess cyclopentadiene or anthracene to C_{60} is reported and the polyadducts are isolated [283] (Fig. 112). Up to six molecules of cyclopentadiene or anthracene could be added to C_{60} moiety. Following this a number of works reported several cycloadducts of C_{60} with anthracene, cyclopentadiene, 1,3-diphenylisobenzofuran, and 1,3-dimethyl butadiene. With anthracene both mono- (**284**) and bis-adducts (**285**) are isolated and characterized by NMR, UV-vis, IR, mass spectroscopy, thermogravimetric analysis, and EPR studies [284]. In all the reactions studied, the cycloadduct

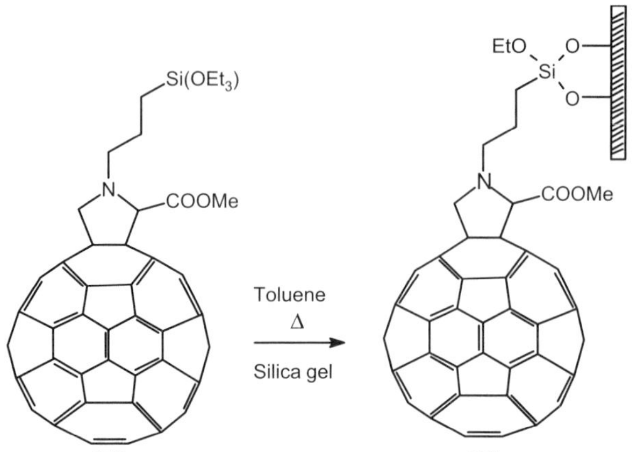

276 277

Figure 107. Anchoring C_{60} to silica surface via the pyrrolidine ring. Adapted from [270].

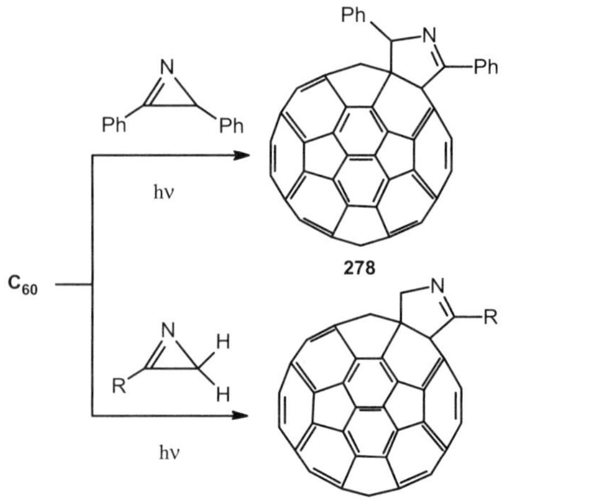

279

Figure 108. Synthesis of fulleropyrrolines by 1,3-dipolar cycloaddition of azirines. Adapted from [272].

formed revert back to C_{60}. Cycloaddition experiments with anthracene are particularly attractive when it is found that at room temperature anthracene reacts with C_{60} under irradiation of the reaction mixture. Irradiation of a solution of equimolar amounts of C_{60} and anthracene in benzene affords mono- and bisadducts along with anthracene dimer. Presence of a methyl group on position 9 of anthracene prevented dimerization of anthracene [285], which suggested a stepwise mechanism for cycloaddition involving photoinduced electron transfer. However, when 9,10-DMA is used, cycloaddition takes place by a concerted mechanism [286]. High-speed vibration milling of the mixture of C_{60} and DMA leads to efficient formation of the mono-adduct. However, when dissolved in a suitable solvent, the cycloadduct undergoes ready dissociation with a half-life of about 2 hours [287].

Higher linear acenes like pentacene also add on to C_{60} at the central C(6), C(13) atoms when the mixture is refluxed in toluene [288]. Under solid phase milling conditions both mono- and bisadducts are formed along with a *trans*-bisfullerene adduct (Fig. 113) [287]. Use of two equivalents of C_{60} increases the yield of the bisfullerene adduct. When positions C(6) and C(13) are substituted with aryl groups, *cis*-bisfullerene adduct is the major product [289].

Cycloaddition of tetracycline involves the initial electrocyclic isomerization into bicyclooctatriene, the diene component on the six-membered ring reacting with C_{60} (Fig. 114) [290]. As in the case of cyclohexadiene, the reaction is reversible, the product reverting back to the starting materials on heating. The product, however, is stable at room

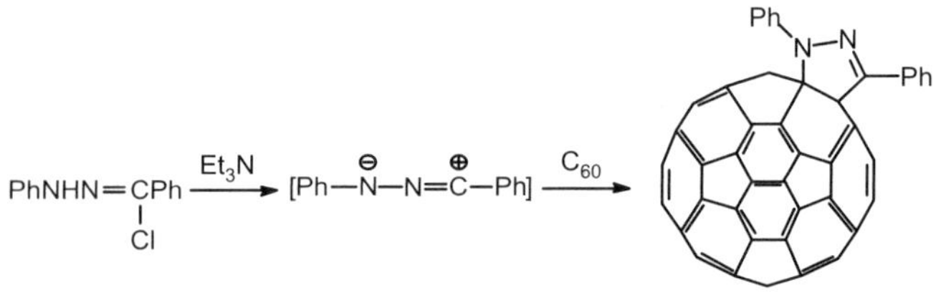

Figure 109. Synthesis of fulleropyrazolines. Adapted from [273].

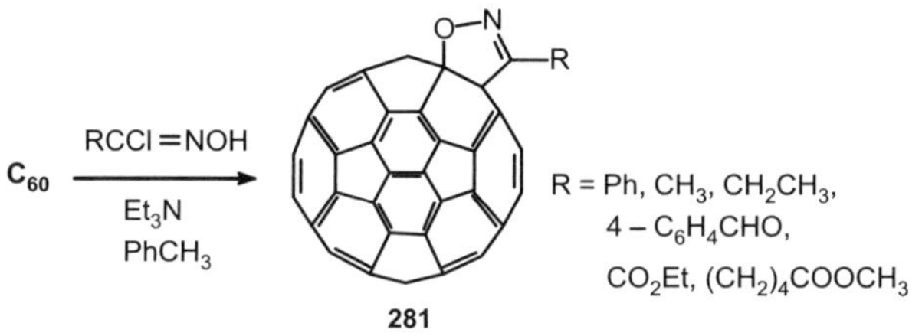

Figure 110. Synthesis of isoxazoline derivative of C_{60}. Adapted from [274].

temperature and hence further reaction on the reactive double bond on the cyclobutenemoiety has been carried out.

8.4.2. Functionalizations via Six-Membered Ring

A variety of Diels–Alder reactions performed on C_{60} lead to introduction of functional groups on an attached six-membered ring. The cycloaddition product **302** resulting from the Diels–Alder reaction of 1,3-butadiene and C_{60} is unstable toward air and light, affording hydroperoxide, **303** (Fig. 115) [291]. This instability arises out of efficient photoexcitation of C_{60} to a triplet excited state and subsequent energy transfer to oxygen from air to generate singlet oxygen ($^1O_2$). Reaction of **302** with $^1O_2$ affords the allylic alcohol **304** after reduction with PPh_3. A number of 1,2-dihydrofullerenes obtained by Diels–Alder reaction of the functionalized 1,3-dienes with C_{60} are known. This strategy leads to the synthesis of a rich variety of oxygenated systems after reduction of the corresponding hydroperoxide with PPh_3 [292].

Diels–Alder reaction of 2-trimethylsilyl-1,3-butadiene with C_{60} is an efficient reaction that leads to further functionalization of C_{60} via the six-membered ring. Cycloaddition of the butadiene with C_{60} leads to the ketone **305** which is reduced to the 1,9-(4-hydroxycyclohexano)fullerene **306**. The compounds **305** and **306** bear the versatile ketone and alcohol functionalities, respectively, which permits the preparation of a variety of other, water-soluble C_{60}-α-amino acid derivatives, **307** and **308** (Fig. 116) [293]. A bromoacetate derivative of **305** is used for a C_{60}-linked deoxy-oligonucleotide [294]. This C_{60}-oligonucleotide conjugate is hybridized to a complementary single-stranded DNA. This oligonucleotide conjugate reacts with light and oxygen to cleave only guanosines in the single stranded region, which are closest to C_{60}, by a single-electron transfer mechanism between guanosine and $^3C_{60}$.

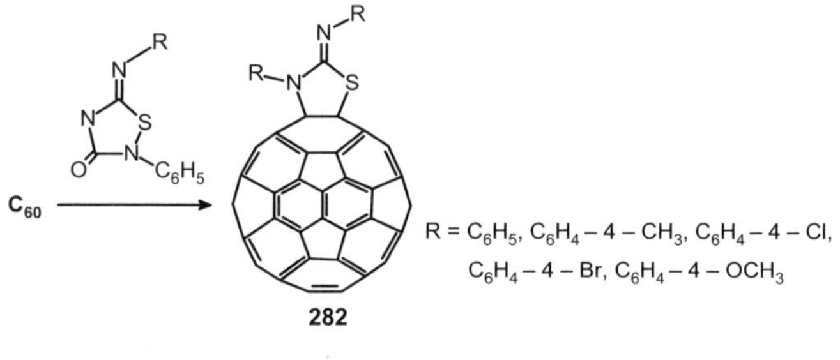

Figure 111. Reaction of C_{60} with 5-imino-1,2,4-thiadiazolidine-3-ones. Adapted from [275].

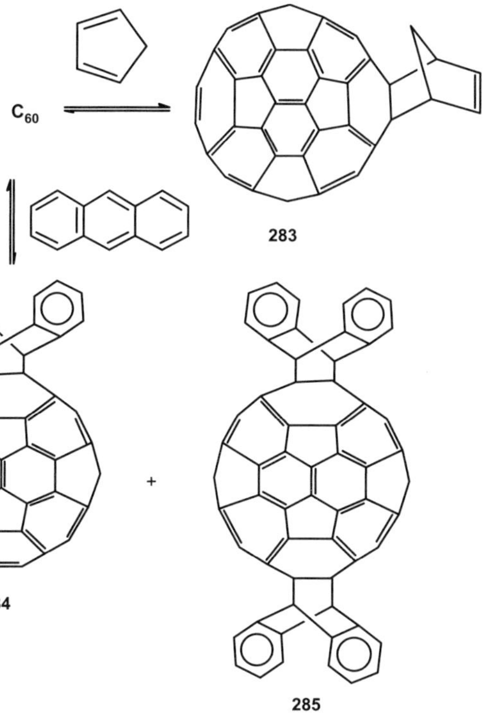

Figure 112. Cycloaddition of cyclopentadiene and anthracene to C_{60}. Adapted from [283].

Several chiral nonracemic C_{60} derivatives are synthesized by cycloaddition of C_{60} with 1-methoxy-3-trimethylsilyloxy-1,3-butadiene and 3-trimethylsiloxy-1,3-pentadiene [295]. The cycloadducts lead to ketones **309** and **312**, respectively, after hydrolysis (Fig. 117). A related cyclohexanone-C_{60} adduct **311** has a temperature dependent NMR spectrum due to slow cyclohexanone ring inversion but such inversion is prevented for substituted ketones **309** and **312**.

A facile photochemically promoted rearrangement of the diene **313** leads to the synthesis of a stable bridged bis-fulleroid **315** (Fig. 118) [296]. This process occurs via the initial [4 + 4] photoadduct **314**, which undergoes a thermally

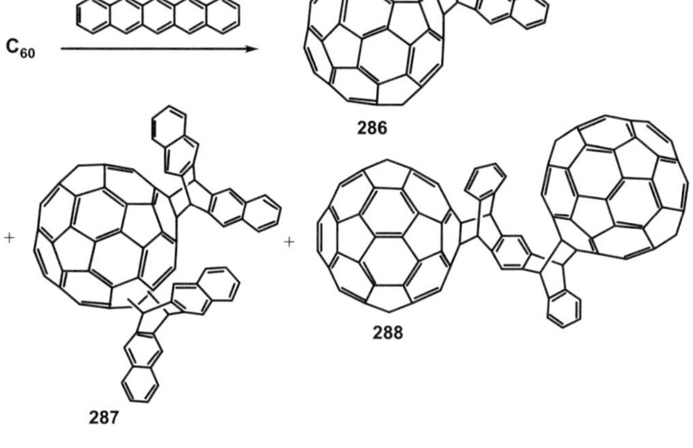

Figure 113. Cycloaddition of pentacene to C_{60}. Adapted from [287].

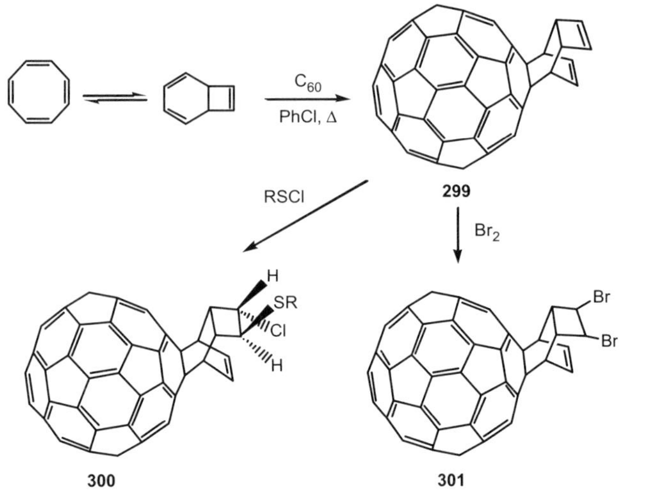

Figure 114. Addition of cyclotetracene to C_{60}. Adapted from [290].

allowed [2 + 2 + 2] cycloreversion to afford a bridged bisfulleroid **315**, which also forms a Co(III) complex readily. In an alternate route to synthesis of **315** in higher yields, a twofold excess of 1-(trimethylsilyl)-1,3-butadiene is added to C_{60} leading to the silly ether **316**, which on acidification furnishes the allylic alcohol **317**. Photolysis of the allylic alcohol affords good yields of the bridged bisfulleroid.

A diastereomeric alcohol **320** is synthesized which is a C_{60}-based inhibitor of HIV-1 protease. The reaction involves a "double-Michael" addition of a dienolate anion to C_{60}. Protonation of **318** results in the ketone **319**, which on reduction furnishes the desired diastereomeric alcohol **320** (Fig. 119) [297]. Synthesis of the cyclohexane ring to C_{60} with a sterically crowded 1,4-position is accomplished. This is a remarkably facile method for the synthesis of sterically crowded fullerene derivatives. Various enones are subjected to the dienolate addition to give bicyclic ketone of C_{60} (**321**–**325**) (Fig. 120) [297].

Crown ether can be appended to C_{60} via diene carrying the crown ether moiety. Thus a crown ether bearing the diene component adds on to C_{60} giving a stable cycloadduct. Again a monoadduct is obtained at room temperature and

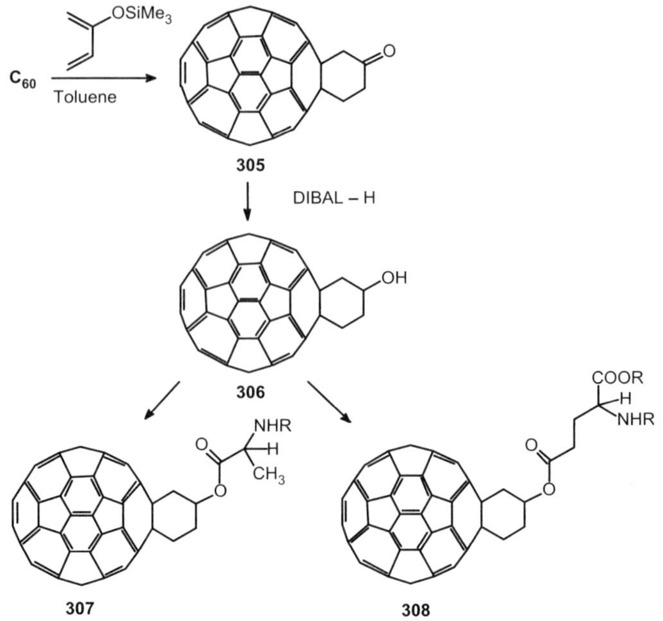

Figure 116. Appending α-amino acids to C_{60} via the hydroxycyclohexano-buckminsterfullerene. Adapted from [293].

the reaction yields a 1:1 mixture of mono- and bisadduct at higher temperature (Fig. 121) [298].

Multiple cycloaddition to C_{60}, as seen in case of cyclopropanation reaction, leads a mixture of regiosymmetric products. Using saccharides as template molecules two boronic acid groups are introduced into C_{60}. A number of saccaride–boronic acid 1:2 complexes are reacted with C_{60}. It is found that the regioselectivity of bisaddition depends on the saccaride structure [299]. Regioselective sixfold [2 + 4] cycloaddition of 1,3-diene to C_{60} is reported (Fig. 122) [300]. A sixfold Diels–Alder reaction of a 1,3-diene at the six symmetry equivalent 6,6-ring junctions results in a chromophore structure of eight completely symmetry-equivalent, strained benzene rings (**327**), which is apparently a multiply bridged cyclophane.

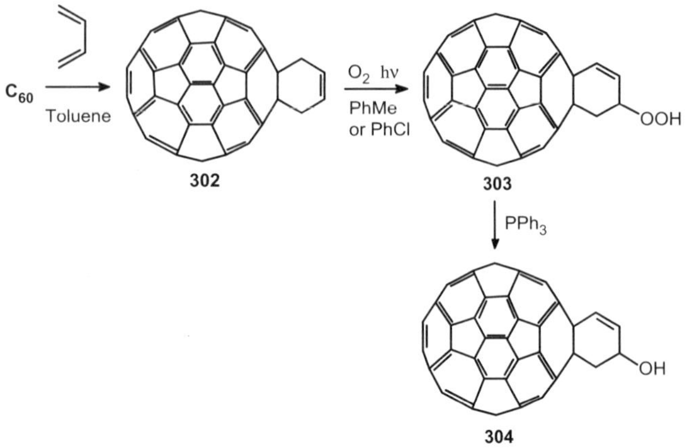

Figure 115. Ene reaction of cyclohexenobuckminsterfullerene. Adapted from [291].

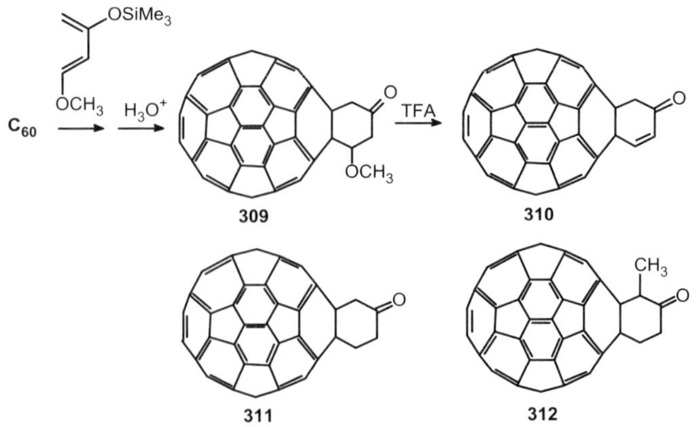

Figure 117. Synthesis of chiral cyclohexanone–C_{60} adducts. Adapted from [295].

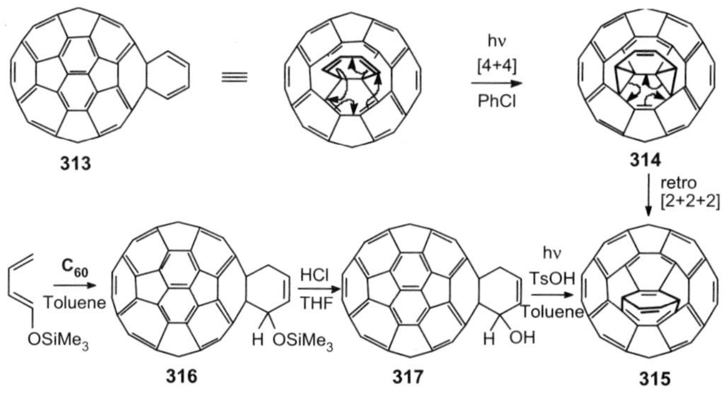

Figure 118. Synthesis of a bridged bisfulleroid. Adapted from [296].

8.4.3. [2 + 4] Cycloaddition via o-Quinodimethane Intermediate

A number of [2 + 4] cycloadditions on C_{60} have been carried out using *o*-quinodimethane precursors. Since *o*-quinodimethane (**328**) is unstable and undergoes first aromatization, it is generated *in-situ* for which various methods are known (Fig. 123). The reaction of *o*-quinodimethane and its derivatives lead to stable products, because the cleavage of adducts is an uphill reaction due to the generation of unstable *o*-quinodimethane species.

Cycloadduct becomes thermally stable and the reaction irreversible when the generated double bond in adduct is a part of the aromatic ring, for example, the cycloaddition of isobenzofuran to C_{60}. When C_{60} reacts with isobenzofuran generated *in-situ* from 1,4-dihydro-1,4-epoxy-3-phenylisoquinoline, the corresponding cycloadduct does not undergo cycloreversion [301]. Synthesis of adduct by [2+4] cycloaddition of C_{60} with a norboranenone leads to a stable adduct **331** by elimination of CO (Fig. 124) generating an aromatic ring. Hence the adduct cannot revert back to the starting materials by retro-Diels–Alder reaction.

o-Quinodimethane is routinely generated by 1,4-elimination of Br_2 from 1,2-bis(bromomethyl)benzenes. From a bromomethyl bicyclooctadiene, the *o*-quinodimethane intermediate is obtained by dehydrohalogenation on treating with potassium *tert*-butoxide. The adduct is unstable and it eliminates a molecule of HBr to produce fullerocyclohexadiene derivative, which is too unstable and decomposes further [302]. However, a number of reactions with C_{60} lead to stable monoadducts [303]. *o*-Quinodimethane can be used as a vehicle for the introduction of a multitude of reactive functional groups. Reaction of 4-amino-*o*-quinodimethane with C_{60} affords a thermally stable and chemically reactive adduct [304]. The synthesis of carboxy-substituted adducts of C_{60} and *o*-quinodimethane is reported and the products are used for subsequent reactions [305]. Following this strategy a variety of C_{60} derivatives are synthesized to study their biological activities and material properties. For example, through the reactive acid moiety a molecule of cholesterol is attached (**334**) for the purpose (Fig. 125).

Heating 1,2-bis(bromomethyl)-4,5-dimethoxybenzene with C_{60} in toluene in the presence of tetrabutylammonium iodide leads to functionalized C_{60} readily (Fig. 126) [306]. As stated earlier, the [2 + 4] cycloaddition reactions involving *o*-quinodimethane intermediate generated *in-situ* are irreversible, because *o*-quinodimethane undergoes fast aromatization.

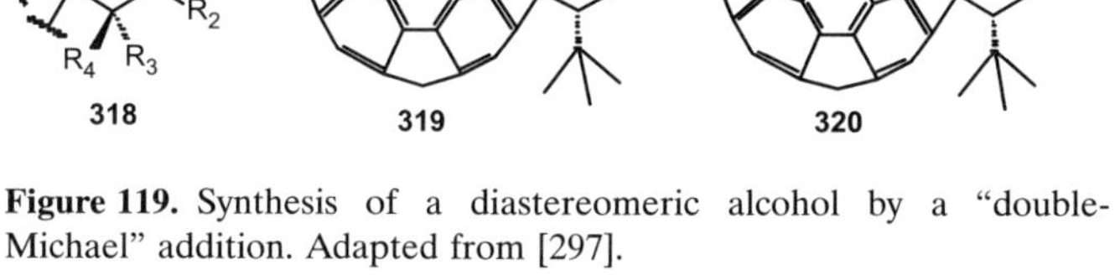

Figure 119. Synthesis of a diastereomeric alcohol by a "double-Michael" addition. Adapted from [297].

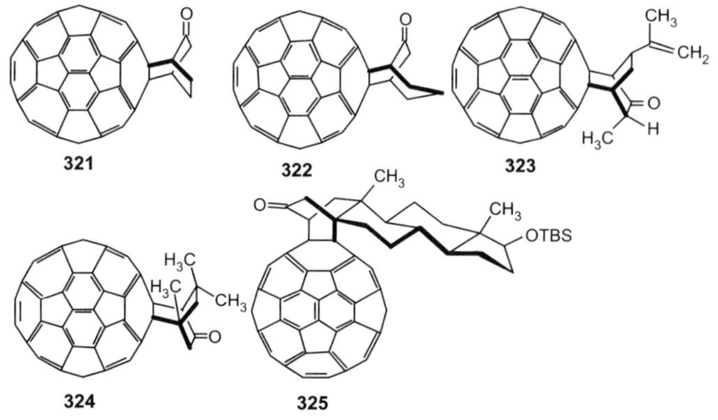

Figure 120. Dienolate addition to C_{60} giving bicyclic ketones. Adapted from [297].

Benzocyclobutene as o-Quinodimethane Precursor

Benzocyclobutenes are interesting precursors for *o*-quinodimethane intermediates. The addition reaction of fullerene with *o*-quinodimethane-type species, using the thermally allowed conrotatory [$\sigma 2 + \pi 2$] electrocyclic ring opening of benzocyclobutane homologs, is reported (Fig. 127) [307].

Thermolysis of the benzocyclobutenes also leads to generation of *o*-quinodimethanes *in-situ*. For example, heating a mixture of C_{60} and a benzocyclobutene derivative leads to facile synthesis of the adduct (Fig. 128) [308]. Both mono- and bisadducts are formed, the major product being the monoadduct.

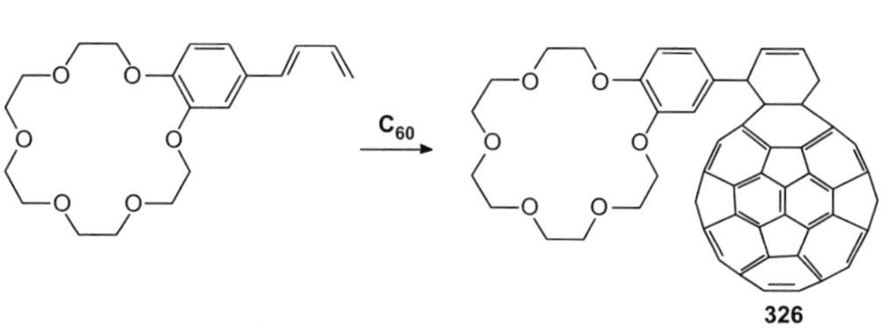

Figure 121. Attaching a crown ether to C_{60}. Adapted from [298].

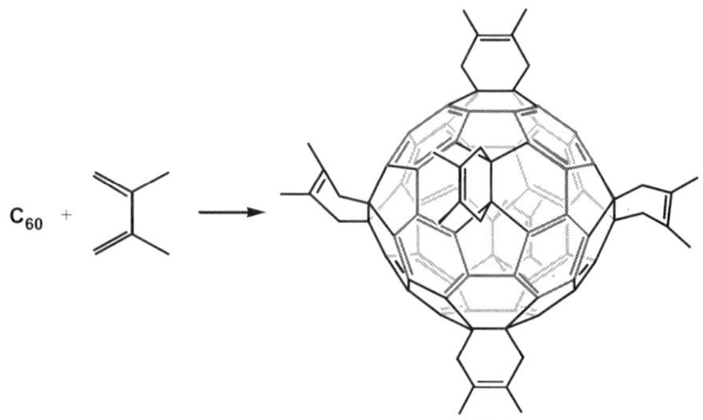

327

Figure 122. Regiosymmetric sixfold [2 + 4] cycloaddition of diene leaving behind an aromatic framework of eight benzene rings. Adapted from [300].

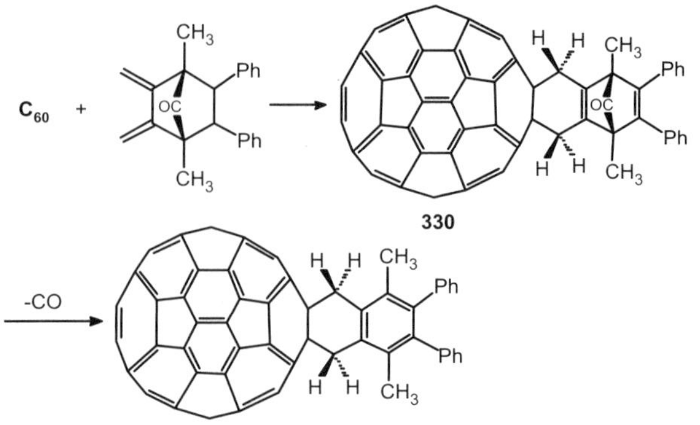

331

Figure 124. Addition of *o*-quinodimethane precursor for synthesis of stable cycloadduct by aromatization of adduct upon CO elimination. Adapted from [282].

Reactive functional groups on benzocyclobutene moiety lead to *o*-quinodimethane carrying a center for further functionalization. Thus, refluxing a mixture of C_{60} and benzocyclobutenol in toluene affords a fullerene derivative with a hydroxy or ether function on the cyclohexane ring (**340–343**) (Fig. 129) [309].

Stable Diels–Alder adducts of C_{60} carrying reactive functions are also synthesized from benzocyclobutenols. For example, heating 3-methoxy-3-phenylbenzocyclobutene with C_{60} yields the cycloadduct stable upon prolonged reflux in toluene (Fig. 130).

The functional group on benzocyclobutene moiety can be further functionalized starting from an adduct, or the starting material itself can be functionalized before cycloaddition. For example, a benzocyclobutene carrying a carboxy group is allowed to react with azacrown ether by 1,3-dicyclohexylcarbodiimide (DCC) coupling and is then reacted with C_{60} (Fig. 131) [310]. Both mono- and bisadducts are formed, the monoadduct being formed as the major product.

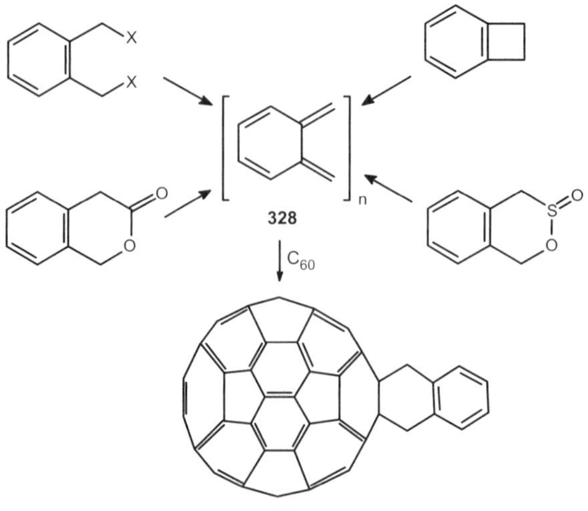

329

Figure 123. Generation of *o*-quinodimethane *in-situ* for cycloaddition to C_{60}.

8.4.4. Donor–Acceptor Systems from [2 + 4] Cycloadditions

A number of donor–acceptor systems from [2 + 4] cycloaddition reactions of fullerenes are known. Fullerene derivatives of quinones as a donor–acceptor system involving fullerene and quinone parts are known. Such a redox system containing fullerene and *p*-benzoquinone (**347**) is synthesized from 3,6-dimethoxybenzocyclobutane and C_{60} (Fig. 132) [311]. Removal of the methyl group leads to the *p*-benzoquinone adducts. The fulleroquinones are efficient redox-active materials. The synthesis of C_{60}-based electron acceptors with attached tetracyano-*p*-quinondimethane and dicyano-*p*-quinondiimine (DCNQI) moieties are reported. These derivatives can accept up to eight electrons in solution [312].

Use of sultines as the precursor for the generation of *o*-quinodimethane by extrusion of SO_2 is quite efficient.

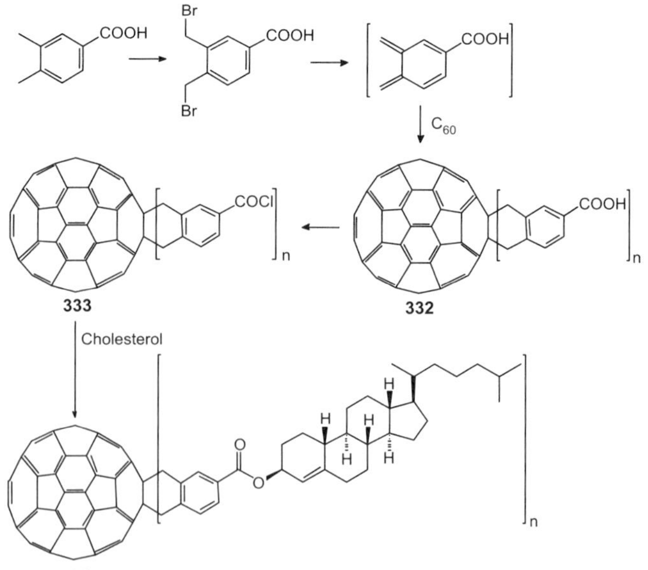

334

Figure 125. *o*-Quinodimethane generation by 1,4-elimination of Br_2 from 1,2-bis(bromomethyl)benzenes, and subsequent attachment of cholesterol. Adapted from [305].

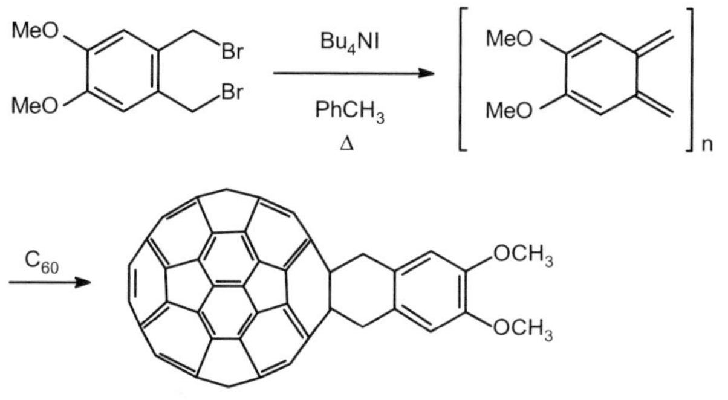

335

Figure 126. Introduction of functional groups via *o*-quinodimethane intermediate. Adapted from [306].

The facile formation of C_{60} adducts from sultines via a Diels–Alder reaction for functionalization of C_{60} is reported. The reaction of various substituted sultines with C_{60} leads to the respective cycloadducts. The substituted 1,4-dialkoxybenzenes are converted into the corresponding *p*-benzoquinone moiety, which are interesting precursors for the preparation of stronger organofullerene acceptors derived from TCNQ and DCNQI (Fig. 133) [313].

Tetrathiafulvalene (TTF) is known as one of the strongest donors used in the design of donor–acceptor systems. Cycloaddition reaction of C_{60} to 2-(thi)oxo-4,5-bis(methylene)-1,3-dithioles is a route to the bislinking of tetrathiafulvalene to C_{60}. The main step to the synthesis of the target molecules lies in the Diels–Alder reaction of C_{60} with the *in-situ* generated *o*-quinodimethanes. To generate the 2-oxo intermediate, two alternative pathways—the thermal rearrangement of *S*-propargyl xanthate and the reductive 1,4-elimination of the corresponding 2-oxo-4,5-bis(bromomethyl)-1,3-dithiole—are adopted (Fig. 134) [314]. The first route furnishes the mono-, bis-, and trisadducts and the second route only a monoadduct.

Through a sultine TTF is attached to C_{60}. The Diels–Alder reaction between C_{60} and tetrathiafulvalene sultine is facile which leads to the synthesis C_{60}-TTF diad, **355** (Fig. 135) [315]. Analogs of tetrathiafulvalene are also synthesized and attached via the *in-situ* generated *o*-quinodimethane moiety. The resulting donor–acceptor systems are photoactive compounds (Fig. 136) [316].

Like sultines, the sulfolene derivatives of oligothiophenes react with C_{60} on heating. A sulfolene derivative of thiophene containing up to five thiophene rings per molecule on reaction with C_{60} furnishes the corresponding fullerene–oligothiophene, **356** (Fig. 137) [317].

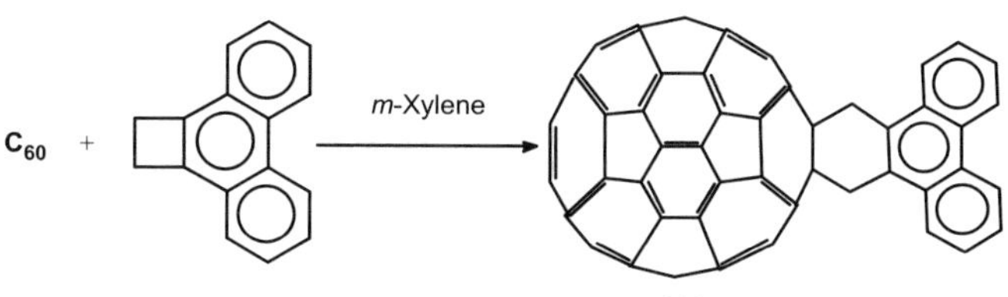

336

Figure 127. Benzocyclobutene as precursor for *o*-quinodimethane. Adapted from [307].

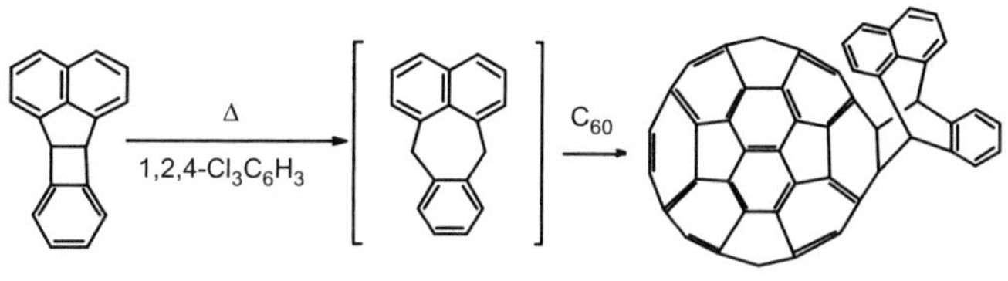

337

Figure 128. Generation of *o*-quinodimethane by thermolysis of a benzocyclobutene. Adapted from [308].

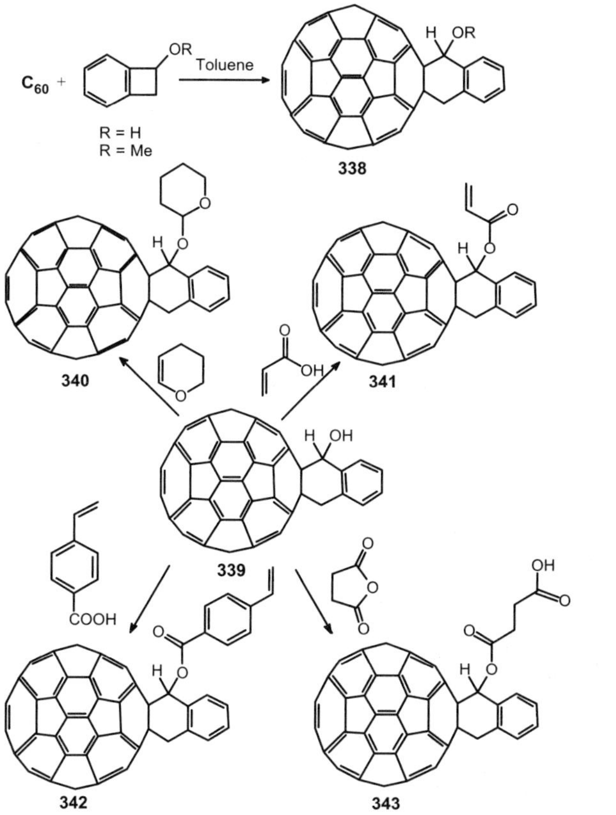

Figure 129. Introduction of a reactive functional group on the cyclohexane ring by reaction of C_{60} with benzocyclobutenol. Adapted from [309].

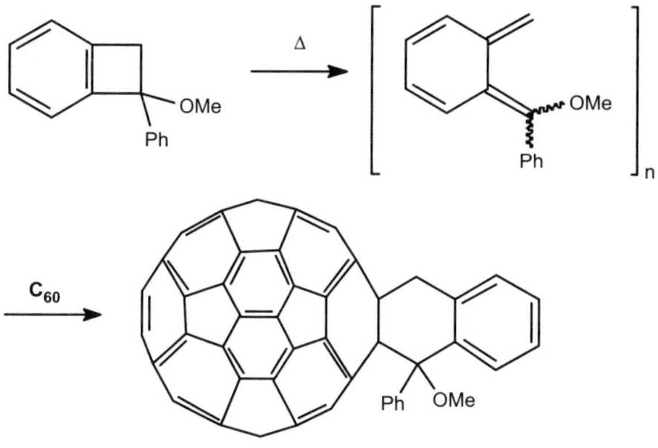

Figure 130. Synthesis of stable cycloadducts from benzocyclobutene carrying reactive functions.

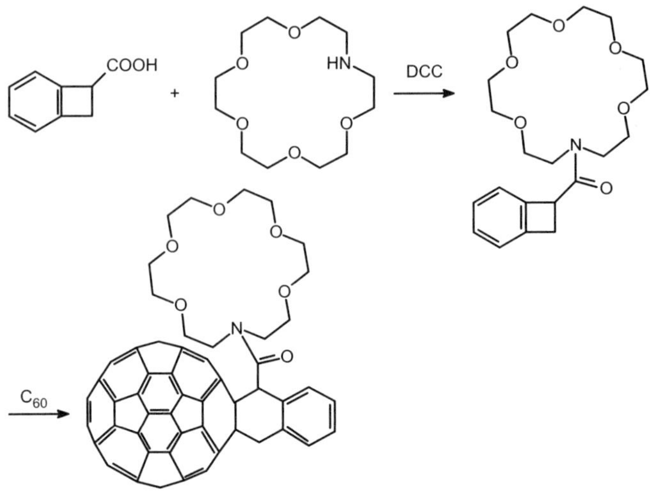

345

Figure 131. Reaction of benzocyclobutene carrying an azacrown ether ring with C_{60}. Adapted from [310].

8.4.5. Attaching Heterocyclic Rings by [2 + 4] Cycloaddition

Heterocycle-fused C_{60} derivatives are conveniently synthesized by the Diels–Alder reaction of C_{60} with suitable dienes carrying the heteroatoms like *N*, *O*, and *S*. The reactions can be irreversible if the diene component is *o*-quinodimethane intermediate. Thus a nitrogen-containing heteroring is fused to C_{60} by the reaction of an *o*-quinodimethane generated by decarboxylation of a benzoxazinone (Fig. 138) [318].

The heterocyclic ring can have more than one heteroatom by a suitable choice of the diene moiety. With a carboethoxy-substituted azabutadiene and thio-substituted azabutadiene C_{60} undergoes hetero-Diels–Alder reaction to give derivatives fused with cyclic amino acid and thiourea, respectively (Fig. 139). When the diene itself is reacted a pyrrolidine derivative is formed [319].

Heterocyclic rings can be appended to fullerenes following reactions other than cycloaddition. For example, a 1-bromoisoquinoline derivative of C_{60} is synthesized by reaction of BrCN with $C_{60}Ph_5Cl$ (Fig. 140). The reaction results in chlorine loss and in the formation of a bromine-substituted CN bridge between the cage and the ortho position of the adjacent phenyl ring [320]. Nucleophilic substitution of the bromine by hydroxide leads to the isoquinolone derivative.

As in the case of introduction of a nitrogen atom, an oxygen atom can be introduced by reacting fullerene with a suitable *o*-quinodimethane derivative. Thermolysis of *o*-hydroxy benzyl alcohol leads to *o*-quinomethide, which reacts with C_{60} giving a oxygen-containing heterocycle, **362** (Fig. 141) [321].

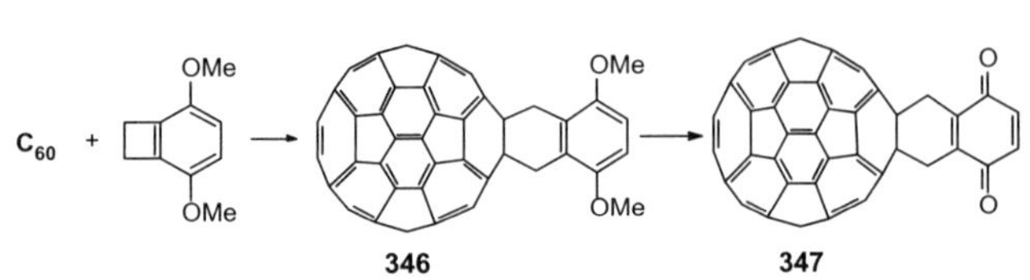

346 347

Figure 132. Synthesis of redox system containing fullerene and *p*-benzoquinone. Adapted from [311].

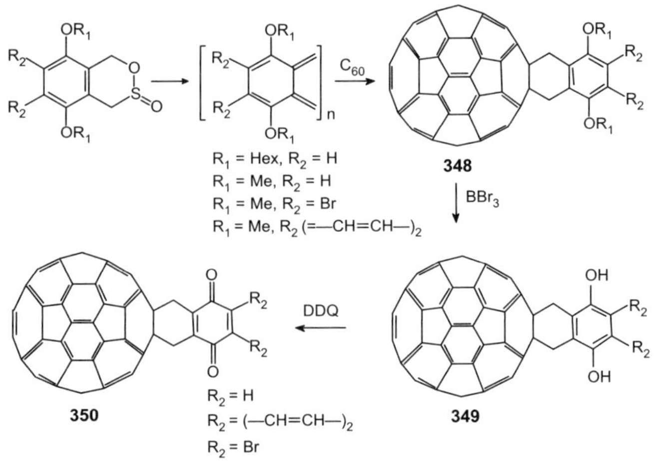

Figure 133. *o*-Quinodimethanes from sultines by extrusion of SO_2. Adapted from [313].

To introduce a sulfur atom, the thio analog of benzocyclobutene is used, which first generates the benzothietane. Later on reaction with C_{60} leads to the fullerothiochroman, **363** (Fig. 142) [322]. Sulfur containing heterorings can be fused to C_{60} by hetero-Diels–Alder reaction with an α,β-unsaturated thiocarbonyl compound, prepared *in-situ* from thioacrylamide and acyl chloride (Fig. 143) [323].

Attaching a Heterocyclic Ring to Fullerene via a Cyclohexane Linker Addition of a cyclobutene derivative carrying a useful functional group is a convenient approach to introduction of a functional group on the generated cyclohexane ring. Following this protocol compound **365** is synthesized, which has a great deal of potential as a starting point to various interesting functionalities. Its diketone is a direct precursor of different 1,4-diazine-fused fullerenes. The synthesis of **365** is easily carried out by

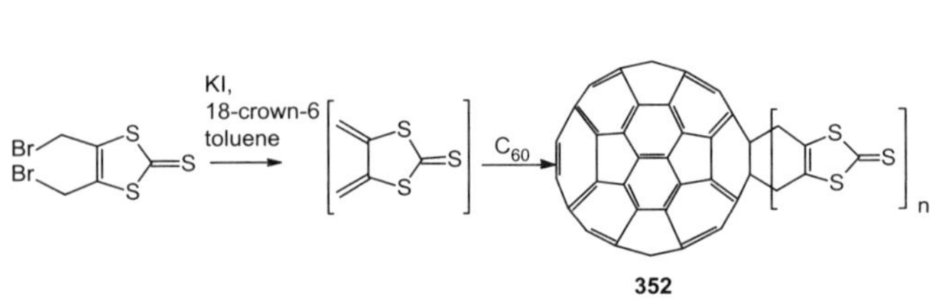

Figure 134. [4 + 2] Cycloaddition of C_{60} to 2-(thi)oxo-4,5-bis(methylene)-1,3-dithioles. Adapted from [314].

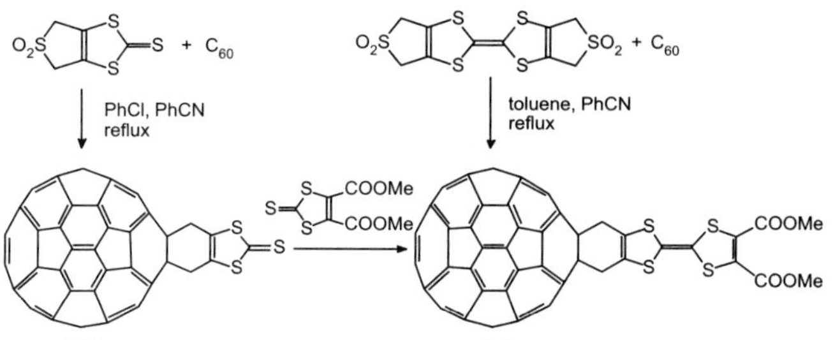

353 354

Figure 135. Synthesis of C_{60}–tetrathiafulvalene diad. Adapted from [315].

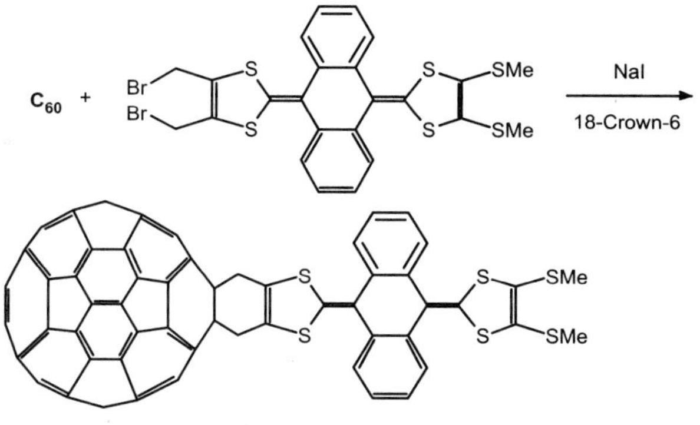

355

Figure 136. Attaching tetrathiafulvalene analogs to C_{60} via *o*-quinodimethane moiety generated from the dibromoderivatives. Adapted from [316].

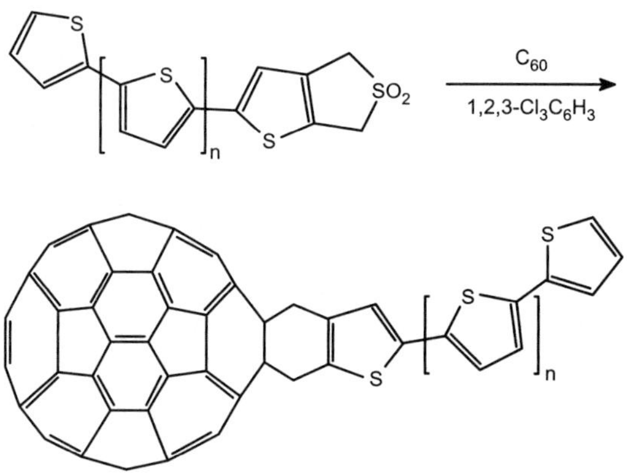

356

Figure 137. Attaching an oligothiophene to C_{60} via the sulfolene. Adapted from [317].

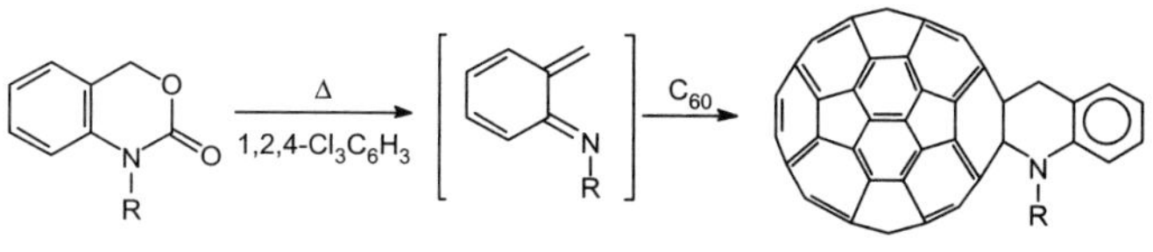

357

Figure 138. Nitrogen heterocycle-fused C_{60}. Adapted from [318].

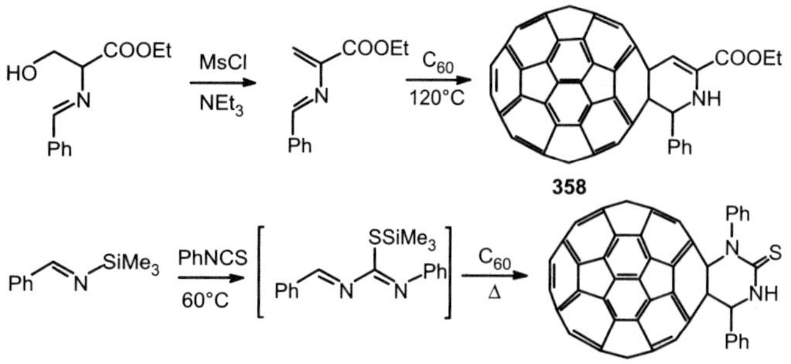

359

Figure 139. C_{60} derivatives fused with cyclic amino acid and thiourea. Adapted from [319].

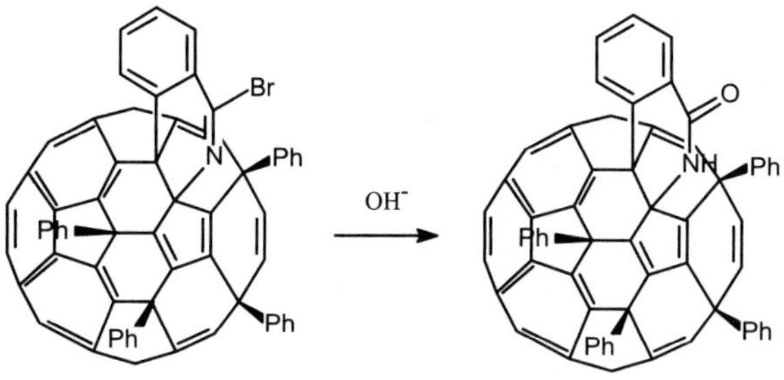

360 361

Figure 140. Formation of a phenylated isoquinoline derivative of C_{60}. Adapted from [320].

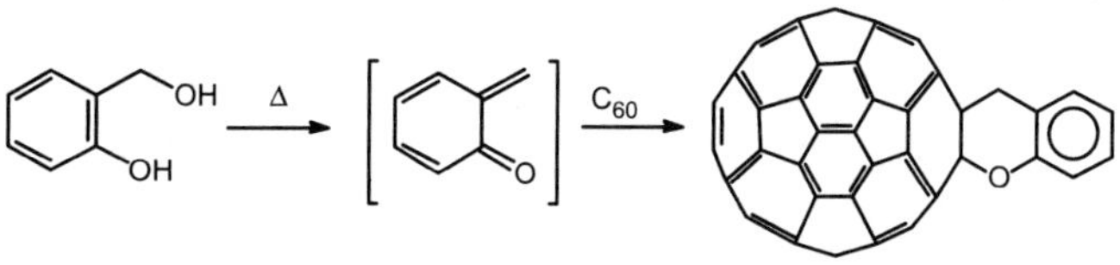

362

Figure 141. Fusing an oxygen-containing heterocyclic ring to C_{60}. Adapted from [321].

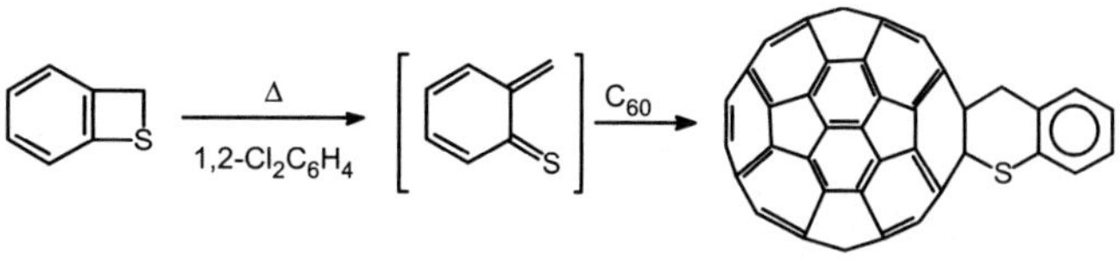

363

Figure 142. Attaching sulfur-containing heterocyclic ring to C_{60}. Adapted from [322].

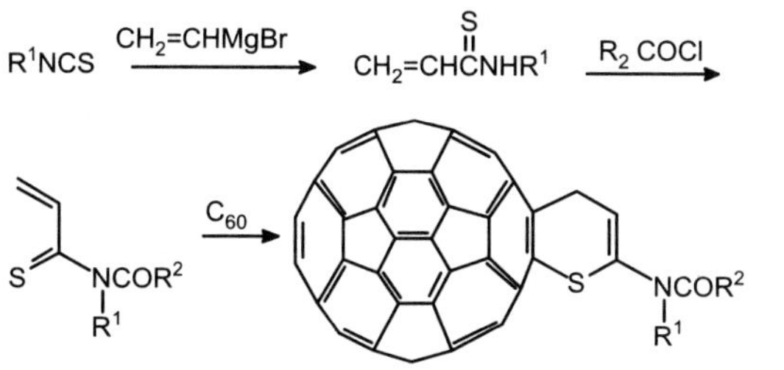

364

Figure 143. Dihydrothiopyran-fused C_{60}. Adapted from [323].

electrocyclic ring opening of 1,2-bis(trimethylsiloxy)-cyclobutene in dichlorobenzene at 180 °C, generating the corresponding butadiene followed by [2 + 4] cycloaddition to C_{60}. Oxidative cleavage of the bis(trimethylsilyl) ether **365** in nonprotic medium followed by low temperature bromination and direct derivatization of the diketone **366** with *O*-methoxyhydroxylamine under basic conditions leads to the isolation of the corresponding dimethoxime **367** in good yield. A number of 1,4-diazines (**368**) are obtained by using the *o*-diaminoarenes. The reactions are carried out in acidic conditions in order to avoid nucleophilic addition to the double bonds of C_{60} (Fig. 144) [324].

The pyrazine-containing cycloadducts **369** are also synthesized by Diels–Alder reaction of C_{60} with the corresponding 2,3-bis(bromomethyl)pyrazine derivatives. The 2,3-bis(bromomethyl)pyrazines are synthesized from condensation of diamines with diketones, from which corresponding *o*-quinodimethane is generated *in-situ*. These reactive intermediates are easily trapped as the respective Diels–Alder adducts by reaction with C_{60} (Fig. 145) [325].

Heterocyclic *o*-quinodimethanes are versatile intermediates for the synthesis of a number of heterocycle-linked C_{60} derivatives. The furan, thiophene, oxazole, thiazole, imidazole, and diazine linked C_{60} derivatives (**370–375**) are obtained by treating the corresponding *o*-quinodimethanes with C_{60} (Fig. 146) [326]. The furan and oxazole derivatives are photosensitive, and hence the products of photooxygenation of these derivatives are isolated.

The Diels–Alder cycloadducts obtained from reactions of C_{60} with electron rich, *cis*-fixed 2,3-dioxy-substituted butadienes are potential precursors of valuable functionalities like acyloin or diketone derivatives, which can be further functionalized (Fig. 147) [327]. The cycloaddition provides the 1:1 cycloadduct as the major product. However, ambient

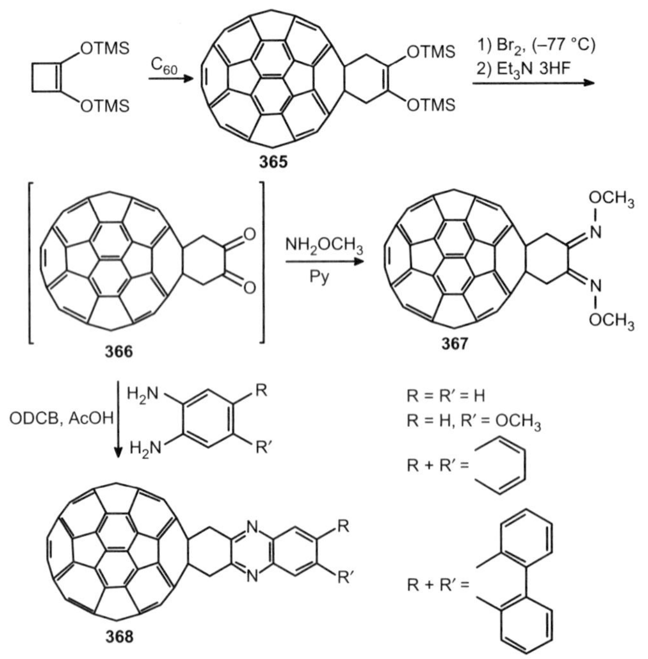

Figure 144. Synthesis of substituted 1,4-diazine-fused fullerenes. Adapted from [324].

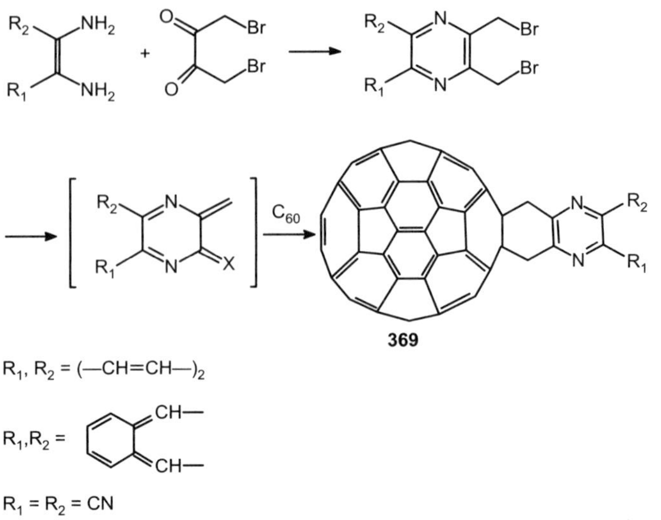

Figure 145. Synthesis of pyrazine-containing cycloadducts by Diels–Alder reaction of C_{60} with pyrazinoquinodimethane intermediates. Adapted from [325].

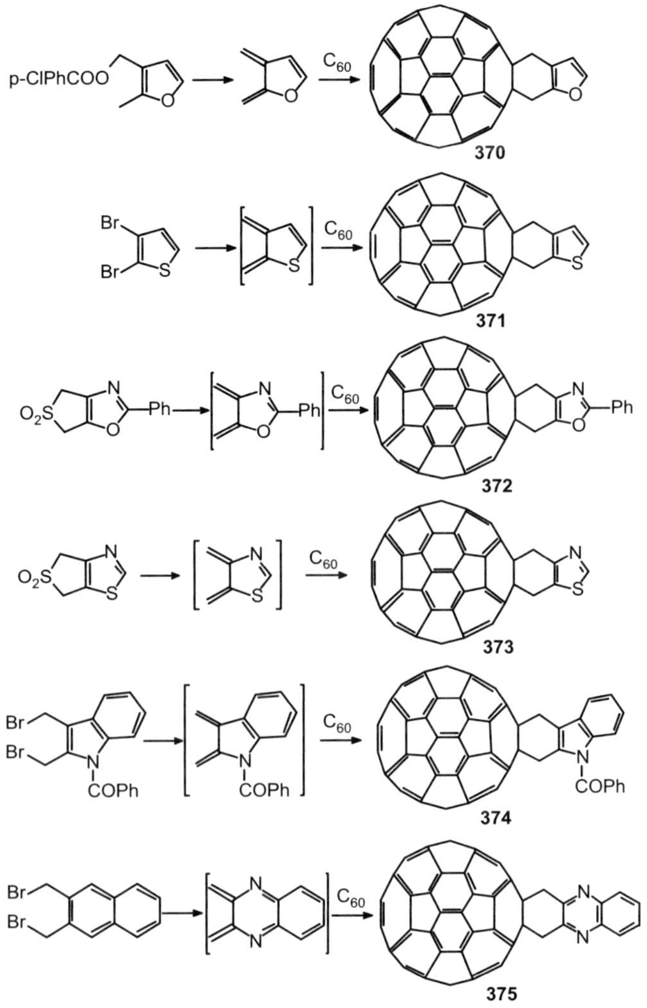

Figure 146. Cycloaddition of heterocyclic *o*-quinodimethane to C_{60} leading to various heterocycle-linked C_{60} derivatives. Adapted from [326].

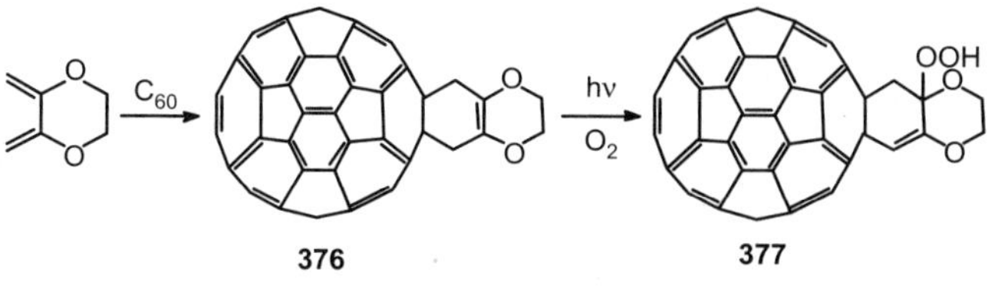

Figure 147. Diels–Alder reaction of C_{60} with electron rich *cis*-fixed butadiene. Adapted from [327].

light caused a rapid decomposition yielding the hydroperoxide derivative **377**. This high reactivity of the electron rich alkene in **376** provides a clue for further derivatization.

Cycloaddition reaction of thieno-*o*-quinodimethane derivatives to C_{60} furnishes the thiophene-linked C_{60} derivatives. Substituted thieno-*o*-quinodimethanes are generated *in-situ* from the respective 2,3-bis(chloromethyl)thiophene and 2,3-bis(bromomethyl[*b*]thiophene) (Fig. 148) [328].

8.4.6. Retro-Diels–Alder Reactions

The preparation of fullerene derivative is often hampered due to their poor solubility. This problem has been resolved through temporary addition of a solubilizing group to a fullerene framework [329]. A solubilizing "tag" is added to C_{60} and the resulting solubilized fullerene is functionalized with the desired group. The solubilizing tag is removed after chemical transformations by retro-Diels–Alder reaction leading to the fullerene derivative. Two illustrative examples of this principle are cited here. The Diels–Alder reaction of 1,3-butadiene and C_{60} leads to a product unstable in air and light affording hydroperoxide **380** (Fig. 149), which after reduction with PPh_3 afforded the allylic alcohol **381**. Soluble derivatives of alcohol **381** are converted to the cyclohexadiene derivative **382**. Diels–Alder reaction of the diene with maleic anhydride and*N*-phenylmaleimide affords bicyclic derivatives. A similar cycloaddition reaction with DMAD leads to the formation of the bicyclo[2.2.2]octadiene intermediate **384**, which readily suffers from retro-Diels–Alder reaction leading to the unreactive dimethyl phthalate and C_{60}. The derivatives of cyclohexadiene can directly be treated with DMAD, for the reversal to occur.

As shown the derivatives of the allylic alcohol can be cleanly removed. This property has been utilized in introducing a solubilizing tag to C_{60}, then carrying out further

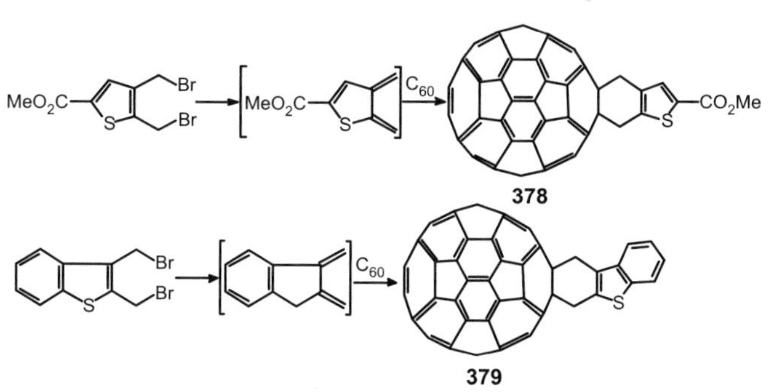

Figure 148. Cycloaddition reaction of thieno-*o*-quinodimethane derivatives to C_{60} furnishing the thiophene-linked C_{60} derivatives. Adapted from [328].

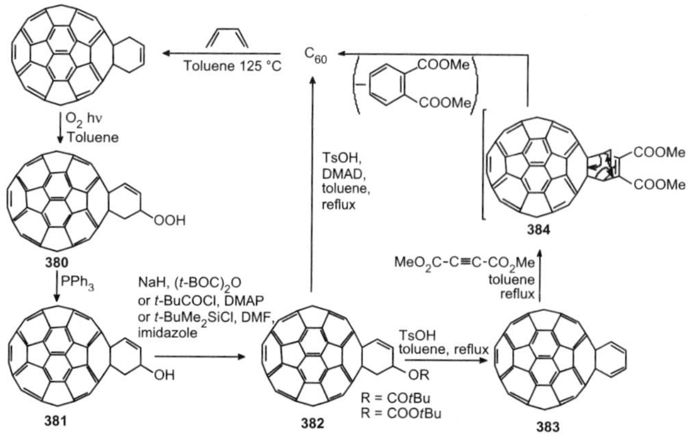

Figure 149. Removing a solubilizing group from C_{60} derivative by retro-Diels–Alder reaction. Adapted from [329].

reaction on the carbon cage, and finally removing the solubilizing moiety. For example, the synthesis of highly insoluble adduct **387** is carried out by this method. For this the allylic alcohol **381** is converted to *t*-BOC-protected alcohol **385** (Fig. 150). Diels–Alder reaction of **385** with the diene gives a diastereomeric mixture of adducts **386**. The solubilizing component, which aided the synthesis of the diastereomers, is cleanly removed by heating **386** under reflux in toluene in the presence of *p*-toluenesulphonic acid and DMAD. The pure adduct **387** is obtained in excellent yield.

Following this strategy, the synthesis of a methanofullerene **390** has been achieved (Fig. 151). Reaction of 3-bromo-1,5-bis(trimethylsilyl)-1,4-pentadiyne with allylic alcohol derivative **385** leads to the solubilized diethynylmethanofullerene **388**, which is obtained after desilylation of the monoadduct. The solubilized diethynylmethanofullerene is obtained as a diastereomeric mixture. Coupling this with phenyl acetylene affords bis(phenylbutadiynyl) derivative **389**, which upon the removal of the solubilizing group leads to the ethanofullerene **390**.

In addition to synthesizing highly insoluble functional fullerenes by introducing easily removable solubilizing groups the reversibility of the Diels–Alder adducts to the parent compounds has been gainfully utilized in purifying C_{60} and separating a mixture of C_{60} and C_{70}. The reaction

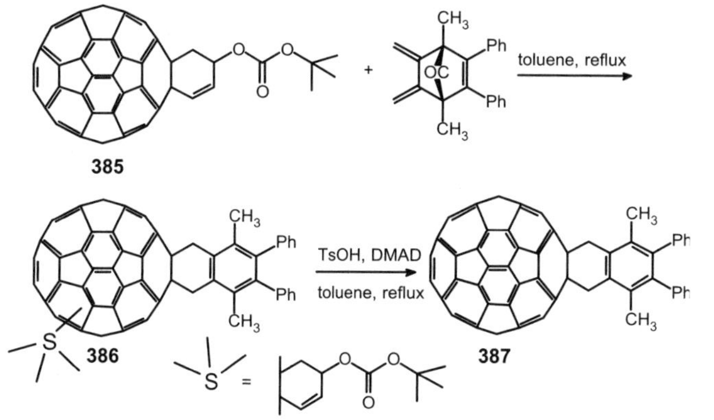

Figure 150. Using retro-Diels–Alder reaction for the synthesis of an insoluble adduct. Adapted from [329].

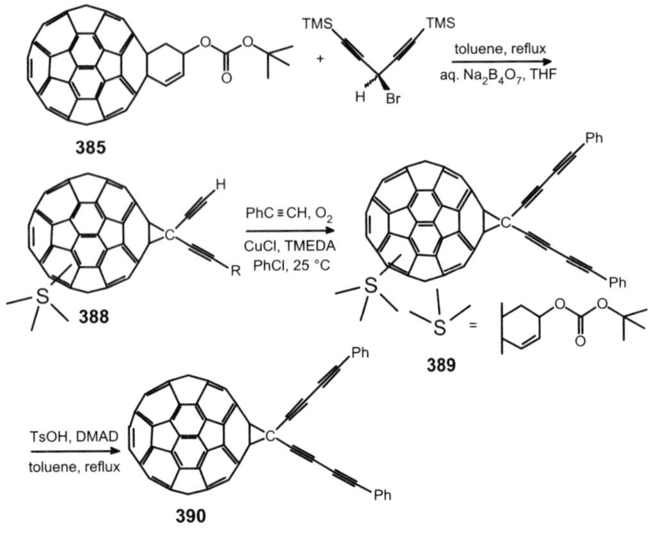

Figure 151. Synthesis of a cyclopropane derivative by removing the solubilizing group. Adapted from [329].

of Merrifield peptide resin in toluene with excess sodium cyclopentadienylide at −20 °C forms a cyclopentadiene-functionalized polymer, which undergoes cycloaddition reaction with C_{60} at room temperature to give fullerene appended to a polymer support (**391**). The product is shown to readily undergo retro-Diels–Alder reaction upon heating, allowing recovery of pure C_{60} (Fig. 152) [330].

The property of reversible cycloaddition polymer supported dienes to C_{60} leads to devise a simple method for the purification of fullerenes [331]. A cyclopentadiene-functionalized silica gel is prepared by the reaction of chloropropyl-functionalized silica gel with lithium cyclopentadienylide (Fig. 153). Diels–Alder reaction of the cyclopentadiene on silica with C_{60} and C_{70} readily leads to silica loaded with C_{60}/C_{70}. This on heating released 93.7% of bound C_{60} and 41.1% of bound C_{70}. The difference in the extent of release of the fullerenes provided a route to purification of C_{60}. The initial C_{60}/C_{70} ratio is 90:10 and the final ratio is 96:4, thus leading to fullerenes significantly enriched in C_{60}.

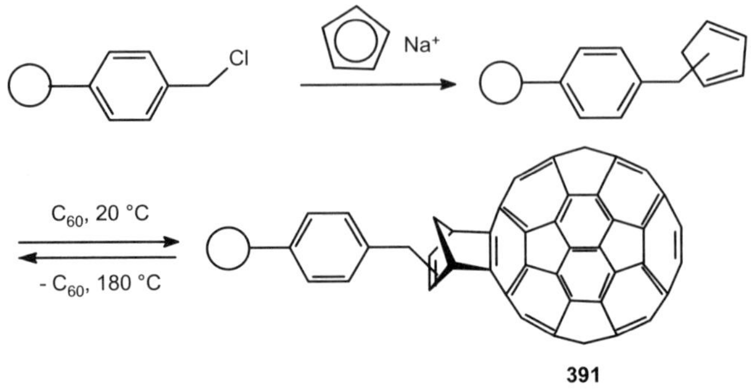

Figure 152. Reversible covalent attachment of C_{60} to a polymeric support. Adapted from [330].

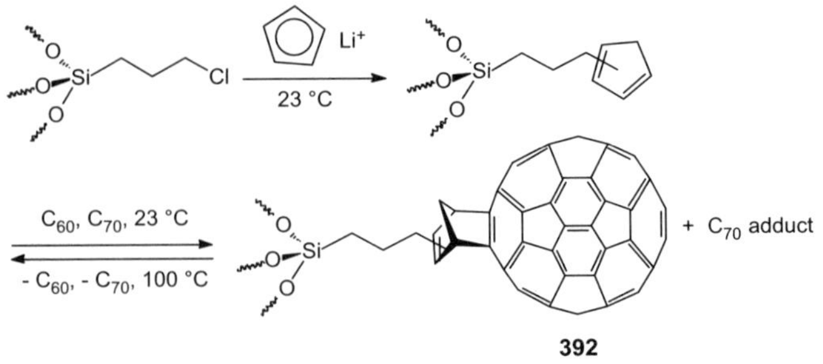

Figure 153. Reversible cycloaddition to silica-supported dienes, a nonchromatographic means to fullerenes significantly enriched in C_{60}. Adapted from [331].

8.5. Higher Cycloaddition Reactions

While majority of the cycloaddition reaction carried out on C_{60} are [2 + 1], [2 + 2], [2 + 3], and [2 + 4] type, a few other higher cycloaddition reactions are known. For example, a [2 + 2 + 2] cycloaddition product **393** is obtained by slow addition of ethyl propiolate to a toluene solution of C_{60} in tricyclohexylphosphine [P(Cy)$_3$] at ambient temperature (Fig. 154) [332]. P(Cy)$_3$ acts as a nucleophile as well as a leaving group. The spectral data strongly support a head-to-tail addition of two ethyl propiolate units to C_{60}.

Another uncommon reaction is the [2 + 5] cycloaddition. The reaction of C_{60} with 1,8-bis(bromomethyl)naphthalene in refluxing benzene in presence of NaI and 18-crown-6 results in a fullerocylcloheptane, **394** (Fig. 155) [333]. A lone example of [2 + 6] cycloaddition on C_{60} is known [334]. In the course of synthesis of fulleroaziridines by reaction between fullerene and ethyl azidoformate, two products are identified, the normal [2 + 4] cycloadduct (**395**) as well as a [2 + 6] cycloadduct (**396**) (Fig. 156). Benzene, used as the solvent, captures the photolytically generated nitrene from ethyl azidoformate to produce *N*-ethoxycarbonylazepine, which on further photoreaction with C_{60} furnishes the cycloadducts.

9. FULLERENE DIMERS

All-carbon fullerene dimers are intriguing molecules due to their unique properties and hence are of considerable interest [335]. The fullerene dimers provide intriguing possibilities as model compounds for nano- and polymer

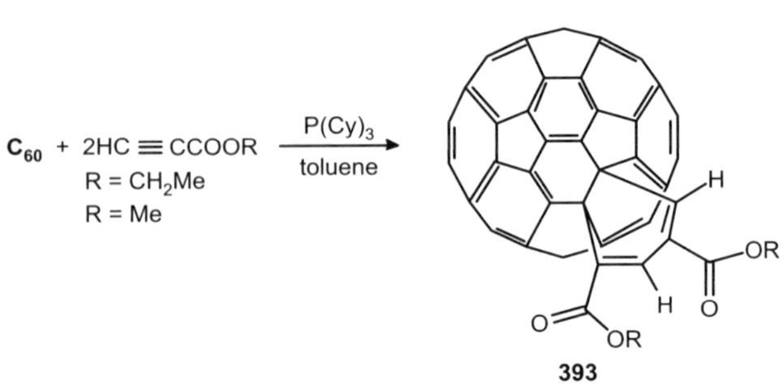

Figure 154. [2 + 2 + 2] Cycloaddition of an alkynes with C_{60}. Adapted from [332].

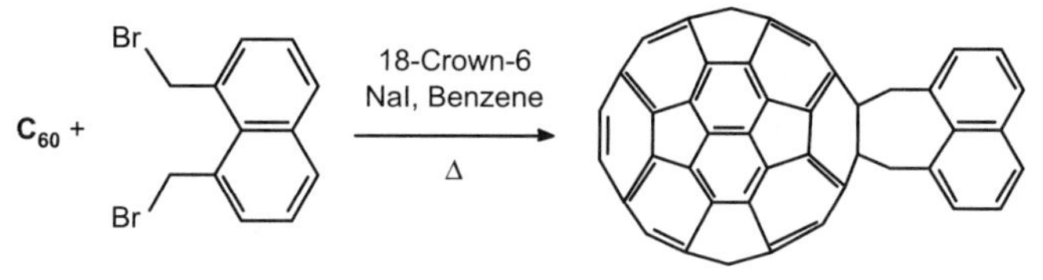

394

Figure 155. An unusual [2 + 5] cycloaddition reaction of C_{60}. Adapted from [333].

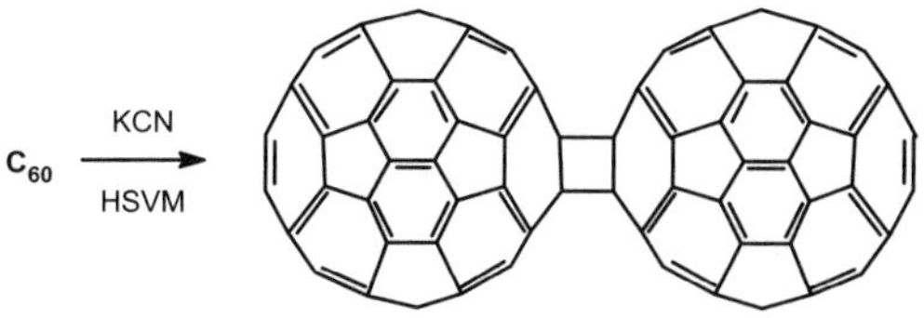

397

Figure 157. Synthesis of C_{120} through a [2 + 2] cycloaddition reaction. Adapted from [338].

science [336, 337]. There are two distinct approaches to synthesis of the fullerene dimers: (i) dimerization of either fullerene itself or a suitably functionalized fullerene and (ii) bifunctional cycloadditions to fullerenes. The simplest such fullerene dimer $(C_{60})_2$, or C_{120} (**397**), has been prepared by solid-state [338] and by chemical [339] methods. The generation of the $(C_{60})_2$ dimer was first accomplished via a mechanochemical reaction with KCN using a high-speed vibration milling (HSVM) technique (Fig. 157) [338a, b]. The structure of the dimer has been determined by X-ray crystallography for a single crystal grown in *o*-dichlorobenzene. The mechanism of dimerization is not clearly known. Dimerization can be via nucleophilic addition or there can be coupling through electron transfer.

A number of theoretical studies have been done to understand the mechanism of dimerization. Orbital-symmetry analysis of the dimerization of C_{60} via [2 + 2] cycloaddition indicates that reactant monomers should approach one another along a pathway in which C_{2h} symmetry is conserved [340]. A mechanism for helium incorporation into one of the C_{60} spheres in the dimer has been examined by modified neglect of differential overlap (MNDO), density functionalization theory, and *ab initio* calculations [341]. The mechanism involves dimerization of two molecules by [2+2] cycloaddition between double bonds at hexagon junctions, formation of a stable closed-shell window, and helium insertion through this window (Fig. 158).

A rigorous proof for photochemical interfullerene [2 + 2] cycloaddition has been provided by direct experimental evidence [342]. In this experiment a *m*-phenylene(methanofullerene) with general structure R–C_{61}–(1,3-Ph)–C_{61}–R is chosen as a synthetic target, since conformational freedom allows the two fullerene cages to be in close contact. For this purpose a bisfulleroid upon heating is first cleanly converted into the methanofullerene. Prolonged irradiation of a deoxygenated solution of the methanofullerene leads to the photodimer by [2 + 2] cycloaddition between the fullerene spheres in close contact, **401** (Fig. 159). Novel dimeric fullerenes incorporated in a 2,3-diazabicyclo[2.2.2]oct-2-ene framework, with and without direct inter-fullerene-cage bonds, are synthesized and fully characterized spectroscopically; the electronic communication between the two fullerene cages is clarified by differential pulse voltammetry [343].

The C_{60} dimer has also been subjected to further reaction. A highly symmetrical decakis-cyclopropanated derivative of C_{120} is synthesized using template activation and the structure is determined by X-ray crystallography (Fig. 160) [344]. The reaction is carried out by addition of excess amounts of diethyl bromomalonate and DBU to a solution of C_{120} and DMA in *o*-dichlorobenzene. Further, the decakisadduct is converted into the icosa-carboxylic acid derivative that is water soluble.

The synthesis of dimeric $(C_{70})_2$ has also been reported. The synthesis is carried out by simply mixing C_{70} and K_2CO_3 in a small mortar and pestle and grinding the mixture by hand for 15 min [345]. Reaction leads to a mixture of five dimers. The isomers are separated and their structures elucidated. The fullerene C_{60}–C_{70} cross-dimer C_{130} is synthesized by the mechanochemical solid-state reaction using the HSVM technique [346]. The reaction is conducted by placing C_{60} and C_{70} in a mixing capsule made of stainless steel, together with 4-aminopyridine and a mixing ball, and treating them by HSVM for 30 min.

Thermolysis of $C_{60}O$ with a 6,6-epoxide structure in the presence of C_{60} affords the corresponding oxygen-bridged C_{60} dimer [347]. By similar experimental condition, the methylene bridged C_{60} dimer is synthesized from thermolysis of fullerene cyclopropane and C_{60} (Fig. 161). The thermolysis of $C_{120}O$ leads to the isolation of $C_{120}O_2$ [348]. The ^{13}C NMR spectra of $C_{120}O_2$ show two groups of four equivalent sp^3-hybridized carbon atoms which leads to the

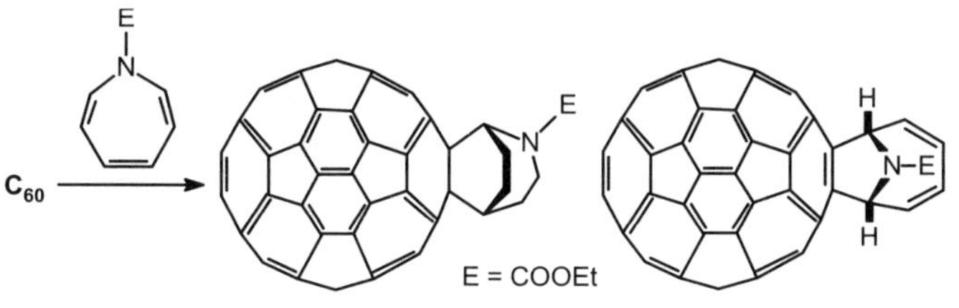

395 396

Figure 156. A [2+6] cycloaddition reaction of C_{60}. Adapted from [334].

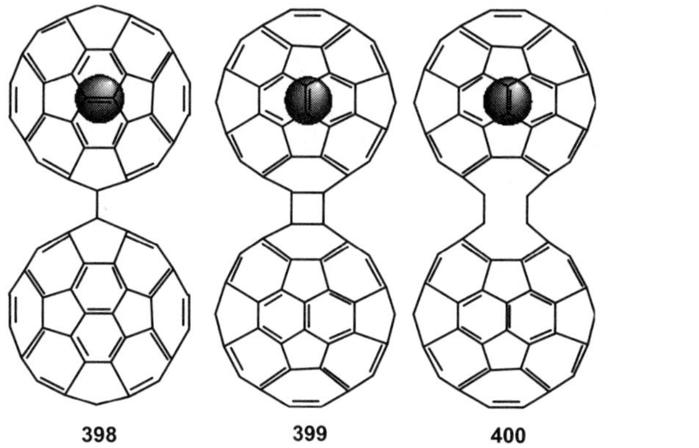
398 399 400

Figure 158. Theoretical studies involving introduction of a helium atom to C_{60} dimers. Adapted from [341].

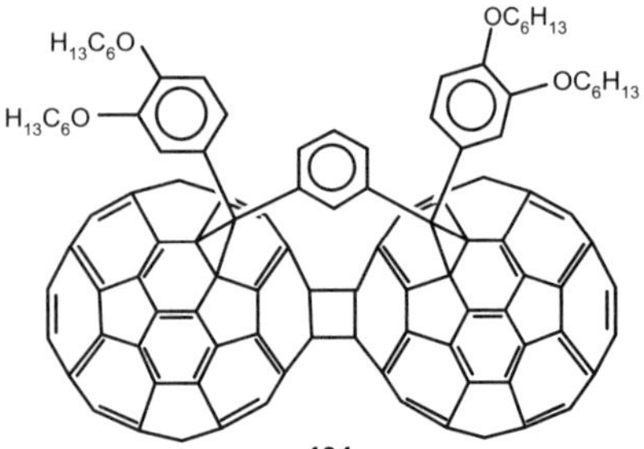

401

Figure 159. Photodimerization of a methanofullerene as proof for interfullerene [2 + 2] cycloaddition. Adapted from [342].

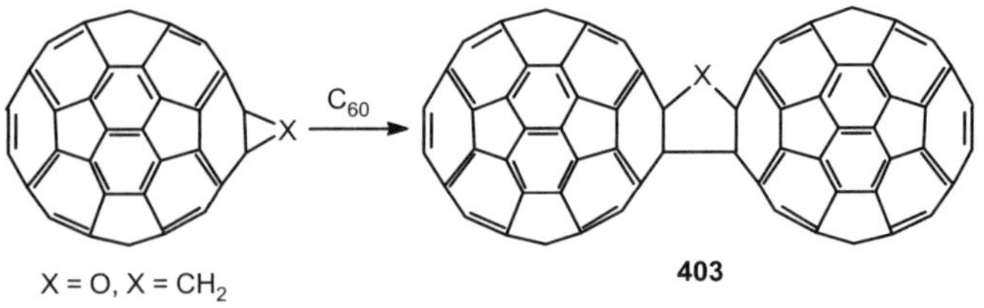

403

Figure 161. Oxygen- and methylne-bridged C_{60} dimers. Adapted from [347].

conclusion that each fullerene cage is bislinked by adjacent furanoid rings. The two furanoid rings link the cages via 6,5-junctions. A 6,6-link results in unfavorable double bonds in pentagons. Semiempirical models confirm the stability of bridged dimer structures assigned in the literature to C_{120}, $C_{120}O$, and $C_{120}O_2$, interpreting all three as additions across formal double bonds of the C_{60} monomers. Analogous structures are proposed and compared for C_{60}/C_{70} dimers C_{130}, $C_{130}O$, $C_{130}O_2$, C_{140}, $C_{140}O$, and $C_{140}O_2$ [349].

There are several reports of C_{60} dimerizing into various singly bonded structures [350]. Some of the best-studied dimers are those of general structure $RC_{60}C_{60}R$ (R = H, halo, alkyl, fluoroalkyl) obtained by dimerization of $RC_{60}^{\bullet}$ radicals. The rotation about the intercage bond of $RC_{60}C_{60}R$ has been studied by molecular mechanics [351]. The radical adducts of C_{60} exist in equilibrium with their dimers and the dimer bond strength depends on the size of the group R (Fig. 162).

A monoprotected diethynylmethanofullerene is homocoupled under Hay coupling conditions to the corresponding butadiynyl-linked dimeric methanofullerene, **405** (Fig. 163) [352]. From cyclic voltametric experiments it is found that both through-bond and through-space electronic interactions are not very efficient. This is in agreement with the UV-vis data as well as X-ray structural data, which indicate that the sp^3 carbon atoms in the two butadiynyl-linked methano bridges act as insulators.

Dimeric fullerenes with acetylinic spacers attached directly to the fullerene core have been reported. Reaction of dilithioacetylene with C_{60} in toluene, followed by dodecyl iodide addition at reflux temperature, leads to a fullerene dimer with an acetylene spacer, **406** (Fig. 164) [353]. The synthesis of a bisfullerene, in which the fullerenes are linked directly by two acetylene units, has been accomplished by homocoupling of the ethynylfullerenes, bearing base stable substituents at their 2-position [354]. In these dimers, even though there is a direct link between the fullerene units, electronic interaction between them is not observed.

Addition of methano-fullerenecarbene, $C_{60}C$:, to C_{60} leads two fullerene units to share the methylene bridge carbon forming a dimer **408** (C_{121}) (Fig. 165) [355]. The double-bonded dimer **409** (C_{122}) obtained by the dimerization of $C_{60}C$: itself has been detected by mass spectroscopy. However, there is a possibility that both **408** and **409** are formed under mass spectrometric conditions. Further, the ^1H NMR spectra of the dimer show a very broad band indicating the presence of a number of isomers.

In all these studies, two fullerene units are reacted either directly or through a linking or spacer group. In an alternative but more versatile approach, fullerene dimers have been prepared by first preparing a suitable bifunctional linking group, which can undergo cycloaddition reactions with C_{60} leading to the desired dimer. A number of interesting products have been prepared following this strategy. However, the reaction invariably leads to formation of a mixture containing regioisomers, which are often difficult to separate. Following the first synthesis of a dimer by reaction of bisdiazo derivative [356], a number of dumbbell-like fullerene dimers have been synthesized by using Diels–Alder reaction of bisdienes with C_{60}. A solid-state [2 + 4] cycloaddition of pentacene to C_{60} using the HSVM technique has led to a C_{60} dimer, the structure of which is suggested to be the less sterically crowded *anti*-isomer (Fig. 166) [357].

Bis-*o*-quinodimethanes with spacer groups provide unique structures upon [2 + 4] cycloaddition reaction with C_{60}. Molecules providing controlled spacing with 1,4-cyclohexadienyl ladders and of different length are reacted with C_{60} by Diels–Alder cycloaddition [358]. Depending on the spacer length a number of dumbbell-like structures could be designed (Fig. 167).

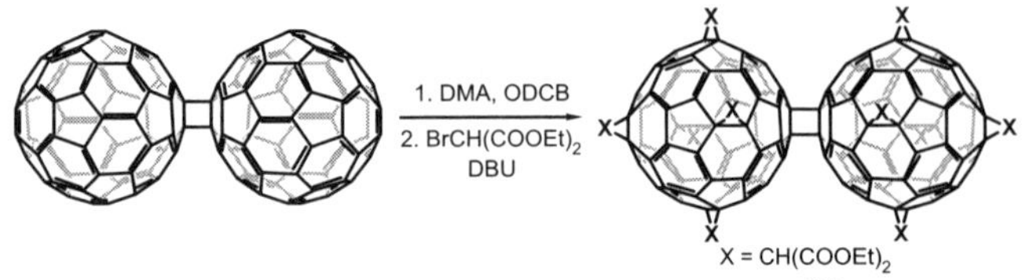

402

Figure 160. A highly symmetrical decakisadduct of fullerene dimer C_{120}. Adapted from [344].

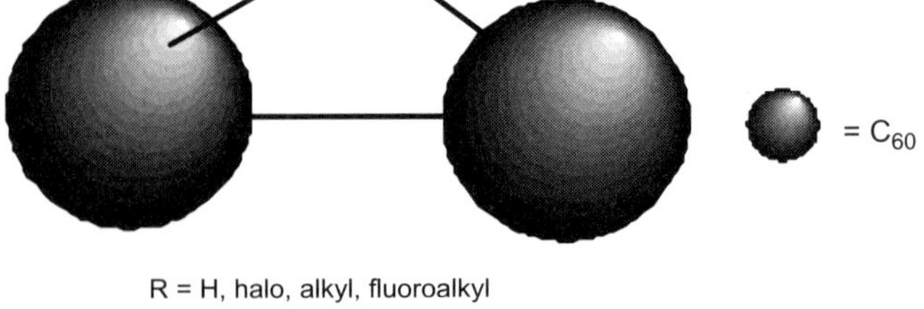

404

Figure 162. Singly bonded dimers of substituted C_{60}. Adapted from [351].

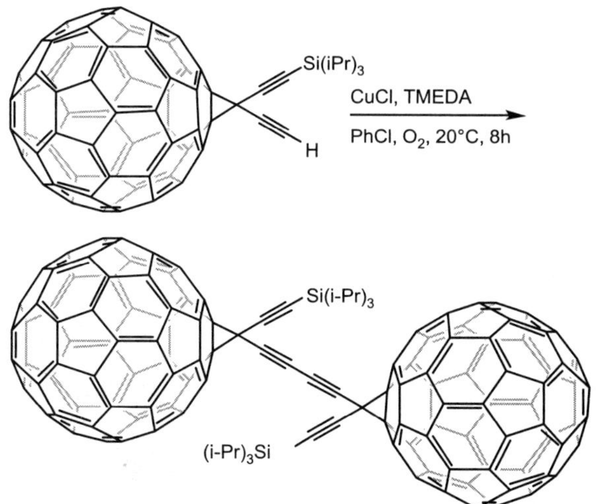

405

Figure 163. Butadiynyl-linked dimeric methanofullerene. Adapted from [352].

The covalently linked, multichromophoric ball-and-chain molecules in which the bridge is a polynorbornane-bicyclo[2.2.0]hexane are synthesized [359]. Analogous reaction with a bisdiene affords the soluble dumbbell system, bearing two C_{60} chromophores (Fig. 168).

The first synthesis of a fullerene dimer by 1,3-dipolar cycloaddition of azomethine ylides to C_{60} is accomplished by treating sarcosine and a dialdehyde in refluxing toluene. The product, a black solid, is insoluble in most organic solvents. The CV experiment showed that there is no significant interaction between both fullerene cages of the dimer (Fig. 169) [360].

Functionalization of C_{60} using the Bingel reaction followed by a multistep reaction pathway leads to a fullerene dimer containing a bipyridine ligand between the fullerene units. Subsequent reaction of the ligand with $Ru(bpy)_2Cl_2$ leads to a Ru(II) complex, which is an electroactive moiety in the fullerene dimer (Fig. 170) [361]. The electronic absorption spectra suggest that there is no charge transfer between Ru(II) and fullerene units.

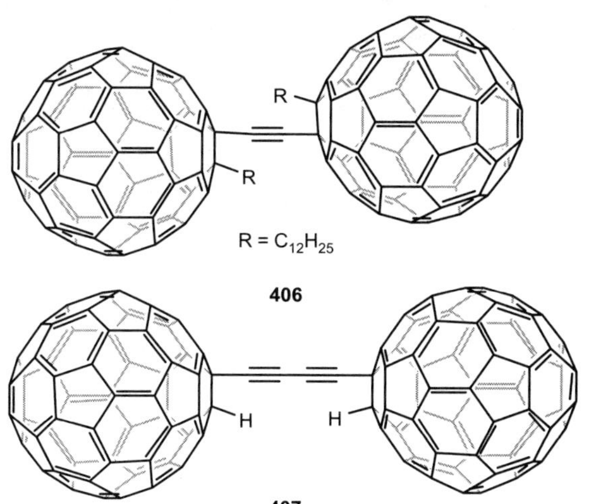

407

Figure 164. Bisfullerenes with acetylene spacer. Adapted from [353].

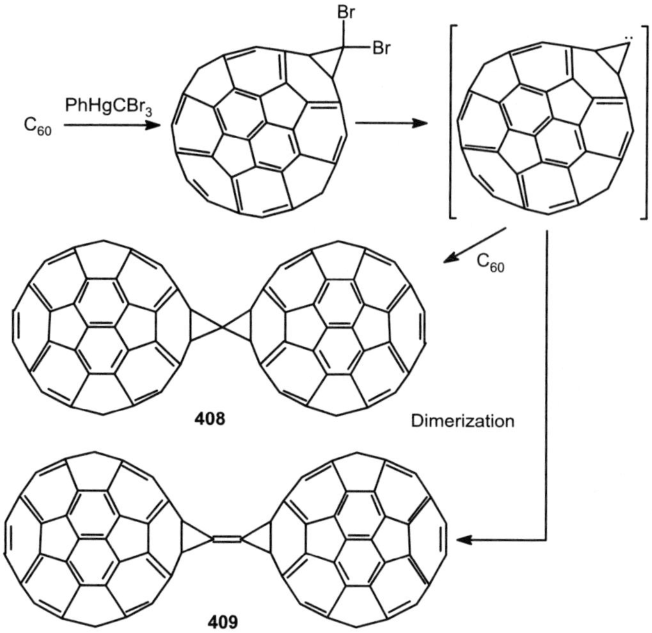

409

Figure 165. Synthesis of C_{121} and C_{122} from methano-fullerenecarbene. Adapted from [355].

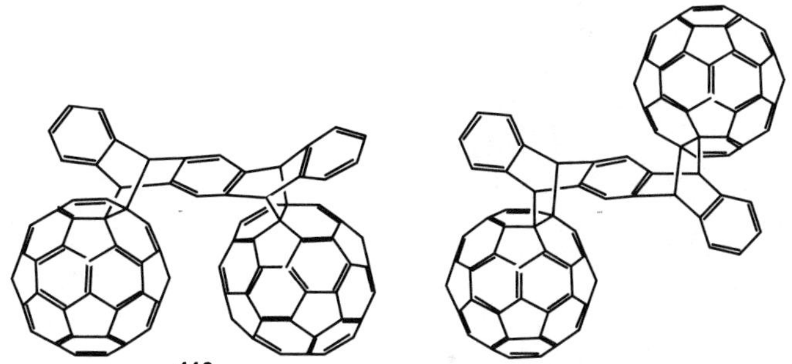

410 411

Figure 166. Addition of pentacene to C_{60} leading to a dimer anti-configuration. Adapted from [357].

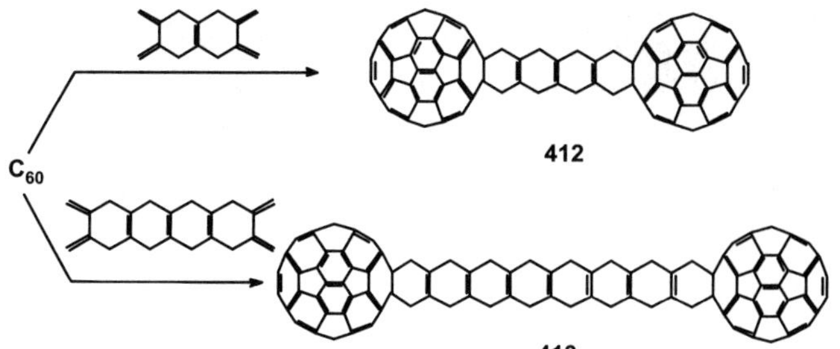

413

Figure 167. Pairwise Diels–Alder cycloaddition to C_{60}. Adapted from [358].

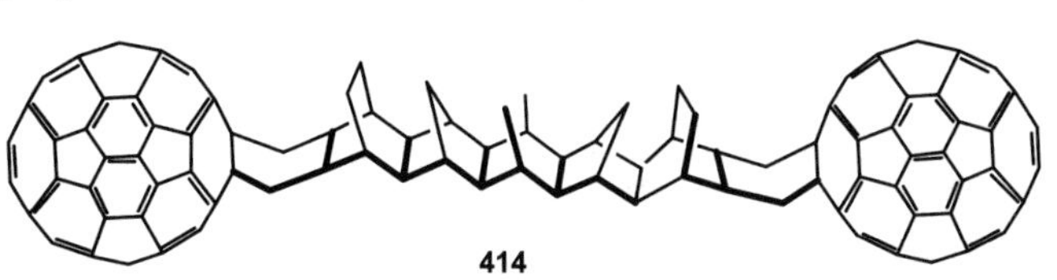

414

Figure 168. A C_{60} dumbbell bearing polynorbornane-bicyclo[2.2.0]-hexane chromophore. Adapted from [359].

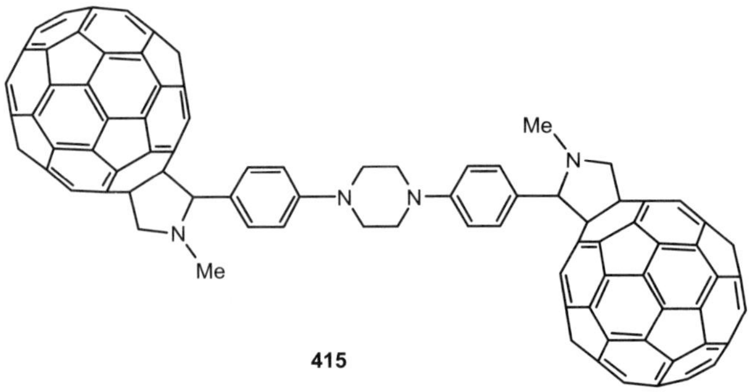

415

Figure 169. Synthesis of a fullerene dimer by 1,3-dipolar cycloaddition of azomethine ylides to C_{60}. Adapted from [360].

Fullerenes covalently attached to TTF lead to interesting species due to their potential applications in photovoltaic devices. The TTF units readily lose en electron, since in the process the radical cation and dication species are aromatic. Since fullerenes are strong electron acceptors, coupling these two species leads to easy electron transfer from TTF to C_{60}. A TTF unit is covalently linked to two or four C_{60} molecules through a flexible bridge (Fig. 171) [362]. These systems are promising candidates for photophysical studies. However, from UV-vis and CV measurements, no interaction between C_{60} and TTF unit has been observed.

The synthesis of a rotaxane, assembled via the copper (I) templated approach bearing two C_{60} units as chemical stoppers, has been reported [363]. The synthesis is based on the oxidative coupling reaction of terminal alkynes functionalized on C_{60} and copper (I) rotaxane in presence of Hay catalyst [CuCl-TMEDA-O_2 (TMEDA = *N*, *N*,*N*′,*N*′-tetramethylethylenediamine)]. A homocoupling reaction leads to a dimeric fullerene derivative as a side product (Fig. 172).

10. HETEROFULLERENES

Apart from exohedral and endohedral fullerene derivatives, heterofullerenes represent the third fundamental group of modified fullerenes. In heterofullerenes, one or more carbon atoms are substituted by heteroatoms, such as trivalent nitrogen or boron atoms. In these cases, substitution of an

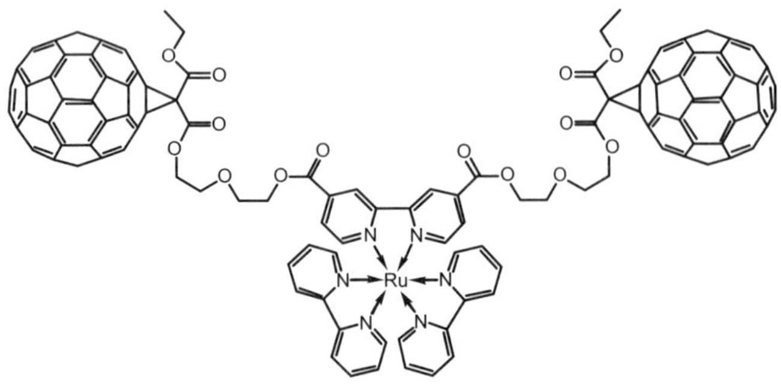

416

Figure 170. Incorporation of a Ru(II) complex in a fullerene dimer. Adapted from [361].

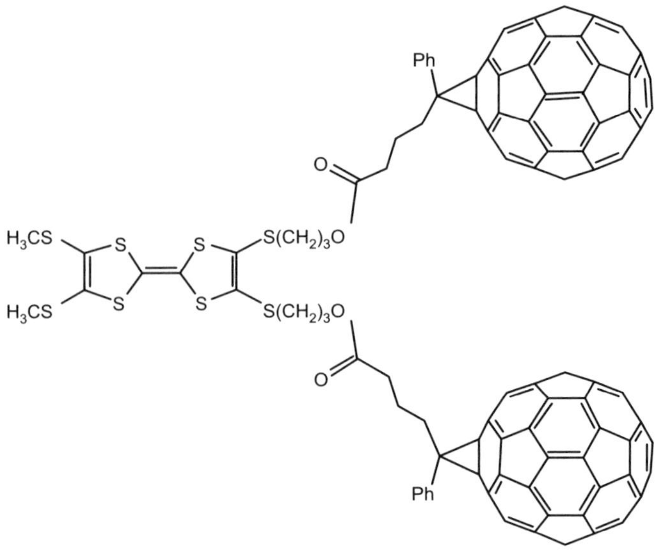

417

Figure 171. Tetrathiafulvalene linked covalently to two C_{60} molecules through a flexible bridge. Adapted from [362].

odd number of C atoms leads to radicals, whereas replacement of an even number of carbon atoms leads to closed-shell systems. In spite of such interesting chemistry, the area of heterofullerenes is still a very young discipline within the scope of synthetic organic chemistry. Nevertheless, there has been some pioneering work that clearly shows a great structural diversity of this compound class [364]. Nitrogen heterofullerene is first synthesized from cage-opened *N*-MEM ketolactam, which is obtained when *N*-MEM-substituted 6,5-azafulleroid reacts with singlet oxygen. Refluxing *N*-MEM ketolactam in *o*-dichlorobenzene in the presence

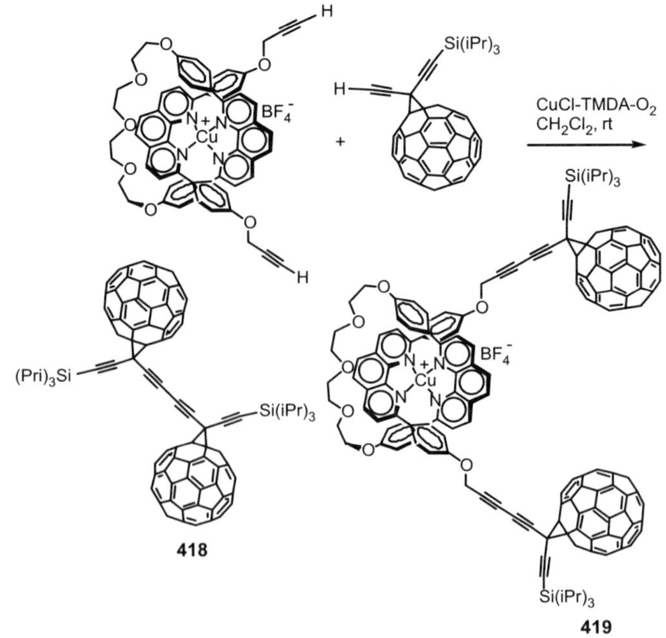

418

419

Figure 172. A copper(I)-complexed rotaxane with two fullerene stoppers. Adapted from [363].

of 12–20-fold excess of *p*-TsOH under an atmosphere of nitrogen leads to the formation of heterofullerene, $C_{59}N$ (Fig. 173) [365]. The suggested mechanism for the formation of the nitrogen heterofullerene dimer involves as a first step the acid-catalyzed cleavage of the MEM group, followed by an intramolecular ring formation.

The synthesis of $(C_{59}N)_2$ and $(C_{69}N)_2$ is also achieved using diazabis-(1,6);(1,9)-homofullerenes as starting materials via their butylamine monoadducts [366]. Heating butylamino adducts of diazabis-(1,6);(1,9)-homofullerenes bearing MEM-protected imino bridges with toluene-*p*-sulphonic acid provides the heterofullerenes and their adducts $RC_{59}N$ and $RC_{69}N$ (Fig. 174) [367]. Starting from a mixture of two different C_{70}-bisazafulleroids, it is possible to isolate three constitutional isomers of $(C_{70}N)_2$. Formation of heterofullerene can occur only if the substituents of the imino bridges are easily removed. 2-MEM is an easily removable protecting group. The C atom of the fullerene cage is eliminated as an isonitrile or carbodiimide species.

Mixed $C_{59}N/C_{69}N$ heterodimers have been also synthesized [368]. Starting with a 1:1 mixture of the ketolactams of C_{60} and C_{70} as well as thermal treatment of equimolar amounts of $(C_{59}N_2)$ and an isomer of $(C_{69}N_2)$, the mixed heterodimers are obtained along with some homodimers (Fig. 175).

A variety of derivatives of the heterofullerenes have been synthesized. Two possible processes of derivatization of $C_{59}N$ are (1) deprotonation of azafullerene $C_{59}HN$ followed by attack of an electrophile, and (2) homolytic dissociation of the interdimer bond of $(C_{59}N)_2$ followed by free radical reactions. Treatment of $(C_{59}N)_2$ with diphenyl methane in refluxing *o*-chlorobenzene provides the substituted azafullerene $C_{59}(CHPh_2)N$ (Fig. 176) [369]. The interdimer bond is relatively weak (18 kcal/mol) and, under photolysis and thermolysis conditions, undergoes facile homolysis. In the presence of a good hydrogen donor or radical source, the resulting azafullerenyl radical undergoes free radical reaction.

Another reaction sequence leads very efficiently to arylated heterofullerenes. The thermal treatment of the heterofullerene dimer with anisole, toluene, and 1-chloronaphthalene in the presence of toluene-*p*-sulphonic acid and air leads to the formation of monoarylated azafullerenes $ArC_{59}N$ in very good yields [370]. The mechanism of this reaction is assumed to be an electrophilic aromatic substitution (S_EAr) by $C_{59}N^+$, which is possibly formed via thermal homolysis of the dimer and subsequent oxidation with O_2. Since arylated aza[60]fullerenes are stable and are

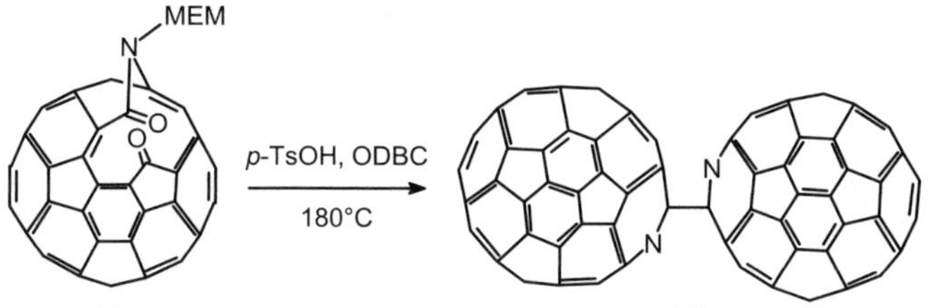

419 **420**

Figure 173. Synthesis of $(C_{59}N)_2$ starting from the cluster opened ketolactam. Adapted from [365].

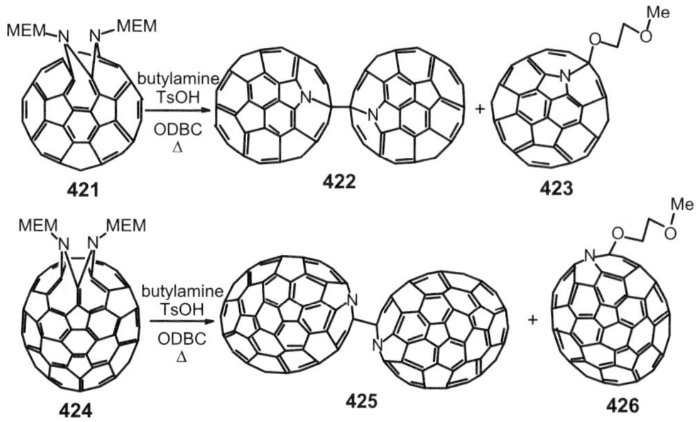

Figure 174. Nitrogen heterofullerenes from diazabishomofullerenes. Adapted from [367].

obtainable in high yields, they represent ideal starting materials for investigation of the behavior of the nitrogen heterofullerene core toward addition reactions. The treatment of monoarylated azafullerenes in CS_2 leads to the exclusive formation of the tetrachlorinated heterofullerene $Cl_4ArC_{59}N$, which contains a pyrrole moiety in the fullerene cage, **430** (Fig. 177) [371]. The pyrrole moiety is decoupled from the conjugated π-system of the fullerene cage. An excess of PPh_3 can easily remove the chlorine addends, leading to reversal of the reaction.

11. OLIGOFULLERENES

Solubilized derivatives of C_{195} and C_{260}, a new class of carbon allotropes, are oligofullerenes having three or four fullerene units per molecule [372]. The smallest number of the methano-bridged all carbon family C_{195} consists of a cyclo-C_{15} core surrounded by three peripheral fullerenes, whereas C_{260} is a cyclo-C_{20} derivative with four fullerenes. The fullerenes are either attached to acetylenic core via methano bridges, or are incorporated into the macrocycle itself (**431**, **432**) (Fig. 178). Although these are expected to be stable compounds, their synthesis by oxidative coupling of 61,61-diethynylmethano[60]fullerene failed; the result is an intractable mixture of insoluble compounds [373]. The synthesis of these oligofullerenes as their derivatives is carried out using the tether directed remote functionalization technique. The starting material synthesized by the template activation method is subjected to nucleophilic

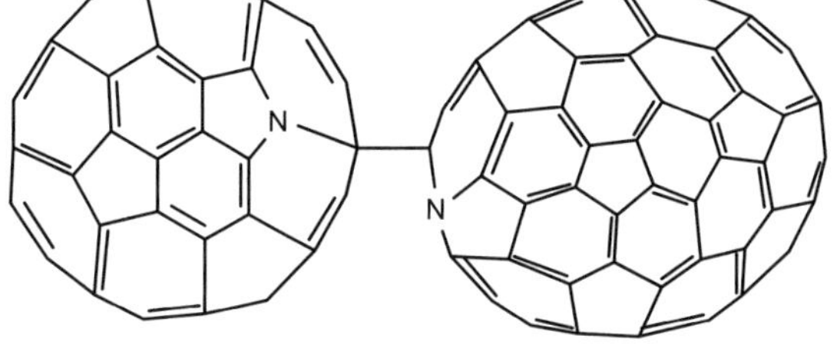

427

Figure 175. $C_{59}N/C_{69}N$ heterodimers. Adapted from [368].

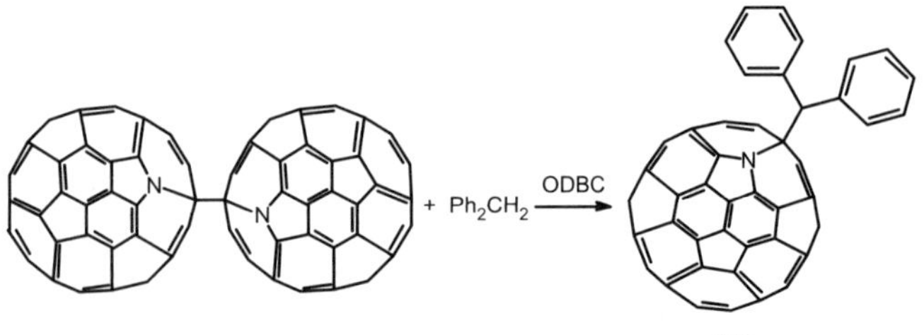

Figure 176. Derivatization of $C_{59}N$ leading to $C_{59}(CHPh_2)N$. Adapted from [369].

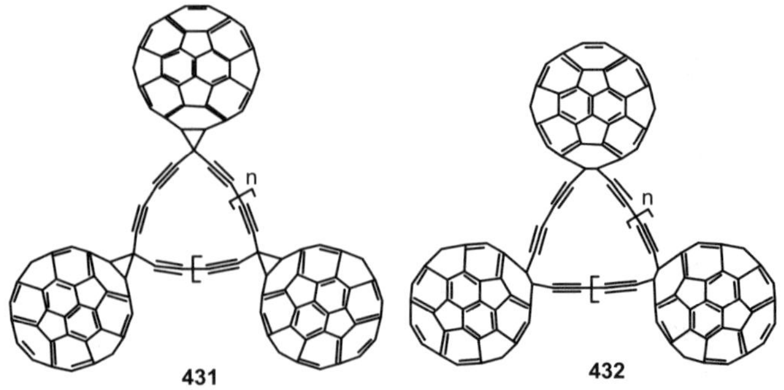

Figure 178. The methano-bridged all carbon family C_{195}, and C_{260} derivative with four fullerenes. Adapted from [372].

addition/elimination of 1,5-bis(trimethylsilyl)-3-bromopenta-1,4-diyne. This introduces diethynylmethano group leading to dialkynylmethanofullerene hexaadduct. Deprotection and subsequent oxidative cyclization leads to two major derivatives of the carbon allotropes C_{195} and C_{260} providing **434** and **435** (Fig. 179).

Cycloaddition reactions are a convenient route to synthesis of fullerene oligomers. For example, herecene, 2,3,5,6,7,8-hexamethylidenebicyclo[2.2.2]octane, reacts readily with C_{60} leading to a fullerene trimer. Herecene dissolved in hot toluene is treated with a toluene solution of C_{60} and the product is generated by threefold cycloaddition [374]. Clusters of bis- and trisfullerenes (**436**) have been prepared following the Prato reaction, in which a mixture of the trialdehyde, fullerene, and sarcosine is refluxed in toluene (Fig. 180) [375]. The fullerene trimers and larger assemblies form a new class of molecular system with potential applications in molecular electronic devices. The photophysical and photochemical behaviors of the cluster forms of bis- and trisderivatives are described. Reaction of C_{60} from addition of dienes generated *in-situ* upon elimination of SO_2 is an efficient approach to synthesis of cycloadducts. Refluxing of the tetrasufolene derivative of zinc meso-tetraarylporphyrinate with 10 equiv. of C_{60} in 1,2-dichlorobenzene produces porphyrins containing four fullerene substitutents (**437**) (see Fig. 180) [376]. Depending on the conditions and duration of the reaction, mono-, bis-, or tris(fullereno)porphyrinates can be obtained as the major product. The photophysical properties of the fullerene tetramer are studied.

12. FULLERENE DENDRIMERS

Starburst dendrimers [377] possess controllable size, shape, topology, surface chemistry, and flexibility. These three-dimensional, highly ordered oligomeric and polymeric compounds formed from smaller molecules by reiterative reaction sequences can mimic certain properties of micelles, liposomes, even those of biomolecules, and still the more complicated, but highly organized, building blocks of biological systems. This branch of "supramolecular chemistry" has sparked new developments in both organic and macromolecular chemistry. Numerous applications of these compounds are conceivable, particularly in mimicking the functions of large biomolecules as drug carriers and immunogens. Combining various dendrimer morphologies would give rise to a wide variety of construction components and combinations.

The fascinating thing about the dendrimeric addition to fullerenes is the polymeric fragment, which, due to its globular three-dimensional structure, leads to greater encapsulation and the ability to prepare monofunctional, monodisperse materials of known molecular mass. The dendritic macromolecules dramatically improve the solubility of the fullerenes and would provide a more compact insulating layer around the clusters than would linear polymers, minimizing or preventing direct fullerene–fullerene interaction. If the dendritic moiety is hydrophilic, the dendritic fullerene (fullerodendrimer) may dissolve in water, which can be useful for biomedical applications. Dendrimers can be attached covalently to fullerenes, or there can be noncovalent com-

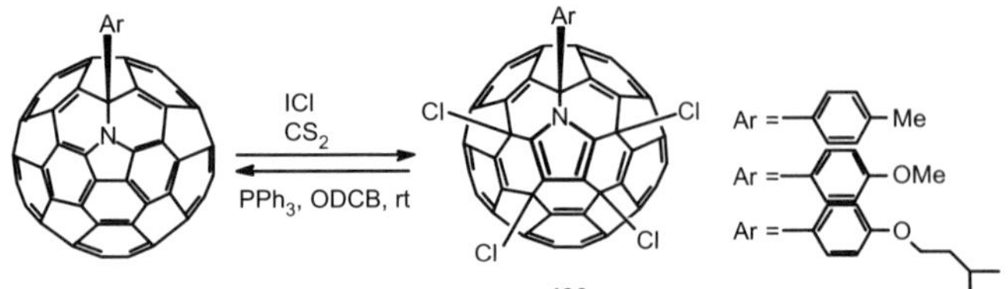

Figure 177. An integral pyrrole moiety decoupled from the π-system of the fullerene cage. Adapted from [371].

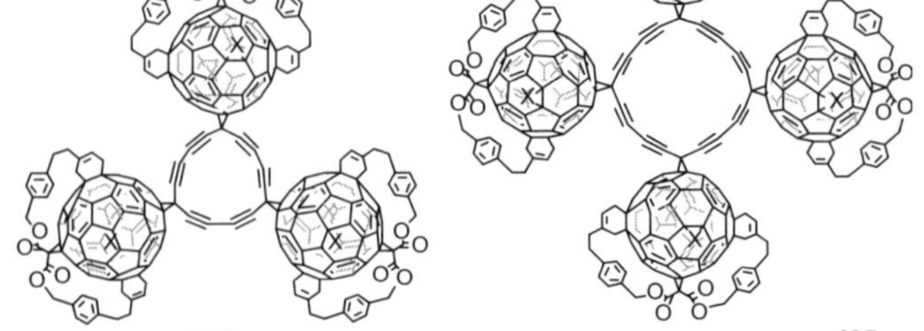

Figure 179. Solubilized derivatives of C_{195} and C_{260}, oligofullerenes having three or four fullerene units per molecule. Adapted from [372].

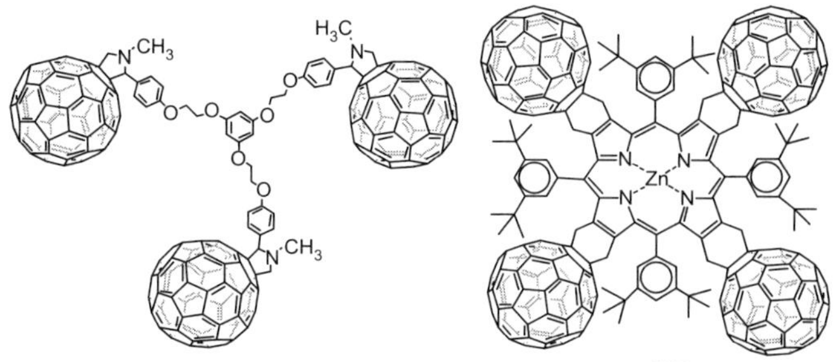

Figure 180. Oligofullerenes prepared by cycloaddition route. Adapted from [375].

plexation of fullerenes with a dendrimer. C_{60} is a convenient core for dendrimers, since its almost spherical shape leads to globular systems even with low generation dendrons. Furthermore, variable degrees of addition within the fullerene core, especially from mono- up to hexaadducts, are possible [378].

The first reported addition of dendrimeric molecules to fullerenes is the one-step cycloaddition reaction of a dendritic azide with C_{60} (Fig. 181) [379]. The dendrimer chosen is the terminally deuterated fourth generation azide, D_{112}-[G-4]-N_3 (**439**), which is prepared by reaction of the corresponding bromide with sodium azide in dimethylsulfoxide. Due to the increase of molecular mass by 3400 for each dendrimer addition, only mono- and diaddition products are observed and no higher addition products. When six equivalents of the dendrimer azide are used per C_{60} the mono- and diaddition compounds are again the major products (2:3 ratio) with little or no observable higher addition products. This inability to obtain significant tri- or higher substitution may be due to steric hindrance around the C_{60} nucleus and/or unfavorable electronic considerations after diaddition. Cyclovoltametry of the fullerodendrimer shows the expected multireduction waves. However, the first three reduction waves occur at lower values than those previously reported. This may be due to the insulating influence of the globular dendritic macromolecule.

The attachment of polyether dendrimers to phenol-functionalized fullerenes is reported [380], which is prepared by reaction of C_{60} with bis(*p*-methoxyphenyl)diazomethane followed by hydrolysis of methyl ethers with BBr_3 (Fig. 182). Polyether dendrimers prepared by the convergent synthesis route are ideally suited for attachment to phenol-functionalized C_{60} since they carry a single electrophilic site at their focal point, and their attachment to phenolic moieties has been studied extensively. The benzylic bromide group at the focal point of a fourth-generation dendrimer [G-4]-Br (**441**) reacts with the phenolic groups located at the surface of the functionalized fullerene in the presence of K_2CO_3. Two dendrimers couple with one fullerene to give the fullerene-bound dendrimer **442** together with a significant amount of higher molecular impurities.

If the dendrimer moiety is hydrophilic, the resulting amphiphilic fullerodendrimers can form monolayers at the air–water interface. However, the hydrophilic headgroups should be large and bulky enough to prevent strong fullerene–fullerene aggregation, while still keeping the fullerene spheres sufficiently close to each other so that

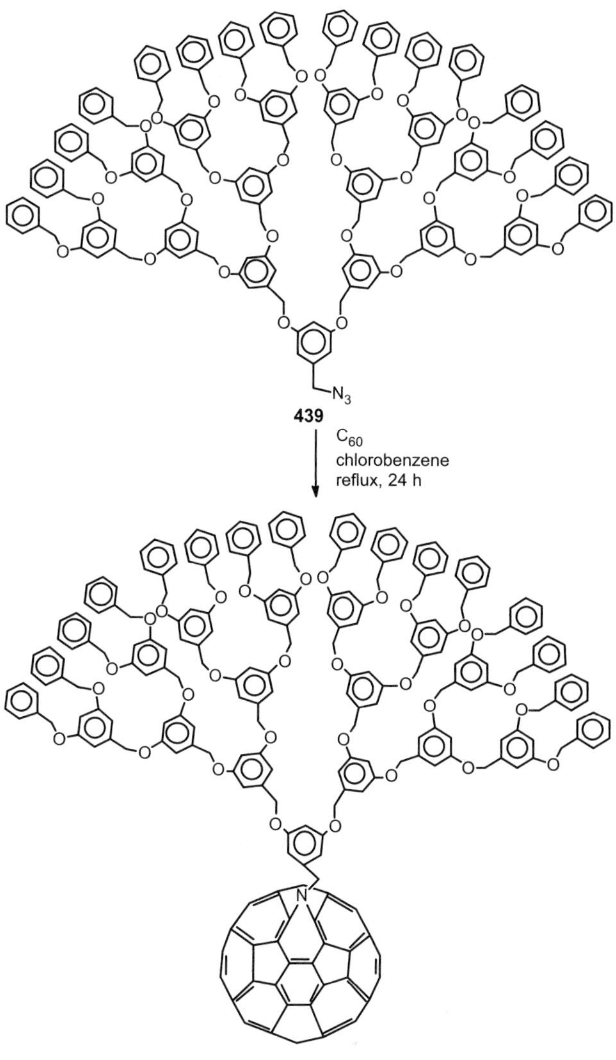

Figure 181. Synthesis of the first reported fullerodendrimer by cycloaddition of a dendritic azide to C_{60}. Adapted from [379].

fullerene–fullerene interactions are not entirely disrupted. Such fullerodendrimers with peripheral acylated glucose units and a fullerene core are prepared and incorporated in Langmuir films (Fig. 183) [381]. Fullerene amphiphiles with their glycoldendron headgroups form stable ordered monolayers at the air–water interface and show reversible behavior in successive compression/expansion cycles. The dendron headgroup is effective in preventing the irreversible aggregation usually observed in amphiphilic fullerene derivatives [382].

The fullerene–dendrimer conjugates **445** and **447** are prepared by DCC(*N*,*N*-dicyclohexylcarbodiimide)-mediated amide bond formation between carboxylic acids **443** and **446** and the *O*-acetylated trisglucoside wedge. Successive compression and expansion cycles are performed on the monolayers of the fullerene-glycodendron conjugates in the air–water interface in order to explore the reversibility of their formation. The results indicate that the formed films are very stable. The hydrophilic carbohydrate dendrons are polar enough to establish a strong amphiphilic character of

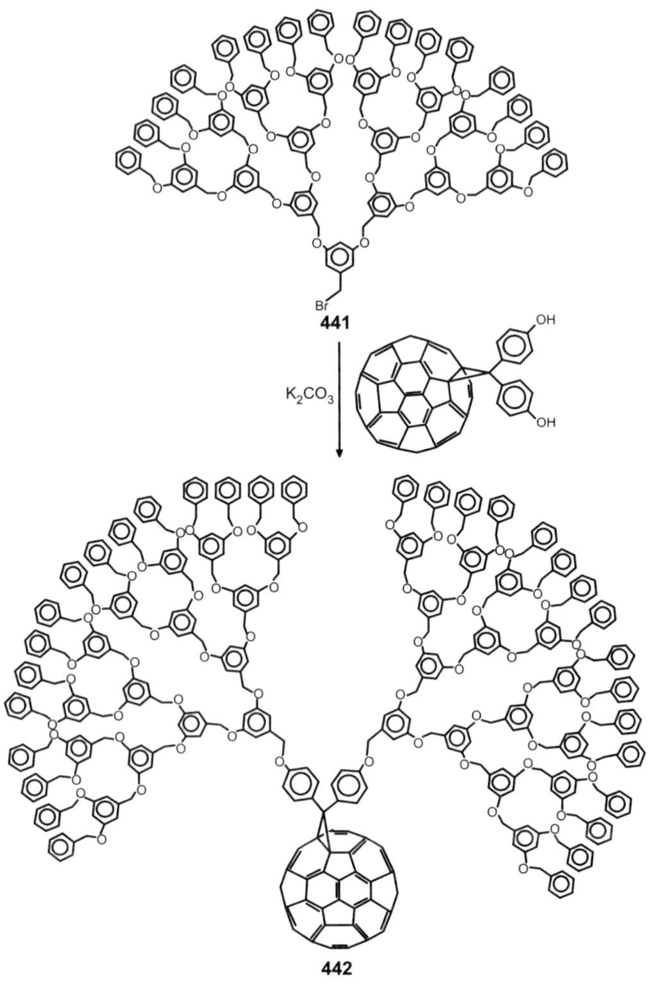

Figure 182. Reaction of a fourth-generation dendrimer with phenolic groups located at the surface of the functionalized fullerene. Adapted from [380].

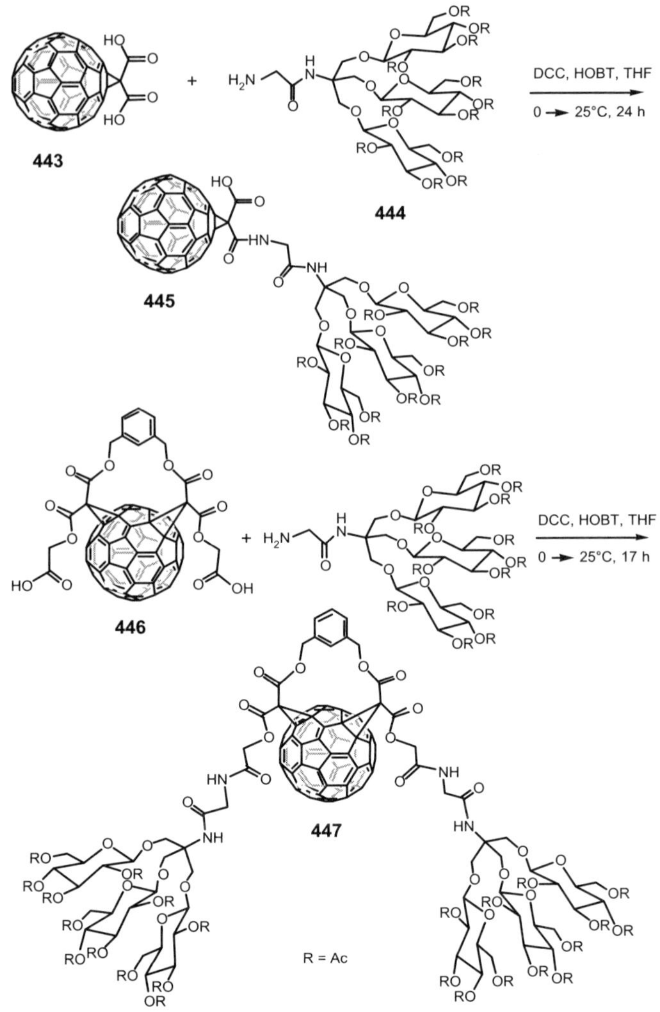

Figure 183. Carbohydrate-containing fullerene–dendrimer conjugates (DCC = *N*,*N*-dicyclohexylcarbodiimide, HOBT = 1-hydroxybenzotriazole) used for LB studies. Adapted from [381].

fullerene derivatives, yet bulky enough to avoid the irreversible aggregation of the fullerene centers.

Dendrimers with a fullerene core are appealing candidates to demonstrate the shielding effects resulting from the presence of the surrounding dendritic shell. Lifetime measurements in different solvents can be used to evaluate the degree of isolation of the central C_{60} moiety from external contacts. With this idea in mind, two series of fullerodendrimers are prepared (Fig. 184) [383]. In the design of these dendrimers, poly(aryl ether) dendritic branches terminated with peripheral triethyleneglycol chains are chosen to enable the compounds to dissolve in a wide variety of solvents. The synthetic approach to prepare compounds **448–451** relies upon the 1,3-dipolar cycloaddition of the dendritic azomethine ylides generated *in-situ* from the corresponding aldehydes and *N*-methylglycine. Dendrimers **452–455** are obtained by regioselective reaction at the fullerene sphere with bismalonate derivatives in a double Bingel cyclopropanation.

The photophysical properties of fullerodendrimers **448–455** have been studied in different solvents like toluene, dichloromethane, and acetonitrile. Interesting trends are obtained in the lifetimes of the lowest triplet excited states in air-equilibrated solutions. A steady increase in lifetime (τ) is found by increasing the dendrimer size in all the solvents (for example, τ in toluene increased from 279 to 374 ns for **448–451** and from 288 to 877 ns for **452–455**. This suggested that the dendritic wedges are able to shield, at least partially, the fullerene core from external contacts with solvent and from quenchers such as molecular oxygen. The triplet lifetimes for **451**, however, are different in the three different solvents, likely reflecting specific solvent fullerene interactions that affect excited state deactivation rates. This suggests that even the largest dendritic wedge is not able to provide a complete shielding of the central fulleropyrrolidine core. In contrast, the triplet lifetime of **455** in the three solvents lead to a similar value. This suggests that the fullerene core is in a similar environment irrespective of the nature of the solvent. In other words, the C_{60} unit is, to a large extent, not surrounded by solvent molecules but substantially buried in the middle of the dendritic structure. This creates a specific site-isolated microenvironment around the fullerene moiety. Molecular modeling calculations for both **451** and **455** reveal that in **451** the dendritic

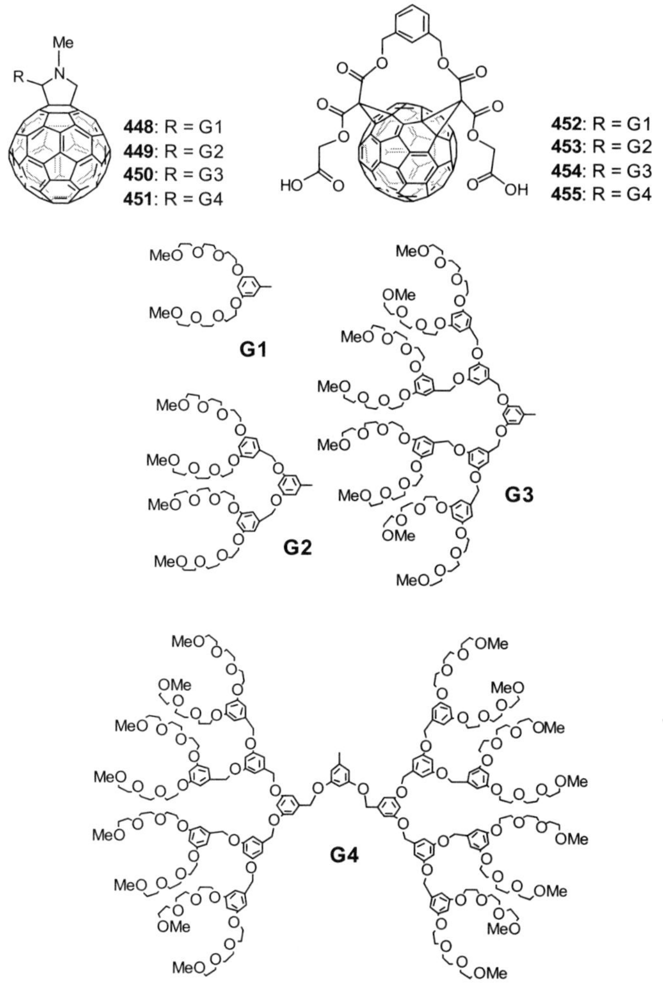

Figure 184. Fullereodendrimers with versatile photoactive core. Adapted from [383].

wedge indeed does not fully cover the fullerene moiety, whereas in **455** dendritic branches are able to fully cover the central fullerene core.

The efficient covering of the C_{60} core is useful to optimize the optical limiting property characteristics of fullerene derivatives. Fullerenes have a larger absorption cross-section in the excited states compared with that of the ground state. Therefore increased triplet lifetimes observed for the largest fullerodendrimers allow for an effective optical limitation. For practical applications, the use of solid devices is largely preferred to solutions and sol–gel glasses loaded with fullerene derivatives have demonstrated good application prospects [384]. Two factors, (i) faster deexcitation dynamics and (ii) reduced triplet yields, however, are limitations for fullerene-doped sol–gel glasses when compared to solutions [385]. The perturbation of the molecular energy levels due to the interactions with the sol–gel matrix and interactions between neighboring fullerene spheres due to aggregation are the possible reasons. The use of the fullerodendrimers prevents such undesirable effects. The incorporation of fullerodendrimers **448–455** in sol–gel glasses is achieved by soaking mesoporous silica glasses with a solution of the dendrimers. Measurements on the resulting doped samples reveal efficient optical limiting properties [383].

Amphiphilic dendrimers with peripheral fullerene units **456–458** are synthesized and their Langmuir and Langmuir–Blodgett (LB) films are studied (Fig. 185) [386]. The compounds form stable Langmuir films at the air–water interface and exhibit reversible compression/decompression behavior. The films can withstand pressures up to $P \approx 20$ mN m^{-1}. In the compression–decompression cycles no hysterisis is observed as long as the collapse pressure is not exceeded. The films of **456** and **458** are transferred onto silicon or glass substrates covered with a monolayer of octadecyltricholorosilane.

A series of fullerodendrimers is investigated in which Fréchet-type dendrons have been connected to a fullerene moiety via an acetylene linker (Fig. 186) [387]. 2,6-Dihydroxyethynylbenzene bearing benzyl ether type dendrons on the oxygens is covalently linked to C_{60} via an intervening triple bond (**459**). The dendritic fullerenes GnC_{60}H (n = 1–4) are highly soluble in common organic solvents. Their structures are characterized as 1,2-addition products with the ethynyl group and a proton added at a 6,6-bond on the fullerene cage. Molecular dynamics simulations with a MM2 forcefield show dense packing of dendron branches around the fullerene cages. The presence of favorable π–π stacking interactions is indicated by the face-to-face arrangement of the fullerene cage with several aromatic rings of the dendrons. Such interactions should have arisen out of intimate wrapping of dendron units with the fullerene core. The proton attached to the fullerene is highly acidic. The electronic properties of the fullerene core greatly change when it is transformed to the anion by the abstraction of this acidic proton. When the THF solution of the dendrimers is treated with one equivalent of Bu^{t}OK, the initial brown solution immediately turns dark green, indicating the formation of fulleryl anions. Oxidation of the anion GnC_{60}^- with iodine leads to the dimer GnC_{60}–GnC_{60}, the structure of which is confirmed by MALDI-TOF MS. ESR studies of the benzene solution of the dimer, however, indicated the presence of an equilibrium between monomeric and dimeric species. The photophysical and photochemical properties of these dendrimers are investigated by time-resolved fluorescence and time-resolved absorption methods, in addition to steady-state spectra [388]. The photophysical properties of the dendrimers such as lifetimes of the singlet and triplet excited states are essentially the same, regardless of the dendrimer generation. However, the rate constants of intermolecular processes such as triplet–triplet annihilation, triplet energy transfer, and electron transfer via the triplet states decrease with the increase in dendrimer generation.

Photophysical investigations of fulleropyrrolidine derivatives substituted with oligophenylenevinylene (OPV) moieties reveal a very efficient singlet–singlet OPV→C_{60} photoinduced energy transfer process. Based on this observation, dendrimers **460–463** with a fullerene core and peripheral OPV subunits are prepared (Fig. 187) [389].

The photophysical properties of fullerodendrimers are investigated in dichoromethane solutions. Upon excitation at the OPV band maximum OPV→C_{60} singlet–singlet

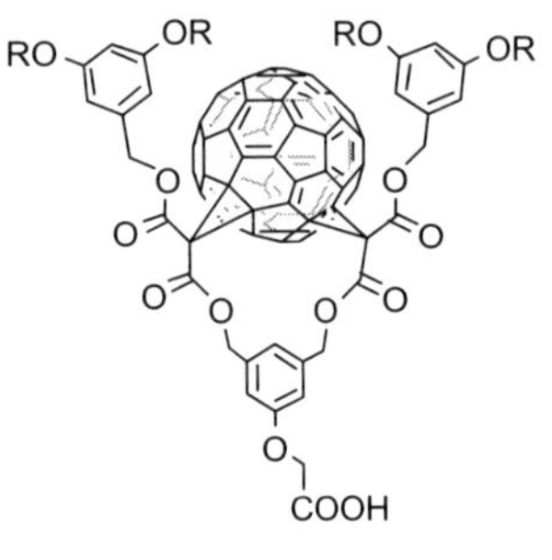

456

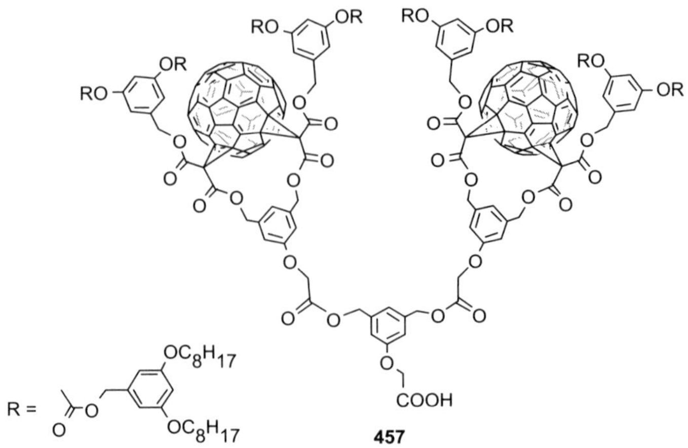

457

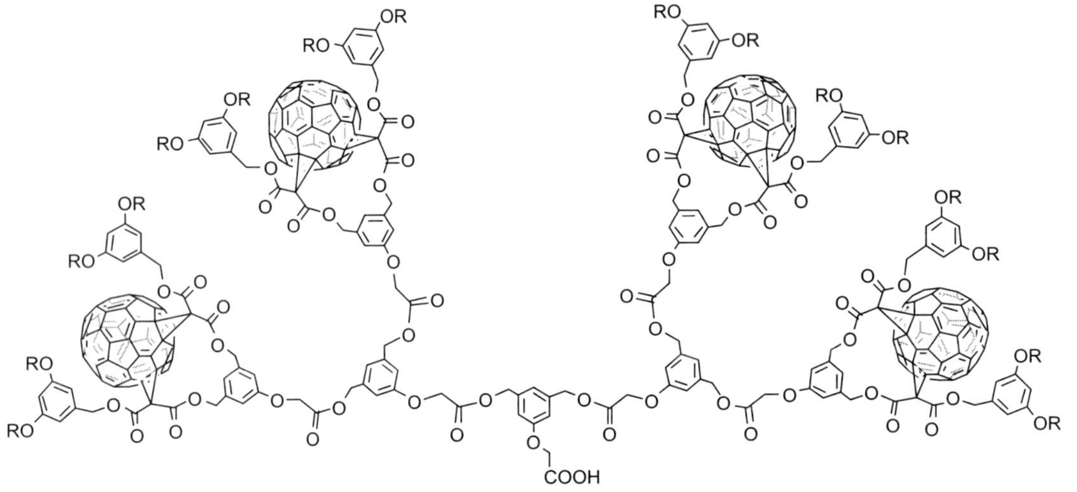

458

Figure 185. Amphiphilic dendrimers bearing peripheral fullerene subunits. Adapted from [386].

quenching of OPV fluorescence is observed for all fullerodendrimers. This indicates that the dendritic wedge is capable of channelling the absorbed energy to the fullerene core. In structurally related dendrimers it is established that the dendritic wedge plays at the same time the role of an antenna capable of channelling the absorbed energy to the fullerene core as well as of an electron-donating unit [390]. Phenylenevenylene dendritic wedge is attached to a pyrazoline[60]fullerene core, and efficient energy transfer from the excited antenna moiety to the C_{60} core is observed [391].

The dendritic wedges containing peripheral fullerene units are used to prepare new dendrimers with a bis(1,10-phenanthroline)copper(I) core [392]. This leads to dendrimers with a fullerene π-chromophore at the periphery (Fig. 188). The largest dendritic copper complex reported contains 16 peripheral C_{60} groups (**466**).

12.1. Dendritic Hexakisadducts of C_{60}

Fullerodendrimers in the foregoing examples have one dendritic wedge bound to the fullerene core. The single dendritic moiety imparts the property desired of a dendrimer. As the fullerene can undergo multiple additions, it can also generate a dendritic species even when the addends themselves are not high generation number dendrimers. The globular dendrimers with a C_{60} core can be structurally varied because C_{60} allows synthesis with T_h (hexakisadduct), C_{2v} (pentakisadduct), C_s (tetrakis and *e*-bisadducts), and C_3 (*e*,*e*,*e*-trisadduct) symmetry. The addends can all be same or different, thus leading to a number of possible regioisomers. Starting from precursors of incomplete octahedral addition pattern, the regioselective formation of a further mixed hexaaddition pattern can be expected. Indeed, this aspect has drawn a lot of attention resulting in many interesting reports which are summarized in a review article [18]. Template-mediated cyclopropanation techniques or tether-directed functionalization allow for straightforward production of hexakisadducts. Two views of macromolecular architectures based on the T_h symmetric motif are shown in Figure 189.

By a suitable choice of the dendritic moiety and the linking chain it is possible to attach dendrimeric units at all the six octahedral sites. A number of hexakisadduct of the type T_h-$C_{66}(COOR)_{12}$, where R is benzyl ether based dendritic unit, are reported [394]. Even if low generation dendra are employed, the products are still spherical dendrimers, some

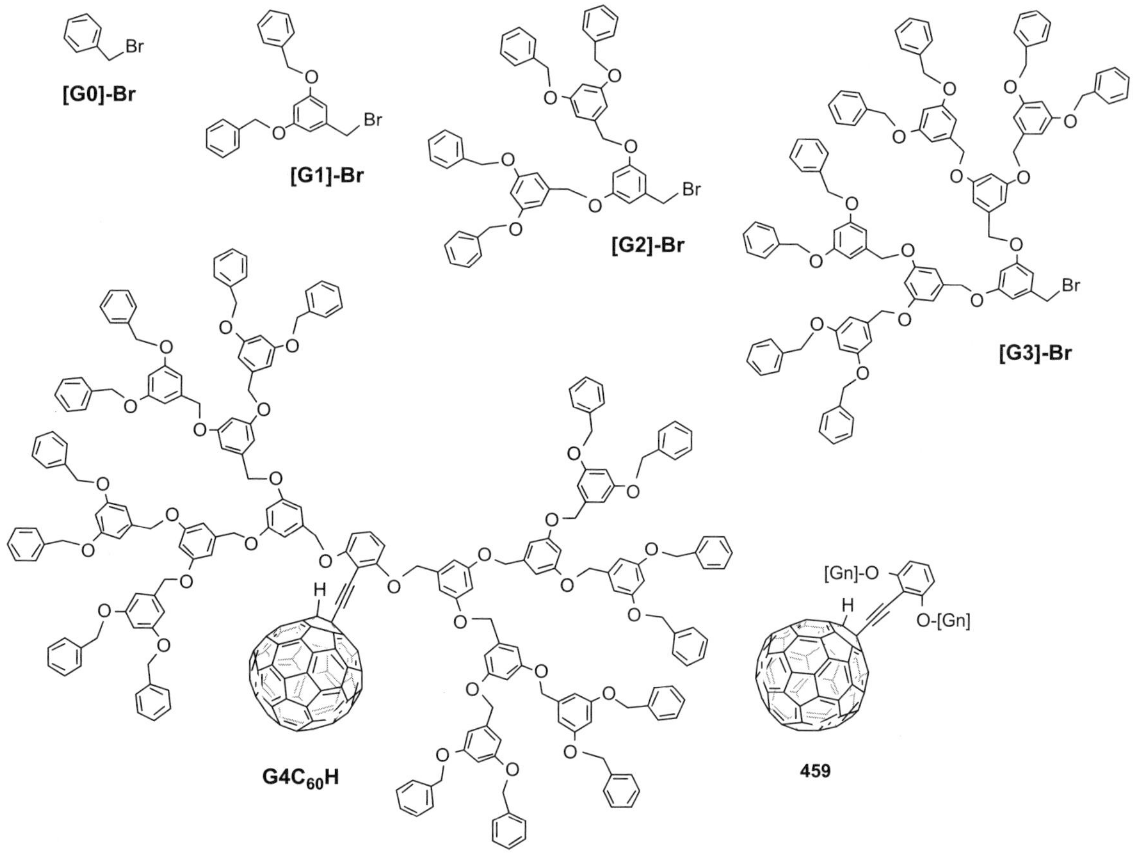

Figure 186. Benzyl-ether type dendrimers connected to C_{60} core by a single covalent bond. Adapted from [387].

of which can have more specific functions. For example, addition of six mesotropic cyanobiphenyl malonate addends produces the spherical thermotropic liquid crystal **467** and a highly water-soluble dendrimer **468** containing 12 carboxylic acid groups at the periphery (Fig. 190) [395].

A number of fullerene precursors with incomplete octahedral addition pattern and dendritic wedges of different generations in the form of their bromomalonates are used (Fig. 191) [396], which on nucleophilic cyclopropanation reaction affords the dendrimeric hexaadducts. For example, reaction of the C_{2v}-symmetrical penatkisadduct leads to the dendrimer **469** (Fig. 192). Similarly dendrimers from the tetrakis-, tris-, and bisadducts are obtained from multiple cyclopropanation of the bromomalonates containing the **G1–G3** dendrons. As seen before, the 6,6-ring junctions in the vacant octahedral sites are preferred for further reaction.

13. POLYFULLERENES AND FULLEROPOLYMERS

In view of the interesting properties of fullerenes, there have been a number of studies exploring the behavior of fullerenes under the routine polymerization conditions. Extensive literature is available on the subject, with a number of review articles [397]. Polyfullerene ideally should be a linear chain of the fullerene molecules joined by one or more covalent bonds. Fullerenes are known to undergo photopolymerization in the solid state [398]. In this reaction a [2 + 2] cycloaddition between two adjacent fullerene molecules involving double bonds at the hexagonal junctions has been suggested. C_{60} can also be polymerized under the combined action of high pressure and temperature [399]. Fullerene polymers revert to pristine monomer upon heating [400]. Photopolymerization of C_{60} in solution using CCl_4, CH_2Cl_2, cyclohexane, and decalin as solvents has been carried out [401]. Irradiation of C_{60} solutions under nitrogen flow produced C_{60} photopolymer, the structure of which is found to be comparable to that of C_{60} photopolymer prepared in solid state. A perfect linear C_{60} polymer has been difficult to comprehend and hence to realize in practice because of lack of specificity of reaction on the fullerene sphere. In order for such a linear polymer to be synthesized, it is necessary that the molecules of C_{60} be linearly arranged. Fullerenes are prone to co-crystallization with other molecules, and in such co-crystals it is possible that the C_{60} molecules could be organized in a linear fashion. One such example has been C_{60}-calixarene co-crystal, in which C_{60} molecules are arranged strictly in a linear manner. Further, such linear chains of C_{60} molecules are well separated. Action of heat and pressure on a co-crystallate of C_{60} and calixarene furnished a linear [2 + 2] addition polymer **470** without cross-linking (Fig. 193) [402].

Though this polyfullerene is interesting and attractive, most studies have been carried out with fullerenes incorporated into a conventional polymer. Fullerene molecules

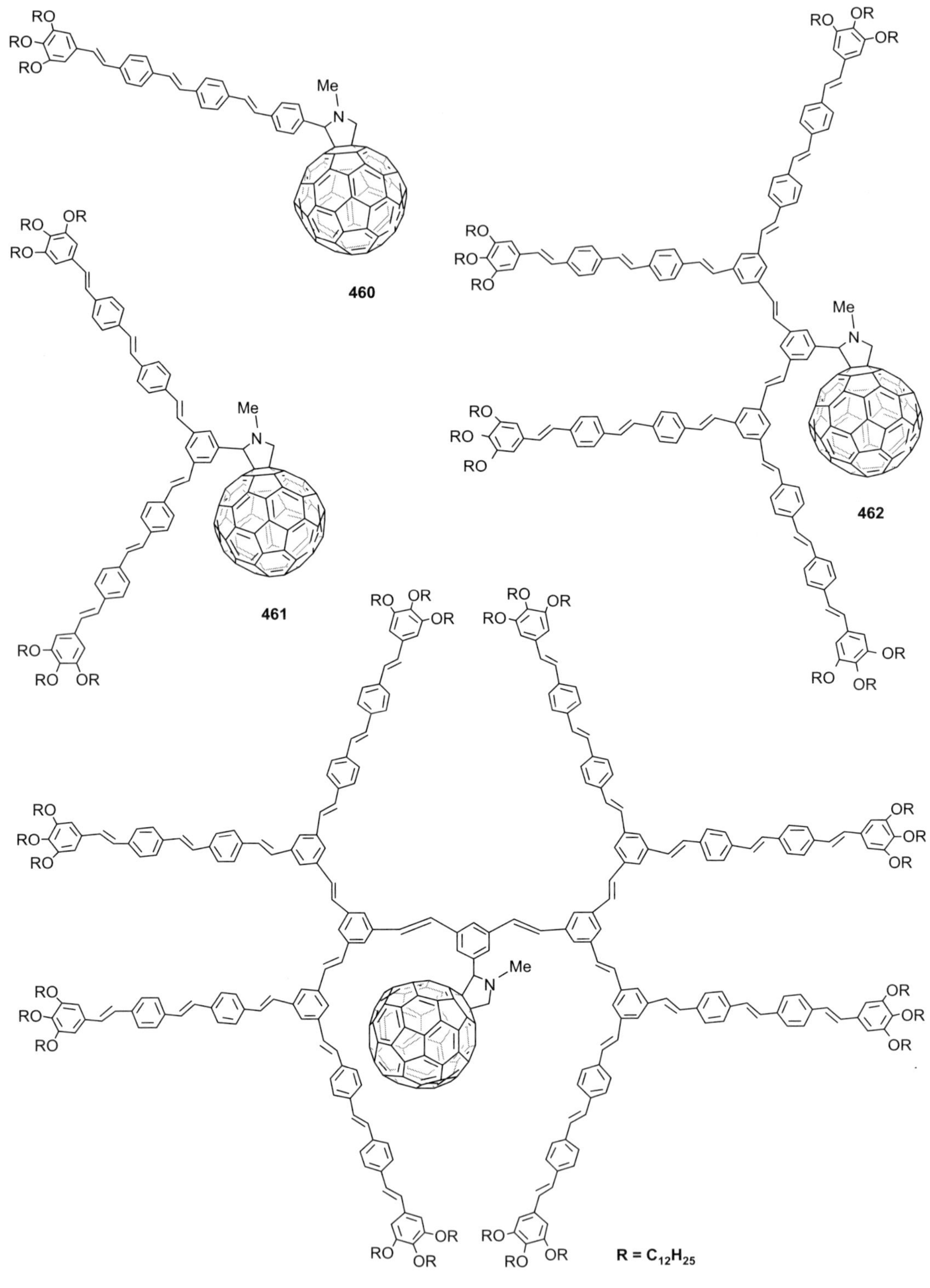

Figure 187. Fullerene–oligophenylenevinylene conjugates. Adapted from [389].

can be incorporated in polymers by several means leading to distinctly different structures of the resulting fullerene polymer. The fullerene moiety can be (1) in the main chains of the polymer, (2) in the core of a starburst polymer, (3) a dendritic polymer, (4) cross-linkers in the polymer network, and (5) grafted onto a solid support. Further, these materials can be obtained by several means, such as free radical, cationic, and anionic polymerizations, condensation, and addition reactions, selected examples of which are discussed here. Since the field of fullerene-dendrimers has significantly grown in volume in recent years, this subject is described in a separate section.

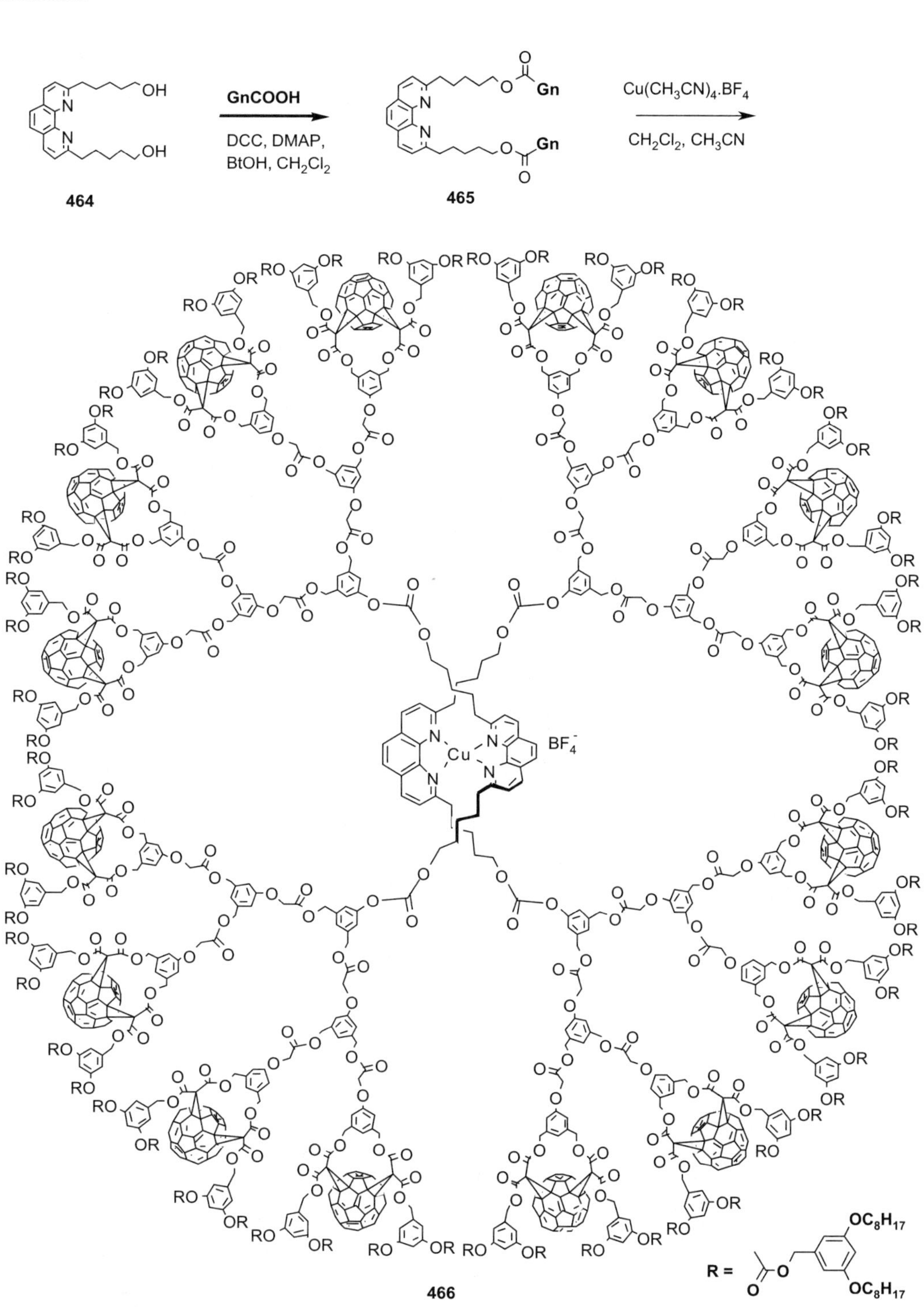

Figure 188. A fullerodendrimer with a bis(phenanthroline)Cu(I) core and peripheral fullerene units. Adapted from [392].

13.1. Radical Polymerization

One of the earliest free radical reactions on C_{60} is the reaction of xylylene diradical leading to C_{60}-*p*-xylylene copolymer [403]. The diradical is obtained when paracyclophane is sublimed at 300 °C into a tube furnace set at 650 °C. It is then swept into a solution of C_{60} in toluene cooled to −78 °C. Such reaction conditions could allow multiple additions of the radicals, leading to a complex product.

Both C_{60} and C_{70} are copolymerized with a number of vinyl monomers, such as styrene, in a standard free radical polymerization either in bulk or co-dissolved in an aromatic solvent using azoisobutyronitrile (AIBN) as the initiator [404]. C_{60}-styrene random copolymers are prepared using bulk and solution polymerization techniques [405]. The resulting polymers are soluble and behave like ordinary polystyrene in solution except for the dark brown color. Multiple additions

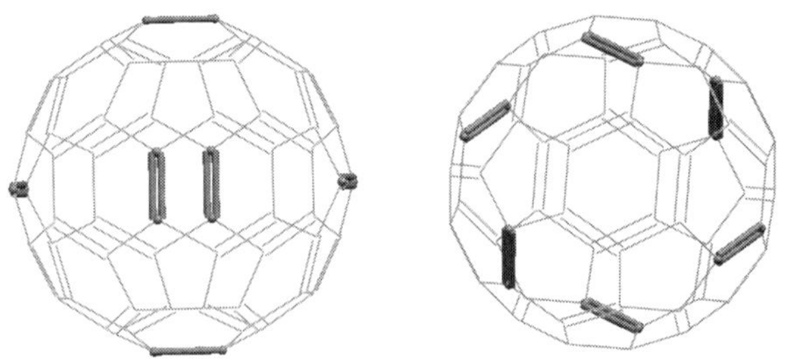

Figure 189. 6,6-Bond positions for a T_h symmetrical octahedral addition pattern leading to a hexakisadduct: tow different views. Adapted from [393].

are commonly observed in radical polymerization of styrene and MMA in the presence of C_{60} leading to a branched structure [406]. Upon copolymerization, the characteristic UV-vis spectra of C_{60} disappears, which has been ascribed to incorporation of substitutenets onto C_{60} molecules as a result of polymerization. Other AIBN-initiated copolymerizations of C_{60} with 4-vinylbenzoic acid, 2-vinylpyridine, or 4-vinylpyridine have been carried out [407]. Laser light scaterring studies of these polymers indicate that the change in the UV-vis pattern is a consequence of scattering rather than true absorbance. Solubility and micellization behavior of C_{60} fullerenes with two well-defined polymer arms leading to materials of the types C_{60}-$(PS)_2$, C_{60}-$(PVP)_2$, and C_{60}-(PS-PVP)$_2$, where PS is polystyrene, PVP is poly(*p*-vinylphenol), and PS-PVP is a diblock copolymer, are prepared by applying the nitroxide controlled free radical polymerization technique [408]. Direct fullerenation of commercially available optical polymer, polycarbonate (PC), is achieved by simply irradiating a solution of PC and C_{60} at room temperature using a conventional UV lamp or by warming a C_{60}/PC solution to a moderate temperature in the presence of AIBN [409]. Radical polymerization of vinyl monomers can be inhibited by fullerenes. The effect of C_{60} on the radical polymerization of vinyl acetate with dimethyl 2,2′-azobisisobutyrate in benzene is investigated kinetically and by means of ESR. C_{60} is found to act as an effective inhibitor of polymerization [410].

13.2. Fullerene–Initiated Polymerization

The dications of C_{60} and C_{70} are found to initiate the polymerization of 1,3-butadiene in gas phase [411]. Reactions of $C_{60}^{\bullet+}$, $C_{70}^{\bullet+}$, C_{60}^{2+}, and C_{70}^{2+} with 1,3-butadiene are conducted. $C_{60}^{\bullet+}$ and $C_{70}^{\bullet+}$ are found to be unreactive toward 1,3-butadiene. However, both C_{60}^{2+} and C_{70}^{2+} are observed to react rapidly with 1,3-butadiene with the sequential addition of up to six molecules of 1,3-butadiene per C_{60}^{2+} and C_{70}^{2+} species. The sequential addition of allenes to C_{60}^{2+} leads to the formation of a chain of six-membered rings. Fullerene dications and trications have been used as initiators of polymerization of allene and propyne in the gas phase leading to observation of a remarkable periodicity in chain growth [412]. In the presence of $AlCl_3$, C_{60} is activated as a cationic species. This activated fullerene is reacted with poly(9-vinylcarbazole) (PVK) leading to grafting of C_{60} onto the polymer chain. The mechanism involves initial cationic activation of a C_{60} double bond by Lewis acid, followed by nucleophilic attack of the arene moiety of PVK [413].

13.3. Fullerenation of Polystyrene

C_{60} can be incorporated into non-cross-linked polystyrene by reaction with $AlCl_3$. The preparation of pendant C_{60}-polystyrene polymers is possible by use of Friedel–Crafts

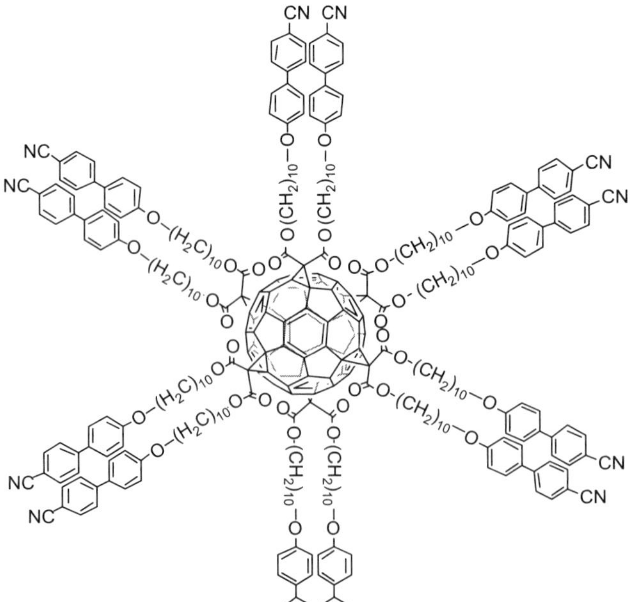

467

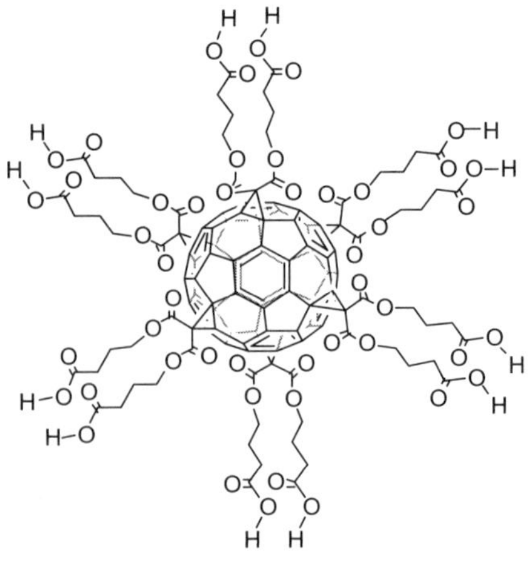

468

Figure 190. A spherical thermotropic liquid crystal and a highly water-soluble C_{60} dendrimer. Adapted from [395].

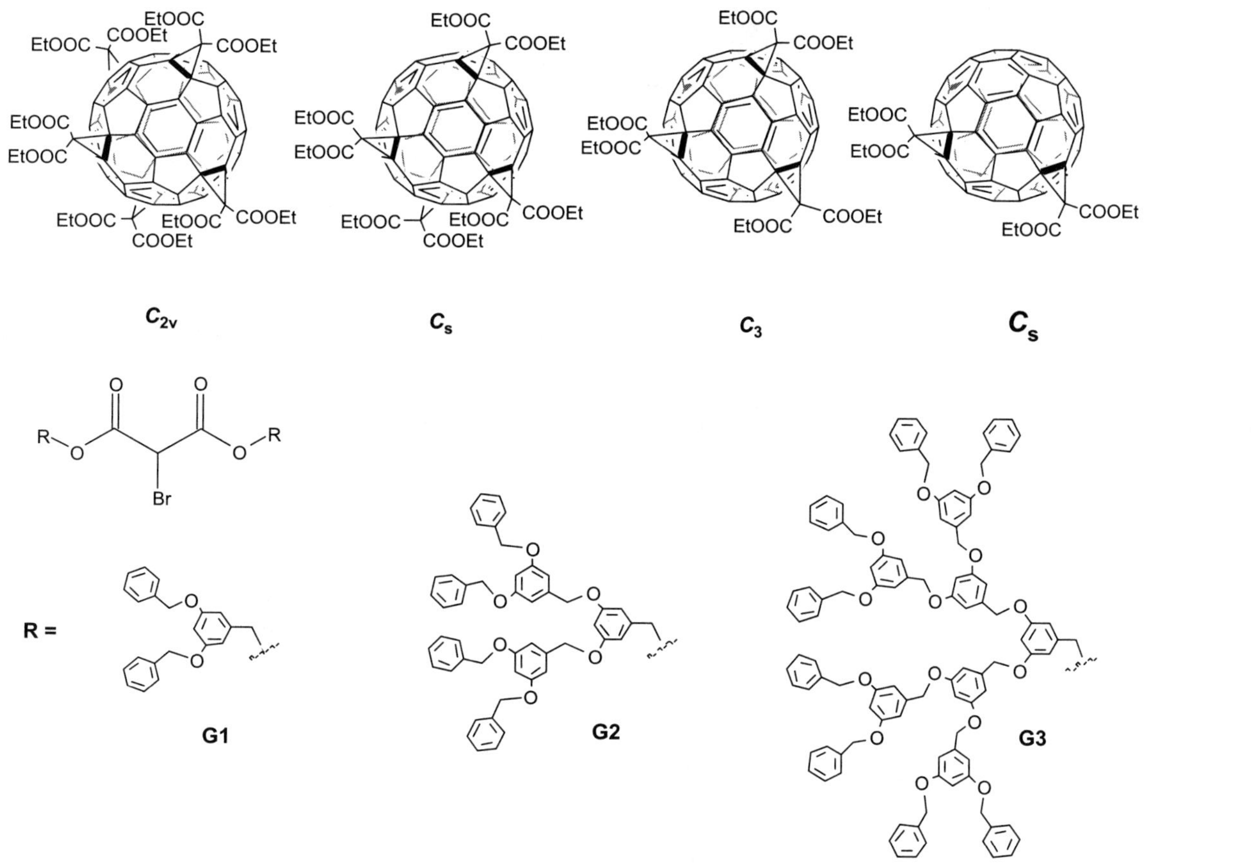

Figure 191. Fullerene derivatives with incomplete octahedral addition patterns and the bromomalonates used for synthesis of dendritic mixed hexakisadducts. Adapted from [396].

type reactions (Fig. 194) [414]. The resulting polymers are soluble in organic solvents such as chloroform. The fullerenation of polystyrene provides a facile method of homogeneously dispersing fullerene molecules into polymer matrices in a structurally somewhat controllable fashion.

13.4. Fulleropolymers by Cycloaddition Reactions

Polymers containing suitable reactive functions can undergo cycloaddition reaction with C_{60} leading to covalent anchoring of the fullerene to the polymer. In the first such example,

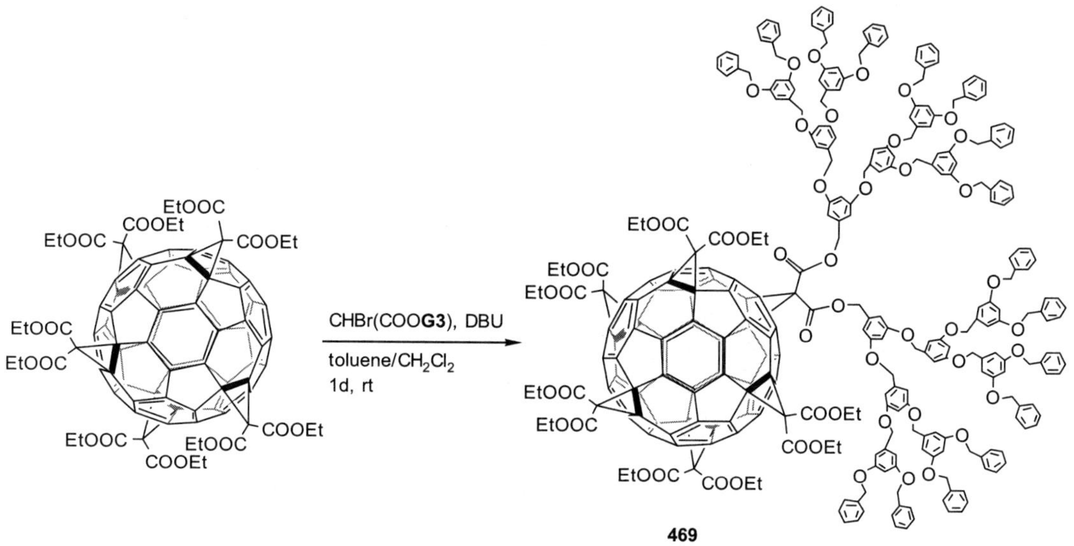

469

Figure 192. The C_{2v}-symmetrical tetrakisadduct as a dendrimer core. Adapted from [396].

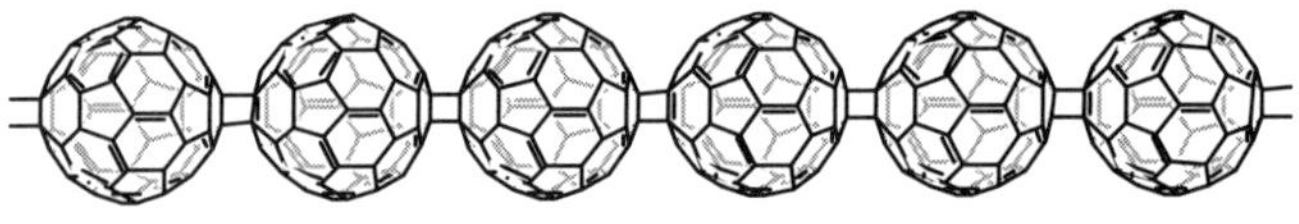

470

Figure 193. A linear polymer of C_{60} via supramolecular preorganization. Adapted from [402].

a simple linear polymer containing azide groups is reacted with C_{60} leading to a polymer with pendant C_{60} connected to the polymer backbone by nitrogen bridging. For this purpose, azidomethyl-substituted polystyrene is prepared. Percentage of azidomethyl units is controlled by the initial feed ratio of styrene and chloromethyl styrene. Reaction of the azimethyl-functionalized polymer with refluxing chlorobenzene furnishes the required polymer in which C_{60} molecules are pendant to the backbone polymer (Fig. 195) [415].

The end functionalization of hexaarm star-shaped polystyrene, prepared via living cationic polymerization by organosilicon compound, has resulted in a star polymer containing C_{60} as end caps [416]. After activation of the chain ends by $TiCl_4$, azido groups are introduced through the use of their trimethylsilyl derivatives, which on cycloaddition reaction with C_{60} furinshes the hexafullerene star polymer, **473** (Fig. 196).

Diels–Alder reaction of bis-*o*-quinodimethanes leads to a fullerene main-chain polymer. For this reason, bis-hexyloxy-substituted bis-sulphone is synthesized, which on heating generates the desired bis-*o*-quinodimethane *in-situ* [417]. Admixing of a mono-*o*-quinodimethane effectively suppressed cross-linking. The main-chain polymer **474** of high molecular weight consisting of 80 C_{60} units on the average is prepared (Fig. 197).

Benzocyclobutenone (BCBO) is known to generate reactive *o*-quinodimethane upon thermal activation. The diene α-oxo-*o*-quinodimethane is found to undergo [2 + 4] cycloaddition reaction with C_{60}. Benzocyclobutene is prepared in large quantities from anthranilic acid, homophthalic anhydride, or *o*-toluoyl chloride. Nitration followed by reduction gives a functionalized BCBO, which is used to prepare a BCBO containing vinyl monomer (Fig. 198) [418].

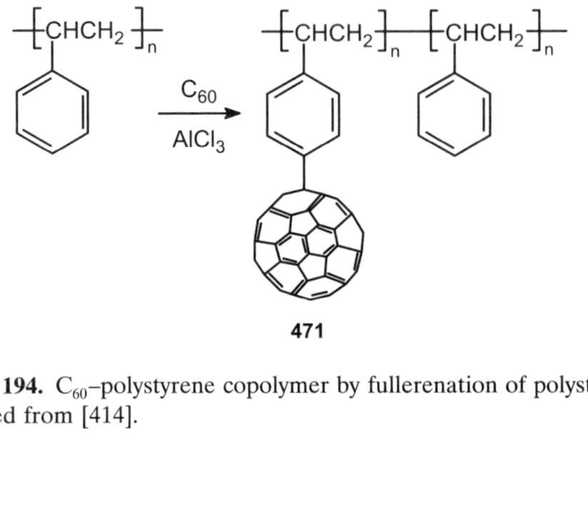

471

Figure 194. C_{60}–polystyrene copolymer by fullerenation of polystyrene. Adapted from [414].

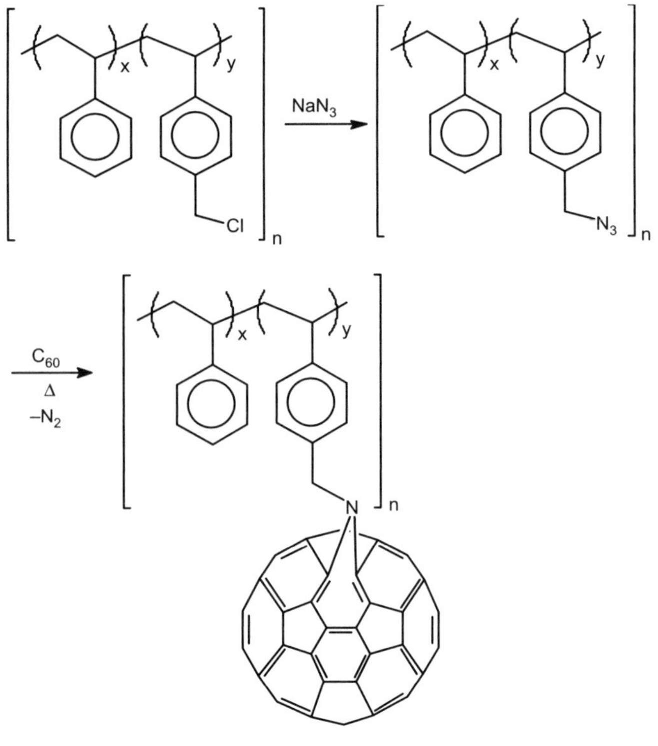

472

Figure 195. Reacting C_{60} with azimethyl-functionalized polymers. Adapted from [415].

13.5. Anionic Living Polymerization

Reactions of preformed polymeric species in the form of anions have been studied following living anionic polymerization route. A starlike polymer with polystyrene units attached to a C_{60} center has been reported [419]. In this method polystyrene anions are allowed to react

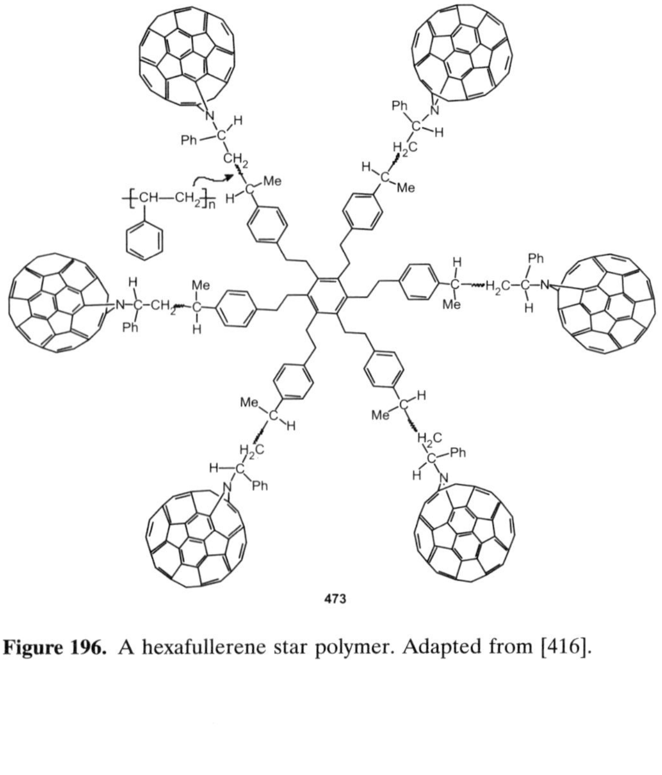

473

Figure 196. A hexafullerene star polymer. Adapted from [416].

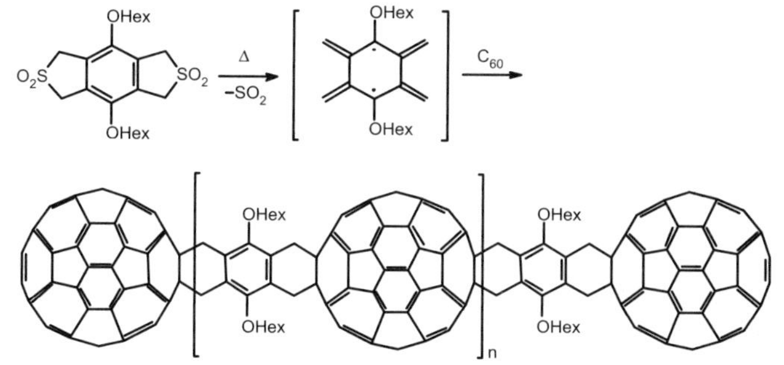

474

Figure 197. Repetitive Diels–Alder reaction of C_{60} and bis-*o*-quinodimethane furnishing a fullerene main-chain polymer. Adapted from [417].

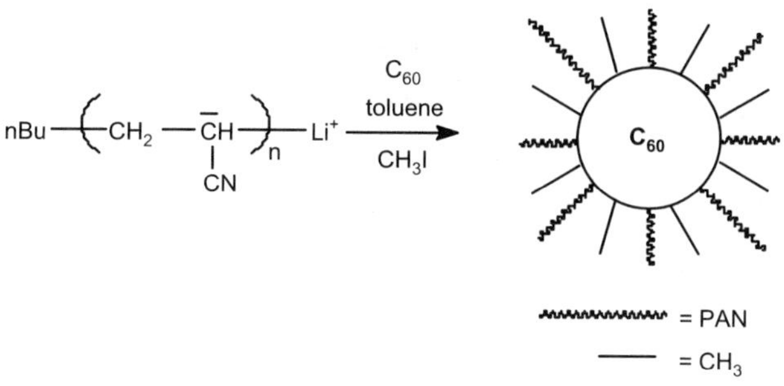

477

Figure 199. C_{60} and polyacrylonitrile copolymer. Adapted from [421].

with C_{60}, followed by an end-capping reaction with CH_3I. One to ten multiple polystyrene arms get attached to C_{60} leading to a star polymer which is insoluble. The key to the development of polymer-modified fullerenes is to form soluble fullerenated polymers in high yield under very mild conditions. Addition of C_{60} to carbanion intermediates of such polymers as polystyrene, poly(bromostyrene), and poly(vinylbenzylchloride) generated by an organometalic reaction leads to soluble polymers [420]. The preparation and structural characterization of novel starlike $C_{60}(CH_3)_3$PAN and $C_{60}(CH_3)_3$PMS copolymers by the reaction of living *n*-butyl-terminated polyacrylonitrile (PAN) and poly(methyl styrene) (PMS) with C_{60} in a heterogeneous medium, followed by a capping reaction with methyl iodide, have been reported (Fig. 199) [421]. Multiple PAN and PMS arms are attached to the C_{60} core. The *n*-butyl-terminated polyacrylonitrile copolymer $C_{60}(CH_3)$PAN has been pyrolyzed under vacuum [422]. As pyrolysis proceeds, evidence for aromatic ring formation is found. The *in-situ* infrared absorption measurements demonstrated that the nitrile groups of the polymers disappeared gradually as heating progressed and conjugated C=N chains appeared.

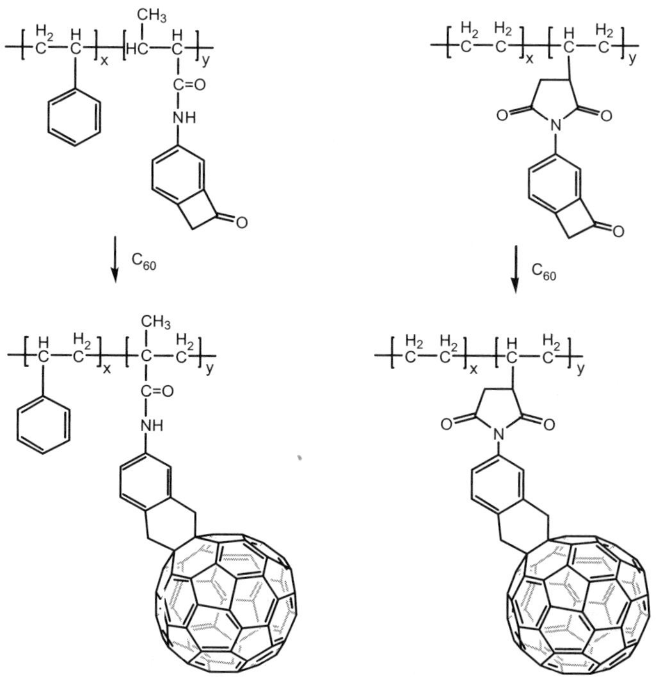

475

476

Figure 198. Incorporation of C_{60} into polymers by cycloaddition reaction with cyclobutenone. Adapted from [418].

Two methods to introduce a negative charge on a C_{60} and the further use of these carbanions as initiators for anionic polymerization of styrene and methyl methacrylate are investigated [423]. Fullerene anion is generated from the potassium salts of C_{60}, obtained by reduction with potassium naphthalenide. The resulting hexaanion only can polymerize MMA. The mechanism of polymerization is by electron transfer so that no fullerene is attached to the PMMA. Alternatively, polystyryllithium is added to C_{60} to form living star molecules with a C_{60} core bearing a well-controlled number of carbanion branches. While hexaanion of C_{60} can only initiate anionic polymerization of styrene by addition onto the monomer, five are enough to initiate the polymerization of MMA. Stars with seven PS branches and heterostars with six PS and two PMMA have been produced. By reacting six PSLi per C_{60} molecule, fairly pure star-shaped polymers with six branches and six carbanions located on the fullerene core are obtained. Only one of these six cabanions is able to initiate the anioninc polymerization of styrene or isoprene leading to well-defined palm-tree-like achitectures $(PS_a)_6C_{60}PS_b$ or $(PS)_6C_{60}PI$, **478** (Fig. 200) [424]. Also the out-growing chain bearing the carbanion has been allowed to react with the coupling agent dibromo-*p*-xylene and dibromohexane to produce the dumbbell-like structures.

Covalent incorporation of C_{60} onto the functionalized polyethylene film is reported by the anionic polymerization route (Fig. 201) [425]. Diphenylmethyl terminated ethylene oligomer is prepared from living oligomers by reaction with 1,1-diphenylethylene followed by deprotonation using *n*-BuLi. First films containing diphenylmethyl terminated ethylene oligomer are prepared by codissolving with HDPE, casting the solution mixture, and then treating with *n*-BuLi–TMEDA. The lithiated film is finally treated with C_{60}, which leads to the grafting of C_{60} onto polyethylene surface (**479**).

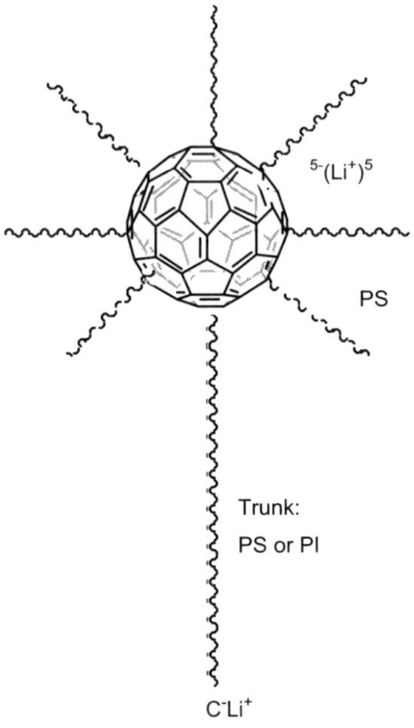

478

Figure 200. Palm-tree-like polymer architecture based on C_{60}. Adapted from [424].

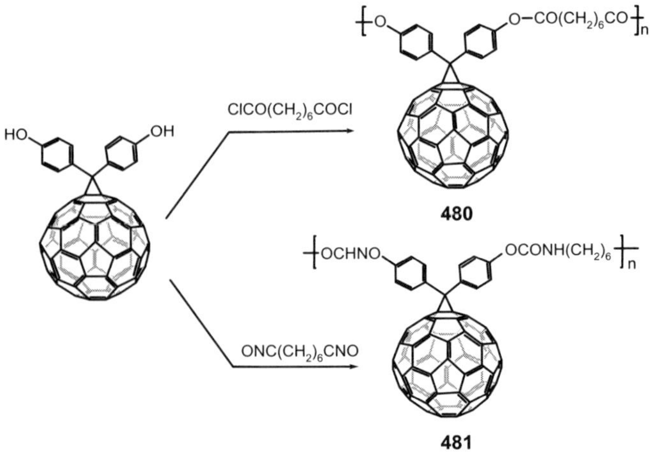

Figure 202. Co-condensation of fullerene bearing carboxy group to yield fullerene polyester and polyurethane. Adapted from [426].

13.6. Condensation Polymerization

One of the early approaches to fullerene-based polymers is to first introduce reactive function to C_{60} and then co-condense it with another reactive monomer. Diphenyl fullerene derivatives bearing hydroxy- and amino-functions are used for C_{60} containing polyesters and polyamides. The diphenyl fullerene derivative containing phenolic OH groups is condensed with sebacryl chloride or hexamethylene diisocyanate to produce two polymers containing C_{60} molecules (**480**, **481**) dangling from the polymer chain (Fig. 202) [426].

Methano[60]fullerene dicarboxylic acid (MFDA) obtained from hydrogenolysis of diethyl methano[60]fullerene carboxylate is a suitable derivative for reaction with a preformed polymer, such as poly(propionylethylenimine-*co*-ethylene). Reaction of MFDA with poly(propionylethylenimine-*co*-ethylene) using 1-ethyl-3-(dimethylaminopropyl)carbodiimide as a coupling agent leads to a highly water-soluble pendant [60]fullerene-polymer, **482** (Fig. 203) [427].

Fullerenols contain multiple polar hydroxy functions and hence are versatile molecules in the prospective design of three-dimensionally stretched polymers. The synthesis of a urethane-connected polyether star utilizing fullerenol as a molecular core is synthesized [428]. The polymer is found to contain six polymer arms per C_{60} on average. Fullerenol is first treated with a polymer of diisocyanated urethane polyether followed by termination reaction with dodecan-1-ol, which affords the corresponding C_{60}-based urethane-connected polyether polymers (Fig. 204). The synthesis of diisocyanate-terminated polyurethane is carried out by the reaction of poly(tetramethylene oxide) glycol and a diisocyanate. This polymer is then allowed to react with fullerenol [429]. The resultant flexible polymers **484** exhibit greatly enhanced thermal stability.

The addition of 2-methylaziridine to C_{60} and subsequent reaction of the adduct with a preformed polymer is investigated. The adduct, suggested to have 10 aziridine rings per molecule of C_{60}, reacts readily with Bisphenol A and Novolac via aziridine ring opening to produce polymers with good thermal stability [430]. The copolymers are three-dimensional polymers containing C_{60} with low coefficients of friction and good wear properties for use as potential solid lubricants. A new approach toward well-defined fullerene-containing poly(dimethylsiloxanes) using an *o*-xylylene derivative method has been described. The polymers **485** exhibited outstanding solubility in common solvents such as chloroform, THF, and benzene, a high thermal stability, and excellent film-forming properties (Fig. 205) [431]. Different strategies for the synthesis of polymers, either containing C_{60} in the polymer backbone or bearing C_{60} as part of the side chains, are described.

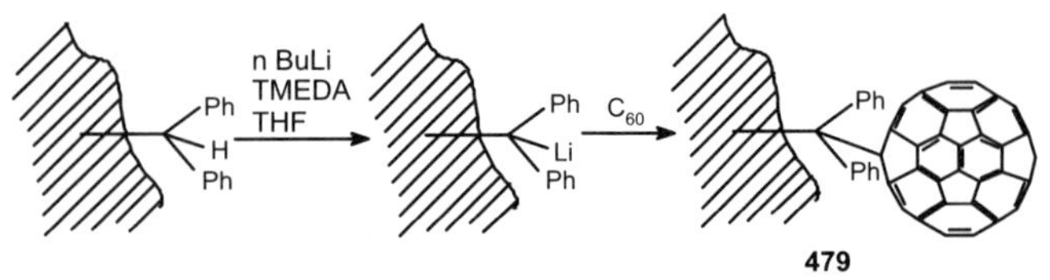

Figure 201. Covalent attachment of fullerene on a polyethylene surface. Adapted from [425].

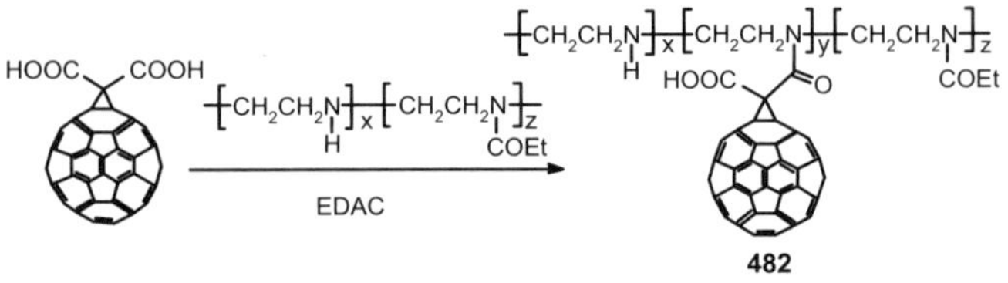

Figure 203. A highly water-soluble [60]fullerene-poly(propionylethylenimine-*co*-ethylene) polymer. Adapted from [427].

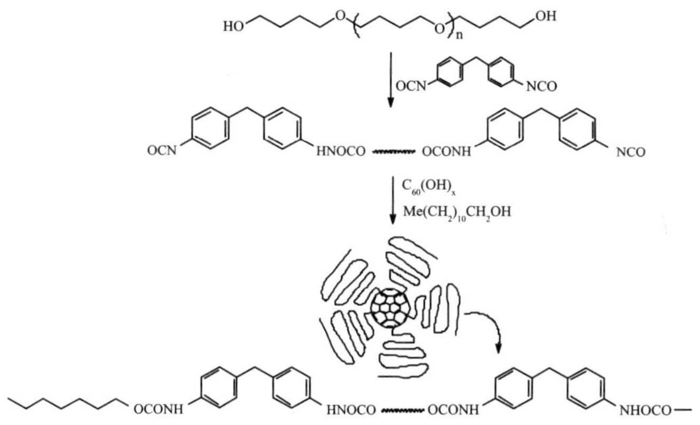

483

Figure 204. Urethane-connected polyether dendritic polymer of C_{60}. Adapted from [428].

13.7. Nucleophilic Reactions

Polymers having amino groups either on the branch or at the termini react readily with C_{60}. Invariable such reactions lead to cross-linked products, which are insoluble. A bifunctional amino-terminated poly(oxyethylene) or poly(oxypropylene) is reacted with C_{60} or C_{70} in toluene [432]. Soluble products formed in the early stages of the reaction later cross-link by reaction between the polyfunctional fullerenes and the bifunctional polyethers. Fullerene functionalized amine containing polymers is capable of withstanding a temperature of at least about 300 °C and in some instances as high as 650 °C. In some cases cross-linking results in enhanced mechanical properties. Polymers with C_{60} pendants on the side chain of polymers soluble in toluene and carbon disulfide are synthesized by titrating toluene solutions of C_{60} with the aminopolymers: poly(ethylene imine), poly[4-[[(2-aminoethyl)imino]methyl]styrene] (**486**), and poly(propylene imine) (**487**) (Fig. 206) [433].

Fullerenes are found to add to amine containing flexible hydrocarbon polymers such as ethylene propylene terpolymer (EPDM-amine) to obtain C_{60}-functionalized polymers, **488** (Fig. 207) [434]. Thermogravimetric analyses show that the fullerene-attached EPDM-amine is thermoxidatively more stable than pure EPDM-amine.

A soluble well-defined fullerene-end-capped polystyrenes is prepared by reaction of amino-terminated polystyrene with C_{60} (**489**). The polymers formed clear homogeneous films; that is, no microphase or nanophase separation is observed (Fig. 208) [435].

A novel strategy to achieve water-soluble polyfullerenes is developed in which a bifunctional monomer is encapsulated in a cyclodextrin cavity and then the inclusion complex is reacted with C_{60}. The result is the first main-chain polyfullerene (**490**), where the cyclodextrin units are fixed at precise positions close to fullerenes without being in direct bond formations with the fullerenes (Fig. 209) [436].

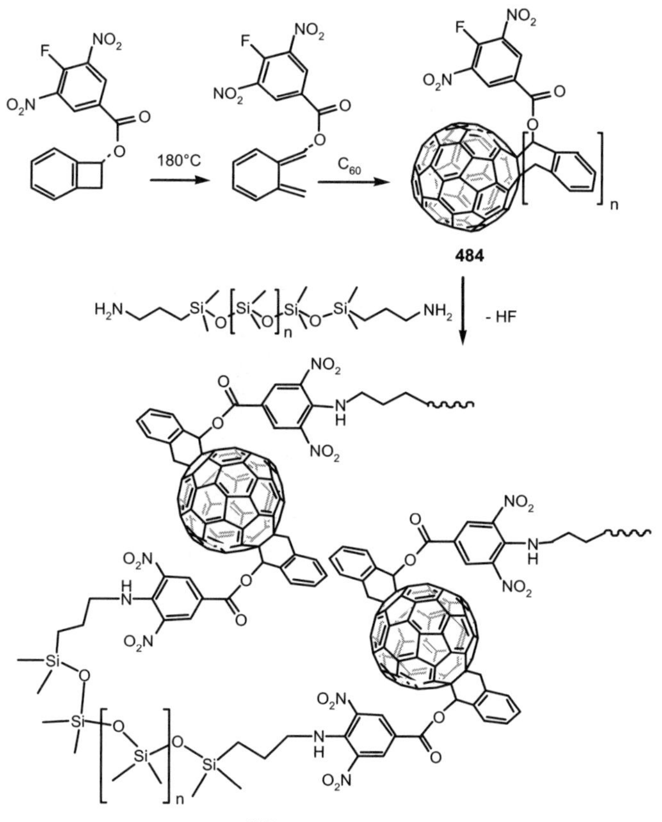

484

485

Figure 205. C_{60}-containing poly(dimethylsiloxanes): soluble polymers with high fullerene content. Adapted from [431].

14. FULLERENE MACROCYCLES

The attachment of macrocyclic compounds such as calixarenes, catenane, crown ethers, cyclophane, cyclotriveratrylene, phthalocyanines, etc. to fullerenes leads to a rich chemistry of supramolecular fullerenes [437], which is interesting because the materials lead to useful applications [438]. These applications stem from the remarkable ability of fullerenes to accept electrons resulting in the formation of donor–acceptor complexes. Fullerenes are known to interact electronically with molecules of organic and inorganic origin like P_4, S_8, I_2, benzene, saturated amines, metallocenes,

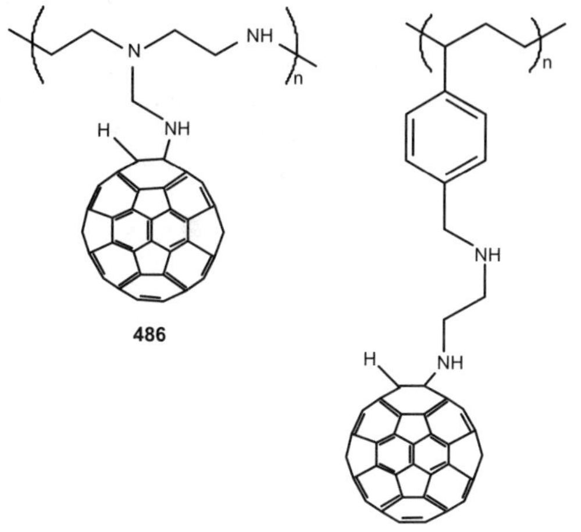

486

487

Figure 206. Binding of C_{60} to amino polymers. Adapted from [433].

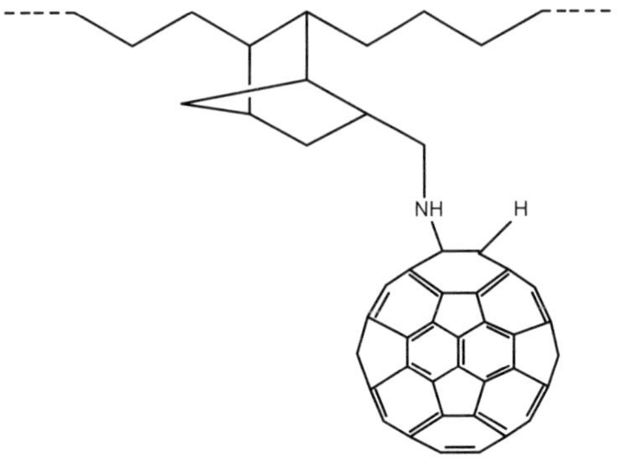

488

Figure 207. Reaction amino-containing hydrocarbon polymers with C_{60}. Adapted from [434].

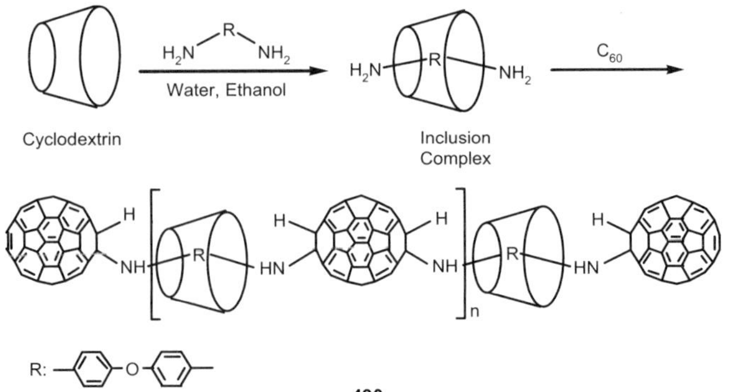

490

Figure 209. The polyfullerocyclodextrins, water-soluble fullerene main-chain polymers. Adapted from [436].

tetrathiafulvalene, and so on [439]. For example, when C_{60} crystallizes from solutions of aromatic hydrocarbons as simple as benzene, the solvent molecules are incorporated into the crystal leading to species such as C_{60}:4(C_6H_6). A solution of hydroquinone and C_{60} in hot benzene on evaporation leads to the black $[C_{60}][HQ]_3$ crystal. Solutions of C_{60} (or C_{70}) and ferrocene deposit black crystals of $[C_{60}][Fc]_2$. In these examples, the π–π interaction between donor and acceptor is the driving force for the complex formation. These interactions are noticed in the solid state though in solution they disappear. The interactions between the fullerenes and the macrocycles, on the other hand, are strong and even exist in solution, often leading to stable complexes.

14.1. Fullerene Calixarenes

Fullerene complexes with calixarenes have been studied extensively [437a]. The reaction of toluene solutions of C_{60} with *p-tert*-butylcalix[8]arene leads to a sparingly soluble yellow–brown precipitate, which is identified as a 1:1 complex [440]. The selective inclusion of C_{60} into the *p-tert*-butylcalix[8]arene cavity leads to a viable method of purification of C_{60} [441]. When carbon soot is allowed to react with *p-tert*-butylcalix[8]arene, the crude product consists of 89% C_{60} and 11% C_{70} complex. This corresponds to a 90% extraction of the C_{60} content of the soot. Treatment of this *p-tert*-butylcalix[8]arene fullerene complex with chloroform dissolved the host that leads the fullerenes to precipitate. This paved the way for a substantial enrichment of the C_{60} content over C_{70} in the fullerene mixture and a repetition of the process leads to the effective separation of C_{60} from C_{70}.

Most of the reported complexes are insoluble in water. A calix[8]arene with a hydrophilic SO_3Na function is reported to form a water-soluble complex [442]. The 1:1 complexes between *p-tert*-butylcalix[8]arene and C_{60} is micelle-like with trimeric aggregates of fullerenes as the core surrounded by three host molecules, each in double cone conformation and spanning two fullerenes along a triangular edge (Fig. 210) [443]. Introduction of electron-rich aniline and 1,3-diaminobenzene derivatives into calix[8]arene creates new C_{60} receptors [444], which gives no indication of complex formation. Deep inclusion of C_{60} into the calix[*n*]arene cavity, as expected for calix[8]arenes, is not a prerequisite for C_{60} inclusion. Calix[*n*]arenes that have a conformation and a proper inclination of the benzene rings such as a calix[5]arene, homocalix[3]arene, and to some extent calix[6]arene do interact with C_{60} in toluene.

Conformational freedom in calix[*n*]aryl esters can be frozen to cone conformation by complexation with appropriate metal cations. Such metal cations appropriately preorganize the calix[*n*]aryl ester derivatives into their cone conformers and can alter them into excellent C_{60} receptors

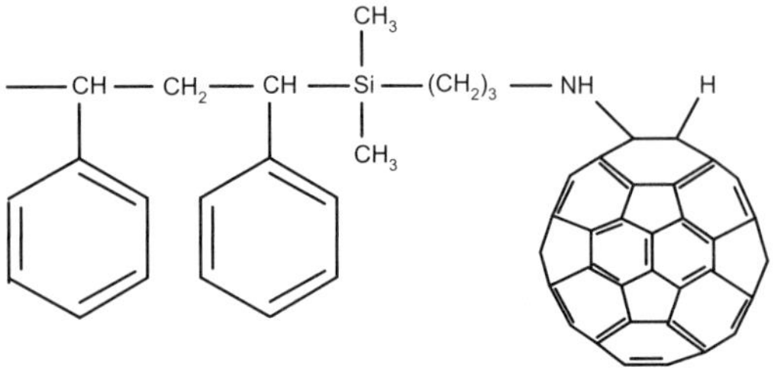

489

Figure 208. A fullerene-end-capped polystyrene. Adapted from [435].

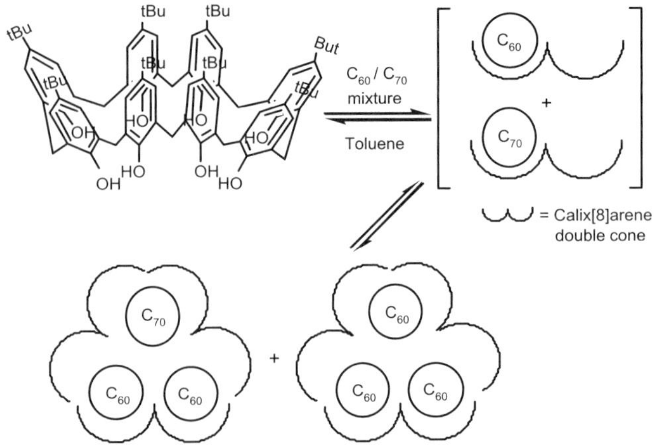

Figure 210. Aggregate of fullerenes surrounded by calixarene host. Adapted from [443].

in solution. A conformationally mobile calix[6]aryl ester does not react with C_{60}. Introduction of bulky ester groups to OH groups leads the phenol units in the calix[6]aryl ester to stand up, which makes the cavity edge smaller and the benzene ring inclination improper. When the calixarene is reacted with a Cs^+ salt, the metal complex could easily include C_{60} into the cavity (**492**). The metal cation formed a complex with the phenolic oxygen converting the calixarene to a symmetrical cone conformation facilitating C_{60} binding to the upper rim (Fig. 211) [445].

Two calixarenes have been covalently connected in order to produce shape-selective receptors, with well-defined cavity sizes (Fig. 212) [446]. With these calixarenes a dramatic increase in association constants for 1:1 complexes with C_{60} in solution has been achieved. The bridged receptor binds to C_{70} even better than C_{60}. The binding constant for C_{60} in toluene at 298 K is 76,000 M^{-1} for C_{60}, whereas for C_{70} in similar conditions it is 163,000 M^{-1}. A bipyridyl ligand attached to a calixarene on reacting with metal ions can generate a suitable receptor for C_{60}. Thus, two calix[5]arene precursor units, each carrying a bipyridine ligand, upon complexation with a metal ion like Ag(I) create a cavity sufficiently large to bind C_{60} and C_{70} (Fig. 213) [447].

A number of covalently bound fullerene calixarenes such as a C_{60}-functionalized calix[8]arene connected by a polyether chain have been designed [448]. An azido-terminated polyether chain attached to the rim of the calix[8]arene is reacted with C_{60}. From UV-vis and fluorescence studies, it is proved that the covalently bound fullerene still interacts with the calixarene in solution. Two fullerene calixarenes are synthesized, one of which had two fullerenes bonded to the lower rim of the same calixarene [449]. From light scattering measurements, it is observed that in chloroform–acetonitrile and chloroform–methanol the compound with two fullerenes aggregates strongly. No aggregation is observed for compounds with a single fullerene. Two compounds, in which C_{60} is linked to a cone-calix[4]arene or to a 1,3-alternate-calix[4]arene, are synthesized (Fig. 214) [450]. The fullerene is used as a "lid" for the ionophoric cavity. Several metal cations such as Li^+, Na^+, Ag^+ influence the electronic spectra of C_{60} in cone-calix[4]arene, indicating the metal cation bound therein interacts with

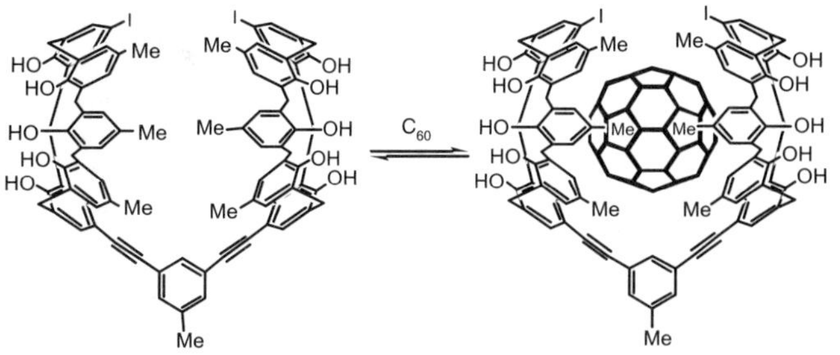
493

Figure 212. A bridged calixarene dimer for encapsulation of C_{60} and C_{70}. Adapted from [446].

the C_{60} surface. No spectral change is noticed when 1,3-alternate-calix[4]arene is treated with the metal ions.

14.2. Fullerene Catenane

For the synthesis of the fullerene containing catenane the bis(*p*-phenylene)-[34]crown-10, which has two π-electron-rich hydroquinone rings, is fused to C_{60} in a regioselective manner (Fig. 215) [451]. The formation of cyclobis(paraquat-*p*-phenylene) around the C_{60}-appended macrocyclic polyether leads to the [2]catenane. ^{1}H NMR investigation indicates the molecule to be intramolecular stacks of donor–acceptor–donor–acceptor, exhibiting spatial interaction between fullerene and bipyridinium units.

14.3. Fullerene Crown Ethers

A fullerene monolayer is attractive because the interesting fullerene properties are transferred to the bulk materials by simple surface coating. However, strong hydrophobic interactions aggregate C_{60} resulting in multilayer films. The aggregation is prevented in the presence of crown ether. However, in spite of the several fullerene crown ether conjugates known, in reality, most lead to multilayer formation at the air–water interface [452]. Apparently, the hydrophobic head groups are not bulky enough to prevent aggregation of the carbon spheres. Amphiphilic azacrown derivatives and the fullerenes could form stable complexes because the substituents on the nitrogen atoms wrapped around the

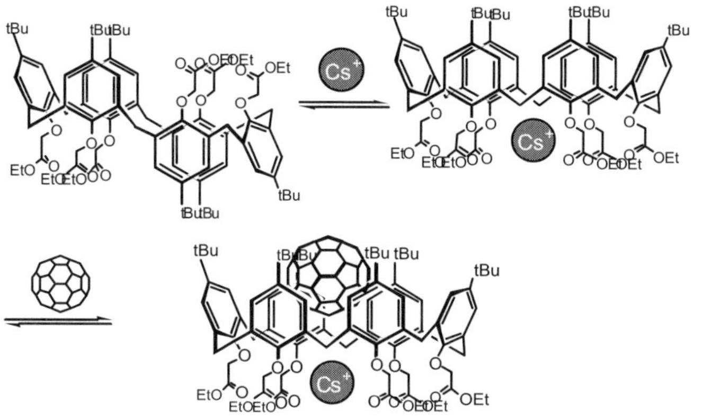
492

Figure 211. Preorganizing a calix[*n*]aryl ester derivative into cone conformer, altering it into an excellent C_{60} receptor. Adapted from [445].

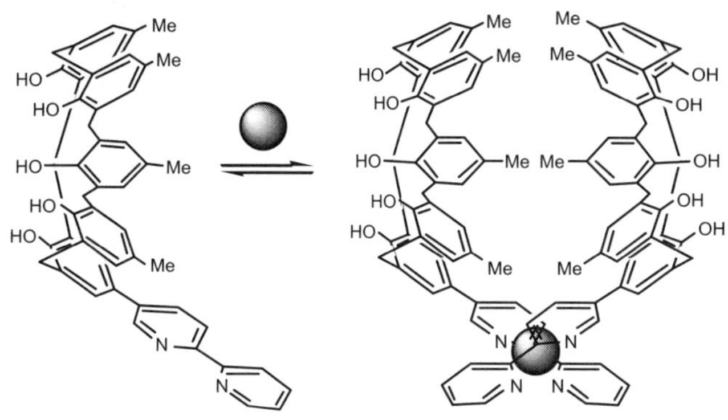
494

Figure 213. Metal ion mediated organization of a C_{60} receptor. Adapted from [447].

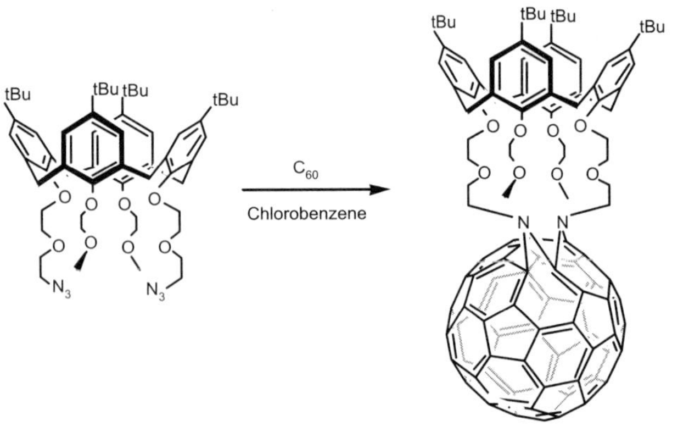

Figure 214. Synthesis of a cone calixarene; the metal ions bound to calixarene moiety interact with C_{60}. Adapted from [450].

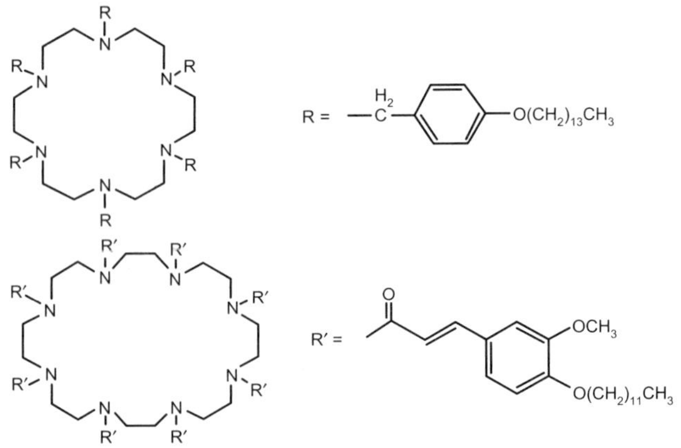

Figure 216. Amphiphilic azacrown derivatives form stable mono- and multilayers of C_{60} and C_{70}. Adapted from [453].

fullerenes like lipophilic arms, thus creating a hydrophobic envelope around the fullerene sphere (Fig. 216) [453]. The binding between the fullerene molecule and the azacrown is a noncovalent interaction. Mono- and multilayers of 1:1 mixtures of C_{60} and C_{70} and the macrocycles could be spread on the air–water interface. The monolayers could be transferred onto solid supports by the Langmuir–Blodgett technique.

Stable Langmuir monolayers and Langmuir–Blodgett films of C_{60} are also reported in a C_{60} benzo-18-crown-6 derivative [454]. This derivative forms a thin film at the air/water interface, especially when K^+ ions are present in the subphase. A pseudo 15-crown-5 macro ring directly attached to C_{60} via a bisaziridine linkage and the subphase containing K^+, Ba^{2+}, and NH_4^+ ions leads to monolayers [455]. The addition of the cations stabilized the monolayers and increased the limiting area. A gold surface is modified using a thiol-terminated ammonium salt (Fig. 217) [456]. When the modified gold layer is immersed in dichloromethane solution of the fullerene derivative a compact monolayer of C_{60} is obtained (**498**).

The crown ether can also be precisely positioned in close proximity to the fullerene surface even when not bound directly (Fig. 218) [437b]. When it is made to complex with the alkali metal ions, a significant perturbation of the electronic structure of the fullerene is observed, indicating that fixing the macrocycle at a precise position on the fullerene surface leads to more effective interaction between the metal ion and the fullerene sphere.

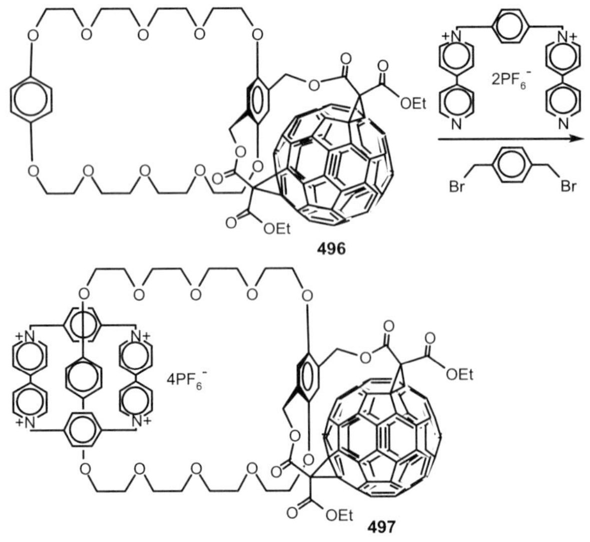

Figure 215. A catanane attached to C_{60} in a regiosymmetric manner. Adapted from [451].

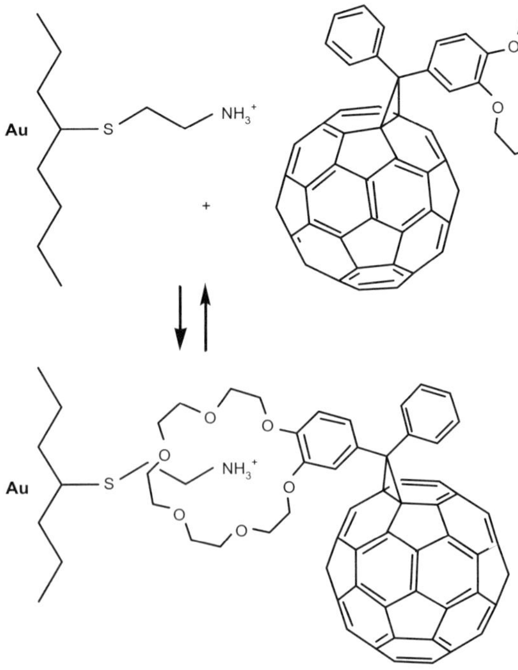

498

Figure 217. Planting a compact monolayer of C_{60} on a gold surface by molecular recognition of 18-crown-6 functionalized fullerene. Adapted from [456].

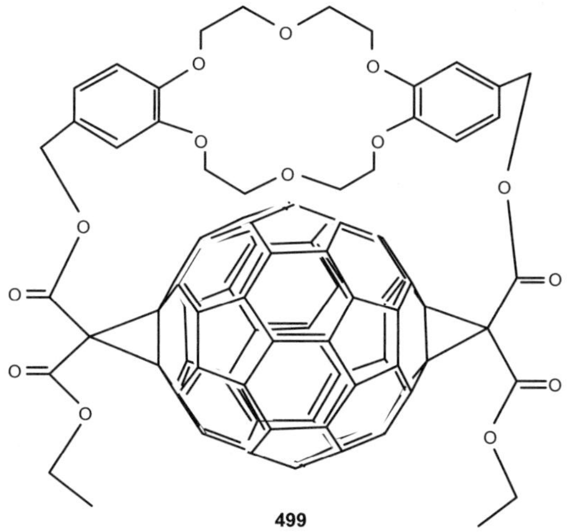

499

Figure 218. Crown ether precisely positioned on C_{60} leads to more effective interaction with metal ion. Adapted from [437b].

14.4. Fullerene Cyclophanes

A cyclophane has been attached to C_{60} leading to a "charming ring" (Fig. 219) [457]. The [2.2]cyclophane had a diazoalkane moiety, through which reaction takes place at the [6,6]-ring junction, leading to a methanofullerene derivative. The ^{13}C NMR signals of aromatic carbons of the cyclophane ring are not markedly affected by substitution of C_{60}. However, changes in the UV-vis spectra indicated intramolecular charge transfer interaction between the cyclophane and fullerene.

14.5. Fullerene Cyclotriveratrylenes

The macrocycle cyclotriveratrylene (CTV) forms an inclusion complex with both C_{60} and C_{70}. It envelops the fullerenes in such a manner that the electron-rich aryl rings of CTV cover three adjacent five-membered rings of C_{60}. The stability of the complex, an outcome of the π–π interactions, leads to changes in the absorption spectra of CTV.

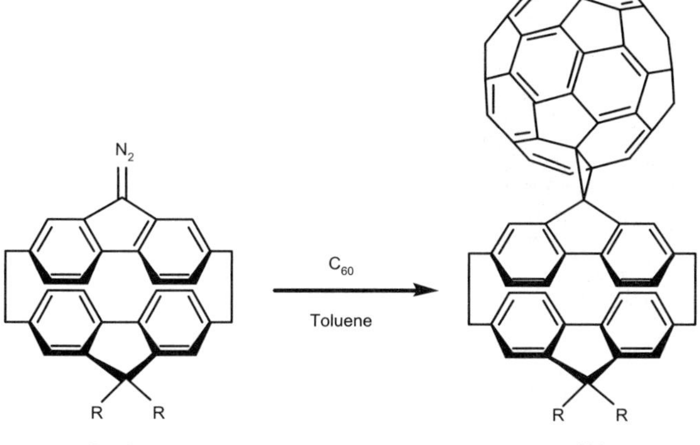

500

Figure 219. Attaching a cyclophane to C_{60} providing a "charming ring." Adapted from [457].

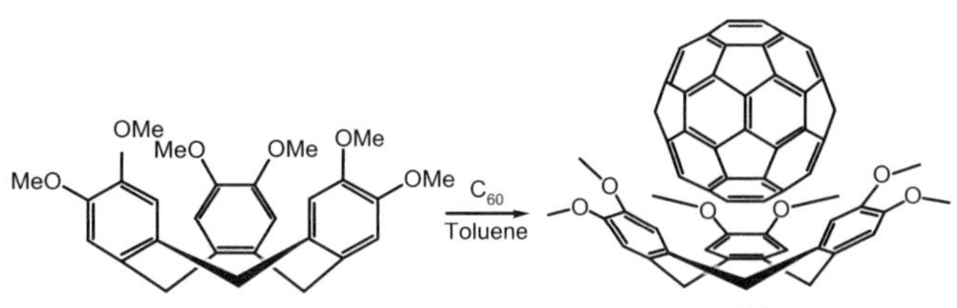

501

Figure 220. Noncovalently bound fullerene cyclotriveratrylene. Adapted from [458].

From a toluene solution of C_{60} and cyclotriveratrylene, solvated 1:1 complex (C_{60})(CTV) is formed and from this solution crystals of $(C_{60})_{1.5}(CTV)(PhMe)_{0.5}$ have been isolated (Fig. 220) [458].

Polybenzyl ether dendritic branches with peripheral triethyleneglycol chains have been attached to a cyclotriveratrylene core and the supramolecular complexes obtained from the resulting macrocyclic derivatives and C_{60} are found to be water-soluble [459]. In the design of the CTV hosts (Fig. 221), the polybenzyl ether dendritic type structure is chosen to provide an internal cavity capable of encapsulating totally the hydrophobic C_{60} guest, thus preventing the fullerene–fullerene interactions and precipitation from the aqueous solutions.

The C_3-symmetrical cyclotriveratrylene-C_{60} trisadducts with a *trans*-3,*trans*-3,*trans*-3 addition pattern (Fig. 222) or with an *e*,*e*,*e*-addition pattern are prepared by regio- and diastereoselctive tether-directed Bingel reaction of C_{60} with the trismalonate appended cyclotriveratrylene derivative [460]. Interchromophoric interactions between the electron-rich cyclotriveratrylene cap and the electron-attracting fullerene moiety have a profound effect on the electrochemical behavior of the carbon sphere.

14.6. Fullerene Phthalocyanines

The absorption and electron-accepting properties of C_{60} and the properties of electro- and photoactive molecules are combined by covalently linking these two types of molecules [461]. This leads to donor–acceptor systems with long-lived charge-separated states for several applications with the ultimate goal of artificial photosynthesis or conversion of solar

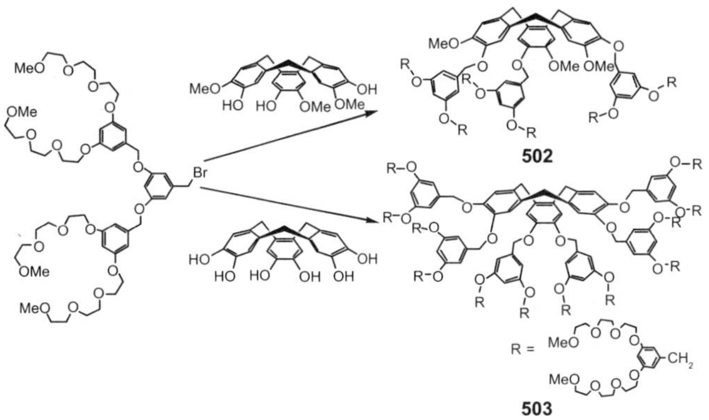

Figure 221. Cyclotriveratrylene hosts for supramolecular complexation with C_{60}. Adapted from [459].

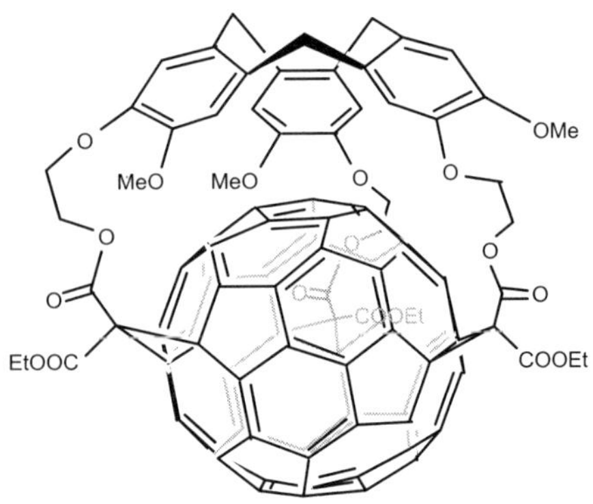

504

Figure 222. Cyclotriveratrylene-C_{60} trisadducts with a *trans*-3,*trans*-3,*trans*-3 addition pattern. Adapted from [460].

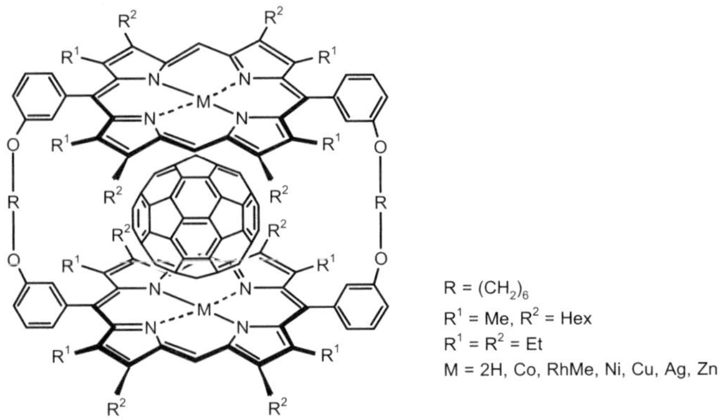

506

Figure 224. Cyclic dimers of metalloporphyrins as tunable hosts for fullerenes. Adapted from [464].

light to electric current. The fullerene phthalocyanines are an interesting class of molecules, because the phthalocyanines themselves are known for their remarkable electronic and optical behavior. A nickel phthalocyanine is synthesized from which fullerene phthalocyanine is obtained by Diels–Alder reaction of the phthalocyanine derivative and C_{60} (Fig. 223) [462]. From cyclovoltametry studies it is found that the reduction of the fullerene moiety has a pronounced influence on the optical properties of the phthalocyanine.

14.7. Fullerene Porphyrins

The porphyrins are ideal donors and hence a number of molecules have been synthesized to covalently link the fullerene sphere with the macrocycle [463]. A large number of porphyrins have been synthesized and their interaction with C_{60} when bound noncovalently or covalently has been explored. For example two derivatives of a cyclic dimer of porphyrin without or with metal ions are shown to complex C_{60} and C_{70} (Fig. 224) [464]. The effect of the metal ions in the porphyrin ring on binding affinity toward the fullerenes has been explored. Extremely high association constants ($\sim 10^8$ M^{-1}) and low dissociation activities of the RhMe/fullerene system should allow construction of more complex supramolecular architectures based on fullerenes.

A fullerene porphyrin conjugate, covalently linked via a pyrrolidine linkage, is prepared [465]. The Ni(II) complex is similarly prepared from the Ni(II) complex of the porphyrin. A porphyrin-linked fullerene, where a C_{60} moiety is covalently linked to the meso position of 5,15-diarylporphyrin with a pyrrolidine spacer, is synthesized [466]. Three pyrrolidine-functionalized C_{60} has been reported [467], in which a simple monomeric methyl pyrrolidine derivative, a pyrrolidine-linked tetraphenyl porphyrin-C_{60} dyad, and a pair of bifullerenes have been investigated. The Diels–Alder reaction has been used to prepare porphyrin-C_{60} dyads. The separation or spatial orientation of the donor and acceptor can influence the extent and rate of charge recombination. By changing the bond connectivity, the spacing and relative orientation of the porphyrin and fullerene moieties in the dyads are systematically changed (Fig. 225) [468]. Such dyads prefer to adopt a conformation in which the porphyrin and fullerene moieties are as close as possible in space as allowed by the structural constraints of the spacer [468, 469]. A few triads **509** have been synthesized and studied in terms of electron and energy transfer (Fig. 226) [470].

Several linear triads have been built around the zinc porphyrin/C_{60} (ZnP/C_{60}) couple. A set of molecular triads **510** in which a fullerene moiety is linked to a ferrocene-ZnP (Fc-ZnP) fragment (Fig. 227) [471]. In this molecule two consecutive electron transfers yield the Fc^+-ZnP–$C_{60}^{\bullet-}$. In oxygen-free benzonitrile, the final charge separated state decays to the singlet ground state. A fullerene moiety is also attached to an array of two prophyrins [472], in which the ZnP moiety performs as an antenna molecule, transferring its singlet excited state energy to the lower lying H_2P. By efficient funneling of light from the antenna chromophore, ZnP, to the H_2P chromophore and injecting charge into

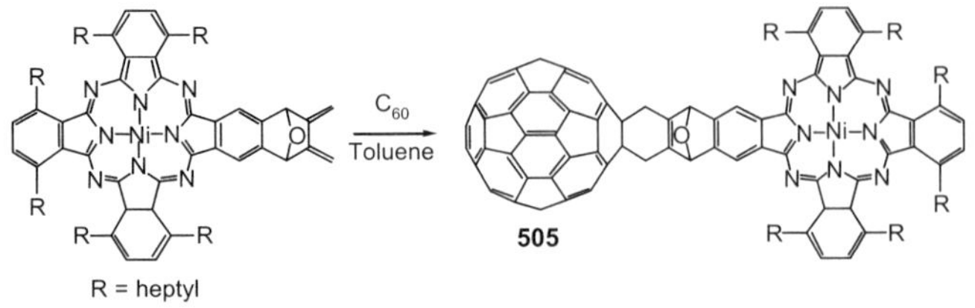

505

Figure 223. Synthesis of fullerene phthalocyanine by Diels–Alder reaction of the phthalocyanine derivative and C_{60}. Adapted from [462].

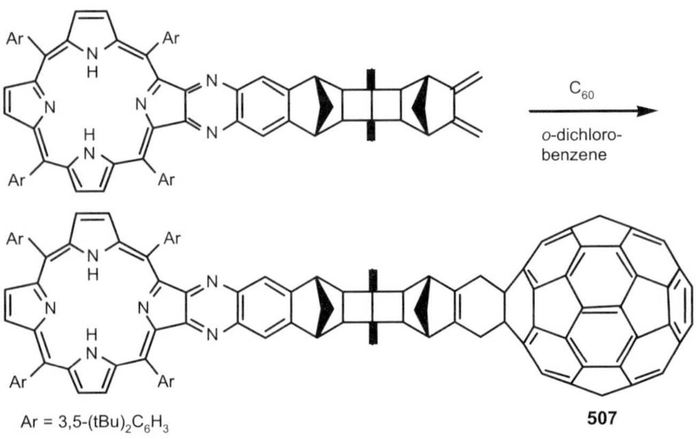

507

Figure 225. Fullerene porphyrin dyads. Adapted from [468].

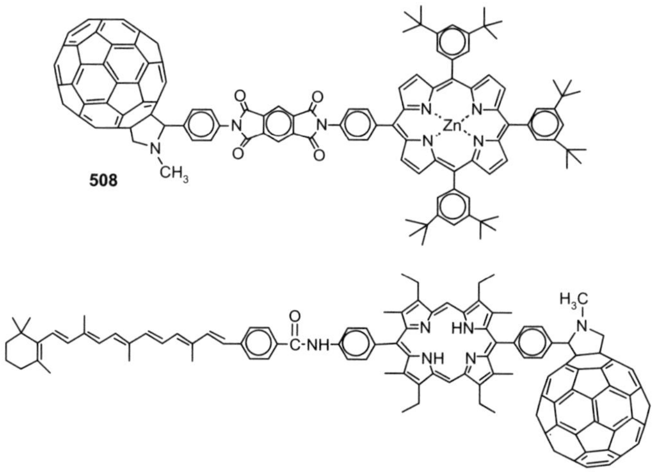

508

509

Figure 226. Fullerene–porphyrin triads. Adapted from [470].

the fullerene core effecting charge shift, this artificial reaction center reproduces the natural system very efficiently. Successful mimicry of the primary event of photosynthesis using these systems leads to combine two porphyrins, a ferrocene and C_{60}, into an integrated single system tetrad, Fc-ZnP–H_2P–C_{60}. The result turned out to be a highly efficient energy transfer system [473]. Indeed, the lifetime of the spatially separated radical pair, the product of a sequence of energy and multistep electron transfer reactions, reaches well beyond milliseconds into a time domain which has never been accomplished so far in an artificial photosynthetic reaction center.

A sophisticated $(ZnP)_3$–ZnP–H_2P–C_{60} hexad **511** comprised of a $(ZnP)_3$–ZnP model antenna system linked to a H_2P–C_{60} reaction center has shown to reveal indeed a cascade of efficient light-driven energy and electron transfer processes (Fig. 228) [474]. The antenna reveals rapid transduction of the singlet excited energy from the peripheral prophyrins to the central ZnP, which then transfers energy further to the ZnP chromophore, finally resulting in electron transfer to C_{60}. At the end of the process a 1.3 ns long-lived charge separated state $(ZnP)_3$–ZnP–$H_2P^{\bullet+}$–$C_{60}^{\bullet-}$ is formed in 70% yield.

A C_{60}–porphyrin interaction is reported in which the macrocycle is not covalently bound to C_{60} but fixed rigidly above the C_{60} leading to a true supramolecular dyad, **512** (Fig. 229) [475]. A pyridine-functionalized fulleropyrrolidone is linked to the Zn(II) ion of a tetraphenylphorphyrin by coordination with high affinity. Using the principle

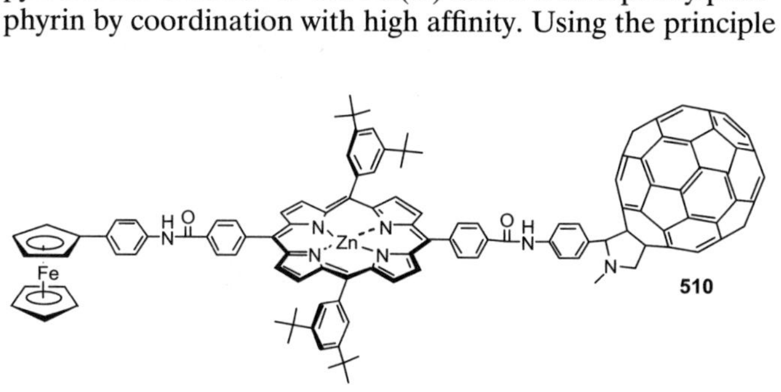

510

Figure 227. A ferrocene–porphyrin–C_{60} triad. Adapted from [471].

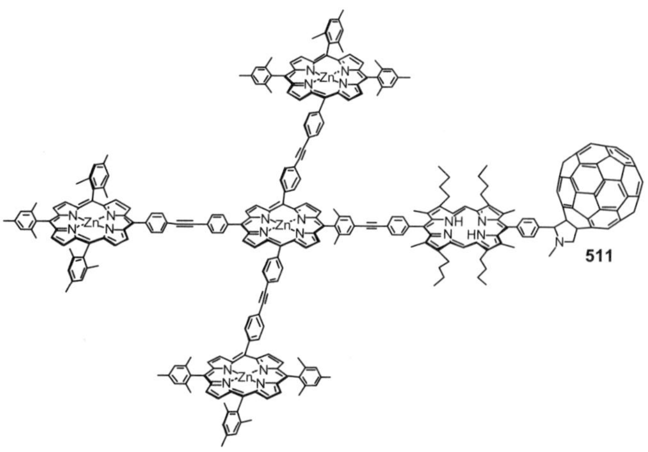

511

Figure 228. A $(ZnP)_3$–ZnP–H_2P–C_{60} hexad. Adapted from [474].

of macrocyclization between C_{60} and bismalonate derivatives it is possible to synthesize a C_{60}–Zn(II) tetraphenyl porphyrin dyad in which the spatial position of the porphyrin ring is fixed above the fullerene [476]. The organic chromophore could be precisely positioned in close proximity to the fullerene surface.

The π–π stacked C_{60}-prophyrin dyad has been used to synthesize rigid electroactive arrays by reacting the dyad with diazabicyclooctane or a pyridine-fullerene ligand, respectively (Fig. 230) [477]. The triad **515** and the pentad **516** are obtained by a careful control of the reactant concentrations. Another triad is formed from a reaction of the C_{60}–Zn(II) porphyrin dyad with the corresponding fulleropyrrolidine.

Two zinc tetraphenylporphyrin (ZnP) chromophores are attached to C_{60} leading to compounds of the type ZnP–C_{60}–ZnP (**517**, **518**) (Fig. 231), a flexible molecule, which leads to rigid, confined model donor–acceptor ensembles by self-organization [478]. The monomeric species is confirmed by molecular modeling using semiempirical methods. Following tether directed bisfunctionalization, a molecule with four appended fullerenes **519** is synthesized, in which the porphyrin ring is rigidly fixed close to the fullerene spheres (Fig. 232).

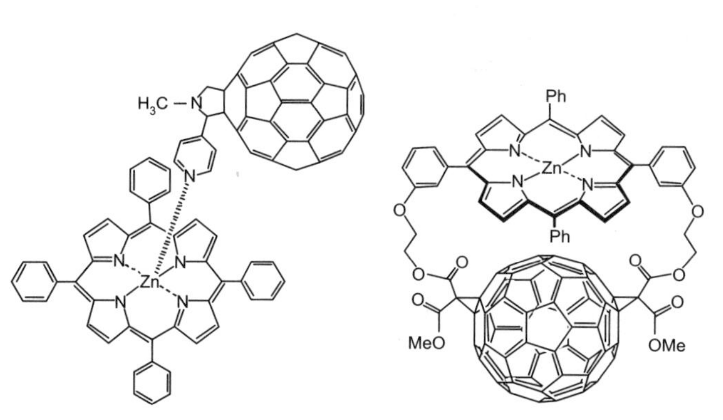

512 513

Figure 229. Supramolecular C_{60}–zinc tetraphenyl prophyrin dyads. Adapted from [475].

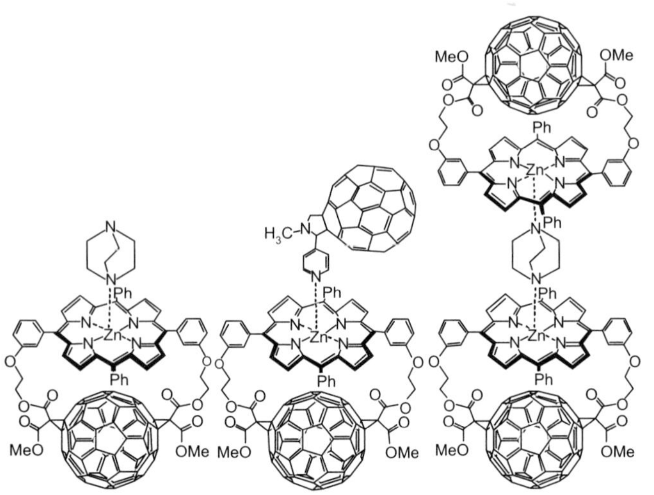

514 515 516

Figure 230. Electroactive arrays of C_{60}–prophyrin traids and a pentad. Adapted from [477].

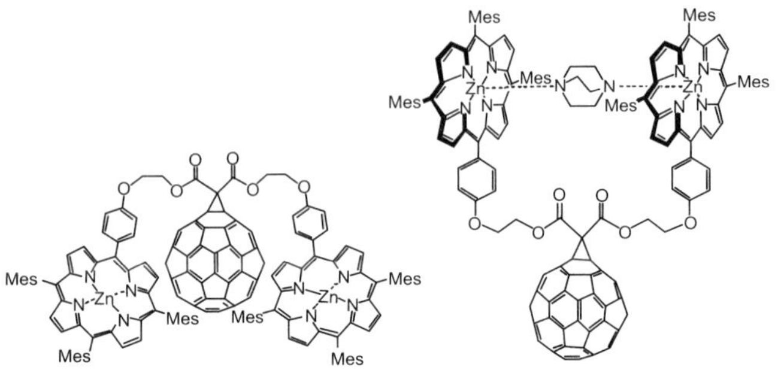

517 518

Figure 231. Flexible or confined zinc tetraphenylporphyrin–C_{60} donor–acceptor ensembles. Adapted from [478].

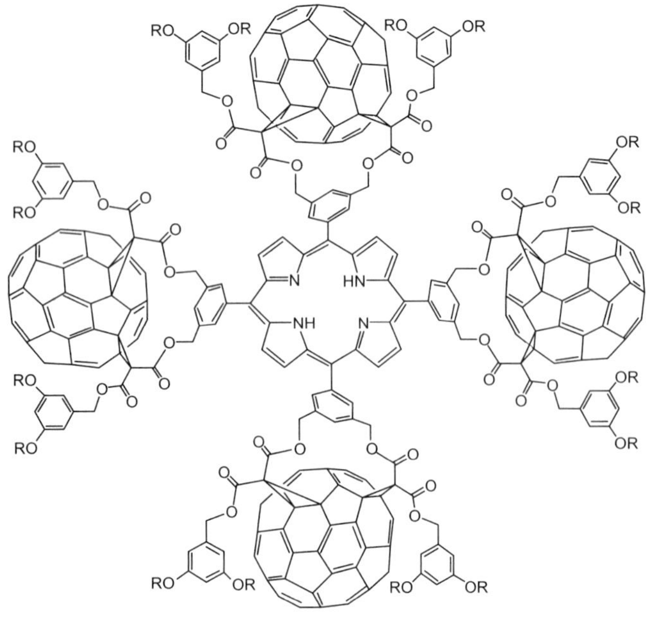

519

Figure 232. A porphyrin appended to four fullerene molecules.

14.8. Fullerene Cyclodextrins

14.8.1. Fullerene Cyclodextrin Inclusion Complexes

Cyclodextrins (CD) are oligoglycosides of six (α-CD), seven (β-CD), or eight (γ-CD) glucose units forming macrocyclic rings and they greatly vary in their aqueous solubility, β-CD being the least soluble. Noncovalent interaction between the cyclodextrin hosts and guest molecules has been effectively used to solubilize the otherwise insoluble hosts in water. The stability constants of the 1:1 inclusion complexes (k_{11}) depend upon the diameter of the guests and the cavity dimensions of the host. In the early stages of the fullerene cyclodextrin inclusion complexation studies, the dimensions of the three CDs, alongside those of C_{60} and C_{70}, are the major considerations that influenced the research of fullerene cyclodextrin inclusion complexation spanning over a decade. The matching cavity size of γ-CD and the diameter of C_{60} as well as its high aqueous solubility leads to synthesis of the host–guest complex and its detailed studies [479, 480]. In this complexation process, n–π donor–acceptor interactions between the n-orbitals of the sugar O atoms and the π-system of the fullerene contribute to the stability of the complex [481]. The cone-shaped cavity of γ-CD (950 pm) and the cavity depth of (780 pm) do not completely encapsulate the C_{60} sphere of van der Waals diameter of $\sim$1000 pm. Hence a complex with a 2:1 (γ-CD:C_{60}) stoichiometry that very effectively covers C_{60} is proposed for the structure of the complex. From photophysical studies it is indicated that γ-CD covered a large portion of the carbon sphere [482]. This preferential complexation leads to extraction of C_{60} from a mixture of C_{60} and C_{70} in solid state on boiling the mixture with an aqueous solution of γ-CD.

Though the unmodified α- and β-CD are reported not to complex C_{60}, their derivatives such as dimethyl-β-cyclodextrin, hydroxypropyl-β-cyclodextrin, and α-, β-, and γ-cyclodextrin–epichlorohydrin prepolymers have been reported to form 1:1 and 2:1 complexes with C_{60} [483]. These complexes have been prepared by kneading. It was recently noticed that β-CD does indeed form a complex with C_{60}, contrary to what is believed to be not probable [484], though under similar reaction conditions α-CD is seen not to form a complex. However, the γ-CD/C_{60} complex is a lot more soluble in water than the β-CD counterpart. Encapsulation of the fullerenes by the cyclodextrins does not necessarily mean that the fullerene sphere is shielded completely from reacting further, although the reactivity of the fullerene in γ-CD/C_{60} complex is noticeably lower than in the unprotected form. The complex is still susceptible for easy reduction by radicals generated in the aqueous phase [485]. A number of electron donors can interact with γ-CD/C_{60} leading to donor–acceptor complexes though the fullerene moiety is embedded between two host molecules of γ-CD. The formation of donor–acceptor complexes with electron-rich molecules, such as triethylamine, pyridine, hexamethylene tetramine, acrylamide, thioglycolic acid, and 2,2′-thiodiethanol, have been reported [486].

14.8.2. Covalently Bound Fullerene Cyclodextrins

The synthesis of water-soluble, covalently bound fullerene cyclodextrin conjugates (**520**, **521**) reacting cyclodextrin monoamine derivatives and fullerenes has been reported (Fig. 233) [487]. Depending upon the macrocycle, the aqueous solubility of the corresponding conjugate varied. Even though α-CD has a very small cavity, which does not facilitate complex formation easily, covalent binding to the fullerenes leads to a material which is even more soluble than the ones from β-CD. Considering the fact that the cyclodextrins are bioamenable, the covalently bound monofunctionalized fullerene cyclodextrins can have a number of biomedical applications. Fullerenes have also been attached to cyclodextrins via a bridging nitrogen atom by treating α- or β-cyclodextrin azides with C_{60} [488]. The extent of their aqueous solubility is again dependent on the solubility of the macrocycle.

A 2:1 (permethylated-β-cyclodextrin)–fullerene conjugate **522** has been synthesized [489], in which two β-cyclodextrin units are attached covalently to C_{60} by Bingel cyclopropanation (Fig. 234). The linking units allow enough flexibility such that the CD rings effectively encapsulate C_{60} leading to high aqueous solubility. Such a complexation strategy is expected to prevent interaction of C_{60} molecules thus minimizing aggregation. However, in the UV-vis spectra of the compound in water, a relative maximum at 430 nm is not seen, which is a sign of aggregation. Nevertheless, the conjugate displays a negative solubility coefficient and has a partition coefficient between octanol and water in the suitable range for biological studies.

15. APPLICATIONS

The fullerenes have three distinct properties that are linked directly to their applications. These are (1) efficient electron transfer in the triplet state, (2) scavenging of free radicals, and (3) generation of singlet oxygen in presence of light. Electron transfer from an electron rich source such as a conducting polymer leads to applications of fullerene materials as photodiodes. Free radical scavenging and singlet oxygen generation are properties associated with the applications of fullerenes and their derivatives in the fields of biology and medicine. A number of fullerene derivatives are known to possess excellent liquid crystalline property. Some of these areas of applications will be discussed here briefly. Besides, there are a number of areas where fullerenes have promising applications. All-carbon ferromagnets of fullerenes are

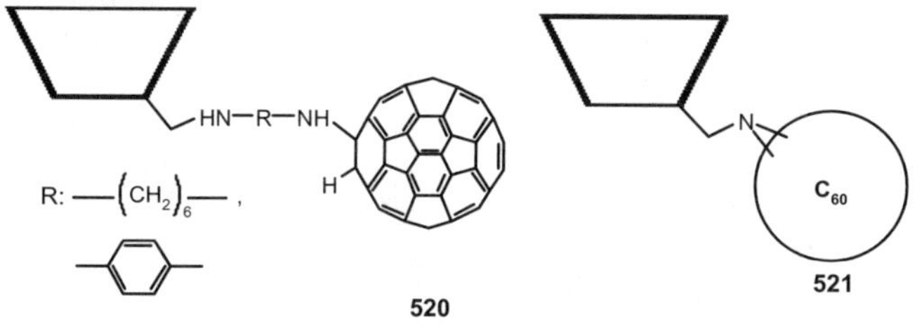

Figure 233. Fullerene–cyclodextrin covalent conjugates. Adapted from [487].

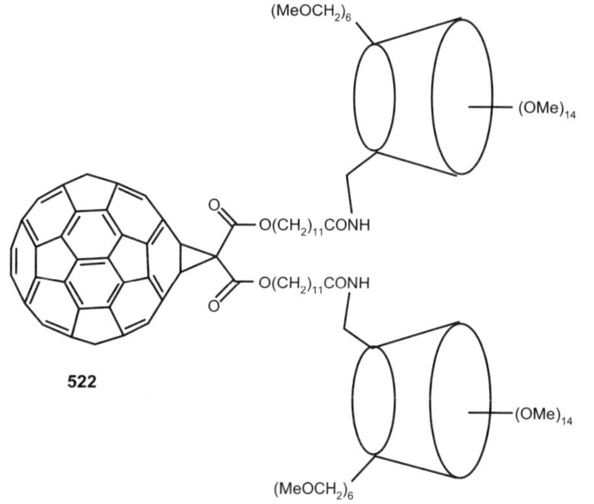

Figure 234. Two β-cyclodextrin units are covalently attached to C_{60} by Bingel cyclopropanation. Adapted from [489].

exciting new materials [490]. The linear and nonlinear optical properties of a number of fullerene derivatives have been thoroughly studied [491].

15.1. Fullerenes and Conjugated Polymers

Fullerenes and conducting polymers with a nondegenerate ground state have been used to design photodiodes [492]. Several such polymers have been doped with C_{60} and C_{70} and their electrical conductivity and photoconductivity have been studied. The preparation of photoconducting films of polyvinylcarbazole (PVK) doped with fullerenes is reported [493], the performance of which is comparable to some of the best photoconductors available commercially. The synthesis of C_{60} and C_{70} chemically modified polymeric photoconductors such as C_{60}-modified PVK and its bromine and iodine modified derivatives is also reported [494]. In these polymers, the charge-transfer interaction between the positively charged carbazole, resulting from charge transfer toward the fullerenes, brought about the enhanced photoconductivity relative to PVK itself. Polymer-based photodetectors and photovoltaic cells have been made by using conjugated polymer soluble derivatives of poly(p-phenylene vinylene) (PPV) and poly(2-methoxy-5-(2′-ethylhexyloxy)-1,4-phenylene vinylene) (MEH-PPV) as donors and C_{60} as acceptor [495]. By optimizing the donor–acceptor ratio, a power conversion efficiency of 2.5% and a collection of efficiency of 26% have been achieved for photovoltaic cells. Doping effects of C_{70} on electrical and optical properties of conducting polymers such as poly(3-alkylthiphene), poly(2,5-dialkoxy-p-phenylene vinylene), and poly(9,9-dialkylfluorene) have been studied and compared with those by C_{60} doping. Fluorescence intensity is quenched and photoconductivity is enhanced upon doping [496]. In order to enhance the intramolecular energy and charge transfer of PPV derivatives and fabricate highly efficient photovoltaic cells, a novel C_{60}–PPV system has been synthesized. Flexible alkyl side groups have been introduced to further improve the processability of the copolymer. Carbazole and triphenylamine moieties are introduced into the copolymer main chains to enhance electrical and photoactive properties [497]. From studies on X-band photoluminescence (PL)-detected magnetic resonance spectra of C_{60}- and C_{70}-doped PPV-based polymers it is concluded that fullerenes quench the PL, and both the polaron and triplet exciton resonances are dramatically enhanced [498]. Photoinduced electron transfer in films of poly(3-decyloxy-1,4-phenylene) (DO-PPP) blended with C_{60} is observed by infrared photoexcitation spectroscopy. The photoinduced IR-active vibrational structure and electronic subgap absorption features of the DO-PPP are greatly enhanced upon adding just 2% of C_{60} [499]. Photoinduced electron-transfer processes from oligothiophenes/polythiophenes to fullerenes have been studied by the nanosecond laser flash photolysis method, observing the transient absorption spectra in the visible and near-IR region. When fullerene is selectively photoexcited in polar solvents, electron transfer from oligothiophene to the excited triplet state of fullerene is confirmed [500].

15.2. Liquid Crystalline Fullerenes

Liquid crystals play a key role in everyday life as the essential component of several display devices. Fullerene-containing liquid crystals are interesting materials for fundamental studies as well as for practical applications. Binding the liquid crystalline moiety to C_{60} leads to better supramolecular organization of the liquid crystal. Further, C_{60} influences the formation and stability of the liquid–crystalline phase. The typical shape can facilitate the formation of molecular devices and switches showing outstanding performance and characteristics. For example, a hexapyrrolidine C_{60} adduct has been found to be suitable in designing a single layer organic light-emitting device [501]. Two concepts have been developed for the design of C_{60}-containing thermotropic liquid crystals: covalent [502] or noncovalent [503] binding of the liquid crystalline moiety with the fullerenes. The work on fullerene containing liquid–crystalline dendrimers with several examples where the dendritic liquid–crystalline moiety is covalently or noncovalently bonded to C_{60} has been reviewed [504]. Covalent linking of the liquid crystal to C_{60} is accomplished through a cyclopropane ring using Bingel reaction or via a pyrrolidine ring generated by reacting an aldehyde derivative and sacrosine with C_{60}. Mesomorphic cyclotriveratrylene (CTV) derivatives are used to noncovalently complex C_{60}.

15.2.1. Liquid Crystals Containing Covalently Bound Fullerenes

Cholesterol is a strong liquid–crystalline promoter. The precursor compound **523** containing cholesterol gives smetic A, chiral nematic phases, and a blue phase. The first fullerene-containing thermotropic liquid crystal **524** is prepared by reacting the malonate precursor **523** with C_{60} (Fig. 235) [502a]. Incorporation of C_{60} leads to a drastic change of the liquid–crystalline properties. For the fullerene-containing compound (**524**) a smetic A phase forms when the sample is cooled from isotropic melt, revealing a monotropic character of the mesophase. Introduction of C_{60} apparently disrupts the liquid–crystalline organization. This could be a consequence of the size and shape of C_{60} molecule.

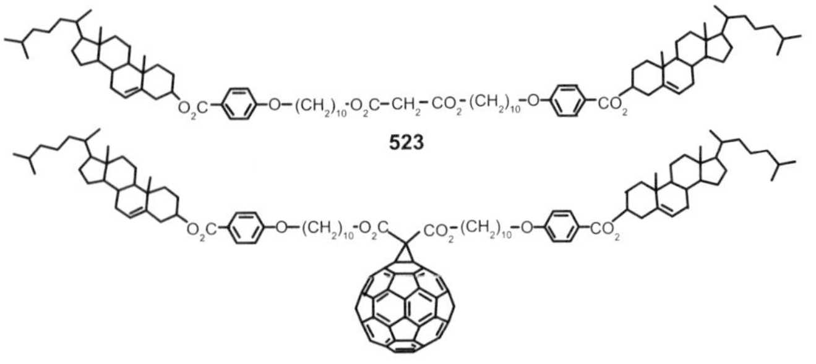

Figure 235. Liquid–crystalline fullerene–cholesterol conjugate. Adapted from [502a].

Liquid–crystalline materials incorporating C_{60} along with electron-donor units such as ferrocene have been reported. Cholesterol-based liquid crystals containing ferrocene have been reported (Fig. 236) [505]. These are multifunctional liquid crystals and are promising candidates for the molecular switches since electron transfer is shown to occur in solution with fullerene–ferrocene dyads and chemical oxidation is applied to generate ferrocenium derivatives from ferrocenes. The photoinduced electron-transfer reaction in fullerene–ferrocene liquid crystals could be used as a switching mechanism because of the presence of either ferrocene (light off) or ferrocinium (light on). Starting from the ferrocene- and cholesterol-containing precursor **525** the fullerene bismalonate derivative **526** is synthesized. As expected, for the fullerene–ferrocene liquid–crystalline material **526**, photoinduced intramolecular electron transfer is observed in organic solvents. Each subunit (i.e., cholesterol, ferrocene and fullerene) plays a role in the supramolecular organization within the liquid–crystalline phases as well as contribute to the liquid–crystalline property and photoinduced electron transfer. In the smectic A phase, a relative order with microsegregation of the different units (ferrocene, fullerene, and cholesterol) is observed. This order might be important to control the electron-transfer processes upon light irradiation. Studies in solution reveal that the ferrocene–fullerene system does indeed give rise to intramolecular electron transfer with relatively long-lived charge-separated species.

Modification of C_{60} with dendritic addends is an elegant approach to prevent aggregation. A hydrophilic dendrimer allows the construction of stable and highly ordered Langmuir–Blodgett films. In such systems C_{60} is buried in the dendrimer. As a consequence, there is a decrease of the unfavorable C_{60}–C_{60} interaction and a beneficial increase of the interactions between the polar groups and the water surface. Functionalization of dendrimers with various mesomorphic groups leads to liquid–crystalline materials (Fig. 237) [506]. Such dendrimers can have tailor-made properties. The size of the dendrimers counterbalances that of C_{60} and minimizes the influence of C_{60} bulk. The second-generation dendrimer **527** represents dendritic structure incorporating the ferrocene. Both the dendrimer and its precursor showed a smetic A phase, indicating that in this case the presence of C_{60} does not have a disrupting influence on the liquid–crystal characteristics.

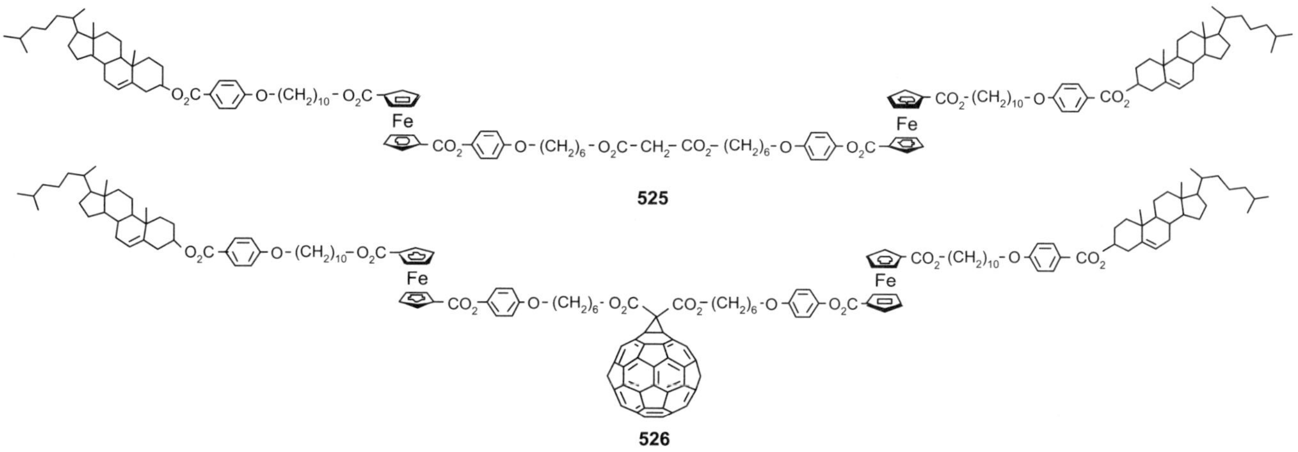

Figure 236. Multifunctional fullerene–ferrocene liquid crystals containing ferrocene and cholesterol. Adapted from [505].

Similar to cholesterol, cyanobiphenyl imparts liquid–crystalline properties to the molecule bearing this function. C_{60} derivatives bearing cyanobiphenyl functions **529** are synthesized starting from the precursor **528** (Fig. 238) [502e]. The *d*-layer spacing determined from X-ray diffraction in the smetic A phase is 76 Å. The layers are viewed as consisting several molecules of C_{60} containing mesogenic groups. The model proposed [504] is that the two arms of the molecule are folded in the same direction and structure consists of a head-to-tail arrangement of these groups (Fig. 239).

The effect of dendrimer size on evolution of mesomorphism is demonstrated by using a series of dendrimers varying in size. A comparison indicates that the stability of the liquid–crystalline phase increases with the increase in the dendrimer generation. The fullerene dendrimers **532–534** (Fig. 240) [502d] and their malonate precursors display the same mesophases, while the precursor of **531** gives rise to a nematic phase. In this case introduction of C_{60} changed the nature of the liquid–crystalline phase.

The layer spacing for **531** is 82 Å, which is a rather high value. Simulation experiments of **531** indicate that the molecule adopts a V shape, the arms of the V being of mesogenic groups. This leaves enough space for mesogenic groups from adjacent layers to be included. Compounds **532–534** have a relatively lower layer spacing compared to **531**. In these dendrimers, the branching parts have large bulk and, to avoid steric crowding, the branches extend laterally rather than coming together to generate the V shape. The smetic A phase for these compounds is stabilized by an efficient lateral arrangement of the mesogenic groups. The dendritic branches with the mesogenic groups connected through intervening aliphatic spacers represent the major part. The fullerene moiety represents a minor part. Therefore the polar cyano groups essentially govern the characteristic properties of these molecules. Molecular modeling shows these groups as pointing to one main direction. This permits the interpenetration of the mesogenic groups from adjacent layers. Thus the central part of the layer is constituted by the fullerene moiety embedded in the dendritic branches. The long aliphatic chains and the ester groups provide the arms with a lot of flexibility that facilitates good space filling. The malonate precursors of **533** and **534** exhibit also smetic A phases similar to the fullerene derivatives. Hence the fullerene moiety does not play a significant role.

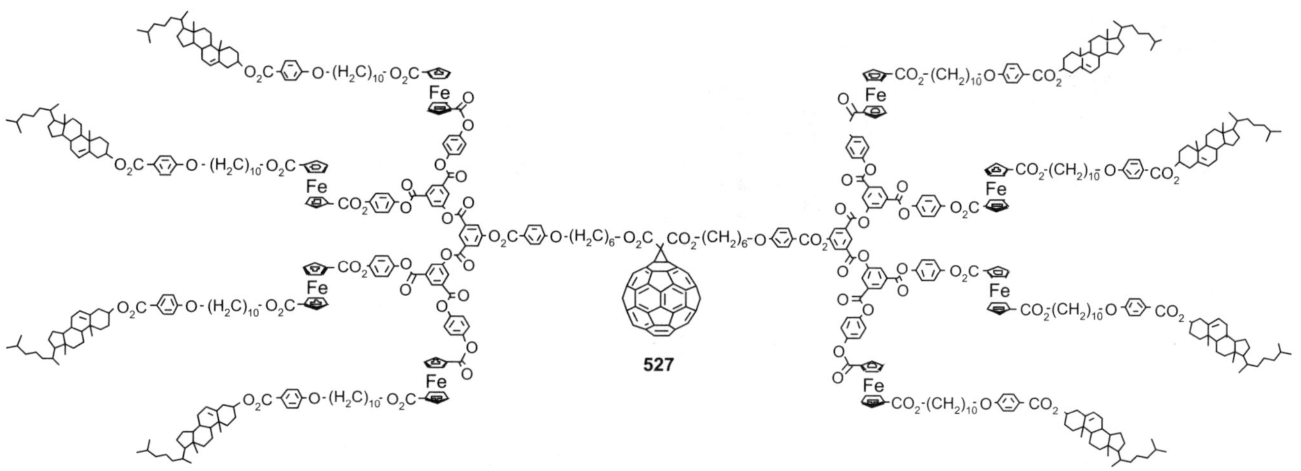

Figure 237. Fullerene dendrimers with ferrocene and cholesterol in the dendritic moiety imparting liquid crystallinity. Adapted from [506].

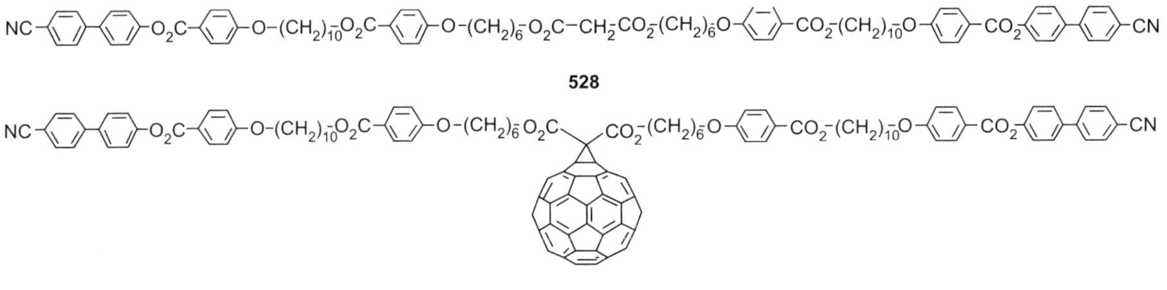

528

529

Figure 238. Liquid–crystalline fullerene derivative containing cyanobiphenyl function. Adapted from [502e].

The first fullerene dendrimer connected to C_{60} via a pyrrolidone ring is **535** (Fig. 241) [502f]. Fulleropyrrolidones are stable to chemical or electrochemical reductions, in contrast to Bingel adducts which undergo retro-Bingel reactions. This property is interesting because the redox-active fulleropyrrolidone liquid crystals remain stable. A smetic phase is observed for its aldehyde precursor as well as for the fullerene derivative. Esterification of a bis(methanofullerene) carrying two carboxylic acid functions with an amine-based dendritic wedge leads to yet another cyanobiphenyl-terminated dendritic liquid–crystalline material **536**. The fullerene derivative displays a mesophase, whereas the precursor shows a nematic phase.

Hexaadducts of C_{60} are attractive materials due to their three-dimensional architecture. Attaching six liquid–crystalline addends to C_{60} leads to materials with remarkable liquid–crystalline properties (**538**) (Fig. 242) [502h]. The organization of 12 cyanobiphenyl units around a central fullerene core generated the required structural anisotropy and intermolecular interactions for mesomorphism to occur. Though the precursor **537** showed a nematic phase, an enantiomeric smectic A phase is observed for **538**. The formation of the smectic A phase indicates that the cyanobiphenyl units are not oriented radially around C_{60} but form a cylinder-like structure with the mesogenic fragments oriented upward and downward. This is a feature demonstrated by the liquid–crystalline materials from dendrimers with a central core functionalized by large number of mesogenic groups.

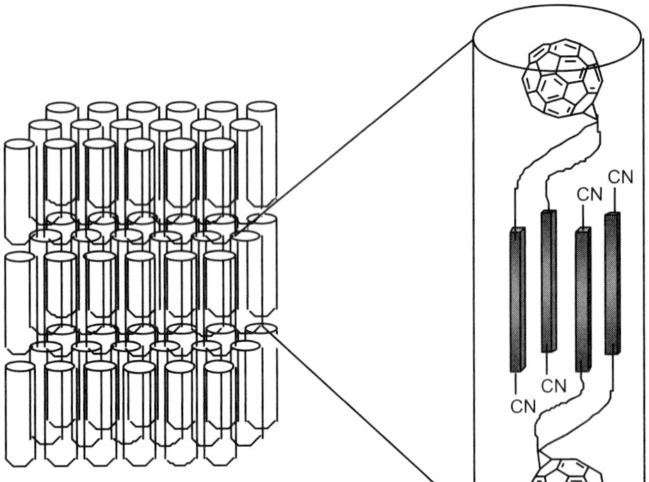

Figure 239. Proposed supramolecular organization of the dendrimer in the liquid–crystalline phase. Adapted from [504].

Attachment of five aromatic groups to one pentagon of a C_{60} fullerene molecule yields deeply conical molecules that stack into polar columnar assemblies (Fig. 243) [507]. The stacking is driven by attractive interactions between the spherical fullerene moiety and the hollow cone formed by the five aromatic side groups of a neighboring molecule in the same column. This packing pattern is maintained when the aromatic groups are extended by attaching flexible aliphatic chains. The compounds possess excellent thermotropic and lyotropic liquid–crystalline properties.

15.2.2. Liquid Crystals of Noncovalently Bound Fullerenes

Cyclotriveratrylenes form supramolecular complexes with C_{60}. Such complexes are easily obtained by evaporation of a benzene solution containing the CTV and C_{60} (Fig. 244) [503]. CTV containing liquid–crystalline units can form a complex with C_{60} in a similar fashion. The noncovalent approach is interesting because it does not require the development of specific reactions to render C_{60} mesomorphic. On the other hand, such a material cannot be used at high temperature when C_{60} is expelled from the complex. Interestingly **540** shows mesomorphic properties at room temperature.

15.3. Fullerene Derivatives for Biological and Biomedical Applications

The biological and biomedical application potentials of fullerenes have received considerable attention [508]. These properties emerge from fullerenes efficiently scavenging free radicals. Additionally fullerenes generate singlet oxygen very efficiently in presence of UV and visible light [509]. In order for fullerenes to be used for these applications, they should be water-soluble. This is possible when hydrophilic groups are appended to the otherwise highly hydrophilic fullerenes.

A number of interesting syntheses have led to fullerene derivatives either soluble in water in an organic cosolvent or soluble in water without the need of a cosolvent.

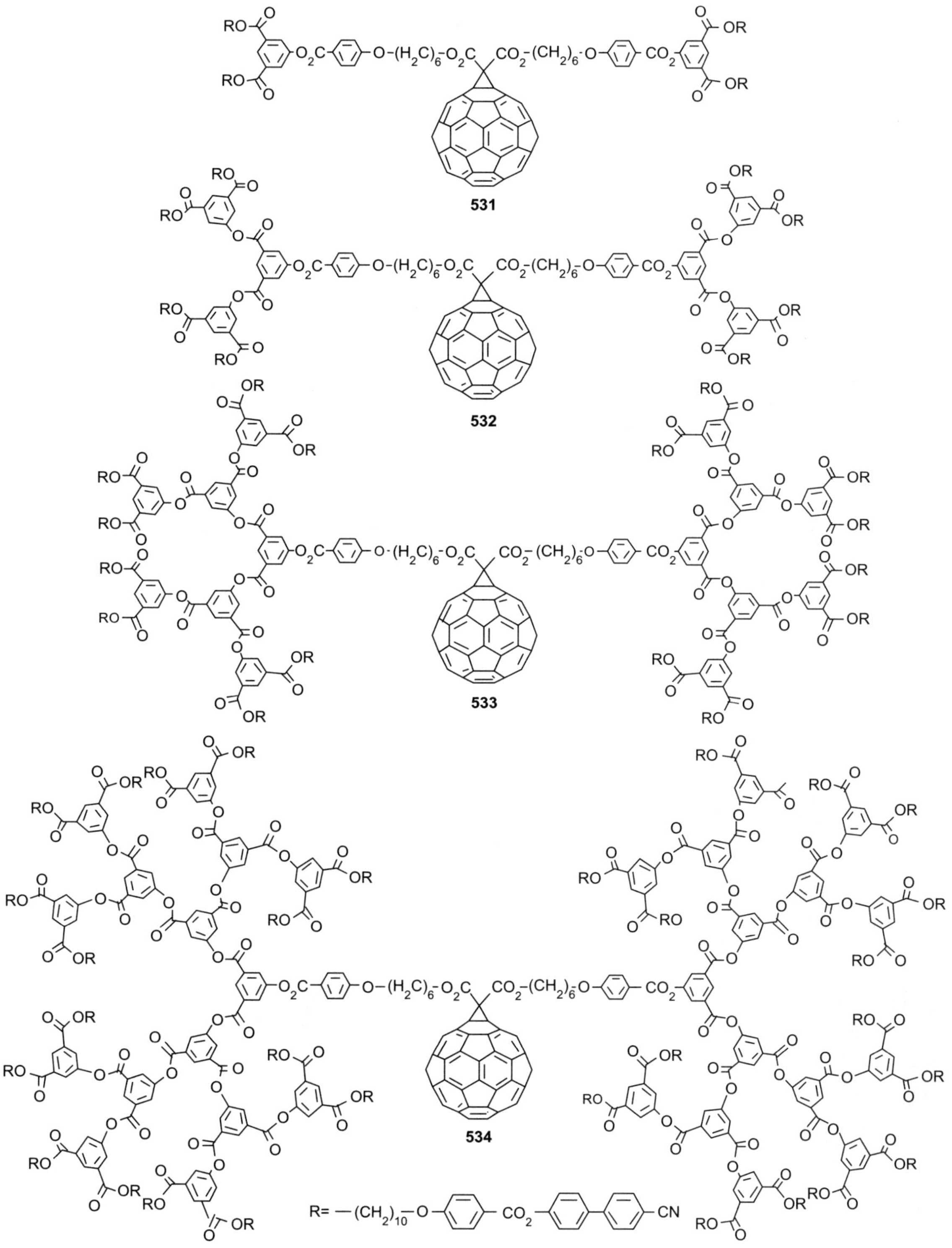

Figure 240. Liquid–crystalline dendrimers of different generation. Adapted from [502d].

Attaching a hydrophilic function usually carrying OH and COOH groups has led to a number of such derivatives. Usually a minimum of 12 OH groups or 6 COOH groups is needed to effectively solubilize C_{60} in water. Thus, polyfunctional fullerene derivatives are more soluble in water than their monofunctional counterparts. Since the radical scavenging property is dependent upon the number of π-bonds left unbroken on fullerene, a compromise is often made between monofunctonalization and polyfunctionalization. Prevention of aggregation is also an equally important aspect. Close contact between fullerene spheres leads to a decrease in triplet lifetime, which then affects the generation of singlet oxygen. The singlet oxygen generated by fullerene in the presence of light is responsible

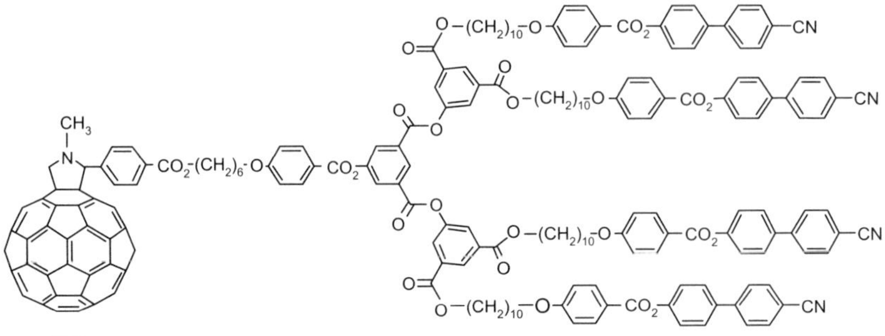

535

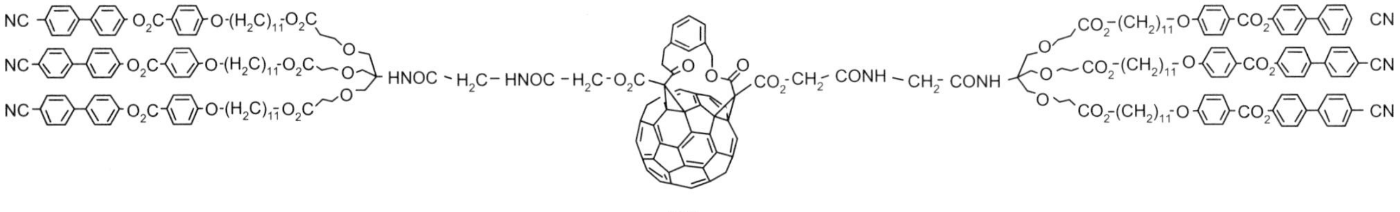

536

Figure 241. Cyanobiphenyl-terminated dendritic liquid–crystalline materials connected to fulleropyrrolidine or a bis(methanofullerene). Adapted from [502f].

for cytotoxic behavior of the fullerenes. Following is a brief outline of the major areas of biology and medicine in which such soluble fullerene derivatives find potential applications.

15.3.1. Antitumor Studies

One of the easily accessible water-soluble fullerenes studied for its anticancer activity is fullerol $C_{60}(OH)_{24}$. This compound affected the growth kinetics of human lymphocyte cultures, epidermal carcinoma cell cultures, efficiently preventing cell growth in a manner similar to known anticancer drugs such as taxol. When water-soluble C_{60}–PEG conjugate is intravenously injected to the tumor-bearing mice, their accumulation in tumor tissue is higher than that in normal tissue [510]. Treatment with C_{60}–PEG conjugate injection coupled with light irradiation is seen to strongly induce tumor necrosis. The photodynamic effect on tumors greatly depends on the C_{60} dose and the power of light irradiation. All the tumor-bearing mice are cured. The synthesis of a highly water-soluble dendro[60]fullerene **541** is carried out (Fig. 245) [511]. Upon hydrolysis of the COOMe groups, the dendritic moiety generates 18 COOH groups, which impart the molecule a very high aqueous solubility. Further, the dendrimer effectively shields the fullerene moiety thus preventing aggregation.

The cytotoxic and photocytotoxic effects of the dendritic C_{60} monoadduct and the malonic acid C_{60} trisadduct are tested [512]. It is found that only the dendritic monoadduct inhibits cell growth, whereas the trismalonic acid adduct has little effect. These fullerene derivatives become toxic when irradiated with UV light. The cell death is mainly caused by membrane damage and it is UV dose-dependent. Trismalonic acid fullerene is found to be more phototoxic than the dendritic derivative.

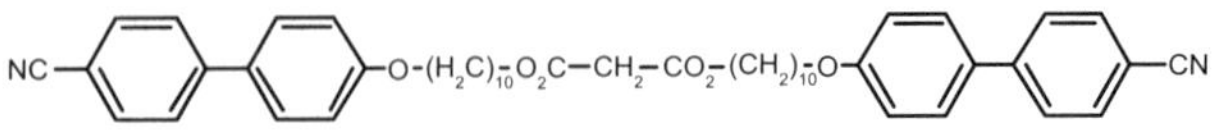

537

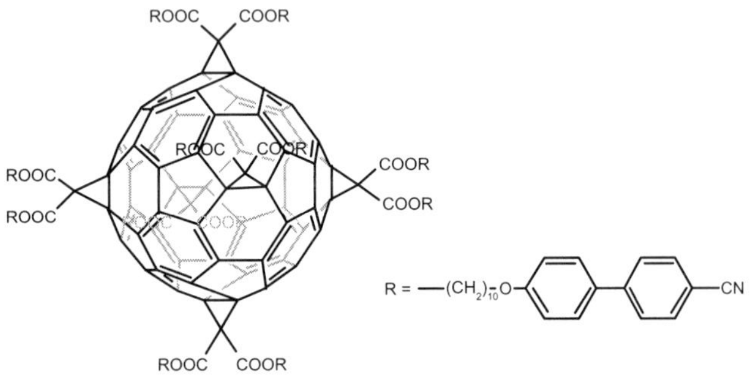

538

Figure 242. Liquid–crystalline hexaadducts of C_{60}. Adapted from [502h].

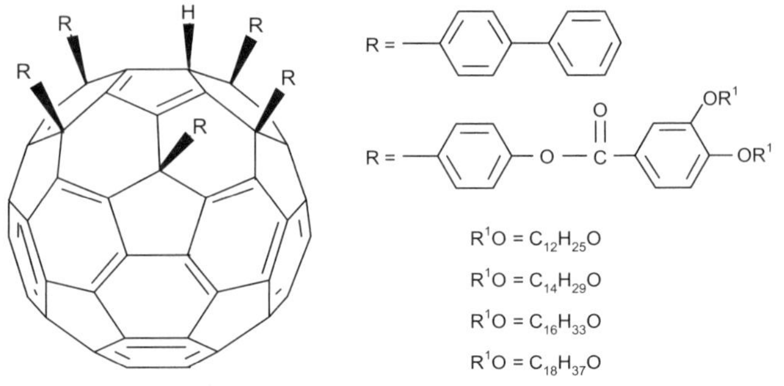

539

Figure 243. The liquid–crystalline pentaarylated fullerenes organize into polar columnar assemblies. Adapted from [507].

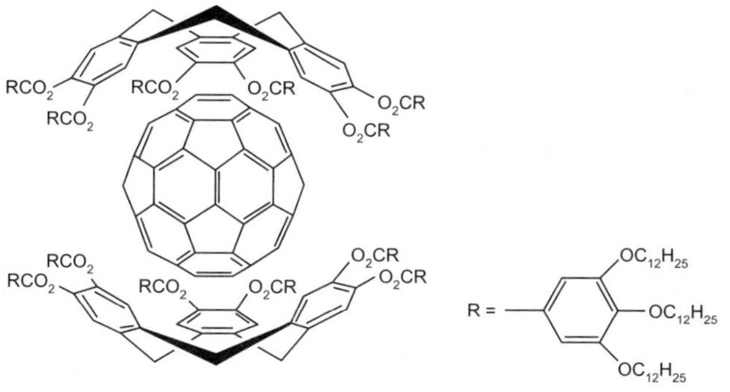

540

Figure 244. A supramolecular fullerene-containing liquid–crystalline material. Adapted from [503].

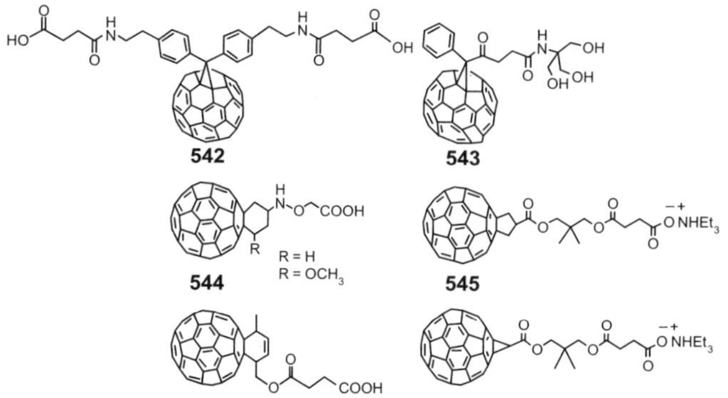

Figure 246. Fullerene carboxylates used for antiviral activity experiments.

15.3.2. HIV Protease Inhibition and Antiviral Activity Studies

Fullerene dicarboxylates (Fig. 246) are amphiphilic. In one of the first reports, methano[60]fullerene dicarboxylate is suggested to show anti-HIV activities based on molecular modeling studies. This study showed that the fullerene derivative and hydrophobic surface of the enzyme cavity could have good van der Waals interaction. Under actual experimental conditions these compounds are found to possess good anti-HIV activity in the μM range [513]. The water soluble fullerene derivative *N*-tris(hydroxymethyl)propylamidomethano[60]fullerene is very active against HIV-1 and exhibited no cytotoxicity [514]. A number of C_{60} and C_{70} derivatives tested as dimethylsulphoxide–water emulsions against human cells infected with HIV showed antiviral activity in low micromolar range [515]. Two C_{70} derivatives are also evaluated for anti-HIV activity. Models analogous to those used for the C_{60} derivative indicate that C_{70} fits in the HIVP hydrophobic cavity like C_{60} or even better [508a]. A series of amphiphilic fullerene derivatives also manifested enzyme inhibitory activity in the micromolar range [516]. Monosubstituted methanofullerene derivatives are found to be more potent compared to the disubstituted methanofullerene derivative. A kinetic study suggested that the monosubstituted derivative becomes hydrophobically bound to the active site of the enzyme.

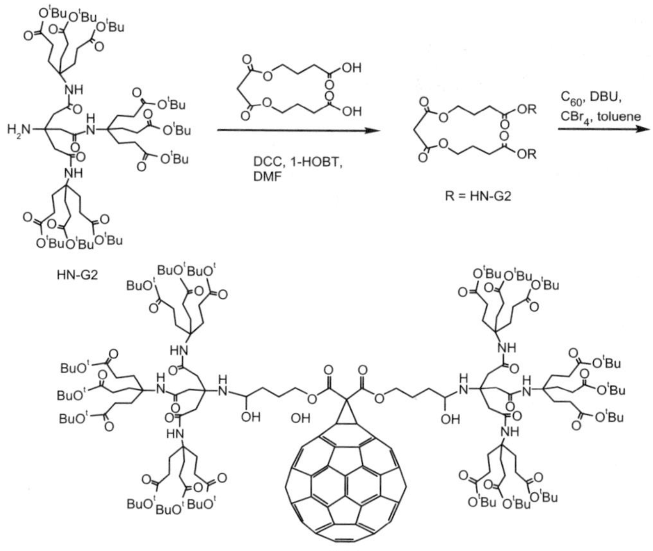

541

Figure 245. A dendritic [60]fullerene used for cytotoxicity and photocytotoxicity studies. Adapted from [511].

15.3.3. DNA Cleaving Studies

A number of studies have been carried out with water-soluble derivatives of fullerenes for DNA cleaving experiments [517]. The photoinduced DNA cleaving activities of C_{60} itself and of a C_{60} derivative with an acridine group are reported. The acridine moiety is also incorporated into a poly(vinyl pyrrolidone) micellar system for the DNA cleaving study. Under visible light irradiation using a supercoiled plasmid (pBR322) the acridine adduct showed stronger DNA cleaving activity than C_{60} itself. Fullerene carboxylic acids also cleaved double stranded DNA with moderate efficiency at 100 μM concentration upon irradiation with visible light. Photolysis of the carboxylic acid in solution containing molecular oxygen generated singlet oxygen, which is responsible for cleaving. The cleaving takes place at the guanine base. A fullerene amine adduct, $C_{60}H_n(NHCH_2CH_2NMe_2)_n$, is highly efficient and it cleanly cut DNA (supercoiled pBR322) into nicked circular one under irradiation with visible light [516]. DNA–fullerene hybrids (**548**, **549**) are

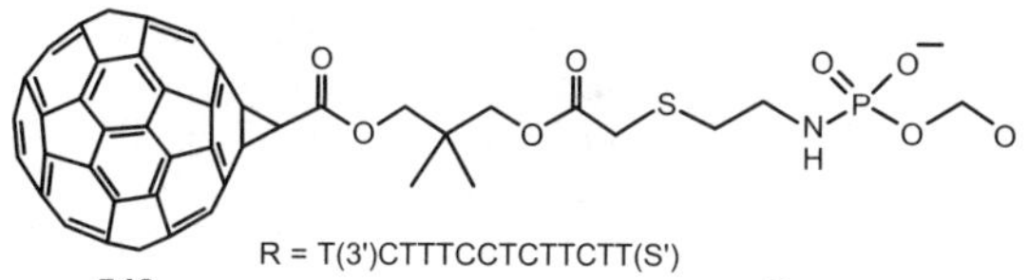

548

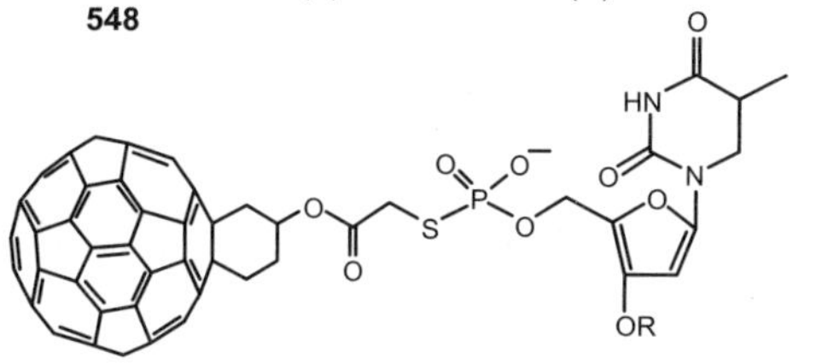

549 R = C(3')TAACGACAATATGTACAAGCCTAATTGTGTAGCATCT

Figure 247. DNA–fullerene hybrid for cleaving studies.

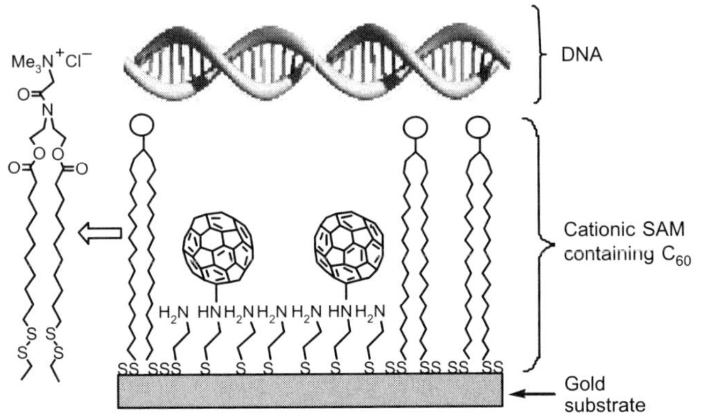

Figure 248. Cationic SAM containing C_{60} on gold for immobilization of DNA. Adapted from [518].

synthesized in which DNA is covalently bound to fullerene via a linking unit (Fig. 247). These hybrids could bind single and double stranded DNA, and upon exposure to light it cleaved DNA specifically at guanosine residues proximal to the fullerene moiety.

Immobilization of DNA on a two-dimensional solid surface is of interest both in studies of DNA itself and in various applications such as biosensors. On a gold surface, a well-ordered monolayer assembly containing quaternary ammonium salts that could interact with the phosphate groups of DNA has been devised. On such a self-assembled monolayer (SAM) C_{60} has been incorporated and then DNA has been immobilized for specific cleaving studies (Fig. 248) [518].

16. SUMMARY

The spherical π-electron surface of C_{60} has provided an unprecedented fundamental building block. The π-bonds in fullerenes are ethylinic and hence undergo a variety of addition reactions. These reactions lead to many derivatives of fullerenes synthesized for fundamental studies as well as for materials science applications. Since the fullerene surface contains a large number of double bonds, multiple additions can take place leading to a mixture of regioisomers. Quite often, addition takes place at the 6,6-ring junction π-bonds of the six pyracyclene units. The driving force for such reactions is the creation of a stable framework of eight benzene rings. By tether directed functionalization, specific regioisomers can be synthesized. The electronic, magnetic, and optical properties of the fullerene derivatives have been investigated for materials science applications. Unmodified fullerenes as well as derivatives are excellent dopants for conducting polymers and very promising for photodiode applications. A number of dyad and triad molecules are synthesized for intramolecular photoinduced charge separation. C_{60} derivatives are studied for optical limiting applications. Water-soluble endohedral fullerenes find applications as magnetic resonance imaging contrast media agents. Fullerenes are powerful radical scavengers. In the presence of visible light they can cleave DNA. Fullerene derivatives soluble in water are efficient neuroprotective agents. A number of such derivatives have been studied in HIV protease inhibition and tumor suppression experiments.

Since 1990, when milligram quantities of fullerenes could be accessible for preparative scale experiments, there has been an unprecedented growth of the chemistry and physics of the fullerenes. Science around these exotic molecules grew at such a rapid pace that, for a time, it was difficult to keep track of the new developments. At a time when it is felt that fullerene research has attained a level of saturation, new and exciting discoveries are being reported. Though functional fullerene chemistry has been thoroughly studied, these are mostly centered around C_{60} and to some extent C_{70}. Eventually such studies involving higher fullerenes will be enthusiastically pursued. Functionalization of endohedral fullerenes is almost at its infancy. Since magnetic, electrical, and optical properties of the endohedral fullerenes are very different from the parent molecules, their derivatives would be expected to possess quite interesting properties. The materials science applications of fullerenes in the broad disciplines of bio-, nano-, and information technology are yet to be thoroughly explored.

GLOSSARY

AM1 A method for semi-empirical molecular mechanics calculation.

Bingel reaction A general method for cyclopropanation of fullerenes by introducing a methylene or substituted methylene bridge onto the fullerene 6,6-ring junction.

B3LYP/6-31G* A hybrid method based on density functional theory (DFT) for predicting chemical reaction mechanism.

Buckminsterfullerene The hollow carbon cage made up of 60 carbon atoms named after R. Buckminster Fuller, American architect, engineer and constructor of geodesic domes.

Dendrimers Three-dimensional, highly ordered oligomeric and polymeric compounds formed from smaller molecules by reiterative reaction sequences.

DNA cleaving A process by which a DNA molecule can be cut into smaller ones.

Endohedral fullerenes Fullerenes with enclosed atom(s) in the cage.

Euler's theorem A applied to fullerenes, each molecule of the fullerene family (C_n) consists of 12 pentagons and m number of hexagons ($m \neq 1$), conforming to the relation $m = (n - 20)/2$.

Exohedral addition reactions of fullerenes Addition reactions on the exterior of fullerene molecule.

Fullerenes A family of large carbon cage molecules about 7–15 Å in diameter.

Fullerene dimers Fullerene derivatives containing two fullerene spheres per molecule.

Fullerene macrocycles Compounds obtained by the covalent or noncovalent attachment of macrocyclic compounds such as calixarenes, catenane, crown ethers, cyclophane, cyclotriveratrylene, phthalocyanines, etc. to fullerenes.

η^5-Fullerene metal complexes Exohedral fullerene-transition metal complexes, in which the fullerene act as an olefinic ligand and bind the metal center.

Heterofullerenes Fullerenes in which one or more carbon atoms are substituted by heteroatoms, such as nitrogen or boron.

HF/3-21G$^{(*)}$ A method for *ab initio* molecular orbital studies.

HIV A retrovirus that causes AIDS (aquired immune deficiency syndrome) by infecting helper T cells of the immune system.

Inclusion complexes Complexes obtained by non-covalent interaction between a host and a suitable guest molecule.

Isolated pentagon rule The rule that signifies that all pentagons in a fullerene be isolated by hexagons, C_{60} being the smallest possible fullerene to obey the isolated pentagon rule.

Macrocyclization A double Bingel reaction between fullerenes and bismalonate derivatives leads to macrocyclization, a method for the preparation of covalent bisadducts of C_{60} with high regio- and diastereoselectivity.

MM2 force field A method in molecular mechanics calculation.

MNDO The semi-empirical molecular orbital (MO) calculations, using a modified neglect of diatomic (differential) overlap approximation.

Periconjugation The interaction between the p_z orbitals of the addends and the "p_z" orbitals of the adjacent fullerene carbon atoms.

Oligofullerene Low molecular weight derivative of fullerenes having three or more fullerenes per molecule.

PM3 A semiempirical molecular mechanics calculation.

Pyramidization angle Defined as $\theta_{\sigma\pi} - 90°$, where $\theta_{\sigma\pi}$ is the angle between the σ and π bonds of a carbon atom.

***o*-Quinodimethane** A intermediate generated by 1,4-elimination of Br_2 from 1,2-bis(bromomethyl)benzenes.

Radical sponge A property of fullerenes to scavenge large concentrations of free radicals.

Ring expansion reactions Reactions leading to the fullerene cage expansion by introducing brigdging atoms.

Schlegel diagrams A convenient way to present the three-dimensional fullerene cage molecules in a two-dimensional plane.

Singlet oxygen The lowest electronically excited singlet state of molecular oxygen, $^1O_2$.

Tether-directed functionalization An approach to the synthesis of fullerene regioisomers defined structure.

ACKNOWLEDGMENTS

The author is deeply indebted to the Kwangju Institute of Science and Technology for providing necessary support to carry out this work through the IITA professorship program of the institute

In the collection and compilation of literature and preparation of this chapter, members of my research group rendered valuable assistance. I record my deep sense of appreciation to all of them. I am specifically indebted to my wife and colleague, Dr. R. R. Das, who contributed significantly through out the preparation of this chapter.

REFERENCES

1. H. W. Kroto, J. R. Heath, S. C. O'Brien, R. F. Curl, and R. E. Smalley, *Nature* 318, 162 (1985).
2. (a) D. E. H. Jones, *New Scientist* 35, 245 (1966); (b) H. Palmer and M. Shelef, "Chemistry and Physics of Carbon" (J. R. Walker, Jr., Ed.), Vol. 4, pp. 85–135. Dekker; New York, 1967.
3. (a) E. Osawa, *Kagaku* 25, 854 (1970) [in Japanese]; *Chem. Abstr.* 74, 75698v (1971). Z. Yoshida and E. Osawa, *Kagakudojin* 174 (1971) [in Japanese]; (b) D. A. Bochvar and E. G. Gal'pern, *Dokl. Acad. Nauk SSSR* 209, 610 (1973). *Proc. Acad. Sci. USSR* 209, 239 (1973) [English translation].
4. W. Krätschmer, L. D. Lamb, K. Fostiropoulos, and D. R. Huffman, *Nature* 347, 354 (1990).
5. R. Taylor, J. P. Hare, A. K. Abdul-Sada, and H. W. Kroto, *Chem. Soc. Chem. Commun.* 1423 (1990).
6. H. W. Kroto, A. W. Allaf, and S. P. Balm, *Chem. Rev.* 91, 1213 (1991).
7. (a) R. Tycko, R. C. Haddon, G. Dabbagh, S. H. Glarum, D. C. Douglass, and A. M. Mujsce, *J. Phys. Chem.* 95, 518 (1991); (b) R. D. Johnson, D. S. Bethume, and C. S. Yannoni, *Acc. Chem. Res.* 25, 169 (1992).
8. A. V. Nilolaev, T. J. S. Dennis, K. Prassides, and A. K. Soper, *Chem. Phys. Lett.* 223, 143 (1994).
9. (a) C. Thilgen and F. Diederich, *Top. Curr. Chem.* 199, 135 (1999); (b) C. Thilgen, A. Herrmann, and F. Diederich, *Angew. Chem. Int. Ed.* 36, 2269 (1997).
10. R. C. Haddon, L. E. Brus, and K. Raghavachari, *Chem. Phys. Lett.* 125, 459 (1986); S. Satpathy, *Chem. Phys. Lett.* 130, 545 (1986); P. D. Hale, *J. Am. Chem. Soc.* 108, 6087 (1986); S. Larsson, A. Volosov, and A. Rosen, *Chem. Phys. Lett.* 137, 501 (1987); A. Rosen and B. Wastberg, *J. Chem. Phys.* 90, 2525 (1989).
11. L. J. Wilson, S. Flanagan, L. P. F. Chibante, and J. M. Alford, Fullerene electrochemistry: Detection, generation, and study of fulleride ions in solution, in "Buckminsterfullerenes" (W. E. Billups and M. A. Ciufolini, Eds.), Ch. 11. VCH, New York, 1993.
12. F. Arias, L. Echegoyen, S. R. Wilson, Q. Lu, and Q. Lu, *J. Am. Chem. Soc.* 117, 1442 (1995).
13. (a) T. G. Schmalz, W. A. Seitz, D. J. Klein, and G. E. Hite, *Chem. Phys. Lett.* 130, 203 (1986); (b) H. W. Kroto, *Nature* 329, 529 (1987).
14. D. J. Klein, T. G. Schmalz, G. E. Hite, and W. A. Seitz, *J. Am. Chem. Soc.* 108, 1301 (1986).
15. Q. L. Zhang, S. C. O'Brien, J. R. Heath, Y. Liu, R. F. Curl, H. W. Kroto, and R. E. Smalley, *J. Phys. Chem.* 90, 525 (1986).
16. (a) M. Saunders, R. J. Cross, H. A. Jiménez-Vázquez, R. Shimshi, and A. Khong, *Science* 271, 1693 (1996); (b) M. Saunders, H. A. Jiménez-Vázquez, R. J. Cross, S. Mroczkowski, D. I. Freedberg, and F. A. L. Anet, *Nature* 367, 256 (1994); (c) M. Saunders, H. A. Jiménez-Vázquez, R. J. Cross, W. E. Billups, C. Gesenberg, A. Gonzales, W. Luo, R. C. Haddon, F. Diederich, and A. Herrmann, *J. Am. Chem. Soc.* 117, 9305 (1995).
17. (a) P. W. Fowler, D. J. Collins, and S. J. Austin, *J. Chem. Soc., Perkin Trans.* 2, 275 (1993); (b) S. J. Austin, P. W. Fowler, P. Hansen, D. E. Manolopoulos, and M. Zheng, *Chem. Phys. Lett.* 228, 478 (1994).
18. M. Bühl and A. Hirsch, *Chem. Rev.* 101, 1153 (2001).
19. R. C. Haddon and K. Raghavachari, Electronic structure of the fullerene carbon allotropes of intermediate hybridization, in "Buckminsterfullerenes" (W. E. Billups and M. A. Ciufolini, Eds.), Ch. 7. VCH, New York, 1993.
20. T. G. Schmalz and D. J. Klein, in "Buckminsterfullerenes" (W. E. Billups and M. A. Ciufolini, Eds.), p. 83. VCH, New York, 1993.
21. H. D. Beckhaus, C. Rüchardt, M. Kao, F. Diederich, and C. S. Foote, *Angew. Chem. Int. Ed. Engl.* 31, 63 (1992).
22. B. Pietzak, M. Waiblinger, T. Almeida Murphy, A. Weidinger, M. Hohne, E. Dietel, and A. Hirsch, *Chem. Phys. Lett.* 279, 259 (1997).

23. H. Mauser, A. Hirsch, N. J. R. van Eikema Hommes, and T. Clark, *J. Mol. Model.* 3, 415 (1992).
24. R. C. Haddon, *Science* 261, 1545 (1993).
25. T. Braun, A. P. Schubert, and R. N. Kostoff, *Chem. Rev.* 100, 23 (2000); *Chem. Rev.* 100, 2475 (2000).
26. F. Wudl, *Acc. Chem. Res.* 25, 157 (1992); G. S. Hammond and V. J. Kuck, Eds., "ACS Symposium Series," Vol. 481. *Am. Chem. Soc.*, Washington, DC, 1992; R. Taylor, D. R. M. Walton, *Nature* 363, 685 (1993); A. Hirsch, *Synthesis* 895 (1995); A. B. Smith III, Ed., Fullerene Chemistry, Tetrahedron Symposia-in-Print Number 60 (1996); S. Samal and S. K. Sahoo, *Bull. Mater. Sci.* 20, 141 (1997); S. Samal and K. E. Geckeler, in "Advanced Functional Molecules and Polymers" H. S. Nalwa, Ed, Vol. 1, Ch. 1, pp. 1–85. Gordon & Breach, New York, 2001.
27. D. W. Murphy, M. J. Rosseinsky, R. M. Fleming, R. Tycko, A. P. Ramirez, R. C. Haddon, T. Siegrist, G. Dabbagh, J. C. Tully, and R. E. Walstedt, *J. Phys. Chem. Solids* 53, 1321 (1993); W. E. Broderick, *J. Am. Chem. Soc.* 116, 5489 (1994).
28. F. Wudl, *Acc. Chem. Res.* 25, 157 (1992).
29. Y. Chai, T. Guo, C. Jin, R. E. Haufler, L. P. F. Chibante, J. Fure, L. Wang, J. M. Alford, and R. E. Smalley, *J. Phys. Chem.* 95, 7564 (1991).
30. D. S. Bethune, R. D. Johnson, J. R. Salem, M. S. de Vries, and C. S. Yanoni, *Nature* 366, 123 (1993); F. T. Edelmann, *Angew. Chem. Int. Ed. Engl.* 34, 981 (1995); S. Nagase, K. Kobayashi, and T. Akasaka, *Bull. Chem. Soc. Jpn.* 69, 2131 (1996); H Shinohara, *Rep. Progr. Phys.* 63, 843 (2000); S. Liu and S. Sun, *J. Organometall. Chem.* 599, 74 (2000).
31. R. Huang, H. Li, W. Lu, and S. Yang, *Chem. Phys. Lett.* 228, 111 (1994).
32. B. Kubler, E. Millon, J. J. Gaumet, and J. F. Muller, *Fuller. Sci. Technol.* 5, 839 (1997).
33. B.-P. Cao, Z.-J. Shi, X.-H. Zhou, Z.-N. Gu, H.-Z. Xiao, and J.-Z. Wang, *Chem. Lett.* 937 (1997).
34. H. Shinohara, H. Yamaguchi, N. Hayashi, H. Sato, M. Ohkohchi, Y. Ando, and Y. Saito, *J. Phys. Chem.* 97, 4259 (1993); K. Kikuchi, S. Suzuki, Y. Nakao, N. Nakahara, T. Wakabayashi, H. Shiromaru, K. Saito, I. Ikemoto, and Y. Achiba, *Chem. Phys. Lett.* 216, 67 (1993).
35. J. Xiao, M. R. Savina, G. B. Martin, A. H. Francis, and M. E. Meyerhoff, *J. Am. Chem. Soc.* 116, 9341 (1994).
36. D.-Y. Sun, Z.-Y. Liu, X.-H. Guo, W.-G. Xu, and S.-Y. Liu, *J. Phys. Chem. B* 101, 3927 (1997); B.-B. Liu, Z.-Y. Liu, W.-G. Xu, H.-B. Yang, C.-X. Gao, J.-S. Lu, S.-Y. Liu, and G.-G. Zou, *Tetrahedron* 54, 11123 (1998); C. Capp, T. D. Wood, A. G. Marshall, and J. V. Coe, *J. Am. Chem. Soc.* 116, 4987 (1994); D.-Y. Sun, Z.-Y. Liu, X.-H. Guo, W.-G. Xu, and S.-Y. Liu, *Fuller. Sci. Technol.* 5, 137 (1997).
37. Y. Kubozono, H. Maeda, Y. Takabayashi, K. Hiraoka, T. Nakai, S. Kashino, S. Emura, S. Ukita, and T. Sogabe, *J. Am. Chem. Soc.* 118, 6998 (1996).
38. M. Hoinkis, C. S. Yannoni, D. S. Bethune, J. R. Salem, R. D. Johnson, M. S. Crowder, and M. S. deVries, *Chem. Phys. Lett.* 198, 461 (1992).
39. N. Okabe, Y. Ohba, S. Suzuki, S. Kawata, K. Kikuchi, Y. Achiba, and M. Iwaizumi, *Chem. Phys. Lett.* 235, 564 (1995).
40. U. Kirbach and L. Dunsch, *Angew. Chem. Int. Ed. Engl.* 35, 2380 (1996); K. Yamamoto, H. Funasaka, T. Takahashi, T. Akasaka, T. Suzuki, and Y. Maruyama, *J. Phys. Chem.* 98, 12831 (1994).
41. M. Takata, B. Umeda, E. Nishlbori, M. Sakata, Y. Salto, M. Ohno, and H. Shinohara, *Nature* 377, 46 (1995); D. M. Poirier, M. Knupfer, J. H. Weaver, W. Andreoni, K. Laasonen, M. Parrinello, D. S. Bethune, K. Kikuchi, and Y. Achiba, *Phys. Rev. B* 49, 17403 (1994); K. Kobayashi and S. Nagase, *Chem. Phys. Lett.* 282, 325 (1998); M. Rübsam, M. Plüschau, P. Schweitzer, K.-P. Dinse, D. Fuchs, H. Rietschel, R. H. Michel, M. Benz, and M. M. Kappes, *Chem. Phys. Lett.* 240, 615 (1995).
42. T. Akasaka, S. Nagase, K. Kobayashi, M. Wälchli, K. Yamamoto, H. Funasaka, M. Kato, T. Hoshino, and T. Erata, *Angew. Chem. Int. Ed. Engl.* 36, 1643 (1997).
43. E. Yamamoto, M. Tansho, T. Tomiyama, H. Shinohara, H. Kawahara, and Y. Kobayashi, *J. Am. Chem. Soc.* 118, 2293 (1996).
44. Y. Miyake, S. Suzuki, Y. Kojima, K. Kikuchi, K. Kobayashi, S. Nagase, M. Kainosho, Y. Achiba, Y. Maniwa, and K. Fisher, *J. Phys. Chem.* 100, 9579 (1996).
45. H. Shinohara, H. Sato, M. Ohchochi, Y. Ando, T. Kodama, T. Shida, T. Kato, and Y. Saito, *Nature* 357, 52 (1992).
46. J. R. Ungerer and T. Hughbanks, *J. Am. Chem. Soc.* 115, 2054 (1993).
47. E. B. Iezzi, J. C. Duchamp, K. Harich, T. E. Glass, H. M. Lee, M. M. Olmstead, A. L. Balch, and H. C. Dorn, *J. Am. Chem. Soc.* 124, 524 (2002).
48. R. D. Johnson, M. S. de Vries, J. Salem, D. S. Bethune, and C. S. Yannoni, *Nature* 355, 239 (1992); S. Hino, H. Takahashi, K. Iwasaki, K. Matsumoto, T. Miyazaki, S. Hasegawa, K. Kikuchi, and Y. Achiba, *Phys. Rev. Lett.* 71, 4261 (1993).
49. T. Takahashi, A. Ito, M. Inakuma, and H. Shinohara, *Phys. Rev. B* 52, 13812 (1995).
50. T. Suzuki, K. Kikuchi, F. Oguri, Y. Nakao, S. Suzuki, Y. Achiba, K. Yamamoto, H. Funasaka, and T. Takahashi, *Tetrahedron* 52, 4973 (1996); T. Suzuki, Y. Maruyama, T. Kato, K. Kikuchi, and Y. Achiba, *J. Am. Chem. Soc.* 115, 11006 (1993); T. Suzuki, Y. Maruyama, T. Kato, K. Kikuchi, Y. Nakao, Y. Achiba, K. Kobayashi, and S. Nagase, *Angew. Chem. Int. Ed. Engl.* 34, 1094 (1995).
51. J. Cioslowski and A. Nanayakkara, *J. Chem. Phys.* 96, 8345 (1992).
52. T. Akasaka, T. Kato, K. Kbayashi, S. Nagase, K. Yamamoto, H. Funasaka, and T. Takahashi, *Nature* 374, 600 (1995); T. Akasaka, S. Nagase, K. Kobayashi, T. Suzuki, T. Kato, K. Yamamoto, H. Funasaka, and K. Takahashi, *J. Chem. Soc.* 1343 (1995); T. Akasaka, S. Nagase, K. Kobayashi, T. Suzuki, T. Kato, K. Kikuchi, Y. Achiba, K. Yamamoto, H. Funasaka, and T. Takahashi, *Angew. Chem. Int. Ed. Engl.* 34, 2139 (1995).
53. T. Akasaka, T. Kato, S. Nagase, K. Kobayashi, K. Yamamoto, H. Funasaka, and T. Takahashi, *Tetrahedron* 52, 5015 (1996).
54. T. Suzuki, Y. Maruyama, T. Kato, T. Akasaka, K. Kobayashi, S. Nagase, K. Yamamoto, H. Funasaka, and T. Takahashi, *J. Am. Chem. Soc.* 117, 9606 (1995).
55. S.-R. Zhang, D.-Y. Sun, X.-Y. Li, F.-K. Pei, and S.-Y. Liu, *Fuller. Sci. Technol.* 5, 1635 (1997); D.-Y. Sun, H.-J. Huang, S.-H. Yang, Z.-Y. Liu, and S.-Y. Liu, *Chem. Mater.* 11, 1003 (1999).
56. K. Kobayashi, S. Nagase, and T. Akasaka, *Chem. Phys. Lett.* 245, 230 (1995); T. Ohtsuki, K. Masumoto, K. Kikuchi, and K. Sueki, *Mater. Sci. Eng. A* 217/218, 38 (1996); D. W. Cagle, T. P. Thrash, L. J. Wilson, G. J. Ehrhardt, J. Alford, and L. P. F. Chibante, "Fullerenes: Recent Advances in the Chemistry and Physics of Fullerenes and Related Materials" (K. Kadish and R. Ruoff, Eds.), Vol. 3, pp. 854–868. Electrochemical Society, Pennington, NJ, 1997.
57. T. Ohtsuki, K. Masumoto, K. Ohno, Y. Maruyama, Y. Kawazoe, K. Sueki, and K. Kikuchi, *Phys. Rev. Lett.* 77, 3522 (1996).
58. K. Kobayashi, M. Kuwano, K. Sueki, K. Kikuchi, Y. Achiba, H. Nakahara, N. Kananishi, M. Watanabe, and K. Tomura, *J. Radio Anal. Nucl. Chem.* 192, 81 (1995).
59. D. W. Cagle, J. M. Alford, J. Tien, and L. J. Wilson, "Fullerenes: Recent Advances in the Chemistry and Physics of Fullerenes and Related Materials" (K. Kadish and R. Ruoff, Eds.), Vol. 4, pp. 361–368. Electrochemical Society, Pennington, NJ, 1997.
60. L. J. Wilson, D. W. Cagle, T. P. Thrash, S. J. Kennel, S. Mirzadeh, J. M. Alford, and G. J. Ehrhardt, *Coord. Chem. Rev.* 190–192, 199 (1999).
61. J. R. Morton, F. Negri, and K. F. Preston, *Acc. Chem. Res.* 31, 63 (1998).

62. P. J. Krusic, E. Wasserman, P. N. Keizer, J. R. Morton, and K. F. Preston, *Science* 254, 1184 (1991).
63. P. J. Krusic, E. Wasserman, B. A. Parkinson, B. Malone, E. R. Holler Jr., P. N. Keizer, J. R. Morton, and K. F. Preston, *J. Am. Chem. Soc.* 113, 6274 (1991).
64. J. R. Morton, K. F. Preston, P. J. Krusic, S. A. Hill, and E. Wasserman, and *J. Phys. Chem.* 96, 3576 (1992).
65. P. J. Krusic, D. C. Roe, E. Johnston, J. R. Morton, and K. F. Preston, *J. Phys. Chem.* 97, 1736 (1993).
66. R. Borghi, L. Lunazzi, G. Placucci, G. Cerioni, and A. Plumitallo, *J. Org. Chem.* 61, 3327 (1996).
67. C. N. McEwen, R. G. McKay, and B. S. Larsen, *J. Am. Chem. Soc.* 114, 4412 (1992).
68. R. Taylor, *Synlett.* 776 (2000).
69. A. Hirsch, *Top. Curr. Chem.* 199, 1 (1998).
70. (a) C. C. Henderson and P. A. Cahill, *Science* 259, 1885 (1993); (b) A. G. Avent, A. D. Darwish, D. K. Heimbach, H. W. Kroto, M. F. Meidine, J. P. Parsons, C. Remars, R. Roers, O. Ohashi, R. Taylor, and D. R. M. Walton, *J. Chem. Soc., Perkin Trans. 2* 15 (1994); (c) S. Ballenweg, R. Gleiter, and W. Krätschmer, *Tetrahedron Lett.* 34, 3737 (1993).
71. A. D. Darwish, A. K. Abdul-Sada, G. J. Langley, H. W. Kroto, R. Taylor, and D. R. M. Walton, *J. Chem. Soc., Perkin Trans. 2*, 2359 (1995); A. D. Darwish, H. W. Kroto, R. Taylor, and D. R. M. Walton, *Synthetic Metals* 1415 (1996).
72. A. D. Darwish, H. W. Kroto, R. Taylor, and D. R. M. Walton, *J. Chem. Soc. Perkin Trans 2* 1415 (1996).
73. R. Taylor, *J. Chem. Soc., Perkin Trans. 2* 1667 (1992).
74. R. E. Haufler, J. Conceicao, L. P. F. Chibante, Y. Chai, N. E. Byrne, S. Flanagan, M. M. Haley, S. C. O'Brian, C. Pan, Z. Xiao, W. E. Billups, M. A. Ciufolini, R. H. Hauge, J. L. Margrave, L. J. Wilson, R. F. Curl, and R. E. Smalley, *J. Phys. Chem.* 94, 8634 (1990).
75. R. Taylor, *Chem. Eur. J.* 7, 4075 (2001).
76. J. H. Holloway, E. G. Hope, R. Taylor, G. J. Langley, A. G. Avent, T. J. Dennis, J. P. Hare, H. W. Kroto, and D. R. M. Walton, *J. Chem. Soc., Chem. Commun.* 966 (1991); A. A. Tuinman, P. Mukherjee, J. L. Adcock, R. L. Hettich, and R. N. Compton, *J. Phys. Chem.* 96, 7584 (1992).
77. A. Hamwi, C. Fabre, P. Chaurand, S. Della-Negra, C. Ciot, D. Djurado, J. Dupois, and A. Rassat, *Fullerene Sci. Technol.* 1, 499 (1993).
78. R. Taylor, G. J. Langley, J. H. Holloway, E. G. Hope, A. K. Brisdon, H. W. Kroto, and D. R. M. Walton, *J. Chem. Soc. Perkin Trans. 2* 181 (1995).
79. A. A. Gakh, A. A. Tuinman, J. L. Adcock, R. A. Sachleben, and R. N. Compton, *J. Am. Chem. Soc.* 116, 819 (1994); O. V. Boltalina, L. N. Sidorov, V. F. Bagryantsev, V. A. Seredenko, A. S. Zapolskii, J. M. Street, and R. Taylor, *J. Chem. Soc. Perkin Trans. 2* 2275 (1996).
80. A. A. Gakh, A. A. Tuinman, J. L. Adcock, and R. N. Compton, *Tetrahedron Lett.* 34, 7167 (1993).
81. O. V. Boltalina, A. Ya. Borschevskii, L. N. Sidorov, J. M. Street, and R. Taylor, *Chem. Commun.* 529 (1996); O. V. Boltalina, J. M. Street, and R. Taylor, *J. Chem. Soc. Perkin Trans. 2* 649 (1998); O. V. Boltalina, M Bühl, A. Khong, M. Saunders, J. M. Street, and R. Taylor, *J. Chem. Soc. Perkin Trans. 2* 1475 (1999).
82. O. V. Boltalina, V. Yu. Markov, R. Taylor, and M. P. Waugh, *Chem. Commun.* 2549 (1996).
83. I. S. Neretin, K. A. Lyssenko, M. Y. Antipin, Y. L. Slovokhotov, O. V. Boltalina, P. A. Troshin, A. Y. Lukonin, L. N. Sidorov, and R. Taylor, *Angew. Chem., Int. Ed. Engl.* 39, 3273 (2000).
84. A. G. Avent, O. V. Boltalina, A. Ya. Lukonin, J. M. Street, and R. Taylor, *J. Chem. Soc. Perkin Trans. 2* 1359 (2000).
85. R. Taylor, A. K. Abdul-Sada, O. V. Boltalina, and J. M. Street, *J. Chem. Soc. Perkin Trans. 2* 1013 (2000).
86. A. K. Abdul-Sada, A. V. Avakyan, O. V. Boltalina, Yu. Markov, J. M. Street, and R. Taylor, *J. Chem. Soc. Perkin Trans. 2* 2659 (1999); O. V. Boltalina, A. K. Abdul-Sada, T. V. Avakyan, T. J. S. Dennis, V. Yu. Markov, and R. Taylor, *J. Phys. Chem. B* 103, 8189 (1999).
87. O. V. Boltalina, B. de La Vaissieŕe, P. W. Fowler, P. B. Hitchcock, J. B. Sandall, P. A. Troshin, and R. Taylor, *Chem. Commum.* 1325 (2000); O. V. Boltalina, B. de La Vaissière, P. W. Fowler, A. Ya. Lukonin, A. K. Abdul-Sada, J. M. Street, and R. Taylor, *J. Chem. Soc. Perkin Trans. 2* 2212 (2000); T. Suzuki, Q. Li, K. C. Fhemani, F. Wudl, and O. Almarsson, *Science* 254, 1186 (1991); M. Prato, Q. Li, F. Wudl, and V. Lucchini, *J. Am. Chem. Soc.* 115, 1148 (1993).
88. O. V. Boltalina, B. de La Vaissière, P. W. Fowler, A. Ya. Lukonin, A. K. Abdul-Sada, J. M. Street, and R. Taylor, *J. Chem. Soc. Perkin Trans. 2* 2212 (2000).
89. G. A. Olah, J. W. Bausch, and G. K. SuryaPrakash, *J. Am. Chem. Soc.* 113, 3205 (1991).
90. P. R. Birkett, A. G. Avent, A. D. Darwish, H. W. Kroto, R. Taylor, and D. R. M. Walton, *J. Chem. Soc., Chem. Commun.* 683 (1995).
91. F. N. Tebbe, R. L. Harlow, D. B. Chase, D. L. Thorn, G. C. Campbell, Jr., J. C. Calabrese, N. Herron, R. J. Yaung, Jr., and E. Wasserman, *Science* 256, 822 (1992).
92. P. R. Birkett, P. B. Hitchcock, H. W. Kroto, R. Taylor, and D. R. M. Walton, *Nature* 357, 479 (1992).
93. L. Y. Chiang, J. B. Bhonsle, L. Wang, S. F. Shu, T. M. Chang, and J. R. Hwu, *Tetrahedron* 52, 4963 (1996).
94. Z. Slanina, X. Zhao, L. Y. Chiang, and E. Osawa, *Int. J. Quantum Chem.* 74, 343 (1999).
95. B.-H. Chen, J.-P. Huang, L. Y. Wang, J. Shiea, T.-L. Chen, and L. Y. Chiang, *J. Chem. Soc., Perkin Trans I* 1171 (1998).
96. D. Sun, H. Huang, S. Yang, Z. Liu, and S. Liu, *Chem. Mater.* 11, 1003 (1999).
97. G. Schick, K. Kampe, and A. Hirsch, *J. Chem. Soc., Chem. Commun.* 2023 (1995).
98. G. A. Olah, I. Busci, C. Lambert, R. Aniszfeld, N. J. Trivedi, D. K. Sensharma, and G. K. SuryaPrakash, *J. Am. Chem. Soc.* 113, 9385 (1991); *J. Am. Chem. Soc.* 113, 9387 (1991).
99. R. Taylor, G. J. Langley, M. F. Meidine, J. P. Parsons, A. K. Abdul-Sada, T. J. Dennis, J. P. Hare, H. W. Kroto, and D. R. M. Walton, *J. Chem. Soc., Chem. Commun.* 667 (1992).
100. A. D. Darwish, P. R. Birkett, G. J. Langley, H. W. Kroto, R. Taylor, and D. R. M. Walton, *Fullerene Sci. Technol.* 5, 705 (1997).
101. A. G. Avent, P. R. Birkett, J. D. Crane, A. D. Darwish, G. J. Langley, H. W. Kroto, R. Taylor, and D. R. M. Walton, *J. Chem. Soc., Chem. Commun.* 1463 (1994).
102. A. G. Avent, P. R. Birkett, H. W. Kroto, R. Taylor, and D. R. M. Walton, *Chem. Commun.* 2153 (1998).
103. A. G. Avent, P. R. Birkett, A. D. Darwish, H. W. Kroto, R. Taylor, and D. R. M. Walton, *Chem. Commun.* 1579 (1997).
104. A. K. Abdul-Sada, A. G. Avent, P. R. Birkett, H. W. Kroto, R. Taylor, D. R. M. Walton, and O. B. Woodhouse, *Chem. Commun.* 307 (1998).
105. A. K. Abdul-Sada, A. G. Avent, P. R. Birkett, H. W. Kroto, R. Taylor, and D. R. M. Walton, *J. Chem. Soc., Perkin Trans. I* 393 (1998).
106. A. G. Avent, P. R. Birkett, A. D. Darwish, H. W. Kroto, R. Taylor, and D. R. M. Walton, *Tetrahedron* 52, 5235 (1996); P. R. Birkett, A. G. Avent, A. D. Darwish, H. W. Kroto, R. Taylor, and D. R. M. Walton, *Chem. Commun.* 1231 (1996).
107. P. R. Birkett, A. G. Avent, A. D. Darwish, H. W. Kroto, R. Taylor, and D. R. M. Walton, *J. Chem. Soc., Chem. Commun.* 1869 (1995).
108. O. V. Boltalina, J. M. Street, and R. Taylor, *Chem. Commun.* 1827 (1998).
109. S. Yamago, M. Yanagawa, H. Mukai, and E. Nakamura, *Tetrahedron* 52, 5091 (1996).
110. M. Sawamura, H. Iikura, and E. Nakamura, *J. Am. Chem. Soc.* 118, 12850 (1996).

111. A. G. Avent, P. R. Birkett, J. D. Crane, A. D. Darwish, G. J. Langley, H. W. Kroto, R. Taylor, and D. R. M. Walton, *J. Chem. Soc., Chem. Commun.* 1463 (1994).
112. H. Ikura, S. Mori, M. Sawamura, and E. Nakamura, *J. Org. Chem.* 62, 7912 (1997).
113. (a) M. Sawamura, H. Iikura, T. Ohama, U. E. Hackler, and E. Nakamura, *J. Organometal. Chem.* 599, 32 (2000). (b) M. Sawamura, M. Toganoh, Y. Kuninobu, S. Kato, and E. Nakamura, *Chem. Lett.* 262 (2000). (c) M. Sawamura, N. Nagahama, M. Toganoh, and E. Nakamura, *J. Organometal. Chem.* 652, 31 (2002).
114. (a) H. Nagashima, H. Terasaki, E. Kimura, K. Nakajima, and K. Itoh, *J. Org. Chem.* 59, 1246 (1994). (b) G. Schick, K.-D. Kampe, and A. Hirsch, *J. Chem. Soc., Chem. Commun.* 2023 (1995). (c) S. Miki, M. Kitao, and K. Fukunishi, *Tetrahedron Lett.* 37, 2049 (1996). (d) G.-W. Wang, Y. Murata, K. Komatsu, and T. S. M. Wan, *J. Chem. Soc., Chem. Commun.* 2059 (1996).
115. M. Sawamura, H. Iikura, A. Hirai, and E. Nakamura, *J. Am. Chem. Soc.* 120, 8285 (1998).
116. M. Sawamura, Y. Kuninobu, M. Toganoh, Y. Matsuo, M. Yamanaka, and E. Nakamura, *J. Am. Chem. Soc.* 124, 9354 (2002).
117. M. A. Yurovskaya and A. A. Ovcharenko, *Khim. Geterotsikl. Soedin.* 291 (1998) [*Chem. Heterocycl. Compd.* (1998)]; E. N. Karaulova and E. I. Bagii, *Usp. Khim.* 68, 979 (1999) [*Russ. Chem. Rev.* 68, 979 (1999)].
118. M. A. Yurovskaya and I. V. Trushkov, *Russ. Chem. Bull., Int. Ed.* 51, 343 (2002).
119. J. D. Crane, P. B. Hitchcock, H. W. Kroto, R. Taylor, and D. R. M. Walton, *J. Chem. Soc., Chem. Commun.* 1764 (1992).
120. L. Hao, M. M. Olmstead, and A. L. Balch, *Abstr. Papers. Am. Chem. Soc.* 166-INOR, 211 (1996).
121. J. D. Crane and P. B. Hitchcock, *J. Chem. Soc., Dalton Trans.* 2537 (1993).
122. P. J. Fagan, J. C. Calabrese, and B. Malone, *Acc. Chem. Res.* 25, 134 (1992); A. Stephens and M. L. H. Green, *Adv. Inorg. Chem.* 44, 1 (1997); P. Mathur, I. J. Mavunkal, and S. B. Umbarkar, *J. Cluster Sci.* 9, 393 (1998); A. L. Balch and M. M. Olmstead, *Chem. Rev.* 98, 2123 (1998); K. Lee, H. Song, and J. T. Park, *Acc. Chem. Res.* 36, 78 (2003).
123. V. K. Cherkasov, Y. F. Rad'kov, M. A. Lopatin, and M. N. Bochkarev, *Russ. Chem. Bull.* 43, 1834 (1994).
124. J. R. Shapley, Y. Du, H.-F. Hsu, and J. J. Way, in "Recent Advances in the Chemistry and Physics of Fullerenes and Related Materials" (K. M. Kadish and R. S. Ruoff, Eds.), Vol. 94-24, p. 1255. Electrochemical Society, Pennington, NJ, 1994.
125. H.-F. Hsu and J. R. Shapley, in "Recent Advances in the Chemistry and Physics of Fullerenes and Related Material" (R. S. Ruoff and K. M. Kadish, Eds.), Vol. 95-10, p. 1087. Electrochemical Society, Pennington, NJ, 1995.
126. S. Zhang, T. L. Brown, Y. Du, and J. R. Shapley, *J. Am. Chem. Soc.* 115, 6705 (1993).
127. J. M. Hawkins, T. A. Lewis, S. D. Loren, A. Meyer, J. R. Heath, Y. Shibato, and R. J. Saykally, *J. Org. Chem.* 55, 6250 (1990); J. M. Hawkins, A. Meyer, T. A. Lewis, S. Loren, and F. J. Hollander, *Science* 252, 312 (1991); J. M. Hawkins, *Acc. Chem. Res.* 25, 150 (1992); J. M. Hawkins, S. Loren, A. Meyer, and R. Nunlist, *J. Am. Chem. Soc.* 113, 7770 (1991).
128. J. M. Hawkins, A. Meyer, and M. A. Solow, *J. Am. Chem. Soc.* 115, 7499 (1993).
129. J. T. Park, J.-J. Cho, and H. Song, *J. Chem. Soc., Chem. Commun.* 15 (1995); J. T. Park, H. Song, J.-J. Cho, M.-K. Chung, J.-H. Lee, and I. H. Shu, *Organometallics* 17, 227 (1998).
130. H.-F. Hsu and J. R. Shapley, *J. Am. Chem. Soc.* 118, 9192 (1996).
131. T. Braun, M. Wohlers, T. Belz, and R. Schlögl, *Catal. Lett.* 43, 175 (1997); T. Braun, M. Wohlers, T. Belz, G. Nowitzke, G. Wortmann, Y. Uchida, N. Pfänder, and R. Schlögl, *Catal. Lett.* 43, 167 (1997).
132. H.-F. Hsu and J. R. Shapley, *J. Chem. Soc., Chem. Commun.* 1125 (1997).
133. K. Lee, H.-F. Hsa, and J. R. Shapley, *Organometallics* 16, 3876 (1997).
134. M.-J. Arce, A. L. Viado, S. I. Khan, and Y. Rubin, *Organometallics* 15, 4340 (1996).
135. M. D. Westmeyer, C. P. Galloway, and T. B. Rauchfuss, *Inorg. Chem.* 33, 4615 (1994); M. D. Westmeyer, T. B. Rauchfuss, and A. K. Verma, *Inorg. Chem.* 35, 7140 (1996).
136. A. L. Balch, V. J. Catalano, and J. W. Lee, *Inorg. Chem.* 30, 3980 (1991).
137. A. L. Balch, D. A. Costa, J. W. Lee, B. C. Noll, and M. M. Olmstead, *Inorg. Chem.* 33, 2071 (1994).
138. A. L. Balch, D. A. Costa, and M. M. Olmstead, *Chem. Commun.* 2449 (1996).
139. R. S. Koefod, M. F. Hudgens, and J. R. Shapley, *J. Am. Chem. Soc.* 113, 8957 (1991).
140. A. L. Balch, J. W. Lee, B. C. Noll, and M. M. Olmstead, *Inorg. Chem.* 32, 55 (1993).
141. M. Iyoda, F. Sultana, S. Sasaki, and H. Butenschön, *Tetrahedron Lett.* 36, 579 (1995).
142. S. A. Lerke, B. A. Parkinson, D. H. Evans, and P. J. Fagan, *J. Am. Chem. Soc.* 114, 7807 (1992); P. J. Fagan, J. C. Calabrese, and B. Malone, *Science* 252, 1160 (1991).
143. V. V. Bashilov, P. V. Petrovskii, V. I. Sokolov, S. V. Lindeman, I. A. Guzey, and Y. T. Struchkov, *Organometallics* 12, 991 (1993).
144. P. J. Fagan, J. C. Calabrese, and B. Malone, *J. Am. Chem. Soc.* 113, 9408 (1991).
145. H. Nagashima, A. Nakaota, Y. Saito, M. Kato, T. Kawanishi, and K. Itoh, *J. Chem. Soc., Chem. Commun.* 377 (1992); H. Nagashima, A. Nakaoka, S. Tajima, Y. Saito, and K. Itoh, *Chem. Lett.* 1361 (1992).
146. J. M. Woods, B. Kahr, S. H. Hoke II, L. Dejarme, R. G. Cooks, and D. Ben-Amotz, *J. Am. Chem. Soc.* 113, 5907 (1991).
147. R. Taylor, J. P. Parsons, A. G. Avent, S. P. Rannard, T. J. Dennis, J. P. Hare, H. W. Kroto, and D. R. M. Walton, *Nature* 351, 277 (1991).
148. F. Diederich, R. Ettl, Y. Rubin, R. L. Whetten, R. Beck, M. Alvarez, S. Anz, D. Sehsharma, F. Wudl, K. C. Khemani, and A. Koch, *Science* 252, 548 (1991).
149. R. Malhotra, S. Kumar, and A. Satyam, *J. Chem. Soc., Chem. Commun.* 1339 (1994).
150. K. M. Creegan, J. L. Robbins, W. K. Robbins, J. M. Millar, R. D. Sherwood, P. J. Tindall, D. M. Cox, A. B. Smith III, J. P. McCauley, Jr., D. R. Jones, and R. T. Gallagher, *J. Am. Chem. Soc.* 114, 1103 (1992).
151. W. A. Kalsbeck and H. H. Thorp, *J. Electroanal. Chem.* 314, 363 (1991).
152. T. Hamano, T. Mashino, and M. Hirobe, *J. Chem. Soc., Chem. Commun.* 1537 (1995).
153. C. Bingel, *Chem. Ber.* 126, 1957 (1993); A. Hirsch, I. Lamparth, and H. R. Karfunkel, *Angew. Chem. Int. Ed. Engl.* 33, 437 (1994).
154. P. Timmerman, H. L. Anderson, R. Faust, J.-F. Nierengarten, T. Habicher, P. Seiler, and F. Diederich, *Tetrahedron* 52, 4925 (1996); F. Cheng, X. Yang, H. Zhu, and Y. Song, *Tetrahedron Lett.* 41, 3947 (2000).
155. H. J. Bestmann, D. Hadawi, T. Roder, and C. Moli, *Tetrahedron Lett.* 35, 9017 (1994); Y. Wang, J. Cao, D. I. Schuster, and S. R. Wilson, *Tetrahedron Lett.* 38, 6843 (1995).
156. P. L. Boulas, Y. Zuo, and L. Echegoyen, *Chem. Commun.* 1547 (1996).
157. H. Yamaguchi, S. Murata, T. Akasaka, and T. Suzuki, *Tetrahedron Lett.* 38, 3529 (1997).
158. T. Suzuki, Q. Li, K. C. Khemini, and F. Wudl, *J. Am. Chem. Soc.* 114, 7301 (1992).
159. A. B. Smith III, R. M. Strongin, L. Brard, G. T. Furst, and W. J. Romanow, *J. Am. Chem. Soc.* 115, 5829 (1993).

160. G. Roth and P. Adelmann, *J. Phys. 1 France* 2, 1541 (1992); D. R. McKenzie, C. A. Davis, D. J. H. Cockayne, D. A. Muller, and A. M. Vassallo, *Nature* 155, 622 (1992); G. E. Scuseria, *Chem. Phys. Lett.* 180, 451 (1991).
161. J. M. Hawkins, A. Meyer, and M. Nambu, *J. Am. Chem. Soc.* 115, 9844 (1993).
162. A. B. Smith III, R. M. Strongin, L. Brard, G. T. Furst, W. J. Romanow, K. G. Owens, and R. J. Goldschmidt, *J. Chem. Soc., Chem. Commun.* 2187 (1994).
163. A. B. Smith III, R. M. Strongin, L. Brard, G. T. Furst, W. J. Romanow, K. G. Owens, R. J. Goldschmidt, and R. C. King, *J. Am. Chem. Soc.* 117, 5492 (1995).
164. F. Wudl, R. A. Janssen, and J. C. Hunmelen, *J. Am. Chem. Soc.* 117, 544 (1995).
165. T. Ohno, N. Martin, B. Knight, F. Wudl, T. Suzuki, and H. Yu, *J. Org. Chem.* 61, 1306 (1996).
166. H. J. Bestmann and C. Moll, *Synlett.* 729 (1996).
167. R. Pilliccani, D. Annibali, G. Costantino, M. Marinozzi, and B. Natalini, *Synlett.* 1196 (1997).
168. A. Vasella, P. Uhimann, C. A. A. Waldraff, F. Diederich, and C. Thilgen, *Angew. Chem. Int. Ed. Engl.* 31, 1388 (1992).
169. J. Osterodt, A. Zett, and F. Vögtle, *Tetrahedron* 52, 4949 (1996).
170. Z. Li, K. H. Bouhadir, and P. B. Shevlin, *Tetrahedron Lett.* 37, 4651 (1996).
171. T. Ohno, K. Moriwaki, and T. Miyata, *J. Org. Chem.* 66, 3397 (2001).
172. L. Shu, G. Wang, S. Wu, and H. Wu, *J. Chem. Soc., Chem. Commun.* 367 (1995).
173. (a) M. Eiermann, R. C. Haddon, B. Knight, Q. C. Li, M. Maggini, N. Martin, T. Ohno, M. Prato, T. Suzuki, and F. Wudl, *Angew. Chem. Int. Ed. Engl.* 34, 1591 (1995). (b) T. Ohno, N. Martin, B. Knight, F. Wudl, T. Suzuki, and H. Yu, *J. Org. Chem.* 61, 1306 (1996).
174. M. Keshavarz-K, B. Knight, R. C. Haddon, and F. Wudl, *Tetrahedron* 52, 5149 (1996).
175. K.-Y. Kay and I. C. Oh, *Tetrahedron Lett.* 40, 1709 (1999); K.-Y. Kay, L. H. Kim, and I. C. Oh, *Tetrahedron Lett.* 41, 1397 (2000); N. Martin, L. Sanchez, and D. M. Guldi, *Chem. Commun.* 113 (2000).
176. T. Habicher, J.-F. Nierengarten, V. Gramlich, and F. Diederich, *Angew. Chem., Int. Ed. Engl.* 37, 1916 (1998).
177. F. Paolucci, M. Marcaccio, S. Roffia, G. Orlandi, F. Zerbetto, M. Prato, M. Maggini, and G. Scorrano, *J. Am. Chem. Soc.* 117, 6572 (1995).
178. D. Armspach, E. C. Constable, F. Diederich, C. E. Housecroft, and J.-F. Nierengarten, *Chem. Commun.* 2009 (1996).
179. F. Diederich and R. Kessinger, *Acc. Chem. Res.* 32, 537 (1999); A. Hirsch and O. Vostrowsky, *Eur. J. Org. Chem.* 829 (2001).
180. A. Hirsch, I. Lamparth, and H. R. Karfunkel, *Angew. Chem., Int. Ed. Engl.* 33, 437 (1994).
181. (a) M. Keshavarz, K. B. Knight, R. C. Haddon, and F. Wudl, *Tetrahedron* 52, 5149 (1996). (b) M. Eiermann, R. C. Haddon, B. Knight, Q. C. Li, M. Maggini, N. Martin, T. Ohno, M. Prato, T. Suzuki, and F. Wudl, *Angew. Chem., Int. Ed. Engl.* 34, 1591 (1995).
182. R. Kessinger, J. Crassous, A. Herrmann, M. Rüttimann, L. Echegoyen, and F. Diederich, *Angew. Chem., Int. Ed.* 37, 1919 (1998).
183. R. Kessinger, M. Gómez-López, C. Boudon, J.-P. M. Gross, L. Echegoyen, and F. Diederich, *J. Am. Chem. Soc.* 120, 8545 (1998).
184. (a) A. Hirsch, I. Lamparth, T. Grösser, and H. R. Karfunkel, *J. Am. Chem. Soc.* 116, 9358 (1994). (b) T. Grösser, M. Prato, V. Lucchini, A. Hirsch, and F. Wudl, *Angew. Chem., Int. Ed. Engl.* 34, 1343 (1995).
185. (a) L. Issacs, R. F. Haldimann, and F. Diederich, *Angew. Chem., Int. Ed. Engl.* 33, 2434 (1994). (b) B. Kräutler and J. Maynallo, *Angew. Chem., Int. Ed. Engl.* 34, 87 (1995). (c) P. Fagan, J. C. Calabrese, and B. Malone, *J. Am. Chem. Soc.* 113, 9408 (1991).
186. I. Lamparth, C. Maichle-Mössner, and A. Hirsch, *Angew. Chem., Int. Ed. Engl.* 34, 1607 (1995).
187. X. Camps and A. Hirsch, *J. Chem. Soc., Perkin Trans. 1* 1595 (1997).
188. P. Timmerman, L. E. Witschel, F. Diederich, C. Boudon, J.-P. Gisselbrecht, and M. Gross, *Helv. Chim. Acta* 79, 6 (1996).
189. T. Habicher, J.-F. Nierengarten, V. Gramlich, and F. Diederich, *Angew. Chem., Int. Ed. Engl.* 37, 1916 (1998).
190. I. Lamparth, C. A. Herzog, and A. Hirsch, *Tetrahedron* 52, 5065 (1996).
191. B. Kräutler, T. Müller, J. Maynollo, K. Gruber, C. Kratky, P. Ochsenbein, D. Schwarzenbach, and H.-B. Bürgi, *Angew. Chem., Int. Ed. Engl.* 35, 1204 (1996).
192. R. Schwenniger, T. Müller, and B. Kräutler, *J. Am. Chem. Soc.* 119, 9317 (1997); F. Cardullo, L. Isaacs, F. Diederich, J.-P. Gisselbrecht, C. Boudon, and M. Gross, *Chem. Commun.* 797 (1996).
193. S. H. Friedman and G. L. Kenyon, *J. Am. Chem. Soc.* 119, 447 (1997).
194. L. Isaacs, R. F. Haldimann, and F. Diederich, *Angew. Chem., Int. Ed. Engl.* 33, 2339 (1994).
195. Y.-Z. An, G. A. Ellis, A. L. Viado, and Y. Rubin, *J. Org. Chem.* 60, 6353 (1995).
196. F. Cardullo, L. Isaacs, F. Diederich, J.-P. Gisselbrecht, C. Boudon, and M. Gross, *Chem. Commun.* 797 (1996); F. Cardullo, P. Seiler, L. Isaacs, J.-F. Nierengarten, R. F. Haldimann, F. Diederich, T. Mordasini-Denti, W. Thiel, C. Boudon, J.-P. Gisselbrecht, and M. Gross, *Helv. Chim. Acta* 80, 343 (1997).
197. T. Hamano, K. Okuda, T. Mashino, M. Hirobe, K. Arakane, A. Ryu, S. Mashiko, and T. Nagano, *Chem. Commun.* 21 (1997).
198. D. M. Guldi and K.-D. Asmus, *J. Phys. Chem. A* 101, 1472 (1997).
199. J.-F. Nierengarten, V. Gramlich, F. Cardullo, and F. Diederich, *Angew. Chem., Int. Ed. Engl.* 35, 2101 (1996); J.-F. Nierengarten, A. Herrmann, R. R. Tykwinski, M. Rüttimann, F. Diederich, C. Boudon, J.-P. Gisselbrecht, and M. Gross, *Helv. Chim. Acta* 80, 293 (1997); J.-F. Nierengarten, T. Habicher, R. Kessinger, F. Cardullo, F. Diederich, V. Gramlich, J.-P. Gisselbrecht, C. Boudon, and M. Gross, *Helv. Chim. Acta* 80, 2238 (1997).
200. R. W. Alder and S. P. East, *Chem. Rev.* 96, 2097 (1996).
201. J.-F. Nierengarten, D. Felder, and J.-F. Nicoud, *Tetrahedron Lett.* 39, 2747 (1998).
202. W. Qian and Y. Rubin, *Angew. Chem., Int. Ed.* 38, 2356 (1999).
203. W. Qian and Y. Rubin, *Angew. Chem., Int. Ed.* 39, 3133 (2000).
204. W. Qian and Y. Rubin, *J. Am. Chem. Soc.* 112, 9564 (2000).
205. C. R. Woods, J.-P. Bourgeois, P. Seiler, and F. Diederich, *Angew. Chem., Int. Ed.* 39, 3813 (2000).
206. F. Diederich and R. Kissinger, in "Templated Organic Synthesis" (F. Diederich and P. J. Stang, Eds.), pp. 189–218. Wiley–VCH, Weinheim, 1999.
207. G. Rapenne, J. Crassous, A. Collet, L. Echegoyen, and F. Diederich, *Chem. Commun.* 1121 (1999).
208. U. Reuther, T. Brandmüller, W. Donaubauer, F. Hampel, and A. Hirsch, *Chem. Eur. J.* 8, 2261 (2002).
209. A. Hirsch and B. Number, *Acc. Chem. Res.* 32, 795 (1999); U. Reuther and A. Hirsch, *Carbon* 38, 1539 (2000).
210. M. Prato, Q. C. Li, F. Wudl, and V. Lucchini, *J. Am. Chem. Soc.* 115, 1148 (1993).
211. I. P. Romanova, G. G. Yusupova, S. G. Fattakhov, A. A. Nafikova, V. I. Kovalenko, V. V. Yanilkin, V. E. Kataev, N. M. Azancheev, V. S. Reznik, and O. G. Sinyashin, *Russ. Chem. Bull., Int. Ed.* 50, 445 (2001).
212. J. Averdung and J. Mattey, *Tetrahedron* 52, 5407 (1996).
213. M. Cases, M. Duran, J. Mesters, N. Martin, and M. Sola, *J. Org. Chem.* 66, 433 (2001).
214. B. Nuber, F. Hampel, and A. Hirsch, *Chem. Commun.* 1799 (1996).

215. M. R. Banks, J. I. G. Cadogen, I. Gosney, P. K. G. Hodgson, P. R. R. Langridge-Smith, and D. W. H. Rankin, *J. Chem. Soc. Chem. Commn.* 1365 (1994).
216. J. Averdung , J. Mattey, D. Jacobi, and W. Abraham, *Tetrahedron* 51, 2543 (1995).
217. M. Yan, S. X. Cai, and F. W. Keana, *J. Org. Chem.* 59, 5951 (1994).
218. N. Jagerovic, J. Elguero, and J. Aubagnac, *Tetrahedron* 52, 6733 (1996).
219. M. R. Banks, J. I. G. Cadogen, I. Gosney, P. K. G. Hodgson, P. R. R. Langridge-Smith, and D. W. H. Rankin, *Synth. Met.* 77, 77 (1996).
220. M. R. Banks, J. I. G. Cadogen, I. Gosney, P. K. G. Hodgson, P. R. R. Langridge-Smith, J. R. A. Millar, J. A. Parkinson, D. W. H. Rankins, and A. T. Taylor, *J. Chem. Soc. Chem. Commun.* 887 (1995).
221. A. B. Smith III and H. Tokuyama, *Tetrahedron* 52, 5257 (1996).
222. M. R. Banks, J. I. G. Cadogan, I. Gosney, A. J. Henderson, K. G. Hodgson, W. G. Kerr, A. Kerth, P. R. R. Langridge-Smith, J. R. A. Millar, A. R. Mount, J. A. Parkinson, A. T. Taylor, and P. Thornburn, *Chem. Commun.* 507 (1996).
223. B. Nuber and A. Hirsch, *Fullerene Sci. Technol.* 4, 715 (1996); C. Bellavia-Lund and F. Wudl, *J. Am. Chem. Soc.* 119, 943 (1997).
224. Y. Yamakoshi, T. Yagami, S. Sueyoshi, and N. Miyata, *J. Org. Chem.* 61, 7236 (1996).
225. H. Murakami, Y. Watanabe, and N. Nakashima, *J. Am. Chem. Soc.* 118, 4484 (1996).
226. J. C. Hummelen, M. Prato, and F. Wudl, *J. Am. Chem. Soc.* 117, 7003 (1995).
227. T. Grösser, M. Prato, V. Lucchini, A. Hirsch, and F. Wudl, *Angew. Chem., Int. Ed. Engl.* 34, 1343 (1995).
228. G. Schick, A. Hirsch, H. Mauser, and T. Clark, *Chem. Eur. J.* 2, 935 (1996).
229. B. Nuber and A. Hirsch, *Fullerene Sci. Technol.* 4, 715 (1996).
230. G. Dong, J. Li, and T. Chan, *J. Chem. Soc., Chem. Commun.* 1725 (1995).
231. L. Shiu, K. Chien, T. Liu, T. Lin, G. Her, and T. Luh, *J. Chem. Soc., Chem. Commun.* 1159 (1995).
232. C. K.-F. Sen, K. Chien, C. Juo, G. Her, and T. Luh, *J. Org. Chem.* 61, 9242 (1996).
233. G. Schick, T. Jarrosson, and Y. Rubin, *Angew. Chem. Int. Ed.* 38, 2360 (1999); J.-F. Nierengarten, *Angew. Chem. Int. Ed.* 40, 2973 (2001).
234. Y. Rubin, T. Jarrosson, G.-W. Wang, M. D. Bartberger, K. N. Houk, G. Schick, M. Saunders, and R. J. Cross, *Angew. Chem. Int. Ed.* 40, 1543 (2001).
235. Y. Murata, M. Murata, and K. Komatsu, *Chem. Eur. J.* 9, 1600 (2003).
236. Y. Murata, M. Murata, and K. Komatsu, *J. Am. Chem. Soc.* 125, 7152 (2003).
237. S. H. Hoke II, J. Molstad, D. Dilettato, M. J. Jay, D. Carlson, B. Kahr, and R. G. Cooks, *J. Org. Chem.* 57, 5069 (1992); A. D. Darwish, A. K. Abdul-Sada, G. J. Langley, H. B. Kroto, R. Taylor, and D. R. M. Walton, *J. Chem. Soc., Chem. Commun.* 2133 (1994).
238. G. Vassilikogiannakis, N. Chronakis, and M. Orfanopoulos, *J. Am. Chem. Soc.* 120, 9911 (1998).
239. G. Vassilikogiannakis and M. Orfanopoulos, *J. Org. Chem.* 64, 3392 (1999); G. Vassilikogiannakis, N. Chronakis, and M. Orfanopoulos, *J. Am. Chem. Soc.* 120, 9911 (1998).
240. R. Bernstein and C. S. Foote, *Tetrahedron Lett.* 39, 7051 (1998).
241. S. Yamago, A. Takeichi, and E. Nakamura, *J. Am. Chem. Soc.* 116, 1123 (1994).
242. X. Zhang, A. Fan, and C. S. Foote, *J. Org. Chem.* 61, 5456 (1996).
243. X. Zhang, A. Romero, and C. S. Foote, *J. Am. Chem. Soc.* 115, 11024 (1993); X. Zhang and C. S. Foote, *J. Am. Chem. Soc.* 117, 4271 (1995).
244. H. Tokuyama, H. Isobe, and E. Nakamura, *J. Chem. Soc., Chem. Commun.* 2753 (1994); S. Yamago, A. Takeichi, and E. Nakamura, *J. Am. Chem. Soc.* 116, 1123 (1994).
245. S. R. Wilson, Y. Wu, N. Kaprinidis, and D. I. Schuster, *J. Am. Chem. Soc.* 115, 8495 (1993); S. R. Wilson, Y. Wu, A. Nikolas, N. Kaprinidis, and D. I. Schuster, *J. Org. Chem.* 58, 6548 (1993).
246. D. I. Schuster, J. Cao, N. Kaprinidis, Y. Wu, A. W. Jensen, Q. Lu, H. Wang, and S. R. Wilson, *J. Am. Chem. Soc.* 118, 5639 (1996); A. W. Jensen, A. Khong, M. Saunders, S. R. Wilson, and D. I. Schuster, *J. Am. Chem. Soc.* 119, 7303 (1997).
247. M. Prato, M. Maggini, G. Scorrano, and V. Lucchini, *J. Org. Chem.* 58, 3613 (1993).
248. R. N. Warrener, G. M. Elsey, and M. A. Houghton, *J. Chem. Soc., Chem. Commun.* 1417 (1995).
249. L. Shiu, T. Lin, S. Peng, G. Her, D. Ju, S. Lin, J. Hwang, C. Mou, and T. Luh, *J. Chem. Soc., Chem. Commun.* 647 (1994).
250. M. Prato, T. Suzuki, H. Forrudian, Q. Li, K. Khemani, F. Wudl, J. Leonetti, R. D. Little, T. White, B. Rickborn, S. Yamago, and E. Nakamura, *J. Am. Chem. Soc.* 115, 1594 (1993).
251. S. Yamago, H. Tokuyama, E. Nakamura, M. Prato, and F. Wudl, *J. Org. Chem.* 58, 4796 (1993).
252. L. Shu, W. Sun, D. Zhang, S. Wu, H. Wu, J. Xu, and X. Lao, *Chem. Commun.* 79 (1997); B. F. O'Donovan, P. B. Hitchcock, M. F. Meidine, H. W. Kroto, R. Taylor, and D. R. M. Walton, *Chem. Commun.* 81 (1997).
253. M. Prato and M. Maggini, *Acc. Chem. Res.* 31, 519 (1998).
254. M. Maggini, G. Scorrano, and M. Prato, *J. Am. Chem. Soc.* 115, 9798 (1993).
255. Q. Lu, D. I. Schuster, and S. R. Wilson, *J. Org. Chem.* 61, 4764 (1996).
256. M. Maggini, G. Scorrano, A. Bianco, C. Toniolo, R. P. Sijbesma, F. Wudl, and M. Prato, *J. Chem. Soc., Chem. Commun.* 305 (1994).
257. X. Zhang, M. Willems, and C. S. Foote, *Tetrahedron Lett.* 51, 8187 (1993).
258. L. Gan, D. Zhou, C. Luo, H. Tan, C. Huang, M. Lu, J. Pan, and Y. Wu, *J. Org. Chem.* 61, 1954 (1996).
259. M. Maggini, A. Dono, G. Scorrano, and M. Prato, *J. Chem. Soc., Chem. Commun.* 845 (1995).
260. K. Liou and C. Cheng, *Chem. Commun.* 1423 (1996).
261. D. M. Guldi, M. Maggini, G. Scorrano, and M. Prato, *J. Am. Chem. Soc.* 119, 974 (1997).
262. M. Prato, M. Maggini, C. Giacometti, G. Scorrano, G. Sandona, and G. Farnia, *Tetrahedron* 52, 5221 (1996).
263. R. M. Williams, J. M. Zwier, and J. W. Verhoeven, *J. Am. Chem. Soc.* 117, 4093 (1995).
264. R. M. Williams, M. Koeberg, J. M. Lawson, Y. An, Y. Rubin, M. N. Paddon-Row, and J. W. Verhoeven, *J. Org. Chem.* 61, 5055 (1996).
265. M. Maggini, A. Dono, G. Scorrano, and M. Prato, *J. Chem. Soc., Chem. Commun.* 845 (1995).
266. B. Illescas, N. Martin, and C. Seoane, *Tetrahedron Lett.* 38, 2015 (1997); M. A. Herranz, B. Illescas, and N. Martin, *J. Org. Chem.* 65, 5728 (2000); B. M. Illescas and N. Martin, *J. Org. Chem.* 65, 5986 (2000).
267. T. Da Ros, M. Prato, F. Novello, M. Maggini, and E. Manfi, *J. Org. Chem.* 61, 9070 (1996).
268. F. Novello, M. Prato, T. Da Ros, M. De Amici, A. Bianco, and C. Toniolo, *Chem. Commun.* 903 (1996).
269. X. Tan, D. I. Schuster, and R. Wilson, *Tetrahedron Lett.* 39, 4187 (1998).
270. A. Bianco, F. Gasparrini, M. Maggini, D. Misiti, A. Polese, M. Prato, G. Scorrano, C. Toniolo, and C. Villani, *J. Am. Chem. Soc.* 119, 7550 (1997).
271. J. Averdung, E. Albrecht, J. Lauterwein, J. Mattay, H. Mohn, W. H. Muller, and H. U. ter Meer, *Chem. Ber.* 127, 787 (1994).
272. J. Averdung and J. Mattay, *Tetrahedron* 52, 5407 (1996).

273. Y. Matsubara, H. Tada, S. Nagase, and Z. Yoshida, *J. Org. Chem.* 60, 5372 (1995).

274. M. S. Meier and M. Poplawaka, *J. Org. Chem.* 60, 5372 (1995); M. S. Meier and M. Poplawaka, *Tetrahedron* 52, 5043 (1996).

275. W. Duczek and H.-J. Niclas, *Tetrahedron Lett.* 36, 2457 (1995); W. Duczek, F. Tittelbach, B. Costisella, and H. Niclas, *Tetrahedron* 52, 8733 (1996).

276. F. Wudl, *Acc. Chem. Res.* 25, 157 (1992).

277. S. R. Wilson and Q. Lu, *Tetrahedron Lett.* 34, 8043 (1993); C. M. A. Alonso, M. G. P. M. S. Neves, A. C. Tome, A. M. S. Silva, and J. A. S. Cavaleiro, *Tetrahedron Lett.* 41, 5679 (2000); M. Ohno, S. Kojima, S. Shirakawa, and S. Eguchi, *Heterocycles* 46, 49 (1997).

278. V. M. Rotello, J. B. Howard, T. Yadav, M. M. Conn, E.Viani, L. M. Giovane, and A. L. Lafluer, *Tetrahedron Lett.* 34, 1561 (1993).

279. M. F. Meidine, A. G. Avent, A. D. Darwish, H. W. Kroto, O. Ohashi, R. Taylor, and D. R. M. Walton, *J. Chem. Soc., Perkin Trans. 2* 1189 (1994).

280. M. F. Meidine, R. Roers, G. J. Langley, A. G. Avent, A. D. Darwish, S. Firth, H. W. Kroto, R. Taylor, and D. R. M. Walton, *J. Chem. Soc., Chem. Commun.* 1342 (1993).

281. B. Kräutler and J. Maynollo, *Tetrahedron* 52, 5033 (1996).

282. Y. Rubin, S. I. Khan, D. I. Freedberg, and C. Yeretzian, *J. Am. Chem. Soc.* 115, 344 (1993).

283. F. Wudl, R. Sijbesma, G. Srdanov, J. A. Castoro, C. Wilkins, S. H. Friedman, D. L. Decamp, and G. L. Kenyon, *J. Am. Chem. Soc.* 115, 6510 (1993).

284. K. Komatsu, Y. Murata, N. Sugita, K. Takeuchi, and T. S. M. Wan, *Tetrahedron Lett.* 34, 8476 (1993); M. Tsuda, T. Ishida, T. Nogami, S. Kurono, and M. Ohashi, *J. Chem. Soc., Chem. Commun.* 1296 (1993); J. A. Schlueter, J. M. Seaman, S. Taha, H. Cohen, K. R. Lykke, H. H. Wang, and J. M. Williams, *J. Chem. Soc., Chem. Commun.* 972 (1993); R. Klemt, E. Roduner, and H. Fischer, *J. Am. Chem. Soc.* 117, 8019 (1995).

285. K. Mikami, S. Matsumoto, T. Tonoi, Y. Okubo, T. Suenobu, and S. Fukuzumi, *Tetrahedron Lett.* 39, 3773 (1998).

286. N. Chronakis and M. Orfanopoulos, *Tetrahedron Lett.* 42, 1201 (2001).

287. Y. Murata, N. Kato, K. Fujiwara, and K. Komatsu, *J. Org. Chem.* 64, 3483 (1999).

288. J. Mack and G. P. Miller, *Fullerene Sci. Technol.* 5, 607 (1997).

289. G. P. Miller and J. Mack, *Org. Lett.* 2, 3979 (2000).

290. H. Ishida, K. Komori, K. Itoh, and M. Ohno, *Tetrahedron Lett.* 41, 9839 (2000).

291. Y. An, G. A. Ellis, A. L. Viado, and Y. Rubin, *J. Org. Chem.* 60, 6353 (1995).

292. Y. An, J. A. L. Viado, M. Arce, and Y. Rubin, *J. Org. Chem.* 60, 8330 (1995).

293. Y. An, J. L. Anderson, and Y. Rubin, *J. Org. Chem.* 58, 4799 (1993).

294. Y. An, C. B. Chen, J. L. Anderson, D. S. Sigman, C. S. Foote, and Y. Rubin, *Tetrahedron* 52, 5179 (1996).

295. S. R. Wilson, Q. Lu, J. Cao, Y. Wu, C. J. Welch, and D. I. Schuster, *Tetrahedron* 52, 5131 (1996).

296. M. Arce, A. L. Viado, Y. An, S. I. Khan, and Y. Rubin, *J. Am. Chem. Soc.* 118, 3775 (1996).

297. P. S. Ganapati, S. H. Friedman, G. L. Kenyon, and Y. Rubin, *J. Org. Chem.* 60, 2954 (1995).

298. S. R. Wilson and Q. Lu, *Tetrahedron Lett.* 34, 8043 (1993).

299. T. Ishi-i, K. Nakashima, and S. Shinkai, *Chem. Commun.* 1047 (1998).

300. B. Kräutler and J. Maynollo, *Angew. Chem., Int. Ed. Engl.* 34, 87 (1995).

301. M. Prato, T. Suzuki, H. Foroudian, Q. Li, K. Khemani, and F. Wudl, *J. Am. Chem. Soc.* 115, 1594 (1993).

302. S. Cossu, G. Cuomo, O. De Lucchi, M. Maggini, and G. Valle, *J. Org. Chem.* 61, 153 (1996).

303. P. Belik, A. Gügel, J. Spickermann, and K. Müllen, *Angew. Chem., Int. Ed. Engl.* 32, 78 (1993).

304. M. Walter, A. Gügel, J. Spickermann, P. Belik, A. Kraus, and K. Müllen, *Fullerene Sci. Technol.* 4, 101 (1996).

305. P. Belik, A Gügel, A. Kraus, M. Walter, and K. Müllen, *J. Org. Chem.* 60, 3307 (1995).

306. F. Diederich, U. Jonas, V. Gramlich, A. Herrmann, H. Ringsdorf, and C. Thilgen, *Helv. Chim. Acta* 76, 2445 (1993).

307. T. Tago, T. Minowa, Y. Okada, and J. Nishimura, *Tetrahedron Lett.* 34, 8461 (1993).

308. A. Kraus, A. Gügel, P. Belik, M. Walter, and K. Müllen, *Tetrahedron* 51, 9927 (1995).

309. X. Zhang and C. S. Foote, *J. Org. Chem.* 59, 5235 (1994).

310. H. Tomioka and K. Yamamoto, *J. Chem. Soc., Chem. Commun.* 1961 (1995).

311. M. Iyoda, F. Sultana, S. Sasaki, and M. Yoshida, *J. Chem. Soc., Chem. Commun.* 1929 (1994).

312. B. M. Illescas, N. Martin, and C. Scoane, *Tetrahedron Lett.* 38, 2015 (1997).

313. B. M. Illescas, N. Martin, C. Scoane, P. de la Cruz, F. Langa, and F. Wudl, *Tetrahedron Lett.* 36, 8307 (1995); B. M. Illescas, N. Martín, C. Scoane, E. Ortí, P. M. Viruela, R. Viruela, and A. de la Hoz, *J. Org. Chem.* 62, 7585 (1997).

314. C. Boulle, M. Cariou, M. Bainville, A. Gorgues, P. Hudhomme, J. Orduna, and J. Garín, *Tetrahedron Lett.* 38, 81 (1997).

315. J. Llacay, M. Mas, E. Molins, J. Veciana, D. Powell, and C. Rovira, *Chem. Commun.* 659 (1997).

316. (a) M. A. Herranz and N. Martin, *Org. Lett.* 1, 2005 (1999). (b) C. Boulle, J. M. Rabreau, P. Hudhomme, M. Cariou, J. Orduna, and J. Garin, *Tetrahedron Lett.* 38, 3909 (1997). (c) J. Llacay, M. Mas, E. Molins, J. Veciana, D. Powell, and C. Rovira, *Chem. Commun.* 659 (1997).

317. F. Effenberger and G. Grube, *Synthesis* 1372 (1998).

318. M. Ohno, H. Sato, and S. Eguchi, *Synlett.* 207 (1999).

319. M. Ohno, S. Kojima, Y. Shirakawa, and S. Eguchi, *Heterocycles* 46, 49 (1997).

320. A. K. Abdul-Sada, A. G. Avent, P. R. Birkett, A. D. Darwish, H. W. Kroto, R. Taylor, D. R. M. Walton, and O. B. Woodhouse, *Chem. Commun.* 307 (1998).

321. M. Ohno, T. Azuma, and S. Eguchi, *Chem. Lett.* 1833 (1993).

322. M. Ohno, S. Kojima, Y. Shirakawa, and S. Eguchi, *Tetrahedron Lett.* 36, 6899 (1995).

323. M. Ohno, S. Kojima, and S. Eguchi, *J. Chem. Soc., Chem. Commun.* 565 (1995).

324. G. Torres-García H. Luftmann, C. Wolff, and J. Mattay, *J. Org. Chem.* 62, 2752 (1997).

325. U. M. Fernandez-Paniagua, B. M. Illescas, N. Martín, C. Seoane, P. de la Cruz, A. de la Hoz, and H. Langa, *J. Org. Chem.* 62, 3705 (1997).

326. M. Ohno, S. Kojima, Y. Shirakawa, and S. Eguchi, *Tetrahedron Lett.* 36, 6899 (1995); M. Ohno, N. Koide, H. Sato, and S. Eguchi, *Tetrahedron* 93, 9075 (1997).

327. G. Torres-García and J. Mattay, *Tetrahedron* 52, 5421 (1996).

328. U. M. Fernandez-Paniagua, B. M. Illescas, N. Martín, and C. Seoane, *J. Chem. Soc., Perkin Trans. 1* 1077 (1996).

329. Y. Z. An, G. A. Ellis, A. L. Viado, and Y. Rubin, *J. Org. Chem.* 60, 6353 (1995); Y. Z. An, A. L. Viado, M. Arce, and Y. Rubin, *J. Org. Chem.* 60, 8330 (1995).

330. K. I. Guhr, M. D. Greaves, and V. M. Rotello, *J. Am. Chem. Soc.* 116, 5997 (1994).

331. B. Nie and V. M. Rotello, *J. Org. Chem.* 61, 1870 (1996).

332. K. Liou and C. Cheng, *J. Chem. Soc., Chem. Commun.* 1603 (1995).

333. S. Pyo, L. Shu, and L. Echegoyen, *Tetrahedron Lett.* 39, 7653 (1998).

334. M. R. Banks, J. I. G. Cadogan, I. Gosney, P. K. G. Hodgson, P. R. R. Langridge-Smith, J. R. A. Millar, J. A. Parkinson, I. H. Sadler, and A. T. Taylor, *J. Chem. Soc., Chem. Commun.* 1171 (1995).
335. J. L. Segura and N. Martin, *Chem. Soc. Rev.* 29, 13 (2000).
336. M. Fujitsuka, C. Luo, O. Ito, Y. Murata, and K. Komatsu, *J. Phys. Chem. A* 103, 7155 (1999).
337. D. Sun and C. A. Reed, *Chem. Commun.* 2391 (2000).
338. (a) G.-W. Wang, K. Komatsu, Y. Murata, and M. Shiro, *Nature* 387, 583 (1997). (b) K. Komatsu, G.-W. Wang, Y. Murata, T. Tanaka, and K. Fujiwara, *J. Org. Chem.* 63, 9358 (1998). (c) K. Komatsu, K. Fujiwara, T. Tanaka, and Y. Murata, *Carbon* 38, 1529 (2000). (d) Y. Iisa, K. Tanoue, T. Mitani, A. Izuoka, T. Sugawara, and T. Yagi, *Chem. Commun.* 1411 (1998).
339. S. Lebedkin, W. Hull, A. Soldatov, B. Renker, and M. Kappes, *J. Phys. Chem. B* 104, 4101 (2000).
340. E. A. Halevi, *Helv. Chim. Acta* 84, 1661 (2001).
341. S. Patchkovskii and W. Thiel, *J. Am. Chem. Soc.* 120, 556 (1998).
342. J. Knol and J. C. Hummelen, *J. Am. Chem. Soc.* 122, 3226 (2000).
343. Y. Murata, M. Suzuki, and K. Komatsu, *Chem. Commun.* 2338 (2001).
344. K. Fujiwara and K. Komatsu, *Chem. Commun.* 1986 (2001).
345. G. S. Forman, N. Tagmatarchis, and H. Shinohara, *J. Am. Chem. Soc.* 124, 178 (2002).
346. K. Komatsu, K. Fujiwara, and Y. Murata, *Chem. Commun.* 1583 (2000).
347. A. B. Smith III, H. Tokuyama, R. M. Strongin, G. T. Furst, and W. J. Romanow, *J. Am. Chem. Soc.* 117, 9359 (1995); S. Lebedkin, S. Ballenweg, J. Gross, R. Taylor, and W. Krätschmer, *Tetrahedron Lett.* 36, 4971 (1995).
348. A. Gromov, S. Lebedkin, S. Ballenweg, A. G. Avent, R. Taylor, and W. Krätschmer, *Chem. Commun.* 209 (1997).
349. P. W. Fowler, D. Mitchell, R. Taylor, and G. Seifert, *J. Chem. Soc. Perkin Trans. 2* 1901 (1997).
350. J. R. Morton, K. F. Preston, P. J. Krusic, S. A. Hill, and E. Isserman, *J. Am. Chem. Soc.* 114, 5454 (1992); J. R. Morton, K. F. Preston, P. J. Krusic, S. A. Hill, and E. Isserman, *J. Phys. Chem.* 96, 3576 (1992); P. N. Keizer, Z. R. Morton, K. F. Preston, and P. J. Krusic, *J. Chem. Soc., Perkin Trans. 2* 1041 (1993). M. Yashida, A. Morishima, Y. Morinaga, and M. Iyoda, *Tetrahedron Lett.* 35, 9045 (1994).
351. S. Osawa, E. Osawa, and M. Harada, *J. Org. Chem.* 61, 257 (1996).
352. H. L. Anderson, R. Faust, Y. Rubin, and F. Diederich, *Angew. Chem., Int. Ed. Engl.* 33, 1366 (1994); H. L. Timmerman, R. F. Anderson, J.-F. Nierengarten, T. Habicher, P. Seiler, and F. Diederich, *Tetrahedron* 52, 4925 (1996).
353. K. Komatsu, N. Takimoto, Y. Murata, T. S. M. Wan, and T. Wong, *Tetrahedron Lett.* 37, 6153 (1996).
354. P. Timmerman, L. E. Witschel, F. Diederich, C. Boudon, J.-P. Gisselbrecht, and M. Gross, *Helv. Chim. Acta* 79, 6 (1996).
355. M. Saunders, H. A. Jiménez-Vázquez, R. J. Cross, and R. J. Poreda, *Science* 259, 1428 (1993).
356. T. Suzuki, Q. Li, K. C. Khemani, and F. Wudl, *J. Am. Chem. Soc.* 114, 7300 (1992).
357. Y. Murata, N. Kato, K. Fujiwara, and K. Komatsu, *J. Org. Chem.* 64, 3483 (1999).
358. L. A. Paquette and R. J. Graham, *J. Org. Chem.* 60, 2958 (1995); L. A. Paquette and W. E. Trego, *Chem. Commun.* 419 (1996).
359. S. I. Khan, A. M. Oliver, M. N. Paddon-Row, and Y. Rubin, *J. Am. Chem. Soc.* 115, 4919 (1993); J. M. Lawson, A. M. Oliver, D. F. Rothenfluh, Y. An, G. A, Ellis, M. G. Ranasinghe, S. I. Khan, A. G. Franz, P. S. Ganapathi, M. J. Shephard, M. N. Paddon-Row, and Y. Rubin, *J. Org. Chem.* 61, 5032 (1996).
360. A. I. de Lucas, N. Martín, L. Sánchez, and C. Seoane, *Tetrahedron Lett.* 52, 9391 (1996).
361. D. Armspach, E. C. Constable, F. Diederich, C. E. Housecroft, and J.-F. Nierengarten, *Chem. Eur. J.* 4, 723 (1998).
362. S. Ravaine, P. Delhaès, P. Leriche, and M. Sallé, *Synth. Met.* 87, 93 (1997).
363. F. Diederich, C. Dietrich-Buchecker, J. Nierengarten, and J. Sauvage, *J. Chem. Soc., Chem. Commun.* 781 (1995).
364. A. Hirsch and B. Nuber, *Acc. Chem. Res.* 32, 795 (1999); U. Reuther and A. Hirsch, *Carbon* 38, 1539 (2000).
365. J. C. Hummelen, B. Knight, J. Pavlovich, R. González, and F. Wudl, *Science* 269, 1554 (1995).
366. I. Lamparth, B. Nuber, G. Schick, A. Skiebe, T. Grösser, and A. Hirsch, *Angew. Chem., Int. Ed. Engl.* 35, 2257 (1995).
367. B. Nuber and A. Hirsch, *Chem. Commun.* 1421 (1996).
368. J. C. Hummelen, C. Bellavia-Lund, and F. Wudl, *Top. Curr. Chem.* 199, 93 (1999).
369. C. Bellavia-Lund, R. González, J. C. Hummelen, R. G. Hicks, A. Sastre, and F. Wudl, *J. Am. Chem. Soc.* 119, 2946 (1997).
370. B. Nuber and A. Hirsch, *Chem. Commun.* 405 (1998).
371. U. Reuther and A. Hirsch, *Chem. Commun.* 1401 (1998).
372. F. Diederich and Y. Rubin, *Angew. Chem., Int. Ed. Engl.* 31, 1101 (1992).
373. F. Diederich, *Nature* 369, 199 (1994); H. L. Anderson, R. Faust, Y. Rubin, F. Diederich, J.-P. Giesselbrecht, M. Gross, and P. Seiler, *Angew. Chem., Int. Engl.* 33, 1366 (1994); H. L. Anderson, C. Boudon, R. Faust, F. Diederich, J.-P. Giesselbrecht, M. Gross, and P. Seiler, *Angew. Chem., Int. Engl.* 33, 1628 (1994).
374. L. A. Paquette and W. E. Trego, *Chem. Commun.* 419 (1996).
375. V. Biju, P. K. Sundeep, K. G. Thomas, and M. V. George, *Langmuir* 18, 1831 (2002).
376. A. Rieder and B. Kräutler, *J. Am. Chem. Soc.* 122, 9050 (2000).
377. G. R. Newkome, C. N. Moorefield, and F. Vögtle, "Dendrimers and Dendrons: Concepts, Syntheses, Applications." VCH, Weinheim, 2001; D. A. Tomalia, A. M. Naylor, and W. A. Goddart III, *Angew. Chem., Int. Engl.* 29, 138 (1990).
378. J.-F. Nierengarten, *Chem. Eur. J.* 6, 3667 (2000); J.-F. Nierengarten, N. Armaroli, G. Accorsi, Y. Rio, and J.-F. Eckert, *Chem. Eur. J.* 9, 37 (2003).
379. C. J. Hawker, K. L. Wooley, and J. M. Frétcht, *J. Chem. Soc., Chem. Commun.* 925 (1994).
380. K. L. Wooley, C. J. Hawker, J. M. J. Frétcht, F. Wudl, G. Srdanov, S. Shi, C. Li, and M. Kao, *J. Am. Chem. Soc.* 115, 9836 (1993).
381. F. Cardullo, F. Diederich, L. Echegoyen, T. Habicher, N. Jayaraman, R. M. Leblanc, J. F. Stoddart, and S. Wang, *Langmuir* 14, 1955 (1998).
382. S. Zhou, C. Burger, B. Chu, M. Sawamura, N. Nagahama, M. Toganoh, U. E. Hackler, H. Isobe, and E. Nakamura, *Science* 291, 1944 (2001); S. Samal and K. E. Geckeler, *Chem. Commun.* 2224 (2001); T. Song, S. Dai, K. C. Tam, S. Y. Lee, and S. H. Goh, *Langmuir* 19, 4798 (2003).
383. Y. Rio, G. Accorsi, H. Nierengarten, J.-L. Rehspringer, B. Hönerlage, G. Kopitkovas, A. Chugreev, A. V. Dorsselaer, N. Armaroli, and J.-F. Nierengarten, *New J. Chem.* 9, 1146 (2002); Y. Rio, J.-F. Nicoud, J.-L. Rehspringer, and J.-F. Nierengarten, *Tetrahedron Lett.* 41, 10207 (2000).
384. D. Felder, D. Guillon, R. Levy, A. Mathis, J.-F. Nicoud, J.-F. Nierengarten, J.-L. Rehspringer, and J. Schell, *J. Mater. Chem.* 10, 887 (2000).
385. J. Schell, D. Felder, J.-F. Nierengarten, J.-L. Rehspringer, R. Lévy, and B. Hönerlage, *J. Sol–Gel Sci. Technol.* 22, 225 (2001).
386. D. Felder, J.-L. Gallani, D. Guillon, B. Heinrich, J.-F. Nicoud, and J.-F. Nierengarten, *Angew. Chem., Int. Ed.* 39, 201 (2000).
387. Y. Murata, M. Ito, and K. Komatsu, *J. Mat. Chem.* 12, 2009 (2002).
388. R. Kunieda, M. Fujitsuka, O. Ito, M. Ito, Y. Murata, and K. Komatsu, *J. Phys. Chem. B* 106, 7193 (2002).
389. N. Armaroli, F. Barigelletti, P. Ceroni, J.-F. Eckert, J.-F. Nicoud, and J.-F. Nierengarten, *Chem. Commun.* 599 (2000).
390. J. L. Segura, R. Gómez, N. Martín, C. Luo, A. Swartz, and D. M. Guldi, *Chem. Commun.* 707 (2001).

391. F. Langa, M. J. Gómez-Escalonillaa, E. Díez-Barrab, J. C. García-Martínezb, A. de la Hozb, J. Rodríguez-Lópezb, A. González-Cortésc, and V. López-Arza, *Tetrahedron Lett.* 42, 3453 (2001).
392. J.-F. Nierengarten, D. Felder, and J.-F. Nicoud, *Tetrahedron Lett.* 40, 273 (1999).
393. A. Hirsch and O. Vostrowsky, *Eur. J. Org. Chem.* 829 (2001).
394. X. Camps, H. Schönberger, and A. Hirsh, *Chem. Eur. J.* 3, 561 (1997).
395. T. Chaud, R. Deschenaux, A. Hirsch, and H. Schönberger, *Chem. Commun.* 2103 (1999).
396. A. Herzog, A. Hirsch, and O. Vostrowsky, *Eur. J. Org. Chem.* 171 (2000).
397. Y. Chen, Z.-E. Huang, R.-F. Cai, and B.-C. Yu, *Eur. Polym. J.* 34, 137 (1998); K. E. Geckeler and S. Samal, *Polym. Int.* 48, 743 (1999); K. E. Geckeler and S. Samal, *J. Macromol. Chem.—Rev. Macromol. Chem. Phys.* 40, 193 (2000); L. Dai, *J. Macromol. Chem.—Rev. Macromol. Chem. Phys.* 39, 273 (1999); L. Dai, *Polym. Adv. Technol.* 10, 357 (1999).
398. A. M. Rao, P. Zhou, K. Wang, G. T. Hager, J. M. Holden, Y. Wang, W. T. Lee, X. Bi, P. C. Eklund, D. S. Cornett, M. A. Duncan, and I. J. Amster, *Science* 259, 955 (1993).
399. Y. Iisa, T. Arima, R. M. Flemming, T. Siegrist, O. Zhou, R. C. Haddon, L. J. Rothberg, K. B. Lyons, H. L. Carter, Jr., A. F. Hebard, R. Tycko, G. Dabbagh, J. J. Krajewski, G. A. Thomas, and T. Yagi, *Science* 264, 1570 (1994); T. Ozaki, Y. Iisa, and T. Mitani, *Chem. Phys. Lett.* 285, 289 (1996).
400. Y. Wang, J. M. Holden, X. Bi, and P. C. Eklund, *Chem. Phys. Lett.* 217, 12705 (1995).
401. F. Cataldo, *Polym. Int.* 48, 143 (1999).
402. D. Sun and C. A. Reed, *Chem. Commun.* 2391 (2000).
403. D. A. Loy and R. A. Assink, *J. Am. Chem. Soc.* 114, 3977 (1992).
404. A. G. Camp, A. Lary, and W. T. Ford, *Macromolecules* 28, 7959 (1995).
405. C. E. Bunker, G. E. Lawson, and Y. Sun, *Macromolecules* 28, 3744 (1995).
406. W. T. Ford, T. D. Graham, and T. H. Mourey, *Macromolecules* 30, 6422 (1997).
407. P. L. Nayak, S. Alva, K. Yang, P. K. Dhal, J. Kumar, and S. K. Tripathy, *Macromolecules* 30, 7351 (1997).
408. H. Okumura, N. Ide, M. Minoda, K. Komatsu, and T. Fukuda, *Macromolecules* 31, 1839 (1998).
409. B. Z. Tang, S. M. Leung, H. Peng, N.-T. Yu, and K. C. Su, *Macromolecules* 30, 2848 (1997).
410. M. Seno, M. Maeda, and T. Sato, *J. Polym. Sci. Part A, Polym. Chem.* 38, 2572 (2000).
411. J. Wang, G. Javahery, S. Petrie, and D. K. Bohme, *J. Am. Chem. Soc.* 114, 9665 (1992).
412. V. Baranov, J. Wang, G. Javahery, S. Petrie, A. C. Hopkinson, and D. K. Bohme, *J. Am. Chem. Soc.* 119, 2040 (1997).
413. R. N. Thomas, *J. Polym. Chem. A, Polym. Chem.* 32, 2727 (1994).
414. B. Liu, C. E. Bunker, and Y. Sun, *Chem. Commun.* 1241 (1996).
415. C. J. Hawker, *Macromolecules* 27, 4836 (1994).
416. E. Cloutet, Y. Gnanou, J.-L. Fillaut, and D. Astruc, *Chem. Commun.* 1565 (1996); E. Cloutet, J.-L. Fillaut, D. Astruc, and Y. Gnanou, *Macromolecules* 32, 1043 (1999).
417. A. Gügel, P. Belik, M. Walter, A. Kraus, E. Harth, M. Wagner, J. Spickermann, and K. Müllern, *Tetrahedron* 52, 5007 (1996).
418. Z. Y. Wang, L. Kuang, X. S. Meng, and J. P. Gao, *Macromolecules* 31, 5556 (1998).
419. E. T. Samulski, J. M. Desimone, M. O. Hunt, Jr., Y. Z. Menceloglu, R. C. Jarnagin, G. A. York, K. B. Labat, and H. Wang, *Chem. Mater.* 4, 1153 (1992).
420. Y. Chen, E. Z. Huang, R. F. Kai, S. Q. Kong, S. Chen, Q. Shao, X. Yan, F. Zhao, and D. Fu, *J. Polym. Sci. A; Polym. Chem.* 34, 3297 (1996); Y. Chen, R. F. Cai, Z. E. Huang, and S. Q. Kong, *Polym. Bull.* 35, 705 (1995); Y. Chen, Z. E. Huang, R. F. Cai, S. Q. Kong, S. M. Chen, Q. F. Shao, X. M. Yan, F. L. Zhao, and D. Z. Fu, *J. Polym. Sci. A, Polym. Chem.* 34, 3297 (1996); Y. Chen, Z. E. Huang, R. F. Cai, B. C. Yu, W. W. Ma, S. M. Chen, Q. F. Shao, X. M. Yan, and Y. F. Huang, *Eur. Polym. J.* 33, 291 (1997).
421. Y. Chen, W. Huang, Z. Huang, R. Cai, H. Yu, S. M. Chen, and X. Yan, *Eur. Polym. J.* 33, 823 (1997); B. Yu, Y. Chen, R. Cai, Z. Huang, and Y. Xiao, *Eur. Polym. J.* 33, 3049 (1997).
422. Y. Chen, Z. Huang, R. Cai, B. Yu, Y. Huang, Y. Su, and Z. Wu, *Eur. Polym. J.* 34, 421 (1998).
423. Y. Ederlé and C. Mathis, *Macromolecules* 30, 4262 (1997).
424. Y. Ederlé and C. Mathis, *Macromolecules* 32, 554 (1999).
425. D. E. Bergbreiter and H. N. Gray, *J. Chem. Soc., Chem. Commun.* 645 (1993).
426. S. Shi, K. C. Khemani, L. Chan, and F. Wudl, *J. Am. Chem. Soc.* 114, 10656 (1992).
427. Y. P. Sun, B. Liu, and D. K. Morton, *Chem. Commun.* 2699 (1996).
428. L. Y. Chiang, L. Y. Wang, S. Tseng, J. Wu, and K. Hsieh, *J. Chem. Soc. Chem.Commun.* 2675 (1994).
429. L. Wang, L. Y. Chiang, C. S. Kuo, J. G. Lin, and C. Y. Huang, *Mater. Res. Soc. Symp. Proc.* 413, 571 (1996).
430. A. Nigam, T. Shekharam, T. Bharadwaj, J. Giovanola, S. Narang, and R. Malhotra, *J. Chem. Soc. Chem. Commun.* 1547 (1995).
431. A. Kraus and K. Müllen, *Macromolecules* 32, 4214 (1999).
432. N. Manalova, I. Rashkov, H. Van Damme, and F. Beguin, *Poym. Bull. (Berlin)* 33, 175 (1994).
433. K. E. Geckeler and A. Hirsch, *J. Am. Chem. Soc.* 115, 3850 (1993).
434. S. Y. Kuishima, M. Kubota, K. Kushida, T. Ishida, M. Ohashi, and T. Nogami, *Tetrahedron Lett.* 35, 4371 (1994).
435. C. Weis, C. Friedrich, R. Mülhaup, and H. Frey, *Macromolecules* 28, 403 (1995).
436. S. Samal, B.-J. Choi, and K. E. Geckeler, *Chem. Commun.* 1373 (2001).
437. (a) S. Shinkai and A. Ikeda, *Gazz. Chim. Ital.* 127, 657 (1997). (b) F. Diederich and M. Gomez-López, *Chem. Soc. Rev.* 28, 263 (1999). (c) F. Diederich and R. Kessinger, *Acc. Chem. Res.* 32, 537 (1999). (d) M. J. Hardie and C. L. Raston, *Chem. Commun.* 1153 (1999). (e) D. M. Guldi, *Chem. Commun.* 321 (2000). (f) J. Rebek, Jr., *Chem. Commun.* 637 (2000).
438. (a) J. L. Atwood, G. A. Koutsantonis, and C. L. Raston, *Nature* 368, 229 (1994). (b) H. Imahori and Y. Sakata, *Adv. Mater.* 9, 537 (1997). (c) N. Martin, J. Segura, and C. Seoane, *J. Mater. Chem.* 7, 1661 (1997). (d) L. Dai, *J. Macromol. Sci., Rev. Macromol. Chem. Phys.* 39, 273 (1999).
439. D. V. Konarev, R. N. Lyubovskaya, N. V. Drichko, E. I. Yudanova, Y. M. Shulga, A. L. Litvinov, V. N. Semkin, and B. P. Tarasov, *J. Mater. Chem.* 10, 803 (2000).
440. T. Suzuki, K. Nakashima, and S. Shinkai, *Chem. Lett.* 699 (1994).
441. E. C. Constable, *Angew. Chem., Int. Ed. Engl.* 33, 2269 (1994).
442. R. M. Williams and J. W. Verhoeven, *Recl. Trav. Chim. Pays-Bas* 111, 531 (1992).
443. C. L. Raston, J. L. Atwood, B. J. Nichols, and I. B. N. Sudria, *Chem. Commun.* 2615 (1996).
444. (a) K. Araki, K. Akao, A. Ikeda, T. Suzuki, and S. Shinkai, *Tetrahedron Lett.* 37, 73 (1996). (b) A. Ikeda, M. Yoshimura, and S. Shinkai, *Tetrahedron Lett.* 38, 2107 (1997).
445. A. Ikeda, Y. Suzuki, M. Yoshimura, and S. Shinkai, *Tetrahedron* 54, 2497 (1998).
446. T. Haino, M. Yanase, and Y. Fukazawa, *Angew. Chem., Int. Ed.* 37, 997 (1998).
447. T. Haino, H. Araki, Y. Yamanaka, and Y. Fukazawa, *Tetrahedron Lett.* 42, 3203 (2001).
448. M. Takeshita, T. Suzuki, and S. Shinkai, *J. Chem. Soc., Chem. Commun.* 2587 (1994).
449. A. Ikeda and S. Shinkai, *Chem. Lett.* 803 (1996).
450. M. Kawaguchi, A. Ikeda, and S. Shinkai, *J. Chem. Soc., Perkin Trans. I* 179 (1998).

451. P. R. Ashton, F. Diederich, M. Gomez-Lopez, J.-F. Nierengarten, J. A. Preece, F. M. Raymo, and J. F. Stoddart, *Angew. Chem., Int. Ed.* 36, 1448 (1997).
452. U. Jonas, F. Cardullo, P. Belik, F. Diederich, A. Gügel, E. Hart, A. Hermann, L. Isaacs, K. Müllen, H. Ringsdorf, C. Thilgen, P. Uhlmann, A. Vasella, C. A. A. Waldraff, and M. Walter, *Chem. Eur. J.* 1, 243 (1995).
453. F. Diederich, J. Effing, U. Jonas, L. Jullien, T. Plesnivy, H. Ringsdorf, C. Thilgen, and D. Weinstein, *Angew. Chem., Int. Ed. Engl.* 31, 1599 (1992).
454. S. Wang, R. M. Leblanc, F. Arias, and I. Echegoyen, *Langmuir* 13, 1672 (1997).
455. S. Wang, R. M. Leblanc, F. Arias, and L. Echegoyen, *Thin Solid Films* 327, 141 (1998).
456. F. Arias, L. A. Godinez, S. R. Wilson, A. E. Kaifer, and L. Echegoyen, *J. Am. Chem. Soc.* 118, 6086 (1996).
457. K.-Y. Kay and I. C. Oh, *Tetrahedron Lett.* 40, 1709 (1999).
458. J. W. Steed, P. C. Junk, J. L. Atwood, M. J. Barnes, C. L. Raston, and R. S. Burkhalter, *J. Am. Chem. Soc.* 116, 10346 (1994).
459. Y. Rio and J.-F. Nierengarten, *Tetrahedron Lett.* 43, 4321 (2002).
460. G. Rapenne, J. Crassous, L. E. Echegoyen, L. Echegoyen, E. Flapan, and F. Diederich, *Helv. Chim. Acta* 83, 1209 (2000).
461. N. Martin, L. Sanchez, B. Illescâs, and I. Pérez, *Chem. Rev.* 98, 2527 (1998).
462. T. G. Linssen, K. Durr, M. Hanack, and A. Hirsch, *J. Chem. Soc., Chem. Commun.* 103 (1995).
463. (a) P. A. Liddell, J. P. Sumida, A. N. Macpherson, L. Noss, G. R. Seely, K. N. Clark, A. L. Moore, T. A. Moore, and D. Gust, *Photochem. Photobiol.* 60, 537 (1994). (b) H. Imahori, T. Hagiwara, T. Akiyama, S. Taniguchi, T. Okada, and Y. Sakata, *Chem. Lett.* 265 (1995). (c) M. G. Ranasinghe, A. M. Oliver, D. F. Ruthenfluh, A. Salek, and M. N. Paddon-Row, *Tetrahedron Lett.* 37, 4797 (1996). (d) D. M. Guldi and N. Martin, *J. Mater. Chem.* 12, 1978 (2002). (e) D. M. Guldi, *Chem. Soc. Rev.* 31, 22 (2002).
464. J.-Y. Zheng, K. Tashiro, Y. Hirabayashi, K. Kinbara, K. Saigo, T. Aida, S. Sakamoto, and K. Yamaguchi, *Angew. Chem., Int. Ed.* 40, 1858 (2001).
465. T. Drovetskaya, C. A. Reed, and P. Boyd, *Tetrahedron Lett.* 36, 7971 (1995).
466. H. Imahori and Y. Sakata, *Chem. Lett.* 199 (1996).
467. Y. Sun, T. Drovetskaya, R. D. Bolskar, R. Bau, P. D. W. Boyd, and C. A. Reed, *J. Org. Chem.* 62, 3642 (1997).
468. H. Imahori, K. Hagiwara, M. Aoki, T. Akiyama, S. Taniguchi, T. Okada, M. Shirakawa, and Y. Sakata, *J. Am. Chem. Soc.* 118, 11771 (1996).
469. R. M. Williams, M. Koeberg, J. M. Lawson, Y.-Z. An, Y. Rubin, M. N. Paddon-Row, and J. W. Verhoeven, *J. Org. Chem.* 61, 5055 (1996).
470. P. A. Liddell, D. Kuciauskas, J. P. Sumida, B. Nash, D. Nguyen, A. L. Moore, T. A. Moore, and D. Gust, *J. Am. Chem. Soc.* 119, 1400 (1997).
471. M. Fijitsuka, O. Ito, H. Imahori, K. Yamada, H. Yamada, and Y. Sakata, *Chem. Lett.* 721 (1999); S. Fukuzumi, H. Imahori, M. E. El-Khouly, M. Fujitsuka, O. Ito, and D. M. Guldi, *J. Am. Chem. Soc.* 123, 2571 (2001).
472. C. Cuo, D. M. Guldi, H. Imahori, K. Tamaki, and Y. Sakata, *J. Am. Chem. Soc.* 122, 6535 (2000).
473. H. Imahori, D. M. Guldi, K. Tamaki, Y. Yoshida, C. Luo, Y. Sakata, and S. Fukuzumi, *J. Am. Chem. Soc.* 123, 6617 (2001).
474. D. Kuciauskas, P. A. Liddel, S. Lin, T. E. Johnson, S. J. Weghorn, J. S. Lindsy, A. L. Moore, T. A. Moore, and D. Gust, *J. Am. Chem. Soc.* 121, 8604 (1999).
475. N. Armaroli, F. Diederich, C. O. Dietrich-Buchecker, L. Flamigni, G. Marconi, and J. F. Nierengarten, *New J. Chem.* 23, 77 (1999).
476. E. Dietel, A. Hirsch, E. Eichhorn, A. Ricker, S. Hackbarth, and B. Röder, *Chem. Commun.* 1981 (1998).
477. D. M. Guldi, C. Luo, T. D. Ros, M. Prato, E. Dietel, and A. Hirsch, *Chem. Commun.* 375 (2000).
478. D. M. Guldi, C. Luo, A. Swartz, M. Scheloske, and A. Hirsch, *Chem. Commun.* 1066 (2001).
479. T. Braun, *Fullerene Sci. Technol.* 5, 615 (1997).
480. K. I. Priyadarsini, H. Mohan, A. K. Tyagi, and J. P. Mittal, *J. Phys. Chem.* 98, 4756 (1994); K. Kanazawa, H. Nakanishi, Y. Ishizuka, T. Nakamura, and M. Matsumoto, *Fullerene Sci. Technol.* 2, 189 (1994).
481. Z.-i. Yoshida, H. Takekuma, S.-i. Takekuma, and Y. Matsubara, *Angew. Chem., Int. Ed. Engl.* 33, 1597 (1994).
482. T. Anderson, K. Nilson, M. Sundhal, G. Westman, and O. Wennerström, *J. Chem. Soc., Chem. Commun.* 604 (1992).
483. (a) M. Wei, N. Li, Z. Gu, and X. Zhou, *Electrochim. Acta* 45, 4441 (2000). (b) M. Wei, N. Li, Z. Gu, and X. Zhou, *Electroanalysis* 12, 442 (2000). (c) H.-C. Hu, Y. Liu, D.-D. Zhang, and L.-F. Wang, *J. Inclusion Phenom. Macrocyclic Chem.* 33, 295 (1999).
484. C. N. Murthy and K. E. Geckeler, *Chem. Commun.* 1194 (2001).
485. V. Ohlendorf, A. Willnow, H. Hungerbuhler, D. M. Guldi, and K.-D. Asmus, *J. Chem. Soc., Chem. Commun.* 759 (1995).
486. K. I. Priyadarsini and H. Mohan, *J. Photochem. Photobiol. A* 85, 63 (1995).
487. S. Samal and K. E. Geckeler, *Chem. Commun.* 1101 (2000).
488. S. Samal and K. E. Geckeler, *Synth. Commun.* 32, 3367 (2000); D.-Q. Yuan, K. Koga, Y. Kourogi, and F. Fujita, *Tetrahedron Lett.* 42, 6727 (2001).
489. S. Filippone, F. Heimann, and A. Rassat, *Chem. Commun.* 1508 (2002).
490. T. L. Makarova, B. Sundqvist, R. Höhne, P. Esquinazi, Y. Kopelevichk, P. Scharff, V. A. Davydov, L. S. Kashevarova, and A. V. Rakhmanina, *Nature* 413, 716 (2001).
491. G. Brusatin and R. Signorini, *J. Mater. Chem.* 12, 1964 (2002).
492. F. Wudl, *J. Mater. Chem.* 12, 1959 (2002).
493. Y. Wang, *Nature* 356, 585 (1992).
494. Y. Chen, Z. E. Huang, and R. F. Cai, *J. Polym. Sci. B, Polym. Phys.* 34, 631 (1996); Y. Chain, R. F. Kai, Z. E. Huang, X. Bai, B. C. Yu, W. Jin, D. C. Pan, and S. T. Wang, *Polym. Bull.* 36, 203 (1996); Z. E. Huang, Y. Chen, R. F. Cai, C. G. Rui, and F. P. Zhang, *J. Appl. Polym. Sci.* 60, 573 (1996); Y. Chen, Z. E. Huang, R. F. Cai, X. M. Yan, S. M. Chen, W. Jin, D. C. Pan, and S. T. Wang, *J. Appl. Polym. Sci.* 61, 2185 (1996); R. F. Cai, X. Bai, Y. Chen, and Z. E. Huang, *Eur. Polym. J.* 34, 7 (1998).
495. J. Gao, F. Hide, and H. Wang, *Synth. Met.* 84, 979 (1997).
496. K. Yoshino, S. Morita, T. Akashi, K. Yoshimoto, M. Yoshida, A. Fujii, A. A. Zakhidov, K. Kikuchi, and Y. Achiba, *Synth. Met.* 70, 1481 (1995).
497. S. Xiao, S. Wang, H. Fang, Y. Li, Z. Shi, C. Du, and D. Zhu, *Macromol. Rapid. Commun.* 22, 1313 (2001).
498. P. A. Lane, J. Shinar, and K. Yoshino, *Phys. Rev.* 54, 9308 (1996).
499. E. K. Miller, K. Lee, J. Cornil, Q. Pei, F. Wudl, J. Heeger, and J. L. Brédas, *Synth. Met.* 84, 631 (1997).
500. K. Matsumoto, M. Fujitsuka, T. Sato, S. Onodera, and O. Ito, *J. Phys. Chem. B* 104, 11632 (2000).
501. K. Hutchison, J. Gao, Y. Rubin, and F. Wudl, *J. Am. Chem. Soc.* 121, 5611 (1999).
502. (a) T. Chuard and R. Deschenaux, *Helv. Chim. Acta* 79, 736 (1996). (b) R. Deschenaux, M. Even, and D. Guillon, *Chem. Commun.* 537 (1998). (c) M. Even, B. Heinrich, D. Guillon, D. M. Guldi, M. Prato, and R. Deschenaux, *Chem. Eur. J.* 7, 2595 (2001). (d) B. Dardel, D. Guillon, B. Heinrich, and R. Deschenaux, *J. Mater. Chem.* 11, 2814 (2001). (e) B. Dardel, R. Deschenaux, M. Even, and E. Serrano, *Macromolecules* 32, 5193 (1999). (f) S. Campidelli and R. Deschenaux, *Helv. Chim. Acta* 84, 589 (2001). (g) N. Tirelli, F. Cardullo, T. Habicher, U. W. Suter, and F. Diederich, *J. Chem. Soc., Perkin Trans. 2* 193 (2000).

(h) T. Chuard, R. Deschenaux, A. Hirsch, and H. Schönberger, *Chem. Commun.* 2103 (1999).

503. D. Felder, B. Heinrich, D. Guillon, J.-F. Nicoud, and J.-F. Nierengarten, *Chem. Eur. J.* 6, 3501 (2000).

504. T. Chuard and R. Deschenaux, *J. Mater. Chem.* 12, 1944 (2002).

505. (a) R. Deschenaux, M. Schweissguth, and A.-M. Levelut, *Chem. Commun.* 1275 (1996). (b) R. Deschenaux, M. Schweissguth, M.-T. Vilches, A.-M. Levelut, D. Hautot, G. J. Long, and D. Luneau, *Organometallics* 18, 5553 (1999); (c) F. Turpin, D. Gullion, and R. Deschenaux, *Mol. Cryst. Lid. Cryst.* 362, 171 (2001).

506. (a) V. Percec, W.-D. Cho, G. Ungar, and D. J. P. Yeardley, *J. Am. Chem. Soc.* 123, 1302 (2001). (b) P. Busson, J. Örtegren, H. Ihre, U. W. Gedde, A. Hult, and G. Andersson, *Macromolecules* 34, 1221 (2001). (c) N. Boiko, X. Zhu, A. Bobrovsky, and V. Shibaev, *Chem. Mater.* 13, 1447 (2001). (d) J. Barberá, B. Donnio, R. Giménez, D. Guillon, M. Marcos, A. Omenat, and J. L. Serrano, *J. Mater. Chem.* 11, 2808 (2001).

507. M. Sawamura, K. Kawai, Y. Matsuo, K. Kanie, T. Kato, and E. Nakamura, *Nature* 419, 702 (2002); C. Tschierske, *Nature* 419, 681 (2002).

508. (a) A. W. Jensen, S. R. Wilson, and D. I. Schuster, *Bioorg. Med. Chem.* 4, 767 (1996). (b) T. D. Ros and M. Prato, *Chem. Commun.* 663 (1999).

509. J. W. Aborgast, A. P. Darmanyan, C. S. Foote, Y. Rubin, F. N. Diedrich, M. M. Alvarez, S. J. Anz, and R. L. Whetten, *J. Phys. Chem.* 95, 11 (1991).

510. Y. Tabata and Y. Ikada, *Pure Appl. Chem.* 71, 2047 (2000).

511. M. Brettreich and A. Hirsch, *Tetrahedron Lett.* 39, 2731 (1998).

512. F. Rancan, S. Rosan, F. Boehm, A. Cantrell, M. Brellreich, H. Schoenberger, A. Hirsch, and F. Moussa, *J. Photochem. Photobiol. B: Biol.* 67, 157 (2002).

513. S. H. Friedmann, D. L. Decamp, R. P. Sijbesma, G. Srdanov, F. Wudl, and G. L. Kenyon, *J. Am. Chem. Soc.* 115, 6506 (1993).

514. R. F. Schinazi, C. Bellavia, R. Gonzalez, C. L. Hill, and F. Wudl, *Proc. Electrochem. Soc.* 95, 696 (1995).

515. D. I. Schuster, S. R. Wilson, and R. F. Schinazi, *Bioorg. Med. Chem. Lett.* 6, 1253 (1996).

516. E. Nakamura, H. Tokuyama, S. Yamago, T. Shiraki, and Y. Sugiura, *Bull. Chem. Soc. Jpn.* 69, 2143 (1996).

517. H. Tokuyama, S. Yamago, E. Nakamura, T. Shiraki, and Y. Sugiura, *J. Am. Chem. Soc.* 115, 7918 (1993); A. S. Bourtorine, H. Tokuyama, M. Takasugi, H. Isobe, E. Nakamura, and C. Helene, *Angew. Chem., Int. Ed. Engl.* 33, 2462 (1994); J. Hirayama, H. Abe, N. Kamo, T. Shinbo, Y. O. Yamada, S. Kurosawa, K. Ikebuchi, and S. Sekicuchi, *Biol. Pharm. Bull.* 22, 1106 (1999).

518. N. Higashi, T. Inoue, and M. Niwa, *Chem. Commun.* 1507 (1997).

Encyclopedia of Nanoscience and Nanotechnology
www.aspbs.com/enn

Fullerenes for Photovoltaics

E. A. Katz
Ben-Gurion University of the Negev, Sede Boqer, Israel

CONTENTS

1. INTRODUCTION

The discovery [1] of C_{60}, a third variety of carbon in addition to the more familiar diamond and graphite forms, together with an easy method [2] to produce macroscopic quantities of the material, have generated enormous interest in many areas of physics, chemistry, and material science. Furthermore, it turns out that C_{60} is only the first of an entire class of fully conjugated, all-carbon molecules—the fullerenes (C_{60}, C_{70}, C_{82}, C_{120}, etc., carbon nanotubes). And yet, brief as the history of research into these materials may be [3], it is already clear from its vast and increasing literature that much complex and fascinating physics lies behind the geometrical simplicity of these structures. In addition to the interesting properties of pristine fullerenes, by doping these materials in an appropriate manner, not only can their electronic properties be "tuned" to coincide with those of semiconductors and conductors, but even superconductivity may result [4]. There are, accordingly, great expectations that fullerenes will find practical use in many advanced technologies. This chapter reviews the prospects of one such application—direct conversion of solar energy to electricity by photovoltaic solar cells.

2. PHOTOVOLTAIC SOLAR CELLS

Photovoltaic solar cells are devices that convert solar radiation directly into electrical energy due to the photovoltaic effect which was first observed by Becquerel in 1839 [5]. A photon can interact with matter by losing its energy, and using this energy to promote an electron to a higher energy level. If a semiconductor solar cell is illuminated with light of photon energy higher than its bandgap energy of E_g, free electron–hole pairs are generated. To separate these photoinduced charge carriers and generate photovoltage, a solar cell should have a junction between two materials across which there is an electrochemical potential difference in equilibrium.

There are at least three main types of such junctions, namely, homo- and hetero-*p–n* junctions, formed between *p* and *n* types of the same or different semiconductors, respectively, and the Schottky barrier, that is, a rectifying interface between a semiconductor and a metal. Other types of solar cells may include a *p–i–n* junction, formed from two doped and one intrinsic semiconductor layer, or a metal–insulator–semiconductor (MIS) configuration.

Most inorganic photovoltaic devices to date have been formed using a *p–n* junction [6]. The different Fermi levels E_F in *p*- and *n*-type regions result in a depletion region at their interface, where this difference is accommodated by a built-in electric field (Fig. 1). Photoinduced carriers, generated within approximately one diffusion length from the depletion region, may diffuse to this region, where they will be separated by the built-in electric field. Electrons and holes move to the *n* and *p* regions, respectively. A potential difference thus appears between the external metal electrodes attached to the *p* and *n* regions, and a current may flow through a connected load resistor.

Figure 2(a) shows the schematic diagram of a homo *p–n* junction solar cell. Ohmic contact is made to the back surface of the base region. The contact to the emitter top layer is made in the form of a grid pattern. An antireflection (AR) coating is then used to reduce the amount of light lost by reflection from the front surface.

A simple circuit for measurements of a solar cell's current–voltage (I–V) characteristics under illumination is shown in Figure 2(b). The dark I–V curve of an ideal *p–n* junction is given by the Shockley equation [6]

$$I_d(V) = I_0(e^{qV/KT} - 1) \quad (1)$$

where I_0 is the reverse saturation current of the *p–n* junction, q is the electron charge, T is the absolute temperature,

ISBN: 1-58883-059-4/$35.00

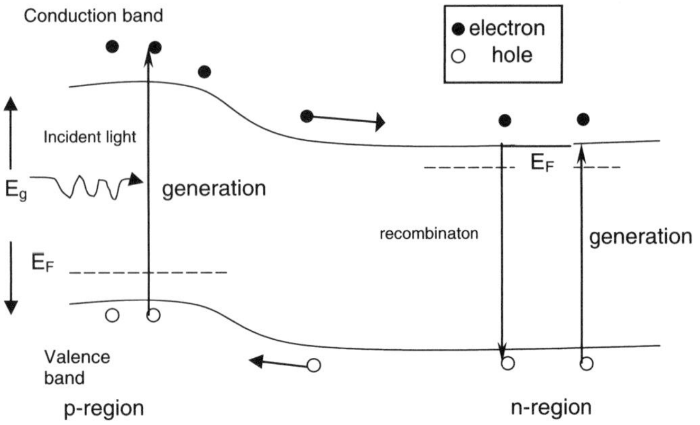

Figure 1. Energy band diagram for a *p–n* junction under illumination.

K is the Boltzmann constant, and *V* is the voltage. The dark *I–V* curve is shown in Figure 3(a). The dark characteristic is measured with an external bias in the forward (first quadrant) and reverse (third quadrant) direction.

Under illumination, the equation for the *I–V* curve becomes

$$I = I_0(e^{qV/KT} - 1) - I_L \tag{2}$$

where I_L is the photocurrent at zero bias, and results from the excitation of excess carriers by solar radiation. A plot of Eq. (2) is given in Figure 3(a) as well. Figure 3(b) represents the part of the irradiated solar cell *I–V* curve from the fourth quadrant just by its inversion around the voltage axis. The quantities $I_{\max}$ and $V_{\max}$, corresponding to the current and voltage respectively, for the maximum power output $P_{\max} = I_{\max} \times V_{\max}$ are also defined in Figure 3(b).

Now, we can define such important solar cell parameters as the short-circuit current I_{sc}, the open-circuit voltage V_{oc}, the fill factor *FF*, and the power conversion efficiency η.

For $V = 0$, $I = -I_L = I_{sc}$, and for $I = 0$, $V = V_{oc}$. The fill factor is defined by the relation

$$FF = \frac{I_{\max} \times V_{\max}}{V_{oc} \times I_{sc}} \tag{3}$$

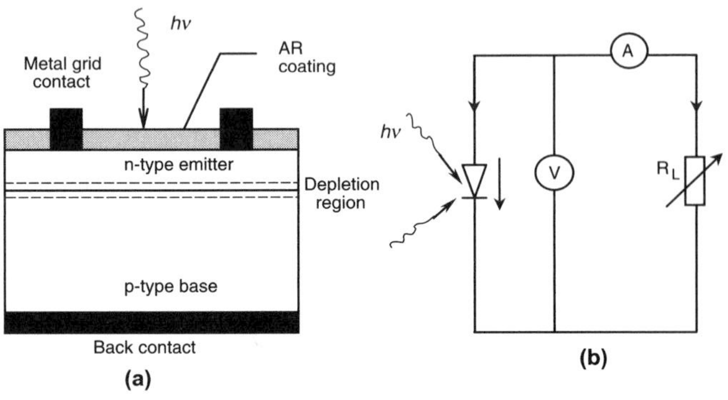

Figure 2. Cross-sectional view of a typical *p–n* junction solar cell (a) and a simplest circuit diagram for the measurements of solar cell *I–V* characteristics under irradiation (b).

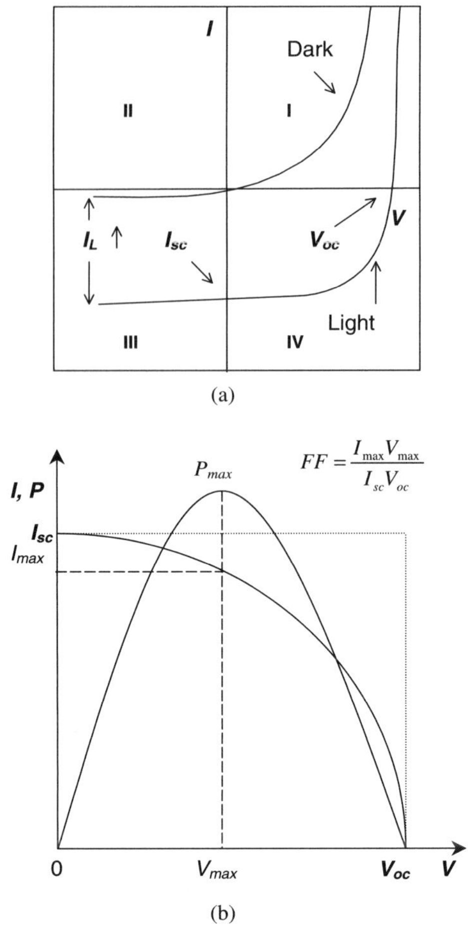

Figure 3. Dark and light *I–V* curves (a) and the general representation of the light *I–V* curve in the first quadrant showing maximum power $P_{\max}$ (b).

Finally, the power conversion efficiency is given by

$$\eta = \frac{I_{\max} \times V_{\max}}{P_{\text{in}}} = FF\frac{V_{oc} \times I_{sc}}{P_{\text{in}}} \tag{4}$$

3. FULLERENES AND PHOTOVOLTAICS

We consider a combination of the following five reasons that make fullerene-based materials unique among all known semiconductors, and promising for photovoltaic applications.

1. Relatively high values of photoconductivity (in comparison with the dark conductivity values) were observed in C_{60} thin films [7].
2. Early studies of solid C_{60} as a material for inorganic solar cells [8–9] were instigated by the theoretical prediction [10] that a C_{60} crystal has a direct bandgap of 1.5 eV. On one hand, this value is close to the experimental value of about 1.6 eV for the fundamental edge in the optical absorption spectra [11] and that of 1.7 eV for the photoconductivity spectra [12] of C_{60} thin films. On the other hand, this is the optimal value for high-efficiency photovoltaic devices of

the single-junction type [13]. Recent experimental studies [14–15] demonstrated that the electronic structure of a C_{60} crystal is more complicated. Specifically, the bandgap value is suggested to be about 2.3 eV (the mobility gap), but the optical absorption extends from the gap energy to the lower energy side (the optical gap is ~1.6 eV). These experimental works stimulated new theoretical calculations, using a "many-body" approach, which yielded a bandgap value of 2.15 eV [16]. However, even if this is true, the electronic structure and optical properties of C_{60} thin films are suitable for the use of this material in efficient heterojunction solar cells (such as C_{60}/Si devices).

3. Solid C_{60} is a molecular crystal in which C_{60} molecules occupy the lattice sites of a face-centered cubic (fcc) structure [17]. If one contrasts a C_{60} molecular crystal unit cell with the corresponding Si cell, the former has 240 atoms to the latter's 8. This allows the possibility of doping to a much greater percentage than would be possible with conventional semiconductors. Furthermore, unlike the case of conventional semiconductors for which doping is only possible via *substitution* of atoms (e.g., P or B for Si), fullerenes may be doped in three additional *distinct* ways: *intercalation* of dopants between the buckyballs, *endohedral doping* within the buckyballs themselves, and *adductive doping* to the outside of the buckyballs [4]. The spectrum of resultant doping energy levels will naturally differ from each manner of doping, thus greatly increasing the probability of finding successful dopants for high-efficiency photovoltaic devices [9].
4. For organic donor–acceptor solar cells, a C_{60} molecule is an ideal candidate due to the strong acceptor properties of the C_{60} molecule, which can accept as many as six electrons [4]. Such cells are now becoming a serious alternative to conventional inorganic photovoltaic devices due to a number of potential advantages, such as their light weight, flexibility, and low-cost fabrication of large areas. The efficient photoresponse of the donor–acceptor cells consisting of conjugated polymers and fullerenes [18–22] originates from an ultrafast (subpicosecond) photoinduced electron transfer from the conjugated polymer (as a donor) to a nearby C_{60} molecule (as an acceptor) [23, 24]. Since this process is faster than any other relaxation mechanism in the conjugated polymers, the quantum efficiency of this process is estimated to be close to unity. Accordingly, the best photovoltaic parameters ever reported for organic solar cells—and in particular, energy conversion efficiencies of 2.5 [25] and 3.6% [26]—were for fullerene-containing devices.
5. There is probably no more environmentally benign semiconductor than C_{60} in that it can be synthesized from graphite using nothing more than a beam of concentrated solar energy, and subsequently purified and crystallized using the same energy source (as described later). To demonstrate this statement, let us review very briefly the state of the art for fullerene production, and in particular, the production of fullerenes and carbon nanotubes by concentrated sunlight.

In 1990, Krätschmer et al. [2] discovered a simple technique to produce fullerenes. The method uses arc discharge between two graphite electrodes in a helium atmosphere to vaporize graphite and produce fullerene-containing soot, the subsequent extraction of soluble fullerenes from the insoluble species of the soot by an organic solvent (toluene, for example), and final chromatographic separation of isolated fullerenes (C_{60}, C_{70}, etc.). At present, the most efficient methods of fullerene production involve different modifications of thermal evaporation of graphite in an inert atmosphere (e.g., ac or dc arc discharge between graphite electrodes, laser ablation of graphite) [4].

The electric-arc process requires use of high-quality, electrically conductive graphite rods. On the other hand, the yield of fullerene production from the collected soot is low (5–10%) due to the photodestruction of fullerene molecules by UV radiation of the electric arc [27]. Furthermore, because the yield is low, excessive amounts of toluene are required to separate the fullerenes from the soot. These factors present serious limitations on the minimum cost that can be obtained by this process.

Even though the yield from the laser ablation process can be high, it is generally accepted that this process will not be cost effective for scaling up to large production levels [28]. The search for a more efficient and environmentally benign method led to the use of concentrated solar energy to evaporate the carbon and efficiently produce fullerenes [27–34] and carbon nanotubes [32, 34–40].

In all reported solar techniques, concentrated sunlight ($\geq$1100 W/cm^2) is focused onto a graphite target, resulting in its heating up to temperatures in excess of 3500 K. Controlling vaporization and condensation conditions leads to the formation of fullerene-rich soot or/and carbon nanotubes (if the target consists of a mixture of graphite and metal catalysts).

The UV component in sunlight is very small (compared with irradiation of an electric arc). Therefore, the photodestruction process is very weak, and the yield of fullerenes is as high as 20% [32]. Increased yield leads to less solvent use, which reduces cost. Furthermore, the relaxation of the requirement to use conductive graphite rods opens the possibility to use less expensive forms of graphite, including mineral graphite powders. A preliminary cost analysis [29] suggested that solar production can be less expensive than the arc process by at least a factor of 4.

It is also noteworthy that fullerenes were demonstrated to be purified and crystallized via differential sublimation [4]. These technological steps are certainly possible to perform in a solar furnace.

4. INORGANIC C_{60}-BASED SOLAR CELLS

4.1. Solar Cells Based on Pristine C_{60}

4.1.1. Metal–Insulator–Semiconductor and Schottky Barrier Devices

In 1992, Hebard et al. patented an idea for the utilization of the photoconductivity and photovoltaic properties of fullerites, and in particular, fabrication of fullerene-based solar cells and "the generation of a current by illumination of appropriate fullerene interfaces" [41]. Such photovoltaic response requires fabrication of a fullerene-based device structure with rectifying properties. However, in that patent publication, the authors described a symmetrical (nonrectifying) Ag/C_{60}/Ag device structure which exhibited photoconductivity, but no photovoltaic effect.

A fullerene-based solid-state device, displaying a remarkable rectifying effect in the dark and photoresponse under illumination, was first demonstrated by Yohehara and Pac [42]. Polycrystalline C_{60} film with a thickness of 100 nm was deposited by the vacuum evaporation technique on a substrate of Al coated glass. This Al layer was used as the front electrode. Sandwich cells were completed by vacuum evaporation of an Al or Au back electrode (0.5 cm^2 in area). The devices without any exposure of the front electrode to oxygen exhibited no rectification and photoresponse. The devices with a front Al electrode exposed to oxygen (before C_{60} film deposition) exhibited a rectification ratio of 66 (at ± 2 V), and greatly enhanced the photocurrent (the quantum yield of the photocurrent for monochromatic irradiation of $\lambda = 400$ nm and 0.1 V forward bias was about 53%). The excitation profile of the photocurrent was observed to follow the optical absorption of a C_{60} film. The authors explained the results by the formation of an oxide layer at the front Al electrode interface and the MIS configuration rather than a Schottky barrier device. MIS devices with a fullerene layer as an active semiconductor contact were also studied by Pichler et al. [43]. Solar cells with a Schottky barrier at the C_{60}/Mg–In [44], C_{70}/Ca [44], and C_{60}/Ag [8, 45] interfaces were demonstrated later.

4.1.2. Fullerene/Inorganic Semiconductor Heterojunction Devices

A heterojunction between a C_{60} thin film and another semiconductor with a high rectifying ratio in the dark ($>10^4$ at ± 2 V) [46] and photovoltage generation [47] were demonstrated, first, for a C_{60}/p–Si interface. The properties of such a heterojunction were then extensively studied in the dark and under irradiation of various light sources [8, 45, 48–53]. Although different groups studied various device configurations (Nb/C_{60}/Si, Ti/C_{60}/Si [46, 48], Au/C_{60}/Si, indium–tin–oxide (ITO)/C_{60}/Si [47, 51], Al/C_{60}/Si [8, 45, 49–50]), all authors seem to agree that potential barrier formation at the C_{60}/p–Si rather than at the metal/C_{60} interface is responsible for the strong rectifying properties of the heterostructures. X-ray photoelectron spectroscopy (XPS) and ultraviolet photoelectron spectroscopy (UPS) [54–55], as well as low-energy electron diffraction (LEED) and Auger electron spectroscopy (AES) [55], have revealed an abrupt character of the C_{60}/Si(111) interface, and no chemical reaction or diffusion between the C_{60} overlayer and Si(111) substrate.

Later, a rectifying heterojunction was reported for the C_{60}/n–Si [44], C_{60}/n–GaAS [56], and C_{60}/n–GaN [57] interfaces as well.

Figure 4 shows a typical dark current–voltage curve of a Ti/C_{60}/p–Si/Al device [48]. Polycrystalline fcc C_{60} film with a thickness of 200 nm was deposited by a vacuum deposition technique onto a chemically terminated (to remove

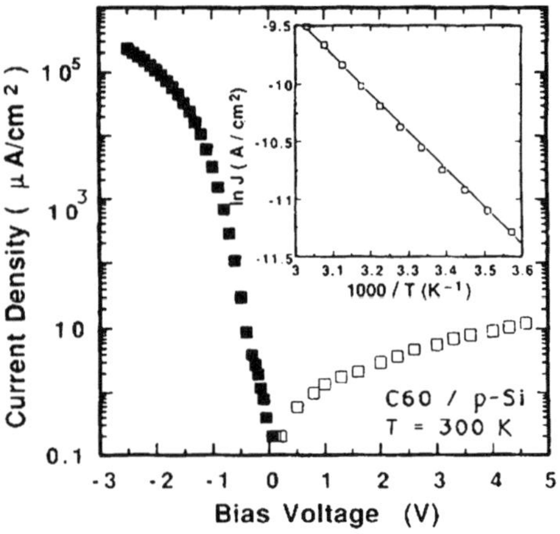

Figure 4. Dark current–voltage characteristics of a Ti/C_{60}/p–Si device. The inset shows data from thermal activation measurements at a fixed forward bias of −0.5 V. Reprinted with permission from [48], K. M. Chen et al., *J. Phys.: Condens. Matter* 7, L201 (1995). © 1995, IOP Publishing Ltd.

any surface oxide) substrate of (111) p-type Si (2–4 $\Omega \cdot$ cm) with a back ohmic contact of Al. Finally, Ti electrode dots of area 5.03×10^{-3} cm^2 were deposited on the C_{60} surface. The device was found to be forward conducting when the Si substrate was positively biased relative to the C_{60} film, the rectification ratio being greater than 10^4 at ± 2 V. Similar results were reported for Nb/C_{60}/p–Si/Al [46] and Al/C_{60}/p–Si/Al [49–51] device configurations. In [48], the forward I–V curve was analyzed using

$$J = J_0\{\exp[q(V - A_e JR_s)/nkT] - 1\} \tag{5}$$

where J is the current density, A_e is the area of the Ti contact, R_s is the series resistance of the sample, and n is the quality factor of a p–n junction.

The inset of Figure 4 displays results of thermal activation measurements over a temperature range of 280–300 K. For a given forward bias at 0.5 V ($A_e JR_s \ll 0.5$ V), the plot of $\ln J$ versus $1/T$ is a straight line, indicating that the current-transport mechanism involves a thermally activated process. After least squares fitting, it was found that J_0 in (1) is an exponential function of $1/T$, that is,

$$J_0 = J_{00}\exp(-q\varphi_{\text{eff}}/kT) \tag{6}$$

where $J_{00} = 1.4 \times 10^5$ μA $\cdot$ cm^{-2} and $q\varphi_{\text{eff}} = 0.48$ eV. $q\varphi_{\text{eff}}$ denotes the effective barrier height of the heterojunction under zero-bias conditions in the dark, which is generally related to the energy band structure of solid C_{60} and Si. For an ideal barrier, the $q\varphi_{\text{eff}}$ value is equal to the difference of the Fermi levels in C_{60} and Si. In a real heterojunction, this value also may be affected by the C_{60}/Si interface states [48, 58, 59].

It was demonstrated by measurements of capacitance–voltage (C–V) characteristics of heterojunctions between an Si substrate and C_{60} films of various thicknesses (from 200 to 900 nm) that the entire C_{60} films and the near-interface region of the Si wafer behave as a depletion region. This means that an internal electric field necessary for a photovoltaic effect exists over the entire thickness of the C_{60} films [58].

Photovoltage generation of a C_{60}/p–Si heterojunction was demonstrated by Wen et al. [47]. C_{60} film with a thickness of 1500 nm was vapor deposited in vacuum on a chemically cleaned p–Si substrate (20 $\Omega \cdot$ cm) with a back ohmic contact of Au. Finally, a front ITO electrode was sputtered on the C_{60} surface. An Au front electrode (transparent to light) was also used for the measurements to confirm that there is no photoelectric effect of the C_{60}/p–Si interface. A difference between the results with ITO and Au electrodes was not observed. Although the rectification ratio in the dark of the ITO/C_{60}/p–Si/Au was not reported, the direction of the rectification was the same as that reported in [8, 46, 48–51]. The V_{oc} value was found to be about 0.2 V under irradiation by a 150 W Xe lamp (longer than 200 nm in wavelength). Unfortunately, other solar cell parameters, as well as the active area of the device, were not reported.

A more detailed study of the photovoltaic properties was performed in [8, 45, 49, 60] for highly crystalline C_{60} thin-film/p–Si heterojunction solar cells. Their dark I–V characteristics exhibited strongly rectifying properties: the rectification ratio being approximately 10^4 at ± 2 V. Under solar irradiation (AM 1.5), the short-circuit current density J_{sc} was found to be 42 μA/cm^2, the open-circuit voltage $V_{oc} = 322$ mV, and $FF = 0.3$. The low values of J_{sc} and FF were attributed to the high resistivity of pristine C_{60} films. SPV spectroscopy characterization of the devices (Fig. 5) suggested that the C_{60} film acts as an active layer in the device, in particular for the conversion of short-wavelength components of sunlight. Indeed, it is evident that the positive SPV signal at 1.1 eV originates from the valence band-to-conduction band transition in p–Si, while the signals at 1.6 and 2.25 eV are attributable to tail-to-tail (optical bandgap) and band-to-band (mobility gap) transitions in the n–C_{60} layer [15]. This conclusion is consistent with the experimental spectral response of similar Al/C_{60}/p–Si/Al solar cells (Fig. 6) [50] and that calculated in [48] using the solar cell property equations [59] and published values of optical absorption coefficient values for C_{60} [62–63] and Si [Fig. 7(a) and (b)]. It should be noted that the experimental and calculated spectral response data qualitatively coincide, except for a surprising sharp peak at about 390 nm. The authors of [50] did not discuss the nature of this peak. A similar peak was observed in the spectral response of a C_{60}-based photoelectrochemical solar cell [64].

The maximum V_{oc} of a heterojunction corresponds to its $q\varphi_{\text{eff}}$ value that depends, as mentioned earlier, on the difference in the Fermi levels of both semiconductors forming the junction, in this case, of C_{60} thin film and Si. The Fermi level of Si can be controlled by changing the type and

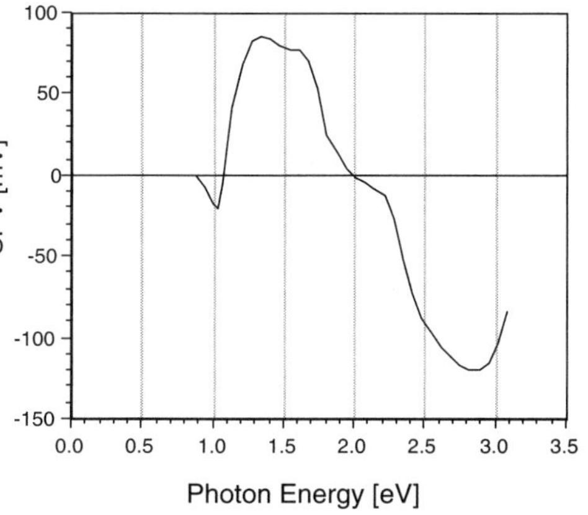

Figure 5. SPV spectrum of an Al/C_{60}/p–Si/Al heterojunction solar cell. Reprinted with permission from [45], E. A. Katz et al., "Thin Films Structures for Photovoltaics," Materials Research Society Series, Vol. 485, p. 113, 1997. © 1997, Materials Research Society.

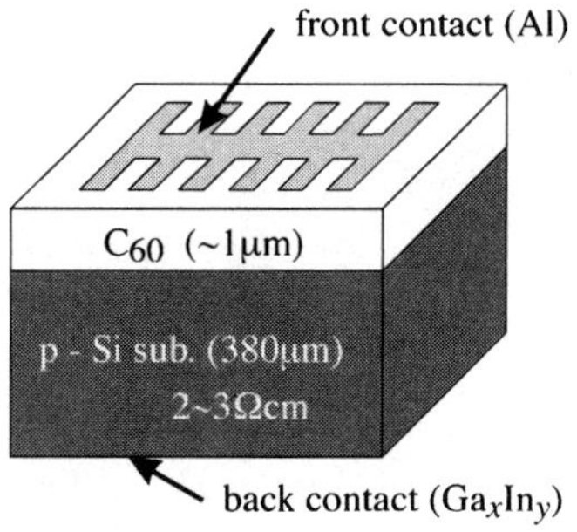

Figure 6. Schematic diagram of the Al/C_{60}/p–Si/Al cell structure. Reprinted with permission from [50], N. Kojima et al., *Jpn. J. Appl. Phys.* 39, 1176 (2000). © 2000, Institute of Pure and Applied Physics.

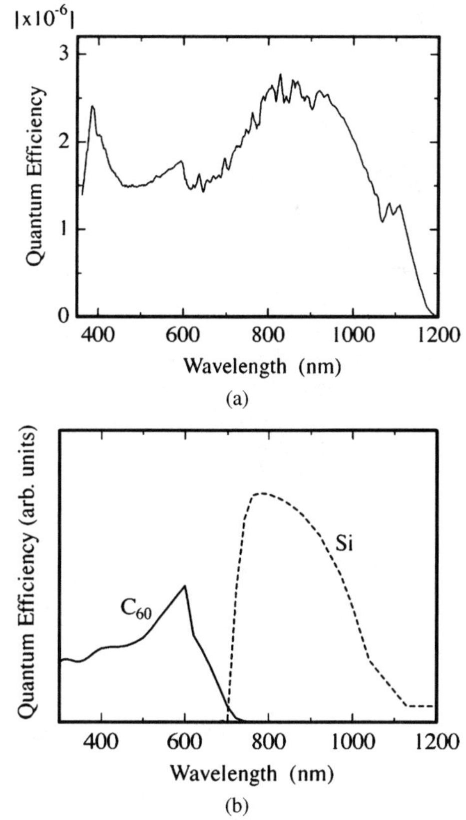

Figure 7. Spectral response of an Al/C_{60}/p–Si/Al solar cell: experimental (a) and calculated (b). Reprinted with permission from [50], N. Kojima et al., *Jpn. J. Appl. Phys.*, 39, 1176 (2000). © 2000, Institute of Pure and Applied Physics.

concentration of dopant. In [50], photovoltage generation by a C_{60}/Si heterojunction was studied with four different types of Si substrates: heavily doped *p*–Si (8×10^{18} cm^{-3} of B), lightly doped *p*–Si (1.5×10^{15} cm^{-3} of B), lightly doped *n*–Si (5×10^{14} cm^{-3} of P), and heavily doped *n*–Si (5×10^{18} cm^{-3} of P). The lightly doped *n*–Si/C_{60} junction was measured to have the highest V_{oc}, 0.4 V, while the lightly doped *p*–Si/C_{60} junction was measured to have the lowest, 0.12 V. This result implies that the Fermi level of C_{60} should be more than 0.4 eV below the Fermi level of lightly doped *n*–Si and more than 0.12 eV above that of lightly doped *p*–Si. The Fermi level of C_{60} was estimated to be about 4.7 eV below the vacuum level. In our opinion, further research is needed to understand the true electronic structure of such heterojunctions.

An increase in the photoactive area of a C_{60}/Si heterojunction may be one of the possible techniques for improvement of the photovoltaic parameters of such solar cells. In order to use this idea, Feng and Miller deposited a C_{60} monolayer on a nanostructured substrate of *p*-type porous silicon [65]. The authors reported an increase in J_{sc} (by 100 times) and V_{oc} in comparison with the corresponding values for a planar configuration of the C_{60}/p–Si heterojunction. Unfortunately, the absolute values of these parameters were not reported.

The very low intrinsic conductivity of C_{60} crystals is considered by most authors to be one of the main limiting factors for solar cell efficiency. Therefore, doping of C_{60} is another research direction for the production of high-efficiency fullerene-based solar cells.

4.2. Devices Based on the Doped C_{60} Layers

4.2.1. Ion Implantation into C_{60} Thin Films

Rectifying device structures based on doped C_{60} films were reported, for the first time, in [52] and [66], where an ion implantation of phosphorus (P) into C_{60} films was demonstrated to result in an increase in their *n*-type conductivity. The same group observed *p*-type conductivity in C_{60} films doped with an Al by its simultaneous sputtering during the film deposition [52]. Photovoltaic parameters of C_{60}/Si solar cells were found to be improved by doping. In our opinion, the authors' conclusion about such *p* doping of C_{60} is quite speculative because of the known fact of the very high electron affinity of the C_{60} molecule.

The idea of ion implantation of P and B in C_{60} films was then developed and applied to solar cell production by Yamaguchi and coauthors [67–71]. Implantation of B^+ (with an energy of 50–80 keV and a dose of 10^{14} cm^{-2}) into C_{60} films, grown on an *n*–Si substrate, was shown to lead to structural changes from the crystalline C_{60} to amorphous carbon, and a dramatic increase in film conductivity. Hall effect measurements of the implanted films indicate *p*-type conduction. Boron implantation significantly improved the parameters of the resultant *p*–C/*n*–Si solar cells [67]. In particular, the cell series resistance R_s was improved by the implantation from 35 kΩ to 370 Ω. J_{sc} for the implanted cells was found to be 0.33 mA/cm^2, $V_{oc} = 0.17$ V, $FF = 0.415$, and efficiency $\eta = 0.023\%$. A typical *I–V* curve for such cells is shown in Figure 8.

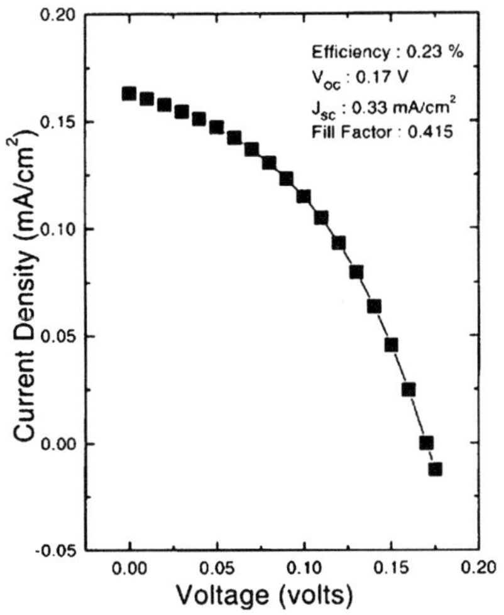

Figure 8. Light-illuminated (AM 1.5) *I–V* curve of a boron ion-implanted C_{60}(α–C)/*n*–Si heterojunction solar cell. Reprinted with permission from [67], K. L. Narayanan and M. Yamaguchi, *Appl. Phys. Lett.* 75, 2106 (1999). © 1999, American Institute of Physics.

The spectral response of these cells (Fig. 9) includes two broad features: 370–600 nm due to the α–C layer, and 700–1150 nm due to Si [67]. This result is in agreement with the experimental finding that the optical gap of the C_{60} films decreased after B^+ implantation [68]. Furthermore, it is possible to control this value by variation of the implantation dose. The optical gap was found to decrease gradually with the ion dose. For example, the gap was reported to be reduced continuously from 1.6 eV for nonimplanted C_{60} films to 0.8 eV after implantation with the dose of 8×10^{14} cm^{-2} [68]. The authors explained this effect by the structural transformation from C_{60} to α–C. We may attribute the intermediate optical gap values to a composite with various concentrations of C_{60} and α–C. In our opinion, this result is even more important than the improvement of the solar cell parameters (which are still not high). Indeed, using this effect, together with the control of the depth-implantation profile by the ion energy variation [69], it might be possible to produce a C_{60}/α–C composite with a nonhomogeneous depth profile optimized for maximum sunlight absorption and cell efficiency. We may predict that the next

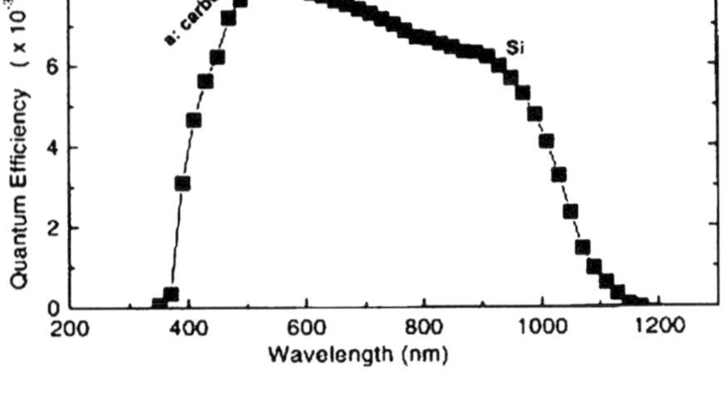

Figure 9. Spectral response of a boron ion-implanted C_{60}(α–C)/*n*–Si heterojunction solar cell. Reprinted with permission from [67], K. L. Narayanan and M. Yamaguchi, *Appl. Phys. Lett.* 75, 2106 (1999). © 1999, American Institute of Physics.

step in this direction will be the production of high efficient multijunction C_{60}/α–C/Si or α–C/C_{60}/Si (p–i–n) solar cells. The possibility of producing α–C/C_{60}/Si (p–i–n) solar cells already has been demonstrated [70, 71]. The efficiency of these devices was found to be 0.1% under AM 1.5 irradiation, which is five times higher than the best cells fabricated using the boron ion-implanted fullerenes without the insulating layer (for comparison, see Table 1). Furthermore, in another recent conference publication [72], the same group has suggested a novel approach for highly efficient quantum-well solar cells [73] with a C_{60}/α–C supperlattice structure produced by an intermittent supply of nitrogen ions during C_{60} deposition (Fig. 10).

4.2.2. Solar Cells with an Oxygenated Fullerene Thin Film

All of the research attempts described in the previous section were aimed at increasing the low intrinsic conductivity of C_{60} crystals by doping. If one uses a fullerene film as an insulator layer in the MIS solar cells, the opposite problem should be solved. Indeed, in this case, a decrease of the fullerene conductivity will increase the solar cell efficiency.

A variety of experiments [7, 74–77] has strongly suggested that the presence of oxygen reduces the dark conductivity and photoconductivity in C_{60} films and single crystals by several orders of magnitude. Yang and Mieno [78] have suggested using this effect in MIS solar cells with an Au/C_{60}/$C_{60}O_x$/Au structure in which C_{60} and $C_{60}O_x$ films act as a semiconductor and an insulator layer, respectively. Highly oxygenated fullerene films ($C_{60}O_x$, with $x = 8, 16, 24, 38, 50$) were obtained using RF O_2 plasma with a cooling system. The experimental results revealed that the highest photocurrent was obtained when the total thickness of the oxygenated fullerene film and the pure fullerene film was about 200 nm. Under this condition, the photocurrent was observed to become 30 times higher than that without the $C_{60}O_x$ layer. Unfortunately, the absolute values of the photovoltaic parameters of these MIS solar cell have not been reported.

5. ORGANIC FULLERENE-BASED PHOTOVOLTAIC DEVICES

Over the last decade, tremendous progress has been achieved in fullerene/conjugated polymer solar cells. A number of detailed reviews are devoted to this topic [20–22, 79, 80]. The operating principles of such solar cells are similar to those for the initial steps of natural photosynthesis, and are based on a very fast (*subpicosecond*) photoinduced electron transfer from a conjugated, semiconducting polymer (as a donor) to a nearby C_{60} molecule (as an acceptor).

5.1. Photoinduced Electron Transfer from Conjugated Polymers onto Fullerenes

The discovery of semiconducting, conjugated polymers and the ability to dope these polymers over the full range from insulator to metal has resulted in the creation of a new class of materials that combines the electronic and optical properties of semiconductors and metals with the attractive mechanical properties and processing advantages of polymers [81–83]. The molecular structures of some semiconducting conjugated polymers are summarized in Figure 11.

Conjugated polymers in their undoped states are electron donors upon photoexcitation (electrons being promoted to the antibonding π^* band). If one uses this property in conjunction with a molecular electron acceptor, long-lived charge separation may be achieved. Once the photoexcited electron is transferred to an acceptor unit, the resulting cation radical (positive polaron) species are known to be highly delocalized and stable. Analogous to the chemical doping process, the photoinduced electron transfer from the conjugated polymer to an acceptor moiety may be considered as "photodoping" [20].

On the other hand, as already mentioned (see Section 3), the C_{60} molecule is an excellent electron acceptor capable of taking on as many as six electrons [4, 84]. In the beginning of the 1990s, independently, Heeger's group [23, 85–86] and Yoshino's group [24, 87–90] reported the discovery of a photoinduced electron transfer from semiconducting conjugated polymers onto C_{60} molecules in solid films. The observed phenomenon is schematically illustrated by Figure 12. The kinetics of this process has recently been time resolved to occur within 40 fs [91]. Since the time scale of this process is more than 10^3 times faster than the radiative or nonradiative decay of photoexcitations, the quantum efficiency for charge transfer and charge separation is estimated to be close to unity [79]. All of these findings provided a new "molecular approach" to low-cost and high-efficiency photovoltaics.

5.2. Bilayered Donor–Acceptor Solar Cells

Immediately after the discovery of photoinduced electron transfer from a conjugated polymer to a C_{60} molecule, this molecular effect was used for the preparation of a heterojunction between poly[2–methoxy, 5–(2′–ethyl–hexyloxy)–1,4–phenylene–vinylene] (MEH–PPV) [Fig. 11(c)] and a C_{60} thin film [18, 92]. Figure 13 displays the dark I–V characteristics of such a device, consisting of successive layers of ITO/MEH–PPV/C_{60}/Au. The rectification ratio at ± 2 V was approximately 10^4. Analysis of this result suggests the formation of a heterojunction at the MEH–PPV/C_{60} interface, with rectifying properties analogous to those of a semiconductor p–n junction. The device demonstrated photovoltaic behavior. The values of J_{sc} were reasonably linear over five decades of light intensity up to approximately 1 W/cm^2, that is, one order of magnitude higher than the terrestrial solar intensity (AM 1.5). The following photovoltaic parameters (measured for a cell area of 0.1 cm^2 under illumination of an argon ion laser with a wavelength $\lambda = 514.5$ nm and a light power of 1 mW/cm^2) were reported: $J_{sc} = 2$ μA/cm^2, $V_{oc} = 0.5$ V, $FF = 0.48$, and $\eta = 0.04\%$.

Later, solar cells with a heterojunction between C_{60} thin films and other organic donor layers, both of various conjugated polymers (poly(paraphenylene) (PPP) [Fig. 11(a)] [93], poly(phenylenevinylene) (PPV) [Fig. 11(b)] [94–95], poly(2,5–dioctyloxy–p–phenylenevinylene) (OOPPV) [Fig. 11(d)] [96], poly(3–octylthiophene) (P3OT) [Fig. 11(e)] [97], poly(isothianaphthene) (PITN) [Fig. 11(f)] [98], poly(3–(4′–(1″,4″,7″–trioxaoctyl)penyl)thiophene) (PEOPT) [Fig. 11(g)] [99]) and small molecules (such as phthalocyanines

Table 1. Photovoltaic parameters of fullerene-based solar cells.

Solar cell design strategy	Cell structure	V_{oc} (mV)	J_{sc} (mA/cm^2)	*FF*	η (%)	Test conditions	Ref.
Pure C_{60}/inorganic semiconductor heterojunction	C_{60}/*p*–Si	306	1.1×10^{-4}	0.55	2.2×10^{-5}	AM 1.5, 100 mW/cm,^{2}	[50]
Ion-implanted C_{60}/inorganic semiconductor heterojunction	B ion-implanted C_{60}/*n*–Si	170	0.33	0.41	0.023	AM 1.5, 100 mW/cm^2	[67]
	α–C/C_{60}/Si *p–i–n* cell	90	4	0.29	0.1	AM 1.5, 100 mW/cm^2	[71]
Conjugated polymer/C_{60} heterojunction	ITO/MEH–PPV/C_{60}/Au	500	2×10^{-3}	0.48	0.04	Monochromatic irradiation, 1 mW/cm^2, $\lambda = 514.5$ nm	[18]
	SnO_2/PPP/C_{60}/Al	300	4×10^{-3}	—	—	Xe arc lamp, 15 mW/cm^2	[93]
	ITO/PPV/C_{60}/Al	800	3.5×10^{-3}	0.48	0.39	Monochromatic irradiation, 0.25 mW/cm^2, $\lambda = 490$ nm	[94]
Phthalocyanine/C_{60} heterojunction	ITO/OTiPc/C_{60}/Al	420	4.1×10^{-4}	0.2	0.35	Monochromatic irradiation, 10 μW/cm^2, $\lambda = 720$ nm	[100]
	ITO/ZnPc/C_{60}/Al	400	0.42	—	—	Xe arc lamp (with IR filter), 100 mW/cm^2	[101]
Donor–acceptor bulk heterojunction between a C_{60} derivative and a conjugated polymer	ITO/MEH–PPV: PCBM/Ca	800	0.5	—	—	Monochromatic irradiation, 20 mW/cm^2 $\lambda = 430$ nm	[19]
	ITO/MDMO–PPV:PCBM/Al	720	0.85	0.3	1.5	Monochromatic irradiation, 10 mW/cm^2, $\lambda = 488$ nm	[111]
	ITO/PEDOT/ MDMO–PPV: PCBM/LiF/Al	820	5.25	0.61	2.5	AM 1.5, 100 mW/cm^2, $T = 50$ °C	[25]
	ITO/PEDOT/ MDMO–PPV: PCBM/LiF/Al	870	4.9	0.60	2.55	STC (AM 1.5G, 100 mW/cm^2, $T = 25$ °C)	[119]
	ITO/PEDOT: PSS/PPV:PCBM/ LiF/Al(Ca)	810	4.9	0.60	3.0	AM 1.5, 80 mW/cm^2	[137]
Screen printing of bulk heterojunction	ITO/PEDOT/ MDMO–PPV: PCBM//Al	841	3.16	0.44	4.3	Monochromatic irradiation, 27 mW/cm^2, $\lambda = 488$ nm, $T = 25$ °C	[185]
Bulk heterojunction between a C_{60} derivative and a low-bandgap polymer	ITO/PEDOT:PSS/ P3HT:PCBM/Ca	580	8.7	0.55	2.8	AM 1.5, 100 mW/cm^2, $T = 25$ °C	[133]
	ITO/PEDOT: PSS/PTPTB: PCBM/LiF/Al(Ca)	720	2.95	0.37	0.1	AM 1.5, 80 mW/cm^2	[137]
	ITO/PEDOT:PSS/ PDTI:PCBM/Al	260	0.5	0.27	0.4	AM 1.5, 80 mW/cm^2	[138]
Bulk heterojunction between carbon nanotubes (CNT) and conjugated polymers	ITO/PPV: multi-walled CNT/Al	900	5.6×10^{-4}	0.23	0.081	Monochromatic irradiation, 37 μW/cm^2, $\lambda = 485$ nm	[169]
	ITO/P3OT: single-walled CNT/Al	750	0.12	0.4	0.04	AM 1.5, 100 mW/cm^2	[170]
"Double-cable" donor–acceptor solar cells	ITO/PEDOT:PSS/ methanofullerene covalently attached to hybrid of PPV and poly (*p*–phenylene ethynylene)/Al	830	0.42	0.29	0.1	AM1.5, 100 mW/cm^2	[153]
C_{60}/CuPc hetero-junction with an exciton-blocking layer	ITO/PEDOT:PSS/ CuPc/C_{60}/BCP/Al	580	18.8	0.52	3.6	AM1.5, 150 mW/cm^2	[26]
Inorganic semiconductor solar cells	Monocrystalline Si	706	42.2	0.83	24.7	STC (AM 1.5G, 100 mW/cm^2, $T = 25$ °C)	[184]
	Amorphous/ nanocrystalline Si	539	24.4	0.77	10.1	STC (AM 1.5G, 100 mW/cm^2, $T = 25$ °C)	[184]

Note: For comparison, the parameters of the best monocrystalline and amorphous/nanocrystalline silicon cells are listed.

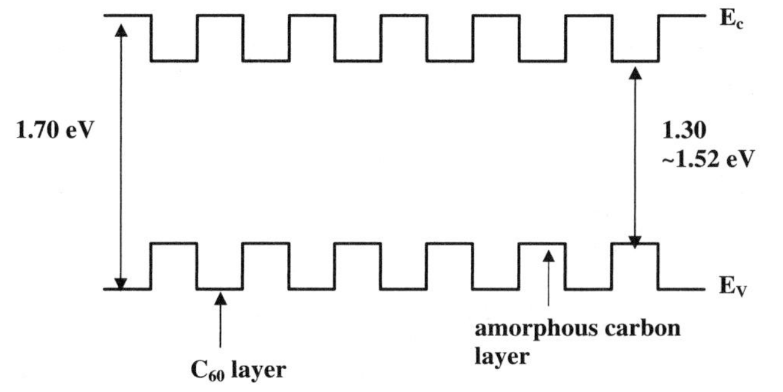

Figure 10. Band diagram of C_{60}/amorphous carbon supperlattice structure. Reprinted with permission from [72], N. Kojima et al., "Proceedings of the 28th IEEE Photovoltaic Specialists Conference," 2000, p. 873. © 2000, IEEE.

(Fig. 14) [100–101]), were demonstrated as well. The photovoltaic parameters of some such devices are summarized in Table 1.

The main factor limiting photovoltaic parameters of these bilayered donor–acceptor solar cells lies in the basic proper-

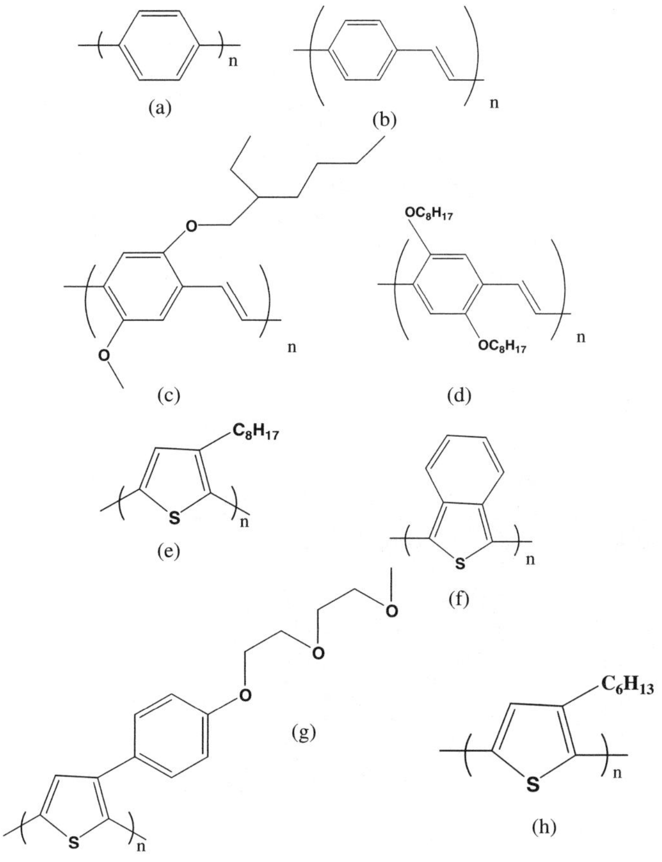

Figure 11. Molecular structures of some semiconducting conjugated polymers used in fullerene-based solar cells. a: poly(paraphenylene) (PPP); b: poly(phenylenevinylene) (PPV); c: poly[2–methoxy, 5–(2′–ethyl–hexyloxy)–1,4–phenylene–vinylene] (MEH–PPV); d: poly(2,5–dioctyloxy–*p*–phenylenevinylene) (OOPPV); e: poly(3–octylthiophene) (P3OT); f: poly(isothianaphthene) (PITN); g: poly[3–(4′–(1″,4″,7″–trioxaoctyl)penyl)thiophene] (PEOPT); h: regioregular poly(3–hexylthiophene) (P3HT).

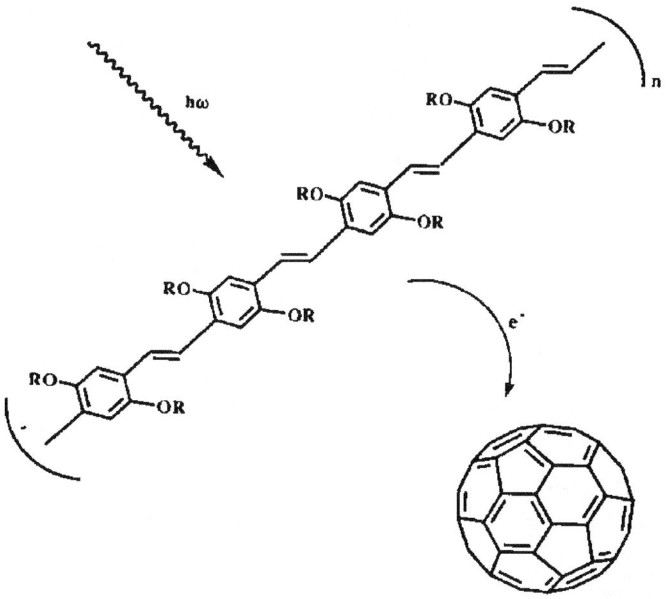

Figure 12. Schematic illustration of photoinduced electron transfer from semiconducting conjugated polymers onto C_{60}. Reprinted with permission from [79], N. S. Sariciftci, *Prog. Quantum Electron.* 19, 131 (1995). © 1995, Elsevier Science.

ties of photogenerated excitations in molecular solids (such as fullerites and polymers), and in the very operation principle of such solar cells. Indeed, intermolecular van der Waals forces in molecular solids are weak compared to bonds in inorganic crystals, and much weaker than the intramolecular bonds. As a consequence, photogenerated excitations ("excitons") are strongly bound, and do not spontaneously dissociate into separate charges. (Dissociation requires an input of energy of ~100 meV compared to a few millielectronvolts for an inorganic semiconductor [102].) This means that carrier generation does not necessarily result from the absorption of light. In other words, contrary to conventional solar cells with an inorganic semiconductor *p–n* junction, as described in Section 2, in a bilayered donor–acceptor molecular solar cell, neutral excitons are first created by light.

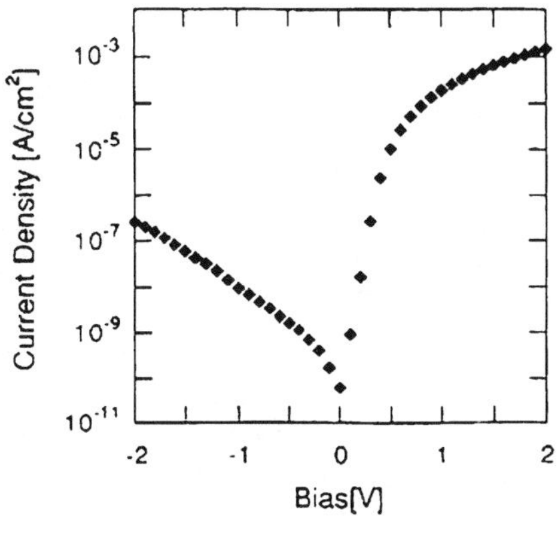

Figure 13. Dark *I–V* characteristics of an ITO/MEH–PPV/C_{60}/Au solar cell. Reprinted with permission from [18], N. S. Sariciftci et al., *Appl. Phys. Lett.* 62, 585 (1993). © 1993, American Institute of Physics.

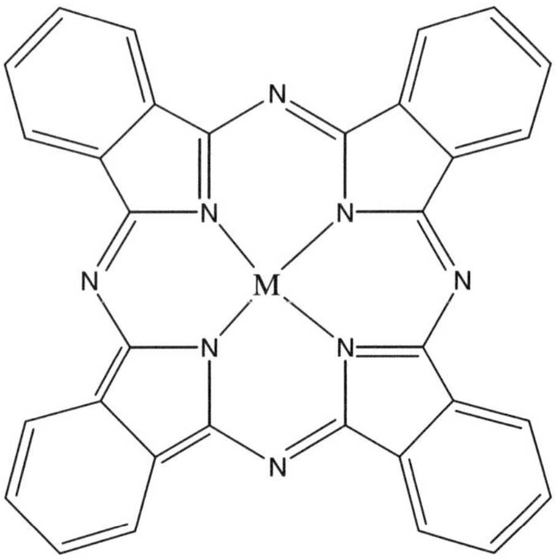

Figure 14. Molecular structures of phthalocyanines (MPc). Fullerene-based solar cells have been demonstrated with MPc, where M = Cu [26], TiO [100], Zn [101, 178].

Finally, charges are separated in the narrow interface region due to donor–acceptor electron transfer interactions.

The above statement is well illustrated by the spectral response of an ITO/MEH–PPV/C_{60}/Au solar cell with illumination through the ITO/MEH–PPV side of the device [18] (Fig. 15). The onset of photocurrent at a photon energy of 1.7 eV follows the absorption of MEH–PPV, which initiates the photoinduced electron transfer. However, the minimum in the photocurrent at a photon energy of 2.5 eV corresponds to the region of maximum absorption in the polymer layer. The MEH–PPV layer, therefore, acts as a filter which reduces the number of photons reaching the MEH–PPV/C_{60} interface. The photoactive region is restricted to a very thin layer adjacent to the heterojunction interface. The thickness of this layer is limited by low values of the exciton diffusion length. In [20], this length in MEH–PPV was estimated to be only a few angstroms. Recent direct measurements of the exciton diffusion length in bilayered solar cells provided values of 4–6 nm for the PPV layer in C_{60}/PPV cells [94–95], as well as 4.7 and 7.7 nm for the PEOPT and C_{60} layers, respectively, in C_{60}/PEOPT cells [99]. These experimental values support the conclusion that, in bilayered molecular solar cells, the only photoexcitations generated in a very thin layer at the heterojunction interface may contribute to the cell photocurrent. This results in a relatively low collection efficiency of such cells.

Significant improvement of the collection efficiency has been achieved using composite material with a network of internal heterojunctions between a fullerene and conjugated polymer, forming so-called *bulk heterojunctions*.

5.3. Fullerene/Polymer Solar Cells with a Network of Internal Heterojunctions

A revolutionary development in organic photovoltaics came in the mid-1990s with the introduction of an *interpenetration network of internal heterojunctions* or the so-called *bulk heterojunction*, where an electron-accepting (fullerenes or fullerene derivatives) and an electron-donating material (conjugated polymers) are blended together, forming a composite. Through control of the morphology of the phase separation into an interpenetration network, one can achieve a high interfacial area within the bulk material. If any point in the bicontinuous network is within an exciton diffusion length from a donor–acceptor interface, then wherever an exciton is photogenerated in either material, it is likely to diffuse to the interface and break up. Then the separated charge carriers may travel to the contacts, and deliver current to an external circuit (Fig. 16).

Using this principle of the material construction, a solar cell based on an MEH–PPV/fullerene derivative composite (without any macrojunction!) was demonstrated [19]. The authors believed that a built-in electric field was created by asymmetric metallic contacts (with different work functions), one of which was transparent ITO. The idea of such internal electric field generation in organic devices was first described by Parker [103], and then developed for organic light-emission diodes (LEDs) [103], photodiodes, and solar cells [104–107].

In [19], uniform films of such soluble fullerene derivative as [6,6]–phenyl C_{61}–butyric acid methyl ester (PCBM) [108] and MEH–PPV were cast from 1,2–dichlorobenzene solutions containing 1:4 weight ratio MEH–PPV:PCBM. The enhanced solubility of PCBM compared to pure C_{60} allows such a high fullerene/conjugated polymer ratio, and strongly

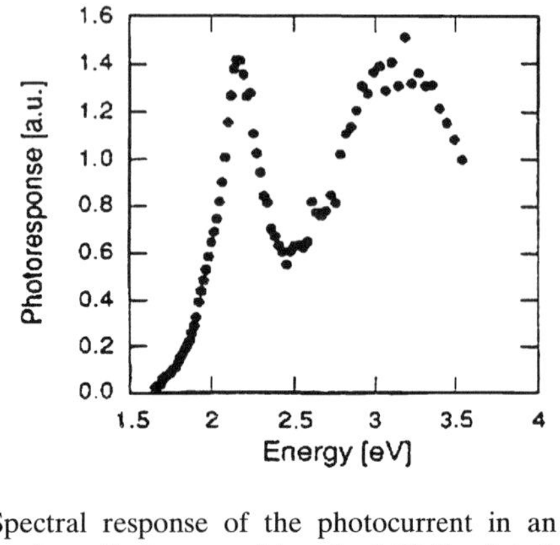

Figure 15. Spectral response of the photocurrent in an ITO/MEH–PPV/C_{60}/Au solar cell at reverse bias of −1 V. Reprinted with permission from [18], N. S. Sariciftci et al., *Appl. Phys. Lett.* 62, 585 (1993). © 1993, American Institute of Physics.

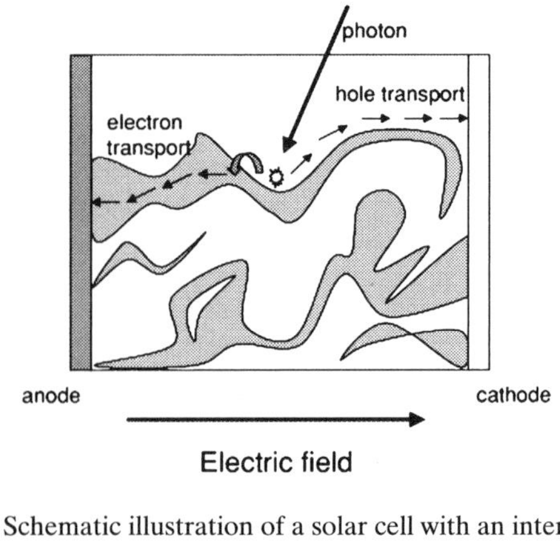

Figure 16. Schematic illustration of a solar cell with an interpenetration network of internal heterojunctions (bulk heterojunction). Reprinted with permission from [102], J. Nelson, *Mater. Today* 20 (2002). © 1995, Elsevier Science.

supports the formation of bulk donor–acceptor heterojunctions. Using this film as a photoactive layer as well as Ca and transparent ITO films as asymmetric electrodes, bulk heterojunction solar cells, with $J_{sc} = 0.5$ mA/cm^2 under 20 mW/cm^2 monochromatic illumination, were demonstrated. This J_{sc} value corresponds to a collection efficiency of 7.4% electrons per incident photon, which is about 2 orders of magnitude higher than that of the bilayered MEH–PPV/C_{60} solar cells described in the previous section.

Starting with [19], substantial progress has been made toward efficient solar cells based on fullerene/conjugated polymer blends [25, 109–120]. In particular, a significant breakthrough has been achieved by realizing that the morphology of the photoactive composite layer plays an important role in charge carrier mobility [25, 121] and solar cell performance [25, 112, 114, 118].

To the best of our knowledge, the highest efficiency of the fullerene/polymer bulk heterojunction solar cells under AM 1.5 irradiation has been reported for the following device configuration [25]: poly [2–methoxy, 5–(3′,7′–dimethyl–octyloxy)]–*p*–phenylene–vinelyne (MDMO–PPV) [125] was used as the electron donor in these cells, while the electron acceptor was (PCBM). The thickness of the spin-cast MDMO–PPV:PCBM active layer with the improved morphology was about 100 nm. As electrodes, a transparent ITO film on one side and an LiF/Al bilayer contact on the other side were used. An LiF/Al electrode was chosen instead of pristine Al in order to guarantee a good ohmic contact between the metal and the organic layer [123–124]. (Recently, it was shown that insertion of a thin LiF interlayer into such cells significantly enhances electron injection [118] due to the fact that such an interlayer between Al and PCBM prevents a chemical reaction of Al with the carboxylic oxygen of PCBM [125].) For improvement of the ITO contact, the ITO was coated with a thin layer of poly(ethylene dioxythiphene) (PEDOT) [113]. The device structure of ITO/PEDOT/MDMO–PPV:PCBM/LiF/Al layered solar cells, together with the molecular structure of the MDMO–PPV and PCBM compounds used for the active layer, are shown in Figure 17. Figure 18(a) displays the molecular structure of PEDOT. The reported photovoltaic parameters of such solar cells under AM 1.5 irradiation, at 50 °C, were $J_{sc} = 5.25$ mA/cm^2, $V_{oc} = 0.82$ V, $FF = 0.61$, and $\eta = 2.5\%$. These high parameters have been approved by accurate indoor [119] and outdoor [120] photovoltaic characterization of such solar cells, the results of which had been appropriately adjusted to standard test conditions (STCs). Such accurate characterization is necessary in order to quantify cell performance in a manner that may be compared from one laboratory to another. In the case of conventional inorganic solar cells, a set of STCs has been defined. These correspond to a radiant intensity of 1000 W/m^2, with a spectral distribution defined as “AM1.5G” (IEC 904-3) and a cell temperature of 25 °C. In spite of the existence of such a standard, all kinds of efficiencies have been reported for organic [126] and, in particular, fullerene-containing solar cells, based on measurements performed under a wide variety of test conditions (Table 1).

The *I–V* characteristics of an ITO/PEDOT/MDMO–PPV:PCBM/LiF/Al solar cell measured under appropriately corrected simulated STC conditions are shown in Figure 19

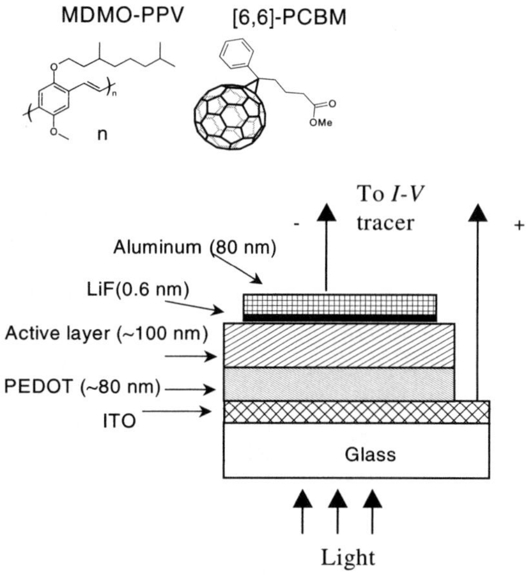

Figure 17. Schematic device structure of an ITO/PEDOT/MDMO–PPV:PCBM/LiF/Al solar cell, together with chemical structure of PCBM and MDMO–PPV compounds used for the cell active layer. Reprinted with permission from [120], E. A. Katz et al., *J. Appl. Phys.* 90, 5343 (2001). © 2001, American Institute of Physics.

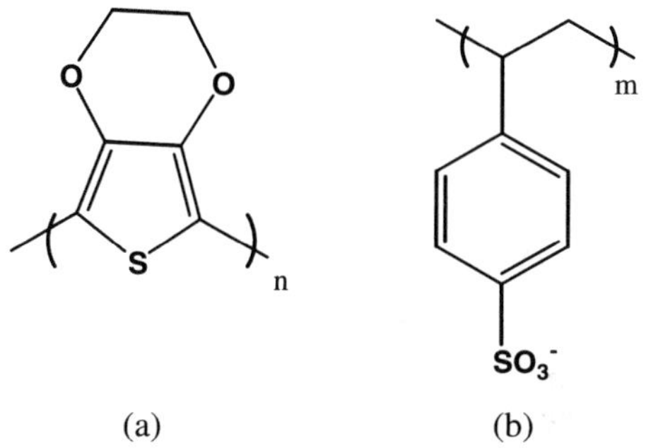

Figure 18. Molecular structures of poly(ethylene dioxythiphene) (PEDOT) (a) and poly(styrenesulfonate) anion (PSS$^-$) (b).

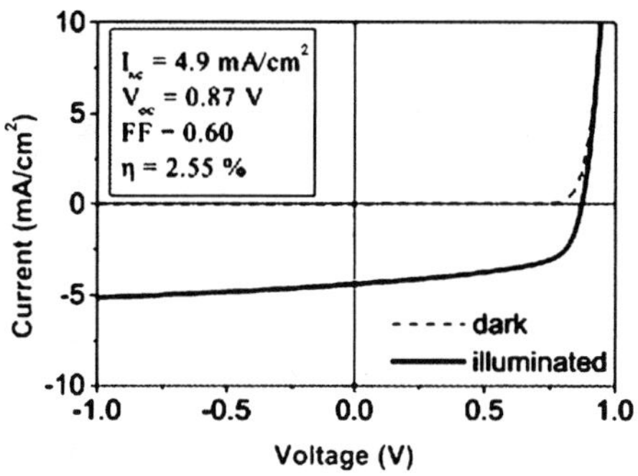

Figure 19. *I–V* characteristics of a 0.1 cm^2 ITO/PEDOT/MDMO–PPV:PCBM/LiF/Al solar cell measured under appropriately corrected simulated STC conditions. Reprinted with permission from [119], J. M. Kroon et al., *Thin Solid Films* 403–404, 223 (2002). © 2002, Elsevier Science.

[119]. The correction [119–120] is based on the determination of the simulator spectral mismatch factor using spectral responses of the cell under test and of the reference cell (Fig. 20), as well as the simulator spectrum and the AM1.5G standard solar spectrum.

The authors of [127] suggested another approach for improvement of the morphology of the solar cell active layer, and the following enhancement of the charge transfer and improvement of the charge transport. In this study, a bilayer system, consisting of spin-cast MEH–PPV and sublimed C_{60} layers, was heated above the MEH–PPV glass transition temperature in order to enhance the diffusion of the fullerene into the polymer, resulting in a concentration gradient structure. With this process, a controlled, *bulk gradient heterojunction* was created. Because the fullerene acceptor was thus distributed throughout the film, exciton recombination was dramatically reduced. This was observed as a decrease in the photoluminescence by an order of magnitude and an increase in the photocurrent by an order of magnitude throughout much of the visible spectrum. Unfortunately, the photovoltaic parameters of such solar cells measured under the proper conditions were not reported.

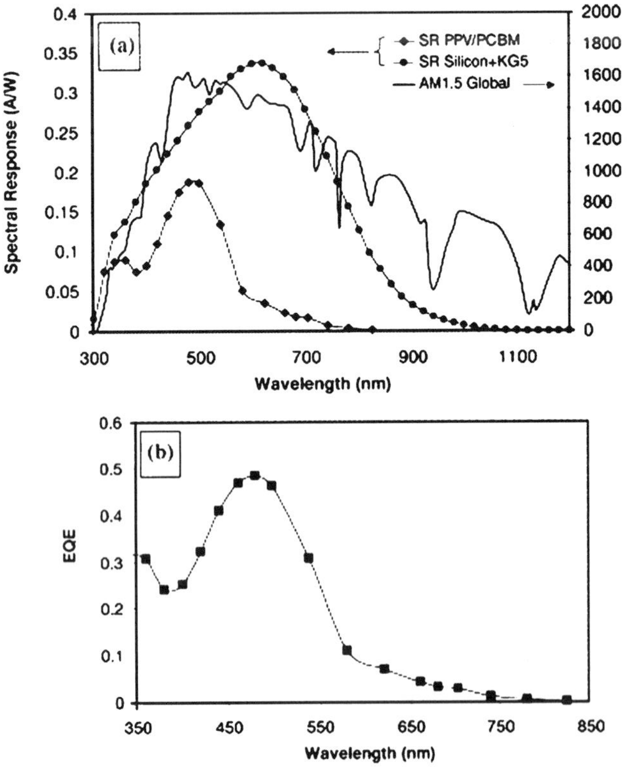

Figure 20. (a) Spectral response (SR) of an ITO/PEDOT/MDMO–PPV:PCBM/LiF/Al solar cell and a filtered (KG5) silicon reference cell, as well as the AM1.5G solar spectrum. (b) External quantum efficiency (EQE) of an ITO/PEDOT/MDMO–PPV:PCBM/LiF/Al solar cell as calculated from the SR. Reprinted with permission from [119], J. M. Kroon et al., *Thin Solid Films* 403–404, 223 (2002). © 2002, Elsevier Science.

5.4. Irradiance and Temperature Dependences of Photovoltaic Parameters of Fullerene/Polymer Solar Cells

To optimize cell efficiency, the electronic processes limiting device performance should be identified. Studies of I–V characteristics of fullerene/conjugated polymer solar cells as a function of temperature and illumination intensity may provide information on photovoltage generation and injection mechanisms, charge transport properties, and the presence of defects (both in the bulk and at the interface).

The irradiance dependence of the photovoltaic parameters of MDMO–PPV:PCBM bulk heterojunction solar cells was studied in the cell temperature and irradiance ranges $T = 283$–306 K and $P_{\text{light}} = 80$–550 W/m^2, respectively [120, 128]. The almost linear irradiance dependence of I_{sc} at all studied temperatures was demonstrated. Very recently [117], it was confirmed that the photocurrent behavior is close to linear at $T = 300$ K, that is, $I_{sc} \propto I_{sc}^{1.01}$, but deviates at lower temperatures ($T \sim 100$ K) and irradiance levels ($P_{\text{light}} \sim 10$ W/m^2). The observed near-linear dependence is a key condition for efficient photoconversion. In simple terms, it indicates that a recombination of the photogenerated charge carriers is sufficiently slow compared to charge transport for all of the photogenerated carriers to be collected at short circuit [129].

Figure 21 summarizes the temperature dependences of the principal photovoltaic parameters of an MDMO–PPV:PCBM bulk heterojunction solar cell, under real sun irradiation in the temperature range 30–60 °C [120].

V_{oc} was found to decrease linearly with increasing temperature [Fig. 21(b)]. For all studied samples, the observed linear decrease had a temperature coefficient in the range $dV_{oc}/dT = -1.40$–1.65 mV/K and extrapolated to $T = 0$ K values of 1.33–1.40 V. It should be noted, however, that recent low-temperature measurements on the current–voltage behavior of conjugated polymer/fullerene bulk heterojunction solar cells in the range 80–300 K showed that this linear temperature dependence of V_{oc} is lost at temperatures below 200 K, and that V_{oc} begins to saturate [117].

In order to try to understand the physical mechanisms which may be responsible for the observed temperature dependence of V_{oc} in the high- and low-temperature ranges, it is instructive to start with an analysis of the V_{oc} behavior of conventional inorganic semiconductor solar cells with a p–n junction (see Section 2). For that situation,

$$V_{oc} = \frac{A\kappa T}{q} \ln\left(\frac{I_{sc}}{I_0} + 1\right) \tag{7}$$

where A is a so-called diode quality factor of the p–n junction and I_0 is the reverse saturation current. According to a geometrically simple model of Shockley [6], I_0 is given by

$$I_0 = qN_vN_c[\exp(-E_g/\kappa T)] \cdot \left(\frac{L_n}{n_n\tau_n} + \frac{L_p}{p_p\tau_p}\right) \tag{8}$$

where N_v and N_c are the effective densities of states in the valence and conduction band, respectively, E_g is the

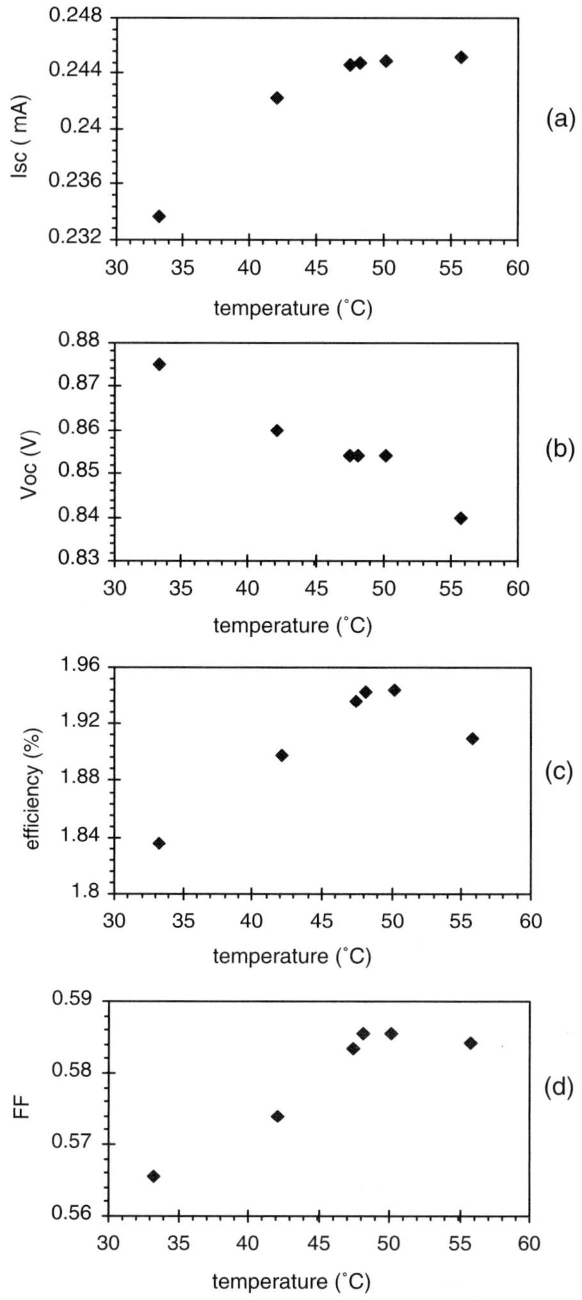

Figure 21. Temperature dependences of the principal photovoltaic parameters for a typical ITO/PEDOT/MDMO–PPV:PCBM/LiF/Al cell derived from outdoor measurements of its I–V curves. Plotted values of efficiency and I_{sc} have been adjusted to the STC irradiance level of 1000 W/m^2. Reprinted with permission from [120], E. A. Katz et al., *J. Appl. Phys.* 90, 5343 (2001). © 2001, American Institute of Physics.

semiconductor bandgap, and L_n, L_p, n_n, p_p, τ_n, τ_p are the diffusion lengths, the carrier densities, and the lifetimes of electrons and holes, respectively. Since $I_{sc} \gg I_0$, by inserting (5) into (4), one obtains

$$V_{oc} = \frac{AE_g}{q} - \frac{A\kappa T}{q} \cdot \ln\left[\frac{1}{I_{sc}} \cdot qN_vN_c\right]\left(\frac{L_n}{n_n\tau_n} + \frac{L_p}{p_p\tau_p}\right)$$
$$= a - bT \qquad (9)$$

where

$$a = V_{oc}(0K) = \frac{AE_g}{q}$$

and

$$b = -dV_{oc}/dT = \frac{A\kappa}{q}\ln\left[\frac{1}{I_{sc}} \cdot qN_vN_c\right]\left(\frac{L_n}{n_n\tau_n} + \frac{L_p}{p_p\tau_p}\right).$$

To the extent to which such a simple model might also be relevant to the organic solar cell situation, that is, in that a linear decrease of V_{oc} with T was also observed for the fullerene/polymer solar cells [117, 120, 128], at least for the high-temperature range ($T > 200$ K), V_{oc} for these solar cells may be described by a diode equation similar to (4) with $I_0 \sim \exp(-E_{DA}/kT)$. Here, E_{DA} is a parameter analogous to E_g for a conventional semiconductor. If we adopt such a model for these fullerene–polymer bulk heterojunction solar cells, E_{DA} corresponds to the energy difference between the lowest unoccupied molecular orbital (LUMO) of the acceptor and the highest occupied molecular orbital (HOMO) of the donor components of the active layer, as will be argued later.

The observed room-temperature values of $V_{oc} \sim 0.8$ V [25, 111, 117, 119–120] are considerably higher than the V_{oc} value of 0.53 V previously reported for bilayer conjugated polymer/fullerene solar cells under intense illumination [18]. They are also two times higher than the work function difference of the two electrodes (Al:4.3 eV, ITO:4.7 eV), which would give a value of 0.4 V. This result strongly supports the conclusion of [109] that photovoltage generation in bulk donor–acceptor heterojunctions cannot be explained by a model of the work function difference of the two electrodes [103], which is generally accepted for single-layer conjugated polymers devices [104].

On the other hand, it has recently been shown that the open-circuit voltage in solar cells based on interpenetrating networks of conjugated polymers with fullerenes is directly related to the acceptor strength of the fullerenes [130–131]. This result fully supports the view that the open-circuit voltage of this type of donor–acceptor bulk-heterojunction cell is related directly to the energy difference between the HOMO level of the donor and the LUMO level of the acceptor components of the active layer. Furthermore, and also in full agreement with this view, it was found that a variation of the negative electrode work function influences the open-circuit voltage only in a minor way. This view on the V_{oc} generation is additionally supported by the fact that the values of the temperature coefficient $dV_{oc}/dT = -1.40$–1.65 mV/K for the cells with bilayered LiF/Al and ITO/PEDOT contacts [120] coincide with those reported for polymer/fullerene bulk heterojunction solar cells of the "previous generation" (i.e., with the same components in the active layer, but without LiF and PEDOT contact layers) [128].

In this picture, the temperature dependence of V_{oc} is directly correlated to the temperature dependence of the quasi-Fermi levels of the components of the active layer under illumination, that is, of the polymer and the fullerene.

Figure 21(a) and (d) shows a relatively large monotonic increase with temperature for I_{sc} and FF, followed by a saturation region. A slight increase in I_{sc} with temperature is also a common feature for inorganic solar cells [132]. However, in the case of fullerene–polymer solar cells, the rate of increase was so dramatic that the increase of the I_{sc} and *FF* product, with temperature, overtook the decrease of V_{oc}. As a result, there was an absolute increase of the power efficiency η with temperature, reaching a maximum value at temperature T_{max} which, for different samples, lay in the range 47–60 °C [Fig. 21(c)]. It should be noted that similar positive temperature coefficients for I_{sc} also have since been observed for fullerene/polymer bulk heterojunction solar cells with other active layers, for example, in PCBM:poly(3–hexylthiophene) (P3HT) solar cells [133]. [The molecular structure of P3HT is shown in Fig. 11(h).] PCBM:polyfluorene solar cells showed even higher values of the I_{sc} temperature coefficient. For such devices, j_{sc} was revealed to increase linearly from 0.8 to 2.2 mA/cm^2 (i.e., by 175%!) as the temperature increased from 5 to 100 °C [134].

A positive temperature dependence of η is a remarkable peculiarity for fullerene–polymer solar cells, which is not observed in most conventional inorganic solar cells [132]. It is significant, therefore, that the heating of solar cells to the reported T_{max} temperatures may be expected to arise naturally by the absorption of solar radiation, that is, without any artificial heating. Therefore, unlike conventional solar cells, we would expect fullerene–polymer solar cells to perform more efficiently under natural (warm climate) operating conditions than under standard test conditions!

Possible reasons for the observed, unusually large, positive temperature coefficient for I_{sc} are still under discussion. It has been attributed to the thermally activated mobility of the charge carriers in the cell's active layer [120], the difference in the temperature dependences of the mobility of electrons and holes within the fullerene and polymer transport paths [117], as well as space-charge effects [129].

The observed temperature dependence of *FF* [Fig. 21(d)] was quite similar to that of I_{sc}. The former, however, can be qualitatively understood in terms of the temperature-dependent series resistance of the solar cell R_s. Fullerene/polymer solar cells have a relatively high resistivity of the organic active layer, but one which *decreases* with increasing temperature.

5.5. Efficiency Optimization of Fullerene/Polymer Solar Cells

5.5.1. Improvement in Short-Circuit Current and Fill Factor

While fullerene/polymer solar cells show quite respectable V_{oc}, the values of I_{sc} and *FF* are much lower than those available from inorganic solar cells. The lower photocurrent is caused by weak light absorption (the reported cells absorb only ~60% of the incident photons [18]) and charge transport, while the low *FF* values are due to poor charge transport and the recombination in these solar cells. Therefore, research activity aimed at I_{sc} improvement should be focused mainly on the two following goals: improving light harvesting, and improving charge transport.

Improving Light Harvesting The optical bandgap of the poly(*p*–phenylenevinelene)s (PPVs) and polythiophenes that are currently used in the construction of bulk heterojunction solar cells lies in the range $E_g \sim 2.0$–2.2 eV, and is not optimized with respect to the solar spectrum (see, for example, Fig. 20). One promising strategy is to replace donor moieties in devices with conducting polymers of lower bandgap ($E_g < 1.8$ eV), which may increase the light absorption in the visible and near-infrared regions of the solar spectrum. In designing new lower bandgap polymers for bulk heterojunction solar cells, various aspects need to be considered. For a conventional homo-*p*–*n* junction solar cell, a reduction of E_g results in an increased light absorption and I_{sc} values, but at the same time lowers V_{oc}. As a consequence, the optimum bandgap of such devices is at ~1.5–1.6 eV [13]. However, as mentioned in the previous section, V_{oc} of fullerene/polymer bulk heterojunction solar cells is likely related to the energy difference between the LUMO level of the acceptor and the HOMO level of the donor components of the active layer. A reduction of the bandgap of the polymer by increasing its HOMO level (i.e., decreasing the polymer oxidation potential) may therefore reduce V_{oc}. On the other hand, a lowering of the bandgap by bringing down the polymer LUMO level (i.e., reducing the reduction potential) will reduce the driving force for the polymer to transfer an electron in the excited state to the fullerene. To counteract such a decreased driving force, a lowering of the bandgap should be accompanied by a concomitant decrease in the oxidation potential of the polymer [135].

The synthetic principles for lowering the bandgap of linear π-conjugated polymers have been reviewed by Roncali [136]. Fullerene/polymer bulk heterojunction solar cells, using novel polypyrolle/thiadiazole (PTPTB) copolymers [Fig. 22(a)] with $E_g \sim 1.5$–1.6 eV [135, 137] and thiophene/isothianaphthene (PDTI) copolymers [Fig. 22(b), (c)] with $E_g \sim 1.2$–1.8 eV [138–139], as a low bandgap donor moiety, have been reported recently. The photovoltaic parameters of such devices are listed in Table 1.

Other strategies for a broad spectral sensitization of such solar cells include introducing a conjugated crystalline dye, such as a perylene derivative [Fig. 22(d)], to a fullerene/polymer blend in the cell active layer [140] or using a coevaporated layer of C_{60} and a small organic molecule, such as zinc phthalocyanine (ZnPc) (Fig. 14), on top of the PCBM:MDMO–PPV blend [141].

In all of these studies [135, 137–141], serious improvements in photoconversion in the visible and near-infrared regions of the solar spectrum have been demonstrated. Although the energy conversion efficiency of these devices is still relatively small (<2.5%), further improvement can be expected. This obviously involves optimization of the synthesis, device architecture, and processing conditions. Multilayered structures of the solar cell active layer combining various bandgap materials, or solar cells with a smooth bandgap gradient in the active layer should be investigated as well. Some solar cells based on new ideas on device architecture will be reviewed in Section 5.5.3.

A different approach is to increase the number of absorbed photons by *light-trapping* structures. This number is limited by a strong requirement that the active layer in

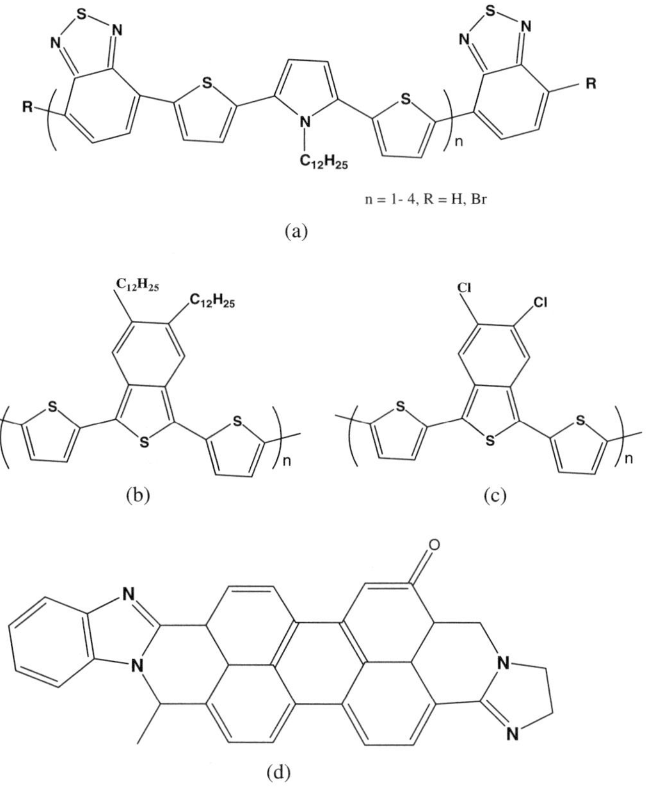

Figure 22. Molecular structures of low-bandgap donor moieties used in fullerene-based solar cells. a: polypyrolle/thiadiazole (PTPTB) copolymer; b, c: thiophene/isothianaphthene (PDTI) copolymers; d: perylene tetracarboxylic derivative.

fullerene/polymer bulk heterojunction solar cells should be rather thin (≤100 nm). By making the active layer thicker, to collect more of the light by absorption, one also reduces the built-in electric field [142] and increases the solar cell series resistance [143], thus reducing the collection efficiency and forcing a compromise on the photocurrent. It is, therefore, desirable to make very thin devices, but to find ways of enhancing the photon path in their active layer. The possibilities for reconciling these requirements by light trapping in the active layer of fullerene/polymer solar cells were discussed by Inganäs et al. [144]. Some of them have been investigated by coupling the incoming light into waveguided modes within the embossed organic thin layers [145]. An elastomeric mold has been used to transfer a submicrometer grating pattern from a commercially available grating template to the active layer of fullerene/polymer bulk heterojunction solar cells. Using spectral response measurements of the grating and reference (flat) cells, with a 90 nm thick active layer, an enhancement in external quantum efficiency all over the spectrum has been demonstrated for the grating cell. At maximum peak position, an enhancement in external quantum efficiency of about 26% has been recorded.

An additional enhancement in light harvesting can be achieved by decreasing the fraction of the incident light reflected by the front surface of a solar cell. Usually, for this purpose, special antireflection coating layers are evaporated on the top of solar cells. It should be noted, however, that all of the solar cells described in the present review had no such layer. The possibility of the discussed improvement has been demonstrated recently just by sandblasting the glass substrate of a fullerene/polymer bulk heterojunction solar cell [146]. Reduced reflection of the incoming light and a corresponding increase in I_{sc} were observed.

Improving Charge Transport Charge transport in fullerene/polymer solar cells is limited mainly by the low intrinsic mobility of charge carriers in molecular solids, and by the charge trapping effects of impurities and defects (in the bulk and at the interfaces).

Recently, several research groups have reported a very strong influence of fullerene concentration in the cell active layer on the carrier mobility, which can be tuned in order to obtain a bipolar transport mechanism [134, 147–149]. The most surprising result of these experiments is a still unexplained fact that an increase in fullerene concentration resulted not only in increased electron mobility, but also in an enhanced hole mobility [134, 149].

As already noted in Section 5.3, in fullerene/polymer bulk heterojunction solar cells, both photocurrent generation and charge transport (in particular, charge carrier mobility) depend upon the morphology of the blended active layer. In the case of the MDMO–PPV/PCBM active layer, an improved morphology and the highest photovoltaic parameters of the cells have been achieved by using chlorobenzene as a solvent [25]. However, a direct transfer of this finding to other systems is not straightforward; therefore, a morphology optimization, in which the choice of solvent is only one of the parameters, should be performed for any combination of materials.

A promising strategy for the vital improvement of morphology and charge transport is the preparation of so-called "double-cable" polymers, that is, supramolecules, in which the polymeric electron donors and the fullerene electron acceptors are covalently linked. The recent developments of the design, characterization, and photovoltaic application of this novel class of fullerene functional materials are reviewed by Cravino and Sariciftci [150]. In this supramolecular photovoltaic approach, the desired nanoscopic bicontinuous network has been achieved by synthesizing novel π-conjugated polymers with pendant fullerenes which prevents any problem arising from interface defects, as observed for fullerene/polymer blends. The molecular structures of some "double-cable" polymers [151–153] are shown in Figure 23. It is worth noting that the performance of the first solar cells based on "double-cable" polymers [153] [Fig. 23(c)], which were not optimized, is already competitive with that of devices prepared according to the p/n heterojunction and "bulk heterojunction" approaches [109, 154]. Photovoltaic measurements of the first "double-cable" solar cells under white-light illumination (100 mW/cm^2) revealed a J_{sc} value of 0.42 mA/cm^2, V_{oc} of 830 mV, and FF of 0.29 [153]. This new synthetic approach also offers great versatility for design tuning the photovoltaic system [153–162]. Figure 24 schematically illustrates one example of such a supramolecular solar cell based on a C_{60}–3PV hybrid compound, in which hole-conducting oligophenylenevinelene (OPV) moiety was chemically linked to an electron-conducting fullerene subunit [155–156]. In such a device, the compound is not only able to generate

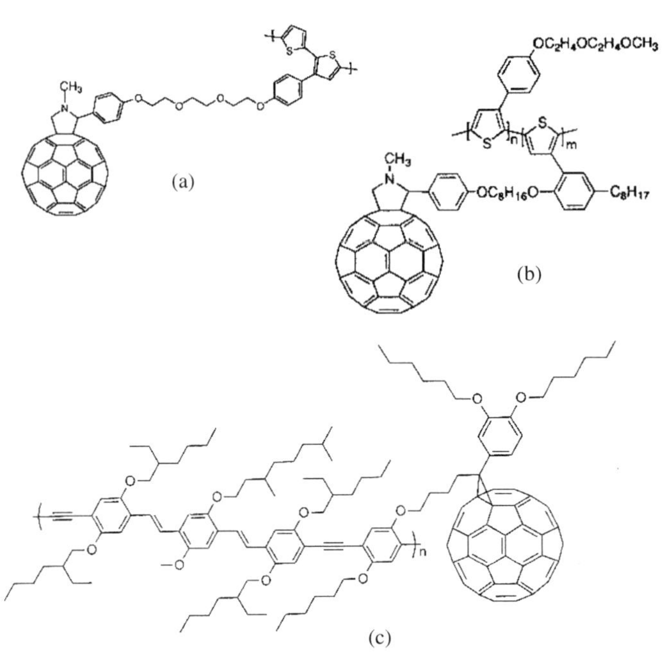

Figure 23. Molecular structures of some "double-cable" polymers. a: Poly(bithiophene–fulleropyrrolidine). Reprinted with permission from [151], H. Neugebauer et al., *Electronic Properties of Molecular Nanostructures*, AIP Conference Proceedings, p. 591, 2001. © 2001, American Institute of Physics. b: Covalent attachment of fullerene derivative to substituted polythiophene (PEOPT). Reprinted with permission from [152], F. Zhang et al., *Adv. Mater.* 13, 1871 (2001). © 2001, Wiley-VCH. c: Covalent attachment of methanofullerene moiety to hybrid of poly(*p*–phenylene vinylene) and poly(*p*–phenylene ethynylene). Reprinted with permission from [153], A. M. Ramos et al., *J. Am. Chem. Soc.* 123, 6714 (2001). © 2001, American Chemical Society.

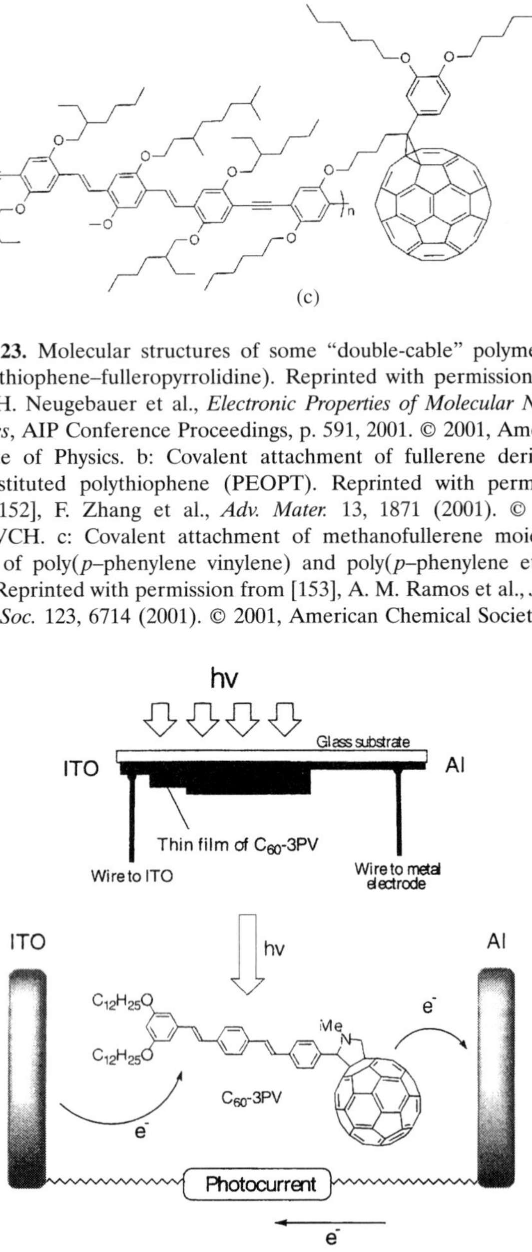

Figure 24. Schematic structure of the supramolecular solar cell based on a C_{60}–3PV hybrid compound, in which hole-conducting oligophenylenevinelene (OPV) moiety was chemically linked to an electron conducting fullerene subunit. Reprinted with permission from [155], J.-F. Nierengarten et al., *Mater. Today* 4, 16 (2001). © 2001, Elsevier Science.

electrons and holes, but it also provides pathways for their subsequent collection at opposite electrodes.

Another research approach for charge transport improvement suggests the use of the longest fullerene molecule, a carbon nanotube (CNT), instead of C_{60} or its derivatives in fullerene/polymer solar cells. Being blended with conjugated polymers, CNTs may not only act as electron acceptors, but also allow the transferred electrons to be efficiently transported along their length, thus providing percolation paths. CNT can be metallic or semiconducting, depending on their diameters and chirality [4]. The electrical and photoelectrical properties of CNT/conjugated polymer composites and interfaces have been investigated since 1996 [163–168]. Recently, bulk heterojunction solar cells based on conjugated polymers blended with multiwalled [169] and single-walled [170] CNTs (Fig. 25) have been reported. The photovoltaic parameters of these CNT-containing solar cells are listed in Table 1.

In addition to this application of CNT, Koprinarov et al. suggested using metallic CNT as a contact material for a wide range of inorganic solar cells [171].

5.5.2. Improvement in Open-Circuit Voltage

Since V_{oc} of fullerene/polymer bulk heterojunction solar cells is related to the energy difference between the LUMO level of the acceptor and the HOMO level of the donor components of the active layer (see Sections 5.4 and 5.5.1.1 and [130–131]), for further increase of the V_{oc}, one should be able to vary the LUMO level of the acceptor, that is, its electron affinity. Brabec et al. synthesized two new fullerene derivatives, ketolactam and azafulleroid, and compared them to C_{60} and PCBM, as acceptors of different strengths in bulk heterojunction solar cells (Fig. 26) [130–131]. It was found that PCBM, which has the lowest electron affinity, that is, acceptor strength, resulted in the highest V_{oc} values. For a further increase of the V_{oc}, new acceptors with lower electron affinity may help.

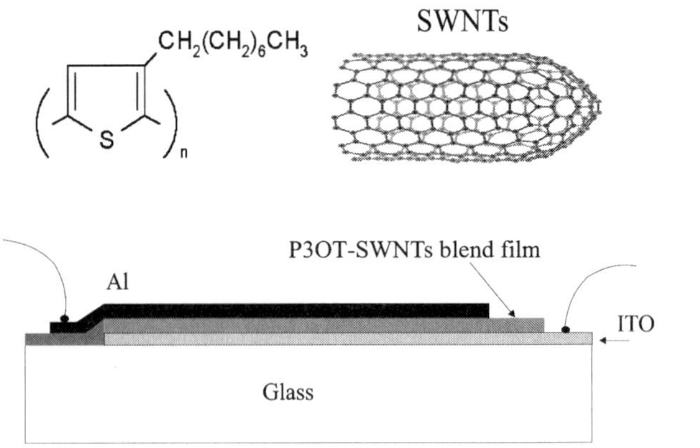

Figure 25. Schematic representation of a bulk heterojunction solar cell based on a conjugated polymer blended with single-walled CNTs, as well as the chemical structures of the compounds in its active layer. Reprinted with permission from [170], E. Kymakis and G. A. J. Amaratunga, *Appl. Phys. Lett.* 80, 112 (2002). © 2002, American Institute of Physics.

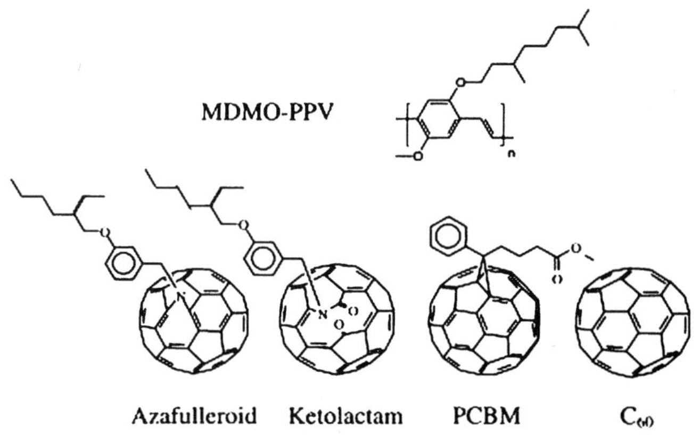

Figure 26. Chemical structure of fullerene derivatives of various acceptor strengths. Reprinted with permission from [131], C. J. Brabec et al., *Thin Solid Films* 403–404, 368 (2002). © 2002, Elsevier Science.

5.5.3. Novel Concepts of Organic Fullerene-Based Solar Cells

Yoshino et al. [21] have suggested the insertion of an additional *middle excitonic layer* between the polymeric donor and fullerene acceptor layers, in a three-layered molecular solar cell, which is reminiscent of a *p–i–n* structure in inorganic solar cells. To test this idea, the authors fabricated a three-layered cell utilizing OOPPV, octaethylporphine (OEP) (Fig. 27), and C_{60} films as the donor, middle excitonic, and acceptor layers, respectively. Photocurrent yield spectra, measured under irradiation from different sides of the cell, were interpreted by light absorption mainly in the OEP layer, exciton migration, and charge generation at both heterojunctions. Both interfacial regions of OOPPV/OEP and OEP/C_{60} were found to contribute to the charge generation by excitonic dissociation. Photocurrent enhancement and broadening of spectral sensitivity were observed for the OOPPV/OEP/C_{60} solar cells, as compared with the OEP single-layer and OOPPV/C_{60} bilayered solar cells. The same research team suggested an idea of fullerene/conjugated polymer multilayered solar cells with a quantum well structure (Fig. 28) [172]. In our opinion, serious experimental research is still needed for further clarification of the advantages of either *the double-heterojunction* or *the multilayered* solar cells, in comparison with the most efficient bulk heterojunction devices.

Meanwhile, Peumans and Forrest [26] have experimentally demonstrated C_{60}/copper phthalocyanine (CuPc) bilayered solar cells, incorporating an *exciton-blocking layer*. The chemical structure of CuPc is shown in Figure 14, where $M = \text{Cu}$. These devices showed an efficiency of $(3.6 \pm 2\%)$ under AM1.5 spectral illumination of 150 mW/cm^2 (1.5 suns) [26]. The cells were fabricated on glass substrates predeposited with an ITO anode contact. The ITO film was then spin coated with an ~320 Å thick film of PEDOT:PSS. The chemical structures of PEDOT and poly(styrenesulfonate) anion (PSS^-) are shown in Figure 18. Then the photoactive layers were grown at room temperature in high vacuum ($\sim 10^{-6}$ torr) in the following sequence: a 50–400 Å thick film of the donor-like CuPc, followed by a 100–400 Å thick film of the acceptor-like C_{60}. Next, a 50–400 Å thick exciton-blocking layer of bathocuproine (BCP) was deposited. Figure 29 displays a molecular structure of BCP. Finally, an Al cathode was deposited by thermal evaporation. (Previously [173], a large-bandgap ($E_g > 3$ eV) and, therefore, a transparent BCP layer had been found to transport electrons to the cathode from the adjoining acceptor layer, while effectively blocking excitons in the lower energy-gap acceptor layer from recombining at the cathode.) The highest photovoltaic parameters were observed for the ITO/PEDOT:PSS/200 Å CuPc/400 Å C_{60}/120 Å BCP/Al solar cells. At an illumination intensity of 150 mW/cm^2, values of $\eta = (3.6 \pm 2\%)$, $V_{oc} = 0.58$ V, $I_{sc} = 18.8$ mA/cm^2, and

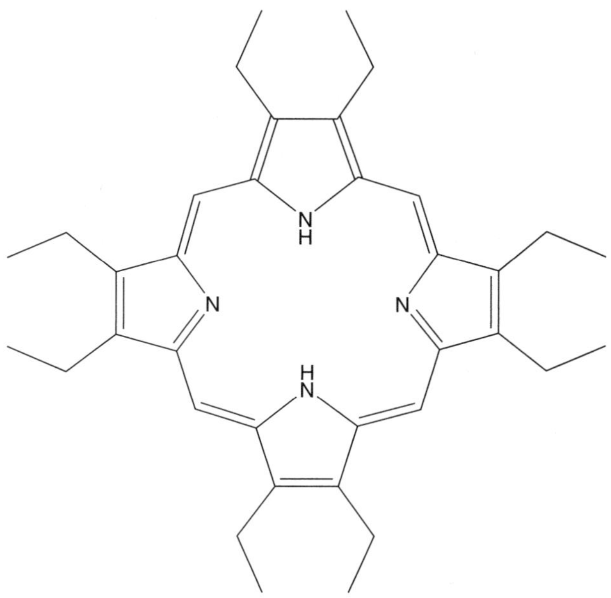

Figure 27. Chemical structure of octaethylporphine (OEP).

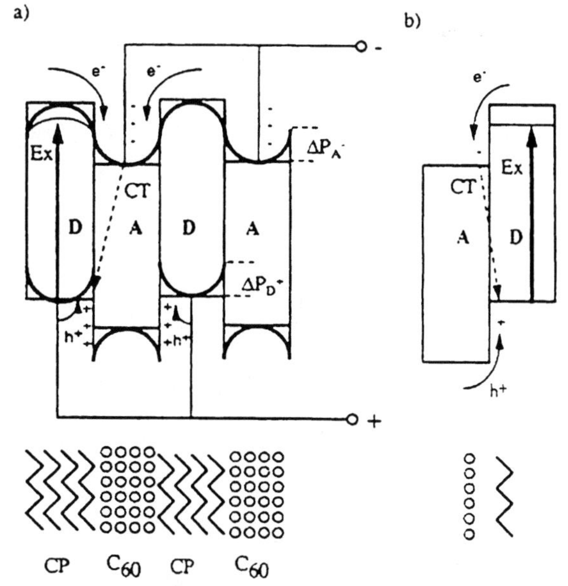

Figure 28. Schematic diagram of fullerene/conjugated polymer multilayered solar cells with quantum well structure. Reprinted with permission from [172], K. Yoshino and A. A. Zahidov, *Synth. Met.* 71, 1875 (1995). © 1995, Elsevier Science.

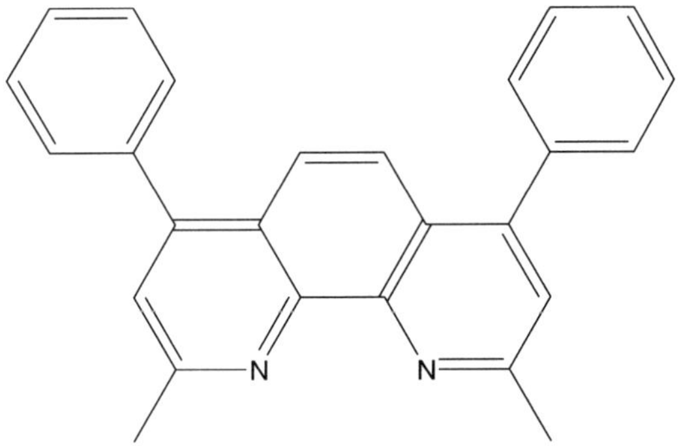

Figure 29. Molecular structure of bathocuproine (BCP).

$FF = 0.52$ were demonstrated. It should be noted that some features of the reported results and their interpretation are still not clear. For example, the reported I_{sc} value is apparently inconsistent with the reported spectral quantum efficiency data (very recently [174], the authors have explained this inconsistency by the degradation of the devices during their exposure to air between the *I–V* and the spectral response measurements). Nevertheless, this approach, of layered solar cells with fullerene and phthalocyanine molecular layers, is very promising, and is now being developed by different research groups [141, 175–176].

Other approaches utilizing the high absorption of phthalocyanines around 1.8 eV include doping of a phthalocyanine layer by C_{60} [175, 177, 178] and making covalently linked molecular systems, like phthalocyanine–fullerene dyads (Fig. 30) [179].

5.6. Stability of Organic Fullerene-Based Solar Cells

The relatively low stability of fullerene/polymer solar cells is a serious problem limiting their industrial application. Under the simultaneous influence of light and exposure to oxygen, a rapid degradation of photogeneration and transport properties of conjugated polymer/fullerene composite material occurs [180]. A faster degradation rate was found for the polymer compared with the fullerene in this material [180–181]. Protection from air and humidity is absolutely necessary to achieve long lifetimes [182].

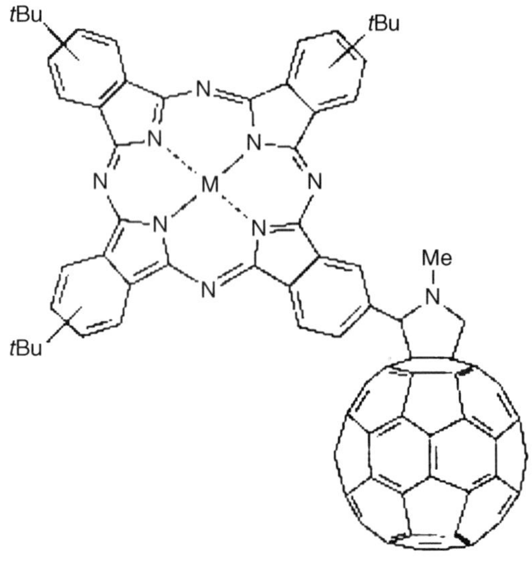

Figure 30. Molecular structure of fulleropyrrolidinphthalocyanine. Reprinted with permission from [179], A. Gouloumis et al., *Chem. Eur. J.* 6, 3600 (2000). © 2000, Wiley-VCH.

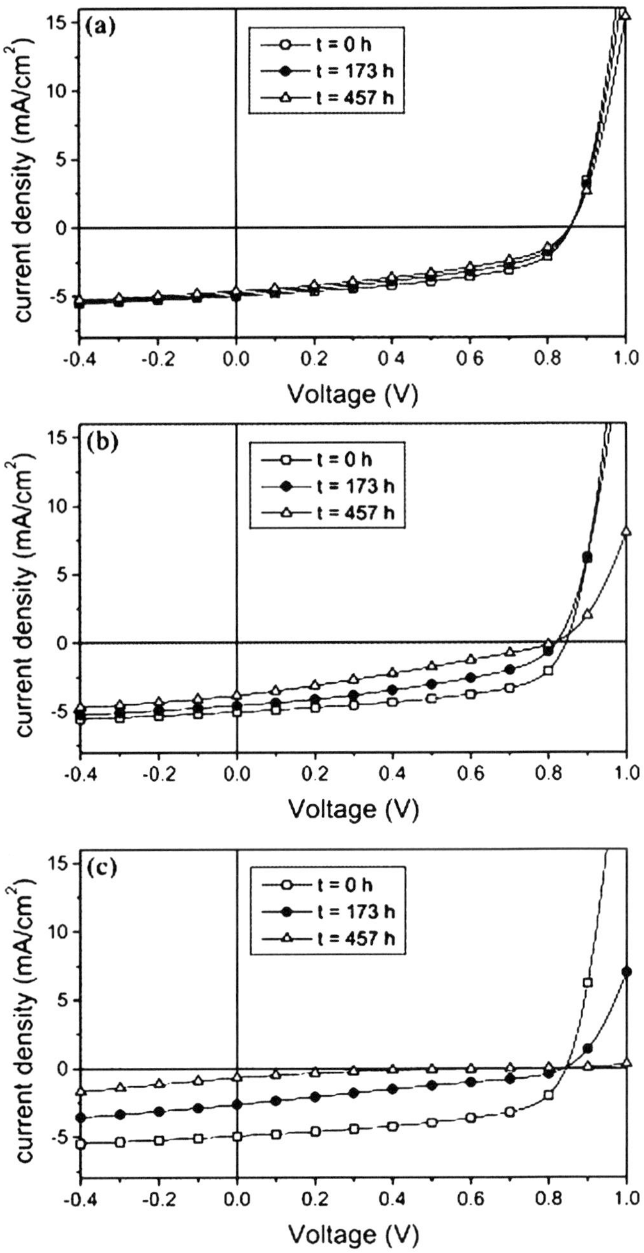

Figure 31. *I–V* curves for 0.1 cm² fullerene/polymer bulk heterojunction solar cell measured at $t = 0$, 173 and 457 h upon aging at (a) room temperature, dark, (b) 40 °C, dark, (c) 50 °C, illumination. Reprinted with permission from [119], J. M. Kroon et al., *Thin Solid Films* 403–404, 223 (2002). © 2002, Elsevier Science.

Stability studies on encapsulated bulk heterojunction fullerene/polymer solar cells were performed under outdoor [120, 183] and indoor [119] operational conditions. During 11 successive days of the outdoor experiment, V_{oc} was found to remain almost constant, while degradation in the I_{sc} and FF values was evident [120, 183]. This V_{oc} stability and degradation of I_{sc} and FF, under operational conditions, was confirmed by the indoor I–V measurements. The latter were performed during 460 h of cell aging, both in the dark, at room and elevated temperatures, and under continuous illumination at 1 sun, equivalent at 50 °C (Fig. 31) [119]. The cells which were kept in the dark, at room temperature, remained stable in terms of V_{oc} and I_{sc}, while FF demonstrated a slight decrease [Fig. 31(a)]. At 40 °C in the dark, a substantial decrease of I_{sc} and FF was observed, while V_{oc} remained constant, resulting in an overall decrease of 30–50% in the efficiency. The combination of thermal stress at 50 °C and visible light soaking resulted in complete degradation of the cell, mostly by I_{sc} and FF, within 460 h of aging [Fig. 31(c)]. The corresponding degradation mechanisms have not yet been clearly identified. Morphological changes in the photoactive layer after thermal aging, as well as photochemical degradation of the polymers, seem to have detrimental effects on the cell performance [119].

6. CONCLUDING REMARKS

In this chapter, we have reviewed the prospects of fullerene-based materials for photovoltaics, and have presented a comprehensive picture of state-of-the-art and future directions of fullerene-based solar cell research. Device structures and photovoltaic parameters of inorganic and organic fullerene-based solar cells were presented.

The very low intrinsic conductivity of C_{60} crystals is one of the main factors limiting the efficiency of inorganic fullerene-based solar cells. As a consequence, development of the appropriate doping processes is ultimately needed. Photovoltaic parameters of C_{60}/Si solar cells were found to be improved by P and B ion implantation in C_{60} thin films. Such implantation resulted in a dramatic increase in film conductivity, partial structural transformation from C_{60} to α–C, and corresponding changes in the optical gap of the C_{60} films. Using this effect, together with the control of the depth-implantation profile by ion energy regulation, it might be possible to produce a C_{60}/α–C composite with a nonhomogeneous depth-profile optimized for maximum sunlight absorption and cell efficiency. We expect that the next step in this direction will be the production of highly efficient multijunction C_{60}/α–C/Si or α–C/C_{60}/Si (p–i–n) solar cells. The ability to produce α–C/C_{60}/Si (p–i–n) solar cells has already been demonstrated. Another approach may include a C_{60}/α–C supperlattice structure for quantum-well solar cells.

Over the last decade, significant progress has been achieved in fullerene-based organic solar cells. The operating principles of such solar cells are similar to those for the initial steps of natural photosynthesis, and are based on very fast (*subpicosecond*) photoinduced electron transfer from a donor molecule (a conjugated, semiconducting polymer, for example) to a nearby C_{60} or other fullerene (or their derivative) acceptor molecule. Such cells are predicted to become a serious alternative to conventional inorganic photovoltaic devices. Efficiencies of over 2% have now been reported for two different types of fullerene-containing organic photovoltaic devices, such as bulk heterojunction fullerene/conjugated polymer solar cells and C_{60}/copper phthalocyanine bilayered solar cells, incorporating an exciton-blocking layer. The performance of these devices is limited by weak light absorption, poor charge transport, and low stability, but improvements may be expected via further optimization of materials and device architecture. Based on current trends, an efficiency of 5% appears to be within reach in the near future [102].

It should be noted that, up to now, only one member of the fullerene family, C_{60}, has been used in inorganic solar cells. In organic photovoltaic devices, pure C_{60}, a variety of C_{60} derivatives, and carbon nanotubes have been already utilized. The photovoltaic parameters of fullerene-based solar cells are summarized in Table 1. For comparison, the parameters of the best monocrystalline and amorphous/nanocrystalline silicon solar cells are also listed. A complete table summarizing the photovoltaic parameters of the best inorganic semiconductor solar cells is documented in [184]. Table 1 shows that the highest efficiency values of 2–3% demonstrated by organic fullerene-based devices are comparable to, but still lower than, those of conventional inorganic semiconductor solar cells [184]. In spite of this, organic fullerene-based devices have a number of advantages, such as their light weight, flexibility, simple processing at relatively low temperatures, and therefore, low fabrication cost. Furthermore, fullerene/polymer solar cells can be scaled up for production of large areas without a significant loss of device efficiency. Recently, a photoactive layer of bulk heterojunction solar cells, consisting of an MDMO–PPV:PCBM blend, was successfully produced by screen printing [185]. The average thickness of the active layer and the root-mean-square surface roughness were 40 and 2.6 nm, respectively. The devices yielded a power conversion efficiency of 4.3% when illuminated by monochromatic light with a wavelength of 488 nm. These results illustrate that screen printing can be a powerful technique for the fast, inexpensive fabrication of roll-to-roll fullerene/polymer solar cells, while retaining nanometer-scale control of film thickness. Using this technique for solar cell fabrication further increases the strong potential that fullerene/polymer solar cells may have practical applications.

Finally, it is worth mentioning that a number of strategies have also been suggested for incorporating fullerene moieties in nanostructured *photoelectrochemical* solar cells [64, 186–201], photosensitive biomaterials, and *artificial photosynthetic centers* [193, 195, 202–218]. Although a detailed description of such devices was not within the scope of the present review, we should point out, for completeness, that various research directions aimed at the efficient harvesting of solar energy by photovoltaic and photoelectrochemical solar cells and artificial photosynthetic systems are closely interrelated. Progress in these directions, as well as a better understanding of natural photosynthesis [219], could well be achieved on the basis of a multidisciplinary effort in physics, chemistry, and biology.

GLOSSARY

C_{60} Fullerene molecule consisting of 60 carbon atoms located at the vertices of a truncated icosahedron.

Carbon nanotube Tubular fullerene molecule, cylindrical in shape, with caps at each end such that two caps can be joined to form a fullerene. The cylindrical portion of a carbon nanotube consists of a single graphene sheet, rolled to form the cylinder.

Conjugated polymers Polymeric molecules containing a regular array of alternating single and double carbon–carbon bonds.

Fullerenes Closed cage molecules consisting of carbon atoms located at the vertices of a polyhedron which contains only hexagonal and pentagonal faces.

Homo- and hetero-*p*–*n* junctions Rectifying interfaces formed between *p* and *n* types of the same or different semiconductors, respectively.

Photovoltaic solar cells Devices that convert solar radiation directly into electrical energy due to the photovoltaic effect.

Schottky barrier Rectifying interface between a semiconductor and a metal.

ACKNOWLEDGMENTS

Dr. C. J. Brabec, Prof. D. Faiman, Dr. J. Kroon, and Dr. J. Nelson are warmly acknowledged for fruitful discussions. The author thanks Prof. D. Faiman and Prof. N. S. Sariciftci for initiating his interest in fullerene solar cells and organic solar cells, respectively. The author also thanks Dr. J. Nelson and Dr. K. L. Narayanan for providing him with the unpublished materials, and Mrs. S. Dann for assistance with manuscript preparation. This work was funded by the Israel Ministry of Energy and National Infrastructures, the European Community, the Israel Ministry of Immigrant Absorption, the Sally Berg Foundation, and the Gensseler Foundation.

REFERENCES

1. H. W. Kroto, J. R. Heath, S. C. O'Brien, R. F. Curl, and R. E. Smalley, *Nature* 318, 162 (1985).
2. W. Krätschmer, L. D. Lamb, K. Fostipoulos, and D. R. Huffman, *Nature* 347, 354 (1990).
3. H. W. Kroto, in "Electronic Properties of Fulerenes" (H. Kuzmany, J. Fink, M. Mering, and S. Roth, Eds.), Vol. 117, p. 1. Springer-Verlag, Berlin–Heidelberg, 1993.
4. M. S. Dresselahaus, G. Dresselahaus, and P. C. Eklund, "Science of Fullerenes and Carbon Nanotubes." Academic, New York, 1996.
5. A. E. Becquerel, *C. R. Acad. Sci.* 9, 561 (1839).
6. W. Shockley, "Electrons and Holes in Semiconductors." van Nostrand, New York, 1950.
7. A. Hamed, in "Organic Conductive Molecules and Polymers" (H. S. Nalwa, Ed.), p. 486. Wiley, Sussex, 1997.
8. D. Faiman, S. Goren, E. Katz, M. Koltun, E. Kunoff, A. Shames, S. Shtutina, and B. Uzan, "Proceedings of the 13th European Photovoltaic Solar Energy Conference" (W. Freiesleben, W. Palz, H. A. Ossenbrink, and P. Helm, Eds.), H. S. Stephens & Associates, 1995, Vol. 2, p. 1284.
9. M. Koltun, D. Faiman, S. Goren, E. A. Katz, E. Kunoff, A. Shames, S. Shtutina, and B. Uzan, *Solar Energy Mater. Solid Cells* 44, 485 (1996).
10. S. Saito and A. Oshiyama, *Phys. Rev. Lett.* 66, 2637 (1991).
11. A. Skumanich, *Chem. Phys. Lett.* 182, 486 (1991).
12. M. Hosoya, K. Ichimura, Z. H. Wang, G. Dresselhaus, and P. C. Eklund, *Phys. Rev. B* 49, 4981 (1994).
13. J. J. Loferski, *J. Appl. Phys.* 27, 777 (1956).
14. R. W. Lof, M. A. van Veenendaal, B. Koopmans, H. T. Jonkman, and G. A. Sawatzky, *Phys. Rev. Lett.* 68, 3924 (1992).
15. B. Mishori, E. A. Katz, D. Faiman, and Y. Shapira, *Solid State Commun.* 102, 489 (1997).
16. E. L. Shirley and S. G. Louie, *Phys. Rev. Lett.* 71, 133 (1993).
17. R. Tycko, G. Dabbagh, R. M. Fleming, R. C. Haddon, A. V. Makhia, and S. M. Zahurak, *Phys. Rev. Lett.* 67, 1886 (1991).
18. N. S. Sariciftci, D. Braun, C. Zhang, V. I. Srdanov, A. J. Heeger, G. Stucky, and F. Wudl, *Appl. Phys. Lett.* 62, 585 (1993).
19. G. Yu, J. Gao, J. C. Hummelen, F. Wudl, and A. J. Heeger, *Science* 270, 1789 (1995).
20. N. S. Sariciftci and A. J. Heeger, in "Handbook of Organic Conductive Molecules and Polymers" (H. S. Nalwa, Ed.), Vol. 1, p. 413. Willey, New York, 1997.
21. K. Yoshino, K. Tada, A. Fujii, E. M. Conwell, and A. A. Zahidov, *IEEE Trans. Electron Devices* 44, 1315 (1997).
22. N. S. Sariciftci, in "Optical and Electronic Properties of Fullerenes and Fullerene-Based Materials" (J. Shinar, Z. V. Vardeny, and Z. H. Kafafi, Eds.), p. 367. Marcel Dekker, New York, 2000.
23. N. S. Sariciftci, L. Smilowitz, A. J. Heeger, and F. Wudl, *Science* 258, 1474 (1992).
24. S. Morita, A. A. Zakhidov, and K. Yoshino, *Solid State Commun.* 82, 249 (1992).
25. S. E. Shaheen, C. J. Brabec, N. S. Sariciftci, F. Padinger, T. Fromherz, and J. C. Hummelen, *Appl. Phys. Lett.* 78, 841 (2001).
26. P. Peumans and S. R. Forrest, *Appl. Phys. Lett.* 79, 126 (2001).
27. L. F. P. Chibante, A. Tess, J. Alford, M. D. Dierner, and R. E. Smalley, *J. Phys. Chem.* 97, 8696 (1993).
28. J. R. Pitts, D. Mischler, C. L. Fields, A. Lewandowski, and C. Bingham, *Solar Chem. News* no. 6, 2 (1998).
29. C. L. Fields, D. H. Parker, J. R. Pitts, M. J. Halle, C. Bingham, A. Lewandowski, and D. E. King, *J. Phys. Chem.* 97, 8701 (1993).
30. D. Laplaze, P. Bernier, G. Flamant, M. Lebrun, A. Brunelle, and S. Della-Negra, *J. Phys. B* 29, 4943 (1996).
31. D. Laplaze, P. Bernier, C. Journet, V. Vie, G. Flamant, and M. Lebrun, *Synth. Met.* 86, 2295 (1997).
32. G. Flamant, A. Ferriere, D. Laplaze, and C. Monty, *Solar Energy* 66, 117 (1999).
33. C. L. Fields, J. R. Pitts, and A. Lewandowski, *Abstr. Papers Am. Chem. Soc.* 212, 26 (1996).
34. T. Guillard, G. Flamant, and D. Laplaze, *J. Solar Energy Eng.* 123, 153 (2001).
35. M. J. Heben, T. A. Bekkedahl, D. L. Schulz, K. M. Jones, A. C. Dillon, C. J. Curtis, C. Bingham, J. R. Pitts, A. Lewandowski, and C. Fields, in "Recent Advances in the Chemistry and Physics of Fullerenes and Related Materials III" (K. M. Kadish, and R. S. Ruoff, Eds.), PV 96-10, p. 803. Electrochemical Society, Pennington, NJ, 1996.
36. D. Laplaze, P. Bernier, W. K. Maser, G. Flamant, T. Guillard, and A. Loiseau, *Carbon* 36, 685 (1998).
37. T. Guillard, G. Flamant, J.-F. Robert, B. Rivoire, J. Giral, and D. Laplaze, *J. Solar Energy Eng.* 124, 22 (2002).
38. L. Alvarez, T. Guillard, J. L. Sauvajol, G. Flamant, and D. Laplaze, *Appl. Phys. A* 70, 169 (2000).
39. D. Laplaze, L. Alvarez, T. Guillard, J. M. Badie, and G. Flamant, *Carbon* 40, 1621 (2002).
40. W. K. Maser, A. M. Benito, and M. T. Martinez, *Carbon* 40, 1685 (2002).

41. A. F. Hebard, B. Miller, J. M. Rosamilia, and W. L. Wilson, U.S. Patent 517, 1373, 1992.
42. H. Yohehara and Ch. Pac, *Appl. Phys. Lett.* 61, 575 (1992).
43. K. Pichler, M. G. Harisson, R. H. Friend, and S. Pekker, *Synth. Met.* 56, 3229 (1993).
44. S. Curran, J. Callaghan, D. Weldon, E. Bourdin, K. Cazini, W. J. Blau, E. Waldron, D. McGovcran, M. Delamesiere, Y. Sarazin, and C. Hogrel, in "Electronic Properties of Fullerenes" (H. Kuzmany, J. Fink, M. Mehring, and S. Roth, Eds.), Vol. 117, p. 427. Springer Series in Solid-State Sciences, 1993.
45. E. A. Katz, D. Faiman, S. Shtutina, A. Shames, S. Goren, B. Mishori, and Y. Shapira, in "Thin Films Structures for Photovoltaics" (E. D. Jones, J. Kalejs, R. Noufi, and B. Sopori, Eds.), Vol. 485, p. 113. Materials Research Society Series, Warrendale, PA, 1997.
46. K. M. Chen, Y. Q. Jia, S. X. Jin, K. Wu, X. D. Zhang, W. B. Zhao, C. Y. Li, and Z. N. Gu, *J. Phys.: Condens. Matter* 6, L367 (1994).
47. C. Wen, T. Aida, I. Honma, H. Komiyama, and K. Yamada, *Denki Kagaku* 62, 264 (1994).
48. K. M. Chen, Y. Q. Jia, S. X. Jin, K. Wu, W. B. Zhao, C. Y. Li, Z. N. Gu, and X. H. Zhou, *J. Phys.: Condens. Matter* 7, L201 (1995).
49. E. A. Katz, D. Faiman, S. Goren, S. Shtutina, B. Mishori, and Y. Shapira, *Fullerene Sci. Technol.* 6, 103 (1998).
50. N. Kojima, M. Yamaguchi, and N. Ishikawa, *Jpn. J. Appl. Phys.* 39, 1176 (2000).
51. K. Kita, C. Wen, M. Ihara, and K. Yamada, *J. Appl. Phys.* 79, 2798 (1996).
52. D. J. Fu, Y. Y. Lei, J. C. Li, M. S. Ye, H. X. Guo, Y. G. Peng, and X. J. Fan, *Appl. Phys. A* 67, 441 (1998).
53. Y. Shi, C. M. Xiong, X. S. Wang, C. H. Lei, H. X. Guo, and X. J. Fan, *Appl. Phys. A* 63, 353 (1996).
54. J. S. Zhu, X. M. Liu, S. H. Xu, J. X. Wu, and X. F. Sun, *Solid State Commun.* 98, 417 (1996).
55. O. Janzen and W. Mönch, *J. Phys.: Condens. Matter* 11, L111 (1999).
56. K. M. Chen, Y. X. Zhang, G. G. Qin, S. X. Jin, K. Wu, C. Y. Li, Z. N. Gu, and X. H. Zhou, *Appl. Phys. Lett.* 69, 3557 (1996).
57. K. M. Chen, W. H. Sun, K. Wu, C. Y. Li, G. G. Qin, Y. X. Zhang, X. H. Zhou, and Z. N. Gu, *J. Appl. Phys.* 85, 6935 (1999).
58. K. Kita, M. Ihara, K. Sakaki, and K. Yamada, *J. Appl. Phys.* 81, 6246 (1997).
59. A. Khan, M. Yamaguchi, and N. Kojima, *Solid-State Electron.* 44, 1471 (2000).
60. E. A. Katz, D. Faiman, S. Goren, S. Shtutina, B. Mishori, and Y. Shapira, in "Proceedings of the 14th European Photovoltaic Solar Energy Conference" (H. A. Ossenbrink, P. Helm, and H. Ehmann, Eds.), H. S. Stephens & Associates, 1997. Vol. 2, p. 1777.
61. H. J. Hovel, in "Semiconductors and Semimetals," Vol. 11, p. 14. Academic Press, New York, 1975.
62. W. Zhao, T. N. Zhao, J. J. Yue, L. Q. Chen, and J. Q. Liu, *Solid State Commun.* 84, 323 (1992).
63. T. Gotoh, S. Nonomura, S. Hirata, and S. Nitta, *Appl. Surf. Sci.* 113/114, 278 (1997).
64. S. Licht, P. A. Ramakrishnan, D. Faiman, E. A. Katz, A. Shames, and S. Goren, *Solar Energy Mater. Solar Cells* 56, 45 (1998).
65. W. Feng and B. Miller, *Electrochem. Solid-State Lett.* 1, 172 (1998).
66. Y. Shi, C. M. Xiong, Y. X. He, H. X. Guo, Y. G. Peng, and X. J. Fan, *Fullerene Sci. Technol.* 4, 963 (1996).
67. K. L. Narayanan and M. Yamaguchi, *Appl. Phys. Lett.* 75, 2106 (1999).
68. N. Dharmasu, T. Sakai, N. Kojima, M. Yamaguchi, Y. Ohshita, and K. L. Narayanan, *J. Appl. Phys.* 89, 318 (2001).
69. K. L. Narayanan, O. Goetzberger, A. Khan, N. Kojima, and M. Yamaguchi, *Solar Energy Mater. Solar Cells* 65, 29 (2001).
70. K. L. Narayanan and M. Yamaguchi, in "Technical Digest of the 12th International Photovoltaic Science and Engineering Conference," Jeju, Korea, 2001, p. 137.
71. K. L. Narayanan and M. Yamaguchi, *Solar Energy Mater. Solar Cells* 75, 345 (2003).
72. N. Kojima, O. Goetzberger, Y. Ohshita, and M. Yamaguchi, in "Proceedings of the 28th IEEE Photovoltaic Specialists Conference" (A. Rohatgi, Ed.), IEEE, 2000, p. 873.
73. K. W. J. Barnham and G. Duggan, *J. Appl. Phys.* 67, 3490 (1990).
74. T. Arai, Y. Muracami, H. Suematsu, K. Kikuchi, Y. Ashiba, and I. Ikemoto, *Solid State Commun.* 84, 827 (1992).
75. S. Kazaoui, R. Ross, and N. Minami, *Solid State Commun.* 90, 623 (1994).
76. B. Pevzner, A. F. Hebard, and M. S. Dresselhaus, *Phys. Rev. B* 55, 16439 (1997).
77. A. Hamed, Y. Y. Sun, Y. K. Tao, R. L. Meng, and P. H. Hor, *Phys. Rev. B* 47, 10873 (1993).
78. S.-C. Yang and T. Mieno, *Jpn. J. Appl. Phys.* 40, 1067 (2001).
79. N. S. Sariciftci, *Prog. Quantum Electron.* 19, 131 (1995).
80. P. F. van Hutten and G. Hadziioannou, *Monatshefte für Chemie* 132, 129 (2001).
81. "Handbook of Conductive Polymers" (T. A. Skotheim, Ed.), Vol. 1 & 2. Marcel Dekker, New York, 1986.
82. A. J. Heeger, S. Kivelson, J. R. Schrieffer, and W.-P. Su, *Rev. Mod. Phys.* 60, 782 (1988).
83. A. J. Heeger and P. Smith, in "Conjugated Polymers" (J. L. Bredas and R. Sibley, Eds.), p. 141. Kluwer Academic, Dordrecht, 1991.
84. M. M. Allemand, A. Koch, F. Wudl, Y. Rubin, F. Diederich, M. M. Alavarez, S. Z. Anz, and R. L. Whetten, *J. Am. Chem. Soc.* 113, 1050 (1991).
85. L. Smilowitz, N. S. Sariciftci, R. Wu, C. Gettinger, A. J. Heeger, and F. Wudl, *Phys. Rev. B* 47, 13835 (1993).
86. C. H. Lee, G. Yu, N. S. Sariciftci, D. Mozes, K. Pakbaz, C. Zhang, A. J. Heeger, and F. Wudl, *Phys. Rev. B* 48, 15425 (1993).
87. K. Yoshino, X. H. Yin, K. Muro, S. Kiomatsu, S. Morita, A. A. Zakhidov, T. Noguchi, and T. Ohnishi, *Jpn. J. Appl. Phys.* 32, L357 (1993).
88. K. Yoshino, X. H. Yin, S. Morita, T. Kawai, and A. A. Zakhidov, *Solid State Commun.* 85, 85 (1993).
89. S. Morita, S. Kiyomatsu, X. H. Yin, A. A. Zakhidov, T. Noguchi, T. Ohnishi, and K. Yoshino, *J. Appl. Phys.* 74, 2860 (1993).
90. S. Morita, S. Kiyomatsu, M. Fukuda, A. A. Zakhidov, K. Yoshino, K. Kikuchi, and Y. Achiba, *Jpn. J. Appl. Phys.* 32, L1173 (1993).
91. C. J. Brabec, G. Zerza, N. S. Sariciftci, G. Cerullo, S. DeSilvestri, S. Luzatti, and J. C. Hummelen, *Chem. Phys. Lett.* 340, 232 (2001).
92. N. S. Sariciftci and A. J. Heeger, U.S. Patent 5,331,183, University of California, 1994.
93. S. B. Lee, P. K. Khabillaev, A. A. Zakhidov, S. Morita, and K. Yoshino, *Synth. Met.* 71, 2247 (1995).
94. J. J. M. Halls, K. Pichler, R. H. Friend, S. C. Moratti, and A. B. Holmes, *Appl. Phys. Lett.* 68, 3120 (1996).
95. J. J. M. Halls, K. Pichler, R. H. Friend, S. C. Moratti, and A. B. Holmes, *Synth. Met.* 77, 277 (1996).
96. K. Yoshino, K. Yoshimoto, K. Tada, H. Araki, T. Kawai, M. Ozaki, and A. A. Zakhidov, in "Fullerenes and Photonics II" (Z. H. Kafafi, Ed.), Vol. 2530, p. 60. SPIE, 1995.
97. S. Morita, A. A. Zakhidov, and K. Yoshino, *Jpn. J. Appl. Phys.* 32, L873 (1993).
98. K. Tada, S. Morita, T. Kawai, M. Onoda, K. Yoshino, and A. A. Zakhidov, *Synth. Met.* 70, 1347 (1995).
99. L. A. A. Petterson, L. S. Roman, and O. Inganäs, *J. Appl. Phys.* 86, 487 (1999).
100. H. Yohehara and Ch. Pac, *Thin Solid Films* 278, 108 (1996).
101. Ch. Pannemann, V. Dyakonov, J. Parisi, O. Hild, and D. Wöhrle, *Synth. Met.* 121, 1585 (2001).
102. J. Nelson, *Mater. Today* 20 (2002).
103. I. D. Parker, *J. Appl. Phys.* 75, 1656 (1994).

104. G. Yu, C. Zhang, and A. J. Heeger, *Appl. Phys. Lett.* 64, 1540 (1994).
105. G. Yu, K. Pakbaz, and A. J. Heeger, *Appl. Phys. Lett.* 64, 3422 (1994).
106. H. Antoniadis, B. R. Hsiesh, M. A. Abkowitz, M. Stolka, and S. A. Jehneke, *Polym. Preprints* 34, 490 (1993).
107. J. J. M. Halls, K. Pichler, R. H. Friend, S. C. Moratti, and A. B. Holmes, *Appl. Phys. Lett.* 68, 311 (1996).
108. J. C. Hummelen, B. W. Knight, F. Lepec, and F. Wudl, *J. Org. Chem.* 60, 532 (1995).
109. L. S. Roman, M. R. Andersson, and O. Inganäs, *Adv. Mater.* 9, 1164 (1997).
110. J. Gao, F. Hide, and H. Wang, *Synth. Met.* 84, 979 (1997).
111. C. J. Brabec, F. Padinger, N. S. Sariciftci, and J. C. Hummelen, *J. Appl. Phys.* 85, 6866 (1999).
112. J. J. M. Halls, A. C. Arias, J. D. MacKenzei, W. Wu, M. Inbasekaran, E. P. Woo, and R. H. Friend, *Adv. Mater. Res.* 12, 498 (2000).
113. T. Fromherz, F. Padinger, D. Gebeyehu, C. Brabec, J. C. Hummelen, and N. S. Sariciftci, *Solar Energy Mater. Solar Cells* 63, 61 (2000).
114. D. Gebeyehu, C. J. Brabec, F. Padinger, T. Fromherz, J. C. Hummelen, D. Badt, H. Schindler, and N. S. Sariciftci, *Synth. Met.* 118, 1 (2001).
115. N. Camaioni, L. Garlaschelli, A. Geri, M. Maggini, G. Possami, and G. Ridolphi, *J. Mater. Chem.* 12, 2065 (2002).
116. T. Aernouts, W. Geens, J. Poortmans, P. Heremans, S. Borghs, and R. Mertens, *Thin Solid Films* 403–404, 297 (2002).
117. V. Dyakonov, *Physica E* 14, 53 (2002).
118. C. J. Brabec, S. E. Shaheen, C. Winder, N. S. Sariciftci, and P. Denk, *Appl. Phys. Lett.* 80, 1288 (2002).
119. J. M. Kroon, M. M. Wienk, W. J. H. Verhees, and J. C. Hummelen, *Thin Solid Films* 403–404, 223 (2002).
120. E. A. Katz, D. Faiman, S. M. Tuladhar, J. M. Kroon, M. M. Wienk, T. Fromherz, F. Padinger, C. J. Brabec, and N. S. Sariciftci, *J. Appl. Phys.* 90, 5343 (2001).
121. W. Geens, S. E. Shaheen, B. Wessling, C. J. Brabec, J. Poortmans, and N. S. Sariciftci, *Organ. Electron.* 3, 105 (2002).
122. G. H. Gelinck, J. Warman, and E. G. J. Staring, *J. Phys. Chem.* 100, 5485 (1996).
123. G. E. Jabbour, B. Kippelen, N. R. Armstrong, and N. Peyghambarian, *Appl. Phys. Lett.* 73, 1185 (1998).
124. L. S. Hung, C. W. Tang, and M. G. Mason, *Appl. Phys. Lett.* 70, 152 (1997).
125. W. J. H. van Gennip, J. K. J van Duren, P. C. Thüne, R. A. J. Janssen, and J. W. Niemantsverdriet, *J. Chem. Phys.* 117, 5031 (2002).
126. J. Rostalski and D. Meissner, *Solar Energy Mater. Solar Cells* 21, 87 (2000).
127. M. Dress, K. Premaratne, W. Graupner, J. R. Heflin, R. M. Davis, D. Marciu, and M. Miller, *Appl. Phys. Lett.* 81, 4607 (2002).
128. E. A. Katz, D. Faiman, Y. Cohen, F. Padinger, C. Brabec, and N. S. Sariciftci, in "Organic Photovoltaics" (Z. H. Kafafi, Ed.), Vol. 4108, p. 117. SPIE, 2000.
129. J. Nelson, *Phys. Rev. B* (in press).
130. C. J. Brabec, A. Cravino, D. Meissner, N. S. Sariciftci, T. Fromherz, M. T. Rispens, L. Sanchez, and J. C. Hummelen, *Adv. Funct. Mater.* 11, 374 (2001).
131. C. J. Brabec, A. Cravino, D. Meissner, N. S. Sariciftci, M. T. Rispens, L. Sanchez, J. C. Hummelen, and T. Fromherz, *Thin Solid Films* 403–404, 368 (2002).
132. K. Emery, J. Burdick, Y. Caiyem, D. Dunlavy, H. Field, B. Kroposki, T. Moriarty, L. Ottoson, S. Rummel, T. Strand, and M. W. Wanlass, "Proceedings of the 25th IEEE Photovoltaic Specialists Conference," IEEE, 1996, p. 1275.
133. P. Schilinsky, C. Waldauf, and C. J. Brabec, *Appl. Phys. Lett.* 81, 3885 (2002).
134. R. Pacios, D. D. C. Bradley, J. Nelson, and C. J. Brabec, *Synth. Met.* (in press).
135. A. Dhanabalan, J. K. J. van Duren, P. A. van Hal, J. L. J. van Dongen, and R. A. J. Janssen, *Adv. Funct. Mater.* 11, 255 (2001).
136. J. Roncali, *Chem. Rev.* 97, 173 (1997).
137. C. Winder, G. Matt, J. C. Hummelen, R. A. J. Janssen, N. S. Sariciftci, and C. J. Brabec, *Thin Solid Films* 403–404, 373 (2002).
138. S. E. Shaheen, D. Vangeneugden, R. Kiebooms, D. Vanderzande, T. Fromherz, F. Padinger, C. J. Brabec, and N. S. Sariciftci, *Synth. Met.* 121, 1583 (2001).
139. D. L. Vangeneugden, D. J. M. Vanderzande, J. Salbeck, P. A. van Hal, R. A. J. Janssen, J. C. Hummelen, C. J. Brabec, S. E. Shaheen, and N. S. Sariciftci, *J. Phys. Chem. B* 105, 11106 (2001).
140. W. Feng, A. Fuji, S. Lee, H. Wu, and K. Yoshino, *J. Appl. Phys.* 88, 7120 (2000).
141. D. Godovsky, L. Chen, L. Pettersson, O. Inganäs, and J. C. Hummelen, *Adv. Mater. Opt. Electron.* 10, 47 (2000).
142. R. S. Ingram, M. J. Hostetler, R. W. Murray, T. P. Bigioni, D. K. Guthrie, and P. N. First, *J. Am. Chem. Soc.* 119, 9279 (1997).
143. T. Aernouts, W. Geens, J. Poortmans, P. Heremans, S. Borghs, and R. Mertens, *Thin Solid Films* 403–404, 297 (2002).
144. O. Inganäs, L. S. Roman, F. Zhang, D. M. Johansson, M. R. Andersson, and J. C. Hummelen, *Synth. Met.* 121, 1525 (2001).
145. L. S. Roman, O. Inganäs, T. Granlund, T. Nyberg, M. Svensson, M. R. Andersson, and J. C. Hummelen, *Adv. Mater.* 12, 189 (2000).
146. W. Geens, T. Aernouts, J. Poortmans, and G. Hadziioannou, *Thin Solid Films* 403–404, 438 (2002).
147. W. Geens, S. E. Shaheen, C. J. Brabec, J. Poortmans, and N. S. Sariciftci, in "Electronic Properties of Novel Materials Molecular Nanostructures" (H. Kuzmany, J. Fink, M. Mehring, and S. Roth, Eds.), AIP Conference Proceedings, Woodbury, NY, 2000, Vol. 544, p. 516.
148. J. Balberg, R. Naidis, M.-K. Lee, J. Shinar, and L. F. Fonseca, *Appl. Phys. Lett.* 79, 197 (2001).
149. W. Geens, S. E. Shaheen, C. J. Brabec, J. Poortmans, and N. S. Sariciftci, *Appl. Phys. Lett.* (submitted).
150. A. Cravino and N. S. Sariciftci, *J. Mater. Chem.* 12, 1931 (2002).
151. H. Neugebauer, A. Cravino, G. Zerza, M. Maggini, S. Bucella, G. Scoranno, M. Svensson, M. R. Andersson, and N. S. Sariciftci, in "Electronic Properties of Molecular Nanostructures" (H. Kuzmany, J. Fink, M. Mehring, and S. Roth, Eds.), AIP Conference Proceedings, Woodbury, NY, 2001, Vol. 591, p. 511.
152. F. Zhang, M. Svensson, M. R. Andersson, M. Maggini, S. Bucella, E. Menna, and O. Inganäs, *Adv. Mater.* 13, 1871 (2001).
153. A. M. Ramos, M. T. Rispens, J. K. J. van Duren, J. C. Hummelen, and R. A. J. Janssen, *J. Am. Chem. Soc.* 123, 6714 (2001).
154. C. J. Brabec, N. S. Sariciftci, and J. C. Hummelen, *Adv. Funct. Mater.* 11, 15 (2001).
155. J.-F. Nierengarten, G. Hadziioannou, and N. Armaroli, *Mater. Today* 4, 16 (2001).
156. J.-F. Nierengarten, J.-F. Eckert, J.-F. Nicoud, L. Ouali, V. V. Krasnikov, and G. Hadziioannou, *Chem. Commun.* 617 (1999).
157. U. Stalmach, B. de Boer, C. Videlot, P. F. van Hutten, and G. Hadziioannou, *J. Am. Chem. Soc.* 122, 5464 (2000).
158. J.-F. Eckert, J.-F. Nicoud, J.-F. Nierengarten, S.-G. Liu, L. Echegoven, F. Barigelletti, N. Armaroli, L. Ouali, V. V. Krasnikov, and G. Hadziioannou, *J. Am. Chem. Soc.* 122, 7467 (2000).
159. E. Peeters, P. A. van Hal, J. Knol, C. J. Brabec, N. S. Sariciftci, J. C. Hummelen, and R. A. J. Janssen, *J. Phys. Chem. B* 104, 10174 (2000).
160. T. Gu and J.-F. Nierengarten, *Tetrahedron Lett.* 42, 3175 (2001).
161. A. Cravino, G. Zerza, H. Neugebauer, M. Maggini, S. Bucella, E. Menna, M. Svensson, M. R. Andersson, C. J. Brabec, and N. S. Sariciftci, *J. Phys. Chem. B* 106, 70 (2002).

162. T. Gu, D. Tsamouras, C. Melzer, V. Krasnikov, J. P. Gisselbrecht, M. Gross, G. Hadziioannou, and J.-F. Nierengarten, *Chem. Phys. Chem.* 3, 124 (2002).
163. D. B. Romero, M. Carrard, W. de Heer, and L. Zuppiroli, *Adv. Mater.* 8, 899 (1996).
164. K. Yoshino, H. Kajii, H. Araki, T. Sonoda, H. Take, and S. Lee, *Fullerene Sci. Technol.* 7, 695 (1999).
165. H. Ago, M. S. P. Shaffer, D. S. Ginger, A. H. Windle, and R. H. Friend, *Phys. Rev. B* 61, 2286 (2000).
166. S. B. Lee, T. Katayama, H. Kajii, H. Araki, and K. Yoshino, *Synth. Met.* 121, 1591 (2001).
167. H. S. Woo, R. Czerw, S. Webster, D. L. Carroll, J. W. Park, and J. H. Lee, *Synth. Met.* 116, 369 (2001).
168. L. Dai and A. W. H. Mau, *Adv. Mater.* 13, 899 (2001).
169. H. Ago, K. Pettrish, M. S. P. Shaffer, A. H. Windle, and R. H. Friend, *Adv. Mater.* 11, 1281 (1999).
170. E. Kymakis and G. A. J. Amaratunga, *Appl. Phys. Lett.* 80, 112 (2002).
171. N. Koprinarov, R. Stefanov, G. Pchelanov, M. Konstantinova, and I. Stambolova, *Synth. Met.* 77, 47 (1996).
172. K. Yoshino and A. A. Zahidov, *Synth. Met.* 71, 1875 (1995).
173. P. Peumans, V. Bulović, and S. R. Forrest, *Appl. Phys. Lett.* 76, 2650 (2000).
174. P. Peumans and S. R. Forrest, *Appl. Phys. Lett.* 80, 338 (2002).
175. Ch. Pannemann, V. Dyakonov, J. Parisi, O. Hild, and D. Wöhrle, *Synth. Met.* 121, 1585 (2001).
176. Y. J. Ahn, G. W. Kang, and C. H. Lee, *Mol. Cryst. Liq. Cryst.* 377, 301 (2002).
177. T. Tsuzuki, Y. Shihora, J. Rostalski, and D. Meissner, *Solar Energy Mater. Solar Cells* 61, 1 (2000).
178. M. Murgia, F. Biscarini, M. Cavalini, C. Taliani, and G. Ruani, *Synth. Met.* 121, 1533 (2001).
179. A. Gouloumis, S.-G. Liu, Á. Sastre, P. Vázques, L. Echegoyen, and T. Torres, *Chem. Eur. J.* 6, 3600 (2000).
180. H. Neugebauer, C. Brabec, J. C. Hummelen, and N. S. Sariciftci, *Solar Energy Mater. Solar Cells* 61, 35 (2000).
181. J. C. Hummelen, J. Knol, and L. Sánchez, in "Organic Photovoltaics" (Z. H. Kafafi, Ed.), Vol. 4108, p. 76. SPIE, 2000.
182. F. Padinger, T. Fromherz, P. Denk, C. J. Brabec, J. Zettner, T. Hierl, and N. S. Sariciftci, *Synth. Met.* 121, 1605 (2001).
183. E. A. Katz, D. Faiman, S. Tuladhar, F. Padinger, C. Brabec, and N. S. Sariciftci, in "Proceedings of the 10th Sede Boqer Symposium on Solar Electricity Production" (D. Faiman, Ed.), 2001, p. 127.
184. M. A. Green, K. Emery, D. L. King, S. Igari, and W. Warta, *Progr. Photovolt: Res. Appl.* 11, 39 (2003).
185. S. E. Shaheen, R. Radspinner, N. Peyghambarian, and G. E. Jabbour, *Appl. Phys. Lett.* 79, 2996 (2001).
186. B. Miller, J. M. Rosamilia, G. Dabbagh, R. Tycko, R. C. Haddon, A. J. Muller, W. Wilson, D. W. Murphy, and A. F. Hebbard, *J. Am. Chem. Soc.* 113, 6291 (1991).
187. P. V. Kamat and K.-D. Asmus, *Interface* 5, 22 (1996).
188. L. L. Larina, O. I. Shevaleevskii, and L. A. Cernozatonskii, in "Polycrystalline Semiconductors IV—Physics, Chemistry and Technology" (S. Pizzini, H. P. Strunk, and J. H. Werner, Eds.), Vol. 51–52, p. 479. Solid State Phenomena, Scitec Publications Ltd., Uetikon-Zuerich, Switzerland, 1996.
189. H. Imahori, T. Azuma, K. Ushida, M. Takahashi, T. Akiyama, M. Hasegawa, T. Okada, and Y. Sakata, in "Fullerenes and Photonics IV" (Z. H. Kafafi, Ed.), Vol. 3142, p. 104. SPIE, 1997.
190. S. Licht, P. A. Ramakrishnan, D. Faiman, E. A. Katz, A. Shames, and S. Goren, *Solar Energy Mater. Solar Cells* 56, 45 (1998).
191. M. Kocher, T. K. Däbler, E. Harth, U. Scherf, A. Gügel, and D. Neher, *Appl. Phys. Lett.* 72, 650 (1998).
192. H. Imahori and Y. Sakata, *Eur. J. Org. Chem.* 10, 2445 (1999).
193. P. V. Kamat, S. Barazzouk, K. G. Thomas, and S. Hotchandani, *J. Phys. Chem. B* 104, 4014 (2000).
194. P. V. Kamat, S. Barazzouk, S. Hotchandani, and K. G. Thomas, *Chem.—A Eur. J.* 6, 3914 (2000).
195. A. Ikeda, T. Hatano, S. Shinkai, T. Akiyama, and S. Yamada, *J. Am. Chem. Soc.* 123, 4855 (2001).
196. H. Imahori, T. Hasobe, H. Yamada, P. V. Kamat, S. Barazzouk, M. Fujitsuka, O. Ito, and S. Fukuzumi, *Chem. Lett.* 8, 784 (2001).
197. V. Biju, S. Barazzouk, K. G. Thomas, M. V. George, and P. V. Kamat, *Langmuir* 17, 2930 (2001).
198. S. F. Yang and S. H. Yang, *J. Phys. Chem. B* 105, 9406 (2001).
199. H. Yamada, H. Imahori, and S. Fukuzumi, *J. Mater. Chem.* 12, 2034 (2002).
200. P. K. Sudeep, B. I. Ipe, K. G. Thomas, M. V. George, S. Barazzouk, S. Hotchandani, and P. V. Kamat, *Nano Lett.* 2, 29 (2002).
201. D. Hirayama, K. Takimiya, Y. Aso, T. Otsubo, T. Hasobe, H. Yamada, H. Imahori, S. Fukuzumi, and Y. Sakata, *J. Am. Chem. Soc.* 124, 532 (2002).
202. D. Gust, T. A. Moore, and L. A. Moore, *Res. Chem. Intermed.* 120, 4398 (1997).
203. D. M. Guldi, M. Maggini, G. Scorrano, and M. Prato, *J. Am. Chem. Soc.* 119, 974 (1997).
204. H. T. Tien, L.-G. Wang, X. Wang, and A. L. Ottova, *Bioelectrochem. Bioenerg.* 42, 161 (1997).
205. P. A. Liddell, D. Kuciauskas, J. P. Sumida, B. Nash, D. Nguyen, A. L. Moore, T. A. Moore, and D. Gust, *J. Am. Chem. Soc.* 119, 1400 (1997).
206. N. Martin, L. Sanchez, B. Illescas, and I. Perez, *Chem. Rev.* 98, 2527 (1998).
207. F. Diederich and M. Gómez-López, *Chem. Soc. Rev.* 263 (1999).
208. K. G. Thomas, *Interface* 8, 30 (1999).
209. Y. Kureishi, H. Tamiaki, H. Shiraishi, and K. Maruyama, *Bioelectrochem. Bioenerg.* 48, 95 (1999).
210. D. M. Guldi, C. P. Luo, T. Da Ros, M. Prato, E. Dietel, and A. Hirsch, *Chem. Commun.* 375 (2000).
211. O. Enger, F. Nuesch, M. Fibbioli, L. Echegoyen, E. Pretsch, and F. Diederich, *J. Mater. Chem.* 10, 2231 (2000).
212. H. Imahori, K. Tamaki, H. Yamada, K. Yamada, Y. Sakata, Y. Nishimura, I. Yamazaki, M. Fujitsuka, and O. Ito, *Carbon* 38, 1599 (2000).
213. D. L. Jiang, J. X. Li, P. Diao, Z. B. Jia, R. T. Tong, H. T. Tien, and A. L. Ottova, *J. Photochem. Photobiol. A Chem.* 132, 219 (2000).
214. H. T. Tien and A. L. Ottova, *J. Membrane Sci.* 189, 83 (2001).
215. M. E. El-Khouly, Y. Araki, M. Fujitsuka, A. Watanabe, and O. Ito, *Photochem. Photobiol.* 74, 22 (2001).
216. H. Imahori, K. Tamaki, D. M. Guldi, C. Luo, M. Fujitsuka, O. Ito, Y. Sakata, and S. Fukuzumi, *J. Am. Chem. Soc.* 123, 6617 (2001).
217. Y. Sakata, H. Imahori, and K.-I. Sugiura, *J. Inclusion Phenomena Macrocyclic Chem.* 41, 31 (2001).
218. H. Murakami, R. Matsumoto, Y. Okusa, T. Sagara, M. Fujitsuka, O. Ito, and N. Nakashima, *J. Mater. Chem.* 12, 2026 (2002).
219. T. Markvart, *Prog. Quantum. Electron* 24, 107 (2000).

ENN Encyclopedia of Nanoscience and Nanotechnology
www.aspbs.com/enn

Functionalities of Dendrimers

Toyoko Imae

Nagoya University, Nagoya, Japan

Katsuya Funayama, Yuko Nakanishi, Kenkichi Yoshii

Nagoya University, Nagoya, Japan

CONTENTS

1. INTRODUCTION

The structures and properties of novel dendritic polymers—dendrimers—have been studied extensively as nanomolecules with the attractive, potential applications in a variety of fields [1–8]. In particular, the investigations are focused on the phenomena based on the concept of "dendritic boxes" or "unimolecular micelles" for the capture of guest molecules and as the compartment of chemical reaction. Host-guest systems, consisting of dendrimers, have enormous demands in the fields of molecular recognition and transport, such as drug delivery, gene therapy, and chemical separation.The interest in dendritic polymers for such applications stems from their distinctive dual structural properties, an external periphery bearing multiple functional groups for solubilization in media and an internal region possessing cavities for the capsulation of small molecules.

Novel dendrimers are prepared by covalent bonding from a functional core through the successive repeating synthesis of a spacer and a branching part (divergent method) or from a conjugation at the conic center of dendrons, units of a dendrimer, (convergent method). When two steps are needed for the extension of repeating units, each step is called a half generation. While the structures of dendrimers at low generations are opened and asymmetric, the structures become crowded and spherical during the increase of the generation [9]. These structural changes influence the properties of dendrimers. Then, dendrimers have a unique surface of plural terminal groups, the number of which can be accurately controlled on a synthesis process. Furthermore, it is possible to bind covalently different functional units in the core, branches, and terminal groups on the divergent synthesis process and to conjugate the dendrons of different chemical constituents on the convergent synthesis process. Thus, dendrimers having various structures and properties can be synthesized.

In this article, the up-to-date investigations concerning the functionalities of dendrimers are reviewed. In particular, dendrimer properties of doping, molecular recognition, adsorption, and hybridization are the focus, and the phenomena of solubilization, complexiation, self-organization, and nanocomposite formation are discussed in relation to the structures and functionalities of dendrimers.

2. DOPING AND SOLUBILIZATION OF GUEST MOLECULES

It is well established that the solubility of hydrophobic molecules in water can be dramatically enhanced in the presence of water-soluble, surface-active agents. For instance, the number of pyrene molecules solubilized in nonionic surfactant micelles of heptaethyleneoxide monoalkyl ether (CnE_7) has been found to be linearly proportional to the aggregation number of the micelles [10]. The solubilization capacity for pyrene is low in spherical micelles of ionic dodecyl sulfate, limited at most to two pyrenes per micelle [11]. Since dendrimers have cavity for maintaining small molecules and may be regarded as unimolecular micelles, water-soluble dendrimers are expected to

ISBN: 1-58883-059-4/$35.00

increase the solubility of water-insoluble or weakly soluble organic compounds. The solubilization of guest molecules in such dendrimers has been investigated mostly by a fluorescence probing agent—pyrene [9, 12–15]. Caminanti et al. [9] have used pyrene as a photoluminescence probe to sense hydrophobic sites in poly(amido amine) (PAMAM) dendrimers possessing sodium carboxylated surfaces. The probe method provided experimental evidence for a surface structural transition between generations 3.5 to 4.5. Pistolis et al. [12, 13] have reported that the solubilizing capability of amine-terminated PAMAM and amine-terminated poly(propylene imine) (PPI) dendrimers in aqueous media increases with increasing their generations. Pyrene fluorescence undergoes significant quenching in the solubilized state, and excimer fluorescence is observed. It has been also concluded, by the findings concerning exciplex formation, that water is progressively excluded from the dendrimer interior, as the generation of PPI dendrimer increases. It has been reported by Schmitzer et al. [14] that glucose-persubstituted PAMAM dendrimers increase solubilizate quantities of pyrene and benzoylcyclohexane in water, depending on the number of microcavities in each dendrimer, although the solubility of pyrene is 3–4 times larger than that of benzoylcyclohexane. Quaternary ammonium chloride PPI dendrimers with both hydrophilic (triethylenoxy methyl ether) and hydrophobic (octyl) chain ends have been synthesized by Pan and Ford [15]. These amphiphilic dendrimers solubilize pyrene in aqueous solution, and the limiting solubility corresponds to one pyrene per dendrimer molecule.

In order to clarify the solubilization capacity of dendrimers, depending on structures and conditions, we have investigated ultraviolet-visible and steady-state fluorescence spectroscopy of aromatic guest molecules in aqueous solutions of fourth generation PAMAM dendrimer with hydroxyl and amine terminals (G4 PAMAM-OH dendrimer, G4 PAMAM-NH_2 dendrimer) and fourth and fifth generation PPI dendrimers with amine terminals (G4 and G5 PPI-NH_2 dendrimers) as host molecules. As seen in Figures 1 and 2, pyrene is soluble in aqueous solutions of dendrimers, although the solubility of pyrene in dendrimers depends on terminal groups, internal chemical structures, and generations of dendrimers as well as pH. On the other hand, a guest molecule larger than pyrene—benzopyrene—is not solubilized because of its larger size. However, the solubility of phenantherene and benzene are also not enhanced in the presence of dendrimers under the same conditions, except the solubility (0.2) of phenanthrene into an aqueous solution of G4 PAMAM-NH_2 dendrimer at pH 11.01. Phenanthrene and benzene are presumably too small, in comparison to the sizes of microcavities in dendrimers studied, and easily pass through the cavities. Therefore, the structural matching of guest molecules noninteracting with dendrimers plays an important role in their doping into the microcavities of dendrimers.

Since dendrimers with tertiary amine groups in interior and primary amine groups in exterior have pKa of 6.65 and 9.20, respectively, the electric charges of dendrimers change with the pH of the aqueous solutions of dendrimers [16]. The solubility of pyrene in aqueous solutions of dendrimers below pH 6.65 is larger than above

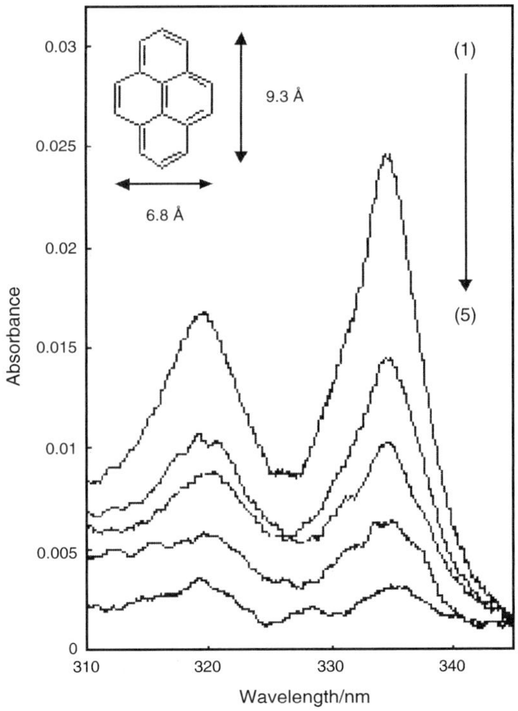

Figure 1. UV-visible spectra of pyrene solubilized in aqueous solutions with and without dendrimers at 25 °C. Powder of guest molecules was added into aqueous solutions of dendrimers at a concentration of 5.0×10^{-7} M, which were adjusted at pH ~6. After the mixtures were stirred overnight, excess guest molecules insoluble in the solutions of dendrimers were filtrated out. Then, UV-visible spectra of the filtrates were measured. From top to bottom: (1) fourth generation hydroxyl-terminated PAMAM dendrimer at pH 5.85; (2) fifth generation amine-terminated PPI dendrimer at pH 6.01; (3) fourth generation amine-terminated PAMAM dendrimer at pH 5.99; (4) fourth generation amine-terminated PPI dendrimer at pH 5.91; (5) without dendrimer.

pH 6.65 (see Fig. 2). This indicates that the protonation of amine groups in interior influences in the doping of pyrene, since the microcavities in the dendrimers are expanded into the preferable sizes by the protonation. Chen et al. [17] have caught the clear evidence for a pH-responsive conformational change of PAMAM dendrimer with amine terminals by means of polarity changes of a polarity-responsive probe, although they could not distinguish between the "denser shell" and "denser core" models. The solubility of pyrene in G4 PAMAM-OH dendrimer is higher than that in G4 PAMAM-NH_2 dendrimer (see Fig. 2). Dendrimers with protonated hydrophilic terminal groups may exclude hydrophobic organic molecules to approach, in contrast to dendrimers with nonionic terminal groups. Incidentally, a positively charged G4 PAMAM-NH_2 dendrimer has demonstrated having "closed" structure [18]. The dimensions of the charged dendrimers have been investigated for carboxylic acid-functionalized PPI dendrimers [19]. The dimension is larger for charged dendrimers and the uncharged dendrimers have the lowest radius. Therefore, the size change of dendrimers occurs with pH.

One must notice that the size change of dendrimers depends not only on the functionality of the branching

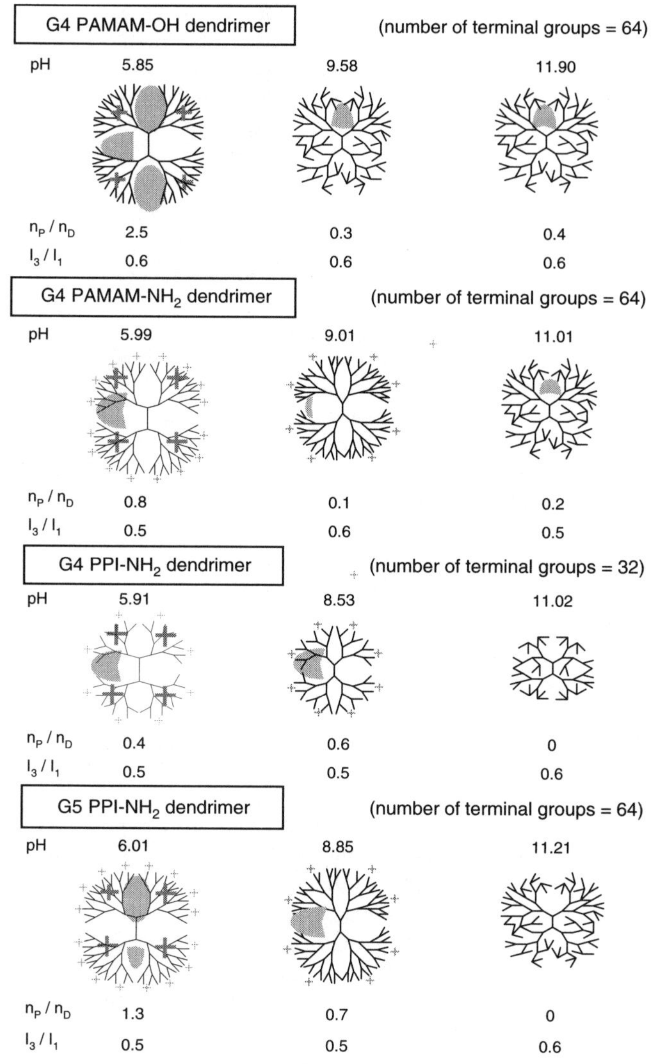

Figure 2. Solubility of pyrene in aqueous solutions of dendrimers at different pHs at 25 °C. The molar fractions of guest molecules per dendrimer were calculated by using molar extinction coefficients of guest molecules (1.7×10^7 cm^{2}mole^{-1} at 337 nm for pyrene, 1.4×10^7 cm^{2}mole^{-1} at 293 nm for phenanthrene). n_P/n_D = number of pyrene molecules solubilized in a dendrimer (solubility). I_3/I_1 = ratio of the third (I_3, 386 nm) to the first (I_1, 375 nm) monomer fluorescence emission intensities of pyrene (polarity index).

points and the functional terminal groups but also on the core, size of spacer, and generation. Although G5 PPI-NH_2 dendrimer and G4 PAMAM-NH_2 dendrimer have the same number and property of terminal groups, the solubility of pyrene in solutions of G5 PPI-NH_2 dendrimer is higher than that of G4 PAMAM-NH_2 dendrimer (see Fig. 2). Since the segment density of a PPI dendrimer is lower than that of a PAMAM dendrimer, due to a longer core alkyl chain [20], guest molecules are doped easier in a PPI dendrimer. The solubility is superior in solutions of G5 PPI-NH_2 dendrimer than in solutions of G4 homologue (see Fig. 2), because of larger volume or microcavity, as found in the solubilization of pyrene in CnE_7 micelles [10].

Pyrene is a useful fluorescence probe for getting micropolarity around it. Micropolarities around solubilization sites of dendrimers can be evaluated as the ratio of the third to the first monomeric fluorescence emission intensities of pyrene, which is well known as a polarity index. As seen in Figure 2, the polarity index of pyrene in aqueous solutions of all dendrimers is close to that in water and independent of the pH. It can be predicted that pyrene exists in the hydrophilic environment in dendrimers, which is invented by the penetration of water [20]. This causes the less solubility of pyrene in dendrimers than in the hydrophobic interior in surfactant micelles, where a larger amount of pyrene is solubilized [10]. Caminati et al. [9] have revealed that pyrene is probably bound in the hydrophilic palisade layer of the PAMAM dendrimer at the small distance from the charged surface groups, supporting the conclusion previously described.

Besides determining the solubilization capacity of dendrimers, extensive works have been reported on the behaviors of solubilized pyrene in dendrimer assemblies and of covalent-bonded pyrene in dendrimers [21–23]. The solubilization of pyrene was used for the determination of micellar formation of amphiphilic linear-dendritic diblock copolymers consisting of hydrophilic linear polyethylene oxide and hydrophobic dendritic carbosilane [21]: The critical micelle concentrations were determined from pyrene fluorescence. It was also found that increasing the size of dendritic block increases the partition equilibrium constants of pyrene in micellar solution. The steady-state fluorescence anisotropy of 1,6-diphenyl-1,3,5-hexatriene in micelles of amphiphilic linear-dendritic diblock copolymers are lower than that of the linear polymeric amphiphiles, suggesting that the microviscosity of the dendritic micellar core is lower than those of linear polymeric micelles. Baker and Crooks [22] have synthesized four generations of PPI dendrimers, which are covalently modified with pyrene moieties, and compared the behavior of pyrene moieties with those of pyrene solubilized in dendrimers. More intensive excimer emission was observed for higher generation dendrimers, while little or no evidence for interdendrimer interactions was observed. Protonation of the tertiary amine units increased monomer fluorescence significantly, while there was only a slight increase in the observed excimer fluorescence. The excitation energy transfer has been investigated between a first generation dendrimer as a donor molecule and a pyrene-containing polymer as an acceptor molecule in Langmuir–Blodgett monolayers [23]. The adjacent donor and accepter layers dramatically increased the fluorescence activity of the pyrene group in a pyrene-containing polymer.

3. MOLECULAR RECOGNITION AND COMPLEXIATION WITH LINEAR POLYMERS

Dendrimers are regularly branched polymers possessing a large number of surface functional groups [24–30]. Their molecular recognition ability is strongly influenced by the nature of the terminal functionalities. The number of terminal groups of dendrimers is increased by the number of branching sites with increasing their generation. Since the terminals behave as functional groups, dendrimers are useful as probes of molecular recognition and carriers of molecular transport. Dendrimers with ionic terminal groups electrostatically interact with counter-ionic groups. Such interaction is useful for catching linear polymers on interpolyelectrolyte complexiation. Many researches for such complexiation were carried out for native polymers [31–43] and synthetic polymers [44–49].

The investigations of the complexation with DNA were performed from the view of gene delivery in mammalian organisms. Since dendrimers are spherical molecules with a size comparable to histone, which is a gene-transporting globular protein, the biomimetic investigation for DNA transfer in the living cells was carried out by using PAMAM dendrimers [31–36]. Plank et al. [31] have examined the complement-activating properties of synthetic cationic molecules and their complexes with DNA. While strong complement activation was seen with a fifth generation PAMAM dendrimer, on the complexiation with DNA it depended on the ratio of polycation and DNA. The ability of PAMAM dendrimers to bind DNAs in a variety of mammalian cells and to enhance their transfer has been investigated by Kukowska-Latallo et al. [32, 33]. Since dendrimers bind various forms of nucleic acids on the basis of electrostatic interaction, the ability of DNA-dendrimer complexes for transfering oligonucleotides and plasmid DNA to mediate antisense inhibition was assessed in an *in-vitro* cell culture system. Although the capability of dendrimers to transfect cells appears to depend on the size and shape of dendrimers and the number of primary amine groups on the surface of the dendrimers, the results indicate that dendrimers function as an effective delivery system. Tang et al. [34] have reported that the transfection activity of the dendrimers is dramatically enhanced by heat treatment. The increased transfection after the heating process is principally due to the increase in flexibility which enables the dendrimer to be compact at the complexiation with DNA and swell at the release from DNA. Gebhart and Kabanov [35] have evaluated nonviral transfection *in vitro*, based on the complexes of DNA and polycations, with respect to their effectiveness, toxicity, and cell-type dependence. The effects of PAMAM and PPI dendrimers were compared to those of other polycations. Luo et al. [36] have chemically modified fifth generation PAMAM dendrimer with biocompatible poly(ethylene glycol) (PEG) chains. This novel conjugate produced a 20-fold increase in transfection efficiency in comparison with partly degraded dendrimer controls. Since the cytotoxicity of PEGylated dendrimer is very low, this kind of compound should be extremely efficient, highly biocompatible, and a low-cost DNA delivery system.

The physicochemical properties, interactions, and structures of DNA-dendrimer complexes have been investigated apart from the gene transport [37–42]. It has been proved by Ottaviani et al. [37] that the concentration dependence of the interaction between polynucleotide and dendrimers is different between small (second generation) and large (sixth generation) dendrimers. The interaction with small dendrimers decreased with an increase in concentration due to self-aggregation of dendrimer molecules. Conversely, the interaction with large dendrimers increased until saturation of the interacting sites occurred. It has also been demonstrated that the supramolecular structure of the complexes changes by varying the mixing ratio of DNA and dendrimers [38]. According to Chen et al. [39], the electrostatic interaction between polynucleotide and PAMAM dendrimer is essential for the effective DNA transfer process. The DNA wraps around seventh generation dendrimers, as illustrated in Figure 3(a), while the wrapping does not occur for fourth and lower generation dendrimers. Kabanov et al. [40] have clarified that electroneutral water-insoluble interpolyelectrolyte complexes of PPI dendrimers interacting with DNA are formed at an equal concentration of amine groups of dendrimer and phosphate groups of DNA. However, the excess addition of fourth- and fifth-generation dendrimers to DNA solution forms positively charged, water-soluble interpolyelectrolyte complexes. Complexed DNA compacts, revealing a wound double-helical structure. These relations may be compared with the Monte Carlo simulations, which were used in the study of the complexation between a polyelectrolyte and an oppositely charged spherical particle in the Debye–Hückel approximation and chain rigidity effects [50, 51]. Bielinska et al. [41] have reported the formation of complicated DNA-dendrimer complexes. Electron microscopic examination of complexes indicated that the majority of plasmid DNA is contracted into isolated toroids, but larger irregular aggregates of polymer and DNA also coexist. This indicates that the binding of plasmid DNA to dendrimer appears to alter its secondary and tertiary structures. Mitra and Imae [42] have revealed the morphology of complicated DNA-dendrimer complexes from atomic force microscopic observation.

Imae et al. [43] have investigated the binding of fourth generation amine-terminated PAMAM dendrimer on native polysaccharide, sodium hyaluronate (NaHA), in an aqueous 0.25 M NaCl solution. The observed variation of molecular weight as a function of mixing ratios of dendrimer to NaHA obeyed the model, where an average number of dendrimers binds to each NaHA chain. However, at a high mixing ratio,

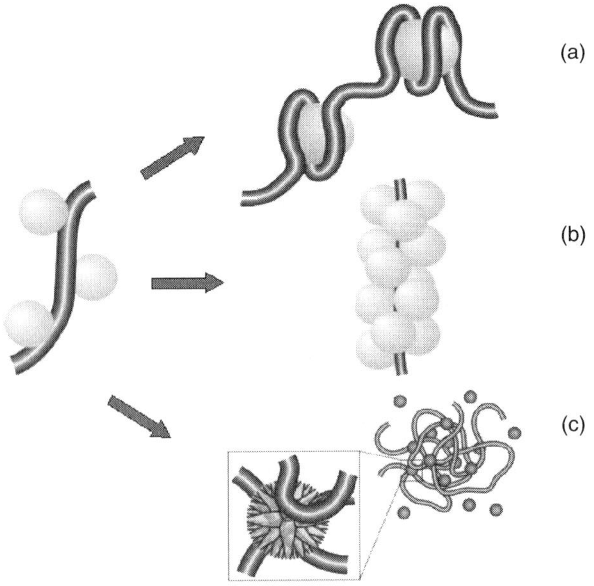

Figure 3. Binding models of cationic dendrimers on linear anionic polymers in solutions: (a) dendrimer-DNA complex; (b) dendrimer-sodium hyaluronate complex; (c) dendrimer-sodium poly(L-glutamate) complex.

the binding is saturated at two dendrimers per three repeating units on a NaHA chain, as illustrated in Figure 3(b). Whereas, at the initial stage of the addition of dendrimer, in a solution, NaHA maintains a worm-like character similar to NaHA without bound dendrimers, NaHA behave like a rigid rod at high dendrimer concentrations. In the NaHA-dendrimer complexiation, the hydrogen-bonding interaction, besides the electrostatic interaction, should play an important role.

The investigation has also been performed for the binding of dendrimers to synthetic linear polyelectrolytes with pH-dependent charge density [44–48]. Dubin et al. [44–47] have reported the complexiation between polycations and carboxylated dendrimers. The complexiation with poly(diallyldimethylammonium chloride) and other polycations occurs most readily for the 7.5th generation with high-charge density and does abruptly at a critical pH. The proposed model is that the polyelectrolyte backbone in the complex is distorted for bending around the contour of the spherical macroions [46]. The interaction of PPI dendrimers to a linear polyanion was compared with that to DNA by Kabanov et al. [48]. That is, dendrimers are penetrable for flexible oppositely charged polyelectrolyte chains. However, rigid negatively charged DNA double helixes apparently bind only to the dendrimer shell. The mixtures of dendrimers with polyanions containing equal amounts of cationic and anionic groups consist of mostly water-insoluble, ion-pair (stoichiometric) complexes but include water-soluble nonstoichiometric interpolyelectrolyte complexes.

The binding of PAMAM dendrimers on sodium poly-L-glutamates (NaPGA) in an aqueous 0.25 M NaCl solution has been reported by Imae and Miura [49]. The apparent molecular weight and the radius of gyration increased rapidly above a certain mixing ratio of dendrimer to NaPGA. The numerical analysis supported the aggregation model that free dendrimers are in equilibrium with dendrimers bound to NaPGA. Dendrimer-NaPGA complexes make aggregates in solution through the junction of dendrimer, and the aggregates consist of i-mer of dendrimer-NaPGA complexes, as shown in Figure 3(c). The aggregates take globular structure with a larger size than that of a single NaPGA without bound dendrimers.

The interesting aspect is the different complexiations of dendrimers with different linear polyelectrolytes such as DNA, NaHA, and NaPGA chains. They depend on the chemical and morphological structures of linear polyelectrolytes. The persistence lengths of the polyelectrolyte chains decrease in the order of rather rigid DNA > semi-flexible NaHA > randomly coiled NaPGA. The common is the electrostatic interaction of charged dendrimers with oppositely charged polyelectrolytes. Additionally, amide or hydroxyl groups in hyaluronate make hydrogen bonding with amide groups in PAMAM dendrimer, and, as a result, many dendrimers are attracted on a NaHA chain, where the steric and electrostatic repulsions between bound guest polyelectrolytes stretch the host polyelectrolyte chain. On the other hand, dendrimers act as a junction connecting between NaPGA chains, and the dendrimer-NaPGA complexes change from "intramolecular" (several guest polyelectrolytes attached to a single host polyelectrolyte) to "intermolecular" (containing more than one host polyelectrolyte), as found in mixtures of polyelectrolytes and surfactant micelles [52]. In the case of DNA, the morphology of complexes with dendrimers is complicated. It seems that the aggregation of complexes occurs simultaneously with the isolated DNA toroid formation including dendrimers and the aggregate sizes are not controlled.

Welch and Muthukumar [53] have examined by the computer simulation based on the theoretical background, the equilibrium and dynamic complexiation behavior of monocentric dendrimer with charged terminal groups to a flexible, oppositely charged polyelectrolyte. They noted three different types of complexes depending upon the ionic strength of solution, the size of the dendrimer, and the length of polymer chain. In three complexes, a dendrimer encapsulates a flexible polyelectrolyte chain, mutually interpenetrates with it, or walks along it. Chodanowski and Stoll [50, 51] have also proposed, as computer-simulated results, the pictures of aggregates of charged dendrimers with oppositely charged polyelectrolytes and the conditions necessary for forming aggregates, such as chain rigidity. Three types of complexes depend upon the ionic strength of solution and the size of the dendrimer or polymer chain, being similar to the results by Welch and Muthukumar [53]. However, both groups did not attach importance to the aggregation of complexes.

We have focused on "intermolecular" dendrimer-NaPGA complexiation. The small-angle X-ray scattering (SAXS) intensity $I(Q)$ at a scattering vector Q for particle solutions is contributed by the number of density particles, the intraparticle form factor depending on the particle geometry, and the interparticle structure factor that is related to the interparticle interactions [20, 54–57]. Figure 4(a) shows SAXS profiles of aqueous 0.25 M NaCl solutions of fourth generation amine-terminated PAMAM dendrimer and its mixture with NaPGA. There is the difference between two $I(Q)$ curves, and both have a peak at $Q = 0.055$ or $0.07\ A^{-1}$. Moreover, SAXS intensities of PAMAM dendrimer decreased with the addition of NaPGA, indicating the interaction between dendrimers and NaPGAs. The complexiation and the aggregation between dendrimer and NaPGA were also supported from dynamic light scattering. Main hydrodynamic diameter (149.4 nm) for a solution of dendrimer-NaPGA mixture was larger than those of free dendrimer (6.2 nm) and free NaPGA.

Normalized intermediate correlation function $I(Q,t)/I(Q,0)$ at a time t from neutron spin echo (NSE) for particle dynamics, which is contributed by two diffusion modes, is described by a double-decaying exponential function [57–60]. As seen in Figure 4(b), the experimental data for a 0.25 M NaCl solution of fourth generation amine-terminated PAMAM dendrimer-NaPGA complex is described by a double-decaying exponential function, although the contribution of the fast mode is small (3–9%) the same as in a case of free dendrimer [60]. This mode may originate in the deformation motion of dendrimer [57]. The major effective diffusion coefficient, that is, the coefficient of the slow mode ($6{\sim}8 \times 10^{-11}\ m^2s^{-1}$), is comparable to the hydrodynamic translational diffusion coefficient ($2{\sim}3 \times 10^{-11}\ m^2s^{-1}$) of the complex obtained from dynamic light scattering. This indicates the translational motion of dendrimers is not disturbed by the complexiation.

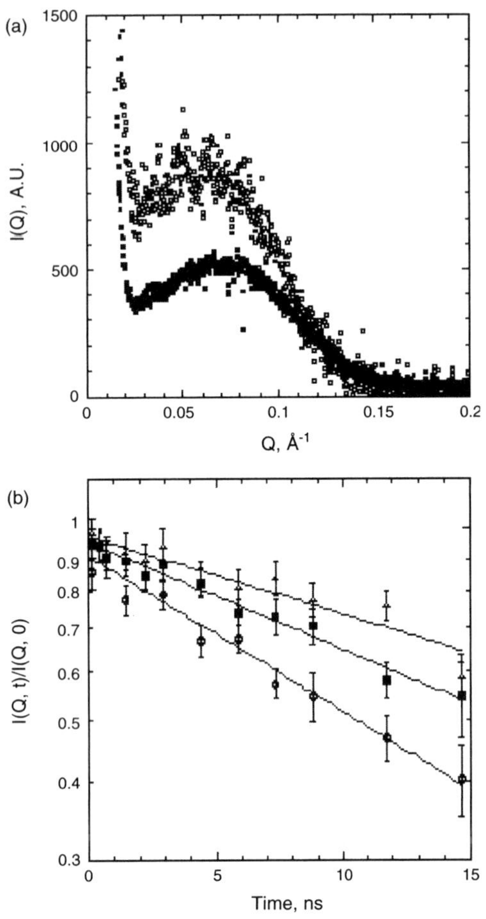

Figure 4. (a) Small-angle X-ray scattering intensities $I(Q)$ as a function of scattering vector Q for aqueous 0.25 M NaCl solutions of fourth generation amine-terminated PAMAM dendrimer (□) and fourth generation amine-terminated PAMAM dendrimer-sodium poly(L-glutamate) mixture (■). Dendrimer and poly(L-glutamate) concentrations are 5 and 0.1 wt%, respectively, and then a number ratio of amine terminal groups of dendrimer against side chains of sodium poly(L-glutamate) is 34.6. (b) Normalized intermediate scattering functions $I(Q, t)/I(Q, 0)$ as a function of time t for a 0.25 M NaCl solution of fourth generation amine-terminated PAMAM dendrimer-sodium poly(L-glutamate) mixture. Q ($Å^{-1}$); △, 0.06: ■, 0.08: ○, 0.1. Solid lines are theoretical ones with a decaying exponential function.

Interpolyelectrolyte complex formation between fourth generation amine-terminated PAMAM dendrimer and NaPGA in aqueous 0.25 M NaCl solution has been studied as a function of pH by Leisner and Imae [61]. Coacervation was observed at pH ~9 and it increased slightly with the dendrimer excess. Three relaxation modes were revealed from dynamic light scattering. The relaxation times of the faster modes are attributed to the hydrodynamic translational diffusions of dendrimer and of the interpolyelectrolyte complex. An additional slow mode dominates the multiexponential dynamic structure factor of the solution of interpolyelectrolyte complex network or the micro-gel inhomogeneity fluctuation. This mode is also associated with a sharp increase of the radius of gyration of mesoscopic particles from ~50 to ~250 nm.

4. ADSORPTION AND SELF-ORGANIZATION AT INTERFACES

4.1. Two-Dimensional Array in Bulk

One of novel utilizations of dendrimers is the construction of self-assemblies and supramolecular assemblies as building blocks, being valuable for the fabrication of higher ordered architectures. The structures of dendritic groups were discussed in relation to their self-assemblies and the liquid crystal formation was predicted [62]. Persec et al. [63, 64] have reported thermotropic nematic liquid crystal formation by dendritic polyethers. Poly(phenylenevinylene)s substituted with dendritic side chains also yielded thermotropic nematic liquid crystals [65]. Hudson et al. [66, 67] have found that flat, tapered, and conical monodendrons, respectively, self-assemble in hexagonal columnar and cubic thermotropic liquid crystal phases with high uniformity. The functionalization of poly(propylene imine) dendrimers with cholesteryl moieties through a carbamate linkage resulted in the formation of smectic A phases at a broad thermal range [68]. A series of polynorbornenes containing second-generation monodendrons as side chain have been synthesized [69]. Although the mesogens are rodlike, the polymers are conformationally flexible overall and displayed liquid-crystal behavior—during cooling, a nematic, smectic, and hexatic phase were observed.

In the cases described above, dendritic polymers with unique structures form the liquid crystals. Now it is assumed that the formation of anisotropically ordered architecture of dendrimers without unique structures is possible, once templates like liquid crystals are used. The lamellar liquid crystal of the cationic surfactant, didodecyldimethylammonium bromide (DDAB) in water, was selected as a template of two-dimensional architecture of anionic PAMAM dendrimers possessing carboxylate terminal groups [70]. The incorporation of dendrimers into the lamellar liquid crystal, at DDAB concentrations in the range of 4–30 wt %, resulted in the transition from a monophasic lamellar structure (L_α) to biphasic lamellar mixtures (L_α and L_α^D), at a similar molar mixing ratio for two dendrimers of 2.5 and 4.5 generations, as illustrated in Figure 5. The anionic dendrimer molecules doped into the water domains stick to the DDAB bilayers in the L_α structure, due to the stronger electrostatic interaction between the cationic surfactants and the oppositely charged dendrimers. In the narrow-spaced L_α^D structure, monolayers of flattened dendrimers are adsorbed between the DDAB bilayers, since the thicknesses of the dendrimer monolayers are smaller than the diameters of the corresponding dendrimers in aqueous solutions.

The investigations on the binding interactions and aggregation processes between dendrimers and surfactants have been carried out at low surfactant concentrations. Those are concerned with the supramolecular assemblies of surfactants with dendrimers [71–74], the dendrimer adsorption onto micelle surface [75, 76], and the structural modification of vesicles upon the interaction with dendrimers [77, 78]. The supramolecular assemblies are constructed by the primary noncooperative binding of monomeric surfactants and the secondary cooperative binding of micelle-like surfactants

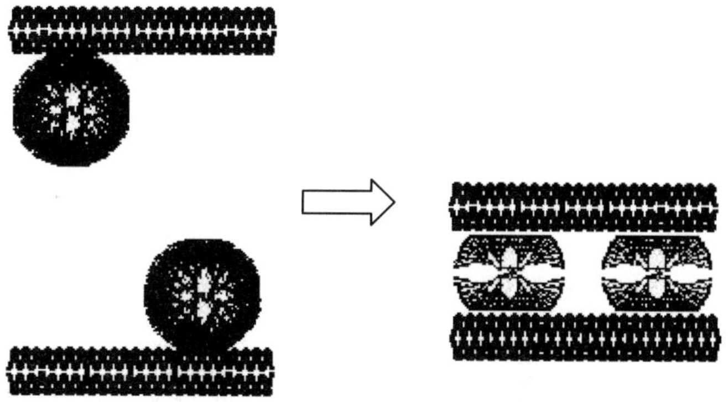

Figure 5. Schematic models of the lamellar L_α structure (left) and the lamellar L_α^D structure (right) in mixtures of didodecyldimethylammonium bromide, anionic PAMAM dendrimer, and water. Reprinted with permission from [70], X. Li et al., *J. Phys. Chem. B* 106, 12170 (2002). © 2002, American Chemical Society.

in/on dendrimer [71, 73]. Conversely, the reports at high surfactant concentrations are very few [79, 80]. Friberg et al. [79] have reported the formation of lamellar liquid crystal by octanoic acid and poly(ethylene imine) dendrimer. A third-generation dendrimer served as a solvent in the liquid crystal, and surfactants ionized and formed ionic pairs with amine terminals of dendrimer. Baars et al. [80] have investigated the scattering on the addition of dendrimers to continuous liquid crystals, but did not refer the morphology of hybrid liquid crystal structures.

4.2. Adlayers

Organization of dendrimers can be attained by the construction of one-, two-, and three-dimensional architectures. Especially, two-dimensional ordering is readily formed at air-liquid, air-solid, solid-liquid, and liquid-liquid interfaces. It is well known that traditional low-molecular-weight surfactants consist of hydrophilic head and lipophilic tail moieties and are characterized as typical amphiphilic molecules owing to their surface activity and association behavior. These kinds of molecules are suitable materials for the organization at the interface. On the process of the stepwise synthesis of dendrimers, central core, spacer, branch, and terminal group can be modified, and a variety of functional moieties or blocks are conjugated in dendrimers. Then, amphiphilic (hydrophilic and hydrophobic) characters can be introduced in dendrimers [81].

Linear polymer-dendrimer hybrids with unique head-tail structure have been synthesized [82–91]. Such copolymers with novel chemical structure possess specific characteristics such as amphiphilicity besides intrinsic dendritic and polymeric characters. They are surface active and arranged as a monolayer at the air-water interface. Amphiphilic dendrimers with a spherical shape have also been synthesized. Core-shell block dendrimers have different amphiphilicity at interior concentric generation layers and exterior generation layers [92–96]. Surface-block dendrimers have a character that the amphiphilicity of two hemispherical parts in dendrimers is different [82, 97–102]. That is, amphiphilic surface-block dendrimers consist of surface blocks with different affinities for solvents, where one is hydrophilic and the other is hydrophobic. Such structural character is dominant not only to associate into micelles, vesicles, microemulsions, liquid crystals, and other self-assemblies but also to adsorb at interfaces. Recently, Imae [81] has reviewed the association behavior of amphiphilic dendritic polymers.

Two dimensionally organized systems of amphiphilic dendrimers are prepared as Langmuir or Gibbs monolayer at an air-water interface and adsorption layer or self-assembled monolayer on solid substrate. Gibbs monolayer formation of amphiphilic dendrimers has been investigated for third and fourth generation amphiphilic surface-block dendrimers with amine and n-hexyl terminals (amine/hexyl), hydroxyl and n-hexyl terminals (hydroxyl/hexyl) and N-acetyl-D-glucosamine and n-hexyl terminals (glucosamine/hexyl) at the air-water interface [81, 99]. The organized adlayer formation and adsorption kinetics have been reported for the same dendrimers on the liquid-solid interfaces [101, 102]. The amine/hexyl and hydroxyl/hexyl dendrimers formed bilayers and their accumulation, owing to their amphiphilic character, which were different from the disordered adsorption of symmetric PAMAM dendrimer, as seen in Figure 6. Dendrimers in the bilayer form the flattened "pancake" structure by pairing between hydrophobic surface blocks. The hydrophilic terminals face to the solution and the solid substrate, since the adsorption film surface is hydrophilic. On the other hand, the adlayer of the glucosamine/hexyl dendrimers was rather flat. The time dependence of the adlayer formation was monitored *in-situ*. It proceeded through fast and slow adsorption steps. The adsorbed amount at the equilibrium decreased in the order of hydroxyl/hexyl > amine/hexyl > glucosamine/hexyl dendrimers and increased linearly with dendrimer concentration. The adsorption was abundant for the third

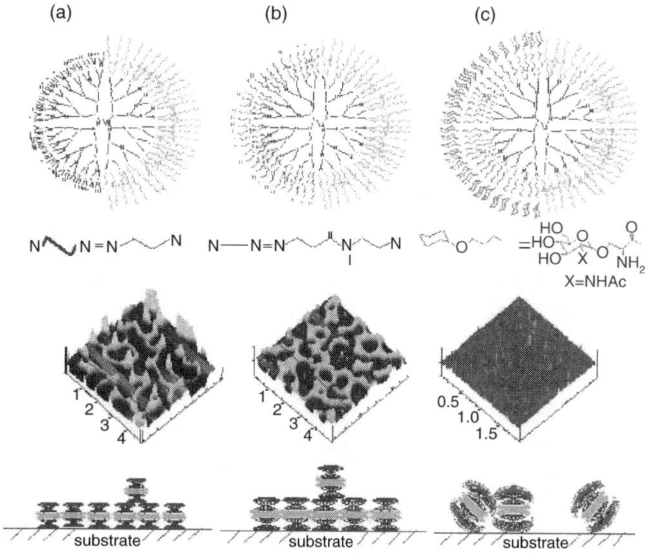

Figure 6. Adlayer formation by amphiphilic surface-block dendrimers on solid substrates. Reprinted with permission from [102], M. Ito et al., *Langmuir* 18, 9757 (2002). © 2002, American Chemical Society. (Top) chemical structure; (middle) atomic force microscopic image; (bottom) Adlayer model. (a) Fourth generation amine/hexyl PAMAM dendrimer; (b) fourth generation hydroxyl/hexyl PAMAM dendrimer; (c) fourth generation glucosamine/hexyl PAMAM dendrimer.

generation dendrimer in comparison to the fourth generation dendrimer and proceeded more for surface-block dendrimers than for PAMAM dendrimer in relation to the hydrophilicity-hydrophobicity balance of dendrimers. The molecular orientation of the hexyl chains in the adlayers was in the order of amine/hexyl > hydroxyl/hexyl > glucosamine/hexyl dendrimers.

Functional dendrimers of symmetrically branched structure are also focused as building blocks of the architectures such as thin adlayer films on solid substrates. Watanabe and Regen [103] have reported multilayer construction by PAMAM dendrimers on solid substrate. Adsorption and aggregation of carbosilane dendrimers and their hydroxyl- or mesogen substitutes on mica and pyrolytic graphite have been investigated by Sheiko et al. [104, 105] and Coen et al. [106]. Bar et al. [107] have prepared dendrimer-modified silicon oxide surfaces and used them as a platform for the deposition of gold and silver colloid monolayers. The thermodynamics and kinetics of adsorption of redox-active dendrimers, containing ferrocenyl moieties on the periphery, have been studied on a Pt electrode surface [108]. Adsorption isotherm depended on the applied potential. At the potential where the ferrocenyl sites are in the reduced form, the adsorption thermodynamics were characterized by the Langmuir adsorption isotherm. When the ferrocenyl sites are oxidized, the electrodeposition of a multilayer of the dendrimer took place. Focally substituted organothiol dendrons were used to construct self-assembled adlayers on a gold surface [109]. As the number of hyperbranches in the dendron increases, the dendron adlayers become more permeable, indicating a tradeoff between size and the packing efficiency. This research suggests the formation of homogeneous but permeable dendron adlayers on the gold surface. Díaz et al. [110] have reported the interfacial reaction of the terpyridyl-pendant dendrimers with Fe^{2+} and Co^{2+} on highly oriented pyrolytic graphite surfaces. The films form highly ordered, two-dimensional hexagonal arrays, which appear to be composed of one-dimensional polymeric strands. The dimensions obtained are consistent with the size of the dendrimer. The ordering in the dendrimer film is dependent on the dendrimer generation. The films are electrochemically active. Sabapathy and Crooks [111] have prepared a hydroxyl-terminated monolayer from fourth generation PAMAM dendrimer and reacted it with heptanoyl chloride to yield ester-coupled bilayers. The interfacial reactivity was discussed in relation to the monolayer structure and the intrinsic properties of the particular coupling reaction studied. The lateral organization of alkyl-substituted polyphenylene dendrimers on the basal plane of graphite has been investigated by Loi et al. [112]. The dendrimers consist of a core of twisted, interlocked benzene rings and an external shell of dodecyl chains. While three kinds of dendrimers show a tetrahedral or a disk-like shape in solution, the dendrimers on graphite spontaneously form a stable, almost pinhole-free monolayer. Complex two-dimensional arrangements and supramolecular ordering were observed in the monolayers prepared by spin-coating. One prominent structure is regions of parallel rows of 6 nm spacing. In addition, pairs of dendrimers formed two-dimensional crystals on graphite. The crystal structure depends sensitively on the structure of the dendrimer, on the solvent, and on the concentration. The interfacial behavior of third and fourth generations of hyperbranched polyesters with hydroxyl-terminal groups was studied by Sidorenko et al. [113]. The molecular adsorption on a bare silicon surface was described in terms of the Langmuir isotherm. A higher adsorption amount was found for a lower generation. The shape of a third generation dendrimer within an adsorbed layer evolved from a pancake within a thickness less than 1 nm for very low surface coverage to densely packed, worm-like bilayer structure with a thickness of about 3 nm for the highest surface coverage. The fourth generation molecules hold a stable close-to-spherical shape with a diameter of 2.5 nm throughout the entire range of surface coverage. The intermolecular flexibility or constrained mobility of dendrimers was considered to be responsible for different surface behavior.

Recently, the morphology of aggregates of dendrimers on the solid substrate was investigated. Sato et al. [114] have reported that dendrimers aggregate in a form of nanometer-sized round dots over an area measured in centimeters on rapid evaporation of a solvent from a thin cast film of an electrolyte solution of poly(amido amine) dendrimers on mica. Chemically specific adsorption, electrostatic interactions, phase transitions, and dewetting instabilities were shown not to be responsible for the dot formation. Changes in temperature and humidity have only little effect. There is a threshold evaporation rate for the formation of dots. Once formed, the dot size and spacing are independent of the evaporation rate. The self-assembly of monodendrons with peripherally attached alkyl substituents has been studied by scanning tunneling microscopy (STM) [115]. The disk-like assembly, a characteristic structure of low-generation dendrimers, was directly observed. The subunits of disk structure can be well resolved as well as the ordered alkyl parts. Xiao et al. [116] have reported organosilane thin films derived on mica surfaces from $SiCl_3$-terminated carbosilane dendrons by means of spin-coating. The morphology of the films was highly dependent upon the generation of the dendrons and the film thickness—mesoscopic ring, disk, or hole structures were observed. These structures were composed of nanoparticles with sizes corresponding to one dendron molecule or the cluster of a few laterally bound molecules. At submonolayer coverage, the molecules tended to flatten and spread out on mica surfaces. The mechanisms for the formation of the observed film morphology were discussed, based on the structure and properties of the dendrons and the effect of the evaporation process.

4.3. Self-Assembled Monolayers

The organized self-assembled monolayers (SAM) of dendrimers were prepared by using a template SAM on substrate. Amine-terminated PAMAM dendrimers were covalent-linked with organized, surface-confined, self-assembled monolayers of mercaptoundecanoic acid [117]. Yoon et al. [118] have also prepared same dendrimer monolayer on acid SAM on gold and then functionalized the dendritic surface amine groups with biotin analogues or desthiobiotin amidocaproate. For testing the association/dissociation reaction cycles at the affinity surface, avidine adlayer was formed onto the biotin analogue-functionalized surface and displaced with free biotin. With the optimized affinity-surface construction steps and reaction conditions, continuous association/dissociation reaction cycles were achieved, resulting in a repeatedly regenerable affinity-sensing surface. Ionized, carboxylate-terminated polyphenylene dendrimer molecules were adsorbed on a carboxylic acid group-terminated, self-assembled monolayer on a gold substrate through the linkage with Cu^{2+} ions [119]. The strong interaction between dendrimer and Cu^{2+} ions, the latter of which is preadsorbed on the COOH SAM, led to a compression or deformation of the SAM and resulted in a decreased height of the dendrimer ions. Fail et al. [120] have immobilized amine-terminated PAMAM dendrimers onto anhydride-functionalized pulsed plasma polymer surfaces via amide linkage formation. The PAMAM dendrimer layers were useful for a variety of surface-related phenomena of fluorination, adhesion, and gas barrier enhancement.

Nagaoka and Imae [121] have reported adsorption structure of amine-terminated PAMAM dendrimer onto 3-mercaptopropionic acid (MPA) SAM on Au island film. The carboxylic acid species of SAM diminished, as the adsorption of PAMAM dendrimer from an aqueous solution proceeded. This indicates that the protonated, amine terminal groups of the dendrimer bind electrostatically to carboxylate groups of SAM. The flattened molecular structure or the imperfect covering of the adsorbed dendrimers was assumed. The dome-shaped, "spread out" configuration of PAMAM dendrimers on substrate has been visually observed by Li et al. [122].

Adsorption kinetics of fourth generation amine-terminated PAMAM and PPI dendrimers from aqueous solutions has been investigated on MPA SAM on Au film [123]. The adsorption kinetics at high dendrimer concentrations obeyed two-step adsorption process [124] but not Langmuir adsorption isotherm. Then, monolayer and additional adsorptions of dendrimers must occur on MPA SAM. There is a difference on the adsorption structure between two dendrimers, as seen in Figure 7. PAMAM dendrimer adsorbs with "the hemi-micellar structure," as illustrated in Figure 7(a), because amine and amido groups of PAMAM dendrimer interact electrostatically and through hydrogen bonding with carboxylate of MPA SAM. The adsorption

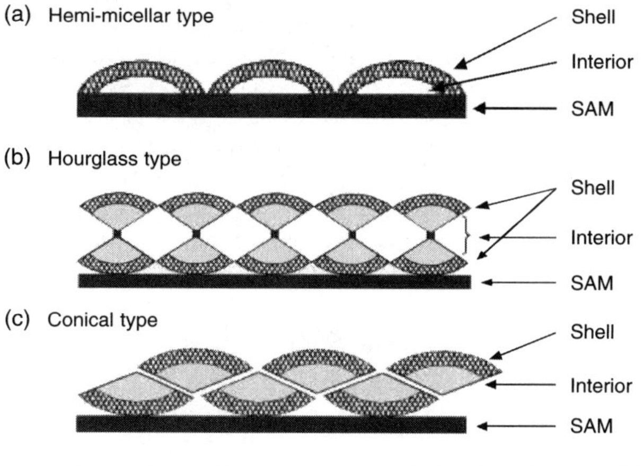

Figure 7. Adsorption models of PAMAM and PPI dendrimers on MPA SAM. Reprinted with permission from [123], H. Nagaoka and T. Imae, *Int. J. Nonlinear Sci. Numer. Simu.* 3, 233 (2002). © 2002, Freund Publishing House Ltd.

structure of PPI dendrimer is "the hour glass type" or "the conical type" (see Fig. 7(b) and (c)), since there is no hydrogen-bonding interaction between dendrimer interior and MPA SAM.

In the case of PAMAM dendrimers, surface-functionalized with ruthenium (II), the adsorption thermodynamics of this redox-active metallodendrimers onto Pt electrodes was well characterized by the Langmuir adsorption isotherm [125]. The adsorption kinetics was found to be activation-controlled with the rate-constant decreasing with decreasing dendrimer generation. The rate of adsorption of positively charged poly(propylene imine) dendrimers on glass, an oppositely charged surface, has been studied as a function of generation and charge [126, 127]. The adsorption kinetics was mostly diffusion/convection-controlled with a linear dependence on the bulk concentration. At high bulk concentration, there is a drop in concentration dependence, and this crossover concentration shifts to higher concentrations with decreasing pH. Bahman et al. [128] have studied *in-situ* adsorption of PAMAM dendrimers from dilute solutions onto cleaned gold. In ethanol, the equilibrium surface coverage corresponds to almost monolayer and increases slightly with generation. The adsorption is likely driven by the weak favorable interaction between the primary amine end groups on PAMAM dendrimer and gold. In aqueous media, under the conditions where the PAMAM amine end groups tend to protonate and bear positive charge, the equilibrium surface coverage grows exponentially with generation, indicating the multilayer formation. The formation of multilayers can be explained by a favorable electrostatic image-charge interaction between the positively charged dendrimers and the gold substrate. At the seventh generation, there was a manifestation of "dense-shell" packing, which lowers the effective charge on the higher generation dendrimers.

Stable ultra-thin self-assembled films were prepared by dendron thiol. Zhang and his collaborators [129, 130] have synthesized polyether dendron with a thiol group at the focal point and used it for the preparation of self-assembled monolayers on metal surface. The dendron SAM covered the gold surface homogeneously and consisted of the ordered stripes by the packing of two flat-cone shaped dendrons. The patterned stripes can be improved by thermal annealing. From the studies of the kinetic process of dendron thiol self-assembling on gold, it was shown that the dendron thiol assembling proceeds with different adsorption rates depending on the assembly time [130]. They have also fabricated ultra-thin organic films via dendritic growth on modified gold, silicon, and quartz substrates [131].

4.4. Composite Films

Investigation was extended to the preparation of composite self-assembled monolayers including dendrimer primarily by Crooks et al. [132–136]. PAMAM dendrimers were surface-immobilized in thiol, self-assembled monolayers by means of two different methods [132], which are the preparation of chemically sensitive dendrimer surfaces as a vapor-phase dosing probe. Zhao et al. [133] have embedded surface-confined dendrimers within self-assembled monolayers of alkane thiol. These films act as ion gates of molecular dimension. The two-dimensional phase behavior of mixed monolayers composed of amine-terminated PAMAM dendrimers and n-alkylthiols has been studied by Lackowski et al. [134]. Mixed monolayers were prepared by sequential immersion of Au substrates in ethanolic solutions of dendrimers and then n-alkylthiols. Time-dependent morphological changes were observed in the mixed monolayers. Dendrimer monolayers immersed in a hexane solution of n-alkylthiols did not phase-separate to an appreciable extent. Moreover, thiol-terminated dendrimers did not phase-separate either, when exposed to ethanolic n-alkylthiol solutions. It was suggested that one of the driving forces for the phase segregation in the mixed monolayers is the difference in adsorption energies of the amine and thiol groups against the Au substrate. Tokuhisa et al. [135] have reported that exposure of high-generation dendrimer monolayers to ethanoic solutions of hexadecanethiol results in a dramatic compression of the dendrimers and causes them to reorient on the surface from an oblate to prolate configuration without desorb. It was shown that the primary driving force for the structural change is the solvation of the dendrimers. Oh et al. [136] have recently described a strategy for constructing a dendrimer-based electrochemical current rectifier permitting current flow in only one direction. The dendrimer-based rectifying layer was prepared by direct adsorption of ferrocene-functionalized poly(propylene imine) dendrimers onto Au surfaces, followed by adsorption of n-hexadecanethiol. The surface coverage of ferrocenyl dendrimers in the two-component monolayers is estimated to be about 50% of maximum coverage. It was revealed that the ferrocenyl dendrimer/n-alkanethiol-modified electrodes exhibit excellent barrier properties, which prevent direct oxidation/reduction of the solution-phase redox molecules. Electrochemical current rectification occurred via mediated electron transfer across the surface-confined ferrocenyl dendrimers. Friggeri et al. [137] have investigated the insertion process of individual, dendrimer sulfide molecules into self-assembled monolayers of 11-mercapto-1-undecanol. The immersion of thiol self-assembled monolayers in a dendrimer solution of increasing concentrations at longer immersion times leads an increase in the dendrimer number inserted into the thiol layer. A mechanism consists of rapid dissociation of surface thiols and follows slow dendrimer adsorption. The rate-determining step of the process is the insertion of the individual dendrimer molecules. On the dissociation step, first-order kinetics was found for the fraction of covered substrate versus time of exposure to solvent.

Layer-by-layer composite films were also prepared. Electrostatic layer-by-layer composite deposition films were fabricated by mutual self-assembly of dendritic macromolecules of two adjacent generations [138]. The formation of layer-by-layer composite films on cationic SAM has carried out by the stepwise adsorption of two PAMAM dendrimers among amine-, carboxylate-, and hydroxyl-terminated species, and is compared with the competitive adsorption of the same paired dendrimers in aqueous solutions [139]. There was apparent conformational change of hydroxyl-terminated dendrimers in the first layer by the accumulated adsorption of amine- or carboxyl-terminated dendrimers, as seen in Figure 8. Normal layer-by-layer accumulation was formed in other combinations of two dendrimers in three species.

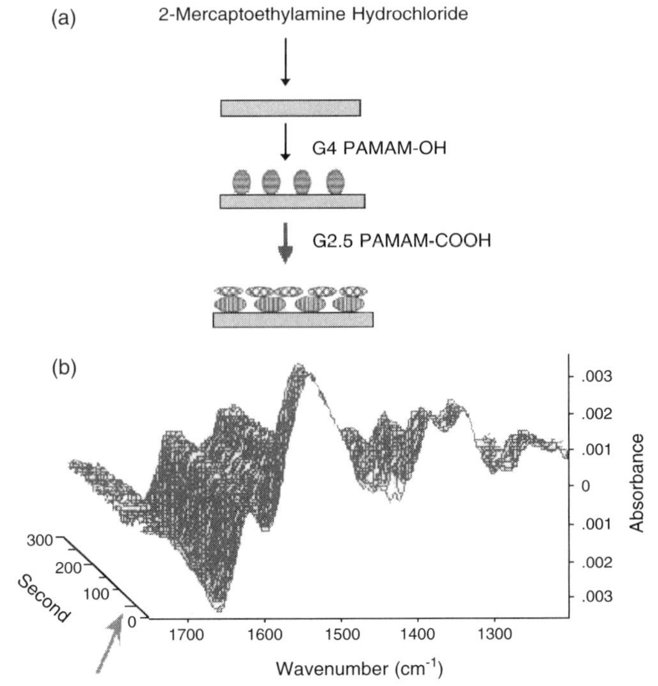

Figure 8. Stepwise adsorption of carboxylate and hydroxyl-terminated dendrimers on 2-mercaptoethylamine hydrochloride SAM. After the adsorption equilibrium of 4th generation hydroxyl-terminated PAMAM dendrimer was attained, the desorption of adlayer in water was equilibrated. The adlayer after the desorption equilibrium was used for the adsorption of 2.5th generation carboxyl-terminated PAMAM dendrimer. (a) Scheme of the stepwise adsorption; (b) the *in-situ* observation of adsorption of 2.5th generation carboxyl terminated PAMAM dendrimer on the adlayer of 4th generation hydroxyl-terminated PAMAM dendrimer. The absorption bands of amide II and OH bending vibrations on attenuated total reflection surface-enhanced infrared absorption spectroscopy decreased in intensity, after the adsorption of carboxylate terminated dendrimer started.

Wang et al. [140] have constructed a photosensitive, layer-by-layered, self-assembled, ultra-thin film based on polyanionic PAMAM dendrimer (with carboxylate terminals) and polycationic nitrocontaining diazoresin via sequential deposition and subsequent UV irradiation, which causes the linkage between the layers to change from ionic to covalent. PAMAM dendrimer/poly(styrenesulfonate) (PSS) multiplayer films were prepared by an iterative electrostatic self-assembly process, namely, the layer-by-layer deposition of PAMAM dendrimer and PSS onto planar supports [141]. The evidence of multi-player film growth was provided. An adsorption-desorption phenomenon was sensitive to the pH and ionic strength of the PSS and dendrimer solutions, the dendrimer generation and concentration, and the PSS molecular weight and concentration.

5. HYBRIDIZATION AND NANOCOMPOSITE FORMATION WITH METAL NANOPARTICLES

Clusters of metals and semiconductors are focused due to their unique mechanical, electronic, optical, magnetic, and chemical properties. Metal nanoparticles are also investigated because of their novel optical, electrical, catalytic, and other properties. Homogeneous, self-assembled monolayers or thin films of alkanethiol, N-hexadecylethylenediamine [142], polymer [143, 144], polymerized methyl methacrylate [145], hydrophobically modified dendrimers [146–150] have been used as stabilizers in the synthesis of stable hydrophobic nanoparticles. On the other hand, the synthesis of water-soluble hydrophilic particles is recently required in relation to the environmental pollution. Then, water-soluble dendrimers were focused as valuable materials for the protection of clusters and nanoparticles.

Zhao et al. [151] and Balogh and Tomalia [152] have introduced a template-synthesis strategy for preparing copper nanoclusters within dendrimer nanoreactors. Transition metal ions are condensed into the interior of a fourth generation PAMAM dendrimer and chemically reduced to zerovalent copper. Nanocomposite clusters prepared ranged in size from 4 to 64 atoms [151]. The surface properties of the host dendrimers seem to determine the solubility of the metal domains [152].

Esumi et al. [153] have prepared gold colloids by the reduction of metal salt by means of ultraviolet irradiation in the presence of zero to fifth generation PAMAM dendrimers with amine terminals. The average particle sizes decreased by increasing the concentration of dendrimers, and gold colloids with a diameter less than 1 nm were obtained in the presence of third to fifth generation dendrimers above the molar ratio of amine group of dendrimer: $HAuCl_4 = 4:1$. Sugar-persubstituted PAMAM dendrimers (sugar balls) were also used for the spontaneous formation of gold nanoparticles [154]. Au^{3+} ions were reduced by the hydroxyl groups of the sugar balls, resulting in the formation of very stable gold nanoparticles. Garcia et al. [155] have stated that large dendrimers encapsulate the colloids and lower-generation dendrimers give rise to larger colloids. Fourth-generation PAMAM dendrimers having terminals functionalized with thiol groups have been synthesized by Chechik and Crooks [156]. While these thiolated dendrimers formed stable monolayers on planar Au substrates, the reduction of $HAuCl_4$ in the presence of thiol-modified dendrimers resulted in the formation of water-soluble, dendrimer-stabilized nanoparticles with 1.5–2.1 nm in diameter. Stable nanoparticles were obtained at Au/dendrimer ratios as large as 120:1.

Esumi et al. [157] have compared the role of generations (third to fifth generation) and terminal groups (amine and carboxyl) of PAMAM dendrimers for preparing nanoparticles of gold, platinum, and silver. Whereas the size of gold and silver nanoparticles decreased with increasing the concentration and the generation of dendrimers, the size of platinum nanoparticles was insensitive against the concentration as well as the generation. Stable nanoparticles were obtained above unity of the ratio of surface group of dendrimer/metal salt for gold and silver nanoparticles and above 40 for platinum nanoparticles. It was suggested that the metal nanoparticles adsorb on the exteriors of the dendrimers.

Manna et al. [158] have synthesized silver and gold nanospherical particles stabilized by a fourth-generation, amine-terminated PAMAM dendrimer by the reduction of $AgNO_3$ and $NaAuCl_4$. The particle size could be controlled by the metal ion-to-dendrimer mixing ratio. Silver nanoparticles are always larger than gold nanoparticles, when they are synthesized by the same condition. A short-ranged hexagonal arrangement of particles was observed in a monolayer onto a carbon-coated copper transmission electron microscopic grid. Then, the terminal amine groups of the dendrimers take part in the stabilization of the nanoparticles, and the dendrimers protecting nanoparticles behave as a spacer packed between the ordered arrangements of the nanoparticles.

PAMAM denderimers with amine terminals possess both internal and external functional groups providing reaction sites for metal ions. Then, it is very important in the application that nanoparticles are either formed in the interior of dendrimers or at the external surface of dendrimers. The previous reports indicated that the hybrid structures of dendrimer-metal nanoparticles depend on the generation, concentration, and terminal group of dendrimers, the type of metals, and the ratio of dendrimers and metals. Gröhn et al. [149, 159, 160] have clarified the influence of reaction conditions and dendrimer generation on the resulting nanostructures of dendrimer-metal hybrid colloids. Second to fourth generation PAMAM dendrimers behave like colloid stabilizers; that is, several dendrimers surround the surface of the metal particle, as shown in Figure 9. On the other hand, sixth to tenth dendrimers act as effective nanotemplates. Then, metal particles are completely formed inside individual dendrimers. Gröhn et al. [160] have prepared hydrophilic polymer networks containing higher generation PAMAM dendrimers, where nanoparticles are located inside. Crooks et al. [161] have used hydroxyl-terminated PAMAM dendrimers for the preparation of metal nanoparticles in the interior of dendrimers, because the dendrimers do not have functional terminal groups as reaction sites for metal ions. They reviewed synthesis and characterization of dendrimer-encapsulated metal nanoparticles and their applications to catalysis. Recently, Ottaviani et al. [162] have achieved the transformation of silver from Ag^+ to Ag^0 in the interior of dendrimer by irradiating the solution of Ag^+-dendrimer complex by means of an excessive dose of X-rays to avoid alternation of the

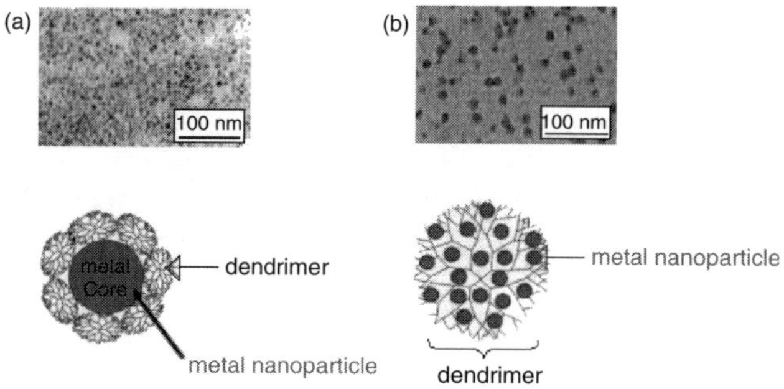

Figure 9. Transmission electron microscopic (TEM) photographs and models of dendrimer-passivated and dendrimer-encapsulated nanoparticles. TEM: (a) deposition on a carbon-coated cupper grid of 4th generation amine-terminated PAMAM dendrimer-passivated gold nanoparticles prepared at the 1:10 ratio for the Au atom:terminal amine group of dendrimer; (b) dendrimer-encapsulated nanoparticles. Reprinted with permission from [159], F. Gröhn et al., *Macromolecules* 33, 6042 (2000). © 2000, American Chemical Society.

structure of the complexes during transformation into nanocomposites.

High stability of dendrimers surrounding the surface of the metal particles is indispensable not only in the protection of particles but also in the utilization of particles. We have investigated the exchange of dendrimers on gold nanoparticles by dodecanethiols at various mixing ratios of nanoparticle and dodecanethiol. The wine-red aqueous solution of dendrimer-passivated gold nanoparticles was vigorously shaken with an n-heptane or diethyl ether solution of dodecanethiols, until the exchange equilibrium was reached. When two phases were separated into the upper organic and lower aqueous phases, wine-red color shifted into organic phase. Figure 10(a) shows the absorption spectra of the water and n-heptane phases after the exchange reaction. The plasmon band (around 510 nm) intensity of nanoparticles in the aqueous phase decreased with the exchange and, at the same time, the band intensity in n-haptane phase increased. The intensity variation of the plasmon band as a function of the mixing ratio of dodecanethiol to gold ions is plotted in Figure 10(b). Although the average particle sizes are maintained even after the exchange reaction, the infrared absorption spectroscopic vibration bands of dendrimers exchanged by absorption bands of dodecanethiol.

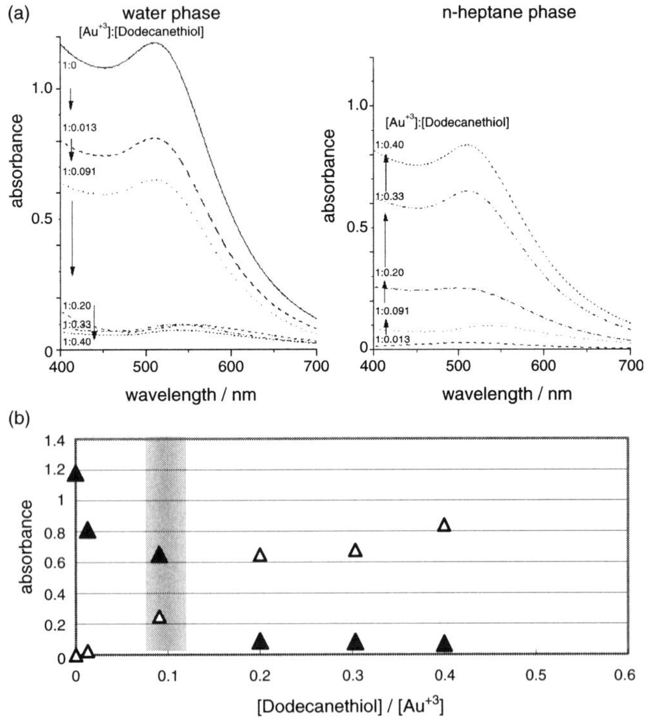

Figure 10. (a) The ultraviolet-visible absorption spectra of hybrid-layer-passivated nanoparticles in water and in n-hexane phases after the ligand replacement reaction at the different ratios of dodecanethiol against Au atom, [dodecanthiol]/[Au^{3+}]. Dendrimer-passivated gold nanoparticles in water and dodecanethiol in n-heptane were shaken until the equilibrium was reached, and the water phase was separated from the n-heptane phase. (b) Plasmon band intensity, as a function of [dodecanthiol]/[Au^{3+}] in water phase (▲) and in n-heptane phase (△) after the ligand replacement reaction. In the shadowed region, solutions of both phases colored.

Although the exchange of dendrimer by dodecanethiol in diethyl ether solution proceeded as well as in n-heptane solution, dendrimer on nanoparticles did not exchange with N-hexadecylethylenediamine in n-heptane solution. On the other hand, N-hexadecylethylenediamine, which was passivated on gold nanoparticles in n-heptane solution, also did not exchange with free dendrimer in water. (N-hexadecylethylenediamine-passivated gold nanoparticles were prepared according to the procedure previously reported for the preparation of silver nanoparticles [142].) The ligand replacement reaction is schematically illustrated in Figure 11. The replacement by the addition of a small amount of dodecanethiol indicates that the interaction of a gold particle surface with amine terminal groups of the dendrimer is weaker than that with thiol. However, the interaction of dendrimer and N-hexadecylethylenediamine on gold nanoparticle surface cannot be discriminated.

In the process of the replacement of dendrimer-passivated nanoparticles by dodecanethiols, gold nanoparticles, surface-confirmed by hybrid self-assembled monolayers, was synthesized in the mixing ratio shadowed in Figure 10(b), where both solutions of water and organic solvent were colored. The formation of the hybrid monolayer is very feasible by the partial replacement of surface-confined dendrimer onto the gold particle surface. The hybrid-layer-passivated nanoparticles separated into domains of small and large spheres [163]. That is, the change of the hydrophilic particle surface to the amphiphilic one induces the segregation into the close-packed monodispersed phases. This indicates that the replacement may be dependent on the particle size. The potentialities of nanoparticles stabilized by hybrid monolayers of fascinating molecules, such as dendritic polymers and surfactants, may be high as well as those of the homogeneous ones. Thus, the hydrophilic/hydrophobic hybrid-layer-protected nanoparticles may stimulate the new development of fundamental surface sciences and new applications of such nanoparticles on chemical sensing, catalysis,

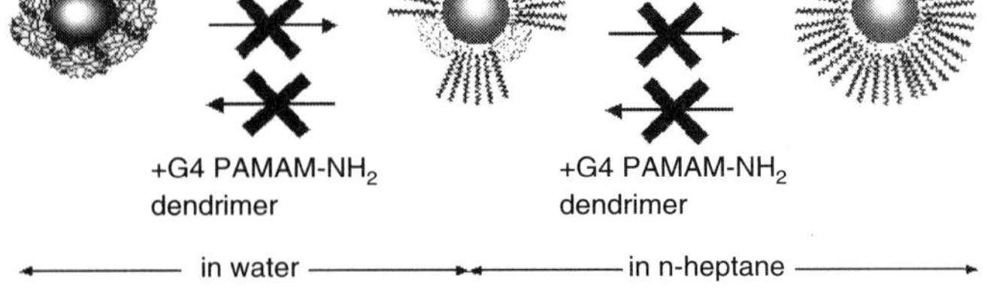

Figure 11. The schematic presentation of ligand replacement reaction. (a) Dendrimer-passivated gold nanoparticles in water + dodecanethiol in n-heptane or diethyl ether. (b) Dendrimer-passivated gold nanoparticles in water + N-hexadecylethylenediamine in n-heptane and N-hexadecylethylenediamine-passivated gold nanoparticles in n-heptane + dendrimer in water.

coating, etc., different from homogeneous hydrophobic or hydrophilic nanoparticles.

Esumi and Goino [164] have reported the adsorption of PAMAM dendrimer on alumina and silica particles at pH 5. The adsorption of dendrimers with surface ionic groups on oppositely charged particles increased with increasing dendrimer generation. While the earlier generation dendrimers behave as ordinary electrolytes, the later generation dendrimers behave as ionic surfactant or polyelectrolyte for the inorganic particle dispersion. The adsorption amounts of dendrimers with carboxylate terminals were greater than those with amine terminals. Esumi et al. [165, 166] have investigated the simultaneous adsorption of PAMAM dendrimers and anionic surfactants on positively charged alumina particles. There was competitive adsorption between the dendrimers with carboxylate terminal groups and sodium dodecyl sulfate at pH 5 [165]. A preferential adsorption of 5.5th generation over 1.5th generation was also observed in the competitive adsorption with anionic surfactant. The adsorption amount of sugar-persubstituted poly(amido amine) dendrimers, sugar balls, at pH 3.5 changes through a maximum, while the adsorption of anionic surfactant increases with increasing it [166]. The enhancement in the simultaneous adsorption is due to the adsorption of complexes consisting of sugar ball and anionic surfactant. In addition, the dispersion stability of alumina suspensions, which was caused by the adsorption of the sugar ball and anionic surfactant, depends on the zeta potential of alumina. The generation of sugar ball and the kind of anionic surfactant did not significantly influence the adsorption characteristics.

There are very few reports of nanocomposites of dendrimers with nanoparticles except metal nanoparticles. Naka et al. [167] have carried out the crystallization of $CaCO_3$ in the presence of PAMAM dendrimer containing carboxylate groups at the external surface. Spherical vaterite crystals were formed, different from the formation of rhombohedral calcite crystals in the absence of additives. CdS in PAMAM dendrimer solutions precipitated as aggregates composed of dendrimer-stabilized nanoclusters with long-term optical stability [168]. Kinetics for the growth of aggregates was discussed—temperature and reactant concentration are dominant variables controlling the aggregation kinetics.

6. CONCLUSIONS

Applications of dendrimers, in connection with the novel characters of dendrimers, are considered in the field of not only industry but also life science. Concerning the doping ability, one of the potentials for the use of PAMAM dendrimers, which are adaptable to native bodies, is biomedical application as magnetic resonance imaging (MRI) contrast agents and drug delivery for diagnoses [5, 7]. Recently, the potentials of water-soluble dendritic unimolecular micelles as drug delivery agents were explored using a model drug—indomethacin [169]. Entrapment of indomethacin in the dendritic micelles was achieved at 11 wt % loading content, and preliminary *in-vitro* release tests showed that sustained release characteristics were achieved. This study supports the possible application of dendritic unimolecular micelles with a hydrophobic core and a hydrophilic shell in drug delivery and suggests the necessary construction of dendrimers with possible structures to optimize chemical, physical, and pharmacological properties.

For the complexiation of PAMAM dendrimers with DNA, which is a mimicry of nucleohistone, the utilization as gene deliverer or vector is examined. Moreover, the complexiation ability of ionic dendrimers with polyelectrolytes may be relevant for applications in polymer separation and viscosity control. The extension of complexiation investigation of dendrimers with linear polyelectrolytes is the formation of gels and networks using dendrimers as building blocks. Amphiphilic hydrogels with highly shaped persistent cross-link junctions were synthesized using linear blocks and perfectly branched (dendritic) macromolecules [170]. The swelling of the gels formed is affected by the relative content of linear polymer, the polarity of the medium, and the temperature. In the preparation of the hydrogels, the influence of various factors on the degree of crystallinity and phase segregation must be considered. Covalently cross-linked, three-dimensional nano-domained networks were prepared from radially layered poly(amido amine-organosilicon) copolymers containing hydrophilic PAMAM dendrimer interiors and reactive organosilicon exteriors [171]. Sizes and shapes of the nanoscopic domains can be controlled by the selection of dendrimer network precursors and by conditions applied to the network conformation.

Architectures of dendrimers designed at the molecular level on interfaces are developed for the purpose of manifold applications such as molecular, chemical, biomimetic, electronic, photochromic, and pH-sensitive sensors or devices. Then, the dendrimer molecules that produce stable, functional, and select interfacial architectures as nanoreactors and nanoreservoirs are exploited as many nanoscopic smart materials or their building blocks, including molecular deliveries, segregations, reaction catalysts, transport agents, and molecular recognitions. The sensitive and selective sensor layers were prepared from polyphenylene dendrimers and used to monitor the concentration of various organic compounds in different environments [172]. The homogeneous, rigid layers of polyphenylene dendrimers allowed the selective detection with high sensitivity and accuracy of polar aromatic target molecules. This is comparable to the benzamido-terminated PAMAM dendrimer-type gas sensor [132]. This sensor probed volatile, organic molecules of heptane, carbon tetrachloride, trichloroethylene, benzene, and 1-buthanol. These types of sensors should be of increasing significance in environmental science. The difference of microcavities between two dendrimers was elucidated using solvatochromic probe, phenol blue [173]. Fluorescence sensing a 2.5th generation PAMAM dendrimer with a naphthalene core unit was prepared, and its potential use as a pH sensor material was examined [174]. It displayed sensitive fluorescence signal amplification towards tetrafluoroacetic acid in comparison to acetic acid. It is then demonstrated that the effects of chemical structural differences of dendrimers are essential as selective sensing probe for different target molecules, and the design and synthesis of more elaborate dendrimers are a fascinating challenge for fabricating the best sensors.

Recently, the preparation of patterned thin films including dendrimers was reported [175–178]. The soft lithographic technique, microcontact-printing method, was used to transfer patterned thin films to surfaces with submicrometer resolution [175]. The fourth-generation amine-terminated PAMAM dendrimers were used as the "ink" molecules. Stable multilayer structures up to roughly 60 nm in height were prepared from dendrimer solutions of high concentrations. The deposition of gold-containing dendrimers on stripe-patterned self-assembled monolayer substrates was studied using the microcontact printing method [176]. The atomic force microscopic (AFM) image showed that the PAMAM dendrimers are selectively deposited on hydrophilic –COOH stripes of the patterned substrates but not on the hydrophobic $-CH_3$ stripes. Arbitrary submicrometer patterns of cobalt were fabricated on a nonconducting surface with wet chemical and microcontact printing methods [177]. A hydroxyl-terminated dendrimer was transferred from a stamp to the surface. The guests, palladium ions, adsorbed into the dendrimer layer as host and acted as nucleation centers for electroless cobalt plating. Patterned dendrimer/gold nanoparticles, immobilized on polymeric substrates, were prepared by masking a patterning method [178]. Polymer substrate, such as poly(dimethylsiloxane) or poly(ethyleneterephthalate) film, was plasma-treated with maleic anhydride. The resulting succinic anhydride groups on the surface were reacted with amine groups of PAMAM dendrimers containing gold salts. The film was irradiated with ultraviolet light through a photomask. The gold particles were well patterned on the surface of the polymer film.

New families of nanocomposite clusters and particles were synthesized. Nanoscale super clusters, assembled around a dendritic core, were synthesized and characterized as quantum dots [179]. Compounds were prepared by binding cluster units $[Ru_5C(CO)_{12}]$ or $[Au_2Ru_6C(CO)_{16}]$ to third-generation dendrimers. Those show a molecular multi-layer structure, which is viewed as a sphere consisting of a nonconducting dendritic core with a sheath of conducting metal clusters over the surface and, in turn, being covered by an additional outer layer of insulating carboxyl ligands. Their properties suggest that they may be of use in a gas delivery/storage system switched thermally or electrochemically, delivering metal particles of precise nuclearity for nanoelctronics and catalysis. Polymerization of styrene in aqueous dispersions of sodium dodecyl sulfate and PPI dendrimer with initiation by potassium persulfide at 80 °C produced latexes [180]. Microscopic observation showed independent particles with diameters of 26–64 nm and small clusters.

GLOSSARY

Dendrimer Spherical polymer prepared by covalent-bonding from a functional core through the successive repeating synthesis of a spacer and a branching part.

Deoxyribonucleic acid (DNA) A biopolymer consisting of two polynucleotide chains with the sugar-phosphate backbones. Two chains forming right-handed helices run in opposite direction.

Lamellar layer One type of lyotropic liquid crystals constructed by surfactant molecules. Molecules are associated into accumulates of bilayers with hydrophobic and hydrophilic spaces.

Linear polymer Molecule with high-molecular-weight which was prepared by the polymerization of bifunctional monomers.

Self-assembled monolayer Monolayer formed by the interaction (covalent bond) of adsorbed molecules with the substrate.

Sodium hyarulonate Native polysaccharide of the glucosamminoglycan family, which is composed of alternating units of D-glucuronic acid and N-acetyl-D-glucosamine.

Sodium poly-L-glutamate Polypeptide composed of L-glutamic acid which is one of components of protein. It takes structures of α-helix, β-form and random coil, depending on the condition.

REFERENCES

1. D. A. Tomalia, A. M. Naylor, and W. A. Goddard, III, *Angew. Chem. Int. Ed.* 29, 138 (1990); *Angew. Chem.* 102, 119 (1990).
2. J. M. J. Fréchet and C. J. Hawker, in "Comprehensive Polymer Science," 2nd suppl. (G. Allen, Ed.). Elsevier Science, Pergamon, Oxford, 1996.
3. G. R. Newkome, C. N. Moorefield, and F. Vögtle, "Dendritic Molecules: Concepts, Synthesis, Perspectives." VCH, Weinheim, 1996.
4. F. Zeng and S. C. Zimmerman, *Chem. Rev.* 97, 1681 (1997).
5. M. Fischer and F. Vögtle, *Angew. Chem. Int. Ed.* 38, 884 (1999).
6. V. Percec and M. N. Holerca, *Biomacromolecules* 1, 6 (2000).
7. R. Esfand and D. A. Tomalia, *DDT* 6, 427 (2001).
8. F. Vögtle, "Dendrimers III Design, Dimension, Function, 212 Topics in Current Chemistry." Springer, Berlin, 2001.
9. G. Caminanti, N. J. Turro, and D. A. Tomalia, *J. Am. Chem. Soc.* 112, 8515 (1990).
10. C. Honda, M. Itagaki, R. Takeda, and K. Endo, *Langmuir* 18, 1999 (2002).
11. J.-H. Kim, M. M. Domach, and R. D. Tilton, *Langmuir* 16, 10037 (2000).
12. G. Pistolis, A. Milliartis, C. M. Paleos, and D. Tsiourvas, *Langmuir* 13, 5870 (1997).
13. G. Pistolis and A. Milliartis, *Langmuir* 18, 246 (2002).
14. A. Schmitzer, E. Perez, I. Rico-Lattes, A. Lattes, and S. Rosca, *Langmuir* 15, 4397 (1999).
15. Y. Pan and W. T. Ford, *Macromolecules* 33, 3731 (2000).
16. D. Leisner and T. Imae, *J. Phys. Chem. B* 197, 8098 (2003).
17. W. Chen, D. A. Tomalia, and J. L. Thomas, *Macromolecules* 33, 9169 (2000).
18. S. Jockusch, J. Ramirez, K. Sanghvi, R. Nociti, N. J. Turro, and D. A. Tomalia, *Macromolecules* 32, 4419 (1999).
19. I. B. Bietveld, W. G. Bouwman, M. W. P. L. Baars, and R. K. Heenan, *Macromolecules* 34, 8380 (2001).
20. K. Funayama, T. Imae, K. Aoi, K. Tsutsumiuchi, M. Okada, M. Furusaka, and M. Nagao, *J. Phys. Chem. B* 107, 1532 (2003).
21. Y. Chang, Y. C. Kwon, S. C. Lee, and C. Kim, *Macromolecules* 33, 4496 (2000).
22. L. A. Baker and R. M. Crooks, *Macromolecules* 33, 9034 (2000).
23. U. Oertel, D. Appelhans, P. Friedel, D. Jehnichen, H. Komber, B. Pilch, B. Hänel, and B. Voit, *Langmuir* 18, 105 (2002).
24. C. J. Hawker and J. M. J. Fréchet, *J. Am. Chem. Soc.* 112, 7638 (1990).
25. N. C. Greenham, S. C. Moratti, D. D. C. Bradley, R. H. Friend, and A. B. Holmes, *Nature* 365, 628 (1993).

26. C. J. Hawker, K. L. Woley, and J. M. J. Fréchet, *Chem. Soc. Perkin. Trans. Part 1*, 1287 (1993).
27. J. M. J. Fréchet, *Science* 263, 1710 (1994).
28. M. Antonietti and C. Goltner, *Angew. Chem. Int. Ed. Engl.* 36, 910 (1997).
29. M. Zhao, L. Sun, and R. M. Crooks, *J. Am. Chem. Soc.* 120, 4877 (1998).
30. L. Balogh and D. A. Tomalia, *J. Am. Chem. Soc.* 120, 7355 (1998).
31. C. Plank, K. Mechtler, F. C. Szoka, Jr., and E. Wagner, *Human Gene Therapy* 7, 1437 (1996).
32. J. F. Kukowska-Latallo, A. U. Bielinska, J. Johnson, R. Spindler, D. A. Tomalia, and J. R. Baker, Jr., *Proc. Natl. Acad. Sci. USA* 93, 4897 (1996).
33. A. U. Bielinska, J. F. Kukowska-Latallo, J. Johnson, D. A. Tomalia, and J. R. Baker, Jr., *Nucl. Acids Res.* 24, 2176 (1996).
34. M. X. Tang, C. T. Redemann, and F. C. Szoka, Jr., *Bioconjugate Chem.* 7, 703 (1996).
35. C. L. Gebhart and A. V. Kabanov, *J. Control Rel.* 73, 401 (2001).
36. D. Luo, K. Haverstick, N. Belcheva, F. Han, and W. M. Saltzman, *Macromolecules* 35, 3456 (2002).
37. M. F. Ottaviani, B. Sacchi, N. J. Turro, W. Chen, S. Jockusch, and D. A. Tomalia, *Macromolecules* 32, 2275 (1999).
38. M. F. Ottaviani, F. Furini, A. Casini, N. J. Turro, S. Jockusch, D. A. Tomalia, and L. Messori, *Macromolecules* 33, 7842 (2000).
39. W. Chen, N. J. Turro, and D. A. Tomalia, *Langmuir* 16, 15 (2000).
40. V. A. Kabanov, V. G. Sergeyev, O. A. Pyshkina, A. A. Zinchenko, A. B. Zezin, J. G. H. Joosten, J. Brackman, and K. Yoshikawa, *Macromolecules* 33, 9587 (2000).
41. A. U. Bielinska, J. F. Kukowska-Latallo, and J. R. Baker, Jr., *Biochim. Biophys. Acta* 1353, 180 (1997).
42. A. Mitra and T. Imae, *Biomacromolecules*, in press.
43. T. Imae, T. Hirota, K. Funayama, K. Aoi, and M. Okada, *J. Coll. Interf. Sci.* 263, 306 (2003).
44. Y. Li, P. L. Dubin, R. Spindler, and D. A. Tomalia, *Macromolecules* 28, 8426 (1995).
45. H. Zhang, P. L. Dubin, R. Spindler, and D. A. Tomalia, *Ber. Bunsenges. Phys. Chem.* 100, 923 (1996).
46. H. Zhang, P. L. Dubin, J. Ray, G. S. Manning, C. N. Moorefield, and G. R. Newkome, *J. Phys. Chem. B* 103, 2347 (1999).
47. N. Miura, P. L. Dubin, C. N. Moorefield, and G. R. Newkome, *Langmuir* 15, 4245 (1999).
48. V. A. Kabanov, A. B. Zezin, V. B. Rogacheva, Zh. G. Gulyaeva, M. F. Zansochova, J. G. H. Joosten, and J. Brackman, *Macromolecules* 32, 1904 (1999).
49. T. Imae and A. Miura, *J. Phys. Chem. B* 107, 8088 (2003).
50. P. Chodanowski and S. Stoll, *Macromolecules* 34, 2320 (2001).
51. S. Stoll and P. Chodanowski, *Macromolecules* 35, 9556 (2002).
52. J. Xia, H. Zhang, D. R. Rigsbee, P. Dubin, and T. Shaikh, *Macromolecules* 26, 2759 (1993).
53. P. Welch and M. Muthukumar, *Macromolecules* 33, 6159 (2000).
54. S. H. Chen and T. L. Lin, *Methods Exp. Phys.* 23, 489 (1987).
55. T. Imae, K. Funayama, K. Aoi, K. Tsutsumiuchi, M. Okada, and M. Furusaka, *Langmuir* 15, 4076 (1999).
56. K. Funayama and T. Imae, *J. Phys. Chem. Solids* 60, 1355 (1999).
57. K. Funayama, T. Imae, K. Aoi, K. Tsutsumiuchi, M. Okada, H. Seto, and M. Nagao, *J. Phys. Soc. Jpn.* 70, 326 (2001).
58. S. Komura, H. Seto, and T. Takeda, *Progr. Colloid Polym. Sci.* 106, 1 (1997).
59. H. Matsuoka, Y. Yamamoto, M. Nakano, H. Endo, H. Yamaoka, R. Zorn, M. Monkenbusch, D. Richter, H. Seto, Y. Kawabata, and M. Nagao, *Langmuir* 16, 9177 (2000).
60. K. Funayama, T. Imae, H. Seto, K. Aoi, K. Tsutsumiuchi, M. Okada, M. Nagao, and M. Furusaka, *J. Phys. Chem. B* 107, 1353 (2003).
61. D. Leisner and T. Imae, *J. Phys. Chem. B*, in press.
62. V. Percec, C.-H. Ahn, G. Ungar, D. J. P. Yeardley, M. Möller, and S. S. Sheiko, *Nature* 391, 161 (1998).
63. V. Percec and M. Kawasumi, *Macromolecules* 25, 3843 (1992).
64. V. Percec, P. Chu, and M. Kawasumi, *Macromolecules* 27, 4441 (1994).
65. Z. Bao, K. R. Amundson, and A. J. Lovinger, *Macromolecules* 31, 8647 (1998).
66. S. D. Hudson, H.-T. Jung, V. Percec, W.-D. Cho, G. Johansson, G. Ungar, and V. S. K. Balagurusamy, *Science* 278, 449 (1997).
67. H.-T. Jung, S. O. Kim, Y. K. Ko, D. K. Yoon, S. D. Hudson, V. Percec, M. N. Holerca, W.-D. Cho, and P. E. Mosier, *Macromolecules* 35, 3717 (2002).
68. D. Tsiourvas, T. Felekis, Z. Sideratou, and C. M. Paleos, *Macromolecules* 35, 6466 (2002).
69. Z. Liu, L. Zhu, Z. Shen, W. Zhou, S. Z. D. Cheng, V. Percec, and G. Ungar, *Macromolecules* 35, 9426 (2002).
70. X. Li, T. Imae, D. Leisner, and M. A. López-Quintela, *J. Phys. Chem. B* 106, 12170 (2002).
71. N. J. Caminati, N. J. Turro, and D. A. Tomalia, *J. Am. Chem. Soc.* 112, 8515 (1990).
72. D. M. Watkins, Y. Sayed-Sweet, J. W. Klimash, N. J. Turro, and D. A. Tomalia, *Langmuir* 13, 3136 (1997).
73. S. M. Ghoreishi, Y. Li, J. F. Holtzwarth, E. Khoshdel, J. Warr, D. M. Bloor, and E. Why-Jones, *Langmuir* 15, 1938 (1999).
74. M. Miyazaki, K. Torigoe, and K. Esumi, *Langmuir* 16, 1522 (2000).
75. X. Li, X. He, A. C. H. Ng, C. Wu, and D. K. P. Ng, *Macromolecules* 33, 2119 (2000).
76. Y. Li, C. A. McMillan, D. M. Bloor, J. Penfold, J. Warr, J. F. Holzwarth, and E. Wyn-Jones, *Langmuir* 16, 7999 (2001).
77. M. F. Ottaviani, R. Daddi, M. Brustolon, N. J. Turro, and D. A. Tomalia, *Langmuir* 15, 1973 (1980).
78. M. F. Ottaviani, P. Favuzza, B. Sacchi, N. J. Turro, S. Jockusch, and D. A. Tomalia, *Langmuir* 18, 2347 (2002).
79. S. E. Friberg, M. Podzimek, D. A. Tomalia, and D. M. Hedstrand, *Mol. Cryst. Liq. Cryst.* 164, 157 (1988).
80. M. W. P. L. Baars, M. C. W. van Boxtel, C. W. M. Bastiaansen, D. J. Broer, S. H. Sontjens, and E. W. Meijer, *Advanced Materials* 12, 715 (2000).
81. T. Imae, in "Structure-Performance Relationships in Surfactants: Second Edition, Revised and Expanded" (K. Esumi, Ed.). Marcel Dekker, New York, 525, 2002.
82. J. M. J. Fréchet, *Science* 263, 1710 (1994).
83. T. M. Chapman, G. L. Hillyer, E. J. Mahan, and K. A. Shaffer, *J. Am. Chem. Soc.* 116, 11195 (1994).
84. J. C. M. van Hest, D. A. P. Delnoye, M. W. P. L. Baars, M. H. P. van Genderen, and E. W. Meijer, *Science* 268, 1592 (1995).
85. J. C. M. van Hest, M. W. P. L. Baars, C. Elissen-Román, M. H. P. van Genderen, and E. W. Meijer, *Macromolecules* 28, 6689 (1995).
86. K. Aoi, A. Motoda, M. Okada, and T. Imae, *Macromol. Rapid Commun.* 18, 945 (1997).
87. J. Iyer, K. Fleming, and P. T. Hammond, *Macromolecules* 31, 8757 (1998).
88. D. Yu, N. Vladimirov, and J. M. J. Fréchet, *Macromolecules* 32, 5186 (1999).
89. C. Román, H. R. Fischer, and M. Meijer, *Macromolecules* 32, 5525 (1999).
90. J. P. Kampf, C. W. Frank, E. E. Malmström, and C. J. Hawker, *Langmuir* 15, 227 (1999).
91. K. Aoi, A. Motoda, M. Ohno, K. Tsutsumiuchi, M. Okada, and T. Imae, *Polymer J.* 31, 1071 (1999).
92. K. Aoi, K. Itoh, and M. Okada, *Macromolecules* 28, 5391 (1995).
93. K. Aoi, K. Tsutsumiuchi, E. Aoki, and M. Okada, *Macromolecules* 29, 4456 (1996).
94. M. C. Coen, K. Lorentz, J. Kressler, H. Frey, and R. Mülhaupt, *Macromolecules* 29, 8069 (1996).
95. S. Stevelmans, J. C. M. van Hest, J. F. G. A. Jansen, D. A. F. J. van Boxtel, E. M. M. de Brabander-van der Berg, and E. W. Meijer, *J. Am. Chem. Soc.* 118, 7398 (1996).

96. A. P. H. J. Schenning, C. Elissen-Román, J.-W. Weener, M. W. P. L. Baars, S. J. van der Gaast, and E. W. Meijer, *J. Am. Chem. Soc.* 120, 8199 (1998).
97. C. J. Hawker, K. L. Wooley, and J. M. J. Fréchet, *J. Chem. Soc. Perkin Trans.* 1, 1287 (1993).
98. K. Aoi, K. Itoh, and M. Okada, *Macromolecules* 30, 8072 (1997).
99. K. Aoi, H. Noda, K. Tsutsumiuchi, and M. Okada, *IUPAC 37th International Symposium on Macromolecules* 765 (1998).
100. T. Imae, K. Funayama, K. Aoi, K. Tsutsumiuchi, and M. Okada, in "Yamada Conference L, Polyelectrolytes" (I. Noda and E. Kokufuta, Eds.), 439 (1999).
101. T. Imae, M. Ito, K. Aoi, K. Tsutsumiuchi, H. Noda, and M. Okada, *Coll. Surf. A: Phys. Eng. Asp.* 175, 225 (2000).
102. M. Ito, T. Imae, K. Aoi, K. Tsutsumiuchi, H. Noda, and M. Okada, *Langmuir* 18, 9757 (2002).
103. S. Watanabe and S. L. Regen, *J. Am. Chem. Soc.* 116, 8855 (1994).
104. S. S. Sheiko, G. Eckert, G. Ignat'eva, A. M. Muzafarov, J. Spickermann, H. J. Räder, and M. Möller, *Macromol. Rapid Commun.* 17, 283 (1996).
105. S. S. Sheiko, A. M. Muzafarov, R. G. Winkler, E. V. Getmanova, G. Echert, and P. Reineker, *Langmuir* 13, 4172 (1997).
106. M. C. Coen, K. Lorenz, J. Kressler, H. Frey, and R. Mülhaupt, *Macromolecules* 29, 8069 (1996).
107. G. Bar, S. Rubin, R. W. Cutts, T. N. Taylor, and T. A. Zawodzinski, *Langmuir* 12, 1172 (1996).
108. K. Takada, D. J. Díaz, H. D. Abruña, I. Cuadrado, C. Casado, B. Alonso, M. Morán, and J. Losada, *J. Am. Chem. Soc.* 119, 10763 (1997).
109. C. B. Gorman, R. L. Miller, K.-Y. Chen, A. R. Bishop, R. T. Haasch, and R. G. Nuzzo, *Langmuir* 14, 3312 (1998).
110. D. J. Díaz, G. D. Storrier, S. Bernhard, K. Takeda, and H. D. Abruña, *Langmuir* 15, 7351 (1999).
111. R. C. Sabapathy and R. M. Crooks, *Langmuir* 16, 1777 (2000).
112. S. Loi, U.-M. Wiesler, H.-J. Butt, and K. Müllen, *Macromolecules* 34, 3661 (2001).
113. A. Sidorenko, X. W. Zhai, S. Peleshanko, A. Greco, V. V. Shevchenko, and V. V. Tsukruk, *Langmuir* 17, 5924 (2001).
114. M. Sato, J. Okamura, A. Ikeda, and S. Shinkai, *Langmuir* 17, 1807 (2000).
115. P. Wu, Q. Fan, G. Deng, Q. Zeng, C. Wang, and C. Bai, *Langmuir* 18, 4342 (2002).
116. Z. Xiao, C. Cai, A. Mayeux, and A. Milenkovic, *Langmuir* 18, 7728 (2002).
117. M. Wells and R. M. Crooks, *J. Am. Chem. Soc.* 118, 3988 (1996).
118. H. C. Yoon, M.-Y. Hong, and H.-S. Kim, *Langmuir* 17, 1234 (2001).
119. H. Zhang, P. C. M. Grim, D. Liu, T. Vosch, S. De Feyter, U.-M. Wiesler, A. J. Berresheim, K. Müllen, C. Van Haesendonck, N. Vandamme, and F. C. De Schryver, *Langmuir* 18, 1801 (2002).
120. C. A. Fail, S. A. Evenson, L. J. Ward, W. C. E. Schofield, and J. P. S. Badyal, *Langmuir* 18, 264 (2002).
121. H. Nagaoka and T. Imae, *Trans. Mater. Res. Soc. Japan* 26, 945 (2001).
122. J. Li, L. T. Piehler, D. Qin, J. R. Baker, Jr., and D. A. Tomalia, *Langmuir* 16, 5613 (2000).
123. H. Nagaoka and T. Imae, *Int. J. Nonlinear Sci. Numer. Simu.* 3, 223 (2002).
124. H. Nagaoka and T. Imae, *J. Coll. Interf. Sci.*, 264, 335 (2003).
125. K. Takeda, G. D. Storrier, M. Morán, and H. D. Abrña, *Langmuir* 15, 7333 (1999).
126. R. C. van Duijvenbode, G. J. M. Koper, and M. R. Böhmer, *Langmuir* 16, 7713 (2000).
127. R. C. van Duijvenbode, I. B. Rietveld, and G. J. M. Koper, *Langmuir* 16, 7720 (2000).
128. K. M. A. Rahman, C. J. Durning, N. J. Turro, and D. A. Tomalia, *Langmuir* 16, 10154 (2000).
129. Z. Bo, L. Zhang, B. Zhao, X. Zhang, J. Shen, S. Höppener, L. Chi, and H. Fuchs, *Chem. Lett.* 1197 (1998).
130. L. Zhang, F. Huo, Z. Wang, L. Wu, X. Zhang, S. Höppener, L. Chi, H. Fuchs, J. Zhao, L. Niu, and S. Dong, *Langmuir* 16, 3813 (2000).
131. L. Zhang, Z. Bo, B. Zhao, Y. Wu, X. Zhang, and J. Shen, *Thin Solid Films* 327–329, 221 (1998).
132. H. Tokuhisa and R. M. Crooks, *Langmuir* 13, 5608 (1997).
133. M. Zhao, H. Tokuhisa, and R. M. Crooks, *Angew. Chem. Int. Ed. Engl.* 36, 2596 (1997).
134. W. M. Lackowski, J. K. Campbell, G. Edwards, V. Chechik, and R. M. Crooks, *Langmuir* 15, 7632 (1999).
135. H. Tokuhisa, M. Zhao, L. A. Baker, Vy T. Phan, D. L. Dermody, M. E. Garcia, R. F. Feez, R. M. Crooks, and T. M. Mayer, *J. Am. Chem. Soc.* 120, 4492 (1998).
136. S.-K. Oh, L. A. Baker, and R. M. Crooks, *Langmuir* 18, 6987 (2002).
137. A. Friggeri, H. Schönherr, H.-J. van Manen, B.-H. Huisman, G. J. Vancso, J. Huskens, F. C. J. M. van Veggel, and D. N. Reinhoudt, *Langmuir* 16, 7757 (2000).
138. V. V. Tsukruk, F. Rinderspacer, and V. N. Bliznyuk, *Langmuir* 13, 2171 (1997).
139. T. Imae and N. Yoshida, unpublished data.
140. J. Wang, J. Chen, X. Jia, W. Cao, and M. Li, *Chem. Commun.* 511 (2000).
141. A. J. Khopade and F. Caruso, *Langmuir* 18, 7669 (2002).
142. A. Manna, T. Imae, M. Iida, and N. Hisamatsu, *Langmuir* 17, 6000 (2001).
143. A. Mayer and M. Antonietti, *Coll. Polym. Sci.* 276, 769 (1998).
144. L. M. Bronstein, S. N. Sidorov, P. M. Valetsky, J. Hartmann, H. Cölfen, and M. Antonietti, *Langmuir* 15, 6256 (1999).
145. N. Yanagihara, K. Uchida, M. Wakabayashi, Y. Uetake, and T. Hara, *Langmuir* 15, 3038 (1999).
146. K. Esumi, T. Hosoy, A. Suzuki, and K. Torigoe, *J. Coll. Interf. Sci.* 229, 303 (2000).
147. K. Esumi, R. Nakamura, A. Suzuki, and K. Torigoe, *Langmuir* 16, 7842 (2000).
148. K. Esumi, A. Kameo, A. Suzuki, and K. Torigoe, *Coll. Surf. A: Physicochem. Eng. Aspects* 189, 155 (2001).
149. F. Gröhn, B. J. Bauer, and E. J. Amis, *Macromolecules* 34, 6701 (2001).
150. Y.-S. Seo, K.-S. Kim, K. Shin, H. White, M. Rafailovich, J. Sokolov, B. Lin, H. J. Kim, C. Zhang, and L. Balogh, *Langmuir* 18, 5927 (2002).
151. M. Zhao, L. Sun, and R. M. Crooks, *J. Am. Chem. Soc.* 120, 4877 (1998).
152. L. Balogh and D. A. Tomalia, *J. Am. Chem. Soc.* 120, 7355 (1998).
153. K. Esumi, A. Suzuki, N. Aihara, K. Usui, and K. Torigoe, *Langmuir* 14, 3157 (1998).
154. K. Esumi, T. Hosoya, A. Suzuki, and K. Torigoe, *Langmuir* 16, 2978 (2000).
155. M. E. Garcia, L. A. Baker, and R. M. Crooks, *Anal. Chem.* 71, 256 (1999).
156. V. Chechik and R. M. Crooks, *Langmuir* 15, 6364 (1999).
157. K. Esumi, A. Suzuki, A. Yamahira, and K. Torigoe, *Langmuir* 16, 2604 (2000).
158. A. Manna, T. Imae, K. Aoi, M. Okada, and T. Yogo, *Chem. Mater.* 13, 1674 (2001).
159. F. Gröhn, B. J. Bauer, V. A. Akpalu, C. L. Jackson, and E. J. Amis, *Macromolecules* 33, 6042 (2000).
160. F. Gröhn, G. Kim, B. J. Bauer, and E. J. Amis, *Macromolecules* 34, 2179 (2001).
161. R. M. Crooks, M. Zhao, L. Sun, V. Chechic, and L. K. Yeung, *Accounts Chem. Res.* 34, 181 (2001).
162. M. F. Ottaviani, R. Valluzzi, and L. Balogh, *Macromolecules* 35, 5105 (2002).

163. A. Manna, T. Imae, K. Aoi, and M. Okazaki, *Molecular Simulation*, 29, 661 (2003).
164. K. Esumi and M. Goino, *Langmuir* 14, 4466 (1998).
165. K. Esumi, N. Fujimoto, and K. Torigoe, *Langmuir* 15, 4613 (1999).
166. K. Esumi, K. Sakagami, S. Kuniyasu, Y. Nagata, K. Sakai, and K. Torigoe, *Langmuir* 16, 10264 (2000).
167. K. Naka, Y. Tanaka, Y. Chujo, and Y. Ito, *Chem. Commun.* 1999, 1931
168. L. H. Hanus, K. Sooklal, C. J. Murphy, and H. J. Ploehn, *Langmuir* 16, 2621 (2000).
169. M. Liu, K. Kono, and J. M. J. Fréchet, *J. Controlled Release* 65, 121 (2000).
170. I. Gitsov and C. Zhu, *Macromolecules* 35, 8418 (2002).
171. P. R. Dvornic, J. Li, A. M. de Leuze-Jallouli, S. D. Reeves, and M. J. Owen, *Macromolecules* 35, 9323 (2002).
172. M. Schlupp, T. Weil, A. J. Berresheim, U. M. Wiesler, J. Bargon, and K. Müllen, *Angew. Chem. Int. Ed.* 40, 4011 (2001).
173. D. L. Richter-Egger, A. Tesfai, and S. A. Tucker, *Anal. Chem.* 73, 5743 (2001).
174. S. Ghosh and A. K. Banthia, *Tetrahedron Lett.* 43, 6457 (2002).
175. D. Arrington, M. Curry, and S. C. Street, *Langmuir* 18, 7788 (2002).
176. F. Gröhn, X. Gu, H. Grüll, J. C. Meredith, G. Nisato, B. J. Bauer, A. Karim, and E. J. Amis, *Macromolecules* 35, 4852 (2002).
177. X. C. Wu, A. M. Bittner, and K. Kern, *Langmuir* 18, 4984 (2002).
178. J. Won, K. J. Ihn, and Y. S. Kang, *Langmuir* 18, 8246 (2002).
179. N. Feeder, J. Geng, P. G. Goh, B. F. G. Johnson, C. M. Martin, D. S. Shephard, and W. Zhou, *Angew. Chem. Int. Ed.* 39, 1661 (2000).
180. Z. Xu and W. T. Ford, *Macromolecules* 35, 7662 (2002).

Encyclopedia of Nanoscience and Nanotechnology
www.aspbs.com/enn

GaAs-Based Nanodevices

Eric Donkor
University of Connecticut, Storrs, Connecticut, USA

CONTENTS

1. INTRODUCTION

Gallium arsenide (GaAs) nanodevices are fabricated by growing ultra-thin heteroepitaxial layers of binary, tertiary, quanternary, or quinary III-V semiconductor alloys, having different energy bandgap, on a GaAs semi-insulating substrate such that one or more of the device feature sizes are comparable with the De Broglie wavelength of carriers in GaAs [1]. As a result, the carriers may be spatially confined into two-, one-, or zero-dimensions giving rise to quantum-well, quantum-wire, and quantum-dot nanodevices. In GaAs quantum-well nanodevices, the carriers are spatially confined into a two-dimensional (2D) space, GaAs quantum-wire nanodevices confine carriers into a one-dimensional (1D) space, and GaAs quantum-dot nanodevices confine carriers into a zero-dimensional (0D) space.

Quantum-well device technology is well established and practical devices have been reported in which the carriers were confined into single-quantum wells (SQWs), double-quantum wells (DQWs), multiple-quantum wells (MQWs), or superlattices (SLs). GaAs quantum-wire and quantum-dot nanodevices are new with fewer practical devices reported to date. Their fabrication is much more challenging since even small fluctuations in feature size or inhomogeneity of the constituents could suppress or destroy the low-dimensional confinement effects that are critical to the device operation.

Progress in GaAs nanodevice development has closely been linked to advances in thin film, or epitaxial growth of GaAs-based semiconductor materials. Epitaxial growth is a process of growing an epi-layer on a crystalline substrate layer. The epi- and substrate layers may be similar or dissimilar materials, resulting in homoepitaxy or heteroepitaxy growth, respectively. In epitaxial growth, it is crucial that the heterojunction be free of interface defects, since such defects render the structure unsuitable for device application. It is easier to achieve this condition with a heterosystem in which the lattice constants of the various epilayers are closely matched. The AlGaAs/GaAs system was the first lattice-matched heterostructure system to be widely grown and used for device applications ranging from tunnel diodes to laser diodes. Continued improvements in epitaxial growth techniques, such as molecular-beam epitaxy, metal-organic, chemical-vapor deposition, and chemical-beam epitaxy have enabled device quality growth of lattice-mismatched heterostructures as well.

Molecular beam epitaxy (MBE) is one of the most common processes for fabricating nanostructures. It can be employed to grow thin films with thicknesses from microns to atomic layer sizes. It has ultra-high vacuum (UHV) multiple chambers for growth, loading, preparation, and *in-situ* analysis. The growth chamber, usually of 450-mm inner diameter, holds a removable source flange, which provides support for effusion or *Knudsen* cells, their shutters, and liquid nitrogen (LN_2) shroud. Constituents of the growth material are evaporated, as atomic beams, from a solid source placed in the Knudsen cells. The cells are oriented such that their beams converge on the substrate in the growth position. The beam sources are thermally isolated from each other by an LN_2-cooled radial vane baffle, which prevents chemical cross-contamination. A second cryopanel surrounds the substrate whose purpose is to reduce contamination due to outgassing of the chamber walls. The sample preparation and loading chamber is connected to the growth chamber via large diameter channels and isolation valves. The analytical chamber allows for *in-situ* and after-growth

ISBN: 1-58883-059-4/$35.00

Encyclopedia of Nanoscience and Nanotechnology
Edited by H. S. Nalwa
Volume 3: Pages (703–717)

surface analysis without exposing the sample to outside environments. The substrate surface must be prepared prior to epitaxial growth. Preparation of the surface entails preloading chemical treatment, as well as *in-situ* cleaning processes. The goal of the preloading chemical treatment is to provide either a clean chemical oxide layer of a few Armstrong thick on the GaAs substrate surface or hydrogen passivation of the GaAs surface. Next, the substrate is loaded into the airlock high-vacuum growth chamber. The pressure of the growth chamber is typically 10^{-9} Torr.

Metal-organic chemical-vapor deposition (MOCVD) is another common process for the growth of GaAs nanostructures. The layers are grown by transporting different precursors or reactants in the vapor phase, under controlled pressure, into a reactor (vertical, horizontal, or barrel) chamber that holds the semi-insulating GaAs substrate. Alkyls of the group III metals and hydrides of the group V elements are typical precursors. The metal-organic sources are contained in stainless steel cylinders equipped with a temperature-controlled bath capable of temperature control over a wide range about −20° to 100 °C at ±0.01% accuracy. Controlling the temperature precisely regulates the partial pressure of the source. Electronic mass flow controllers control the exact amount of carrier gas flowing through the bubblers, and at the same time maintain a constant vapor pressure of the source. A heating system heats the substrate during growth. The heater can raise the temperature of the substrate to about 1000 °C. Heating can be achieved by radio frequency (RF) induction heating, radiative heating, and resistance heating. Radio frequency heating is common in large commercial systems. In this type of heating, the substrate susceptor is inductively coupled to the RF coil. In radiative heating, the heat energy from a resistance element is transformed into radiant energy. The susceptor absorbs the radiant energy and converts it back to heat energy. In the resistance heating method, current flows through an electrically conductive layer mounted to the susceptor to generate the required heat energy. Safety precautions include gas leak detectors for the carrier gas, the purifier, and toxic gases. Commercial MOCVD are computer controlled, and can be used for defining the growth process and for monitoring reactor conditions.

Nanodevices can be classified according to a) device structure, b) mode of operation, and c) circuit functionality. When described according to device, structure nanodevices are classified into quantum-wells, quantum-wires, and quantum-dots. Quantum-well nanodevices may be divided further into SQW, DQW, MQW, and SL. Quantum-wire and quantum-dot nanodevices may also be subdivided based upon the mode of carrier confinement, namely, mechanical confinement (strain, selective growth, and etching effects), electrostatic-field confinement, and self-organized (or self-assembly) devices.

A second method of classification is based on the quantum-mechanical transport phenomena underlying the mode operation of nanodevices. Accordingly, devices are classified into hot-electron and ballistic devices, resonant tunneling devices, quantum-interference, and spin precess devices. A third method classifies nanodevices, based upon their circuit application, into two-terminal diodes, three-terminal transistors, and multi-terminal integrated circuits.

Figure 1 depicts characteristic lengths of electronic material, namely, relaxation length, mean-free path, coherence length, and De Broglie wavelength. These characteristic lengths are used to illustrate and compare two nomenclatures of electronic device scaling. These nomenclatures are indicated at the top and bottom of the figure. In one such historical development shown at the top, scaling progressed from classical, hot-electron, ballistic, mesoscopic, and quantum-effect devices. Classical electronic devices have feature sizes between 0.5 μm to over 10 μm and are comparable with relaxation length of materials. The next scaled-down devices are the hot-electron devices and they have feature sizes of between 0.1 μm and 0.5 μm. Such feature sizes fall between the characteristic mean-free path and relaxation length. The carriers in hot-electron devices may have temperatures that exceed the thermodynamic equilibrium temperature. The next scaled-down devices are ballistic devices with feature sizes in the range 0.05 μm to 0.1 μm, corresponding to feature size between the coherence length and the free-mean path. Ballistic devices suffer little scattering, and therefore exhibit very high speed and high-frequency performance. The devices described thus far (classical, hot-electron, ballistic) can be characterized by classical or semi-classical physical laws. The electrical characteristics of classical devices are described adequately by the classical drift-diffusion theory. The electrical characteristics of hot-electron and ballistic devices are described by semi-classical laws. The semi-classical model is also based upon the classical drift-diffusion theory. However, transport parameters, such as relaxation time and conductivity, are determined based upon quantum-mechanical models incorporating additional effects arising from size quantization.

The two remaining scaled-down devices of the top nomenclature, namely, mesoscopic and quantum devices, require quantum-mechanical laws for their characterization. Mesoscopic devices have feature sizes that lie between 0.007 μm to about 0.015 μm (7 nm to 15 nm). Devices whose feature sizes are truly comparable with atomic feature size,

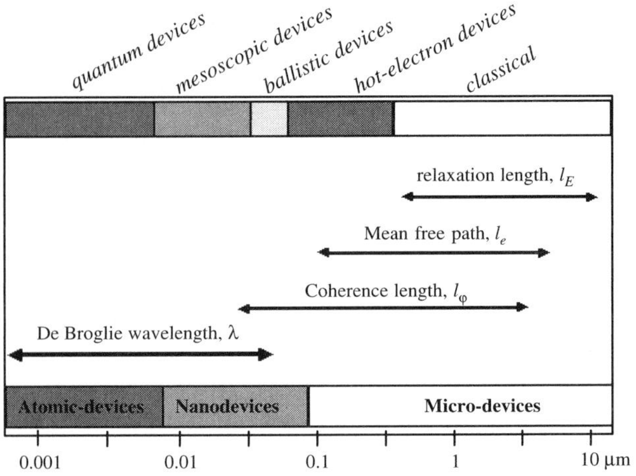

Figure 1. Comparison of two historical evolution of miniaturization of electronic devices in terms of material characteristics lengths (De Broglie wavelength, coherence length, mean-free path, relaxation length).

which for GaAs is of the order of 6 nm or less, are called quantum devices. They require full quantum-mechanical treatment to analyze or characterize their performance. Invariably, mesoscopic and quantum devices require low temperature operation.

The second type of nomenclature for device scaling is shown at the bottom of Figure 1, namely, microdevices, nanodevices, and atomic devices. Historically, this nomenclature for device scaling evolved correspondingly with advances in semiconductor material growth and device fabrication technology. Microelectronics material growth and device fabrication are based on the self-aligned technology. Self-aligned technology has the capabilities for scaling device feature size in the range of 0.1 μm to 10 μm. An effort to find a solution to the difficulty of applying self-aligned technology to fabricating scaled-down devices with feature sizes less than 0.1 μm gave rise to heteroepitaxial technology. Concomitant with heteroepitaxial growth came nanodevice technology, whereby one or more of the device feature size is a few nanometers.

This work treats GaAs nanodevices. The treatment includes emerging as well as existing quantum-well, quantum-wire, and quantum-dot devices. Electronic and transport properties in GaAs nanostructures are described in Section 2. In particular, we emphasize how size effects impact key electronic and transport parameters. Quantum-well and superlattice devices are treated in Section 3. In Section 4, GaAs resonant tunneling devices are discussed. In Section 5, quantum-wire and quantum-dot devices are discussed. GaAs spintronic devices are introduced in Section 6.

The rapid manner in which GaAs nanotechnology is evolving makes it impossible to include all devices reported to date. However, the devices covered here were carefully selected to adequately span the field of GaAs nanodevice technology reported in the literature between 1980 and 2002. The references are only representative and by no means exhaustive.

2. ELECTRONIC AND TRANSPORT PROPERTIES OF GaAs NANOSTRUCTURES

The key features of the electronic properties of GaAs nanostructures for characterizing nanodevices are quantization of energy levels, density of states, excitonic formation, and spin processing. These properties are observed as a consequence of quantum-mechanical carrier confinement induced by size effects, mechanical effect, electrostatic potential, or magnetic fields. A theoretical determination of these properties starts with a solution of the Schrödinger equation for the electronic wave function and the energy band. Many physical realities complicate solution of the Schrödinger equation. First, the electrostatic interaction that exists between the carriers in the confined system requires a many-bodied model to the problem. Second, band mixing requires a solution of couple nonlinear second-order differential equations. Third, compositional variations, material disorder, or statistical fluctuations influence the solutions. The effective mass, or k.p. method, offers a semi-classical approach to determining the electronic properties of nanostructures.

Scattering, high field, tunneling processes, hot electron, localization, and quantum interference effects dominate carrier transport in nanostructures. Scattering processes may be elastic or inelastic. The main sources of elastic scattering are ionized impurities, interface roughness, and electron screening. Inelastic scattering is primarily due to carrier-phonon interactions. Scattering in nanostructures causes transitions in momentum, energy, or quantum state. These transitions ultimately lead to momentum loss, energy loss, intersubband transitions, or change in spin number and are characterized in terms of transition probabilities from an initial state to a final state. Scattering limits the device operation speed and also introduces noise, and therefore its elimination or minimization is crucial. Elastic scattering can be minimized using selective doping in which some selected layers of a nanostructure are doped while others are undoped. The choice of materials and their arrangement is carefully designed to introduce real-space transfer of carriers from the doped layers to the undoped layers, so that the ionized impurities located in the doped layers are separated from the carriers transferred to the undoped layers. A spacer layer may be added to better separate the ionized impurities in the undoped layer. Also, minimization of elastic scattering can be achieved by eliminating interface nonuniformities during growth of the nanostructures.

High electric field inherent in nanostructures causes the mean energy and momentum of carriers to exceed their thermal equilibrium values. The most noted effects are velocity overshoot and saturation. The saturation velocity in GaAs nanolayers has a value between $1\text{–}2 \times 10^7$ cm/s and occurs at electric fields exceeding 10 kV/cm. Velocity overshoot is observable at low field of between 2–5 kV/cm, and depending on the composition of the nanolayer, the overshoot velocity could be as high as 5 times the saturation velocity. Velocity overshoot is also observed in the transient state on a time scale of 1–2 ps. The steady state velocity overshoot occurs at a critical electrical field, whereas the transient state velocity overshoot increases with increasing electric field strength.

3. GaAs QUANTUM-WELL AND SUPERLATTICE DEVICES

GaAs quantum-well and superlattice nanodevices may be classified based upon the physical phenomena employed to control current flow in the devices. Current flow may be controlled by an electric field, electric potential, or by quantum-mechanical tunneling. Accordingly, GaAs quantum-well and superlattice devices may be classified into a) heterostructure field-effect, b) heterostructure potential-effect, and c) resonant tunneling devices. These devices may be implemented as two-terminal diodes, three-terminal transistors, or multi-terminal integrated circuits.

3.1. GaAs Heterostructure Field Effect Transistors

GaAs heterostructure field effect transistor (HFET) is one of the earliest quantum-well field-effect devices. A family tree of GaAs HFETs is illustrated in Figure 2. The three

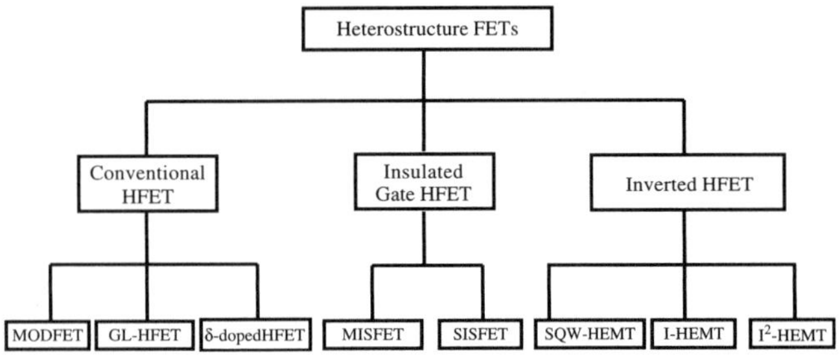

Figure 2. Family tree of heterostructure field-effect transistor (HFETs). Reprinted with permission from [3], E. Donkor, "Handbook of Advanced Electronic and Photonic Materials and Devices," 2001. © 2001, Academic Press.

main branches are conventional, insulated gate, and inverted HFET [2]. A quantum-well structure confines carriers into two-dimensional electron gas (2DEG). In a conventional HFET, the 2DEG resides in the smaller bandgap layer of the quantum-well and are separated from the gate by the wider bandgap material. The 2DEG is formed as a result of real-space charge transfer from the doped wide bandgap material to the undoped smaller bandgap material. MODulation-doped FETs (MODFETs) form a subclass of conventional HFET. In these HFETs, the wide bandgap material is uniformly doped. Another subclass of conventional HFET is the GL-MODFET in which the $Al_xGa_{1-x}As$ layer between the 2DEG and the gate is gradually graded in composition by varying the Al molar fraction from about $x = 0.3$ at the 2DEG interface to $x = 0$ at the gate interface. A third subclass of the conventional HFET is the SL-MODFET in which the quantum-wells are replaced by superlattices. The HFET may be doped in a narrow region called δ-doped to cause separation of the 2DEG from ionized impurities.

Insulated gated HFETs have an insulated layer separating the gate electrode from the quantum-wells that confine the 2DEG. Metal-Insulator-FET (MISFET) and semiconductor-insulator-semiconductor-FET (SISFET) form the two basic variations of the insulated gate HFETs. The motivation for developing these structures is to form complimentary HFET for low-power logic circuits analogous to silicon CMOS circuits. The challenge in growing insulated-gate FET is due to the fact that GaAs has no native oxide. In the inverted HFET, an undoped, small bandgap material is grown on a doped wide bandgap material. Also, the gate is grown on the small bandgap material. This structure is thus inverted with respect to the conventional FET. Inverted HFET may be further classified into single quantum-well, high-electron mobility transistor (SQW-HEMT), insulated-gate, high-electron mobility transistor (I-HEMT) and inverted-insulated gate, high-electron mobility transistor (I^2-HEMT).

3.1.1. Conventional HFET

Modulation-Doped HFETs Figure 3 is a cross-sectional view of an $GaAs/Al_xGa_{1-x}As$ single quantum-well MODFET. It shows the various GaAs and AlGaAs layers. Due to the higher bandgap of AlGaAs compared to the adjacent GaAs region, free electrons diffuse from the AlGaAs into the GaAs and form a 2DEG at the heterojunction. A potential barrier, arising from the band discontinuities, confines the free electrons in the GaAs layer to a very thin sheet. Due to the absence of ionized donors in the channel of the MODFET, electrons forming the 2DEG suffer little Coulomb scattering and enjoy high mobility. A thin "spacer" layer of undoped AlGaAs placed between the doped AlGaAs and the undoped GaAs, further separates the 2DEG from the ionized donors at the interface, increasing electron mobility at the cost of a total charge transferred to the interface. The heavily doped GaAs "cap" layer facilitates formation of ohmic contacts to the device.

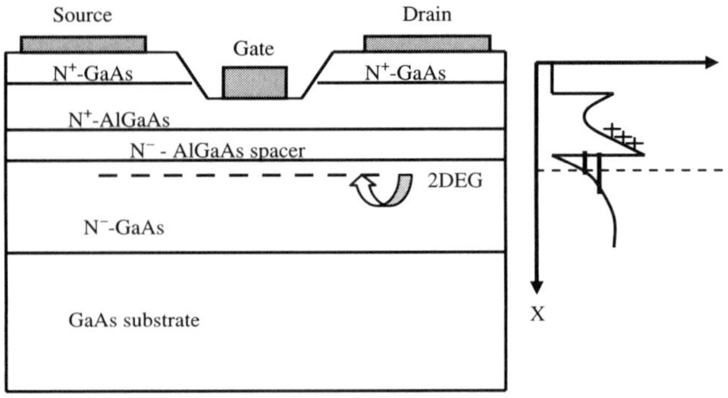

Figure 3. Cross-sectional view of a conventional HFET. An energy band diagram Is also shown. The two-dimensional electron gas (2DEG) is formed due to real-Space charge transfer. Due to size effects the energy of the 2DEG is quantized. Reprinted with permission from [3], E. Donkor, "Handbook of Advanced Electronic and Photonic Materials and Devices," 2001. © 2001, Academic Press.

An energy band diagram of the MODFET is also shown in Figure 3. The difference in electron affinities translates into a discontinuity in the conduction band at the heterojunction. Electrons from the doped AlGaAs diffuse or tunnel to the GaAs, causing the formation of a dipole layer with ionized donors on one side and electrons on the other. Equilibrium is established when the potential difference equals the conduction band offset. Charge transfer across the heterointerface requires only that a discontinuity of greater than $2kT$ appear in the conduction band, not a difference in bandgap. Electrons in the 2DEG are confined to a potential well at the heterojunction, which is often approximated as a triangular well. Electron energies are quantized perpendicular to the heterointerface. Motion in the other two directions is unconstrained.

A simplified explanation for the formation of the energy band structure is to consider the impurities of the doped (AlGaAs) region as isolated with binding energy E_b. At very low temperatures $E_b \gg K_BT$, the carriers remain frozen onto the impurity sites. When $E_b \cong K_BT$, carrier detrapping takes place. To establish thermal equilibrium, the electrons escape, either by diffusion or quantum mechanical tunneling into the small bandgap (GaAs) material. The electrons quickly lose energy by emitting phonons. The reverse process, that is, recapturing of the electrons by the ionized impurities, is prevented by the built-in potential. The residual impurities however, affect the charge transfer. A band bending takes place due to the formation of a dipole between the ionized impurities and the electrons.

Graded-Layer and Delta-Doped HFET An improvement on the MODFET is obtained by grading the epilayers forming the heterointerface. The goal is to lower the effective interfacial resistance. In this case, the wide bandgap $Al_xGa_{1-x}As$ donor layer is graded in composition from $Al_xGa_{1-x}As$ (with the Al molar fraction $x = 0.3$ approximately) to GaAs at the cap interface to facilitate ohmic contact formation.

The delta-doped, or pulsed-doped HFET [4, 5] is so called because it has well-defined, high-peak, narrow doping carrier concentration. Delta-doped HFETs have many attractive characteristics relative to conventional HFETs, including high drain current, high intrinsic transconductance, high breakdown voltage, and better threshold controllability [6]. Delta-doped HFETs find applications in high-power microwave circuits and high-speed digital circuits.

Superlattice-Doped HFETs Another modification of the MODFET incorporates multiple quantum-well layers [7]. The goal here is to increase the 2DEG sheet density. This allows for increased device current and improved power performance. The disadvantage is lower frequency and speed performance [8] compared to single quantum-well MODFET. DX centers and doped impurities in heavily doped layers degrade device performance. One method of resolving this problem is to incorporate a superlattice donor layer in place of the doped layer [9, 10]. In the SL-doped HFET, only the thin GaAs layers are n-doped within the undoped AlGaAs. As a result, the donors are separated from the AlGaAs lattice and should not induce DX centers.

Pseudomorphic HFETs A pseudomorphic, or strained-layer HFET (PHEFT) can also improve upon the basic HFET performance by replacing the GaAs active region with a different material, such as InGaAs, which has relatively low DX centers and higher mobility [11, 12]. The most studied PHEFT is the n-AlGaAs/InGaAs/GaAs structure. By using a strained layer InGaAs instead of GaAs for the active region, higher band discontinuity, ΔE_c, is obtained so that an n-$Al_xGa_{1-x}As$ with $x < 0.2$ can be used as an electron supply layer and prevent DX centers. Pseudomorphic HFETs have been designed for low-noise amplification [13], high frequency, and high-power operations [14].

3.1.2. Insulated-Gate HFETs

Heterostructure insulated-gate field effect transistors (HIGFET) has two main subclasses: the SISFET [15, 16] and MISFET [17–19]. The SISFET has a semiconductor gate and a heterojunction barrier in contrast with a metal gate and Schottky barrier for the conventional HFET. The SISFET exhibits good thermal stability—its threshold voltage is near zero and less dependent on material characteristics [20]. The transconductance versus gate voltage shows a distinctive sharp turn at low temperatures near its knee region and has a higher slope in the linear region than conventional HFETs. Metal-insulator semiconductor HFETs offer the integration of n-channel and p-channel HFET to form complimentary circuits, which is the basic building block for integrated circuits.

Figure 4 illustrates a complimentary MIS-HFET. The substrate is a semi-insulating GaAs. An n^+-IMPLANT isolates the n-channel MISHFET, likewise a p^+-IMPLANT isolates the p-channel MISHFET. The n-channel incorporates an n^+-Ge under the gate electrode to maintain a low threshold voltage and for nonalloy contact to the gate metal.

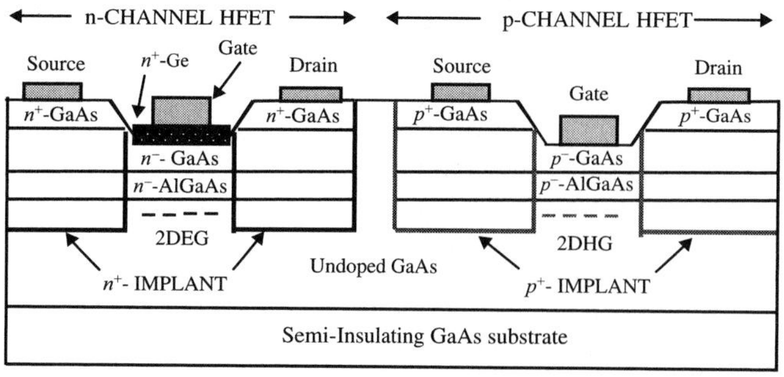

Figure 4. Cross-sections of a complimentary MIS-HFETs. The n-channel shown on the left incorporates a heavily doped Ge under the gate for low threshold voltage and for non-alloy contact with the gate metal.

3.1.3. Inverted HFETs

In inverted AlGaAs/GaAs HFETs, an undoped GaAs layer is grown between the gate metal and the AlGaAs layer. The advantages of the inverted HFET compared with the normal MODFET are lower contact resistance to the 2DEG, better carrier confinement, high transconductance, and higher carrier concentration. Inverted HFETs, however, suffer from large gate current, parasitic MESFET effects under the gate region, and performance degradation due to substrate effects.

3.1.4. HFET Electrical Characteristic

The most widely used analytical approaches for describing the charge distribution of the 2DEG are the self-consistent method [21] and the variational method [22]. An expression for the sheet carrier density of the 2DEG is given by

$$n_s = \left(\sqrt{\frac{2\varepsilon_2 N_d}{q}\left[V_{20} - \frac{\Delta E_{FO}(T)}{q} + \delta\right] + N_d^2(d_i + \Delta d)^2}\right) - N_d(d_i + \Delta d) \tag{1}$$

Here, ε_2 is the permittivity of the doped layer, N_d is the ionized donor density in the doped layer, d_i is the thickness of the undoped spacer layer, δ is a correction factor due to the use of the depletion approximation in the doped layer. For a donor binding energy of 15 meV, δ is 25 meV at 300 K and 50 meV at 77 K. The position of the Fermi level measured from the bottom of the potential well, represented by ΔE_{FO}, at the heterointerface is a function of the sheet carrier density [23].

Control of the sheet carriers in the HFETs is achieved through a Schottky gate. Accounting for the gate voltage, the sheet carrier density expression of Eq. (1) can be modified as

$$n_s = \frac{\varepsilon_2}{q}\frac{V_G - V_{TH}}{d + \Delta d} \tag{2}$$

$$V_{TH} = \phi_b - V_p - \frac{\Delta E_c + \Delta E_{Fo}}{q} \tag{3}$$

where $d = d_d + d_i$, d_d is the thickness of the doped AlGaAs beneath the Schottky gate, ϕ_b is the Schottky barrier height, and V_p is the pinch-off voltage.

Current voltage, transconductance voltage, and capacitance voltage are the main d.c. characteristics of interest. A simplified analytical model for these characteristics is derived from the charge control method [24]. The main difference between various analytical models is due to the choice of the functional relationship between the drift velocity and the applied electric field. The import of these relationships is to account for velocity overshoot, and other nonequilibrium effects, while still providing a tractable or closed-form functional relationship [25, 26].

A multiregion model, whereby the linear and saturation region are analyzed separately and then integrated into a unified model, is often employed to obtain analytical closed-form expressions for the I-V characteristics. The linear region corresponds to low-applied electric field, such that the velocity of carriers in the channel is directly proportional to the applied field. The saturation region is taken as corresponding to the saturation (or constant) velocity of the carriers with applied voltage. The I-V relation for the respective linear and saturation regions are of the form

$$I_d = \frac{w\mu\varepsilon_2}{2(d+\Delta d)} \frac{(V_G - V_{TH})^2 - (V_G - V_{TO} - V_d)^2}{L_g + V_d/F_c} \tag{4}$$

$$I_d = \frac{\beta V_{s1}^2 (1 + 2\beta R_s V_g' + V_g'/V_{s1}^2)^{\frac{1}{2}}}{1 - \beta^2 R_s^2 V_{s1}^2} - \frac{\beta V_{s1}^2 (1 + \beta R_s V_g')}{1 - \beta^2 R_s^2 V_{s1}^2} \tag{5}$$

Here $V_g' = V_G - V_{TH}$, $V_{s1} = F_c L_g$, $\beta = \varepsilon_2 \mu W/(d + \Delta d)$, L_g is the gate length, w is the gate width, and R_s is the source series resistance. A comparison between these models and experimental results show good agreement [27].

The ac model attempts to determine nonlinear elements such as gate capacitance and other parasitic capacitances, found in the equivalent circuit [28] of the HFET. The relationship between the maximum frequency, $f_{\max}$, and the cut-off frequency, f_t, in terms of parasitics, have been widely reported and takes the form

$$f_{\max} = \frac{f_t}{\sqrt{4\frac{g_o}{g_m}\left(g_m R_i + \frac{R_s+R_g}{1/g_m+R_s}\right)} + \frac{4}{5}\frac{C_{gd}}{C_{gs}}\left(1+\frac{2.5C_{gd}}{C_{gs}}\right)(1+g_m R_s)^2} \tag{6}$$

$$f_t = \frac{g_m}{2\pi C_{gs}} \tag{7}$$

Here g_m is the transconductance and the other parameters are as defined in the literature [29].

The dc characteristics turn to depend on device structure and material systems used to design the HFET. As the gate length shrinks, the dc characteristics generally improve [30]. Improvement in device characteristics is also observed at low temperatures [31]. The growth of pseudomorphic HFETs has new approaches to achieving enhancement in device characteristics opening up. Dickmann et al. [32] achieved transconductance of 368 mS/mm and a drain saturation current of 326 mA/mm in an $(Al_{0.7}Ga_{0.3})_{0.5}In_{0.5}P/In_{0.15}Ga_{0.85}As/GaAs$ pseudomorphic HFET. Likewise, Pereiaslavets et al. [33] reported drain saturation current of 800 mA/mm and transconductance of 600 mS/mm for a narrow channel (50–80 Å) GaInP/InGaAs/GaAs pseudomorphic HFET. The a.c. characteristics show similar trends of improvement in short channel devices, and in pseudomorphic structures that incorporate devices with high mobility and high velocity overshoot effects.

3.2. GaAs Potential-Effect Transistors (HPETs)

Heterostructure potential-effect transistors (HPETs) are current-controlled devices and are classified into heterostructure bipolar transistors (HBTs), ballistic-injection transistors and real-space transfer transistors as shown in Figure 5. The base or emitter structure of HBTs can be made to be single-well (SW-HBT), double-well (DW-HBT), or superlattice (SL-HBT). Ballistic injection transistors hot electrons are injected over or through an emitter barrier into the base region. The injected carriers travel through the barrier with little or no scattering. Current control is maintained by changing the collector barrier height. An induced-base, ballistic-injection transistor (or induced-base transistor) was proposed in 1985 [34] and experimentally implemented by Chang et al. [35]. Fabrication of this device is complicated albeit it has shown no remarkable improvement over the basic HBT. The resonant tunneling hot-electron transistor (RHET) and the tunneling hot-electron transfer amplifier (THETA) are two other ballistic injection transistors. The charge injection transistor [36] (CHINT) is an example of a real-space charge transfer transistor. The structure has a source, high-mobility channel, drain, and an additional collector terminal. Thus, it may be considered an integrated HFET/HBT.

3.2.1. Heterostructure Bipolar Transistor

Figure 6 depicts a cross-sectional view of an npn AlGaAs/GaAs single heterojunction bipolar transistor (SHBT). The structure consists of a lightly (n^--) doped wide bandgap $Al_xGa_{1-x}As$ emitter, a heavily (p^+-) doped GaAs

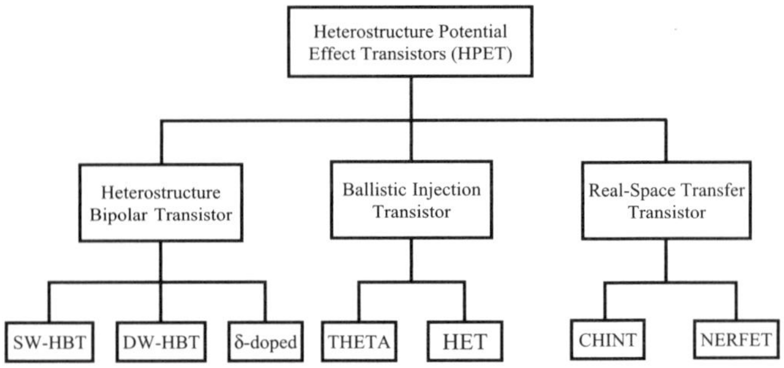

Figure 5. Family tree of heterostructure potential-effect transistor. Reprinted with permission from [3], E. Donkor, "Handbook of Advanced Electronic and Photonic Materials and Devices," 2001. © 2001, Academic Press.

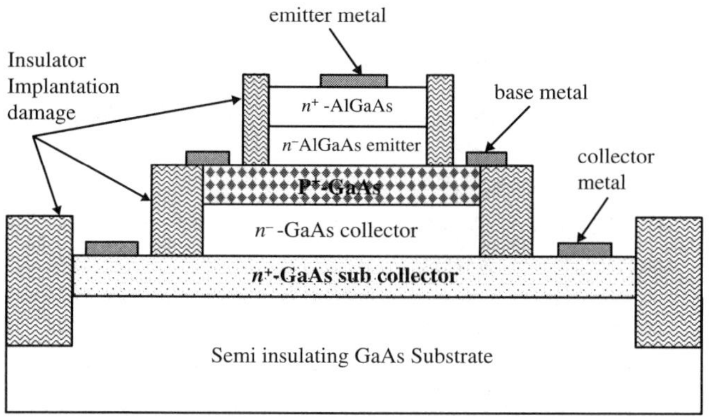

Figure 6. Cross-sectional view of an AlGaAs/GaAs SW-HBT. Reprinted with permission from [3], E. Donkor, "Handbook of Advanced Electronic and Photonic Materials and Devices," 2001. © 2001, Academic Press.

base, and a n^--doped GaAs collector. A double heterostructure bipolar transistor (DHBT) is formed by replacing the GaAs collector with another wide bandgap material such as AlGaAs. The performance enhancement that an HBT offers over a homojunction bipolar transistor can be deduced from the current gain figure of merits which is given by [37]:

$$\beta_{\max} = \frac{N_e}{P_b}\frac{v_{nb}}{v_{pe}} \exp(\Delta\varepsilon_g/kT) \tag{8}$$

Here N_e and P_b are the emitter and the base doping levels. v_{nb} and v_{pe} are the mean speeds of electrons at the emitter end of the base and holes at the base end of the emitter, respectively. Also, the energy gap difference is $\Delta\varepsilon_g = q(V_p - V_n)$, where V_p and V_n are the potential barriers of holes and electrons. The energy gap difference of HBTs is several kTs; as a result, very high $\beta_{\max} > 1000$ can be achieved [38].

Improvement in high frequency performance can be realized in an HBT, since the base can be highly doped to reduce base resistance without deteriorating the current gain. However, the process of heavily doping the base can be impeded due to an increasing diffusion coefficient of the base acceptors at higher concentrations. Thus, the choice of dopant becomes very important. Carbon appears to be a promising acceptor in MBE, MOMBE, and MOCVD growth. Carbon diffuses less in comparison with Be or Zn, allowing for high temperature processing and high doping concentrations. The discontinuity in the energy bands may be abrupt or graded, and their advantages and disadvantages have been discussed [39].

The lattice-matched $Ga_{0.52}In_{0.48}P$/GaAs HBT [40–44] has evolved as an alternative to the more conventional AlGaAs/GaAs HBT. Some of its attractive features include near ideal current-voltage characteristics, high current gain [45, 46], and constant current gain for a wide range of temperatures [47]. An etching solution for GaInP does not attack GaAs [48], making selective etching possible for material processing. Also, GaInP is less susceptible to oxidation, whereas the AlGaAs is readily oxidized due to the acute reactivity of Al. Also DX centers are relatively less compared to AlGaAs.

3.2.2. Ballistic Injection Heterostructure Bipolar Transistor

Tunneling Hot-Electron Transfer Amplifier (THETA) Figure 7 is an illustration of a tunneling hot-electron transistor. The structure consists of heteroepitaxially grown layers of heavily doped 200 nm-thick GaAs (emitter) layer, a 10 nm-thick intrinsic AlGaAs, a 30 nm-thick heavily doped GaAs (base) layer, a 6 nm-thick grade AlGaAs layer, a 100 nm-thick intrinsic AlGaAs, and a heavily doped thick GaAs (collector) layer. The tunneling hot-electron transistor is a unipolar potential effect device and forming an ohmic contact to the base is quite challenging. One approach is by reactive ion etching to expose the base. However, etching should not penetrate deep into the base; otherwise, the base-collector junction could be short-circuited.

The emitter-base junction forms an injector for hot-electron tunneling, from the emitter into the base, through the thin 10 nm *i*-AlGaAs layer. The kinetic energy of the injected carrier depends on the emitter-base voltage and the difference between the quasi-Fermi level and the bottom of the conduction band in the base region. The base is made thinner than the mean-free path of the carriers so that the emitted electrons travel ballistically through the base. The base collector forms a barrier and only hot electrons with energy in excess of the barrier height can reach the collector. The RHET can be biased to operate as an amplifier, if the emitter is negatively biased and the collector positively biased with respect to the base. A typical base-to-emitter voltage, $V_{\mathrm{be}} > 0.6$ V. However, such high voltages can cause electrons in GaAs to scatter from the Γ-valley to the upper L and X valleys resulting in low-tunneling current arriving at the collector of the devices. For ballistic transport to occur $V_{\mathrm{be}} < 0.3$ V. This can be achieved by higher doping of the base and emitter regions.

Hot Electron Transistor (HET) The HET is so called because at normal biased voltages, the electronic temperature, T_e, of the current carrying electrons is higher than their thermal equilibrium temperature T in the device. Such a situation occurs if the device feature size is smaller than the energy relaxation length of electrons for the material. The energy band diagram of the basic GaAs HET is shown in Figure 8. It depicts three heavily doped GaAs layers forming the emitter (200 nm), base (30 nm), and collector

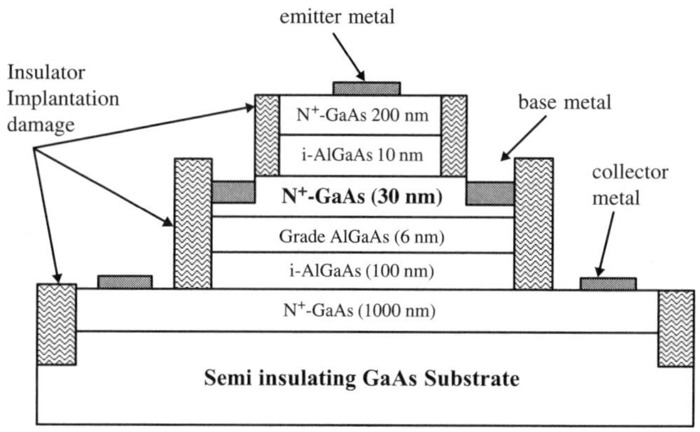

Figure 7. Cross sectional view of a tunneling hot electron transistor.

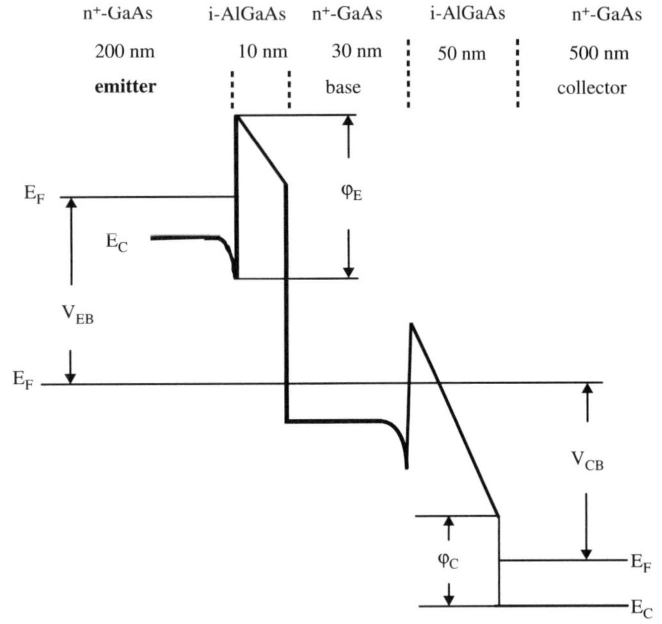

Figure 8. Energy band diagram of GaAs hot-electron transistor.

(500 nm) regions, respectively, and separated by two intrinsic AlGaAs regions of thickness 10 nm and 50 nm. The first AlGaAs/GaAs heterojunction at the emitter-base junction is a ballistic injector of carriers, and the second AlGaAs/GaAs heterojunction at the base-collector junction forms a barrier with height φ_c allowing only hot electrons with kinetic energy higher than the barrier to surmount. By applying a base-emitter voltage, V_{be}, across the base-emitter junction, electrons are injected ballistically into the base region. A base-collector voltage, V_{bc}, of a few volts enables the carriers to be collected.

Two variations for the HET are the V-grooved barrier transistor (VBT) and the injected-induced barrier transistor (IBT). The VBT has an ultra-thin heavily doped base region. The base can be p^+-layer in which case a bipolar HET is formed or can be n^+-doped layer to form a monopolar HET. Unlike the VBT, the base of an IBT is undoped or rather the base carriers are induced and injected. As a result, it has both higher common base current gain and operating speed than the basic HET.

3.2.3. Real-Space, Charge-Transfer Transistors

Real-space, charge-transfer transistors are three-terminal devices in which hot electrons are transferred between two conducting layers separated by a potential barrier. There are two main types, namely, the CHINT and the negative differential resistance, field-effect transistor (NERFET). Hess first suggested the device applications of real-space hot-electron transfer [49]. In layered heterostructures, hot electrons can move between adjacent layers. If the layers have different mobilities, the charge transfer between these layers results in negative differential resistance (NDR) effects. A real-space, charge-transfer device has source, drain, and collector terminals. The source-drain configuration has the basic HFET structure and is designed to inject hot electrons to be eventually collected at the collector terminal. The source-drain, hot-electron injector configuration is separated from the collector terminal by an insulator layer.

3.2.4. HBT Modeling and Characteristics

A procedure for modeling HBTs is based on a modification of the basic theory of homojunction BJTs to account for variation in semiconductor composition. In the simplest approximation, carrier fluxes are taken to be linear functions of the applied external biases, which lead to the drift-diffusion theory. In the drift-diffusion approximation, the electron current density, J_n, and the hole current density, J_p, are expressed as the sum of two terms—a drift one due to the electrical field and a diffusion term due to the inhomogeneous carrier densities. For the HBT, they are given by

$$J_n = nq\mu_n\left(\frac{-dV}{dx} + \frac{1}{q}\frac{dE_c}{dx}\right) + qD_n\left(\frac{dn}{dx} - \frac{n}{N_c}\frac{dN_c}{dx}\right) \quad (9)$$

$$J_p = pq\mu_p\left(\frac{dV}{dx} + \frac{1}{q}\frac{dE_v}{dx}\right) - qD_p\left(\frac{dp}{dx} - \frac{p}{N_v}\frac{dN_v}{dx}\right) \quad (10)$$

Here μ_n, μ_p, D_n, D_p are, respectively, the mobility and diffusion coefficients for electrons and holes and q is the electronic charge. In general, these parameters are dependent on the electric field. E_c, N_c, E_v, N_v are, respectively, the energy and effective density of states in the conduction and valence bands. The terms that involve the derivatives of the conduction and valence bands act as quasi-electric field that act on the electrons and holes, respectively. Similarly, the derivatives of the densities of states act as diffusion like terms for electrons and holes. The Poisson's equation has to be solved simultaneously with the drift-diffusion equation to obtain the behavior of carrier transport in the HBT. If we assume that the conduction and valence band bending are the same, then the Poisson's equation can be written simply as

$$div(\varepsilon_{\text{eff}}\, grad(V)) = q(n - p + N_a^- - N_d^+) \quad (11)$$

where ε_{eff} is the effective permittivity, N_a^- and N_d^+ are the ionized acceptors and donors, respectively. The solution to coupled Eqs. (9–11) is given by [50]:

$$J_n = qn\mu_n\frac{dF_n}{dx} \quad (12)$$

$$J_p = qp\mu_p\frac{dF_p}{dx} \quad (13)$$

where F_n and F_p are the quasi-Fermi levels for the conduction and valence bands, respectively. Numerical techniques, such as the Monte Carlo method, offers an alternative approach to solving the governing equations for the HBT. The Monte Carlo method is more appropriate if generation-recombination processes are to be accounted for [51]. Refinements of the basic drift-diffusion model are the thermionic-diffusion and the thermionic-field models [52].

The DC modeling of HBT entails the determination of the emitter current density (J_E), collector current density

(J_C), and base current density (J_B), the transconductance (g_m), and the current gain factor (β). These terminal currents can be obtained from the electron and hole current densities using the Ebers–Moll approach. The AC model determines device parasitics [53], the unity power-gain frequency, $f_{\max}$, the unity current gain frequency, f_t, and the high frequency and switching performance of the HBT.

4. GaAs RESONANT-TUNNELING DEVICES

The superlattice resonant-tunneling structure consists of a number of alternating layers of different energy bandgap materials with the layer thickness in the nanometer range, such that the electronic wave functions in adjacent layers overlap. Energy minibands and minigaps are formed which extend throughout the entire superlattice structure as a result of the overlap of the carrier wave functions. Doping and compositional superlattices form the two basic types. The *n-i-p-i* superlattice is an example of a doping superlattice. In a *n-i-p-i* superlattice, a periodic carrier concentration is introduced by selective doping a semiconductor material with acceptors and donors alternatively. The ionized impurities create repeated layers of negative (n) and positive (p) space charge separated by intrinsic (i) layers. As a result of such doping profile, a periodically repeating electric field is created in the semiconductor structure.

The first *n-i-p-i* structure was reported by Dohler in 1971 [55, 56]. Later in 1980, a GaAs *n-i-p-i* superlattice was grown by Ploog et al. [57]. These earlier structures were homoepitaxially grown. A schematic diagram of a homoepitaxial doping superlattice is shown in Figure 9. It consists of alternately p- and n-doped layers. Ohmic contact is established in all the n-layers by alloying Sn-metal, which is blocking to the p-layers, through all the layers. Likewise, ohmic contacts are established to the p-layer by alloying Sn-Zn metal through all the layers. Thus a bias voltage, V_{np}, can be applied between all n-layers and p-layers of the superlattice. Space charge is formed at the pn-junctions as a result of real-space, charge transfer across the junctions. This produces a modulation potential that bends both the valence and conduction bands. In effect, the p- and n-layers of the doping superlattice have thickness corresponding to the width of the space charges created. Figure 10 shows the energy band diagram of a doping superlattice. Electrons and holes are confined in small layers by the confining potential, but are separated from each other by the depletion regions. The electronic wavefunction, ψ_n, and the energy level, E_n, of the nth subband are given by

$$\psi_{c,n} = (2^n a_c n! \sqrt{\pi})^{-\frac{1}{2}} \exp\left(\frac{-z^2}{2a_c^2}\right) \times H_n\left(\frac{z}{a_c}\right) \quad n = 1, 2, 3, \ldots \tag{14}$$

$$a_c = \sqrt{\frac{\hbar}{m_c \omega_c}} \quad \omega_c = \sqrt{\frac{4q^2 \pi N_D}{\varepsilon_o \varepsilon_r m_c}}$$

$$E_{c,n} = \hbar\omega_c\left(n + \frac{1}{2}\right) \tag{15}$$

Here N_D is the doping concentration for electrons, m_c, the effective mass of electrons, ε_o is the permittivity of free

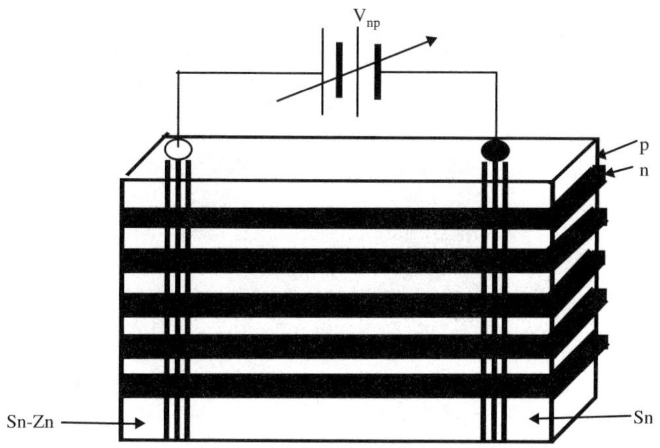

Figure 9. Homoepitaxial *n-i-p-i* structure fabricated with alternating p- and n-layers. The Sn metal makes contact with all the n-layers but blocks contact to the p-layers. Similarly, the Sn-Zn alloy makes contact to all p-layers but blocks the n-layers. Reprinted with permission from [54], G. H. Dohler, "Properties of III–V Quantum Wells and Superlattices," 1996. © 1996, IEEE.

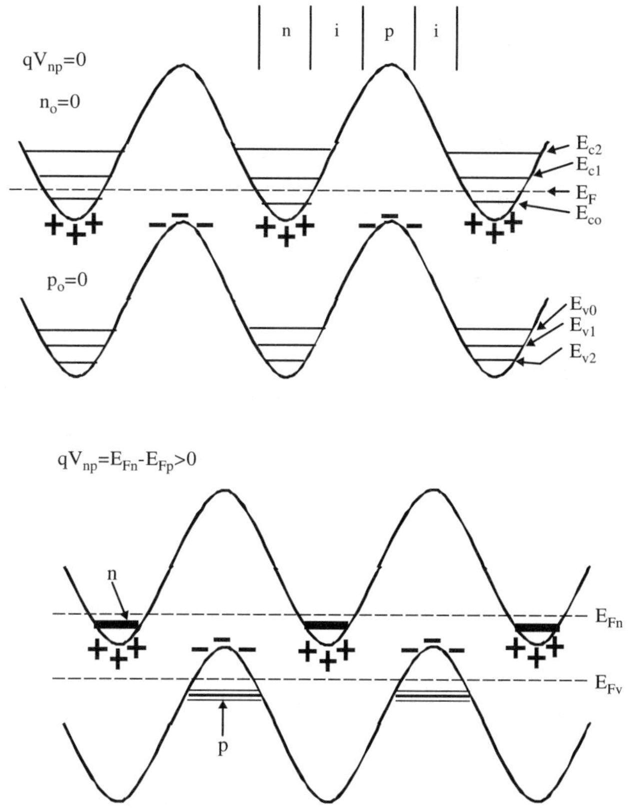

Figure 10. Energy band diagram of homoepitaxy *nipi* superlattice. Electrons and holes are confined in small layers by the confining potential, but are separated from each other by the depletion regions. Reprinted with permission from [54], G. H. Dohler, "Properties of III–V Quantum Wells and Superlattices," 1996. © 1996, IEEE.

space, ε_r is the relative permittivity, $\hbar$ is the Planck's constant, and q is the electronic charge. Equation 15 indicates that the energies are quantized and characterized for the nth subband by an integer, n, analogous to any 2D system. The observation of such reduced dimensionality of doping superlattice has been demonstrated experimentally [58–60].

Compositional superlattices are grown as lattice-matched or strained-layer structures consisting of alternating layers of wide bandgap and small bandgap materials with a layer thickness of a few nanometers, which allow tunneling of the electronic wave function across the barriers. The wave function, energy states, and density of states in a compositional superlattice with quantum number j and subband index μ assuming infinite well approximation are as follows:

$$\phi_{j,\mu,k_x} = \sqrt{\frac{1}{N}} \sum_{l=1} \chi_j(z - ld) \exp i(k_x ld) \exp i(k_x x + k_y y) \tag{16}$$

$$E_{j,\mu,k_x,k_z} = \varepsilon_{j,\mu} + s_j + \frac{\hbar^2}{2m_c^*}(k_x^2 + k_y^2) + 2t_j \cos(k_z d) \tag{17}$$

$$\rho_{SL}(E) = \frac{L_x L_y m^*}{\pi\pi\hbar^2} \sum_{j,k_z} \Theta(E - E_j - s_j + 2t_j \cos(k_z d)) \tag{18}$$

The heteroepitaxy n-i-p-i superlattice structure, introduced in the 1980's [61], is designed as an integrated structure of doping and compositional superlattices. They are of two main types, namely, Type-I and Type-II. In the Type-I doping heteroepitaxial superlattice, the center of the n- and/or the p-layer consists of undoped quantum wells of a lower bandgap material, separated by an undoped spacer layer to reduce impurity scattering [62]. The Type-II doping heteroexpitaxial superlattice [63] also has intrinsic layers (i-layers) that consist of a quantum-well of a lower bandgap. In addition, the electrons and holes are spatially separated from their parent dopants. Figures 11 and 12 show their respective energy band diagrams.

Carriers may be transported in superlattices either parallel or perpendicular to the growth layer surfaces. Perpendicular transport is dominated by tunneling processes and is the predominant mechanism underlying superlattice devices such as SL-tunneling diodes, SL-bipolar transistor, and SL-hot-electron transistor. On the other hand, carrier transport in superlattice devices such as SL-FET, RT-FET, and gated quantum-well SL-transistor are parallel to the growth surface.

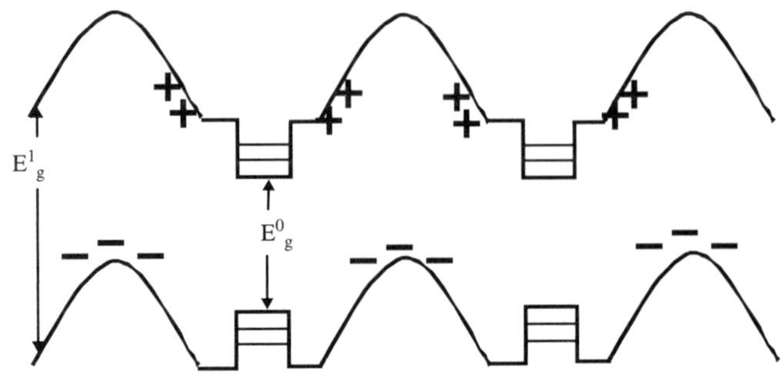

Figure 11. Energy band diagram of TYPE I heteroepitaxial *nipi* structure. Reprinted with permission from [54], G. H. Dohler, "Properties of III–V Quantum Wells and Superlattices," 1996. © 1996, IEEE.

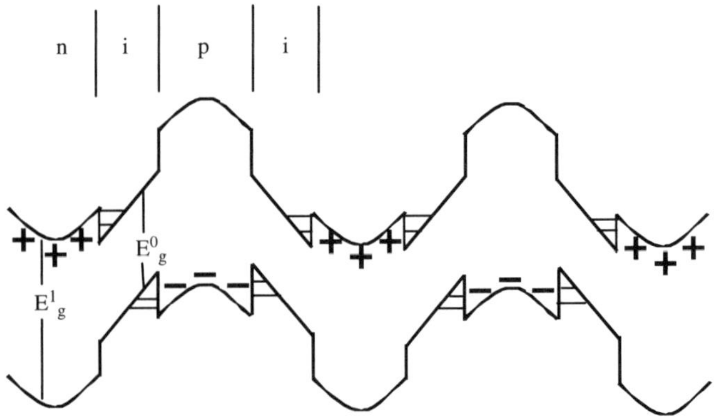

Figure 12. Energy band diagram of a TYPE-II heteroepitaxial *nipi* structure. Reprinted with permission from [54], G. H. Dohler, "Properties of III–V Quantum Wells and Superlattices," 1996. © 1996, IEEE.

Superlattice and quantum-well resonant-tunneling devices exhibit current-voltage characteristics that uniquely exhibit negative differential resistance. Such negative differential resistance has been realized in double quantum-well, multiple quantum-well, *nipi*-superlattice and heteroepitaxial-superlattice diodes and transistors. In these devices, the peak-to-valley ratio of the negative differential resistance determines their performance. This ratio, especially for the superlattice-based resonant tunneling devices, is enhanced at low temperatures.

4.1. GaAs Quantum-Well and Superlattice Tunneling Diode

Quantum-well and superlattice tunneling diodes with a current-voltage relationship that exhibit multiple peaks have a wide range of applications, including implementing high-speed multiple-value logic circuits, high-frequency oscillators, and amplifier circuits. In order to achieve the desired multiple peak I-V characteristics, a number of superlattices are integrated in a series or in parallel. Figure 13 illustrates a GaAs/AlGaAs superlattice tunneling diode. The structure consists of two superlattice sections separated by a tunneling barrier. The period of the superlattice is 4 nm and the tunnel barrier is 20 nm and is therefore thicker than each superlattice section. The heavily doped GaAs layers under the Au contact ensure better ohmic contacts. The operation of the diode is described with respect to its energy band diagram. Consider the application of a bias voltage across the devices. At low voltage biasing, the device current is due to tunneling of electrons from filled states of a miniband in the top section (TOP-SL) to empty states in the corresponding bottom section (BOT-SL) miniband on the other side of the barrier.

Increasing the bias voltage causes the current to rise steadily until the Fermi level on the top superlattice section side of the barrier falls within the lower minigap. When the second miniband on the topside becomes accessible to the tunneling electrons, the current rises again. A typical current-voltage characteristics of a four-section series diode

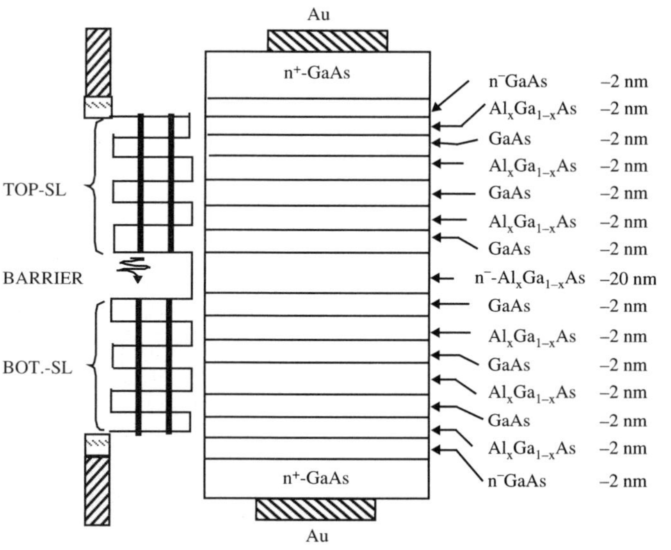

Figure 13. Superlattice tunneling diode.

arrangement, is shown in Figure 14. The curve depicts four peaks symmetrical about the current axis.

Resonant-tunneling (RT) diodes may be integrated to form more complex circuits. They can be integrated either vertically or horizontally. The integrated diodes turn to have multiple peaks current-voltage characteristics. Such integrated RT diodes have been used as frequency multipliers, multiple logic circuits, and as memory elements.

4.2. GaAs Resonant Tunneling Transistors

4.2.1. Quantum-Well and Superlattice-Base HBT

A superlattice structure can be incorporated in both a potential-effect and field-effect heterostructure transistor. Such hybrid structures have been used to design circuits with reduced transistor count. Both unipolar and bipolar potential-effect superlattice transistor operation is possible. A superlattice-base bipolar transistor is shown in Figure 15.

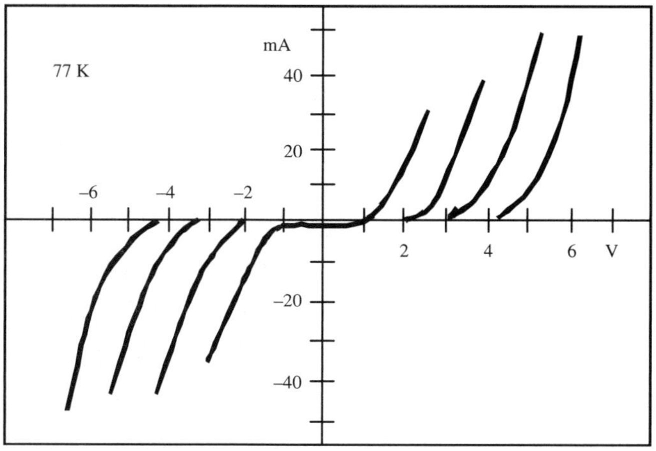

Figure 14. Current-voltage characteristics of a four-section superlattice diode. There are four distinct peaks that are symmetrical about the current axis.

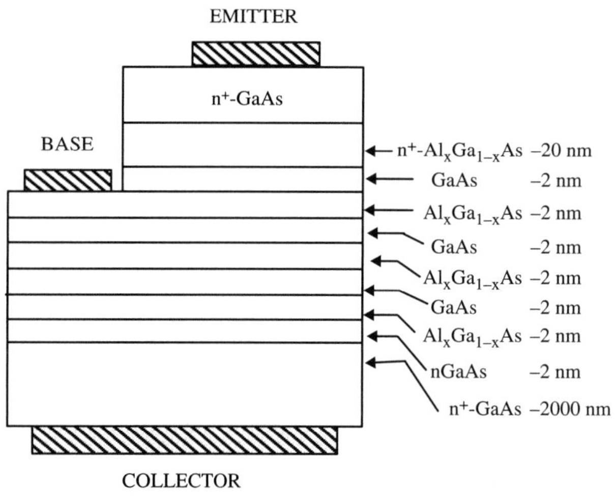

Figure 15. AlGaAs/GaAs superlattice-based HBT.

The emitter forms an abrupt junction with the base and is heavily doped to allow for tunneling into the GaAs/AlGaAs superlattice base. The superlattice base is grown as 20-section alternating layers of GaAs and $Al_xGa_{1-x}As$ with $x = 0.3$. Each layer has a thickness of 2 nm. The base is p-doped and the emitter is a 3–5 μm thick n^+-GaAs and is separated from the base by an undoped GaAs layer, thereby forming an *npn* structure. A typical input transfer characteristics (I_C vs. V_{BE}) are shown together with the output characteristics (I_C vs. V_{CE}) for a common-emitter transistor configuration in Figure 16.

4.2.2. Quantum-Well and Superlattice-Emitter Resonant Tunneling HBT

A GaAs superlattice-emitter resonant tunneling heterostructure bipolar transistor has been reported in InGaP/GaAs structures [64]. The insertion of a superlattice in the emitter layer, and the large valence band discontinuity ΔE_V of the InGaP/GaAs material system serves to provide an enhancement in the negative differential resistance effect at room temperature in both the saturation and linear regimes of device characteristics. The device also gives higher current gain due to better confinement of minority carriers in the base region.

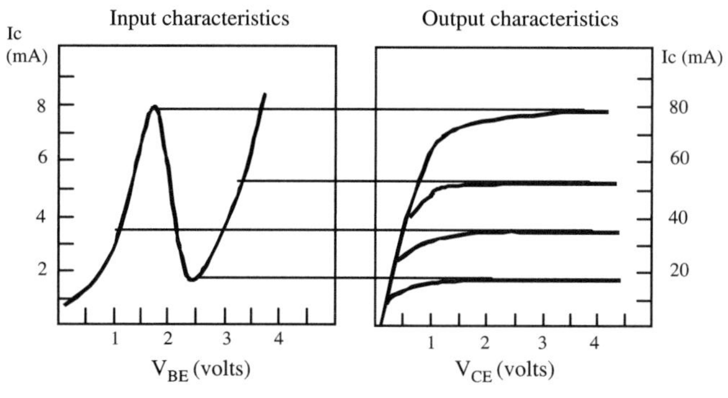

Figure 16. Input characteristics and output characteristics of AlGaAs/GaAs superlattice-based HBT.

5. GaAs QUANTUM-WIRE AND QUANTUM-DOT DEVICES

5.1. Single-Electron Transistor

A single-electron transistor (SET) is a device whose operation relies on single electron tunneling through a nanoscale junction. An SET can be considered as a field effect transistor whosc channcl consists of two tunncling junctions forming a quantum-dot (QD) island. The quantum-dot island is channeled to the source and drain so current can flow under the influence of the source and drain bias voltages V_s and V_d, respectively. The island is capacitive-coupled to a gate voltage V_g. An equivalent circuit representation of the SET is shown in Figure 17. The amount of charge that can occupy the quantum-dot island under the influence of the gate bias is discrete even though the bias voltage may change continuously. Thus, the charging energy of the system depends on the number of electron n on the quantum-dot island and the gate voltage. The charging energy, E_{ch}, is given by

$$E_{ch}(n, Q_G) = \frac{(ne - Q_G)^2}{2C} \tag{19}$$

$$Q_G = C_s V_s + C_G V_G + C_D V_D \tag{20}$$

$$C = C_s + C_G + C_D \tag{21}$$

The charging energy required to increase the charge on the quantum-dot island from n to $n+1$ as a consequence of tunneling is given by

$$E_{ch}(n+1, Q_G) - E_{ch}(n, Q_G) = \left(n + \frac{1}{2} - \frac{Q_G}{e}\right)\frac{e^2}{C} \tag{22}$$

The energy differences are equally spaced and can be tuned by the gate voltage. A single electron can tunnel from the source RTD1 to the drain RTD2 through the quantum-dot island (QD) if

$$eV_s > E_{ch}(n+1, Q_G) - E_{ch}(n, Q_G) > eV_d \tag{23}$$

The current on a macroscopic conductor is often assumed to flow without any blockage in charging a capacitor. This is because the capacitance of such macroscopic wires is large, and therefore the energy $E_{ch} = e^2/2C$ required to charge an electron being transported on the wire is often negligible compared with the thermal energy of the electron. The situation is drastically different for nanoscale conductors or at extremely low electron temperatures. For instance, the capacitance of a quantum dot is of the order of 10^{-18} F or less. Thus, transporting a single electron through the quantum-dot requires energy of the order of 100 meV, which is comparable with the thermal energy of an electron at room temperature $kT = 25$ meV. As a result, the transport of the electron is blocked until the applied bias provides an energy in excess of the required charging energy. This phenomenon is an example of Coulomb blockade in quantum transport. The quantum-dot junction of the SET is in the nanometer scale and electron tunneling through the junction, and, consequently, the current-voltage relationship of the SET can be expected to be governed by Coulomb blockade processes. The I–V relationship of the SET can be expressed as [65]

$$I = \frac{1}{4R}\left[V - \frac{4e^2}{C^2 V}\left(\frac{Q_c}{e} - n - \frac{1}{2}\right)^2\right]$$
$$\text{for} \quad -\frac{VC}{2e} \le \frac{Q_c}{e} - n - \frac{1}{2} \le \frac{VC}{2e} \tag{24}$$

Here, R is the tunneling resistance and assumed equal at the source and drain ends.

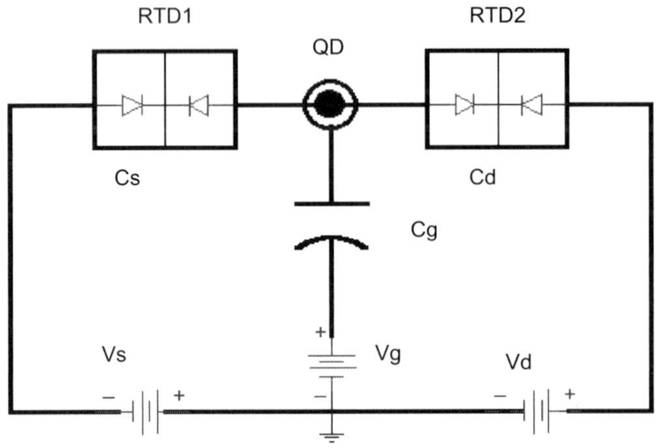

Figure 17. Equivalent circuit of a single electron transistor.

5.2. Quantum Interference Transistor

The quantum interference transistor is an electron wave interferometer. One configuration is the Mach–Zehnder interferometer as shown in Figure 18. It has an input port or source, an output port or drain, and a control terminal or gate that is placed on one arm of the interferometer. Another nomenclature for this structure is the gated-ring, quantum interference transistor. The gate forms a Schottky barrier such that by applying voltage to the gate electrode the electron density under the gate region can be reduced. In a 1D electron gas, with carrier concentration n, the electron wavelength, λ_F, at the Fermi energy is related to the concentration n as $\lambda_F = 2/n$. Thus, an electron wavelength can be controlled to produce constructive or destructive interference at the output (or drain port).

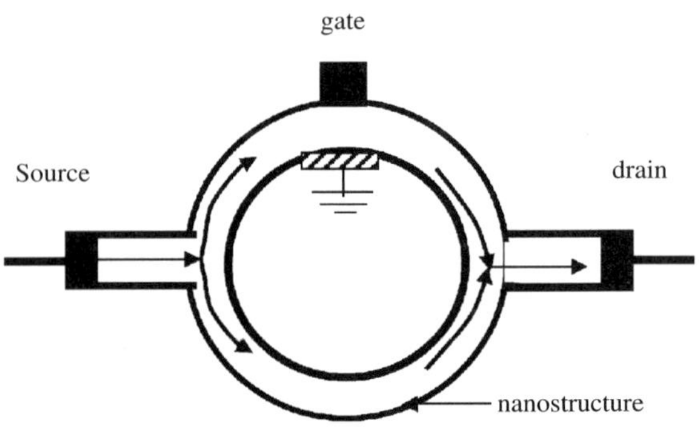

Figure 18. Gated-ring quantum interference transistor.

An alternative structure of the quantum interference transistor is the electron wave stub tuner [66, 67] illustrated in Figure 19. The structure has source, drain, and gate terminals. Electron waves emanate from the source and travel towards the drain. A fraction travel directly to the drain, while a porting reaches the drain after reflection at the gate electrode. This causes a path length difference between the waves, which can be tuned by changing the length of the gate stub by using fringing fields under the gate. The strength of this field is controlled by the bias on the gate.

6. GaAs SPINTRONIC DEVICES

Spin transport electronics or spintronics [68] combine electronics and ferromagnetics technologies to store, amplify, and process information. In comparison to conventional electronic device, spintronic devices consume less power and are also nonvolatile. Spintronic devices that show promise for practical application are spin-diodes, spin-FETs, and spin valve transistor (SVT). Two basic requirement for the successful operation of spintronic devices are a) a method to inject spin-polarized current from a ferromagnetic material into a semiconductor material and vice versa, and b) detection of spin-polarized current. Many schemes may be employed to effect spin injection [69]. The simplest method is to inject spin-polarized current through an ohmic contact between the ferromagnet and the semiconductor materials. To form a good ohmic contact the semiconductor must be heavily doped. However, this leads to spin-flip scattering losses. An alternative approach is to inject spin-polarized current through a ferromagnetic-insulator-semiconductor, or ferromagnetic-insulator-ferromagnetic tunneling junction. Other techniques are ballistic-electron spin-polarized injection, and hot-electron, spin-polarized injection. Spin-polarized current can be detected using electroluminescence, ferromagnetic electrodes, or by potentiometric methods.

6.1. Spintronic Diode

A cross-sectional view of a spintronic diode is shown in Figure 20. It consists of a ferromagnetic p^+-(Ga,Mn)As and nonmagnetic n^+-GaAs epitaxially grown on top of a nonmagnetic n-GaAs/(In,Ga)As/p-GaAs quantum-well LED [70]. A spin-polarized current injection is achieved through

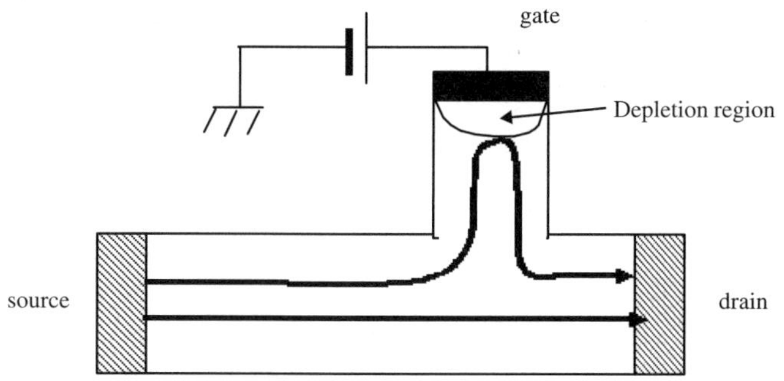

Figure 19. An electron wave stub tuner quantum interference transistor.

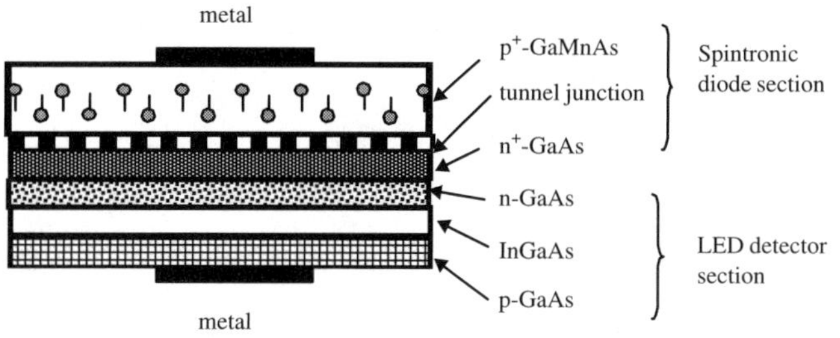

Figure 20. Schematic diagram of a spintronic diode/LED integrated structure. Spin polarized electrons from the GaMnAs ferromagnetic tunnel to the heavily doped GaAs and detected as electronuminescence of the LED.

interband tunneling between the valence band of the ferromagnetic layer and the conduction band of the nonmagnetic layer under reverse bias. The measurement scheme utilizes LED to detect spin by circular polarization of electroluminescense (EL). This detection state is based on the physical principle that all spin-polarized current in dilute magnetic semiconductors shows distinct circular polarization of the EL signal. The degree of polarization depends on the thickness of the spin aligner [71].

6.2. Spin Field Effect Transistor (SPINFET)

A cross-sectional view of spin-FET is shown in Figure 21. Its source and drain are dilute magnetic semiconductor materials. The density of states of electrons with one spin greatly exceeds that of the other spin at the Fermi level of the source and drain ferromagnetic material. Thus, the source and drain contacts preferentially inject and detect spin electrons. The channel of the spin FET consists of a narrow gap InGaAs quantum-well structure capable of splitting between up-spin and down-spin electrons even in the absence of a magnetic field. Suppose the magnetization of the injected electrons from the source is along the x-direction ($+x$ pol.); then the splitting between the spins will result in a positive z-polarized ($+z$ pol.) and negative z-polarized ($-z$ pol.) electrons, that is,

$$\underset{(+x \text{ pol.})}{\begin{pmatrix}1\\1\end{pmatrix}} = \underset{(+z \text{ pol.})}{\begin{pmatrix}1\\0\end{pmatrix}} + \underset{(-z \text{ pol.})}{\begin{pmatrix}1\\0\end{pmatrix}} \qquad (25)$$

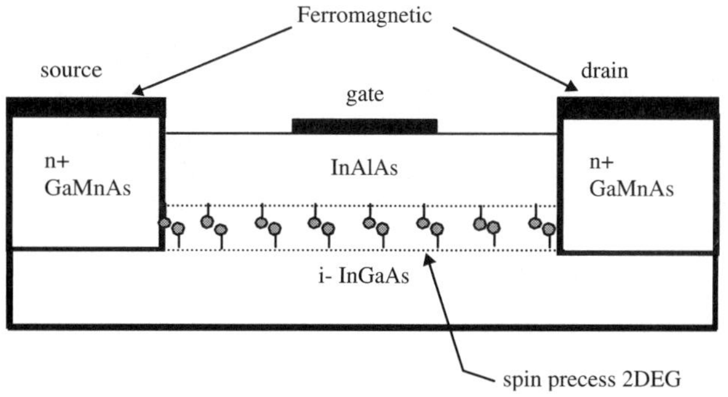

Figure 21. Cross-sectional view of a spinFET. The source and drain consists of highly doped dilute magnetic semiconductor in contact with ferromagnet source and drain metal. A spin precess 2DEG is formed at the interface of the InAlAs/InGaAs heterointerface.

The energies of the positive z-polarized and the negative z-polarized electrons are given by

$$E(+z \text{ pol.}) = \frac{\hbar^2 k_{x1}^2}{2m^*} - \eta k_{x1} \tag{26}$$

$$E(-z \text{ pol.}) = \frac{\hbar^2 k_{x2}^2}{2m^*} + \eta k_{x2} \tag{27}$$

Here, k_{x1}, k_{x2} are the electron wave vectors, η is the spin-orbit coupling constant, m^* is the effective mass of electrons, and $\hbar$ is Plank's constant. In the absence of inelastic scattering, the energy is invariant so that a differential phase shift is introduced between the up- and down-spin (or $+z$ pol. and $-z$ pol.), which is given for a spin FET with channel length L as

$$\Delta\theta = (k_{x1} - k_{x2})L = \frac{2m^*\eta L}{\hbar^2} \tag{28}$$

The transistor action entails controlling or changing the spin-orbit coupling constant, which in turn changes the phase shift. This will effectively rotate or precess the spin of the electron and modulate the current. The spin-coupling constant is controlled by a gate voltage.

GLOSSARY

Ballistic device An electronic device with the feature size of its active medium being smaller than the mean-free path of electrons in the host material. Electrons transit the device with little or no scattering, thereby reaching ballistic speeds.

Gallium arsenide (GaAs) noanodevices Quantum-well, quantum-wire, and quantum-dot electronic and optoelectronic devices grown on semi-insulating GaAs substrate, with GaAs compounds serving as the active medium.

Hot-electron device A nanodevice in which the temperature of electrons, or carriers, exceed the temperature of the host material during its operation.

Nanodevice An electronic device with at least one of its feature size having dimensions of the order of a nanometer.

Quantum dot Heteroepitaxial thin-film structure with all three of its feature sizes of the order of a nanometer.

Quantum-dot device A nanodevice in which electrons are confined from moving freely along any of its dimensions. Electrons move by hopping from one *dot* to another.

Quantum-effect device Atomic scale device in which quantum mechanical phenomena such as diffraction, interference and electron hopping describe its behavior and performance.

Quantum interference device An electronic device whose operation is based on the constructive and destructive interference of electron waves.

Quantum well Epitaxially grown thin-film material system in which a narrow bandgap material, such as gallium arsenide (GaAs) is sandwiched between two wide-bandgap materials such as aluminum gallium arsenide (AlGaAs).

Quantum-well device A nanodevice in which electrons are confined from moving freely in one dimensions.

Quantum wire A heteroepitaxial thin-film structure with two of its feature sizes of the order of a nanometer.

Quantum-wire device A nanodevice in which electrons are confined from moving freely in two-dimensions.

Resonant tunneling The transport of electrons across a metallurgical interface of an electronic device or structure, when the electron energies are lower than the potential energy barrier arising at the interface.

Resonant tunneling device A device in which current is controlled by manipulating the tunneling of electrons across a potential energy barrier.

Single electron transistor A device whose operation relies on single-electron tunneling through a nanoscale junction.

Spintronic devices Quantum devices that combine electronics and ferromagnetic effects for their operation.

Superlattices Thin-film material system consisting of a large number of heterostructures.

REFERENCES

1. E. Donkor, "Gallium Arsenide Heterostructures" in Semiconductors and Semimetal" (N. Willardson, Ed.). Academic Press, San Diego, Vol. 73. Chapter 1, 2001.
2. W.-C. Liu, W.-L. Chang, W.-S. Lour, K.-H. Yu, K.-W. Lin, C.-C. Cheng, and S.-Y. Cheng, *IEEE Trans Electron. Dev.* 48, 1290 (2001).
3. E. Donkor, in "Handbook of Advanced Electronic and Photonic Materials and Devices" (H. S. Nalwa, Ed.). Academic Press, San Diego, Vol. 1, Chap. 2, p. 81, 2001.
4. M. Zaknoune, B. Bonte, C. Gaquiere, Y. Cordier, Y. Druelle, D. Theron, and Y. Crosnier, *IEEE Trans. Electron. Dev. Lett.* 19, 345 (1998).
5. S. E. Rosenbaum, B. K. Kormanyos, L. M. Jelian, M. Matloubian, A. S. Brown, L. E. Larson, L. D. Nguyen, M. A. Thompson, L. P. B. Katehi, and G. M. Rebeiz, *IEEE Micro. Theory Tech.* 43, 927 (1995).
6. H.-Y. Lee, I.-J. Lin, H.-M. Shieh, and C.-T. Lee, *Solid State Electron.* 46, 1075 (2002).
7. P. Saunier and J. W. Lee. *IEEE Electron. Dev. Lett.* EdL-7, 503–505 (1986).
8. E. Sovero, A. K. Gupta, J. A. Higgins, and W. A. Hill, *IEEE Trans. Electron. Dev.* ED-33, 1434–1438 (1986).
9. T. Baba, T. Mizutani, and M. Ogawa, *J. Appl. Phys.* 59, 526–532 (1986).
10. C. W. Tu, W. L. Jones, R. F. Kopf, L. D. Urbanek, and S.-S. Pei, *IEEE Electron. Dev. Lett.* EDL-7, 552–554 (1986).
11. H. Morkoc and H. Unlu, in "Semiconductor and Semimetals," Vol. 24 (R. Dingle, Ed.). Academic Press, New York, Chapter 2, 1987.
12. H.-Y. Lee, I.-J. Lin, H.-M. Shieh, and C.-T. Lee, *Solid State Electron.* 46, 1075 (2002).
13. Y. Itoh, Y. Horiie, K. Nakahara, N. Yoshida, T. Katoh, and T. Takagi, *IEEE Micro. Guided Wave Lett.* 5, 48 (1995).
14. Y. L. Lai, E. Y. Chang, C. Y. Chang, T. K. Chen, T. H. Lui, S. P. Wang, T. H. Chen, and C. T. Lee, *IEEE Electron. Dev. Lett.* 17, 229–231 (1996).
15. P. M. Solomon, C. M. Knoeder, and S. L. Wright, *IEEE Electron. Dev. Lett.* EDL-5, 379 (1984).
16. M. K. Matsumoto, M. Ogura, T. Wada, N. Hashizume, and Y. Hayashi, *Electron. Lett.* 20, 462–463 (1984).
17. K. Remashan and K. N. Bhat, *IEEE Trans. Electron. Dev.* 49(3), 343 (2002).
18. Z. Chen and S. N. Mohammad, *Electron. Lett.* 1906 (1997).
19. H. Baratte, D. C. LaTulipe, C. M. Knoedler, T. N. Jackson, D. J. Frank, P. M. Solomon, and S. L. Wright, *IEDM Tech. Dig.* 444–447 (1986).
20. R. A. Kiehl, P. M. Solomon, and D. J. Frank, *IBM J. Res. Develop.* 34, 506 (1990).

21. T. J. Drummond, W. T. Masselink, and H. Morkoc, *IEEE Proc.* 74, 773–882 (1986).
22. G. Bastard, "Wave Mechanics Applied to Semiconductor Heterostructures." John Wiley, New York, Chapter 5, 1988.
23. K. Lee, M. S. Shur, T. J. Drummond, and H. Morkoc, *IEEE. Trans. Electron. Dev.* ED-30, 207–212 (1983).
24. K. Lee, M. S. Shur, T. J. Drummond, and H. Morkoc, *IEEE Trans. Electron. Dev.* ED-30, 207–212 (1983).
25. H. Hida, T. Itoh, and K. Ohata, *IEEE Trans. Electron. Dev.* ED-33, 1580–1586 (1986).
26. K. Yokoyama and H. Sakaki, *IEEE Electron. Dev. Lett.* EDL-8, 73075 (1987).
27. M. H. Weiler and Y. Ayasli, in "Microwave and Millimeter Wave Heterostructure Transistors and Their Application" (F. Ali, I. Bahl, and A. Gupta, Eds.). Artech, Norwood, 133–140, 1989.
28. L. F. Lester, P. M. Smith, P. Ho, P. C. Chao, R. C. Tiberio, K. H. Duh, and E. D. Wolf, in "Microwave and Millimeter Wave Heterostructure Transistors and Their Application" (F. Ali, I. Bahl, and A. Gupta, Eds.). Artech, Norwood, 254–257, 1989.
29. M. B. Das, *IEEE Trans. Electron. Dev.* 32, 11 (1985).
30. L. F. Lester, P. M. Smith, P. Ho, P. C. Chao, R. C. Tiberio, K. H. Duh, and E. D. Wolf, in "Microwave and Millimeter Wave Heterostructure Transistors and Their Application" (F. Ali, I. Bahl, and A. Gupta, Eds.). Artech, Norwood, 254–257, 1989.
31. R. A. Kiehl, P. M. Solomon, and D. J. Frank, *IBM J. Res. Develop.* 34, 506 (1990).
32. J. Dickmann, M. Berg, A. Geyer, H. Daembkes, F. Scholz, and M. Moser, *IEEE Trans. Electron. Devices* 42, 2 (1995).
33. B. Pereiaslavets, G. H. Martin, L. F. Eastman, R. W. Yanka, J. M. Ballingall, J. Braunstein, K. H. Bachem, and B. K. Ridley, *IEEE Trans. Electron. Devices* 44, 1341 (1997).
34. S. Luryi, *IEEE Electron. Dev. Lett.* EDL-6, 178–180 (1985).
35. C. Y. Chang, W. C. Liu, M. S. Jame, Y. H. Wang, S. Luryi, and S. M. Sze, *IEEE Electron. Dev. Lett.* EDL-7, 497–499 (1986).
36. S. Luryi, A. Kastalsky, A. C. Gossard, and R. Hendel, *IEEE Trans. Electron. Dev.* ED-31, 32 (1984).
37. H. Kroemer, in "Microwave and Millimeter-Wave Heterostructure Transistors and their Applications" (F. Ali, I. Bahl, and A. Gupta, Eds.). Artech House, Norwood, MA, 78, 1989.
38. M. Konagai, K. Katsukawa, and K. Takahashi, *J. Appl. Phys.* 48, 4389–4394 (1977).
39. H. Kroemer, in "Microwave and Millimeter-Wave Heterostructure Transistors and Their Applications" (F. Ali, I. Bahl, and A. Gupta, Eds.). Artech House, Norwood, MA, 78, 1989.
40. M. J. Mondry and H. Kromer, *IEEE Electron. Dev. Lett.* 6, 175–177 (1985).
41. K. Mochizuki, R. J. Welty, P. M. Asbeck, C. R. Lutz, R. E. Welser, S. J. Whitney, and N. Pan, *IEEE Trans. Electron. Dev.* 47, 2277 (2000).
42. C. R. Abernathy, F. Ren, P. W. Wisk, S. J. Pearton, and R. Esagui, *Appl. Phys. Lett.* 61, 1092–1094 (1992).
43. B.-C. Lye, P. A. Houston, H.-K. Yow, and C. C. Button, *IEEE Trans. Electron. Dev.* 45, 2417 (1998).
44. J. W. Park, D. Pavlidis, S. Mohammadi, J. L. Guyaux, and J.-C. Garcia, *IEEE Trans. Electron. Dev.* 48, 1297 (2001).
45. W. Liu, E. Beam, T. Henderson, and S. K. Fan, *IEEE Electron. Dev. Lett.* 14, 301–303 (1993).
46. W. Liu and S. K. Fan, *IEEE Electron. Dev. Lett.* 13, 510–512 (1992).
47. W. Liu, S. K. Fan, T. Henderson, and D. Davito, *IEEE Trans. Electron. Dev.* 40, 1351–1353 (1993).
48. J. R. Lothian, J. M. Kuo, F. Ren, and S. J. Pearton, *J. Electron. Mater.* 21, 441–445 (1992).
49. K. Hess, *Physica* 117B, 723 (1983).
50. D. Ankri, R. Azoulay, E. Caquot, J. Dangla, C. Dubon, and J. F. Palmier, in "Microwave and Millimeter-Wave Heterostructrure Transistors and Their Applications" (F. Ali, I. Bahl, and A. Gupta, Eds.). Artech House, Norwood, 179, 1989.
51. K. Tomizawa, "Numerical Simulation of Submicron Semiconductor Devices." Artech, Boston, Chapter 4, 1993.
52. J. J. Liou, "Advanced Semiconductor Device Physics and Modeling." Artech, Boston, Chapter 3, 1994.
53. M. B. Das, in "Microwave and Millimeter-Wave Heterostructrure Transistors and Their Applications" (F. Ali, I. Bahl, and A. Gupta, Eds.). Artech House, Norwood, 227–237, 1989.
54. G. H. Dohler in "Prop. of III–V Quantum Wells and Superlattices" (P. Bhattachara, Ed.). IEE, UK. Chapter 1, 1996.
55. G. H. Dohler, *Phys. Status Solidi B* 52, 79–92 (1972).
56. G. H. Dohler, *Phys. Status Solidi B* 52, 533–545 (1972).
57. K. Ploog, A. Fischer, G. H. Dohler, and H. Kunzel, *Inst. Phys. Conf. Ser.* 59, 721 (1981).
58. R. C. Miller, A. C. Gossard, D. A. Kleinman, and O. Munteanu, *Phys. Rev.* B29, 3740 (1984).
59. S. P. Kowalczyk, W. J. Schaffer, E. A. Kraut, and R. W. Grant, *J. Vac. Sci. Technol.* 20, 705 (1982).
60. H. Sakaki, L. L. Chang, R. Ludeke, C. A. Chang, C. A. Sai-Halasz, and L. Esaki, *Appl. Phys. Lett.* 31, 211 (1977).
61. P. Ruden and G. H. Dohler, *Surface Sci.* 132, 540 (1983).
62. H. Kunzel, A. Fischer, J. Knecht, and K. Ploog, *Appl. Phys. A.* 30, 73–81 (1983).
63. G. H. Dohler and P. Ruden, *Surf. Sci.* 142, 474–485 (1984).
64. C.-Y. Chen, W.-C. Wang, W.-H. Chiou, C.-K. Wang, H.-M. Chuang, S.-Y. Cheng, and W.-C. Liu, *Solid State Electron.* 46, 1289 (2002).
65. T. Dittrich, P. Hanggi, G.-L. Ingold, B. Kramer, G. Schon, and W. Zwerger, "Quantum Transport." Wiley-VCH, New York, 159, 1998.
66. F. Sols, M. Macucci, U. Ravaioli, and K. Hess, *Appl. Phys. Lett.* 54, 350 (1989).
67. S. Datta, *Superlatt. Microstructures* 6, 83 (1989).
68. J. F. Gregg, I. Petej, and C. Dennis, *J. Appl. Phys. D* 35, R121 (2002).
69. M. Johnson, *Semicond. Sci. Technol.* 17, 298 (2002).
70. M. Kohda, Y. Ohno, K. Takamura, F. Matsukura, and H. Ohno, "A Spin Esaki Diode," *Jpn. J. Appl. Phys.* 40 (2001).
71. G. Schmidt and L. W. Molenkamp, in "Semiconductor and Spintronics and Quantum Computation" (D. D. Awschalom, D. Loss, and N. Samarth, Eds.). Springer, New York, Chapter 3, 2002.

Encyclopedia of Nanoscience and Nanotechnology
www.aspbs.com/enn

GaInNAs Quantum Well Lasers

W. Li, M. Pessa, T. Jouhti, C. S. Peng, E.-M. Pavelescu

Tampere University of Technology, Tampere, Finland

CONTENTS

1. INTRODUCTION

When atoms are brought together to form a crystal, the energy levels of the electrons will merge to form bands of energy, in which electrons can only occupy the allowed bands separated by a forbidden bandgap. The band that is normally filled with electrons in the absence of thermal excitations (at $T = 0$ K) in semiconductors is called the valence band, while the upper unfilled band is called the conduction band [1]. The optical processes in semiconductors are mediated by transitions involving electrons in the conduction band and holes (the empty states) in the valence band. Electrons and holes can be injected into a semiconductor either by a voltage applied to electrical contacts at the surface of the crystal or by absorption of light.

The operating characteristics of semiconductor devices (e.g., diode lasers) depend critically on the physical properties of the constituent materials. At present, III–V compound semiconductors (which consist of elements from columns III and V of the periodic table), such as GaAs and InP, provide the material basis for a number of well-established technologies of electronic and optoelectronic devices and systems. Since the electronic and optical properties of semiconductors are completely determined by their band structures or crystal structures, in order to tailor the band structure one method is to alloy two or more binary compounds via an appropriate growth technique. Another method is to fabricate artificial heterostructures—quantum wells (QWs) and superlattices—at dimensions comparable to the electron or hole wavelength in the semiconductor material (typically around 10 nm), in which a layer of one semiconductor is sandwiched between two layers of a larger bandgap material. As a result, the electrons and holes are unable to move freely in the crystal growth (confinement) direction. This confinement changes the density of the electron states and their specific energy levels, resulting in a totally new generation of quantum semiconductor devices (see Fig. 1).

Semiconductor diode lasers are small, efficient devices with typical dimensions of less than 1 mm. They are used in a wide variety of applications, ranging from the readout sources in compact disc players to transmitters in optical fiber communication systems. Similar to all lasers, a semiconductor diode laser is made of a gain medium providing the light amplification inserted between two reflective mirrors. It essentially exploits a forward-biased p–n junction to inject electrons and holes to generate light (photons). The device also consists of an optical cavity, which can guide the photons generated to achieve stimulated emission. The simplest way to make a mirror in a semiconductor is simply to cleave the crystal, and the facets will function as mirrors (often the facets are coated with antireflection and high-reflection layers to improve the laser performance and reliability).

The operation wavelength of semiconductor QW laser diodes is largely determined by the semiconductor bandgap. Figure 2 plots the bandgap vs. lattice constant for several families of III–V compound semiconductors. These direct bandgap semiconductors have emerged as the materials of choice for lasers that emit in the 0.7- to 1.6-μm wavelength range. For a long time, the alloy semiconductor $In_xGa_{1-x}As_yP_{1-y}$ grown on an InP substrate was the only material system for 1.3- to 1.55-μm-wavelength lasers used in optical fiber communications. Although InP-based lasers have proven successful for applications in optical communications, they have some drawbacks [2]: (1) The GaInAsP/InP laser has higher temperature sensitivity compared to shorter wavelength lasers that are grown on GaAs substrates, primarily due to the relatively small conduction band offset between InGaAsP and InP, $\Delta E_c/\Delta E_g = 0.4$ (about 0.6 for the GaAs/AlGaAs system), which allows electrons to escape from the QW when they have enough

ISBN: 1-58883-059-4/$35.00

Encyclopedia of Nanoscience and Nanotechnology
Edited by H. S. Nalwa
Volume 3: Pages (719–730)

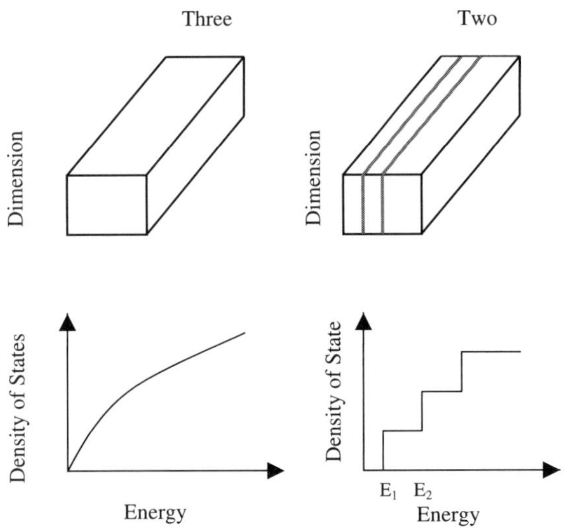

Figure 1. Variation in the energy dependence of the density of states in three-dimensional and two-dimensional (quantum well) systems.

thermal energy; (2) it is difficult to achieve single-step epitaxy in vertical-cavity surface-emitting lasers (VCSELs) in which high-quality distributed Bragg reflector (DBR) mirrors are required, since the refractive index difference of InP/InGaAsP is insufficient and the thermal conductivity is too poor to realize low-cost and high-performance long-wavelength VCSELs [3].

After the successful development of short-wavelength (850 nm) GaAs/AlGaAs VCSEL technology, the extension of light emission to the even more important 1.3-μm transmission window has been pursued to meet the demands of metro area network and optical switching and routers. One approach is using self-organized InAs-based quantum dots [4], in which the electrons are confined in three dimensions. Although 1.3-μm-wavelength quantum dot lasers have been demonstrated, they suffer from high temperature sensitivity and relatively low modulation speed. The second approach applies GaAsSb/GaAs type II QWs, however, with only modest success because of the gain blueshift with increasing current injection [5]. The last candidate for the 1.3-μm VCSELs is the GaInNAs/GaAs QW, which is discussed in this review.

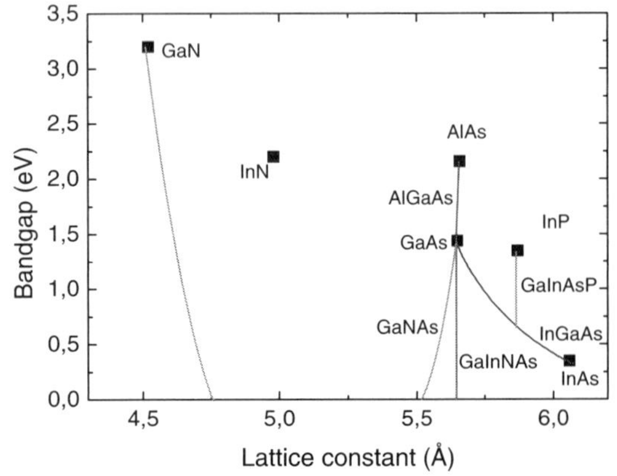

Figure 2. Selection of ternary and quaternary III–V compounds (alloys) plotted as a function of energy gap and lattice constant.

Ga(In)N_yAs_{1-y} alloys have unique properties: Incorporation of nitrogen into Ga(In)As simultaneously decreases the lattice constant and the bandgap [6]. With 1% N incorporated, the bandgap of Ga(In)NAs is reduced by about 150 meV, exhibiting a giant bowing effect, which arises from the large difference in size and electronegativity of the As and N atoms. Therefore, with less than 2% of nitrogen incorporated into GaInAs, the emission wavelength of GaInNAs can be extended to 1.3 μm. Compared to the conventional GaInAsP/InP system, the GaInNAs/GaAs heterostructure offers several advantages. First, the GaInNAs/GaAs conduction band well is deeper and the electron effective mass is larger [7], thus providing better confinement for electrons and better match of the valence and conduction band densities of states. This should lead to a diode laser with higher operating temperature, higher efficiency, and higher output power [2]. Second, monolithic VCSELs can be straightforwardly fabricated using the well-developed GaAs/AlAs mirrors and selective wet oxidation of AlAs for current and optical confinement (the lattice mismatch between AlAs and GaAs is only 1592 ppm). Third, as compared to the control of group V composition in the $InGaAs_yP_{1-y}$ alloy, growth of the Ga(In)N_yAs_{1-y} alloy by molecular beam epitaxy (MBE) is relatively easy because atomic nitrogen is very active, resulting in the group V composition only determined by the N flux.

We review the present knowledge of dilute nitride Ga(In)NAs as well as the progress in the growth of high-quality GaInNAs materials and in realizing high-performance edge-emitting lasers and VCSELs. The review is divided into three main sections, which deal with the physical properties of Ga(In)NAs alloys, epitaxial growth, and state-of-the-art GaInNAs lasers.

2. GaInNAs MATERIALS

2.1. Ga(In)NAs Alloy

In the dilute regime of N $< 10^{18}$ cm^{-3}, the nitrogen atom acts as an isoelectronic electron trap in GaAs [8, 9]. The crystal potential is deformed at the N site such that a highly localized N-related state is formed due to the large difference in size and electronegativity of As and N. This isoelectronic N level is resonant with the conduction band about 180 meV above the conduction band edge [8]. The existence of this N-related state and the strong perturbation due to the introduction of N cause the dramatic band structure changes observed in Ga(In)N_yAs_{1-y} alloy [10]. The transition from nitrogen acting as an isoelectronic impurity to the formation of N-induced bands takes place in GaN_xAs_{1-x} grown on GaAs at $x \approx 0.21\%$. The corresponding mean distance between N atoms is about 30 Å.

2.1.1. Band Structure of GaNAs

The band structure of GaAsN is difficult to calculate, since alloys are not perfect crystals even if they have a perfect lattice. This is because the atoms are placed randomly and not in any periodic manner. The bandgap of the alloy compound can usually be described by

$$E_g(A_{1-x}B_x) = (1-x)E_g(A) + xE_g(B) - x(1-x)C$$

where the so-called bowing parameter C accounts for the deviation from a linear interpolation between the two binaries A and B. The bowing parameter for conventional III–V alloys is typically a fraction of an electronvolt. However, it is surprisingly huge and composition dependent in GaN_xAs_{1-x} (~26 eV for $x < 1\%$ and ~16 eV for $x > 1\%$) [11]. Due to the giant bowing parameter, GaAsN alloys provide the opportunity to simultaneously decrease both the lattice constant and the bandgap with increasing N content. Unfortunately, it appears that when the N concentration increases, the alloy quality deteriorates dramatically as reflected by the very low photoluminescence efficiency and very short carrier diffusion length, impeding the potential application of Ga(In)Nas alloys in electronic and optoelectronic devices [2, 6, 12].

The carrier effective masses are fundamentally important in semiconductor lasers, as they determine the density of states (in a quantum well structure, the density of states in any subband is linearly proportional to the effective mass). The electron effective masses are greater in GaN_xAs_{1-x} alloy than in GaAs, whereas the hole effective masses are similar to those of GaAs. Values of $0.12m_0$ and $0.19m_0$ were reported for $x = 1.2\%$ and 2.0%, respectively (the electron effective mass of GaAs is $0.067m_0$) [13]. The observed trend of an increasing mass value with N content is in agreement with theoretical predictions [14, 15].

From absorption and photomodulated reflectance measurements [10, 16], the GaN_xAs_{1-x} alloy has been proven to have a direct bandgap and exhibits a decrease in the bandgap energy with increasing N composition. This can be explained by two main models: One is the two-level band-anticrossing model [15], and the other one is based on pseudopotential local density approximation (LDA) calculations [17, 18]. Recently, the empirical pseudopotential method has been adopted to yield detailed theories of many electronic anomalies and puzzling phenomena of GaAsN alloys [19].

2.2. Metastability of GaInNAs Alloys

Since Ga and N are smaller atoms than In and As, respectively, the bond configuration of Ga–As + In–N will be better lattice matched (thus, possess less strain) than that of Ga–N + In–As. On the other hand, the cohesive energies of the respective binary zincblende solids follow the sequence GaN > InN > GaAs > InAs (being, respectively, 2.24, 1.93, 1.63, and 1.55 eV per bond), so the (highly strained) Ga–N + In–As configuration is preferred in terms of bond energy [20]. Thus, in equilibrium, N prefers to be surrounded by In atoms ("local indium enrichment"), whereas As prefers to be surrounded by Ga. This type of short-range order is a novel feature of quaternary isovalent systems, leading to a significant blueshift of the bandgap (i.e., reduced bowing) with respect to random alloys. The interplay of growth conditions and metastability of GaInNAs alloys can explain many experimental findings [21].

2.3. Band Alignment in Ga(In)NAs/GaAs Heterostructures

One extremely important parameter in Ga(In)NAs/GaAs heterostructures is the band edge discontinuity produced when they are brought together, which has a large impact on the charge transport and recombination process in Ga(In)NAs/GaAs heterostructures and their potential device applications. Unfortunately, the experimental results are quite controversial, and both type I (the electrons and holes are located in the same layer) and type II (the electrons and holes are in different layers) band alignment were reported. Recent experimental results strongly favor a type I band alignment for this material system [22–24]. The studies of band alignment in Ga(In)NAs/GaAs have been based mainly on the effect of quantum confinement on the optical transition energies from strained Ga(In)NAs/GaAs QWs [25, 26].

The photoluminescence (PL) and PL excitation measurements of GaInNAs/GaAs QWs of different thickness have yielded important results: The conduction band discontinuity amounts to 80% of the bandgap difference and a strongly increased electron mass compared to InGaAs with the same In content is observed [27].

3. EPITAXIAL GROWTH

3.1. MBE/MOCVD

The key to growing semiconductor nanostructures is epitaxial growth techniques such as MBE and metal–organic chemical vapor deposition (MOCVD), which have had a tremendous impact on semiconductor physics and technology in the past decades. MBE and MOCVD are the techniques that allow monolayer (about 3 Å) control in the growth of the desired crystals. In MBE, atoms or molecules of the species (groups III and V) to be grown impinge upon the heated substrate in high vacuum (~10^{-11} torr). The crystal growth is governed predominantly by a kinetic process. Achieving a stoichiometric III–V epitaxial growth lies in the surface chemical dependence of the sticking coefficient of the group V elements. Under proper growth conditions, the growth rate is only determined by the arrival rate of the group III elements. In MOCVD, the growth process is much more complex than that of MBE. The source materials in MOCVD are complex molecules (precursors) that are transported to the heated substrate in the reactor cell via a carrier gas stream. These reactants decompose in the heated zone above the substrate and growth takes place. Unlike MBE growth, the MOCVD process is mass transport limited, and the growth rate is dependent mainly on the partial pressure of the metal–organic reactants present in the reactor. A limiting factor of MOCVD, with respect to MBE, is carbon incorporation into the films, resulting from the organic nature of the precursors.

3.2. Characterization of Epitaxial Structures

X-ray diffraction is a powerful characterization technique (nondestructive) for providing very detailed information on epitaxial structures, which ranges from the determination of the lattice constant of single epitaxial layers to studies of the structural parameters of superlattices and QW interfaces [28]. X-ray diffraction can probe large areas of the sample to a depth on the order of 10 μm, and X-ray rocking curves can be computer simulated to provide information on an atomic scale.

Another very useful characterization technique in studies of single QWs and superlattices is transmission electron microscopy (TEM). The TEM cross sections reveal the well and barrier thickness and provide some measure of sample uniformity. Cross sections imaged using (002) reflection yield detailed information on the strain distribution in the sample. The well–barrier contrast allows for evaluation of the interface abruptness [29].

The optical properties of epitaxial structures can be investigated by PL measurement, in which the carrier population is photoexcited with a pump laser and the recombination light is usually evaluated via a spectrum analyzer. It is fast and nondestructive and can provide valuable information of the bandgap of a material and QW spontaneous recombination as well as the crystalline quality of QWs via the shape of the PL spectrum (intensity and linewidth).

3.3. Growth of Ga(In)NAs/GaAs Heterostructure

The growth of high-quality Ga(In)NAs material on GaAs is not straightforward. It is essential that an alloy not be clustered or phase separated; otherwise, the material will not be very useful for optoelectronic devices and cannot be used in any reliable device process. Due to the large difference in size and bonding energy between N and As, it is difficult to achieve a good "mixing" in the Ga(In)NAs alloy formation process. Thus, nonequilibrium growth has to be applied to overcome the "natural dislike" of such materials. The most effective way is growth of GaInNAs at lower temperatures, which is one of the advantages of MBE over MOCVD (low-temperature growth would be a problem for MOCVD where decomposition reactions at the surface are difficult).

3.3.1. Growth of Ga(In)NAs/GaAs by MBE

Several nitrogen sources (e.g., NH_3, monomethylhydrazine, dimethylhydrazine, N plasma source, etc.) have been developed for MBE growth of dilute nitrides on a GaAs substrate. It has been shown that the best choice for growth of Ga(In)NAs by MBE is using a radio frequency (rf) plasma cell to generate an atomic nitrogen flux. In principle, the plasma (including ions and electrons) is produced by passing an rf electrical discharge through nitrogen gas at a low pressure. Due to the presence of very active atomic N, the group V composition of Ga(In)N_yAs_{1-y} is only determined by the arriving N flux at substrate temperatures below about 500 °C. In other words, the nitrogen content in the Ga(In)N_yAs_{1-y} epilayer is inversely proportional to the group III (Ga, In) growth rate, implying the sticking coefficient of atomic N is near unity [2]. These results are quite different from those of any other mixed column V, III–V alloy semiconductor (e.g., GaInAsP, GaInAsSb, GaInPSb), indicating that MBE-grown GaInNAs has certain advantages over InGaAsP in which the group V composition is sensitive to changes in the growth conditions (temperature and group V fluxes) [30, 31], making control of the group V composition somewhat complicated.

3.3.2. Growth of Ga(In)NAs/GaAs by MOCVD

For growth of GaInNAs/GaAs by MOCVD, group III source materials of triethylgallium (TEGa) and trimethylindium (TMIn) have been used in combination with tertiarybutylarsine (TBAs) and dimethylhydrazine (DMHy) [or 1,1-dimethylhydrazine (UDMHy)] as group V sources which decompose at a low temperature compared with that of often-used materials such as AsH_3 and NH_3 [32]. With DMHy as the N source, the MOCVD-grown samples showed a strong superlinearity in nitrogen composition against the nitrogen source flow, and the nitrogen composition increased with increasing growth rate, but decreased significantly by incorporating indium into GaNAs or by increasing the growth temperature [32].

The growth of GaInNAs using UDMHy as the N source is also quite complicated. Both the TBAs/UDMHy partial pressure ratio and the growth rate drastically influence N incorporation in the GaInNAs layers. It was shown that the key parameter for nitrogen incorporation in the GaInNAs/GaAs material system was the partial pressure of TBAs, and N incorporation was reduced with decreasing growth rate of GaInNAs. In addition, the N incorporation was also influenced by the In composition of the epitaxial layer [33].

3.4. Material Quality of Ga(In)NAs/GaAs Heterostructure

Since the growth window of GaAs is quite different from that of GaN, achieving Ga(In)NAs of high crystalline quality is very challenging. The growth of Ga(In)NAs at too high a temperature may result in surface segregation of nitrogen and clustering (or phase separation), which will deteriorate the crystalline quality [34, 35]. On the other hand, growth at too low a temperature leads to a high density of point defects (Ga vacancies, As_{Ga}, defect complexes) due to insufficient cation migration [36–38]. In addition, to reduce ion-induced damage of the growing layer, the plasma operating condition must be optimized to maximize the amount of atomic nitrogen in the beam flux.

Considering the large lattice mismatch between GaAs and GaN, even incorporating small amounts of N will give rise to a substantial tensile strain in pseudomorphic GaNAs on GaAs. The most promising range of N composition in Ga(In)NAs is around 2%, where the material quality is quite acceptable for device applications. Incorporation of more N atoms leads to a large amount of N interstitials and N clustering due to the large size mismatch between the solute and the solvent atoms [34, 37, 39]. Also, the photoluminescence (PL) efficiency degrades severely [2, 35, 40–42].

For the determination of the nitrogen content of Ga(In)NAs epilayers on GaAs, the X-ray diffraction (XRD) technique is often used. However, by XRD, the lattice spacings are determined and not directly the N content of the Ga(In)N_yAs_{1-y} alloys. Thus, the calculation of these values requires assumptions on the variation of the lattice parameters and of the elastic constants with *y*. A linear interpolation (Vegard's law) between the values of GaAs and cubic-type GaN has been commonly assumed to be justified. On the other hand, although most N atoms are believed to occupy predominantly the As sublattice, the N atoms might energetically prefer moving out from their substitutional sites to neighboring interstitial sites [36, 39, 43]. It is shown that the formation of isolated interstitial nitrogen in GaNAs is unlikely because of its high formation energy in the lattice. Instead, N complexes such as As–N and N–N split interstitials are energetically favored to form, occupying a single lattice site [34, 37]. The formation of split interstitial N–As complexes induces a compressive strain in the epilayer, while N–N complexes cause less tensile strain as compared to the substitutional N_{As} atom [34].

Postgrowth annealing can significantly improve the PL efficiency of Ga(In)NAs alloys (the PL intensity typically increases by a factor of 10–100), which has been attributed to a significant reduction in the concentration of competing nonradiative defects [35, 36, 40, 44]. Both positron annihilation measurements and nuclear reaction analysis in conjunction with Rutherford backscattering spectrometry have been used to study the point defects present in MBE-grown Ga(In)NAs. The experimental data reveal a high density of nitrogen interstitials [$(2.3 \pm 0.2) \times 10^{19}$ cm^{-3}] and Ga vacancies [$(4.2 \pm 0.2) \times 10^{16}$ cm^{-3}]. The existence of a high density of N interstitials and Ga vacancies (part of the Ga vacancies may be converted into an arsenic antisite [38]) in the as-grown nitride–arsenide materials will significantly affect the device performance, since they introduce bandgap levels affecting the carrier capture and recombination as well as transport process. Therefore, Ga(In)NAs must be annealed to obtain device-quality material. It is also important to reduce the ion-induced damage and metastable neutral molecules N_2^* produced from the rf plasma source or to incorporate more indium into the epilayer, inhibiting the formation of the N interstitial. The role of indium is athreefold: (1) Adding indium reduces the bandgap; (2) highly compressive strain in the epilayer induced by In makes N atoms prefer the As sublattice rather than the interstitial sites [36]; and (3) incorporation of In can inhibit the formation of N-related clusters [45].

The annealing of GaInNAs/GaAs QWs also leads to a reduction in the PL linewidth, which is related to the more homogeneous In distribution [46, 47], as well as a blueshift of the PL spectra (typically in the range of 30–70 meV, depending on the growth and annealing conditions) as shown in Figure 3. The blueshift can be attributed in part to enhanced interdiffusion (As–N interdiffusion through a kick-out mechanism and/or vacancy assistant group III interdiffusion) [2, 36, 40, 47]. Another source of the blueshift is the rearrangement of the nitrogen nearest neighbor environments from Ga-rich to In-rich environments [20, 21], which reflects an interplay of growth conditions and metastability of the alloy. During the growth of GaInNAs, chemical bonding aspects dominate at the surface, which favor Ga–N bonds instead of In–N bonds. This surface state is frozen in during the nonequilibrium growth process. In contrast to the

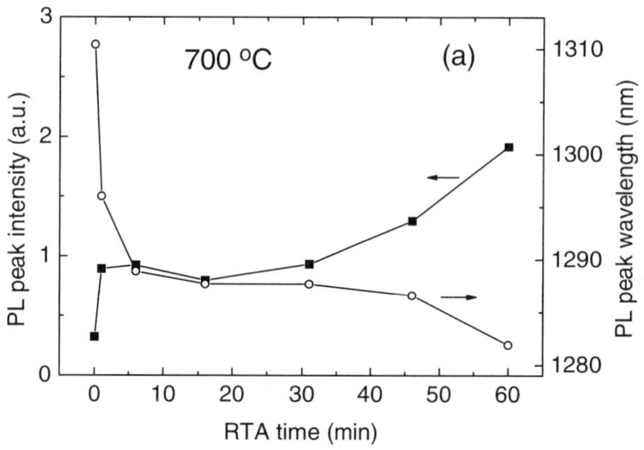

Figure 3. Effect of rapid thermal annealing on the PL properties as a function of annealing time at 700 °C.

surface, In-rich nearest neighbor configurations of N ("local indium enrichment") are favored in bulk at equilibrium due to the dominance of local strain effects. Therefore, the frozen non-equilibrium bulk state can be transformed into the equilibrium bulk state by postgrowth annealing under appropriate conditions [20, 21]. Recently, we have used computer simulations of double crystal X-ray diffraction curves for as-grown and annealed 6-nm-thick Ga0.7In0.3N0.03As 0.97/GaAs multiple-quantum well (MQW) structures in conjunction with the photoluminescence experiments to clarify the origin of annealing-induced photoluminescence blueshift in GaInNAs/GaAs QWs. For samples annealed at 700 °C for 6 min, the atomic interdiffusion was negligible where the blueshift was significant. After annealing for 1 h, atomic interdiffusion was indeed observed, and the diffusion length of In and N were determined to be 0.6 nm and 1.0 nm, respectively. Our results indicate that short range atomic rearrangements in GaInNAs well layer dominate in the initial stage of annealing where the blueshift is significant; as annealing continues ($>=$ 700 °C), the atomic interdiffusion across the QW interfaces takes place, which further blueshifts the emission.

In general, 7-nm-thick GaInNAs/GaAs QWs with In and N contents of about 35% and 1.7%, respectively allow us to achieve 300-K luminescence at 1.3 μm. To extend the emission of annealed GaInNAs/GaAs QWs to 1300 nm and beyond without deteriorating the material quality, GaNAs barriers have been grown instead of GaAs barriers. There are several advantages to GaNAs barriers: (i) The GaNAs barrier results in longer wavelength emission due to the reduced barrier potential; (ii) GaNAs acts as a source of nitrogen, which is believed to reduce out-diffusion of N from the QW; and (iii) the presence of tensile strain in the GaNAs barrier balances the highly compressive strain in the GaInNAs well layer, making the whole structure more stable [2, 48].

In attempts to further improve the structural and optical properties of the active region, a GaInNAs strain-mediating layer was inserted between the GaInNAs QW and the GaNAs barrier. It was found that the combination of strain-compensating GaNAs barriers and GaInNAs strain-mediating layers redshifted the PL emission due to the broadening of the effective QW width and the reduction of In out-diffusion from the QW [48, 49].

3.5. Role of Sb in GaInNAs

The linewidth of the PL emission from a GaInNAs/GaAs heterostructure is extremely important from the physical and technological points of view. In general, there are two inhomogeneous broadening sources in these alloys: interface roughness and alloy potential fluctuations. It has been confirmed that the roughening of the growth front of GaInNAs, particularly at high In and N composition, can be reduced by incorporating Sb (<1%), which acts as a surface surfactant [50]. The incorporation of Sb in GaInNAs also improves the luminescence efficiency. Furthermore, it has been reported that the N incorporation rate is enhanced in GaAsSbN at a higher Sb content [51, 52]. Recently, a room-temperature lasing up to about 1.5 μm has been achieved by GaInNAsSb/GaNAsSb QW lasers (Sb incorporated 6–7%) [53].

3.6. Role of Hydrogen in GaInNAs

Since hydrogen is present during the growth (except for solid-source MBE) and the processing of semiconductor materials, considerable interest has been focused on the high chemical activity of hydrogen to passivate electrically active defects (including both n- and p-type dopants), which is likely due to the formation of a neutral complex with no dangling bond left [54]. Recently, the role of hydrogen in altering the electronic properties of InGaNAs/GaAs QWs has been investigated. The most remarkable effects observed are a decrease in PL efficiency and a blueshift in emission upon hydrogen irradiation. The optical properties are fully recovered by thermal annealing [55]. These phenomena can be accounted for by the formation of N–H bonds, which effectively passivate the N atoms in the lattice [55].

The incorporation of H also affects the electrical properties of GaInNAs materials [57]. It has been shown that H acts as an isolated donor in InGaNAs and makes the as-grown undoped samples slightly n type. Annealing above 700 °C reduces the hydrogen concentration and renders the samples p type [56].

4. QUANTUM WELL HETEROSTRUCTURE LASERS

The typical semiconductor laser diode consists of a thin slab of undoped active material sandwiched between p- and n-type cladding layers with larger bandgaps. The lower bandgap active region has a higher refractive index than the cladding, so that a transverse dielectric optical waveguide is formed. Modern semiconductor diode lasers utilize so-called separate-confinement QW heterostructures, which use a thinner QW carrier-confining active region (~10 nm) and a surrounding intermediate-bandgap separate-confinement region to confine the photons. Typically, QW lasers have threshold currents that are an order of magnitude below those of conventional double-heterojunction laser diodes fabricated from bulklike materials, because quantum confinement reduces the density of states in the active region and thereby lowers the number of injected carriers required to invert the population. The other major attraction of QW lasers is their advantages in dynamic characteristics, which include a very high modulation speed, a low-frequency chirp, and a narrow linewidth.

Practical semiconductor diode lasers come in two basic configurations: those with in-plane cavities and those with vertical cavities (as illustrated schematically in Figs. 4 and 5, respectively). The planar geometry of the edge-emitting laser results in a beam emitted parallel to the surface of the substrate, while the vertical-cavity laser emits perpendicular to the substrate. Edge-emitting diode lasers have been in existence since the late 1960s. Vertical-cavity surface-emitting lasers (VCSELs) have been available from the beginning of the 1990s.

4.1. Broad-Area and Ridge Waveguide Edge-Emitting Lasers

For the assessment of material quality, broad-area lasers are often fabricated. Standard photolithographic processing is used to fabricate approximately 100-μm-wide oxide-defined stripes on the sample. Then the substrates are lapped and polished, n-type contacts are performed by alloyed Ni/Au/Ge/Au, and Ti/Pt/Au contacts are evaporated on the p side. Finally, various cavity lengths are formed by cleaving, and the resulting bars are cleaved into individual diodes for testing [58].

Another widely produced semiconductor laser is the index-guided single-mode laser. It is designed to propagate the lowest order ("fundamental") optical mode that has a *near*-Gaussian shape in both lateral and transverse dimensions [59]. A popular index-guided single-mode laser structure is the ridge waveguide laser, in which the waveguide in the plane of a p–n junction is provided by etching almost all of the p-side cladding region away from a narrow ridge (less than 5 μm in width). The two-dimensional optical mode is largely confined to the layer beneath the ridge. Contact metallization is fabricated on top of the ridge to channel light through the ridge waveguide and enable lasing.

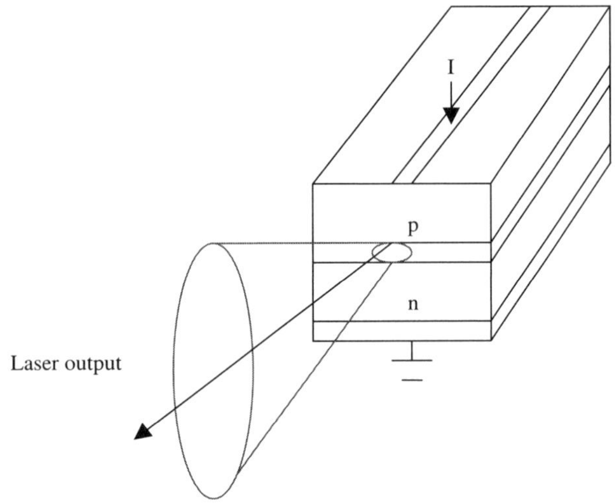

Figure 4. Schematic of an in-plane (edge-emitting) diode laser.

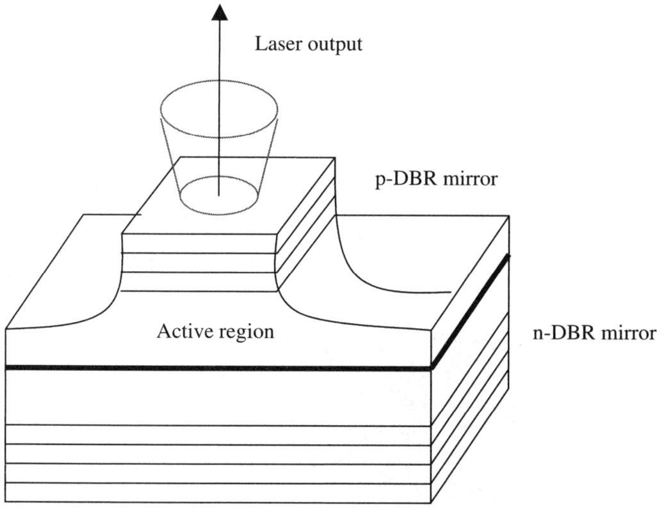

Figure 5. Schematic of a vertical-cavity surface-emitting laser.

4.2. Implanted and Selectively Oxidized VCSELs

VCSEL fabrication generally involves establishing electrical contact to the anode and cathode of the diode and defining the transverse extent of the optical cavity. A method to achieve electrical confinement in a planar VCSEL topology is to utilize ion implantation [60]. Implantation of ions into the top DBR mirror can be used to render the material around the laser cavity nonconductive and thus to concentrate the injected current into the active medium. Fabrication of implanted VCSELs usually begins with the deposition of electrical contacts. Next, an implantation mask of photoresist or plated metal can be used to block the ions. Deep proton implants near the active region are used to define the transverse extent of the laser cavity.

Another method to achieve both electrical and optical confinement in VCSELs is to utilize selective oxidation of AlGaAs. Fabrication of selectively oxidized VCSELs begins with the deposition of electrical contacts. A silicon nitride mask is deposited and patterned on the wafer top to encapsulate the metal contact and to form an etch mask. Dry etching, such as reactive ion etching, is used to define mesas to expose the oxidation layers. The lateral oxidation extent of a layer within the etched mesa is controlled by the composition of the layer and the oxidation time. After oxidation, the nitride mask is removed to permit laser testing.

4.3. Characterization of Semiconductor Lasers

The semiconductor laser is unique in that the population inversion required to reach lasing threshold results from electrical charge injection into the active region. The accumulating electrical charge provides gain for the spontaneously emitted photons recirculating within the laser resonator. As the injection current is increased above its threshold value, the oscillation intensity builds up, and lasing results. Lasing is evidenced by both a sharp increase in output power and a narrowing of the broad spontaneous emission spectrum into distinct lasing lines. From the light–current curve of a laser, its threshold current and slope efficiency can be obtained, while the current–voltage curve is principally characterized by its junction voltage and series resistance. The useful output of the laser is limited by multimode operation at the kink in its light–current curve, as witnessed by the broadening and steering of its lateral far-field pattern [59].

The temperature dependence of the laser characteristics is critical to device performance at high output powers, where the heat generation is highest. Some applications may require laser operation over a wide temperature range. The thermal characteristic is a result of the temperature dependence of the lasing threshold current [described by $J_{\text{th}}(T) = J_{\text{th}}^0 \exp(T/T_0)$], the thermal resistance of the laser and heat sink, and the conversion efficiency of the laser. In addition, the maximum temperature, T_{max}, at which continuous-wave (cw) operation can still be achieved, is determined by the heat resistance and the dissipation on the one hand and by the temperature dependence of the threshold current on the other.

Apart from the dc characteristics of QW lasers mentioned above, their dynamic characteristics also need to be evaluated to meet the requirements of optical communication. To examine the dynamic response of a semiconductor laser, small signal modulation is often adopted, in which the laser is biased at a point well above threshold and a small ac signal is applied to it. This approach provides important insight into the physics of the laser and how the laser design can be improved. In addition, the spectral properties of lasers are very important for high-speed transmission. In order to have no additional noise due to mode partition, single-mode lasers with side-mode suppression ratios, better than 30 dB, are preferred.

The reliability or lifetime of a semiconductor laser is an important factor to be considered, especially for its application in optical communication. The reliability of a laser is determined by dissipated heat, thermal stresses, current density, handling, and power density at the facet. All of these parameters are dependent on the waveguide structure, resonator configuration, thermal resistance, and packaging.

4.4. GaInNAs/GaAs Edge-Emitting Laser

Semiconductor lasers (see Fig. 4) emitting at 1.3 and 1.55 μm are key components for optical fiber communication systems. Currently, these lasers are predominantly based on GaInAsP alloys grown on InP substrates. They are quite sensitive to variations in temperature. This high temperature sensitivity is primarily due to Auger recombination and the weak electron confinement resulting from the small conduction band offset in GaInAsP/InP. The GaInNAs/GaAs heterostructures, on the other hand, with a deeper conduction band QW and a larger electron effective mass, will thus provide better confinement for electrons and better match of the valence and conduction band densities of state, which lead to a higher characteristic temperature (T_0) and higher operating temperature, higher efficiency, and higher output power [2, 61].

The first long-wavelength GaInNAs/GaAs laser operating at room temperature in the cw mode was reported in

1996 [62]. The laser emits at 1.18 μm with a threshold current density (J_{th}) of 1.83 kA/cm^2 and a T_0 of 126 K. Its structure consists of an active region of a single 6-nm-thick $Ga_{0.7}In_{0.3}N_{0.02}As_{0.98}$ well layer, a waveguide layer of 0.3-μm-thick GaAs, and 1.5-μm-thick n- and p-type $Al_{0.3}Ga_{0.7}As$ cladding layers doped to 5×10^{17} and 9×10^{17} cm^{-3}, respectively. Finally, a 0.2-μm-thick p$^+$-GaAs contact layer is heavily doped to 3×10^{19} cm^{-3} in order to minimize the contact resistance. After growth of the laser structure, it may not be necessary to carry out annealing if the cladding layers are grown at a temperature high enough to anneal out most of the nonradiative centers present in the GaInNAs QW.

The laser performance has also been found to depend strongly on the growth temperature of the active QW region [53]. A single-QW (SQW) laser grown under optimized conditions yields threshold current densities ($W = 20$ μm, $L = 800$ μm) as low as 450 A/cm^2 with wavelengths in the range of 1230–1250 nm. The laser thresholds decrease as the growth temperature of the active region increases up to 440 °C; beyond 440 °C, the threshold currents increase.

By introducing a strain-compensated barrier to GaInNAs/GaAs, it is possible to grow highly strained GaInNAs wells free of misfit dislocations and to increase the number of QWs in the laser structures. Growth of strain-compensated GaInNAs/GaAsP/GaAs/GaInP SQW laser structures has been proposed [40]. The active region of the laser consists of a single 6-nm-thick GaInNAs/GaAsP quantum well with two 100-nm-thick GaAs space layers sandwiched between $Ga_{0.49}In_{0.51}P$ cladding layers. Recently, high-performance strain-compensated $In_{0.4}Ga_{0.6}As_{0.995}N_{0.005}$ QW lasers, with a lasing emission wavelength of 1.295 μm, have been achieved by MOCVD utilizing AsH_3 and U-dimethylhydrazine (U-DMHy) as the group V sources. The threshold (J_{th}) current densities of the GaInNAs QW lasers have been measured to be 211 A/cm^2 ($L = 2000$ μm).

Truly high power GaInNAs/GaAs (emitting at 1.3 μm) SQW lasers have been achieved with a threshold current density down to 405 A/cm^2 for a 1000-μm-long cavity (at 10 °C) and 8-W cw output (10 °C) [63]. No catastrophic optical damage is observed at a facet power density of 30 mW/cm^2. Preliminary life tests carried out at $P_{out} = 1.5$ W and $T = 35$ °C for 1000 h indicate that the lasers are robust; no power degradation is found. For GaInNAs 3QW lasers, in which three $Ga_{0.65}In_{0.35}N_{0.017}As_{0.983}$ QWs with a thickness of 6–7 nm, separated by 20-nm barrier layers, are symmetrically inserted into the 300-nm-thick GaAs cavity, the laser ($L = 1120$ μm, $W = 100$ μm) exhibits very low threshold current density down to 680 A/cm^2 and a slope efficiency of 0.59 W/A (output per two facets) with wavelength around 1.29 μm. Using heat sinking at 10 °C, cw operation is demonstrated with output power of 2.4 W and 4 W under cw and pulsed operation, respectively [64]. The internal efficiency and losses are estimated to be 81% and 10 cm^{-1}, respectively.

To extend the lasing wavelength to 1.31 μm, increasing the nitrogen content in the GaInNAs quantum well has often been adopted, which results in much higher threshold current densities mainly due to the formation of the N-related nonradiative centers [36]. A GaInNAs SQW laser emitting at 1.3045 μm exhibits a threshold current density as high as 2.86 kA/cm^2 [65]. In our group, a GaInNAs/GaNAs/GaAs SQW laser structure has been designed and grown by solid-source molecular beam epitaxy using an rf plasma source. A threshold current density as low as 546 A/cm^2 was achieved for 1600-μm-long, as-cleaved shallow ridge stripe lasers emitting at 1.32 μm (20 °C). The characteristic temperature is 97–133 K in the temperature range from 20 to 80 °C. The internal quantum efficiency for these lasers is 80%, while the material losses are 7.0 cm^{-1}. Under cw operation, optical output up to 40 mW per facet was achieved at room temperature for these uncoated lasers [66, 67].

To achieve low threshold current and single-mode lasing, GaInNAs ridge waveguide lasers have been developed [68]. The emission wavelength is around 1290 nm at room temperature; the active region consists of two QWs. Lasers with as-cleaved facets show threshold currents as low as 21 mA with an efficiency of 0.25 W/A per facet. By high-reflection coating of one side (with a reflectivity of about 75%), the threshold is lowered to 16 mA and the efficiency increases to 0.35 W/A. The best laser achieved could be operated under cw conditions up to 100 °C and exhibits a T_0 of 110 K [44]. These values compare favorably to those of 1.3-μm InGaAsP/InP ridge waveguide lasers, where the typical T_0 is around 60 K.

To fully utilize GaInNAs edge-emitting lasers as Raman pumps or for nonlinear optical sources in channel switching, one must extend the wavelength over 1.4 μm and produce high power (>500 mW) in the single lateral mode. This is a key requirement for GaAs-based lasers to be competitive with their InP counterparts. With approximately 5% N incorporated in the QWs, a GaInNAs/GaAs laser with emission at 1.5 μm was achieved, while the threshold current density was extremely high (50 kA/cm^2) [69]. Recently, ridge waveguide lasers with cleaved facets and three GaInNAsSb QWs and GaNAsSb barriers have been fabricated. A maximum output power in excess of 70 mW and a slope efficiency of 0.21 W/A have been demonstrated. The emission wavelength is at 1.465 μm, and the minimum threshold current density is 2.8 kA/cm^2 [70].

4.5. Vertical-Cavity Surface-Emitting Lasers

Fiberoptic components commonly use two types of semiconductor lasers: edge-emitting lasers and VCSELs (see Fig. 5). The most prevalent gigabit-speed diode lasers at 1.55 μm in data communications and telecommunications are distributed feedback (DFB) edge-emitting lasers, which are based on the propagation of optical waves in a periodic grating to create strong mode selectivity. However, the edge emitters have numerous drawbacks in cost, manufacturability, and reliability. In contrast, VCSELs can be on-wafer tested. Large arrays of devices can also be fabricated to be used for the interconnection of data links in network computing and massive parallel computing. In the telecommunications industry, the VCSEL's uniform, single-mode beam profile (due to its short cavity length) is desirable for coupling into optical fibers. As a whole, VCSELs retain the following advantages over edge-emitting lasers: lower cost, lower power consumption, higher reliability, smaller size, easier testing, easier manufacturing and packaging, and more efficient coupling to fiber. The first VCSELs were made by combining bulk active regions with metal mirrors. Subsequently, other approaches were tried, including dielectric mirrors, semiconductor distributed Bragg reflector (DBR) mirrors, and air–semiconductor DBR mirrors.

For a VCSEL, since the round-trip gain path length is very short, the reflectivities of the mirrors need to be very high. The only mirrors that can provide over 99% reflectivity are dielectric stacks created by growing quarter-wavelength-thick layers of semiconductors with alternating refractive indices, which are referred to as distributed Bragg reflectors (DBRs). For electrical-pump monolithic VCSELs, the DBR mirrors represent the most unique and challenging aspect of VCSELs. They serve a dual role, defining the laser cavity as highly reflective mirrors, as well as providing the pathway for current injected from the contact layers. Thus, the principal design issues associated with the DBRs are a trade-off between providing sufficiently high reflectivity for light generated in the cavity along with the lowest possible resistance for both electrons and holes transported to the active region [60].

In a typical VCSEL, as many as 60 individual semiconductor layers may be stacked within a structure that is only about 10 μm thick. After growth, conventional semiconductor processing techniques are used to define the laser and complete the device. VCSELs are most often fabricated with a circular aperture between 5 and 25 μm in diameter to enable single- or multiple-mode operation. Due to the high gain and small volume of the VCSEL structure, the lasers are very efficient and have low turn-on voltages, high modulation frequencies (can be directly modulated at speeds beyond 10 Gbps), and single longitudinal mode operation.

4.5.1. GaInNAs/GaAs Vertical-Cavity Surface-Emitting Lasers

As the Internet and data transmission continue to grow, a concurrent demand for higher bandwidth and decreasing costs makes the prospect of using VCSEL-based fiberoptic links very attractive. Presently, all commercial VCSELs are epitaxially grown on GaAs substrates, for which the standardized wavelength is 850 nm. At 1310 nm, however, due to the optical characteristics of a single-mode fiber (relatively low attenuation and minimal dispersion), the optical signal is capable of traveling much greater distances than can be achieved with 850-nm wavelengths. Therefore, 1310-nm VCSELs are the ideal solution for short to intermediate distances (less than 20 km) for telecommunications and data communications applications.

Despite the strong demand for 1310-nm VCSELs, the ability to grow an optimal combination of a high-quality active region and high-reflectivity mirrors with low electrical drop is technologically challenging. Furthermore, considering the higher free-carrier absorption of the p-type DBR mirrors at 1.3 μm than at 850 nm (the free-carrier absorption increases as λ^2), the mirror design for 1.3-μm VCSELs is far more challenging than for 850-nm VCSELs. The above-mentioned problems suggest that modifications to the conventional VCSEL structure are needed for achieving high-performance, long-wavelength devices.

The first VCSELs with an active region of GaInNAs/GaAs QWs, emitting around 1.2 μm, were realized in 1998 [71]. The etched-pillar VCSEL was designed for substrate emission to avoid optical losses. The epitaxial layers were grown by gas source MBE (As_2 was produced from cracked AsH_3). The bottom mirror was a 25.5-period Si-doped GaAs–AlAs n-type DBR. A single 7-nm-thick $Ga_{0.7}In_{0.3}N_{0.004}As_{0.996}$ QW was located at the center of the GaAs cavity. Above the cavity was a 21-period Be-doped GaAs–AlAs p-DBR with parabolically graded interfaces to reduce series resistance. A Ti–Au contact electrode, combined with a p-phase-matching layer and cap, enhanced the reflectance of the p-DBR. Electrical contact to the n substrate was made using indium solder. Room-temperature pulsed operation was achieved with a threshold current density of 3.1 kA/cm^2, a slope efficiency of 0.04 W/A, and an output power above 5 mW for 45-μm-diameter devices. Laser oscillation was observed for temperatures as high as 95 °C.

With a similar design, room-temperature cw operation of oxide-confined GaInNAs/GaAs VCSELs emitting at a wavelength of 1.2 μm was demonstrated later, in which the active region was embedded in a one-wavelength-long cavity and consisted of three 7-nm $Ga_{0.7}In_{0.3}N_{0.02}As_{0.98}$ QWs separated by 20-nm GaAs barriers [72]. The VCSELs were grown by solid-source MBE and the threshold current was as low as 1 mA and the slope efficiency above 0.045 W/A. Pulsed threshold current varied from 0.89 mA for 3.6-μm devices to approximately 21 mA for 29-μm sizes. The threshold current density was within the range of 2–2.5 kA/cm^2 at larger sizes, and a slope efficiency above 0.09 W/A was achieved. The slope efficiency dropped for aperture sizes below 7 μm, which was likely due to optical scattering, and the current density rose to nearly 7 kA/cm^2 for the smallest (3.6 μm) devices.

GaInNAs/GaAs VCSELs with a conventional p–n structure were also demonstrated. The growth was done in two separate MOCVD systems. The devices achieved cw room-temperature operation with a threshold current of approximately 7.6 kA/cm^2, a slope efficiency of 0.1 W/A, and an output power of 100 μW at 1.26 μm from a 10 × 10 μm^2 aperture with a threshold voltage of 2.8 V [73].

Truly 1300-nm GaInNAs VCSELs were demonstrated much later, in which a p^{++}/n^{++} tunnel junction was inserted at the first optical node above the microcavity to convert electrons into holes and then uses a second n-type DBR mirrors [74, 75]. The VCSELs were grown by MBE. The threshold current was 0.9 mA (5 kA/cm^2), and the threshold was 4.3 V. A maximum output power of 0.77 mW was obtained, and the peak differential efficiency was 0.24 W/A. The device with a large aperture (15 × 18 μm^2) exhibited a maximum cw output power of 2.1 mW. For a 5 × 5-μm^2 device, cw operation at temperatures as high as 105 °C was achieved. Devices with the more conventional p–n structure exhibited output characteristics similar to the tunnel junction devices. The threshold current density was 4.4 kA/cm^2, and the threshold voltage of 2.7 V was significantly smaller than that for the tunnel junction devices. Undoubtedly, the tunnel junction in that device contributes significantly to the voltage drop. Further improvements in the tunnel diode design may enable significantly lower device voltages without substantial increases in absorption losses.

Table 1. VCSELs on GaAs substrates.

Research group	λ (nm)	Active region	Growth method and design	Single-/ multimode	P_{out} (mW)	I_{th} (mA) V_{th} (V) @RT	J_{th} (kA/cm^2)	Remarks	Ref.
Infineon	1293	GaInNAs	MOCVD, intracavity	SM	1.40@25 °C 0.50@85 °C	1.25 1.06	3	cw, 10 Gb/s, SMSR > 40 dB	[80]
Infineon	1302	GaInNAs	MBE, intracavity	SM	~1.0@RT	1.90 —	7.9	cw, f-3 dB 5.4 GHz, SMSR > 35 dB	[81]
NEC	1295	GaAsSb	MOCVD, p–n	SM	0.06@RT	1.2 2.1	—	cw, T_{max} 70 °C	[82]
NEC	1270	GaAsSb	MOCVD, p–n	SM	0.46@RT	1.1 2.0	—	cw, f-3 dB 4.5 GHz, T_{max} 80 °C, SMSR > 30 dB	[83]
Sandia	1288	GaInNAs	MBE, tunnel junction	SM	0.77@RT	0.9 4.3	5.0	cw, T_{max} 105 °C, 3 × 6 μm^2	[84]
Sandia	1288	GaInNAs	MBE, tunnel junction	MM	2.10@RT	0.9 4.3	—	cw, T_{max} 105 °C, 15 × 18 μm^2	[84]
Cielo	1289	GaInNAs	MBE, p–n	SM	0.98@20 °C	2.9 2.2	—	cw, f-3 dB 7 GHz, SMSR > 50 dB	[85]
Ioffe Inst., Tech. Univ. Berlin	1300	InAs QDs	MBE, intracavity	MM	0.65@25 °C	1.2 1.5	1.5	cw, P_{out} < 0.1 mW@ 45 °C	[86]

To completely avoid the problems of optical losses and voltage drop in the mirrors, a double-intracavity contact design was proposed [76–78]. The MBE-grown VCSELs (oxide aperture 4 × 6 μm^2) exhibit an output power of more than 1 mW at room temperature with a wavelength of 1306 nm. Lasing in cw operation can be maintained up to 80 °C. The threshold current at room temperature is 2.2 mA, and the side-mode suppression ratio at a typical drive current of 5 mA (for 2.5 Gbit s^{-1}) is better than 30 dB, but near the thermal rollover it drops to about 10 dB. For MOCVD-grown GaInNAs VCSELs with the same structure, a record P_{out} of 1.4 mW (cw single mode with a side-mode suppression ratio greater than 40 dB) at room temperature and 0.5 mW at 85 °C have been obtained. The cw threshold current is 1.25 mA, and the wavelength is 1.293 μm. The threshold voltage is 1.06 V, which is remarkably lower than the voltage drop in tunnel junction VCSELs. This approach appears to be the most suitable for preparing low-threshold VCSELs of small sizes, while disadvantages are current crowding and nonuniform injection at power levels significantly higher than 1 mW. Modifications to the device structure could be a single-intracavity contact design, in which the top DBR mirrors are undoped and current driven through the bottom DBR mirrors [79].

The emission dynamics and the optical gain of 1.3-μm GaInNAs/GaAs VCSELs have been investigated both experimentally and theoretically [78]. An ultrabroad temperature operation range between 30 and 388 K with picosecond emission dynamics has been demonstrated. The shapes and widths of the gain spectra are similar to those of GaInAsP/InP at moderate carrier densities, while at higher densities the gain spectra broaden considerably. These results can be well described by the microscopic theory.

The data transmission properties of GaInNAs VCSELs are encouraging. It has been demonstrated that transmission rates up to 10 Gbit/s in a back-to-back configuration and 2.5 Gbit/s for transmission over 20.5 km using a standard single-mode silica fiber with a power penalty of only 0.5 dB over this distance [77, 80]. Clearly, the 1.3-μm VCSELs possess excellent capabilities for high-speed communications, incomparably better than those of their 850-nm counterparts.

Table 1 gives a selected list of performance characteristics of various 1.3-μm VCSELs, as reported in the literature. There are clearly many parameters not yet optimized for these devices, but the results to date demonstrate that the GaInNAs/GaAs material system is the most promising candidate for 1.3-μm VCSELs. Still, further improvement is desirable, since the performance of these devices lags that of the well-developed 850-nm VCSELs considerably.

5. SUMMARY

We have reviewed the physical properties of Ga(In)NAs alloys as well as their application in quantum well lasers for optical communications. Although substantial improvements have been made in the material quality and in the laser characteristics in recent years, there are concerns about the reliability of GaInNAs/GaAs lasers. Therefore, further improvements of both materials and device designs will be required to realize high-performance, reliable 1.3-μm GaInNAs lasers for optical communications.

GLOSSARY

Distributed Bragg reflector (DBR) Many-layered structure consisting of semiconductor material with contrasting refractive index and thickness such that the reflected components of the electromagnetic wave from the layers interfaces arrive in phase with each other and superpose to create a highly reflective mirror.

Electronegativity The tendency (or power) of an atom in a molecule to attract electrons.

Epitaxial structures Thin films, grown on single-crystal substrates, with the same crystal structure and orientation as the substrate.

Photolithography Pattern definition method that uses UV radiation to expose the resist.

p–n junction It consists of two semiconductor regions of opposite type and shows a pronounced rectifying behavior.

Rocking curve Intensity versus angle plot for X-ray scattering.

Stimulated emission The photons emitted are coherent with the photons already present and the rate is proportional to the existing photon density.

Superlattice Repeated sequence of layers forming a period structure.

REFERENCES

1. J. Singh, "Optoelectronics: An Introduction to Materials and Devices." McGraw–Hill, New York, 1996.
2. J. S. Harris, *IEEE J. Select. Topics Quantum Electron.* 6, 1145 (2000).
3. J. Joos, F. Mederer, M. Kicherer, I. Ecker, R. Jäger, W. Schmid, M. Grab-herr, and K. J. Ebeling, *IEEE Photon. Technol. Lett.* 12, 344 (2000).
4. N. Kirstaedter, N. N. Ledentsov, M. Grundmann, D. Bimberg, V. M. Ustinov, S. S. Ruminov, M. V. Maximov, P. S. Kop'ev, Z. I. Alferov, U. Richter, P. Werner, U. Gosele, and J. Heydenreich, *Electron. Lett.* 30, 1416 (1994).
5. T. Anan, M. Yamada, K. Nishi, K. Kurihara, K. Tokutome, A. Kamei, and S. Sugou, *Electron. Lett.* 37, 566 (2001).
6. M. Kondow, K. Uomi, A. Niwa, T. Kitatani, S. Watahiki, and Y. Yazawa, *Jpn. J. Appl. Phys.* 35, 1273 (1996).
7. M. Hetterich, M. D. Dawson, A. Y. Egorov, D. Bernklau, and H. Riechert, *Appl. Phys. Lett.* 76, 1030 (2000).
8. D. J. Wolford, J. A. Bradley, K. Fry, and J. Thompson, "Proceedings of the 17th Conference on the Physics of Semiconductors," 1984, p. 627.
9. X. Liu, M. E. Pistol, L. Samuelson, S. Schwetlick, and W. Seifert, *Appl. Phys. Lett.* 56, 1451 (1990).
10. P. J. Klar, H. Grüning, W. Heimbrodt, J. Koch, F. Höhnsdorf, W. Stolz, P. M. A. Vicente, and J. Camassel, *Appl. Phys. Lett.* 76, 3439 (2000).
11. K. Uesugi, N. Morooka, and I. Suemune, *Appl. Phys. Lett.* 74, 1254 (1999).
12. S. R. Kurtz, A. A. Allerman, E. D. Jones, J. M. Gee, J. J. Banas, and B. E. Hammons, *Appl. Phys. Lett.* 74, 729 (1999).
13. P. N. Hai, W. M. Chen, I. A. Buyanova, H. P. Xin, and C. W. Tu, *Appl. Phys. Lett.* 77, 843 (2000).
14. C. Skierbiszewski, P. Perlin, P. Wisniewski, W. Knap, T. Suski, W. Walukiewicz, W. Shan, K. M. Yu, J. W. Ager, E. E. Haller, J. F. Geisz, and J. M. Olson, *Appl. Phys. Lett.* 76, 2409 (2000).
15. W. Shan, W. Walukiewicz, J. W. Ager, E. E. Haller, J. F. Geisz, D. J. Friedman, J. M. Olson, and S. R. Kurtz, *Phys. Rev. Lett.* 82, 1221 (1999).
16. M. Weyers, M. Sato, and H. Ando, *Jpn. J. Appl. Phys.* 31, L853 (1992).
17. E. D. Jones, N. A. Modine, A. A. Allerman, S. R. Kurtz, A. F. Wright, S. T. Torez, and X. Wei, *Phys. Rev. B* 60, 4430 (1999).
18. T. Mattila, S.-H. Wei, and A. Zunger, *Phys. Rev. B* 60, R11245 (1999).
19. P. R. C. Kent, L. Bellaiche, and A. Zunger, *Semicond. Sci. Technol.* 17, 851 (2002).
20. K. Kim and A. Zunger, *Phys. Rev. Lett.* 86, 2609 (2001).
21. P. J. Klar, H. Grüning, J. Koch, S. Schäfer, K. Volz, W. Stolz, W. Heimbrodt, A. M. Kamal Saadi, A. Lindsay, and E. P. O'Reilly, *Phys. Rev. B* 64, 121203R (2001).
22. P. Krispin, S. G. Spruytte, J. S. Harris, and K. H. Ploog, *J. Appl. Phys.* 88, 4153 (2000).
23. P. J. Klar, H. Grüning, W. Heimbrodt, J. Koch, W. Stolz, P. M. A. Vicente, A. M. Kamal Saadi, A. Lindsay, and E. P. O'Reilly, *Phys. Status Solidi B* 223, 163 (2001).
24. I. A. Buyanova, G. Pozina, P. N. Hai, W. M. Chen, H. P. Xin, and C. W. Tu, *Phys. Rev. B* 63, 033303 (2001).
25. H. P. Xin and C. W. Tu, *Appl. Phys. Lett.* 72, 2442 (1998).
26. T. Miyamoto, K. Takeuchi, T. Kageyama, F. Koyama, and K. Iga, *J. Cryst. Growth* 197, 67 (1999).
27. M. Hetterich, M. D. Dawson, A. Yu. Egorov, D. Bernklau, and H. Riechert, *Appl. Phys. Lett.* 76, 1030 (2000).
28. L. Tapfer, "X-Ray Investigations of Superlattices and Quantum Wells." EMIS Datareviews Series Number 15 (P. Bhattacharya, Ed.). England, 1996.
29. M. B. Panish and H. Temkin, "Gas Source Molecular Beam Epitaxy: Growth and Properties of Phosphorus Containing III–V Heterostructures." Springer-Verlag, Berlin, 1993.
30. G. B. Stringfellow, "Organometallic Vapor-Phase Epitaxy: Theory and Practice." Academic Press, Boston, 1989.
31. K. Tappura, T. Hakkarainen, K. Rakennus, and H. Asonen, *J. Cryst. Growth* 120, 145 (1992).
32. T. Miyamoto, T. Kageyama, S. Makino, D. Schlenker, F. Koyama, and K. Iga, *J. Cryst. Growth* 209, 339 (2000).
33. F. Höhnsdorf, J. Koch, C. Agert, and W. Stolz, *J. Cryst. Growth* 195, 391 (1998).
34. W. Li, M. Pessa, and J. Likonen, *Appl. Phys. Lett.* 78, 2864 (2001).
35. S. G. Spruytte, M. C. Larson, W. Wampler, C. Coldren, P. Krispin, H. E. Petersen, S. Picraux, K. Ploog, and J. S. Harris, *J. Cryst. Growth* 227–228, 506 (2001).
36. W. Li, M. Pessa, T. Ahlgren, and J. Dekker, *Appl. Phys. Lett.* 79, 1094 (2001).
37. S. B. Zhang and S. H. Wei, *Phys. Rev. Lett.* 86, 1789 (2001).
38. N. Q. Thinh, I. A. Buyanova, W. M. Chen, H. P. Xin, and C. W. Tu, *Appl. Phys. Lett.* 79, 3089 (2001).
39. S. G. Spruytte, C. W. Coldren, J. S. Harris, W. Wampler, P. Krispin, K. Ploog, and M. C. Larson, *J. Appl. Phys.* 89, 4401 (2001).
40. W. Li, J. Turpeinen, P. Melanen, P. Savolainen, P. Uusimaa, and M. Pessa, *Appl. Phys. Lett.* 78, 91 (2001).
41. S. Sato and S. Satoh, *J. Cryst. Growth* 192, 381 (1998).
42. M. Fischer, M. Reinhardt, and A. Forchel, *Electron. Lett.* 36, 1208 (2000).
43. W. J. Fan, S. F. Yoon, T. K. Ng, S. Z. Wang, W. K. Loke, R. Liu, and A. Wee, *Appl. Phys. Lett.* 80, 4136 (2002).
44. H. Riechert, A. Yu. Egorov, B. Borchert, and S. Illek, *Compd. Semicond.* 6, 71 (2000).
45. B. Q. Sun, D. S. Jiang, Z. Pan, L. H. Li, and R. H. Wu, *Appl. Phys. Lett.* 77, 4148 (2000).
46. Z. Pan, L. H. Li, W. Zhang, Y. W. Lin, and R. H. Wu, *Appl. Phys. Lett.* 77, 1280 (2000).
47. M. Albrecht, V. Grillo, T. Remmele, H. P. Strunk, A. Yu. Egorov, Gh. Dumitras, H. Riechert, A. Kaschner, R. Heitz, and A. Hoffmann, *Appl. Phys. Lett.* 81, 2719 (2002).
48. W. Li, J. Konttinen, C. S. Peng, T. Jouhti, E. Pavelescu, M. Suominen, and M. Pessa, "14th Indium Phosphide and Related Materials Conference," 2002, p. 23.
49. E.-M. Pavelescu, C. S. Peng, T. Jouhti, J. Konttinen, W. Li, M. Pessa, M. Dumitrescu, and S. Spânulescu, *Appl. Phys. Lett.* 80, 3054 (2002).
50. X. Yang, M. J. Jurkovic, J. B. Heroux, and W. I. Wang, *Appl. Phys. Lett.* 75, 178 (1999).
51. J. C. Harmand, G. Ungaro, L. Largeau, and G. Le Roux, *Appl. Phys. Lett.* 77, 2482 (2000).

52. J. C. Harmand, A. Caliman, E. V. K. Rao, L. Largeau, J. Ramos, R. Teissier, L. Travers, G. Ungaro, B. Theys, and I. F. L. Dias, *Semicond. Sci. Technol.* 17, 778 (2002).
53. J. S. Harris, *Semicond. Sci. Technol.* 17, 880 (2002).
54. J. I. Pankove and N. M. Johnson, "Hydrogen in Semiconductors," Semiconductors and Semimetals Vol. 34. Academic Press, New York, 1991.
55. A. Polimeni, G. Baldassarri Höger von Högersthal, M. Bissiri, M. Capizzi, A. Frova, M. Fischer, M. Reinhardt, and A. Forchel, *Semicond. Sci. Technol.* 17, 797 (2002).
56. A. Janotti, S. B. Zhang, S.-H. Wei, and C. G. Van de Walle, *Phys. Rev. Lett.* 89, 086403 (2002).
57. H. P. Xin, C. W. Tu, and M. Geva, *Appl. Phys. Lett.* 75, 1416 (1999).
58. P. S. Zory, Jr. Ed., "Quantum Well Lasers." Academic Press, New York, 1993.
59. E. Kapon, Ed., "Semiconductor Lasers II: Materials and Structures." Academic Press, New York, 1999.
60. C. Wilmsen, H. H. Temkin, and L. A. Coldren, Eds., "Vertical-Cavity Surface-Emitting Lasers: Design, Fabrication, Characterization, and Applications." Cambridge Univ. Press, Cambridge, UK, 1999.
61. M. Kondow, T. Kitatani, S. Nakatsuka, M. C. Larson, K. Nakahara, Y. Yazawa, and M. Okai, *IEEE J. Select. Topics Quantum Electron.* 3, 719 (1997).
62. M. Kondow, S. Nakatsuka, T. Kitatani, Y. Yazawa, and M. Okai, *Electron. Lett.* 32, 2244 (1996).
63. D. A. Livshits, A. Yu. Egorov, and H. Riechert, *Electron. Lett.* 36, 1381 (2000).
64. A. Yu. Egorov, D. Bernklau, D. Livshits, V. Ustinov, Zh. I. Alferov, and H. Riechert, *Electron. Lett.* 35, 1643 (1999).
65. K. Nakahara, M. Kondow, T. Kitatani, M. C. Larson, and K. Uomi, *IEEE Photon. Technol. Lett.* 10, 487 (1998).
66. W. Li, T. Jouhti, C. S. Peng, J. Konttinen, P. Laukkanen, E.-M. Pavelescu, M. Dumitrescu, and M. Pessa, *Appl. Phys. Lett.* 79, 3386 (2001).
67. C. S. Peng, T. Jouhti, P. Laukkanen, E.-M. Pavelescu, J. Konttinen, W. Li, and M. Pessa, *IEEE Photon. Technol. Lett.* 12, 777 (2000).
68. B. Borchert, A. Yu. Egorov, S. Illek, M. Komainda, and H. Riechert, *Electron. Lett.* 35, 2204 (1999).
69. M. Fischer, M. Reinhardt, and A. Forchel, *Electron. Lett.* 36, 1208 (2000).
70. W. Ha, V. Gambin, M. Wistey, S. Bank, H. Yuen, S. Kim, and J. S. Harris, *Electron. Lett.* 38, 277 (2002).
71. M. C. Larson, M. Kondow, T. Kitatani, K. Nakahara, K. Tamura, H. Inoue, and K. Uomi, *IEEE Photon. Technol. Lett.* 10, 188 (1998).
72. M. C. Larson, C. W. Coldren, S. G. Spruytte, H. E. Petersen, and J. S. Harris, *IEEE Photon. Technol. Lett.* 12, 1598 (2000).
73. S. Sato, N. Nishiyama, T. Miyamoto, T. Takahashi, N. Jikutani, M. Arai, A. Matsutani, F. Koyama, and K. Iga, *Electron. Lett.* 36, 2018 (2000).
74. K. D. Choquette, J. F. Klem, A. J. Fischer, O. Blum, A. A. Allerman, I. J. Fritz, S. R. Kurtz, W. G. Breiland, R. Sieg, K. M. Geib, J. W. Scott, and R. L. Naone, *Electron. Lett.* 36, 1388 (2000).
75. J. F. Klem, D. K. Serkland, and K. M. Geib, *Proc. SPIE* 4646, 137 (2002).
76. G. Steinle, A. Y. Egorov, and H. Riechert, *Electron. Lett.* 37, 93 (2001).
77. H. Riechert, A. Ramakrishnan, and G. Steinle, *Semicond. Sci. Technol.* 17, 892 (2002).
78. M. Hoffman, N. Gerhardt, Am. Wagner, C. Ellmers, F. Hönsdorf, J. Koch, W. Stolz, S. W. Koch, W. W. Ruhle, J. Hader, J. V. Moloney, E. P. O'Reilly, B. Borchert, A. Y. Egorov, H. Riechert, H. C. Schneider, and W. W. Chow, *IEEE J. Quantum Electron.* 38, 213 (2002).
79. M. Pessa, C. Peng, T. Jouhti, E.-M. Pavelescu, W. Li, S. Karirinne, and O. Okhotnikov, "International Workshop on the Physics and Technology of Diluted Nitrides for Optical Communications," Battalimani, Istanbul 2002.
80. A. Ramakrishnan, G. Steinle, C. Degen, and G. Ebbinghaus, *Electron. Lett.* 38, 322 (2002).
81. F. Mederer, I. Ecker, R. Michalzik, G. Steinle, H. Riechert, B. Lunitz, J. Moisel, and D. Wiedenmann, "Proceedings of the 52nd Electronic Components and Technology Conference," 2002, p. 5.
82. T. Anan, M. Yamada, K. Nishi, K. Kurihara, K. Tokutome, A. Kamei, and S. Sugou, *Electron. Lett.* 37, (566) 2001.
83. M. Yamada, T. Anan, K. Kurihara, K. Nishi, K. Tokutome, A. Kamer, and S. Sugou, "LEOS 2001, 14th Annual Meeting of the IEEE," 2001, Vol. 2, p. 598.
84. J. F. Klem, D. K. Serkland, and K. M. Geib, "LEOS 2001, 14th Annual Meeting of the IEEE," 2001, Vol. 2, p. 596.
85. R. L. Naone, A. W. Jackson, S. A. Feld, D. Galt, K. J. Malone, and J. J. Hindi, "CLEO '01. Technical Digest. Summaries," 2001, p. 657.
86. N. N. Ledentsov, D. Bimberg, V. M. Ustinov, Zh. I. Alferov, and J. A. Lott, "IEEE International Conference on Indium Phosphide and Related Materials," 2001, p. 5.

Encyclopedia of Nanoscience and Nanotechnology
www.aspbs.com/enn

γ-Irradiation Preparation of Nanomaterials

Yi Xie, Yujie Xiong
University of Science and Technology of China, Hefei, Anhui, People's Republic of China

CONTENTS

1. INTRODUCTION

1.1. Importance of Nanomaterials

In 1959 Richard Feynman, the future Nobel Laureate, suggested in his famous lecture, entitled "There's Plenty of Room at the Bottom," a variety of tests that might be achieved at very small scales. Three decades later, Feynman's vision has become the greatest scientific frontier of the century. It opened a new field of "nanoscale materials" having microstructures with characteristic length scales of the order of a few (typically 1–100) nanometers. Nanoscale materials can often display chemical, optical, electric, magnetic, and thermal properties, which are both quantitatively and qualitatively different from their respective bulk materials and from the discrete atomic or molecular species from which they are derived. The intriguing alteration of these properties allows nanoscale materials to have numerous potential applications in various areas such as materials and biomedical sciences, optics, magnetism, electronics, catalysis, ceramics, energy storage, gas sensors, and electrochemistry. In almost all applications of nanoscale materials, fabrication represents the first and one of the most significant challenges to their realization. Many methods for the fabrication of nanoscale materials have been developed, ranging from lithographic techniques to sputtering, from the sol–gel method to chemical vapor deposition (CVD), from grinding to a template-based method, and from the microemulsion method to the γ-irradiation method, etc. Among these methods, the γ-irradiation technique has some special advantages such as processing at room temperature and ambient pressure, high yield, as well as easy control of particle size and size distribution. Therefore, the γ-irradiation technique has been more and more widely used to prepare nanocrystalline metals, alloys, metal oxides, metal sulfides, and nanocomposites since it can provide high energy and an evenly distributed irradiation field. The intent of this chapter is to provide an overview of the γ-irradiation technique as a promising method for fabricating nanoscale materials.

1.2. Phylogeny of γ-Irradiation Technique

Above all, it is necessary to illustrate the phylogeny of the γ-irradiation technique. In 1962, Fujita et al. began the synthesis of metal aggregates by the radiolytic reduction of metal cations in solution [1]. In 1982, Belloni et al. reported microaggregates of non-noble metals and bimetallic alloys obtained by radiation-induced reduction for the first time and used a magnet to separate the products, which were carefully examined by electron microscopy [2]. In recent decades, this method has been developed further [3–5]. The technique of pulse radiolysis has been used to study the yield of short-lived clusters and their optical spectra in dilute aqueous solutions of about 10^{-4} M metallic ions [6–8]. By this method, silver cluster and colloidal silver could form [9, 10]. The formation of other colloidal metals from corresponding ions such as Cu^{2+} [11–13], Ti^{4+} [14, 15], Pb^{2+} [16, 17], and Pd^{2+} [18] has also been studied. By high-energy electron irradiation of 1.7×10^{-4} M $NaAu(CN)_4$ solution, Mosseri et al. [19] prepared ultrafine Au particles. Marignier et al. [20] prepared ultrafine Ni and Cu–Pd alloy particles by separate γ-irradiation of $NiSO_4$ and $CuSO_4$–$PdCl_2$ solution. Several years later, the radiolytic generation and control of extremely small CdS particles [21] were first described by

ISBN: 1-58883-059-4/$35.00

Hayes and co-workers. However, the solutions used in the aforementioned studies are about 10^{-4} M metal ions, and the ultrafine particles produced are in the colloidal soluble state. Later, study of the γ-irradiation of more concentrated solutions containing 10^{-2}–10^{-3} M metal ions produces nanoscale metal, which thus makes the γ-irradiation technique become the new focus of nanoscience and extends to the fabrication of nonmetals, alloys, metal oxides, metal sulfides, and nanocomposites with all kinds of nanostructures.

2. γ-IRRADIATION TECHNIQUE TO INORGANIC NANOMATERIALS

2.1. Noble Metals and Their Alloys

2.1.1. γ-Irradiation–Hydrothermal Technique

Noble metals are the first objective class studied by the γ-irradiation technique to prepare nanomaterials. On the principle of γ-irradiation reactions, during γ-irradiation, the formation of hydrated electrons can be shown as follows [22, 23]:

$$H_2O \sim\sim\longrightarrow H^\bullet, {}^\bullet OH, e^-_{aq}, H_3O^+, H_2, H_2O_2, HO_2, H_2O^* \quad (1)$$

Then, the radiation reduction of metal ions by hydrated electrons leads to the formation of metal nanoparticles:

$$e^-_{aq} + M^{m+} \rightarrow M^{(m-1)+} \quad (2)$$

$$e^-_{aq} + M^{+} \rightarrow M^{0} \quad (3)$$

$$nM^0 \rightarrow M_2 \rightarrow M_n \rightarrow M_{agg} \quad (4)$$

where n is a number of aggregations of a few units, M_n is cluster colloid, and M_{agg} is the aggregate in the final stable state [12].

The typical experimental process can be described as follows: A solution containing metal salts of appropriate concentration (10^{-2}–10^{-3} M) was irradiated in a ^{60}Co γ-source at appropriate dose for a long time to precipitate powders, then filtered out and washed. It should be mentioned that appropriate $(CH_3)_2CHOH$ as scavenger to eliminate hydroxyl radicals needs to be added to the reaction system.

$${}^\bullet OH + (CH_3)_2CHOH \rightarrow H_2O + (CH_3)_2COH^\bullet \quad (5)$$

In the experimental process, salt concentration, irradiation time, and dose can be adjusted.

Based on the absorption spectrum studies on after-irradiated solution [24], it was found that not only were the above three parameters crucial to the formation of noble metal clusters and colloids, but also appropriate surfactant should be added to stabilize the formed small clusters and colloids. It seems reasonable [25] to attribute the stabilizing effect of the surfactant molecules to the adherence of their long hydrocarbon chains to the surface of colloidal silver particles, the anionic head groups sticking out into the aqueous surroundings and thus keeping the aggregates dispersed as colloidal silver in solution, as Scheme 1 shows.

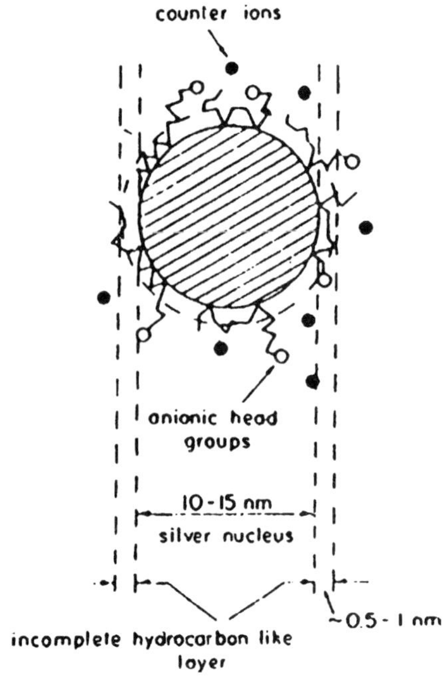

Scheme 1. Schematic presentation of the cross section through a silver–surfactant aggregate.

With increasing surfactant concentration, the hydrocarbon-like layer at the surface of silver particle may be expected to become more complete and thicker; this favors the formation and stabilization of small clusters and colloidal silver.

In the experimental process, noble metal clusters were found to exist for more than one month in γ-irradiated concentrated saltsolution in air. This is much longer than that in dilute metal ion solutions previously reported [26–32]. Thus, the hydrothermal treatment was used to aggregate the colloid to metal powders. This process to prepare nanoscale noble metals was called the γ-irradiation technique. For example, in the case of preparing silver nanocrystals, the as-obtained silver solution was then hydrothermal-treated for several hours at about boiling point. As a result of transmission electron microscope (TEM) studies, it was found that the shape of the silver particles produced by γ-irradiation depended on the surfactant used. For example, when sodium dodecyl sulfate is used as a surfactant, the silver particles are uniform quasispherical (Fig. 1a). Under appropriate conditions, some silver dendrites can be obtained (Fig. 1b). However, when poly(vinyl alcohaol) is used, the silver particles have various shapes.

From these results, it can be concluded that the γ-irradiation technique is an excellent method to prepare nanoscale noble metals with as-desired nanostructures, combined with additional methods such as surfactant-assisted process. However, to get over the shortcoming that the produced metal colloids are difficult to aggregate and the as-obtained metals have poor crystallinity by γ-irradiation, the hydrothermal treatment should be used. This technique can be applied to synthesize nanoscale Ag [24], Cu [33, 34], Pd, Pt, and Au [35].

By adding a complex agent such as $NH_3{\cdot}H_2O$ into the reagent mixture, this technique can be extended to prepare their alloys [36, 37]. The complex agent is used to adjust

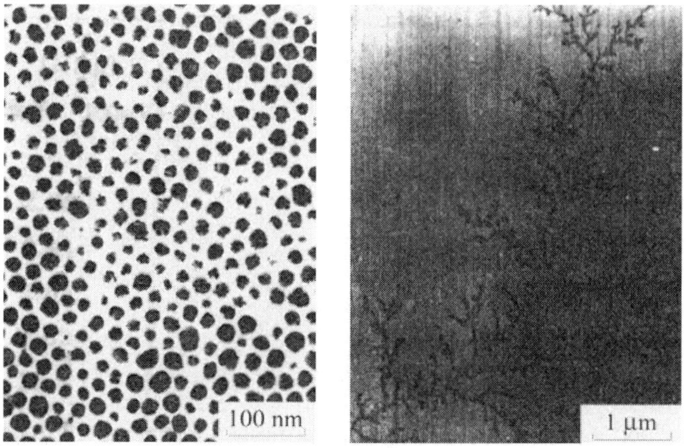

(a) (b)

Figure 1. TEM images of silver nanocrystals prepared by the γ-irradiation method. Solution: (a) 0.05 M $AgNO_3$, 0.01 M $C_{12}H_{25}NaSO_4$, 1.0 M $(CH_2)_2CHOH$; (b) 0.1 M $AgNO_3$, 0.1 M $C_{12}H_{25}NaSO_4$, 1.0 M $(CH_2)_2CHOH$. Dose rate (Gy min^{-1}) and irradiation time: (a) 77, 5.5 h; (b) 74.2, 12 h.

the condition of the solution, which may be caused by the change in the rate of reduction reactions of noble metal ions during γ-irradiation resulting from the complexation of metal ions with the complex agent. On the other hand, the ligand on metal ions may act as a bridge for electron transfer from the solution to the metal ions.

2.1.2. Fractal and Dendritic Growth of Metallic Silver Studies under γ-Irradiation

Many physical, chemical, technological, and biological processes exhibit noninteger (or fractal) dimensionality as was recognized by Mandelbrot in his studies on complex geometrical shapes [38]. Fractals are generally observed in far-from-equilibrium growth phenomena; hence they provide a natural framework for the study of disordered systems. There are many reports on the study of fractal structures and their formation in both theoretical and experimental aspects in the past decade. Electrochemical metallic deposition [39–41], fluid and air displacement in other media [42, 43], and dielectric breakdown [44] are a few experimental examples that are important for understanding general aggregation processes and establishing model simulation. The diffusion limited aggregation (DLA) model [45, 46] and cluster–cluster aggregation (CCA) model [47, 48] are widely used to explain and analyze these phenomena. The aggregating process of particles in a Laplacian field can be described by the DLA model, which involves cluster formation by the adhesion of a particle with random walk to a selected seed on contact. The resulting structure of the DLA model has a fractal dimension $D \sim 1.7$. Aggregation of small particles in a fluid medium, for example, metal colloids, can be depicted by the CCA model. Two limiting regimes of CCA aggregation have been accepted [49]: fast diffusion-limited cluster aggregation (DLCA) and slow reaction-limited cluster aggregation. In the DLCA regime, the fractal dimension D of the grown structures is about 1.45 in two dimensions ($d = 2$) and may increase to 1.75 ± 0.07 in a system of high initial particle density as the system approaches the sol–gel transition [50].

Since it was found that some silver dendrites could form via the γ-irradiation technique in the presence of surfactant more studies on their aggregating process should be put forward to reveal their mechanism [51]. With γ-irradiation on $AgNO_3$ dissolved in deionized water (a), ethanol (b), and $C_{12}H_{25}NaSO_4$aqueous solution (c), the as-obtained products were determined by TEM images (Fig. 2). For measuring the dimensions of Ag fractals, TEM images were digitized by the use of an OPTON image analyzer at 512×512 pixels. A threshold criterion was then applied to separate pixels in the fractal regions from the background. The fractal dimension (D) was obtained by using the box-counting algorithm [52]. In this method, the digitized images were covered with a grid of squares.

The aggregating morphology of sample A is regular dendritic patterns along three different directions and with a fractal dimension of 1.81. Due to the restraining effect of ethanol and surfactant, the produced smaller silver particles of samples B and C aggregated to fractal patterns with center with a fractal dimension of 1.73 and without center with dimension of 1.70, respectively. The former two cases share common characteristics with the DLA model.

On the basis of these dimension calculations, the aggregating process of fractal and dendritic growth was given. First, silver particles formed in γ-irradiation, and ethanol and surfactant might prevent them from aggregating. Then,

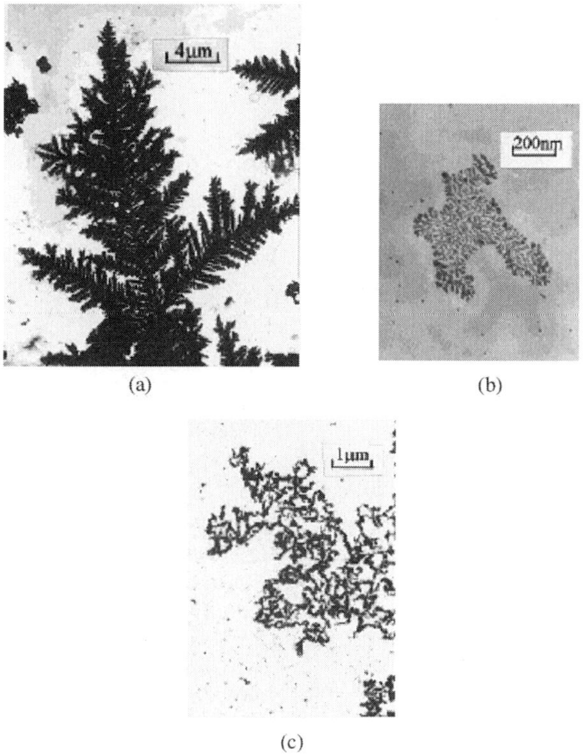

(a) (b)

(c)

Figure 2. TEM images of Ag colloids prepared (a) in aqueous solution; (b) in ethanol; (c) in $C_{12}H_{25}NaSO_4$ aqueous solution. Reprinted with permission from [51], S. Z. Wang and H. W. Xin, *J. Phys. Chem. B* 104, 5681 (2000). © 2000, American Chemical Society.

the aggregation of Ag particles on the C-coated Cu grid could cause van der Waals attraction between metallic Ag particles. By γ-irradiation on $AgNO_3$ aqueous solution (solution A), the Ag particles obtained were relatively large and these Ag particles were single crystals as determined from X-ray diffraction (XRD) patterns. When the turbid liquid was dripped onto the C-film of the Cu grid, the Ag particles crystallized on the edge of the tiny irregular edge of the Cu grid and involved low-angle grain boundaries. The other Ag particles, once absorbed on the surface of the aggregate from the turbid fluid, diffused with a random walk until they irreversibly stuck to the perimeter of the growing aggregate. According to the former reports, the dendritic patterns are formed only when the anisotropy dominates noise [53]. Anisotropy, which is essential for the formation of dendritic pattern, can come from the interaction between the background and aggregating particles [54] or the anisotropy of the growing crystalline itself [55]. In the present experimental system, the aggregation of Ag particles took place in a two-dimensional C-film-coated Cu grid. The substrate of C-film was amorphous and the medium was also isotropic; no noise was induced into the growth system. The anisotropy, which causes dendrite patterns of the aggregate, comes from the Ag particles themselves. Because the single crystals of Ag particles intrinsically contain a strong crystalline anisotropy, the aggregation of Ag particles was mainly along the preferred growth direction and, accordingly, brought about a dendrite pattern. On the other hand, the large Ag particles have lower mobility; they cannot change their position or growth direction freely under the agent of anisotropy and grow strictly along the preferred growth direction. Consequently, the final aggregation pattern was not very regular shaped branches, or oriented on definite directions.

The metallic Ag prepared in ethanol has a smaller particle size and formed fractal patterns as aggregation took place. This can be assumed for the following reasons: first, nanoscale Ag grains contain a significantly higher percentage of their atoms at the grain boundary and the distortion of lattice in the surface layer. These nanocrystalline characteristics decreased the crystalline anisotropy of Ag particles. Second, solvation occurred on the surface of Ag particles. The interaction of metallic surfaces with the solvent makes the surfaces become homogeneous; thus, Ag particles lost the anisotropy. When these Ag particles were absorbed on the C-film from the solution, they walked and stuck to the surface of aggregate randomly and formed typically fractal patterns.

The occurrence of Ag colloid by γ-irradiation on solution C was due to $C_{12}H_{25}NaSO_4$, which was absorbed on the surface of the Ag particles and prevented the Ag particles from aggregating. Following the standard theory of colloids, the stability of a colloid is governed by the balance between van der Waals attraction and Coulombic repulsion of charged particles [56]. Enhancing the van der Waals attraction force or weakening the Coulombic repulsion force induces rapid coagulation of colloids. When this Ag colloid was dripped onto the C-coated Cu grid, the evaporation of solvent resulted in the break of the balance and the aggregation took place. The final aggregates were random, tortuous structures and have no center (i.e., there were no seeds for the aggregation of Ag particles). This process is analogous to the two-dimensional DLCA simulation.

The above analyses indicate that producing small silver particles in a surfactant-assisted system is crucial to form fractal patterns. The γ-irradiation technique has characteristics of dispersive energy and mild conditions, which are suitable for building a surfactant-assisted system. This further confirms that the γ-irradiation technique is an excellent pathway to form fractal patterns.

2.1.3. γ-Irradiation with Organic Capping Technique

Although the surfactant-assisted process is suitable for obtaining better-dispersed noble metal nanoparticles, it is difficult to control their sizes by adjusting some parameters. Thus, an organic capping technique was substituted for the surfactant-assisted process. For example, a γ-irradiation reducing route with citrate capping [57] was designed to prepare silver nanoparticles with different sizes: appropriate sodium citrate and N_2O were added to the irradiated reaction system. By adjusting the citrate concentration, the as-obtained products were well-dispersed colloidal nanoparticles with different sizes (Fig. 3). The reaction mechanism was identified as

$$^{\bullet}OH(^{\bullet}H) + (CH_3)_2CHOH \rightarrow H_2O(H_2) + (CH_3)_2C^{\bullet}OH \tag{6}$$

$$e^-_{aq} + N_2O + H_2O \rightarrow N_2 + OH^- + OH \tag{7}$$

$$Ag^+ + (CH_3)_2C^{\bullet}OH \rightarrow Ag^0 + (CH_3)_2CO + H^+ \tag{8}$$

$$Ag_x + \text{citrate} \leftrightarrow Ag_{x(\text{citr})} \tag{9}$$

$$Ag_{x(\text{citr})} + Ag_{x(\text{citr})} \rightarrow Ag_{2x(\text{citr})}, \text{ etc.} \tag{10}$$

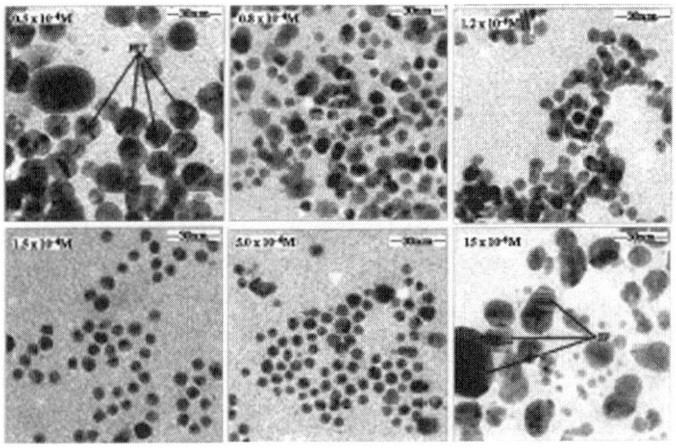

Figure 3. TEM images of silver colloidal nanoparticles obtained at various citrate concentrations. MT: multiple twinned particles. SP: particles with staple faults. Reprinted with permission from [57], A. Henglein and M. Giersig, *J. Phys. Chem. B* 103, 9533 (1999). © 1999, American Chemical Society.

$$\mathrm{Ag}_{m(\mathrm{citr})} + n(\mathrm{CH}_3)_2\mathrm{C}^{\bullet}(\mathrm{OH}) \rightarrow Ag^{n-}_{m(citr)} + n(\mathrm{CH}_3)_2\mathrm{CO} + n\mathrm{H}^+ \quad (11)$$

$$\mathrm{Ag}^{n-}_{m(\mathrm{citr})} + \mathrm{Ag}^+ \rightarrow \mathrm{Ag}^{(\mathrm{n}-1)-}_{(\mathrm{m}+1)(\mathrm{citr})} \quad (12)$$

By studying the size and size distribution of the silver particles formed at various citrate concentrations, the formation process of nanoparticles can be described as in Scheme 2.

This technique has also been applied to prepare copper nanocrystals [58] with different sizes.

2.2. Active Metals and Nonmetals

The preparation of monodispersed active metals and nonmetals is always a challenge due to their high reaction activity and low melting point, respectively. Since the γ-irradiation technique has been successfully developed to prepare nanoscale noble metals, it can be extended to the active metals and nonmetals field, which overcomes the above hindrances.

Active metals, for example, nickel ions, can be reduced under γ-irradiation and form small clusters. However, the zero valence state of nickel, which resulted from the two-electron transfer, could also undergo oxidation:

$$\mathrm{Ni}_n + \mathrm{H}_3\mathrm{O}^+ \rightarrow \mathrm{Ni}_{n-1} + \mathrm{Ni}^+ + 1/2\mathrm{H}_2 + \mathrm{H}_2\mathrm{O} \quad (13)$$

Thus, selecting the appropriate pH condition is crucial to the preparation of active metals. For example, in the presence of $NH_3 \cdot H_2O$ [59], which acts as an alkalizing agent, the pH value of the solution is kept in the range of 10 to 11, so the reoxidation of atoms or aggregates of nickel in solution is greatly suppressed. It is obvious that the γ-irradiation technique can remain in a reductive environment and hold back the reoxidation of products under appropriate pH condition, which provides a pathway to prepare active metals (Ni [59] and Co [60]).

Some nanocrystalline metals (Cd [61], Sn [62], Sb [63], and In [64]) with low melting points can also be prepared by similar routes. Compared with other methods such as electrolysis and reduction with more active metals [65–67], the particle size of as-prepared nanoparticles by γ-irradiation was smaller.

In the case of nonmetals, selecting an appropriate precursor is the key problem. For example, to prepare nanoscale selenium, the most familiar selenium source is SeO_2 (IV). How can we transform it to a form that can contain hydrated electron best? It is well known that selenium (IV) exists as Se^{4+} ions and SeO_3^{2-} ions in hydrochloric acid and NaOH solutions, respectively. The reduction of SeO_3^{2-} ions was more difficult than that of Se^{4+} ions, which is due to the stability of SeO_3^{2-} ions caused by the strong covalent bonding between selenium and oxygen. So a hydrochloric acid system should be chosen to prepare selenium nanoparticles [68] (Fig. 4a) from SeO_2.

Nanoscale Te [69] and As [70] uniform particles (Fig. 4) have also been successfully prepared by γ-irradiation of an aqueous solution of TeO_2 and As_2O_3, respectively. From the above process, it is found that the γ-irradiation technique avoids thermal treatment and overcomes the difficulty of nonmetals' low melting point.

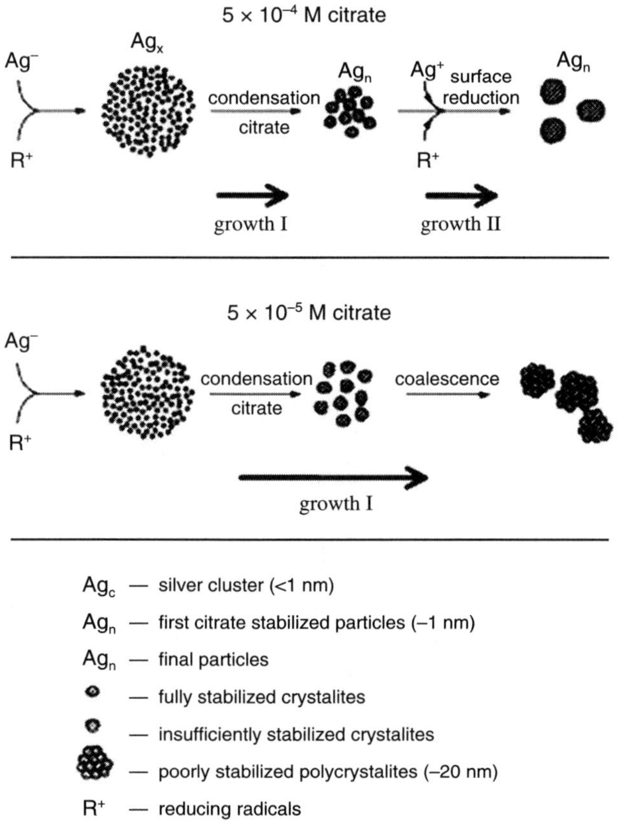

Scheme 2. Illustration of important routes of particle growth at two citrate concentrations. At the higher citrate concentration, the stages of nucleation and growth are well separated. At the lower citrate concentration, coalescence reactions of growing particles contribute to particle enlargement. Reprinted with permission from [57], A. Henglein and M. Giersig, *J. Phys. Chem. B* 103, 9533 (1999). © 1999, American Chemical Society.

2.3. Metal Oxides

Many metal oxides have poor stability, resulting from the fact that they are easily oxidized in air. For example, cuprous oxide nanocrystals are difficult to prepare via regular methods. The γ-irradiation technique is advantageous for the maintainance of reductive environment and can prevent as-prepared products from oxidization. However, cupous ions will partially be reduced to elemental copper in the reagent solution with a pH value of about 3.0–3.5 and the final product is a mixture of copper and cuprous oxide. In this case, the adjustment of pH value is crucial to the value state of final products. When the solution contains a CH_3COOH/CH_3COONa buffer pair that keeps the pH value in the range of 4.0 to 4.5, the reduction and dismutation of cuprous ions are completely suppressed and the final product is pure cuprous oxide. Thus, γ-irradiation in solution of appropriate pH value has more application in the preparation of unstable metal oxides.

For instance, in the case of preparing nanosized Cr_2O_3 powders [72], there are two common chromium sources,

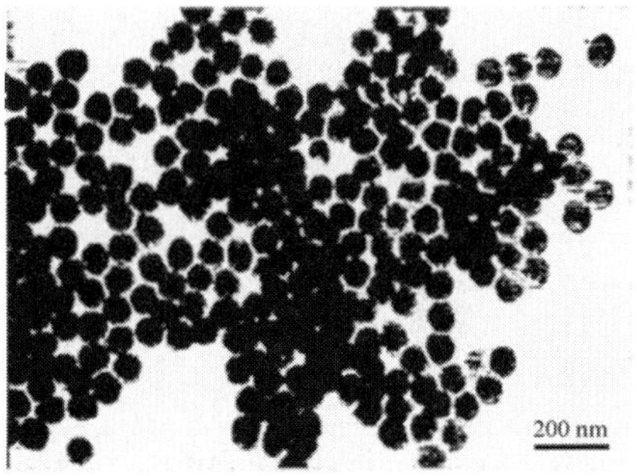

(a)

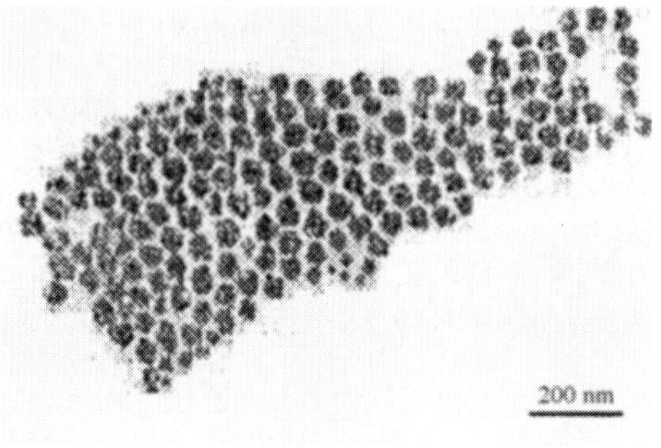

(b)

Figure 4. (a) TEM image of selenium. Reprinted with permission from [68], Y. J. Zhu et al., *Mater. Lett.* 28, 119 (1996). © 1996, Elsevier Science. (b) TEM image of arsenic uniform nanopaticles prepared by the γ-irradiation method. Reprinted with permission from [70], Y. J. Zhu and Y. T. Qian, *Mater. Sci. Eng. B* 47, 184 (1997). © 1997, Elsevier Science.

CrO_4^{2-} and $Cr_2O_7^{2-}$ ions. In the CrO_4^{2-} ion, there exists a tetrahedral arrangement in which a chromium atom is located at the center. In the dichromate ion, there are two tetrahedral units linked together by one oxygen atom. The Cr–O distance for the bridging oxygen is greater than that for the other oxygen atoms. The σ–π donor properties of the bridging oxygen atom are much less compared with the terminal oxygen atoms. Thus, the Cr–O bond for the bridging oxygen in the $Cr_2O_7^{2-}$ ion could break down to form a CrO_3 radical in the process of γ-irradiation. The CrO_3 radical is unstable and is reduced rapidly by hydrated electrons to form Cr_2O_3. Thus, $Cr_2O_7^{2-}$ ions are more suitable for the precursor to nanoscale Cr_2O_3 under γ-irradiation.

Nanoscale γ-Fe_2O_3 [73], MoO_2 [74, 75], and Mn_2O_3 [76] have also been successfully prepared by γ-irradiation of an aqueous solution of $FeSO_4$, $[(NH_4)_6Mo_7O_{24}\cdot 4H_2O]$, and $KMnO_4$, respectively.

From the above example, it can be concluded that γ-irradiation can easily reduce metal ions of a high value state to as-desired metal oxides nanocrystals under the effect of hydrated electrons.

2.4. Metal Chalcogenides

In recent years, the synthesis and characterization of chalcogenides of different groups have attracted considerable attention due to their important nonlinear properties [77], luminescent properties [78], quantum size effects [79], and other important physical and chemical properties [80].

2.4.1. γ-Irradiation Technique

Considerable progress has been made in the synthesis of metal chalcogenides powders. Conventionally, metal chalcogenides could be prepared by a variety of methods, including gas-phase, solid/vapor, and aqueous solution reactions [81, 82]. However, as-prepared metal chalcogenides have irregular crystallites with a wide size distribution ranging from 25 to 200 nm. Using organometallic precursors was another route for synthesizing metal chalcogenides, but most of the organometallic precursors were toxic, easy to hydrolyze and oxidize, and hard to deal with. To control better morphologies, soft-chemical techniques are necessary as assisted methods. Thus, seeking a mild method that can be well combined with other soft-chemical techniques is a challenge for the preparation of metal chalcogenides. The development of γ-irradiation on the preparation of metal and nonmetal nanomaterials provides a new pathway to synthesize metal chalcogenides.

It is clear that hydrated electrons produced in solution by γ-irradiation can reduce metal ions to lower valence [83]:

$$H_2O \rightarrow e^-_{aq}, H_3O^+, H^\bullet, H_2, {}^\bullet OH, H_2O_2 \quad (14)$$

$$e^-_{aq} + M^{2+} \rightarrow M^+ \quad (15)$$

As for the sulfur source, there are five common available compounds: carbon disulfide, thiourea, sodium thiosulfate, mercaptoethanol, and thiol. Research on γ-irradiation of carbon disulfide and thiourea indicates that the formation of nonionic intermediates, possibly as $S^\bullet$ radicals and free sulfur, might be liberated when such $S^\bullet$ radicals collide in solution [84, 85]:

$$CS_2 \rightarrow S^\bullet + CS^\bullet \quad (16)$$

The precipitate of MS probably prevents the formation of sulfur from $S^\bullet$ radical in the irradiation. This was identified by the experimental fact that no sulfur was observed in the products.

$$S^\bullet + M^+ \rightarrow MS \quad (17)$$

In the case of sodium thiosulfate as the sulfur source, S^{2-} was homogeneously released from its decomposition [86, 87]. Research of γ-irradiation on sodium thiosulfate indicates the following radiolytic reaction [88]:

$$2S_2O_3^{2-} + 4OH^- + 2H^+ \rightarrow HS^- + 3SO_3^{2-} + 3H^+ + H_2O \quad (18)$$

Hayes et al. [21] report that aqueous thiol irradiated by gray releases sulfide ions into the solution. Another sulfur source, mercaptoethanol, which also has a -SH group, may also release HS^- ions under γ-irradiation. In both cases, HS^- can be formed in the γ-irradiation of the solution:

$$RSH + e^-_{aq} \rightarrow R^\bullet + HS^- \quad (19)$$

$$HS^- \rightarrow S^{2-} + H^+ \quad (20)$$

Thus, metal sulfides can be formed in the following way:

$$M^{2+} + S^{2-} \rightarrow MS \quad (21)$$

From the above theoretic analyses, γ-irradiation may provide a mild pathway for the synthesis of metal sulfides.

Thus, the metal sulfides [89–91] can be prepared by following method: The solutions' dissolving metal sulfate (or acetate), various sulfur sources (i.e., sodium thiosulfate, carbon disulfide, thiourea, and mercaptoethanol), and a scavenger isopropanol in distilled water (or absolute ethanol) were irradiated for some time.

In the case of preparing ZnS, it is found from the XRD study that for sulfur sources with strong polarity such as sodium thiosulfate, the released energy of reactions (20) and (21) between ions is probably only sufficient to allow the formation of the sphalerite phase. When the sulfur source is mercaptoethanol, the polarity of which is in between that of sodium thiosulfate and carbon disulfide, the solvent plays an important role in the structure of the product: sphalerite in water and wurtzite in anhydrous ethanol. Thus, selecting an appropriate sulfur source is crucial to control the phase of products under γ-irradiation. Similar methods have been developed to synthesize CdS [92], ZnS [93], MoS_2 [94], NiS [95], CdSe, and $Cu_{2-x}Se$ [96].

Considering ethylenediamine's property of dissolving sulfur powder and forming a homogeneous solution, γ-irradiation can be extended in ethylenediamine systems. It was reported that sulfur–amine solutions contain S^{2-}, S_6^{2-}, S_4^-, S_4N^-, and other sulfur polyanions [97, 98]. Thus, sulfur dissolving in ethylenediamine may easily be reduced to S^{2-} by solvated electrons upon γ-irradiation.

On the other hand, ethylenediamine has a strong ability for *N*-chelation to many kinds of metal ions and can form a stable complex, which can well inhibit the reduction of many high-valence metal ions to lower valence metal ions or metal colloids by solvated electrons and can ensure the slower growth of sulfide nanoparticles.

Ag_2S, PbS [99], CdS [99, 100], CdSe [101], and PbSe [102] have been synthesized in nonaqueous phase by this route. Thus, γ-irradiation can prepare metal chalcogenide nanocrystals in either aqueous or nonaqueous phase. This development helps to build up soft-chemical systems among several solvents.

2.4.2. γ-Irradiation Soft-Template Technique

Recently, the preparation and properties of nanowires or nanorods of various materials have received considerable attention, and the potential wide-ranging applications of these kinds of nanomaterials have been predicted [103–108]. For the fabrication of nanowires or nanorods of desired materials, the template method has been demonstrated to be very effective; examples include the use of carbon nanotubes [103–105] and nanochannels in nuclear track polycarbonate membranes [106] or porous alumina [107, 108] which act as hard templates within which desired materials are synthesized. In addition, certain soft template methods [109] may be applied although soft templates cannot always regulate the size and shape of the products; for example, rodlike cationic surfactant micelles have been used as templates for the preparation of rodlike gold particles [110]. It will be interesting to explore the possibility of the use of solution-based synthesis of one-dimensional nanomaterials without the presence of preformed templates.

However, soft-chemical environments are easy to destroy under thermal treatment. Thus, the γ-irradiation method can be combined with the soft template method to prepare one-dimensional metal chalcogenides. For example, there are strong hydrogen bonds among molecules of the alcohols. Therefore, long chains can be easily formed by molecules of hexadecanol through hydrogen bonds. These long chains can act as new soft templates. When metal ions meet with S^{2-} (S^{2-} ions can be from the decomposition of sodium thiosulfate ($Na_2S_2O_3$)) upon γ-irradiation [111, 112]), metal sulfides may be grown along these long chains, leading to the formation of metal sulfides nanofibrils [113]. CdS and ZnS nanofibrils have been prepared via this route with hexadecanol as soft template.

2.4.3. γ-Irradiation Liquid Crystal-Template Technique

It is quite noteworthy that a "soft" liquid crystal phase could directly template a hard mineral phase. The liquid crystal-template-assisted route to group IIB–VI semiconductor nanowires has received more attention for its special properties, such as convenience in operation and ease in control of morphology. Li [114] also reported the synthesis of parallel nanowires of CdS via a hexagonal liquid crystal soft template. However, Braun and co-workers [115] prepared ZnS, CdS, and CdSe superlattices templated by a hexagonal liquid crystal using H_2S or H_2Se gas rather than semiconductor nanowires.

In the case of liquid crystal-template to sulfide nanowires, γ-irradiation is obviously appropriate as a reaction source: First, it does not disrupt the texture of the liquid crystal template. Second, the irradiation energy is so high that the reactants produce some radicals that play an important role in the expected reactions. Li [114] prepared CdS nanowires in a hexagonal liquid crystal template via a flow of H_2S gas at room temperature. Braun [115] prepared ZnS, CdS, and CdSe superlattices templated by hexagonal liquid crystal also under H_2S or H_2Se gas. The gaseous H_2S and H_2Se are toxic and should be used carefully. Except for them, thiourea and CS_2 were commonly used as sulfur sources. However, thiourea does not react with the metal salt at room temperature. When the temperature of this hexagonal mesophase is raised to 80 °C (at which point the texture of hexagonal liquid crystal was partly disrupted), metal sulfides nanowires cannot be obtained. Similarly, CS_2 was added to the hexagonal mesophase and no products were formed at room temperature without γ-irradiation. On the basis of the above research, γ-irradiation can activate the sulfur source well.

Thus, combining γ-irradiation with the liquid crystal-template method, a novel simultaneous *in-situ* formation (SISF) technique [116] for the synthesis of ZnS nanowires via γ-irradiation at room temperature has been developed. The so-called SISF technique is based on a mechanism of simultaneous occurrence of the decomposition of thiourea

(Tu) or carbon disulfide (CS_2) under γ-irradiation and the formation of ZnS nanowires in an inverted hexagonal liquid crystal phase. The hexagonal mesophase was formed by mixing 30 vol% $(EO)_{10}$-oleyl, 30 vol% *n*-hexane, 10 vol% *n*-hexanol/*i*-propanol (2:1), and 30 vol% water. It is found that the ZnS nanowires (Fig. 5) formed are well dispersed before and after removal of the liquid crystal template. The probable process of the formation of ZnS nanowires is shown in Scheme 3.

The SISF technique assisted by γ-irradiation can be applied to the synthesis of a wide variety of technologically important one-dimensional semiconductor nanomaterials.

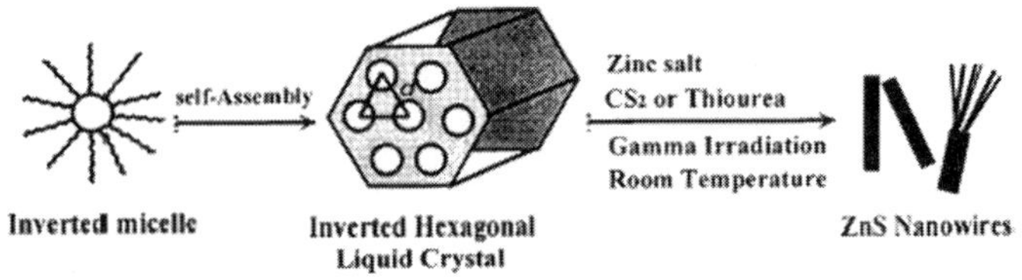

Scheme 3. Illustration of the formation process of ZnS nanowires in liquid crystal template. Reprinted with permission from [116], X. C. Jiang et al., *Chem. Mater.* 13, 1213 (2001). © 2001, American Chemical Society.

2.4.4. γ-Irradiation with Ultrasonic Radiation Technique

Wang et al. [117] have reported the preparation of PbSe nanowires in ethylenediamine using KBH_4 as the reductant and proposed that the formation of PbSe nanowires may be attributed to the existence of active H_2 in KBH_4, and the special structure and *N*-chelation properties of ethylenediamine. γ-Irradiation can play the same role as KBH_4. It is thought that, upon γ-irradiation, solvated electrons (e_s^-) with strong reductive properties can be formed in ethylenediamine [118], hydrogen radicals ($H^\bullet$) and hydrated electrons can also be generated due to the presence of traces of water [23], and these reductive species can activate the reaction and favor the formation of nanorods in ethylenediamine.

However, the transformation from Se powder to Se^{2-} requires a lot of energy. It was reported that, under the sonolysis, Se powder tends to disproportionate in liquid ethylenediamine with basic, *N*-chelation properties and form divalent selenium. Hydroxyl (OH^-) ion may exist in ethylenediamine with traces of water.

$$3Se + 6OH^- \rightarrow 2Se^{2-} + SeO_3^{2-} + 3H_2O$$

In addition, the divalent selenium can further interact with N atoms in organic amines to form a pseudo-high-valent selenium [119], although there is no report about the direct reaction between Se powder and amines. Therefore, sonication can favor the formation of Se^{2-} and the nonvalence interaction between Se^2 and ethylenediamine.

Prompted by these studies, an ultrasonic and γ-irradiation technique has been developed to prepare well crystallized PbSe nanorods in an ethylenediamine solution at room temperature [120]: An appropriate amount of selenium powder was added to 100 mL ethylenediamine in a vessel at room temperature. Then the tank was sealed and sonicated in a commercial ultrasonic cleaner under ambient conditions for more than an hour. Following this step, a stoichiometric amount of lead acetate was added to the sonicated slurry, which was then irradiated in γ-ray. Thus, the γ-irradiation and ultrasonic technique not only makes use of the reductive species in γ-irradiation but also obtains much reaction energy from the ultrasonic process.

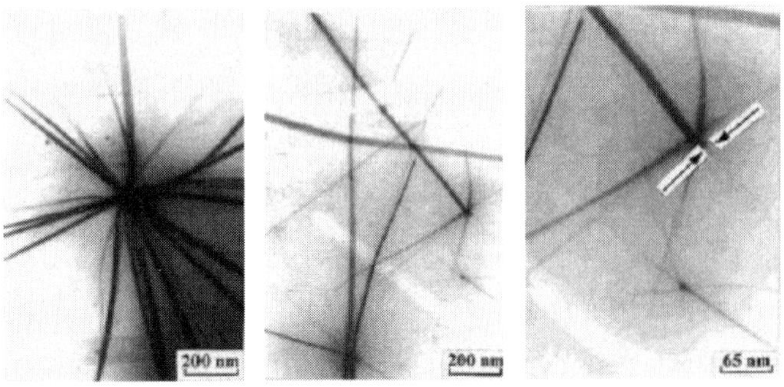

(a) (b) (c)

Figure 5. TEM images of (a) ZnS nanowires before the liquid crystal template is removed; (b) ZnS nanowires after the liquid crystal template is removed; (c) one magnified part of a ZnS bundle composed of ZnS nanowires. Reprinted with permission from [116], X. C. Jiang et al., *Chem. Mater.* 13, 1213 (2001). © 2001, American Chemical Society.

3. γ-IRRADIATION TECHNIQUE TO POLYMER NANOMATERIALS

3.1. Polymerization under γ-Irradiation

During the past decade microemulsion polymerization, as a new branch of emulsion polymerization, has been extensively studied. However, most attention was paid to the polymerization rate. Only a few papers focus on the growth of polymer particles and the variation of the distribution of particles during polymerization. This may be due in part to the difficulties indirectly obtaining the particle size using the most conventional analytic techniques, because of the high emulsifier concentration and the various complications associated with microemulsions. It was reported that in W/O microemulsion, the nucleation is continuous throughout the polymerization [121]. The early work on styrene O/W microemulsion polymerization assumed that the nucleation occurs in the microdroplets and that nucleation was limited below 20% conversion. After reexamination of the same system, Guo et al. [122, 123] postulated that particle nucleation is continuous. Gan et al. [124] investigated the formation of polystyrene nanoparticles in ternary cationic microemulsion through photon correlation spectroscopy and TEM. It was reported that the initiation occurred mainly in the styrene swollen microemulsion droplets. The average hydrodynamic radius of swollen particles first increased rapidly to a maximum at 4–7% conversion. Due to the continuous nucleation, the number of polymer particles and the polydispersity of particle size increased continuously throughout the polymerization process.

The growth of poly(methyl methacrylate) (PMMA) particles in ternary cationic microemulsion had also been studied [125]. In contrast to that of the styrene system, the sizes of methyl methacrylate (MMA)-swollen polymer particles increased continuously during polymerization, and no maximum of hydrodynamic radius of swollen particles was observed. In a similar system [126], a two-stage process of polymerization is observed. The first stage, described by a very slow increase in conversion, is attributed mainly to homogenous nucleation. The second stage, characterized by a much higher rate of conversion, involves continuous nucleation and is governed mainly by a micelle-entry mechanism.

Research studies indicate that the polymerization under γ-irradiation in aqueous solution is considered a radical mechanism, meaning that the polymerization reaction is absolutely caused by the radicals produced by the decomposition of water under irradiation. During γ-irradiation, the formation of $H^\bullet$ and $OH^\bullet$ can be shown as follows:

$$H_2O \sim\sim\longrightarrow R^\bullet(H^\bullet \text{ or } OH^\bullet) \quad (22)$$

$$\text{chain solicitation: } R^\bullet + M \rightarrow RM\bullet \quad (23)$$

$$\text{chain increase: } RM^\bullet + M \rightarrow RM_2^\bullet \quad (24)$$

$$RM_{n-1}^\bullet + M \rightarrow RM_n^\bullet \quad (25)$$

$$\text{chain expiration: } RM_n^\bullet + RM_m^\bullet \rightarrow P_{(n+m)} \quad (26)$$

or

$$RM_n^\bullet + RM_m^\bullet \rightarrow P_n + P_m \quad (27)$$

The polymerization can be initiated with γ-radiation, while it is very easy to stop at the desired conversion by removing the γ-ray source.

In this way, variation of the size and its distribution of monomer-swollen polymer nanoparticles with conversion can be obtained in a microemulsion environment. On the basis of this strategy, the polymerization [127] of three different monomers, styrene, butyl acrylate, and methyl methacrylate, has been carried out in appropriate microemulsion [128]. In all these three microemulsion polymerizations, the evidence of continuous nucleation was observed. It was found that the size of polymer particles rapidly increases up to their maximum at the early stage. With the increase of conversion, the large particles supply their monomer to newly formed particles and become smaller.

From these studies, we can see that γ-irradiation is a convenient way to polymerize monomers without any solicitation reagent.

3.2. Multilayer Nanoparticles

One of the most prominent properties of amphiphilic block copolymers is their ability to form micelles [129–132]. The micelles consist of a dense core and a protective corona. The core formed by insoluble blocks can serve as a host for insoluble low-molecular-weight compounds, such as drugs. Numerous realized or potential applications are based on this property. The possibility of controlling the transport into or from the micelle core and its solubilization capacity is of great importance for applications of micelles.

Three-layer (onion-type) micelles rank among promising candidates for the controlled uptake/release processes. It has been shown quantitatively [133] that the rate of the uptake of cyclohexane into the poly(2-ethylhexyl acrylate) cores of micelles with poly(acrylic acid) shells can be slowed by the presence of a PMMA middle layer in the micelles. The three-layer micelles have been prepared using a pair of diblock copolymers of types AB and BC [134–136], a heteroarm star copolymer (A_nB_n) with a BC copolymer [137], or an ABC block copolymer [133, 138]. The common drawback of these methods of preparing onion-type micelles is that the parameters of the resulting micelles are to a great extent predetermined by the properties of the copolymers used and cannot easily be varied.

Another possible technique of preparing multiplayer particles is polymerization of a suitable monomer dispersed in a micellar solution. Liu et al. [139] studied polymerization of MMA and butyl acrylate in aqueous micellar solutions of poly(methyl methacrylate)-block-poly(methacrylic acid). Hirzinger et al. [140] prepared PMMA particles using micelles of the copolymer polystyrene-block-poly(etheneco-propene) in *n*-decane. In these works, light-scattering techniques were used for the structure characterization. These methods could not provide direct information on the inner structure of the particles.

On the basis of the previous research on γ-radiation to polymerization, three-layer nanoparticles [141] were prepared by γ-radiation-induced polymerization of MMA in polystyrene-block-poly(methacrylic acid) aqueous micellar solution, and their structure was studied by small-angle neutron scattering (SANS). The scattering data were fitted using the bare-core approximation for a two-component core to elucidate the distribution of MMA before polymerization and the structure of nanoparticles with polymerized MMA. The newly formed polymer (PMMA) is deposited on the surface of polystyrene cores of the original micelles. The thickness of the PMMA layer depends on the monomer concentration. The SANS results indicate that the PMMA layer is penetrated by the poly(methacrylic acid) corona chains forming channels accessible to water.

From these results, it can be concluded that γ-irradiation combined with micelle science provides a new pathway to the preparation and assembly of polymer.

4. γ-IRRADIATION TECHNIQUE TO NANOCOMPOSITES

Quantum states in the nanocrystalline inorganic materials are size-dependent, leading to new electronic, optical, magnetic, mechanical, and catalytic properties [142–145]. The areas of application, which can be foreseen to benefit from the small size and organization of nanocrystallites, include nanoelectronics, nonlinear optics, catalysis, high-density information storage, and hybrid materials [146–149]. Studies relating to these well-defined inorganic particles have brought about not only a deep understanding of quantum confinement effects but also the development of new and useful spectroscopic methods and detection technologies [150, 151]. Nanocomposite techniques provide convenient routes to improve the processability and stability of nanocrystals with intriguing novel electronic, optical, and magnetic capacities [152, 153]. Recently, special attention has concentrated on developing methods for assembling nanoscale building blocks [154, 155]. These methods usually rely on both building block components and assembly strategies. For example, a variety of methods have been developed for preparing monodispersed samples of CdS, CdSe, and Au particles with diameters ranging from 1 to 40 nm [147, 156–159]. The general principle in the construction of nanocomposites involves the intimate mixing of nanocrystallites with procurable matrices. The nanoscale matrices usually provide spatially localized sites for nucleation, which minimizes the degree of particle aggregation and imposes an upper limit on the size [160–162]. In the view of materials science, nanocomposites are usually valuable materials, for example potentially useful in improving the photocatalytic conversion efficiency of photoelectrochemical systems [163]. These properties are largely determined by matrix materials and the size of metallic or semiconducting nanocrystallites within the matrices. Therefore, it will be significant to select suitable reaction routes to obtain worthwhile matrix materials and control the size of nanocrystallites within the matrices.

4.1. Inorganic–Inorganic Nanocomposites

It is well known that nanoscaled metal oxides have potential or demonstrated applications in many technologies, such as solar energy conversion, batteries, catalysis, and ductile ceramics [164–170]. Thus, inorganic–inorganic nanocomposites, such as metal/oxide nanocomposites, have been of interest recently [171–179].

Various methods are used to prepare nanocomposite materials, the most common being gas evaporation [171, 172], radio frequency sputtering [173], ion exchange–reduction [174, 175], and the sol–gel process [176–179]. In these methods, heat treatment or hydrogen reduction at relative high temperature is necessary. However, high temperature readily leads to the growth or oxidation of nanocrystalline particles. Thus, a relatively low temperature is favorable for preparation of nanocomposite materials. Based on the successful preparation of inorganic nanocrystals by the γ-irradiation technique, a γ-irradiation sol–gel method has been developed to synthesize metal/oxide nanocomposites. In this method, hydrogels containing a homogeneous dispersion of metal ions were first prepared by the sol–gel process and then oxide–metal nanocomposites were prepared by γ-irradiating the hydrogels instead of the solutions of metal salts.

Ag/TiO_2 [180] and Ag/SiO_2 [181] nanocomposites have been prepared via this route.

4.2. Inorganic–Polymer Nanocomposites

Recently, a large amount of scientific and technological research has been devoted to hybrid inorganic/polymer composites [182–184], in which the particle sizes, morphology, and orientation can be confined by the reaction space [185]. Also such hybrid composites can frequently provide properties that are combinations of those of the original inorganic and polymer materials. The key step in the confinement of inorganic particle size employed in these works is the initial isolation of a space. Then, under controlled conditions, an inorganic component is induced to form within the limited space. This class of totally innovative materials is considered to be novel and functional, with a wide range of potential applications in nonlinear optical materials [186, 187], conductive composites [188], nanoscaled electronics [189], electromagnetic interference shielding materials, and the like. Since γ-irradiation is convenient for the polymerization of monomers and the preparation of inorganic particles, it can be extended to synthesize inorganic–polymer nanocomposites.

4.2.1. Embedded Metal–Polymer Nanocomposites

Nonlinear optical properties of metal colloids prepared by various methods were studied [190, 191]. In general, the polymer–metal nanocomposites were synthesized in two steps: the polymer was first synthesized, and then the metal ions introduced into the polymer matrix were reduced to zero valent state by a reducing agent or by postheating. This caused a wide distribution of metal particle size and poor dispersion of metal particles in the polymer. As the material to be used in nonlinear optical devices, the chemical stability and optical homogeneity of nanocomposites of inorganic particles dispersed in matrix are required, and for mass production a simple fabrication method for processing at relatively low temperatures is sought.

Thus, it was necessary that the γ-irradiation technique was introduced to the preparation of metal–polymer nanocomposites. For example, nanosilver particles embedded in polyacrylamide matrix [192] were synthesized *in-situ* by using radiation–reduction of silver ions and radiation–polymerization of acrylamide simultaneously with γ-rays.

This route has also been extended to the preparation of Ag/polyacrylamide [193], Ag/poly(butyl acrylate-co-styrene) [194], Ag/polyacrylonitrile [195], Cu/polyacrylamide [196], and Ni/polyacrylamide [197].

4.2.2. Embedded Metal Sulfide–Polymer Nanocomposites

Metal sulfide semiconductors/polymer nanocomposites are considered to be highly functional materials with many applications, such as in photoluminescent, photoelectric, and nonlinear optical materials. The flexibility and processability of polymer matrices can provide good mechanical properties. As optical materials applications are expanding, the need for novel optically functional and transparent materials increases. These needs range from high performance, all optical switches for future use in optical computing to hard transparent coatings as protective or barrier layers. In addition to optical needs such as switching and amplification, the materials must be integrated into existing structures such as waveguides and optical fibers. As such, films and fibers are of great interest as the final form of these novel materials. Nanocomposite materials show great promise as they can provide the necessary stability and processability for these important applications.

In general, two basic routes have been used to synthesize the composites [198–201]. One is to carry the inorganic reaction between metal ion sources with H_2S or organometallic sulfur sources inside the polymer matrices. The other is to polymerize the monomer after doping the inorganic semiconductor components inside. In recent years, scientists have turned their attention to devising different strategies to control the morphology of inorganic components in the polymer matrices. However, in the above traditional methods, the dispersion of metal sulfide is difficult to control since the inorganic reaction and polymerization were performed separately, and semiconductors could not be well dispersed in the polymer matrix.

Thus, the progress on the extension of the γ-irradiation method to the preparation of semiconductor/polymer nanocomposites in a single step, in which the inorganic component displays nearly monodispersion and uniformity, was necessary. The synthetic factors optimized in order ensure the efficient dispersion of synthetic composites were (1) good "solvation" of the inorganic components in the polymer matrix and (2) an ordered, regular polymer environment in which to induce nucleation (matrix preorganization).

For example, CdSe/polyacrylamide nanocomposites [202] were prepared by this method. Being a material with good thermoplastic and water soluble monomers, polyacrylamide was selected to be the polymer matrix in this work. In addition, cross-linked polyacrylamide can provide a spongelike system in water, which requires that the nanoparticles are well distributed within the polyacrylamide network. CdSe, which shows quantum size effect by the shift of the optical absorption peak toward higher energies as the crystallite size approaches values comparable with or below its exciton size (9 nm for CdSe) and has unusual photoluminescence, etc., was selected to be the inorganic phase [203–205]. Using this approach, the metal ion source $CdSO_4$, selenium source Na_2SeSO_3, and organic monomer acrylamide were mixed homogeneously in the solution; the formation of nanocrystalline metal selenide particles and polymerization of monomer occurred almost simultaneously under γ-irradiation, leading to a good distribution of CdSe nanoparticles in the polymer matrix.

To ensure the formation of the polyacrylamide network, which precedes the further growth of inorganic particles, the concentration of the monomer was selected at the point with the highest rate of polymerization [206], judging from the concentration–polymerization velocity relationship (Fig. 6).

From the ultraviolet-visible spectrum analyses on products with different doses, it can be concluded that the as-prepared polyacrylamide surrounding, in which an amount of holes among the polymer chains fill the whole system and provide an ordered, regular network, is in fact used as a nanoreactor to carry out chemical reactions in restricted geometry.

The same route has been used to prepare CdS/polyacrylamide [207, 208], PbS/poly(vinyl acetate) [209], CdS/ polyacrylonitrile [210], and PbS/polyacrylonitrile [211].

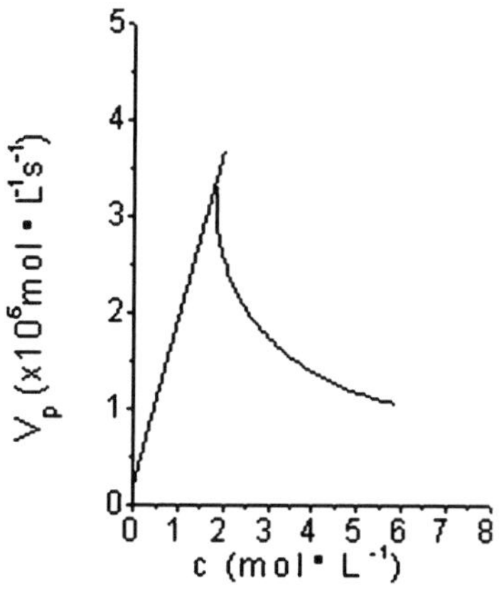

Figure 6. The concentration–polymerization velocity relationship.

4.2.3. Metal Sulfide–Polymer Composite Nanofibers

Researchers are trying to explore novel templates for the synthesis of functional materials. For example, polycationic poly(*N*-vinylcarbazole) (PNVK) is applied as templates to fabricate regular composite nanowire arrays of thallic oxide (Tl_2O_3) and polycationic PNVK at the molecular level [212].

Thus, polymers with long chains can provide templates for the growth of nanowires. Particularly, maleic anhydride (MA) and styrene (St) are excellent candidates for templates. Under γ-irradiation, St monomers are activated by the radicals generated in solution upon irradiation and are converted into St radicals. These St radicals rapidly initiate copolymerization of St and MA, and copolymer chains are propagated and provide attachment sites for metal sulfide nanoparticles owing to the presence of highly polar groups along the chains. It has been reported that the copolymerization of St and MA gives a copolymer in which the monomers alternate with near perfect regularity along the chain [213]. This is due to the fact that MA with an electron-attracting substituent has electron-accepting character ($e = 2.25$) while St with an electron-releasing substituent has electron-donating character ($e = 20.80$), which leads to a very small value of the product of monomer reactivity ratios ($r_1 \times r_2 = 0.006$).

Part of the copolymer undergoes esterification in ethanol with the acid anhydride groups being esterified. In addition, the two carbonyl groups alternate with the aromatic ring, leading to a homogeneous dispersion of carbonyl groups. In most previous studies on inorganic/copolymer composites, the polymer matrices were block copolymers, and particle growth was expected to be restricted to ionic or highly polar microdomains so providing a mechanism for both size control and particle stabilization [214, 215]. However, such ionic or highly polar microdomains were not necessarily evenly distributed in the block copolymers in contrast to the highly polar carbonyl groups in this work that lead to better control over size and shape of the nanosize particles.

Thus, after introducing γ-irradiation to this field, CdS nanofibers [216] were prepared in an alternate copolymer of MA and St via a one-pot procedure using γ-irradiation. The ordered and regular polymer environment induces nucleation of CdS (matrix preorganization), and initially formed highly polar groups evenly distributed along the chains can effectively control the size and shape of the CdS particles.

4.2.4. Assemblies of Metal Sulfide–Polymer Nanocomposites

In the traditional methods, the dispersion of metal sulfide is difficult to control since the inorganic reaction and polymerization were performed separately, and semiconductors could not disperse well in the polymer matrix. Thus, utilizing the hydrophilic and hydrophibic properties of monomers is an efficient way to solve this problem.

It is known that an organic monomer with polar groups can self-organize into micellelike aggregates with an inner hydrophilic region, utilizing the difference of the solubility of different fragments in the monomer molecule. Such micellelike aggregates can polymerize to polymer spheres with an inner hydrophilic region under γ-irradiation; the latter can act as both the template and nanoreactor for the following growth of inorganic semiconductor nanoparticles or nanofibers. In this approach, γ-irradiation offers an ideal means by which both polymerization of the monomer aggregates and the further growth of the inorganic nanofibers occur at room temperature and under ambient pressure. In a typical heterogeneous system of poly(vinyl acetate) (PVAc), $CdCl_2$, and CS_2, the mutual solubility between VAc and water is about 2% [217]. So before the polymerization, there is a small amount of water in the upper VAc layer. With the existence of water, the hydrophilic groups (C=O) in the VAc molecules tend to shrink together to form micellelike cores in order to minimize the free energy, and the other parts of vinyl acetate molecules then form the extended micellelike shell in its entity [218–220]. Such inner hydrophilic micellelike aggregates result in the inner hydrophilic cores in PVAc after γ-irradiation. In addition, under γ-irradiation, polymerization is faster than the formation of inorganic components, and the density of the PVAc spheres increases and the formed PVAc spheres drop into the under water layer. Water-soluble $CdCl_2$ will automatically distribute in the favorite inner hydrophilic region of the PVAc spheres and react further with CS_2 taken along with the PVAc spheres. The following nucleation and growth of acicular CdS are confined in the limited reactor region; thus the resulting CdS spreads homogeneously in the spherical polymer matrix. Based on the above strategy, spherical assemblies [221] of CdS nanofibers in PVA were successfully synthesized from a heterogeneous system of VAc monomer, cadmium chlorite, and carbon disulfide under γ-irradiation at room temperature and under ambient pressure.

By appropriate control of the experimental parameters, this technique could be extended to the preparation of a variety of spherical assemblies of semiconductor in polymer in heterogeneous systems under γ-irradiation at room temperature and in ambient pressure.

4.2.5. Metal Sulfide–Polymer Nanocables

Special attention has been paid to nanotubes and nanowires due to their interesting properties and potential applications. Since the discovery of carbon nanotubes [222] various nanotubes such as WS_2, MoS_2, BN, and BC_2N and semiconductor nanowires such as CdTe and ZnCdSe have been synthesized [223–228]. The template technique is an important method of nanowire fabrication. Usually, there are two types of template: "track-etch" polymer membranes, such as the polycarbonate membrane with 60 nm diameter pores, and the porous aluminum oxide membrane, with a pore diameter of 70 nm [229, 230]. However, the challenges of fabricating semiconductor nanowires with a quantum confinement effect are great because technically useful quantum wires will require lateral dimensions less than 10 nm, and it is difficult to fabricate quantum wires on a scale lower than 10 nm [231].

Nanocables with a wire/sheath structure are another kind of potentially useful one-dimensional nanostructure. There have been a few reports on the preparation of semiconductor/insulator nanocables. For example, Si/SiO_2 nanocable has been prepared by combining laser-ablation cluster formation with vapor–liquid–solid growth [232], and β-SiC/SiO_2 nanocable has been obtained by the carbothermal reduction of sol–gel derived silica xerogels containing carbon nanoparticles at 1650 °C [233]. In both cases, the outer layers of SiO_2 in nanocables are induced by the reaction atmosphere.

As described in the former section, an organic monomer with polar groups can self-organize into amphiphilic supramolecules, utilizing the difference in solubility of different fragments in the monomer molecule. Such supramolecules can polymerize to a preorganized polymer tubule with a hydrophilic core and a hydrophobic sheath. Then the polymer tubule acts as both template and nanoreactor for the following growth of inorganic semiconductor nanowires in the hydrophilic cores from various water-soluble sources. Thus, a nanocable with semiconductor wire in a polymer sheath can be obtained in a heterogeneous solution system. In this approach, γ-irradiation offers an ideal means by which the supramolecules can be polymerized and the tubular structure can be solidified with the desired diameter at room temperature under ambient pressure.

For example, the PVAc tubule can be formed through a continuous process from monomer to supramolecule then to polymer in a heterogeneous system under γ-irradiation. As proposed in Scheme 4, the heterogeneous system includes the upper VAc region and the lower water region. The mutual solubility between VAc and water is about 2% [217]. In the presence of a small amount of water, the hydrophilic groups (C=O) in the VAc molecules tend to shrink together to form micellelike cores in order to minimize the free energy, and the other hydrophobic parts of vinyl acetate molecules then form the extended micellelike shell in its entity. Then the enthalpic contributions of hydrogen bonding interactions drive the self-assembly of the supramolecular structure [218, 234, 235] as shown in the inset of Scheme 4. Such a hydrogen bonding interaction model is widely used to explain the process of molecular self-assembly into a well-defined supramolecular structure [236, 237]. A similar process can also be found in the formation of macroporous polymer monoliths with large pores that are obtained in a poor solvent due to an earlier onset of phase separation [238]. Obviously, the self-assembly structure consisting

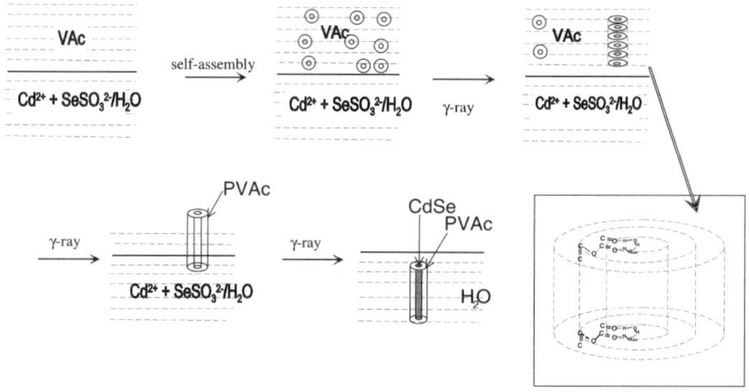

Scheme 4. Schematic illustration of the process of formation of CdSe/PVAc nanocables; inset: schematic illustration of the self-assembly of VAc. Reprinted with permission from [239], Y. Xie et al., *Adv. Mater.* 11, 1512 (1999). © 1999, Wiley-VCH.

of a hydrophilic core and extended shells in the massive monomer is related to the hydrogen-bonding bridge of the adjacent $C = O$ groups in VAc monomer in the presence of water.

In this approach, the entire preparation process is carried out by γ-irradiation at room temperature and under ambient pressure. A key of the method is irradiation of the heterogeneous system at a rather low dose rate for a relatively long time, to make sure there is enough time for the self-organization of the monomer into supramolecules. Then polymerization of the self-assembled supramolecules proceeds; the tubular structure forms and solidifies. Then the PVAc tubules drop into the lower water layer since the density of as-formed PVAc is greater than water. Water-soluble inorganic compounds will automatically distribute in the favorable hydrophilic cores of the PVAc tubules, and further irradiation will cause inorganic semiconductor to grow.

Based on the above strategy, a CdSe/PVAc nanocable with a 6 nm core and an 80 nm diameter sheath [239] (Fig. 7a) was successfully synthesized from a heterogeneous system of VAc monomer, cadmium sulfate, and sodium selenosulfate under γ-irradiation at room temperature and ambient pressure.

The successful approach is confirmed by the coexistence of some hollow tubules in the products, which can be clearly observed under the TEM, as shown in Figure 7b, and the ED pattern (Fig. 7b, inset), without the CdSe diffraction dots or circles, confirmed that it is only amorphous polymer. This route can also be used to synthesize CdS/PVA nanocables with $Na_2S_2O_3$ as reagent. If substituting hydrophibic sulfur source CS_2 [240] with hydrophilic $Na_2S_2O_3$, the size of the cores (Fig. 7c) would be adjusted as desired.

It is reasonable that, under thermal initiation conditions, the violent thermal motion breaks the preorganized supramolecular structure while the hydrophilic groups (C=O) make the polymer shrink to a sphere in order to minimize the free energy. This point was identified by the experiment under solvothermal conditions [241]. Some composite nanorods (Fig. 8) rather than nanocables were found in the as-obtained product. It further confirms that γ-irradiation is a better way to soft-chemical synthesis.

Compared with the preparation of embodied nanocomposites, it is found that the key to the formation of CdSe/(polyvinyl acetate) nanocable is slow polymerization under mild conditions that ensures that there is sufficient time for self-organization of the monomer into the supramolecular, maintenance of the tubular structure throughout the polymerization process, and the growth of inorganic crystallites in good orientation. The irradiation process carried out at a rather low dose rate (20.3 Gy/min) for a relatively long time was designed in our approach. Another tick of slow polymerization is the adding of $(CH_3)_2CHOH$, which is present in a much higher concentration than metal salts. OH and H radicals, which are generated in the radiolysis of water and induce the polymerization of vinyl acetate, were scavenged by abstracting an H atom from $(CH_3)_2CHOH$, leading to the rate of chain inducement and thus the rate of whole polymerization is decreased.

This technique could be extended to the preparation of a variety of semiconductor/polymer or metal/polymer nanocables in a heterogeneous system, by an appropriate choice of the inorganic/organic system and the experimental conditions. The diameter of the inner hole in the polymer tubule, which will control the diameter of the semiconductor or metal core, can be adjusted by choosing a suitable organic monomer with different solubility of different groups in selective solvents. As well as the hydrophilic cores, hydrophobic cores can also be obtained in a selective solvent and growth of inorganic nanowires can be carried out in such hydrophobic holes by choosing suitable precursors. In addition, semiconductor or metal nanowires with a desired

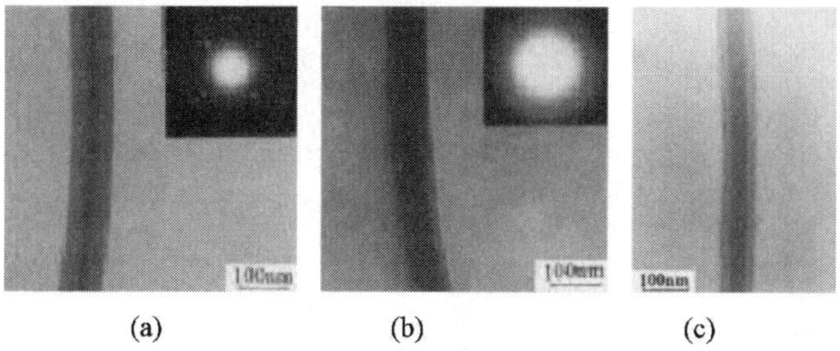

Figure 7. (a) TEM micrograph at higher magnification of part of the nanocable; inset: ED pattern of this area; (b) TEM micrograph of the coexisting PVAc empty tubules; inset: ED pattern of this area; (c) TEM micrograph of the nanocables with thicker cores. Reprinted with permission from [239], Y. Xie et al., *Adv. Mater.* 11, 1512 (1999). © 1999, Wiley-VCH.

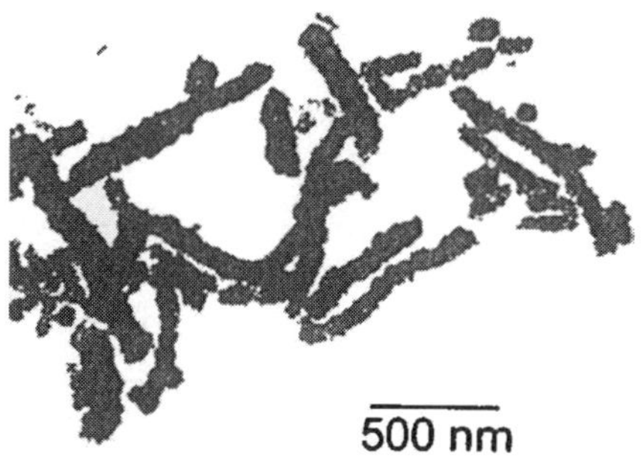

Figure 8. TEM image of the CdS/PVA composite nanorods obtained under hydrothermal treatment. Reprinted with permission from [241], J. H. Zeng et al., *Chem. Commun.* 1332 (2001). © 2001, Royal Society of Chemistry.

diameter can be prepared by this method, since the polymer sheath can be easily removed by dissolving in suitable organic solvents.

5. CONCLUSION

In the preceding sections, a summary of recent reports on γ-irradiation preparation of nanomaterials has been presented. The γ-irradiated-synthesized nanomaterials encompass metals, alloys, nonmetals, metal oxides, metal chalcogenide nanocrystals, and inorganic–inorganic and inorganic–polymer nanocomposites, with as-desired morphologies, including nanoparticles, nanofibers, nanorods, nanocables, and some assemblies. The motivation for use of γ-irradiation obviously has been to produce hydrated electrons and reduce metal ions, avoiding using any additional reducing agents in reactions. It should be mentioned that appropriate $(CH_3)_2CHOH$ as scavenger to eliminate hydroxyl radicals needs to be added into reaction system. While the preparation of intended nanomaterials has been achieved, an understanding of the reaction mechanism in γ-irradiation has been investigated. Since it is a relatively new activity in the nanomaterial and nanochemistry field, there is a growing volume of empirical knowledge about γ-ray–material interaction.

On of the most important issues in γ-irradiation applications for syntheses of nanomaterials is that this technique with relatively high yield of product and low cost can process under normal pressure at room temperature. Thus, it can be conveniently combined with other methods, such as hydrothermal, hard/soft template, and sonochemical methods, or other chemistry branches, such as coordination chemistry, surface and colloidal chemistry, and polymer science. In particular, it is advantageous for the stability of some soft-chemical environments (for example, micelles and liquid crystals) and the polymerization of monomers, although inorganic compounds can act as photoinitiators to initiate the polymerization sometimes. Therefore, the γ-irradiation technique is especially suitable for the soft-chemical synthesis of one-dimensional nanomaterials and their assemblies of inorganic nanomaterials and inorganic–polymer nanocomposites.

However, the reported literature suggests that the shortcoming of the γ-irradiation technique is obvious in that its energy is too dispersive. Therefore, it is unsuitable for the preparation of materials (for example, III–V compounds) which need high temperature and energy. On the other hand, the products prepared by γ-irradiation are usually crystalline-poor. Thus, sometimes the combination of the γ-irradiation technique with the hydrothermal method is necessary.

By the combination with many other fields, the γ-irradiation technique is expected to focus on the synthesis of metals, metal oxides, metal chalcogenides, and their polymer composites with as-desired nanostructure.

GLOSSARY

Fractal A geometric pattern that is repeated at ever smaller scales to produce irregular shapes and surfaces that cannot be represented by classical geometry. Fractals are used especially in computer modeling of irregular patterns and structures in nature.

Gamma radiation Emission and propagation of energy in the form of a gamma-ray.

Hydrated electron Transferring electron in aqueous phase.

Liquid crystal Any of various liquids in which the atoms or molecules are regularly arrayed in either one dimension or two dimensions, the order giving rise to optical properties, such as anisotropic scattering, associated with the crystals.

Micelle A submicroscopic aggregation of molecules, as a droplet in a colloidal system.

Nanocable One-dimensional nanomaterial with wire/sheath structure.

Nanocomposite Metallic or semiconducting nanocrystallites within the matirces.

Nanomaterial Nanoscale material having a microstructure with a characteristic length scale on the order of a few (typically 1–100) nanometers and displaying a quantum-confined effect.

Nanorod One-dimensional rodlike nanomaterial.

Nanoscale Scale ranging from 1 to 100 nm (one billionth of a meter).

Nanowire One-dimensional wirelike nanomaterial.

Polymer Any of numerous natural and synthetic compounds of usually high molecular weight consisting of up to millions of repeated linked units, each a relatively light and simple molecule.

Ultrasonic Of or relating to acoustic frequencies above the range audible to the human ear, or above approximately 20,000 hertz.

ACKNOWLEDGMENTS

This work was supported by the National Natural Science Foundation of China and the Chinese Ministry of Education.

REFERENCES

1. H. Fujita, M. Izawa, and H. Yamazaki, *Nature* 196, 666 (1962).
2. J. Belloni, M. O. Delcourt, and C. Leclere, *Nouv. J. Chim.* 6, 507 (1982).
3. M. O. Delcourt, N. Keghouche, and J. Belloni, *Nouv. J. Chim.* 7, 131 (1983).
4. M. O. Delcourt, J. Belloni, J. L. Marigrder, C. Mory, and C. Couex, *Radiat. Phys. Chem.* 23, 485 (1984).
5. Z. Chen, B. Chen, Y. Qian, M. Zhang, L. Yang, and C. Fan, *Acta Metall. Sinica B* 5, 407 (1992).
6. R. Tausch-Trekml, A. Henglein, and J. Lilie, *Ber. Bunsenges. Phys. Chem.* 82, 1335 (1978).
7. B. O. Ershov and N. L. Sukhov, *Radiat. Phys. Chem.* 36, 93 (1990).
8. J. H. Baxendale, E. M. Fielden, and J. P. Keene, *Proc. Roy. Soc. London Ser. A* 286, 320 (1965).
9. A. Henglein and R. Tausch-Trekml, *Colloid J. Interface Sci.* 80, 84 (1981).
10. M. Mostafavi, J. L. Marignier, J. Amblard, and J. Belloni, *Radiat. Phys. Chem.* 34, 605 (1989).
11. Y. Uan, Y. A. Ran, and G. Czapski, *Biochem. Biophys. Acta* 503, 339 (1978).
12. N. L. Sukhov, M. A. Aldnshin, and B. O. Ershov, *Khim. Vysokikh Energii* 20, 392 (1986).

13. B. O. Ershov, E. Janata, M. Michaelis, and A. Henglein, *J. Phys. Chem.* 95, 8996 (1991).
14. B. Cercek, M. Ebert, and A. J. Swallow, *J. Chem. Soc. A* 612 (1966).
15. J. Butler and A. Henglein, *Radiat. Phys. Chem.* 15, 603 (1980).
16. M. Breitenkamp, A. Henglein, and J. Lille, *Ber. Bunsenges. Phys. Chem.* 80, 973 (1976).
17. A. Henglein, E. Janata, and A. Fojtik, *J. Phys. Chem.* 96, 4734 (1992).
18. M. Michaelis and A. Henglein, *J. Phys. Chem.* 96, 4719 (1992).
19. S. Mossed, A. Henglein, and E. Janata, *J. Phys. Chem.* 93, 6791 (1989).
20. J. L. Marigrder, J. Belloni, M. O. Delcourt, and J. P. Chevalier, *Nature* 317, 344 (1985).
21. D. Hayes, O. I. Micic, M. T. Nenedovie, V. Swayambunathan, and D. Meidel, *J. Phys. Chem.* 93, 4603 (1989).
22. M. Anbar, M. Bambeneck, and A. B. Ross, *Natl. Bur. Stand. Ref. Data Set.* 43 (1973).
23. A. B. Ross, *Natl. Bur. Stand. Ref. Data Set.* 43, Suppl. (1975).
24. Y. Zhu, Y. Qian, M. Zhang, Z. Chen, and C. Fan, *Chin. J. Chem. Phys.* 8, 435 (1995).
25. B. B. Damaski and A. N. Frumkin, "Reaction of Molecules at Electrodes" (N. S. Bush, Ed.). Interscience, New York, 1971.
26. A. Henglein, *Chem. Phys. Lett.* 154, 473 (1989).
27. A. Henglein, *Chem. Rev.* 89, 1861 (1989).
28. M. Mostafavi, N. Keghouche, M. O. Delcourt, and J. Belloni, *Chem. Phys. Lett.* 167, 193 (1990).
29. A. Henglein, T. Linnert, and P. Mulvaney, *Ber. Bunsenges. Phys. Chem.* 94, 1449 (1990).
30. T. Linnert, P. Mulyaney, A. Henglein, and H. Weller, *J. Am. Chem. Soc.* 112, 4657 (1990).
31. A. Henglein, in "Modern Trends of Colloid Science in Chemistry and Biology" (H. F. Eicke, Ed.), p. 132. Birkhauser Verlag, Basel, 1985.
32. B. G. Ershov, J. Janata, and A. Henglein, *J. Phys. Chem.* 97, 339 (1993).
33. Y. J. Zhu, Y. T. Qian, M. W. Zhang, Z. Y. Chen, B. Liu, and G. E. Zhou, *Mater. Sci. Eng. B* 23, 116 (1994).
34. Y. J. Zhu, Y. T. Qian, M. W. Zhang, and Z. Y. Chen, *Acta Metall. Sinica B* 30, 259 (1994).
35. Y. J. Zhu, Y. T. Qian, H. Huang, M. W. Zhang, and L. Yang, *J. Mater. Sci. Lett.* 15, 1346 (1996).
36. Y. J. Zhu, Y. T. Qian, M. W. Zhang, Y. J. Li, Z. Y. Chenand, and G. E. Zhou, *J. Alloys Compd.* 221, L4 (1995).
37. Z. Y. Chen, B. Chen, Y. T. Qian, M. W. Zhang, L. Yang, and C. G. Fan, *Acta Metall. Sinica B* 28, 169 (1992).
38. B. B. Mandelbrot, "The Fractal Geometry of Nature." Freeman, New York, 1982.
39. M. Matsushita, M. Sano, Y. Hayakawa, H. Honjo, and Y. Sawada, *Phys. Rev. Lett.* 53, 286 (1984).
40. F. Argoul, A. Arneodo, G. Grasseau, and H. L. Swinney, *Phys. Rev. Lett.* 61, 2558 (1988).
41. V. J. Fleury, *Mater. Res.* 6, 1169 (1991).
42. J. Nittmann, G. Daccord, and H. E. Stanley, *Nature* 314, 141 (1984).
43. G. Daccord, J. Nittmann, and H. E. Stanley, "On Growth and Form: Fractal and Non-Fractal Patterns in Physics" (H. E. Stanley and N. Ostrowsky, Eds.). Martinus Nijhoff, Dordrecht, 1986.
44. L. Niemeyer, L. Pietronero, and H. J. Wiesmann, *Phys. Rev. Lett.* 52, 1033 (1984).
45. T. A. Witten and L. M. Sander, *Phys. Rev. Lett.* 47, 1400 (1981).
46. S. R. Forrest and T. A. Witten, *J. Phys. A* 12, L109 (1979).
47. P. Meakin, *Phys. Rev. Lett.* 51, 1119 (1983).
48. M. Kolb, R. Botetand, and R. Jullien, *Phys. Rev. Lett.* 51, 1123 (1983).
49. M. Y. Lin, H. M. Lindesay, D. A. Weitz, R. C. Ball, R. Klein, and P. Meakin, *Nature* 339, 360 (1989).
50. H. J. Herrmann and M. Kolb, *J. Phys. A* 19, L1027 (1986).
51. S. Z. Wang and H. W. Xin, *J. Phys. Chem. B* 104, 5681 (2000).
52. D. A. Russel, J. D. Hanson et al., *Phys. Rev. Lett.* 45, 1175 (1980).
53. E. Ben-Jacob, R. Godbey, N. Goldenfeld, J. Koplik, H. Levine, T. Mueller, and L. M. Sander, *Phys. Rev. Lett.* 55, 1315 (1985).
54. H. Brune, C. Romainczyk, H. Roder, and K. Kern, *Nature* 369, 469 (1994).
55. H. Honjo and S. Ohat, *Phys. Rev. A* 9, 4555 (1987).
56. D. H. Everett, "Basic Principles of Colloid Science," p. 130. Royal Society of Chemistry, London, 1988.
57. A. Henglein and M. Giersig, *J. Phys. Chem. B* 103, 9533 (1999).
58. A. Henglein, *J. Phys. Chem. B* 104, 1206 (2000).
59. Y. J. Zhu, Y. T. Qian, and M. W. Zhang, *Chinese Sci. Bull.* 42, 644 (1997).
60. Y. J. Zhu, Y. T. Qian, M. W. Zhang, Z. Y. Chen, M. Chen, and G. E. Zhou, *J. Mater. Sci. Lett.* 13, 1243 (1994).
61. Y. J. Zhu, Y. T. Qian, M. W. Zhang, Y. J. Li, W. Z. Wang, and Z. Y. Chen, *Mater. Trans.* 36, 80 (1995).
62. Y. J. Zhu, Y. T. Qian, L. Yang, W. Z. Wang, M. W. Zhang, Z. Y. Chen, and S. Tan, *Nanostruct. Mater.* 4, 915 (1994).
63. Y. P. Liu, Y. T. Qian, C. Wang, M. W. Zhang, and Y. H. Zhang, *Mater. Res. Bull.* 31, 973 (1996).
64. Y. P. Liu, Y. T. Qian, M. W. Zhang, Z. Y. Chen, and C. S. Wang, *Mater. Lett.* 26, 81 (1996).
65. Y. Kato, W. Sckiguchi, and M. Ishida, Jpn., Kokai, Tokkyo, Koho, JP 05, 86, 491 [93, 86, 491] (C1, C25 C5/02) (1993).
66. H. Senzaki, M. Sumita, M. Onoe, and H. Hata, Jpn., Kokai, Tokkyo, Koho, JP 05, 209, 208 [93, 209, 208] (C1, B22 F9/12) (1993).
67. M. Figlarz, F. Fievet, Lagier, and P. Jean, Fr. Demande. FR 2537898 (C1, B22 F9/18) (1984).
68. Y. J. Zhu, Y. T. Qian, H. Huang, and M. W. Zhang, *Mater. Lett.* 28, 119 (1996).
69. Y. J. Zhu, Y. T. Qian, H. Huang, and M. W. Zhang, *J. Mater. Sci. Lett.* 15, 1700 (1996).
70. Y. J. Zhu and Y. T. Qian, *Mater. Sci. Eng. B* 47, 184 (1997).
71. Y. J. Zhu, Y. T. Qian, M. W. Zhang, Z. Y. Chen, D. F. Xu, L. Yang, and G. E. Zhou, *Mater. Res. Bull.* 29, 377 (1994).
72. Y. J. Zhu, Y. T. Qian, and M. W. Zhang, *Mater. Sci. Eng. B* 41, 294 (1996).
73. Y. H. Ni, X. W. Ge, Z. C. Zhang, and Q. Ye, *Chem. Mater.* 14, 1048 (2002).
74. Y. P. Liu, Y. T. Qian, M. W. Zhang, Z. Y. Chen, and C. Wang, *Mater. Res. Bull.* 31, 1029 (1996).
75. G. S. Chu, G. Z. Bian, Z. C. Zhang, and J. F. Chen, *Acta Phys. Chim. Sin.* 15, 365 (1999).
76. Y. P. Liu, Y. T. Qian, Y. H. Chen, M. W. Zhang, Z. Y. Chen, L. Yang, C. S. Wang, and Z. C. Zhang, *Mater. Lett.* 28, 357 (1996).
77. L. E. Brus, *Appl. Phys. A* 53, 465 (1991).
78. W. Hoheisel, V. L. Colvin, C. S. Johnson, and A. P. Alivisatos, *J. Chem. Phys.* 101, 8555 (1994).
79. C. B. Murray, C. R. Kagan, and M. G. Bawendi, *Science* 270, 1335 (1995).
80. H. Weller, *Angew. Chem. Chem. Int. Ed. Engl.* 32, 41 (1993).
81. D. M. Wilhelmy and E. Matijevic, *J. Chem. Soc. Faraday Trans.* 80, 563 (1984).
82. R. Williams, P. N. Yocom, and F. S. Sotofko, *J. Colloid Interface Sci.* 106, 388 (1985).
83. P. N. Moorthy and J. J. Weiss, *Nature* 201, 1317 (1964).
84. F. Johnston, *J. Phys. Chem.* 79, 419 (1975).
85. W. M. Dale and J. V. Davies, *Radiat. Res.* 7, 35 (1957).
86. A. J. Elliot, D. R. McCracken, G. V. Buxton, and N. D. Wood, *J. Chem. Soc. Faraday Trans.* 86, 1539 (1990).
87. G. V. Buxton and D. C. Walker, *Radiat. Phys. Chem.* 23, 207 (1984).
88. R. L. Eager and D. S. Mahadevappa, *Can. J. Chem.* 43, 614 (1965).
89. Z. P. Qiao, Y. Xie, X. J. Li, C. Wang, Y. J. Zhu, and Y. T. Qian, *J. Mater. Chem.* 9, 735 (1999).

90. Z. P. Qiao, Y. Xie, J. G. Xu, Y. J. Zhu, and Y. T. Qian, *J. Colloid Interface Sci.* 214, 459 (1999).
91. Z. P. Qiao, Y. Xie, Y. T. Qian, and Y. J. Xu, *Mater. Chem. Phys.* 62, 88 (2000).
92. Y. D. Yin, X. L. Xu, X. W. Ge, C. J. Xia, and Z. C. Zhang, *Chem. Commun.* 1641 (1998).
93. Y. D. Yin, X. L. Xu, X. W. Ge, Y. Lu, and Z. C. Zhang, *Radiat. Phys. Chem.* 55, 353 (1999).
94. G. S. Chu, G. Z. Bian, Y. L. Fu, and Z. C. Zhang, *Mater. Lett.* 43, 81 (2000).
95. G. Z. Bian, Y. D. Yin, Y. L. Fu, Z. H. Wu, T. D. Hu, and T. Liu, *Acta Phys. Chim. Sin.* 16, 55 (2000).
96. Z. P. Qiao, Y. Xie, J. G. Xu, X. M. Liu, Y. J. Zhu, and Y. T. Qian, 78, 1143 (2000).
97. Y. D. Li, Y. Ding, H. W. Liao, and Y. T. Qian, *J. Phys. Chem. Solids* 60, 965 (1999).
98. P. Belton, I. P. Parkin, and J. D. Woollins, *J. Chem. Soc. Dalton Trans.* 511 (1989).
99. M. Chen, Y. Xie, H. Y. Chen, Z. P. Qiao, and Y. T. Qian, *J. Colloid Interface Sci.* 237, 47 (2001).
100. Y. H. Ni, X. W. Ge, H. R. Liu, X. L. Xu, and Z. C. Zhang, *Radiat. Phys. Chem.* 61, 61 (2001).
101. X. W. Ge, Y. H. Ni, H. R. Liu, Q. Ye, and Z. C. Zhang, *Mater. Res. Bull.* 36, 1609 (2001).
102. Y. Xie, Z. P. Qiao, M. Chen, Y. J. Zhu, and Y. T. Qian, *Chem. Lett.* 875 (1999).
103. Ruoff, *Nature* 372, 731 (1994).
104. P. M. Ajayan, O. Stephan, P. Redlich, and C. Colliex, *Nature* 375, 564 (1995).
105. Y. K. Chen, M. L. H. Green, and S. C. Tsang, *Chem. Commun.* 2489 (1996).
106. C. R. Martin, *Science* 266, 1961 (1994).
107. C. Huber, M. Sadoqi, T. Huber, and D. Chacko, *Adv. Mater.* 7, 316 (1995).
108. D. Routkevitch, T. Bigioni, M. Moskovits, and J. M. Xu, *J. Phys. Chem.* 100, 14037 (1996).
109. K. Torigoe and K. Esumi, *Langmuir* 11, 4199 (1995).
110. K. Esumi, K. Matsuhisa, and K. Torigoe, *Langmuir* 11, 3285 (1995).
111. A. J. Elliot, D. R. Mccracken, G. V. Buxton, and N. D. Wood, *J. Chem. Soc. Faraday Trans.* 86, 1539 (1990).
112. G. V. Buxton and D. C. Walker, *Radiat. Phys. Chem.* 23, 207 (1984).
113. B. Cheng, W. Q. Jiang, Y. R. Zhu, and Z. Y. Chen, *J. Mater. Sci. Lett.* 19, 503 (2000).
114. Y. Li, J. Wan, and Z. Gu, *Acta Phys. Chim. Sin.* 15, 1 (1999).
115. P. V. Braun, P. Osenar, and S. I. Stupp, *Nature* 380, 325 (1996).
116. X. C. Jiang, Y. Xie, J. Lu, L. Y. Zhu, W. He, and Y. T. Qian, *Chem. Mater.* 13, 1213 (2001).
117. W. Wang, Y. Geng, Y. Qian, M. Ji, and X. Liu, *Adv. Mater.* 10, 1479 (1998).
118. M. Anbar and E. Hart, *J. Phys. Chem.* 69, 1244 (1965).
119. W. Nakanishi, S. Hayashi, A. Sakaue, G. Ono, and Y. Kawada, *J. Am. Chem. Soc.* 120, 3635 (1998).
120. M. Chen, Y. Xie, J. C. Lu, Y. J. Zhu, and Y. T. Qian, *J. Mater. Chem.* 11, 518 (2001).
121. F. Candau, Y. S. Leong, and J. R. Fitch, *J. Polym. Sci. Polym. Chem. Ed.* 23, 193 (1985).
122. J. S. Guo, M. S. El-Aasser, and J. W. Vanderhoff, *J. Polym. Sci. Polym. Chem. Ed.* 27, 691 (1989).
123. J. S. Guo, E. D. Sudol, J. W. Vanderhoff, and M. S. El-Aasser, *J. Polym. Sci. Polym. Chem. Ed.* 30, 691 (1992).
124. L. M. Gan, C. H. Chew, K. C. Lee, and S. C. Ng, *Polymer* 35, 2659 (1994).
125. L. M. Gan, K. C. Lee, C. H. Chew, E. S. Tok, and S. C. Ng, *J. Polym. Sci. Part A: Polym. Chem.* 33, 1161 (1995).
126. F. Bleger, A. K. Murthy, F. Pla, and E. W. Kaler, *Macromolecules* 27, 2559 (1994).
127. X. L. Xu, X. W. Ge, Q. Ye, Z. C. Zhang, J. Zuo, A. Z. Niu, and M. W. Zhang, *Radiat. Phys. Chem.* 49, 469 (1997).
128. X. L. Xu, X. W. Ge, Z. C. Zhang, Z. C. Wu, and Z. W. Zhang, *Radiat. Phys. Chem.* 49, 469 (1997).
129. Z. Tuzar and P. Kratochvíl, "Surface and Colloid Science" (E. Matijevic, Ed.), Vol. 15. Plenum Press, New York, 1993.
130. "Solvents and Self-Organization of Polymers" (S. E. Webber, P. Munk, and Z. Tuzar, Eds.). Kluwer Academic, Dordrecht, 1996.
131. K. Mortensen, *Curr. Opin. Colloid Interface Sci.* 3, 12 (1998).
132. "Amphiphilic Block Copolymers: Self-Assembly and Applications" (P. Alexandridis and B. Lindman, Eds.). Elsevier Science, Amsterdam, 2000.
133. J. Kříž, B. Masař, J. Pleštil, Z. Tuzar, H. Pospíšil, and D. Doskočilová, *Macromolecules* 31, 41 (1998).
134. K. Procházka, T. J. Martin, S. E. Webber, and P. Munk, *Macromolecules* 29, 6526 (1996).
135. M. R. Talingting, P. Munk, S. E. Webber, and Z. Tuzar, *Macromolecules* 32, 1593 (1999).
136. J. Pleštil, J. Kříž, Z. Tuzar, K. Procházka, Yu. B. Melnichenko, G. D. Wignall, R. Talingting, P. Munk, and S. E. Webber, *Macromol. Chem. Phys.* 202, 553 (2001).
137. C. Tsitsilianis, D. Voulgaris, M. Štěpánek, K. Podhájecká, K. Procha'zka, Z. Tuzar, and W. Brown, *Langmuir* 16, 6868 (2000).
138. J. Pleštil, H. Pospíšil, B. Masař, and M. A. Kiselev, *Annual Report 1998, FLNP*, Joint Institute for Nuclear Research, Dubna, Russia, 1998, pp. 63–66.
139. T. Liu, H. Schuch, M. Gerst, and B. Chu, *Macromolecules* 32, 6031 (1999).
140. B. Hirzinger, M. Helmstedt, and J. Stejskal, *Polymer* 41, 2883 (2000).
141. J. Pleštil, H. Pospíšil, J. Kříž, Z. Tuzar, and R. Cubitt, *Langmuir* 17, 6699 (2001).
142. J. H. Fendler and F. C. Meldrum, *Adv. Mater.* 7, 607 (1995).
143. B. O. Dabbousi, C. B. Murray, M. F. Rubner, and M. G. Bawendi, *Chem. Mater.* 6, 216 (1994).
144. C. B. Murray, D. J. Norris, and M. G. Bawendi, *J. Am. Chem. Soc.* 115, 8706 (1993).
145. G. A. Ozin, *Adv. Mater.* 4, 612 (1992).
146. A. P. Alivisatos, *Science* 271, 933 (1996).
147. H. Weller, *Adv. Mater.* 5, 88 (1993).
148. Y. Wang and N. Herron, *J. Phys. Chem.* 95, 525 (1991).
149. L. Brus, *J. Phys. Chem.* 90, 2555 (1986).
150. R. S. Kane and R. E. Cohen, *Chem. Mater.* 11, 90 (1999).
151. R. N. Bhargava, D. Gallagher, X. Hong, and A. Nurmikko, *Phys. Rev. Lett.* 72, 416 (1994).
152. D. Wang, Y. Cao, X. Zhang, Z. Liu, X. Qian, X. Ai, F. Liu, D. Wang, Y. Bai, T. Li, and X. Tang, *Chem. Mater.* 11, 392 (1999).
153. L. L. Beecroft and C. K. Ober, *Chem. Mater.* 9, 1302 (1997).
154. L. C. Brousseau, Q. Zhao, D. A. Shultz, and D. L. Feldheim, *J. Am. Chem. Soc.* 120, 7645 (1998).
155. J. J. Storhoff, R. C. Mucic, and C. A. Mirkin, *J. Cluster Sci.* 8, 179 (1997).
156. K. C. Grabar, R. G. Freeman, M. B. Hommer, and M. Natan, *J. Anal. Chem.* 67, 735 (1995).
157. C. B. Murray, D. J. Norris, and M. G. Bawendi, *J. Am. Chem. Soc.* 115, 8706 (1993).
158. H. Weller, *Angew Chem. Int. Ed. Engl.* 32, 41 (1993).
159. M. A. Hayat, "Colloidal Gold: Principles, Methods, and Applications." Academic Press, San Diego, 1991.
160. V. Sankaran, T. Yue, R. E. Cohen, R. R. Schrock, and R. S. Sillbey, *Chem. Mater.* 5, 1133 (1993).
161. C. C. Cummins, R. R. Schrock, and R. E. Cohen, *Chem. Mater.* 4, 27 (1992).
162. V. Sankara, C. C. Cummins, R. R. Schrock, R. E. Cohen, and R. J. Silbey, *J. Am. Chem. Soc.* 112, 6858 (1990).
163. K. R. Gopidas, M. Bohorquez, and P. V. Kamat, *J. Phys. Chem.* 96, 6435 (1990).

164. W. Li, H. Osora, L. Otero, D. C. Duncan, and M. A. Fox, *J. Phys. Chem. A* 102, 5333 (1998).
165. L. Bedja, P. V. Kamat, A. G. Lapin, and S. Hotchandani, *Langmuir* 13, 2398 (1997).
166. A. Hagfeldt and M. Gratzel, *Chem. Rev.* 95, 49 (1995).
167. L. Kavan, K. Kratochvilova, and M. Grätzel, *J. Electroanal. Chem.* 394, 93 (1995).
168. B. O'Regan and M. Grätzel, *Nature* 353, 737 (1991).
169. M. J. Mayo, R. W. Siegel, A. Narayanasamy, and W. D. Nix, *J. Mater. Res.* 5, 1073 (1990).
170. J. Karch, R. Birringer, and H. Gleiter, *Nature* 330, 556 (1987).
171. C. G. Granqvist and O. Hunderi, *Phys. Rev. B* 16, 3513 (1977).
172. G. A. Niklasson, *J. Appl. Phys.* 62, 258 (1987).
173. B. Abeles, P. Sheng, M. D. Coutts, and Y. Arie, *Adv. Phys.* 24, 407 (1975).
174. D. Chakravorty, *J. Non-Cryst. Solids* 15, 191 (1974).
175. D. Chakravorty, A. Shuttleworth, and P. H. Gaskell, *J. Mater. Sci.* 10, 799 (1975).
176. S. Datta, S. S. Mitra, D. Chakravorty, S. Ram, and D. Bahadur, *J. Mater. Sci. Lett.* 5, 89 (1986).
177. A. Chatterjee and D. Chakravorty, *J. Phys. D* 22, 1386 (1986).
178. S. Roy, A. Chatterjee, and D. Chakravorty, *J. Mater. Res.* 8, 689 (1993).
179. H. L. Su, Y. Xie, P. Gao, Y. J. Xiong, and Y. T. Qian, *J. Mater. Chem.* 11, 684 (2000).
180. Y. J. Zhu, Y. T. Qian, H. Huang, M. W. Zhang, and S. X. Liu, *Mater. Lett.* 28, 259 (1996).
181. Y. J. Zhu, Y. T. Qian, M. W. Zhang, Z. Y. Chen, and G. E. Zhou, *J. Mater. Chem.* 4, 1619 (1994).
182. C. B. Murry, D. J. Norris, and M. G. Bawendi, *J. Am. Chem. Soc.* 115, 8706 (1993).
183. X. Li, J. R. Fryer, and D. J. Cole-Hamilton, *J. Chem. Soc. Chem. Commun.* 1715 (1994).
184. D. S. Kim and T. Bonhoeffer, *Nature* 370, 370 (1994).
185. J. T. Mcdevitt, D. C. Jurbergs, and S. G. Haupt, *Chemtech* 24, 24 (1994).
186. H. S. Zhou, T. Wada, and H. Sasabe, *Appl. Phys. Lett.* 68, 1288 (1996).
187. S. Ogawa, Y. Hayashi, N. Kobayashi, T. Tokizaki, and A. Nakamura, *Jpn. J. Appl. Phys.* 33, L331 (1994).
188. C. C. Yen and T. C. Chang, *J. Appl. Polym. Sci.* 40, 53 (1990).
189. R. P. Andres, J. D. Bielefeld, J. I. Henderson, D. B. Janes, V. R. Kolagunta, C. P. Kubiak, W. J. Mahoney, and R. G. Osifchin, *Science* 273, 1690 (1996).
190. D. Ricard, P. Roussignol, and C. Flytzanis, *Opt. Lett.* 10, 511 (1985).
191. K. Uchida, S. Kaneko, S. Omi, C. Hata, H. Tanji, Y. Asahara, A. J. Ikushima, T. Tokizaki, and A. Nakamura, *J. Opt. Soc. Am. B* 11, 1236 (1994).
192. Y. J. Zhu, Y. T. Qian, X. J. Li, and M. W. Zhang, *Chem. Commun.* 1081 (1997).
193. Y. J. Zhu, Y. T. Qian, X. J. Li, and M. W. Zhang, *Nanostruct. Mater.* 10, 673 (1998).
194. Y. D. Yin, X. L. Xu, C. J. Xia, X. W. Ge, and Z. C. Zhang, *Chem. Commun.* 941 (1998).
195. H. R. Liu, X. W. Ge, Y. H. Ni, Q. Ye, and Z. C. Zhang, *Radiat. Phys. Chem.* 61, 89 (2001).
196. Y. H. Ni, X. W. Ge, H. R. Liu, Z. C. Zhang, Q. Ye, and F. Wang, *Chem. Lett.* 458 (2001).
197. H. R. Liu, X. W. Ge, Y. J. Zhu, X. L. Xu, Z. C. Zhang, and M. W. Zhang, *Mater. Lett.* 46, 205 (2000).
198. A. Sellinger, P. M. Weiss, A. Nguyen, Y. Lu, R. A. Assink, W. Gong, and C. J. Brinker, *Nature* 394, 256 (1998).
199. A. Blaaderen Van, R. Ruel, and P. Wiltzius, *Nature* 385, 321 (1997).
200. M. Antonietti, E. Wenz, L. Bronstein, and M. Seregina, *Adv. Mater.* 7, 1000 (1995).
201. M. Moffitt, L. McMahon, V. Pessel, and A. Eisenberg, *Chem. Mater.* 7, 1185 (1995).
202. Z. P. Qiao, Y. Xie, J. X. Huang, Y. J. Zhu, and Y. T. Qian, *Radiat. Phys. Chem.* 58, 287 (2000).
203. T. Inkuma, T. Arai, and M. Ishikawa, *Phys. Rev. B* 42, 11093 (1990).
204. M. G. Bawendi, P. J. Carroll, W. L. Wilson, and L. E. Brus, *J. Chem. Phys.* 96, 946 (1992).
205. W. Hoheisel, V. L. Colvin, C. S. Johnson, and A. P. Alivisatos, *J. Chem. Phys.* 101, 8455 (1994).
206. A. Chapiro, "Radiation Chemistry of Polymeric Systems." Wiley, New York, 1962.
207. Y. H. Ni, X. W. Ge, H. R. Liu, Z. C. Zhang, and Q. Ye, *Chem. Lett.* 924 (2001).
208. Z. P. Qiao, Y. Xie, G. Li, Y. J. Zhu, and Y. T. Qian, *J. Mater. Sci.* 35, 285 (2000).
209. Z. P. Qiao, Y. Xie, Y. J. Zhu, and Y. T. Qian, *Mater. Sci. Eng. B* 77, 144 (2000).
210. Z. P. Qiao, Y. Xie, J. G. Xu, Y. J. Zhu, and Y. T. Qian, *Mater. Res. Bull.* 35, 1355 (2000).
211. Z. P. Qiao, Y. Xie, Y. J. Zhu, and Y. T. Qian, *J. Mater. Chem.* 9, 1001 (1999).
212. J. F. Liu, K. Z. Yang, and Z. H. Lu, *J. Am. Chem. Soc.* 119, 11061 (1997).
213. P. J. Flory, "Principles of Polymer Chemistry." Cornell Univ. Press, New York, 1953.
214. M. Moffitt, H. Vali, and A. Eisenberg, *Chem. Mater.* 10, 1021 (1998).
215. S. W. Haggata, D. J. Cole-Hamilton, and J. R. Fryer, *J. Mater. Chem.* 7, 1967 (1997).
216. M. Chen, Y. Xie, H. Y. Chen, Z. P. Qiao, Y. J. Zhu, and Y. T. Qian, *J. Colloid Interface Sci.* 229, 217 (2000).
217. "Lange's Handbook of Chemistry" (J. A. Dean, Eds.). McGraw–Hill, New York, 1985.
218. S. Mann, *Nature* 265, 499 (1993).
219. G. M. Whitesides, J. P. Mathias, and C. T. Seto, *Science* 254, 1312 (1991).
220. M. R. Ghadiri, J. R. Granja, and L. K. Buehler, *Nature* 369, 301 (1994).
221. Y. Xie, Z. P. Qiao, M. Chen, Y. J. Zhu, and Y. T. Qian, *Nanostruct. Mater.* 11, 1165 (1999).
222. S. Iijima, *Nature* 354, 56 (1991).
223. R. Tenne, L. Margulis, M. Genut, and G. Hodes, *Nature* 360, 444 (1992).
224. Y. Feldman, E. Wasserman, D. J. Srolovitt, and R. Tenne, *Science* 267, 222 (1995).
225. N. G. Chopra, R. J. Luyken, K. Cherrey, V. H. Crespi, M. L. Cohen, S. G. Souie, and A. Zettl, *Science* 269, 966 (1995).
226. Z. W. Sieh, K. Cherrey, N. G. Chopra, X. Blase, Y. Miyamoto, A. Rubio, M. L. Cohen, S. G. Louie, A. Zettl, and R. Gronsky, *Phys. Rev. B* 51, 11229 (1995).
227. D. Brinkmann, G. Fishman, C. Gourgon, L. S. Dang, A. L.ffler, and H. Mariette, *Phys. Rev. B* 54, 1872 (1996).
228. B. P. Zhang, W. X. Wang, T. Yasuda, Y. Segawa, H. Yaguchi, K. Onabe, K. Edamatsu, and T. Itoh, *Mater. Sci. Eng. B* 51, 224 (1998).
229. C. M. Zelenski and P. K. Dorhout, *J. Am. Chem. Soc.* 120, 734 (1998).
230. C. R. Martin, *Science* 266, 1961 (1994).
231. M. Sundaram, S. A. Chalmers, P. F. Hopkins, and A. C. Gossard, *Science* 254, 1326 (1991).
232. A. M. Morales and C. M. Lieber, *Science* 279, 208 (1998).
233. G. W. Meng, L. D. Zhang, C. M. Mo, S. Y. Zhang, Y. Qin, S. P. Feng, and H. J. Li, *J. Mater. Res.* 13, 2533 (1998).
234. M. Whitesides, J. P. Mathias, and C. T. Seto, *Science* 254, 1312 (1991).

235. M. R. Ghadiri, J. R. Granja, and L. K. Buehler, *Nature* 369, 301 (1994).
236. P. R. Sijbesma, F. H. Beijer, L. Brunsveld, B. J. B. Folmer, J. H. K. Hirschberg, R. F. M. Lange, J. K. L. Lowe, and E. W. Meijer, *Science* 278, 1601 (1997).
237. V. A. Russell, C. C. Evans, W. Li, and M. D. Ward, *Science* 276, 575 (1997).
238. C. Viklund, F. Svec, and J. M. J. Fr.chet, *Chem. Mater.* 8, 744 (1996).
239. Y. Xie, Z. P. Qiao, M. Chen, Y. J. Zhu, and Y. T. Qian, *Adv. Mater.* 11, 1512 (1999).
240. Z. P. Qiao, Y. Xie, J. G. Xu, Y. J. Zhu, and Y. T. Qian, *Chem. Phys. Lett.* 321, 504 (2000).
241. J. H. Zeng, J. Yang, Y. Zhu, Y. F. Liu, Y. T. Qian, and H. G. Zheng, *Chem. Commun.* 1332 (2001).

ENN Encyclopedia of Nanoscience and Nanotechnology
www.aspbs.com/enn

Gas Adsorption on Carbon Nanotubes

A. D. Migone, S. Talapatra

Southern Illinois University at Carbondale, Carbondale, Illinois, USA

CONTENTS

1. INTRODUCTION

The study of gas adsorption on carbon nanotube substrates is an ever-expanding field of research. This chapter summarizes much of the experimental work done on this subject over the past five years. We review results obtained on a variety of systems, including the rare gases, and simple molecular systems such as nitrogen, oxygen, methane, and other hydrocarbons. The most important topic that, while mentioned throughout the chapter, will not be discussed in any great detail is the work on hydrogen adsorption on nanotubes. The choice to omit a detailed review of this subject was made because there are several comprehensive, and fairly recent, reviews of hydrogen adsorption available [1–3]. The present work is the first review devoted to summarizing the findings on all other adsorbates on carbon nanotube substrates.

We briefly describe the main experimental techniques that have been used to study gas adsorption in general, and adsorption on carbon nanotubes in particular. We then review separately work done on multiwall nanotube substrates and that conducted on single-wall carbon nanotubes. While the focus is on experimental work, occasional reference will also be made to especially relevant theoretical and simulational results; in each case, we try to provide the broader context in which the experimental research was conducted, pointing out the fundamental theoretical issues as well as the practical ones that the researchers were addressing with their experiments.

Adsorption is the process by which atoms or molecules of one chemical species (the adsorbate) are attracted, and stay on the surface of a solid of a different chemical species (the substrate). Depending on the strength of the interaction between the adsorbate and the substrate, we can have physisorption (weakly bound adsorbates, having binding energies typically below about 0.3 eV) or chemisorption (more strongly bound adsorbates) [4]. In physisorption, the adsorbate retains its chemical identity.

The study of gas adsorption on carbon nanotubes is of considerable interest, both from a fundamental point of view as well as from a practical perspective. Gases adsorbed on nanotubes can be used to provide us with one of the few readily available physical realizations of matter in one dimension [5–10]. As such, these systems are a testing ground for theoretical predictions specifically made for 1-D matter. In addition, because they permit the determination of the gas–substrate binding energy, gases adsorbed on nanotubes provide us with an experimental test for the level of accuracy achieved for the potentials used to describe the interaction between different adsorbate species and carbonaceous substrates [11].

From a practical perspective, carbon nanotubes, and other related carbon nanomaterials, hold significant promise for the development of an entirely new approach to gas storage technology [12]. The implications that these new approaches may have for the storage of hydrogen have been discussed extensively [1, 2, 12]. However, the development of new methods of gas storage would have an impact that would be felt well beyond hydrogen-related applications. In addition to gas storage applications, the significant levels of differential adsorption that can be achieved with these novel carbon materials can be effectively used to develop new methods for gas mixture separation and purification [13].

2. GAS ADSORPTION MEASUREMENTS

A number of the standard adsorption techniques have been utilized to study gases physisorbed on nanotubes. These include: adsorption isotherms [9, 14–35], thermal programmed desorption [6–8, 12, 36–40], neutron scattering [41, 42], X-ray diffraction [32], X-ray spectroscopy [43], Raman spectroscopy [44], thermogravimetric analysis [37], FTIR spectroscopy [37], and calorimetry [24, 45]. Several of these techniques have been summarized in a review article

ISBN: 1-58883-059-4/$35.00

[46] on adsorbed films by Vilches. In addition to techniques traditionally associated with the study of adsorbed systems, it has been found that some electrical transport properties [47–53] of carbon nanotubes are sensitive to the presence of adsorbed species on the tubes, and thus provide information on adsorption on these substrates.

2.1. Adsorption Isotherm Measurements

By far the most common approach used to investigate gas adsorption on any substrate is to perform adsorption isotherms. In this type of measurement, the amount of gas adsorbed onto a substrate is determined as a function of the chemical potential (or, equivalently, of the pressure) of the coexisting adsorbate that remains present in the vapor phase at a fixed temperature. The method by which the determination of the amount adsorbed is made varies.

In volumetric isotherms, the quantity of gas adsorbed on the substrate is obtained as the difference between the amount of gas initially dosed into the experimental cell which contains the substrate and the amount of gas that remains present in the vapor phase in the experimental cell after the new equilibrium conditions have been reached. Computer-controlled volumetric adsorption setups are able to produce isotherms of sufficiently high quality to allow for the meaningful determination of pointwise derivatives of the experimental data. The isothermal compressibilities of the adsorbed films can then be obtained directly from these derivatives.

In gravimetric adsorption isotherms, the amount of gas adsorbed on the substrate is determined by measuring the weight of gas adsorbed. The gravimetric approach is well suited for the performance of adsorption isotherms at and above room temperature, where in many cases the large fraction of the gas present in the vapor phase makes volumetric determinations impractical.

While other approaches for measuring adsorption isotherms (such as ellipsometric measurements or quartz crystal microbalance frequency determinations) have been applied to the study of gases adsorbed on planar substrates, their use has not yet been extended to the investigation of gases adsorbed on carbon nanotube substrates.

Collections of adsorption isotherms measured at a number of different temperatures (see Fig. 1, which displays such data, as presented in [23]) allow the determination of the isosteric heat of adsorption of an adsorbate on a substrate [54]. The isosteric heat of adsorption is the amount of heat released (adsorption is an exothermic process) when an atom or molecule adsorbs onto the substrate [4]. In terms of adsorption isotherms, the isosteric heat of adsorption q_{st}, is defined as [54]

$$q_{st} = kT^2(d \ln P/dT)_n$$

Here, n is the fixed value of the coverage at which the isosteric heat is calculated, dT is the difference between the two temperatures used in the calculation of the isosteric heat, $d \ln P$ is the difference in the logarithms of the pressures at these two temperatures, k is Boltzmann's constant, and T is the average of the two temperatures. When the isosteric heat of adsorption is determined at sufficiently low coverages [4], it can be used to provide an experimental value for the binding energy of the adsorbed species on the substrate.

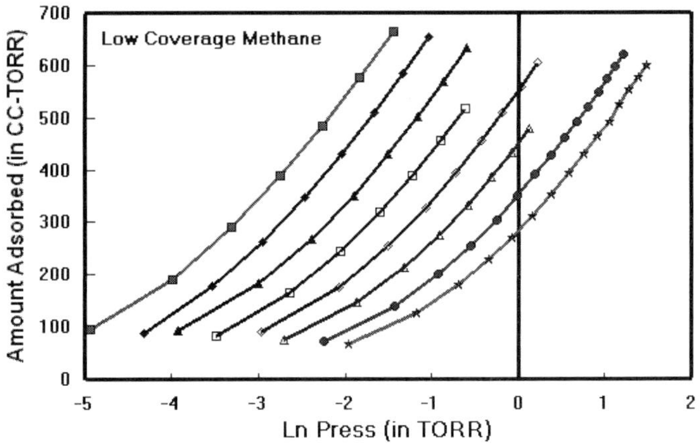

Figure 1. Set of very low coverage adsorption data, for various temperatures, for CH_4 adsorbed on bundles of SWNTs. The coverages (which are given in $cm^3 \cdot$ torr (labeled "cc-torr") in this figure; 1 cc $\cdot$ torr corresponds to 58.8 nmol) are plotted as a function of the logarithm of the pressure (in torr). The coverages displayed in this figure range between 1 and 10% of one full layer of methane on the nanotube bundles. The temperatures are, from left to right, 159.88, 164.82, 174.82, 179.84, 184.80, 189.85, and 194.68 K. The isosteric heat of adsorption can be determined from these data in the manner described in the text. Reprinted with permission from [23], S. E. Weber et al., *Phys. Rev. B* 61, 13150 (2000). © 2000, American Physical Society.

2.2. Temperature-Programmed Desorption

In thermal-programmed desorption measurements (TPD), gas is adsorbed on a substrate at a low temperature. The adsorbate remaining in the vapor phase is then pumped out of the experimental cell, so that equilibrium is established at a low value of the pressure at the beginning of the measurements, and so that most of the gas present inside the experimental cell is adsorbed onto the substrate. The temperature of the cell is then allowed to increase at some preset rate. The amount of gas desorbing from the substrate is monitored (for example, with a mass spectrometer) as a function of the increasing temperature. Thermal-programmed desorption can be used for the determination of the binding energy of a gas on a substrate because the temperature at which the desorption rate reaches a maximum is related to the binding energy of the adsorbate. While this technique is most often used for studying chemisorbed systems (for example, it has been used to determine the thermal evolution of chemical functional groups chemisorbed on the nanotubes) [10, 38], thermal-programmed desorption also has been used for the determination of the binding energy of physisorbed adsorbates such as helium (Fig. 2) [6–8], and other simple adsorbates such as xenon, methane, and SF_6 (see Fig. 3) [36, 40].

2.3. Thermogravimetric Analysis

In thermogravimetric analysis, the change in the mass of the film adsorbed on a substrate, either as a result of adsorption or desorption, is studied under preselected conditions.

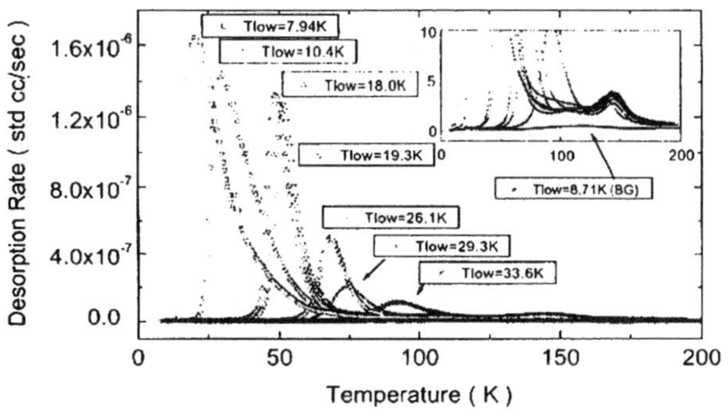

Figure 2. Temperature-programmed desorption data for helium on SWNT. The figure presents the helium desorption rates from the nanotube substrate as a function of temperature. Reprinted with permission from [8], Y. H. Kahng et al., *J. Low Temp. Phys.* 126, 223 (2002). © 2002, Kluwer Academic/Plenum.

For example, the mass change can be determined when a gas mixture containing the adsorbate of interest, together with an inert (nonadsorbing) species, is made to flow at a constant rate over the substrate under isothermal conditions [37]. These determinations provide information as to the effectiveness of a substrate for differential adsorption processes for specific adsorbates. Experiments can also be conducted in a manner entirely analogous to TPD measurements, but instead of monitoring the amount of gas desorbed with a mass spectrometer, it is the mass change that is monitored as a function of the temperature of the system.

2.4. Scattering Techniques

Scattering techniques, in particular X-ray diffraction [32] and neutron scattering [42], have been applied to study the structure of the adsorbed films. Typically, the diffraction pattern corresponding to the bare substrate (i.e., the nanotube bundles before the introduction of the adsorbate species) is measured. Gas is then introduced into the experimental cell, and the diffraction pattern is measured again. In its simplest approach, the difference between these two diffraction patterns is taken to correspond to the diffraction pattern from the adsorbed film. By monitoring the diffraction peaks of the substrate, these structural techniques can also be used to determine whether there are any changes in the structure of the substrate that occur as a result of the adsorption process [32, 42].

In addition to structural determinations, neutron scattering can be used to probe the dynamics of adsorbed films. Quasielastic neutron scattering measurements have been used to investigate the mobility of films adsorbed on carbon nanotubes (Fig. 4) [41].

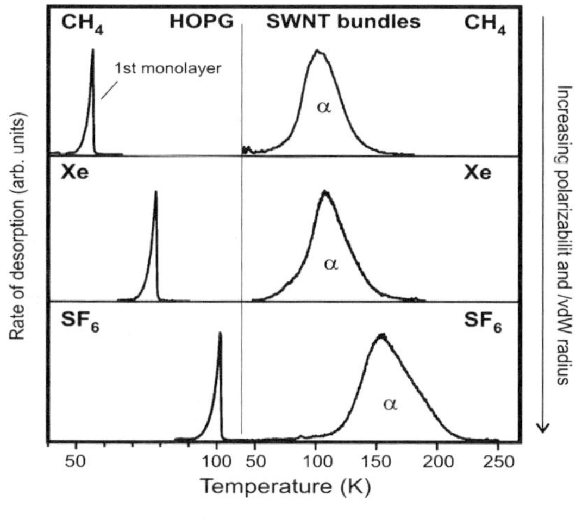

Figure 3. Thermal-programmed desorption traces for CH_4, Xe, and SF_6 on HOPG (left panels) and on SWNTs (right panels). The higher values of the desorption temperature on the SWNTs indicate stronger binding than on HOPG. These data are somewhat different form previously published data from this same group (see [36]) in that lower temperature multilayer peaks are not present. This is probably due to the fact that the older samples were rougher, and diffusion in them was more inhibited than in the sample used here. The unpublished data presented in this figure were provided by T. Hertel.

2.5. Spectroscopic Techniques

Raman spectroscopy has been utilized to study the adsorption of hydrogen on single-wall nanotubes (Fig. 5) [44]. This approach uses shifts in the Raman frequencies of the adsorbate species relative to the values of the frequencies in the bulk phase (which occur as a consequence of adsorption) to obtain information on the adsorbed system. The shifts are compared with theoretical estimates of the shifts that would result under different adsorption scenarios. Inferences can

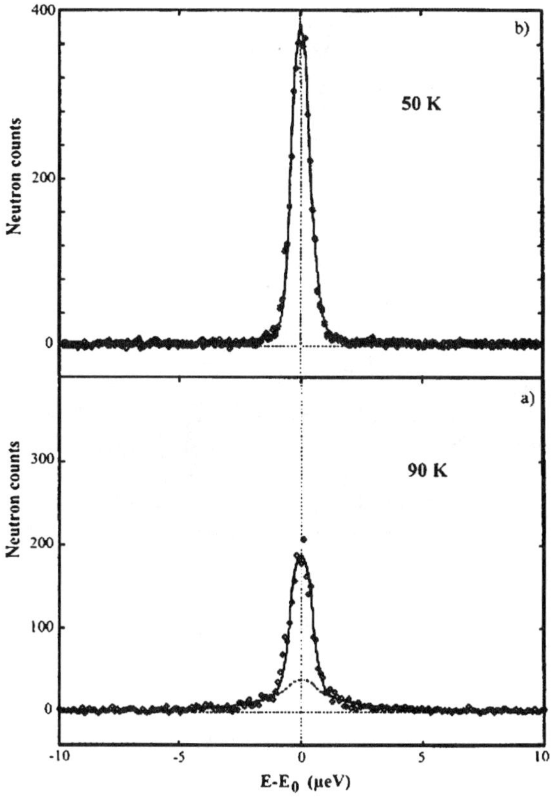

Figure 4. QENS spectra of 20 cm^3 STP of methane adsorbed on 600 mg of SWNT (the contributions from the background have been subtracted) at: (a) 90 K and (b) 50 K for $Q = 0.91$ Å^{-1}. The 50 K data are characteristic of a solid phase, while the broad wings at the base of the 90 K data indicate the presence of a liquid phase. Reprinted with permission from [41], M. Bienfait et al., *Surf. Sci.* 460, 243 (2000). © 2000, Elsevier Science.

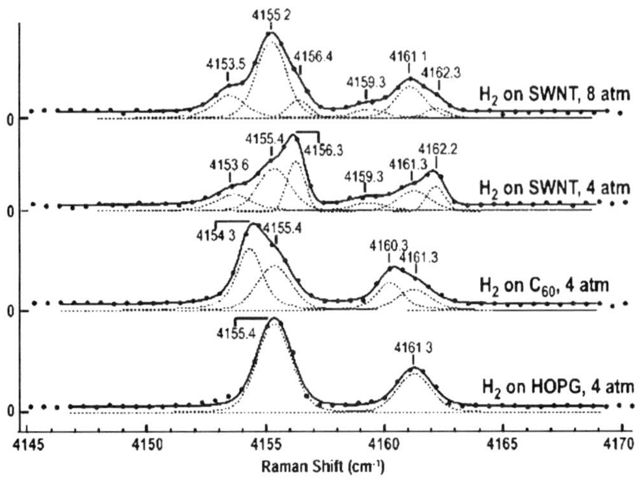

Figure 5. Raman spectroscopy data for hydrogen adsorbed on SWNT at 8 and 4 atm, together with data for C60 and HOPG at 4 atm. Reprinted with permission from [44], K. A. Williams et al., *Phys. Rev. Lett.* 88, 165502 (2002). © 2002, American Physical Society.

be drawn as to the locations of the adsorption sites, and their strength, as a result of comparing experiments and theoretical calculations.

FTIR spectroscopy has been used to look at the growth of films on carbon nanotube substrates [37]. The change in intensity of specific IR peaks of the adsorbate as a function of exposure time can be monitored to determine the increase in the amount of adsorbate present on the substrate.

2.6. Calorimetry

Calorimetry has been used to investigate gas adsorption on bundles of single-wall carbon nanotubes [24, 45]. Two different quantities have been determined calorimetrically: the heat of adsorption, and the specific heat of the adsorbed film. Measurement of the heat released when adsorption occurs under essentially isothermal conditions permits the obtainment of the isosteric heat of adsorption of a given adsorbate on the nanotube substrate [24]. The specific heat of an adsorbed film, on the other hand, is measured by substracting the heat capacity of the bare substrate from the heat capacity of the loaded substrate (i.e., the substrate that already has a fixed amount of gas adsorbed onto it) [45]. The film's contribution to the heat capacity is taken as the difference between these two values.

2.7. Adsorption Isotherm Simulations

In addition to these strictly speaking experimental tools, computer simulation "experiments" (in particular, simulated adsorption isotherms) [5] have been used to model the behavior of a variety of adsorbed species on carbon nanotubes. Combining the contributions from actual experiments with those from simulated experiments has proven to be a powerful approach to further our understanding of the behavior of films adsorbed on carbon nanotubes.

3. ADSORPTION ON CARBON NANOTUBES

3.1. Adsorption on Multiwalled Nanotubes

While some experimental techniques can be implemented with microscopic-sized samples, the nature of adsorption experiments is such that they almost universally require macroscopic amounts of substrate (on the order of milligrams or greater) on which to perform the measurements. Methods for the production of nanotubes in sufficiently large amounts to make adsorption studies possible were first developed for multiwalled tubes [55], and later for single-walled tubes [56, 57]. As a result, some of the first studies involving adsorption on nanotubes were conducted on multiwalled tubes [14].

Unlike the case with single-walled nanotubes (which, upon production, spontaneously arrange themselves into "bundles," that is, a triangular ordered phase) [57], multiwalled nanotubes do not self-organize into ordered arrays at production. There are just two groups of adsorption sites on samples of multiwalled nanotubes: the external surface of the tubes, and, if the tubes have at least one of their two end caps removed, the hollow space at the interior of the tube. This does not imply that multiwalled tubes cannot be grown on especially prepared substrates so that they form what, at the micron level, appears to be an ordered mat (see, for example, the electron micrographs shown in [19]). However, the arrangement in a mat is far from that which would give rise to new adsorption behavior, as occurs when SWNTs assemble into bundles, because the separation between the parallel multiwalled tubes in the mat is too large to give rise to a new set of adsorption sites.

Initial reports on the adsorption characteristics of multiwalled carbon nanotubes focused on the determination of the specific surface areas of these materials. In a study to establish a method for uncapping the ends of multiwalled nanotubes, Green's group reported on the effect of CO_2 treatment on the specific surface area of a sample of tubes as determined by applying the BET equation to nitrogen adsorption isotherms at liquid nitrogen temperatures [14]. They found that the specific area before the treatment was 21.0 m^2/g, while after the treatment it increased to 37.0 m^2/g. The increase in the value of the specific surface area was taken as evidence of the uncapping of the tubes after the CO_2 treatment [14].

The storage of Ar at high pressures, in the hollow interior of capped catalytically produced multiwalled nanotubes, has been investigated [43]. The study was performed by hot isostatically pressing catalytically produced multiwalled tubes with outer diameters between 20 and 150 nm and inner diameters between 10 and 60 nm under 170 MPa of argon gas at 650 °C for 48 h. The results, which identified Ar inside a sizable fraction of the close-ended multiwalled tubes, were obtained by performing energy dispersive X-ray maps of the tubes. They were explained in terms of the high pressure helping the gas penetrate into the hollow center of the tubes, and the amorphous outer layer of the tube that formed as a result of the high temperature helping seal the gas at the interior of the tube. However, Kuznetsova et al. [10] noted that the interpretation of the above experiments is complicated by the presence of large amounts of impurities that can incorporate gas in regions other than the interior of the nanotubes. There has been one other study that explored the encapsulation of gas [58]. In this case, however, the encapsulation process was quite different. The more recent study investigated the simultaneous formation of CN_x nanotubes and the encapsulation of N_2 gas inside the tubes. The encapsulation process takes place in a single step, at the time in which the MWNTs are produced. The presence of the gas inside the tubes was determined from the existence of a peak corresponding to molecular nitrogen in the EELS spectrum of the sample.

Gaucher et al. [15] investigated the adsorption of nitrogen and argon on catalytically produced multiwalled carbon nanotubes at liquid nitrogen temperature. In this study, the ends of the nanotubes were always open. The diameters of the hollow interiors of the tubes investigated were about 4 nm, while the outer diameters of the tubes ranged typically between 20 and 40 nm. Adsorption–desorption cycles displayed hysteresis loops, indicating the presence of mesopores in the samples. The mesopore volume and pore radius were determined from the adsorption data by using the Barrett–Joyner–Halenda (BJH) approach [59]. The maximum pore radius, at about 2 nm, coincided with the diameter of the hollow core of the tubes seen in electron micrographic studies. Upon heating the samples in an argon atmosphere to 2800 °C, a majority of the tubes developed closed ends, and BJH analysis of the adsorption–desorption data on these samples yielded much smaller mesopore volumes.

Gravimetric adsorption isotherms at 77.3 K were used to investigate the adsorption of N_2 on multiwalled nanotubes, produced by Hyperion [16]. A majority of the nanotubes that were used in this study reportedly were open at one end. The tubes had diameters on the order of 4 nm and wall thicknesses on the order of 3 nm (yielding an average overall tube diameter of 10 nm). The study undertook to determine the values of the interior and exterior surfaces of the tubes, and the presence or absence of capillary condensation for this system. The determination of the adsorption areas on the interior and exterior surfaces was accomplished by using an alpha-plot comparison in which the nonporous reference isotherm was measured on nonporous carbon black. An alpha plot is the curve that results when one displays the amount adsorbed (in fractions of a layer) at a given value of the reduced pressure P/P_o on a test substrate (Y axis) as a function of the amount adsorbed, at the same value of P/P_o, on a nonporous reference substrate of similar adsorption characteristics (X axis). For a nonporous test sample, such a plot should yield a straight line. Deviations from a straight line are interpreted in terms of the presence of porosity on the test sample. The overall specific area determined for the tubes employed in this study was 268 m^2/g, of which 90 m^2/g were attributed to the inner surface. No hysteresis was seen in the adsorption–desorption cycle; this result was expected for tubes that are open only at one end. (It should be noted that very recently, however, on especially prepared porous Si substrates, it has been shown that even tubes with one end open show adsorption hysteresis [60].)

Mackie et al. [17], in a set of volumetric measurements performed at 77.3 K, studied methane adsorption/desorption cycles on catalytically produced multiwalled nanotubes. Studies were performed on two groups of tubes: one of these groups of tubes had interior diameters ranging between 10 and 100 nm (with an estimated 70% falling between 40 and 70 nm), and the other group consisted of larger tubes, with interior diameters between 1 and 2 μm. The narrower tubes were split into three subgroups that received different treatments before being used as a substrate in the adsorption studies. One subset was treated with nitric acid, another was subjected to high temperatures (2400 °C) under vacuum, and the final subset was used in an as-produced fashion. This study focused on three things: determining the wetting characteristics of the methane films on the tubes, ascertaining the presence or absence of hysteresis loops indicative of capillary condensation on adsorption–desorption cycles for these substrates, and determining the specific areas of the different sets of tubes. The study found that solid methane forms very thin, incompletely wetting films on the nanotubes. By contrast, thick solid methane films, of at least 8–11 layers in thickness, can be formed on planar graphite before reaching saturation [61, 62]. This difference in behavior is due to the additional strain on the film growing on the tubes, which results in a reduction of the number of layer that are present at saturation. Hysteresis loops were present in the chemically treated tubes (which were supposed to have both of their ends uncapped), while they were not observed on the as-produced tubes. The specific surface areas for the untreated tubes were on the order of 17 m^2/g, while they were 23 m^2/g for the acid-treated tubes. On the vacuum-heated tubes (which have more highly graphitized outer surfaces), a number of steps were obtained in the adsorption data, indicative of layer-by-layer adsorption. The pressures at which these steps were observed, relative to the saturated vapor pressure, were similar to those observed for planar graphite (not a very surprising result, in view of the relatively large thickness of the walls of the tubes used in these measurements).

Bougrine et al. [18], who studied the adsorption of Kr films on catalytically produced multiwalled tubes, conducted a somewhat similar investigation. The majority of the tubes used in this investigation were closed. The diameters of the tubes fell into three groups within the same batch: around 10 nm, between 20 and 30 nm, and around 50 nm. The smaller diameter tubes had walls that were more uniform than those in the larger diameter tubes. The diameter of the hollow center of the tube was observed to be inversely correlated to the outer diameter of the tube. This study focused on determining the value of the relative pressure at which the first and second steps appeared on the tubes, and compared these features to the corresponding ones on planar graphite. The steps occurred at higher pressures on the nanotubes than the corresponding first- and second-layer steps for Kr on planar graphite. The authors explain this behavior as resulting from the curvature exhibited by the nanotubes. Only two separate steps were seen in the adsorption data before the saturated vapor pressure was reached. This indicates that the solid film incompletely wets the tubes, in contrast to the behavior on planar graphite, which is completely wetted by solid Kr [62]. No hysteresis was observed between the adsorption and desorption curves. The specific area estimated for the tubes used in this study was 105 m^2/g.

Hilding et al. [19] investigated the adsorption of butane on multiwalled nanotubes by performing gravimetric isotherms at room temperature. The tubes used were grown through chemical vapor deposition, which resulted in the formation of an aligned mat with very few contact points between nanotubes. TEM studies were used to determine that, in many cases, one end of the tubes was open. Three batches of tubes were produced, all having diameters for the hollow interior of the nanotubes on the order of 2.6 nm, but having outer tube diameters of about 24, 49, and 53 nm. Adsorption took place primarily on the external surface of the tubes, with tubes with smaller outer diameters having greater specific adsorption capacities. The smaller tubes were determined to have a surface area of 38 m^2/g, the next smaller tubes 14.8 m^2/g, and the largest diameter tubes 11.2 m^2/g. A small amount of adsorption (corresponding to between 0.97 and 0.07 m^2/g) occurred in the hollow interior of the tubes. No hysteresis loops were observed in adsorption/desorption cycles for these samples.

Yang et al. [20] investigated adsorption of N_2 on pristine and purified, catalytically grown carbon nanotubes. As a result of the purification process, some of the tips of the nanotubes appear, upon TEM inspection, to have been opened. Adsorption isotherms were conducted volumetrically at 77.3 K. A hysteresis loop was observed in the adsorption–desorption cycles for the purified tubes, while measurements made on pristine samples show no sign of hysteresis. The largest fraction of the hysteresis loop observed on the purified tubes (close to 80%) was attributed to aggregated pores, that is, condensation in the space between tubes, while a much smaller fraction of the capillarity observed was attributed to adsorption in the space at the interior of the nanotubes, which were uncapped as a result of the purification process.

Very recently, Masenelli-Verlot et al. [21] reported on the layering behavior of ethylene films adsorbed on multiwalled carbon nanotubes. From measurements performed between 90 and 100.5 K, these authors infer the existence of a layering transition, for the second layer of ethylene, at a temperature in the vicinity of 82 K. The layering transition temperature for the *n*th layer of an adsorbed film is the temperature at which the number of layers present in the film in equilibrium at the saturated vapor pressure (i.e., in coexistence with the bulk adsorbate) increases from $n - 1$ to n. The layering transition temperature for the second layer of ethylene on planar graphite is 78.9 K [63]. The authors also report that the value of the layering temperature increases for decreasing tube diameters. In other words, there is a temperature interval over which, at the same temperature, thinner solid ethylene films grow on nanotube substrates than on planar graphite. This effect is more pronounced, the narrower the tubes used.

A summary of adsorption results for various adsorbates on multiwalled carbon nanotubes is shown in Table 1.

3.2. Adsorption on Single-Wall Nanotubes

After methods were developed for the production of macroscopic amounts of single-walled carbon nanotubes [56, 57], adsorption studies on these systems became possible. While theoretical activity had provided some initial stimulus for these experimental investigations (specifically, calculations for the adsorption of helium and hydrogen on carbon nanotube bundles) [66], there is little doubt that a great deal of the impetus for performing adsorption studies came from the very remarkable results from Heben's group regarding hydrogen adsorption [12]. Using thermal-programmed desorption on a sample that contained approximately 0.2% of SWNTs, Heben's group reported that this new carbon material was able to adsorb relatively large amounts of hydrogen at room temperature [12]. This study estimated a value for the binding energy for hydrogen on the SWNTs of 19.6 kJ/mol, and it estimated the H_2 uptake by the SWNTs to be somewhere between 5 and 10% by weight. These

Table 1. Summary of adsorption results for various gases on multiwalled carbon nanotubes.

Adsorbate	Production/treatment	Tube diameter (nm)	Temp. (K)	Specific surface area (m^2/g)	Ref.
N_2	Arc discharge/as produced		77.3	21.0	[14]
	Arc discharge/CO_2 at 850 °C		77.3	37.0	[14]
	Manufactured by Hyperion/heated in vacuum at 373 K for 2 h	4 (interior) 3 (wall thickness)	77.3	268 (total) 90 (tube interior)	[16]
Kr	Catalytic/heated in vacuum to 773 for 6 h	10–50	77.3	105	[18]
CH_4	Catalytic/evacuated at room temp. for 12 h	10–100	77.3	17	[17]
	Catalytic/HNO_3 treated, then evacuated at room temp. for 12 h	10–100	77.3	23	[17]
Butane	CVD grown/heating 473.15 K at atmospheric pressure	2.6 (interior)/24 (exterior)	298.15	37.9	[19]
	CVD grown/heating 473.15 K at atmospheric pressure	2.6 (interior)/49 (exterior)	298.15	14.8	[19]
	CVD grown/heating 473.15 K at atmospheric pressure	2.6 (interior)/53 (exterior)	298.15	11.2	[19]

Note: The columns include a listing of the adsorbate used, a brief description of the method of production of the nanotubes and the treatment to which they were subjected, the diameters of the tubes, the range of temperatures at which the measurements were conducted, the specific surface areas, and the source.

results opened the possibility that this novel carbon material could be used for the room-temperature, low-pressure storage of H_2. A number of other studies of H_2 storage and H_2 adsorption followed this initial, exciting report [1–3]. Some of the more recent reports are not able to confirm the initial claims of very high H_2 storage capacity by the SWNTs (see, for example, [3]).

Single-walled nanotubes form bundles [56, 57]. The van der Waals interaction among the individual tubes causes them to order in a triangular array. For (10, 10) tubes, which have a diameter of 1.37 nm, the intertube separation is 1.7 nm. Typically, bundles are formed by between 40 and 200 tubes [56, 57]. The formation of bundles has important consequences for the adsorption characteristics of these materials.

Three different groups of sites for adsorption have been predicted to exist for bundles formed by close-ended, single-walled carbon nanotubes [11] (see Fig. 6). These are: the sites present on the surface of individual nanotubes which sit on the perimeter or periphery of the bundle, which we will call outer surface sites; the sites present in the convex valleys formed by the space in which two of these tubes on the periphery of the bundle come together, called groove sites or grooves; and, finally, the open spaces at the interior of the bundles which are left by three tubes coming as close together as they can, called the interstitial channels or ICs. In case the nanotubes are open or cut, in addition to these three groups of sites, the hollow interior of the open tubes will also be potentially available for adsorption.

On close-ended tubes, the outer surface sites and the grooves are always accessible for adsorption. Whether the interstitial channels are or are not available for adsorption will depend on the relative sizes of the adsorbate species and of the channels [11]. A purely geometrical calculation for (10, 10) nanotubes which considers the tube diameters and the intertube separation gives a diameter of 5.2 Å for the interstitial channels. This simple calculation, however, grossly overestimates the actual space available for the ICs because it neglects the effect of considering the electronic clouds present on each of the tubes. When a reasonable guess is made regarding the size of this electronic cloud, the diameter of the interstitial channels becomes approximately 2.3 Å [26]. A more sophisticated analysis, one which takes into consideration both the attractive and repulsive portions of the adsorption potential [11], concludes that only the smallest adsorbates (helium, neon, and hydrogen) can adsorb in the interstitial channels. For the first two of these adsorbates, in fact, the adsorption potential on the interstitials is most favorable, while it is less so for hydrogen; all other adsorbates are too large to fit in. Experimentally, the question whether adsorption occurs in the interstitial channels for the smallest adsorbates is one that has not yet been completely settled [25, 31, 32].

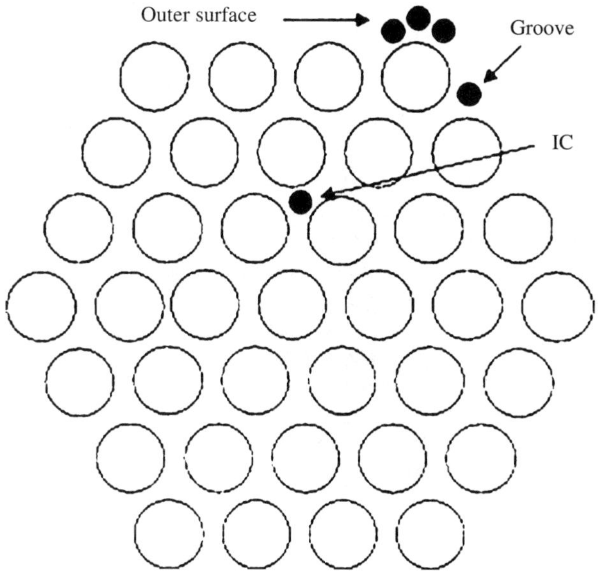

Figure 6. Schematic drawing depicting the three groups of sites that are potentially available for adsorption on bundles of closed-ended single-wall carbon nanotubes. The groove sites and the outer surface sites are available for adsorption by all gases, regardless of the size of the adsorbate. Whether adsorption occurs or not in the interstitial sites is currently a matter actively being investigated. Reprinted with permission from [26], A. J. Zambano et al., *Phys. Rev. B* 64, 075415 (2001). © 2001, American Physical Society.

Adsorption studies on SWNTs have dealt with a variety of questions. Some of these include: determining which of the various sets of sites present in the nanotube bundles are actually available for adsorption by a given adsorbate; establishing the values of the binding energies, or of the isosteric heats of adsorption, of different species adsorbed on the SWNT bundles; and identifying which new phases, if any, are present in the adsorbed films. Many studies have also determined the area available for adsorption on specific samples of SWNTs as part of their characterization of the substrates used.

3.2.1. Adsorption on As-Produced and Purified SWNT Bundles

Eswaramoorthy et al. [22] investigated the adsorption of N_2, benzene, and methanol on as-produced arc discharge SWNTs, as well as on SWNTs treated with hydrochloric and with nitric acids. No characterization for the samples was provided (e.g., no value for the estimated purity was given), and no details were provided on the adsorption measurement process. The report dealt with overall features of adsorption, such as the determination of the specific surface areas of the samples (which presumably was obtained by using the BET equation) and the determination of the pore surface area (the latter was determined by means of alpha plots with the comparison substrate being nonporous carbon black). Large microporosity was found in all cases. The Horvath–Kawazoe (HK) equation [65] was applied to the data in order to obtain an estimate of the pore size distribution, and a maximum was found at the expected tube diameter of about 1.0 nm in all cases, indicating that, according to this analysis, even a large number of the as-produced tubes are open. Adsorption/desorption cycles were investigated for the as-produced sample, and a hysteresis loop was found.

From a temperature-programmed desorption investigation of helium, Hallock and his collaborators determined the value of the binding energy of this rare gas on SWNT bundles at low temperatures to be 340 K [6]. This value was in very good agreement with calculations for helium adsorbed in the interstitial channels of SWNTs, which predicted a value for this adsorbate [11, 66]. The numerical agreement between the measured binding energy and the theoretically calculated binding energy for the interstitial channels, together with the large amount of gas desorbed, led the authors to conclude that the interstitial channels were the sites on which helium was adsorbing. Since the atoms adsorbed on the interstitial channels form a long column that is 1 atom thick, the phase was identified as one dimensional. In a subsequent paper, the authors revised downwards the value of the binding energy measured by TPD to 210 K, as a result of a correction of their temperature scale [7]. However, since the amount of gas desorbed was unaffected by the correction, they still interpreted their data in terms of adsorption in the ICs. A recent set of measurements, and a new analysis by the same group, found that the TPD data for helium indicate that the binding energy of this gas on SWNTs is coverage dependent [8]. The binding energy decreases for increasing coverage until a plateau region is reached, in which this quantity remains constant with coverage. This more recent report (Fig. 2) tentatively concluded that, for the higher coverage regime explored in the study, adsorption occurred in the grooves, and that the much higher binding energy sites observed for the lower coverages investigated may originate from surface inhomogeneity or corrugation.

Weber et al. [23] investigated the binding energy of methane on a sample of unpurified single-walled nanotubes produced in Prof. P. Bernier's laboratory by the arc-discharge method. The binding energy was determined from the isosteric heat of adsorption measurements. In turn, the isosteric heat values were obtained from a series of adsorption isotherms performed at different temperatures, following the procedure described in Section 2 (see Fig. 1). The adsorption isotherms were performed at relatively high temperatures and at very low coverages (only coverages of less than 8% of one full layer were used in the analysis). Thus, in this experiment, only the binding energy of the highest energy sites was determined. The measured value was 1.76 times larger than the corresponding binding energy for methane on planar graphite [67]. This study did not univocally identify the sites on which adsorption occurred. Rather, it provided two exclusive possible alternative interpretations for the data. The authors noted that methane was larger than neon, helium, or hydrogen, gases for which very favorable adsorption in the interstitial channels had been calculated [11, 66]. For He and Ne, the values of the binding energy were greater by about 150% than the corresponding values on planar graphite [11]. They suggested that one possibility was that the higher binding energy sites for methane could be a set of adsorption sites on the ICs, on which methane would fit less favorably than either helium or neon (and, hence, would have a lower relative value of the binding energy increase than these other gases). The alternative possibility suggested by these authors was that methane was too big to fit in the interstitial channels, and that the higher energy sites corresponded to adsorption on the grooves.

Muris et al. [24] reported on the adsorption of methane (and krypton) on single-walled carbon nanotubes. The sample used in their measurements was unpurified nanotubes produced by the arc-discharge method, also from Bernier's laboratory. They reported on volumetric adsorption isotherm measurements and microcalorimetric results. The adsorption isotherm data corresponding to the formation of the first layer on the carbon nanotube bundle surface displayed two well-defined, but broad, substeps. This indicated the presence of two groups of sites, each with a fairly uniform binding energy, in the first layer. From adsorption isotherm measurements performed at different temperatures, they extracted the two values of the first-layer binding energies. The higher binding energy value, determined from a coverage corresponding to the midpoint of the lower pressure substep, was approximately 1.2 times larger than the binding energy for methane adsorbed on planar graphite. On the other hand, the binding energy of the lower energy sites was lower than that for the first layer of methane on graphite, but larger than that for the second layer of methane on graphite [68]. The values derived from the adsorption isotherm measurements were confirmed by microcalorimetric determinations performed at the same coverage. The lower pressure step was identified as resulting from adsorption on the interstitial channels, while the higher pressure (lower binding energy) step was identified as adsorption occurring on the outer surface sites. The possibility of adsorption on the grooves was not considered in this study.

Bienfait's group reported on the mobility of the methane films adsorbed on single-wall nanotube bundles [42]. They used quasielastic neutron scattering to determine the diffusion coefficients of the adsorbed gas (Fig. 4). The results indicated that, on the more strongly binding sites, the adsorbed film was solid-like for temperatures below 120 K. For the less strongly binding sites, a very viscous liquid-like behavior was observed for temperatures between 70 and 120 K. At 50 K, the film in the lower energy binding sites was solid.

Talapatra et al. [25] reported on the binding energy determinations for the highest energy binding sites for neon, methane, and xenon on SWNT bundles. The measurements were performed on the same sample of SWNTs. The binding energy of the highest energy sites was determined from isosteric heat measurements performed at very low coverages. In all three cases, the values of the binding energies were higher than those found on planar graphite by a factor of about 1.75. Since the amount of the increase was the same in all three cases, the authors concluded that the different gases were all occupying the same type of sites. Theoretical calculations provide the values of the binding energies for these gases for the three different groups of sites: ICs, outer surface sites, and grooves [11]. For adsorption on the interstitial channels, the changes in the binding energies relative to the values on exfoliated graphite are very different for neon, methane, and xenon. On the other hand, the changes on the grooves, relative to planar graphite, are very similar for these three gases, and in all cases, they are greater than those for graphite. Additionally, the size of the Xe atom is too large to allow it to fit in the ICs. Taking these different factors into account, the study concluded that none of the gases studied was able to penetrate in the ICs, and that the high energy-binding sites present on the nanotube bundles were on the grooves. The same conclusion was reached from specific surface area measurements, which found essentially the same values of the specific surface area for the same sample of SWNTs using Ne and Xe as the adsorbates. It should be noted, however, that in [25], incorrect values for the monolayer capacities were used in these determinations, leading to incorrect estimates for the surface area. When the correct values for monolayer completion are used, the specific surface areas are found to be 161.1 m^2/g using Xe and 173.1 m^2/g using Ne. Since these results are very similar, the validity of the argument made in [25], leading to the conclusion that the ICs are not accessible for adsorption by any one of the three gases studied, remains unaffected.

Very similar binding energy increases to those reported by Talapatra et al. [25] were found in a temperature-programmed desorption study performed by Hertel et al. [36], who investigated thermal desorption spectra from methane, Xe, and SF_6 (Fig. 5). The nanotube sample used in this study came from tubes@rice. A very interesting aspect of this study is that the treatment to which the nanotubes were subjected to was significantly different from that used by Talapatra et al. [25]. While in the Talapatra study [25] the as-produced sample was just subjected to extended pumping at room temperature, the sample investigated by Hertel's group was subjected to heating to 1200 K in a Ta foil in an UHV chamber, and cycled repeatedly to anneal it [36]. The study determined that, for methane, xenon, and SF_6, adsorption at lower coverages occurs principally in the groove sites. From the peak desorption temperatures, they concluded that the value of the binding energies for these gases on the groove sites were between 190 and 160% greater than on the surface of HOPG for methane, and between 170 and 130% greater than on the surface of HOPG for Xe. The ranges of values quoted for each species correspond to the range of coverages studied. The higher percentage increases correspond to the lower coverages studied, while the lower percentage increases correspond to the higher coverages studied for each gas.

Zambano et al. [26] reported on adsorption isotherm results for xenon on SWNT bundles. The measurements for this study were conducted on arc-discharge, as-produced tubes produced in Prof. P. Bernier's laboratory by Journet. The study involved the performance of complete sets of monolayer adsorption isotherms at a variety of temperatures. As had been the case for the report by Muris et al. [24] for methane adsorbed on SWNTs, two substeps were found in the adsorption data. The top of the second substep was identified as monolayer completion [26] (as had been done by Muris et al.) [24]. In order to identify the nature of the adsorption sites responsible for the lower pressure substep, the authors performed a geometrical calculation. They estimated the number of Xe atoms that would adsorb on the outer surface sites and on the grooves; they then considered two cases: that the ICs were available for adsorption, and that they were not available. Comparison of the experimental results with these calculations led to the conclusion that only adsorption on the grooves and the outer surface sites was needed to explain the results, and that, thus, no adsorption occurred in the ICs. A comparison was also made between the values of the isosteric heat of adsorption and binding energy for Xe determined experimentally and theoretically calculated values for the binding energy on the grooves [11]. Very good agreement was found between the two values. As a result of comparing both the geometric calculations and the theoretically calculated value of the binding energy on the grooves with experiment, the authors concluded that the lower pressure substep corresponded to adsorption on the grooves.

Talapatra and Migone [27] investigated in some detail the adsorption of methane on closed-ended SWNTs. Their results largely confirmed those of Muris et al. [24]: two substeps were observed in first-layer adsorption isotherms, corresponding to the presence of two groups of adsorption sites with different binding energies, one greater and the other smaller than those for planar graphite. The interpretation of these results, however, differed from that of Muris et al. [24] as the more strongly binding sites were identified in this study as corresponding to the grooves. Both studies agreed in the identification of the weaker sites as corresponding to the outer surface sites. Talapatra and Migone [27], were able to explain why previous studies had reported different increases in the strength of the binding of methane on SWNTs bundles relative to that on graphite: (Talapatra et al. [25] had found that the binding energy of methane on the most strongly binding sites on the bundles was 1.75 times that found on graphite; while Muris et al. [24] had found an increase of only 1.2 times in the value of the isosteric heat of adsorption relative to that on graphite). They noted that the values reported by Muris et al. [24] were measured at higher fractional coverages on the nanotube surface than those used by Talapatra et al. [25] in their determinations, and that the isosteric heat was a strongly dependent function of the coverage at low coverages on the SWNT substrates [27].

Muris et al. [31] performed a comparative adsorption study of Xe, SF_6, and CF_4, adsorbed on nanotubes produced by arc discharge. They combined their results for these gases with those already available, from measurements performed by the same group [24], for CH_4 to determine the location of the sites on which these different gases adsorb on the bundles of single-walled nanotubes. Based on the size of the substeps present in the data for the first layer for the different gases (Fig. 7), the authors concluded that Xe, SF_6, and CF_4 adsorb only on the grooves and outer surface sites, and that they do not adsorb in the interstitial channels. In the case of methane, the adsorption data from these authors display a relatively larger lower pressure substep than that which is present for Xe (and for the other gases studied). The authors concluded that the larger low-pressure adsorption substep present for methane is due to adsorption occurring on some of the largest interstitial channel sites present in this substrate, in addition to the adsorption occurring on grooves for this adsorbate.

Fujiwara et al. [32] performed adsorption isotherm and X-ray diffraction measurements on as-grown nanotube samples, as well as on samples that had been heat treated in air at 350 °C (the latter were estimated to be open ended). They found that the specific surface area increased from about 245 m^2/g on the as-grown samples to about 515 m^2/g for the samples heat treated in air, indicating that there were additional sites available in the heat-treated sample. The authors looked at one single diffraction peak for the as-produced samples, and found that the peak shifted to a

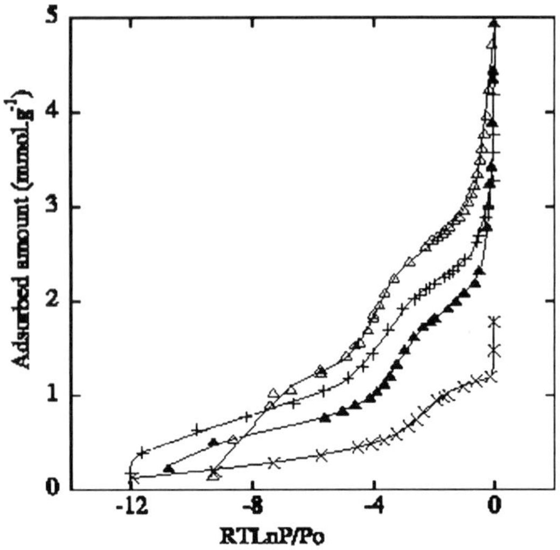

Figure 7. Adsorption isotherms for (from left) methane at 78 K, xenon at 110.6 K, CF4 at 96 K, and SF6 at 142.6 K on single-walled carbon nanotubes. Reprinted with permission from [31], M. Muris et al., *Surf. Sci.* 492, 67 (2001). © 2001, Elsevier Science.

lower Q (Å^{-1}) value after adsorption. In contrast to all other experimental and theoretical results for adsorption on nanotubes, these authors interpreted their adsorption and X-ray results for nitrogen and oxygen on closed-ended nanotubes exclusively in terms of adsorption in the interstitial channels. Thus, these authors entirely neglected to consider adsorption on either the grooves or the outer surface sites that are present and available in the bundles [32]. In their results for the heat-treated samples, these authors interpret the data as indicating adsorption in the interstitial channels and on the interior of the tubes, once again neglecting the outer surface sites and the grooves. As the two groups of sites that these authors neglected to consider in their interpretation (i.e., the grooves and the outer sites) are always present and available for adsorption by adsorbates of the molecular sizes used here, and as the area measured in the closed tubes can be very effectively explained in terms of adsorption occurring only on grooves and outer sites (and, for the heat-treated tubes, in terms of adsorption occurring only on grooves, outer sites, and the inside of the open tubes), the interpretation put forth by Fujiwara et al. [32] has to be viewed with a great deal of skepticism.

Bienfait's group investigated the adsorption of methane and deuterium on bundles of as-grown SWNTs using neutron scattering [42]. They observed, for the case of CD_4, a 5% shift toward smaller scattering vectors for the 0.4 Å^{-1} peak, indicating an apparent expansion of the lattice. They offer two alternative explanations for this shift, both of which can account for this experimental feature. The first one is that some CD_4 is adsorbing in the interstitial channels, and that this adsorption results in a deformation (expansion) of the nanotube lattice. The second explanation is that the lattice expansion is only apparent; it is the result of CD_4 adsorbing on the grooves, at a position that is displaced by 1 Å outwardly above the lattice sites, that is, 1 Å above the surface of the bundle, thus appearing as an expansion of the lattice. Combining the results of their diffraction experiments with prior adsorption isotherm measurements, they proposed a scenario in which the formation of the first layer of CD_4 on SWNT bundles starts with the adsorbate filling the groove sites and the widest available ICs (these two groups are the highest binding energy sites present) [42]. After these sites are filled, monolayer completion proceeds by filling the less attractive sites in the first layer, that is, the outer surface sites.

3.2.2. Adsorbate Size Effects on Phase Formation

The existence of new phases in the systems formed by gases adsorbed on closed-ended SWNTs has been explored in an adsorption isotherm study by Talapatra and Migone [9]. This study focused on first-layer features of the adsorption of Xe and Ar films on the nanotubes. It provided a detailed comparison with results obtained in computer simulations of adsorption isotherms for these same systems [5]. The simulations predicted that monolayer completion for these gases proceeds in a series of stages [5]. The number and characteristics of the stages vary from one adsorbate to another, depending on the size of the adsorbate. For the case of Xe, simulations predict that adsorption takes place in two stages. In the first stage, formation of the first layer begins with the filling of the grooves, at the lowest pressures. The adsorbate forms 1 atom thick long lines of atoms on the grooves; in the simulation study, these lines were termed the "one-channel" phase [5]. After the grooves are filled, monolayer completion follows with the filling up of the outer surface sites, with what in the computer simulation study was called the "five-channel" phase. The adsorption isotherms obtained in the simulations [see Fig. 8(a)] closely match the features obtained by Talapatra and Migone in their experimental work [some of the experimental results are displayed in Fig. 8(b)] [9]. To provide a quantitative comparison, the pressures at which the two first-layer features occurred in the simulations were plotted in a $\log P$ versus $1/T$ curve, as were the pressures determined from the experiment for the two first-layer substeps [9]. The experimental points fell on the same line as those from the simulations, indicating that the agreement between them was excellent [see Fig. 8(c)]. This permitted the identification of the lower pressure substep in the experiments as a one-dimensional phase formed by adsorption on the grooves. For the case of Ar, the simulations predict that monolayer completion proceeds in three stages [5]. The first one is the formation of the one-channel phase on the grooves. This phase is followed, at higher pressures, by the formation of two additional lines of atoms, one on each side of the atoms present on the grooves, in what was called the "three channel" phase in the simulations. Finally, at pressures only slightly higher than those needed to form the three-channel phase, monolayer completion occurs in what was identified as a "six-channel" phase. Because the difference in chemical potential between the three-channel and six-channel phases is not too large, and because the surface offered by the SWNT substrates is not ideal and exhibits heterogeneities that broaden and blur adsorption features, it is difficult to extract evidence for the existence of the three-channel phase from the adsorption data. In order to search for evidence of the existence of the three-channel phase in the experimental data, the isothermal

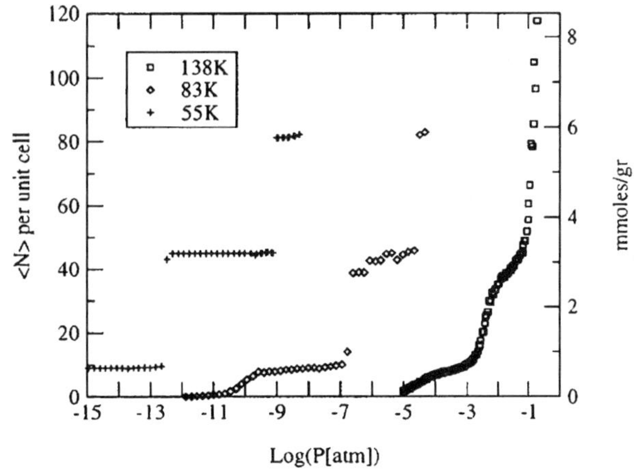

(a)

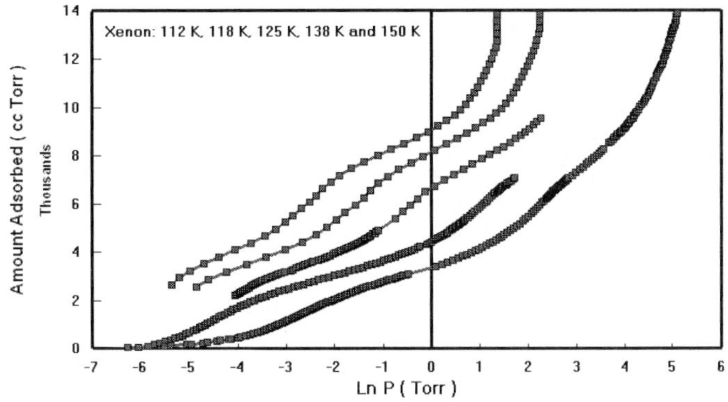

(b)

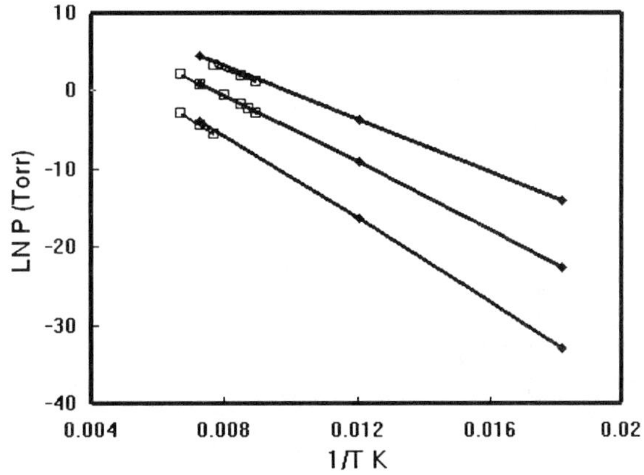

(c)

Figure 8. (a) Grand canonical Monte Carlo simulations for adsorption isotherms for Xe on the outside surface of closed-ended single-wall carbon nanotubes. The amount adsorbed, in arbitrary units, is presented as a function of the logarithm of the pressure (in atmospheres). One full layer on the surface of the bundle corresponds to a coverage of approximately 50 in these units. The temperatures shown are, from left to right, 55, 83, and 138 K. The lower pressure substep (0–10 in coverage) corresponds to the filling of the grooves, and the second substep (10–50 in coverage) to the filling of the outer surface sites. Reprinted with permission from [5], M. M. Calbi et al., *J. Chem. Phys.* 115, 9975 (2001). © 2001, American Institute of Physics. (b) Adsorption isotherms measured for Xe on the outer surface of bundles of closed-ended carbon nanotubes. The temperatures displayed are, from left to right, 112.07, 118.06, 125.07, 138.05, and 150.06 K. The coverage (vertical axis) is in $cm^3 \cdot$ torr; the logarithm of the pressure (in torr) is given in the horizontal axis. The lowest coverage substep corresponds to adsorption on the grooves, the middle coverage substep corresponds to adsorption on the outer surface, and the highest coverage substep is the formation of a second layer on the outside surface of the bundles. Reprinted with permission from [9], S. Talapatra and A. D. Migone, *Phys. Rev. Lett.* 87, 206106 (2001). © 2001, American Physical Society. (c) Logarithm of the pressure at the midpoint of the adsorption substeps (corresponding to filling the grooves, the outer surface sites, and the second layer) for Xe on closed-ended single-wall carbon nanotube bundles, plotted as a function of the inverse of the temperature. Filled symbols are the theoretical results [see (a)], while the closed symbols are the experimental data points [see (b)]. Excellent agreement is found between the two sets of points. Reprinted with permission from [9], S. Talapatra and A. D. Migone, *Phys. Rev. Lett.* 87, 206106 (2001). © 2001, American Physical Society.

compressibility of the adsorbed films was obtained [9]. Computing the isothermal compressibility of a film essentially involves performing a derivative of the adsorption data. Features that are hard to resolve in the adsorption data appear with greater clarity in the compressibility (in particular, a step or a substep in the isotherm will appear as a peak in the isothermal compressibility). For Ar, the higher pressure substep in the isotherm that results from all of the adsorption that occurs in the monolayer after the filling of the grooves gives rise to two peaks in the isothermal compressibility [9]. The lower pressure peak in this structure was identified as corresponding to the three-channel phase. The pressures at which the one-channel, three-channel, and six-channel phases were observed in the simulations were plotted in a $\log P$ versus $1/T$ curve, as were the experimental results for the pressures of the lower coverage substep, and those for the two peaks in the isothermal compressibility present at higher coverages. Again, very good agreement was found between experiments and simulations [9]. Thus, by conducting a detailed comparison with computer simulation results, this adsorption isotherm study was able to identify one-dimensional phases for Ar and Xe (the one-channel phases) and a new phase (the three-channel phase) for Ar adsorbed on the outer surfaces of closed-ended SWNTs.

3.2.3. Coverage Dependence on Isosteric Heats of Adsorption

A number of quantum [28, 29] and classical [28] adsorbates have been explored by adsorption isotherm measurements (and, in the case of helium, also by heat capacity) [45] by Vilches' group. They studied adsorption isotherms on arc-discharge produced nanotubes for H_2, D_2, N_2, Ar [28] and, more recently, for ^4He [29]. The monolayer data investigated for H_2, D_2, N_2, and Ar display the characteristic two substeps that have been observed by other authors for several other adsorbates. The low-pressure substep (high energy-binding sites) was attributed to adsorption on the grooves and on the widest interstitial channels present in the bundles, and the higher pressure substep (lower binding-energy sites) to adsorption on the outer surface sites. The isosteric heat of adsorption was determined for a wide range of coverages in the first layer for H_2, D_2, N_2, and Ar [28]. This quantity is found to be higher for lower coverages, and to reach a plateau as coverage increases (Fig. 9); the plateau starts roughly at coverage of one third of a complete monolayer. The behavior of the isosteric heat is consistent with that seen in other experiments for this quantity. The helium adsorption isotherm measurements were conducted between 2.1 and 12 K. This study was able to resolve, for the first time for any adsorbed system on SWNT bundles, up to three distinct layers in the data measured at the lower temperatures [29]. The isosteric heat of adsorption was calculated from the lowest coverages in the first layer up to the third layer. The first-layer isosteric heat increases as the coverages decreases, and then it plateaus. The values of the isosteric heat are consistent with an interpretation in which the groove sites get filled first at the lower coverages, and then the external surface sites get filled as monolayer completion is approached [29]. There are two additional plateaus, at even lower values of the isosteric heat, as the coverage

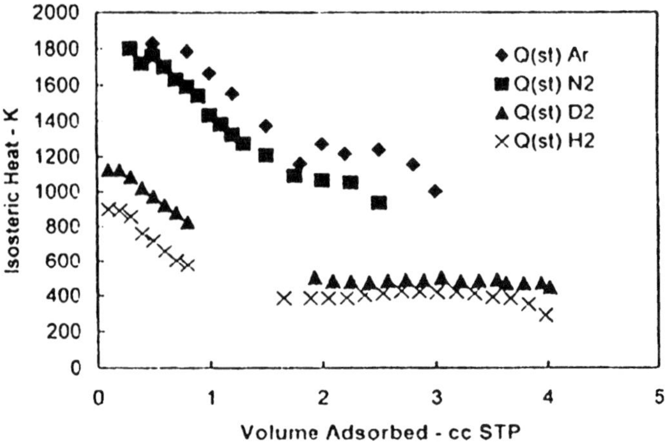

Figure 9. Coverage dependence of the isosteric heat for N_2, Ar, D_2, and H_2 adsorbed on SWNT bundles. Reprinted with permission from [28], T. Wilson et al., *J. Low Temp. Phys.* 126, 403 (2002). © 2002, Kluwer Academic/Plenum.

goes through the second and third layers. The heat capacity measurements for helium films on the SWNTs are currently in progress [45].

The adsorption of N_2 on SWNTs has been explored by Kim's group with adsorption isotherms [30]. Their results indicate that the isosteric heat is higher for lower coverages, and that this quantity decreases with increasing coverage, in a manner that is entirely consistent with the observations of other experimental groups. These measurements were performed on pulsed-laser-produced nanotubes that were purchased from tubes@rice (this company's present name is Carbon Nanotechnologies Inc. or CNI). Interestingly, the two substeps that are present in the arc-discharge tubes investigated by other groups are not present in these data. This suggests that there is a possibility that nanotube bundles produced by different methods may result in significantly different adsorption characteristics. The question of the dependence of the adsorption characteristics on the production and treatment of the nanotube substrates used in adsorption measurements is one which has not yet been addressed in a systematic fashion.

3.2.4. Adsorption Results on Chemically and Mechanically Opened SWNTs

Stepanek et al. [33] investigated the effects on adsorption of opening the ends of the tubes, and making the interior space of the nanotubes available to the adsorbate species. These authors employed a mechanical process, involving the use of abrasive diamond powder, to cut open single-wall nanotubes. This mechanical approach obviated the need for the relatively harsh chemical treatments that are the norm in opening processes for tubes. The nanotubes preserved a well-organized structure after the mechanical cutting. Evidence that the tubes were open was obtained from high-resolution transmission electron micrographs, and evidence that the process did not disturb the order in the tubes was obtained from X-ray diffraction measurements. Adsorption measurements were performed before and after the cutting treatment. The adsorption results were analyzed in terms of the Horvath–Kawazoe equation [65]; this analysis revealed significant differences in the data taken before and after the treatment. In particular, the HK analysis showed evidence of the presence of nanoporosity, in the appropriate range for open tubes (1–2 nm), only for the nanotubes that were cut open mechanically [33].

Two recent studies have investigated the effects of purification on the pore structure and adsorption characteristics of HiPCO single-walled carbon nanotube bundles [34, 35]. Samples from two different batches of as-produced HiPCO SWNTs, manufactured by Carbon Nanotechnologies Inc., were investigated by Simard's group [34]. Studies on the samples were conducted before and after subjecting them to acid purification treatment. The samples were investigated with X-ray diffraction (for the verification of the ordered nature of the bundles), Raman spectroscopy (for independent confirmation of the tube diameters), and with argon and nitrogen adsorption isotherm measurements (at 87 and 77 K, respectively). The pore structure was estimated using the HK equation [65]. The results, both for the specific surface areas as well as for the HK pore distribution, on the samples from the two batches of as-produced samples were significantly different from one another. This indicates that the adsorption characteristics of the same type of material can vary significantly from batch to batch. The pore structure analysis of the low-pressure portion of the data obtained from the HK analysis for the untreated tubes yielded two peaks in the pore size distribution. These peaks were interpreted in terms of adsorption occurring on the grooves and on open tubes. The fraction of open tubes in the as-produced samples, as estimated from the HK approach, was observed to vary from an upper limit of 38% in one batch, down to 15% in another. After the purification treatment, the HK analysis yielded a smaller proportion of open tubes. This result was interpreted in terms of blocking of the tube openings by the functional groups that result from the chemical purification process [34].

HiPCO nanotubes produced by Carbon Nanotechnologies Inc. were also investigated by Kaneko's group [35]. The nanotubes were subjected to two different purification treatments (one involving attack by concentrated HCl for 48 h, and the other involving heating in air at 623 K, followed by immersion in concentrated hydrochloric acid for 48 h). These cleaning procedures have been shown to be effective in removing the metallic catalyst impurities present on HiPCO nanotube samples. The adsorption characteristics of the nanotube samples before and after the purification processes were investigated by means of volumetric N_2 adsorption isotherms at liquid nitrogen temperature. The pore structure of the samples was inferred from the determination of alpha plots on the materials investigated. There is clear evidence of the presence of hysteresis loops after subjecting the tubes to either one of the purification treatments, indicating the formation of mesopores as the result of the purification processes. The adsorption at low values of P/P_o is considerably enhanced for the tubes subjected to both air oxidation and HCl treatment, indicating the production of micropores on these samples, in addition to the mesopores produced by either treatment. The alpha-plot analysis yields an external surface area for pristine HiPCO tubes on the order of 436 m^2/g, which was attributed to the presence of amorphous carbon on the untreated material; the total external area of the tubes is estimated at 524 m^2/g. After treatment with just HCl, the total area increases to 587 m^2/g, while the external area is estimated at 364 m^2/g. After treatment with both oxidation in air and attack by HCl, the total area is considerably increased, to 861 m^2/g, while the external area in this case is estimated at 334 m^2/g [35].

The values of the pore-size distribution and the internal and external areas determined in the three previous sets of studies depend on the applicability of the Horvath–Kawazoe analysis [65] and on that of the alpha plots, respectively, as techniques for interpreting adsorption data for bundles of carbon nanotubes. The validity of these approaches is not automatically assured in these cases since there are uniform high binding energy surfaces on the nanotube bundles that are not the result of pores, and since there will be considerable adsorption occurring on them at low pressures. However, these three studies [33–35] make it very clear that the purification and tube-opening processes (either mechanical [33] or chemical [34, 35]) have a dramatic effect on the adsorption characteristics of the resulting material.

The effects of the functional groups that result from purification and opening treatments with nitric and sulfuric acids on the adsorption characteristics of the SWNT bundles has been investigated by Yates's group [10, 38]. Kuznetsova et al. studied the adsorption of Xe on both closed-ended and open-ended nanotubes by means of temperature-programmed desorption and other analytical techniques [10, 38]. The closed-ended nanotubes were purified by the nitric/sulphuric acid treatment. The open tubes were sonicated in a mixture of nitric and sulfuric acid, followed by treatment with oxygen peroxide/sulfuric acid. The resulting samples were placed inside an ultrahigh-vacuum apparatus, and heated *in-situ* to either 623 or 1073 K prior to exposure to Xe. A mass of 45 μg of tubes was used for both the open- and closed-ended tubes. After a heat treatment to 623 K for closed tubes, a small desorption peak was observed near 120 K in temperature-programmed desorption measurements for Xe. For the open tubes heated to 623 K, at least two temperature-programmed desorption peaks, and a much greater desorbed amount than for the closed tube case, were observed (Fig. 10). There was a factor of 12 larger amount of Xe adsorbed on the open tubes than on the closed ones. During the heating process, the evolution of CO, CO_2, H_2, and CH_4, was observed; this evolution started at about 500 K [10, 38]. These gases result from chemical decomposition processes occurring in the nanotubes. Similar gas evolution characteristics were observed for open and for closed nanotubes. Xe thermal desorption spectra were also measured after heat treatment to 1073 K. The intake of Xe on both closed and open tubes increased very considerably after the high-temperature heating relative to the intake of Xe exhibited for tubes heated to 623 K [10, 38]. The closed tubes show two weak desorption peaks at about 105 and 120 K. For the open tubes, the 120 K feature shifts to higher temperature as the initial Xe coverage of the sample is increased. These experiments showed that the chemical treatment with nitric and sulfuric acids leaves functional groups containing carbon, oxygen, and hydrogen on the nanotubes, and that these groups block the entry ports to the nanotube interior. The high-temperature heating treatment of the sample leads to the removal of the blocks on

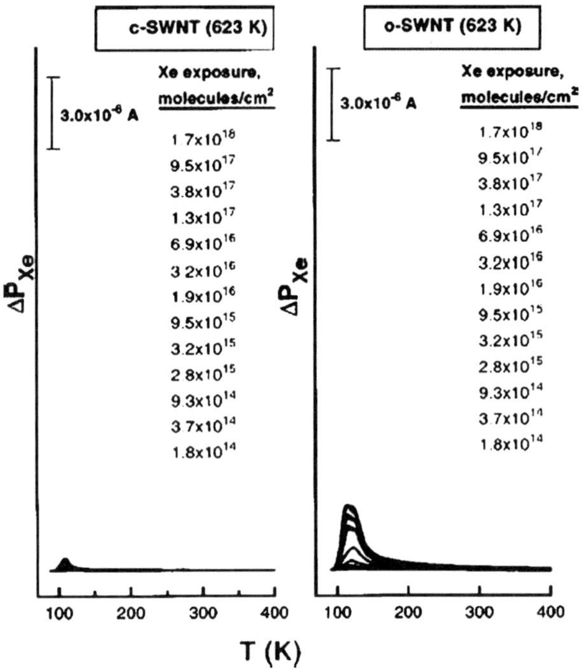

Figure 10. Comparison of programmed desorption rates for xenon on bundles of closed (c-SWNT) and open (o-SWNT) after activation at 623 K at a variety of different Xe exposures. Reprinted with permission from [10], A. Kuznetsova et al., *J. Chem. Phys.* 112, 9590 (2000). © 2000, American Institute of Physics.

the entry ports of the tubes, and hence results in increased adsorptive capacity. From the analysis of the desorption kinetics observed for Xe from the high-temperature treated open tubes, this study concluded that the Xe inside the tubes is quasi one- dimensional. The value for the binding energy measured at the interior of the tubes agrees well with calculated values for this quantity obtained by Cole's group [11].

The same group also investigated the effects of successive etching of purified SWNT with ozone on the adsorption of Xe gas [39]. A sample of purified nanotubes was heated in vacuum to 1073 K, exposed to a monitored amount of ozone, had the ozone removed, was heated to 973 K, and had Xe temperature-programmed desorption measurements performed. The sample was then reexposed to ozone, and the same sequence was repeated. The effect of ozone exposure on the tubes was initially to increase the Xe uptake. After a certain point, repeated exposure to ozone resulted in a decrease in the Xe uptake. The decrease became progressively worse as exposure to ozone continued. The comparison of the experimental results with simulations for this system led to the conclusion that there was an optimal etching of the nanotubes that corresponded to openings at the ends and at the side walls of the tubes on the order of 0.5–0.7 nm, and that etching beyond this point results in a decrease in the level of Xe uptake because further removal of C atoms from the tubes reduces the surface area available for adsorption [39].

A summary of adsorption results for various adsorbates on single-walled carbon nanotubes is shown in Table 2.

3.3. Adsorption Effects on Electrical Transport Properties of SWNTs

The effect of gas adsorption on the electrical properties of carbon nanotubes has been the object of considerable attention [47–53].

The possibility of using semiconducting SWNTs for chemical sensor applications has been explored by Kong et al. [47]. These studies were based on changes on the electrical resistivity of the semiconducting SWNTs upon exposure to gaseous NO_2 and NH_3, which were present at dilute concentrations (2–200 ppm for NO_2 and 0.1–1.0% concentrations for ammonia, respectively) in a carrier gas (air or Ar). The conductance of the nanotubes was observed to change by about three orders of magnitude upon exposure to NO_2, and by about two orders of magnitude for ammonia. The effects of the change were slowly reversed (over a 12 h period) upon continued exposure to the flowing inert gas or more quickly by heating.

Zettl's group reported that exposure of carbon nanotubes to air or oxygen resulted in dramatic changes in the electrical resistance and thermopower [48, 49]. The electrical resistance at room temperature changed between 10 and 15% as a result of cycling between an air atmosphere and vacuum. At 350 K in an initial atmosphere of pure oxygen, the thermopower was +20 μV/K. As the oxygen was removed from the space surrounding the tubes, the thermopower switched gradually to lower values, passing through 0, and eventually settling on an equilibrium value while under vacuum of −10 μV/K. This group also investigated the temperature dependence of the reversibility of these changes when the gas pressure surrounding the samples was changed. Switching between vacuum values and air-exposed values occurred very fast at 380 K. At 350 K, reverting back to the equilibrium values under both sets of conditions took a period of several minutes. At 300 K, these changes were much slower, taking periods of hundreds of hours. The explanation proposed for these observations was that oxygen dopes the tubes so that they go from *n* type in vacuum to *p* type when in air or oxygen [48]. The study also concluded that previous measurements of electrical properties of nanotubes that had not taken into account the effects of air exposure (and consequently, exposure to oxygen) would yield "air-doped" and not intrinsic nanotube values for the quantities measured [48].

The values of the electrical resistance of mats of SWNTs in vacuum and upon exposure to different adsorbates (including N_2, H_2, O_2, Ar, CO_2 , H_2O) were studied by Marliere and co-workers [50, 51]. For N_2, H_2, O_2, and Ar, an increase in the electrical resistivity was observed upon introduction of gas. For CO_2, after the initial increase in resistivity, there was a decrease in the resistivity measured. A decrease in the resistivity was also observed with the introduction of water. The interpretation was that water and carbon dioxide act as electron donors upon adsorption on the SWNT mat [50, 51].

The effect of gas adsorption on the thermolelectric power and on the electrical resistivity on SWNTs was investigated by Eklund's group [52]. The thermopower went from positive to negative as a sample prepared in air was degassed under vacuum at 500 K. The electrical resistivity decreased

Table 2. Summary of adsorption results for various gases on single-walled carbon nanotubes.

Adsorbate	Production/treatment	Temp. range	Isost. heat/ bind. energy	Specific surface area (m^2/g)	Ref.	Source
He	Laser ablation/purified by boiling in 70% HCl at 120 °C for 4 h and degassed at 500 °C and at ~5×10^{-6} torr for 4 h	7.94–33.6 K	274 K (G)		[8]	E
	Arc discharge/unpurified; baked under vacuum 5×10^{-7} torr at 500 K	2–14 K	240 K (G)	80	[29]	E
	Potential calculations		386 K (IC)		[11]	T
			270 K (G)		[11]	T
			166 K (OS)		[11]	T
Ne	Arc discharge/unpurified; evacuated to 1×10^{-6} torr at room temp.	37–58 K	603 K (G)		[25]	E
	Potential calculations		902 K (IC)		[11]	T
			666 K (G)		[11]	T
			398 K (OS)		[11]	T
H_2	Arc discharge/unpurified; evacuated below 5×10^{-8} torr at 970 K	100–500 K	2365 K (I/IC)		[12]	E
	Arc discharge/unpurified; baked under vacuum 5×10^{-7} torr at 500 K	77.3–91.3 K	*~900 K (G/IC)*		[28]	E
		20–37 K	*~499 K (OS)*		[28]	E
	Laser ablation/unpurified, heated to 350 K for 12 h at 10^{-3} torr	25–65 K	~724 K (G/IC)		[69]	E
	Model rope consisting of seven SWNTs	77 K	1443 K (IC)		[72]	S
			1088 K (G)		[72]	S
			758 K (I)		[72]	S
			603 K (OS)		[72]	S
	Potential calculations		292 K (IC)		[11]	T
	Potential calculations		618 K (G)		[11]	T
	Potential calculations		383 K (OS)		[11]	T
D_2	Arc discharge/unpurified; baked under vacuum 5×10^{-7} torr at 500 K	~85 K	*~1150 K (G/IC)*		[28]	E
		~33 K	*~535 K (OS)*		[28]	E
Ar	Arc discharge/unpurified; baked under vacuum 5×10^{-7} torr at 500 K	~90 K	*~1850 K (G/IC)*		[28]	E
		~86 K	*~1250 K (OS)*		[28]	E
	Arc discharge/unpurified; evacuated to 1×10^{-6} torr at room temp.	110–160 K	1949 K (G)	~170	[70]	E
	Laser ablation/chemically opened tubes annealed at 1073 K for 12 h in vacuum after acid treatment	120–133 K	2090 K (I)		[71]	E
	Potential calculation		934 K (OS)		[11]	T
	Potential calculation		1550 K (G)		[11]	T
N_2	Arc discharge/unpurified, baked under vacuum 5×10^{-7} torr at 500 K	~100 K	*~1750 K (G/IC)*		[28]	E
		~90 K	*~1100 K (OS)*		[28]	E
	Laser ablation/evacuated for 24 h at 350 K	93.7–81.2 K	998 (G)		[30]	E
	Arc discharge/untreated	77.3 K		240	[32]	E
	Arc discharge/heated in air at 350 °C for 1 h	77.3 K		501	[32]	E
	HiPCO/pristine, heated under vacuum at 423 K for 2 h	77.3 K		436	[35]	E
	HiPCO/attacked with concentrated HCl for 48 h; heated under vacuum for 2 h at 423 K	77.3 K		587	[35]	E
	HiPCO/heated in air at 623 K, followed by immersion in concentrated HCl for 48 h, evacuated at 423 K for 2 h	77.3 K		861	[35]	E
O_2	Arc discharge/untreated	77.3 K		255	[32]	E
	Arc discharge/heated in air at 350 °C for 1 h	77.3 K		534	[32]	E

continued

during the same process. The effect of introducing and pumping out N_2 and then He was, first, to produce small increases in the thermopower and resistance (introduction of the gas), and then, to decrease both quantities (pumping out) in a reversible manner. The observations for the air-saturated samples were explained in terms of weak electron transfer from the O_2 . The smaller effects observed for nitrogen and helium were attributed to changes in the

Table 2. Continued

Adsorbate	Production/treatment	Temp. range	Isost. heat/ bind. energy	Specific surface area (m^2/g)	Ref.	Source
CH_4	Arc discharge/unpurified; evacuated to 1×10^{-6} torr at room temp.	160–195 K	2575 K (G)		[25]	E
	Arc discharge/outgassed in vacuum ($\sim 10^{-5}$ torr) at 773 K	95.3–109.9 K 78.6–87.6 K	*2200 K (IC)* *1338 K (OS)*	204	[24]	E
	Potential calculation		1714 K (G)		[11]	T
	Potential calculation		1036 K (OS)		[11]	T
Kr	Potential calculation		1981 K (G)		[11]	T
	Potential calculation		1814 K (I)		[11]	T
Xe	SWNT-paper/outgassed by repeated heating and annealing cycles under UHV with peak temperatures to 1200 K.		3257 K (G)		[40]	E
	Molecular mechanics		2823 K (G)		[40]	T
	Molecular mechanics		3137 K (I)		[40]	T
	Arc discharge/unpurified; evacuated to 1×10^{-6} torr at room temp.	220–295 K	*3271 K (G)*		[25]	E
	Arc discharge/unpurified, outgassed in vacuum ($\sim 10^{-4}$ Pa) at 773 K	90–130 K	1894 K (OS)		[31]	E
	Chemically cut using H_2SO_4/HNO_3 mixture, followed by treatment with H_2SO_4/H_2O_2 accompanied by sonication and suspension in methanol dimethyl formaldehyde		3233 K (I)		[38]	E
	Potential calculation		2580 K (G)		[11]	T
	Potential calculation		2503 K (I)		[11]	T
	Potential calculation		1573 K (OS)		[11]	T
CF_4	Arc discharge/unpurified outgassed in vacuum ($\sim 10^{-4}$ Pa) at 773 K	92–103 K	*1872 K (OS)*		[31]	E
	Potential calculation		2433 K (G)		[11]	T
	Potential calculation		1498 K (OS)		[11]	T
SF_6	Arc discharge/unpurified outgassed in vacuum ($\sim 10^{-4}$ Pa) at 773 K	130–150 K	*2067 K (OS)*		[31]	E
	Potential calculation		3272 K (G)		[11]	T
	Potential calculation		2037 K (OS)		[11]	T

Note: The columns include a listing of the adsorbate used, a brief description of the method of production of the nanotubes and the treatment to which they were subjected, the range of temperatures at which the measurements were conducted, the values for the isosteric heats (in italicized numerals) or of the binding energies (standard numerals) both in K, the specific surface areas (when they were available), the source, and a letter to indicate whether the result is experimental (E), theoretical calculation (T), or computer simulation (S). The entries for the isosteric heat of adsorption and binding energies are reported for the different binding energy sites, namely, the interstitial channels (IC), grooves (G), outer surface of the outer tubes in the nanotube bundles (OS), and the inside of the open nanotubes (I).

electron and hole lifetimes due to collisions with nonthermal phonons generated by collisions of the gas molecules with the tube walls [52].

The IBM group headed by Avouris very recently [53] explored whether the *p*-to-*n* conversion of vacuum-annealed nanotubes is the result of doping of the nanotubes by O_2, as has been suggested [48]. This group investigated the characteristics of two different approaches for converting the nanotubes from *p* type to *n* type in carbon nanotube field-effect transistors (CN–FETs) [53]. Results for conventional doping of the tubes with an electron donor (K) were compared to those obtained for tubes in which the conversion was achieved by annealing the nanotube–metal electrodes, and partially exposing them to different oxygen atmospheres. The evolution of the transfer characteristics of the FETs as a function of gradual introduction of oxygen in the case of the annealed tubes was quite different from the evolution observed for the conventionally doped tubes as a function of K doping. This indicates that there are fundamental differences in the mechanisms of the transformation for each one of these cases. This group proposed a different explanation for the effect of oxygen adsorption on the nanotubes [53]. This study concluded that oxygen did not dope the bulk of the nanotubes, but that, rather, it modified the Schottky barriers at the metal electrode–semiconducting nanotube contacts. That the contacts were central to the observed behavior, as was proposed in this explanation, was verified by performing a test in which the characteristics of the nanotube FET were investigated for a case in which only one of the two contacts of a nanotube FET was annealed by local heating (by passing a large current in vacuum through only one of the two metal-semiconducting contacts). The drain current curve characteristics of the FET went from being symmetrical (before the heating) to completely asymmetrical (after heating just one of the contacts), verifying the conclusion that the observations were the result of changes at the metal–nanotube contact.

4. APPLICATIONS

Multiwalled carbon nanotubes have been investigated for potential use in differential adsorption. The adsorption of NO, CO_2, and SO_2 by multiwalled nanotubes from a gas stream which consisted of a mixture of He and O_2 has been investigated. Long and Yang [37] produced the multiwalled tubes used in these measurements by catalytic decomposition of acetylene on a metal catalysts; the tubes were purified prior to performing the measurements. The differential adsorption of the three different gases tested was investigated by means of thermogravimetric analysis, temperature-programmed desorption, and FTIR spectroscopy. It was found that the tubes were quite effective adsorbers for NO at room temperature, but were not equally effective for adsorbing CO_2 or SO_2.

Multiwalled carbon nanotubes have been used to produce capacitive and resistive prototypes of gas-sensing devices [64]. In the capacitive device, a silicon dioxide/multiwalled nanotube composite was used as the active sensing element of the capacitor, while in the resistive device, the active sensing elements were multiwalled nanotubes grown on a serpentine silicon dioxide pattern. The basis of the operation of such devices is that the impedance of the nanotubes changes as a result of the adsorption of different chemical species on them. The prototypes proved especially effective in sensing water and ammonia. The effects of water on the sensors were reversible, while those of ammonia were not.

Calculations for molecular sieving, using the SWNT bundles for separating hydrogen isotopes have found that the interstitial spaces between the nanotubes provide suitable pores for such process [13]. Specifically, the calculations found that the ICs of (10, 10) nanotube bundles can act as a quantum molecular sieve for separating hydrogen from tritium. If it turns out that such separation can be achieved with nanotubes, it would be of significant technological importance, since the present isotope separation techniques are expensive and complicated. However, one should note that this potential application depends on hydrogen and tritium being able to access the ICs. Whether hydrogen, or any other gas smaller than Xe, can access the ICs is a question that, at present, remains unresolved and the subject of some controversy [6–8, 24–29].

The high binding energy offered by the SWNT surface can be used to simplify, in some cases, the separation of gas mixtures, for example, xenon–krypton. Processes to achieve the separation of this mixture could have commercial implications because xenon is being increasingly used in the automobile industry, in headlights of automobiles; this provides a strong economic incentive for finding new and efficient ways of distilling this rare gas from air. Xenon's extraction from air is carried out in several consecutive steps [73]. In the last one of these steps, krypton (Kr) and Xe are separated by solidifying an Xe–Kr mixture at liquid nitrogen temperatures, and using the difference between the respective vapor pressures of these two rare gases to extract what is essentially pure Kr from the vapor, leaving an Xe-rich solid as residue. Since xenon has a sufficiently larger binding energy than Kr on SWNT bundles, a room-temperature xenon-capturing filter could be designed to replace this last purification step.

5. SUMMARY AND FUTURE DIRECTION

There is a considerable degree of consistency among the different sets of adsorption results obtained on the multiwalled nanotubes described above. There is a general trend in the data that shows an inverse correlation between the external tube diameters and the values of the specific surface area of the samples. The different sets of data also indicate that the greater the number of walls in the tubes, the closer the adsorption pressures for the formation of different layers on the tubes are to the values observed on planar graphite. All of the studied solid films adsorbed on multiwalled nanotubes are found to incompletely wet these substrates, even in those cases in which solid films of the same adsorbate completely wet planar graphite. Finally, the data on hysteresis in adsorption/desorption cycles indicate that such a feature is less likely to be present in pristine or as-produced tubes, while they are generally present for multiwalled tubes that have been subjected to some purification or activation treatment.

None of the trends in the data reported by the different authors cited above is unexpected. Thicker tubes, which involve a greater number of walls, have interaction potentials that resemble more closely those of planar graphite (which is the limiting case of an infinitely thick tube wall). On the other hand, the effects of curvature, in particular, the strain on a solid adsorbed film that results from it growing on a curved surface, will result in incomplete wetting. The presence of hysteresis loops in adsorption–desorption curves for nanotube substrates subjected to purification or activation treatments is probably the result of damage on the tube walls produced by the treatments.

Unlike the case with multiwalled carbon nanotubes, the degree of consensus that has been reached as a result of investigations on adsorption on single-walled nanotubes is more limited. Regarding the determination of the sites where adsorption takes place on bundles of closed-ended single-wall nanotubes, most authors now agree that the grooves and the outer surface sites play an important role, and that these sites account for either the majority or the totality of the adsorption that occurs on the bundles. All of the measurements performed at low coverages indicate that the isosteric heat of adsorption increases with decreasing coverage. The issue of whether or not various gases are able to adsorb on the ICs is one that is not completely settled yet, although it is generally accepted that gases as large or larger than Xe do not go into the ICs. Several new phases have been identified on the nanotube bundles, and it is clear that some of them are essentially one-dimensional entities. The potential for discovering a greater variety of phases when more systems are investigated in detail is considerable. The effect of contaminants on the tubes as a result of the purification processes to remove metal catalysts and amorphous carbon impurities is one that has been partly explored. It is clear that, on chemically cut tubes, the cutting agents have a pore-blocking effect that requires their removal at high temperatures to make the space at the interior of the tubes available for adsorption. The effect of adsorbates on the electrical properties of SWNTs has been explored; rather dramatic changes in the resistivity and other electrical properties have been found as a result of adsorption. The source of this dramatic change is a subject of current investigation, but it now appears clear that adsorption at the nanotube–metal electrode contact plays a major role.

A number of issues remain to be investigated or more fully explored.

- No systematic study of the differences between the adsorbed systems formed by gases on carbon nanotubes produced by different methods has yet been undertaken. A significant fraction of the work discussed here has been conducted on samples of less than ideal purity. As purer samples become available as a result of gentler purification processes, it would be of great interest to compare the results of adsorption on these samples with the results summarized in this chapter.
- Although exploration of some of the effects of the chemical purification treatments on adsorption has began, a more complete investigation of this subject remains to be undertaken.
- The question of access to the interior of cut tubes is one that requires additional exploration. It is not clear, for example, whether the availability of this interior space is limited to treated tubes maintained in ultrahigh-vacuum conditions or whether the space would be available under less stringent environments.

All in all, the study of adsorption on carbon nanotubes has been an exciting area of research. With the open questions that need to be answered, it is likely to remain such for some time to come.

GLOSSARY

Adsorption A process by which atoms or molecules of one chemical species (the adsorbate) are attracted, and stay on the surface of a solid of a different chemical species (the substrate).

BET method A simple theoretical model proposed by Brunauer, Emmett, and Teller in 1938 (*J. Am. Chem. Soc.* 60, 309) that arrives at an equation for the determination of the monolayer capacity of a substrate from volumetric adsorption isotherm measurements.

Chemical potential Total free energy per molecule.

Heat of adsorption Amount of heat released when an atom or molecule adsorbs onto a substrate.

Isostere Function that relates the equilibrium value of the pressure to the temperature at a constant value of the amount of a gas adsorbed by a given amount of substrate.

ACKNOWLEDGMENTS

A. D. M. acknowledges the Research Corporation for partial support of this work, the donors of the Petroleum Research Fund of the American Chemical Society for partial support of this work, and the National Science Foundation for supporting this work through grant DMR-0089713. S. T. acknowledges the Link Foundation Energy Fellowship Program for their financial support through a scholarship. The authors are grateful to V. Krungleviciute for assistance with the preparation of this manuscript, and M. W. Cole for reading the manuscript and providing useful and insightful comments. The authors thank Prof. T. Hertel for allowing them to reproduce some of his results prior to publication.

REFERENCES

1. A. C. Dillon and M. J. Heben, *Appl. Phys. A* 72, 133 (2001).
2. M. S. Dresselhaus, K. A. Williams, and P. C. Eklund, *Mater. Res. Soc. Bull.* 24, 45 (1999).
3. M. Hirscher, M. Becher, M. Haluska, U. Dettlaff-Weglikowska, A. Quintel, G. S. Duesberg, Y.-M. Choi, P. Downes, M. Hulman, S. Roth, I. Stepanek, and P. Bernier, *Appl. Phys. A* 72, 129 (2001).
4. L. W. Bruch, M. W. Cole, and E. Zaremba, "Physical Adsorption: Forces and Phenomena." Oxford University Press, Oxford, 1997.
5. S. M. Gatica, M. J. Bojan, G. Stan, and M. W. Cole, *J. Chem. Phys.* 114, 3765 (2001); M. M. Calbi, S. M. Gatica, M. J. Bojan, and M. W. Cole, *J. Chem. Phys.* 115, 9975 (2001).
6. W. Teizer, R. B. Hallock, E. Dujardin, and T. W. Ebbesen, *Phys. Rev. Lett.* 82, 5305 (1999).
7. W. Teizer, R. B. Hallock, E. Dujardin, and T. W. Ebbesen, *Phys. Rev. Lett.* 84, 1844 (2000).
8. Y. H. Kahng, R. B. Hallock, E. Dujardin, and T. W. Ebbesen, *J. Low Temp. Phys.* 126, 223 (2002).
9. S. Talapatra and A. D. Migone, *Phys. Rev. Lett.* 87, 206106 (2001).
10. A. Kuznetsova, J. T. Yates, J. Liu, and R. E. Smalley, *J. Chem. Phys.* 112, 9590 (2000).
11. G. Stan, M. J. Bojan, S. Curtarolo, S. M. Gatica, and M. W. Cole, *Phys. Rev. B* 62, 2173 (2000).
12. A. C. Dillon, K. M. Jones, T. A. Bekkedahl, C. H. Kiang, D. S. Bethune, and M. J. Heben, *Nature* 367, 377 (1997).
13. S. R. Challa, D. S. Sholl, and J. K. Johnson, *Phys. Rev. B* 63, 245419 (2001); Q. Wang, S. R. Challa, D. S. Sholl, and J. K. Johnson, *Phys. Rev. Lett.* 82, 956 (1999).
14. S. C. Tsang, P. J. F. Harris, and M. L. H. Green, *Nature* 362, 520 (1993).
15. H. Gaucher, R. J. M. Pellenq, Y. Grillet, S. Bonnamy, and F. Beguin, "Carbon '97—Extended Abstracts of the 23rd Biennial Conference on Carbon," 1997, Vol. II, p. 388.
16. S. Inoue, N. Ichikuni, T. Suzuki, T. Uematsu, and K. Kaneko, *J. Phys. Chem. B* 102, 4689, (1998).
17. E. B. Mackie, R. A. Wolfson, L. M. Arnold, K. Lafdi, and A. D. Migone, *Langmuir* 13, 7197 (1997).
18. A. Bougrine, N. Dupont-Pavlovsky, J. Ghanbaja, D. Billaud, and F. Beguin, *Surf. Sci.* 506, 137 (2002).
19. J. Hilding, E. Grulke, S. B. Sinnott, D. Qian, R. Andrews, and M. Jagtoyen, *Langmuir* 17, 7540 (2001).
20. Q.-H. Yang, P.-X. Hou, S. Bai, M.-Z. Wang, and H.-M. Cheng, *Chem. Phys. Lett.* 384, 18 (2001).
21. K. Masenelli-Varlot, E. McRae, and N. Dupont-Pavlovsky, *Appl. Surf. Sci.* (2002).
22. M. Eswaramoorthy, R. Sen, and C. N. R. Rao, *Chem. Phys. Lett.* 304, 207 (1999).
23. S. E. Weber, S. Talapatra, C. Journet, A. J. Zambano, and A. D. Migone, *Phys. Rev. B* 61, 13150 (2000).
24. M. Muris, N. Dufau, M. Bienfait, N. Dupont-Pavlovsky, Y. Grillet, and J. P. Palmari, *Langmuir* 16, 7019 (2000).
25. S. Talapatra, A. J. Zambano, S. E. Weber, and A. D. Migone, *Phys. Rev. Lett.* 85, 138 (2000).
26. A. J. Zambano, S. Talapatra, and A. D. Migone, *Phys. Rev. B* 64, 075415 (2001).
27. S. Talapatra and A. D. Migone, *Phys. Rev. B* 65, 045416 (2002).
28. T. Wilson, A. Tyburski, M. R. DePies, O. E. Vilches, D. Becquet, and M. Bienfait, *J. Low Temp. Phys.* 126, 403 (2002).
29. T. Wilson and O. E. Vilches, *J. Low Temp. Phys.* (in print).

30. D.-H. Yoo, G.-H. Rue, Y.-H. Hwang, and H.-K. Kim, *J. Phys. Chem. B* 106, 3371 (2002).
31. M. Muris, N. Dupont-Pavlovsky, M. Bienfait, and P. Zeppenfeld, *Surf. Sci.* 492, 67 (2001).
32. A. Fujiwara, K. Ishii, H. Suematsu, H. Kataura, Y. Maniwa, S. Suzuki, and Y. Achiba, *Chem. Phys. Lett.* 336, 205 (2001).
33. I. Stepanek, G. Maurin, P. Bernier, J. Gavillet, and A. Loiseau, *Mater. Res. Soc. Symp. Proc.* 593, 119 (2000).
34. W.-F. Du, L. Wilson, J. Ripmeester, R. Dutrisac, B. Simard, and S. Denommee, *Nano Lett.* 2, 343 (2002).
35. C.-M. Yang, K. Kaneko, M. Yudasaka, and S. Iijima, *Nano Lett.* 2, 385 (2002).
36. T. Hertel, J. Kriebel, G. Moos, and R. Fasel, in "Nanonetworks Materials, Fullerenes Nanotubes and Related Systems" (S. Saito, T. Ando, Y. Iwasa, K. Kikuchi, M. Kobayashi, and Y. Saito, Eds.), AIP Conference Proceedings, Kamkura, Japan, 2001, Vol. 590, p. 181. (2001).
37. R. Q. Long and R. T. Yang, *Ind. Eng. Chem. Res.* 40, 4286 (2001).
38. A. Kuznetsova, D. B. Mawhinney, V. Naumenko, J. T. Yates, Jr., J. Liu, and R. E. Smalley, *Chem. Phys. Lett.* 321, 292 (2000).
39. A. Kuznetsova, J. T. Yates, Jr., V. V. Simonyan, J. K. Johnson, C. B. Huffman, and R. E. Smalley, *J. Chem. Phys.* 115, 6691 (2001).
40. T. Hertel, private communication. Note that the more recent data, presented in Figure 3, have some differences with respect to similar data presented in [6]: the low-temperature multilayer-like desorption peaks are not present. The low-temperature features are interpreted as resulting from higher roughness and more inhibited diffusion in the older samples used in the work reported in [36].
41. M. Bienfait, B. Asmussen, M. Johnson, M. P. Zeppenfeld, *Surf. Sci.* 460, 243 (2000).
42. M. M. Muris, M. Bienfait, P. Zeppenfeld, N. Dupont-Pavlovsky, M. Johnson, O. E. Vilches, and T. Wilson, *Appl. Phys. A* (in press).
43. G. E. Gadd, M. Blackford, S. Moricca, N. Webb, P. J. Evans, A. M. Smith, G. Jacobsen, S. Leung, A. Day, and Q. Hua, *Science* 277, 993 (1997).
44. K. A. Williams, B. K. Pradhan, P. C. Eklund, M. K. Kostov, and M. W. Cole, *Phys. Rev. Lett.* 88, 165502 (2002).
45. O. E. Vilches, personal communication.
46. O. E. Vilches, *Ann. Rev. Phys. Chem.* 31, 463 (1980).
47. J. Kong, N. R. Franklin, C. Zhou, M. G. Chapline, S. Peng, K. Cho, and H. Dai, *Science* 287, 622 (2000).
48. P. Collins, K. Bradley, M. Ishigami, and A. Zettl, *Science* 287, 1801 (2000).
49. K. Bradley, S. H. Jhi, P. G. Collins, J. Hone, M. L. Cohen, S. G. Louie, and A. Zettl, *Phys. Rev. Lett.* 85, 4361 (2000).
50. C. Marliere, P. Poncharal, L. Vaccarini, and A. Zahab, *Mater. Res. Soc. Symp. Proc.* 593, 173 (2000).
51. A. Zahab, L. Spina, P. Pncharal, and C. Marliere, *Phys. Rev. B* 62, 10000 (2000).
52. G. U. Sumanasekera, C. K. W. Adu, S. Fang, and P. C. Eklund, *Phys. Rev. Lett.* 85, 1096 (2000).
53. V. Derycke, R. Martel, J. Appenzeller, and Ph. Avouris, *Appl. Phys. Lett.* 60, 2773 (2002).
54. J. G. Dash, "Films on Solid Surfaces." Academic, New York, 1975.
55. T. W. Ebbesen and P. M. Ajayan, *Nature* 358, 221 (1992).
56. C. Journet, W. K. Maser, P. Bernier, A. Loiseau, M. Lemy de la Chappele, S. Lefrant, P. Deniard, R. Lee, and J. E. Fischer, *Nature* 388, 756 (1997).
57. A. Thess, R. Lee, P. Nikolaev, H. Dai, P. Petit, J. Robert, C. Xu, Y. H. Lee, S. G. Kim, A. G. Rinzler, D. T. Colbert, G. E. Scuseria, D. Tomanek, J. E. Fischer, and R. E. Smalley, *Science* 273, 483 (1996).
58. M. Terrones, R. Kamalakaran, T. Seeger, and M. Ruhle, *Chem. Commun.* 2335 (2000).
59. E. P. Barrett, L. G. Joyner, and P. P. Halenda, *J. Am. Chem. Soc.* 73, 373 (1951).
60. B. Coasne, A. Grosman, C. Ortega, and M. Simon, *Phys. Rev. Lett.* 88, 256102 (2002).
61. H. S. Nham and G. B. Hess, *Langmuir* 5, 575 (1989).
62. G. B. Hess, in "Phase Transitions in Surface Films 2" (H. Taub, G. Torzo, H. J. Lauter, and S. C. Fain, Jr., Eds.), NATO ASI Series B: Physics, Vol. 267, p. 357. Plenum, New York, 1991.
63. J. Menaucourt, A. Thomy, and X. Duval, *J. de Phys. (Paris)* 38, C4-195 (1977).
64. O. K. Varghese, P. D. Kichamore, D. Gong, K. G. Ong, E. C. Dickey, and C. A. Grimes, *Sensors Actuators B* 81, 32 (2001).
65. G. Horvath and K. Kawazoe, *J. Chem. Eng. Jpn.* 18, 470 (1983).
66. G. Stan and M. W. Cole, *Surf. Sci.* 395 280 (1998); G. Stan, V. H. Crespi, M. W. Cole, and M. Bonninsegni, *J. Low Temp. Phys.* 113, 447 (1998).
67. G. Vidali, G. Ihm, H.-Y. Kim, and M. W. Cole, *Surf. Sci. Rep.* 12, 136 (1991).
68. A. Thomy and X. Duval, *J. Chim. Phys. (Paris)* 67, 1101 (1970).
69. C. M. Brown, T. Yildirim, D. A. Neumann, M. J. Heben, T. Gennett, A. C. Dillion, J. L. Alleman, and J. E. Fisher, *Chem. Phys. Lett.* 329, 311 (2001).
70. S. Talapatra, D. S. Rawat, and A. D. Migone, *J. Nanosci. Nanotechnol.* 5, 1 (2002).
71. D.-H. Yoo, G.-H. Rue, J.-Y. Seo, Y.-H. Hwang, M. H. W. Chan, and H.-K. Kim, *J. Phys. Chem. B* 106, 9000 (2002).
72. K. A. Williams and P. C. Eklund, *Chem Phys. Lett.* 320, 352 (2000).
73. N. A. Downie, "Industrial Gases." Blackie Academic & Professional, London, 1997.

Encyclopedia of Nanoscience and Nanotechnology
www.aspbs.com/enn

Gas Sensors from Nanostructured Metal Oxides

Simonetta Capone, Pietro Siciliano

Istituto per la Microelettronica e Microsistemi (I.M.M) del C.N.R., Sezione di Lecce, via per Arnesano—Campus Universitario, 73100 Lecce, Italy

CONTENTS

1. INTRODUCTION

1.1. Nanomaterials and Applications

Most properties of solids depend on the microstructure, that is, the chemical composition, the arrangement of the atoms in the structural lattice and the size of the solid in one, two, or three dimensions. This basic result of material science implies that a change of one or several of these parameters causes a variation of the material properties [1]. In particular, when the size of the crystallites in a polycrystalline material is reduced from macroscopic dimensions down to the scale of a few nanometers, some special effects, such as interfacial effects, size effects, and macroquantum tunnel effects, occur so that a new class of nanostructured materials, with different (electronic, optical, electrical, magnetic, chemical, and mechanical) properties with respect to single crystalline materials (or coarse-grained polycrystals) and/or glasses with the same average chemical composition, can be distinguished [3, 24]. The term "nanoparticles" is generally used in the material science community to indicate particles with diameters smaller than 100 nm [2, 3, 7]. However, in many applications, the term nanoparticles is used when the particle sizes are less than 50 nm [4, 5] or even 10 nm [1, 6].

The explosion in both academic and industrial interest in nanostructured materials (a broad spectrum of which includes insulators, semiconductors, superconductors, metals and alloys, nanocomposites) in the last ten years arises just from the extremely fascinating and useful properties of such materials. Some examples of these properties are: an increased electrical conductivity of ceramics and magnetic nanocomposites, an increased electrical resistivity in metals, an increased magnetic coercivity down to a critical particle size in the nanoscale regime, and below this critical particle size, the occurrence of supermagnetic behavior, an increased hardness and strength of metals and alloys, and an enhanced ductility of structural and functional ceramics [9]. A blue shift in the optical spectra of quantum-confined particles [10, 11], an increase in the luminescent efficiency of some porous semiconductors [12–14], and a lower melting temperature [15] are also observed.

In general, the nanomaterial properties, incompletely understood today, are the results of the small size and/or dimensionality of the nanoparticles, as well as of the large fraction of interfaces between adjacent crystallites (grain boundaries), and hence of the increased percentage of surface atoms. In fact, one important feature for nanocrystalline materials is the large surface area. Since the average size of an atom is on the order of 1–2 Å in radius and a nanometer comprises 10 Å, in 1 nm, there will be 3–5 atoms, depending on the atomic radius, and a grain with a size ranging from 1 to 100 nm will have a number of atoms varying from several to a million. Thus, as a schematic example, we can notice that, for a 4096-atom cubic grain, the relative percentage of surface atoms N_t/N_s (N_t: bulk atoms, N_s: surface atoms) is 33%; it becomes 57% by splitting the cube into eight cubes, and increases up to about 87% after splitting each cube into eight cubes (Fig. 1) [8].

Crystallites and grain boundaries in a nanocrystalline material have comparable volume fractions, and are both of the dimension of a few nanometers. Hence, the overall properties of the solid will be strongly influenced by the contribution coming from grain boundaries. The grain boundaries are complex and highly disordered structures consisting of a few layers of atoms with different coordination numbers compared with the corresponding atoms in the grain; they generally contain high densities of interface states that may induce an intrinsic chemical reactivity property toward the environment chemical compounds. The

ISBN: 1-58883-059-4/$35.00

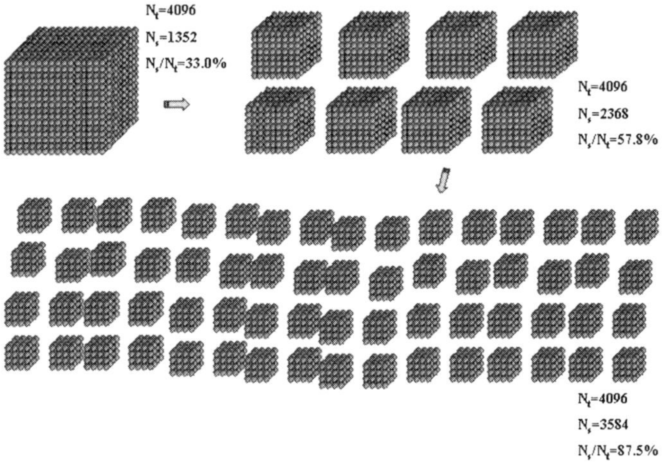

Figure 1. Schematic example that shows how the ratio N_t/N_s (N_t: bulk atoms, N_s: surface atoms) increases for a 4096-atom cubic grain by splitting the cube into eight cubes, and next each cube into eight cubes [8].

study of grain boundaries and their effect on the properties of nanocrystalline materials has been the subject of extensive discussion among researchers working in the field of nanostructured material science [20–23]. Moreover, the atomic structure in the interior of crystallites is modified through the introduction of defects, strain fields, or short-range correlated static or dynamic displacements of atoms from their ideal crystal lattice positions. Many physical properties of nanocrystalline materials will be modified as a consequence of the combined effects of these structural changes, that is, the reduction of the length scale on which there is coherency in the atomic arrangement (long-range order) and the introduction of interfaces with an atomic short-range order different from the one in the crystal lattice [19].

Because of their unique properties, nanostructured materials can be exploited for a wide variety of functional applications, most of which use nanocrystalline films supported by a suitable substrate [2–7, 27–36]. Nanoscale materials have a promising future in microelectronic and optoelectronic applications (quantum nanostructures: wells, wire, dots, tubes, belts) [25, 26]; electroluminescent silicon-based devices; and next-generation computer chips, single-electron devices, and so on. They also can be used in the development of new photocells, very small active elements in magnetic recording, electrochromic display devices, high-energy density batteries, scanning nanoprobes, physic filters to separate different specimens, and "aerogels" for insulation, heating, and cooling bills, in addition to longer lasting biocompatible medical implants, biomimetics, smart structures, microdevices for telemedicine, are further examples of the broad spectrum of nanomaterial applications. In applications to bulk devices, the small size and the homogeneity of nanomaterials give rise to particular solids with very special mechanical properties; they are exceptionally strong, hard, ductile at high temperature, wear resistant, erosion resistant, and corrosion resistant.

Moreover, a high catalytic and a strong reactivity with gaseous species are fundamental properties of nanocrystalline semiconducting metal oxides that make them particularly suitable for applications in catalyzers [16] and semiconductor gas sensors [17, 18, 37, 38].

The field of nanostructured materials is thus really very broad, and the scientific research (nanoscience) that studies material and system properties in the nanoscale range involves an interdisciplinary approach based on solid-state physics, chemistry, biology, and material science whose cross fertilization will lead in the future to new avenues in material science. Gleiter reported a simple and exhaustive classification schema for nanostructured materials (NsM) according to their chemical composition (families) and the dimensionality and shape of the crystallites forming the NsM (categories) [1, 11] (Fig. 2). The cluster of novel technologies, related to the synthesis of materials and/or the development of devices with new properties by means of the controlled manipulation of their microstructure on a nanometer scale, is well known as nanotechnology, and it no doubt constitutes the most innovative and emerging technology of the 21st century [5].

1.2. Nanocrystalline Oxides in Gas Sensors

In this chapter, the main research lines of nanoscience and nanotechnology applied to semiconductor gas sensors (SGS) will be reviewed. They cover new materials, new concept and device types, and a basic understanding of the gas-sensing function. The application of nanocrystalline metal oxides films in semiconductor gas sensors will be extensively discussed, with the aim of better understanding the new opportunities offered by such nanostructured materials for enhancing the performance of this kind of gas sensor.

In the field of semiconductor gas sensors, recent applied studies and product releases have shown some significant trends in nanotechnologies and gas-sensing layers to be employed. One of these trends aims to implement low cost,

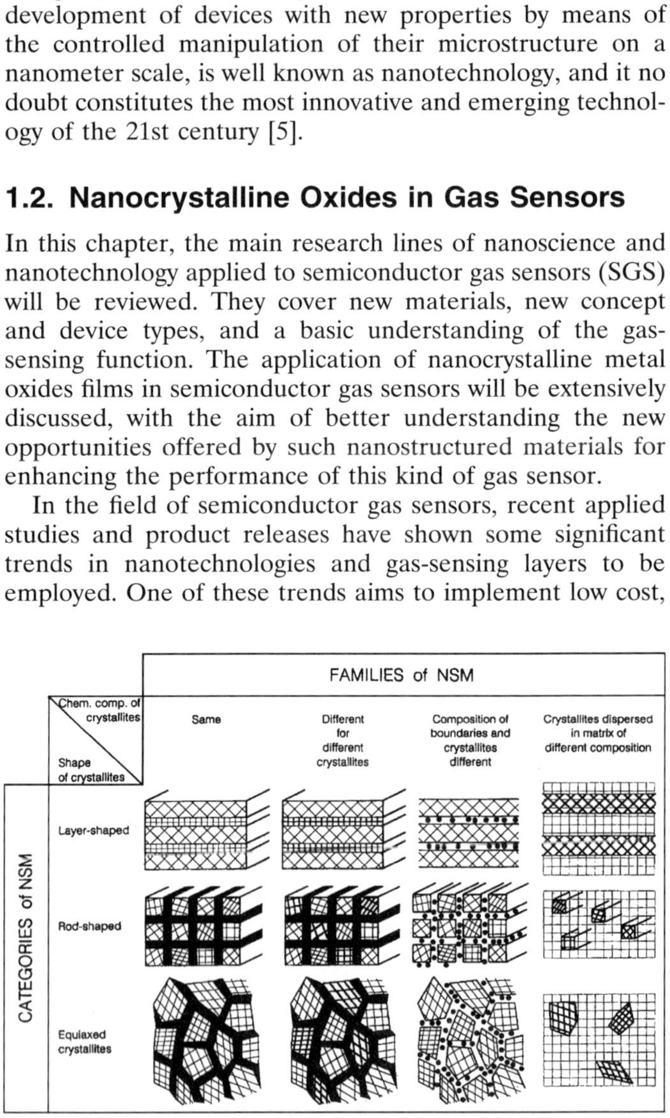

Figure 2. Classification schema for NsM according to their composition and the dimensionality (shape) of the crystallites (structural elements) forming the NsM. The boundary regions of the first and second families of NsM are indicated in black to emphasize the different atomic arrangements in the crystallites and in the boundaries. The chemical composition of the (black) boundary regions and the crystallites is identical in the first family. In the second family, the (black) boundaries are the region crystallites where two crystals of different chemical compositions are joined together, causing a steep concentration gradient. Reprinted with permission from [1], H. Gleiter, *Acta Mater.* 48, 1 (2000). © 2000, Elsevier Science.

low-power consumption, reliable, smart, and miniaturized sensing devices, and it shows the decisive advantage of using a micromachined silicon platform as a substrate for the sensitive layers [39–42]. On the other hand, several theoretical and applied articles have shown the advantage of reducing the metal-oxide grain size down to nanometer scale in order to improve the sensing properties (mainly sensitivity and selectivity), as well as stability over time of the oxide layer [43–54]. Nanocrystalline semiconducting metal oxides with controlled composition are indeed of increasing interest in gas sensing, and also constitute a new and exciting subject of fundamental research. In fact, it is well known that the response of semiconductor gas sensors to gases, the gas sensitivity and selectivity, and the degradation of performance with time of the sensing layers depend on the grain size, the surface morphology, and the internal porosity. The microstructure of the sensing layers is determined by the preparation method and the deposition parameters, as well as by postannealing treatments at high temperature for material stabilization, and in general, both an empirical and systematic study of the influence of all of these parameters is needed to obtain improved sensor devices by better control of the microstructure of their active layers. Some examples of sensing films of different morphologics are given in Figures 3–5.

Thanks to their small grain size, high density of grain boundaries, and interfaces, nanostructured metal oxides show a strong reactivity with gaseous species. The high reaction rates of these fine-grained semiconducting oxides with chemical species in air are a direct consequence of their high specific area. This specific feature, together with the potential to control both the microstructure and the chemistry due to the high technological levels reached in the preparation of the sensing layers, make nanocrystalline semiconducting oxides very attractive materials for the development of high-sensitivity fast-response semiconducting gas sensors, in which surface interaction processes play the key role in the formation of the sensor signal [47, 55, 56]. The gas/semiconductor surface interactions on which is based the gas-sensing mechanism of SGS occur at the grain boundaries of the polycrystalline oxide film. They generally include reduction/oxidation processes of the semiconductor, adsorption of the chemical species directly on the semiconductor and/or adsorption by reaction with surface states associated with preadsorbed ambient oxygen, electronic

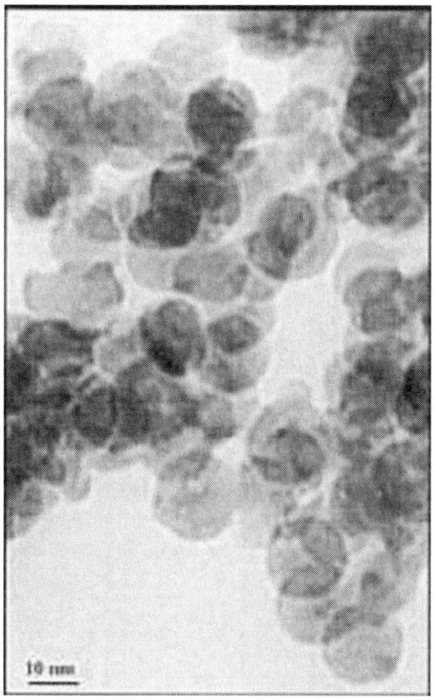

Figure 3. TEM images of tin-oxide nanoparticles obtained by thermal oxidation of tin/tin-oxide nanoparticles. Reprinted with permission from [205], C. Nayral et al., *Appl. Surf. Sci.* 164, 219 (2000). © 2000, Elsevier Science.

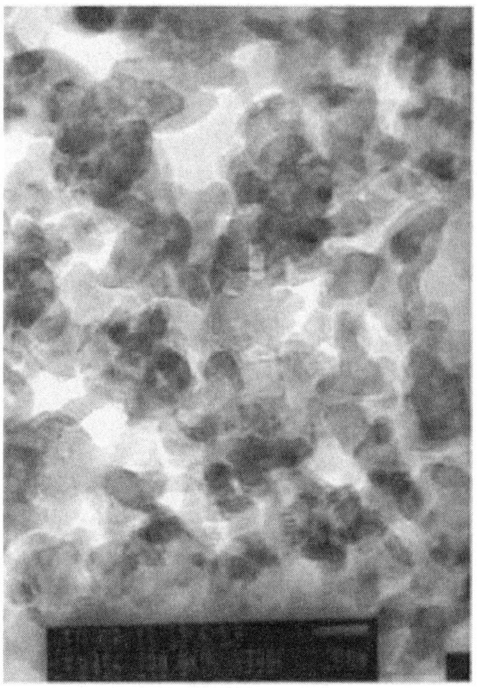

Figure 4. TEM micrographs of W/TiO_2 thin film (W : Ti ratio of 1 : 33) (bar 20 nm). Reprinted with permission from [195], C. Garzella et al., *Sensors Actuators B* 83, 230 (2002). © 2002, Elsevier Science.

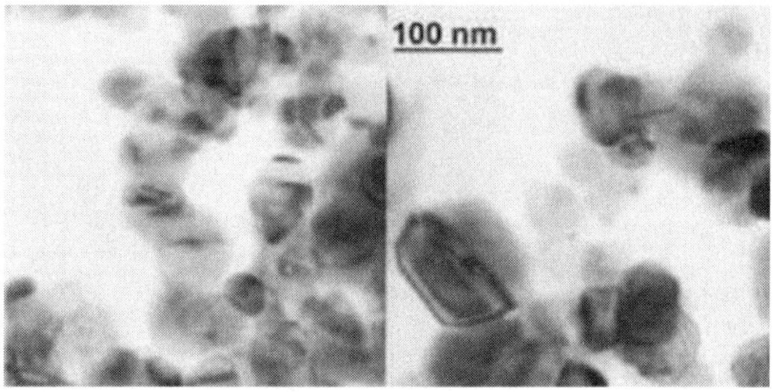

(a) (b)

Figure 5. TEM micrograph of a microwave pure tin oxide conventionally stabilized at 1000 °C for 8 h (a) and 0.2% Pd *in-situ* catalyzed (b). Reprinted with permission from [181], A. Cirera et al., *Mater. Sci. Eng. C* 15, 203 (2001). © 2001, Elsevier Science.

transfer of delocalized conduction-band electrons to localized surface states and vice versa, catalytic effects, and in general, complex surface chemical reactions between the different adsorbed chemical species [17, 18, 69]. The effect of these surface phenomena is a reversible and significant change in electrical resistance (i.e., a resistance increase or decrease under exposure to oxidizing, respectively reducing, gases referring as an example to an *n*-type semiconductor oxide). This resistance variation can be easily observed and used to detect chemical species in the ambient. The gas/solid interactions exploited in SGS can involve changes in the surface conductance or changes in the bulk conductance; correspondingly, two principal types of metal-oxide-based gas sensors can be distinguished [69, 78]. Since a semiconductor oxide is, in general, nonstoichiometric and the oxygen vacancies are the main bulk defects, changes in the bulk conductance are due to changes in the oxygen partial pressure at operating temperatures so high (600–1000 °C) that the oxygen vacancies can quickly diffuse from the interior of the grains to the surface and vice versa, and the bulk of the oxide has to reach an equilibrium state with ambient oxygen. The main application of this first kind of sensor is the measurement of oxygen partial pressure as required in combustion control systems, in particular, in the feedback control of the air/fuel ratio of automobile engine exhaust gases near the λ point in order to improve the fuel economy efficiency and reduce the harmful emission of gases as CO, NO_x, and hydrocarbons [79–81]. The second principal type of semiconductor gas sensors is based only on changes in surface conductivity at lower temperatures (<600 °C) and at quasiconstant oxygen partial pressure. In this condition, the sensor detects small concentrations of reactive gases in air by a displacement from the constant oxygen pressure equilibrium state, induced by gas-interfering effects at the surface of the sensor. Also in this operation mode, there is a sensitivity of the sensor to changes in the oxygen partial pressure, but in this case, the defect oxide chemistry has a minimum importance as far as the gas sensitivity is concerned. This chapter will be devoted especially to this latter class of resistive-type gas sensors.

The working temperature at which these devices work varies, depending on the specific target gas in the ambient, and on the selected sensor material in conjunction with its properties in every case. As this working temperature usually ranges from 200 to 400 °C, it is necessary to implement a heating element in a sensor device. A simple SGS is thus basically composed of a substrate in alumina or silicon (on which the sensing layer is deposited), the electrodes (to measure the resistance changes of the sensing film), and the heater (commonly a Pt resistive-type heater) to reach the optimum sensing temperature. A simple scheme of a sensor on alumina substrate is shown in Figure 6.

A gas sensor has three principal functions [40, 57]: (1) the *receptor function*, that is, the ability to detect a particular gas, which depends on the surface chemical processes; (2) the *transducer function*, which translates the gas-induced changes at the semiconductor surface into an electric output signal, and which depends on the microstructure of the sensing material and on the electrical conduction mechanism through the grain network; and (3) the *operation mode*, the

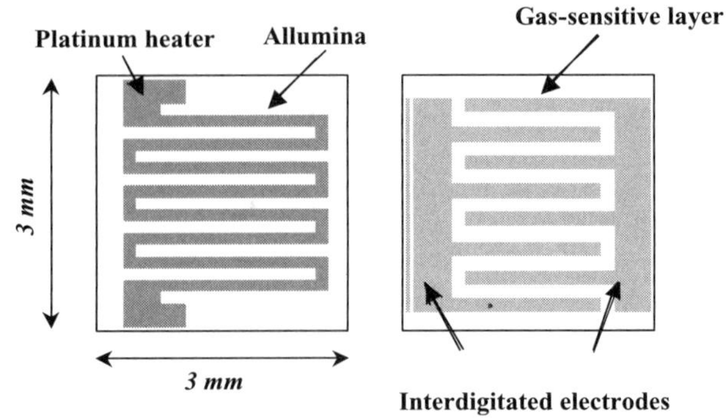

Figure 6. Simple scheme of a gas sensor on alumina substrate.

most used one based on the monitoring of the sensor resistance at constant temperature operation. Other operation modes are the modulated operation temperature [58–61], the single and/or simultaneous measurement of sensor resistance and thermovoltage [62, 63], and the single and/or simultaneous measurement of sensor resistance and temperature [64–68].

In this chapter, we will first consider the background and the gas-sensing principles of semiconductor gas sensors. Next, since the most important factors affecting the sensing properties of these devices via the transducer function are the microstructure and the morphology of the sensing element, the models for electric transport in polycrystalline semiconductors will be reported, discussing the modifications of these models for nanocrystalline semiconductors. The aim was to gain a deeper insight into the nature of the gas-sensing properties of semiconducting nanostructured metal oxides by correlating the macroscopic characteristics of semiconductor gas sensors to the microscopic characteristics of their sensing elements [70]. Great efforts are in fact devoted to the basic understanding of the real grain boundaries and size effect. A survey of the most used nanostructured metal-oxide films and deposition techniques for semiconductor gas sensors also will be reported. Next, particular interest will be devoted to the influence of dopants and the use of mixed oxides or heterolayers, the latter representing a new design concept of a semiconductor gas sensor. Application examples of SGS to the detection of toxic and/or pollutant gases will be looked into on the basis of experimental results reported in recent literature. Moreover, the concept of electronic nose will be introduced, and the most recent research developments in multisensor systems for specific functional applications in the food industry, environmental monitoring, analysis of different ambients, and so on, will be reviewed. Finally, the future of nanostructured oxides and the trends of scientific research in the gas-sensing field will be discussed in the last section.

2. GAS-DETECTION PRINCIPLES

2.1. General Features

The principle of operation of gas sensors based on semiconducting metal oxides relies on gas-induced variations of the surface chemistry that cause large and reversible changes in the electrical conductivity of the sensing layers when exposed to oxidizing or reducing gases. The discovery of an electrical response of a semiconductor to variations in the chemical composition of the ambient is not recent. Gas-sensitive effects on semiconductors were first discovered by Brattain and Bardeen [71] (1953) on Ge. Soon after, Seiyama et al. (1962) [72] and Taguchi (1971) [73] applied this discovery to gas detection by producing the first chemoresistive gas sensors. Seiyama et al. demonstrated that the conductivity of thin films of ZnO, when heated up to 300 °C, was strongly modulated by the presence of traces of reactive gases in air. Taguchi also demonstrated similar properties for sintered compressed and polycrystalline SnO_2 powder, with the advantage of greater stability and high sensitivity at a lower operating temperature, and established Figaro Engineering Inc., which today remains one of the major manufacturers of gas-sensor devices.

Since then, a huge literature on the characterization and use of semiconducting oxides as gas sensors has been produced. Various books that report the principles and applications of such a kind of gas sensor also have been published [17, 18, 74–77]. Nowadays, an appreciable number of manufacturers of gas sensors and multisensor systems (often referred to as *electronic noses*) are on the market. Section 4 will be completely devoted to the development of these latter multisensor devices. The major international conferences on gas sensors and electronic nose of such devices are listed in Table 1.

The urgent demand for gas-sensor systems is clearly explained by the growing number of their potential applications. Two important groups of applications are the detection of single gases (such as NO_2, O_3, CO, CH_4, etc.) and the discrimination of odors or generally the monitoring of changes in the ambient. Single-gas sensors can, for example, be used as fire detectors, leakage detectors, controllers of ventilation in cars and planes, and alarm devices warning the overcoming of threshold concentration values of hazardous gases in workplaces. Next, we will report some of the most noteworthy examples of applications of gas sensors and electronic noses.

Table 1. List of the main gas-sensor manufacturers.

Gas detectors
Capteur (http://www.capteur.co.uk)
Figaro Engineering Inc. (http://www.figarosensor.com)
Fisinc (http://www.fisinc.jp)
Microsens (http://www.microsens.ch)
Microchemical (http://www.microchemical.com)
Th (http://www.th-online.de.firmen/ust/webpages/gas.html)
Multisensor systems
Motech (http://www.motech.de)
Nordic Sensor (http://www.nordicsensor.se)
Airsense (http://www.airsense.com)
Alpha-Mos (http://www.alpha-mos.com)
AromaScan (http://www.aromascan.com)
Marconi Tech. (http://www.marconitech.com)
RST—Rostock (http://www.rst-rostock.de)
Lennartz Electronics (http://www.lennartz-electronic.de)

The great interest of the industrial and scientific world in SGS comes from their numerous advantages, like small size, high sensitivity in detecting very low concentrations (at levels of ppm or even ppb) of a wide range of gaseous chemical compounds, the possibility of on-line operation, and due to possible bench production, low cost. Besides, it should be considered that semiconductor gas sensors belong to the class of solid-state gas-sensor devices that furnish information about the ambient composition directly in the form of an electric signal. The possibility of easily combining in the same device the functions of a sensitive element and a signal converter and control electronics markedly simplifies the design of a sensor, and constitutes the main advantage of chemoresistive-type sensors over biochemical, optical, acoustic, and other gas-sensing devices. This real simplicity in function is also the main advantage over analytical techniques as gas chromatography (GC) and mass spectroscopy (MS), which are complex, expensive, difficult to operate, and off-line apparatus.

In spite of the numerous advantages of resistive-type gas sensors, they show different disadvantages, such as poor reproducibility and long-time instability due to aging. There are two undesired aging effects that may appear when the sensor works for a long period: a *drift* of the baseline signal (defined as the conductance in air or in a reference gas) or a *drift* in the sensor response. The sensor response is usually defined as

$$X = \begin{cases} x_s/x_0, & \text{if } x_s > x_0 \\ x_0/x_s, & \text{if } x_s < x_0 \end{cases} \tag{1}$$

or

$$X = \begin{cases} x_s - x_0/x_0, & \text{if } x_s > x_0 \\ x_0 - x_s/x_s, & \text{if } x_s < x_0 \end{cases} \tag{2}$$

(x_s and x_0 are the electrical signal values at equilibrium in the presence of a gas/odor and in a reference gas, respectively), depending on the reducing or oxidizing character of the chemical species interacting with the sensor surface. A general trend, which has been experimentally verified, is that the sensor response to a particular gas increases up to a maximum corresponding to an optimum working temperature, then falls off toward zero at higher temperatures. As the sensor response is used to characterize the change in the sensor signal due to a certain stimulus, on the other hand, the (partial) sensitivity S describes the change in the sensor response due to a specific change in the stimulus. In particular, if y_k is the concentration of the chemical species k, the (partial) sensitivity S_k of the sensor to detect the species

k is defined by the slope of the sensor calibration curve at concentration y_{k_0}:

$$S_k = \left(\frac{\partial X}{\partial y_k}\right)_{y_k=y_{k_0},\, y_{j\neq k}\ \text{const.}} \tag{3}$$

Metal oxides are nonlinear sensors; the change in sensor response due to a defined change in gas concentration depends generally on the concentration of the gas to be monitored y_k, and also on the concentration of other gases $y_{j\neq k}$ (cross-sensitivity effect). In this case, more than one sensor or more than one operation mode of the same sensor are required to determine the concentration level y_k. In the worst case, S_k varies with the aging of the sensor due to drifts or contamination effects.

Long-time instabilities are of considerable importance for the practical use of the sensor; preaging thermal treatments and cycle calibration checks have to be carried out in order to avoid incorrect use of the device. The causes of instabilities are mainly microstructural and morphological changes (change in size, number, and distribution of grains and intergranular boundaries) of the sensing elements, but also irreversible reactions with chemical species in the ambient, modifications of the sensor heating element, or of the electrodes have to be taken into account.

Other important disadvantages are the sensitivity to water vapor and the lack of selectivity. Metal-oxide-based gas sensors are normally sensitive to more than one chemical species in air, and usually show cross sensitivities. This nonspecificity of the response to a chemical species whose presence, identity, and concentration in air have to be determined is by now considered an intrinsic property of metal-oxide-based gas sensors. It is easy to understand how this disadvantage represents a real problem when different reactive gases are present simultaneously in the same atmosphere and interference effects between them can occur. Unselectivity cannot be eliminated completely, but it can be improved in different ways, such as: (1) the use of filters [82, 83] or chromatographic columns to discriminate between gases on the basis of their molecular size or other physical properties [84], (2) the use of catalysts and promoters or more specific surface additives [85–88], (3) the selection of the material for the sensing layer [89, 38] and its physical preparation, (4) the analysis of the transient sensor response [90], (5) the selection of a fixed temperature to maximize sensitivity to a particular analyte gas [91], or (6) the use of a temperature-modulated operation mode [58–61]. A different approach to the problem of unselectivity is based on the development of the above-mentioned *electronic nose*, which consists of an array of different sensing elements with partially overlapping sensitivity and a *pattern recognition* system [92]. Basically, the idea of an electronic nose is to exploit the unavoidable cross sensitivity of the sensors instead of trying to eliminate it, by linking the sensors in an array configuration, and by analyzing the responses of the sensors in a subsequent data-processing step in order to perform a qualitative and/or quantitative analysis of the ambient under examination.

The basic part of the efforts made nowadays by the scientific research community in the SGS field are devoted to solving the problems that sensing elements can have. Sensitivity, selectivity, and stability are the three principal parameters that have to be improved in order to develop more accurate devices. These parameters are strongly dependent both on the surface chemical processes of the specific material used as a sensing element and on its structural and morphological properties. Both aspects have to be considered in order to gain a deeper insight into the receptor and transducer function of the gas sensor.

In general, the characterization and development of chemoresistive gas sensors has been based on an empirical approach. This approach consists principally of systematic experimental conductivity measurements in a controlled atmosphere by using the direct-current (dc) or alternating-current (ac) method. dc measurements are the most used ones in gas-sensing tests, and constitute the most common operation mode in commercial gas sensors. However, the ac method is a powerful tool to examine the nature of conduction processes. As processes of different time constants can be distinguished by varying the frequency, ac measurements compared with corresponding dc measurements allow us to separate contributions from contacts, bulk, surface, and grain boundaries, all of which may contribute to the overall sensor response; equivalent circuits with different resistance/capacitance (RC) units describe formally the frequency behavior [113–115, 118]. Gas-sensing tests are carried out by using suitable and flexible gas-mixing benches [94] that allow mixing many gases in a wide range of concentrations and combinations, and adjust all of the conditions that are essential for the sensor's behavior (temperature, gas flow, humidity, etc.). The aim of gas-sensing tests in a controlled atmosphere is to analyze the sensor response and characterize the sensor in terms of a few phenomenological parameters (sensitivity, selectivity, response and recovery time, optimum working temperature, and repeatability of the measure and stability) [95].

Moreover, as the reactivity of the sensor surface is critically dependent on the condition of the surface, that is, on its elemental composition including any alloying, doping, or impurity constituents, on its electronic and defect structure, and on its morphology, a good general characterization of a gas sensor should also include a spectroscopic and microscopic characterization of the sensor surface. A large number of techniques, suitable for application to the analysis of gas-sensing layers, have been developed to study various surface properties, including the structure, composition, oxidation states, morphology, and changes of chemical or electronic properties by utilizing scattering, absorption, or emission of photons, electrons, atoms, and ions [96, 97]. Advances in modern surface instrumentations allow the detection of finer detail: atomic spatial resolution, ever-smaller energy resolution, and shorter time scales. Moreover, the ability to monitor surface conditions during exposure of the sensor surface to reactants could be fundamental for the understanding of sensor mechanisms and behavior. Hence, a great interest has been devoted to surface analytical techniques that can be applied under an *in-situ* real operating state of the sensor [98, 107, 111, 112]. Table 2 reports some of the most commonly used surface-science techniques.

Thus, in the above-described experimental approach, mainly chosen by *users* of gas sensors, the commercially

Table 2. Most commonly used surface-science techniques for application to the study of gas-sensor surfaces.

Name	Acronym	Primary surface information
Transmission electron microscopy	TEM	Surface morphology
High-resolution transmission electron microscopy [97–101, 109, 116]	HRTEM	Surface morphology with atomic scale resolution
Scanning electron microscopy [97, 98, 100]	SEM	Surface morphology
Atomic force microscopy [97, 101]	AFM	Surface morphology
Scanning tunneling microscopy [105, 107, 108]	STM	Surface morphology with atomic scale resolution
Fourier transform infrared spectroscopy [109–112]	FTIR	Bonding geometry and strength, surface reactions, and surface species
Surface photovoltage spectroscopy	SPS	Energy position of surface state, surface charge distribution,
Electric-field-induced surface photovoltage spectroscopy [103]	EFISPS	bandgap, photosensitization in semiconductors
X-ray photoemission spectroscopy [96, 105, 115–117]	XPS	Composition, oxidation state
Synchrotron radiation photoelectron spectroscopy [102]	SRPS	
Ultraviolet photoemission spectroscopy [96, 102]	UPS	Valence band structure
X-ray diffraction [96, 115, 116, 118]	XRD	Surface structure
Selected area electron diffraction [53, 97]	SAED	Surface structure
Electron energy loss spectroscopy [96]	ELS	Electronic structure and surface structure
Auger electron spectroscopy [96, 106]	AES	Chemical composition
Scanning Auger microscopy [96, 105]	SAM	Map of the distribution of an element over the surface
Raman spectroscopy [104, 116]		Surface structure
Electron paramagnetic resonance [119]	EPR	Surface structure
Electrical spectroscopy for chemical analysis [120]	ESCA	Composition and oxidation state
Thermal desorption spectroscopy [115]	TDS	Composition, heat of adsorption, and surface structure
Energy dispersive X-ray spectroscopy	EDX	Composition, oxidation state

available sensors or sensors from research labs are tested and used without changing their conditions of preparation. The results of the experimental approach are necessary for optimizing the sensor parameters, and they constitute a useful practical guidance for the *developers*. In production, during the sensor-processing phase, the approach is different because the efforts are devoted to developing sensors that have been optimized by systematically changing all of the system parameters, that is, preparation of gas-sensitive films, doping and aging procedures, device structure (substrates, electrodes, heater, packaging), materials for filters, modulation conditions during sensor operation, and so on, for different applications. In spite of extensive activities in the scientific research and development of chemoresistive gas sensors based both on "trial and error" or on an empirical systematic research approach, our present comprehension of the elementary steps in gas-sensing phenomena is still poor. A basic science understanding of the processes underlying sensor operation is a fundamental prerequisite for further improvements of these types of gas sensors. The processing of the sensors in production, the experimental characterization procedures, and the theoretical studies on sensor behavior have to be closely linked, and the advances of one field have to be functional for the advances of the others. A better connection among theoretical, experimental, and technological aspects of semiconductor gas sensors is essential for a complete description of gas-sensing mechanisms [56, 93]. Basic research on gas sensors provides a framework for experimental results, and attempts to define atomistic models that relate the behavior of a sensor system as a whole to the elementary mechanisms of chemical sensing, and to the assumed electrical properties and grain structure of sensor (*modeling*) [121, 122]. In such a way, the macroscopic response of a sensor, consisting of a modulation of its surface resistance, is explained from a microscopic point of view based on gas/semiconductor interaction processes and on the electrical transport properties of the sensing layers. To this purpose, the basic scientific approach, by interpreting experimental data provided by gas-sensing tests and surface microscopic and spectroscopic analysis techniques, performs quantum-mechanics-based calculations, considers thermodynamic and kinetics aspects of gas-sensing mechanisms, and applies semiconductor physics theory. By using quantum-mechanical calculations, the energy levels related to surface and adsorbed molecules can be determined, while by using thermodynamics and kinetics, these calculated electronic levels can be linked to the surface coverage of adsorbed species corresponding to the real experimental condition (dependent on sensor working temperature, gas concentration, etc.). Finally, changes in the concentration and mobility of free charge carriers and in the sensor resistance are correlated on the basis of specific models for electrical conduction mechanisms [56].

The receptor function by which a sensor recognizes a chemical species at the surface of semiconducting metal-oxide film includes adsorption processes (physisorption, chemisorption, and ionosorption), chemical and catalytic reactions at surface adsorption sites, and at grain boundaries and similar reactions at electrode/semiconductor interfaces or at surface metallic clusters. The influence of these surface chemistry phenomena on the sensor response may be understood on the basis of the charge-transfer model (CTM) and the modified band model of semiconducting metal-oxide sensor devices that also take into account the effects of additives, dopants, grain size, as well as contacts [18, 122, 123]. According to these models, the changes in the electrical resistance of the sensor are described by the formation of depletion space-charge layers at the surface and around the grains, with upwards bending of the energy bands. Surface-energy barriers for conduction electrons result, whose height and width are variable, depending on the occupancy of surface states related to adsorbed species. Other models, such as the Volkenstein model (VM) [124], have been defined to describe the dependence of conductivity variations on gas interaction.

However, a complete comprehensive theory on the gas-sensing mechanism is still missing due to the real complexity of the specific reactions between gas species and a semiconducting polycrystalline metal-oxide film when the latter is exposed to a reactive ambient. Moreover, the microstructure of the sensing material transduces such gas reactions into a measurable sensor signal. For a basic understanding of the gas-sensing mechanisms in a resistive chemical gas sensor, it is useful to separate the receptor function from the transducer one. By a combination of various transducers and operation modes, a large variety of sensor signals results. Next, in this chapter we will present a review of the necessary theoretical background in gas-sensing mechanisms by considering both receptor and transducer functions. Finally, particular interest will be devoted to the grain-size effect on the gas-sensing properties.

2.2. Receptor Function

2.2.1. Intrinsic and Gas-Induced Surface States

The surface of a solid, even when ideally perfectly "clean," that is, without impurities, deposited metals, or adsorbed species, is itself an imperfection for the crystalline lattice because the periodicity of the crystal is roughly broken. The atoms or ions on the semiconductor surface are not surrounded by the exact number of their nearest neighbors like the corresponding atoms in the bulk; hence, they have to rearrange themselves, producing substantial deformation and a large local electrical field. This condition gives to the surface itself a reactivity toward the environmental components. Quantum-mechanics calculations, that take into account only the interruption of lattice periodicity, demonstrate that a solid surface leads to the formation of "intrinsic" localized energy levels that may lie in the forbidden energy region of a semiconductor material [18, 125]. In this case, these surface states can capture or inject charge carriers to crystal bulk—they can be acceptors or donors or both.

In highly ionic semiconductors, any surface ion has an incomplete coordination; hence, they can electrostatically interact with molecules in the environment. Most metal oxides are ionic semiconductors, with the metal atoms as cations and the oxygen atoms as anions; the surface metal cations tend to capture extra electrons (acting as acceptors), while the surface oxygen anions tend to give up electrons (acting as donors). In a semiclassical approach, Madelung calculated, in a surface site of interest, the potential due to all of the point charges in the crystal, and he showed how this electrostatic effect gives rise to intrinsic surface states.

In covalent semiconductors, on the other hand, there are covalent bonds resulting from the overlap of valence orbitals along preferential directions, with each atom contributing an electron; hence, for each covalent bond, there is an electron pair shared by the atoms. At the surface, a surface atom finds no neighbor to pair its electron with, that is, a valence orbital of a surface atom facing the ambient finds no neighboring orbital with which to be overlapped. The orbital with an unpaired electron and extending out of the surface forms the so-called "dangling bond." Now, this dangling bond can either accept another electron from an external or it can be given to the bulk of the crystal, leaving the surface energy level (the "surface state") unoccupied. Again, both acceptor and donor electronic levels may be available at a clean surface. Shockley first calculated the intrinsic surface states for a covalent semiconductor.

Although it is really impossible to obtain a perfectly clean surface, that has no practical application, the study of the electrical effects of surface states at a clean surface constitutes the first fundamental step toward the comprehension of a more important and realistic case where gases and/or odorous molecules are adsorbed onto the semiconductor surface and complex adsorbates may be formed.

The effects of impurities and chemical compounds adsorbed from the environment are very similar to the effects of the "intrinsic" surface states at a clean surface, although they are complicated by adsorption/desorption processes, relocations in the neighboring active sites, and other chemical reactions. Both kinds of surface states are dealt with in the same way by using quantum-mechanical or semiclassical methods.

In quantum mechanics, two approaches are generally applied: the rigid band and the molecular approach. In the rigid-band model, the electrons are described by nonlocalized Bloch wave functions, and localized surface states arise when the electron affinity of the surface species (adsorbed atoms or ions) is very different from that of bulk atoms or ions. The surface electronic levels due to the adsorbates can act as acceptors or donors, but local bonding of the adsorbate to one or a few atoms of the solid can be neglected, the adsorbate being treated simply as a perturbation of the electronic band scheme of the semiconductor. This model is commonly used in the treatment of adsorption with charge transfer (ionosorption) on ionic solids, such as ambient oxygen on metal oxides. Contrary to the rigid-band model, the molecular-orbital model, used especially in the case of strong interaction between the adsorbate and the solid, considers the local bonding of the adsorbate and the solid, neglecting the energy band structure of the solid.

A semiclassical approach is a useful method based on the Madelung potential, and applied in particular to ionic surface states of metal-oxide semiconductors. It considers the energy affinity of any cation or anion in the lattice as the electron affinity at infinity, slightly modified by the electrostatic potential of the ionic lattice (Madelung potential). Details of the approaches used by the two methods are given in [69, 126].

2.2.2. Physisorption and Chemisorption

Due to the reactivity of surface atoms, which lack binding partners, molecules from the gas phase will be adsorbed at the surface, leading to the previously discussed surface states. Before discussing how intrinsic and gas-induced surface states cause the formation of surface-charge layers, and consequently the modulation of the surface conductivity of semiconducting gas-sensitive films, it is worthwhile emphasizing some aspects of the adsorption processes [18, 69, 124, 126, 127].

In general, one distinguishes two major types of adsorption with respect to the underlying adsorptive forces: physical and chemical adsorption. Physisorption is a weak adsorption caused by van der Waals' forces, whereas chemisorption is based on the stronger covalent forces (overlapping between the adsorbate and adsorbent wave functions), and hence is connected with a partial electron transfer between adsorbent and adsorbate.

Starting from the physical picture that physical adsorption arises at large distances R from the adsorbent ($R \gg$ lattice constant), a molecule approaching the surface becomes slightly polarized, and induces an equivalent dipole in the adsorbent, that typically leads to 0.10–0.12 eV (i.e., a physisorption heat of about 2–5 kcal $\cdot$ mol^{-1}) dipole–dipole binding interaction of the adsorbate to the surface. According to quantum mechanics (the Volkenstein model [127, 124]), the overlap between the Bloch function of the conduction-band electron and the wave function of the ad electron tends to zero, and therefore it is sufficient to restrict the perturbation theory to second-order approximation of the dipole–dipole interaction to explain the van der Waals interaction ($\Delta E \propto R^{-6}$). As the ad particle approaches closer to the surface (R tends to zero), repulsive forces arise, so that a weak minimum in the curve of the energy potential $E(R)$ arises, and the ad particle will be physisorbed at a certain distance R_{phys} from the support surface. According to the VM model, chemisorption will take place depending on the "quantum-mechanical nature" of the system; other wave functions which allow stronger attractive forces like antiparallel spin orientation result in covalent binding. Hence, physisorption should be treated separately from chemisorption. The latter can have different degrees of chemisorption: the neutral weak-chemisorbed form and the charged strong-chemisorbed form (if charge transfer occurs). Due to the different electronic systems (wave functions) being involved in physisorption and chemisorption, the Fermi–Dirac statistics cannot describe the distribution on both states. According to the VM, this is the point at which the charge-transfer model fails because the latter uses Fermi–Dirac statistics for both physisorbed and chemisorbed states, while it is valid only for the same electronic system.

In the CTM, physisorption is a first step of association of gas species with a solid surface; then the physisorbed species can be chemisorbed/ionosorbed if they exchange electrons with the semiconductor surface. As in CTM, in the Lennard–Jones representation of adsorption processes, physisorption and chemisorption are closely linked. Moreover, the model also takes into account that, for stronger bonding of chemisorption, normally a gas species must dissociate into atoms. Figure 7 shows the energy of the system for physisorption (E_{phys}) and chemisorption (E_{chem}) as a function of the distance R from the adsorbent surface. In the case of physisorption, the system is at zero energy (defined) with infinite R, while in the case of chemisorption, it has substantial energy, even at infinite distance R, to account for the dissociation energy (E_{dis}). Physisorption, with its small heat of adsorption (Q_{phys}), is characterized by a high fractional surface coverage θ at a low temperature and a low coverage at a high temperature. The fractional

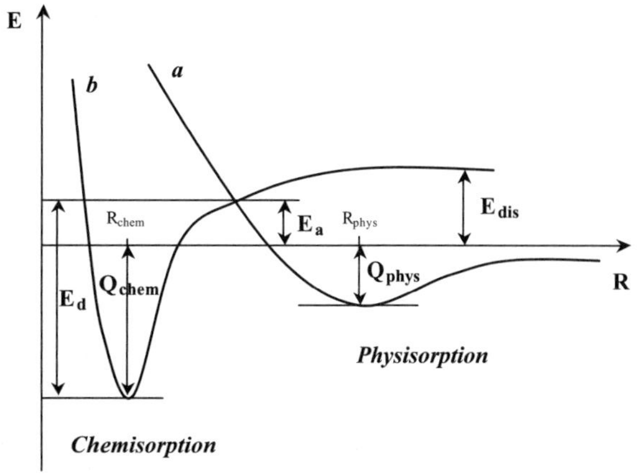

Figure 7. Lennard–Jones representation of physisorption and chemisorption: energy of the adsorbent (semiconductor surface)/adsorbate system versus the distance R between the adsorbate and the surface for physisorption (a) and for chemisorption with dissociation (b).

surface coverage θ, as long as no more than a monolayer is adsorbed, is defined as

$$\theta = N/N_t \tag{4}$$

where N is the number of adsorbed molecules per unit area, and N_t is the total surface density of adsorption sites available.

The chemisorption energy curve shows a deep minimum corresponding to the higher heat of chemisorption (Q_{chem}) ($Q_{\text{chem}} > 15$ kcal $\cdot$ mol^{-1}). The intersection of curves a and b yields an activation energy (E_a) for chemisorption. This is the energy that has to be supplied to a molecule so that it can be dissociated, and hence chemisorbed, starting from its physisorbed state, while, starting from if the molecule is in the gas phase, the total dissociation energy (E_{dis}) needs for its chemisorption. A lesser energy is thus sufficient to promote chemisorption starting from already physisorbed molecules. However, direct chemisorption is also encountered, as asserted by the Volkenstein model.

Also, desorption, like chemisorption, requires an energy (E_d) to be promoted ($E_d = Q_{\text{chem}} + E_a$). So, in contrast to physisorption, which is a slightly exothermic, unactived process, chemisorption and desorption are activated processes. The activation energies can be supplied either thermally or by a nonequilibrium process such as illumination.

Assuming a uniform distribution of identical adsorption sites, the net rate of adsorption is given by the algebraic sum of an adsorption term and a desorption term:

$$\frac{d\theta}{dt} = k_{\text{ads}}(1-\theta)P_g - k_{\text{des}}\theta \tag{5}$$

where $k_{\text{ads}} = A\exp(-E_a/kT)$ is the rate constant for adsorption and $k_{\text{des}} = B\exp(-E_d/kT)$ is the rate constant for desorption.

Setting $d\theta/dt = 0$ for the steady state leads to an equilibrium coverage θ:

$$\theta = \frac{P_g}{P_g + k_{\text{des}}/k_{\text{ads}}} \tag{6}$$

that represents the Langmuir isotherm. At very low pressures, the Langmuir isotherm reduces to the simple Henry's law:

$$\theta = \frac{k_{\text{des}}}{k_{\text{ads}}} P_g = C \exp(Q_{\text{chem}}/kT) \cdot P_g \tag{7}$$

in which the fractional coverage is simply proportional to the gas pressure. The preexponential term C is given by $C = A\sigma\tau(2\pi RTM)^{-1/2}$, where A is Avogadro's number, σ is the cross-sectional area of the adsorbed molecule, τ is the molecular vibration time, R is the universal constant of gases, and M is the molecular weight. In most real cases, Henry's isotherm is a very useful, but oversimplified approximation, inadequate to describe the more complex dependence of coverage on the pressure of the species in the gas phase. Many expressions, both analytical and empirical, have been proposed [128]. We report, as an example, the isotherm of the Volkenstein model [124, 127]:

$$\theta = \frac{\beta P_g}{\beta P_g + 1} \tag{8}$$

with

$$\beta = b\left\{f^0\left[1 + \frac{\nu^-}{2\nu^0}\exp\left(\frac{E_f - E_c}{kT}\right)\right]\right\}^{-1},$$

$$f^0 = \frac{1}{\frac{1}{2}\exp[(E_f - E_s)/kT] + 1}$$

occupation probability of the neutral weak-chemisorbed state, $E_s = E^-_{\text{ads}} - E^0_{\text{ads}}$, ν^0 and ν^- typical phonon frequencies on the order of 10^{-13} s^{-1}, E_f the Fermi level, and E_c the energy of the bottom of conduction band. In contradiction to the Langmuir isotherm, where $b = k_{\text{des}}/k_{\text{ads}}$ is a function of only the temperature, here the coefficient β additionally depends on the difference of E_f and the surface energy E_s; hence, according to VM in chemical adsorption, the fractional coverage depends not only on the external parameters gas pressure P_g and temperature T, but also on the electronic state of the system as a whole.

In Figure 8, a typical adsorption isobar, reporting the temperature dependence of the volume adsorbed, is sketched.

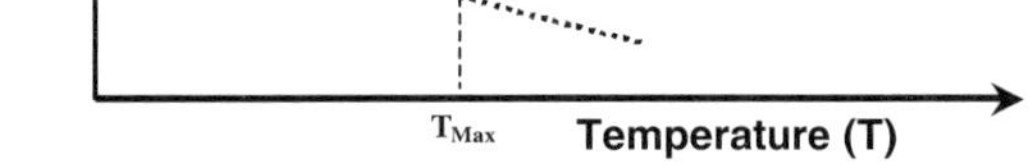

Figure 8. Typical dependence of the volume adsorbed V_{ads} on the temperature (adsorption isobar).

Physisorption, which is an unactivated process, starts from the lowest temperatures, and decreases exponentially with increasing temperature. At very high temperatures, there is equilibrium chemisorption, again decreasing with increasing temperature following the dependence of $b = k_{\text{des}}/k_{\text{ads}}$ in Eq. (7). In the intermediate temperatures, the adsorption isobar indicates an irreversible chemisorption, so that a maximum adsorption is observed at some temperature T_{max}. This can be explained by taking into account the increase of the activation energy E_a and the decrease of the heat of chemisorption Q_{chem} when increasing the fractional coverage θ. Moreover, due to the opposite variation with θ of E_a and Q_{chem}, the activation energy of desorption $E_{\text{des}} = (E_a + Q_{\text{chem}})$ will not vary too much with the coverage, and thus also the desorption rate. So in the intermediate temperature range, the higher the temperature, the higher the value of E_a reached, and thus the higher the θ value reached; in opposition to this trend, as T increases, the progressive increase of E_a with θ limits the volume adsorbed, and on the other hand, the rate of desorption is negligible because of its high activation energy E_{des}. The adsorption rate in the range of irreversible chemisorption can be expressed by the empirical Elovich equation: $dN/dt = a\exp(-bN)$, with a and b experimental adjustable parameters.

Since metal-oxide-based sensors should be reversible devices, gas molecules have to be coupled retrievably on the surface of the sensing elements so that adsorbed species can leave the sensor surface. At the same time, for a change of the electrical properties of semiconducting metal oxides, it is essential that formation of the charged adsorbed species, that is, a charge transfer between the adsorbed species and the semiconductor, takes place (ionosorption). Thus, the careful adjustment of the operation temperature and ionosorption are the most basic parameters for gas-sensor functioning.

Next, we will analyze the formation of surface space-charge layers and of surface energy barriers by surface states associated with the chemisorption of active species, in particular oxygen, whose presence is fundamental for the basic processes of sensor behavior.

2.2.3. Surface Space-Charge Layers and Charge-Transfer Model

For simplicity, we will take as a reference the case when there is no net excess charge at the surface, and the energy bands of a semiconductor have the same levels from the bulk up to the surface (ideal flat-band state); this ideal situation neglects the presence of "intrinsic" surface electronic states in the bandgap. However, in the presence of surface states introduced by chemisorbed species, the electronic band scheme near the surface has to be modified by reaching a new equilibrium. Figure 9 shows the electronic band scheme of an n-type semiconductor near its surface in the ideal flat-band state (A) and after chemisorption and charge transfer from the semiconductor to the acceptor-type surface state E_{SS} (B).

According to the just-mentioned charge-transfer model, the connection between the partial pressure of the gas phase and the concentration of chemisorbed species is interposed by physical adsorption, by considering the "kinetic" interaction between the gas phase and the physisorbed state,

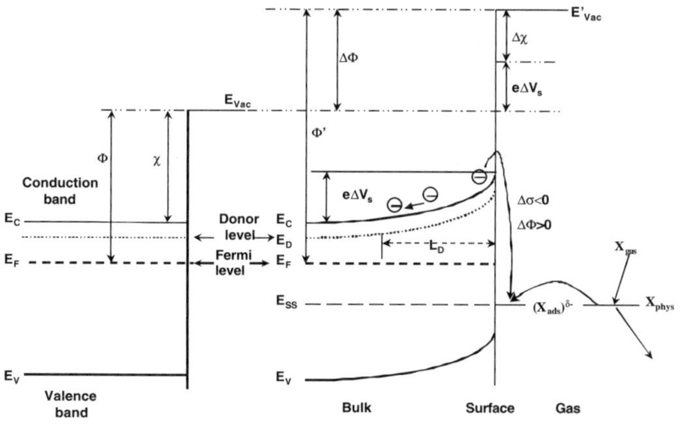

(a) (b)

Figure 9. Schematic representation of chemisorption and charge transfer at an n-type semiconductor surface in the electronic band scheme of the surface. (A) situation before the chemisorption (ideal flat-band case), and (B) band bending after charge transfer from the semiconductor to the surface state. The gas-phase molecule X_{gas} is first physisorbed in a precursor state X_{phys}. Acceptor-type states due to molecules (e.g., $X = O_2$, NO_2) with an electron affinity higher than the work function of the semiconductor ($E_{SS} < E_F$) capture electrons from the conduction band, leading to a negatively charged surface with the countercharge the positively charged donors below the surface ("double layer") and to surface energy barriers for the conduction electrons. In the sketch, the work functions Φ before chemisorption and Φ' after chemisorption, corresponding to electron affinity χ and band bending $-e\Delta V_S$, are also indicated. Chemisorption can be measured by monitoring changes of the work function ($\Delta\Phi = \Phi' - \Phi$) and changes of the surface conductivity ($\Delta\sigma$) upon occupation of surface states E_{SS} by forming $(X_{ads})^{\delta-}$.

where the physisorbed state acts as the precursor state of the chemisorbed state. Hence, as a molecule X from the gas phase approaches the semiconductor surface, it is first physisorbed, then it can be chemisorbed, and it can exchange electrons with the semiconductor. If the molecule is an acceptor-type molecule (e.g., $X = O_2$, NO_2) with an electron affinity higher than the work function of the semiconductor ($E_{SS} < E_F$) (as in Fig. 9), it attracts electrons from the conduction band, whereas if it is a donor-type molecule (e.g., $X = H_2$) with an electron affinity lower than the work function of the semiconductor ($E_{SS} > E_F$), it injects electrons into the conduction band. When such a transition occurs (specifically, Fig. 9), after the electron has moved from the immobile ionized donors to the acceptor-type surface states, a "double layer" is formed: on one side, the positively charged donor ions below the semiconductor surface as a "space-charge layer" compensated, on the other side, by the negatively charged surface states as a surface sheet of charges. An electric field develops between the double layer, and raises the surface band energies with respect to those of the surface; in such a way, the surface energy bands result bent upwards, building up energy barriers for the conduction electrons. The depth of the "depletion layer" (x_0) and the surface potential, or Schottky barrier height (ΔV_S), both depend on the electronic properties of the semiconductor and on the density of surface states.

Depending on the type of surface states (acceptor or donor type) and on the conductivity type (n or p) of the semiconductor, three types of space-charge layers may appear below the surface: (1) a depletion layer, (2) an accumulation layer, or (3) an inversion layer.

The depth of the "space-charge layer" (x_0) and the surface potential, or Schottky barrier height (ΔV_S), can be calculated by solving the one-dimensional Poisson's equation:

$$\frac{d^2V}{dx^2} = -\frac{\rho}{\varepsilon\varepsilon_0} \tag{9}$$

where V is the electric potential, x is the distance from the surface, ε is the dielectric constant of the material, ε_0 is the permittivity of free space, and ρ is the charge density in the space-charge layer, consisting of electron (n), holes (p), ionized donors (N_D^+), and ionized acceptors (N_A^-):

$$\rho = (N_D^+ - N_A^- + p - n)e \tag{10}$$

The boundary conditions are

$$V = V_S \quad \text{at } x = 0 \quad \text{and} \quad V = V_b \quad \text{at } x = x_0 \tag{11}$$

and, since at $x \geq x_0$ the semiconductor is uncharged, we can used the further boundary condition $dV/dx = 0$ at $x = x_0$.

For the condition of charge neutrality, the surface sheet of charges stored in the surface states (Q_S) has to be compensated by the space charges contained in the space-charge layer of thickness x_0 (Q_{SC}): $Q_S = -Q_{SC} = \int_0^{x_0} \rho(x)\,dx$. Assuming that all donors are fully ionized, and neglecting the electron and hole concentrations in the space-charge layer (Shockley approximation of exhaustion layer), the Poisson's equation can be solved analytically. In this case, in fact, the charge density in the space-charge layer ρ is constant:

$$\rho = eN_D^+ = eN_D, \qquad \text{for } 0 \leq x \leq x_0 \tag{12}$$

Therefore, the thickness of the depletion layer results to be

$$x_0 = -Q_S/eN_D. \tag{13}$$

By twice integrating the Poisson's equation reduced to the simpler form

$$\frac{d^2V}{dx^2} = -\frac{eN_D}{\varepsilon\varepsilon_0} \tag{14}$$

we obtain

$$V(x) = -\frac{eN_D}{2\varepsilon\varepsilon_0}(x - x_0)^2 + V_b \tag{15}$$

So the height of the Schottky barrier $e\Delta V_S = e[V(0) - V_b]$ is given by

$$e\Delta V_S = -\frac{eN_D x_0^2}{2\varepsilon\varepsilon_0} = -\frac{Q_S^2}{2eN_D\varepsilon\varepsilon_0} = \frac{e\left(\sum_j N_{S,j}\alpha_j\theta_j\right)}{2N_D\varepsilon\varepsilon_0} \tag{16}$$

where N_{Sj} is the number of surface states of the adsorbed species j with charge α_j electronic units per unit area, and θ_j is the fractional coverage of surface states (occupancy) of the type j.

A rate equation simulation with mobile donors has been considered by Lantto and co-workers; the solution $e\Delta V_s = -kT\ln\lfloor 1-(Q_s^2/2eN_{Ds}\varepsilon\varepsilon_0 kT)\rfloor$ gives much smaller, and also more realistic values for $e\Delta V_S$ compared to those calculated with a constant distribution of donors [129, 130]. It seems that the grain geometry has a marked relative effect on the $e\Delta V_S$ values only in grains with $L < 10$ nm, as results from the mathematical analysis reported in [133].

Of course, the presence of Schottky barriers at surface and at grain boundaries directly influences the electrical conductance of the gas-sensing layer. However, the real situation for a gas-sensing layer that consists of numerous interconnected metal-oxide nanocrystals with different intergranular contacts is very complex. The heights of these potential energy barriers may vary due to grain size, or the size and shape of the necks between grains [134]. The microstructure is a fundamental in the electrical conduction of the sensing materials, as will be discussed in Section 2.3.

2.2.4. Gas Interaction

The Role of Oxygen In general, a semiconducting metal oxide is a nonstoichiometric oxide, and more commonly, an oxygen-deficient oxide containing oxygen vacancies. As surface defects, they may influence chemical reactions with adsorbates; on the other hand, they behave as donors in the bulk. Indeed, bulk oxygen vacancies are known to be the origin of the n-type behavior of tin oxide (SnO_2)—the most used material in semiconductor gas sensors—but also of other oxides [135]. SnO_2 is a wide-bandgap ($E_{gap} = 3.6$ eV) semiconductor with a thermodynamically stable rutile structure. Two donor levels E_{D_1} and E_{D_2} due to singly and doubly ionized oxygen vacancies are located at 0.03 and 0.15 eV below the Fermi level [131, 136]. However, according to theoretical calculations by Rantala et al., it seems that surface vacancies have no influence on the conductance, but they have to move toward the bulk in order to become electrically active [132]. In fact, at high temperature (700–1100 °C), oxygen vacancies diffuse quickly from the interior of grains to the surface or from the surface to the interior of the grains, and the basic reaction between oxygen molecules in the gas phase $O_{2(gas)}$, and the oxygen vacancy $V^{2+}_{O(bulk)}$ in the bulk can be expressed as

$$O_{lattice} \leftrightarrow V^{2+}_{O(bulk)} + 2e^- + \frac{1}{2}\cdot O_{2(gas)} \quad (17)$$

In this high-temperature range, semiconducting metal oxides are used as oxygen sensors due to their strong dependencies of electrical conductivity σ on oxygen partial pressure P_{O_2}:

$$\sigma = A\exp(-E_A/kT)\cdot P_{O_2}^{1/m} \quad (18)$$

where E_A is the activation energy which represents the sensitivity of the electrical conductivity to temperature changes, and the absolute value of $1/m$ depends on the dominating defects involved in the reaction between oxygen gas and the sensor [81].

At low and medium temperatures (about 200–500 °C), that is, in the working temperature range of resistive-type gas sensors, chemisorption of oxygen at the surface of oxide and at the grain boundaries leads to the formation of surface oxygen ions and Schottky barriers between grains. According to the above-discussed model, the height of the Schottky barriers depends on the density of the ionosorbed gas molecules. Since semiconducting metal-oxide-based gas sensors usually work in air, and oxygen is known as a main active component of the air participating in charge-transfer processes at the oxide surfaces, the analysis of the interaction of oxygen species with the semiconductor surface and with other adsorbed species is crucial.

It has been shown, by using different experimental methods (infrared analysis IR [137], temperature-programmed desorption TPD [138–140], and electron paramagnetic resonance EPR [119, 141]) that oxygen is adsorbed onto the surface of semiconducting n-type metal oxides in the molecular neutral form (O_2 physisorbed) and in different charged forms: mainly $O^-_{2,ads}$ and O^-_{ads}, and seldom O^{2-}_{ads}. This last form, doubly charged adsorbed oxygen, is unstable, and its stabilization should signify lattice incorporation.

Oxygen chemisorption is an activated process; the activation energy is lower for $O^-_{2,ads}$ than for O^-_{ads}; therefore, $O^-_{2,ads}$ is found on the surface at temperatures below 200 °C at which no O^-_{ads} is found. The energy level for O^-_{ads} is deeper than the level for $O^-_{2,ads}$, which first desorbs when the temperature increases further. So below 200 °C, the molecular species ($O^-_{2,ads}$) dominate, and above this temperature, the atomic species (O^-_{ads}, O^{2-}_{ads}) dominate. The adsorption sites for oxygen are the surface oxygen vacancies. We can describe in a simple model the adsorption of oxygen using the quasichemical reaction steps:

$$O_{2,gas} \longleftrightarrow O_{2,phys} \qquad \text{physisorption} \quad (19a)$$

$$O_{2,phys} + e^- + S \longleftrightarrow O^-_{2,ads} \qquad \text{chemisorption} \quad (19b)$$

$$O_{2,phys} + 2e^- + 2S \longleftrightarrow 2O^-_{ads} \quad (19c)$$

or

$$O^-_{2,ads} + e^- + S \longleftrightarrow 2O^-_{ads} \quad (19d)$$

$$O^-_{ads} + e^- \longleftrightarrow O^{2-}_{ads} \longleftrightarrow O^{2-}_{lat} \qquad \text{lattice incorporation} \quad (19e)$$

where S is an adsorption site for oxygen (i.e., an oxygen vacancy $V_{O(surf)}$).

As a result of this ionosorption of oxygen, conduction electrons are depleted from the surface regions, and potential energy barriers are developed, as discussed in the previous paragraph. So, in air, the most common operating ambient for gas sensors, adsorbed oxygen is the dominant contributor to the negatively charged surface states referred to in Eq. (16), providing the main surface coverage θ with ionosorbed oxygen. It has been experimentally observed that the coverage with ionosorbed oxygen can increase up to a saturation level. The concentration of negative charges on the metal oxide surface (in the present case, ionosorbed oxygen) is limited, as is directly seen from Eq. (16), when a reasonable limit (say 0.5–1 eV) is placed on ΔV_S. A maximum of 10^{12}–10^{13} cm^{-2} surface charges is expected (Weisz limitation), while a simple evaluation of depletion layer thickness x_0 [Eq. (13)] gives $x_0 = 1$–100 nm. It is equally important to note the corollary that about 10^{12}–10^{13} cm^{-2} surface states will normally completely control the surface barrier. With about this density of surface states, the Fermi level of the semiconductor will equalize the electrochemical potential of electrons in the surface states; indeed, since the system semiconductor–adsorbent/adsorbate is at equilibrium, the Fermi level is constant through all phases. This limitation on the amount of ionosorbed species, and the resulting ability to control the depletion layer with this density of states, was first pointed out by Weisz [142]. The term for such surface control is *Fermi energy pinning*, that is, the Fermi energy in the semiconductor is pinned at the value determined by the surface states.

The effect of adsorbed oxygen on the electrical properties of the semiconductor is to cause a high resistance in *n*-type semiconductors (as they are mostly metal oxides) [144]. Kačiulis et al. studied the effect of surface oxygen species on the chemoresistance by XPS of the surface chemical composition before and after different sample treatments [143]. Reducing gaseous species (e.g., CO) react with oxygen at the surface of the semiconductor; they remove a fraction of the ionosorbed oxygen, restoring, in such a way, electrons to the conduction band, and decreasing the resistance of the semiconductor. This sensitivity of resistance to the coverage of adsorbed oxygen leads to a close connection between the resistance and the concentration of the reducing agent in the gas phase, and this is very schematically the way in which standard semiconductor gas sensors work to detect traces of reducing gases in air.

Next, some gas interactions will be reported in more detail.

Water Vapor In many applications, water vapor is present as an interfering compound, hence, the interaction of a metal-oxide semiconductor with water vapor is of great interest in gas sensing. It is well-known experimental evidence that water vapor increases the conductance of semiconductor gas sensors, in a similar way to reducing gases [153]. At low temperatures, water vapor is physisorbed as a molecule, but this specie, completely desorbed below 400 K, does not make any contribution to electrical conduction, while at higher temperatures, H_2O is chemisorbed after dissociation in an hydroxyl group OH^-, that shares its electronic pair with a lattice tin atom Sn_{lat}, and in a weakly bonded proton H^+, that reacts easily with lattice oxygen or with adsorbed oxygen [140, 144, 145].

Two surface interaction mechanisms between SnO_2 and water vapors have been suggested [146, 147] to explain the conductance increase. In the first mechanism, the hydrogen atom H, resulting from a homolytic dissociation of an H_2O molecule, binds itself to a lattice oxygen O_{lat} in a hydroxyl group $O_{lat}H^-$, creating a free electron:

$$H_2O + Sn_{lat} + O_{lat} \longrightarrow (HO - Sn_{lat}) + O_{lat}H^- + e^- \quad (20)$$

In the second mechanism, a "rooted" specie, consisting of two hydroxyl groups bound to two Sn atoms, is formed. One derives from the dissociation of a water molecule, and the other from the reaction between the hydrogen and the oxygen lattice. The result is the formation of surface oxygen vacancies:

$$\begin{aligned} H_2O_{gas} + 2Sn_{lat} + O_{lat} &\longrightarrow 2(OH - Sn_{lat}) + V_{O(surf)} \\ V_{O(surf)} &\longrightarrow V^{2+}_{O(bulk)} + 2e^- \end{aligned} \quad (21)$$

These surface oxygen vacancies can diffuse to the bulk, where they can be ionized, hence acting as an electron donor.

It can be observed that, according to these two interaction mechanisms, the adsorption sites for H_2O are proportional to O_{lat} and Sn_{lat}, while the adsorption sites for gaseous oxygen are the oxygen vacancies V_O. However, water vapor adsorption creates oxygen vacancies that can be occupied by chemisorbed oxygen species (O^-_{2ads}, O^-_{ads}), compensating for the resistance decrease due to water adsorption. Since the number of adsorption sites O_{lat} and Sn_{lat} is normally quite a lot more than the adsorption sites $V_{O(surf)}$, this explains the predominant resistance-decrease effect in humid air compared with the resistance-increase effect due to the only oxygen adsorption in dry air.

Morrison [126], as well as Henrich and Cox [148], considered an indirect effect, that is, the interaction between either OH^- or H^+ with an acidic or basic group, which are also acceptor surface states. The coadsorption of water with another adsorbate, which could be an electron acceptor, may change the electron affinity of the latter. Henrich and Cox suggested that preadsorbed oxygen can be displaced by water adsorption. Also, Caldararu et al. [149] assumed a blocking of the adsorption sites for oxygen by water. In fact, if there are many surface oxygen vacancies, it is difficult for water to find enough lattice oxygen to be adsorbed. On the contrary, if the level of humidity is too high, so that there is a large coverage with hydroxyl groups OH^-, the adsorption of oxygen can be limited. In this condition, the presence of water vapor can interfere with oxygen adsorption, not allowing gaseous oxygen to be adsorbed onto the surface vacancies. The relative surface distribution and coverage of hydroxyl groups and oxygen species may be related to the history and thermal treatments of the samples. This is why the effect (enhancing or interfering) of humidity on the response to reducing gases is often ambiguous and controversial [150–152, 162].

CO Carbon monoxide is generally accepted to act as a reducing gas for a metal-oxide semiconducting film [145, 154–158], even if some anomalous responses have been observed [159, 160]. All experimental studies in air in a temperature range of 150–450 °C reported a decrease of the surface resistance in the presence of CO. CO molecules react with ionosorbed oxygen species adsorbed or lattice oxygens [148]. IR studies identified CO-related species, that is, unidentate and bidentate carbonate between 150 and 400 °C and carboxylate between 250 and 400 °C [161]. Moreover, the formation of CO_2 as a reaction product between 200 and 400 °C was identified [137].

$$\text{a)}\ CO_{gas} + 2O^-_{ads} \longrightarrow CO_3^{2-} \begin{cases} \longrightarrow CO_{2gas} + O^{2-}_{ads} \\ \longrightarrow CO_{2gas} + O^-_{ads} + e^- \\ \longrightarrow CO_{2gas} + 1/2 \cdot O_{2gas} + 2e^- + S \end{cases} \quad (22)$$

$$\text{b)}\ CO_{gas} + O^-_{ads} \longrightarrow CO_2^- \ (\uparrow O^-_{ads}) \downarrow CO_{2gas} + e^- + S$$

CH_4 Experimental studies report a decrease in the sensor resistance in the presence of methane. Since methane is thermodynamically more stable than most other reducing gases, it is relatively difficult to detect it with high sensitivity at temperatures lower than 350–400 °C. Only a few works report the detection of CH_4 at low temperature [165]. The CH_4 detection mechanism involves the oxidation reactions of methane with the adsorbed oxygen species, but also with the lattice oxygen, especially at high temperatures, producing CO_2 and H_2O via CH_n or CH_nO intermediates ($0 < n < 4$) [163, 164].

Kohl assumed, based on TDS, two different pathways for methane [144, 147]. The first involves the reactions with ionosorbed oxygen, and the second involves the reaction with lattice oxygen (Fig. 10). The decrease of the resistance of a semiconducting gas sensor as a consequence of the interaction with methane is due to the formation of surface oxygen vacancies and/or electrons.

For the influence of water on methane, several models have been proposed [145, 151, 166]. Egashira et al. assumed that water blocks the adsorption sites for methane to explain the decreasing interaction of CH_4 with oxygen decrease in the presence of water [166]. Ionescu et al. proposed a model, which is based on the measurement of conductance transients during and after rapid transitions from dry into humid air containing CH_4 and CO, respectively [151]. They assumed that CH_4 reacts with lattice oxygen, thereby competing with water for the same oxygen site, whereas CO reacts mainly on other oxygen sites, which are not involved in the reactions with water vapor.

Moreover, in considering the gas/solid interaction phenomena at the surface of a gas-sensor device, we have to remember the role of the metal electrodes, which give two different contributions: an electrical and a chemical contribution. An electrical contribution of the semiconducting sensing layer/electrode interface to the overall sensor resistance has, in fact, to be considered. Moreover, a chemical influence of, for example, the catalytic activity of the contact material in the region close to the contact may occur. The electrode geometry and position effects are also important for semiconductor gas sensors. Several works that took into consideration the role of electrical contacts have been produced [156, 162, 167–174].

Figure 11 shows schematically a typical situation for a conductometric gas sensor based on polycrystalline film, including the current path through, the role of adsorption, charge-transfer processes, surface chemical reactions, the effect of contacts, and also possible catalytic effects at the contact area.

2.3. Transducer Function

2.3.1. Electrical Conduction Mechanisms

The microstructure of the sensing layers together with the chosen operation mode determine how the above-discussed surface reactions, which allow us to detect a chemical species, are transduced in the output signal. The microstructure is strongly dependent above all on the deposition technique, but also on the addition of dopants, promoters, as well as the chosen burning-in cycles. Both thin-film and thick-film deposition techniques are available for the production of metal-oxide layers for semiconducting gas sensors [175]. Most of the chemical (CVD) and physical (PVD) vapor deposition techniques provide thin films with microstructures and transducer functions which are very different from the one obtained by thick-film deposition.

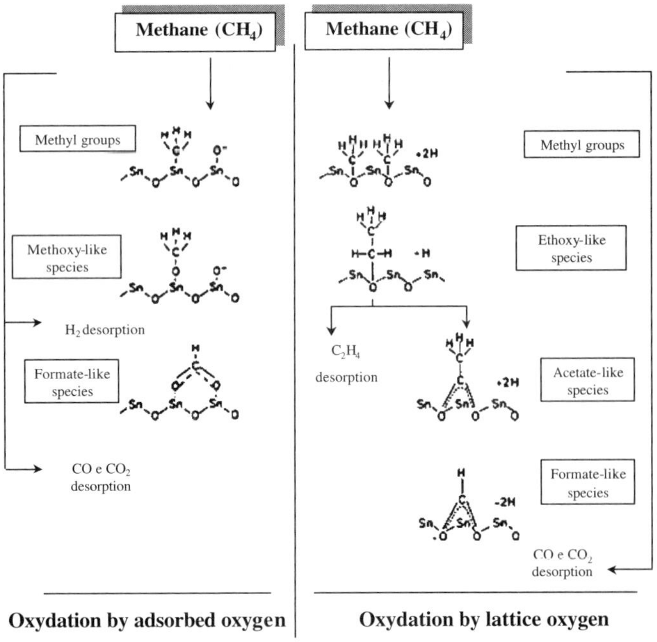

Figure 10. Diagram of the methane oxidation process by lattice oxygen (right side) and adsorbed oxygen (left side).

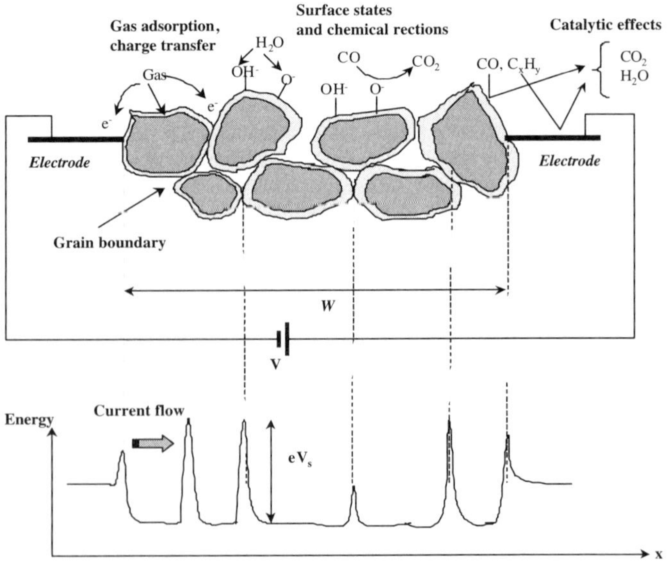

Figure 11. Scheme of a conductometric gas sensor with electrodes placed high up in the sensor face. The figure shows the energy profile of the semiconductor along the path between a pair of electrodes with the energy barriers at grain boundaries and at the metal/semiconductor interface, the species (O_2, H_2O, CO) adsorption, the charge transfer from the gas to sensor or to electrodes, the surface chemical reaction between CO and chemisorbed oxygen species, and possible catalytic effects at the contact area.

According to Bârsan et al., a simple distinction between two different types of layers can be made [56, 156].

- Compact oxide layers, in which the interaction with chemical species in the gas phase takes place only at the geometric surface. The electron flow is parallel to the surface, that is, to the depletion layer.
- Porous polycrystalline layer, in which the volume of the layer is also accessible to the gases, and in this case, the active surface is much higher than the geometric one.

The former is obtained by most CVD and PVD techniques, and the latter is characteristic of thick-film deposition techniques and thin-film deposition techniques such as RGTO[1] [175, 176] and sol–gel process [194–200], which result in porous layers with a large surface area. In Section 3.2, a brief survey of some the most used depositions will be given.

For a compact layer, two kinds of behavior can be distinguished, depending on the ratio between the layer thickness t and the width of the depletion layer x_0 (or, equivalently, the Debye length[2] L_D).

1. In the first case, the layer is completed depleted ($t \sim L_D$); the conductance is limited by surface traps, that is, the conductance is determined by the electrons excited from the surface states, and their activation energy equals the surface-state energy E_{SS} (*surface-trap-limited conduction*).
2. In the second case, the layer is partly depleted ($t > L_D$), and the surface reactions do not influence the conduction in the entire layer. The conduction takes place in a bulk layer of variable width ($t - x_0$) (much more conductive than the depleted layer) by means of electrons provided by donor states, with the energy of the bulk donor E_D as the activation energy. The effect of surface states and surface gas reactions on the conduction is to modulate the conduction channel width ($t - x_0$). This behavior is known as *bulk-trap-limited conduction* (see Fig. 12). Based on simple geometric considerations, the conductance G of a thin single crystal layer of thickness t, width W, and length L is given by

$$G = G_0\left(\frac{Wt}{L}\right)\left(1 - \frac{x_0}{t}\right) \quad \text{with} \quad x_0 = \sqrt{\frac{2e\Delta V_S \varepsilon\varepsilon_0}{e^2 N_D}} \tag{23}$$

For the case of a completely depleted layer in the absence of reducing gases, it is possible that exposure to reducing gases acts as a switch to the partly depleted layer case (due to the injection of extra free electrons). It is also possible that exposure to oxidizing gases acts as a switch from the partly depleted layer to the completely depleted layer. However, it is important to emphasize that, in this situation, the sensitivity to gases is poor, and the sample has to be very thin in order to obtain appreciable sensitivity values.

For porous nanocrystalline layers, the conduction mechanism is more complex, depending on the presence of necks

[1] Rheotaxial growth and thermal oxidation of tin layers.

[2] $L_D = \sqrt{(\varepsilon\varepsilon_0 kT/n_0 e^2)}$.

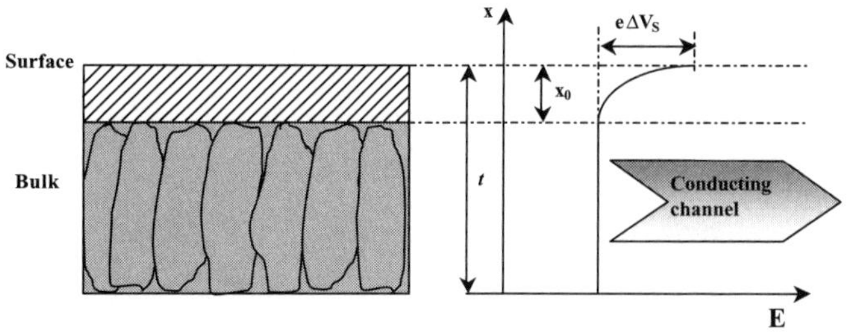

Figure 12. Schematic representation of a compact sensing with geometry and energy band representation. x_0 is the thickness of the depleted surface layer, t is the layer thickness, and $e\Delta V_S$ is the height of the energy barrier.

between grains, on Schottky barriers, and on the grain size. The nanocrystalline structure implies lower resistance paths through the bulk of crystallites alternating with higher resistance constrictions at the points of contacts. Three general cases may be considered, according to the degree of sintering of the material. They are illustrated in Figure 13.

Figure 13a shows the case of a well-sintered material with a fully open "neck" between adjacent grains. Surface states cause depletion layers extending below the grains to a depth marked by the dashed lines on both sides of the neck. In this case, the conduction mechanism takes place in a bulk layer of variable width ($z_n - 2z_0$) (much more conductive than the depleted layer) by means of electrons provided by donor states, with the energy of the bulk donor E_D as the activation energy. The effect of surface states and surface-gas reactions

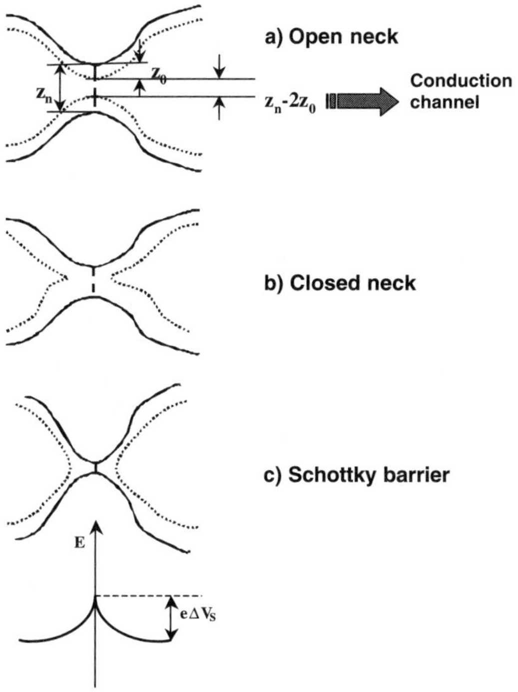

Figure 13. Three models for conductance limited by intergrain connections. a) Open neck (*bulk-trap-limited conduction*), b) closed neck (*surface-trap-limited conduction*), c) Schottky barrier. In all of the cases, we considered the diameter of the grain l higher than the Debye length L_D ($l > L_D$).

on the conduction is to modulate the conduction channel width $(z_n - 2z_0)$. This behavior is known as *bulk-trap-limited conduction*.

Figure 13b shows the case of a "closed" neck in which the depletion zones from the surfaces of the adjacent grains overlap, leaving a higher resistance path through the center. This geometry corresponds to a less complete sintering that left a narrower neck or to an "open" neck that has been constricted by surface reactions to a change in the depletion layer thickness. In this case, the conductance is directly influenced by the gaseous ambient through the occupancy of the surface states; the conductance is determined by the activation of the electrons from the surface states into the conduction band (*surface-trap-limited conduction*).

Figure 13c shows a third situation where a Schottky barrier is formed at the boundaries of two adjacent grains. In this case, the conduction mechanism between grains is a thermionic emission, the conductance G depending on the barrier height $e\Delta V_S$:

$$G = G_b \exp\left(-\frac{e\Delta V_S}{kT}\right) \tag{24}$$

This conduction mechanism (*barrier-limited conduction*) is clearly different from the other two cases, $e\Phi_b$ being the activation energy for conductance directly related to the square of the surface charge. The changes of Schottky barriers as a function of the ambient gaseous composition lead to a high gas sensitivity.

According to the grain boundary carrier-trapping model, in this latter situation, the presence of intergrain boundaries results in band bending, which sets up potentials barriers to current flow (effectively reducing the carrier mobility μ) or in a severe carrier depletion which reduces the carrier density n considerably below the bulk doping level N (or, in some conditions, to a combination of these two effects). The density n_t (per unit area) of charges trapped in extrinsic interface states depends on the environmental composition via the adsorption of active gaseous species. The density N_t of surface states determines the maximum amount of charge which can be trapped (Weisz limitation), while the trap energy E_t may determine the activation energy for free-carrier freeze out.

The standard treatment of the electrical transport phenomenon in polycrystalline thin films is due to Petritz [207]. The model is based on a thermionic emission mechanism between adjacent grains. Assuming that the electrical resistivity in the grain is negligible with respect to the electrical resistivity in the boundary regions ($\rho_{\text{grain}} \ll \rho_{\text{boundary}}$), he showed that the total electrical resistivity is controlled by the energy barriers at the grain boundaries:

$$\sigma_g = eN\mu_0 \exp\left(\frac{-e\Phi_b}{kT}\right) \tag{25}$$

where N is the bulk doping level, $\mu_0 = (M/n_c kT)$, n_c is the number of crystallites per unit length along the film, and M is a factor that is dependent on the barrier, but independent of Φ_b (i.e., $e\Delta V_S$). Since $\rho_{\text{grain}} \ll \rho_{\text{boundary}}$, Petritz also assumed that the carrier concentration was not reduced by the exponential term, but rather, that all carriers took part in the conduction process, but with reduced mobility. The effective carrier mobility is so thermally activated:

$$\mu_g = \mu_0 \exp\left(\frac{-e\Phi_b}{kT}\right) \tag{26}$$

Seto [208] developed a more comprehensive analysis of the grain boundary-trapping model based on some assumptions. He modified the preexponential term in Eq. (26):

$$\mu_0 = el\left(\frac{1}{2\pi m^* kT}\right)^{1/2} \tag{27}$$

where l is the average grain size and m^* is the electron effective mass.

Thermionic emission results from those carriers with high enough energy to overcome the Schottky barrier at the grain boundary, but a contribution to current across the grain boundary also comes from tunneling from those carriers with less energy than the barrier height that go through the barrier by quantum-mechanical tunneling.

The Hall effect measurements are fundamental tools in order to obtain information about the electrical transport properties of polycrystalline gas-sensing films [210]. Since both μ and n can control the conductivity σ, the Hall effect allows us to distinguish if σ is mainly controlled by intergrain depletion barriers (modulation of μ) or by the carrier depletion in the grains (modulation of n) or by a combination of the two effects. However, more care has to be taken in the interpretation of the results of Hall effect measurements, depending on the relative magnitude of n_t compared to Nl and on the Debye length L_D compared to the grain size l. According to the Orton and Powell treatment of the Hall effect in polycrystalline thin films, three situations can be distinguished [209]. Figure 14 summarizes the three different conduction band profiles that can occur for an n-type film, and Table 3 shows the corresponding expressions for n and μ for the three regimes.

a) $\boldsymbol{n_t < Nl}$: The resulting depletion layer width W is less than $l/2$, and the grains are partially depleted. In this case, the carrier concentration is quite high compared with n_t, and the depletion layer is just located at the surface of the grains, so that the Hall effect measures the carrier concentration N in the bulk of the grains. The conductivity σ will be mainly controlled by intergrain depletion barriers with a thermally activated Hall mobility given by Eq. (26), where μ_0 is related to the grain diameter l according to Eq. (27). Hence, the activation energy for Hall mobility is $E_\mu = e\Phi_b$.

The height of Schottky barriers, calculated by solving the Poisson's equation and by equating the interface charge to the fixed charge in the depletion layer (i.e., $n_t = 2NW$), is given by

$$e\Phi_b = \frac{e^2 n_t^2}{8\varepsilon N} \tag{28}$$

b) $\boldsymbol{n_t \approx Nl,\ L_D < l/2}$: The depletion layer extends right through the grain. In this case, the carrier concentration n may be very much lower than the bulk doping level N; in particular, it is given by the following relationship:

$$n = \sqrt{2\pi}\frac{L_D}{l}\frac{NN_c}{(N_t/l) - N}\exp\left(-\frac{(E_c - E_t) - e\Phi_b}{kT}\right) \tag{29}$$

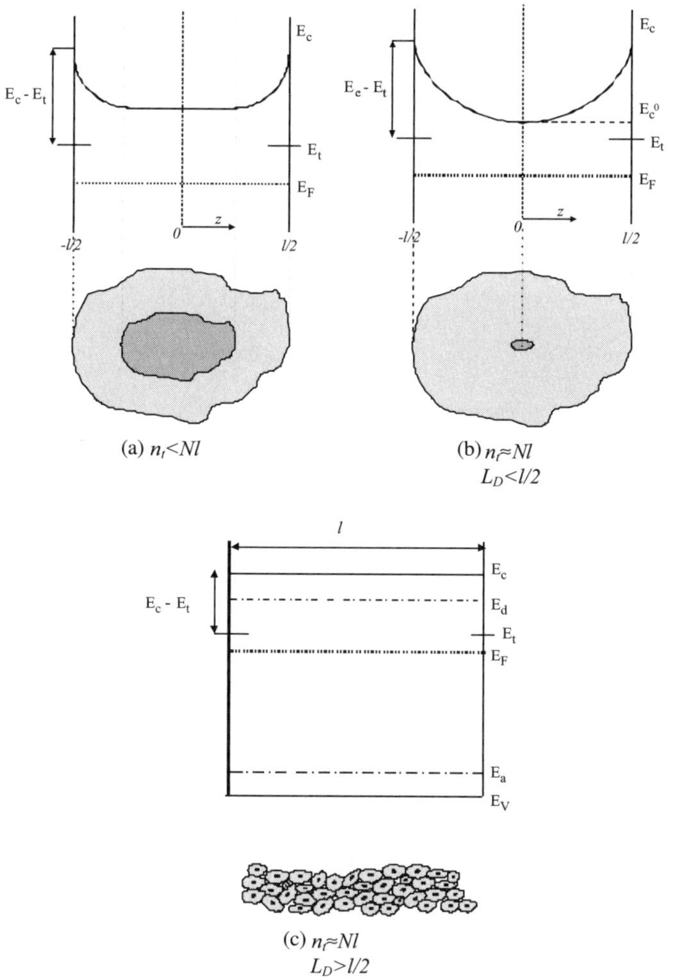

Figure 14. Conduction band profiles through single grains of an n-type film for the three cases discussed by Orton and Powell [209]: a) $n_t < Nl$, b) $n_t \approx Nl$, $L_D < l/2$, c) $n_t \approx Nl$, $L_D > l/2$.

and μ is still barrier limited (26). Thus, in this regime, both n and μ control the conductivity σ. Next, we will use the following general relation for n:

$$n = n_0 \exp\left(-\frac{E_n}{kT}\right) \tag{30}$$

One can notice that, since the activation energy for the mobility is $E_\mu = e\Phi_b$ and for the carrier density $E_n = (E_c - E_t) - e\Phi_b$, and since the conductivity $\sigma \propto n_{av}\mu \propto \exp[(E_c - E_t)/kT]$, the activation energy for the conductivity is $E_\sigma = E_n + E_\mu = (E_c - E_t)$.

Table 3. Expression of n and μ for the three regimes.

Regime	n	μ
a) $n_t < Nl$	N	$\mu_0 \exp(-\frac{e\Phi_b}{kT})$
b) $n_t \approx Nl$, $L_D < l/2$	$\sqrt{2\pi}\frac{L_D}{l}\frac{NN_c}{(N_t/l)-N}\exp(-\frac{(E_c-E_t)-e\Phi_b}{kT})$	$\mu_0 \exp(-\frac{e\Phi_b}{kT})$
c) $n_t \approx Nl$, $L_D > l_1/2$	$\frac{NN_c}{(N_t/l)}\exp(-\frac{(E_c-E_t)}{kT})$	μ_0 or μ_{bulk}

In this regime, the Schottky barrier is given by the following relationship:

$$e\Phi_b = \frac{e^2Nl^2}{8\varepsilon} \tag{31}$$

Band bending is related to the doping level N and the interface trap density N_t. Comparing the case $Nl < N_t$ where $e\Phi_b$ varies linearly with N in this regime (31) with the case $Nl > N_t$ where $e\Phi_b$ varies with N^{-1} (28), it follows that there is a maximum barrier height that occurs when $Nl = N_t$ (Weisz limit), and correspondingly μ reaches a minimum.

c) $\boldsymbol{n_t \approx Nl}$, $\boldsymbol{L_D > l/2}$: Since almost all of the conduction electrons are trapped in the available interface states on the grain boundaries, and since the Debye length L_D is larger than the grain size, the grains will be almost fully depleted of free carriers, and the energy bands will be essentially flat through each grain. This always occurs below a critical grain size so that it is now clear how the grain size is a fundamental parameter in determining the electrical properties of polycrystalline, and why it should be measured and controlled. In this regime, there are no barriers to current flow, and the mobility will not be activated, although the carrier concentration may be thermally activated because of the excitation of the carriers from the trap level E_t to the conduction band. In particular, n is given by the following relationship:

$$n = \frac{NN_c}{(N_t/l)}\exp\left(-\frac{(E_c - E_t)}{kT}\right) \tag{32}$$

Thus, in this regime, only n controls the conductivity σ with an activation energy $E_n = E_c - E_t$. Small variations in n may quickly change the grains from highly depleted to nearly fully undepleted; this conductivity modulation by carrier depletion in the grains is enhanced when $L_D > l/2$, and could explain the better sensitivity toward gas compounds of nanocrystalline films.

Finally, it has to be remembered that, for porous oxide layers, since the volume of the sensing layer is accessible to gas, a diffusion theory has to be considered as an electrical transport mechanism, together with thermoelectronic emission theory. We refer to [93, 156, 245] for details.

2.3.2. Conductance–Concentration Relationships

Many different formulas for describing the dependence of the sensor conductance on the concentration of the gaseous species which has to be detected can be found in the literature. These different equations, expressing the conductance G as a function of the gas partial pressure or concentration P_g, are mostly power laws ($G \propto P_g^{-m}$). Several both empirical and analytical approaches have been used for deriving them. The variety of conductance–concentration relationships proposed is due to the different conduction mechanism assumed, to the different surface catalytic reactions assumed for the same adsorbate–adsorbent system, to the different adsorption isobar curve assumed, as well as to the real complexity of the numerous parameters that have to be controlled in the sensor device. Table 4 shows typical equations for describing the calibration curves of SnO_2-based sensors for reducing gases.

Table 4. Equations describing the dependence of conductance on the concentration of reducing gases as derived empirically or from theoretical calculations.

No.	Formula	Procedure	Comments	Ref.
1	$G = A_1 c^{\beta}$	Empirical	Large database, deviation at high concentration	[211]
2	$G = G_0 + A_1 \cdot p_{CO}^{1/2}$	Rate equations and SC physics	Reaction of CO and chemisorbed oxygen	[212]
3	$G = G_0 + A_1 \cdot p^{1/2}$	Empirical	Thin film	[213]
4	$R = R_0 p_{O_2}^{\beta}(1 + K_{CO}[H_2O][CO])^{-\beta}$	Empirical	For CO, CH_4, H_2O, CH_4/H_2 constant and modulated T	[145]
7	$G/G_0 = x_i \sim (1 + K_{ik} p_k)^{\beta ik}$	Empirical	i denotes the sensor, k the gas	[214]
8	$G \sim p_{O_2}^{-m/2.18}(A_0 + A_1 p_R^m)^{0.92}$	Rate equations and SC physics	Oxygen species (O_m^-) specified by $m = 0.5, 1, 2$	[135]
9	$G = A_0 + A_1 c^{\beta} + 1/\{A_2 \cdot c^{\beta} + 1/(G_0 - A_0)\}$	Williams' band bending and SC physics	Account for hole contribution in surface conductance	[154]
10	$R = p_R^{\beta}, \beta = -0.5(1 + 1/(1 - 2 \cdot a \cdot R/b))$	Rate equations and SC physics	Oxygen species O^-, exponent depends on the actual resistance	[18]
11	$G^2 - G_0^2 \sim p_{CO}$	Rate equations and SC physics	Fully depleted layer at the surface	[158]
12	$G/G_0 = \Pi(p_i/p_{i,0})^{\beta i}$	Empirical, Clifford '83	In gas mixtures, normalized to a reference	[215, 216]
13	$(G^{\beta} - G_0^{\beta}) \sim p_{CO}, \beta \geq 2$	Rate equations and SC physics		[217]
14	$(G^2 - G_0^2)/G_0 \sim p_{CO}$	Empirical	Procedure for reduced humidity interference	[218]
15	$G/G_0 = (p_{CO}/p_{0,CO})^{\beta CO} \cdot (p_{H_2O}/p_{0,H_2O})^{\beta H_2O}$	Rate equations and SC physics	Interaction of water vapor V_O- and dipoles associated with hydroxyl groups uses a normalization	[152]
16	$G = p_R^{2\beta/(\alpha+1)}(L_D > r), G = p_R^{\beta/(\alpha+1)}\ (L_D < r)$	Rate equations and SC physics	Interacting oxygen species ($O_{\beta}^{-\alpha}$), α and $\beta = 1, 2$, mobility effects	[155, 156]
17	$G = A_0(A_1 \cdot p_R)^{\beta}$	Empirical	Long-term properties by calibration with reference gas	[219]

2.4. Grain-Size Effect on Gas Response

In Section 2.3, we discussed some important aspects related to the electrical conduction properties of a metal-oxide semiconductor layer, which are also closely linked to its microstructure and to the gas-sensing mechanisms. The grain size, the type of intergrain connections, and the distribution of the grains in the grain network that builds the sensing layer are fundamental parameters of a semiconductor gas sensor. However, it should be considered that there is no easy description of the electrical transport properties of polycrystalline films, and a simple distinction between the different conduction models (surface trap-, bulk trap-, or Schottky barrier-limited conduction) is not realistic. Even in the case of Schottky barrier-limited conduction, different regimes can be distinguished according to the above discussion. A real three-dimensional, incompletely sintered material normally has an inhomogeneous structure, and it likely contains more than one single type of intergranular contact ("neck" or "barrier"). Moreover, as the sensor works for a long time at a relatively high temperature, the proportion between the different types of intergranular contacts may change with time. In fact, small adjacent grains can sinter to a larger grain, so that an "aging" effect of the sensor is usually observed. Hence, experimental results could be misinterpreted if only one of the two connection mechanisms between grains is considered.

Wang et al. [48] defined a model that combines both the grain-boundary and neck mechanisms, which are formulated on the basis of the thermoionic emission and junction modulation theory. The neck-conduction mechanism alone has a very low sensitivity, while the grain-boundary resistance is more affected by the gas concentration variation. According to Wang et al., at a large grain size, grain-boundary resistance dominates at low gas concentration, while at high gas concentration, the neck resistance is dominant. However, the neck resistance is much increased over the entire concentration range for a much smaller grain size. They discussed quantitatively the gas response as a function of grain size and chemisorbed gas concentration. When the grain size decreases uniformly, the sensor response increases more and more quickly. According to their model, only for very small grains ($l \leq 40$ nm) does the sensor response have a threshold, linear, and saturation region over a large range of adsorbed gas concentrations N_{ads}. This conclusion, on the other hand, is well supported by experimental evidence. In the threshold region, the gas response increases slowly because the neck resistance is still large, while as the adsorbed gas concentration N_{ads} further increases, the increase of the sensor response becomes quicker because the N_{ads} modulation of the neck cross section becomes more significant (linear region). According to the model, gas response saturates because, at high gas concentrations, N_{ads} is proportionally high; hence, most of the adsorbed oxygen is consumed to react with the gases, and the depletion width becomes very small. Consequently, the diameter of the neck cross section approaches its maximum, and the height of grain-boundary Schottky barriers approaches its minimum.

As emerges from the discussion above, the dimension of the grains in metal-oxide layers hence has a strong influence on the electrical conduction mechanism, and it is fundamental in determining the gas-sensing properties. At present, great efforts are being devoted to developing gas sensors based on nanocrystalline metal oxides, with a precise control of the sensing film nanostructure. Scientific research is also addressed to study the effect of grain size on the sensor response to specific gases. It is well known, in fact, that the gas response of a polycrystalline metal oxide with a very small average grain size may be even orders of magnitude higher than that of a metal oxide with a large grain size. As an example, Figure 15 reports the sensor response at 250 °C to 1 ppm NO_2 versus grain size for an In_2O_3 thin film deposited by the sol–gel process.

The grain size effect can be reasonably considered a very promising approach to the development of high-sensitivity metal-oxide gas sensors. However, the real role of grain size in the gas detection mechanism is not well understood yet. Fundamental studies are needed in order to define a comprehensive model that better explains the gas-sensing properties of a metal-oxide semiconductor layer on the basis of its microstructure and the surface chemical reactions. A theoretical investigation of the dependence of gas sensitivity of nanostructured semiconductor gas sensors on cluster size was presented by Vlachos and Xenoulis [46]. They developed a model in which spherical grains and a uniform surface-state density are considered, and the current between grains is calculated, taking into account the thermal emission and the tunneling of carriers. The sensitivity is calculated taking the derivative of the logarithm of the conductivity over the surface-state density. The results of this model show that there is a critical grain size, which is material dependent, at which the sensitivity is maximal. This grain size depends mainly on free-carrier concentration, and thus on temperature too. Figure 16 shows the theoretical calculated sensitivity versus grain size for some metal oxides. Moreover, for metal-oxide gas sensors, this optimum size depends on the degree of oxidation, the presence of surface additives, and the poisoning of surface parameters, which are known to control the carrier concentration.

As just discussed in the paragraph above relative to the regime c) ($n_t \approx Nl$, $L_D > l/2$), under standard operating conditions of the sensor in air, the grains will be almost fully depleted of free carriers. The significance of the Debye length L_D can no longer be ignored, and the energy bands will be essentially flat through each grain. A flat-band condition is, in fact, the most frequently assumed hypothesis for nanostructured semiconductor films [108, 155, 156, 215, 220], although there is not exact agreement about the critical grain size value below which this condition may be justified. According to Bârsan, the flat-band condition is fulfilled in the range of grain sizes below 50 nm [155]. The flat-band approximation allows us to neglect both the exact electron distribution in the oxide crystallites and the change of the electron mobility at grain boundaries [221].

Recently, Castro et al. provided some experimental proofs of the flat-band condition by scanning-tunneling microscopy and spectroscopy on SnO_2 thick films whose average grain size is about 100 nm [108]. They found a quite symmetric behavior of the tunneling current–voltage characteristics with current flow for positive and negative voltages. This indicates that the SnO_2 nanocrystallites present a nonrectifying behavior. They also explained the observed changes in resistance with temperature in oxygen atmosphere by including oxygen adsorption and diffusion into the grains.

It is easy to understand how, for small grains ($l < L_D$), the effects of, for example, reducing gases can no more be traditionally described as for large grains, that is, as a decrease of chemisorbed oxygen species which is accompanied by a reduction of the band bending and the depletion layers. This is why the depletion layers can increase up to the point for which the grain depletion layers cover the whole grains. This happens when the grains are small enough for intergrain barriers to overlap, and then the entire

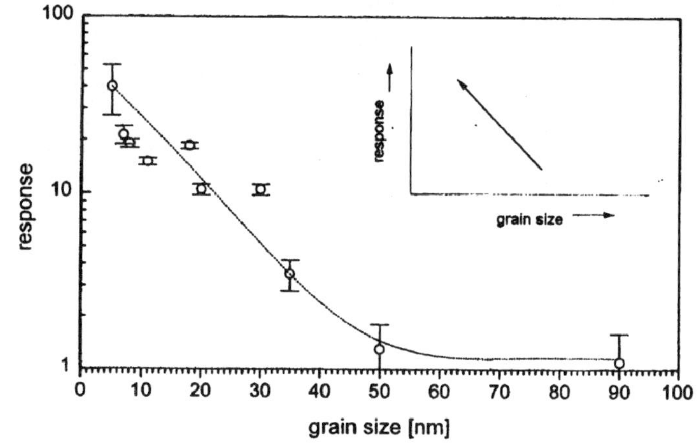

Figure 15. Sensor response at 250 °C to 1 ppm NO_2 versus grain size for an In_2O_3 thin film deposited by sol–gel process. Reprinted with permission from [50], A. Gurlo et al., *Sensors Actuators B* 47, 92. (1998). © 1998, Elsevier Science.

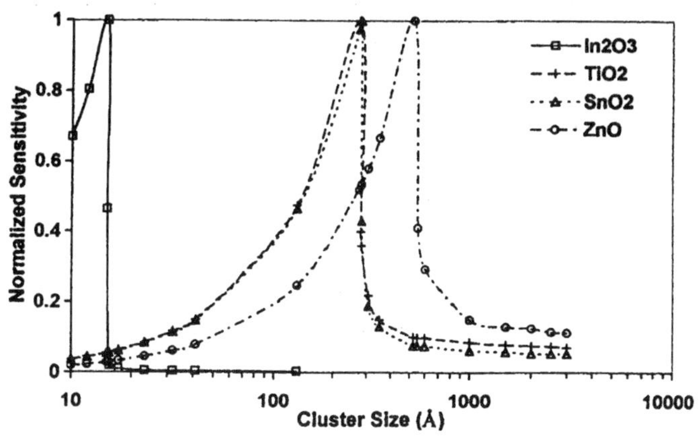

Figure 16. Theoretical calculated sensitivity versus grain size for some metal oxides: ZnO (circles), SnO_2 (triangles), TiO_2 (crosses), In_2O_3 (squares) at 500 K.

grains become electron depleted. All conduction electrons are trapped in the surface states at intergrains. The electron concentration changes dramatically and homogeneously within the grains upon changes in the occupation of surface states. The expected exponential dependence of the conductance on the temperature is

$$G = P\sigma_0 \exp[-(E_C - E_F)/kT] \tag{33}$$

where σ_0 denotes the temperature-dependent bulk conductivity, which is determined by the concentration and ionization energies of the bulk donor and acceptor states, as well as the mobilities of conduction electrons. The proportionality constant P characterizes the sample geometry and the optimum percolation paths expected theoretically for $T \to \infty$ for the conduction of electrons through the sample [215]. The value of G in Eq. (33) is modulated by changes in the difference between the Fermi level E_F and the conduction band edge E_C in the bulk upon exposure to gases. Indeed, the effect of gases can be approximated by a shift in energy bands relative to the Fermi level and the related shift in electron concentration (Fig. 17). Here, no difference exists between the bulk and surface position of E_C, that is, $\Delta(E_C - E_F)_{\text{surf}} = \Delta(E_C - E_F)_{\text{bulk}}$, due to the flat-band approximation [122, 215].

The relationship between the G and gas concentration is a delicate matter. Kissine et al. derived some expressions for the electron concentration n in the film by assuming a flat-band condition [220]. Using the condition of charge neutrality and the adsorption–desorption balance rule for acceptors and surface donors, they expressed n as a function of partial normalized gas pressures p_a, p_d:

$$n = N_D + \frac{n_i}{n} + \Delta n(N_a, N_D, n_s, p_a, p_d) \tag{34a}$$

where

$$\Delta n = \frac{n_s n_d p_d}{n + (n + n_d)p_d} \quad \text{for donor-like gas} \tag{34b}$$

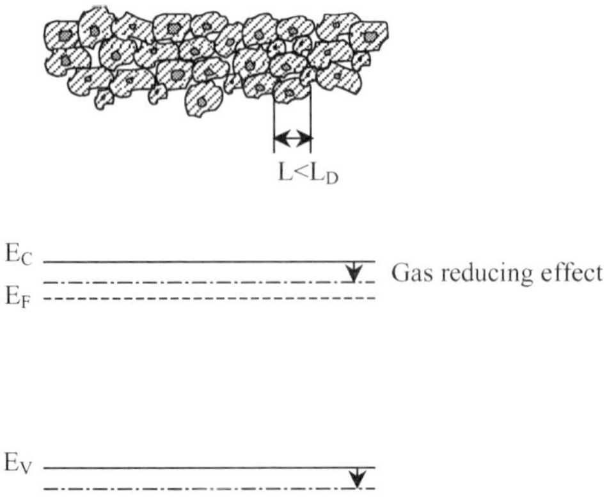

Figure 17. Characteristic representation of nanocrystal ($l < L_D$) with its surface electronic band scheme. A homogeneous shift of the energy bands occurs across the grains upon charge-transfer reactions at their surfaces and interfaces. Reprinted with permission from [46], D. S. Vlachos and A. C. Xenoulis, *Nanostruct. Mater.* 10, 1355 (1998). © 1998, Elsevier Science.

$$\Delta n = -\frac{n_s n_a p_a}{n_a + (n + n_a)p_a} \quad \text{for acceptor-like gas} \tag{34c}$$

$$\Delta n = n_s \frac{n_a n_d p_d - n^2 p_a}{n_a(n + n_d)p_d + n(n + n_a)p_a + n_a n} \tag{34d}$$

when donor- and acceptor-like gas are simultaneously adsorbed at the film surface.

Bârsan proposed a model for conduction in gas-sensing SnO_2 layers with a grain size below 50 nm [155]. Bârsan et al. also presented a modeling approach for the receptor and transducer functions, with particular attention to the microstructure influence and the role of grain size. He emphasized that, for small grains, since the mean-free path of electrons becomes comparable to the dimension of the grains, the number of collisions experimented by a free electron in the grain bulk becomes comparable to the number of collisions at the grain surface; the latter also may be influenced by adsorbed species acting as additional scattering centers. Using the formula derived by Many et al. [222] for the influence of surface scattering, adapted to a simple spherical grain geometry, one gets

$$\mu = \frac{\mu_b}{1 + W \cdot \lambda/2r} \tag{35}$$

where W, which represents the probability of a diffuse (inelastic) scattering process, is assumed equal to surface coverage θ. In the case of very small grains, the ratio $\lambda/2r$ is not negligibly small, and the influence of the surface on the electron mobility may become relevant; hence, it should be taken into account. The dependence of G on the gas concentration of a reducing gas takes the form

$$\begin{aligned} G &= p_R^{2\beta/(\alpha+1)}, \quad & L_D > r \\ G &= p_R^{\beta/(\alpha+1)}, \quad & L_D < r \end{aligned} \tag{36}$$

with $\alpha = 1$ for O_2^- and O^-, $\alpha = 2$ for O^{2-}, $\beta = 1$ for O^- and O^{2-}, and $\beta = 2$ for O_2^-.

A particularly attractive feature, that is beginning to strongly interest the basic science on nanostructured materials used as gas sensors, is the possible effect of the surface and quantum confinement of electrons in nanograins. However, scarce scientific literature on the matter is currently present. Li et al. studied this effect on a nano-$LaFeO_3$ surface by the vacuum adsorptive surface photovoltage technique (SPV) [103]. They verified a quantum confinement effect which depends exponentially on the gas pressure. They also found that there are basically two different methods of charge-transfer transition for a nano-$LaFeO_3$ surface adsorbed by ethanol molecules.

Finally, it is worthless for some aspects of the electrical transport properties of a gas-sensitive layer which can be considered as a large cluster of interconnected nanocrystalline grains. In real nanocrystalline films, the electrical current is constrained to flow in nonuniform patterns for purely geometric reasons, following a preferred path of least resistance, and the effective mobility may be limited by such effects, even in the absence of band bending (*percolation effect*) [209]. As long as each grain contributes to the conductivity of the layer, the size and shape of the conducting network do not change, and the gas sensitivity depends only on the change in the conductivity at single-grain transitions. In this case, the morphology leads to a constant prefactor in the expression of conductivity, and the whole sample can be described approximately by a single mean-grain transition. On the contrary, when some grains switch between a conducting and an insulating state, only the conductive part of the sample will contribute to the conductivity. The size and shape of the conducting part depend on the morphology of the sensing layer and, of course, on the gas concentration.

Ulrich et al. reported a percolation model for a nanocrystalline gas-sensitive layer that takes into account the possibility of having ultrasmall compensated grains that are insulating [45]. The probability that a nanograin is insulating depends on the diameter and number of next neighbors (the coordination number k). According to the percolation theory, the concentration of conducting grains must exceed a critical concentration p_c, the percolation threshold, in order to form a conducting path spanning the sample (Fig. 18). The conductivity is given by a power law just above threshold:

$$\sigma \propto (p - p_c)^{\mu}, \qquad p > p_c \tag{37}$$

This leads to a large dependence of σ on the gas concentration which triggers p. The critical exponent μ depends only on the dimension of the network, and not on its internal structure. The percolation threshold also leads to a gas-detection limit [44]. In fact, experimental studies revealed that the gas response does not change until a critical gas concentration is reached; above the detection limit, the gas response rapidly increases. Ulrich et al. attributed the enhanced gas response, when the grain size is decreased, to the presence of percolation effects. They also observed that the percolation threshold for a sample should be adjustable

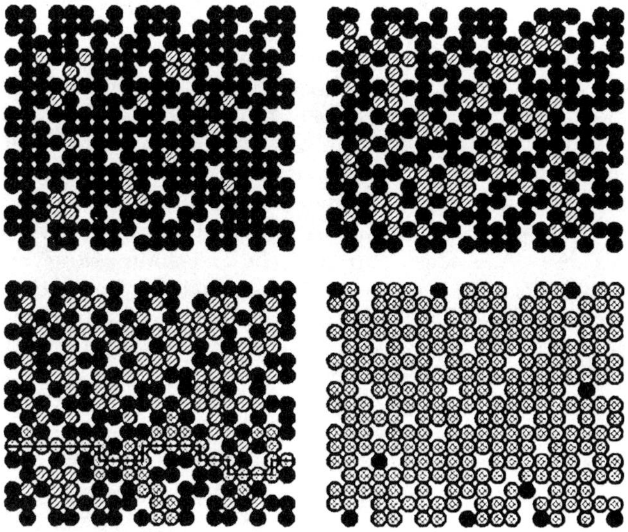

Figure 18. Small grain cluster below and above the percolation threshold p_c. For simplicity, a two-dimensional cluster on a rectangular lattice is drawn. Open, shaded circles represent conducting grains. Above p_c, the first conducting path is drawn as a solid line. The side spanning conducting cluster of the sublattice is shaded with a cross mesh. Reprinted with permission from [45], M. Ulrich et al., *Thin Solid Films* 391, 299 (2001). © 2001, Elsevier Science.

with the grain size distribution and the average coordination k_{mean}. The first, together with the mean diameter, can be controlled by the synthesis and deposition parameters of the sensing layer; the latter is nearly random, and results from the sintering process.

3. CURRENT STATUS OF RESEARCH

3.1. Materials

SnO_2-based gas sensors represent the first prototypes of chemoresistive gas sensors. Tin oxide is the most known and used material in this type of gas sensor. However, SnO_2 is not the only material with acceptable gas-detection properties, but in general, metal oxides are the most suitable semiconducting materials for gas-sensing applications [146]. In fact, it results that materials with predominantly ionic bonding (such as ZnO, SnO_2) are much more sensitive to changes in the surface coverage by adsorbed molecules and, consequently, to changes in the gas-phase composition with respect to materials with covalent bonding (Si, Ge, GaAs) [47]. Moreover, by using metal oxides as semiconducting sensing elements, it is possible to avoid the problem of instability of other semiconductors (such as germanium), mainly due to the formation of stable surface oxide layers under cycled thermal annealing in air. On the contrary, metal oxides bind atmospheric oxygen at their surface in a reversible way [18]. Of course, oxides characterized by exceedingly high resistivity, such as SiO_2 and Al_2O_3, are not used in resistive gas sensors because of the difficulties encountered in electrical conductivity measurements.

A screening of a large group of oxides was originally carried out. It showed that the sensitivity of conductivity to low concentrations of reactive gases in air was a phenomenon common to metal oxides, and not specific to SnO_2 or to a few special cases [38, 223]. Also, at present, an important part of the scientific research is continuously involved in the systematic study of new metal oxides for gas sensors. The reason for the research on novel gas-sensitive materials is that, although it is obvious that other gas-sensitive oxides should exist, there is no a priori basis for identifying them or for determining which, if any, materials may have more advantageous properties than SnO_2, and particularly a selectivity of response.

A huge amount of studies on SnO_2-based gas sensors have been published. Table 5 shows an incomplete list of literature about SnO_2 sensors for specific target gases; it informs us about the preparation process and the doping/catalytic element [122]. The aim of these extensive research activities was to gain a better understanding of gas-sensing mechanisms [74–77, 224–226], and naturally, to improve the properties of SnO_2 sensors [49, 91].

Simple binary metal oxides, such as Ga_2O_3, WO_3, In_2O_3, and TiO_2, are the most investigated semiconducting materials. Additionally, there are several ternary (e.g., ITO, $SrTiO_3$, and $BaSnO_3$), quaternary, and complex metal oxides, including heterometallates, that are promising candidates [49]. A list of new gas-sensitive metal oxides, alternative to tin oxide, is reported in Table 6. The concept of n-type and p-type materials has long been established in

Table 5. Incomplete list of studies performed on SnO_2.

	Doping/ catalyst	Fabrication technology	Operation temperature	Ref.
H_2	SO_2	Ceramic	~380 °C	[281]
	—	Thick film	~320 °C	[282]
	Sb	Thick film	~240 °C	[283]
	—	Whiskers	~520 °C	[284]
	Pd	—	—	[285]
	Sb_2O_3, Au	Thin film	150–510 °C	[286]
	Ag	Ceramic	25–150 °C	[287]
	—	Thin film	—	[289]
NH_3	Pd, Bi, $AlSiO_3$	—	—	[290]
SO_2	SO_2	Ceramic	~220 °C	[281]
CO	—	Thick film	~280 °C	[282]
	ThO_2, SiO_2	Thick film	~220 °C	[291]
	Pd, ThO_2	—	—	[292]
	Pt, Sb	Ceramic film	25–50 °C	[293]
	Pd, Cu	Thick film	150–280	[294]
	PdO, MgO, ThO_2	—	—	[295]
	Pt	Thick film	100–250 °C	[296]
	Pt, In	Thin film	100–330	[297]
	Sb, Pt	Thin film	25–330	[298]
	Pd	Thin film	200–250	[199, 299]
	Pd, Pt	Thick film	—	[300, 66, 39, 116, 185]
CH_3COOH	Pd, Bi, $AlSiO_3$	—	—	[290]
	— , Pd	—	100–480	[301]
C_2H_5OH	—	Thick film	~250	[282]
	—	Thin film	250–500	[302]
CH_4	—	Thick film	—	[282]
	Pd	Thick film	100–500	[303]
	Pd	Thick film	250–350	[185]
	Os	Thin film	250–300	[165, 199, 299]
AsH_3	—	—	~420	[304]
H_2S	SO_2	Ceramic	~420 °C	[281]
	—	Thick film	~100	[305]
NO_2, NO_x	—	Thick film	180–250	[306]
	—	Thin film	—	[228, 299, 190, 201]
	—	Thin film	100–200	[307]
	Bi_2O_3	Thick film	200–400	[308]
	Pd	Thick film	200–300	[309]
	Cd	Thin film	250–400	[310]
	In, Al, Pt	Thin film	25–400	[311]
	—	Thick film	250–330	[312]
	V, In	Thin film	100–350	[313]
	In	Thick film	~300	[314]
CCl_4	Pd	Thick film	~200	[315]
CO_2	La_2O_3	—	~400	[316]

oxide semiconductors. In the gas-sensing context, the sign of response (resistance increase or decrease) leads to a simple and useful classification. Based on the *p*- or *n*-type conduction of the oxide, gases can be classified as oxidizing or reducing according to the sign of the response: *n*-type oxides show a resistance increase in the presence of oxidizing gases (like O_2, NO_2, etc.) and a resistance decrease to reducing gases (like CO, CH_4, hydrocarbons, etc.), while *p*-type oxides have the opposite behavior.

Table 6. Incomplete list of studies on semiconducting metal oxides.

Semiconductor material	Doping material	Detectable gas	Ref.
$YBa_2Cu_3O_{7-\delta}$	—	NO	[317]
In_2O_3	—	NO_2, O_3	[193, 318]
WO_3	—	NO_x	[272, 322]
$LnCoO_{3-x}$ Ln = Nd, Sm, Eu, La	—	CO	[319]
Co_3O_4	—	CO	[320, 321]
MoO_3	—	NO_2	[323]
$Bi_xMo_yO_z$	—	Alcohols, ketones, H_2, CO	[324]
$Cr_{1.8}Ti_{0.2}O_3$	—	NH_3	[325]
ZnO	Al_2O_3, In_2O_3	NH_3, H_2, CH_4, C_4H_{10}, C_2H_5OH	[326]
ITO		NO_2, NO, CO, O_3	[327–329]
$CdIn_2O_4$		CO	[330]
In_2O_3	CeO_2	O_3	[331]
$Sn_{1-x}Fe_xO_y$		CO, ethanol	[332, 334]

3.1.1. Influence of Dopants and Surface Additives

Noble metals (e.g., Ag, Pd, Pt, etc.), uniformly distributed and finely dispersed as catalytic clusters on the surface of the oxide-sensing layer or added as dopants into the oxide bulk, are often used in semiconductor gas sensors. It is well known and experimentally proven that catalysts added to the sensor may influence its response to specific gases, may speed up the surface reaction, or may impart selectivity to the reaction so that some reaction processes are favored over others [18, 227, 86, 87].

Because semiconductor gas sensors depend on surface chemical reactions, they are strongly influenced by the activation energy needed to initiate the reaction. As discussed in Section 2.2.2, the activation of a species involved in a surface reaction may be the dissociation of a molecule, the ionization of the species, or some other intermediate reaction that presents the species in a form ready for an exothermic reaction on the sensor surface. The activation energy necessary to promote a reaction, usually provided as thermal energy, depends on the route of the reaction. If the activation energy using a particular route is high, the rate of the reaction will be low, and vice versa, if a route for the reaction can be found with a low activation energy, the reaction will be fast. Energetically, the effect of catalysis is to provide a more favorable reaction path. In such a way, a lower operating temperature for the sensor may be needed, and the rate of only some reactions also can be greatly enhanced in contrast to the rate of others, so that selectivity is provided to the system.

Thus, the use of catalysts/dopants is a feature of great importance in semiconductor sensor design. Although the promoting effect of modifiers in the semiconductor bulk is not well understood yet, the way in which surface-dispersed metal additives influence the gas response characteristics could be explained in terms of the *spillover* and/or the *Fermi-level* mechanism [18, 88].

Spillover is a chemical interaction in which the metal catalyst dissociates the molecule, the atoms of which can "spill over" onto the surface of the semiconductor support, and eventually react with ionosorbed oxygen species, thus controlling intergranular resistance. Hydrogen and oxygen spillover has been extensively studied. Hydrogen spillover is important for hydrogen-containing combustible gases. Oxygen spillover, favoring its adsorption and dissociation in the atomic species O^-_{ads}, which is more reactive than the molecular species $O^-_{2,ads}$, accelerates the oxidation of reducing gases, and consequently promotes their detection in air.

The *Fermi-level* mechanism is an electronic interaction in which gas adsorption on the catalyst removes electrons from the catalyst, and the catalyst, in turn, removes electrons from the supporting semiconductor. The supported catalyst pins the Fermi level of the semiconductor support at the Fermi level of the catalyst, thus controlling the band bending in the semiconductor, and consequently, the intergranular resistance, in the same way as adsorbed oxygen or another adsorbate would do in a normal situation without a catalyst. Of course, the Fermi level of the catalyst at the catalyst/semiconductor interface has to depend on oxygen adsorption, and the small catalyst particles have to be highly dispersed (distance between clusters less than about 500 Å) so that the space-charge regions of neighboring catalysts overlap.

Platinum and palladium are the most commonly used dopants to improve the performance of SnO_2 for CO detection; it seems that Pt promotes CO oxidation via the spillover mechanism, while it seems that Pd acts via electronic interaction (Fermi-level control) [49], even if an electronic interaction between Pt and tin oxide also has been proposed [86].

The most active metal catalysts are known to be the transition metals (Pt, Pd, Rh, Ir, Os) that, having partially filled *d* orbitals of valence-band electrons, allow a strong gas–solid interaction [18].

A huge amount of experimental work on metal oxides doped with metallic elements or modified with surface metallic additives has been published. We mention, just as an example, only some of the most significant experimental works on the effect of dopants and/or surface additives on the gas-sensing properties of metal oxides. Capone and co-workers studied the gas-sensing properties of Pd-, Pt-, Os-, and Ni-doped SnO_2 thin films deposited by the sol–gel method [165, 198, 199, 228–233]. The films are prepared by spin coating, starting from a hydrolyzed alcoholic solution of tin precursors. The addition of inorganic transition metal salts in the solution enables the activation of the resulting film. The presence of modifiers is evidenced both in the form of nanoparticles (Pd and Pt) and doping atoms (Os, Ni). The dopants increase the sensitivity to certain gases, hence improving the selectivity of the sensing material. In particular, it has been shown that Pd and Os act as activators, respectively, for CO and CH_4, enhancing the sensitivity and simultaneously decreasing the optimum operating temperature. The good sensitivity to methane at a low working temperature makes the Os–SnO_2 thin films very attractive materials to be used in domestic CH_4 leak detectors. An SnO_2-based sensor array also has been developed for application in the electronic nose. Good results have been obtained for the discrimination of air pollutants and different food odors.

A detailed analysis of the role and activity of three of the most common catalytic additives (such as Pd, Pt, and Au introduced in an SnO_2 nanopowder) in the gas-sensing mechanism have been reported by Cabot et al. [116, 234, 235]. Ruiz et al. also found that the efficiency in catalytic activity improves by incorporating Pt nanoclusters in anatase TiO_2, whose surface reactivity is better [236].

Palladium is the most used dopant/additive to improve the detection of CO and CH_4 [185, 188, 237–241]. Platinum is also a widely used metallic catalyst for gas-sensor applications. Wöllenstein et al. found that Pt nanoclusters, covering the surface of highly dense sputtered polycrystalline SnO_2 layers, increases the sensitivity to CO with lower temperatures without a distinctive influence on the dynamic response constants; on the other hand, the Pt-surface doping eliminates the reaction to O_3 and NO_2 at working temperature around 340 °C [242]. A simple method to achieve a layer of noble metal, such as Pd, Ag, Pt, or Au, nanoparticles on oxide surfaces was proposed by Mizsei et al. [243]. It consists of a first sputtering deposition of an amorphous metal nanofilm of about 5 nm thickness; the film is agglomerated to nanoparticles either via direct crystallization by heating, in the case of nonreacting atmosphere, or by the growth of oxide crystallites and the subsequent reduction of these in a reducing atmosphere. Finally, we mention a work on ruthenium-doped tin-oxide thin films prepared by spray pyrolysis; the films showed an unsually high sensitivity to hydrocarbon (LPG) at a lower operating temperature of 250 °C [244].

3.1.2. Mixed Oxides and Heterojunctions

The search for novel materials suited to gas sensing has led the scientific community working in the field to consider a new concept design for gas sensors based on the use of nanostructured mixed oxides and heterojunctions. The idea originated from some observations related to the working principles of metal-oxide-based gas sensors. As is well known, the resistance variation of the sensing layer involves two important functions, that is, the recognition and the transducer functions. Gas/solid interaction phenomena are involved in the receptor function, while the microstructure of the oxide determines the trasduction of the chemical stimulus in air into an electrical signal. Generally speaking, if a single-oxide system is adopted, these two functions cannot be optimized independently. Instead, by introducing in the system a foreign material, which is very reactive to a target gas and acts as an "antenna" material, both functions may be optimized simultaneously, and the sensor may become more sensitive, even to low reactive gas concentrations. In these cases, the material acting as a unique receptor (antenna material) should be interfaced electronically to the transducer material, and its chemical change should sensitively modulate the semiconducting properties of the transducer oxide through the heterojunction [246]. Layered-type sensors and composite-type sensors containing heterocontacts between the two phases fulfill this novel concept of gas sensors. They constitute a promising class of gas sensors with enhanced gas-sensing properties.

Heterojunction gas sensors attempt to overcome the problems of existing monolithic gas sensors through the introduction of tunable selectivity, self-cleaning properties, and a built-in failure detection mechanism. Geometrically, heterojunction gas sensors consist of two semiconducting oxides in contact, and the active region is in the interface between the two materials, which behave like a *p–n* junction. Gas adsorbed on either side of the heterocontact is oxidized at varying rates at the heterojunction interface, and influences its rectifying characteristics [247]. Since a loss of rectifying characteristics implies a breakdown of the *p–n* junction, there is a built-in failure detection mechanism [248, 249]. Many works on the gas-sensing properties of junctions based on an *n*-type and a *p*-type oxide have been published. The most common *n*-type material used in these sensors is zinc oxide (ZnO) or tin dioxide (SnO_2), and the *p*-type component is usually a doped copper oxide (CuO) or nickel oxide (NiO) or a perovskite phase such as La_2CuO_4 [250–257].

Since the layered-type sensors are difficult to fabricate, mixed oxides were suggested as suitable materials to accomplish this new concept design of gas sensors. A mixed-oxide sensing layer should consist of nanograins of the two oxides connected by many heterojunctions. The efforts to achieve a mixed metal-oxide film also arise from the need for a stable and nanostructured material. There are, in fact, many effects that can derive from these mixed systems, such as the stabilization of the sensing material, ensuring, for example, the maintenance of its nanosized character. Furthermore, the presence of a second phase can maintain the grain size within dimensions comparable to the Debye length. Moreover, one of the two phases could act as a filter, and the other one as a sensing material. It seems that the oxide at lower concentration restrains the growth of the other oxide and stabilizes it, avoiding the well-known grain growth that occurs when the sensor is heated [258].

A mixed oxide can be represented as $M_1O–M_2O$, where M_1O is the nanocrystalline matrix and M_2O is the doping oxide distributed between the surface and the bulk of the M_1O nanocrystalline grain. According to Gas'kov et al. [47], the advantage of mixed oxides over simple nanocrystalline oxides is associated with the redistribution of M_2 between the bulk and the surface of the M_1O grain, depending on the redox properties of the gas phase. Low oxidation levels of the M_2 cation, corresponding to a larger ionic radius and predominant distribution of M_2 over the surface of M_1O grains, occur in the reducing atmosphere. On the contrary, cations in higher oxidation states are formed in an oxidizing atmosphere, with increasing probability that M_2 cations occupy regular cationic positions in M_1O. The lability of the chemical state of the M_2 cation in the nanocrystalline system may result in a dramatic change in the state of grain boundaries, and in a modification of the electronic properties of the material in the presence of even trace amounts (0.1–10 ppm) of reducing or oxidizing gas molecules in the gas phase. The oxides to be used in creating nanocomposites are chosen taking into account the ionic radii of the elements.

Numerous studies on the gas-sensing properties of mixed oxides have been published. The most common mixed oxides which have been considered for application in gas sensors are: $MoO_3–TiO_2$, $MoO_3–WO_3$ [195, 258–263], $In_2O_3–SnO_2$, $In_2O_3–NiO$ [264], $MoO_3–In_2O_3$ [50], $TiO_2–Fe_2O_3$ [265], $SnO_2–TiO_2$ [266], $LaOCl–SnO_2$ [267], and *n–p* TiO_2 [268].

3.2. Synthesis and Deposition Techniques

A variety of different preparation techniques can be used for the deposition of the nanocrystalline metal-oxide layers. The sensor preparation methods can be divided into two classes: preparation by thick- or thin-film technology. A summary of typical deposition methods is reported in Table 7. As mentioned in Section 2.3.1, thick-film sensors can be characterized by being rather thick (a few to several hundred micrometers) and porous, so that a high active surface area is exposed to the gas interactions. Thin-film sensors, on the contrary, can be prepared by established preparation methods from thin-film technology (semiconductor technology). They are rather compact, with a typical thickness in the range of a few to several hundred nanometers, and therefore comparably thin. In this case, gas interactions are mainly restricted to the surface layer. Comparisons of thin- and thick-film metal-oxide-based sensors showed that the changes in the electrical resistance of the sensors are typically much more significant for thick-film sensors than for thin-film sensors.

Table 7. Typical deposition techniques used for the preparation of sensitive layers for metal-oxide gas sensors.

Chemical vapor deposition (CVD)	Physical vapor deposition (PVD) Sputtering	Physical vapor deposition (PVD) Evaporation	Powder preparation	Other techniques
Thermal CVD	Sputtering	Molecular beam epitaxy	Sol–gel from precursors	Sol–gel thin film
Plasma-activated CVD	Reactive sputtering			
Laser-induced CVD	Cathode sputtering with bias voltage	Thermal evaporation	Precipitation of precursors	Rheotaxial grow and thermal oxidation (RGTO)
Electroless plating		Reactive evaporation	Laser pyrolysis	
Spray pyrolysis	Ion beam deposition	Ion plating		
Melt dipping	Ionized cluster beam (reactive)	Reactive ion plating		
Liquid quenching		Arc evaporation		
Deposition of organic polymers	Plasma decomposition	Laser evaporation		
Deposition of emulsions				

The first metal-oxide sensors were thick (or ceramic) sensors. A heater coil and an electrode wire were pressed in the powder material. Then the so-called Taguchi sensors were introduced, where the sensitive material was painted as a paste on an alumina tube in which the heater was separated from the layer. Currently, SnO_2 is the most used material in ceramic sensors. SnO_2 thick films are usually prepared starting from commercial tin oxide, tin, tin salt (mostly $SnCl_4$), or organo–tin compounds (mostly tin alkoxides). The starting materials are used to prepare powders, colloid solutions (gels), or sprayable solutions. The sensitive layer is then fabricated based on these. The solutions of the base material can be used to prepare thin-film sensors.

The often-used term sol–gel is referred to this special preparation route: here, an aqueous tin salt solution forms a sol with poly SnO_2 clusters. They can be precipitated by basic solutions, or they undergo a gelation step when clusters grow, but still dissolve as a colloid. A critical step in this process is the calcinations of the powders because this will define the grain size of the material.

Different methods are used to deposit the powders or solutions onto electrode-equipped substrates with a heater at the back side. The most common methods are dipping, painting, screen printing, and drop coating for the deposition of thick films, and spraying and spin coating for the deposition of thin films. These methods use a thermal treatment after deposition to create the sensitive layer.

The sensors can be optimized by chemically modifying the SnO_2 material by adding dopants. Various methods have been developed for doping. These methods can be divided into two groups; in the first, a compound containing the dopant is added to the precursor; in the second, the SnO_2 powders are treated with such a compound. Typical examples for the first method are the addition of organometallic compounds to alkoxides or of metal chlorides to $SnCl_4$. If, in the latter case, the $SnCl_4$–metal chloride mixture is afterwards precipitated by means of adding NH_3, this is called coprecipitation. Typical examples for the second group are: the colloid method, the impregnation of SnO_2 with metal chlorides, the fixation of metallic particles on SnO_2, and the treatment of SnO_2 with nitric/hydrochloric acid solutions of the dopant [98].

Thus, the sol–gel technique is a wet chemical deposition method, that is well suited for the preparation both of thick and thin films. It offers several advantages with respect to conventional processing technologies.

1. It is simple.
2. It is inexpensive.
3. It is easy to control the morphology and to modify the composition with uniformly dispersed dopants and modifiers.
4. It produces films with a high porosity and a large surface area that improves the efficiency of the sensors in the mechanism dominated by surface phenomena.

Nowadays, the sol–gel method is mostly applied for the preparation of thick films through a dip- or drop-coated deposition procedure of the base material paste onto the transducer structure.

However, great interest is devoted to the preparation of sol–gel metal-oxide thin films because thin-film devices are perfectly compatible with silicon technology and Si-micromachined heater substrates. Most of the commercially available gas sensors based on thick-film metal oxides use ceramic heater substrates (mainly alumina substrates). A common shortcoming of such thick-film devices is their high level of heating power consumption (0.5–1 W). This level can be reduced by about one order of magnitude using Si-micromachined heater substrates. A complicating feature in employing such miniaturization techniques, however, is that these are not compatible with standard screen-printing technologies, and hence, it is difficult to form miniaturized low-power-consumption devices with commercially established thick-film sensing layers. In this context, novel approaches such as drop coating have been proven to be fruitful and advantageous thick-film deposition techniques for the coating of Si microhotplates [39, 184, 185, 198].

Apart from the sol–gel thin-film deposition which gives porous layers with a large surface area [194–200], another technique that provides a thin film with these characteristics is the RGTO [78, 176]. This technique, called by Sberveglieri [175] rheotaxial growth and thermal oxidation, allows the preparation of a thin film with a high surface area and crystallites with a nanometric size. Referring, for example, to SnO_2, the technique consists of the deposition of tin on a substrate with the temperature just below the melting point of tin, and subsequent oxidation (reactive growth) of the grains with a well-defined temperature program in an oxidizing atmosphere. Due to the melting point of the tin, spheres are formed. First, the spheres are not totally oxidized, and the contact among them is small; the conductance of this film is low, although it is partly metallic. Only an annealing step of the layer in an oxygen-containing atmosphere leads to an increase of the conductance because of the growth of the spheres (approximately 35% in volume [175]). We find cauliflower-like grains from a high surface area and good contacts among each other.

We do not describe here the CVD and PVD techniques, well known in semiconductor technology, but we mention some of the most interesting novel approaches to the deposition of sensitive layers for semiconductor gas sensors, for which we refer to [98, 123, 175–206].

3.3. Micromachined Substrates for Gas Sensors

Microtechnologies have been developed for the production of silicon semiconductors. Thanks to the batch production, the growing size of silicon wafers, and the miniaturization of each microdevice, an ever-increasing number of elements can be processed in parallel, resulting in a considerable reduction of costs. The silicon revolution has enabled us to produce small, reliable processors in the form of integrated circuits (ICs) using microelectronic technology. Silicon processing technology is no doubt the most advanced technology in the world, and may not be easily surpassed by other technologies.

All of the facilities offered by silicon microtechnology make it an excellent tool for the production of microsensors [269, 270]. Additionally, since the electrical sensor signals have to be acquired, amplified, and evaluated, the sensor technology has to be compatible with modern electronics.

Thus, another cost reduction may be achieved by the integration of the sensing part and electronic circuitry on the same chip, and this provides a second argument for second fabrication by microelectronic technologies.

In this context, micromachined gas-sensing devices are acquiring an ever-growing importance in the field of gas sensors [39–42, 271–273]. The heart of this structure is the active area that is comprised of a heater, sensor electrodes, and the gas-sensitive layer situated at the center of a thin membrane which itself is supported by an outer frame of silicon. Basically, there are two different structures for micromachined gas sensors: the *closed-membrane type* and the *suspended-membrane type* of gas sensors. The latter are also called *spider-type* gas sensors. This resistively heated dielectric membrane (a–SiON or LPCVD Si_3N_4) provides the thermal insulation between the active area heated up to high temperatures and the silicon frame that remains at room temperature.

The attractive features and advantages offered by using such membranes in gas-sensing devices are numerous.

- They can be handled easily in the electronic industry, and enable integration with electronics; in particular, they need a voltage source both for resistance measurement and for the necessary heater compatible with standard digital electronics (e.g., CMOS).
- They allow a low power consumption not exceeding 100 mW for operation. This small amount of heating power is caused by the reduction in the heated surface area, as well as by the excellent thermal isolation provided by the thin dielectric membranes.
- They allow the production of tiny sensor devices of small size and lower weight (greater portability).
- They allow the reduction of manufacturing costs.
- They can be easily grouped into battery-operated arrays (i.e., sensor arrays integrated on a sensing chip).

Figure 19 schematically shows a top view (a) and a cross section of a single micromachined structure drop coated by the sensing layer (b). In particular, this model of microhotplates was fabricated at the Institute of Microtechnology (IMT) of Neuchâtel in Switzerland [39, 198].

3.4. Electronic Nose

The previous sections of this review have described the science and technology underlying the present resistive-type semiconductor gas sensors based on nanostructured metal-oxide films. One of the main problems related to this kind of gas sensor is a still unsatisfactory understanding of the details of device function; in particular, the effective role of grain size, when it decreases to a nanometer scale, needs more intensive studies. Moreover, there is a significant development effort to turn even a laboratory device into a commercial useful sensor. In spite of the advantages reviewed in the sections above, our ability to design marketable sensors to respond selectively and sensitively to specific gases at an appropriate temperature is still limited. In particular, problems of drift and reproducibility are difficult to overcome. On the contrary, the poor selectivity of chemoresistive gas sensors can be overcome by the design concept of an *electronic nose*.

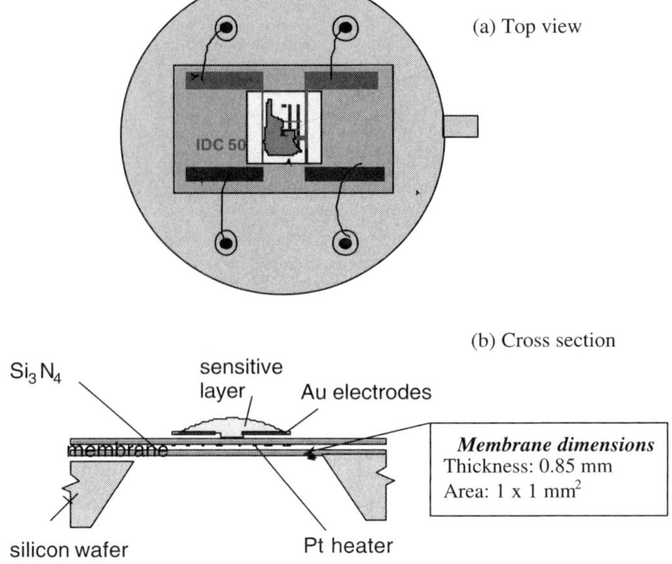

Figure 19. Scheme of a closed-membrane-type gas sensor. In particular, the model of microhotplates was fabricated at the Institute of Microtechnology (IMT) of Neuchâtel, Switzerland.

The performance of nonideal gas sensors with partially overlapping sensitivities can be improved by linking a number of different nonselective chemoresistive sensors in an array configuration. The term "array" is not intended as a particular spatial arrangement of the sensors in the device; on the contrary, it is in the sense of a vectorial organization of the output of the single sensors under exposure to the same environment, so that the contribution of each array element is a component of a multidimensional general problem. The signals of such a sensor array can be evaluated by means of pattern recognition (PARC) and multicomponent analysis (MCA) methods which, exploiting the cross correlations between the sensor responses, extract information contained in the sensor-output ensemble [274–278].

In particular, a PARC procedure performs a qualitative analysis of the environment; it evaluates the multidimensional dataset from a gas-sensor array, seeking the underlying main relationships in the dataset itself in order to analyze the data structure and discriminate between different data classes (*clusters*) belonging to different "chemical patterns" (*classification*). Figure 20 illustrates the basic principles of a pattern recognition procedure in the chemical gas-sensing scenario. Moreover, a PARC procedure has to assign a sensor array output from an unknown gas/odor to a class, recognizing in such a way the occurrence of a particular chemical pattern (*identification*). An MCA procedure, based on multivariate techniques of regression, performs a quantitative analysis of the environment, determining the concentration of one or more compounds in a mixture. These analytical regression methods provide a general correlation between the analytical information (composition of the chemical pattern) and the output of the sensor array.

Of course, such multisensor systems need calibration; by presenting many different chemicals to the sensor array, a database of patterns is built up. This database of labeled

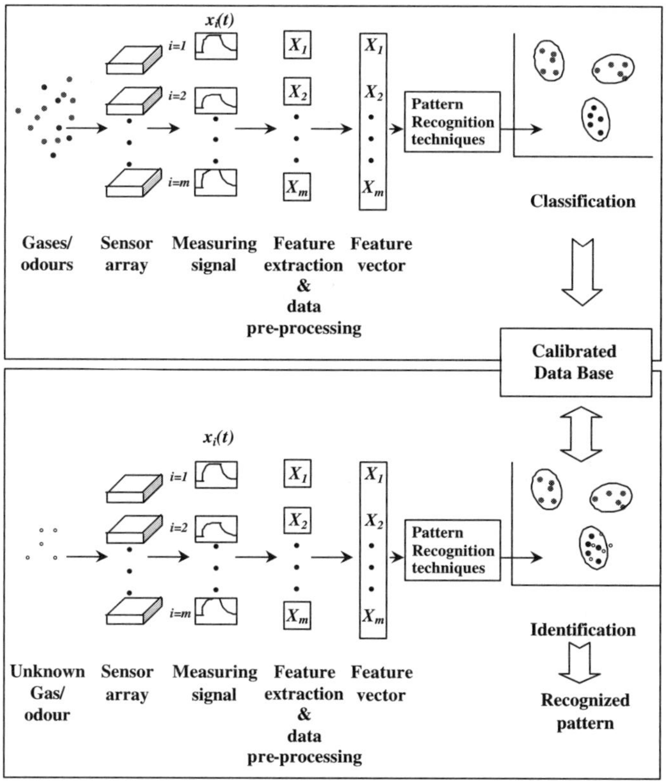

Figure 20. Basic principles for applying pattern recognition methods to gas/odor analysis. The sensor array is exposed to the ambient of interest, the measuring signals are preprocessed, and the resulting individual features X_i are extracted. The complete signal vector X containing all of the sensors' information is used as input of the pattern recognition algorithm. The method may be either unsupervised if it seeks to discriminate between different gases/odors, or supervised in the sense that an unknown gas/odor is analyzed using a dataset built up a priori from a set of known gases/odors in an initial calibration, learning, or training stage.

patterns is used to train the pattern recognition system. The goal of this training process is to configure the recognition system so as to classify and eventually quantify each chemical compound in a gas mixture.

Data processing techniques adapted to the data analysis from an electronic nose have a long history, and in parallel with the first well-established approach provided by *chemometrics*, a new approach based on *artificial neural networks* (ANNs) has been developed [276].

Chemometrics, which has been in existence since the 1960s, is the chemical discipline that uses mathematical and statistical methods to provide the maximum chemical information by analyzing chemical data. These chemometric techniques, applied to data from a sensor array, seek to determine the distribution of the data and the underlying relationships between one set of independent variables (i.e., the preprocessed sensor responses from a sensor array) and another set of dependent variables (i.e., the gases/odors classes or the component concentrations). This statistical approach is sometimes referred to as *parametric* in that it assumes that the data can be described by a probability density function, such as the multinormal distribution. In chemometric techniques, the correlation between the measuring data and the composition or concentrations of the analytes in the chemical pattern during the calibration procedure may be described by a calibration function using an exactly defined set of parameters. These parameters normally are related to physical variables (such as partial sensitivity or slope of the calibration function). Most chemometric techniques use linear transformations so that, by explicitly determining the inverse calibration function, it is possible to calculate the concentrations of unknown samples from the measured sensor signals.

The second approach to data analysis in an electronic nose, based on artificial neural networks, is a relatively new one that arose in the 1980s. It seeks to solve multivariate problems in a manner similar to the human cognitive process by using a biologically inspired neural architecture and rules based upon human reasoning. This approach is also referred to as nonparametric because it uses an implicit representation of the correlation sensor signal/analyte concentrations without a definition of parameters which can be identified as physical variables.

For both the chemometrics- and ANN-based approaches, pattern recognition for qualitative analysis and the multicomponent analysis techniques for quantitative analysis are available.

An important distinction in data analysis techniques concerns their *unsupervised* or *supervised* character. Unsupervised methods makes no a priori assumption about the sample classes, but they try to discriminate between clusters of unknown gases/odors by enhancing the differences between their associated input vectors; on the contrary, the task of a supervised technique is to classify unknown gases/odors as known ones that have been learned during an earlier calibration procedure.

Numerous data analysis techniques are commonly employed to analyze data from electronic nose instrumentation. Multiple linear regression (MLR), principal component regression (PCR), and partial least squares (PLS) are quantitative data processing techniques which are well known in analytical chemistry and in the sensor array area, while the most used conventional statistical pattern analysis techniques are principal component analysis (PCA), cluster analysis (CA), and discriminant function analysis (DCA). Among the neural- network-based techniques, the most important ones are back propagation (BP), radial basis function (RBF), Kohonen self-organizing map (SOM), and learning vector quantization (LVQ). Moreover, there are other more recent dynamical methods, that seek to improve the performance of the pattern recognition method through the use of the dynamical (transitory) response $x_{ri}(t)$ of the sensors, or through the use of models that evolve with time. The latter type of technique is known as adaptive, and deals with nonstationary sources of data, for example, a sensor output that, say, systematically ages or becomes poisoned with time. Adaptive techniques are not well established, but currently are an important topic of research; we mention among them the methods based on genetic algorithms (GAs) or the adaptive resonance theory (ART). Finally, we mention the fuzzy neural networks (FNNs) that apply fuzzy logic to electronic nose data.

3.5. Applications

Since early work in the area of electronic noses arose in response to the problem of nonselectivity in toxic and contaminant gas mixtures, environmental monitoring remains one of the most interesting application fields for such sensing systems. It includes the analysis of fuel mixtures, the detection of oil leaks, the identification of household odors and toxic wastes, air-quality monitoring, and the monitoring of factory emissions.

But the analysis of gaseous environments is not the only task that an electronic nose is required to fulfill; an electronic nose can also be used to analyze and classify different kinds of odors. In this sense, the term "electronic nose" comes from the analogy of these electronic devices with the natural olfaction, generalizing odors as chemical patterns. Although the electronic nose is far from replacing the human olfactory system, the possible uses from this technology are endless. Currently, the largest market for electronic noses is the food industry. Applications of electronic noses in the food industry include (just to mention some of them): quality assessment in food production, inspection of food quality by odor, control of food-cooking processes, inspection of fish, monitoring the fermentation process, checking the rancidity of mayonnaise, verifying if orange juice is natural, monitoring food and beverage odors, grading whiskey, and inspection of beverage containers. In some instances, electronic noses can be used to augment or replace panels of humans experts. In other cases, electronic noses can be used to reduce the amount of analytical chemistry that is performed in food production, especially when only qualitative results are needed.

Moreover, because the sense of smell is an important sense to the doctor, an electronic nose also has applicability as a diagnostic tool. An electronic nose can examine odors from the body (e.g., breath, wounds) and identify possible problems. Odors in the breath can be indicative of gastrointestinal problems, sinus problems, infections, diabetes, liver, and even cancer problems.

In Table 8, examples of gas sensors and gas-sensor system applications are listed. Here, we report, as an example, some results related to metal-oxide gas sensors for applications in an electronic nose [198].

The application of a multisensor system to the analysis of binary gas mixtures of carbon monoxide and methane (CO/CH_4) in air at different relative humidity levels (0, 30, and 50% R.H.) have been considered by Capone et al. [198, 185]. The sensor array consisted of eight Pd-doped SnO_2-based sensors deposited by drop coating onto Si-micromachined hotplates used for this experimental work. They selected CO and CH_4 as target gases because both are toxic and hazardous gases which may be present in a domestic and/or industrial environment. Moreover, they chose Pd as the dopant in SnO_2 thick-film-based sensors to promote the gas detection mechanisms based on surface chemical reactions between the adsorbed oxygen species and the considered gases. In fact, it is known that Pd (as Pt) is a catalytically active metal that improves the sensitivity to reducing gases, such as carbon monoxide and methane, through chemical or electronic interaction. The sensing elements of the array were diversified by varying the Pd-doping content (0.2, 2%wt. of Pd) and the geometry of the Pt contact electrodes in comb-like and gap configurations with differently spaced fingers. The idea was to modulate the selectivity of each sensor of the array, both by different Pd doping and by simply changing the geometric arrangement of the electrodes since it leads to a different Schottky barrier modulation at the SnO_2 semiconductor/metal contact interface, and so to a different overall resistance modulation [167].

Table 8. Examples of gas sensors and gas-sensor system applications.

Automobiles
- Car ventilation control
- Filter control
- Gasoline vapor detection
- Alcohol breath tests

Safety
- Fire detection
- Leak detection
- Toxic/flammable/explosive gas detectors
- Boiler control
- Personal gas monitor

Indoor air quality
- Air purifiers
- Ventilation control
- Cooking control

Environmental control
- Weather stations
- Pollution monitoring

Food
- Food quality control
- Process control
- Packaging quality control (off-odors)

Industrial production
- Fermentation control
- Process control

Medicine
- Breath analysis
- Disease detection

Figure 21 shows, as an example, the dynamic response of the sensors to a sequence of different CO concentrations at 350 °C and 50% R.H. The sensors resulted as very sensitive both to CO and CH_4. Humidity enhances the response to CO and decreases the response to CH_4; however, in humid air, there is always an enhancement of the response of the SnO_2/Pd sensors when carbon monoxide and methane are present simultaneously.

Principal component analysis gave good discrimination results for the different binary CO/CH_4 mixtures. In particular, we found that data are distributed in PC space according to a well-defined geometric structure, with data clusters oriented along preferential directions (Fig. 22). This is a useful tool for a subsequent classification process between the different classes of CO/CH_4 mixtures, based on other pattern recognition methods like discriminant analysis (DA) or cluster analysis. Principal component regression (PCR) gave discrete results for the prediction of the real level of concentration of each component (CO and CH_4) in a mixture, especially for medium–high concentrations of the two gases.

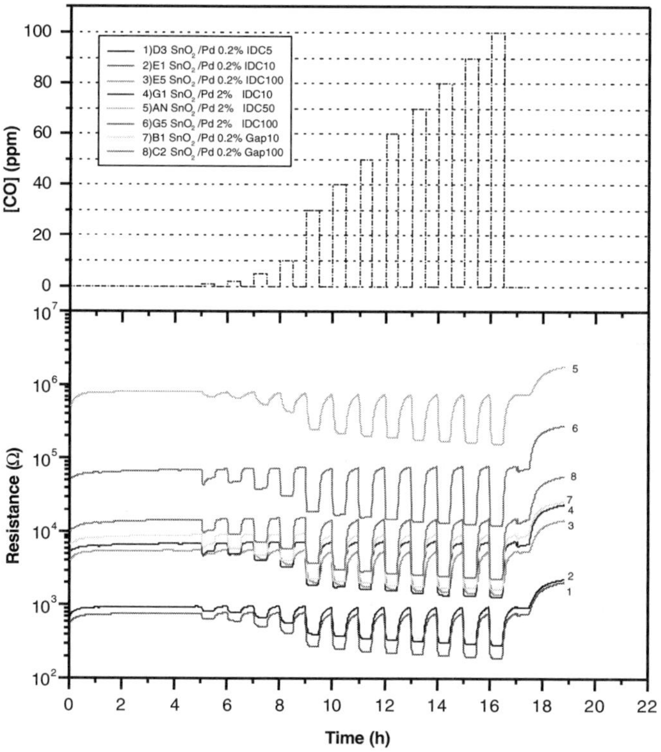

Figure 21. Dynamic response of SnO_2–Pd thick-film-based sensor array to a sequence of different CO concentrations at 350 °C and 50% R.H. [198].

An array of differently doped SnO_2 sol–gel thin films (i.e., undoped SnO_2 and Pd-, Pt-, and Os-doped SnO_2) have been characterized in a controlled atmosphere toward NO_2, CO, CH_4, C_2H_5OH, and CH_3OH to test the ability of the sensor array to be applied in the discrimination of these air pollutants. The evaluation and analysis of the data coming from the array were performed by means of principal component analysis. As usually reported for visualizing clusters, Figure 23a shows the score plot as a function of the first two principal components. As one can see from the figure, the orthogonal projection of the scores onto the PC1–PC2 plane allow five gas clusters (CO, CH_4, NO_2, C_2H_5OH, and CH_3OH) to be well identified. A complete visualization of the discrimination process is also possible using the plot of the first and third principal components (Fig. 23b), where a better separation between the CO and the CH_4 groups is evident.

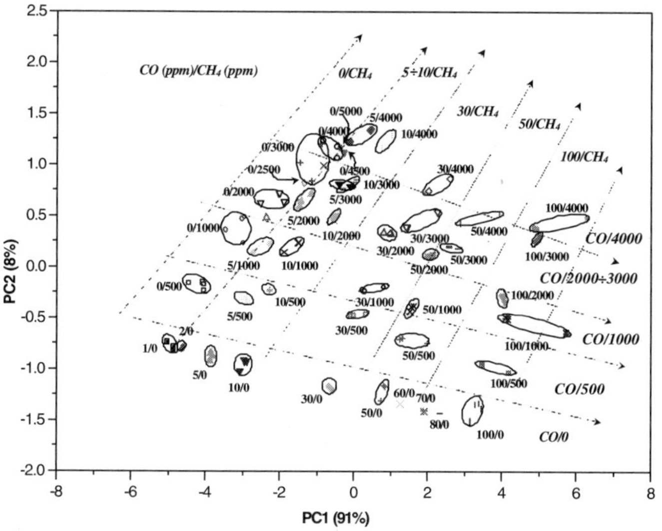

Figure 22. PCA results in the plane of the first two principal components for the dataset related to 30% R.H.

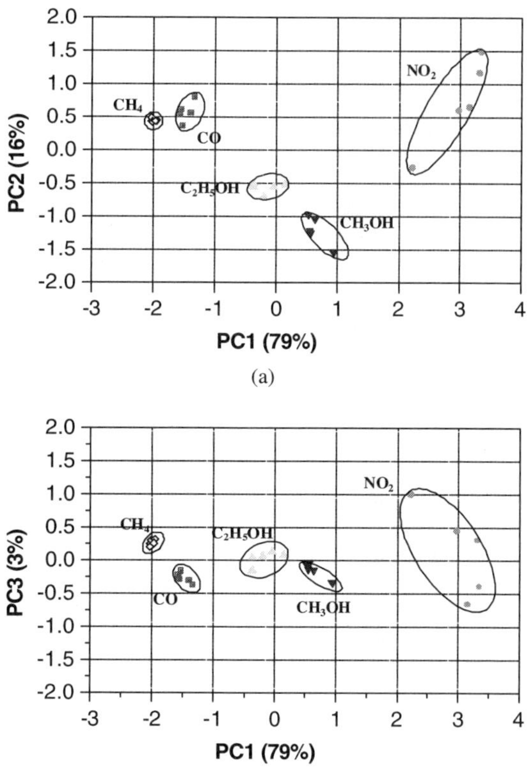

(a)

(b)

Figure 23. Results of the principal component analysis for some air contaminants. a) Score plot in the PC1–PC2 plane. b) Score plot in the PC1–PC3 plane.

An important field of application of the electronic nose is also the food industry. Since the European community has defined strict standards for food quality, significant efforts have been devoted to the development of new techniques that could complement the traditional sensorial and analytical analysis of foodstuffs. The attention has been turned to electronic noses which, in mimicking the human nose, offer an objective way of detecting aromatic fingerprints. In particular, due to the fundamental importance of olive oil in the "Mediterranean diet" and in the marketplace, there is a great interest in all of the techniques able to provide an analysis of olive oil quality. Olive oil was the first foodstuff to be classified by both chemical and sensorial analysis according to the EU Normative. This means that an olive oil can be labeled as extra virgin only if both its chemical and sensorial characteristics are within certain standards established by law. Thus, the market requires urgently reproducible, reliable, inexpensive, easy to train and to use, objective "sniffing" electronic devices dedicated to olive oil for different applications, mainly for classification and degradation

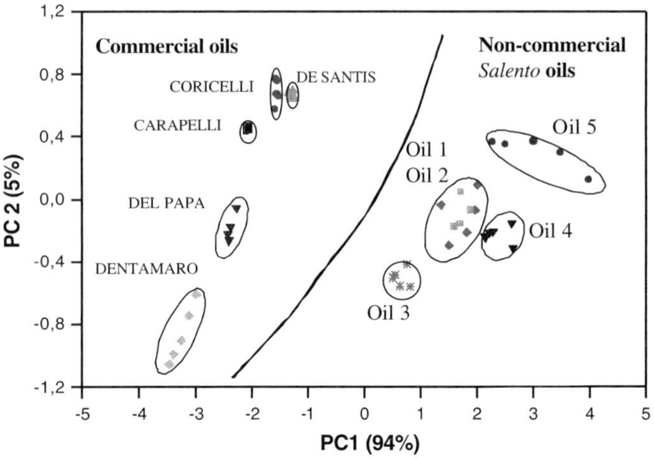

Figure 24. PCA plot related to commercial and noncommercial olive oils from Salento, Italy. The commercial olive oils are labeled with the name of the trademark.

studies. Authenticity is an issue of major concern across the oil industry. Olive oil is marketed based on acidity grade in "olive-husk oil, olive oil, virgin olive oil, extra-virgin olive." Olive oil is also marketed based upon the country and region of production and upon olive type. There is a very large difference in price between, for example, an Italian and a Spanish or Greek olive oil.

The authors developed an electronic nose at the Institute IMM–CNR for testing a number of different commercial olive oils of various qualities produced in different areas in Italy [198, 232, 233]. Furthermore, the analyzed olive oils were compared with local oils of the Salento region in Apulia, Italy. Indeed, one of the most important goals in the field of olive-oil characterization is the classification of them on the basis of their region of origin. This is most important for the regions that have the so called DOP (protect origin denomination) label, which is an index of the quality of the product.

A significant result was obtained by analyzing the array response to the volatile compounds of some commercial and noncommercial local olive oils from Salento by PCA. In

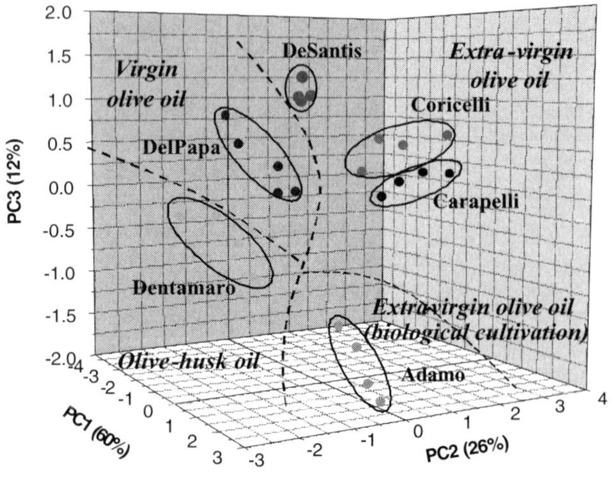

Figure 25. PCA score plot in the space of the first three PCs for all six commercial olive oils.

the score plot, we can clearly distinguish all of the different classes of olive oils, and a discrimination between the commercial and noncommercial local olive oils (Fig. 24). In Figure 25, the PCA score plot related to the same commercial olive oils included an extra-virgin olive oil derived from a Salento biological cultivation.

4. SUMMARY

In this chapter, the authors wanted to offer an overview of the basic science, the technology and the applications of one of the principal types of solid-state gas sensor, that is, resistive-type gas sensors based on nanocrystalline metal-oxide films. The nanostructured metal oxide has been considered as a particular group of the nanomaterials whose general properties and applications were mentioned in Section 1. The background and the gas-sensing principles of semiconductor gas sensors based on nanostructured metal oxides were given in Section 2. The understanding of the device working principles have focused on the description of the receptor and the transducer functions. Particular attention has been devoted to the importance of the microstructure of the sensitive layers, and to the role of grain size when the latter decreases down to the nanometer scale. In Section 3, we described the current status of semiconductor gas-sensor research and development. This section dealt with the most relevant metal oxides reported in the literature as gas-sensing materials, the synthesis and deposition techniques, the opportunity to improve the sensor properties related to the use of Si-micromachined substrates, the concept design of an electronic nose, and finally, some applications of gas-sensor and multisensor systems.

In the future, there is no doubt that nanocrystalline metal oxides will constitute the key for the development of semiconducting gas sensors with improved gas-sensing properties. Nanoscience and nanotechnology are also devoting great efforts to the development of novel materials for gas-sensor applications. We mention, just as an example, the interest in inorganic–organic hybrid nanocomposite-containing conducting polymers as the organic part and metal oxides as the inorganic part [279], and in ribbon-like *nanobelts* made of semiconducting metal oxides. The latter are chemically pure, structurally uniform, and largely defect free, with clean surfaces not requiring protection against oxidation; each nanobelt is made up of a single crystal with specific surface planes and shape. This new class of nanostructures offers a great potential for applications in ultrasmall sensors because the conductivity of these materials changes dramatically when gas or liquid molecules attach their surfaces [280]. Driving elements in the gas-sensor field are also the miniaturization of the devices, the use of silicon microfabrication techniques in sensor production, the development of specific electronic noses trained for specific applications, and the optimization of smarter pattern recognition techniques.

GLOSSARY

Adsorption A process involving the separation of a substance from one phase accompanied by its accumulation or concentration at the surface of another. The adsorbing phase is the adsorbent, and the material concentrated or adsorbed at the surface of that phase is the adsorbate. In semiconductor gas sensors the adsorbent is the semiconducting gas sensitive film and the adsorbate the gaseous molecules. Adsorption is thus different from absorption, a process in which material transferred from one phase to another (e.g. liquid) interpenetrates the second phase to form a "solution". The term sorption is a general expression encompassing both processes. According to the bond of the adsorbate to the adsorbent different types of adsorption processes can be distinguished: physisorption, chemisorption and ionosorption.

Electronic nose A device which comprises an array of electronic chemical sensors with partially specificity and an appropriate pattern recognition system, capable of recognizing simple or complex odors and performing qualitative and quantitative analysis of gaseous mixtures.

Energy bands They consist of a large number of closely spaced energy levels existing in crystalline materials. The bands can be thought of as the collection of the individual energy levels of electrons surrounding each atom. The wavefunctions of the individual electrons, however, overlap with those of electrons confined to neighboring atoms. The Pauli exclusion principle does not allow the electron energy levels to be the same so that one obtains a set of closely spaced energy levels, forming an energy band. The energy band model is crucial to any detailed treatment of semiconductor devices. It provides the framework needed to understand the concept of an energy bandgap and that of conduction in an almost filled band as described by the empty states.

Nanomaterials Materials characterized by crystalline grains ranging in size from 0.1 to 100 nanometers (1 nanometer equal to 10^{-9} meters).

Semiconductor gas sensor A gas detector whose gas sensing element is a thick or thin material with semiconducting electrical properties.

REFERENCES

1. H. Gleiter, *Acta Mater.* 48, 1 (2000).
2. J. Schooman, *Solid State Ionics* 135, 5 (2000).
3. P. Moriarty, *Rep. Prog. Phys.* 64, 297 (2001).
4. F. Einar Kruis, H. Fissan, and A. Peled, *J. Aerosol Sci.* 29, 511 (1998).
5. *J. Aerosol Sci.* 29, 749 (1998).
6. H. Gleiter, J. Weissmüller, O. Wollersheim, and R. Würschum, *Acta Mater.* 49, 737 (2001).
7. http://www.nanomat.com/nanoint.htm.
8. J. R. Morante, Personal Communication, Workshop ENVSENS, S. Cesarea Terme, Lecce, Italy, May 2002.
9. A. J. A. Winnubst and M. M. R. Boutz, *J. Eur. Ceramic Soc.* 18, 2101 (1998).
10. I. Kosacki, V. Petrovsky, and H. U. Anderson, *Appl. Phys. Lett.* 74, 341 (1999).
11. H. Gleiter, *Nanostruct. Mater.* 6, 3 (1995).
12. Y. Kanemitsu, T. Ogawa, K. Shiraishi, and K. Takeda, *Phys. Rev. B* 48, 4883 (1993).
13. Y. Kanemitsu, *Phys. Rep.* 263, 1 (1995).
14. S. M. Prokes, in "Nanomaterials: Synthesis, Properties and Applications" (A. S. Edelstein and R. C. Cammarata, Eds.), pp. 439–457. Institute of Physics Publishing, Bristol, U.K., 1996.
15. A. N. Goldstein, C. M. Echer, and A. P. Alivisatos, *Science* 256, 1425 (1992).
16. H. W. Sarkas, S. T. Arnold, J. H. Hendricks, J. H. Kidder, C. A. Jones, and K. H. Bowen, *Z. Phys. D* 26, 46 (1993).
17. P. T. Moseley and B. C. Tofield (Eds.), "Solid State Gas Sensors." Adam Hilger, Bristol, U.K. and Philadelphia, 1987.
18. M. J. Madou and S. R. Morrison (Eds.), "Chemical Sensing with Solid State Devices." Academic, New York, 1989.
19. J. Weissmüller, in "Synthesis and Processing of Nanocrystalline Powder" (D. L. Bourel, Ed.), p. 3. TMS, 1996.
20. J. Weissmüller, in "Nanomaterials: Synthesis, Properties and Applications" (A. S. Edelstein and R. C. Cammarata, Eds.), p. 219. Institute of Physics Publishing, Bristol, U.K., 1996.
21. S. R. Phillpot, D. Wolf, and H. Gleiter, *Scripta Metall. Mater.* 33, 1245 (1995).
22. X. Xiao, X. Hu, and X. Chen, *Thin Solid Films* 375, 151 (2000).
23. L. L. Kazmerski, in "Polycrystalline and Amorphous Thin Films and Devices" (L. L. Kazmerski, Ed.), Chap. 3, p. 59. Academic, 1980.
24. S. Vepřek, *Thin Solid Films* 297, 145 (1997).
25. C. M. Lieber, *Solid State Commun.* 117, 607 (1998).
26. A. P. Alivisatos, *Science* 271, 933 (1996).
27. B. H. Kear and P. R. Strutt, *Nanostruct. Mater.* 6, 227 (1995).
28. A. Aviram (Ed.), "Molecular Electronics—Science and Technology, AIP Conference Proceedings," 1992, Vol. 262.
29. G. C. Hadjipanayis, K. J. Klabunde, and C. M. Sorensen, in "Nanomaterials: Synthesis, Properties and Applications" (A. S. Edelstein and R. C. Cammarata, Eds.), p. 375. Institute of Physics Publishing, Bristol, U.K., 1996.
30. X. Chen, X. Hu, and J. Feng, *Nanostruct. Mater.* 6, 309 (1995).
31. A. Bezryadin, C. Dekker, and G. Schmid, *Appl. Phys. Lett.* 71, 1273 (1997).
32. Frontiers in cluster science: Nanoclusters, *J. Cluster Sci.* (Special Issue), 8 (1997).
33. Y.-M. Chiang (Ed.), Nanostructured materials for energy applications, 3, *J. Electroceram.* (Special Issue), 1 (1997).
34. H. Maleki, D. Illa, R. L. Zimmerman, G. M. Jenkins, and D. Poker, *Mater. Res. Soc. Symp. Proc.* (C. M. Cotell, A. E. Meyer, S. M. Gorbatkin, and G. L. Grobe, III, Eds.), 414, (1996).
35. A. ten Wolde (Ed.), "Nanotechnology, Towards a Molecular Construction Kit," The Netherlands Study Centre for Technology Trends (STT), The Hague, 1998.
36. K. Nihara, A. Nakahira, and T. Sekino, in "Nanophase and Nanocomposite Materials" (D. C. Van Aken, G. S. Was, and K. Ghosh, Eds.), p. 27. TMS, 1991.
37. D. Kohl, *J. Phys. D: Appl. Phys.* 34, 125 (2001).
38. H. Meixner and U. Lampe, *Sensors Actuators B* 33, 198 (1996).
39. D. Briand, A. Krauss, B. van der Schoot, U. Weimar, N. Bârsan, W. Göpel, and N. F. de Rooij, *Sensors Actuators B* 68, 223 (2000).
40. I. Simon, N. Bârsan, M. Bauer, and U. Weimar, *Sensors Actuators B* 73, 1 (2001).
41. S. Semancik, R. E. Cavicchi, M. C. Wheeler, J. E. Tiffany, G. E. Poirier, R. M. Walton, J. S. Suehle, B. Panchapakesan, and D. L. DeVoe, *Sensors Actuators B* 77, 579 (2001).
42. P. Fau, M. Sauvan, S. Trautweiler, C. Nayral, L. Erades, A. Maisonnat, and B. Chaudret, *Sensors Appl. B* 78, 83 (2001).
43. C. Xu, J. Tamaki, N. Miura, and N. Yamazoe, *Sensors Actuators B* 3, 147 (1991).
44. G. Bläser, Th. Rühl, C. Diehl, M. Ulrich, and D. Kohl, *Phys. A* 266, 218 (1999).
45. M. Ulrich, C.-D. Kohl, and A. Bunde, *Thin Solid Films* 391, 299 (2001).
46. D. S. Vlachos and A. C. Xenoulis, *Nanostruct. Mater.* 10, 1355 (1998).
47. A. M. Gas'kov and M. N. Rumyantseva, *Russian J. Appl. Chem.* 74, 440 (2001).

48. X. Wang, S. S. Yee, and W. P. Carey, *Sensors Actuators B* 24–25, 454 (1995).
49. N. Yamazoe, *Sensors Actuators B* 51, 7 (1991).
50. A. Gurlo, N. Bârsan, M. Ivanovskaya, U. Weimar, and W. Göpel, *Sensors Actuators B* 47, 92 (1998).
51. D. E. Williams and K. F. E. Pratt, *Sensors Actuators B* 70, 214 (2000).
52. G. Zhang and M. Liu, *Sensors Actuators B* 69, 144 (2000).
53. M. C. Horrillo, A. Serventi, D. Rickerby, and J. Gutiérrez, *Sensors Actuators B* 58, 474 (1999).
54. A. M. Mazzone, *Philos. Mag. Lett.* 82, 99 (2002).
55. G. Chow and N. I. Noskova, "Nanostructured Materials, Science and Technology," NATO ASI Series, 1998, p. 335.
56. N. Bârsan, M. Schweizer-Berberich, and W. Göpel, *Fresenius J. Anal. Chem.* 365, 287 (1999).
57. S. Yamauchi, *Chem. Sensor Technol.* 4, Kodansha, Tokio, 20 (1992).
58. A. P. Lee and B. J. Reedy, *Sensors Actuators B* 60, 35 (1999).
59. R. Cavicchi, J. Suehle, P. Chaparala, G. Poirier, K. Kreider, and M. Gaitan, *Sensors Actuators B* 33, 142 (1996).
60. A. Heilig, N. Bârsan, U. Weimar, M. Schweizer-Berberich, J. Gardner, and W. Göpel, *Sensors Actuators B* 43, 45 (1997).
61. S. Nakata and N. Ojima, *Sensors Actuators B* 56, 79 (1999).
62. K. Široký, *Sensors Actuators B* 17, 13 (1993).
63. R. Ionescu, *Sensors Actuators* 48, 392 (1998).
64. P. Krebs and A. Grisel, *Sensors Actuators B* 13/14, 155 (1993).
65. G. Williams and G. Coles, *Sensors Actuators B* 57, 108 (1999).
66. A. Heilig, N. Bârsan, U. Weimar, and W. Göpel, *Sensors Actuators B* 58, 302 (1999).
67. T. Takada, "Proceedings of the IMCS 98," 1998, p. 434.
68. H. Shin, C. Lloyd, and J. Gardner, "Proceedings of Transducers 97," 1997, p. 935.
69. A. Vancu, R. Ionescu, and N. Bârsan, in "Thin Film Resistive Sensors" (P. Ciureany and S. Middelhoek), Chap. 6, p. 437. IOP Publishing Ltd., 1992.
70. A. Lopes, E. Fortunato, P. Nunes, P. Vilarinho, and R. Martins, *Int. J. Inorgan. Mater.* 3, 1349 (2000).
71. W. H. Brattain and J. Bardeen, *Bell Syst. Tech. J.* 32, 1 (1953).
72. T. Seiyama, A. Kato, K. Fujushi, and M. Nagatani, *Anal. Chem.* 34, 1502 (1962).
73. N. Taguchi, U.S. Patent 3, 644, 795.
74. W. Göpel, J. Hesse, and Z. N. Zemel (Eds.), "Sensors—A Comprehensive Survey, Chemical and Biochemical Sensors," Vol. 2, Parts I and II. VCH, Weinheim, 1991.
75. P. T. Moseley, J. O. W. Norris, and D. E. Williams, "Techniques and Mechanisms in Gas Sensing." Adam Hilger, Bristol, U.K., 1991.
76. G. Sberveglieri (Ed.), "Gas sensors—Principles, Operation and Developments." Kluwer Academic, Dordrecht, 1992.
77. K. Ihokura and J. Watson, "The Stannic Oxide Gas Sensor—Principles and Applications." CRC Press, Boca Raton, FL, 1994.
78. G. Sberveglieri, *Sensors Actuators B* 23, 103 (1995).
79. U. Lampe, M. Fleischer, and H. Meixner, *Sensors Actuators B* 17, 187 (1994).
80. M. Fleischer and H. Meixner, *Sensors Actuators B* 43, 1 (1997).
81. Y. Xu, X. Zhou, and O. Toft Sorensen, *Sensors Actuators B* 65, 2 (2000).
82. C. O. Park, S. A. Akbar, and J. Hwang, *Mater. Chem. Phys.* 75, 56 (2002).
83. M. Fleischer, S. Kornely, T. Weh, J. Frank, and H. Meixner, *Sensors Actuators B* 69, 205 (2000).
84. I. Heberle, A. Liebminger, U. Weimar, and W. Göpel, *Sensors Actuators B* 68, 53 (2000).
85. J. Arbiol i Cobos, Ph.D. Dissertation, Facultat de Física, Departament d'Electrònica, Universitat de Barcelona, Chap. 1, p. 17, 2001.
86. J. O. W. Norris, in "Solid State Sensors" (P. T. Moseley and B. C. Tofield, Eds.), p. 124, Adam Hilger, Bristol, U.K. and Philadelphia, 1987.
87. D. Kohl, *Sensors Actuators B* 1, 158 (1990).
88. D. S. Vlachos, C. A. Papadopolos, and J. A. Avaritsiotis, *Sensors Actuators B* 44, 458 (1997).
89. P. T. Moseley, *Sensors Actuators B* 6, 149 (1992).
90. C. Distante, M. Leo, P. Siciliano, and K. Persaud, *Sensors Actuators B* 87, 274 (2002).
91. N. Yamazoe and N. Miura, in "Chemical Sensor Technology" (S. Yamauchi, Ed.), Vol. 4, p. 19. Kodansha, Tokyo, 1992.
92. J. W. Gardner and P. Bartlett, *Sensors Actuators B* 18–19, 211 (1994).
93. D. E. Williams, *Sensors Actuators B* 57, 1 (1999).
94. H.-E. Endres, H. D. Jander, and W. Göttler, *Sensors Actuators B* 23, 163 (1995).
95. T. A. Jones, in "Solid State Gas Sensors" (P. T. Moseley and B. C. Tofield, Eds.), p. 51. Adam Hilger, Bristol, U.K. and Philadelphia, 1987.
96. J. C. Rivière, in "Solid State Gas Sensors" (P. T. Moseley and B. C. Tofield, Eds.), p. 169. Adam Hilger, Bristol, U.K. and Philadelphia, 1987.
97. J. Arbiol i Cobos, Ph.D. Dissertation, Facultat de Física, Departament d'Electrònica, Universitat de Barcelona, Chap. 2, p. 61, 2001.
98. J. Kappler, Ph.D. Dissertation, University of Tübigen, 2001.
99. A. M. Serventi, D. G. Rickerby, M. C. Horrillo, R. G. Saint-Jacques, and J. Gutiérrez, *Nanostruct. Mater.* 11, 813 (1999).
100. D. G. Rickerby, M. C. Horrillo, J. P. Santos, and P. Serrini, *Nanostruct. Mater.* 9, 43 (1997).
101. K.-I. Fukui, H. Onishi, and Y. Iwasawa, *Appl. Surf. Sci.* 140, 259 (1999).
102. M. Sinner-Hettenbach, N. Bârsan, U. Weimar, T. Weiß, H. von Schenck, M. Göthelid, L. Giovanelli, and G. Le Lay, *Thin Solid Films* 391, 192 (2001).
103. K. Li, D. Wang, F. Wu, T. Xie, and T. Li, *Mater. Chem. Phys.* 60, 226 (1999).
104. M. Boulova, A. Gaskov, and G. Lucazeau, *Sensors Actuators B* 81, 99 (2001).
105. M. P. Casaletto, S. Kaciulis, G. Mattogno, L. Pandolfi, G. Scavia, L. Dori, S. Nicoletti, M. Severi, and S. Zampolli, *Appl. Surf. Sci.* 189, 39 (2002).
106. R. Schlesinger and M. Bruns, *Thin Solid Films* 366, 265 (2000).
107. J. Arbiol, P. Gorostiza, A. Cirera, A. Cornet, and J. R. Morante, *Sensors Actuators B* 78, 57 (2001).
108. M. S. Castro, M. P. Suárez, and C. M. Aldao, *J. Eur. Ceram. Soc.* 21, 1115 (2001).
109. G. Ghiotti, A. Chiorino, F. Prinetto, M. C. Carotta, G. Martinelli, and M. Merli, *Sensors Actuators B* 44, 474 (1997).
110. G. Ghiotti, A. Chiorino, G. Martinelli, and M. C. Carotta, *Sensors Actuators B* 25, 520 (1995).
111. S. Emiroglu, N. Bârsan, U. Weimar, and V. Hoffmann, *Thin Solid Films B* 391, 176 (2001).
112. R. Pohle, M. Fleischer, and H. Meixner, *Sensors Actuators B* 78, 133 (2001).
113. J. R. MacDonald, "Impedance Spectroscopy." Wiley, New York, 1987.
114. U. Weimar and W. Göpel, *Sensors Actuators B* 26, 13 (1995).
115. K.-D. Schierbaum, *Sensors Actuators B* 24–25, 239 (1995).
116. A. Cabot, J. Arbiol, J. R. Morante, U. Weimar, N. Bârsan, and W. Göpel, *Sensors Actuators B* 70, 87 (2000).
117. T. Kawabe, S. Shimomura, T. Karasuda, K. Tabata, E. Suzuki, and Y. Yamaguchi, *Surf. Sci.* 448, 101 (2000).
118. G. Kiss, Z. Pintér, I. V. Perczel, Z. Sassi, and F. Réti, *Thin Solid Films* 391, 216 (2001).
119. S.-C. Chang, *J. Vac. Sci. Technol.* 71, 366 (1980).
120. R. Sanjinés, D. Rosenfeld, F. Gozzo, Ph. Alméras, L. Perez, F. Lévy, G. Margaritondo, and H. Schreiner, *Surf. Interface Anal.* 22, 372 (1994).

121. A. M. Stoneham, in "Solid State Gas Sensors" (P. T. Moseley and B. C. Tofield, Eds.). Adam Hilger, Bristol, U.K. and Philadelphia, 1987.
122. W. Göpel and K. D. Schierbaum, *Sensors Actuators B* 26–27, 1 (1995).
123. M. Schweizer-Berberich, Ph.D. Dissertation, University of Tübingen, Germany, 1998.
124. H. Geistlinger, *Sensors Actuators B* 17, 47 (1993).
125. H. Lüth, "Surfaces and Interfaces of Solids," 2nd ed. Springer, Heidelberg, 1993.
126. S. R. Morrison, "The Chemical Physics of Surfaces," 2nd ed., Plenum, New York, 1990.
127. T. Volkenstein, "Electronic Processes on Semiconductor Surfaces During Chemisorption." Consultants Bureau, New York, 1991.
128. S. Brunauer, L. S. Deming, W. E. Deming, and E. Teller, *J. Am. Chem. Soc.* 62, 142 (1940).
129. T. S. Rantala, V. Lantto, and T. T. Rantala, *Sensors Actuators B* 13/14, 234 (1993).
130. T. S. Rantala and V. Lantto, *Sensors Actuators B* 18/19, 711 (1994).
131. D. F. Cox, T. B. Fryberger, and S. Semancik, *Phys. Rev. B* 38, 2072 (1988).
132. T. S. Rantala, V. Lantto, and T. T. Rantala, *Sensors Actuators B* 18–19, 716 (1994).
133. T. S. Rantala and V. Lantto, *Surf. Sci.* 352–354, 765 (1996).
134. P. Romppainen and V. Lantto, *J. Appl. Phys.* 63, 5159 (1988).
135. D. E. Williams, in "Solid State Gas Sensors" (P. T. Moseley and B. C. Tofield, Eds.). Adam Hilger, Bristol, U.K. and Philadelphia, 1987.
136. T. S. Rantala, P. Pirttiaho, and V. Lantto, *Sensors Actuators B* 15/16, 323 (1993).
137. S. Lenaerts, J. Roggen, and G. Maes, *Spectrochim. Acta Part A* 51, 883 (1995).
138. J. P. Joly, L. Gonzalez-Cruz, and Y. Arnaud, *Bull. Soc. Chim. Fr.* 11 (1986).
139. B. Gillot, C. Fey, and D. Delafosse, *J. Chem. Phys.* 73, 19 (1976).
140. N. Yamazoe, J. Fuchigami, and T. Seiyama, *Surf. Sci.* 86, 335 (1979).
141. A. M. Volodin and A. E. Cherkasin, *Reac. Kinet. Catal. Lett.* 17, 329 (1981).
142. P. B. Weisz, *J. Chem. Phys.* 21, 1531 (1953).
143. S. Kačiulis, G. Mattogno, A. Galdikas, A. Mironas, and A. Šetkus, *J. Vac. Sci. Technol. A* 14, 3164 (1996).
144. C. D. Kohl, in "Gas Sensors" (G. Sberveglieri, Ed.), Chap. 2, p. 43. Kluwer, Dordrecht, 1992.
145. P. K. Clifford and D. T. Tuma, *Sensors Actuator* 3, 233 (1982/1983).
146. G. Heiland and D. Kohl, in "Chemical Sensor Technology" (T. Seiyama, Ed.), Vol. 1, Chap. 2, p. 15. Kodansha, Tokyo, Elsevier, Amsterdam, 1988.
147. D. Kohl, *Sensors Actuators* 18, 71 (1989).
148. V. A. Henrich and P. A. Cox, "The Surface Science of Metal Oxides," p. 312. University Press, Cambridge, 1994.
149. M. Caldararu, D. Sprînceana, V. T. Popa, and N. I. Ionescu, *Sensors Actuators B* 30, 35 (1996).
150. D. S. Vlachos, P. D. Skafidas, and J. N. Avaritsiotis, *Sensors Actuators B* 24–25, 491 (1995).
151. R. Ionescu, A. Vancu, C. Moise, and A. Tomescu, *Sensors Actuators B* 61, 39 (1999).
152. N. Bârsan and R. Ionescu, *Sensors Actuators B* 12, 71 (1993).
153. F. Réti, M. Fleischer, J. Gerblinger, U. Lampe, E. B. Várhegyi, I. V Perczel, H. Meixner, and J. Giber, *Sensors Actuators B* 26–27, 106 (1995).
154. N. Bârsan, R. Grigorovici, R. Ionescu, M. Motronea, and A. Vancu, *Thin Solid Films* 171, 53 (1989).
155. N. Bârsan, *Sensors Actuators B* 17, 241 (1994).
156. N. Bârsan and U. Weimar, *J. Electroceram.* 7, 143 (2001).
157. S. Strässler and A. Reis, *Sensors Actuators* 4, 465 (1983).
158. M. Ippomatsu, H. Sazaki, and Yanajida, *J. Mater. Sci.* 25, 259 (1990).
159. P. D. Skafidas, D. S. Vlachos, and J. N. Avaritsiotis, *Sensors Actuators B* 21, 109 (1994).
160. H. Teterycz, B. W. Licznerski, K. Nitsch, K. Wiśniewski, and L. J. Golonka, *Sensors Actuators B* 47, 153 (1998).
161. M. J. Willett, in "Techniques and Mechanisms in Gas Sensing" (P. T. Moseley, J. Norris, and D. E. Williams), Vol. 3, p. 61. Adam Hilger, Bristol, U.K., 1991.
162. S. Capone, P. Siciliano, F. Quaranta, R. Rella, M. Epifani, and L. Vasanelli, *Sensors Actuators B* 77, 503 (2001).
163. D. D. Lee, W. Y. Chung, and B. K. Sohn, *Sensors Actuators B* 13–14, 252 (1993).
164. C. A. Papadopoulus, D. S. Vlachos, and J. N. Avaritsiotis, *Sensors Actuators B* 32, 61 (1996).
165. F. Quaranta, R. Rella, P. Siciliano, S. Capone, M. Epifani, L. Vasanelli, A. Licciulli, and A. Zocco, *Sensors Actuators B* 58, 350 (1999).
166. M. Egashira, M. Nakashima, and S. Kawasumi, "Conference Proceedings of IMCS," Fukuoka, Japan, 1983.
167. U. Jain, A. H. Harker, M. Stoneham, and D. E. Williams, *Sensors Actuators B* 2, 111 (1990).
168. X. Vilanova, E. Llobet, J. Brezmes, J. Calderer, and X. Correig, *Sensors Actuators B* 48, 425 (1998).
169. J. W. Gardner, *Sensors Actuators B* 26–27, 261 (1995).
170. J. W. Gardner, H. V. Shurmer, and P. Corcoran, *Sensors Actuators B* 4, 117 (1991).
171. M. Schweizer-Berberich, N. Bârsan, U. Weimar, J. R. Morante, and W. Göpel, "Proceedings of the 11th European Conference on Solid State Transducers, EUROSENSORS XI," Warsaw, Poland, Sept. 1997, p. 1377.
172. A. Ylinampa, V. Lantto, and S. Leppävuori, *Sensors Actuators B* 13–14, 602 (1993).
173. K. Fukui and M. Nakane, *Sensors Actuators B* 13–14, 589 (1993).
174. H. Gourari, M. Lumbreras, R. Van Landschoot, and J. Schoonman, "Proceedings of the 12th European Conference on Solid State Transducers, EUROSENSORS XII," Southampton, U.K., Sept. 1998, p. 677.
175. G. Sberveglieri, *Sensors Actuators B* 6, 239 (1992).
176. W. Hellmich, Ch. Bosch-v. Braunmühl, G. Müller, G. Sberveglieri, M. Berti, and C. Perego, *Thin Solid Films* 263, 231 (1995).
177. Th. Becker, S. Ahlers, Ch. Bosch-v. Braunmühl, G. Müller, and O. Kiesewetter, *Sensors Actuators B* 7, 55 (2001).
178. V. Guidi, M. A. Butteri, M. C. Carotta, B. Cavicchi, M. Ferroni, C. Malaga, G. Martinelli, D. Vincenzi, M. Sacerdoti, and M. Zen, *Sensors Actuators B* 84, 72 (2002).
179. A. Diéguez, A. Romano-Rodríguez, J. R. Morante, U. Weimar, M. Schweizer-Berberich, and W. Göpel, *Sensors Actuators B* 31, 1 (1996).
180. A. Cirera, A. Vilà, A. Diéguez, A. Cabot, A. Cornet, and J. R. Morante, *Sensors Actuators B* 64, 65 (2000).
181. A. Cirera, A. Vilà, A. Cornet, and J. R. Morante, *Mater. Sci. Eng. C* 15, 203 (2001).
182. S. C. Ray, M. K. Karanjai, and D. Dasgupta, *Thin Solid Films* 247, 162 (1994).
183. V. Golovanov, J. L. Solis, V. Lantto, and S. Leppävuori, *Sensors Actuators B* 34, 401 (1996).
184. D. Briand, B. van der Schoot, and N. F. de Rooij, "Proceedings of the 13th European Conference on Solid State Transducers, EUROSENSORS XI," The Hague, The Netherlands, Sept. 1999, p. 703.
185. S. Capone, P. Siciliano, N. Bârsan, U. Weimar, and L. Vasanelli, *Sensors Actuators B* 78, 40 (2001).
186. G. Micocci, A. Serra, A. Tepore, S. Capone, R. Rella, and P. Siciliano, *J. Vac. Sci. Technol. A* 15, 34 (1997).
187. L. M. Cukrov, P. G. McCormick, K. Galatsis, and W. Wlodarski, *Sensors Actuators B* 77, 491 (2001).

188. D. Briand, M. Labeau, J. F. Currie, and G. Delabouglise, *Sensors Actuators B* 48, 395 (1998).
189. J. R. Brown, P. W. Haycock, L. M. Smith, A. C. Jones, and E. W. Williams, *Sensors Actuators B* 63, 109 (2000).
190. G. Leo, R. Rella, P. Siciliano, S. Capone, J. C. Alonso, V. Pankov, and A. Ortiz, *Sensors Actuators B* 58, 370 (1999).
191. M. de la Olvera and R. Asomoza, *Sensors Actuators B* 45, 49 (1997).
192. M. A. El Khakani, R. Dolbec, A. M. Serventi, M. C. Horrillo, M. Trudeau, R. G. Saint-Jacques, D. G. Rickerby, and I. Sayago, *Sensors Actuators B* 77, 383 (2001).
193. T. K. H. Starke, G. S. V. Coles, and H. Ferkel, *Sensors Actuators B* 85, 239 (2002).
194. C. Cantalini, W. Wlodarski, H. T. Sun, M. Z. Atashbar, M. Passacantando, and S. Cantucci, *Sensors Actuators B* 65, 101 (2000).
195. C. Garzella, E. Comini, E. Bontempi, L. E. Depero, C. Frigeri, and G. Sberveglieri, *Sensors Actuators B* 83, 230 (2002).
196. T. M. Racheva and G. W. Critchlow, *Thin Solid Films* 292, 299 (1997).
197. S.-S. Park and J. D. Mackenzie, *Thin Solid Films* 274, 154 (1996).
198. S. Capone, Ph.D. Dissertation, University of Lecce, 2001.
199. S. Capone, R. Rella, P. Siciliano, L. Vasanelli, A. Licciulli, and M. Epifani, "Proceedings CIMTEC'98, Advances in Science and Technology, Solid State Chemical and Biochemical Sensors (T. Srl, Ed.), 1999, Vol. 26, p. 161.
200. P. Siciliano, *Sensors Actuators B* 70, 153 (2000).
201. C. Cobianu, C. Savaniu, A. Arnautu, R. Iorgulescu, D. Dascalu, S. Capone, G. Leo, M. Mazzer, R. Rella, P. Siciliano, and L. Vasanelli, *Sensors Actuators B* 58, 552 (1999).
202. C. Cobianu, C. Savaniu, P. Siciliano, S. Capone, M. Utriainen, and L. Niinisto, *Sensors Actuators B* 77, 496 (2001).
203. I. Hotovy, J. Huran, P. Siciliano, S. Capone, L. Spiess, and V. Rehacek, *Sensors Actuators B* 78, 126 (2001).
204. G. Neri, A. Bonavita, S. Galvagno, P. Siciliano, and S. Capone, *Sensors Actuators B* 82, 40 (2002).
205. C. Nayral, E. Viala, V. Collière, P. Fau, F. Senocq, A. Maisonnat, and B. Chaudret, *Appl. Surf. Sci.* 164, 219 (2000).
206. I. Jiménez, A. Cirera, J. Folch, A. Cornet, and J. R. Morante, *Sensors Actuators B* 78, 78 (2001).
207. R. L. Petritz, *Phys. Rev.* 110, 1508 (1956).
208. J. Y. W. Seto, *J. Appl. Phys.* 46, 5247 (1975).
209. J. W. Orton and M. J. Powell, *Rep. Prog. Phys.* 43, 1267 (1980).
210. M. Epifani, A. Forleo, S. Capone, F. Quaranta, R. Rella, P. Siciliano, and L. Vasanelli, Hall effect measurements in gas sensors based on nanosized Os-doped sol–gel derived SnO_2 thin films, accepted by "IEEE Sensors," 2002.
211. Figaro Engineering Inc., *Figaro Gas Sensor, Product Catalogue* (1995).
212. H. Windischmann and P. Mark, *J. Electrochem. Soc.: Solid-State Sci. Technol.* 126, 672 (1979).
213. H. Pink, L. Treitinger, and L. Vite, *Jpn. J. Appl. Phys.* 19, 513 (1980).
214. A. Ikegami and M. Kaneyasu, "Transducer 85," Philadelphia, PA, 1985, p. 136.
215. K. D. Schierbaum, U. Weimar, W. Göpel, and R. Kowalkowski, *Sensors Actuators B* 3, 205 (1991).
216. K. D. Schierbaum, U. Weimar, and W. Göpel, *Sensors Actuators B* 2, 71 (1990).
217. R. Ionescu, V. Vasilesgu, A. Vancu, and N. Bârsan, *Rev. Roum. Phys.* 36, 885 (1991).
218. R. Ionescu, V. Vasilesgu, and A. Vancu, *Sensors Actuators B* 8, 151 (1992).
219. N. Bârsan and A. Tomescu, *Sensors Actuators B* 26–27, 45 (1995).
220. V. V. Kissine, V. V. Sysoev, and S. A. Voroshilov, *Sensors Actuators B* 79, 163 (2001).
221. V. Demarne and R. Sanjines, in "Gas Sensors—Principles, Operation and Developments" (G. Sberveglieri, Ed.), p. 89. Kluwer Academic, Dordrecht, 1992.
222. A. Many, Y. Goldstein, and N. B. Grover, "Semiconductor Surfaces," p. 308. Interscience, New York, 1965.
223. P. T. Moseley, A. M. Stoneham, and D. E. Williams, in "Techniques and Mechanisms in Gas Sensing" (P. T. Moseley, J. O. W. Norris, and D. E. Williams, Eds.), p. 108. Adam Hilger, Bristol, U.K., 1991.
224. T. S. Rantala, V. Lantto, and T. T. Rantala, *Sensors Actuators B* 47, 59 (1998).
225. V. Brinzari, G. Korotchenkov, and S. Dmitriev, *Sensors Actuators B* 61, 143 (1999).
226. V. M. Aroutiounian and G. S. Aghababian, *Sensors Actuators B* 50, 80 (1998).
227. N. Yamazoe, K. Kurokawa, and T. Seymana, *Sensors Actuators* 4, 283 (1983).
228. R. Rella, P. Siciliano, S. Capone, M. Epifani, L. Vasanelli, and A. Licciulli, *Sensors Actuators B* 58, 283 (1999).
229. S. Capone, P. Siciliano, F. Quaranta, R. Rella, M. Epifani, and L. Vasanelli, *Sensors Actuators B* 69, 230 (2000).
230. S. Capone, M. Epifani, F. Quaranta, P. Siciliano, and L. Vasanelli, *Thin Solid Films* 391, 314 (2001).
231. S. Capone, M. Epifani, F. Quaranta, P. Siciliano, A. Taurino, and L. Vasanelli, *Sensors Actuators B* 78, 174 (2001).
232. A. Taurino, S. Capone, C. Distante, M. Epifani, R. Rella, and P. Siciliano, *Thin Solid Films* 418, 59 (2002).
233. F. Quaranta, R. Rella, P. Siciliano, S. Capone, C. Distante, M. Epifani, and A. Taurino, *Sensors Actuators B* 84, 55 (2002).
234. A. Cabot, A. Diéguez, A. Romano-Rodríguez, J. R. Morante, and N. Bârsan, *Sensors Actuators B* 79, 98 (2001).
235. A. Cabot, A. Vilà, and J. R. Morante, *Sensors Actuators B* 84, 12 (2002).
236. A. Ruiz, J. Arbiol, A. Cirera, A. Cornet, and J. R. Morante, *Mater. Sci. Eng. C* 19, 105 (2002).
237. G. Tournier, C. Pijolat, R. Lalauze, and B. Patissier, *Sensors Actuators B* 26–27, 24 (1995).
238. O. Renault, A. V. Tadeev, G. Delabouglise, and M. Labeau, *Sensors Actuators B* 59, 260 (1999).
239. M. Gaidi, B. Chenevier, and M. Labeau, *Sensors Actuators B* 62, 43 (2000).
240. M. I. Ivanovskaya, P. A. Bogdanov, D. R. Orlik, A. Ch. Gurlo, and V. V. Romanovskaya, *Thin Solid Films* 296, 41 (1997).
241. J. C. Kim, H. K. Jun, J.-S. Huh, and D. D. Lee, *Sensors Actuators B* 45, 271 (1997).
242. J. Wöllenstein, H. Böttner, M. Jaegle, W. J. Becker, and E. Wagner, *Sensors Actuators B* 70, 196 (2000).
243. J. Mizsei, L. Pirttiaho, M. Karppinen, and V. Lantto, *Sensors Actuators B* 65, 195 (2000).
244. R. S. Niranjan, S. R. Sainkar, K. Vijayamohanan, and I. S. Mulla, *Sensors Actuators B* 82, 82 (2002).
245. G. Sakai, N. Matsunaga, K. Shimanoe, and N. Yamazoe, *Sensors Actuators B* 80, 125 (2001).
246. N. Yamazoe, J. Tamaki, and N. Miura, *Mater. Sci. Eng. B* 41, 178 (1996).
247. K. Hikita, M. Miyayama, and H. Yanagida, *J. Am. Ceram. Soc.* 78, 865 (1995).
248. E. Traversa, *J. Am. Ceram. Soc.* 78, 2625 (1995).
249. S. J. Jung, H. Ohsawa, Y. Nakamura, H. Yanagida, K. Hasumik, and O. Okada, *J. Electrochem. Soc.* 141, 53 (1994).
250. R. B. Vasiliev, M. N. Rumyantseva, N. V. Yakovlev, and A. M. Gaskov, *Sensors Actuators B* 50, 186 (1998).
251. R. B. Vasiliev, M. N. Rumyantseva, S. E. Podguzova, A. S. Ryzhikov, L. I. Ryabova, N. V. Yakovlev, and A. M. Gaskov, *Mater. Sci. Eng. B* 57, 241 (1999).
252. K. Suga, N. Koshizaki, K. Yasumoto, and E. Smela, *Sensors Actuators B* 13–14, 598 (1993).

253. Y. Ushio, M. Miyayama, and H. Yanagida, *Sensors Actuators B* 17, 221 (1994).
254. J. H. Yu and G. M. Choi, *Sensors Actuators B* 61, 59 (1999).
255. Z. Ling, C. Leach, and R. Freer, *J. Eur. Ceram. Soc.* 21, 1977 (2001).
256. C. Leach, Z. Ling, and R. Freer, *Scripta Mater.* 42, 1083 (2000).
257. T. Hyodo, T. Mori, A. Kawahara, H. Katsuki, and Y. Shimizu, *Sensors Actuators B* 77, 41 (2001).
258. E. Comini, M. Ferroni, V. Guidi, G. Faglia, G. Martinelli, and G. Sberveglieri, *Sensors Actuators B* 84, 26 (2002).
259. K. Galatsis, Y. X. Li, W. Wlodarski, E. Comini, G. Faglia, and G. Sberveglieri, *Sensors Actuators B* 77, 472 (2001).
260. M. Ferroni, D. Boscarino, E. Comini, D. Gnani, V. Guidi, G. Martinelli, P. Nelli, V. Rigato, and G. Sberveglieri, *Sensors Actuators B* 58, 289 (1999).
261. K. Galatsis, Y. X. Li, W. Wlodarski, and K. Kalantar-Zatech, *Sensors Actuators B* 77, 478 (2001).
262. V. Guidi, D. Boscarino, E. Comini, G. Faglia, M. Ferroni, C. Malaga, G. Martinelli, V. Rigato, and G. Sberveglieri, *Sensors Actuators B* 65, 264 (2000).
263. K. Galatsis, Y. X. Li, W. Wlodarski, E. Comini, G. Sberveglieri, C. Canalini, S. Cantucci, and M. Passacantando, *Sensors Actuators B* 83, 276 (2002).
264. M. Ivanovskaya, P. Bogdanov, G. Faglia, and G. Sberveglieri, *Sensors Actuators B* 68, 344 (2000).
265. E. Comini, V. Guidi, C. Frigeri, I. Riccò, and G. Sberveglieri, *Sensors Actuators B* 77, 16 (2001).
266. M. Radecka, K. Zakrzewska, and M. Rekas, *Sensors Actuators B* 47, 194 (1998).
267. E. H. A. Diagne and M. Lumbreras, *Sensors Actuators B* 78, 98 (2001).
268. N. Savage, B. Chwieroth, A. Ginwalla, B. R. Patton, S. A. Akbar, and P. K. Dutta, *Sensors Actuators B* 79, 17 (2001).
269. J. W. Gardner, "Microsensors—Principles and Applications." Wiley, New York, 1994.
270. U. Dibbern, in "Sensory and Sensory Systems for an Electronic Nose" (J. W. Gardner and P. N. Bartlett, Eds.), NATO ASI Series, 1992, p. 147.
271. D. G. Rickerby, N. Wächter, M. C. Horrillo, J. Gutiérrez, I. Gràcia, and C. Cané, *Sensors Actuators B* 69, 314 (2000).
272. D.-S. Lee, J.-W. Lim, S.-M. Lee, J.-S. Huh, and D.-D. Lee, *Sensors Actuators B* 64, 31 (2000).
273. R. E. Cavicchi, R. M. Walton, M. Equino-Class, J. D. Allen, and B. Panchapakesan, *Sensors Actuators B* 77, 145 (2001).
274. S. Vaihinger and W. Göpel, in "Sensors—A Comprehensive Survey," Vol. 2, Part II, p. 192. VCH, Weinheim, 1991.
275. A. Hierlemann, M. Schweizer-Berberich, U. Weimar, G. Kraus, A. Pfau, and W. Göpel, in "Sensors Update," Vol. 5, p. 121. VCH, Weinheim, 1991.
276. J. W. Gardner and P. N. Bartlett, "Electronic Noses—Principles and Applications." Oxford University Press, Oxford, 1999.
277. C. Di Natale, F. Davide, and A. D'Amico, *Sensors Actuators B* 23, 111 (1995).
278. J. W. Gardner and P. N. Bartlett, in "Techniques and Mechanisms in Gas Sensing" (P. T. Moseley, J. O. W. Norris, and D. E. Williams, Eds.), p. 347. Adam Hilger, Bristol, U.K., 1991.
279. K. Suri, S. Annapoorni, A. K. Sarkar, and R. P. Tandon, *Sensors Actuators B* 81, 277 (2002).
280. Z. W. Pan, Z. R. Dai, and Z. L. Wang. "Nanobelts of Semiconducting Oxides." *Science* 291, 1947 (2001).
281. R. Lalauze, N. Bui, and C. Pijolat, *Sensors Actuators* 6, 119 (1984).
282. G. Heiland and D. Kohl, *Sensors Actuators* 8, 227 (1985).
283. L. N. Yannopoulos, *Sensors Actuators* 12, 77 (1987).
284. M. Egashira, T. Matsumoto, Y. Shimizu, and H. Iwanaga, *Sensors Actuators* 14, 205 (1988).
285. T. Oyabu, *J. Appl. Phys.* 53, 2785 (1982).
286. G. N. Advani, Y. Komem, J. Hasenkopf, and A. G. Jordan, *Sensors Actuators* 2, 139 (1981/1982).
287. N. Yamazoe and Y. Kurokawa, "Proceedings of the International Meeting on Chemical Sensors," Fukuoka, Japan, 1983, p. 35.
288. V. N. Mishra and R. P. Agarwal, "Sensitivity, Response and Recovery Time of SnO_2 Based Thick-Film Sensor Array for H_2, CO, CH_4 and LPG," *Microelectronics Journal (Incorporating Journal of Semicustom ICs)*, 29, 861 (1998).
289. C. Cobianu, R. Iorgulescu, C. Savaniu, A. Dima, P. Siciliano, S. Capone, R. Rella, F. Quaranta, and L. Vasanelli, *Proc. SPIE* 3680, 1151 (1999).
290. G. S. V. Coles, K. J. Gallacher, and J. Watson, *Sensors Actuators* 7, 89 (1985).
291. M. Prudenziati and B. Morten, *Sensors Actuators* 10, 65 (1986).
292. M. Nitta and M. Haradome, *J. Electron. Mater.* 8, 571 (1979).
293. Y. Okayama, H. Fukaya, K. Kojima, Y. Terasama, and T. Handa, "Proceedings of the International Meeting on Chemical Sensors," Fukuoka, Japan, 1983.
294. R. Lambrich, W. Hagen, and J. Lagois, "Proceedings of the International Meeting on Chemical Sensors," Fukuoka, Japan, 1983.
295. M. Nitta and M. Haradome, *IEEE Trans. Electron Devices* ED-26, 219 (1979).
296. M. J. Esper, E. M. Logothetis, and J. C. Chu, "SAE Automotive Engineering Congress," Ser. 790, 1979, p. 140.
297. G. Sulz, G. Kühner, H. Reiter, G. Uptmoor, W. Schweizer, H. Löw, M. Locher, and K. Steiner, *Sensors Actuators B* 15–16, 390 (1993).
298. V. Ambrazeviciene, A. Kaldikas, G. Grebinskij, A. Mironas, and H. Tvardauskas, *Sensors Actuators B* 17, 27 (1993).
299. P. Siciliano, *Sensors Actuators B* 70, 153 (2000).
300. M. Schweizer-Berberich, J. G. Zheng, U. Weimar, W. Göpel, N. Bârsan, E. Pentia, and A. Tomescu, *Sensors Actuators B* 31, 71 (1996).
301. M. Labeau, B. Toutheron, G. Delabouglise, J. Pena, V. Ragel, A. Varela, J. Roman, J. Martinez, J. M. Gonzalez-Calbet, and M. Vallet-Regi, *Sensors Actuators B* 15–16, 379 (1993).
302. G. Sberveglieri, G. Faglia, S. Groppeli, P. Nelli, and A. Taroni, *Sensors Actuators B* 7, 721 (1992).
303. G. Martinelli and M. C. Carotta, *Sensors Actuators B* 15–16, 363 (1993).
304. W. Mokwa, D. Kohl, and G. Heiland, *Sensors Actuators* 8, 101 (1985).
305. V. Lantto, P. Romppainen, T. S. Rantela, and S. Lepparvuori, *Sensors Actuators B* 4, 451 (1991).
306. A. Diéguez, A. Romano-Rodríguez, J. R. Morante, J. Kappler, N. Bârsan, and W. Göpel, *Sensors Actuators B* 60, 125 (1999).
307. K. D. Schierbaum, S. Vaihinger, W. Göpel, H. H. Van den Vlekkert, B. Kloeck, and N. F. Rooij, *Sensors Actuators B* 1, 171 (1990).
308. G. Williams and G. S. V. Coles, *Sensors Actuators B* 15–16, 349 (1993).
309. H. Torrela, C. Pijolet, and R. Lalaouze, *Sensors Actuators B* 4, 445 (1991).
310. G. Sberveglieri, S. Groppelli, and P. Nelli, *Sensors Actuators B* 4, 457 (1991).
311. F. J. Gutiérrez, L. Arés, J. I. Robla, M. C. Horrillo, I. Sayago, J. M. Getino, and J. A. de Agapito, *Sensors Actuators B* 15–16, 354 (1993).
312. G. D. Barbi and D. S. Blanco, *Sensors Actuators B* 15–16, 372 (1993).
313. G. Sulz, G. Kühner, H. Reiter, G. Uptmoor, W. Schweizer, H. Löw, M. Locher, and K. Steiner, *Sensors Actuators B* 15–16, 390 (1993).
314. G. Wiegleb and J. Heitbaum, *Sensors Actuators B* 17, 93 (1994).
315. H. Torrela, C. Pijolet, and R. Lalaouze, *Sensors Actuators B* 4, 445 (1991).
316. N. Mizuna, T. Yoshioka, K. Kazo, and M. Iwamato, *Sensors Actuators B* 13–14, 473 (1993).

317. S. Kudo, H. Ohnishi, T. Matsumoto, and M. Ippomatsu, *Sensors Actuators B* 23, 219 (1995).
318. M. Ivanovskaya, A. Gurlo, and P. Bogdanov, *Sensors Actuators B* 77, 264 (2001).
319. T. Arakawa, K. Takada, Y. Tsunemine, and J. Shiokawa, "Proceedings of the 2nd International Meeting on Chemical Sensors," Bordeaux, France, July 1986, p. 115.
320. J. R. Stetter, *J. Colloid Interface Sci.* 65, 432 (1978).
321. K. W. Kirby and H. Kimura, *Sensors Actuators B* 32, 49 (1996).
322. S. Capone, R. Rella, P. Siciliano, and L. Vasanelli, *Thin Solid Films* 350, 264 (1999).
323. V. Guidi, G. C. Cardinali, L. Dori, G. Faglia, M. Ferroni, G. Martinelli, P. Nelli, and G. Sberveglieri, *Sensors Actuators B* 49, 88 (1998).
324. N. Hykaway, W. M. Sears, R. F. Frindt, and S. R. Morrison, *Sensors Actuators* 15, 105 (1988).
325. P. T. Moseley and D. E. Williams, *Sensors Actuators B* 1, 113 (1990).
326. H. Nanto, T. Minami, and S. Takata, *J. Appl. Phys.* 60, 482 (1986).
327. G. Sberveglieri, S. Groppelli, and G. Coccoli, *Sensors Actuators* 15, 235 (1988).
328. G. Sberveglieri, P. Benussi, G. Coccoli, S. Groppelli, and P. Nelli, *Thin Solid Films* 186, 349 (1990).
329. A. Hattori, H. Tachibana, N. Yoshiike, and A. Yoshida, *Sensors Actuators B* 77, 120 (1999).
330. Z. Szlarski, K. Zakrzewska, and M. Rekas, *Thin Solid Films* 174, 269 (1989).
331. T. Takada, in "Chemical Sensor Technology" (T. Seiyama, Ed.), Vol. 2, p. 59. Elsevier, Amsterdam/Kodansha, Tokyo, 1989.
332. G. Sberveglieri, C. Perego, F. Parmigiani, and G. Quattrini, *Sensors Actuators B* 20 163 (1994).
333. Y. Dawei, M. Xiaocui, W. Zongchang, and W. Jun, *Thin Solid Films* 224, 257 (1993).

ENN Encyclopedia of Nanoscience and Nanotechnology
www.aspbs.com/enn

Germanium Nanocrystals

Shinji Nozaki, Hiroshi Morisaki

University of Electro-Communications, Chofu-shi, Tokyo, Japan

Seiichi Sato

Himeji Institute of Technology, Kamigori-cho, Ako-gun, Hyogo, Japan

CONTENTS

1. INTRODUCTION

Since the discoveries of visible luminescence at room temperature from porous silicon (Si) [1] and the other forms of Si nanostructures [2, 3], there has been an increasing interest in the physical properties of semiconductor nanostructures distinct from those of bulk semiconductors. When the size of a material is scaled down to the order of nanometers, the number of atoms in the material becomes as low as 1000. Such a small material exhibits chemical and physical properties distinct from those of the bulk material. This is called the size effect. There are three major phenomena induced by the size effect: confinement of carriers and phonons, surface-related phenomena, and phase transformation. The visible luminescence from the Si nanostructures was often attributed to the confinement of carriers, and the luminescence mechanism was well reviewed by Kanemitsu [4].

The visible luminescence from the Si nanocrystals was rather surprising to us, because such a "simple" technique to form Si nanocrystals made Si more attractive as a light-emitting material. Here, a nanocrystal refers to a crystalline nanoparticle formed by various techniques and differs from the nanostructures formed by epitaxial growth or lithography. The most popular and "simple" techniques to form Si nanocrystals are electrochemical etching, co-sputtering of SiO_2 and Si, and gas evaporation [4]. These techniques, except for electrochemical etching, have been employed to form germanium (Ge) nanocrystals, from which visible luminescence is also expected for the same reason as Si nanocrystals. Because of its lower melting point, the formation of Ge nanocrystals is easier than that of Si nanocrystals. In addition to the ease of fabrication, the quantum confinement is expected to appear more conspicuously for Ge than for Si, because the Bohr radius of the excitons in Ge of 24.3 nm [5] is much larger than that in Si of 4.9 nm [6]. The Ge nanocrystals formed by co-sputtering of SiO_2 and Ge, gas evaporation of Ge, and cluster-beam evaporation are included in this review. Although the last technique, cluster-beam evaporation, was developed by the authors and not commonly used to form Ge nanocrystals, it should be described in detail because of the manifestation of the third effect of the aforementioned size effects, that is, the phase transformation. While the first two size effects are well known, the phase transformation is less known to most readers and will be discussed in detail.

When the size of a material becomes smaller, its characterization becomes more difficult, simply because the intensity of the physical or chemical signal to assess the structural or physical properties becomes smaller with respect to the background noise. Although many methods used to characterize bulk materials have been applied to semiconductor nanocrystals, there is not well-established technique suitable for such extremely small materials. Even with a transmission electron microscope (TEM), great care must be taken so that the beam irradiation does not change the crystal structure of the nanocrystals, particularly the metastable structure. Various characterization techniques that have been applied to Ge nanocrystals are reviewed along with the characterization results.

One of the attractive features of Ge nanocrystals is visible-light emission. Like Si, bulk Ge with an indirect bandgap is less attractive for light-emitting device applications.

ISBN: 1-58883-059-4/$35.00

However, many photoluminescence colors have been observed from Ge nanocrystals. The mechanism of visible-light luminescence has been a subject of debate. Some support the radiative recombination in Ge nanocrystals, while others support the radiative defect center associated with the native Ge oxide. The arguments on both sides are presented and compared.

The application of Ge nanocrystals is still limited. Although an electroluminescence device was suggested, the light emission efficiency is far below those of the devices made of a direct-bandgap material. One promising application is a nanocrystal memory. To integrate them on a large scale, good control of the positioning and size of the Ge nanocrystals embedded in SiO_2 is mandatory. Recent progress in nanocrystal memory fabrication is reviewed.

This chapter is organized as follows. A discussion of the methods to form Ge nanocrystals, including the cluster-beam evaporation technique, is presented in Section 2, and various characterization techniques, which have been employed to study the properties of Ge nanocrystals, are described along with the characterization results in Section 3. Section 4 concerns the physical properties of the Ge nanocrystals. Among the physical properties, the phase transformation is an important size effect but unfamiliar to most readers. It is discussed in great detail. Section 4.3 is allocated for the phase transformation in Ge nanocrystals. This section is mainly based on our own work, because, except for our pioneering work, there are not many reports on the phase transformation in semiconductor nanocrystals. Section 5 describes the potential applications of Ge nanocrystals. In this review, we included a minimum number of figures and have selected several unpublished ones in order to supplement the discussion. Readers are encouraged to refer to the published figures in the articles listed in the References if they consider that the descriptions or discussions are not clear without the figures.

2. FABRICATION OF Ge NANOCRYSTALS

Several techniques used to form Ge nanocrystals are reviewed. The formed nanocrystals are either directly deposited on a substrate or embedded in the matrices, which are Si [7], porous Si [8], zeolites [9], and SiO_2. The Ge nanocrystals embedded in SiO_2 have been the most intensively studied because of good confinement of the electrons with a large potential barrier of the oxide. The as-deposited Ge nanoparticles formed by some methods are amorphous and become crystallized only after high-temperature annealing. The material properties of the Ge nanocrystals are also affected by the fabrication technique and are briefly presented for each technique in this section. More details of the properties are discussed in Section 4.

2.1. Chemical Synthesis

Chemical synthesis is the simplest technique to form Ge nanocrystals. The sol–gel method has been used to fabricate semiconductor and metal nanocrystals such as CdS CuCl, CdTe, PbS, Au, and Ag dispersed in SiO_2 glasses [10–14]. Nogami and Abe first fabricated Ge nanocrystals embedded in SiO_2 containing 7 wt% Ge by the sol–gel method using $Si(OC_2H_5)_4$(TEOS) and $GeCl_4$ as the starting materials [15]. However, it was very difficult to control the reaction and form nanocrystals with different sizes because the hydrolysis of $GeCl_4$ was extremely fast. Yang et al. synthesized the gels through the hydrolysis of $Si(OC_2H_5)_4$ and 3-trichlorogermaniumpropanoic acid (Cl_3–Ge–C_2H_4–COOH). The gels were then heated at 600 °C for 10 h in air to form the GeO_2–SiO_2 glasses [16]. The gels were further heated at 500–700 °C in a H_2 gas atmosphere in which the Ge^{4+} ions were reduced to precipitate cubic Ge nanocrystals. The sizes of the Ge nanocrystals ranged from 1 to 13 nm.

Highly crystalline Ge nanocrystals in the size range of 2–10 nm were grown in inverse micelles and purified and size-separated by high-pressure liquid chromatography [17]. Controlled nucleation and growth occurred in the interior of the nanosize surfactant aggregates called inverse micelles. An anhydrous ionic salt (e.g., GeX_4, where X = Cl, Br, or I) was dissolved in the hydrophilic interior of a solution of micelles. Since the ionic salts were completely insoluble in the continuous oil medium used (e.g., octane), the nucleation and growth of Ge was restricted to the micelle interior. The interior volume of the micelles could be varied over the range of 1–10 nm. Control over the final nanocrystal size was achieved by a variation in the micelle size, intermicellar interactions, and reaction chemistry.

Kauzlarich et al. synthesized Ge nanocrystals by reacting the Zintl phase precursors with Ge tetrachloride [18]. The surface of the Ge nanocrystals was passivated with alkyl groups through the reactions with RLi and MgBrR (R = alkyl). The surface was kept from oxidizing and was stable. These nanocrystals emitted light over a wide range in the visible range depending on the size.

2.2. Gas Evaporation of Ge

The gas evaporation technique has been conventionally employed to form ultrafine metal and semiconductor particles. Although this technique is similar to vacuum evaporation, the source material in a boat is evaporated in an inert-gas atmosphere. During the inert-gas evaporation of Ge, the Ge atoms that effused from the source will rapidly lose their energy by collision with gas atoms, that is, macroscopically, the Ge vapor is cooled in the gas. The collision mean free path is very short; for Ge atoms in 1 torr of argon (Ar), for example, it is approximately 10^{-7} m. This efficient cooling locally produces a high supersaturation of Ge vapor, which leads to a homogeneous nucleation. The size of the formed Ge nanocrystal increases with the increasing evaporation rate and shortening of the mean free path. In other words, for a higher evaporation temperature of the source material and higher gas pressure, a larger size is expected. Furthermore, some of the Ge nanocrystals collide with others and coalesce during travel to the substrate after the nucleation.

An earlier study of the Ge nanocrystals formed by the gas evaporation technique was reported by Saito [19]. He heated Ge powder in a tungsten boat at 1500–1900 °C in an argon atmosphere. The gas pressure was varied from 3 to

40 torr. In the transmission electron diffraction (TED) pattern, some Ge nanocrystals showed a diamond structure, while the others showed a tetragonal structure, which will be discussed in detail later.

Han et al. [20] observed a blueshift from the bulk's optical bandgap as large as 2 eV for 10-nm Ge nanocrystals in the optical absorption spectra and attributed the shift to quantum confinement. However, the luminescence from the Ge nanocrystals formed by the gas evaporation of Ge is generally weak due to nonradiative recombination via surface states. The surface dangling bonds of the Ge nanocrystal must be terminated by surface passivation. Si et al. made an attempt to passivate the surface of Ge nanocrystals with Si by evaporation of Si and Ge [21]. Two methods were employed in the gas evaporation technique to form Ge nanocrystals whose surface was passivated with Si. One used a boat with a SiGe alloy as the source, and the other used two boats each with Si and Ge. As a result of the characterization by X-ray diffraction (XRD), Raman spectroscopy, X-ray photoelectron spectroscopy (XPS), and TEM, it was found that the Ge nanocrystals with a Si-passivated surface were formed by the co-evaporation of Si and Ge from the two boats, while the SiGe alloy nanocrystals were formed by evaporation of the SiGe alloy source from one boat. However, even the Ge nanocrystals passivated with Si failed to exhibit visible luminescence at room temperature.

2.3. Co-sputtering of SiO_2 and Ge

An ideal Ge quantum dot, where carriers are confined by a high potential barrier, can be made by surrounding the Ge nanocrystal with SiO_2. The surface of the Ge nanocrystals is unstable and becomes oxidized or contaminated in air, while the surface of the Ge nanocrystals embedded in SiO_2 is passivated with SiO_2 and is expected to be more stable. Such a structure with Ge nanocrystals embedded in SiO_2 is also called Ge-doped glass. The thin film of Ge nanocrystals most often studied for size effects is the Ge-doped glass.

One of the most popular methods to make Ge-doped glass is the co-sputtering of SiO_2 and Ge. In this method, Ge chips are placed on a Si target and co-sputtered in argon gas [22–26]. The Ge nanocrystals are obtained by annealing the as-deposited Ge-doped glass film. In this technique, the size of the Ge nanocrystals can be controlled by changing the volume fraction of Ge in the as-deposited films or by varying the annealing profile [25–29]. Takeoka et al. obtained average diameters of 0.9–5.3 nm by changing the volume fraction of Ge from 0.2 to 7.2% [27] and modeled the relationship between the volume fraction, $f_{Ge}(\%)$, and the average diameter, d_{ave}, by

$$\frac{f_{Ge}}{100} = \frac{(4/3)\pi\left(d_{ave}/2\right)^3}{(s + d_{ave})^3} \tag{1}$$

where s is the distance between two adjacent Ge nanocrystals [27]. The size distribution of the Ge nanocrystals is often characterized by the log-normal function [30].

The growth process of Ge nanocrystals in a SiO_x matrix was studied by an X-ray diffraction (XRD) technique using the grazing incidence of X-rays [24]. The conclusion drawn from the XRD analysis was consistent with those from Raman spectroscopy and X-ray photoelectron spectroscopy. The as-deposited Ge-doped glass film consisted of Ge and GeO_x. Since the content of GeO_x decreased with the increasing annealing temperature, the GeO_x appeared to be thermally decomposed into Ge and O, which then oxidized the SiO_x matrix, as proposed by Dutta [31]: $SiO_x + GeO_x \rightarrow SiO_2 + Ge$. Two stages of crystallization were found in the temperature range studied for annealing; the first stage was the nucleation of Ge nanocrystals at 600 °C, and the second stage was the growth of the nanocrystals at 800 °C [24]. During the first stage, the Ge nanocrystals were formed by crystallization from the amorphous Ge, which was contained in the as-deposited samples. During the second stage, the decomposition of GeO_x supplied additional Ge and enabled the nanocrystals to grow faster.

The sizes of the Ge nanocrystals in the Ge-doped glass film can be made smaller by oxidation. Rath et al. oxidized the Ge-doped glass film by wet oxidation [32]. After annealing the as-deposited film in an argon atmosphere, the nanocrystals were found to be oxidized. When the as-deposited film was annealed and oxidized at 800 °C, the sizes estimated from the phonon confinement model in the Raman spectra were 12 and 6 nm before and after oxidation, respectively. The size reduction was also confirmed by optical absorption measurements.

The film deposited by the co-sputtering of Ge and SiO_2 is amorphous $Si_xGe_yO_z$. Amorphous $Si_xGe_yO_z$ can also be deposited by co-sputtering Si and Ge in an oxygen-containing atmosphere. In the experiment reported by Zacharias et al. [33, 34], the water vapor added to Ar was the oxygen source. The oxygen content in the as-deposited $Si_xGe_yO_z$ depended on both the deposition rate and the partial water pressure. With an increase in the sputtering power, the oxygen content decreased. The sputtering yield depended on the energy of the sputtering Ar ions and is 1.55 (Ge) compared with 1.0 (Si) for Ar ions of 1 keV. The formation of the Ge nanocrystals in the SiO_2 matrix was induced by annealing at high temperature. The average size of the Ge nanocrystals formed by annealing at 800 °C decreased with an increase in the oxygen content or sputtering power.

2.4. Ion Implantation of Ge in SiO_2

The implantation of Ge in SiO_2 is another way to form amorphous $Si_xGe_yO_z$, which turns into Ge-doped glass by annealing of the Ge-implanted SiO_2 at high temperature. Kawamura et al. [35] and Hosono et al. [36], however, implanted protons in $9SiO_2$–$1GeO_2$ glasses at 1.5 MeV without heating the substrate and formed Ge-doped glass, where Ge nanocrystals with diameters of 6–10 nm were embedded in SiO_2 without postthermal annealing. The fact that Ge nanocrystals were not formed by He^+ but by H^+ implantation was strong evidence that some chemical process was involved in the nanocrystal formation. The overall chemical reaction might be the reduction of GeO_2 by H_2, that is, $GeO_2 + 2H_2 \rightarrow Ge + 2H_2O$.

There are numerous reports on the fabrication of Ge-doped glass by the ion implantation of Ge in SiO_2 [37–45]. In this method, Ge is implanted in SiO_2 thermally grown on Si substrates, and then the samples are mostly annealed in a nitrogen atmosphere at high temperature. Zhu et al. found that the Ge dose was the controlling factor for the size distribution [37]. With a decrease in the Ge dose, the fraction of large particles in the size distribution decreased. In contrast to the finding by Zhu et al., Dowd et al. concluded that the size distribution was independent of the Ge dose [41]. Their size distribution was characterized by the log-normal function. In both studies, the range of the Ge dose was the same but the implant energy was different, that is, 500 keV and 1 MeV for Zhu et al. and Dowd et al., respectively.

The Ge redistribution and nanocrystal formation in Ge-implanted SiO_2 during annealing were systematically studied by Heinig et al [39]. The Ge redistribution and nanocrystal evolution could be strongly influenced by the annealing ambient [38]. The annealing behavior was explained by the in-diffusion of an oxidant from the annealing atmosphere, which significantly changed the Ge depth profile and nanocrystal distribution. Annealing in N_2 or Ar leads to a dramatic change in the as-implanted Gaussian-like Ge depth distribution to a bimodal profile and, additionally, an accumulation of Ge at the Si/SiO_2 interface. No Ge nanocrystals were found in the subsurface region, while nanocrystals were found in the center of the SiO_2 layer. Annealing in dry O_2 increases the surface concentration of the oxidant, and noncrystalline GeO_2 particles were formed.

The Ge nanocrystals formed in SiO_2 by ion implantation and postannealing are usually distributed in the depth direction and form the band in SiO_2 [39]. However, in some device applications such as nanocrystal memories, which will be discussed later, a monolayer of Ge nanocrystals should be formed in SiO_2. A thin layer of Ge nanocrystals with a thickness of 3–4 nm was successfully formed by lowering the implant energy to 3 keV [44].

For size uniformity, multienergy ion implantation [43] was proposed. In this method, the Ge ions were implanted with energies of 50, 80, 120, 200, and 360 keV, and with doses of 1×10, 1.2×10, 2×10, 4×10, and 1×10, respectively (a total dose of 1.82×10 cm^{-2}). The sizes of the Ge nanocrystals were more uniform than those obtained by single-energy ion implantation or co-sputtering of Ge and SiO_2 due to the improved uniformity of the Ge concentration across the depth of the as-implanted sample. The average size of the Ge nanocrystals after annealing at 800 °C was 3.7 nm. However, a further increase in the annealing temperature to 1000 °C resulted in an increased size and wider size distribution.

2.5. Cluster-Beam Evaporation

A unique method to deposit Ge nanocrystalline films, which is called the cluster-beam evaporation technique, was developed by the authors [46]. This technique resembles the well-known ionized cluster-beam (ICB) technique. The ICB technique was developed to deposit high-quality thin films by Takagi [47]. The major difference between the cluster-beam evaporation and ICB techniques is whether the formed cluster is ionized or not. In the ICB technique, the clusters are intentionally ionized and accelerated by an electric field. The clusters with sufficient kinetic energies become resolved into individual atoms when they land on the substrate. They migrate on the substrate surface and form an epitaxial film even at low substrate temperature. In contrast to the ICB technique, the clusters make a "soft" landing on the substrate during cluster-beam evaporation. Because they are not ionized or accelerated, they are not resolved. The clusters may migrate on the surface after their "soft" landing.

In both techniques, the clusters are formed by supersonic expansion through the nozzle of a crucible. The classical nucleation theory predicts that the sizes of the Ge nanocrystals formed by these techniques will be much smaller than those formed by the gas evaporation technique, which is conventionally used to deposit metal or semiconductor micro- and nanoparticles. The cluster-beam evaporator apparatus is a vacuum evaporator with a capped carbon crucible. The cap is 1 mm thick and has a nozzle hole with a diameter of 1 mm. The crucible loaded with small pieces of undoped Ge is heated by electron bombardment.

During the deposition, the temperature of the crucible is kept at 2000 K, and the pressure in the chamber is 5×10^{-6} torr. The substrate temperature is either at room or liquid-nitrogen temperature. The typical film thickness is 100 nm after a 60-min deposition. The TEM micrographs of the nanocrystalline films showed a smooth morphology without any features and a rough morphology consisting of Ge nanocrystals with diameters ranging from 4 to 40 nm for substrate temperatures of 300 and 77 K, respectively [46]. However, even a larger nanocrystal in the latter film may be due to agglomeration of the smaller nanocrystals. The clusters in the former film seem to have migrated on the surface even at room temperature. They were also found to be hardly oxidized in air [46]. The depth profile of Ge3d indicated the presence of Ge oxide, GeO_x, and GeO_2 only on the surface of the Ge nanocrystalline film and did not change even after a long exposure to air. However, the nanocrystalline film deposited on the substrate whose temperature was 77 K was subject to oxidation and exhibited a unique oxidation process called photooxidation [46, 48], which will be discussed in detail in Section 4.1.2.

2.6. Oxidation of Ge

As mentioned earlier, forming a monolyear or a sheet of the Ge nanocrystals in SiO_2 is required for device applications such as nanocrystal memories. A clever and simple method to form a thin layer of the Ge nanocrystals embedded in SiO_2 is to oxidize a SiGe or SiGeC alloy layer [49–54]. King et al. formed a thin epitaxial layer of $Si_{1-x}Ge_x$ on the surface of a Si wafer by ion implantation of Ge and subsequent high-temperature wet oxidation, which caused the implanted Ge atoms to pile up at the Si/SiO_2 interface. The Ge segregated out of the growing oxide under these oxidation conditions. After the high-temperature oxide film was etched off, the oxide was grown in dry O_2. Next, a $Si_{1-x}Ge_xO_2$ layer was grown in wet O_2 in order to incorporate the Ge into the oxide. The top oxide was then grown in dry O_2. Finally, a high-temperature annealing step was cmployed to cause the Ge within the oxide to precipitate and form nanocrystals. The cross-sectional transmission electron micrograph clearly showed one monolayer-like Ge nanocrystals.

Instead of thermal oxidation of a SiGe layer, photooxidation can form Ge nanocrystals from the SiGe oxides. Craciun et al. exposed a 15-nm-thick SiGe strained layer grown by molecular-beam epitaxy (MBE) to ultraviolet (UV) light from a low-pressure Hg lamp in the presence of oxygen [51]. The sample temperature was kept at 550 °C. After a 5-h photooxidation, the multilayer structure of the pure and Ge-doped SiO_2 layers was formed. The oxidation temperature was significantly lower than that usually used for the formation of Ge nanocrystals by the reduction of SiGe oxides.

The structures with Ge nanocrystals embedded in SiO_2 were also formed by oxidation of a Ge film on Si or the SiO_2/Ge/SiO_2 structure [55–57]. However, there was no conclusive evidence presented to show that the Ge nanocrystals were embedded in the SiO_2 without being pushed down to the SiO_2/Si interface.

Nevertheless, the oxidation of a Ge layer on Si to form Ge nanocrystals embedded in SiO_2 should be further explored, because it can form a two-dimensional (2D) array of Ge nanocrystals with controlled positioning. One possible way to obtain a well-ordered structure of Ge nanocrystals in SiO_2 is to oxidize the MBE-grown structure of Ge dots embedded in Si. The Ge dots can be ordered by making the best use of the self-assembly during the MBE growth. Oxidation of the MBE-grown structure was attempted in order to form the Ge-doped glass [58]. Although the oxide was formed on the layer of Ge dots, it was found that the Ge layer was pushed down to the Si substrate without oxidizing the Si layer underneath the Ge layer. This may have occurred because the MBE-grown Ge dots were not well separated from each other. There is plenty of room for improvement.

2.7. Miscellaneous

Other techniques to form Ge nanocrystals are pulsed-laser ablation (PLA) [7, 59] and the spark process [60, 61]. The apparatus for the PLA synthesis of Ge nanocrystals in the gas phase developed by Ngiam et al. [7] is similar to that for the cluster-beam evaporation technique. A target rod was ablated by a frequency-doubled pulsed Nd:YAG laser. The ablated Ge plasma was cooled by a pulse of He carrier gas released into the growth channel and started to condense in the growth channel before expanding into a vacuum. The Ge nanocrystals passed into the deposition chamber and were deposited on a substrate.

The spark process was performed in stagnant air by applying a pulsating unipolar high voltage (15 kV) and a low current between a tungsten-tip electrode (anode) and a p-type (100) Ge wafer (cathode) for 20 min [60, 61]. The surface of the Ge nanocrystals was found to be oxidized.

Since the nanocrystals were formed by gas condensation in both methods, they may be regarded as modified gas evaporation methods.

3. CHARACTERIZATION METHODS

The characterization method to directly observe Ge nanocrystals and determine their sizes is transmission electron microscopy (TEM). However, the specimen preparation requires time and great care in order not to damage the samples. Although various spectroscopies are also used, the signals are usually weak, and their analyses are more involved. In this section, the characterization methods often used to obtain information on the structure and properties of the Ge nanocrystals are reviewed.

3.1. TEM

TEM is a direct method to observe Ge nanocrystals and requires a high resolution to obtain the lattice image of a Ge nanocrystal. A high-resolution TEM (HRTEM) image provides information on the crystal structure, crystal size, lattice constant, and defects of each nanocrystal.

The Ge nanocrystals or the Ge-doped glass thin film can be deposited either directly on a substrate or on the carbon-coated TEM grids as a plan-view TEM specimen. Plan-view and cross-sectional TEM specimens of the Ge nanocrystals deposited directly on a substrate should be prepared from the thin film containing Ge nanocrystals by the standard procedures for the preparation of TEM specimens described in a TEM textbook [62].

The HRTEM image of the as-deposited Ge nanocrystal fabricated by cluster-beam evaporation is shown as an example in Figure 1. The Ge nanocrystal is circled in the figure. The diameter of the nanocrystal is 3.6 nm.

The crystal structure of the Ge nanocrystal was studied in detail by using the ultrahigh vacuum (UHV)–MBE–HRTEM system [63]. Ge nanocrystals were deposited on thinned Si substrates in a UHV–MBE chamber and transferred into a UHV-field emission HRTEM through a UHV-transferring system without exposure to air. The developed UHV–MBE–HRTEM system keeps the samples from contamination or oxidation and allows observation of an individual nanocrystal without their overlapping. The real-time TEM images were collected using a video camera with a time resolution of 60 frames/s. Such real-time TEM was mandatory to observe the Ge nanocrystals, which are unstable under electron-beam irradiation.

The attachments of the TEM instrument such as transmission electron diffraction (TED) and energy-dispersive

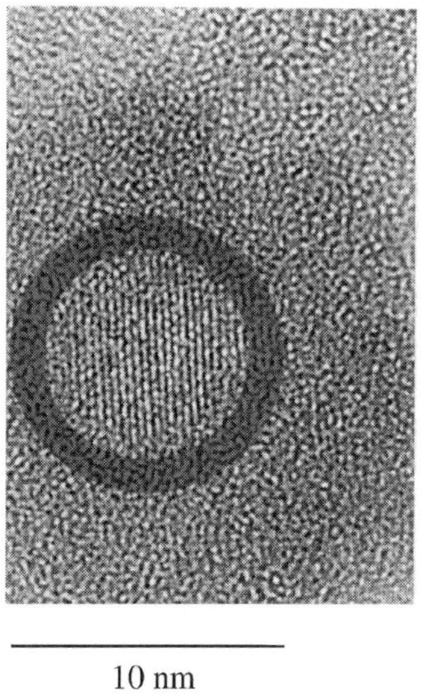

Figure 1. High-resolution transmission electron micrograph (HRTEM) of the Ge nanocrystalline film deposited by the cluster-beam evaporation technique. The Ge nanocrystal is circled.

X-ray spectroscopy (EDS) supplement the TEM analysis. The spacings between the lattice planes calculated from the TED pattern provide useful information on the crystal structure of a Ge nanocrystal, and Sato et al. identified the ST-12 structure in the Ge nanocrystalline film deposited by cluster-beam evaporation based on comparison of the obtained spacings with those of the ST-12 and diamond structures [64]. The TED also provides assessment of the crystallinity of the Ge nanocrystals, for example, single crystal, polycrystal, or amorphous. The EDS provides information on the chemical composition. The element maps of Si and Ge in the Ge nanocrystals embedded in SiO_2 or Si can determine whether Ge is localized in a nanocrystal or mixed with Si near the surface of a nanocrystal.

3.2. XRD

Conventional X-ray diffraction (XRD) may not be very useful for determining the crystallinity and dimension of small-sized Ge nanocrystals because of the lower intensity and broader width of the reflection peak [65, 66]. A meaningful analysis usually requires a thick sample including many nanocrystals. A glazing incidence XRD, which is sometimes called the thin-film diffraction (TFD) method, is, however, very effective for the measurement of thin films less than 100 nm thick because the X-ray penetration depth becomes shallow and the diffracted intensity is expected to be a few times greater than that of the conventional $\theta/2\theta$ method. It was found that the TFD was useful for characterization of the Ge nanocrystals in Ge-doped glass thin films [24].

The particle size of a Ge nanocrystal, D, is calculated from the full width at half-maximum (FWHM) of the XRD peaks at $2\theta_B$ using the Scherrer formula:

$$\frac{0.89\lambda}{\beta(2\theta_B)\cos\theta_B} \tag{2}$$

where λ is the X-ray wavelength (0.154 nm for Cu Kα radiation) and $\beta(2\theta_B)$ is the FWHM in radians. In the TFD, although the spatial width of the diffracted beams is fairly wide, the accuracy of $\beta(2\theta_B)$ is within 0.1° because the planar monochromator is set directly before the detector, providing a better angular accuracy. In the TFD on the films deposited on a Si(100) substrate, the samples are set at a glancing angle of 2° and to the (100) azimuth to avoid the 311 reflection peak of the substrate, since there is a strong reflection peak from the {311} planes of the substrate that appears for the glancing angle of 2.8°, which must be avoided. Even if this method is used, the XRD pattern still contains a small tail attributable to the 311 reflection peak of the Si substrate but the intensity of this tail is small compared to the reflection peak of the Si substrate using the ordinary method. For Ge nanocrystals with a good crystallinity, the Ge (111), (220), and (311) Bragg reflection peaks should be observed at $2\theta_B$ of 27.3°, 45.3°, and 53.7°, respectively.

3.3. XPS

The detection and energy analysis of photoelectrons produced by radiation whose energy exceeds their binding energies is the subject of an extensively used technique known as photoelectron (PE) spectroscopy. This technique can be conveniently divided into two broad areas, the first employing ultraviolet radiation and hence called ultraviolet photoelectron spectroscopy (UPS), and the second using X-rays, termed X-ray photoelectron spectroscopy (XPS). The details of XPS can be found in many textbooks on materials or surface science [67, 68]. XPS is also known as electron spectroscopy for chemical analysis (ESCA).

One of the most important applications of XPS is the accurate measurement of core-level binding energies. Figure 2 shows typical Ge3d XPS spectra of Ge nanocrystalline film deposited by cluster-beam evaporation. The peak for the as-deposited sample is fitted with three pure Gaussian curves (solid curves) [69, 70]. From the peak energies, these three peaks are associated with Ge (~29.3 eV), GeO_x (~31 eV), and GeO_2 (~32.6 eV). In the figure, the XPS spectrum of the film after annealing at 500 °C in N_2 flow is also shown. After annealing, the peak is fitted with two curves, which are for Ge and GeO_2. GeO_x is thermally decomposed into pure Ge and oxygen, which oxidizes Ge, for example, $2GeO_x \rightarrow xGeO_2 + (2-x)Ge$. Since x is less than 2, pure Ge atoms are supplied as a result of the thermal decomposition, and the Ge nanocrystal grows larger.

XPS is also able to probe the electronic structure of the valence band of metals, semiconductors, and insulators [71–73]. Joannopoulos and Cohen reported good agreement between the calculated density of states (DOS) and the XPS spectrum, which were related to the DOS for Si and Ge in amorphous structures [71]. We also studied the DOS of the ST-12 of the Ge nanocrystals by XPS [64] and these details will be discussed later. It was clear from a comparison with the calculated DOS curve that the XPS spectrum of the Ge nanocrystalline film was in good agreement with the theoretical DOS of the ST-12 structure; both showed two peaks at about 1 and 8 eV.

Another application of XPS for the Ge nanocrystalline film is to study the plasmon loss features [74]. After removing the oxide from the surface of the Ge nanocrystalline film, we collected the XPS signals in the range of the binding energy from 25 to 70 eV and transformed the data into the electron energy loss (EEL) spectra. Every zero-loss peak corresponds to the Ge3d XPS peak and the loss peak due to the bulk plasmon is seen at 16–18 eV. The plasmon energy of the Ge nanocrystalline film was observed at 17.1–17.4 eV, which was about 1 eV higher than the value of the bulk Ge

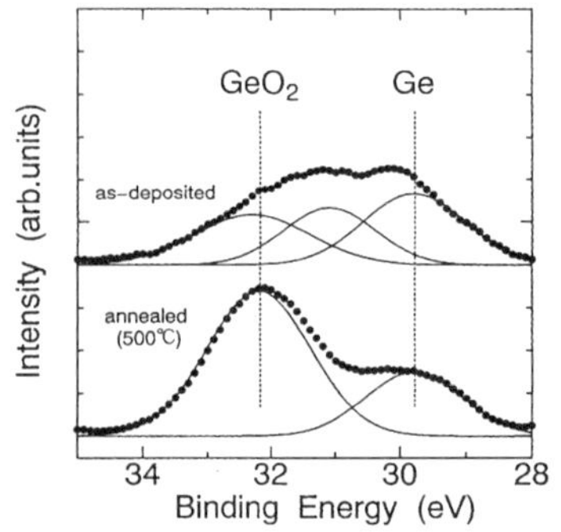

Figure 2. X-ray photoelectron spectroscopy (XPS) spectra of the Ge nanocrystalline films deposited by cluster-beam evaporation before and after annealing at 500 °C.

(16.3 eV). The plasmon energies of both the Ge nanocrystalline film and bulk Ge were well explained by the classical model, and we attributed the observed energy difference to a difference in the crystal structure.

3.4. Photoluminescence and Optical Absorption

Like Si nanocrystals, Ge nanocrystals have been mostly studied for light emission because of their potential application to light-emitting devices. Photoluminescence (PL) spectroscopy is widely used to assess the efficiency of light emission from Ge nanocrystals. It is important to physically understand the luminescence mechanism and the size dependence of the PL peak energy. Photoluminescence spectroscopy of crystalline semiconductors was reviewed in detail by Gilliland [75]. However, a simple PL spectroscopy alone is not sufficient to provide conclusive answers to questions on the luminescence mechanism and size dependence of the PL peak energy. The analysis of the PL spectrum should include effects of the surface oxide and nonhomogeneous sizes.

The measured peak is usually very broad and can be deconvoluted into a few peaks, which are still broad [61, 69, 76]. Because of the broadness of the peak, it does not seem meaningful to say that the PL peak energy increases with a decrease in the average size. The broadening of the PL peak is associated with the nonhomogeneous sizes of the Ge nanocrystals. The nanocrystals with different sizes contribute to the PL spectra with different peak energies. If each PL originates from the radiative recombination of carriers confined in a nanocrystal, the peak becomes much narrower in the plot of the PL intensity at a selected emission energy as a function of the PL excitation energy. Such a PL measurement is called photoluminescence excitation (PLE). In the PLE spectra, the nonhomogeneous broadening is suppressed. If the PLE peak energy is the same as the selected luminescence energy, the luminescence is most likely to originate from the band-to-band recombination or exciton recombination in the Ge nanocrystals. If there is a difference in the energies, which is associated with the Stokes shift between the absorption-edge energy and the PL energy, the luminescence is primarily due to Ge surface defects or defects present in the Ge nanocrystals or surface oxides. The PLE has been employed to study the luminescence mechanism of the Ge nanocrystals in great detail [77, 78].

Time-resolved PL, which measures the decay of the PL intensity for a selected wavelength, provides more profound information on the recombination of the carriers [61, 77, 79]. The internal luminescence efficiency is generally described by

$$\eta = \frac{1}{1 + \tau_r/\tau_n} \tag{3}$$

where τ_r and τ_n are the lifetimes for radiative and nonradiative recombinations, respectively. The PL decay time constant is a combination of the radiative and nonradiative constants and is given by

$$\tau_{PL}^{-1} = \tau_r^{-1} + \tau_n^{-1} \tag{4}$$

Therefore, if there is a large difference between the radiative and nonradiative recombination lifetimes, the PL decay becomes single exponential; otherwise, it becomes double exponential. It is clear from (4) that if the nonradiative lifetime is extremely small, the PL decay becomes fast. This suggests that a fast decay may not be necessarily due to radiative recombination but due to a fast nonradiative recombination [77].

Optical absorption measurements are often carried out to estimate the bandgaps of bulk semiconductors and to determine whether they are direct or indirect. They are also employed to characterize Ge nanocrystals [17, 20, 22, 77, 80–82]. The optical absorption coefficients of bulk semiconductors are mostly described by the following three expressions [83, 84]:

$$\alpha(h\nu) \propto (h\nu - E_g)^{1/2} \quad \text{for direct-bandgap semiconductors} \tag{5}$$

$$\alpha(h\nu) \propto (h\nu - E_g - E_p) \quad \text{for indirect-bandgap semiconductors} \tag{6}$$

and

$$h\nu\alpha(h\nu) \propto (h\nu - E_g)^2 \quad \text{for amorphous semiconductors} \tag{7}$$

where $h\nu$, E_g, and E_p are the photon energy, bandgap, and phonon energies, respectively.

Although some have claimed a good fit to (7) [20, 82], in most cases, the absorption spectra of the Ge nanocrystals do not follow any of the above equations. This is due to the following complications in the optical absorption of the Ge nanocrystals. First, both the optical absorption and the light scattering affect the transmission spectrum, which is directly measured to obtain the optical absorption spectra. To calculate the optical absorption from the transmission, one should not ignore the effect of light scattering. Second, the nonhomogeneous size distribution complicates the analysis of the optical absorption of the Ge nanocrystals. Third, one must consider the quantum confinement effect. If the size distribution is uniform, and the measurement temperature is low enough to suppress thermal fluctuation of the electron energy, the energy of three-dimensionally confined electrons should exhibit discrete levels and not follow the absorption expressions for bulk semiconductors. An analysis of the absorption spectra of Ge nanocrystals, taking into account all of the above complications, has not yet been undertaken.

3.5. Raman Scattering

Raman scattering is a very powerful characterization technique to detect a signal from as little as a few monolayers and has been employed to characterize Ge nanocrystals formed by various methods. Figure 3 shows the typical Raman spectra of two types of Ge nanocrystals: one with the ordinary diamond structure and the other with the ST-12 as a result of a phase transformation, which will be discussed in detail later. In the spectrum of Ge nanocrystals with the

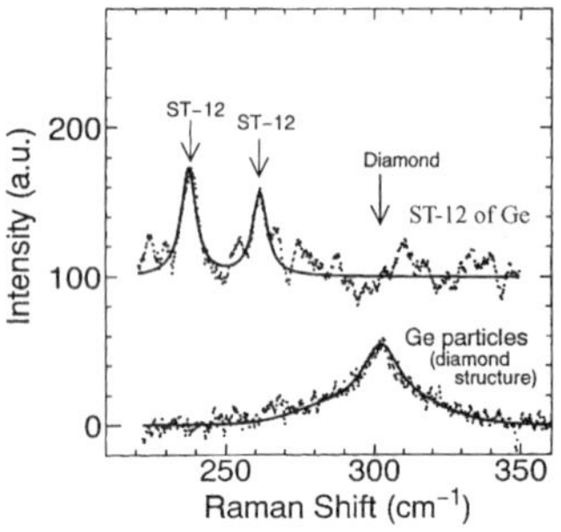

Figure 3. Raman spectrum of the Ge nanocrystalline film deposited by the cluster-beam evaporation technique. The Raman spectrum of Ge particles with the diamond structure is also shown for comparison.

diamond structure, a broad peak is seen at the Raman shift of 300 cm^{-1}.

The Raman line shapes of the low-dimensional materials have been derived based on the phonon confinement model. A quantitative model was developed by Richter et al. [85] and was later improved by Campbell and Fauchet [86] to estimate the average size of the nanocrystals. Using this model, the first-order Raman spectrum of a spherical crystal can be written as

$$\int \frac{|C(0,\, q)|^2}{[\omega - \omega(q)]^2 + (\Gamma/2)^2}\, d^3q \tag{8}$$

where q is expressed in units of $2\pi/a$; $C(0, q)$ is the Fourier coefficient, which describes the phonon confinement; $w(q)$ is the phonon dispersion curve; and Γ is the natural linewidth of the Raman spectrum of bulk Ge($\sim$3 cm^{-1}). The integral is taken over the entire Brillouin zone. Sato et al. used a Gaussian function for the Fourier coefficient, which is written as

$$|C(0,\, q)| = \exp(-q^2L^2/4) \tag{9}$$

where L is the diameter in units of the lattice constant, a, and an adjusting parameter to obtain the best fit to the measured line shape, and found a good match of the diameter with that measured by TEM [87].

The Raman scattering has been used to study the formation of Ge nanocrystals by high-temperature annealing [88–90]. Ninomiya et al. characterized the Ge-doped SiO_2 thin films annealed at various temperatures by Raman scattering and concluded that there were two stages of crystallization: crystallization of amorphous Ge contained in the as-deposited film at 600 °C and decomposition of GeO_x at 800 °C [24]. The conclusion based on the Raman scattering was in good agreement with that from XRD and XPS.

For probing electronic transitions well above the fundamental bandgap, resonant Raman scattering (RRS) is superior to both optical absorption and emission, which are the standard techniques for studying the confinement of excitons [91, 92]. In addition to electronic transitions, RRS can provide information on phonons and their interactions with electrons. The latter ability is significant since it has been established that confinement effects can modify the vibrational modes and hence the electron–phonon interaction of semiconductor nanocrystals. Recently, RRS has been employed to study the E_1 transitions in nanometer-sized Ge nanocrystals embedded in SiO_2 [92]. The E_1 transitions in bulk Ge occur around 2.2 eV, well above the indirect and direct bandgaps with energies of 0.6 and 0.9 eV, respectively. Most semiconductors possess higher energy excitons with quite large oscillator strengths. Teo et al. studied Ge nanocrystals with diameters ranging from 4 to 10 nm by RRS [92]. The samples were formed by the ion implantation of Ge^+ ions into SiO_2 films followed by annealing. The observed size-dependent blueshifts of the E_1 exciton energy were found to agree with the theoretical calculations based on the effective mass approximation.

3.6. STM/AFM

Although scanning probe microscopy (SPM), such as scanning tunneling microscopy (STM) and atomic force microscopy (AFM), is known as a powerful tool to study nanostructures, its ability to image atoms has not been demonstrated for Ge nanocrystals. A difficulty in obtaining the atomic image of an individual nanocrystal comes from its spherical shape, which does not have an atomically flat area large enough to obtain an atomic image.

Sato et al. obtained the surface image of the Ge nanocrystalline film using an ultrahigh vacuum scanning tunneling microscope (UHV–STM) [64]. It clearly showed individual nanocrystals with diameters of 3–4 nm. The size distribution was narrow, and they were densely positioned so that the electron wave functions in two adjacent nanocrystals could overlap. Caragianis-Broadbridge et al. made a cross-sectional AFM study on Ge-doped SiO_2 thin films and obtained the sizes of the Ge nanocrystals and their distribution [53]. The cross-sectional AFM samples were prepared using a preparation technique based on the TEM tripod sample preparation method. This method allows imaging of nanocrystals without inducing electron-beam damage or beam-induced phase transformation. It is clear, however, that the resolution is quite limited by the radius of the AFM tip ($\sim$10–20 nm).

4. PHYSICAL PROPERTIES OF Ge NANOCRYSTALS

Like Si nanocrystals, Ge nanocrystals have been mostly studied for luminescence. The luminescence characteristic strongly depends on the methods used to form the Ge nanocrystals. For instance, the Ge nanocrystalline film deposited at 77 K by cluster-beam evaporation exhibited a unique property referred to as photooxidation and showed blue-light emission after photooxidation. Besides luminescence, the optical nonlinearity found in Ge-doped SiO_2 is another interesting optical property. In addition to the optical poperties, there is recent increasing interest in single-electron tunneling in Ge-doped glass films. In this section, the optical and electrical properties of Ge nanocrystals are reviewed.

4.1. Optical Properties

4.1.1. PL and EL

Various forms of Ge nanocrystals with a well-passivated surface exhibit strong visible-light photoluminescence (PL). The photoluminescence of Ge nanocrystals has been mostly studied in Ge-doped glass, where photogenerated electrons are expected to be confined in the Ge nanocrystals surrounded by a high potential barrier of SiO_2. A strong PL associated with Ge was observed only after annealing to form the Ge nanocrystals. Nevertheless, the mechanism of light emission is still a subject of debate.

Takeoka et al. observed strong size-dependent PL spectra from Ge-doped glass fabricated by the co-sputtering of Ge and SiO_2 and concluded that the PL originated from the recombination of the photogenerated electron–hole pairs confined in the Ge nanocrystals [27]. The PL peak energy was 0.88 eV for an average diameter of 5.3 nm and blueshifted to 1.54 eV with decreasing size to an average diameter of 0.9 nm. Broad PL spectra were observed in the visible wavelength range by Maeda [69]. The peak energy showed a blueshift from 2.1 to 2.5 eV with a decreasing average diameter from 5 to 2 nm, respectively. He, however, did not see any change in the PL peak energy for sizes greater than 5 nm. His experimental data for the sizes less than 5 nm were consistent with a quantum confinement model. Although both studies include the PL of Ge nanocrystals with the same sizes, and the fabrication methods are the same, the PL energies are significantly different.

Zacharias and co-workers observed PL with a peak around 3.1 eV regardless of the size of the Ge nanocrystals embedded in SiO_2, and they attributed the origin of the PL to a Ge/O-related defect [79, 93]. Okamoto and Kanemitsu also observed PL from the Ge-doped SiO_2 thin films fabricated by the co-sputtering of Ge and SiO_2 [77]. The surfaces of the Ge nanocrystals with an average diameter of 3.7 nm were covered with a native oxide. The nanocrystals had the characteristics of direct-bandgap semiconductors, but the PL spectra were similar to those of isolated small molecules. Excitons were localized near the surface of the nanocrystals, and the strong coupling of excitons and the stretch vibrations of the Ge oxides caused such PL spectra.

The surface oxides of the Ge nanocrystals also play an important role in PL from the Ge-doped glass thin films fabricated by the ion implantation of Ge into SiO_2. Several PL peaks have been reported. The PL peaks in the blue [65, 78] and infrared [94] wavelength regions are associated with the oxide covering the Ge nanocrystals, while that in the red [63] wavelength region was attributed to the radiative recombination of excitons confined in the Ge nanocrystals.

The electronic structure of the Ge nanocrystals was studied using the sp^3 tight binding description by Niquet et al. [95]. Their results with ab initio calculations in the local density approximation for smaller clusters indicated that the blue–green PL above 2 eV was associated with deep traps in the oxide, while the size-dependent PL in the near infrared of less than 1 eV involved deep traps in the gap of the nanocrystals.

Min et al. compared the PL from the SiO_2 films implanted with Ge and Xe [96, 97]. The peak was located around 1.8 eV for all the samples. Observation of luminescence in the Xe-implanted samples similar to that in the Ge-implanted samples suggested that the radiative defect centers in the SiO_2 matrix were responsible for the luminescence in the Ge-doped glass. The PL was quenched by implantation of deuterium into the Ge-doped glass thin film and recovered by removing deuterium by thermal annealing in a vacuum. These results also suggest that the observed PL is indeed defect related.

The Ge nanocrystals fabricated by the spark process [61], gas evaporation [98], and cluster-beam evaporation [48, 99] also exhibited visible PL, which was associated with the oxide defects. In particular, those by the cluster-beam evaporation at a substrate temperature of 77 K showed a strong blue PL at room temperature. The luminescence was seen only after photooxidation of the Ge nanocrystals, which will be discussed later. Since the PL intensity increases with the photooxidation time, and the peak energy of 2.8 eV was independent of the oxidation time, the PL was associated with the radiative centers in GeO_x on the surface of the Ge nanocrystals [48].

Tunable luminescence over a broad range in the visible region of the spectrum was demonstrated from the chemically synthesized Ge nanocrystals terminated with alkyl groups [18]. The PL peak wavelength was varied from 450 to 600 nm by controlling the size of the nanocrystals. The shift in the spectrum indicated size-dependent energy levels consistent with the quantum confinement model.

Electroluminescence (EL) was observed from a MOS-like capacitor with the Ge-doped glass thin film deposited on a Si substrate [100, 101]. Rectifying behavior was seen in the I–V characteristics. The EL was observed by Shcheglov et al. [100] at a reverse-bias voltage, resulting in an electrical field of 10^6 V/cm across the oxide, but by Zhang et al. under a forward-bias condition. Shcheglov et al. compared the PL and EL spectra and found similar results. The mechanism for the EL excitation was related to impact ionization by "hot" carriers in the oxide layer. However, the origin of the EL was not clarified. Zhang et al. observed a broad peak around 600 nm at low and moderate voltages and a sharp peak around 400 nm at high voltages. The broad peak around 600 nm was attributed to the recombination in the nanocrystals, while the sharp peak was attributed to the Ge oxygen-deficient center excited by impact ionization of the "hot" electrons.

4.1.2. Photooxidation

Photooxidation of the Ge microstructures was first reported by Harshavardhan and Hedge [102]. When the Ge nanocrystals are exposed to UV light, they are found to be rapidly oxidized even at room temperature. According to the model of photonically stimulated oxidation enhancement of Si proposed by Young [103], the rate determination step of the oxidation is the dissociation of molecular oxygen into its atoms. In his model, electrons in Si need to be excited to the conduction band of SiO_2 in order to dissociate molecular oxygen. The photooxidation of Ge nanocrystals was recently observed in those formed by the gas evaporation [98] and cluster-beam evaporation [48] techniques.

In the Ge nanocrystals made by cluster-beam evaporation, the photooxidation is evidenced by the optical absorption spectra in Figure 4. Although it is difficult to estimate the optical bandgaps from the spectra because of a gradual increase in the absorption near the band edge, it is clear from a comparison of both spectra that the photooxidation increases the optical bandgap. If we assume that the shift is due to the quantum confinement of carriers, the observed blueshift of the absorption spectrum can be explained by a decrease in the Ge core size after photooxidation.

Our systematic study [48] of the photooxidation of Ge nanocrystals deposited at a substrate temperature of 77 K by the cluster-beam evaporation technique supported Young's electron-active oxidation model. Since the dissociation energy of molecular oxygen is 5.1 eV and much higher than the photon energies of UV light, it is unlikely to be directly dissociated by the UV light. The schematic band diagram shown in Figure 5 depicts our model of the photooxidation of the Ge nanocrystals. As shown in Figure 2, a Ge core is surrounded by GeO_x. The surface of the GeO_x shell, where a supply of oxygen is sufficient, may be GeO_2. The bandgap of GeO_x increases with an increase in the value of x. When the Ge nanocrystal is exposed to UV light, the light is absorbed in the Ge core, where electron–hole pairs are generated. In the Ge nanocrystals, the electron injection into GeO_x was facilitated much more by the lower potential barrier due to an increase in the electron energy resulting from the quantum confinement. The electrons injected in GeO_x meet oxygen molecules during their diffusion in GeO_x and assist them in breaking up into smaller atomic components as discussed in the electron-active oxidation model [103]. On the other hand, the holes are not injected into the oxide but are confined in the Ge core because of its greater effective mass. The remaining holes make it easier for Ge atoms to bond to oxygen atoms.

If the preceding photooxidation model is correct, the size distribution of the Ge cores becomes smaller with the irradiation time, and the final size is determined by the photon energy of the incident light. The oxidation appears to stop when the size of the Ge core decreases and the apparent bandgap reaches the photon energy. A further increase in the apparent bandgap will result in the transmission of the incident light. Therefore, after prolonged photooxidation, the sizes of the Ge cores would become uniform. It is also

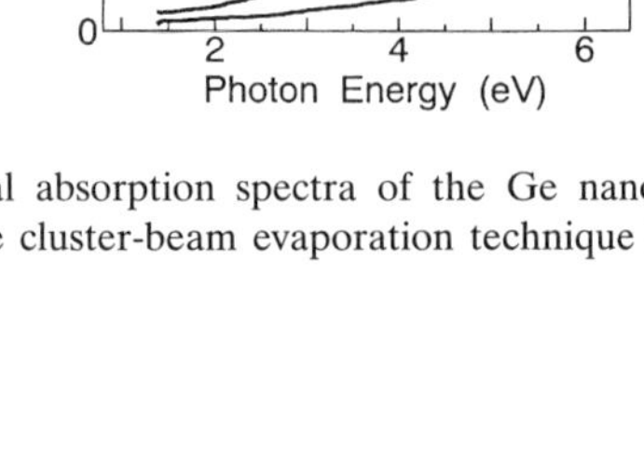

Figure 4. Optical absorption spectra of the Ge nanocrystalline films deposited by the cluster-beam evaporation technique before and after UV exposure.

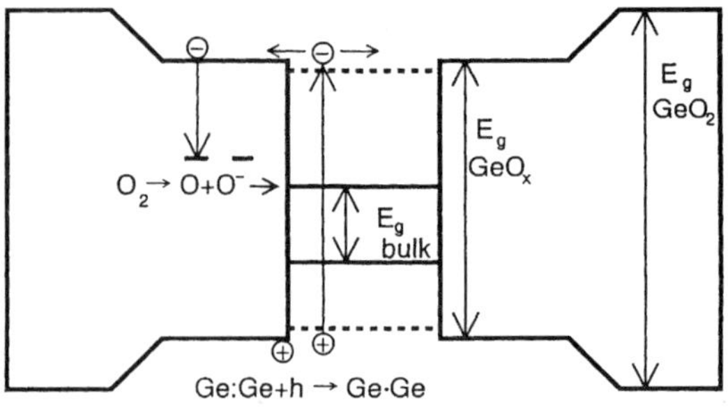

Figure 5. Schematic band diagram of the Ge nanocrystal to explain the photooxidation mechanism.

predicted that the higher the incident photon energy is, the smaller the final size of the Ge cores is.

In contrast to the Ge nanocrystals deposited at a substrate temperature of 77 K, those deposited at a substrate temperature of 300 K did not exhibit photooxidation. The difference was explained by the density of the nanocrystals, as discussed in Section 2. Since they were resistant to oxidation, no luminescence was observed from the samples.

4.1.3. Nonlinearity

Ge-doped silica is commonly employed for controlling the refractive index profile of optical fibers and exhibits other useful optical properties such as photorefraction [104] and second-harmonic generation [105]. In most of these applications, the Ge is present as a mixed oxide (SiO_2:GeO_2). However, recent experiments have shown that the Ge-doped glass exhibits a significant third-order optical nonlinearity [106–108].

The nonlinear response, intensity-dependent refractive index of the Ge-doped glass thin film with a Ge fluence of 3×10^{17} cm^{-2} implanted in SiO_2 was found to be large and more than three orders of magnitude greater than that of pure silica: 2.7×10^{-13} and 6.9×10^{-13} cm^2W^{-1} for the 100- and 650-fs pulses, respectively; that is, the nonlinear response increased with the increasing pulse duration [106]. It also linearly increased with the increasing Ge fluence implanted in SiO_2. The optical absorption in the films linearly increased with the Ge fluence, suggesting that the nonlinearity was absorptive rather than refractive.

Relaxation of the nonlinear response had two characteristic time constants, one less than 100 fs and the other about 1 ps. The carrier lifetimes are typically on the order of 10^{-3} s in pure Si and Ge, which is much longer than those observed in Ge-doped glass. However, the presence of impurities and defects can significantly reduce the carrier lifetimes. Therefore, the measured relaxation time of about 1 ps is characteristic of the defect-mediated recombination of excited carriers. The instantaneous response is believed to result from the field-induced polarization of the material.

4.2. Electrical Properties

Although the three-dimensional (3D) quantum confinement of an electron–hole pair is theoretically expected in the Ge nanocrystals, there has not been clear evidence reported until now. The difficulty in experimentally observing the physical phenomena indicating the 3D quantum confinement is mainly due to the nonhomogeneous size distributions of the Ge nanocrystals.

The effective Bohr radius, a_B, is given by

$$a_B = \frac{\varepsilon \hbar^2}{\mu q^2}, \tag{10}$$

where ε is the dielectric constant, $\hbar$ is the reduced Planck constant, μ is the reduced mass of an electron–hole pair, and q is the electronic charge. In the case of bulk Ge, ε is 15.8 [109], and μ is the reduced mass obtained from $1/\mu = 1/m_e + 1/m_h$. The lightest electron effective mass, m_e, and the lightest hole effective mass, m_h, at the L and Γ points are $0.082m_0$ and $0.043m_0$, respectively. We calculate $\mu = 0.028m_0$ and obtain $a_B = 24.3$ nm. This effective Bohr radius is very large due to the small reduced mass of the electron–hole pair and the large dielectric constant.

If the nanocrystal size is much smaller than the effective Bohr radius, the electrons and holes can be independently confined into the infinite spherical potential. In this case, the lowest energy of the electron–hole pair E_1 is given by [110]

$$E_1 = E_g + \frac{\pi^2 \hbar^2}{2\mu R^2} \tag{11}$$

where E_g is the bandgap of bulk Ge. A more detailed calculation of the confinement energy was presented with the experimental data by Teo et al. [92].

Another interesting electrical feature associated with the Ge nanocrystals is Coulomb blockade and single-electron tunneling. Inoue et al. observed Coulomb gaps and steplike features in the low-temperature I–V characteristics of Ge-doped glass formed by co-sputtering and attributed them to the combined effects of resonant tunneling of electrons through the discrete electronic state of Ge nanocrystals and single-electron tunneling [111].

The Coulomb blockade and single-electron tunneling are also observed in Ge nanocrystalline films besides Ge-doped glass as long as there is a thin insulator between the Ge nanocrystals. The I–V characteristics of a 30-nm-thick Ge nanocrystalline film deposited at 77 K by the cluster-beam evaporation technique exhibited a distinct Coulomb gap and a steplike feature at 100 K [112–114]. They were also observed even at room temperature after photooxidation, which reduced the size of the nanocrystals and increased the inter-nanocrystal spacings. The observed Coulomb blockade effect results from a very highly selective electron transport process where the electron transport is dominated by one local well-defined current path with the largest conductance.

The carrier transport properties across Ge nanocrystalline films deposited by the cluster-beam evaporation technique have been thoroughly studied. There is a difference in the electron transport characteristics of the films deposited at substrate temperatures of 77 and 300 K. The difference is explained by a structural difference, which is mentioned in Section 2.5; that is, the Ge nanocrystals deposited at a substrate temperature of 300 K are hardly oxidized and are physically in contact with each other, whereas Ge nanocrystals deposited at a substrate temperature of 77 K often agglomerate and are oxidized. It was found that, under relatively low electric fields, the Ge nanocrystalline film deposited at 77 K exhibited a $T^{-1/2}$ conductivity dependence, which could be explained by the variable range hopping model of Efros and Shklovskii [115] due to a weak overlap of electron wave functions, while in that deposited at 300 K, where the nanocrystals are in physical contact with each other, causing a significant overlap of electron wave functions, the conductivity–temperature dependence was explained by the variable range hopping model proposed by Mott and Davis [116] and Greaves [117].

4.3. Phase Transformation

4.3.1. ST-12 of Ge

One of the conventional methods to transform the crystal structure to a more closely packed one is the application of hydrostatic pressure. The high-pressure forms of a material are also experimentally obtained by reducing the size of a material [118–120]. The phase-transformation mechanism of a material due to size reduction is explained as follows. A strong surface tension due to many dangling bonds is expected to favor compact structures when the size decreases. A possible phase transformation of silicon nanocrystals from the diamond to the body-centered cubic (bcc) structure for sizes of 100–1000 atoms was theoretically predicted by Tomanek and Schluter [121].

For instance, Si and Ge, which are normally found in the diamond structure, are known to have several crystal structures, the so-called high-pressure forms: (1) hexagonal structure with four atoms per unit cell (2H-4), (2) body-centered cubic structure with eight atoms per unit cell (BC-8), and (3) simple tetragonal structure with 12 atoms per unit cell (ST-12). Figure 6 shows these four crystal structures with their densities, bandgaps, and types of bandgaps (direct or indirect) for Ge.

Among the high-pressure forms of Ge, the ST-12 structure is the most interesting from the viewpoint of device application. Joannopoulous and Cohen [71, 122] made a theoretical calculation and predicted that Ge with the ST-12 structure was a semiconductor with a direct bandgap of 1.47 eV. Bulk Ge with the diamond structure has been less attractive for device application because of its indirect and narrow bandgap. The earliest experimental observation of the ST-12 structure in Ge nanocrystals was reported by Saito [19], who found the ST-12 structure in Ge nanocrystals with diameters less than 20 nm formed by the conventional gas evaporation technique. Later Kanemitsu et al. found from high-resolution electron microscopy (HREM) that the crystal structure of Ge nanocrystals embedded in SiO_2 for diameters less than 4 nm was not like the ordinary diamond structure, although they could not identify the structure to be ST-12 [30]. They also briefly mentioned that the Ge nanocrystals with a new crystal structure had the characteristic of direct optical transition.

Recently, Wu et al. studied Ge nanocrystals smaller than 5 nm in diameter which were deposited in the MBE chamber with an ultrahigh vacuum field emission transmission electron microscope (UHV–FE–TEM) [63]. After deposition, the samples were transferred into the UHV–FE–TEM through a UHV-transferring system. They did not find the

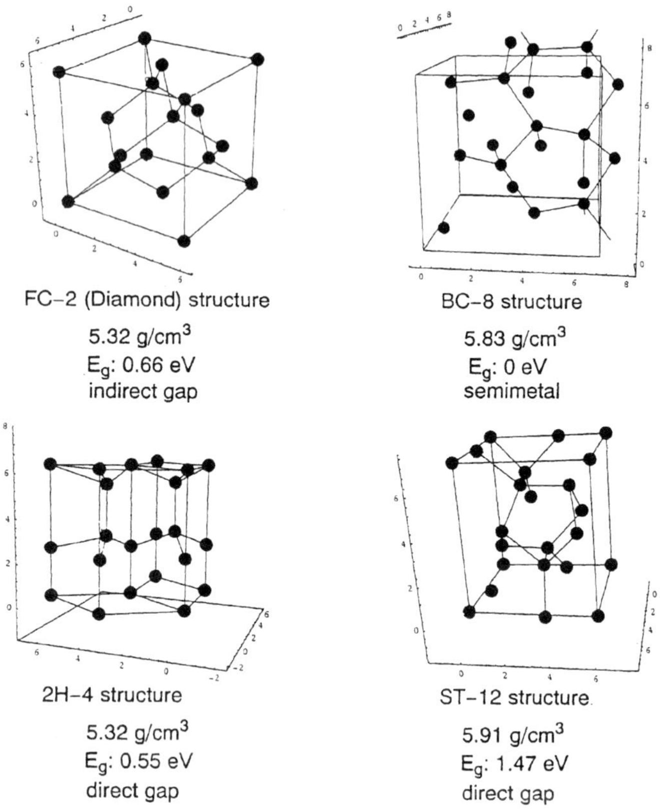

Figure 6. Possible crystal structures of Ge.

ST-12 structure but rather β-tin structure. The β-tin structure was theoretically predicted as a high-pressure form of Ge [123] and experimentally observed by Oh-Ishi et al. [124]. Although the β-tin structure was obtained under high pressure, it was transformed to the ST-12 during pressure release [125].

We have successfully formed the thermally stable ST-12 structure of Ge by the cluster-beam evaporation technique and identified the structure of the formed Ge nanocrystals as the ST-12 structure by various characterization methods [64, 74]. Although the Ge nanocrystals for both substrate temperatures of 77 and 300 K had the crystal structure of ST-12, in this section, we limit our discussion to the ST-12 found in the Ge nanocrystalline film deposited at 300 K. A similar discussion can be applied to that at 77 K.

The Raman spectrum of Ge nanocrystalline film deposited by the cluster-beam evaporation technique at a substrate temperature of 300 K is shown in Figure 3. Two peaks were observed around 240 and 265 cm^{-1} in the Raman spectrum of the Ge nanocrystalline film. Kobliska et al. reported the peaks at 273 $\pm$ 3 and 246 $\pm$ 3 cm^{-1} for the ST-12 Ge obtained by applying a high pressure [126]. In Figure 3, the Raman spectrum of the Ge particles with the diamond structure is also shown for comparison. The peak wavenumber is 300 cm^{-1}, the same as that of bulk Ge. The Raman spectrum of the Ge nanocrystalline film resembles that of the ST-12 Ge obtained by applying high pressure and suggests that the crystal structure in the Ge nanocrystalline film is not the diamond but the ST-12. We also identified the ST-12 by TED, XPS, and optical absorption measurements [64, 74].

4.3.2. Thermal Stability of ST-12 of Ge

The ST-12 of Ge is known to be metastable [127], and the ST-12 structure in the Ge nanocrystals disappeared after annealing at 300 °C [19]. We confirmed the metastability of the ST-12 of Ge by studying the ST-12 phase obtained by applying high pressure onto a small piece of bulk Ge placed in a diamond anvil cell [124]. The Raman spectrum suggested the presence of the diamond, wurzite, and ST-12 phases in the Ge piece after applying 6 GPa for a short time and then slowly removing the pressure. However, the peak for the ST-12 phase became weaker with time and finally disappeared after 4 h, while the other peaks remain unchanged.

In contrast to the ST-12 Ge formed by applying high pressure, the ST-12 Ge in the Ge nanocrystalline films deposited by the cluster-beam evaporation technique is surprisingly thermally stable. The DOS and optical absorption of the Ge nanocrystalline films annealed at 500, 700, and 800 °C were studied [64]. The sample annealed at 800 °C showed a DOS curve and absorption spectrum significantly different from those of the as-deposited and samples annealed at 500 and 700 °C. They resemble those of bulk Ge. From these studies, the critical temperature for the phase transformation from the ST-12 to the diamond structure was estimated to be between 700 and 800 °C. This result was consistent with those of the Raman [128] and plasmon-loss measurements [74]. It is concluded that the ST-12 structure of the Ge nanocrystals deposited by the cluster-beam evaporation technique is far more stable than that obtained by the other methods.

5. APPLICATIONS

Although light-emitting devices may be the first choice among the device applications of Ge nanocrystals similar to Si nanocrystals, the short-term development of such devices is hindered by a low luminescence efficiency and the broad peak of the luminescence spectra. These problems may be solved by improving the size uniformity of the Ge nanocrystals. Nevertheless, it is questionable that the light-emitting devices made of Ge nanocrystals can compete with conventional light-emitting devices made of compound semiconductor materials with direct bandgaps. Once the discovery of luminescence from Si and Ge nanocrystals attracted many researchers and engineers engaged in Si technology, it gave them hope that light-emitting devices such as light-emitting diodes (LEDs) and lasers could even be made of semiconductor materials with indirect bandgaps. Researchers and engineers in the Si technology area can easily accept the use of Si- or Ge-based materials rather than compound semiconductors. However, recent progress in blue-light-emitting devices made of GaN-based III–V compound semiconductors has lowered the value of Si or Ge nanocrystals as a light-emitting device material, because all colors are now commercially available with light-emitting devices made of compound semiconductors.

Like the other semiconductor nanocrystals, a new application that makes the best use of the properties of Ge nanocrystals and that cannot be realized with any other material must be explored for Ge nanocrystals to become more industrially important.

One of the promising electron device applications of Ge nanocrystals is a nanocrystal memory. Currently, flash electrically erasable and programmable read only memory (EEPROM) is the most popular form of nonvolatile memory. Figure 7 compares the structures of the conventional EEPROM and the nanocrystal memory. During the operation of the conventional EEPROM, electrons are injected by tunneling into a continuous floating gate to cause a threshold voltage shift. To achieve nonvolatility in the flash EEPROM, the tunneling oxide thickness is maintained high enough (~7 nm) in order to prevent floating gate charge loss under normal read and retention. The use of a large oxide thickness, however, necessitates the injection of a charge into the floating gate at large voltages, which result in hot-carrier degradation and large power consumption. To solve the problem of the flash EEPROM, a nanocrystal memory was proposed by Tiwari et al. [129]. In the nanocrystal memory, the charge loss through lateral paths is suppressed by use of a large inter-nanocrystal spacing. Thus, a smaller oxide thickness becomes possible without sacrificing the retention and makes the charge injection into the floating gate at small voltages possible. These nanocrystals can be made of either Si or Ge.

Nanocrystal memories have been demonstrated with Si [129, 130] and Ge [49, 131] nanocrystals. The electron storage characteristics of the nanocrystal memories with Si and Ge nanocrystals studied by computer simulation showed that the retention time for Ge nanocrystals was several orders longer than that for Si nanocrystals because of a higher potential barrier at the Ge/SiO_2 interface [132].

The capacitor with a Ge-doped glass film as a gate dielectric is a good test structure to quickly study programming and retention without forming a transistor structure [55, 56, 90, 133]. The capacitance–voltage (C–V) curve of the MOS capacitor should be shifted by injection of the charge into Ge nanocrystals, and the flatband voltage shift for a single electron per nanocrystal will be [129]:

$$\Delta V_{\mathrm{FB}} = \frac{qn}{\varepsilon_{\mathrm{ox}}}\left(t_{\mathrm{cntl}} + \frac{\varepsilon_{\mathrm{ox}} D}{2\varepsilon_{\mathrm{Si}}}\right) \tag{12}$$

where t_{cntl} is the thickness of the control oxide between the gate and the nanocrystal, D is the diameter of the nanocrystal, ε is the dielectric constant, q is the magnitude of the electronic charge, and n is the density of the nanocrystals. For the 5-nm-diameter Ge nanocrystals with a density of 1×10^{12} cm^{-2}, the flatband voltage shift is calculated as 1.0 V.

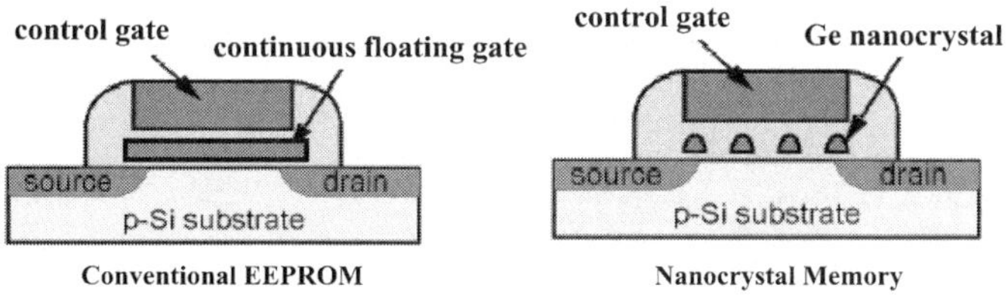

Figure 7. Comparison of the electrically erasable and programmable read only memory (EEPROM) and the nanocrystal memory.

The Coulomb blockade effect can also be significant for this nanocrystal size. The Coulomb charging energy ($q^2/2C$, where C is the nanocrystal capacitance) of a 5-nm-diameter nanocrystal is approximately 74 meV, larger than the thermal energy at room temperature, and limits the injection of additional electrons into the nanocrystal.

The process flow to fabricate Ge nanocrystal memories developed by King et al., which is described in Section 2.6, is easily incorporated in the present Si technology. In order to serve in a large-scale memory array, however, the distribution of the threshold voltage of the memory cells must be small. For high-density memory application, the memory devices have to be scaled to a deep submicrometer size. The variation in the number of electrons trapped, that is, the number of nanocrystals in the memory cell, has to be within a certain tolerance determined by the design of the sense amplifier. If there is nonuniformity in the size and density of the Ge nanocrystals, the variation in the threshold voltage in the programmed states increases as the size of the memory cell decreases. Therefore, the uniformity of the size and density of the Ge nanocrystals must be well controlled for application in high-density and large-scale memory arrays.

6. SUMMARY

We have reviewed the fabrication techniques, characterization methods, and properties of Ge nanocrystals. The most popular form of Ge nanocrystals is the Ge nanocrystals embedded in SiO_2, which is often called Ge-doped glass. The Ge nanocrystals in Ge-doped glass films are ideal quantum dots because of the high potential barrier formed by the surrounding oxide. Although the photoluminescence spectra reported for Ge-doped glass films are not conclusive evidence of quantum confinement, a recent RRS study supports quantum confinement in the Ge nanocrystals of the Ge-doped glass film.

Although light-emitting devices are the first choice among the device applications, it is not certain that the light-emitting devices made of Ge nanocrystals can compete with those made of compound semiconductors. In order for Ge nanocrystals to become an important electronic or photonic material, an application should make the best use of the properties of the Ge nanocrystals and should not be realized with any other material. One promising application of the Ge nanocrystals is a nanocrystal memory, which is to replace the conventional EEPROM flash memory. However, good control of the size and density of Ge nanocrystals is required for application in high-density and large-scale memory arrays.

Besides good control of the size and density, the challenge is to control the positioning of the Ge nanocrystals in SiO_2. The oxidation of the MBE-grown structure of the Ge dots embedded in Si is a good way to control the positioning of the Ge nanocrystals in the oxide. The Ge dots can be well ordered in an array by controlling the strain, resulting in a lattice mismatch between Si and Ge.

One of the unique deposition techniques of Ge nanocrystals is cluster-beam evaporation. The as-deposited Ge nanocrystalline thin films consist of the ST-12 Ge instead of the ordinary diamond crystal structure. The ST-12 structure is one of the high-pressure forms of Ge with optical properties of a direct-bandgap semiconductor and is known to be metastable. In contrast to the ST-12 Ge formed by applying high pressure and the gas evaporation technique, the ST-12 Ge in the Ge nanocrystalline films is thermally stable. They also exhibited the unique property of photooxidation. The Ge nanocrystals were oxidized under UV irradiation and showed blue luminescence after photooxidation. The luminescence was attributed to defects in the oxide. The ST-12 Ge has a direct bandgap of 1.47 eV, which may lead to a wide variety of device applications in contrast to the diamond-structured Ge.

No matter what the device application might be, uniformity in the size and density and a good control of positioning may always be required. Future research on Ge nanocrystals must focus on control of the positioning, size, and density. Most techniques that have been used to fabricate Ge nanocrystals do not have a good control of positioning, and we are all longing for the emergence of a new technique.

GLOSSARY

Bohr radius Radius within which the electron and hole are bound to each other, as in the case of the hydrogen atom. The electron and the hole are correlated in their motion in a way that can be described as the simple Coulomb attraction of the electron and the hole.

Coulomb blockade Physical phenomenon that the electron tunneling is forbidden in one small capacitance tunneling junction under a certain bias condition, $-e/2C < V < e/2C$. This results from a significant change in the electrostatic energy of the junction by adding or removing a single electron for extremely small C.

Direct (indirect) bandgap The bandgap is defined as the energy difference between the conduction-band minimum and the valence-band maximum. If the conduction-band minimum occurs at the same point in k space or momentum space as the valence-band maximum, then the bandgap is direct. If not, it is indirect.

Electrically erasable programmable read only memory (EEPROM) NMOS memory utilizing Fowler–Nordheim tunneling to charge and discharge the floating gate through the thin oxide as the tunneling medium.

Flatband voltage Gate voltage required to bring about the flatband condition (no charge induced in the semiconductor) in the MOS capacitor. Although the flatband voltage of the ideal MOS capacitor is 0, it can be shifted by a difference in the work functions of the gate (M) and semiconductor (S) and electronic charge in the oxide or at the semiconductor/oxide interface.

Ge-doped glass Silicon dioxide layer with Ge nanoparticles or nanocrystals dispersed.

Nanocrystal Crystalline particle with a size of several nanometers or less. It differs from the nanostructures formed by epitaxial growth or lithography in this chapter.

Photooxidation Oxidation process enhanced under illumination. The light used for photooxidation is usually ultraviolet light with high-energy photons.

ACKNOWLEDGMENTS

The studies on the Ge nanocrystals deposited by the cluster-beam evaporation technique have been supported in part by research grants from the Japan Securities Scholarship Foundation, the Mechanical Industry Development and Assistance Foundation, and the Yazaki Science Foundation; by a Grant-in-Aid for Science Research (B) from the Ministry of Education, Science and Culture; and by Special Coordination Funds for Promoting Science and Technology titled "Research on New Material Development by Nanospace Lab."

REFERENCES

1. L. T. Canham, *Appl. Phys. Lett.* 57, 1046 (1991).
2. S. Furukawa and T. Miyasato, *Phys. Rev. B* 38, 5726 (1988).
3. H. Takagi, H. Ogawa, Y. Yamazaki, A. Ishizaki, and T. Nakagiri, *Appl. Phys. Lett.* 59, 2349 (1990).
4. Y. Kanemitsu, *Phys. Rep.* 263, 1 (1995).
5. Y. Maeda, N. Tsukamoto, Y. Yazawa, Y. Kanemitsu, and Y. Masumoto, *Appl. Phys. Lett.* 59, 3168 (1991).
6. A. G. Cullis, L. T. Canham, and P. D. J. Calcott, *J. Appl. Phys.* 82, 909 (1997).
7. S.-T. Ngiam, K. F. Jensen, and K. D. Kolenbrander, *J. Appl. Phys.* 76, 8201 (1994).
8. X. L. Wu, Y. Gu, G. G. Siu, E. Fu, N. Tang, T. Gao, and X. M. Bao, *J. Appl. Phys.* 86, 707 (1999).
9. H. Migues, V. Fornes, F. Meseguer, F. Marquez, and C. Lopez, *Appl. Phys. Lett.* 69, 2347 (1996).
10. M. Nogami, K. Nagasaka, and E. Kato, *J. Am. Ceram. Soc.* 73, 2097 (1990).
11. M. Nogami, Y. Q. Zhu, Y. Tohyama, K. Nagasaka, T. Tokizaki, and A. Nakamura, *J. Am. Ceram. Soc.* 74, 238 (1991).
12. M. Nogami, K. Nagasaka, and T. Suzuki, *J. Am. Ceram. Soc.* 75, 220 (1992).
13. L. Z. Yao, C. H. Ye, C. M. Mo, W. L. Cai, and L. D. Zhang, *J. Cryst. Growth* 216, 147 (2000).
14. S. T. Selvan, M. Nogami, A. Nakamura, and Y. Hamanaka, *J. Non-Cryst. Solids* 255, 254 (1999).
15. M. Nogami and Y. Abe, *Appl. Phys. Lett.* 65, 2545 (1994).
16. H. Yang, X. Wang, H. Shi, F. Wang, X. Gu, and X. Yao, *J. Cryst. Growth* 236, 371 (2002).
17. J. P. Wilcoxon, P. P. Provencio, and G. A. Samara, *Phys. Rev. B* 64, 035417 (2001).
18. S. M. Kauzlarich, Q. Liu, S.-C. Yin, H. W. H. Lee, and B. Taylor, *Mater. Res. Soc. Symp. Proc.* 638, F6.5.1 (2001).
19. Y. Saito, *J. Cryst. Growth* 47, 61 (1979).
20. M. Han, Y. Gong, J. Ma, F. Liu, and G. Wang, *Jpn. J. Appl. Phys.* 34, 49 (1995).
21. J. Si, H. Ono, K. Uchida, S. Nozaki, and H. Morisaki, *Mater. Res. Soc. Symp. Proc.* 638, F14.4.1 (2001).
22. S. Hayashi, M. Fujii, and K. Yamamoto, *Jpn. J. Appl. Phys.* 28, L1464 (1989).
23. Y. Maeda, N. Tsukamoto, Y. Yazawa, Y. Kanemitsu, and Y. Masumoto, *Appl. Phys. Lett.* 59, 3168 (1991).
24. H. Ninomiya, N. Itoh, S. Rath, S. Nozaki, and H. Morisaki, *J. Vac. Sci. Technol., B* 17, 1903 (1999).
25. S. Takeoka, M. Fujii, S. Hayashi, and K. Yamamoto, *Appl. Phys. Lett.* 74, 1558 (1999).
26. W. K. Choi, V. Ng, Y. W. Ho, T. B. Chen, and V. Ho, *Mater. Res. Soc. Symp. Proc.* 638, F14.1.1 (2001).
27. S. Takeoka, M. Fujii, S. Hayashi, and K. Yamamoto, *Phys. Rev. B* 58, 7921 (1998).
28. M. Fujii, S. Hayashi, and K. Yamamoto, *Appl. Phys. Lett.* 57, 2692 (1990).

29. M. Fujii, S. Hayashi, and K. Yamamoto, *Jpn. J. Appl. Phys., Part 1* 30, 687 (1991).
30. Y. Kanemitsu, H. Uto, Y. Masumoto, and Y. Maeda, *Appl. Phys. Lett.* 61, 2187 (1992).
31. A. K. Dutta, *Appl. Phys. Lett.* 68, 1189 (1996).
32. S. Rath, H. Ono, S. Nozaki, and H. Morisaki, *J. Surf. Anal.* 3, 251 (1998).
33. M. Zacharias, R. Weigand, J. Blasing, and J. Christen, *Mater. Res. Soc. Symp. Proc.* 452, 117 (1997).
34. M. Zacharias, R. Weigand, B. Dietrich, F. Stolze, J. Blasing, P. Veit, T. Drusedau, and J. Christen, *J. Appl. Phys.* 81, 2384 (1997).
35. K. Kawamura, H. Hosono, and H. Kawazoe, *J. Appl. Phys.* 80, 1357 (1996).
36. H. Hosono, N. Matsunami, A. Kudo, and T. Ohtsuka, *Appl. Phys. Lett.* 65, 1632 (1994).
37. J. G. Zhu, C. W. White, J. D. Budai, S. P. Withrow, and Y. Chen, *J. Appl. Phys.* 78, 4386 (1995).
38. J. von Borany, R. Grotzschel, K. H. Heinig, A. Markwitz, W. Matz, B. Schmidt, and W. Skorupa, *Appl. Phys. Lett.* 71, 3215 (1997).
39. K. H. Heinig, B. Schmidt, A. Markwitz, R. Grotzchel, M. Strobel, and S. Oswald, *Nucl. Instrum. Methods Phys. Res. B* 148, 969 (1999).
40. C. Bonafos, B. Garrido, M. Lopez, A. Perez-Rodriguez, J. R. Morante, Y. Kihn, G. Ben Assayag, and A. Claverie, *Appl. Phys. Lett.* 76, 3962 (2000).
41. A. R. Dowd, D. Llewellyn, J. D. Fiz Gerald, and T. G. Elliman, *Mater. Res. Soc. Symp. Proc.* 650, R3.39.1 (2001).
42. E. S. Marstein, A. E. Gunnaes, U. Serincan, R. Turan, A. Olsen, and T. G. Finstad, *Surf. Coat. Technol.* 158–159, 544 (2002).
43. M. Yamamoto, T. Koshikawa, T. Yasue, H. Harima, and K. Kajiyama, *Thin Solid Films* 369, 100 (2000).
44. P. Normand, K. Beltsios, E. Kapetanakis, D. Tsoukalas, T. Travlos, J. Stoemenos, J. Van Den Berg, S. Zhang, C. Vieu, H. Launois, J. Gautier, F. Jourdan, and L. Palun, *Nucl. Instrum. Methods Phys. Res. B* 178, 74 (2001).
45. K. Masuda, M. Yamamoto, M. Kanaya, and Y. Kanemitsu, *J. Non-Cryst. Solids* 299, 1079 (2002).
46. S. Sato, S. Nozaki, H. Ono, H. Morisaki, and M. Iwase, *Appl. Phys. Lett.* 66, 3176 (1995).
47. H. Takagi, "Ionized-Cluster Beam Deposition and Epitaxy." Noyes, Park Ridge, NJ, 1988.
48. S. Sato, S. Nozaki, and H. Morisaki, *J. Appl. Phys.* 81, 1518 (1997).
49. Y.-C. King, T.-J. King, and C. Hu, *IEEE Trans. Electron Devices* 48, 696 (2001).
50. A. E. Bair, Z. Atzmon, T. L. Alford, and D. J. Smith, *J. Appl. Phys.* 83, 2835 (1998).
51. V. Craciun, I. W. Boyd, A. H. Reader, and D. E. W. Vandenhoudt, *Appl. Phys. Lett.* 65, 3233 (1994).
52. T.-S. Yoon and K.-B. Kim, *J. Vac. Sci. Technol., B* 20, 631 (2002).
53. C. Caragianis-Broadbridge, J. M. Blaser, and D. C. Paine, *J. Appl. Phys.* 82, 1626 (1997).
54. D. C. Paine, C. Caragianis, T. Y. Kim, Y. Shigesato, and T. Ishihara, *Appl. Phys. Lettt.* 62, 2842 (1993).
55. T. Kobayashi, T. Endoh, H. Fukuda, S. Nomura, A. Sakai, and Y. Ueda, *Appl. Phys. Lett.* 71, 1195 (1997).
56. H. Fukuda, T. Kobayashi, T. Endoh, S. Nomura, A. Sakai, and Y. Ueda, *Appl. Surf. Sci.* 130–132, 776 (1998).
57. W. K. Choi, V. Ng, V. S. L. Swee, C. S. Ong, M. B. Yu, Rusli, and S. F. Yoon, *Scripta Mater.* 44, 1873 (2001).
58. Y. S. Lim, F. Bassani, A. Portavoce, A. Ronda, S. Nozaki, and I. Berbezier, "European Materials Research Society Spring Meeting, Strasbourg, France, June 18–21, 2001."
59. T. Gao, S. Tong, X. Q. Zheng, X. L. Wu, L. M. Wang, and X. M. Bao, *Appl. Phys. Lett.* 72, 3312 (1998).
60. M. H. Ludqig, R. E. Hummel, and S.-S. Chang, *J. Vac. Sci. Technol., B* 12, 3023 (1994).
61. S.-S. Chang, G. J. Choi, and R. E. Hummel, *Mater. Sci. Eng., B* 76, 237 (2000).
62. D. B. Williams and C. B. Carter, "Transmission Electron Microscopy I." Plenum, New York, 1996.
63. Y. Wu, M. Takeguchi, and K. Furuya, *Jpn. J. Appl. Phys.* 38, 7241 (1999).
64. S. Sato, S. Nozaki, and H. Morisaki, *Appl. Phys. Lett.* 72, 2460 (1998).
65. J.-Y. Zhang, X.-M. Bao, Y.-H. Ye, and X.-L. Tan, *Appl. Phys. Lett.* 73, 1790 (1998).
66. A. G. Rolo, O. Conde, M. J. M. Gomes, and M. P. dos Santos, *J. Mater. Process. Technol.* 92–93, 269 (1999).
67. D. J. O'Connor, B. A. Sexton, and R. St. C. Smart, Eds., "Surface Analysis Methods in Materials Science," Springer Series in Surface Science, Vol. 23. Springer-Verlag, New York, 1992.
68. J. M. Walls, Ed., "Methods of Surface Analysis." Cambridge University Press, Cambridge, UK, 1989.
69. Y. Maeda, *Phys. Rev. B* 51, 1658 (1995).
70. W. K. Choi, Y. W. Ho, S. P. Ng, and V. Ng, *J. Appl. Phys.* 89, 2168 (2001).
71. J. D. Joannopoulos and M. L. Cohen, *Phys. Rev. B* 7, 2644 (1973).
72. S. Hufner, "Photoelectron Spectroscopy." Springer-Verlag, Berlin, 1995.
73. M. L. Cohen and J. R. Chelikowsky, "Electronic Structure and Optical Properties of Semiconductors." Springer-Verlag, Berlin, 1988.
74. S. Sato, S. Nozaki, and H. Morisaki, *Thin Solid Films* 343–444, 481 (1999).
75. G. D. Gilliland, *Mater. Sci. Eng.* R18, 99 (1997).
76. X. L. Wu, G. G. Siu, Y. Gu, N. Tang, T. Gao, and X. M. Bao, *Appl. Phys. Lett.* 74, 827 (1999).
77. S. Okamoto and Y. Kanemitsu, *Phys. Rev. B* 54, 16421 (1996).
78. J.-Y. Zhang, Y.-H. Ye, X.-L. Tan, and X.-M. Bao, *J. Appl. Phys.* 86, 6139 (1999).
79. M. Zacharias and P. M. Fauchet, *Appl. Phys. Lett.* 71, 380 (1997).
80. R. Hayashi, M. Yamamoto, K. Tsunetomo, K. Kohno, Y. Osaka, and H. Nasu, *Jpn. J. Appl. Phys.* 29, 756 (1990).
81. P. Tognini, L. C. Andreani, M. Geddo, A. Stella, P. Cheyssac, R. Kofman, and A. Migliori, *Phys. Rev. B* 53, 6992 (1996).
82. A. Dowd, R. G. Elliman, and B. Luther-Davies, *Appl. Phys. Lett.* 79, 2327 (2001).
83. J. I. Pankove, "Optical Processes in Semiconductors." Dover, New York, 1971.
84. R. H. Klazes, M. H. L. M. van den Broek, J. Bezemer, and S. Radelaar, *Philos. Mag. B* 45, 377 (1982).
85. H. Richter, Z. P. Wang, and L. Ley, *Solid State Commun.* 39, 625 (1981).
86. I. H. Campbell and P. M. Fauchet, *Solid State Commun.* 58, 739 (1986).
87. S. Sato, S. Rath, S. Akiyama, S. Nozaki, and H. Morisaki, *J. Appl. Phys.* 86, 1774 (1999).
88. A. V. Kolobov, Y. Maeda, and K. Tanaka, *J. Appl. Phys.* 88, 3285 (2000).
89. Y. W. Ho, V. Ng, W. K. Choi, S. P. Ng, T. Osipowicz, H. L. Seng, W. W. Tjui, and K. Li, *Scripta Mater.* 44, 1291 (2001).
90. H. Fukuda, S. Sakuma, T. Yamada, S. Nomura, M. Nishino, T. Higuchi, and S. Ohshima, *J. Appl. Phys.* 90, 3524 (2001).
91. S. H. Kwok, P. Y. Yu, C. H. Tung, Y. H. Zhang, M. F. Li, C. S. Peng, and J. M. Zhou, *Phys. Rev. B* 59, 4980 (1999).
92. K. L. Teo, S. H. Kwok, P. Y. Yu, and S. Guha, *Phys. Rev. B* 62, 1584 (2000).
93. M. Zacharias, M. Schmidt, and P. M. Fauchet, *Electrochem. Soc. Proc.* 97-11, 287 (1997).
94. X. L. Wu, T. Gao, G. G. Siu, S. Tong, and X. M. Bao, *Appl. Phys. Lett.* 74, 2420 (1999).
95. Y. M. Niquet, G. Allan, C. Delerue, and M. Lanno, *Appl. Phys. Lett.* 77, 1182 (2000).

96. K. S. Min, K. V. Shcheglov, C. M. Yang, R. P. Camata, H. Atwater, M. L. Brongersma, and A. Polman, *Mater. Res. Soc. Symp. Proc.* 405, 247 (1996).
97. K. S. Min, K. V. Shcheglov, C. M. Yang, H. Atwater, M. L. Brongersma, and A. Polman, *Appl. Phys. Lett.* 68, 2511 (1996).
98. S. Nozaki, S. Sato, A. Denda, H. Ono, and H. Morisaki, *Mater. Res. Soc. Symp. Proc.* 358, 133 (1995).
99. S. Nozaki, S. Sato, S. Rath, H. Ono, and H. Morisaki, *Bull. Mater. Sci.* 22, 377 (1999).
100. K. V. Shcheglov, C. M. Yang, K. J. Vahala, and H. A. Atwater, *Appl. Phys. Lett.* 66, 745 (1995).
101. J.-Y. Zhang, Y.-H. Ye, and X.-L. Tan, *Appl. Phys. Lett.* 74, 2459 (1999).
102. K. Harshavardhan and M. S. Hedge, *Solid State Commun.* 69, 117 (1989).
103. E. M. Young, *Appl. Phys. A* 47, 259 (1988).
104. M. V. Bazylenko, D. Moss, and J. Canning, *Opt. Lett.* 23, 697 (1998).
105. A. J. Ikushima, T. Fujiwara, and K. Saito, *J. Appl. Phys.* 88, 1201 (2000).
106. A. Dowd, R. G. Elliman, M. Samoc, and B. Luther-Davies, *Appl. Phys. Lett.* 74, 239 (1999).
107. Y. X. Jie, Y. N. Xiong, A. T. S. Wee, C. H. A. Huan, and W. Ji, *Appl. Phys. Lett.* 77, 3926 (2000).
108. A. Dowd, D. Llewellyn, R. G. Elliman, B. Luther-Davies, M. Samoc, and J. D. Fitz Gerald, *Nucl. Instrum. Methods Phys. Res. B* 175–177, 637 (2001).
109. C. Kittel, "Introduction to Solid State Physics," 5th ed., p. 233. Wiley, New York, 1976.
110. L. E. Brus, *IEEE J. Quantum Electron.* 22, 1909 (1986).
111. Y. Inoue, M. Fujii, S. Hayashi, and K. Yamamoto, *Solid-State Electron.* 42, 1605 (1998).
112. S. Banerjee, S. Nozaki, and H. Morisaki, *Mater. Res. Soc. Symp. Proc.* 536, 551 (1999).
113. S. Banerjee, S. Nozaki, and H. Morisaki, *Appl. Phys. Lett.* 76, 445 (2000).
114. S. Banerjee, S. Nozaki, and H. Morisaki, *J. Appl. Phys.* 91, 4307 (2002).
115. A. L. Efros and B. I. Shklovskii, *J. Phys. C* 8, L49 (1975).
116. N. F. Mott and E. A. Davis, "Electron Process in Non-Crystalline Materials." Clarendon, Oxford, 1979.
117. G. N. Greaves, *J. Non-Cryst. Solids* 11, 427 (1973).
118. F. Farges, M. F. de Fereudy, B. Raoult, and G. Torchet, *Adv. Chem. Phys.* 70, 45 (1989).
119. W. Miehle, O. Kandler, T. Leisner, and O. Echt, *J. Chem. Phys.* 91, 5940 (1989).
120. T. P. Martin, T. Bergmann, H. Gohlich, and T. Lange, *J. Phys. Chem.* 95, 6421 (1991).
121. D. Tomanek and M. A. Schluter, *Phys. Rev. B* 36, 1208 (1987).
122. J. D. Joannopoulos and M. L. Cohen, *Phys. Rev. B* 8, 2733 (1973).
123. M. T. Yin and M. L. Cohen, *Phys. Rev. B* 26, 5668 (1982).
124. M. Oh-Ishi, S. Akiyama, K. Uchida, S. Nozaki, and H. Morisaki, *Phys. Status Solidi B* 223, 391 (2001).
125. C. S. Menoni, J. Z. Hu, and I. L. Spain, *Phys. Rev. B* 34, 362 (1986).
126. R. J. Kobliska, S. A. Solin, M. Selders, R. K. Chang, R. Alben, M. F. Thorpe, and D. Weaire, *Phys. Rev. Lett.* 29, 725 (1972).
127. F. P. Bundy and J. S. Kasper, *Science* 139, 340 (1963).
128. S. Nozaki, S. Sato, H. Ono, H. Morisaki, and M. Iwase, *Mater. Res. Soc. Symp. Proc.* 405, 223 (1996).
129. S. Tiwari, F. Rana, H. Hanafi, A. Hartstein, E. F. Crabbe, and K. Chan, *Appl. Phys. Lett.* 68, 1377 (1996).
130. J. J. Welser, S. Tiwari, S. Rishton, K. Y. Lee, and Y. Lee, *IEEE Electron Device Lett.* 18, 278 (1997).
131. H. I. Hanafi, S. Tiwari, and I. Khan, *IEEE Trans. Electron Devices* 43, 1553 (1996).
132. H. G. Yang, Y. Shi, H. M. Bu, J. Wu, B. Zhao, X. L. Yuan, B. Shen, P. Han, R. Zhang, and Y. D. Zheng, *Solid-State Electron.* 45, 767 (2001).
133. W. K. Choi, W. K. Chim, C. L. Heng, L. W. Teo, V. Ho, V. Ng, D. A. Antoniadis, and E. A. Fitzgerald, *Appl. Phys. Lett.* 80, 2014 (2002).

ENN Encyclopedia of Nanoscience and Nanotechnology
www.aspbs.com/enn

Germanium Nanocrystals Structure

A. V. Kolobov
National Institute of Advanced Industrial Science and Technology, Tsukuba, Ibaraki, Japan

CONTENTS

1. INTRODUCTION

Silicon (Si) technology clearly dominates the semiconductor industry. However, most optoelectronic devices still rely on the use of III-V semiconductors. The reason is that direct band structure in III-V semiconductors allows one to fabricate efficient light emitters while indirect band structure of Si and Ge puts a serious barrier in the way to achieve this goal.

It has been suggested that use of nanosize Ge and Si structures should result in the band structure modification, namely, making it direct. Several reports of intense photoluminescence (PL) from nanosize Si and Ge have generated increased interest in these materials.

There are two major ways to fabricate Ge nanostructures. One is epitaxial growth on Si when Ge nanostructures are formed as a result of self-assembly process and the other one is fabrication of nanocrystals confined in various matrixes. In this entry the latter process is discussed. We start with methods of Ge nanocrystal fabrication. This section is followed by description of their average and local structure. Optical and electrical properties of Ge nanocrystal structures are further discussed. The entry is concluded by a summary of possible advanced application of Ge nanocrystals.

2. FABRICATION TECHNIQUES

Attempts to incorporate the Ge nanocrystals into industrial Si technology are likely to be responsible for the fact that in most cases the Ge nanocrystals are formed embedded in a SiO_2 matrix, the substrate usually being silicon. Various methods of the nanocrystal fabrication have been successfully applied. These techniques include (co-)sputtering [1–7], sol–gel processing [8, 9], and ion implantation [10–13] followed by annealing at temperatures in the 300–1000 °C range (either conventional annealing or rapid thermal annealing (RTA) can be used). Each of the techniques has its own merits and disadvantages. Thus sputtering allows one to vary the Ge content in a very broad range, sol–gel seems to produce nanocrystals of a smaller size, and ion implantation allows to incorporate more Ge species into the SiO_2 host compared to the sol–gel technique and is compatible with the conventional Si technology.

Other methods include H_2-reduction of Ge-oxide [14], atmospheric pressure chemical vapor deposition (CVD) [15], epitaxial growth of Ge on oxidized Si substrates [16], low-temperature UV oxidation [17], and deposition of Ge on SiO_2 and subsequent high-temperature oxidation [18]. It should be noticed that the evidence for the formation of the Ge nanocrystals in the latter case is rather ambiguous.

Host materials different from SiO_2, such as silicon when Ge nanoparticles were produced by pulsed laser ablation and co-deposited in a Si film simultaneously grown by chemical beam epitaxy using silane [19], AlN (by pulsed laser deposition) [20], and SiC (by ion implantation and subsequent annealing) [21], have been used.

An interest has been observed in fabrication of Ge nanocrystals inside nanoporous materials, in particular in zeolite cages [22–26] and in opals [27, 28]. In the opal case, after filling the opal pores with Ge, the host matrix is etched away and the so-called "inverse opal" structure results. The interest in this case is mainly driven by photonic bandgap properties of three-dimensional periodic structures.

Arrays of Ge nanocrystals which are not embedded into any matrix have also been produced from colloids [29, 30] and using a Ge cluster source evaporation technique [31].

3. NANOCRYSTAL GEOMETRY

The nanocrystal geometry, that is, their size and shape as well as their spatial distribution within the host matrix, is determined by the preparation conditions. Most of

ISBN: 1-58883-059-4/$35.00

these data have been obtained by direct observation of the nanocrystals using high-resolution transmission electron microscopy (HR-TEM). While this technique is very convenient, it has some restrictions with respect to the sample preparation. Depending on the sample thickness, for example, certain nanocrystals may or may not be observed, occasionally resulting in poor reproducibility.

Some of the experimental problems related to TEM studies of the nanostructures are discussed in [32]. While there are some differences between the results reported by various groups, general trends can be summarized as follows.

- The minimum experimentally observed nanocrystal size reported in the literature is 1 nm [5].
- The size of the nanocrystals increases with the annealing time.
- The size of the nanocrystals increases with the annealing temperature.
- The maximum nanocrystal size reported is ~30 nm [7].

It should also be noted that the nanocrystals generally possess a rather spherical shape and the typical size of the nanocrystals is smaller than the typical size of Ge/Si self-assembled quantum dots prepared by epitaxial growth through the Stranski–Krastanov mode. In the latter case the lateral size of the Ge nanostructures is usually on the order of 100 nm with the height on the order of 20 nm [33].

The geometrical parameters of the nanocrystals obtained by various groups are summarized in Table 1 together with the preparation conditions. The size of the nanocrystals also depends on the concentration of Ge in the sample, being larger for larger concentrations (Table 2).

It should also be noticed that the size of the nanocrystals strongly depends on the substrate material. Thus, in our experiments, when samples containing 60 mol.% Ge were simultaneously deposited onto Si(100) and quartz-glass substrates and then annealed in Ar atmosphere at 800 °C for 1 hour, the size of the nanocrystals formed on Si(100) was 20 nm while the size of the nanocrystals formed on a quartz-glass substrate was much smaller, namely, 5 nm [37].

The smallest observed size of the nanocrystals either may be determined by the TEM resolution or can be limited by the thermodynamical-energy balance. An estimate based on the latter assumption gives a value of 2–3 nm, which is in good agreement with the experimental observations.

The minimum nanocrystal size was also calculated in [38]. These authors have shown that Ge clusters with the diamond structure are stable for clusters containing more than 40 atoms, which also falls into the same size range.

In [4] it was suggested that double annealing at 300 °C and 600 °C for nucleation and growth results in a better size control and more uniform size distribution than annealing at a single temperature.

Annealing of a sample with a larger Ge content (60 mol.%) results in a bimodal Ge nanocrystal formation [37]. In a direct vicinity of the Si(100) substrate, spherical and rather densely placed nanocrystals were formed with a typical diameter of ~20 nm. It is worth noticing that the film–Si(100) interface was no longer flat, with dips in the immediate vicinity of the nanocrystals. In the area away from the substrate, well-faceted nanocrystals of comparable size were formed. It is worth noting that whilst the faceted nanocrystals in the bulk of the SiO_2 matrix were usually single-domain single crystals, those located in the vicinity of the interface possess multiple-twin structure [37].

Spherical Ge multiple-twin structured nanocrystals located at the Si–SiO_2 interface were also reported in [7] where rapid thermal annealing has been used. The size of the nanocrystals varied from 6 nm for 800 °C annealing to 20–30 nm for the 1000 °C annealing.

It has been suggested [39] that a higher concentration of the Ge nanocrystals close to the interface can be due to the fact that the solubility of Ge is different at the Si–SiO_2 interface and in the bulk of the oxide with the concentration gradient leading the diffusion flux and subsequent accumulation

Table 1. Nanocrystal size dependence on the growth conditions.

Fabrication method	Annealing temperature, °C and time	Nanocrystal size, nm	Ref.
Sputtering	300, 30 min	2–3	[4]
	800, 30 min	18	[4]
	800	1–5	[5]
	800, RTA	6	[7]
	1000, RTA	20–30	[7]
	700	4–5	[34]
	800	5.5–7	[34]
	900	>15	[34]
	800	6	[35]
Ion implantation	800	4–10	[12]
	600	4–5	[11]
	800	25	[11]
	1100, 1 hour	5–6	[13]
(in SiC host)	1600, RTA	2–10	[21]
Low-temperature UV oxidation	550	2–8	[17]
sol–gel	700, 2 hours	4–6	[8]
	800	10	[8]
Pulsed laser deposition (on AlN)	500 (substrate temperature)	15–20	[20]
Epitaxy on an oxidized Si wafer	300–400 (substrate temperature)	6–7	[16]
Laser ablation + CVD		>2	[19]
From colloids	270	6–20	[29]
		2–10	[30]
Cluster source		3–4	[31]
Plasma-enhanced CVD		5	[36]

Table 2. Nanocrystal size dependence on the Ge concentration.

Fabrication technique	Ge concentration	Annealing temperature	Size, nm	Ref.
Sputtering	25 mol.%	800	not detected	[37]
	60	800	20	[37]
Ion Implantation	1×10^{17} ions/cm^2	800	4	[12]
	2×10^{17}	800	7	[12]
	3×10^{17}	800	10	[12]

of Ge at the interface. Alternatively [14], it has been argued that the outdiffusion of Si from the substrate at elevated temperatures should enhance the nucleation and growth of the nanocrystals at the interface.

While in [7, 37] the nanocrystals formed in the vicinity of the Si substrate are clearly separated from the substrate by an amorphous medium, in some cases Ge nanoislands can grow at the interface epitaxially [17].

In most cases annealing of a Ge-SiO_2 composite results in the formation of Ge nanocrystals. However, in the case of a higher Si concentration in the composite, GeSi nanocrystals can also be produced using this technique [6]. In this case it was found that the size of the nanocrystals, typically in the 10-nm range, was gradually decreasing with the increasing distance from the Si(100) interface.

4. NANOCRYSTAL STRUCTURE

The TEM results described above provide direct evidence on the formation of the nanocrystals. They can also in certain cases provide information on the structure of the nanocrystals. Alternative, or rather complementary, techniques for the investigation of the average and local structure of the nanocrystals are electron and X-ray diffraction, Raman scattering, and X-ray absorption fine structure and X-ray photoemission spectroscopy (XPS). The latter technique provides mainly the information on the oxidation state of Ge species [4, 7].

It should be noted that diffraction techniques rely on the long-range order in the nanocrystals and, as a result, measure the average structure of the nanocrystals; the same is true for the TEM measurements. Raman scattering and especially X-ray absorption fine structure are sensitive to local bonding and provide information on the local structure. The latter may be different from the average structure, as will be seen below. In this section we review the results obtained by different techniques and discuss the merits and disadvantages of each approach.

4.1. Average Structure

In most embedded Ge nanocrystal samples the interplane distances measured from HR-TEM images have led the authors to a conclusion that the nanostructures possess diamond structure [4, 17, 40]. In [2, 4] it was mentioned that the structure was size-dependent. While in the nanocrystals with a size >5 nm the (111) interplane spacing of 0.326 nm was observed which corresponds to the diamond structure, in the nanocrystals whose diameter was $\leq$4 nm the spacing was 0.298 nm which could not be derived from the diamond structure. It has been suggested [41] that small nanocrystals have tetragonal structure with the spacing of 0.292 nm. This value agrees with the interplane spacing observed in small embedded crystals and the authors of [2] suggest that smaller nanocrystals possess tetragonal structure.

For the nanocrystals embedded in AlN the observed lattice fringes correspond to (111) planes of the diamond structure (0.326 nm) [20]. A similar conclusion was drawn for the Ge nanocrystals in a Si host matrix [19] even though the size of the nanocrystals was rather small ($\geq$2 nm).

X-ray diffraction studies on Ge nanocrystallites prepared by plasma-enhanced CVD revealed the presence of the crystalline plane spacing of 0.345 nm in the crystals with the diameter of 5 nm. This value is different from the diamond structure and the authors concluded that a new Ge crystalline phase was formed.

For the Ge nanocrystals grown from a cluster source [31], it has been argued, based on Raman scattering, XRD, TEM, and transmission electron diffraction (TED), that the nanocrystals possess the ST-12 structure [42, 43]. This high-pressure form possesses a direct gap of about 1.5 eV [44, 45]. The size of the nanocrystals in this case was 3–4 nm so that this result is in agreement with the previous suggestions.

On the other hand, the colloidal Ge nanocrystals retain the bulk-like crystalline structure down to the size of 2 nm [30].

What may be the reasons for the nanocrystal-size dependence of the crystal structure? A possible and likely explanation would be an increased surface tension for small-size crystals. It is known that a high-pressure form of a material can be produced by reducing the size [46]. The SP-12 structure is a high-pressure (tetragonal) form of Ge which is metastable and is transformed into the diamond structure upon annealing in the 200–300 °C range [41, 44]. Theoretical simulations also confirmed that a transition from tetragonal to diamond structure should occur on Ge clusters when the number of atoms in the cluster reaches 40 and more, that is, for the size exceeding 1–1.5 nm [38]. The host matrix and the preparation method may (and do) influence the interface energy and hence the crystal structure. For example, it has been shown [38] that chemical preparation methods should always yield diamond structures.

To summarize the known data, we can say that the nanocrystals with the size of 5 nm and larger possess the diamond structure while smaller nanocrystals may have the tetragonal structure. It was suggested that by tuning the surface properties, the pressure exerted on the dot core may be varied and used to tailor the atomic structure [38].

4.2. Local Structure

4.2.1. Raman Scattering: Merits and Pitfalls

Raman scattering is a technique used to characterize the Ge nanostructures most often. While it is a very powerful technique, in the particular case of Ge nanostructures and Si substrate it has severe restrictions, often not taken into account. Before discussing the positive knowledge provided by the Raman scattering studies, we shall discuss the pitfalls of this technique.

Ge possesses a Raman peak around 300 cm^{-1}. The exact peak position, its line shape, and intensity are often used to discuss the degree of phonon confinement, stress, composition, and size of the nanostructure as well as the amount of Ge-Ge bonds [47]. It should be kept in mind that comparison is usually made with the Raman peak position of bulk Ge. The latter has been measured in various samples under various conditions, for example, temperatures, and the reported values vary from 298.5 cm^{-1} (calculated from Eq. (3) of [47]) to 300.5 cm^{-1} [48] to 305 cm^{-1} [49]. If this fact is neglected, one can draw quite different conclusions depending on the choice of reference.

The most serious problem arises from a fact which is often neglected, viz., a Raman spectrum of pure Si also exhibits a peak located at almost exactly the same position, the so-called two-phonon acoustic peak. The latter has been known since the 1970s and has been well documented [50, 51]. It has a characteristic shape with three features located at 225 cm^{-1}, 301 cm^{-1}, and 435 cm^{-1}. This peak intensity may be quite strong and is often mistaken for the Ge peak. A detailed summary of this has been given in [52–55].

Another problem is a similarity in positions of the 435 cm^{-1} peak of Si with the peak due to Ge-Si vibrations and the so-called local Si-Si vibrations. Their misinterpretation can result in significant mistakes in determining the composition of the nanostructures from the peak positions.

Finally, it should be kept in mind that in the case of nanostructures the peak intensity is not proportional to the number of the corresponding bonds. Thus annealing or capping of self-assembled Ge/Si(100) quantum dots results in simultaneous decrease of the Ge-Ge and Ge-Si peak intensities [56, 57]. This fact has been interpreted in [57] as intermixing of Ge with Si, but this should have caused an increase in the Ge-Si peak intensity, which is not what has been observed experimentally.

Possible ways to distinguish between the true Ge Raman peak and the two-phonon acoustic Si peak include use of resonant Raman scattering [12, 54], use of polarized Raman scattering [54], and measurements under pressure [58]. The former two techniques allow to suppress the relative contribution from the Si peak, while the latter one makes the two peaks shift in the opposite directions. Use of pure Ge isotopes is another possible technique [56], although the isotopes may not be easily available.

Now that the reader has been provided with crucial information on Raman scattering in the particular case of Ge nanostructure on Si substrates and can make his own judgment of the original papers, we shall proceed to discuss some of the positive results obtained through Raman scattering studies.

In cases of small-size nanocrystals, the observed Raman peaks are asymmetrically broadened [12] over the low-frequency range which is believed to be due to confinement of phonons [59] since the confinement results in the breakdown of the wave vector selection rules. Therefore, phonons at $q \neq 0$ also contribute to the Raman spectrum, resulting in the shift of the peak position from its bulk value and asymmetric broadening [60].

The size of the nanocrystals, determined from the line-shape fitting [60] using the approach developed in [61], agrees quite well with the size measured by TEM. The observed disagreement between theoretical and experimentally observed peak positions may be due to stress, intermixing, or inhomogeneous Ge distribution. Each of these reasons may be dominant in a particular case. Small nanocrystals whose presence can be detected by HR-TEM may exhibit completely amorphous-like Raman spectra [62].

In addition to the size-dependent line shape and peak position, the Raman scattering efficiency was also demonstrated to change (increase) considerably for the nanocrystals with smaller sizes [12, 63].

Use of polarized Raman scattering demonstrated that (in samples containing 60 mol.% Ge in SiO_2) any Ge-Si alloying is limited to the interface region and takes place on the Si substrate surface rather than in the nanocrystals. On the other hand, in samples containing excess Si (compared to SiO_2), the process is more complicated. The Ge phase has been found to be formed first, followed by the formation of GeSi outer shells [6]. This result may be explained by the fact that Ge oxide is thermodynamically less stable than SiO_2 [64].

While in most cases Raman scattering has been successfully used for characterization of the Ge nanocrystals embedded in SiO_2 [4, 5, 7, 12, 40], it has also been used to detect the Ge phase in zeolites [24, 26] and for the nanocrystals prepared by evaporation using a cluster source. In the latter case it was concluded [42] that the nanocrystals possessed ST-12 structure.

4.2.2. X-Ray Absorption Fine Structure

X-ray absorption coefficient of a solid above the absorption edge exhibits oscillations that are caused by back-scattering of the created photoelectron wave on the nearest neighbors. These oscillations are called extended X-ray absorption fine structure (EXAFS). Analysis of EXAFS spectra allows one to obtain information on the coordination numbers, distances to the nearest neighbors, and chemical nature of the neighbors, as well as a mean-square relative displacement, both in crystalline and amorphous phases.

The absorption edge itself is also strongly modified by the chemical surrounding and can be a source of additional information. Recent progress in theoretical simulation of XAFS makes it possible to simulate both the edge structure and EXAFS oscillations with high accuracy.

Use of XAFS has allowed one to measure the Ge-Ge bond length (usually with the accuracy of ± 0.005 Å). For example, in embedded 20-nm nanocrystals the bond length was found to be the same as in bulk Ge; that is, the nanocrystals are unstrained [37]. (In self-assembled Ge/Si(100) quantum dots the Ge nanostructures exhibit clear bond shortening [65].) The strain information obtained from EXAFS combined with the Raman peak position allows the best-to-date determination of the degree of the phonon confinement [6].

In zeolite-confined Ge clusters Ge-Ge bond length was found to vary from 2.41 to 2.55 Å depending on the CVD precursor and the host matrix [22].

Of interest is combined use of EXAFS and diffraction anomalous fine structure (DAFS). The latter technique, which exhibits similar oscillations in the Bragg peak intensity as a function of X-ray energy, is based on the long-range order, that is, only probes the nanocrystalline phase. EXAFS provides information on both phases. Combined use of both techniques allows one to investigate both phases of the same material independently; that is, the details of the crystallization process can be studied on an atomic scale [66].

5. PROPERTIES

5.1. Optical Properties

5.1.1. Electronic Band Structure

As already mentioned in the Introduction, one of the major reasons pushing the interest in Ge nanocrystals is a conjecture that they should possess direct bandgap structure and, as a result, can be a basis for fabrication of efficient light emitters perfectly compatible with Si, technology. This conjecture was first made in [67]. Since the effective Bohr radius for Wannier excitons in Ge (24 nm) is larger than in Si, it is likely to be easier to modify the electronic structure of Ge than that of Si [4].

Optical properties of Ge nanocrystals have been studied both experimentally and theoretically [29, 35, 68, 69]. Optical absorption of embedded and colloidal nanocrystals has been studied in detail in [13] and [30], respectively, and it was suggested [13] that scattering makes a significant contribution to the extinction spectra damping the spectral features.

In [35] optical nonlinearity of Ge nanocrystals embedded in silicon oxide thin films using Z scan and pump-and-probe technique with 532-nm ps laser pulses. The nonlinear refractive index and absorption coefficient are Ge-concentration-dependent and range between 0.0026 and 0.0082 cm^2/GW and 190 and 760 cm/GW, respectively, when the Ge atomic fraction increases from 8.5% to 27.6%. It was concluded that the nonlinearity mainly comes from excited carriers and two-photon absorption makes a small contribution. The excited carrier lifetime was determined to be $\sim$70 ps and independent of the laser intensity up to 19.2 GW/cm^2.

Extinction spectroscopy of the colloidal Ge nanocrystals [29] over the 0.5- to 5-eV energy range revealed a number of features attributable to transitions both near and away from $k = 0$. Strong size effects were observed for the direct transitions near $k = 0$, while those away from $k = 0$ seem to be unaffected by confinement. The authors interpret these results within the context of bulk Ge electronic structure.

In [69] *ab initio* calculations were performed that provided evidence for a direct gap in the center of a two-dimensional Brillouin zone depending on the particular crystallographic orientation of the Ge structures. Strong optical absorption and emission of photons without phonon coupling is theoretically predicted for [100]-oriented Ge quantum films. This result is associated with the presence of a direct gap band structure, with a large mixing of the bulk Γ derived states into the conduction band minimum of the films, and with the particular localization of the band edges wave functions in the real space. The [110]- and [111]-oriented Ge films seem to not possess such a pronounced optical activity in spite of the fact that the bandgap in the [111]-oriented films is also direct. The authors concluded that the [100]-oriented Ge quantum films could be a good candidate for applications in silicon-compatible light-emitting devices.

Electronic structure of Ge nanocrystals has been studied in [68] using sp^3 tight-binding description. The authors claim that the blue-green photoluminescence of Ge nanocrystals comes from defects in the oxide and that the size-dependent photoluminescence in the near-infrared probably involves a deep trap in the gap of the nanocrystals. They also predicted that radiative lifetimes remain long in spite of the small difference, $\sim$0.14 eV, between direct and indirect gaps of bulk Ge.

The authors of [70] have used a tight-binding approximation to calculate the electronic states in silicon nanostructures. They have found that amorphous Si nanostructures (rather than those with special symmetries) are most promising for light-emitting devices. The same may or may not be true for Ge.

5.1.2. Photoluminescence

While intense photoluminescence (PL) in the visible range from Ge nanocrystals has been reported by numerous groups and is an established fact, its mechanism remains a subject of severe debates. Before addressing this controversial issue, we shall summarize the established experimental results.

- Ge nanocrystals exhibit strong PL in the visible range, the most often observed peak position being slightly above 2 eV [4, 5, 16, 30, 59, 71–77]. What "strong" PL means is a matter of the authors' perception. Regrettably, the PL intensities in the original papers are given in arbitrary units. For this reason it is difficult to compare these results with those for light-emitting III-V structures.
- The peak position does not exhibit strong size-dependence [2, 4, 8, 9, 14, 15, 59, 74, 75] (with the exception of [78] where it does).
- Strong size-dependence has been observed [78] for near-IR PL. These authors have also observed increased PL efficiency for smaller nanocrystals.
- In some cases (when the excitation energy is high enough), two peaks have been detected [9, 71, 76].

The dependence of the peak position on the nanocrystal size is summarized in Figure 1. One can see that there

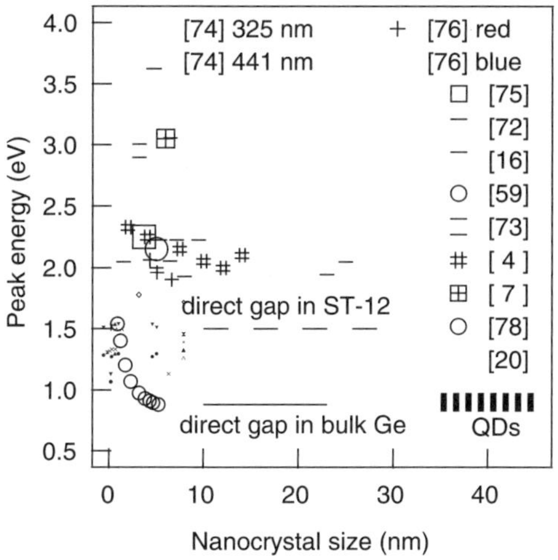

Figure 1. Nanocrystal size dependence of the PL peak position for Ge nanocrystals embedded in SiO_2 (summary of the literature data; sources are indicated in the figure). For comparison, PL peak positions for Ge nanocrystals embedded in AlN and for self-assembled Ge/Si quantum dots are also shown. The lines indicate the direct optical gap in the bulk diamond Ge (dashed line) and in the high-pressure ST-12 structure.

the experimental data peak around three different energies, namely, ~1 eV, ~2 eV, and ~3 eV. Most often the 2-eV peak is reported and we shall start the discussion with this.

Regarding the mechanism of the photoluminescence (in this paragraph we discuss the peak located above 2 eV), the opinions are strongly polarized. Thus in [4, 73, 76, 77] the authors argue, based on the observed weak size-dependence of the PL peak, that the PL is due to quantum confinement. PL intensity-dependence on the nanocrystal size is another argument [4, 73]. The observed size-dependence of the peak position is much weaker than predicted theoretically [67]. The supporters of the confinement approach explain this in terms of shortening of the Ge-Ge bond length in the nanocrystals which modifies the electronic structure of Ge [76] or by presence of fast nonradiative relaxation channels in nanocrystalline Ge taking part at the surface [4].

The opponents of the quantum confinement model argue that the size-dependence of the peak position is negligible [71, 72, 74, 79] and that the PL is due to emission from some centers in SiO_x layer surrounding the nanocrystals [5, 72], in the matrix itself [74], or is associated with Ge-related defects at the interfaces [71]. A similar conclusion has also been drawn from tight-binding calculations [69].

In the case of AlN host, the PL peak was observed at ~1.0 eV [20]. We would like to mention here that PL from self-assembled Ge/Si quantum dots (which can be considered as Ge nanocrystals epitaxially embedded in Si) is also characterized by a peak located around 0.8 eV, the exact peak position being a function of Ge coverage (see, e.g., Fig. 19 of [33]). The dependence of the visible PL peak position on the host material is yet another indication that the interface states play a very important role in the process.

Of special interest is the PL reported in [78, 80]. These are the only reports where strong size-dependence of both the peak position and the intensity has been detected. The observed peak was located in the near-IR and did change its position from 0.88 eV to 1.55 eV as the nanocrystal size decreased from 5.3 to 0.9 nm. (It should be noticed that the authors were not able to directly detect the nanocrystals with the size less than 2.3 nm and the size of the smaller nanocrystals was estimated from the volume fraction of Ge. This is a disappointing fact because the strongest size-dependence was reported exactly for the range where the nanocrystals could not be observed by HR-TEM.)

It is worth noting that the observed peak position for the largest nanocrystals agrees very well with the value of direct gap in bulk Ge (0.88 eV) [81], which makes it reasonable that the observed PL is, indeed, due to the direct optical transition in small nanocrystals. It should be noted that despite the size-dependence being much stronger in this case than in the majority of other reports, it is still not as strong as theoretically predicted. Thus the prediction requires the peak position of 2.3 eV for the 4-nm Ge nanocrystals [67, 75], while the experimentally observed value is ~0.9 eV.

Finally, there are reports of a peak located at ~3 eV. Most authors seem to agree that its origin is related to defects, either Ge/O related defects [74] or implantation-induced defects [76].

As follows from the above discussion, the situation with the PL mechanism is very complex and it is impossible to draw a definite conclusion about the PL mechanism at this stage. In [38] it was argued that in understanding the properties of the small semiconductor dots, both quantum confinement and surface effects have to be taken into account.

It seems likely that PL around 1 eV is indeed mainly due to the quantum confinement because the peak value agrees with the direct gap of bulk Ge and the peak position does not depend on the host matrix. The peaks at ~2 eV and ~3 eV are likely to be due to the SiO_2 matrix. It may be possible, however, that the ~2-eV peak is also a manifestation of the confinement in the case when the structure of the Ge of small nanocrystals embedded in SiO_2 changes from diamond structure to ST-12 structure.

The 3-eV feature is likely to be caused by defects. Further studies are needed in order to elucidate the true mechanism of photoluminescence in Ge nanocrystals.

5.2. Electrical Properties

Studies of electric properties of Ge nanocrystals are quite limited. Temperature-dependent current–voltage (I–V) characteristics have been studied across the thickness of Ge nanocrystalline films prepared by the cluster beam evaporation technique [82, 83]. It was found that in a film with a thickness of 30 nm, deposited on a substrate kept at 77 K, step-like features in the I–V characteristics were observed at low temperatures which are characteristic of single-electron transport.

6. ADVANCED APPLICATIONS

The present state of research of the Ge nanocrystals is still at a stage when basic properties of the nanocrystals are studied and the technological conditions optimized. One can only discuss potential advanced applications of the Ge nanocrystals which may be the following.

- Light emitters (based on numerous PL studies).
- Single-electron transistors. Since the pioneering paper [84], a great deal of interest has been generated around the detection of single-electron transport. Recent reports of step-like features in I–V curves of Ge nanocluster samples [82, 83] make the latter a possible candidate for the realization of the single-electron transistor ideas.
- Memories. A memory effect was observed in Ge nanocrystals embedded in an amorphous SiO_2 matrix of a metal-insulator-semiconductor structure [85].
- Photonic structures. Creation of high-quality three-dimensional periodic structures inside zeolites and opals [86] allows for fabrication of Ge-based photonic crystals.
- Nonlinear optical media [35], and photosensitive materials [87].
- Quantum computers [88].

GLOSSARY

Direct-gap structure Electronic structure of bands in a semiconductor when the minimum of the conduction band is located above the maximum of the valence band, i.e., at the same wavevector. Optical absorption and emission due to direct transitions possess much higher efficiency than those due to indirect ones.

Inverse opal The structure resulting after incorporation of foreign species into the pores of an opal and subsequent etching of the host matrix.

Local structure Arrangement of atoms on a scale of a few Ångstroms; includes the coordination number, bond lengths and bond angles, and chemical nature of the first-nearest neighbor species.

Nanocrystals Crystals with a characteristic size in the 1- to 100-nm range.

Quantum confinement effect Manifestation of the small size of nanoparticles. When a charge carrier is confined in a small volume, its energy levels shift with respect to those in the bulk material, hence, size dependence of the photoluminescence peak.

ST-12 structure A high-pressure form of germanium. It is believed that Ge nanocrystals of very small size possess this form as opposed to the diamond structure of bulk germanium.

ACKNOWLEDGMENTS

The original experiments of the author referred to in this article have been performed at Joint Research Center for Atom Technology (JRCAT). The author is very grateful to K. Tanaka, H. Oyanagi, Y. Maeda, A. Frenkel, and S. Wei for their valuable contribution to the original experiments.

REFERENCES

1. S. Hayashi, M. Fujii, and K. Yamamoto, *Jpn. J. Appl. Phys., Part 2* 28, L1464 (1989).
2. Y. Kanemitsu, H. Uto, Y. Matsumoto, and Y. Maeda, *Appl. Phys. Lett.* 61, 2187 (1992).
3. Y. Maeda, N. Tsukamoto, Y. Yazawa, Y. Kanemitsu, and Y. Matsumoto, *Appl. Phys. Lett.* 59, 3168 (1992).
4. Y. Maeda, *Phys. Rev. B* 51, 1658 (1995).
5. M. Zacharias, J. Blasing, J. Christen, P. Veit, B. Dietrich, and D. Bimberg, *Superlattices and Microstructures* 18, 139 (1995).
6. A. Kolobov, H. Oyanagi, N. Usami, S. Tokumitsu, T. Hattori, S. Yamasaki, K. Tanaka, S. Ohtake, and Y. Shiraki, *Appl. Phys. Lett.* 80, 488 (2002).
7. W. K. Choi, Y. W. Ho, S. P. Ng, and V. Ng, *J. Appl. Phys.* 89, 2168 (2001).
8. M. Nogami and Y. Abe, *Appl. Phys. Lett.* 65, 2545 (1994).
9. A. Saito and T. Suemoto, *Phys. Rev. B* 56, R1688 (1997).
10. X. L. Wu, T. Gao, X. M. Bao, F. Yan, S. S. Jiang, and D. Feng, *J. Appl. Phys.* 82, 2704 (1997).
11. J. G. Zhu, C. W. White, J. D. Budai, S. P. Withrow, and Y. Chen, *J. Appl. Phys.* 78, 4386 (1995).
12. K. L. Teo, S. H. Kwok, P. Y. Yu, and S. Guha, *Phys. Rev. B* 62, 1584 (2000).
13. A. Dowd, R. G. Ellimann, and B. Luther-Davies, *Appl. Phys. Lett.* 79, 2327 (2001).
14. D. C. Paine, C. Caragianis, T. Y. Kim, Y. Shigesato, and T. Ishahara, *Appl. Phys. Lett.* 62 (1993).
15. A. K. Dutta, *Appl. Phys. Lett.* 68, 1189 (1996).
16. A. Shklyaev and M. Ichikawa, *Appl. Phys. Lett.* 80, 1432 (2002).
17. V. Craciun, W. Boyd, A. H. Reader, and D. E. W. Vanderhoudt, *Appl. Phys. Lett.* 65, 3233 (1994).
18. T. Kobayashi, T. Ehdoh, H. Fukuda, S. Nomura, A. Sakai, and Y. Ueda, *Appl. Phys. Lett.* 71, 1195 (1997).
19. S. T. Ngiam, K. F. Jensen, and K. D. Kolenbrander, *J. Appl. Phys.* 76, 8201 (1994).
20. K. M. Hassan, A. K. Sharma, J. Narayan, J. F. Muth, C. W. Teng, and R. M. Kolbas, *Appl. Phys. Lett.* 75, 1222 (1999).
21. C. Shubert, U. Kaiser, A. Hedler, W. Wesch, T. Gorelik, U. Glatzel, J. Krausslich, B. Wunderlich, G. Hess, and K. Goetz, *J. Appl. Phys.* 91, 1520 (2002).
22. O. Dag, A. Kuperman, and G. A. Ozin, *Adv. Mater.* 6, 147 (1994).
23. A. Zunger, S. Wagner, and P. M. Petroff, *J. Electronic Mater.* 22, 3 (1993).
24. A. Aparisi, V. Fornes, F. Marquez, R. Moreno, C. Lopez, and F. Meseguer, *Solid State Electronics* 40 (1996).
25. C. L. Bowes, A. Malek, and G. A. Ozin, *Chem. Vapor. Depos.* 2, 97 (1996).
26. H. Miguez, V. Fornes, F. Meseguer, F. Marquez, and C. Lopez, *Appl. Phys. Lett.* 69, 2347 (1996).
27. H. Miguez, E. Chomski, F. Garcia-Santamaria, M. Ibisate, S. John, C. Lopez, F. Meseguer, J. P. Mondia, G. A. Ozin, O. Toader, and H. M. van Driel, *Adv. Mater.* 13, 1634 (2001).
28. S. John and K. Busch, *J. Lightwave Tech.* 17, 1931 (1999).
29. J. R. Heath, J. J. Shiang, and A. P. Alivisatos, *J. Chem. Phys.* 101, 1607 (1994).
30. J. P. Wilcoxon, P. P. Provencio, and G. A. Samara, *Phys. Rev. B* 64, 035417 (2001).
31. S. Sato, S. Nozaki, H. Morisaki, and M. Iwase, *Appl. Phys. Lett.* 66, 3176 (1995).
32. C. Bonafos, B. Garido, M. Lopez, A. Perez-Rodriguez, J. R. Morante, Y. Kihn, G. Ben Assayag, and A. Claverie, *Appl. Phys. Lett.* 76, 3962 (2000).
33. K. Brunner, *Rep. Prog. Phys.* 65, 27 (2002).
34. M. Zacharias, R. Weigand, B. Dietrich, F. Stoltze, J. Blasing, P. Veit, T. Drusedau, and J. Christen, *J. Appl. Phys.* 81, 2384 (1997).
35. Y. E. Jie, Y. N. Xiong, A. T. S. Wee, C. H. A. Huan, and W. Ji, *Appl. Phys. Lett.* 77, 3926 (2000).
36. J. Jiang, K. Chen, X. Huang, Z. Li, and D. Feng, *Appl. Phys. Lett.* 65, 1799 (1994).
37. A. Kolobov, S. Wei, W. S. Yan, H. Oyanagi, Y. Maeda, and K. Tanaka, *Phys. Rev. B*, submitted (2002).
38. L. Pizzagalli, G. Gali, J. E. Klepeis, and F. Gygi, *Phys. Rev. B* 63, 165324 (2001).
39. K. H. Heining, B. Schmidt, A. Markwitz, R. Grotzschel, M. Strobel, and S. Oswald, *Nucl. Instrum. Methods. Phys. Res. B* 148, 969 (1999).
40. M. Fujii, S. Hayashi, and K. Yamamoto, *Appl. Phys. Lett.* 57, 2692 (1990).
41. Y. Saito, *J. Cryst. Growth* 47, 61 (1979).
42. S. Nozaki, S. Sato, H. Ono, H. Morisaki, and M. Iwase, *Mater. Res. Soc. Symp. Proc.* 405, 223 (1996).
43. S. Sato, S. Nozaki, and H. Morisaki, *Appl. Phys. Lett.* 77, 2460 (1998).
44. F. P. Bundy and J. S. Kasper, *Science* 139, 340 (1963).
45. J. D. Joannopoulos and M. L. Cohen, *Phys. Rev. B* 7, 2644 (1973).
46. T. P. Martin, T. Bergmann, H. Gohlich, and T. Lange, *J. Chem. Phys.* 95, 6421 (1991).
47. Z. F. Krasil'nik, P. Lytvyn, D. N. Lobanov, N. Mestres, A. V. Novikov, J. Pascual, M. Y. Valakh, and V. A. Yakhumchuk, *Nanotechnology* 13, 81 (2002).
48. J. H. Parker, D. W. Feldman, and M. Ashkin, *Phys. Rev.* 155, 712 (1967).
49. J. M. Zhang, M. Gieler, A. Goebel, M. Cardona, E. E. Haller, and K. Itoh, *Phys. Rev. B* 57, 1348 (1998).
50. P. A. Temple and C. E. Hathaway, *Phys. Rev. B* 7, 3685 (1973).
51. K. Uchinokura, T. Sekine, and E. Matsuura, *J. Phys. Chem. Solids* 35, 171 (1974).
52. A. Kolobov and K. Tanaka, *Appl. Phys. Lett.* 75, 3572 (1999).
53. A. V. Kolobov, *J. Vac. Sci. Tech. A* 20, 1116 (2002).

54. A. V. Kolobov, *J. Appl. Phys.* 87, 2926 (2000).
55. A. Kolobov and K. Tanaka, *Appl. Phys. Lett.* 78, 3550 (2001).
56. A. Kolobov, K. Morita, K. M. Itoh, and E. E. Haller, *Appl. Phys. Lett.* 81, 3855 (2002).
57. J. Wan, Y. H. Luo, Z. M. Jiang, G. Jin, K. L. Wang, X. Z. Liao, and J. Zou, *J. Appl. Phys.* 90, 4290 (2001).
58. K. L. Teo, L. Qin, Z. X. Shen, and O. G. Schmidt, *Appl. Phys. Lett.* 80, 2919 (2002).
59. V. Craciun, C. Boulmer-Leborgne, E. J. Nicolls, and I. W. Boyd, *Appl. Phys. Lett.* 69, 1506 (1996).
60. W. K. Choi, V. Ng, S. P. Ng, H. H. Thio, Z. X. Shen, and W. S. Li, *J. Appl. Phys.* 86, 1398 (1999).
61. P. M. Fauchet and I. H. Campbell, *Crit. Rev. Solid State Mater. Sci.* 14, S79 (1988).
62. S. Hayashi and H. Abe, *Jpn. J. Appl. Phys.* 23, L824 (1984).
63. D. V. Murphy and S. R. J. Brueck, *Optics Lett.* 8, 494 (1983).
64. F. K. LeGoues, R. Rosenberg, T. Nguyen, F. Himpsel, and B. S. Meyerson, *J. Appl. Phys.* 65, 1724 (1989).
65. A. V. Kolobov, H. Oyanagi, S. Wei, K. Brunner, G. Abstreiter, and K. Tanaka, *Phys. Rev. B* 66, 075319 (2002).
66. A. Kolobov, H. Oyanagi, A. Frenkel, J. Robinson, J. Cross, S. Wei, K. Brunner, G. Abstreiter, A. Shklyaev, M. Ichikawa, Y. Maeda, S. Yamasaki, and K. Tanaka, "Proceedings of the 3rd International Conference on Synchrotron Radiation in Materials Science," 2002, in print.
67. T. Takagahara and K. Takeda, *Phys. Rev. B* 46, 15578 (1992).
68. H. C. Weissker, J. Furthmuller, and F. Bechstedt, *Phys. Rev. B* 65, 155327 (2002).
69. A. N. Kholod, S. Ossicini, V. E. Borisenko, and F. A. d'Avitaya, *Phys. Rev. B* 65, 115315 (2002).
70. F. Yonezawa, K. Nishio, J. Koga, and T. Yamaguchi, *J. Optoelectronics Adv. Mater.* 4, 569 (2002).
71. X. L. Wu, Y. Gu, G. G. Siu, E. Fu, N. Tang, T. Gao, and X. M. Bao, *J. Appl. Phys.* 86, 707 (1999).
72. S. Y. Ma, Z. C. Ma, W. H. Zong, H. X. Han, Z. P. Wang, G. H. Li, G. Qin, and G. G. Qin, *J. Appl. Phys.* 84, 559 (1998).
73. L. Yue and Y. He, *J. Appl. Phys.* 81, 2910 (1997).
74. M. Zacharias and P. M. Fauchet, *Appl. Phys. Lett.* 71, 380 (1997).
75. S. Okamoto and Y. Kanemitsu, *Phys. Rev. B* 54, 16421 (1996).
76. J. Y. Zhang, X. M. Bao, Y. H. Ye, and X. L. Tan, *Appl. Phys. Lett.* 73, 1790 (1998).
77. F. G. Liu, Z. G. Wang, G. H. Li, and G. H. Wang, *J. Appl. Phys.* 83, 3435 (1998).
78. S. Takeoka, M. Fujii, S. Hayashi, and K. Yamamoto, *Phys. Rev. B* 58, 7921 (1998).
79. Y. M. Niquet, G. Allan, C. Delerue, and M. Lannoo, *Appl. Phys. Lett.* 77, 1182 (2000).
80. S. Takeoka, M. Fujii, S. Hayashi, and K. Yamamoto, *Appl. Phys. Lett.* 74, 1558 (1999).
81. J. C. Phillips, "Bonds and Bands in Semiconductors." Academic, New York, 1973.
82. S. Banerjee, S. Nozaki, and H. Morisaki, *Appl. Phys. Lett.* 76, 445 (2000).
83. S. Banerjee, S. Nozaki, and H. Morisaki, *J. Appl. Phys.* 91, 4307 (2002).
84. D. V. Averin and K. K. Likharev, *J. Low Temp. Phys.* 62, 345 (1986).
85. W. K. Choi, W. K. Chim, C. L. Heng, L. W. Teo, V. Ho, V. Ng, D. A. Antoniadis, and E. A. Fitzgerald, *Appl. Phys. Lett.* 80, 2014 (2002).
86. F. Meseguer, A. Blanco, H. Miguez, F. Garcia-Santamaria, M. Ibisate, and C. Lopez, *Colloids and Surfaces* 202, 281 (2002).
87. J. Nishii, K. Kintaka, H. Hosono, H. Kawazoe, M. Kato, and K. Muta, *Phys. Rev. B* 60, 7166 (1999).
88. I. Shlimak, I. Vagner, and V. I. Safarov, "Proceedings of the 25th International Conference on Physics of Semiconductors," 2000, p. 1717.

ENN Encyclopedia of Nanoscience and Nanotechnology
www.aspbs.com/enn

Germanium-on-Silicon Infrared Detectors

G. Masini, L. Colace, G. Assanto
University Roma Tre, Roma, Italy

CONTENTS

1. INTRODUCTION

Infrared (IR), that is, light of wavelengths from 700 nm to a few hundred micrometers, is of paramount relevance in several applications ranging from communications to defense, and airborne and environmental sensing. IR is commonly subdivided into three roughly defined regions: the near-infrared (NIR) or short-wavelength infrared (SWIR) from 0.7 to 2 μm, the mid-infrared (MIR) or mid-wavelength infrared (MWIR) from 2 to 10 μm, and the far-infrared or long-wavelength infrared (LWIR) beyond 10 μm. With regards to the NIR, despite the recent contraction of the related market, optical fiber communications remains the dominant application area, with spectral "windows" located around 900 nm (first), 1300 nm (second), and 1550 nm (third), where silica fibers exhibit minima in propagation losses and/or (chromatic) dispersion. For this reason long-haul systems operate in the second and/or third windows [1], while the first is employed in short-reach and low-cost systems employing plastic fibers, GaAs emitters, and silicon photodiodes [2]. The near- and mid-IR are the best option for remote sensing and imaging owing to a reduced Raleigh scattering (compared to the visible) and to atmospheric transparency around 1.6 μm, from 3 to 5 μm and from 8 to 14 μm, respectively. The latter is of great relevance for thermal imaging, since the blackbody radiation of an object at 300 K peaks at about 10 μm. Airborne or satellite-acquired SWIR maps are commonly used (often in conjunction with visible ones) to identify the water content in the flora and, thereby, assess the risk of fire occurrence in regions subject to dry and/or hot weather [3]. In the field of art restoration, near-infrared imaging can bring to evidence the preparatory drawing lying under a painting [4]. All of these applications, and many others in medicine, biology, spectroscopy, etc., rely on a sensitive, sometime fast, detection of IR light. The technology, the performance, and the use of infrared detectors based on a specific material system, namely germanium on silicon, are the subject of this overview.

This chapter is organized as follows: in Section 1, after the Introduction, we review the basic concepts of detection of near- and mid-infrared light and the relevant parameters characterizing sensors at these wavelengths. Section 2, examines the techniques used for the growth of SiGe and pure Ge on silicon. Finally, Sections 3 and 4 summarize the major achievements with NIR and MIR photodetectors and arrays based on Ge and SiGe-on-Si technologies.

1.1. Near-Infrared Detectors

Detectors for fiber optic communications must be very fast (speeds as high as 10 Gb/s are common today, and 40-Gb/s systems are entering the arena [5]), sensitive, and efficiently coupled to the fiber tip. Wavelength division multiplexing (WDM) efficiently exploits the enormous transmission bandwidth offered by silica fibers. In such systems, signals carried at different "colors" (i.e., wavelengths) travel in the same waveguide and are demultiplexed at the receiver before detection. Wavelength demultiplexers, usually arrayed waveguide gratings (AWG) [6], provide their output via a set of closely spaced waveguides to be matched to an array of detectors. The possibility to fabricate and integrate NIR detector arrays with AWG can therefore improve the compactness of WDM systems. As will be clear in the following, this is one of the key advantages of the Ge-on-Si technology for NIR photodetection. Furthermore, as mentioned above, near-infrared light around 1.55 μm is

ISBN: 1-58883-059-4/$35.00

Encyclopedia of Nanoscience and Nanotechnology
Edited by H. S. Nalwa
Volume 3: Pages (829–842)

also used in free-space communications owing to the transparency of the atmosphere and the eye-safety characteristic of this spectral region [7].

Near-infrared detection is commonly based on the promotion of an electron from the valence to the conduction band of a semiconductor, through the absorption of a photon [8]. The unbound electron and the corresponding hole created in the valence band, in the presence of an electric field, generate a current flow through the contacts realized on the sample. A very common device implementing this concept is the *p-i-n* photodiode, sketched in Figure 1 [9]. Its electric characteristic in dark is almost identical to that of a simple *p-n* junction diode; that is, the dependence of current I on applied voltage V obeys the Shockley equation

$$I = I_0\left(e^{V/V_T} - 1\right) \tag{1}$$

with V_T the threshold voltage and I_0 the saturation (reverse) current. When light of appropriate wavelength shines on the device, the photocarriers add up to the thermally generated ones to originate an extra current term flowing in the same direction as the reverse dark current, that is,

$$I = I_0\left(e^{V/V_T} - 1\right) - I_L \tag{2}$$

Typical current–voltage characteristics are plotted in Figure 2.

Due to the planar technology used to fabricate electronic devices in Si [10], the light to be detected enters the device through one of the doped layers. The intrinsic layer, sandwiched between the two oppositely doped ones providing the built-in electric field, is responsible for the light absorption and the transport of the photogenerated carriers. Therefore, in the spectral range of operation, the *i*-semiconductor must completely absorb the light in a limited thickness; at the same time it must exhibit high electronic quality to grant a sufficient drift-length for the generated carriers to reach the contacts before recombining. Absorption is nonvanishing if the material bandgap energy (cut-off energy) is lower than the energy of the photons to be absorbed; moreover, moving away from the cutoff, it increases more steeply in direct (e.g., GaAs) than in indirect (e.g., Si, Ge) bandgap semiconductors. From this point of view, the former ones are preferable. In choosing the best semiconductor for a given wavelength range, another important consideration is that, as explained above, photogenerated and dark currents add to one another and both originate shot noise [11] with power

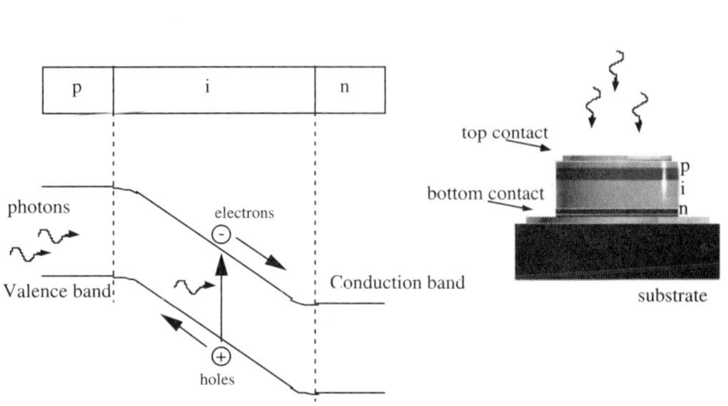

Figure 1. The *p-i-n* photodiode: band structure and sketch.

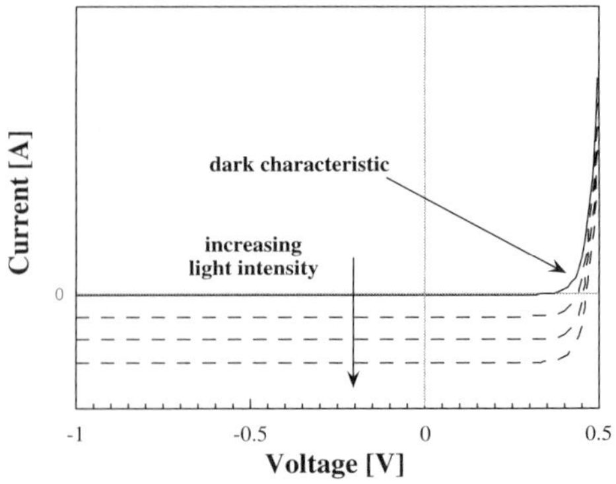

Figure 2. Typical current–voltage characteristics of a *p-i-n* photodiode in the dark and under illumination.

spectral density proportional to the current intensity. Therefore, if sensitivity is a concern, care has to be taken to minimize the dark contribution. The latter, depending on the thermal carrier generation rate, decreases exponentially with the bandgap energy, suggesting the selection of bandgap values not much lower than the photon energies. This is further advantageous in order to prevent a too large absorption in the doped layer, where a lower electric field and poorer transport properties (than in the *i*-layer) would facilitate carrier loss by recombination. In Figure 3 we show the bandgap energy of various semiconductor crystals and alloys versus lattice parameter. Based on the preceding discussion, from Figure 3 it is apparent that suitable semiconductors for NIR detection are Ge, InGaAs, and InGaSb alloys, the latter two favored by a direct bandgap. Due to the commercial availability of InP substrates which are lattice-matched to InGaAs, InGaAs detectors are the most employed for the NIR.

To characterize NIR detectors, the most relevant and commonly used figures are their spectral range of operation, the responsivity, the dark current or the noise equivalent power, and the speed. The responsivity R_λ is measured in A/W and defined as the photocurrent generated per unity of impinging power at wavelength λ onto the device. It relates to the external quantum efficiency η_{ext}, that is, the number of electrons flowing in the circuit for each incident photon

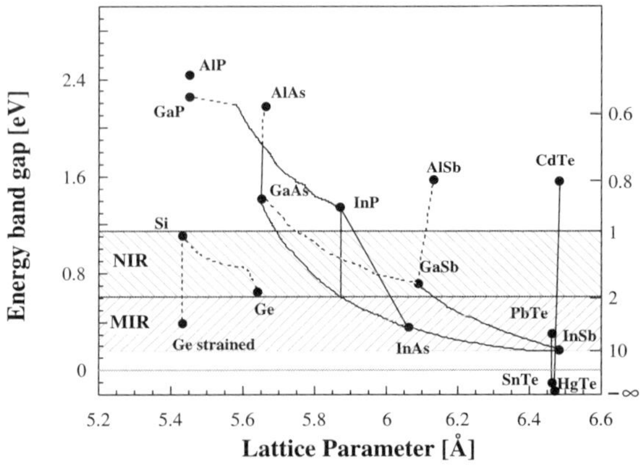

Figure 3. The bandgap of various semiconductors as a function of the lattice parameter.

of wavelength λ:

$$R_\lambda = \eta_{ext} q\lambda / hc \quad (3)$$

with q the electron charge, h the Planck constant, and c the speed of light in vacuum. The external quantum efficiency is limited by the Fresnel reflection Θ_r at the photodiode surface, by the finite thickness d of the absorbing layer, and by the losses R due to recombination:

$$\eta_{ext} = (1 - R)(1 - \Theta_r)(1 - e^{-\alpha d}) \quad (4)$$

where α is the absorption coefficient. Owing to the dependence (3), the spectral responsivity typically increases with wavelength, up to the cutoff corresponding to the bandgap. At $\lambda = 1.24$ μm, for a unity quantum efficiency, $R_\lambda = 1$ A/W.

The dark current, for the reasons outlined above, is orders of magnitude larger in InGaAs than in Si photodiodes. It is proportional to the device area for diameters larger than a few tens of microns, with densities J_d in the range 1–100 μA/cm^2. At smaller diameters, the leakage current (which scales with the perimeter rather than the area of the device) becomes dominant and the dark current density increases. Responsivity and dark current can be combined into the noise equivalent power (NEP) which, expressed in WHz$^{-1/2}$, is the optical power able to produce a current signal equal to the noise, for a unity bandwidth.

Finally, another important figure is the photodetector speed of response. The speed of a *p-i-n* photodiode is limited by two main mechanisms: the transit time and the *RC* time constant. The first is the time required by the photogenerated carriers to travel to the doped layers; it can be optimized by reducing the thickness of the intrinsic layer to the minimum required for complete absorption of the light, and by making the carriers move at the saturation velocity by increasing the electric field with an applied (reverse) bias. The second stems from the low pass characteristic of the load resistance (or feedback resistance when the detector is connected to a transimpedance preamplifier [12]) with the photodiode capacitance. Since the capacitance is proportional to the photodiode area, the speed can be increased by reducing the device size. InGaAs photodiodes for 40-Gb/s applications, for instance, have a diameter of about 10 μm [13].

1.2. Mid-Infrared Detectors

Going towards longer wavelengths (2–10 μm), semiconductors of progressively smaller bandgaps must be employed for infrared detection. Here, the most used material is the ternary II–VI alloy $Hg_{1-x}Cd_xTe$ (or MCT). By increasing the Cd molar fraction (x), the bandgap energy of this compound can be varied from -0.14 ($x = 0$) to about 1.5 eV ($x = 1$) with only a 0.3% variation in lattice parameter [14]. This allows great flexibility in the fabrication of heterostructures where a wider bandgap material in one of the doped layers helps reduce absorption losses.

A more recent, and quite successful, approach to the realization of MIR detectors is based on *intra*band transitions (instead of *inter*band as in the previous paragraphs). In this case the absorption of a photon promotes a carrier (either an electron or a hole) from a bound state in a quantum well (QW) to the continuum, where it can drift away under an electric field (see Fig. 4) [15]. The barrier height from the bound state within the QW to the continuum, corresponding to the cutoff wavelength of the device, can be tailored by a careful design of the well width. Quantum-well infrared photodetectors (QWIPs) have been demonstrated using various semiconductor families: InGaAs/AlGaAs [16], InGaAs/InP [17], and GaAs/AlGaAs [18], the last being the most common thanks to technological maturity and despite short-wavelength limitations due to a relatively low band-offset. Advantages and disadvantages of QWIP versus MCT devices have been the subject of extensive investigations [19–24]. QWIPs with different cutoff wavelengths can be stacked on the same substrate to obtain multicolor detectors [25]; both MCT [26] and QWIP [27] focal plane arrays (FPAs) of detectors have been demonstrated and are used in commercial (rather expensive) cameras [28]. Due to the lack of an appropriate technology enabling the monolithical integration with electronics for pixel addressing and read-out, in these devices a Si read-out integrated circuit is often bump-bonded to the array [29]. Very recently, QWIPs have been suggested for THz detection at wavelengths longer than 30 μm [30].

Mid-infrared detector performance can be quantified again in terms of spectral responsivity, dark current (or NEP), and speed. However, in the MIR it is common practice to refer to the specific peak detectivity (D*), which is the NEP inverse normalized to the square root of the detector area. In this wavelength range an important contribution to shot noise comes from the photocurrent generated by background radiation (i.e., emitted by the "scene" at temperatures higher than 0 K). This contribution is unavoidable and limits the detector sensitivity. To reduce the shot noise stemming from the dark current and, eventually, reach the background-limited infrared performance (BLIP), MIR detectors are commonly operated at low temperatures. A widely used figure is the maximum temperature at which BLIP operation is attained.

1.3. SiGe: A Silicon-Compatible Alternative?

The monolithic integration of detectors (and FPAs) with Si electronics for amplification, I–V conversion, and processing of the photogenerated signal has been intensively pursued aiming at an increase in compactness, a reduction in costs, and an improvement in reliability [31, 32]. More

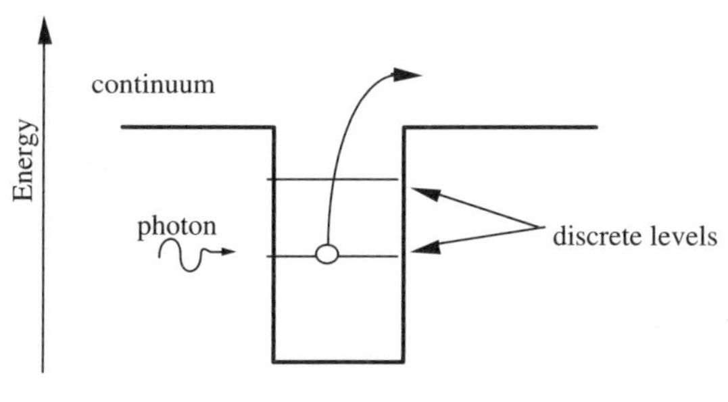

Figure 4. Schematic operation of a QWIP: photoexcitation of a carrier from a bound state (in a quantum well) to the continuum.

recently, the ultra-high speed required by optical communications and, at the same time, the lower cost sought for optical receivers in distributed FTTH (fiber-to-the-home) networks, have prompted extra efforts in this field [33]. Despite various interesting attempts (epitaxy [34,35], including the recent claim by Motorola [36], wafer fusion [37,38], wafer bonding [39–41]), however, the growth of III–V semiconductors on silicon has never reached the quality (in terms of defect density and reliability) required by optoelectronic devices. Moreover, even when detectors of reasonable performance have been integrated with silicon complementary metal-oxide-semiconductor (CMOS) electronics, the latter had to be fabricated using a modified process, thus giving up the advantages of a standardized technology [42].

Silicon-based MIR detection can be achieved with metal/semiconductor barrier detectors employing PtSi/Si or IrSi/Si Schottky junctions [43]. In such devices, free carriers in the silicide absorb infrared photons, thereby gaining the energy necessary to overcome the Schottky barrier, be injected in silicon, and drift under the electric field. In CMOS technology, silicides are often used to lower the resistance of the polycrystalline Si gates in MOS transistors; therefore, their compatibility with standard IC fabrication process is well established. Arrays of Schottky-barrier MIR detectors have been demonstrated and commercialized, despite the fact that their low quantum efficiency and limited spectral response (up to a few micrometers) cannot compare to MCTs in top-notch applications.

SiGe offers a valuable alternative to III–V, MCTs, and silicides for fabricating near- and mid-infrared detectors [44, 45]. The SiGe technology has gained a large popularity in recent years thanks to its success in heterojunction bipolar transistors (HBTs) [46, 47]. Si HBTs with a SiGe base outperform conventional homojunction bipolar transistors in terms of speed, reaching—or even exceeding—that of III–V devices; in addition, they can be seamlessly integrated with standard CMOS electronics to yield highly performing and dense integrated circuits. SiGe electronics is largely used in wideband circuitry with applications to wireless and optical communication systems [48].

SiGe, being a lower bandgap material than silicon, can extend the spectral responsivity of Si detectors to the NIR, while Si/SiGe heterostructures and QW can be used to detect the MIR. By choosing the appropriate molar fraction of Ge in a Si_xGe_{1-x} alloy, the bandgap can vary from that of silicon (1.16 eV; $x = 1$) to that of pure Ge (0.66 eV; $x = 0$). Unfortunately, contrary to MCTs, the full excursion from $x = 1$ to $x = 0$ implies a 4.2% change in the lattice parameter [49]. In the following section we review both the problems caused by such lattice mismatch in growing SiGe on Si and the methods to minimize or eliminate them.

2. THE GROWTH OF SiGe ON SILICON

Epitaxy, or the extension (i.e., thickening) of a crystal by growing on its surface layers replicating the substrate structure, has been a key in the development of microelectronics [50]. When the grown material differs from the substrate, the process is named "heteroepitaxy." Heteroepitaxy allows combination of materials with different electrical and optical properties in the same structure, thereby introducing a great design flexibility. At the beginning of the microelectronics age, Shockley [51] and Kroemer [52] foresaw the enormous potential of heterojunction transistors. Later, the amazing physical properties of quantum wells and superlattices [53, 54] were exploited and mastered to improve existing devices (e.g., semiconductor lasers) or create completely new ones (quantum cascade devices [55], QWIPs [56]).

2.1. Films of SiGe

A successful heteroepitaxy relies on a substrate and a growing layer which share similar properties; in particular, they should exhibit the same crystalline structure and lattice parameter, as in the cases of AlGaAs deposited on GaAs or MCT on CdTe. Conversely, in the growth of pure Ge on Si (same crystalline structure but 4.2% difference in lattice parameter), the Ge layer tends to initially assume the Si lattice parameter, undergoing compression (tensile strain) in the growth plane (along the normal). This nonequilibrium persists until the elastic energy in the film builds up beyond threshold; the film eventually relaxes and its own lattice parameter is restored via the creation of dislocations [57]. These are line defects (see Fig. 5) which lay in the growth plane (misfit dislocations) as well as perpendicularly to it (threading dislocations); they act as scattering centers for the free carriers and introduce electronic states in the otherwise forbidden semiconductor bandgap [58], thus impairing its electronic quality. The maximum thickness which can be pseudomorphically (i.e., with the same lattice parameter of the substrate) grown is called critical thickness (h_c). For a SiGe alloy grown on silicon, the critical thickness is plotted in Figure 6 as a function of Ge content. It is apparent from Figure 6 that only a few monolayers of pure Ge can be pseudomorphically deposited on Si, larger thicknesses being allowed for lower Ge molar fractions. This limitation imposed by lattice mismatch has severe consequences in a number of applications. For NIR detectors, a high absorption is required in the range 1.3–1.55 μm; as shown in Figure 7, this can only be obtained at Ge fractions close or equal to 1. Even in this case, however, a reasonable efficiency with an absorption coefficient of about 10^4 cm^{-1} would require a thickness of a few micrometers (see Eq. (4)), well beyond the critical value [59–61].

A fundamental parameter for QW design is the conduction and valence band line-up at the Si/SiGe heterointerface. For strained SiGe films grown on unstrained Si, the alignment is of type I (Fig. 8a), with a larger barrier at

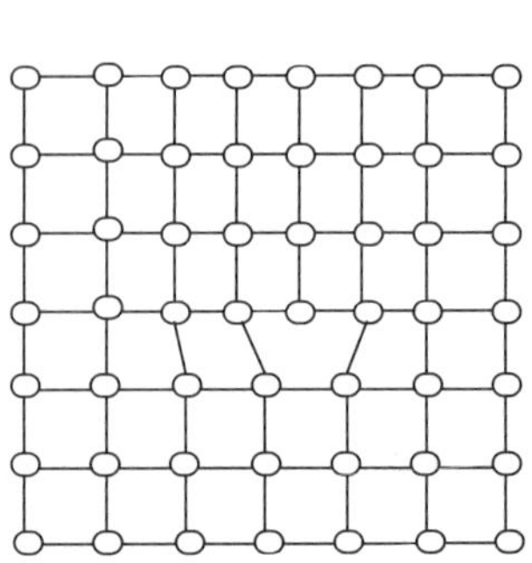

Figure 5. Schematic representation of a dislocation in a 2D lattice.

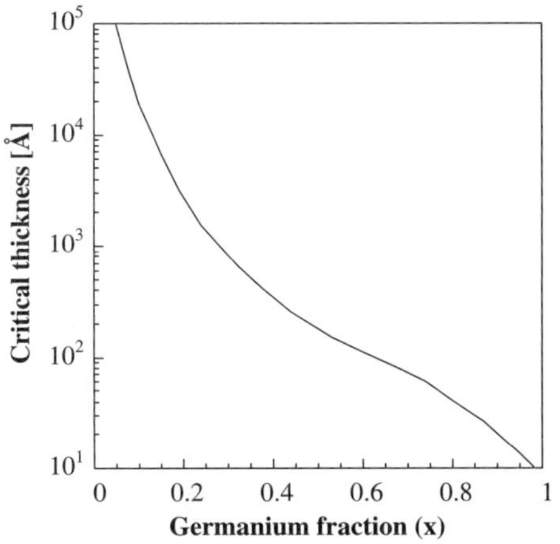

Figure 6. Critical layer thickness of SiGe grown on a silicon substrate as a function of Ge molar fraction.

the valence band discontinuity (Fig. 8b) up to a Ge concentration of about 60% [62–64]. The alignment changes to type II (Fig. 8a) for higher Ge molar fractions or when SiGe is grown on a strained Si layer (e.g., Si deposited on an entirely relaxed SiGe film).

Besides determining a critical thickness, the lattice mismatch affects the morphology of the grown layer. Epitaxy, in fact, can occur in three basic "growth modes": a two-dimensional or layer-by-layer mode ("Frank Van der Merwe"), a three-dimensional or island mode ("Volmer–Weber"), and a hybrid mode where islands grow on a few-monolayer-thick wetting layer ("Stransky–Krastanov") [65]. The latter is usually observed in the deposition of pure Ge (or high-Ge-fraction SiGe) on Si. A three-dimensional growth can give rise to strained/ wavy layers [66] or islands [67].

SiGe layers are obtained by either molecular beam epitaxy (MBE) or ultra- high vacuum chemical vapor deposition (UHV-CVD). In MBE the substrate is heated in a vacuum chamber with pressure below 10^{-9} Torr. The materials to be deposited (grains or rods of Si and Ge in our case) are evaporated by electron beams and/or electric heaters, a shutter in front of the sources allows control of the molecular flux, and a piezoelectric balance monitors the film thickness. Additionally, reflection high-energy electron diffraction (RHEED) is often available to monitor the crystallinity of the growing layer. MBE allows a very tight control of film growth, down to the atomic layer scale; for this reason, it is preferred whenever sharp and abrupt interfaces are needed, as in QW fabrication. On the other hand, UHV-CVD, being capable of processing a large batch of wafers in parallel, is best suited for large-scale production. In the case of CVD, the film precursors are in a gaseous state (typically SiH_4-silane and GeH_4-germane for SiGe) and are thermally decomposed. Two regimes are possible: a low-temperature regime in which the growth is limited by the reaction rate, and a high-temperature one dominated by the transport of chemical species (either precursors or reaction by-products) in the gas.

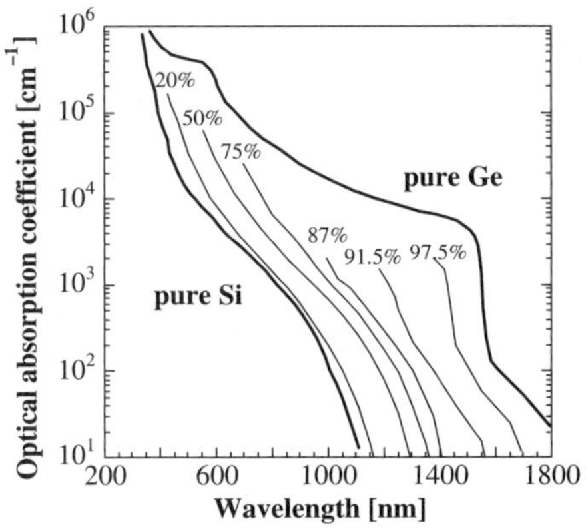

Figure 7. Absorption coefficients of bulk SiGe alloys versus wavelength. Data from [60] for pure Si and Ge, [61] for 20, 50, and 75% Ge, and [62] for 87, 91.5, and 97.5% Ge.

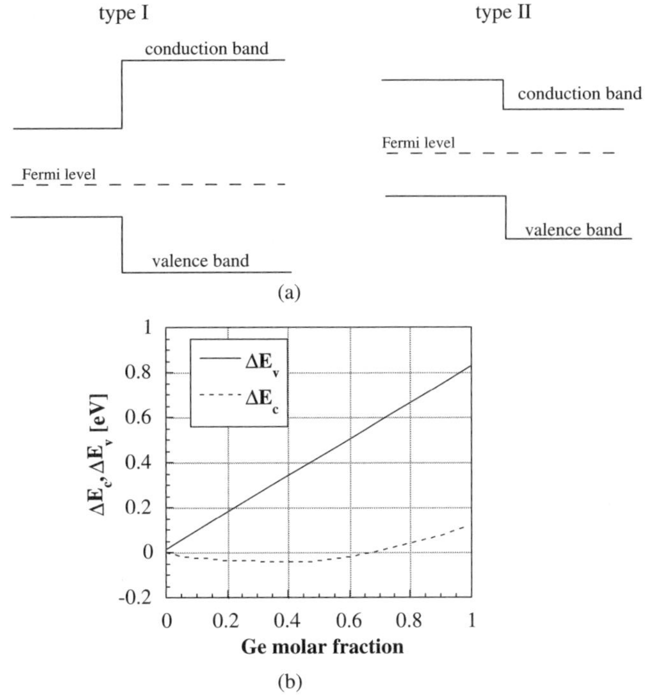

Figure 8. (a) Band alignment for type I and II heterostructures. (b) Conduction- and valence-band offsets at the SiGe/Si heterointerface versus Ge molar fraction in the alloy. We refer to strained SiGe on Si.

2.2. Relaxed Films of Pure Ge

As discussed above, the growth of thick pure-Ge films for the active layer in NIR detectors is hindered by the large lattice mismatch with Si. The direct growth of Ge-rich alloys with molar fraction exceeding 30% (corresponding to a lattice mismatch larger than 1%) yields films with threading dislocation densities exceeding 10^{11} cm^{-2} [68]. This figure must be compared with a maximum tolerable density of about 10^3 cm^{-2} for minority carrier devices (such as bipolar transistors, where a long diffusion length is a major requirement), or about 10^6 cm^{-2} for majority carrier devices (such as MOSFETs, which mainly suffer from mobility depression induced by defects). Various approaches have been developed to minimize the defect density in Ge films: the use of gradual buffers, the incorporation of carbon, and the deposition of a low-temperature buffer.

2.2.1. Gradual Buffers

A slow, gradual transition in lattice constant from Si substrate to Ge reduces the strain accumulated in the film by a continuous nucleation of dislocations. The density of threading dislocations can be estimated to be much lower in presence of this gradual buffer than in the case of an abrupt change of the lattice parameter [69]. The process begins with the deposition of a low-Ge-content SiGe layer up to the thickness needed to achieve complete relaxation. Then the Ge percentage is raised to retrieve the same lattice mismatch. The cycle is repeated until the Ge content is 100%. Based on this idea, digital (i.e., stepwise) and linearly graded buffers have been attempted. Despite some initial failure [70], the advantages of such an approach were clearly demonstrated by Fitzgerald et al. using MBE [71, 72]. Their work outlined the importance of a complete relaxation at each stage in the realization of the gradual buffer, such that the strain is always low due to the small lattice mismatch between two successive steps. Record-low dislocation densities of $4.4 \cdot 10^5$, $1.7 \cdot 10^6$, and $3 \cdot 10^6$ cm^{-2} were obtained for Ge percentages of 23, 32, and 50% in the resulting film, respectively. This technique was applied to the CVD growth of pure-Ge films with the addition of chemimechanical polishing (CMP) at a 50% Ge concentration [73]. CMP is standard in planarizing wafer surfaces during IC fabrication; the wafer is pressed and spun over a velvet pad with an abrasive slurry in solution with chemical etchants [74]. The grading rates were 10 and 5% Ge/μm and the resulting buffers were 10 and 20 μm thick, respectively [73]. The threading dislocations in the Ge layer (of thickness 1.5 to 3 μm) had a density of about $2 \cdot 10^6$ cm^{-2}, an order of magnitude smaller than in samples without CMP. Two other groups have tried to grow Ge-rich alloys with the graded buffer approach; LeGoues et al. obtained similar results with step graded buffers, sometimes combined with linear grading [68]. Presting et al. grew Si_mGe_m short-period superlattices on a buffer graded up to 50% [75, 76]. Another technique often used in conjunction with graded buffers consists in the use of surface-active species (surfactants) [77]. The role of surfactants, deposited on the substrate before film growth, is that of reducing the surface free energy. This entails the successive atoms to accommodate on a bidimensional film instead of piling up into islands. Remarkable results have been obtained by Liu et al., who fabricated a 0.3-μm-thick, $Si_{0.5}Ge_{0.5}$ layer with $1.5 \cdot 10^4$ cm^{-2} threading dislocation density on a 25% Ge/μm graded buffer and a single layer of Sb [78]. Antimony, as well as other group-V elements used as surfactants, tends to be partially incorporated in the growing film, thus resulting in an undesired n-type doping [79]. A valid alternative, both in MBE and CVD, is the use of hydrogen [80–82]. Despite the encouraging results, however, the use of graded layers is not completely adequate for industrial processing, due to the long times required and the resulting thick (tens of micrometers) structure, the latter hindering a seamless integration of the detector (to be fabricated on the buffer) with standard MOS devices.

2.2.2. SiGeC Alloys

Carbon can be incorporated into SiGe to change its lattice parameter and obtain a better match to the substrate [83, 84]. The ternary compound $Si_{1-x-y}Ge_xC_y$ has a lattice constant given by the Vegard law:

$$a_{Si_{1-x-y}Ge_xC_y} = a_{Si} + (a_{Ge} - a_{Si})x + (a_C - a_{Si})y \tag{5}$$

with $a_{Si} = 5.43$, $a_{Ge} = 5.64$, and $a_C = 3.56$ Å. For a given Ge percentage, from (5) one can calculate the amount of carbon needed for an alloy constant as in Si. Unfortunately, carbon does not only reduce the lattice parameter, but widens the bandgap, thus counteracting the Ge-induced reduction. Nevertheless, this by-phenomenon is not as effective as the lowering of lattice constant, and carbon incorporation can be used either to reduce the stress in the film maintaining the same bandgap, or to lower the bandgap at a constant stress [85]. Threading dislocation densities of 10^5 cm^{-2} have been measured in films with up to 30% Ge [86]. When its concentration in SiGeC films exceeds 4%, however, carbon tends to precipitate by forming SiC [87]. This hampers its use in the growth of high Ge-content alloys: from (5), in fact, a >5% carbon concentration is required to compensate the lattice mismatch with Si when the Ge molar faction reaches 50%.

2.2.3. Low-Temperature Buffers

The direct deposition of pure Ge on silicon results in islands: after a few atomic layers, Ge atoms find it energywise more convenient to pile up and build three-dimensional structures on the surface (Stransky–Krastanov growth). However, if the substrate temperature is low enough (typically below 400 °C) to slow down the diffusion of precursors on the surface, a layer-by-layer growth can be achieved [88]. The latter is favored in CVD due to the presence of hydrogen as a surfactant [89]. Under such conditions, the deposition of a few tens of nanometers of pure Ge on Si results in a flat, completely relaxed film where stress has been released by dislocations. The realization of a thin buffer, however, is time consuming due to the low deposition rates at moderate temperatures. Nevertheless, once the buffer reaches the thickness required for complete relaxation, the temperature can be raised to the usual values (~600 °C) and the growth can proceed at higher rates (above 10 nm/min) as in homoepitaxy. Such an approach was proposed by Colace et al., who demonstrated a material quality well suited for sensitive and fast metal-semiconductor-metal Ge-on-Si NIR photodiodes [90]. Later, the same technique was adopted by Luan et al., with the addition of a postgrowth cyclic thermal annealing [91]. The latter was very effective in further reducing the dislocation density from $9.5 \pm 0.4 \cdot 10^8$ cm^{-2} (in as-grown films) to $7.9 \pm 0.6 \cdot 10^7$ and $5.2 \pm 0.6 \cdot 10^7$ cm^{-2} after one and 10 thermal cycles between 900 and 100 °C, respectively. Thermal cycling, causing stress through the mismatch between Si and Ge expansion coefficients, enhances the drift of dislocations and thereby their probability to collide and annihilate [92, 93]. If Ge is deposited on a patterned substrate (e.g., oxidized silicon with small windows opened in the SiO_2), the dislocation density can be further reduced to the extent that several small (10 × 10 μm^2) dislocation-free mesas exist. In this case, the mesa sidewalls act as sinkholes and allow dislocation to disappear upon reaching them [84]. As a final remark, it should be noted that the determination of dislocation densities can be affected by large errors due to the limited field of view in cross-section images by transmission electron microscopy (TEM). To alleviate this problem, Luan et al. combined two different techniques: TEM analysis and etch-pit density count (EPD) [94]. The latter consists in counting the threads in a plane-view SEM or TEM picture, after etching with a selective solution (CH_3COOH: 67 ml, HNO_3: 20 ml, HF: 10 ml, and I_2: 30 mg) acting faster near the defects [94].

2.3. Films of Polycrystalline Ge

The problems with the substrate/film mismatch can be avoided if the film is nonepitaxial: a polycrystalline film, while consisting of a material with inferior electronic quality, can be obtained at low temperatures, increasing the compatibility of the process with standard silicon technology. Polycrystalline SiGe, with Ge molar fraction lower than 60%, was proposed as an appealing alternative to polycrystalline Si in various integrated electronics components, among them the gate electrodes in CMOS transistors [95], and the active layer of thin-film transistors for liquid-crystal displays [96]. Poly-SiGe requires temperatures lower than poly-Si and exhibits higher conductivity and grain size. Poly-Ge was suggested as a buffer for the growth of GaAs (with nearly the same lattice parameter of Ge) on inexpensive substrates (glass) for solar cell applications [97]. Ge can be deposited in polycrystallites by evaporation at temperatures as low as 300 °C in a vacuum with base pressure of about 10^{-7} Torr [98]. Raman spectroscopy proves that amorphous and polycrystalline (poly-Ge) phases are obtained below and above about 250 °C, respectively [99].

3. NEAR-INFRARED DETECTORS AND ARRAYS

3.1. SiGe-Based Detectors

3.1.1. Waveguide Detectors

Due to the low absorption afforded by SiGe alloys at the wavelengths of interest (1.3 and 1.55 μm), a guided-wave configuration is to be preferred for detectors based on this material. For an adequate sensitivity under normal incidence, in fact, the thickness d of the active layer must exceed the absorption length (the inverse of the absorption coefficient α) to ensure efficient collection of incident photons, and be shorter than the drift length (the product of saturation velocity and free carrier lifetime) to avoid recombination losses. Conversely, in waveguide detectors, the light propagates and gets absorbed along a direction perpendicular to the path of photogenerated carriers (see Fig. 9), the latter being entirely collected in a properly designed active thickness. On the other hand, the device must be longer than the absorption length, and this can result in speed limitations owing to the detector capacitance. Waveguides can be integrated on silicon with various approaches: NIR light can be confined in the overlayer of silicon-on-insulator (SOI) wafers [100], or by way of a doping-induced refractive index increase in Si [101], or alloying Si with Ge [102]; moreover, silica or Si-nitride [103, 104] waveguides can be fabricated on silicon substrates using standard IC technologies. Among the above, SOI and nitride approaches are the most promising. SiGe waveguide detectors were first demonstrated in 1986 by Temkin et al [105]. A coherently strained Ge_xSi_{1-x}/Si superlattice (SLS), as in [106], was embedded in a *p-n* junction and became the active medium of the detectors, fabricated with various Ge fractions in it. Larger Ge contents resulted in a photoresponse extended towards longer wavelengths, the sample with 60% Ge showing a maximum external quantum efficiency of 10.2% at 1.3 μm, a dark current of 7.1 mA/cm^2, and a fast response of 312 ps for a reverse bias of 10 V. As expected from the drop in SiGe absorption, however, the photoresponse at 1.55 μm was much lower. This first achievement spurred a substantial interest, and was followed by several results in Europe and the United States. In [107] the device consisted of a *p*-type 2.5-μm-thick $Si_{0.98}Ge_{0.02}$ waveguide with a 5-μm-wide rib for lateral confinement, and the NIR sensitive region was made of 20 periods of 5-nm-thick $Si_{0.55}Ge_{0.45}$ quantum wells separated by 30-nm Si spacers. The detector was grown on top of the waveguide, with light reaching it through the higher refractive index of the active region. A maximum external quantum efficiency of 11% was measured at 1.282 μm and a reverse bias of 7 V, while the best dark current (at the same bias but for a shorter device with about 70% lower efficiency) was 1 mA/cm^2. The speed was investigated in both time and frequency domains measuring the response to a light step with a fast oscilloscope and a microwave network analyzer. The latter provided the "small signal" transfer function with a bandwidth well in the GHz range, while the former allowed to measure a 50-ns long-lasting photocurrent tail. The latter is usually associated to traps and results in a time-drift of the diode operating point and in a shrinkage of the eye-diagram at the receiver. Similar performances (1.32 μm external quantum efficiency of 7% and dark current of 2.7 mA/cm^2 at 14 V reverse bias) were demonstrated in [108] with a multiple quantum well with a 71% Ge content (in the alloy used for the wells). Wide open eye-diagrams were reported detecting 0.5- and 1.5-Gb/s pseudorandom

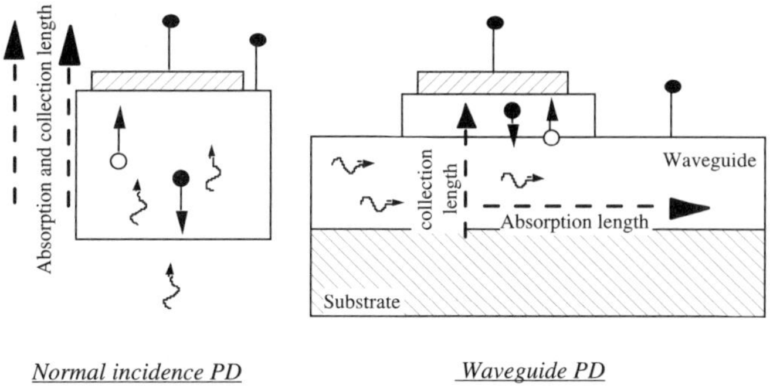

Figure 9. Sketch of normal incidence (left) and guided-wave photodetectors (right). Absorption and collection lengths are defined along parallel or orthogonal directions in the two cases, respectively. Black (white) circles represent electrons (holes), and incident photons are pictured as wormlike arrows.

nonreturn to zero (NRZ) signals. The best performances of a SiGe-strained layer waveguide photodetector were theoretically explored in [109]. A number of constraints were accounted for in the simulation, among them, a maximum 60% molar Ge (to avoid ultrathin films), a 1-mm waveguide length, and a 10-μm width. The external efficiency included the Fresnel reflection at the fiber–waveguide interface, the modal mismatch, and the overlap between the guided mode and the absorbing region. By referring to two different models for the critical thickness, namely Dodson–Tsao [110] and People–Bean [111], efficiencies as high as 13 and 16%, respectively, were calculated for a (realistic) 1-μm thickness limit. It should be stressed that, when adopting the People–Bean model, the most important loss contribution is the fiber-to-waveguide coupling: neglecting it, a 60% quantum efficiency can be expected. This clearly suggests the need to use optimized waveguide couplers, rather than the commonly employed end-fire.

3.1.2. Normal Incidence Detectors

A few attempts have been made to fabricate normal-incidence photodiodes based on strained SiGe. Reference [112] describes a device employing a 500-nm SLS with 10-nm-thick wells of $Si_{0.5}Ge_{0.5}$ in a *p-i-n* structure. The detector was antireflection coated to optimize light in-coupling at 1.3 μm, and the external quantum efficiency was 1% for a reverse bias of 4 V, although a further increase was reported at voltages close to 20 V (breakdown). The dark current density was 3 mA/cm^2 (at 4 V), comparable to similar devices based on SiGe. More recently, undulating, Ge-rich SiGe multilayers have extended the spectral response towards 1.55 μm [113, 114]. A vertical MSM structure with a buried $CoSi_2$ bottom contact exhibited quantum efficiencies of 5.2% at 1.3 μm and 1% at 1.55 μm, respectively.

3.1.3. Avalanche Detectors

Aiming at an enhancement of the sensitivity of SiGe detectors, avalanche photodiodes (APDs) have been designed and demonstrated, as well. APDs are appealing for optical communications, since their large gain (easily in excess of 10^3) can eliminate the need for front-end amplifiers and the related noise. Moreover, owing to the large field applied to induce avalanche multiplication, they are also fast. In terms of excess noise produced by carrier multiplication, APDs perform better if the avalanche is promoted by only one type of photocarriers; for example, if electrons' and holes' ionization coefficients are very different from one another [115, 116]. Silicon is well suited for APDs, with a ratio between electrons' (α_n) and holes' (α_p) ionization coefficients varying from 5 to 500 (depending on the electric field). Such ratio is lower than 2 in bulk germanium. The dependence of the ionization coefficients on the percentage of Ge in a SiGe alloy on Si has been investigated in [117]. In this work, strongly absorbed light was shined on either the *p*- or the *n*-side of a SiGe *p-i-n* diode, thus originating a single-carrier current in the intrinsic layer: electrons for absorption near the *p*-layer and holes near the *n*-layer. A ratio α_n/α_p ranging from 10 to 0.75 was estimated for Ge percentages from 0 to 100% with an electric field of 330 KV/cm.

The good noise characteristic of Si can be combined with the nonvanishing NIR absorption of SiGe in "separate absorption and multiplication" APDs (SAM-APD): a relatively low-field region (lower than needed to start avalanche) where absorption occurs is adjacent to a Si high-field region devoted to photocarriers multiplication. SAM-APDs were first fabricated with III–V semiconductors [118] and later designed [119] and demonstrated [120] using SiGe-on-Si. As in the previous case, the SiGe SAM-APD was realized in a waveguide configuration and exhibited a maximum responsivity of 1.1 A/W at 1.3 μm when biased at about 30 V. Soon after, the same team reported an optimized SAM-APD [121] with a responsivity of 4 A/W at 1.3 μm and 30 V, and a 100-ps response. In terms of receiver sensitivity, that is, the minimum optical power required at the output of the fiber to grant a bit error rate (BER) better than 10^{-9} [122], the SAM-APD above yielded -29.4 dBm at a data rate of 800 Mb/s, comparable to commercial InGaAs *p-i-n* detectors [123].

3.1.4. Photoconductive Detectors

A different approach to obtain photocurrent gain in a semiconductor device is the use of photoconductivity [124]. Photoconductivity can be observed both in a semiconductor with ohmic contacts and in forward biased *p-n* junctions: conductivity is enhanced due to photon absorption and the related increase of free carrier density in conduction and valence bands. Since electrons and holes are swept at different speeds by the electric field in the semiconductor, however, a net charge builds up. Assuming that electrons move much faster than holes (i.e., the drift time τ_e for electrons is much lower than τ_h for holes), the semiconductor will remain positively charged. To reestablish charge neutrality, extra electrons are injected by the contact and so provide a "secondary" current. Therefore, from a single photon, several photocarriers contribute to the current and a net gain is obtained. The photoconductive gain is given by the ratio between the drift times

$$G_{ph} = \frac{\tau_h}{\tau_e} \tag{6}$$

It can be much larger than unity due to the presence of traps, which can slow down the motion of the holes. Despite the high achievable gain, however, photoconductors exhibit a lower sensitivity than photodiodes. Dark current densities in these devices and the associated shot noise are, indeed, typically much larger owing to the lack of injection barriers at the contacts. The speed is limited by the turn-off time, that is, the time required to sweep all the slower carriers (holes in our example) out of the active region. To increase the speed, it is therefore preferable to reduce τ_h, although this reduces the gain. A SiGe SLS waveguide detector operating in the photoconductive mode is described in [125], where the large difference in drift times was attributed to hole trapping in the wells. The sensitivity at 1.3 μm was about -16 dBm at 180 Mb/s.

3.1.5. SiGeC Detectors

Carbon was incorporated in a NIR detector to compensate for the lattice mismatch due to Ge in a SiGe alloy, in 1996 [126]. The active region of this normal-incidence *p-i-n* diode was an 80-nm-thick $Si_{0.25}Ge_{0.6}C_{0.15}$ alloy. But, despite the use of carbon, this first device exhibited a quantum efficiency of only 1% at 1.3 μm, with a large dark current of 7 mA/cm^2 at 0.5 V reverse bias. Slightly better results were obtained by the same team using a waveguide [127], yielding 8 and 0.2% external quantum efficiencies at 1.3 and 1.55 μm, respectively, for a reverse bias of just 0.3 V. The dark current density was 4 mA/cm^2. Two years later, in order to raise the absorption in the active area, a normal-incidence detector was fabricated based on pure-GeC alloy [128]. Various C compositions from 0 up to 2% (due to the poor solid solubility of C in Ge: 10^8 atoms/cm^3) were attempted, well below the near 10% molar fraction required to compensate the lattice mismatch between pure Ge and Si. The films ended up being undercompensated and with a high (10^{10} cm^{-2}) dislocation density. The device exhibited a dark current in the mA/cm^2 range with responsivity extending up to 1.55 μm and a maximum of 35 mA/W at 1.3 μm.

3.2. Pure-Ge-on-Si Detectors

Ge detectors on Si were first grown in the 1960s by Milnes and co-workers [129, 130]. They exhibited an extended photoresponse in the near-infrared, with characteristics strongly affected by the large amount of defects at the Si/Ge heterointerface. It was only after two decades, in 1984, that Luryi et al. reported on graded buffers for the growth of Ge on Si [131]. Their *p-i-n* diode exhibited a 40% quantum efficiency at 1.3 μm; regretfully, the significant dislocation density resulted in a dark current larger than 50 mA/cm^2. To this extent, using a superlattice embedded in the buffer layer, the same group managed to reduce the defect density and the dark current [132], but at the expense of a smaller 3% quantum efficiency (at 1.3 μm). Research in pure-Ge detectors was therefore at stake compared to strained SiGe alloys and superlattices until 1994 when, in an effort to produce sensitive devices with a spectral coverage including the third window of optical communications (1.55 μm), Sutter et al. realized an MBE-grown Ge-on-Si *p-i-n* photodiode [133]. To minimize the dislocation count in the unavoidably relaxed thick Ge, a 50-nm-thick Ge buffer was deposited at a relatively low temperature (420 °C) and rate (0.1 nm/s) to prevent the formation of islands, while maintaining good crystalline quality. The devices underwent annealing steps at 700 °C before the lithographic processes for the definition of the mesas and the metallization. The dark current was still large (51 mA/cm^2), but the response extended to 1.6 μm with a peak 43% external quantum efficiency at 1.55 μm. Four years later, Samavedam et al. realized a *p-n* Ge photodiode on a graded buffer with a CMP step at an intermediate $Si_{0.5}Ge_{0.5}$ composition [134]. This device, despite the limited responsivity of 133 mA/W at 1.3 μm due to the thinness (240 nm) of the absorbing layer, exhibited a record low dark current of 0.2 mA/cm^2 thanks to the high quality of the epitaxial layer. Contemporarily and independently, Colace et al. introduced the concept of a low-temperature buffer in the CVD growth of Ge on Si [135, 136]. Metal-semiconductor-metal detectors fabricated using the latter approach provided high quantum efficiencies at both 1.3 and 1.55 μm and response times of 2.2 ns, but the incomplete rectification at the Ag/Ge contacts of the interdigitated structure resulted in large dark currents. Substantial improvements were brought about by complementing the low-temperature buffer with the introduction of postgrowth annealing steps in a *p-i-n* structure [137–140]. This allowed to reach high responsivities of 0.89 and 0.75 A/W at 1.3 and 1.55 μm, respectively, pulse responses shorter than 200 ps at 1.3 μm [141], and dark currents as low as 15 mA/cm^2 at 1 V (reverse bias).

3.3. Polycrystalline Ge Detectors and Arrays

The main advantage offered by SiGe with respect to III–V semiconductors is the integrability of detectors with standard silicon CMOS electronics. The relatively high temperatures used for the epitaxy of Ge-rich SiGe on Si, and the cleaning procedures required by high-quality "epi-ready" surfaces, however, impose to insert the Ge growth after the realization of the MOS oxide and before the deposition of the first metal contact. This would alter the standard CMOS process, thus resulting in impracticalities in silicon IC foundries. To alleviate this problem, the detectors have to be realized with a process conducted at a low temperature and less dependent on surface preparation, such that it can be performed at the very end of the IC fabrication, just before chip dicing and packaging. Polycrystalline Ge appears as one such viable solution and has been actively pursued by us [142]. Mesa poly-Ge on Si heterojunction photodiodes have been demonstrated [142] with a responsivity of 16 mA/W at 1.3 μm, fast (650 ps) pulse response, and low dark current (1 mA/cm^2 at 1 V reverse bias). A few functional devices have also been obtained in poly-Ge on Si: a linear detector array (16 pixels) [143], a wavemeter encompassing a set of photodiodes with different spectral responses [144, 145], a detector integrated onto a SOI waveguide [146]. More recently, the effective compatibility of poly-Ge deposition with standard IC technology has been demonstrated by integrating an 8-pixel array with the addressing/amplifying electronics in standard—2 μm—CMOS technology [147]. While the low responsivity remains the main problem of this approach, its simplicity and prospective low cost qualify it as one of the most promising candidates for future exploitation in consumer-oriented commercial devices.

In Table 1, a synoptic view of the most relevant SiGe-on-Si-based detectors for the NIR is reported.

4. MID-INFRARED DETECTORS AND ARRAYS

4.1. SiGe QWIPs

SiGe quantum wells for MIR photodetectors were first proposed in 1998 by Yang et al. [148, 149]. They investigated the advantages of an indirect-gap semiconductor QW for normal-incidence absorption of infrared light through intersubband transitions. In direct-gap semiconductor QWIPs with an isotropic band structure at the Γ point, in fact, intersubband transitions are forbidden at normal incidence, that is, when the electric field of the incident light is parallel to the walls of the quantum well. To be efficiently detected, then, the light must couple at a nonzero angle with the normal, as actually done by means of diffraction gratings [150].

Table 1. Synoptic view of SiGe on Si detectors for the near-infrared.

Material/deposition technique	Device structure	Responsivity, quantum efficiency	Speed	Dark current density	Ref.
SiGe/MBE	waveguide, *p-n*	10.2% @ 1.3 μm	312 ps	7.1 mA/cm^2	[105]
SiGe/MBE	waveguide, *p-i-n*	11% @ 1.28 μm	400 ps	1 mA/cm^2	[107]
SiGe/CVD	waveguide, *p-i-n*	7% @ 1.3 μm	1.5 Gb/s	2.7 mA/cm^2	[108]
SiGe/MBE	*p-i-n*	1% @ 1.3 μm		3 mA/cm^2	[112]
SiGe/MBE	m-s-m	5.2% @ 1.3 μm, 1% @ 1.55 μm			[114]
SiGe/MBE	waveguide SAM-APD	4 A/W @ 1.3 μm	100 ps		[121]
SiGeC/MBE	waveguide, *p-i-n*	8% @ 1.3 μm, 0.2% @ 1.55 μm		4 mA/cm^2	[127]
SiGeC/MBE	*p-i-n*	35 mA/W @ 1.3 μm			[128]
Ge/MBE	*p-i-n*	40% @ 1.3 μm		50 mA/cm^2	[131]
Ge/MBE	*p-i-n*	43% @ 1.3 μm		51 mA/cm^2	[133]
Ge/CVD	*p-n*	133 mA/W @ 1.3 μm		0.2 mA/cm^2	[134]
Ge/CVD	m-s-m	240 mA/W @ 1.3 μm	2 ns		[135]
Ge/CVD	*p-i-n*	0.89 A/W @ 1.3 μm, 0.75 A/W @ 1.55 μm	200 ps	15 mA/cm^2	[140]
poly-Ge/PVD	*p-n* (heterojunction)	16 mA/W @ 1.3 μm	600 ps	1 mA/cm^2	[142]
poly-Ge/PVD	waveguide, *p-n* (heterojunction)	8 mA/W @ 1.3 μm		0.8 mA/cm^2	[146]
poly-Ge/PVD	8 pixel array integrated on a standard IC	43 V/W @ 1.3 μm		1 mA/cm^2	[147]

Conversely, using Si or Ge, owing to the anisotropy of their effective mass tensor and, therefore, to the possibility of carrier acceleration perpendicular to the electric field, the absorption of normally incident photons is nonvanishing. This consideration, and the forthcoming integration of detectors with readout electronics on a single chip, has stimulated the investigation of SiGe QWIPs. While the first theoretical studies, in route with the understanding developed for III–Vs, focused on the use of conduction-band QWs [151–153], the control of the valence-band line-up between strained SiGe and Si has steered the research towards the use of holes rather than electrons. The use of strained Si (111) grown on relaxed SiGe layers was suggested in [154]. Despite the calculations, however, no related experimental results are available yet. In 1990 Lin et al. successfully fabricated the first SiGe-based MIR detectors [155, 156]. They departed from standard intraband transitions in QWIPs, exploiting photon-induced emission of holes from a heavily doped SiGe epilayer into the Si substrate at the heterointerface. For this reason these so-called heterojunction internal photoemission (HIP) detectors exhibited quantum efficiencies around 1% for wavelengths from 2 to 10 μm, with 30% Ge in the 40-nm-thick SiGe layer. Tailoring of the spectral response was also demonstrated by lifting the valence-band barrier through an increase of the Ge molar fraction from 20 to 40%. Following these results, a 400 $\times$ 400 focal plane array of HIP detectors was demonstrated, each pixel (28 $\times$ 28 μm^2) consisting of a $Ge_{0.44}Si_{0.56}$/Si heterojunction with an MOS transistor used as a transfer gate [157, 158]. In this array the photogenerated charges moved to the chip periphery through a charge coupled device (CCD), and a good image quality was obtained at 53 K (nominal temperature) in both LWIR and MWIR bands.

The devices described above were fabricated by MBE, and the degenerate *p*-doping of the SiGe layer, required to get a large hole density and a large free carrier absorption, was obtained by co-evaporating HBO_2. While the latter helped the deposition (HBO_2 has a relatively high vapor pressure), the high substrate temperature needed to initiate the reaction with silicon facilitated SiGe relaxation and, additionally, let oxygen from HBO_2 incorporate in the film. Both such effects lower the electronic quality of the SiGe layer. An alternative approach based on elemental boron is discussed in [159]: the authors largely improved the material quality at comparable doping levels (2×10^{20} cm^{-3}), which resulted in smaller dark currents of 50 μA/cm^2 at 38 K. A model explaining the photocurrent spectra exhibited by such a device can be found in [160]. Elemental boron for doping the SiGe layer in HIP detectors was also reported in [161], where the authors introduced an undoped SiGe spacer to improve the SiGe/Si heterointerface. The first truly SiGe-based QWIP was demonstrated by Karunasiri et al. in 1991 [162, 163]. The absorbing layer consisted of 50 periods of 3 nm-thick $Si_{0.85}Ge_{0.15}$ wells *p*-doped at 10^{19} cm^{-3} and separated by 50-nm-thick Si barriers. The SiGe composition and well thickness were adjusted to obtain a single energy level separated by about 120 meV from the extended states, thus providing an absorption peak close to 10 μm. The device exhibited high responsivities both for light coupling through a facet at a 45° angle with the surface ($D^* = 10^9$ cmHz$^{1/2}$/W at 77 K) and at normal incidence [164]. By increasing the Ge in the QW, these authors also demonstrated a shift of the photoresponse peak to shorter wavelengths (2 and 3 μm for 60 and 30% Ge, respectively) [165]. Longer-wavelength QWIP operation with broadband response from 8 to 14 μm was demonstrated by People et al. with a

10-period, 4-nm-thick $Ge_{0.25}Si_{0.75}$ MQW [166]. The spectral response measured at 77 K showed three broad peaks centered at 8.2, 10.8, and 13 μm, respectively. In 1986 a detector with spectral response peaked at 5 μm and a background limited performance maintained up to 85 K was demonstrated [167]. In recent years, a systematic attempt to optimize SiGe MIR photodetectors has been carried out, testing a number of single and multiple QW structures [168, 169], and introducing some novel concepts [170]. In particular, the important role of the doping setback layer (i.e., the undoped SiGe layer at the Si/SiGe interface) in reducing the dark current, and, therefore, improving the detectivity has been outlined. Moreover, a gradient in the doping and in the Ge content of the SiGe layer was implemented to ease the photocarrier drift towards the heterointerface.

While free carrier absorption is commonly considered the excitation mechanism in *p*-type SiGe HIPs, the origin of the strong normal-incidence photoresponse of *p*-type SiGe QWIPs is quite debated. In [167], based on the theoretical work by Chang and James [171], researchers attributed the observed 8.2-μm response peak to inter-subband transitions from heavy hole 1 (HH1) to heavy hole 2 (HH2) levels, normally forbidden for normal-incidence light [172]. The second peak at 10.8 μm, instead, was associated to intervalence transitions between HH1 and the split-off band (SO) [173]. On the other hand, free carrier absorption was recognized as the fundamental excitation mechanism in *p*-type SiGe QWIPs by Robbins et al. [174], Strong et al. [175–177], and Corbin et al. [178]. In [179], a detailed comparison was carried out between absorption and photocurrent spectra of similar structures for different polarizations of the incoming light. A close-to-ideal agreement was found between the measured absorption spectra and the predictions of a model accounting for intervalence and inter-subband transitions with appropriate selection rules. The photocurrent spectra, conversely, gave a larger-than-expected contribution for light polarized in the quantum well plane. This was ascribed to a transport efficiency (and, therefore, photoconductive gain) higher in the light hole (LH) and SO (states excited by normal-incidence light) than in the HH2 band.

4.2. Quantum Dot Detectors

Zero-dimensional structures or *quantum dots* have been proposed to overcome the main limitation of conduction-band III–V QWIPs, namely the lack of photoresponse at normal incidence. In a classical view, in fact, light with electric field polarized in the growth plane can accelerate a carrier in a quantum dot, thus allowing its injection in the surrounding semiconductor. The same is not possible in conduction-band QWs, since an acceleration perpendicular to the barriers (i.e., a field polarization perpendicular to the growth plane) is required. In addition, reduced phonon scattering is expected in quantum dots compared to quantum wells, with an increased probability of photocarrier emission before relaxation to the ground state [180]. Several quantum dot IR photodetectors (QDIP) have been already demonstrated based on III–V semiconductors [181–184], and Ge-on-Si QDIPs have been introduced [185]. These encompass an active region realized as a multiple stack of layers of Ge islands interleaved with silicon spacers. The Ge is *p*-type doped through co-evaporation of boron, and the average island-size ranges from 7.5 (height) $\times$ 75 nm (width) with a surface density of 4.4×10^9 cm^{-2} [186] to 1.5 $\times$ 15 nm with density 3×10^{11} cm^{-2} [187]. Photoresponse at normal incidence has been demonstrated in the mid-infrared [188], with detectivity peaking near 20 μm (1.7×10^8 cm Hz$^{1/2}$/W) and 9 μm (7×10^7 cm Hz$^{1/2}$/W) at room temperature.

5. CONCLUSIONS

Near- and mid-infrared light detectors with Ge and SiGe compounds on silicon are challenging III–V and II–VI based devices in a number of applications, from optical communications to remote sensing. High responsivity, detectivity, and speed have been demonstrated in different wavelength ranges and temperatures. Different approaches have been proposed and investigated to get rid of the lattice mismatch existing between SiGe and Si. Thanks to a mature and well-mastered fabrication technology, new devices based on nanoscale Ge islands grown on silicon are entering the arena. The integrability of these devices and standard silicon electronics into functional microsystems appears to be the fundamental advantage of SiGe, SiGeC, Ge, and poly-Ge, and is expected to drive research and development of such technology towards affordable and reliable components for the consumer market.

GLOSSARY

Bandgap The range of energy levels forbidden to electrons in the band structure of a semiconductor.

Mid-infrared (MIR) Light of wavelength in the spectral range from 2 to 10 μm.

Near-infrared (NIR) Light of wavelength in the spectral range from 700 mm to 2 μm.

Photodetector Electronic device used to transform a lightwave signal into an electrical one usually proportional to the power of the former.

Quantum well (QW) A potential well in a semiconductor heterostructure whose small thickness induces quantization of the allowed levels for electrons.

REFERENCES

1. J. M. Senior, "Optical Fiber Communications," pp. 88–110. Prentice-Hall, Englewood Cliffs, NJ, 1992.
2. T. Kaino, *Appl. Phys. Lett.* 48, 757 (1998).
3. P. Ceccato, N. Gobron, S. Flasse, B. Pinty, and S. Tarantola, *Remote Sensing of Environment* 82, 188 (2002).
4. http://www.ino.it/~luca/rifle/rifle_en.html.
5. M. Sauter and W. Sitter, *Lightwave* 1, 16 (2002).
6. H. Takashi, S. Suzuki, K. Kato, and I. Nishi, *Electron. Lett.* 26, 87 (1990).
7. S. Esener and P. Marchand, *Mat. Sci. in Semicon. Proc.* 3, 433 (2000).
8. J. I. Pankove, "Optical Processes in Semiconductors," p. 34. Dover Publications, New York, 1975.
9. A. Yariv, "Optical Electronics," pp. 422–432. Saunders College Publishing, New York, 1971.
10. R. S. Muller and T. I. Kamins, "Device Electronics for Integrated Circuits," pp. 57–60. Wiley, New York, 1986.

11. P. R. Gray and R. G. Meyer, "Analog Integrated Circuits," p. 635. Wiley, New York, 1984.
12. J. M. Senior, "Optical Fiber Communications," pp. 489–492. Prentice-Hall, Englewood Cliffs, NJ, 1992.
13. G. Masini, L. Colace, and G. Assanto, in "Handbook of Thin Film Materials," (H. S. Nalwa, Ed.), Vol. 4, p. 331. Academic Press, New York, 2002.
14. M. A. Herman and M. Pessa, *J. Appl. Phys.* 57, 2671 (1985).
15. B. F. Levine, *J. Appl. Phys.* 74, R1 (1993).
16. K. K. Choi, S. V. Bandara, S. D. Gunapala, W. K. Liu, and J. M. Fastenau, *J. Appl. Phys.* 91, 551 (2002).
17. D. K. Sengupta, S. L. Jackson, D. Ahmari, H. C. Kuo, J. I. Malin, S. Thomas, M. Feng, G. E. Stillman, Y. C. Chang, L. Li, and H. C. Liu, *Appl. Phys. Lett.* 69, 3209 (1996).
18. K. K. Choi, *J. Appl. Phys.* 73, 5230 (1993).
19. M. A. Kinch and A. Yariv, *Appl. Phys. Lett.* 55, 2093 (1989).
20. P. M. Young, C. H. Grein, H. Ehrenreich, and R. H. Miles, *J. Appl. Phys.* 74, 4774 (1993).
21. G. M. Williams, *J. Appl. Phys.* 77, 4153 (1995).
22. C. H. Grein, M. E. Flatté, H. Ehrenreich, and R. H. Miles, *J. Appl. Phys.* 77, 4156 (1995).
23. J. Piotrowski and A. Rogalski, *J. Appl. Phys.* 80, 2542 (1996).
24. H. Ehrenreich, C. H. Grein, R. H. Miles, and M. E. Flatté, *J. Appl. Phys.* 80, 2545 (1996).
25. M. Z. Tidrow, X. Jiang, S. S. Li, and K. Bacher, *Appl. Phys. Lett.* 74, 1335 (1999).
26. W. E. Tennant, C. A. Cockrun, J. B. Gilpin, M. A. Kinch, M. B. Reine, and R. P. Ruth, *J. Vac. Sci. Technol. B* 10, 1359 (1992).
27. T. R. Schimert, S. L. Barnes, A. J. Brouns, F. C. Case, P. Mitra, and L. T. Claiborne, *Appl. Phys. Lett.* 68, 2846 (1996).
28. http://www.indigosystems.com/s_merlin_long.html.
29. N. Y. Aziz, G. T. Kincaid, J. L. Heat, W. J. Parrish, and J. T. Woolaway, *Proc. SPIE* 3360, 80 (1998).
30. M. A. Gadir, P. Harrison, and R. A. Soref, *J. Appl. Phys.* 91, 5820 (2002).
31. R. A. Soref, *Proc. IEEE* 81, 1687 (1993).
32. H. Zimmermann, "Integrated Silicon Opto-Electronics." Springer-Verlag, Berlin, 2000.
33. G. Masini, L. Colace, and G. Assanto, *Mat. Sci. Eng. B* 89, 2 (2002).
34. Y. J. Mill, R. P. G. Karunasiri, K. L. Wang, and G. Bai, *J. Vac. Sci. Technol. B* 7, 340 (1989).
35. E. Droge, R. F. Schnabel, E. H. Bottcher, M. Grundmann, A. Krost, and D. Bimberg, *Electron. Lett.* 30, 1348 (1994).
36. http://www.thinfilmmfg.com/Noteworthy/Noteworthy01/GaAs20Sept01.htm.
37. A. R. Hawkins, T. E. Reynolds, D. R. England, D. I. Babic, M. J. Mondry, K. Streubel, and J. E. Bowers, *Appl. Phys. Lett.* 68, 3692 (1996).
38. A. R. Hawkins, W. Wu, P. Abraham, K. Streubel, and J. E. Bowers, *Appl. Phys. Lett.* 70, 303 (1997).
39. F. E. Ejeckam, C. L. Chua, Z. H. Zhu, Y. H. Lo, M. Hong, and R. Bath, *Appl. Phys. Lett.* 67, 3936 (1995).
40. Z. H. Zhu, F. E. Ejeckam, Y. Qian, J. Zhang, Z. Zhang, G. L. Christenson, and Y. H. Lo, *IEEE J. Sel. Topics Quantum Electron.* 3, 927 (1997).
41. B. F. Levine, A. R. Hawkins, S. Hiu, B. J. Tseng, J. P. Reilley, C. A. King, L. A. Gruezke, R. W. Johnson, D. R. Zolnowski, and J. E. Bowers, *Appl. Phys. Lett.* 71, 1507 (1997).
42. S. Ting, M. Bulsara, V. Yang, M. Groenert, S. Samavedam, M. Currie, T. Langdo, E. A. Fitzgerald, A. M. Joshi, R. Brown, X. Wang, R. M. Sieg, and S. A. Ringel, *Proc. SPIE* 3630, 19 (1999).
43. H. Zimmermann, "Integrated Silicon Opto-Electronics," p. 86. Springer-Verlag, Berlin, 2000.
44. T. P. Pearsall, *Mat. Sci Eng. B* 9, 225 (1991).
45. K. L. Wang and R. P. G. Karunasiri, *J. Vac. Sci. Technol. B* 11, 1159 (1993).
46. B. S. Meyerson, *Sci. American*, p. 42, March (1994).
47. R. A. Metzger, *Comp. Semicon.* 1, 2 (1995).
48. S. Subbanna, G. Freeman, J.-S. Rieh, D. Ahlgren, K. Stein, C. Dickey, J. Mecke, P. Bacon, R. Groves, M. Meghelli, M. Soyuer, B. Jagannathan, K. Schonenberg, S.-J. Jeng, A. Joseph, D. Coolbaugh, R. Volant, D. Greenberg, H. Chen, K. Brelsford, D. Harame, J. Dunn, L. Larson, D. Herman Jr., and B. Meyerson, *Jpn. J. Appl. Phys.* 41, 1111 (2002).
49. M. Levinshtein, S. Rumyantsev, and M. Shur, Eds., "Handbook Series on Semiconductor Parameters," pp. 1–33. World Scientific, Singapore, 2000.
50. C. W. Pearce, in "VLSI Technology" (S. M. Sze, Ed.), p. 55. McGraw-Hill, New York, 1988.
51. W. Shockley, U.S. Patent 2, 569, 347, 1951.
52. H. Kroemer, *Proc. IRE* 45, 1535 (1957).
53. L. Esaki, *IEEE J. Quantum Electron.* 22, 1611 (1986).
54. Z. I. Alferov, *Rev. Mod. Phys.* 73, 767 (2001).
55. F. Capasso and J. Faist, *Opt. Photon. News* 5, 15 (1994).
56. J. S. Smith, L. C. Chiu, S. Margalit, A. Yariv, and A. Y. Cho, *J. Vac. Sci. Technol. B* 1, 376 (1983).
57. F. K. LeGoues, *MRS Bulletin* 21, 38 (1998).
58. P. M. Mooney, L. Tilly, C. P. D'Emic, J. O. Chu, F. Cardone, F. K. LeGoues, and B. S. Meyerson, *J. Appl. Phys.* 82, 688 (1997).
59. S. M. Sze, "Physics of Semiconductor Devices," p. 750. Wiley, New York, 1981.
60. R. Potter, "Handbook of Optical Constants of Solids," p. 465. Academic Press, New York, 1985.
61. R. Braunstein, A. R. Moore, and F. Herman, *Phys. Rev.* 109, 695 (1958).
62. S. C. Jain and W. Hayes, *Semicon. Sci. Technol.* 6, 547 (1991).
63. J. C. Bean, *Proc. IEEE* 80, 571 (1992).
64. R. People, *IEEE J. Quantum Electron.* 22, 1696 (1986).
65. E. Bauer and H. Poppa, *Thin Solid Films* 12, 167 (1974).
66. D. C. Houghton, D. D. Perovic, J.-M. Baribeau, and G. C. Weatherly, *J. Appl. Phys.* 67, 1850 (1990).
67. O. P. Pchelyakov, Y. B. Bolkhovityanov, A. V. Dvurechenskii, L. V. Sokolov, A .I. Nikiforov, A. I. Yakimov, and B. Voigtlander, *Semiconductors* 34, 1229 (2000).
68. F. K. LeGoues, B. S. Meyerson, J. F. Morar, and P. D. Kirchner, *J. Appl. Phys.* 71, 4230 (1992).
69. G. Heigl, G. Span, and E. Kasper, *Thin Solid Films* 222, 184 (1992).
70. J. M. Baribeau, T. E. Jackman, P. Maigne, D. C. Houghton, and M. W. Denhoff, *J. Vac. Sci. Technol. A* 5, 1898 (1987).
71. E. A. Fitzgerald, Y. H. Xie, M. L. Green, D. Brasen, A. R. Kortan, J. Michel, Y. J. Mii, and B. E. Weir, *Appl. Phys. Lett.* 59, 811 (1991).
72. Y. H. Xie, E. A. Fitzgerald, P. J. Silverman, A. R. Kortan, and B. E. Weir, *Mat. Sci. Eng. B* 14, 332 (1992).
73. M. T. Currie, S. B. Samavedam, T. A. Langdo, C. W. Leitz, and E. A. Fitzgerald, *Appl. Phys. Lett.* 72, 1718 (1998).
74. S. K. Gandhi, "VLSI Fabrication Principles," p. 726. Wiley, New York, 1994.
75. H. Presting and H. Kibbel, *Thin Solid Films* 222, 215 (1992).
76. W. Jager, D. Stenkamp, P. Ehrhart, K. Leifer, W. Sybertz, H. Kibbel, H. Presting, and E. Kasper, *Thin Solid Films* 222, 221 (1992).
77. M. Copel, M. C. Reuter, E. Kaxiras, and R. M. Tromp, *Phys. Rev. Lett.* 63, 632 (1989).
78. J. L. Liu, C. D. Moore, G. D. U'Ren, Y. H. Luo, Y. Lu, G. Jin, S. G. Thomas, M. S. Goorsky, and K. L. Wang, *Appl. Phys. Lett.* 75, 1586 (1999).
79. D. Reinking, M. Kammler, M. Horn-von-Hoegen, and K. R. Hofmann, *Appl. Phys. Lett.* 71, 924 (1997).
80. M. Copel and R. M. Tromp, *Appl. Phys. Lett.* 58, 2648 (1991).
81. A. Sakai, T. Tatsumi, and K. Aoyama, *Appl. Phys. Lett.* 71, 3510 (1997).
82. A. Sakai and T. Tatsumi, *Appl. Phys. Lett.* 64, 52 (1994).

83. K. Eberl, S. S. Iyer, S. Zollner, J. C. Tsang, and F. K. LeGoues, *Appl. Phys. Lett.* 60, 3033 (1992).
84. J. Mi, P. Warren, P. Letourneau, M. Judelewicz, M. Gailhanou, M. Dutoit, C. Dubois, and J. C. Dupuy, *Appl. Phys. Lett.* 67, 259 (1995).
85. A. St. Amour, C. W. Liu, J. C. Sturm, Y. Lacroix, and M. L. W. Thewalt, *Appl. Phys. Lett.* 67, 3915 (1995).
86. H. J. Osten and E. Bugiel, *Appl. Phys. Lett.* 70, 2813 (1997).
87. R. A. Soref, *J. Vac. Sci. Technol. A* 14, 913 (1996).
88. B. Cunningham, J. O. Chu, and S. Akbar, *Appl. Phys. Lett.* 59, 3574 (1991).
89. I. Daruka, J. Tersoff, and A. L. Barabasi, *Phys. Rev. Lett.* 82, 2753 (1999).
90. L. Colace, G. Masini, F. Galluzzi, G. Assanto, G. Capellini, L. Di Gaspare, and F. Evangelisti, *Solid State Phenom.* 54, 55 (1997).
91. H.-C. Luan, D. R. Lim, K. K. Lee, K. M. Chen, J. G. Sandland, K. Wada, and L. C. Kimerling, *Appl. Phys. Lett.* 75, 2909 (1999).
92. E. Kasper, Ed., "Properties of Strained and Relaxed Silicon Germanium;" Vol. 12. INSPEC, The Institution of Electrical Engineers, London, 1995.
93. H.-C. Luan, D. R. Lim, L. Colace, G. Masini, G. Assanto, K. Wada, and L. C. Kimerling, "Proceedings of the Material Research Society," Vol. 607, 279 1999.
94. S. K. Ghandhi, "VLSI Fabrication Principles," p. 647. Wiley Interscience, New York, 1994.
95. T. J. King, J. R. Pfiester, J. D. Shott, J. P. Mc Vittie, and K. C. Saraswat, *IEDM Tech. Dig.* p. 253 (1990).
96. T. J. King and K. C. Saraswat, *IEDM Tech. Dig.* p. 567 (1991).
97. R. Venkatasubramanian, D. P. Malta, M. L. Timmons, J. B. Posthill, J. A. Hutchby, R. Ahrenkiel, B. Keyes, and T. Wangensteen, "Proceedings of the 1st World Conference on Photovoltaic Energy Conversion (WCPEC)," 1994, p. 1692.
98. L. Colace, G. Masini, F. Galluzzi, and G. Assanto, *Proc. Mat. Res. Soc.* 536, 469 (1998).
99. F. Evangelisti, M. Garozzo, and G. Conte, *J. Appl. Phys.* 53, 7390 (1982).
100. B. Jalali, S. Yegnanarayanan, T. Yoon, T. Yoshimoto, I. Rendina, and F. Coppinger, *IEEE J. Sel. Top. Quantum Electron.* 4, 938 (1998).
101. R. Soref and R. Bennet, *IEEE J. Quantum Electron.* 23, 123 (1987).
102. S. P. Pogossian, L. Vescan, and A. Vonsovici, *Appl. Phys. Lett.* 75, 1440 (1999).
103. S. Yokoyama, T. Nagata, Y. Kuroda, T. Doi, T. Namba, K. Miyake, T. Miyamoto, S. Miyazaki, M. Koyanagi, and M. Hirose, *J. Vac. Sci. Technol. A* 13, 629 (1995).
104. K. Wörhoff, L. T. H. Hilderink, A. Driessen, and P. V. Lambeck, *J. Electrochem. Soc.* 149, 85 (2002).
105. H. Temkin, T. P. Pearsall, J. C. Bean, R. A. Logan, and S. Luryi, *Appl. Phys. Lett.* 48, 963 (1986).
106. J. C. Bean, L. C. Feldman, A. T. Fiory, S. Nakahara, and I. K. Robinson, *J. Vac. Sci. Tech. A.* 2, 436 (1984).
107. A. Splett, T. Zinke, K. Petermann, E. Kasper, H. Kibbel, H.-J. Herzog, and H. Presting, *IEEE Photon. Techn. Lett.* 6, 59 (1994).
108. B. Jalali, L. Naval, and A. F. J. Levi, *J. Lightw. Techn.* 12, 930 (1994).
109. L. Naval, B. Jalali, L. Gomelsky, and J. M. Liu, *J. Lightw. Techn.* 14, 787 (1996).
110. B. W. Dodson and J. Y. Tsao, "Proc. 2nd Int. Symp. on Silicon Molecular Beam Epitaxy" (J. C. Bean and L. J. Scholwater, Eds.), pp. 105–113. Electrochemical Soc., Pennington, NJ 1988.
111. R. People and J. C. Bean, *Appl. Phys. Lett.* 47, 322 (1985).
112. F. Y. Huang, X. Zhu, M. O. Tanner, and K. L. Wang, *Appl. Phys. Lett.* 67, 566 (1995).
113. H. Lafontaine, N. L. Rowell, S. Janz, and D. X. Xu, *J. Appl. Phys.* 86, 1287 (1999).
114. S. Winnerl, D. Buca, S. Lenk, C. Buchal, S. Mantl, and D. X. Xu, *Mat. Sci. Eng. B* 89, 73 (2002).
115. F. Capasso, in "Lightwave Communication Technology, Semiconductors and Semimetal Series" (W. T. Tsang, Ed.). Academic Press, New York, 1985.
116. R. J. McIntyre, *IEEE Trans. Electron Dev.* 13, 164 (1966).
117. J. Lee, A. L. Gutierrez-Aitken, S. H. Li, and P. K. Battacharya, *IEEE Trans. Electron Dev.* 43, 977 (1996).
118. J. C. Campbell, A. G. Dentai, W. S. Holden, and B. L. Kasper, *Electron. Lett.* 19, 818 (1983).
119. S. Luryi, T. P. Pearsall, H. Temkin, and J. C. Bean, *IEEE Electron Dev. Lett.* 7, 104 (1986).
120. T. P. Pearsall, H. Temkin, J. C. Bean, and S. Luryi, *IEEE Electron Dev. Lett.* 7, 330 (1986).
121. H. Temkin, A. Antreasyan, N. A. Olsson, T. P. Pearsall, and J. C. Bean, *Appl. Phys. Lett.* 49, 809 (1986).
122. J. M. Senior, "Optical Fiber Communications," p. 471. Prentice-Hall, Englewood Cliffs, NJ, 1992.
123. S. Forrest, *J. Lightw. Technol.* 3, 347 (1985).
124. J. M. Senior, "Optical Fiber Communications," p. 458. Prentice-Hall, Englewood Cliffs, NJ, 1992.
125. H. Temkin, J. C. Bean, T. P. Pearsall, N. A. Olsson, and D. V. Lang, *Appl. Phys. Lett.* 49, 155 (1986).
126. F. Y. Huang and K. L. Wang, *Appl. Phys. Lett.* 69, 2330 (1996).
127. F. Y. Huang, K. Sakamoto, K. L. Wang, P. Trinh, and B. Jalali, *IEEE Phot. Techn. Lett.* 9, 229 (1997).
128. X. Shao, S. L. Rommel, B. A. Orner, H. Feng, M. W. Dashiell, R. T. Troeger, J. Kolodzey, P. R. Berger, and T. Laursen, *Appl. Phys. Lett.* 72, 1860 (1998).
129. W. G. Oldham, A. R. Riben, D. L. Feucht, and A. G. Milnes, *J. Electrochem. Soc.* 110, 53c (1963).
130. J. P. Donnely and A. G. Milnes, *Sol. State Electron.* 9, 174 (1966).
131. S. Luryi, A. Kastalsky, and J. C. Bean, *IEEE Trans. Electron. Dev.* 31, 1135 (1984).
132. A. Kastalsky, S. Luryi, J. C. Bean, and T. T. Sheng, *Proc. Electrochem. Soc.* PV85-7, 406 (1985).
133. P. Sutter, U. Kafader, and H. Von Känel, *Solar Energy Mat. Solar Cells* 31, 541 (1994).
134. S. B. Samavedam, M. T. Currie, T. A. Langdo, and E. A. Fitzgerald, *Appl. Phys. Lett.* 73, 2125 (1998).
135. L. Colace, G. Masini, F. Galluzzi, G. Assanto, G. Capellini, L. Di Gaspare, E. Palange, and F. Evangelisti, *Appl. Phys. Lett.* 72, 3175 (1998).
136. L. Colace, G. Masini, F. Galluzzi, G. Assanto, G. Capellini, L. Di Gaspare, E. Palange, and F. Evangelisti, *Proc. Mat. Res. Soc.* 486, 193 (1998).
137. G. Masini, L. Colace, G. Assanto, H.-C. Luan, K. Wada, and L. C. Kimerling, *Electron. Lett.* 35, 1 (1999).
138. L. Colace, G. Masini, G. Assanto, H.-C. Luan, K. Wada, and L. C. Kimerling, *Appl. Phys. Lett.* 76, 1231 (2000).
139. G. Masini, L. Colace, G. Assanto, H.-C. Luan, and L. C. Kimerling, *Electron. Lett.* 36, 2095 (2000).
140. G. Masini, L. Colace, G. Assanto, H.-C. Luan, and L. C. Kimerling, *IEEE Trans. Electron Dev.* 48, 1092 (2001).
141. S. Famà, L. Colace, G. Masini, G. Assanto, and H.-C. Luan, *Appl. Phys. Lett.* 81, 586 (2002).
142. G. Masini, L. Colace, and G. Assanto, *Mat. Sci. Eng. B* 69, 257 (2000).
143. L. Colace, G. Masini, G. Assanto, and F. Galluzzi, *Electron. Lett.* 34, 1968 (1998).
144. G. Masini, L. Colace, and G. Assanto, *Electron. Lett.* 35, 1549 (1999).
145. L. Colace, G. Masini, F. Galluzzi, and G. Assanto, *Mat. Sci. Semicon. Proc.* 3, 545 (2000).
146. G. Masini, L. Colace, and G. Assanto, *Opt. Mat.* 17, 243 (2001).
147. G. Masini, V. Cencelli, L. Colace, F. Denotaristefani, and G. Assanto, *Appl. Phys. Lett.* 80, 3268 (2002).
148. C.-L. Yang and D.-S. Pan, *J. Appl. Phys.* 64, 1573 (1988).

149. C.-L. Yang, D.-S. Pan, and R. Somoano, *J. Appl. Phys.* 65, 3253 (1989).
150. Y. Fu, M. Willander, W. Lu, and W. Xu, *J. Appl. Phys.* 84, 5750 (1998).
151. V. D. Shadrin, V. T. Coon, and F. L. Serzhenko, *Appl. Phys. Lett.* 62, 2679 (1993).
152. V. D. Shadrin and V. T. Coon, *J. Appl. Phys.* 73, 7747 (1993).
153. V. D. Shadrin, *Appl. Phys. Lett.* 65, 71 (1994).
154. K. Fujita, S. Fukatsu, Y. Shiraki, H. Yaguchi, and R. Ito, *Appl. Phys. Lett.* 61, 210 (1992).
155. T. L. Lin and J. Maserjian, *Appl. Phys. Lett.* 57, 1422 (1990).
156. T. L. Lin, A. Ksendzov, S. M. Dejewski, E. W. Jones, R. W. Fathauer, T. N. Krabach, and J. Maserjian, *IEEE Trans. Electron. Dev.* 38, 1141 (1991).
157. B. Y. Tsaur, C. K. Chen, and S. A. Marino, *IEEE Electron. Dev. Lett.* 12, 293 (1991).
158. B.-Y. Tsaur, C. K. Chen, and S. A. Marino, *Opt. Eng.* 33, 72 (1994).
159. T. L. Lin, T. George, E. W. Jones, A. Ksendzov, and M. L. Huberman, *Appl. Phys. Lett.* 60, 380 (1991).
160. B. S. M. Lin and J. Hwang, *J. Appl. Phys.* 75, 388 (1994).
161. H. C. Liu, L. Li, J.-M. Baribeau, M. Buchanan, and J. G. Simmons, *J. Appl. Phys.* 71, 2039 (1992).
162. R. P. G. Karunasiri, J. S. Park, and K. L. Wang, *Appl. Phys. Lett.* 59, 2588 (1991).
163. R. P. G. Karunasiri, J. S. Park, K. L. Wang, and S. K. Chun, *Opt. Eng.* 33, 1468 (1994).
164. J. S. Park, R. P. G. Karunasiri, and K. L. Wang, *Appl. Phys. Lett.* 60, 103 (1992).
165. R. P. G. Karunasiri, J. S. Park, and K. L. Wang, *Appl. Phys. Lett.* 61, 2434 (1992).
166. R. People, J. C. Bean, C. G. Bethea, S. K. Sputz, and L. J. Peticolas, *Appl. Phys. Lett.* 61, 1122 (1992).
167. P. Kruck, M. Helm, T. Fromherz, G. Bauer, J. F. Nutzel, and G. Abstreiter, *Appl. Phys. Lett.* 69, 3372 (1996).
168. J. Uschmann, H. Presting, H. Kibbel, K. Thonke, R. Sauer, W. Cabanski, and M. Jaros, *Thin. Solid Films* 294, 340 (1997).
169. H. Presting, M. Hepp, H. Kibbel, K. Thonke, R. Sauer, M. Mahlein, W. Cabanski, and M. Jaros, *J. Vac. Sci. Technol. B* 16, 1520 (1998).
170. H. Presting, J. Konle, M. Hepp, H. Kibbel, K. Thonke, R. Sauer, E. Corbin, and M. Jaros, *Opt. Eng.* 39, 2624 (2000).
171. Y.-C. Chang and R. B. James, *Phys. Rev. B* 39, 12672 (1989).
172. R. P. G. Karunasiri, J. S. Park, Y. J. Mii, and K. L. Wang, *Appl. Phys. Lett.* 57, 2585 (1990).
173. T. P. Pearsall, *Progr. Quantum Electron.* 18, 97 (1994).
174. D. J. Robbins, M. B. Stanaway, W. Y. Leong, R. T. Carline, and N. T. Gordon, *Appl. Phys. Lett.* 66, 1512 (1995).
175. R. Strong, R. Misra, D. W. Greve, and P. C. Zalm, *J. Appl. Phys.* 82, 5191 (1997).
176. R. Strong, R. Misra, D. W. Greve, and P. C. Zalm, *J. Appl. Phys.* 82, 5199 (1997).
177. R. Strong, D. W. Greve, R. Misra, M. Weeks, and P. Pellegrini, *Thin Solid Films* 294, 343 (1997).
178. E. Corbin, M. J. Shaw, M. R. Kitchin, M. Jaros, J. Konle, and H. Presting, *Opt. Eng.* 40, 2753 (2001).
179. T. Fromherz, P. Kruck, M. Helm, G. Bauer, J. F. Nutzel, and G. Abstreiter, *Appl. Phys. Lett.* 68, 3611 (1996).
180. H. Benisty, C. M. Sotomayor-Torres, and C. Weisbuch, *Phys. Rev. B* 44, 10945 (1991).
181. S. J. Xu, S. J. Chua, T. Mei, X. C. Wang, X. H. Zhang, G. Karunasiri, W. J. Fan, C. H. Wang, J. Jiang, S. Wang, and X. G. Xie, *Appl. Phys. Lett.* 73, 3153 (1998).
182. Z. Chen, O. Baklenov, E. T. Kim, I. Mukhametzhanov, J. Tie, A. Madhukar, Z. Ye, and J. C. Campbell, *J. Appl. Phys.* 89, 4558 (2001).
183. E.-T. Kim, Z. Chen, M. Ho, and A. Madhucar, *J. Vac. Sci. Technol. B* 20, 1188 (2002).
184. S.-W. Lee, K. Hirakawa, and Y. Shimada, *Appl. Phys. Lett.* 75, 1428 (1999).
185. P. Schittenhelm, C. Engel, F. Findeis, G. Abstreiter, A. A. Darhubber, G. Bauer, A. O. Kosogov, and P. Werner, *J. Vac. Sci. Technol. B* 16, 1575 (1998).
186. C. Miesner, O. Rothig, K. Brunner, and G. Abstreiter, *Physica E* 7, 146 (2000).
187. A. I. Yakimov, A. V. Dvurechenskii, Y. Y. Proskuryakov, A. I. Nikiforov, O. P. Pchelyakov, S. A. Teys, and A. K. Gutakovskii, *Appl. Phys. Lett.* 75, 1413 (1999).
188. A. I. Yakimov, A. V. Dvurechenskii, A. I. Nikiforov, and Y. Y. Proskuryakov, *J. Appl. Phys.* 89, 5676 (2001).

Encyclopedia of Nanoscience and Nanotechnology
www.aspbs.com/enn

Giant Magnetoresistance in Cobalt-Based Multilayers

D. H. Mosca, D. K. Lottis
Departamento de Física–UFPR, 81531-990 Curitiba PR, Brazil

CONTENTS

1. INTRODUCTION

Co-based multilayers with nanometer spacings exhibit a wide range of fascinating phenomena, such as quantum well states, oscillatory magnetic coupling, and giant magnetoresistance (GMR). The presence of these phenomena in Co-based multilayered structures has led to extensive studies and wide use in various applications due to their high magnetoresistance effect at room temperature and moderate saturation fields. The most important of these is their use in magnetoresistive readout heads in the latest generation of high-density recording heads for the computer industry. Technological applications of GMR devices also include switching and logic elements as well as memory devices. The ensemble of revolutionary ideas and technology was initially termed magnetoelectronics [1], but currently the term *spintronics* has become widely used for all GMR materials as well as all systems where spin-polarized electron transport arising from band structure and spin scattering occurs: tunneling magnetoresistance (TMR) materials, dilute magnetic semiconductors (DMSs), and hybrid ferromagnetic metal–semiconductor structures.

From the perspective of fundamental physics, GMR systems are equally interesting. Phenomena such as the oscillatory interlayer exchange coupling and the GMR effect (with its dependence on sample structure, composition, and geometry) have provided challenges on both the experimental and the theoretical fronts, with new techniques and approaches being developed to solve new families of problems. In addition to extending and testing our knowledge by applying well-established physical models to new systems, advances such as these have frequently resulted in new insights into that knowledge.

The variety of devices and transducers available on the market might raise questions about continued progress: have researchers exhausted the possibilities for advances in this field? To the contrary, scientific conferences on magnetism devoted specifically to magnetic sensors and actuators still bring revolutionary news such as conceptually novel nonvolatile magnetoresistive random access memories.

The aim of this review is to survey the state of the art in physics and technology concerning the GMR effect in Co-based nanostructures. This implies good coverage of GMR in general, since the vast majority of systems displaying the effect are based on either iron or cobalt. Theoretical calculations and models reflecting current understanding of GMR in transition metal multilayers have been reported in many reviews [1–7]. The emphasis of this review will be on experimental physics. The physics of Co-based multilayers is presented comprehensively, and sophisticated GMR technology is described and discussed. The review is organized into six sections: (1) Introduction, (2) Fabrication of Cobalt-Based Structures, (3) Magnetic Coupling, (4) Charge and Spin Transport, (5) Technological Applications, and (6) Final Remarks.

1.1. Giant Magnetoresistance Phenomena

Many readers are already familiar with the most outstanding aspects of the phenomenon that has come to be known as giant magnetoresistance (GMR). They will be content to skip over this section, which introduces some of the underlying physical concepts and facts prior to the detailed discussions offered in Sections 2–5.

A well-known feature of electrical conduction in nonmagnetic metals is the increase in resistivity that is observed when a magnetic field is applied. Roughly speaking, the effect is due to the Lorentz force acting on the conduction

ISBN: 1-58883-059-4/$35.00

electrons, and it is not surprising that resistance tends to increase most dramatically when the field is applied in a direction perpendicular to the current flow. This "ordinary" magnetoresistance (MR) is quite small (under 1%) for most metals at room temperature and in moderate fields—a tesla or less [8].

Ferromagnetic materials, on the other hand, display qualitatively different properties. When a current flows in a bulk ferromagnetic alloy, resistance variations in the presence of applied magnetic fields are strongly dominated by the field's influence on the magnetization. For samples in a saturated, single-domain state, there is a dependence of the resistivity on the angle between the current and the magnetization that can be quite significant in some alloys: changes of 2 or 3% in small fields (under 10 mT) are achieved in many systems at room temperature. This effect, known as anisotropic magnetoresistance (AMR) or spontaneous anisotropy of the resistivity (SAR), has physical origins that are quite different from the ordinary MR.

One strong hint at the difference in mechanisms is the fact that bulk polycrystalline ferromagnets show a resistance *maximum* when the magnetization is parallel to the current, and a *minimum* when the current and the magnetization are perpendicular. Furthermore, at room temperature a ferromagnet's resistance continues to change even after the magnetization is saturated—forced into a single-domain state: the resistance is observed to *decrease* with increasing field. As we will see in Section 4, these effects are understood largely in terms of the interaction between the s electrons and the spin-polarized d electrons responsible for the ferromagnetism in transition metals and their alloys.

In 1986, Grunberg and co-workers [9] discovered antiferromagnetic interlayer exchange coupling in Fe/Cr(0.9 nm)/Fe sandwiches produced by molecular beam epitaxy. In the absence of applied fields, the magnetization in neighboring Fe layers points in opposite directions. This finding happened at the same time as Majkrzak and co-workers [10] were conducting research that resulted in the discovery of antiferromagnetic coupling in the Gd–Y superlattices. Two years later, while working as a postdoctoral research associate with the research group led by A. Fert in Orsay, Baibich [11] performed MR measurements on antiferromagnetically coupled Fe(3 nm)/Cr(X nm) superlattices with $X = 1.8, 1.2$, and 0.9 nm. The results were astonishing: the sample's resistance showed a decrease of about 50% in the resistance when the magnetic field applied was strong enough to magnetically saturate the system, bringing the magnetization of all layers into alignment. The magnitude of the effect motivated the coining of a term that has been almost universally adopted: giant magnetoresistance.

The discovery reported by Baibich and collaborators [11], along with a qualitatively similar report by Binasch and co-workers [12], quickly prompted the establishment of research projects in numerous well-known laboratories. In addition to replicating the initially reported effects, efforts were aimed at identifying new systems, particularly ones displaying large effects at room temperature. The best results for exchange-coupled multilayered structures were simultaneously obtained by Mosca et al. [13] and Parkin et al. [14] with an MR effect as high as 50% at room temperature (an unpublished work announced results better than 110% at room temperature [15]). Parkin [16] also demonstrated that GMR can be produced in multilayered systems with various transition metal spacers and that the epitaxial stacking was not an essential characteristic.

Within a time span of only three or four years, research in GMR had taken on a position of substantial significance, with hundreds of accumulated publications and over a dozen well-established programs at world-class research laboratories. A number of fascinating and controversial physical phenomena were being studied, with the potential for applications rising at the same dizzying rate as the number of interesting fundamental physical issues.

1.2. Cobalt-Based Nanostructures: Multilayers, Pillars, Nanowires, Cluster Arrays, Dots

The preparation and characterization of materials at the beginning of this century makes it possible to fabricate materials with control on the atomic scale. The interest in low-dimensional materials is fundamentally driven by access to properties (electric, magnetic, optical, mechanical, and chemical) which are a function of the dimensions of the materials. Magnetic nanostructured systems are a particularly interesting subclass of atomically engineered structures, since the cooperative phenomenon itself (magnetic order) and the connections between magnetic properties and the underlying electronic states lead to a wealth of opportunities for developing magnetic devices.

Confinement and quantization phenomena, already visible in semiconductors at room temperature at dimensions greater than 100 nm [17], are typically seen in metals at dimensions of a few nanometers owing to their high electron density [4]. A particularly interesting example is the observation of confined standing waves in a "quantum corral" consisting of Fe atoms on top of a (001) surface of Cu single crystals [18].

Among the nanostructured materials described in the current literature, we can mention examples ranging from three-dimensional to zero-dimensional geometries: multilayers, stripes, pillars, nanowires, cluster arrays, and dots [4, 19]. Figure 1 shows different types of GMR nanostructured materials.

From the point of view of preparation methods, the most versatile class of GMR material is cluster arrays or granular systems consisting of nanosized granules of magnetic materials dispersed in a nonmagnetic host material which can be fabricated by co-deposition of immiscible ferromagnetic and noble transition metals from sputtering [20, 21], ultrahigh vacuum (UHV) evaporation [22], electrodeposition [23], melt spinning [24, 25], chemical synthesizing [26, 27], and ball milling [28] or by co-deposition of discontinuous multilayered structures [29, 30].

It is also worth noting that very interesting features have been recently observed in zero- and one-dimensional Co structures. Let us give two examples. First, a study of a monolayer-thick array of Co nanodots on Cu(001) and Au(111) indicated that Co dots have a spontaneous magnetization at a finite Curie temperature and the transition temperature to a long-range two-dimensional (2D)

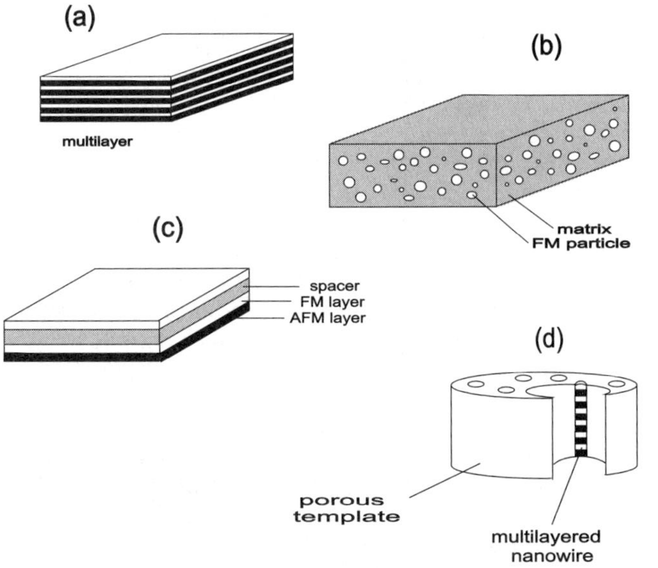

Figure 1. Various types of GMR structures: (a) multilayer, (b) granular film, (c) spin valve, and (d) multilayered nanowires.

ferromagnetic state is higher than that for a Co film directly deposited on a Cu(001) surface [31]. Second, large magnetoresistance (described as *ballistic magnetoresistance*) has recently been observed in nanocontacts [32] and in an array of Co dots [33, 34].

The fabrication of early generations of magnetic nanostructured materials required very sophisticated and refined techniques such as molecular beam epitaxy (MBE) and ultrahigh vacuum sputtering. These materials can now be obtained using simpler and less expensive techniques such as electrodeposition [35, 36]. Typical methods include a variety of processes based on vacuum evaporation, electrodeposition, sputtering, ion plating, chemical deposition, thermochemical reactions, and mechanical alloying. Some of these methods will be described in Section 3.

2. FABRICATION OF COBALT-BASED STRUCTURES

Many compilations have been devoted to the physical and chemical properties of metallic cobalt. The most recent and complete of these dates back to 1979 [37]. First-principles theories of magnetic phenomena in Co nanostructures have been recently reviewed by Wu and Freeman [38]. Some important structural and magnetic properties of cobalt are reviewed later. Next, we briefly present the most important Co-based nanostructures and introduce the techniques employed in their preparation.

2.1. Metallic Cobalt: Stable and Metastable Phases

Bulk Co appears in two crystalline structures, that is, a hexagonally close packed (hcp) form which is stable below about 673 K and a face-centered cubic (fcc) form stable at higher temperatures up to the melting point. The lattice parameters of cobalt at ambient temperature are $a = 2.5074$ Å and $c = 4.0699$ Å for hexagonal Co and $a = 3.5442$ Å for cubic Co, respectively. The major part of the binding energy in the cobalt is associated with the directional homopolar interaction of d electrons, which implies that the stacking-fault energy must be small [39]. A low allotropic transformation energy favors a martensitic phase transformation from hcp to fcc (470 to 510 J/mol) and phase coexistence of the cobalt near the ambient temperature in aggregates and thin films [37]. The relative stability of the two phases has long been known to depend on the grain size [40] and film thickness [41]. For very small grain size or small film thickness, the fcc structure becomes stable down to room temperature. Co exhibits a relatively high surface energy ($\gamma = 2.7$ J m^{-2}), owing to the partially filled d shell. Thus, smooth Co layers are, in general, difficult to obtain on top of noble metals [42] or insulating [43] substrates that have smaller surface energies.

To overcome the surface energy mismatch problems, nonequilibrium growth conditions are necessary, having sometimes been established by using a surfactant to promote a layer-by-layer growth mode [44, 45]. The self-diffusion of Co at constant temperature is markedly dependent on sample metallurgy due to the high diffusion rates along the grain boundaries [46] and are anomalously affected by the magnetic transformation [47].

Metastable phases were obtained for thin films epitaxially grown onto suitable single-crystal substrates. In general, the thickness to which a metastable cubic Co film will grow before transforming into hcp Co depends on the growth kinetics (substrate temperature and incoming Co flux rate) and also on the condition of the initial substrate surface.

Early in the history of epitaxially grown ferromagnetic films, cubic Co films were grown on the cleaved faces of NaCl and KCl [48] and the {111} faces of Cu single crystals [37]. Epitaxial cubic Co films have also been grown by electrodeposition on Cu and Ni substrates [49]. The structure of these films and the phase transformations depend on the film thickness.

More recently, metastable cubic Co was epitaxially grown on several substrates by the sputtering [50] and MBE [51, 52] techniques.

The match between the lattice spacing of body-centered cubic (bcc) Co and the (110) GaAs substrate is sufficiently close to enable Co epitaxy up to 37 nm [53, 54]. Ultrathin fcc Co on Cu(001) [55] and fcc Co(001)/Cu(001) [56] and Co(111)/Cu(111) [57] multilayers were also epitaxially grown by UHV evaporation. The influence of preparation techniques (UHV evaporation, direct-current (dc) magnetron sputtering, radio frequency (rf) diode sputtering) on different buffer layers and the coexistence of fcc/hcp Co phases were also investigated [58–60].

According to first-principles theoretical calculations by Moruzzi et al. [61], bcc Co is ferromagnetic for any reasonable value of unit cell volume, while fcc Co can exist in either a nonmagnetic or a ferromagnetic phase, and there is a range of volumes where the two phases can coexist. Experimental results confirm the ferromagnetic behavior of cubic Co, corroborating the theoretical predictions.

2.2. Magnetism and Magnetotransport of Cobalt

The spontaneous magnetization of cobalt at absolute zero is equal to 1.72 Bohr magnetons (μ_B), which corresponds to an absolute saturation magnetization of 162.5 G/g (16.25 T/kg) [62]. No significant change in the saturation magnetization is apparent at the allotropic transformation. As in the case of other ferromagnetic materials, the temperature dependence of the spontaneous magnetization of Co is weak at low temperatures. The magnetization decreases very slowly, with the slope increasing with temperature until it vanishes abruptly at the ferromagnetic Curie temperature T_c near 1121 °C (1394 K). Single crystals of Co are magnetically anisotropic.

For cobalt, the hexagonal axis $\langle 0001 \rangle$ is the direction of easiest magnetization at room temperature. On heating hexagonal Co, the magnetic anisotropy reduces; it becomes nearly isotropic around 250 °C (523 K). For cubic Co (above about 673 K), the crystal directions in order of increasing difficulty of magnetization are $\langle 111 \rangle$, $\langle 110 \rangle$, and $\langle 100 \rangle$, respectively. With increasing temperature, cobalt again becomes magnetically isotropic ($\sim$1000 °C). Single crystals of hexagonal Co exhibit a rather large volume magnetostriction with the ratio $\Delta V/V$ varying from contractions as large as 100×10^{-6} to expansions of up to 150×10^{-6}, depending on the crystallographic direction [63].

From the point of view of electronic transport, it is important to remember that the resistivity ρ at room temperature of single crystals of hexagonal Co is anisotropic. In the $\langle 0001 \rangle$ direction, the resistivity is almost twice the value observed along the $\langle 10\bar{1}0 \rangle$ direction. The magnetoresistance effect of single crystals of 99.8% cobalt along the $\langle 0001 \rangle$ direction has a saturation value $\Delta\rho/\rho$ of about 0.4% for a field strength of only about 200 Oe [64]. The anisotropic magnetoresistance (AMR) of single-crystal cobalt whiskers has been examined with current flow in different crystalline directions from temperatures below 4.2 K to far above room temperature [65]. It should be emphasized that for single crystals the dependence of the AMR on the current and field directions is much more complex than is the case for polycrystalline samples. More relevant to our discussion of GMR in Co-based nanostructures is the presence of a peak in the resistance versus applied field curves, near zero field, for the high-quality single-crystal samples of Coleman and Morris. This "GMR-like" feature is associated with the transition between multidomain and single-domain states and is observed at low temperatures. In the large field limit, the general behavior of the field dependence of magnetoresistance in hcp single-crystal Co is described by the functional form $\Delta\rho/\rho_0 = aB^n$, where the exponent n is less than 1 and Shubnikov–de Haas oscillations are observed [65].

2.3. Molecular Beam Epitaxy

MBE was originally developed for preparing compound semiconductors with improved structural and electrical qualities to allow the fabrication of devices such as lasers and modulated doped-channel FETs and HEMTs. In the 1990s, MBE was extensively applied to epitaxial insulators, magnetic metal multilayers, buried epitaxial metals in semiconductors, and dilute magnetic semiconductors. Co/Au, Co/Pd, Co/Pt, and Co/Cu were the first examples of MBE-grown magnetic superlattices with non-bulk-like properties [66].

An interesting property observed in MBE Co-based multilayers is the perpendicular magnetic anisotropy. We recall that an isotropic ferromagnetic material formed into a thin-film geometry shows "shape anisotropy": the magnetization tends to remain aligned in the plane of the film. This is due to the powerful demagnetizing field generated by the magnetization when it is perpendicular to the surface of the film. For Co/Au, Co/Pd, and Co/Pt structures with sufficiently thin cobalt layers, perpendicular magnetic anisotropy was observed persisting well above room temperature [67]. Co/Cr and Co/Pt multilayers as well as Co–Cr and Co–Pt alloys have been investigated for technological applications in perpendicular magnetic recording, which has enormous potential for ultrahigh density recording, but several breakthroughs will be required before their practical use [68]. Also, for Co/Pt multilayers a significant Kerr rotation of the plane of polarization of the light by the magnetization of the sample was also observed when interfacial Pt itself becomes magnetic due to its proximity to the Co [69].

Arrays of magnetic Co dots and wires are other interesting examples of MBE Co-based nanostructures. By demanding a higher degree of sophistication in the growth and characterization, Co nanometer-scaled dots self-arranged on a Cu(001) surface [31] and Co one- and two-dimensional nanostructures on a Au(111) surface [70] have been investigated. Both monolayer-thick arrays of Co dots on Cu(001) and Au(111) indicate that there exists a spontaneous magnetization and a superparamagnetic behavior below the finite Curie temperature (<50 K). An interesting feature observed was an oscillatory variation of the average surface in-plane lattice spacing with increasing Co monolayer coverage on Cu(001) which has been associated with relaxation of edge atoms [71].

2.4. Sputtering

Sputtering deposition methods (including diode sputtering, ion beam sputtering, and magnetron sputtering) emerged in the 1940s as a commercial coating process. It was only in the 1970s that the magnetically enhanced variant of diode sputtering (magnetron sputtering) emerged as a competitive alternative to other vacuum coating techniques such as thermal evaporation and electron beam evaporation. In the last two decades, along with applications such as metallizing for microelectronic circuits and chip carriers and depositing electrical resistance films, the industry has turned to sputter deposition techniques as a method for producing materials, samples, and devices. Particularly relevant here is the manufacture of magnetic thin films of Co and Co alloys for general magnetic storage devices, floppy disks, and thin-film magnetic heads. A wide variety of magnetic multilayered structures have been produced by sputtering. An extensive list of such structures was recently reviewed by Schuller et al. [72]. The low cost of the sputtering process and its compatibility with large-scale production with entirely automatic machines encouraged research activity.

Perpendicular recording materials such as Co–Cr with columnar microstructure and strong perpendicular anisotropy are good examples of Co-based films produced by sputtering. It is also worth noting that Co alloys and refractory metals cause problems because of differences in alloy constituent vapor pressures and their high melting points when conventional thermal evaporation techniques are used.

Magnetic multilayered films produced by sputtering with a ferromagnetic layer consisting of body-centered cubic Co–Fe alloys have been extensively studied. We recall that among all possible pairwise combinations of the three ferromagnetic 3d transition metals (Co, Fe, and Ni), Co–Fe displays the largest values of the Curie temperature and saturation induction (a room temperature value of 24.5 kG is obtained at 35 at% Co). The large magnetization of Co–Fe alloys correlates with the transition between bcc and hcp structures as the concentration varies.

In addition to single films and multilayers, researchers have used sputter deposition to produce increasingly sophisticated structures with new geometries. Lithographically patterned pillars for the study of magnetization reversal by spin-polarized current have been obtained from sputtered Co/Cu nanostructures [73]. Finally, it is remarkable that the fabrication of planar tunnel junctions is commonly carried out in multitarget sputtering systems by means of a lithographic process [74].

2.5. Electrodeposition

Electrodeposition is one of the simpler and cheaper processes available for the fabrication of metallic films and nanostructures [36]. The electrodeposition of metal, often referred to as electroplating, is a highly versatile technique that has recently been used to fabricate metallic alloys, compositionally modulated alloys, heterogeneous alloys, nanowires, multilayered nanowires, pillars, point contacts, and nanodots.

Compositionally modulated alloys with an (FM/I/...) nanostructure composed of Co-based alloys have been made using a single-bath method using potentiostatic control and a pulsed deposition technique [75] exhibiting oscillatory exchange coupling and a GMR effect at room temperature [30, 76–79].

Heterogeneous alloy films consisting of ferromagnetic particles or clusters in a nonmagnetic matrix exhibiting the GMR effect were also prepared using an electrochemical method [80, 81].

Nanowires and multilayered nanowires can be fabricated by electrodeposition into templates [82–84]. In particular, isolated magnetic nanowires and arrays of magnetic nanowires have recently attracted considerable interest motivated by ultrahigh-density magnetic recording applications, understanding of the CPP-GMR, and the determination of spin diffusion lengths [84–87]. As an example, pure Co nanowires grown by electrodeposition appear essentially single crystalline and possess their normal equilibrium phase (hcp) with a preferred orientation along the $\langle 1010 \rangle$ direction, that is, with close-packed planes almost parallel to the wires [88].

Other types of nanostructures have been grown by electrodeposition such as Ni/NiO/Co pillars [89], dot arrays [34], point contacts [90–93], and nanodots [19].

3. MAGNETIC COUPLING

As mentioned in Section 1.1, the GMR effect was originally found in transition metal–based exchange-coupled multilayers. Subsequently, the effect was found to oscillate with increasing thickness of the spacer layer and was quickly correlated to the change of interlayer exchange coupling between the ferromagnetic and antiferromagnetic cases. The connection between GMR and interlayer exchange coupling (IEC) was soon understood as follows: antiparallel orientations of consecutive magnetic layers resulted in higher resistance than the magnetically saturated configuration with all magnetic layers having parallel moments. The IEC "causes" the GMR by ensuring field-driven transitions between the high-resistance and low-resistance states.

More recently, evidence has accumulated suggesting a connection at a more fundamental level. Both phenomena can be understood as arising at least in part because of quantum mechanical effects altering the wave functions of conduction electrons in the multilayer. The connections between oscillatory interlayer coupling and GMR, along with the intrinsically interesting nature of the coupling, motivated the inclusion of this section on magnetic coupling in the present review.

3.1. Background: Magnetostatic Coupling

The existence of magnetic coupling in a simple "sandwich" structure, consisting of two magnetic slabs separated by a nonmagnetic spacer layer, is rather intuitive. The magnetization in the two slabs will tend to assume a configuration such that the moments in each layer are aligned with the field lines produced by the neighboring layer. Since the moments in each layer also tend to align with the field lines produced by the remainder of the same layer, the moments have a tendency to lie in the plane of the layers (shape anisotropy). Consequently, an antiparallel configuration for the magnetizations in a sandwich structure is expected on the basis of simple magnetostatic (dipolar) interactions.

In the thin-film limit of a sandwich structure, however, the magnitude of the magnetostatic coupling (for ideal, smooth layers) is vanishingly small. In other words, as the transverse dimensions of the sandwich structure become large, the energy associated with the interlayer coupling becomes vanishingly small. Actual sandwich structures produced with typical deposition processes also tend to exhibit in-plane anisotropy: the magnetization has one or more preferred orientations. This anisotropy is frequently much stronger than the magnetostatic coupling expected for smooth, thin sandwich structures.

Even in the thin-film limit, magnetostatic coupling can become significant if surface roughness is present: steps, ripples, and other features in the surfaces of the magnetic layers result in "stray" magnetic field lines that couple with nearby magnetic moments in the neighboring layer. This type of mechanism for producing interlayer coupling in the thin-film limit has been evaluated with quantitative models and is expected to produce small but measurable effects [3].

Rather complicated models have been developed to take into account the various structural influences (such as interfacial roughness, pinholes, strain, misfit dislocations) [2–4, 94–97].

3.2. Interlayer Exchange Coupling

The magnitude of the antiferromagnetic coupling observed in Fe/Cr/Fe nanostructures by Grunberg et al. [9] was much larger than what could be expected on the basis of magnetostatic coupling. Instead, the interactions between moments in the neighboring magnetic layers seemed to arise from quantum mechanical mechanisms analogous to the exchange coupling responsible for ferromagnetism in metals. The presence and thickness of the intervening layers ruled out direct coupling of the moments: indirect exchange coupling involving the electrons in the spacer layer must be present.

The antiferromagnetic coupling was directly confirmed using neutron scattering [98]. Magnetometric methods and the magnetooptical Kerr effect (MOKE) were used to directly measure the coupling in magnetic multilayers [3, 6, 7]. It was clear that dipolar interactions could not be responsible for the strength of the antiferromagnetic interlayer coupling ($J \sim 1$ mJ/m^2) observed in the Fe/Cr multilayers.

In 1990, Parkin et al. [99] reported that GMR in sputtered Fe/Cr multilayers oscillates when the thickness of the Cr increases beyond 1.8 nm, indicating that the coupling changes with the thickness of the Cr layers. These oscillations correspond to a sequence of zero-field magnetizations along the multilayered film aligned either parallel or antiparallel as a result of an effective ferromagnetic and antiferromagnetic exchange coupling. In contrast with Gd–Y multilayers, the oscillation in Fe/Cr superlattices has a period of 1.8 nm, which is far higher than the value given by a direct and simplistic application of the Rudderman–Kittel, Kasuya–Yoshida (RKKY) theory [3]. The unique magnetic properties of Cr were also ruled out as providing a mechanism for the interlayer coupling for various reasons, including the lack of correlation between the oscillation period and the characteristic structural and magnetic lengths for bulk Cr, along with the persistence of significant coupling strength well above the bulk Néel temperature. Furthermore, Parkin et al. [99] soon determined that the Co/Ru multilayer shows an oscillatory IEC somewhat similar to what is observed in Fe/Cr, suggesting that other mechanisms could be present. The coupling strength (and GMR) of the Co/Cu multilayer described by Mosca et al. [13] and Parkin et al. [14] also oscillated with spacer thickness, and later work by Parkin [16] showed that the oscillatory IEC is indeed present in multilayers with a wide variety of compositions.

3.3. Long- and Short-Period Oscillations

In general, the period of oscillation depends on the material of the spacer, the crystal structure, and the growth direction of both ferromagnetic and nonferromagnetic layers [100, 101].

More precise investigations of exchange interlayer coupling have revealed that, besides the oscillations mentioned previously (long-period oscillations), faster oscillations (short-period oscillations) also existed. In fact, two different periods of oscillation were revealed [101–103]. Short-period oscillations are quite hard to detect since they are strongly dependent on the roughness of the interfaces. Oscillations of the GMR ratio reveal the oscillatory exchange coupling as shown in Figure 2.

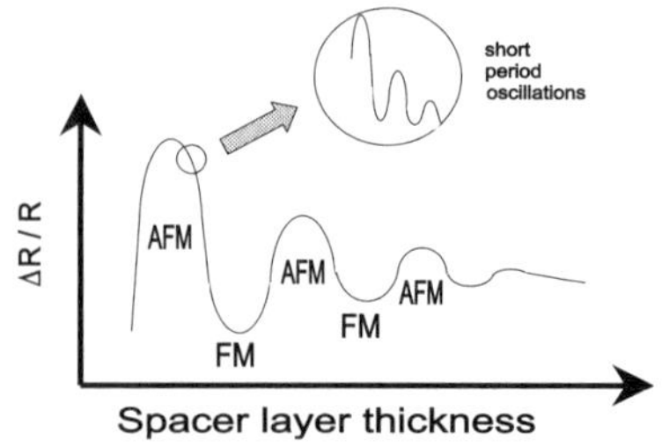

Figure 2. Schematic drawing of oscillations in the magnetoresistance ratio. The character of coupling between consecutive FM layers changes from AFM to FM with some period when the spacer thickness is varied. Short-period oscillations in the zoom.

Grunberg and co-workers [104] and Purcell et al. [105, 106] used a wedge-shaped spacer to investigate the local value of the interlayer coupling for precisely determined thicknesses of the spacer. Similarly, a deeper insight into the complex behavior of the coupling oscillations with short-range and long-range periods resulted from studies conducted with epitaxial sandwiches of Fe/Cr/Fe [102] and Co/Cu/Co [107] with a wedge-shaped spacer layer. The same results were essentially obtained for sputtered Co/Cu multilayers with wedge-shaped Cu layers [108]. Various other methods have been used and new ones developed, such as Brillouin light scattering (BLS) [109], ferromagnetic resonance (FMR) [109], spin-polarized low-energy electron diffraction (SPLEED) [110], and scanning electron microscopy with polarization analysis (SEMPA) [102].

The magnetic interlayer coupling in various Co-based structures textured (111) and (110) for fcc and bcc, respectively, such as Co/Cu [13, 14, 56, 111], Co/Au [112], Co/Ru, Co/Mo, Co/V, Co/Ir [16], Co/Ag [113], CoFe/Cu [114], and Co/Cr [115], exhibits a period of oscillation Λ varying between 0.9 and 1.1 nm, except for the Cr spacer with 1.8 nm. It is also worth noting that for fcc Co/Cu(111) multilayers $\Lambda =$ 1.0–1.1 nm [13, 14], whereas for fcc Co/Cu(100) multilayers $\Lambda =$ 2.1–2.8 nm [107, 116]. A short-period oscillation is present in the case of fcc Co/Cu(100) multilayers where $\Lambda = 0.9$ nm [116].

These results corroborate the generalized RKKY theory [3]. This theoretical approach employs concepts and calculations employed in the original RKKY theory for magnetic impurities interacting by means of a free-electron gas. In the generalized form, the model must reflect the participation of the large numbers of magnetic moments in the interacting ferromagnetic layers, as well as incorporating the nonspherical shape of the Fermi surface of the spacer layer. The latter consideration is essential for accounting for experimental observations; the results obtained using the generalized RKKY model allow prediction in an essentially correct manner of the periods of oscillatory coupling. In particular, in the case of multilayers with noble metals as the spacer layer, two different oscillation periods are expected from Fermi surface nesting along the (100) and (110) orientations. In fact, two different periods of oscillation were revealed in other (100) textured multilayers such as Fe/Cr/Fe [105], Fe/Au/Fe [117], and Fe/Ag/Fe [118].

Although the theoretical results obtained from the generalized RKKY model for the periods of oscillation correspond fairly well with the experimental ones, results of modeling the magnitude of the oscillations and phase are not as satisfactory. Furthermore, the magnitude of the oscillations is found to depend on certain properties of the ferromagnetic layer, once again departing from predictions obtained from the generalized RKKY model. Other theoretical models have been proposed to explain the interlayer exchange coupling. Jones and others [2–4, 94–97] have published comparative reviews of these models.

3.4. Other Coupling Effects

Besides the coupling effects mentioned previously, several others have been observed. We have selected a few of these for discussion in this section.

The magnetic coupling in spin valve structures is very complex relative to a multilayered structure since the exchange bias, RKKY-like, and orange peel or Néel couplings intervene with similar magnitudes.

Another type of coupling that may also occur is the so-called biquadratic coupling. This kind of magnetic coupling is represented by a term proportional to the square of the internal product of the unit vectors (i.e., cosine of angle) in the direction of magnetization of interacting adjacent layers. Biquadratic coupling tends to align the adjacent layers perpendicular to each other, and its origin is not fully clear. Ordinary, antiferromagnetic IEC is proportional to the same cosine, and it has been found that the biquadratic coupling term must be included in phenomenological expressions describing the behavior of certain systems. A comprehensive analysis of biquadratic coupling in magnetic trilayers was reported by Rezende and co-workers [119–123] using various techniques: MOKE, BLS, and FMR. In brief, they concluded that biquadratic coupling could not be attributed exclusively to intrinsic mechanisms.

A new direction in the research of the interlayer exchange coupling was opened, addressing the effect of the magnetic layer on the strength and phase of the oscillatory coupling. Bruno [124, 125] and Barnas [126] have predicted that the strength of the coupling should oscillate with the ferromagnetic layer thickness. Bloemen et al. [81] have confirmed that the oscillatory interlayer exchange coupling across Cu(100) depends on the Co thickness, strongly supporting the model of Bruno. To understand this effect, it has been necessary to consider the effect of the nanostructure on the electronic structure of the metals in each layer, reflecting the alteration (compared to the bulk) in the quantum mechanical states available to the conduction electrons.

3.5. Quantum Well States

The discovery that a quantum well state (QWS) exists in a metal overlayer grown on a ferromagnetic substrate promoted intense research on the intrinsic relation between the QWS and the oscillatory coupling and GMR magnetic multilayers [4]. The QWS nature of the interlayer coupling was identified by magnetic measurements [127, 128] and photoemission experiments [129].

The QWS can be regarded as a consequence of the nanometer-scale interlayer spacing, which tends to confine certain electron wave functions to dimensions comparable with their wavelengths. A more fundamental way of describing QWS is to recognize that conduction electrons in the layers of a GMR structure can "experience" potential wells in the vertical direction associated with the material's layered structure. The energy contribution associated with their motion in this direction is quantized into bands with structure on the same energy scale as the band structure for the bulk metals. This results in changes to the shape of the density of states (DOS) function and to the value of the density of states at the Fermi level.

An essential physical aspect of (FM/N/...) layered magnetic nanostructures is the formation of the spin-polarized QWS due to spin-dependent electron confinement due to the reflection of electrons at the interfaces. These reflections are spin dependent. Figure 3 shows QWS formation by reflections of electrons at the interfaces of ferromagnetic layers.

Himpsel et al. [4] have recently achieved significant progress by modeling exchange coupling with the assumption that quantum well states tailor the electronic properties of a nanoscaled material.

The quantum interference model proposed by Bruno [96, 130] describes the exchange coupling as spin-dependent reflections of Bloch waves at the multilayer interfaces. It has been found that the antiferromagnet (AFM) coupling strength oscillates as a function of the ferromagnet (FM) layer thickness because of a density of polarized states modulation which, in turn, is due to multiple electron reflections within the FM layer.

The experimental observation of oscillations in the exchange coupling as a function of the thickness of the ferromagnetic layer corroborates the quantum interference model [127, 128]. Also, the strong temperature dependence of IEC strength in Co(hcp)/Cu sandwiches, implying temperature dependence which is not only governed by the spacer Fermi surface, supports the QSW models [131].

4. CHARGE AND SPIN TRANSPORT

The theoretical explanation for the GMR effect is relatively well established and covered in a large number of papers and books [3, 6, 7]. On the other hand, TMR systems continue to attract considerable research [132]. In this section, we will describe the most important milestones and ideas in the development of the GMR and TMR theories.

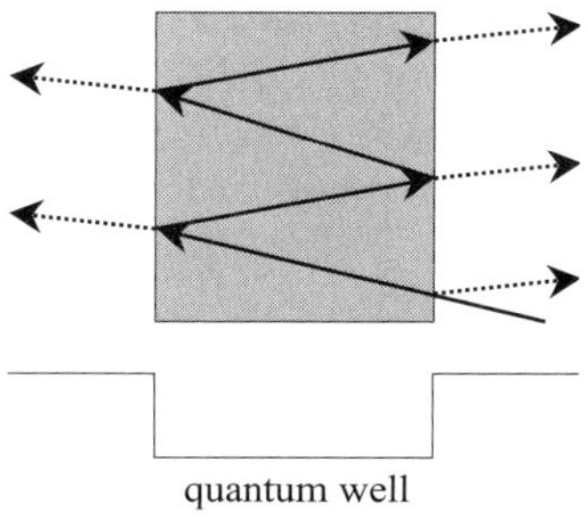

Figure 3. Quantum well states formed by reflections of electrons at the interfaces of a ferromagnet with a nonmagnetic metal.

4.1. GMR Systems

A brief introduction to GMR was offered in Section 1. In this section, we offer a more detailed discussion of the wealth of phenomena associated with GMR in a variety of systems. Initially, we shall broaden our scope of GMR systems by describing some of the nanostructures developed during the first few years of research on the magnetoresistance of layered systems. Next, we shall offer an overview of key advances in theoretical descriptions of this class of phenomena. The section concludes by describing some of the more interesting Co-based GMR nanostructures developed in recent years and by highlighting unresolved issues in theoretical models.

The spectacular magnitude of the negative MR reported for carefully prepared multilayers of Fe/Cr—and, within less than two years, Co/Cu—attracted a great deal of attention (see Fig. 4 for an example of typical GMR data for Co/Cu multilayers). Another less spectacular result reported at about the same time led to similarly important physical conclusions. The research group led by Renard reported "enhanced magnetoresistance" in $(Co/Au)_N$ multilayers with perpendicular anisotropy [133]. The resistance was found to pass through a maximum value during the reversal of the structure's magnetization, with an effect substantially larger than what would be expected based on AMR within the Co layers.

The same group subsequently reported similar effects in conceptually simpler sandwich-type structures, where the top and bottom Co layers had different thicknesses, which was known to result in different coercive fields. Since the Co layers experience magnetization reversal at different values of a reverse field, the structure has an antiparallel magnetic configuration for certain values of the applied field during a hysteresis cycle. The enhanced MR was observed for these ranges of applied field. Although the effect was only a few percent, it suggested that a mechanism similar to the one present in the AF-coupled Fe/Cr system could cause increased MR even in the absence of interlayer exchange coupling. The correlation between the magnetic configuration (parallel and antiparallel orientations of consecutive ferromagnetic layers) and the electrical resistance (low-resistance and high-resistance states) was further documented with the discovery, described in the previous section, of oscillations in the IEC between ferromagnetic and antiferromagnetic alignment. As described in the previous section, the GMR was found to oscillate in phase with the coupling strength, with maxima for the AFM-coupled and minima for the FM-coupled structures.

The presence of a "GMR-like" effect in uncoupled multilayers with ferromagnetic layers having equal thicknesses was hypothesized to arise due to a "partially antiparallel" configuration during the magnetization reversal process, since the resistance maxima were at or near the coercive fields.

Many other researchers later described the GMR effect in various uncoupled multilayered systems with asymmetric magnetic layers FM1 and FM2 in (FM1/N/FM2/...) structures [134–137], where FM1 and FM2 are similar ferromagnetic metals with distinct thicknesses, qualitatively similar to the Co/Au structure reported by Renard's group. The other way of ensuring that alternating ferromagnetic layers would have different coercivities was to employ dissimilar ferromagnetic metals (such as cobalt and Permalloy, a magnetically soft nickel–iron alloy) for FM1 and FM2 as defined previously. Figure 5 shows the principle of GMR in uncoupled FM1/NM/FM2 sandwiches.

Subsequent work showed that dissimilar metals (Permalloy and cobalt) can display *oscillatory interlayer exchange coupling* through copper spacer layers, resulting in two physically distinct mechanisms for driving transitions between the high-resistance (antiparallel) and the low-resistance (parallel) configurations of the layer-by-layer magnetizations.

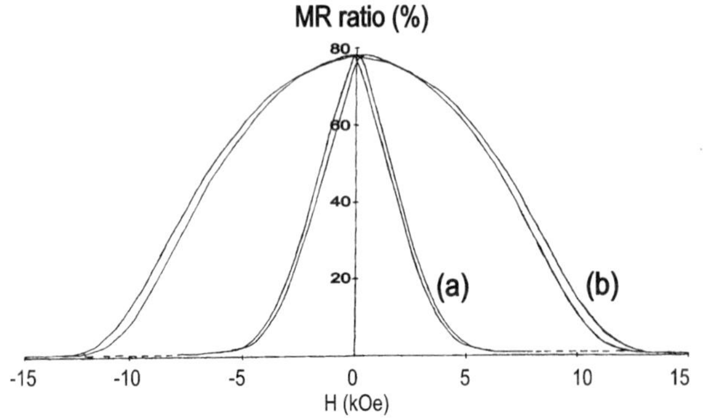

Figure 4. Magnetoresistance curve of a (Co 15 Å/Cu 9 Å)$_{30}$ multilayer measured at 4.2 K for a magnetic field applied (a) parallel to the current in the film plane and (b) perpendicular to the film plane. The magnitude of the MR ratio, given in terms of the resistances as $1 - R(H_s)/R(0)$, is independent of the field direction; only the saturation field H_s is changed due to the large demagnetizing effect when the external field is applied perpendicular to the film plane [13].

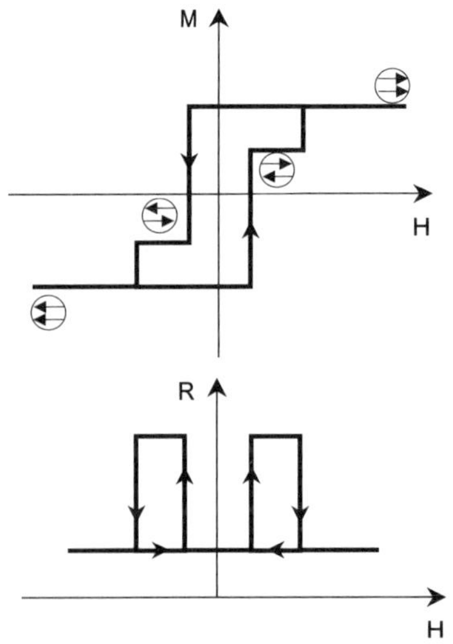

Figure 5. Principle of the asymmetric spin valve mechanism for an uncoupled structure having magnetic layers with different coercive fields. For large magnetic field strength H, both layers are parallel magnetized. When H decreases, the layer of the smaller coercive field switches its state of magnetization M. The second layer remains magnetized antiparallel with respect to the first layer until H is sufficient to reverse its magnetization. The resistance R correspondingly follows these changes in magnetization alignment.

The interplay between the two mechanisms results in an intricate variation of the measured GMR with the various layer thicknesses [138, 139].

Dieny et al. [140–142] proposed a new exchange-biased coupled sandwich structure they called a *spin valve*, where an antiferromagnetic AFM layer (CoO, NiO, FeMn) is used for pinning the magnetization of one of the ferromagnetic FM layers in (FM/N/FM/AFM) structures. The pinned layer maintains a fixed magnetization in the presence of weak to moderate applied fields, whereas the free layer, particularly when magnetically soft, is easily reversed with small applied fields. These structures are therefore characterized by transitions between the high-resistance and the low-resistance states that are extremely abrupt compared to AF-coupled Fe/Cr and Co/Cu structures. This, in turn, can imply, in some cases, a very high *sensitivity,* or percentage change in resistance per unit applied field, at the expense of a narrow range of field values over which the resistance changes are produced.

The first generation of spin valves was constructed using the magnetically soft Ni–Fe alloy known as Permalloy [6, 140–142], thus optimizing the abruptness of the low-field transition. Later work investigated the increase in the GMR and better temperature stability produced by inserting a thin layer of cobalt at the interfaces between the Permalloy and the Cu layers [143].

An impressive compilation of data on various polycrystalline Co-based multilayers with magnetic layer composed of Co and spacer layers of the transition metals (T, V, Cr, Mn, Zr, Nb, Mo, Tc, Ru, Rh, Pd, Hf, Ta, W, Re, Os, Ir, Pt) and noble metals (Cu, Ag, Au) was reported by Parkin and co-workers at IBM [3, 16].

The starting point for theoretical research into the mechanisms responsible for the resistance variations between parallel and antiparallel configurations of ferromagnetic layer magnetizations in GMR systems was, of course, existing knowledge about conduction in ferromagnetic metals. In the case of the main ferromagnetic transition metals, the electrons responsible for the magnetic moments are in d bands that are near the Fermi energy and that are intermediate between the very broad s bands and the very narrow bands associated with localized core electrons. In other words, these metals are itinerant ferromagnets: the magnetic moments arise from electrons that participate in conduction, although in a much less significant manner than their s-band counterparts. The presence of d states at the Fermi energy implies that s–d scattering processes can be quite significant in these metals. Indeed, there is substantial experimental evidence supporting the importance of s–d scattering in Fe, Co, Ni and their alloys.

For these conductors, the majority band is almost entirely below the Fermi energy, while the minority band is displaced in such a manner that a larger portion of the band is empty. Given the shape of the bands (density of states versus energy), this tends to result in the s electrons having spin-dependent scattering probabilities for s–d processes. For example, the s electrons in the minority band usually "see" a larger number of available d states at the Fermi energy, increasing the probability of a collision (transition between k states) taking place.

In many cases, particularly at low temperatures, collisions involving a reversal of the electron's spin direction are extremely rare compared to spin-preserving collisions. The spin-up and spin-down conduction electrons behave essentially as separate currents, visualized as separate "wires" or resistors.

To account for properties of magnetic alloy resistivities, the preceding concepts were developed into a phenomenological model based on the classical two-current model introduced by Mott [144, 145], originally used to describe the contributions to the measured current from nearly free s electrons and the less mobile narrow-band d electrons in nonmagnetic transition metals. In adapting that model to ferromagnets, the two currents corresponded to majority- and minority-spin conduction electrons. The model focused on the s-electron contribution to the conductivity, with d electrons appearing only through their contribution to spin-dependent scattering rates and mean-free paths. This "spin-polarized" version of the two-current model was probed and substantiated by Campbell and Fert for bulk alloys with research conducted largely in the 1960s and 1970s [8]. These systematic studies allowed the determination of the resistivities associated with the spin-up and spin-down channels, as well as the corresponding asymmetries, for the ferromagnetic transition metals and a wide range of ferromagnetic alloys based on Fe, Co and/or Ni. Figure 6 shows the schematic diagram of the two-current mechanism using a density of states picture.

With the two-current picture of conduction in ferromagnets, a simple conceptual interpretation was formulated by Fert's group and included in the first report on GMR in Fe/Cr. For the ferromagnetic configuration, one spin direction is systematically less scattered than the other one. These electrons dominate the contribution to the total current, in a sort of "short-circuit" effect (see Fig. 7). For the antiparallel configuration, the short-circuit effect vanishes, since all of the electrons will experience substantial scattering for either one or the other magnetic alignments.

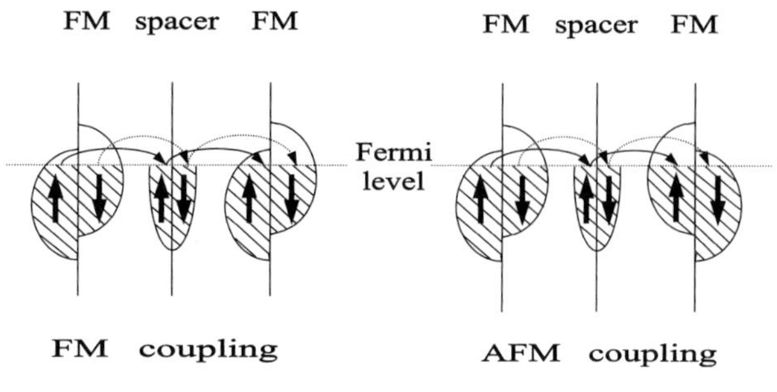

Figure 6. Schematic diagram of the density of states in nonmagnetic spacer and ferromagnetic metals. Energy bands for nonmagnetic metals are the same for electrons with different spins, whereas, for ferromagnetic metals, they split into two subbands, one in which electrons have their spin parallel to the magnetization (spin-up electrons) and the other in which the spin is antiparallel to the magnetization (spin-down electrons). The difference in the density of states at the Fermi level for spin-up and spin-down d electrons leads to different scattering rates with different mean-free paths and resistivities. The two-current mechanism corresponding to spin-up and spin-down electrons is represented in the AFM- and FM-coupled configuration of magnetizations.

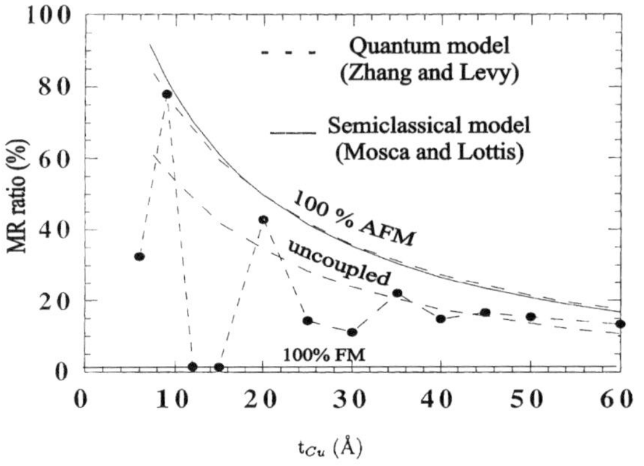

Figure 7. MR ratio as a function of the Cu spacer layer thickness in (Co 15 Å/Cu t_{Cu} Å)$_{30}$ multilayer series. The dashed lines are the best fit of quantum calculations for fully antiferromagnetic and uncoupled magnetization configurations. The solid lines are the best fit of semiclassical calculations for fully antiferromagnetic and fully ferromagnetic magnetization configurations. Experimental data are joined by a short-dashed line only to guide the eyes [13].

Therefore, the physics of GMR is quite similar to the spin-dependent conduction in ferromagnetic metals where the two-current model is applied with current carried by spin-up and spin-down electrons flowing in two independent "channels." Due to the spin splitting of the electronic structure in ferromagnets, the number of spin states and spin diffusion length (length characterizing the spatial extent of the spin polarization) are different. There exists a spin asymmetry coefficient $\alpha = \rho_\uparrow/\rho_\downarrow$ defined in terms of the spin-up and spin-down resistivities, $\rho_\uparrow$ and $\rho_\downarrow$, respectively. The ordinary magnetoresistance of ferromagnetic metals and their alloys is proportional to this spin asymmetry coefficient. In GMR materials, spin asymmetry coefficients can also be defined, but the spin-up and spin-down resistivities differ due to the intrinsic and extrinsic spin-dependent scattering in the "bulk" and interfaces of the GMR structure. In Section 5.3, the bulk and interface contributions to GMR will be discussed in more detail.

In GMR materials, the magnetoresistance effect, which is traditionally defined by the variation in the electrical resistivity ρ as a function of the applied field as $(\rho(H)-\rho(0))/\rho(0)$, can be defined as

$$\mathrm{GMR} = (\rho_{\mathrm{AP}} - \rho_{\mathrm{P}})/\rho_{\mathrm{AP}}$$

where ρ_{AP} and ρ_{P} are, respectively, the resistivities in the ideal antiparallel and parallel magnetization configurations in an antiferromagnetically coupled multilayer. A similar relation is also used in (FM1/NM/FM2/...)-coupled or -uncoupled sandwich structures when the ferromagnetic layers FM1 and FM2 show fully parallel and antiparallel magnetization configurations in a certain applied field interval. It is worth noting that the definition of GMR using ρ_{P} in the denominator allows a magnetoresistance effect larger than 100%; for example, the GMR record using the definition $(\rho_{\mathrm{AP}} - \rho_{\mathrm{P}})/\rho_{\mathrm{P}}$ is 220% in antiferromagnetically coupled Fe/Cr multilayers at low temperature [146].

Departing from the low-temperature limit, the transfer of momentum between the two channels occurs by spin-flip electron–magnon scattering. Therefore, the so-called spin-mixing effect tends to equalize the two currents due to spin-up (spin-down) scattering to spin-down (spin-up) states by the annihilation or creation of magnons. Since the shunting by the less resistive channel is reduced, the overall resistivity increases and the magnetoresistance decreases.

A simplified but quantitative approach to describing the GMR was reported early on, based on the semiclassical model for conduction in metals. To apply the model to GMR systems, the two-current model was associated with the semiclassical formalism originally developed by Fuchs [147] and Sondheimer [148] for describing conduction in nonmagnetic thin films. The Fuchs–Sondheimer theory had already been applied to nonmagnetic, metallic superlattice systems by Carcia and Suna [149].

For the GMR systems, separate spin-up and spin-down contributions to the current profile were computed by applying the Boltzmann equation and inserting spin-dependent mean-free paths in the ferromagnetic layers, as well as (in some cases) spin-dependent scattering at the interfaces and outer surfaces of the multilayers. The first results obtained using this approach were the Camley–Barnas [150] and Barthélémy–Fert [151] calculations. A simple albeit somewhat restricted version of the calculation was developed by Mosca and Lottis that employed the analytical expression for the conductivity of a metallic superlattice obtained by Carcia and Suna and allowed an analytical expression to be obtained for the GMR in hypothetical superlattices where the conduction electrons in the nonmagnetic spacer layer have the same mean-free path as one of the spin channels in the magnetic layers. This approach was used for the comparison between semiclassical and quantum calculations for the Co/Cu multilayers displayed in Figure 8 [152].

Early models using a Boltzmann equation incorporated, among others, the assumption that the two current channels do not mix and that scattering within the layers is isotropic with respect to conduction electron wave vectors. The first assumption corresponds to low temperatures, since collisions involving the reversal of a conduction electron's spin commonly involve thermally excited spin waves within the ferromagnet. The inclusion of spin-flip collisions couples the otherwise independent differential equations for the spin-up and spin-down electron distribution functions. Solutions of the coupled equations, thus incorporating spin

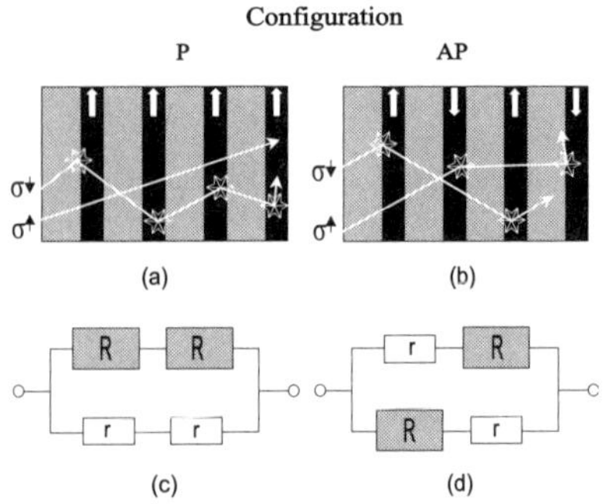

Figure 8. Spin-dependent scattering of the conduction electrons in a multilayered structure with magnetization aligned (a) parallel and (b) antiparallel. Network model of spin-dependent scattering for (c) parallel and (d) antiparallel magnetization configurations.

mixing, were obtained numerically by Fert's group in the early 1990s [153], with an analytical solution being reported more recently [154]. Calculations including anisotropic scattering within the ferromagnetic layers [155] and angle-dependent interfacial scattering and reflection coefficients [150] were also introduced fairly early on. The early models using a Boltzmann equation approach were followed by the resistor network of Edwards [156] and scattering in superlattices by White [157].

A quantum mechanical approach, based on the Kubo formula and surface scattering introduced by Tesanovic and co-workers [158, 159], was developed in a long series of papers by Levy [2, 160]. Results of the semiclassical calculations deviate from those of the quantum models when the film thickness is much smaller than the electron mean-free path. But the major open problem is the microscopic origin of the spin-dependent scattering by interfaces. Inoe and Maekawa [161] have introduced a theory of the spin-dependent scattering by interfaces in superlattices with nonmagnetic spacer layers by reducing the problem to the classical case of scattering by magnetic impurities. They have considered the scattering of the s electrons by the interface magnetic atoms and the formation of virtual bound states to describe the hybridization of the s- and d-band states at the interfaces. However, this type of calculation should be extended in order to describe rough interfaces and their scattering effects.

4.2. CIP and CPP Experiments

The usual measurements of GMR are made in a straightforward manner with the current flowing in the layer planes (CIP measurements). However, in this geometry, it is difficult to separate and quantify bulk and interface scattering. An interesting development was the breakthrough that allowed measuring GMR in magnetic multilayers with current flowing perpendicular to the multilayer planes (CPP measurements) [162, 163]. Current flowing in the plane of a multilayered structure is expected to be less sensitive to the magnetic configuration of the layers than current flowing *perpendicular* to the layers. Aware of this fact, researchers at Michigan State University (MSU) succeeded in implementing a technique that allowed them to perform an astonishing measurement of the resistance of a multilayer with current flowing from patterned electrodes deposited on the bare substrate (beneath the multilayer) to electrodes deposited on the top surface of the multilayers [164, 165].

The CPP (current perpendicular to the plane) geometry produced the expected enhancement of the GMR relative to the in-plane (CIP) geometry. To overcome the daunting obstacle of measuring the extremely small resistance of macroscopic regions of thin multilayered samples (typically 10–1000 nΩ for 1 mm^2 at 4.2 K), the MSU group employed a fascinating technique involving superconducting metals for the current/voltage leads and a SQUID-based detection system for registering the small voltages and corresponding variations.

The experimental requirements for detecting the GMR in the CPP geometry, at least using the approach developed at MSU, did not offer short-term potential for applications. Near the end of this section, we shall discuss novel systems where the enhancement of the GMR in the CPP geometry has been brought much closer to future incorporation in devices of various kinds.

Much of the excitement over the development of a technique capable of performing CPP experiments was related to the theoretical models that were being employed to describe transport in GMR systems. Many of the models depend strongly on phenomenological parameters, and some of these are nearly impossible to determine without CPP measurements, which allow direct determination of the crucial bulk and interface spin-dependent parameters [166]. Besides, CPP measurements eliminate the shunting effects caused by buffers, capping or biasing layers present with CIP measurements.

Grooved substrates arise as structures intermediate between CIP and CPP structures to probe the GMR effect [167]. Figure 9 shows the basic geometries of current flowing in GMR systems.

In the current notation of CPP-GMR multilayered films, the resistivities of the magnetic layers for the spin-up and spin-down electrons are written as

$$\rho_{\downarrow(\uparrow)} = 2\rho_F(1 \pm \beta)$$

Similarly, the interface resistance for the unit area is written as

$$r_{\downarrow(\uparrow)} = 2r_b(1 \pm \gamma)$$

where β and γ are the spin asymmetry coefficients of bulk and interface scattering, respectively, and the "average" or zero-asymmetry-limit quantities ρ_F and r_b are given detailed definitions in the Valet–Fert model [166]. The determination of the spin asymmetry coefficients, β and γ, allows the separate identification of bulk and interface scattering contributions to the GMR.

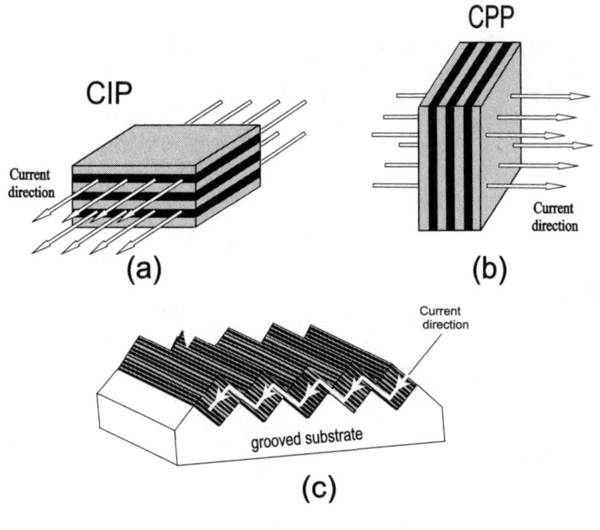

Figure 9. Three basic geometries of current flowing in GMR systems: (a) current in the plane of the film (CIP), (b) current perpendicular to the plane of the film (CPP), and (c) mixed CIP and CPP geometry using the deposition of the CPP structures on the grooved substrate.

4.3. Spin-Dependent Scattering: Bulk and Interface Contributions

The spin asymmetry coefficients β and γ for bulk and interface scattering of Co and several Co-based alloys were measured in CPP multilayered structures by the group at Michigan State University [168–170] and multilayered nanowires by the Louvain group [171, 172].

Table 1 presents the spin asymmetry coefficients β and γ for bulk and interface scattering results obtained for Co and some Co-based alloys. Both spin asymmetry coefficients are positive and the interface contribution is higher than the bulk contribution for Co, in agreement with previous measurements performed on dilute alloys [8] and theoretical calculations in the dilute limit [179].

It is worth noting that Co/Cu and Co/Ag multilayers have positive values for β and γ, and γ is larger than β independently of the temperature and CPP-GMR measurements employed to derive these parameters.

Therefore, it is interesting to note that in Co/Cu, Co/Ag, and CoFe/Cu multilayers the bulk scattering prevails at larger thicknesses, and there is no compensation thickness where GMR passes through 0 as observed for Fe-based structures where bulk and interface scattering may compete, pulling in opposite ways and in some cases canceling out ($\beta > 0$ and $\gamma > 0$) [180].

That Co–Cr and Co–Mn alloys give rise to negative β is in accordance with measurements and theoretical arguments based on the electronic structure of the bulk materials.

Also, whereas the introduction of Cu in the spacer layer gives rise to positive γ, the introduction of Cr in the spacer layer gives rise to negative γ. The interfacial contribution is a more delicate subject since it is connected to extrinsic potentials (associated with diffusion centers at the interfaces) as well as intrinsic potentials (associated with periodicity of the multilayer). In a general way, the sign of the spin asymmetry coefficients of alloys and interfaces reflects the matching of the d states in the spin direction of majority electrons. Interfacial doping and dusting investigations may shed more light on the extrinsic and intrinsic contributions, but roughness effects make it difficult to clarify this problem.

Mertig et al. [179] have demonstrated that in multilayers, such as Co/Cu, the types of electrons in CIP electronic transport differ from those in CPP. Most of the electronic states involved in the CIP conduction are confined to the nonmagnetic layer probing the scattering potentials within the layer and at their interfaces, whereas the electronic states in CPP conduction are nonlocalized.

There is still another factor that contributes to the "channeling" of the current in CIP as compared to that in CPP: in CIP, the characteristic length scale of the scattering process is the mean-free path, which is, in general, of the same order or smaller than the layer thicknesses. Thus, the influence of the spin-dependent scattering inside the layers is limited to a length scale on the order of the mean-free path around the interfaces. As a consequence, the scattering processes in CIP and CPP systems are distinct with bulk scattering contributing more weakly in CIP than in CPP. Figure 10 shows the influence of QWS in CIP and CPP conduction.

Nowadays, there are at least two important open questions concerning the origin of the GMR: (i) is the GMR mainly an extrinsic effect resulting from the spin-dependent scattering by imperfections and impurities or is it due to the intrinsic nanoscaled potential of a multilayered structure without any need for spin-dependent scattering? (ii) can the spin asymmetry coefficients from bulk and interface scattering, adopted in the GMR models, be derived from electronic structure calculations? An important contribution to the discussion of these questions is given by Vouille et al. [169].

These questions remain open. There is evidence that extrinsic potentials effectively contribute to the GMR. For instance, the doping of the ferromagnetic layers with a few percent of impurities can strongly influence the GMR and even change its sign from negative (normal GMR) to positive (inverse GMR), in keeping with electronic structure calculations. Inverse GMR, where smaller resistance is attained in the antiparallel configuration of the magnetization of consecutive ferromagnetic layers, has been observed in both CIP [181, 182] and CPP [180, 183] geometries when magnetic layers FM1 and FM2 with opposite spin asymmetries alternate in (FM1/N/FM2/N/...) structures. Inverse GMR is

Table 1. Spin asymmetry coefficients for bulk and interface scattering obtained from CPP measurements at low temperatures (4.2–77 K) on selected (FM/Cu/Co/Cu) multilayers, where FM=Co or Co-based alloys [169], Co/Cu grooved substrates [173], Co/Cu multilayered nanowires prepared by electrodeposition [171], Co/Ag multilayers [174], Co/Cr multilayers [169], and CoFe/Cr [175].

System	Bulk scattering spin asymmetry coefficient, β	Interface scattering spin asymmetry coefficient, γ	Ref.
Co/Cu at 4.2 K	+0.46	+0.75	[176, 177]
Co/Cu at 77 K	+0.36	+0.86	[171]
Co/Cu (multilayers)	+0.51	+0.76	[169]
Co/Cu (grooved)	+0.27	+0.76	[178]
Co/Cu (nanowires)	+0.36	+0.85	[171]
Co/Ag	+0.48	+0.83	[174]
Co/Cr	+0.46	−0.24	[175]
$Co_{91}Fe_9$/Cr	+0.65	−0.14	[175]
$Co_{91}Fe_9$/Cu	+0.65	+0.76	[169]
$Co_{90}Cr_{10}$/Cu	−0.12	+0.35	[169]
$Co_{95}Mn_5$/Cu	−0.03	+0.96	[169]

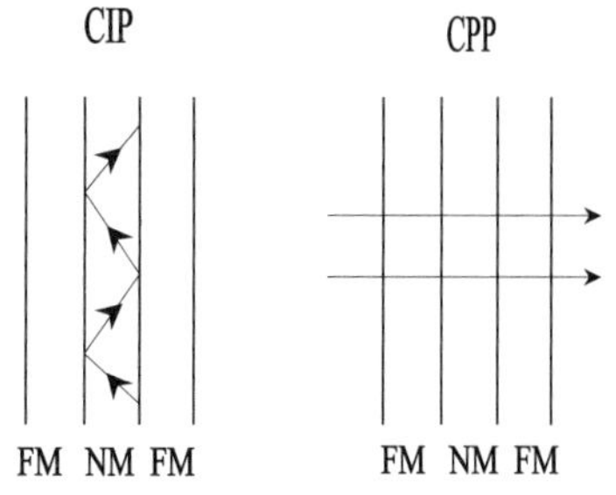

Figure 10. Scheme of the CIP and CPP conduction in FM/NM multilayers. The conduction in CIP weakly probes the scattering process in the ferromagnetic layers since the electronic states responsible are partially confined in the nonmagnetic layer (quantum well states), whereas the conduction in CPP is mainly nonlocalized, probing the bulk as well as the interfaces of the multilayered structure.

strong evidence that extrinsic effects are important in many systems and that the two-current model with its spin asymmetry coefficients is useful for describing GMR phenomena.

Before closing this section, we will address some experimental findings on the influence of growth condition parameters on the GMR effect in Co-based nanostructures which demonstrate the complexity of the problem.

Resistivity is not a truly intrinsic property of metals determined exclusively by band structure. It depends on extrinsic properties such as concentrations of defects, impurities, and other scattering centers (zero resistivity is expected in an ideally perfect crystal or superlattice). Therefore, minor details introduced at the beginning or during the growth stage of Co-based GMR materials may have a significant impact on the strength of the GMR effect due to the interplay between bulk and interface spin-dependent scattering. Indeed, the field dependence of the resistance as a whole could significantly change.

In particular, several experiments indicate that the GMR is strongly dependent on the interface structure (interfacial effects are tremendously enhanced in a multilayer), which, in turn, can be influenced by the deposition parameters such as residual pressure in the growth chamber, the presence of surfactant agents, or intentional mixing of materials at the interface involving thermal treatment during growth.

As an example, let us consider how a specific surfactant modifies the magnetic properties of nanostructures. Thin Co films and Co/Cu superlattices grown with Pb exhibit an induced layer-by-layer growth of Co which delays the fcc to hcp structural transition and strongly reduces the twinning formation in the superlattices. As a consequence, the easy axis of magnetization remains perpendicular to the film plane, allowing the observation of complete antiferromagnetic coupling for both in-plane and out-of-plane magnetization [44].

Let us give two other examples, now involving the deposition conditions. First, the strong influence of the purity of the base pressure on the GMR ratio of Co/Cu multilayers was described by Egelhoff and co-workers [184]. By reducing the partial pressure of water below 10^{-7} Pa, an increase in the GMR ratio has been observed, but a further reduction to 10^{-8} Pa leads to a decrease of about 20% in the GMR ratio. According to Egelhoff et al., the oxygen in the gas acts as a surfactant on the growth, suppressing the intermixing between Co and Cu and influencing the surface roughness. Second, the strong dependence of the GMR ratio of Co/Cu multilayers on the argon acceleration voltage U_B has been reported by Saito and co-workers [185]. By increasing the U_B from 400 to 600 V, the GMR ratio increases from approximately 5% to 50%. However, a further increase in U_B to 1000 V reduces the GMR effect to 30%. As demonstrated by nuclear magnetic resonance (NMR) measurements, the largest GMR ratio corresponds to the sharpest interface concomitant with predominance of the fcc Co component relative to hcp Co.

Parkin et al. [186], Greig et al. [187], and Tomlinson et al. [188] have discussed the integrity of the layers (pinholes and other accidental bridging) and emphasized the importance of the buffer layer in Co-based multilayers. George et al. [189], Parkin and co-workers [3], and Kataoka et al. [114] have discussed the results of planar doping or dusting in Co-based multilayers where a few atomic layers of any ferromagnetic or nonmagnetic metal is inserted at the interfaces which gives rise to an increase or decrease in spin-dependent scattering or, equivalently, resistivity.

4.4. Current-Driven Magnetization Reversal by Spin Injection

Berger [190] and Slonczewski [191] have argued that a sufficiently large current flowing perpendicular to a magnetic multilayer composed of alternating ferromagnetic FM and nonmagnetic N metals can accumulate and transfer spin between FM layers, stimulate spin wave excitations, and even excite precession of the FM layer magnetizations. Such current-driven magnetization reversal and spin wave excitation phenomena were first confirmed in Co/Cu multilayers at current densities of about 10^8 A/cm^2 [90–93]. To achieve the required high current densities, point contacts (with an area of about 10^2 nm^2) and electron beam lithographically patterned pillars (with an area of approximately 2500 nm^2) were used with contact resistance combining ballistic (Shaving) and diffuse (Maxwell) scattering. Magnetization precession on the femtosecond time scale and even precessional magnetization reversal in nanosized ferromagnetic bits can also be obtained by hot spin injection [192].

Recent experiments performed by Katine et al. [193] and Fert et al. [194] demonstrate clearly the feasibility of control of the magnetization configuration of Co/Cu/Co trilayers by the current intensity. Such experiments have potential applications in high-speed, high-density storage and memory.

4.5. Spin Polarization, Accumulation, and Relaxation

The current understanding of GMR illustrates an important basic point: spin electronic nanostructures work by transferring spin information from one part of the structure to another. This information is mediated by the charge carriers at the Fermi level or above (hot spin electrons), and it decays on a characteristic length scale (the spin diffusion length), which is the average distance diffused by a carrier spin before flipping.

A spin-polarized electron injected into a diamagnetic or paramagnetic medium undergoes many collisions that modify its momentum before it eventually spin flips. The spin diffusion length depends both on the mean-free path and on the spin-flip time and is very sensitive to the presence of impurities. Therefore, material composition and purity are crucial engineering parameters in GMR materials.

Spin accumulation is simply an excess number of up-spins over down-spins (or vice versa), which can be seen as a splitting of the electrochemical potential of the two spin channels. Spin dynamics essentially controls the accumulation and both longitudinal and transverse spin relaxation.

Typical values of the spin diffusion length l_s in Co-based multilayers, where both interfacial and bulk contributions must be taken into account, are 500 nm for Cu and 60 nm for spin-up electrons and 1 nm for spin-down electrons for Co.

Figure 11 shows the spin accumulation around an FM/NM/FM structure.

A crucial difference between the CPP and CIP geometries is the existence of spin accumulation in CPP-GMR systems with the appearance of the spin diffusion length as the unique damping length for the currents. The spin diffusion length is fixed by spin orbit scattering in the low-temperature limit and is expected to be relatively long in GMR systems composed of 3d and noble metals.

The spin accumulation effects at interfaces between ferromagnetic FM and nonferromagnetic NM metals have been described by Johnson and Silsbee [195], van Son et al. [196], and Valet and Fert [166] for isolated interfaces separating two semiinfinite media. Due to the spin accumulation around the interfaces with the NM metal, a spin-coupled interface resistance r_I appears at each FM/NM interface. The physical origin of the r_I is that, to balance spin accumulation by spin relaxation, the chemical potentials of the spin-up and spin-down directions are shifted in opposite directions, which gives rise to spin-dependent pseudo electric fields. As a consequence, there is an additional potential drop proportional to the product between r_I and the current density J. The case of multilayered structures consists of a sort of addition of interface resistance calculated for isolated interfaces.

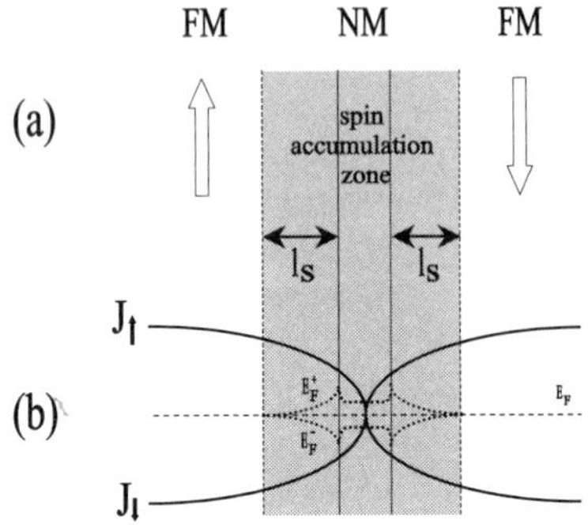

Figure 11. Schematic drawing of spin accumulation in the case of two semiinfinite layers of ferromagnets FM with opposite magnetization spaced by a nonmagnetic NM metal: (a) the spin accumulation zone extends over a distance on the order of the spin diffusion length l_s at each side of the interface between NM and FM, and (b) variation of the spin-up and spin-down current densities and spin splitting of the Fermi level.

The same assumptions of the Valet–Fert model applied to the FM/semiconductor interface indicate that the main obstacle to spin transfer directly from an FM electrode into a semiconductor, which is associated with the large conductivity mismatch of these materials [197, 198], may be overcome by insertion of a tunnel contact [199]. A concomitant FM resistance comparable to the magnitude of the tunnel contact resistance is needed to induce a *spin splitting* of electrochemical potential in the FM for spin-polarized emission to occur. Recent experimental results corroborate these theoretical predictions [200–202].

Hybrid structures involving the integration of ferromagnetic metals onto semiconductors have been probed and studied in order to utilize the spin degree of freedom in semiconductors. Kikkawa and Awschalom [203] have recently demonstrated that the spin polarization and coherence in semiconductors can be sustained at a time close to a microsecond and transported over a 100-μm scale. A special issue of *Semiconductor Science and Technology* [204] covers the state of the art of semiconductor spintronics to date.

4.6. TMR Systems

In 1995, two research groups [74, 205] independently developed another kind of GMR multilayered material which had been proposed six years later by Slonczewski [206] from the seminal work of Jullière [207]. In these magnetic tunnel junctions, the magnetic layers FM are separated by a thin insulator I barrier in (FM/I/FM) structures. The GMR phenomenon in these tunnel structures is called TMR—tunneling magnetoresistance.

Although some applications of TMR are already in development, the fundamental physics of spin polarization of tunneling electrons is still not completely understood. For example, whereas the polarization of the probability of tunneling from a ferromagnetic metal is expected to be the polarization of the density of states at the Fermi level, the positive polarization found for electrons tunneling from most ferromagnetic transition metals, such as Co, across alumina disagrees with the negative polarization of their density of states [208]. The positive polarizations measured were exclusively attributed to the s character of the tunneling electrons. The state of the art of magnetotransport in tunnel junctions has been recently reviewed by Moodera and Mathon [74] and Tsymbal et al. [132].

Table 2 presents spin polarization measurements for selected ferromagnetic FM transition metals and their alloys, which were measured in superconductor/Al_2O_3/FM planar junctions by three different research groups.

Table 2. Experimental values of spin polarization of selected ferromagnetic transition metals and their alloys.

	Research group		
Ferromagnetic electrode	Meservey and Tedrow [208]	Moodera and Mathon [74]	Monsma and Parkin [209]
Fe	+40%	+44%	+45%
Co	+35%	+45%	+42%
Ni	+23%	+33%	+31%
$Ni_{80}Fe_{20}$	+32%	+48%	+45%
$Co_{50}Fe_{50}$	+47%	+51%	+50%

In spite of being distinct phenomena, GMR and TMR are essentially connected by the spin polarization states in ferromagnetic materials. In lithographically patterned magnetic tunnel junctions with planar geometry involving two ferromagnetic metals FM separated by an insulating barrier layer I (FM/I/FM), the tunneling conductivity varies as a function of the angle between the magnetization of the FM electrodes. Similar to those observed in GMR systems, normal TMR (negative) and inverse TMR (positive) have been observed.

Essentially, the physics of TMR is based on Jullière's work [207] using the spin polarization defined as

$$P = \frac{[D_\uparrow(E_F) - D_\downarrow(E_F)]}{[D_\uparrow(E_F) - D_\downarrow(E_F)]}$$

where $D_\uparrow(E_F)$ and $D_\downarrow(E_F)$ are the density of states at the Fermi level E_F for spin-up and spin-down electrons in the ferromagnetic FM electrodes of an FM1/I/FM2 planar junction with an insulating barrier layer I. In a planar junction, the tunneling magnetoresistance (TMR) effect is usually defined as

$$\text{TMR} = (\rho_{AP} - \rho_P)/\rho_P = 2P_1P_2/(1 - P_1P_2)$$

where ρ_{AP} and ρ_P are, respectively, the resistivities in the antiparallel and parallel magnetization configurations of the ferromagnetic electrodes, FM1 and FM2, in a certain applied field interval and P_1 and P_2 are the spin polarizations of the ferromagnetic electrodes at each side of the barrier layer.

In a review of recent results obtained by the Orsay group on the spin-polarized tunneling in magnetic tunnel junctions combining electrodes of ferromagnetic transition metal and half-metallic oxide ($La_{0.7}Sr_{0.3}MnO_3$—LSMO), it is demonstrated that the spin polarization of tunneling electrons is governed by the specific type of electronic bonding at the electrode/barrier interface [194].

In magnetic tunnel junctions Co/I/LSMO with I = $SrTiO_3$ (STO), $Ce_{1-x}La_xO_y$ (CLO), Al_2O_3 (ALO), and ALO/STO, a normal TMR confirms the positive polarization of the tunneling probability from Co when the barrier is ALO. In contrast, when the ALO barrier is replaced by STO or CLO the TMR is inverse. This means that when tunneling is through STO or CLO the polarization of Co is negative, therefore, with the same sign for the polarization of the density of states of the d band of Co at the Fermi level. The role of the nature of the barrier in determining the sign of the spin polarization is consistent with several recent *ab initio* calculations [210–212].

The covalent bonding between Co and O at the interface of Co with O-terminated ALO or hybridization between Co and Al at the interface of Co with Al-terminated ALO is identified as the mechanism responsible for electronic bonding at the interface. Studies of Co/ALO/STO/LSMO junctions with thicker STO exhibit normal TMR. Therefore, the influence of the barrier on the sign of the polarization is mainly due to interface effects.

There is a TMR of about 5% at 300 K in Co/STO/LSMO tunnel junctions. This means that junctions with half-metallic manganites are promising candidates for technological applications in the future.

5. TECHNOLOGICAL APPLICATIONS

MR sensors are based on materials for which the resistance depends on the magnetization state. MR sensors can be divided into two classes of sensors: anisotropic magnetoresistance (AMR) sensors and giant magnetoresistance (GMR) sensors. GMR sensors can be further classified according to the corresponding mechanisms and features: GMR, TMR, and spin valve structured materials. The AMR sensors are currently regarded as being essentially obsolete, whereas members of the GMR class of sensors are largely in development.

Thin-film MR sensors can be used in several kinds of applications, such as magnetic measurements (including magnetometers and gradiometers), electrical measurements (as current transducers as well as switching and logic elements), magnetic data storage (as magnetoresistive heads, magnetic random access memory devices, and magnetic card readers), mechanical transducers (for linear displacements or measurements of rotational speed), and magnetic field imaging.

In this section, we will only address the major improvements and new concepts resulting from the utilization of Co-based GMR sensors. A detailed and complete description of applications of MR sensors can be found in articles dedicated specifically to the subject such as the recent reviews published by Tumanski [6] or Prinz [213].

All devices using GMR, spin valve, and TMR structures are rather complex systems with properties and performance determined by various parameters. Technological factors related to sensors or reading heads and affecting their performance include the following: large MR ratio, large sensitivity, small change of parameters with temperature, and good repeatability and reliability. These features depend on the interface quality (roughness, interfacial mixing), crystal quality (texture, grain size), and material quality (purity of composition and structure definition). The technological parameters also depend on the deposition conditions intrinsic to each deposition method employed such as sputtering, electron beam evaporation, molecular beam epitaxy, and electrodeposition. Important deposition conditions include substrate temperature, quality of the substrate, base pressure, and bias voltage. Also, the performance of patterned sensors may be different from the as-deposited structure.

The most important technological applications of the GMR, spin valve, and TMR structures that we described here are magnetic sensors, read heads, spin valve transistors, and magnetic random access memories (MRAMs). Each of these will be discussed. Finally, we briefly discuss the implications of spin-dependent transport for quantum computing.

5.1. Magnetic Field Sensors

The first large-scale commercial application of GMR has been as magnetic field sensors in the read heads of magnetic recording media for digital information storage.

In comparison with other magnetic sensors, the MR sensors offer some very competitive behavior [214] such as the largest sensitivity per unit area; that is, extremely small sensors still retain good sensitivity. This advantage is appreciated in data storage reading applications [215]. The following attributes of MR sensors can be emphasized: small dimensions, high reliability, low cost, and wide-frequency bandwidth (from 100 to 10,000 Hz), including dc fields [216].

The magnetic field range recommended for Co-based MR sensors is presented in Figure 12. According to comparative analyses of Co-based GMR systems performed by Tumanski [6], the higher GMR ratio and the higher magnetic field sensitivity for magnetic field sensor applications is found in Co/Cu and CoFe/Cu antiferromagnetic-coupled multilayers.

Despite the large MR effect, the room temperature sensitivity ($\Delta R/R/H_s$) is rather small in AFM-coupled multilayers. Best results are obtained with uncoupled structures and spin valve structures. The parameters responsible for these characteristics for selected systems are presented in Table 3.

A disadvantage of GMR sensors is that the fabrication technology is much more sophisticated, and, therefore, they are more expensive than AMR sensors. Furthermore, GMR sensors are not suitable at high temperatures, while AMR sensors can work without problems at temperatures well above room temperature (up to 190 °C). Therefore, it is not expected that GMR and TMR read heads will completely replace AMR heads.

5.2. Magnetoresistance Readouts

It may be stated that magnetoresistive heads are the main and most important application area of MR sensors due to their market price. The introduction of AMR-based MR heads by Hunt [225] caused a revolution in high-density recording in the beginning of the 1970s. The areal recording density is conventionally expressed in terms of the number of bits stored per unit area (1 Gb/in^2 = 1.55 Mb/mm^2). In the last 10 years, the areal bit density has increased from 0.1 Gb/in^2 to about 10 Gb/in^2 with a bit length of 0.1 μm [226, 227]. As a comparison, we note that a compact disc (CD) exhibits a recording density of about 0.7 Gb/in^2, and a digital video disc (DVD) exhibits a recording density of about 3.3 Gb/in^2.

The operating principle of the unshielded MR head is very simple. An MR head (vertical or horizontal type) in the form of a stripe detects the magnetic field above moving media such as disks and tapes.

Unshielded MR heads have poor performance for high-density recording. The requirements of higher resolution, areal density, and data transfer speed mean that unshielded MR heads are commonly used for long-wavelength signals such as magnetic card readers and detection of ample magnetic field sources.

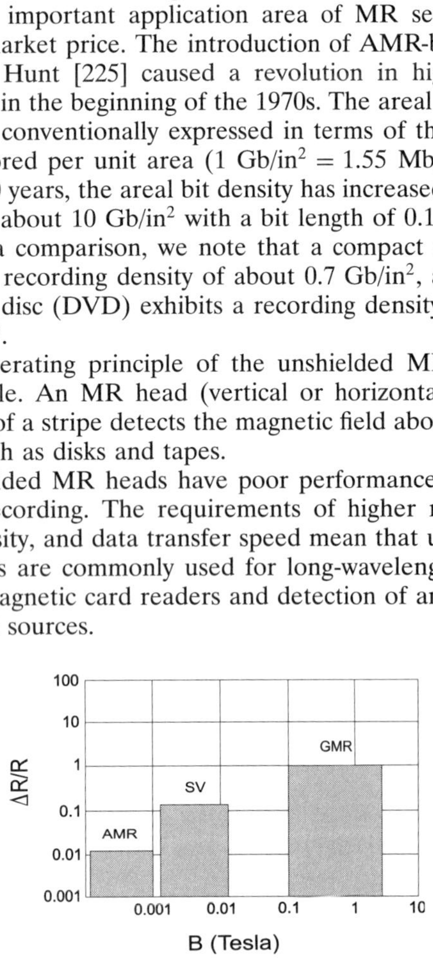

Figure 12. Typical magnitude and field range of anisotropic magnetoresistance (AMR), spin valve (SV), and giant magnetoresistance (GMR) systems.

Table 3. Selected parameters of Co-based AFM-coupled multilayers, uncoupled structures, and spin valve structures at room temperature, according to Tumanski [6]. (1 kA/m = 12.57 Oe.)

Structure	$\Delta R/R$ (%)	H_S (kA/m)	$\Delta R/R/H_S$ (%:kA/m)	Ref.
(Co1.0/Cu0.8) × 30[a]	38	880	0.08	[217]
(FeCo1.5/Cu0.9) × 30[a]	32	160	0.20	[218]
Au/Co/Au/Co/Au[b]	1.5	8	0.19	[219]
Fe/Cu/Co/Cu[b]	3	8	0.37	[220]
NiFe/Co/Cu/Co[b]	0.5	0.08	3.10	[221]
(Co/Cu) × 3/(Fe/Pd) × 3[b]	1.7	0.3	5.70	[222]
Co/Cu/Co[c]	9.5	1.2	7.92	[223]
Co/Cu/Co/FeMn[d]	6.9	0.8	8.60	[224]
NiO/Co/Cu/Co/Cu/Co/NiO[e]	24.8	3.3	7.51	[184]
AMR sensor[f]	1.6	1	1.60	[6]

[a] Antiferromagnetically coupled multilayer.
[b] Uncoupled structure.
[c] Spin valve structure.
[d] Exchange-biased spin valve.
[e] Double-exchange-biased spin valve.
[f] Typical anisotropic magnetoresistive sensor.

Usually, MR heads integrated with inductive writing heads require a shield. Thus, dual-shielded heads are widely used for very high density recording applications [228]. The shield on each side of an MR element can significantly improve its spatial resolution since the shields change the distribution of the magnetic field above the medium concentrating the flux in the gap over the MR element. The spatial resolution depends mainly on the gap length, sensor height (one-track or multitrack heads), and sensor permeability.

A typical commercial MR head consists of two magnetic layers ($Ni_{81}Fe_{19}$ and Co) with a few angstroms of thickness separated by a Cu spacer layer. The moments of one of the magnetic layers are pinned by an FeMn antiferromagnetic layer. The other layer is free to rotate under the influence of the fringing fields from the magnetized media.

Fontana [229] and Tsang [230] have reported the parameters of extremely small heads for gigabit density recording (over 40 Gb/in^2). Submicrometer-wide spin valve and GMR read heads have also been reported [231].

Practical limitations on electronics at very high densities arise from heat dissipation in the head and thermal instability of the medium, and comparisons between read heads containing both a TMR element and a GMR element with sense current in the plane (CIP-GMR) or perpendicular to the plane (CPP-GMR) of the sensor films, as well as the expected parameters of future 100 Gb/in^2 until 300 Gb/in^2 recording, are discussed by Ruigrok et al. [232].

5.3. Spin Valve Transistor

An important concept associated with spin-dependent transport is called spin accumulation, which is simply an excess number of up-spins over down-spins at the electrochemical potential level.

Datta and Das have proposed a spin-based device they called "spin-polarized field effect high transistor" or spin-FET [233].

Johnson [234] has proposed a three-terminal spin transistor, which includes a nonmagnetic base layer sandwiched between a ferromagnetic emitter layer and a ferromagnetic collector layer. The spin accumulation effect makes the base layer have a different chemical potential than the collector layer and will pump the current either into or out of the collector, depending on the magnetization direction of the collector layer.

The first attempt to realize a GMR metal-based transistor was reported in 1995 [235] and is illustrated in Figure 13. This more sophisticated spin valve transistor uses two Si layers as emitter and collector and a GMR Co-based multilayer as the base [235]. The inelastic mean-free path of hot electrons is spin dependent in ferromagnets, but in spin valve transistors the hot electrons at a mean excess energy of about 0.9 eV became relevant for magnetic sensors and spintronics [237].

The source and drain are different and form Schottky barriers at the metal/semiconductor interface with the GMR metal multilayer base. When one switches the Co/Cu sandwich from the aligned to the antialigned state, a lower drain current will result due to the shorter electron path length (and higher base resistance). More work is needed on this interesting new device for future applications.

5.4. Magnetoresistive Memories

The MR readout scheme consists of the storage of data in a magnetic medium, which, in turn, produces a stray magnetic field that can be detected by a separate MR sensing element. Initially, this concept did not lead to high-density data storage because it was difficult to get a sufficiently large external stray field from a small magnetic storage cell. In the mid 1980s, a new concept of magnetoresistive memory (MRAM) was developed at Honeywell by Daughton and co-workers which included a writing operation using magnetic hysteresis and reading using magnetoresistance of the same components where data are stored. The new class of MRAM

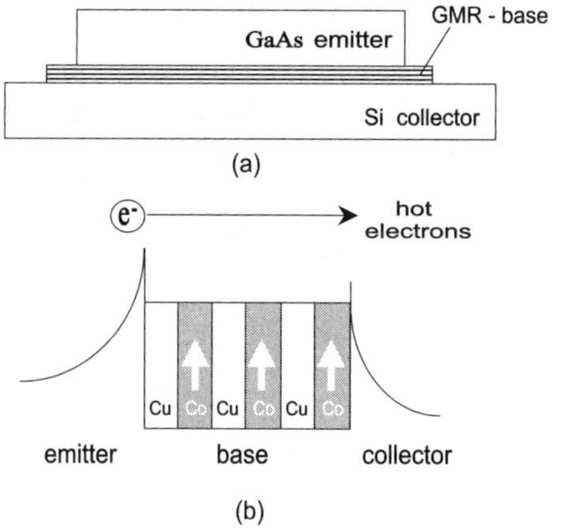

Figure 13. (a) Schematic cross section of a spin valve transistor with Co/Cu nanostructure sandwiched between Si and GaAs substrates. (b) Schematic energy band diagram of the spin valve transistor. The emitter Schottky barrier is slightly higher than the collector Schottky barrier, reducing the quantum mechanical reflections of hot electrons. A 215% change in the collector current was observed at liquid nitrogen temperature between the parallel and antiparallel configurations of magnetization, but only $10^{-4}\%$ of the emitter current was seen at the collector, which seriously limits the applicability of the device. The leakage current from base to collector also precludes operation at room temperature at reverse bias [236].

devices offered revolutionary advances for high-density, random access, nonvolatile memory that was later adapted for use with GMR and TMR structures as magnetoresistive elements. Indeed, the discovery of GMR materials in 1989 [11, 12] immediately prompted higher output signals (factor of about 3) and faster read access time (about 9 times) improvements.

The MRAM concept is illustrated in Figure 14. The magnetic field produced by a current through the stripe magnetized the data cell clockwise or counterclockwise when aided by a current (field) from an orthogonal strip line. Current in either strip by itself would not change the storage state. The original AMR version featured a readout process where the clockwise and counterclockwise configurations were distinguished by their differential resistance (dR/dH) in the presence of a modulated bias current in the orthogonal strip line. For GMR- and TMR-based memory cells, the logical states consist of parallel and antiparallel orientations. The resistance of the cell is higher when the magnetization of the soft layer is antiparallel to the magnetization of the hard layer, and lower for the parallel orientation. It is worth noting that a single MRAM cell within a two-dimensional array could be selectively written. The maximum difference in resistance of the cell between a "1" and a "0" when it is read is about 0.5% with differential sense signals of 0.5–1.0 mV. These MRAMs operate with a read access time of about 250 ns [238] while the write time was 100 ns [239].

It was initially believed that, as the memory cells approached the dimensions of a domain wall width, there would be no more problems with multidomain magnetization in the cells. Recently, a vortices magnetization storage mode in round MRAM cells has been observed [240, 241]. The practical lower limit to MRAM storage area would be about 0.1 μm on a side. The utilization of circumferential remnant magnetization structures may circumvent this problem. This storage method does not create vortices for cell sizes down to less than 100 nm diameter (approximately 30 nm inside diameter).

Another challenge for high-density MRAM is cell stability at nanometer dimensions. As the cell size shrinks and the volume of magnetic material becomes smaller, thermal agitation can cause a cell to lose data, that is, the so-called superparamagnetic limit. If a spin valve cell is used as the storage cell, the heating effects of the word and digit currents can heat the cell above the Néel temperature of the antiferromagnetic pinning layer, erasing the storage information.

The stability of the MRAM cell is an energy problem, involving thermal activation energy estimated as $k_B T$, where k_B is the Boltzmann constant and T is the temperature, and the magnetic energy associated with storage estimated as

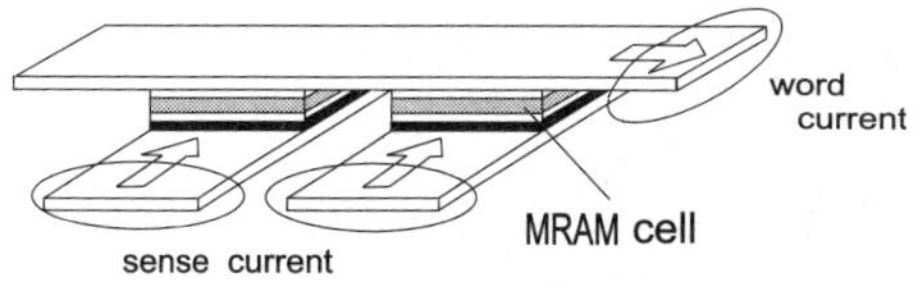

Figure 14. Design of an MRAM cell which has radiation hardened and which has faster switching than the conventional SRAM. Comparison of performance and parameters for various types of volatile and nonvolatile memory chip alternatives can be found in [236].

MH_cV, where H_c is a critical field that prevents magnetization reversal, M is the saturation magnetization, and V is the volume of the magnetic material in the cell. If the volume is reduced, the magnetic energy approaches a value of k_BT and the error rate in the memory becomes unacceptable. It is interesting to note that making H_c ever higher does not work because of the current required to write and the resultant heating of the cell (raising kT).

Besides, GMR materials had serious limitations in comparison with semiconductor memories (SRAM and DRAM), which are still faster because of the low MRAM sense signal. Recently, MRAM as a nonvolatile, solid-state memory has been tested to replace flash memory and EPROM.

TMR devices provide a higher percentage of magnetoresistance than GMR sandwiches or spin valve structures and thus have the potential for higher signals and higher speed. Recent results indicate TMR cells giving magnetoresistance about 4 times higher than good GMR cells [242].

Promising results in the construction of an MRAM with tunnel junction cells arranged in the cross-point architecture were reported by Parkin et al. [243]. The design of the cell is quite similar to that shown in Figure 14. The tunnel junction consists of a bottom electrode and a top FM electrode of Permalloy separated by a tunnel junction of oxidized Al. The bottom electrode is composed of a Co/Ru/Co/FeMn exchange-biased sandwich. The Co-based exchange-biased magnetic tunnel junction devices might be used as magnetic nonvolatile storage elements. The advantages of TMR-based devices compared to GMR-based devices are the following: higher nominal resistances (typically 10 Ω in GMR becomes 1000 Ω in TMR), which reduce the required operation power; higher resistance change (typically 1 Ω in GMR becomes 100 Ω in TMR), which gives a better sensitivity, and an output signal about 100 times larger than those obtained in GMR-based devices and about 1000 times those obtained in conventional AMR-based devices.

5.5. Quantum Computation

The current standard of electronic devices and data storage media has reached a level of miniaturization and storage density such that magnetic materials have to be fabricated on a nanometer scale. The emerging concept of spintronics, which is based on the fact that current carriers have not only charge but also spin, requires the assembling of nanometer-sized magnetic structures with the desired magnetic properties. This is an important part of the background that motivates scientists and engineers to attempt to grow and characterize magnetic objects at smaller and smaller length scales, ranging from 2D films and multilayers to 1D wires and eventually to 0D dots.

In particular, Shen and Kirschner [244] recently addressed the impact of tailoring magnetic phenomena in artificially structured Co-based material in the spintronics field. They review, for example, the most significant progress in recent years in the effort of growing artificially structured magnetic materials and the new structural and magnetic properties of these materials are discussed.

An important point is that the quantum transport properties of charge carriers, such as an electron or hole, are not only determined by the quantum motion of the particle as a whole, but can involve internal degrees of freedom such as spin. Spintronics is the fastest growing area of quantum electronics due largely to the possible applications to conventional information processing and storage [245]. The possible application to quantum computing has been noted by many research groups [246].

6. FINAL REMARKS

The existence of ferromagnetic order and large magnetoresistance in nanoscaled structures obviously is encouraging and important news for scientists who are interested in low-dimensional magnetism. Since the magnetization direction can be oriented along any desired direction with high accuracy, magnetic structures can be efficiently used in data storage and magnetic switching and sensoring devices.

The field of spintronics has been growing dramatically in recent years. The central idea of spintronics is the fact that conduction electrons (or holes) carry not only charge, but spin. The recognition of this additional degree of freedom is the key that promises a new generation of spin-based devices whose operation relies on differentiated manipulation of independent families of spin-polarized current carriers.

The emerging field of spin-based electronics promises to have a substantial impact on the future technology. Major challenges in this field include (i) the efficient injection of spin polarization as well as spin-polarized currents, (ii) the transport of spin-polarized carriers across relevant length scales and heterointerfaces, and (iii) the control and manipulation of electron spins on sufficiently fast time scales. As demonstrated in this review, Co-based nanostructures are archetypes and intrinsically favorable as materials for spintronics applications because Co provides an effective source of electron spin polarization at room temperature for ohmic injection, tunnel injection, hot electron injection, and ballistic electron injection. Co-based GMR magnetic field sensors for general purposes and GMR spin valve read heads for hard disk drives are currently commercially available. Co-based MRAM prototypes are not available, but production is anticipated in the next few years. Co-based multilayers are also potential candidates for GMR isolator elements, which combine a planar coil and a GMR sensor on an integrated circuit chip to perform a function similar to that of opto-isolators, which provide electrical isolation of grounds between electric circuits.

Considerable advances in our basic understanding of spin interactions in the solid state as well as in new materials engineering, lithography, and device fabrication are still necessary, but certainly Co-based nanostructures will continue to be an important part of this future.

GLOSSARY

Exchange bias Exchange interaction across the interface between antiferromagnets and ferromagnets with pinning of the ferromagnet magnetization.

Giant magnetoresistance Appreciable change in the resistance of nanostructured systems when an applied magnetic field, strong enough to magnetically saturate the system, brings the magnetization of all magnetic entities into alignment.

Interlayer exchange coupling Magnetic interaction via quantum mechanical mechanisms that aligns parallel or antiparallel the magnetic moments of ferromagnetic layers spaced by an intervening layer.

Magnetoresistance effect Phenomenon in which the resistance of a material depends on its state of magnetization.

Quantum well state Quantum mechanical state of nanostructured systems that comes from the confinement of the electron wave functions to dimensions comparable with their wavelengths.

Spin accumulation Excess number of up-spins over down-spins (or vice versa) and hence also of magnetization in a certain spatial zone.

Spin injection Spin information transfer from one part to another mediated by the electrical carriers near the Fermi energy.

Spin polarization Excess number of spin-up over spin-down electrons (or vice versa) at the Fermi level of ferromagnetic materials or in an electric current flow.

ACKNOWLEDGMENTS

DHM acknowledges support from CNPq. DKL acknowledges support from the Fundação Araucária of the state of Paraná.

REFERENCES

1. G. Prinz, *Science* 282, 1660 (1998).
2. P. M. Levy, *Solid State Phys.* 47, 367 (1994).
3. K. B. Hathaway, A. Fert, P. Bruno, D. T. Pierce, J. Unguris, R. J. Celotta, and S. S. P. Parkin, in "Ultrathin Magnetic Structures I" (B. Heinrich and J. A. C. Bland, Eds.), Vol. 2, p. 45. Springer-Verlag, Berlin, 1994.
4. F. J. Himpsel, J. E. Ortega, G. J. Mankey, and R. F. Willis, *Adv. Phys.* 47, 511 (1998).
5. F. J. Himpsel, K. N. Altmann, G. J. Mankey, J. E. Ortega, and D. Y. Petrovykh, *J. Magn. Magn. Mater.* 200, 456 (1999).
6. S. Tumanski, "Thin Film Magnetoresistive Sensors." IOP Publishing, Bristol, 2001.
7. K. H. J. Bushow, Ed., "Handbook of Magnetic Materials," Vol. 12. Elsevier, Amsterdam, 1999.
8. I. A. Campbell and A. Fert, in "Ferromagnetic Materials" (E. P. Wolfarth, Ed.), p. 769. North-Holland, Amsterdam, 1982.
9. P. Grunberg, R. Schreiber, Y. Pang, M. B. Brodsky, and H. Sowers, *Phys. Rev. Lett.* 57, 2442 (1986).
10. C. F. Majkrzak, *Phys. Rev. Lett.* 56, 2700 (1986).
11. M. N. Baibich, J. M. Broto, A. Fert, F. Nguyen Van Dau, F. Petroff, P. Etienne, G. Creuzet, A. Frieddrich, and J. Chazelas, *Phys. Rev. Lett.* 61, 2472 (1988).
12. G. Binasch, P. Grunberg, F. Saurenbach, and W. Zinn, *Phys. Rev. B* 39, 4828 (1989).
13. D. H. Mosca, F. Petroff, A. Fert, P. A. Schroeder, W. P. Pratt, Jr., and R. Loloee, *J. Magn. Magn. Mater.* 94, L1 (1991).
14. S. S. P. Parkin, R. Bhadra, and K. P. Roche, *Phys. Rev. Lett.* 66, 2152 (1991).
15. S. S. P. Parkin, *IBM J. Res. Dev.* 42, 3 (1998).
16. S. S. P. Parkin, *Phys. Rev. Lett.* 67, 3598 (1991).
17. L. Esaki and L. L. Chang, *Phys. Rev. Lett.* 33, 495 (1974).
18. M. F. Crommie, C. P. Lutz, and D. M. Eigler, *Science* 62, 218 (1993).
19. X. G. Liu, L. Fu, S. H. Hong, V. P. Dravid, and C. A. Mirkin, *Adv. Mater.* 14, 231 (2002).
20. J. Q. Xiao, J. S. Jiang, and C. L. Chien, *Phys. Rev. Lett.* 68, 3749 (1992).
21. A. E. Berkowitz, J. R. Mitchell, M. J. Carey, A. P. Young, S. Zhang, F. E. Spada, F. T. Parker, A. Hutten, and G. Thomas, *Phys. Rev. Lett.* 68, 3745 (1992).
22. S. S. P. Parkin, *Phys. Rev. Lett.* 71, 1641 (1993).
23. H. Zaman, S. Ikeda, and Y. Ueda, *IEEE Trans. Magn. Mater.* 33, 3517 (1997).
24. B. Dieny, *J. Magn. Magn. Mater.* 126, 433 (1993).
25. J. Wecker, R. von Helmolt, L. Schutz, and K. Samwer, *Appl. Phys. Lett.* 62, 1985 (1993).
26. K. A. Tarasov, V. P. Isupov, B. B. Bokhonov, Yu. A. Gaponov, B. P. Tolochko, M. R. Sharafutdinov, and S. S. Shatskaya, *J. Mater. Synth. Proc.* 8, 21 (2000).
27. O. Santini, D. H. Mosca, W. H. Schreiner, R. Marangoni, J. L. Guimarães, F. Wypych, and A. J. A. de Oliveira, *J. Phys. D: Appl. Phys.* 36, 1 (2003).
28. S. M. Thompson, J. F. Gregg, C. R. Staddon, D. Daniel, S. J. Dawson, K. Ounadjela, J. Hammann, C. Fermon, G. Saux, K. O'Grady, S. J. Grieves, J. M. D. Coey, and A. Fagan, *Philos. Mag. B* 68, 923 (1993).
29. B. Dieny, S. Sankar, M. R. McCartney, D. J. Smith, P. Bayle-Guillemaud, and A. E. Berkowitz, *J. Magn. Magn. Mater.* 185, 283 (1998).
30. E. M. Kakuno, I. Mazzaro, N. Mattoso, W. H. Schreiner, D. H. Mosca, M. A. B. Gomes, and M. Cantão, *J. Electrochem. Soc.* 144, 3222 (1997).
31. F. Komori, K. D. Lee, K. Nakatsuji, T. Iimori, and Y. Q. Cai, *Phys. Rev. B* 63, 214420 (2001).
32. J. J. Versluijs, M. A. Bari, and J. D. M. Coey, *Phys. Rev. Lett.* 87, 026601 (2001).
33. O. M. Stoll, M. I. Montero, J. Guimpel, J. J. Akerman, and I. K. Schuller, *Phys. Rev. B* 65, 104518 (2002).
34. N. Garcia, G. G. Qiang, and I. G. Saveliev, *Appl. Phys. Lett.* 80, 1785 (2002).
35. J. H. Fendler, in "Nanoparticles and Nanostructured Films: Preparation, Characterization, and Application" (J. H. Fendler, Ed.). Wiley–VCH, London, 1998.
36. W. Schwarzacher and D. S. Lashmore, *IEEE Trans. Magn.* 32, 3133 (1996).
37. W. Betteridge, *Prog. Mater. Sci.* 24, 51 (1979).
38. R. Wu and A. J. Freeman, *J. Magn. Magn. Mater.* 200, 498 (1999).
39. T. Ericsson, *Acta Metall.* 14, 853 (1966).
40. E. A. Owe and D. M. Jones, *Proc. Phys. Soc. B* 67, 456 (1954).
41. E. Votava, *Acta Metall.* 8, 901 (1960).
42. L. Z. Mezey and J. Giber, *Jpn. J. Appl. Phys.* 21, 1569 (1982).
43. R. F. Cook, *J. Mater. Res.* 1, 852 (1986).
44. J. Camarero, T. Graf, J. J. de Miguel, R. Miranda, W. Kuch, M. Zharnikov, A. Dittschar, C. M. Schneider, and J. Kirschner, *Phys. Rev. Lett.* 76, 4428 (1996).
45. W. F. Egelhoff, P. J. Chen, C. J. Powell, M. D. Stiles, R. D. McMichael, C. L. Lin, J. M. Sivertsen, J. H. Judy, K. Takano, and A. E. Berkowitz, *J. Appl. Phys.* 80, 5183 (1996).
46. W. Lange, A. Hässner, and W. Krug, *Phys. Status Solidi* 8, K11 (1965).
47. K. Hirano, R. P. Agarwala, B. L. Averbach, and M. Cohen, *J. Appl. Phys.* 33, 3049 (1962).
48. T. Honma and Wayman, *J. Appl. Phys.* 36, 2791 (1965).
49. J. G. Wright, *Thin Solid Films* 22, 197 (1974).

50. Ch. Morawe, A. Stierle, N. Metoki, K. Bröhl, and H. Zabel, *J. Magn. Magn. Mater.* 102, 223 (1991).
51. G. A. Prinz, *Mater. Res. Soc. Symp. Proc.* 56, 139 (1986).
52. P. C. Reidi, T. Dumelow, M. Rubinstein, G. A. Prinz, and S. B. Qadri, *Phys. Rev.* 36, 4595 (1987).
53. P. C. Riedi, T. Dumelow, M. Rubinstein, G. A. Prinz, and S. B. Quadri, *Phys. Rev. B* 36, 4595 (1987).
54. J. J. Krebs, *Appl. Phys. A* 49, 513 (1989).
55. C. M. Schneider, P. Bressler, P. Schuster, J. Kirschner, J. J. de Miguel, and R. Miranda, *Phys. Rev. Lett.* 64, 1059 (1990).
56. A. Cebollada, J. L. Martinez, J. M. Gallego, J. J. de Miguel, R. Miranda, S. Ferrer, F. Battalan, G. Fillion, and J. P. Rebouillat, *Phys. Rev. B* 39, 9726 (1989).
57. J. P. Renard, P. Beavullain, C. Dupas, K. Le Dang, P. Veillet, E. Velu, C. Marlière, and D. Renard, *J. Magn. Magn. Mater.* 115, L147 (1992).
58. K. Le Dang, P. Veillet, P. Beauvillain, C. Chapert, H. He, F. J. Lamelas, C. H. Lee, and R. Clarke, *Phys. Rev. B* 16, 13228 (1991).
59. H. A. M. de Gronckel, K. Kopinga, W. J. M. de Jonge, P. Panissod, J. P. Schillé, and F. J. A. den Broeder, *Phys. Rev. B* 44, 9100 (1991).
60. C. Mény, P. Panissod, and R. Loloee, *Phys. Rev. B* 45, 12269 (1992).
61. V. L. Moruzzi, P. M. Marcus, K. Schwarz, and P. Mohn, *Phys. Rev. B* 34, 1784 (1986).
62. R. M. Bozorth, "Ferromagnetism." Van Nostrand, New York, 1951.
63. R. M. Bozorth, *Phys. Rev.* 96, 311 (1954).
64. H. Masumoto, H. Saito, and M. Kikuchi, *Nippon Kinzoku Gakkaishi* 30, 885 (1966).
65. R. V. Coleman and R. C. Morris, *Phys. Rev. B* 8, 317 (1973).
66. R. F. C. Farrow, *J. Cryst. Growth* 104, 556 (1990).
67. J. F. Cochran, B. Hillebrands, and G. Günterodt, in "Ultrathin Magnetic Structures" (B. Heinrich and J. A. C. Bland, Eds.), Vol. 2, p. 222. Springer-Verlag, Berlin, 1994.
68. Y. Nakamura, *J. Magn. Magn. Mater.* 200, 634 (1999).
69. S. D. Bader and J. L. Erskine, in "Ultrathin Magnetic Structures" (B. Heinrich and J. A. C. Bland, Eds.). Vol. 2, p. 320. Springer-Verlag, Berlin, 1994.
70. S. Padovani, I. Chado, F. Scheurer, and J. P. Bucher, *Phys. Rev. B* 59, 11865 (1993).
71. J. Fassbender, U. May, B. Schirmer, R. M. Jungblut, B. Hillebrands, and G. Güntherodt, *Phys. Rev. Lett.* 75, 1176 (1995).
72. I. K. Schuller, S. Kim, and C. Leighton, *J. Magn. Magn. Mater.* 200, 571 (1999).
73. F. J. Albert, N. C. Emley, E. B. Myers, D. C. Ralph, and R. A. Burhman, *Phys. Rev. Lett.* 89, 226802 (2002).
74. J. Moodera and G. Mathon, *J. Magn. Magn. Mater.* 200, 248 (1999).
75. M. Alper, K. Attenborough, R. Hart, S. L. Lane, D. S. Lashmore, C. Younes, and W. Schwarzacher, *Appl. Phys. Lett.* 63, 2144 (1993).
76. J. W. Chang and L. T. Romankiw, in "Magnetic Materials, Processes and Devices" (L. T. Romankiw and D. A. Herman, Eds.), p. 223. Electrochemical Society, Pennington, NJ, 1994.
77. S. Z. Hua, D. S. Lashmore, L. Salamanca-Riba, W. Schwarzacher, L. J. Swartzendruber, R. D. McMichael, L. H. Bennett, and R. Hart, *J. Appl. Phys.* 76, 6519 (1994).
78. K. D. Bird and M. Schlesinger, *J. Electrochem. Soc.* 142, L65 (1995).
79. E. M. Kakuno, R. C. da Silva, N. Mattoso, W. H. Schreiner, D. H. Mosca, and S. R. Teixeira, *J. Phys. D: Appl. Phys.* 32, 1209 (1999).
80. Y. Ueda and M. Ito, *Jpn. J. Appl. Phys.* 33, L1403 (1994).
81. P. J. H. Bloemen, M. T. Johnson, M. T. H. van de Vorst, R. Coehoorn, A. Reinders, J. aan de Stegge, R. Jungblut, and W. J. M. de Jonge, *J. Magn. Magn. Mater.* 148, 193 (1995).
82. C. R. Matin, *J. Mater. Chem.* 7, 1075 (1997).
83. J. E. Wegrowe, D. Kelly, A. Franck, S. E. Gilbert, and J. Ph. Ansermet, *Phys. Rev. Lett.* 82, 3681 (1999).
84. A. Fert and L. Piraux, *J. Magn. Magn. Mater.* 200, 338 (1999).
85. T. M. Whitney, J. S. Jiang, P. Searson, and C. Chien, *Science* 261, 1316 (1993).
86. L. Piraux, J. M. George, J. F. Despres, C. Leroy, E. Ferain, R. Legras, K. Ounadjela, and A. Fert, *Appl. Phys. Lett.* 65, 2484 (1994).
87. A. Blondel, J. P. Meir, B. Doudin, and J. Ph. Ansermet, *Appl. Phys. Lett.* 65, 3020 (1994).
88. J. L. Maurice, D. Imhoff, P. Etienne, O. Durand, S. Dubois, L. Piraux, J. L. George, P. Galtier, and A. Fert, *J. Magn. Magn. Mater.* 184, 1 (1998).
89. B. Doudin, G. Redmond, S. E. Gilbert, and J. P. Ansermet, *Phys. Rev. Lett.* 79, 933 (1997).
90. M. Tsoi, A. G. M. Jansen, J. Bass, W. C. Chiang, M. Seck, V. Tsoi, and P. Wyder, *Phys. Rev. Lett.* 80, 4281 (1998).
91. M. Tsoi, A. G. M. Jansen, J. Bass, W. C. Chiang, M. Seck, V. Tsoi, and P. Wyder, *Phys. Rev. Lett.* 81, 493 (1998).
92. E. B. Myers, F. J. Albert, J. C. Sankey, E. Bonet, R. A. Buhrman, and D. C. Ralph, *Phys. Rev. Lett.* 89, 196801 (2002).
93. E. B. Myers, D. C. Ralph, J. A. Katine, R. N. Louie, and R. A. Buhrman, *Science* 285, 867 (1999).
94. B. A. Jones, *IBM J. Res. Dev.* 42, 25 (1998).
95. J. Mathon, *Phys. Rev. Lett.* 74, 3696 (1995).
96. P. Bruno, *Phys. Rev. B* 52, 411 (1994).
97. J. E. Ortega, F. J. Himpsel, G. J. Mankey, and R. F. Willis, *Phys. Rev. B* 47, 1540 (1993).
98. A. Barthelemy, A. Fert, M. N. Baibich, S. Hadjoudj, F. Petroff, P. Etienne, R. Cabanel, S. Lequien, F. N. Vandau, and G. Creuzet, *J. Appl. Phys.* 67, 5908 (1990).
99. S. S. P. Parkin, N. More, and K. P. Roche, *Phys. Rev. Lett.* 64, 2304 (1990).
100. R. Coehoorn, *Phys. Rev. B* 44, 9331 (1991).
101. M. D. Stiles, *J. Magn. Magn. Mater.* 200, 322 (1999).
102. J. Unguris, R. J. Celotta, and D. T. Pierce, *Phys. Rev. Lett.* 67, 140 (1991).
103. R. J. Celotta, D. T. Pierce, and J. Unguris, *MRS Bull.* 20, 30 (1995).
104. S. Demokritov, J. A. Wolf, P. Grunberg, and A. Zinn, *Mater. Res. Soc. Symp. Proc.* 231, 133 (1992).
105. S. T. Purcell, W. Folkerts, M. T. Johnson, N. W. E. McGee, K. Jager, J. aan de Stegge, W. B. Zeper, W. Hoving, and P. Grunberg, *Phys. Rev. Lett.* 67, 903 (1991).
106. S. T. Purcell, M. T. Johnson, N. W. E. McGee, R. Coehoorn, and W. Hoving, *Phys. Rev. B* 45, 13064 (1992).
107. M. T. Johnson, S. T. Purcell, N. W. E. McGee, R. Coehoorn, J. aan de Stegge, and W. Hoving, *Phys. Rev. Lett.* 68, 2688 (1992).
108. R. Mattheis, W. Andrä, Jr., L. Fritzsch, A. Hubert, M. Rührig, and F. Thrum, *J. Magn. Magn. Mater.* 121, 424 (1993).
109. B. Heinrich, J. F. Cochran, B. Hillebrands, and G. Günterodt, in "Ultrathin Magnetic Structures" (B. Heinrich and J. A. C. Bland, Eds.), Vol. 2, p. 222. Springer-Verlag, Berlin, 1994.
110. D. Pescia et al., *Phys. Rev. B* 78, 475 (1990).
111. F. Petroff, A. Barthélémy, D. H. Mosca, D. K. Lottis, A. Fert, P. A. Schroeder, W. P. Pratt, R. Loloee, and S. Lequien, *Phys. Rev. B* 44, 5355 (1991).
112. V. Grolier, D. Renard, B. Barthélémy, P. Beauvillain, C. Chapert, C. Dupas, J. Ferré, M. Galtier, E. Kolb, M. Mulloy, J. P. Renard, and P. Veillet, *Phys. Rev. Lett.* 71, 3023 (1993).
113. G. Tosin, L. F. Schelp, M. Carara, J. E. Schmidt, A. A. Gomes, and M. N. Baibich, *J. Magn. Magn. Mater.* 121, 399 (1993).
114. N. Kataoka, K. Saito, and H. Fujimori, *J. Magn. Magn. Mater.* 121, 383 (1993).
115. N. Metoki et al., *J. Magn. Magn. Mater.* 126, 397 (1993).
116. J. J. de Miguel, A. Cebollada, J. M. Gallego, R. Miranda, C. M. Schneider, P. Schuster, and J. Kirschner, *J. Magn. Magn. Mater.* 93, 1 (1991).
117. A. Fuss, S. Demokritov, P. Grunberg, and W. Zinn, *J. Magn. Magn. Mater.* 103, L221 (1992).

118. J. Unguris, R. J. Cellota, and D. T. Pierce, *J. Magn. Magn. Mater.* 127, 205 (1993).
119. S. M. Rezende, A. Azevedo, F. M. de Aguiar, J. R. Fermin, W. F. Egelhoff, and S. S. P. Parkin, *Phys. Rev. B* 66, 064109 (2002).
120. J. R. L. de Almeida and S. M. Rezende, *Phys. Rev. B* 65, 092412 (2002).
121. S. M. Rezende, C. Chesman, M. A. Lucena, A. Azevedo, F. M. de Aguiar, and S. S. P. Parkin, *J. Appl. Phys.* 84, 958 (1998).
122. S. M. Rezende, C. Chesman, M. A. Lucena, A. Azevedo, F. M. de Aguiar, and S. S. P. Parkin, *J. Magn. Magn. Mater.* 177, 1213 (1998).
123. A. Azevedo, C. Chesman, S. M. Rezende, F. M. de Aguiar, X. Bian, and S. S. P. Parkin, *Phys. Rev. Lett.* 76, 4837 (1996).
124. P. Bruno, *Europhys. Lett.* 23, 615 (1993).
125. P. Bruno, *J. Magn. Magn. Mater.* 121, 248 (1993).
126. J. Barnas, *J. Magn. Magn. Mater.* 128, 171 (1994).
127. P. J. H. Bloemen, M. T. Johnson, M. T. H. van de Vorst, R. Coehoorn, J. J. de Vries, R. Jungblut, J. aan de Stegge, A. Reinders, and W. J. M. de Jonge, *Phys. Rev. Lett.* 72, 764 (1994).
128. S. N. Okuno and K. Inomata, *Phys. Rev. Lett.* 72, 1553 (1994).
129. R. K. Kawakami, E. Rotenberg, E. J. Escorcia-Aparicio, H. J. Choi, J. H. Wolfe, N. V. Smith, and Z. Q. Qiu, *Phys. Rev. Lett.* 82, 4098 (1999).
130. P. Bruno, *Phys. Rev. B* 52, 411 (1995).
131. N. Persat and A. Dinia, *Phys. Rev. B* 56, 2676 (1997).
132. E. Y. Tsymbal, O. N. Mryasov, and P. R. LeClair, *J. Phys. Condens. Matter* 15, R109 (2003).
133. E. Velu, C. Dupas, D. Renard, J. P. Renard, and J. Seiden, *Phys. Rev. B* 37, 668 (1988).
134. T. Shinjo and H. Yamamoto, *J. Phys. Soc. Jpn.* 59, 3061 (1990).
135. C. Dupas et al., *J. Appl. Phys.* 67, 5680 (1991).
136. A. Chaiken et al., *Appl. Phys. Lett.* 59, 240 (1991).
137. D. H. Mosca, A. Barthelemy, F. Petroff, A. Fert, P. A. Schroeder, W. P. Pratt, R. Lolloe, and R. J. Cabanel, *J. Magn. Magn. Mater.* 93, 480 (1991).
138. T. Valet, J. C. Jacquet, J. M. Coutellier, L. G. Pereira, R. Morel, D. Lottis, and A. Fert, *Appl. Phys. Lett.* 61, 3187 (1992).
139. D. Lottis, A. Fert, R. Morel, L. G. Pereira, J. C. Jacquet, P. Galtier, J. M. Coutellier, and T. Valet, *J. Appl. Phys.* 73, 5515 (1993).
140. B. Dieny, V. S. Speriosu, S. S. P. Parkin, B. A. Gurney, D. R. Wilhoit, and D. Mauri, *Phys. Rev. B* 43, 1297 (1991).
141. B. Dieny, V. S. Speriosu, B. A. Gurney, S. S. P. Parkin, D. D. Wilhoit, K. P. Roche, S. Metin, D. T. Peterson, and S. Nadimi, *J. Magn. Magn. Mater.* 93, 101 (1991).
142. V. S. Speriosu, B. Dieny, P. Humpert, B. A. Gurney, and H. Lefakis, *Phys. Rev. B* 44, 5358 (1991).
143. J. Szucs, T. O'Brien, D. K. Lottis, S. Gangopadhyay, S. Mao, and E. Murdock, *J. Appl. Phys.* 81, 4014 (1997) and references therein.
144. N. F. Mott, *Proc. R. Soc. London, Ser. A* 153, 699 (1936).
145. N. F. Mott, *Adv. Phys.* 13, 325 (1964).
146. Schad et al., *Appl. Phys. Lett.* 64, 3500 (1994).
147. K. Fuchs, *Proc. Cambridge Philos. Soc.* 34, 100 (1938).
148. H. Sondheimer, *Adv. Mater.* 1, 1 (1952).
149. P. F. Carcia and A. Suna, *J. Appl. Phys.* 54, 2000 (1983).
150. R. E. Camley and J. Barnas, *Phys. Rev. Lett.* 63, 664 (1989).
151. A. Barthélémy and A. Fert, *Phys. Rev. B* 43, 13124 (1991).
152. D. H. Mosca, Ph.D. Thesis, IF-UFRGS, Porto Alegre, 1992.
153. J. L. Duvail, A. Fert, L. G. Pereira, and D. K. Lottis, *J. Appl. Phys.* 75, 7070 (1994).
154. L. G. Pereira, J. L. Duvail, and D. K. Lottis, *J. Appl. Phys.* 88, 4772 (2000).
155. B. Dieny, *Phys. Rev. B* 45, 806 (1992).
156. D. M. Edwards, *IEEE Trans. Magn.* 27, 3548 (1991).
157. R. L. White, *IEEE Trans. Magn.* 28, 2482 (1992).
158. Z. Tesanovic, M. V. Jaric, and S. Maekawa, *Phys. Rev. Lett.* 57, 2760 (1986).
159. A. Vedyayev, B. Dieny, and N. Ryhanova, *Europhys. Lett.* 19, 329 (1992).
160. P. M. Levy, *Phys. Rev. Lett.* 65, 1643 (1990).
161. J. Inoe and S. Maekawa, *Prog. Teor. Phys.* 106, 184 (1991).
162. W. P. Pratt, Jr., S. F. Lee, P. Holody, Q. Yang, R. Loloee, J. Bass, and P. A. Schroeder, *J. Magn. Magn. Mater.* 126, 406 (1993).
163. M. A. M. Gijs et al., *Appl. Phys. Lett.* 63, 111 (1993).
164. W. P. Pratt et al., *J. Magn. Magn. Mater.* 126, 406 (1993).
165. W. P. Pratt et al., *J. Appl. Phys.* 73, 5326 (1993).
166. T. Valet and A. Fert, *J. Magn. Magn. Mater.* 121, 378 (1993).
167. M. A. M. Gijs, M. T. Johnson, A. Reinders, and P. E. Huisman, *Appl. Phys. Lett.* 66, 1839 (1995).
168. Q. Yang, P. Holody, S. F. Lee, L. L. Henry, R. Loloee, P. A. Schroeder, W. P. Pratt, Jr., and J. Bass, *Phys. Rev. Lett.* 72, 3274 (1994).
169. C. Vouille, A. Barthélémy, F. Elokan Mpondo, A. Fert, P. A. Schroeder, S. Y. Hsu, A. Reilly, and B. Loloee, *Phys. Rev. B* 60, 6710 (1999).
170. J. Bass and W. P. Pratt, Jr., *J. Magn. Magn. Mater.* 200, 274 (1999).
171. L. Piraux, S. Dubois, C. Marchal, J. M. Beuken, L. Fillipozzi, J. F. Depres, K. Ounadjela, and A. Fert, *J. Magn. Magn. Mater.* 156, 317 (1996).
172. L. Piraux, S. Dubois, C. Marchal, J. M. Beuken, L. Fillipozzi, J. F. Depres, K. Ounadjela, and A. Fert, *J. Magn. Magn. Mater.* 159, L287 (1996).
173. K. Ounadjela and A. Fert, *J. Magn. Magn. Mater.* 156, 317 (1996).
174. A. Fert, A. Barthélémy, P. Galtier, P. Holody, R. Loloee, R. Morel, F. Petroff, P. Schroeder, L. B. Steren, and T. Valet, *Mater. Sci. Eng., B* 31, 1 (1995).
175. C. Vouille, Ph.D. Thesis, Université Paris XI, Orsay, 1998.
176. Q. Yang et al., *Phys. Rev. B* 51, 3226 (1995).
177. J. Bass, P. A. Schroeder, W. P. Pratt, S. F. Lee, Q. Yang, P. Holody, L. L. Henry, and R. Loloee, *Mater. Sci. Eng., B* 31, 77 (1995).
178. W. Oeps et al., *Phys. Rev. B* 53, 14024 (1996).
179. I. Mertig et al., *J. Magn. Magn. Mater.* 151, 363 (1995).
180. S. Y. Hsu, A. Barthélémy, P. Holody, R. Loloee, P. A. Schroeder, and A. Fert, *Phys. Rev. Lett.* 78, 2652 (1997).
181. J. M. George, L. G. Pereira, A. Barthélémy, F. Petroff, L. B. Steren, J. L. Duvail, A. Fert, R. Loloee, P. Holody, and P. A. Schroeder, *Phys. Rev. Lett.* 72, 408 (1994).
182. J. P. Renard, P. Bruno, R. Megy, B. Bartenlian, P. Beauvillain, C. Chappert, C. Dupas, E. Kolb, M. Mulloy, P. Veilet, and E. Velu, *Phys. Rev. B* 51, 12821 (1995).
183. C. Vouille, A. Fert, A. Barthélémy, S. Y. Hsu, R. Loloee, and P. A. Schroeder, *J. Appl. Phys.* 81, 4573 (1997).
184. W. F. Egelhoff et al., *J. Appl. Phys.* 82, 6142 (1997).
185. Y. Saito et al., *J. Magn. Magn. Mater.* 126, 466 (1993).
186. S. S. P. Parkin, Z. G. Li, and D. J. Smith, *Appl. Phys. Lett.* 58, 2710 (1991).
187. D. Greig, M. J. Hall, C. Hammand, B. J. Hickey, H. P. Ho, M. A. Howson, M. J. Walker, and D. G. Wright, *J. Magn. Magn. Mater.* 110, 239 (1992).
188. M. E. Tomlinson, R. J. Pollard, D. G. Lord, and P. J. Grundy, *J. Magn. Magn. Mater.* 111, 79 (1992).
189. J. M. George, A. Barthélemy, F. Petroff, T. Valet, and A. Fert, *Mater. Res. Soc. Symp. Proc.* 313, 737 (1993).
190. L. Berger, *Phys. Rev. B* 54, 9353 (1996).
191. J. C. Slonczewski, *J. Magn. Magn. Mater.* 159, L1 (1996).
192. W. Weber, S. Riesen, and H. C. Siegmann, *Science* 291, 1015 (2001).
193. J. A. Katine, F. J. Albert, R. A. Buhrman, E. B. Myers, and D. C. Ralph, *Phys. Rev.* 84, 3149 (2000).
194. A. Fert, A. Barthélémy, J. Ben Youssef, J. P. Contour, V. Cros, J. M. de Teresa, A. Hamzic, J. M. George, G. Faini, J. Grollier, H. Jaffrès, H. Le Gall, F. Montaigne, F. Pailloux, and F. Petroff, *Mater. Sci. Eng., B* 84, 1 (2001).

195. M. Johnson and R. H. Silsbee, *Phys. Rev. B* 35, 4959 (1991); M. Johnson, *Phys. Rev. Lett.* 67, 3594 (1991).

196. P. C. van Son, H. van Kempen, and P. Wyder, *Phys. Rev. Lett.* 58, 2271 (1987).

197. E. I. Rashba, *Phys. Rev. B* 62, R1667 (2000).

198. G. S. Schmidt and L. W. Molenkamp, *Semicond. Sci. Technol.* 17, 310 (2002).

199. A. Fert and H. Jaffrès, *Phys. Rev. B* 64, 184420 (2001).

200. H. J. Zhu, M. Ramsteiner, H. Kostial, M. Wassermeier, H. P. Schonher, and K. H. Ploog, *Phys. Rev. Lett.* 87, 016601 (2001).

201. F. Gustavsson, J. M. George, V. H. Etgens, and M. Mahmoud, *Phys. Rev. B* 64, 184422 (2001).

202. D. H. Mosca, J. M. George, J. L. Maurice, A. Fert, M. Eddrief, and V. H. Etgens, *J. Magn. Magn. Mater.* 226, 917 (2001).

203. J. M. Kikkawa and D. D. Awschalom, *Phys. Rev. Lett.* 80, 4313 (1998).

204. See article collection in *Semicond. Sci. Technol.* 17, 275 (2002).

205. J. S. Moodera, L. R. Kinder, T. M. Wong, and R. Meservey, *Phys. Rev. Lett.* 74, 3273 (1995).

206. J. C. Slonczewski, *Phys. Rev. B* 39, 6995 (1989).

207. M. Jullière, *Phys. Lett. A* 54, 225 (1975).

208. R. Meservey and P. M. Tedrow, *Phys. Rep.* 238, 173 (1994).

209. D. J. Monsma and S. S. P. Parkin, *Appl. Phys. Lett.* 77, 720 (2000).

210. P. K. de Boer, G. A. de Wijs, and R. A. de Groot, *Phys. Rev. B* 58, 15422 (1998).

211. I. I. Oleinik, E. Y. Tsymbal, and D. G. Pettifor, *Phys. Rev. B* 62, 3952 (2000).

212. E. Y. Tsymbal, I. I. Oleinik, and D. G. Pettifor, *J. Appl. Phys.* 87, 5230 (2000).

213. G. A. Prinz, *J. Magn. Magn. Mater.* 200, 57 (1999).

214. V. Popovic, *Sen. Actuators, A* 56, 39 (1996).

215. F. B. Shelledy and J. L. Nix, *IEEE Trans. Magn.* 28, 2283 (1992).

216. T. Hermann, W. Black, and S. Hui, *IEEE Trans. Magn.* 33, 4029 (1997).

217. P. J. Grundy, *J. Magn. Magn. Mater.* 126, 516 (1993).

218. M. Jimbo et al., *J. Magn. Magn. Mater.* 126, 422 (1993).

219. C. Dupas, P. Beauvilain, C. Chapert, J. P. Renard, F. Trigui, P. Veillet, and E. Velu, *J. Appl. Phys.* 67, 5680 (1990).

220. A. Chaiken, P. Lubitz, J. J. Krebs, G. A. Prinz, and M. Z. Harford, *J. Appl. Phys.* 70, 5864 (1991).

221. K. Matsuyama et al., *IEEE Trans. Magn.* 32, 4612 (1996).

222. A. Schul et al., *J. Appl. Phys.* 75, 7061 (1994).

223. B. Dieny, *J. Magn. Magn. Mater.* 136, 335 (1994).

224. B. Dieny, A. Granovsky, A. Vedyaev, N. Ryzhanova, C. Cowache, and L. G. Pereira, *J. Magn. Magn. Mater.* 151, 378 (1995).

225. R. P. Hunt, *IEEE Trans. Magn.* 7, 150 (1971).

226. F. Tomiyama, *IEEE Trans. Magn.* 34, 1970 (1998).

227. K. O'Grady and H. Laidler, *J. Magn. Magn. Mater.* 200, 616 (1999).

228. R. H. Dee and J. C. Cates, *IEEE Trans. Magn.* 34, 1866 (1998).

229. R. E. Fontana, *IEEE Trans. Magn.* 35, 806 (1999).

230. C. Tsang, *IEEE Trans. Magn.* 35, 689 (1999).

231. J. A. Brug, *J. Appl. Phys.* 79, 4491 (1996).

232. J. J. M. Ruigrok, R. Coehoorn, S. R. Cumpson, and H. W. Kesteren, *J. Appl. Phys.* 87, 5398 (2000).

233. S. Datta and B. Das, *Appl. Phys. Lett.* 56, 665 (1990).

234. M. Johnson, *Science* 26, 32 (1993).

235. D. J. Monsma, J. C. Lodder, Th. J. A. Popma, and B. Dieny, *Phys. Rev. Lett.* 74, 5260 (1995).

236. J. de Boek, W. Van Roy, J. Das, V. Motsnyi, Z. Liu, L. Lagae, H. Boeve, K. Dessein, and G. Borghs, *Semicond. Sci. Technol.* 17, 342 (2002).

237. R. Jansen, P. S. Anil Kumar, O. M. J. van't Erve, R. Vlutters, P. de Haan, and J. C. Lodder, *Phys. Rev. Lett.* 85, 3277 (2000).

238. A. Pohm, C. Comstock, and A. Hurst, *J. Appl. Phys.* 67, 4881 (1990).

239. J. Daughton, *Thin Solid Films* 216, 162 (1992).

240. J. Gadbois and J. Zhu, *IEEE Trans. Magn.* 34, 1066 (1998).

241. J. G. Zhu, Y. Zheng, and G. Prinz, *J. Appl. Phys.* 87, 6668 (2000).

242. J. Daughton, *J. Appl. Phys.* 81, 3758 (1997).

243. S. S. P. Parkin, *J. Appl. Phys.* 85, 5828 (1999).

244. J. Shen and J. Kirschner, *Surf. Sci.* 500, 300 (2002).

245. S. A. Wolf, D. D. Awschalom, R. A. Buhrman, J. M. Daughton, S. von Molnar, M. L. Roukes, A. Y. Chtchelkanova, and D. M. Treger, *Science* 294, 1488 (2001).

246. J. A. Gupta et al., *Science* 292, 2458 (2001).

Encyclopedia of Nanoscience and Nanotechnology
www.aspbs.com/enn

Growth Simulations of Nanoclusters

F. Baletto, R. Ferrando, A. C. Levi
Università di Genova, Genova, Italy

CONTENTS

1. INTRODUCTION

As this whole encyclopedia shows, the fabrication and study of nanostructures are presently central subjects of both science and technology. These structures typically form by growth (or by erosion), either on a substrate or in free space. The growth process occurs in the laboratory, but also may be simulated in a computer; in the latter case (provided that the simulation is realistic enough) the properties of the growing structure can be followed from the beginning and more deeply understood at the atomistic or molecular level.

In the 20th century, the study of how matter behaves has followed two streams: one focused on the properties and interactions of individual atoms, and later of subatomic components (which is the field of nuclear and particle physics); the other emphasized the properties of very many atoms or molecules together, so many that the aggregates they make can be treated as infinite. The realm between these two limits concerned only a few independent souls until about the late 1970s.

Although the birth date of nanotechnology can be fixed on December 29, 1959, when Richard Feynman in a famous speech said: "What I want to talk about is the problem of manipulating and controlling things on a small scale.... In the distant future we will be able to arrange the atoms the way we want.... There is plenty of room at the bottom." Very little happened until some time later. We recall that the term *nanotechnology* was introduced for the first time in 1974 by N. Taniguchi (of Tokyo University) to define the engineering of submicroscopic systems on the atomic scale (that of nanometres), and to separate it from the more conventional *microtechnology*. Since then, interest has increased explosively in the study of nano-objects, both deposited or grown on a substrate, and free in space (nanoscience), as well as in their fabrication for useful purposes (nanotechnology). For the latter, it is necessary to work with systems of sizes between 1 and 100 nm, and these systems have to be produced via well defined methods in order to get a good definition of the final structure and a good control on it.

In the 1980s two fundamental discoveries in the field of nanoscience were made, allowing a wider study of the nanoworld and opening new horizons. At the IBM of Zurich, G. Binnig and H. Rohrer (Nobel prize 1986) built the scanning tunneling microscope (STM), which is of basic importance in order to visualize and study a surface from an atomic point of view. At the Rice University of Houston, Texas, R. Smalley (Nobel prize 1996, with H. Kroto and R. Curl) found a new molecule, C_{60}, which became the progenitor of a new class of materials (called *fullerenes* in the honor of a famous American architect), including the so-called nanotubes.

What can we expect from this nanoworld? Introducing the nanotechnology into our macroworld, it is possible to improve the present technology, in order to fabricate more and more efficient microchips, able to elaborate data faster and faster; to accumulate terabytes in few square centimeters of magnetic substrates; to create new materials with new structural properties; to produce new sensors for medical purposes.... up to nanorobots and nanomachines: the new science fiction.

Of central importance in this context are what have come to be called *clusters*, which can be considered as the building blocks of nanoscience and nanotechnology. Clusters are aggregates of atoms (or molecules), generally intermediate in size between individual atoms and objects large enough to be considered as *bulk matter*. When we speak of *small* clusters we mean something of size $d \approx$ 1–3 nm, containing no more than a few hundreds or possibly a thousand particles (atoms or molecules); *large* clusters, on the other hand, consist of many thousands of particles ($d \approx$ tens of nm). We can grow clusters either on a substrate or in free space. In both cases, clusters can be 2-D or 3-D.

ISBN: 1-58883-059-4/$35.00

The nanoworld is at the boundary between the world of atoms and molecules, which is ruled by quantum mechanics, and the macroworld ruled by classical mechanics. A basic aim of nanoscience is to understand the properties of nano-objects, grown either on a substrate or in free space: Since their behavior is intermediate between quantum and classical, it is not trivial to find a universal modeling to describe their properties.

In this sense, a good way to study both free and supported clusters (the latter corresponding to nanostructured surfaces) is to simulate their growth processes from the beginning, and to compare the results obtained with the experimental data, in order to understand more deeply the properties we wish to investigate. In this chapter, we will discuss such simulations in two geometries:

- Surface growth on a substrate, which may also be of the same nature as the growing structure.
- Cluster growth in free space (and then deposited on a substrate).

First of all, we want to describe the computational methods that can be used to analyze nanostructure growth processes, focusing our attention on a ticklish question in this field: the time scale. In fact, an important point that must be discussed before dealing with the simulation of materials growth is the problem of reaching the correct length and time scales. These scales depend on the specific features of the growth experiments, but there are typical ranges which can be easily singled out.

In the case of the epitaxial growth on a substrate [1], the typical length scale is determined by the size of the crystal terraces, which can be of 10^3 Å or more. This means that each atomic layer in the growing crystal contains at least several thousands of atoms. On the other hand, the experimental fluxes range from a few monolayers per second to fractions of a monolayer per minute.

On the contrary, the growth of free clusters of a few hundreds of atoms in gas phase takes place on the time scale of fractions of a millisecond. In this case, the length scale, which corresponds to the diameter of the cluster, is a few tens of Å [2–4].

Because of the above differences in the typical length and time scales, and depending on the quantities that have to be calculated, different methods are appropriate in the simulation of these systems. The typical tools range from *ab initio* calculations of the electronic structure to molecular dynamics (MD) with interatomic potentials to kinetic Monte Carlo (KMC) simulations. Let us analyze these methods in relation to their possible applications to the simulation of actual crystal or cluster growth.

The *ab initio* calculations [5] are the most reliable and least approximate, but are computationally very expensive. At the moment, they are limited to sizes around a hundred atoms, and they can extend to times of a few tens of picoseconds. Therefore, they cannot be used at all for simulating any real growth process, neither for crystals nor for clusters. However by *ab initio* methods, the energy barriers for many diffusion mechanisms can be computed, and these barriers may serve as an input for KMC simulations. Unfortunately, it is quite expensive to implement methods for the automatic search of saddle points into an *ab initio* scheme. In the case of semiconductors, longer time scales (some nanoseconds, but again on small sizes) can be reached by tight-binding simulations [6]. These methods are thus better suited to investigate the elementary diffusion processes than for simulating crystal growth.

The MD [7] technique relies on the description of the element under study by means of semiempirical interatomic potentials. Reliable interatomic potentials have been developed for simple metals and transition metals, and for some semiconductors [8]. In the MD simulations, a detailed microscopic description of the system is achieved. In a standard MD simulation, the classical equations of motion are solved by the discretization of time in steps δt. In order to have a precise solution of the equations of motion, δt must be much shorter than the fastest phononic periods (which are in the order of 10^{-12} s). Therefore, typical δt are in the range of 10^{-14}–10^{-15} s. The number of steps that can be accumulated during a simulation depends on the system size; at present, systems with $N \sim 10^4$ atoms can be simulated on about 10^7 steps. This means that a simulation of epitaxial crystal growth on experimental sizes by standard MD can reach a time scale of 10^{-7} s, which is at least seven orders of magnitude faster than the real experiments. The situation is different in the case of free nanocluster growth. The experimental time scales are much shorter (10^{-3}–10^{-4} s), and many fewer atoms must be simulated. This allows one to extend the number of time steps up to 10^{10}, thus reaching a simulation time that is only one or two orders of magnitude faster than the experimental time, and this allows a direct comparison between the simulations and the experimental data.

A recent method for extending the time scale of simulations maintaining the detailed microscopic description of the MD technique is accelerated molecular dynamics (AMD) [9], for which different methods have been developed, mostly based on the assumption that the transition state theory (TST) holds [10]. The AMD simulations can reach the actual experimental time scales of epitaxial growth but still on small sizes (say a few hundred atoms) and only at low temperatures, where the acceleration boost is much more effective.

However, the most effective tool for simulating the epitaxial growth is still the KMC method [11, 12]. In these simulations, space is discretized on a lattice. A set of possible moves (a move corresponding to the displacement of an atom from an equilibrium site to another) is chosen, and time is coarse grained, so that the elementary time scale is the rate of the fastest movement. Rates are assumed usually to have an activated form, so that the rate of a given process a is $\Gamma_a = \Gamma_a^0 \exp[-E_a/(k_B T)]$, where E_a is the activation barrier and Γ_a^0 is a temperature-independent prefactor. The probability of choosing a move is proportional to its rate. In this way, both experimental length and time scales can be reached, at the expense of a much less detailed description of the system under study. In fact, it is practically impossible to take into account all possible atomistic processes in a KMC simulation. Therefore, one usually chooses a subset of processes, with the hope that it will be sufficient to grasp the essential physics of the growth process. On the other hand the KMC method is not well-suited to study free nanocluster growth, because nanoclusters can assume both crystalline and noncrystalline structures, and this forbids the use of a fixed lattice of coordinates. Moreover, the rearrangement processes are so complicated that they cannot be fitted into a simple table of elementary moves.

This chapter is subdivided into two parts. In the first part, of general character, which corresponds to Section 2, we shall deal with a description of the three above mentioned computational/simulational techniques, describing the most important algorithms of specific interest for nanogrowth simulations. In the second part, we shall treat examples that constitute successful applications of the methods described in Section 2. Specifically, in Section 3 we shall describe applications to thin films and cluster growth on surfaces, whereas in Section 4 we shall deal with the growth, solidification, and coalescence of free nanoclusters. Finally in Section 5 we present the conclusions.

2. SIMULATION METHODS

2.1. *Ab Initio*

There are different kinds of *ab initio* simulations, depending on the degree of how *ab initio* they really are. In all cases, nowadays, density functional theory (DFT) is used. We briefly recall the main principles of the theory, as described (in the context of surface physics) for example in Refs. [5, 13].

According to the Hohenberg–Kohn theorems [14, 15], the specification of an electronic ground state density $n(\mathbf{r})$ determines the corresponding external potential $v_{ext}(\mathbf{r})$ uniquely: In other words, the density $n(\mathbf{r})$ contains all the relevant information for the calculation of the electronic energy and of all the other relevant properties of the system (and the explicit calculation of the many-body ground-state wave function is not necessary). The external potential v_{ext} in principle is nothing but the Coulomb potential due to the nuclei, assumed fixed according to the Born–Oppenheimer (BO) approximation (see below, however). The electronic energy is thus a density functional $E_v[n]$, that has to be minimized.

The functional is

$$E_v[n] = T[n] + \int v_{ext}(\mathbf{r})n(\mathbf{r})\, d^3r + 1/2 \int u(\mathbf{r})n(\mathbf{r})\, d^3r + E_{xc}[n] \quad (1)$$

where: $T[n]$ is the kinetic energy functional; the second term describes the interaction with the external potential; the third term describes the Coulombic repulsion

$$u(\mathbf{r}) = e^2 \int \frac{n(\mathbf{r}')}{|\mathbf{r}' - \mathbf{r}|}\, d^3r' \quad (2)$$

of the electrons among themselves; and the fourth, *difficult* term $E_{xc}[n]$ contains all the effects of exchange and correlation.

Because of the last term, evaluating the energy functional may seem hopeless. However, it must be recalled that for a *uniform* electron gas (where there is no external potential and the density is the only variable) the functional reduces to a *function*, which is known with considerable precision, such as from the work of Ceperley and Alder [16].

Minimizing the functional implies solving a system of effective single-particle Schrödinger equations (the Kohn–Sham equations [17]), the potential occurring in the equation being an appropriate functional derivative with respect to the density. In the local density approximation (LDA) the functional used is computed at constant density, using the above well-known results for the uniform electron gas, while more advanced approximations take into account approximately the effect of the density gradients for a nonuniform electron gas.

It must be observed that in real calculations the electrons studied are nearly always the valence electrons only. The core electrons are taken into account replacing the nuclear potential by an appropriate pseudopotential, which plays then the role of the "external potential" of the theory.

True *ab initio* simulations would imply treating the electrons and the nuclei together. For decades this was believed to be an impossible feat, not only because of mathematical complications, but also because of the very different time scales involved. From the BO (or, more generally, adiabatic) point of view, the rapidly moving electrons adapt at each instant of time to the nuclear positions, and their energy constitutes, in turn, the potential in which the nuclei move. To each electronic state corresponds a potential energy surface (PES), or rather hypersurface, in the space of nuclear positions and energy.

This point of view is still valid, of course (apart from special points on hypercurves where two hypersurfaces cross and the adiabatic approximation does not apply). But some modern calculations based on density-functional theory, while retaining the BO approximation, manage to follow both the nuclear and the electronic motions.

The most extreme case is the Car–Parrinello approach [18], where the electronic behavior (quantum-mechanical, of course) is simulated by replacing the electron variables by a (large) set of fictitious classical variables. If the latter are chosen properly, they mimic the electron behavior very closely while at the same time allowing the use of a classical simulation based on Newton equations.

A simple use of the Car–Parrinello method is to let the surface atoms move and relax until they find different (partial) equilibrium positions, whose energies are then compared to find the most favorable one, as in the true equilibrium situation. A very recent, instructive example is the work by Sonnet et al., where the different arrangements for coadsorption of C and H on a Si(100) are discussed [19]. (The most favorable arrangement, in particular, is found to be that where H is bound to C and C is imbedded in the Si surface: not surprisingly, since, as the authors point out, the C-H bond is considerably stronger than the Si-H bond.)

The Car–Parrinello method, however, may be used, more daringly, to follow the real motions of atoms in time: i.e., kinetically. This approach is completely correct, because by the Hellmann–Feynman theorem the forces follow exactly (within the BO approximation) from the simulation. The theorem [20] asserts, indeed, that (except for the small forces that arise from deviations from the BO approximation, and apart from magnetic or relativistic effects) the quantum-mechanical forces on the nuclei are just exactly the classical Coulombic forces arising from the charges on the nuclei interacting with the charge density of the electron cloud.

In this case a new parameter, the total duration t_{total} of the simulation (in microscopic time), becomes relevant.

Here an appropriate example is the simulated diffusion of a proton within a silicon matrix by Buda et al. (even though this outstanding work does not refer specifically to surfaces or clusters, we cite it here for its paradigmatic nature) [21]. Trajectories for the H^+ motion through both crystalline and amorphous Si are obtained. An inherent limitation, however, is that the proton dynamics is treated classically, while in principle H^+ should be treated quantum-mechanically along with the electrons. In the crystalline case, diffusion proceeds via jumps between states of high symmetry, while very different paths are followed in the amorphous case. This study is also remarkable for its rather long t_{total}, but in less fortunate cases the latter can be quite short.

To briefly explain the Car–Parrinello method, we recall that in DFT [15] the electron density (which can be written in terms of occupied single-particle orbitals) minimizes the energy functional for given nuclear positions and external constraints. Instead of writing down the equations corresponding to such minimization (i.e., the Kohn–Sham equations [17], which, for relatively large systems, become rapidly difficult to solve), Car and Parrinello introduced a Lagrangian equation including in the kinetic part, beyond the nuclear velocities with their real masses, the time derivatives of electronic orbitals with fictitious masses, and solve the problem by "classical" molecular dynamics, using *simulated annealing* to avoid being trapped in local minima. An illuminating discussion of the numerical implementation of the Car–Parrinello method was given by Remler and Madden [22].

The Car–Parrinello method is by no means the only relevant *ab initio* method for simulating surfaces, clusters, etc. The quantum Monte Carlo methods, in their various versions, are equally important. In particular, the *path-integral Monte Carlo* method (essentially due to Pollock and Ceperley [23]) allows the study of specifically quantum-mechanical properties. One such application is the remarkable study by Sindzingre et al. of the superfluidity of hydrogen clusters, which we shall quote again in the chapter dealing with nanoclusters [24].

The *ab initio* method proposed and used by Gross and Scheffler is quite different [25]. Since (if the electronic structure, total energy, and forces acting on the nuclei are to be evaluated during the journey of the particles) computing each trajectory is a heavy computational task to be performed over and over again, Gross and Scheffler prefer to avoid all this by first evaluating the PES once for all, then studying (usually classically, but in the case of hydrogen, quantum-mechanically [26]) the nuclear motion on such PES. This method reverts in a sense to a more traditional, adiabatic treatment, which is, however, quite sufficient in many cases where nonadiabatic effects are inessential. Surface chemistry usually involves nonadiabatic effects in a mild fashion; for example, the dissociative adsorption of a diatomic molecule, although a chemical reaction, can be satisfactorily described, as is well known, by a single PES where the relevant coordinates are the distance of the molecular center of mass from the surface and the distance between the nuclei of the two atoms. This is precisely the case studied by Gross and Scheffler in their seminal article [25]; moreover, they consider the dissociative adsorption of *hydrogen* on Pd(100) in order to show that the quantum-mechanical properties of the nuclei do not hinder a satisfactory treatment according to their method.

A less delicate case (and also more relevant for the present review) is that of diffusion. For many diffusion problems, there is little doubt that Scheffler's methods apply: calculating correctly the PES may be a difficult problem, and studying the diffusion process on such PES may be another, but nonadiabatic effects (with the possible exception of diffusion on the surfaces of semiconductors and valence crystals, where bonds are continuously broken and reformed) can hardly be a difficulty.

In a previous article [27] Stumpf and Scheffler treated in great detail the self-diffusion of Al atoms on several aluminum surfaces, showing how density-functional theory affords considerable precision in the computation of the barriers for diffusion.

Using DFT, Stumpf and Scheffler investigated properties of Al(111), Al(100), Al(110), and stepped Al(111) surfaces. In particular, in the case of Al(111), for adsorption and diffusion of Al on flat regions, the hcp site was found to be slightly preferred energetically over the fcc site. The energy barrier for diffusion is very low (0.04 eV). Close to either of the two sorts of close-packed monatomic steps on Al(111), Al adatoms feel an electronic attraction toward the step edge, which has a range of several lattice spacings. Stumpf and Scheffler also discussed island shapes: in equilibrium the islands are hexagonal; during growth the shapes may be fractal, triangular, or hexagonal depending on the kinetics [27].

Similar to the methods of Scheffler's group are those of Feibelman [28, 29]. A general discussion of the theoretical methods used is given in the first paper cited, while for example in the second paper cited Feibelman computes from first principles the energetics of steps on Pt(111). Platinum is complicated because of the *d*-electrons (previous calculations, including those of Stumpf and Scheffler, had been mostly done on *sp*-bonded materials), and Feibelman stresses the new problems that arise for Pt. The pseudopotential is essential (Feibelman uses that of Hamann [30]), but equally essential is a correct use of the LDA where the energies of a uniform electron gas are those calculated with great accuracy by Ceperley and Alder [16] and parametrized by Perdew and Zunger [31]. The step energies computed by Feibelman, however, are not in complete agreement with experiments, probably showing that the LDA is not accurate enough.

Indeed, more recent work (too vast to be reviewed here) tries to achieve a better agreement by going beyond the LDA. For example, the generalized gradient approximation (GGA) goes beyond the LDA by including, besides the properties of the uniform electron gas, terms related in a rather sophisticated way to the density gradients. The GGA is believed to be a definite improvement over the LDA (contrary to older attempts to include the gradients, which failed, the reason being that those *naive* gradient corrections violated some fundamental physical rules of the electron gas, such as that the exchange hole density must be strictly negative and its integral must be exactly -1 [5]).

But even with the GGA the results are sometimes disappointing. An interesting paper by Lorensen et al. on the mechanisms of self-diffusion on Pt(110) compares, both in the LDA and in two versions of the GGA (PW91, RPBE), two different diffusion paths along the grooves of the missing-row reconstructed (110) surface [32]: the *direct* path and a more winding *facet* path exploring the flanks of the groove. While the LDA favors the direct path, the GGA is ambiguous; the RPBE version favors the facet path, but for the PW91 version both paths are equally probable.

Recent papers by Feibelman consider various interesting problems relevant for surface diffusion [33–35]. *Concerted substitution*, i.e., the process where growth occurs by implanting adsorbed atoms into the outermost layer of a metal, was studied by Feibelman and Stumpf [33], who proved that such a process occurs more easily on those transition metal surfaces where the relaxation around an adatom is greater, as had previously been conjectured by Kellogg et al. [36]. Feibelman then considers diffusion along the bottom side of a step, where another kind of *concerted substitution* is possible, but proves that (at least on Pt(111)) this process has to overcome a higher energy barrier than hopping diffusion [34]. In the last paper cited above, two different diffusion modes of dimers on Pt(110) (dimer dissociation and recombination, and the leapfrog mechanism suggested by Montalenti and Ferrando [37] where the back atom overcomes the front atom) are compared [35]. Feibelman concludes that leapfrog should be by far the most probable mechanism. Since this is not what is observed experimentally, Feibelman suggests that the discrepancy may be due to CO contamination, frequent on Pt surfaces.

Finally we remark that calculations from first principles are desirable, but rather heavy computationally. As a consequence, the size of the systems that can be simulated is usually rather small: A typical piece of surface may contain, for example, $10 \times 10 \times 2$ atoms (the third number referring of course to the direction into the bulk). The situation is not catastrophic, however, since the cell is repeated on the surface and no spurious boundary effects appear. Concerning free clusters, the *ab initio* calculations are limited to sizes of about one hundred atoms.

2.2. Molecular Dynamics Simulations

Numerical simulations are good tools to study the properties of the many-body systems. They allow one to compute both instantaneous quantities and averages on time or the number of particles. The latter are of fundamental importance since they are comparable with experimental data. To estimate correctly these quantities it is necessary to introduce statistical mechanics into a computational simulation.

Molecular dynamics is the method used to study equilibrium and transport properties of classical many-body systems. *Classical* means that the motion of particles can be well described by Newtonian laws. This approximation is valid for a great variety of materials and in a wide range of temperatures. In fact, let λ_T be the thermal De Broglie wavelength given by $\lambda_T = \frac{h}{\sqrt{2\pi m k_B T}}$. For typical values of the lattice constant $a \sim 2$–$3\,\text{Å}$, we have that $\lambda_T < a$ for $k_B T > \frac{h^2}{2\pi m a^2}$, which corresponds to consider $T > 20$ K for heavy metals. The aim of an MD simulation is not to predict precisely what will happen to a system that has been prepared in a precisely initial condition: we are always interested in statistical predictions. We wish to predict the average behavior of a system that was prepared in an initial state about which we know something (e.g., the total energy).

We deal with the techniques used to solve the classical equations of motion for a system of N particles (atoms or molecules) interacting via a potential $\mathcal{V}$, which can be written in terms depending on coordinates of individual particles, pairs, triplets, and so on:

$$\mathcal{V} = \sum_i v_1(\mathbf{r}_i) + \sum_i \sum_{j>i} v_2(\mathbf{r}_i, \mathbf{r}_j) + \sum_i \sum_{j>i} \sum_{k>j>i} v_3(\mathbf{r}_i, \mathbf{r}_j, \mathbf{r}_k) + \cdots \tag{3}$$

We can simulate microcanonical systems to analyze the diffusive motion of adparticles on a substrate and canonical systems to study growth processes.

We consider a classical system Γ consisting of N particles in a volume V in the phase space $\Omega(\mathbf{p}, \mathbf{q})$: its microscopic state is described in term of positions p_i and momentum q_i of each element of our system. Working in the Born–Oppenheimer approximation, it is possible to write the Hamiltonian $\mathcal{H}$ of the system as $\mathcal{H} = \mathcal{T} + \mathcal{V}$, where $\mathcal{V}$ is the potential term and $\mathcal{T}$ is the kinetic energy, and it is usually given by

$$\mathcal{T} = \sum_{i=1}^{N} \sum_{\alpha=1}^{3} \frac{p_{i\alpha}^2}{2m_i} \tag{4}$$

where m_i is the mass of the particles and the index α runs on the components of the momentum.

If the potential is conservative, we can write the motion equations in the Hamilton–Jacobi formalism. In the configurational space $\mathcal{Q}(\mathbf{q}, \dot{\mathbf{q}})$, the time evolution of a holonomic system is given by

$$\begin{aligned} -\dot{p}_k &= \frac{\partial \mathcal{H}}{\partial q_k} \\ \dot{q}_k &= \frac{\partial \mathcal{H}}{\partial p_k} \end{aligned} \tag{5}$$

If we consider a system of atoms, with Cartesian coordinates $\mathbf{r}_i$ and the previous definitions of $\mathcal{V}$ and $\mathcal{T}$, the motion equation can be written as

$$m_i \ddot{\mathbf{r}}_i = \mathbf{f}_i \tag{6}$$

where m_i is the mass of particles i and $\mathbf{f}_i = -\nabla_{\mathbf{r}_i} \mathcal{H} = -\nabla_{\mathbf{r}_i} \mathcal{V}$ is the force on particles i. For systems described by a potential that is independent of time and velocities, we have that $\mathcal{H}$ is automatically equal to the total energy; and thus we have

$$\begin{aligned} \dot{\mathbf{r}}_i &= \frac{\mathbf{p}_i}{m_i} \\ \dot{\mathbf{p}}_i &= \mathbf{f}_i \end{aligned} \tag{7}$$

Computing center-of-mass trajectories, then, involves solving either a system of $3N$ second-order differential equations or an equivalent set of $6N$ first-order equations. Before considering how to do this, we can make just one general remark concerning the equations of motion: Newton's equations are time reversible, and so should be our algorithms. By changing the signs of all velocities or momenta, we will cause the particles to retrace their trajectories. The computer-generated trajectories will also have this property.

If we simulate a microcanonical system of N particles in a given volume V, its Hamiltonian $\mathcal{H}$ satisfies the condition $E < \mathcal{H} < E + \Delta E$ with $(\Delta E/E) \ll 1$; in other words in a microcanonical system N, V is constant and E is well conserved. The locus of points in Ω describing $\mathcal{H}(\mathbf{p}, \mathbf{q}) \equiv E$ defines a hypersurface of energy E. If the system evolves in time following the Hamilton equations, the representative point runs on this hypersurface. We require the system to preserve its energy in the range between E and $E + \Delta E$. We may also speak about a Gibbs ensemble, which is represented by a point distribution in the phase space characterized by a density function $\rho = \rho(t, \mathbf{p}, \mathbf{q})$. In the Gibbs scheme, the Hamiltonian satisfies the Liouville theorem $(\partial\rho/\partial t) + \{\mathcal{H}, \rho\} = 0$. If we want to study equilibrium properties we can focus our attention on those systems whose equilibrium density ρ_{eq} depends only on $\mathcal{H}$, i.e. on the energy. If ergodicity is assumed, time averages are equal to ensemble averages, the latter being defined by

$$\bar{f} = \frac{1}{N!h^{3N}} \frac{\int f(\mathbf{q}, \mathbf{p})\rho_{eq}(\mathbf{q}, \mathbf{p})\, d^{3N}p\, d^{3N}q}{\int \rho_{eq}(\mathbf{p}, \mathbf{q})\, d^{3N}p\, d^{3N}q} \tag{8}$$

where f is a macroscopic function in phase space. In particular, letting γ_i be a generic coordinate $(\mathbf{p}_i, \mathbf{q}_i)$, its mean value in the microcanonical ensemble is given by

$$\begin{aligned}\overline{\gamma_i \frac{\partial \mathcal{H}}{\partial \gamma_j}} &= \frac{\int_{E<\mathcal{H}<E+\Delta E} \left(\gamma_i \frac{\partial \mathcal{H}}{\partial \gamma_i}\right) d^{3N}p\, d^{3N}q}{\int_{E<\mathcal{H}<E+\Delta E} d^{3N}p\, d^{3N}q} \\ &= \cdots \text{to first order} \cdots \delta_{ij} k_B T\end{aligned} \tag{9}$$

which is the analytic form of the generalized energy equipartition theorem.

The limit of considering only the microcanonical ensemble is that we have to study only a given region of phase space and so it is impossible to describe directly the thermal equilibrium. We will show in the following that it is possible to simulate a canonical ensemble by a change in the Hamiltonian, i.e. adding a thermal contact with a thermostat. The most remarkable difference between the two approaches is that in a canonical simulation the energy fluctuates around its average and its variance is given by $\sigma_{\mathcal{H}}^2 = \overline{\mathcal{H}^2} - (\overline{\mathcal{H}})^2 = k_B T^2 c_V$, where c_V is the heat capacity at costant volume, which is proportional to the number N of particles, thus for $N \to \infty$ the limit $\sigma_{\mathcal{H}}/\overline{\mathcal{H}} \to 0$ will hold.

2.2.1. Force Calculation

A brief comment about the calculation of the forces among the N particles in the system: This is the most time-consuming part of almost all MD simulations. Here we deal with the computational time necessary to calculate the forces at fixed system size to study diffusion properties, and varying the size of the system to study growth processes.

In the case of a pairwise additive potential, we have to consider only the interaction between a particle and its neighbors, which implies that, for N particles, we must evaluate $N\frac{N-1}{2}$ pair distances. Thus, in MD simulations, the computational time t_{mac} scales as $t_{mac} \propto cN + dN^2$ where c and d are constants. We can introduce some tricks to reduce this time a bit. The most common technique is to introduce a cut-off radius r_{cut} to make an evaluation of all pair distances every m time steps. It means that in this way, the t_{mac} scale as $cN + d'N^2$ where $d' = d/m$. For a very large system, the computational times scale are always as N^2. The introduction of a cut-off is possible only if the potential is short ranged, which means that it decays asymptotically faster than r^{-d}, where d is the dimensionality of the system. In a three-dimensional space, it means that a short-ranged $\mathcal{V}$ decreases faster than r^{-3}.

In the case of a growth simulation, the number of particles is not a constant. Let us evolve a system of $N = N_{in} + N_{dep}$ particles, where N_{dep} is the number of deposited particles and N_{in} is the initial size of the system, for a time t_{mac} depending on N_{dep} by

$$\begin{aligned} t_{mac}(N_{dep}) &= cN + dN^2 \\ &= c(N_{in} + N_{dep}) + d(N_{in} + N_{dep})^2 \end{aligned} \tag{10}$$

then each $\Delta t = n\delta t$ we increase N_{dep}, so the total computational time is given by $t_{mac}^{TOT} = \sum_{N_{dep}} t_{mac}(N_{dep}) \propto N^3$.

Of course, the calculation of the force is different depending on the choice of the interparticle potential, whose specific form (which is not necessarily analytic) depends on the particular system that is under study. Typical examples are the Lennard–Jones potential for rare gases, the embedded atom potential for metals, and the Girifalco potential for fullerites [38, 39].

2.2.2. Integration Method: Verlet Algorithms

A standard method for the solution of ordinary differential equations is the finite difference approach. The general idea is as follows: Given the molecular positions, velocities, and other dynamic information at time t, we attempt to obtain the positions, velocities ... at time $t + \delta t$ to a sufficient degree of accuracy. The equations are solved on a step-by-step basis; the choice of the time interval δt (time step) will depend somewhat on the method of solution, but δt will be significantly smaller than the typical time taken for a molecule to travel its own length. If the classical trajectory is continuous, then an estimate of the positions at time $t + \delta t$ may be obtained by Taylor expansion about time t

$$\mathbf{r}(t + \delta t) = \mathbf{r}(t) + \mathbf{v}(t)\delta t + \frac{\mathbf{f}(t)}{2m}\delta t^2 + \frac{\delta t^3}{3!}\mathbf{b}(t) + \cdots \tag{11}$$

similarly,

$$\mathbf{r}(t - \delta t) = \mathbf{r}(t) - \mathbf{v}(t)\delta t + \frac{\mathbf{f}(t)}{2m}\delta t^2 - \frac{\delta t^3}{3!}\mathbf{b}(t) + \cdots \tag{12}$$

Summing these two equations, we have the standard Verlet algorithm

$$\mathbf{r}(t+\delta t) \sim 2\mathbf{r}(t) - \mathbf{r}(t-\delta t) + \frac{\mathbf{f}(t)}{m}\delta t^2 \tag{13}$$

The estimate of the new position contains an error of the order of δt^4. We can derive the velocity from a knowledge of the trajectory, but in this way the expression for the velocity is only accurate to order δt^2, which is not sufficient to calculate accurately the kinetic energy and thus the temperature T.

It is more useful to introduce the velocity-Verlet algorithm, which is equivalent to the standard Verlet [7] but involves explicitly the velocity variables in the propagation of the trajectories; at time $t + \delta t$ we have

$$\begin{aligned}\mathbf{r}(t+\delta t) &= \mathbf{r}(t) + \mathbf{v}(t)\delta t + \frac{\mathbf{f}(t)}{2m}\delta t^2 + \cdots \\ \mathbf{v}(t+\delta t) &= \mathbf{v}(t) + \frac{\mathbf{f}(t)+\mathbf{f}(t+\delta t)}{2m}\delta t^2 + \cdots\end{aligned} \tag{14}$$

The velocity-Verlet algorithm is fast, even if this is relatively unimportant; it is not particularly accurate for long time steps. Hence, we should expect to compute the forces on all particles rather frequently. Also, it requires about as little memory as is at all possible. Plus, its short-time energy conservation is fair, but it exhibits little long-term energy drift. This is related to the fact that Verlet algorithms are area-preserving and time-reversible [7]. In fact, although Verlet algorithms do not conserve the total energy of the whole system exactly, strong evidence indicates that they conserve a pseudo-Hamiltonian approaching the true Hamiltonian in the limit of infinitely short time steps. The accuracy of the trajectories generated by the Verlet algorithms is not impressive. But then, it would hardly help to use a better algorithm. Such an algorithm may only postpone the unavoidable exponential growth of the error in the trajectories close to the true trajectories for a time comparable to the duration of a typical MD run.

Just a few comments about the choice of the time step. Obviously, if it is longer, fewer evaluations of the forces are needed per unit of simulation time. This would suggest that it is advantageous to use a sophisticated algorithm that allows the use of a long time step. Algorithms that allow the use of a large time step achieve this by requiring more memory storage. Moreover, it is necessary to integrate the motion equations on realistic time scales: The time step must be much less than the typical amount of time that a particle spends in a single oscillation in its well. The typical phonon frequency is of the order of 10^{12} s^{-1}; thus time steps in the range of some femtoseconds are small enough to ensure solving the Newton equations correctly.

2.2.3. MD at Constant Temperature

Before describing briefly how it is possible to perform canonical MD simulations, we want to underline that they are necessary for studying growth processes. In fact, when a particle is deposited on a surface, the release of its adsorption energy causes an increase of the temperature of the system, which is huge for the rather small systems used in MD simulations (for example, several tens of degrees for a system of several hundreds of particles). Therefore subsequent depositions would cause the melting of the system, and this is not in agreement with the experimental observations of growth processes.

Even if at first sight it would seem impossible to perform canonical MD simulations, fortunately it turns out that there are two different solutions to this problem. One is based on the idea that dynamical simulation of other ensembles (e.g., $[N, V, T]$, $[N, P, T]$) is possible by mixing Newtonian MD with certain MC moves. The second approach is completely dynamical in origin. It is based on a reformulation of the Lagrangian equations of motion of the system. The latter, introduced by Andersen [40] in the context of constant-pressure MD simulations, has become one of the most important tricks to extend the applicability of MD simulations. Before describing briefly the Andersen approach, we want to discuss a bit the meaning of "constant temperature". From a statistical point of view, there is no ambiguity; we can impose a temperature on a system by bringing it into a thermal contact with a large heat bath. Under these conditions, the probability to find the system in a given energy state is given by the Boltzmann distribution. In the constant temperature method proposed by Andersen [40], the system is indeed coupled to a heat bath that imposes the desired T. The coupling to the heat bath is represented by stochastic impulsive forces that act occasionally on randomly selected particles. These stochastic collisions with the bath can be considered as MC moves that transport the system from one constant-energy shell to another. Between stochastic collisions, the system evolves at constant energy according to the standard Newtonian equations of motion. The collisions ensure that all accessible constant-energy shells are visited according to their Boltzmann weights. Before starting such a constant-temperature simulation, we should first select the strength of the coupling to the bath. This coupling strength is determined by the frequency of stochastic collisions. Let ν be this frequency. If subsequent collisions are uncorrelated, then the distribution of time intervals between two subsequent stochastic collisions, $P(t; \nu)$, is of the Poisson form

$$P(t; \nu) = \nu e^{-\nu t} \tag{15}$$

Thus in a constant-temperature simulation, we follow the scheme: initialization of the system with given positions and momenta, integration of the Newton equations, and stochastic collisions with the heat bath with a probability of $\nu\delta t$ in a time step δt. For the particles that undergo a collision, the new velocity is drawn from a Maxwell-Boltzmann distribution corresponding to T.

The mixing of Newtonian dynamics a with stochastic collision model turns the MD simulation into a Markov process. As shown by Andersen [40], a canonical distribution in phase space is invariant under repeated application of the Andersen algorithm. Combined with the fact that the Markov chain is also irreducible and aperiodic, this implies that the Andersen algorithm does, indeed, generate a canonical distribution.

2.3. Kinetic Monte Carlo

Standard thermodynamical Monte Carlo is a method for the calculation of average values in a given *equilibrium* thermodynamical ensemble. States in a space of configurations are generated and used for the calculation of quantities of interest [41–43]. The focus is on the convergence of series for the calculated quantities, and one is not interested in the generation of a sequence of states that properly corresponds to the dynamics of a system (in many cases the opposite is true [44]).

About 20 years ago, MC started to be used also for the study of kinetic processes [45, 46]. The aim of KMC is to reproduce faithfully *nonequilibrium*, or *relaxation* processes. This time the emphasis is on the correctness of time evolution in the simulation.

In order to fix ideas, let us denote the space of all possible configurations, $\mathcal{C}$, in a statistical-mechanical model by $\mathcal{S} = \{\mathcal{C}\}$, and by $P(\mathcal{C}, t)$, a time-dependent distribution of configurations at time t.

The MC technique can be also viewed as a method of solving the master equation

$$\frac{\partial P(\mathcal{C}, t)}{\partial t} = -\sum_{\mathcal{C}'} W(\mathcal{C} \to \mathcal{C}')P(\mathcal{C}, t) + \sum_{\mathcal{C}'} W(\mathcal{C}' \to \mathcal{C})P(\mathcal{C}', t) \tag{16}$$

associated with a matrix of *transition probabilities*, $W(\mathcal{C} \to \mathcal{C}')$, between two states $\mathcal{C}$ and $\mathcal{C}'$. Equation (16) describes a stochastic process in the Markovian approximation [47].

The transition probabilities in standard thermodynamical MC do not need to have any relation to the dynamics of the system. They are not considered as given a priori, but they are constructed in a way that guarantees that the distribution of generated states converges as quickly as possible to the desired static distribution $P(\mathcal{C})$. At each MC step a trial configuration is generated, which is then either accepted or rejected. A sequence of configurations, $\mathcal{C}_k$, $k = 1, \ldots, M$, form a Markovian chain, which is used for the construction of series for the quantities to be calculated.

Let us now describe the principle of kinetic Monte Carlo. KMC is a method for solving the master equation (16), which describes the stochastic evolution of the system. The way of solving is essentially as in thermodynamical MC: making random choices so that a Markov chain is generated. Now, however, in contrast to standard thermodynamical MC, this chain has to represent a possible evolution of the simulated system. Configuration changes have to correspond to real events in the stochastic system. Each of the events in the real system can happen with some probability per unit time (rate). To be specific, let $N(\mathcal{C})$ be the number of possible events in a given configuration $\mathcal{C}$; in the case of crystal growth they are, for example, the hopping of an adatom, the adsorption of an atom, and so on. Let the rates of these events be Γ_a, $a = 1, \ldots, N(\mathcal{C})$. Both $N(\mathcal{C})$ and the set $\{\Gamma_a\}$ depend on the configuration $\mathcal{C}$. Let us define the total rate

$$Q = Q(\mathcal{C}) = \sum_{a=1}^{N(\mathcal{C})} \Gamma_a \tag{17}$$

The transition probabilities can now be formally written as

$$W(\mathcal{C} \to \mathcal{C}') = \sum_{a=1}^{N(\mathcal{C})} \Gamma_a V^a(\mathcal{C} \to \mathcal{C}') \tag{18}$$

where $V^a(\mathcal{C} \to \mathcal{C}')$ is again an "underlying" stochastic matrix for an event a, specifying whether the transition $\mathcal{C} \to \mathcal{C}'$ is possible by this event. In the simulation event a should occur with probability $(\Gamma_a/Q(\mathcal{C}))$. It is realized by selecting possible events with probabilities proportional to their physical rates. Therefore, the key point to obtain a good model is a proper choice of the transition rates. These rates belong either to deposition or diffusion processes. The rate of deposition processes is determined by the deposition flux F, usually given in units of adatoms per unit time per adsorption site. In simulations of growth by molecular beam epitaxy (MBE), adatoms land on any site with equal rate; if the site where the adatom lands is a stable adsorption site, the adatom stops there; on the contrary, a cascade process may begin to look for a stable site in the neighborhood [48, 49].

The rates of diffusion processes must have an activated form [10, 50]; for a process a, with energy barrier E_a (the latter is the energy difference between the saddle point and the initial minimum related to process a), the rate Γ_a is given by

$$\Gamma_a = \Gamma_a^0 \exp\left(-\frac{E_a}{k_B T}\right) \tag{19}$$

where the prefactor Γ_a^0 can be evaluated for example in the framework of the TST [10], which gives

$$\Gamma_a^0 = \frac{\prod_{i=1}^{3N} \nu_i^M}{\prod_{i=1}^{3N-1} \nu_i^S} \tag{20}$$

where the ν_i^M is the normal-mode frequency at the minimum, and the ν_i^S is the stable normal mode frequency at the saddle point.

A fast way to implement a KMC simulation without unsuccessful attempts (also called N-fold way algorithm) was formulated by Bortz, Kalos, and Lebowitz (BKL) [51] for the Ising model, and is often used in KMC simulations of crystal growth [52]. The basic idea is that at each MC step, one process is selected with its corresponding probability and then also realized, instead of attempting a generic process (whose probability may be much lower). We shall describe first the simplest variant of this algorithm. Let us consider the kth time step:

1. Choose a random number z_1 with uniform distribution in the range $[0, Q(\mathcal{C}_k)]$.
2. Find the corresponding event. This is done by the choice of the first index s for which $\sum_{a=1}^{s} \Gamma_a(\mathcal{C}) \geq z_1$.
3. Carry out the event s leading to a new configuration, $\mathcal{C}_{k+1}$.
4. Update those Γ_a that have changed as a result of event s; update Q and any data structure being used.

This algorithm is usually not applied in this form but in a modified, more effective form (see below). Let us consider the dependence of the computer time needed on N, which is related to the system size. Steps 1 and 3 require a time independent of N, but Step 2 is time-consuming. If linear search is used, the search time is $\mathcal{O}(N)$. Since the growth rules are usually local the updating in Step 4 does not have to cost too much computer time, although careful programming is needed. One has to identify events that are not possible any more in a new configuration, they have to be removed from the list, and new events that become possible have to be added to the list. Depending on the data structure used, time needed in Step 4 is $\mathcal{O}(N)$ at the most.

A faster algorithm can be obtained if one considers, instead of individual events, groups of events. Let us group events into n groups, labeled by $\alpha = 1, \ldots, n$. This can be done either formally by forming groups with the same number of events, which allows maximal effectivity of the algorithm, or in a way that keeps physics clear, forming groups of the same kind of events that correspond to a certain kind of process (diffusion of an adatom over the specific energy barrier, desorption of an atom with the specific binding energy, and so on). Let us consider explicitly the second case; each group will represent a certain kind of process, all processes in a group have the same rate Γ_α.

In a given configuration, $\mathcal{C}$, there are some possible processes, and each kind of possible process can be realized in one or more ways, by one or more event. Assume that a process α can be realized in $n_\alpha(\mathcal{C})$ ways, in the configuration $\mathcal{C}$. We shall call *multiplicities* the quantities $n_\alpha(\mathcal{C})$. For example, there may be $n_{adat}(\mathcal{C})$ adatoms with the same surroundings that can diffuse, or there are $n_{dep}(\mathcal{C})$ sites where a new particle can be deposited. Some particles can take part in more processes, and some processes may not be possible in the given configuration. To each kind of process we assign a partial rate, $q_\alpha(\mathcal{C}) = n_\alpha(\mathcal{C})\Gamma_\alpha$, and a relative probability, $p_\alpha(\mathcal{C}) = (q_\alpha(\mathcal{C})/Q(\mathcal{C}))$, which are conditional to the given configuration. The total transition rate in a configuration $\mathcal{C}$ is now $Q(\mathcal{C}) = \sum_{\alpha=1}^{n} n_\alpha(\mathcal{C})\Gamma_\alpha$. In each step of the simulation (in the given configuration) the multiplicities of processes are known. The algorithm in the kth step of the simulation proceeds as follows:

1. Choose a random number z_1 in the range $[0, Q(\mathcal{C}_k)]$.
2. Decide which kind of process will take place choosing the first index σ for which $\sum_{\alpha=1}^{\sigma} q_\alpha(\mathcal{C}) \geq z_1$.
3. Select a realization of the process σ. Technically this can be done with the help of a list of coordinates for each kind of movement, and an integer random number z_2 in the range $[1, n_\sigma(\mathcal{C}_k)]$; z_2 is generated and the corresponding member from the list is selected.
4. Perform the selected movement.
5. Update multiplicities n_α, relative rates q_α, total rate Q, and any data structure being used.

To estimate computer time demands, let us suppose that the multiplicities are approximately the same, $n_\alpha \approx (N/n)$. The search has two parts: searching for group, which takes time $\mathcal{O}(N/n)$, and searching within the group, which takes time $\mathcal{O}(n)$. Minimizing the total time leads to an optimal number of groups $n \propto N^{1/2}$, and the computer time then scales as $\mathcal{O}(N^{1/2})$. The time for Step 5 may still be $\mathcal{O}(N)$ at the most, but it is quite short in many cases.

An even faster algorithm for large N can be obtained if we, instead of the two-level search scheme, use a $K > 2$ level search scheme [53], i.e., if we divide the groups into subgroups, and these subgroups again into smaller subgroups, and so on down to the level K. Then the total search time scales are as $\mathcal{O}(KN^{1/K})$. The best asymptotic behavior $\mathcal{O}(\ln_2 N)$ is obtained by using the largest possible K, for which there are only two events in the lowest level subgroups. In practice, for typical simulation sizes, $K = 4$ or $K = 5$, can give a faster scheme than the $\mathcal{O}(\ln_2 N)$ method [53].

We shall conclude this subsection on the problem of introducing the physical time into the KMC simulation. This can be done [54], provided that all physical processes can be separated so that in any time instance only one event takes place, and the events are Poisson processes. The time interval between two successive events (waiting time), τ, is a random variable with the distribution $P(\tau) = Qe^{-Q\tau}$, and the average value is $\langle\tau\rangle = \frac{1}{Q}$ [47], Q being the total rate as defined before. In the algorithms described above, another random number, z_3, uniformly distributed between 0 and 1, will be generated, thus obtaining the time interval, $\Delta t_k = -\frac{1}{Q(\mathcal{C}_k)} \ln z_3$, spent before the passage to a new configuration $\mathcal{C}_{k+1}$. Note that $Q(\mathcal{C}_k)$ depends on the configuration $\mathcal{C}_k$. In an approximation, one can only suppose that the system stays in the state $\mathcal{C}_k$ for a time inversely proportional to the total transition rate $Q(\mathcal{C}_k)$ [54]. Mean quantities during the growth are then calculated as time averages over M time points

$$\langle A\rangle = t^{-1} \sum_{k=1}^{M} \Delta t_k A(\mathcal{C}_k) \tag{21}$$

where the total time is $t = \sum_{k=1}^{M} \Delta t_k$. It was tested that this procedure agrees with the results of the exact solution of the master equation [54].

3. MODELING AND SIMULATION OF CLUSTER GROWTH ON SURFACES

In typical growth experiments, atoms of molecules are deposited on a crystal surface by an atomic or molecular beam, which impinges normally on to the surface (shadowing effects due to non-normal incidence have been studied in [55]). Three parameters are usually controlled in the experiment:

- **Deposition flux** F, measured in number of atoms per second per adsorption site, which can be as slow as a small fraction of a monolayer per minute.
- **Substrate temperature** T, which is usually kept constant.
- **Coverage** θ (or equivalently, growth time t_g, since $\theta = Ft_g$).

The temperature of the beam source generally plays a minor role since the kinetic energy of the beam is usually a small fraction of the adsorption energy of the atoms or molecules on the crystal surface.

From the point of view of the simulations, the KMC method is almost universally adopted, and crystal growth involves the modeling of the following physical processes (see Fig. 1):

- **Deposition:** Adparticles can land at any site of the surface. As they approach the surface, they acquire a considerable amount of kinetic energy, which comes out from the adsorption energy. This kinetic energy is directed toward the local normal to the crystal surface, and it is mostly dissipated in phonons upon impact [56]. However, a part of this kinetic energy can be transformed into translational energy on the surface plane, giving thus rise to what is called *transient mobility*, because it takes place before the adparticle is thermalized on the surface. For an excellent review on this topic see Ref. [57]. Transient mobility amounts usually to a few (one or two) lattice spacings in most systems (for example, in metal-on-metal epitaxy) and is usually not taken into account in simulations, even if in some low-T cases it can be the main source of atomic mobility, playing thus an important role. Another effect that can be important at deposition is *funneling* [48, 49]. This amounts to a way of effectively taking into account the different time scales of the mobility on different crystal facets. For example, consider the growth of an fcc(100) surface. There, adatom mobility is slow. During the growth of a thin film of this orientation, also (111) facets are formed. There, the mobility is much faster, so that a new adatom impinging on a (111) facet will hop there many times in a short period. This is very time-consuming to simulate. How to avoid this problem? Usually, when one adatom is deposited on (111) facet, it is treated as if it was in a metastable configuration, and a search of the nearest adsorption sites on other kinds of facets is automatically performed, so that the adatom is placed on one of them at the end of the deposition process. A possible choice is to treat all sites where the adparticle has a coordination lower than four as metastable, and all the other sites as stable.
- **Diffusion:** The adparticle makes a random walk on the surface, the movements of which have a frequency that depends on the local environment. Diffusion can be on a flat terrace (*intralayer diffusion*) or between two different terraces (*interlayer diffusion*). It can lead the adparticle to attach to existing steps (*aggregation process*) or to detach from them (*detachment process*). Once the particle is attached at a step it can move along it (*periphery diffusion*), and cross corners toward a nearby step (*corner crossing*). All these processes are usually modeled in KMC simulations with TST rates (see Eq. (20)) depending on the local environment (number of neighbors in initial and final positions) of the diffusing particle. A special effort must be devoted to the calculation of the energy barriers pertaining to these processes, and (with less emphasis) of the prefactors. Energy barriers calculations can be performed either by *ab initio* or by semiempirical methods.
- **Desorption:** Adparticles may, in principle, leave the surface. However, the desorption energy is usually much larger than the energy barriers for diffusion, so that experiments usually can be performed in temperature ranges where desorption takes place on such a long time scale as to be practically negligible. This is certainly the case of metal-on-metal systems. In modeling these systems, adparticle desorption is simply not considered as a possible move in the KMC simulations.

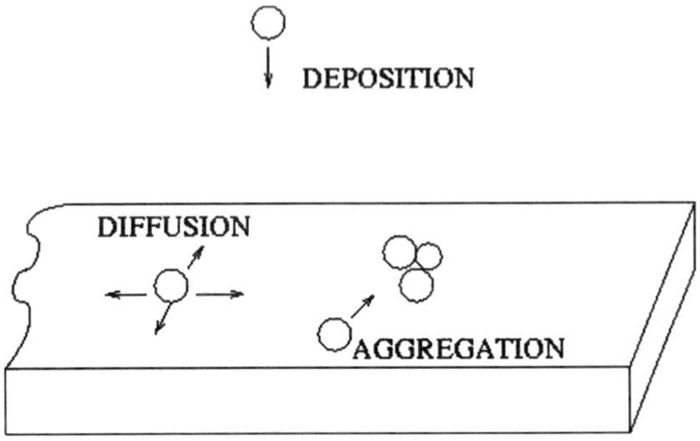

Figure 1. Representation of the elementary diffusion processes taking place during growth on a crystal surface: deposition, diffusion, and aggregation. These processes are common to free and to supported growth. In usual epitaxial growth conditions for supported cluster growth desorption (not represented in the figures) is negligible; it must be taken into account in the case of the high-temperature growth of noble-gas or C_{60} nanoclusters.

In this section we want to illustrate, by three different examples, how the computational techniques described in the previous section can be used in the modeling of experiments involving the growth of aggregates or thin films on surfaces. Specifically, we shall treat the epitaxial growth of Pt on Pt(111), comparing the experimental results with those following from the density functional calculations; the epitaxial growth of Ag on Ag(110), comparing the experimental results with those of KMC simulations; and the growth of lipid aggregates at the air–water interface, simulated again by Monte Carlo techniques.

3.1. Three-Dimensional Cluster Growth in Pt/Pt(111)

This system gives a very nice example of the interplay among experiments and first-principle DFT calculations, showing that the experiments can simulate the calculations of the barriers of specific elementary processes, which in turn suggest further experiments.

The (111) surface has a hexagonal symmetry and high-symmetry steps on this surface can have two different symmetries (see Fig. 2); steps of type A, which present square nanofacets on the step riser, and steps of type B, presenting triangular nanofacets on the step riser. At sufficiently high temperatures, Pt islands epitaxially grown on Pt(111) are compact and limited by these kinds of steps [58]. Triangular islands are limited either by A steps or by B steps, while hexagonal islands are limited by steps of both kinds. When depositing further adatoms on these compact islands, growth proceeds in a 3-D mode [59], and mounds are formed.

The cause of the onset of 3-D growth is the inability of the adatoms deposited above the islands to descend down to the lower level, due to the existence of an additional energy barrier, the Ehrlich–Schwoebel (ES) barrier [60, 61], for downward crossing of the step (see Fig. 3). This extra barrier is

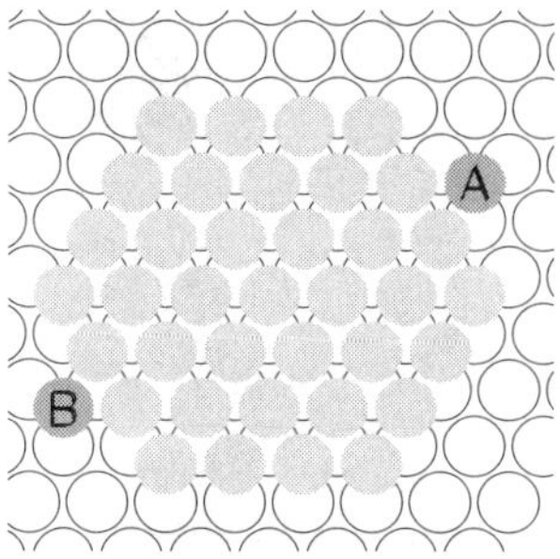

Figure 2. Schematic representation of a (111) surface with a hexagonal island above it. The island is limited by three A steps (square faces on the step riser, atom A is place along one of them) and by three B steps (triangular facets on the step riser, atom B is placed along one of them). These steps are alternating along the island perimeter. On this surface, hexagonal islands must be limited by both kinds of steps, while triangular islands can be limited by steps of a single kind. This is equally true for islands placed either on fcc or on hcp stacking.

due to the reduced coordination of the adatom at the saddle point position of the downward crossing process, and can be very low when exchange processes are possible (see for example [50]). The ES barrier is in general dependent on the step orientation, and therefore one could expect that A and B steps present different ES barriers. This has important consequences on the growth shape of the mounds. For example, if the ES barrier is small at A steps and large at B steps, deposited adatoms will descend down preferentially at A steps, and will accumulate there. As a result, A facets will tend to disappear during growth, and the mounds will present mostly facets with B steps.

The first experimental observation (by the group of S. Esch) of Pt mounds grown Pt(111) have the shape of triangular pyramids up to a temperature of 450 K, and these pyramids are mostly bounded by A steps [62]. This observation is consistent with a small ES barrier at B steps, and with a large ES barrier at A steps. This experimental result stimulated DFT calculations by P. J. Feibelman [34, 63], which were however in complete disagreement with the previous experimental finding. According to different DFT schemes,

Table 1. Energy barriers (in eV) calculated by *ab initio* methods (in LDA and GGA for diffusion of Pt adatoms on Pt(111)).

Diffusion process	LDA	GGA
Hop on (111) terrace	0.29	—
Exchange down A step	0.31	—
Exchange down B step	0.64	—
Hop down A step	0.53	—
Hop down B step	0.80	—
Hop along A step	0.84	0.71
Hop along B step	0.90	0.77

Note: All data are taken from Refs. [34, 63].

the ES barrier is large at B steps and small at A steps (see Table 1).

The solution of this discrepancy came out of new experiments [64], which showed that the older experiments were affected by CO contamination. Removing the CO contamination, mounds on Pt(111) grow bounded by B steps, as predicted by the DFT calculations. This result shows the predictive power of state-of-the-art DFT calculations.

3.2. Homoepitaxial Growth of Ag(110)

The (110) surface of Ag is of rectangular symmetry (see Fig. 4), with channels along the $[1\bar{1}0]$ direction, which are separated by atomic rows. In-channel diffusion of adatoms occurs mostly by jumps, while cross-channel diffusion occurs by exchange and is more difficult [65].

Due to the anisotropy of the surface, one may expect that a rather peculiar surface nanostructuring can be achieved by growth on this surface. This is indeed the case, as shown by the experiments performed in the group of Valbusa, both in the submonolayer and in the multilayer regimes. At submonolayer coverages, different kinds of islands are grown depending on T [66]. At low temperatures, the islands are small and present preferentially a cross-channel orientation. At intermediate temperatures, islands change their orientation by 90 degrees, being well-defined one-dimensional strips in the in-channel direction. Finally, at high temperatures, the in-channel orientation is still preserved, but islands

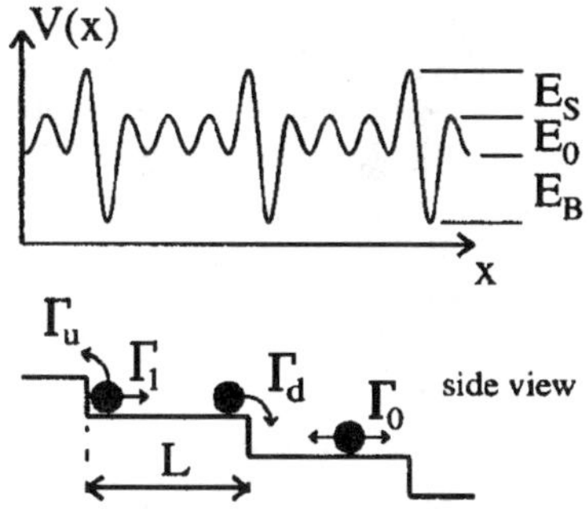

Figure 3. Potential energy profile (*upper panel*) for an adatom diffusing on an array of equally spaced terraces (*lower panel*). The Ehrlich–Schwoebel additional barrier E_s hinders the descent of adatoms at the lower level, so that the descent rate Γ_d is smaller than the diffusion rate on the flat terrace Γ_0, since $\Gamma_0 \propto \exp[-E_0/(k_B T)]$, while $\Gamma_d \propto \exp[-(E_0 + E_s)/(k_B T)]$. At the lower side of the step, coordination is increased so that adatoms are trapped there for long times.

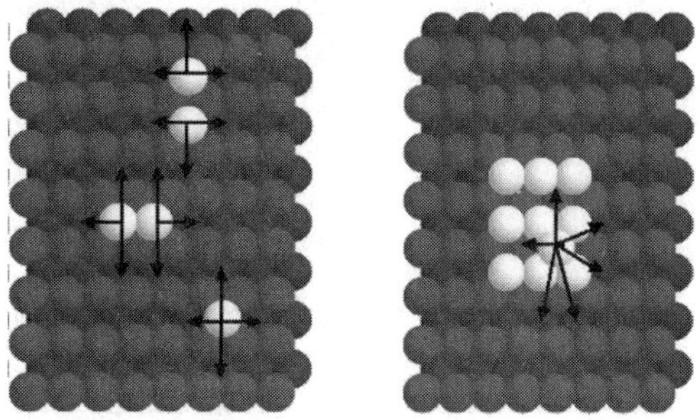

Figure 4. Left panel: fcc(110) surface with adatoms, and dimers adsorbed on it (adsorbed atoms are represented in white). The possible intralayer diffusion moves [75] are indicated by arrows. Right panel: fcc(110) surface with a 3×3 island adsorbed on it. Above the island there is a further atom, whose possible diffusion moves (intralayer and interlayer moves) are indicated by arrows. In both panels, the easy-diffusion in-channel $[1\bar{1}0]$ is horizontal.

are two-dimensional. In the case of multilayer growth, ripples are formed, and in analogy with the submonolayer islands, these ripples rotate from the cross-channel direction at low temperatures to the in-channel direction at intermediate and high temperatures [67]. The time scale of the experiments is several minutes (adatom fluxes are of 1 ML/min), and the size of the terraces is several hundreds of nanometers, while temperatures range from 100 K to room temperature.

To study these time scales and terrace sizes, it is necessary to perform KMC simulations, that aim to single out a small number of energetic parameters, which are able to describe accurately the growth process [68].

The first step in the KMC modeling is thus the choice of the important processes and the calculations of their activation barriers. To this end, Ag is described by means of semiempirical interaction potentials [69], and we calculate the barriers of several diffusion processes by means of quenche MD simulations [70]. The full barrier set can be rationalized by the following simple formula depending on a rather small set of energetic parameters [71]

$$E_{p,n} = E^0_{p,n} + n_p E^b_p + n_n E^b_n + n^S E^S_{p,n} \tag{22}$$

where $E^0_{p,n}$ are the barriers for diffusion on the bare surface in the parallel (in-channel) or normal (cross-channel) directions. $E^b_{p,n}$ and $n_{p,n}$ are the strengths and numbers of in-channel and cross-channel bonds, respectively. $E^S_{p,n}$ are anisotropic Ehrlich–Schwoebel barriers, and n^S is 1 for interlayer processes and 0 for intralayer processes. The resulting model is referred to as the anisotropic bond-breaking model (ABBM).

The values of these energetic parameters are given in Table 2, and the different diffusion moves are schematically represented in Figure 4. For simplicity, we set all prefactors to the common value of 10^{12} s^{-1}.

In a model with full fcc geometry, special care must be given to the way by which adatoms are deposited on the surface, and to the way by which the final position is chosen after an interlayer process.

For example, a deposited atom can land on any site of the surface. However, in a given surface configuration, not all sites fulfill the restricted solid-on-solid (RSOS) conditions: depositing on such *unstable* sites (which indeed can be either really unstable positions, like the on-top positions, or metastable sites; see Fig. 5) would create overhangs. In our model, adatoms can be deposited on any site. If the latter is *unstable*, a cascade follows to look for *stable* sites (i.e., the sites that fulfill the RSOS condition) in its neighborhood. This procedure is justified for example when the *unstable* site is located on a (111) facet, where mobility is much faster compared to other facets [50, 72], and temperature is high enough. If T is so low that mobility on the (111) facets is frozen, overhangs are easily created [73]. In the case of silver, the diffusion barrier is of 0.10 eV or even less [50, 72], so that diffusion is activated well below 100 K on the experimental time scales. Our deposition procedure does not take into account the fact that adatoms are attracted by the mounds on a rough surface, as has been recently demonstrated by MD simulations [74]. However, this effect should be important only at very low temperatures, where the growing surface is found to be rough ($T <$ 100 K). A growth model that includes such an effect may be only tractable by MD simulations. This investigation is left for future work. Concerning the search for the arrival site after an interlayer move, we proceed as follows. If the nearest arrival site is *stable* (compare Figs. 4 and 5), the diffusing adatom stops there. If not, a cascade follows to look for a *stable* site in the neighborhood. Again, this procedure is fully justified on (111) facets. On the other facets, this procedure can *overestimate* interlayer mobility. In order to check this point, we made simulations also by a second procedure, letting the interlayer move to be executed only if the nearest arrival site is *stable*; on the contrary, we suppress the move. This procedure *underestimates* interlayer mobility. We have verified that this second procedure leads to very minor changes in our results, showing that growth morphologies are not influenced by the way the arrival site is searched for in interlayer moves. This could indicate that also the choice of the cascade procedure at deposition is not a crucial point for the results contained in the following.

Let us now briefly describe how the KMC algorithm decides to make a movement (either deposition or diffusion) process in a given configuration C of the system. The total probability of having a deposition event, $\Gamma_{dep}(C)$, is given by the total number of sites N in the surface multiplied by the deposition flux F (in ML/s). On the other hand, the total probability of having a diffusion process is calculated within a binary tree. The lowest level of the binary tree numbers

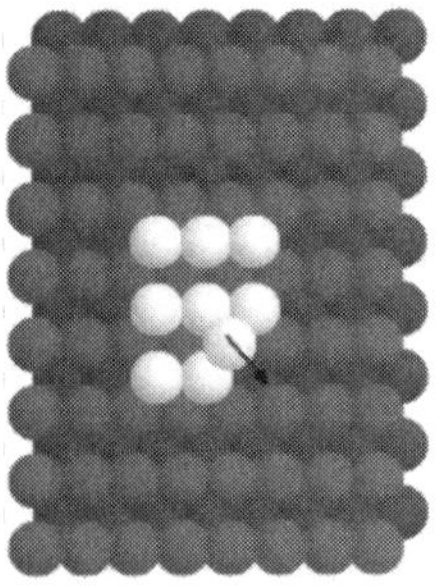

Figure 5. fcc(110) surface with an island of 8 atoms above it. Compare this island with the one of the bottom panel of Figure 4. Above the island there is a further adatom, which is placed in a short-lived metastable site. In KMC simulations, this atom searches for the nearest sites of higher coordination and chooses one of them as a stable site [75]. The path to the closest stable site is indicated by the arrow.

Table 2. Energetic parameters (in eV) of the ABBM model for the growth of Ag/Ag(110) [75].

Parameter	Value	Variability range
E^0_p	0.27	0.24–0.30
E^0_n	0.34	0.31–0.37
E^b_p	0.18	—
E^b_n	0.04	0.02–0.06
E^S_p	0.07	0.05–0.09
E^S_n	0.18	—

Note: In the last column we give the interval of variability of the parameters. Within these limits a good agreement with experiments is obtained.

N elements. (For simplicity, we choose the size of the simulation box in such a way that $N = 2^p$ for some integer p.) Each element is a diffusion rate associated with a given site i, and its value $\Gamma_0(i)$ is the sum of the four frequencies of the diffusion processes in the four diffusion directions. The rates of thc level $m+1$ of the binary tree are obtained by combining the rates at level m as follows: $\Gamma_{m+1}(i) = \Gamma_m(2i-1) + \Gamma_m(2i)$, so that level $m+1$ has half of the elements of level m. At level p, we have a single element, which is the total diffusion rate $\Gamma_{diff}(C)$. The total rate associated with C is $\Gamma_{dep}(C) + \Gamma_{diff}(C)$. For choosing the process to be performed, a random number z, in the range $[0, \Gamma_{dep}(C) + \Gamma_{diff}(C)]$, is extracted. If $z \leq \Gamma_{dep}(C)$, a deposition process starts; otherwise the algorithm chooses a diffusion process. If deposition is to be performed, a site is randomly extracted; if diffusion is to be performed, we go downward in the binary tree by extracting random numbers at each level to choose which branch to follow. In this way, at level 0 in the tree, a site is singled out. Now a further random number is extracted in order to choose one of the possible directions for diffusion, and the move is performed.

As shown in Figures 6 and 7, this model reproduces qualitatively both the island and the ripple rotation depending on temperature. Moreover, the agreement is quantitative since the rotation takes place in the same T range as in the experiments, and the experimental ripple wavelengths are well reproduced as shown in Figure 8 [75].

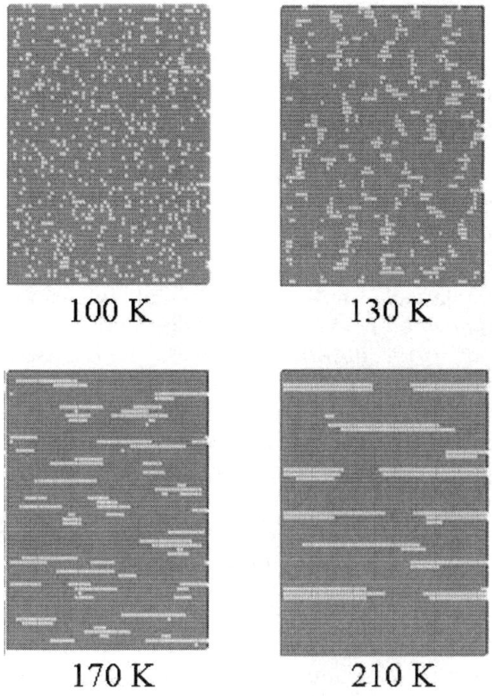

Figure 6. Snapshots from simulations of the growth of Ag/Ag(110) [75]; the in-channel direction is horizontal. The snapshots are taken after the deposition of 0.15 monolayers (submonolayer growth) at a flux of 1 ML/min at different temperatures: 100 (top left), 130 (top right), 170 (bottom left), and 210 K (bottom right). At 100 K the islands are small and irregular; at 130 K, the islands are elongated in the cross-channel direction; at 170 K, the islands are mostly monatomic strips oriented in the in-channel direction; at 210 K, the islands are elongated in the in-channel direction but present a more compact shape.

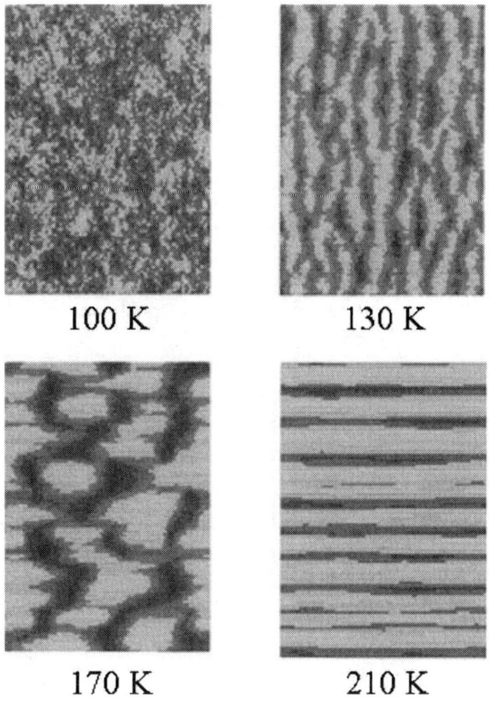

Figure 7. Snapshots from simulations of the growth of Ag/Ag(110) [75]; the in-channel direction is horizontal. The snapshots are taken after the deposition of 30 monolayers (multilayer growth) at a flux of 1 ML/min at different temperatures: 100 (top left), 130 (top right), 170 (bottom left), and 210 K (bottom right). The top-layer atoms are represented in white; lower layers in darker colors. At 100 K the surface is rough with no regular features; at 130 K, ripples are formed, with a cross-channel orientation (the same orientation as the islands grown at this temperature, see Fig. 6, left panel); at 170 K, mounds are grown with no special orientation; at 210 K, ripples are grown with an in-channel orientation, which is the same orientation of the islands grown at 210 K (see Fig. 6, right panel). These results about the ripple rotation with temperature are in excellent agreement with the experiments [67].

The behavior of the cross-channel ripple wavelength at low temperatures can be rationalized into a simple analytical expression by the following argument. Indeed, at sufficiently low temperatures (to be conservative, $T \leq 150$ K at the experimental fluxes), cross-channel mobility is practically inhibited; interlayer in-channel mobility is practically absent, so that the only possible process is intralayer in-channel diffusion. If one could neglect the existence of the cross-channel bonds, and the fact that the full fcc geometry induces some cross-channel correlations in the deposition process, one could consider adatoms in different channels as completely independent, and reduce the problem to a simple one-dimensional model (with no interlayer mobility) along the in-channel direction. In this one-dimensional model, the only characteristic length is the in-channel diffusion length ℓ of the adatoms on terraces [1]. It is known that [1, 76]

$$\ell \propto \left(\frac{D_p}{F}\right)^{\chi} \tag{23}$$

where D_p is the in-channel jump frequency, and $\chi = 1/4$ for one-dimensional systems. In this framework, we expect that

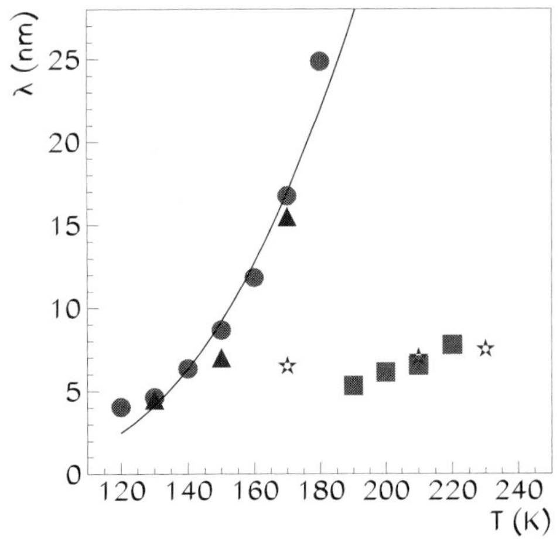

Figure 8. Ripple wavelength λ as a function of T. The triangles and the circles are the experimental [67] and the simulation [75] results for the low-T cross-channel ripples respectively. The stars and the squares are the experimental [67] and the simulation [75] results for the high-T in-channel ripples respectively. The agreement between experiments and simulations is remarkable. The full line is the best fit to the cross-channel ripple wavelength using Eq. (24). This equation assumes that the cross-channel ripple wavelength is proportional to the in-channel diffusion length of adatoms, because in-channel diffusion is the only process possible at low temperatures. The agreement Eq. (24) with both experiments and simulations is very good. Reprinted with permission from [75], A. Videcoq et al., *Surf. Sci.* 515, 575 (2002). © 2002, Elsevier Science.

λ is proportional to ℓ, and therefore

$$\lambda = a_0 \exp\left(-\frac{E_p^0}{4k_B T}\right) \tag{24}$$

where a_0 is treated as a fitting parameter, and the best fit is obtained for the value $a_0 = 1.7 \times 10^4$ Å. In Figure 8 we compare the results from Eq. (24) with the simulation data. As can be seen from Figure 8, the agreement is in general good, showing that the wavelength of cross channel ripples follows to a reasonable approximation an activated behavior with a barrier that is a quarter of the in-channel intralayer diffusion barrier, but with some discrepancies at low and high temperatures. The high T discrepancy is readily explained by noticing that the assumptions of the absence of cross-channel intralayer diffusion and of in-channel interlayer diffusion break up at increasing temperature. The low-T discrepancy might be due to the fact that cross-channel bonds are not negligible there. We have thus made a series of simulations with $E_n^b = 0$, comparing the results with Eq. (24). Even if there is some improvement in the linearity of the Arrhenius plot, the low-T deviation still remains, showing that it is not due to cross-channel bonds.

The analysis of the simulations allows us to single out the key microscopic mechanisms causing island and ripple rotation. At low temperatures, interlayer diffusion is completely inhibited, and intralayer diffusion takes place only in the in-channel direction. Both in-channel and cross-channel bonds are stable. In this way, islands are formed preferentially with cross-channel orientation, and the ripple instability develops in the cross-channel direction because there is only an in-channel uphill surface current [67, 68]. At higher temperatures, cross-channel bonds are no more stable, while in-channel bonds are still stable, so that islands assume the shape of in-channel strips. On the other hand, cross-channel intralayer diffusion and in-channel interlayer diffusion become active, and begin to cause the rotation of 90 degrees in relation to the uphill surface current, and therefore to the ripples. Varying the energetic parameters of the model, and comparing the results with the experiments, one can set limits of variability of the parameter themselves. These limits are given in Table 2.

3.3. Lipid Cluster Growth at the Air–Water Interface

Phospholipids and fatty acids on the water surface can form Langmuir monolayers, which exist in a number of different phases. In Langmuir monolayers, one of the most important transitions is that between the liquid-expanded and the liquid-condensed phases, very often called the "main transition" [77]. When the equilibrium phase is liquid-condensed, but the system finds itself in the liquid-expanded phase, clusters of the liquid-condensed phase are seen to grow within the liquid-expanded phase with peculiar shapes, ranging from compact to fractal, according to the overall physical parameters. The simulations described in the present section generate clusters that bear a striking similarity with the experimental ones.

Since Witten and Sander introduced their fortunate diffusion-limited aggregation (DLA) model [78], the literature on the subject has developed enormously. The main reason is that DLA is the simplest tool to generate random fractals. Important connections were also discovered with clusters occurring in critical phenomena or percolation problems (although the properties were different in each case). The more important contribution of DLA was probably to furnish a *paradigm*, within which the fractal properties of natural or computer-generated objects could be studied, from the relatively simple overall fractal dimension to quantities obtainable only via much subtler analysis.

It is out of the question to discuss these problems here (see, for example, [79] or [80] for a thorough treatment). We wish, however, to consider cluster growth from the same point of view as we have considered crystal growth in the remainder of this paper: in terms of physical (thermodynamical) parameters, temperature T and disequilibrium $\Delta\mu$ in the first place. The clusters that we are going to consider, therefore, must in some limiting case ($\Delta\mu = 0$) be in equilibrium with the environment; hence they must be able to compensate growth by evaporation or dissolution, so a double DLA-like process must be involved. Such a double DLA process was first introduced (as many other things in statistical mechanics) by Kadanoff [81] (see below).

Let us first recall ordinary DLA, as was first proposed in [78]. DLA is a growth process, taking place in a computer. The growth starts from a preexisting *germ*, placed at the origin. In the simplest case the simulation is two-dimensional. *Atoms* are sent in sequentially from the circumference of a large circle, having the germ at its center. Each atom performs a *random walk* in the plane; finally either it reaches a site neighboring the growing cluster, in which case it sticks irreversibly to the cluster, or it reaches the outer circumference again, in which case it is lost. In this way clusters are generated, possessing a characteristically ramified appearance. The fractal dimension of DLA clusters, as obtained numerically, is 1.71 ± 0.01 [82]. Precise and convincing theoretical results, in agreement with the above, were obtained in [83].

DLA growth is irreversible and, from our point of view, corresponds to $\Delta\mu \to \infty$. In order to allow for finite disequilibria, Kadanoff [81] introduced a very simple modification of DLA, which he calls a *pedestrian model*. Two competing processes are taking place. One process is ordinary DLA; in the other process atoms detach from the cluster, perform again a random walk, and either stick to the cluster at another point or are lost. A modified version of this model was studied in detail in [84]. The work of Gliozzi et al. was addressed to study the temperature and disequilibrium dependence of the *growth morphology* of two-dimensional clusters. The morphology changes in terms of physically significant parameters that were studied, in particular, by Gliozzi et al. [84], who specifically applied the analysis to the growth of clusters of the gel phase at the expense of the fluid phase in Langmuir monolayers. In these works the physical parameters are T and $\Delta\mu$. The role of disequilibrium is obvious; temperature, on the other hand, occurs indirectly, via the T-dependence of *anisotropy*.

The algorithm of [84] describes a competition between growth and rearrangement and is a modified version of Kadanoff's "pedestrian" model [81]. Particles are continuously sent in from the outside, in DLA fashion. They perform a random walk, then either stick to the growing cluster or are lost. At the same time other particles detach from the periphery of the cluster and, again, perform a random walk and either stick to the growing cluster at another place or are lost (*evaporate*).

The energy E of the cluster is given by the lattice-gas formula

$$E = -J \sum_{i,j} n_i n_j \tag{25}$$

where n_i is 1 if site i is occupied, J is the pair interaction energy, and the sum is over pairs of nearest neighbor sites. A particle is added or removed from the cluster with a probability $\mathcal{P}$ given by the Glauber rule

$$\mathcal{P} = \frac{1}{\exp[\beta(\Delta E - \mu_{eq}\Delta N)] + 1} \tag{26}$$

where μ_{eq} is the chemical potential for solid-vapor coexistence (for the square lattice $\mu_{eq} = -2J$, for the triangular lattice $\mu_{eq} = -3J$), $\beta = 1/k_B T$, and ΔE and ΔN ($= 1$ or -1) are the variations of energy and particle number.

The two random-walk processes, originating at long distances and at the periphery of the cluster, respectively, alternate in such a way that the ratio between sticking and evaporating particles is $\exp(\beta\Delta\mu)$ to 1, where $\Delta\mu$ is the disequilibrium between the two phases. Thus the growth and rearrangement processes compete with the correct rates. By construction, the resulting growth rate is $\exp(\beta\Delta\mu) - 1$, corresponding to the so-called Wilson-Frenkel law [85].

Gliozzi et al. found, at a rather well-defined temperature (or anisotropy), a morphology transition of the *tip-splitting* type [86], from the dendritic to the dense-branching morphology (DBM). Interestingly, even for large disequilibria and (relatively) high temperatures, the clusters generated in this way (contrary to what happens for simple DLA) are not ordinary fractals. Rather, they belong to the typology of *fat fractals* [79], i.e. their measure, when measured with rulers of length l, is given by

$$M(l) = M(0) + Al^{\gamma} \tag{27}$$

(For ordinary, or *thin* fractals, $M(0)$ vanishes.) A further development in terms of the description of cluster growth in Langmuir monolayers took into account the important effects of long-range interactions [87–89], which have a destabilizing influence even in thermodynamic equilibrium [90, 91].

In the repulsive case [87], a remarkable agreement is found between the simulated clusters and those observed in experiments performed on clusters growing in Langmuir monolayers [92], as can be seen in Figure 9. The same shapes (compact in some cases, highly ramified in others) are found, showing that the model correctly captures the competition between random and ordering effects, although in the simulation the intensity of random behavior is determined by temperature, while in the experiments

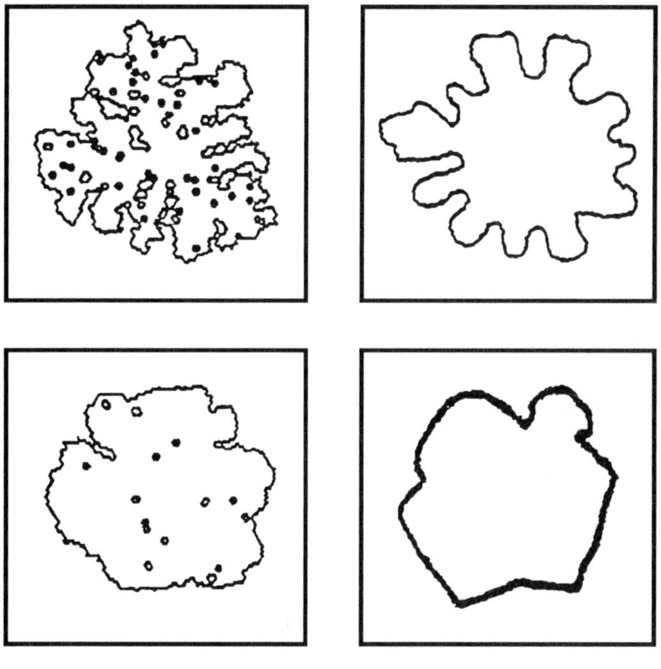

Figure 9. Comparison between simulations and experiments [87] of lipid cluster growth at the air–water interface. The left panels show simulated clusters, while the right panels reproduce experimental clusters [92], which were made of two different lipids. The cluster of the top right panel is made of triple-chain lecithin with the third chain attached to the 1-position. The cluster in the bottom right panel is made of triple-chain lecithin with the third chain attached to the 2-position. The simulated clusters, grown by the procedure outlined in the text, are remarkably similar to the experimental ones. Reprinted with permission from [87], G. Indiveri et al., *Thin Solid Films*. 284/285, 106 (1996). © 1996, Elsevier Science.

the same effects are obtained depending on the chemistry of the amphiphilic molecules, such as on the slight difference between two kinds of triple-chain lecithin (Fig. 9).

In the attractive case, where the simulation is appropriately performed over a lattice [88] a more surprising behavior is found. Assuming a potential energy of the form

$$U = p \sum_{i,j} R_{ij}^{-\alpha} \tag{28}$$

for small $p/k_B T$ the clusters have a typical DBM, but for very large $p/k_B T$ they take the shape of a star with branches corresponding to the lattice symmetry (e.g., a sixfold star on a triangular lattice), so that a characteristic morphology transition takes place as a function of p. Interesting results are also found as a function of the exponent α (when $p \to 0$ and $\alpha \to \infty$ the model becomes equivalent to ordinary DLA) (Fig. 10). In this case the algorithm had to be appropriately modified, the random walks taking place in a circular corona. Indeed, in the presence of long-range attractive forces it is not possible to choose the starting (or end) points for diffusion very close to the aggregate (which is permitted for DLA) since the morphology would depend on the initial conditions. The width of the corona must be chosen carefully; it turns out that a ratio of 5 to 4 between outer and inner radii gives good results.

A fairly complete analysis of morphology transitions in the (disequilibrium, anisotropy)-plane was presented by Brener et al. [93], leading to the identification of four distinct morphologies, which they call compact dendrite, fractal dendrite,

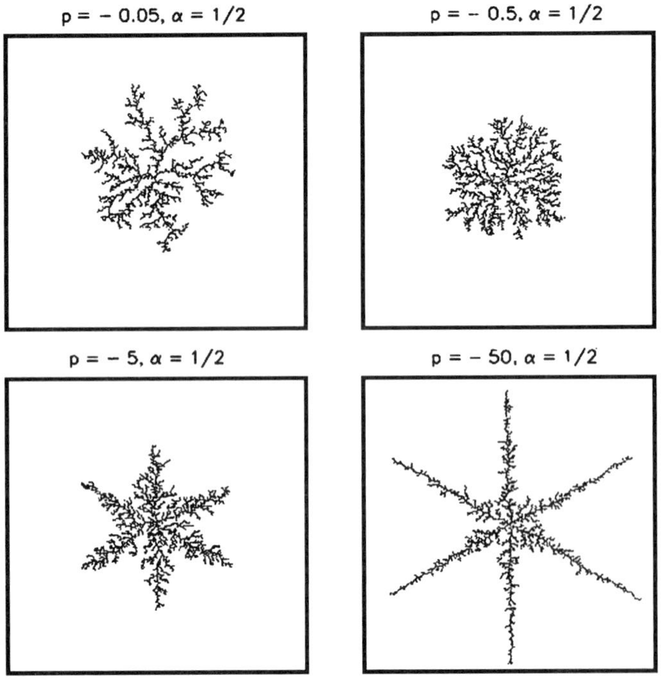

Figure 10. Snapshots from simulations [88] of the aggregation model with repulsive long-range interactions, where the interaction energy between two particles i and j is given by $U_{ij} = p/R_{ij}^{\alpha}$, α is positive, and negative p correspond to attractive interactions [see Eq. (28)]. Here α is kept fixed at the value 0.5, and p is varied. Reprinted with permission from [88], G. Indiveri et al., *Physica A* 237, 217 (1999). © 1999, Elsevier Science.

compact seaweed, and fractal seaweed (where *seaweed* indicates the DBMs).

The simulations considered in the present section aim at reproducing clusters in Langmuir monolayers. However, equally important potentially will be the application of the above mentioned methods to the growth of metal clusters on metals. Such clusters exhibit very diverse forms, from compact to ramified, fat-fractal shapes [96], depending, as Bartelt and Evans [97] have shown, on the ratio between the rate of arrival of new atoms from the outside and the rate of diffusion along the periphery of the cluster. An intermediate case between Langmuir monolayers and metal adsorbates is that of self-assembled structures of, for example, thiols on gold [98], which have been the object of very interesting studies.

4. SIMULATION OF THE GROWTH OF FREE NANOCLUSTERS

Nanoclusters—nearly monodispersed particles that are generally less than 10 nm in diameter—have generated intense interest over the past decade. One reason for this is the belief that nanoclusters will have unique properties, derived in part from tha fact that these particles and their properties lie somewhere between those of bulk and single-particle species. Such "strange morsels of matter" have many fascinating potential uses, including quantum dots or computers and electronic devices, chemical sensors, light-emitting diodes, "ferrofluids" for cell separations, industrial lithography, and photochemical pattern applications such as flat-panel display. Nanoclusters also have significant potential as new types of catalysts of higher activity and selectivity. In the following section, we shall treat the growth of free nanoclusters: They can grow by condensation of a vapor, then the formed clusters can collide and coalesce. These free nanoclusters have peculiar properties, which can be preserved when they are deposited, for example by soft landing, on inert substrates. The structure of nanoclusters is one of the most fundamental properties and plays an important role to understand all aspects of their physical and chemical behaviors. Much experimental and theoretical effort has been devoted to find how the most favorable structure depends on the cluster size N. Now it is known that cluster arrangements may be different from those of the bulk: Also structures that are noncrystalline symmetries (containing for example symmetries of order five) are possible, because the constraint of translational invariance does not apply to these finite structures.

Progress in cluster physics closely follows the development of new experimental methods. Important advances in production and detection techniques for metal clusters in molecular beams provide the possibility of studying clusters in an interaction-free environment. All molecular-beam experiments on clusters need a cluster source, which is chosen on the basis of the material, and a cluster detector. Here we report a brief description of the six most important ways to produce nanoclusters [2]:

- *Seeded supersonic nozzle* sources are used mostly to produce intense cluster beams of low-boiling-point metals. These sources produce continuous beams with reasonably narrow speed distributions. Clusters with up to several hundreds (and, under suitable conditions, with thousands) of atoms per cluster may be produced in adequate abundance. The cluster temperatures are not well known; however, larger clusters are often assumed to be near the evaporation limit.

- *Gas-aggregation* sources are particulary efficient in the production of large clusters ($N < 10,000$). The intensities are generally much lower than for nozzle sources. The overall cluster size distributions can be adjusted within rather broad limits. These sources are used for low-to-medium boiling-point materials (<2,000 K). Low cluster temperatures (<100 K) can be achieved.
- *Laser vaporization* sources produce clusters in the size range from one atom to typically several hundreds of particles per cluster. These sources are pulsed, and, although the time-averaged flux is low compared with the nozzle source, intensities within a pulse are much higher. In principle, these sources can be used for all metals to produce neutral and positively and negatively charged ionized clusters. The cluster temperatures are expected to be near or below the source temperature, depending on the supersonic expansion conditions. Then it is possible to cool down or anneal these clusters by letting them into an inert-gas atmosphere.
- *Pulsed-arc cluster-ion* sources are basically similar to laser vaporization ones. Here, however, an intense electrical discharge rather than a laser is used to produce the clusters, resulting in intense cluster-ion beams.
- *Ion sputtering* sources are used primarily to produce intense continuous beams of small singly ionized clusters of most metals. The clusters are hot, typically near the evaporation limit.
- *Liquid-metal ion* sources produce singly and multiply ionized clusters of low-melting-point metals. These clusters are hot, since in-flight evaporation and fission are observed.

The detection of a slow neutral cluster seems to be difficult, but by diffraction methods we can obtain information about cluster structure. Then clusters can be ionized for an efficient mass selective detection. An atomic ion has a detection efficiency of near unity, if it is accelerated to a kinetic energy of some keV and impinges onto a metal surface. The secondary electrons emitted can be used for further amplification. The same method is used universally to detect cluster ions. Then we can analyze the cluster beam by a mass-filter (or a quadrupole) in order to obtain mass spectra. A typical experiment consists of a well collimated electron beam with an energy of 30 to 50 keV crossing a supersonic cluster beam. The fast electrons are scattered from the atoms in the cluster, and the diffraction pattern is recorded usually on photographic film. A series of diffraction rings around the position of the primary electron beam is recorded. For the interpretation, a geometrical model is constructed and the diffraction pattern calculated, remembering that the scattered intensity is the squared modulus of the Fourier transform (FT) of the scatterers with respect to the momentum transfer. In this way three pieces of information can be extracted.

- The mean geometry of the cluster.
- The mean cluster size.
- The cluster temperature.

The geometry is obtained directly from the fit of the assumed model to the experimental data. It is plausible that an icosahedron (see next section) has a different FT than a part of an fcc lattice, and so on. The larger the cluster, the narrower are the diffraction rings, such as in light diffraction from a grating with few or many grooves. The more the scatterers add their contributions coherently, the sharper the resulting pattern is. The position of an atom is described by the vector

$$\mathbf{r}(t) = \mathbf{r}_0 + \mathbf{u}(t) \tag{29}$$

where $\mathbf{r}_0(t)$ is a constant vector pointing to an atom in the cluster, and the time dependent vector $\mathbf{u}(t)$ describes the thermally activated vibrations. An average over the thermal motion gives for the temperature dependence of the diffracted intensity

$$I(T) = I_{(T=0)}e^{-2W} \tag{30}$$

with

$$W = \frac{1}{3}\langle u^2\rangle q^2 \tag{31}$$

where $\langle u^2\rangle$ is the mean square thermal motion of the atoms from Eq. (29), and q^2 is the square of the momentum experienced by the electron beam upon scattering. The value of $\langle u^2\rangle$ increases with temperature. The diffraction lines do not broaden but decrease exponentially in intensity according to Eq. (31).

4.1. Cluster Structures

Many kinds of nanoclusters have as their most stable forms structures based on polyhedra. They can be crystalline structures (fcc for the elements that we will consider in the following) in the sense that we can build an infinite, periodic array of them; but, since for a finite system there is no translational symmetry constraint, also noncrystalline geometries are found. The latter can show very often fivefold symmetries as those found in decahedra (Dh) or icosahedra (Ih). Here we present a brief description of the most common morphologies.

Icosahedra—They consist of 20 tetrahedra sharing a common vertex (see Fig. 11). Tetrahedra are not space-filling. Ino [99] proposed that there were small gaps in the regions separating the tetrahedra. For this reason, icosahedral particles are often called multitwinned particles since they can be thought of as tetrahedra each having twin boundaries with its neighbor. Alternatively, the tetrahedra might distort, as they are packed together so that they fill the entire volume of the icosahedron. A Mackay [100] Ih is structured in shells, as in an onion-like structure. An Ih with k shells has

$$N_{Ih}(k) = \frac{10}{3}k^3 - 5k^2 + \frac{11}{3}k - 1 \tag{32}$$

Figure 11. Icosahedra are noncrystalline structures with six fivefold symmetry axes, having two fivefold symmetry vertices each. Around these vertices, 20 close-packed triangular facets are arranged. The icosahedral shape is quasispherical with a close-packed surface, and optimizes the surface/volume ratio very well. In nanoclusters, icosahedra can be obtained by arranging subsequent shells of atoms around a central one (Mackay icosahedra). Here we represent the fifth-shell icosahedron, which has 309 atoms: from left to right this icosahedron is given in top view, side view, and along a fivefold symmetry axis.

atoms (so that the series of magic numbers is 1, 13, 55, 147...) and presents 20 triangular facets of side k. Each facet has a (111) geometry. Icosahedral clusters, as we said before, if they continue to grow, will never become crystals. They possess 12 fivefold symmetry axes. This is a symmetry operation not consistent with the translational symmetry found in crystals. The icosahedral packing has many favorable attributes. The Ih has the highest symmetry of all discrete point groups. Each of 20 triangular faces of an Ih can be constructed with close-packed spheres [(111)-facets]. But the icosahedron is a highly strained structure. The interatomic distances between shells are smaller than those within a shell. For this reason, icosahedral packing is found only in relatively small clusters having a high surface-to-volume ratio.

Marks decahedra—They are the other very common noncrystalline structures with a fivefold symmetry axis. A decahedron consists of a pentagonal bipyramid where some vertices are cut in order to have a more spherical structure (see Fig. 12). Alternatively, a Dh can be thought of as five tetrahedra sharing a common edge. It is possible to arrange spheres in decahedral arrangement having a shell structure. The next Dh is formed by placing a large overlapping umbrella on top of the previous member of the set. Since it is energetically favorable to truncate a decahedron in order to make it more spherical, these truncations expose (100) facets that are not close-packed and that have a correspondingly high surface energy. The degree of truncation can be varied in order to minimize the total energy for a given material. The best solution to this problem was given by Marks [101]. He introduced three indices (m, n, p), which characterize unambiguosly decahedra. m and n are the length of the sides of the (100) facets of the truncation, perpendicular and parallel to the fivefold axis, respectively; p is the depth of the Marks reentrance ($p = 1$ corresponds to no reentrance, i.e., to the Ino decahedron [102]). A (m, n, p) Marks Dh has $h = m + n + 2p - 3$ atoms along its symmetry axis and a total number of atoms $N_{Dh}(m, n, p)$ given by

$$N_{Dh}(m,n,p) = \frac{1}{6}\{30p^3 - 135p^2 + 207p - 102 + [5m^3 + (30p-45)m^2 + (60(p^2-3p)+136)m] + n[15m^2 + (60p-75)m + 3(10p^2-30p)+66]\} - 1 \quad (33)$$

Figure 12. As icosahedra, also decahedra are noncrystalline structures with fivefold symmetries. Decahedra present a single fivefold axis, corresponding to two fivefold vertices. A decahedron is obtained by joining two pentagonal pyramids at their basis. This kind of decahedron has ten triangular close-packed facets, but its shape is far from being spherical. A better structure can be obtained by truncating the sides in such a way that rectangular facets are exposed, and forming reentrances exposing further close-packed facets. In this way a Marks decahedron is obtained. The Marks decahedron of $N = 146$ atoms is shown: from left to right, top view (along the fivefold axis), side view, and a different representation of the top view, which shows that atoms are perfectly arranged in columns from this perspective. This decahedron has $(m, n, p) = (3, 2, 2)$: m is the length of the horizontal side of the rectangular facets (see the middle panel) in number of atoms; n is the length of the vertical side of the rectangular facets; p is the depth of the reentrance (see the left panel), with $p = 1$ corresponding to no reentrance at all.

fcc polyhedra—Starting from the octahedron, better fcc polyhedra are obtained truncating symmetrically the six vertexes, thus obtaining square and hexagonal (triangular in the case of cuboctahedra) facets (see Fig. 13). In the following we characterize a given truncated octahedron (TO) by two indexes: n_l is the length of the edge of the complete octahedron; n_{cut} is the number of layers cut at each vertex. A TO has a number of atoms

$$N_{TO}(n_l, n_{cut}) = \frac{1}{3}(2n_l^3 + n_l) - 2n_{cut}^3 - 3n_{cut}^2 - n_{cut} \quad (34)$$

and square facets with edges of $n_{cut} + 1$ atoms. Concerning the (111) hexagonal facets, they are not in general regular hexagons. In fact, three edges of the hexagons are in common with square facets, having thus $n_{cut} + 1$ atoms, while the remaining three edges have $n_l - 2n_{cut}$ atoms. Regular hexagons are thus possible if $n_l = 3n_{cut} + 1$; in the following the TO with regular hexagonal facets will be referred to as

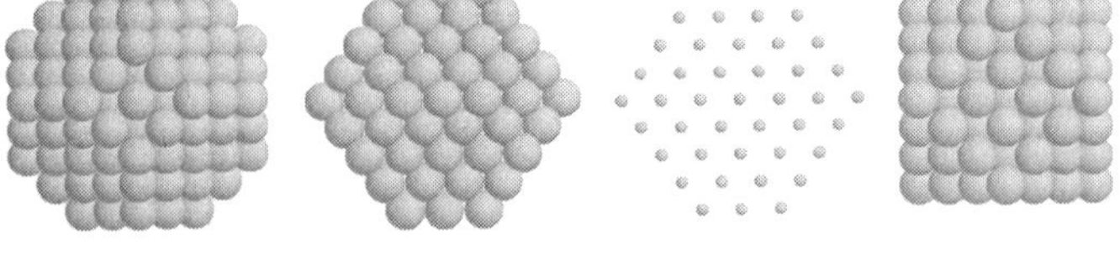

Figure 13. Truncated octahedra (TO). These are fcc polyhedra that are obtained truncating the vertices of an octahedron, which is simply a square-basis bipyramid. When the truncation is not too deep, the TO present hexagonal and square facets, as in the first, second, and third figure from the left. When the truncation gets deeper and deeper, the hexagonal facets decrease in size at the expense of the square facets, until the hexagonal facets are transformed into triangular facets. In this case, the structure is called cuboctahedron (at right).

regular TO. On the other hand, cuboctahedra are characterized by $n_l = 2n_{cut} + 1$, which gives energetically unfavorable clusters with large (100) facets and triangular (111) facets; by substituting this relation into Eq. (34), and comparing with Eq. (32), one finds that a cuboctahedron with a given n_{cut} has the same number of atoms as an Ih with $k = n_{cut} + 1$. The most favorable TO can be found by means of the geometrical Wulff construction [1], which minimizes the surface energy, as shown in several metallic systems in [103].

In order to distinguish the different cluster structures, we have implemented the common neighbor analysis (CNA) method [104]. Briefly, in this method each pair of nearest neighbors (NN) is classified by a triplet of integers (r, s, t) where r is the number of common nearest neighbors, s is the number of nearest-neighbor bonds among the r common nearest neighbors, and t is the length of the longest chain that can be formed with the s bonds. Monitoring the signatures $(r, s, t) = (5, 5, 5), (4, 2, 1), (4, 2, 2)$, we can distinguish whether the cluster is still molten and how its structure is classified. (In Table 3, one can find typical values of these CNA-signatures for molten, Ih, Dh, and fcc clusters.)

4.2. Growth Models

Before dealing with the modeling of growth processes, we wish to spend some words to justify the use of MD simulations to study the growth of nanoaggregates. In typical experiments performed in inert gas aggregation sources, one has

- experimental time scale: $t^C_{exp} \sim$ ms and
- sizes: $d^C \sim$ several nm,

while in the case of epitaxial crystal growth the same parameters are $t^{MBE}_{exp} \sim$ minutes (hours!) and $d^{MBE} \sim \mu$m. In the case of small systems (N is up to several hundred particles) with the present computational resources it is possible to simulate up to $t_{sim} \sim 10^{-5}$ s, which is comparable with experimental time scales; while in the case of MBE only $t_{sim} \sim 10^{-8}$ s can be reached, eight order of magnitude faster than the experimental time scale. Thus, in the case of free clusters, since the system size is small and the time scale is short, MD is a very good tool. If the interparticle potential is reliable, MD simulations offer an accurate and detailed description of the growth process, which allows us to single out the key microscopic mechanisms leading to structural transformations. Contrary to KMC, MD does not imply restrictive choices about the possible processes or the assumption of some rate theory, such as TST, to hold.

The first step to analyzing in detail the growth of free nanoaggregates is to study their lowest energy sequences. Different groups [105–107] suggested several methods, from genetic algorithms to Basin hopping to searching the global minimum (or very good local minima) at a given size N. This often is not sufficient to understand the structures found in experiments. Then it is necessary to include the effect of temperature on clusters. Thus, the problem is to find the minimum free-energy structures. But sometimes taking into account entropic effects is not enough, because the growth time is finite and the system is not able to explore all its configurational space. In other words, since the experimental time of building up of nanosystems is short, the kinetics of the growth process are of fundamental importance to understanding the experimental data. There are two alternative models of obtaining solid clusters, built in contact with a thermal bath, at the end of the growth process. Clusters can grow in the solid state by subsequent additions of single particles on a small seed. Alternatively, clusters can be grown as liquid droplets and then solidify after the growth is completed. From the point of view of simulations, the first approach corresponds to a growth at fixed T and increasing N; while in the second case N is kept fixed and T is lowered to freeze the liquid droplet. In typical conditions, solid-state growth is strongly out of equilibrium.

Table 3. Comparison of several nanocluster structures, pertaining to the three different structural motifs (icosahedral, decahedral and fcc), with respect to the most significant CNA signatures [104].

N	Structure	N_5	$P(5,5,5)$	$P(4,2,1)$	$P(4,2,2)$
75	(2,2,2) m-Dh	5	1.25	28.2	20.4
100	(3,1,2) m-Dh	5	0.92	27.6	19.6
101	(2,3,2) m-Dh	6	1.12	33.7	20.2
146	(3,2,2) m-Dh	6	0.75	39.6	18.7
192	(3,3,2) m-Dh	7	0.66	45.3	18.2
212	(2,2,3) m-Dh	7	0.60	44.7	16.4
238	(3,4,2) m-Dh	8	0.61	48.6	18.0
247	(4,2,2) m-Dh	7	0.51	47.2	16.9
268	(2,3,3) m-Dh	8	0.54	48.6	15.8
318	(4,3,2) m-Dh	8	0.45	52.5	16.3
55	Ih	5	10.3	0.00	38.5
147	Ih	7	5.17	17.2	38.8
309	Ih	9	3.10	31.0	34.9
561	Ih	11	2.06	41.2	30.9
116	TO	—	0.00	60.5	0.00
201	TO	—	0.00	64.6	0.00
225	TO	—	0.00	61.8	0.00
314	TO	—	0.00	68.0	0.00

Note: Size N, type of structure, length of the fivefold symmetry axes N_5 (in atoms), and percentages P of the occurrences of the CNA signatures (5, 5, 5), (4, 2, 1), and (4, 2, 2).

Source: The data are taken from [111].

4.3. Solid-State Growth

The modeling of a growth process through solid states is very close to the DLA method. In fact, we can image that the building up of a cluster is due to the deposition of adatoms onto a small solidified seed, which is in contact with an inert gas, which thermalizes and carries the seeds. Then, when the incoming adatom adsorbs on a cluster, it is free to move on the surface, to reach and attach on islands or steps (as happens on a flat surface), as seen in Figure 1. Desorption is allowed, even if in the case of metallic cluster has never been seen, while for clusters of C_{60} molecules it can take place at high temperatures [108], because these clusters do not melt but sublimate. In order to simulate the presence of the inert gas, canonical MD simulations are needed, introducing for example an Andersen thermostat, as described previously (see [7] for details). The average collision frequency of the thermostat ν must be chosen according to two opposite needs: ν has to be high enough in order to have an efficient thermostat, but it must not be too large because the thermostat must not influence the diffusive properties [110]. Free nanoclusters growth simulations are started from a small seed (seven atoms in the case of silver below), and further particles are deposited one by one at intervals τ_{sim}, by choosing the incoming particle on a random point on a large sphere centered around the cluster center of mass and directing its velocity towards the cluster. The initial position of the adatom is far enough to ensure that there are no interactions with the atoms of the aggregate. The initial velocity of the deposited adatom is chosen as the average velocity at T_{ev} corresponding to a typical evaporation temperature of the material; but this is not so important, since the kinetic energy has to be compared with the adsorption energy, which is about 1–2 eV. In between two subsequent depositions, all the atoms of the cluster are thermalized at a given temperature T, and they are free to move following Newton's equations.

The parameters of this model are the temperature T of the thermal bath and the time interval τ between two subsequent depositions, which is the inverse of the adatom flux ϕ^M. An estimate of the experimental adatom flux ϕ^M on the cluster is given by

$$\phi^M = \frac{PA_{eff}}{\sqrt{2\pi m_M k_B T}} \tag{35}$$

where m_M is the mass of the atom, A_{eff} is the effective area of the cluster, P is the pressure of the vapor, and T is the experimental growth temperature. Below we summarize the growth simulation results for different systems.

4.3.1. Silver Clusters

In the case of silver the simulations were performed by Baletto et al. [4, 111]; the parameters were chosen in this way [4]: time interval $\tau = 1/\phi^M$ in the range 2.1–98 ns; simulation temperature T between 350 and 650 K. The experimental time scale in IGA sources is of the order of one atom every $\sim 10^2$ ns, as seen by putting in Eq. (35) typical values for Ag such as $P \sim 3$ mbar, $m_M = m_{Ag}$, with $R \sim$ 2–3 nm, $T \in$ 300–500 K (for details see [109]). simulations are started from a seed of 7 atoms. The experimental data by Reinhard et al. [3] were in disagreement with any thermodynamics calculations, showing a great abundance of small decahedra and few small icosahedra, while icosahedra were (very surprisingly) the most frequent structures at large sizes. fcc clusters were found if the growth took place at high T. The growth simulations by Baletto et al. [110, 4, 111] have shown that this can be explained in terms of kinetic effects. Small (2 nm of diameter) metastable decahedra are grown at intermediate temperatures because of a kinetic trapping phenomenon in stable smaller Dh (reentrant morphology transition [4]). The large metastable icosahedra are then obtained by growth above the small decahedra (see the sequences in Fig. 14). In fact, a small decahedron is a fragment of a larger icosahedron. The latter grows readily if islands on hcp stacking are nucleated on the starting decahedron [111, 112]. This solid–solid transformation of a Dh into an Ih (sstDI) takes place because it is always more convenient to adsorb an island onto a decahedron on hcp stacking than on fcc stacking, since fcc adsorption sites between two neighboring (111) facets are unstable for single adatoms.

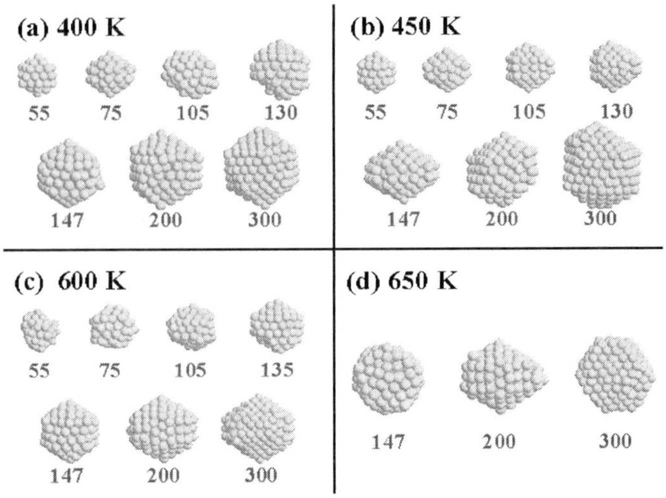

Figure 14. Snapshots at different sizes (given in the figure) from simulations of Ag clusters growth, with a flux of one atom each 7 ns, but at different temperatures. Sequence (a) is at low T (400 K). At 55 atoms a perfect Ih is obtained; then the Ih transforms into a Dh, which is a perfect (2, 2, 2) Marks Dh at 75 atoms. The deposition of further atoms causes the nucleation of islands in hcp stacking and the formation of the Ih at 147 atoms, which grows shell by shell to form larger metastable Ih. Sequence (b) is at intermediate T (450 K). Up to 75 atoms it is analogous to (a), but the addition of further atoms continues the growth of a Dh structure up to 147 atoms. Dh are the most stable structures up to about 100 atoms, and, at this T, kinetic energy is high enough to achieve structural optimization up to this size. After this size, there is a kinetic trapping phenomenon in Dh structures, which ends up above 150 atoms, again because of the transformation into an Ih, which is metastable in this case and then grows shell by shell. Sequence (b) is in complete agreement with the experiments in [3], where small Dh and large Ih were produced. Sequence (c) is at high T (600 K). The cluster is liquid up to 135 atoms, and then solidifies into the most stable structure in this size range, which is an Ih. After that, the growth keeps on following the thermodynamics predictions, giving Dh at larger sizes. Sequence (d) is at even higher T (650 K) and shows the formation of an fcc cluster at large sizes. We underline that at sizes around 150 atoms, there is an intermediate T range where metastable Dh are grown (reentrant morphology transition [4]).

The formation of an hcp island breaks the decahedral symmetry and so the Dh grows towards an Ih. The sstDI is the key ingredient to explain the experimental results: from small decahedra, larger icosahedra are grown, which can then grow further in a shell-by-shell mode preserving their symmetry. The sstDI is a very peculiar example of complete structural transformation, since it has a clear kinetic origin and occurs only through solid states. On the contrary, the reverse transformation of an Ih into a Dh takes place only via the melting of the Ih. Finally, the growth simulations have shown that fcc clusters can be obtained at high temperatures, again in agreement with the experimental results.

4.3.2. Aluminum Clusters

The growth of aluminum clusters has been studied by Valkealahti and Manninen [113] by means of MD simulations, with the aim of explaining the observation of clusters of octahedral shape (see Fig. 13), thus exposing only (111) facets [114–116]. In their simulations, the starting cluster is a

TO of 586 atoms. After having calculated the energy barriers for the most relevant diffusion processes of adatoms and dimers above the cluster surface, Valkealahti and Manninen performed growth simulations by depositing atoms one by one (up to a total of 201), with a much faster deposition rate (one atom each 0.5 ns) than the rates of the above-discussed silver cluster growth simulations [4]. The cluster temperature was kept constant by a thermostat, and simulations were performed in a range around 400 K. Even such a fast deposition rate was however sufficient to demonstrate the possibility of transforming a TO into an octahedron by a growth process. In fact, the deposited adatoms can diffuse fast on the (111) facets, and when they reach the border close to a (100) facet, they can exchange easily with an edge atom, which is then trapped on the (100) facet. The reverse process (exchange from a (100) to a (111) facet) is very difficult because the adsorption energy is much more favorable on (100) than on (111) facets. The accumulation of atoms on (100) facets leads naturally to the formation of an octahedron exposing (111) facets only.

4.3.3. Gold Clusters

A simulation procedure of the same kind as the one of Valkealahti and Manninen for aluminum (see previous section) was applied to the growth of gold TO by Baletto et al. [110], starting from a TO of 201 atoms and depositing with rates of one atom each ns and of one atom each 25 ns. Also in the case of gold, the transformation of the TO into an octahedron can take place easily, because the diffusion processes lead the adatoms to reach the (100) facets, where adsorption is by far more favorable. In comparison, the growth of silver octahedra starting from a TO is still possible, but somewhat more difficult than for gold [110].

4.3.4. Core-Shell Silver/Copper and Silver/Palladium Clusters

The growth of bimetallic clusters in the core-shell geometry has been studied by F. Baletto et al. [117] by the same procedure described above [4, 111]. The systems silver/copper and silver/palladium were considered. In both cases, silver has the tendency to segregate at the surface [118]. Thus, starting either with a palladium or with a copper cluster as a core, and depositing silver atoms above it, there is hope to cover the core by a well-defined silver shell. The aim is to build up a single-layer silver shell. This layer will be strained, due to the size mismatch with the core atoms. Strained overlayers are interesting for catalytic applications [119], especially when obtained above nanoclusters that have a large surface/volume ratio. The simulations by Baletto et al. have indeed shown that sharply defined silver shells with a thickness of a single layer can be grown in a wide range of temperatures. Intermixing between silver and palladium or copper atoms is weak. There is however a difference between the two systems: On palladium cores, the best layers are grown at low T, because the increase of T causes the appearance of some intermixing; on copper cores, the tendency of silver to segregate is so strong that intermixing is always negligible, and the increase of temperature leads to the growth of better shells because diffusion processes are more efficient in eliminating growth defects.

4.3.5. Clusters of C_{60} Molecules

The growth of clusters of C_{60} molecules was studied experimentally several years ago [120], and the mass spectra showed magic numbers sequences related to clusters of icosahedral symmetry, up to sizes around 100 molecules. Subsequent calculations by different methods (see [121] for a summary) were however in complete disagreement with the experimental results, since the lowest energy clusters are icosahedra up to sizes of about 15 molecules, and then, either decahedral or close-packed clusters are preferred. This discrepancy was attributed to the fact that the experiments were performed on charged clusters, while the calculations considered neutral clusters. However, the repetition of the experiments with neutral clusters [122] has given again the growth of icosahedral clusters. Close-packed and decahedral clusters can be obtained only after annealing at high T [122]. The simulations by F. Baletto et al. [123] explained the experimental results by showing that the growth of large metastable icosahedra is due to kinetic trapping effects. During the growth process, small fragments of larger icosahedra are formed around sizes of 20–25 molecules. These fragments are still the lowest-energy structures. Adding further molecules, the cluster remains trapped in a pathway which leads to the completion of the large icosahedron. This is metastable, but with a very long lifetime. Due to a problem of time scales, these simulations were however not able to obtain the close-packed and the decahedral clusters after annealing. Further simulations by Branz et al. [124] showed that the transformation of metastable icosahedra into the most stable structures should be possible, but they have forced their simulated clusters to keep their size constant, not allowing cluster dissolution by evaporation.

4.4. Liquid-State Growth Followed by Freezing

Liquid-state growth implies that solidification takes place after growth. Simulations are performed starting from a liquid droplet of N particles, and then, after a thermalization at such high temperature to ensure that the cluster is liquid, T is lowered. In order to mimick a thermal contact with a cold inert gas, the cluster should be cooled down at a slow rate r_c. For example, the cluster can be cooled down is small steps δT at each time interval δt so $r_c = \delta T/\delta t$. How does one choose the cooling rate? In an inert-gas atmosphere, one can estimate that a cluster of radius R and area $A_{eff} = 4\pi R^2$ collides with gas atoms with a frequency ϕ^G_{exp} given by

$$\phi^G_{exp} \sim \frac{P_G A_{eff}}{\sqrt{2\pi m_G k_B T_G}} \tag{36}$$

where P_G and T_G are, respectively, the pressure and the temperature of the inert gas of mass m_g, and k_B is the Boltzmann constant. In the harmonic approximation, the energy loss is given by

$$\frac{\delta T}{\delta t} \sim \phi^G_{exp} \frac{\delta E}{3Nk_B} \tag{37}$$

where δE is the energy transfer at each collision, and since $N \sim R^3$ we obtain $(\delta T/\delta t) \sim (1/R)$. From [125], we can estimate that the loss for each collision with a helium atom is $\delta E \sim$ 5–10 meV. Using typical parameters for the gas ($P_g \sim 3$ mbar, $T_g \sim 300$ K) and considering a cluster of $R \sim$ 1 nm, we have $(\delta T/\delta t) \sim 10^{-1}$ K/ns or even less. This is the procedure adopted in [130], where slow rates, close to the experimental ones, were adopted. However, in many cases the cooling rates of simulations are much faster than the experimental ones (see [126–129]).

4.4.1. Silver Clusters

In Ref. [130], Baletto et al. considered several free silver clusters with size N in the range 130–310 atoms and, in addition, at icosahedral magic numbers 147, 309, 561, and 923. Different slow cooling rates were employed (choosing $\delta T = 2.5$ K and the cooling rate between 0.1 and 5 K/ns), and, independently of that, particles presented always a well-defined structure. At small sizes (around 2–3 nm), Ih clusters are very common, around the magic size of 147 atoms and in the wide range $N \sim$ 245–310. On the other hand, decahedral particles are the most frequent structures at sizes $N \sim$ 170–240. fcc clusters are found, at intermediate cooling rates, at $N > 200$ atoms, especially around 280, but their percentage decreases if the cooling rate is decreased. By freezing a silver nanodroplet of a diameter around 2 nm, the probability to solidify as an icosahedron is quite high. At magic icosahedral numbers, we find that Mackay icosahedra are not always obtained. In fact, already at 561 atoms nonicosahedral particles are preferentially obtained, and at $N = 923$ it is very unlikely to find icosahedral structures, especially if the cooling rate is slow. Indeed, at $N = 147$ there is 100% of Ih structures, while at $N = 923$, there is only ~10%. On increasing the cluster size, it is more and more difficult to grow large icosahedra by freezing on a liquid droplet. From these results, it follows that entropic effects [131] probably cause the appearance of some icosahedra after freezing the droplets of 923 atoms, but they are not sufficient to make the icosahedron the most favorable structure at that size. These findings are not in agreement with the experimental data of [3], where Dh structures are dominant at small sizes ($d \sim 2$ nm) and a great abundance of large icosahedra ($d \sim 10$ nm) is found. Therefore, the experimental data are in much better agreement with the previously reported simulations of the solid-state growth [110, 4, 111, 112] than with the results of liquid state growth followed by freezing. The conclusion is that in an inert-gas atmosphere, the silver vapor condenses, some seeds are formed, and the particles grow as solid from these seeds because of collisions between them, or by deposition of atoms from the metallic vapor, but it is more unlikely that silver nanoparticles are grown as liquid nanodroplets.

4.4.2. Gold Clusters

The solidification of gold nanoclusters has been simulated by Chushak and Bartell [126] by a rather different method with respect to Baletto et al. [130]. In their MD simulations, they start from high T liquid clusters of sizes of 459, 1,157, and 3,943 atoms. The clusters are then cooled very fast (5×10^4 K/ns) down to about 700 K, were production runs of 1 ns are performed. After that, clusters are further cooled down to 300 K, again with a fast rate (300 K/ns). The result of this procedure is that icosahedral clusters are preferentially produced, even if close-packed or decahedral clusters should be more stable in these sizes (see for example [103]). The production of icosahedra during freezing is due to the width of the icosahedral funnel [107] with respect of the funnel leading to close-packed structures on the potential-energy surface. This preference for icosahedral structures is especially seen at fast cooling rates (compare with the above results for silver [130]), and again, is more evident for the smallest cluster than for the others.

4.4.3. Nickel Clusters

Yue Qi et al. [129] studied the freezing of several Ni clusters of different sizes (from 336 to 8,007 atoms) by MD simulations. They performed heating and cooling cycles: Starting from a solid cluster at low T they raised the temperature until the cluster melted, and after that, they cooled down to solidify the cluster again. During these cycles, the temperature was varied at a fast rate, of 4×10^3 K/ns. The freezing of the clusters produced icosahedral clusters at sizes below 500 atoms and fcc clusters at larger sizes. Due to the fast cooling rate, these resolidified clusters contained a rather large number of defects, with extended region with a disordered character. Also the results of Yue Qi et al. [129] confirm the tendency to form icosahedral clusters only at small sizes after freezing a liquid droplet, in agreement with the previous cases.

4.4.4. Lead Clusters

The solidification of lead clusters by MD simulations was studied first by Lim et al. [132]; they considered a single cluster of about 8,000 atoms, obtaining an icosahedron after fast quenching, in disagreement with the energetic calculations that indicated cuboctahedra as the most favorable clusters [133]. Hendy and Hall [128] performed a more systematic study of the solidification of liquid lead droplets. In their procedure, the droplet is quenched suddenly below the cluster melting point at this size, and then equilibrated for 10 ns. In the size range between 600 and 4,000 atoms, their freezing simulations produced icosahedra with a reconstructed surface, which were shown to be lower in energy than any other known structure of the same size. (We remark however that global optimization at these sizes is unfeasible.) The freezing of larger clusters (around 6,000 atoms) produced fcc structures, even though with several stacking faults forming a narrow band with hcp structure. The production of lead Ih clusters is in agreement with the experimental observation of icosahedral particles of diameter between 3 and 6 nm by Hyslop et al. [134].

4.4.5. Lennard–Jones Clusters

Lennard–Jones clusters are good models for noble-gas nanoparticles, for example for argon clusters, which have been extensively studied in experiments [135]. Their solidification has been studied by Ikeshoji et al. [127] by MD simulations, taking into account two possible ways to cool down the cluster: by a thermostat or by evaporation of atoms. This second possibility must be taken into account for these systems. In fact, Lennard–Jones systems exhibit a rather narrow temperature range for the existence of liquid structures.

Therefore, the evaporation of atoms is still likely at temperatures that are slightly above the melting temperature of the cluster. For metallic systems, the temperature range where the liquid is stable is so wide that solidification by evaporation is practically negligible. Ikeshoji et al. considered 380 clusters in the size range between 160 and 2,200 atoms, and found that there is a transition at increasing size from icosahedra to a variety of different structures (Dh, fcc, hcp, and Ih). The transition does not depend on the cooling method, either evaporation or thermostat, and takes place around size 450. This is lower than the transition size (750 atoms) observed in experiments on argon clusters [135], but it is closer than any estimate based on total energy optimization.

4.4.6. Parahydrogen Clusters

An important issue concerning the freezing of a cluster concerns the way by which solidification is taking place inside the cluster itself. An interesting case for studying this phenomenon is related to parahydrogen nanoclusters, which are at present widely studied because they could be superfluid [136, 24, 137]. In such aggregates, the principal antagonist of superfluidity is solidification. This was recently studied by Levi and Mazzarello [138, 139], who investigated under what conditions solidification either cannot occur at all, or would occur only in times much longer than the lifetime of the cluster. Due to surface melting effects (enhanced by van der Waals forces, via the Hamaker contribution to the free energy [140]), the exterior layers of the cluster do not solidify. As far as the inner core is concerned, the solidification times depend strongly on the exact values of the parameters; they also depend strongly on the thickness of the molten layer. Nucleation implies a barrier, which in principle can be overcome either by thermal fluctuations (at relatively high T) or by tunneling (at very low T). Tunneling in a system with an infinite, or very large, number of degrees of freedom (which can be treated in terms of a small number of macroscopic variables) was introduced in the context of phase transitions, by Lifshitz and Kagan [141]. Two terms in the macroscopic kinetic energy are relevant here. One refers to the liquid motions when a portion of liquid is replaced by the denser solid phase, the other to the motion of the individual molecules across the interface. In both cases the nucleation times can be exceedingly long. More recently, the solidification of parahydrogen clusters containing a foreign, nonspherical molecule at their centers (e.g., OCS, as in the Göttingen experiment [137]), was studied [139]. The foreign molecule a) partly replaces the solid phase, b) imposes a nonspherical symmetry to the solid–liquid interface, c) furnishes a template for solidification, and d) promotes solidification via its van der Waals attraction, which is the most important effect. As a result, the solidification times are drastically reduced with respect to the case of pure hydrogen clusters. This study was used to discuss, from the point of view of solidification, the experiment of the Göttingen group [137], where strong indications in favor of hydrogen superfluidity were found by studying the spectrum of an OCS molecule, surrounded by hydrogen molecules at the center of a helium cluster. The disappearance at 0.15 K of the Q-branch in the spectrum was interpreted as disappearance of the drag of hydrogen by the OCS rotations at low temperatures, hence as superfluidity of the former.

4.5. Coalescence of Nanoclusters

Understanding coalescence is of primary importance for understanding the structure of cluster-assembled materials. These materials can be grown by the low-energy deposition of preformed clusters containing hundreds or thousands of particles on a surface. The structure of the resulting film depends crucially on the diffusion properties of the deposited clusters. By thermodynamic considerations, it is evident that clusters deposited on a surface will tend to coalesce and to form larger drops. The kinetics of this process is not known in detail, but there are some theoretical studies about the coalescence of unsupported clusters. Constant-temperature MD simulations allow us to study the coalescence of free clusters, but at the same time the coalescence events here can be relevant to supported clusters in the case that they are loosely bound to the substrate. The coalescence process can present differences if it takes place between two liquid clusters, one liquid and one solid cluster, or two solid clusters. In order to simulate the process, the starting point configuration consists of two fully equilibrated clusters, which are placed in contact with one another (i.e., along the z axis) at a distance of approach roughly equal to the nearest-neighbor distance for the material. The initial angular momentum is set to zero. For like clusters, the initial positions are chosen in order to allow rotations of clusters with respect to one another by a random angle; indeed, it is expected that coalescence will proceed differently if two facets are in contact than if a facet of one cluster is in contact with a vertex of the other. In the following, the results of simulations by Lewis et al. [142] of coalescence of gold clusters are summarized.

- **Coalescence of liquid–liquid clusters**. In this case, the two small clusters rapidly coalesce into a single essentially spherical cluster. The radii of gyration of the coalescing clusters of N atoms are $\rho_{g\alpha} = (1/N)\sum_{i=1}^{N}(\alpha_i - \alpha_{cm})$, where $\alpha = x, y, z$, and α_{cm} are the coordinates of the center of mass of the cluster, and their average is

$$\rho_g = \sqrt{\sum_{\alpha} \rho_{g\alpha}^2} \tag{38}$$

  Analyzing the evolution in time of all three radii of gyration, it is found that they converge to the overall average on a time scale of several tens of ps. It is important to note that coalescence into a spherical cluster proceeds by deformation of the two clusters in such a way as to optimize the contact surface, i.e. without interdiffusion of one cluster into—or onto—the other. The coalescence of two liquid clusters is essentially a collective phenomenon, involving hydrodynamics flow driven by surface tension forces. Of course, on longer time scales, since the cluster is liquid, diffusion takes over and there is intermixing of the two initial clusters.
- **Coalescence of liquid–solid clusters**. The analysis of the radii for the system (liquid cluster of 767 against a solid cluster of 1505) reveals that coalescence proceeds in two stages. First maximizing the contact surface, an extremely rapid approach of the two clusters is observed, taking place on a time of about 100 ps, which is of the same order of magnitude for coalescence of

two liquid clusters. The cluster is far from spherical but has a faceted ovoidal shape. This rapid approach is followed by a slow “sphericization” of the system driven by surface diffusion on time scales of several hundreds of ns, which is much longer than the sintering time predicted by macroscopic theories of sintering for two soft spheres. The rapid changes seen at short times are due to elastic and plastic deformations; at long times the presence of facets slows down the diffusion [110].

- **Coalescence of solid–solid clusters**. The case of two solid clusters of 1055 atoms is considered. At the beginning, the coalescence seems to proceeds faster here than it did for the liquid–solid case, but the sphericization is slower. However, we can expect that the overall coalescence is faster in the case of liquid–solid clusters. The internal structure seems to be a single domain. On the other hand, considering clusters different in size (i.e., 767 and 1,505) it is found that the internal structure is complex and presents grains. Thus we can conclude that the coalescence of two solid clusters depends strongly on their sizes, while this should not be observed in previous cases.

5. CONCLUSIONS

The study of the nanoworld presently takes enormous benefit from the interplay of experiments and computer simulations. Indeed, when tackling real problems, with the aim to explain quantitatively the experimental outcomes about specific systems, analytical techniques are very often not feasible, and the only way to interpret the experimental data relies on simulations. Thus experiments are interpreted with the help of simulations, which can also suggest new experiments, as in the case of Pt/Pt(111) growth, which in turn stimulate further simulations.

In this article we have given an overview of the most widely employed computational techniques for studying the growth of nanosystems. We have also shown specific examples where these techniques have been employed successfully to explain puzzling experimental data.

Ab initio calculations can provide activation energy barriers for complicated processes in a large variety of systems. The advantages of this technique are its versatility (almost all systems can be studied in the same framework) and its accuracy. The drawback is its computational costs, which allow neither time-dependent simulations on a scale which can be significant for growth processes, nor the study of systems with thousands of atoms. Moreover, since the calculation of each energy barrier is cumbersome, one must decide in advance which processes are significant and which are not. If this task is accomplished, the set of *ab initio* barriers is a very good basis to build up a kinetic Monte Carlo simulation.

Molecular dynamics simulations with interatomic potentials can reach much longer time scales, and on larger sizes. MD simulations give a detailed microscopic description of diffusive and growth processes, and, apart from the choice of an appropriate interatomic potential, they do not require any specific assumption about the kinetics of the system under study. Therefore, when they can be used, they are the most appropriate tool for studying growth phenomena at the microscopic level. MD simulations can be employed to study free nanocluster growth in realistic conditions, following for example complicated transformations between noncrystalline structures. Moreover, they allow a complete study of diffusion processes without any a priori assumption about the choice of the significant processes. This technique requires the development of reliable interatomic potentials. There is no general recipe for this, but accurate atom–atom potentials have been developed for many systems, for example noble gases, several simple and transition metals, and noble metals. MD simulations are however not well-suited to simulate epitaxial crystal growth, because the time scale is too long and terraces are too large. Accelerated MD techniques are promising for bridging the time scale gap. However, at the moment they are useful only for rather small systems at very low temperatures. Moreover, they rely on the assumption of some specific rate theory, which is not universally valid.

Kinetic Monte Carlo simulations can reach the correct sizes and the time scales of epitaxial crystal growth. KMC modeling is based on a large set of assumptions (coarse graining in time and space, choice of a limited set of possible processes, choice of their rates), so that its predictive power may in some cases be rather weak. In particular, the choice of the set of the significant processes may require some skill. However, in the literature, there are several examples of successful application of KMC simulations to explain experimental data.

We have shown that computer simulations have already played a very important role in the investigation of the nanoworld. Since computer power is increasing at a fast rate, we expect that simulation methods will extend their range of applicability in the near future, so that their help in explaining experiments and in designing new materials at the nanoscale will become more and more important.

GLOSSARY

ab initio A calculation or simulation that computes the atomic motions from first principles, following quantum mechanics and avoiding the introduction of fitting parameters.

Coalescence The process by which two clusters unite to form one larger cluster.

Growth The process by which a cluster, starting from a small seed, increases in size by incorporating additional atoms or molecules.

Molecular dynamics A simulation that computes the atomic motions within a given (possibly many-body) potential, solving exactly the Lagrange equations.

Monte Carlo A simulation that treats a physical system in terms of a Markovian stochastic process. In equilibrium Monte Carlo, only the minimal energy and the corresponding atomic positions are sought; kinetic Monte Carlo, on the contrary, attempts to really simulate the atomic motions.

Nanoclusters Aggregates of a small number of atoms or molecules (typically less than 10,000) held together by metallic, chemical, or van der Waals forces.

Solidification (also called freezing) The process by which a cluster, formerly liquid, goes over, totally or partially, to the solid phase.

REFERENCES

1. A. Pimpinelli and J. Villain, "Physics of Crystal Growth." Cambridge University Press, Cambridge, 1998.
2. W. de Heer, *Rev. Mod. Phys.* 65, 611 (1993).
3. D. Reinhard, B. D. Hall, D. Ugarte, and R. Monot, *Phys. Rev. B* 55, 7868 (1997).
4. F. Baletto, C. Mottet, and R. Ferrando, *Phys. Rev. Lett.* 84, 5544 (2000).
5. G. P. Brivio and M. I. Trioni, *Rev. Mod. Phys.* 71, 231 (1999).
6. K. M. Ho et al., *Nature* 392, 582 (1998).
7. D. Frenkel and B. Smit, "Understanding Molecular Simulations" Academic Press, London, 1996.
8. S. Erkoc, *Phys. Rep.* 278, 79 (1997).
9. A. F. Voter, F. Montalenti, and T. C. Germann, *Annu. Rev. Mater. Res.* 32, 321 (2002).
10. P. Hänggi, P. Talkner, and M. Borkovec, *Rev. Mod. Phys.* 62, 251 (1990).
11. A. F. Voter, *Phys. Rev. B* 34, 6819 (1986).
12. M. Kotrla, *Comput. Phys. Commun.* 96, 82 (1996).
13. C. Ratsch, P. Ruggerone, and M. Scheffler, in " Morphological Organization in Epitaxial Growth and Removal" (Zhenyu Zhang and M. G. Lagally, Eds.), p. 3. World Scientific, Singapore 1998.
14. P. Hohenberg and W. Kohn, *Phys. Rev. B* 136, 864 (1964).
15. See, e.g., W. Kohn, in "Highlights of Condensed Matter Theory," Varenna School, Course 89, 1985, p. 1.
16. D. M. Ceperley and B. J. Alder, *Phys. Rev. Lett.* 45, 566 (1980).
17. W. Kohn and L. J. Sham, *Phys. Rev. A* 140, 1133 (1965).
18. R. Car and M. Parrinello, *Phys. Rev. Lett.* 55, 2471 (1985).
19. P. Sonnet, A. Selloni, L. Stauffer, A. DeVita, and R. Car, *Phys. Rev. B* 65, 085322 (2002).
20. J. O. Hirschfelder and W. J. Meath, *Adv. Chem. Phys.* 12, 3 (1967).
21. F. Buda, G. L. Chiarotti, R. Car, and M. Parrinello, *Physica B* 119, 98 (1991).
22. D. K. Remler and P. A. Madden, *Mol. Phys.* 70, 921 (1990).
23. E. L. Pollock and D. M. Ceperley, *Phys. Rev. B* 30, 2555 (1984).
24. P. Sindzingre, D. M. Ceperley, and M. L. Klein, *Phys. Rev. Lett.* 67, 1871 (1991).
25. A. Gross and M. Scheffler, *Phys. Rev. B* 57, 2493 (1998).
26. A. Gross, S. Wilke, and M. Scheffler, *Phys. Rev. Lett.* 75, 2718 (1995).
27. R. Stumpf and M. Scheffler, *Phys. Rev. B* 53, 4958 (1996).
28. P. J. Feibelman, *Phys. Rev. B* 35, 2626 (1986).
29. P. J. Feibelman, *Phys. Rev. B* 52, 16845 (1995).
30. D. R. Hamann, *Phys. Rev. B* 40, 2980 (1989).
31. J. Perdew and A. Zunger, *Phys. Rev. B* 23, 5048 (1981).
32. H. T. Lorensen, J. K. Nørskov, and K. W. Jacobsen, *Phys. Rev. B* 60, R5149 (1999).
33. P. J. Feibelman and R. Stumpf, *Phys. Rev. B* 59, 5892 (1999).
34. P. J. Feibelman, *Phys. Rev. B* 60, 4972 (1999).
35. P. J. Feibelman, *Phys. Rev. B* 61, R2452 (1999).
36. G. L. Kellogg, A. F. Wright, and M. S. Daw, *J. Vac. Sci. Technol., A* 9, 1757 (1991).
37. F. Montalenti and R. Ferrando, *Phys. Rev. Lett.* 82, 1498 (1999).
38. M. S. Daw and M. I. Baskes, *Phys. Rev. B* 29, 6443 (1984).
39. L. A. Girifalco, *J. Phys. Chem.* 96, 858 (1992).
40. H. C. Andersen, *J. Chem. Phys.* 72, 2384 (1980).
41. K. Binder and D. W. Heermann, "Monte Carlo Simulations in Statistical Physics. An Introduction," vol. 80 of Springer Series in Solid–State Sciences. Springer, Berlin, 1988.
42. D. W. Heermann, "Computer Simulation Methods in Theoretical Physics," 2nd ed. Springer, Berlin, 1990.
43. J. P. Valleau, in "Computer Simulations in Material Science: Interatomic Potentials, Simulation Techniques and Applications" (M. Meyer and V. Pontikis, Eds.), number 205 in NATO ASI, Series E: Applied Science, p. 67, Kluwer, Dordrecht, 1991.
44. H. C. Kang and W. H. Weinberg, *J. Chem. Phys.* 90, 2824 (1989).
45. K. Binder (1979), in "Monte Carlo Methods in Statistical Physics" (K. Binder, Ed.), Vol. 7 of Topics in Current Physics, p. 1. Springer, Berlin, 2nd ed., 1986.
46. K. Binder and M. H. Kalos (1979), in "Monte Carlo Methods in Statistical Physics," Vol. 7 of Topics in Current Physics, p. 225. Springer, Berlin, 2nd ed., 1986.
47. N. G. van Kampen, "*Stochastic Processes in Physics and Chemistry*," North–Holland, Amsterdam, 1981.
48. J. G. Amar and F. Family, *Phys. Rev. B* 54, 4742 (1996).
49. M. C. Bartelt and J. W. Evans, *Phys. Rev. Lett.* 75, 4250 (1995).
50. T. Ala-Nissila, R. Ferrando, and S. C. Ying, *Adv. Phys.* 51, 949 (2002).
51. A. B. Bortz, M. H. Kalos, and J. L. Lebowitz, *J. Comp. Phys.* 17, 10 (1975).
52. S. Clarke, M. Wilby, and D. D. Vvedensky, *Surf. Sci.* 255, 91 (1991).
53. J. L. Blue, I. Beichl, and F. Sullivan, *Phys. Rev. E* 51, R867 (1995).
54. M. Kotrla and A. C. Levi, *J. Stat. Phys.* 64, 579 (1991).
55. S. van Dijken, L. C. Jorritsma, and B. Poelsema, *Phys. Rev. Lett.* 82, 4038 (1999).
56. G. De Lorenzi and G. Ehrlich, *Surf. Sci.* 293, L900 (1993).
57. J. V. Barth, *Surf. Sci. Rep.* 40, 75 (2000).
58. T. Michely, M. Hohage, M. Bott, and G. Comsa, *Phys. Rev. Lett.* 70, 3943 (1993).
59. R. Kunkel, B. Poelsema, L. K. Verheij, and G. Comsa, *Phys. Rev. Lett.* 65, 733 (1990).
60. G. Ehrlich and F. G. Hudda, *J. Chem. Phys.* 44, 1039 (1966).
61. R. L. Schwoebel, *J. Appl. Phys.* 40, 614 (1968).
62. S. Esch, M. Hohage, T. Michely, and G. Comsa, *Phys. Rev. Lett.* 72, 518 (1994).
63. P. J. Feibelman, *Phys. Rev. Lett.* 81, 168 (1998).
64. M. Kalff, G. Comsa, and T. Michely, *Phys. Rev. Lett.* 81, 1255 (1998).
65. R. Ferrando, *Phys. Rev. Lett.* 76, 4195 (1996).
66. C. De Giorgi, P. Aihemaiti, F. Buatier de Mongeot, C. Boragno, R. Ferrando, and U. Valbusa, *Surf. Sci.* 487, 49 (2001).
67. F. Buatier de Mongeot, G. Costantini, C. Boragno, and U. Valbusa, *Phys. Rev. Lett.* 84, 2445 (2000).
68. F. Hontinfinde and R. Ferrando, *Phys. Rev. B* 63, 121403(R) (2001).
69. M. Guillopé and B. Legrand, *Surf. Sci.* 215, 577 (1989).
70. C. H. Bennett, in "Diffusion in Solids, Recent Developments" (A. S. Nowick and J. J. Burton, Eds.), p. 73. Academic Press, New York, 1975.
71. F. Hontinfinde, R. Ferrando, and A. C. Levi, *Surf. Sci.* 366, 306 (1996).
72. F. Montalenti and R. Ferrando, *Phys. Rev. B* 59, 5881 (1999).
73. K. J. Caspersen and J. W. Evans, *Phys. Rev. B* 64, 075401 (2001).
74. F. Montalenti and A. F. Voter, *Phys. Rev. B* 64, 081401(R) (2001).
75. A. Videcoq, F. Hontinfinde, and R. Ferrando, *Surf. Sci.* 515, 575 (2002).
76. M. C. Bartelt and J. W. Evans, *Europhys. Lett.* 21, 99 (1993).
77. H. Möhwald, *Rep. Progr. Phys.* 56, 653 (1993).
78. T. A. Witten and L. M. Sander, *Phys. Rev. Lett.* 47, 1400 (1981).
79. T. Vicsek, "Fractal Growth Phenomena," World Scientific, Singapore, second ed., 1992.
80. P. Meakin, in "Phase Transitions and Critical Phenomena" (C. Domb and M. S. Green, Eds.), Vol. 12, p. 335. Academic Press, London, 1988.
81. L. P. Kadanoff, *J. Stat. Phys.* 39, 267 (1985).

82. P. Meakin, *Phys. Rev. A* 27, 604 (1983); *ibid* A27 1495 (1983).
83. A. Erzan, L. Pietronero, and A. Vespignani, *Rev. Mod. Phys.* 67 545 (1995).
84. A. Gliozzi, A. C. Levi, M. Menessini, and E. Scalas, *Physica A* 203, 347 (1994).
85. A. C. Levi and M. Kotrla, *J. Phys. Chem.* 9, 299 (1997).
86. O. Shochet, K. Kassner, E. Ben-Jacob, S. G. Lipson, and H. Müller-Krumbhaar, *Physica A* 181, 186; *ibid* A187, 87 (1992).
87. G. Indiveri, A. C. Levi, A. Gliozzi, E. Scalas, and H. Möhwald, *Thin Solid Films* 284/285, 106 (1996).
88. G. Indiveri, E. Scalas, A. C. Levi, and A. Gliozzi, *Physica A* 273, 217 (1999).
89. R. Pastor-Satorras and J. M. Rubí, *Phys. Rev. E* 51, 5994 (1995).
90. H. McConnell, *Annu. Rev. Phys. Chem.* 42, 171 (1991).
91. T. Vanderlick and H. Möhwald, *J. Phys. Chem.* 94, 886 (1989).
92. A. Dietrich, H. Möhwald, W. Rettig, and G. Brezezinski, *Langmuir* 7, 539 (1991).
93. E. Brener, H. Müller-Krumbhaar, K. Kassner, and D. Tëmkin, *International Journal of Modern Physics C* 3, 825 (1992).
94. R. Rietz, G. Brezesinski, and H. Möhwald, *Ber. Bunsen-Ges. Phys. Chem.* 97, 1394 (1993).
95. Y. Pomeau, *Europhys. Lett.* 3, 1201 (1987).
96. R. Q. Hwang, J. Schröder, C. Günther, and R. J. Behm, *Phys. Rev. Lett.* 67, 3279 (1991).
97. M. C. Bartelt and J. W. Evans, *Surf. Sci.* 314, L385 (1994).
98. N. Camillone III, P. Eisenberger, T. Y. B. Leung, P. Schwartz, and G. Scoles, *J. Chem. Phys.* 101, 11031 (1994).
99. S. Ino, *J. Phys. Soc. Jpn.* 27, 941 (1969).
100. A. L. Mackay, *Acta Crystallogr.* 15, 916 (1962).
101. L. D. Marks, *Rep. Prog. Phys.* 57, 603 (1994).
102. T. P. Martin, *Phys. Rep.* 273, 199 (1996).
103. F. Baletto, R. Ferrando, A. Fortunelli, F. Montalenti, and C. Mottet, *J. Chem. Phys.* 116, 3856 (2002).
104. D. Faken and H. Jónsson, *Comput. Mater. Sci.* 2, 279 (1994).
105. K. Michaelian, N. Rendón, and I. L. Garzón, *Phys. Rev. B* 60, 2000 (1999).
106. C. L. Cleveland et al., *Phys. Rev. Lett.* 79, 1873 (1997).
107. D. J. Wales, J. P. K. Doye, M. A. Miller, P. N. Mortenson, and T. R. Walsh, *Adv. Chem. Phys.* 111, 1 (2000).
108. M. H. J. Hagen, E. J. Meijer, G. C. A. M. Mooij, D. Frenkel, and H. N. W. Lekkerkerker, *Nature* 365, 425 (1993).
109. B. D. Hall, M. Flüeli, R. Monot, and J.-P. Borel, *Phys. Rev. B* 43, 3906 (1991).
110. F. Baletto, C. Mottet, and R. Ferrando, *Surf. Sci.* 446, 31 (2000).
111. F. Baletto, C. Mottet, and R. Ferrando, *Phys. Rev. B* 63, 155408 (2001).
112. F. Baletto and R. Ferrando, *Surf. Sci.* 490, 361 (2001).
113. S. Valkealahti and M. Manninen, *Phys. Rev. B* 57, 15533 (1998).
114. J. Lerme, M. Pellarin, J. L. Vialle, B. Baguenard, and M. Broyer, *Phys. Rev. Lett.* 68, 2818 (1992).
115. T. P. Martin, U. Näher, and H. Schaber, *Chem. Phys. Lett.* 199, 470 (1992).
116. S. Valkealahti, U. Näher, and M. Manninen, *Phys. Rev. B* 51, 11039 (1995).
117. F. Baletto, C. Mottet, and R. Ferrando, *Phys. Rev. B* 66, (2002).
118. G. Tréglia, B. Legrand, J. Eugène, B. Aufray, and F. Cabané, *Phys. Rev. B* 44, 5842 (1991); S. Ouannasser, L. T. Wille, and H. Dreyssé, *Phys. Rev. B* 55, 14245 (1997).
119. M. Mavrikakis, B. Hammer, and J. K. Nørskov, *Phys. Rev. Lett.* 81, 2819.
120. T. P. Martin, U. Näher, H. Schaber, and U. Zimmermann, *Phys. Rev. Lett.* 70, 3079 (1993).
121. J. P. K. Doye, D. J. Wales, W. Branz, and F. Calvo, *Phys. Rev. B* 64, 235409 (2001).
122. W. Branz, N. Malinowski, H. Schaber, and T. P. Martin, *Chem. Phys. Lett.* 328, 245 (2000).
123. F. Baletto, J. P. K. Doye, and R. Ferrando, *Phys. Rev. Lett.* 88, 075503 (2002).
124. W. Branz, N. Malinowski, A. Enders, and T. P. Martin, *Phys. Rev. B* 66, 094107 (2002).
125. J. Westergren, H. Gronbeck, A. Rosen, and S. Nordholm, *Nanostruct. Mater.* 12, 281 (1999).
126. Y. G. Chushak and L. S. Bartell, *J. Phys. Chem. B* 105, 11605 (2001).
127. T. Ikeshoji, G. Torchet, M.-F. De Feraudy, and K. Koga, *Phys. Rev. E* 63, 031101 (2001).
128. S. C. Hendy and B. D. Hall, *Phys. Rev. B* 64, 085725 (2001).
129. Y. Qy, T. Cagin, W. L. Johnson, and W. A. Goddard, *J. Chem. Phys.* 115, 385 (2001).
130. F. Baletto, C. Mottet, and R. Ferrando, *Chem. Phys. Lett.* 354, 82 (2002).
131. J. P. K. Doye and F. Calvo, *Phys. Rev. Lett.* 86, 3570 (2001).
132. H. S. Lim, C. K. Ong, and F. Ercolessi, *Comput. Mater. Sci.* 2, 495 (1994).
133. H. S. Lim, C. K. Ong, and F. Ercolessi, *Surf. Sci.* 269/270, 1109 (1992).
134. M. Hyslop, A. Wurl, S. A. Brown, B. D. Hall, and R. Monot, *Eur. Phys. J. D* 16, 233 (2001).
135. J. Farges, M.-F. de Feraudy, B. Raoult, and G. Torchet, *J. Chem. Phys.* 84, 3491 (1986).
136. V. L. Ginsburg and A. A. Sobyanin, *JETP Lett.* 15, 242 (1972).
137. S. Grebenev, B. Sartakov, J. P. Toennies, and A. F. Vilesov, *Science* 289, 1532 (2000).
138. A. C. Levi and R. Mazzarello, *J. Low Temp. Phys.* 122, 75 (2001).
139. R. Mazzarello and A. C. Levi, *J. Low Temp. Phys.* 127, 259 (2002).
140. A. C. Levi, in "Phase Transitions in Surface Films" (H. Taub, G. Torzo, H. J. Lauter, and S. C. Fain, Eds.), p. 327. Plenum, New York, 1990.
141. I. M. Lifshitz and Yu. Kagan, *Sov. Phys. JETP* 35, 206 (1972).
142. L. J. Lewis, P. Jensen, and J.-L. Barrat, *Phys. Rev. B* 56, 2248 (1997).

Encyclopedia of Nanoscience and Nanotechnology
www.aspbs.com/enn

High-Field Conduction in Nanostructures

R. D. Gould
Keele University, Keele, UK

CONTENTS

1. INTRODUCTION

We can conceive, in the not too distant future, the fabrication of integrated electronic circuits where the feature size of individual components is reduced to the order of nanometers. These nanostructures will require fabrication methods that are either currently only under the initial stages of investigation or even totally unknown. Already nanosized objects may be manipulated using self-assembly and scanning probe microscopy, but the problem remains of how to build functional devices [1]. The mode of operation of such devices is also questionable at the present time, and the assumption that bulk properties will transfer to the nanoscale is very unlikely to be justified. Quantum-mechanical interactions will inevitably be of paramount importance. However, whatever the fabrication methods used and the details of the electronic interactions within the individual nanostructures, it is clear that contacts will be required in order to provide biasing and device control. Furthermore, the one type of nanostructure (admittedly with nanometer dimensions in only one direction) that has been extensively investigated to date is based on thin films. A film of thickness 100 nm having a potential difference of 1 V applied between its surfaces would experience an electric field of 10^7 V m^{-1}, approaching the dielectric breakdown strength of many materials. As we move into the realm of nanostructures, dimensions of only a few nanometers will become commonplace, consequently with even higher electric fields developed. It is likely that voltage levels will have to be restricted considerably in order to avoid damage and premature failure. Nevertheless, it is evident that high-field conduction mechanisms, which have previously been identified in many thin film structures, will become the norm rather than the exception as at present. High-field conduction processes have been previously reviewed by several authors in the context of metal-dielectric-metal layers [2–4]. Of particular interest in the present context is that phthalocyanine films, one of the earliest investigated organic semiconductors, exhibit a wide variety of such high-field conduction processes [5]. The basic phthalocyanine molecule is exceptionally stable for an organic structure, is planar, and is well suited to an important role in future organic nanostructures. High-field conduction processes are also well-known in many conventional inorganic, thin-film semiconductor materials, particularly the cadmium compounds [6], which are beginning to be investigated in the field on nanotechnology. Layer-by-layer growth of CdSe-based nanocrystalline LEDs has already been reported [7].

In keeping with these comments, the review covers the physics of the established high-field conduction processes which occur under steady direct current (DC) bias, plus where appropriate, some speculation concerning conduction processes that may become increasingly important as nanotechnology progresses. Alternating current (AC) processes are excluded from the review on the grounds of simplicity. There is no detailed description of thin film deposition, nanostructure fabrication, and assembly processes or current prototype devices, most of which are fully described elsewhere in this encyclopedia. The major emphasis is on the physics of the high-field processes that will underpin the operation of future nanodevices; although at present, these relate primarily to one-dimensional cases in metal-dielectric-metal systems, it is extremely likely that in a nanostructure, a full three-dimensional theoretical analysis of these processes will be required. The theory for this has not been developed at the present time, but it is clear that the starting point will be the solution of one of the four Maxwell equations

ISBN: 1-58883-059-4/$35.00

Encyclopedia of Nanoscience and Nanotechnology
Edited by H. S. Nalwa
Volume 3: Pages (891–915)

of electromagnetism, namely, Gauss's law, which in principle allows us to calculate the electric field F at any point in the structure. This is implicit in Poisson's equation, which relates the gradient of the electric field to the local volume charge density. In conjunction with the current-density equations, which also include the diffusion component of current density resulting from carrier concentration gradients and the carrier continuity equations, a derivation of the theoretical current density-voltage (J-V) characteristics should be possible. These three sets of equations are the basis for present day semiconductor device operation [4] and in a suitably modified form, which takes into account effects such as the inclusion of microscopic fields and quantum-mechanical behavior, will also be important in the development of the theory of nanoelectronic devices.

Following this introduction, Section 2 briefly covers the basic electromagnetic equations mentioned above. They are presented in the general three-dimensional form using vector operator notation; these are the fundamental equations that will be the basis of the applications of nanostructures in electronics. The equation for the electric field at a distance x from the center of a charged sphere of radius a is discussed as the basis for speculation that the electric field in nanostructures will be extremely high, especially in localized regions. The simpler, one-dimensional form of Poisson's equation, as conventionally used in deriving the J-V characteristics for thin film structures, is then presented. Section 3 describes the types of contact commonly presented to dielectric films, particularly, but not exclusively in metal-dielectric-metal systems. This is particularly germane, since the conduction processes that occur in the dielectric are heavily influenced by the external contacts, in terms of factors such as the barrier height at the interface, and the variation of space charge density within the dielectric. This discussion is restricted to metal-dielectric contacts, and does not cover internal barriers such as semiconductor-semiconductor junctions and grain boundaries. Speculation on the form of these in nanostructures is unlikely to be of value, considering the current state of the art, and can await experimental results on nanostructures with internal barriers.

The main body of the discussion focuses on the various high-field conduction mechanisms previously identified in thin film structures, and which are likely to carry over in some form to nanostructures. These are covered in Sections 4 to 7. The origin of the mechanisms, and a derivation of the expected J-V characteristics, are given where appropriate, together with a small selection of experimental results from the literature specifically chosen to illustrate the processes. Neither the theoretical derivations or the experimental results described are expected to be typical of what may in due course be observed in three-dimensional nanostructures, but both will offer hints of how progress may be achieved in this field. In Section 4, the thermionic emission process is described. This process results from the fact that in a metal at a given temperature, due to electron energies following a Fermi-Dirac distribution, there will be some electrons with energies sufficient to escape over the surface potential barrier. In the presence of an electric field, the potential barrier is effectively lowered, and the resulting current collected at a second electrode is voltage-dependent. This is known as the Schottky effect. For very thin dielectric regions, typically up to about 10 nm only, the possibility of direct quantum-mechanical tunneling between the two electrodes is significant. Various aspects of the tunneling process are discussed in Section 5. In Section 6, the important mechanism of space-charge-limited conduction (SCLC) is described. This arises from the presence of a space charge reservoir at the interfacial region resulting from the use of ohmic contacts. Various different J-V characteristics can result, depending on whether such a charge is condensed in traps, and if so on their specific energy distribution. Depending primarily on the dielectric film thickness, the space charge is present either only at the contacts (in thicker films) or over the entire thickness (in thinner films). In nanostructures, the latter is likely to be the more important case. The Poole–Frenkel effect, a process similar in some ways to the Schottky effect, is discussed in Section 7. This results from the lowering of internal potential barriers due to the applied field, which results in an increased number of carriers entering the conduction (or valence) band. The field-lowering properties of the Schottky and Poole–Frenkel effects are very similar, including the basic dependencies of the J-V characteristics; these similarities and some differences are outlined. Impurity conduction (or hopping) is not addressed in this discussion, as activation energies are very low and it is rarely considered as a high-field process. This occurs when impurity centers are sufficiently close to each other to enable hopping of carriers between individual sites. In Section 8, some speculations concerning filamentary conduction are given. Although large scale filaments, of the type known to exist in various thin film structures, would almost certainly be impossible to control in nanostructures, smaller filaments of the type proposed by Dearnaley et al. [8], could possibly be exploited in a new class of nanodevices. Finally, in Section 9, these various mechanisms are briefly compared and their relevance in nanostructures is assessed.

For comparability between different works, most quantities have been quoted in the SI system of units, but other convenient units such as the electron-Volt (eV) are also used where this aids clarity. When work was originally presented using non-SI units, these have normally been converted. The main emphasis in this review is on the physics of the high-field processes that dominate the electrical conductivity of dielectric films and how such processes may prove important in nanoelectronic devices.

2. ELECTROMAGNETIC FUNDAMENTALS

The most important electromagnetic consideration in the possible functioning of nanodevices appears at present to be Gauss's law and the related Poisson's equation. Gauss's law allows us to calculate the electric field for a given charge distribution, although the equation is often very difficult or even impossible to solve for nontrivial situations. Nevertheless, it gives us an insight into the intensity of the electric field that may be expected within nanostructures. Poisson's equation directly relates the potential to the charge distribution in the structure which, in conjunction with the current density equations and the carrier continuity equations mentioned in Section 1, allows the J-V characteristics to be calculated in simple cases with known boundary conditions. We should note that the following discussion is based on the conventional macroscopic description of electromagnetism. The delineation between the macroscopic and microscopic electromagnetic approach clearly depends on the details of the situation considered, but a distance of 10 nm has been considered plausible for the absolute lower limit of the macroscopic domain [9]. For smaller distances, the microscopic approach is generally necessary. Jackson [9] has shown how the macroscopic Maxwell equations can be derived by considering a spacial average of the microscopic equations. He has also discussed the need for a quantum-mechanical treatment for calculations of the electric field at the surface of a conductor, quoting the results of such a calculation involving the solution of the many-electron problem in which the ions of the conductor are approximated by a continuous constant charge density. The excess electron charge is seen to be confined within a region of ± 0.2 nm of the surface of the ionic distribution. This demonstrates the ultimate need for a sophisticated, quantum-mechanical treatment for the smallest of nanodevices. However, in this discussion we restrict ourselves to the macroscopic equations of electromagnetism, with the proviso that below 10 nm distances or so, these will probably be inadequate.

We first consider Gauss's law in a dielectric

$$\oint \mathbf{D} \cdot d\mathbf{S} = Q \tag{1}$$

where $d\mathbf{S}$ is a surface element perpendicular to the electric displacement field $\mathbf{D}$ at any point on a closed surface $\mathbf{S}$ enclosing a total charge Q. The relationship between $\mathbf{D}$ and the electric field $\mathbf{F}$ may be taken as

$$\mathbf{D} = \varepsilon_r \varepsilon_0 \mathbf{F} \tag{2}$$

where ε_r is the relative permittivity of the material and $\varepsilon_0 = 8.85 \times 10^{-12}$ F m^{-1} is the permittivity of free space; Eq. (2) is a constitutive equation of the medium. This applies particularly for linear, isotropic, homogeneous media, and neglects electric quadrupole and higher moment densities in the medium. Furthermore, $\mathbf{D}$ is always parallel to $\mathbf{F}$. For our estimates of electric fields in nanostructures, this approximation is perfectly adequate. In our medium, Eq. (1) then reduces to

$$\oint \mathbf{F} \cdot d\mathbf{S} = \frac{Q}{\varepsilon_r \varepsilon_0} \tag{3}$$

We then consider a conducting sphere of radius a embedded in the dielectric. The treatment below is based on that of Duffin [10] for a conducting sphere in vacuum, but similar treatments are given in many books on electromagnetism. The conducting sphere will carry all its charge Q on its surface, and from the symmetry of the sphere there will be a uniform charge per unit area (area charge density) σ over the surface of the sphere, where $\sigma = Q/4\pi a^2$. To find the electric field $\mathbf{F}$ at a distance r from the center of the sphere, we consider a second sphere of radius r ($r \geq a$) concentric with the charged sphere. From the symmetry of the system, the electric field F is the same at all points on the surface of the second sphere and is directed radially outwards. The outward flux over the surface of the sphere will then be $\oint \mathbf{F} \cdot d\mathbf{S} = 4\pi r^2 F$. When substituted into the Gauss's law expression of Eq. (3), this yields

$$F = \frac{Q}{4\pi \varepsilon_r \varepsilon_0 r^2} \tag{4}$$

for the electric field at any point a distance $r \geq a$ from the center of the charged sphere. In particular, the field at the surface of the charged sphere F_s, where $r = a$, is

$$F_s = \frac{Q}{4\pi \varepsilon_r \varepsilon_0 a^2} = \frac{\sigma}{\varepsilon_r \varepsilon_0} \tag{5}$$

where we have also used the earlier charge density expression. The potential V can be obtained from the electric field F of Eq. (4) using $V = -\int F\,dr$, and the surface potential V_s of the charged sphere is then given by

$$V_s = \frac{Q}{4\pi \varepsilon_r \varepsilon_0 a} = aF_s = \frac{a\sigma}{\varepsilon_r \varepsilon_0} \tag{6}$$

This equation demonstrates that if we were to have a series of spheres having varying radii, but all charged to the same potential, the electric field just outside the surface and also the surface area charge density are inversely proportional to a, the radius of curvature. This result carries over to the more realistic situation, where we have a single conductor with a varying radius of curvature. The electric field and the area charge density are higher in regions having a small radius of curvature, and this leads to the conclusion that because of the small size of nanostructures, and therefore the smaller radii of curvature, the electric field will be particularly high in certain regions and able to sustain a high-field conduction process. These regions are as likely to be present by accident (e.g., asperities and voids) as by design.

Using some of the above results, the field can be estimated in some nanostructures. The calculations are, of course, at best order of magnitude estimates of the field values, in view of the assumptions made, and the fact that the above theory is based on macroscopic rather than microscopic fields. Consider first a conductive region in contact with a dielectric. The conductive region is at a potential of 1 V and the radius of curvature of the conductor $a = 1$ μm. Both of these values are likely to be large in the context of nanostructures, but are typical of values in very-large-scale integration (VLSI) technology. The dielectric is assumed to have a value of relative permittivity $\varepsilon_r = 3.9$, which corresponds to silicon dioxide, SiO_2 [11]. Using these values, Eq. (6) predicts an electric field at the boundary of the conductor with the dielectric of 10^6 V m^{-1}. At a distance of 1 μm away from the radius of curvature in the dielectric ($r = 2$ μm), Eq. (4) predicts an electric field of 2.5×10^5 V m^{-1}. Although these values are not as high as the field of 10^7 V m^{-1} in the 100 nm-thick film mentioned in Section 1, they are bordering on the field intensity necessary for high-field conduction processes. Conversely, if we consider a nanostructure where the radius of curvature is 10 nm, fields at the conductor/dielectric interface of 10^8 V m^{-1} and 10^7 V m^{-1} are obtained for potentials of 1 V and 0.1 V, respectively. These and smaller radii of curvature lead to local high fields that would clearly support the operation of high-field conduction processes. Thus, in nanostructures, control of the radius of curvature of conducting regions could determine whether or not high-field conduction processes occur. As smaller structures are contemplated, the existence of high fields in the structures would become inevitable.

Having laid the foundations for the inevitability of high fields in nanostructures, we now turn our attention to Poisson's equation, which is the starting point for deriving the J-V characteristics in such regions. Gauss's law of Eq. (1), which is presented in the integral form, can by the application of Gauss's theorem be written in the corresponding differential form as

$$\nabla \cdot \mathbf{D} = \rho(x, y, z) \tag{7}$$

where $\rho(x, y, z)$ represents the charge per unit volume (or the volume charge density) at the point (x, y, z). The differential form is in general more easy to work with, and encapsulates the behavior of the field at a *point* in space, rather than over a closed surface. Furthermore, the integral form of this and the other Maxwell equations are only solvable for highly symmetric situations. The differential forms, although they may not be solvable analytically for a general shape, lend themselves to computational numerical solutions. If we make the assumption that Eq. (2) is still valid, Eq. (7) becomes

$$\nabla \cdot \mathbf{F} = \frac{\rho(x, y, z)}{\varepsilon_r \varepsilon_0} \tag{8}$$

We now consider the definition of the electric field $\mathbf{F}$ as the negative gradient of the potential V, or

$$\mathbf{F} = -\nabla V \tag{9}$$

Substitution of this into Eq. (8) gives

$$\nabla \cdot \nabla V = \nabla^2 V = -\frac{\rho(x, y, z)}{\varepsilon_r \varepsilon_0} \tag{10}$$

which is the three-dimensional form of Poisson's equation. In the absence of a charge density when $\rho = 0$, Poisson's equation reduces to $\nabla^2 V = 0$, which is Laplace's equation. The vector operator ∇^2 is known as the Laplacian operator, and in Cartesian coordinates is

$$\nabla^2 = \frac{\partial^2}{\partial x^2} + \frac{\partial^2}{\partial y^2} + \frac{\partial^2}{\partial z^2} \tag{11}$$

If the charge density varies in one direction only, say the x-direction, Poisson's equation itself reduces to

$$\frac{d^2 V}{dx^2} = \frac{dF}{dx} = -\frac{\rho(x)}{\varepsilon_r \varepsilon_0} \tag{12}$$

This is the form that is used in the discussion on contacts and in the derivation of some of the J-V characteristics in the following sections.

In this section, the basic macroscopic equations of electromagnetism relevant to nanostructures have been presented, and the case made for the existence of very high fields in nanostructures even at relatively low voltage levels. Although Maxwell's equations, the current density equations and the carrier continuity equations, are likely to be the basis for a suitable theoretical treatment of currents in nanostructures, it is almost certain that a solution of the microscopic Maxwell's equations will be required in any rigorous theory, and that a full quantum-mechanical treatment will be required in the limit. For dimensions greater than about 10 nm, it is assumed that the macroscopic approach will hold reasonably well. Most of the work to date concerning contacts and high-field conduction that is reviewed in the following sections falls into this category, although quantum mechanics form the basis of most of the mechanisms described. In particular, tunneling is a purely quantum-mechanical process and can take place across a maximum distance of about 10 nm.

3. CONTACTS

The electrical contact made to a dielectric (either semiconductor or insulator) can drastically influence the dominant type of conductivity that occurs. Simmons [2] has classified the conduction processes within the dielectric as either bulk-limited or electrode-limited. For bulk-limited conductivity, the electrodes are largely irrelevant, in that the charge carriers are generated in the bulk, normally by thermal or optical excitation. In this case, the purpose of the electrodes is merely to supply a potential leading to a drift current in the bulk. In electrode-limited conductivity, the electrode/dielectric interface acts as a potential barrier to charge injection from the contacts, and becomes the limiting impedance to charge flow. Contacts are normally classified as ohmic or blocking, with an intermediate type known as a neutral contact. An ohmic contact has been defined by Rhoderick [12] as a contact, which has a sufficiently low resistance for the current to be determined by the resistance of the bulk semiconductor (or current-voltage characteristic of a device), rather than by the properties of the contact. In its most rudimentary form, the type of contact formed at a metal/dielectric interface depends on the relative values of the metal work function ψ_m and that of the dielectric (semiconductor or insulator) ψ_i. In equilibrium, the vacuum and Fermi levels of the metal and the dielectric must be continuous across the interface. If no potential is applied across the interface, the Fermi level must not only be continuous but also flat; otherwise a current would flow across the junction, contrary to experience. At a sufficient distance away from the interface within the bulk of the dielectric, the energy difference between the vacuum and Fermi levels must be equal to the bulk dielectric work function ψ_i. In the absence of surface states, the height of the potential barrier at the interface φ_0 is given by

$$\varphi_0 = \psi_m - \chi \tag{13}$$

where χ is the dielectric electron affinity. This expression was first stated explicitly by Mott [13], although Schottky [14] gave the first model for what has now become known as a Schottky barrier. The expression represents the difference between the Fermi level on the metal side of the junction and the bottom of the conduction band on the dielectric side. It is therefore a measure of the energy required to excite an electron in the metal into the dielectric conduction band. Although this expression is valid for both ohmic and blocking contact, we shall see later that φ_0 is only relevant in limiting the current in the latter case. In Figure 1 [2], energy band diagrams of both ohmic and blocking contacts are shown, in which the relationship of Eq. (13) is clearly evident.

It has been previously pointed out by Simmons [2] that the class of conductivity (bulk- or electrode-limited) is primarily governed by the type of contact, and not by the value of φ_0. We mainly consider conduction by electrons rather than by holes. In insulators [2] and in wide bandgap semiconductors [6], this is almost universally true owing to the lower hole mobilities and their immobilization by trapping effects. In any case, in the occasional cases of p-type conductivity, the conditions relating to the various workfunction values are reversed [12].

The various types of contact are illustrated in Figure 1 [2], together with the basic conditions for their formation in terms of the metal and dielectric workfunctions for the case of electron injection. Ohmic contacts occur in this case when $\psi_m < \psi_i$. Electrons are injected from the electrode into the dielectric in order to comply with thermal equilibrium requirements (equality of the Fermi levels). This is illustrated in Figures 1(a) and 1(b). Figure 1(a) shows the situation before the metal is in contact with the dielectric, where the vacuum levels of both materials are aligned. F represents the Fermi levels, and since the workfunction represents the difference between the vacuum level and the Fermi level for each material, the dielectric Fermi level is below that of the metal. The injected electrons penetrate a distance λ_0 into the dielectric conduction band, and constitute a negative space charge region, termed an accumulation region. An equal amount of positive charge accumulates on the surface of the electrode in order to satisfy the requirement of charge neutrality, and the interaction between these two sets of charges establishes a varying electric field within the accumulation region. Because the Fermi level represents an occupation probability, this must remain constant for any junction in the absence of an applied voltage. However, since the band structure of the dielectric must remain effectively unaltered in the bulk region, which is shielded from the electrode by the accumulation region, it therefore results in upward band bending of the conduction band edges. In this region, the accumulation region acts as a reservoir of charge, which can supply electrons to the dielectric as required by the bias conditions. The conduction process is limited by the rate at which electrons can flow into the bulk of the dielectric and not by the rate at which they are supplied by the electrode [2] and is therefore bulk-limited for ohmic contacts.

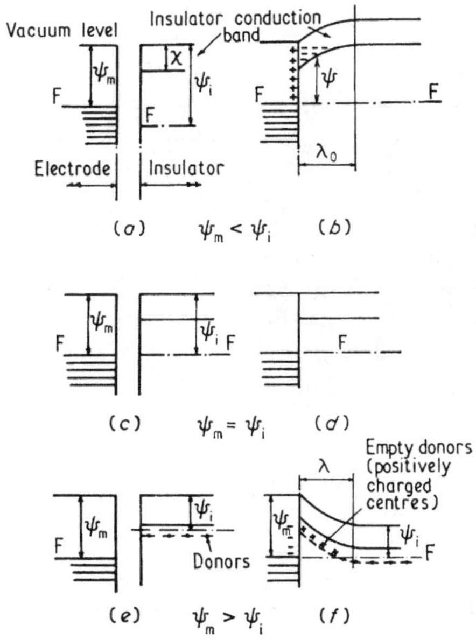

Figure 1. Contacts. Energy band diagrams showing the requirements and type of contacts for the case of an ohmic contact (a), (b), a neutral contact (c), (d) and a blocking contact (e), (f). F indicates the Fermi level. Reprinted with permission from [2], J. G. Simmons, *J. Phys. D: Appl. Phys.* 4, 613 (1971). © 1971, Institute of Physics Publishing Ltd.

Various workers have attempted to calculate the thickness of the space-charge region and derive expressions for the potential or electric field in situations where $\psi_m < \psi_i$. Perhaps the earliest was that of Mott and Gurney [15], who solved Poisson's equation (Eq. (12)) in one direction for the case of an electron density which decayed exponentially with the potential ψ, subject to boundary conditions including a constant, fixed electron density at the metal electrode, and a zero field in the bulk of the dielectric. This work established that for the space-charge region to be of a thickness up to 10 nm, the value of $\psi_m - \chi$ must be 0.1 eV or less. A later and extremely useful theoretical treatment was also provided by Simmons [16]. This assumed that the charge density in the space-charge region $\rho(x)$ decreases exponentially with the potential energy of the dielectric conduction band edge above the electrode Fermi level ψ such that

$$\rho(x) = eN_c \exp\left(-\frac{\psi}{kT}\right) \qquad (14)$$

where ψ is shown in Figure 1(b) [2], N_c represents the effective density of states in the dielectric conduction band, e is the electronic charge, k is Boltzmann's constant, and T is the absolute temperature. This is the starting point for the solution of Poisson's equation given in Eq. (12), which becomes

$$\frac{dF}{dx} = \frac{1}{e}\frac{d^2\psi}{dx^2} = -\frac{eN_c}{\varepsilon_r\varepsilon_0}\exp\left(-\frac{\psi}{kT}\right) \qquad (15)$$

Integrating with respect to x, subject to the boundary conditions that $d\psi/dx = 0$ at $\psi = \psi_i - \chi$ (*i.e.*, the electric field becomes zero at the boundary of the accumulation region) yields

$$\frac{d\psi}{dx} = \left(\frac{2e^2kTN_c}{\varepsilon_r\varepsilon_0}\right)^{1/2}\left[\exp\left(-\frac{\psi}{kT}\right) - \exp\left(-\frac{\psi_i - \chi}{kT}\right)\right]^{1/2} \qquad (16)$$

Integrating again, and using the boundary conditions that $\psi = \psi_m - \chi$ at $x = 0$ according to Eq. (13) and $\psi = \psi_i - \chi$ at $x = \lambda_0$ gives

$$\lambda_0 = \left(\frac{2kT\varepsilon_r\varepsilon_0}{e^2N_c}\right)^{1/2} \times \left\{\frac{\pi}{2} - \sin^{-1}\left[\exp\left(-\frac{\psi_i - \psi_m}{2kT}\right)\right]\right\}\exp\left(\frac{\psi_i - \chi}{2kT}\right) \qquad (17)$$

which reduces in the usual case where $\psi_i - \psi_m > 4kT$, to

$$\lambda_0 \approx \frac{\pi}{2}\left(\frac{2kT\varepsilon_r\varepsilon_0}{e^2 N_c}\right)^{1/2} \exp\left(\frac{\psi_i - \chi}{2kT}\right) \tag{18}$$

Expressions for the field at the interface ($x = 0$) and for the total space charge per unit area in the accumulation region are also derived in [16]. For a good ohmic contact, a value of $\lambda_0 = 16$ nm was calculated for $\psi_m - \chi = 0.1$ eV, reasonably close to the value predicted earlier [15]. For the case of an insulator with N_t shallow electron traps positioned at an energy E_t below the bottom of the conduction band, this expression is modified to [2]

$$\lambda_0 = \frac{\pi}{2}\left(\frac{2kT\varepsilon_r\varepsilon_0}{e^2 N_t}\right)^{1/2} \exp\left(\frac{\psi_i - \chi - E_t}{2kT}\right) \tag{19}$$

Calculations show that $\lambda_0 \geq 10$ nm for virtually all situations, except for very high trap concentrations. This implies that in nanostructures, where traps are very unlikely to be easily controllable, the accumulation region established at ohmic contacts will probably envelop the entire structure. This is likely to encourage SCLC as discussed in Section 6.

Blocking contacts occur in principle when $\psi_m > \psi_i$. Electrons flow from the dielectric to the metal electrode on contact, which is the opposite situation to that which takes place with ohmic contacts. The region where the electrons have been removed becomes a positive space-charge region, which is termed a depletion region because of the absence of majority carriers (electrons in this case). A corresponding negative charge is located on the electrode, and the interaction of these charges leads to an electric field in the depletion region. This situation is illustrated in Figure 1(e) and 1(f), and results in the conduction band edge in the dielectric bending downwards until it reaches a point where the Fermi level in the bulk of the dielectric lies ψ_i below the vacuum level, such that the region beyond the depletion region retains the bulk dielectric band structure. In this case, the current through the dielectric is limited by the rate of flow of electrons over the interfacial potential barrier whose height is given by Eq. (13). Note that there is no reservoir of mobile charge in the depletion region, as there is in the case of an accumulation region in an ohmic contact, and so the excitation of electrons over the potential barrier becomes the rate-limiting process. In this case, therefore, the conduction is electrode-limited. Such contacts are also frequently termed Schottky barriers, as much of the early theory concerning them was performed by Schottky [14, 17].

The analysis for blocking contacts, where $\psi_m > \psi_i$, is similar to the familiar analysis of *p-n* junctions in semiconductor theory. Mott and Gurney [15] derived an expression for the depletion region width, which is essentially the same as that currently used today, using what is known as the depletion approximation [12]. More sophisticated treatments of the depletion region at a blocking contact have been performed by various workers. A general expression for the barrier height was obtained by Cowley and Sze [4, 18], assuming an intimate contact and an interfacial layer between the metal and a semiconductor. The interfacial layer was supposed to be transparent to electrons and able to withstand a potential across it. Surface states were included in the analysis, whose density is a property of the semiconductor surface, but independent of the metal. Rhoderick [12] has given an exact analysis of the electric field in a Schottky barrier. Copious information concerning blocking contacts is given in the standard text on rectifying semiconducting contacts by Henisch [19]. For the purposes of the present discussion, the depletion approximation is sufficient, and we follow the treatment followed by Simmons [16]. It is assumed that the dielectric contains a concentration of donors N_d below the conduction band edge, as shown in Figure 1(f) [2]. Assuming the donors are uniformly distributed and completely ionized from the interface to the full extent of the depletion region, a depth λ_0, the space charge density in the depletion region is constant, that is, $\rho(x) = eN_d$, and Poisson's Eq. (12) becomes

$$\frac{d^2\psi}{dx^2} = \frac{e^2 N_d}{\varepsilon_r\varepsilon_0} \tag{20}$$

Integrating with respect to x gives

$$\frac{d\psi}{dx} = \frac{e^2 N_d}{\varepsilon_r\varepsilon_0}(x - \lambda_0) \tag{21}$$

where the integration constant is determined by the fact that $d\psi/dx = 0$ at $x = \lambda_0$. Integrating again and using the fact that $\psi = \psi_m - \chi$ at $x = 0$, as in the case of the ohmic contact, we obtain

$$\psi = \frac{e^2 N_d}{\varepsilon_r\varepsilon_0}\left(\frac{x^2}{2} - \lambda_0 x\right) + (\psi_m - \chi) \tag{22}$$

Finally, using $\psi = \psi_i - \chi$ at $x = \lambda_0$ we obtain the thickness of the depletion region

$$\lambda_0 = \left[\frac{2(\psi_m - \psi_i)\varepsilon_r\varepsilon_0}{e^2 N_d}\right]^{1/2} \tag{23}$$

Calculations using this expression [2] for $\psi_m - \psi_i = 2$ eV and $\varepsilon_r = 5$ show that N_d needs to be of the order of 10^{25} m^{-3} to reduce λ_0 to the order of 10 nm, and therefore the depletion region is likely to cover the whole nanostructure for undoped and moderately doped dielectrics.

A neutral contact is intermediate between an ohmic and a blocking contact. In this case, $\psi_m = \psi_i$ and the Fermi levels line up automatically on contact, with no flow of charge across the interface. In this case, neither an accumulation region nor a depletion region is established, and there is no band bending as shown in Figures 1(c) and 1(d) [2]. At low applied biases, such a contact behaves as if it is ohmic; however, at higher biases the maximum current that the electrode can supply is limited by the saturated thermionic (Richardson) current over the potential barrier, and ohmic conduction ceases.

The previous discussion has shown that, in principle, the type of contact formed depends on the relative values of ψ_m and ψ_i. A seriously complicating factor, however, is the existence of surface states. These are additional energy levels located at the surfaces, that is at interfaces, that are absent in the bulk of the dielectric. They are due to the effects of dangling bonds or impurities at the interface and to the curtailment of long-range order in this region [3]. Equation (13) does not apply directly in the presence of a significant number of surface states. In nanoelectronic structures, control of surface states will be of paramount importance. The existence of impurities may well be controlled using advanced assembly techniques, but the presence of dangling bonds will depend on the detailed surface chemistry, and the absence of long-range order will be felt internally as well as at interfaces, owing to the nanoscale dimensions. It is clear that surface state control, or even exploitation, will be a key feature in the development of nanoscale devices.

Such considerations are complicated by the fact that electronic devices (nano or otherwise) generally require at least two contacts. The interaction of the two contact regions has previously been considered for metal-dielectric-metal thin film systems in terms of the width of the accumulation or depletion regions and the dielectric layer thickness [2, 16]. When two ohmic contacts are provided, and the film thickness d is less than the total extent of the two accumulation regions (*i.e.*, $d < 2\lambda_0$ for zero-applied bias), there is insufficient charge within the accumulation regions to screen the interior of the dielectric from the conditions at the interfaces. In this case, the dielectric conduction band does not become flat in the interior as shown in Figure 1(b) [2], remaining curved throughout the dielectric; thus, an electric field exists throughout the entire dielectric, even in the absence of a bias. In contrast to the situation shown in Figure 1(b) [2], the highest point of the bottom of the conduction band is less than $\psi_i - \chi$ above the Fermi level. A space charge is therefore present throughout the dielectric and swamps the indigenous charge. Space-charge-limited conductivity, as described in Section 6, is thus a strong contender for the dominant conduction process under these circumstances. Indeed, it has been previously stressed that overlapping accumulation regions are required in order to observe good SCLC [2]. Such a situation is highly likely in the case of nanodevices, where the dimensions are small and the accumulation regions are likely to penetrate throughout the dielectric. Conversely, if $d > 2\lambda_0$, the accumulation regions do not overlap—the bulk of the dielectric is screened from the interfaces and absent of space charge. There is no electric field in the interior region, and it is the indigenous charge that keeps the bottom of the conduction band at a height $\psi_i - \chi$ above the Fermi level. However, as noted earlier, this situation is unlikely to occur in nanoscale devices. When two blocking contacts are provided, overlap of depletion (rather than accumulation) regions occurs if $d < 2\lambda_0$. When this takes place, the conduction band also bends, but in the opposite sense to that for two ohmic contacts. In this case, there is a positive space charge in the interior of the dielectric—an electric field is also present throughout and the bottom of the conduction band is higher than $\psi_i - \chi$ above the Fermi level. The dielectric is depleted of electrons throughout its extent and an electrode-limited conduction process will occur. As with the case of two ohmic contacts, the situation where $d > 2\lambda_0$ is improbable in nanodevices, although if it should occur the bottom of the conduction band would be $\psi_i - \chi$ above the Fermi level in the interior, and would be effectively screened from the surfaces.

An interesting and nontrivial effect takes place when two dissimilar neutral contacts are used with an insulator. If the workfunctions of the two electrodes are denoted by ψ_{m1} and ψ_{m2}, respectively ($\psi_{m2} > \psi_{m1}$), the difference in the potential barrier heights at the two electrodes is $\psi_{m2} - \psi_{m1}$. This difference gives rise to an intrinsic electric field F_{in} given by

$$F_{in} = \frac{\psi_{m2} - \psi_{m1}}{ed} \tag{24}$$

which is the result of a transfer of charge between the two electrodes. Electrons from the electrode of lower workfunction (electrode 1) transfer to the electrode with the higher workfunction (electrode 2), which gives a positive surface charge on electrode 1 and a negative surface charge on electrode 2. The intensity of this field can be substantial. For instance, if $(\psi_{m2} - \psi_{m1})/e = 1$ V and $d = 2$ nm, the value of $F_{in} = 5 \times 10^8$ V m^{-1} [2], and even for our limiting assumption that $d = 10$ nm, results in a field value of 10^8 V m^{-1}. When a voltage V is applied between the electrodes, the total field F in the dielectric is given by

$$F = F_{in} + \frac{V}{d} = \frac{(\psi_{m2} - \psi_{m1})/e + V}{d} \tag{25}$$

Thus, depending on the polarity of the applied voltage, the total field is either increased above or reduced below F_{in}. If the electrode of higher workfunction is negatively biased, the insulator will be able to withstand greater applied voltages before dielectric breakdown. This effect could be useful in the avoidance of dielectric breakdown in nanodevices. The intrinsic field can be determined by investigating the breakdown voltage of metal-insulator-metal junctions as a function of voltage bias [20].

It has been argued that in metal-dielectric-metal structures, the type of contact is given in principle by the relative values of the metal and dielectric workfunctions, while the potential barrier between the electrode Fermi level and the conduction band of the dielectric is given by Eq. (13). Where present, surface states can override these considerations and entirely determine the contact behavior. A space-charge region of negative (ohmic contact) or positive (blocking contact) charge is established within the dielectric. For the case of ohmic contacts, the accumulation region is likely to extend to a distance of at least 10 nm, and may be the dominant effect in determining the conduction process observed. Overlapping accumulation regions will lead to SCLC. For blocking contacts, the depletion region is also likely to extend across entire nanostructures, particularly for undoped and moderately doped dielectrics, and will lead to an electrode-limited process. Even neutral contacts can give rise to extremely high internal fields, without the application of applied voltages. It may be possible to exploit this effect to mitigate dielectric breakdown and failure mechanisms.

Within the last few years, several experiments have been performed concerned with the ultimate prospects of nanoscale devices, and in particular, the quantization of conductance in such devices. Of particular relevance is the use of quantum point contacts, which are normally metallic structures just a few nanometers wide; such contacts are formed by using a scanning tunneling microscope (STM) tip, which is gradually retracted from a metallic surface. Ohnishi et al. [21] have used a gold STM tip which was slowly withdrawn from a gold island. A gold bridge formed between the substrate and tip during withdrawal of the tip;

electron microscope images showed rows of gold atoms in this region. Parallel with the imaging, electrical measurements showed that the conductance was quantized in units of $2e^2/h$ ($\sim$13 kΩ^{-1}), where e is the electronic charge and h is Planck's constant. This value was observed to correspond to a single strand of gold atoms, while that of a double strand was twice as large. Yanson et al. [22] performed similar electrical measurements and determined the same quantized conductance values by making measurements of conductance as a function of electrode displacement. They were able to stretch a metallic wire of single gold atoms up to a distance of 2 nm. Contacts made to nanodevices will probably be larger than this, consisting of several parallel metallic chains, although quantum behavior is still probable. Particularly interesting is the fact that chains of metallic atoms, that is, what are presently termed nanowires, were originally proposed many years ago by Dearnaley et al. [8] in metal-insulator-metal structures which showed electroforming as described in Section 8. From this point of view, electroforming may be ascribed to a solid-state analogue of the nano-contact experiments, in which the metallic chains are established during the electroforming process owing to extremely high electric fields at asperities with a small radius of curvature, as suggested by Eq. (6). However, as yet there are no convincing images of such metallic filaments of nanometer dimensions in metal-insulator-metal structures. The stability of filaments (nanowires) in solid-state structures should be greater than in vacuum, owing to the rigidity of the structures and effective heat-conduction mechanisms. The quantum-mechanical effects described will inevitably become of overwhelming importance as nanostructures decrease in size towards their theoretical limit. Nevertheless, it is likely that the traditional contact properties discussed earlier will still remain relevant in the larger structures, while allowing for the fact that accumulation and depletion regions will dominate the conductivity.

4. THE SCHOTTKY EFFECT

The initial approach to deriving the J-V characteristics of metal-semiconductor or insulator-metal tunnel structures has been given in the form of a generalized current density equation by several workers [23–25]. Simmons [2] has used the form of this equation for the parabolic electron energy-momentum relation and the free-electron mass m in each of the three regions (*i.e.*, the cathode, the insulator or semiconductor, and the anode) as a starting point to obtain both the Schottky-effect current density and the tunneling current density. The form used is [2, 25]

$$J = \frac{4\pi em}{h^3}\int_0^\infty dE(f_c(E) - f_a(E))\int_0^E P(E_x)\,dE_x \quad (26)$$

where e is the electronic charge, h is Planck's constant, $P(E_x)$ is the electron transmission probability across the dielectric, and E_x is the "x-directed" electron energy. f_c and f_a represent the Fermi-Dirac distribution function for electrons in the cathode and the anode, respectively:

$$f_c = \frac{1}{1+\exp\left(\frac{E-E_f}{kT}\right)} \quad \text{and} \quad f_a = \frac{1}{1+\exp\left(\frac{E+eV-E_f}{kT}\right)} \quad (27)$$

where k is Boltzmann's constant, E is the electron energy, E_f is the Fermi energy of the cathode, and V is the voltage difference between the cathode and anode.

Providing the potential barrier is too thick to allow tunneling (Section 5) to occur, and also that the concentration of allowed states in the forbidden gap of the dielectric is too small to allow impurity conduction (or hopping) to occur, the current that flows through the dielectric is limited by the rate at which electrons are thermally excited over the interfacial potential barrier from the Fermi level in the cathode to the bottom of the dielectric conduction band, φ. Simmons [2] has argued that the current density expression in this case can be derived from Eq. (26) by replacing the lower limit in the integral over dE_x by φ and the transmission function $P(E_x)$ by 1. On performing the integration, the following expression is obtained:

$$J = \frac{4\pi mek^2T^2}{h^3}\exp\left(-\frac{\varphi}{kT}\right) \equiv AT^2\exp\left(-\frac{\varphi}{kT}\right) \quad (28)$$

This expression is known as the Richardson saturated thermionic emission equation when φ is independent of the applied voltage, so J depends only on the barrier height and the temperature T. A is called the Richardson constant, whose theoretical value is 1.2×10^6 A m^{-2} K^{-2}.

Equation (28) can also be derived using slightly different (but equivalent) approaches [3, 4, 12], often in connection with the behavior of Schottky-barrier diodes. An example of the classic derivation is given by Levine [26], and considers electrons to be emitted only in the x-direction. Assuming an electron's energy to be entirely kinetic, this leads to the inequality

$$v_x > \left(\frac{2(E_f+\varphi)}{m}\right)^{1/2} \quad (29)$$

where v_x is the velocity component along the positive x-direction. The problem is basically that of calculating the number of electrons at a given temperature which have the required velocities. The derivation starts from the general expression for the number of electrons dn with a velocity in the interval v to $v + dv$

$$dn = \rho(v)f(v)\,dv_x\,dv_y\,dv_z \quad (30)$$

where $\rho(v) = (2m/h)^3$ is the density of states for electrons, $f(v)$ is the Fermi-Dirac occupation probability which is f_c of Eq. (27), and v_x, v_y, and v_z are the three velocity components. The total energy of an electron E is

$$E = \frac{1}{2}m\left(v_x^2 + v_y^2 + v_z^2\right) \quad (31)$$

The total current density is obtained by multiplying Eq. (30) by ev_x, and integrating over v_x from the lower limit given by Eq. (29) to ∞, and over v_y and v_z from $-\infty$ to $+\infty$; Eq. (28) results from this integration.

The above discussion and Eq. (28) imply that the current density emitted into the dielectric is not field-dependent. However, this is not true, owing primarily to the image-force effect. When an electron is emitted, the metal surface at the interface becomes positively charged (polarized) as a result of the loss of the electron. Therefore, there is an attractive force between the electron and the polarized electrode, with a corresponding electric field established. The electric field is directed perpendicular to the electrode, and under these conditions, the field and the potential can be calculated as if there were a positive charge located at a mirror image position behind the electrode. If the electron is a distance x in front of the electrode, the positive image charge is a distance x behind it, so that charges are separated by a perpendicular distance $2x$. Coulomb's law then allows us to calculate the magnitude of force on the electron as $e^2/4\pi\varepsilon_r\varepsilon_0(2x)^2 = e^2/16\pi\varepsilon_r\varepsilon_0 x^2$ towards the surface of the electrode, where ε_r is the high-frequency value since the electron spends only a very short time near the cathode during emission. Since this force is attractive, its sign is negative. The potential energy φ_{im}, relative to that of an electron at infinity due to the image force, then becomes

$$\varphi_{im} = -\frac{e^2}{16\pi\varepsilon_r\varepsilon_0 x} \tag{32}$$

This expression is not valid *at* the electrode surface, since it becomes $-\infty$ at that point, but it is considered to be applicable when $x > x_0$, and the potential is assumed to be linear for $x < x_0$ [16, 27]. This technicality need not concern us here. Thus, when the image force is taken into consideration, the potential barrier that an electron needs to overcome in order to be emitted into the dielectric conduction band is, for $x > x_0$, given by $\varphi_0 - \varphi_{im}$. When an electric field F is applied, it conspires with the image force to lower the potential barrier by adding a further third term $-eFx$ to the potential barrier term. The potential step $\varphi(x)$ with an electric field applied, and in the presence of an image force, then becomes

$$\varphi(x) = \varphi_0 - \frac{e^2}{16\pi\varepsilon_r\varepsilon_0 x} - eFx \tag{33}$$

for a neutral contact. The case for a blocking contact has been covered by Simmons [2], but will not be discussed further here. This expression clearly depends on x and has a maximum value. Providing an electron has sufficient energy to surmount this maximum, it is emitted into the dielectric. Differentiation shows that this maximum occurs at

$$x_m = \left(\frac{e}{16\pi\varepsilon_r\varepsilon_0 F}\right)^{1/2} \tag{34}$$

and that the maximum value of $\varphi(x) = \varphi_0 - (e^3/4\pi\varepsilon_r\varepsilon_0)^{1/2}F^{1/2}$. The effective barrier height φ is then field-dependent and given by

$$\varphi = \varphi_0 - \Delta\varphi_S \tag{35}$$

where

$$\Delta\varphi_S = \beta_S F^{1/2} \tag{36}$$

represents the amount of field lowering of the potential barrier, and β_S is called the Schottky field-lowering coefficient given by

$$\beta_S = \left(\frac{e^3}{4\pi\varepsilon_r\varepsilon_0}\right)^{1/2} \tag{37}$$

The current density in the presence of a field is then obtained by substituting Eqs. (35) to (37) into Eq. (28) to yield

$$J = AT^2 \exp\left(-\frac{\varphi_0}{kT}\right)\exp\left(\frac{\beta_S F^{1/2}}{kT}\right) \tag{38}$$

This expression was first obtained by Schottky in 1914 and is known as the Richardson-Schottky equation [28]. Providing the electric field applied across the dielectric is reasonably uniform, the J-V characteristic is then

$$J = AT^2 \exp\left(-\frac{\varphi_0}{kT}\right)\exp\left(\frac{\beta_S V^{1/2}}{kTd^{1/2}}\right) \tag{39}$$

where d represents the dielectric thickness. Thus, Schottky emission is indicated by a linear plot of $\ln J$ against $V^{1/2}$ and the gradient of such a plot is determined by β_S, which in turn depends on the permittivity of the dielectric.

The Schottky effect has been observed in a wide range of insulators and semiconductors, both organic and inorganic. Emtage and Tantraporn [29] observed this mechanism in Al-Al_2O_3-Al and Al-GeO_2-Al structures, although some of their work was also performed on gold films in contact with polymerized silicon oil. Pollack [30] observed Schottky emission through 30 nm-thick reactively evaporated Al_2O_3 layers having Pb electrodes. In addition to observing a $\ln I \propto V^{1/2}$ dependence at fixed temperatures of 241 K and 273 K, the temperature dependence of the phenomenon was also investigated, and it was shown that gradients of the curves over a range of temperatures increased linearly with reciprocal temperature $1/T$ as also predicted by Eq. (39). Below 235 K, the temperature dependence of the gradient was minimal, and it was therefore suggested that tunneling was dominant at lower temperatures. In Ta-Ta_2O_5-Au samples prepared by electron beam evaporation and plasma oxidation [31], Schottky emission was observed when the Ta electrode was biased negatively, and tunneling when it was biased positively. More recently, Schottky emission has been observed in many insulating films having various different metal contacts.

The Schottky effect has also been reported in organic metal phthalocyanine (MPc) films, which are invariably p-type semiconductors [5]; the Schottky effect occurs for hole injection at the blocking contact. Examples of such systems are Al-H_2Pc-Al [32] and Al-CuPc-Al [33]. In CuPc films [34] and in ZnPc films [35] with at least one Al electrode, the Schottky effect was also identified, but was attributed to an aluminum oxide interfacial layer at the interface. Wilson and Collins [36] found behavior consistent with Eq. (39) for planar CuPc films with Al electrodes, deducing a depletion region width of 96 nm. Hassan and Gould [37] found experimental values of the field-lowering coefficient several times higher than the theoretical value of $\beta_S = 2 \times 10^{-5}$ eV m$^{1/2}$ V$^{-1/2}$ in CuPc films with a positively biased Al electrode. However, this was reconciled by adopting the explanation of Wilson and Collins [36] that most of the applied voltage is dropped over a depletion region of thickness d_S and assuming the theoretical value of β_S. Values of $d_S = 120$ nm and $\varphi_0 = 0.88$ eV were determined for an annealed sample. For hole injection from the Pb electrode in Au-CuPc-Pb structures, enhanced values of the field-lowering coefficient were similarly found, and were satisfactorily explained in terms of a Schottky barrier region at the injecting electrode [38]. In triclinic PbPc films with one Al electrode and one Au electrode, both Ahmad and Collins [39] and Shafai and Gould [40] used a similar Schottky barrier analysis to derive consistent values of the Schottky barrier width of 50 nm with slightly different barrier heights of 1.11 eV [39] and 1.0 eV [40].

In films of the compound semiconductor CdS, Zuleeg [41] used dissimilar ohmic In contacts and Au blocking contacts. Ohmic conductivity was observed for both polarities at low voltages, but under reverse bias (Au injecting electrode), a field-lowering conduction process was observed, as shown in Figure 2 [41], where the logarithm of the current is proportional to the square root of the applied voltage. This figure is typical of a sample showing the Schottky effect. The behavior could, in principle, be either the Schottky effect or the Poole–Frenkel effect (Section 7), but Schottky emission was indicated in this case using differential capacitance measurements, giving a Schottky barrier height of 0.88 eV, which is close to that obtained for gold contacts to grown CdS single crystals. Field-lowering behavior has also been observed in cadmium arsenide (Cd_3As_2) films having a bottom Ag electrode and a top electrode of either Ag, Al, or Au [42]. The simple barrier-height model [13, 14] is unlikely to be applicable in this case, owing to the very low bandgap of Cd_3As_2 [6], but there was a good correlation with the experimental field-lowering coefficient and the value of β_S predicted by Eq. (37) for the relative permittivity value of 12 for Cd_3As_2. There appeared to be a gradual transition from the Schottky effect to the Poole–Frenkel effect as the film thickness increased, and the effect of the injecting electrode/Cd_3As_2 interface became less significant.

The basic Schottky emission process, under the influence of an applied field as given by Eq. (39), has been observed in many metal/dielectric systems. As will be seen in Section 7, a similar J-V dependence is also observed for the Poole–Frenkel effect, although in principle it is possible to distinguish between these on the basis of the different expressions for the field-lowering coefficient. However, in practice, this is often inconclusive, and it is normally better to first determine the type of contact; if blocking, then the Schottky effect is probably responsible and if ohmic the conductivity is likely to be via the Poole–Frenkel effect. The existence of a Schottky barrier can be determined using differential capacitance measurements [41], and Poole–Frenkel conduction, being bulk-limited, is not sensitive to the electrode species. Hence, experiments using a range of contacts with different workfunctions can be used to distinguish between these two effects. In some cases, the Schottky effect is influenced by an interfacial oxide layer [34, 35], while in some organic materials the field-lowering coefficient is somewhat higher than the theoretical value, implying that most of the applied voltage is dropped over a depletion region [36–40].

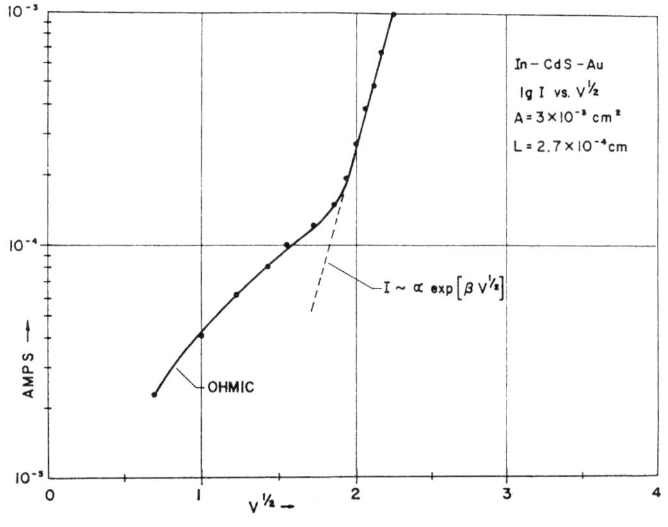

Figure 2. The Schottky effect. Reverse current measurements (Au negative) in an evaporated Au-CdS-In diode of thickness 2.7 μm showing the Schottky effect above 4 V. Reproduced with permission from [41], R. Zuleeg, *Solid State Electron.* 6, 645 (1963). © 1963, Elsevier Science.

A good indication as to whether the Schottky effect may be observed in nanoscale devices may be obtained from Eq. (34), which gives the dependence of the position of the maximum in the potential barrier height x_m below the metal/dielectric interface. Simple calculations show that for $\varepsilon_r = 4$ (a typical insulator), this will occur at a depth of approximately 3 nm for a field of 10^7 V m^{-1} and 1 nm for a field of 10^8 V m^{-1}. Higher values of ε_r, such as those applicable to most semiconductors, reduces the value of x_m still further. For all practical purposes, x_m will generally be less than 10 nm from the interface and the Schottky effect should therefore be observable in most metal/dielectric combinations having the appropriate workfunction/electron affinity combinations. However, tunneling through the potential barrier, rather than field-assisted emission over it, may well be more probable, particularly in structures with dimensions less than 10 nm.

5. TUNNELING

Tunneling is an extremely important conduction process that follows directly from the quantum-mechanical description of particles, such as electrons, in terms of a wave-function $\psi(x)$. The product $\psi(x)\psi^*(x)\,dx$ represents the probability of finding an electron with wave-function $\psi(x)$, whose complex conjugate is $\psi^*(x)$, in the range x to $x + dx$. Effectively, this means that an electron can tunnel through a potential barrier, without the need to surmount the potential barrier as in the case of the Schottky effect. However, for practical purposes, the value of $\psi(x)\psi^*(x)\,dx$ is nonzero only for very small values of x, that is, those less than a maximum of about 10 nm. Consequently, the tunneling process only takes place across very narrow potential barriers.

Early work on tunneling continued on from the initial development of quantum mechanics in the 1920s. The first work of relevance in the present context is that of Fowler and Nordheim [43] who derived an expression (the Fowler–Nordheim equation) for the tunnel emission current from a metal in a high electric field. This equation is still directly relevant to tunnel currents for relatively large applied voltages, and appears as the limiting high-voltage case in the comprehensive model of tunneling developed by Simmons [44]. Expressions for the tunnel current between two metal electrodes were first derived by Frenkel [45]. Other major contributions to the theory of tunneling were by Sommerfield and Bethe [46] and Holm [47], who derived expressions for low and high voltages, and for intermediate voltages, respectively. Stratton [25] expressed the tunneling current density in terms of Eq. (26) and derived expressions for an arbitrary-shaped potential barrier, while Simmons [44] derived a series of expressions applicable over four alternative voltage ranges.

The approach given below follows that of the Simmons model. The starting point is effectively Eq. (26) and the Wentzel–Kramer–Brillouin (WKB) approximation for the tunneling probability $P(E_x)$, an approach also followed by Stratton [25]. The WKB approximation is [44, 48]

$$P(E_x) = \exp\left\{-\frac{4\pi}{h}\int_{s_1}^{s_2}[2m(\varphi(x) - E_x)]^{1/2}\,dx\right\} \tag{40}$$

where $\varphi(x)$ represents the varying height of the potential barrier and s_1 and s_2 represent the limits of the barrier at the Fermi level, with the width of the barrier $\Delta s = s_2 - s_1$. In view of the complexity of the mathematical derivation, it is not appropriate to give it fully here. The full derivation of the J-V characteristics is given by Simmons [44], and summaries have also been given elsewhere [2, 3, 16]. The generalized tunneling characteristic, applicable for all voltage ranges and barrier shapes, becomes

$$J = J_0\{\bar{\varphi}\exp(-A\bar{\varphi}^{1/2}) - (\bar{\varphi} + eV)\exp[-A(\bar{\varphi} + eV)^{1/2}]\} \tag{41}$$

where

$$A = \frac{4\pi\beta\Delta s(2m)^{1/2}}{h} \quad \text{and} \quad J_0 = \frac{e}{2\pi h(\beta\Delta s)^2} \tag{42}$$

In these expressions, $\bar{\varphi}$ represents the mean barrier height above the Fermi level of the negatively biased electrode and β is a function of barrier shape, which is approximately unity.

Equation (41) is particularly useful in that it can be applied to a potential barrier of any shape, provided the mean barrier height $\bar{\varphi}$ and the barrier width Δs are known. It may be interpreted as the difference between two currents, one flowing from the first to the second electrode, and the other in the reverse direction. Simmons applied Eqs. (41) and (42) to the case of a rectangular potential barrier, as considered in earlier work [46, 47]. It is thus relatively simple to calculate $\bar{\varphi}$ and Δs in this case. For very low applied voltage ($V \approx 0$), $\bar{\varphi} \approx \varphi_0$, the energy difference between the Fermi level of the negative electrode and the bottom of the dielectric conduction band as given by Eq. (13), and $\Delta s = d$, the electrode spacing. The J-V characteristic is then described by

$$J = \left(\frac{3e^2}{2h^2 d}\right)(2m\varphi_0)^{1/2} V \exp\left[-\frac{4\pi d}{h}(2m\varphi_0)^{1/2}\right] \tag{43}$$

Thus, at very low voltages, Ohm's law is followed. For intermediate voltages, ($V < \varphi_0/e$), $\bar{\varphi} \approx \varphi_0 - eV/2$, $\Delta s = d$ and we obtain

$$J = \left(\frac{e}{2\pi h d^2}\right) \times\left\{\left(\varphi_0 - \frac{eV}{2}\right)\exp\left[-\frac{4\pi d}{h}(2m)^{1/2}\left(\varphi_0 - \frac{eV}{2}\right)^{1/2}\right] - \left(\varphi_0 + \frac{eV}{2}\right)\exp\left[-\frac{4\pi d}{h}(2m)^{1/2}\left(\varphi_0 + \frac{eV}{2}\right)^{1/2}\right]\right\} \tag{44}$$

For high voltages ($V > \varphi_0/e$), $\bar{\varphi} \approx \varphi_0/2$ and $\Delta s = d\varphi_0/eV$. In this case, the value of the factor $\beta = 23/24$ and the J-V characteristic becomes

$$J = \left(\frac{2.2e^3V^2}{8\pi h\varphi_0 d^2}\right) \times\left\{\exp\left[-\frac{8\pi d}{2.96heV}(2m)^{1/2}\varphi_0^{3/2}\right] - \left(1 + \frac{2eV}{\varphi_0}\right) \times\exp\left[-\frac{8\pi d}{2.96heV}(2m)^{1/2}\varphi_0^{3/2}\left(1 + \frac{2eV}{\varphi_0}\right)^{1/2}\right]\right\} \tag{45}$$

For very high voltages ($V > (\varphi_0 + E_f)/kT$), the Fermi level of the second electrode lies below the bottom of the conduction band of the first electrode, and electrons cannot tunnel from the second to the first electrode. The second term in Eq. (45) becomes negligible and we obtain

$$J = \frac{2.2e^3V^2}{8\pi h\varphi_0 d^2}\exp\left[-\frac{8\pi d}{2.96heV}(2m)^{1/2}\varphi_0^{3/2}\right] \tag{46}$$

The last equation is similar to the Sommerfield–Bethe [46] relationship for high voltages, except for the factor of 2.2 and the replacement of 3 by 2.96 in the denominator of the exponential. These differences were considered negligible, in consideration of the dominant influence of the exponential term. Equation (46) is also substantially equivalent to the well-known Fowler–Nordheim equation [43], which has been observed many times in the literature.

An interesting point regarding tunneling is that there is no significant temperature-dependence in any of the four voltage ranges. This is a distinctive feature of tunneling, which allows it to be distinguished from other conduction processes, such as the Schottky effect. There is, in fact, a slight quadratic dependence, with the current density at a temperature T, $J(T)$, related to its value at 0 K, $J(0)$, by the approximate expression [2] $J(T) = J(0)(1 + \alpha T^2)$. For typical values of $\bar{\varphi} = 2$ eV and $\Delta s = 2$ nm, the value of $\alpha = 6 \times 10^{-7}$, and there is little difference between the current density at absolute zero and at room temperature.

Tunneling has been observed in a wide variety of structures. Fisher and Giaever [49] made measurements of currents in Al-Al_2O_3-Al structures produced by oxidation, and compared their J-V characteristics with the tunneling model of Holm [47]. There was good qualitative agreement, with the films showing ohmic conduction at low voltages and an exponential dependence of current on voltage at higher voltages. However, they were only able to obtain quantitative agreement by assuming the electron effective mass in the insulator to be about 1/9 of the free electron mass; later, Mead [50] showed that the results were actually influenced by electric field penetration into the electrodes. Mayerhofer and Ochs [51] investigated very thin films of Al_2O_3 and BeO, again finding ohmic and faster-than-exponential dependences of current on voltage at lower and higher voltages, respectively. Tunneling was thought to be directly from metal to metal electrode at low voltages and directly into the insulator conduction band at higher voltages. Pollack and Morris [52] also made measurements on thermally grown Al_2O_3 films on Al; however, they used a wide variety of different evaporated counter electrodes. They compared their measurements with the model of Simmons [44] described above for similar metal electrodes, and in view of their varying counter electrodes, also with an extended model version of the Simmons model for dissimilar electrodes [53]. They found a remarkable correlation with the model at room temperature over nine decades of current, thus substantially verifying the model. Lenzlinger and Snow [54] later studied tunneling in thermally grown SiO_2 films on Si of a thickness 65–500 nm. The electrodes were of either Al or Ag. Since the films were relatively thick, it was necessary to apply high voltages to reduce the Fermi-level barrier thickness to that suitable for tunneling. Thus, the conditions for Eq. (46) were essentially fulfilled. Data were plotted in the form of a Fowler–Nordheim plot, $\log(J/V^2)$ against $1/F$ (or d/V) and were shown to give excellent linear behavior over five orders of magnitude of current. The measured currents were somewhat lower than predicted by the theory, probably due to trapping effects, and the slopes of the curves implied effective mass values of $0.48m$ for Ag, $0.39m$ for Al, and $0.42m$ for Si, where m is the free electron mass. Since these early experimental verifications of the tunneling models, it has been observed in many insulating systems.

Although tunneling is normally associated with current flow through insulators, it may also be observed in semiconductors. Perhaps the prime example of this at present is the tunnel diode, which consists of a *p-n* junction in which both semiconducting regions are degenerate, and the Fermi level is located within one of the allowed bands, rather than in the forbidden band. Tunneling can occur across the depletion region at any point where there are occupied energy levels on one side of the junction and unoccupied levels on the other side. The varying alignment of such regions as the applied voltage varies, changes the probability of the tunneling process, and leads to a differential negative resistance region in the *J-V* characteristic, which is useful in oscillators and other high-frequency applications. The theory and experimental characteristics of contemporary tunnel diodes is fully described in the literature [4]. Due to the requirement for semiconductor doping, *p-n* junctions of a similar type are unlikely to be fabricated as nanostructures, and will not be further discussed. In copper phthalocyanine (CuPc) thin films with indium electrodes, it was observed that a very sharp increase in the current density occurred when an applied voltage of approximately 1 V was exceeded [55]. However, this feature only occurred when the substrate electrode was biased positive, notwithstanding the fact that both electrodes were of the same metal. It had been suggested earlier by Wilson and Collins [36] that oxidation of indium contacts during deposition was likely to occur, and that tunneling might therefore take place. Although it was considered unlikely that such an oxide layer would exceed a thickness of about 1.5 nm, and therefore would be unlikely to dominate the overall conductivity, it was suggested that such a layer may modify the character of the In/CuPc interface, allowing tunneling to occur to the valence band, since CuPc is *p*-type. Simmons [56] had proposed that a modified Fowler–Nordheim expression may be applicable for tunneling through the interfacial barrier when the electric field was sufficiently high to reduce the barrier width to about 5 nm. Data in this region were therefore analyzed in terms of a Fowler–Nordheim plot ($\log(I/V^2)$ against $1/V$), which according to Eq. (46) gives a linear dependence. This is illustrated in Figure 3 [55], and is typical of plots obtained with Fowler–Nordheim tunneling. The figure shows two curves, the first for a simple In-CuPc-In sample (A) and the second for a sample that also incorporated an "edge-thickening" insulating layer to avoid localized high fields and a potential dielectric breakdown originating at the edge of the bottom electrode (B). Assuming a tunneling thickness of 5 nm, barrier heights of 0.36 eV and 0.27 eV were determined for the samples of type A and B, respectively. The current density drawn was, however, considerably lower than that predicted by Eq. (46). This may be due to two effects: first, the effective tunneling area may be a factor of as low as 10^{-5} times the geometric area [57]; second, Geppart [58] has shown that in the presence of space charge and traps there is a

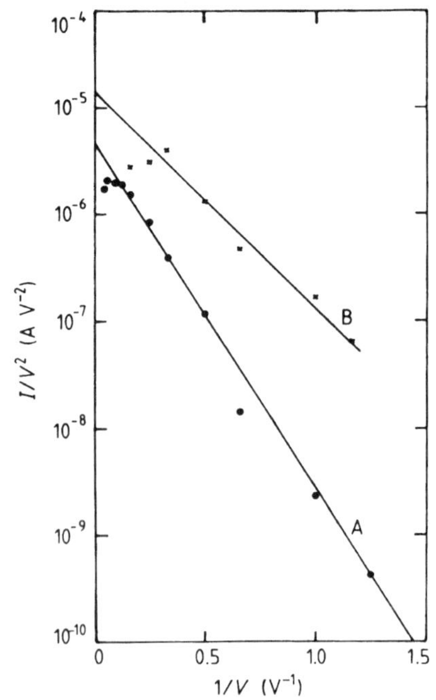

Figure 3. Tunneling. Fowler–Nordheim plots for evaporated In-copper phthalocyanine-In sandwich structures of thickness 1.2 μm. Type A samples were simple sandwich structures, whereas type B samples had an additional insulating "edge-thickening" layer to avoid preferential dielectric breakdown. The curves yield barrier heights of 0.36 eV (type A) and 0.27 eV (type B) assuming tunneling thicknesses of 5 nm. Reproduced with permission from [55], A. K. Hassan and R. D. Gould, *J. Phys. D: Appl. Phys.* 22, 1162 (1989). © 1989, Institute of Physics Publishing Ltd.

significant decrease in the current density. A further example of tunneling in semiconductors is given by Canali et al. [59] for CdTe who found a linear dependence of the logarithm of the drift mobility on inverse electric field ($1/F$) for $F > 1.2 \times 10^6$ V m^{-1}, which was consistent with tunneling.

The theory of tunneling was developed following the initial development of quantum mechanics. Since the Fowler–Nordheim expression was first derived [43], the tunneling theory has been extended to cover a wide range of potential barriers, including those of arbitrary shape with similar [44] and dissimilar [53] electrodes. There is excellent agreement between experiment and theory for insulators, and tunneling behavior is also observed in both organic and inorganic semiconductors, in addition to the well-known case of tunneling across the depletion region in the degenerate semiconductor junctions of tunnel diodes.

In nanostructures, it is unlikely that semiconductor doping control will initially prove sufficient for the fabrication of useful devices. However, thin insulators of nanoscale dimensions are likely to be utilized in nanoelectronic devices as carrier tunnel injectors. In view of the low thickness, dielectric breakdown will probably be a problem, unless voltage levels are restricted, and Ohm's law behavior as predicted by Eq. (43) is probable. The type of tunneling observed is unlikely to be well-described by a rectangular potential barrier, and the theory needs to encompass nanoscale dimensions in more than one direction. As devices become smaller, the probability of direct tunneling into the allowed band of a dielectric will probably be greater than that of Schottky emission, particularly for grown oxides, which may be only a few nanometers thick.

6. SPACE-CHARGE-LIMITED CONDUCTION

When there is an appreciable space-charge concentration in a dielectric, carriers passing through the space-charge region are affected by it in a similar way to electrons emitted from the cathode in a vacuum tube (Child's Law). The first derivation of the J-V characteristic for space-charge limited currents in solids was performed by Mott and Gurney [14] in 1940. They assumed that the space-charge region extended throughout the dielectric; furthermore, the effects of traps were not considered in the original theory. In Mott and Gurney's derivation, the effect of carrier diffusion is considered, but discarded later in the derivation as negligible. Lamb [3] followed the earlier derivation, but included the effect of dielectric permittivity. The equivalent, but simpler, derivation of Simmons [16] ignores diffusion from the outset, but includes the permittivity of the dielectric. This approach is also followed here.

The total current density is given by

$$J = \rho(x)\mu F \tag{47}$$

where $\rho(x)$ is the space-charge density, μ is the carrier mobility, and F is the electric field. If the effects of diffusion currents are also included, there is an additional term $-De(dn/dx)$ added to the right-hand side of the equation, where D is the diffusion coefficient and n is the number of carriers per unit volume. Using Poisson's equation to substitute from Eq. (12) for $\rho(x)$, we obtain

$$J = -\mu\varepsilon_r\varepsilon_0 F\frac{dF}{dx} \tag{48}$$

Integrating, subject to the boundary conditions that $F = 0$ at $x = 0$, gives

$$F = -\frac{dV}{dx} = \left(\frac{2Jx}{\mu\varepsilon_r\varepsilon_0}\right)^{1/2} \tag{49}$$

Further integration, subject to the boundary conditions that V is zero at $x = 0$ and V at $x = d$ (at the opposite side of the dielectric), yields

$$J = \frac{9}{8}\varepsilon_r\varepsilon_0\mu\frac{V^2}{d^3} \tag{50}$$

This classic SCLC expression predicts that the current density is proportional to the square of the applied voltage and also inversely proportional to the cube of the dielectric thickness. Both these variations have been observed in structures showing SCLC. However, Eq. (50) also predicts that there is no temperature-dependence, and implies values of the current density far in excess of those observed experimentally. This discrepancy can be resolved by discounting the trap-free SCLC expression and considering the situation where traps are present in the dielectric. This was first considered by Rose [60], who considered the case of a set of shallow traps of concentration N_t positioned at an energy E_t below the bottom of the conduction band. Using Fermi-Dirac statistics, it is simple to show that the ratio of free-to-trapped charge is given by

$$\theta = \frac{N_c}{N_t}\exp\left(-\frac{E_t}{kT}\right) \tag{51}$$

The current is then carried only by the free charge and the current density in Eq. (50) is reduced by the factor θ to give [16, 60]

$$J = \frac{9}{8}\varepsilon_r\varepsilon_0\mu\theta\frac{V^2}{d^3} \tag{52}$$

This has essentially the same form as Eq. (50). The factor θ, however, has the effect of reducing the current density value below that for the trap-free case. Depending on the circumstances, θ may have a value of less than 10^{-5}. A second feature that becomes apparent is that there is an exponential dependence of θ on $1/T$, that is, θ is thermally activated, and therefore in the presence of shallow traps, there is a $J \propto \exp(-E_t/kT)$-dependence. Thus, the trap depth may be determined from a plot of $\ln J$ against $1/T$ at a constant voltage and the trap concentration calculated from Eq. (51).

The conductivity in samples containing traps distributed in energy was also tackled by Rose [60], who considered both traps distributed exponentially and uniformly. The first case was shown to follow a $J \propto V^n$-dependence, where $n \geq 2$, and the second to follow a $J \propto \exp(\alpha V)$-dependence, where α is a constant. For the exponential trap distribution, where the trap concentration per unit energy range at an energy E

below the conduction band edge $N(E)$, may be represented by

$$N(E) = N_0 \exp\left(-\frac{E}{kT_t}\right) \quad (53)$$

where N_0 is the trap concentration at the conduction band edge and T_t is a temperature parameter describing the distribution, Lampert [61] has derived a J-V characteristic given by

$$J \sim e\mu N_c \left(\frac{\varepsilon_r \varepsilon_0 V}{eN_0 kT_t}\right)^l \frac{V^{l+1}}{d^{2l+1}} \quad (54)$$

for n-type conductivity, where l represents the ratio T_t/T. A similar expression is applicable for p-type conductivity [5]. There is, therefore, a power-law dependence of J on V, with $J \propto V^n$, where $n = l+1$. The dependence on thickness is $J \propto d^{-(2l+1)}$ and the temperature-dependence is via the power law, since $l = T_t/T$. Such dependencies have been widely observed in the literature [61]. J-V characteristics corresponding to a uniform trap distribution have been rarely observed, and will not therefore be further discussed.

It should be noted that at low voltages, Ohm's law is almost universally obeyed, with a transition to SCLC at a transition voltage V_t. Derivations of expressions for V_t, both for the trap-free case and for the case of a single shallow trap level are given by Lampert [61]. The transition voltage between ohmic conduction and SCLC for the case of an exponential trap distribution has also been derived [62]. Lampert [61] has discussed the situation where all the traps become filled, and the J-V characteristic then follows the trap-free law above a voltage V_{TFL}, corresponding to the trap-filled limit. It has also been shown that for SCLC dominated by an exponential trap distribution, measurements of J as a function of temperature at constant applied voltage in the SCLC region may be used to determine the mobility and trap concentration [63]. This is a useful method where the mobility is low and Hall-effect measurements are difficult to perform. The case of double injection (holes from one electrode and electrons from the other) has been considered, and leads in some cases to negative resistance behavior [3, 61]. However, these references to the theory and practice of SCLC currents are beyond the scope of this discussion, and will not be referred to further. Additional information on many of these aspects of SCLC is given in the standard text of Lampert and Mark [64].

Space-charge-limited conduction has been observed in some insulators and in many different semiconductors. Much of the earlier work was performed on CdS, which has been described as the workhorse of the SCLC field [61]. Much of this was performed on thin single crystals of the material [65–67]. In this early work, there was clear evidence of a shallow-trap square-law, followed by a rapid current rise over several orders of magnitude at the trap-filled limit. Amorphous selenium was shown to exhibit SCLC when provided with hole-injecting Te contacts [68]. In Sb_2S_3 dielectric films with bismuth ohmic contacts, Budinas et al. [69] observed SCLC having square-law behavior. In accordance with the theory, the current was initially ohmic before showing the square-law dependence. At high voltages, a rapidly increasing current was evident, suggestive of trap filling. Furthermore, V_t was found to be proportional to d^2, and J was found to be proportional to d^{-3} at constant voltage. All of these observed dependences implied SCLC and have since been confirmed in many different materials.

The phthalocyanines have been particularly fruitful in exhibiting SCLC. Perhaps the earliest measurements showing SCLC in these organic materials are those of Delacote et al. [70] on Al-CuPc-Au structures, where at low voltage an ohmic hole current was observed, followed by exponential behavior implying a uniform trap distribution as suggested by Lampert [61]. Sussman [71] made an extensive series of measurements on CuPc films with Au electrodes. It was proven that the material was p-type, and after an ohmic region a power law with exponent n in the range 2.6–4.0 was observed. The value of n decreased with increasing temperature T, since $n = 1 + (T_t/T)$. He also reported an inverse power-law dependence of current on thickness, as was also predicted by Eq. (54). Since the publication of this work SCLC, dominated by both discrete trap levels (Eqs. (51) and (52)) and by an exponential trap distribution (Eqs. (53) and (54)), has been observed in many phthalocyanine materials. Their properties have been tabulated elsewhere [5]. In general, these materials are p-type, having a thermally generated hole concentration in the range of 10^{16}–10^{18} m^{-3}. Trap concentrations for a single trap level are in the range 5×10^{16}–7.1×10^{25} m^{-3} and for an exponential distribution 6×10^{20}–9.3×10^{26} m^{-3}. Activation energies for discrete trap levels vary between the different phthalocyanines, having typical values of 0.77 eV [38] and 0.62 eV [72] in CuPc, 0.42 eV in PbPc [40], 0.58 eV in NiPc [73], and 0.45 eV (fresh) and 0.69 eV (annealed) in CoPc [74]. Such values are probably a function of the central metal atom in the molecule and the oxygen content. Trap concentrations and carrier mobility have been measured in CuPc films as a function of the ratio of evaporation background pressure to deposition rate [75] as suggested previously [63].

In CdS thin films, like the single crystals mentioned earlier, SCLC has also been observed. In samples having an ohmic In and a blocking Au contact, Zuleeg [41] observed SCLC under forward bias (In injecting electrode) with a single trap level, although at higher voltage levels, a $J \propto V^3$ dependence suggested that an exponential trap distribution was also present. Zuleeg and Muller [76] also noted a $J \propto d^{-3}$ dependence in agreement with that expected from Eq. (52). Space-charge-limited conduction dominated by a single trap level in coevaporated CdS films, also with In electrodes, has been investigated by Pizzarello [77], who used measurements of V_{TFL} to determine a trap concentration of 2.5×10^{20} m^{-3}. CdTe films similarly show SCLC behavior. In n-type CdTe films, ohmic conduction was observed at low voltages, whereas at higher voltages the current followed an exponential dependence on voltage [78, 79]. The theoretical model of Rose [60] for a uniform trap distribution was modified, and substantial agreement between this and the experimental results was obtained. In Ni-CdTe-Au structures, a $J \propto V^2$ dependence was observed when the Ni electrode was biased negative, indicative of a single trap level [80] and consistent with Eq. (52). These results are shown in Figure 4 [80], and are typical of those for samples showing

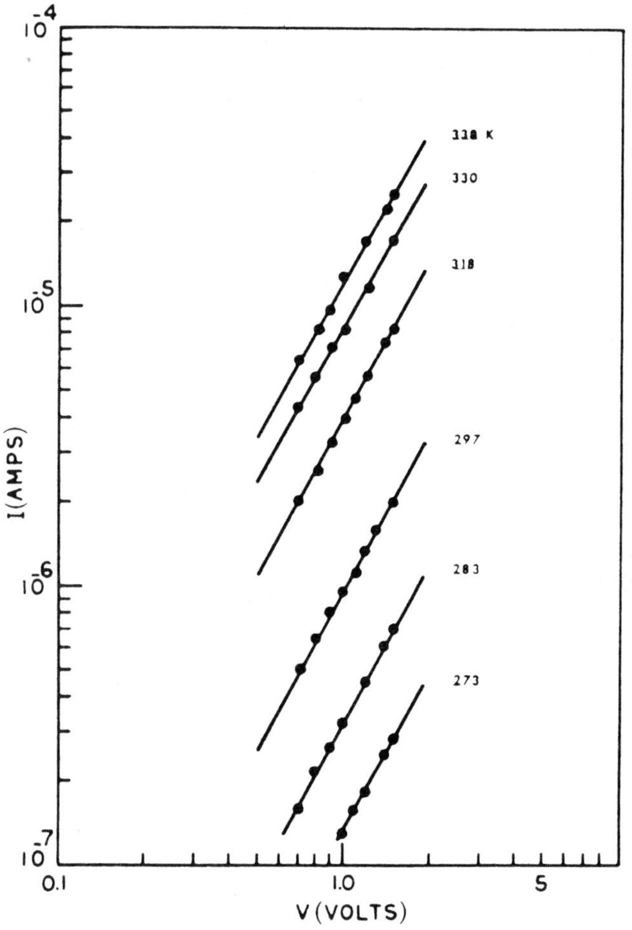

Figure 4. Space-charge-limited conduction. V^2 dependence of current on voltage in an electrodeposited Ni-CdTe-Au sandwich structure of thickness approximately 1 μm showing SCLC for temperatures of 273–338 K. Reproduced with permission from [80], B. M. Basol and O. M. Stafsudd, *Solid State Electron.* 24, 121 (1981). © 1981, Elsevier Science.

SCLC. They are plotted logarithmically, and all have similar gradients of approximately 2. Furthermore, the figure also shows that the current density increases systematically with temperature over the whole temperature range of 273–338 K. The same workers observed a $J \propto d^{-3}$ dependence, thus confirming SCLC. In evaporated CdTe films, with both electrodes of Al, similar ohmic and single trap-level behavior was observed [81]. However, at higher voltages, there was a $J \propto d^n$ dependence, with n in the range 6.0–6.5, changing to 11–12, suggesting an exponential trap distribution may also be apparent. Similar studies have also shown ohmic conduction and SCLC dominated by an exponential trap distribution, without the intermediate single trap level behavior [82]. Both the J-V and J-d dependencies were as predicted by Eq. (54). Measurements on similar samples [83] using a method proposed previously [63], where the dependence of J on $1/T$ was measured for several different applied voltages, yielded values of the mobility of (4–8) $\times 10^{-2}$ m^2 V^{-1} s^{-1} and the trap concentration of (1.4–4.2) $\times 10^{24}$ m^{-3}. CdSe films similarly show SCLC. Some samples have shown behavior dominated by a single trap level only [84, 85], while others additionally showed behavior dominated by an exponential trap distribution [86–88]. The differences appear to depend on the differing deposition methods and processing used.

Space-charge-limited conduction is observed in many thin film systems, but it is restricted to structures having at least one ohmic contact. For n-type conduction, this implies that $\psi_m < \psi_i$, although the influence of surface states cannot be neglected. It has also been argued that the accumulation regions in nanostructures are likely to envelop the entire structure (Eq. (18)) and is likely to be a dominant influence on the conductivity. The type of SCLC observed will clearly depend on the nanostructure architecture. Trap-free behavior is clearly possible for perfectly assembled nanostructures with no impurities. The addition of impurities in a controlled manner could be sufficient to immobilize nearly all of the injected charge by trapping effects. Nanostructures are unlikely to be large enough to accommodate a sufficient number of impurities to approximate an exponential trap distribution, and therefore a single trap level giving a $J \propto V^2$ dependence, is more probable. However, the influence of added impurities will probably be disproportionate to their number, as only one impurity atom in a structure containing 100 atoms would correspond to a trap concentration of about 10^{28} m^{-3}. As nanostructure sizes inevitably decrease, quantum effects will probably become apparent, considerably modifying any observed SCLC behavior.

7. THE POOLE–FRENKEL EFFECT

The Poole–Frenkel effect is a bulk-limited process, which may be considered as the bulk analogue of the Schottky effect [16]. Since it is a bulk property, at low fields the current density is determined only by the conductivity and the applied field, and may be written as

$$J = \sigma_0 F \tag{55}$$

where σ_0 represents the low-field conductivity. In a bulk insulator or wide bandgap semiconductor without impurities, the contribution of holes to σ_0 may be considered negligible, and

$$\sigma_0 = e\mu N_c \exp\left(-\frac{E_g}{2kT}\right) \tag{56}$$

where μ is the mobility and N_c is the effective density of states in the conduction band. It is clear that if these two equations only are considered, the current density is proportional only to the voltage at a given temperature, and that therefore Ohm's law is followed.

Frenkel [89] proposed that in the presence of a high electric field, the ionization potential of the atoms in the solid E_g would be lowered by an amount depending on the electric field, in the same way as the potential barrier at a contact may be lowered by the Schottky effect. The potential barrier $\varphi(x)$ for thermal excitation is assumed to follow a coulombic form, and is lowered by application of the high field.

The potential barrier for excitation into the conduction band is then given by

$$\varphi(x) = \varphi_0 - \frac{e^2}{4\pi\varepsilon_r\varepsilon_0 x} - eFx \qquad (57)$$

where φ_0 is the ionization energy. This expression is similar to Eq. (33) for the Schottky effect. However, in the case of the Poole–Frenkel effect, an image force with characteristic distance $2x$ is not present, and the x in the above equation represents the distance of the electron from the emitting center. A maximum in this expression occurs at

$$x_m = \left(\frac{e}{4\pi\varepsilon_r\varepsilon_0 F}\right)^{1/2} \qquad (58)$$

which on substituting into Eq. (57), leads to a reduced ionization energy given by

$$\varphi = \varphi_0 - \Delta\varphi_{PF} \qquad (59)$$

where

$$\Delta\varphi_{PF} = \beta_{PF} F^{1/2} \qquad (60)$$

represents the amount of field lowering of the barrier at the center, and β_{PF} is the Poole–Frenkel field-lowering coefficient given by

$$\beta_{PF} = \left(\frac{e^3}{\pi\varepsilon_r\varepsilon_0}\right)^{1/2} = 2\beta_S \qquad (61)$$

In principle, therefore, the Poole–Frenkel field-lowering coefficient is twice that of the Schottky value given by Eq. (37). If we then replace E_g by $E_g - \Delta\varphi_{PF}$ in Eq. (56), Eq. (55) becomes

$$J = J_0 \exp\left(\frac{\beta_{PF} F^{1/2}}{2kT}\right) = J_0 \exp\left(\frac{\beta_{PF} V^{1/2}}{2kTd^{1/2}}\right) \qquad (62)$$

where $J_0 = \sigma_0 F$ is the low field current density. Simmons [2] has pointed out that although $\Delta\varphi_{PF} = 2\Delta\varphi_S$, the coefficient of $F^{1/2}$ in Eqs. (38) and (62) is the same for both effects, owing to the additional factor of 2 in the denominator of Eq. (62). In thin films, however, Mead [90] has argued that since, for the case of an insulator, traps are almost universally present, the current density in thin films containing shallow traps is given by

$$J = J_0 \exp\left(\frac{\beta_{PF} F^{1/2}}{kT}\right) = J_0 \exp\left(\frac{\beta_{PF} V^{1/2}}{kTd^{1/2}}\right) \qquad (63)$$

In thin films, Eq. (63) is therefore normally associated with the Poole–Frenkel effect rather than Eq. (62). In principle, if Eq. (63) is followed, this may be distinguished from the Schottky effect of Eq. (39) by the value of the field-lowering coefficient derived from a $\ln J \propto V^{1/2}$ plot. However, this is not true if Eq. (62) is followed. Furthermore, many modified Poole–Frenkel expressions may be found in the literature, having different coefficients of $F^{1/2}$, which makes the assignment of the type of conductivity even more problematic. For instance, when a material contains donor levels below the Fermi level and shallow neutral traps [91], the coefficient of $F^{1/2}/kT$ is one half the value given by Eq. (61). Moreover, in the presence of a nonuniform electric field, having a maximum value of $\alpha^2 F$ where $\alpha > 1$, the Poole–Frenkel field-lowering coefficient is enhanced by the factor α [92]. The fact remains, however, that like the Schottky effect the Poole–Frenkel effect is indicated by a linear $\ln J \propto V^{1/2}$ dependence, although distinguishing between the two effects is particularly difficult.

Poole–Frenkel conductivity has been observed in many wide bandgap oxide films, including anodized Ta_2O_5 and silicon monoxide (SiO) [90, 93–95]; the interpretation of some of these results has been questioned, and alternatively classified as Schottky emission when the correct high-frequency permittivity values are used in the analysis [91]. In SiO films, good agreement with Eq. (63) was obtained, but the value of β_{PF} was greater than the theoretical value [96]; such discrepancies have been discussed by Jonscher [97]. Further work on SiO films established that this is indeed Poole–Frenkel behavior [98], as the characteristics were symmetric over five orders of applied bias, irrespective of different combinations of metal electrodes, and therefore inconsistent with the Schottky mechanism. Another very important microelectronic device material in which Poole–Frenkel conductivity has been observed is silicon nitride (Si_3N_4) [99]. Metal-insulator-semiconductor (MIS) diodes were fabricated on a degenerate Si substrate, by reacting $SiCl_4$ and NH_3 at 1000 °C to produce a Si_3N_4 film. An evaporated gold layer completed the structure. At low values of electric field, the J-V characteristic was asymmetric, with a higher current carried when the Au electrode was positive. This asymmetry was attributed to differences in the barrier heights at the gold-nitride and the nitride-silicon interfaces. At high fields, exceeding approximately 4×10^8 V m^{-1}, typical Poole–Frenkel behavior was observed with linear $\ln J - V^{1/2}$ behavior apparent over nearly four decades of current. For a given field, it was found that the characteristics of current density versus field were essentially independent of the film thickness, device area, electrode materials, and polarity, strongly indicating that the conductivity was bulk controlled, rather than electrode controlled as in the Schottky effect.

Poole–Frenkel conduction in a structure incorporating a phthalocyanine was reported in In-CuPc-SiO_x-In structures [55], but was proven to be dominated by the insulating SiO_x region. In oxygen-doped CuPc films, with a positively biased Al electrode and a negatively biased gold electrode, a slightly enhanced value of β_{PF} was determined [37] and attributed to a nonuniform electric field distribution as proposed previously [92]. In annealed films, the measured value of the field-lowering coefficient was in reasonably good agreement with the theoretical value. Similar agreement was observed in Au-CuPc-Pb samples [38]. Au-PbPc-Al samples showed two different values of the field-lowering coefficient [40], that at high voltages was consistent with Poole–Frenkel conductivity in the presence of a nonuniform field, while the lower voltage value was consistent with the Schottky effect. Similar results on the same system had, however, been interpreted in terms of the Schottky effect with two different barrier widths [39], although the presence of the Poole–Frenkel effect was not totally ruled out. Both Schottky and Poole–Frenkel behavior were observed in PbPc films containing oxygen in which both electrodes were gold [100].

Poole–Frenkel conductivity has been reported in radio frequency (rf)-sputtered CdS films sandwiched between Al and Au electrodes [101, 102]. A clear dependence of $\ln I$ on $V^{1/2}$ was observed, and this was symmetric with respect to the applied field, and therefore tended to discount the possibility of Schottky emission. However, detailed calculations of the value of the field-lowering coefficient were not performed. The process has also been observed by several groups in CdTe films [59, 92, 103, 104], although in one case the material was intentionally doped [104]. The earliest work was performed on compensated bulk samples [59], in which the mobility was found to vary linearly with $F^{1/2}$. The experimental value of β_{PF} was reasonably close to the theoretical value, although at fields greater than 1.2×10^6 V m^{-1}, the Poole–Frenkel effect gave way to tunneling. In Al-CdTe-Al thin film samples, Poole–Frenkel type behavior was observed consistently over a temperature range from 0 °C to 108 °C. However, the value of the field-lowering coefficient increased from 4.95×10^{-5} eV m$^{1/2}$ V$^{-1/2}$ at 0 °C to 7.70×10^{-5} eV m$^{1/2}$ V$^{-1/2}$ at 108 °C, and was of the order of twice that expected for the Poole–Frenkel effect in CdTe. However, the Schottky effect was discounted, since the field lowering did not appear to depend on the electrode workfunction, when the injecting electrode was replaced by either In or Ag. These results, which appear typical of those associated with the Poole–Frenkel effect, are shown in Figure 5 [103]. The apparent discrepancy between the experimental and theoretical values of β_{PF} was resolved by adopting the suggestion of Jonscher and Ansari [98] that electrons produced by thermal ionization at donor-like centers could hop between the sites as the result of thermal activation. When the data were reinterpreted according to this model, there was good agreement between the theoretical and experimental values of β_{PF}, and a barrier of height 0.15 eV between hopping sites was deduced, which was attributed to Cd vacancies. Similar linear $\ln J \propto V^{1/2}$ characteristics were obtained on the same system in which the experimental β_{PF} value also exceeded the theoretical value [92]; measurements of the gradients of these characteristics as a function of $1/T$ also generated experimental β_{PF} values, which were consistent with those determined from the J-V characteristics. It was suggested that the similarity between the results of this work [92] and that of the earlier work [103] could be related to individual centers experiencing an electric field greater than the mean field in samples where the electric field was nonuniform, and an approximate expression to account for such characteristics was presented. It was also observed that p-type CdTe films, which had been deliberately doped with $PbCl_2$, showed a change in conductivity type from SCLC to Poole–Frenkel [104], which may be related either to the inclusion of Poole–Frenkel centers owing to the doping process or to the establishment of a nonuniform electric field. Evidence of Poole–Frenkel conductivity has also been reported in CdSe films [88] with either two Au electrodes or one Al and one Au electrode. Again, measured values of β_{PF} were somewhat in excess of the theoretical value and were associated with a nonuniform electric field. In Cd_3As_2 films [42] with various electrode combinations, there was also evidence for Poole–Frenkel conductivity in thicker films, although the Schottky effect was thought to be responsible for conductivity in thinner films.

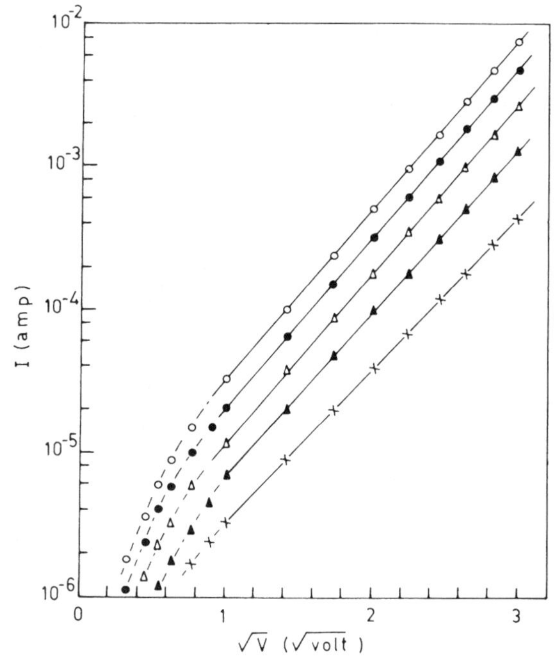

Figure 5. The Poole–Frenkel effect. Linear dependence of the logarithm of current on the square root of the applied voltage in an evaporated Al-CdTe-Al sandwich structure of thickness 722.5 nm at 0 °C (bottom curve), 30 °C, 55 °C, 80 °C, and 108 °C (top curve). The behavior was identified with a form of the Poole–Frenkel effect. Reproduced with permission from [103], S. Gogoi and K. Barua, *Thin Solid Films* 92, 227 (1982). © 1982, Elsevier Science.

There is thus a considerable amount of evidence for Poole–Frenkel conductivity in many different insulators and semiconductors. The basic J-V characteristic expected for this type of conductivity may, in principle, be distinguished from that for the Schottky effect in thin film systems which contain shallow traps by the differing values of the field-lowering coefficients, where $\beta_{PF} = 2\beta_S$. However, there are several complicating factors. Even in such films that contain, for instance, donor levels below the Fermi level [91] or which have a nonuniform electric field [92], the effective value of the field-lowering coefficient is modified. Thus, distinguishing Poole–Frenkel from Schottky emission is not a simple task, and a more meaningful assignment needs to include a consideration of whether the conductivity is bulk- or electrode-limited. This may be informed by determining whether or not the J-V characteristics are symmetric using dissimilar electrodes, or by using capacitance measurements to obtain Schottky barrier heights as a controlling feature of the conductivity.

In the context of nanoelectronics, it is likely that in practice the type of Poole–Frenkel conductivity will follow Eq. (62) rather than Eq. (63), since at least in the foreseeable future most samples are likely to be assembled from pure elemental semiconductors. Samples of nanosized dimensions consisting of compound semiconductors are very unlikely to show simple behavior owing to discontinuities at the surface boundary and other crystal imperfections. An estimate of the lateral extent of the region across which field-assisted ionization might occur can be obtained from Eq. (58). By comparison with Eq. (34), this will be twice the value obtained for the Schottky effect at the same field with the same dielectric, or about 6 nm for a field of 10^7 V m^{-1} and 2 nm for 10^8 V m^{-1}, where $\varepsilon_r = 4$. For silicon, a lower value would result, owing to the higher value of ε_r, and would allow several Poole–Frenkel emission centers to operate within the volume of a nanostructure. It is unclear, however, whether these would dominate the electrical conductivity, or what the ultimate form of the J-V characteristic might be; to answer this would require a quantum-mechanical analysis.

8. FILAMENTARY CONDUCTION

For some considerable time, negative resistance and electron emission effects have been observed in metal-insulator-metal (MIM) sandwich systems that have undergone an electroforming process. Providing the insulator (or semiconductor in many cases) is reasonably thin (typically a few tens of nanometers), the application of a voltage bias of the order of several volts under vacuum conditions results in the structure undergoing an essentially irreversible electroforming process. Prior to electroforming, such samples normally exhibit one of the high-field mechanisms described in Sections 4–7, but following it the conductivity is enhanced by several orders of magnitude, and voltage-controlled differential negative resistance (VCNR) behavior is observed. Early experiments were performed on insulators, such as SiO_x sandwiched between Al and Au electrodes [105] or on anodized Al_2O_3 with an evaporated Au counter electrode [106]. Initial models of the electroforming process and conductivity in such samples involved, for instance, SCLC in an impurity band of localized states introduced during electroforming and located near the center of the bandgap [107], or the injection of ions from the anode (Au in this case) to form a broad impurity band [105]. Neither of these models involved localized conduction in the form of filaments, a mechanism which was originally proposed by Dearnaley et al. [108]. This model assumed that microscopic filaments were established during the electroforming process that spanned the insulating region between the two electrodes, and that the subsequent negative resistance behavior was a consequence of filament rupture owing to heat dissipation. Since this early work, electroforming has been observed in an extremely wide variety of materials, and has been extensively reviewed in the literature [8, 109–111].

Electrical conductivity through narrow channels was first suggested by Dearnaley [112] in chemically unsaturated oxides. This was developed into a theory of the oxide-coated cathode [113] and its efficient emission properties. Generally, it was suggested that at favorable points on the metal-insulator interface, the high electric field will encourage filament growth in which oxygen vacancies migrate to the filament tip, enabling it to grow. Equation (6) demonstrates that at a region of small radius of curvature, for example, at structural imperfections or at the ends of existing filaments, the electric field will be particularly intense and therefore conducive to filament growth. Thus, we have a simple mechanism both for the initiation of filaments and for their propagation through the insulator. It is clear, however, that such a process may also occur in other materials where oxygen vacancies are not present, and in such cases the filaments are likely to contain metal originating from the electrodes [111].

The original model of filamentary conduction as first proposed by Dearnaley et al. [108] remains the basis of current understanding, although modifications to the model and various refinements have been proposed. It was assumed that conduction through individual filaments follows Ohm's law; furthermore, in any particular electroformed structure there are a very large number of filaments (roughly estimated at 5×10^6 cm^{-2} or 5×10^{10} m^{-2}). Such filaments were considered to be essentially noninteracting, except via thermal processes. Since the conduction in the individual filaments follows Ohm's law, the individual electrical behavior of a filament is governed by the filament resistance, ρ. However, ρ varies between individual filaments, and for a given electroformed sample, the assembly of filaments may be described by a probability distribution of filament resistances $P(\rho)$, where

$$\int_0^\infty P(\rho)\, d\rho = 1 \tag{64}$$

The range of integration is clearly from $0 \to \infty$ in the general case, although the lower and upper limits may be replaced by other values depending on the physical properties of the assumed filament distribution. The first calculated distribution of filament resistances was approximately triangular, as shown in Figure 6 [108]. ρ_t and ρ_u represent, respectively, the minimum and maximum values of ρ in the distribution.

Since the filaments are effectively connected in parallel between the two electrodes, the total current through the

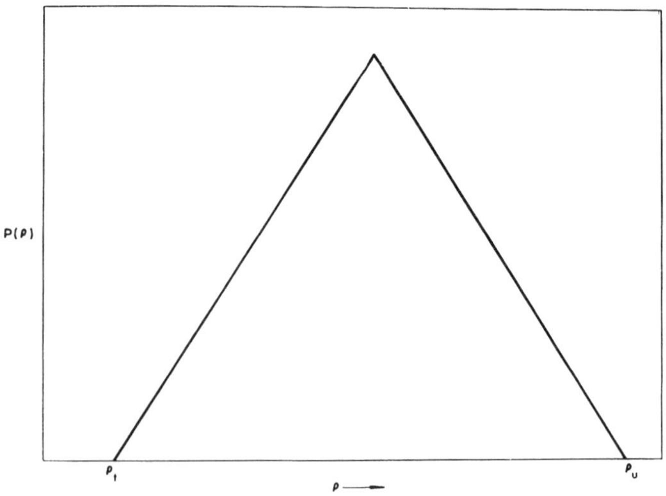

Figure 6. Filamentary conduction. The simple triangular probability distribution of filament resistances, ρ, assumed as an approximation to the calculated distribution deduced from experimental current-voltage behavior. Reproduced with permission from [108], G. Dearnaley, D. V. Morgan, and A. M. Stoneham, *J. Non-Cryst. Solids* 4, 593 (1970). © 1970, Elsevier Science.

electroformed sample can be obtained by summing the currents through the individual filaments and is given by

$$I = \int_{\rho_{\min}}^{\infty} \frac{V}{\rho} P(\rho)\, d\rho \tag{65}$$

where $\rho_{\min}$ represents the minimum resistance of a filament remaining conductive at the voltage V. The crucial part of the theory, which causes the negative resistance behavior, is the fact that as the voltage applied increases the value of $\rho_{\min}$ also increases, that is, individual filaments either cease to conduct or fracture, as a response to the applied voltage. The exact mechanism for this is of secondary importance, as long as lower resistance filaments cease to conduct at lower voltages than higher resistance filaments. In the original model, and in much of the subsequent work, it is assumed that the process is thermal, and that filaments cease to conduct when a local temperature $T_{\max}$ is reached. Solution of the steady state heat flow equation for the filament, where α characterizes the Joule heating of the filaments and τ_c characterizes the cooling by conduction, yields an expression for $\rho_{\min}$ as

$$\rho_{\min} = \frac{\tau_c \alpha V^2}{T_{\max} - T_0} \tag{66}$$

where T_0 represents the temperature of the environment surrounding the filaments. For the case of cylindrical filaments of radius d and length L cooling to a cylindrical sink of radius D and temperature T_0, the product $\tau\alpha_c$ is given by

$$\tau_c \alpha = \frac{\ln(D/d)}{2\pi K L} \tag{67}$$

where K (W m^{-2} K^{-1}) is the thermal conductivity of the insulator.

Using simple arguments and typical values of thermal conductivity, it was concluded that $(T_{\max} - T_0) \sim 1000$ K, $d \sim 1$ nm, and the mean resistance per filament is about 2.5×10^7 Ω. Since this initial work, many variations to the original assumptions have been considered. A wide variety of different filament probability distributions $P(\rho)$ have been investigated, including uniform, parabolic, and bimodal triangular distributions [114], with rather good agreement observed for the parabolic distribution. Radial, as well as cylindrical symmetry in the heat flow, was also considered [115], and it was shown that in both cases the voltage V_p, at which filaments of resistance ρ will cease to conduct, is given by

$$V_p = \beta \rho^{1/2} \tag{68}$$

where β is determined by the heat flow symmetry. For the case of radial heat flow symmetry, for instance, where the filament resistance is assumed to be dominated by weak spots, the filaments are effectively isolated from each other. It was early recognized that the filament conductivity was unlikely to be ohmic, and Sutherland [116] investigated the case where the current $I \propto I_0 \sinh(V/V_0)$, where I_0 and V_0 are constants. It has also been suggested that cooling in electroformed structures may take place through the top electrode [117], when biased at the higher potential, since the thermal conductivity of the metal electrode is much higher than that of the unformed insulator regions. Nevertheless, even with this departure from the more localized heat conduction symmetry, an appropriate value of β in Eq. (68) was derived to account for the situation. These workers also suggested that a normal probability distribution of filament resistances was more fundamental than those investigated previously, and that both the triangular and the parabolic distributions were merely approximations to this [118]. Normal probability distributions of filament radii [119] and cross-sectional area [120] have also been investigated. It has been shown recently [121] that Monte Carlo techniques offer a good method for investigating more complex situations, such as those requiring multivariate probability distributions. This technique has been tested for the case of the normal probability density distributions of filament resistances and radii, and is anticipated to become ubiquitous for investigation of filaments in the case of nonohmic filamentary conduction.

Although the above discussion of filamentary conduction assumes that it is a thermal process that causes the filaments to fracture, since individual filaments of this type have not been unequivocally observed, this is by no means certain. Dearnaley et al. [108] considered that the localized regions in which filaments are initiated to be of size $R < 2$ nm, and that filaments were therefore unlikely to be observed microscopically. Bräuer et al. [122] also concluded more recently that direct observation of filaments having nanometer dimensions, where the length is much greater than the thickness or width, would not be possible with existing experimental methods. Only when filaments can be isolated from each other, such that a single or a small number of filaments can be investigated directly, will such questions be answered. It is not impossible with conventional technology to fabricate MIM samples of area $A = 2 \times 10^{-11}$ m^2, which would correspond on average to the estimated filament area density of 5×10^{10} m^{-2}, and may be expected to contain typically one filament. Required feature size at the metallic contact would be of the order of $\sqrt{A}$, or a few micrometers. It is possible that by electroforming such structures, individual filaments could be formed and their electrical characteristics investigated. Owing to the estimated filament radius $d \sim 1$ nm, it is entirely possible that they may show quantized conductance effects, as have recently been observed in metallic chain structures investigated using a STM [21, 22], and discussed in Section 3.

Electroforming may well be a suitable method to fabricate nanowires (or quantum wires) within a solid structure. For ballistic electron transport along a narrow channel, the conductance G_0 is proportional to the transmission probability T for current injected from an electron reservoir at one end of the channel, and is given by the Landauer formula [123]:

$$G_0 = \frac{e^2}{h} T \tag{69}$$

Size quantization is expected to set in when the width of the conducting channel becomes less than the Fermi wavelength of the electrons λ_F, which is of the order of 0.5 nm for metals. However, for semiconductors, and in particular for the material comprising the proposed filaments, λ_F can be larger than this because of the smaller electron density [123]. We may therefore reasonably expect quantum behavior in filaments established during electroforming, providing their radius is of about 1 nm as estimated by Dearnaley et al. [108].

Electroforming is well known in many thin film sandwich structures, usually giving rise to negative resistance behavior. If the current is indeed carried by an assembly of filaments with radii of the order of 1 nm, it is very likely that the individual filaments will exhibit quantum behavior and act as nanowires (or quantum wires). It is therefore feasible that the electroforming process may be a relatively simple route to the fabrication of nanowires, which will facilitate investigation of their electrical properties and perhaps direct observation using one of the tunneling microscopy techniques.

9. CONCLUDING REMARKS

In this review, it has been demonstrated that in nanostructures there is a high probability of the existence of high electric fields, whether intentionally by design or as a consequence of other factors. Precautionary measures, as used in present day metal-oxide-semiconductor (MOS) devices, will be necessary to protect structures, as the existence of unwanted high electric fields, resulting from charge inhomogenieties, need to be neutralized. Such considerations will be as necessary in nanostructures having nonelectronic applications as in electronic nanodevices.

It is a direct consequence of Maxwell's equations of electromagnetism, and Gauss's law in particular, that very high electric fields are generated at regions of a small radius of curvature. In three-dimensional structures of nanoscale dimensions, there may in principle be several regions where such considerations apply and high fields are present. The existence of high fields in nanostructures therefore warrants a consideration of high field conduction processes in the design of all electronic nanostructures. In deriving the current-voltage characteristics of most of the established high-field conduction processes described in this review, Poisson's equation, which relates the localized electric field to the charge density, must be solved subject to suitable boundary conditions. At the time of writing this article, this is usually performed only in the one-dimensional case, and although this is satisfactory in establishing the current-voltage characteristic equations for potential differences applied across thin films, when three-dimensional nanostructures are seriously considered this will be inadequate and may require a full three-dimensional analysis. Furthermore, the implications of the microscopic set of Maxwell's equations will need to be assessed, and in many cases a quantum-mechanical description will be required.

Many of the conduction models described in this review involve parameters such as the carrier concentrations at the band edges, and impurity and trap concentrations. Familiar concepts, such as the assumption of an effective density of states of approximately constant energy at the conduction and valence band edges, may not be applicable in nanostructures, where the inclusion of a small number of impurities may lead to degenerate behavior. For similar reasons, in nanostructures trap concentrations resulting from imperfections may be exceptionally high, and impurity doping of semiconductors is likely to be tactically avoided when not essential.

In all nanoelectronic devices, as with their conventional counterparts, electrical contacts will be required to introduce electric charge into the structure. Estimates derived from the conventional expressions for the widths of the accumulation region or the depletion region in the cases of ohmic and blocking contacts, respectively, show that such regions are likely to extend to a depth of at least 10 nm, and are therefore capable of encompassing entire nanostructures. Thus, the contact region cannot be considered separately from the remaining nanostructure, but becomes integral with it, and can in principle wholly determine the nanostructure conductivity. Consideration of the contact parameters, as well as doping and the other effects mentioned earlier, will be of paramount importance in the development of nanoelectronic devices. The basic features of the known high-field conduction processes in thin films are summarized below, together with some implications of their possible appearance in nanostructures.

For metal-dielectric-metal structures, a generalized current density expression is applicable, which includes the Fermi-Dirac probability distribution functions for electrons in the two electrodes and the electron transmission probability across the dielectric. In essence, the dielectric region is considered as providing a barrier to the flow of electrons between the electrodes. For relatively thick dielectrics, the current is determined by the rate at which electrons are thermally excited over the interfacial potential barrier at the injecting electrode. Due to the interplay between the contributions to the overall potential from the metal/dielectric barrier height, the image force, and the applied field, there is a maximum in the potential function inside the dielectric. The effective barrier height is lowered as the applied field is increased, and thus the current increases with the applied voltage. The location of the maximum in the potential depends on the dielectric permittivity and the applied field. For all practical purposes, this is located within 10 nm of the electrode/dielectric interface and would, in principle, permit Schottky emission providing the interfacial workfunctions are suitable. For thinner films, it is more probable that tunneling will take place through the dielectric, either directly or via excitation to the conduction or valence bands. Such structures may well find applications as injectors of carriers (electrons) in nanostructures.

For structures having an ohmic contact, space-charge-limited conductivity is a strong possibility. Since the accumulation region is likely to penetrate throughout entire nanostructures, the primary condition for SCLC is fulfilled. The type of current-voltage dependence will depend on the presence or lack of trapping centers and on their distribution in energy. For assembled nanostructures with no impurities, it is possible that behavior similar to the classic Mott–Gurney trap-free behavior will be observed, permitting high current densities. Trapping effects may be used to immobilize injected charge if it is not required. Similarly Poole–Frenkel bulk conductivity will presumably be observed, but in the absence of traps, and exhibit a current density dependence differing from that normally observed in thin films with shallow traps. Field-assisted ionization of Poole–Frenkel centers would occur over a region of approximately twice that for the equivalent Schottky effect in the same material, and would allow the operation of several Poole–Frenkel centers in a given nanostructure. It is, however, unclear whether this would then constitute the dominating conduction mechanism, or whether an electrode-limited process would prevail.

The final high-field process that has been considered in this review is observed only after an electroforming process. Among other theories of electroforming, it has been widely argued that during electroforming a considerable number of conducting filaments are formed between the electrodes. There is considerable evidence that quantized conduction occurs in metallic chains, which may be formed by retracting an STM tip from a metallic surface. This represents a quantum point contact. There is a reasonable correlation between the dimensions predicted for conducting filaments in electroformed structures, those predicted for quantum conduction in metallic chains and experimental observations in quantum point contacts. This poses the question that individual filaments may exhibit quantum behavior. The electroforming process, which with its subsequent conductivity has been seen as a curiosity in solid state physics, may well be a suitable mechanism for forming nanowires in solids and investigating any quantum conduction behavior. Applications of electroforming in nanostuctures, whether or not electroforming involves conducting filaments, are unlikely owing to the unpredictable and time-varying currents normally observed.

To conclude, the major high field conduction processes observed in thin film structures have been reviewed, and typical examples have been cited for each of these. Owing to the decreased size of nanostructures, high fields will be an almost inevitable consequence of nanoelectronics. Both the experimental and theoretical aspects of nanostructures are in their infancy, although this encyclopedia shows that remarkable progress has been made in many areas. It is clear that some of the experimental and theoretical work concerned with thin films (nanostructures in one dimension) will carry over to the three-dimensional situation; for structures having nanodimensions in all directions, quantum effects are to be expected and the theory developed will need to be based on this. Although nanoelectronics is still in its infancy, it is based on many years of earlier work on thin films. High field conductivity is as likely to be found in nanostructures as in thin films and will play a crucial role in nanoelectronics.

GLOSSARY

Accumulation region A space-charge region at a semiconductor or insulator surface where majority carriers are accumulated following a flow of carriers into the region from the contact in order to equalize Fermi levels.

Barrier height The quantity of energy required to surmount a potential barrier.

Blocking contact A contact which can only supply carriers to a semiconductor or insulator after they have overcome a potential barrier. The contact resistance is high, since the applied potential falls largely over a low conductivity depletion region. Essentially a Schottky barrier.

Bulk-limited Determined by bulk properties of the semiconductor or insulator.

Conduction band A partially filled energy band in a semiconductor or insulator in which electrons can move freely to establish an electric current.

Constitutive equation An equation relating the electric displacement field **D** and the electric field intensity **F**, or the magnetic flux density **B** and the magnetic field intensity **H** in a particular medium.

Continuity equations Equations relating the rate of change of the concentration of charge carriers (electrons and holes) in a semiconductor or insulator to the carrier generation and recombination rates and the divergence of the current density.

Coulomb's law The force between two electric point charges is directed along the line connecting them and its magnitude is proportional to the product of the charges and inversely proportional to the permittivity of the medium in which they are situated and to the square of their separation.

Current density equations Equations relating the electron and hole current densities in a semiconductor or insulator to their concentrations, including the influences of both an electric field and carrier concentration gradients.

Dangling bond A chemical bond associated with an atom at the surface of a solid that is not linked with a second atom, but extends towards the exterior of the solid.

Depletion approximation An approximation where the majority free carrier concentration is assumed to fall *abruptly* from the value in the bulk of a semiconductor or insulator to a negligible value at the boundary of the depletion region.

Depletion region A space-charge region at a semiconductor or insulator surface where majority carriers are depleted following a flow of carriers out of the region to the contact in order to equalize Fermi levels.

Dielectric breakdown The permanent failure of a dielectric to act as an insulator, arising for instance from heat generated or electric discharges at high voltages. Following dielectric breakdown the dielectric may be short circuited or open circuited.

Differential form The expression of one of Maxwell's four equations of the electromagnetic field in terms of the vector operators $\nabla\cdot$ (divergence) or $\nabla\times$ (curl), which encapsulates the behavior of the field at a point in space.

Donor An impurity or imperfection in a semiconductor or insulator which donates electrons to the conduction band, thus increasing the electron conductivity.

Effective density of states The number of energy levels (states) per unit volume required to be placed at a specific energy within a band of allowed states which, when multiplied by the probability of occupation by the relevant type of carriers, gives the concentration of carriers in the band, e.g. the effective density of states of electrons at the bottom of the conduction band E_c or of holes at the top of the valence band E_v.

Effective mass A parameter having the dimensions of mass that is assigned to charge carriers in a solid which allows them to be considered as free carriers in an applied electric field. Effective masses are not restricted in terms of magnitude or sign.

Electrode-limited Determined by the conditions at the injecting electrode.

Electroforming A process whereby some dielectrics undergo voltage-induced structural changes, after which the electrical conductivity is drastically increased and they exhibit characterisitic VCNR, electron emission and other features.

Electron affinity The difference in energy between an electron at the bottom of the conduction band and an electron at rest outside the surface of a semiconductor or insulator, i.e. at the vacuum level.

Fermi-Dirac distribution A statistical density function specifying the probability that a member of an assembly of independent fermions, such as electrons in a semiconductor, insulator or metal, occupies a particular energy level when in a state of thermal equilibrium.

Fermi level The energy level in a solid at which the probability of occupation is 1/2, e.g. where the value of the Fermi-Dirac distribution function is 1/2 for electrons in a solid.

Field-lowering Effective lowering of a potential barrier height as a result of modifications to the potential barrier in response to an applied electric field, e.g. lowering of an image force potential barrier in the Schottky effect, or a coulombic potential barrier in the Poole–Frenkel effect.

Field-lowering coefficient A coefficient β quantifying the amount of field-lowering, e.g. the Schottky field-lowering coefficient β_S, or the Poole–Frenkel field-lowering coefficient β_{PF}.

Filament A region of relatively high electrical conductivity embedded in a medium of high resistivity, through which electrical current may flow preferentially; in particular long narrow regions connecting the electrodes in metal-insulator or semiconductor-metal structures, proposed to be established during an electroforming process.

Filamentary conduction Electrical conduction preferentially through localized high-conductivity filaments, particularly those proposed to be established during an electroforming process.

Fowler-Nordheim Pertaining to a type of quantum-mechanical tunneling behavior observed at high electric fields F, where the current density is proportional to $F^2 \exp(-1/F)$, as originally derived by Fowler and Nordheim.

Gauss's law The outward flux of the electric field intensity **F** over a closed surface **S** is equal to the algebraic sum of the conduction and polarization charges enclosed by **S** divided by the permittivity of free space ε_0; or equivalently the outward flux of the electric displacement field **D** is equal to the algebraic sum of the conduction charges enclosed by **S**.

High-field conduction Electrical conductivity operating under high-field conditions, where the concentration and/or mobility of charge carriers may be field dependent. Under these conditions the current density-voltage characteristics are distinguished by the type of conduction process, and do not generally obey Ohm's law.

Hopping An electrical conduction process sustained by hopping between impurity centers which have energy levels within the forbidden band of a compensated semiconductor or insulator. Electrons in occupied states below the Fermi level obtain energy from phonon interactions enabling them to move to unoccupied states above the Fermi level. A type of impurity conduction.

Image force The electrostatic force on an electric charge due to the polarization charge that it induces on a neighbouring conductor, which may be considered as equivalent to the attraction between the charge and its electrical image in the conductor.

Impurity conduction An electrical conduction process sustained by transfer between impurity centers which have energy levels within the forbidden band gap of a semiconductor or insulator. Hopping is usually classified as impurity conduction.

Integral form The expression of one of Maxwell's four equations of the electromagnetic field in terms of volume or surface integrals bounded by closed surfaces and paths respectively, which encapsulates the behavior of the field over the volume or surface.

Landauer formula An expression relating the conductance to the transmission probability for current injected in the case of ballistic transport in a narrow conducting region.

Laplace's equation A partial differential equation describing the variation in electric field **F** or potential V with position in the absence of an electric charge distribution, and subject to suitable boundary conditions. A special case of Poisson's equation.

Macroscopic electromagnetism This is described by the set of four classical Maxwell's equations described below. These may be derived from the equations of microscopic electromagnetism using suitable spatial and time averaging of the microscopic fields, where the derived macroscopic fields **D** and **H** are identified with contributions relating to bulk properties of the macroscopic medium.

Maxwell's equations A set of four self-consistent equations summarizing the classical space and time dependence of the electromagnetic field, and relating the electric displacement field **D**, the electric field intensity **F**, the magnetic flux density **B**, the magnetic field intensity **H**, the volume electric charge density ρ and the current density **J**. These may be expressed in either the differential form or the integral form.

Microscopic electromagnetism This is described by a set of four Maxwell's equations applying on a microscopic scale, relating the microscopic electric field intensity **f**, the microscopic magnetic flux density **b**, the microsopic charge density η and the microscopic current density **j**. At the microscopic scale there is no corresponding microscopic electric displacement field **d** or microscopic magnetic field intensity **h** as for the macroscopic description, because all the charges are included in η and **j**.

Mobile charge Charges which can move freely as a response to an electric field, i.e. electrons in the conduction band of a semiconductor or insulator. Conversely, ionized impurities are immobile charges.

Nanowire A long narrow chain of atomic dimensions.

Negative resistance The condition where the current I decreases with increasing voltage V. Strictly *differential* negative resistance, with $dI/dV < 0$ on a current-voltage characteristic.

Neutral contact A contact intermediate between ohmic and blocking contacts, possessing neither accumulation or depletion space-charge regions.

N-type Pertaining to an extrinsic semiconductor or insulator having a concentration of negative electron charge carriers greater than that of positive hole charge carriers.

Ohmic contact A contact which can readily supply carriers to a semiconductor or insulator from a reservoir of mobile charge in an accumulation region. The contact resistance is sufficiently low for the current to be determined by the bulk material and not by the contact.

Ohm's law This is obeyed when the current flowing through a specimen is directly proportional to the voltage applied across it. It is not generally applicable at high electric fields, or across junctions between different materials.

Poisson's equation A partial differential equation which describes the variation in electric field **F** or potential V with position produced by a given distribution of charge density, and subject to suitable boundary conditions.

Poole–Frenkel Pertaining to the Poole–Frenkel effect described below.

Poole–Frenkel effect Lowering of an internal potential barrier in a semiconductor or insulator between an impurity centre and the conduction or valence band edges, when it interacts with an electric field. This leads to field-assisted thermal ionization which causes an increase in the conductivity.

Potential barrier A region where a difference between potentials exists which needs to be overcome to enable carriers to pass from one region to another.

P-type Pertaining to an extrinsic semiconductor or insulator having a concentration of positive hole charge carriers greater than that of negative electron charge carriers.

Quantized conductance Conductance which only varies in discrete quantum steps.

Quantum The minimum amount by which a particular property of a system can change; pertaining to such a discrete change in the value of a property.

Quantum-mechanical Pertaining to the theory of quantum mechanics.

Quantum mechanics The contemporary theory of matter, electromagnetic radiation and their interactions, which generalizes and supercedes classical physics, particularly in the atomic and subatomic domains.

Quantum point contact A structure in which a 'neck' of atoms of width only a few atomic diameters, and comparable to the Fermi wavelength of the conduction electrons, bridges two electric contacts. They are normally metallic and formed by retraction of the tip in a scanning tunneling microscope (STM).

Quantum wire A long narrow chain, whose dimensions in two orthogonal directions are less than some appropriate property, such as the mean-free path or the electron Fermi wavelength. The latter constitutes a one-dimensional conductor.

Richardson–Schottky equation An equation relating the current density of electrons emitted from a conductor to the temperature, potential barrier height for electron emission and the applied electric field. The potential barrier height from a metal to vacuum is equal to the metal work function, and from a metal to a semiconductor or insulator by the metal work function minus the relevant electron affinity.

Richardson's thermionic emission equation (also known as the Richardson-Dushman equation) An equation relating the current density emitted from a conductor, in the absence of an applied electric field, to the temperature and the potential barrier height for electron emission. The potential barrier height from a metal to vacuum is equal to the metal work function, and from a metal to a semiconductor or insulator by the metal work function minus the relevant electron affinity.

Scanning tunneling microscope (STM) A type of microscope in which a fine conducting tip is located close to the surface of a sample and scanned across the surface. Electrons tunnel between the tip and the sample producing an electrical signal, which is used to control the distance between the tip and the sample and produces a contour map of the surface. The method is capable of resolving individual atoms.

Schottky Pertaining to a Schottky barrier or the Schottky effect.

Schottky barrier A potential barrier at a metal/ semiconductor or insulator interface which needs to be overcome for carriers to pass between them. There is no potential barrier for the same type of carrier in the reverse direction. Essentially a blocking contact.

Schottky-barrier diode An electronic device based on a junction between a metal and a semiconductor, which allows a high current in one direction and only a small current in the reverse direction. Unlike the *pn* junction this is a unipolar device.

Schottky effect Lowering of a potential barrier between the Fermi level of a metal and the conduction or valence band edges of a semiconductor or insulator in the presence of an electric field. This leads to enhanced field-dependent thermionic emission and causes an increase in the conductivity.

Schottky emission Emission of a carrier over a Schottky barrier.

Shallow trap A trap located at an energy level above the Fermi level.

SI system of units An internationally recognized system of units (Système International d'Unités) with the metre (m), kilogram (kg), second (s), ampere (A), kelvin (K), candela (cd) and mole (mol) as base units.

Size quantization Quantization of a property owing to restriction in size in one or more dimensions, e.g. a quantum wire showing quantized conductance suffers size quantization in two dimensions when the Fermi wavelength is less than that of its diameter, and is therefore a one-dimensional conductor.

Space charge Charge existing in a region of space (e.g. within a solid) arising from a distribution of positively or negatively charged entities. The charge may be mobile (e.g. associated with electrons or holes) or immobile (e.g. associated with impurities or traps) and gives rise to it own electric field as described by Poisson's equation.

Space-charge-limited conduction (SCLC) A conduction process dominated by the effects of a space-charge region, which is injected into a semiconductor or insulator. The level of current drawn is strongly influenced by the presence and occupation of traps.

Surface state An electron state in a semiconductor or insulator whose wavefunction is localized in a region near the surface, and which arises from the lack of periodicity and the presence of dangling bonds at the surface.

Thermal equilibrium A state in which no net energy flow takes place between two regions; this is achieved by equilization of the Fermi levels.

Thermionic emission The emission of electrons from a conductor by thermal excitation over a potential barrier. The emitted current density is given by Richardson's thermionic emission equation in the absence of an applied electric field and by the Richardson-Schottky equation when an electric field is applied.

Thin film A thin layer of material, normally deposited on the surface of another material (the substrate) by processes such as evaporation or sputtering. In this context 'thin' means of a thickness below that at which a normal bulk property is observed, and can therefore depend on the property of interest; however it is usually below 1 μm.

Trap An impurity or imperfection in a semiconductor or insulator at which an electron or hole can be immobilized until released by thermal excitation.

Trap-filled limit The situation in materials showing space-charge-limited conduction where all the traps are filled, and therefore all additional injected charge contributes to the current. The applied voltage at which this occurs is denoted by V_{TFL}.

Tunneling A conduction process dominated by the quantum-mechanical tunnel effect, in which electrons are able to pass across a narrow region which on classical theory would require more energy than the electron possesses.

Vacuum level The energy of an electron at rest just outside the surface of a solid.

Valence band The highest electronic energy band in a semiconductor or insulator which can be filled with electrons and is located below the conduction band. If partially filled an electric current can be established.

Very-large-scale integration (VLSI) A level of integration having typically more than 10^5 devices per chip.

Voltage-controlled differential negative resistance (VCNR) A type of differential negative resistance in which the current is single-valued for a given applied voltage, with the current-voltage characteristic having an N-shaped curve. In contrast, for current-controlled differential negative resistance (CCNR) the voltage is single-valued for a given current and the current-voltage characteristic is S-shaped.

Wentzel-Kramer-Brillouin (WKB) approximation An approximate method of solving linear differential equations with slowly varying coefficients, which is useful in handling wavefunctions and energy levels, and in particular the tunneling probability.

Work function The difference in energy between the Fermi level and the energy of an electron at rest outside the surface of a solid, i.e. at the vacuum level.

REFERENCES

1. J. Gimzewski, *Physics World* 11, 29 (1998).
2. J. G. Simmons, *J. Phys. D: Appl. Phys.* 4, 613 (1971).
3. D. R. Lamb, "Electrical Conduction Mechanisms in Thin Insulating Films." Methuen, London, 1967.
4. S. M. Sze, "Physics of Semiconductor Devices," 2nd Ed. Wiley, New York, 1981.
5. R. D. Gould, *Co-ordination Chemistry Reviews* 156, 237 (1996).
6. R. D. Gould, in "Handbook of Thin Film Materials" (H. S. Nalwa, Ed.), Vol. 4, Ch. 4, p. 187. Academic Press, New York, 2001.
7. J. Lee, M. Mathai, F. Jain, and F. Papadimitrakopoulos, *J. Nanoscience Nanotech.* 1, 59 (2001).
8. G. Dearnaley, A. M. Stoneham, and D. V. Morgan, *Rep. Prog. Phys.* 33, 1129 (1970).
9. J. D. Jackson, "Classical Electrodynamics," 3rd Ed. Wiley, New York, 1998.
10. W. J. Duffin, "Electricity and Magnetism," 4th Ed. McGraw–Hill, London, 1990.
11. W. S. Ruska, "Microelectronic Processing." McGraw–Hill, New York, 1987.
12. E. H. Rhoderick, "Metal-Semiconductor Contacts." Clarendon Press, Oxford, 1978.
13. N. F. Mott, *Proc. Cambridge Phil. Soc.* 34, 568 (1938).
14. W. Schottky, *Naturwiss.* 26, 843 (1938).
15. N. F. Mott and R. W Gurney, "Electronic Processes in Ionic Crystals," 2nd Ed. Clarendon Press, Oxford, 1948.
16. J. G. Simmons, in "Handbook of Thin Film Technology" (L. I. Maissel and R. Glang, Eds.). McGraw–Hill, New York, 1970.
17. W. Schottky, *Z. Physik* 113, 367 (1939).
18. A. M. Cowley and S. M. Sze, *J. Appl. Phys.* 36, 3212 (1965).
19. H. K. Henisch, "Rectifying Semiconducting Contacts." Clarendon Press, Oxford, 1957.
20. J. G. Simmons, *Phys. Rev. Letters* 10, 10 (1963).
21. H. Ohnishi, Y. Kondo, and K. Takayanagi, *Nature* 395, 780 (1998).
22. A. I. Yanson, G. R. Bollinger, H. E. van den Brom, N. Agraït, and J. M. van Ruitenbeek, *Nature* 395, 783 (1998).
23. P. J. Price and J. M. Radcliffe, *IBM J. Res. Dev.* 3, 364 (1959).
24. W. A. Harrison, *Phys. Rev.* 123, 85 (1961).
25. R. Stratton, *J. Phys. Chem. Solids* 23, 1177 (1962).
26. S. Levine, in "The Encyclopedia of Physics" (R. M. Besançon, Ed.), p. 1229. Van Nostrand Reinhold, New York, 1985.
27. A. J. Dekker, "Solid State Physics." Macmillan, London, 1958.
28. W. Schottky, *Z. Phys.* 15, 872 (1914).
29. P. R. Emtage and W. Tantraporn, *Phys. Rev. Letters* 8, 267 (1962).
30. S. R. Pollack, *J. Appl. Phys.* 34, 877 (1963).
31. W. E. Flannery and S. R. Pollack, *J. Appl. Phys.* 37, 4417 (1966).
32. M. Fusstöss-Wegner, *Thin Solid Films* 36, 89 (1976).
33. B. Sh. Barkhalov and Yu. A. Vidadi, *Thin Solid Films* 40, L5 (1977).
34. C. Hamann and C. Tantzscher, *J. Chem. Phys.* 36, 81 (1976).
35. F.-R. Fan and L. R. Faulkner, *J. Chem. Phys.* 69, 3334 (1978).
36. A. Wilson and R. A. Collins, *Sensors and Actuators* 12, 389 (1987).
37. A. K. Hassan and R. D. Gould, *Int. J. Electron.* 69, 11 (1990).
38. R. D. Gould and A. K. Hassan, *Thin Solid Films* 193/194, 895 (1990).
39. A. Ahmad and R. A. Collins, *Phys. Stat. Sol. (A)* 126, 411 (1991).
40. T. S. Shafai and R. D. Gould, *Int J. Electron.* 73, 307 (1992).
41. R. Zuleeg, *Solid State Electron.* 6, 645 (1963).
42. M. Din and R. D. Gould, *Thin Solid Films* 340, 28 (1999).
43. R. H. Fowler and L. W. Nordheim, *Proc. Roy. Soc. London* A119, 173 (1928).
44. J. G. Simmons, *J. Appl. Phys.* 34, 1793 (1963).
45. J. Frenkel, *Phys. Rev.* 36, 1604 (1930).
46. A. Sommerfield and H. Bethe, in "Handbuch der Physik" (H. Geiger and K. Scheel, Eds.), Vol. 244/2, p. 450. Springer, Berlin, 1933.

47. R. Holm, *J. Appl. Phys.* 22, 569 (1951).
48. N. F. Mott, "Elements of Wave Mechanics." Cambridge University Press, Cambridge, 1952.
49. J. C. Fisher and I. Giaever, *J. Appl. Phys.* 32, 172 (1961).
50. C. A. Mead, *Phys. Rev. Letters* 6, 545 (1961).
51. D. Mayerhofer and S. A. Ochs, *J. Appl. Phys.* 34, 2535 (1963).
52. S. R. Pollack and C. E. Morris, *J. Appl. Phys.* 35, 1503 (1964).
53. J. G. Simmons, *J. Appl. Phys.* 34, 2581 (1964).
54. M. Lenzlinger and E. H. Snow, *J. Appl. Phys.* 40, 278 (1969).
55. A. K. Hassan and R. D. Gould, *J. Phys. D: Appl. Phys.* 22, 1162 (1989).
56. J. G. Simmons, *Phys. Rev.* 166, 912 (1968).
57. K. L. Chopra, "Thin Film Phenomena." Kreiger, New York, 1979; (first published by McGraw–Hill, New York, 1969).
58. D. V. Geppert, *J. Appl. Phys.* 33, 2993 (1962).
59. C. Canali, F. Nava, G. Ottaviani, and K. Zanio, *Solid State Commun.* 13, 1255 (1973).
60. A. Rose, *Phys. Rev.* 97, 1538 (1955).
61. M. A. Lampert, *Rep. Prog. Phys.* 27, 329 (1964).
62. R. D. Gould and B. A. Carter, *J. Phys. D: Appl. Phys.* 16, L201 (1983).
63. R. D. Gould, *J. Appl. Phys.* 53, 3353 (1982).
64. M. A. Lampert and P. Mark, "Current Injection in Solids." Academic Press, New York, 1970.
65. R. W. Smith, *Phys. Rev.* 97, 1525 (1955).
66. R. W. Smith and A. Rose, *Phys. Rev.* 97, 1531 (1955).
67. R. W. Smith, *R.C.A. Rev.* 20, 69 (1959).
68. J. L. Hartke, *Phys. Rev.* 125, 1177 (1963).
69. T. Budinas, P. Mackus, A. Smilga, and J. Vivvakas, *Phys. Stat. Sol.* 31, 375 (1969).
70. G. M. Delacote, J. P. Fillard, and F. J. Marco, *Solid State Commun.* 2, 373 (1964).
71. A. Sussman, *J. Appl. Phys.* 38, 2738 (1967).
72. A. K. Hassan and R. D. Gould, *J. Phys: Condens. Matter* 1, 6679 (1989).
73. A. K. Hassan and R. D. Gould, *Int. J. Electron.* 74, 59 (1993).
74. S. Gravano, A. K. Hassan, and R. D. Gould, *Int. J. Electron.* 70, 477 (1991).
75. R. D. Gould, *J. Phys. D: Appl. Phys.* 9, 1785 (1986).
76. R. Zuleeg and R. S. Muller, *Solid State Electron.* 7, 575 (1964).
77. F. A. Pizzarello, *J. Appl. Phys.* 35, 2730 (1964).
78. C. Lhermitte, D. Carles, and C. Vautier, *Thin Solid Films* 28, 269 (1975).
79. C. Lhermitte and C. Vautier, *Thin Solid Films* 58, 83 (1979).
80. B. M. Basol and O. M. Staffsudd, *Solid State Electron.* 24, 121 (1981).
81. V. S. Dharmadhikari, *Int. J. Electron.* 54, 787 (1983).
82. B. B. Ismail and R. D. Gould, *Phys. Stat. Sol. (A)* 115, 237 (1989).
83. R. D. Gould and B. B. Ismail, *Int. J. Electron.* 69, 19 (1990).
84. R. Glew, *Thin Solid Films* 46, 59 (1977).
85. R. K. Pandey, R. B. Gore, and A. J. N. Rooz, *J. Phys. D: Appl. Phys.* 20, 1059 (1987).
86. K. N. Sharma and K. Barua, *J. Phys. D: Appl. Phys.* 12, 1729 (1979).
87. A. O. Oduor and R. D. Gould, *Thin Solid Films* 270, 387 (1995).
88. A. O. Oduor and R. D. Gould, *Thin Solid Films* 317, 409 (1998).
89. J. Frenkel, *Phys. Rev.* 54, 647 (1938).
90. C. A. Mead, *Phys. Rev.* 128, 2088 (1962).
91. J. G. Simmons, *Phys. Rev.* 155, 657 (1967).
92. R. D. Gould and C. J. Bowler, *Thin Solid Films* 164, 281 (1988).
93. H. Hirose and Y. Wada, *Japan. J. Appl. Phys.* 3, 179 (1964).
94. I. T. Johansen, *J. Appl. Phys.* 37, 449 (1966).
95. T. E. Hartman, J. C. Blair, and R. Bauer, *J. Appl. Phys.* 37, 2488 (1966).
96. A. Servini and A. K. Jonscher, *Thin Solid Films* 3, 341 (1969).
97. A. K. Jonscher, *Thin Solid Films* 1, 213 (1967).
98. A. K. Jonscher and A. A. Ansari, *Philos. Mag.* 23, 205 (1971).
99. S. M. Sze, *J. Appl. Phys.* 38, 2951 (1967).
100. J. Kašpar, I. Emmer, and R. A. Collins, *Int. J. Electron.* 76, 793 (1994).
101. H. Murray and A. Tosser, *Thin Solid Films* 36, 247 (1976).
102. A. Piel and H. Murray, *Thin Solid Films* 44, 65 (1977).
103. S. Gogoi and K. Barua, *Thin Solid Films* 92, 227 (1982).
104. R. D. Gould and B. B. Ismail, *Vacuum* 50, 99 (1998).
105. J. G. Simmons and R. R. Verderber, *Proc. Roy. Soc. A* 301, 77 (1967).
106. T. W. Hickmott, *J. Appl. Phys.* 33, 2669 (1962).
107. T. W. Hickmott, *J. Appl. Phys.* 35, 2679 (1964).
108. G. Dearnaley, D. V. Morgan, and A. M. Stoneham, *J. Non-Cryst. Solids* 4, 593 (1970).
109. A. K. Ray and C. A. Hogarth, *Int. J. Electron.* 57, 1 (1984).
110. H. Pagnia and N. Sotnik, *Phys. Stat. Sol. (A)* 108, 11 (1988).
111. A. K. Ray and C. A. Hogarth, *Int. J. Electron.* 69, 97 (1990).
112. G. Dearnaley, *Phys. Letters* 25A, 760 (1967).
113. G. Dearnaley, *Thin Solid Films* 3, 161 (1969).
114. R. D. Gould, *Thin Solid Films* 57, 33 (1979).
115. R. D. Gould, *J. Non-Cryst. Solids* 55, 363 (1983).
116. R. R. Sutherland, *J. Phys. D: Appl. Phys.* 4, 468 (1971).
117. A. K. Ray and C. A. Hogarth, *J. Phys. D: Appl. Phys.* 20, 552 (1987).
118. A. K. Ray and C. A. Hogarth, *Thin Solid Films* 127, 69 (1985).
119. S. Gravano and R. D. Gould, *Int. J. Electron.* 73, 315 (1992).
120. S. Gravano and R. D. Gould, *Thin Solid Films* 248, 263 (1994).
121. S. Gravano, E. Amr, R. D. Gould, and M. Abu Samra, *Thin Solid Films*, 433, 321 (2003).
122. S. Bräuer, H. Pagnia, and N. Sotnik, *Int. J. Electron.* 76, 707 (1994).
123. S. Elliott, "The Physics and Chemistry of Solids." Wiley, New York, 1998.

Encyclopedia of Nanoscience and Nanotechnology
www.aspbs.com/enn

HiPco Process for Single-Wall Carbon Nanotube Production

Pavel Nikolaev
G.B. Tech/NASA Johnson Space Center, Houston, Texas, USA

CONTENTS

1. INTRODUCTION

Single-walled carbon nanotubes (SWCNTs) exhibit many unique and useful physical and chemical properties [1]. Demonstrated methods for producing SWCNTs involve laser vaporization of metal-doped carbon targets [2], arc evaporation of metal-doped carbon electrodes [3], and chemical vapor deposition (CVD) of carbon-containing molecules such as C_2H_4, CH_4, and CO [4–6] on supported nanometer-sized metal particles that serve as catalysts for SWCNT growth [7]. All of these methods produce SWCNTs in milligram to gram quantities in a few hours. However, many potential applications of SWCNTs require kilogram to ton quantities. Laser vaporization and arc discharge are essentially batch techniques with rather limited potential for bulk production. Traditional CVD techniques employ catalyst supported on alumina or silica particles, which need to be separated from the nanotubes afterward, requiring time-consuming purification. Also, CVD techniques are better geared toward high-yield production of multiwall tubes or vapor-grown carbon fibers (VGCF), and yields of SWCNTs produced by CVD on supported catalyst are rather low.

Many groups have investigated gas-phase continuous-flow production of carbon fibers. These studies typically involve passing a mixture of carbon source gas and organometallic catalyst precursor molecules through a heated furnace. The organometallics decompose and react, forming metal clusters on which carbon fibers nucleate and grow. Tibbets et al. have reported gas-phase synthesis of carbon fibers (VGCF) in heated flowing mixtures of methane or hexane with organometallics such as iron pentacarbonyl [$Fe(CO)_5$] [8] and ferrocene [dicyclopentadienyl iron, $Fe(Cp)_2$] [9]. Sen and co-workers [10–12] have investigated the catalytic growth of carbon nanotubes in gas phase, both by direct pyrolysis of ferrocene and other metallocenes and by catalytic decomposition of hydrocarbons or carbon monoxide in the presence of metallocenes or $Fe(CO)_5$. In two of these studies, this group reported production of SWCNTs from mixtures of $Fe(CO)_5$ with acetylene [10] and benzene [12]. Cheng and co-workers have reported the production of SWCNT in a heated flow of benzene and ferrocene [13, 14]. Recently Ploenjes et al. reported SWCNT production in low-pressure vibrationally nonequilibrium CO excited by a continuous wave CO laser in the presence of a small amount of $Fe(CO)_5$ [15].

To date, all but one [15] investigation reporting gas-phase continuous-flow production of SWCNT relied on hydrocarbons as the carbon source molecule for nanotube growth. However, hydrocarbons pyrolize readily on most surfaces heated above 600 to 700 °C (temperatures required for growth of SWCNT), forming graphitic deposits or amorphous carbon. Nanotubes grown in heated flows of gaseous hydrocarbons often show substantial amorphous carbon overcoating, which would need to be removed in subsequent processing steps. One may anticipate that such effects would complicate efforts to scale up production methods using hydrocarbon feedstocks.

HiPco is a new gas-phase process for single-wall carbon nanotube production discovered at Rice University in early 1998. HiPco stands for "high-pressure CO." In the HiPco process nanotubes grow in high-pressure, high-temperature flowing CO on catalytic clusters of iron. Catalyst is formed *in-situ* by thermal decomposition of iron pentacarbonyl, which is delivered intact within a cold CO flow and then rapidly mixed with hot CO in the reaction zone. Upon heating, the $Fe(CO)_5$ decomposes into atoms which condense into larger clusters. SWCNTs nucleate and grow on these particles in gas phase via CO disproportionation

ISBN: 1-58883-059-4/$35.00

(Boudouard reaction): $CO + CO \xrightarrow{Fe} CO_2 + C_{(SWCNT)}$, catalyzed by the Fe surface. The rate of the Boudouard reaction scales as a square of CO pressure, hence the importance of high pressure of CO for efficient SWCNT production.

The HiPco process is an attractive alternative to other methods of nanotube production for several reasons. It is a gas-phase process which does not employ premade catalyst particles, like other varieties of CVD techniques. It is a continuous rather than a batch process, and the current HiPco apparatus is able to run continuously for up to 2 weeks at a time, limited by the capacity of the nanotube collector. Nanotubes produced by HiPco have significant amount of Fe impurity, but Fe particles are not encased in heavy graphitic shells like in the case of arc or laser and are relatively easy to remove. Most importantly, HiPco is amenable for relatively easy scale-up by chemical engineers, and this undertaking is currently underway at Carbon Nanotechnologies Inc. in Houston, TX.

The main limitation of the HiPco process is the low rate of the Boudouard reaction. Even at 30 atm pressure, only about 1 out of 10,000 CO molecules entering the reactor is converted into nanotubes. The rest are recirculated through the reactor over and over again. Since the amount of $Fe(CO)_5$ entering the reactor is constant (~1.5 ppm) and all of it decomposes into iron particles, this results in ~5–6 at% of iron in nanotubes.

2. HIGH-PRESSURE CO (HiPco) DISCOVERY AND EARLY EXPERIMENTS

Nanotube production in gas phase in high-pressure CO was discovered and initially studied in a simple flow cell reactor consisting of a 1 in. Outside diameter thick-walled quartz flow tube contained within a tube furnace, through which reactant gases are flowed. It could withstand pressures up to 10 atm. The tube section inside the furnace was maintained between 800 and 1200 °C, while the tube entrance and exit were maintained at room temperature. The rate at which the reactant gases were heated had substantial effects on the amount and quality of SWCNT produced. In some experiments the CO and $Fe(CO)_5$ were introduced through a water-cooled injector positioned inside the quartz tube, which maintained the gases at low temperature until they were injected into the furnace, resulting in rapid heating. Around the exit of this injector could be positioned a circle of four narrow-gauge steel needles through which preheated CO was passed at high flow rate to mix with the cool flow emerging from the injector, further increasing the heating rate of the injected gas. The CO sprayed from this "showerhead" mixer was preheated by passing through a helical heat exchanger positioned within the furnace.

2.1. Feedstock and Catalyst

The gas mixture consisted of CO (1–10 atm flowing at 1-2 slm) dosed with a small amount (0–25 mTorr) of gaseous $Fe(CO)_5$. Dosing was accomplished by passing a measured fraction of the CO stream through a $Fe(CO)_5$-filled bubbler maintained at 0 °C. This produced a partial pressure of $Fe(CO)_5$ equal to its vapor pressure of ~7 Torr, and the partial pressure of $Fe(CO)_5$ was subsequently reduced by diluting with additional CO before its introduction into the flow cell. Commercial CO will always contain a few ppm of $Fe(CO)_5$ as a contaminant, which was filtered out using an in-line catalytic purifier (Oxyzorb, Matheson) prior to introduction into flow system. Alternately, this small "background" concentration of $Fe(CO)_5$ was used directly in experiments with high CO pressure and/or low $Fe(CO)_5$ concentration because the bubbler was found to be unreliable under these conditions. The concentration of background $Fe(CO)_5$ in the CO stream, typically 2–10 ppm, was determined by measuring the total amount of iron deposited in the reactor.

Flow of $CO/Fe(CO)_5$ mixtures through the heated reactor resulted in black deposits on the walls of the quartz tube outside the furnace. These deposits consisted of SWCNT, non-nanotube carbon impurities and iron particles apparently overcoated with carbon. In control experiments using no $Fe(CO)_5$ (with the catalytic purifier in use), no iron or carbon reaction products were observed.

2.2. Temperature and Flow Velocity Study

Gas-phase iron pentacarbonyl will decompose rapidly at 250–300 °C [17]. Boudouard reaction occurs at a significant rate only at temperatures above 500 °C [18]. One may thus anticipate that the rate at which the gas mixture is heated through the temperature range of 200–500 °C will be important in determining the outcome of this process: too slow a heating rate could allow the Fe clusters to grow too large before they reach the temperature required for CO disproportionation to proceed, thus preventing SWCNT nucleation and growth in the gas phase. This was confirmed in initial experiments conducted without the cooled injector. In these experiments, 6.5 atm of unpurified CO was flowed directly through the quartz tube heated to 850 °C. It was determined that the CO contained 4–5 ppm of $Fe(CO)_5$, giving a $Fe(CO)_5$ partial pressure of 20–25 mTorr. When the flow velocity in the region of transition from room temperature to SWCNT growth temperature (about 10 cm long) was around 10–20 cm/sec, it was found that SWCNTs were produced and deposited as thin mats or films on the cold parts of the quartz tube. Figure 1 shows transmission electron microscope (TEM) and scanning electron microscope (SEM) images of the SWCNT material from these experiments. When the flow velocity was slower,[1] 0.5–1 cm/sec, no SWCNTs were formed, and instead multiwall fibers were found deposited at the oven entrance. Approximately 3.6 mg of material was produced in 16 hours of gas flow at 1500 sccm. EDS and TGA a showed that the material contained 45% of iron atoms, or 75% iron by weight. Such a level of purity and production rate is very poor, but nevertheless this experiment demonstrated the feasibility of the gas-phase approach.

[1] This experiment was performed with the furnace positioned vertically, which forced a uniform gas flow. In the former experiment the furnace was positioned horizontally, and gas was drawn into the furnace much faster (10–20 cm/sec) due to formation of a convection cell.

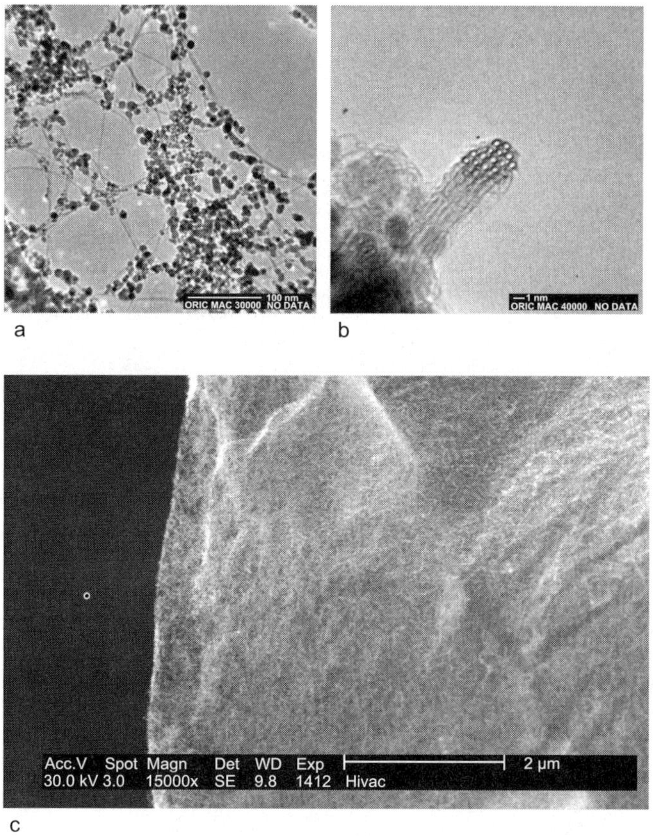

Figure 1. TEM (a, b) and SEM (c) images of SWCNT ropes produced in a first experiment [6.5 atm CO with 5 ppm $Fe(CO)_5$ at 850 °C]. (b) and (c) Reprinted with permission from [16], P. Nikolaev et al., *Chem. Phys. Lett.*, 313, 91 (1999). © 1999, Elsevier Science.

Much more rapid heating of the $CO/Fe(CO)_5$ mixture was achieved using the cooled injector, and this setup was used to study the dependence of SWCNT production on temperature and pressure. In these experiments, CO containing 5 ppm of $Fe(CO)_5$ was flowed through the heated reactor tube at 1000 sccm for 5–15 hours. Nanotube containing material was deposited on the walls of the quartz tube and on a filter positioned downstream of the reactor. This material was collected and weighed, and the mass fraction of carbon was determined by TGA. TEM observations of the product material suggest that most of the carbon was present as SWCNT, rather than other forms such as amorphous carbon. TEM images also gave the diameter distributions of the product SWCNT. Very few tube ends were observed, implying that the tubes are long compared to their width: average tube length was estimated to be approximately one micrometer, based on the number of tube ends observed.

2.3. Process Optimization

Table 1 lists the rate of SWCNT production vs temperature at 10 atm. The highest production rate was achieved at the highest available temperature, 1200 °C. Next CO pressure was varied from 1 to 10 atm while the relative concentration of $Fe(CO)_5$ was maintained at 5 ppm and oven temperature at 1200 °C. It was found that production rate increased with increasing pressure up to 10 atm, the maximum pressure that could be achieved. Table 2 lists the SWCNT production rate vs CO pressure. It was also observed that nanotubes consistently showed smaller diameters as pressure was increased. Figure 2 shows the diameter distributions observed at various pressures. Reaction at 10 atm gave tubes as small as 0.7 nm in diameter, which is about the diameter of a C_{60} molecule.

Table 1. Nanotube production rate and yield vs temperature at 10 atm pressure.

Temperature (°C)	Production rate (mg/hr)	SWCNT yield (mol%)	SWCNT yield (wt%)
850	0.68	73%	37%
1000	1.00	71%	34%
1200	1.24	79%	44%

Source: Reprinted with permission from [16], P. Nikolaev et al., *Chem. Phys. Lett.* 313, 91 (1999). © 1999, from Elsevier Science.

Experiments with the cooled injector alone using higher $Fe(CO)_5$ concentrations did not give additional nanotube material. Instead, iron in excess of 5 ppm merely formed additional iron particles mixed in with the product material. This inefficient use of available iron could result from too small a heating rate of the $CO/Fe(CO)_5$ mixture. Therefore the "showerhead" mixer was implemented to achieve even faster heating of the injected gases. In experiments with this mixer the CO pressure was maintained at 3 atm rather than 10 atm in order to achieve higher gas flow velocities, since the net flow rate was limited by the mass-flow meter's capacity. The $CO/Fe(CO)_5$ mixture was flowed through the cooled injector at 1000 sccm, while hot gas (pure CO) was sprayed from the mixer at 5000 sccm. All showerhead mixer experiments were carried out at 1200 °C.

The showerhead allowed utilization of somewhat higher $Fe(CO)_5$ concentrations in the CO. $Fe(CO)_5$ concentrations of 8 ppm gave SWCNT production rates of up to 15 mg/hr. But more importantly, the showerhead led to much more efficient use of the available iron: 8 ppm of $Fe(CO)_5$ gave product material containing only 7 at% iron, compared to 31 at% obtained using 5 ppm $Fe(CO)_5$ with the cooled injector alone. The diameter distributions of the SWCNT from the showerhead configuration were similar to those from the cooled injector alone (Fig. 2). Clearly, the rate of heating of the $CO/Fe(CO)_5$ mixture has a substantial effect on the SWCNT material produced. Figure 3 shows TEM images of the SWCNT material produced using the cooled injector with the showerhead mixer. Note from the TEM images in both Figures 1 and 3 that SWCNTs produced

Table 2. Nanotube production rate and yield vs CO pressure at 1200 °C temperature.

Pressure (atm)	Production rate (mg/hr)	SWCNT yield (mol%)	SWCNT yield (wt%)
1	1.16	61%	25%
3	1.38	67%	30%
10	1.24	79%	44%

Source: Reprinted with permission from [16], P. Nikolaev et al., *Chem. Phys. Lett.* 313, 91 (1999). © 1999, Elsevier Science.

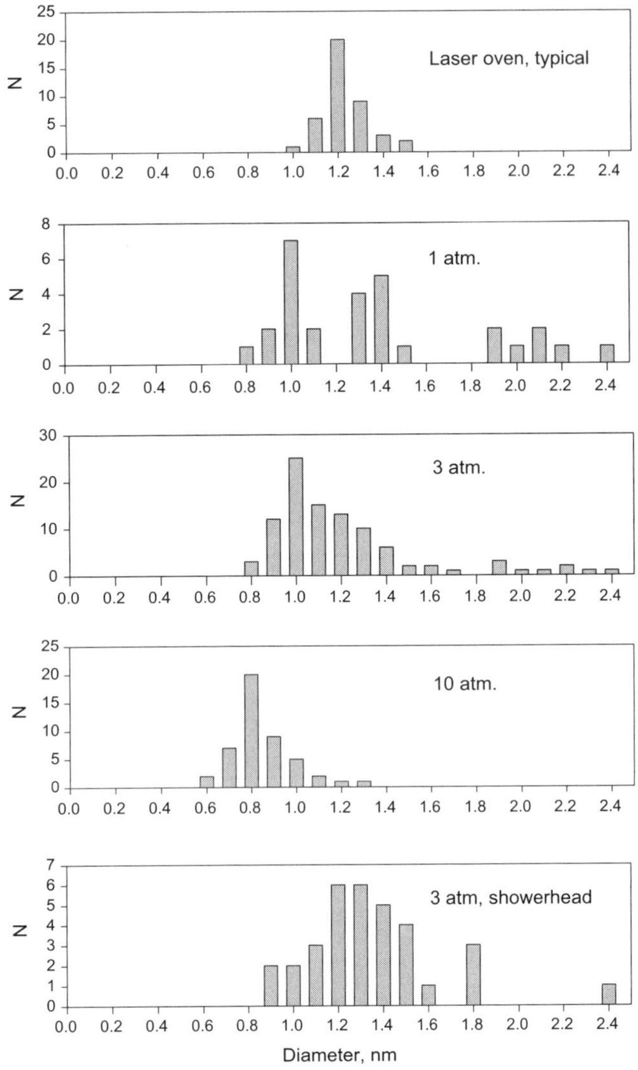

Figure 2. Diameter distributions of SWCNTs produced in the first-generation reactor at 1200 °C and varying CO pressures. Reprinted with permission from [16], P. Nikolaev et al., *Chem. Phys. Lett.* 313, 91 (1999). © 1999, Elsevier Science.

by this process are essentially free of amorphous carbon overcoating.

3. SECOND-GENERATION HiPco APPARATUS

The experimental setup described has four major drawbacks. First, flow rates are limited by the capacity of mass-flow meters. Second, carbon monoxide is exhausted into the atmosphere, which is quite wasteful, since only about 1 out of 10,000 carbon atoms entering reactor in the form of CO is converted into nanotubes, and all this CO has to be heated to 1200 °C. Third, the maximum pressure is limited by the strength of the quartz tube. Fourth, stainless steel parts which are not water-cooled (showerhead mixer and spiral heat exchanger) were literally falling apart in the carbon-rich, reducing atmosphere in the reactor (CO

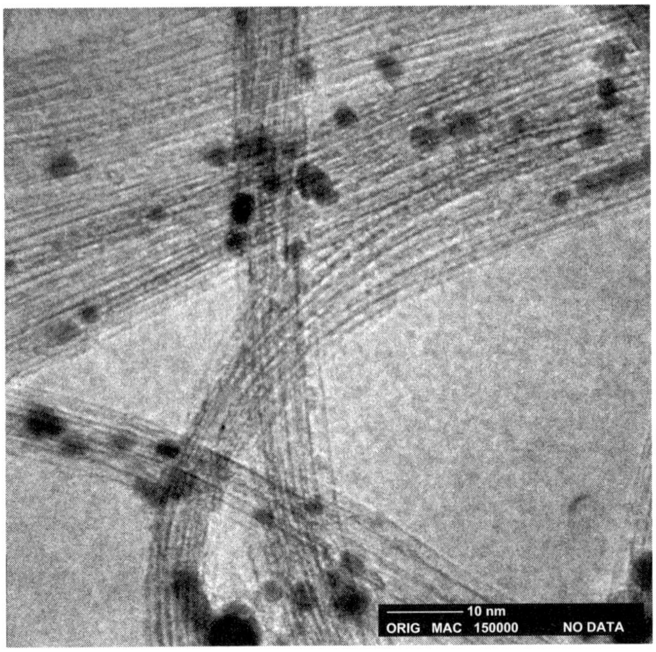

Figure 3. TEM image of SWCNT produced in the first-generation reactor using the cooled injector with showerhead mixer [3 atm CO with 8 ppm of $Fe(CO)_5$ at 1200 °C]. Reprinted with permission from [16], P. Nikolaev et al., *Chem. Phys. Lett.* 313, 91 (1999). © 1999, Elsevier Science.

disproportionates appreciably on the steel surfaces, and the resulting carbon reacts with steel making it brittle).

3.1. Closed-Loop Reactor for High-Pressure Operation

To address these issues, a second-generation reactor was built in 1999 [19]. It is a closed-loop reactor; that is, CO is recirculated through the system using a compressor (rather than being exhausted into the atmosphere). The mass-flow meters capacities were increased to match that of the compressor, allowing net flow rate of up to 300 slm (compared to 6 slm in first-generation reactor). The pressure vessel, instead of being a quartz tube, was constructed out of aluminum, allowing pressures up to 100 atm. The materials problem was solved by having all hot parts made out of graphite or quartz.

This reactor (Fig. 4) consists of a 3-in.-diameter thin-walled quartz tube surrounded by an electrical heating element, both of which are contained within a thick-walled aluminum cylinder. The heating element itself and the space between the quartz tube and the aluminum cylinder's inner wall reside under an atmosphere of argon maintained at slightly higher pressure than that of the CO inside the quartz tube. Mixtures of $Fe(CO)_5$ and CO are injected into the reactor through an insulated, air- or water-cooled stainless steel injector tipped with a copper nozzle. Around the end of the injector was positioned a circle of orifices (the "showerhead"), through which preheated CO was sprayed to mix with and heat the cold flow emerging from the catalyst injector nozzle (Fig. 5). The showerhead CO preheater consists of a thick-walled graphite tube with six longitudinally bored, equally spaced channels. Each of the six channels contains

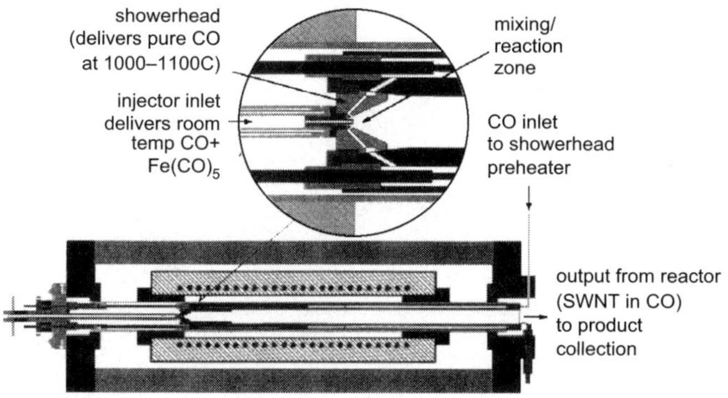

Figure 4. 4. Layout of the second-generation HiPco reactor. Reprinted with permission from [19], M. J. Bronikowski et al., *J. Vac. Sci. Technol. A* 19, 1800 (2001). © 2001, American Vacuum Society.

a resistively heated graphite rod. As shown in Figure 4, CO for the showerhead flows through these channels and is heated by contact with graphite rods. The heated CO passes then into the showerhead orifice circle (also machined from graphite) and is delivered into the reaction zone through the showerhead orifices. There, it collides and mixes with the cold injector flow emerging from the copper nozzle, which protrudes through the showerhead and into this reaction zone. The temperature of the heated CO passing through the showerhead, T_{SH}, was controlled by varying the current passed through the heating rods and was measured by thermocouples inserted into thermal wells in the front face of the graphite CO preheater. Note that because the hot showerhead CO mixes with the room temperature injector flow, the actual reaction temperature will be lower than T_{SH} by a calculable amount.

The highest SWCNT production rate was achieved with a showerhead consisting of a circle of orifices distributed equally around a circle, positioned with the orifices about 1 mm away from the injector nozzle, and set at an angle of 30° with respect to the injector flow (the quartz tube axis). Three, six, and eight orifices were tried, all with similar results. It should be noted here that due to the very large number of possible geometries for collision/mixing of gas jets, it is unlikely that the current showerhead/injector geometry represents the best design to accomplish $Fe(CO)_5$ heating and SWCNT formation. Attempts to optimize the showerhead design will be discussed in Section 5.1.

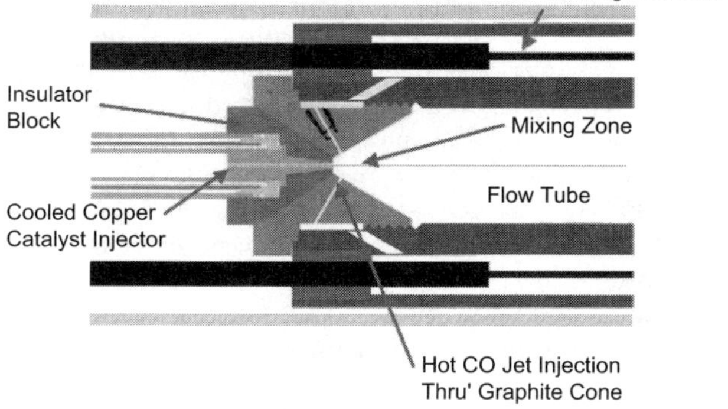

Figure 5. Close-up of the injector nozzle.

$Fe(CO)_5$ is introduced into the injector flow by passing a measured fraction of the CO flow through a liquid $Fe(CO)_5$-filled bubbler maintained at room temperature. The SWCNTs and iron particles are carried out of the reactor and pass through a series of filters and cooled surfaces, upon which the SWCNTs condense. Once purged of SWCNTs the CO gas passes through absorption beds containing NaOH and a molecular sieve (type 3A), which remove CO_2 and H_2O, respectively. The CO is then recirculated back through the gas flow system and reactor using a compressor (Fluitron, Y5-200); the reactor flow system thus constitutes a closed loop through which CO is continuously recycled. The gas flowing in the system can be analyzed with a gas chromatography/mass spectrometry analysis system (Varian, Saturn 2000), which is used to measure product CO_2 concentration in the reactor exhaust stream, carbonyl concentration in the reactant stream, and concentrations of trace gases such as H_2.

3.2. Standard Running Conditions

The standard running conditions are 450 psi (30 atm) of CO pressure and 1050 °C reactor (showerhead) temperature. CO is flowed through the showerhead preheater at 250 slm mass flow, while CO containing 0.25 Torr of $Fe(CO)_5$ is flowed through the injector at 42 slm. The total flow of CO through the reactor is thus 292 slm, and the ratio of showerhead flow to injector flow is 6:1. Typical run times are 24–72 h. Under these conditions, the reactor produces SWCNT material at a rate of approximately 450 mg/h. Figure 6 shows a TEM image of the typical product.

3.3. CO_2 Concentration as a Measure of SWCNT Production

The primary emphasis of the research involving a second-generation setup has been on optimizing the mass yield of SWCNTs per unit time with respect to available experimental variables. The behavior of the SWCNT production with

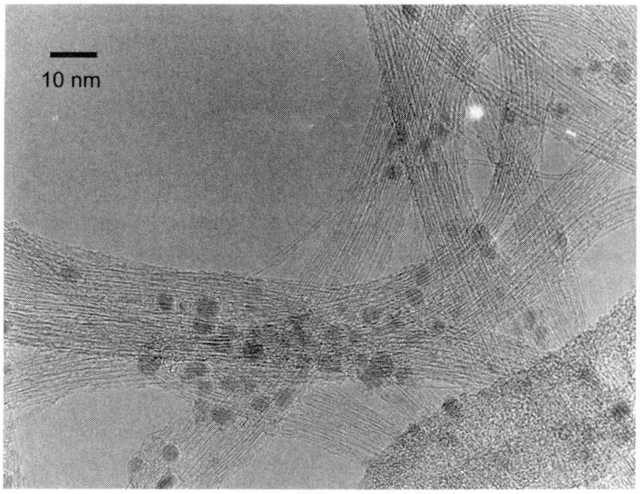

Figure 6. TEM image of the typical nanotubes produced in the second-generation HiPco reactor.

respect to such parameters also yields insight into the details of the SWCNT production mechanism.

For maximizing SWCNT production, it was found that real-time measurement of the CO_2 concentration in the gas flow exiting the reactor is a valuable tool. By varying experimental parameters while monitoring CO_2 production, it is possible to maximize CO_2 (and, hence, carbon) production with respect to any parameter. The maximum CO_2 production will correspond to the maximum production of nanotubes provided that the carbon product is in the form of nanotubes (which can be verified by microscopy), since for each CO_2 molecule there is one carbon molecule produced. Simultaneous measurements of CO_2 and $Fe(CO)_5$ pressures allow calculation of atomic carbon concentration (yield) in the resulting mixture of nanotubes and iron, according to eq. (1). $Fe(CO)_5$ in the cooled injector is diluted by a factor of 7 in the showerhead, hence the 1/7 coefficient.

$$\text{yield} = \frac{P(CO_2)}{P(CO_2) + \frac{1}{7}P(Fe(CO)_5)} \tag{1}$$

Unfortunately, the published parametric study [19] lacks such analysis and relies instead on TGA measurements exclusively.

3.4. Parametric Study of SWCNT Production

First SWCNT production was studied as a function of reaction temperature. Figure 7 gives the concentrations of CO_2 in the exhaust stream measured for various showerhead temperatures. The production of SWCNTs is very small at showerhead temperatures below 800 °C and appears to start between 850 and 900 °C. By 950 °C the SWCNT production rate is beginning to level out, reaches a maximum at 1050 °C, and then begins to fall off at higher temperatures.

CO_2 production was also investigated as a function of the concentration of $Fe(CO)_5$ in the injector gas flow. Figure 8 shows a typical result: the relative concentration of CO_2 produced as $Fe(CO)_5$ concentration was varied, at a pressure of 45 atm. It was found that CO_2 production first increases approximately linearly with increasing $Fe(CO)_5$ concentration, but that it eventually levels off. As the $Fe(CO)_5$ concentration is increased further, the CO_2 production actually

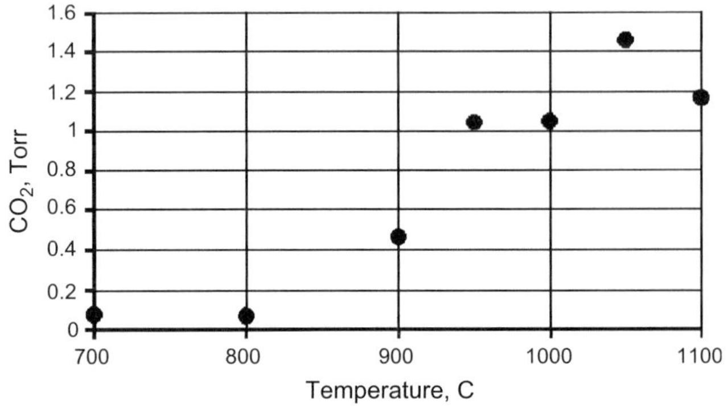

Figure 7. CO_2 yield vs reactor temperature at 30 atm pressure. Reprinted with permission from [19], M. J. Bronikowski et al., *J. Vac. Sci. Technol. A* 19, 1800 (2001). © 2001, American Vacuum Society.

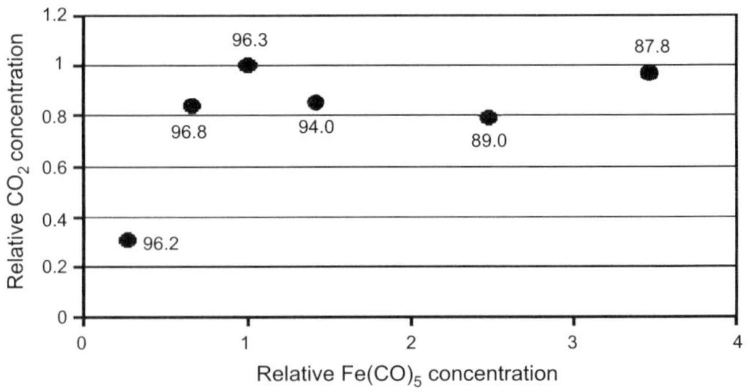

Figure 8. CO_2 yield vs $Fe(CO)_5$ concentration in the injector flow at 45 atm. Values are shown relative to $Fe(CO)_5$ and CO_2 concentrations at maximum SWCNT yield (0.38 and 1.4 Torr, respectively). Numbers indicate calculated atomic fraction of carbon in the resulting mix of SWCNT and iron. Reprinted with permission from [19], M. J. Bronikowski et al., *J. Vac. Sci. Technol. A* 19, 1800 (2001). © 2001, American Vacuum Society.

decreases and only increases again at much higher concentrations. This qualitative behavior was observed for all temperature, pressure, gas flow, and heating/mixing conditions investigated and appears quite general. SEM and TEM analysis of the SWCNTs produced at very high $Fe(CO)_5$ concentrations indicated that a much larger fraction of the carbon was present as amorphous carbon or as graphitic, fullerene-like partial shells associated with the larger metal particles, rather than as additional SWCNTs. Thus, it appears that the production of SWCNTs is optimum at the initial maximum ("turnover point") in the curve of CO_2 vs $Fe(CO)_5$; excess iron over this maximum does not give more SWCNTs but instead appears to retard the growth of SWCNTs and eventually gives much carbon in undesirable forms.

Note that operating at the turnover point serves to maximize the production of SWCNTs, without consideration of Fe in the resulting mix. Superimposed on Figure 8 are atomic fractions of carbon in the resulting mix, calculated according to the expression (1). It is obvious that nanotubes with the least amount of iron (96.8 at% of C) are produced at $Fe(CO)_5$ concentration of 0.25 Torr, at the point where the CO_2 vs $Fe(CO)_5$ curve just begins to deviate from the linear dependence. Therefore $Fe(CO)_5$ pressure can be selected to produce the maximum amount of nanotubes with slightly higher Fe content (3.7 at% Fe), or a slightly lower amount of nanotubes with minimal Fe content (3.2 at% Fe). (These numbers are for 45 atm pressure and will be different for other reactor pressures.)

SWCNT production was also studied as a function of reaction pressure. At any given pressure, the CO_2 vs $Fe(CO)_5$ curve has qualitatively the same appearance as that shown in Figure 8. The maximum CO_2 concentration (at the turnover point) and the corresponding level of $Fe(CO)_5$ were found to increase as pressure was increased up to the maximum (45 atm) in an approximately linear fashion. These results are shown in Figure 9 (normalized to standard running condition, 30 atm). Superimposed on Figure 9 are atomic fractions of carbon in the resulting mix, calculated according to the expression (1). Because of the small deviations from the linear dependence in both CO_2 and $Fe(CO)_5$ curves, nanotubes with the least amount of iron

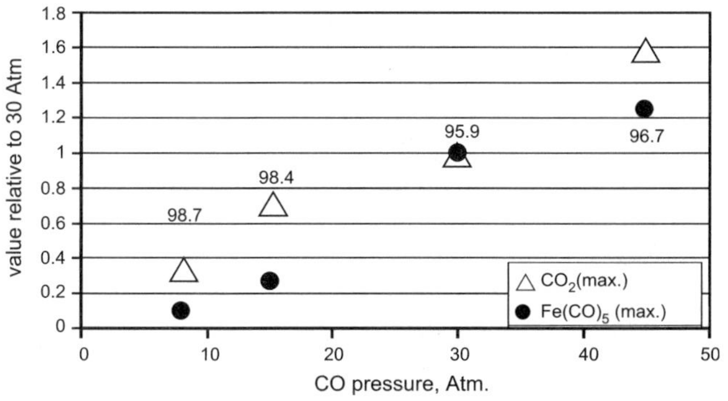

Figure 9. Maximum CO_2 produced and corresponding $Fe(CO)_5$ concentrations vs CO pressure. Values are shown relative to $Fe(CO)_5$ and CO_2 concentrations at maximum SWCNT yield at 30 atm (0.3 and 1 Torr, respectively). Numbers indicate calculated atomic fraction of carbon in the resulting mix of SWCNT and iron. Reprinted with permission from [19], M. J. Bronikowski et al., *J. Vac. Sci. Technol. A* 19, 1800 (2001). © 2001, American Vacuum Society.

are actually produced at 8 and 15 atm. Running at 30 atm produces nanotubes with the highest amount of iron. Maximum nanotube production rate was achieved at 45 atm, with the amount of iron still lower than that at 30 atm. Based on this analysis, it is not clear why 30 atm was chosen as a standard pressure, since maximum nanotube production rate is achieved at 45 atm, and nanotubes with the least amount of iron are produced at 15 atm.

4. MECHANISMS OF SWCNT GROWTH

Carbon SWCNTs have been observed to nucleate and grow from small metal clusters on supported catalysts [4, 20]. It was proposed in [19] that gas-phase nucleation and growth of SWCNTs occur in the HiPco process by an analogous mechanism: the iron carbonyl molecules decompose at high temperature to release their Fe atoms, which cluster together to form small catalytic particles from which SWCNTs nucleate and grow, all in the gas phase. Gas-phase iron pentacarbonyl decomposes rapidly above 300 °C [17]. CO disproportionation occurs at a significant rate only at temperatures above 500 °C [18]. Thus, the rate at which the gas mixture is heated through the temperature range of 300–500 °C will be important in determining the outcome of this process. If this rate is too slow, the clusters may grow too big to nucleate nanotubes. Instead, they would simply overcoat with carbon [4]. On the other hand, the per-atom binding energy for small iron clusters (2–10 atoms) is typically 1.5–2.5 eV, much less than the bulk value of 4.28 eV [21]. At temperatures where SWCNT formation occurs, such small clusters would tend to evaporate quickly. Thus, heating the gas mixture too quickly might curtail the formation of SWCNTs.

4.1. Cluster Formation and Inactivation

If one assumes that all carbonyl molecules are first stripped of CO to form atomic iron vapor and then begin to form Fe_n clusters, there appears to be a bottleneck in the cluster formation: the $Fe + Fe \rightarrow Fe_2$ reaction. Moreover, once a few Fe_2 clusters are formed, the rest of Fe atoms would react much easier with Fe_2 than with each other. This would lead to preferred formation of large Fe clusters not suitable for SWCNT nucleation. Scott et al. have attempted to simulate numerically the relevant iron–carbon cluster kinetics in the nucleation zone of the HiPco reactor [22]. One of the interesting results which emerged from these simulations is that the Fe dimer formation occurs through other, much more effective channels:

$$Fe(CO) + Fe \rightarrow Fe_2 + CO \quad \text{and}$$
$$Fe(CO) + Fe(CO) \rightarrow Fe_2 + 2CO \qquad (2)$$

At the same time, in addition to direct evaporation of small Fe clusters ($Fe_n \rightarrow Fe_{n-1} + Fe$), they can evaporate via carbonyl formation: $Fe_n + CO \rightarrow Fe_{n-1} + Fe(CO)$. This is relevant to the mechanism by which nanotube growth ceases. Since the per-atom binding energy of small Fe clusters is smaller than that of larger clusters, any evaporation of smaller clusters which are suitable for SWCNT nucleation will lead to growth of larger clusters, which presumably have already nucleated nanotubes. As a catalytic particle grows larger, a more and more catalytically active surface area is created, and eventually a graphitic shell or other structures not associated with the growing nanotube will begin to form on the particle. Eventually the particle will become covered with carbon, preventing the diffusion of additional CO to the particle's surface and terminating further nanotube growth. Alternately, if a growing nanotube is attached to the small cluster, it may evaporate completely too early, ceasing nanotube growth. The cessation of growth of the nucleated nanotubes would thus derive from a combination of these two effects: some nanotubes would stop growing when their attached catalyst particle evaporates or becomes too small, and some would stop when their catalyst cluster grows too large.

4.2. Discussion

The observed temperature dependence of SWCNT production is consistent with this picture. Both the decomposition of $Fe(CO)_5$ and the rate of CO disproportionation become appreciable only at high temperatures, so one expects that the SWCNT production rate will initially increase as the temperature is increased. At temperatures above 1050 °C the SWCNT production rate begins to decline. Apparently, at higher temperatures the rate of evaporation of small, active catalytic clusters is fast enough that growth of SWCNTs is strongly curtailed, overcoming gains in growth arising from faster $Fe(CO)_5$ decomposition and a greater rate of CO disproportionation.

SWCNT yield as a function of $Fe(CO)_5$ concentration can also be understood in terms of this SWCNT growth model. At low concentrations of iron, accretion of Fe onto already growing nanotubes is slow enough that the faster death of growing nanotubes is more than offset by the nucleation of additional nanotubes, as Fe concentration increases. The addition of more iron thus results simply in the formation of proportionally more nanotubes. As $Fe(CO)_5$ concentration is increased further, we see a slowing down in the rate of

SWCNT formation increase, and at sufficiently high concentrations the differential SWCNT yield can actually be negative. This behavior could indicate a concentration regime in which diminished nanotube production due to Fe accretion onto growing nanotubes exceeds increased production from formation of additional catalytic particles. Above a critical "turnover" $Fe(CO)_5$ concentration, the rate of Fe accretion becomes large enough, and the resultant shortening of a nanotube growth time severe enough, that the overall yield of SWCNTs decreases with increasing $Fe(CO)_5$ concentration. Eventually, at very high iron concentrations, the yield of carbon turns around and begins increasing again, but as mentioned, the extra carbon is in forms other than nanotubes. This behavior is qualitatively independent of pressure. Quantitatively, the turnover concentration increases as pressure increases (Fig. 9). Higher CO pressure may simply cause a greater SWCNT growth rate from a catalytic cluster, allowing the SWCNT to grow longer before its attached particle is deactivated. Alternately, higher CO pressure may give a higher nucleation rate of nanotubes from very small clusters, so that more small clusters nucleate nanotubes and fewer collide with and adsorb onto previously nucleated tubes. Results of the parametric study [19] suggest that limits on SWCNT production arise primarily from inactivation of Fe catalyst particles once they become too large. There could be other mechanisms of catalyst inactivation.

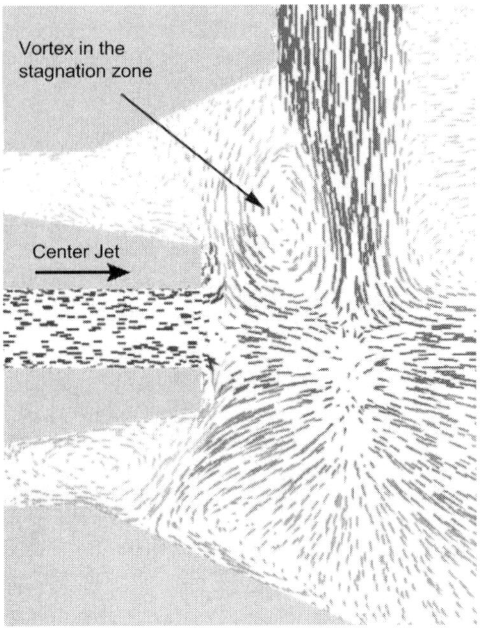

Figure 10. Flow field in the x–z section for the nozzle configuration with three peripheral jets positioned perpendicular to the central jet. Velocity vectors are colored with local temperature. Reprinted with permission from [24], A. Povitsky and M. D. Salas, ICASE Report 2001-4, February 2001.

5. NUMERICAL MODELING OF HiPco PROCESS

5.1. Nozzle Design

Povitsky and Salas [23, 24] have attempted to optimize the showerhead geometry by modeling the gas flow in the nozzle numerically using the FLUENT code (three-dimensional, steady–state Navier-Stokes equations combined with RNG k–ε turbulence model). They report that for the showerhead geometry with three peripheral jets described previously, the central part of the central cold jet penetrates the peripheral jets, while the peripheral part of the central jet forms a vortex in the stagnation zone between jets (Fig. 10). There are generally three types of trajectories originating in the central cold jet (Fig. 11). The first type deflects toward the nozzle axis and provide relatively slow and monotonic heating. These originate close to the nozzle axis and in-plane with peripheral jets. The second type deflects away from the nozzle axis and provide faster and monotonic heating. These originate close to the nozzle axis and in between the peripheral jets planes. The third type represents trajectories in the vortex. These originate on the periphery of the central jet and provide the fastest heating, but temperature profiles exhibit large oscillations. The heating rate is within $(0.8–1.5)\times 10^5$ K/s for the geometry described previously (30° angle between central and peripheral jets) and can be increased to $(1–2.5\times10^5$ K/s if the angle is increased to 45°. What is probably most important is not the heating rate itself (it can be adjusted by changing mass flow rates in the nozzle), but the variations in the heating rates for different trajectory types and temperature oscillations of the third type trajectories. It is clear that gas flowing through the showerhead along different types of trajectories experiences widely varying temperature histories. It is possible that just one type of temperature history is optimal for nanotube production, while others simply produce unwanted iron particles. Determining the optimal temperature history and the relevant nozzle design computationally is a formidable task.

Povitsky and Salas have proposed a nozzle design with three (instead of one) central jets and three peripheral jets spaced by 60°. It all but eliminates type one trajectories and increases the heating rate to $(3.5–4.5)\times10^5$ K/s but has not been explored experimentally yet. Another design studied in [24] is a showerhead with two peripheral jets which are

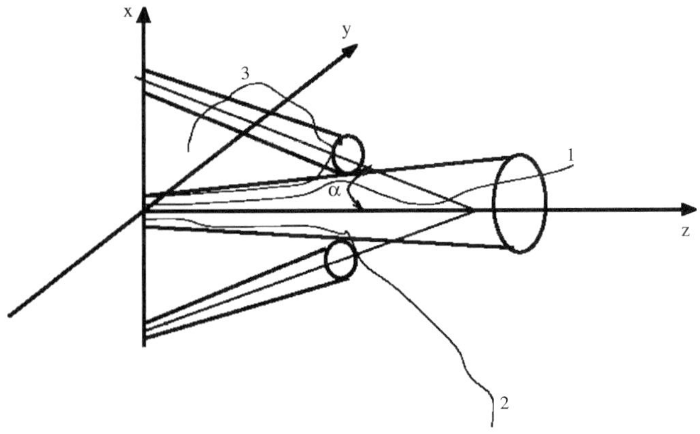

Figure 11. Three types of trajectories: (1) Trajectory bends inward in the central jet. (2) Trajectory bends outward. (3) Trajectory involved in the vortex formed upstream of the intersection of the jets. Reprinted with permission from [24], A. Povitsky and M. D. Salas, ICASE Report 2001-4, February 2001.

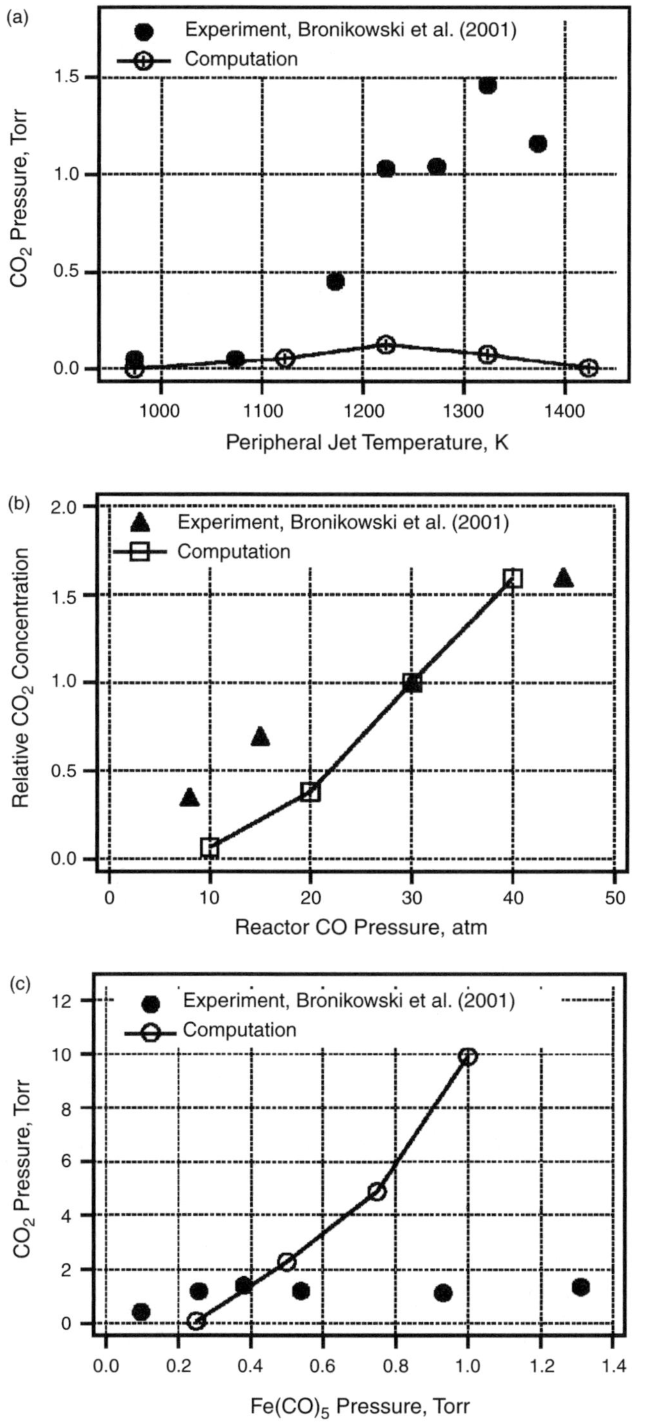

Figure 12. Computed CO_2 production compared to experiment. (a) CO_2 production vs temperature. (b) CO_2 production vs CO pressure. (c) CO_2 production vs $Fe(CO)_5$ concentration. Reprinted with permission from [26], T. Gökçen et al., *J. Nanosci. Nanotech.* 2, 535 (2002). © 2002, American Scientific Publishers.

positioned perpendicular to the central jet, which produces mostly type three trajectories. This design was tried in the HiPco reactor [25] and produced very low nanotube yields, indicating that very fast heating combined with large temperature oscillations and large spread in the heating rates does not constitute optimal conditions.

Another attempt to model the gas dynamics in the HiPco nozzle has been published recently [26] by Gokcen et al. It does not take the turbulence into account. Typical Reynolds numbers are around 1000–2000, but jets are still likely to be nonlaminar due to nonparallel interaction between them. The authors approximated three-dimensional (3D) nozzle geometry with 2D axisymmetrical design; that is, peripheral hot CO is injected through an axisymmetrical circular gap, with size determined so that mass and momentum ratios of central and peripheral flows match the experimental values. This eliminates the type two trajectories mentioned and therefore the calculated flow field is probably less realistic.

5.2. Chemical Kinetics

The flow field calculated in [26] was fully coupled to the chemical kinetic model developed by Dateo et al. [27]. Authors then attempted to reproduce the results of the parametric study discussed previously [19], that is, CO_2 production vs temperature, overall pressure, and carbonyl partial pressure [26] (Fig. 12). The calculated CO_2 production vs temperature is qualitatively similar to the experiment, but it reaches maximum at 1225 K (compared to the 1325 K in experiment) and significantly underestimates the overall CO_2 production. A large percentage of iron exits the reactor

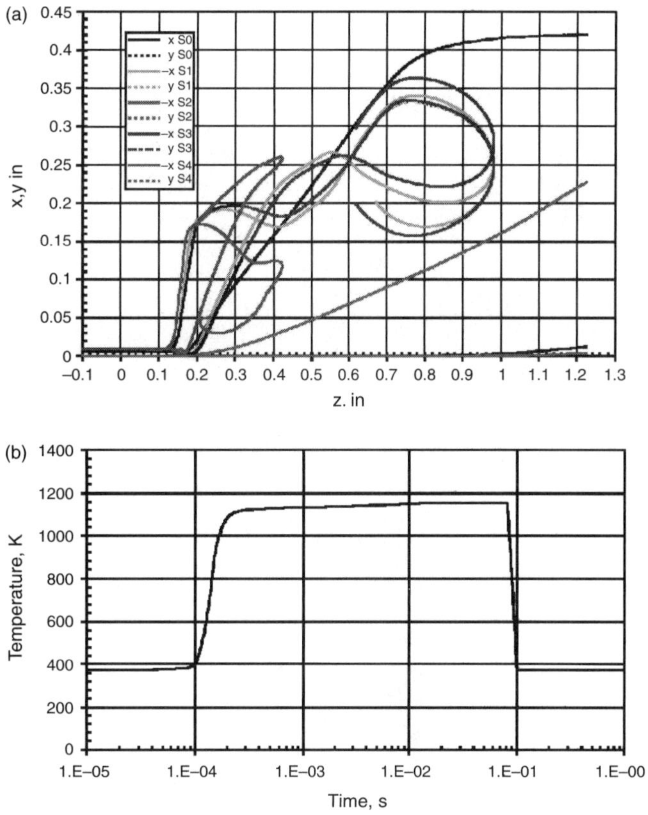

Figure 13. (a) Computed trajectories (x and y coordinates vs axial distance) in the standard nozzle geometry. S2 is type 1, S3 is type 2, and S0, S5, and S5, are type 3. (b) Temperature history along type 1 trajectory (S2). Reprinted with permission from [22], C. D. Scott et al., *J. Nanosci. Nanotech.* 3, 63 (2003). © 2003, American Scientific Publishers.

in the form of Fe atoms, because apparently cluster evaporation becomes significant above this temperature and competes with cluster formation and nucleation of nanotubes. This suggests that dimer dissociation and cluster evaporation rates used in the chemical kinetics modes are too high. It is also possible that the overall computed heating rates are slower than in experiment due to the choice of the nozzle geometry and failure to take turbulence into account. The qualitative behavior of CO_2 production vs CO pressure follows the experiment, except that again it is underestimated by about an order of magnitude. The computed CO_2 production vs $Fe(CO)_5$ concentration dependence does not exhibit the turnover effect seen in the experiment. Bronikowski et al. argue in [19] that CO_2 (and nanotube) production slows above turnover concentration because of Fe accretion onto existing clusters followed by inactivation. Since the kinetic model developed in [26, 27] does not take such mechanisms into account, this result is quite expected.

Scott et al. [22] have also attempted to simulate numerically the relevant iron–carbon cluster kinetics in the nucleation zone of the HiPco reactor. Several chemical kinetics models of varying degrees of complexity were developed and solved with the CHEMKIN package utilizing several sets of rate coefficients, which are mostly similar to [26, 27]. Some coefficients were adjusted to reduce the unrealistically high Fe atom concentration at the exit in [26, 27]. The chemical model, unlike in [26], was coupled to only one type one trajectory taken from [24], which means that computed cluster distributions and CO_2 production cannot reflect the experimental values. The trajectory and its temperature history are shown in Figure 13. The resulting CO_2 and iron concentrations are shown in Figure 14. All models significantly underestimate Fe and overestimate CO_2 concentrations, which is probably due to the trajectory choice (type one trajectories provide the slowest heating rate). This may also be due to the choice of the rate coefficients for nanotube formation.

The effect of the carbonyl bond strength on the nanotube nucleation was studied computationally in [28]. The authors compared nickel and iron carbonyls as catalyst precursors using the same models as in [22]. It appears that much higher Ni–CO bond strength (170 kJ/mole) compared to that of Fe(CO) (23.9 kJ/mole) prevents dimer formation via reaction (2), thus decreasing CO_2 production by 12–22 orders of magnitude. Indeed, no nanotube formation was observed with pure $Ni(CO)_4$ as a catalyst [25].

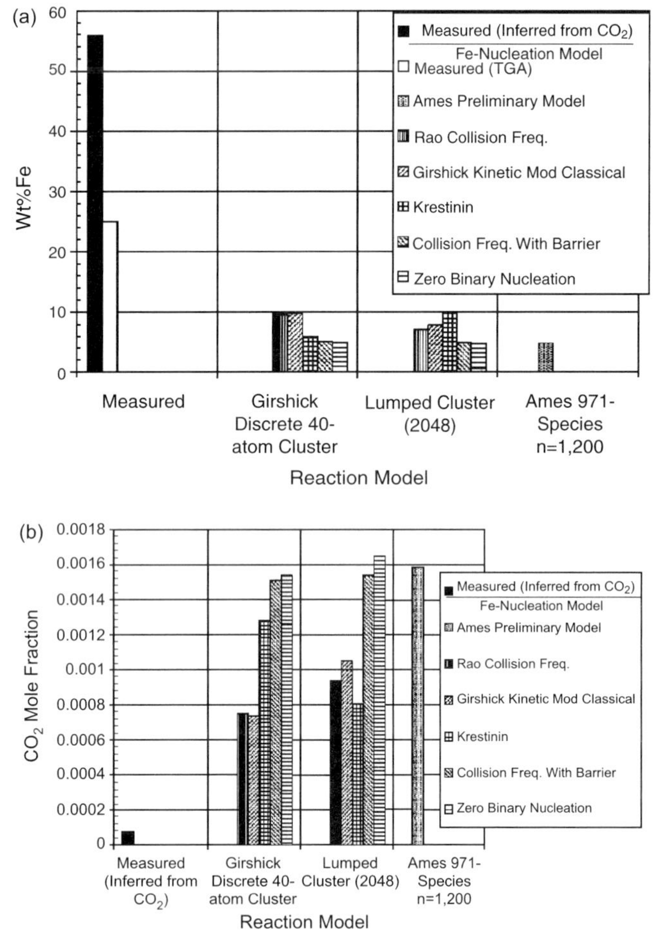

Figure 14. Iron (a) and CO_2 (b) concentration at the reactor exit, computed along single type 1 trajectory. Reprinted with permission from [22], C. D. Scott et al., *J. Nanosci. Nanotech.* 3, 63 (2003). © 2003, American Scientific Publishers.

6. CONCLUSIONS

The HiPco process has demonstrated its potential to produce single-wall carbon nanotubes inexpensively and on a large scale. Nanotubes contain significant amounts of Fe impurity (3–6 at%), which fortunately is much easier to remove compared to SWCNTs produced by other methods. The amount of nontubular carbon impurities seems to be low. The prevalence of iron in the nanotubes underscores the low efficiency of CO conversion into nanotubes. Nanotube yield can be improved by more efficient nucleation and prevention of the catalyst deactivation in the later stages of SWCNT growth. More effort is necessary to better understand the gas dynamics and cluster kinetics of the process.

GLOSSARY

Chemical vapor deposition Chemical process based on catalytically assisted decomposition of gas-phase precursor molecules.

Energy dispersive spectroscopy (EDS) A microanalytical technique that is based on the characteristic X-ray peaks that are generated when the high energy beam of the electron microscope interacts with the specimen. Each element yields a characteristic spectral fingerprint that may be used to identify the presence and relative concentration of that element within the sample.

Thermogravimetry analysis (TGA) An analytical technique that in which the mass of a substance (and/or its reaction products) is measured as a function of temperature whilst the substance is subjected to a controlled temperature program.

ACKNOWLEDGMENTS

The author thanks Carl Scott, Sivaram Arepalli, and Richard Smalley for helpful discussions. Financial help from NASA grant NAS9-19100 is greatly appreciated.

REFERENCES

1. B. I. Yakobson and R. E. Smalley, *Am. Sci.* 85, 324 (1997).
2. A. Thess, P. Nikolaev, H. Dai, P. Petit, J. Robert, C. H. Xu, Y. H. Lee, S. G. Kim, A. G. Rinzler, D. T. Colbert, G. E. Scuseria, D. Tomanek, J. E. Fischer, and R. E. Smalley, *Science* 273, 483 (1996).
3. C. Journet, W. K. Maser, P. Bernier, A. Loiseau, M. L. Delachapelle, S. Lefrant, P. Deniard, R. Lee, and J. E. Fischer, *Nature* 388, 756 (1997).
4. J. H. Hafner, M. J. Bronikowski, B. R. Azamian, P. Nikolaev, A. G. Rinzler, D. T. Colbert, K. A. Smith, and R. E. Smalley, *Chem. Phys. Lett.* 296, 195 (1998).
5. M. Su and J. Liu, *Chem. Phys. Lett.* 322, 321 (2000).
6. W. E. Alvarez, F. Pompeo, L. Balzano, J. E. Herrera, B. Kitiyanan, A. Borgna, and D. E. Resasco, *J. Nanopart, Res.* 4, 131 (2002).
7. J. Kong, A. M. Cassel, and H. Dai, *Chem. Phys. Lett.* 292, 567 (1998).
8. G. G. Tibbetts, C. A. Bernardo, D. W. Gorkiewitz, and R. L. Alig, *Carbon* 32, 569 (1994).
9. G. G. Tibbetts, D. W. Gorkiewicz, and R. L. Alig, *Carbon* 31, 809 (1993).
10. B. C. Satishkumar, A. Govindaraj, R. Sen, and C. N. R. Rao, *Chem. Phys. Lett.* 293, 47 (1998).
11. R. A. Sen, A. Govindaraj, and C. N. R. Rao, *Chem. Phys. Lett.* 267, 276 (1998).
12. R. A. Sen, A. Govindaraj, and C. N. R. Rao, *Chem. Mater.* 9, 2078 (1997).
13. H. M. Cheng, F. Li, G. Su, H. Y. Pan, L. L. He, X. Sun, and M. S. Dresselhaus, *Appl. Phys. Lett.* 72, 3282 (1998).
14. H. M. Cheng, F. Li, X. Sun, S. D. M. Brown, M. A. Pimenta, A. Marucci, G. Dresselhaus, and M. S. Dresselhaus, *Chem. Phys. Lett.* 289, 602 (1998).
15. E. Ploenjes, P. Palm, G. B. Viswanathan, V. V. Subramaniam, I. V. Adamovich, W. R. Lempert, H. L. Frasier, and J. W. Rich, *Chem. Phys. Lett.* 352, 342 (2002).
16. P. Nikolaev, M. J. Bronikowski, R. K. Bradley, F. Rohmund, D. T. Colbert, K. A. Smith, and R. E. Smalley, *Chem. Phys. Lett.* 313, 91 (1999).
17. D. Nicholls, in "Comprehensive Inorganic Chemistry" (J. C. Bailar Jr., H. J. Emeleus, R. Nyholm, and A. F. Trotman-Dickenson, eds.), Vol. 3, P. 990. Pergamon, Oxford, 1973.
18. G. D. Renshaw, C. Roscoe, and P. L. Walker, Jr., *J. Catal.* 18, 164 (1970).
19. M. J. Bronikowski, P. A. Willis, D. T. Colbert, K. A. Smith, and R. E. Smalley, *J. Vac. Sci. Technol. A* 19, 1800 (2001).
20. H. J. Dai, A. G. Rinzler, P. Nikolaev, A. Thess, D. T. Colbert, and R. E. Smalley, *Chem. Phys. Lett.* 260, 471 (1996).
21. L. Lian, C.-X. Su, and P. B. Armentrout, *J. Chem. Phys.* 97, 4072 (1992).
22. C. D. Scott, A. Povitsky, C. Dateo, T. Gökçen, P. A. Willis, and R. E. Smalley, *J. Nanosci. Nanotech.* 3, 63 (2003).
23. A. Povitsky, ICASE Report 2000-18, April 2000.
24. A. Povitsky and M. D. Salas, ICASE Report 2001-4, February 2001.
25. R. E. Smalley, personal communication.
26. T. Gökçen, C. E. Dateo, and M. Meyyappan, *J. Nanosci. Nanotech.* 2, 535 (2002).
27. C. E. Dateo, T. Gökçen, and M. Meyyappan, *J. Nanosci. Nanotech.* 2, 523 (2002).
28. C. D. Scott and R. E. Smalley, *J. Nanosci. Nanotech.* 3, 75 (2003).

Encyclopedia of Nanoscience and Nanotechnology
www.aspbs.com/enn

Hybrid Solar Cells

Elif Arici, N. Serdar Sariciftci
Johannes Kepler University, Linz, Austria

Dieter Meissner
Fachhochschule Wels, Wels, Austria

CONTENTS

1. INTRODUCTION

Ever since the invention of the silicon solar cell in the 1940s, people have acknowledged the enormous potential that photovoltaic systems have for a large-scale electricity production. But silicon wafers are expensive, so great efforts have been focused on the development of cheaper thin-film solar cells. Such films may be purely inorganic or contain organic materials as an essential part of the device. Examples of the latter are junctions consisting of thin layers of organic donor and acceptor molecules; bulk heterojunctions in which the donor and acceptor phases exist as an interpenetrating network [1]; or dye-sensitized photoelectrochemical solar cells [2, 3], in which light absorbance occurs in metallo-organic dye molecules, but generated electrons are injected into the conduction band of a wide-bandgap semiconductor, mainly highly porous nanocrystalline TiO_2 (the so-called Grätzel cells).

Hybrid materials combine the unique properties of one or more kinds of inorganic nanoparticles with the film-forming properties of polymers. The availability of preparing thin organic/inorganic hybrid materials with various semiconducting polymers has opened the door to the construction of a new class of devices, in which both components are photovoltaically active. Most of the polymers can be processed from solution at room temperature, enabling the manufacturing of large area, flexible, and lightweight devices. A convenient way to fabricate p-n junctions is shown to use a p-doping semiconducting polymer (polyhexylthiophenes) and n-type II-VI semiconductor nanoparticle CdSe [4].

1.1. Bulk Heterojunction Concept

In organic semiconductors the absorbed photons generate electron-hole pairs, that can be regarded as excitons with a binding energy (0.05–1.0 eV) much higher than that of most inorganic semiconductors. In solar cells based on pristine semiconducting polymers, charge collection requires that the neutral excited states produced by photoexcitation can be separated into free charge carriers first, the so-called exciton dissociation. The generated free charges are then transported through the device to the electrodes. The charge collection is dependent on how the carriers can reach the electrodes without recombining with oppositely charged carriers in the photoactive layer.

Exciton dissociation is known to occur very efficiently at interfaces between the two organic semiconductors mixed together in a composite thin film such as a conjugated polymer and fullerene mixtures [5–7]. The photoexcited electrons will then be accepted by the material with the higher electron affinity (electron acceptor, in this case fullerene or its derivatives), and the hole by the material with the lower ionization potential, which also acts as electron donor. Many conjugated polymers are electron donors upon photoexcitation. The idea using this property in conjunction with a molecular electron acceptor to achieve long-living charge separation was based on the stability of the photoinduced excitations on the conjugated polymer. The energy gain by electron transfer from the polymer to the electron acceptor must be sufficient to overcome the coulombic exciton binding energy in the polymer to produce charge separation.

Examples of further electron acceptor materials include also cyano-substituted conjugated polymers [8, 9], tetracyanoanthraquinodimethanes [10], and various oligomers [11].

ISBN: 1-58883-059-4/$35.00

Encyclopedia of Nanoscience and Nanotechnology
Edited by H. S. Nalwa
Volume 3: Pages (929–944)

Efficient charge separation has been realized in two different cell configurations:

- bilayer configurations of n-type layer/ p-type polymer (vice versa)
- an interconnected network structure of p- and n-type materials

In a bilayer configuration, excitons, which are generated in the polymer layer, need to reach the donor/acceptor interface, where the charge separation occurs. Since the exciton diffusion length in organic semiconductors is usually only about ~10 nm [12, 13], the photoactive layer in bilayer heterojunction structures is rarely thicker than 20 nm.

In the interconnected network structure, blending an electron-accepting and a hole-accepting material can result in the formation of domains close to the desired 20 nm length scale. Thus, most excitons can reach a donor-acceptor interface even in thicker devices. Replacing the flat interface by an interconnected network structure of p- and n-type materials leads to a high surface area in the whole volume for charge separation, that is, the "bulk heterojunction" (Fig. 1).

Based on interconnected networks of hole-conducting polymers (p-type) with percolating electron-conducting Buckminsterfullerene (C_{60}) derivatives (n-type), recently more than 2.5% solar power conversion efficiency in devices of less than 100 nm thickness of the absorber layer, in which only a relatively small portion of the solar light is absorbed, have been achieved [14].

In bulk heterojunction cells, the transport of the carriers to the electrodes without recombination is a more complicated problem to solve, since it requires that once the electrons and holes are separated onto different materials, each carrier type has a pathway to the appropriate electrode without needing to pass through a region of the other material. The nanoscopic morphology of the composite layers plays in this manner a fundamental role for efficient solar cell performance.

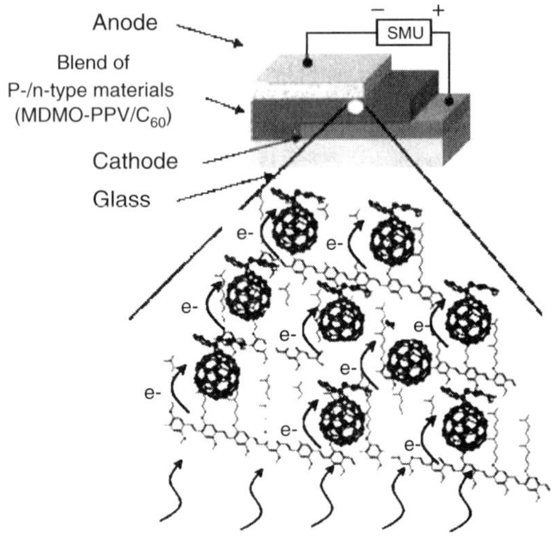

Figure 1. Formation of a bulk heterojunction in an interpenetrating donor/acceptor network.

An additional fundamental phenomenon which determines the behavior of conductivity in bulk heterojunction devices is the percolation threshold. Percolation takes place when the electron acceptor and donor concentrations reach a value high enough to provide the conduction along the connected network of each type of materials. The solubility of the n-type and p-type components in the same solvent is an important parameter for the construction of hybrid solar cells.

1.2. Nanocrystalline Semiconductors

Nanoclusters of inorganic semiconductor materials, known as nanocrystals, have been the subject of intense research in the 1990s. The primary motivation for this increased activity is related to the size-dependent new properties of these materials. For example, magnetic and semiconductor nanocrystals exhibit strongly size-dependent properties that may find applications in magnetic storage and microelectronics, key components in information technology [15–17]. The emission and absorption of semiconductor nanocrystals are tunable by size as a result of the quantum confinement effect. Also, the use of quantized nanocrystals as fluorescent or absorbing materials in optoelectronic and photovoltaic devices is an interesting and quickly developing area of research [18, 19].

There are in principle two distinct synthetic routes to create semiconductor nanostructures: physical or chemical procedure. In the "physical" approach, the particles are grown by lithographic or molecular beam techniques, whereas in the "chemical" approach, they are formed either by particle growth in various matrices or by the methods of colloidal chemistry. In this contribution, we will focus on the latter approach, which allows us to produce stable particles, which can be handled like ordinary chemical substances.

Motivated chemists have spent many years developing synthetic strategies for the preparation of nanocrystalline materials, which should have desirable chemical properties such as high crystallinity, stability against degradation, but also flexible handling properties for specific arrangement of the nanocrystals on various substrates. Nowadays, series of nanocrystals with different organic/inorganic shells, shapes, and sizes are found in the literature. Many of them are commercially available.

To exploit the full potential for the technological applications of the nanocrystalline materials, it is very important to endow them with good processing attributes. This has ultimately guided scientists toward using conventional polymers as matrix material, resulting in inorganic/organic guest-host systems generally termed "hybrid" or "composite" materials.

An important strategy was to create nanoparticles shielded by an organic ligand, the so-called "core-shell" particles. The shell can alter the dispersion characteristic of the particle by initiating attractive forces with the polymer chains, in which the particles should be homogeneously arranged. But also the stability against oxidation and against aggregation can be inhibited by tailoring the surface modification.

1.3. Hybrid Solar Cells

An effective strategy of the hybrid solar cell investigations is to fabricate blends of nanocrystals with semiconductive polymers as a photovoltaic layer. The basics of this heterojunction concept have been already described.

The investigations of inorganic semiconductor nanoparticles embedded into semiconducting polymer blends are promising materials for several reasons. The classical inorganic semiconductor materials for photovoltaic applications can have a higher absorbance coefficient and photoconductivity than organic materials mentioned earlier. Furthermore, the n- or p-type doping level of these nanocrystalline semiconductors can easily be varied by synthesis. So we can study charge transfer in composites of n-type and p-type conducting polymers with corresponding inorganic counterparts.

The first requirement in inorganic/organic hybrid solar cells is to blend a high concentration of inorganic nanoparticles into the polymer matrix, to form an interpenetrating network. Phase separation in macroscopic scale should be avoided. As discussed in Section 1.1, for an effective charge separation in electron donor/acceptor composites, the average distance between formed excitons in the two materials should be in the range of 20 nm, which determines the size of domains. For a profitable morphology of the blends, nanoscale inorganic particles should be used. To give an idea, the fullerene molecules embedded in the polymer are about <1 nm in diameter; the cluster size, however, is heavily dependent on the processing conditions. It has been shown that composite films cast from toluene solutions result in much bigger (50–500 nm scale) fullerene clusters as compared to chlorobenzene-cast films with much finer grain sizes (below 50 nm) [20, 21].

Typically, inorganic semiconductors in macroscopic dimensions, irrespective of their size, will absorb all electromagnetic radiation with an energy greater than the bandgap. However, if the particles become smaller than that of the exciton in the bulk semiconductor (typically about 10 nm), their electronic structure changes. The electronic properties of such small particles are hence more like those of a molecule than an extended solid. The electronic and optical properties of such small particles will depend not only on the material of which they are composed, but also on their size, the so-called quantum confinement effect [22–26]. The lowest energy of optical transition, among others, will increase significantly due to the quantum confinement with decreasing size of the inorganic clusters. For example, the quantized particles of $CuInS_2$ dispersed in acetonitrile can have an energy gap about 2.3 eV. In comparison to the energy gap of the bulk material, the increase is about 0.9 eV [27].

Bandgap tuning in nanoparticles can be used in realization of device architectures, such as tandem solar cells in which the different bandgaps can be obtained by modifying only one chemical compound [28, 29]. Since the energy levels of the polymers can be tuned by chemical modification of the backbone chain and the energy levels of the nanoparticles can be tuned through the size-dependent quantum confinement effects, blends of the two materials offer the possibility of tailoring optimal conditions for a solar cell, including energy gain from charge transfer for the efficient charge separation and the spectral range of the absorbing light. In contrast to C_{60} derivatives, the bandgap of the nanoparticles lies conveniently in the visible region. Furthermore, the possibility is given, to study not only electron transfer from p-type polymer to n-type nanoparticle, but also hole transfer from the nanoparticle to the polymer after excitation in the nanoparticle.

Also, the interaction between the inorganic particle and polymer should be good enough to disperse high nanoparticle concentrations in the polymer solution and to prepare high-quality films from this solution. Contrary to the organic materials, the nanocrystals are not easily dispersed in polymer matrices. Because of the small size of the cluster, they have a high surface energy, which leads to aggregation or oxidation. Because of this reason, the synthesis of nanoparticles shielded with various organic ligands is important in determining the morphology of the polymer/nanocrystal composites.

On the other hand, the performance of nanocrystal-based devices is often limited by poor charge transport through the nanoparticle-polymer interface, since the nanocrystals are surrounded by a stabilizing shell of capping molecules. In these cases, it is possible to remove the capping molecules partly by repeated washing with organic solvents such as pyridine or cyclohexane, providing a better charge carrier exchange between nanocrystals and the conducting polymer.

Another promising technique is the modification of the nanocrystal surface with monomers or oligomers of semiconducting polymers. Subsequent co-polymerization of the nanocrystals with the matrix facilitates incorporation of the inorganic component into the semiconducting polymer and charge transport through the nanocrystal-polymer heterojunction. Details on the methods used to synthesize polymers on the nanoparticle surface can be found elsewhere [30–35]. State-of-the-art reviews for engineering particle surfaces are also given [36, 37].

In-situ formation of nanoparticles in a high-dielectric-constant polymeric medium has been realized for CuO and Cu_nS [38]. To our knowledge, no studies in photovoltaic devices have yet been reported using these methods.

The existence of concentration gradient of the n-type nanoparticles within the p-type polymer matrix may allow optimization of the topology of the heterojunction network. Introducing the gradient device morphology leads to an increase in the active area for charge separation while decreasing the charge carrier recombination during their transfer to electrodes. To obtain concentration gradient of nanoparticles, the colloid solution has to be placed on top of a polymer film. Diffusion of nanoparticles into the polymer matrix takes place with the penetration depth controlled by temperature, swelling of the polymer layer, and not at least by the size and shape of the nanocrystals. The control over the size and shape of nanoparticles is thereby an additional aspect to outline in the section below.

Because of the enormous importance of the nature of the shell molecules, the form of the particles, and the interactions with the chosen polymer matrix to construct efficient solar cells, special emphasis is given to the synthesis of the "core-shell" particles. Several strategies will be described in more detail in Section 2.

The photovoltaic performance of various inorganic/organic hybrid solar cells, prepared via spin-coating, will be the focus of this chapter. The other two common thin-film techniques, namely, electrodeposition and layer-by-layer self-assembly, are described very briefly. References are given to these techniques and to the performance of the cells prepared with these techniques for detailed information.

An efficient charge transfer from the nanocrystalline inorganic semiconductors to the conduction band of wide-bandgap semiconductors (such as TiO_2, ZnO, Ta_2O_5) in combination with high extinction coefficients of nanocrystals in the visible range makes them very attractive for dye-sensitized solar cells [39–42]. The special case of dye-sensitized photoelectrochemical solar cells would go beyond the scope of this review. We just point out the latest developments in dye-sensitized solar cells formed with semiconducting polymers.

2. METHODS

The three most common techniques [43] used for characterizing the semiconductor nanoparticles are:

1. *Optical absorption* (*or transmission*). Since the most apparent effect of quantized nanocrystals is the increasing of the bandgap, hypsochromic shift of the absorption spectra is a fast and simple indicator of the particle size.
2. *Transmission electron microscopy* (*TEM*) *together with electron diffraction* (*ED*). TEM is used to image the size, shape, and size distribution of the nanocrystals. Blended in polymer matrix, the images give information about the orientation of the particle in the polymer film and roughly about the morphology of the film. ED provides information about the stoichiometry of the particle and crystal phase.
3. *X-ray powder diffraction* (*XRD*). One very important use of XRD when dealing with nanocrystals is to estimate crystal dimensions L through the Debye–Scherrer relationship:

$$L \approx \frac{\lambda}{\Delta(2\Theta) \cdot \cos\Theta}$$

where λ is the X-ray wavelength (0.15418 nm for CuK_α radiation), $\Delta(2\Theta)$ is the peak full width at half maximum (FWHM) in radians, and Θ is the peak position. This relationship is valid only for a spherical crystal.

The size is inversely proportional to the FWHM. To be more precise, the value of FWHM depends on the length over which the periodicity of the crystal is complete (coherence length). That means, also the presence of defects in the crystal, which destroy the periodicity, can lead to broadening of the X-ray diffractions. Also a broad distribution in the crystal particle size can lead to wrong interpretation of the systems: Even if the amount of large crystals is less, they may dominate the diffraction pattern, since peak heights increase with increasing particle size. (Constant peak area for same amount of the material independent of the size, but decreasing FWHM, leads to increase of the peak heights.) Thus, the interpretation of the XRD peak broadening should be carried out with complementary methods.

2.1. Synthesis of Core-Shell Particles for Hybrid Solar Cell Applications

Nanoparticles with modified surfaces can easily be mixed with polymer materials to form thin films by several techniques such as spin-coating, doctor blade, or layer-by-layer self-assembly. The modification of the surface with an organic surfactant depends on the method by which the particle has been prepared. The nature of the organic ligand not only influences the nanoparticle distribution in the polymer matrix but also prevents more or less the aggregation and reduces the oxidative degradation. Core-shell materials have been synthesized using decomposition of organometallic compounds, colloidal chemistry, and solvothermal route.

2.1.1. Decomposition of Organometallic Compounds

CdS and CdSe have been widely studied as photoactive nanomaterials, due to the availability of precursors and ease of crystallization. The highly cited method for making CdE (E = S, Se, Te) nanocrystals is that of Murray et al. [44]. A general scheme for preparing monodisperse nanocrystals via organometallic route requires a single temporally short nucleation event followed by slower growth on the existing nuclei. This has been realized first by the pyrolysis of organometallic reagent (precursor) alkyl (dimethylcadmium) and a chalcogen source TOPSe (tri-n-octylphosphine selenid) in a hot tri-octylphosphine oxide (TOPO) matrix, which is a polar coordinating Lewis base.

The two precursors were dissolved in a liquid phase similar in nature to the matrix. The used liquid phase was tri-octylphosphine (TOP). The nucleation of CdSe nanocrystals was achieved in an inert atmosphere by the sudden introduction of the concentrated reagents in the TOPO matrix. The temperature of the TOPO is sufficient to decompose the reagents (200–350 °C), forming a supersaturation of species in solution that is relieved by nucleation of nanocrystals. Upon nucleation, the concentration of these species in solution drops below the critical concentration. Also the temperature decreases by addition of the precursors, which have been held at room temperature. Further material can only add to the existing nuclei.

During the growth of nanocrystals in the matrix, the surfactants dynamically adsorb to the growing crystals, allowing the atoms to further adsorb or de-adsorb, which controls the particle size and shape, while suppressing the aggregation. Each nanoparticle in the sample is surrounded by a TOPO layer.

Considerable efforts have been reported on the mild synthesis conditions by use of a single-molecular precursor in which the metal-chalcogen bond is already in place [45, 46]. The mild-pyrolyse method has been successfully applied to the growth of II-VI [47–51], III-V [52–57], and IV-VI [58, 59] semiconductor nanocrystals.

It is important to note that the shape of the particles depends strongly on the growth kinetics. For example, working with highly diluted suspensions leads to small nanocrystals first. Systems with mostly small nanocrystals, on the other hand, exhibit a second growth stage: the high surface energy of small nanocrystals promotes their dissolution at the expense of the larger ones. This growth stage is known as Ostwald ripening. Such slow growth conditions favor the formation of particles with least surface energy. The resulting particle shape is spherical [60]. At higher reagent concentration, relative differences between the growth rates can lead to anisotropic shapes. Controlled growth of crystalline arrays can be further achieved by manipulation of growth kinetics. Combining surfactants with different strength of the adsorption on the particle surface can be used to vary the growth rates and synthesize anisotropic particle shapes.

The breakthrough in the morphological preparation was that the ability to assemble these colloidal particles into crystalline arrays [61–65] can be used to increase the photoconductivity of the polymer/nanoparticle blends.

Controlled formation of nanoparticles with spherical and rod shapes was realized by CdSe nanocrystals. This is possible because CdSe crystals with the wurtzite structure are intrinsically anisotropic, with a unique c-axis. If the overall growth rate by an extremely high monomer concentration is fast, growth is generally faster along the c-axis. Rods form at high growth rates by unidirectional growth of the c-axis.

The experience showed that TOPO is not a suitable surfactant for controlled growth of rodlike CdSe structures. The resulting rodlike particles are too big in three dimensions and mostly insoluble in common solvents. The blends of these particles in polymers are characterized by large-scale phase separation, which is a disadvantage for photovoltaic applications. Addition of a molecule that coordinates more strongly to cadmium has been used to control the growth rate and shape of the particles. An important discovery was that hexylphosphonic acid (HPA), which is almost always present as impurities in technical grade TOPO, accentuates the differences in the growth rates among various faces [66]. Controlled formation of CdSe nanocrystals with spherical and rod shapes was realized by the variation of the surfactant ratio of TOPO and HPA. The role of HPA seems to increase the growth rate along the c-axis. At low concentrations of HPA, spherical-shaped nanocrystals form. Intermediate amounts of HPA favor the formation of rods. Higher HPA ratios lead to the formation of arrow-shaped nanocrystals. More complex shapes such as teardrop or tetrapod-shaped crystals are produced by varying the ratio of surfactants, injection volume, and time-dependent monomer concentration [67].

Another strategy to fabricate superstructures from TOPO capped particles with a narrow size distribution is gentle destabilization of the dispersion. Attractive surface forces of nanoparticles drive self-organization and superlattice formation [68].

The crystallization of TOPO capped CdSe particles, 3.5 nm in diameter, was carried out in vertically positioned glass tubes. The toluene dispersion of the particles was slowly destabilized by diffusion of methanol, resulting in aggregation of the particles. A buffer layer of propane-2-ol between solvent-nonsolvent contact was placed, to decrease the growing rate of the particles. Single crystals grew in the form of perfectly faceted hexagonal plates. The optical behavior of these three-dimensional (3D) ordered systems, 100 μm in lateral dimensions and 20 μm in depth, was very similar to the optical behavior of individual nanoparticles.

Theoretical aspects for the kinetics of growth of solid-state nanocrystals in a liquid medium have been also reported [69].

Nanoparticles of TOPO capped II-VI materials are found to degrade in the presence of light and oxygen, forming unstable chalcogene oxide species [70]. Alternatively, TOPO has been also replaced by other organic groups such as pyridine, 4-picoline, tris (2-ethylhexyl) phosphate, and 4-(trifluoromethyl) thiophenol [71, 72]. Another way to eliminate the degradation is coating the nanoparticle surface by a layer of a higher bandgap semiconductor [73–76]. The relative stability of CdSe (CdS) nanocrystals compared to CdSe is drawn from energetic offsets between the core and shell valence bands. When the CdS shell is present, there is a possibility given, that a hole on the core nanoparticle will recombine with electrons of CdS but not degrade in the presence of oxygen. CdS capped CdSe nanocrystals have been introduced as a new class of electroluminescent materials. The light-emitting diodes of ITO/PPV (40 nm)/CdSe (CdS) (24 nm)/Mg/Ag show low operating voltages, a factor of 8 increase in quantum efficiencies, and a factor of 10 increase in lifetimes in comparison to those made by organically capped CdSe nanocrystals [77].

2.1.2. Colloidal Route

An alternative synthetic approach involves mixing the reagents at a temperature low enough to preclude any appreciable reaction. A controlled dropping of nucleation reagent initiates the reaction. As long as the temperature is adjusted to keep the rate at which the reagents react less than or equal to the dropping rate, controlled reaction conditions are given.

Because the energy of the surface is higher than in the volume, the tendency of nature is to reduce the surface-to-volume ratio, which is also the driving force of particle growth in a colloidal route after nucleation.

To stabilize the smallest particle in a dispersion, the material exchange between the particles should be hindered. This can be realized by adsorbance of charged (electrostatic forces) or bulkier surfactants (greater steric hindrance). Tailoring the ratio of the reagents concentration to that of the surfactants provides a control over the particle size. Higher surfactant-to-reagent ratios favor the formation of a smaller particle size. Surfactants in solution are only reversibly adsorbed to the surfaces of the reagents and control the particle growth. Also the chemistry of the surfactant plays an important role. For example, bulkier trioctylphosphines provide greater steric hindrance than more compact tributylphosphines, slowing the growing step [78]. The stability of the particle dispersions depends on the temperature and on the solvent. If the interaction between the surfactants and the solvent is favorable, the nanocrystals are then isolated. Introducing an additional solvent, which is miscible with the first solvent but has an unfavorable interaction with the capping group, destabilizes the particle dispersion. A controlled destabilization may lead to formation of 3D arrays, but a fast growing of the particles cause just their aggregation.

Synthesis of $CuInS_2$ (CIS) nanoparticles was performed by a colloidal route at room temperature [79]. In this synthesis, the surfactant used was triphenyl phosphite (TPP). In the first step, $InCl_3$ and CuI were suspended in a mixture of acetonitrile and TPP. The solutions were refluxed in an argon atmosphere to form TPP complexed $InCl_3$ and CuI solutions. Bis(trimethylsilyl) sulfide, which was diluted with TPP, was added for the nucleation, drop by drop to the mixture at room temperature. A remarkable dynamic color change of the solution occurs from colorless through yellow to orange and red depending on the amount of added bis(trimethylsilyl) sulfide and on the size of the formed CIS particles. The average particle size increases over time with a compensating decrease of the nanoparticle number due to the Ostwald ripening. Under cooling conditions, the dispersion is stable for ca. one month. X-ray diffraction studies performed on the product are consistent with those of the chalcopyrite structure of CIS.

2.1.3. Solvothermal Route

Using the solvothermal approach, the synthesis of III-V semiconductors has been demonstrated [80]. An advantage of this method is that the particles formed are usually crystalline and do not need post-treatment at high temperatures. Also, no organometallic precursor, which can complicate the system, and no organic stabilizers are used in this route. The nature of the solvent affects the reaction process and product quality. The synthetic route was extended to the synthesis of nanocrystalline GaN [81] and $CuInSe_2$ [82].

We synthesized $CuInSe_2$ (CISe) by the solvothermal route described by Y. Qian and co-workers. A stoichiometric mixture of elementary Se, $CuCl_2 \cdot 2H_2O$, and $InCl_3 . \cdot 4H_2O$ was loaded into a 50 ml autoclave, which was then filled with anhydrous ethylenediamine up to 80% of the total volume. The experiment was performed at 180 °C for 15 h. The product was washed with water and ethanol several times in order to remove the by-product, a crystalline, undefined mixture of Cu, In, and Se beige in color. The product was black in color, which predicts the nonquantized nature of the particle. The CISe particles were elongated in shape with dimensions of about 15–60 nm [130].

Due to its strong basic capacity, an important role of the ethylenediamine is to increase the reactivity of elemental selenium forming Se^{2-} [83] and to promote the reaction between $InCl_3$ and Se forming the anions $(InSe_2)^-$. Since ethylenediamine dissolves $CuCl_2$, it can act as a bidentate ligand that could facilitate the formation of chelate complexes of $[Cu(en)_2]^+$ and likely force the $CuInSe_2$ grains to grow in one direction. When diethylamine, also a strong base ($pKa = 10.98$) but without double N-chelation, was used instead of ethylenediamine, spherical $CuInSe_2$ particles have been obtained [82]. Similar observations between the particle shape and solvent were made also for the synthesis of PbS nanoclusters [84]. These results indicate that only a bidentate solvent sustains whisker-like shapes. A major limitation of this method has been that it often leads to aggregated particles embedded in a polymer matrix. Further wet chemical synthesis methods for nanocrystalline chalcopyrites are given in [85, 86].

2.2. Thin-Film Preparation Techniques

There are three common methods to prepare films of hybrid materials: spin-coating from a solution, electrodeposition, and layer-by-layer assembly. Electrodeposition and layer-by-layer (LBL) self-assembly will be described only very briefly. References are given to these techniques and to the performance of the cells prepared with these techniques, for more information.

Hybrid solar cells fabricated via spin-coating from dispersions of CdS (CdSe) nanocrystals and semiconducting polymers are the most extensively explored systems in general. So we will describe these systems in more detail. $CuInS_2$ ($CuInSe_2$) embedded in semiconducting polymer matrices will be discussed for different photovoltaic solar cell configurations.

2.2.1. Spin-Coating

Most incorporation techniques of hybrid systems require mixing of the components in a good solvent/dispersion and spin-coating of the solution to form thin films on selected substrates. Since using semiconducting polymers as an optoelectronically active component in such hybrid systems requires optimal morphology of the film, an important issue is how to assemble nanocrystals within films. The nanocrystals consist of organic shells readily dispersible in a variety of organic solvents, and they easily form uniform films of closely packed nanocrystals by spin-casting and spin-coating. To tailor the right components for an ideal morphology of a polymer/inorganic composite film is still a challenge, which will be the target of experimental work in future.

2.2.2. Electrochemical Deposition

Films of the quantized particles can be prepared by using electrochemical deposition from nonaqueous solutions such as DMSO containing, for example, elementary chalcogen (S, Se) and a metal salt [87]. In the second step, semiconducting polymers have been electrolytically grown onto this nanocrystalline inorganic layer. Classical photovoltaic materials like CdTe [88, 89], CdS [90, 91], Cd/ZnSe [92], TiO_2 [93], $CuInS_2$ [94, 95], and $CuInSe_2$ [96] have also been nanoscale structured onto different substrates by using this technique. The experimental details can be found in [90, 91, 97, 98]. Horowitz and Garnier, who studied a polythiophene/GaAs p-n junction, reported on a hybrid solar cell fabricated by electrodeposition [99].

Schottky-type photovoltaic cells, based on polythiophene derivatives and CdTe [100] or CdS [101–103], are also reported. In these cells, the polymer is in its highly doped state and behaves like a metal. The active area for charge generation and separation may be in the inorganic layer.

Heterojunction cells consisting of n-type $CuInSe_2$ and p-type polypyrrole multilayers have been fabricated electrochemically [104].

2.2.3. Layer-by-Layer Assembly

Self-assembly is defined as the spontaneous liquid-phase adsorption of materials onto a surface. Driving forces are mostly electrostatic in nature, but they can also involve hydrogen bonding, van der Waals, or hydrophobic interactions. The substrates, for example, ITO, silica, metal electrodes used for electro-optical applications, have surfaces with negatively charged oxides. A well-cleaned substrate is immersed into a dilute aqueous solution of an oppositely charged solution, for a time optimized for adsorption of a monolayer, rinsed, and dried. Rinsing with a solvent between the adsorption steps removes the excess solution and leaves a thin monolayer of the polyelectrolyte on the surface. The next step is the immersion of the monolayer covered substrate into a dilute dispersion of oppositely charged species (cationic/anionic polyelectrolytes or nanoparticles), also for a time of optimized adsorption of the second layer, rinsing, and drying. The cycle can be repeated to reach the desirable thickness of multilayers. Strong interpenetration of the used polymer chains makes impossible a description of the border between oppositely charged layers. In the resulting laminated structure, the two different types of semiconductors can diffuse into each other, forming a mixed layer at the interface, while still providing a percolation path to the respective electrodes. Detailed information for the fabrication techniques can be found in [105–110]. Such growth of multilayers was demonstrated for single layers of oppositely charged polyelectrodes [111] and semiconductor polymers such as water-soluble poly(p-phenylene)s [112–115] and poly(pyrrole) [115] .

The inorganic particles were usually shielded by organic layers with functional groups of dithiols, in which covalent metal-S bonds can be formed. LBL film-coating studies for nanoparticles such as CdTe, HgTe, CdSe, PbS, TiO_2, ZrO_2, Ag, Au, SiO_2, and MoO_2 have been also investigated [116–119].

The CdTe/polyaniline and CdTe/polypyrrole composites have also been produced in a second step via the treatment of electrochemically prepared polyaniline films with aqueous colloidal solutions of thiol capped nanocrystals [120] and via the electrochemical polymerization of pyrrole into the porous CdTe films [121] .

Approaches to create an electroactive interface included the incorporation of CdSe nanoparticles with 1,6-hexadecanethiol (HDT) onto a film of poly(p-phenylene-vinylene). The diodes had a good rectification and enhanced electroluminescence [122, 123]. However, the current-voltage characteristics of these cells did not originate in a p-n junction, but in a Schottky diode junction.

Composite diodes have been fabricated by self-assembling TOPO capped n-type, 2- to 4-nm CdSe particles and HDT (as a binder) onto the p-doped semiconducting polymers poly(pyrrole) (Ppy) or poly(3-methylthiophene) (PMeT). The semiconducting polymers, on the other hand, have been either layered by electrochemical deposition (PMeT) or layer by layer chemically assembled onto derivatized conducting substrates [119]. These films have shown p-n junctions in an ITO/polymer/$(HDT/CdSe)_3$/Al configuration.

Layer-by-layer assembly of hybrid poly(vinylpyridine)/CdSe multilayer thin films has been reported, whereas the driving force for the formation of PVP/CdSe layer was H-bond interaction between the pyridine group of the polymer and the carboxylic acid group of the surfactant surrounding the nanoparticle [124].

3. SOLAR CELLS

3.1. Hybrid Solar Cells based on CdSe-CdS Nanocrystals

Photovoltaic characteristics of CdSe and CdS core/shell particles have been most extensively explored by the group of Alivisatos et al. According to the photoluminescence quenching measurements in composite films of poly(2-methoxy,5-(2′-ethyl-hexyloxy)-p-phenylene-vinylene (MEH-PPV) and TOPO capped CdSe nanoparticles (5 nm), a charge separation at the interface between the polymer and CdSe nanoparticles has been reported [125].

Replacing the surfactant TOPO by pyridine leads to a more effective charge separation, which has been concluded because of the decrease of MEH-PPV photoluminescence efficiency to a value of ca. 0.01 by mixing CdSe in a composite film. The increase of short circuit photocurrent with increasing amount of CdSe was consistent with this picture. Photoluminescence quenching experiments have been confirmed by the observation of charge on the polymer chain in photoinduced absorption measurements [126].

By monochromatic illumination at 514 nm of an ITO/CdSe-MEH-PPV/Al cell (95 weight % CdSe) with a power intensity of 5 W/m^2, they received an open circuit voltage, V_{oc}, of 500 mV and a short circuit current of about 1 μA for a 7.3 mm^2 active area. There was a rectification ratio of 100 at $+/-3$ V in the dark and no rectification under illumination. A power conversion efficiency of 0.25% was obtained.

The low quantum efficiency is attributed to the poor transport of charges. TEM images of the films show phase separation for all composite films, unfortunately more pronounced for pyridine capped CdSe particles. Because of the structural traps in the form of incomplete pathways in the percolation network, the separated charges on different materials recombine with opposite charges before reaching the electrodes. To improve the charge collection, the film morphology should be optimized.

An idea to improve the film morphology was replacing the spherical nanocrystal particles with elongated ones. Van der Waals interactions become stronger as the contact area between two particles increases [127]. Because of this, the tendency for elongated particles to agglomerate will be stronger than that for spherical particles of the same volume. In agreement with this argument, TEM images of particles in polymer matrices show that the agglomeration domain sizes increase with increasing length of the particles (Fig. 2).

The efficient packing of larger nanocrystals may result in a decrease of charge recombination possibilities in a bulk heterojunction hybrid solar cell. Indeed, the use of nanocrystalline rods in conjunction with a regioregular poly(3-hexylthiophene) (P3HT) has resulted in almost an order of magnitude increase in power conversion efficiency, in comparison to the cell discussed above [128]. P3HT is an effective hole transport material in its regioregular form, with field effect hole mobilities up to 0.1 $cm^2 V^{-1} s^{-1}$ [129].

Solar cells were fabricated by spin-casting a solution of 90 weight % CdSe nanorods (7×60 nm) in P3HT onto an indium tin oxide glass substrate coated with

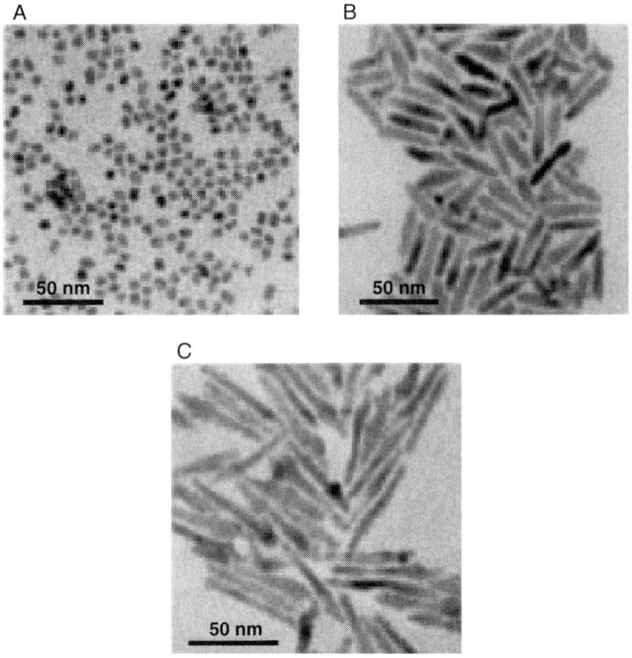

Figure 2. Cadmium selenide nanocrystals with aspect ratios ranging from 1 to 10. The samples, shown by transmission electron micrographs at the same scale, have dimensions (A) 7 nm × 7 nm, (B) 7 nm × 30 nm, and (C) 7 nm × 60 nm. Reprinted with permission from [4], W. Tu. Huynh et al., *Science* 295, 2425 (2002). © 2002, American Association for the Advancement of Science.

poly(ethylene dioxythiophene) doped with polystyrene sulfonic acid (PEDOT:PSS) with aluminum as counter electrode. PEDOT:PSS is a highly p-doped polymer, usually coated on the ITO electrode, to smooth the electrode surface and increase the adhesion of organic materials to form good-quality layers via spin-coating.

By monochromatic illumination at 515 nm of an ITO/PEDOT:PSS/CdSe-P3HT/Al cell with a power intensity of 0.1 mW/cm^2, they received an open circuit voltage, V_{oc}, of ca. 500 mV and a short circuit current of about 0.02 mA/cm^2. There was a rectification ratio of 10^5 in the dark. A power conversion efficiency of 6.9% is obtained (Fig. 3B). Solar cell characteristics of this 7 × 60-nm nanorod device illuminated with AM 1.5 global light, include a short circuit current of 5.7 mA/cm^2 and an open circuit voltage of 680 mV (Fig. 3C). Due to the quantization effect, the form of the photocurrent spectra changes depending on the size of the nanorods (Fig. 3D). The external quantum efficiency (IPCE), which is the percentage of electrons collected per incident photon, increases with increasing length of the nanorods from ca. 20% for 7 × 7 nm up to ca. 55% for 7 × 60 nm particles (Fig. 3A) [3].

3.2. Hybrid Solar Cells based on $CuInS_2$-$CuInSe_2$ Nanocrystals

We investigated first blends of poly (2-methoxy-5-(3′,7′-dimethyloctyloxy)-1,4-phenylene vinylene) (MDMO-PPV) and of regioregular poly (3-hexylthiophene) (P3HT) with copper indium diselenide (CISe) and copper indium disulfide (CIS) nanocrystals [130].

The crystallographic structure of the particles was determined by powder X-ray diffraction using Cu K_α radiation of 1.5418 Å. Figure 4 shows the X-ray diffraction pattern of the CISe and CIS. An intense peak at $2\theta = 26.6°$ oriented along the (112) direction and other prominent peaks observed at 44.3° ((220)/(204)) and 52.3° ((312)/(116)) indicate the chalcopyrite structure of CISe. These patterns are in good agreement with JCPDS data [162].

In contrast, the upper plot shows the results of X-ray diffraction investigations performed on the CIS. The three

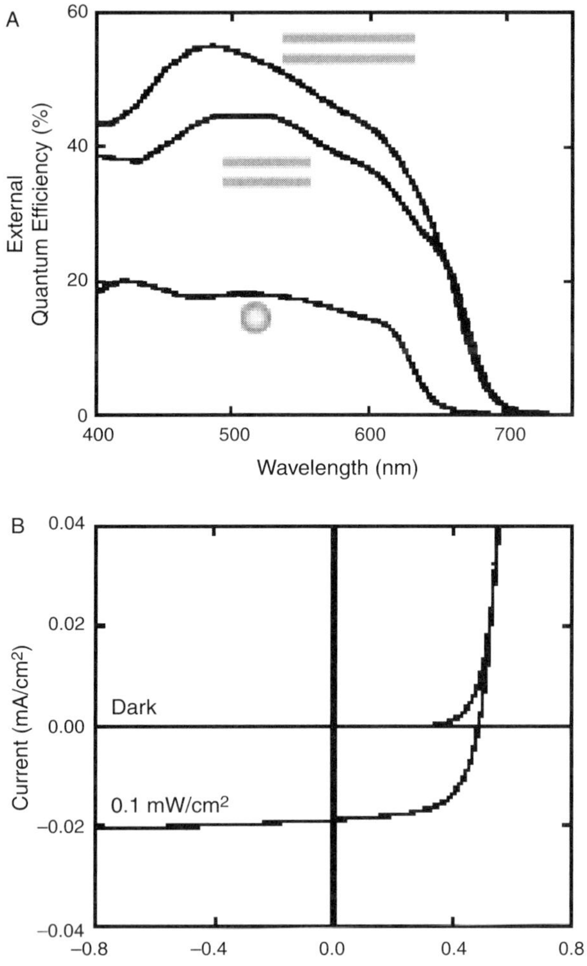

Figure 3. (A) External quantum efficiencies of 7-nm-diameter nanorods with lengths 7, 30, and 60 nm. The device structure consists of film, which was spin-cast from a pyridine-chloroform solution of 90 wt% CdSe nanorods in P3HT. The intensity is 0.084 mW/cm^2 at 515 nm. (B) The current-voltage characteristics of the 7 nm × 60 nm nanorod device exhibit rectification ratios of 10^5 in the dark and a short circuit current of 0.019 mA/cm^2 under illumination of 0.084 mW/cm^2 and at 515 nm. (C) Solar cell characteristics of this 7 nm × 60 nm nanorod device illuminated with simulated AM 1.5 global light, include a short circuit current of 5.7 mA/cm^2. (D) Photocurrent spectra for two devices with 60-nm-long nanorods with diameters 7 and 3 nm. Reprinted with permission from [4], W. U. Huynh et al., *Science* 295, 2425 (2002). © 2002, American Association for the Advancement of Science.

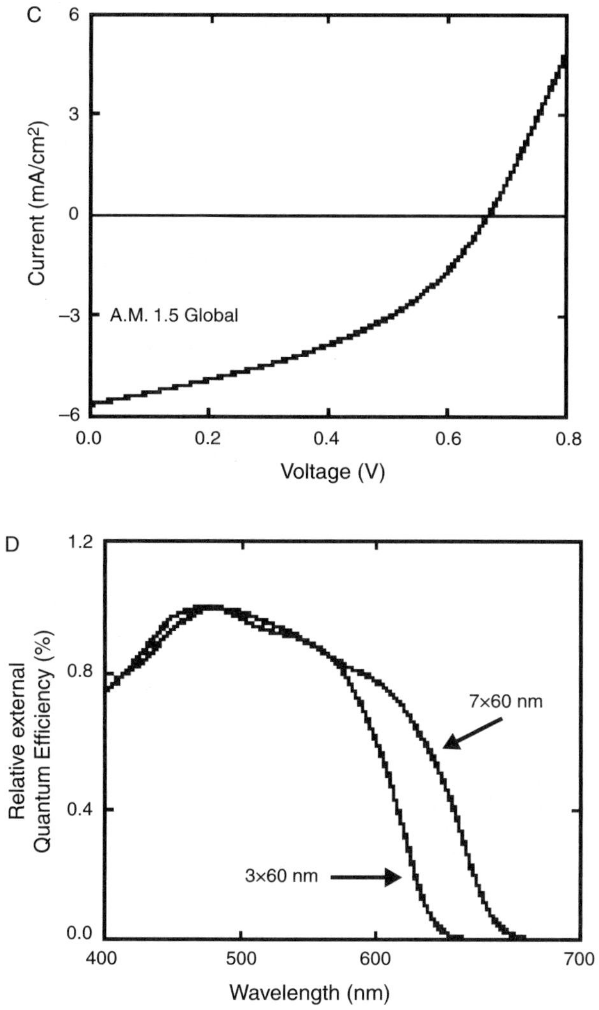

Figure 3. Continued.

now broad reflections at 2θ values of 27.9 (112), 46.5 (220)/(204), and 55.0 (312)/(116) are those of the chalcopyrite structure of CIS [163]. However, in the range between 15° and 35° a broad halo appears, probably due to an amorphous contribution. This may indicate that still TPP is present, which may not have been removed completely.

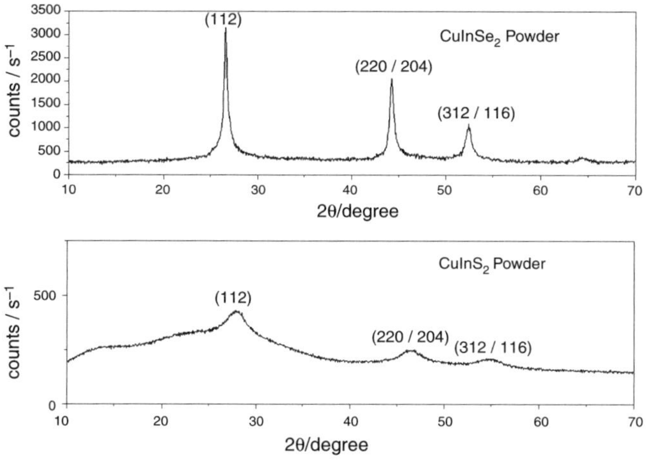

Figure 4. X-ray diffractograms of CISe and CIS.

Apart from this, a comparison of the diffraction patterns clearly shows a pronounced broadening of the CIS diffraction patterns due to the Debye–Scherrer relationship (Section 2). TEM is used to image the size, shape, and size distribution of the nanocrystals [130]. CISe samples in MDMO-PPV appear to display whisker shaped particles with dimensions of at least 15 × 60 nm. CIS particles in P3HT matrix have also elongated forms with dimensions on the order of 5 × 20 nm (Fig. 5A). The preparation of the particles has been mentioned in Sections 2.1.2 and 2.1.3.

In the next step, films of $CuInS_2$ particles have been fabricated by spin-coating on the ITO electrode, which was first covered with (PEDOT:PSS). The films were homogeneous, with a roughness of about 40 nm (Fig. 5B). The CIS layer exhibits due to the quantum confinement effect, a significant hypsochromic shift of absorption relative to the bulk

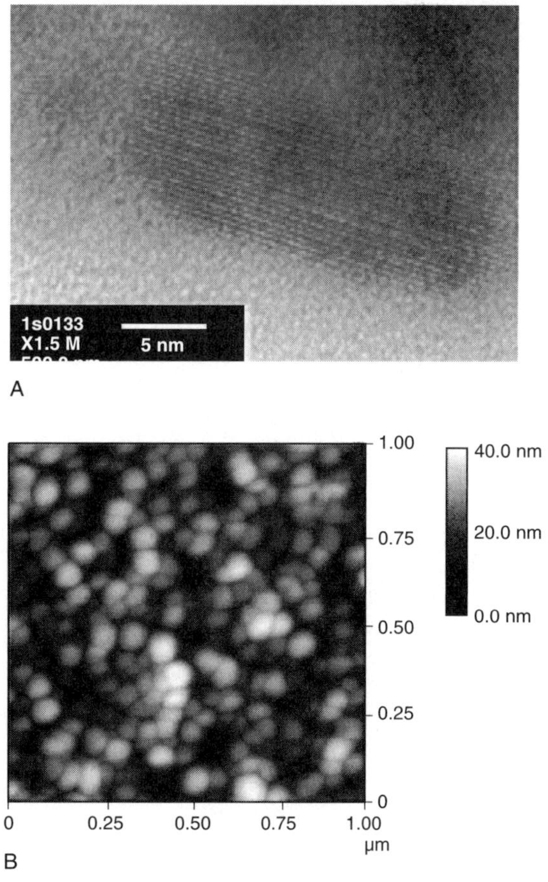

Figure 5. (A) Purely CIS nanocrystal shown by transmission electron micrographs with dimensions of ca. 5 × 12 nm. The particle size distribution in general depends strongly on the preparation parameters; we received particles with dimensions starting with 5 × 5 up to tens of nanometers. (B) Photocurrent and absorbance spectra for CIS layers show similar form with an edge at 700 nm, indicating the size quantization. (C) Atomic force micrographs of the sample, characterized by (B), show a homogeneous film with an agglomerate size of ca. 50 nm. (D) The current-voltage characteristics of the CIS device exhibit a short circuit current of 20 $\mu A/cm^2$ and 450 mV under illumination with simulated AM 1.5 global light (semilogarithmic plot).

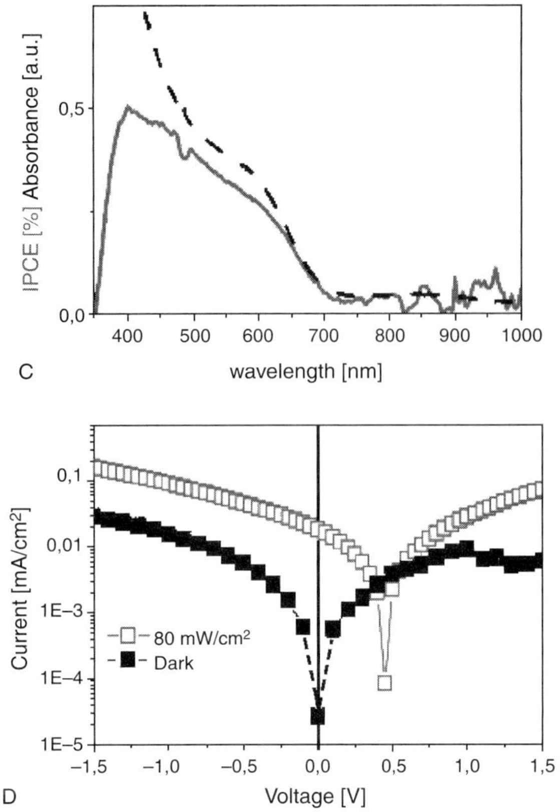

Figure 5. Continued.

material absorption at 825 nm [131]. The absorption spectrum of films, prepared via spin-coating from a 0.06 M solution of CIS dispersions, exhibits a high shoulder at 700 nm and a weak absorbance at longer wavelengths (Fig. 5C). The IPCE onset is more or less identical with the onsets of the corresponding absorbance spectra. It is obvious that quantized CIS particles are photovoltaic active.

In the case of an ITO/PEDOT:PSS/CIS/LiF/Al photovoltaic device, a short circuit density I_{sc} of 20 μA/cm^2 and an open circuit voltage V_{oc} of 450 mV have been reported. A fill factor of about 0.25 indicates high serial resistance. However, a low rectification ratio RR at $+/-2$ V in dark conditions of 2.5 and under illumination of 6.5 indicates nonoptimized diode characteristics (Fig. 5D). Repeating the measurements and drying the films under argon leads to an improvement of the cell.

Flat interfaces formed in bilayer structures of the CIS and the nicely soluble methanofullerene, 1-(3-methoxy carbonyl)-propyl-1-1-phenyl-(6,6) C_{61} (PCBM), have shown a significant improvement in the photovoltaic performance [27]. PCBM is the electron-transporting and hole-blocking "gate" material in this bilayer configuration [132]. An additional advantage of PCBM is its solubility in toluene. TPP capped CIS particles are also slightly soluble in toluene. However, when PCBM is deposited onto a previously prepared CIS film, some swelling of the CIS layer may occur and some roughening of the formed interfaces between two layers.

Thc CIS//PCBM double layer exhibits a reasonable diode behavior with a typical current versus voltage I-V response (Fig. 6). Illuminating the cell with 80 mW/cm^2 white light from a solar simulator produces an open circuit voltage, V_{oc}, of 710–790 mV and a short circuit current density of ca. 0.26 mA/cm^2. The calculated fill factor of the diode is 0.44. Conversion efficiency of 0.086% is obtained. The rectification ratio low at $+/-2$ V is 2.5 in the dark and 3.5 under illumination.

The dependence of the short circuit current on the incident light intensity, which follows a power law $I_{sc} \sim F^{\alpha}$ with a scaling exponent α, has been fitted around 0.7 for single layers of CIS. On the other hand, for the bilayer configurations CIS//PCBM, the dependence of the I_{sc} follows a power law $I_{sc} \sim F^{\alpha}$ with $\alpha \sim 0.9$.

It is obvious that in layers of TPP capped CIS, contrary to the layers consisting of TOPO capped CdS nanocrystals, a photovoltaic response can be detected without removing the organic ligand. TPP seems to be weakly bound on the inorganic core and not totally insulating the particles from each other.

The fact that particle growth occurs already by changing the viscosity of the dispersion solution leads us to the assumption that the organic shell TPP is adsorbed only reversibly on the CIS core. There might be a cooperative interaction between core, shell, and solvent. Tailoring the ratio of the concentration of organic shell to that of the correctly chosen solvent, acetonitrile provides control over the nanoparticle size, since low shell-to-solvent ratio favors the formation of bigger CIS particles. Reducing the amount of acetonitrile leads to weakening of the complexation capability of the TPP and results in a loss of size control. The absorption maximum shifts to the longer wavelengths like the bulk material. The existence of large domains in CIS and MDMO-PPV or P3HT layers, which were spin-cast from a toluene suspension, are in agreement with this argument.

We investigated a bulk heterojunction device prepared from CIS and highly p-doped poly(3,4-ethylenedioxythiophene):poly(4-styrenesulfonate) complex (PEDOT:PSS). The main point thereby was to prepare a well-percolating film, while at the same time to avoid the CIS particle-growth during the preparation [132]. The advantage of PEDOT-PSS is in this aspect its solubility in acetonitrile/TPP mixtures. PEDOT:PSS does not absorb in the visible range. Therefore it is easy to investigate the size distribution of the CIS particles by studying the absorbance behavior of CIS depending on the preparation. The aim was to form an interconnected network of the nanoparticle and of the polymer, so that the charge separation can occur at the very large interfacial area of the nanoparticle-polymer contacts. The optical and microscopic investigations on the layers of CIS-PEDOT:PSS blends showed that the nanoparticles do not grow during spin-coating from an acetonitrile dispersion of CIS and PEDOT:PSS.

In our systems, due to the high HOMO level of PEDOT:PSS ($E_{HOMO} \sim -5.0$ eV) not far from the work function of the Al electrode (~ -4.3 eV), an ohmic contact can be formed even to this electrode short circuiting the cell (Fig. 7).

By white-light illumination of an ITO/CIS-PEDOT:PSS/LiF/Al cell with a power intensity of 80 mW/cm^2, we received an open circuit voltage, V_{oc}, of 150 mV and a short circuit current density of 4 μA/cm^2. The calculated fill factor

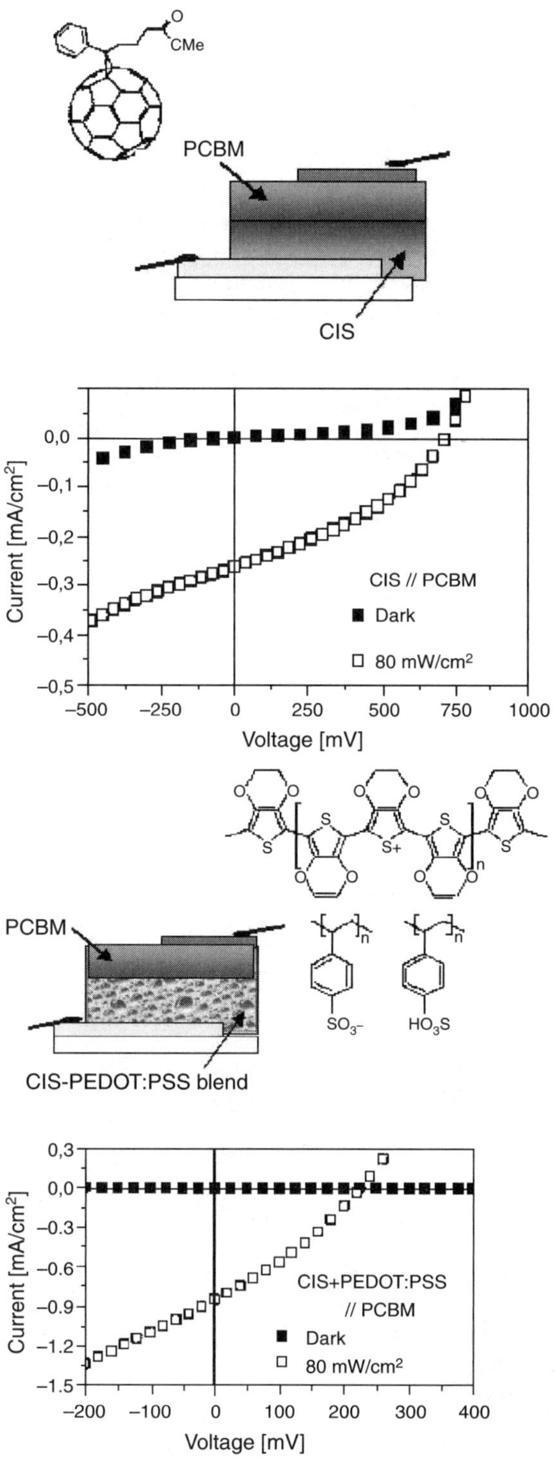

Figure 6. (left) Schematic structure of bilayer devices ITO/CIS/PCBM/Al and their *I-V* characteristics in dark and under illumination with simulated AM 1.5 global light (Inset: The molecular structure of PCBM.) (right) Schematic structure of "optimized" bulk heterojunction devices ITO/CIS-PEDOT:PSS/PCBM/Al and their *I-V* characteristic in dark and under illumination with simulated AM 1.5 global light (Inset: The molecular structure of the polymer PEDOT:PSS.)

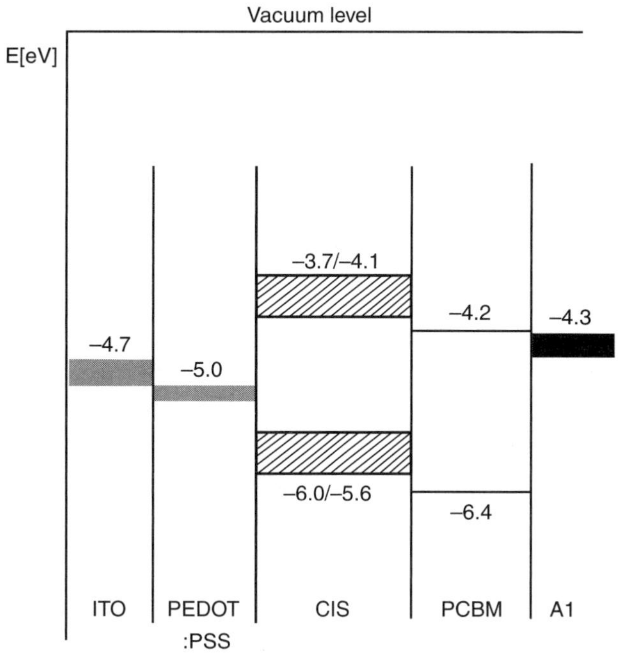

Figure 7. Energy levels of investigated materials. The HOMO-LUMO levels of macroscopic CIS are about −5.6 eV and −4.1 eV. Because of the size distribution of the quantized CIS particles, a distribution of their energy gaps must be expected. The anticipated maximal broadening of HOMO and LUMO levels from the smallest particle up to macroscopic CIS is calculated by adding the maximal hypsochromic shift of CIS absorption in acetonitrile dispersion to the literature data of bulk CIS. However, for relatively large particles, the bandgap might be smaller than shown here. Reprinted with permission from [132], E. Arici et al., *Adv. Funct. Mater.* 13, 165 (2003). © 2003, Wiley.

of the diode is 0.1. A conversion efficiency of only 0.003% is obtained. There is no rectification in the dark and a rectification ratio of 2.2 at +/−2 V under illumination [132].

In order to prevent the short circuiting contact between the hole conductor PEDOT:PSS and the electron-collecting Al electrode, a PCBM interlayer was introduced (HOMO = −6.4 eV, LUMO = −4.2 eV). Thereby a white-light illumination of the cell with a power intensity of 80 mW/cm^2 produced an open circuit voltage, V_{oc}, of 220 mV and a short circuit current density of 0.84 mA/cm^2 (Fig. 6b). The rectification ratio in the dark is 13.1 at +/−2 V and 11.3 under illumination.

The most significant differences between the different cell types investigated can be seen from the incident-photon-to-current-conversion efficiencies (IPCE) spectra (Fig. 8). The IPCE onsets are more or less identical with the onsets of the corresponding absorbance spectra. The photocurrent response is not limited by the absorption depth of the film. The obtained photocurrent for the CIS single layer is very poor; the photogenerated excitons mostly recombine before they can reach the electrodes. However, current generation can be significantly improved by adding PCBM as a strong electron acceptor.

Because of the ohmic contact between PEDOT:PSS and the Al electrode, the IPCE value is decreased for CIS-PEDOT:PSS monolayers as compared to the monolayer of nanocrystalline CIS. The IPCE value of CIS-PEDOT:PSS/

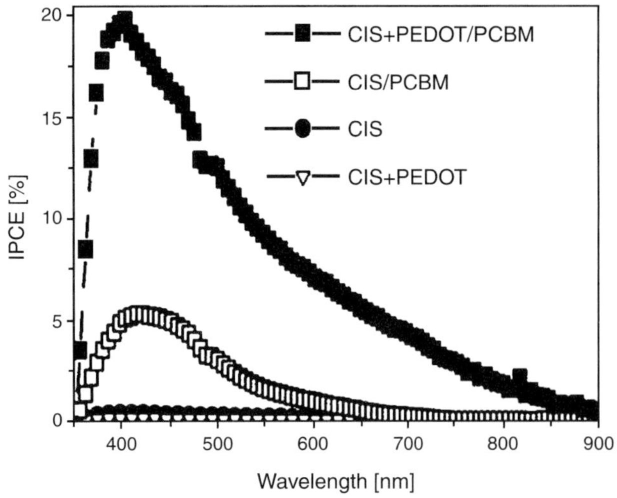

Figure 8. IPCE spectra for photovoltaic devices using different device structures. Reprinted with permission from [132], E. Arici et al., *Adv. Funct. Mater.* 13, 165 (2003). © 2003, Wiley.

PCBM bilayer gives by far the best results. The holes move along the PEDOT:PSS network towards the high work function electrode ITO and the electrons, in turn, can move through the PCBM layer, where they can easily be transported to the Al electrode, whereas the holes are blocked by the high ionization energy of the PCBM HOMO level.

3.3. Hybrid Solar Cells based on TiO_2 Nanocrystals

Dye-sensitized nanocrystalline TiO_2 solar cells [2, 161] are based on a fundamentally different working principle than solar cells described in Sections 3.1 and 3.2. The semiconductor used in these devices is not photoexcited directly, in contrast to the case for the p/n photovoltaic cells. Briefly speaking, these cells consist of a nanoporous working electrode formed by a sintered film of TiO_2 semiconductor nanoparticles (10–30 nm in diameter), which serves as electron acceptor and transport layer. The TiO_2 layer is coated with a single layer of a Ru-bipyridl-based dye for light absorption and electron injection into the TiO_2 conduction band. An electrolyte solution containing the redox couple like iodide/triiodide (I^-/I_3^-) serves as redox medium to regenerate the photoexcited dye molecules by reduction. Charge separation occurs at the semiconductor/electrolyte interface. For the operation of the photoelectrochemical device, the counter electrode (Au or Pt) is sealed to the working photoelectrode with a spacer, whereas the volume between the electrodes is filled with the electrolyte solution. Photovoltaic energy conversion efficiencies as high as 10–11% at AM 1.5 standard solar spectrum have been achieved [134].

Recent efforts in dye-sensitized nanocrystalline TiO_2 solar cell research are focusing more on replacing the liquid electrolyte with solid-state analogous to eliminate problems with sealing and degradation. Early work was focused on use of solid-state p-type conductors CuI [135, 136] and CuSCN [137, 138]. Cells made in this way gave solar efficiencies of several percents, but they were not stable in air and under illumination.

Various procedures have been employed in the fabrication of solid-state dye-sensitized solar cells with p-type organic materials, that accept holes from the Ru-bipyridl-based dye cation. Triphenyldiamine, for example, is a hole conductor with high intrinsic charge carrier mobilities of about 10^{-3} cm^2 V^{-1} s^{-1} at room temperature. Nevertheless, replacing the liquid electrolyte by a triphenyldiamine layer via evaporation led to low external quantum efficiencies (IPCE) up to 0.2% [139].

For an efficient cell, organic materials placed on the top of the working electrode should penetrate into the nanoporous TiO_2 and should form a good contact to the adsorbed dye. Crystallization of the organic material is undesirable, as it would impair the formation of a good contact between the electrode and the hole conductor. Triphenyldiamine and semiconducting polymers based on triphenyldiamine have been investigated in order to understand the relationship between the film-forming properties, dye stability, and cell performance [140, 141].

The hole-conductor material containing a spiro center (a tetrahedral carbon linking two aromatic moieties), namely, 2,2′7,7′-tetrakis (N,N-di-p-methoxyphenyl-amine) 9,9′-spirobifluorene (spiro-OMeTAD), has been introduced as an alternative because of its relatively high glass transition temperature and amorphous film-forming properties via spin-coating. In these systems, a better electronic contact between the dye molecules and the hole conductor accompanied by increasing penetration of the solid hole conductors, leads to IPCE values up to 5% by illuminating through the working electrode. In the presence of additives, which oxidize partly the hole-conductive material and increase the amount of the free charge carriers, the maximum value of the IPCE is 33% [142]. This value is only a factor of ca. 2 lower than with liquid electrodes for the same working electrode/dye system [134].

An attractive idea is to sensitize the high-surface-area p-n heterojunction between TiO_2 and spiro-OMeTAD by using PbS quantum dots. PbS nanoparticles were formed *in-situ* on the porous TiO_2 electrode using chemical deposition techniques. These cells showed IPCE values of up to 45% [143].

An alternative approach to low molecular hole conductors is the use of polymer gel electrodes. In this case, plasticizers are used as additive during the preparation of the electrolyte to increase the chain mobility and the ionic conductivity of the electrolyte. The existence of the polymer gelling agents in the liquid electrolyte promotes its solidification afterwards [144–146]. These approaches have yielded "quasi-solid-state" devices with energy conversion efficiencies similar to those achievable with liquid electrolytes. However, the sealing of such cells remains still a problem.

The use of polymer electrolytes in such dye-sensitized photovoltaic cells exploits the advantages of ionic conduction and easy processing [147]. Polymer electrolytes comprising a poly(epichlorohydrin-co-ethylen oxide) elastomer, NaI and I_2 as mobile redox carriers, have been reported as an effective hole transporter in solid-state dye-sensitized devices. The maximum value of the IPCE reported is about 52% by illuminating through the working electrode [148, 149]. Devices have been fabricated also using a polymer electrolyte consisting of high-molecular-mass poly(ethylene oxide) filled with titanium oxide spherical particles (20 nm in diameter), LiI, and I_2. These cells present IPCE values as high as 40% [150].

Another method to form a good penetration between the dye-sensitized working electrode and solid hole conductor is *in-situ* photoelectrochemical polymerization. This concept is realized for polypyrrole [151] and poly(3,4-ethylene-dioxythiophene) [152] based devices.

In principle, the dye can also be replaced by the semiconducting polymer layer, bringing together the functions of light absorption and charge transport in a single material. Photovoltaic studies on junctions consisting of MEH-PPV on TiO_2 have shown that on excitation of the MEH-PPV layer, electrons are injected into the conduction band of TiO_2 [153]. The maximum value of the IPCE is about 6% [154]. Comparative studies using poly(3-octylthiophene) and its derivatives have been also investigated [155–157].

One of the problems associated with conjugated polymers and nanocrystalline TiO_2 bilayers is the inefficient filling of the TiO_2 pores, which leads to charge transport limited devices. Increasing the size of the TiO_2 pores, on the other hand, leads to devices with a smaller interface for charge generation. An alternative idea is to prepare bulk heterojunction devices in which a continuous network of TiO_2 is created inside a MDMO-PPV film [158].

Using the bulk heterojunction concept, the resulting IPCE value is up to 11%. For such devices, additional processes and parameters contribute to the efficiency, including the exciton diffusion length and efficiency of the electron transfer from the polymer layer into TiO_2 [159, 160].

4. OUTLOOK

Research efforts over the last decade have led to a number of new strategies for the preparation of organic/inorganic hybrid solar cells. Earlier methods have primarily focused on achieving nanocrystalline layers of well-defined inorganic materials and covering them in a second step with a semiconductive polymer layer. A new field of scientific research has grown out of the new strategy of creating composite layers of organic/inorganic blends with huge surface areas for charge separation. More recent approaches, particularly those based on core-shell particles, have been shown to be effective in producing composite blends by easy processing techniques like spin-coating. These techniques permit remarkable control over the layer thickness, and thus the assembled components and experimental conditions can be easily manipulated to create the favorable morphology for a better performance of a hybrid solar cell. The synthetic approaches outlined provide a high level of flexibility, thus allowing the cell design to be variable. On the other hand, the morphology control of organic/inorganic blends is still in its infancy.

A number of major scientific advances in the construction of bulk heterojunction hybrid solar cells have occurred recently and many more are expected. It is envisioned that future research will, to a large extent, continue to be focused on optimizing existing approaches, as well as developing new procedures, in order to modify the surface properties of nanoparticles. An interesting extension of the core/shell nanoparticles may be the manufacturing of semiconductive polymer layers on the surfaces of nanoparticles instead of using insulating organic shells between them. Accordingly, assembling of nanoparticles into colloidal crystals is expected to be an additional strategy for the morphology control. Finally, a significant goal is the development of the long-term stability.

GLOSSARY

Bulk heterojunction Interconnected network structure created by blending of an electron-accepting and a hole-accepting material, whereas the exciton dissociation can occur at a donor-acceptor interface.

Excitons Electron-hole pairs generated in the material upon absorption of photons are the so-called excitons. Thereby an electron will be promoted from the valence band to the conduction band, leaving a positively charged "hole" behind. This hole can be thought of as the absence of an electron and acts as a particle with its own effective mass and charge in the solid, when exciton dissociation occurs.

Ostwald ripening The growth mechanism where the smaller particles dissolve and the monomer released thereby is consumed by the larger particles. As a result the average nanocrystal size increases with time and the particle concentration decreases.

Percolation Percolation takes place when the electron acceptor and donor concentrations reach the value high enough to provide the conduction along the interconnected network of each type of materials in a bulk heterojunction structure.

Size quantization The bandgap of a quantum confined semiconductor is greater than that of the parent bulk material from which it is derived. This effect is a consequence of the confinement of the electron in a three-dimensional "box" with discrete energy levels rather than in the quasi-continuous band of a bulk material. The electronic and optical properties of such small particles will depend not only on the material of which they are composed, but also on their size.

ACKNOWLEDGMENTS

Financial support by the Bundesministerium für Bildung und Forschung in Germany, Grant 01SF0026, is gratefully acknowledged. We also thank the European Commission for partial support of this work. We express our sincere thanks to Professor F. Schäffler, Institute for Semiconductor and Solid-State Physics, and Dr. Neugebauer, LIOS, Physical Chemistry, Johannes Kepler University Linz.

REFERENCES

1. C. Brabec, N. S. Sariciftci, and J. C. Hummelen, *Adv. Func. Mat.* 11, 15 (2001).
2. B. O'Regan and M. Grätzel, *Nature* 353, 737 (1991).
3. J. Krüger, R. Plass, M. Grätzel, and H. J. Matthieu, *Appl. Phys. Lett.* 81, 367 (2002).
4. W. U. Huynh, W. W. Dittmer, and A. P. Alivisatos, *Science* 295, 2425 (2002).

5. N. S. Sariciftci, L. Smilowitz, A. J. Heeger, and F. Wudl, *Science* 258, 1474 (1992).
6. D. Braun, C. Zhang, V. Srdanov, A. J. Heeger, G. Stucky, and F. Wudl, *Appl. Phys. Lett.* 62, 585 (1993).
7. G. Yu, J. Gao, J. C. Hummelen, F. Wudl, and A. J. Heeger, *Science* 270, 1474 (1995).
8. G. Yu and A. J. Heeger, *J. Appl. Phys.* 78, 4510 (1995).
9. J. M. Halls, C. A. Walsch, N. C. Greenham, E. A. Marseglia, R. H. Friend, S. C. Moratti, and A. B. Holmes, *Nature* 376, 498 (1995).
10. G. Zerza, A. Cravino, H. Neugebauer, N. S. Sariciftci, R. Gomez, J. L. Segura, N. Martin, M. Svensson, and M. R. Andersson, *J. Phys. Chem. A* 105, 4173 (2001).
11. R. A. Janssen, M. P. T. Cristians, C. Hare, N. Martin, N. S. Sariciftci, A. J. Heeger, and F. Wudl, *J. Chem. Phys.* 103, 8840 (1995).
12. K. E. Ziemelis, A. T. Hussain, D. D. C. Bradley, R. H. Friend, J. Rühe, and G. Wagner, *Phys. Rev. Lett.* 66, 2231 (1991).
13. J. J. M. Halls, K. Picher, R. H. Fried, S. C. Moratti, and A. B. Holmes, *Appl. Phys. Lett.* 68, 3120 (1996).
14. S. Shaheen, C. J. Brabec, F. Padinger, T. Frommherz, J. C. Hummelen, and N. S. Sariciftci, *Appl. Phys. Lett.* 78, 841 (2001).
15. M. H. Huang, S. Mao, H, Feick, H. Yan, Y. Wu, H. Kind, E. Weber, R. Russo, and P. Yang, *Science* 292, 1897 (2001).
16. S. Sun, C. B. Murray, D. Weller, L. Folks, and A. Moser, *Science* 287, 1989 (2000).
17. S. A. Empodocles and M. G. Bawendi, *Science* 278, 2114 (1997).
18. M. Gao, C. Lesser, S. Kirstein, H. Möhwald, A. L. Rogach, and H. Weller, *J. Appl. Phys.* 87, 2297 (2000).
19. M. Gao, B. Richter, S. Kirstein, and H. Möhwald, *J. Phys. Chem. B* 102, 4096 (1998).
20. S. E. Shaheen, C. J. Brabec, N. S. Sariciftci, F. Padinger, T. Fromherz, and J. C. Hummelen, *Appl. Phys. Lett.* 78, 841 (2001).
21. T. Munters, T. Martens, L. Goris, V. Vrindts, J. Manca, L. Lutsen, W. de Ceuninck, D. Vanderzande, L. De Schepper, J. Gelan, N. S. Sariciftci, and C. J. Brabec, *Thin Solid Films* 403–404, 247 (2002).
22. H. Weller, *Angew. Chem. Int. Ed. Engl.* 32, 41 (1993).
23. M. L. Steigerwald and L. E. Brus, *Acc. Chem. Res.* 23, 283 (1990).
24. A. P. Alivisatos, *Science* 217, 933 (1996).
25. S. A. Empedocles and M. G. Bawendi, *Acc. Chem. Res.* 32, 389 (1999).
26. C. J. Murphy and J. L. Coffer, *Appl. Spectros.* 56, 16 (2002).
27. E. Arici, N. S. Sariciftci, and D. Meissner, *Mol. Cryst. Liq. Cryst.* 383, 129 (2002).
28. M. Green, *Physica E* 14, 65 (2002).
29. M. Green, *Prog. Photovoltaics* 9, 123 (2001).
30. H. T. Oyama, R. Sprycha, Y. Xie, R. E. Partch, and E. Matijevic, *J. Colloid Interface Sci.* 160, 298 (1993).
31. S. M. Marinakos, D. A. Schultz, and D. L. Feldheim, *Adv. Mater.* 34, 11 (1999).
32. S. M. Marinakos, J. P. Novak, L. C. Brousseau, A. B. House, E. M. Edeki, J. C. Feldhaus, and D. L. Feldheim, *J. Am. Chem. Soc.* 121, 8518 (1999).
33. Y. Yang, J. Huang, S. Liu, and J. Shen, *J. Mater. Chem.* 7, 131 (1997).
34. S. W. Haggata, D. J. Cole-Hamilton, and J. R. Fryer, *J. Mater. Chem.* 7, 1969 (1997).
35. F. M. Pavel and R. A. Mackay, *Langmuir* 16, 8568 (2000).
36. F. Caruso, *Adv. Mater.* 13, 11 (2001).
37. K. Yoshinaga, *Surfactant Sci. Ser.* 92, 626 (2000).
38. Weitz, J. Worrall, and F. Wudl, *Adv. Mater.* 12, 106 (2000).
39. U. Bach, D. Lupo, P. Comte, J. E. Moser, F. Weissörtel, J. Salbecj, H. Spreitzer, and M. Grätzel, *Nature* 395, 583 (1998).
40. R. Vogel, P. Hoyer, and H. Weller, *J. Phys. Chem.* 98, 2183 (1994).
41. S. Hotchandani and P. V. Kamat, *J. Phys. Chem.* 96, 6834 (1992).
42. M. Grätzel, *Nature* 414, 338 (2001).
43. Gary Hodes and I. Rubinstein, in "Electrochemistry of Nanomaterials" (G. Hodes, Ed.), p. 27. Wiley-VCH, Weinheim, 2001.
44. C. B. Murray, D. J. Norris, and M. G. Bawendi, *J. Am. Chem. Soc.* 115, 8706 (1993).
45. T. Trindade and P. O'Brien, *Adv. Mater.* 8, 61 (1996).
46. T. Trindade, P. O'Brien, and X. Zhang, *Chem. Mater.* 9, 523 (1997).
47. J. E. Katari, V. L. Colvin, and A. P. Alivisatos, *J. Phys. Chem.* 98, 4109 (1994).
48. B. O. Dabboussi, J. Rodriguez-Viejo, and F. V. Mikulec, *J. Phys. Chem. B* 101, 9463 (1997).
49. T. Cassagneau, T. E. Mallouk, and J. H. Fendler, *J. Am. Chem. Soc.* 120, 7647 (1998).
50. X. Peng, J. Wickham, and A. P. Alivisatos, *J. Am. Chem. Soc.* 120, 5343 (1998).
51. M. A. Hines and P. Guyot-Sionnest, *J. Phys. Chem.* 100, 468 (1996).
52. R. L. Wells, C. G. Pitt, A. T. McPhail, A. P. Purdy, S. Shafieezad, and H. B. Hallock, *Chem. Mater.* 1, 4 (1989).
53. S. S. Kher and R. L. Wells, *Chem. Mater.* 6, 2056 (1994).
54. O. I. Micic, J. R. Sprague, C. J. Curtis, K. M. Jones, J. L. Machol, A. J. Nozik, H. Giessen, B. Fluegel, G. Mohs, and N. Peyghambarian, *J. Phys. Chem.* 99, 7754 (1995).
55. Guzelian, U. Banin, A. V. Kadavanich, X. Peng, and A. P. Alivisatos, *Appl. Phys. Lett.* 69, 1432 (1996).
56. O. I. Micic, J. R. Sprague, and C. J. Curtis, *J. Phys. Chem.* 100, 7212 (1996).
57. M. Green and P. O'Brien, *Adv. Mater.* 10, 527 (1998).
58. M. Green and P. O'Brien, *J. Mater. Chem.* 9, 243 (1999).
59. T. Trindade, P. O'Brien, X. Zhang, and M. Motevalli, *J. Mater. Chem.* 7, 1011 (1997).
60. X. G. Peng, J. Wickham, and A. P. Alivisatos, *J. Am. Chem. Soc.* 120, 5343 (1998).
61. X. Duan and C. M. Lieber, *Adv. Mater.* 12, 298 (2000).
62. Y. Cui, X. Duan, J. Hu, and C. M. Lieber, *J. Phys. Chem. B* 104, 5213 (2000).
63. M. S. Gudiksen, L. J. Lauhon, J. Wang, D. Smith, and C. M. Lieber, *Nature* 415, 617 (2002).
64. X. Duan, J. Wang, and C. M. Lieber, *Appl. Phys. Lett.* 76, 1116 (2000).
65. X. Duan and C. M. Lieber, *J. Am. Chem. Soc.* 122, 188 (2000).
66. X. G. Peng, L. Manna, W. D. Yang, J. Wickham, E. Scher, A. Kadavanich, and A. P. Alivisatos, *Nature* 404, 59 (2000).
67. L. Manna, E. Scher, A. Kadavanich, and A. P. Alivisatos, *J. Am. Chem. Soc.* 122, 12700 (2000).
68. D. V. Talapin, E. V. Shevchenko, A. Kornowski, N. Gaponik, M. Haase, A. L. Rogach, and H. Weller, *Adv. Mater.* 13, 1868 (2001).
69. C. D. Dushkin, S. Saita, K. Yoshie, and Y. Yamaguchi, *Adv. Colloid Interface Sci.* 88, 37 (2000).
70. J. E. Bowen-Katari, V. L. Colvin, and A. P. Alivisatos, *J. Phys. Chem.* 98, 4109 (1994).
71. K. Yoshinaga, *Surfactant Sci. Ser.* 92, 626 (2000).
72. M. Kuno, J. K. Lee, B. O. Dabbousi, F. V. Mikulec, and M. C. Bawendi, *J. Chem. Phys.* 106, 9869 (1997).
73. X. Peng, M. C. Schlamp, A. V. Kadavanich, and A. P. Alivisatos, *J. Am. Chem. Soc.* 119, 7019 (1997).
74. Y. Cao and U. Banin, *J. Am. Chem. Soc.* 122, 9692 (2000).
75. Y. Cao and U. Banin, *Angew. Chem. Int. Ed.* 38, 3692 (1999).
76. M. T. Harrison, S. V. Kershaw, A. L. Rogach, A. Kornowski, A. Eychmüller, and H. Weller, *Adv. Mater.* 12, 123 (2000).
77. M. C. Schlamp, X. Peng, and A. P. Alivisatos, *J. Appl. Phys.* 82, 5837 (1997).
78. C. T. Black, C. B. Murray, R. L. Sandstrom, and S. Sun, *Science* 290, 1131 (2000).
79. C. Czekelius, M. Hilgendorff, L. Spanhel, I. Bedja, M. Lench, G. Müller, U. Bloeck, D. Su, and M. Giersig, *Adv. Mater.* 11, 643 (1999).
80. Y. Qian, *Adv. Mater.* 11, 1101 (1999).

81. Y. Xie, Y. Qian, W. Wang, S. Zhang, and Y. Zhang, *Science* 272, 1926 (1996).
82. B. Li, Y. Xie, J. Huang, and Y. Qian, *Adv. Mater.* 11, 1456 (1999).
83. T. Jiang, G. A. Ozin, and R. L. Bedard, *Adv. Mater.* 6, 860 (1994).
84. W. Wang, Y. Geng, Y. Qian, M. Ji, and X. Liu, *Adv. Mater.* 10, 1479 (1998).
85. J. Xiao, Y. Xie, R. Tang, and Y. Qian, *J. Solid State Chem.* 161, 179 (2001).
86. A. B. Mandale, S. D. Sathaye, and K. R. Patil, *Mater. Lett.* 55, 30 (2002).
87. S. Baranski and W. R. Fawcett, *J. Electrochem. Soc.* 130, 579 (1983).
88. A. C. Rastogi and K. S. Balakrishnan, *J. Electrochem. Soc.* 440, 134 (1989).
89. J. von Windheim and M. Cocivera, *J. Electrochem. Soc.* 136, 1502 (1989).
90. A. S. Baranski, W. R. Facett, and A. C. McDonald, *J. Electroanal. Chem.* 160, 271 (1984).
91. A. S. Baranski and W. R. Facett, *J. Electrochem. Soc.* 131, 2509 (1984).
92. G. Hodes, Y. Golan, D. Behar, Y. Zhang, B. Alperson, and I. Rubenstein, in "Nanoparticles and Nanostructured Films" (J. H. Fendler, Ed.), pp. 1–22. Wiley-VCH, Weinheim, W. Germany, 1998.
93. L. Kavan, T. Stoto, M. Grätzel, D. J. Fitzmaurice, and V. Shklover, *J. Phys. Chem.* 97, 9493 (1993).
94. G. Hodes and D. Cahen, *Sol. Cells* 16, 245 (1986).
95. S. Nakamura and A. Yamamoto, *Sol. Energy Mater. Sol. Cells* 75, 81 (2002).
96. S. Menezes, *Electrochem. Solid State Lett.* 5, C79 (2002).
97. A. S. Baranski, W. R. Fawcett, K. Gatner, A. C. McDonald, J. R. MacDonald, and M. Selen, *J. Electrochem. Soc.* 130, 579 (1983).
98. Y. Golan, G. Hodes, and I. Rubenstein, *J. Phys. Chem.* 100, 2220 (1996).
99. G. Horowitz and F. Garnier, *Sol. Energy Mater. Sol. Cells* 13, 47 (1986).
100. S. A. Gamboa, P. J. Sebastian, X. Mathew, H. Nguyen-Cong, and P. Chartier, *Sol. Energy Mater. Sol. Cells* 59, 115 (1999).
101. A. J. Franck, S. Glenis, and A. S. Nelson, *J. Phys. Chem.* 93, 3818 (1989).
102. C. H. Nguyen, C. Sene, and P. Chartier, *Sol. Energy Mater. Sol. Cells* 52, 413 (1998).
103. C. H. Nguyen, M. Dieng, C. Sene, and P. Chartier, *Sol. Energy Mater. Sol. Cells* 63, 23 (2000).
104. S. Bereznev, J. Kois, E. Mellikov, A. Öpik, and D. Meissner, "Proceedings of the 17th European Photovoltaic Solar Energy Conference," 2001.
105. C. Luo, D. M. Guldi, M. Maggini, E. Menna, S. Mondini, N. A. Kotov, and M. Prato, *Angew. Chem. Int. Ed. Engl.* 39, 3905 (2000).
106. M. Fang, C. H. Kim, G. B. Saupe, H. N. Kim, C. C. Waraksa, T. Miwa, A. Fujishima, and T. E. Mallouk, *Chem. Mater.* 11, 1526 (1999).
107. A. A. Mamedov, A. Belov, M. Giersig, N. Mamedova, and N. A. Kotov, *J. Am. Chem. Soc.* 123, 7738 (2001).
108. R. K. Iler, *J. Colloid Interface Sci.* 21, 569 (1996).
109. H. Lee, L. J. Kepley, H. G. Hong, S. Akhter, and T. E. Mallouk, *J. Phys. Chem.* 92, 2597 (1988).
110. G. Decher and J. D. Hong, *Makromol. Chem. Macromol. Symp.* 46, 321 (1991).
111. G. Decher, *Science* 277, 1232 (1997).
112. P. K. Ho, M. Granström, R. H. Friend, and N. C. Greenham, *Adv. Mater.* 10, 769 (1998).
113. C. Fou, O. Onitsika, M. Ferreira, M. F. Rubner, and B. R. Hsieh, *J. Appl. Phys.* 80, 4067 (1996).
114. J. W. Baur, S. Kim, P. B. Balanda, J. R. Reynolds, and M. F. Rubner, *Adv. Mater.* 10, 1452 (1998).
115. T. Cassagneau, T. E. Mallouk, and J. H. Fendler, *J. Am. Chem. Soc.* 120, 7848 (1998).
116. M. Y. Gao, C. Lesser, S. Kirstein, H. Möhwald, A. L. Rogach, and H. Weller, *J. Appl. Phys.* 87, 2297 (2000).
117. A. L. Rogach, D. S. Koktysh, M. Harrison, and N. A. Kotov, *Chem. Mater.* 12, 1526 (2000).
118. N. A. Kotov, I. Dekany, and J. H. Fendler, *J. Phys. Chem.* 99, 13065 (1995).
119. T. Cassagneau, T. E. Mallouk, and J. H. Fendler, *J. Am. Chem. Soc.* 120, 7647 (1998).
120. N. P. Gaponik, D. V. Talapin, and A. L. Rogach, *Phys. Chem. Chem. Phys.* 1, 1787 (1999).
121. N. P. Gaponik, D. V. Talapin, A. L. Rogach, and A. Eychmüller, *J. Mater. Chem.* 10, 2163 (2000).
122. V. L. Colvin and A. P. Alivisatos, *Nature* 354, 370 (1994).
123. N. S. Kumar, M. P. Joshi, C. S. Friend, P. N. Prasad, and R. Burzynski, *Appl. Phys. Lett.* 71, 1388 (1997).
124. E. Hao and T. Lian, *Langmuir* 16, 7879 (2000).
125. N. C. Greenham, X. Peng, and A. P. Alivisatos, *Phys. Rev. B* 54, 17628 (1996).
126. D. S. Ginger and N. C. Greenham, *Synth. Met.* 101, 425 (1999).
127. M. R. Böhmer, *J. Colloid. Interface Sci.* 197, 251 (1998).
128. U. Wendy, X. P. Huynh, and A. P. Alivisatos, *Adv. Mater.* 11, 923 (1999).
129. H. Sirringhaus, N. Tessler, and R. H. Friend, *Science* 280, 1741 (1998).
130. E. Arici, H. Hoppe, A. Reuning, N. S. Sariciftci, and D. Meissner, "Proceedings of the 17th European Photovoltaic Solar Energy Conference," 2001.
131. H. W. Schock and D. Meissner, in "Solarzellen—Physikalische Grundlagen und Anwendungen in der Photovoltaik." Vieweg & Sohn, Wiesbaden, 1993.
132. E. Arici, N. S. Sariciftci, and D. Meissner, *Adv. Funct. Mater.* 13, ??? (2003).
133. T. Fromherz, F. Padinger, D. Gebeyahu, C. Brabec, J. C. Hummelen, and N. S. Sariciftci, *Sol. Energy Mater. Sol. Cells* 63, 61 (2000).
134. M. K. Nazeeruddin, A. Kay, I. Rodicio, R. Humphry-Baker, E. Müller, R. Liska, N. Vlachopoulos, and M. Grätzel, *J. Am. Chem. Soc.* 115, 6382 (1993).
135. K. Tennakone, G. R. R. A. Kumara, A. R. Kumarasinghe, K. G. U. Wijayantha, and P. M. Sirimanne, *Sci. Technol.* 10, 1689 (1995).
136. K. Tennakone, G. R. R. A. Kumara, I. R. M. Kottegada, K. G. U. Wijayantha, and V. P. S. Perera, *J. Phys. D: Appl. Phys.* 32, 347 (1999).
137. B. O'Regan and D. T. Schwartz, *Chem. Mater.* 9, 439 (1997).
138. B. O'Regan, D. T. Schwartz, S. M. Zakeeruddin, and M. Grätzel, *Adv. Mater.* 12, 1263 (2000).
139. J. Hagen, W. Schaffarth, P. Otschik, R. Fink, A. Bacher, H. W. Schmidt, and D. Haarer, *Synth. Met.* 89, 215 (1997).
140. C. Jäger, R. Bilke, M. Heim, D. Haarer, H. Karickal, and M. Thelakkat, *Synth. Met.* 121, 1543 (2001).
141. K. R. Haridas, J. Ostrauskaite, M. Thelakkat, M. Heim, R. Bilke, and D. Haarer, *Synth. Met.* 121, 1573 (2001).
142. U. Bach, D. Lupo, P. Comte, J. E. Moser, F. Weissörtel, J. Salbeck, H. Spreitzer, and M. Grätzel, *Nature* 395, 583 (1998).
143. R. Plass, S. Pelet, J. Krüger, U. Bach, and M. Grätzel, *J. Phys. Chem. B* 106, 7578 (2002).
144. F. Cao, G. Oskam, and P. C. Searson, *J. Phys. Chem.* 99, 17071 (1995).
145. M. Matsumoto, H. Miyasaki, K. Matsuhiro, Y. Kumashiro, and Y. Takaoka, *Solid State Ionics* 89, 263 (1996).
146. K. Tennakone, G. K. R. Senadeera, V. R. S. Perera, I. R. M. Kottegoda, and L. A. A. De Silva, *Chem. Mater.* 11, 2474 (1999).
147. F. Cao, G. Oskam, and P. C. Searson, *J. Phys. Chem.* 99, 17071 (1995).

148. A. F. Nogueira, J. R. Durrant, and M. A. De Paoli, *Adv. Mater.* 13, 826 (2001).
149. A. F. Nogueira and M. A. De Paoli, *Sol. Energy Mater. Sol. Cells* 61, 135 (2000).
150. G. Katsaros, T. Stergiopoulos, I. M. Arabatzis, K. G. Papadokostaki, and P. Falaras, *J. Photochem. Photobiol. A: Chem.* 149, 191 (2002).
151. K. Murakoshi, R. Kogure, Y. Wada, and S. Yanagida, *Chem. Lett.* 5, 471 (1997).
152. Y. Saito, T. Kitamura, Y. Wada, and S. Yanagida, *Synth. Met.* 131, 185 (2002).
153. T. J. Savenije, J. M. Warman, and A. Goossens, *Chem. Phys. Lett.* 287, 148 (1998).
154. T. J. Savenije, M. J. W. Vermeulen, M. P. De Haas, and J. M. Warman, *Sol. Energy Mater. Sol. Cells* 61, 9 (2000).
155. D. Gebeyehu, C. J. Brabec, and N. S. Sariciftci, *Thin Solid Films* 403–404, 271 (2002).
156. D. Gebeyehu, C. J. Brabec, and N. S. Sariciftci, D. Vangeneugden, R. Kiebooms, D. Vanderzande, F. Kienberger, and H. Schindler, *Synth. Met.* 125, 279 (2002).
157. S. Luzatti, M. Basso, M. Castellani, C. J. Brabec, and N. S. Sariciftci, *Thin Solid Films* 403–404, 52 (2002).
158. P. A. van Hal, M. M. Wienk, J. M. Kroon, W. J. H. Verhees, L. H. Slooff, W. J. H. van Gennip, P. Jonkheijm, and R. A. J. Janssen, in press.
159. A. C. Arango, S. A. Carter, and P. J. Brock, *Appl. Phys. Lett.* 74, 1698 (1999).
160. J. S. Salafsky, H. Kerp, and R. E. I. Schropp, *Synth. Met.* 102, 1256 (1999).
161. J. Desilvestro, M. Grätzel, L. Kavan, and *J. Moser, J. Am. Chem. Soc.* 107, 2988 (1985).
162. H. Hwang et al., *Thin Solid Films* 67, 83 (1980).
163. D. Suri, K. Nagpal, and G. Chadha, *J. Appl. Crystallogr.* 22, 578 (1989).